Ross
MATH 263
MONTEREY PENINSULA COLLEGE LIBRARY
3 4262 00127 1648
1 Hour
No Overnight

The Algebra Pyramid

Equations and Inequalities
$2 + 5(7) = 37$ or $x + 2y > 5$

Expressions
$2 + 5(7)$ or $x + 2y$

Constants and Variables
$2, 5, 7, x, y$

The Algebra Pyramid illustrates how variables, constants, expressions, equations, and inequalities relate. At the foundation of algebra, and this pyramid, are constants and variables, which are used to build expressions, which, in turn, are used to build equations and inequalities.

Complex Numbers
Examples: $5 + 2i$, $15 - 3i\sqrt{2}$, etc.

Real Numbers

Irrational Numbers
$-5\sqrt{3}$, $\sqrt{2}$, π, etc.

Rational Numbers $-3, -2.4, -1\frac{4}{5}, 0, 0.\overline{6}, 1$, etc.

Integers . . . $-3, -2, -1, 0, 1, 2, 3, \ldots$

Whole Numbers $0, 1, 2, 3, \ldots$

Natural Numbers $1, 2, 3, \ldots$

Imaginary Numbers
$6i$, $5i\sqrt{3}$, etc.

Arithmetic Summary

Each operation has an inverse operation. In the diagram below, the operations build from the top down. Addition leads to multiplication, which leads to exponents. Subtraction leads to division, which leads to roots.

PROPERTIES OF ARITHMETIC

In each of the following, a, b, and c represent real numbers.

Additive Identity

$a + 0 = a$

Commutative Property of Addition

$a + b = b + a$

Associative Property of Addition

$(a + b) + c = a + (b + c)$

Multiplicative Identity

$a \cdot 1 = a$

Multiplicative Property of 0

$a \cdot 0 = 0$

Commutative Property of Multiplication

$ab = ba$

Associative Property of Multiplication

$(ab)c = a(bc)$

Distributive Property

$a(b + c) = ab + ac$

INVERSE OPERATIONS

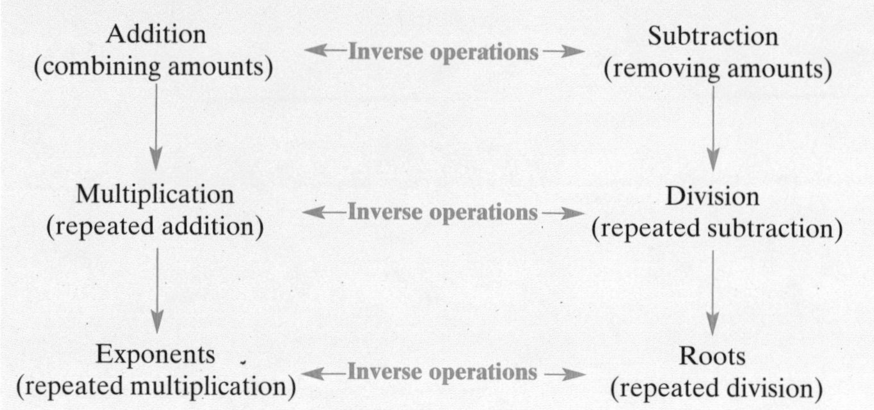

ORDER OF OPERATIONS

1. Grouping symbols
2. Exponents or roots from left to right
3. Multiply or divide from left to right
4. Add or subtract from left to right

SOLVING LINEAR EQUATIONS

To solve linear equations in one variable:

1. Simplify both sides of the equation as needed.
   a. Distribute to clear parentheses.
   b. Clear fractions or decimals by multiplying through by the LCD. In the case of decimals, the LCD is a power of 10 determined by the decimal number with the most places. (Clearing fractions and decimals is optional.)
   c. Combine like terms.
2. Use the addition principle so that all variable terms are on one side of the equation and all constants are on the other side. (Clear the variable term with the lesser coefficient.) Then combine like terms.
3. Use the multiplication principle to clear the remaining coefficient.

USING THE QUADRATIC FORMULA

To solve a quadratic equation in the form $ax^2 + bx + c = 0$, where $a \neq 0$, use the quadratic formula:

$$x = \frac{-b \pm \sqrt{b^2 - 4ac}}{2a}$$

FACTORING A POLYNOMIAL

To factor a polynomial, first factor out any monomial GCF, then consider the number of terms in the polynomial. If the polynomial has:

I. Four terms, then try to factor by grouping.

II. Three terms, then determine if the trinomial is a perfect square or not.

A. If the trinomial is a perfect square, then consider its form.
  1. If in the form $a^2 + 2ab + b^2$, then the factored form is $(a + b)^2$.
  2. If in the form $a^2 - 2ab + b^2$, then the factored form is $(a - b)^2$.

B. If the trinomial is not a perfect square, then consider its form.
  1. If in the form $x^2 + bx + c$, then find two factors of c whose sum is b, and write the factored form as $(x +$ first number$)$ $(x +$ second number$)$.
  2. If in the form $ax^2 + bx + c$, where $a \neq 1$, then use trial and error. Or, find two factors of ac whose sum is b; write these factors as coefficients of two like terms that, when combined, equal bx; and then factor by grouping.

III. Two terms, then determine if the binomial is a difference of squares, a sum of cubes, or a difference of cubes.

A. If given a binomial that is a difference of squares $a^2 - b^2$, then the factors are conjugates and the factored form is $(a + b)(a - b)$. Note that a sum of squares cannot be factored.

B. If given a binomial that is a sum of cubes, $a^3 + b^3$, then the factored form is $(a + b)(a^2 - ab + b^2)$.

C. If given a binomial that is a difference of cubes, $a^3 - b^3$, then the factored form is $(a - b)(a^2 + ab + b^2)$.

Note: Always check to see if any of the factors can be factored.

Intermediate Algebra

Second Edition

TOM CARSON
Midlands Technical College

ELLYN GILLESPIE
Midlands Technical College

BILL JORDAN
Seminole Community College

Boston San Francisco New York
London Sydney Tokyo Singapore Madrid
Mexico City Paris Cape Town Hong Kong Montreal

Publisher: Greg Tobin
Editor in Chief: Maureen O'Connor
Executive Editor: Jennifer Crum
Executive Project Manager: Kari Heen
Project Editor: Lauren Morse
Editorial Assistants: Elizabeth Bernardi and Emily Ragsdale
Managing Editor: Ron Hampton
Senior Designer: Dennis Schaefer
Cover Designer: Dennis Schaefer
Photo Researcher: Beth Anderson
Supplements Supervisor: Emily Portwood
Media Producer: Sharon Smith
Software Development: Rebecca Williams, MathXL; Marty Wright, TestGen
Marketing Manager: Jay Jenkins
Marketing Coordinator: Alexandra Waibel
Senior Prepress Supervisor: Caroline Fell
Senior Manufacturing Buyer: Carol Melville
Production Coordination: Pre-Press Company, Inc.
Composition: Pre-Press Company, Inc.
Artwork: Pre-Press Company, Inc.

Cover photo: © R. Ian Lloyd/Masterfile—The Baha'i Temple at dusk, Delhi, India

Photo credits can be found on page I-14 in the back of the book.

Many of the designations used by manufacturers and sellers to distinguish their products are claimed as trademarks. Where those designations appear in this book, and Addison-Wesley was aware of a trademark claim, the designations have been printed in initial caps or all caps.

Library of Congress Cataloging-in-Publication Data

Carson, Tom, 1967–
Intermediate Algebra—2nd ed./Tom Carson, Ellyn Gillespie, Bill E. Jordan.
p. cm.
Includes index.
ISBN 0-321-35835-X (Student's Edition)
1. Albegra I. Gillespie, Ellyn. II. Jordan, Bill E. III. Title.

QA154.3.C37 2005b
512.9—dc22 2005050956

Copyright © 2007 Pearson Education, Inc. All rights reserved. No part of this publication may be reproduced, stored in a retrieval system, or transmitted, in any form or by any means, electronic, mechanical, photocopying, recording, or otherwise, without the prior written permission of the publisher. Printed in the United States of America. For information on obtaining permission for use of material in this work, please submit a written request to Pearson Education, Inc., Rights and Contracts Department, 75 Arlington Street, Suite 300, Boston, MA 02116.

1 2 3 4 5 6 7 8 9 10—VH—10 09 08 07 06

Contents

Preface

Welcome to the second edition of *Intermediate Algebra* by Carson, Gillespie, and Jordan! Revising this series has been both exciting and rewarding. It has given us the opportunity to respond to valuable instructor and student feedback and suggestions for improvement. It is with great pride that we share with you both the improvements and additions to this edition as well as the hallmark features and style of the Carson/Gillespie/Jordan series.

Intermediate Algebra, Second Edition, is the fifth book in a series that includes *Prealgebra*, Second Edition, *Elementary Algebra*, Second Edition, *Elementary Algebra with Early Systems of Equations*, and *Elementary and Intermediate Algebra*, Second Edition. This text is designed to be versatile enough for use in a standard lecture format, a self-paced lab, or even in an independent study format. Written in a relaxed, nonthreatening style, *Intermediate Algebra* takes great care to ensure that students who have struggled with math in the past will be comfortable with the subject matter. Explanations are carefully developed to provide a sense of why an algebraic process works the way it does, instead of just an explanation of how to follow the process. In addition, problems from science, engineering, accounting, health, the arts, and everyday life link algebra to the real world. A complete study system beginning with a Learning Styles Inventory and supported by frequent Learning Strategy boxes, is also provided to give students extra guidance and to help them be successful. (See page xxiii.)

Changes to the Second Edition

This revision includes refinements to the presentation of the material as well as the addition of many more examples and applications throughout the text. However, the primary focus of this revision is the exercise sets. The section-level exercise sets have been scrutinized and reworked to create a gradation that slowly progresses from easy to more difficult. There is also better pairing between odd and even exercises, and many more midlevel problems have been added.

In addition to the exercise sets, the Learning Strategy boxes and Algebra Pyramid references have been enhanced and increased in number to provide students with even more guidance.

We have added section number references to the review exercises at the end of each section as well as in the Chapter Review Exercises and Cumulative Review Exercises.

Small versions of the Algebra Pyramid have been added to the Chapter Review Exercises and the Cumulative Review Exercises to help students distinguish groups of expression exercises from groups of equation or inequality exercises.

Finally, the number of exercises included in MyMathLab and MathXL has been increased dramatically for an even stronger correlation between the book and the technology that supports it.

Key Features

Study System A study system is presented in the *To the Student* section on pages xvii–xxii. This system is then reinforced throughout the text. The system recommends color codes for taking notes. The color codes are consistent in the text itself: red for definitions, blue for procedures and rules, and black for notes and examples. In addition, the study system presents strategies for succeeding in the course. These learning strategies have been expanded and are revisited in the chapter openers and throughout the body of the text.

Learning Styles Inventory A Learning Styles Inventory is presented on page xxiii to help students assess their particular learning style. Learning Strategy boxes are then presented throughout the book with different learning styles in mind.

> ## Learning Styles Inventory
>
> **What is your personal learning style?**
>
> A learning style is the way in which a person processes new information. Knowing your learning style can help you make choices in the way you study and focus on new material. Below are fifteen statements that will help you assess your learning style. After reading each statement, rate your response to the statement using the scale below. There are no right or wrong answers.
>
> 3 = Often applies   2 = Sometimes applies   1 = Never or almost never applies
>
> ____ **1.** I remember information better if I write it down or draw a picture of it.
>
> ____ **2.** I remember things better when I hear them instead of just reading or seeing them.
>
> ____ **3.** When I receive something that has to be assembled, I just start doing it. I don't read the directions.
>
> ____ **4.** If I am taking a test, I can "visualize" the page of text or lecture notes where the answer is located.

> **Learning Strategy**
>
> The summaries in this textbook are like the study sheet suggested in the To the Student section. Recall that your study sheet is a list of the rules, procedures, and formulas that you need to know. If you are a tactile or visual learner, spend a lot of time reviewing and writing the rules or procedures. Try to get to the point where you can write the essence of each rule and procedure from memory. If you are an auditory learner, record yourself saying each rule and procedure, then listen to the recording over and over. Also, consider developing clever rhymes or songs for each rule or procedure to help you remember them.

Learning Strategy Boxes Learning Strategy boxes appear where appropriate in the text to offer advice on how to effectively use the study system and how to study specific topics based on a student's individual learning style (see pages 2, 121, 154, and 394).

The Algebra Pyramid An Algebra Pyramid is used throughout the text to help students see how the topic they are learning relates to the big picture of algebra—particularly focusing on the relationship between constants, variables, expressions, and equations (see pages 56, 326, and 417). In Chapter Review Exercises and Cumulative Review Exercises, an Algebra Pyramid icon indicates the level of the pyramid that correlates to a particular group of exercises to help students determine what actions are appropriate with these exercises, for example, whether to "simplify" or "solve" (see pages 50, 378, and 527).

The Algebra Pyramid

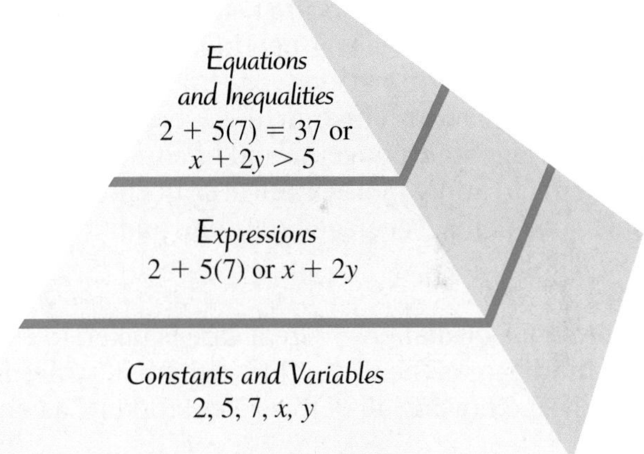

CHAPTER 3

Equations and Inequalities in Two Variables and Functions

"Science begins with the world we have to live in, accepting its data and trying to explain its laws. From there, it moves towards the imagination: it becomes a mental construct, a model of a possible way of interpreting experience. The further it goes in this direction, the more it tends to speak the language of mathematics, which is really one of the languages of imagination, along with literature and music."
—Northrop Frye, Canadian educator and critic

3.1 Graphing Linear Equations
3.2 The Slope of a Line
3.3 The Equation of a Line
3.4 Graphing Linear Inequalities
3.5 Introduction to Functions and Function Notation

In Chapter 2, we learned to solve equations and inequalities in one variable. We also graphed solution sets on a number line. In this chapter, we will find solution sets for equations and inequalities that have two variables. We will also see that the solution sets for these equations can be graphed. However, instead of a single number line, we will use two number lines, one for each variable, in what is called the *coordinate plane*. We will finish the chapter by exploring a generalized relationship between two sets of data called a *function*.

135

Chapter Openers Like the Algebra Pyramid, chapter openers are designed to help students see how the topics in the upcoming chapter relate to the big picture of the entire course. The chapter openers give information about the importance of the topics in each chapter and how they fit into the overall structure of the course (see pages 1, 55, and 135).

Connection Boxes Connection boxes bridge concepts and ideas that students have learned elsewhere in the text so they see how the concepts are interrelated and build on each other (see pages 151, 194, and 343).

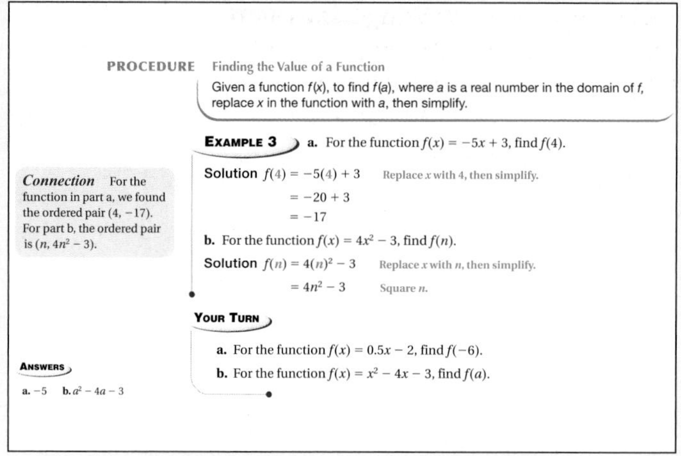
PROCEDURE Finding the Value of a Function

Given a function $f(x)$, to find $f(a)$, where a is a real number in the domain of f, replace x in the function with a, then simplify.

EXAMPLE 3 a. For the function $f(x) = -5x + 3$, find $f(4)$.

Solution $f(4) = -5(4) + 3$ Replace x with 4, then simplify.

$= -20 + 3$

$= -17$

b. For the function $f(x) = 4x^2 - 3$, find $f(n)$.

Solution $f(n) = 4(n)^2 - 3$ Replace x with n, then simplify.

$= 4n^2 - 3$ Square n.

Connection For the function in part a, we found the ordered pair $(4, -17)$. For part b, the ordered pair is $(n, 4n^2 - 3)$.

YOUR TURN

a. For the function $f(x) = 0.5x - 2$, find $f(-6)$.

b. For the function $f(x) = x^2 - 4x - 3$, find $f(a)$.

ANSWERS

a. -5 b. $a^2 - 4a - 3$

Your Turn Practice Exercises Your Turn practice exercises are found after most examples to give students an opportunity to work problems similar to the examples they have just seen. This practice step makes the text more interactive and provides immediate feedback so students can build confidence in what they are learning (see pages 36, 83, and 190).

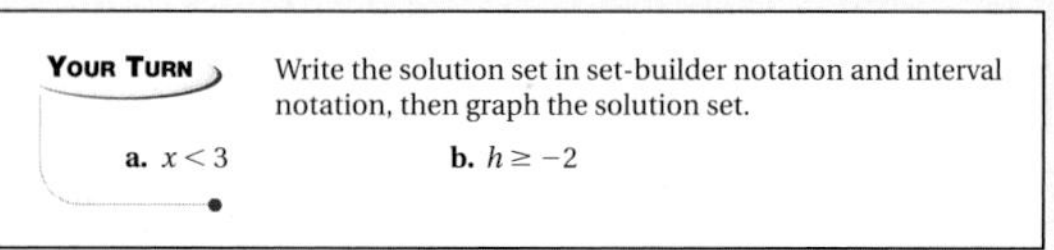
YOUR TURN Write the solution set in set-builder notation and interval notation, then graph the solution set.

a. $x < 3$ b. $h \geq -2$

Real, Relevant, and Interesting Applications A large portion of application problems in examples and exercise sets are taken from real situations in science, engineering, health, finance, the arts, or just everyday life. The real-world applications illustrate the everyday use of basic algebraic concepts and encourage students to apply mathematical concepts to solve problems (see pages 199, 205, and 259–266).

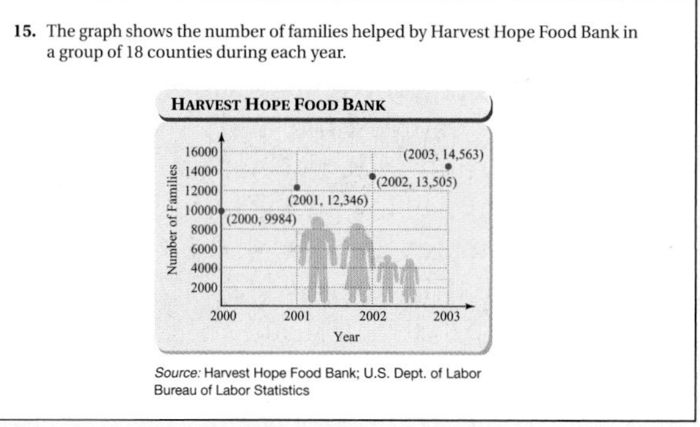
15. The graph shows the number of families helped by Harvest Hope Food Bank in a group of 18 counties during each year.

Source: Harvest Hope Food Bank; U.S. Dept. of Labor Bureau of Labor Statistics

Thorough Explanations Great care is taken to explain not only how to do the math, but also why the math works the way it does, where it comes from, and how it is relevant to students' everyday lives. Knowing all of this gives students a context in which to remember the concept.

Problem-Solving Outline A five-step problem-solving outline is introduced on page 66 of Section 2.2 with the following headings:

1. Understand
2. Plan
3. Execute
4. Answer
5. Check

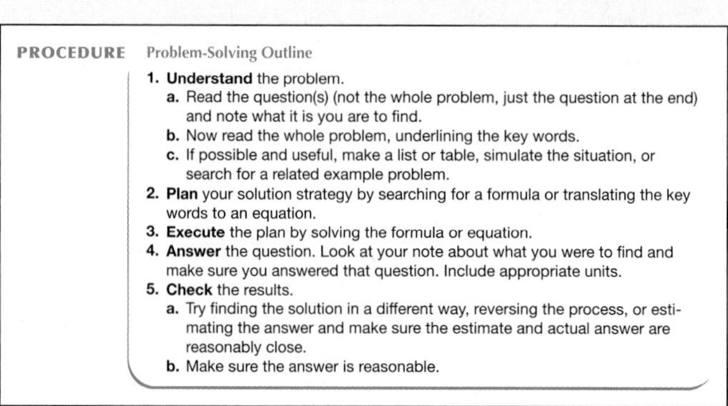
PROCEDURE Problem-Solving Outline

1. **Understand** the problem.
   a. Read the question(s) (not the whole problem, just the question at the end) and note what it is you are to find.
   b. Now read the whole problem, underlining the key words.
   c. If possible and useful, make a list or table, simulate the situation, or search for a related example problem.
2. **Plan** your solution strategy by searching for a formula or translating the key words to an equation.
3. **Execute** the plan by solving the formula or equation.
4. **Answer** the question. Look at your note about what you were to find and make sure you answered that question. Include appropriate units.
5. **Check** the results.
   a. Try finding the solution in a different way, reversing the process, or estimating the answer and make sure the estimate and actual answer are reasonably close.
   b. Make sure the answer is reasonable.

Application examples throughout the rest of the text follow the steps given in this outline, presenting the headings to model the thinking process clearly (see pages 74, 254, and 505).

Warning Boxes Warning boxes alert students to common mistakes and false assumptions that students often make and explain *why* these are incorrect (see pages 28, 62, and 227).

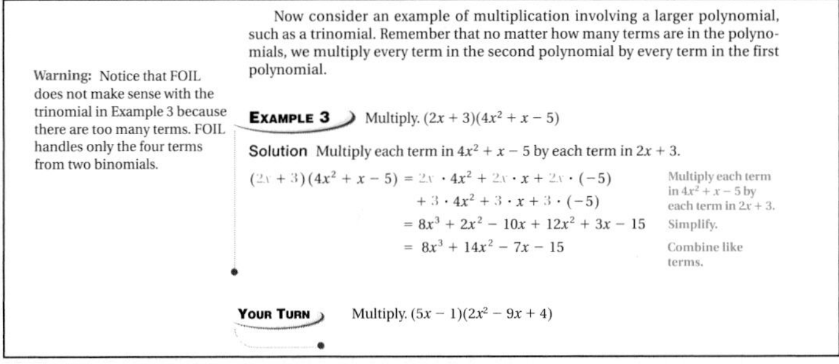

Now consider an example of multiplication involving a larger polynomial, such as a trinomial. Remember that no matter how many terms are in the polynomials, we multiply every term in the second polynomial by every term in the first polynomial.

Warning: Notice that FOIL does not make sense with the trinomial in Example 3 because there are too many terms. FOIL handles only the four terms from two binomials.

EXAMPLE 3 Multiply. $(2x + 3)(4x^2 + x - 5)$

Solution Multiply each term in $4x^2 + x - 5$ by each term in $2x + 3$.

$$\begin{aligned}(2x + 3)(4x^2 + x - 5) &= 2x \cdot 4x^2 + 2x \cdot x + 2x \cdot (-5) + 3 \cdot 4x^2 + 3 \cdot x + 3 \cdot (-5) && \text{Multiply each term in } 4x^2 + x - 5 \text{ by each term in } 2x + 3.\\ &= 8x^3 + 2x^2 - 10x + 12x^2 + 3x - 15 && \text{Simplify.}\\ &= 8x^3 + 14x^2 - 7x - 15 && \text{Combine like terms.}\end{aligned}$$

YOUR TURN Multiply. $(5x - 1)(2x^2 - 9x + 4)$

Of Interest

René Descartes was born in 1596 at La Haye, near Tours, France. While serving in the military, Descartes had a series of dreams in which he became aware of a new way of viewing geometry using algebra. In 1637, after some urging by his friends, he reluctantly allowed one work known as the *Method* to be printed. It was in this book that the rectangular coordinate system and analytical geometry was given to the world.

Of Interest Boxes Of Interest boxes are positioned throughout the text to offer a unique perspective on content that some students might otherwise consider to be ho-hum mathematics. Sometimes containing trivia and other times historical notes, Of Interest boxes are designed to enhance the learning process by making concepts fun, interesting, and memorable (see pages 45, 136, and 513).

Puzzle Problems These mathematical brainteasers, often solved without a formulaic approach, appear at the end of selected exercise sets to encourage critical thinking (see pages 107, 266, and 405).

PUZZLE PROBLEM

A cyclist is involved in a multiple-day race. She feels she needs to complete today's 40 kilometers part in somewhere between 1 hour and 45 minutes and 2 hours. Find the range of values her rate can be to complete this leg of the race in her desired time frame.

Collaborative Exercises OPTICAL ILLUSION OR CONFUSION?

An optical shop at a local mall advertises the sale shown to the right. The total cost for a pair of glasses is the sum of the costs of the frame and lenses.

Le Optical Shoppe
All lenses
40% Off

1. Let F represent the regular price of a frame and L the regular price for lenses. Write an expression that describes the total cost of a pair of glasses at regular price.
2. Does the expression $F + 0.60L$ give the cost of the glasses during the advertised 40% off sale? Explain why or why not.
3. The regular price for Anna's lenses is \$90. Anna chooses frames listed at \$120. How much did Anna save by buying her glasses during the sale?
4. Write and solve an inequality to determine the price of the most expensive frame Anna can choose during the sale if she wishes to keep the cost of her glasses to at most \$125.
5. The optical shop gives a discount of 25% on frames and lenses every day to seniors. Pat has chosen a \$140 frame. She wears bifocals, so her lenses are \$260. The shop will apply only one of the discounts, so which would be better for Pat, the 25% senior discount or the advertised 40% off sale?
6. The optical shop has a complete series of economy frames for \$60. Write and solve an inequality showing for which lens prices the 40% off sale would be more economical than the senior discount if an economy frame is used.
7. College students are eligible for a 15% discount on glasses that cost over \$100. Using F to represent the regular price of a frame and L for the regular price of lenses, write an expression that describes the total cost of a pair of glasses with the student discount.
8. If a student uses the economy \$80 frames, write and solve an inequality showing for which lens prices the 40% off sale would be better than the everyday student discount.

Collaborative Exercises These exercises, which appear once per chapter, encourage students to work in groups to discuss mathematics and use the topics from a particular section or group of sections to solve a problem (see pages 23, 43, and 163).

Calculator Tips The relevant functions of calculators (scientific or graphing, depending on the topic) are explained and illustrated throughout the text in the optional Calculator Tips feature. In addition, an occasional calculator icon in the exercise sets indicates that the problem is designed to be solved using a calculator, though one is not required (see pages 143, 183, and 319).

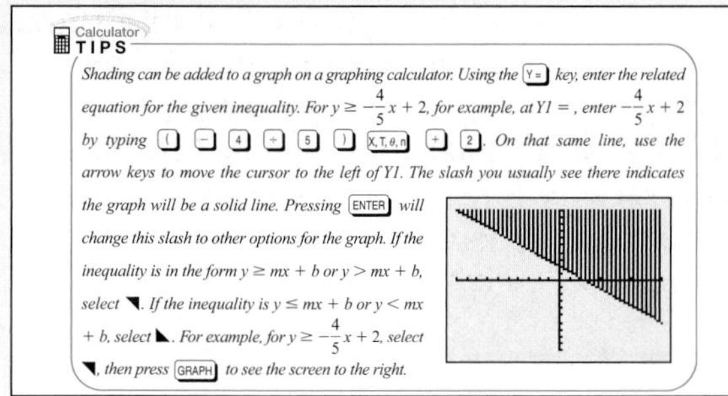

Calculator TIPS

Shading can be added to a graph on a graphing calculator. Using the Y= *key, enter the related equation for the given inequality. For* $y \geq -\frac{4}{5}x + 2$, *for example, at Y1 = , enter* $-\frac{4}{5}x + 2$ *by typing* (– 4 ÷ 5) X,T,θ,n + 2. *On that same line, use the arrow keys to move the cursor to the left of Y1. The slash you usually see there indicates the graph will be a solid line. Pressing* ENTER *will change this slash to other options for the graph. If the inequality is in the form* $y \geq mx + b$ *or* $y > mx + b$, *select* ◥. *If the inequality is* $y \leq mx + b$ *or* $y < mx + b$, *select* ◣. *For example, for* $y \geq -\frac{4}{5}x + 2$, *select* ◥, *then press* GRAPH *to see the screen to the right.*

REVIEW EXERCISES

For Exercises 1 and 2, graph the solution set.

[2.6] **1.** $|2x - 5| \geq 1$

[4.6] **2.** $\begin{cases} x + y < 5 \\ 2x - y \geq 4 \end{cases}$

[5.1] **3.** Write 0.0004203 in scientific notation.

[5.4] **4.** Divide: $\dfrac{x^3 + x^2 - 10x + 11}{x + 4}$

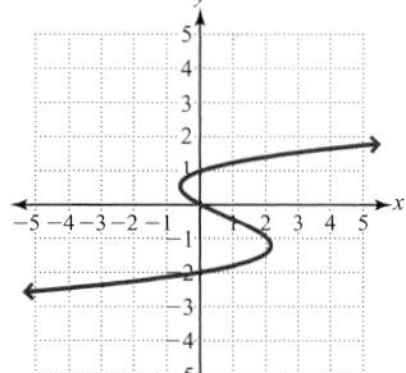

[3.5] **5.** Is the relation graphed to the right a function?

[5.2] **6.** If $f(x) = 5x^3 + 6x - 19$ and $g(x) = 3x^2 - 6x - 8$, find $(f + g)(x)$.

Review Exercises Since continuous review is important in any mathematics course, this text includes Review Exercises at the end of each exercise set. These exercises review previously learned concepts not only to keep the material fresh for students, but also to serve as a foundational review for the discussion in the upcoming section (see pages 107, 432, and 499).

Chapter Summaries and Review Exercises An extensive Summary at the end of each chapter provides a list of defined terms referenced by section and page number, a two-column summary of key concepts, and a list of important formulas appearing in that chapter. A set of Review Exercises is also provided with answers to all Review Exercises provided in the back of the book (see pages 127–131, 212–216, and 297–306).

Chapter 3 Summary

Defined Terms

Section 3.1
Linear equation in two variables (p. 137)
x-intercept (p. 139)
y-intercept (p. 139)

Section 3.2
Slope (p. 152)

Section 3.4
Linear inequality in two variables (p. 179)

Section 3.5
Relation (p. 189)
Domain (p. 189)
Range (p. 189)
Function (p. 189)

Procedures, Rules, and Key Examples

Procedures/Rules	Key Examples
Linear Equations ... an equation in two variables, ...e variables with a chosen number ...n for the value of the other variable.	**Example 1:** Find two solutions for the equation $y = 3x - 1$. First solution: Let $x = 0$: $y = 3(0) - 1$, $y = -1$, Solution: $(0, -1)$. Second solution: Let $x = 1$: $y = 3(1) - 1$, $y = 3 - 1$, $y = 2$, Solution: $(1, 2)$

Chapter 3 Review Exercises

For Exercises 1–6, answer true or false.

[3.1] **1.** When writing coordinates, the vertical coordinate is written first.

[3.1] **2.** There are an infinite number of solutions to every linear equation in two variables.

[3.2] **3.** The slope-intercept form of a linear equation is $y = mx + b$.

[3.1] **4.** Both coordinates in Quadrant I are positive.

[3.3] **5.** We can find the equation of a line given any two points on that line.

[3.1] **6.** We must have at least three points to correctly graph a straight line.

For Exercises 7–10, complete the rule.

[3.2] **7.** The slope of a line connecting two points is found by the equation ______.

[3.5] **8.** A function must pass the ______ line test.

Chapter Practice Tests A Practice Test follows each set of chapter review exercises. The problem types in the practice tests correlate to the short-answer tests in the *Printed Test Bank.* This is especially comforting for students who have math anxiety or who experience test anxiety (see pages 54, 132, and 217).

Chapter 3 Practice Test

1. Determine the coordinates for each point in the graph to the right.

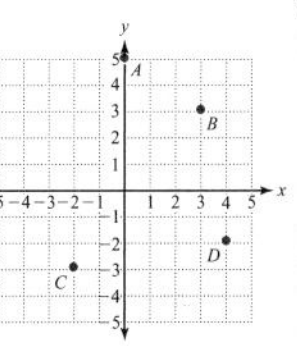

2. State the quadrant in which $\left(3.6, -501\frac{2}{3}\right)$ is located.

3. Determine whether $(-3, 5)$ is a solution for $y = -\frac{1}{4}x + 3$.

For Exercises 4 and 5, determine the slope and the coordinates of the y-intercept, then graph.

4. $y = -\frac{4}{3}x + 5$

5. $x - 2y = -8$

For Exercises 6 and 7, determine the slope of the line through the given points.

6. $(-1, -4), (-2, -4)$

7. $(-3, -9), (4, -1)$

8. Write the equation of a line in slope-intercept form with y-intercept $(0, 5)$ and slope $\frac{2}{7}$.

9. Write the equation of a line in slope-intercept form that passes through the points $(4, 2), (-5, -1)$.

10. Write the equation of a line through the points $(1, 4)$ and $(-3, -1)$ in the form $Ax + By = C$, where A, B, and C are integers and $A > 0$.

11. Are the graphs of $y = \frac{3}{4}x - 2$ and $y = \frac{4}{3}x + 2$ parallel, perpendicular, or neither?

For Exercises 12 and 13, graph the linear inequality.

12. $y \geq 3x - 1$

13. $2x - 3y < 6$

Practice Test 217

Cumulative Reviews Cumulative Review Exercises appear after Chapters 3, 6, 9, and 11. These exercises help students stay current with all the material they have learned and help prepare them for midterm and final exams (see pages 219 and 443).

Chapters 1–3 Cumulative Review Exercises

For Exercises 1–6, answer true or false.

[3.1] 1. The first coordinate of an ordered pair is the x-coordinate.

[2.3] 2. The inequality $x > 0$ can also be expressed as $[0, \infty)$ in interval notation.

[3.3] 3. Parallel lines have the same slope.

[1.3] 4. If the base is a negative number and the exponent is odd, then the product is negative.

[1.2] 5. $a(b + c) = ab + ac$

[3.1] 6. $y = 3x^2 + 2$ is a linear equation.

For Exercises 7–10, fill in the blank.

[2.1] 7. To clear decimal numbers in an equation, we multiply by an appropriate power of ______ as determined by the decimal number with the most decimal places.

[1.3] 8. The radical symbol $\sqrt{\ }$ denotes only the ______ square root.

[1.3] 9. List the order in which we perform operations of arithmetic.

[1.4] 10. Two expressions, like $5x$ and $7x$, that have the same variable raised to the same exponent are called ______ terms.

[1.1] 11. Write a set containing the vowels.

[2.4] 12. Find the intersection and union of the given sets. $A = \{w, e, l, o, v\}$ $B = \{m, a, t, h\}$

[1.3] *For Exercises 13–16, evaluate.*

13. $(-3)^2$

14. $\sqrt[3]{27}$

15. $\sqrt{-25}$

16. $\sqrt[4]{\frac{16}{81}}$

[1.3] *For Exercises 17–18, simplify.*

17. $-|4 + 3| - 8(1 - 5)^2$

18. $5\{2 - [2 - (3 - 1)]\} + 5 \cdot 3$

[1.4] *For Exercises 19 and 20, evaluate the expression using the given values.*

19. $\sqrt{x + y}$; $x = 9, y = 16$

20. $\frac{3x}{x - 5}$; $x = 5$

Cumulative Review Exercises 219

Supplements for *Intermediate Algebra,* Second Edition

Student Supplements

STUDENT'S SOLUTIONS MANUAL

- By Doreen Kelly, *Mesa Community College.*
- Contains complete solutions to the odd-numbered section exercises and solutions to all of the section-level Review Exercises, Chapter Review Exercises, Practice Tests, and Cumulative Review Exercises.
  ISBN: 0-321-37512-2

DIGITAL VIDEO TUTOR

- Complete set of digitized videos on CD-ROM for student use at home or on campus.
- Ideal for distance learning or supplemental instruction.
  ISBN: 0-321-37513-0

MATHXL® TUTORIALS ON CD

- Provides algorithmically generated practice exercises that correlate to the exercises in the textbook.
- Every exercise is accompanied by an example and a guided solution, and selected exercises may also include a video clip to help students visualize concepts.
- The software provides helpful feedback for incorrect answers and can generate printed summaries of students' progress.
  ISBN: 0-321-37515-7

ADDISON-WESLEY MATH TUTOR CENTER

- Staffed by qualified mathematics instructors.
- Provides tutoring on examples and odd-numbered exercises from the textbook through a registration number with a new textbook or purchased separately.
- Accessible via toll-free telephone, toll-free fax, e-mail, or the Internet.
- White Board technology allows tutors and students to actually see problems worked while they "talk" in real time over the Internet during tutoring sessions.
  www.aw-bc.com/tutorcenter

Instructor Supplements

ANNOTATED INSTRUCTOR'S EDITION

- Includes answers to all exercises, including Puzzle Problems and Collaborative Exercises, printed in bright blue near the corresponding problems.
- Useful teaching tips are printed in the margin.
- A ★ icon, found in the AIE only, indicates especially challenging exercises in the exercise sets.
  ISBN: 0-321-35836-8

INSTRUCTOR'S SOLUTIONS MANUAL

- By Doreen Kelly, *Mesa Community College.*
- Contains complete solutions to all even-numbered section exercises, Puzzle Problems, and Collaborative Exercises.
  ISBN: 0-321-37519-X

INSTRUCTOR AND ADJUNCT SUPPORT MANUAL

- Includes resources designed to help both new and adjunct faculty with course preparation and classroom management.
- Offers helpful teaching tips specific to the sections of the text.
  ISBN: 0-321-37520-3

PRINTED TEST BANK

- By Laura Hoye, *Trident Technical College.*
- Contains one diagnostic test per chapter, four free-response tests per chapter, one multiple-choice test per chapter, a mid-chapter check-up for each chapter, one midterm exam, and two final exams.
  ISBN: 0-321-37518-1

VIDEOTAPES

- A series of lectures presented by author-team member Ellyn Gillespie correlated directly to the chapter content of the text. A video symbol at the beginning of each exercise set references the videotape or CD (*see also* Digital Video Tutor).
- Include a pause-the-video feature that encourages students to stop the videotape, work through an example, and resume play to watch the video instructor work through the same example.
  ISBN: 0-321-37514-9

Instructor Supplements

TestGen®

- Enables instructors to build, edit, print, and administer tests.
- Features a computerized bank of questions developed to cover all text objectives.
- Alogrithmically based, allowing instructors to create multiple, but equivalent, verisions of the same questions or test with the click of a button.
- Instructors can also modify test bank questions or add new questions.
- Tests can be printed or administered online.
- Available on a dual-platform Windows/Macintosh CD-ROM.
  ISBN: 0-321-37517-3

MathXL®

MathXL® is a powerful online homework, tutorial, and assessment system that accompanies your Addison-Wesley textbook in mathematics or statistics. With MathXL, instructors can create, edit, and assign online homework and tests using algorithmically generated exercises correlated at the objective level to your textbook. They can also create and assign their own online exercises and import TestGen tests for added flexibility. All student work is tracked in MathXL's online gradebook. Students can take chapter tests in MathXL and receive personalized study plans based on their test results. The study plan diagnoses weaknesses and links students directly to tutorial exercises for the objectives they need to study and retest. Students can also access supplemental video clips directly from selected exercises. MathXL® is available to qualified adopters. For more information visit our Web site at www.mathxl.com, or contact your Addison-Wesley sales representative.

MyMathLab

MyMathLab is a series of text-specific, easily customizable online courses for Addison-Wesley textbooks in mathematics and statistics. MyMathLab is powered by CourseCompass™—Pearson Education's online teaching and learning environment—and by MathXL®—our online homework, tutorial, and assessment system. MyMathLab gives instructors the tools they need to deliver all or a portion of their course online, whether students are in a lab setting or working from home. MyMathLab provides a rich and flexible set of course materials, featuring free-response exercises that are algorithmically generated for unlimited practice and mastery. Students can also use online tools, such as video lectures, animations, and a multimedia textbook, to independently improve their understanding and performance. Instructors can use MyMathLab's homework and test managers to select and assign online exercises correlated directly to the textbook, and they can also create and assign their own online exercises and import TestGen tests for added flexibility. MyMathLab's online gradebook—designed specifically for mathematics and statistics—automatically tracks students' homework and test results and gives the instructor control over how to calculate final grades. Instructors can also add offline (paper-and-pencil) grades to the gradebook. MyMathLab is available to qualified adopters. For more information, visit our Web site at www.mymathlab.com or contact your Addison-Wesley sales representative.

Acknowledgments

Many people gave of themselves in so many ways during the development of this text. Mere words cannot contain the fullness of our gratitude. Though the words of thanks that follow may be few, please know that our gratitude is great.

We would like to thank the following people who gave of their time in reviewing the text. Their thoughtful input was vital to the development of the text.

Marwan Abu-Sawwa, *Florida Community College at Jacksonville*
Daniel Bacon, *Massasoit Community College*
Sandra Belcher, *Midwestern State University*
Nancy Carpenter, *Johnson County Community College*
Sharon Edgmon, *Bakersfield College*
Kathy Garrison, *Clayton College and State University*
Haile Haile, *Minneapolis Community and Technical College*
Nancy R. Johnson, *Manatee Community College*
Tracey L. Johnson, *University of Georgia*
Jeffrey Kroll, *Brazosport College*
Peter Lampe, *University of Wisconsin–Whitewater*
Sandra Lofstock, *California Lutheran University*
Stephanie Logan, *Lower Columbia College*
John Long, *Jefferson Community College*
Janis Orinson, *Central Piedmont Community College*
Merrel Pepper, *Southeast Technical Institute*
Patrick Riley, *Hopkinsville Community College*
Reynaldo Rivera, *Estrella Mountain Community College*
Rebecca Schantz, *Prairie State College*
Kay Stroope, *Phillips County Community College*
John Thoo, *Yuba College*
Bettie Truitt, *Black Hawk College*
Judith A. Wells, *University of Southern Indiana*
Joe Westfall, *Carl Albert State College*
Peter Willett, *Diablo Valley College*
Tom Williams, *Rowan-Cabarrus Community College*

We would like to extend a heartfelt thank-you to everyone at Addison-Wesley for giving so much to this project. We would like to offer special thanks to Jennifer Crum and Greg Tobin, who believed in us and gave us the opportunity; to Elizabeth Bernardi, Emily Ragsdale, Lauren Morse, and Kari Heen, for keeping us on track; and also to Jay Jenkins, Tracy Rabinowitz, and Alexandra Waibel for the encouragement and working so hard to get us "out there."

A very special thank-you to Dennis Schaefer, who created the beautiful, student-friendly text design and cover; to Ron Hampton, whose keen eyes and editorial sense were invaluable during production; and to Gordon Laws, Sam Blake, and all of the fabulous people at Pre-Press Company, Inc. for working so hard to put together the finished pages.

To Sharon Smith, Ruth Berry, Mary Ann Perry, and all the people involved in developing the media supplements package, we are so grateful for all that you do. A special thank-you to Laura Hoye, who created the excellent *Printed Test Bank,* and to Doreen Kelly for her work on the solutions manuals. Thank you to Sharon O'Donnell, Cheryl Cantwell, Elizabeth Morrison, and Vince Koehler for their wonderful job of accuracy checking the manuscript and page proofs. A big thank-you goes to Lisa Sims, Cheryl Cantwell, and Laura Wheel for their help keeping the application problems fresh and up to date.

Finally, we'd like to thank our families for their support and encouragement during the process of developing and revising this text.

Tom Carson

Ellyn Gillespie Stewart

Bill Jordan

To the Student

Why do I have to take this course?

Often this is one of the first questions students ask when they find out they must take an algebra course, especially when they believe that they will never use the math again. You may think that you will not use algebra directly in daily life, and you may assume that you can get by knowing enough arithmetic to balance a checkbook. So, what is the real point of education? Why don't colleges just train students for the jobs they want? The purpose of education is not just job training but also exercise—mental exercise. An analogy that illustrates this quite well is the physical training of athletes.

During the off-season, athletes usually develop an exercise routine that may involve weight lifting, running, swimming, aerobics, or maybe even dance lessons. Athletes often seek out a professional trainer to push them further than they might push themselves. The trainer's job is not to teach an athlete better technique in his or her sport, but to develop the athlete's raw material—to work the body for more strength, stamina, balance, etc. Educators are like physical trainers, and going to college is like going to the gym. An educator's job is to push students mentally and work the "muscle" of the mind. A college program is designed to develop the raw material of the intellect so the student can be competitive in the job market. After the athlete completes the off-season exercise program, he or she returns to the coach and receives specific technique training. Similarly, when students complete their college education and begin a job, they receive specific training to do that job. If the trainer or teacher has done a good job with hardworking clients, the coaching or job training should be absorbed easily.

Taking this analogy a step further, a good physical trainer finds the athlete's weaknesses and designs exercises that the athlete has never performed before, and then pushes him or her accordingly. Teachers do the same thing—their assignments are difficult in order to work the mind effectively. If you feel "brain-strained" as you go through your courses, that's a good sign that you are making progress, and you should keep up the effort.

The following study system is designed to help you in your academic workouts. As teachers, we find that most students who struggle with mathematics have never really *studied* math. A student may think, "Paying attention in class is all I need to do." However, when you watch a teacher do math, keep in mind that you are watching a pro. Going back to the sports analogy, you can't expect to shoot a score of 68 in golf by watching Tiger Woods. You have to practice golf yourself in order to learn and improve. The study system outlined in the following pages will help you get organized and make efficient use of your time so that you can maximize the benefits of your course work.

What do I need to do to succeed?

We believe there are four prerequisites one must have or acquire in order to succeed in college:

1. **Positive Attitude**
2. **Commitment**
3. **Discipline**
4. **Time**

A **Positive Attitude** is most important because commitment and discipline flow naturally from it. Consider Thomas Edison, inventor of the lightbulb. He tried more than 2000 different combinations of materials for the filament before he found the successful combination. When asked by a reporter about all his failed attempts, Edison replied, "I didn't fail once, I invented the lightbulb. It was just a 2000-step process." Recognize that learning can be uncomfortable and difficult, and mistakes are part of the process. So, embrace the learning process with its discomforts and difficulties, and you'll see how easy it is to be committed and disciplined.

Commitment means giving everything you've got with no turning back. Consider Edison again. Imagine the doubts and frustrations he must have felt trying material after material for the filament of his lightbulb without success. Yet he forged ahead. In Edison's own words, "Our greatest weakness lies in giving up. The most certain way to succeed is always to try just one more time."

Discipline means doing things you should be doing even when you don't want to. According to author W. K. Hope, "Self-discipline is when your conscience tells you to do something and you don't talk back." Staying disciplined can be difficult given all the distractions in our society. The best way to develop discipline is to create a schedule and stick to it.

Make sure you have enough **Time** to study properly, and make sure that you manage that time wisely. Too often, students try to fit school into an already full schedule. Take a moment to complete the exercise that follows under "How do I do it all?" to make sure you haven't overcommitted yourself. Once you have a sense of how much time school requires, read on about the study system that will help you maximize the benefits of your study time.

How do I do it all?

Now that we know a little about what it takes to be successful, let's make sure that you have enough time for school. In general, humans have a maximum of 60 hours of productivity per week. Therefore, as a guide, let's set the maximum number of work hours, which means time spent at your job(s) and at school combined, at 60 hours per week. Use the following exercise to determine the time you commit to your job and to school.

Exercise: Calculate the time that you spend at your job and at school.

1. Calculate the total hours you work in one week.
2. Calculate the number of hours you are in class each week.
3. Estimate the number of hours you should expect to spend outside of class studying. *A general rule is to double the number of hours spent in class.*
4. Add your work hours, in-class hours, and estimated out-of-class hours to get your total time commitment.
5. Evaluate the results. *See below.*

Evaluating the Results

a. If your total is greater than 60 hours, you will probably find yourself feeling overwhelmed. This feeling may not occur at first, but doing that much for an extended period of time will eventually catch up with you, and something may suffer. It is in your best interest to cut back on work or school until you reduce your time commitment to under 60 hours per week.
b. If your total is under 60 hours, good. Be sure you consider other elements in your life, such as your family's needs, health problems, commuting, or anything that could make demands on your time. Make sure that you have enough time for everything you put in your life. If you do not have enough time for everything, consider what can be cut back. It is important to note that it is far better to pass fewer classes than to fail many.

How do I make the best use of my time? How should I study?

We've seen many students who had been making D's and F's in mathematics transform their grades to A's and B's by using the study system that follows.

The Study System

Your Notebook

1. Get a loose-leaf binder so that you can put papers in and take them out without ripping any pages.
2. Organize the notebook into four parts:
   a. Class notes
   b. Homework
   c. Study sheets (a single piece of paper for each chapter onto which you will transfer procedures from your notes)
   d. Practice tests

In Class

Involve your mind completely.

1. **Take good notes.** Use three different colors. Most students like using red, blue, and black (pencil).
   - Use the red pen to write *definitions.* Also, use this color to mark problems or items that the instructor indicates will be covered on a test.
   - Use the blue pen to write procedures and rules.
   - Use the pencil to write problems and explanations.

When taking notes, don't just write the solutions to the problems that the instructor works out, but write the explanations as well. To the side of the problem, make notes about each step so that you remember the significance of the steps. Pay attention to examples or issues the instructor emphasizes: they will usually appear on a test, so make an effort to include them in your notes. Include common errors that the instructor points out or any words of caution. If you find it is difficult to write and pay attention at the same time, ask your instructor if you can record the lectures with a tape recorder. If your instructor follows the text closely, when he or she points out definitions or procedures in the text, highlight them or write a page reference in your notes. You can then write these referenced items in their proper place in your notes after class.

2. **Answer the instructor's questions.** This does not mean you have to answer every question verbally, but you should think through every question and answer in your mind, write an answer in your notes, or answer out loud.
3. **Ask questions.** You may find it uncomfortable to ask questions in front of other people, but keep in mind that if you have a question, then it is very likely that someone else has the same question. If you still don't feel like asking in class, then be sure to ask as soon as class is over. The main thing is to get that question answered as soon as possible because in mathematics, one misconception can grow and cause confusion in the future.

After Class

Prepare for the next class meeting as if you were going to have a test on everything covered so far. To make the most of your time, set aside a specific time that is reserved for math. Since there are often many distractions at home, study math while on campus in a quiet place such as the library or tutorial lab. Staying on campus also allows you to visit your instructor or tutorial services if you have a question that you cannot resolve. Here is a systematic approach to organizing your math study time outside of class:

1. As soon as possible, go over your notes. Clarify any sentences that weren't quite complete. Fill in any page-referenced material.
2. Read through the relevant section(s) in the text again, and make sure you understand all the examples.
3. Transfer each new procedure or rule to your study sheet for that chapter. You might also write down important terms and their definitions. Make headings for each objective in the section(s) you covered that day. Write the procedures and definitions in your own words.
4. Study the examples worked in class. Transfer each example (without the solution) to the practice test section of your notebook, leaving room to work it out later.
5. Use your study sheet to do the assigned practice problems. As soon as you finish each problem, check your answer in the back of the book or in the *Student's Solutions Manual.* If you did not get it correct, then immediately revisit the problem to determine your error (see the box on troubleshooting). If you are asked to do even-numbered problems, then work odd-numbered problems that mirror the even problems. This way you can check your answers for the odd-numbered problems and then work the even-numbered problems with confidence.
6. After completing the homework, prepare a quiz for yourself. Select one of each type of homework problem. Don't just pick the easy ones! Set the quiz aside for later.
7. After making the quiz, study your study sheet. To test your understanding, write the rules and procedures in your own words. Do not focus on memorizing the wording in the textbook.
8. Now it is time to begin preparing for the next class meeting. Read the next section(s) to be covered. Don't worry if you do not understand everything. The idea is to get some feeling for the topics to be discussed so that the class discussion will actually be the second time you encounter the material, not the first. While reading, you might mark points that you find difficult so that if the instructor does not clear them up, you can ask about them. Also, attempt to work through the examples. The idea is for you to do as much as possible on your own before class so that the in-class discussion merely ties together loose ends and solidifies the material.
9. After you have finished preparing for the next day, go back and take the quiz that you made. If you get all the answers correct, then you have mastered the material. If you have difficulty, return to your study sheet and repeat the exercise of writing explanations for each objective.

Troubleshooting: For the problems that you do not get correct, first look for simple arithmetic errors. If you find no arithmetic errors, then make sure you followed the procedure or rules correctly. If you followed the or rules correctly, then you have likely interpreted something incorrectly, either with the problem or the rules. Read the instructions again carefully and try to find similar examples in your notes or in the book. If you still can't find the mistake, go on to something else for a while. Often after taking a fresh look you will see the mistake right away. If all these tips fail to resolve the problem, then mark it as a question for the next class meeting.

How do I ace the test?

Preparing for a Test

If you have followed all of the preceding suggestions, then preparing for a test should be quite easy.

1. **Read.** In one sitting, read through all of your notes on the material to be tested. In the same sitting, read through the book, observing what the instructor has highlighted in class. To guide your studies, look at any information or documents provided by your instructor that address what will be on the test. The examples given by the instructor will usually reflect the test content.
2. **Study.** Compare your study sheet to the summary in the book at the end of the chapter. Use both to guide you in your preparation, but keep in mind that the sheet you made from your notes reflects what the instructor has emphasized. Make sure you understand everything on your study sheet. Write explanations of the objectives until you eliminate all hesitation about how to approach an objective. The rules and procedures should become second nature.
3. **Practice.** Create a game plan for the test by writing the rule, definition, or procedure that corresponds to each problem on your practice tests. (Remember, one practice test is in your book and the other you made from your notes.) Next, work through the practice tests without referring to your study sheet or game plan.
4. **Evaluate.** Once you have completed the practice tests, check them. The answers to the practice tests in the book are in the answer section in the back of the book. Check the practice test that you made using your notes.
5. **Repeat.** Keep repeating steps 2, 3, and 4 until you get the right answer for every problem on the practice tests.

Taking a Test

1. When the test hits your desk, don't look at it. Instead, do a memory dump. On paper, dump everything you think you might forget. Write out rules, procedures, notes to yourself, things to watch out for, special instructions from the instructor, and so on. This will help you relax while taking the test.
2. If you get to a problem that you cannot figure out, skip it and move on to another problem. First do all the problems you are certain of and then return to the ones that are more difficult.
3. Use all the time given. If you finish early, check to make sure you have answered every problem. Even if you cannot figure out a problem, at least guess. Use any remaining time to check as many problems as possible by doing them over on separate paper.

If You Are Not Getting Good Results

Evaluate the situation. What are you doing or not doing in the course? Are you doing all the homework and taking the time to prepare as suggested? Sometimes people misjudge how well they have prepared. Just like an athlete, to excel, you will need to prepare beyond the minimum requirements.

Here are some suggestions:

1. **Go to your instructor.** Ask your instructor for help to evaluate what is wrong. Make use of your instructor's office hours, because this is your opportunity for individual attention.
2. **Get a tutor.** If your school has a tutorial service, and most do, do your homework there so you can get immediate help when you have a question.

3. **Use Addison-Wesley's support materials.** Use the support materials that are available with your text, which include a *Student's Solutions Manual,* the Addison-Wesley Math Tutor Center, videotapes (available on CD-ROM as well), and tutorial software available on CD and online. Full descriptions of these supplements are provided on page xii of this book.
4. **Join a study group.** Meet regularly with a few people from class and go over material together. Quiz each other and answer questions. Meet with the group only after you have done your own preparation so you can then compare notes and discuss any issues that came up in your own work. If you have to miss class, ask the study group for the assignments and notes.

We hope you find this study plan helpful. Be sure to take the Learning Styles Inventory that follows to help determine your primary learning style. Good luck!

Learning Styles Inventory

What is your personal learning style?

A learning style is the way in which a person processes new information. Knowing your learning style can help you make choices in the way you study and focus on new material. Below are fifteen statements that will help you assess your learning style. After reading each statement, rate your response to the statement using the scale below. There are no right or wrong answers.

3 = Often applies **2** = Sometimes applies **1** = Never or almost never applies

____ **1.** I remember information better if I write it down or draw a picture of it.

____ **2.** I remember things better when I hear them instead of just reading or seeing them.

____ **3.** When I receive something that has to be assembled, I just start doing it. I don't read the directions.

____ **4.** If I am taking a test, I can "visualize" the page of text or lecture notes where the answer is located.

____ **5.** I would rather the professor explain a graph, chart, or diagram to me instead of just showing it to me.

____ **6.** When learning new things, I want to do it rather than hear about it.

____ **7.** I would rather the instructor write the information on the board or overhead instead of just lecturing.

____ **8.** I would rather listen to a book on tape than read it.

____ **9.** I enjoy making things, putting things together, and working with my hands.

____ **10.** I am able to conceptualize quickly and visualize information.

____ **11.** I learn best by hearing words.

____ **12.** I have been called hyperactive by my parents, spouse, partner, or professor.

____ **13.** I have no trouble reading maps, charts, or diagrams.

____ **14.** I can usually pick up on small sounds like bells, crickets, frogs, or distant sounds like train whistles.

____ **15.** I use my hands and gesture a lot when I speak to others.

Write your score for each statement beside the appropriate statement number below.

Then add the scores in each column to get a total score for that column.

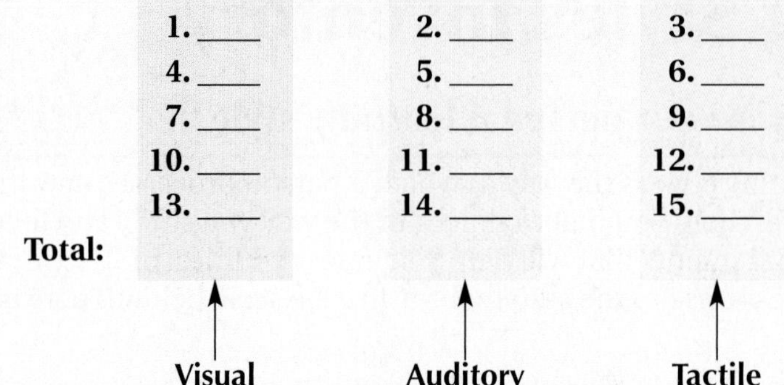

	Visual	Auditory	Tactile
	1. ____	2. ____	3. ____
	4. ____	5. ____	6. ____
	7. ____	8. ____	9. ____
	10. ____	11. ____	12. ____
	13. ____	14. ____	15. ____
Total:			

The largest total of the three columns indicates your dominant learning style.

Visual learners learn best by seeing. If this is your dominant learning style, you should focus on learning strategies that involve seeing. The color coding in the study system (see page xix) will be especially important. The same color coding is used in the text. Draw lots of diagrams, arrows, and pictures in your notes to help you see what is happening. Reading your notes, study sheets, and text repeatedly will be an important strategy.

Auditory learners learn best by hearing. If this is your dominant learning style, you should use learning strategies that involve hearing. After getting permission from your instructor, bring a tape recorder to class to record the discussion. When you study your notes, play back the tape. Also, when you learn rules, say the rule over and over. As you work problems, say the rule before you do the problem. You may also find the video tapes beneficial in that you can hear explanations of problems taken from the text.

Tactile (also known as kinesthetic) learners learn best by touching or doing. If this is your dominant learning style, you should use learning strategies that involve doing. Doing lots of practice problems will be important. Make use of the Your Turn exercises in the text. These are designed to give you an opportunity to do problems like the examples as soon as the topic is discussed. Writing out your study sheets and doing your practice tests repeatedly will be important strategies for you.

Note that the study system developed in this text is for all learners. Your learning style will help you decide what aspects and strategies in the study system to focus on, but being predominantly an auditory learner does not mean that you shouldn't read the textbook, do lots of practice problems, or use the color-coding system in your notes. Auditory learners can benefit from seeing and doing, and tactile learners can benefit from seeing and hearing. In other words, do not use your dominant learning style as a reason for not doing things that are beneficial to the learning process. Also, remember that the Learning Strategy boxes presented throughout the text provide tips to help you use your personal learning style to your advantage.

This Learning Styles Inventory is adapted from *Cornerstone: Building on Your Best,* Third Edition, by Montgomery/Moody/Sherfield © 2002. Reprinted by permission of Prentice-Hall, Inc., Upper Saddle River, NJ.

CHAPTER 1

Real Numbers and Expressions

"The artist in me cries out for design."

—Robert Frost, U.S. poet

"Always design a thing by considering its next larger context—a chair in a room, a room in a house, a house in an environment, an environment in a city plan."

—Eliel Saarinen, Finnish-American architect

"It affords me no satisfaction to commence to spring an arch before I have got a solid foundation."

—Henry David Thoreau, U.S. author, philosopher, naturalist

Algebra is like a building. It has a foundation and various topics build upon that foundation and connect to one another to form a beautiful structure. In this chapter, we review the foundation, which is arithmetic. It is important to note that our review is not meant to be a complete exploration. Rather, we are assuming that the foundation has already been built, and we are merely doing a brief tour so that we can quickly move on to the higher structures.

1.1 Sets and the Structure of Algebra

OBJECTIVES

1. Understand the structure of algebra.
2. Classify number sets.
3. Determine the absolute value of a number.
4. Compare numbers.

OBJECTIVE 1. Understand the structure of algebra. In algebra, our basic components are **variables** and **constants**, which we use to build **expressions**, **equations**, and **inequalities**.

DEFINITIONS

Variable: A symbol varying in value.
Constant: A symbol that does not vary in value.
Expression: A collection of constants, variables, and arithmetic symbols.
Equation: Two expressions set equal to each other.
Inequality: Two expressions separated by $\neq$, $<$, $>$, $\leq$, or $\geq$.

Learning Strategy

Developing a good study system and understanding how you best learn is essential to academic success. If you haven't already done so, read the study system in the To the Student section at the beginning of the text. Also, take a moment to complete the Learning Styles Inventory found on page xix to discover your learning style. In these learning strategy boxes, we will offer tips on how to connect the study system and your learning style to be successful in the course.

Variables are usually letters of the alphabet, like x or y. Constants are symbols for specific values like 1, 2, $\frac{3}{4}$, 6.74, or π. Following are some examples of expressions:

$$2 + 6 \qquad 4x - 5 \qquad \frac{1}{3}\pi r^2 h$$

Note: *Expressions do not contain an equal sign or inequality symbols.*

Examples of equations are

$$2 + 6 = 8 \qquad 4x - 5 = 12 \qquad V = \frac{1}{3}\pi r^2 h$$

Connection Think of expressions as phrases and equations as complete sentences. The expression $2 + 6$ is read "two plus six," which is not a complete sentence. The equation $2 + 6 = 8$ can be read as "Two plus six equals eight." Notice the equal sign translates to the verb "equals," so we have a complete sentence.

The following table contains examples of inequalities and their verbal translations.

Inequality Symbols and Their Translations

Symbolic Form	Translation
$8 \neq 3$	Eight is not equal to three.
$5 < 7$	Five is less than seven.
$7 > 5$	Seven is greater than five.
$x \leq 3$	x is less than or equal to three.
$y \geq 2$	y is greater than or equal to two.

The following Algebra Pyramid illustrates how variables, constants, expressions, equations, and inequalities relate to one another. Constants and variables form the foundation on which we build expressions, which in turn form equations and inequalities.

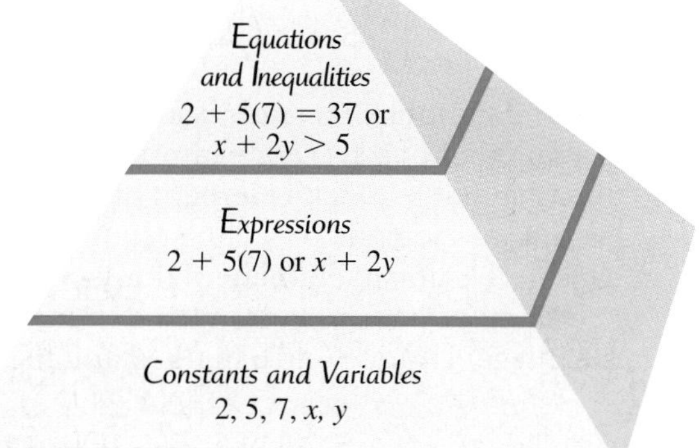

Note: *During this course, we will move back and forth between expressions, equations, and inequalities. When we change topics, we will use this Algebra Pyramid as a visual aid so that it's clear what we are working on.*

OBJECTIVE 2. Classify number sets. In mathematics, we often group numbers, variables, or other objects in **sets**.

DEFINITION ***Set:*** A collection of objects.

Braces are used to indicate a set. For example, the set containing the numbers 1, 2, 3, and 4 can be written {1, 2, 3, 4}. The numbers 1, 2, 3, and 4 are called the *members* or *elements* of this set.

EXAMPLE 1 Write the set containing the last five letters of the alphabet.

Answer {V, W, X, Y, Z}

YOUR TURN Write the set containing the first four months of the year.

Sets can contain a finite number of elements, an infinite number of elements, or no elements at all. A set with no elements is an *empty set*, which we write using empty braces { } or the symbol $\varnothing$. The set in Example 1 is a finite set with five elements. Some examples of infinite sets are the **set of natural numbers**, the **set of whole numbers**, and the **set of integers**.

Note: *The three dots are known as* ellipses *and indicate that the numbers continue forever in the same pattern.*

DEFINITIONS
The set of natural numbers: {1, 2, 3, . . . }
The set of whole numbers: {0, 1, 2, 3, . . . }
The set of integers: {. . . , −3, −2, −1, 0, 1, 2, 3, . . . }

ANSWER

{January, February, March, April}

So far, we have written sets in *roster form*, which means we have listed each element or, in the case of infinite sets, the pattern of the set. We can also write sets in *set-builder notation*, where we describe what the set contains. Here, we write the whole numbers from 0 to 4 using both notations.

Note: *The set-builder form is read "the set of all x such that x is a whole number and x is less than or equal to 4."*

Roster form: $\{0, 1, 2, 3, 4\}$

Set builder: $\{x \mid x \text{ is a whole number and } x \leq 4\}$

The symbol $\in$ is read "is an element of" and is used to indicate that an object is a member of a set. For example, $-3 \in \{x \mid x \text{ is an integer}\}$ is a true statement indicating that -3 "is an element of" the set of integers. The symbol $\notin$ is translated "is not an element of." The statement $-3 \notin \{x \mid x \text{ is a whole number}\}$ is true because -3 is not an element of the set of whole numbers.

Number lines are often used in mathematics. The following number line has the integers from -6 to 6 marked and the integer 2 plotted:

Of Interest

Symbols are sometimes used for the number sets. For example, the set of integers is often written as ***Z***, and the set of real numbers is often written as $\mathbb{R}$ or ***R***.

Although we can view only a portion of a number line, the arrows at the ends indicate that the line and the numbers on it continue forever in both directions. If we were to travel along the number line forever in both directions we would encounter every number in the set of *real numbers*.

Every integer is in the set of real numbers. Consequently, we say that the set of integers is a **subset** of the set of real numbers.

DEFINITION ***Subset:*** If every element of a set B is an element of a set A, then B is a subset of A.

The symbol $\subseteq$ is used to indicate a subset. For example, to indicate that the set of integers is a subset of the set of real numbers, we write $\boldsymbol{Z} \subseteq \boldsymbol{R}$.

The set of real numbers also contains numbers like fractions that are not integers. For example, the following number line has the numbers $\frac{5}{4}$ and -0.3 plotted, which are not integers:

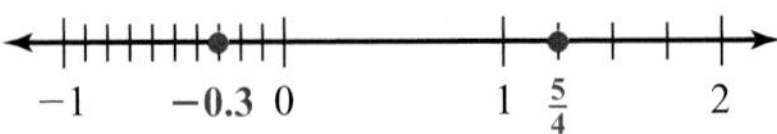

Numbers like $\frac{5}{4}$ and -0.3, along with the integers, are in a subset of the real numbers called the set of **rational numbers**.

DEFINITION ***Rational number:*** Any real number that can be expressed in the form $\frac{a}{b}$, where a and b are integers and $b \neq 0$.

Note: *In the definition for a rational number, the notation $b \neq 0$ is important because if the denominator b were to equal 0, then the fraction would be undefined. The reason it is undefined will be explained later.*

Note: *We will learn how to write numbers with repeating decimal digits as fractions in Appendix B. For now, you just need to be able to recognize that these types of decimal numbers are rational numbers.*

The number $\frac{5}{4}$ is a rational number because 5 and 4 are integers. Similarly, -0.3 is rational because it can be written as $-\frac{3}{10}$. All the integers are rational numbers because they can be written in the form $\frac{a}{b}$, where b will be 1. For example, $7 = \frac{7}{1}$. Some less obvious rational numbers are decimal numbers like $0.\overline{6}$. The repeat bar in $0.\overline{6}$ means the digit 6 repeats forever, as in 0.6666 . . . . It is a rational number because $0.\overline{6}$ can be written as $\frac{2}{3}$.

Some real numbers are not rational numbers. For example, the exact value of the real number π cannot be expressed as a ratio of integers. Real numbers like π are called **irrational numbers**.

DEFINITION ***Irrational number:*** Any real number that is not rational.

Some other irrational numbers are $\sqrt{2}$ and $\sqrt{3}$. Square roots will be explained in more detail in Section 1.3. Because an irrational number cannot be written as a ratio of integers, if a calculation involves an irrational number, we must leave it in symbolic form or use a rational number approximation. We can approximate π with rational numbers like 3.14 or $\frac{22}{7}$.

The following diagram illustrates how the set of real numbers and all its subsets are organized. Note the letters used to indicate each set.

Another set of numbers that we often use in mathematics is the set of **prime numbers**, which is a subset of the natural numbers.

DEFINITION ***Prime number:*** A natural number with exactly two different factors, 1 and the number itself.

There are an infinite number of prime numbers. The first ten prime numbers are 2, 3, 5, 7, 11, 13, 17, 19, 23, 29.

EXAMPLE 2 Determine whether the statement is true or false.

a. $-5 \in \{x \mid x \text{ is an integer}\}$

Answer True, because -5 is a member of the set of integers.

b. $\frac{2}{3} \in \boldsymbol{W}$

Answer False, because $\frac{2}{3}$ is not a whole number.

c. November $\notin \{n \mid n \text{ is a month}\}$

Answer False, because November is a member of the set containing the months.

d. Given $A = \{1, 2, 3, 4, 5\}$ and $B = \{2, 3, 5\}$, then $B \subseteq A$.

Answer True, B is a subset of A because every element of B is in A.

e. $0.4 \in \boldsymbol{Q}$

Answer True, because 0.4 is the decimal equivalent of $\frac{2}{5}$.

YOUR TURN Determine whether the statement is true or false.

a. Friday $\in \{n \mid n \text{ is a day of the week}\}$

b. $0.2 \notin \boldsymbol{I}$

c. Given $A = \{1, 2, 3, 4, 5\}$ and $B = \{2, 3, 5\}$, then $A \subseteq B$.

OBJECTIVE 3. Determine the absolute value of a number. In mathematics, we often discuss the size, or **absolute value**, of a number.

DEFINITION ***Absolute value:*** A number's distance from zero on a number line.

For example, the absolute value of 5, written $|5|$, is 5 because it is 5 units from 0 on a number line. Similarly, $|-5| = 5$ because -5 is also 5 units from zero.

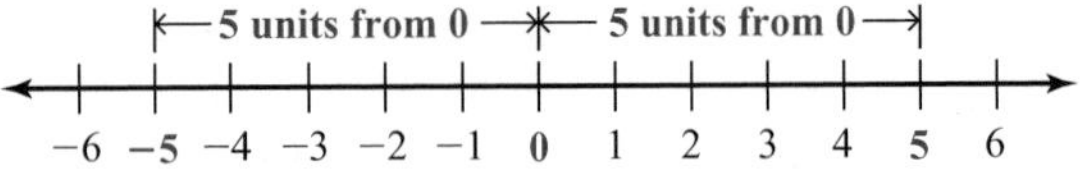

Consider $|0|$. Since 0 is zero units from itself, $|0| = 0$.

RULE **Absolute Value**

The absolute value of every real number is either positive or 0 according to the following rule:

$$|n| = \begin{cases} n \text{ when } n \geq 0 \\ -n \text{ when } n < 0 \end{cases}$$

ANSWERS

a. true
b. true
c. false

EXAMPLE 3 Determine the absolute value.

a. $|-2.3|$

Answer $|-2.3| = 2.3$

b. $\left|\frac{1}{4}\right|$

Answer $\left|\frac{1}{4}\right| = \frac{1}{4}$

YOUR TURN Determine the absolute value.

a. $\left|-4\frac{1}{2}\right|$

b. $|9.8|$

OBJECTIVE 4. Compare numbers. We can also use a number line to determine which of two numbers is greater. Because numbers increase from left to right on a number line, the number farthest to the right will be the greater of two numbers.

RULE **Comparing Numbers**

For any two real numbers *a* and *b*, *a* is greater than *b* if *a* is to the right of *b* on a number line.

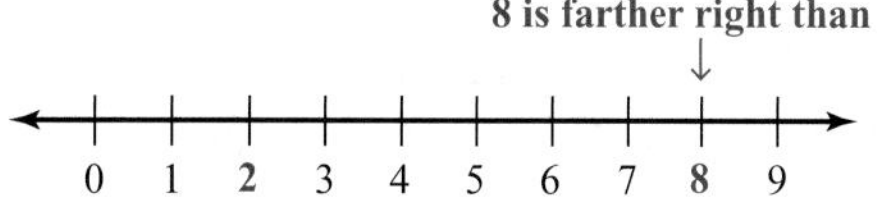

Because the number 8 is farther to the right on a number line than the number 2, we say that 8 is greater than 2, or, in symbols, $8 > 2$.

EXAMPLE 4 Use $=$, $<$, or $>$ to write a true statement.

a. $6 \ \square \ -8$

Answer $6 > -8$ because 6 is farther right on a number line than -8.

b. $-3.2 \ \square \ -3.1$

Answer As we see on the following number line, $-3.2 < -3.1$ because -3.1 is farther right than -3.2.

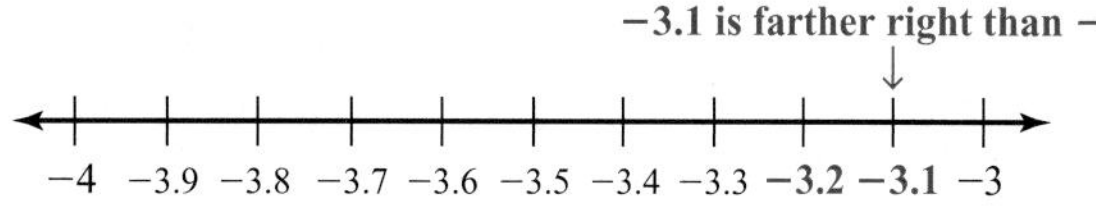

c. $\left|-2\frac{3}{4}\right| \ \square \ 2.75$

Answer $\left|-2\frac{3}{4}\right| = 2.75$ because the absolute value of $-2\frac{3}{4}$ is $2\frac{3}{4}$, which is 2.75 when written as a decimal number.

ANSWERS

a. $4\frac{1}{2}$

b. 9.8

ANSWERS

a. $-14 > -25$

b. $-2.4 < 0$

c. $3\frac{5}{6} > 3\frac{1}{4}$

d. $\left|-9\frac{1}{2}\right| = 9.5$

YOUR TURN Use =, <, or > to write a true statement.

a. -14 ▢ <-25 b. -2.4 ▢ 0 c. $3\frac{5}{6}$ ▢ $3\frac{1}{4}$ d. $\left|-9\frac{1}{2}\right|$ ▢ 9.5

1.1 Exercises

For Extra Help MyMathLab  Videotape/DVT  InterAct Math 

Math Tutor Center  Math XL.com

1. Define *set* in your own words.

2. What is the difference between a finite set and an infinite set?

3. What is a subset of a set?

4. We can indicate an empty set using { } or ∅, but not {∅}. Why is {∅} incorrect?

5. Explain the difference between a rational number and an irrational number.

6. Explain why it is incorrect to say that absolute value is always positive.

For Exercises 7–20, write a set representing each.

7. The days in a weekend

8. The vowels

9. The months that begin with J

10. The months containing 31 days

11. The states in the United States whose names begin with the word New

12. The states in the United States whose names begin with *W*

13. The whole numbers less than 5

14. The prime numbers less than 10

15. The natural-number multiples of 3

16. The natural-number multiples of 5

17. The integers greater than $-\frac{5}{4}$ and less than $\frac{4}{3}$

18. The even prime numbers

19. The integers greater than $-\frac{3}{4}$ and less than $-\frac{1}{4}$

20. The integers greater than 2 and less than -3

For Exercises 21–28, write the set in set-builder notation.

21. $\{2, 4, 6, \ldots\}$

22. $\{1, 2, 3, \ldots\}$

23. $\{a, b, c, \ldots, x, y, z\}$

24. $\{a, e, i, o, u\}$

25. {Monday, Tuesday, Wednesday, Thursday, Friday, Saturday, Sunday}

26. {January, February, March, April, May, June, July, August, September, October, November, December}

27. $\{5, 10, 15, \ldots\}$

28. $\{1, 2, 3, 4, 5, 6, 7, 8, 9, 10\}$

***For Exercises 29–54, answer* true *or* false.**

29. $n \in \{a, e, i, o, u\}$

30. $m \in \{l, m, n, o, p\}$

31. "go" $\in \{n \mid n$ is a verb$\}$

32. "leave" $\in \{n \mid n$ is a noun$\}$

33. $4.5 \in \{x \mid x$ is a rational number$\}$

34. $\pi \in \{x \mid x$ is an irrational number$\}$

35. "Garfield" $\notin \{n \mid n$ is the last name of a U.S. president$\}$

36. "Florida" $\notin \{n \mid n$ is the name of a state in the United States$\}$

37. $-0.6 \notin \boldsymbol{R}$

38. $\sqrt{3} \notin \boldsymbol{Q}$

39. $\{a, e, i, o, u\} \subseteq \{n \mid n$ is a letter of the English alphabet$\}$

40. {red, blue, yellow} $\subseteq \{n \mid n$ is a color$\}$

41. $\{1, 3, 5, 7, \ldots\} \subseteq \boldsymbol{Z}$

42. $\{2, 4, 6, 8, \ldots\} \subseteq \boldsymbol{Q}$

43. $\boldsymbol{Q} \subseteq \boldsymbol{Z}$

44. $\boldsymbol{Z} \subseteq \boldsymbol{Q}$

45. $\boldsymbol{I} \subseteq \boldsymbol{R}$

46. $\boldsymbol{N} \subseteq \boldsymbol{Q}$

47. A number exists that is both rational and irrational.

48. Some integers are irrational numbers.

49. The only difference between the set of whole numbers and the set of natural numbers is that the set of whole numbers contains 0.

50. The set of integers contains every negative number.

51. Every rational number can be written as a fraction.

52. A number expressed as a repeating decimal cannot be written in the form $\frac{a}{b}$, where a and b are integers and b is not equal to zero.

53. The set of rational numbers contains more numbers than the set of irrational numbers.

54. All real numbers are either positive or negative.

For Exercises 55–62, graph each number on a number line.

55. $\frac{2}{3}$

56. $-\frac{3}{4}$

57. $1\frac{3}{8}$

58. $-8\frac{1}{5}$

59. -2.1

60. 5.8

61. 3.62

62. -1.22

For Exercises 63–66, round the given decimal approximation to the nearest tenth, then graph the rounded value on a number line.

63. $\sqrt{13} \approx 3.6056$

64. $\sqrt{60} \approx 7.7460$

65. $-\sqrt{12} \approx -3.4641$

66. $-\sqrt{42} \approx -6.4807$

For Exercises 67–74, simplify each number.

67. $|2.6|$ **68.** $|-2.8|$ **69.** $\left|-1\frac{2}{5}\right|$ **70.** $\left|1\frac{1}{4}\right|$ **71.** $|-1|$ **72.** $|0|$ **73.** $|-8.75|$ **74.** $|18|$

For Exercises 75–86, use =, <, or > to write a true statement.

75. $-6 \;\square\; -8$

76. $-19 \;\square\; -7$

77. $0 \;\square\; 1.8$

78. $-1.7 \;\square\; -1.6$

79. $3\frac{4}{5} \;\square\; 3\frac{3}{4}$

80. $2\frac{3}{5} \;\square\; 2\frac{1}{4}$

81. $|-3| \;\square\; |3|$

82. $6.2 \;\square\; |-6.2|$

83. $6.7 \;\square\; |6.7|$

84. $|10.4| \;\square\; 10.4$

85. $\left|-\frac{2}{3}\right| \;\square\; \left|-\frac{4}{3}\right|$

86. $\left|-\frac{7}{4}\right| \;\square\; \left|-\frac{5}{3}\right|$

For Exercises 87–90, list the given numbers in order from least to greatest.

87. $0.4, -0.6, 0, 3\frac{1}{4}, |-0.02|, -0.44, \left|1\frac{2}{3}\right|$

88. $-2.56, 5.4, |8.3|, \left|-7\frac{1}{2}\right|, -4.7$

89. $2.9, 1, \left|-2\frac{3}{4}\right|, -12.6, |-1.3|, -9.6$

90. $-1, \pi, |-0.05|, -1.3, \left|4\frac{2}{3}\right|, 0.\overline{4}$

For Exercises 91–94, use the following graph.

91. Write a set containing the three years from 1994 to 2003 that had the greatest travel volume.

92. Write a set containing the three years from 1994 to 2003 that had the least travel volume.

93. Write a set containing the years from 1994 to 2003 that had a travel volume less than 850 million person-trips.

94. Write a set containing the years from 1994 to 2003 that had a travel volume greater than 930 million person-trips.

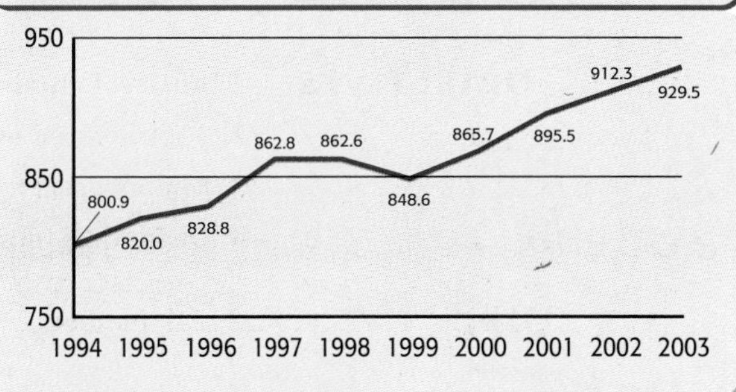

Source: Travel Industry Assn. of America, Travel Scope

For Exercises 95–98, use the following graph.

95. Write a set that contains the three most popular types.

96. Write a set containing the three least popular types.

97. Write a set containing those types whose portion of sales is 20% or higher.

98. Write a set containing those types that make up exactly 10% of sales.

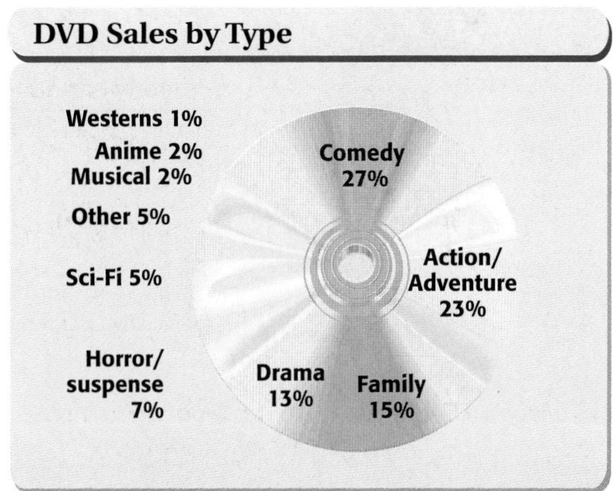

Source: Video Store magazine, based on DVD units sold in 2003

For Exercises 99–102, use the following data.

99. Write a set containing the amount of oil in millions of barrels imported from Canada, Mexico, and Saudi Arabia.

100. Write a set containing the amount of oil imported from Nigeria and Venezuela.

101. Write a set containing those countries that contributed more than 600 million barrels of crude oil.

102. Write a set containing those countries that contributed more than 450 million barrels of crude oil.

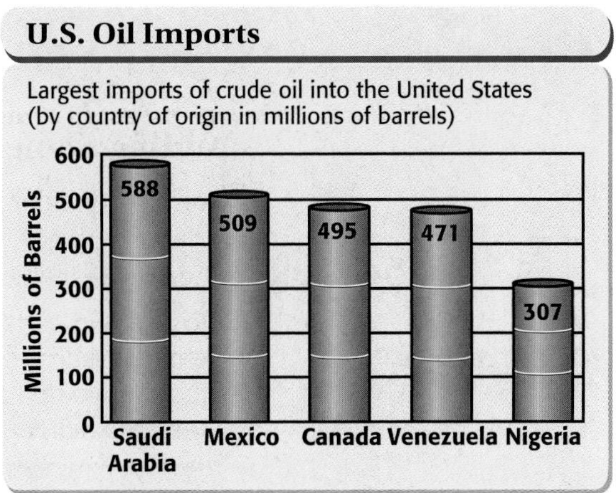

Source: Energy Information Administration

1.2 Operations with Real Numbers; Properties of Real Numbers

OBJECTIVES

1. Add real numbers.
2. Subtract real numbers.
3. Multiply real numbers.
4. Divide real numbers.

OBJECTIVE 1. Add real numbers.

Addition Properties of Real Numbers

First, let's consider some properties of addition that are true for all real numbers. The following table summarizes these properties:

Property of Addition	Symbolic Form	Word Form
Identity property of addition	$a + 0 = a$	The sum of a number and 0, the additive identity, is that number.
Commutative property of addition	$a + b = b + a$	Changing the order of addends does not affect the sum.
Associative property of addition	$a + (b + c) = (a + b) + c$	Changing the grouping of three or more addends does not affect the sum.

YOUR TURN Identify the property of addition that is illustrated in the given equation.

a. $x + 2 = 2 + x$ **b.** $(3 + m) + n = 3 + (m + n)$ **c.** $t + 0 = t$

Adding Signed Numbers

Now we consider how to add signed numbers

PROCEDURE **Adding Real Numbers**

To add two numbers that have the *same sign*, add their absolute values and keep the same sign.
To add two numbers that have *different signs*, subtract their absolute values and keep the sign of the number with the greater absolute value.

ANSWERS

a. commutative property
b. associative property
c. identity property

Note: *We can relate the addition of signed numbers to money. Example 1(a) illustrates a sum of two debts, which increases the amount of debt. Example 1(b) illustrates a credit with a debt where the value of the credit is more than the debt, so the result is a credit. In Example 1(c), we see the value of the debt is more than the credit, so the result is a debt.*

EXAMPLE 1 Add.

a. $-15 + (-21)$

Solution $-15 + (-21) = -36$

b. $26 + (-17)$

Solution $26 + (-17) = 9$

c. $-28 + 5$

Solution $-28 + 5 = -23$

d. $-\frac{4}{5} + \frac{2}{3}$

Solution $-\frac{4}{5} + \frac{2}{3} = -\frac{4(3)}{5(3)} + \frac{2(5)}{3(5)}$ Write equivalent fractions with their least common denominator (LCD), 15.

$= -\frac{12}{15} + \frac{10}{15}$

$= \frac{-12 + 10}{15}$ Add numerators and keep the common denominator. Because the addends have different signs, we subtract and keep the sign of the number with the greater absolute value.

$= -\frac{2}{15}$

e. $-12.5 + (-16.4)$

Solution $-12.5 + (-16.4) = -28.9$

YOUR TURN Add.

a. $-52 + (-16)$ **b.** $37 + (-42)$ **c.** $-29 + 12$

d. $-\frac{1}{6} + \left(-\frac{5}{8}\right)$ **e.** $-21.6 + 15.2$

ANSWERS

a. -68 **b.** -5 **c.** -17 **d.** $-\frac{19}{24}$ **e.** -6.4

Note: *Additive inverses add to equal the additive identity.*

Note that adding two numbers that have the same absolute value but different signs gives us 0. For example, $6 + (-6) = 0$. We say 6 and -6 are **additive inverses**, or opposites.

DEFINITION ***Additive inverses:*** Two numbers whose sum is 0.

YOUR TURN Find the additive inverse of the given number.

a. -7 **b.** 0.8 **c.** $-\frac{3}{4}$

ANSWERS

a. 7 **b.** -0.8 **c.** $\frac{3}{4}$

We sometimes indicate the additive inverse, or opposite, of a number using a minus sign. For example, the additive inverse of -7 can be written symbolically as $-(-7)$, which when simplified is 7.

RULE **Double Negative Property**

For any real number n, $-(-n) = n$.

EXAMPLE 2 Simplify.

a. $-(-(-6))$

Solution $-(-(-6)) = -(6)$ Simplify $-(-6)$, which is 6.

$= -6$ Find the additive inverse of 6.

b. $-|-15|$

Solution $-|-15| = -(15)$ Simplify $|-15|$, which is 15.

$= -15$ Find the additive inverse of 15.

Warning: Don't confuse $-|-15|$ with $-(-15)$. The expression $-|-15|$ indicates the additive inverse of the absolute value of -15, which is -15. The expression $-(-15)$ indicates the additive inverse of -15, which is 15.

YOUR TURN Simplify.

a. $-(-(5))$ **b.** $-|13|$

OBJECTIVE 2. Subtract real numbers. Recall that when we add two numbers that have different signs we actually subtract, which suggests that subtraction and addition are related. In fact, every subtraction statement can be written as an equivalent addition statement. For example, the subtraction statement $8 - 5 = 3$ and the addition statement $8 + (-5) = 3$ are equivalent. Note that in writing $8 - 5$ as $8 + (-5)$, we change the operation sign and also change the subtrahend 5 to its additive inverse.

$$8 - 5$$

Change the operation from minus to plus. Change the subtrahend to its additive inverse

$$= 8 + (-5)$$
$$= 3$$

PROCEDURE **Rewriting Subtraction**

To write a subtraction statement as an equivalent addition statement, change the operation symbol from a minus sign to a plus sign and change the subtrahend to its additive inverse.

ANSWERS

a. 5 **b.** -13

EXAMPLE 3 Subtract.

Connection Notice the use of the double negative property when changing $-(-6)$ to $+6$.

a. $-15 - (-6)$

Solution $-15 - (-6) = -15 + 6$ Write as an equivalent addition.
$= -9$

b. $-\frac{2}{3} - \frac{3}{4}$

Solution
$$-\frac{2}{3} - \frac{3}{4} = -\frac{2}{3} + \left(-\frac{3}{4}\right)$$ Write as an equivalent addition.
$$= -\frac{2(4)}{3(4)} + \left(-\frac{3(3)}{4(3)}\right)$$
$$= -\frac{8}{12} + \left(-\frac{9}{12}\right)$$
$$= -\frac{17}{12}$$

Note: *Subtraction is neither commutative nor associative.*

Note: *Just because we* can *rewrite subtraction does not mean we* **have** *to rewrite a subtraction as an equivalent addition. For example, it is easy enough to determine the result of $9 - 2$ without rewriting it as $9 + (-2)$.*

c. $5.04 - 8.01$

Solution $5.04 - 8.01 = 5.04 + (-8.01)$
$= -2.97$

YOUR TURN Subtract.

a. $25 - (-17)$ **b.** $-\frac{3}{8} - \left(-\frac{1}{3}\right)$ **c.** $0.06 - 4.02$

OBJECTIVE 3. Multiply real numbers.

Properties of Multiplication

Like addition, multiplication has properties, which we list in the following table:

Property of Multiplication	Symbolic Form	Word Form
Multiplicative property of 0	$0 \cdot a = 0$	The product of a number and 0 is 0.
Identity property of multiplication	$1 \cdot a = a$	The product of a number and 1, the multiplicative identity, is the number.
Commutative property of multiplication	$ab = ba$	Changing the order of factors does not affect the product.
Associative property of multiplication	$a(bc) = (ab)c$	Changing the grouping of three or more factors does not affect the product.
Distributive property of multiplication over addition	$a(b + c) = ab + ac$	A sum multiplied by a factor is equal to the sum of that factor multiplied by each addend.

ANSWERS

a. 42 **b.** $-\frac{1}{24}$ **c.** -3.96

ANSWERS

a. commutative property
b. multiplicative property of 0
c. distributive property

YOUR TURN Identify the property of multiplication that is illustrated in the given equation.

a. $xyz = yxz$ **b.** $5 \cdot 0 \cdot n = 0$ **c.** $2(x + 3) = 2x + 6$

Multiplying Signed Numbers

Now, let's consider multiplying signed numbers.

RULES Multiplying Two Signed Numbers

The product of two numbers that have the *same sign* is positive.
The product of two numbers that have *different signs* is negative.

EXAMPLE 4 Multiply.

a. $(-8)(-4)$

Solution $(-8)(-4) = 32$

b. $6(-7)$

Solution $6(-7) = -42$

Warning: Don't confuse $6(-7)$ with $6 - 7$. The expression $6(-7)$ indicates multiplication, whereas $6 - 7$ indicates subtraction.

c. $-\frac{3}{4} \cdot \left(\frac{5}{6}\right)$

Solution $-\frac{3}{4} \cdot \left(\frac{5}{6}\right) = -\frac{3 \cdot 5}{4 \cdot 6} = -\frac{15}{24} = -\frac{5}{8}$

d. $(-2.3)(-0.07)$

Solution $(-2.3)(-0.07) = 0.161$

YOUR TURN Multiply.

a. $-12(-6)$ **b.** $-4(9)$ **c.** $\left(-\frac{5}{8}\right)\left(-\frac{7}{10}\right)$ **d.** $(-0.6)(7.82)$

When multiplying more than two factors, we can multiply from left to right. However, the following examples suggest that to determine the sign of the answer, we can simply consider the number of negative factors.

$\underbrace{(-2)(-3)(-4)}_{\text{Odd number of negative factors}} = 6(-4) = \underset{\text{Negative result}}{-24}$

$\underbrace{(-2)(-3)(4)}_{\text{Even number of negative factors}} = 6(4) = \underset{\text{Positive result}}{24}$

ANSWERS

a. 72 **b.** −36 **c.** $\frac{7}{16}$
d. −4.692

RULE Multiplying with Negative Factors

The product of an even number of negative factors is positive, whereas the product of an odd number of negative factors is negative.

EXAMPLE 5 Multiply.

a. $(-1)(-1)(-3)(8)$

Solution $(-1)(-1)(-3)(8) = -24$ Because there are an odd number of negative factors, the result is negative.

b. $(-1)(-2)(-5)(3)(-4)$

Solution $(-1)(-2)(-5)(3)(-4) = 120$ Because there are an even number of negative factors, the result is positive.

YOUR TURN Multiply.

a. $(-1)(6)(-4)(2)$ **b.** $(7)(-2)(-1)(3)(-1)$

ANSWERS

a. 48 **b.** -42

Multiplication also has a **multiplicative inverse**.

DEFINITION ***Multiplicative inverses:*** Two numbers whose product is 1.

For example, $\frac{2}{3}$ and $\frac{3}{2}$ are multiplicative inverses because their product is 1.

$$\frac{2}{3} \cdot \frac{3}{2} = \frac{6}{6} = 1$$

Connection Additive inverses *add* to equal the additive identity, 0, and multiplicative inverses *multiply* to equal the multiplicative identity, 1.

Notice that to write a number's multiplicative inverse, we simply invert the numerator and denominator. Multiplicative inverses are also known as *reciprocals*.

YOUR TURN Find the multiplicative inverse.

a. $\frac{4}{5}$ **b.** $-\frac{1}{7}$ **c.** 9

OBJECTIVE 4. Divide real numbers. Recall that we can write a subtraction statement as an equivalent addition statement using the additive inverse. Similarly, we can write a division statement as an equivalent multiplication statement using the multiplicative inverse. We use this rule when dividing fractions.

Change the operation from multiplication to division.

$$\frac{1}{3} \div \frac{2}{5} = \frac{1}{3} \cdot \frac{5}{2} = \frac{5}{6}$$

Change the divisor to its multiplicative inverse

ANSWERS

a. $\frac{5}{4}$ **b.** -7 **c.** $\frac{1}{9}$

The rules for determining the sign of a quotient, which is the result of a division problem, are the same as for determining the sign of a product.

RULES Dividing Signed Numbers

The quotient of two numbers that have the *same sign* is positive.
The quotient of two numbers that have *different signs* is negative.

EXAMPLE 6 Divide.

a. $-45 \div (-5)$

Solution $-45 \div (-5) = 9$

b. $56 \div (-8)$

Solution $56 \div (-8) = -7$

c. $-\frac{2}{3} \div \frac{5}{6}$

Solution
$$-\frac{2}{3} \div \frac{5}{6} = -\frac{2}{3} \cdot \frac{6}{5} \quad \text{Write an equivalent multiplication.}$$
$$= -\frac{12}{15} \quad \text{Multiply.}$$
$$= -\frac{4}{5} \quad \text{Simplify.}$$

Note: *Division is neither commutative nor associative.*

d. $-12.6 \div (-0.5)$

Solution $-12.6 \div (-0.5) = 25.2$

YOUR TURN Divide.

a. $(-60) \div 12$ **b.** $-42 \div (-6)$ **c.** $\frac{5}{6} \div \left(-\frac{3}{4}\right)$

Division Involving 0

Calculator **TIP**

Try dividing any number by 0 on your calculator. When you press the [=] *or* [ENTER] *key the screen will say* error, *which is your calculator's way of indicating undefined or indeterminate operations.*

If 0 is involved in division, we must be careful how we evaluate the result. If we divide 0 by a nonzero number, the quotient is 0. For example $0 \div 9 = 0$, which we can check using multiplication: $0 \cdot 9 = 0$.

Now consider division by 0, as in $8 \div 0$. To check this division, the quotient must multiply by 0 to equal 8, which is impossible because any number multiplied by 0 would equal 0, not 8. Consequently, we say that dividing a nonzero number by 0 is *undefined.*

Now consider $0 \div 0$. To check this division, the quotient must multiply by 0 to equal 0. Notice the quotient could be any number because any number times 0 equals 0. Because we cannot determine a unique quotient, we say that $0 \div 0$ is *indeterminate.*

Following is a summary of these rules:

RULES Division Involving 0

$0 \div n = 0$ when $n \neq 0$
$n \div 0$ is undefined when $n \neq 0$
$0 \div 0$ is indeterminate

ANSWERS

a. -5 **b.** 7 **c.** $-\frac{10}{9}$

1.2 Exercises

For Extra Help

MyMathLab Videotape/DVT InterAct Math Math Tutor Center Math XL.com

1. Explain the difference between the commutative property of addition and the associative property of addition.

2. Why is 0 called the additive identity and 1 the multiplicative identity?

3. Explain why 8 and -8 are additive inverses.

4. What are multiplicative inverses?

5. Explain how to add two numbers that have the same sign.

6. Explain how to add two numbers that have different signs.

7. Explain how to write a subtraction statement as an equivalent addition statement.

8. What is the sign of the product or quotient of two numbers with different signs?

For Exercises 9–20, indicate whether the given equation illustrates the additive identity property, multiplicative identity property, additive inverses, or multiplicative inverses.

9. $3 + (-3) = 0$

10. $4 \cdot \frac{1}{4} = 1$

11. $-6.1 + 0 = -6.1$

12. $-8\frac{1}{2} + 8\frac{1}{2} = 0$

13. $(-8.1)1 = -8.1$

14. $-\frac{2}{3}\left(-\frac{3}{2}\right) = 1$

15. $-\frac{1}{5}(-5) = 1$

16. $2\frac{3}{5} + 0 = 2\frac{3}{5}$

17. $-5.2(1) = -5.2$

18. $9\frac{1}{7} + 0 = 9\frac{1}{7}$

19. $-\frac{2}{3} \cdot 1 = -\frac{2}{3}$

20. $-6 + 6 = 0$

For Exercises 21–28, find the additive inverse and multiplicative inverse.

21. 8

22. 12

23. -7

24. -9

25. $-\frac{5}{8}$

26. $\frac{7}{2}$

27. 0.3

28. -2.5

For Exercises 29–40, indicate whether the equation illustrates the commutative property of addition, commutative property of multiplication, associative property of addition, associative property of multiplication, or the distributive property.

29. $3 + 2 = 2 + 3$

30. $2 \cdot n \cdot 5 = 2 \cdot 5 \cdot n$

31. $5(x + y) = 5x + 5y$

32. $-\frac{1}{3} + \frac{2}{5} = \frac{2}{5} + \left(-\frac{1}{3}\right)$

33. $\frac{3}{4}\left(\frac{2}{9} \cdot \frac{5}{7}\right) = \left(\frac{3}{4} \cdot \frac{2}{9}\right)\frac{5}{7}$

34. $0.5 + (-8.1 + (-9)) = (0.5 + (-8.1)) + (-9)$

35. $-3(mn) = (-3m)\, n$

36. $6t + 6u = 6(t + u)$

37. $2x + (3y + 5x) = 2x + (5x + 3y)$

38. $x(5 + y) = (5 + y)x$

39. $(x - 3)(x + 4) = (x + 4)(x - 3)$

40. $m + (2m + 5) = (m + 2m) + 5$

For Exercises 41–56, add.

41. $9 + (-16)$

42. $23 + (-29)$

43. $-27 + (-13)$

44. $-8 + (-10)$

45. $-15 + 9$

46. $-21 + 14$

47. $14 + (-19)$

48. $32 + (-16)$

49. $-\frac{3}{4} + \frac{1}{6}$

50. $\frac{1}{4} + \left(-\frac{5}{6}\right)$

51. $-\frac{1}{8} + \left(-\frac{2}{3}\right)$

52. $-\frac{2}{5} + \left(-\frac{3}{4}\right)$

53. $-0.18 + 6.7$

54. $-15.81 + 4.28$

55. $-0.28 + (-4.1)$

56. $-7.8 + (-9.16)$

For Exercises 57–72, subtract.

57. $2 - (-3)$

58. $-4 - (-9)$

59. $0 - (-2)$

60. $8 - (-1)$

61. $7 - 11$

62. $6 - 21$

63. $8 - 3$

64. $17 - 13$

65. $\frac{7}{10} - \left(-\frac{3}{5}\right)$

66. $-\frac{1}{2} - \left(-\frac{1}{3}\right)$

67. $-\frac{1}{5} - \left(-\frac{1}{5}\right)$

68. $-\frac{2}{5} - \left(-\frac{4}{10}\right)$

69. $4.01 - 3.65$

70. $8.1 - 4.76$

71. $-6.1 - (-4.5)$

72. $-7.1 - (-2.3)$

For Exercises 73–88, multiply or divide.

73. $4(-3)$

74. $5(-2)$

75. $(-2)\,(-1)$

76. $(-5)\,(-1)$

77. $-2 \cdot \frac{1}{4}$

78. $-3 \cdot \left(-\frac{2}{5}\right)$

79. $-25 \div (-5)$

80. $-28 \div (-4)$

81. $-\frac{1}{4} \div \frac{2}{3}$

82. $\frac{3}{5} \div \left(-\frac{1}{5}\right)$

83. $-12 \div 0.3$

84. $-1.4 \div (-0.7)$

85. $-1\,(-2)(-3)$

86. $-2(5)(-2)(-3)$

87. $2.7(-0.1)(-2)$

88. $(-0.6)(-2.5)(-0.1)$

For Exercises 89–92, solve.

89. On April 15, 2005, the NASDAQ index closed at 1908.15, which was a change of −38.56 from the previous day's closing value. What was the closing value on April 14, 2005? (*Source:* Motley Fool)

90. The temperature at which molecular motion is at a minimum is −273.15°C. A piece of metal is cooled to −256.5°C. Write an equation that describes the difference in the temperatures, then calculate the difference.

Of Interest

The temperature −273.15°C is also known as *absolute zero* on the Kelvin temperature scale. The Kelvin scale is the scale most scientists use to measure temperatures. Increments on the Kelvin scale are equal to increments on the Celsius scale, so a change of 1 K corresponds to a change of 1°C.

91. A district manager has a balance of −$1475.84 on her business credit card. In one week, the following transactions occurred. Find the new balance.

Description	Amount
Payment	$1200
Italian Café	−$124.75
Starbucks	−$12.50
Office Maxx	−$225.65
Kinko's	−$175.92

92. A small company has a balance of $10,450.75 in the bank. The following table lists income and expenses for one week. Find the new balance.

Description	Income	Description	Expense
Sales	$3400	Electricity	−$234.45
		Sewer	−$150.00
		Payroll	−$8500
		Materials	−$2400
		Office Supplies	−$142.75
		Phone Service	−$225.80

For Exercises 93 and 94, find the resultant force on each object. The resultant force on an object is the sum of all the forces acting on the object.

93.

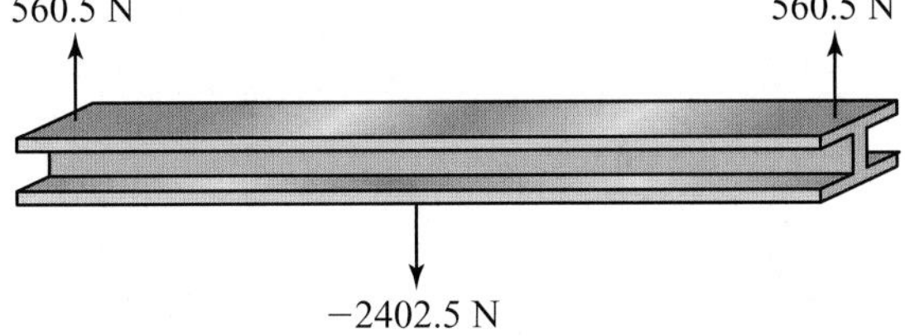

Of Interest

N is an abbreviation for the metric unit of force called the newton. The unit is named after Sir Isaac Newton (1642–1727) in honor of his development of a theory of forces.

94.

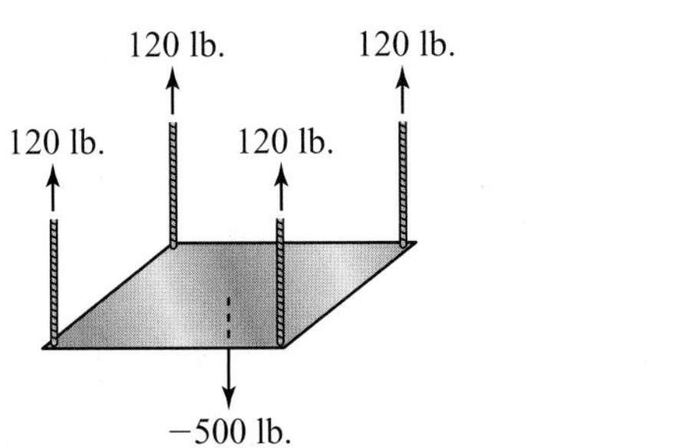

95. The budget for a company has a −3.5% change from last year's budget. If last year's budget was $4.8 million, what is the new budget?

96. A company experiences a −2.5% change in profit from the previous year. If the profit for the previous year was $1.4 million, what is this year's profit?

For Exercises 97 and 98, use the fact that an object's weight is a downward force due to gravity and is calculated by multiplying the object's mass by the acceleration due to gravity, which is a constant (force = mass · acceleration). The following table lists the gravitational acceleration constants in American units for various bodies in our solar system. American mass units are slugs (s) *and force units are pounds* (lb.).

Earth	-32.2 ft./sec.^{2}
The Moon	-5.5 ft./sec.^{2}
Mars	-12.3 ft./sec.^{2}

Of Interest

On July 20, 1969, during the *Apollo 11* mission, Neil Armstrong and Buzz Aldrin became the first people to walk on the surface of the Moon. Armstrong was the first out of the lunar lander and stated the now famous words, "That's one small step for man, one giant leap for mankind."

97. a. Find the force due to gravity on a person with a mass of 5.5 slugs on Earth, the Moon, and Mars.

b. On which of the Earth, the Moon, or Mars does a person experience the greatest force due to gravity?

98. a. During the NASA Apollo missions, the lunar lander had a mass of about 271 slugs (without fuel and equipment). Find the force due to gravity on the lander on Earth and on the Moon.

b. What gravitational force would that same lander have experienced on Mars?

For Exercises 99 and 100, use the fact that in an electrical circuit, voltage is equal to the product of the current, measured in amperes (A)*, and the resistance of the circuit, measured in ohms* (Ω) *(voltage = current × resistance).*

99. Suppose the current in a circuit is -6.4 A and the resistance is 8 Ω. Find the voltage.

100. An electrical technician measures the voltage in a circuit to be -15 V and the current to be -8 A. What is the resistance of the circuit?

PUZZLE PROBLEM

Two books in an algebra series, Elementary Algebra *and* Intermediate Algebra, *are placed on a bookshelf as shown. Suppose the hardback covers are each $\frac{1}{8}$ of an inch thick and each book without the covers is 2 inches thick. If a bookworm begins at the first page of* Elementary Algebra *and bores straight through to the last page of* Intermediate Algebra, *how far has it traveled?*

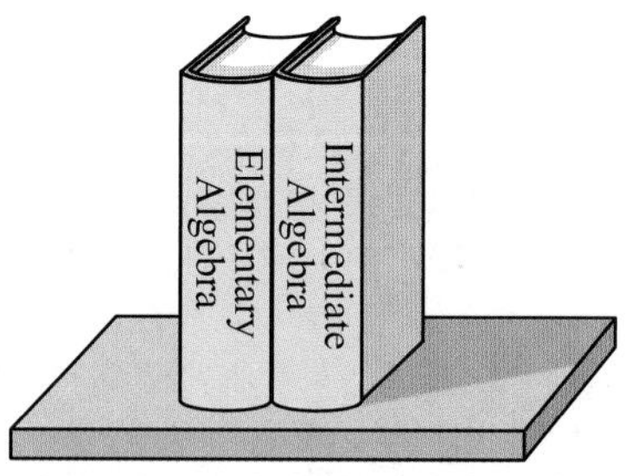

Collaborative Exercises WINDCHILL

Since wind helps exposed portions of your body to lose heat, how cold you feel outside depends not only on the temperature of the air but also on how windy it is. For the last half of the 20th century, the accompanying table was used to determine a combination of air temperature (in °F) and wind speed called* windchill. *For example, with a wind speed of 15 miles per hour and an air temperature of 30°F, the windchill is 9°F. That is, the effect on loss of body heat would be similar to 9°F in still air. Use the table to answer the questions that follow.

Air Temperature (degrees Fahrenheit)

Wind Speed (mph)	45	40	35	30	25	20	15	10	5	0	−5	−10	−15	−20	−25	−30
5	43	37	32	27	22	16	11	6	0	−5	−10	−15	−21	−26	−31	−36
10	34	28	22	16	10	3	−3	−9	−15	−21	−27	−34	−40	−46	−52	−58
15	29	22	15	9	2	−5	−12	−18	−25	−32	−38	−45	−52	−59	−65	−72
20	25	18	11	4	−3	−11	−18	−25	−32	−39	−46	−53	−60	−68	−75	−82
25	22	15	8	0	−7	−15	−22	−30	−37	−44	−52	−59	−67	−74	−82	−89
30	20	13	5	−3	−10	−18	−25	−33	−41	−48	−56	−64	−71	−79	−87	−94
35	19	11	3	−5	−12	−20	−28	−36	−44	−51	−59	−67	−75	−83	−90	−98
40	18	10	2	−6	−14	−22	−30	−38	−46	−53	−61	−69	−77	−85	−93	−101
45	17	9	1	−7	−15	−23	−31	−39	−47	−55	−63	−71	−79	−87	−95	−103
50	17	9	1	−7	−15	−23	−31	−40	−48	−56	−64	−72	−80	−88	−96	−104

Source: National Center for Atmospheric Research

1. For an air temperature of 15°F and a wind speed of 5 miles per hour, what is the windchill?

2. What other temperature/wind speed combinations give the same windchill as in number 1?

3. For an air temperature of 45°F, verify that the expression **20 − 43** shows the change in windchill as the wind speed *increases* from 5 to 30 miles per hour. Simplify the expression **20 − 43** and comment on the sign of your answer.

4. For an air temperature of 25°F, write and simplify a mathematical expression for the change in windchill if the wind speed *decreases* from 35 miles per hour to 15 miles per hour. Comment on the sign of your answer.

5. For an air temperature of 10°F, what is the difference in windchill when the wind speed decreases from 35 miles per hour to 10 miles per hour?

6. Water freezes at 32°F. If the air temperature is 45°F, estimate how fast the wind must be blowing so that the windchill is 32°F.

7. Examine the row in the table for a 5 mile per hour wind speed. For an air temperature of 45°F, the windchill temperature is 43°F, or 2 degrees less than the air temperature.

a. Is this difference the same for all air temperatures at a wind speed of 5 miles per hour?

b. How about for other wind speeds?

c. Describe the trend in the difference between air temperature and windchill as you move from left to right across any row in the table. Interpret your answer in terms of how wind affects how cold you feel.

REVIEW EXERCISES

1. Write a set containing the last names of the first four presidents of the United States.

2. If $A = \{1, 3, 5, 7, 9\}$ and $B = \{3, 5, 6\}$, is B a subset of A? Why?

3. Is the set of whole numbers a finite set or an infinite set?

4. Explain why -6 is a rational number.

5. Evaluate: $|-25|$

6. Use $<$, $>$, or $=$ to make a true statement. $-\frac{5}{6} \;\square\; -\frac{15}{18}$

1.3 Exponents, Roots, and Order of Operations

OBJECTIVES

1. **Evaluate numbers in exponential form.**
2. **Evaluate roots.**
3. **Use the order of operations agreement to simplify numerical expressions.**

OBJECTIVE 1. Evaluate numbers in exponential form. An expression like 2^4 is in *exponential form* where the number 2 is called the *base* and 4 is an *exponent*. The expression 2^4 is read "two to the fourth power" or simply "two to the fourth." To evaluate 2^4, we write 2 as a factor four times, then multiply.

$$2^4 = \underbrace{2 \cdot 2 \cdot 2 \cdot 2}_{\text{Four 2s}} = 16$$

Base (2), Exponent (4)

RULE Evaluating an Exponential Form

For any real number b and any natural number n, the nth power of b, or b^n, is found by multiplying b as a factor n times.

$$b^n = \underbrace{b \cdot b \cdot b \cdot \cdots \cdot b}_{b \text{ used as a factor } n \text{ times}}$$

EXAMPLE 1 Evaluate.

a. $(-9)^2$

Solution $(-9)^2 = (-9)(-9) = 81$

b. $\left(-\frac{3}{4}\right)^3$

Solution $\left(-\frac{3}{4}\right)^3 = \left(-\frac{3}{4}\right)\left(-\frac{3}{4}\right)\left(-\frac{3}{4}\right) = -\frac{27}{64}$

Note: *A base raised to the second power, as in $(-9)^2$, is said to be* **squared**. *A base raised to the third power, as in $\left(-\frac{3}{4}\right)^3$, is said to be* ***cubed.***

Example 1 suggests the following sign rules:

RULES **Evaluating Exponential Forms with Negative Bases**

If the base of an exponential form is a negative number and the exponent is even, then the product is positive.

If the base is a negative number and the exponent is odd, then the product is negative.

YOUR TURN Evaluate.

a. $\left(-\frac{1}{2}\right)^5$

b. $(-0.3)^4$

Calculator TIPS

Use the [^] *key to indicate an exponent. For example, to evaluate* $\left(-\frac{2}{3}\right)^3$, *type:*

[(] [(−)] [2] [÷] [3] [)] [^] [3] [ENTER]

The answer will be given as a decimal. To change the decimal answer to a fraction, press the [MATH] *key, select 1:Frac, then press* [ENTER].

OBJECTIVE 2. Evaluate roots.

Square Roots

The square root of a number is a number whose square is the given number. For example, one square root of 25 is 5 because $5^2 = 25$. Another square root of 25 is -5 because $(-5)^2 = 25$. In fact, every positive real number has *two* real-number square roots, a *positive* root and a *negative* root. For convenience, we can write both the positive and negative roots in a compact expression, ± 5. The number 0 has only one square root, 0.

ANSWERS

a. $-\frac{1}{32}$ **b.** 0.0081

EXAMPLE 2 Find all real-number square roots of the given number.

a. 64

Answer ± 8

b. -36

Answer No real-number square roots exist.

Explanation The square of a real number is always positive or zero.

Note: *We do not say that this situation is undefined or indeterminate because we will eventually define the square root of a negative number. Such roots will be defined using imaginary numbers.*

YOUR TURN Find all real-number square roots of the given number.

a. 121 **b.** -49

The Principal Square Root

The symbol $\sqrt{}$ is called a *radical* and indicates the *principal* (positive) square root of a number.

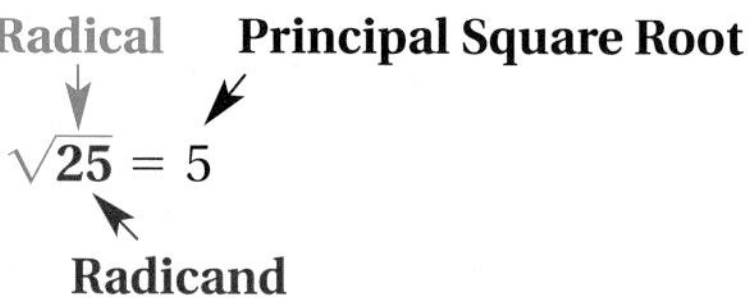

EXAMPLE 3 Evaluate.

Note: *This suggests that* $\sqrt{\frac{25}{81}} = \frac{\sqrt{25}}{\sqrt{81}} = \frac{5}{9}$.

a. $\sqrt{49}$

Answer $\sqrt{49} = 7$

b. $\sqrt{\frac{25}{81}}$

Answer Since $\left(\frac{5}{9}\right)^2 = \frac{25}{81}$, we can say $\sqrt{\frac{25}{81}} = \frac{5}{9}$.

c. $\sqrt{0.36}$

Answer $\sqrt{0.36} = 0.6$

d. $\sqrt{-100}$

Answer not a real number

Note: *For now, we will consider only simple rational roots. In Chapter 8, we will consider irrational roots.*

Calculator TIPS

To find a square root, use the $\sqrt{}$ *function, which may require pressing*  2nd *and then the* x^2 *key. For example, to find the square root of 144, press:*

We can now summarize what we have learned so far with the following rules for square roots.

ANSWERS

a. ± 11

b. No real-number roots exist.

RULES Square Roots

Every positive number has two square roots, a positive root and a negative root.
The square root of 0 is 0.
Negative numbers have no real-number square roots.
The radical symbol $\sqrt{\ }$ denotes only the positive (principal) square root.
$\sqrt{\frac{a}{b}} = \frac{\sqrt{a}}{\sqrt{b}}$, where $a \geq 0$ and $b > 0$

YOUR TURN Evaluate.

a. $\sqrt{121}$ b. $\sqrt{\frac{81}{169}}$ c. $\sqrt{-0.09}$

ANSWERS

a. 11 **b.** $\frac{9}{13}$
c. not a real number

Other Roots

Each natural number exponent has a corresponding root, which we indicate as an *index* in a radical. For example, to indicate the cube root of a number *a*, we write $\sqrt[3]{a}$. To find a cube root, we find a number that when cubed, equals the radicand.

Note: *If no index is shown, as in* $\sqrt{16}$, *the radical is understood to mean a square root.*

The index is 3. → $\sqrt[3]{64} = 4$ because $4^3 = 64$

Higher roots follow a similar pattern.

EXAMPLE 4 Evaluate.

a. $\sqrt[4]{81}$

Solution $\sqrt[4]{81} = 3$ because $3^4 = 81$

b. $\sqrt[5]{32}$

Solution $\sqrt[5]{32} = 2$ because $2^5 = 32$

YOUR TURN Evaluate.

a. $\sqrt[3]{125}$ b. $\sqrt[4]{16}$ c. $\sqrt[5]{243}$

OBJECTIVE 3. Use the order of operations agreement to simplify numerical expressions. If an expression contains a mixture of operations, we agree to perform the operations in a specific order because using other orders can result in different answers.

ANSWERS

a. 5 **b.** 2 **c.** 3

PROCEDURE Order of Operations Agreement

Perform operations in the following order:

1. Grouping symbols: parentheses ( ), brackets [ ], braces { }, absolute value | |, and radicals $\sqrt{\ }$.
2. Exponents/Roots from left to right, in order as they occur.
3. Multiplication/Division from left to right, in order as they occur.
4. Addition/Subtraction from left to right, in order as they occur.

EXAMPLE 5 Simplify.

Warning: The expressions $(-5)^2$ and -5^2 are different. The expression $(-5)^2$ means the square of the number -5, whereas -5^2 means the additive inverse of the square of 5.

$(-5)^2 = (-5)(-5) = 25$

$-5^2 = -(5 \cdot 5) = -25$

a. $18 - 12 \div 3(-2) + (-5)^2$

Solution $= 18 - 12 \div 3(-2) + 25$ Simplify the exponential form.

$= 18 - 4(-2) + 25$ Multiply or divide from left to right: $-12 \div 3$ is first.

$= 18 + 8 + 25$ Multiply: $-4(-2) = 8$.

$= 51$ Add.

b. $(-2)^3 - 3|5 - (4 + 3)| - \sqrt{16 + 9}$

Solution When grouping symbols are embedded one within another, as in $|5 - (4 + 3)|$, we work from the innermost set of grouping symbols outward.

$= (-2)^3 - 3|5-7| - \sqrt{25}$ Simplify the innermost parentheses: $4 + 3 = 7$; add inside the radical: $16 + 9 = 25$.

$= -8 - 3|-2| - 5$ Evaluate the exponential form, subtract within the absolute value, and find the square root.

$= -8 - 3(2) - 5$ Simplify the absolute value: $|-2| = 2$.

$= -8 - 6 - 5$ Multiply: $-3(2) = -6$.

$= -19$ Subtract.

c. $\dfrac{(-3)^4 - 3(-5)}{-\sqrt{4 \cdot 9}}$

Solution Simplify the numerator and denominator separately, then divide the results.

$= \dfrac{81 - 3(-5)}{-\sqrt{36}}$ Evaluate the exponential form in the numerator and simplify within the radical in the denominator.

$= \dfrac{81 + 15}{-6}$ Multiply in the numerator, evaluate the square root in the denominator.

$= \dfrac{96}{-6}$ Add in the numerator.

$= -16$ Divide.

Connection The expression in Example 5(c) is equivalent to the expression $\left[(-3)^4 - 3(-5)\right] \div \left[-\sqrt{4 \cdot 9}\right]$.

YOUR TURN Simplify.

a. $15 - [3 - 5]^2 - \sqrt{25 - 9}$

b. $24 \div (-3)(-4) + |9 - 16| + (-4)^3$

c. $\dfrac{8(4 - 5)^5 - 1}{15 - 3(12)}$

ANSWERS

a. 7 **b.** -25 **c.** $\dfrac{3}{7}$

Now, let's consider some applications that require the use of the order of operations agreement.

EXAMPLE 6 A cumulative grade point average (GPA) is found by dividing the total number of grade points by the total number of course credits. Grade points are calculated by multiplying the numerical value of the letter grade (A = 4, B = 3, C = 2, D = 1, F = 0) by the number of credits for the course. For the following student's grade report, write a numerical expression that describes the GPA, then calculate the GPA.

Course	Course Credits	Grade
ENG 102	3	A
MAT 102	3	B
BIO 110	4	C
SOC 101	3	F

Solution $\text{GPA} = \dfrac{3 \cdot 4 + 3 \cdot 3 + 4 \cdot 2 + 3 \cdot 0}{3 + 3 + 4 + 3}$

$= \dfrac{12 + 9 + 8 + 0}{13}$ Multiply in the numerator; add in the denominator.

$= \dfrac{29}{13}$ Add in the numerator.

≈ 2.231 Divide and round the quotient to the nearest thousandth.

Counting problems often generate expressions with exponents that require using order of operations. Consider counting the number of different numbers that can be expressed using two LED displays if a single digit (0–9) is shown in each display.

Note: *Each display can contain 0–9, so there are ten different digits that are possible in each display.*

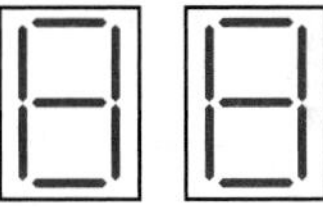

We could list every possible number using two digits—00, 01, 02, 03, 04, . . . , 99—to see that there are 100 possible numbers. Or, since each of the ten digits in the first display is paired with ten digits in the second display, we could multiply $10 \cdot 10$ (or 10^2) to find out that there are 100 possible numbers. In general, if a problem involves counting the total number of possibilities with items in a set of positions, the result will be the product of the number of items possible in each position.

Connection We will study these types of counting problems in more detail in Appendix D.

EXAMPLE 7 How many license plates are possible if each license plate has three capital letters (A–Z) followed by three digits (0–9)?

Solution Multiply the number of letters or digits that are possible in each position.

Note: *Each of the three letter positions has 26 different letter possibilities. Each of the three digit positions has 10 different digit possibilities.*

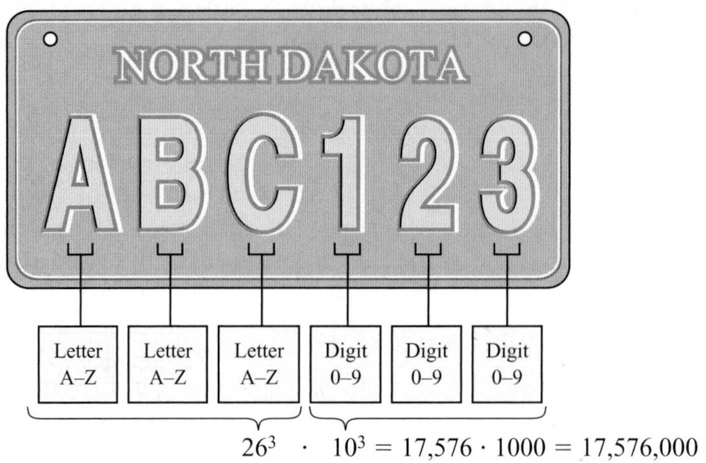

$26^3 \cdot 10^3 = 17{,}576 \cdot 1000 = 17{,}576{,}000$

YOUR TURN How many different combinations are possible if a combination lock has four dials, one dial with the letters A–L and the other three dials each with the digits 0–9 on them?

ANSWER

12,000

1.3 Exercises

For Extra Help MyMathLab · Videotape/DVT · InterAct Math · Math Tutor Center · Math XL.com

1. Explain how to evaluate a^b, where a is any real number and b is a natural number.

2. What is the difference between a number's square and its square root?

3. Why is there no real-number square root of a negative number?

4. If a numerical expression contains embedded grouping symbols, such as $\{15 - 4[3 - (5 + 2)]\}$, explain the order in which you simplify the grouping symbols.

5. Discuss all possible ways to simplify $\sqrt{9 \cdot 4}$.

6. Discuss all possible ways to simplify $\sqrt{169 - 25}$.

For Exercises 7–10, identify the base and the exponent, then translate the expression into words.

7. $(-4)^3$

8. 5^2

9. -1^7

10. -2^8

For Exercises 11–26, evaluate.

11. 5^4 **12.** 2^7 **13.** $(-3)^3$ **14.** $(-7)^2$ **15.** -6^2 **16.** -6^3

17. $(-1)^{10}$ **18.** $(-1)^7$ **19.** $\left(-\frac{3}{8}\right)^2$ **20.** $\left(-\frac{1}{9}\right)^2$ **21.** $\left(-\frac{5}{6}\right)^3$ **22.** $\left(-\frac{2}{3}\right)^4$

23. $(0.4)^3$ **24.** $(0.1)^4$ **25.** $(-2.1)^3$ **26.** $(-0.3)^4$

For Exercises 27–34, find all the square roots of the given number.

27. 225 **28.** 36 **29.** 256 **30.** 121 **31.** 100 **32.** 49 **33.** -25 **34.** -400

For Exercises 35–48, evaluate the root.

35. $\sqrt{25}$ **36.** $\sqrt{100}$ **37.** $\sqrt[5]{32}$ **38.** $\sqrt[4]{81}$ **39.** $\sqrt{0.36}$ **40.** $\sqrt{0.01}$ **41.** $\sqrt[3]{-27}$

42. $-\sqrt[3]{-64}$ **43.** $\sqrt[3]{\frac{8}{27}}$ **44.** $\sqrt[4]{\frac{1}{16}}$ **45.** $\sqrt{-36}$ **46.** $\sqrt{-16}$ **47.** $\sqrt[3]{\frac{24}{3}}$ **48.** $\sqrt{\frac{50}{2}}$

For Exercises 49–72, evaluate using the order of operations.

49. $-4 + 3(-1)^4 + 18 \div 3$

50. $(-2)^3 + 8 - 7(4 - 3)$

51. $-8^2 + 36 \div (9 - 5)$

52. $-7^2 - 25 \div (8 - 3)$

53. $\sqrt[3]{31 - 4} + 2^3 - 9$

54. $6 + 4^2 \div \sqrt{13 + 3}$

55. $-2|-8 - 2| \div (-5)(2)$

56. $|-6 - 8| \div 2(-3)$

57. $-24 \div (-6)(2) + \sqrt{169 - 25}$

58. $\sqrt{100 - 64} + 18 \div (-3)(-2)$

59. $-18 \cdot \frac{2}{9} \div (-2) + |9 - 5(-2)|$

60. $24\left(-\frac{3}{8}\right) \div (-3) + |-6 + 2(-3)|$

61. $13.02 \div (-3.1) + 6^2 - \sqrt{25}$

62. $4^2 + \sqrt{49} - 9.03 \div (-4.3)$

63. $(1 - 0.8)^2 + 2.4 \div (0.3)(-0.5)$

64. $(0.1)^3 - (-6)(3) + 5(2 - 8)$

65. $\frac{4}{5} \div \left(-\frac{1}{10}\right) \cdot (-2) + \sqrt[5]{16 + 16}$

66. $-\frac{3}{4} \div \frac{1}{8} + \left(-\frac{2}{5}\right)(-3)(-4)$

67. $\frac{9}{8} \cdot \left(-\frac{2}{3}\right) + \left(\frac{1}{5} - \frac{2}{3}\right) \div \sqrt{\frac{125}{5}}$

68. $\frac{6}{5} \cdot \left(-\frac{2}{3}\right) + \left(\frac{2}{5} - \frac{2}{5}\right) \div \sqrt{\frac{36}{4}}$

69. $\frac{12 - 2^3}{5 - 3 \cdot 2}$

70. $\frac{6^2 + 4}{8 \div (-2) \cdot 5}$

71. $\frac{6^2 - 3(4 + 2^5)}{5 + 20 - (2 + 3)^2}$

72. $\frac{4[5 - 8(2 + 1)]}{3 - 6 - (-4)^2}$

In Exercises 73–76, a property of arithmetic was used as an alternative to the order of operations. Determine what property of arithmetic was applied and explain how it is different from the order of operations agreement.

73. $13 - 1 \cdot 3 \cdot 3 + 8^2$
$= 13 - 1 \cdot 9 + 64$
$= 13 - 9 + 64$
$= 68$

74. $-3(2 + 5) - \sqrt{36}$
$= -6 - 15 - 6$
$= -27$

75. $-6[-1 + 6^2] - \sqrt{14 + 11}$
$= -6[-1 + 36] - \sqrt{25}$
$= 6 + (-216) - 5$
$= -215$

76. $(-3)^3 + 2[-11 + 8 + (-2)]$
$= -27 + 2[-13 + 8]$
$= -27 + 2[-5]$
$= -27 + (-10)$
$= -37$

For Exercises 77–80, explain the mistake. Then simplify correctly.

77. $12 \div 4 \cdot 3 - 11$
$= 12 \div 12 - 11$
$= 1 - 11$
$= -10$

78. $25 - 3(1 - 8)$
$= 25 - 3(-7)$
$= 22(-7)$
$= -154$

79. $30 \div 2 + \sqrt{16 + 9}$
$= 15 + \sqrt{16 + 9}$
$= 15 + 4 + 3$
$= 22$

80. $-2^4 + 16 \div 4 - (5 - 7)$
$= 16 + 4 - (-2)$
$= 20 + 2$
$= 22$

81. Jennifer is preparing for a 10-mile minimarathon. To qualify, she must have an average time of 72 minutes or less on four 10-mile runs. She records her time on each of four practice runs, which are listed here.

April 2, 2005	76.5 minutes	April 16, 2005	71.4 minutes
April 9, 2005	74.5 minutes	April 23, 2005	69.2 minutes

a. Find her average time for the four 10-mile practice runs.
b. Does she qualify?

82. Harold receives an offer from his electric and gas provider where he can pay a fixed rate of \$100 each month to avoid drastic variations in his bill from month to month. To determine if the offer is reasonable, he decides to compare the \$100 fixed monthly rate with his average monthly cost from the previous year. The following list contains his monthly charges from 2004.

January	\$210	July	\$20
February	\$224	August	\$22
March	\$110	September	\$25
April	\$85	October	\$88
May	\$60	November	\$148
June	\$28	December	\$198

a. Find the average of the monthly bills from 2004.
b. Is the \$100 fixed monthly rate a reasonable deal?

For Exercises 83–84, use the following bar graph that shows the number of Georgia HOPE scholarship recipients each school year from 1993–94 through 2002–2003.

83. Find the average number of HOPE scholarship recipients from 1993–94 through 1997–98.

84. Find the average number of HOPE scholarship recipients from 1998–99 through 2002–2003.

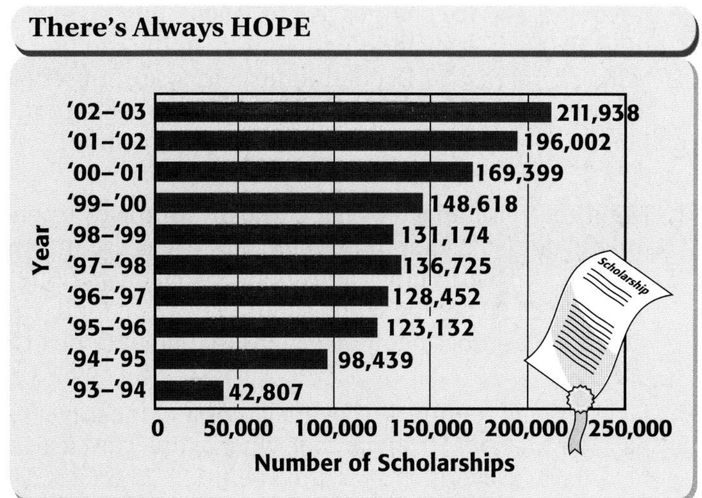

Source: Chronicle of Higher Education, 11/21/03

For Exercises 85 and 86, use the following line graph, which shows the closing value of the S&P 500 index on each day from June 20, 2005 through July 1, 2005.

85. Find the average closing value for the S&P 500 for the week of June 20–June 24.

86. Find the average closing value for the S&P 500 for the week of June 27–July 1.

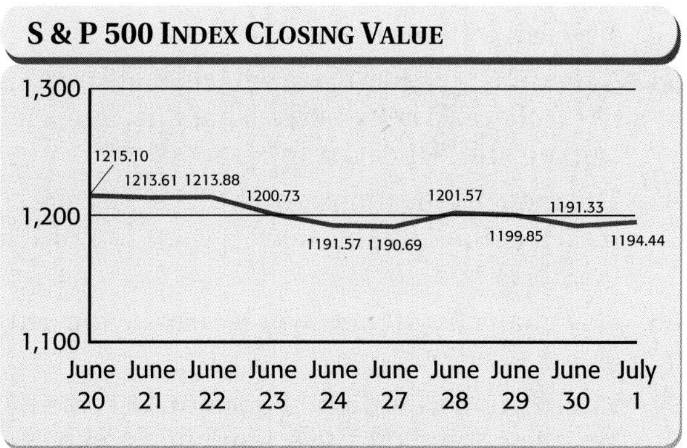

Source: Fool.com

For Exercises 87 and 88, find the GPA.

87.

Course	Course Credits	Grade
CHM 110	4	B
MAT 140	4	A
PSY 201	3	C
ECO 101	3	D

88.

Course	Course Credits	Grade
MAT 110	3	B
ENG 201	3	A
PHI 101	3	C
ART 101	3	A
PHY 101	4	D

For Exercises 89–98, solve.

89. A wireless service charges \$35.50 for 500 minutes in the primary calling zone, then \$0.10 for each additional minute in the calling zone. Calls outside of the calling zone cost \$0.12 per minute. Suppose a subscriber uses 658 minutes in the calling zone and 45 minutes outside the calling zone. Write a numerical expression that describes the total cost for the month, then calculate the total.

90. A wireless service charges \$49 for 1000 minutes in the primary calling zone, then \$0.05 for each additional minute in the calling zone. Calls outside of the calling zone cost \$0.15 per minute. Suppose a subscriber uses 1258 minutes in the calling zone and 72 minutes outside the calling zone. Write a numerical expression that describes the total cost for the month, then calculate the total.

91. Dona gets reimbursed by her company for travel expenses. Her company pays for fuel and \$0.35 per mile when employees use their own vehicle, which she did. She drove 814 miles and spent \$54.50 on gas. She spent three nights in a hotel at \$89.90 per night and spent \$112.45 on food. Write a numerical expression that describes her total expenses, then find the total expenses.

92. Jackson gets reimbursed by his company for travel expenses. His company pays for fuel and \$0.30 per mile when employees use their own vehicle, which he did. He drove 127 miles and spent \$21.26 on gas. He spent two nights in a hotel at \$112.80 per night and spent \$131.08 on food. Write a numerical expression that describes his total expenses, then find the total expenses.

93. How many combinations are possible on a combination lock that has four dials, each with the letters A–H?

94. A slot machine has three wheels that spin. Each wheel has an orange, an apple, a cherry, a strawberry, a banana, and a joker. In how many different ways can the three wheels stop?

95. The memory locations on a MIDI device are coded using a 7-bit memory chip. If each of the 7 bits can have a value of 0 or 1, how many memory locations are possible?

96. How many five-letter "words" can be formed using the five lowercase vowels (a, e, i, o, u)?

97. How many seven-digit phone numbers are possible if the first three digits can be 7, 8, or 9 and the other four digits can be any numeral 0–9?

98. How many ten-digit phone numbers are possible with the area codes 801, 802, or 803?

REVIEW EXERCISES

[1.1] **1.** Is $5x^2 - 4y$ an expression or an equation? Explain.

[1.1] **2.** Write a set containing the whole numbers less than 9.

[1.1] **3.** What property is being applied in $4(9 + 8) = (9 + 8)4$?

[1.2] ***For Exercises 4–6, simplify.***

4. $19 - (-54)$

5. $-2(-5)(-6)$

6. $\dfrac{-38}{2}$

1.4 Evaluating and Rewriting Expressions

OBJECTIVES

1. **Translate word phrases to variable expressions.**
2. **Evaluate variable expressions.**
3. **Rewrite expressions using the distributive property.**
4. **Rewrite expressions by combining like terms.**

We have reviewed how to simplify numeric expressions. Now, we will review basic concepts of variable expressions.

OBJECTIVE 1. Translate word phrases to variable expressions. To translate a word phrase, select a variable to represent the unknown amount, then use the key words to translate to the appropriate operation. The following table contains some basic phrases and their translations:

Addition	Translation
The sum of x and 3	$x + 3$
h plus k	$h + k$
7 added to t	$t + 7$
3 more than a number	$n + 3$
y increased by 2	$y + 2$

Note: *Since addition is a commutative operation, it does not matter in what order we write the translation.*

For "the sum of x and 3," we can write

$$x + 3 \text{ or } 3 + x.$$

Subtraction	Translation
The difference of x and 3	$x - 3$
h minus k	$h - k$
7 subtracted from t	$t - 7$
3 less than a number	$n - 3$
y decreased by 2	$y - 2$

Note: *Subtraction is not a commutative operation; therefore, the order in which we write the translation matters. We must translate each key phrase exactly as shown. Notice when we translate "less than" or "subtracted from," the translation is reverse order from what we read.*

Multiplication	Translation
The product of x and 3	$3x$
h times k	hk
twice a number	$2n$
triple the number	$3n$
$\frac{2}{3}$ of a number	$\frac{2}{3}n$

Note: *Multiplication is a commutative operation, so it does not matter in what order we write the translation.*

$$h \text{ times } k \text{ can be } hk \text{ or } kh$$

Division	Translation
The quotient of x and 3	$x \div 3$
h divided by k	$h \div k$
h divided into k	$k \div h$
the ratio of a to b	$a \div b$

Note: *Division is not a commutative operation; therefore, we must translate division phrases exactly as shown. Notice how "divided into" is translated in reverse order of what we read.*

Exponents	Translation
c squared	c^2
the square of *b*	b^2
k cubed	k^3
the cube of *b*	b^3
n to the fourth power	n^4
y raised to the fifth power	y^5

Roots	Translation
The square root of *x*	$\sqrt{x}$
The cube root of *y*	$\sqrt[3]{y}$
The fifth root of *n*	$\sqrt[5]{n}$

EXAMPLE 1 Translate the phrase to an algebraic expression.

Note: *The commutative property of addition allows us to write the expression in either order.*

a. four more than three times a number

Translation $4 + 3n$ or $3n + 4$

Note: *Because subtraction is not commutative, we must translate its key words carefully. The phrase "six* less than*" indicates that the 6 comes* after *the subtraction sign.*

b. six less than the square of a number

Translation $n^2 - 6$

c. the sum of h raised to the fifth power and fifteen

Translation $h^5 + 15$

d. the ratio of m times n to r cubed

Translation $\dfrac{mn}{r^3}$

Note: *When coupled with the word* ratio, *the word* to *translates into the fraction line. The amount to the left of the word* to *goes in the numerator and the amount to the right of the word* to *goes in the denominator.*

e. one-half of v divided by the square root of t

Translation $\frac{1}{2}v \div \sqrt{t}$

Note: *When the word* of *is preceded by a fraction, it means multiply.*

Note: *Without the parentheses, the expression is $5x + y$, which is "the sum of five times x and y."*

f. five times the sum of x and y

Translation $5(x + y)$

g. the square root of the difference of the square of x and the square of y

Note: *"The square root of the difference" indicates that the entire subtraction is the radicand.*

Translation $\sqrt{x^2 - y^2}$

h. the product of x and y divided by the sum of x^2 and five

Translation $xy \div (x^2 + 5)$ or $\dfrac{xy}{x^2 + 5}$

YOUR TURN Translate the phrase to an algebraic expression.

a. two-thirds subtracted from the product of nine and a number

b. -6.2 increased by 9.8 times a number

c. a number minus twice the sum of the number and seven

d. the difference of m and n, all raised to the fourth power

ANSWERS

a. $9n - \frac{2}{3}$ **b.** $-6.2 + 9.8y$

c. $x - 2(x + 7)$ **d.** $(m - n)^4$

OBJECTIVE 2. Evaluate variable expressions. Recall that variables represent numbers. When we replace the variables in an expression with numbers, we are *evaluating* the expression.

PROCEDURE Evaluating a Variable Expression

1. Replace each variable with its corresponding given value.
2. Simplify the resulting numerical expression.

EXAMPLE 2 Evaluate $\frac{4x^2}{x-6}$ when

a. $x = -3$

Solution $\frac{4(-3)^2}{-3-6}$ Replace x with -3.

$= \frac{36}{-9} = -4$ Simplify.

b. $x = 6$

Solution $\frac{4(6)^2}{6-6}$ Replace x with 6, then simplify.

$= \frac{144}{0}$, which is undefined.

Connection We evaluate expressions when we check solutions to equations, as we will see in Chapter 2.

YOUR TURN

a. Evaluate $2x^3 - 6y$ when $x = -3$ and $y = 1$.

b. Evaluate $\frac{2n}{n+5}$ when $n = -5$.

In Example 2(b), we found that $\frac{4x^2}{x-6}$ was undefined when $x = 6$. With any variable expression, it is important to identify values for the variable that cause the expression to be undefined. Remember that if a denominator is 0, the expression is undefined.

EXAMPLE 3 Determine all values for the variable that causes the expression to be undefined.

a. $\frac{x-3}{x+2}$

Answer If $x = -2$, this expression is undefined because the denominator is 0.

b. $\frac{y}{y-7}$

Answer If $y = 7$, this expression is undefined.

ANSWERS

a. -60 **b.** undefined

Answer

8

Your Turn Determine all values for the variable that causes $\frac{k+2}{k-8}$ to be undefined.

OBJECTIVE 3. Rewrite expressions using the distributive property. Recall from Section 1.2 the distributive property of multiplication over addition.

RULE **The Distributive Property of Multiplication over Addition**

$$a(b + c) = ab + ac$$

We can use the distributive property to rewrite an expression in another form that is equivalent to the original form.

Example 4 Use the distributive property to write an equivalent expression.

a. $3(x + 4)$

Solution $3(x + 4) = 3 \cdot x + 3 \cdot 4$ — Distribute 3.

$= 3x + 12$ — Multiply.

b. $-5(t - 2)$

Solution $-5(t - 2) = -5 \cdot t - (-5) \cdot 2$ — Distribute -5.

$= -5t - (-10)$ — Multiply.

$= -5t + 10$ — Rewrite $-(-10)$ as $+10$.

Note: *You might find it helpful to think of $-5(t - 2)$ as $-5(t + (-2))$ so that multiplying -5 and -2 gives positive 10 directly without having to use the double negative property.*

c. $\frac{3}{8}\left(2m + \frac{4}{5}\right)$

Solution $\frac{3}{8}\left(2m + \frac{4}{5}\right) = \frac{3}{8} \cdot 2m + \frac{3}{8} \cdot \frac{4}{5}$ — Distribute $\frac{3}{8}$.

$= \frac{3}{4}m + \frac{3}{10}$ — Multiply and simplify.

Your Turn Use the distributive property to write an equivalent expression.

a. $-4(6 - 5y)$

b. $\frac{4}{5}\left(\frac{1}{2}y - 10\right)$

OBJECTIVE 4. Rewrite expressions by combining like terms. Some expressions like $5x^2 + 9x + 7$ contain addition of expressions called **terms**.

DEFINITION ***Term:*** An expression that is separated by addition.

Answers

a. $-24 + 20y$ **b.** $\frac{2}{5}y - 8$

For example, the terms in $5x^2 + 9x + 7$ are $5x^2$, $9x$, and 7. An expression containing subtraction, like $3x^2 - 9x + 2$, can be rewritten as a sum to identify its terms. So $3x^2 - 9x + 2$ becomes $3x^2 + (-9x) + 2$ and we see that its terms are $3x^2$, $-9x$, and 2. A term that has no variable, like 2, is called a *constant term*.

The numerical factor in a term is called the **numerical coefficient**, or simply the coefficient of the term.

DEFINITION ***Numerical coefficient:*** The numerical factor in a term.

For example, the coefficients of the terms in $\frac{2}{3}x - 7x + 2$ are $\frac{2}{3}$, -7, and 2.

Sometimes expressions contain **like terms.**

DEFINITION ***Like terms:*** Variable terms that have the same variable(s) raised to the same exponents, or constant terms.

Examples of like terms:	Examples of unlike terms:	
$3x$ and $5x$	$0.2x$ and $8y$	(different variables)
$\frac{3}{4}y^2$ and $9y^2$	$4t^2$ and $4t^3$	(different exponents)
$7xy$ and $3xy$	x^2y and xy^2	(different exponents)
6 and 15	12 and $12x$	(different variables)

Combining Like Terms

We can rewrite expressions containing like terms so that they have fewer terms by combining the like terms. The process of rewriting an expression with fewer symbols is called *simplifying*, and when the expression is written with the fewest symbols possible, we say it is in *simplest form.*

To simplify an expression by combining like terms, we use the distributive property. Consider the expression $3x + 5x$. Notice there is a common factor of x in both terms. Using the distributive property, we write the common factor x outside the parentheses and write the remaining factors, which are the coefficients 3 and 5, as a sum inside the parentheses.

$$\begin{aligned} 3x + 5x &= (3 + 5)x \\ &= 8x \end{aligned}$$

Because the parentheses contain a sum of two numbers, we can simplify by adding them.

Notice that when combining like terms we simply add the coefficients and keep the variable the same.

PROCEDURE **Combining Like Terms**

To combine like terms, add or subtract the coefficients and keep the variables and their exponents the same.

EXAMPLE 5 Combine like terms.

a. $8x + 7x$

Solution $8x + 7x = 15x$

Note: *Think "eight x's plus seven x's is fifteen x's."*

b. $15k^2 - k^2$

Solution $15k^2 - k^2 = 14k^2$

Note: *The coefficient of $-k^2$ is -1.*

YOUR TURN Combine like terms.

a. $2.4n + 6.9n$

b. $\frac{1}{4}y^3 - \frac{2}{3}y^3$

Sometimes, expressions are more complex and contain different sets of like terms or may require using the distributive property to simplify parentheses before like terms can be combined.

EXAMPLE 6 Simplify.

Note: *When we use the commutative property to rearrange the expressions so that like terms are together, we say we are* **collecting** *the like terms.*

a. $10x + 7y - x + 6y$

Solution

$$10x + 7y - x + 6y$$

$$= \underbrace{10x - x} + \underbrace{7y + 6y}$$ Use the commutative property to rearrange the terms.

$$= 9x + 13y$$ Combine like terms: $10x - x = 9x$ and $7y + 6y = 13y$.

b. $0.2m - 1.5mn - 2.5m + 9 - 0.3mn$

Solution

$$0.2m - 1.5mn - 2.5m + 9 - 0.3mn$$

$$= 0.2m - 2.5m - 1.5mn - 0.3mn + 9$$ Use the commutative property to collect the like terms.

$$= -2.3m - 1.8mn + 9$$ Combine like terms.

c. $3h + k - 2(5h + 3k) - 9 + 5k$

Solution

$$3h + k - 2(5h + 3k) - 9 + 5k$$

$$= 3h + k - 10h - 6k - 9 + 5k$$ Distribute -2.

$$= 3h - 10h + k - 6k + 5k - 9$$ Collect like terms.

$$= -7h - 9$$ Combine like terms.

ANSWERS

a. $9.3n$ **b.** $-\frac{5}{12}y^3$

YOUR TURN Combine like terms.

a. $3.2x^2 - 9x + 12 - 0.4x - 12$

b. $\frac{5}{6}h - k + 4 - \frac{3}{4}h + 5k$

c. $3(t + 5) - (5t - 3u) - 1$

ANSWERS

a. $3.2x^2 - 9.4x$ **b.** $\frac{1}{12}h + 4k + 4$ **c.** $-2t + 3u + 14$

1.4 Exercises

For Extra Help

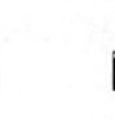

MyMathLab Videotape/DVT InterAct Math Math Tutor Center Math XL.com

1. Why can we translate a phrase that indicates addition such as "eight more than x" as $8 + x$ or $x + 8$?

2. The phrase "nine subtracted from n" translates to $n - 9$, which is in reverse order of what we read. What other key words for subtraction translate in reverse order?

3. In your own words, explain how to evaluate a variable expression.

4. Explain the difference between evaluating and rewriting an expression.

5. What are like terms?

6. Explain how to combine like terms.

For Exercises 7–28, translate the phrase to an algebraic expression.

7. five times a number

8. the product of a number and eight

9. the sum of twice a number and two

10. the difference of five times a number and four

11. five less p

12. seven less than m

13. the product of a number to the fourth power and 8

14. the quotient of the square of y and 9

15. twenty less than the product of a number and two

16. a number subtracted from twice another number

17. fifteen divided by r squared

18. the product of x cubed and y

19. one-half subtracted from the quotient of p and q

20. two-thirds less than the quotient of m and n.

21. m minus three times the sum of a number and 5

22. the difference of seven and a number, decreased by nine

23. the difference of four and t, all raised to the fifth power

24. negative four increased by the sum of x and y

25. six-sevenths of a number multiplied by seven

26. seven-tenths of a number added to 12

27. the sum of x and y less than the difference of m and n.

28. the difference of a and b subtracted from the sum of c and d

For Exercises 29–32, explain the mistake. Then translate the phrase correctly.

29. six less than y; Translation $6 - y$

30. four times the sum of r and seven; Translation: $4r + 7$

31. four subtracted from the square root of m; Translation: $4 - \sqrt{m}$

32. seven divided by the product of two and r; Translation: $2r \div 7$

For Exercises 33–42, translate the indicated phrase.

33. The length of a rectangle is five times the width. If the width is represented by w, then write an expression that describes the length.

34. The width of a rectangle is twice the length. If the length is represented by l, then write an expression that describes the width.

35. The length of a rectangle is three times the width subtracted from two. If the width is represented by the variable w, then write an expression that describes the length.

36. The width of a rectangle is one-fifth the length. If the length is represented by l, then write an expression that describes the width.

37. The diameter of a circle is twice the radius. If r represents the radius, then write an expression for the diameter.

38. The radius of a circle is half the diameter. If d represents the diameter, then write an expression for the radius.

39. Mickie has seventeen coins in her change purse that are all either nickels or quarters. If n represents the number of nickels she has, then write an expression in terms of n that describes the number of quarters.

40. Millie split $2500 between two savings accounts. If n represents the amount in dollars in one account, then write an expression in terms of n for the amount in the other account.

41. Zelda passes a rest stop on the highway. One-half of an hour later, Scott, traveling in the same direction, passes the same rest stop. If t represents the amount of time in hours it takes Scott to catch up to Zelda, write an expression in terms of t that describes the amount of time Zelda has traveled since passing the rest stop.

42. Renee is on her way home and passes a sign. One-fourth of an hour later, Gary, who is traveling in the same direction, passes the same sign. If t represents the amount of time in hours it takes Gary to catch up to Renee, write an expression in terms of t that describes the amount of time Renee has traveled since passing the sign.

Exercises 43–54 contain word descriptions of expressions from mathematics and physics. Translate each description to a variable expression.

43. The circumference of a circle can be found by multiplying the diameter by π. Write an expression for finding the circumference.

44. The perimeter of a rectangle is twice the width plus twice the length. Write an expression for finding the perimeter of a rectangle.

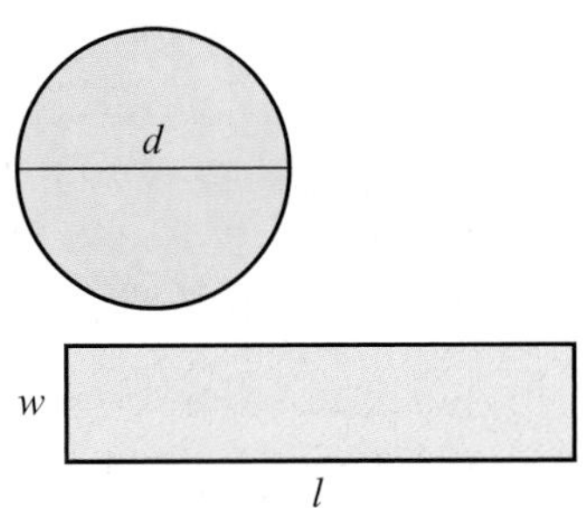

45. The area of a trapezoid is one-half of the product of the height and the sum of the lengths of the sides a and b. Write an expression for finding the area of a trapezoid.

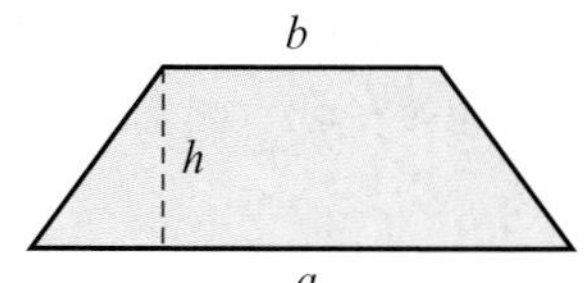

46. The volume of a cylinder is found by multiplying the product of the height and the square of the radius by π. Write an expression for finding the volume of a cylinder.

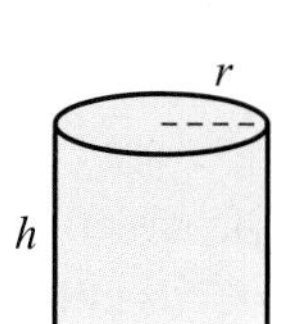

47. The volume of a cone is one-third of the product of π, the square of the radius, and the height of the cone. Write an expression for finding the volume of a cone.

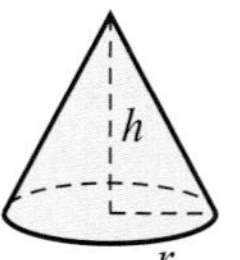

48. The volume of a sphere is four-thirds of the product of π and the cube of the radius. Write an expression for finding the volume of a sphere.

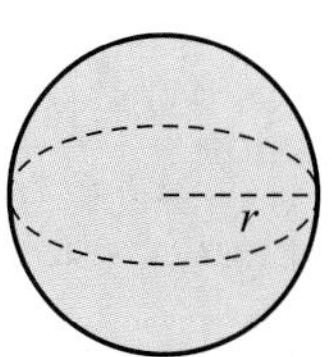

49. In physics, energy is calculated by multiplying the product of the mass and the square of the velocity by one-half. Write an expression for finding energy.

50. When accelerating from rest for an amount of time, the distance an object travels is the product of one-half of the acceleration, a, and the square of the time, t. Write an expression for finding the distance.

51. Isaac Newton discovered that the relationship for gravitational attraction between two bodies is the product of their masses M and m, divided by the square of the distance, d, between them. Write an expression for finding the gravitational attraction between two bodies.

52. Albert Einstein developed an expression that describes the energy of a particle at rest, which is the product of the mass, m, of the particle and the square of the speed of light, which is represented by c. Write an expression for finding the rest energy.

53. Albert Einstein's theory of relativity includes a mathematical expression that is the square root of the difference of 1 and the ratio of the square of the velocity, v, of an object in motion to the square of the speed of light, c. Translate to an algebraic expression.

54. René Descartes developed an expression for the distance between two points in the coordinate plane. The distance between two points is the square root of the sum of the square of the difference of x_2 and x_1 and the square of the difference of y_2 and y_1. Write an expression for finding the distance between two points in the coordinate plane.

Of Interest

Isaac Newton is considered one of the greatest mathematicians ever. He received a B.A. from Trinity College, Cambridge, in 1664. Over the next two years, he developed calculus, discovered the universal law of gravitation, and proved that white light is made up of all colors in the spectrum (all before he was 25). Newton's mathematical model for gravity held until the early 1900s when Albert Einstein revolutionized physics with his theory of relativity.

For Exercises 55–62, evaluate the expression using the given values.

55. $-0.4(x+2)-5;\ x=3$

56. $1.6(y-3)+4;\ y=-2$

57. $-3m^2+5m+1;\ m=-\dfrac{2}{3}$

58. $y^2-2y+3;\ y=-\dfrac{1}{2}$

59. $-|5-x|+|2+y^3|;\ x=5,\ y=-1$

60. $|8m^2-3m|+2;\ m=-0.4$

61. $\sqrt[3]{c}-2ab^2;\ a=-1,\ b=-2,\ c=8$

62. $\sqrt{m-9}+3n^2;\ m=13,\ n=-2$

63. The expression $ad-bc$ is used to calculate the determinant of a matrix, which we will study in detail in Section 4.5. Find the determinant given the values:

a. $a=5,\ b=0.2,\ c=-3,\ d=7$

b. $a=-8,\ b=\dfrac{2}{3},\ c=2,\ d=-\dfrac{5}{6}$

64. The expression b^2-4ac is called the *discriminant* and is used to determine the types of solutions for quadratic equations. We will study the discriminant in detail in Section 9.2. Find the value of the discriminant of each set of values:

a. $a=-2,\ b=4,\ c=3$

b. $a=-1,\ b=2,\ c=-4$

65. The expression $\dfrac{y_2-y_1}{x_2-x_1}$ is used to calculate the slope of a line, which we will discuss fully in Section 3.2. Find the slope given each set of values:

a. $x_1=3,\ y_1=-1,\ x_2=5,\ y_2=-7$

b. $x_1=3,\ y_1=-1,\ x_2=-1,\ y_2=-2$

66. The expression $\sqrt{(x_2-x_1)^2+(y_2-y_1)^2}$ is used to calculate the distance between two points in the coordinate plane. Evaluate the expression using the given values:

a. $x_1=5,\ y_1=-2,\ x_2=-7,\ y_2=3$

b. $x_1=1,\ y_1=-2,\ x_2=9,\ y_2=4$

Connection We will discuss this distance formula more completely in Section 11.1.

For Exercises 67–74, determine all values for the variable that cause the expression to be undefined.

67. $\dfrac{-7}{6-y}$

68. $\dfrac{8}{m+2}$

69. $\dfrac{-5y}{(y+5)(y-1)}$

70. $\dfrac{2m}{(m+2)(m-1)}$

71. $\dfrac{5+x^2}{x}$

72. $\dfrac{6-u^2}{u}$

73. $\dfrac{x+1}{4x+1}$

74. $\dfrac{3y}{3-5y}$

For Exercises 75–82, use the distributive property to write an equivalent expression.

75. $9(3x-5)$

76. $4(x+2)$

77. $-5(m+2)$

78. $-8(2x-7)$

79. $\dfrac{3}{8}\left(\dfrac{2}{9}x-24\right)$

80. $-\dfrac{4}{7}\left(-14k-\dfrac{1}{8}\right)$

81. $-2.1(3x+2.4)$

82. $4.2(2.1x-5)$

For Exercises 83–92, simplify by combining like terms.

83. $2x-13x$

84. $5p+8p$

85. $\dfrac{6}{7}b^2-\dfrac{8}{7}b^2$

86. $\dfrac{1}{2}y-\dfrac{5}{6}y$

87. $4x-9y-12+y+3x$

88. $n-2m+7+2m+8n$

89. $1.5x+y-2.8x+0.3-y-0.7$

90. $0.4t^2+t-2.8-t^2+0.9t-4$

91. $2.6h^2+\dfrac{5}{3}h-\dfrac{2}{5}h^2+h+7$

92. $-0.3t+\dfrac{3}{4}t^2-8+t+\dfrac{1}{2}t-\dfrac{1}{3}t^2$

For Example 93–96, simplify.

93. $2(5n - 6) + 4(n + 1) - 8$

94. $3(x - 7) + 6(2x - 3) + 9$

95. $7a + 5b - 3(4a + 2b) - 12 + 8b$

96. $t - 2(t + 9u) + 7t - 9 + 18u$

97. **a.** Translate to an algebraic expression: fourteen plus the difference of six times a number and eight times the same number.

b. Simplify the expression.

c. Evaluate the expression when the number is -3.

98. **a.** Translate to an algebraic expression: the sum of negative five times a number and eight minus two times the same number.

b. Simplify the expression.

c. Evaluate the expression when the number is 0.2.

REVIEW EXERCISES

[1.1] **1.** Write the set containing all integers greater than or equal to -2 in set-builder notation.

[1.2] **2.** What property of addition is represented by $-9 + (6 + 2) = -9 + (2 + 6)$?

[1.2] **3.** What property is represented by $-1(5 - 2) = -5 + 2$?

[1.3] ***For Exercises 4–6, evaluate.***

4. $-5^2 + 3(6 - 8) - \sqrt{16}$

5. $(20 - 24)^3 - 4|3 - 8|$

6. -7^2

Chapter 1 Summary

Defined Terms

Section 1.1
Variable (p. 2)
Constant (p. 2)
Expression (p. 2)
Equation (p. 2)
Inequality (p. 2)
Set (p. 3)
Natural numbers (p. 3)
Whole numbers (p. 3)
Integers (p. 3)
Subset (p. 4)
Rational number (p. 4)
Irrational number (p. 5)
Prime number (p. 5)
Absolute value (p. 6)

Section 1.2
Additive inverses (p. 13)
Multiplicative inverses (p. 17)

Section 1.3
Exponential form (p. 24)
Base (p. 24)
Exponent (p. 24)
Principal square root (p. 26)

Section 1.4
Term (p. 38)
Numerical coefficient (p. 39)
Like terms (p. 39)

The Real-Number System

Real Numbers (R)

Irrational Numbers (I): Any real number that is not rational, such as $-\sqrt{2}$, $-\sqrt{3}$, $\sqrt{0.8}$, and π.

Rational Numbers (Q): Real numbers that can be expressed in the form $\frac{a}{b}$, where a and b are integers and $b \neq 0$, such as $-4\frac{3}{4}$, $-\frac{2}{3}$, 0.018, $0.\overline{6}$, and $\frac{5}{8}$.

Integers (Z): $\ldots, -3, -2, -1, 0, 1, 2, 3, \ldots$

Whole Numbers (W): $0, 1, 2, 3, \ldots$

Natural Numbers (N): $1, 2, 3, \ldots$

Arithmetic Summary Diagram

Each operation has an inverse operation. In the diagram, the operations build from the top down. Addition leads to multiplication, which leads to exponents. Subtraction leads to division, which leads to roots.

Properties of arithmetic

In each of the following *a*, *b*, and *c* represent real numbers.

Additive Identity
$a + 0 = a$

Commutative Property of Addition
$a + b = b + a$

Associative Property of Addition
$(a + b) + c = a + (b + c)$

Multiplicative Identity
$a \cdot 1 = a$

Multiplicative Property of 0
$a \cdot 0 = 0$

Commutative Property of Multiplication
$ab = ba$

Associative Property of Multiplication
$(ab)c = a(bc)$

Distributive Property
$a(b + c) = ab + ac$

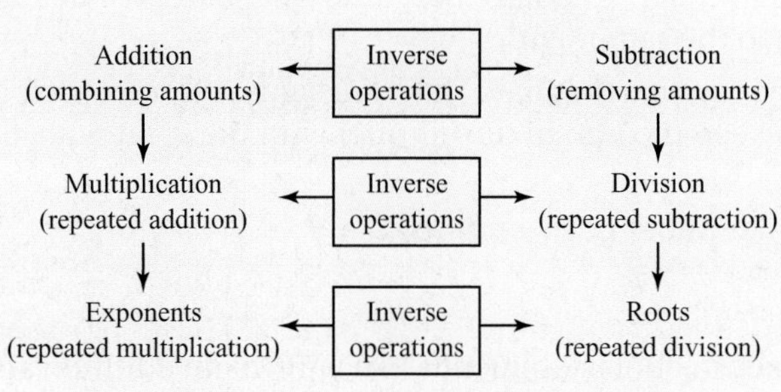

Order of Operations
1. Grouping symbols
2. Exponents and roots
3. Multiply or divide from left to right
4. Add or subtract from left to right

Procedures, Rules, and Key Examples

Procedures/Rules

Section 1.1 Sets and the Structure of Algebra

To write a set, write the elements or members of the set within braces.

The absolute value of a number is positive or zero.

For any two real numbers *a* and *b*, *a* is greater than *b* if *a* is to the right of *b* on a number line.

Key Examples

Example 1: Write the set containing odd natural numbers.

Answer: $\{1, 3, 5, 7, \ldots\}$

Example 2: Find the absolute value.
a. $|12| = 12$ **b.** $|-9| = 9$ **c.** $|0| = 0$

Example 3: Use $<$ or $>$ to make a true sentence.
a. -15 ▒ -10 **b.** 8 ▒ -5

Answers:
a. $-15 < -10$ **b.** $8 > -5$

continued

Procedures/Rules	Key Examples
Section 1.2 Operations with Real Numbers; Properties of Real Numbers	
To add two numbers that have the same sign, add their absolute values and keep the same sign. To add two numbers that have different signs, subtract their absolute values and keep the sign of the number with the greater absolute value.	**Example 1:** Add. **a.** $3 + 9 = 12$ **b.** $-3 + (-9) = -12$ **c.** $-3 + 9 = 6$ **d.** $3 + (-9) = -6$
For any real number n, $-(-n) = n$.	**Example 2:** Simplify $-(-6)$. Answer: $-(-6) = 6$
To write a subtraction statement as an equivalent addition statement, change the operation symbol from a minus sign to a plus sign and change the subtrahend to its additive inverse.	**Example 3:** Subtract. **a.** $3 - 9 = 3 + (-9) = -6$ **b.** $-3 - 9 = -3 + (-9) = -12$ **c.** $3 - (-9) = 3 + 9 = 12$ **d.** $-3 - (-9) = -3 + 9 = 6$
The product of two numbers that have the same sign is positive. The product of two numbers that have different signs is negative. The product of an even number of negative factors is positive, whereas the product of an odd number of negative factors is negative.	**Example 4:** Multiply. **a.** $(3)(4) = 12$ **b.** $(-3)(-4) = 12$ **c.** $(3)(-4) = -12$ **d.** $(-3)(4) = -12$ **e.** $(-1)(-2)(-3)(-4) = 24$ **f.** $(-2)(-3)(-4) = -24$
The quotient of two numbers that have the same sign is positive. The quotient of two numbers that have different signs is negative. $0 \div n = 0$ when $n \neq 0$ $n \div 0$ is undefined when $n \neq 0$ $0 \div 0$ is indeterminate	**Example 5:** Divide. **a.** $24 \div 8 = 3$ **b.** $-24 \div (-8) = 3$ **c.** $-24 \div 8 = -3$ **d.** $24 \div (-8) = -3$ **e.** $0 \div 7 = 0$ **f.** $35 \div 0$ is undefined.
Section 1.3 Exponents, Roots, and Order of Operations	
For any real number b and any natural number n, the nth power of b, or b^n, is found by multiplying b as a factor n times. $b^n = \underbrace{b \cdot b \cdot b \cdot \cdots \cdot b}_{b \text{ used as a factor } n \text{ times}}$ If the base of an exponential form is a negative number and the exponent is even, then the product is positive. If the base is a negative number and the exponent is odd, then the product is negative.	**Example 1:** Evaluate. **a.** $3^4 = 3 \cdot 3 \cdot 3 \cdot 3 = 81$ **b.** $(-5)^4 = (-5)(-5)(-5)(-5) = 625$ **c.** $(-2)^5 = (-2)(-2)(-2)(-2)(-2) = -32$

continued

Procedures/Rules	Key Examples

Section 1.3 Exponents, Roots and Order of Operations (continued)

Every positive number has two square roots, a positive root and a negative root.

Example 2: Find all square roots of 36.

Answer: ± 6

The square root of 0 is 0.

Negative numbers have no real-number square roots.

The radical symbol $\sqrt{}$ denotes only the positive (principal) square root.

$\sqrt{\dfrac{a}{b}} = \dfrac{\sqrt{a}}{\sqrt{b}}$, where $a \geq 0$ and $b > 0$

Example 3: Simplify.

a. $\sqrt{100} = 10$

b. $\sqrt{-64}$ is not a real number.

c. $\sqrt{\dfrac{25}{81}} = \dfrac{\sqrt{25}}{\sqrt{81}} = \dfrac{5}{9}$

d. $\sqrt[3]{64} = 4$

Perform operations in the following order:

1. Grouping symbols: parentheses ( ), brackets [ ], braces { }, absolute value | |, and radicals $\sqrt{}$.
2. Exponents/Roots from left to right, in order as they occur.
3. Multiplication/Division from left to right, in order as they occur.
4. Addition/Subtraction from left to right, in order as they occur.

Example 4: Simplify:

$$\begin{aligned} &-4^3 - 3|16 - (4 + 2 \cdot 9)| + \sqrt{49} \\ &= -64 - 3|16 - (4 + 18)| + 7 \\ &= -64 - 3|16 - 22| + 7 \\ &= -64 - 3|-6| + 7 \\ &= -64 - 3(6) + 7 \\ &= -64 - 18 + 7 \\ &= -75 \end{aligned}$$

Section 1.4 Evaluating and Rewriting Expressions

To translate a word phrase to an expression, identify the variables, constants, and key words, then write the corresponding symbolic form.

To evaluate an algebraic expression,

1. Replace each variable with its corresponding given number.
2. Simplify the resulting numerical expression.

To combine like terms, add or subtract the coefficients and keep the variables and their exponents the same.

Example 1: Translate the phrase to an expression:

a. Three less than four times a number

Answer: $4n - 3$

b. Eight times the difference of a number and five

Answer: $8(n - 5)$

Example 2:

Evaluate $5x^3 - 4x$ when $x = -2$.

$$\begin{aligned} 5(-2)^3 - 4(-2) &= 5(-8) - 4(-2) \\ &= -40 + 8 \\ &= -32 \end{aligned}$$

Example 3: Distribute.

$$\begin{aligned} -3(x - 5) &= -3 \cdot x - (-3) \cdot 5 \\ &= -3x + 15 \end{aligned}$$

Example 4: Combine like terms.

$$\begin{aligned} &6x^2 - 9x + x^2 - 12 + 9x + 7 \\ &= 6x^2 + x^2 - 9x + 9x - 12 + 7 \\ &= 7x^2 - 5 \end{aligned}$$

Chapter 1 Review Exercises

For Exercises 1–6, answer true or false.

[1.1] **1.** $\sqrt{3}$ is an irrational number.

[1.2] **2.** When using the order of operations, multiplication always comes before division.

[1.1] **3.** The absolute value of every number is positive.

[1.3] **4.** $-3^2 = -9$

[1.3] **5.** $(-3)^2 = 9$

[1.1] **6.** All whole numbers are rational numbers.

For Exercises 7–10, fill in the blank.

[1.1] **7.** A variable is a symbol that can ______________ in value.

[1.2] **8.** The product of an even number of negative factors is ______________, whereas the product of an odd number of negative factors is ______________.

[1.3] **9.** If the base of an exponential form is a negative number and the exponent is even, then the product is ______________.

[1.4] **10.** To evaluate a variable expression, (1) ______________ each variable with its corresponding given value; (2) simplify the resulting numerical expression.

[1.1] ***For Exercises 11–14, write a set representing each description.***

11. The states outside the 48 contiguous states of the United States

12. The odd integers

13. The natural number multiples of 5

14. The letters in the word *simplify*

[1.1] ***For Exercises 15–18, write the set in set-builder notation.***

15. $\{3, 6, 9, \ldots\}$

16. $\{0, 1, 2, 3, \ldots\}$

17. $\{2, 3, 5, 7, 11, \ldots\}$

18. {Monday, Tuesday, Wednesday, Thursday, Friday, Saturday, Sunday}

[1.1] ***For Exercises 19–22, answer true or false.***

19. March $\in \{n \mid n$ is a day of the week$\}$

20. If $A = \{3, 4, 5\}$ and $B = \{2, 3, 5\}$, then $B \subseteq A$.

21. $\frac{2}{3} \in I$

22. $\sqrt{2} \notin I$

[1.1] ***For Exercises 23–26, use =, <, or > to write a true statement.***

23. $-(-5) \; \square \; |-5|$

24. $\left|-3\frac{1}{4}\right| \; \square \; -3\frac{1}{4}$

25. $0.5 \; \square \; \frac{1}{2}$

26. $-|-3| \; \square \; -4$

[1.2] ***For Exercises 27–30, indicate whether the given equation illustrates the additive identity, multiplicative identity, additive inverses, or multiplicative inverses.***

27. $5.8 + (-5.8) = 0$

28. $-\frac{2}{5} \cdot -\frac{5}{2} = 1$

29. $0 + (-9) = -9$

30. $8.7(1) = 8.7$

[1.2] ***For Exercises 31–36, indicate whether the given equation illustrates the commutative property of addition, commutative property of multiplication, associative property of addition, associative property of multiplication, or distributive property.***

31. $6(7 + 5) = 6 \cdot 7 + 6 \cdot 5$

32. $-3(4 \cdot 5) = (-3 \cdot 4)5$

33. $-6 + 5 = 5 + (-6)$

34. $3 \cdot (2 \cdot 4) = 3 \cdot (4 \cdot 2)$

35. $-7 + (1 + 8) = (-7 + 1) + 8$

36. $-2(3 + 5) = -2 \cdot 3 - 2 \cdot 5$

[1.2] ***For Exercises 37–40, add.***

37. $6 + (-7)$

38. $-4 + 9$

39. $-15 + (-2)$

40. $-2 + (-5)$

[1.2] ***For Exercises 41–44, subtract.***

41. $7 - 9$

42. $-2 - 8$

43. $-15 - (-2)$

44. $8 - (-1)$

[1.2] ***For Exercises 45–52, multiply or divide.***

45. $-2(4)$

46. $-3(-5)$

47. $7(-8)$

48. $25 \div (-5)$

49. $-10 \div (-5)$

50. $-8 \div 4$

51. $-1(-2)(-3)$

52. $-50 \div (4 \div (-2))$

[1.3] ***For Exercises 53–54, identify the base and the exponent, then translate the expression to words, and then simplify.***

53. -2^7

54. $(-1)^4$

[1.3] ***For Exercises 55–58, evaluate.***

55. -3^2

56. -2^3

57. $(-4)^2$

58. $\left(-\frac{2}{5}\right)^3$

[1.3] *For Exercises 59–62, evaluate the root.*

59. $\sqrt{121}$

60. $\sqrt[4]{81}$

61. $\sqrt[3]{27}$

62. $\sqrt[8]{1}$

[1.3] *For Exercises 63–68, evaluate using the order of operations.*

63. $-7|8 - 2^4| + 7 - 3^2$

64. $8(1 - 3^2) + \sqrt{16 - 7}$

65. $\sqrt[3]{8} - 7 \cdot 3^2$

66. $\sqrt{16} + \sqrt{9} - 3(2 - 7)$

67. $5^2(3 - 8)^2$

68. $\dfrac{3^4 - 5[3 - 4(-2)]}{16 \div 8(-5)}$

[1.4] *For Exercises 69–74, translate the phrase to an algebraic expression.*

69. fourteen minus a number times eight

70. twice the sum of a number and two

71. the sum of a number and one-third of the difference of the number and four

72. the ratio of m and the square root of n

73. sixteen subtracted from half of the difference of a number and eight

74. twenty less than the sum of a number and five

[1.4] *For Exercises 75–76, translate the indicated phrase.*

75. The length of a rectangle is twice the width. If the width is represented by w, then write an expression that describes the length.

76. Laura is traveling south on a highway and passes a grocery store. One-third of an hour later, Tom, who is traveling in the same direction on the highway, passes the same store. If t represents the amount of time in hours it takes Tom to catch up to Laura, write an expression in terms of t that describes the amount of time Laura has traveled since passing the store.

[1.4] *For Exercises 77–80, evaluate the expression using the given values.*

77. $3a^2 - 4a + 2;\ a = -1$

78. $-4|-b + 2ac|;\ a = 1, b = -2, c = -1$

79. $15(x + y)^2 + x^2;\ x = 1, y = -2$

80. $\sqrt{a - b} + 3^0 - a;\ a = 25, b = 16$

[1.4] *For Exercises 81 and 82, determine all values for the variable that cause the expression to be undefined.*

81. $\dfrac{2}{x - 3}$

82. $\dfrac{5x - 3}{3x + 5}$

[1.4] *For Exercises 83 and 84, use the distributive property to write an equivalent expression.*

83. $-2(5x + 1)$

84. $4(2a + 3b - 4)$

[1.4] *For Exercises 85–88, simplify by combining like terms.*

85. $3x + 2x^2 - 4x - x - 3x^2$

86. $5m^5 + 3mn - 2mn^2 - mn - 4m^5$

87. $-7ab + 3ab^2 + 2ab + 3a - 7a^2 - 8$

88. $6r - 3 - r - 2r - 7$

For Exercises 89–92, solve.

Equations and Inequalities

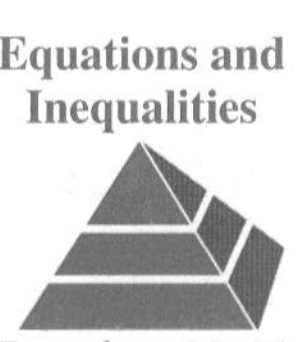

Exercises 89–92

[1.2] 89. Leanne has a balance of $-\$245.85$ on her credit card. Find her new balance if she makes the following transactions:

Transaction	Amount
Payment	\$125.00
Wal-Mart	−\$72.34
Starbucks	−\$12.50
Krispy Kreme	−\$14.75

[1.3] 90. Find the GPA.

Course	Course Credits	Grade
NUR 110	2	B
MAT 140	4	A
PSY 201	3	B
ECO 101	3	D

[1.3] 91. A wireless service charges \$59 for 2000 minutes in the primary calling zone, then \$0.10 for each additional minute in the calling zone. Calls outside of the calling zone cost \$0.15 per minute. Suppose a subscriber uses 1568 minutes in the calling zone and 37 minutes outside the calling zone. Write a numerical expression that describes the total cost for the month, then calculate the total.

[1.3] 92. A store owner wants to use a code system to label his inventory. How many different codes are possible if each code label has three letters (A–E) and five digits (0–9)?

Chapter 1 Practice Test

For Exercises 1–5, simplify.

1. $|8.1|$
2. $-\left|-\frac{11}{4}\right|$
3. $\sqrt{169}$
4. $\sqrt[3]{125}$
5. $\sqrt[5]{\frac{1}{32}}$

For Exercises 6 and 7, indicate whether the expression illustrates the commutative property of addition, the associative property of addition, the commutative property of multiplication, the associative property of multiplication, or the distributive property.

6. $8(1 + 3) = 8(3 + 1)$
7. $2(5 \cdot 3) = (2 \cdot 5)3$

For Exercises 8–19, simplify.

8. $9 + (-1)$
9. $\frac{2}{3} - \left(-\frac{1}{4}\right)$
10. $(-3)(2.5)$
11. $(-5)^2$
12. $-\frac{2}{5} \div \frac{5}{2}$
13. $\sqrt[4]{16}$
14. $8 \div 4 \cdot 2$
15. $-3^2 + 7 - 2(5 - 1)$
16. $\sqrt[4]{16} + 8 - (3 + 1)^2$
17. $4 \div |8 - 6| + 2^4$
18. $(5 - 4)^5 + (2 - 3)^3$
19. $\sqrt{9 + 16} + [-2^2 + 3(2 - 5)]$

20. Collin has a balance of $-\$423.75$ on a credit card. Find his balance after the following transactions:

Description	Amount
Outback Steakhouse	−\$84.50
Texaco	−\$24.80
Payment	\$500
Best Buy	−\$356.45

21. Listed are the top Internet reference web sites. Find the average number of visitors per site. Each site is listed with the number of unique visitors in July, in millions.

Is That a Fact?

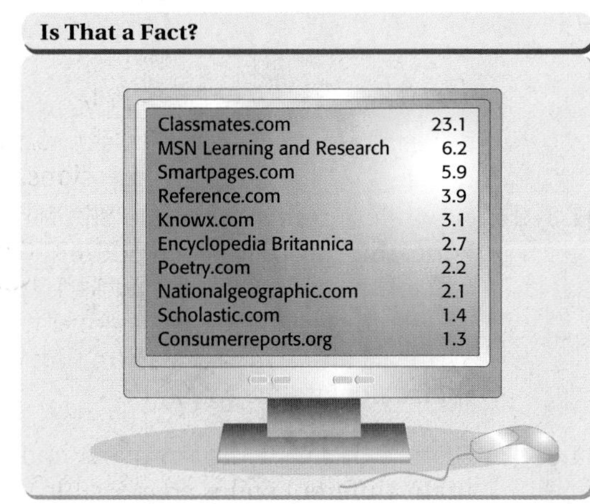

Site	Visitors
Classmates.com	23.1
MSN Learning and Research	6.2
Smartpages.com	5.9
Reference.com	3.9
Knowx.com	3.1
Encyclopedia Britannica	2.7
Poetry.com	2.2
Nationalgeographic.com	2.1
Scholastic.com	1.4
Consumerreports.org	1.3

Source: Comscore Media Metrix

22. Evaluate: $-2|3 - 4xy^2|$ when $x = 2$ and $y = -3$
23. Evaluate: $\frac{a}{b} - \sqrt{a} + \sqrt{b}$ when $a = 16$ and $b = 4$
24. Use the distributive property to write an equivalent expression: $-7(3x + 5)$
25. Simplify: $\frac{2}{5}x + \frac{3}{4}y - 5x + 6y + 2.7$

CHAPTER 2

Linear Equations and Inequalities in One Variable

"The world looks like a multiplication table or a mathematical equation, which, turn it how you will, balances itself."

"All love is mathematically just, as much as the two sides of an algebraic equation."

—Ralph Waldo Emerson (1803–1882)
American essayist, poet, and philosopher

"Inequality is a fact. Equality is a value."

—Mason Cooley U.S. aphorist

In Chapter 1, we reviewed number sets, arithmetic operations, properties, and how to evaluate and rewrite expressions. In this chapter, we build upon the foundation we put in place in Chapter 1 and learn to solve equations and inequalities that have one variable. We will also discuss (in Section 2.2) a general approach to solving application problems, which we will use throughout the text.

2.1 Linear Equations and Formulas

OBJECTIVES

1. **Solve linear equations in one variable.**
2. **Identify identities and contradictions.**
3. **Solve formulas for specified variables.**

In Chapter 1, we reviewed real numbers and their properties and also defined constants, variables, expressions, equations, and inequalities. We now move up the algebra pyramid and consider equations. Recall that an equation has an equal sign.

Learning Strategy

As suggested in the To the Student section on page xix, arrange your notebook with four sections: Notes, Homework, Study Sheets, and Practice Tests. Remember to use color in your notes: red for definitions and blue for rules and procedures. Organization and color coding will help you locate important items faster.

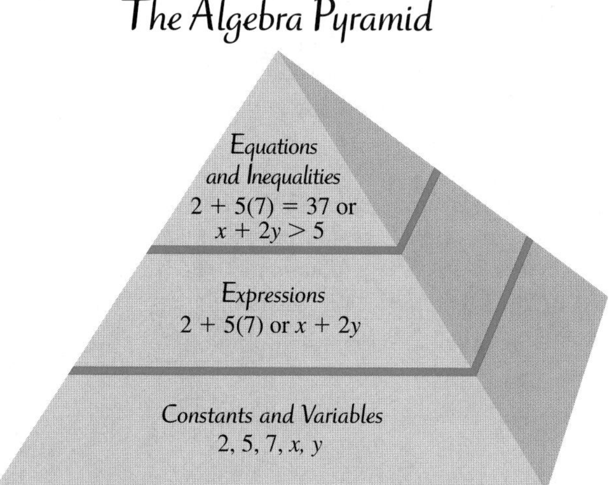

An equation can be either true or false. For example, the equation $4 + 1 = 5$ is true, and the equation $8 - 2 = 7$ is false. If an equation contains a variable for a missing number, such as $x + 3 = 9$, our goal is to find its **solution(s)**. We can write those solutions in a **solution set**.

DEFINITIONS ***Solution:*** A number that makes an equation true when it replaces the variable in the equation.
Solution set: A set containing all the solutions for a given equation.

For example, 6 is a solution to the equation $x + 3 = 9$ because replacing x with 6 makes the equation true.

$$x + 3 = 9$$
$$\downarrow$$
$$6 + 3 = 9 \quad \text{True}$$

Note: *By showing that the equation $x + 3 = 9$ is a true statement when x is replaced with 6, we have checked that 6 is the solution for the equation.*

The solution set for this equation is $\{6\}$ or, in set-builder form, $\{x \mid x = 6\}$. We will focus on writing solution sets when we learn about inequalities in Section 2.3.

OBJECTIVE 1. Solve linear equations in one variable. In this chapter, we will solve **linear equations in one variable.**

DEFINITION ***Linear equation in one variable:*** An equation that can be written in the form $ax + b = c$, where a, b, and c are real numbers and $a \neq 0$.

The equation $2x + 3 = 7$ is a linear equation in one variable. Notice that the exponent of x is 1, which is an important feature of linear equations. If an equation in one variable has a variable raised to an exponent other than 1, then the equation is not linear. For example, $x^2 - 2x + 5 = 8$ is not linear because it contains x^2.

Two principles of equality are used to solve linear equations in one variable: the *addition principle of equality* and the *multiplication principle of equality.*

RULES **The Addition Principle of Equality**

If $a = b$, then $a + c = b + c$ is true for all real numbers a, b, and c.

The Multiplication Principle of Equality

If $a = b$, then $ac = bc$ is true for all real numbers a, b, and c, where $c \neq 0$.

The addition principle of equality says that adding the same number to both sides of an equation gives an equivalent equation, meaning its solution(s) is the same. Likewise, the multiplication principle says that multiplying both sides of an equation by the same nonzero number gives an equivalent equation.

When we solve an equation using these principles, our goal is to write an equation equivalent to the original that has the variable isolated. In other words, we want the variable alone on one side of the equation.

EXAMPLE 1 Solve and check. $2x - 9 = 5$

Solution Use the addition principle to isolate the variable term, then use the multiplication principle to isolate the variable.

$2x - 9 + 9 = 5 + 9$ Add 9 to both sides to isolate $2x$.

$2x = 14$ Simplify.

$\frac{1}{\cancel{2}_1} \cdot \frac{\cancel{2}^1 x}{1} = \frac{\cancel{14}^7}{1} \cdot \frac{1}{\cancel{2}_1}$ Multiply both sides by $\frac{1}{2}$ to isolate x.

$x = 7$

Check $2(7) - 9 \stackrel{?}{=} 5$ Replace x in the original equation with 7 and verify that the equation is true.

$14 - 9 \stackrel{?}{=} 5$

$5 = 5$ True, therefore, 7 is correct.

Note: *Dividing by 2 is equivalent to multiplying by $\frac{1}{2}$. Also, dividing both sides by 2 looks cleaner.*

$$\frac{2x}{2} = \frac{14}{2}$$
$$x = 7$$

From now on, if the coefficient is an integer, we will divide both sides by that integer instead of multiplying by its reciprocal.

YOUR TURN Solve and check.

a. $4y - 19 = 1$

b. $9 = -12t + 3$

Solving Equations with Variable Terms on Both Sides

If variable terms appear on *both* sides of the equal sign, we use the addition principle to get the variable terms on one side of the equal sign and the constant terms on the other side.

EXAMPLE 2 Solve and check. $5y - 6 = 2y - 30$

Note: *Had we added $-5y$ to both sides, after simplifying, the right side of the equation would contain $-3y$, which has a negative coefficient. By adding $-2y$, we avoided a negative coefficient after simplifying.*

Solution $5y - 2y - 6 = 2y - 2y - 30$ — Add $-2y$ to both sides to get the variable terms together.

$3y - 6 + 6 = -30 + 6$ — Add 6 to both sides to isolate $3y$.

$\frac{3y}{3} = \frac{-24}{3}$ — Divide both sides by 3 to isolate y.

$y = -8$

Check $5(-8) - 6 \stackrel{?}{=} 2(-8) - 30$ — Replace y in the original equation with -8, and verify that the equation is true.

$-40 - 6 \stackrel{?}{=} -16 - 30$

$-46 = -46$ — True; therefore, -8 is correct.

Simplifying First

If an equation contains parentheses or like terms that appear on the same side of the equal sign, we first clear the parentheses and combine like terms.

EXAMPLE 3 Solve and check. $18 - (3n - 5) = 11n - 4(n + 3)$

Solution $18 - 3n + 5 = 11n - 4n - 12$ — Distribute to clear parentheses.

$23 - 3n = 7n - 12$ — Combine like terms.

$23 - 3n + 3n = 7n + 3n - 12$ — Since $-3n$ has the lesser coefficient, we add $3n$ to both sides.

$23 + 12 = 10n - 12 + 12$ — Add 12 to both sides to isolate $10n$.

$\frac{35}{10} = \frac{10n}{10}$ — Divide both sides by 10 to isolate n.

$3.5 = n$

Check $18 - (3(3.5) - 5) \stackrel{?}{=} 11(3.5) - 4(3.5 + 3)$ — Replace n in the original equation with 3.5 and verify that the equation is true.

$18 - (10.5 - 5) \stackrel{?}{=} 38.5 - 4(6.5)$

$18 - 5.5 \stackrel{?}{=} 38.5 - 26$

$12.5 = 12.5$ — True; therefore, 3.5 is correct.

ANSWERS

a. 5 b. $-\frac{1}{2}$

YOUR TURN Solve and check.

a. $6y - 12 + 3y = 15 + 4y - 7$ **b.** $11 + 5(k - 1) = 3k - (6k + 2)$

Equations with Fractions or Decimals

We can use the multiplication principle of equality to clear fractions and decimals from an equation so that it contains only integers. Although equations can be solved without clearing fractions or decimals, most people find equations that contain only integers easier to solve.

If the equation contains fractions, we multiply both sides by a number that will eliminate all the denominators. The easiest number to use is the least common denominator (LCD).

EXAMPLE 4 Solve and check. $\frac{1}{5}x - \frac{3}{4} = \frac{1}{2}x + 1$

Solution

$20\left(\frac{1}{5}x - \frac{3}{4}\right) = 20\left(\frac{1}{2}x + 1\right)$ Clear the fractions by multiplying both sides by the LCD, 20.

$20 \cdot \frac{1}{5}x - 20 \cdot \frac{3}{4} = 20 \cdot \frac{1}{2}x + 20 \cdot 1$ Distribute.

$4x - 15 = 10x + 20$ Simplify.

$4x - 4x - 15 = 10x - 4x + 20$ Subtract $4x$ from both sides.

$-15 - 20 = 6x + 20 - 20$ Subtract 20 from both sides.

$\frac{-35}{6} = \frac{6x}{6}$ Divide both sides by 6 to isolate x.

$-\frac{35}{6} = x$

Check

$\frac{1}{5}\left(-\frac{35}{6}\right) - \frac{3}{4} \stackrel{?}{=} \frac{1}{2}\left(-\frac{35}{6}\right) + 1$ Replace x in the original equation with $-\frac{35}{6}$ and verify that the equation is true.

$-\frac{7}{6} - \frac{3}{4} \stackrel{?}{=} -\frac{35}{12} + 1$ Multiply.

$-\frac{14}{12} - \frac{9}{12} \stackrel{?}{=} -\frac{35}{12} + \frac{12}{12}$ Write equivalent fractions with their LCD.

$-\frac{23}{12} = -\frac{23}{12}$ True; therefore, $-\frac{35}{6}$ is correct.

Clearing Decimals

To clear decimal numbers in an equation, we multiply by an appropriate power of 10 as determined by the decimal number with the most places.

ANSWERS

a. 4 **b.** −1

Note: *Since 0.25 has more places than any of the other decimal numbers, it determines the size of the power of 10. Since it has two places, we multiply by 10^2, which is 100.*

Connection Since decimal numbers represent fractions with denominators that are powers of 10, we are still multiplying by the LCD when we clear them. For example,

$$0.25 = \frac{25}{100} \quad \text{and} \quad 0.4 = \frac{4}{10}$$

The LCD for these fractions is 100.

EXAMPLE 5 Solve and check. $0.4(n - 3) = 0.25n + 6$

Solution

$100 \cdot 0.4(n - 3) = 100(0.25n + 6)$	Multiply both sides by 100.
$40(n - 3) = 100 \cdot 0.25n + 100 \cdot 6$	Multiply $100 \cdot 0.4$ and distribute 100.
$40n - 120 = 25n + 600$	Distribute 40 and multiply.
$40n - 25n - 120 = 25n - 25n + 600$	Subtract $25n$ from both sides.
$15n - 120 + 120 = 600 + 120$	Add 120 to both sides.
$\frac{15n}{15} = \frac{720}{15}$	Divide both sides by 15.
$n = 48$	

Check

$0.4(48 - 3) \stackrel{?}{=} 0.25(48) + 6$	Replace n in the original equation with 48 and verify that the equation is true.
$0.4(45) \stackrel{?}{=} 12 + 6$	Simplify.
$18 = 18$	True; therefore, 48 is correct.

The following outline summarizes the process of solving linear equations in one variable.

PROCEDURE **Solving Linear Equations**

To solve linear equations,

1. Simplify both sides of the equation as needed.
   a. Distribute to clear parentheses.
   b. Clear fractions or decimals by multiplying through by the LCD. In the case of decimals, the LCD is the power of 10 with the same number of 0 digits as decimal places in the number with the most decimal places. (Clearing fractions and decimals is optional.)
   c. Combine like terms.
2. Use the addition principle so that all variable terms are on one side of the equation and all constants are on the other side. (Clear the variable term with the lesser coefficient.) Then combine like terms.
3. Use the multiplication principle to clear the remaining coefficient.

YOUR TURN Solve and check.

a. $\frac{1}{3}(x - 1) = \frac{3}{4}x + \frac{1}{6}$ **b.** $4.8t - 2.46 = 0.3t - 14.16$

ANSWERS

a. $-\frac{6}{5}$ **b.** -2.6

OBJECTIVE 2. Identify identities and contradictions. In general, a linear equation in one variable has only one real-number solution. We say such an equation is a **conditional equation.**

DEFINITION ***Conditional linear equation in one variable:*** An equation with exactly one solution.

However, there are two special cases that we need to consider.

Identities

If the left and right sides of an equation are *identical*, as in $2x + 1 = 2x + 1$, then it is an **identity**. Since the expressions are identical, we can replace x with any real number and the equation will check, so every real number is a solution for an identity.

DEFINITION ***Identity:*** An equation in which every real number (for which the equation is defined) is a solution.

Sometimes, we may have to simplify the expressions in an equation to see that it is an identity.

EXAMPLE 6 Solve and check. $4(3x - 2) + 5 = 9x - 3(1 - x)$

Solution

$$12x - 8 + 5 = 9x - 3 + 3x \quad \text{Distribute.}$$

$$12x - 3 = 12x - 3 \quad \text{Combine like terms.}$$

Because the left and right sides are identical, the equation is an identity and every real number is a solution. The solution set is $\{x \mid x \text{ is a real number}\}$, or $\mathbb{R}$.

Although we stopped solving the equation at $12x - 3 = 12x - 3$, we could have continued the process. Each step would affirm that the equation is an identity.

$$12x - 12x - 3 = 12x - 12x - 3 \quad \text{Subtract } 12x \text{ from both sides.}$$

$$-3 = -3 \quad \text{It is still clear that the equation is an identity.}$$

Contradictions

After simplifying both sides of an equation, if the variable terms are identical but the constant terms are not identical, as in $2x + 1 = 2x + 5$, then the equation is a **contradiction**. Replacing x in $2x + 1$ and $2x + 5$ with the same real number will always yield different results, suggesting that contradictions have no solution.

DEFINITION ***Contradiction:*** An equation that has no solution.

EXAMPLE 7 Solve and check. $4x + 2x - 7 = 5 + 6x - 8$

Solution $6x - 7 = 6x - 3$ Combine like terms.

Because the variable terms are identical and the constant terms are not, the equation is a contradiction and has no solution. The solution set is empty, which we indicate by { } or ∅.

As we saw with the identity in Example 6, we can continue the process of solving a contradiction and each step will affirm that it is a contradiction.

$$6x - 6x - 7 = 6x - 6x - 3$$ Subtract $6x$ from both sides.

$$-7 = -3$$ This false equation with no variable terms affirms that the equation is a contradiction.

Warning: Do not write the symbol for empty set within braces, as in {∅}. This no longer indicates an empty set because the braces now contain something, so they are not truly empty.

OBJECTIVE 3. Solve formulas for specified variables. If an equation contains more than one variable, such as in a formula, we can often use the methods we have learned to solve for one of the variables. For example, suppose we are asked to solve for w in the formula for the area of a rectangle, $A = lw$. Since w is multiplied by l, we view l as a coefficient of w, so to solve for w, we divide both sides by l.

$$\frac{A}{l} = \frac{\cancel{l}w}{\cancel{l}}$$

$$\frac{A}{l} = w$$

Our example suggests the following procedure.

PROCEDURE **Isolating a Variable in a Formula**

To isolate a particular variable in a formula,

1. Treat all other variables like constants.
2. Isolate the desired variable using the outline for solving equations.

EXAMPLE 8 Solve the given formula for the specified variable.

a. $P = R - C$; solve for R.

Solution $P + C = R - C + C$ Add C to both sides to isolate R.

$P + C = R$ Simplify.

b. $I = Prt$, solve for r.

Solution $$\frac{I}{Pt} = \frac{\cancel{P}r\cancel{t}}{\cancel{P}\cancel{t}}$$ The product Pt is a coefficient of r, so we divide both sides by Pt to isolate r.

$$\frac{I}{Pt} = r$$ Simplify.

Of Interest

The formula $P = R - C$ is used to calculate profit, given revenue R and cost C. The formula $I = Prt$ is used to calculate interest I given principal P at a rate r over a time t.

c. $P = 2l + 2w$; solve for w.

Solution $P - 2l = 2l - 2l + 2w$ Subtract $2l$ from both sides to isolate $2w$.

$\frac{P - 2l}{2} = \frac{2w}{2}$ Divide both sides by 2 to isolate w.

$\frac{P - 2l}{2} = w$ Simplify.

Note: *We could rewrite* $\frac{P - 2l}{2}$.

$w = \frac{P - 2l}{2} = \frac{P}{2} - \frac{2l}{2} = \frac{P}{2} - l$

d. $C = \frac{5}{9}(F - 32)$; solve for F

Solution $\frac{9}{5} \cdot C = \frac{9}{5} \cdot \frac{5}{9}(F - 32)$ Multiply both sides by $\frac{9}{5}$ to isolate $F - 32$.

$\frac{9}{5}C + 32 = F - 32 + 32$ Add 32 to both sides.

$\frac{9}{5}C + 32 = F$ Simplify.

Of Interest

The formula

$$C = \frac{5}{9}(F - 32)$$

is used to convert degrees Fahrenheit to degrees Celsius.

Your Turn Solve the formula for the indicated variable.

a. $V = ir$; solve for i.

b. $A = \frac{1}{2}h(a + b)$; solve for h.

c. $B = P + Prt$; solve for r.

Answers

a. $i = \frac{V}{r}$

b. $h = \frac{2A}{a + b}$

c. $r = \frac{B - P}{Pt}$ or $r = \frac{B}{Pt} - \frac{1}{t}$

2.1 Exercises

1. What is a solution for an equation?

2. What is a solution set?

3. What is the solution set for an identity?

4. What is the solution set for a contradiction?

5. Explain how you would isolate t in the distance formula, $d = rt$.

6. What is the first step for isolating x in the slope formula, $y = mx + b$?

For Exercises 7–24, solve and check.

7. $2x - 2 = 10$

8. $3x + 3 = 15$

9. $9 - 3m = 12$

10. $7 - 5u = 13$

11. $2x + 2 = x + 3$

12. $3p + 4 = 4p - 4$

13. $4t + 3 = 2t + 9$

14. $6q - 5 = 3q + 4$

15. $2(3z + 1) = 8$

16. $3(h + 2) = 12$

17. $n + 2(3n + 1) = 9$

18. $2b + 3(b - 2) = 4$

19. $3(w + 2) - 3 = 3(w - 1)$

20. $2(l + 5) - 2 = 2 + 2(l + 4)$

21. $\frac{1}{2}z - 1 = 4z - 3 - 3z$

22. $\frac{1}{3}y - 5 = 7y + 4 - 5y$

23. $0.5x + 0.95 = 0.2x - 1$

24. $0.2p - 0.7 = 0.7p - 3.6$

For Exercises 25–44, solve and check. Note that some equations may be identities or contradictions.

25. $17a - 5 - 7a = 2a + 19$

26. $7d - 42 - 27d = -64 - 9d$

27. $8r - 2(r + 5) = 5(2r - 6) - 8$

28. $12c + 2(c - 3) = 7 + 5(2c - 1)$

29. $8 - 5(3x - 2) = 38 - (x - 8)$

30. $18 - 8(3 - s) = -5(s + 4) + 27$

31. $8h - 27 + 5(2h - 3) = 62 - (3h + 8)$

32. $5m + 3 + 3(4m - 5) = 5(m + 5) - 5$

33. $\frac{2}{3}q - 4 = \frac{1}{5}q + 10$

34. $\frac{4}{5}k - 6 = 7 - \frac{1}{2}k$

35. $\frac{2}{7}(x - 5) = \frac{1}{5}(x + 5)$

36. $\frac{2}{3}(d - 6) = 3 + \frac{3}{4}d$

37. $0.5(x - 2) + 1.76 = 0.3x + 0.8$

38. $3.3 - 0.6a = 1.1(4 - a) - 2$

39. $2x - 3.24 + 2.4x = 6.2 + 0.08x + 3.52$

40. $x - 2.28 + 1.6x = 4.6 + 0.05x - 0.25$

41. $6(m + 3) - 5 + 2m = 3(3m + 1) - m$

42. $16(q - 3) + 2q - 1 = 5(q + 1) + 13q + 2$

43. $7(n + 2) - 3n = 4 + 4n + 10$

44. $-5(p + 2) - 3p + 5 = -5 - 8p$

For Exercises 45–48, explain the mistake, then solve correctly.

45.
$$\begin{aligned} 2(x + 3) &= 7x - 1 \\ 2x + 3 &= 7x - 1 \\ 3 &= 5x - 1 \\ 4 &= 5x \\ \frac{4}{5} &= x \end{aligned}$$

46.
$$\begin{aligned} 4 + 3(p - 1) &= 14 \\ 4 + 3p - 1 &= 14 \\ 3 + 3p &= 14 \\ 3p &= 11 \\ p &= \frac{11}{3} \end{aligned}$$

47. $$\begin{aligned} 4 - 5(x+1) + 4x &= 7 - 11 \\ -1(x+1) + 4x &= 7 - 11 \\ -x - 1 + 4x &= -4 \\ 3x - 1 &= -4 \\ 3x &= -3 \\ x &= -1 \end{aligned}$$

48. $$\begin{aligned} 7 - 2(d+2) - 3 &= 5 \\ 5(d+2) - 3 &= 5 \\ 5d + 10 - 3 &= 5 \\ 5d + 7 &= 5 \\ 5d &= -2 \\ d &= -\frac{2}{5} \end{aligned}$$

For Exercises 49–66, solve the formula for the indicated variable.

49. $P = R - C$; solve for C

50. $c^2 = a^2 + b^2$; solve for b^2

51. $A = bh$; solve for b

52. $I = Prt$; solve for t

53. $A = 2\pi pw$; solve for w

54. $A = 2\pi rh$; solve for r

55. $A = \dfrac{\theta r^2}{2}$; solve for r^2

56. $V = \dfrac{1}{3}\pi r^2 h$; solve for h

57. $F = \dfrac{kMm}{d^2}$; solve for M

58. $t = -\dfrac{2\omega}{\alpha}$; solve for ω

59. $A = \pi s(R + r)$; solve for s

60. $B = P(1 + rt)$; solve for P

61. $P = 2l + 2w$; solve for l

62. $P = 2\pi r + 2d$; solve for d

63. $F = \dfrac{9}{5}C + 32$; solve for C

64. $h = -16t^2 + h_0$; solve for t^2

65. $x = vt + \dfrac{1}{2}at^2$; solve for a

66. $E = \dfrac{1}{2}mv^2 + mgy$; solve for v^2

For Exercises 67–70, explain the mistake, then solve correctly.

67. $V = lwh$; solve for h
$$\begin{aligned} V - lw &= lwh - lw \\ V - lw &= h \end{aligned}$$

68. $A = bh$; solve for h
$$\begin{aligned} A - b &= bh - b \\ A - b &= h \end{aligned}$$

69. $P = 2l + 2w$; solve for l

$$P - 2w = 2l + 2w - 2w$$
$$P - 2w = 2l$$
$$P - 2w - 2 = 2l - 2$$
$$P - 2w - 2 = l$$

70. $A = \frac{1}{2}bh$; solve for b

$$A = 2 \cdot \frac{1}{2}bh$$
$$A = bh$$
$$\frac{A}{h} = b$$

REVIEW EXERCISES

[1.1] **1.** Write a set containing the odd natural numbers less than 14.

[1.2] **2.** Simplify: $14 - 12[6 - 8(3 + 2^5)] + \sqrt{9 \cdot 16}$

[1.4] ***For Exercises 3 and 4, translate to an expression.***

3. nine less than the product of seven and n

4. negative three times the sum of a number and eight

5. Combine like terms: $8x - y - 10x + 9 - 5y - 4$

6. Distribute: $-6(9m - 4)$

2.2 Solving Problems

OBJECTIVES

1. **Use formulas to solve problems.**
2. **Translate words to equations.**
3. **Solve problems involving two unknowns.**
4. **Solve motion problems.**
5. **Solve mixture problems.**

A primary purpose of studying mathematics is to develop and improve problem-solving skills. George Polya formulated a problem-solving process that follows a four-step outline: (1) understand the problem, (2) devise a plan for solving the problem, (3) execute the plan, and (4) check the results. The following outline for problem solving is based on Polya's four-step process. We'll see Polya's process illustrated throughout the rest of the text in application problems.

PROCEDURE Problem-Solving Outline

1. **Understand** the problem.
   a. Read the question(s) (not the whole problem, just the question at the end) and note what it is you are to find.
   b. Now read the whole problem, underlining the key words.
   c. If possible and useful, make a list or table, simulate the situation, or search for a related example problem.
2. **Plan** your solution strategy by searching for a formula or translating the key words to an equation.
3. **Execute** the plan by solving the formula or equation.
4. **Answer** the question. Look at your note about what you were to find and make sure you answered that question. Include appropriate units.
5. **Check** the results.
   a. Try finding the solution in a different way, reversing the process, or estimating the answer and make sure the estimate and actual answer are reasonably close.
   b. Make sure the answer is reasonable.

OBJECTIVE 1. Use formulas to solve problems. We will explore the various strategies listed in the outline. First, we focus on using formulas to solve problems.

EXAMPLE 1 The average temperature on Mars is $-81°F$. Use $F = \frac{9}{5}C + 32$ to convert this temperature to degrees Celsius.

Understand We are given a temperature in degrees Fahrenheit and the formula $F = \frac{9}{5}C + 32$.

Plan We can replace F with -81 and solve for C.

Execute

$$-81 = \frac{9}{5}C + 32 \quad \text{Replace } F \text{ with } -81.$$

$$-81 - 32 = \frac{9}{5}C + 32 - 32 \quad \text{Subtract 32 from both sides.}$$

$$\frac{5}{9} \cdot (-113) = \frac{5}{9} \cdot \frac{9}{5}C \quad \text{Simplify and then multiply both sides by } \frac{5}{9}.$$

$$-62.\overline{7} = C$$

Answer The average temperature on Mars in degrees Celsius is $-62.\overline{7}°C$.

Check To check, we could convert the answer back to degrees Fahrenheit, which we will leave to the reader.

Connection We could have first solved $F = \frac{9}{5}C + 32$ for C as we did in Section 2.1, which would give the formula $C = \frac{5}{9}(F - 32)$. If we had to convert a lot of temperatures, this would be a more efficient approach.

YOUR TURN The formula $B = P + Prt$ can be used to calculate the balance in an account if simple interest is added to a principal P at an interest rate of r after t years. Suppose that an account earns 4% interest and after one year the balance is \$4368. Find the principal that was invested.

ANSWER

\$4200

EXAMPLE 2 The face of one wing of a new house is to be covered with brick. To order the correct amount of bricks, the builder needs to know the area to be covered. Calculate the area that will be covered.

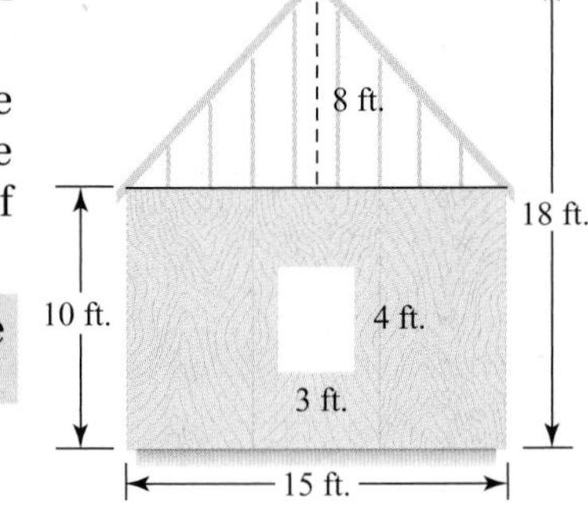

Understand We are to calculate the area that will be covered with brick.

Plan Since the window will not be covered, the area in question is a combination of a large rectangle and a triangle, minus the area of the window.

Note: *Uppercase L and lowercase l indicate different values. Similarly, uppercase W and lowercase w indicate different values.*

$A =$ Area of the large rectangle $+$ Area of the triangle $-$ Area of the window

$$A = LW + \frac{1}{2}bh - lw$$

Execute Replace the variables with the corresponding values and calculate.

$$A = LW + \frac{1}{2}bh - lw$$

$$A = (15)(10) + \frac{1}{2}(15)(8) - (3)(4)$$

$$A = 150 + 60 - 12$$

$$A = 198 \text{ ft.}^2$$

Answer The area to be covered by brick is 198 ft.^{2}

Check Verify the calculations. We will leave the check to the reader.

YOUR TURN A house is to be built on a lot as shown in the figure to the right. Once the house is built, the remaining area will be landscaped. Calculate the area to be landscaped.

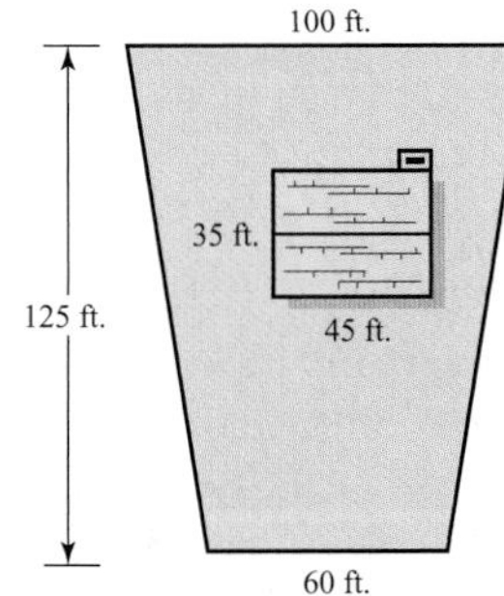

ANSWER

8425 ft.^{2}

OBJECTIVE 2. Translate words to equations. In Section 1.4, we translated word phrases to expressions. Now we will translate sentences to equations. The key words to look for that indicate an equal sign are listed here.

Note: *You may want to review the key words for various operations in Section 1.4 before proceeding with this material.*

Key words for an equal sign:	is equal to	is	yields
	is the same as	produces	results in

PROCEDURE Translating Word Sentences

To translate a word sentence to an equation, identify the variable(s), constants, and key words, then write the corresponding symbolic form.

EXAMPLE 3 Three-fourths of the sum of a number and five is equal to one minus one-third of the number. Translate to an equation, then solve.

Understand The phrase "three-fourths of the sum" indicates to multiply "the sum of the number and five" by $\frac{3}{4}$. Because the sum is being multiplied, we write it in parentheses. "Is equal to" means an equal sign, "minus" means subtract, and "one-third of the number" indicates multiplication.

Plan Let n represent the unknown number, translate the key words to an equation, then solve.

Execute Translation:

Three-fourths of the sum of a number and five is equal to one minus one-third of the number.

$$\frac{3}{4} \cdot \quad (n+5) \quad = \quad 1 \quad - \quad \frac{1}{3} \cdot n$$

$$\frac{3}{4}(n+5) = 1 - \frac{1}{3}n$$

Solve: $$12 \cdot \frac{3}{4}(n+5) = 12 \cdot \left(1 - \frac{1}{3}n\right)$$ Multiply both sides by the LCD, 12, to clear the fractions.

$$9(n+5) = 12 \cdot 1 - 12 \cdot \frac{1}{3}n$$ Multiply $12 \cdot \frac{3}{4}$ and distribute 12.

$$9n + 45 = 12 - 4n$$ Distribute 9.

$$9n + 4n + 45 = 12 - 4n + 4n$$ Add $4n$ to both sides.

$$13n + 45 - 45 = 12 - 45$$ Subtract 45 from both sides.

$$\frac{13n}{13} = \frac{-33}{13}$$ Divide both sides by 13.

Answer $$n = -\frac{33}{13}$$

Check Verify that $\frac{3}{4}$ of the sum of $-\frac{33}{13}$ and 5 is equal to 1 minus $\frac{1}{3}$ of $-\frac{33}{13}$. We will leave this check to the reader.

YOUR TURN Translate to an equation, then solve. One-half of the difference of a number and six is equal to two-fifths plus three-fourths of the number.

ANSWERS

$\frac{1}{2}(n-6) = \frac{2}{5} + \frac{3}{4}n;$

$n = -\frac{68}{5}$ or $-13\frac{3}{5}$

EXAMPLE 4 A clothing store determines retail prices by adding 20% of the wholesale price to the wholesale price. If the retail price of a suit is \$84.90, find the wholesale price that the store paid.

Understand Since \$84.90 is a retail price, it is the result of adding 20% of the wholesale price to the wholesale price.

Plan Let p represent the wholesale price, translate the key words to an equation, then solve.

Execute Translation:

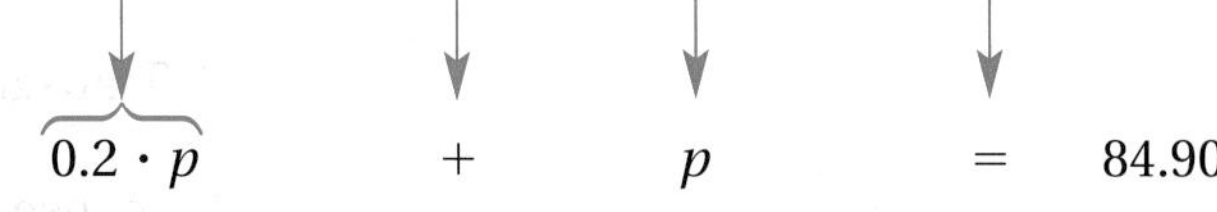

$$0.2 \cdot p \quad + \quad p \quad = \quad 84.90$$

Note: *Because addition is commutative, we can translate "Adding 20% of the wholesale price to the wholesale price" as* $p + 0.2p$ *or* $0.2p + p$.

$$0.2p + p = 84.90$$

Solve: $1.2p = 84.90$ **Combine like terms.**

$$\frac{1.2p}{1.2} = \frac{84.90}{1.2}$$ **Divide both sides by 1.2.**

$$p = 70.75$$

Answer The wholesale price is \$70.75.

Check Verify that adding 20% of \$70.75 to \$70.75 results in a price of \$84.90.

$$0.2(70.75) + 70.75 = 14.15 + 70.75 = 84.90$$

YOUR TURN A telemarketer selling children's books indicates that after a 15% discount for ordering over the phone, the price will be \$5.27 per book. What was the original price for each book?

OBJECTIVE 3. Solve problems involving two unknowns. Some problems involve two unknown amounts and two relationships. Our approach to solving these problems will be to use the relationships to write an equation that we can solve. The following outline gives a more specific approach.

PROCEDURE **Solving Problems with Two or More Unknowns**

To solve problems with two or more unknowns,

1. Determine which unknown will be represented by a variable.
2. Use one of the relationships to describe the other unknown(s) in terms of the variable.
3. Use the other relationship to write an equation.
4. Solve the equation.
5. Check.

ANSWER

\$6.20

First, we will consider problems in which we translate key words.

EXAMPLE 5 One positive number is three times another. The greater number minus the smaller number is 26. What are the two numbers?

Understand We must find two unknown positive numbers. We are given two relationships that we can translate.

Plan Translate the relationships to an equation and then solve.

Execute Relationship 1: One number is three times another.

Translation: one number $= 3n$

Note: *n represents the smaller number.*

Now we use the second relationship to write an equation that we can solve.

Relationship 2: The greater number minus the smaller number is 26.

Translation: greater number − smaller number = 26

$$3n - n = 26$$

$$2n = 26 \quad \text{Combine like terms.}$$

$$\frac{2n}{2} = \frac{26}{2} \quad \text{Divide both sides by 2.}$$

$$n = 13 \quad \text{Simplify.}$$

Answer The smaller number is 13. The greater number, $3n$, is $3(13) = 39$.

Check Verify both relationships: 39 is 3 times 13 and 39 minus 13 is 26.

YOUR TURN One positive number is one-fourth of another number. The sum of the two numbers is 31.

Consecutive Integers

Sometimes, problems do not give all the needed relationships in an obvious manner, as in problems involving consecutive integers.

EXAMPLE 6 The sum of three consecutive odd integers is 75. Find the integers.

Understand If we let n represent the smallest of the odd integers, then the pattern for three consecutive odd integers is

Smallest odd integer: n

Next odd integer: $n + 2$

Third odd integer: $n + 4$

Plan Translate to an equation and then solve.

ANSWER

24.8 and 6.2

Execute The sum of three consecutive odd integers is 75.

$$\boxed{\text{Smallest odd integer}} + \boxed{\text{Next odd integer}} + \boxed{\text{Third odd integer}} = 75$$

$$n \quad + \quad n+2 \quad + \quad n+4 \quad = 75$$

$$3n + 6 = 75 \quad \text{Combine like terms.}$$

$$3n + 6 - 6 = 75 - 6 \quad \text{Subtract 6 from both sides.}$$

$$\frac{3n}{3} = \frac{69}{3} \quad \text{Divide both sides by 3.}$$

$$n = 23$$

Answer The smallest of the three odd integers, n, is 23. The next odd integer, $n + 2$, is $23 + 2 = 25$. And the third odd integer, $n + 4$, is $23 + 4 = 27$.

Check The numbers 23, 25, and 27 are three consecutive odd integers and their sum is $23 + 25 + 27 = 75$.

The sum of three consecutive even integers is 102. What are the integers?

Geometry Problems

In geometry problems, the definition of a geometry term often provides the needed relationship. For example, a problem might involve angles that are **complementary** or **supplementary**.

DEFINITIONS ***Complementary angles:*** Two angles are complementary if the sum of their measures is 90°.

Supplementary angles: Two angles are supplementary if the sum of their measures is 180°.

In the accompanying figure, $\angle ABD$ and $\angle DBC$ are complementary because $32° + 58° = 90°$. Also, $\angle EFH$ and $\angle HFG$ are supplementary because $20° + 160° = 180°$.

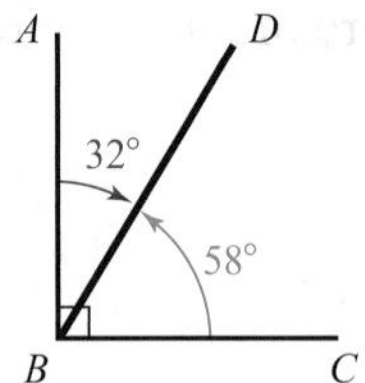

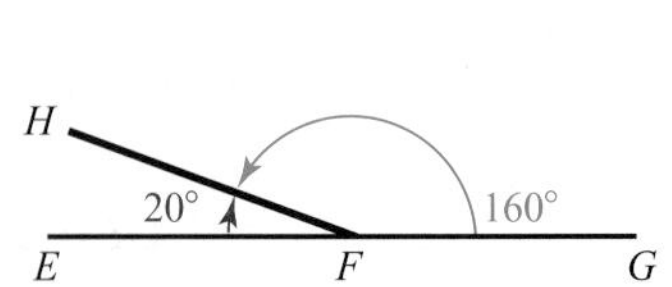

ANSWER

32, 34, 36

Now let's consider a problem containing one of these terms.

EXAMPLE 7 A steel cable is connected to a suspension bridge, creating two angles. If the larger angle is 30° less than twice the smaller angle, then what are the angle measurements?

Understand We must find the two angle measurements. We need two relationships. First, the larger angle is 30° less than twice the smaller angle. A sketch shows our second relationship, that the two angles are supplementary.

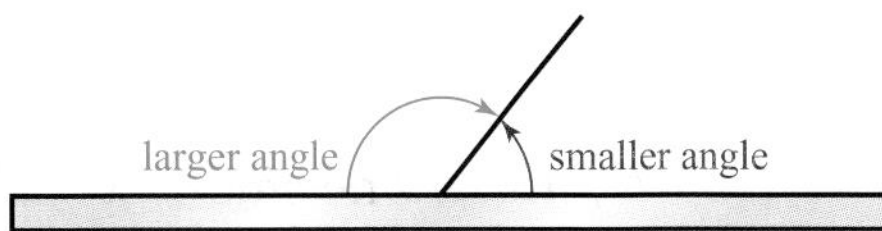

Plan Translate to an equation, and then solve.

Execute Relationship 1: The larger angle is 30° less than twice the smaller angle.

Translation: larger angle $= 2a - 30$

Note: *a represents the smaller angle.*

Relationship 2: The two angles are supplementary.

Translation: larger angle + smaller angle = 180

$$2a - 30 + a = 180$$

$$3a - 30 = 180 \quad \text{Combine like terms.}$$

$$3a - 30 + 30 = 180 + 30 \quad \text{Add 30 to both sides.}$$

$$\frac{3a}{3} = \frac{210}{3} \quad \text{Divide both sides by 3.}$$

$$a = 70$$

Answer The smaller angle is 70°. The larger angle, $2a - 30$, is $2(70) - 30 = 140 - 30 = 110°$.

Note: *We could have also found the larger angle by subtracting 70° from 180° to get 110°.*

Check The two angles are supplementary because 110° + 70° = 180° and 30° less than twice 70° is 110°.

YOUR TURN Wood joists in a roof are connected as shown in the figure. The measure of $\angle DBC$ is 9° more than twice the measure of $\angle ABD$. Find the measure of $\angle DBC$ and $\angle ABD$.

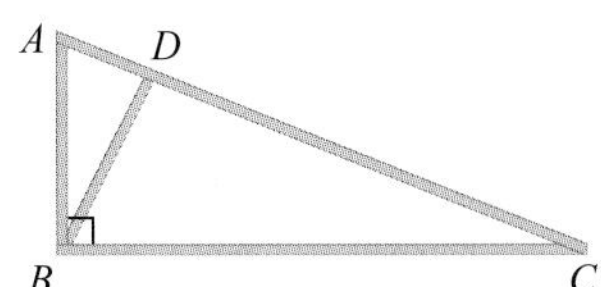

Using a Table

It is sometimes helpful to organize the information in a problem using a table.

ANSWER

27° and 63°

EXAMPLE 8 H&R Block sells two versions of its TaxCut software: a standard version for \$19.95 and a deluxe version for \$29.95. In one day, the company sold 2700 copies of the software over the Internet and took in a total revenue of \$61,365. How many copies of each version were sold?

Understand We are given the price of each version of software. Note that the price times the number of copies sold gives the revenue from the sale of that version. We are also told that there were 2700 copies of software sold. We can organize this information in a table and describe the revenue for each version.

Version	Price	Number of Copies	Revenue
Deluxe	\$29.95	n	$29.95n$
Standard	\$19.95	$2700 - n$	$19.95(2700 - n)$

We let n represent the number of copies of the deluxe version. Since 2700 copies of software were sold, $2700 - n$ describes the number of copies of the standard version.

To describe revenue, we multiply the price and the number of copies.

Plan The last column of our table shows the revenue for each version of the software. We can use these expressions with the fact that the total revenue was \$61,365 to write an equation.

Execute

$$\text{Revenue from the deluxe version} + \text{Revenue from the standard version} = 61{,}365$$

$$29.95n + 19.95(2700 - n) = 61{,}365$$

$$29.95n + 53{,}865 - 19.95n = 61{,}365 \quad \text{Distribute 19.95.}$$

$$10n + 53{,}865 = 61{,}365 \quad \text{Combine like terms.}$$

$$10n + 53{,}865 - 53{,}865 = 61{,}365 - 53{,}865 \quad \text{Subtract 53,865 from both sides.}$$

$$\frac{10n}{10} = \frac{7500}{10} \quad \text{Divide both sides by 10.}$$

$$n = 750$$

Answer H&R Block sold 750 copies of the deluxe version and $2700 - 750 = 1950$ copies of the standard version.

Check The total number of copies is correct because $750 + 1950 = 2700$. The total revenue is correct because $750(\$29.95) + 1950(\$19.95) = \$22{,}462.50 + \$38{,}902.50 = \$61{,}365$.

YOUR TURN An administrative assistant orders day minders and desk calendars for the people in his department. Day minders cost \$14.95 and desk calendars cost \$8.95. If he ordered three times as many calendars as day minders at a total cost of \$334.40, how many of each did he order?

ANSWER

8 day minders, 24 desk calendars

OBJECTIVE 4. Solve motion problems. We can use a table similar to the one in Example 8 with problems involving two objects in motion.

EXAMPLE 9 Two cars are traveling toward each other on the same highway. One car is traveling at 65 miles per hour and the other at 60 miles per hour. If the two cars are 20 miles apart, how long will it be until they meet?

Understand Draw a picture of the situation:

Note: *Car 1 will travel the greater distance because it is going faster.*

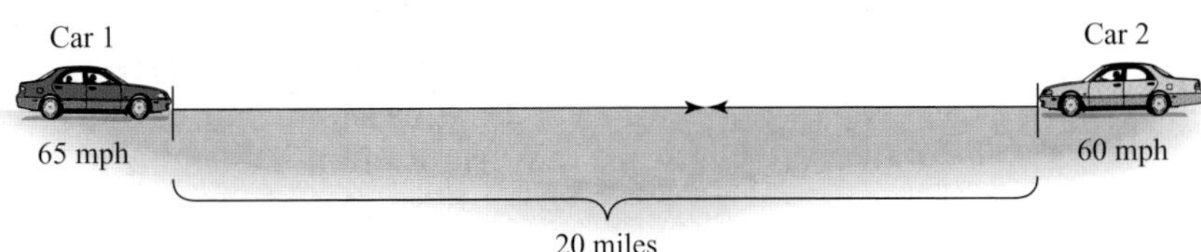

From the diagram we see that the sum of the individual distances traveled will be the total distance separating the two cars, which is 20 miles. Since we are given their rates, we can use $d = rt$ to write expressions for their individual distances.

Car	Rate	Time	Distance
Car 1	65 mph	t	$65t$
Car 2	60 mph	t	$60t$

We were given these rates.

Both cars start at the same time and meet at the same moment in time, so they traveled the same amount of time, t.

To describe each distance, we use the fact that $d = rt$ and multiply the rate and time.

Learning Strategy

If you are a tactile learner, you may find it helpful to simulate the situations in rate problems with toy cars, pencils, or small pieces of paper.

Plan Use the sum of the individual distances traveled (20 miles) to write an equation.

Execute Car 1's distance + Car 2's distance = 20

$$65t + 60t = 20$$
$$125t = 20$$
$$\frac{125t}{125} = \frac{20}{125}$$
$$t = 0.16$$

Note: *If solving for time, the time unit will match the unit of time in the rate. In this problem, the rate is in miles per hour, so the time unit is hours.*

Answer The cars will meet in 0.16 hours, which is 9.6 minutes or 9 minutes and 36 seconds.

Check Use $d = rt$ to verify that in 0.16 hours, the cars will travel a combined distance of 20 miles.

$$\text{Car 1: } d = 65(0.16) = 10.4 \text{ miles}$$
$$\text{Car 2: } d = 60(0.16) = 9.6 \text{ miles}$$

The combined distance is $10.4 + 9.6 = 20$ miles.

YOUR TURN Deon and Kayla are jogging along the same trail. Kayla crosses a bridge at 5:00 P.M. Deon crosses the same bridge at 5:10 P.M. Kayla is traveling at 4 miles per hour while Deon is traveling at 6 miles per hour. What time will Deon catch up to Kayla?

ANSWER

5:30 P.M.

OBJECTIVE 5. Solve mixture problems. In chemistry, *concentration* refers to the amount of a solution that is pure solute. For example, a 60-milliliter solution that has a 5% concentration of hydrochloric acid (HCl) has $0.05(60) = 3$ milliliters of pure HCl and the rest is another chemical, usually water. Solutions with different concentrations can be mixed to produce a desired concentration. We can use algebra to determine how much of each solution to mix so that the desired concentration is achieved.

EXAMPLE 10 A chemist has a bottle containing 80 milliliters of 10% HCl solution and a bottle of 25% HCl solution. She wants a solution that is 20% HCl. How much of the 25% solution should be added to the 10% solution so that a 20% concentration is created?

Understand We can use a table to organize the information and describe the volume of HCl in each of the three solutions.

Solution	Concentration	Solution Volume	Volume of HCl
10% Solution	0.10	80	0.10(80)
25% Solution	0.25	n	$0.25n$
20% Solution	0.20	$80 + n$	$0.20(80 + n)$

These are the concentration percentages written in decimal form.

The 10% solution and an unknown amount of the 25% solution are combined to form the 20% solution.

The volume of HCl in each solution is found by multiplying its concentration and volume.

Plan Write an equation that describes the mixture. Then solve for n.

Execute The 10% solution and 25% solution combine to equal the 20% solution. We can say:

Volume of HCl in the 10% solution	+	Volume of HCl in the 25% solution	=	Volume of HCl in the 20% solution
$0.10(80)$	+	$0.25n$	=	$0.20(80 + n)$

$$8 + 0.25n = 16 + 0.20n$$

$$8 + 0.25n - 0.20n = 16 + 0.20n - 0.20n$$ Subtract $0.20n$ from both sides.

$$8 - 8 + 0.05n = 16 - 8$$ Subtract 8 from both sides.

$$\frac{0.05n}{0.05} = \frac{8}{0.05}$$ Divide both sides by 0.05.

$$n = 160$$

Answer Adding 160 milliliters of the 25% solution to 80 milliliters of the 10% solution produces a 20% HCl solution.

Check The volume of HCl in the 25% solution is $0.25(160) = 40$ milliliters, so the total volume of HCl in the mixture is 48 milliliters. Notice this is verified by the 20% concentration: $0.20(240) = 48$ milliliters.

120 ml

A chemical engineer has 60 milliliters of a 5% acid solution. How much of a 20% acid solution must be added to create a solution that is 15% acid?

2.2 Exercises

1. What are the four steps to problem solving according to George Polya?

2. List four strategies suggested as aids to understanding a problem.

3. How do you translate a problem involving two relationships and two unknowns to a single equation with one variable?

4. Suppose a problem involves three consecutive even integers. If the first even integer is represented by n, then what expressions describe the second and third even integers?

5. What are supplementary angles?

6. What is a solution's concentration?

For Exercises 7–10, solve.

7. If you cool carbon dioxide down to $-109.3°F$ at sea level, you get "dry ice." Use $F = \frac{9}{5}C + 32$ to convert this temperature to degrees Celsius.

8. The coldest spot on the Earth was found in Antartica at $-89°C$. Use $C = \frac{5}{9}(F - 32)$ to convert this temperature to degrees Fahrenheit.

9. The figure to the right shows the wall of a den that is to be painted. The two windows are 4 feet by 5 feet. In order to purchase the correct amount of paint, the homeowner needs to calculate the area to be painted. Calculate the area to be covered.

10. The figure to the right shows a lot plan for a new house. Once the house is built, the remainder of the lot will be landscaped. Find the area to be landscaped.

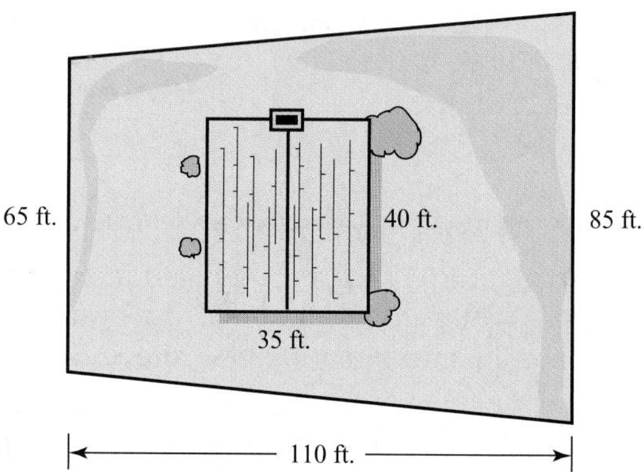

For Exercises 11 and 12, use the formula $B = P + Prt$, which is used to calculate the balance in an account if simple interest is added to a principal P at an interest rate of r after t years.

11. Suppose that an account earns 5.5% interest and after two years the balance is \$3330. Find the principal that was invested.

12. Suppose that an account earns 3% interest and after five years the balance is \$1725. Find the principal that was invested.

For Exercises 13–32, solve.

13. Five subtracted from three times a number is forty. What is the number?

14. Fifteen is equal to six less than seven times a number. What is the number?

15. Three times the sum of a number and four is negative six. What is the number?

16. Twice the difference of a number and three is negative five. What is the number?

17. Eight less than the product of five and the difference of a number and two is twice the number. What is the number?

18. Six minus three times the difference of a number and five is equal to four times the number. What is the number?

19. A bookstore determines the retail price by marking up the wholesale price 25%. The retail price of a chemistry book is \$110.90. What was the wholesale price?

20. The MSRP (manufacturer's suggested retail price) of a particular car is \$22,678. If the MSRP represents a markup of 15% of the wholesale price, what is the wholesale price?

21. A computer store is selling older stock computers at a discount of 30% off the original price. If the price after the discount is \$699.95, what was the original price?

22. A clothing store has a special sale with every item discounted 40%. If the price of a dress after the discount is \$24.50, what was the original price?

23. An annual report for a company shows that 45,000 units of a particular product were sold, which is an increase of 20% over the previous year. How many units of that product were sold the previous year?

24. A report shows that revenue for one month was \$24,360, which is a decrease of 12.5% from the previous month. What was the previous month's revenue?

25. The sum of three consecutive even integers is 78. What are the integers?

26. The sum of two consecutive integers is 191. What are the integers?

27. The sum of three consecutive odd integers is 165. What are the integers?

28. The sum of four consecutive integers is 34. What are the integers?

29. Two angles are supplementary. If one angle is twice the other, what are the angles?

30. Two angles are complementary. If one angle is 15° more than twice the other angle, what are the angles?

31. Two angles are complementary. If one angle is 5° less than three times the other angle, what are the angles?

32. Two angles are supplementary. If one angle is 15° less than twice the other angle, what are the angles?

For Exercises 33–38, complete the table. Do not solve the problem.

33. A vending machine has 12-ounce and 16-ounce drinks. The 12-ounce drinks are \$0.50 and the 16-ounce drinks are \$1.00. If 3600 total drinks were sold in one month, and the total sales were \$2225, how many of each size drink were sold?

Categories	Value	Number	Amount
12 oz.	0.50	$3600 - x$	
16 oz.	1	x	

34. Kelly is purchasing plants from a garden center where the prices are based on the size of the pot. She is purchasing 13 plants in one-gallon pots and 7 plants in half-gallon pots. A single 1-gallon and single half-gallon have a combined price of \$14.50. The total for all the plants she is buying is \$155.50. What is the individual price of each size pot?

Categories	Value	Number	Amount
1 gallon	x	13	
1/2 gallon	$14.50 - x$	7	

35. At 6 A.M., a freight train leaves Washington, DC, traveling at 50 miles per hour. At 9 A.M., a passenger train leaves the same station traveling in the same direction at 75 miles per hour. How long will it take the passenger train to overtake the freight train? How far will they be from the station at this time?

Categories	Rate	Time	Distance
Freight	50	t	
Passenger	75	$t - 3$	

36. Janet and Paul were both traveling west from Charleston, West Virginia, to the Rupp Arena in Lexington, Kentucky, for a concert. At 6:00 P.M., Janet, traveling at 45 miles per hour, was 10 miles west of Paul. A little later, Paul, traveling at 50 miles per hour, passed Janet. What time did Paul pass Janet?

Categories	Rate	Time	Distance
Janet	45		d
Paul	50		$d + 10$

37. How many liters of a 40% solution of HCl are added to 2000 liters of a 20% solution to obtain a 35% solution?

Solutions	Concentrate	Vol. of Sol.	Vol. of HCl
40%	0.40	x	$0.40x$
20%	0.20	2000	
35%	0.35		$0.35(x + 2000)$

38. A pharmacist has a 45% acid solution and a 35% acid solution. How many liters of each must be mixed to form 80 liters of a 40% acid solution?

Solutions	Concentrate	Vol. of Sol.	Vol. of Acid
45%	0.45	x	$0.45x$
35%	0.35		$0.35(80 - x)$
40%	0.40	80	

For Exercises 39–44, solve Exercises 33–38 using your completed table.

39. Solve Exercise 33.

40. Solve Exercise 34.

41. Solve Exercise 35.

42. Solve Exercise 36.

43. Solve Exercise 37.

44. Solve Exercise 38.

For Exercises 45–56, solve.

45. Laquanda ordered 16 desktop computers and 6 laptop computers from Dell for her department at a total cost of $33,100. If the laptops cost $200 more than the desktops, how much did each type of computer cost?

46. Wade ordered computers for a different department but decided to go with Gateway. He ordered 10 desktop computers and 4 laptop computers at a total cost of $20,080. If the laptops cost $260 more than the desktops, how much did each type of computer cost?

47. An administrative assistant orders cellular phones for the people in his department. The Motorola phones cost $74.95 and the Nokia phones cost $49.95. If he ordered twice as many Nokia phones as Motorola phones at a total cost of $1398.80, how many of each did he order?

48. A manager at a Toys "R" Us orders board games for a sales promotion. Monopoly games cost \$8.95 and Trivial Pursuit games cost \$15.95. If he ordered half as many Trivial Pursuit games as Monopoly games at a total cost of \$575.45, how many of each did he order?

49. Two cars are traveling towards each other on the same road. One car is traveling 50 miles per hour and the other at 45 miles per hour. If the two cars are 190 miles apart, how much time will it take for them to meet?

50. Two cars traveling in opposite directions meet and then pass each other. The eastbound car maintains an average speed of 60 miles per hour and the westbound car averages 48 miles per hour. If they maintain those average rates, how long will it take them to be 270 miles apart?

51. Andrea passes under a bridge at 55 miles per hour. One-half hour later, her husband, Paul, traveling in the same direction passes under the same bridge at 65 miles per hour. How long will it take Paul to catch up?

52. Bonnie and Bob are traveling south in separate cars on the same highway. Bonnie is traveling at 65 miles per hour and Bob at 70 miles per hour. Bonnie passes Exit 64 at 1:30 P.M. Bob passes the same exit at 1:45 P.M. At what time will Bob catch up to Bonnie?

53. A mechanic has 24 ounces of a mixture that is 20% antifreeze and another mixture that is 10% antifreeze. She wants a 16% antifreeze mixture. How much of the 10% solution should be added to the 20% solution to form a 16% antifreeze solution?

54. A gardener has 15 ounces of a 50% plant fertilizer solution and some 40% plant fertilizer solution. He wants a 46% plant fertilizer solution. How much of the 40% solution should be added to the 50% solution to form a 46% solution?

55. A farmer has 10 gallons of 40% pesticide mixture and some 10% pesticide mixture. He wants to make a 20% pesticide mixture. How much of the 10% mixture should he combine with the 40% mixture to form a 20% mixture?

56. Janice has 50 milliliters of 10% HCl solution and some 25% HCl solution. How much of the 25% solution must be added to the 10% solution to create a 20% concentration?

REVIEW EXERCISES

[1.1] **1.** Place the following values in order from least to greatest:

$$0.02,\ -\frac{1}{6},\ 4.5\%,\ |-15.8|,\ \sqrt{48},\ -5\frac{3}{8}$$

[1.1] **2.** What is the smallest prime number?

[1.1] ***For Exercises 3 and 4, use <, >, or = to write a true statement.***

3. $-|5.8| \ \blacksquare\ -|-5.8|$

4. $-(-6)\ \blacksquare\ -(-(-8))$

[2.1] ***For Exercises 5 and 6, solve.***

5. $6x - 19 = 4x - 31$

6. $-\frac{4}{9}n = \frac{5}{6}$

2.3 Solving Linear Inequalities

OBJECTIVES

1. **Represent solutions to linear inequalities in one variable using set-builder notation, interval notation and graphs.**
2. **Solve linear inequalities in one variable.**
3. **Solve problems involving linear inequalities.**

OBJECTIVE 1. Represent solutions to linear inequalities in one variable using set-builder notation, interval notation, and graphs. Not all problems translate to equations. Sometimes a problem can have a range of values as solutions. We can often use inequalities to describe situations where a range of solutions is possible. Inequality symbols and their meanings are

$<$ is less than
$>$ is greater than
$\leq$ is less than or equal to
$\geq$ is greater than or equal to

In this section, we will focus on how to solve **linear inequalities**.

DEFINITION ***Linear inequality in one variable:*** An inequality that can be written in the form $ax + b \square c$, where $\square$ is an inequality symbol.

Some examples of linear inequalities in one variable:

$$x > 5 \qquad n + 2 < 6 \qquad 2(y - 3) \leq 5y - 9$$

A solution for an inequality is any number that can replace the variable(s) in the inequality and make it true. Since inequalities can have a range of solutions, we often write those solutions in set-builder notation, which we introduced in Section 1.1. For example, the solution set for the inequality $x > 5$ is $\{x \mid x > 5\}$.

Another popular notation used to indicate ranges of values is *interval notation.* With interval notation, parentheses are used for end values that are not included in the interval and brackets are used for end values that are included. For example, since the solution set for $x > 5$ does not include 5, we would write $(5, \infty)$, whereas the solution set for $x \geq 5$ would be written as $[5, \infty)$. The symbol ∞ means infinity and is never included as an end value.

We can represent these solution sets graphically using a number line. We use parentheses or brackets from interval notation.

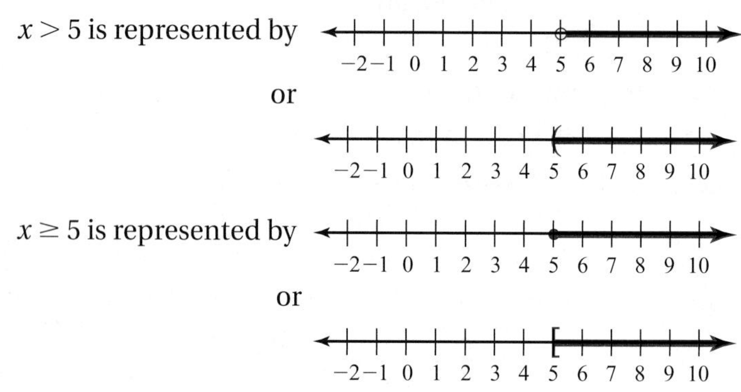

Note: *This form reinforces the interval notation* $(5, \infty)$.

Note: *This form reinforces the interval notation* $[5, \infty)$.

Your instructor may have a preference of style. We will use the parentheses and brackets because they reinforce interval notation.

PROCEDURE **Graphing Inequalities**

To graph an inequality in one variable on a number line,

1. If the symbol is $\le$ or $\ge$, then draw a bracket (or solid circle) on the number line at the indicated number. If the symbol is $<$ or $>$ then draw a parenthesis (or open circle) on the number line at the indicated number.
2. If the variable is greater than the indicated number, then shade to the right of the indicated number. If the variable is less than the indicated number, then shade to the left of the indicated number.

EXAMPLE 1 Write the solution set in set-builder notation and interval notation, then graph the solution set.

a. $x \le -4$
Set-builder notation: $\{x \mid x \le -4\}$
Interval notation: $(-\infty, -4]$

Graph:

−9 −8 −7 −6 −5 −4 −3 −2 −1 0 1 2 3

Note: *When writing interval notation, we write from the left-most end value to the right-most end value.*

b. $n < 0$
Set-builder notation: $\{n \mid n < 0\}$
Interval notation: $(-\infty, 0)$

Graph:

−6 −5 −4 −3 −2 −1 0 1 2 3 4 5 6

Connection The set $\{n \mid n < 0\}$ is a way of expressing the set of all negative real numbers.

YOUR TURN Write the solution set in set-builder notation and interval notation, then graph the solution set.

a. $x < 3$ **b.** $h \ge -2$

ANSWERS

a. Set-builder notation: $\{x \mid x < 3\}$
Interval notation: $(-\infty, 3)$
Graph:

−2 −1 0 1 2 3 4 5

b. Set-builder notation: $\{h \mid h \ge -2\}$
Interval notation: $[-2, \infty)$
Graph:

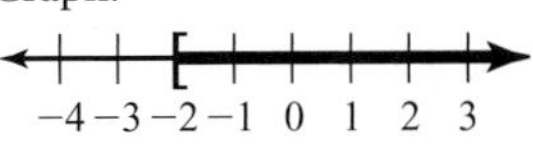

OBJECTIVE 2. Solve linear inequalities in one variable. To solve linear inequalities like $n + 2 < 6$ and $2(x - 3) \le 5x - 9$, we follow essentially the same process as for solving linear equations. The principles for inequalities, summarized next, are very similar to the principles for equations. Be careful to note the special condition in the multiplication principle of inequality.

RULES **The Addition Principle of Inequality**

If $a < b$, then $a + c < b + c$ is true for all real numbers a, b, and c.

The Multiplication Principle of Inequality

If a and b are real numbers where $a < b$, then $ac < bc$ is true if c is a *positive* real number and $ac > bc$ is true if c is a *negative* real number.

Note: Although we have written the principles in terms of the $<$ symbol, the principles are true for any inequality symbol.

Notice that the addition principle of inequality indicates that adding the same amount to (or subtracting the same amount from) both sides of an inequality does not affect the inequality. Similarly, multiplying (or dividing) both sides of an inequality by the same *positive* real number does not affect an inequality. But if we multiply (or divide) both sides by the same *negative* real number, then we must *reverse the direction of the inequality symbol* to keep the inequality true. The following procedure summarizes the process of solving a linear inequality.

PROCEDURE **Solving Linear Inequalities in One Variable**

To solve a linear inequality in one variable,

1. Simplify both sides of the inequality as needed.
   a. Distribute to clear parentheses.
   b. Clear fractions or decimals by multiplying through by the LCD just as we did for equations. (This step is optional.)
   c. Combine like terms.
2. Use the addition principle so that all variable terms are on one side of the inequality and all constants are on the other side. Then combine like terms.
3. Use the multiplication principle to clear any remaining coefficient. If you multiply (or divide) both sides by a negative number, then reverse the direction of the inequality symbol.

EXAMPLE 2 For each inequality, solve and write the solution set in set-builder notation and in interval notation. Then graph the solution set.

a. $-6x > 18$

Solution $\dfrac{-6x}{-6} < \dfrac{18}{-6}$ **Divide both sides by −6 and change the inequality.**

$x < -3$

Note: *Because we divided both sides by a negative number, we reversed the direction of the inequality symbol.*

Set notation: $\{x \mid x < -3\}$
Interval notation: $(-\infty, -3)$

Graph: −6 −5 −4 −3 −2 −1 0 1 2 3 4 5 6

b. $-10 \le 3m + 2$

Solution $-10 - 2 \le 3m + 2 - 2$ **Subtract 2 from both sides.**

$\dfrac{-12}{3} \le \dfrac{3m}{3}$ **Divide both sides by 3.**

$-4 \le m$

Set notation: $\{m \mid m \ge -4\}$
Interval notation: $[-4, \infty)$

Graph: −9 −8 −7 −6 −5 −4 −3 −2 −1 0 1 2 3

Note: *Dividing both sides by a positive number does not affect the inequality. Also, $-4 \le m$ and $m \ge -4$ are the same.*

c. $\frac{1}{3}(9h + 5) - 2h > \frac{1}{4}h + 1$

Solution

$$3h + \frac{5}{3} - 2h > \frac{1}{4}h + 1 \qquad \text{Distribute } \frac{1}{3}.$$

$$h + \frac{5}{3} > \frac{1}{4}h + 1 \qquad \text{Combine like terms.}$$

$$12 \cdot h + 12 \cdot \frac{5}{3} > 12 \cdot \frac{1}{4}h + 12 \cdot 1 \qquad \text{Multiply both sides by the LCD, 12, to clear the fractions.}$$

$$12h + 20 > 3h + 12$$

$$12h - 3h + 20 > 3h - 3h + 12 \qquad \text{Subtract } 3h \text{ from both sides.}$$

$$9h + 20 - 20 > 12 - 20 \qquad \text{Subtract 20 from both sides.}$$

$$\frac{9h}{9} > \frac{-8}{9} \qquad \text{Divide both sides by 9.}$$

$$h > -\frac{8}{9}$$

Note: *By subtracting $3h$ from both sides, we get a positive 9 coefficient after combining like terms, so we won't have to reverse the inequality when we divide both sides by 9.*

Set-builder notation: $\left\{h \mid h > -\frac{8}{9}\right\}$

Interval notation: $\left(-\frac{8}{9}, \infty\right)$

Graph:

Note: *We could have multiplied by the LCD, 12, first.*

$$12\left[\frac{1}{3}(9h + 5) - 2h\right] > 12\left(\frac{1}{4}h + 1\right)$$

$$4(9h + 5) - 24h > 3h + 12$$

Continuing the solution process will yield the same solution, $h > -\frac{8}{9}$.

YOUR TURN For each inequality, solve and write the solution set in set-builder notation and in interval notation. Then graph the solution set.

a. $-0.2x > 0.5$ **b.** $\frac{3}{4}n - 1 \ge \frac{1}{2}n + \frac{1}{5}$ **c.** $2(k - 5) + 4k \le 7k + 6 - 5k$

Linear inequalities can also have no solution or every real number as a solution.

EXAMPLE 3 For each inequality, solve and write the solution set in set-builder notation and in interval notation. Then graph the solution set.

a. $9x - 8 < 4(x + 3) + 5x$

Solution

$$9x - 8 < 4x + 12 + 5x \qquad \text{Distribute 4.}$$

$$9x - 8 < 9x + 12 \qquad \text{Combine like terms.}$$

$$9x - 9x - 8 < 9x - 9x + 12 \qquad \text{Subtract } 9x \text{ from both sides.}$$

$$-8 < 12$$

Since there are no variable terms left in the inequality and $-8 < 12$ is true, every real number is a solution for the original inequality.

ANSWERS

a. $x < -2.5$

Set-builder notation: $\{x \mid x < -2.5\}$

Interval notation: $(-\infty, -2.5)$

Graph:

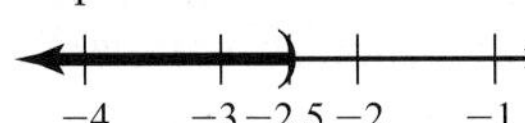

b. $n \ge \frac{24}{5}$ or $4\frac{4}{5}$

Set-builder notation: $\left\{n \mid n \ge \frac{24}{5}\right\}$

Interval notation: $\left[\frac{24}{5}, \infty\right)$

Graph:

c. $k \le 4$

Set-builder notation: $\{k \mid k \le 4\}$

Interval notation: $(-\infty, 4]$

Graph:

Graph: (number line from −6 to 6)

−6 −5 −4 −3 −2 −1 0 1 2 3 4 5 6

Set-builder notation: $\{x \mid x \text{ is a real number}\}$, or $\mathbb{R}$
Interval notation: $(-\infty, \infty)$

b. $5(t-3) \geq 7-(t+4)+6t$

Solution

$5t - 15 \geq 7 - t - 4 + 6t$ Distribute.

$5t - 15 \geq 5t + 3$ Combine like terms.

$5t - 5t - 15 \geq 5t - 5t + 3$ Subtract $5t$ from both sides.

$-15 \geq 3$

Since there are no variable terms left in the inequality and $-15 \geq 3$ is false, there are no solutions for the original inequality.

Graph: (number line from −6 to 6)

−6 −5 −4 −3 −2 −1 0 1 2 3 4 5 6

Set-builder notation: $\{\ \}$ or $\varnothing$
Interval notation: We do not write interval notation if there are no solutions.

OBJECTIVE 3. Solve problems involving linear inequalities. Problems requiring inequalities can be translated using key words much like we used key words to translate sentences to equations. The following table lists some common key words that indicate inequalities.

Words That Mean $\geq$	Words That Mean $>$	Words That Mean $\leq$	Words That Mean $<$
at least	is greater than	at most	is less than
minimum of	more than	maximum of	smaller than
no less than		no more than	

EXAMPLE 4 Sasha's grade in her math course is calculated by the average of four tests. To receive an A for the course, she needs an average of at least 89.5. If her current test scores are 84, 92, and 94, what range of scores can she make on the last test to receive an A for the course?

Understand We are given three out of four test scores and must find the range of values for the fourth score that will give her an average for the four scores of 89.5 or higher.

Plan Since we are to find a range of scores, we will write an inequality with n representing the score on the last test, then solve.

Execute Sasha's average must be at least 89.5.

$$\frac{84+92+94+n}{4} \geq 89.5$$

Note: *n represents Sasha's score on the fourth test.*

$\dfrac{270+n}{4} \geq 89.5$ Simplify in the numerator.

$270 + n \geq 358$ Multiply both sides by 4.

$n \geq 88$ Subtract 270 from both sides.

Answer A score of 88 or greater on the last test will give her an average of at least 89.5.

Check Verify that a score of 88 will give her an average of 89.5, then choose a value greater than 88 and verify that the average will be greater than 89.5. We will leave this check to the reader.

EXAMPLE 5 A painter charges \$80 plus \$1.50 per square foot. If a family is willing to spend no more than \$500, then what is the range of square footage they can afford?

Understand We are given the painter's prices and the maximum amount of money.

Plan Since we are looking for a range of square footage, we will write an inequality with n representing the number of square feet, and then solve.

Execute Translate: The total cost has to be less than or equal to \$500.

$$80 + 1.5n \leq 500$$

Solve: $80 - 80 + 1.5n \leq 500 - 80$ Subtract 80 from both sides.

$$\frac{1.5n}{1.5} \leq \frac{420}{1.5}$$ Divide both sides by 1.5.

$$n \leq 280$$

Answer They can afford up to 280 square feet.

Check Verify that if they paid for exactly 280 square feet, then the painter would charge exactly \$500.

EXAMPLE 6 A business breaks even if its revenue and costs are equal and makes a profit if the revenue exceeds costs. Suppose the expression $30n + 2000$ describes the monthly revenue for a company, where n is the number of units of product sold. The expression $10n + 18{,}000$ describes the monthly cost of producing those n units of the product. How many units must be sold to break even or make a profit?

Understand We are given expressions for revenue and cost and are to find the number of units that the company must sell in order to break even or make a profit.

Plan Write an inequality and then solve.

Execute Translate: To break even or make a profit, the revenue has to be greater than or equal to the cost.

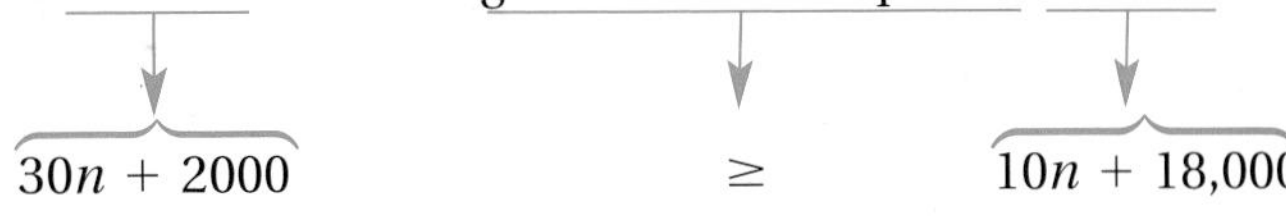

$$30n - 10n + 2000 \geq 10n - 10n + 18{,}000$$ Subtract $10n$ from both sides.

Solve:

$$20n + 2000 \geq 18{,}000$$

$$20n + 2000 - 2000 \geq 18{,}000 - 2000$$ Subtract 2000 from both sides.

$$\frac{20n}{20} \geq \frac{16{,}000}{20}$$ Divide both sides by 20.

$$n \geq 800$$

Answer The company would need to sell 800 units of its product to break even and more than 800 units to make a profit.

Check Verify that selling exactly 800 units causes the company to break even and that selling more than 800 units causes the company to make a profit. We will leave this to the reader.

YOUR TURN A wireless company offers a calling plan that charges \$35 for the first 500 minutes and \$0.06 for each additional minute. Find the range of total minutes Polina could use with this plan if she budgets a maximum of \$50 per month for her wireless charges.

ANSWER

$t \leq 750$ minutes (500 minutes plus 250 additional minutes)

2.3 Exercises

1. What is a solution for an inequality?

2. Explain the difference between $x < a$ and $x \leq a$, where a is a real number.

3. The following graph is a solution set for an inequality. What does it show?

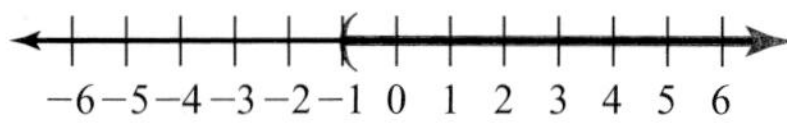

4. Explain what $(-\infty, 4]$ means.

5. How do you read $\{x \mid x > 2\}$? What does it mean?

6. What action causes the direction of an inequality symbol to change?

For Exercises 7–14: ***a. Write in set-builder notation.***
b. Write in interval notation.
c. Graph the solution set.

7. $x \geq 5$

8. $n \leq 7$

9. $q < -1$

10. $x > -3$

11. $p < \frac{1}{5}$

12. $a \geq -\frac{2}{3}$

13. $r \leq 1.9$

14. $s \leq -3.2$

For Exercises 15–40, solve: ***a. Write the solution set using set-builder notation.***
b. Write the solution set using interval notation.
c. Graph the solution set.

15. $r - 6 < -12$

16. $x + 6 \geq -1$

17. $-4y \leq -16$

18. $-3x < -18$

19. $2p + 9 < 17$

20. $9y - 7 > 11$

21. $2 - 5x > 25$

22. $-3c + 4 < 17$

23. $\frac{a}{8} + 1 < \frac{3}{8}$

24. $4 > \frac{1}{5} - \frac{q}{5}$

25. $7 - x < 2 + x$

26. $5b - 3 \geq 13 + 4b$

27. $-11k - 8 > -16 - 9k$

28. $4 - 9k < -4k + 19$

29. $3(2w - 4) - 5 \leq 13$

30. $6(3a + 2) - 10 \geq 2$

31. $5(y + 4) + 9 > 3(3y - 1)$

32. $4(v - 10) \leq 17(v + 3) + 13$

33. $\frac{1}{6}(5x + 1) < -\frac{1}{6}x - \frac{5}{3}$

34. $\frac{1}{7}(3x - 28) > -\frac{4}{7}x + \frac{1}{4}$

35. $\frac{1}{6}(7m + 2) - \frac{1}{6}(11m - 7) \geq 0$

36. $\frac{1}{5}(p + 10) - \frac{1}{2}(p + 5) > 0$

37. $0.7x - 0.3 \leq 0.8x + 0.7$

38. $1.2b - 1.4 \geq 1.5b - 0.5$

39. $0.09z + 20.34 < 3(1.4z - 1.5) + 2.1z$

40. $3.2f + 3.6 - 1.8f \leq 0.3(f + 6) - 0.4$

For Exercises 41–48, translate to an inequality, then solve.

41. Three-fourths of a number is less than negative six.

42. Two-thirds of a number is greater than negative eight.

43. Five times a number increased by one is greater than sixteen.

44. Two less than three times a number is less than ten.

45. One more than four times a number is at most twenty-five.

46. Three minus two times a number is at least seventeen.

47. Six more than twice the difference of a number and five will not exceed twelve.

48. Eight more than three times the difference of a number and four is at least negative sixteen.

For Exercises 49–64, solve.

49. To earn a B in her math course, Yvonne needs the average of five tests to be at least 80. Her current test scores are 82, 92, 73, and 69. What range of scores on the fifth test would earn her at least a B in the course?

50. In an introduction to theater course, the final grade is determined by the average of four papers. The department requires any student whose average falls below 70 to repeat the course. Eric's scores on the first three papers are 70, 62, 75. What range of scores on the fourth paper would cause him to have to repeat the course?

51. For an upcoming golf tournament, Blake must show his scores for his most recent four rounds of golf. He wants the average of those scores to be 84 or less. If he has played three rounds and scored 86, 82, and 88, what range of scores must he have on his fourth round to meet his goal?

52. Paulette wants to average at least 180 during a bowling tournament in which bowlers roll three games. So far she has scored 186 and 178. What range of values must she score on the third game to meet her goal?

53. A homeowners association is planning a rectangular common garden. The area of the garden may not exceed 338 square feet. If the length is 26 feet, what range of values can the width have?

54. Dante needs to make a rectangular spa cover that is at least 14 square feet. If the spa length is 4 feet, what range of values can the width have?

55. Susan makes ceramic flower pots. One particular flower pot requires her to glue a strip of mosaic around the circumference of the top. She is designing a new flower pot and has enough mosaic to cover 150 inches. What range of values can the radius have?

56. Frank is designing a drain from a storage tank. He has a gasket that can vary in size up to a maximum circumference of 12 centimeters. Find the range of values for the diameter of the pipe.

57. Juan is trying to meet his girlfriend at the airport before she gets on a plane. If the airport is 210 miles away and she is leaving in 3 hours, what is the minimum average rate that he must drive to see her?

58. Therese does not drive over 65 miles per hour on the highway. If she plans a trip of 338 miles, find the range of values that describes her time to complete the trip.

59. A company produces lamps. The expression $12.5n + 3000$ describes the monthly revenue if n lamps are sold. The expression $8n + 21{,}000$ describes the monthly cost of producing those n lamps. How many lamps must be sold per month to break even or make a profit?

60. The expression $65n + 10{,}000$ describes the monthly revenue for a cleaning service, where n is the number of homes cleaned. The expression $25n + 15{,}000$ describes the monthly cost of operating the service. How many homes must be cleaned per month to break even or make a profit?

61. Gold is a solid up to a temperature of 1064.58°C. At 1064.58°C or hotter, gold is liquid.

    a. Find the range of temperatures in degrees Fahrenheit that gold remains solid.

    b. Find the range of temperatures in degrees Fahrenheit that gold is liquid.

62. Iron is a solid up to a temperature of 2795°F. At 2795°F or hotter, iron is liquid.

    a. Find the range of temperatures in degrees Celsius that iron remains solid.

    b. Find the range of temperatures in degrees Celsius that iron is liquid.

63. The design of a circuit specifies that the voltage cannot exceed 15 V. If the resistance of the circuit is 6 Ω, find the range of values that the current, in amps, can have. (Use the formula voltage = current × resistance.)

64. A label on a forklift indicates that the maximum safe load is 24,500 N, that is, the maximum downward force is −24,500 N. If the acceleration due to gravity is $-9.8\ \text{m/sec.}^2$, find the range of mass, in kilograms, that can be safely lifted. (Use the formula force = mass × acceleration.)

PUZZLE PROBLEM

A lady and a gentleman are sister and brother. We do not know who is older.

Someone asked them: Who is older?

The sister said: I am older.

The brother said: I am younger.

At least one of them was lying. Who is older?

Collaborative Exercises *Optical Illusion or Confusion?*

An optical shop at a local mall advertises the sale shown to the right. The total cost for a pair of glasses is the sum of the costs of the frame and lenses.

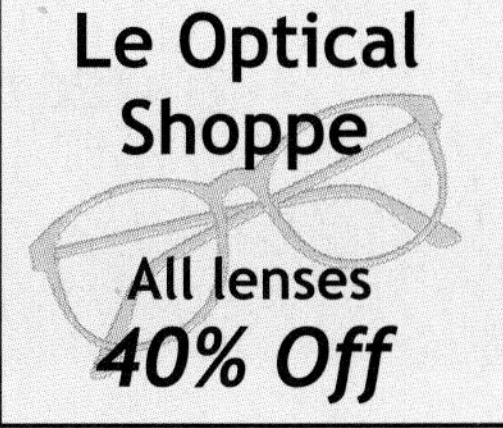

1. Let F represent the regular price of a frame and L the regular price for lenses. Write an expression that describes the total cost of a pair of glasses at regular price.
2. Does the expression $F + 0.60L$ give the cost of the glasses during the advertised 40% off sale? Explain why or why not.
3. The regular price for Anna's lenses is $90. Anna chooses frames listed at $120. How much did Anna save by buying her glasses during the sale?
4. Write and solve an inequality to determine the price of the most expensive frame Anna can choose during the sale if she wishes to keep the cost of her glasses to at most $125.
5. The optical shop gives a discount of 25% on frames and lenses every day to seniors. Pat has chosen a $140 frame. She wears bifocals, so her lenses are $260. The shop will apply only one of the discounts, so which would be better for Pat, the 25% senior discount or the advertised 40% off sale?
6. The optical shop has a complete series of economy frames for $60. Write and solve an inequality showing for which lens prices the 40% off sale would be more economical than the senior discount if an economy frame is used.
7. College students are eligible for a 15% discount on glasses that cost over $100. Using F to represent the regular price of a frame and L for the regular price of lenses, write an expression that describes the total cost of a pair of glasses with the student discount.
8. If a student uses the economy $80 frames, write and solve an inequality showing for which lens prices the 40% off sale would be better than the everyday student discount.

REVIEW EXERCISES

[1.1] **1.** If $A = \{1, 2, 3, 4\}$ and $B = \{0, 1, 2, 3, 4, 5, 6\}$, is $A \subseteq B$?

[1.2] **2.** What property is illustrated by $-5(2 + 7) = -5(7 + 2)$?

[1.4] **3.** Evaluate $0.5r^2 - t$ when $r = -6$, and $t = 8$.

[1.4] **4.** Evaluate $|x^3 - y|$ when $x = -4$ and $y = -19$.

[2.1] ***For Exercises 5 and 6, solve and check.***

5. $6x + 9 = 13 + 4(3x + 2)$

6. $\frac{3}{4}n - 1 = \frac{1}{5}(2n + 3)$

2.4 Compound Inequalities

OBJECTIVES

1. Solve compound inequalities involving "and."
2. Solve compound inequalities involving "or."

In Section 2.3, we learned to solve linear inequalities and represent their solution sets with set notation, interval notation, and graphically on number lines. For example, the solution set for $x > 3$ is represented in those three ways here.

Set-builder notation: $\{x \mid x > 3\}$

Interval notation: $(3, \infty)$

Graph:

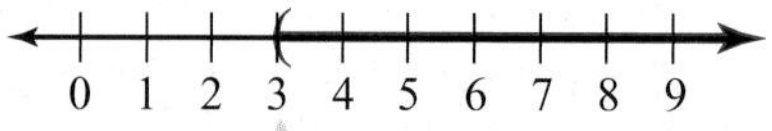

Note: *Remember that the parenthesis indicates that 3 is not included in the solution set.*

In this section, we will build on this foundation and explore **compound inequalities.**

The Algebra Pyramid

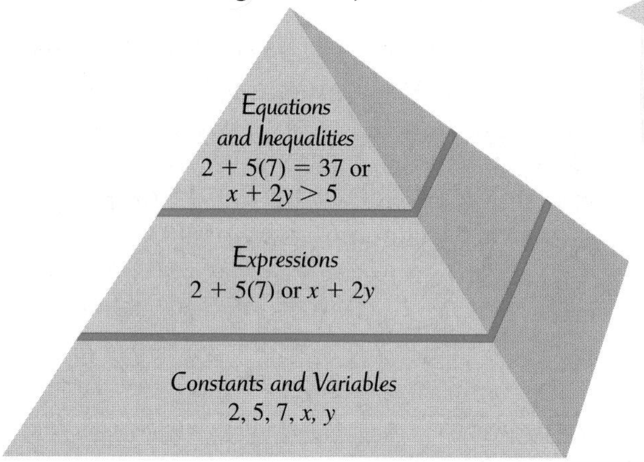

Note: *In this chapter, we explore inequalities and equations, so we are at the top of our Algebra Pyramid.*

DEFINITION ***Compound inequality:*** Two inequalities joined by either "and" or "or."

Some examples of compound inequalities are

$$x > 3 \text{ and } x \le 8 \qquad -2 \ge x \text{ or } x > 4$$

OBJECTIVE 1. Solve compound inequalities involving "and."

Interpreting Compound Inequalities with "and"

First, let's consider inequalities involving "and," such as $x > 3$ *and* $x \leq 8$. The word *and* indicates that the solution set contains only values that satisfy *both* inequalities. Therefore, the solution set is the **intersection**, or overlap, of the two inequalities' solution sets.

DEFINITION ***Intersection:*** For two sets A and B, the intersection of A and B, symbolized by $A \cap B$, is a set containing only elements that are in both A and B.

For example, if $A = \{1, 2, 3, 4, 5\}$ and $B = \{3, 4, 5, 6, 7\}$, then $A \cap B = \{3, 4, 5\}$ because 3, 4, and 5 are the only elements in both A and B. Let's look at our compound inequality $x > 3$ and $x \leq 8$. First we will graph the two inequalities separately, then consider their intersection.

Note: *We use the blue field to indicate the intersection and the dashed lines to indicate the intersection boundaries.*

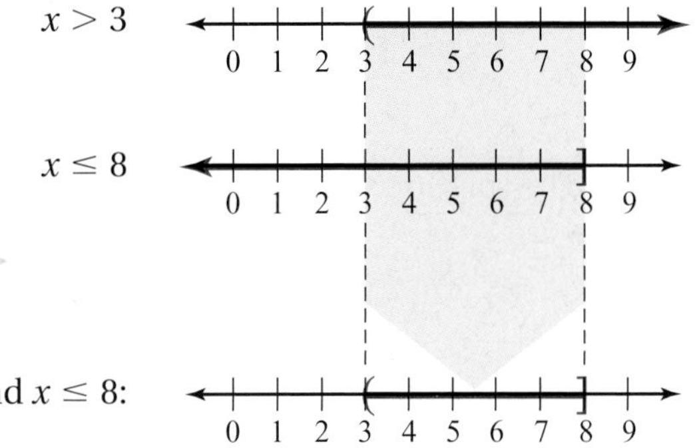

Note: *The intersection of the two inequalities contains elements that are only in* **both** *of their solution sets. Notice that 3 is excluded in the intersection because it was excluded in one of the individual graphs.*

So $x > 3$ and $x \leq 8$ means that x is any number greater than 3 *and* less than or equal to 8. Since this inequality indicates x is between two values, we can write it without the word *and*: $3 < x \leq 8$. In set-builder notation, we write $\{x | 3 < x \leq 8\}$. Using interval notation, we write $(3, 8]$.

EXAMPLE 1 For the compound inequality $x > -2$ and $x < 3$, graph the solution set and write the compound inequality without "and," if possible. Then write in set-builder notation and in interval notation.

Solution The solution set is the region of intersection.

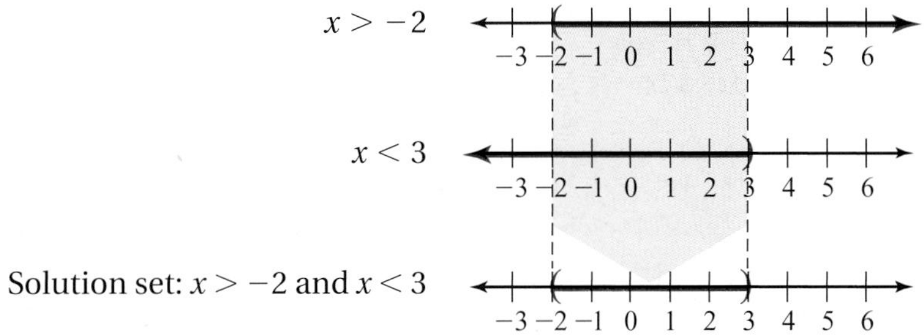

Without "and": $-2 < x < 3$

Set-builder notation: $\{x | -2 < x < 3\}$

Interval notation: $(-2, 3)$

Warning: Be careful not to confuse the interval notation $(-2, 3)$ with an ordered pair.

YOUR TURN For each compound inequality, graph the solution set and write the compound inequality without "and," if possible. Then write in set-builder notation and in interval notation.

a. $x > 4$ and $x < 7$ **b.** $x \geq -8$ and $x \leq -2$

Solving Compound Inequalities with "and"

Now let's consider solution sets for more complex inequalities involving "and."

EXAMPLE 2 For the inequality $3x - 1 > 2$ and $2x + 4 \leq 14$, graph the solution set. Then write the solution set in set-builder notation and in interval notation.

Solution First, we solve each inequality in the compound inequality.

$3x - 1 > 2$ and $2x + 4 \leq 14$

$3x > 3$ and $2x \leq 10$ Use the addition principle to isolate the x term.

$x > 1$ and $x \leq 5$ Divide out each coefficient.

The solution set is the intersection of the two individual solution sets.

$x > 1$ 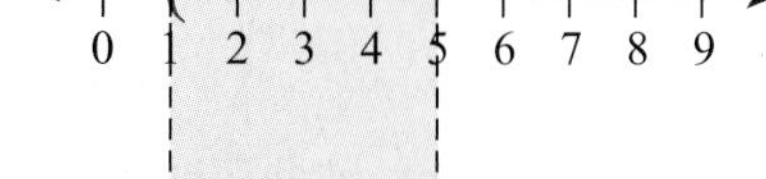

$x \leq 5$ 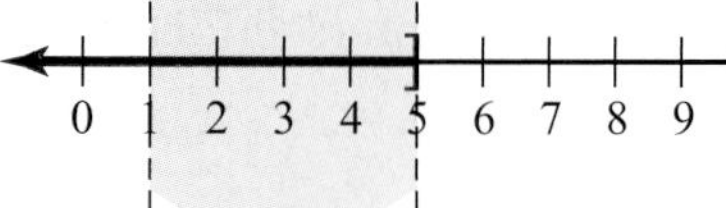

Solution set: $x > 1$ and $x \leq 5$ 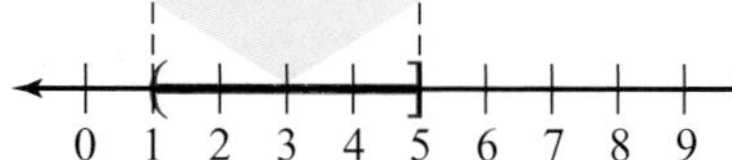

Set-builder notation: $\{x \mid 1 < x \leq 5\}$
Interval notation: $(1, 5]$

Example 2 suggests the following procedure.

PROCEDURE **Solving Compound Inequalities Involving "and"**

To solve a compound inequality involving "and,"

1. Solve each inequality in the compound inequality.
2. The solution set will be the intersection of the individual solution sets.

Finally, we consider some special situations with compound inequalities involving "and."

ANSWERS

a. 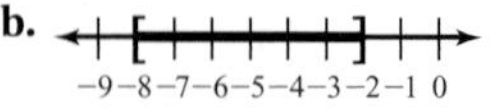

Without "and": $4 < x < 7$
Set-builder: $\{x \mid 4 < x < 7\}$
Interval: $(4, 7)$

b. −9 −8 −7 −6 −5 −4 −3 −2 −1 0

Without "and": $-8 \leq x \leq -2$
Set-builder: $\{x \mid -8 \leq x \leq -2\}$
Interval: $[-8, -2]$

EXAMPLE 3 For the compound inequality, graph the solution set. Then write the solution set in set-builder notation and in interval notation.

a. $-2 \le -3x + 7 \le 1$

Solution Note that $-2 \le -3x + 7 \le 1$ means $-3x + 7 \ge -2$ and $-3x + 7 \le 1$. Solving the two inequalities gives us

> **Note:** *Remember, when we multiply or divide both sides of an inequality by a negative number, we must change the direction of the inequality.*

$-3x + 7 \ge -2$	and	$-3x + 7 \le 1$	
$-3x \ge -9$	and	$-3x \le -6$	Subtract 7 from both sides of each inequality.
$x \le 3$	and	$x \ge 2$	Divide both sides of each inequality by -3.

Remember, we can write $x \le 3$ and $x \ge 2$ as $2 \le x \le 3$. Graphing this solution set we have

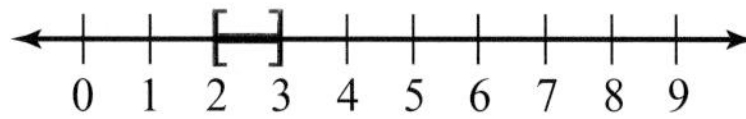

We could have also solved the compound inequality in its original form.

$$-2 \le -3x + 7 \le 1$$
$$-9 \le -3x \le -6 \quad \text{Subtract 7 from all three parts of the compound inequality.}$$
$$3 \ge x \ge 2 \quad \text{Divide all three parts of the compound inequality by } -3.$$

Note that $3 \ge x \ge 2$ is the same as $2 \le x \le 3$.

Set-builder notation: $\{x \mid 2 \le x \le 3\}$

Interval notation: $[2, 3]$

b. $2x + 11 \ge 5$ and $2x + 11 > 3$

Solution To graph the solution set, we first solve each inequality.

$2x + 11 \ge 5$	and	$2x + 11 > 3$	
$2x \ge -6$	and	$2x > -8$	Subtract 11 from both sides of each inequality.
$x \ge -3$	and	$x > -4$	Divide both sides of each inequality by 2.

$x \ge -3$

−5 −4 −3 −2 −1 0 1 2 3 4

$x > -4$

−5 −4 −3 −2 −1 0 1 2 3 4

> **Note:** *The region of intersection for these graphs is from -3 to ∞.*

Solution set:

−5 −4 −3 −2 −1 0 1 2 3 4

Set-builder notation: $\{x \mid x \ge -3\}$

Interval notation: $[-3, \infty)$

c. $-4x - 3 > 1$ and $-4x - 3 < -11$

Solution First, we solve each inequality.

$-4x - 3 > 1$	and	$-4x - 3 < -11$	
$-4x > 4$	and	$-4x < -8$	Add 3 to both sides of each inequality.
$x < -1$	and	$x > 2$	Divide both sides of each inequality by -4.

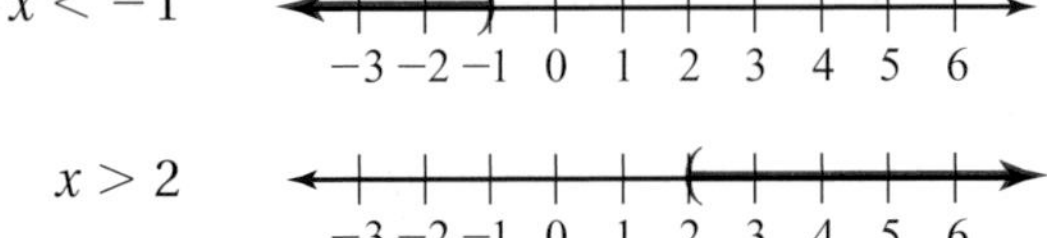

Note: *There is no region of intersection for these graphs, so the solution set contains no values and is said to be empty.*

To graph an empty set, we draw a number line with no shading.

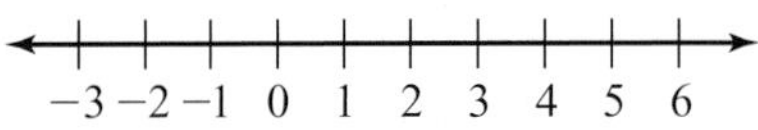

Set-builder notation: { } or $\varnothing$

Interval notation: We do not write interval notation because there are no values in the solution set.

Your Turn For each compound inequality, graph the solution set and write the compound inequality without "and," if possible. Then write in set-builder notation and in interval notation.

a. $-9 \le -5x + 11 < 21$

b. $-4x - 3 > 1$ and $-4x - 3 < -5$

OBJECTIVE 2. Solve compound inequalities involving "or."

Interpreting Compound Inequalities with "or"

In a compound inequality such as $x > 2$ or $x \le -1$, the word *or* indicates that the solution set contains values that satisfy *either* inequality. Therefore, the solution set is the **union**, or joining, of the two inequalities' solution sets.

DEFINITION ***Union:*** For two sets A and B, the union of A and B, symbolized by $A \cup B$, is a set containing every element in A or in B.

Answers

a. 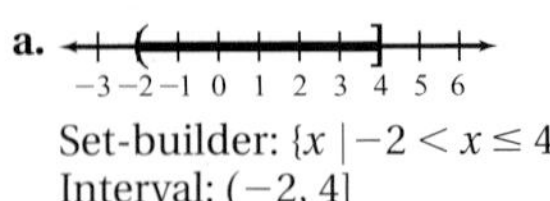

Set-builder: $\{x \mid -2 < x \le 4\}$
Interval: $(-2, 4]$

b. −3 −2 −1 0 1 2 3 4 5 6

Set-builder: { } or $\varnothing$
no interval notation

Learning Strategy

If you are a visual learner, imagine placing one graph on top of the other to form their union.

For example, if $A = \{1, 2, 3, 4, 5\}$ and $B = \{3, 4, 5, 6, 7\}$, then $A \cup B = \{1, 2, 3, 4, 5, 6, 7\}$. Consider our compound inequality $x > 2$ or $x \leq -1$. Let's graph the two inequalities separately, then consider their union.

$x > 2$

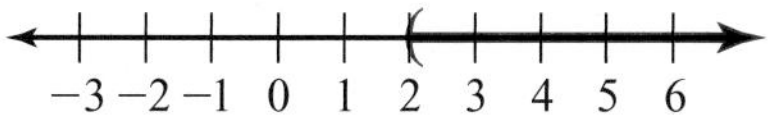

$x \leq -1$

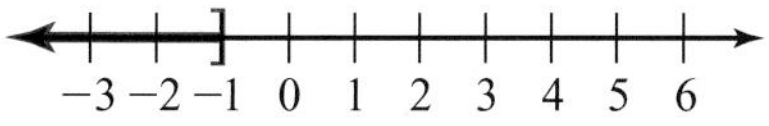

Solution set:
$x > 2$ or $x \leq -1$

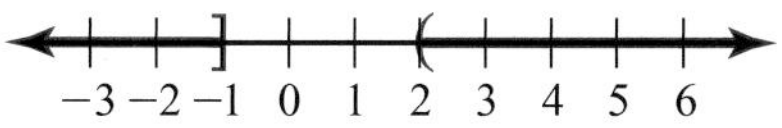

Note: *This graph is the union of the two individual solution sets because it contains every element in **either** of the individual solution sets.*

Using set-builder notation, we write the solution set as $\{x \mid x > 2 \text{ or } x \leq -1\}$. Interval notation is a little more challenging for compound inequalities like $x > 2$ or $x \leq -1$ because the solution set is a combination of two intervals. As we scan from left to right, the first interval is $(-\infty, -1]$. The second interval is $(2, \infty)$. To indicate the union of these two intervals, we write $(-\infty, -1] \cup (2, \infty)$.

Solving Compound Inequalities with "or"

Now let's solve more complex inequalities involving "or." The process is similar to the process we used for solving compound inequalities involving "and."

PROCEDURE Solving Compound Inequalities Involving "or"

To solve a compound inequality involving "or,"

1. Solve each inequality in the compound inequality.
2. The solution set will be the union of the individual solution sets.

EXAMPLE 4 For each compound inequality, graph the solution set. Then write the solution set in set-builder notation and in interval notation.

a. $\frac{2}{3}x - 5 \leq -7 \quad \text{or} \quad \frac{2}{3}x - 5 \geq -1$

Solution First, we solve each inequality in the compound inequality.

$$\frac{2}{3}x - 5 \leq -7 \quad \text{or} \quad \frac{2}{3}x - 5 \geq -1$$

$$\frac{2}{3}x \leq -2 \quad \text{or} \quad \frac{2}{3}x \geq 4 \qquad \text{Add 5 to both sides of each inequality.}$$

$$x \leq -3 \quad \text{or} \quad x \geq 6 \qquad \text{Multiply both sides by } \frac{3}{2} \text{ in each inequality.}$$

The solution set is the union of the two individual solution sets.

$x \le -3$

$x \ge 6$

Solution set:
$x \le -3$ or $x \ge 6$

Set-builder notation: $\{x \mid x \le -3 \text{ or } x \ge 6\}$
Interval notation: $(-\infty, -3] \cup [6, \infty)$

b. $-3x + 8 < -1$ or $-3x + 8 \le 20$

Solution

$-3x + 8 < -1$ or $-3x + 8 \le 20$

$-3x < -9$ or $-3x \le 12$ **Subtract 8 from both sides of each inequality.**

$x > 3$ or $x \ge -4$ **Divide both sides of each inequality by -3.**

Now we graph the individual solution sets and consider their union.

$x > 3$

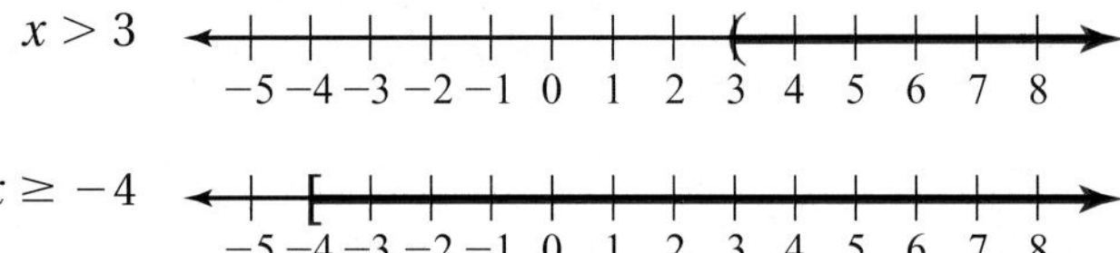

$x \ge -4$

Solution set:
$x > 3$ or $x \ge -4$

Note: *When we join the two graphs, the graph of $x \ge -4$ covers all of the graph of $x > 3$, so the union of the two graphs is actually $x \ge -4$.*

Set-builder notation: $\{x \mid x \ge -4\}$
Interval notation: $[-4, \infty)$

c. $9x - 16 > -34$ or $9x - 16 \le 29$

Solution

$9x - 16 > -34$ or $9x - 16 \le 29$

$9x > -18$ or $9x \le 45$ **Add 16 to both sides of each inequality.**

$x > -2$ or $x \le 5$ **Divide both sides of each inequality by 9.**

Now we graph the individual solution sets and consider their union.

$x > -2$

−5 −4 −3 −2 −1 0 1 2 3 4 5 6 7 8

$x \leq 5$

−5 −4 −3 −2 −1 0 1 2 3 4 5 6 7 8

Solution set:
$x > -2$ or $x \leq 5$

−5 −4 −3 −2 −1 0 1 2 3 4 5 6 7 8

Note: *When we join the two graphs, the entire number line is covered, so the union of the two graphs is the set of real numbers.*

Set-builder notation: $\{x \mid x \text{ is a real number}\}$, or $\mathbb{R}$.
Interval notation: $(-\infty, \infty)$

YOUR TURN For each compound inequality, graph the solution set. Then write the solution set in set-builder notation and in interval notation.

a. $-5x - 8 > 12$ or $-5x - 8 \leq -3$ **b.** $\frac{3}{4}x + 5 < -1$ or $\frac{3}{4}x + 5 < 8$

c. $5 - 3x \geq 11$ or $5 - 3x < 17$

ANSWERS

a. −6 −5 −4 −3 −2 −1 0 1 2 3
Set-builder: $\{x \mid x \leq -4$ or $x \geq -1\}$
Interval: $(-\infty, -4) \cup [-1, \infty)$

b. −3 −2 −1 0 1 2 3 4 5 6
Set-builder: $\{x \mid x < 4\}$
Interval: $(-\infty, 4)$

c. −4 −3 −2 −1 0 1 2 3 4
Set-builder: $\{x \mid x \text{ is a real number}\}$, or $\mathbb{R}$
Interval: $(-\infty, \infty)$

2.4 Exercises

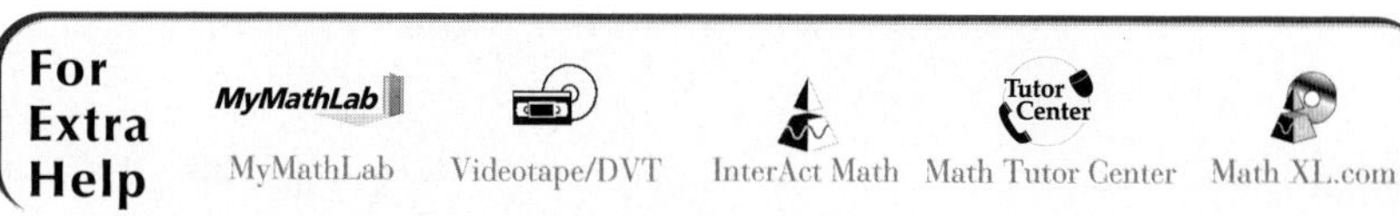

1. What is a compound inequality?

2. Compound inequalities involve what two key words?

3. Describe the intersection of two sets.

4. Describe the union of two sets.

5. Describe how to graph a compound inequality involving "and."

6. Describe how to graph a compound inequality involving "or."

***For Exercises 7–14:* a. Find the intersection of the given sets.**
b. Find the union of the given sets.

7. $A = \{1, 3, 5\}$
$B = \{1, 3, 5, 7, 9\}$

8. $A = \{2, 4, 6\}$
$B = \{2, 4, 6, 8, 10\}$

9. $A = \{5, 6, 7\}$
$B = \{7, 8, 9\}$

10. $A = \{8, 10, 12, 14\}$
$B = \{14, 16, 18\}$

11. $A = \{c, a, t\}$
$B = \{d, o, g\}$

12. $A = \{l, o, v, e\}$
$B = \{m, a, t, h\}$

13. $A = \{x, y, z\}$
$B = \{w, x, y, z\}$

14. $A = \{m, n, o, p, q\}$
$B = \{a, b, c, \ldots x, y, z\}$

For Exercises 15–22, write each inequality without "and."

15. $x > -4$ and $x < 5$

16. $n \geq 3$ and $n < 7$

17. $y > -2$ and $y \leq 0$

18. $m \geq 0$ and $m < 3$

19. $w > -7$ and $w < 3$

20. $r > -1$ and $r \leq 1$

21. $u \geq 0$ and $u \leq 2$

22. $t > 7$ and $t < 15$

For Exercises 23–30, graph the compound inequality.

23. $x > 2$ and $x < 7$

24. $x > 3$ and $x < 9$

25. $x > -1$ and $x \leq 5$

26. $x \geq -4$ and $x < 0$

27. $1 \leq x \leq 10$

28. $-2 < x < -1$

29. $-3 \leq x < 4$

30. $0 < x \leq 8$

***For Exercises 31–50:* a. Graph the solution set.**
b. Write the solution set in set-builder notation.
c. Write the solution set in interval notation.

31. $x > -3$ and $x < -1$

32. $x > 4$ and $x < 7$

33. $x + 2 > 5$ and $x - 4 \leq 2$

34. $x - 3 \geq 1$ and $x + 2 < 10$

35. $-x > 1$ and $-2x \leq -10$

36. $-4x \geq 8$ and $3x \leq 15$

37. $2x + 6 \geq -4$ and $3x - 1 < 8$

38. $3x + 5 \geq -1$ and $5x - 2 < 13$

39. $-3x - 8 > 1$ and $-4x + 5 \leq -3$

40. $-6x + 4 < -14$ and $-x - 3 \geq -2$

41. $-3 < x + 4 < 1$

42. $-2 \leq x - 2 \leq 2$

43. $-2 \leq 5x + 3 \leq 13$

44. $7 < 2x - 1 < 11$

45. $0 \leq 2 + 3x < 8$

46. $0 \leq -2 + 5x \leq 13$

47. $-6 < -3x + 3 \leq 3$

48. $-3 < 5 - 2x < 1$

49. $3 \leq 6 - x \leq 6$

50. $-4 \leq -4 - x \leq 2$

For Exercises 51–58, graph each inequality.

51. $x < -2$ or $x > 6$

52. $n \leq 3$ or $n > 4$

53. $y < -3$ or $y \geq 0$

54. $m \leq 2$ or $m > 8$

55. $w > -3$ or $w > 2$

56. $r > -3$ or $r \geq -1$

57. $u \geq 0$ or $u \leq 2$

58. $t > 6$ or $t < 13$

For Exercises 59–76: ***a. Graph the solution set.***
b. Write the solution set in set-builder notation.
c. Write the solution set in interval notation.

59. $y + 2 < -7$ or $y + 2 > 7$

60. $a - 4 \le -2$ or $a - 4 \ge 2$

61. $3r + 2 < -4$ or $3r - 3 > 0$

62. $4t + 5 \le -7$ or $4t + 5 \ge -3$

63. $-w + 2 \le -5$ or $-w + 2 \ge 3$

64. $-2x - 3 \le -1$ or $-2x - 3 \ge 5$

65. $7 - 3k \le -2$ or $7 - 3k \ge 1$

66. $8 - 2q \le 2$ or $8 - 2q \ge 4$

67. $5x + 2 \le -1$ or $5x + 2 \ge 3$

68. $2m - 1 < 0$ or $2m - 1 > 5$

69. $-2c + 2 < -4$ or $-2c + 4 < -6$

70. $5d + 3 < 8$ or $5d + 3 < 18$

71. $6 - 3m \le -2$ or $6 - 3m \ge 3$

72. $3 - 4n > 8$ or $3 - 4n < -3$

73. $2x + 9 \le 1$ or $2x + 9 \le -3$

74. $5 + y \ge 4$ or $5 + y \ge 6$

75. $-3x + 2 \leq -1$ or $-3x + 2 \geq -2$

76. $4k - 7 \leq 1$ or $4k - 7 \geq -5$

For Exercises 77–88: a. Graph the solution set.
b. Write the solution set in set-builder notation.
c. Write the solution set in interval notation.

77. $-3 < x + 4$ and $x + 4 < 7$

78. $0 < x - 1$ and $x - 1 \leq 3$

79. $2x + 3 < -1$ or $2x + 3 > 7$

80. $3x - 2 < -5$ or $3x - 2 > 7$

81. $4 < -2x < 6$

82. $9 \leq -3x \leq 15$

83. $1 \leq 2x - 5 \leq 7$

84. $-7 < 3x - 4 < 2$

85. $-5 < 1 - 2x < -1$

86. $4 \leq -3x + 1 < 7$

87. $x - 3 < 2x + 1 < 3x$

88. $2x - 2 \leq x + 1 \leq 2x + 5$

For Exercises 89–98, solve. Then: a. Graph the solution set.
b. Write the solution set in set-builder notation.
c. Write the solution set in interval notation.

89. If Andrea's current long-distance bill is between \$30 and \$45 in a month, then she can save money by switching to another company. Her current rate is \$0.09 per minute. What is the range of minutes that Andrea must use long distance to justify switching to the other company?

90. A mail-order music club offers one bonus CD if the total of an order is from \$75 to \$100. Dayle decides to buy the lowest-priced CDs, which are \$12.50 each. In what range would the number of CDs he orders have to be in order to receive a bonus CD?

91. Juan has taken four of the five tests in his history course. To get a B in the course, his average needs to be at least 80 and less than 90. If his scores on the first four tests are 95, 80, 82, 88, in what range of values can his score on the fifth test be so that he has a B average?

92. Students in a chemistry course receive a C if the average of four tests is at least 70 and less than 80. Suppose a student has the following scores: 60, 72, and 70. What range of scores on the fourth test would cause the student to receive a C?

93. To conserve energy, it is recommended that a home's thermostat be set at 5° above 73°F during summer or 5° below 73° during winter. If the heat pump does not run when the temperature is at or between those values, in what range of temperatures is the heat pump off?

94. The house thermostat is set so that the heat pump/air conditioner comes on if the temperature is 2° or more above or below 70°. In what range of temperatures is the heat pump off?

95. To maintain a saltwater aquarium, the temperature should be within 4° of 76°. What range of temperatures would be acceptable for saltwater fish? (*Source: The Conscientious Marine Aquarist*, Robert M. Fenner © 2001.)

96. When driving in the state of Mississippi, motorists risk a ticket if they drive more than 15 miles per hour over or under the 55 miles per hour speed limit on a state highway. In what range of speeds would a motorist not risk receiving a ticket? (*Source:* Mississippi State Code 63-3-509.)

97. A building is to be designed in the shape of a trapezoid as shown. Find the range of values for the length of the back side of the building so that the square footage is from 6000 to 8000 square feet. (Use $A = \frac{1}{2}h(a + b)$.)

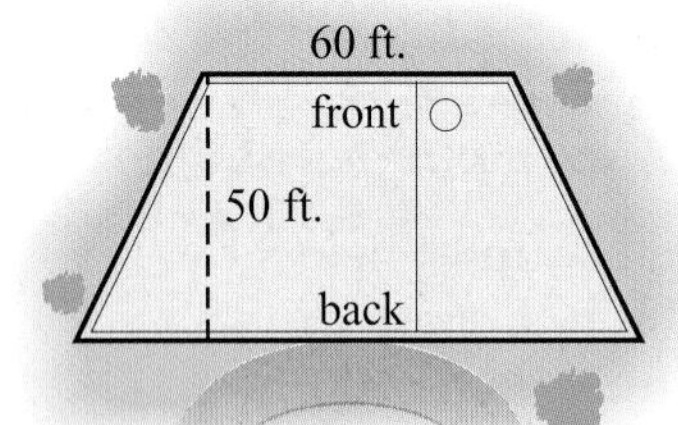

98. At 32°F and below, water is a solid (ice). At 212°F and above, water is a gas (steam). Use a compound inequality to describe the range of values in degrees Celsius for water in its liquid state. (Use $C = \frac{5}{9}(F - 32)$.)

For Exercises 99–104, use the following graph and write each solution set using set-builder notation and interval notation.

99. If x represents time, write a solution set that describes the range of years in which there were at least 20,000 drug defendants and 15,000 or more property defendants.

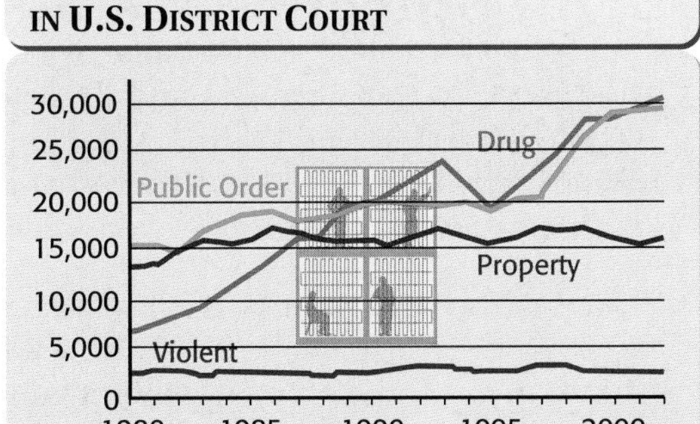

Source: Compendium of Federal Justice Statistics.

100. If x represents time, write a solution set that describes the range of years in which there were at least 5,000 and at most 10,000 drug defendants.

101. If x represents time, write a solution set that describes the range of years in which there were fewer drug defendants than property and public order defendants.

102. If x represents time, write a solution set that describes the range of years in which there were fewer drug defendants than property defendants or fewer drug defendants than public order defendants.

103. If x represents time, write a solution set that describes the range of years in which there were fewer public order defendants than drug defendants but more public order defendants than property defendants.

104. If x represents time, write a solution set that describes the range of years in which there were at most 10,000 drug defendants or fewer property defendants than drug defendants.

PUZZLE PROBLEM

A cyclist is involved in a multiple-day race. She feels she needs to complete today's 40 kilometers part in somewhere between 1 hour and 45 minutes and 2 hours. Find the range of values her rate can be to complete this leg of the race in her desired time frame.

REVIEW EXERCISES

[1.1] **1.** Is the absolute value of a number always positive? Explain.

[1.3] ***For Exercises 2 and 3, simplify.***

2. $-|2 - 3^2| - |-4|$

3. $-|-|-16||$

[2.1] ***For Exercises 4 and 5, solve.***

4. $3x - 14x = 17 - 12x - 11$

5. $\frac{3}{5}(25 - 5x) = 15 - \frac{3}{5}$

[2.2] **6.** One-half the sum of a number and 2 is zero. Find the number.

2.5 Equations Involving Absolute Value

OBJECTIVE 1. Solve equations involving absolute value. We learned in Section 1.1 that the absolute value of a number is its distance from zero on a number line. For example, $|4| = 4$ and $|-4| = 4$ because both 4 and -4 are 4 units from 0 on a number line.

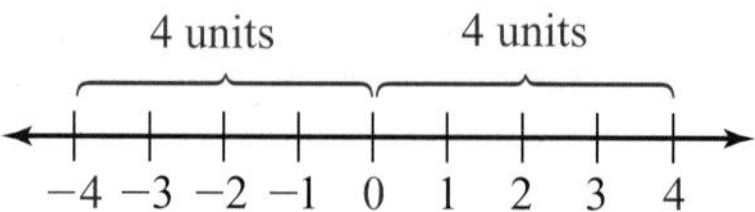

Now we will solve equations in which a variable appears within the absolute value symbols, as in $|x| = 4$. Notice the solutions are 4 and -4, which suggests the following rule.

RULE **Absolute Value Property**

If $|x| = a$, where x is a variable or an expression and $a \geq 0$, then $x = a$ or $x = -a$.

EXAMPLE 1 Solve.

a. $|2x - 5| = 9$

Solution Using the absolute value property, we separate the equation into a positive case and a negative case, then solve the two cases separately.

$$\begin{aligned} 2x - 5 &= 9 \quad &\text{or}\quad 2x - 5 &= -9 \\ 2x &= 14 \quad &\text{or}\quad 2x &= -4 \\ x &= 7 \quad &\text{or}\quad x &= -2 \end{aligned}$$

The solutions are 7 and -2.

b. $|4x + 1| = -11$

Solution This equation has the absolute value equal to a negative number, -11. Because the absolute value of every real number is a positive number or zero, this equation has no solution.

From Example 1b we can say that an absolute value equation in the form $|x| = a$, where $a < 0$, has no solution.

YOUR TURN Solve.

a. $|3x + 4| = 8$ **b.** $|2x - 1| = 0$ **c.** $|5x - 2| = -3$

ANSWERS

a. $\frac{4}{3}$ and -4 **b.** $\frac{1}{2}$

c. no solution

To use the absolute value property, we need the absolute value isolated.

EXAMPLE 2 Solve. $|4x - 3| + 9 = 16$

Solution $|4x - 3| = 7$ Subtract 9 from both sides to isolate the absolute value.

$4x - 3 = 7$ or $4x - 3 = -7$ Use the absolute value property.

$4x = 10$ or $4x = -4$

$x = \frac{5}{2}$ or $x = -1$

The solutions are $\frac{5}{2}$ and -1.

Examples 1 and 2 suggest the following procedure.

PROCEDURE **Solving Equations Containing a Single Absolute Value**

To solve an equation containing a single absolute value,

1. Isolate the absolute value so that the equation is in the form $|ax + b| = c$. If $c > 0$, proceed to steps 2 and 3. If $c < 0$, the equation has no solution.
2. Separate the absolute value into two equations, $ax + b = c$ and $ax + b = -c$.
3. Solve both equations.

YOUR TURN Solve.

a. $|6x - 3| - 5 = 4$ **b.** $|2x + 1| + 6 = 4$

More Than One Absolute Value

Some equations contain more than one absolute value expression, such as $|x + 3| = |2x - 8|$. If two absolute values are equal, they must contain expressions that are *equal* or *opposites*. Some simple examples follow:

Equal		**Opposites**
$\|7\| = \|7\|$	or	$\|7\| = \|-7\|$
$\|-5\| = \|-5\|$	or	$\|-5\| = \|5\|$

For the equation $|x + 3| = |2x - 8|$, therefore, the expressions $x + 3$ and $2x - 8$ must be equal or opposites.

Equal		**Opposites**
$x + 3 = 2x - 8$	or	$x + 3 = -(2x - 8)$

This suggests the following procedure.

ANSWERS

a. 2 and -1 **b.** no solution

PROCEDURE **Solving Equations in the Form $|ax + b| = |cx + d|$**

To solve an equation in the form $|ax + b| = |cx + d|$,

1. Separate the absolute value equation into two equations: $ax + b = cx + d$ and $ax + b = -(cx + d)$.
2. Solve both equations.

EXAMPLE 3 Solve. $|3x + 5| = |x - 9|$

Solution Separate the absolute value equation into two equations.

Equal		**Opposites**
$3x + 5 = x - 9$	or	$3x + 5 = -(x - 9)$
$2x + 5 = -9$	or	$3x + 5 = -x + 9$
$2x = -14$	or	$4x + 5 = 9$
$x = -7$	or	$4x = 4$
		$x = 1$

Note: *Think of $-(x - 9)$ as $-1(x - 9)$.*

The solutions are -7 and 1.

Sometimes when we separate an equation containing two absolute values into its two equivalent equations, we find that only one of them yields a solution.

EXAMPLE 4 Solve. $|3x - 7| = |5 - 3x|$

Solution

Equal		**Opposites**	
$3x - 7 = 5 - 3x$	or	$3x - 7 = -(5 - 3x)$	Separate into two equations.
$6x - 7 = 5$	or	$3x - 7 = -5 + 3x$	
$6x = 12$	or	$-7 = -5$	
$x = 2$			

Note: *Subtracting $3x$ from both sides gives us a false equation, so the "opposites" equation has no solution.*

This absolute value equation has only one solution, 2.

YOUR TURN Solve.

a. $|9 + 4x| = |1 - 4x|$ **b.** $|5 - 2x| = -|x + 3|$

ANSWERS

a. -1 **b.** no solution

2.5 Exercises

For Extra Help

MyMathLab

Videotape/DVT

InterAct Math

Math Tutor Center

Math XL.com

1. What does absolute value mean in terms of a number line?

2. How do you interpret $|x| = 5$?

3. State the absolute value property in your own words.

4. If given an equation in the form $|ax + b| = c$ and $c < 0$, what can you conclude about its solution(s)?

5. Explain the first step in solving an equation containing two absolute values, such as $|2x + 1| = |3x - 5|$.

6. How can you tell that an absolute value equation containing two absolute values has only one solution?

For Exercises 7–22, solve using the absolute value property.

7. $|x| = 2$

8. $|y| = 5$

9. $|a| = -4$

10. $|r| = -1$

11. $|x + 3| = 8$

12. $|w - 1| = 4$

13. $|2m - 5| = 1$

14. $|3s + 7| = 10$

15. $|6 - 5x| = 1$

16. $|3x - 2| = 5$

17. $|4 - 3w| = 6$

18. $|2x - 1| = 2$

19. $|4m - 2| = -5$

20. $|6p + 5| = -1$

21. $|4w - 3| = 0$

22. $|6 - 3m| = 0$

For Exercises 23–38, isolate the absolute value, then use the absolute value property.

23. $|2y| - 3 = 5$

24. $|3r| + 7 = 10$

25. $|y - 1| + 2 = 4$

26. $|x + 3| - 1 = 5$

27. $|b + 4| - 6 = 2$

28. $|v - 2| + 2 = 4$

29. $3 + |5x - 1| = 7$

30. $2 + |3t - 2| = 5$

31. $1 - |2k + 3| = -4$

32. $-3 = 1 - |4u - 2|$

33. $4 - 3|z - 2| = -8$

34. $3 - 2|x - 5| = -7$

35. $6 - 2|3 - 2w| = -18$

36. $15 - 2|5 - 2x| = -5$

37. $|3x - 2(x + 5)| = 10$

38. $|4y + 2(7 - y)| = 5$

For Exercises 39–48, solve by separating the two absolute values into equal and opposite cases.

39. $|2x + 1| = |x + 5|$

40. $|2p + 5| = |3p + 10|$

41. $|x + 3| = |2x - 4|$

42. $|p - 1| = |2p + 8|$

43. $|3v + 4| = |1 - 2v|$

44. $|3 - 6c| = |2c - 3|$

45. $|n - 3| = |3 - n|$

46. $|2r - 1| = |1 - 2r|$

47. $|2k + 1| = |2k - 5|$

48. $|4 + 2q| = |2q + 8|$

For Exercises 49–56, solve.

49. $|10 - (5 + h)| = 8$

50. $|7 - (2 - n)| = 17$

51. $\left|\frac{b}{2} - 1\right| = 4$

52. $\left|\frac{u}{3} + 2\right| = 3$

53. $\left|\frac{4 - 3x}{2}\right| = \frac{3}{4}$

54. $\left|\frac{6 - 5w}{6}\right| = \frac{2}{3}$

55. $\left|2y + \frac{3}{2}\right| - \frac{1}{2} = 5$

56. $\left|p + \frac{2}{3}\right| - 1 = 7$

REVIEW EXERCISES

[2.3] ***For Exercises 1 and 2, solve.***

1. $6 - 4x > -5x - 1$

2. $-4x \le 12$

[2.3] **3.** Translate to a linear inequality and solve: Two less than a number is less than five.

[2.4] **4.** Use interval notation to represent the solution set shown on the graph to the right.

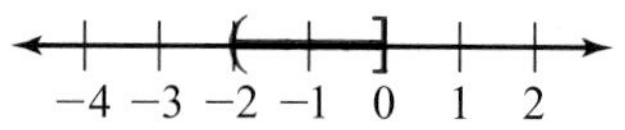

[2.4] **5.** Graph the compound inequality: $-1 < x \le 0$

[2.4] **6.** Solve the compound inequality: $-1 < \frac{3x + 2}{4} < 4$

2.6 Inequalities Involving Absolute Value

OBJECTIVES

1. **Solve absolute value inequalities involving less than.**
2. **Solve absolute value inequalities involving greater than.**

OBJECTIVE 1. Solve absolute value inequalities involving less than. Now let's solve inequalities that contain absolute value. Think about $|x| \le 3$. A solution for this inequality is any number whose absolute value is less than or equal to 3. This means the solutions are a distance of 3 units or less from 0 on a number line.

Note: *Solutions for the equal to part of $|x| \le 3$ are 3 and -3. The less than part of $|x| \le 3$ means all numbers in between 3 and -3 because their absolute values are less than 3.*

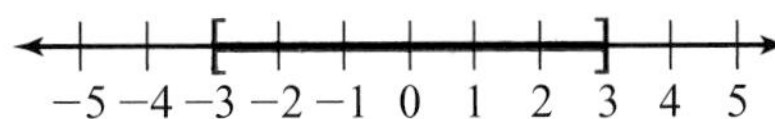

Notice the solution region corresponds to the compound inequality $x \ge -3$ and $x \le 3$, which we can write as $-3 \le x \le 3$. In set-builder notation, the solution is $\{x \mid -3 \le x \le 3\}$ and in interval notation, we write $[-3, 3]$.

Our examples suggest the following procedure.

PROCEDURE **Solving Inequalities in the Form $|x| < a$, where $a > 0$**

To solve an inequality in the form $|x| < a$, where $a > 0$,

1. Rewrite as a compound inequality involving "and": $x > -a$ and $x < a$. (We can also use $-a < x < a$.)
2. Solve the compound inequality.

Similarly, to solve $|x| \le a$, we would write $x \ge -a$ and $x \le a$ (or $-a \le x \le a$).

EXAMPLE 1 For each inequality, solve, graph the solution set, and write the solution set in both set-builder and interval notation.

a. $|x - 4| \le 3$

Solution $x - 4 \ge -3$ and $x - 4 \le 3$ Rewrite as a compound inequality.

$x \ge 1$ and $x \le 7$ Add 4 to both sides of each inequality.

Recall that $x \ge 1$ and $x \le 7$ means $1 \le x \le 7$, so our graph is as follows:

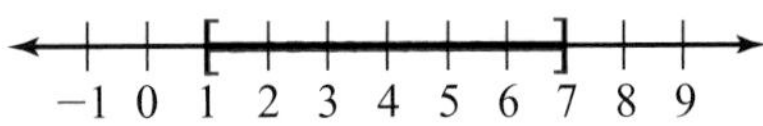

Note: *The solution set for $|x - 4| \le 3$ contains every number whose distance from 4 is 3 units or less.*

Set-builder notation: $\{x \mid 1 \le x \le 7\}$

Interval notation: $[1, 7]$

b. $|3x + 1| < 5$

Solution Instead of $3x + 1 > -5$ and $3x + 1 < 5$, we will use the more compact form.

$-5 < 3x + 1 < 5$ Rewrite as a compound inequality.

$-6 < 3x < 4$ Subtract 1 from all parts of the inequality.

$-2 < x < \frac{4}{3}$ Divide all parts of the inequality by 3 to isolate x.

Set-builder notation: $\left\{x \mid -2 < x < \frac{4}{3}\right\}$

Interval notation: $\left(-2, \frac{4}{3}\right)$

c. $|-0.5x + 1| - 2 < 1$

Solution Notice that the equation is not in the form $|x| < a$, so our first step is to isolate the absolute value.

$\lvert -0.5x + 1 \rvert < 3$	Add 2 to both sides to isolate the absolute value.
$-3 < -0.5x + 1 < 3$	Rewrite as a compound inequality.
$-4 < -0.5x < 2$	Subtract 1 from all parts of the inequality.
$8 > x > -4$	Divide all parts of the inequality by -0.5 to isolate x.

Note: *Because we divided by a negative, we changed the direction of the inequalities.*

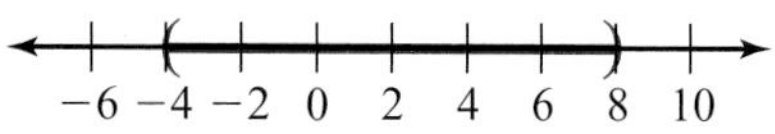

Set-builder notation: $\{x \mid -4 < x < 8\}$

Interval notation: $(-4, 8)$

d. $\left|\frac{1}{3}x - 4\right| + 6 < -1$

Solution $\left|\frac{1}{3}x - 4\right| < -7$ Subtract 6 from both sides to isolate the absolute value.

Because absolute values cannot be negative, this inequality has no solution, so the solution set is empty.

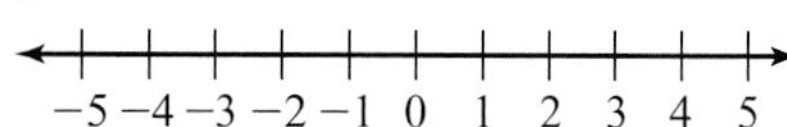

Set-builder notation: { } or $\varnothing$

Interval notation: We do not write interval notation because there are no values in the solution set.

YOUR TURN For each inequality, solve, graph the solution set, and write the solution set in both set-builder notation and interval notation.

a. $|x + 5| < 2$

b. $|2x - 3| \le 7$

c. $\left|-\frac{1}{3}x + 1\right| - 2 \le 1$

d. $\left|\frac{1}{2}x - 1\right| + 3 < -2$

ANSWERS

a. Set-builder: $\{x \mid -7 < x < -3\}$
Interval: $(-7, -3)$

b. Set-builder: $\{x \mid -2 \le x \le 5\}$
Interval: $[-2, 5]$

c. Set-builder: $\{x \mid -6 \le x \le 12\}$
Interval: $[-6, 12]$

d. Set-builder: { } or $\varnothing$
no interval notation

OBJECTIVE 2. Solve absolute value inequalities involving greater than. Now we consider inequalities with greater than, such as $|x| \geq 5$. A solution for $|x| \geq 5$ is any number whose absolute value is greater than or equal to 5. As the graph shows, the solution set contains all values that are a distance of 5 units or more from 0.

Note: *Solutions for the* ***equal to*** *part of* $|x| \geq 5$ *are 5 and* -5*. The* ***greater than*** *part means all values that are farther from 0 than 5 and* -5 *because their absolute values are greater than 5.*

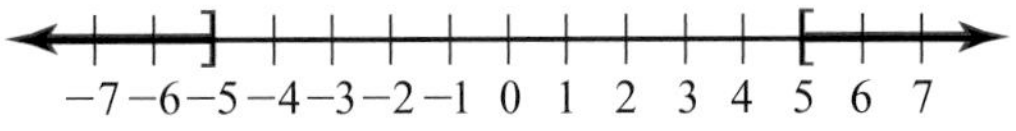

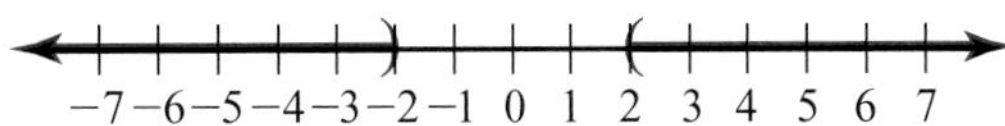

Notice that the solutions are equivalent to $x \leq -5$ or $x \geq 5$.

An inequality such as $|x| > 2$ uses parentheses:

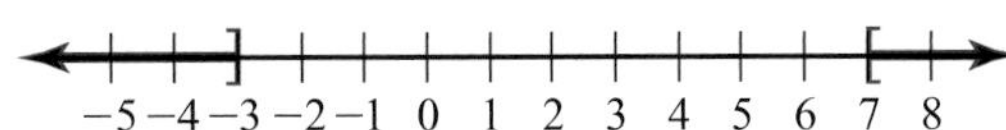

The solutions are equivalent to $x < -2$ or $x > 2$. Our examples suggest that we can split these inequalities into a compound inequality involving "or."

PROCEDURE **Solving Inequalities in the Form $|x| > a$, where $a > 0$**

To solve an inequality in the form $|x| > a$, where $a > 0$,

1. Rewrite as a compound inequality involving "or": $x < -a$ or $x > a$.
2. Solve the compound inequality.

Similarly, to solve $|x| \geq a$, we would write $x \leq -a$ or $x \geq a$.

EXAMPLE 2 For each inequality, solve, graph the solution set, and write the solution set in both set-builder notation and interval notation.

a. $|x - 2| \geq 5$

Solution $x - 2 \leq -5$ or $x - 2 \geq 5$ — Rewrite as a compound inequality.

$x \leq -3$ or $x \geq 7$ — Add 2 to both sides of each inequality.

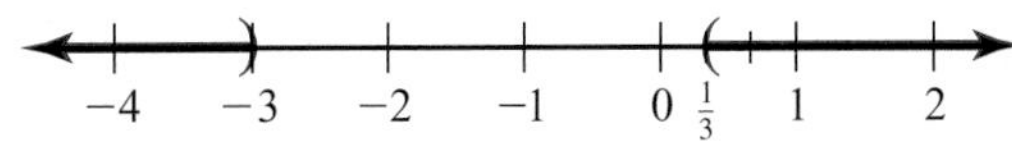

Set-builder notation: $\{x \mid x \leq -3 \text{ or } x \geq 7\}$

Interval notation: $(-\infty, -3] \cup [7, \infty)$

b. $|-3x - 4| > 5$

Solution $-3x - 4 < -5$ or $-3x - 4 > 5$ — Rewrite as a compound inequality.

$-3x < -1$ or $-3x > 9$ — Add 4 to both sides of each inequality.

$x > \frac{1}{3}$ or $x < -3$ — Divide both sides of each inequality by -3.

Set-builder notation: $\left\{x \mid x < -3 \text{ or } x > \frac{1}{3}\right\}$

Interval notation: $(-\infty, -3) \cup \left(\frac{1}{3}, \infty\right)$

c. $\left|\frac{3}{4}x - 2\right| + 1 \geq 6$

Solution $\left|\frac{3}{4}x - 2\right| \geq 5$ — Subtract 1 from both sides to isolate the absolute value.

$\frac{3}{4}x - 2 \leq -5$ or $\frac{3}{4}x - 2 \geq 5$ — Rewrite as a compound inequality.

$\frac{3}{4}x \leq -3$ or $\frac{3}{4}x \geq 7$ — Add 2 to both sides of each inequality.

$x \leq -4$ or $x \geq \frac{28}{3}$ — Multiply both sides of each inequality by $\frac{4}{3}$.

Set-builder notation: $\left\{x \mid x \leq -4 \text{ or } x \geq \frac{28}{3}\right\}$

Interval notation: $(-\infty, -4] \cup \left[\frac{28}{3}, \infty\right)$

d. $|0.2x + 1| - 3 > -7$

Solution $|0.2x + 1| > -4$ — Add 3 to both sides to isolate the absolute value.

This equation indicates that the absolute value is greater than a negative number. Since the absolute value of every real number is either positive or 0, the solution set is $\mathbb{R}$. The graph is the entire number line.

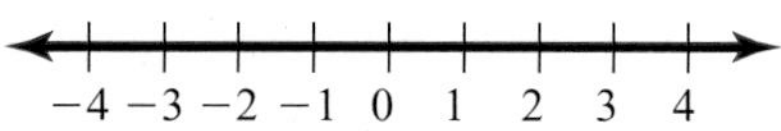

Set-builder notation: $\{x \mid x \text{ is a real number}\}$

Interval notation: $(-\infty, \infty)$

Your Turn For each inequality, solve, graph the solution set, and write the solution set in both set-builder notation and interval notation.

a. $|x - 3| > 2$

b. $|4x + 1| \geq 9$

c. $|-0.4x + 1| + 2 \geq 3$

d. $\left|\frac{1}{3}x - 4\right| - 2 > -9$

Answers

a. Set-builder: $\{x \mid x < 1 \text{ or } x > 5\}$

Interval: $(-\infty, 1) \cup (5, \infty)$

b. Set-builder: $\left\{x \mid x \leq -\frac{5}{2} \text{ or } x \geq 2\right\}$

Interval: $\left(-\infty, -\frac{5}{2}\right] \cup [2, \infty)$

c. Set-builder: $\{x \mid x \leq 0 \text{ or } x \geq 5\}$

Interval: $(-\infty, 0] \cup [5, \infty)$

d. 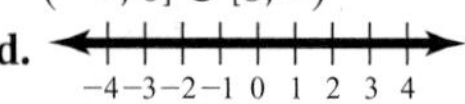

Set-builder: $\{x \mid x \text{ is a real number}\}$

Interval: $(-\infty, \infty)$

2.6 Exercises

For Extra Help MyMathLab | Videotape/DVT | InterAct Math | Math Tutor Center | Math XL.com

For Exercises 1–4, we assume* a *is a positive number.

1. What compound inequality is related to $|x| \le a$, where $a > 0$?

2. What compound inequality is related to $|x| \ge a$, where $a > 0$?

3. How would you characterize the graph of $|x| < a$, where $a > 0$?

4. How would you characterize the graph of $|x| > a$, where $a > 0$?

5. Under what conditions does $|x| < a$ have no solution?

6. Under what conditions does the solution set for $|x| > a$ contain all real numbers?

For Exercises 7–20, solve the inequality. Then: a. Graph the solution set.
b. Write the solution set in set-builder notation.
c. Write the solution set in interval notation.

7. $|x| < 5$

8. $|y| \le 3$

9. $|x + 3| \le 7$

10. $|m - 5| < 2$

11. $|s + 3| + 2 < 5$

12. $|p - 3| + 4 \le 8$

13. $|2m - 5| + 1 < 10$

14. $|4x + 7| + 2 < 5$

15. $|-3k + 5| + 7 \le 8$

16. $|-5h - 1| - 6 < 8$

17. $2|x| - 7 \le 3$

18. $3|u| + 2 < 8$

19. $2|w - 3| + 4 < 10$

20. $4|n + 2| - 3 \le 9$

For Exercises 21–34, solve the inequality. Then: a. Graph the solution set.
b. Write the solution set in set-builder notation.
c. Write the solution set in interval notation.

21. $|c| > 12$

22. $|h| \ge 6$

23. $|y + 2| \ge 7$

24. $|a - 4| > 3$

25. $|p - 6| + 2 > 10$

26. $|x + 5| - 3 \ge 6$

27. $|3x + 6| - 3 \ge 9$

28. $|2x - 5| - 1 \ge 10$

29. $|-4n - 5| + 3 > 8$

30. $|-6h + 1| - 7 > 4$

31. $4|v| - 3 \ge 1$

32. $5|m| + 2 > 12$

33. $4|y + 2| - 1 > 3$

34. $3|x - 2| + 5 \ge 11$

For Exercises 35–54, solve the inequality. Then: a. Graph the solution set.
b. Write the solution set in set-builder notation.
c. Write the solution set in interval notation.

35. $|4m + 8| - 2 > 10$

36. $|2x + 4| - 2 \geq 10$

37. $|-3x + 6| < 6$

38. $|-2y + 3| \leq 3$

39. $|2r - 3| > -3$

40. $|3b + 7| > -2$

41. $2 - |x + 3| > 1$

42. $5 - 2|u + 4| \geq 1$

43. $6|2x - 1| + 3 < 9$

44. $4|3p + 6| - 2 < 22$

45. $5 - |w + 4| > 10$

46. $4 - |5 + k| > 7$

47. $\left|2 - \frac{3}{2}k\right| \leq 5$

48. $\left|3 - \frac{1}{2}x\right| \geq 7$

49. $|0.25x - 3| + 2 > 4$

50. $|0.5y - 3| + 4 < 5$

51. $\left|2.4 - \frac{3}{4}y\right| \leq 7.2$

52. $\left|5.3 - \frac{2}{3}w\right| \geq 5.3$

53. $|2b - 8| + 5 < 1$

54. $|8p + 7| + 4 > 3$

For Exercises 55–62, write an inequality involving absolute value that describes the graph shown.

55. 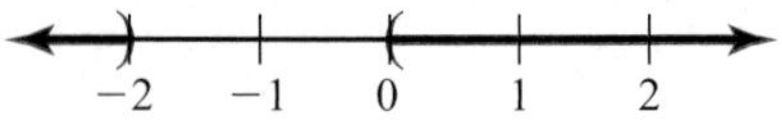

56. 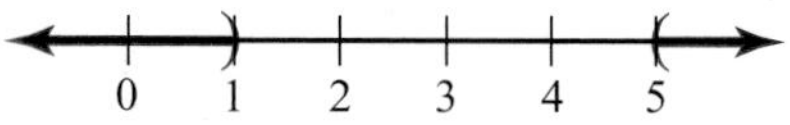

57. 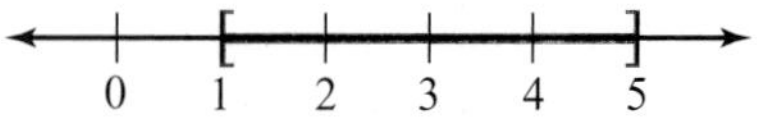

58. 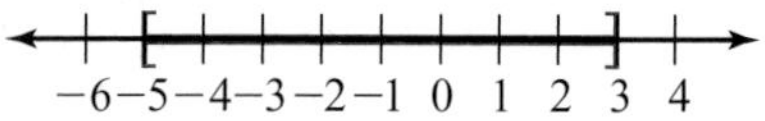

59. 

60. 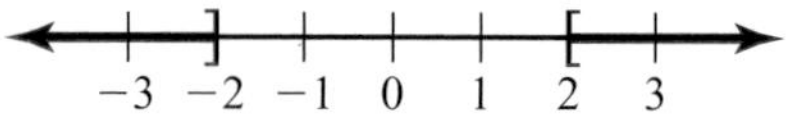

61. 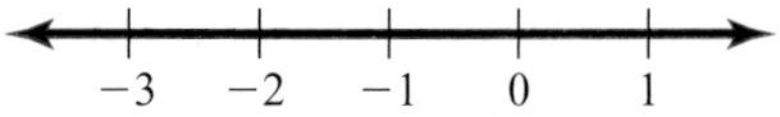

62. 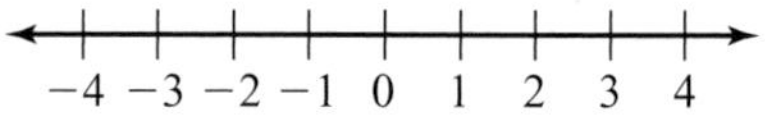

REVIEW EXERCISES

[1.1] ***For Exercises 1 and 2, indicate whether the statement is true or false.***

1. $-\frac{2}{3} \in \{x | x \text{ is a rational number}\}$

2. $\{1, 3, 5, 7\} \subseteq N$

[2.1] ***For Exercises 3 and 4, solve.***

3. $23 - 5(2 - 3x) = 12x - (4x + 1)$

4. $\frac{5}{6}x - \frac{3}{4} = \frac{2}{3}x + \frac{1}{2}$

[2.4] **5.** Solve $Ax + By = C$ for y.

[2.2] **6.** The sum of three consecutive integers is 141. Find the three integers.

Chapter 2 Summary

Defined Terms

Section 2.1
Solution (p. 56)
Solution set (p. 56)
Linear equation in one variable (p. 57)
Conditional linear equation in one variable (p. 61)
Identity (p. 61)
Contradiction (p. 61)

Section 2.2
Complementary angles (p. 72)
Supplementary angles (p. 72)

Section 2.3
Linear inequality in one variable (p. 82)

Section 2.4
Compound inequality (p. 94)
Intersection (p. 95)
Union (p. 98)

Learning Strategy

The summaries in this textbook are like the study sheet suggested in the To the Student section. Recall that your study sheet is a list of the rules, procedures, and formulas that you need to know. If you are a tactile or visual learner, spend a lot of time reviewing and writing the rules or procedures. Try to get to the point where you can write the essence of each rule and procedure from memory. If you are an auditory learner, record yourself saying each rule and procedure, then listen to the recording over and over. Also, consider developing clever rhymes or songs for each rule or procedure to help you remember them.

Procedures, Rules, and Key Examples

Procedures/Rules

Section 2.1 Linear Equations and Formulas

The Addition Principle of Equality

If $a = b$, then $a + c = b + c$ is true for all real numbers a, b, and c.

The Multiplication Principle of Equality

If $a = b$, then $ac = bc$ is true for all real numbers a, b, and c where $c \neq 0$.

To solve linear equations,

1. Simplify both sides of the equation as needed.
   a. Distribute to clear parentheses.
   b. Clear fractions or decimals by multiplying through by the LCD. In the case of decimals, the LCD is the power of 10 with the same number of 0 digits as decimal places in the number with the most decimal places. (Clearing fractions and decimals is optional.)
   c. Combine like terms.
2. Use the addition principle so that all variable terms are on one side of the equation and all constants are on the other side. (Clear the variable term with the lesser coefficient.) Then combine like terms.
3. Use the multiplication principle to clear the remaining coefficient.

Key Examples

Example 1: Solve and check: $\frac{2}{3}x - \frac{1}{2}(x + 1) = \frac{5}{6}x - 2$.

$$6 \cdot \frac{2}{3}x - 6 \cdot \frac{1}{2}(x + 1) = 6 \cdot \frac{5}{6}x - 6 \cdot 2 \quad \text{Multiply both sides by 6.}$$

$$4x - 3(x + 1) = 5x - 12 \quad \text{Simplify.}$$

$$4x - 3x - 3 = 5x - 12 \quad \text{Distribute } -3.$$

$$x - 3 = 5x - 12 \quad \text{Combine like terms.}$$

$$x - x - 3 = 5x - x - 12 \quad \text{Subtract } x \text{ from both sides.}$$

$$-3 + 12 = 4x - 12 + 12 \quad \text{Add 12 to both sides.}$$

$$\frac{9}{4} = \frac{4x}{4} \quad \text{Divide both sides by 4.}$$

$$\frac{9}{4} = x$$

Check: Substitute $\frac{9}{4}$ for x in the original equation, then verify that the equation is true. We will leave this to the reader.

continued

Procedures/Rules	Key Examples
Section 2.1 Linear Equations and Formulas (Continued) To isolate a particular variable in a formula, **1.** Treat all other variables like constants. **2.** Isolate the desired variable using the outline for solving equations.	**Example 2:** Isolate t in the formula $d = vt + x$. $d - x = vt + x - x$ Subtract x from both sides. $\frac{d-x}{v} = \frac{vt}{v}$ Divide both sides by v. $\frac{d-x}{v} = t$
Section 2.2 Solving Problems To translate a word sentence to an equation, identify the variable(s), constants, and key words, then write the corresponding symbolic form.	**Example 1:** Translate to an equation. **a.** Three less than eight times a number is thirteen. Answer: $8n - 3 = 13$ **b.** Four times the sum of a number and five is equal to negative three times the difference of the number and two. Answer: $4(n + 5) = -3(n - 2)$
Problem-Solving Outline **1. Understand** the problem. **a.** Read the question(s) (not the whole problem, just the question at the end) and note what it is you are to find. **b.** Now read the whole problem, underlining key words. **c.** If possible and useful, make a list or table, simulate the situation, or search for a related example problem. **2. Plan** your solution strategy by searching for a formula or translating the key words to an equation. **3. Execute** the plan by solving the formula or equation. **4. Answer** the question. Look at your note about what you were to find and make sure you answered that question. Include appropriate units. **5. Check** results. **a.** Try finding the solution in a different way, reversing the process, or estimating the answer and make sure the estimate and actual answer are reasonably close. **b.** Make sure the answer is reasonable.	**Example 2:** Two angles are supplementary. One angle measures 16° less than three times the other angle. Find the two angle measurements. Solution: If we let x represent the measure of the smaller angle, then the larger angle measures $3x - 16$. Since the angles are supplementary, the sum of their measures is 180°. $3x - 16 + x = 180$ $4x - 16 = 180$ $4x - 16 + 16 = 180 + 16$ $\frac{4x}{4} = \frac{196}{4}$ $x = 49$ Smaller angle: $x = 49$ Larger angle: $3x - 16 = 3(49) - 16 = 131$
To solve problems with two or more unknowns, **1.** Determine which unknown will be represented by a variable. **2.** Use one of the relationships to describe the other unknown(s) in terms of the variable. **3.** Use the other relationship to write an equation. **4.** Solve the equation. **5.** Check.	**Example 3:** A chemist has 200 milliliters of a 15% nitric acid solution. How many milliliters of a 30% nitric acid solution should she add to make a solution that is 20% nitric acid? Solution: Use a table to organize the information. *continued*

Procedures/Rules

Key Examples

Section 2.2 Solving Problems (Continued)

Solution	Concentration	Solution Volume	Volume of Nitric Acid
15%	0.15	200	0.15(200)
30%	0.3	n	$0.3n$
20%	0.2	$200 + n$	$0.2(200 + n)$

Now write an equation relating the volumes of the three solutions.

$$\begin{aligned} 0.15(200) + 0.3n &= 0.2(200 + n) \\ 30 + 0.3n &= 40 + 0.2n \\ 30 + 0.3n - 0.2n &= 40 + 0.2n - 0.2n \\ 30 - 30 + 0.1n &= 40 - 30 \\ 0.1n &= 10 \\ n &= 100 \end{aligned}$$

She needs to add 100 milliliters of the 30% solution.

Section 2.3 Solving Linear Inequalities

To graph an inequality in one variable on a number line,

1. If the symbol is $\le$ or $\ge$, then draw a bracket (or solid circle) on the number line at the indicated number. If the symbol is $<$ or $>$ then draw a parenthesis (or open circle) on the number line at the indicated number.
2. If the variable is greater than the indicated number, then shade to the right of the indicated number. If the variable is less than the indicated number, then shade to the left of the indicated number.

The Addition Principle of Inequality

If $a < b$, then $a + c < b + c$ is true for all real numbers a, b, and c.

The Multiplication Principle of Inequality

If a and b are real numbers where $a < b$, then $ac < bc$ is true if c is a *positive* real number, and $ac > bc$ is true if c is a *negative* real number.

Note: Though we have written the principles in terms of the $<$ symbol, the principles are true for any inequality symbol.

Example 1: Graph $x \le 3$ on a number line.

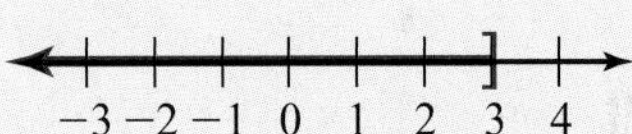

Example 2: Graph $x > -4$ on a number line.

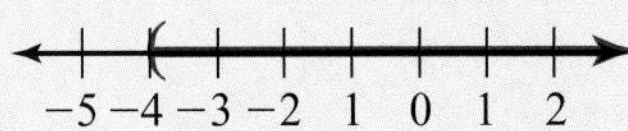

Example 3: Solve. Then write the solution set in set-builder notation and in interval notation.

a. $-3x \ge 12$

$$\frac{-3x}{-3} \le \frac{12}{-3}$$ Divide both sides by -3, which changes the direction of the inequality.

$$x \le -4$$

Set-builder notation: $\{x \mid x \le -4\}$

Interval notation: $(-\infty, -4]$

continued

Procedures/Rules	Key Examples
Section 2.3 Solving Linear Inequalities (Continued) To solve a linear inequality in one variable, **1.** Simplify both sides of the inequality as needed. **a.** Distribute to clear parentheses. **b.** Clear fractions or decimals by multiplying through by the LCD just as we did for equations. (This step is optional.) **c.** Combine like terms. **2.** Use the addition principle so that all variable terms are on one side of the inequality and all constants are on the other side. Then combine like terms. (Remember, moving the term with the lesser coefficient results in a positive coefficient.) **3.** Use the multiplication principle to clear any remaining coefficient. If you multiply (or divide) both sides by a negative number, then reverse the direction of the inequality symbol.	**b.** $12m - (2m - 9) > 5m - 7 + 3m$ $12m - 2m + 9 > 8m - 7$ **Distribute.** $10m - 8m + 9 > 8m - 8m - 7$ **Subtract 8*m* from both sides.** $2m + 9 - 9 > -7 - 9$ **Subtract 9 from both sides.** $\frac{2m}{2} > \frac{-16}{2}$ **Divide both sides by 2.** $m > -8$ Set-builder notation: $\{m \mid m > -8\}$ Interval notation: $(-8, \infty)$
Section 2.4 Compound Inequalities To solve a compound inequality involving "and," **1.** Solve each inequality in the compound inequality. **2.** The solution set will be the intersection of the individual solution sets.	For each example, solve the compound inequality, then **a.** Graph the solution set. **b.** Write the solution set in set-builder notation. **c.** Write the solution set in interval notation. **Example 1:** $-3 < -2x - 5 \le 1$ Solution: **a.** $-2x - 5 > -3$ and $-2x - 5 \le 1$ $-2x > 2$ and $-2x \le 6$ $x < -1$ and $x \ge -3$ Graph: −6 −5 −4 −3 −2 −1 0 1 **b.** Set-builder notation: $\{x \mid -3 \le x < -1\}$ **c.** Interval notation: $[-3, -1)$
To solve a compound inequality involving "or," **1.** Solve each inequality in the compound inequality. **2.** The solution set will be the union of the individual solution sets.	**Example 2:** $5x - 1 \le -16$ or $5x - 1 > 9$ Solution: **a.** $5x - 1 \le -16$ or $5x - 1 > 9$ $5x \le -15$ or $5x > 10$ $x \le -3$ or $x > 2$ Graph: −5 −4 −3 −2 −1 0 1 2 3 4

continued

Procedures/Rules | Key Examples

Section 2.4 Compound Inequalities (Continued)

b. Set-builder notation:
$\{x \mid x \leq -3 \text{ or } x > 2\}$

c. Interval notation: $(-\infty, -3] \cup (2, \infty)$

Section 2.5 Equations Involving Absolute Value

If $|x| = a$, where x is a variable or an expression and $a \geq 0$, then $x = a$ or $x = -a$.

To solve an equation containing a single absolute value,

1. Isolate the absolute value so that the equation is in the form $|ax + b| = c$.
2. Separate the absolute value into two equations, $ax + b = c$ and $ax + b = -c$.
3. Solve both equations.

To solve an equation in the form $|ax + b| = |cx + d|$,

1. Separate the absolute value equation into two equations:
$ax + b = cx + d \quad \text{and} \quad ax + b = -(cx + d)$
2. Solve both equations.

Example 1: Solve $|4x + 1| - 3 = 2$.

Solution:

$|4x + 1| = 5$ Isolate the absolute value.

$4x + 1 = 5$ or $4x + 1 = -5$ Use the absolute value property.

$4x = 4$ or $4x = -6$

$x = 1$ or $x = -\frac{3}{2}$

Example 2: Solve $|5x - 7| = |3x + 1|$.

Solution: Separate into two equations.

Equal		Opposites
$5x - 7 = 3x + 1$	or	$5x - 7 = -(3x + 1)$
$2x - 7 = 1$	or	$5x - 7 = -3x - 1$
$2x = 8$	or	$8x = 6$
$x = 4$	or	$x = \frac{3}{4}$

Section 2.6 Inequalities Involving Absolute Value

For each example, solve the inequality, then

a. Graph the solution set.
b. Write the solution set in set-builder notation.
c. Write the solution set in interval notation.

To solve an inequality in the form $|x| < a$ where $a > 0$,

1. Rewrite as a compound inequality involving "and": $x > -a$ and $x < a$.
2. Solve the compound inequality.

Similarly, to solve $|x| \leq a$, we would write $x \geq -a$ and $x \leq a$.

Example 1: $|2x - 3| - 1 < 4$

Solution:

a. $|2x - 3| < 5$ Isolate the absolute value.

$2x - 3 > -5$ and $2x - 3 < 5$ Separate using "and."

$2x > -2$ and $2x < 8$

$x > -1$ and $x < 4$

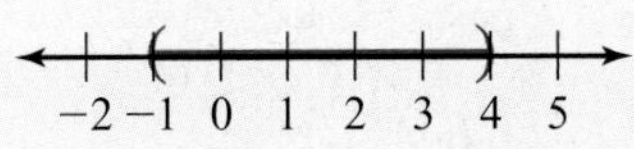

b. Set-builder notation. $\{x \mid -1 < x < 4\}$

c. Interval notation: $(-1, 4)$

continued

Procedures/Rules

Section 2.6 Inequalities Involving Absolute Value (Continued)

To solve an inequality in the form $|x| > a$ where $a > 0$,

1. Rewrite as a compound inequality involving "or": $x < -a$ or $x > a$.
2. Solve the compound inequality.

Similarly, to solve $|x| \geq a$, we would write $x \leq -a$ or $x \geq a$.

Key Examples

Example 2: $|3x - 5| - 7 \geq 4$

Solution:

a. $|3x - 5| \geq 11$ **Isolate the absolute value**

$3x - 5 \leq -11$ or $3x - 5 \geq 11$

$3x \leq -6$ or $3x \geq 16$ **Separate using "or."**

$x \leq -2$ or $x \geq \frac{16}{3}$

b. Set-builder notation:

$$\left\{x \mid x \leq -2 \text{ or } x \geq \frac{16}{3}\right\}$$

c. Interval notation: $(-\infty, -2] \cup \left[\frac{16}{3}, \infty\right)$

Chapter 2 Review Exercises

For Exercises 1–6, answer true or false.

[2.1] **1.** Both { } and ∅ indicate an empty set.

[2.1] **2.** $3x + 5 = 3x - 5$ is an identity.

[2.2] **3.** Supplementary angles have a sum of 180°.

[2.3] **4.** The interval notation (2, 5) represents all numbers between 2 and 5.

[2.5] **5.** Absolute value equations always have two solutions.

[2.5] **6.** To solve absolute value equations, the absolute value must be isolated.

For Exercises 7–10, fill in the blank.

[2.1] **7.** An equation that has no solution is called a __________.

[2.2] **8.** Two angles are complementary if the sum of their measures is ___.

[2.5] **9.** To solve an equation containing a single absolute value,

a. Isolate the absolute value so that the equation is in standard form, $|ax + b| = c$.

b. Separate the absolute value into two equations: ________ and __________.

c. Solve both equations.

[2.6] **10.** To solve an inequality in the form $|x| < a$,

a. Rewrite as a compound inequality involving *and*: _______ and ______.

b. Solve the compound inequality.

[2.1] ***For Exercises 11–22, solve and check. Note that some equations may be identities or contradictions.***

Equations and Inequalities

Exercises 11–82

11. $3x - 9 = 12$

12. $7m - 3 = 4m + 9$

13. $6(3a - 4) + (3a + 4) = 5(a - 2) - 10$

14. $2 - 2y - 6(y + 3) = 5 - 7y$

15. $\frac{1}{2}k + \frac{5}{8} = \frac{1}{4}$

16. $\frac{1}{5}d - 2 = -3 + d$

17. $2w + 2(7 - w) = 2(5w + 7)$

18. $5r - 2(5r - 3) = 6 - 5r$

19. $0.53 - 0.2z = 0.2(2z - 13) - 0.47$

20. $18(v + 2) - 6v = 30 + 12v - 2$

21. $\frac{2}{5}p - \frac{3}{20} = \frac{13}{20} - \frac{3}{10}(6 - 2p)$

22. $0.5(l + 4) = 0.3(3l - 1) - 0.9$

[2.1] ***For Exercises 23–26, solve the formula for the indicated variable.***

23. $I = Prt$; solve for t

24. $P = 2l + 2w$; solve for w

25. $A = \frac{1}{2}bh$; solve for b

26. $P = a + b + c$; solve for a

[2.3] ***For Exercises 27–32, solve and then: a. Write the solution set in set-builder notation.***
b. Write the solution set using interval notation.
c. Graph the solution set.

27. $-8n \geq -32$

28. $3x + 13 > 7$

29. $9 - 2m > 15$

30. $3h - 13 \geq 7h + 7$

31. $\frac{2}{3}t - 1 \leq \frac{1}{4}t + \frac{1}{2}$

32. $12 - (u + 7) < 3(u - 1)$

[2.4] ***For Exercises 33 and 34, find the intersection and union of the given sets.***

33. $A = \{2,4,6\}$
$B = \{1,2,3,4,5\}$

34. $A = \{a,b,c,d\}$
$B = \{a,e,i,o,u\}$

[2.4] ***For Exercises 35–46, solve and then: a. Graph the solution set.***
b. Write the solution set in set-builder notation.
c. Write the solution set in interval notation.

35. $4 < -2x < 6$

36. $9 \leq -3x \leq 15$

37. $-3 < x + 4 < 7$

38. $0 < x - 1 \leq 3$

39. $-1 \leq 2x + 3 < 3$

40. $-5 < 3x - 2 < 7$

41. $w + 4 \leq -2$ or $w + 4 \geq 2$

42. $4w - 3 < 1$ or $4w - 3 > 0$

43. $2m - 5 < 0$ or $2m - 5 > 5$

44. $3x + 2 \leq -2$ or $3x + 2 \geq 8$

45. $-x - 6 \leq -2$ or $-x - 6 \geq 3$

46. $-4w + 1 \leq -3$ or $-4w + 1 \geq 5$

[2.5] ***For Exercises 47–56, solve.***

47. $|x| = 4$

48. $|x - 4| = 7$

49. $|2w - 1| = 3$

50. $|3u - 4| = 2$

51. $|5r + 8| = -3$

52. $|q - 4| - 3 = 8$

53. $|5w| - 2 = 13$

54. $2|3x - 4| = 8$

55. $1 - |2x + 3| = -3$

56. $4 - 2|r - 5| = -8$

[2.6] ***For Exercises 57–66, solve and then:*** ***a. Graph the solution set.***
b. Write the solution set in set-builder notation.
c. Write the solution set in interval notation.

57. $|x| < 5$

58. $|p| \geq 4$

59. $|x - 3| > 7$

60. $|2m + 6| < 4$

61. $|3s - 1| \geq -2$

62. $5|b| - 2 > 3$

63. $7|m + 3| \le 21$

64. $-2|t - 5| < -10$

65. $5 - 2|2k - 3| \le -15$

66. $3 - 7|2p + 4| \le 24$

[2.2] ***For Exercises 67–79, solve.***

67. Liquid nitrogen is often used to quickly freeze objects. It is so cold that when a room-temperature object is immersed, the liquid nitrogen will immediately boil at a temperature of about $-196°C$. Using $C = \frac{5}{9}(F - 32)$, convert this temperature to degrees Fahrenheit.

68. The formula $B = P + Prt$ can be used to determine the final balance, B, in an account if a principal P earns interest at a rate r for t years. If the balance in an account is \$4368 after earning interest at a rate of 4% for one year, find the principal that was invested.

69. Two-thirds of a number is equal to two plus one-half of the number. Find the number.

70. Twice the difference of a number and seven is negative six. What is the number?

71. A clothing store has a special sale with every item discounted 30%. If the price of a dress after the discount is \$44.94, what was the original price?

72. An annual report for a company shows that 37,791 units of their top-selling product were sold during the year, which is an increase of 10.5% over the previous year's sales. How many units of that product were sold the previous year?

73. The sum of two consecutive integers is 125. Find the integers.

74. The sum of three consecutive even integers is 108. Find the integers.

75. Two angles are complementary. If one angle is $6°$ more than triple the other angle, what are the angles?

76. Two pieces of steel are welded together, forming two angles as shown. If one angle is $15°$ less than twice the other angle, what are the angles?

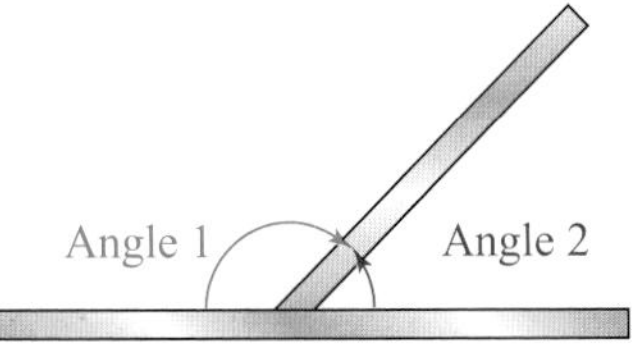

77. A music store manager purchases 400 CDs from a particular vendor at two different prices. Some of the CDs cost \$10.50 and others cost \$12.00. If she spent a total of \$4320, how many CDs did she purchase at each price?

78. Two cars are traveling toward each other on the same road. One car is traveling at 50 miles per hour and the other at 55 miles per hour. If the two cars are 315 miles apart, how long until they meet?

79. Juan has a bottle containing 45 milliliters of 15% HCl solution and a bottle of 35% HCl solution. He wants a 25% solution. How much of the 35% solution must be added to the 15% solution so that a 25% concentration is created?

For Exercises 80–82, solve using inequalities.

[2.3] 80. A homeowner is planning to put a pool in the backyard. The area of a rectangular pool may not exceed 300 square feet. If the width is 12 feet, what range of values can the length have?

[2.4] 81. Students in a biology course receive a B if the average of four tests is at least 80 and less than 90. Suppose a student has the following scores: 70, 89, and 83. What range of scores on the fourth test would allow the student to receive a B?

[2.4] 82. Water is in its liquid form when the temperature is between 32°F and 212°F.

a. Write a compound inequality that describes the range of temperatures at which water is liquid.

b. Write a compound inequality that describes the range of temperatures at which water is not liquid.

c. Write a compound inequality that describes the range of temperatures for which water is liquid in degrees Celsius.

d. Write a compound inequality that describes the range of temperatures for which water is not liquid in degrees Celsius.

Chapter 2 Practice Test

1. Find the intersection and union of the given sets.

   $A = \{h, o, m, e\}$

   $B = \{h, o, u, s, e\}$

For Exercises 2–7, solve.

2. $2(w + 2) - 3 = 3(w - 1)$
3. $\frac{2}{3}q - 4 = \frac{1}{5}q + 10$
4. $|x + 3| = 5$
5. $3 - |2x - 3| = -6$
6. $|2x + 3| = |x - 5|$
7. $|5x - 4| = -3$
8. Solve for b; $A = \frac{1}{2}h(b + B)$

For Exercises 9–16, solve. Then: a. Graph the solution set.
b. Write the solution set in set-builder notation.
c. Write the solution set in interval notation.

9. $-3 < x + 4 \leq 7$
10. $4 < -2x \leq 6$
11. $|x + 4| < 9$
12. $2|x - 1| > 4$
13. $3 - 2|x + 4| > -3$
14. $|3y - 2| < -2$
15. $|8t + 4| \geq -12$
16. $2|3x - 4| \leq 10$

17. Five decreased by three times a number is forty-one. What is the number?

18. A clothing store has a special sale in which items are discounted 30%. If the price after discount for a dress is $31.36, what was the original price?

19. A pharmaceutical salesperson meets with two different clients and sells a total of 500 boxes of a particular medication. The first client paid full price for each box, which is $250.00, whereas the second client, who has a preferred customer status, got a discount and paid $225.00. If the combined amount of the sale was $120,000, how many boxes were sold to each client?

20. A mail-order book club offers one bonus book if the total of an order is between $40 and $50. Johnson decides to buy the least expensive books, which are $8 each. What range would the number of books he orders have to be in order to receive a bonus book?

CHAPTER 3

Equations and Inequalities in Two Variables and Functions

"Science begins with the world we have to live in, accepting its data and trying to explain its laws. From there, it moves towards the imagination: it becomes a mental construct, a model of a possible way of interpreting experience. The further it goes in this direction, the more it tends to speak the language of mathematics, which is really one of the languages of imagination, along with literature and music."

—Northrop Frye, Canadian educator and critic

In Chapter 2, we learned to solve equations and inequalities in one variable. We also graphed solution sets on a number line. In this chapter, we will find solution sets for equations and inequalities that have two variables. We will also see that the solution sets for these equations can be graphed. However, instead of a single number line, we will use two number lines, one for each variable, in what is called the *coordinate plane*. We will finish the chapter by exploring a generalized relationship between two sets of data called a *function*.

3.1 Graphing Linear Equations

OBJECTIVES

1. **Plot points in the coordinate plane.**
2. **Graph linear equations.**

OBJECTIVE 1. Plot points in the coordinate plane. In 1619, René Descartes, the French philosopher and mathematician, recognized that positions of points in a plane could be described using two perpendicular number lines, called *axes*. The axes form what is called a *rectangular coordinate system*, or the *Cartesian* coordinate system, named in honor of René Descartes. Usually, we call the horizontal axis the *x-axis* and the vertical axis the *y-axis*.

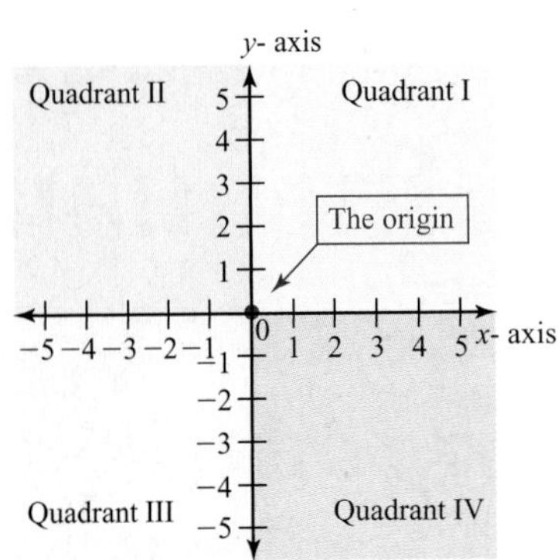

The point where the axes intersect is called the *origin* and has the value 0 for both the x-axis and y-axis. Also, the axes divide the plane into four regions, which we call *quadrants*. The quadrants are numbered using Roman numerals, as shown in the figure to the left.

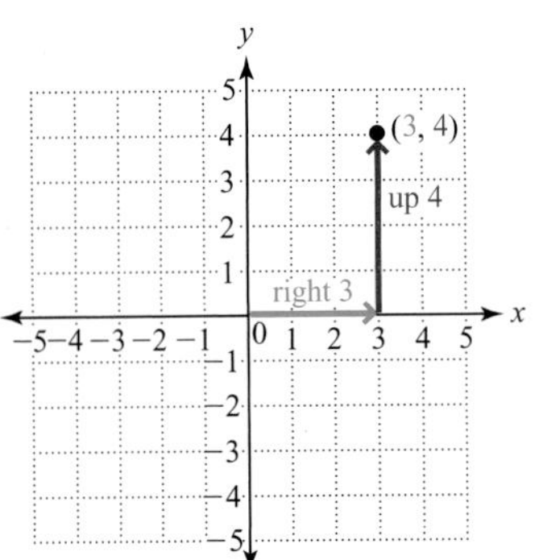

Any point in the plane can be described using one number from each axis written in a specific order. The horizontal axis value is given first and then the vertical axis value. Because the order is specific, we say that the two numbers form an *ordered pair*. Each number in an ordered pair is called a *coordinate* of the ordered pair.

The notation for writing ordered pairs is (horizontal coordinate, vertical coordinate). For example, the ordered pair for the point shown in the coordinate plane to the left is (3, 4) because it is 3 units to the right and 4 units up from the origin.

Of Interest

René Descartes was born in 1596 at La Haye, near Tours, France. While serving in the military, Descartes had a series of dreams in which he became aware of a new way of viewing geometry using algebra. In 1637, after some urging by his friends, he reluctantly allowed one work known as the *Method* to be printed. It was in this book that the rectangular coordinate system and analytical geometry was given to the world.

EXAMPLE 1 Plot the point described by the coordinates and identify the point's quadrant.

a. $(-3, -2)$ **b.** $(0, 4)$ **c.** $(2, -4)$

Solution

Note: *The ordered pair* $(-3, -2)$ *means 3 units to the left of the origin and 2 units down, placing the point in Quadrant III.*

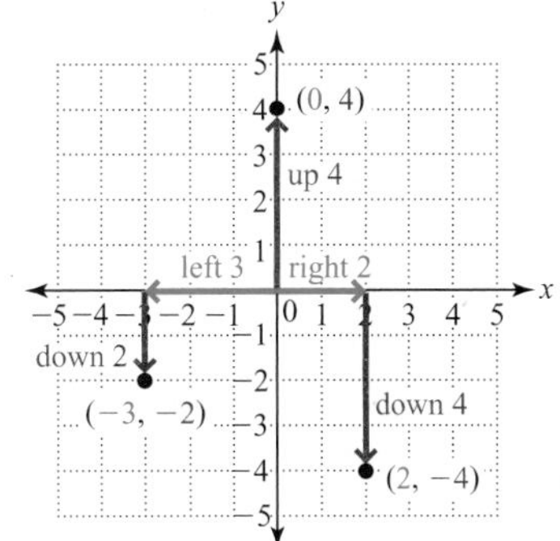

Note: *The ordered pair* $(0, 4)$ *means the point is 0 units from the origin along the x-axis and 4 units up the y-axis. It is not in a quadrant.*

Note: *The ordered pair* $(2, -4)$ *means 2 units to the right of the origin and 4 units down, placing the point in Quadrant IV.*

Learning Strategy

If you are a tactile learner, when you plot a point, move your pencil from the origin along the x-axis first, then move up or down to the point location, just as we've done with the arrows.

YOUR TURN Plot the point described by the coordinates and identify the point's quadrant.

a. $A: (4, 3)$ **b.** $B: (-3, 2)$ **c.** $C: (-5, -2)$ **d.** $D: (2, 0)$

OBJECTIVE 2. Graph linear equations. We can graph the solution set of an equation or inequality in the coordinate plane. In this section, we will focus on linear equations. Since the coordinate plane has an x- and a y-axis, we need to discuss solutions to **linear equations in two variables**, like $x + y = 3$ and $y = 2x - 3$.

DEFINITION ***Linear equation in two variables:*** An equation that can be written in the form $Ax + By = C$, where A and B are not both 0.

An equation written in the form $Ax + By = C$, like $x + y = 3$, is said to be in *standard form*. The equation $y = 2x - 3$ is not in standard form.

A solution for an equation with two variables is a pair of numbers, one number for each variable, that can replace the corresponding variables and make the equation true. Note that a solution for an equation with two variables is an ordered pair. For example, $(1, 2)$ is a solution for $x + y = 3$, which we can verify by replacing x with 1 and y with 2:

$$x + y = 3$$
$$1 + 2 = 3 \quad \text{This is true, so the ordered pair is a solution.}$$

Note that $(1, 2)$ is just one solution out of an infinite number of possible solutions for $x + y = 3$. An infinite number of solutions are possible because every

ANSWERS

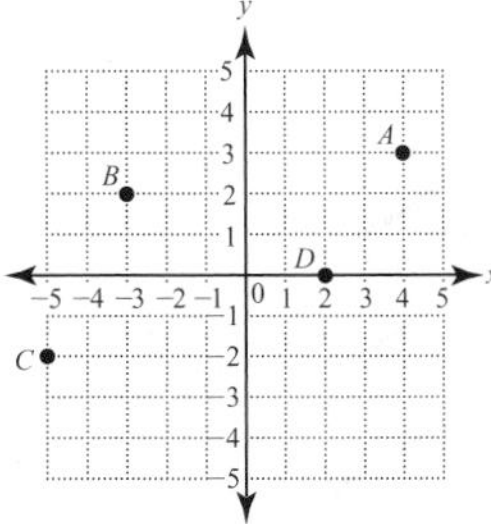

A: Quadrant I
B: Quadrant II
C: Quadrant III
D: not in a quadrant (on the x-axis)

x-value has a corresponding y-value that will add to the x-value to equal 3, and vice versa. We list some other solutions in the following table:

x	y	Ordered Pair
−2	5	(−2, 5)
−1	4	(−1, 4)
0	3	(0, 3)
1	2	(1, 2)
2	1	(2, 1)
3	0	(3, 0)
4	−1	(4, −1)
5	−2	(5, −2)

Note: *To find these solutions, we choose a value for x or y, then use the equation to find the corresponding value of the other variable. For example, if we choose x to be 2, then y has to be 1 so that the sum is 3.*

Our exploration suggests the following procedure for finding solutions to equations with two variables.

PROCEDURE **Finding Solutions to Equations with Two Variables**

To find a solution to an equation in two variables,

1. Replace one of the variables with a chosen number (any number).
2. Solve the equation for the other variable.

Since these equations have an infinite number of solutions, a graph offers a nice way to represent the entire solution set. For example, on the accompanying grid we plot each of the solutions that we listed for $x + y = 3$.

Note: *The arrows on either end of the line indicate that the solutions continue indefinitely beyond our grid.*

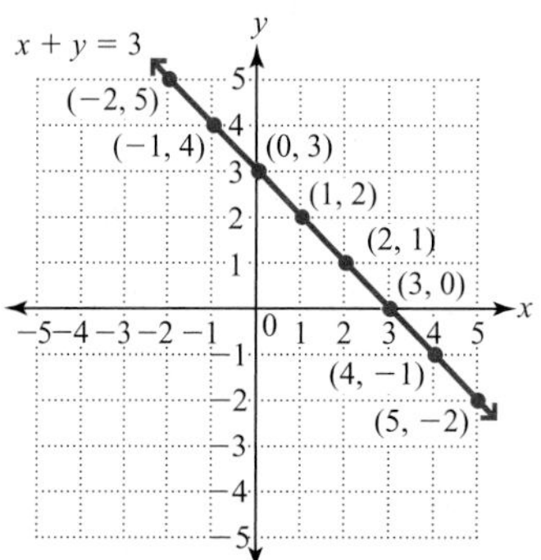

Notice that the ordered pair solutions lie on a straight line. In fact, every ordered pair along that line is a solution for $x + y = 3$, so the line represents the solution set for the equation.

The graph of the solutions of any linear equation is a straight line. Since two points determine a line in a plane, we need at least two ordered pair solutions to draw the graph of a linear equation. It is wise, however, to find three solutions, using the third solution as a check. If we plot all three points and they cannot be connected with a straight line, then we know something is wrong.

PROCEDURE **Graphing Linear Equations**

To graph a linear equation,

1. Find at least two solutions to the equation.
2. Plot the solutions as points in the rectangular coordinate system.
3. Connect the points to form a straight line.

EXAMPLE 2 Graph. $2x + y = 4$

Solution We will find three solutions.

If $x = 0$,	If $x = 1$,	If $x = 2$,	Choose a value for x.
$2(0) + y = 4$	$2(1) + y = 4$	$2(2) + y = 4$	In the equation, replace x with the chosen value. Then solve for y.
$y = 4$	$2 + y = 4$	$4 + y = 4$	
First solution	$y = 2$	$y = 0$	
$(0, 4)$	Second solution	Third solution	
	$(1, 2)$	$(2, 0)$	

A table helps organize the solutions. To graph, we plot our solutions and connect the points to form a straight line.

x	y	Ordered Pair
0	4	(0, 4)
1	2	(1, 2)
2	0	(2, 0)

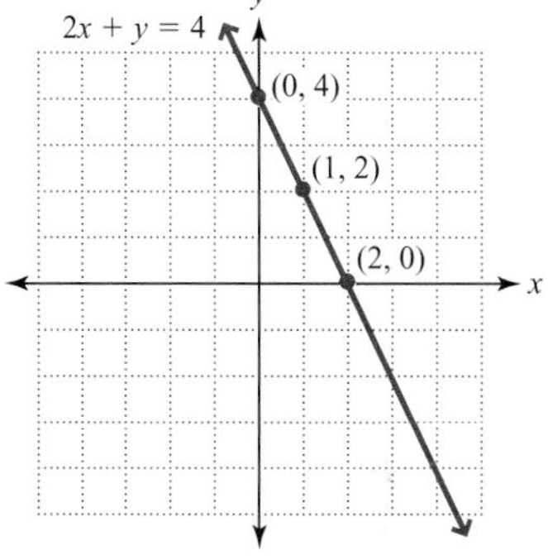

Notice that the graph of $2x + y = 4$ intersects the x-axis at $(2, 0)$ and the y-axis at $(0, 4)$. Points where a graph intersects an axis are called **intercepts**.

DEFINITIONS

***x*-intercept:** A point where a graph intersects the x-axis.
***y*-intercept:** A point where a graph intersects the y-axis.

Note that x-intercepts will always have the form $(a, 0)$ and y-intercepts will have the form $(0, b)$, where a and b are real numbers.

PROCEDURE **Finding the x- and y-intercepts**

To find an x-intercept,

1. Replace y with 0 in the given equation.
2. Solve for x.

To find a y-intercept,

1. Replace x with 0 in the given equation.
2. Solve for y.

Since equations are usually simpler when x or y is replaced with 0, many people find it easy to graph equations by finding the intercepts.

EXAMPLE 3 Find the x- and y-intercepts, then graph.

a. $y = -\frac{1}{3}x + 2$

Solution $0 = -\frac{1}{3}x + 2$ For the x-intercept, replace y with 0.

$-2 = -\frac{1}{3}x$ Subtract 2 from both sides.

$6 = x$ Multiply both sides by -3.

x-intercept: $(6, 0)$

$y = -\frac{1}{3}(0) + 2$ For the y-intercept, replace x with 0.

$y = 2$ Simplify.

y-intercept: $(0, 2)$

Note: *In the equation $y = -\frac{1}{3}x + 2$, we can see that 2 is the y-coordinate of the y-intercept because replacing x with zero eliminates the x term leaving $y = 2$.*

Note: *Remember that although two points determine a line, it is wise to find a third point as a check. To find a third point, we will let $x = 3$.*

$y = -\frac{1}{3}(3) + 2$ Replace x with 3.

$y = 1$ Simplify.

Notice that (3, 1) does indeed lie on the line.

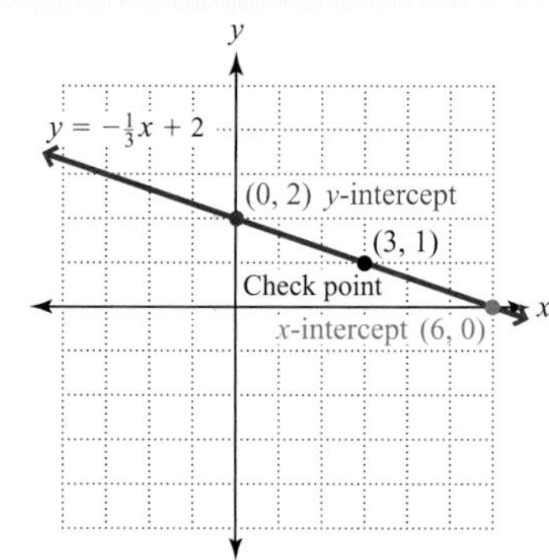

b. $y = 2x$

Solution $0 = 2x$ For the x-intercept, replace y with 0.

$0 = x$ Divide both sides by 2.

$y = 2(0)$ For the y-intercept, replace x with 0.

$y = 0$ Simplify.

Note: *The origin is the only point in the coordinate plane that can be both the x- and y-intercept.*

The x- and y-intercepts are the same point: $(0, 0)$. To graph, we need at least one more point. We will find another point by choosing a value for x and solving for y.

Let $x = 1$:

$y = 2(1)$ Replace x with 1.

$y = 2$ Simplify.

Second solution: $(1, 2)$

Note: *We will find a third point as a check.*

Let $x = 2$:

$y = 2(2)$ Replace x with 2.

$y = 4$ Simplify.

Notice that (2, 4) is on the line.

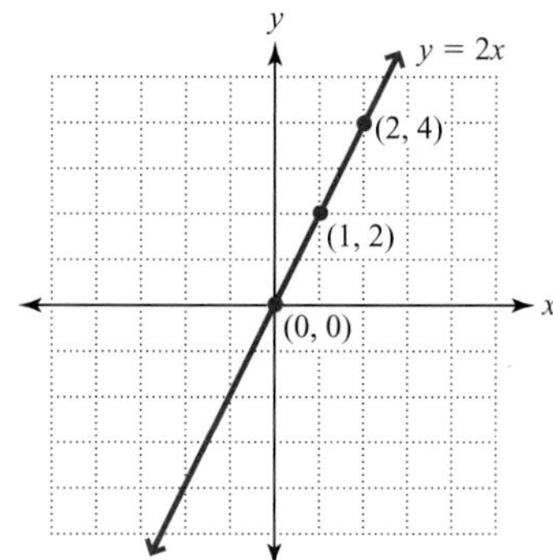

The equations $y = -\frac{1}{3}x + 2$ and $y = 2x$ from Example 3 are in the form $y = mx + b$. A nice feature of equations in the form $y = mx + b$ is that b is the y-coordinate of the y-intercept. If we replace x with 0 in $y = mx + b$, we have $y = m(0) + b = b$, so the y-intercept is $(0, b)$. We will learn more about equations of the form $y = mx + b$ in Section 3.2.

RULE The y-intercept of $y = mx + b$

Given an equation in the form $y = mx + b$, the coordinates of the y-intercept are $(0, b)$.

YOUR TURN Find the x- and y-intercepts, then graph.

a. $x + 3y = 3$ **b.** $y = 2x - 1$ **c.** $y = -3x$

ANSWERS

a. x-intercept: $(3, 0)$, y-intercept: $(0, 1)$

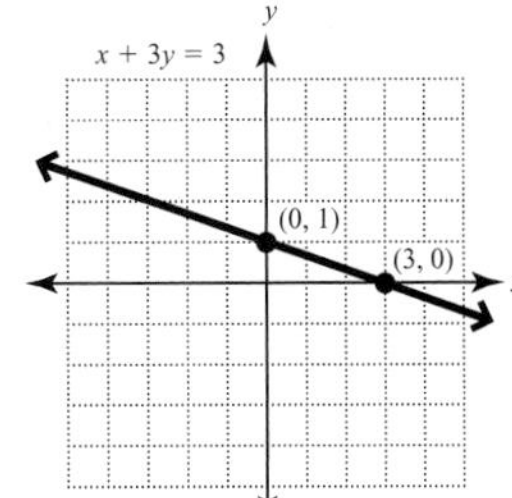

b. x-intercept: $\left(\frac{1}{2}, 0\right)$, y-intercept: $(0, -1)$

c. x- and y-intercepts: $(0, 0)$

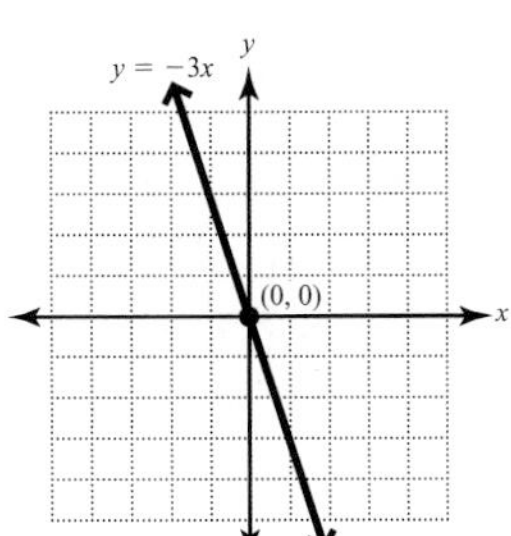

We have graphed linear equations in two variables. Now let's consider equations with one variable, such as $y = 3$ or $x = 4$, in which the variable is equal to a constant.

EXAMPLE 4 Graph.

a. $y = 3$

Solution To find solutions, we can think of $y = 3$ as $0x + y = 3$. Since the coefficient of x is 0, y is always 3 no matter what we choose for x.

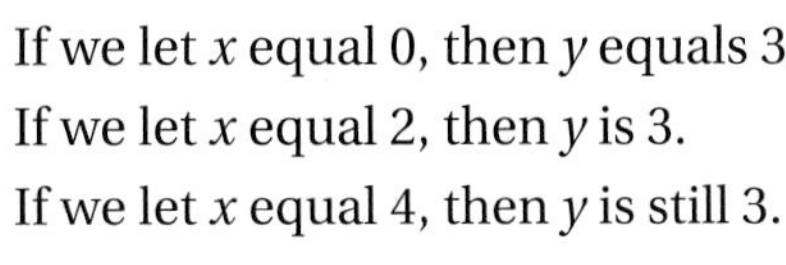

If we let x equal 0, then y equals 3.
If we let x equal 2, then y is 3.
If we let x equal 4, then y is still 3.

x	y	Ordered Pair
0	3	(0, 3)
2	3	(2, 3)
4	3	(4, 3)

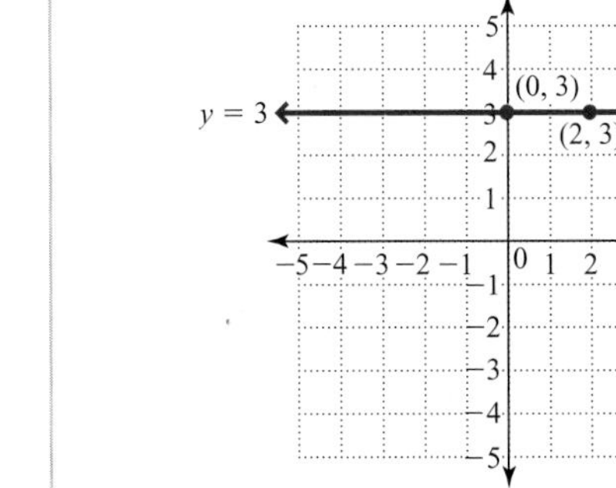

Note: *The graph of $y = 3$ is a horizontal line parallel to the x-axis. It has no x-intercept and its y-intercept is $(0, 3)$.*

b. $x = -4$

Solution The equation $x = -4$ indicates that x is equal to a constant, -4. We can write $x = -4$ as $x + 0y = -4$ and complete a table of solutions.

If we let y equal 0, then x equals -4.
If we let y equal 1, then x is -4.
If we let y equal 2, then x is still -4.

x	y	Ordered Pair
−4	0	(−4, 0)
−4	1	(−4, 1)
−4	2	(−4, 2)

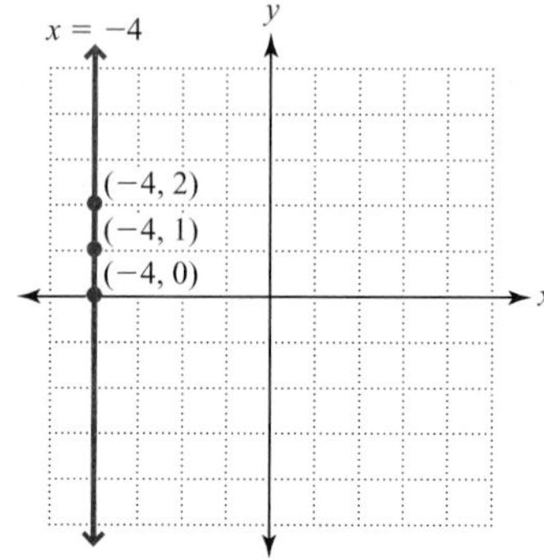

Note: *The graph of $x = -4$ is a vertical line parallel to the y-axis. It has no y-intercept and its x-intercept is $(-4, 0)$.*

RULES Horizontal Lines and Vertical Lines

The graph of $y = c$, where c is a real-number constant, is a horizontal line parallel to the x-axis with a y-intercept at $(0, c)$.

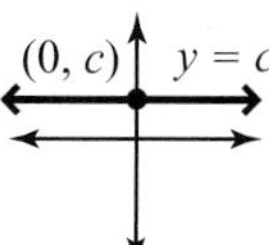

The graph of $x = c$, where c is a real-number constant, is a vertical line parallel to the y-axis with an x-intercept at $(c, 0)$.

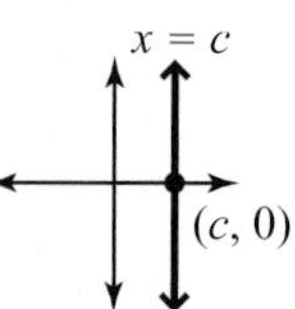

ANSWERS

a.

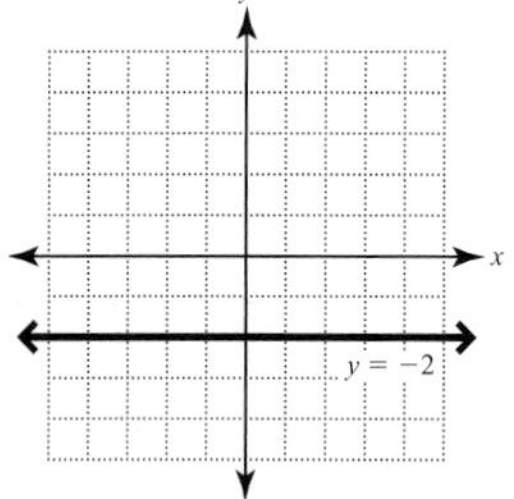

b.

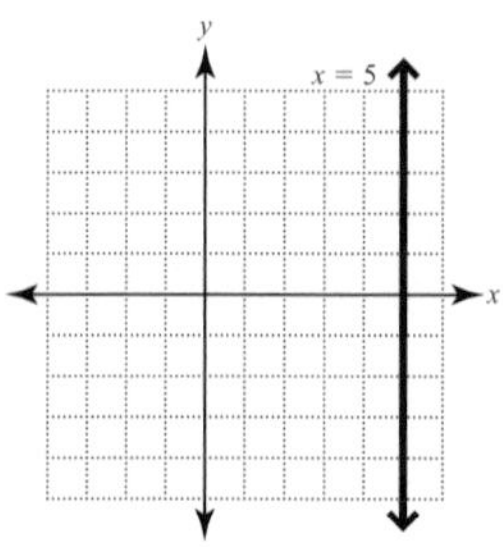

YOUR TURN Graph.

a. $y = -2$ **b.** $x = 5$

Calculator **TIPS**

To graph an equation on a graphing calculator, first press the [Y=] *key, which gives a list: Y1, Y2, Y3, etc. Each of the Ys can be used to enter an equation, which means you can graph multiple equations on the same grid if desired.*

To enter the equation $y = 2x + 3$, press [Y=] *and at Y1, type:* [2] [X, T, θ, n] [+] [3]

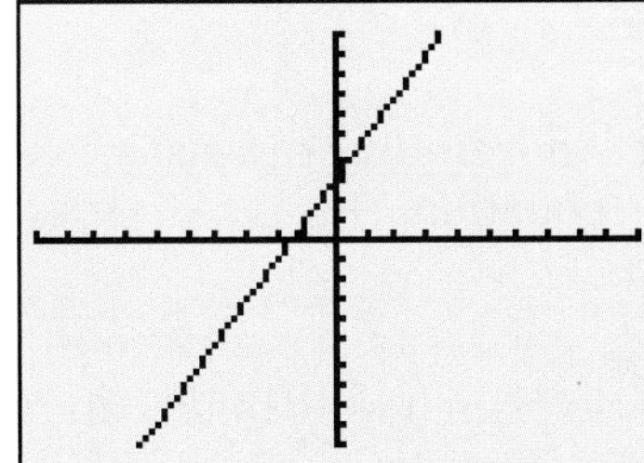

Notice that the [X, T, θ, n] *key is used to type the variable x. To see the graph, press* [GRAPH]. *You should see a screen like the one to the right.*

Sometimes, you may need to adjust the viewing window to see the graph. For example, if you want the x-axis to range from -10 to 10, press [WINDOW], *then at Xmin, enter -10 and at Xmax, enter 10. Xscl allows you to change the distance between each tick mark on the x-axis. If the Xscl is 1, the distance between each mark is 1. Similarly, Ymax and Ymin change the range on the y-axis and Yscl affects the tick marks on the y-axis.*

Ordered pairs can also be determined. One way is to graph the equation, then press [TRACE]. *You will see a small cursor appear on the graph with coordinates of the corresponding ordered pair at the bottom of the screen. You can trace along the graph using the arrow keys to see other ordered pairs on the graph.*

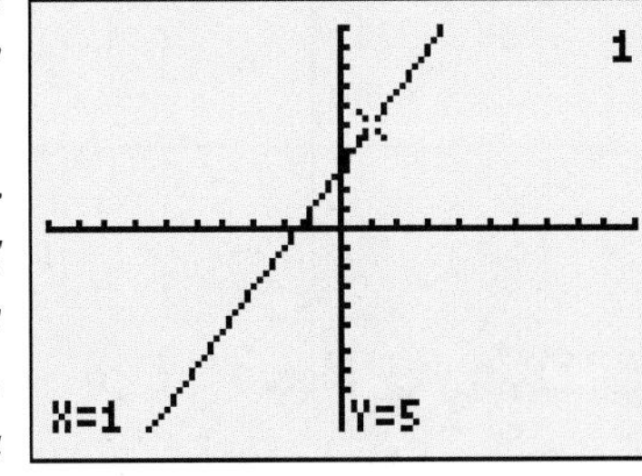

You can also input an x-value and have the calculator determine the corresponding y-value using the VALUE function, which is found in the CALC menu. Press [2nd] [TRACE], *then select 1:VALUE. At the X = prompt, enter a value, such as 1. After pressing* [ENTER], *you should see the screen to the right.*

3.1 Exercises

For Extra Help MyMathLab | Videotape/DVT | InterAct Math | Math Tutor Center | Math XL.com

1. Describe how to locate the point described by $(-4, 3)$ in the rectangular coordinate system.

2. What signs would the coordinates of a point in Quadrant I have? Quadrant II? Quadrant III? Quadrant IV?

3. How do you determine if an ordered pair is a solution for a given equation?

4. How do you find a solution for a given equation with two unknowns?

5. What does the graph of an equation represent?

6. What is the minimum number of points needed to draw a straight line? Explain.

7. After finding at least two solutions, explain the process of graphing a linear equation.

8. Explain how to find x- and y-intercepts.

For Exercises 9–10, write the coordinates for each point.

9.

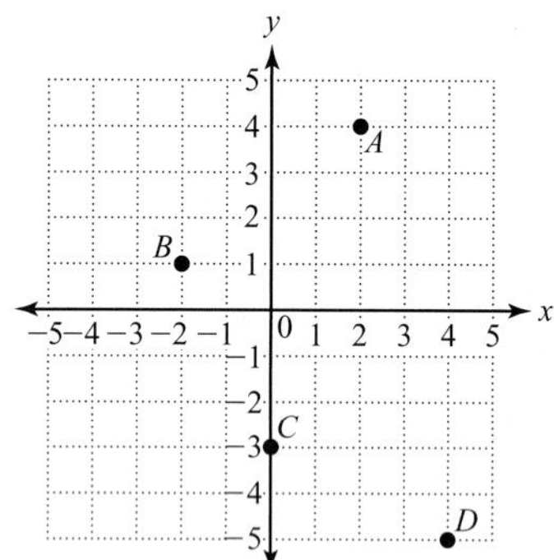

10.

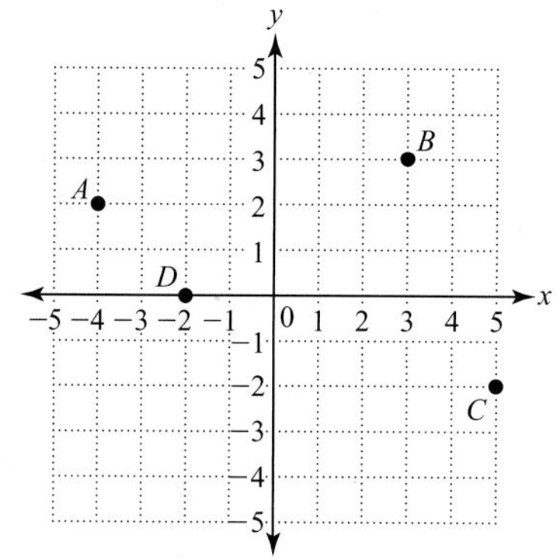

For Exercises 11–14, plot and label the points indicated by the coordinate pairs.

11. $(4, 1), (3, -1), (-5, -2), (0, 2)$

12. $(-1, 4), (2, 5), (-2, 0), (4, -1)$

13. $(-3, 0), (5, -3), (-2, -4), (1, 1)$

14. $(2, -3), (0, -4), (5, -1), (-3, 1)$

For Exercises 15–22, state the quadrant in which the point is located.

15. $(-4.2, 81.6)$

16. $\left(-6\frac{1}{3}, -5\frac{1}{2}\right)$

17. $(72, 98)$

18. $(103, -27)$

19. $(-58.6, -92.7)$

20. $(0.78, 0.09)$

21. $\left(45\frac{1}{2}, -12\frac{3}{4}\right)$

22. $(-47, 501)$

23. The ordered pairs $(-4, 2)$, $(-1, 2)$, $(-1, 4)$, $(2, 4)$, $(2, -3)$, and $(-4, -3)$ form the vertices of a figure.

a. Plot the points and connect them to form the figure.
b. Find the perimeter of the figure.

c. Find the area of the figure.

24. The ordered pairs $(-5, 2)$, $(-2, 2)$, $(-2, 5)$, $(1, 5)$, $(1, 3)$, $(4, 3)$, $(4, -3)$, and $(-5, -3)$ form the vertices of a figure.

a. Plot the points and connect them to form the figure.
b. Find the perimeter of the figure.

c. Find the area of the figure.

For Exercises 25–32, determine whether the ordered pair is a solution for the equation.

25. $(4, 2)$; $2x + 3y = 14$

26. $(-1, -2)$; $3x - 2y = 1$

27. $\left(-\frac{1}{3}, \frac{2}{3}\right)$; $x - y = -1$

28. $\left(\frac{3}{5}, -\frac{1}{4}\right)$; $2x - y = 7$

29. $(5, -1)$; $\frac{2}{3}x - y = 8$

30. $(-2, -4)$; $-y = \frac{1}{2}x + 3$

31. $(3.5, -2.1)$; $5.1x = 2 - 3y$

32. $(-2.1, -0.2)$; $-3.4 - 2x = 4y$

For Exercises 33–40, graph. Note that these equations are in standard form.

33. $x - y = 4$

34. $x + y = 5$

35. $2x + y = 6$

36. $3x - y = 9$

37. $x - 2y = 10$

38. $x + 4y = 12$

39. $2x - 5y = 10$

40. $3x + 4y = 12$

For Exercises 41–64, graph. Note that these equations are not in standard form.

41. $y = x$

42. $y = 2x$

43. $y = 3x$

44. $y = 5x$

45. $y = -x$

46. $y = -2x$

47. $y = -3x$

48. $y = -5x$

49. $y = \frac{1}{3}x$

50. $y = \frac{1}{2}x$

51. $y = -\frac{1}{3}x$

52. $y = -\frac{2}{5}x$

53. $y = 3x + 4$

54. $y = x - 1$

55. $y = -2x + 1$

56. $y = -5x - 3$

57. $y = \frac{3}{4}x + 2$

58. $y = \frac{1}{4}x + 2$

59. $y = -\frac{1}{3}x - 1$

60. $y = -\frac{3}{5}x - 4$

61. $y = 4$

62. $y = -5$

63. $x = 2$

64. $x = -3$

For Exercises 65–76, find the x- and y-intercepts, then graph.

65. $y = x - 3$

66. $y = 2x + 1$

67. $y = -x + 2$

68. $y = -3x - 2$

69. $y = -3x$

70. $y = 2x$

71. $y = -\frac{3}{4}x + 1$

72. $y = \frac{2}{3}x - 2$

73. $2x + y = 4$

74. $3x + 4y = 12$

75. $2x - 3y = 6$

76. $x - 2y = 6$

77. Compare the graphs of $y = x$, $y = 2x$, $y = 3x$, and $y = 5x$ from Exercises 41–44. For an equation in the form $y = mx$, what effect does increasing m seem to have on the graph?

78. Compare the graphs of and $y = x$ and $y = -x$ from Exercises 41 and 45, then compare the graphs of $y = 2x$ and $y = -2x$ from Exercises 42 and 46. For an equation in the form $y = mx$, what can you conclude about the graph when m is positive versus when m is negative?

79. Compare the graphs of $y = 3x$ and $y = 3x + 4$ from Exercises 43 and 53, then compare the graphs of $y = -2x$ and $y = -2x + 1$ from Exercises 46 and 55. For an equation in the form $y = mx + b$, what effect does adding b (where $b > 0$) to mx seem to have on the graph of $y = mx + b$?

80. Compare the graphs of $y = x$ and $y = x - 1$ from Exercises 41 and 54, then compare the graphs of $y = -5x$ and $y = -5x - 3$ from Exercises 48 and 56. For an equation in the form $y = mx - b$, what effect does subtracting b (where $b > 0$) from mx seem to have on the graph of $y = mx - b$?

For Exercises 81 and 82, a polygon has been moved from its original position, shown with the dashed lines, to a new position, shown with solid lines.
a. List the coordinates of each vertex of the original position.
b. List the coordinates of each vertex of the new position.
c. Write a rule that describes in mathematical terms how to move any point on the polygon from its original position to any point on the polygon in its new position.

81.

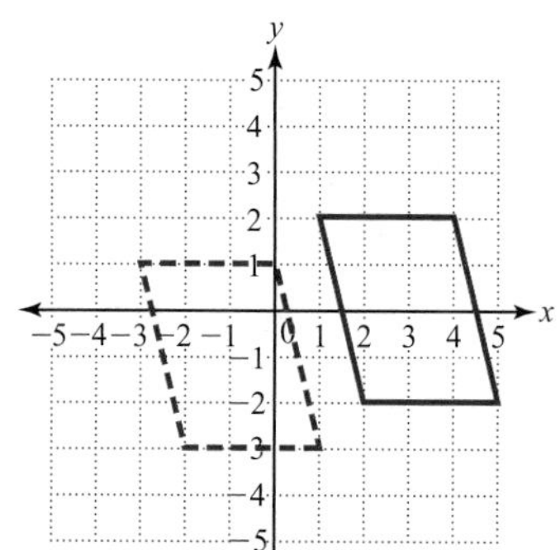

82.

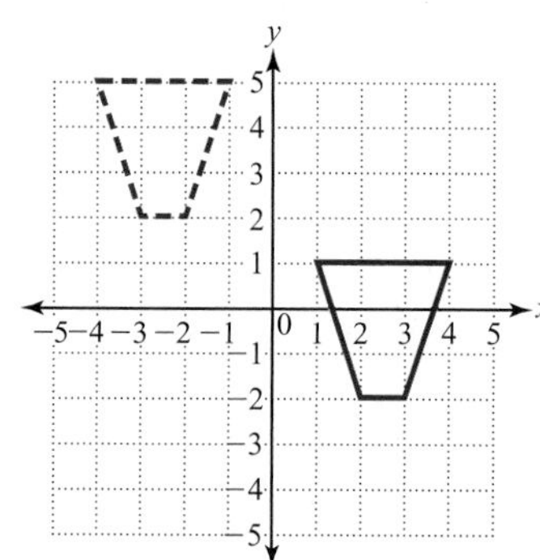

For Exercises 83–88, solve. In each situation, $n \geq 0$ and $c \geq 0$.

83. An academic tutor charges \$60 per month plus \$5 per hour of tutoring. The equation $c = 5n + 60$ describes the total that he would charge for tutoring, where n represents the number of hours of tutoring and c is the total cost.
   a. Find the total cost if the tutor works 7 hours.
   b. If the tutor's total charges are \$90, how many hours of labor is the tutor charging for?
   c. Graph the equation with n along the horizontal axis and c along the vertical axis.

84. A plumber charges \$100 plus \$40 per hour of labor. The equation $c = 40n + 100$ describes the total that he would charge for a service visit, where n represents the number of hours of labor and c is the total cost.

a. Find the total cost if labor is 2.5 hours.

b. If the plumber's total charges are \$180, how many hours of labor is the plumber charging for?

c. Graph the equation with n along the horizontal axis and c along the vertical axis.

85. A wireless company charges \$40 per month for 450 anytime minutes plus \$0.05 per minute above 450. The equation $c = 0.05n + 40$ describes the total that the company would charge for a monthly bill, where n represents the number of minutes used above 450.

a. Find the total cost if a customer uses 600 anytime minutes.

b. If the customer's total charges are \$41.75, how many additional minutes was the customer charged for?

c. Graph the equation with n along the horizontal axis and c along the vertical axis.

86. A stock begins a 10-day decline in price at \$47 per share. The equation $c = -n + 47$ describes the total that the stockbroker could sell the stock for, where n represents the number of days into the decline and c is the total cost.

a. Find the total cost if the stock is at day 4.

b. If the stock is worth \$39 per share, how many days into the decline is the price?

c. Graph the equation with n along the horizontal axis and c along the vertical axis.

87. A copy center charges \$5 for the first 200 copies plus \$0.03 per copy after that. The equation $c = 0.03n + 5$ describes the charges, c, for copying, where n represents the number of copies after the first 200 copies.

a. Find the total cost for 300 copies.

b. If a customer is charged a total of \$11, how many copies did the customer request?

c. Graph the equation with n along the horizontal axis and c along the vertical axis.

88. A mechanic charges \$125 plus \$25 per hour of labor. The equation $c = 25n + 125$ describes the total that she would charge for a service visit, where n represents the number of hours of labor and c is the total cost.

a. Find the total cost if labor is 1.5 hours.

b. If the mechanic's total charges are \$150, how many hours of labor is the mechanic charging for?

c. Graph the equation with n along the horizontal axis and c along the vertical axis.

REVIEW EXERCISES

[1.2] **1.** What property is illustrated by $3x + 5 - 6x + 9 = 5 + 3x - 6x + 9$?

[1.3] **2.** Simplify: $\dfrac{3(5 - 8) + 21}{8 - 3(6)}$

[1.4] ***For Exercises 3 and 4, evaluate the expression*** $\dfrac{y_2 - y_1}{x_2 - x_1}$ ***using the given values.***

3. $x_1 = 4$, $x_2 = 1$, $y_1 = -8$, and $y_2 = -2$

4. $x_1 = -10$, $x_2 = 2$, $y_1 = 13$, and $y_2 = 5$

[2.1] **5.** Solve: $12 - 3(n - 2) = 6n - (2n + 1)$

[2.3] **6.** Solve $7x - 5 > 4x + 13$ and write the solution set in set-builder notation and in interval notation. Then graph the solution set.

3.2 The Slope of a Line

OBJECTIVES

1. **Compare lines with different slopes.**
2. **Graph equations in slope-intercept form.**
3. **Find the slope of a line given two points on the line.**

In Section 3.1, we stated that the standard form for a linear equation in two variables is $Ax + By = C$. We also graphed equations in the form $y = mx + b$, which are not in standard form, and discovered that their y-intercept is $(0, b)$. In this section, we explore equations in the form $y = mx + b$ further and consider how the coefficient m affects the graph.

OBJECTIVE 1. Compare lines with different slopes. First, consider graphs of equations in which b is 0. If $b = 0$, the equation $y = mx + b$ becomes $y = mx$ and the graph has the origin $(0, 0)$ as its x- and y-intercepts.

Connection When an equation is written with y isolated, like those in Example 1, we can think of x-values as inputs and the corresponding y-values as outputs. In Example 1, we chose x-input values of 0, 1, and 2 and used the equations to find the corresponding y-output values. This is the fundamental way of thinking about a mathematical concept called a *function*, which we will discuss in Section 3.5.

EXAMPLE 1 Graph on the same grid. $y = x$, $y = 3x$, and $y = \frac{1}{3}x$

Solution In the following table, the same choice for x has been substituted into each equation and the corresponding y-coordinate has been found:

x	$y = x$	$y = 3x$	$y = \frac{1}{3}x$
0	0	0	0
1	1	3	$\frac{1}{3}$
2	2	6	$\frac{2}{3}$

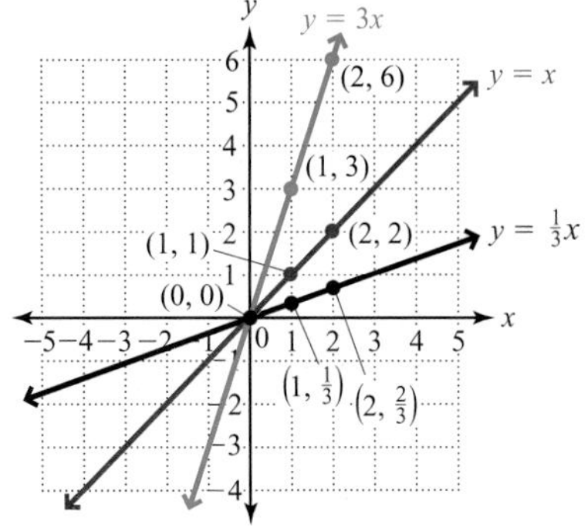

Example 1 suggests that for equations of the form $y = mx$, the coefficient m affects the incline of the line. Consequently, m is called the *slope* of the line.

For $y = x$, the slope is 1 ($m = 1$).
For $y = 3x$, the slope is 3 ($m = 3$).
For $y = \frac{1}{3}x$, the slope is $\frac{1}{3}$ ($m = \frac{1}{3}$).

So far, we have considered only positive slope values, which incline upward from left to right. Now let's graph equations with negative slopes.

EXAMPLE 2 Graph on the same grid. $y = -x$, $y = -2x$, and $y = -\frac{1}{2}x$

Solution

x	$y = -x$	$y = -2x$	$y = -\frac{1}{2}x$
0	0	0	0
1	−1	−2	$-\frac{1}{2}$
2	−2	−4	−1

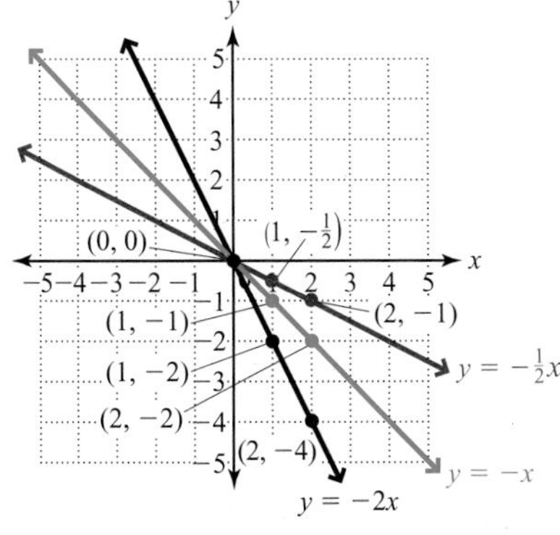

Note: *In math, the convention is to visually scan a graph from left to right just as we would read a sentence.*

Notice that lines with negative slope incline downward from left to right. Also notice that regardless of whether the incline is upward or downward, the greater the absolute value of the slope, the steeper the incline. We will summarize our discoveries about slope in Objective 2.

ANSWERS

a. The graph of $y = 4x$ is steeper than $y = 2x$ because a slope of 4 is greater than a slope of 2.

b. The graph's incline is downward from left to right because the slope is negative.

YOUR TURN

a. Which has a steeper graph, $y = 4x$, or $y = 2x$? Why?

b. Does the graph of $y = -\frac{1}{3}x$ incline upward or downward from left to right? Why?

OBJECTIVE 2. Graph equations in slope-intercept form. Let's explore more specifically how slope determines the incline of a graph. Notice if we isolate m in $y = mx$, we get $m = \frac{y}{x}$, which suggests that **slope** is a ratio of the amount of vertical change (y) to the amount of horizontal change (x).

Note: *A precise formula for the slope of a line will be given in Objective 3.*

DEFINITION ***Slope:*** The ratio of the vertical change between any two points on a line to the horizontal change between those points.

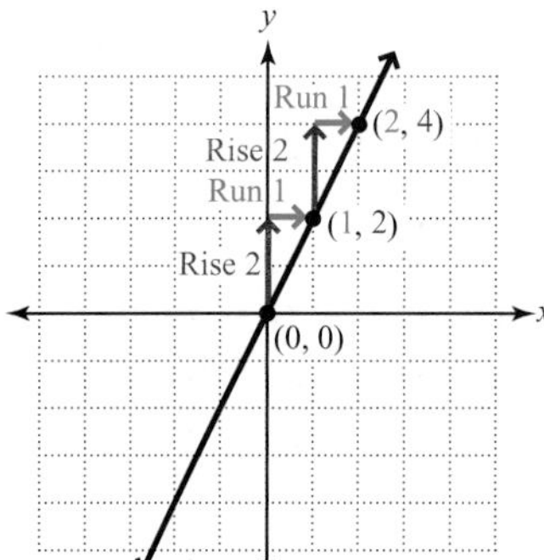

For example, the slope of $y = 2x$ is 2, which can be written as $\frac{2}{1}$. This slope value means that rising vertically 2 units and then running horizontally 1 unit to the right from any point on the line will end up at another point on the same line. For example, rising 2 units and running 1 unit from (0, 0) gives us (1, 2). Similarly, rising 2 units and running 1 unit from (1, 2) gives us (2, 4).

This discovery about slope suggests another approach to graphing. In an equation of the form $y = mx + b$, the slope is m and the y-intercept is $(0, b)$. We can use the y-intercept as a starting point and then find other points on the line using the slope.

EXAMPLE 3 For the equation $y = -\frac{2}{3}x + 1$, determine the slope and the y-intercept. Then graph the equation.

Solution The slope is $-\frac{2}{3}$ and the y-intercept is (0, 1). To graph the line using the slope, we plot the y-intercept, then rise −2 (move down two) and run 3 (move right three), arriving at (3, −1); or rise 2 and run −3, arriving at (−3, 3).

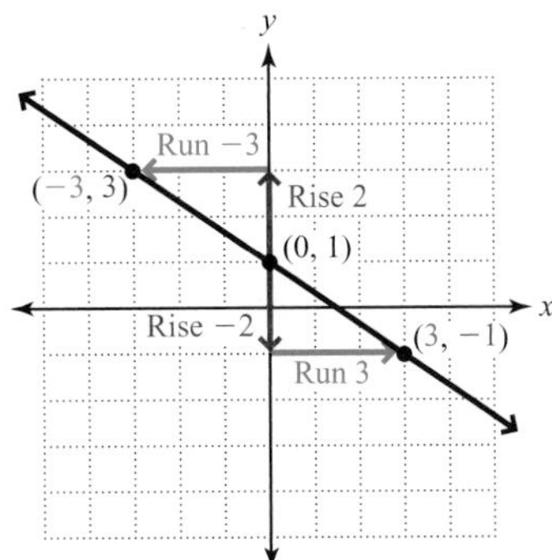

The points we found using the slope can be verified in the equation. We will show the check for (−3, 3) and leave the check for (3, −1) to the reader.

$$y = -\frac{2}{3}x + 1$$

$$3 \stackrel{?}{=} -\frac{2}{3}(-3) + 1$$

$$3 \stackrel{?}{=} 2 + 1$$

$$3 = 3$$ This checks.

The equation $y = -\frac{2}{3}x + 1$ has the form $y = mx + b$, where m is the slope and b is the y-coordinate of the y-intercept. Because we can easily determine the slope and y-intercept of an equation in the form $y = mx + b$, we call it *slope-intercept form.*

RULES Graphs of Equations in Slope-Intercept Form

The graph of an equation in the form $y = mx + b$ (slope-intercept form) is a line with slope m and y-intercept $(0, b)$. The following rules indicate how m affects the graph.

$y = mx+b,\ m > 0$

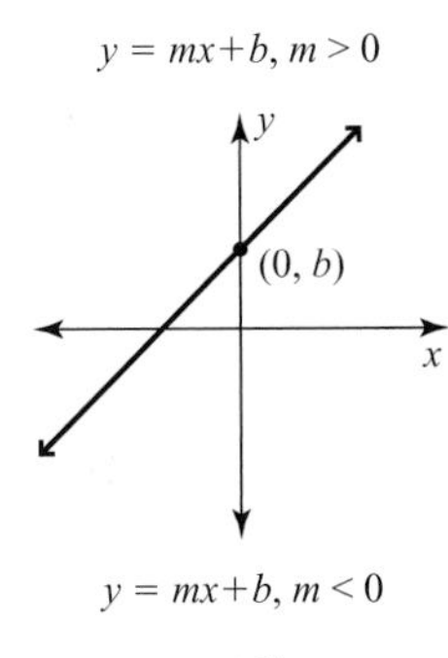

If $m > 0$, then the line slants upward from left to right.

$y = mx+b,\ m < 0$

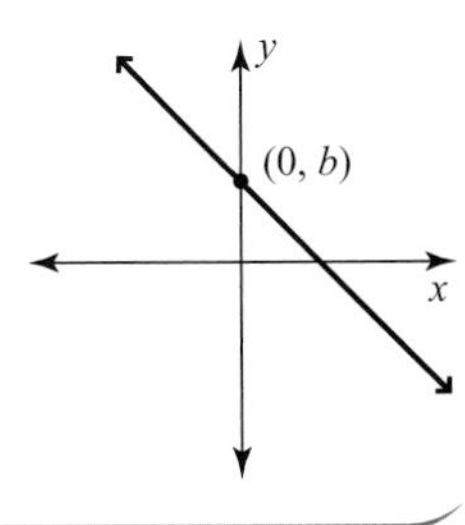

If $m < 0$, then the line slants downward from left to right.

The greater the absolute value of m, the steeper the line.

ANSWER

$m = \frac{2}{5}$, y-intercept: (0, −2)

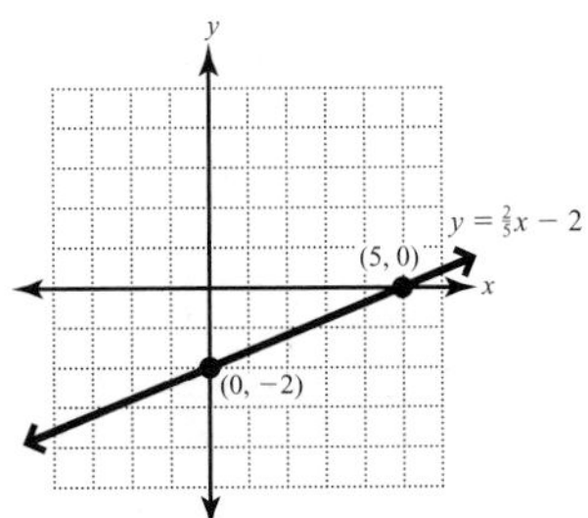

YOUR TURN For the equation $y = \frac{2}{5}x - 2$, determine the slope and the y-intercept. Then graph the equation.

If an equation is not in slope-intercept form and we need to determine the slope and y-intercept, then we can write the equation in slope-intercept form by solving for y.

Learning Strategy

If you are a visual learner, when finding an intercept, cover the term containing the other variable and then solve the remaining portion of the equation mentally. For example, to find the y-intercept of $5x - 2y = 8$, cover $5x$ and then solve for y.

$$5x - 2y = 8$$
$$y = -4$$

EXAMPLE 4 For the equation $5x - 2y = 8$, determine the slope and the y-intercept. Then graph the equation.

Solution Write the equation in slope-intercept form by isolating y.

$$-2y = -5x + 8 \quad \text{Subtract } 5x \text{ from both sides.}$$

$$y = \frac{-5}{-2}x + \frac{8}{-2} \quad \text{Divide both sides by } -2 \text{ to isolate } y.$$

$$y = \frac{5}{2}x - 4 \quad \text{Simplify.}$$

The slope is $\frac{5}{2}$ and the y-intercept is $(0, -4)$. To graph the line we begin at $(0, -4)$ and then rise 5 and run 2, which gives the point $(2, 1)$. Rising 5 and running 2 from $(2, 1)$ gives the point $(4, 6)$.

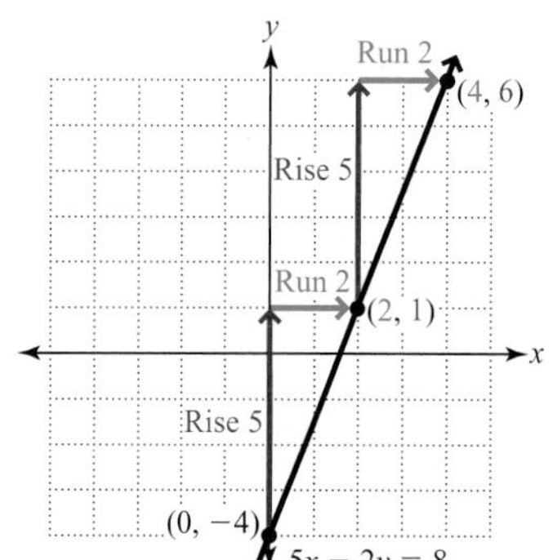

We can check the ordered pairs in the equation:

For (2, 1):	For (4, 6):
$5(2) - 2(1) \stackrel{?}{=} 8$	$5(4) - 2(6) \stackrel{?}{=} 8$
$10 - 2 \stackrel{?}{=} 8$	$20 - 12 \stackrel{?}{=} 8$
$8 = 8$	$8 = 8$

Both ordered pairs check.

YOUR TURN For the equation $3x + 4y = 12$, determine the slope and the y-intercept. Then graph the equation.

ANSWER

$m = -\frac{3}{4}$, y-intercept: $(0, 3)$

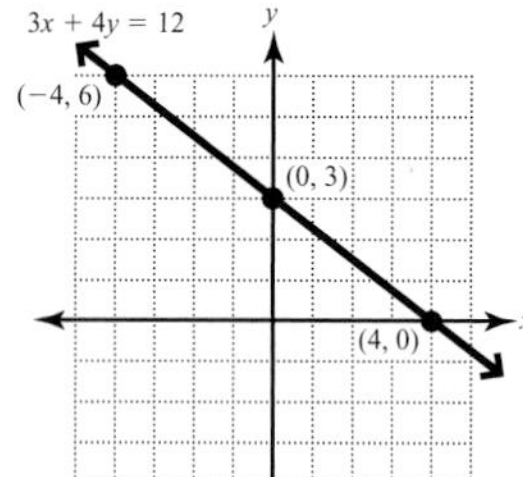

OBJECTIVE 3. Find the slope of a line given two points on the line. Given two points on a line, we can determine the slope of the line. Consider the points $(4, 6)$ and $(2, 1)$ on the graph of $5x - 2y = 8$ from Example 4. Remember that we found the slope of this line to be $\frac{5}{2}$.

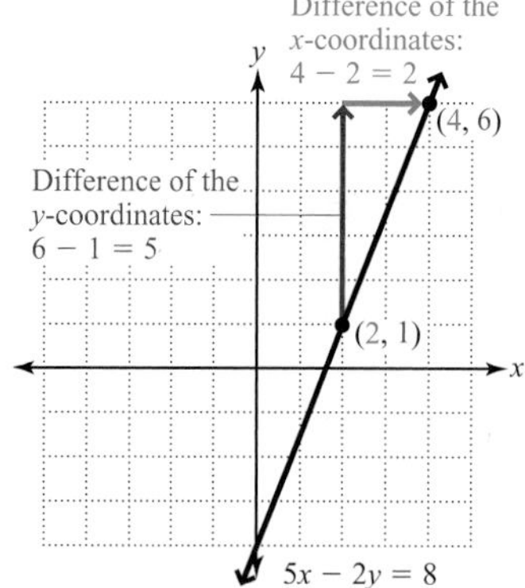

Notice that the amount of rise is equal to the difference in the y-coordinates of the two points, and the amount of run is the difference of the x-coordinates.

$$m = \frac{6-1}{4-2} = \frac{5}{2}$$

In general, we can write a formula for slope.

RULE **The Slope Formula**

Given two points (x_1, y_1) and (x_2, y_2), where $x_2 \neq x_1$, the slope of the line connecting the two points is given by the formula $m = \frac{y_2 - y_1}{x_2 - x_1}$.

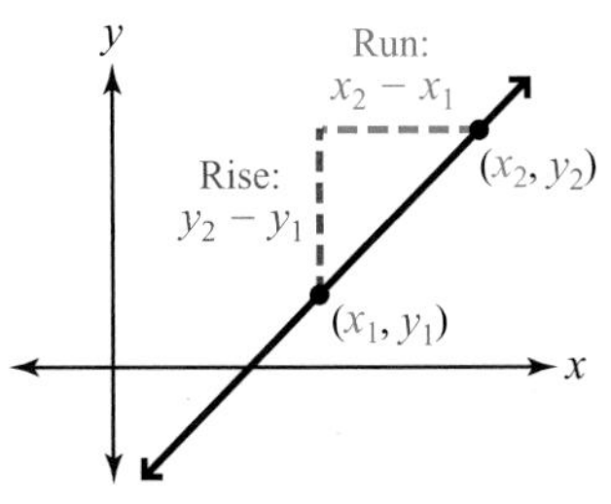

EXAMPLE 5 Find the slope of the line connecting $(-1, 7)$ and $(5, 3)$.

Solution Using $m = \frac{y_2 - y_1}{x_2 - x_1}$, replace the variables with their corresponding values, and then simplify.

$$m = \frac{3 - 7}{5 - (-1)} = \frac{-4}{6} = -\frac{2}{3}$$

Note: *Some people find it helpful to first label the coordinates.*

$(-1, 7)$	$(5, 3)$
(x_1, y_1)	(x_2, y_2)

It does not matter which ordered pair is (x_1, y_1) and which is (x_2, y_2). If we let $(5, 3)$ be (x_1, y_1) and $(-1, 7)$ be (x_2, y_2), we get the same slope.

$$m = \frac{7 - 3}{-1 - 5} = \frac{4}{-6} = -\frac{2}{3}$$

EXAMPLE 6 Graph the line connecting the given points and find its slope.

a. $(-2, -3)$ and $(4, -3)$

Solution Because the y-coordinates are the same, the graph is a horizontal line.

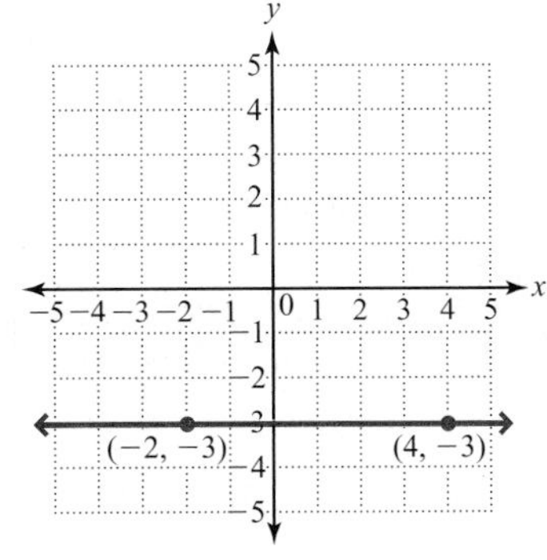

Calculating the slope:

$$m = \frac{-3 - (-3)}{4 - (-2)} = \frac{0}{6} = 0$$

Note: *The equation of this line is* $y = -3$.

b. (3, 4) and (3, −2)

Solution Because the x-coordinates are the same, the graph is a vertical line.

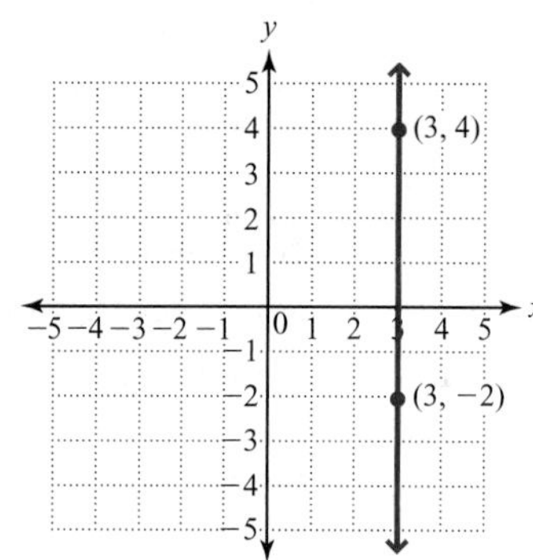

Calculating the slope:

$$m = \frac{4 - (-2)}{3 - 3} = \frac{6}{0}, \text{ which is undefined.}$$

Note: *The equation of this line is* $x = 3$.

We can summarize what we learned in Example 6 with the following rules:

RULES **Slopes of Horizontal and Vertical Lines**

Two points with different x-coordinates and the same y-coordinates, (x_1, c) and (x_2, c), will form a horizontal line with slope 0 and equation $y = c$.

Two points with the same x-coordinates and different y-coordinates, (c, y_1) and (c, y_2) will form a vertical line with undefined slope and equation $x = c$.

YOUR TURN Find the slope of the line connecting the points with the given coordinates.

a. (2, 5) and (6, 1) **b.** (−5, −2) and (7, −5) **c.** (4, −1) and (−3, −1)

EXAMPLE 7 The following graph shows the number of product units a company produced and sold each year from 1995 to 2005. An analyst determines the red line shown reasonably describes the trend shown in the data. Find the slope of that line.

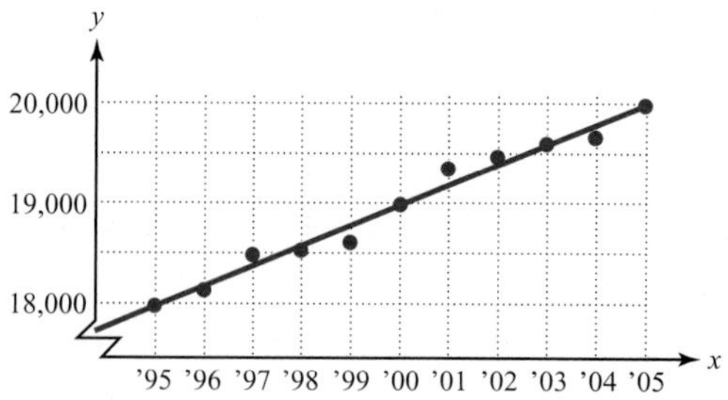

Understand We are to find the slope of a line that passes through or near a set of data points.

ANSWERS

a. −1 **b.** $-\frac{1}{4}$ **c.** 0

Plan To find the slope of the line, we use the slope formula $m = \frac{y_2 - y_1}{x_2 - x_1}$ with two data points that are on the line. We will use (1995, 18,000) and (2000, 19,000).

Execute $m = \frac{19{,}000 - 18{,}000}{2000 - 1995} = \frac{1000}{5} = 200$

Answer The slope of the line is 200, which means demand for the product increased by about 200 units each year.

Check We could use a different pair of data points on the line and see if we get the same slope. We will leave this to the reader.

3.2 Exercises

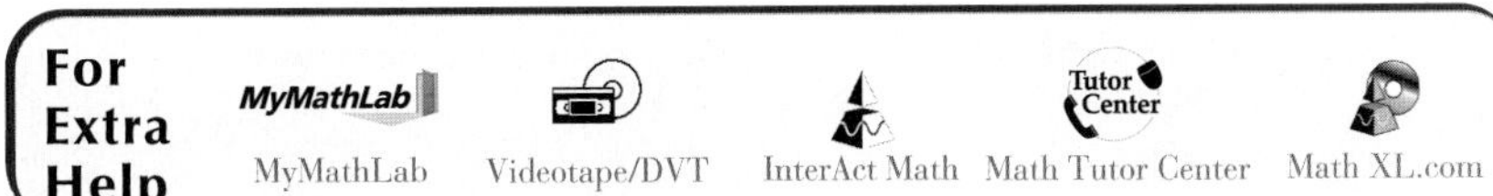

1. In your own words, what is the slope of a line?

2. Explain how you would use the slope and y-intercept to graph $y = \frac{3}{4}x - 2$.

3. Given an equation in the slope-intercept form $y = mx + b$, if $m > 0$, what does that indicate about the incline of the graph?

4. Given an equation in the slope-intercept form $y = mx + b$, if $m < 0$, what does that indicate about the incline of the graph?

5. Describe a line with slope 0. Explain why a slope of 0 produces this type of line.

6. Describe a line with undefined slope. Explain why undefined slope produces this type of line.

For Exercises 7–10, which equation's graph has the steeper slope?

7. $y = 3x + 1$ or $y = 2x - 7$

8. $y = x - 2$ or $y = 4x + 3$

9. $y = 0.2x + 3$ or $y = x - 4$

10. $y = \frac{3}{4}x - 1$ or $y = \frac{2}{3}x + 5$

For Exercises 11–14, indicate whether the graph of the equation inclines upward or downward from left to right.

11. $y = 3x + 2$

12. $y = -4x + 1$

13. $y = -\frac{4}{5}x - 3$

14. $y = \frac{2}{3}x - 5$

For Exercises 15–18, determine the slope and the y-intercept of the line.

15.

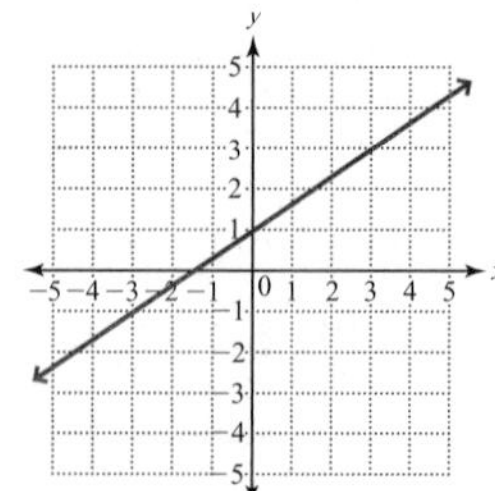

16.

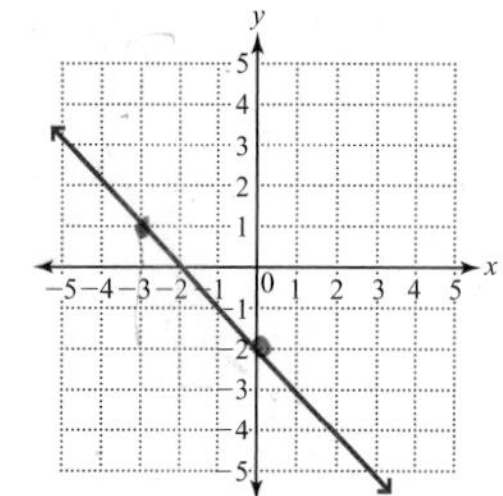

17.

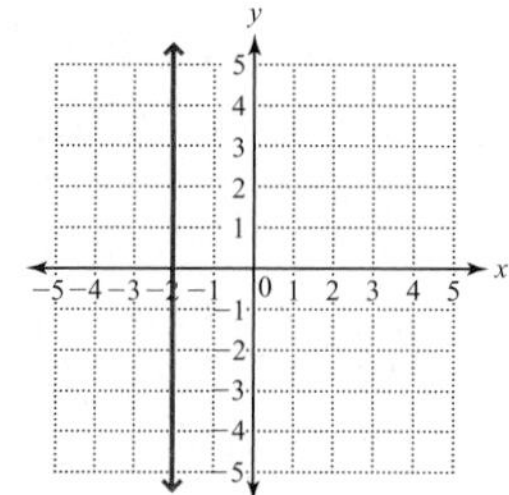

18.

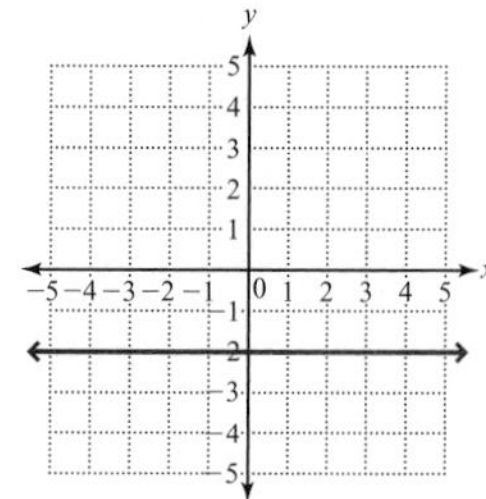

For Exercises 19–26, match the equation with its graph.

19. $y = 4x - 1$

20. $y = 2x + 3$

21. $y = -2x + 3$

22. $y = -4x + 2$

23. $y = \frac{2}{3}x - 1$

24. $y = \frac{3}{4}x + 2$

25. $y = -\frac{1}{2}x + 3$

26. $y = -\frac{3}{4}x - 1$

a.

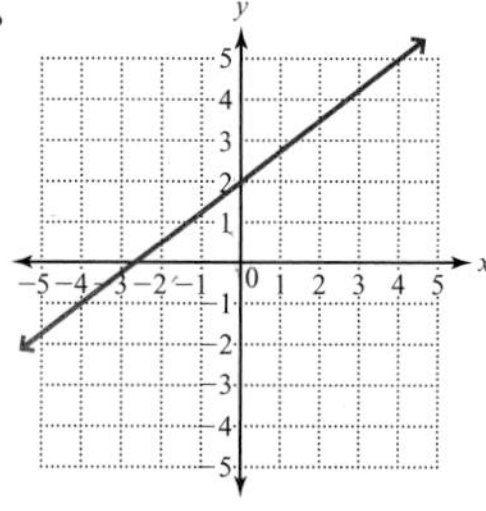

b.

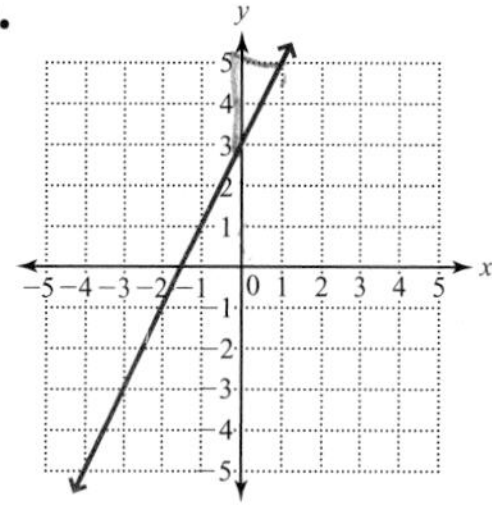

c.

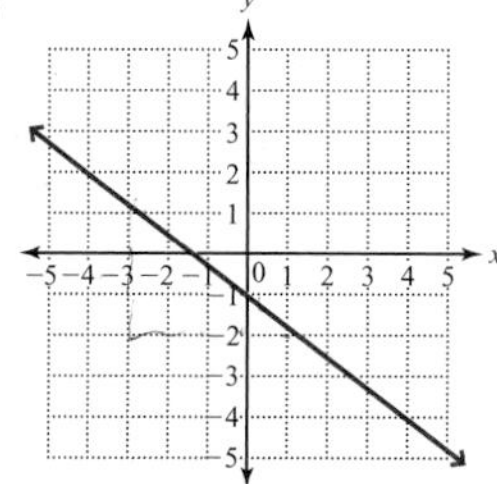

d.

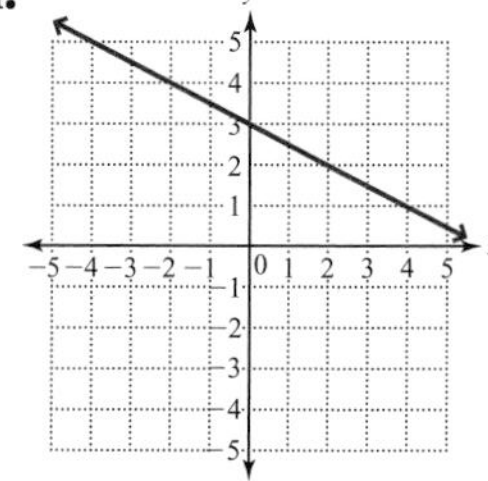

e.

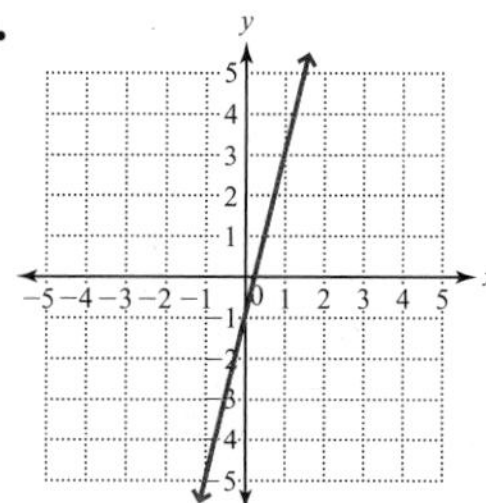

f.

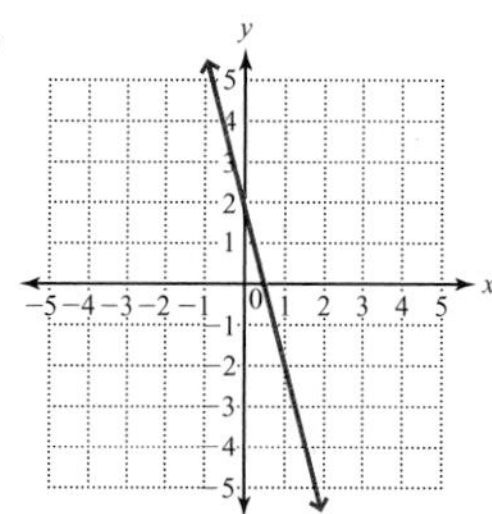

g.

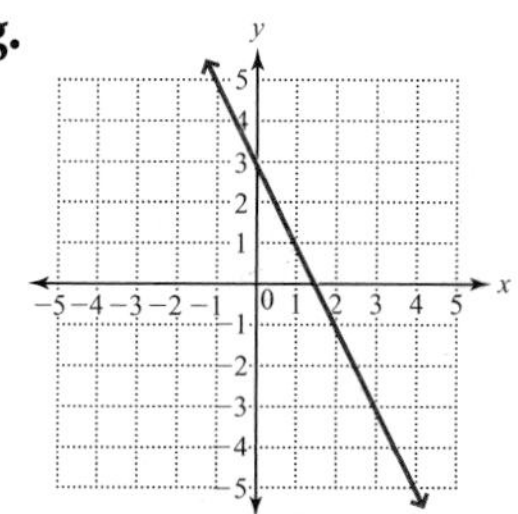

h.

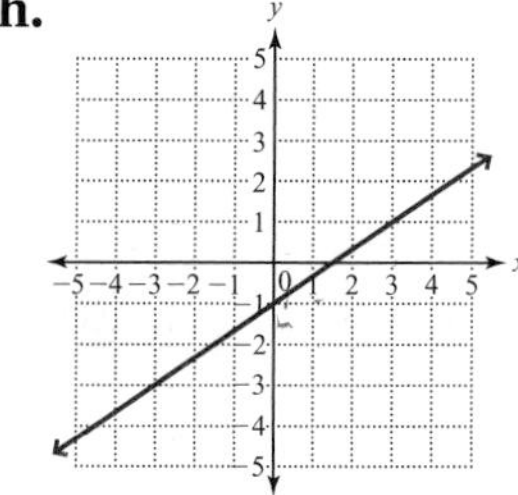

For Exercises 27–42, determine the slope and the y-intercept. Then graph the equation.

27. $y = \frac{2}{3}x + 5$

28. $y = \frac{3}{5}x - 8$

29. $y = -\frac{1}{5}x - 8$

30. $y = -\frac{4}{5}x - 1$

31. $y = x + 4$

32. $y = x - 3$

33. $y = -5x + \frac{2}{3}$

34. $y = -3x - \frac{4}{5}$

35. $2x + 3y = 6$

36. $5x + 3y = 6$

37. $x + 2y = -7$

38. $3x - 2y = 9$

39. $2x - 7y = 8$

40. $5x - 4y = -9$

41. $-x + y = 0$

42. $2x + y = 0$

For Exercises 43–54, find the slope of the line through the given points.

43. $(4, 2), (2, 6)$

44. $(1, 7), (3, 1)$

45. $(-1, -3), (4, 6)$

46. $(6, -7), (-2, 4)$

47. $(1, 4), (-3, 10)$

48. $(-7, 9), (2, 3)$

49. $(8, 2), (8, -5)$

50. $(-3, 1), (-3, -7)$

51. $(5, 12), (-1, 12)$

52. $(-5, -1), (-3, -1)$

53. $(0, 0), (10, -8)$

54. $(9, 11), (-1, -1)$

For Exercises 55 and 56, solve.

55. A parallelogram has vertices at $(-1, 1)$, $(1, 6)$, $(5, 1)$, and $(7, 6)$.
   - **a.** Plot the vertices in a coordinate plane, then connect them to form the parallelogram.
   - **b.** Find the slope of each side of the parallelogram.
   - **c.** What do you notice about the slopes of the parallel sides?

56. A right triangle has vertices at $(-2, 1)$, $(-4, 4)$, and $(-1, 6)$.

a. Plot the vertices in a coordinate plane, then connect them to form the triangle.

b. Find the slope of each side of the triangle.

c. What do you notice about the slopes of the perpendicular sides?

For Exercises 57–60, find the indicated slope. Disregard the object's orientation so that each slope is given as a positive number.

57. A wheelchair ramp is to be built with a height of 29 inches and a length of 348 inches. Find the slope of the ramp.

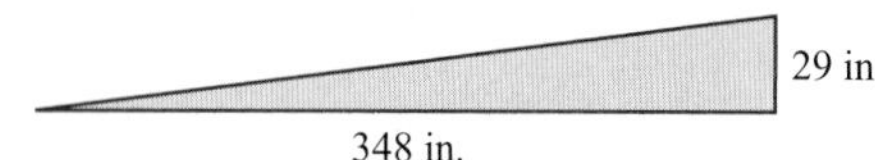

58. A wood frame for a roof is shown. Find the slope of the roof.

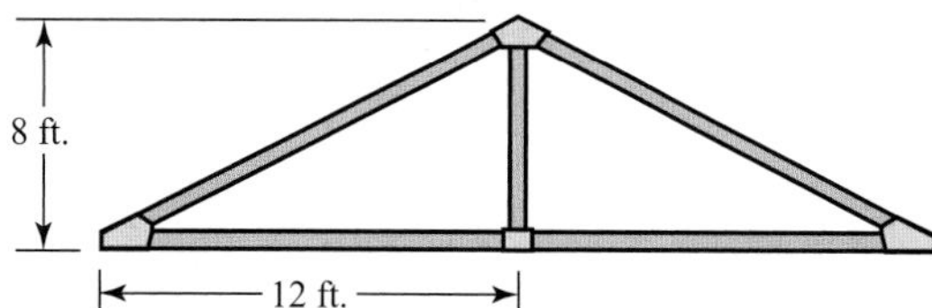

Of Interest

In construction, the slope of a roof is called its *pitch*.

59. An architect consults an elevation plan to determine the slope of a driveway for a new home. The bottom of the driveway will be at an elevation of 210 feet and the top will be at an elevation of 215 feet. The drive will cover a horizontal distance of 20 feet. Find the slope of the driveway. Write your answer as a decimal number.

60. The four faces of the Great Pyramid in Egypt originally had a smooth stone surface. It was originally 481 feet tall and each side along the square base was 754 feet long. If the top of the pyramid was centered over the base, find the slope of its original faces. Write your answer as a decimal number rounded to the nearest hundredth.

Of Interest

After losing the stones that formed its smooth outer surface, the pyramid is now 449 feet tall and each side is 745 feet long.

61. A roller coaster begins on flat elevated track that is 15 feet above the ground, as shown (point B). After traveling 50 feet on the flat stretch of track to point C, it begins climbing a hill. The top of the hill (point D) is 80 feet above the ground and 100 feet horizontally from the beginning position of the ride.

a. If we were to graph the line representing the hill in the coordinate plane with the origin at point A, what would be the coordinates of points C and D?

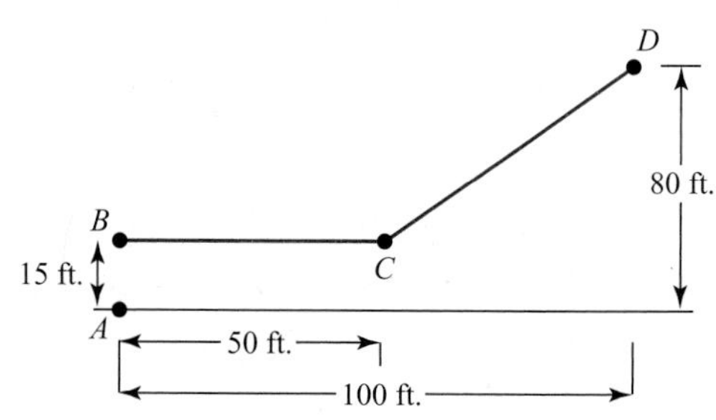

b. Find the slope of the hill that the coaster climbs.

c. What does the slope indicate?

62. Body mass index, BMI, assesses the amount of fat in a person's body. At age 20, June weighed 125 and her BMI was 16.9. Now at age 38, she weighs 150 and has a BMI of 27.4. Suppose the increase in weight and BMI was roughly linear from age 20 to age 38. Plotting her weight along the x-axis and BMI along the y-axis gives the graph to the right.

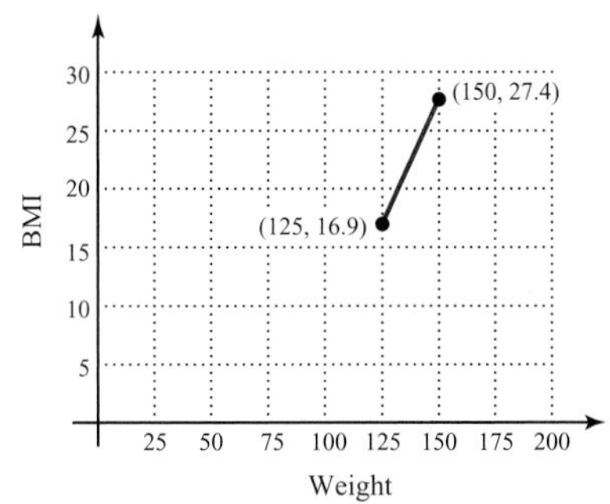

a. Find the slope of the line shown. Write your answer as a decimal number rounded to the nearest hundredth.

b. What does the slope indicate?

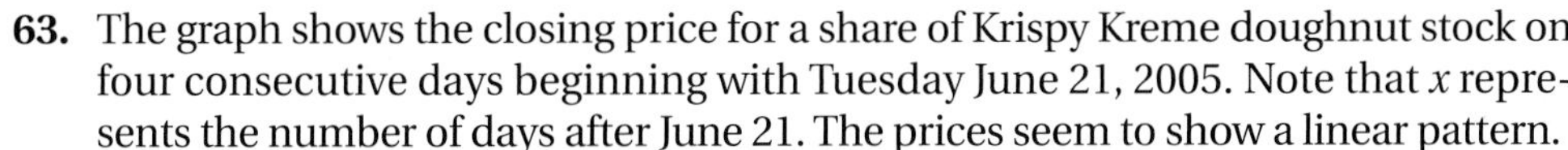

63. The graph shows the closing price for a share of Krispy Kreme doughnut stock on four consecutive days beginning with Tuesday June 21, 2005. Note that x represents the number of days after June 21. The prices seem to show a linear pattern.

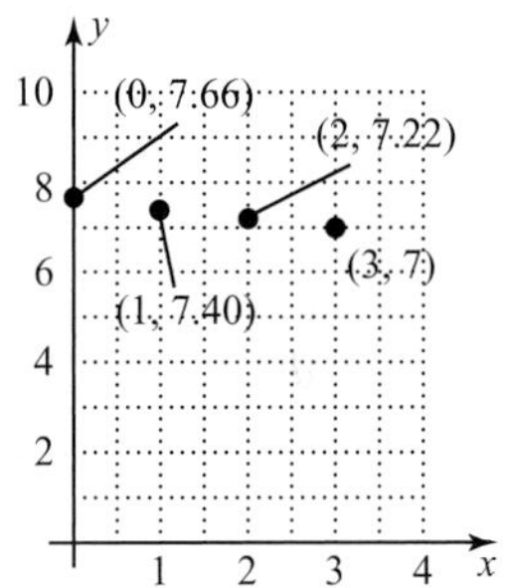

a. Draw a straight line from the point on the graph at June 21 ($x = 0$) to the point on the graph at June 24 ($x = 3$).

b. Find the slope of this line.

c. If the same trend continued, predict the price of a share of Krispy Kreme stock on the next day of trading (Monday June 27).

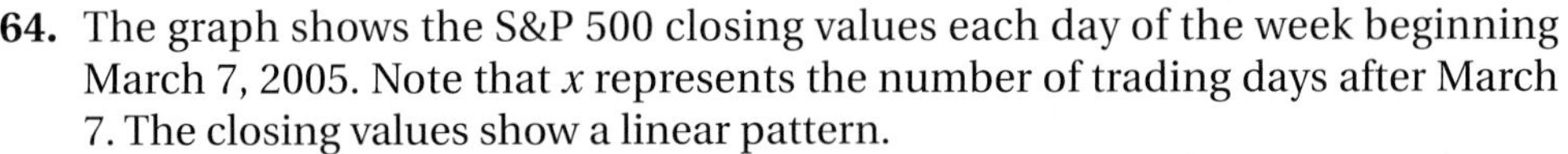

64. The graph shows the S&P 500 closing values each day of the week beginning March 7, 2005. Note that x represents the number of trading days after March 7. The closing values show a linear pattern.

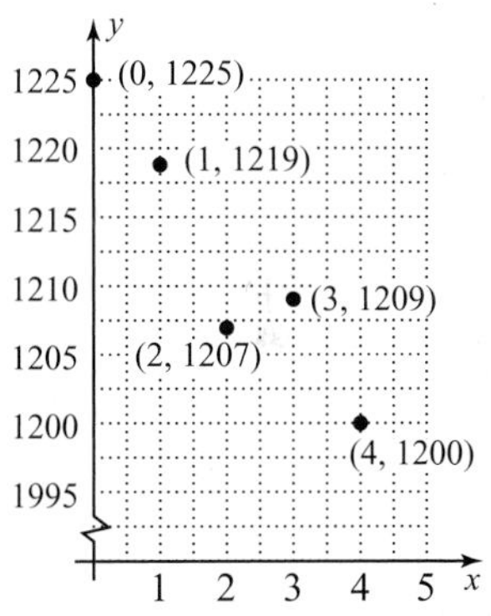

a. Draw a line from the point on the graph at March 7 to the point on the graph at March 11.

b. Find the slope of this line.

c. If the same trend continued, find the closing value on March 14, which was the next day of trading.

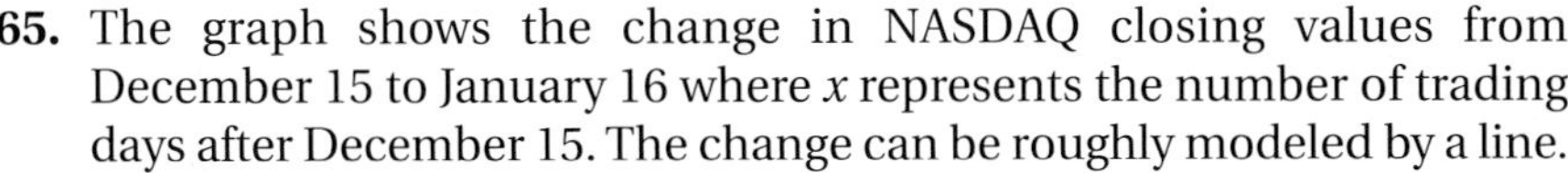

65. The graph shows the change in NASDAQ closing values from December 15 to January 16 where x represents the number of trading days after December 15. The change can be roughly modeled by a line.

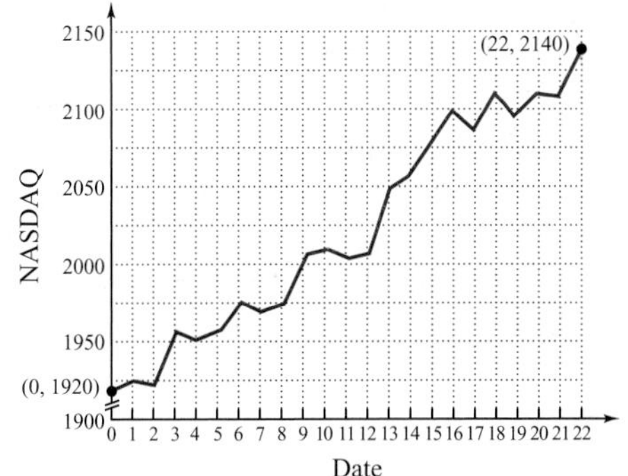

a. Draw a straight line connecting the NASDAQ closing value on December 15 to the value on January 16.

b. Find the slope of this line.

c. If the same trend continued, what was the NASDAQ closing value on January 19, which was the next day the market opened?

66. The graph shows the change in NASDAQ from May 16, 2005 through June 17, 2005 where x represents the number of trading days after May 16. The change can be roughly modeled by a line.

a. Draw a straight line connecting the NASDAQ value on May 16 to the value on June 17.

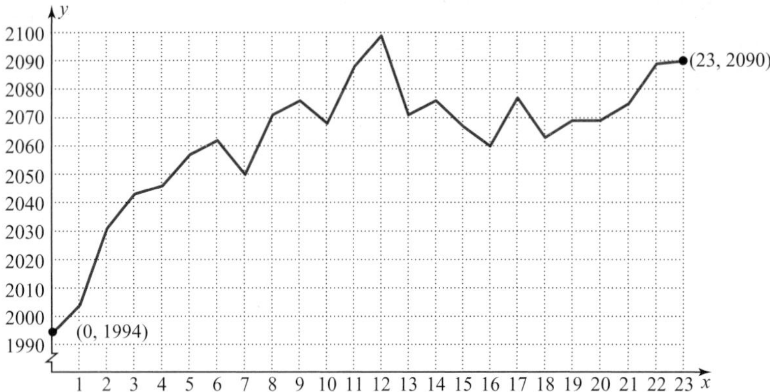

b. Find the slope of this line rounded to the nearest whole number.

c. If the same trend continued, what was the NASDAQ closing value on June 20, which was the next day the market opened?

67. In 1990, 32 million U.S. residents spoke a native language other than English. In 2003, that number had increased to 47 million. (*Source:* U.S. Bureau of the Census)

a. If x represents the number of years after 1990 (with 1990 being $x = 0$), and y represents the number of residents who spoke a native language other than English, plot the two data points given, then connect them with a straight line.

b. Find the slope of the line connecting the two data points given.

68. In 2000, the average earnings per hour for U.S. production workers was \$14.00. The average earnings increased steadily each year so that in 2003 the average hourly earnings was \$15.35. (*Source:* Bureau of Labor Statistics, U.S. Department of Labor)

a. If x represents the number of years after 2000 (with 2000 being $x = 0$), and y represents the average hourly earnings plot the two data points given, then connect them with a straight line.

b. Find the slope of the line connecting the two data points given.

69. In 2000, the average number of hours that production workers in the U.S. worked per week was 34.3. The number of hours declined each year so that in 2003 the average number of hours was 33.7. (*Source:* Bureau of Labor Statistics, U.S. Department of Labor)

a. If x represents the number of years after 2000 (with 2000 being $x = 0$), and y represents the average number of hours per week, plot the two data points given, then connect them with a straight line.

b. Find the slope of the line connecting the two data points given.

70. In 2000, union members represented 13.5% of the U.S. labor force. This percentage decreased steadily each year so that in 2003, union members made up 12.9% of the labor force. (*Source:* Bureau of Labor Statistics, U.S. Department of Labor)

a. If x represents the number of years after 2000 (with 2000 being $x = 0$), and y represents the percent, plot the two data points given, then connect them with a straight line.

b. Find the slope of the line connecting the two data points given.

Collaborative Exercises TABLE MATH

Tables, mental math, and visual techniques can serve as tools for understanding and calculating slope and finding ordered pairs for less complex problems.

1. Consider the ordered pairs in the following table. Suppose we know that the points form a straight line and we wish to find the slope of the line through the points. The lines outside the table show that as x increases by 2 from 0 to 2, y increases by 7 from 2 to 9. Does this pattern hold for all ordered pairs in the table?

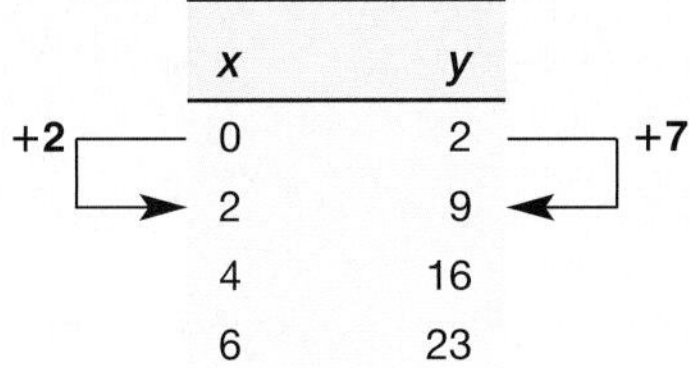

x	y
0	2
2	9
4	16
6	23

2. a. Use the slope formula $m = \dfrac{y_2 - y_1}{x_2 - x_1}$ to find the slope of the line formed by the ordered pairs in the table.

b. Discuss the connection between the slope you calculated and the pattern shown in Question 1.

3. a. Following is the same table as in Question 1. Find the two missing values a and b.

x	y
0	2
2	9
4	16
6	23

+a (from x = 0 to x = 6)   +b (from y = 2 to y = 23)

b. Write the ratio of b to a in simplest form. What is this value?

c. What does your calculation in part b suggest about how to find the slope?

(continued)

4. Note that the y-intercept is given in the preceding table. Write the equation of the line (in slope-intercept form) that fits the ordered pairs in the table.

5. Use the ratio method and the following table to find the slope of the line through the points (1, 7) and (4, −4). Verify your answer with the slope formula.

x	y
1	7
4	−4

6. In graphing $y = -\frac{3}{4}x + 5$, we already know that the y-intercept is (0, 5) and that the slope is $-\frac{3}{4}$. We can use a table to find a second point (and even a third checkpoint). Find those points.

x	y
0	5
+4 →	← −3

REVIEW EXERCISES

[1.4] ***For Exercises 1 and 2, evaluate $4n^2 - m(n + 2)$ using the given values.***

1. $m = 7, n = -2$

2. $m = -5, n = 3$

3. Simplify: $16x - 3y - 15 - x + 3y - 12$

4. Distribute: $-5(3x - 2)$

[2.1] ***For Exercises 5 and 6, solve the equation for the indicated variable.***

5. $V = lwh$; solve for h

6. $Ax + By = C$; solve for y

3.3 The Equation of a Line

OBJECTIVES

1. **Use slope-intercept form to write the equation of a line.**
2. **Use point-slope form to write the equation of a line.**
3. **Write the equation of a line parallel to a given line.**
4. **Write the equation of a line perpendicular to a given line.**

In this section, we will explore how to write an equation of a line given information about the line, such as its slope and the coordinates of a point on the line.

OBJECTIVE 1. Use slope-intercept form to write the equation of a line. If we are given the y-intercept $(0, b)$ of a line, to write the equation of the line we will need either the slope or another point so that we can calculate the slope. First, consider a situation in which we are given the y-intercept and the slope of a line.

EXAMPLE 1 A line has a slope of $-\frac{3}{5}$. If the y-intercept is $(0, 1)$, write the equation of the line in slope-intercept form.

Solution We use $y = mx + b$, the slope-intercept form of the equation, replacing m with the slope, $-\frac{3}{5}$, and b with the y-coordinate of the y-intercept, 1.

$$y = -\frac{3}{5}x + 1$$

YOUR TURN A line has a slope of -5. If the y-intercept is $(0, -2)$, write an equation of the line in slope-intercept form.

Now suppose we are given the y-intercept and a second point on the line. In order to write the equation in slope-intercept form, we need the slope, so our first step is to calculate the slope using the formula from Section 3.2, $m = \frac{y_2 - y_1}{x_2 - x_1}$, where the two points, (x_1, y_1) and (x_2, y_2), are the y-intercept and the other given point.

ANSWER

$y = -5x - 2$

EXAMPLE 2 The graph shows the relationship between Celsius and Fahrenheit temperatures. The table lists two ordered pairs of equivalent temperatures on the two scales.

a. Write the equation for converting degrees Celsius to degrees Fahrenheit.

b. Use the equation to convert 30°C to degrees Fahrenheit.

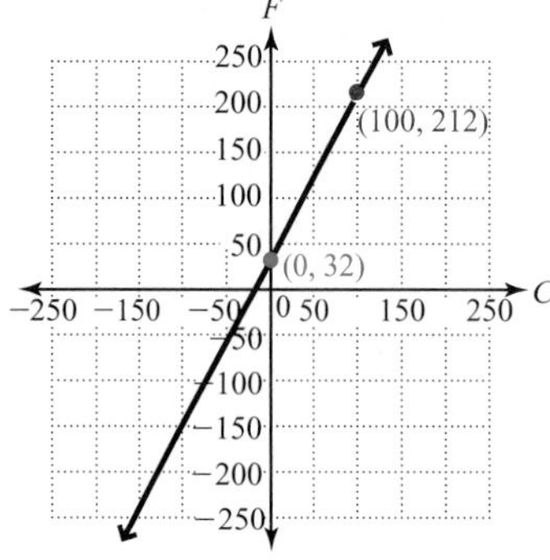

	Celsius (C)	Fahrenheit (F)
Freezing point of water	0	32
Boiling point of water	100	212

Solution a. We are given (0, 32), which is like the y-intercept, and a second point, (100, 212). To use the slope-intercept equation, we first find the slope using the slope formula.

$$m = \frac{212 - 32}{100 - 0} = \frac{180}{100} = \frac{9}{5}$$

Connection The slope $\frac{9}{5}$ means that for every 9° of change on the Fahrenheit scale there is a 5° change on the Celsius scale.

Note: *In this case, the y-coordinates are degrees Fahrenheit and the x-coordinates are degrees Celsius.*

Since the F-intercept is (0, 32) and we know that the slope is $\frac{9}{5}$, we can write an equation in slope-intercept form. We let C (degrees Celsius) take the place of the x-variable, and F (degrees Fahrenheit) take the place of the y-variable.

$$F = \frac{9}{5}C + 32$$

b. Now we can use this equation to convert Celsius temperatures to degrees Fahrenheit. We are to convert 30°C to degrees Fahrenheit.

$$F = \frac{9}{5}(30) + 32 = 54 + 32 = 86°\text{F}$$

Examples 1 and 2 suggest the following procedure:

PROCEDURE **Equation of a Line Given Its y-intercept**

To write the equation of a line given its y-intercept, $(0, b)$, and its slope, m, use the slope-intercept form of the equation, $y = mx + b$. If given a second point and not the slope, we must first calculate the slope using $m = \frac{y_2 - y_1}{x_2 - x_1}$, then use $y = mx + b$.

Your Turn Write the equation of the line passing through (0, 3) and (4, −2) in slope-intercept form.

Answer

$y = -\frac{5}{4}x + 3$

OBJECTIVE 2. Use point-slope form to write the equation of a line. Now let's see how to write the equation of a line given *any* two points on the line. We can use the slope formula to derive the *point-slope* form of the equation of a line. Recall the slope formula:

Connection We can also view the formula for slope as a proportion and cross multiply to get the point-slope form:

$$\frac{m}{1} = \frac{y_2 - y_1}{x_2 - x_1}$$

$(x_2 - x_1)m = y_2 - y_1$

$$m = \frac{y_2 - y_1}{x_2 - x_1}$$

$$(x_2 - x_1) \cdot m = \left(\frac{y_2 - y_1}{x_2 - x_1}\right) \cdot (x_2 - x_1)$$ Multiply both sides by $x_2 - x_1$ to isolate the y's.

$$(x_2 - x_1)m = (y_2 - y_1)$$

$$y_2 - y_1 = m(x_2 - x_1)$$ Rewrite with y's on the left side to resemble slope-intercept form.

With this point-slope form, we can write the equation of a line by substituting the slope of the line for m and the coordinates of any point on the line for x_1 and y_1, leaving x_2 and y_2 as variables. To indicate that x_2 and y_2 remain variables, we remove their subscripts so that we have $y - y_1 = m(x - x_1)$, which is called the point-slope form of the equation of a line.

PROCEDURE Using the Point-Slope Form of the Equation of a Line

To write the equation of a line given its slope and any point, (x_1, y_1), on the line, use the point-slope form of the equation of a line, $y - y_1 = m(x - x_1)$. If given a second point, (x_2, y_2), and not the slope, we first calculate the slope using $m = \frac{y_2 - y_1}{x_2 - x_1}$, then use $y - y_1 = m(x - x_1)$.

Example 3 Write the equation of a line with a slope of −3 and passing through the point (2, −4). Write the equation in slope-intercept form.

Solution Because we are given the coordinates of a point and the slope of a line passing through the point, we use the point-slope formula, $y - y_1 = m(x - x_1)$.

Note: *We leave x and y as variables and substitute only for x_1 and y_1.*

$$y - (-4) = -3(x - 2)$$ Replace m with −3, x_1 with 2, and y_1 with −4.

$$y + 4 = -3x + 6$$ Simplify.

$$y = -3x + 2$$ Subtract 4 from both sides so that we have slope-intercept form.

Check Verify that the point (2, −4) is a solution for the equation.

$$-4 \stackrel{?}{=} -3(2) + 2$$

$$-4 = -4$$ True

EXAMPLE 4 Write the equation of a line passing through the points $(-3, 6)$ and $(9, -2)$. Write the equation in slope-intercept form.

Solution Since we do not have the slope, we first calculate it using $m = \dfrac{y_2 - y_1}{x_2 - x_1}$.

$$m = \frac{-2-6}{9-(-3)} = \frac{-8}{12} = -\frac{2}{3}$$ **Replace x_1 with -3, y_1 with 6, x_2 with 9, and y_2 with -2.**

Now we can use the point-slope form. We can use either of the two given points for (x_1, y_1) in the point-slope equation. We will use $(-3, 6)$ for (x_1, y_1).

$$y - 6 = -\frac{2}{3}(x - (-3))$$ **Using $y - y_1 = m(x - x_1)$, replace m with $-\frac{2}{3}$, x_1 with -3, and y_1 with 6.**

$$y - 6 = -\frac{2}{3}x - 2$$ **Distribute and simplify.**

$$y = -\frac{2}{3}x + 4$$ **Add 6 to both sides to get slope-intercept form.**

YOUR TURN

a. Write the equation of a line with a slope of -0.4 and passing through the point $(-2, 3)$ in slope-intercept form.

b. Write the equation of a line passing through $(-3, -5)$ and $(3, -1)$ in slope-intercept form.

Writing Linear Equations in Standard Form

We may also write equations in standard form, which is $Ax + By = C$, where A, B, and C are real numbers. In standard form, it is customary to write the equation so that the x term is first with a positive coefficient and, if possible, A, B, and C are all integers.

EXAMPLE 5 A line connects the points $(8, 4)$ and $(2, -1)$. Write the equation of the line in the form $Ax + By = C$, where A, B, and C are integers and $A > 0$.

Solution Since we are given two points, we first find the slope using $m = \dfrac{y_2 - y_1}{x_2 - x_1}$.

$$m = \frac{-1-4}{2-8} = \frac{-5}{-6} = \frac{5}{6}$$ **Replace x_1 with 8, y_1 with 4, x_2 with 2, and y_2 with -1.**

Because we were not given the y-intercept, we use the point-slope form of the linear equation, $y - y_1 = m(x - x_1)$.

$$y - (-1) = \frac{5}{6}(x - 2)$$ **Use the slope and the point $(2, -1)$.**

$$y + 1 = \frac{5}{6}(x - 2)$$ **Simplify the signs.**

ANSWERS

a. $y = -0.4x + 2.2$

b. $y = \frac{2}{3}x - 3$

Now manipulate the equation to get the form $Ax + By = C$, where A, B, and C are integers and $A > 0$.

$6(y + 1) = 6 \cdot \frac{5}{6}(x - 2)$	**Multiply both sides by the LCD, 6.**
$6y + 6 = 5x - 10$	**Simplify and distribute.**
$6y - 5x = -16$	**Subtract $5x$ and 6 from both sides to separate the variable terms and constant terms.**
$-5x + 6y = -16$	**Use the commutative property of addition so that the x term is the first term.**
$5x - 6y = 16$	**Multiply both sides by -1 so that the coefficient of the x term is positive. Now A, B, and C are integers and $A > 0$.**

EXAMPLE 6 The following data points show the amount a city charges for the number of cubic feet of water consumed. When plotted, the points all lie on a straight line. Write the equation of the line in standard form with A, B, and C integers and $A > 0$.

x (consumption in cubic feet)	y (price in \$)
0	5.00
100	7.50
200	10.00
300	12.50

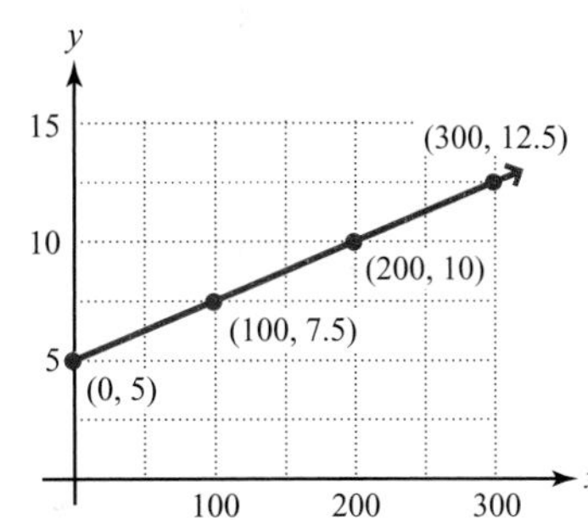

Solution First, we find the slope of the line using any two ordered pairs in the slope formula. We will use (0, 5) and (100, 7.5).

$$m = \frac{7.5 - 5}{100 - 0} = \frac{2.5}{100} = 0.025$$

Using the y-intercept, (0, 5), we can write the equation in slope-intercept form.

$$y = 0.025x + 5 \qquad \textbf{Slope-intercept form.}$$

Now we write the equation in standard form with A, B, and C integers and $A > 0$.

$40y = x + 200$	**Multiply both sides by 40 to eliminate the decimal number 0.025.**
$40y - x = 200$	**Subtract x from both sides.**
$-x + 40y = 200$	**Use the commutative property to write the x term first.**
$x - 40y = -200$	**Multiply both sides by -1. We now have standard form with A, B, and C integers and $A > 0$.**

Note: *We chose to multiply both sides by 40 because $0.025 = \frac{1}{40}$. Other choices are possible, all of which lead to equations equivalent to $x - 40y = -200$. For example multiplying both sides by 200 leads to $5x - 200y = -1000$.*

Your Turn A line connects the points with coordinates $(-1, -2)$ and $(4, 4)$. Write the equation of the line in the form $Ax + By = C$, where A, B, and C are integers and $A > 0$.

Answer

$6x - 5y = 4$

OBJECTIVE 3. Write the equation of a line parallel to a given line. Consider the graphs of $y = 2x + 1$ and $y = 2x - 3$.

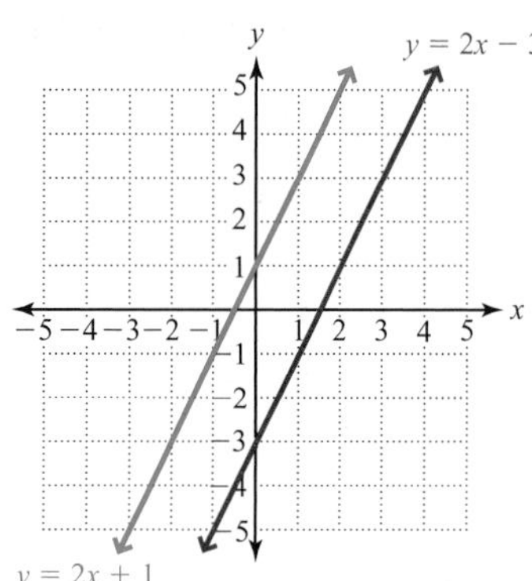

These two lines are parallel, which means they do not intersect at any point. We see in the equations that they have the same slope, 2, but different y-intercepts, which suggests the following rule.

RULE **Parallel Lines**

The slopes of parallel lines are equal.

Example 7 Write the equation of the line in slope-intercept form that passes through $(4, -1)$ and is parallel to the graph of $y = -\frac{1}{2}x + 3$.

Solution The slope of the given equation is $-\frac{1}{2}$, so the slope of the parallel line must also be $-\frac{1}{2}$. We can now write the equation of the parallel line.

$$y - (-1) = -\frac{1}{2}(x - 4)$$

Using $y - y_1 = m(x - x_1)$, replace m with $-\frac{1}{2}$, x_1 with 4, and y_1 with -1.

$$y + 1 = -\frac{1}{2}x + 2$$

Distribute and simplify.

$$y = -\frac{1}{2}x + 1$$

Subtract 1 from both sides to isolate y.

Your Turn Write the equation of a line in slope-intercept form that passes through $(-3, 2)$ and is parallel to the graph of $y = \frac{2}{3}x - 5$.

Answer

$y = \frac{2}{3}x + 4$

OBJECTIVE 4. Write the equation of a line perpendicular to a given line. Consider the graphs of $y = \frac{2}{3}x - 4$ and $y = -\frac{3}{2}x + 1$.

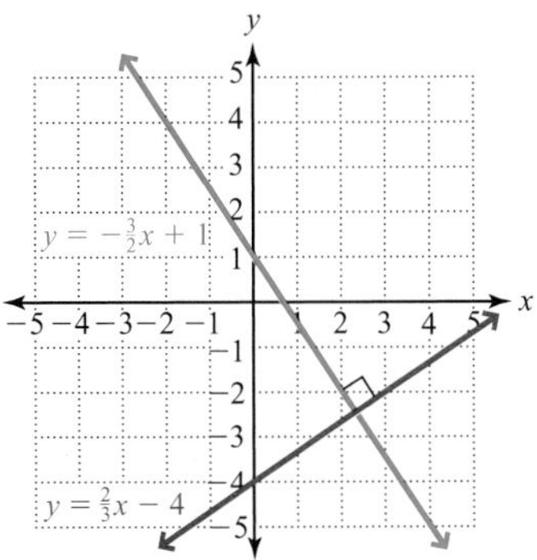

These two lines are perpendicular, which means that they intersect at a 90° angle. In the equations, we see that their slopes, $\frac{2}{3}$ and $-\frac{3}{2}$, are reciprocals with opposite signs, which suggests the following rule.

RULE **Perpendicular Lines**

The slope of a line perpendicular to a line with a slope of $\frac{a}{b}$ is $-\frac{b}{a}$.

EXAMPLE 8 Write the equation of a line in standard form that passes through (3, 1) and is perpendicular to the graph of $3x + 5y = 10$.

Solution First, we need to determine the slope of the given line, so we will write the given equation in slope-intercept form.

$$5y = -3x + 10 \quad \text{Subtract } 3x \text{ from both sides.}$$

$$y = -\frac{3}{5}x + 2 \quad \text{Divide both sides by 5 to get slope-intercept form.}$$

We see that the slope of the given line is $-\frac{3}{5}$, so the slope of the perpendicular line must be $\frac{5}{3}$. We can now write the equation of the perpendicular line.

$$y - 1 = \frac{5}{3}(x - 3) \quad \text{Using } y - y_1 = m(x - x_1)\text{, replace } m \text{ with } \frac{5}{3}, x_1 \text{ with 3, and } y_1 \text{ with 1.}$$

$$3y - 3 = 5x - 15 \quad \text{Multiply both sides by 3, simplify, and distribute.}$$

$$3y - 5x = -12 \quad \text{Subtract } 5x \text{ from and add 3 to both sides.}$$

$$-5x + 3y = -12 \quad \text{Rearrange the } x \text{ and } y \text{ terms.}$$

$$5x - 3y = 12 \quad \text{Multiply both sides by } -1.$$

YOUR TURN Write the equation of a line in standard form that passes through (−6, 2) and is perpendicular to the graph of $3x + y = -1$.

ANSWER

$x - 3y = -12$

3.3 Exercises

For Extra Help

MyMathLab

Videotape/DVT

InterAct Math

Math Tutor Center

Math XL.com

1. If you are given the slope and the y-intercept of a line and asked to write its equation, which form of the equation would be easier to use?

2. Describe the general process of writing the equation of a line if you are given two ordered pairs on it neither of which is its y-intercept.

3. How do you rewrite an equation that is in standard form in slope-intercept form?

4. How do you rewrite an equation that is in point-slope form in standard form?

5. Given two linear equations in two variables, how can you verify that their graphs are parallel?

6. What is the slope of a line perpendicular to a line with a slope of $\frac{a}{b}$?

For Exercises 7–12, use the given slope and coordinate of the y-intercept to write the equation of the line in slope-intercept form.

7. $m = -4; (0, 3)$

8. $m = 3; (0, -1)$

9. $m = \frac{3}{5}; (0, -2)$

10. $m = -\frac{1}{4}; \left(0, \frac{1}{2}\right)$

11. $m = -0.2; (0, -1.5)$

12. $m = 1.6; (0, 4.5)$

For Exercises 13–16, determine the slope and the y-intercept and write the equation of the line in slope-intercept form.

13.

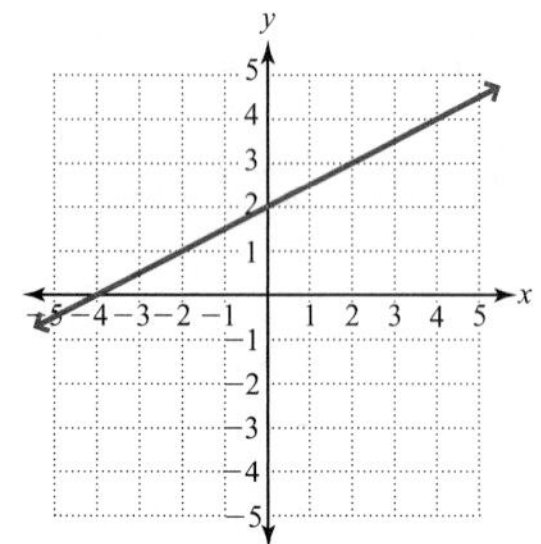

14.

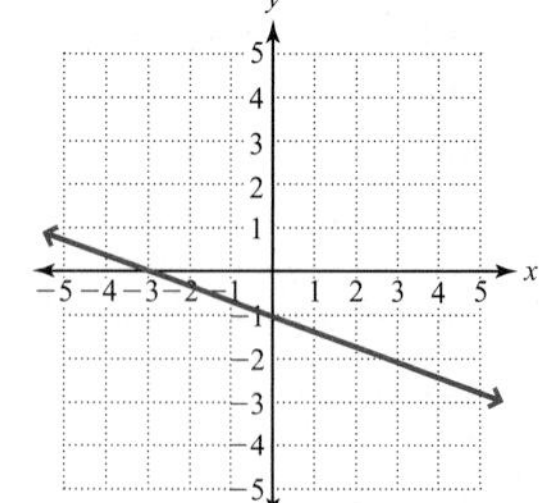

15.

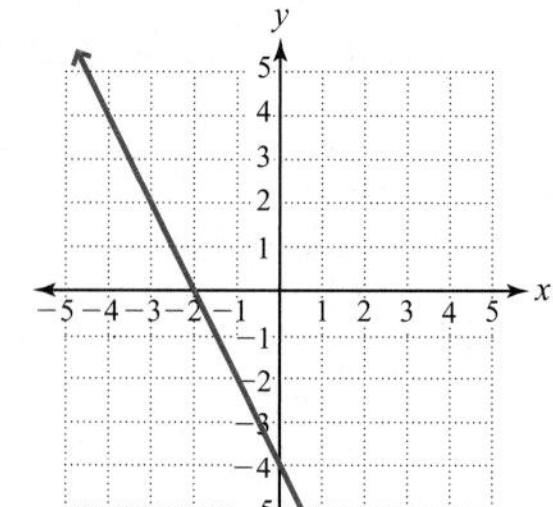

16.

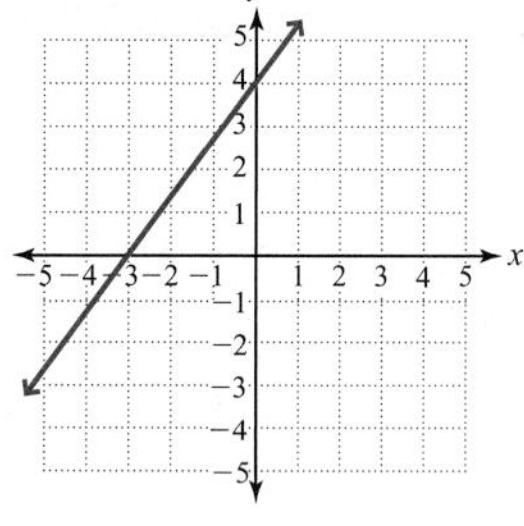

For Exercises 17–20, write the equation of the line with the given y-intercept and passing through the given point. Leave the answer in slope-intercept form.

17. $(0, 2), (3, 4)$

18. $(0, 4), (5, 7)$

19. $(0, -6), (-2, -3)$

20. $(0, -3), (4, -6)$

For Exercises 21–32, write the equation of a line in slope-intercept form with the given slope and passing through the given point.

21. $m = 2; (3, 1)$

22. $m = 3; (2, -5)$

23. $m = -2; (-5, 0)$

24. $m = -4; (-3, 0)$

25. $m = -1; (3, -4)$

26. $m = -1; (0, 0)$

27. $m = \frac{2}{5}; (0, -2)$

28. $m = \frac{2}{3}; (0, 3)$

29. $m = \frac{4}{5}; (-2, -2)$

30. $m = \frac{4}{3}; (-5, -9)$

31. $m = -\frac{3}{2}; (-1, 3)$

32. $m = -\frac{5}{2}; (9, 4)$

For Exercises 33–44: a. Write the equation of a line through the given points in slope-intercept form.
b. Write the equation in standard form, $Ax + By = C$, where A, B, and C are integers and $A > 0$.

33. $(4, 1), (1, -2)$

34. $(3, -4), (-1, 8)$

35. $(-1, -4), (2, -7)$

36. $(1, 3), (-2, -9)$

37. $(-6, -6), (3, 0)$

38. $(-4, -7), (2, 8)$

39. $(-5, 0), (0, 6)$

40. $(5, 2), (-5, -2)$

41. $(4, 2), (-9, 3)$

42. $(-9, 0), (4, -1)$

43. $(-2, 2), (-5, 9)$

44. $(-5, -2), (6, -4)$

For Exercises 45–56, determine whether the given lines are parallel, perpendicular, or neither.

45. $y = \frac{1}{3}x - 5$
$y = \frac{1}{3}x + 2$

46. $y = \frac{2}{5}x - 2$
$y = \frac{2}{5}x + 2$

47. $y = \frac{3}{4}x - 2$
$y = -\frac{4}{3}x + 4$

48. $y = -\frac{2}{5}x + 4$
$y = \frac{5}{2}x - 6$

49. $y = 5x + 1$
$y = -5x - 6$

50. $y = -\frac{1}{4}x + 5$
$y = -4x$

51. $2x + 7y = 8$
$4x + 14y = -9$

52. $3x - 4y = 7$
$-9x + 12y = -2$

53. $3x + 5y = 4$
$5x - 3y = 2$

54. $6x - 2y = 7$
$x + 3y = -5$

55. $x = 4$
$y = -2$

56. $y = 3$
$x = 3$

For Exercises 57–66: ***a. Write the equation in slope-intercept form of a line that passes through the given point and is parallel to the graph of the given equation.***
b. Write the equation in the form $Ax + By = C$, where A, B, and C are integers and $A > 0$.

57. $(2, -6); y = -5x + 3$

58. $(-1, 1); y = -3x + 1$

59. $(-5, -2); y = 4x - 2$

60. $(5, 7); y = 2x - 5$

61. $(3, -4); y = \frac{2}{3}x - 5$

62. $(-4, 2); y = \frac{5}{2}x - 4$

63. $(4, -3); 4x + 6y = 3$

64. $(1, 1); 3x - 4y = 8$

65. $(-3, 7); 2x + 5y - 30 = 0$

66. $(-2, -2); x - 3y + 9 = 0$

For Exercises 67–76: ***a. Write the equation of a line in slope-intercept form that passes through the given point and is perpendicular to the graph of the given equation.***
b. Write the equation in the form $Ax + By = C$, where A, B, and C are integers and $A > 0$.

67. $(2, -1); y = -\frac{1}{3}x + 6$

68. $(-1, 5); y = \frac{1}{2}x - 7$

69. $(0, -3); y = \frac{2}{5}x - 8$

70. $(3, 2); y = -\frac{3}{4}x + 1$

71. $(-2, -9); y = -3x + 4$

72. $(-3, -3); y = 2x - 5$

73. $(-2, -3); x + 4y = -10$

74. $(2, 8); 2y + x = 9$

75. $(1, 4); 2x - 3y = 15$

76. $(-1, -5); 3x + 2y = 1$

77. What is the equation of a horizontal line through $(0, -4)$?

78. What is the equation of a vertical line through $\left(\frac{3}{4}, 0\right)$?

79. What equation describes the x-axis?

80. What equation describes the y-axis?

81. On the following graph, p represents the U.S. resident population, in millions of people, and t represents the number of years after 2000. The graph shows that the U.S. resident population has increased from 282.4 million people in 2000 (year 0) to 295.5 million people in 2005 (year 5). (*Source:* U.S. Bureau of the Census)

a. Find the slope of the line.

b. Write the equation of the line in slope-intercept form.

c. If this trend continues, predict the U.S. resident population in 2012.

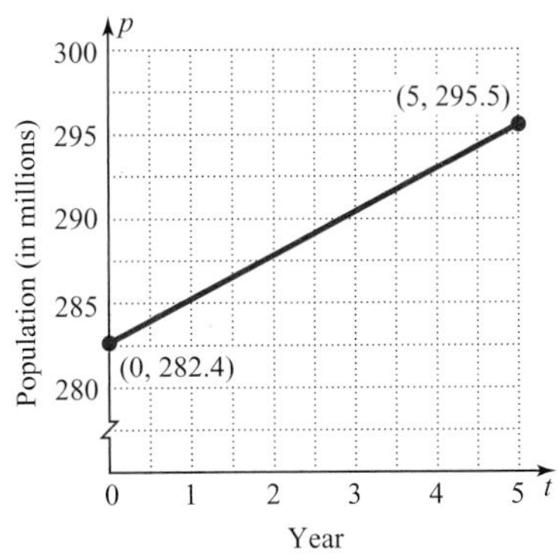

82. On the following graph, c represents the average monthly cell phone bill, in dollars, and t represents the number of years after 1999. The graph shows that the average monthly bill for U.S. cell phone subscribers increased from \$41.24 in 1999 (year 0) to \$50.64 in 2004 (year 5). (*Source:* Cellular Telecommunications and Internet Association)

a. Find the slope of the line.

b. Write the equation of the line in slope-intercept form.

c. If this trend continues, predict the average monthly bill for U.S. cell phone subscribers in 2011.

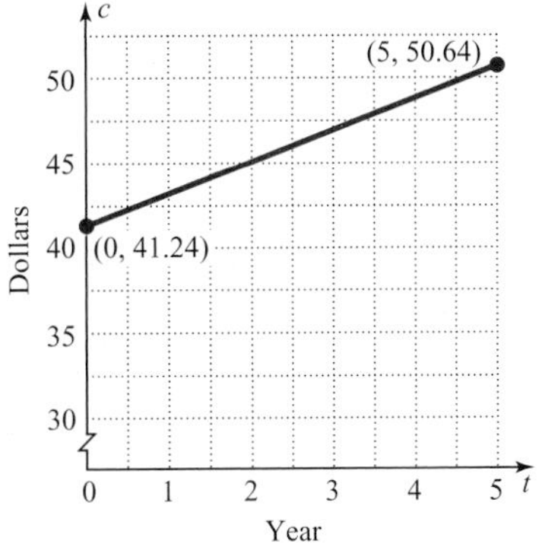

For Exercises 83–90, solve.

83. On the following graph, b represents the average number of barrels of oil produced by the U.S. per day, in thousands, and t represents the number of years after 1996. The graph shows that the U.S. production of crude oil decreased from an average of 6465 thousand barrels per day in 1996 to 5419 thousand barrels per day in 2004. (*Source:* Energy Information Administration, *Monthly Energy Review* July 2005)

a. Find the slope of the line.

b. Write the equation of the line in slope-intercept form.

c. If this trend continues, predict the average number of barrels of crude oil expected to be produced per day in the U.S. in 2012.

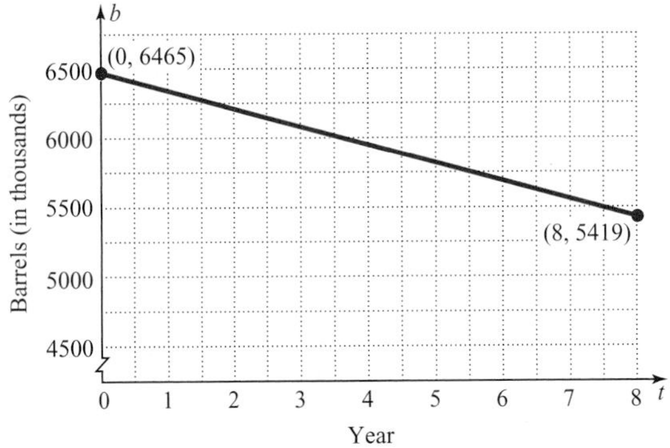

84. On the following graph, n represents the number of music CDs manufactured, in millions, and t represents the number of years after 1998. The graph shows that the number of music CDs manufactured decreased from 847 million in 1998 to 766.9 million in 2004. (*Source:* Recording Industry Association of America)

a. Find the slope of the line.

b. Write the equation of the line in slope-intercept form.

c. If this trend continues, predict the number of music CDs manufacturers are expected to ship in 2012.

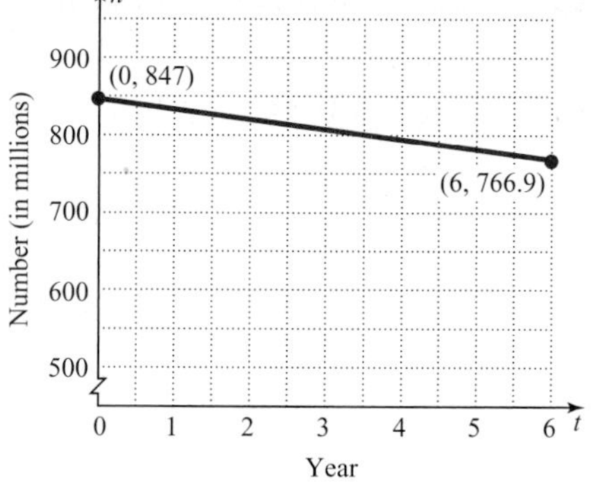

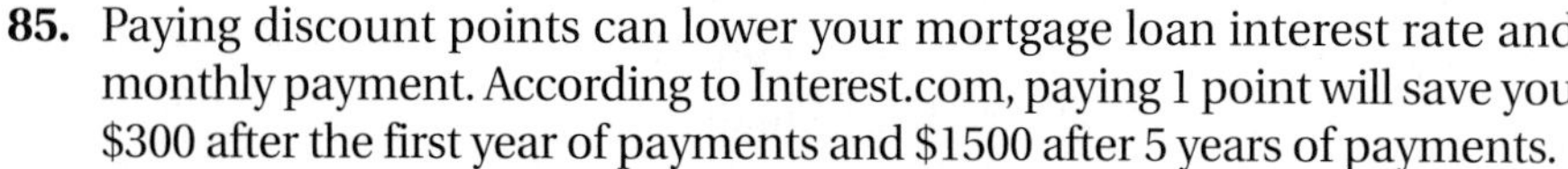

85. Paying discount points can lower your mortgage loan interest rate and monthly payment. According to Interest.com, paying 1 point will save you \$300 after the first year of payments and \$1500 after 5 years of payments.

a. Make a grid with the horizontal axis labeled t and the vertical axis labeled s so that t represents the number of years payments have been completed with 1 point and s represents the amount of money saved. Plot the two given data points and then connect them with a straight line. (Hint: $t = 1$ means payments have been made for 1 year.)

b. Find the slope of the line.

c. Write an equation of the line in slope-intercept form.

d. If the discount savings follow the same trend, predict the savings after 10 years.

86. The Consumer Price Index (CPI) compares the current value of a dollar against what the dollar would buy in a past year. Since 1967, the CPI has steadily risen in a linear pattern. Items that cost \$1.00 in 1967 cost \$5.62 in 2004. (*Source:* Bureau of Labor Statistics, Department of Labor)

a. Make a grid with the horizontal axis labeled t and the vertical axis labeled p so that t represents the number of years after 1967 ($t = 0$) and p represents the CPI. Plot the two given data points and then connect them with a straight line.

b. Find the slope of the line connecting the two data points given.

c. Write the equation of the line in slope-intercept form that connects the two data points given.

d. Using the equation from part b, predict the CPI in 2010.

87. The percentage of men 65 years of age or older who are in the labor force has steadily increased from 16.3% in 1990 to 18.6% in 2003. (*Source:* Bureau of Labor Statistics, Department of Labor)

a. Make a grid with the horizontal axis labeled t and the vertical axis labeled p so that t represents the number of years after 1990 ($t = 0$) and p represents the percentage of men 65 years of age or older who are in the labor force. Plot the two given data points and then connect them with a straight line.

b. Find the slope of the line connecting the two data points given.

c. Write the equation of the line in slope-intercept form that connects the two data points given.

d. Using the equation from part b, predict the percentage of men 65 years of age or older in the labor force in 2010.

88. In 1993, 22.8% of U.S. households had a computer. In 2003, the percent increased to 61.8% of U.S. households. (*Source:* U.S. Bureau of the Census)

a. Make a grid with the horizontal axis labeled t and the vertical axis labeled p so that t represents the number of years after 1993 and p represents the percent of U.S. households that had a computer. Plot the two given data points and then connect them with a straight line. (Hint: $t = 0$ means 1993 so that $t = 10$ means 2003.)

b. Find the slope of the line.

c. Write an equation of the line in slope-intercept form.

d. If this trend continues, predict the percent of U.S. households expected to have a computer in 2010.

89. In 1997, 24.7% of adults ages 18 and over in the United States were cigarette smokers. In 2004, the percent decreased to 20.9% of U.S. adults. (*Source:* Centers for Disease Control, National Center for Health Statistics)

a. Make a grid with the horizontal axis labeled t and the vertical axis labeled p so that t represents the number of years after 1997 ($t = 0$) and p represents the percentage of adults ages 18 and over in the U.S. who smoke. Plot the two given data points and then connect them with a straight line.

b. Find the slope of the line.

c. Write an equation of the line in slope-intercept form.

d. If this trend continues, predict the percent of the U.S. adults ages 18 and over expected to be cigarette smokers in 2015.

90. In 1990, 89.5% of high school seniors in the United States had consumed alcohol within 30 days of participating in the survey. In 2003, the percent decreased to 76.6%. (*Source:* Monitoring the Future, Univ. of Michigan Inst. for Social Research and National Inst. of Drug Abuse)

a. Make a grid with the horizontal axis labeled t and the vertical axis labeled p so that t represents the number of years after 1990 ($t = 0$) and p represents the percentage of high school seniors in the U.S. who consumed alcohol within 30 days of participating in the survey. Plot the two given data points and then connect them with a straight line.

b. Find the slope of the line.

c. Write an equation of the line in slope-intercept form.

d. If this trend continues, predict the percent of high school seniors that will consume alcohol in 2015.

91. Martha Stewart Living Omnimedia stock closed at \$14.46 on Wednesday, March 3, 2004. On Tuesday, March 9, 2004, its stock closed at a price of \$9.55. (*Source:* fool.com)

a. Make a grid with the horizontal axis labeled t and the vertical axis labeled p so that t represents the number of days of trading after March 3 ($t = 0$) and p represents the closing price of the stock in dollars. Plot the two given data points and then connect them with a straight line. (Hint: When numbering the days, remember that no trading occurs on Saturdays or Sundays.)

b. Find the slope of the line.

c. Write an equation of the line in slope-intercept form.

d. If the stock price continued falling along this line, what would be the stock price on March 12?

92. Nike, Inc. stock closed at \$88.50 on Friday, July 15, 2005 and at \$84.50 on Thursday, July 28, 2005. (*Source:* moneycentral.msn.com)

a. Make a grid with the horizontal axis labeled t and the vertical axis labeled p so that t represents the number of days of trading after July 15 ($t = 0$) and p represents the closing price of the stock in dollars. Plot the two given data points and then connect them with a straight line. (Hint: When numbering the days, remember that no trading occurs on Saturdays or Sundays.)

b. Find the slope of the line.

c. Write an equation of the line in slope-intercept form.

d. If the stock price continued falling along this line, what would be the stock price on August 5, 2005?

REVIEW EXERCISES

[1.1] **1.** Use <, >, or = to write a true statement: $-0.8 \blacksquare -\frac{1}{8}$

[2.1] ***For Exercises 2 and 3, solve and check.***

2. $4(2x + 3) - 2(x - 6) = 7 - (3x + 1)$

3. $\frac{1}{4}x - \frac{5}{6} = \frac{3}{2}x - 1$

[2.3] ***For Exercises 4 and 5, solve and graph the solution set. Then write the solution set in set-builder notation and in interval notation.***

4. $-4x - 7 \geq 13$

5. $6x - 2(x - 1) \geq 9x - 8$

6. It is suggested that a certain breed of dog have at least 2000 square feet of space to get adequate exercise. If the length of a fence must be fixed at 50 feet, what range of values can the width have so that the area is at least 2000 square feet?

3.4 Graphing Linear Inequalities

OBJECTIVE 1. Graph linear inequalities in the coordinate plane.

Now let's graph **linear inequalities in two variables**, such as $3x + 2y > 6$ or $y \leq x - 2$ in the coordinate plane.

DEFINITION ***Linear inequality in two variables:*** An inequality that can be written in the form $Ax + By > C$, where the inequality could also be $<$, $\leq$, or $\geq$.

Note: *Linear inequalities have the same form as linear equations except that they contain an inequality symbol instead of an equal sign.*

A solution to a linear inequality is an ordered pair that makes the inequality true. For example, the ordered pair (2, 4) is a solution for $3x + 2y > 6$:

$$3x + 2y > 6$$

$$3(2) + 2(4) \overset{?}{>} 6 \quad \text{Replace } x \text{ with 2 and } y \text{ with 4.}$$

$$14 > 6 \quad \text{This is true, so (2, 4) is a solution.}$$

Recall that the graph of an equation in the coordinate plane is its solution set. This is also true for inequalities. The solution set for a linear inequality, such as $3x + 2y \geq 6$, contains ordered pairs that satisfy $3x + 2y > 6$ *or* $3x + 2y = 6$. The graph of the equation $3x + 2y = 6$ is a line, which we can graph using intercepts (2, 0) and (0, 3). That line is called a *boundary* for the graph of the inequality and it separates the coordinate plane into two half planes. One of those half planes is the solution region for $3x + 2y > 6$.

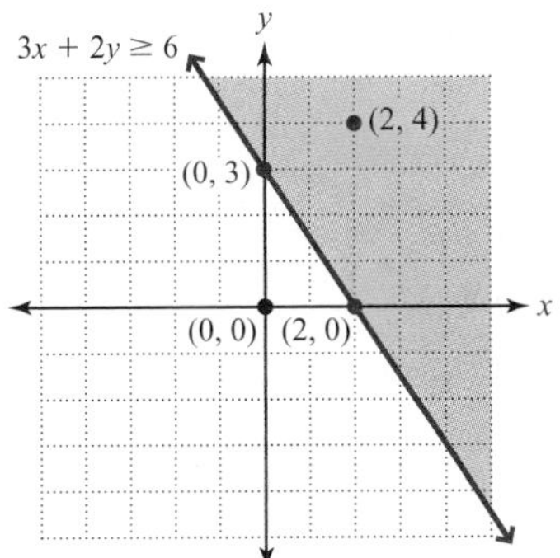

Earlier we showed that (2, 4) is a solution for $3x + 2y > 6$. In fact, every ordered pair in the half plane with (2, 4) is a solution for $3x + 2y > 6$, so we shade that region. Therefore, the solution set for $3x + 3y \geq 6$ contains every ordered pair on the line $3x + 2y = 6$ along with every ordered pair in the half plane containing (2, 4).

Ordered pairs in the other half plane do not satisfy the inequality, which is why it is not shaded. For example, consider (0, 0).

$$3(0) + 2(0) \overset{?}{\geq} 6$$

$$0 \geq 6 \quad \text{This inequality is false, so (0, 0) is not a solution.}$$

If the inequality were $3x + 2y > 6$, we shade the same region, but since the ordered pairs on the line $3x + 2y = 6$ are *not* in the solution set, we draw a dashed line instead of a solid line as shown to the right.

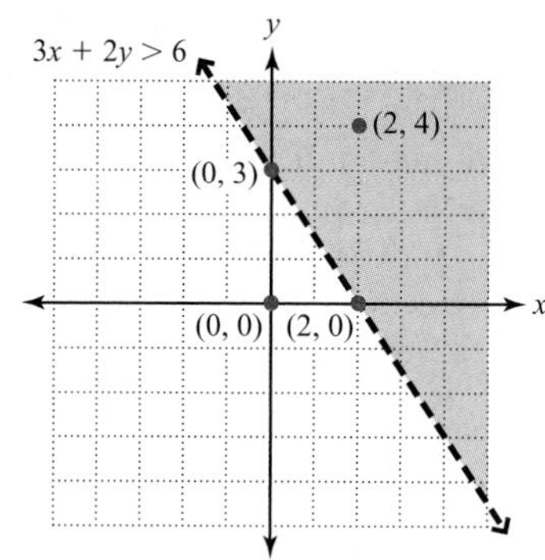

For the inequalities $3x + 2y \leq 6$ and $3x + 2y < 6$, we shade the half plane on the other side of the boundary line, which contains (0, 0).

The graph of $3x + 2y \leq 6$:

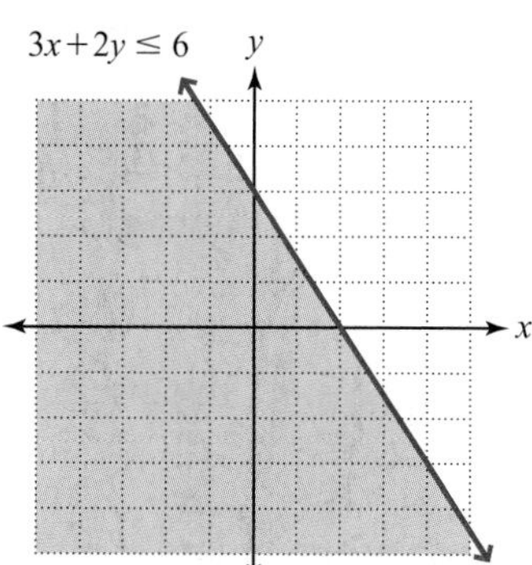

The graph of $3x + 2y < 6$:

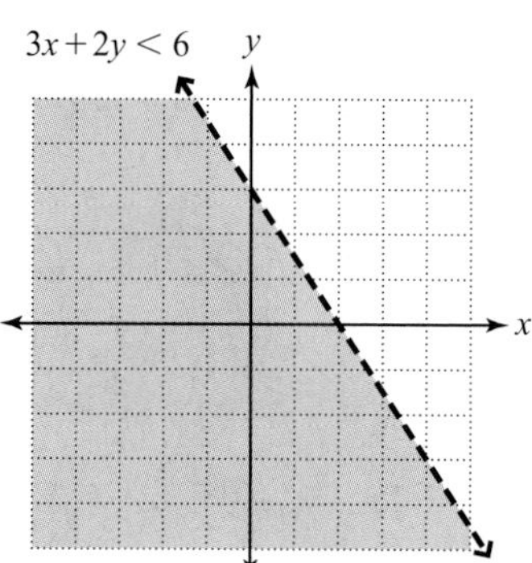

Connection When graphing a solution set on a number line for inequalities involving $\leq$ or $\geq$, we used brackets for end values. For inequalities involving $<$ or $>$, we used parentheses to show that the end values are not part of the solution set. When graphing solution sets for inequalities in the coordinate plane, a solid line is like a bracket, whereas a dashed line is like a parenthesis.

PROCEDURE **Graphing Linear Inequalities**

To graph a linear inequality,

1. Graph the related equation. The related equation has an equal sign in place of the inequality symbol. If the inequality symbol is $\leq$ or $\geq$, draw a solid boundary line. If the inequality symbol is $<$ or $>$, draw a dashed boundary line.
2. Choose an ordered pair on one side of the boundary line and test this ordered pair in the inequality. If the ordered pair satisfies the inequality, then shade the region that contains it. If the ordered pair does not satisfy the inequality, then shade the region on the other side of the boundary line.

EXAMPLE 1 Graph.

a. $y < -2x$

Solution First, graph the related equation $y = -2x$.

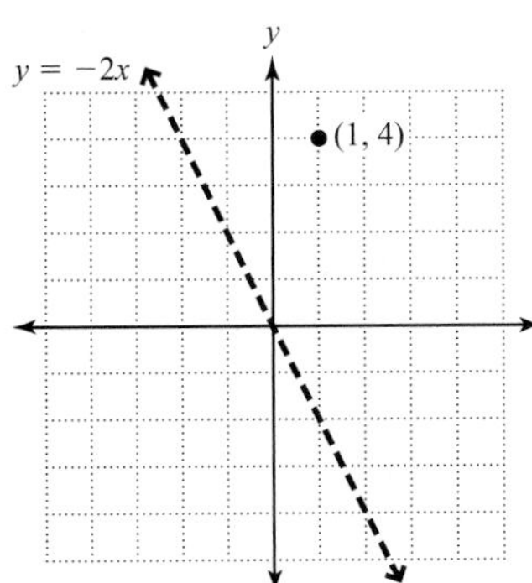

Since the inequality is $<$, we draw a dashed line to indicate that ordered pairs along the boundary line are not in the solution set.

Next we choose a point, (1, 4), to see if it satisfies the inequality.

$4 \stackrel{?}{<} -2(1)$ **Replace x with 1 and y with 4.**

$4 < -2$ **This inequality is false, so (1, 4) is not a solution.**

Since our chosen point (1, 4) did not satisfy the inequality, we shade the region on the other side of the boundary line.

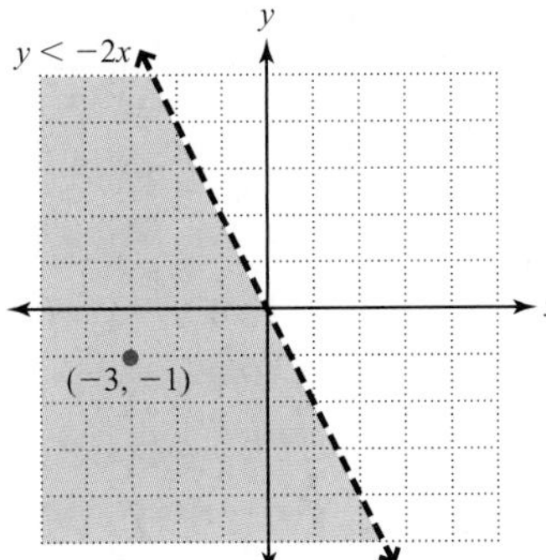

We can confirm our shading by checking an ordered pair in that region. We will choose $(-3, -1)$:

$-1 \stackrel{?}{<} -2(-3)$ **Replace x with -3 and y with -1.**

$-1 < 6$ **This inequality is true, confirming that we have shaded the correct half plane.**

b. $x - 3y \leq 6$

Solution

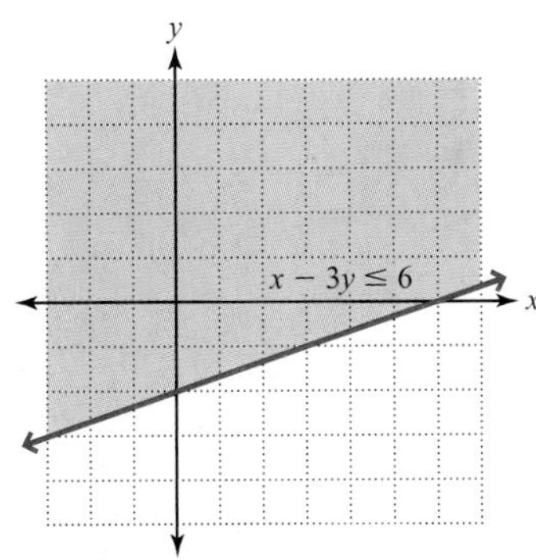

Since the inequality is $\leq$, when we graph the related equation $x - 3y = 6$, we draw a solid line to indicate that ordered pairs along the boundary line are in the solution set.

We will choose the origin (0, 0) as our test point.

$$0 - 3(0) \stackrel{?}{\leq} 6$$
$$0 \leq 6$$

This is true, so (0, 0) is a solution and we shade the half plane containing it.

YOUR TURN Graph.

$y < -\frac{2}{5}x + 3$

EXAMPLE 2 A drink stand in an amusement park sells two sizes of drink. The large size sells for \$3.00 and the smaller for \$2.00. The park management feels that the stand needs to have a total revenue from drink sales of at least \$600 each day to be profitable.

a. Write an inequality that describes the amount of revenue the stand must make to be profitable.
b. Graph the inequality.
c. Find two combinations of the number of large and small drinks that must be sold to be profitable.

a. We can use a table like those from Chapter 2.

Category	Price	Number Sold	Revenue
Large	3.00	x	$3x$
Small	2.00	y	$2y$

The total revenue would be found by the expression $3x + 2y$. If that total revenue must be at least 600, then we can write the following inequality:

$$3x + 2y \geq 600$$

b.

$3x + 2y \geq 600$

Note: *We show only the first quadrant because it is impossible to sell a negative number of drinks. The line represents every combination of numbers of large and small drinks sold that would produce exactly \$600 in revenue. The shaded region represents all combinations that produce revenue exceeding \$600.*

c. We assume that fractions of a particular size are not sold, so we will give only whole number combinations that are profitable. One combination is selling 200 large drinks and 0 small drinks, which gives exactly \$600. A second combination is selling 100 large drinks and 200 small drinks, which gives a total revenue of \$700.

ANSWERS

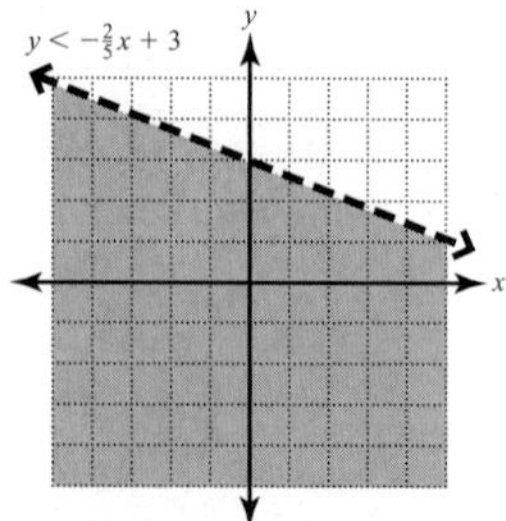

Calculator TIPS

Shading can be added to a graph on a graphing calculator. Using the [Y=] *key, enter the related equation for the given inequality. For* $y \geq -\frac{4}{5}x + 2$*, for example, at Y1 = , enter* $-\frac{4}{5}x + 2$ *by typing* [(] [−] [4] [÷] [5] [)] [X, T, θ, n] [+] [2]*. On that same line, use the arrow keys to move the cursor to the left of Y1. The slash you usually see there indicates the graph will be a solid line. Pressing* [ENTER] *will change this slash to other options for the graph. If the inequality is in the form* $y \geq mx + b$ *or* $y > mx + b$*, select* ◥*. If the inequality is* $y \leq mx + b$ *or* $y < mx + b$*, select* ◣*. For example, for* $y \geq -\frac{4}{5}x + 2$*, select* ◥*, then press* [GRAPH] *to see the screen to the right.*

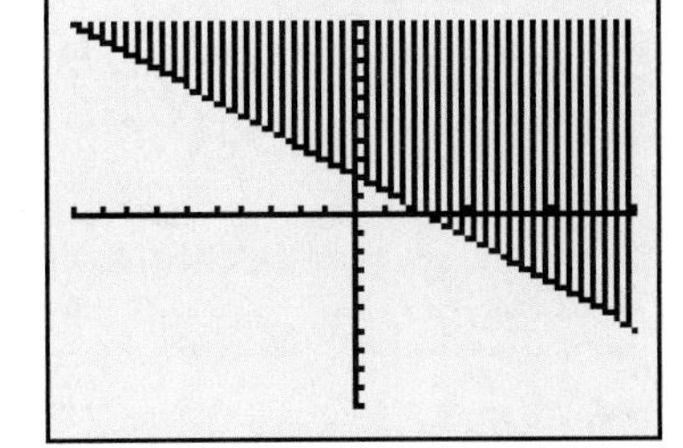

3.4 Exercises

For Extra Help: MyMathLab | Videotape/DVT | InterAct Math | Math Tutor Center | Math XL.com

1. When graphing a linear inequality in two variables, what is the first step?

2. How do you determine the related equation for a given inequality?

3. In graphing a linear inequality, when do you draw a dashed line and when do you draw a solid line?

4. When graphing a linear inequality, how do you determine which side of the boundary line to shade?

For Exercises 5–12, determine whether the ordered pair is a solution for the linear inequality.

5. $(-1, 3)$; $y \geq -x + 7$

6. $(-1, -1)$; $y > x + 15$

7. $(0, 0)$; $2x + 3y \geq 0$

8. $(7, 2)$; $3x - 2y > 4$

9. $(3, 4); 3y - x < 8$

10. $(0, -3); x - y \geq 5$

11. $(5, 3); y > \frac{3}{4}x - 5$

12. $(0, 1); y > \frac{2}{3}x + 4$

For Exercises 13–38, graph the linear inequality.

13. $y \leq -x + 3$

14. $y \leq x - 1$

15. $y \geq -4x + 6$

16. $y \geq -3x + 8$

17. $y > x$

18. $y > -x$

19. $y > -3x$

20. $y < 2x$

21. $y > \frac{2}{5}x$

22. $y < \frac{1}{4}x$

23. $x - y < 2$

24. $x - y < 5$

25. $x + 3y > -9$

26. $x + 4y < -8$

27. $x - 2y \geq -6$

28. $5x + 3y \geq -15$

29. $3x - 2y > 6$

30. $4x + y < 4$

31. $5x - y \leq 0$

32. $x + 7y \geq -14$

33. $4x + 2y \leq 3$

34. $5x + 2y < 9$

35. $x > 6$

36. $x \leq -1$

37. $y \leq 7$

38. $y > 5$

39. A company produces two versions of a game. The board version costs $10 per unit to produce and the video version costs $20 per unit to produce. Management plans for the cost of production to be a maximum of $250,000 for the first quarter. The inequality $10x + 20y \leq 250{,}000$ describes the total cost as prescribed by management.

a. What do x and y represent?

b. Graph the inequality. Since x and y represent positive numbers, the graph should be in the first quadrant only ($x \geq 0$ and $y \geq 0$).

c. What does the line represent?

d. What does the shaded region represent?

e. List two combinations of numbers of units produced of each version of the game that yield a total cost of $250,000.

f. List two combinations of numbers of units produced of each version of the game that yield a total cost less than $250,000.

g. In reality, is every combination of units produced represented by the line and shaded region a possibility? Explain.

40. A cosmetics company sells two different sizes of foundation. The revenue for each unit of the large bottle is \$5.00. The revenue for each unit of the small bottle is \$3.50. For the company to break even during the first quarter, the company must generate \$400,000 in revenue. The inequality $5x + 3.5y \geq 400{,}000$ describes the amount of revenue that must be generated from each size bottle in order to break even or turn a profit.

a. What do x and y represent?

b. Graph the inequality. Since x and y represent positive numbers, the graph should be in the first quadrant only ($x \geq 0$ and $y \geq 0$).

c. What does the boundary line represent?

d. What does the shaded region represent?

e. List two combinations of sales numbers of each size that allow the company to break even.

f. List two combinations of sales numbers of each size that allow the company to turn a profit.

g. In reality, is every combination of units sold represented by the line and shaded region a possibility? Explain.

41. A rectangular fence is to be designed so that the maximum perimeter is 200 feet.

a. Letting l represent the length and w the width, write an inequality in which the maximum perimeter is 200 feet.

b. Graph the inequality. Note that $l > 0$ and $w > 0$.

c. What does the line represent?

d. What does the shaded region represent?

e. Find two combinations of length and width that yield a perimeter of exactly 200 feet.

f. Find two combinations of length and width that yield a perimeter of less than 200 feet.

42. A landscaper plans a rectangular garden so that the maximum perimeter is 250 feet.

a. Letting l represent the length and w the width, write an inequality in which the maximum perimeter is 250 feet.

b. Graph the inequality. Note that $l > 0$ and $w > 0$.

c. What does the line represent?

d. What does the shaded region represent?

e. Find two combinations of length and width that yield a perimeter of exactly 250 feet.

f. Find two combinations of length and width that yield a perimeter of less than 250 feet.

43. Roberto is coordinating a fund-raiser for a high school marching band in which students sell two different boxes of fruit. A box of oranges sells for $12 and a box of grapefruits for $15. The goal is to raise at least $18,000.

a. Letting x represent the number of boxes of oranges sold and y represent the number of boxes of grapefruits sold, write an inequality in which the total sales is at least $18,000.

b. Graph the inequality. Note that $x \geq 0$ and $y \geq 0$.

c. What does the line represent?

d. What does the shaded region represent?

e. Find two combinations of fruit boxes sold that yield exactly $18,000 in total sales.

f. Find two combinations of fruit boxes sold that yield more than $18,000.

44. Veronica visits a home center to purchase some new border edging for her garden. She has a gift certificate for $75. Edging comes in two styles: straight edge pieces, which cost $3.50 each, and scalloped edge pieces which cost $5.50 each.

a. Letting x represent the number of straight edge pieces and y represent the number of scalloped edge pieces, write an inequality that has her total cost within the amount of the gift certificate.

b. Graph the inequality. Note that $x \geq 0$ and $y \geq 0$.

c. What does the line represent?

d. What does the shaded region represent?

e. Find two combinations of containers she could purchase.

f. In reality, are there any combinations that she could purchase that would yield a total in the exact amount of the gift certificate?

REVIEW EXERCISES

[1.4] **1.** Evaluate $x^2 - 4x + 1$ when $x = -2$.

[1.4] **2.** Find every value for x that makes $\frac{x}{x-3}$ undefined.

[2.1] **3.** Solve and check: $3x - 4(x + 8) = 5(x - 1) + 21$

[3.1] **4.** Find the x- and y-intercepts for $4x - 5y = 8$.

[1.1] ***For Exercises 5 and 6, use the ordered pairs:* $(-2, 7), (0, 1), (3, 5)$, *and* $(4, 2)$.**

5. Write a set containing all the x-coordinates.

6. Write a set containing all the y-coordinates.

3.5 Introduction to Functions and Function Notation

OBJECTIVES

1. **Identify the domain and range of a relation and determine whether a relation is a function.**
2. **Find the value of a function.**
3. **Graph functions.**

OBJECTIVE 1. Identify the domain and range of a relation and determine whether a relation is a function. We have used ordered pairs to generate graphs of equations and inequalities in two variables. Ordered pairs are the foundation of two important concepts in mathematics, *relations* and *functions*. First, we consider **relations**.

DEFINITION ***Relation:*** A set of ordered pairs.

For example, the following relation contains ordered pairs that track the price of a share of a particular stock at the close of the stock market each day for a week.

Day	Closing Price
1	\$6.50
2	\$8.00
3	\$10.00
4	\$9.25
5	\$8.00

Because a relation involves ordered pairs such as (4, 9.25), it has two sets of values called the **domain** and **range**.

DEFINITIONS ***Domain:*** A set containing initial values of a relation; its input values; first coordinates in ordered pairs.
Range: A set containing all values that are paired to domain values in a relation; its output values; second coordinates in ordered pairs.

For example, the domain for our stock price relation contains the number of each day, so the domain is {1, 2, 3, 4, 5}. The range contains all the prices, so the range is {6.50, 8.00, 9.25, 10.00}.

Note that each day has exactly one closing price, which makes this relation a **function**.

DEFINITION ***Function:*** A relation in which each value in the domain is assigned to exactly one value in the range.

Note: *Every input in a function has exactly one output.*

From our stock price example, we can see that a function can have different values in the domain assigned to the same value in the range. For example, domain values 2 and 5 each correspond to the same range value, 8.00.

Using arrows to map the domain values to the range values can be a helpful technique in determining whether a relation is a function. Such a map might look like this:

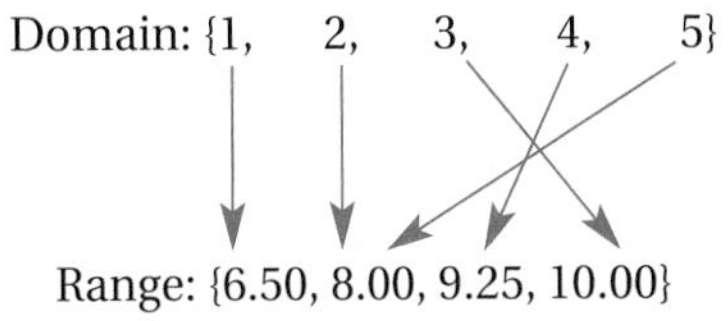

Domain: {1, 2, 3, 4, 5}

Range: {6.50, 8.00, 9.25, 10.00}

Note: *Each element in the domain has a single arrow pointing to an element in the range.*

A relation is not a function if any value in the domain is assigned to more than one value in the range. For example, suppose we have a domain that contains some players on the 2005 University of Southern California football team and a range containing the positions they play. The following relation shows their position assignments:

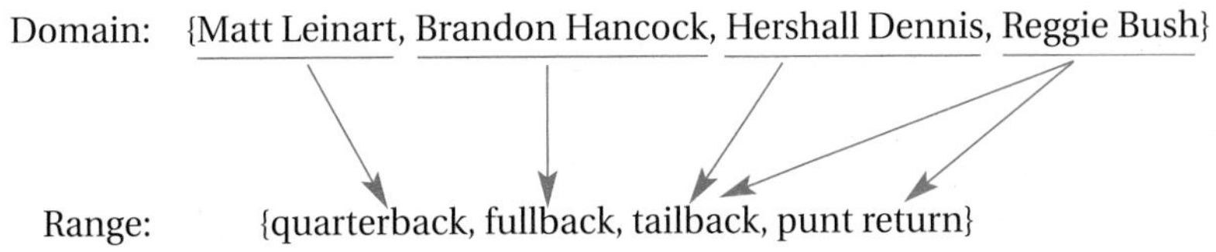

Domain: {Matt Leinart, Brandon Hancock, Hershall Dennis, Reggie Bush}

Range: {quarterback, fullback, tailback, punt return}

Note: *We now have an element in the domain, Reggie Bush, assigned to two elements in the range, so the relation is not a function.*

Conclusion If any value in the domain is assigned to more than one value in the range, then the relation is not a function.

EXAMPLE 1 Identify the domain and range of the relation, then determine if it is a function.

Note: *In this text, relations in table form will always have the domain in the left column and the range in the right column.*

a. Birthrate in the United States for each year:

Year	Rate per 1000 People
1998	14.6
1999	14.5
2000	14.7
2001	14.1
2002	13.9
2003	14.1

Source: Centers for Disease Control and Prevention, National Center for Health Statistics

Solution Domain: {1998, 1999, 2000, 2001, 2002, 2003}
Range: {13.9, 14.1, 14.5, 14.6, 14.7}

The relation is a function because every element in the domain is paired with exactly one element in the range.

b. Results of the 2005 U.S. Open Golf Tournament:

Place	Player
1	Michael Campbell
2	Tiger Woods
3	Sergio Garcia, Tim Clarke, Mark Hensby (three-way tie)

Solution Domain: {1, 2, 3}
Range: {Michael Campbell, Tiger Woods, Sergio Garcia, Tim Clarke, Mark Hensby}

The relation is not a function because an element in the domain, 3, is paired with more than one element in the range: Sergio Garcia, Tim Clarke, *and* Mark Hensby.

YOUR TURN Identify the domain and range of the relation and determine if it is a function.

a. Unemployment rate as a percent of the population from January to June of 2005:

Month	Unemployment Rate
1	5.2
2	5.4
3	5.2
4	5.2
5	5.1
6	5.0

Source: U.S. Department of Labor, Bureau of Labor Statistics

b. Player assignments for each position on the 2005 St. Louis Cardinals roster:

Position	Player
Left field	John Mabry
Center field	John Edmonds, So Taguchi
Right field	Larry Walker, So Taguchi

ANSWERS

a. Domain: {1, 2, 3, 4, 5, 6}
Range: {5.0, 5.1, 5.2, 5.4}
It is a function.

b. Domain: {left field, center field, right field}
Range: {John Mabry, John Edmonds, So Taguchi, Larry Walker}
It is not a function.

Since relations involve ordered pairs, we sometimes see them as graphs in the coordinate plane. Consider a graph of the stock price relation that we explored earlier.

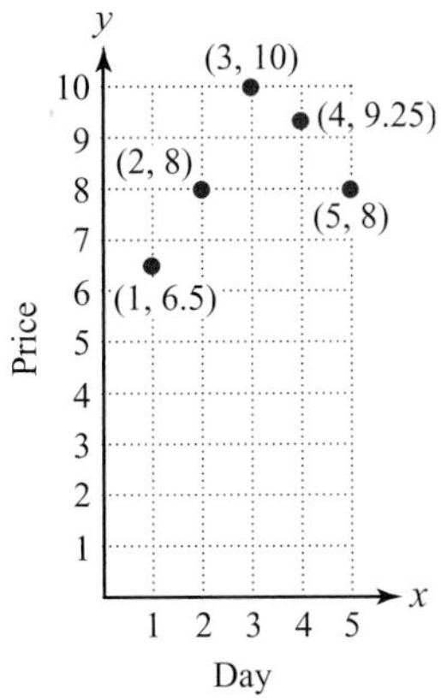

Notice the domain is plotted along the horizontal axis and the range is plotted along the vertical axis. On a graph in the standard x- and y-coordinate plane, the domain will be the x-coordinates and the range will be the y-coordinates.

PROCEDURE **Determining the Domain and Range of a Graph**

The domain is a set containing the first coordinate (x-coordinate) of every point on the graph.
The range is a set containing the second coordinate (y-coordinate) of every point on the graph.

To determine whether a graphed relation is a function, we can perform the *vertical line test.* Since every value in the domain of a function corresponds to exactly one value in the range, on a function's graph every x-coordinate will correspond to exactly one y-coordinate. This means a vertical line drawn through each x-coordinate in the domain of a function will intersect the graph at only one point.

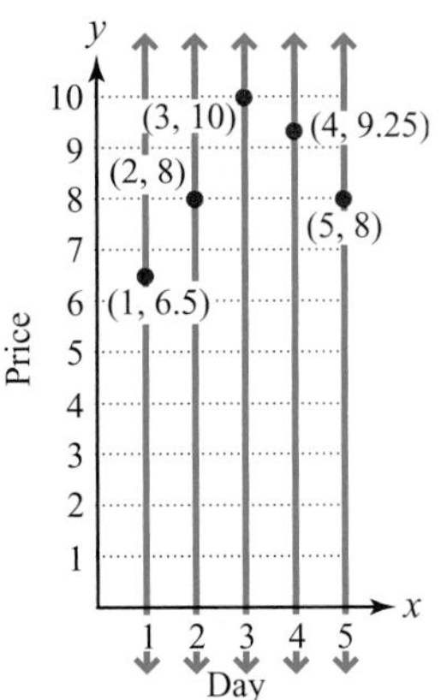

For example, in our graph of the stock price relation, we have drawn vertical lines, shown in blue, through each x-coordinate in the domain. Notice every one of those vertical lines intersects only one point, so the relation is a function.

PROCEDURE **Vertical Line Test**

To determine whether a graphical relation is a function, draw or imagine vertical lines through each value in the domain. If each vertical line intersects the graph at only one point, the relation is a function. If any vertical line intersects the graph more than once, the relation is not a function.

EXAMPLE 2 Identify the domain and range of the relation, then determine if it is a function.

a. Seattle's rain accumulation for January:

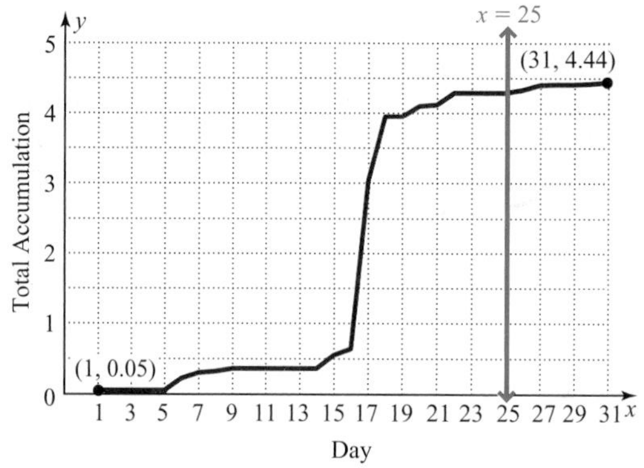

Source: National Weather Service, Sea-Tac International Airport Station

Solution Domain: $\{t | 1 \le t \le 31\}$
Range: $\{r | 0.05 \le r \le 4.44\}$

This relation is a function because a vertical line placed at each t-value would intersect the graph at only one point. We have shown one such vertical line passing through day 25 in blue.

b.

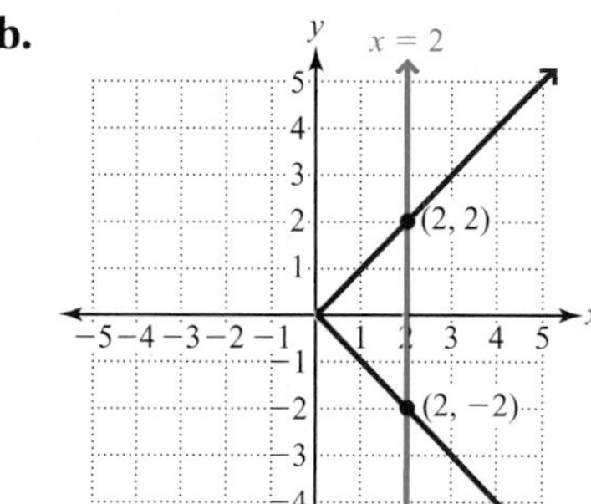

Solution Domain: $\{x | 0 \le x < \infty\}$
Range: $\mathbb{R}$

This relation is not a function because there are values in the domain that correspond to two values in the range. We can see this with the vertical line test. For example, $x = 2$ (shown in blue) passes through two different points on the graph, (2, 2) and (2, −2).

c. The following relation is from a survey that shows the percent of travelers that choose to use their personal vehicle according to trip distance.

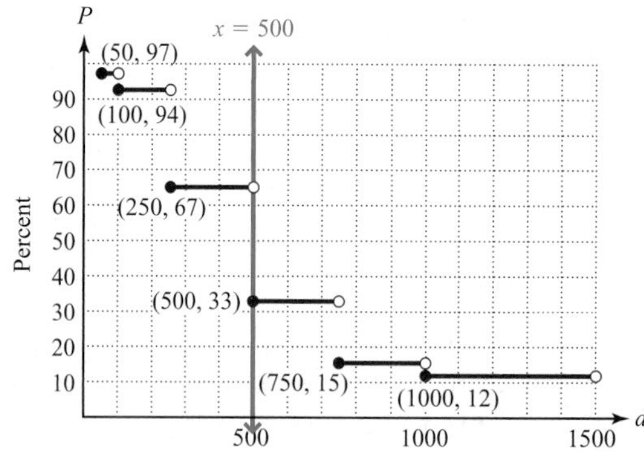

Solution

Domain: $\{x | 50 \le x < 1500\}$
Range: {12%, 15%, 33%, 67%, 94%, 97%}

This relation is a function. Remember that solid circles indicate values that are in the set, whereas open circles are not. Notice that each open circle and solid circle align vertically so that a vertical line through them only intersects the graph at one point, the solid circle. For example, $x = 500$ (shown in blue) intersects the graph only at the solid circle at 33%.

YOUR TURN Identify the domain and range of the relation, then determine if it is a function.

a.

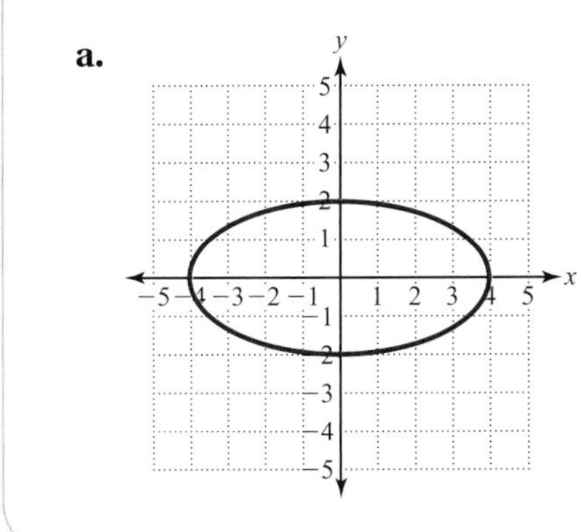

b.

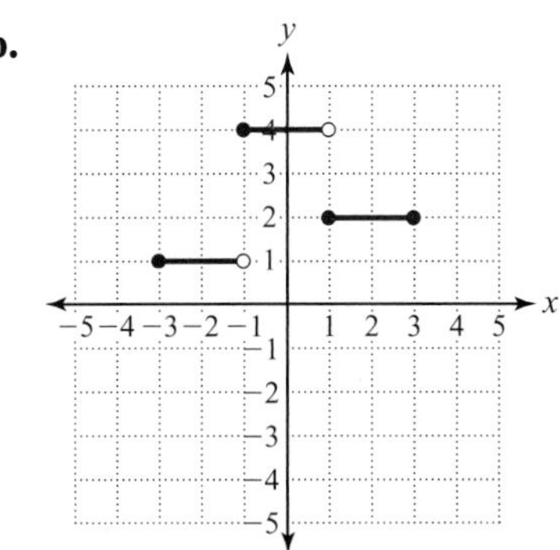

ANSWERS

a. Domain: $\{x \mid -4 \le x \le 4\}$; Range: $\{y \mid -2 \le y \le 2\}$

It is not a function.

b. Domain: $\{x \mid -3 \le x \le 3\}$; Range: $\{1, 2, 4\}$

It is a function.

OBJECTIVE 2. Find the value of a function. We have seen functions as paired sets and graphs, but they can also take the form of equations. In equation form, functions are essentially the same as equations in two variables. For example, we can write the equation $y = 3x + 1$ as $f(x) = 3x + 1$ in function notation. Notice that the notation $f(x)$, which is read "a function in terms of x," or "f of x," takes the place of the variable y. The following table shows some equations in two variables and their corresponding function form:

Equation in two variables:	Function notation:
$y = -2x + 3$	$f(x) = -2x + 3$
$y = \frac{3}{4}x - 1$	$f(x) = \frac{3}{4}x - 1$

We can think of the x-values in equations like $y = 3x + 1$ as inputs and the y-values as outputs. Similarly, the domain's x-values in $f(x) = 3x + 1$ are the inputs and the resulting function values, $f(x)$, are its outputs (elements of the range). Compare the following tables:

x	y
0	$3(0) + 1 = 1$
1	$3(1) + 1 = 4$
2	$3(2) + 1 = 7$

x	$f(x)$
0	$3(0) + 1 = 1$
1	$3(1) + 1 = 4$
2	$3(2) + 1 = 7$

Both tables list the same information. Finding the value of a function is the same as finding the y-coordinate of an equation in two variables that is solved for y. In both cases, we input a value for x and calculate the output value, which is either y or $f(x)$, depending on the form.

Function notation offers a clever way to indicate that a specific x-value is to be used. For the function $f(x) = 3x + 1$, we can indicate that we want the value of the function when $x = 2$ by writing "find $f(2)$."

PROCEDURE **Finding the Value of a Function**

Given a function $f(x)$, to find $f(a)$, where a is a real number in the domain of f, replace x in the function with a, then simplify.

EXAMPLE 3 **a.** For the function $f(x) = -5x + 3$, find $f(4)$.

Connection For the function in part a, we found the ordered pair $(4, -17)$. For part b, the ordered pair is $(n, 4n^2 - 3)$.

Solution $f(4) = -5(4) + 3$ Replace x with 4, then simplify.

$= -20 + 3$

$= -17$

b. For the function $f(x) = 4x^2 - 3$, find $f(n)$.

Solution $f(n) = 4(n)^2 - 3$ Replace x with n, then simplify.

$= 4n^2 - 3$ Square n.

YOUR TURN

a. For the function $f(x) = 0.5x - 2$, find $f(-6)$.

b. For the function $f(x) = x^2 - 4x - 3$, find $f(a)$.

ANSWERS

a. -5 **b.** $a^2 - 4a - 3$

Connection Finding the value of a function is like evaluating an expression. Following are function and expression forms that indicate essentially the same procedure:

Function language:	**Expression language:**
If $f(x) = 3x$, find $f(2)$.	Evaluate the expression $3x$ when $x = 2$.

We can also find the value of a function given its graph. For a given value in the domain (x-value), we look on the graph for the corresponding value in the range (y-value).

EXAMPLE 4 Use the graph to find the indicated value of the function.

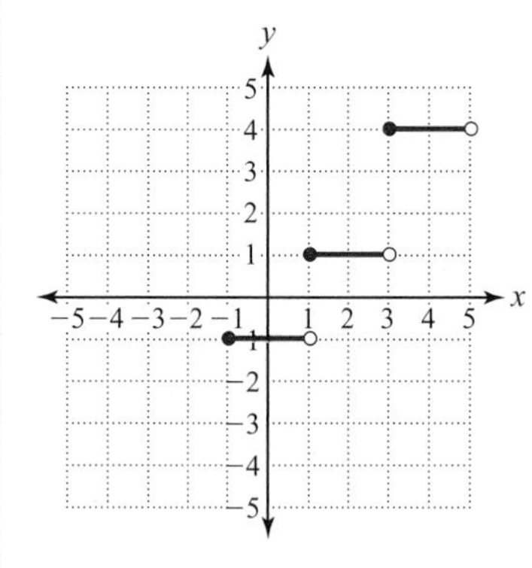

a. $f(0)$

Solution The notation $f(0)$ means to find the value of the function (y-value) when $x = 0$. On the graph we see that when $x = 0$, the corresponding y-value is -1, so we say $f(0) = -1$.

b. $f(3)$

Solution When $x = 3$, we see that $y = 4$, so $f(3) = 4$.

c. $f(-4)$

Solution The domain for this function is $\{x \mid -1 \le x < 5\}$. Since -4 is not in the domain, we say that $f(-4)$ is undefined.

YOUR TURN Use the graph to find the indicated value of the function.

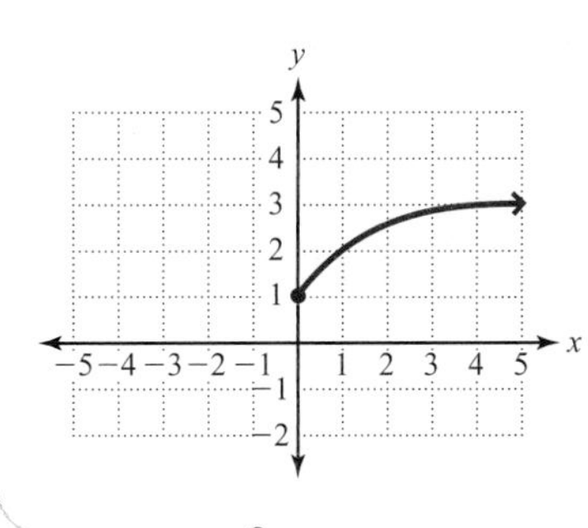

a. $f(-2)$ **b.** $f(0)$ **c.** $f(5)$

ANSWERS

a. undefined **b.** 1 **c.** 3

OBJECTIVE 3. Graph functions. We create the graph of a function the same way that we create the graph of an equation in two variables. First we consider linear functions.

Linear Functions

Linear functions are similar to linear equations in slope-intercept form.

Slope-intercept form: $y = mx + b$
Linear function: $f(x) = mx + b$

The graph of a linear function is a straight line with slope m and y-intercept $(0, b)$.

EXAMPLE 5 Graph. $f(x) = -2x + 3$

Solution Think of the function as the equation $y = -2x + 3$. We could make a table of ordered pairs or use the fact that the slope is -2 and the y-intercept is 3.

Table of ordered pairs:

x	f(x)
0	3
1	1
2	−1

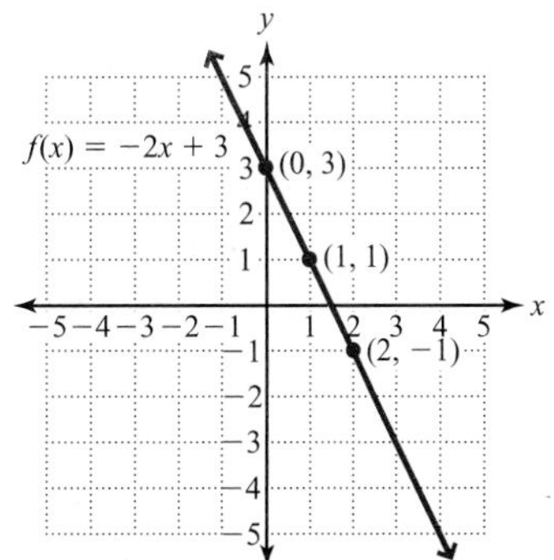

ANSWER

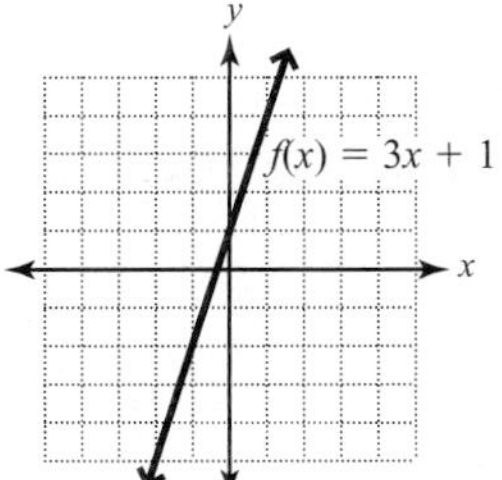

YOUR TURN Graph. $f(x) = 3x + 1$

EXAMPLE 6 An electric and gas company charges \$5 per month plus \$1.1245 per cubic foot of natural gas used. The function $c(v) = 1.1245v + 5$ describes a customer's monthly cost for natural gas, where v represents the volume in cubic feet of natural gas used during the month.

a. Graph the function.

b. Find the cost if a customer uses 24 cubic feet of natural gas.

Solution

a.

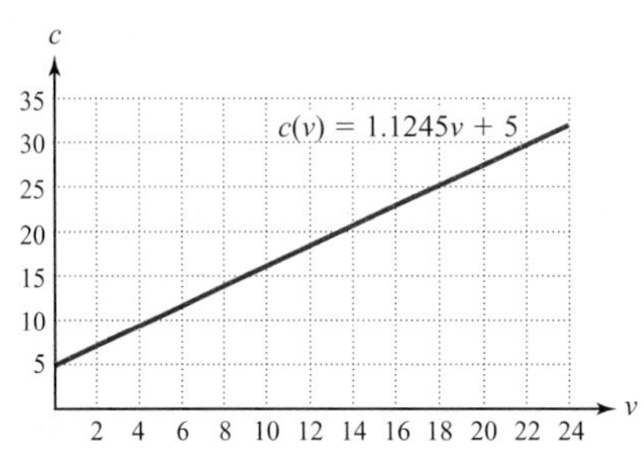

Note: *The slope of the line, 1.1245, is the cost per cubic foot of natural gas used each month. The y-intercept (or c-intercept in this case) is the \$5 flat fee.*

b. Finding the cost for using 24 cubic feet means we are evaluating $c(24)$.

$$c(24) = 1.1245(24) + 5 = 31.988$$

Rounding to the nearest cent, the total cost is \$31.99.

Nonlinear Functions

Graphs of nonlinear functions are not straight lines. We will explore nonlinear functions in greater detail later. However, we can graph some simple nonlinear functions by finding enough ordered pairs to generate a rough sketch of the graph.

EXAMPLE 7 Graph. $f(x) = x^2$

Solution We create a table of ordered pairs, plot the points, and connect them with a smooth curve.

x	f(x)
−2	4
−1	1
0	0
1	1
2	4

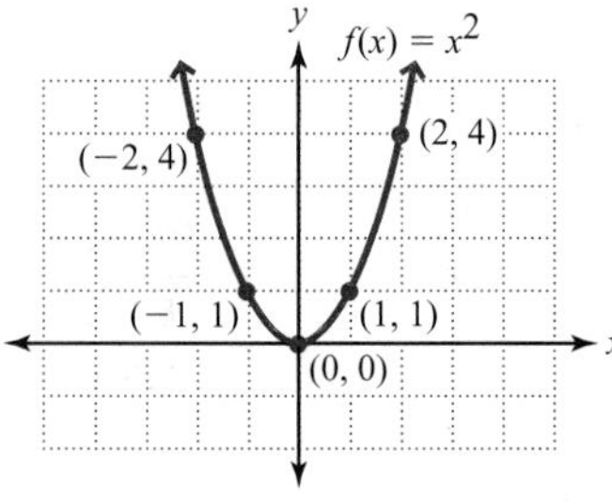

Note: *The graph of $f(x) = x^2$ is called a parabola.*

YOUR TURN Graph. $f(x) = x^2 + 1$

ANSWER

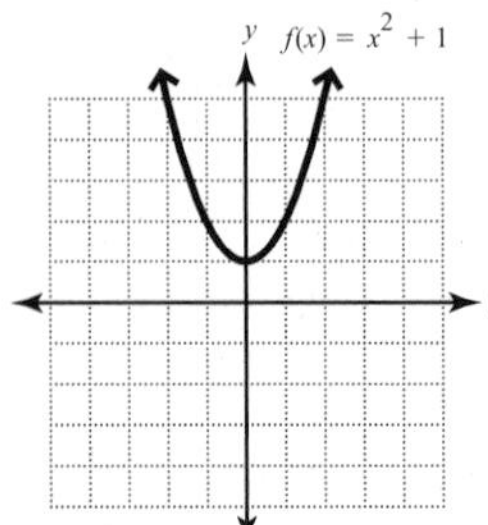

3.5 Exercises

For Extra Help

MyMathLab

Videotape/DVT

InterAct Math

Math Tutor Center

Math XL.com

1. What is the domain of a relation?

2. What is the range of a relation?

3. What makes a relation a function?

4. When using the vertical line test on a graph, how do you determine if the graph is the graph of a function?

5. Explain how to find the value of a function.

6. Is every linear equation of the form $y = mx + b$ a function?

For Exercises 7–24, identify the domain and range of the relation, then determine if it is a function.

7. The five most expensive college tuitions according to the *Chronicle of Higher Education*:

Institution	Tuition
Landmark College	$36,750
Sarah Lawrence College	$32,416
Kenyon College	$32,170
Trinity College	$31,940
George Washington University	$31,710

8. The five most populated cities in 2003:

City	Population
New York	8,085,742
Los Angeles	3,819,951
Chicago	2,869,121
Houston	2,009,690
Philadelphia	1,479,339

Source: Bureau of the Census, U.S. Dept. of Commerce

9. Superbowl wins as of February 2005: (*Source:* nfl.com)

Number of Wins	NFL Team
5	San Francisco 49ers, Dallas Cowboys
4	Pittsburgh Steelers
3	Green Bay Packers, New England Patriots, Oakland/L. A. Raiders, Washington Redskins
2	New York Giants, Miami Dolphins, Denver Broncos
1	Baltimore Ravens, Chicago Bears, New York Jets, Tampa Bay Buccaneers, Baltimore Colts, Kansas City Chiefs, St. Louis/L. A. Rams

10. College football national champions from 2000 to 2004:

Year	Team
2000	Oklahoma
2001	Miami (FL)
2002	Ohio State
2003	Louisiana State, Southern California (split title)
2004	Southern California

11. Money sources for purchase of second homes:

Percentage	Where the Money Comes From
41%	Savings
16%	Sale of stock or bonds
5%	Equity from other homes
5%	Financial institution loan
4%	Inheritance

Source: National Association of Realtors

12. Network prime-time averages for the week of 7/25/05–7/31/05:

Rank	Show
1	*CSI*
2	*Without a Trace*
3	*CSI:Miami*
4	*Two and a Half Men*
5	*Law and Order, NCIS* (two-way tie)
7	*Law and Order: Criminal Intent*
8	*Law and Order: SVU, Primetime* (two-way tie)
10	*Everybody Loves Raymond*

Source: Nielsen MediaResearch

13. The top five hardcover fiction bestsellers as of 7/31/05 (*Source: The New York Times*):

1. *Lifeguard* by James Patterson
2. *The Historian* by Elizabeth Kostova
3. *The Da Vinci Code* by Dan Brown
4. *The Interruption of Everything* by Terry McMillan
5. *Until I Find You* by John Irving

14. The top movies as of 6/19/05 (*Source:* Box Office Mojo):

1. *Batman Begins*
2. *Mr. and Mrs. Smith*
3. *Madagascar*
4. *Star Wars Episode III: Revenge of the Sith*

15. The graph shows the number of families helped by Harvest Hope Food Bank in a group of 18 counties during each year.

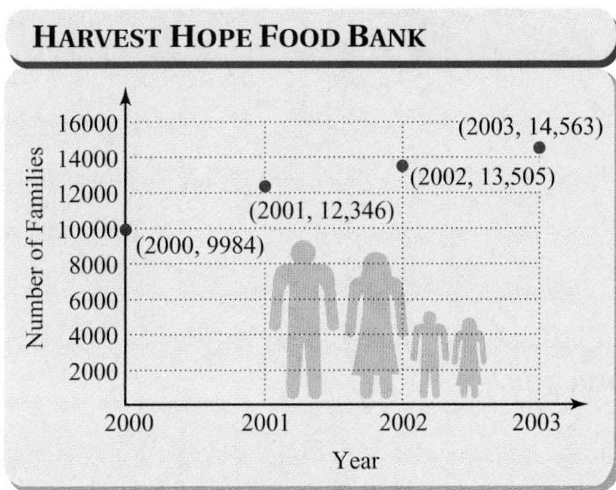

Source: Harvest Hope Food Bank; U.S. Dept. of Labor Bureau of Labor Statistics

16. The graph shows the number of motorcyclist fatalities in the United States for each year.

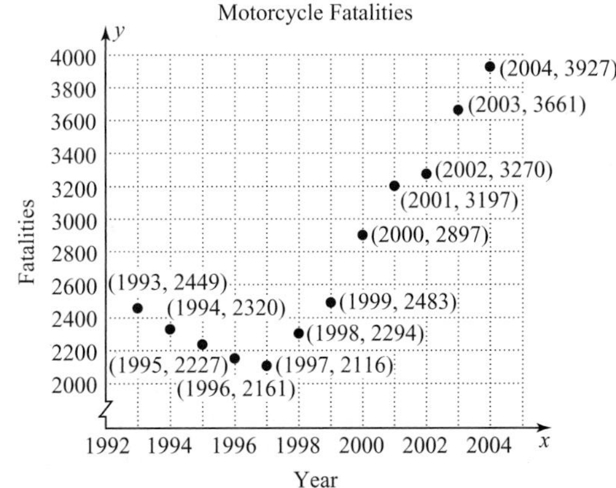

Source: U.S. Department of Transportation, National Highway Traffic Safety Administration

17.

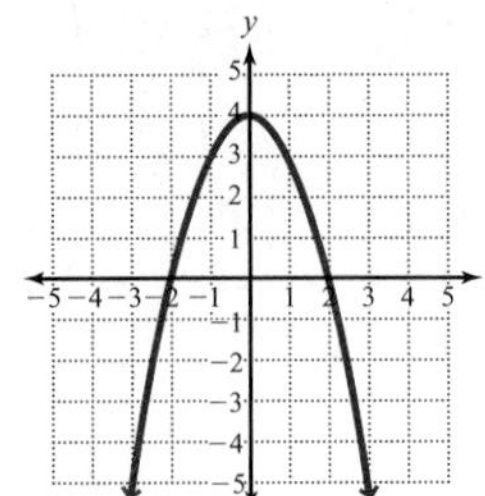

18.

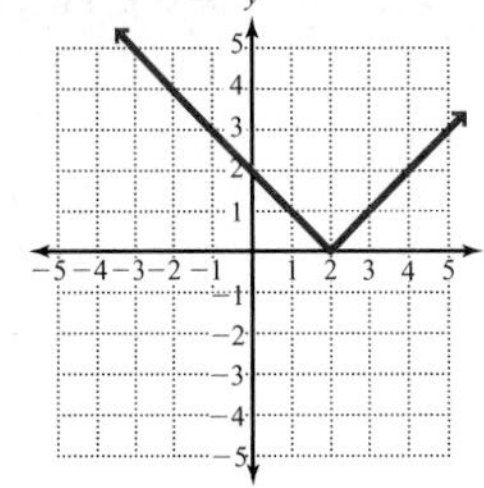

19.

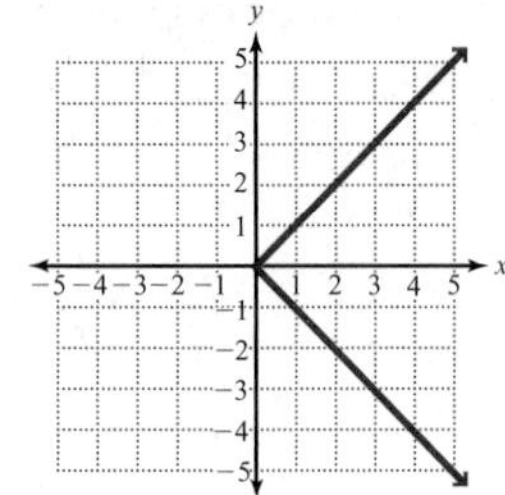

20.

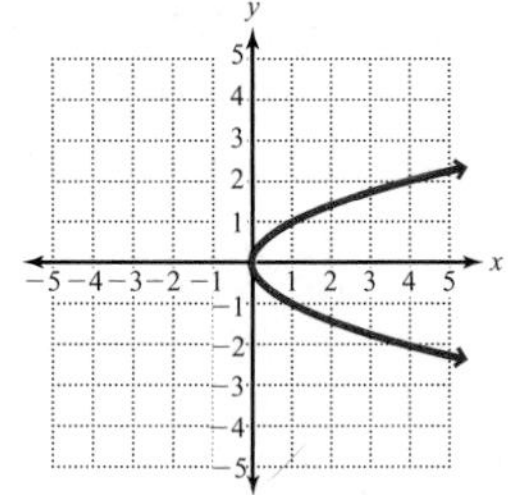

21.

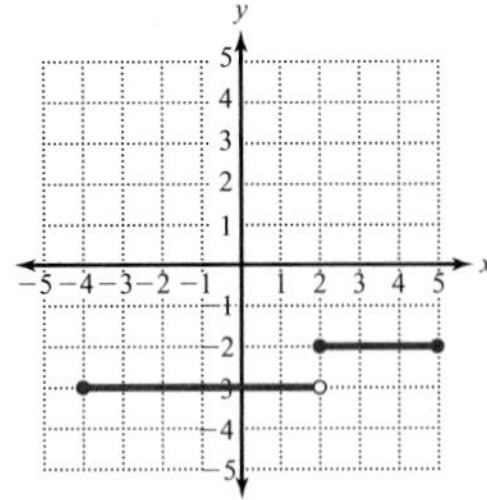

22.

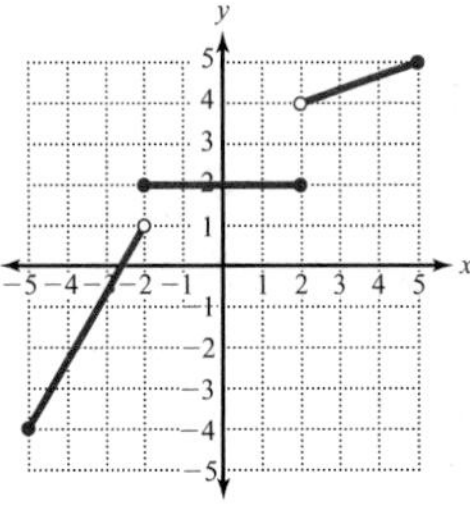

23.

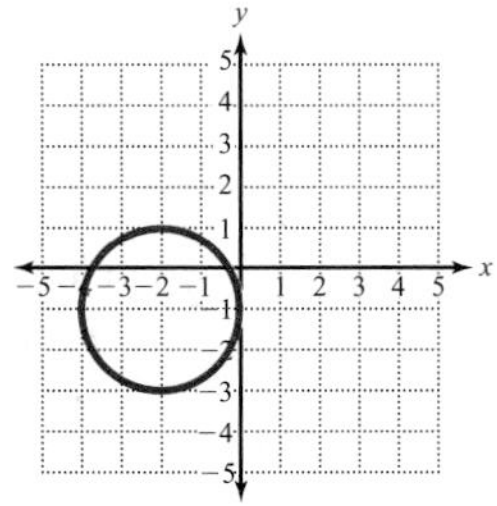

24.

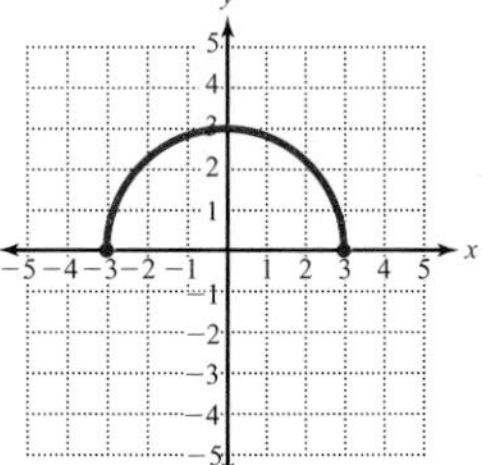

For Exercises 25–44, find the indicated value of the function.

25. $f(x) = -2x - 9$

a. $f(0)$

b. $f(1)$

c. $f(-1)$

d. $f(a + 1)$

26. $f(x) = 6x - 2$

a. $f(1)$

b. $f(0)$

c. $f(-1)$

d. $f(t + 3)$

27. $f(x) = 2x^2 - x + 7$

a. $f(0)$

b. $f(1)$

c. $f(-1)$

d. $f(a)$

28. $f(x) = -3x^2 + 2x$

a. $f(-2)$

b. $f(-1)$

c. $f(3)$

d. $f(a)$

29. $f(x) = \sqrt{3 - x}$

a. $f(-1)$

b. $f(12)$

c. $f(-2)$

d. $f(t)$

30. $f(x) = \sqrt{4x - 2}$

a. $f(-1)$

b. $f(1.5)$

c. $f(3)$

d. $f(a)$

31. $f(x) = \frac{2}{5}x + 1$

a. $f(0)$

b. $f(-5)$

c. $f(-1)$

d. $f(r)$

32. $f(x) = \frac{3}{4}x - 3$

a. $f(4)$

b. $f(0)$

c. $f(3)$

d. $f(t)$

33. $f(x) = x^2 - 2.1x - 3$

a. $f(0)$

b. $f(-2.2)$

c. $f\left(\frac{2}{3}\right)$

d. $f(a)$

34. $f(x) = 3.1x^2 + 2x$
 a. $f(-1)$
 b. $f(0.1)$
 c. $f\left(-\frac{1}{2}\right)$
 d. $f(a)$

35. $f(x) = \sqrt{x^2 - 4x}$
 a. $f(4)$
 b. $f(7)$
 c. $f(2)$
 d. $f(n)$

36. $f(x) = \sqrt{-x^2 + 5x^3}$
 a. $f(-2)$
 b. $f(1)$
 c. $f(0)$
 d. $f(w)$

37. $f(x) = |3x^2 + 1|$
 a. $f(0)$
 b. $f\left(\frac{2}{3}\right)$
 c. $f(-1)$
 d. $f(-2)$

38. $f(x) = |x - 3x^2|$
 a. $f(-2)$
 b. $f(-1)$
 c. $f\left(\frac{1}{3}\right)$
 d. $f(3)$

39. $f(x) = \frac{3 - x}{x - 4}$
 a. $f(5)$
 b. $f(4)$
 c. $f(-2)$
 d. $f(3)$

40. $f(x) = \frac{2x - 1}{x + 2}$
 a. $f(-3)$
 b. $f(-2)$
 c. $f(3)$
 d. $f(-1)$

41. $f(x) = \frac{x}{x^2 - 1}$
 a. $f(0)$
 b. $f(1)$
 c. $f(2)$
 d. $f(m)$

42. $f(x) = \frac{x - 2}{x^2 - 9}$
 a. $f(4)$
 b. $f(-1)$
 c. $f(3)$
 d. $f(a)$

43. $f(x) = \frac{1}{\sqrt{2 - x}}$
 a. $f(-2)$
 b. $f(1)$
 c. $f(2)$
 d. $f(6)$

44. $f(x) = \frac{4x}{\sqrt{2x - 5}}$
 a. $f(3)$
 b. $f\left(\frac{5}{2}\right)$
 c. $f(-2)$
 d. $f(a)$

For Exercises 45–50, use the graph to find the indicated value of the function.

45. a. $f(-4)$ b. $f(0)$ c. $f(2)$

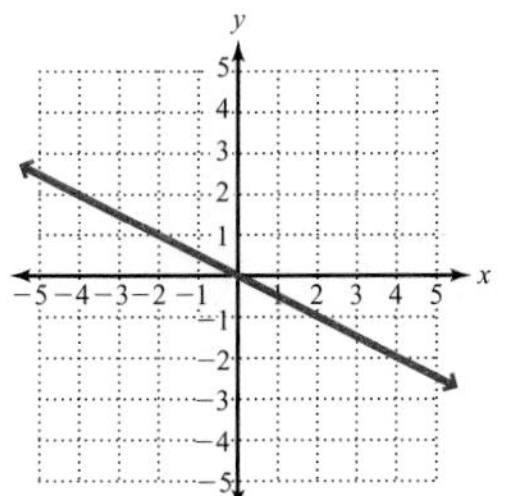

46. a. $f(-4)$ b. $f(-2)$ c. $f(1)$

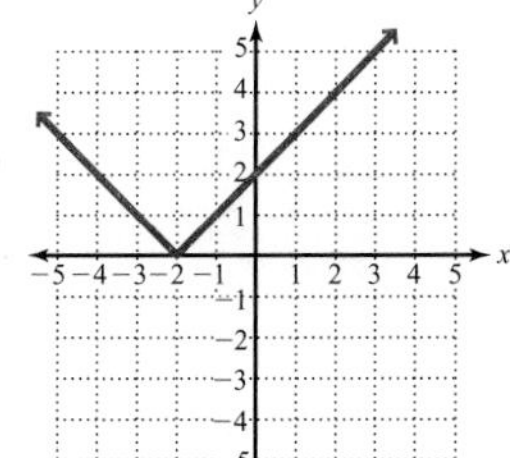

47. a. $f(-2)$ b. $f(0)$ c. $f(2)$

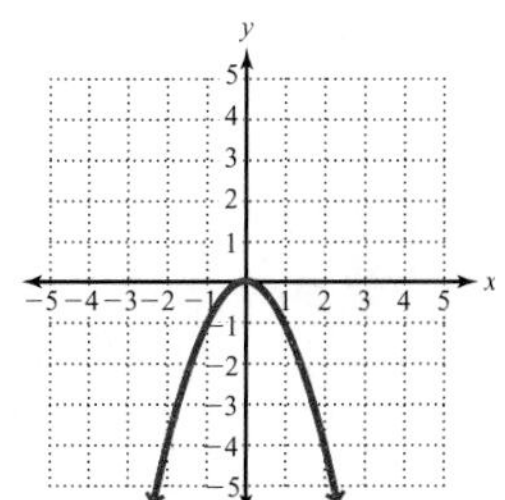

48. **a.** $f(-2)$ **b.** $f(-1)$ **c.** $f(1)$

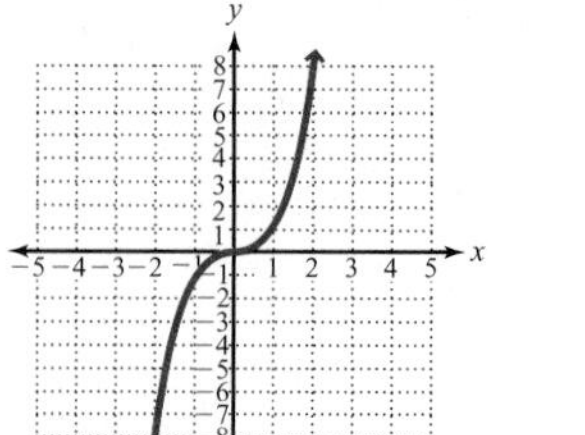

49. **a.** $f(-4)$ **b.** $f(0)$ **c.** $f(3)$

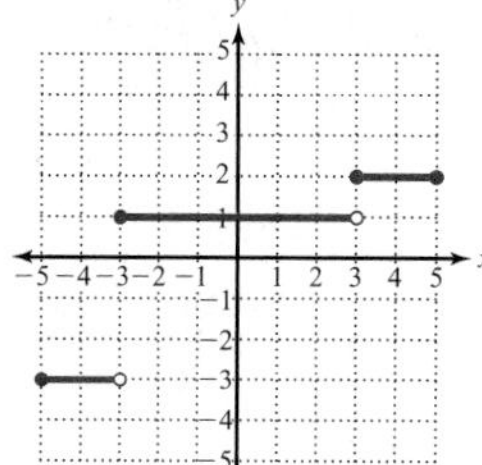

50. **a.** $f(-5)$ **b.** $f(-1)$ **c.** $f(2)$

For Exercises 51–60, graph.

51. $f(x) = -3x + 2$

52. $f(x) = 4x - 1$

53. $f(x) = \frac{2}{3}x - 1$

54. $f(x) = -\frac{3}{2}x - 5$

55. $f(x) = -5x$

56. $f(x) = \frac{4}{3}x$

57. $f(x) = x^2 - 1$

58. $f(x) = x^2 + 2$

59. $f(x) = -x^2 + 3$

60. $f(x) = -x^2 - 2$

61. A factory employee gets paid \$50 per day plus \$10 per hour. The employee's time is kept by a special clock that records when the employee clocks in and out. The function $w(t) = 10t + 50$ describes an employee's daily earnings for working t hours.

a. Graph the function. Note that $t \geq 0$ and $w \geq 0$.

b. Find the daily earnings for an employee who works 7.5 hours.

62. The function $F(C) = \frac{9}{5}C + 32$ is used to convert degrees Celsius to degrees Fahrenheit.

a. Graph the function.

b. Find the intercepts.

63. A city charges \$6 per month plus \$1.225 per cubic foot of water used.

a. If C represents cost in dollars and V represents the volume of water used in cubic feet, write a function for C in terms of V that describes a customer's monthly cost.

b. Graph the function. Note that $V \geq 0$ and $C \geq 0$.

c. Find the monthly cost if 40 cubic feet of water are used.

64. A trucker is paid \$200 plus \$0.85 per mile traveled.

a. If p represents the total amount the trucker is paid in dollars, and d represents the distance traveled in miles, write a function for p in terms of d that describes the trucker's total pay for a trip.

b. Graph the function. Note that $d \geq 0$ and $p \geq 0$.

c. Find the total pay if the trucker travels 250 miles.

65. The formula for the area of a circle, A, can be expressed in function form as $A(r) = \pi r^2$, where r represents the radius of the circle.

a. Use a graphing calculator to graph the function. Set your window so that the range on the x-axis is from 0 to 3 with a scale of 1 and the range on the y-axis is from 0 to 20 with a scale of 5.

b. Is it a linear function?

c. In the notation $A(1.5)$, what does the 1.5 refer to?

d. Calculate $A(1.5)$. Round to the nearest hundredth.

66. The formula for the volume of a cylinder, V, with a height of 8 centimeters can be expressed in function form as $V(r) = 8\pi r^2$, where r represents the radius of the cylinder in centimeters.

a. Use a graphing calculator to graph the function. Set your window so that the range on the x-axis is from 0 to 3 with a scale of 1 and the range on the y-axis is from 0 to 50 with a scale of 5.

b. Is it a linear function?

c. In the notation $V(1.2)$, what does the 1.2 refer to?

d. Calculate $V(1.2)$. Round to the nearest tenth.

67. On the following graph, m represents the amount of meat produced worldwide in kilograms per person, and t represents the number of years after 1993. Notice that the amount produced has increased in roughly a linear pattern.

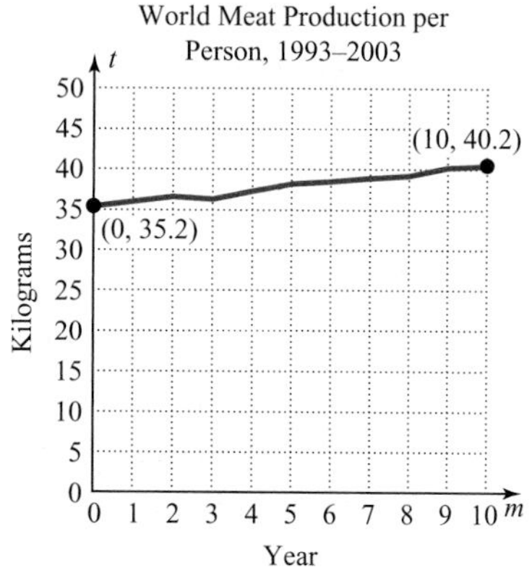

Source: Food and Agricultural Organization of the United Nations

a. Draw a straight line connecting the endpoints of the graph, then find the slope of the line.

b. Write an equation in slope-intercept form that describes the line.

c. Write the equation in function form.

d. Predict the amount of meat that will be produced per person in the year 2013.

68. On the following graph, g represents the amount of natural gas consumed worldwide in trillions of cubic feet and t represents the number of years after 1993. Notice that the amount consumed has increased in roughly a linear pattern.

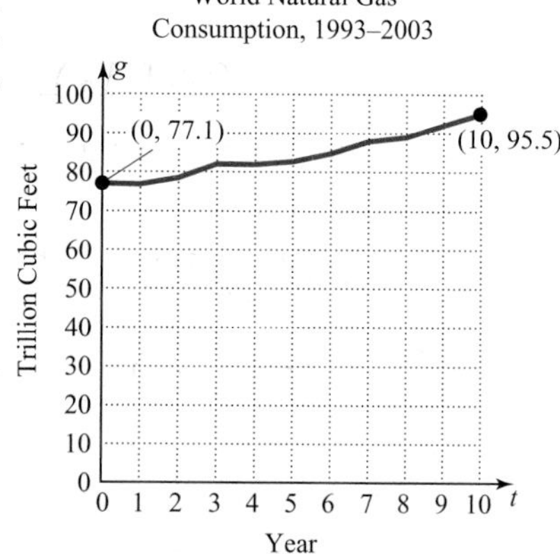

Source: Energy Information Administration, *International Energy Annual,* 2003

a. Draw a straight line connecting the endpoints of the graph, then find the slope of the line.

b. Write an equation in slope-intercept form that describes the line.

c. Write the equation in function form.

d. Predict the amount of natural gas that will be consumed in the year 2013.

69. The table to the right shows the distance traveled by U.S. drivers in trillions of miles for each year.

Year	Amount
1995	2.42
1996	2.49
1997	2.56
1998	2.63
1999	2.69
2000	2.75
2001	2.80
2002	2.86
2003	2.88

Source: National Highway Traffic Safety Administration

a. Make a grid with the horizontal axis labeled t and the vertical axis labeled d so that t represents the number of years after 1995 and d represents the distance traveled in trillions of miles. Plot the ordered pairs in the coordinate plane. (Hint: $t = 0$ means 1995.)

b. Do the points follow a linear pattern? If so, find the slope of the line that passes through them.

c. Write the equation of this line in function notation, $d(t)$.

d. Find $d(20)$, which is the predicted number of miles driven in 2015.

70. The table to the right shows the average movie ticket price in the United States.

a. Make a grid with the horizontal axis labeled t and the vertical axis labeled p so that t represents the number of years after 1999 and p represents the average price of a movie ticket in the United States. Plot the ordered pairs in the coordinate plane. (Hint: $t = 0$ means 1999.)

b. Do the points follow a linear pattern? If so, find the slope of the line that passes through them.

c. Write the equation of this line in function notation, $p(t)$.

d. Find $p(16)$, which is the predicted average movie ticket price in 2015.

Year	Price
1999	$5.06
2000	$5.39
2001	$5.65
2002	$5.80
2003	$6.03
2004	$6.21

Source: National Association of Theater Owners

REVIEW EXERCISES

[1.4] 1. Simplify: $x + 9y - 3x + 7 - 3y - 9$

[2.1] 2. Solve: $5(x + 1) - x = 3x - (x + 1)$

[2.3] 3. Solve $4x - (6x + 3) \geq 8x - 23$. Write the solution set in set-builder notation and in interval notation. Then graph the solution set.

[2.2] 4. The sum of two consecutive even integers is 74. Find the integers.

[2.2] 5. The perimeter of a rectangle is 108 inches. If the length is 6 inches more than the width, find the length and width.

[2.2] 6. An ice-cream shop has two sizes of cups, large and small. The small sells for $1.50 and the large sells for $2.25. In 1 hour a salesperson sells 40 cups for a total of $69.00. Find the number of each size sold.

Chapter 3 Summary

Defined Terms

Section 3.1
Linear equation in two variables (p. 137)
x-intercept (p. 139)
y-intercept (p. 139)

Section 3.2
Slope (p. 152)

Section 3.4
Linear inequality in two variables (p. 179)

Section 3.5
Relation (p. 189)
Domain (p. 189)
Range (p. 189)
Function (p. 189)

Procedures, Rules, and Key Examples

Procedures/Rules	Key Examples
Section 3.1 Graphing Linear Equations To find a solution to an equation in two variables, 1. Replace one of the variables with a chosen number (any number). 2. Solve the equation for the value of the other variable.	**Example 1:** Find two solutions for the equation $y = 3x - 1$. First solution: Let $x = 0$: $y = 3(0) - 1$, $y = -1$. Solution: $(0, -1)$ Second solution: Let $x = 1$: $y = 3(1) - 1$, $y = 3 - 1$, $y = 2$. Solution: $(1, 2)$
To graph a linear equation, 1. Find at least two solutions to the equation. 2. Plot the solutions as points in the rectangular coordinate system. 3. Connect the points to form a straight line.	**Example 2:** Graph $y = 3x - 1$. We found two solutions above: $(0, -1)$ and $(1, 2)$. 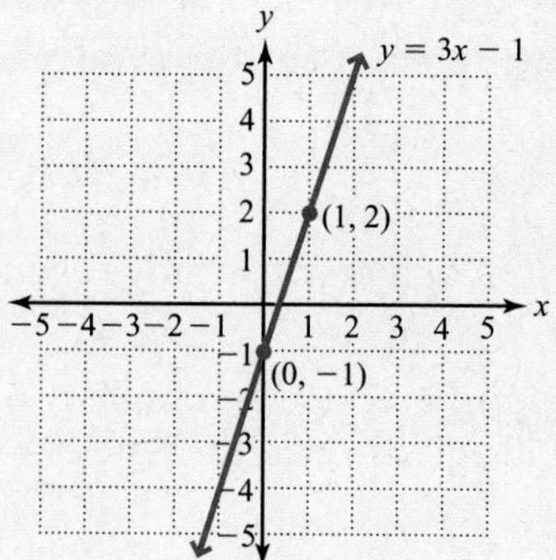

continued

Procedures/Rules	Key Examples

Section 3.1 Graphing Linear Equations (continued)

To find an x-intercept,

1. Replace y with 0 in the given equation.
2. Solve for x.

To find a y-intercept,

1. Replace x with 0 in the given equation.
2. Solve for y.

Given an equation in the form $y = mx + b$, the coordinates of the y-intercept are $(0, b)$

Example 3: Find the x- and y-intercepts for $4x - 3y = 24$.

Solution:

x-intercept:
$$4x - 3(0) = 24$$
$$4x = 24$$
$$x = 6$$
x-intercept: $(6, 0)$

y-intercept:
$$4(0) - 3y = 24$$
$$-3y = 24$$
$$y = -8$$
y-intercept: $(0, -8)$

The graph of $y = c$, where c is a real-number constant, is a horizontal line parallel to the x-axis with a y-intercept at $(0, c)$.

The graph of $x = c$, where c is a real-number constant, is a vertical line parallel to the y-axis with an x-intercept at $(c, 0)$.

Example 4: Graph $y = -2$ and $x = 3$.

Solution:

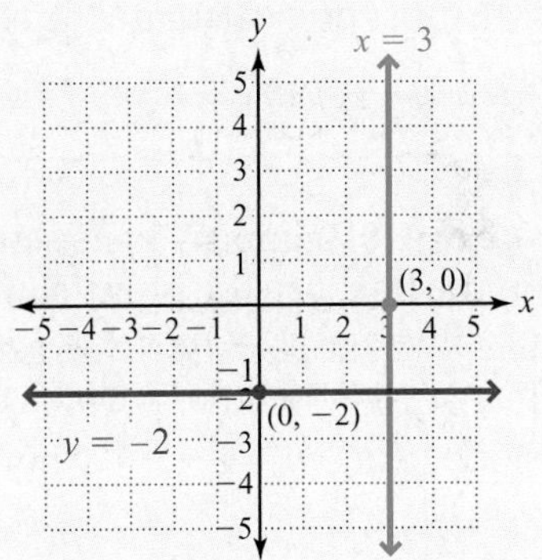

Section 3.2 The Slope of a Line

The graph of an equation in the form $y = mx + b$ (slope-intercept form) is a line with slope m and y-intercept $(0, b)$. The following rules indicate how m affects the graph.

If $m > 0$, then the line slants upward from left to right.

If $m < 0$, then the line slants downward from left to right.

The greater the absolute value of m, the steeper the line.

Example 1: Graph $y = \frac{1}{2}x + 1$ and $y = -\frac{1}{2}x + 1$.

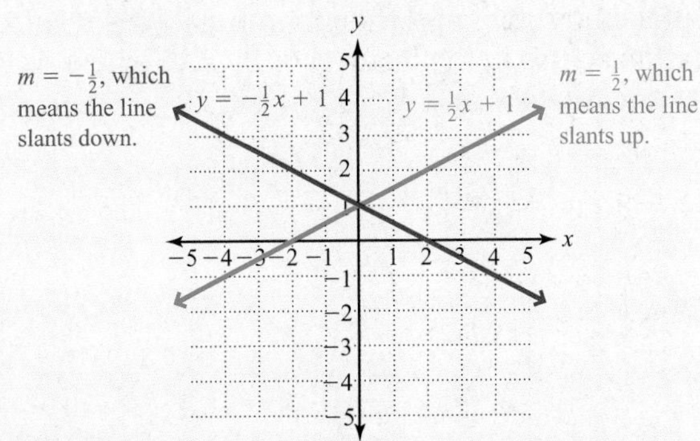

Given two points (x_1, y_1) and (x_2, y_2), where $x_2 \neq x_1$, the slope of the line connecting the two points is given by the formula $m = \frac{y_2 - y_1}{x_2 - x_1}$.

Example 2: Find the slope of a line passing through the points $(4, -2)$ and $(-1, -3)$.

Solution: $m = \frac{-3 - (-2)}{-1 - 4} = \frac{-1}{-5} = \frac{1}{5}$

continued

Procedures/Rules	Key Examples
Section 3.2 The Slope of a Line (continued) Two points with different x-coordinates and the same y-coordinates, (x_1, c) and (x_2, c), will form a horizontal line with a slope of 0 and equation $y = c$.	**Example 3:** The slope of a line connecting the points $(1, 3)$ and $(-2, 3)$ is 0 and the equation of the line is $y = 3$.
Two points with the same x-coordinates and different y-coordinates, (c, y_1) and (c, y_2), will form a vertical line with undefined slope and equation $x = c$.	**Example 4:** The slope of a line connecting the points $(-1, 6)$ and $(-1, -2)$ is undefined and the equation of the line is $x = -1$.
Section 3.3 The Equation of a Line To write the equation of a line given its y-intercept, $(0, b)$, and its slope, m, use the slope-intercept form of the equation, $y = mx + b$. If given a second point and not the slope, we must first calculate the slope using $m = \frac{y_2 - y_1}{x_2 - x_1}$, then use $y = mx + b$.	**Example 1:** Write the slope-intercept form of the equation of a line passing through $(0, 3)$ with slope $-\frac{5}{6}$. Solution: $y = -\frac{5}{6}x + 3$
To write the equation of a line given its slope and any point (x_1, y_1) on the line, use the point-slope form of the equation of a line, $y - y_1 = m(x - x_1)$. If given a second point, (x_2, y_2), and not the slope, we first calculate the slope using $m = \frac{y_2 - y_1}{x_2 - x_1}$, then use $y - y_1 = m(x - x_1)$.	**Example 2:** Write the slope-intercept form of the equation of a line passing through $(-4, 1)$ with slope -2. Solution: $y - y_1 = m(x - x_1)$ $y - 1 = -2(x - (-4))$ **Replace m with -2, x_1 with -4, and y_1 with 1.** $y - 1 = -2x - 8$ **Distribute -2.** $y = -2x - 7$ **Add 1 to both sides.**
To write the equation of a line in standard form, use the preceding rules and then manipulate the resulting equation so it is in the form $Ax + By = C$, where A, B, and C are integers and $A > 0$.	**Example 3:** Write an equation of a line connecting $(1, 6)$ and $(-5, 2)$ in the form $Ax + By = C$, where A, B, and C are integers and $A > 0$. Solution: Find the slope: $m = \frac{2 - 6}{-5 - 1} = \frac{-4}{-6} = \frac{2}{3}$ Write the equation: $y - 6 = \frac{2}{3}(x - 1)$ $3(y - 6) = 3 \cdot \frac{2}{3}(x - 1)$ **Multiply by 3 to clear the fraction.** $3y - 18 = 2x - 2$ **Simplify and distribute.** $-2x + 3y = 16$ **Subtract $2x$ and add 18 on both sides.** $2x - 3y = -16$ **Multiply by -1 so that the x-term is positive.** ***continued***

Procedures/Rules	Key Examples

Section 3.3 The Equation of a Line (continued)

The slopes of parallel lines are equal.

The slope of a line perpendicular to a line with a slope of $\frac{a}{b}$ is $-\frac{b}{a}$.

Example 4: Find the slopes of the lines parallel to and perpendicular to the graph of $4x - 5y = -10$.

Solution: Write $4x - 5y = -10$ in slope-intercept form.

$$-5y = -4x - 10 \qquad \text{Subtract } 4x \text{ from both sides.}$$

$$y = \frac{4}{5}x + 2 \qquad \text{Divide both sides by } -5.$$

The slope of $4x - 5y = -10$ is $\frac{4}{5}$.

The slope of a line parallel to $4x - 5y = -10$ is $\frac{4}{5}$.

The slope of a line perpendicular to $4x - 5y = -10$ is $-\frac{5}{4}$.

Section 3.4 Graphing Linear Inequalities

To graph a linear inequality,

1. Graph the related equation. The related equation has an equal sign in place of the inequality symbol. If the inequality symbol is $\leq$ or $\geq$, draw a solid line. If the inequality symbol is $<$ or $>$, draw a dashed line.
2. Choose an ordered pair on one side of the boundary line and test this ordered pair in the inequality. If the ordered pair satisfies the inequality, then shade the region that contains it. If the ordered pair does not satisfy the inequality, then shade the region on the other side of the boundary line.

Example 1: Graph $3x + 5y > 15$.

Solution: Since the inequality is $>$, we graph the related equation $3x + 5y = 15$ using a dashed line. We then choose a point on one side of that line to see if it satisfies the inequality. We will choose (0, 0).

$$3(0) + 5(0) > 15$$

$$0 > 15$$

This inequality is false, so (0, 0) is not a solution. Therefore, we shade the region on the other side of the line.

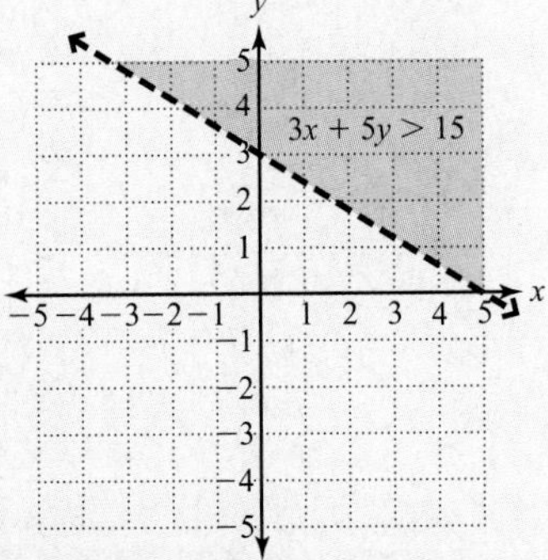

continued

Procedures/Rules	Key Examples

Section 3.5 Introduction to Functions and Function Notation

The domain of a relation is a set that contains all its input values. If given the graph of a relation, the domain contains the first coordinate (x-coordinate) of every point on the graph.

The range of a relation is a set that contains all its ouput values. If given the graph of a relation, the range contains the second coordinate (y-coordinate) of every point on the graph.

Vertical Line Test

To determine whether a graphical relation is a function, draw or imagine vertical lines through each value in the domain. If each vertical line intersects the graph at only one point, the relation is a function. If any vertical line intersects the graph more than once, the relation is not a function.

Example 1: Identify the domain and range of the relation, then determine if it is a function.

a.

Date of Birth	Person
March 1	Candice, Ricky
April 7	Sheri
May 19	Berry

Domain: {March 1, April 7, May 19}
Range: {Candice, Ricky, Sheri, Berry}
It is not a function because one element in the domain corresponds to two elements in the range.

b.

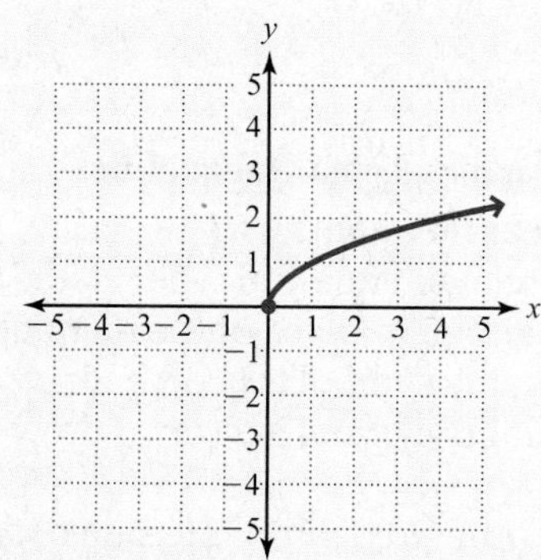

Domain: $\{x \mid x \geq 0\}$
Range: $\{y \mid y \geq 0\}$
It is a function (it passes the vertical line test).

c.

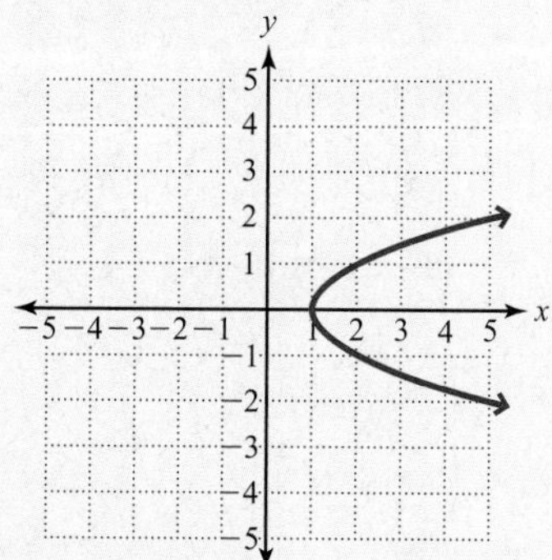

Domain: $\{x \mid x \geq 1\}$
Range: $\mathbb{R}$
It is not a function (it fails the vertical line test).

continued

Procedures/Rules	Key Examples
Section 3.5 Introduction to Functions and Function Notation (continued) Given a function $f(x)$, to find $f(a)$, where a is a real number in the domain of f, replace x in the function with a, then simplify.	**Example 2:** For the function $f(x) = x^3 - 2x$, find $f(-4)$. $f(-4) = (-4)^3 - 2(-4)$ $= -64 + 8$ $= -56$

Summary of the forms of a linear equation:

Slope-intercept form: $y = mx + b$

Point-slope form: $y - y_1 = m(x - x_1)$

Standard form: $Ax + By = C$

Chapter 3 Review Exercises

For Exercises 1–6, answer true or false.

[3.1] 1. When writing coordinates, the vertical coordinate is written first.

[3.1] 2. There are an infinite number of solutions to every linear equation in two variables.

[3.2] 3. The slope-intercept form of a linear equation is $y = mx + b$.

[3.1] 4. Both coordinates in Quadrant I are positive.

[3.3] 5. We can find the equation of a line given any two points on that line.

[3.1] 6. We must have at least three points to correctly graph a straight line.

For Exercises 7–10, complete the rule.

[3.2] 7. The slope of a line connecting two points is found by the equation ______.

[3.5] 8. A function must pass the ______ line test.

[3.2] 9. Given an equation of the form $y = mx + b$,
If $m > 0$, then its graph is a line that slants ______ from left to right.
If $m < 0$, then its graph is a line that slants ______ from left to right.

[3.5] 10. Define a function.

[3.1] 11. Write the coordinates for each point.

[3.1] 12. Plot and label the points indicated by the ordered pairs $(4, -1)$, $(-5, 0)$ $(0, 3)$, $(-2, -5)$.

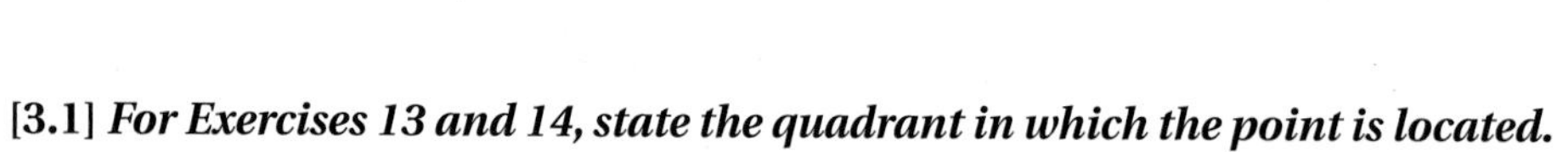

[3.1] ***For Exercises 13 and 14, state the quadrant in which the point is located.***

13. $(-305, 12.7)$

14. $\left(-2\frac{2}{3}, -0.6\right)$

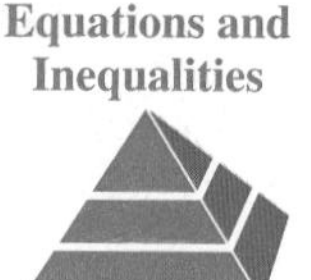

[3.1] ***For Exercises 15–18, determine whether the given pair of coordinates is a solution for the given equation.***

15. $(2, 1)$; $2x - y = -3$

16. $\left(\frac{4}{5}, 2\right)$; $5x + y = 6$

17. $(0.4, 1.2)$; $y + 2x = 4.1$

18. $(5, 2)$; $y = \frac{2}{5}x$

[3.1] *For Exercises 19–24, graph.*

19. $y = x - 4$

20. $y = 5x$

21. $y = \frac{2}{3}x - 3$

22. $y = -\frac{2}{7}x + 1$

23. $2x - 3y = 6$

24. $3x + 4y = 28$

[3.2] *For Exercises 25–28, determine the slope and the y-intercept, then graph.*

25. $y = -3x + 2$

26. $y = \frac{5}{2}x + 3$

27. $2x + 5y = 15$

28. $y - 4x + 3 = 0$

[3.2] *For Exercises 29–32, find the slope through the given points.*

29. $(1, 8), (4, -1)$

30. $(3, -1), (3, 2)$

31. $(7, -4), (2, -9)$

32. $(-1, -1), (3, -4)$

[3.3] *For Exercises 33–36, write the equation of the line in slope-intercept form given the slope and the coordinates of the y-intercept.*

33. $m = 2; (0, -4)$

34. $m = -\frac{2}{5}; (0, 4)$

35. $m = -0.3; (0, -1)$

36. $m = -3; (0, 0)$

[3.3] *For Exercises 37–38, determine the slope and the y-intercept. Then write the equation of the line in slope-intercept form.*

37.

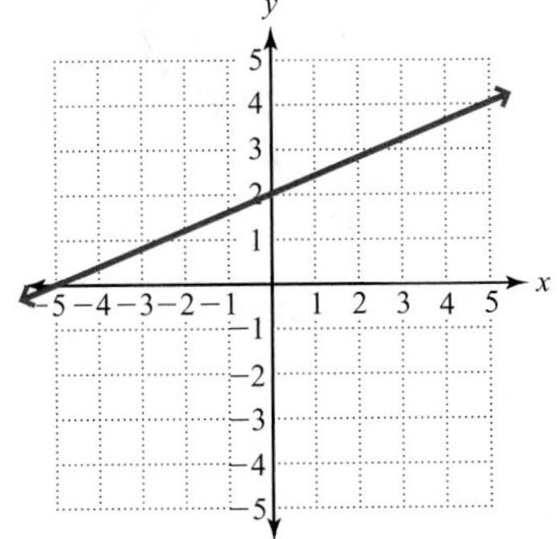

38.

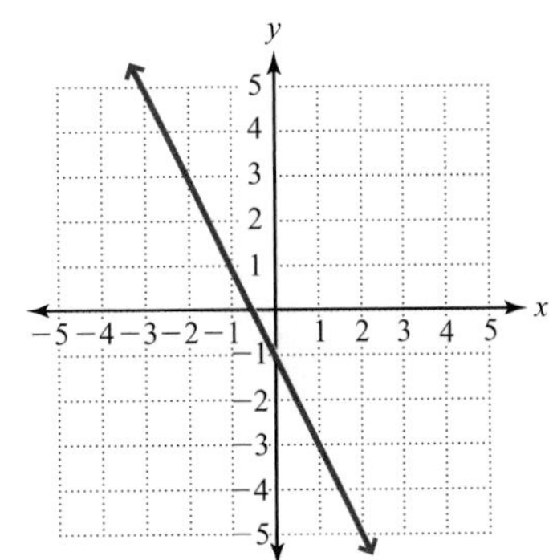

[3.3] ***For Exercises 39–42: a. Write the equation of a line through the given points in slope-intercept form. b. Write the equation in standard form, $Ax + By = C$, where A, B, and C are integers and $A > 0$.***

39. $(3, 3), (5, 9)$

40. $(-2, 2), (-4, -5)$

41. $(-9, -3), (4, -2)$

42. $(6, 0), (0, 6)$

[3.3] ***For Exercises 43–46, determine whether the graphs of the equations are parallel, perpendicular, or neither.***

43. $y = \frac{2}{3}x$
$y = \frac{2}{3}x - 4$

44. $y = -7x - 9$
$y = 7x + 6$

45. $y = \frac{4}{3}x - 1$
$y = -\frac{3}{4}x + 4$

46. $y = -x - 1$
$y = x - 5$

[3.3] ***For Exercises 47–50, write the equation of a line in slope-intercept form with the given slope and passing through the given point.***

47. $m = -1; (2, 8)$

48. $m = 6.2; (3, -5)$

49. $m = \frac{2}{3}; (4, 0)$

50. $m = -1; (-2, -2)$

[3.3] ***For Exercises 51–54, write the equation of the line in slope-intercept form.***

51. Find the equation of a line passing through $(0, -2)$ and parallel to the line $y = -\frac{3}{5}x + 2$.

52. Find the equation of a line passing through $(2, 5)$ and parallel to the line $x - 5y = 10$.

53. Find the equation of a line passing through $(-1, -2)$ and perpendicular to the line $y = -\frac{2}{3}x + 5$.

54. Find the equation of a line passing through $(-3, -5)$ and perpendicular to the line $y = \frac{1}{3}x + 4$.

[3.4] ***For Exercises 55–56, determine whether the ordered pair is a solution for the linear inequality.***

55. $(-3, -1); 3x + 5y > 1$

56. $(-4, 2); y \geq 3x + 1$

For Exercises 57–62, graph the linear inequality.

57. $y < -2x + 5$

58. $3x - 4y > 1$

59. $-2x - 5y \leq -10$

60. $y > \frac{4}{5}x$

61. $x + y \geq 3$

62. $y \geq -3$

[3.5] ***For Exercises 63–68, find the domain and range of the relation, then determine if it is a function.***

63. Tallest peaks in North America:

Peak	Height (ft.)
McKinley	20,320
Logan	19,551
Pico de Orizaba	18,555
St. Elias	18,008
Popocatépetl	17,930

64. The number of drive-in theaters in each state:

Number of Drive-ins	State
21	California
23	Indiana
32	New York
35	Ohio, Pennsylvania

Source: United Drive-in Theater Owners Association

65.

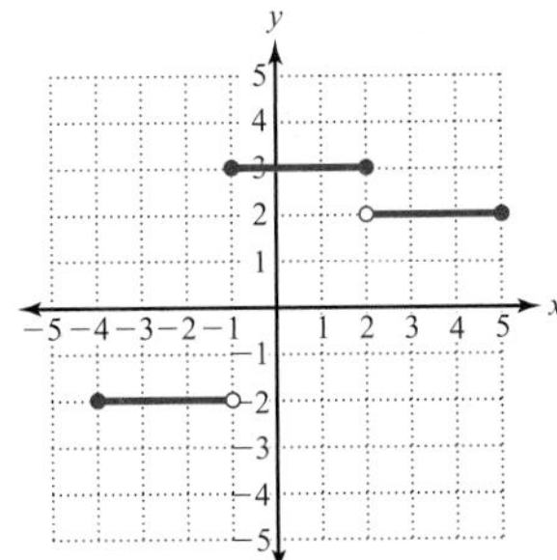

66.

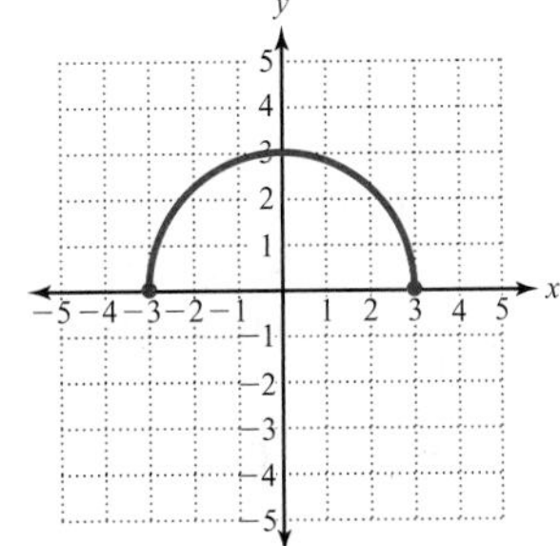

67.

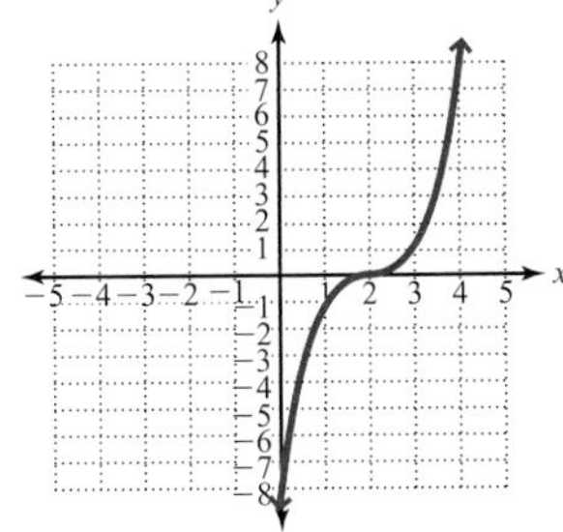

68.

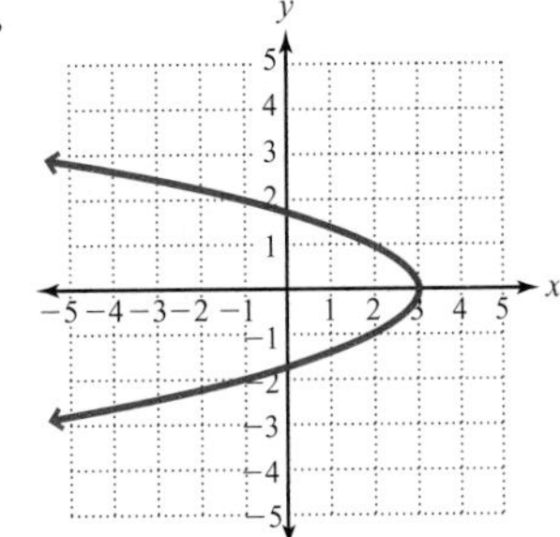

[3.5] **69.** Use the graph in Exercise 65 to find the indicated value of the function.

a. $f(-3)$ **b.** $f(2)$ **c.** $f(0)$ **d.** $f(5)$

[3.5] **70.** Use the graph in Exercise 66 to find the indicated value of the function.

a. $f(-3)$ **b.** $f(0)$ **c.** $f(3)$ **d.** $f(4)$

[3.5] **71.** Find the indicated value of the function $f(x) = x^2 - 4$.

a. $f(2)$ **b.** $f(0)$ **c.** $f(-3)$ **d.** $f(n)$

[3.5] **72.** Find the indicated value of the function $g(x) = \dfrac{x - 3}{x + 5}$.

a. $g(2)$ **b.** $g(3)$ **c.** $g(-5)$

[3.2] **73.** A builder is putting together a staircase where the base of each step is 12 inches and the rise is 10 inches. What is the slope of the staircase?

[2.3] **74.** In 1999, 23.1% of 12th-grade students smoked daily. In 2003, the rate had declined to 15.3%. (*Source: Monitoring the Future,* University of Michigan Institute for Social Research and National Institute on Drug Abuse)

a. If the years are plotted along the x-axis so that 1999 is year 0 and 2003 is year 4, and the percent is plotted along the y-axis, find the slope of the line passing through the two points given.

b. Write the equation of the line in slope-intercept form that connects the two points.

c. Assuming the trend continues, use the equation from part b to determine the percent of 12th-grade students that will smoke daily in 2010.

[3.4] **75.** A furniture company sells two different types of dining room chairs. A basic chair sells for \$35.00 and an armchair sells for \$50.00. For the company to break even or make a profit, the combined revenue from those two chairs must be at least \$70,000.

a. Letting x represent the number of basic chairs sold and y represent the number of armchairs sold, write an inequality that describes the revenue requirement for the company to break even or make a profit.

b. Graph the inequality. Note that $x \geq 0$ and $y \geq 0$.

c. What does the boundary line represent?

d. What does the shaded region represent?

e. List two combinations of sales numbers of each chair that allow the company to break even.

f. List two combinations of sales numbers that give the company a profit.

[3.5] **76.** A landscaper charges \$75 per visit plus \$25 per hour of labor.

a. If c represents the total cost of a visit and t represents the time in hours that she works, write an equation using function notation that describes the total cost.

b. Find the total cost if labor is 1.5 hours.

c. If a client's total charges are \$150, how many hours of labor was the client charged?

d. Graph the function. Note that $t \geq 0$ and $c \geq 0$.

Chapter 3 Practice Test

1. Determine the coordinates for each point in the graph to the right.

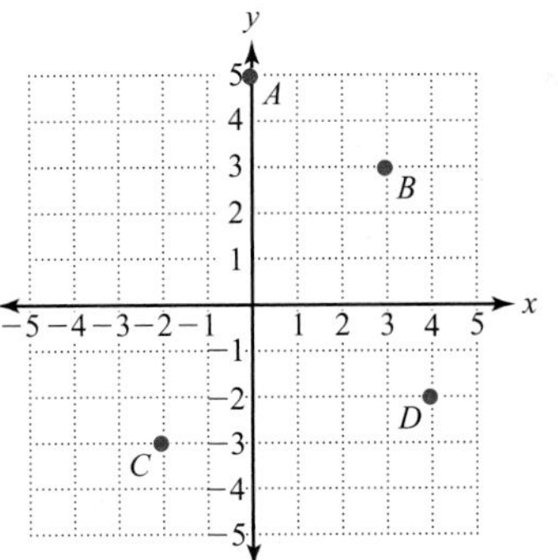

2. State the quadrant in which $\left(3.6, -501\frac{2}{3}\right)$ is located.

3. Determine whether $(-3, 5)$ is a solution for $y = -\frac{1}{4}x + 3$.

For Exercises 4 and 5, determine the slope and the coordinates of the y-intercept, then graph.

4. $y = -\frac{4}{3}x + 5$

5. $x - 2y = -8$

For Exercises 6 and 7, determine the slope of the line through the given points.

6. $(-1, -4), (-2, -4)$

7. $(-3, -9), (4, -1)$

8. Write the equation of a line in slope-intercept form with y-intercept $(0, 5)$ and slope $\frac{2}{7}$.

9. Write the equation of a line in slope-intercept form that passes through the points $(4, 2), (-5, -1)$.

10. Write the equation of a line through the points $(1, 4)$ and $(-3, -1)$ in the form $Ax + By = C$, where A, B, and C are integers and $A > 0$.

11. Are the graphs of $y = \frac{3}{4}x - 2$ and $y = \frac{4}{3}x + 2$ parallel, perpendicular, or neither?

For Exercises 12 and 13, graph the linear inequality.

12. $y \geq 3x - 1$

13. $2x - 3y < 6$

For Exercises 14 and 15, identify the domain and range of the relation, then determine if it is a function.

14. The number of U.S. residents who are not native English speakers:

Language	U.S. Residents in Millions
Spanish	28.1
Chinese	2.0
French	1.6
German	1.4
Tagalog	1.2

Source: U.S. Bureau of the Census

15.

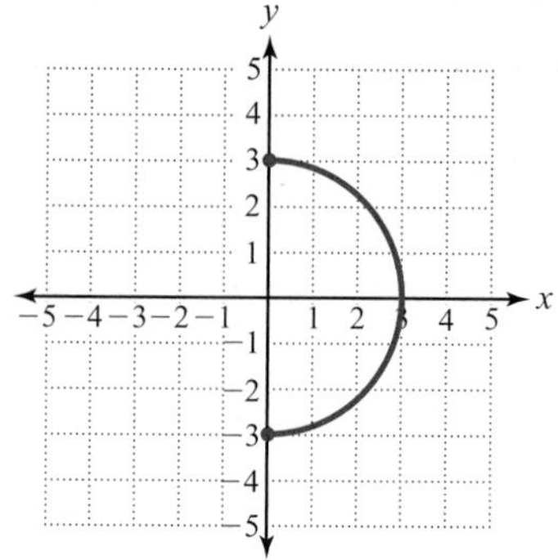

For Exercises 16 and 17, find the indicated value of the function.

16. $f(x) = 2x^2 - 7$

a. $f(-2)$ **b.** $f(3)$ **c.** $f(t)$

17. $f(x) = \dfrac{4x}{x + 5}$

a. $f(-1)$ **b.** $f(5)$ **c.** $f(-5)$

18. a. Give the domain and range of the function shown in the graph.

b. Use the graph to find $f(1)$.

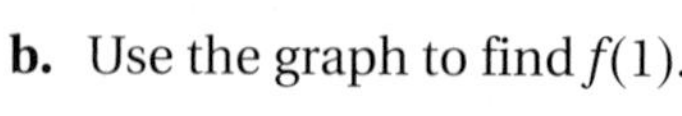

19. Part of a farmer's land is to be enclosed for pasture. Materials limit the perimeter to 1000 feet.

a. Letting l represent the length in feet and w represent the width in feet, write an inequality in which the maximum perimeter is 1000 feet.

b. Graph the inequality. Note that $l > 0$ and $w > 0$.

c. Find two combinations of length and width that satisfy the inequality.

20. A shipping company charges \$3.00 to use its overnight shipping service plus \$0.45 per ounce of weight of the package being shipped.

a. If c represents the total cost in dollars and w represents the package weight in ounces, write an equation using function notation that describes the cost of shipping the package overnight.

b. Graph the function. Note that $w \geq 0$ and $c \geq 0$.

c. Find the cost of shipping a 32-ounce package.

Chapters 1–3 Cumulative Review Exercises

For Exercises 1–6, answer true or false.

[3.1] **1.** The first coordinate of an ordered pair is the x-coordinate.

[2.3] **2.** The inequality $x > 0$ can also be expressed as $[0, \infty)$ in interval notation.

[3.3] **3.** Parallel lines have the same slope.

[1.3] **4.** If the base is a negative number and the exponent is odd, then the product is negative.

[1.2] **5.** $a(b + c) = ab + ac$

[3.1] **6.** $y = 3x^2 + 2$ is a linear equation.

For Exercises 7–10, fill in the blank.

[2.1] **7.** To clear decimal numbers in an equation, we multiply by an appropriate power of ________ as determined by the decimal number with the most decimal places.

[1.3] **8.** The radical symbol $\sqrt{\ }$ denotes only the ________ square root.

[1.3] **9.** List the order in which we perform operations of arithmetic.

[1.4] **10.** Two expressions, like $5x$ and $7x$, that have the same variable raised to the same exponent are called ________ terms.

[1.1] **11.** Write a set containing the vowels.

[2.4] **12.** Find the intersection and union of the given sets.
$A = \{w, e, l, o, v\}$ $B = \{m, a, t, h\}$

[1.3] ***For Exercises 13–16, evaluate.***

13. $(-3)^2$

14. $\sqrt[3]{27}$

15. $\sqrt{-25}$

16. $\sqrt[4]{\frac{16}{81}}$

[1.3] ***For Exercises 17–18, simplify.***

17. $-|4 + 3| - 8(1 - 5)^2$

18. $5\{2 - [2 - (3 - 1)]\} + 5 \cdot 3$

[1.4] ***For Exercises 19 and 20, evaluate the expression using the given values.***

19. $\sqrt{x + y}; x = 9, y = 16$

20. $\frac{3x}{x - 5}; x = 5$

[1.4] **21.** Translate the expression to symbols: five times the sum of six and a number.

[1.4] **22. a.** Translate to an algebraic expression: the product of five and n is subtracted from the sum of eight and negative four times n.

b. Simplify the expression.

c. Evaluate the expression when n is 4.

[1.2] ***For Exercises 23 and 24, indicate whether the given equation illustrates the additive identity, multiplicative identity, additive inverse, commutative property of addition, commutative property of multiplication, associative property of addition, associative property of multiplication, distributive property, or multiplicative inverse.***

23. $-\frac{1}{4} \cdot -4 = 1$

24. $6(5 + 2) = 6 \cdot 5 + 6 \cdot 2$

For Exercises 25–27, solve.

[2.1] **25.** $6(5 + y) - 3y = 2y + 1$

[2.1] **26.** $5(x + 1) = 10 + 5x - 5$

[2.5] **27.** $|2x - 1| = |x + 8|$

For Exercises 28–30, solve and then a. Write the solution set in set-builder notation.
b. Write the solution set using interval notation.
c. Graph the solution set.

[2.3] **28.** $3 - 2x \geq 8 + 4x$

[2.6] **29.** $|3x - 1| \leq 5$

[2.6] **30.** $3|x + 1| - 7 > 2$

[2.1] **31.** Solve $d = rt$ for t.

[3.2] **32.** Find the slope of a line passing through $(3, -1)$ and $(-2, -1)$.

[3.2] **33.** Determine the slope and y-intercept for $4y - 3x = 8$.

[3.3] **34.** Determine whether the graphs of $2x + y = 6$ and $y = -\frac{1}{2}x + 1$ are parallel, perpendicular, or neither.

[3.3] 35. Find the equation of a line passing through $(1, 8)$ and parallel to the line $2x - y = 3$.

[3.4] 36. Determine whether $(2, 1)$ is a solution for $1 - 4x + y > 7$.

[3.4] ***For Exercises 37–38, graph the linear inequality.***

37. $x - y < 4$

38. $2x - y \geq 6$

[3.5] 39. Determine whether the relation is a function. The largest academic libraries:

University	Number of Volumes in Millions
Harvard	14.7
Yale	10.7
University of Illinois	9.6
University of California	9.3
University of Texas	8.1

Source: Association of Public Libraries

[3.5] 40. Given $f(x) = 3x^2 + 2$, find $f(-1)$.

[3.5] 41. Graph $f(x) = \frac{2}{5}x - 1$.

For Exercises 42–50, solve.

[1.3] 42. The Oklahoma wheat crop produced 104 million bushels in 2002, 179 million bushels in 2003, and 164 million bushels in 2004. What is the average wheat crop produced for those years?

[1.3] 43. Find the GPA.

Course	Course Credits	Grade
Math 111	3	A
Eng 212	3	A
Phys 201	4	B

[2.2] 44. The sum of two consecutive even integers is 186. What are the integers?

[2.2] 45. Two angles are supplementary. If the measure of one angle is twice the measure of the other angle, what are the angle measurements?

[2.2] 46. In 2004, there were 175,725 cell sites in the United States, which represented an increase of 881% from the number of cell sites in 1994. Find the number of cell sites that were in the United States in 1994.
(*Source:* Cellular Telecommunications and Internet Association)

[2.2] 47. Two cars are traveling toward each other on the same road. One car is traveling 50 miles per hour and the other at 45 miles per hour. If the two cars are 190 miles apart, how long until they meet?

[2.2] 48. An administrative assistant orders new phones for the people in his department. The standard phones cost $35 and the deluxe phones cost $75. If he ordered two times as many standard phones as deluxe phones at a total cost of $580, how many of each did he order?

[2.2] 49. Janet has 50 milliliters of a 10% HCl solution and some 25% HCl solution. If she wants a 15% HCl solution, how much of the 25% solution should be added to the 10% solution?

[2.3] 50. A wireless company has a plan that charges $40 for 500 minutes, then $0.04 for each additional minute. Find the range of total minutes a person could use with this plan if the budget is a maximum of $60 per month for wireless charges.

CHAPTER 4

Systems of Linear Equations and Inequalities

"Mathematics takes us still further from what is human, into the region of absolute necessity, to which not only the actual world, but every possible world, must conform."

—Bertrand Russell, mathematician (1872–1970)

In Chapter 2, we solved linear equations and inequalities in one variable. In Chapter 3, we found solutions to linear equations and inequalities in two variables and graphed those solutions in the coordinate plane. In this chapter, we learn how to solve *systems* of linear equations and inequalities.

We will begin in Section 4.1 with systems of two linear equations and learn three techniques for solving these systems: graphing, substitution, and elimination. In Section 4.2, we will solve systems of three linear equations. We will then solve problems (4.3) and learn more advanced techniques for solving systems of equations: matrices (4.4) and Cramer's Rule (4.5). Finally, in Section 4.6, we will solve systems of linear inequalities, which will involve graphing.

4.1 Solving Systems of Linear Equations in Two Variables

OBJECTIVES

1. Determine if an ordered pair is a solution for a system of equations.
2. Solve systems of linear equations graphically and classify systems.
3. Solve systems of linear equations using substitution.
4. Solve systems of linear equations using elimination.

OBJECTIVE 1. Determine if an ordered pair is a solution for a system of equations. A problem involving multiple unknowns can be represented by a group of equations called a **system of equations**.

DEFINITION ***System of equations:*** A group of two or more equations.

Consider the following problem: The sum of two numbers is 3. Twice the first number plus three times the second number is 8. What are the two numbers? If x represents the first number and y represents the second number, then we can translate each sentence into an equation. The two equations together form a system of equations that describe the problem.

$$\text{System of equations } \begin{cases} x + y = 3 & \text{(Equation 1)} \\ 2x + 3y = 8 & \text{(Equation 2)} \end{cases}$$

Our goal will be to find the **solution for the system of equations**.

DEFINITION ***Solution for a system of equations:*** An ordered set of numbers that makes all equations in the system true.

For example, in the preceding system of equations, if $x = 1$ and $y = 2$, both equations are true, so they form a solution to the system of equations. We can write the solution as an ordered pair: $(1, 2)$. We can check by substituting those values in place of the corresponding variables.

PROCEDURE **Checking a Solution to a System of Equations**

To verify or check a solution to a system of equations,

1. Replace each variable in each equation with its corresponding value.
2. Verify that each equation is true.

EXAMPLE 1 Determine whether each ordered pair is a solution to the system of equations.

$$\begin{cases} x - 2y = 3 & \text{(Equation 1)} \\ y = 3x + 1 & \text{(Equation 2)} \end{cases}$$

a. (5, 1)

Solution $x - 2y = 3$ (Equation 1) $\qquad y = 3x + 1$ (Equation 2)

$$5 - 2(1) \stackrel{?}{=} 3 \qquad 1 \stackrel{?}{=} 3(5) + 1$$
$$5 - 2 \stackrel{?}{=} 3 \qquad 1 \stackrel{?}{=} 15 + 1$$
$$3 = 3 \quad \text{True} \qquad 1 = 16 \quad \text{False}$$

In both equations, replace x with 5 and y with 1.

Because (5, 1) does not satisfy both equations, it is not a solution for the system.

b. $(-1, -2)$

Solution $x - 2y = 3$ (Equation 1) $\qquad y = 3x + 1$ (Equation 2)

$$-1 - 2(-2) \stackrel{?}{=} 3 \qquad -2 \stackrel{?}{=} 3(-1) + 1$$
$$-1 + 4 \stackrel{?}{=} 3 \qquad -2 \stackrel{?}{=} -3 + 1$$
$$3 = 3 \quad \text{True} \qquad -2 = -2 \quad \text{True}$$

In both equations, replace x with -1 and y with -2.

Because $(-1, -2)$ satisfies both equations, it is a solution for the system.

YOUR TURN Determine whether each ordered pair is a solution to the system of equations.

$$\begin{cases} 3x + y = 1 & \text{(Equation 1)} \\ y = 1 - 2x & \text{(Equation 2)} \end{cases}$$

a. $(2, -5)$ **b.** $(0, 1)$

OBJECTIVE 2. Solve systems of linear equations graphically and classify systems. Let's graph the equations in the system from Example 1: $\begin{cases} x - 2y = 3 & \text{(Equation 1)} \\ y = 3x + 1 & \text{(Equation 2)} \end{cases}$.

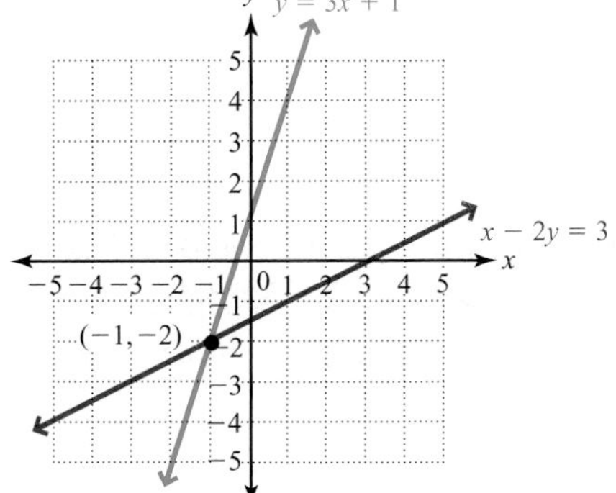

Notice that the graphs of $x - 2y = 3$ and $y = 3x + 1$ intersect at the point $(-1, -2)$. Since $(-1, -2)$ lies on both lines, its coordinates satisfy both equations, so it is the solution for the system. If two linear graphs intersect at a single point, then the system has a *single solution* at the point of intersection. Also, notice that these lines have different slopes, which is always the case when a system of two linear equations has a single solution.

ANSWERS

a. not a solution
b. solution

A system can also have no solution. Look at the graphs of the equations in the system $\begin{cases} y = 2x + 1 & \text{(Equation 1)} \\ y = 2x - 3 & \text{(Equation 2)} \end{cases}$.

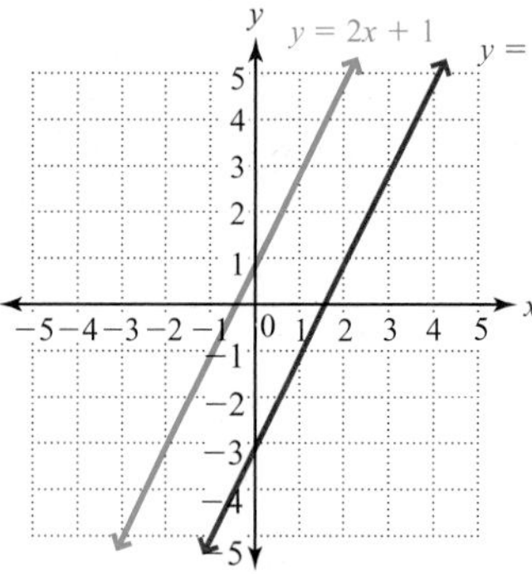

Notice the equations $y = 2x - 3$ and $y = 2x + 1$ have the same slope, 2 and different y-intercepts, so their graphs are parallel lines. Since the graphs of the equations in this system have no point of intersection, the system has no solution.

Some systems have an infinite number of solutions. Consider the system $\begin{cases} 4x + 8y = 16 & \text{(Equation 1)} \\ 2x + 4y = 8 & \text{(Equation 2)} \end{cases}$.

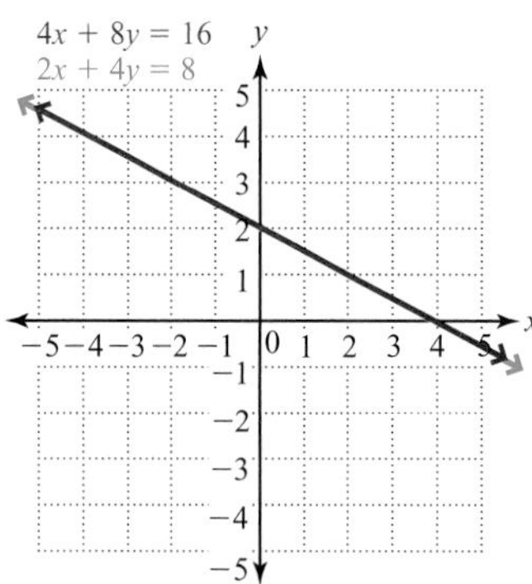

Notice that the graphs of the two equations are identical. Consequently, an infinite number of solutions to this system of equations lie on that line.

We can see without graphing that the two equations describe the same line by writing them in slope-intercept form.

	$4x + 8y = 16$ (Equation 1)	$2x + 4y = 8$ (Equation 2)	
Subtract $4x$ from both sides.	$8y = -4x + 16$	$4y = -2x + 8$	**Subtract $2x$ from both sides.**
Divide both sides by 8 to isolate y.	$y = -\frac{4}{8}x + \frac{16}{8}$	$y = -\frac{2}{4}x + \frac{8}{4}$	**Divide both sides by 4 to isolate y.**
Simplify the fractions.	$y = -\frac{1}{2}x + 2$	$y = -\frac{1}{2}x + 2$	**Simplify the fractions.**

Since the equations are identical, their graphs are the same line.

Our work suggests the following graphical method for solving systems of equations:

PROCEDURE **Solving Systems of Equations Graphically**

To solve a system of linear equations graphically,

1. Graph each equation.
   a. If the lines intersect at a single point, then the coordinates of that point form the solution.
   b. If the lines are parallel, then there is no solution.
   c. If the lines are identical, then there are an infinite number of solutions, which are the coordinates of all the points on that line.
2. Check your solution.

EXAMPLE 2 Solve the system of equations graphically.

a. $\begin{cases} y = -2x + 3 & \text{(Equation 1)} \\ x - 2y = 4 & \text{(Equation 2)} \end{cases}$

Solution Graph each equation.

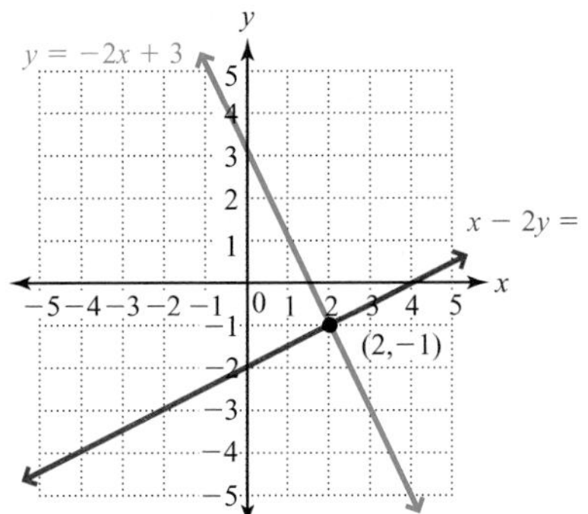

The lines intersect at a single point, which appears to be $(2, -1)$. We can verify that it is the solution by substituting into the equations.

$y = -2x + 3$		$x - 2y = 4$	
$-1 \stackrel{?}{=} -2(2) + 3$		$2 - 2(-1) \stackrel{?}{=} 4$	
$-1 \stackrel{?}{=} -4 + 3$		$2 + 2 \stackrel{?}{=} 4$	
$-1 = -1$	True	$4 = 4$	True

Warning: Graphing by hand can be imprecise and not all solutions have integer coordinates, so you should always check your solutions by substituting them into the original equations.

Answer Because $(2, -1)$ makes both equations true, it is the solution.

b. $\begin{cases} 3x + 4y = 8 & \text{(Equation 1)} \\ y = -\dfrac{3}{4}x - 3 & \text{(Equation 2)} \end{cases}$

Solution Graph each equation.

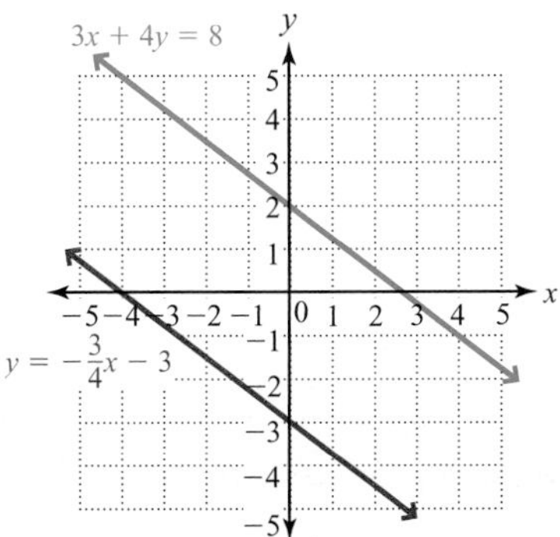

The lines appear to be parallel, which we can verify by comparing the slopes. The slope of Equation 2 is $-\frac{3}{4}$. To determine the slope of Equation 1, we rewrite it in slope-intercept form.

$$3x + 4y = 8$$

$$4y = -3x + 8 \quad \text{Subtract } 3x \text{ from both sides.}$$

$$y = -\frac{3}{4}x + 2 \quad \text{Divide both sides by 4 to isolate } y.$$

Answer Since the slopes are equal and the y-intercepts are different, the lines are parallel, so the system has no solution.

c. $\begin{cases} 6x + 2y = 4 & \text{(Equation 1)} \\ y = -3x + 2 & \text{(Equation 2)} \end{cases}$

Solution Graph each equation.

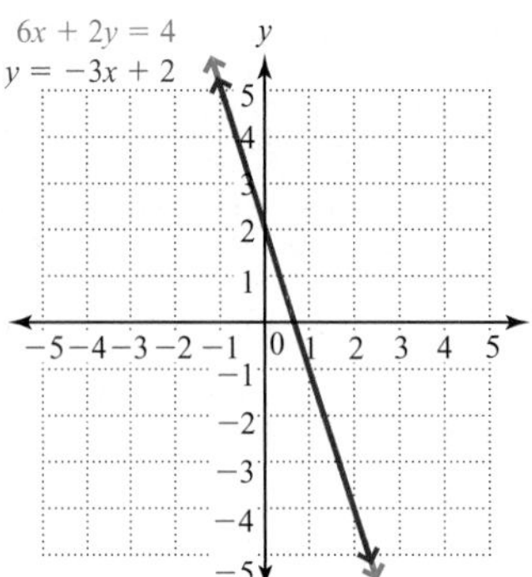

The lines appear to be identical, which we can verify by rewriting $6x + 2y = 4$ in slope-intercept form.

$$6x + 2y = 4$$

$$2y = -6x + 4 \quad \text{Substract } 6x \text{ from both sides.}$$

$$y = -3x + 2 \quad \text{Divide both sides by 2 to isolate } y.$$

Answer Since the equations are the same, the lines coincide and there are an infinite number of solutions to the system, which are the coordinates of all the points on the line.

YOUR TURN Solve the system of equations graphically.

$$\begin{cases} 3x + y = -2 \\ x + 2y = 6 \end{cases}$$

Calculator TIPS

To solve a system of equations using a graphing calculator, begin by graphing each equation. Remember that you must write the equations in slope-intercept form in order to enter them in the calculator. Using the Y= *key, enter one equation in Y1 and the second equation in Y2, then press* GRAPH *to see both graphs.*

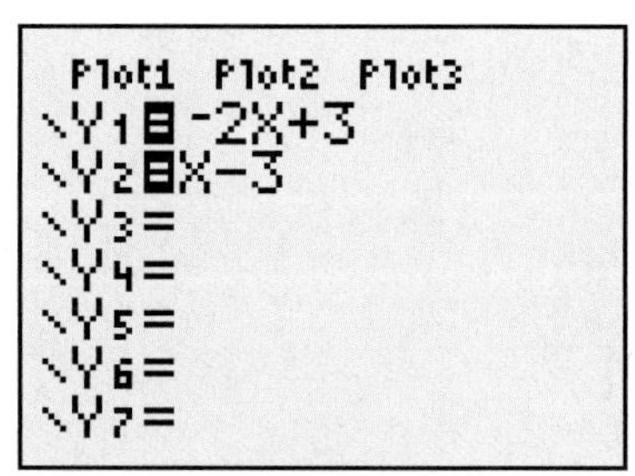

You can find the coordinates of the point of intersection by using the CALC feature found by pressing 2nd TRACE. *Select INTERSECT from the menu. The calculator will then prompt you to verify which lines you want it to find the intersection of, so press* ENTER *for each line. It will then prompt you to indicate which point of intersection you want. Because these straight lines will have only one point of intersection, you can simply press* ENTER. *The coordinates of the point of intersection will appear at the bottom of the screen.*

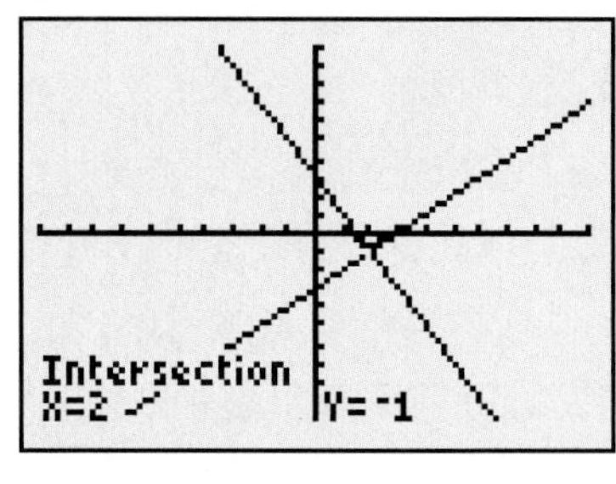

ANSWER

(−2, 4)

x + 2y = 6

3x + y = −2

The solution is $(-2, 4)$.

Classifying Systems of Equations

Systems can be classified as **consistent** or **inconsistent**.

DEFINITIONS ***Consistent system of equations:*** A system of equations that has at least one solution.
Inconsistent system of equations: A system of equations that has no solution.

We can further classify the equations in a system as **dependent** or **independent**. Complete definitions of dependent and independent equations are beyond the scope of this course. For our purposes, we will supply the following definitions, which apply for two linear equations in two unknowns.

DEFINITIONS ***Dependent linear equations in two unknowns:*** Equations with identical graphs.
Independent linear equations in two unknowns: Equations that have different graphs.

We summarize our discussion of classifying systems and equations in those systems with the following procedure:

PROCEDURE **Classifying Systems of Two Equations in Two Unknowns**

To classify a system of two equations in two unknowns, write the equations in slope-intercept form and compare the slopes and y-intercepts.

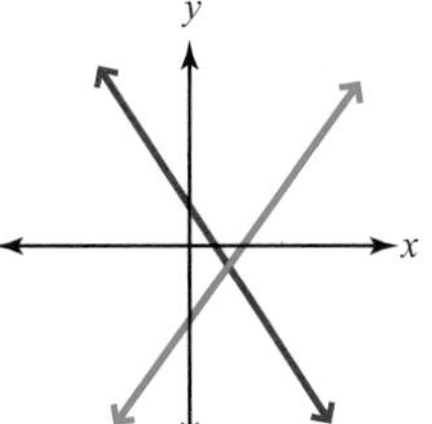

Consistent system: The system has a single solution at the point of intersection.
Independent equations: The graphs are different and intersect at one point. They have different slopes.

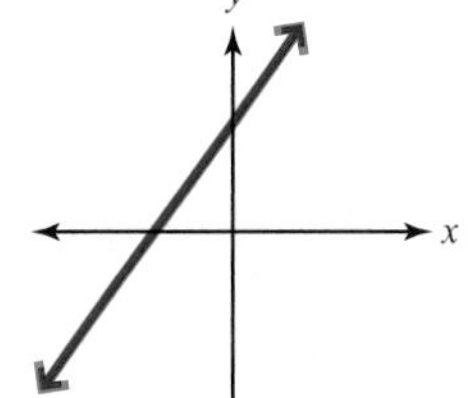

Consistent system: The system has an infinite number of solutions.
Dependent equations: The graphs are identical. They have the same slope and same y-intercept.

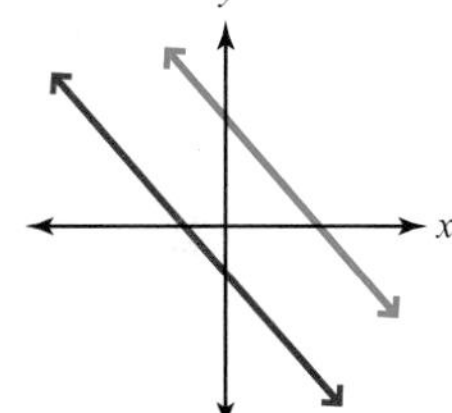

Inconsistent system: The system has no solution.
Independent equations: The graphs are different and are parallel. Though they have the same slope, they have different y-intercepts.

EXAMPLE 3 For each of the systems of equations in Example 2, is the system consistent or inconsistent, are the equations dependent or independent, and how many solutions does the system have?

Solution The system $\begin{cases} y = -2x + 3 \\ x - 2y = 4 \end{cases}$ from Example 2(a) is consistent with independent equations and one solution $(2, -1)$.

The system $\begin{cases} 3x + 4y = 8 \\ y = -\frac{3}{4}x - 3 \end{cases}$ from Example 2(b) is inconsistent with independent equations and no solution.

The system $\begin{cases} 6x + 2y = 4 \\ y = -3x + 2 \end{cases}$ from Example 2(c) is consistent with dependent equations and an infinite number of solutions.

YOUR TURN

a. Determine whether the following system is consistent with independent equations, consistent with dependent equations, or inconsistent with independent equations.

b. How many solutions does the following system have?

$$\begin{cases} x - 2y = 1 \\ x - 4y = 2 \end{cases}$$

OBJECTIVE 3. Solve systems of linear equations using substitution. Using the graphing method to find the solution to a system of equations can be difficult if the solution contains fractions or decimals. A more practical method is the *substitution* method. Consider the following system:

$$\begin{cases} x + 3y = 10 & \text{(Equation 1)} \\ y = x + 4 & \text{(Equation 2)} \end{cases}$$

Remember that in a solution of a system, the same x and y values satisfy both equations. In the preceding system, notice Equation 2 indicates that y is equal to $x + 4$, so y must be equal to $x + 4$ in Equation 1 as well. Therefore, we can substitute $x + 4$ for y in Equation 1.

$$x + 3y = 10$$

$$x + 3(x + 4) = 10 \quad \text{Substitute } x + 4 \text{ for } y.$$

Now we have an equation in terms of a single variable, x, which allows us to solve for x.

$$x + 3x + 12 = 10 \quad \text{Distribute 3.}$$

$$4x + 12 = 10 \quad \text{Combine like terms.}$$

$$4x = -2 \quad \text{Subtract 12 from both sides.}$$

$$x = -\frac{1}{2} \quad \text{Divide both sides by 4 and simplify.}$$

ANSWERS

a. consistent with independent equations

b. one solution

We can find the y-value by substituting $-\frac{1}{2}$ for x in either of the original equations. Equation 2 is easier because y is isolated.

$$y = x + 4$$

$$y = -\frac{1}{2} + 4 \qquad \text{Substitute } -\frac{1}{2} \text{ for } x.$$

$$y = 3\frac{1}{2}$$

The solution to the system of equations is $\left(-\frac{1}{2}, 3\frac{1}{2}\right)$.

PROCEDURE **Solving Systems of Equations Using Substitution**

To find the solution of a system of two linear equations using the substitution method,

1. Isolate one of the variables in one of the equations.
2. In the other equation, substitute the expression you found in Step 1 for that variable.
3. Solve this new equation. (It should now have only one variable.)
4. Using one of the equations containing both variables, substitute the value you found in Step 3 for that variable and solve for the value of the other variable.
5. Check the solution in the original equations.

In step 1, if no variable is isolated in any of the given equations, then we select an equation and isolate one of its variables. Isolating a variable that has a coefficient of 1 or -1 will avoid fractions.

EXAMPLE 4 Solve the system using substitution.

$$\begin{cases} x + y = 6 & \text{(Equation 1)} \\ 5x + 2y = 8 & \text{(Equation 2)} \end{cases}$$

Solution *Step 1* In Equation 1, both x and y have a coefficient of 1, so isolating either variable is easy. We will isolate y.

$$y = 6 - x \qquad \text{Subtract } x \text{ from both sides to isolate } y.$$

Step 2 Substitute $6 - x$ for y in the second equation.

$$5x + 2y = 8$$

$$5x + 2\overbrace{(6 - x)} = 8$$

Step 3 Solve for x using the equation we found in step 2.

$$5x + 12 - 2x = 8 \qquad \text{Distribute 2.}$$

$$3x + 12 = 8 \qquad \text{Combine like terms.}$$

$$3x = -4 \qquad \text{Subtract 12 from both sides.}$$

$$x = -\frac{4}{3} \qquad \text{Divide both sides by 3.}$$

Step 4 Find the value of y by substituting $-\frac{4}{3}$ for x in one of the equations containing both variables. Since we isolated y in $y = 6 - x$, we will use this equation.

$$y = 6 - \left(-\frac{4}{3}\right) \quad \text{Substitute } -\frac{4}{3} \text{ for } x \text{ in } y = 6 - x.$$

$$y = \frac{18}{3} + \frac{4}{3} \quad \text{Write as addition and find a common denominator.}$$

$$y = \frac{22}{3}$$

The solution is $\left(-\frac{4}{3}, \frac{22}{3}\right)$.

Step 5 We will leave the check to the reader.

YOUR TURN Solve the system using substitution.

$$\begin{cases} x - y = 1 \\ 3x + y = 11 \end{cases}$$

If none of the variables in the system of equations has a coefficient of 1, then for Step 1, select the equation that seems easiest to work with and the variable in that equation that seems easiest to isolate. When selecting that variable, recognize that you'll eventually divide by its coefficient, so choose the variable whose coefficient will divide evenly into most, if not all, of the other numbers in its equation.

EXAMPLE 5 Solve the system of equations using substitution. $\begin{cases} 3x + 4y = 6 \\ 2x + 6y = -1 \end{cases}$

Solution *Step 1* We will isolate x in $3x + 4y = 6$.

$$3x = 6 - 4y \quad \text{Subtract } 4y \text{ from both sides.}$$

$$x = 2 - \frac{4}{3}y \quad \text{Divide both sides by 3 to isolate } x.$$

Note: *We chose to isolate x because its coefficient, 3, goes evenly into one of the other numbers in the equation, 6, whereas y's coefficient, 4, does not go evenly into any of the other numbers.*

Step 2 Substitute $2 - \frac{4}{3}y$ for x in $2x + 6y = -1$.

$$2x + 6y = -1$$

$$2\left(2 - \frac{4}{3}y\right) + 6y = -1$$

ANSWER

(3, 2)

Step 3 Solve for y using the equation we found in Step 2.

$$4 - \frac{8}{3}y + 6y = -1$$ Distribute to clear the parentheses.

$$3 \cdot 4 - 3 \cdot \frac{8}{3}y + 3 \cdot 6y = 3 \cdot (-1)$$ Multiply both sides by 3 to clear the fraction.

$$12 - 8y + 18y = -3$$ Simplify.

$$12 + 10y = -3$$ Combine like terms.

$$10y = -15$$ Subtract 12 from both sides.

$$y = -\frac{3}{2}$$ Divide both sides by 10 and simplify.

Step 4 Find the value of x by substituting $-\frac{3}{2}$ for y in one of the equations containing both variables. We will use $x = 2 - \frac{4}{3}y$.

$$x = 2 - \frac{4}{3}\left(-\frac{3}{2}\right)$$ Substitute $-\frac{3}{2}$ for y in $x = 2 - \frac{4}{3}y$.

$$x = 2 + 2$$ Simplify.

$$x = 4$$

The solution is $\left(4, -\frac{3}{2}\right)$.

Step 5 We will leave the check to the reader.

YOUR TURN Solve the system of equations using substitution.

$$\begin{cases} 5x - 2y = 10 \\ 3x - 6y = 2 \end{cases}$$

ANSWER

$\left(\frac{7}{3}, \frac{5}{6}\right)$

OBJECTIVE 4. Solve systems of linear equations using elimination. Because substitution can be tedious when no coefficients are 1, we turn to a third method, the *elimination* method. In this method, we use the addition principle of equality to add equations so that a new equation emerges with one of the variables eliminated. We will work through some examples to get a sense of the method before writing a formal procedure.

EXAMPLE 6 Solve the system using elimination. $\begin{cases} x + y = 5 & \text{(Equation 1)} \\ 2x - y = 7 & \text{(Equation 2)} \end{cases}$

Solution If we add the two equations, the resulting equation has one variable.

$$x + y = 5$$ (Equation 1)

$$2x - y = 7$$ (Equation 2)

$$3x + 0 = 12$$ Add equation 1 to equation 2

$$3x = 12$$ Simplify.

$$x = 4$$ Divide both sides by 3 to isolate x.

Note: *Since y and $-y$ are additive inverses, their sum is 0. Since y is eliminated in the new equation, we can easily solve for the value of x.*

Connection The addition principle of equality (Chapter 2) says that adding the same amount to both sides of an equation will not affect its solution(s). Since Equation 1 indicates that $x + y$ and 5 are the same amount, if we add $x + y$ to the left side of Equation 2 and 5 to the right side of Equation 2, then we are applying the addition principle of equality.

Now we can find the value of y by substituting 4 for x in one of the original equations. We will use $x + y = 5$.

$$4 + y = 5 \quad \text{Substitute 4 for } x \text{ in } x + y = 5.$$
$$y = 1$$

The solution is (4, 1). We can check by verifying that (4, 1) makes both of the original equations true. We will leave this to the reader.

YOUR TURN Solve the system of equations using elimination.

$$\begin{cases} 4x + 3y = 8 \\ x - 3y = 7 \end{cases}$$

ANSWER $\left(3, -\frac{4}{3}\right)$

Connection The multiplication principle of equality (Chapter 2) says that we can multiply both sides of an equation by the same nonzero amount without affecting its solution(s).

Multiplying One Equation by a Number to Create Additive Inverses

In Example 6, you may have noted that the expressions $x + y$ and $2x - y$ conveniently had additive inverses y and $-y$. If no such pairs of additive inverses appear in a system of equations, we can use the multiplication principle of equality to multiply both sides of one or both equations by a number in order to create additive inverse pairs.

EXAMPLE 7 Solve the system using elimination.

$$\begin{cases} x + y = 6 & \text{(Equation 1)} \\ 2x - 5y = -16 & \text{(Equation 2)} \end{cases}$$

Solution We will rewrite one of the equations so that it has a term that is the additive inverse of one of the terms in the other equation. We will multiply both sides of Equation 1 by 5 so the y terms will be additive inverses.

Note: *We could have multiplied both sides of Equation 1 by −2. We would get a −2x term in Equation 1 that is the opposite of the 2x term in Equation 2.*

$$5 \cdot x + 5 \cdot y = 5 \cdot 6 \quad \text{Multiply both sides of Equation 1 by 5.}$$
$$5x + 5y = 30 \quad \text{(Equation 1 rewritten)}$$

Now we add the rewritten Equation 1 to Equation 2 to eliminate the y term.

Note: *Multiplying Equation 1 by 5 made the elimination of the y terms possible.*

$$\begin{aligned} 5x + 5y &= 30 && \text{(Equation 1 rewritten)} \\ 2x - 5y &= -16 && \text{(Equation 2)} \\ \hline 7x + 0 &= 14 && \text{Add rewritten Equation 1 to Equation 2 to eliminate } y. \\ 7x &= 14 && \text{Simplify.} \\ x &= 2 && \text{Divide both sides by 7 to isolate } x. \end{aligned}$$

To finish, we substitute 2 for x in an equation containing both variables. We will use $x + y = 6$.

$$\begin{aligned} 2 + y &= 6 && \text{Substitute 2 for } x \text{ in } x + y = 6. \\ y &= 4 \end{aligned}$$

The solution is (2, 4). We will leave the check to the reader.

YOUR TURN Solve the system using elimination.

$$\begin{cases} -3x - 5y = 6 \\ x + 2y = -1 \end{cases}$$

Multiplying Each Equation by a Number to Create Additive Inverses

If no coefficient in a system of equations is 1, then we may have to multiply each equation by a number to generate a pair of additive inverses.

EXAMPLE 8 Solve the system using elimination.

$$\begin{cases} 4x - 3y = -2 & \text{(Equation 1)} \\ 6x - 7y = 7 & \text{(Equation 2)} \end{cases}$$

Solution We will choose to eliminate x, so we will multiply Equation 1 by 3 and Equation 2 by -2, which makes the x terms additive inverses.

$$\begin{aligned} 4x - 3y = -2 &\xrightarrow{\text{Multiply by 3}} 12x - 9y = -6 \\ 6x - 7y = 7 &\xrightarrow[\text{Multiply by } -2]{} -12x + 14y = -14 \end{aligned}$$

Note: *The x terms are now additive inverses, $12x$ and $-12x$.*

Now we can add the rewritten equations to eliminate the x term.

$$\begin{aligned} 12x - 9y &= -6 \\ -12x + 14y &= -14 \\ \hline 0 + 5y &= -20 && \text{Add the equations.} \\ y &= -4 && \text{Solve for } y \text{ by dividing both sides by 5.} \end{aligned}$$

To finish, we substitute -4 for y in one of the equations and solve for x. We will use $4x - 3y = -2$.

$$\begin{aligned} 4x - 3(-4) &= -2 && \text{Substitute } -4 \text{ for } y \text{ in } 4x - 3y = -2. \\ 4x + 12 &= -2 && \text{Simplify.} \\ 4x &= -14 && \text{Subtract 12 from both sides.} \\ x &= -\frac{7}{2} && \text{Divide both sides by 4 and simplify.} \end{aligned}$$

Note: *It does not matter which equation we use because x has the same value in any equation in the system.*

The solution is $\left(-\frac{7}{2}, -4\right)$. We will leave the check to the reader.

ANSWER

$(-7, 3)$

Fractions or Decimals in a System

We can use the multiplication principle to clear any fractions or decimals from the equations in a system.

EXAMPLE 9 Solve the system using elimination.

$$\begin{cases} \frac{1}{2}x - y = \frac{3}{4} & \text{(Equation 1)} \\ 0.4x - 0.3y = 1 & \text{(Equation 2)} \end{cases}$$

Solution To clear the fractions in Equation 1, we multiply both sides by the LCD, 4. To clear the decimals in Equation 2, we multiply both sides by 10.

$$\frac{1}{2}x - y = \frac{3}{4} \xrightarrow{\text{Multiply by 4}} 2x - 4y = 3$$

$$0.4x - 0.3y = 1 \xrightarrow[\text{Multiply by 10}]{} 4x - 3y = 10$$

Now that both equations contain only integers, the system is easier to solve. We will eliminate x by multiplying the first equation by -2, which makes the x terms additive inverses, then add the equations.

$$\begin{aligned} -4x + 8y &= -6 \\ 4x - 3y &= 10 \\ \hline 0 + 5y &= 4 && \text{Add the equations.} \\ y &= \frac{4}{5}, \text{ or } 0.8 && \text{Solve for } y \text{ by dividing both sides by 5.} \end{aligned}$$

To finish, we substitute 0.8 for y in one of the equations and solve for x.

$$\begin{aligned} 0.4x - 0.3(0.8) &= 1 && \text{Substitute 0.8 for } y \text{ in Equation 2.} \\ 0.4x - 0.24 &= 1 && \text{Simplify.} \\ 0.4x &= 1.24 && \text{Add 0.24 to both sides.} \\ x &= 3.1 && \text{Divide both sides by 0.4.} \end{aligned}$$

The solution is (3.1, 0.8). We will leave the check to the reader.

YOUR TURN Solve the system using elimination.

$$\begin{cases} 0.6x + y = 3 \\ \frac{2}{5}x + \frac{1}{4}y = 1 \end{cases}$$

Rewriting the Equations in the form $Ax + By = C$

In Examples 6–9, all the equations were in standard form, which we learned in Chapter 3 to be $Ax + By = C$. To use the elimination method, the equations need to be written in standard form. Consider the following system:

$$\begin{cases} y = 2 - x & \text{(Equation 1)} \\ 3x - y = 1 & \text{(Equation 2)} \end{cases}$$

ANSWER

$\left(1, \frac{12}{5}\right)$, or (1, 2.4)

We need to rewrite Equation 1 in standard form before using elimination.

$$\begin{cases} x + y = 2 & \text{(Equation 1)} \\ 3x - y = 1 & \text{(Equation 2)} \end{cases}$$

We can now summarize the elimination method with the following procedure:

PROCEDURE **Solving Systems of Equations Using Elimination**

To solve a system of two linear equations using the elimination method,

1. Write the equations in standard form ($Ax + By = C$).
2. Use the multiplication principle to clear fractions or decimals.
3. Multiply one or both equations by a number (or numbers) so that they have a pair of terms that are additive inverses.
4. Add the equations. The result should be an equation in terms of one variable.
5. Solve the equation from step 4 for the value of that variable.
6. Using an equation containing both variables, substitute the value you found in step 5 for the corresponding variable and solve for the value of the other variable.
7. Check your solution in the original equations.

Inconsistent Systems and Dependent Equations

How would we recognize an inconsistent system or a system with dependent equations using the elimination method?

EXAMPLE 10 Solve the system using elimination. $\begin{cases} 2x - y = 1 \\ 2x - y = -3 \end{cases}$

Solution Notice that the left sides of the equations match. Multiplying one of the equations by -1, then adding the equations, will eliminate both variables.

Connection If we were to graph the equations in Example 10, we would see that the lines are parallel, confirming there is no solution for the system.

$$\begin{aligned} 2x - y &= 1 \xrightarrow{\text{Multiply by } -1} & -2x + y &= -1 \\ 2x - y &= -3 & 2x - y &= -3 \\ & & 0 &= -4 \end{aligned}$$

Both variables have been eliminated and the resulting equation, $0 = -4$, is false; therefore, there is no solution. This system of equations is inconsistent.

EXAMPLE 11 Solve the system using elimination. $\begin{cases} 3x + 4y = 5 \\ 9x + 12y = 15 \end{cases}$

Solution To eliminate x, we could multiply the first equation by -3, then combine the equations.

Connection If graphed, both equations in Example 11 generate the same line, indicating the equations are dependent.

$$\begin{aligned} 3x + 4y &= 5 \xrightarrow{\text{Multiply by } -3} & -9x - 12y &= -15 \\ 9x + 12y &= 15 & 9x + 12y &= 15 \\ & & 0 &= 0 \end{aligned}$$

Both variables have been eliminated and the resulting equation, $0 = 0$, is true. This means the equations are dependent, so there are an infinite number of solutions. We can say the solution set contains every ordered pair that satisfies $3x + 4y = 5$ (or $9x + 12y = 15$).

ANSWERS

a. all ordered pairs along $x - 4y = 2$ (dependent equations)

b. no solution (inconsistent)

YOUR TURN Solve the system of equations.

a. $\begin{cases} x - 4y = 2 \\ 5x - 20y = 10 \end{cases}$

b. $\begin{cases} x + 2y = 3 \\ x + 2y = 1 \end{cases}$

4.1 Exercises

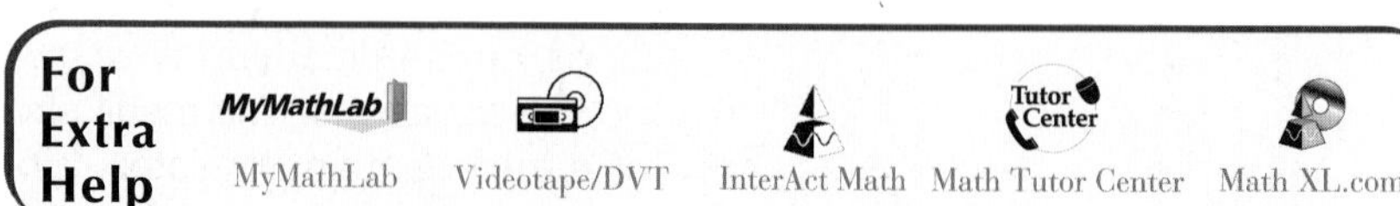

1. How do you check a solution for a system of equations?

2. If a system of linear equations in two variables has one solution, what does this indicate about the graphs of the equations?

3. If a system of linear equations in two variables has no solution (inconsistent), what does this indicate about the graphs of the equations?

4. If two linear equations in two variables are dependent, what does this mean about their graphs?

5. Suppose you plan to use substitution to solve the system $\begin{cases} y = x + 2 \\ 3x - 4y = -9 \end{cases}$.

a. Which variable would you replace in the substitution process?

b. What expression would replace that variable?

6. Suppose you plan to use elimination to solve the system $\begin{cases} 4x + 2y = 10 \\ 3x - 2y = 4 \end{cases}$.

Which variable would you eliminate? Why?

For Exercises 7–14, determine whether the given ordered pair is a solution to the given system of equations.

7. $(-1, 2)$; $\begin{cases} x - y = -3 \\ x + y = 1 \end{cases}$

8. $(1, -2)$; $\begin{cases} 4x - 3y = 10 \\ 3x - 2y = 7 \end{cases}$

9. $(-2, 3)$; $\begin{cases} 3x + 4y = 6 \\ x - 4y = 8 \end{cases}$

10. $(-3, 2)$; $\begin{cases} 2x + 3y = -12 \\ 4x - 5y = 2 \end{cases}$

11. $\left(-\frac{3}{4}, -\frac{2}{3}\right)$; $\begin{cases} 4x + 2y = 9 \\ x - 7y = 3 \end{cases}$

12. $\left(\frac{3}{4}, \frac{2}{3}\right)$; $\begin{cases} 5x + y = 1 \\ x + y = 6 \end{cases}$

13. $(3, -4)$; $\begin{cases} 3x + 2y = 1 \\ \frac{1}{3}x - \frac{1}{2}y = 3 \end{cases}$

14. $(4, -5)$; $\begin{cases} \frac{1}{4}x - \frac{3}{5}y = 4 \\ x + 2y = -6 \end{cases}$

For Exercises 15–18, translate into a system of equations. Do not solve.

15. The sum of two numbers is five. The difference of the same numbers is three. What are the numbers?

16. The difference of two numbers is eight. The sum of the same numbers is twelve. What are the numbers?

17. The perimeter of a rectangle is fifty. If the width is two less than the length, find the dimensions of the rectangle.

18. Jon is three years older than his sister. If the sum of their ages is twelve, how old are they?

For Exercises 19–24: a. Determine whether the system of equations is consistent with independent equations, inconsistent with independent equations, or consistent with dependent equations.
b. How many solutions does the system have?

19.

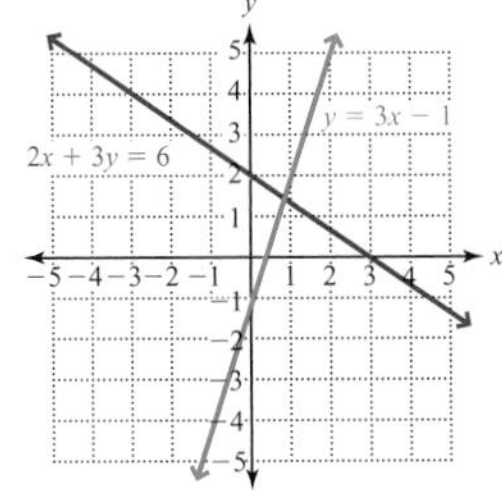

20.

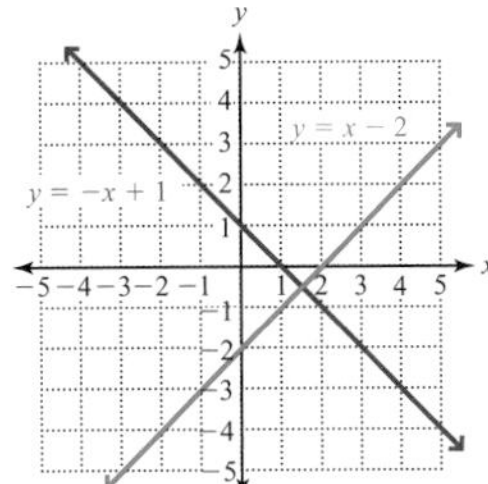

21.

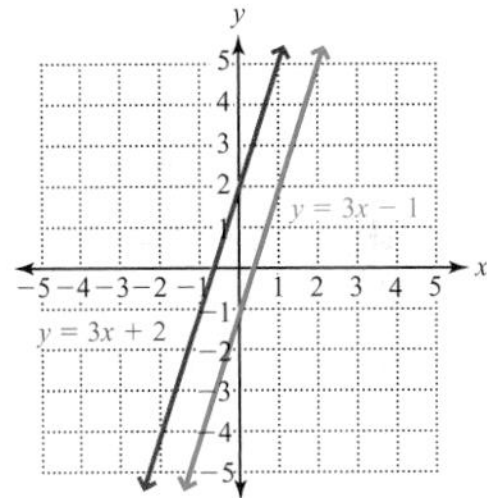

22.

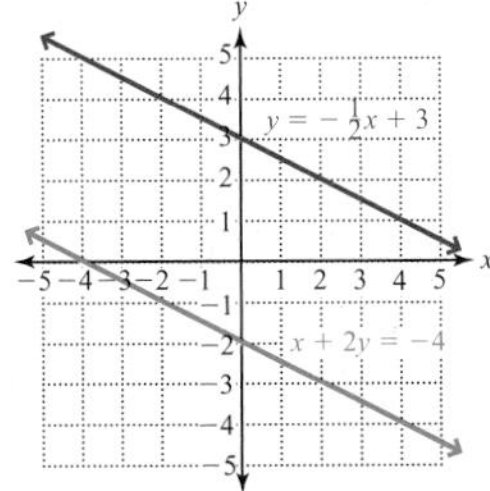

23.

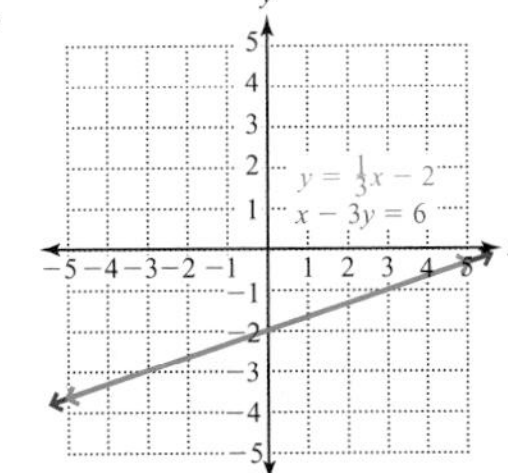

24.

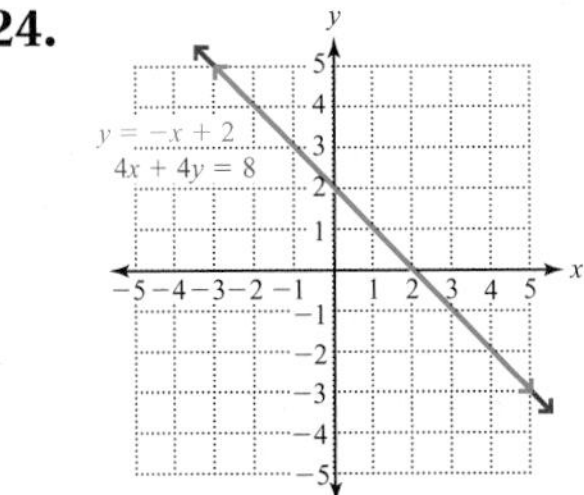

For Exercises 25–30, determine whether the system of equations is consistent with independent equations, inconsistent with independent equations, or consistent with dependent equations.

25. $\begin{cases} y = -x \\ y - x = 6 \end{cases}$

26. $\begin{cases} y = x + 4 \\ 4x + y = 10 \end{cases}$

27. $\begin{cases} x + 3y = 1 \\ 2x + 6y = 2 \end{cases}$

28. $\begin{cases} 6x + 9y = 6 \\ 2x + 3y = 2 \end{cases}$

29. $\begin{cases} 3x + 2y = 12 \\ 6x + 4y = -12 \end{cases}$

30. $\begin{cases} 2x + y = 1 \\ 4x + 2y = 1 \end{cases}$

For Exercises 31–40, solve the system graphically.

31. $\begin{cases} x + y = 5 \\ x - y = 3 \end{cases}$

32. $\begin{cases} x - y = 1 \\ 2x + y = 8 \end{cases}$

33. $\begin{cases} y = 2x + 5 \\ y = -x - 4 \end{cases}$

34. $\begin{cases} y = -3x - 1 \\ y = 2x - 6 \end{cases}$

35. $\begin{cases} 2x - y = 3 \\ 2x - y = 8 \end{cases}$

36. $\begin{cases} 3x + 2y = 6 \\ 6x + 4y = 18 \end{cases}$

37. $\begin{cases} 3x + y = 4 \\ 6x + 2y = 8 \end{cases}$

38. $\begin{cases} 2x + y = -2 \\ 8x + 4y = -8 \end{cases}$

39. $\begin{cases} x = 4 \\ y = -2 \end{cases}$

40. $\begin{cases} x = -3 \\ y = 4 \end{cases}$

For Exercises 41–56, solve the system by substitution.

41. $\begin{cases} 3x - y = 2 \\ y = 2x \end{cases}$

42. $\begin{cases} 2x - y = 5 \\ y = -3x \end{cases}$

43. $\begin{cases} x + y = 1 \\ y = 2x - 5 \end{cases}$

44. $\begin{cases} x = y + 5 \\ x + y = 1 \end{cases}$

45. $\begin{cases} y = -\dfrac{3}{4}x \\ x - 8y = -7 \end{cases}$

46. $\begin{cases} y = \dfrac{2}{3}x \\ 6x + 3y = 4 \end{cases}$

47. $\begin{cases} 3x + 4y = 11 \\ x + 2y = 5 \end{cases}$

48. $\begin{cases} 2x + y = 7 \\ 3x + 2y = 12 \end{cases}$

49. $\begin{cases} 4x - y = -11 \\ 3x - 4y = -5 \end{cases}$

50. $\begin{cases} 5x - 3y = 2 \\ 3x - y = -2 \end{cases}$

51. $\begin{cases} 4x + 3y = -2 \\ 8x - 2y = 12 \end{cases}$

52. $\begin{cases} 2x - 5y = 18 \\ 3x + 2y = 8 \end{cases}$

53. $\begin{cases} x - 3y = -4 \\ 5x - 15y = -6 \end{cases}$

54. $\begin{cases} 2x - y = 1 \\ 4x - 2y = 5 \end{cases}$

55. $\begin{cases} x - 2y = 6 \\ -2x + 4y = -12 \end{cases}$

56. $\begin{cases} x + y = -5 \\ -2x - 2y = 10 \end{cases}$

For Exercises 57 and 58, explain the mistake, then find the correct solution.

57. $\begin{cases} 2x + y = 8 \\ x = y + 1 \end{cases}$

Substitute:

$$2(y + 1) + y = 8$$
$$2y + 1 + y = 8$$
$$3y + 1 = 8$$
$$3y = 7$$
$$y = \frac{7}{3}$$

Solve for x:

$$x = \frac{7}{3} + 1$$
$$x = \frac{10}{3}$$

Solution: $\left(\dfrac{10}{3}, \dfrac{7}{3}\right)$

58. $\begin{cases} 3x + 7y = 8 \\ x - y = 6 \end{cases}$

Rewrite: $x = 6 + y$

Substitute: $6 + y - y = 6$

$6 = 6$

Solution: Infinite solutions along $x - y = 6$

For Exercises 59–78, solve the system using elimination.

59. $\begin{cases} x + y = 1 \\ 2x - y = 2 \end{cases}$

60. $\begin{cases} x - y = 7 \\ x + y = 5 \end{cases}$

61. $\begin{cases} 2x + 3y = 4 \\ 4x + 3y = 5 \end{cases}$

62. $\begin{cases} 3x + 2y = 6 \\ 5x + 2y = 14 \end{cases}$

63. $\begin{cases} 4x + y = 8 \\ 5x + 3y = 3 \end{cases}$

64. $\begin{cases} 5x + y = 12 \\ 5x - 6y = -2 \end{cases}$

65. $\begin{cases} x + 3y = -1 \\ 3x + 6y = -1 \end{cases}$

66. $\begin{cases} x + y = 68 \\ 2x + 3y = 160 \end{cases}$

67. $\begin{cases} 3x = 2y + 7 \\ 4x - 3y = 10 \end{cases}$

68. $\begin{cases} 4x = 3y - 23 \\ 5x - 4y = -30 \end{cases}$

69. $\begin{cases} 12x + 18y = 17 \\ 6x + 10y = 9 \end{cases}$

70. $\begin{cases} 4x - 4y = -1 \\ 12x - 8y = 3 \end{cases}$

71. $\begin{cases} \frac{1}{5}x + \frac{1}{2}y = \frac{1}{5} \\ \frac{1}{2}x + \frac{1}{3}y = -\frac{4}{3} \end{cases}$

72. $\begin{cases} \frac{1}{4}x - \frac{1}{2}y = -\frac{1}{4} \\ \frac{1}{3}x + \frac{1}{6}y = \frac{4}{3} \end{cases}$

73. $\begin{cases} 0.4x - 0.3y = 1.3 \\ 0.3x + 0.5y = 1.7 \end{cases}$

74. $\begin{cases} 0.5x - 0.2y = -2.1 \\ 0.2x + 0.5y = 0.9 \end{cases}$

75. $\begin{cases} 2x + 3y = 6 \\ 4x - 18 = -6y \end{cases}$

76. $\begin{cases} 4x = 2y + 7 \\ 2x - y = 4 \end{cases}$

77. $\begin{cases} x + 2y = 4 \\ x - 4 = -2y \end{cases}$

78. $\begin{cases} 4x + 2y = 6 \\ 6x = 9 - 3y \end{cases}$

For Exercises 79 and 80, explain the mistake, then find the correct solution.

79. $\begin{cases} x + y = 8 \\ \underline{x - y = 7} \end{cases}$

$2x = 1$

$x = \frac{1}{2}$

$\frac{1}{2} + y = 8$

$y = \frac{15}{2}$

Solution: $\left(\frac{1}{2}, \frac{15}{2}\right)$

80. $\begin{cases} x + y = 3 \\ x + 2y = 4 \end{cases}$

$\begin{cases} x + y = 3 \\ -1(x + 2y = 4) \end{cases}$

$\begin{cases} x + y = 3 \\ \underline{-x - 2y = 4} \end{cases}$

$-y = 7$

$y = -7$

$x - 7 = 3$

$x = 10$

Solution: $(10, -7)$

For Exercises 81–84, solve.

81. A business breaks even when its cost and revenue are equal. The following graph shows the cost of a product based on the number of units produced and the revenue based on the number of units sold.

   a. How many units must be sold in order to break even?

   b. What is the amount of revenue and cost for the business to break even?

   c. Write an inequality that describes the number of units that must be sold for the business to make a profit.

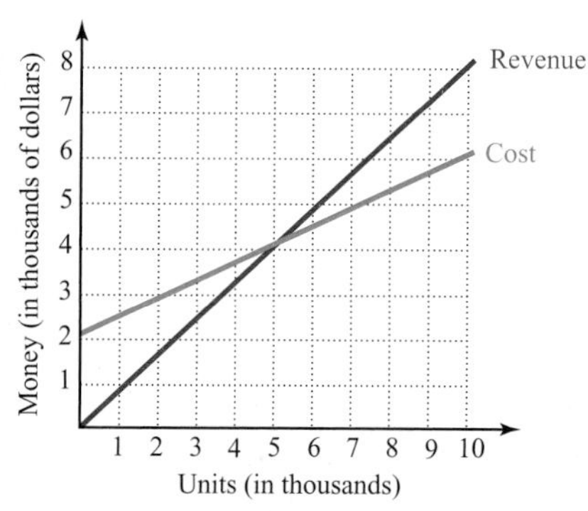

82. The graph shows two long-distance plans. Plan 1 costs \$7.95 per month plus \$0.03 per minute. This plan can be used to call anyone in the country. Plan 2 costs a flat \$19.95 per month with unlimited calls, but you may call only other customers using AT&T at that rate.

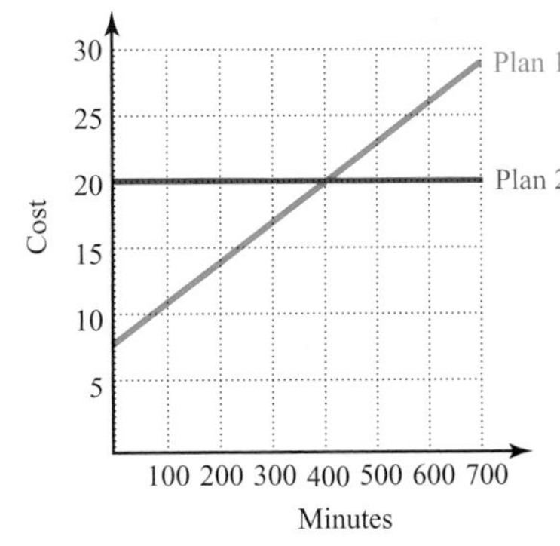

a. How many minutes would a person have to talk in one month using the first plan to equal the monthly cost of the second plan? (Assume all customers are AT&T customers.)

b. If a person considering the first plan were to use 500 minutes calling only AT&T customers, is the first plan the best plan?

c. What other factors need to be considered in choosing one of the two plans?

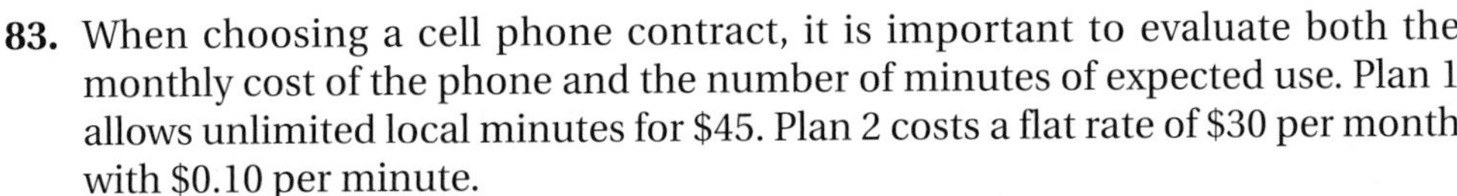

83. When choosing a cell phone contract, it is important to evaluate both the monthly cost of the phone and the number of minutes of expected use. Plan 1 allows unlimited local minutes for \$45. Plan 2 costs a flat rate of \$30 per month with \$0.10 per minute.

a. Write a system of equations so that one equation describes the cost of plan 1 and the other equation describes the total cost of plan 2.

b. On one grid, graph both equations in the system.

c. How many minutes must be used in order for the two plans to cost the same per month?

d. Which plan is cheaper if the customer plans to use more than 200 minutes per month?

e. If the phone is only for emergencies, which plan is the less-expensive choice?

84. Deciding whether to purchase or lease a car can be a challenging decision. Suppose a dealership offers two deals for a particular car. The first deal is to purchase the car with \$2700 down plus a monthly payment of \$550 for 36 months. The second deal is to lease the car with \$1500 down plus a lease payment of \$350 per month for 36 months. Upon returning the vehicle in good condition, the lease agreement can be transferred to a new vehicle at the same payment (\$350) and no additional down payment.

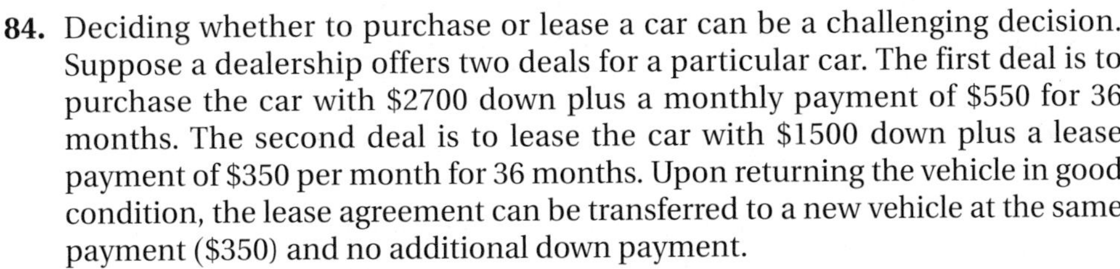

a. Write a system of equations so that one equation describes the total cost of buying the vehicle and the other equation describes the total cost of leasing the car indefinitely.

b. On one grid, graph both equations in the system.

c. After how many months is the total cost of leasing the same as that of purchasing this car?

d. If the customer wants to drive a new vehicle every three years, is it better to go with the purchase or the lease deal?

e. What other factors should a person consider when deciding whether to purchase or lease a car?

REVIEW EXERCISES

[2.1] **1.** Solve for y: $x + 3y = 6$

[2.1] **2.** Solve for x: $\frac{1}{3}x - 2y = 10$

[1.4] **3.** Use the distributive property: $4(5x + 7y - z)$

[1.4] **4.** Simplify: $3x - 4y + z - 2x + y - z$

[1.4] **5.** Given $x + y + 2z = 7$, find x if $y = -1$ and $z = 3$.

[1.4] **6.** Given $x - 3y + 2z = 6$, find y if $x = -2$ and $z = 4$.

4.2 Solving Systems of Linear Equations in Three Variables

OBJECTIVES

1. **Determine if an ordered triple is a solution for a system of equations.**
2. **Understand the types of solution sets for systems of three equations.**
3. **Solve a system of three linear equations using the elimination method.**

OBJECTIVE 1. Determine if an ordered triple is a solution for a system of equations. In this section, we will solve systems of three linear equations with three unknowns. Solutions of these systems will be *ordered triples* with the form (x, y, z). We check solutions for these systems just as we did in Section 4.1: Replace each variable with its corresponding value and verify that each equation is true.

EXAMPLE 1 Determine whether each ordered triple is a solution to the system of equations.

$$\begin{cases} x + y + z = 3 & \text{(Equation 1)} \\ 2x + 3y + 2z = 7 & \text{(Equation 2)} \\ 3x - 4y + z = 4 & \text{(Equation 3)} \end{cases}$$

a. (2, 1, 0)

Solution In all three equations, replace x with **2**, y with **1**, and z with **0**.

Equation 1:

$$x + y + z = 3$$
$$2 + 1 + 0 \stackrel{?}{=} 3$$
$$3 = 3 \quad \text{True}$$

Equation 2:

$$2x + 3y + 2z = 7$$
$$2(2) + 3(1) + 2(0) \stackrel{?}{=} 7$$
$$4 + 3 + 0 \stackrel{?}{=} 7$$
$$7 = 7 \quad \text{True}$$

Equation 3:

$$3x - 4y + z = 4$$
$$3(2) - 4(1) + 0 \stackrel{?}{=} 4$$
$$6 - 4 + 0 \stackrel{?}{=} 4$$
$$2 = 4 \quad \text{False}$$

Because (2, 1, 0) does not satisfy all three equations in the system, it is not a solution for the system.

b. (3, 1, −1)

Solution

Equation 1:

$$x + y + z = 3$$
$$3 + 1 + (-1) \stackrel{?}{=} 3$$
$$3 = 3 \quad \text{True}$$

Equation 2:

$$2x + 3y + 2z = 7$$
$$2(3) + 3(1) + 2(-1) \stackrel{?}{=} 7$$
$$6 + 3 - 2 \stackrel{?}{=} 7$$
$$7 = 7 \quad \text{True}$$

Equation 3:

$$3x - 4y + z = 4$$
$$3(3) - 4(1) + (-1) \stackrel{?}{=} 4$$
$$9 - 4 - 1 \stackrel{?}{=} 4$$
$$4 = 4 \quad \text{True}$$

Because (3, 1, −1) satisfies all three equations in the system, it is a solution for the system.

YOUR TURN Determine whether the following ordered triples are solutions to the system of equations.

a. (4, 1, 1) **b.** (−1, 5, 2)

$$\begin{cases} x + y + z = 6 & \text{(Equation 1)} \\ 3x - 2y + 3z = -7 & \text{(Equation 2)} \\ 4x - 2y + z = -12 & \text{(Equation 3)} \end{cases}$$

OBJECTIVE 2. Understand the types of solution sets for systems of three equations. Recall that we plot an ordered pair using two axes: x and y. Similarly, we plot an ordered triple like (3, 4, 5) using three axes: x, y, and z, each of which is perpendicular to the other two, as shown.

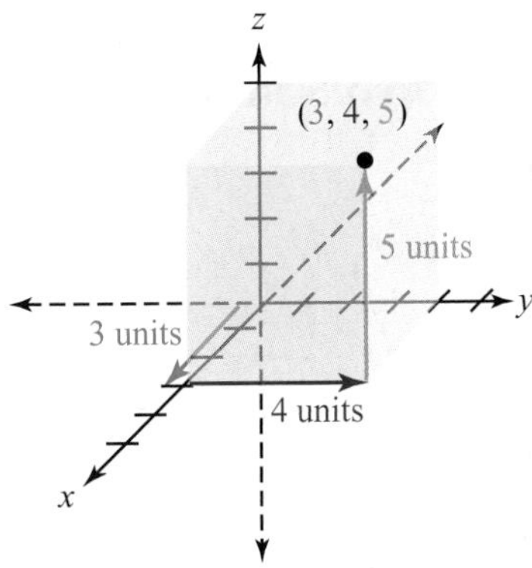

The graph of a linear equation in three variables is a *plane*, which is a flat surface much like a sheet of paper with infinite length and width. We draw a plane as a parallelogram (shown to the right), but remember that the length and width are infinite.

In Section 4.1, we found that two lines could intersect in one point (consistent system), no points (inconsistent system), or an infinite number of points (consistent system with dependent equations). Similarly, the intersection of the planes of a three-variable system can tell us about the number of solutions for the system.

ANSWERS

a. not a solution
b. solution

A Single Solution

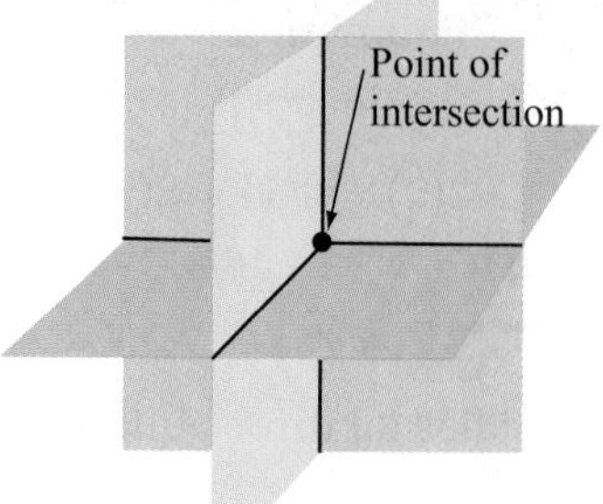

If the planes intersect at a single point, then that ordered triple is the solution to the system.

Connection Recall from Section 4.1 that a system of equations that has a solution (single or infinite) is consistent and a system that has no solution is inconsistent. This is true for systems of three equations as well. We also discussed dependent and independent equations. However, for systems of three equations, defining dependent and independent equations and discussing their graphical representations is too complicated for this course. Here, we will simply say that systems with an infinite number of solutions have dependent equations. As we will see, we can identify these systems when solving them using the elimination method.

Infinite Number of Solutions

If the three planes intersect along a line, then the system has an infinite number of solutions, which are the coordinates of any point along that line.

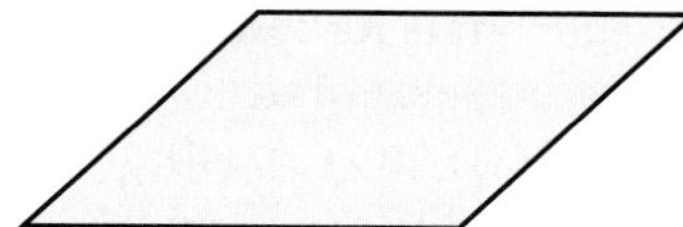

If all three graphs are the same plane, then the system has an infinite number of solutions, which are the coordinates of any point in the plane.

No Solution

If all the planes are parallel, then the system has no solution.

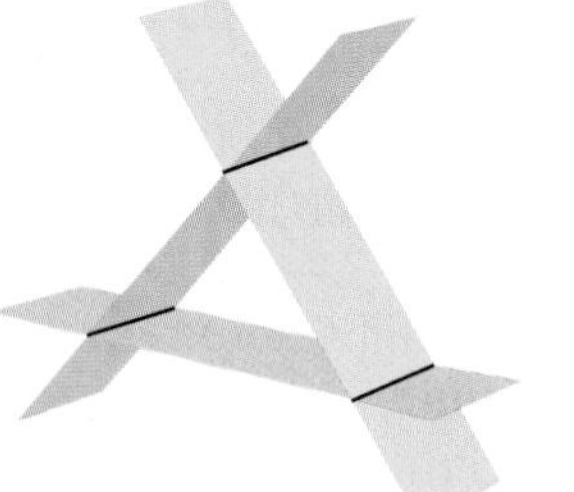

Pairs of planes could also intersect, as shown. However, since all three planes do not have a common intersection, the system has no solution.

Note: *Planes can be configured in still more ways so that they do not have a common intersection.*

OBJECTIVE 3. Solve a system of three linear equations using the elimination method. Since graphing systems of three linear equations is usually impractical, we will solve them using elimination. The process is much like what we learned in Section 4.1.

EXAMPLE 2 Solve the system of equations using elimination.

$$\begin{cases} x + y + z = 2 & \text{(Equation 1)} \\ 2x + y + 2z = 1 & \text{(Equation 2)} \\ 3x + 2y + z = 1 & \text{(Equation 3)} \end{cases}$$

Solution First, we choose a variable and eliminate that variable from two pairs of equations. Let's eliminate y using Equations 1 and 2. We multiply Equation 1 by -1, then add the equations.

Note: *Our initial goal is to generate two equations with two variables that are the same.*

$$\begin{array}{llcrl} \text{(Eq. 1)} & x + y + z = 2 & \xrightarrow{\text{Multiply by } -1} & -x - y - z = -2 & \\ \text{(Eq. 2)} & 2x + y + 2z = 1 & & \underline{2x + y + 2z = 1} & \text{Add the equations.} \\ & & & x \quad + z = -1 & \text{(Equation 4)} \end{array}$$

Now we need to eliminate y from another pair of equations. We will choose Equations 1 and 3.

$$\begin{array}{llcrl} \text{(Eq. 1)} & x + y + z = 2 & \xrightarrow{\text{Multiply by } -2} & -2x - 2y - 2z = -4 & \\ \text{(Eq. 3)} & 3x + 2y + z = 1 & & \underline{3x + 2y + \quad z = 1} & \text{Add the equations.} \\ & & & x \qquad - z = -3 & \text{(Equation 5)} \end{array}$$

Equations 4 and 5 form a system of two equations with two variables x and z. We can eliminate z by adding the equations together.

$$\begin{array}{lrl} \text{(Eq. 4)} & x + z = -1 & \\ \text{(Eq. 5)} & \underline{x - z = -3} & \text{Add the equations.} \\ & 2x \quad = -4 & \\ & x = -2 & \text{Divide both sides by 2 to isolate } x. \end{array}$$

To find z, we substitute -2 for x in Equation 4 or 5. We will use Equation 4.

$$\begin{array}{rl} x + z = -1 & \\ -2 + z = -1 & \text{Substitute } -2 \text{ for } x. \\ z = 1 & \text{Add 2 to both sides.} \end{array}$$

To find y, substitute -2 for x and 1 for z in any of the original equations. We will use Equation 1.

$$\begin{array}{rl} x + y + z = 2 & \\ -2 + y + 1 = 2 & \text{Substitute } -2 \text{ for } x \text{ and 1 for } z. \\ y - 1 = 2 & \text{Simplify the left side.} \\ y = 3 & \text{Add 1 to both sides to isolate } y. \end{array}$$

The solution is $(-2, 3, 1)$. We can check the solution by verifying that the ordered triple satisfies each of the three original equations. We will leave the check to the reader.

In some systems, one or more variables may be missing from one or more of the equations, which can simplify solving the system.

EXAMPLE 3 Solve the system using elimination.

$$\begin{cases} 2x + 3y = 3 & \text{(Equation 1)} \\ 2y - 3z = -8 & \text{(Equation 2)} \\ 4x - z = 10 & \text{(Equation 3)} \end{cases}$$

Solution Notice that a variable is missing in each of the equations. We will eliminate z from Equations 2 and 3.

Note: *We chose Equation 3 because z has a coefficient of -1.*

$$\begin{array}{llcrl} \text{(Eq. 2)} & 2y - 3z = -8 & & 2y - 3z = -8 & \\ \text{(Eq. 3)} & 4x - z = 10 & \xrightarrow{\text{Multiply by } -3} & -12x + 3z = -30 & \text{Add the equations.} \\ & & & -12x + 2y = -38 & \text{(Equation 4)} \end{array}$$

Because Equations 1 and 4 do not have z terms, we do not need to eliminate z from another pair of equations. We will now use Equations 1 and 4 to eliminate x.

$$\begin{array}{llcrl} \text{(Eq. 1)} & 2x + 3y = 3 & \xrightarrow{\text{Multiply by } 6} & 12x + 18y = 18 & \\ \text{(Eq. 4)} & -12x + 2y = -38 & & -12x + 2y = -38 & \text{Add the equations.} \\ & & & 20y = -20 & \\ & & & y = -1 & \text{Divide both sides by 20 to isolate } y. \end{array}$$

To find x, substitute -1 for y in Equation 1 or 4. We will use Equation 1.

$$\begin{aligned} 2x + 3(-1) &= 3 && \text{Substitute } -1 \text{ for } y \text{ in Equation 1.} \\ 2x - 3 &= 3 && \text{Simplify.} \\ 2x &= 6 && \text{Add 3 to both sides.} \\ x &= 3 && \text{Divide both sides by 2 to isolate } x. \end{aligned}$$

To find z, substitute in either Equation 2 or 3 (Equation 1 has no z term). We will use Equation 2.

$$\begin{aligned} 2(-1) - 3z &= -8 && \text{Substitute } -1 \text{ for } y \text{ in Equation 2.} \\ -2 - 3z &= -8 && \text{Simplify.} \\ -3z &= -6 && \text{Add 2 to both sides.} \\ z &= 2 && \text{Divide both sides by } -3 \text{ to isolate } z. \end{aligned}$$

The solution is $(3, -1, 2)$. We will leave the check to the reader.

Following is a summary of the process of solving a system of three linear equations:

PROCEDURE **Solving Systems of Three Equations Using Elimination**

To solve a system of three equations with three unknowns using elimination,

1. Write each equation in the form $Ax + By + Cz = D$.
2. Eliminate one variable from one pair of equations using the elimination method.
3. If necessary, eliminate the same variable from another pair of equations.
4. Steps 2 and 3 result in two equations with the same two variables. Solve these equations using the elimination method.
5. To find the third variable, substitute the values of the variables found in step 4 into any of the three original equations that contain the third variable.
6. Check the ordered triple in all three of the original equations.

Let's see what happens when a system of three linear equations has no solution (inconsistent).

EXAMPLE 4 Solve the system using elimination.

$$\begin{cases} 2x + y + 2z = -1 & \text{(Equation 1)} \\ -3x + 2y + 3z = -13 & \text{(Equation 2)} \\ 4x + 2y + 4z = 5 & \text{(Equation 3)} \end{cases}$$

Solution We need to eliminate a variable from two pairs of equations. Let's eliminate y using Equations 1 and 2.

$$\begin{array}{lll} \text{(Eq. 1)}\ \ 2x + y + 2z = -1 & \xrightarrow{\text{Multiply by } -2} & -4x - 2y - 4z = 2 \\ \text{(Eq. 2)}\ -3x + 2y + 3z = -13 & & \underline{-3x + 2y + 3z = -13} \quad \text{Add the equations.} \\ & & -7x \qquad\quad -z = -11 \quad \text{(Equation 4)} \end{array}$$

Now we eliminate y from Equations 2 and 3.

$$\begin{array}{lll} \text{(Eq. 2)}\ -3x + 2y + 3z = -13 & \xrightarrow{\text{Multiply by } -1} & 3x - 2y - 3z = 13 \\ \text{(Eq. 3)}\ \ 4x + 2y + 4z = 5 & & \underline{4x + 2y + 4z = 5} \quad \text{Add the equations.} \\ & & 7x \qquad\quad + z = 18 \quad \text{(Equation 5)} \end{array}$$

Equations 4 and 5 form a system in x and z.

$$\begin{array}{ll} \text{(Eq. 4)} & -7x - z = -11 \\ \text{(Eq. 5)} & \underline{\ \ 7x + z = 18} \quad \text{Add the equations.} \\ & \quad\ \ 0 = 7 \end{array}$$

All variables are eliminated and the resulting equation is false, which means this system has no solution; it is inconsistent.

Note: *We mentioned in the Connection box on p. 246 that we can identify systems of dependent equations when using the elimination method. A system of three equations is dependent if during the solution process we add two equations and get $0 = 0$ (we saw this with two equations in Example 11 of Section 4.1). For example, if Equation 4 of Example 4 were $-7x - z = -18$, the sum of Equations 4 and 5 would be $0 = 0$.*

$$\begin{array}{ll} \text{(Eq. 4)} & -7x - z = -18 \\ \text{(Eq. 5)} & \underline{\ \ 7x + z = 18} \\ & \qquad\ 0 = 0 \end{array}$$

In such a case, we say the equations in the system are dependent and the system has an infinite number of solutions.

YOUR TURN Solve the following systems of equations.

a. $\begin{cases} x + 3y + 2z = 6 \\ 2x - 3y + z = -18 \\ -3x + 2y + z = 12 \end{cases}$ **b.** $\begin{cases} 2x + y + 2z = 1 \\ x + y + z = 2 \\ 4x + 2y + 4z = 6 \end{cases}$ **c.** $\begin{cases} x + 3y - 6z = 9 \\ -7y + 6z = -3 \\ -2x + y + 6z = -15 \end{cases}$

ANSWERS

a. $(-2, 4, -2)$

b. no solution

c. infinite number of solutions (dependent equations)

4.2 Exercises

For Extra Help

MyMathLab

Videotape/DVT

InterAct Math

Math Tutor Center

Math XL.com

1. When solving a system of three equations with three variables, you eliminate a variable from one pair of equations, then eliminate the same variable from a second pair of equations. Does it matter which equations you choose?

2. When solving a system of three equations with three variables, you eliminate a variable from one pair of equations, then eliminate the same variable from a second pair of equations. Why do we eliminate the same variable from two pairs of equations?

3. Suppose you are given the system of equations $\begin{cases} x + y + z = 0 & \text{(Equation 1)} \\ 2x + 4y + 3z = 5 & \text{(Equation 2)} \\ 4x - 2y + 3z = -13 & \text{(Equation 3)} \end{cases}$.

   Which variable would you choose to eliminate, using which pairs of equations? Why?

4. When using elimination to solve a system of three equations with three variables, how do you know if it has no solution?

5. Two drawings were given in the text of inconsistent systems of equations with three unknowns. Make another drawing of an inconsistent system other than the ones given.

6. Where would a point be located if it solves two equations of a system of three equations in three variables but does not solve the third equation?

For Exercises 7–12, determine if the given point is a solution of the system.

7. $(3, -1, 1)$
$$\begin{cases} x + y + z = 3 \\ 2x - 2y - z = 4 \\ 2x + y - 2z = 3 \end{cases}$$

8. $(2, -2, 1)$
$$\begin{cases} 3x + 2y + z = 3 \\ 2x - 3y - 2z = 8 \\ -2x + 4y + 3z = -9 \end{cases}$$

9. $(1, 0, 2)$
$$\begin{cases} 2x + 3y - 3z = -4 \\ -2x + 4y - z = -4 \\ 3x - 4y + 2z = 5 \end{cases}$$

10. $(3, 4, 0)$
$$\begin{cases} x + 2y + 5z = 11 \\ 3x - 2y - 4z = 1 \\ 2x + 2y + 3z = 12 \end{cases}$$

11. $(2, -2, 4)$
$$\begin{cases} x + 2y - z = -6 \\ 2x - 3y + 4z = 26 \\ -x + 2y - 3z = -18 \end{cases}$$

12. $(0, -2, 4)$
$$\begin{cases} 3x + 2y - 3z = -16 \\ 2x - 4y - z = 4 \\ -3x + 4y - 2z = -16 \end{cases}$$

For Exercises 13–34, solve the systems of equations.

13. $\begin{cases} x + y + z = 5 \\ 2x + y - 2z = -5 \\ x - 2y + z = 8 \end{cases}$

14. $\begin{cases} x + y + z = 4 \\ 3x + y - z = -4 \\ 2x - 2y + 3z = 3 \end{cases}$

15. $\begin{cases} x + y + z = 2 \\ 4x - 3y + 2z = 2 \\ 2x + 3y - 2z = -8 \end{cases}$

16. $\begin{cases} x + y - z = 7 \\ -2x + 2y - z = -5 \\ 3x - 3y + 2z = 6 \end{cases}$

17. $\begin{cases} 2x + y + 2z = 5 \\ 3x - 2y + 3z = 4 \\ -2x + 3y + z = 8 \end{cases}$

18. $\begin{cases} 2x + 3y - 2z = -4 \\ 4x - 3y + z = 25 \\ x + 2y - 4z = -12 \end{cases}$

19. $\begin{cases} x + 2y - z = 1 \\ 2x + 4y - 2z = -8 \\ 3x + y - 4z = 6 \end{cases}$

20. $\begin{cases} 3x - 2y + z = 5 \\ 4x - 5y + 2z = 7 \\ 9x - 6y + 3z = 7 \end{cases}$

21. $\begin{cases} 4x - 2y + 3z = 6 \\ 6x - 3y + 4.5z = 9 \\ 12x - 6y + 9z = 18 \end{cases}$

22. $\begin{cases} -8x + 4y + 6z = -18 \\ 2x - y - 1.5z = 4.5 \\ 4x - 2y - 3z = 9 \end{cases}$

23. $\begin{cases} y = 2x + z - 8 \\ x = 2y - 3z + 11 \\ 2x + 3y - z = -6 \end{cases}$

24. $\begin{cases} z = -3x + y - 10 \\ 2x + 3y - 2z = 5 \\ x = 3y - 3z - 14 \end{cases}$

25. $\begin{cases} x = 4y - z + 1 \\ 3x + 2y - z = -8 \\ x + 6y + 2z = -3 \end{cases}$

26. $\begin{cases} 2x - 3y - 4z = 3 \\ y = -4x + 8z - 1 \\ x = -5y - 2z - 4 \end{cases}$

27. $\begin{cases} 4x + 2y + 3z = 9 \\ 2x - 4y - z = 7 \\ 3x - 2z = 4 \end{cases}$

28. $\begin{cases} 4x + 2y - 3z = 20 \\ 2x + 5z = -4 \\ 3x - 3y + 2z = 2 \end{cases}$

29. $\begin{cases} 2x - 3z = 6 \\ 5x + 4y = -7 \\ y + 2z = -6 \end{cases}$

30. $\begin{cases} 4y + 3z = 2 \\ 3x + 4y = 2 \\ 2x - 5z = -6 \end{cases}$

31. $\begin{cases} 3x + 2y = -2 \\ 2x - 3z = 1 \\ 0.4y - 0.5z = -2.1 \end{cases}$

32. $\begin{cases} 0.2x - 0.3z = -1.8 \\ 3x + 2y = 5 \\ 3x + 2z = -1 \end{cases}$

33. $\begin{cases} \frac{3}{2}x + y - z = 0 \\ 4y - 3z = -22 \\ -0.2x + 0.3y = -2 \end{cases}$

34. $\begin{cases} -0.2x - 0.3y + 0.1z = -0.3 \\ -3x + 2y = 13 \\ \frac{1}{4}y - \frac{1}{2}z = 2 \end{cases}$

REVIEW EXERCISES

[2.2] ***For Exercises 1–6, solve using the methods from Chapter 2.***

1. One number exceeds another number by 62. If five times the smaller is decreased by two times the larger, the difference is 155. What are the numbers?

2. A company has 400 feet of fencing material to enclose a rectangular space. If it were to make the length 50 feet longer than the width, what would be the dimensions of the rectangle?

3. The sum of two consecutive odd integers is 124. Find the integers.

4. Two boards are joined to form two angles as shown. The greater angle measures 15° more than twice the lesser angle. Find the measure of both angles.

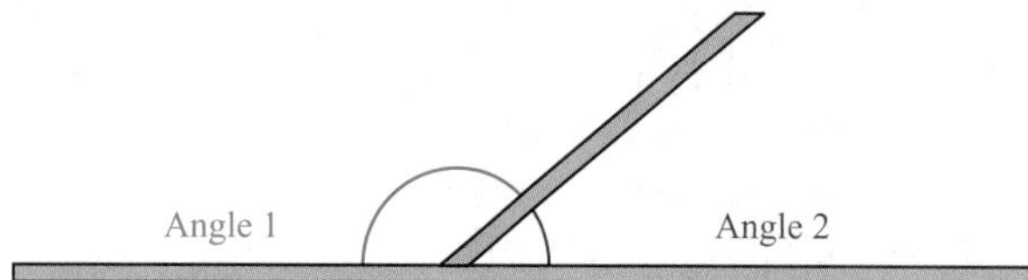

5. Dora is jogging on a trail at an average rate of 5 miles per hour. Jose, who is riding his bike in the opposite direction, passes her at an average rate of 13 miles per hour. If they maintain those average rates, how long will it take them to be 3 miles apart?

6. How many milliliters of 10% HCl solution and 30% HCl solution must be mixed to make 200 milliliters of 20% HCl solution?

4.3 Solving Applications Using Systems of Equations

OBJECTIVES

1. Solve application problems that translate to a system of two linear equations.

2. Solve application problems that translate to a system of three linear equations.

In this section, we use systems of equations to solve a variety of application problems. The following procedure describes our general approach to solving problems using systems of equations:

PROCEDURE **Solving Problems Using Systems of Equations**

To solve problems using a system of equations,

1. Select a variable to represent each unknown.
2. Write a system of equations.
3. Solve the system.

Note: *The graphing method is the least desirable method because it can be time-consuming and inaccurate. The substitution method is simpler than elimination if one of the equations has an isolated variable. If all equations in the system are in standard form, then elimination is better.*

OBJECTIVE 1. Solve application problems that translate to a system of two linear equations.

Business and Currency Problems

EXAMPLE 1 An artist produces two types of ornaments: small and large. In reviewing her records for October, she notes that the total revenue was \$1582, but she did not record how many of each size she sold. She remembers selling 75 more large ornaments than small. If the small sells for \$3.50 and the large for \$5.00, how many of each size did she sell in October?

Understand We are given the total revenue and the price of each ornament. We also know that she sold 75 more large ornaments than small. We are to find the number of each size sold.

Plan and Execute Let x represent the number of small ornaments sold and y represent the number of large ornaments sold.

Categories	Selling Price	Number Sold	Revenue
Small	\$3.50	x	$3.50x$
Large	\$5.00	y	$5.00y$

Relationship 1: She sold 75 more large than small.

Translation: $y = x + 75$

Relationship 2: The total revenue is \$1582.

Translation: $3.50x + 5.00y = 1582$

$$\text{Our system: } \begin{cases} y = x + 75 & \text{(Equation 1)} \\ 3.50x + 5.00y = 1582 & \text{(Equation 2)} \end{cases}$$

Since y is isolated in Equation 1, the substitution method seems easier.

$$3.50x + 5.00y = 1582$$

$$3.50x + 5.00(x + 75) = 1582 \quad \text{Substitute } x + 75 \text{ in place of } y.$$

$$3.50x + 5.00x + 375 = 1582 \quad \text{Distribute.}$$

$$8.50x = 1207 \quad \text{Combine like terms and subtract 375 from both sides.}$$

$$x = 142 \quad \text{Divide both sides by 8.50.}$$

Now we can find the value of y by substituting 142 for x in Equation 1.

$$y = 142 + 75 = 217$$

Answer She sold 142 small and 217 large ornaments.

Check Verify the solution in both given relationships.

YOUR TURN A broker sells a combined total of 40 shares of two different stocks. The first stock sold for \$12.25 per share and the second stock for \$15.75 per share. If the total sale was \$539, how many shares of each stock were sold?

ANSWER

26 shares at \$12.25, 14 shares at \$15.75

Rate Problems

EXAMPLE 2 To gain strength, a rowing crew practices in a stream with a fairly quick current. When rowing against the stream, the team takes 15 minutes to row 1 mile, whereas with the stream, they row the same mile in 6 minutes. Find the team's speed in still water and how much the current changes its speed.

Understand We assume that the amount the current adds to the speed is the same as the amount subtracted, depending on which way the team is traveling.

Plan and Execute Let x represent the team's speed in still water. Let y represent the amount that the current changes the speed. We can use a table to organize the information.

Direction	Rate	Time (in hours)	Distance
Upstream	$x - y$	0.25	$0.25(x - y)$
Downstream	$x + y$	0.1	$0.1(x + y)$

Relationship 1: The distance upstream is 1 mile.

Translation: $0.25(x - y) = 1$

$x - y = 4$ **Divide both sides by 0.25 to clear the decimal (or multiply both sides by 4).**

Relationship 2: The distance downstream is 1 mile.

Translation: $0.1(x + y) = 1$

$x + y = 10$ **Divide both sides by 0.1 to clear the decimal (or multiply both sides by 10).**

Our system: $\begin{cases} x - y = 4 & \text{(Equation 1)} \\ x + y = 10 & \text{(Equation 2)} \end{cases}$

Since the equations are in standard form, we will use the elimination method.

$$x - y = 4$$
$$\underline{x + y = 10}$$
$$2x + 0 = 14 \quad \text{Add the equations to eliminate } y.$$
$$2x = 14$$
$$x = 7 \quad \text{Divide both sides by 2.}$$

Now we can find the value of y by substituting 7 for x in one of the equations. We will use $x + y = 10$.

$7 + y = 10$ **Substitute 7 in place of x.**

$y = 3$ **Subtract 7 from both sides.**

Answer The team can row 7 miles per hour in still water. The stream's current changes the speed by 3 miles per hour depending on the direction of travel.

Check Verify the time of travel for each part of the round trip.

Going upstream: $7 - 3 = 4$ miles per hour

$$\text{Time: } t = \frac{d}{r} = \frac{1}{4} = 0.25 \text{ hours, which is 15 minutes}$$

Going downstream: $7 + 3 = 10$ miles per hour

$$\text{Time: } t = \frac{d}{r} = \frac{1}{10} = 0.1 \text{ hours, which is 6 minutes}$$

YOUR TURN A plane traveling east with the jet stream travels from Salt Lake City to Atlanta in 3 hours. From Atlanta to Salt Lake City, flying against the jet stream, the plane takes 3.75 hours. If it is 1590 miles from Salt Lake City to Atlanta, find the plane's speed in still air.

EXAMPLE 3 Suppose Angel is running in a marathon. He passes the first checkpoint at an average speed of 6 miles per hour. Two minutes later, Jay passes the first checkpoint at an average speed of 10 miles per hour. If both runners maintain their average speeds, how long will it take Jay to catch up to Angel?

Understand We are to find the amount of time it takes for Jay to catch Angel given their individual rates and the amount of time separating them.

Plan and Execute Let x represent Angel's travel time and y represent Jay's travel time. We can use a table to organize the information.

Direction	Rate	Time (in hours)	Distance
Angel	6	x	$6x$
Jay	10	y	$10y$

Relationship 1: When Jay catches up to Angel, Angel will have run for 2 minutes $\left(\frac{1}{30} \text{ of an hour}\right)$ longer than Jay since passing the checkpoint.

$$\text{Translation: } x = y + \frac{1}{30}$$

ANSWER

477 mph

Relationship 2: When Jay catches up, they will have traveled the same distance from the start of the race.

Translation: $6x = 10y$

$$\text{Our system: } \begin{cases} x = y + \dfrac{1}{30} & \text{(Equation 1)} \\ 6x = 10y & \text{(Equation 2)} \end{cases}$$

Equation 1 is solved for x, so we will use substitution.

$$6\left(y + \frac{1}{30}\right) = 10y \quad \textbf{Substitute } y + \frac{1}{30} \textbf{ for } x \textbf{ in Equation 2.}$$

$$6y + \frac{1}{5} = 10y \quad \textbf{Multiply.}$$

$$\frac{1}{5} = 4y \quad \textbf{Subtract } 6y \textbf{ from both sides.}$$

$$\frac{1}{20} = y \quad \textbf{Divide both sides by 4.}$$

Answer Jay will catch up in $\frac{1}{20}$, or 0.05, hours, which is 3 minutes.

Check Notice that when Jay catches Angel, Angel will have run for 5 minutes $(3 + 2)$ after passing the checkpoint. In that 5 minutes $\left(\frac{1}{12} \text{ of an hour}\right)$ after passing the checkpoint, Angel travels $6\left(\frac{1}{12}\right) = \frac{1}{2}$ mile. In the 3 minutes after the checkpoint, Jay travels $10\left(\frac{1}{20}\right) = \frac{1}{2}$ mile, indicating that Jay is side-by-side with Angel.

YOUR TURN Ian and Lindsey are riding bikes in the same direction. Ian is traveling at an average speed of 12 miles per hour and passes a sign at 3:30 P.M. Lindsey, who is traveling at an average speed of 16 miles per hour, passes the same sign at 3:40 P.M. At what time will Lindsey catch up to Ian?

Mixture Problems

EXAMPLE 4 How many milliliters of a 10% HCl solution and 30% HCl solution must be mixed together to make 200 milliliters of 15% HCl solution?

Understand The two unknowns are the volumes of 10% solution and 30% solution that are mixed. One relationship involves concentrations of each solution in the mixture and the other relationship involves the total volume of the final mixture (200 ml).

ANSWER

4:10 P.M.

Plan and Execute Let x and y represent the two amounts to be mixed. We can use a table to organize the information.

Solution	Concentration	Volume of Solution	Volume of HCl
10% HCl	0.10	x	$0.10x$
30% HCl	0.30	y	$0.30y$
15% HCl	0.15	200	0.15(200)

Relationship 1: The total volume is 200 ml.

Translation: $x + y = 200$

Relationship 2: The combined volumes of HCl in the two mixed solutions is equal to the total volume of HCl in the mixture.

Translation: $0.10x + 0.30y = 0.15(200)$

Our system: $\begin{cases} x + y = 200 \\ 0.10x + 0.30y = 30 \end{cases}$

Since the equations are in standard form, elimination is the better method. We will eliminate x.

$x + y = 200 \xrightarrow{\text{Multiply by } -0.10} -0.10x - 0.10y = -20$

$0.10x + 0.30y = 30 \qquad 0.10x + 0.30y = 30$ **Add the equations.**

$0.20y = 10$

$y = 50$ **Divide both sides by 0.20.**

Now we can find x by substituting 50 for y in one of the equations. We will use $x + y = 200$.

$x + \mathbf{50} = 200$ **Substitute 50 for y.**

$x = 150$ **Subtract 50 from both sides.**

Answer Mixing 150 milliliters of the 10% solution with 50 milliliters of the 30% solution gives 200 milliliters of the 15% solution.

Check The mixture volume is $150 + 50 = 200$ milliliters. The volume of HCl is $0.10(150) + 0.30(50) = 15 + 15 = 30$, which means 30 milliliters out of the 200 milliliters is pure HCl, and $\frac{30}{200}$ is 0.15 or 15%.

YOUR TURN How much 20% solution and 40% solution must be mixed together to make 400 milliliters of 35% HCl solution?

ANSWER

300 ml of 40% solution with 100 ml of 20% solution

OBJECTIVE 2. Solve application problems that translate to a system of three linear equations. We follow the same procedure to solve applications involving three unknowns as we did when solving applications with two unknowns.

EXAMPLE 5 At a movie theater, John buys one popcorn, one soft drink, and one candy bar, all for \$7. Fred buys two popcorns, three soft drinks, and two candy bars for \$16. Carla buys one popcorn, two soft drinks, and three candy bars for \$12. Find the price of one popcorn, one soft drink, and one candy bar.

Understand We have three unknowns and three relationships, and we are to find the cost of each.

Plan Select a variable for each unknown, translate the relationships to a system of three equations, and then solve the system.

Execute Let x represent the cost of one popcorn, y represent the cost of one soft drink, and z represent the cost of one candy bar.

Relationship 1: One popcorn, one soft drink, and one candy bar cost \$7.

Translation: $x + y + z = 7$

Relationship 2: Two popcorns, three soft drinks, and two candy bars cost \$16.

Translation: $2x + 3y + 2z = 16$

Relationship 3: One popcorn, two soft drinks, and three candy bars cost \$12.

Translation: $x + 2y + 3z = 12$

$$\text{Our system: } \begin{cases} x + y + z = 7 & \text{(Equation 1)} \\ 2x + 3y + 2z = 16 & \text{(Equation 2)} \\ x + 2y + 3z = 12 & \text{(Equation 3)} \end{cases}$$

We will choose to eliminate z from two pairs of equations. We will start with Equations 1 and 2.

$$\begin{array}{lll} \text{(Eq. 1)}\ \ x + y + z = 7 & \xrightarrow{\text{Multiply by } -2} & -2x - 2y - 2z = -14 \\ \text{(Eq. 2)}\ \ 2x + 3y + 2z = 16 & & \underline{\ \ 2x + 3y + 2z = 16\ \ } \quad \text{Add the equations.} \\ & & \qquad\quad y \qquad\ = 2 \quad \text{(Equation 4)} \end{array}$$

Note: *Although we were trying to eliminate z, we ended up eliminating both x and z.*

Equation 4 gives us the value of y, indicating that a soft drink costs \$2.00. Now, we choose another pair of equations and eliminate z again. We will use Equations 1 and 3.

$$\begin{array}{lll} \text{(Eq. 1)}\ \ x + y + z = 7 & \xrightarrow{\text{Multiply by } -3} & -3x - 3y - 3z = -21 \\ \text{(Eq. 3)}\ \ x + 2y + 3z = 12 & & \underline{\ \ x + 2y + 3z = 12\ \ } \quad \text{Add the equations.} \\ & & -2x - y \qquad = -9 \quad \text{(Equation 5)} \end{array}$$

Since we already know $y = 2$, we can substitute for y in $-2x - y = -9$.

$$-2x - 2 = -9 \quad \text{Substitute 2 for } y.$$
$$-2x = -7 \quad \text{Add 2 to both sides.}$$
$$x = 3.5 \quad \text{Divide both sides by } -2 \text{ to isolate } x.$$

Since x represents the cost of one popcorn, a popcorn costs \$3.50. To find z, substitute for x and y into one of the original equations. We will use Equation 1.

$$3.5 + 2 + z = 7 \quad \text{Substitute 3.5 for } x \text{ and 2 for } y.$$
$$5.5 + z = 7 \quad \text{Simplify.}$$
$$z = 1.5 \quad \text{Subtract 5.5 from both sides to isolate } z.$$

Since z represents the cost of one candy bar, a candy bar costs \$1.50.

Answer Popcorn costs \$3.50, a soft drink costs \$2.00, and a candy bar costs \$1.50.

Check Verify that at these prices, the amounts of money spent by John, Fred, and Carla are correct. We will leave the check to the reader.

YOUR TURN A small aquarium is in the shape of a rectangular solid. The sum of the length, width, and height is 46 inches. The sum of twice the length, three times the width, and the height is 90 inches. The sum of the length, twice the width, and three times the height is 86 inches. Find the dimensions of the aquarium.

ANSWER

length: 20 in.,
width: 12 in.,
height: 14 in.

4.3 Exercises

For Extra Help MyMathLab Videotape/DVT InterAct Math Math Tutor Center Math XL.com

1. Suppose there are four more \$1 bills than \$5 bills in a wallet. If x represents the number of \$5 bills and y represents the number of \$1 bills, write an equation relating the number of bills.

2. If a plane's speed in still air is x miles per hour and the amount the jet stream changes the plane's speed is y miles per hour, write an expression that describes the plane's speed with the jet stream and another expression describing the plane's speed against the jet stream.

3. If a problem has three unknowns, then how many equations would you expect to be in the corresponding system of equations?

4. A problem involves the price of three different sizes of drink: small, medium, and large. If the large sells for \$0.20 less than the small and medium combined, write an equation describing the pricing.

For Exercises 5–38, translate to a system of two equations, then solve.

5. Erica has only \$5 bills and \$10 bills in her wallet. If she has 23 bills totaling \$190, how many of each bill does she have?

6. Mark has 15 total bills, all ones and fives. If he has \$67, how many of each bill does he have?

7. As of August 2005, *Shrek* and *Shrek II* grossed a combined total of \$705 million. If *Shrek II* grossed \$169 million more than *Shrek*, how much did each movie gross? (*Source:* MovieWeb)

8. As of March 2004, the U.S. Senate and House of Representatives had a combined payroll of \$85 million. If the payroll for the House of Representatives was \$19 million more than that of the Senate, how much was allocated to each branch of Congress?

9. Of the 11.6 million students enrolled in community colleges, there are 1.8 million more women than men. How many women are enrolled in a community college? (*Source:* American Association of Community Colleges)

10. In 2003, there were 2927 thousand people in health care occupations. If there were 2305 thousand more women than men, how many men were in health care occupations? (*Source:* U.S. Department of Labor Statistics)

11. The average annual salary for a post-secondary math teacher is \$21,760 less than that of a mathematician in industry. Together, the average salaries total \$134,720. Find each average annual salary. (*Source:* U.S. Department of Labor)

12. The civilian labor force in March 2004 was 1420 thousand less than in March 2005. The total for the two years was 294,894 thousand. How many civilians were in the labor force in March of each year? (*Source:* U.S. Department of Labor)

13. Shark attacks are almost six times as likely to occur in deep water (greater than 30 feet) as in other depths. Find the percentage of attacks occurring in deep water. (*Hint:* Since all attacks occur in the ocean, the combined percent is 100%.) (*Source:* International Shark Attack File; Florida Museum of Natural History, Knight Ridder Tribune, Associated Press)

14. During 2002 and 2003, a total of 9903 people were infected with the West Nile Virus. In 2003, 2.36 times as many people were infected as in 2002. How many people were infected each year? (*Source:* Centers for Disease Control and Prevention)

15. The Old-Fashioned Christmas Store sells two sizes of fresh wreaths. An 18-inch wreath costs \$20 and a 22-inch wreath costs \$25. In one day, the number of 22-inch wreaths sold was six more than triple the number of 18-inch wreaths, for a total of \$1100. How many of each were sold?

16. Pier One sells two sizes of pillar candles. The larger one sells for \$15 and the smaller one for \$10. One day the number of small candles sold was four more than twice the number of larger candles, for a combined total of \$845. How many of each size were sold?

17. In the Civil War, the Union and Confederate forces lost a combined total of about 498,000 soldiers. If the Union lost 38,000 less than three times the number of soldiers lost by the Confederate forces, how many soldiers died on each side? (*Source:* U.S. Department of Defense)

Of Interest

Of the 498,000 soldiers who died in the Civil War, only about 215,000 died in battle (140,000 Union and 75,000 Confederate). The rest died from disease and malnutrition.

18. The unemployment rate in December 1999 was 1.4% less than the rate in December 2004. The rate in 2004 was 2.6% less than twice the rate in 1999. Find the 2004 unemployment rate. (*Source:* U.S. Department of Labor)

19. The perimeter of the base of the Washington Monument is 220.5 feet. If the width is equal to its length, what are the dimensions of the base?

20. The perimeter of the playing lines for a doubles tennis match is 228 feet. If the width is 42 feet less than the length, what are the dimensions?

21. Soccer field dimensions vary according to age and location. The maximum perimeter of a field for 8-year-olds is 130 yards. If the length is 10 yards less than twice the width, what are the dimensions?

22. The perimeter of a dollar bill is 17 inches. If the length is $\frac{1}{2}$ inch more than twice the width, what are the dimensions of a dollar bill?

23. A support beam is attached to a wall and to the bottom of a ceiling truss as shown in the figure to the right. The angle made with the truss on one side of the support beam is two times the angle on the other side of the support beam. Find the two angles.

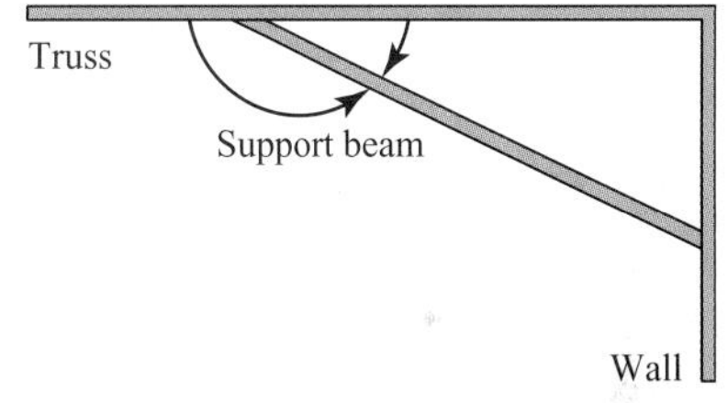

24. A laser beam is aimed at a flat photocell at an angle as shown in the figure to the right. The angle on one side of the beam is 30° more than the angle on the other side of the beam. Find the two angles.

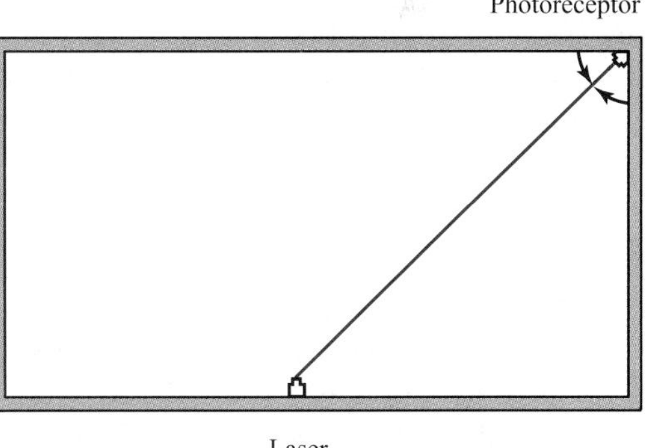

25. A security camera is placed in the corner of a room near the ceiling. The camera is aimed so that in that corner the angle made with the wall on one side of the camera is 20° less than the angle made with the wall on the other side of the camera. Find the two angles.

26. Used as a security device, a laser is placed a few inches above the floor on one wall of a room and aimed at a photoreceptor in the corner of the room. At the receptor, the angle made by the beam with the wall on one side of the receptor is 15° more than the angle made on the other side of the receptor. Find the two angles.

27. In a cross-country skiing event, Josh passes a checkpoint averaging 12 miles per hour. Three minutes later, Dolph passes the same checkpoint averaging 15 miles per hour. If they maintain their average speeds, how long will it take Dolph to catch up?

28. In a NASCAR event, Dale Earnhardt Jr. finds himself 4 seconds behind the leader, who is averaging 180 miles per hour (264 feet per second). If Dale can pick up his average speed to 183 miles per hour (268.4 feet per second), how long will it take him to catch up to the leader?

29. Francis and Poloma are traveling south in separate cars on the same interstate. Poloma is traveling at 65 miles per hour and Francis at 70 miles per hour. Poloma passes a rest stop at 3:00 P.M. Francis passes the same rest stop at 3:30 P.M. At what time will Francis catch up to Poloma?

30. Hale and Austin are jogging in the same direction, Hale passes a waterfall at 9 A.M. Austin passes the same waterfall at 9:30 A.M. Austin is jogging at 5 miles per hour and Hale is jogging at 4 miles per hour. When will Austin catch up with Hale?

31. A boat travels 36 miles upstream in 3 hours. Going downstream, it can travel 48 miles in the same amount of time. Find the speed of the current and the speed of the boat in still water.

32. Going upstream, it takes 2 hours to travel 36 miles in a boat. Downstream, the same distance takes only 1.5 hours. Find the speed of the current and the speed of the boat in still water.

33. If a plane can travel 650 miles per hour with the wind and only 550 miles per hour against the wind, find the speed of the wind and the speed of the plane in still air.

34. If a plane can travel 320 miles per hour with the wind and only 280 miles per hour against the wind, find the speed of the wind and the speed of the plane in still air.

35. How many milliliters of a 5% HCl mixture must be combined with a 20% HCl mixture to get 10 milliliters of 12.5% HCl mixture?

36. How much of a 45% saline solution must be mixed with a 30% solution to produce 20 liters of a 39% saline solution?

37. Julio is investing some money in a money market account that pays 9% and the rest in a savings account at 5%. He invested four times as much in the money market account as in the savings account. If the combined interest from both investments was $1435, how much was invested in each?

38. Marta invested twice as much money in an account paying 7% interest than she did in an account paying 4% interest. If the total interest paid was $720, how much did she invest in each?

For Exercises 39–70, translate to a system of three equations, then solve.

39. The sum of three numbers is 16. One of the numbers is 2 more than twice a second number, and 2 less than the third number. Find the numbers.

40. The sum of three numbers is 18. Twice the first number is 1 less than the third number, and twice the second number is 1 more than the third number. Find the numbers.

41. The sum of the measures of the angles in every triangle is 180°. Suppose the measure of one angle of a triangle is three times that of a second angle, and the measure of the second angle is 5° less than the measure of the third. Find the measure of each angle of the triangle.

42. The measure of one angle of a triangle is 25° less than the measure of a second angle, and twice the measure of the second angle is 65° more than the measure of the third angle. Find the measure of each angle of the triangle.

43. The perimeter of a triangle is 33 inches. The sum of the length of the longest side and twice the length of the shortest side is 31 inches. Twice the length of the longest side minus both the other side lengths is 12 inches. Find the side lengths.

44. The perimeter of the figure shown is 118 centimeters. Twice x minus y is 1 centimeter and y minus z is 10 centimeters. Find the side lengths.

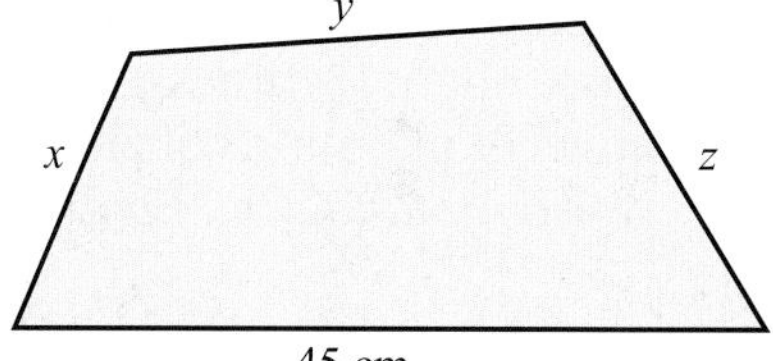

45. At a fast-food restaurant, one burger, one order of fries, and one drink cost \$5; three burgers, two orders of fries, and two drinks cost \$12.50; and two burgers, four orders of fries, and three drinks cost \$14. Find the individual cost of one burger, one order of fries, and one drink.

46. James went to the college bookstore and purchased two pens, two erasers, and one pack of paper for \$6. Tamika purchased four pens, three erasers, and two packs of paper for \$11.50. Jermaine purchased three pens, one eraser, and three packs of paper for \$9.50. Find the individual costs of one pen, one eraser, and one pack of paper.

47. Tickets for a high school band concert were \$3 for children, \$5 for students, and \$8 for adults. There was a total of 500 tickets sold and the total money received from the sale of the tickets was \$2500. There were 150 fewer adult tickets sold than student tickets. Find the number of each type of ticket sold.

48. Tickets to a play were \$2 for children, \$4 for students, and \$6 for adults. There was a total of 300 tickets sold and the total money received from the sale of the tickets was \$1250. There were 35 more student tickets sold than children's tickets. Find the number of each type of ticket sold.

49. In the opening game of the 2005 NBA Western Conference finals in a 121–114 victory over the Phoenix Suns, Tony Parker of the San Antonio Spurs scored 29 points with a combination of 3-point field goals, 2-point field goals, and free throws (1 point each). If he scored a total of 14 times and made twice as many 3-point field goals as free throws, find the number of each that he made.

50. At a track meet, 10 points are awarded for each first-place finish, 5 points for each second, and 1 point for each third. Suppose a track team scored a total of 71 points and had two more firsts than seconds and one less first than third. Find the number of first-, second-, and third-place finishes for the team.

51. Tanisha has 25 coins consisting of nickels, dimes, and quarters, with a total value of \$3.60. If the sum of the number of nickels and dimes is five more than the number of quarters, how many of each type of coin does she have?

52. A vending machine accepts nickels, dimes, and quarters. When the owner checks the machine, there is a total of 56 coins worth \$7.55. If the sum of the number of nickels and dimes is 2 more than twice the number of quarters, find the number of each type of coin.

53. Tommy empties his piggy bank and finds that he has 25 coins consisting of pennies, nickels, and dimes that are worth \$1.36. If there are twice as many nickels as there are pennies, find the number of each type of coin.

54. At the end of each day, Josh puts all the change in his pockets into a jar. After one week he has 25 coins consisting of pennies, dimes, and quarters that are worth \$2.17. If there are 3 more dimes than quarters, find the number of each type of coin.

55. Bronze is an alloy that is made of zinc, tin, and copper in a specified proportion. Suppose an order is placed for 1000 pounds of bronze. The amount of tin in this order of bronze is three times the amount of zinc, and the amount of copper is 20 pounds more than 15 times the amount of tin. Find the number of pounds of zinc, tin, and copper. (*Source: World Almanac and Book of Facts*, 2002)

56. Fresh Roasted Almond Company sells nuts in bulk quantities. When bought in bulk, peanuts sell for \$1.34 per pound, almonds for \$4.36 per pound, and pecans for \$5.88 per pound. Suppose a local specialty shop wants a mixture of 200 pounds that will cost \$2.70 per pound. Find the number of pounds of each type of nut if the sum of the number of pounds of peanuts and almonds is nine times the number of pounds of pecans.

57. The Wunderbar Delicatessen sells Black Forest ham for \$11.96 per pound, turkey breast for \$8.76 per pound, and roast beef for \$9.16 per pound. Suppose they make a 10-pound party tray of these three meats such that the average cost is \$9.80 per pound. Find the number of pounds of each if the number of pounds of breast of turkey is equal to the sum of the number of pounds of Black Forest ham and roast beef.

58. Barnie's coffee shop sells Jamaican Blue Mountain coffee for \$45.99 per pound, Hawaiian Kona for \$36.99 per pound, and Sulawesi Kalossi for \$12.99 per pound. A 25-pound mixture of these three coffees sells for \$30.27 per pound. Find the number of pounds of each if the sum of the number of pounds of Jamaica Blue Mountain and Hawaiian Kona is 5 pounds more than the number of pounds of Sulawesi Kalossi. (*Source:* Barnie's, Orlando, Florida)

59. A total of \$8000 is invested in three stocks, which paid 4%, 6%, and 7% dividends, respectively, in one year. The amount invested in the stock that returned 7% dividends is \$1500 more than the amount that returned 4% dividends. If the total dividends in one year were \$475, find the amount invested in each stock.

60. A total of \$5000 is invested in three funds. The money market fund pays 5% annually, the income fund pays 6% annually, and the growth fund pays 8% annually. The total earnings for one year from the three funds are \$340. If the amount invested in the growth fund is \$500 less than twice the amount invested in the money market fund, find the amount in each of the funds.

61. Nina inherited \$100,000, which she invested in stocks, bonds, and certificates of deposit. After the first year, she had lost 6% in the stocks, earned 5% on the bonds, and earned 4% on the certificates of deposit. The sum of twice the principal in certificates of deposit and the principal in stocks is \$20,000 more than the principal in bonds. Find the principal in each investment if the total income for the year was \$1500.

62. An executive received a \$30,000 end-of-the-year bonus. He invests the bonus in stocks, municipal bonds, and certificates of deposit. At the end of the first year, he had lost 5% in the stocks, earned 5% on the bonds, and earned 4% on the certificates of deposit, which resulted in a net return of \$220. Find the principal in each of the investments if the principal in stocks is \$6000 less than the sum of the principals in bonds and certificates.

63. Eggs are classified by the weight of a dozen. The three largest sizes are jumbo, extra large, and large. The total weight of one dozen of each size is 81 ounces. Three dozen jumbo eggs weigh 39 ounces more than the sum of the weight of one dozen extra large and one dozen large. The sum of the weights of two dozen extra large and one dozen large is 18 ounces more than the weight of two dozen jumbo. Find the weight of one dozen of each type of egg. (*Source: Numbers: How Many, How Long, How Far, How Much*)

64. The intensity of sound is measured in decibels. The decibel reading of a rock concert is 60 less than three times the reading of normal conversation and 10 less than that of a jet at takeoff. The decibel reading of a jet takeoff is 10 more than twice that of normal conversation. Find the decibel reading of each. (*Source: Numbers: How Many, How Long, How Far, How Much*)

65. The number of calories burned per hour bicycling is 120 more than the number burned from brisk walking. The number of calories burned per hour in climbing stairs is 180 more than from bicycling and is twice that from brisk walking. Find the number of calories burned per hour for each of the three activities. (*Source: Numbers: How Many, How Long, How Far, How Much*)

66. The human body is composed of many elements. About 98% of the human body is made up of carbon, oxygen, and hydrogen. The percent of the human body that is hydrogen is the sum of twice the percentage of oxygen and the percentage of carbon. The percent of oxygen is 6% more than twice the percentage of carbon. Find the percent of the human body that is composed of each of these three elements. (*Source: Numbers: How Many, How Long, How Far, How Much*)

67. The three countries in the world with the highest death rates per 1000 are Angola, Mozambique, and Botswana. The sum of the death rates of all three countries is 75 per 1000. The sum of the death rate per 1000 of Angola and Mozambique is twice the death rate of Botswana, and the death rate per 1000 for Mozambique is 2 more than the rate for Angola. Find the death rate per 1000 for each country. (*Source: Phrase Base*)

68. The three countries in the world with the highest birthrates per 1000 are Chad, Mali, and Niger. The sum of the birthrates is 145 per 1000. The sum of the birthrates per 1000 of Chad and Mali is 5 less than twice the birthrate of Niger. The birthrate per 1000 of Chad is 3 less than the birthrate of Niger. Find the birthrate per 1000 for each of these countries. (*Source: Phrase Base*)

69. An object is thrown upward with an initial velocity of v_0 from an initial height of h_0. The height of the object, h, is described by an equation of the form $h = at^2 + v_0t + h_0$, where h and h_0 are in feet and t is in seconds. The table shows the height for different values of t. Find the values of a, v_0, and h_0 and write the equation for h.

t	h
1	234
3	306
6	174

70. Use the same general equation from Exercise 69, $h = at^2 + v_0t + h_0$, where h and h_0 are in feet and t is in seconds. The table shows the distance above the ground for different values of t. Find the values of a, v_0, and h_0 and write the equation for h.

t	h
2	136
3	106
4	44

PUZZLE PROBLEM

The sum of four positive integers a, b, c, and d is 228 and a < b < c < d. The greatest number subtracted from twice the smallest number gives 18. If twice the sum of the third number and fourth number is subtracted from the product of 6 and the second number, the result is 10. If the smallest two numbers are consecutive odd integers, find all numbers.

REVIEW EXERCISES

[1.4] **1.** Use the distributive property to rewrite $-3(2x - 4y - z + 9)$.

[2.1] **2.** Given the equation $x - 0.5y = 8$, let $x = 6$, then solve for y.

For Exercises 3–6, use the function $f(x) = -2x + 1$.

[3.5] **3.** Find $f(1)$.

[3.1] **4.** Find the y-intercept.

[3.3] **5.** Find the slope.

[3.5] **6.** Graph the function.

4.4 Solving Systems of Linear Equations Using Matrices

OBJECTIVES

1. **Write a system of equations as an augmented matrix.**
2. **Solve a system of linear equations by transforming its augmented matrix to echelon form.**

OBJECTIVE 1. Write a system of equations as an augmented matrix. Although the elimination method that we learned in Section 4.1 is effective for solving systems of linear equations, we can streamline the method by manipulating just the coefficients and constants in a **matrix**.

DEFINITION ***Matrix:*** A rectangular array of numbers.

Following are some examples of matrices (plural of matrix):

$$\begin{bmatrix} 1 & -2 \\ -3 & 4 \end{bmatrix} \quad \begin{bmatrix} 1 & -5 & 6 \\ -2 & 4 & 0 \end{bmatrix} \quad \begin{bmatrix} -3 & 0 & 9 \\ -2 & 4 & 7 \\ 9 & -2 & 0 \end{bmatrix}$$

Matrices are made up of horizontal rows and vertical columns.

Column 1 Column 2 Column 3

$$\begin{matrix} \text{Row 1} \rightarrow \\ \text{Row 2} \rightarrow \end{matrix} \begin{bmatrix} 1 & 4 & -2 \\ 5 & -1 & 3 \end{bmatrix}$$

Column 1 Column 2

$$\begin{matrix} \text{Row 1} \rightarrow \\ \text{Row 2} \rightarrow \\ \text{Row 3} \rightarrow \end{matrix} \begin{bmatrix} 3 & -4 \\ -1 & 2 \\ 5 & -3 \end{bmatrix}$$

Note: *One way to remember that columns are vertical is to imagine the columns on a building, which are vertical.*

We call a matrix like $\begin{bmatrix} 2 & 5 & -1 \\ 5 & 4 & 3 \end{bmatrix}$ a 2×3 (read "2 by 3") matrix because it has two rows and three columns; $\begin{bmatrix} 2 & 5 \\ -4 & 6 \\ 3 & -2 \end{bmatrix}$ is a 3×2 matrix because it has three rows and two columns. Each number in a matrix is called an *element*. To solve a system of equations using matrices, we first rewrite the system as an **augmented matrix**.

DEFINITION ***Augmented matrix:*** A matrix made up of the coefficients and the constant terms of a system. The constant terms are separated from the coefficients by a dashed vertical line.

For example, to write the system $\begin{cases} 3x - 2y = 7 \\ 4x - 3y = 10 \end{cases}$ as an augmented matrix, we omit the variables and write $\left[\begin{array}{cc|c} 3 & -2 & 7 \\ 4 & -3 & 10 \end{array}\right]$.

EXAMPLE 1 Write $\begin{cases} 2x + 3y - 4z = 1 \\ 3x - 5y + z = -4 \\ 2x - 6y + 3z = 2 \end{cases}$ as an augmented matrix.

Solution $\left[\begin{array}{ccc|c} 2 & 3 & -4 & 1 \\ 3 & -5 & 1 & -4 \\ 2 & -6 & 3 & 2 \end{array}\right]$

Note: *It is helpful to think of the dashes as the equal sign of each equation.*

YOUR TURN Write the augmented matrix for $\begin{cases} x + 3y = -2 \\ -4x - 5y = 7 \end{cases}$.

OBJECTIVE 2. Solve a system of linear equations by transforming its augmented matrix to echelon form. Recall that in the elimination method, we could interchange equations, add equations, or multiply equations by a number and then add the rewritten equations. Since each row of an augmented matrix contains the constants and coefficients of each equation in the system, we can interchange rows, add rows, or multiply rows by a number and then add the rewritten rows.

RULE **Row Operations**

The solution of a system is not affected by the following row operations in its augmented matrix:

1. Any two rows may be interchanged.
2. The elements of any row may be multiplied (or divided) by any nonzero real number.
3. Any row may be replaced by a row resulting from adding the elements of that row (or multiples of that row) to a multiple of the elements of any other row.

We use these row operations to solve a system of linear equations by transforming the augmented matrix of the system into an equivalent matrix that is in **echelon form**.

DEFINITION ***Echelon form:*** An augmented matrix whose coefficient portion has 1s on the diagonal from upper left to lower right and 0s below the 1s.

For example, $\left[\begin{array}{cc|c} 1 & 3 & 3 \\ 0 & 1 & 4 \end{array}\right]$ and $\left[\begin{array}{ccc|c} 1 & 2 & -4 & 5 \\ 0 & 1 & -5 & 7 \\ 0 & 0 & 1 & 5 \end{array}\right]$ are in echelon form.

In the following examples we will let R_1 mean row 1, R_2 mean row 2, and so on.

ANSWER

$\left[\begin{array}{cc|c} 1 & 3 & -2 \\ -4 & -5 & 7 \end{array}\right]$

EXAMPLE 2 Solve the following linear system by transforming its augmented matrix into echelon form.

$$\begin{cases} x - 2y = -4 & \text{(Equation 1)} \\ -2x + 5y = 9 & \text{(Equation 2)} \end{cases}$$

Solution First, we write the augmented matrix: $\left[\begin{array}{cc|c} 1 & -2 & -4 \\ -2 & 5 & 9 \end{array}\right]$.

Now, we perform row operations to transform the matrix into echelon form. The element in the first row, first column is already 1, which is what we want. So we need to rewrite the matrix so that -2 in the second row, first column becomes 0. To do this, we multiply the first row by 2 and add it to the second row.

$$2R_1 + R_2 \longrightarrow \left[\begin{array}{cc|c} 1 & -2 & -4 \\ 0 & 1 & 1 \end{array}\right]$$

Note: *The result of the row operations replaces the affected row in the matrix. Other rows are rewritten unchanged.*

The resulting matrix represents the system $\begin{cases} x - 2y = -4 \\ y = 1 \end{cases}$.

Since $y = 1$, we can solve for x using substitution.

$$x - 2(1) = -4 \qquad \text{Substitute 1 for } y \text{ in } x - 2y = -4.$$
$$x - 2 = -4 \qquad \text{Simplify.}$$
$$x = -2 \qquad \text{Add 2 to both sides.}$$

The solution is $(-2, 1)$.

EXAMPLE 3 Solve the following linear system by transforming its augmented matrix into echelon form:

$$\begin{cases} 2x + 6y = 1 & \text{(Equation 1)} \\ x + 8y = 3 & \text{(Equation 2)} \end{cases}$$

Solution First, we write the augmented matrix: $\left[\begin{array}{cc|c} 2 & 6 & 1 \\ 1 & 8 & 3 \end{array}\right]$.

Now we perform row operations to get echelon form.

$$\frac{1}{2} \cdot R_1 \longrightarrow \left[\begin{array}{cc|c} 1 & 3 & \frac{1}{2} \\ 1 & 8 & 3 \end{array}\right] \qquad \text{We need the 2 in row 1, column 1 to be a 1, so multiply each number in row 1 by } \frac{1}{2}.$$

$$-1 \cdot R_1 + R_2 \longrightarrow \left[\begin{array}{cc|c} 1 & 3 & \frac{1}{2} \\ 0 & 5 & \frac{5}{2} \end{array}\right] \qquad \text{We need the 1 in row 2, column 1 to be 0, so multiply row 1 by } -1 \text{ and add it to row 2.}$$

$$\frac{1}{5} \cdot R_2 \longrightarrow \left[\begin{array}{cc|c} 1 & 3 & \frac{1}{2} \\ 0 & 1 & \frac{1}{2} \end{array}\right] \qquad \text{We need the 5 in row 2, column 2 to be 1, so multiply row 2 by } \frac{1}{5}.$$

Connection Row operations correspond to equation manipulations in the elimination method. For example, $-1 \cdot R_1 + R_2$ corresponds to multiplying $x + 3y = \frac{1}{2}$ by -1 and then adding the resulting equation to $x + 8y = 3$ so that we get $5y = \frac{5}{2}$.

This matrix represents the system $\begin{cases} x + 3y = \dfrac{1}{2} \\ y = \dfrac{1}{2} \end{cases}$.

Since $y = \frac{1}{2}$, we can solve for x using substitution.

$$x + 3\left(\frac{1}{2}\right) = \frac{1}{2} \qquad \text{Substitute } \frac{1}{2} \text{ for } y \text{ in } x + 3y = \frac{1}{2}.$$

$$x + \frac{3}{2} = \frac{1}{2} \qquad \text{Simplify.}$$

$$x = -1 \qquad \text{Subtract } \frac{3}{2} \text{ from both sides.}$$

The solution is $\left(-1, \frac{1}{2}\right)$.

YOUR TURN Solve the following system by transforming its augmented matrix to echelon form.

$$\begin{cases} 3x + 4y = 6 \\ x - 3y = -11 \end{cases}$$

Now let's use row operations to solve systems of three equations.

EXAMPLE 4 Use the echelon method to solve

$$\begin{cases} 2x + y - 2z = -3 & \text{(Equation 1)} \\ x + y + 2z = 7 & \text{(Equation 2)} \\ 4y + 3z = 5 & \text{(Equation 3)} \end{cases}$$

Solution Write the augmented matrix: $\left[\begin{array}{ccc|c} 2 & 1 & -2 & -3 \\ 1 & 1 & 2 & 7 \\ 0 & 4 & 3 & 5 \end{array}\right]$.

Now we perform row operations to get echelon form

$$\begin{array}{r} R_2 \longrightarrow \\ R_1 \longrightarrow \\ \end{array} \left[\begin{array}{ccc|c} 1 & 1 & 2 & 7 \\ 2 & 1 & -2 & -3 \\ 0 & 4 & 3 & 5 \end{array}\right]$$

Because we need a 1 in row 1, column 1, we will interchange rows 1 and 2.

$$-2 \cdot R_1 + R_2 \longrightarrow \left[\begin{array}{ccc|c} 1 & 1 & 2 & 7 \\ 0 & -1 & -6 & -17 \\ 0 & 4 & 3 & 5 \end{array}\right]$$

To get a 0 in row 2, column 1, we multiply the new row 1 by -2 and add it to row 2.

$$4 \cdot R_2 + R_3 \longrightarrow \left[\begin{array}{cccc} 1 & 1 & 2 & 7 \\ 0 & -1 & -6 & -17 \\ 0 & 0 & -21 & -63 \end{array}\right]$$

Row 3 only needs a 0 in the second column, so we multiply row 2 by 4 and add it to row 3.

$$\begin{array}{r} \\ -1 \cdot R_2 \longrightarrow \\ -\frac{1}{21} \cdot R_3 \longrightarrow \end{array} \left[\begin{array}{ccc|c} 1 & 1 & 2 & 7 \\ 0 & 1 & 6 & 17 \\ 0 & 0 & 1 & 3 \end{array}\right]$$

To get a 1 in row 2, column 2, we multiply row 2 by -1.

To get a 1 in row 3, column 3, we multiply row 3 by $-\frac{1}{21}$.

ANSWER

$(-2, 3)$

The resulting matrix represents the system $\begin{cases} x + y + 2z = 7 \\ \quad\;\; y + 6z = 17. \\ \qquad\quad\; z = 3 \end{cases}$

To find y, substitute 3 for z in $y + 6z = 17$

$$y + 6(3) = 17 \quad \text{Substitute 3 for } z.$$
$$y + 18 = 17 \quad \text{Multiply 6 and 3.}$$
$$y = -1 \quad \text{Subtract 18 from both sides.}$$

To find x, substitute -1 for y and 3 for z in $x + y + 2z = 7$.

$$x + (-1) + 2(3) = 7 \quad \text{Substitute } -1 \text{ for } y \text{ and 3 for } z.$$
$$x + 5 = 7 \quad \text{Simplify the left side of the equation.}$$
$$x = 2 \quad \text{Subtract 5 from both sides of the equation.}$$

The solution is $(2, -1, 3)$. We can check by verifying that the ordered triple satisfies all three of the original equations.

YOUR TURN Solve the following linear system by transforming its augmented matrix into echelon form.

$$\begin{cases} 3x + y + z = 2 \\ x + 2y - z = -4 \\ 2x - 2y + 3z = 9 \end{cases}$$

Calculator TIPS

To use a graphing calculator to solve the system in Example 4, press the MATRX *key, then use the arrow keys to highlight* ***EDIT****. You will see a list of matrices and their current sizes. Press 1 to select the first matrix, named [A]. We need our matrix to be 3 × 4, so press* 3 ENTER*, then* 4 ENTER*. The cursor will now be in row 1, column 1 of a 3 × 4 matrix. You can now enter each number. Use the arrow keys to move the cursor to any position in the matrix. Note that the cursor position's row number, column number, and current value are displayed at the bottom of the screen. After entering each number for Example 4, your screen should look like this:*

MATRX [A] 3 × 4
[2 1 −2 −3]
[1 1 2 7]
[0 4 3 5]

When you finish entering all the numbers, press 2nd MODE *to quit editing.*

ANSWER

(1, −2, 1)

Now, to put the matrix into row echelon form, press MATRX *again and move the cursor to highlight **MATH** at the top of the screen. From the menu, select **A: ref (** and press* ENTER. *You will be prompted to enter the name of the desired matrix. Press* MATRX *again and select **1:[A] 3 × 4.** You will now see ref ([A]. Use the*) *key to close the parentheses and press* ENTER. *You now have the row echelon form and your screen should look like this:*

ref ([A])
[[1 .5 −1 −1.5]
[0 1 .75 1.25]
[0 0 1 3]]

Note: The function **ref** stands for *row echelon form.*

Notice the last row indicates that $z = 3$ and we can use substitution to find the other values.

*Your calculator can also show the complete solution. Instead of selecting **A: ref (**, select **B: rref (** from the math menu. Your screen will look like this:*

rref ([A])
[[1 0 0 2]
[0 1 0 −1]
[0 0 1 3]]

Note: This form is called *reduced row echelon form*, which is why the function is **rref**.

Note that the first row of this form indicates that $x = 2$, the second row indicates that $y = -1$, and the third row indicates that $z = 3$, which is the solution for our system.

4.4 Exercises

1. How many rows and columns are in a 4×2 matrix?

2. How do you write a system of equations as an augmented matrix?

3. What does the dashed line in an augmented matrix correspond to in a system of equations?

4. Explain how solving a system of equations using row operations is similar to solving the system using elimination.

5. What is echelon form?

6. Once an augmented matrix is in echelon form, how do you find the values of the variables?

For Exercises 7–14, write the augmented matrix for the system of equations.

7. $\begin{cases} 14x + 7y = 6 \\ 7x + 6y = 8 \end{cases}$

8. $\begin{cases} 5x + 6y = 2 \\ 10x + 3y = -2 \end{cases}$

9. $\begin{cases} 7x - 6y = 1 \\ 8x - 12y = 6 \end{cases}$

10. $\begin{cases} 3x - y = 6 \\ 9x + 2y = -2 \end{cases}$

11. $\begin{cases} x - 3y + z = 4 \\ 2x - 4y + 2z = -4 \\ 6x - 2y + 5z = -4 \end{cases}$

12. $\begin{cases} 3x + 2y - 3z = 2 \\ 2x - 4y + 5z = -10 \\ 5x - 4y + z = 0 \end{cases}$

13. $\begin{cases} 4x + 6y - 2z = -1 \\ 8x + 3y = -12 \\ -y + 2z = 4 \end{cases}$

14. $\begin{cases} x - 3y = 3 \\ -3x + 4y + 9z = 3 \\ 2x + 7z = -9 \end{cases}$

For Exercises 15–18, given the matrices in echelon form, find the solution for the system.

15. $\left[\begin{array}{cc|c} 1 & 5 & -6 \\ 0 & 1 & -2 \end{array}\right]$

16. $\left[\begin{array}{cc|c} 1 & -3 & 7 \\ 0 & 1 & -5 \end{array}\right]$

17. $\left[\begin{array}{ccc|c} 1 & -4 & -8 & 6 \\ 0 & 1 & -2 & -7 \\ 0 & 0 & 1 & 1 \end{array}\right]$

18. $\left[\begin{array}{ccc|c} 1 & -2 & 6 & 16 \\ 0 & 1 & -4 & -13 \\ 0 & 0 & 1 & 3 \end{array}\right]$

For Exercises 19–24, complete the indicated row operation.

19. Replace R_2 in $\left[\begin{array}{cc|c} 1 & 3 & -1 \\ -2 & 5 & 6 \end{array}\right]$ with $2R_1 + R_2$.

20. Replace R_2 in $\left[\begin{array}{cc|c} 1 & 2 & -2 \\ 3 & 8 & -4 \end{array}\right]$ with $-3R_1 + R_2$.

21. Replace R_3 in $\left[\begin{array}{ccc|c} 1 & -2 & 4 & 6 \\ 0 & 2 & -1 & -5 \\ 0 & 8 & -6 & -3 \end{array}\right]$ with $-4R_2 + R_3$.

22. Replace R_2 in $\left[\begin{array}{ccc|c} 1 & 5 & -3 & 8 \\ -3 & 2 & -4 & -6 \\ 0 & 1 & -2 & 9 \end{array}\right]$ with $3R_1 + R_2$.

23. Replace R_1 in $\left[\begin{array}{cc|c} 4 & 8 & -10 \\ -1 & 3 & 2 \end{array}\right]$ with $\frac{1}{4}R_1$.

24. Replace R_3 in $\left[\begin{array}{ccc|c} 1 & 3 & 4 & 8 \\ 0 & 1 & -2 & -6 \\ 0 & 0 & -6 & 24 \end{array}\right]$ with $-\frac{1}{6}R_3$.

For Exercises 25–28, describe the row operation that should be performed to make the matrix closer to echelon form.

25. $\left[\begin{array}{rr|r} 1 & 2 & -1 \\ -3 & 4 & 5 \end{array}\right]$

26. $\left[\begin{array}{rr|r} 1 & -3 & 2 \\ 5 & 6 & -4 \end{array}\right]$

27. $\left[\begin{array}{rrr|r} 1 & -4 & 3 & 6 \\ 0 & 1 & -2 & 4 \\ 0 & 2 & -5 & 1 \end{array}\right]$

28. $\left[\begin{array}{rrr|r} 1 & -2 & 4 & 7 \\ 0 & 1 & -2 & 3 \\ 0 & -5 & 4 & -9 \end{array}\right]$

For Exercises 29–52, solve by transforming the augmented matrix into echelon form.

29. $\begin{cases} x - y = 2 \\ x + y = 2 \end{cases}$

30. $\begin{cases} x + y = 2 \\ x - y = -8 \end{cases}$

31. $\begin{cases} x + y = 3 \\ 3x - y = 1 \end{cases}$

32. $\begin{cases} x + 3y = -9 \\ -x + 2y = -11 \end{cases}$

33. $\begin{cases} x + 2y = -7 \\ 2x - 4y = 2 \end{cases}$

34. $\begin{cases} x - 3y = -13 \\ 3x + 2y = 5 \end{cases}$

35. $\begin{cases} -2x + 5y = 9 \\ x - 2y = -4 \end{cases}$

36. $\begin{cases} 2x + y = 12 \\ x - 3y = 6 \end{cases}$

37. $\begin{cases} 4x - 3y = -2 \\ 2x - 3y = -10 \end{cases}$

38. $\begin{cases} 6x - 3y = 0 \\ 2x + 4y = 0 \end{cases}$

39. $\begin{cases} 5x + 2y = 12 \\ 2x + 3y = -4 \end{cases}$

40. $\begin{cases} 4x + 3y = 14 \\ 3x - 2y = 2 \end{cases}$

41. $\begin{cases} x + y + z = 6 \\ 2x + y - 3z = -3 \\ 2x + 2y + z = 6 \end{cases}$

42. $\begin{cases} x + 2y + z = 2 \\ 3x + y - z = 3 \\ 2x + y + 2z = 7 \end{cases}$

43. $\begin{cases} x - y + z = -1 \\ 2x - 2y + z = 0 \\ x + 3y + 2z = 1 \end{cases}$

44. $\begin{cases} x + 3y - 2z = 4 \\ 2x - 3y + 2z = -7 \\ 3x + 2y - 2z = -1 \end{cases}$

45. $\begin{cases} 2x - y + z = 8 \\ x - 2y + 3z = 11 \\ 2x + 3y - z = -6 \end{cases}$

46. $\begin{cases} 2x + 3y - 2z = -21 \\ 2x - 4y + 3z = 15 \\ 3x + 2y - 3z = -24 \end{cases}$

47. $\begin{cases} 3x + 2y - 3z = 1 \\ -2x + 3y - 4z = 7 \\ 5x - 2y + z = -5 \end{cases}$

48. $\begin{cases} 2x - 2y + z = 6 \\ -2x + 4y + 3z = 4 \\ 4x - 3y - 2z = -7 \end{cases}$

49. $\begin{cases} 3x - 6y + z = -10 \\ 7y - z = 2 \\ 2x + 4z = 14 \end{cases}$

50. $\begin{cases} 6x - y + 5z = -28 \\ 4x + 2y = 10 \\ 5x + 6z = -23 \end{cases}$

51. $\begin{cases} 2x + 5y - z = 10 \\ x - y - z = 14 \\ x - 6y = 20 \end{cases}$

52. $\begin{cases} 2x + 4y - 5z = 16 \\ x - 2y - 11z = 1 \\ 4x + 5y = -6 \end{cases}$

For Exercises 53–60, solve using a matrix on a graphing calculator.

53. $\begin{cases} 2x + 5y = -14 \\ 6x + 7y = -10 \end{cases}$

54. $\begin{cases} 4x - 3y = 11 \\ 5x + 4y = 68 \end{cases}$

55. $\begin{cases} 4x - 7y = -80 \\ 9x + 5y = -14 \end{cases}$

56. $\begin{cases} 7x - 3y = 23 \\ 4x - 9y = -67 \end{cases}$

57. $\begin{cases} 3x + 2y - 5z = -19 \\ 4x - 7y + 6z = 67 \\ 5x - 6y - 4z = 18 \end{cases}$

58. $\begin{cases} 6x - 7y + 2z = -52 \\ 8x + 5y - 6z = 132 \\ 12x + 4y - 7z = 150 \end{cases}$

59. $\begin{cases} 8x - 7y + 2z = 108 \\ 6x + 11y - 10z = -74 \\ 9x - 2y + 13z = 140 \end{cases}$

60. $\begin{cases} 8x - 11y + 14z = 40 \\ 5x + 15y - 5z = 35 \\ 17x - 9y + 2z = -133 \end{cases}$

For Exercises 61–62, explain the mistake, then find the correct solution.

61. $\begin{cases} x + 3y = 13 \\ -4x - y = -26 \end{cases}$

$$\left[\begin{array}{cc|c} 1 & 3 & 13 \\ -4 & -1 & -26 \end{array}\right]$$

$$4 \cdot R_1 + R_2 \rightarrow \left[\begin{array}{cc|c} 1 & 3 & 13 \\ 0 & 11 & -13 \end{array}\right]$$

$$\frac{1}{11} \cdot R_2 \rightarrow \left[\begin{array}{cc|c} 1 & 3 & 13 \\ 0 & 1 & -\frac{13}{11} \end{array}\right]$$

Solution: $\left(-\frac{182}{11}, -\frac{13}{11}\right)$

62. $\begin{cases} x - y + z = 8 \\ 3x - z = -9 \\ 4y + z = -6 \end{cases}$

$$\left[\begin{array}{ccc|c} 1 & -1 & 1 & 8 \\ 3 & 0 & -1 & -9 \\ 4 & 0 & 1 & -6 \end{array}\right]$$

$$-3 \cdot R_1 + R_2 \rightarrow \left[\begin{array}{ccc|c} 1 & -1 & 1 & 8 \\ 0 & 3 & -4 & -33 \\ 4 & 0 & 1 & -6 \end{array}\right]$$

$$\begin{array}{l} \\ \frac{1}{3} \cdot R_2 \rightarrow \\ -4 \cdot R_1 + R_3 \rightarrow \end{array} \left[\begin{array}{ccc|c} 1 & -1 & 1 & 8 \\ 0 & 1 & -\frac{4}{3} & -11 \\ 0 & 4 & -3 & -38 \end{array}\right]$$

$$-4 \cdot R_2 + R_3 \rightarrow \left[\begin{array}{ccc|c} 1 & -1 & 1 & 8 \\ 0 & 1 & -\frac{4}{3} & -11 \\ 0 & 0 & \frac{7}{3} & 6 \end{array}\right]$$

$$\frac{3}{7} \cdot R_3 \rightarrow \left[\begin{array}{ccc|c} 1 & -1 & 1 & 8 \\ 0 & 1 & -\frac{4}{3} & -11 \\ 0 & 0 & 1 & \frac{18}{7} \end{array}\right]$$

Solution: $\left(-\frac{15}{7}, -\frac{53}{7}, \frac{18}{7}\right)$

For Exercises 63–76, translate the problem to a system of equations, then solve using matrices.

63. Brad purchased three grilled chicken sandwiches and two drinks for \$9.90, and Angel purchased seven grilled chicken sandwiches and four drinks for \$22.30. Find the price of one grilled chicken sandwich and one drink.

64. Sharika purchased three general admission tickets and two student tickets to a college play for \$55, and Yo Chen purchased two general admission tickets and four student tickets for \$50. Find the cost of one general admission and one student ticket.

65. A right triangle has one angle whose measure is 90°. One of the remaining angles is 6° less than twice the measure of the other. Find the measure of each of the remaining angles.

66. The measure of the largest angle of a triangle is 16° less than the sum of the measures of the other two. Twice the measure of the middle angle is 40° more than the sum of the largest and smallest. Find the measure of each of the angles.

Of Interest

During the new and full Moon, a small tidal wave, called a tidal bore, sweeps up the Amazon. The tidal bore can travel 450 miles upstream at speeds in excess of 40 miles per hour, causing waves of 16 feet or more along the riverbank.

67. The two longest rivers in the world, the Nile and the Amazon, have a combined length of 8050 miles. The Nile is 250 miles longer than the Amazon. Find the length of both rivers.

68. The longest vehicular tunnel in the world is the Saint Gotthard Tunnel in Switzerland and the second longest is the Arlberg Tunnel in Austria. The total length of the two tunnels is 18.8 miles. The Saint Gotthard tunnel is 1.4 miles longer than the Arlberg. Find the length of each tunnel. (*Source: Webster's New World Book of Facts*)

69. Nikita invested a total of \$10,000 in certificates of deposit that pay 5% annually and in a money market account that pays 6% annually. If the total interest earned in one year from the two investments is \$536, find the principal that was invested in each.

70. An athlete received a \$60,000 signing bonus, which he invested in three funds—a money market fund paying 5% annually, an income fund paying 5.5% annually, and a growth fund paying 7% annually. The sum of the principals invested in the money market and the growth fund equals the principal in the income fund. The total annual interest received from the three investments is \$3400. Find the principal invested in each fund.

71. John bought two CDs, four tapes, and three DVDs for \$164, and Tanelle bought five CDs, two tapes, and two DVDs for \$160. If the sum of the cost of one CD and one tape equals the cost of one DVD, find the cost of one of each.

72. Angel bought 3 pounds of salmon, 2 pounds of tuna, and 5 pounds of cod for \$63. Sara bought 6 pounds of salmon, 3 pounds of tuna, and 5 pounds of cod for \$94. The cost of 1 pound of salmon and 1 pound of tuna is the same as the cost of 3 pounds of cod. Find the cost of 1 pound of each.

73. In New England's 24–21 victory over Philadelphia in Super Bowl XXXIX, New England scored touchdowns (6 points), extra points (1 point), and field goals (3 points). The number of touchdowns equaled the number of extra points. Also, the number of touchdowns was one less than four times the number of field goals. Find how many of each type of score New England had. (*Source:* Super Bowl.com)

74. In the 2005, NCAA Women's Championship Basketball game, Baylor player Emily Niemann scored eight times for a total of 19 points using a combination of 3-point field goals, 2-point field goals, and free throws (1 point). If the sum of the number of 3-point and 2-point field goals was three times the number of free throws, find the number of each type. (*Source: Orlando Sentinel,* Apr. 16, 2005)

75. An electrical circuit has three points of connection, with different voltage measurements at each of the three connections. An engineer has written the following equations to describe the voltages. Find each voltage.

$$\begin{aligned} 4v_1 - v_2 &= 30 \\ -2v_1 + 5v_2 - v_3 &= 10 \\ -v_2 + 5v_3 &= 4 \end{aligned}$$

76. An engineer has written the following system of equations to describe the forces in pounds acting on a steel structure. Find the forces.

$$\begin{aligned} 6F_1 - F_2 &= 350 \\ 9F_1 - 2F_2 - F_3 &= -100 \\ 3F_2 - F_3 &= 250 \end{aligned}$$

REVIEW EXERCISES

[1.3] ***For Exercises 1–6, simplify.***

1. $(-3)(4) - (-4)(5)$

2. $2(2 - 8) + 3[-1 - (-6)] - 4(4 - 6)$

3. $\dfrac{(-5)(-1) - (9)(3)}{(2)(-1) - (3)(3)}$

4. $\dfrac{2\left(\frac{5}{4}\right) - 8\left(\frac{1}{4}\right)}{\left(\frac{1}{2}\right)\left(\frac{5}{4}\right) - \left(\frac{3}{2}\right)\left(\frac{1}{4}\right)}$

5. $\dfrac{(1.1)(-0.2) - (1.7)(-0.5)}{(0.2)(-0.2) - (0.5)(-0.5)}$

6. $\dfrac{1[-45 - (-9)] + 7(-9 - 2) - 2(-27 - 30)}{1(6 - 3) - 2(-9 - 2) - 2[9 - (-4)]}$

4.5 Solving Systems of Linear Equations Using Cramer's Rule

OBJECTIVES

1. **Evaluate determinants of 2 × 2 matrices.**
2. **Evaluate determinants of 3 × 3 matrices.**
3. **Solve systems of equations using Cramer's Rule.**

OBJECTIVE 1. Evaluate determinants of 2 × 2 matrices. In this section, we will explore another way of solving systems of linear equations using a special type of matrix called a **square matrix**.

DEFINITION ***Square matrix:*** A matrix that has the same number of rows and columns.

For example, $\begin{bmatrix} 2 & 3 \\ 1 & -4 \end{bmatrix}$ and $\begin{bmatrix} 1 & -3 & 5 \\ -7 & 2 & 9 \\ 0 & 4 & -6 \end{bmatrix}$ are square matrices.

Every square matrix has a *determinant*. We write the determinant of a matrix, A, as $\det(A)$ or $|A|$. The method used to find the determinant of a matrix depends upon its size.

RULE Determinant of a 2 × 2 Matrix

If $A = \begin{bmatrix} a_1 & b_1 \\ a_2 & b_2 \end{bmatrix}$, then $\det(A) = \begin{vmatrix} a_1 & b_1 \\ a_2 & b_2 \end{vmatrix} = a_1b_2 - a_2b_1$.

Warning: Be careful to note the difference between $[A]$ and $|A|$. The notation $[A]$ means "the matrix A" whereas $|A|$ means "the determinant of the matrix A."

Notice that the determinant contains diagonal products, as illustrated by the following:

$$\begin{vmatrix} a_1 & b_1 \\ a_2 & b_2 \end{vmatrix} = a_1b_2 - a_2b_1$$

Note: *Since subtraction is not commutative, be sure to note the order of the two products.*

Learning Strategies

The products in the determinant are arranged with the downward cross product subtracting the upward cross product. An easy way to remember this order is that you have to "fall down before you can get up."

EXAMPLE 1 Find the determinants of the following matrices.

a. $A = \begin{bmatrix} 3 & -2 \\ 2 & 4 \end{bmatrix}$

Solution $\det(A) = \begin{vmatrix} 3 & -2 \\ 2 & 4 \end{vmatrix} = (3)(4) - (2)(-2) = 12 + 4 = 16$

b. $B = \begin{bmatrix} -3 & 5 \\ 4 & -2 \end{bmatrix}$

Solution $\det(B) = (-3)(-2) - (4)(5) = 6 - 20 = -14$

c. $M = \begin{bmatrix} 1 & 3 \\ 3 & 9 \end{bmatrix}$

Solution $\det(M) = (1)(9) - (3)(3) = 9 - 9 = 0$

YOUR TURN Find the determinant of $\begin{bmatrix} 1 & -3 \\ -4 & 2 \end{bmatrix}$.

ANSWER

-10

OBJECTIVE 2. Evaluate determinants of 3 × 3 matrices. There are various methods of evaluating the determinant of a 3 × 3 matrix. One of the most common methods is *expanding by minors*. Each element of a square matrix has a number called the **minor** for that element.

DEFINITION ***Minor:*** The determinant of the remaining matrix when the row and column in which the element is located are ignored.

EXAMPLE 2 Find the minor of 2 in $\begin{bmatrix} 2 & -3 & -6 \\ -1 & 5 & -2 \\ 3 & -4 & 1 \end{bmatrix}$.

Solution To find the minor of 2, we ignore its row and column (shown in blue) and evaluate the determinant of the remaining matrix (shown in red).

$$\begin{bmatrix} 2 & -3 & -6 \\ -1 & 5 & -2 \\ 3 & -4 & 1 \end{bmatrix}$$

$$\begin{vmatrix} 5 & -2 \\ -4 & 1 \end{vmatrix} = (5)(1) - (-4)(-2) = 5 - 8 = -3$$

YOUR TURN Find the minor of 6 in $\begin{bmatrix} 1 & -3 & 6 \\ -2 & 2 & 0 \\ 4 & -1 & 5 \end{bmatrix}$.

ANSWER

$\begin{vmatrix} -2 & 2 \\ 4 & -1 \end{vmatrix} = -6$

Note: *We can actually expand by minors along any row or column to find the determinant of a 3×3 matrix. For simplicity, we have chosen to show expanding by minors only along the first column. You may learn how to expand by minors along other rows or columns in future courses.*

To evaluate the determinant of a 3×3 matrix, we will expand by minors along the first column.

RULE **Evaluating the Determinant of a 3 × 3 Matrix**

$$\begin{vmatrix} a_1 & b_1 & c_1 \\ a_2 & b_2 & c_2 \\ a_3 & b_3 & c_3 \end{vmatrix} = a_1\begin{pmatrix}\text{minor}\\ \text{of } a_1\end{pmatrix} - a_2\begin{pmatrix}\text{minor}\\ \text{of } a_2\end{pmatrix} + a_3\begin{pmatrix}\text{minor}\\ \text{of } a_3\end{pmatrix}$$

$$= a_1\begin{vmatrix} b_2 & c_2 \\ b_3 & c_3 \end{vmatrix} - a_2\begin{vmatrix} b_1 & c_1 \\ b_3 & c_3 \end{vmatrix} + a_3\begin{vmatrix} b_1 & c_1 \\ b_2 & c_2 \end{vmatrix}$$

EXAMPLE 3 Find the determinant of $\begin{bmatrix} 2 & -3 & -4 \\ -1 & 2 & -2 \\ 3 & -4 & 1 \end{bmatrix}$.

Solution Using the rule for expanding by minors along the first column, we have

$$\begin{vmatrix} 2 & -3 & -4 \\ -1 & 2 & -2 \\ 3 & -4 & 1 \end{vmatrix} = 2\begin{pmatrix}\text{minor}\\ \text{of } 2\end{pmatrix} - (-1)\begin{pmatrix}\text{minor}\\ \text{of } -1\end{pmatrix} + 3\begin{pmatrix}\text{minor}\\ \text{of } 3\end{pmatrix}$$

$$= 2\begin{vmatrix} 2 & -2 \\ -4 & 1 \end{vmatrix} - (-1)\begin{vmatrix} -3 & -4 \\ -4 & 1 \end{vmatrix} + 3\begin{vmatrix} -3 & -4 \\ 2 & -2 \end{vmatrix}$$

$$= 2(2 - 8) + 1(-3 - 16) + 3(6 + 8)$$

$$= -12 + (-19) + 42$$

$$= 11$$

YOUR TURN Find the determinant of $\begin{bmatrix} 3 & -2 & 4 \\ -2 & 3 & 1 \\ 2 & -4 & 2 \end{bmatrix}$.

Calculator TIP

The determinant of a matrix can be found using a graphing calculator. After entering the matrix (see p. 271) we use the ***det (*** *function. Press the* MATRX *key and select the* ***det (*** *function from the* ***MATH*** *menu. Now indicate the matrix you want to find the determinant of by pressing* MATRX *and selecting the desired matrix from your list. After indicating the desired matrix, press* ENTER.

ANSWER

26

OBJECTIVE 3. Solve systems of equations using Cramer's Rule. Now we can use **Cramer's Rule**, which uses determinants to solve systems of equations. To derive Cramer's Rule, we solve a general system of equations using the elimination method. We will show the derivation for a system of two equations in two unknowns.

$$\begin{cases} a_1x + b_1y = c_1 & \text{(Equation1)} \\ a_2x + b_2y = c_2 & \text{(Equation 2)} \end{cases}$$

We will eliminate y by multiplying equation 1 by b_2 and equation 2 by $-b_1$.

$$\begin{aligned} a_1b_2x + b_1b_2y &= b_2c_1 && \text{Multiply equation 1 by } b_2. \\ -a_2b_1x - b_1b_2y &= -b_1c_2 && \text{Multiply equation 2 by } -b_1. \\ a_1b_2x - a_2b_1x \quad &= b_2c_1 - b_1c_2 && \text{Add the equations.} \\ (a_1b_2 - a_2b_1)x &= b_2c_1 - b_1c_2 && \text{Factor out } x \text{ from the left side.} \\ x &= \frac{b_2c_1 - b_1c_2}{a_1b_2 - a_2b_1} && \text{Divide by } a_1b_2 - a_2b_1. \end{aligned}$$

Notice that the numerator is $\begin{vmatrix} c_1 & b_1 \\ c_2 & b_2 \end{vmatrix}$ and the denominator is $\begin{vmatrix} a_1 & b_1 \\ a_2 & b_2 \end{vmatrix}$, so $x = \dfrac{\begin{vmatrix} c_1 & b_1 \\ c_2 & b_2 \end{vmatrix}}{\begin{vmatrix} a_1 & b_1 \\ a_2 & b_2 \end{vmatrix}}$.

If we repeat the same process and solve for y, we get $y = \dfrac{\begin{vmatrix} a_1 & c_1 \\ a_2 & c_2 \end{vmatrix}}{\begin{vmatrix} a_1 & b_1 \\ a_2 & b_2 \end{vmatrix}}$.

A similar approach is used to derive the rule for a system of three equations in three unknowns.

DEFINITION ***Cramer's Rule***

The solution to the system of linear equations $\begin{cases} a_1x + b_1y = c_1 \\ a_2x + b_2y = c_2 \end{cases}$ is

$$x = \frac{\begin{vmatrix} c_1 & b_1 \\ c_2 & b_2 \end{vmatrix}}{\begin{vmatrix} a_1 & b_1 \\ a_2 & b_2 \end{vmatrix}} = \frac{D_x}{D} \quad \text{and} \quad y = \frac{\begin{vmatrix} a_1 & c_1 \\ a_2 & c_2 \end{vmatrix}}{\begin{vmatrix} a_1 & b_1 \\ a_2 & b_2 \end{vmatrix}} = \frac{D_y}{D}$$

The solution to the system of linear equations $\begin{cases} a_1x + b_1y + c_1z = d_1 \\ a_2x + b_2y + c_2z = d_2 \\ a_3x + b_3y + c_3z = d_3 \end{cases}$ is

$$x = \frac{\begin{vmatrix} d_1 & b_1 & c_1 \\ d_2 & b_2 & c_2 \\ d_3 & b_3 & c_3 \end{vmatrix}}{\begin{vmatrix} a_1 & b_1 & c_1 \\ a_2 & b_2 & c_2 \\ a_3 & b_3 & c_3 \end{vmatrix}} = \frac{D_x}{D}, \quad y = \frac{\begin{vmatrix} a_1 & d_1 & c_1 \\ a_2 & d_2 & c_2 \\ a_3 & d_3 & c_3 \end{vmatrix}}{\begin{vmatrix} a_1 & b_1 & c_1 \\ a_2 & b_2 & c_2 \\ a_3 & b_3 & c_3 \end{vmatrix}} = \frac{D_y}{D}, \quad \text{and} \quad z = \frac{\begin{vmatrix} a_1 & b_1 & d_1 \\ a_2 & b_2 & d_2 \\ a_3 & b_3 & d_3 \end{vmatrix}}{\begin{vmatrix} a_1 & b_1 & c_1 \\ a_2 & b_2 & c_2 \\ a_3 & b_3 & c_3 \end{vmatrix}} = \frac{D_z}{D}$$

Note: *Each denominator, D, is the determinant of a matrix containing only the coefficients in the system. To find D_x, we replace the column of x-coefficients in the coefficient matrix with the constants from the system. To find D_y, we replace the column of y-coefficients in the coefficient matrix with the constant terms, and do likewise to find D_z.*

Of Interest

Cramer's Rule is named after Gabriel Cramer (1704–1752), who was chair of the mathematics department at Geneva, Switzerland. In one of his books, he gave an example that required finding an equation of degree two whose graph passed through five given points. The solution led to a system of five linear equations in five unknowns. In order to solve the system, Cramer referred readers to an appendix, which explained what we now call Cramer's Rule.

EXAMPLE 4 Use Cramer's Rule to solve $\begin{cases} 2x + 3y = -5 \\ 3x - y = 9 \end{cases}$.

Solution First, we find D, D_x, and D_y.

$$D = \begin{vmatrix} 2 & 3 \\ 3 & -1 \end{vmatrix} = (2)(-1) - (3)(3) = -2 - 9 = -11$$

$$D_x = \begin{vmatrix} -5 & 3 \\ 9 & -1 \end{vmatrix} = (-5)(-1) - (9)(3) = 5 - 27 = -22$$

$$D_y = \begin{vmatrix} 2 & -5 \\ 3 & 9 \end{vmatrix} = (2)(9) - (3)(-5) = 18 + 15 = 33$$

Now we can find x and y.

$$x = \frac{D_x}{D} = \frac{-22}{-11} = 2 \qquad y = \frac{D_y}{D} = \frac{33}{-11} = -3$$

The solution is $(2, -3)$, which we can check by verifying that it satisfies both equations in the system. We will leave the check to the reader.

YOUR TURN Use Cramer's Rule to solve $\begin{cases} 3x - 2y = -16 \\ x + 3y = 2 \end{cases}$.

EXAMPLE 5 Use Cramer's Rule to solve $\begin{cases} x + 2y - 2x = -7 \\ 3x - 2y + z = 15 \\ 2x + 3y - 3z = -9 \end{cases}$.

Note: *Recall that we find the determinant of a 3×3 matrix by expanding by minors along the first column.*

Solution We need to find D, D_x, D_y, and D_z.

$$D = \begin{vmatrix} 1 & 2 & -2 \\ 3 & -2 & 1 \\ 2 & 3 & -3 \end{vmatrix} = (1)\begin{vmatrix} -2 & 1 \\ 3 & -3 \end{vmatrix} - (3)\begin{vmatrix} 2 & -2 \\ 3 & -3 \end{vmatrix} + (2)\begin{vmatrix} 2 & -2 \\ -2 & 1 \end{vmatrix}$$

$$= 1(6 - 3) - 3(-6 + 6) + 2(2 - 4)$$

$$= 3 - 0 + (-4)$$

$$= -1$$

$$D_x = \begin{vmatrix} -7 & 2 & -2 \\ 15 & -2 & 1 \\ -9 & 3 & -3 \end{vmatrix} = (-7)\begin{vmatrix} -2 & 1 \\ 3 & -3 \end{vmatrix} - (15)\begin{vmatrix} 2 & -2 \\ 3 & -3 \end{vmatrix} + (-9)\begin{vmatrix} 2 & -2 \\ -2 & 1 \end{vmatrix}$$

$$= -7(6 - 3) - 15(-6 + 6) + (-9)(2 - 4)$$

$$= -21 - 0 + 18$$

$$= -3$$

ANSWER

$(-4, 2)$

$$D_y = \begin{vmatrix} 1 & -7 & -2 \\ 3 & 15 & 1 \\ 2 & -9 & -3 \end{vmatrix} = (1)\begin{vmatrix} 15 & 1 \\ -9 & -3 \end{vmatrix} - (3)\begin{vmatrix} -7 & -2 \\ -9 & -3 \end{vmatrix} + (2)\begin{vmatrix} -7 & -2 \\ 15 & 1 \end{vmatrix}$$

$$= 1(-45 + 9) - 3(21 - 18) + 2(-7 + 30)$$

$$= -36 - 9 + 46$$

$$= 1$$

$$D_z = \begin{vmatrix} 1 & 2 & -7 \\ 3 & -2 & 15 \\ 2 & 3 & -9 \end{vmatrix} = (1)\begin{vmatrix} -2 & 15 \\ 3 & -9 \end{vmatrix} - (3)\begin{vmatrix} 2 & -7 \\ 3 & -9 \end{vmatrix} + (2)\begin{vmatrix} 2 & -7 \\ -2 & 15 \end{vmatrix}$$

$$= 1(18 - 45) - 3(-18 + 21) + 2(30 - 14)$$

$$= -27 - 9 + 32$$

$$= -4$$

Note: *After finding the values of two of the variables, you could find the value of the third variable by substituting these values back into any one of the original equations.*

$$x = \frac{D_x}{D} = \frac{-3}{-1} = 3, \qquad y = \frac{D_y}{D} = \frac{1}{-1} = -1, \qquad z = \frac{D_z}{D} = \frac{-4}{-1} = 4$$

The solution is $(3, -1, 4)$. We will leave the check to the reader.

YOUR TURN Use Cramer's Rule to solve $\begin{cases} 2x + 3y - 2z = 11 \\ 3x + y + 4z = -8. \\ x - 3y - 2z = 4 \end{cases}$

ANSWER

$(1, 1, -3)$

4.5 Exercises

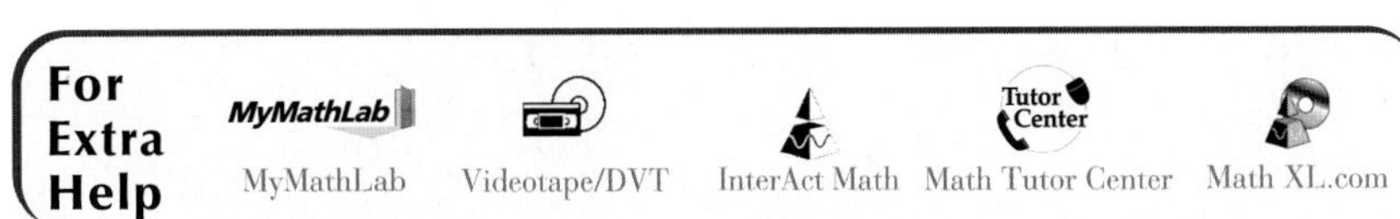

1. Is it possible to find the determinant of $\begin{bmatrix} 1 & 2 & 5 \\ 3 & 6 & 2 \end{bmatrix}$? Why or why not?

2. Explain the difference between a matrix and a determinant.

3. How do you find the minor for an element of a 3×3 matrix?

4. How do you find D_y when solving a system of equations using Cramer's Rule?

For Exercises 5–32, find the determinant.

5. $\begin{bmatrix} 3 & 2 \\ 1 & 5 \end{bmatrix}$

6. $\begin{bmatrix} 4 & 1 \\ 3 & 7 \end{bmatrix}$

7. $\begin{bmatrix} -3 & 5 \\ 2 & 4 \end{bmatrix}$

8. $\begin{bmatrix} -5 & 4 \\ 2 & 6 \end{bmatrix}$

9. $\begin{bmatrix} 3 & 6 \\ -2 & 4 \end{bmatrix}$

10. $\begin{bmatrix} 2 & 8 \\ -3 & 2 \end{bmatrix}$

11. $\begin{bmatrix} -3 & -4 \\ 2 & 5 \end{bmatrix}$

12. $\begin{bmatrix} -3 & -5 \\ 4 & 6 \end{bmatrix}$

13. $\begin{bmatrix} -2 & -3 \\ -4 & 5 \end{bmatrix}$

14. $\begin{bmatrix} -6 & -2 \\ 3 & -4 \end{bmatrix}$

15. $\begin{bmatrix} 0 & 3 \\ -5 & 7 \end{bmatrix}$

16. $\begin{bmatrix} -5 & 0 \\ -4 & 2 \end{bmatrix}$

17. $\begin{bmatrix} 1 & 2 & 1 \\ 3 & 1 & 4 \\ 2 & 3 & 2 \end{bmatrix}$

18. $\begin{bmatrix} 3 & 1 & 4 \\ 2 & 2 & 3 \\ 1 & 4 & 3 \end{bmatrix}$

19. $\begin{bmatrix} -1 & 2 & 0 \\ -3 & 2 & 4 \\ -4 & 2 & 3 \end{bmatrix}$

20. $\begin{bmatrix} -2 & 0 & -1 \\ 3 & -2 & 4 \\ -3 & 2 & 1 \end{bmatrix}$

21. $\begin{bmatrix} 2 & 1 & -3 \\ 0 & -3 & 2 \\ 4 & 1 & -3 \end{bmatrix}$

22. $\begin{bmatrix} 3 & -2 & 4 \\ 3 & 0 & 2 \\ -4 & -2 & 2 \end{bmatrix}$

23. $\begin{bmatrix} 1 & 4 & -2 \\ 3 & 2 & 0 \\ -1 & 4 & 3 \end{bmatrix}$

24. $\begin{bmatrix} 3 & -5 & 0 \\ 2 & -4 & 1 \\ -2 & -1 & -3 \end{bmatrix}$

25. $\begin{bmatrix} 0.3 & -0.5 \\ 1.3 & -0.6 \end{bmatrix}$

26. $\begin{bmatrix} -0.4 & 1.6 \\ -4.7 & 3.1 \end{bmatrix}$

27. $\begin{bmatrix} -0.4 & 0.7 & -1.2 \\ 3.1 & 1.5 & -3.2 \\ 1.6 & -2.2 & -1.5 \end{bmatrix}$

28. $\begin{bmatrix} 1.7 & -3.2 & 4.1 \\ 5.3 & -6.2 & -1.1 \\ -1.3 & 2.3 & -4.5 \end{bmatrix}$

29. $\begin{bmatrix} \frac{1}{2} & -\frac{1}{3} \\ \frac{2}{5} & \frac{3}{5} \end{bmatrix}$

30. $\begin{bmatrix} -\frac{3}{4} & -\frac{3}{5} \\ \frac{3}{2} & \frac{2}{5} \end{bmatrix}$

31. $\begin{bmatrix} \frac{1}{2} & -\frac{3}{4} & \frac{2}{5} \\ \frac{1}{3} & \frac{1}{5} & -\frac{3}{2} \\ -\frac{3}{4} & \frac{1}{2} & \frac{3}{5} \end{bmatrix}$

32. $\begin{bmatrix} -\frac{1}{4} & -\frac{3}{2} & \frac{4}{3} \\ \frac{1}{5} & -\frac{5}{4} & \frac{1}{2} \\ -\frac{5}{3} & \frac{1}{4} & -\frac{4}{5} \end{bmatrix}$

For Exercises 33 and 34, expand by minors along the first column.

33. $\begin{bmatrix} x & y & 1 \\ 2 & -1 & 3 \\ -2 & 0 & 1 \end{bmatrix}$

34. $\begin{bmatrix} x & y & 1 \\ -3 & -2 & 4 \\ 3 & -2 & 2 \end{bmatrix}$

For Exercises 35–48, solve using Cramer's Rule.

35. $\begin{cases} x + y = -5 \\ x - 2y = -2 \end{cases}$

36. $\begin{cases} x + 3y = 1 \\ x + y = -3 \end{cases}$

37. $\begin{cases} 2x - 3y = -6 \\ x - y = -1 \end{cases}$

38. $\begin{cases} 2x + 5y = -7 \\ x - y = -7 \end{cases}$

39. $\begin{cases} -x + 2y = -12 \\ 2x - 3y = 20 \end{cases}$

40. $\begin{cases} -x + 2y = -9 \\ 4x + 5y = -3 \end{cases}$

41. $\begin{cases} 2x - y = -4 \\ -x + 3y = -3 \end{cases}$

42. $\begin{cases} x - 3y = -19 \\ 2x + y = -3 \end{cases}$

43. $\begin{cases} 8x - 3y = 10 \\ 4x + 3y = 14 \end{cases}$

44. $\begin{cases} 2x - y = -4 \\ 4x - 5y = -51 \end{cases}$

45. $\begin{cases} \frac{1}{2}x - \frac{1}{4}y = 0 \\ \frac{3}{4}x + \frac{5}{2}y = \frac{23}{2} \end{cases}$

46. $\begin{cases} \frac{2}{3}x + \frac{1}{4}y = \frac{1}{2} \\ \frac{3}{4}x + \frac{4}{3}y = -\frac{23}{4} \end{cases}$

47. $\begin{cases} 0.2x + 0.5y = 3.4 \\ 0.7x - 0.3y = -0.4 \end{cases}$

48. $\begin{cases} 1.2x - 0.6y = -2.4 \\ 3.1x + 1.3y = -11.9 \end{cases}$

For Exercises 49–60, solve using Cramer's Rule.

49. $\begin{cases} x + y + z = 6 \\ 2x - 4y + 2z = 6 \\ 3x + 2y + z = 11 \end{cases}$

50. $\begin{cases} 2x + y - 3z = -1 \\ x + 2y - 2z = -3 \\ -3x - 4y + z = -3 \end{cases}$

51. $\begin{cases} 3x + y - z = -4 \\ 2x - y + 2z = -7 \\ x - 3y + z = -6 \end{cases}$

52. $\begin{cases} x - y + 3z = -10 \\ 5x + 4y - z = -7 \\ 2x + y - z = -4 \end{cases}$

53. $\begin{cases} 4x + 2y + 3z = 9 \\ 2x - 4y - z = 7 \\ 3x - 2z = 4 \end{cases}$

54. $\begin{cases} 2x - y = -1 \\ 5x - 3y + 2z = 0 \\ 3x + 2y - 3z = -8 \end{cases}$

55. $\begin{cases} 3x + 2y = -12 \\ 3y + 10z = -16 \\ 6x - 2z = 3 \end{cases}$

56. $\begin{cases} 4x + 2z = 7 \\ 8x - 2y = -7 \\ 10y - 2z = -5 \end{cases}$

57. $\begin{cases} \frac{1}{2}x + \frac{1}{3}y + \frac{3}{4}z = \frac{25}{12} \\ \frac{2}{3}x + \frac{3}{2}y - \frac{1}{4}z = -\frac{37}{12} \\ \frac{3}{4}x + \frac{1}{4}y - \frac{2}{3}z = -\frac{7}{4} \end{cases}$

58. $\begin{cases} \frac{3}{4}x - \frac{1}{2}y + \frac{1}{2}z = \frac{1}{4} \\ \frac{2}{5}x + \frac{5}{2}y - \frac{4}{3}z = -\frac{51}{5} \\ \frac{7}{4}x + \frac{5}{3}y - \frac{3}{5}z = -\frac{623}{60} \end{cases}$

59. $\begin{cases} 0.3x + 0.4y - 0.6z = 2.6 \\ 0.5x - 0.2y + 0.7z = -0.8 \\ 1.4x + 1.3y - 2.2z = 9.8 \end{cases}$

60. $\begin{cases} 1.2x + 2.1y - 0.5z = 7.3 \\ 3.2x - 2.4y + 1.3z = 6.1 \\ 2.5x + 1.3y - 1.7z = 8.4 \end{cases}$

For Exercises 61–64, find x.

61. $\begin{vmatrix} 9 & x \\ -6 & 5 \end{vmatrix} = 21$

62. $\begin{vmatrix} 12 & 2 \\ x & -3 \end{vmatrix} = -22$

63. $\begin{vmatrix} 2 & -1 & 0 \\ 0 & x & 1 \\ -2 & 0 & 4 \end{vmatrix} = -38$

64. $\begin{vmatrix} 1 & -2 & 4 \\ -1 & 0 & 2 \\ 0 & x & 5 \end{vmatrix} = -28$

For Exercises 65–70, use the following. Suppose a triangle has vertices of (x_1, y_1), (x_2, y_2), and (x_3, y_3) as shown in the graph to the right. The area of the triangle is given by $A = \frac{1}{2}\left|\det\begin{bmatrix} x_1 & y_1 & 1 \\ x_2 & y_2 & 1 \\ x_3 & y_3 & 1 \end{bmatrix}\right|$. *For example, the area of the triangle whose vertices are* $(1, 2), (3, -2)$, *and* $(-2, 5)$ *is* $A = \frac{1}{2}\left|\det\begin{bmatrix} 1 & 2 & 1 \\ 3 & -2 & 1 \\ -2 & 5 & 1 \end{bmatrix}\right|$. *Expanding about the first column we have:*

Note: This notation indicates half of the absolute value of the determinant.

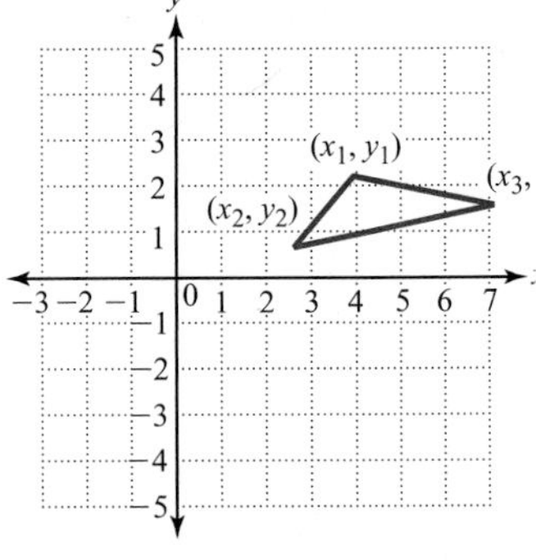

$$A = \frac{1}{2}\left|1\det\begin{bmatrix} -2 & 1 \\ 5 & 1 \end{bmatrix} - 3\det\begin{bmatrix} 2 & 1 \\ 5 & 1 \end{bmatrix} - 2\det\begin{bmatrix} 2 & 1 \\ -2 & 1 \end{bmatrix}\right|$$

$$= \frac{1}{2}|(-2 - 5) - 3(2 - 5) - 2(2 + 2)|$$

$$= \frac{1}{2}|-7 + 9 - 8| = \frac{1}{2}|-6| = 3 \textit{ square units}$$

Find the area of the triangles with vertices at the given points.

65. (2, 4) (4, 0) (6, 5)

66. (0, 2) (3, −2) (5, 5)

67. (−3, 1) (2, −3) (4, 4)

68. (−3, −2) (−1, 4) (3, −4)

69. (−4, −1) (1, 3) (3, −3)

70. (−3, −3) (2, 2) (4, −1)

For Exercises 71–76, translate the problem to a system of equations, then solve using Cramer's Rule.

71. The two heaviest known meteorites to be found on Earth's surface are the Hoba West, which was found in Namibia, and the Ahnighito, which was found in Greenland. The total weight of the two meteorites is 90 tons. The Hoba West is twice as heavy as the Ahnighito. Find the weight of each. (*Source: Webster's New World Book of Facts*)

72. The two largest expenses for the average American family are federal taxes and housing, including household expenses. Together these two items total 43.3% of the average family's income. The amount spent on taxes is 12.1% more than the amount spent on housing. Find the percent spent on each. (*Source: Numbers: How Many, How Long, How Far, How Much*)

73. A restaurant makes a soup that includes garbanzo and black turtle beans. The manager purchased 10 pounds of beans at a cost of \$8.80. If the garbanzo beans cost \$1.00 per pound and the black turtle beans cost \$0.70 per pound, how many pounds of each did he purchase?

74. The perimeter of a triangle is 21 inches and two sides are of equal length. The length of the third side is 3 inches less than the length of the two equal sides. Find the length of each side of the triangle.

75. Coinage bronze is made up of zinc, tin, and copper. The percent of tin is four times the percent of zinc. The percent of copper is nineteen times the sum of the percents of zinc and tin. Find the percent of zinc, tin, and copper in coinage bronze. (*Source: Webster's New World Book of Facts*)

76. In winning the 2005 NCAA basketball championship, North Carolina scored a total of 39 times in their 75 to 70 victory over Illinois. The sum of the number of 3-point field goals and 2-point field goals was three more than twice the number of free throws (1 point each). How many of each did they score? (*Source:* CBS Sports Line)

REVIEW EXERCISES

[2.3] ***For Exercises 1 and 2, solve and graph the solution set.***

1. $x > 2x - 3$

2. $2x + 3 \geq 10 - 5x$

[2.2] **3.** Find two consecutive integers whose sum is 39.

[3.4] ***For Exercises 4 and 5, graph.***

4. $y > 3x + 2$

5. $x + y \leq -2$

[4.1] **6.** Use substitution to solve $\begin{cases} x = 3 - y \\ 5x + 3y = 5 \end{cases}$.

4.6 Solving Systems of Linear Inequalities

OBJECTIVES

1. Graph the solution set of a system of linear inequalities.
2. Solve applications involving systems of linear inequalities.

OBJECTIVE 1. Graph the solution set of a system of linear inequalities. In Section 2.3, we learned to solve linear inequalities in terms of one variable. In Section 3.4, we learned how to graph linear inequalities. In this section, we will develop a graphical approach to solving *systems of linear inequalities.*

Consider the system of linear inequalities $\begin{cases} 3x + 2y \geq 6 \\ x - y < 2 \end{cases}$

First, let's graph each inequality separately.

Note: *We graphed* $3x + 2y \geq 6$ *at the beginning of Section 3.4. We learned that we use a dashed line with* $<$ *or* $>$ *and a solid line with* $\leq$ *or* $\geq$. *Also, we chose a test point to determine which side of the line to shade. Ordered pairs on solid, not dashed, lines are part of the solution set.*

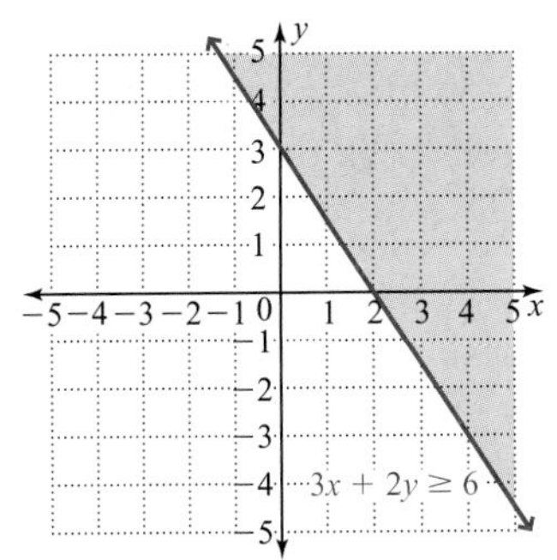

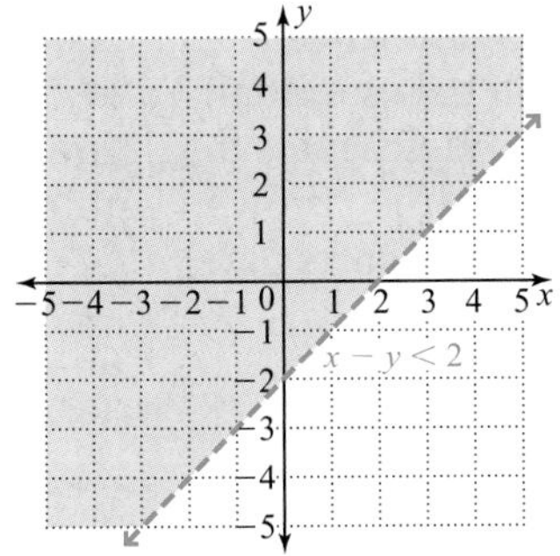

The solution set for $3x + 2y \geq 6$ contains ordered pairs on the line $3x + 2y = 6$ or in the red shaded region defined by $3x + 2y > 6$. The solution set for $x - y < 2$ contains ordered pairs in the blue shaded region only. We place the two graphs together on the same grid to determine the solution set for the system. A solution for a system of inequalities is an ordered pair that makes every inequality in the system true.

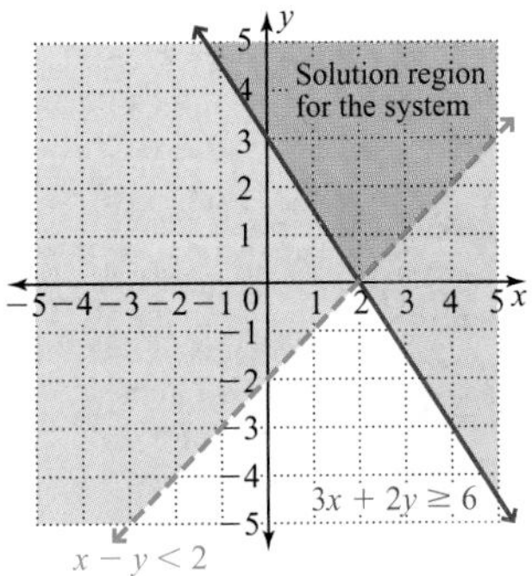

On our graph containing both $3x + 2y \geq 6$ and $x - y < 2$, the region where the shading overlaps contains ordered pairs that make both inequalities true, so all ordered pairs in this region are in the solution set for the system. Also, ordered pairs on the solid line for $3x + 2y \geq 6$ where it touches the region of overlap are in the solution set for the system, whereas ordered pairs on the dashed line for $x - y < 2$ are not. Our example suggests the following procedure.

PROCEDURE **Solving a System of Linear Inequalities**

To solve a system of linear inequalities, graph all of the inequalities on the same grid. The solution set for the system contains all ordered pairs in the region where the inequalities' solution sets overlap along with ordered pairs on the portion of any solid line that touches the region of overlap.

EXAMPLE 1 Graph the solution set for the system of inequalities.

Connection Remember that for a system of two linear equations, a solution is a point of intersection of the graphs of the two equations. Similarly, for a system of linear inequalities, the solution region is the region of intersection of the graphs of the two inequalities.

a. $\begin{cases} x + 2y < 4 \\ y > 3x - 5 \end{cases}$

Solution Graph the inequalities on the same grid. Because both lines are dashed, the solution set contains only those ordered pairs in the region of overlap (the purple shaded region).

Note: *Ordered pairs on dashed lines are* not *part of the solution region.*

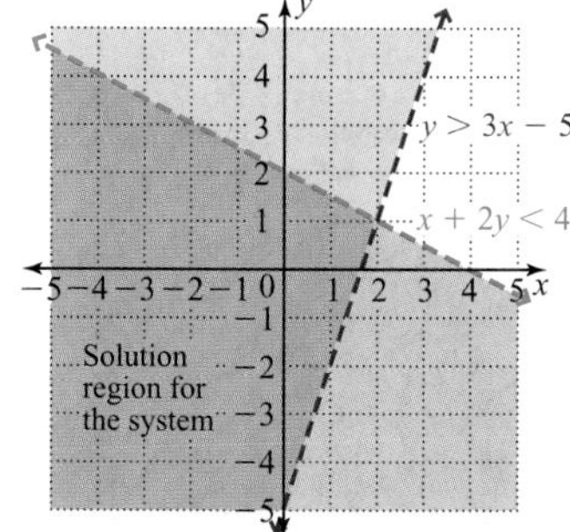

b. $\begin{cases} 2x + y \geq 1 \\ y > -1 \end{cases}$

Solution Graph the inequalities on the same grid. The solution set for this system contains all ordered pairs in the region of overlap (purple shaded region) together with all ordered pairs on the portion of the solid line that touches the purple shaded region.

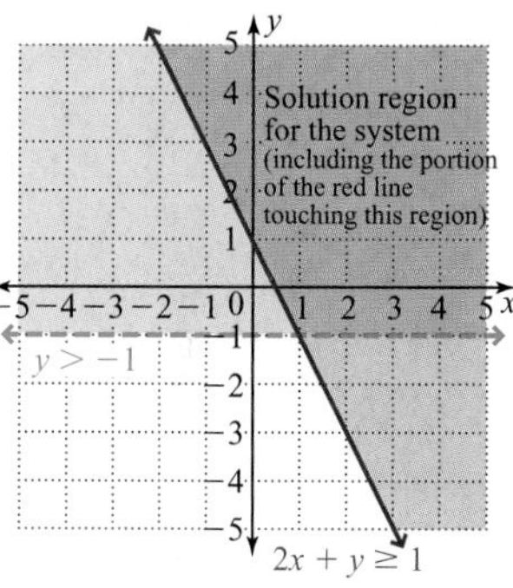

YOUR TURN Graph the solution set for the system of inequalities.

a. $\begin{cases} 3x - y > 6 \\ x + y < 4 \end{cases}$ **b.** $\begin{cases} x \leq 1 \\ 2x - 5y < 10 \end{cases}$

ANSWER

a.

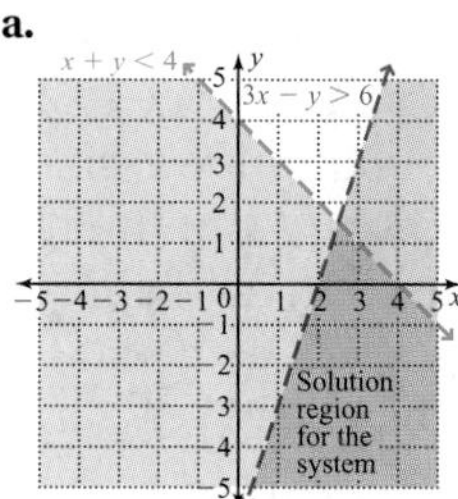

b.

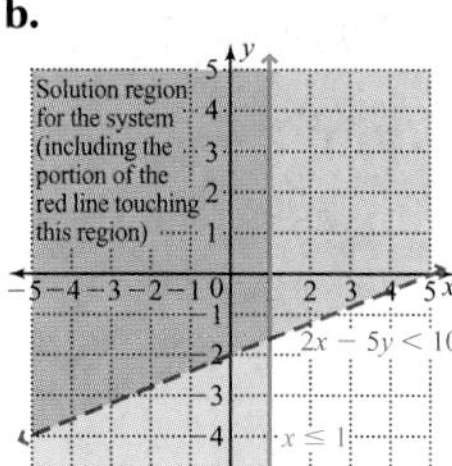

Inconsistent Systems

Some systems of linear inequalities have no solution. We say these systems are *inconsistent.*

EXAMPLE 2 Graph the solution set of the system of inequalities.

$$\begin{cases} y \geq -\frac{1}{2}x + 1 \\ x + 2y \leq -4 \end{cases}$$

Solution Graph each inequality.

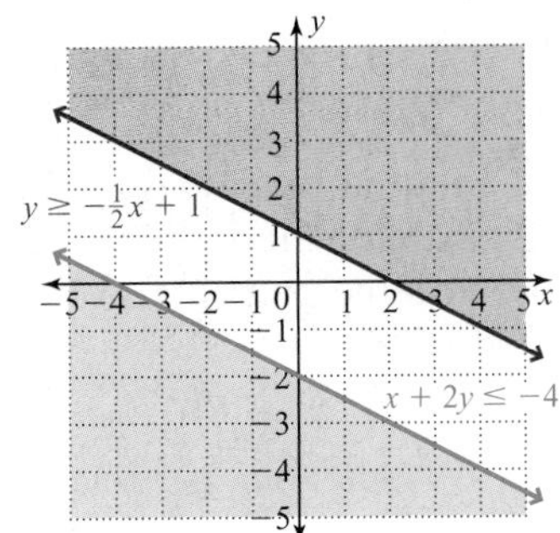

The lines appear to be parallel. We can verify this by writing $x + 2y \leq -4$ in slope-intercept form and comparing the slopes.

$$x + 2y \leq -4$$
$$2y \leq -x - 4 \quad \text{Subtract } x \text{ from both sides.}$$
$$y \leq -\frac{1}{2}x - 2 \quad \text{Divide both sides by 2.}$$

The lines have the same slope, $-\frac{1}{2}$, so they are in fact parallel. Since the lines are parallel and the shaded regions do not overlap, there is no solution region for this system. The system is inconsistent.

Connection Notice that the lines in Example 2 are parallel. Recall that in a system of linear equations, if the lines are parallel, the system is inconsistent because there is no point of intersection. With linear inequalities, the lines must be parallel *and* the shaded regions must not overlap for the system to be inconsistent.

OBJECTIVE 2. Solve applications involving systems of linear inequalities.

EXAMPLE 3 A company uses low-quality paper in photocopiers and high-quality paper in laser printers. The low-quality paper costs \$25 per case and the high-quality paper costs \$50 per case. The administrative assistant keeps a total of at least 15 boxes of paper in inventory at all times. Also, the total monthly cost cannot exceed \$1000. Write a system of inequalities that describes the number of boxes of paper that she could order in one month.

Understand We must translate to a system of inequalities, then solve the system.

Plan and Execute Let x represent the number of boxes of low-quality paper and y represent the number of boxes of high-quality paper.

Relationship 1: The total number of boxes must be at least 15.

The words "at least" indicate that the combined number of boxes is to be greater than or equal to 15, so $x + y \geq 15$.

Relationship 2: The total monthly cost cannot exceed \$1000.

The low-quality paper costs \$25 per case, so $25x$ describes the amount spent purchasing cases of low-quality paper. Similarly, $50y$ describes the amount spent on cases of high-quality paper. The total cannot exceed \$1000, so $25x + 50y \leq 1000$.

$$\text{Our system: } \begin{cases} x + y \geq 15 \\ 25x + 50y \leq 1000 \end{cases}$$

Answer See graph. Since negative numbers of cases cannot be purchased, the solution set is confined to Quadrant I. Any ordered pair in the shaded region or on a portion of either line touching the shaded region is a solution for the system. However, assuming that only whole cases can be purchased, only ordered pairs of whole numbers in the solution set, such as (15, 5) or (20, 10), are realistic.

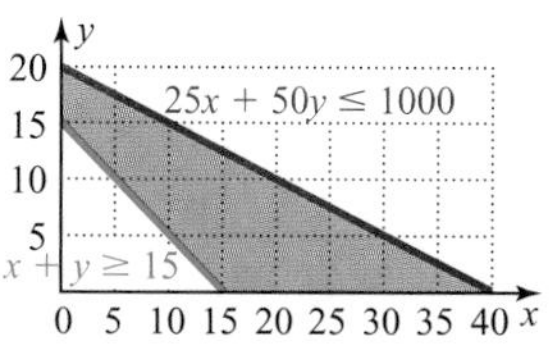

Check We will check one ordered pair in the solution region. The ordered pair (15, 5) indicates a purchase of 15 cases of low-quality paper and 5 cases of high-quality paper, so the company has 20 cases of paper (which is more than 15 cases) that cost a total of $25(15) + 50(5) = \$675$ (which is less than \$1000).

Your Turn A landscaper offers two packages for maintaining quarter-acre lots. He charges \$40 per visit for basic maintenance or \$60 per visit for deluxe maintenance. His crews can maintain a maximum of 70 quarter-acre lots per week. He needs to make at least \$2400 per week to make ends meet.

a. Write a system of inequalities that describes his business's needs.

b. Solve the system by graphing.

c. Give a combination of packages that his crews could maintain and allow him to make ends meet.

Answers

a. $\begin{cases} x + y \le 70 \\ 40x + 60y \ge 2400 \end{cases}$

b. Since x and y must be positive, the solution set is confined to Quadrant I.

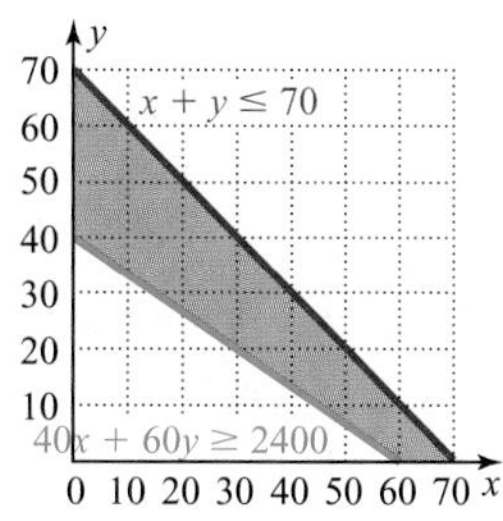

c. (20, 40), which means 20 basic and 40 deluxe

4.6 Exercises

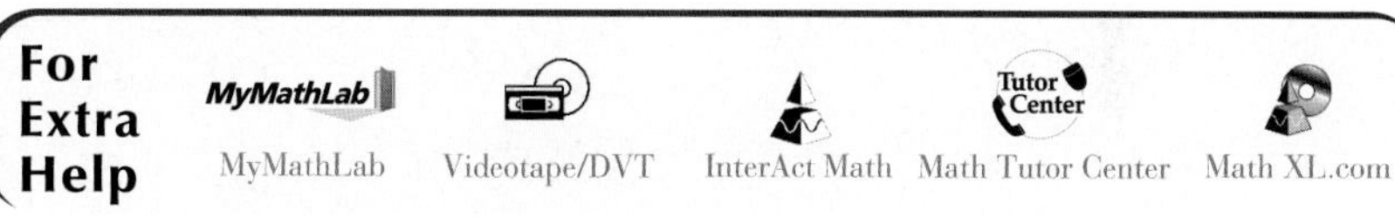

1. How do you determine whether or not a given ordered pair is a solution for a system of linear inequalities?

2. When solving a system of linear inequalities, after graphing all inequalities on the same grid, how do you determine the solution set?

3. What must be true about the boundary lines in a system of two linear inequalities that has no solutions?

4. What is the solution region described by the system $\begin{cases} x > 0 \\ y > 0 \end{cases}$?

5. Write a system of linear inequalities whose solution set is the entire third quadrant.

6. Write a system of linear inequalities whose solution set is the entire fourth quadrant.

For Exercises 7–38, graph the solution set for the system of inequalities.

7. $\begin{cases} y > 2x \\ y < -2x + 4 \end{cases}$

8. $\begin{cases} y < -3x \\ y > x + 3 \end{cases}$

9. $\begin{cases} x - y < -5 \\ x + y < 3 \end{cases}$

10. $\begin{cases} x + y > -2 \\ x - y > 5 \end{cases}$

11. $\begin{cases} 2x + 3y \leq 12 \\ 2x + y \leq 8 \end{cases}$

12. $\begin{cases} 2x - y \leq 6 \\ x + 2y \geq 8 \end{cases}$

13. $\begin{cases} x + 3y \geq 11 \\ 2x - y \leq 1 \end{cases}$

14. $\begin{cases} 2x + y \leq -1 \\ x - y \leq 5 \end{cases}$

15. $\begin{cases} 2x + 3y < 1 \\ x - 4y \geq 3 \end{cases}$

16. $\begin{cases} 4x - y > -10 \\ 4x + 2y \leq 1 \end{cases}$

17. $\begin{cases} 2x + 5y \leq 7 \\ 3x + y > -9 \end{cases}$

18. $\begin{cases} x + 2y > 3 \\ 2x - 4y \leq 10 \end{cases}$

19. $\begin{cases} x + 2y < 6 \\ 2x + 4y \geq -4 \end{cases}$

20. $\begin{cases} x - 3y < 6 \\ 2x - 6y \geq -6 \end{cases}$

21. $\begin{cases} 4x - 2y < 8 \\ 2x - y < -4 \end{cases}$

22. $\begin{cases} 3x + y > 6 \\ 6x + 2y \geq -6 \end{cases}$

23. $\begin{cases} x + 2y > 4 \\ 3x + 6y \le -2 \end{cases}$

24. $\begin{cases} 4x - y \le -5 \\ 8x - 2y \ge -1 \end{cases}$

25. $\begin{cases} x + y \ge 4 \\ y \ge 2 \end{cases}$

26. $\begin{cases} x - 2y > 0 \\ x < 0 \end{cases}$

27. $\begin{cases} 2x + y \ge 0 \\ x < 3 \end{cases}$

28. $\begin{cases} 5x - 6y \ge -12 \\ y > 2 \end{cases}$

29. $\begin{cases} 5x + 3y < -4 \\ 2y \ge 4 \end{cases}$

30. $\begin{cases} 3x - y \ge -13 \\ 4x < 12 \end{cases}$

31. $\begin{cases} x \ge -2 \\ y < 4 \end{cases}$

32. $\begin{cases} y \ge -4 \\ x < 3 \end{cases}$

33. $\begin{cases} x < 1 \\ y \ge 0 \end{cases}$

34. $\begin{cases} x \ge -2 \\ y \ge 1 \end{cases}$

35. $\begin{cases} x + 2y \le 4 \\ 3x + 2y \ge 6 \\ 2x - 4y < 8 \end{cases}$

36. $\begin{cases} x - 2y < 6 \\ 4x + 2y \ge -8 \\ 2x - 3y \ge -9 \end{cases}$

37. $\begin{cases} x + y \le 1 \\ y - x > -5 \\ y > -3 \end{cases}$

38. $\begin{cases} 2x + y \ge 1 \\ y - x < -1 \\ x < 3 \end{cases}$

For Exercises 39–42, explain the mistake.

39. $\begin{cases} x - y > -1 \\ x + y \leq 2 \end{cases}$

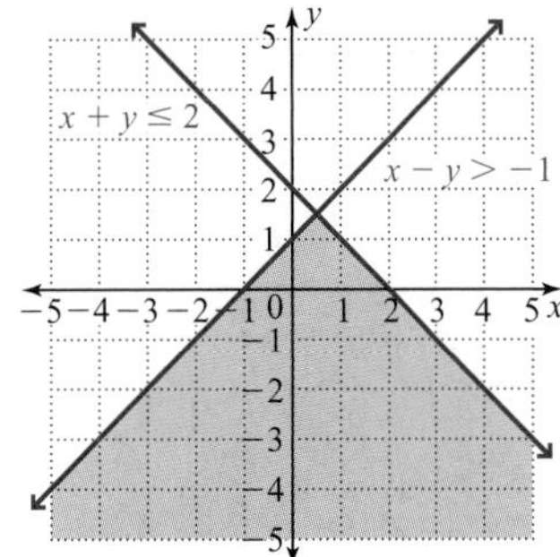

40. $\begin{cases} y < x - 2 \\ x - 3y \geq 2 \end{cases}$

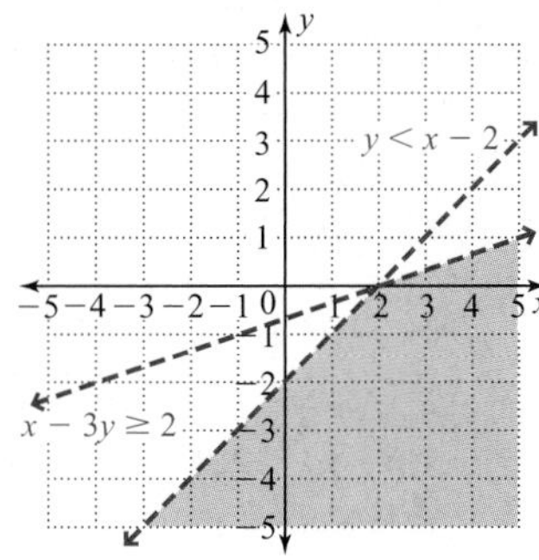

41. $\begin{cases} x - y > 3 \\ x + 2y > 1 \end{cases}$

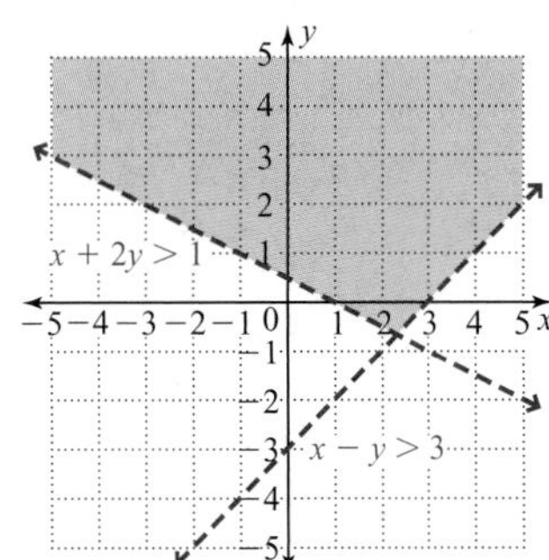

42. $\begin{cases} 3x - y \leq 6 \\ x + 2y \geq 5 \end{cases}$

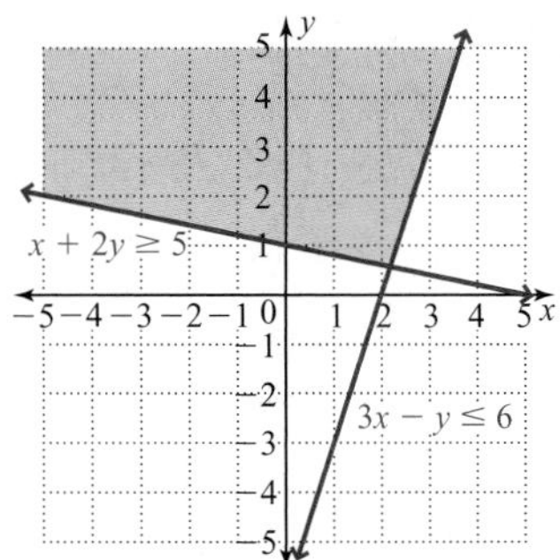

43. A furniture company makes two different benches, the standard bench and the regal bench. The standard bench sells for \$150 and the regal bench sells for \$200. To keep everyone busy and make a profit during any given week, they must sell at least 75 benches and receive at least \$12,000 in sales.

a. Write a system of inequalities that describes the number of benches that must be made during any given week to keep everyone busy and make a profit.

b. Solve the system by graphing.

c. Give two different combinations of number of benches that must be sold in order to keep everyone busy and make a profit.

44. Crane's Roost public park is designing a rectangular floating stage for live band performances. The length must be at least 25 feet more than the width and the perimeter must not exceed 500 feet.

a. Write a system of inequalities that describes the specifications for the stage.

b. Solve the system by graphing.

c. Give two different combinations of length and width that satisfy the requirements for the stage.

45. A sales person will receive at most a $5000 bonus for exceeding his sales quota for the year. He plans to invest the money in certificates of deposit paying 4% annually and municipal bonds that pay 5% annually. He wants his total annual interest to be at least $100.

a. Write a system of inequalities that describes the amount in the two investments and the annual interest received.

b. Solve the system by graphing.

c. Give two different combinations of investment amounts that return at least $400.

46. To be admitted to a certain university, prospective students must have a combined verbal and math score of at least 1150 in the SAT. To be admitted to the college of arts and sciences, students must also have a verbal score of at least 550.

a. Write a system of inequalities that describes the requirements to be admitted to the college of arts and sciences.

b. The maximum score on each part of the SAT is 800. Write two inequalities describing these maximum scores and add them to the system.

c. Solve the system by graphing.

d. Give two different combinations of verbal and math scores that satisfy the admission requirements to the college of arts and sciences.

Collaborative Exercises MAXIMIZING THE PROFIT

An aspiring-artists group is preparing for a fund-raising sale. They design and construct two types of antique-reproduction tables: a semicircular foyer table and a side table. The materials cost $100 for each foyer table and $200 for each side table. Both types of table require 2 hours of cutting and carving. The foyer table requires 3.5 hours to assemble, sand, and finish; and the side table requires 1.75 hours to assemble, sand, and finish. The group has a total of $3300 to purchase all materials; 40 hours available for cutting and carving; and 63 hours available for assembling, sanding, and finishing. For each foyer table that they sell, their profit will be $350. For each side table, their profit will be $215. How many of each should they make to maximize their profit?

Linear programming *is an area of mathematics that solves problems like the one just described. One of the fundamental elements of a linear programming model is the constraint. A* **constraint** *is simply an inequality that describes some limited resource required in the problem. For example, suppose you make two types of bicycles, style* A *and style* B. *Also, suppose that style* A *requires 40 minutes of welding for each bicycle and style* B *requires 25 minutes of welding. If you only have 600 minutes of welding time available,*

then the welding-time constraint would be $40A + 25B \le 600$, where A represents the number of style A produced and B represents the number of style B produced.

1. For the artists fund-raising problem there are three basic constraints: (1) cost of materials, (2) cutting and carving, and (3) assembling-sanding-finishing. For each of these constraints, write an inequality similar to the one in the bicycle example. Let x represent the number of foyer tables and y the number of side tables.

2. Since x and y represent numbers of tables, they cannot represent negative values. Write inequalities for these two additional constraints.

3. Now graph the system of inequalities described by these five constraints.
4. The region that is the the solution of the system of inequalities is called the *feasible region* and includes all points that satisfy the system. The goal is to find the optimal solution which, according to linear programming, is one of the corner points of the feasible region. There should be five corner points: one at the origin, one on each axis, and two more in quadrant I. Determine the coordinates of each corner point. Note that a corner point lying on an axis is the x- or y-intercept for the line passing through that point. Since two lines intersect to form a corner point that is not on an axis, make a system out of their two equations, then solve the system to find their point of intersection.

5. Next develop an algebraic expression describing the profit gained for selling x of the foyer tables and y of the side tables. This is called the *objective function.*

6. Using the objective function, test each of the points found in step 4 to see which one yields the maximum profit.

7. Using your solution, answer the following questions.
   a. How many of each type of table should the group produce?

   b. How many hours will be spent in each phase of the production?

   c. How much money will they need to purchase the materials?

   d. What amount of profit will the group receive?

REVIEW EXERCISES

[1.3] ***For Exercises 1 and 2, evaluate.***

1. 3^4

2. -10^2

[1.4] **3.** Distribute: $-6(5x - 4)$

[1.4] **4.** Simplify: $3x - 2(2x + 7)$

[2.1] **5.** Solve: $4x - 9 = x + 3$

[2.1] **6.** Solve: $3(x - 5) = 10 - (x + 1)$

Chapter 4 Summary

Defined Terms

Section 4.1
System of equations (p. 224)
Solution for a system of equations (p. 224)
Consistent system (p. 229)
Inconsistent system (p. 229)
Dependent linear equations in two unknowns (p. 229)
Independent linear equations in two unknowns (p. 229)

Section 4.4
Matrix (p. 267)
Augmented matrix (p. 267)
Echelon form (p. 267)

Section 4.5
Square matrix (p. 278)
Minor (p. 279)
Cramer's Rule (p. 281)

Procedures, Rules, and Key Examples

Procedures/Rules	Key Examples
Section 4.1 Solving Systems of Linear Equations in Two Variables To verify or check a solution to a system of equations, **1.** Replace each variable in each equation with its corresponding value. **2.** Verify that each equation is true.	**Example 1:** Determine whether the ordered pair is a solution to the system. $\begin{cases} 2x + 3y = 8 \\ y = 5x - 3 \end{cases}$ **a.** $(-2, 4)$ $2x + 3y = 8$; $2(-2) + 3(4) \stackrel{?}{=} 8$; $8 = 8$ $y = 5x - 3$; $4 \stackrel{?}{=} 5(-2) - 3$; $4 \neq -13$ $(-2, 4)$ is not a solution to the system. **b.** $(1, 2)$ $2x + 3y = 8$; $2(1) + 3(2) \stackrel{?}{=} 8$; $8 = 8$ $y = 5x - 3$; $2 \stackrel{?}{=} 5(1) - 3$; $2 = 2$ $(1, 2)$ is a solution to the system.
To solve a system of linear equations graphically, **1.** Graph each equation. **a.** If the lines intersect at a single point, then the coordinates of that point form the solution. **b.** If the lines are parallel, then there is no solution. **c.** If the lines are identical, then there are an infinite number of solutions, which are the coordinates of all the points on that line. **2.** In any case, check your solution.	**Example 2:** Solve $\begin{cases} 2x + 3y = 8 \\ y = 5x - 3 \end{cases}$ graphically. Solution: 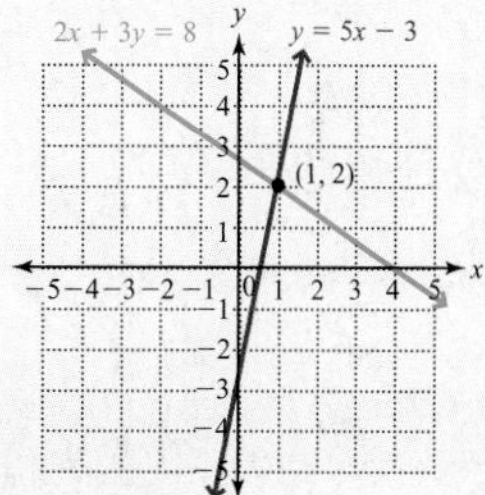 The solution is $(1, 2)$.

continued

Procedures/Rules	Key Examples
Section 4.1 Solving Systems of Linear Equations in Two Variables (Continued) To classify a system of equations, write the equations in slope-intercept form and compare the slopes and y-intercepts. **1.** If the slopes are different, then you have a **consistent system** with **independent equations**; it has a single solution. **2.** If the slopes are equal and the y-intercepts are also equal, then you have a **consistent system** with **dependent equations**; it has an infinite number of solutions. **3.** If the slopes are equal with different y-intercepts, then you have an **inconsistent system** with **independent equations**; it has no solution.	**Example 3:** Classify $\begin{cases} x + 3y = 6 \\ 2x + 6y = -5 \end{cases}$ and discuss the number of solutions. Solution: Write the equations in slope-intercept form. $x + 3y = 6$ becomes $y = -\frac{1}{3}x + 2$ $2x + 6y = -5$ becomes $y = -\frac{1}{3}x - \frac{5}{6}$ Since the slopes are the same $\left(-\frac{1}{3}\right)$, but the y-intercepts are different, this system is inconsistent with independent equations; it has no solution.
To find the solution of a system of two linear equations using the substitution method, **1.** Isolate one of the variables in one of the equations. **2.** In the other equation, substitute the expression you found in step 1 for that variable. **3.** Solve this new equation. (It will now have only one variable.) **4.** Using one of the equations containing both variables, substitute the value you found in step 3 for that variable and solve for the value of the other variable. **5.** Check the solution in the original equations.	**Example 4:** Solve $\begin{cases} y - 4x = -1 \\ 5x - y = 2 \end{cases}$ using substitution. Solution: We will isolate y in the first equation. $y = 4x - 1$ **Add $4x$ on both sides.** Now we substitute $4x - 1$ in place of y in the second equation. $5x - y = 2$ $5x - (4x - 1) = 2$ **Substitute $4x - 1$ for y.** $5x - 4x + 1 = 2$ **Distribute.** $x + 1 = 2$ **Combine like terms.** $x = 1$ **Subtract 1 from both sides.** Now substitute 1 for x in one of the equations. We will use $y = 4x - 1$. $y = 4x - 1$ $y = 4(1) - 1$ $y = 3$ The solution is (1, 3). ***continued***

Procedures/Rules	Key Examples

Section 4.1 Solving Systems of Linear Equations in Two Variables (Continued)

To solve a system of two linear equations using the elimination method,

1. Write the equations in standard form ($Ax + By = C$).
2. Use the multiplication principle of equality to clear fractions or decimals.
3. Multiply one or both equations by a number (or numbers) so that they have a pair of terms that are additive inverses.
4. Add the equations. The result should be an equation in terms of one variable.
5. Solve the equation from step 4 for the value of that variable.
6. Using an equation containing both variables, substitute the value you found in step 5 for the corresponding variable and solve for the value of the other variable.
7. Check your solution in the original equations.

Note: If, after step 4, the result is an equation that is false, such as $0 = 4$, then the system has no solution. If the result is an identity, such as $0 = 0$, then the system has an infinite number of solutions.

Example 5: Solve $\begin{cases} x + 2y = -4 \\ 4x + y = 5 \end{cases}$ using elimination.

Solution: We will eliminate y.

$$\begin{array}{ll} x + 2y = -4 & x + 2y = -4 \\ 4x + y = 5 \xrightarrow{\text{Multiply by } -2.} & -8x - 2y = -10 \\ & -7x + 0 = -14 \quad \text{Add equations.} \\ & x = 2 \quad \text{Divide both sides by } -7. \end{array}$$

Now solve for the value of y.

$$4(2) + y = 5 \quad \text{Substitute 2 in place of } x.$$
$$y = -3$$

The solution is $(2, -3)$.

Section 4.2 Solving Systems of Linear Equations in Three Variables

To solve a system of three equations with three unknowns,

1. Write each equation in the form $Ax + By + Cz = D$.
2. Eliminate one variable from one pair of equations using the elimination method.
3. Eliminate the same variable from another pair of equations.
4. Steps 2 and 3 result in two equations with the same two variables. Solve these equations using the elimination method.
5. To find the third variable, substitute the values of the variables found in step 4 into any of the three original equations that contain the third variable.
6. Check the ordered triple in all three original equations.

Example 1: Solve the following system of equations:

$$\begin{cases} x + 2y - z = -6 & \text{(Equation 1)} \\ 2x - 3y + 4z = 26 & \text{(Equation 2)} \\ -x + 2y - 3z = -18 & \text{(Equation 3)} \end{cases}$$

Eliminate x by adding Equation 1 and Equation 3.

$$\begin{array}{ll} x + 2y - z = -6 & \\ -x + 2y - 3z = -18 & \\ 4y - 4z = -24 & \text{Divide both sides by 4.} \\ y - z = -6 & \text{(Eq. 4)} \end{array}$$

Eliminate x again using Equations 2 and 3.

$$\begin{array}{ll} 2x - 3y + 4z = 26 & 2x - 3y + 4z = 26 \\ -x + 2y - 3z = -18 \xrightarrow{\text{Multiply by 2.}} & -2x + 4y - 6z = -36 \\ & y - 2z = -10 \quad \text{(Eq. 5)} \end{array}$$

Use Equations 4 and 5 to solve for z.

$$\begin{array}{ll} y - z = -6 & y - z = -6 \\ y - 2z = -10 \xrightarrow{\text{Multiply by } -1.} & -y + 2z = 10 \\ & z = 4 \end{array}$$

Substitute 4 for z in Equation 4 and solve for y.

$$y - 4 = -6$$
$$y = -2$$

continued

Procedures/Rules	Key Examples
Section 4.2 Solving Systems of Linear Equations in Three Variables (Continued)	Substitute -2 for y and 4 for z in Equation 1 and solve for x. $$\begin{aligned} x + 2(-2) - 4 &= -6 \\ x - 4 - 4 &= -6 \\ x &= 2 \end{aligned}$$ The solution is $(2, -2, 4)$.
Section 4.3 Solving Applications Using Systems of Equations To solve problems using systems of equations, **1.** Select a variable to represent each unknown. **2.** Write a system of equations. **3.** Solve the system. **Note:** The substitution method is simpler than elimination if one of the equations has an isolated variable. If all equations in the system are in standard form, then elimination is simpler.	**Example 1:** A vendor sells small and large drinks for \$2.00 and \$3.00, respectively. If the vendor sold 2400 drinks for a total of \$5400, how many of each size were sold? Let $x =$ the number of small drinks sold. Let $y =$ the number of large drinks sold. The total number of drinks sold, 2400, translates to $x + y = 2400$. The total revenue is \$5400, which translates to $2x + 3y = 5400$. System: $\begin{cases} x + y = 2400 \\ 2x + 3y = 5400 \end{cases}$ To solve the system, we will use the elimination method because there are no isolated variables. $x + y = 2400 \xrightarrow{\text{Multiply by } -3} -3x - 3y = -7200$ $2x + 3y = 5400 \qquad\qquad 2x + 3y = 5400$ We eliminated y. $\quad -x + 0 = -1800$ Solve for x. $\quad x = 1800$ Substitute 1800 for x in an equation: $1800 + y = 2400$ We chose $x + y = 2400$. $y = 600$ Isolate y. The vendor sold 1800 small and 600 large drinks. *continued*

Procedures/Rules	Key Examples

Section 4.4 Solving Systems of Linear Equations Using Matrices

Row operations:

1. Any two rows may be interchanged.
2. The elements of any row may be multiplied (or divided) by any nonzero real number.
3. Any row may be replaced by a row resulting from adding the elements of that row (or multiples of that row) to a multiple of the elements of any other row.

Row operations can be used to solve a system of linear equations by transforming the augmented matrix of the system into a matrix that is in echelon form. A matrix is in echelon form if the coefficient portion of the augmented matrix has 1s on the diagonal from upper left to lower right and 0s below the 1s.

Example 1: Solve $\begin{cases} 2x + 3y = -5 \\ x + 2y = -4 \end{cases}$ using row operations.

The augmented matrix is $\left[\begin{array}{cc|c} 2 & 3 & -5 \\ 1 & 2 & -4 \end{array}\right]$.

$\left[\begin{array}{cc|c} 1 & 2 & -4 \\ 2 & 3 & -5 \end{array}\right]$ We need row 1, column 1 to be a 1, so interchange the two rows.

$-2 \cdot R_1 + R_2 \longrightarrow \left[\begin{array}{cc|c} 1 & 2 & -4 \\ 0 & -1 & 3 \end{array}\right]$ We need row 2, column 1 to be 0, so multiply row 1 by -2 and add it to row 2.

$-1 \cdot R_2 \longrightarrow \left[\begin{array}{cc|c} 1 & 2 & -4 \\ 0 & 1 & -3 \end{array}\right]$ Multiply row 2 by -1 to get echelon form.

Row 2 means $y = -3$ and row 1 means $x + 2y = -4$.

$x + 2(-3) = -4$ Substitute -3 for y and solve for x

$x = 2$

The solution is $(2, -3)$.

Section 4.5 Cramer's Rule

If $A = \begin{bmatrix} a_1 & b_1 \\ a_2 & b_2 \end{bmatrix}$, then $\det(A) = \begin{vmatrix} a_1 & b_1 \\ a_2 & b_2 \end{vmatrix} = a_1b_2 - a_2b_1$.

Example 1: Find the determinant of $\begin{bmatrix} 3 & -4 \\ 2 & -5 \end{bmatrix}$.

Solution: $\begin{vmatrix} 3 & -4 \\ 2 & -5 \end{vmatrix} = 3(-5) - 2(-4) = -15 + 8 = -7$

The determinant of a 3×3 matrix is found by

$$\begin{vmatrix} a_1 & b_1 & c_1 \\ a_2 & b_2 & c_2 \\ a_3 & b_3 & c_3 \end{vmatrix} = a_1\begin{pmatrix}\text{minor} \\ \text{of } a_1\end{pmatrix} - a_2\begin{pmatrix}\text{minor} \\ \text{of } a_2\end{pmatrix} + a_3\begin{pmatrix}\text{minor} \\ \text{of } a_3\end{pmatrix}$$

$$= a_1\begin{vmatrix} b_2 & c_2 \\ b_3 & c_3 \end{vmatrix} - a_2\begin{vmatrix} b_1 & c_1 \\ b_3 & c_3 \end{vmatrix} + a_3\begin{vmatrix} b_1 & c_1 \\ b_2 & c_2 \end{vmatrix}$$

Example 2: Find the determinant of $\begin{bmatrix} -1 & -2 & 3 \\ 1 & -3 & 4 \\ -2 & 3 & -1 \end{bmatrix}$.

Solution:

$$\begin{vmatrix} -1 & -2 & 3 \\ 1 & -3 & 4 \\ -2 & 3 & -1 \end{vmatrix} = -1\begin{vmatrix} -3 & 4 \\ 3 & -1 \end{vmatrix} - 1\begin{vmatrix} -2 & 3 \\ 3 & -1 \end{vmatrix} + (-2)\begin{vmatrix} -2 & 3 \\ -3 & 4 \end{vmatrix}$$

$$= -1(3 - 12) - 1(2 - 9) + (-2)(-8 + 9)$$
$$= 9 + 7 - 2$$
$$= 14$$

continued

Procedures/Rules	Key Examples

Section 4.5 Cramer's Rule (Continued)

Cramer's Rule

The solution to $\begin{cases} a_1x + b_1y = c_1 \\ a_2x + b_2y = c_2 \end{cases}$ is

$$x = \frac{\begin{vmatrix} c_1 & b_1 \\ c_2 & b_2 \end{vmatrix}}{\begin{vmatrix} a_1 & b_1 \\ a_2 & b_2 \end{vmatrix}} = \frac{D_x}{D} \text{ and } y = \frac{\begin{vmatrix} a_1 & c_1 \\ a_2 & c_2 \end{vmatrix}}{\begin{vmatrix} a_1 & b_1 \\ a_2 & b_2 \end{vmatrix}} = \frac{D_y}{D}$$

Example 3: Solve $\begin{cases} 3x + 2y = -2 \\ 5x + 2y = 2 \end{cases}$.

Solution: $D = \begin{vmatrix} 3 & 2 \\ 5 & 2 \end{vmatrix} = -4$

$$x = \frac{\begin{vmatrix} -2 & 2 \\ 2 & 2 \end{vmatrix}}{-4} = \frac{-4-4}{-4} = \frac{-8}{-4} = 2$$

$$y = \frac{\begin{vmatrix} 3 & -2 \\ 5 & 2 \end{vmatrix}}{-4} = \frac{6+10}{-4} = \frac{16}{-4} = -4$$

The solution is (2, −4).

The solution to $\begin{cases} a_1x + b_1y + c_1z = d_1 \\ a_2x + b_2y + c_2z = d_2 \\ a_3x + b_3y + c_3z = d_3 \end{cases}$ is

$$x = \frac{\begin{vmatrix} d_1 & b_1 & c_1 \\ d_2 & b_2 & c_2 \\ d_3 & b_3 & c_3 \end{vmatrix}}{\begin{vmatrix} a_1 & b_1 & c_1 \\ a_2 & b_2 & c_2 \\ a_3 & b_3 & c_3 \end{vmatrix}} = \frac{D_x}{D},\ y = \frac{\begin{vmatrix} a_1 & d_1 & c_1 \\ a_2 & d_2 & c_2 \\ a_3 & d_3 & c_3 \end{vmatrix}}{\begin{vmatrix} a_1 & b_1 & c_1 \\ a_2 & b_2 & c_2 \\ a_3 & b_3 & c_3 \end{vmatrix}} = \frac{D_y}{D},$$

$$\text{and } z = \frac{\begin{vmatrix} a_1 & b_1 & d_1 \\ a_2 & b_2 & d_2 \\ a_3 & b_3 & d_3 \end{vmatrix}}{\begin{vmatrix} a_1 & b_1 & c_1 \\ a_2 & b_2 & c_2 \\ a_3 & b_3 & c_3 \end{vmatrix}} = \frac{D_z}{D}$$

Example 4: Solve $\begin{cases} x + 2y - z = -4 \\ 2x - 3y - 2z = 13 \\ 2x - 2y + 3z = 20 \end{cases}$.

Solution: $D = \begin{vmatrix} 1 & 2 & -1 \\ 2 & -3 & -2 \\ 2 & -2 & 3 \end{vmatrix} = -35$

$$x = \frac{\begin{vmatrix} -4 & 2 & -1 \\ 13 & -3 & -2 \\ 20 & -2 & 3 \end{vmatrix}}{-35} = \frac{-140}{-35} = 4$$

$$y = \frac{\begin{vmatrix} 1 & -4 & -1 \\ 2 & 13 & -2 \\ 2 & 20 & 3 \end{vmatrix}}{-35} = \frac{105}{-35} = -3$$

$$z = \frac{\begin{vmatrix} 1 & 2 & -4 \\ 2 & -3 & 13 \\ 2 & -2 & 20 \end{vmatrix}}{-35} = \frac{-70}{-35} = 2$$

The solution is (4, −3, 2).

Section 4.6 Solving Systems of Linear Inequalities

To solve a system of linear inequalities, graph all of the inequalities on the same grid. The solution set for the system contains all ordered pairs in the region where the inequalities' solution sets overlap along with ordered pairs on the portion of any solid line that touches the region of overlap.

Example 1: Solve $\begin{cases} x + y > -2 \\ y \le 3x - 1 \end{cases}$.

Solution:

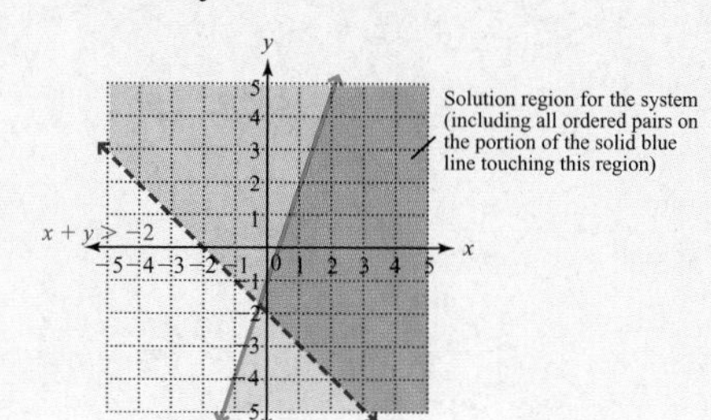

Chapter 4 Review Exercises

For Exercises 1–5, answer true or false.

[4.1] **1.** The solution for a system of two linear equations is always an ordered pair that satisfies all the equations of the system.

[4.1] **2.** Given a system of two linear equations with two variables, if there is no solution then the graphs are parallel lines.

[4.1] **3.** Any solvable system of linear equations can be solved using the elimination method.

[4.4] **4.** Solving a system of linear equations using the echelon method is like solving a system using the elimination method except the echelon method uses the coefficients and constant terms only.

[4.5] **5.** The determinant of a matrix is another matrix.

For Exercises 6–10, complete the rule.

[4.1] **6.** To verify or check a solution to a system of equations,

1. ________ each variable in each equation with its corresponding value.
2. Verify that each equation is true.

[4.1] **7.** An inconsistent system has no solution. The graphs have the same ________ but different ____________.

[4.6] **8.** To solve a system of two linear inequalities,

1. Graph each inequality.
2. The solution of the system is ______________________________ __________.

[4.4] **9.** A matrix is in echelon form if ______________________________ ____.

[4.5] **10.** When solving a system of linear equations using Cramer's Rule, the denominator is the ____________ of the matrix containing only the coefficients of the system.

[4.1] ***For Exercises 11–12, determine whether the given ordered pair is a solution to the given system of equations.***

11. $(4, 3)$; $\begin{aligned} x - y &= 1 \\ -x + y &= -1 \end{aligned}$

12. $(1, 1)$; $\begin{aligned} 2x - y &= 7 \\ x + y &= 8 \end{aligned}$

[4.1] For Exercises 13–16: ***a.*** ***Determine whether the system of equations is consistent with independent equations, inconsistent with independent equations, or consistent with dependent equations.***
b. ***How many solutions does the system have?***

13.

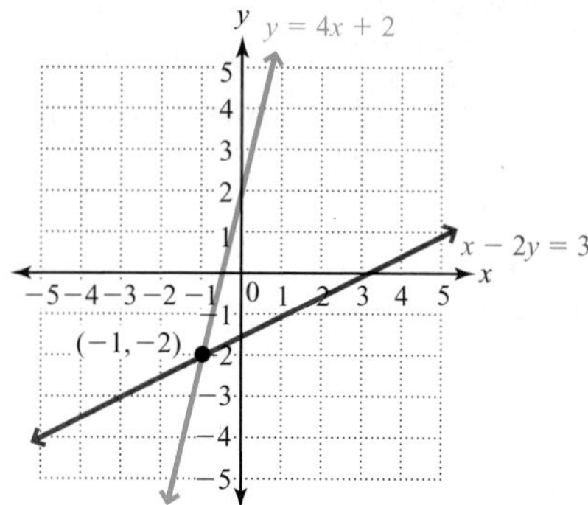

14.

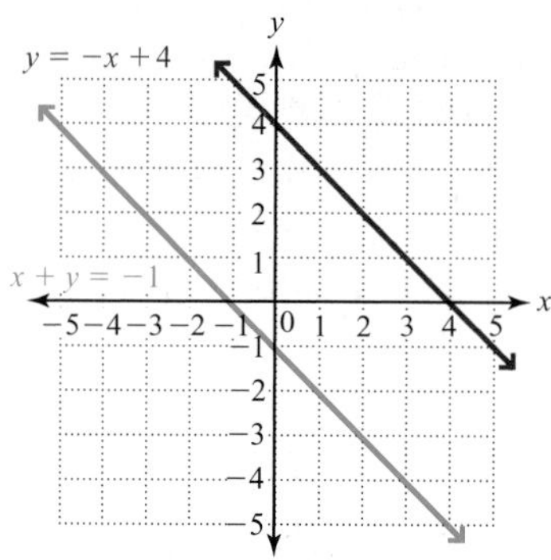

15. $\begin{cases} x - 3y = 10 \\ 2x + 3y = 5 \end{cases}$

16. $\begin{cases} x - 5y = 10 \\ 2x - 10y = 20 \end{cases}$

[4.1] *For Exercises 17–20, solve the system graphically.*

17. $\begin{cases} 4x = y \\ 3x + y = -7 \end{cases}$

18. $\begin{cases} x - y = 4 \\ 2x + 3y = 3 \end{cases}$

19. $\begin{cases} y = -2x - 3 \\ 4x + 2y = 6 \end{cases}$

20. $\begin{cases} 3x - 2y = 6 \\ y = \frac{3}{2}x - 3 \end{cases}$

[4.1] *For Exercises 21–24, solve the system of equations using substitution. Note that some systems may be inconsistent or consistent with dependent equations.*

21. $\begin{cases} 3x + 10y = 2 \\ x - 2y = 6 \end{cases}$

22. $\begin{cases} 2x + 5y = 8 \\ x - 10y = 9 \end{cases}$

23. $\begin{cases} 3y - 4x = 6 \\ y = \frac{4}{3}x + 2 \end{cases}$

24. $\begin{cases} 8x - 2y = 12 \\ y - 4x = 3 \end{cases}$

[4.1] *For Exercises 25–28, solve the system of equations using elimination.*

25. $\begin{cases} x + y = 4 \\ x - y = -2 \end{cases}$

26. $\begin{cases} 3x + 2y = 4 \\ 2x - 3y = 7 \end{cases}$

27. $\begin{cases} 0.25x + 0.75y = 4 \\ x - y = -4 \end{cases}$

28. $\begin{cases} \frac{1}{5}x - \frac{1}{3}y = 2 \\ x + y = 2 \end{cases}$

[4.2] *For Exercises 29–32, solve the systems using the elimination method.*

29. $\begin{cases} x + y + z = -2 \\ 2x + 3y + 4z = -10 \\ 3x - 2y - 3z = 12 \end{cases}$

30. $\begin{cases} x + y + z = 0 \\ 2x - 4y + 3z = -12 \\ 3x - 3y + 4z = 2 \end{cases}$

31. $\begin{cases} 3x + 4y = -3 \\ -2y + 3z = 12 \\ 4x - 3z = 6 \end{cases}$

32. $\begin{cases} 2x + 2y - 3z = 3 \\ x = -3y + 4z - 3 \\ z = -x - 2y + 6 \end{cases}$

[4.4] *For Exercises 33–36, solve the system using the echelon method.*

33. $\begin{cases} x - 4y = 8 \\ x + 2y = 2 \end{cases}$

34. $\begin{cases} 2x - y = -4 \\ 2x - 3y = 0 \end{cases}$

35. $\begin{cases} x + y + z = 2 \\ 2x + y + 2z = 1 \\ 3x + 2y + z = 1 \end{cases}$

36. $\begin{cases} x + 2y + z = -1 \\ 2x - 2y + z = -10 \\ -x + 3y - 2z = 15 \end{cases}$

[4.5] *For Exercises 37–40, solve the system of equations using Cramer's Rule.*

37. $\begin{cases} 9x + 4y = 18 \\ 5x - 2y = -28 \end{cases}$

38. $\begin{cases} 5x - y = 3 \\ 10x - 2y = 5 \end{cases}$

39. $\begin{cases} x + y + z = 5 \\ 3x - 2y + z = -3 \\ x + 3y + 4z = 10 \end{cases}$

40. $\begin{cases} x + 3y + 2z = 6 \\ 2x - 3y + z = -18 \\ -3x + 2y + z = 12 \end{cases}$

[4.6] *For Exercises 41–44, graph the solution set for the system of inequalities.*

41. $\begin{cases} x + y > -5 \\ x - y \geq -1 \end{cases}$

42. $\begin{cases} 2x + y \geq -4 \\ -3x + y \geq -1 \end{cases}$

43. $\begin{cases} 2x + y > 1 \\ 3x - y \leq -1 \end{cases}$

44. $\begin{cases} -3x + 4y < 12 \\ 2x - y \leq -3 \end{cases}$

[4.3] *For Exercises 45–55, solve.*

45. The sum of two numbers is 16. Their difference is 4. What are the two numbers?

46. A movie theater charges $7.00 for adults and $5.50 for children. If 139 tickets were sold for a total of $835, how many of each ticket were sold?

47. In 2003, domestic and international travelers spent a combined total of $556 billion in the United States. If domestic travelers spent $424 billion more than international travelers, how much was spent by each group of travelers? (*Source:* Tourism Industries, International Trade Administration, Department of Commerce)

48. In 2002, Western Europe and North America contributed a combined total of 43% of the world's carbon dioxide emissions. If North America contributed 4% less than twice the percentage contributed by Western Europe, what percent did each of those regions contribute?

> ***Of Interest***
>
> The United States alone contributes 23% of the world's carbon dioxide emissions.

49. The smaller of two complementary angles is 10 degrees less than one-third of the larger angle. Find the measure of the two angles.

50. Jose invested part of $5000 in a fund that earned 7% in interest and the rest in a fund that earned 9% in interest. If he earned $380 in interest during the year, how much was invested in each fund?

51. Yolanda and Dee are traveling west in separate cars on the same highway. Yolanda is traveling at 60 miles per hour and Dee at 70 miles per hour. Yolanda passes the I-95 exit at 4:00 P.M. Dee passes the same exit at 4:15 P.M. At what time will Dee catch up to Yolanda?

52. A plane travels 450 miles in 3 hours with the jet stream. When the plane returns, it takes 5 hours to travel the same distance against the jet stream. What is the speed of the plane in still air?

53. How much of a 10% saline solution is needed to combine with a 60% saline solution to obtain 50 milliliters of a 30% saline solution?

54. John bought some vitamin supplements that cost $8 each, some film that cost $4 per roll, and some bags of candy that cost $3 per bag. He bought a total of eight items that cost $32. The number of rolls of film was one less than the number of bags of candy. Find the number of each type of item that he bought.

55. At a swim meet, 5 points are awarded for each first-place finish, 3 points for each second, and 1 point for each third. Shawnee Mission South High School scored a total of 38 points. The number of first-place finishes was one more than the number of second-place finishes. The number of third-place finishes was three times the number of second-place finishes. Find the number of first-, second-, and third-place finishes for the school.

[**4.6**] ***Solve using a system of linear inequalities.***

56. A real estate course is offered in a continuing education program at a two-year college. The course is limited to 16 students. Also, to run the course, the college must receive at least $2750 in tuition. The college charges $250 per student if they live in the same county as the college and $400 if they live in any other county (or out-of-state).

a. Write a system of inequalities that describes the number of in-county students enrolled in the course versus the number of out-of-county students and how much they paid.

b. Solve the system by graphing.

c. Give a combination of students that could enroll in the course so that it runs.

Chapter 4 Practice Test

For Exercises 1 and 2, determine whether the given ordered pair is a solution to the given system of equations.

1. $(-1, 3)$; $\begin{cases} 3x + 2y = 3 \\ 4x - y = -7 \end{cases}$

2. $(2, 2, 3)$; $\begin{cases} 2x - 3y + z = 1 \\ -2x + y - 3z = 11 \\ 3x + y + 3z = 14 \end{cases}$

3. Solve by graphing. $\begin{cases} x + 3y = 1 \\ -2x + y = 5 \end{cases}$

For Exercises 4–9, solve the system of equations using substitution or elimination. Note that some systems may be inconsistent or consistent with dependent equations.

4. $\begin{cases} 2x + y = 15 \\ y = 7 - x \end{cases}$

5. $\begin{cases} x - 2y = 1 \\ 3x - 5y = 4 \end{cases}$

6. $\begin{cases} 3x - 2y = -8 \\ 2x + 3y = -14 \end{cases}$

7. $\begin{cases} 4x + 6y = 2 \\ 6x + 9y = 3 \end{cases}$

8. $\begin{cases} x + 2y + z = 2 \\ x + 4y - z = 12 \\ 3x - 3y - 2z = -11 \end{cases}$

9. $\begin{cases} x + 2y - 3z = -9 \\ 3x - y + 2z = -8 \\ 4x - 3y + 3z = -13 \end{cases}$

For Exercises 10 and 11, solve the equations using the echelon method.

10. $\begin{cases} x + 2y = -6 \\ 3x + 4y = -10 \end{cases}$

11. $\begin{cases} x + y + z = 6 \\ 3x - 2y + 3z = -7 \\ 4x - 2y + z = -12 \end{cases}$

For Exercises 12 and 13, solve the system of equations using Cramer's Rule.

12. $\begin{cases} 3x + 4y = 14 \\ 2x - 3y = -19 \end{cases}$

13. $\begin{cases} x + y + z = -1 \\ 3x - 2y + 4z = 0 \\ 2x + 5y - z = -11 \end{cases}$

14. Graph the solution set for the system of inequalities.
$\begin{cases} 2x - 3y < 1 \\ x + 2y \le -2 \end{cases}$

For Exercises 15–20, solve.

15. Excedrin surveyed workers in various professions who get headaches at least once a year on the job. There were 9 more accountants than there were waiters/waitresses in the survey who got a headache. If the combined number of accountants and waiters/waitresses that got a headache was 163, how many people in each of those professions got a headache? (*Source:* the *State* newspaper)

16. When asked what the Internet most resembled, three times as many people said a library as opposed to a highway. If 240 people were polled, how many people considered the Internet to be a library? (*Source: bLINK* magazine, winter 2001)

17. A boat traveling with the current took 3 hours to go 30 miles. The same boat went 12 miles in 3 hours against the current. What is the rate of the boat in still water?

18. Janice invested $12,000 into two funds. One of the funds returned 6% interest and the other returned 8% interest after one year. If the total interest for the year was $880, how much did she invest in each fund?

19. Tickets for the senior play at Apopka High School cost $3 for children, $5 for students, and $8 for adults. There were 800 tickets sold for a total of $4750. The number of adult tickets sold was 50 more than two times the number of children's tickets. Find the number of each type of ticket.

20. A landscaper wants to plan a bordered rectangular garden area in a yard. Since she currently has 200 feet of border materials, the perimeter needs to be at most 200 feet. She thinks it would look best if the length is at least 10 feet more than the width.

a. Write a system of inequalities to describe the situation.

b. Solve the system by graphing.

c. Give a combination of length and width that satisfy the requirements for the garden.

CHAPTER 5

Polynomials

"Art, it seems to me, should simplify. That, indeed, is very nearly the whole of the higher artistic process; finding what conventions of form and what detail one can do without and yet preserve the spirit of the whole—so that all one has suppressed and cut away is there to the reader's consciousness as much as if it were in type on the page."

—Willa Cather, U.S. novelist

In Chapters 2–4, we focused on linear equations and functions. In this chapter, we will learn about a specific group of expressions, called *polynomials,* that we will use to build more complicated types of equations and functions. You may have noticed that in every linear equation or function, the exponent of every variable is 1. Polynomial expressions, equations, and functions can involve variables with exponents of 1 or greater. Because polynomials involve exponents to a greater extent, we will first discuss the rules of exponents in Section 5.1. Then, in Section 5.2, we will define polynomials and polynomial functions. The rest of the chapter will be devoted to learning how to add, subtract, multiply, and divide polynomials.

5.1 Exponents and Scientific Notation

OBJECTIVES

1. Use the product rule of exponents to simplify expressions.
2. Use the quotient rule of exponents to simplify expressions.
3. Use the power rule of exponents to simplify expressions.
4. Convert between scientific notation and standard form.
5. Simplify products, quotients, and powers of numbers in scientific notation.

Note: *When the base is negative, the sign of the product depends on whether the exponent is odd or even.*

Recall from Section 1.3 that natural-number exponents indicate repeated multiplication of a base number.

$$2^4 = \underbrace{2 \cdot 2 \cdot 2 \cdot 2}_{\text{four factors of 2}} = 16 \qquad (-5)^3 = \underbrace{(-5)(-5)(-5)}_{\text{three factors of } -5} = -125$$

$$\left(\frac{2}{3}\right)^4 = \underbrace{\frac{2}{3} \cdot \frac{2}{3} \cdot \frac{2}{3} \cdot \frac{2}{3}}_{\text{four factors of } \frac{2}{3}} = \frac{16}{81}$$

Note: *Raising both the numerator and denominator to the power gives the same result.*

Our examples suggest the following rules.

RULES Evaluating Exponential Forms

If n is a natural number and a is a real number, then $a^n = \underbrace{a \cdot a \cdot \cdots \cdot a}_{n \text{ factors of } a}$.

If $a < 0$ and n is even, then a^n is positive.

If $a < 0$ and n is odd, then a^n is negative.

If a and b are real numbers with $b \neq 0$ and n is a whole number, then $\left(\frac{a}{b}\right)^n = \frac{a^n}{b^n}$.

Warning: Remember, expressions like $(-2)^4$ and -2^4 mean different things. In $(-2)^4$, the base is -2, whereas in -2^4, the base is 2. So, the minus sign in -2^4 means to find the additive inverse of 2^4.

$$(-2)^4 = (-2)(-2)(-2)(-2) = 16$$
$$-2^4 = -[2 \cdot 2 \cdot 2 \cdot 2] = -16$$

OBJECTIVE 1. Use the product rule of exponents to simplify expressions. So far we have developed rules only for exponents that are natural numbers (positive integers). However, exponents can be positive, zero, or negative. To show what an exponent of 0 means, we need another rule of exponents called the *product rule*.

Consider $a^2 \cdot a^3$. To find the product, we can expand to individual factors:

$$a^2 \cdot a^3 = a \cdot a \cdot a \cdot a \cdot a = a^5$$

Note: *The product, a^5, could also be found by adding the exponents:*

$$a^2 \cdot a^3 = a^{2+3} = a^5$$

Our example suggests the following product rule of exponents.

RULE **Product Rule of Exponents**

If a is a real number and m and n are natural numbers, then $a^m \cdot a^n = a^{m+n}$.

EXAMPLE 1 Simplify using the product rule of exponents.

a. $3^2 \cdot 3^4$

Solution $3^2 \cdot 3^4 = 3^{2+4} = 3^6$

b. $n^5 \cdot n^7$

Solution $n^5 \cdot n^7 = n^{5+7} = n^{12}$

c. $7x \cdot 2x^3$

Solution $7x \cdot 2x^3 = 7 \cdot 2 \cdot x \cdot x^3$ Use the commutative property to group like bases together.

$= 14x^{1+3}$ Multiply the coefficients and use the product rule of exponents.

$= 14x^4$ Simplify the exponent.

d. $(-9m^2n)(5m^3n^7p)$

Solution $(-9m^2n)(5m^3n^7p)$

$= -9 \cdot 5 \cdot m^2 \cdot m^3 \cdot n \cdot n^7 \cdot p$ Use the commutative property to group like bases together.

$= -45m^{2+3}n^{1+7}p$ Multiply the coefficients and use the product rule of exponents.

$= -45m^5n^8p$ Simplify the exponents.

YOUR TURN Simplify using the product rule of exponents.

a. $x^3 \cdot x^6$ **b.** $4y^2 \cdot 8y^5$ **c.** $(-6ab^3)(7abc^2)$

ANSWERS

a. x^9 **b.** $32y^7$ **c.** $-42a^2b^4c^2$

We can now use the product rule to show what 0 as an exponent means. Consider $a^0 \cdot a^n$, where $a \neq 0$. Using the product rule of exponents we can simplify.

$$a^0 \cdot a^n = a^{0+n} = a^n$$

Conclusion: Since the product of a^0 and a^n is a^n, we can say that a^0 must be equal to 1.

EXAMPLE 2 Simplify. Assume variables do not equal 0.

a. $(-32)^0$

Solution $(-32)^0 = 1$

b. $9n^0$

Solution $9n^0 = 9(1)$ Replace n^0 with 1.

$= 9$ Multiply.

RULE Zero as an Exponent

If a is a real number and $a \neq 0$, then $a^0 = 1$.

YOUR TURN Simplify. Assume variables do not equal 0.

a. $(-28)^0$ **b.** y^0 **c.** $5x^0$ **d.** $-4m^0$

OBJECTIVE 2. Use the quotient rule of exponents to simplify expressions. Now, let's consider dividing exponential forms with the same base, as in $\frac{a^5}{a^3}$, where $a \neq 0$. To determine the rule, we can expand the exponential expressions and then divide out three of the common factors of a.

$$\frac{a^5}{a^3} = \frac{a \cdot a \cdot a \cdot a \cdot a}{a \cdot a \cdot a} = a \cdot a = a^2$$

Note: *The quotient, a^2, could also be found by subtracting the exponents.*

$$\frac{a^5}{a^3} = a^{5-3} = a^2$$

Our example suggests the following quotient rule of exponents.

RULE Quotient Rule for Exponents

If m and n are integers and a is a real number where $a \neq 0$, then $\frac{a^m}{a^n} = a^{m-n}$.

Note: We cannot let a equal 0 because if a were replaced with 0, we would have $\frac{0^m}{0^n} = \frac{0}{0}$, which is indeterminate.

Connection We have seen that $a^2 \cdot a^3 = a^{2+3} = a^5$. Since multiplication and division are inverse operations, it makes sense that when dividing the same base we subtract the exponents.

Note that the divisor's exponent is subtracted from the dividend's exponent. This order is important because subtraction is not a commutative operation.

ANSWERS

a. 1 **b.** 1 **c.** 5 **d.** −4

Connection In Objective 1 we used the product rule of exponents to see that $y^0 = 1$. We can expand the exponential forms and divide out all the common factors to affirm this fact.

$$\frac{y^5}{y^5} = \frac{y \cdot y \cdot y \cdot y \cdot y}{y \cdot y \cdot y \cdot y \cdot y} = 1$$

When we divide exponential forms that have the same base and the same exponent, the result is 1.

EXAMPLE 3 Divide using the quotient rule of exponents. Assume that variables in denominators are not equal to 0.

a. $\dfrac{x^{11}}{x^5}$

Solution $\dfrac{x^{11}}{x^5} = x^{11-5} = x^6$

b. $\dfrac{y^5}{y^5}$

Solution $\dfrac{y^5}{y^5} = y^{5-5} = y^0 = 1$

c. $\dfrac{36x^8}{4x^2}$

Solution $\dfrac{36x^8}{4x^2} = \dfrac{36}{4} \cdot \dfrac{x^8}{x^2}$ Separate the coefficients and variables.

$= 9x^{8-2}$ Divide the coefficients and use the quotient rule for the variables.

$= 9x^6$ Simplify.

Note: *After this example, we will no longer show this step.*

d. $\dfrac{18t^2u^8v}{30t^2u}$

Solution $\dfrac{18t^2u^8v}{30t^2u} = \dfrac{18}{30}t^{2-2}u^{8-1}v$ Divide the coefficients and use the quotient rule for the variables.

$= \dfrac{3}{5}u^7v$ Simplify.

Learning Strategy

If you are a visual learner, look at the fraction line as a subtraction sign between the numerator's exponent and the denominator's exponent.

YOUR TURN Simplify using the quotient rule for exponents. Assume variables in the denominator are not equal to 0.

a. $\dfrac{y^{13}}{y^8}$ **b.** $\dfrac{15t^4}{3t^4}$ **c.** $\dfrac{-12m^5n^3p}{28mn^3}$

We can use the quotient rule to deduce what a negative integer exponent means. Suppose we have a division of exponential expressions in which the divisor's exponent is greater than the dividend's exponent, as in $\dfrac{a^2}{a^5}$. We can use two approaches to simplify this expression.

Using the quotient rule: $\dfrac{a^2}{a^5} = a^{2-5} = a^{-3}$

Expanding the exponential forms: $\dfrac{a^2}{a^5} = \dfrac{a \cdot a}{a \cdot a \cdot a \cdot a \cdot a} = \dfrac{1}{a \cdot a \cdot a} = \dfrac{1}{a^3}$

Both results are correct, indicating that $a^{-3} = \dfrac{1}{a^3}$, which suggests the following rule for negative exponents.

ANSWERS

a. y^5 **b.** 5 **c.** $-\dfrac{3}{7}m^4p$

RULE If a is a real number where $a \neq 0$ and n is a natural number, then $a^{-n} = \frac{1}{a^n}$.

EXAMPLE 4 Evaluate the exponential form.

a. 10^{-3}

Warning: Notice that $10^{-3} \neq -1000$.

Solution $10^{-3} = \frac{1}{10^3} = \frac{1}{1000}$

b. $(-2)^{-4}$

Solution $(-2)^{-4} = \frac{1}{(-2)^4} = \frac{1}{(-2)(-2)(-2)(-2)} = \frac{1}{16}$

What if a negative exponent is in the denominator of a fraction?

EXAMPLE 5 Evaluate. $\frac{1}{3^{-2}}$

Solution Because the denominator is raised to a negative exponent, we write the reciprocal of the denominator in the denominator.

$$\frac{1}{3^{-2}} = \frac{1}{\frac{1}{3^2}} = 1 \cdot \frac{3^2}{1} = 3^2 = 9$$

If we remove the intermediate steps in our solution to Example 5, we see that $\frac{1}{3^{-2}} = 3^2$, which suggests the following rule.

RULE If a is a real number where $a \neq 0$ and n is a natural number, then $\frac{1}{a^{-n}} = a^n$.

What if a fraction is raised to a negative exponent?

EXAMPLE 6 Evaluate. $\left(\frac{3}{4}\right)^{-2}$

Solution $\left(\frac{3}{4}\right)^{-2} = \frac{1}{\left(\frac{3}{4}\right)^2} = \frac{1}{\frac{9}{16}} = \frac{16}{9}$

If we remove the intermediate steps in our solution to Example 6, we see that $\left(\frac{3}{4}\right)^{-2} = \frac{16}{9}$. Since $\frac{16}{9} = \left(\frac{4}{3}\right)^2$, we can conclude that $\left(\frac{3}{4}\right)^{-2} = \left(\frac{4}{3}\right)^2$, which suggests the following rule.

RULE If a and b are real numbers, where $a \neq 0$ and $b \neq 0$ and n is a natural number, then $\left(\frac{a}{b}\right)^{-n} = \left(\frac{b}{a}\right)^{n}$.

YOUR TURN Evaluate the exponential form.

a. $(-4)^{-3}$ **b.** $\frac{1}{5^{-2}}$ **c.** $\left(\frac{2}{5}\right)^{-4}$

ANSWERS

a. $-\frac{1}{64}$ b. 25 c. $\frac{625}{16}$

After using the rules of exponents to simplify, if the exponent in the result is negative, then we rewrite the exponential form so that the exponent is positive.

EXAMPLE 7 Simplify and write the result with a positive exponent.

a. $t^3 \cdot t^{-5}$

Solution $t^3 \cdot t^{-5} = t^{3+(-5)}$ Use the product rule of exponents.

$= t^{-2}$ Simplify the exponent.

$= \frac{1}{t^2}$ Write the expression with a positive exponent.

b. $\frac{12x^3}{9x^7}$

Solution $\frac{12x^3}{9x^7} = \frac{12}{9}x^{3-7}$ Use the quotient rule for exponents.

$= \frac{4}{3}x^{-4}$ Simplify.

$= \frac{4}{3} \cdot \frac{1}{x^4}$ Write with a positive exponent.

$= \frac{4}{3x^4}$ Simplify.

Connection In Chapter 7, we will name expressions like $\frac{12x^3}{9x^7}$ *rational expressions* and use this same technique to simplify them.

YOUR TURN Simplify and write the result with a positive exponent.

a. $x^5 \cdot x^{-6}$ **b.** $\frac{5y^2}{20y^8}$ **c.** $\frac{k^{-4}}{k^3}$

OBJECTIVE 3. Use the power rule of exponents to simplify expressions. We now consider raising a power to a power, as in $(a^2)^3$. To simplify this expression, we need to determine how many factors of a the expression really indicates. The exponent 3 means to multiply three factors of a^2. Since each a^2 means two factors of a, there are a total of six factors of a.

$$(a^2)^3 = a^2 \cdot a^2 \cdot a^2 = a^{2+2+2} = a^6$$

Note: *Using the product rule, we see that we add three 2's, which means we could multiply 2 by 3 to get 6.*

ANSWERS

a. $\frac{1}{x}$ b. $\frac{1}{4y^6}$ c. $\frac{1}{k^7}$

Our example suggests the following power rule of exponents.

RULE **A Power Raised to a Power**

If a is a real number and m and n are integers, then $(a^m)^n = a^{mn}$.

EXAMPLE 8 Simplify using the power rule and write the result with a positive exponent.

a. $(t^3)^4$

Solution $(t^3)^4 = t^{3\cdot 4} = t^{12}$

b. $(x^{-5})^2$

Solution $(x^{-5})^2 = x^{-5\cdot 2} = x^{-10} = \dfrac{1}{x^{10}}$

Raising a Product to a Power

Now consider a product raised to a power, as in $(ab)^2$. The exponent 2 indicates two factors of ab.

$$\begin{aligned} (ab)^2 &= ab \cdot ab \\ &= a \cdot a \cdot b \cdot b && \text{Use the commutative property to rearrange like bases.} \\ &= a^2b^2 && \text{Use the product rule to simplify.} \end{aligned}$$

Notice each factor in the parentheses is raised to the power outside the parentheses, which suggests the following rule.

RULE **Raising a Product to a Power**

If a and b are real numbers and n is an integer, then $(ab)^n = a^nb^n$.

EXAMPLE 9 Simplify.

a. $(3a^4)^2$

Solution
$$\begin{aligned} (3a^4)^2 &= 3^2(a^4)^2 && \text{Use the rule for raising a product to a power.} \\ &= 9a^{4\cdot 2} && \text{Use the power rule.} \\ &= 9a^8 && \text{Simplify the exponent.} \end{aligned}$$

b. $\left(-\dfrac{4}{5}xy^2z^6\right)^3$

Solution
$$\begin{aligned} \left(-\frac{4}{5}xy^2z^6\right)^3 &= \left(-\frac{4}{5}\right)^3(x)^3(y^2)^3(z^6)^3 && \text{Use the rule for raising a product to a power.} \\ &= -\frac{64}{125}x^3y^6z^{18} && \text{Use the power rule.} \end{aligned}$$

YOUR TURN Simplify.

a. $(4y^5)^3$

b. $\left(-\dfrac{2}{3}t^5u^2v\right)^4$

ANSWERS

a. $64y^{15}$ **b.** $\dfrac{16}{81}t^{20}u^8v^4$

OBJECTIVE 4. Convert between scientific notation and standard form. Sometimes, we use very large or very small numbers, such as when we describe the vast distances in space or the tiny size of cells. For example, the distance from the Sun to the next nearest star, Proxima Centauri, is about 24,700,000,000,000 miles. Or, a single streptococcus bacterium is about 0.00000075 meters in diameter. The large number of zero digits in these numbers makes them tedious to write. **Scientific notation** allows us to write such numbers more concisely.

DEFINITION ***Scientific Notation:*** A number expressed in the form $a \times 10^n$, where a is a decimal number with $1 \le |a| < 10$ and n is an integer.

The number 2.65×10^4 is in scientific notation because $1 \le |2.65| < 10$ and the exponent 4 is an integer. Although 26.5×10^3 names the same number, it is not in scientific notation because $|26.5| > 10$. Similarly, 0.265×10^5 is also not in scientific notation because $|0.265| < 1$. We can multiply out 2.65×10^4 to discover its meaning.

$$2.65 \times 10^4 = 2.65 \times 10{,}000 = 26{,}500$$

Note: *Multiplying by 10^4 causes the decimal point to move 4 places to the right.*

Now consider a number in scientific notation with a negative exponent, such as 3.24×10^{-6}. To write this number in standard form, we can rewrite the power of 10 with a positive exponent.

$$3.24 \times 10^{-6} = 3.24 \times \frac{1}{10^6} = \frac{3.24}{1{,}000{,}000} = 0.00000324$$

Note: *Multiplying by 10^{-6} causes the decimal point to move 6 places to the left.*

Our example suggests that the exponent of the power of 10 determines the number of places that the decimal point moves.

PROCEDURE **Changing Scientific Notation to Standard Form**

To change a number from scientific notation, $a \times 10^n$, where $1 \le |a| < 10$ and n is an integer, if $n > 0$, move the decimal point to the right n places. If $n < 0$, move the decimal point to the left $|n|$ places.

EXAMPLE 10 Write each number in standard form.

a. -4.35×10^5

Solution $-4.35 \times 10^5 = -435{,}000$ — **Since $n > 0$, we move the decimal right 5 places.**

b. 9.4×10^{-8}

Solution $9.4 \times 10^{-8} = 0.000000094$ — **Since $n < 0$, we move the decimal left 8 places.**

YOUR TURN Write each number in standard form.

a. 5.403×10^6 **b.** 7×10^{-4}

Now let's convert numbers like 65,000,000 or 0.000045 from standard form to scientific notation. Since scientific notation begins with a decimal number whose absolute value is greater than or equal to 1 but less than 10, our first step will be to determine the position of the decimal point.

65,000,000

Decimal goes here because $1 \le |6.5| < 10$.

0.000045

Decimal goes here because $1 \le |4.5| < 10$.

Now we need to establish the power of 10.

$65{,}000{,}000 = 6.5 \times 10^7$

Since 65,000,000 is greater than 1, the 7 places between the old decimal position and the new position are expressed as a positive power of 10.

$0.000045 = 4.5 \times 10^{-5}$

Since 0.000045 is less than 1, the 5 places between the old decimal position and the new position are expressed as a negative power of 10.

We can summarize the process in the following procedure.

PROCEDURE **Changing Standard Form to Scientific Notation**

To write a number in scientific notation,

1. Locate the new decimal position, which will be to the right of the first nonzero digit in the number.
2. Determine the power of 10.
   a. If the number's absolute value is greater than 1, the power is the number of digits between the old decimal position and new decimal position expressed as a positive power.
   b. If the number's absolute value is less than 1, the power is the number of digits between the old decimal position and new decimal position expressed as a negative power.
3. Delete unnecessary 0's.
   a. If the number's absolute value is greater than 1, delete the zeroes to the right of the last nonzero digit.
   b. If the number's absolute value is less than 1, delete the zeroes to the left of the first nonzero digit.

ANSWERS

a. 5,403,000 **b.** 0.0007

EXAMPLE 11 Write the number in scientific notation.

a. $-46{,}000{,}000$

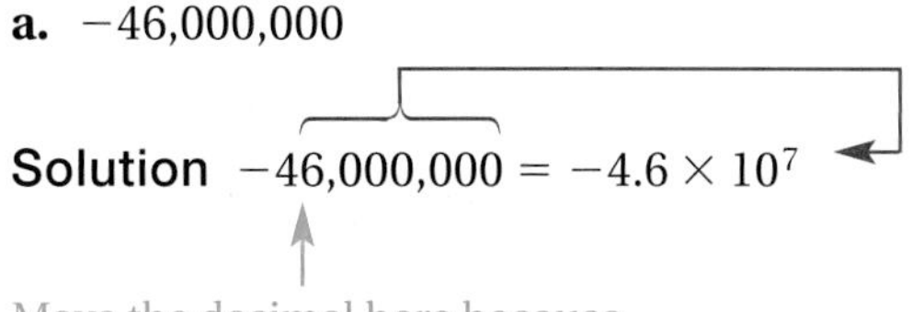

Solution $-46{,}000{,}000 = -4.6 \times 10^7$

Move the decimal here because $1 \le |-4.6| < 10$.

There are 7 places between the new decimal position and the original position. Since $|-46{,}000{,}000| > 1$, the exponent is positive. After placing the decimal point, the zeros to the right of 6 can be deleted.

b. 0.000038

Solution $0.000038 = 3.8 \times 10^{-5}$

Move the decimal here because $1 \le |3.8| < 10$.

There are 5 places between the new decimal position and the original. Since $|0.000038| < 1$, the exponent is negative. After placing the decimal point, the zeroes to the left of 3 can be deleted.

YOUR TURN Write the number in scientific notation.

a. $-206{,}000{,}000$ **b.** 0.00000052

ANSWERS

a. -2.06×10^8 **b.** 5.2×10^{-7}

Calculator TIPS

To enter scientific notation on a graphing or scientific calculator, we can use the EE function. For example, to enter 4.2×10^{-5}, type:

4.2 [EE] [(−)] 5 (*You may have to press* [2nd] *to get to the EE function.*)

Notice that the base 10 is understood and is not entered when using the EE function. You can also enter scientific notation without using EE, like this:

4.2 [×] 10 [^] [(−)] 5

After entering scientific notation, if you press [ENTER]*, the calculator will change the number to standard form. However, if the number has too many digits for the calculator to display, then it will leave it in scientific form. Similarly, if the answer to a calculation is too large for the calculator to display, then it will show the result in scientific form.*

OBJECTIVE 5. Simplify products, quotients, and powers of numbers in scientific notation. We can use the rules of exponents to simplify products, quotients, or powers of numbers in scientific notation.

EXAMPLE 12 Simplify.

a. $(4.8 \times 10^5)(6.4 \times 10^7)$

Solution $(4.8 \times 10^5)(6.4 \times 10^7)$

$= 4.8 \times 6.4 \times 10^5 \times 10^7$ Use the commutative property.

$= 30.72 \times 10^{5+7}$ Multiply the decimal numbers and use the product rule for exponents.

$= 30.72 \times 10^{12}$ Simplify the exponent.

$= 3.072 \times 10^{13}$ Adjust the decimal position and exponent so that we have the same number in scientific notation.

Note: *This number is not in scientific notation because $30.72 \ge 10$. Moving the decimal point 1 place to the left gives the proper position. We add 1 to the exponent to account for this additional place.*

Note: 0.52×10^{-9} *is not in scientific notation because* $0.52 < 1$. *Moving the decimal point 1 place to the right gives the proper position. We subtract 1 from the exponent to account for the new position.*

b. $\dfrac{4.368 \times 10^{-4}}{8.4 \times 10^{5}}$

Solution $\dfrac{4.368 \times 10^{-4}}{8.4 \times 10^{5}} = \dfrac{4.368}{8.4} \times \dfrac{10^{-4}}{10^{5}}$ Separate the decimal numbers and powers.

$= 0.52 \times 10^{-4-5}$ Divide the decimal numbers and use the quotient rule for exponents.

$= 0.52 \times 10^{-9}$ Simplify the exponent.

$= 5.2 \times 10^{-10}$ Adjust the decimal position and exponent so that we have the same number in scientific notation.

c. $(-2 \times 10^5)^3$

Solution $(-2 \times 10^5)^3 = (-2)^3 \times (10^5)^3$ Use the rule for raising a product to a power.

$= -8 \times 10^{15}$ Simplify $(-2)^3$ and use the power rule for exponents.

ANSWERS

a. 2.325×10^5

b. 4.5×10^{-7}

c. 2.56×10^{14}

YOUR TURN Simplify.

a. $(7.5 \times 10^9)(3.1 \times 10^{-5})$ **b.** $\dfrac{2.88 \times 10^4}{6.4 \times 10^{10}}$ **c.** $(-4 \times 10^3)^4$

5.1 Exercises

For Extra Help

MyMathLab

Videotape/DVT

InterAct Math

Math Tutor Center

Math XL.com

1. How do you simplify an expression like $x^3 \cdot x^4$?

2. How do you simplify an expression like $\dfrac{y^7}{y^5}$?

3. After simplifying 5^{-2}, is the result positive or negative? Explain.

4. Why is 42.5×10^6 not in scientific notation?

5. How do you write a number that is greater than 1 in scientific notation?

6. How do you write a number greater than 0 but less than 1 in scientific notation?

For Exercises 7–24, evaluate the exponential expression.

7. 15^0 **8.** 6^0 **9.** $4x^0$ **10.** $-3y^0$ **11.** $(3xy^2)^0$ **12.** $(-6m^3n^2)^0$

13. -4^2 **14.** -3^3 **15.** 2^{-3} **16.** 4^{-2} **17.** -3^{-4} **18.** -5^{-2}

19. $\frac{1}{6^{-3}}$ **20.** $\frac{1}{4^{-2}}$ **21.** $\frac{4}{x^{-3}}$ **22.** $\frac{6}{b^{-5}}$ **23.** $\frac{1}{4a^{-6}}$ **24.** $\frac{1}{9c^{-3}}$

For Exercises 25–28, explain the mistake, then find the correct answer.

25. $-2^4 = (-2)(-2)(-2)(-2) = 16$

26. $(-2)^3 = -6$

27. $4^{-1} = -4$

28. $\left(\frac{2}{3}\right)^{-2} = -\frac{4}{9}$

For Exercises 29–40, simplify using the product rule of exponents.

29. $mn^2 \cdot m^3n$ **30.** $x^3y^2 \cdot xy^4$ **31.** $3^{10} \cdot 3^2$ **32.** $2^5 \cdot 2^4$

33. $(-3)^5(-3)^3$ **34.** $(-4)^4(-4)^6$ **35.** $(-4p^4q^3)(3p^2q^2)$ **36.** $(2m^3n^2)(-3mn^9)$

37. $(6r^2st^4)(-3r^4s^3t^8)$ **38.** $(-8a^3b^2c^7)(-2ab^8c^2)$ **39.** $(1.2u^3t^9)(3.1u^4t^2)$ **40.** $(3.1x^2y^4z)(-x^8y^2z^7)$

For Exercises 41–52, simplify using the quotient rule of exponents.

41. $\frac{h^5}{h^3}$ **42.** $\frac{m^7}{m^3}$ **43.** $\frac{a^2}{a^7}$ **44.** $\frac{u^3}{u^8}$ **45.** $\frac{6x^2y^6}{3xy^3}$ **46.** $\frac{-12j^5k^9}{3j^3k^6}$

47. $\frac{18r^3s^7t}{-12r^4s^4t}$ **48.** $\frac{21m^4n^9}{14m^7n^2}$ **49.** $\frac{15u^{-8}v^3w^4}{-21u^3v^3w^{-2}}$ **50.** $\frac{18q^{-7}r^9s^3}{81qr^{-7}s^3}$ **51.** $\frac{8a^{-2}b^3c^8}{2a^3b^{-4}c^5}$ **52.** $\frac{3x^7y^{-3}z^{-4}}{9x^4y^{-5}z}$

For Exercises 53–62, simplify using the power rule of exponents.

53. $(x^3)^4$

54. $(t^4)^2$

55. $(2x^5)^4$

56. $(4n^6)^3$

57. $(-5x^3y)^2$

58. $(-2m^6n^2)^5$

59. $\left(\frac{3}{4}a^2b^4\right)^3$

60. $\left(-\frac{1}{2}ab^3\right)^4$

61. $(-0.3r^2t^4u)^3$

62. $(0.4ab^5c^6)^4$

For Exercises 63–74, simplify.

63. $(4x^2y^3)^2(-2xy^4)^2$

64. $(-3w^2v^2)^3(2w^8v^3)^2$

65. $\dfrac{(3q^4p^2)^2}{(5q^2p^2)^3}$

66. $\dfrac{(2u^3v^6)^3}{(-4u^6v^2)^2}$

67. $(3h^3t^5)^{-2}(9h^2t^4)^2$

68. $(3p^3q^8)^2(4p^2q^6)^{-3}$

69. $\dfrac{(9u^3v^2)^{-1}}{(2u^4v^2)^4}$

70. $\dfrac{(6x^2y^3z)^{-2}}{(3x^6y^2)^3}$

71. $\dfrac{(2u^2v^3)^{-3}(4u^{-2}v^3)}{(3u^2v^{-3})^{-2}}$

72. $\dfrac{(-3m^2n^3)^{-2}(m^4n^8)^3}{(-2m^{-2}n^{-3})^{-3}}$

73. $\dfrac{(-4a^{-2}b^3c^2)^{-1}(2abc)^{-2}}{(3a^5b^2)^3(2a^{-1}b^{-3}c)^{-3}}$

74. $\dfrac{(9r^6s^2)^{-2}(3r^{-3}s^2)^3}{(6rst^{-2})^{-1}(3r^2s^3t)^{-2}}$

For problems 75–86, write the number in standard form.

75. The Andromeda galaxy is about 2.9×10^6 light-years from Earth.

76. The Large Magellanic Cloud is a small galaxy about 1.79×10^5 light-years from Earth.

77. The Andromeda galaxy contains at least 2×10^{11} stars.

78. Some scientists estimate that there may be as many as 4×10^{11} stars in our galaxy, the Milky Way.

79. *Forbes* magazine lists Bill Gates's net worth as $\$4.65 \times 10^{10}$. (*Source:* "Forbes World's Richest People," 2005)

Andromeda Galaxy

80. In 2003, the United States exported $\$3.154 \times 10^9$ in toys, games, and sporting goods. (*Source:* Office of Trade and Economic Analysis, U.S. Dept. of Commerce)

81. Blonde hair has a smaller diameter than darker hair and can be as thin as 1.7×10^{-7} meters in diameter.

82. The mass of a dust particle is 7.53×10^{-10} kg.

83. The mass of a neutron is about 1.675×10^{-24} g.

84. The mass of a proton is about 1.673×10^{-24} g.

85. The radioactive decay of plutonium-239 emits an alpha particle weighing 6.645×10^{-27} kg.

86. The mass of an electron is about 9.109×10^{-28} g.

For Exercises 87–98, write the number in scientific notation.

87. The all-time highest-rated television program is the last episode of *M*A*S*H*, which aired on February 28, 1983 and had 50,150,000 households viewing it. (*Source:* Nielson Media Research, January 1961–February 2004)

88. The universe is estimated to be 16,500,000,000 years old.

89. In July of 2005, the U.S. national debt reached (rounded to the nearest billion) \$7,879,000,000,000. (*Source:* Bureau of the Public Debt)

90. The star Vega is approximately 155,200,000,000,000 miles from Earth.

91. As of August 2005, the top-grossing film of all time was *Titanic*, which earned a total of \$600,000,000,000 in box office sales. (*Source:* MovieWeb)

92. A gram of hydrogen contains 6,000,000,000,000,000,000 protons.

93. Light with a wavelength of 0.00000055 meters is green in color.

94. The size of a plant cell is 0.00001276 meters wide.

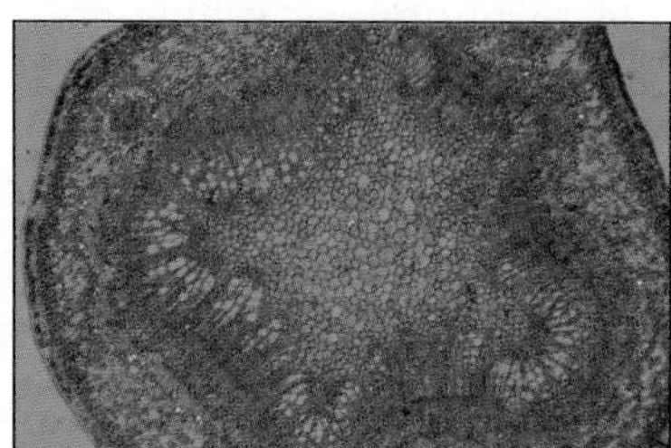

Plant cell

95. The size of the HIV virus that causes AIDS is about 0.0000001 meters.

96. The time light takes to travel 1 meter is 0.000000003 seconds.

97. In June of 2005, the fastest supercomputer was the Blue Gene/L system, which could complete a single calculation in about 0.0000000000000073 seconds. (*Source:* 25th Edition of the *Top 500 List of the World's Fastest Supercomputers*)

98. The size of an atomic nucleus of the lead atom is 0.0000000000000071 meters.

For Exercises 99 and 100, write the numbers in order from smallest to largest.

99. $8.3 \times 10^6, 1.2 \times 10^7, 6 \times 10^5, 7.4 \times 10^6, 2.4 \times 10^8$

100. $6.1 \times 10^{-3}, 7.2 \times 10^{-2}, 9.3 \times 10^{-4}, 3.1 \times 10^{-2}, 4.5 \times 10^{-6}$

For Exercises 101–120, simplify.

101. $(2.1 \times 10^3)(3 \times 10^4)$

102. $(6 \times 10^5)(2.1 \times 10^4)$

103. $(-3.2 \times 10^4)(4.1 \times 10^5)$

104. $(5.2 \times 10^5)(-3.1 \times 10^4)$

105. $(8.1 \times 10^{-4})(2.2 \times 10^5)$

106. $(3.2 \times 10^{-3})(7.4 \times 10^8)$

107. $\dfrac{8.4 \times 10^6}{2.1 \times 10^2}$

108. $\dfrac{8.4 \times 10^{10}}{4.2 \times 10^3}$

109. $\dfrac{9.3 \times 10^{14}}{-3 \times 10^6}$

110. $\dfrac{-8.12 \times 10^8}{4 \times 10^5}$

111. $\dfrac{5.7 \times 10^{-3}}{9.5 \times 10^{-6}}$

112. $\dfrac{9.4 \times 10^{-5}}{4.5 \times 10^{-1}}$

113. $(3 \times 10^4)^3$

114. $(6 \times 10^2)^2$

115. $(2 \times 10^7)^3$

116. $(4 \times 10^3)^3$

117. $(5 \times 10^{-4})^2$

118. $(8 \times 10^{-3})^2$

119. $(-4 \times 10^{-6})^{-2}$

120. $(-2 \times 10^{-4})^{-3}$

For Exercises 121–124, use the following information to solve. The energy in joules of a single photon of light can be determined by $E = hf$, where h is a constant 6.626×10^{-34} joule-seconds and f is the frequency of the light in hertz. Write each answer in scientific notation with the decimal number rounded to two decimal places.

121. Find the energy of red light with a frequency of 4.2×10^{14} Hz.

122. Find the energy of blue light with a frequency of 6.1×10^{14} Hz.

123. A photon of light is measured to have 1.4×10^{-19} joules of energy. Find its frequency.

124. A photon of yellow-green light is measured to have 3.6×10^{-19} joules of energy. Find its frequency.

Of Interest

The color of light is determined by its frequency. The human eye is capable of distinguishing frequencies of light from about 3.92×10^{14} Hz (boundary of infrared) up to about 7.89×10^{14} Hz (boundary of ultraviolet). (*Source:* Emiliani, *The Scientific Companion*)

For Exercises 125 and 126, use the following information. Albert Einstein discovered that if a mass of m kilograms is converted to pure energy (say, in a nuclear reaction), the amount of energy in joules that is released is described by $E = mc^2$, where c represents the constant speed of light, which is approximately 3×10^8 meters per second.

125. Suppose 4.2×10^{-12} kilograms of plutonium is converted to energy in a reactor. How much energy is released?

126. Suppose 4.2×10^{-8} kilograms of uranium is converted to energy in a nuclear reaction. How much energy is released?

PUZZLE PROBLEM

Upon being asked for her house number, a mathematician responded, "It is a three-digit perfect square between 100 and 400, and if you rotate the number so that it is upside down, that number is a perfect square also." What is her house number?

REVIEW EXERCISES

[1.2] **1.** Explain how to rewrite $19 - (-6)$ as addition.

[1.2] **2.** What property is illustrated by $x + (y + z) = (x + y) + z$?

[1.4] **3.** Use the distributive property to rewrite the expression $-8(n - 4)$.

[1.4] ***For Exercises 4–6, simplify by combining like terms.***

4. $7x + 8y - 4x - 12 + 13y - 5$

5. $10x^2 - x + 9 - 12x + 3x^2$

6. $2(u + 8) - (5u - 3) - 7$

5.2 Polynomials and Polynomial Functions

OBJECTIVES

1. Determine the coefficient and degree of a monomial.
2. Determine the degree of a polynomial and write polynomials in descending order of degree.
3. Add polynomials.
4. Subtract polynomials.
5. Classify and graph polynomial functions.
6. Add and subtract polynomial functions.
7. Solve application problems using polynomial functions.

In Section 5.1 we learned about exponents so that we could discuss a class of expressions called polynomials. Notice that we are on the expression level of the algebra pyramid.

Note: *Inequalities are also in the top level of the Algebra Pyramid, but we will not be exploring polynomial inequalities at this point.*

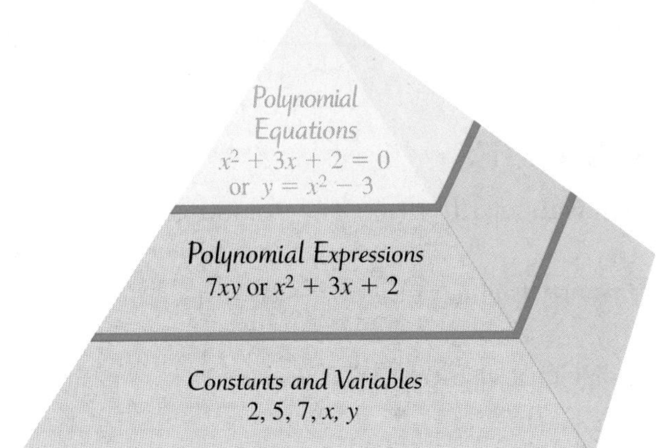

OBJECTIVE 1. Determine the coefficient and degree of a monomial. In Section 1.4 we defined terms and discussed like terms. In order to define a polynomial, we will first discuss a special type of term called a **monomial**.

DEFINITION ***Monomial:*** An expression that is a constant, or a product of a constant and variables that are raised to whole number powers.

Examples of monomials: 1.6 $\quad 4x^3 \quad -\frac{3}{4}tu^2$

In our discussion of terms in Section 1.4, we defined a **coefficient** to be the numerical factor in a term. This holds for monomials as well. Another piece of information about a monomial is its **degree**, which relates to the exponents of the variables.

DEFINITIONS ***Coefficient:*** The numerical factor in a monomial.
Degree of a Monomial: The sum of the exponents of all variables in the monomial.

For example, the coefficient of $-9xy^3$ is -9 and the degree is 4, which is the sum of the variables' exponents: $1 + 3 = 4$. Note that the degree of a constant, such as 8, is 0 because we can write any constant as a product with a variable raised to the 0 power. For example, $8 = 8x^0$ because $x^0 = 1$.

ANSWERS

a. coefficient: -8; degree: 4
b. coefficient: -0.5; degree: 9
c. coefficient: 1; degree: 2
d. coefficient: $\frac{4}{5}$; degree: 0

YOUR TURN Identify the coefficient and degree of each monomial.

a. $-8x^4$ **b.** $-0.5x^4y^5$ **c.** tu **d.** $\frac{4}{5}$

OBJECTIVE 2. Determine the degree of a polynomial and write polynomials in descending order of degree. Now we are ready to formally define a **polynomial**.

DEFINITION ***Polynomial:*** A monomial or an expression that can be written as a sum of monomials.

Examples of polynomials: $5x$ $3x + 7$ $x^2 + 6xy + 3y$ $4x^3 - x^2 + 2x - 12$

Notice that the polynomial $4x^3 - x^2 + 2x - 12$ has the same variable x in each variable term, whereas $x^2 + 6xy + 3y$ has a mixture of two variables, x and y. A polynomial like $4x^3 - x^2 + 2x - 12$ is said to be a **polynomial in one variable**.

DEFINITION ***Polynomial in One Variable:*** A polynomial in which every variable term has the same variable.

Special names have been given to some polynomials. We have already discussed monomials, like $5x$, which are single-term polynomials. Notice the prefix *mono-* in monomial indicates *one* term. Continuing with the prefixes, a two-term polynomial, like $4x + 8$, is called a **binomial** and a three-term polynomial, like $2y^2 + 5y + 8$, is called a **trinomial**. No special names are given to polynomials with more than three terms.

DEFINITIONS ***Binomial:*** A polynomial containing two terms.
Trinomial: A polynomial containing three terms.

Recall that we defined the degree of a monomial to be the sum of the exponents of the variables. We can also identify the **degree of a polynomial**.

DEFINITION ***Degree of a Polynomial:*** The greatest degree of any of the terms in the polynomial.

EXAMPLE 1 Identify the degree of each polynomial, then indicate whether the polynomial is a monomial, binomial, trinomial, or none of these names.

a. $5x^6 + x^3 - 10x^2$

Answer The degree is 6, because it is the greatest degree of any of the terms. The polynomial is a trinomial.

Note: *Remember that the degree of a monomial like $3x^3y^4$ is the sum of the variables' exponents.*

b. $x^5 + 3x^3y^4 - 2xy^3 - y^2 - 6$

Answer The degree is 7, because $3x^3y^4$ has the greatest degree of the terms. None of the listed names apply to this polynomial.

YOUR TURN Identify the degree of each polynomial, then indicate whether the polynomial is a monomial, binomial, trinomial, or none of these names.

a. $-8y^9 + 4y$ **b.** $2x^5 - 9x^6y + x^3y^3 - 15y - 7$

ANSWERS

a. degree = 9; binomial

b. degree = 7; none of these names

Notice the terms in the polynomial $5x^6 + x^3 - 10x^2$ are written in order from the greatest degree to the least degree. A polynomial written in this way is said to be in *descending order of degree.* We call the first term in such a polynomial the *leading term* and that term's coefficient is the *leading coefficient.* For example, in the polynomial $5x^6 + x^3 - 10x^2$, the leading term is $5x^6$ and the leading coefficient is 5. Because it can be complicated to write polynomials in more than one variable in descending order, we will only require polynomials in one variable to be written in descending order.

YOUR TURN Write each polynomial in descending order.

a. $2n^3 + n^5 - 9n^2 + 8 - 3n^4 + 4n$ **b.** $-4t^2 + 2t + 16 + 5t^3 - 13t^6$

OBJECTIVE 3. Add polynomials. We can perform arithmetic operations with polynomials. First, we consider adding polynomials.

EXAMPLE 2 Add. $(3x^2 - 6x + 1) + (4x^2 + x - 7)$

Solution If we remove the parentheses, we have a single polynomial that we can simplify by combining like terms.

$$(3x^2 - 6x + 1) + (4x^2 + x - 7) = 3x^2 - 6x + 1 + 4x^2 + x - 7 \quad \text{Remove parentheses.}$$

$$= 3x^2 + 4x^2 - 6x + x + 1 - 7 \quad \text{Collect like terms.}$$

$$= 7x^2 - 5x - 6 \quad \text{Combine like terms.}$$

Note: *Combining the like terms in order of degree places the resulting polynomial in descending order of degree.*

ANSWERS

a. $n^5 - 3n^4 + 2n^3 - 9n^2 + 4n + 8$

b. $-13t^6 + 5t^3 - 4t^2 + 2t + 16$

From Example 2 we can summarize how to add polynomials.

PROCEDURE Adding Polynomials

To add polynomials, combine like terms.

Many people prefer to combine the like terms without collecting them or even removing the parentheses.

EXAMPLE 3 $(t^4 + 9t^3 - 2t - 9) + (t^4 - 4t^3 + 2t - 11)$

Solution $(t^4 + 9t^3 - 2t - 9) + (t^4 - 4t^3 + 2t - 11)$

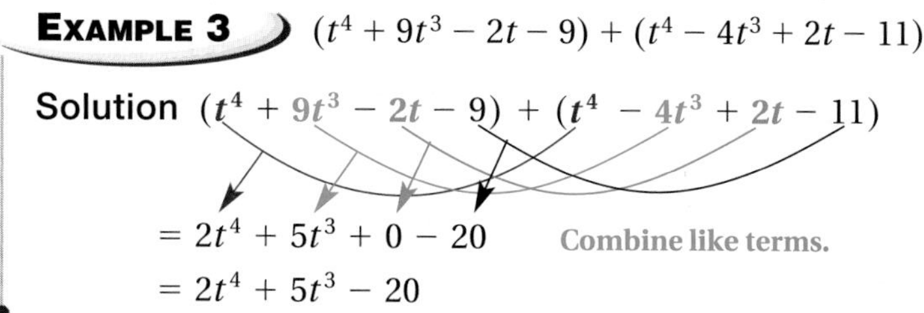

$= 2t^4 + 5t^3 + 0 - 20$ Combine like terms.

$= 2t^4 + 5t^3 - 20$

Sometimes the polynomials may contain terms that have several variables, or coefficients that are fractions or decimals.

EXAMPLE 4 $\left(2x^4 + 0.3xy^2 - \frac{3}{4}xy + 5\right) + \left(-6x^4 - 0.5xy^2 + \frac{1}{3}xy + 4\right)$

Solution $\left(2x^4 + 0.3xy^2 - \frac{3}{4}xy + 5\right) + \left(-6x^4 - 0.5xy^2 + \frac{1}{3}xy + 4\right)$ Combine like terms.

$= -4x^4 - 0.2xy^2 - \frac{5}{12}xy + 9$

YOUR TURN Add.

a. $(9x^2 - 7x - 5) + (3x^2 + 6x - 3)$

b. $(4n^4 - n^3 + 8n - 15) + (2n^4 - 8n^3 - 8n + 2)$

c. $\left(2.5a^4 - \frac{4}{5}ab^3 + \frac{2}{3}ab - 6\right) + \left(a^4 + \frac{1}{3}ab^3 + ab - 1\right)$

ANSWERS

a. $12x^2 - x - 8$
b. $6n^4 - 9n^3 - 13$
c. $3.5a^4 - \frac{7}{15}ab^3 + \frac{5}{3}ab - 7$

OBJECTIVE 4. Subtract polynomials. Recall that we can write subtraction as equivalent addition by *adding* the opposite or *additive inverse* of the subtrahend. For example, $8 - (-2) = 8 + 2$. We can apply this principle to polynomial subtraction.

The subtraction sign is changed to an addition sign.

$$(9x^2 + 7x + 5) - (6x^2 + 2x + 1) = (9x^2 + 7x + 5) + (-6x^2 - 2x - 1)$$

The subtrahend changes to its additive inverse.

Note: *To write the additive inverse of a polynomial, we change the sign of every term. We can also view this process as an application of the distributive property.*

$$\begin{aligned} -(6x^2 + 2x + 1) &= -1(6x^2 + 2x + 1) \\ &= -1 \cdot 6x^2 + (-1) \cdot 2x + (-1) \cdot 1 \\ &= -6x^2 - 2x - 1 \end{aligned}$$

Our exploration suggests the following procedure for subtracting polynomials.

PROCEDURE Subtracting Polynomials

To subtract polynomials,

1. Write the subtraction statement as an equivalent addition statement.
   a. Change the operation symbol from a minus sign to a plus sign.
   b. Change the subtrahend (second polynomial) to its additive inverse. To get the additive inverse, we change the sign of each term in the polynomial.
2. Combine like terms.

EXAMPLE 5 Subtract.

a. $(7x^3 + 13x^2 + 9x + 6) - (2x^3 - 5x^2 + 9x - 6)$

Solution $(7x^3 + 13x^2 + 9x + 6) - (2x^3 - 5x^2 + 9x - 6)$

Change the minus sign to a plus sign. Change all signs in the subtrahend.

$= (7x^3 + 13x^2 + 9x + 6) + (-2x^3 + 5x^2 - 9x + 6)$

$= 5x^3 + 18x^2 + 12$ Combine like terms.

b. $(0.4x^5 - 3x^2y - 0.6xy - y^2) - (x^5 - xy^2 + 1.3xy - 5y^2)$

Solution $(0.4x^5 - 3x^2y - 0.6xy - y^2) - (x^5 - xy^2 + 1.3xy - 5y^2)$

$= (0.4x^5 - 3x^2y - 0.6xy - y^2) + (-x^5 + xy^2 - 1.3xy + 5y^2)$ Write equivalent addition.

$= -0.6x^5 - 3x^2y + xy^2 - 1.9xy + 4y^2$ Combine like terms.

Learning Strategy

When changing the signs, try using a color pen so that you can clearly see the sign changes.

YOUR TURN Subtract.

a. $(6u^4 + u^2 - 12u - 8) - (7u^4 - 5u^2 - u + 4)$

b. $(0.6x^5 - 2xy^2 + y - 1.5) - (1.5x^5 + x^3y^2 - 2xy^2 + 4y - 1.3)$

OBJECTIVE 5. Classify and graph polynomial functions. Polynomials in one variable can be used to make **polynomial functions**.

DEFINITION ***Polynomial Function:*** A function of the form $f(x) = ax^m + bx^n + \cdots$ with a finite number of terms, where each coefficient is a real number and each exponent is a whole number.

Examples of polynomial functions: $f(x) = x^2 + 3x + 1$ $\quad g(x) = x^3 - 2$

Polynomial functions are classified by the degree of the polynomial. A degree 0 polynomial makes a **constant function**, a degree 1 polynomial makes a **linear function**, a degree 2 polynomial makes a **quadratic function**, and a degree 3 polynomial makes a **cubic function**. For now we will only consider these four classifications.

DEFINITION ***Constant Function:*** A function of the form $f(x) = c$, where c is a real number.

The graph of a constant function is a horizontal line through $(0, c)$.

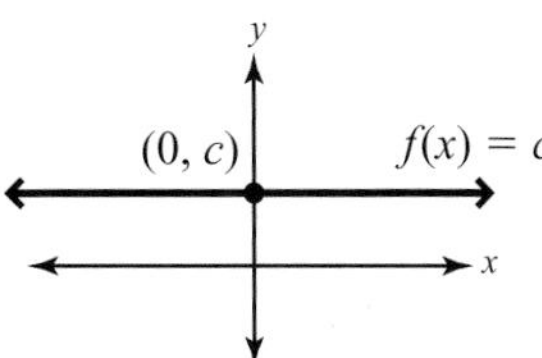

DEFINITION ***Linear Function:*** A function of the form $f(x) = mx + b$, where m and b are real numbers and $m \neq 0$.

Connection Constant functions are special linear functions with $m = 0$.

Graphs of linear functions are lines with slope m and y-intercept $(0, b)$.

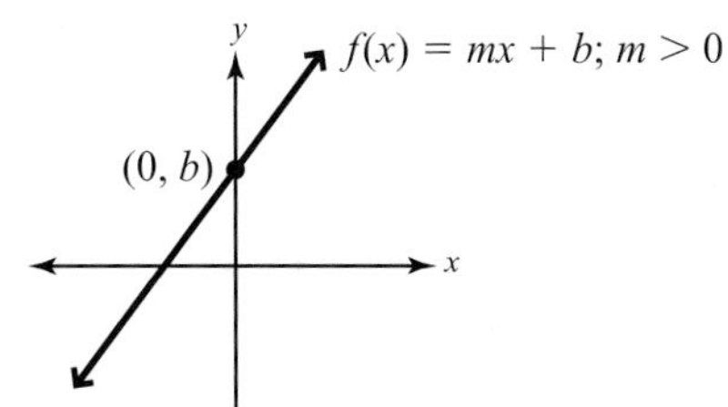

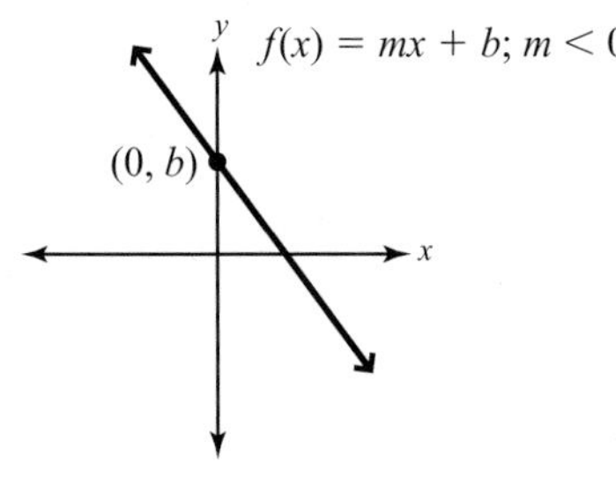

ANSWERS

a. $-u^4 + 6u^2 - 11u - 12$

b. $-0.9x^5 - x^3y^2 - 3y - 0.2$

DEFINITION ***Quadratic Function:*** A function of the form $f(x) = ax^2 + bx + c$, where a, b, and c are real numbers and $a \neq 0$.

Graphs of quadratic functions are parabolas with y-intercept at $(0, c)$.

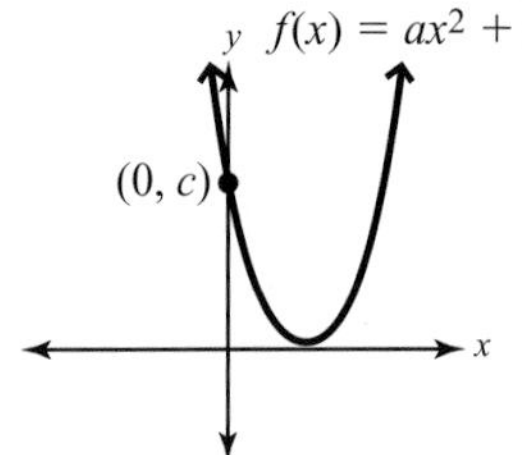

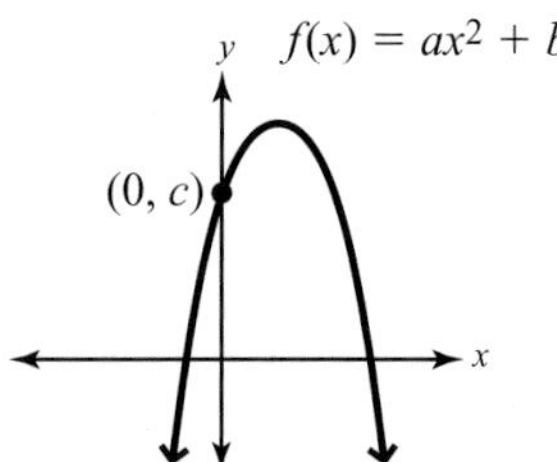

DEFINITION ***Cubic Function:*** A function of the form $f(x) = ax^3 + bx^2 + cx + d$, where a, b, c, and d are real numbers and $a \neq 0$.

Graphs of cubic functions resemble an S-shape with y-intercept at $(0, d)$.

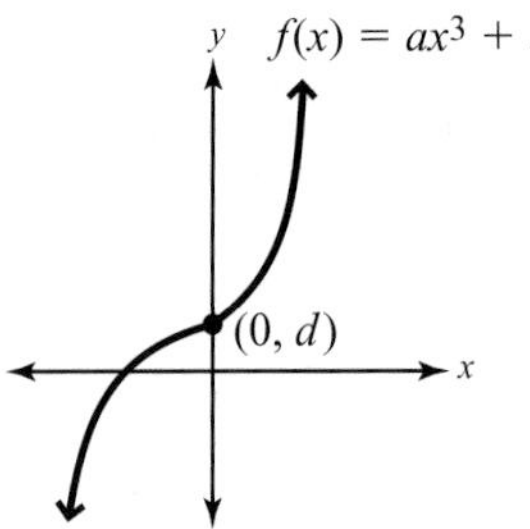

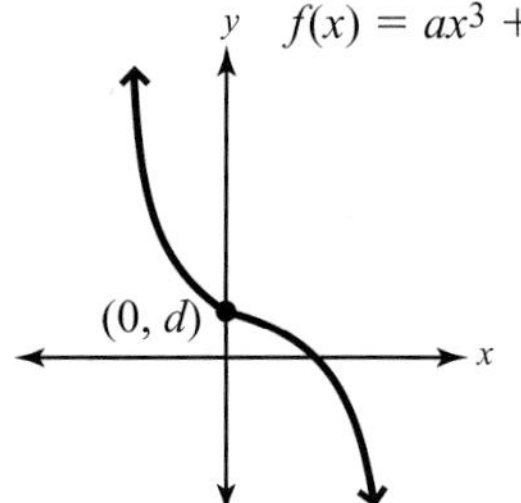

Connection The vertical line test, which we learned in Chapter 3, confirms that all of these are functions.

EXAMPLE 6 Graph. $f(x) = x^3 + 1$

Solution Find enough ordered pairs to generate the graph. Recall that we find an ordered pair by evaluating the function using a chosen value for x. For example, we will show the calculation of $f(-2)$.

$$\begin{aligned} f(-2) &= (-2)^3 + 1 && \text{Replace } x \text{ with } -2. \\ &= -8 + 1 && \text{Simplify.} \\ &= -7 \end{aligned}$$

The ordered pair is $(-2, -7)$.

Connection Finding $f(x)$ is the same as finding the y-coordinate given an x-coordinate.

Additional ordered pairs are found similarly. We list several pairs in the following table.

x	$f(x)$
−2	−7
−1	0
0	1
1	2
2	9

To graph the function, we plot our ordered pairs and then draw a smooth curve. Since this is a cubic, we know the graph will resemble an S-shape.

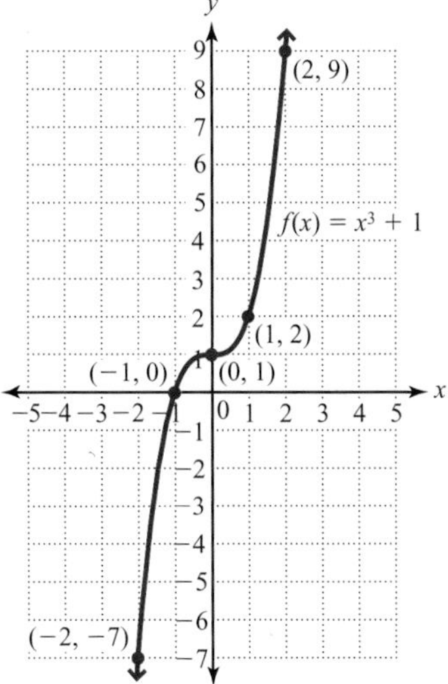

YOUR TURN Graph. $f(x) = \frac{1}{2}x^3 - 1$

OBJECTIVE 6. Add and subtract polynomial functions. We can find the sum or difference of two polynomial functions.

RULE Adding or Subtracting Functions

The sum of two functions, $\boldsymbol{f + g}$, is found by $(f + g)(x) = f(x) + g(x)$.

The difference of two functions, $\boldsymbol{f - g}$, is found by $(f - g)(x) = f(x) - g(x)$.

EXAMPLE 7 Given $f(x) = 2x + 1$ and $g(x) = 7x + 3$, find

a. $f + g$ **b.** $f - g$ **c.** $(f - g)(-4)$

Solution

a. $(f + g)(x) = (2x + 1) + (7x + 3)$ Use the rule $(f + g)(x) = f(x) + g(x)$.
$= 9x + 4$

b. $(f - g)(x) = (2x + 1) - (7x + 3)$ Use the rule $(f - g)(x) = f(x) - g(x)$.
$= -5x - 2$

c. In part b, we found $(f - g)(x) = -5x - 2$.

$(f - g)(-4) = -5(-4) - 2$ Replace x with -4 in $(f - g)(x) = -5x - 2$.
$= 20 - 2$ Simplify.
$= 18$

ANSWER

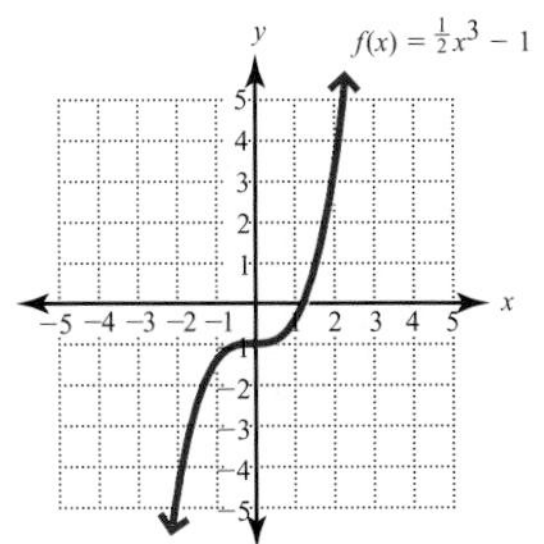

ANSWERS

a. $2x^2 - 5x - 4$

b. $-x + 14$

c. 14

YOUR TURN Given $h(x) = x^2 - 3x + 5$ and $k(x) = x^2 - 2x - 9$, find

a. $h + k$ **b.** $h - k$ **c.** $(h + k)(-2)$

OBJECTIVE 7. Solve application problems using polynomial functions. One common application of polynomial functions is in determining the net profit a business makes given functions for the revenue and the cost.

EXAMPLE 8 A software company produces an accounting program. The function $R(x) = 0.2x^2 + x + 500$ describes the revenue the company makes from sales, where x represents the number of units that it sold. The function $C(x) = 10x + 4000$ describes the cost of producing those x units of software.

a. Find a function, $P(x)$, that describes the profit.

b. Find the profit if the company sells 1000 copies of the software.

Understand We are given functions for the revenue and cost and must determine the profit function and the profit if 1000 copies are sold.

Plan We will first find the profit function, then use that function to determine the profit if 1000 copies are sold.

Execute Since the net profit is the money left after a business deducts costs, the relationship for finding profit is $P(x) = R(x) - C(x)$.

$$P(x) = (0.2x^2 + x + 500) - (10x + 4000)$$

$$P(x) = (0.2x^2 + x + 500) + (-10x - 4000) \quad \text{Write the equivalent addition.}$$

$$P(x) = 0.2x^2 - 9x - 3500 \quad \text{Combine like terms.}$$

Now to find the profit for the sale of 1000 copies, we find $P(1000)$.

$$P(1000) = 0.2(1000)^2 - 9(1000) - 3500 = 187{,}500$$

Answer The profit function is $P(x) = 0.2x^2 - 9x - 3500$, and the profit for the sale of 1000 copies is \$187,500.

Check We can add the cost and profit functions to see if we get the revenue function.

$$R(x) = (0.2x^2 - 9x - 3500) + (10x + 4000)$$

$$R(x) = 0.2x^2 + x + 500 \quad \text{Combine like terms.}$$

YOUR TURN A small publishing company produces books for counseling teenagers. The function $R(x) = 0.1x^2 + 2x + 250$ describes the revenue a company makes from sales, where x represents the number of books sold. The function $C(x) = 5x + 1200$ describes the cost of producing those books.

a. Find a function, $P(x)$, that describes the profit.

b. Find the profit if the company sells 1000 books.

ANSWERS

a. $P(x) = 0.1x^2 - 3x - 950$

b. \$96,050

5.2 Exercises

For Extra Help

MyMathLab

Videotape/DVT

InterAct Math

Math Tutor Center

Math XL.com

1. How do you determine the degree of a monomial?

2. Explain the difference between a monomial, a binomial, and a trinomial.

3. How do you determine the degree of a polynomial?

4. Given a polynomial in one variable, what does *descending order of degree* mean?

5. How do you add polynomials?

6. When subtracting one polynomial from another, after the operation symbol is changed from a minus sign to a plus sign, what is the next step?

7. What is the degree of the polynomial in a quadratic function? What is the graph of a quadratic function?

8. Given two polynomial functions $f(x)$ and $g(x)$, how do you determine $(f+g)(x)$?

For Exercises 9–22, identify the degree of each polynomial, then indicate whether the polynomial is a monomial, binomial, trinomial, or none of these names.

9. $-7uv^3$

10. $-\frac{2}{5}a^3b^5c$

11. $25 - x^2$

12. $x + 4$

13. $5p^3 + 2p^2 - 1$

14. $1.5r^4 - 3r^2 + 8r$

15. $4g^2 - 5g - 11g^3 - 8g^4 + 7$

16. $-7.1k + 2.3k^3 - 8k^2 - 1$

17. $7x + 5x^3 - 19$

18. $-16m^4 + 5m - 7m^2$

19. -7

20. 4

21. $\frac{1}{3}pq - 3p^2q$

22. $3m^2n^2 + 6m^4n$

For Exercises 23–32, add and write the resulting polynomial in descending order.

23. $(5x^2 - 3x + 1) + (2x^2 + 7x - 3)$

24. $(3y^2 + 7y - 3) + (4y^2 + 3y + 1)$

25. $(p^4 - 3p^3 + 4p - 1) + (p^4 - 2p^3 - 4p + 7)$

26. $(12r^4 - 5r^2 + 8r - 15) + (7r^4 + 3r^2 + 2r - 9)$

27. $(4u^3 - 6u^2 + u + 11) + (-5u^3 - 3u^2 + u - 5)$

28. $(7p^3 - 9p^2 + 5p - 1) + (-4p^3 + 8p^2 + 2p + 10)$

29. $\left(\frac{2}{3}u^4 + \frac{3}{4}u^3 - u^2 - u + 3\right) + \left(\frac{2}{3}u^4 - \frac{1}{4}u^3 + 3u^2 + 4u - 1\right)$

30. $\left(\frac{7}{8}w^4 - 5w^3 + 3w^2 - 8\right) + \left(\frac{3}{4}w^4 - \frac{1}{2}w^3 + 4w^2 - 7w + 9\right)$

31. $(3.1t^4 - 2.1t^3 + 7t^2 + 5.8t + 4) + (4.2t^4 + 3.6t^3 - 8t^2 - 3.1t + 3)$

32. $(7.3h^4 - 3.1h^3 - 7.6h^2 + 3.5h + 2.7) + (0.4h^4 - 2.5h^3 + 1.2h^2 - 1.6h - 3.1)$

For Exercises 33–42, subtract and write the resulting polynomial in descending order.

33. $(7a^3 - a^2 - a) - (8a^3 - 3a + 1)$

34. $(4x^3 - 3x + 4) - (6x^3 - 3x^2 + 5)$

35. $(3m^4 + 2m^3 - m^2 + 1) - (6m^3 - 2m^2 - m + 3)$

36. $(5n^4 + 9n^3 - n^2 - 3) - (n^4 - 9n^3 + n^2 + 5)$

37. $(-7r^3 - 7r) - (-5r^3 + 2r^2 - 6r - 1)$

38. $(-y^5 + 6y^3 + 7y^2 - y - 3) - (-3y^2 + 7y - 1)$

39. $\left(\frac{4}{5}y^3 - \frac{1}{3}y^2 + 7y - \frac{2}{7}\right) - \left(\frac{1}{5}y^3 + \frac{2}{3}y^2 + \frac{3}{4}y - \frac{1}{7}\right)$

40. $\left(\frac{9}{10}m^5 - \frac{2}{3}m^4 + \frac{3}{5}m^3 - m^2 + \frac{1}{7}m + 8\right) - \left(\frac{11}{10}m^5 + \frac{4}{3}m^4 - \frac{2}{5}m^3 - m^2 + \frac{2}{7}m + 1\right)$

41. $(-4.5w^3 - 5.1w^2 + 2.7w + 4.1) - (3.8w^3 - 1.4w^2 + 3.4w - 2.6)$

42. $(9.1m^4 - 2.5m^3 + 3m^2 + 2.1m - 1) - (-4.2m^4 + 3.2m^3 + 2.1m^2 - 1.1m - 3)$

For Exercises 43–52, perform the indicated operation and write the resulting polynomial in descending order.

43. $(-5w^4 - 3w^3 - 8w^2 + w - 14) + (-3w^4 + 6w^3 + w^2 + 12w + 5)$

44. $(6h^9 - 7h^5 - 3h^4 - 2h^3 + 3) + (-h^8 + 6h^6 - 8h^5 + 2h^4 - h^2 - 1)$

45. $(6x^4 + 4x^3 - 3x + 5) - (-2x^4 - 5x^2 + 9x - 8)$

46. $(-9y^5 - y^2 + 13y - 16) - (-4y^3 + 15y^2 - y + 11)$

47. $\left(-\frac{7}{3}g^4 + \frac{4}{5}g^3 + \frac{7}{5}\right) + \left(\frac{4}{3}g^4 - \frac{2}{3}g^2 + \frac{2}{5}\right)$

48. $\left(\frac{1}{3}a^4 + 2.1a^3 - a^2 - 2\right) + \left(\frac{3}{4}a^4 + 3.1a^2 + 5\right)$

49. $(4y^2 - 8y + 1) + (5y^2 + 3y + 2) - (6y^2 - 9y + 2)$

50. $(3k^3 - 5k^2 - k - 1) - (2k^3 - 3k^2 - k - 7) + (4k^3 + 4k^2 + k + 7)$

51. $(-6a^3 - 5a^2 + 10) - (9a^3 + 5a^2 - 3a - 1) + (-4a^3 - 2a^2 + 6a - 2)$

52. $-(-r^4 + 3r^2 - 12r - 14) - (3r^4 + 6r^3 + 3r - 8) + (5r^4 - 2r^3 - 3r^2 + 10r - 5)$

For Exercises 53–58, add or subtract.

53. $(8a^2 + 5ab - 2b^2) - (10ab - 6a^2 - 8b^2)$

54. $(3a^2 - 4ab + b^2) - (-2a^2 + 4ab + b^2)$

55. $(7x^3y^4 - 2x^2y^3 - xy + 4) + (-5x^3y^4 + 7x^2y^3 + 8xy - 12)$

56. $(-13a^6 + a^3b^2 + 3ab^3 - 8b^3) + (11a^6 - 3a^2b^3 - 4ab^3 + 10ab^2 + 4b^3)$

57. $(x^2y^2 + 8xy^2 - 12x^2y - 4xy + 7y^2 - 9) - (-3x^2y^2 + 2xy^2 + 4x^2y - xy + 6y^2 + 4)$

58. $(15a^3y^4 + a^2y^2 + 5ay^3 - 7ay + 12y^2 - 11) - (-a^3y^4 + a^2y^2 + 2ay^3 - ay + 9)$

For Exercises 59–62, write an expression for the perimeter in simplest form.

59.

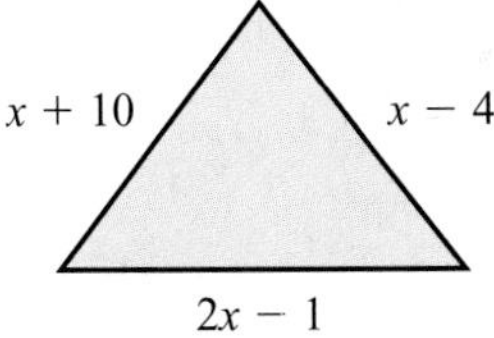

60.

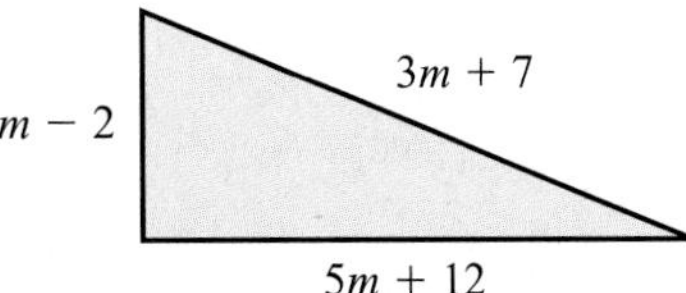

61.

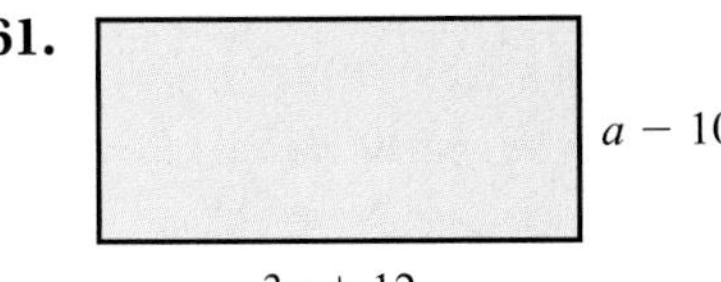

62.

$3x - 2$

$5x + 7$

For Exercises 63–70, indicate whether the function is a constant function, a linear function, a quadratic function, or a cubic function.

63. $g(x) = -5$

64. $g(x) = 4$

65. $f(x) = 3x^2 - 2x + 1$

66. $h(x) = x^2 + 4x - 9$

67. $c(x) = 4x - \frac{3}{5}$

68. $h(x) = -3x - 1$

69. $r(x) = -4x^3 + 7x^2 - x + 3$

70. $f(x) = x^3 - 2x^2 + 1$

For Exercises 71–78, graph the function.

71. $h(x) = -2$

72. $f(x) = 4$

73. $c(x) = -5x + 1$

74. $h(x) = -3x + 4$

75. $g(x) = x^2 + 3x - 4$

76. $g(x) = 3x^2 - 1$

77. $f(x) = x^3 - 2$

78. $c(x) = 2x^3 - 1$

For Exercises 79–86, find: a. $f + g$
b. $f - g$

79. $f(x) = -4; g(x) = 3$

80. $f(x) = 7; g(x) = -1$

81. $f(x) = -4x + 3; g(x) = -x - 1$

82. $f(x) = 8x + 3; g(x) = -x + 3$

83. $f(x) = x^2 - 1; g(x) = 3x^2 + 2$

84. $f(x) = 3x^2 + 2; g(x) = -2x^2 - 5x + 9$

85. $f(x) = x^3 + 2x + 5; g(x) = -2x^3 + 2$

86. $f(x) = x^3 - 8; g(x) = -2x^3 + x^2 + 3$

87. A pharmaceutical company produces an antibiotic. The function $R(x) = 0.2x^2 + x + 3000$ describes the revenue the company makes from sales, where x represents the number of bottles sold. The function $C(x) = 18x + 2000$ describes the cost of producing those bottles.

a. Find a function, $P(x)$, that describes the profit.

b. Find the profit if the company sells 1000 bottles.

88. The function $l(x) = 2x + 5$ describes the length of a rectangle, and the function $w(x) = x - 5$ describes its width.

a. Write a function, $P(x)$, that describes the perimeter.

b. Find $P(8)$.

For Exercises 89 and 90, use the following information. If we neglect air resistance, the polynomial function $h(x) = -16t^2 + 200$ describes the height of a falling object after falling for t seconds from an initial height of 200 feet.

89. What is its height after 0.5 seconds?

90. What is its height after 1.2 seconds?

For Exercises 91 and 92, use the following information. The polynomial function $V(r) = 1.5r^2 - 0.8r + 9.5$ describes the voltage in a circuit, where r represents the resistance in the circuit in ohms.

91. Find the voltage if the resistance is 6 ohms.

92. Find the voltage if the resistance is 8 ohms.

93. The function $F(x) = 0.38x^2 - 4.5x + 33.9$ approximately models the fatality rate of motorcyclists involved in crashes per 100 million vehicle miles traveled each year since 1990. (*Source:* National Highway Traffic Safety Administration, *Traffic Safety Facts,* 2003)

a. Complete the table of ordered pairs.

	Years since 1990	Fatalities
1990		
1992		
1994		
1996		
1998		
2000		
2002		
2004		

b. Graph the function. Note that $x \geq 0$ and $y \geq 0$.

c. Using the model, predict the motorcycle fatality rate in 2010.

94. The function $S(x) = -0.3x^2 + 3.2x + 27.5$ approximately models the percentage of high school students who smoke each year since 1991. (*Source:* Centers for Disease Control, National Center for Health Statistics, *Health, United States,* 2004)

a. Complete the table of ordered pairs.

	Years since 1991	Students Who Smoked
1991		
1993		
1995		
1997		
1999		
2001		
2003		

b. Graph the function. Note that $x \geq 0$ and $y \geq 0$.

c. Using the model, predict the percentage of high school students who smoke in 2007.

95. The function $P(x) = 0.0035x^3 - 0.2x^2 - 7.5x + 915$ approximately models the population of Cleveland, Ohio in millions from 1950 to 2000. (*Source:* U.S. Census Bureau, *Statistical Abstract*, 1900–2003)

a. Complete the table of ordered pairs.

	Years since 1950	Population
1950		
1960		
1970		
1980		
1990		
2000		

b. Graph the function. Note that $x \geq 0$ and $y \geq 0$.

c. Using the model, predict the population of Cleveland in 2010.

96. The function $G(x) = 0.0031x^3 - 0.1x^2 + 0.94x + 3.68$ approximately models the price of natural gas per 1000 cubic feet since 1980. (*Source:* Energy Information Administration)

a. Complete the table of ordered pairs.

	Years since 1980	Price
1980		
1984		
1988		
1992		
1996		
2000		
2004		

b. Graph the function. Note that $x \geq 0$ and $y \geq 0$.

c. Using the model, predict the price of natural gas in 2012.

REVIEW EXERCISES

[1.4] ***For Exercises 1 and 2, distribute.***

1. $8(3x - 9)$

2. $-6(2y + 7)$

For Exercises 3 and 4, solve.

[2.1] **3.** $5x - 3(x + 2) = 7x - 8$

[2.3] **4.** $12x + 9 > 4(2x - 5)$

For Exercises 5 and 6, solve. a. Write the solution set using set-builder notation.
b. Write the solution set using interval notation.
c. Graph the solution set.

[2.4] **5.** $-11 < 2x - 3 \leq 9$

[2.6] **6.** $|2x - 7| \geq 5$

5.3 Multiplying Polynomials

OBJECTIVES

1. Multiply a polynomial by a monomial.
2. Multiply two or more polynomials with multiple terms.
3. Determine the product when given special polynomial factors.
4. Multiply polynomial functions.

In Section 5.1, though we had not formally defined monomials, we used the product rule to multiply them. For example, in Example 1(d) of Section 5.1, we multiplied $(-9m^2n)(5m^3n^7p)$.

$$(-9m^2n)(5m^3n^7p) = -45m^{2+3}n^{1+7}p$$ **Multiply the coefficients and use the product rule of exponents.**

$$= -45m^5n^8p$$ **Simplify the exponents.**

In this section, we will explore how to multiply multiple-term polynomials.

OBJECTIVE 1. Multiply a polynomial by a monomial. First, we consider multiplying a polynomial by a monomial as in $x(x + 4)$. Notice the distributive property applies.

$$x(x + 4) = x \cdot x + x \cdot 4$$ **Use the distributive property.**

$$= x^2 + 4x$$ **Simplify.**

Note: *We used the product rule of exponents to simplify $x \cdot x$.*

Our example suggests the following procedure.

PROCEDURE Multiplying a Polynomial by a Monomial

To multiply a polynomial by a monomial, use the distributive property to multiply each term in the polynomial by the monomial.

EXAMPLE 1 Multiply.

a. $-5x^2(2x^2 + 5x - 1)$

Solution $-5x^2(2x^2 + 5x - 1) = -5x^2 \cdot 2x^2 - 5x^2 \cdot 5x - 5x^2 \cdot (-1)$ **Distribute $-5x^2$.**

$= -10x^4 - 25x^3 + 5x^2$ **Simplify.**

Learning Strategy

If you are a visual learner, try drawing lines connecting each product.

$-5x^2 \quad (2x^2 + 5x^2 - 1)$

$-5x^2 \cdot 2x^2$

$-5x^2 \cdot 5x$

$-5x^2 \cdot -1$

b. $0.2a^2b(3a^3b - ab^2 + 0.5ac)$

Solution

$$0.2a^2b(3a^3b - ab^2 + 0.5ac) = 0.2a^2b \cdot 3a^3b + 0.2a^2b \cdot (-ab^2) + 0.2a^2b \cdot 0.5ac$$ Distribute $0.2a^2b$.

$$= 0.6a^5b^2 - 0.2a^3b^3 + 0.1a^3bc$$ Simplify.

Note: *When multiplying multivariable terms, it is helpful to multiply the coefficients first, then the variables in alphabetical order.*
For $0.2a^2b \cdot 3a^3b$, *think:* $0.2 \cdot 3 = 0.6;$ $a^2 \cdot a^3 = a^5;$ $b \cdot b = b^2$

YOUR TURN Multiply.

a. $2x^3(7x^2 - x + 4)$

b. $-4xy^2z(6x^3z - 5yz^2 + 9xz)$

OBJECTIVE 2. Multiply two or more polynomials with multiple terms. We again use the distributive property to multiply two polynomials with multiple terms, such as $(x + 3)(x + 2)$. Now, however, we distribute the entire $x + 3$ expression.

$$(x + 3)(x + 2) = (x + 3) \cdot x + (x + 3) \cdot 2$$ Distribute $x + 3$.

To complete the multiplication, we need to apply the distributive property again.

Note: *We could skip to this line by multiplying every term in the second polynomial by every term in the first polynomial.*

$$= x \cdot x + 3 \cdot x + x \cdot 2 + 3 \cdot 2$$ Distribute x in $(x + 3) \cdot x$ and 2 in $(x + 3) \cdot 2$.

$$= x^2 + 3x + 2x + 6$$ Simplify.

$$= x^2 + 5x + 6$$ Combine like terms.

Our example suggests the following procedure.

PROCEDURE **Multiplying Polynomials**

To multiply two polynomials,

1. Multiply every term in the second polynomial by every term in the first polynomial.
2. Combine like terms.

ANSWERS

a. $14x^5 - 2x^4 + 8x^3$
b. $-24x^4y^2z^2 + 20xy^3z^3 - 36x^2y^2z^2$

Connection The process of multiplying polynomials is actually the same as for multiplying numbers. In fact, many people multiply polynomials in the same vertical fashion as for numbers. For example, compare (12)(13) with $(x + 2)(x + 3)$.

$$\begin{array}{r} 12 \\ \times\ 13 \\ \hline 36 \\ +\ 12 \\ \hline 156 \end{array} \qquad \begin{array}{r} x + 2 \\ x + 3 \\ \hline 3x + 6 \\ +\ x^2 + 2x \\ \hline x^2 + 5x + 6 \end{array}$$

Again, the distributive property is the governing principle. We multiply each digit in one number by each digit in the other number, then add digits in the like place values. For polynomials, we multiply each term in one polynomial by each term in the other, then combine like terms.

Example 2 Multiply.

a. $(2x + 4)(3x - 5)$

Learning Strategy

As we mentioned earlier, try drawing lines connecting the terms that you multiply so that you can better see the products.

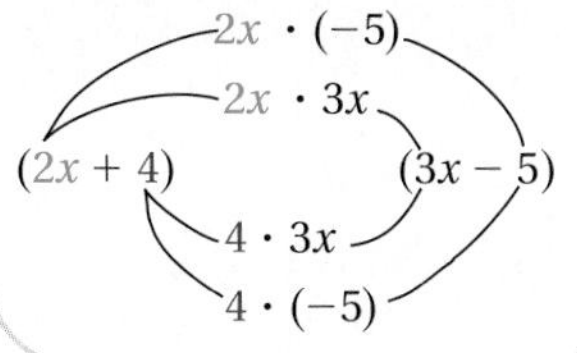

Solution
$$\begin{aligned}(2x + 4)(3x - 5) &= 2x \cdot 3x + 2x \cdot (-5) + 4 \cdot 3x + 4 \cdot (-5) && \text{Multiply each term in } 3x - 5 \text{ by each term in } 2x + 4. \\ &= 6x^2 - 10x + 12x - 20 && \text{Simplify.} \\ &= 6x^2 + 2x - 20 && \text{Combine like terms.}\end{aligned}$$

b. $(4m - 3n)(2m - n)$

Solution
$$\begin{aligned}(4m - 3n)(2m - n) &= 4m \cdot 2m + 4m \cdot (-n) \\ &\quad + (-3n) \cdot 2m + (-3n) \cdot (-n) && \text{Multiply each term in } 2m - n \text{ by each term in } 4m - 3n. \\ &= 8m^2 - 4mn - 6mn + 3n^2 && \text{Simplify.} \\ &= 8m^2 - 10mn + 3n^2 && \text{Combine like terms.}\end{aligned}$$

c. $(x^2 - 3)(x^2 + 4)$

Solution
$$\begin{aligned}(x^2 - 3)(x^2 + 4) &= x^2 \cdot x^2 + x^2 \cdot 4 \\ &\quad + (-3) \cdot x^2 + (-3) \cdot 4 && \text{Multiply each term in } x^2 + 4 \text{ by each term in } x^2 - 3. \\ &= x^4 + 4x^2 - 3x^2 - 12 && \text{Simplify.} \\ &= x^4 + x^2 - 12 && \text{Combine like terms.}\end{aligned}$$

d. $(x^2 + 4)(y + 6)$

Solution
$$\begin{aligned}(x^2 + 4)(y + 6) &= x^2 \cdot y + x^2 \cdot 6 + \\ &\quad 4 \cdot y + 4 \cdot 6 && \text{Multiply each term in } y + 6 \text{ by each term in } x^2 + 4. \\ &= x^2y + 6x^2 + 4y + 24 && \text{Simplify.}\end{aligned}$$

Note: *There are no like terms to combine.*

The word FOIL, which stands for First Outer Inner Last, is a popular way to remember the process of multiplying two binomials. We will use the binomials from Example 2(a) to demonstrate.

Warning: FOIL only helps keep track of products when multiplying two binomials. When multiplying larger polynomials, just remember to multiply every term in the second polynomial by every term in the first polynomial.

$(2x + 4)(3x - 5)$ **First terms:** $2x \cdot 3x = 6x^2$

$(2x + 4)(3x - 5)$ **Outer terms:** $2x \cdot (-5) = -10x$

$(2x + 4)(3x - 5)$ **Inner terms:** $4 \cdot 3x = 12x$

$(2x + 4)(3x - 5)$ **Last terms:** $4 \cdot (-5) = -20$

Connection The product of two binomials can be illustrated geometrically. The following rectangle has a length of $x + 7$ and a width of $x + 5$. We can describe its area in two ways:

1. The product of the length and width, $(x + 7)(x + 5)$; or
2. The sum of the areas of the four internal rectangles, $x^2 + 5x + 7x + 35$. Since these two approaches describe the same area, the two expressions must be equal.

	x	7
x	x^2	$7x$
5	$5x$	35

Length $\cdot$ Width = Sum of the areas of the four internal rectangles

$$(x + 7)(x + 5) = x^2 + 5x + 7x + 35$$

$$= x^2 + 12x + 35 \quad \text{Combine like terms.}$$

YOUR TURN Multiply.

a. $(x - 9)(x + 3)$ **b.** $(5y - 2)(6y - 7)$ **c.** $(4t^2 - u)(5t^2 + u)$

Now consider an example of multiplication involving a larger polynomial, such as a trinomial. Remember that no matter how many terms are in the polynomials, we multiply every term in the second polynomial by every term in the first polynomial.

Warning: Notice that FOIL does not make sense with the trinomial in Example 3 because there are too many terms. FOIL handles only the four terms from two binomials.

EXAMPLE 3 Multiply. $(2x + 3)(4x^2 + x - 5)$

Solution Multiply each term in $4x^2 + x - 5$ by each term in $2x + 3$.

$$(2x + 3)(4x^2 + x - 5) = 2x \cdot 4x^2 + 2x \cdot x + 2x \cdot (-5) + 3 \cdot 4x^2 + 3 \cdot x + 3 \cdot (-5)$$ Multiply each term in $4x^2 + x - 5$ by each term in $2x + 3$.

$$= 8x^3 + 2x^2 - 10x + 12x^2 + 3x - 15$$ Simplify.

$$= 8x^3 + 14x^2 - 7x - 15$$ Combine like terms.

ANSWERS

a. $x^2 - 6x - 27$
b. $30y^2 - 47y + 14$
c. $20t^4 - t^2u - u^2$

ANSWER

$10x^3 - 47x^2 + 29x - 4$

YOUR TURN Multiply. $(5x - 1)(2x^2 - 9x + 4)$

OBJECTIVE 3. Determine the product when given special polynomial factors. We can use formulas to quickly find the product of some special polynomial factors.

Squaring a Binomial

First, let's look for a pattern that we can use to help us square binomials. There are two cases to consider: $(a + b)^2$ and $(a - b)^2$. Recall that squaring a number or expression means we multiply the number or expression by itself.

$$(a + b)^2 = (a + b)(a + b) \qquad (a - b)^2 = (a - b)(a - b)$$
$$= a^2 + ab + ab + b^2 \qquad = a^2 - ab - ab + b^2$$
$$= a^2 + 2ab + b^2 \qquad = a^2 - 2ab + b^2$$

RULE Squaring a Binomial

If a and b are real numbers, variables, or expressions, then $(a + b)^2 = a^2 + 2ab + b^2$
$(a - b)^2 = a^2 - 2ab + b^2$

EXAMPLE 4 Multiply.

a. $(x + 7)^2$

Solution $(a + b)^2 = a^2 + 2\,a\,b + b^2$

$(x + 7)^2 = (x)^2 + 2(x)(7) + (7)^2$ Use $(a + b)^2 = a^2 + 2ab + b^2$, where a is x and b is 7.

$= x^2 + 14x + 49$ Simplify.

Warning: Notice that $(x + 7)^2$ does *not* equal $x^2 + 49$.

b. $(6r - 5)^2$

Solution $(6r - 5)^2 = (6r)^2 - 2(6r)(5) + (5)^2$ Use $(a - b)^2 = a^2 - 2ab + b^2$.

$= 36r^2 - 60r + 25$ Simplify.

c. $(h^3 - 4)^2$

Solution $(h^3 - 4)^2 = (h^3)^2 - 2(h^3)(4) + (4)^2$ Use $(a - b)^2 = a^2 - 2ab + b^2$.

$= h^6 - 8h^3 + 16$ Simplify.

d. $(4x + 5y^2)^2$

Solution $(4x + 5y^2)^2 = (4x)^2 + 2(4x)(5y^2) + (5y^2)^2$ Use $(a + b)^2 = a^2 + 2ab + b^2$.

$= 16x^2 + 40xy^2 + 25y^4$ Simplify.

e. $[(x + 6) - y]^2$

Solution If we view $x + 6$ as a and y as b, the expression has the form $(a - b)^2$.

$$(a - b)^2 = a^2 - 2\,a\,b + b^2$$

$[(x + 6) - y]^2 = (x + 6)^2 - 2(x + 6)(y) + (y)^2$ Use $(a - b)^2 = a^2 - 2ab + b^2$.

$= x^2 + 2(x)(6) + 6^2 - 2xy - 12y + y^2$ Use $(a + b)^2 = a^2 + 2ab + b^2$ to square $x + 6$.

$= x^2 + 12x + 36 - 2xy - 12y + y^2$ Simplify.

$= x^2 + y^2 - 2xy + 12x - 12y + 36$ Write in descending order.

Answers

a. $k^2 - 8k + 16$
b. $9y^2 + 6y + 1$
c. $x^4 + 10x^2 + 25$
d. $n^6 - 8n^3 + 16$
e. $25t^2 + u^2 + 10tu - 30t - 6u + 9$

Your Turn Multiply.

a. $(k - 4)^2$ **b.** $(3y + 1)^2$ **c.** $(x^2 + 5)^2$

d. $(4 - n^3)^2$ **e.** $[5t + (u - 3)]^2$

Multiplying Conjugates

Another special product is the result of multiplying **conjugates**.

DEFINITION ***Conjugates:*** Binomials that differ only in the sign separating the terms.

The following binomial pairs are conjugates:

$x + 6$ and $x - 6$ $\quad$ $3t + 5$ and $3t - 5$ $\quad$ $-2c + 9d$ and $-2c - 9d$

Let's multiply general conjugates $a + b$ and $a - b$ to determine the pattern.

Note: *The like terms $-ab$ and ab are additive inverses, so their sum is 0.*

$$\begin{aligned}(a+b)(a-b) &= a \cdot a + a \cdot (-b) + b \cdot a + b \cdot (-b) && \text{Use FOIL.}\\ &= a^2 - ab + ab - b^2 && \text{Simplify.}\\ &= a^2 - b^2 && \text{Combine like terms.}\end{aligned}$$

The product, $a^2 - b^2$, is called a *difference of squares*. Our example suggests the following rule.

RULE **Multiplying Conjugates**

If a and b are real numbers, variables, or expressions, then $(a + b)(a - b) = a^2 - b^2$.

Example 5 Multiply.

a. $(3t + 5)(3t - 5)$

Solution $(a + b)(a - b) = a^2 - b^2$

$$\begin{aligned}(3t+5)(3t-5) &= (3t)^2 - (5)^2 && \text{Use } (a+b)(a-b) = a^2 - b^2\text{, where } a \text{ is } 3t \text{ and } b \text{ is } 5.\\ &= 9t^2 - 25 && \text{Simplify.}\end{aligned}$$

Note: *The commutative property tells us that $(t^2 - 5u)(t^2 + 5u)$ is the same as $(t^2 + 5u)(t^2 - 5u)$, so this expression is of the form $(a + b)(a - b)$.*

b. $(t^2 - 5u)(t^2 + 5u)$

Solution
$$\begin{aligned}(t^2 - 5u)(t^2 + 5u) &= (t^2)^2 - (5u)^2 && \text{Use } (a+b)(a-b) = a^2 - b^2.\\ &= t^4 - 25u^2 && \text{Simplify.}\end{aligned}$$

c. $[(x + 3) + y][(x + 3) - y]$

Solution If we view $x + 3$ as a and y as b, the expression has the form $(a + b)(a - b)$.

$$\begin{aligned}
(a \quad + \quad b)\ (a \quad - \quad b) &= \quad a^2 \quad - \quad b^2 \\
[(x + 3) + y][(x + 3) - y] &= (x + 3)^2 - y^2 \\
&= x^2 + 2(x)(3) + 3^2 - y^2 && \text{Use } (a+b)^2 = a^2 + 2ab + b^2 \text{ to square } x + 3. \\
&= x^2 + 6x + 9 - y^2 && \text{Simplify.} \\
&= x^2 - y^2 + 6x + 9 && \text{Write in descending order.}
\end{aligned}$$

Your Turn Multiply.

a. $(9y - 2)(9y + 2)$

b. $(3h^2 + k^2)(3h^2 - k^2)$

c. $[2m + (n - 4)][(2m - (n - 4)]$

Answers

a. $81y^2 - 4$ **b.** $9h^4 - k^4$
c. $4m^2 - n^2 + 8n - 16$

OBJECTIVE 4. Multiply polynomial functions. We can find the product of two polynomial functions using the following rule.

RULE **Multiplying Functions**

The product of two functions, $f \cdot g$, is found by $(f \cdot g)(x) = f(x)\,g(x)$.

Example 6 **a.** If $f(x) = 4x + 7$ and $g(x) = 2x^2 - x - 5$, find $(f \cdot g)(x)$.
b. Find $(f \cdot g)(-2)$.

Solution **a.**

$$\begin{aligned}
(f \cdot g)(x) &= (4x + 7)(2x^2 - x - 5) && \text{Use } (f \cdot g)(x) = f(x)g(x). \\
&= 8x^3 - 4x^2 - 20x + 14x^2 - 7x - 35 && \text{Multiply each term in } 2x^2 - x - 5 \text{ by each term in } 4x + 7. \\
&= 8x^3 + 10x^2 - 27x - 35 && \text{Combine like terms.}
\end{aligned}$$

b.

$$\begin{aligned}
(f \cdot g)(-2) &= 8(-2)^3 + 10(-2)^2 - 27(-2) - 35 && \text{In } (f \cdot g)(x), \text{ replace } x \text{ with } -2. \\
&= -64 + 40 + 54 - 35 && \text{Simplify.} \\
&= -5
\end{aligned}$$

Your Turn

a. If $f(x) = 3x - 5$ and $g(x) = x^2 - 6x + 2$, find $(f \cdot g)(x)$.

b. Find $(f \cdot g)(-1)$.

Answers

a. $(f \cdot g)(x) = 3x^3 - 23x^2 + 36x - 10$
b. -72

5.3 Exercises

For Extra Help — MyMathLab, Videotape/DVT, InterAct Math, Math Tutor Center, Math XL.com

1. What mathematical property do we apply when multiplying polynomials?

2. What is the general process for multiplying two multiple-term polynomials?

3. Explain how to multiply two binomials.

4. Explain how to recognize conjugates.

5. What type of expression is the product of a conjugate pair?

6. Show why $(a + b)^2 \neq a^2 + b^2$.

For Exercises 7–26, multiply the polynomial by the monomial.

7. $5x^3(x^2 + 3x - 2)$

8. $4y^5(-3y^3 - 5y^2 + y)$

9. $-6x^3(2x^2 + 4x - 1)$

10. $-2y^4(y^2 + 7y - 8)$

11. $4n^4(n^3 + 7n^2 - 2n - 3)$

12. $6n^6(n^3 + 4n^2 - n + 3)$

13. $9a^2b^4(3a^4b^2 - 2ab^8)$

14. $3yz^3(y^4z - 3y^4z^2)$

15. $\frac{1}{4}m^2np^3(2mn^5 - 5m^2n^2 + 8m^2n^3p^2)$

16. $\frac{2}{3}k^2l^3m(2k^3l^2 - 3lm^4 + 6k^2l^7m^2)$

17. $-0.3p^2q^2(8p^3q^7 - 2q + 7p^6)$

18. $0.4a^6c^2(2ac^7 - ac^2 - 5a^2c^9)$

19. $-5x^5y^2(4x^2y + 2xy^4 - 5x^8y^5)$

20. $-5a^2b(3ab^3 - 2a^2b^2 + 3ab^5)$

21. $-3r^2s(2r^4s^3 - 6r^2s^2 - 3rs + 3)$

22. $-x^2y^3(5xy^2 - 3x^2y^5 + 1 - y^3)$

23. $-0.2a^2b^2(2.1a - 6ab + 3a^2b - a^2b^2)$

24. $-0.5h^2k^3(6k^3 - 2h^2k^3 - 3hk^4 + 4h^8)$

25. $\frac{1}{3}abc^6(9a^2b^2c^2 - 4a^2bc + 12a)$

26. $\frac{2}{5}xy^2z^3(20x^2y^3 - 5x^2z^2 + 10x^2yz)$

For Exercises 27 and 28, a larger rectangle is formed out of smaller rectangles:

a. Write an expression in simplest form for the length (along the top).
b. Write an expression in simplest form for the width (along the side).
c. Write an expression that is the product of the length and width that you found in parts a and b.
d. Write an expression in simplest form that is the sum of the areas of each of the smaller rectangles.
e. Explain why the expressions in parts c and d are equivalent.

27.

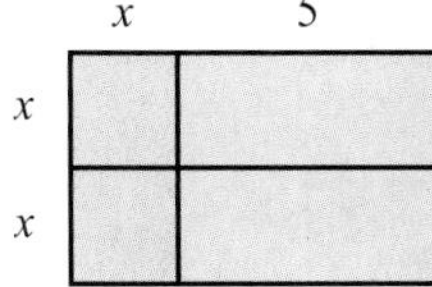

28.

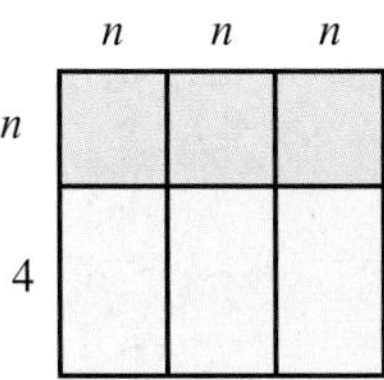

For Exercises 29–40, multiply the binomials (use FOIL).

29. $(2x + 3)(3x + 4)$

30. $(2a + 4)(7a + 9)$

31. $(3x - 1)(5x + 2)$

32. $(4n + 7)(9n - 2)$

33. $(2y - 5x)(3y - 2x)$

34. $(2x - 4y)(5x - 3y)$

35. $(5m - 3n)(3m + 4n)$

36. $(7x + 3y)(3x - 4y)$

37. $(t^2 - 5)(t^2 - 2)$

38. $(m^2 + 7)(m^2 - 1)$

39. $(a^2 + 6b^2)(a^2 - b^2)$

40. $(3x^2 + 4y^2)(x^2 - 5y^2)$

For Exercises 41 and 42, a larger rectangle is formed out of smaller rectangles:

a. Write an expression in simplest form for the length (along the top).
b. Write an expression in simplest form for the width (along the side).
c. Write an expression that is the product of the length and width that you found in parts a and b.
d. Write an expression in simplest form that is the sum of the areas of each of the smaller rectangles.
e. Explain why the expressions in parts c and d are equivalent.

41.

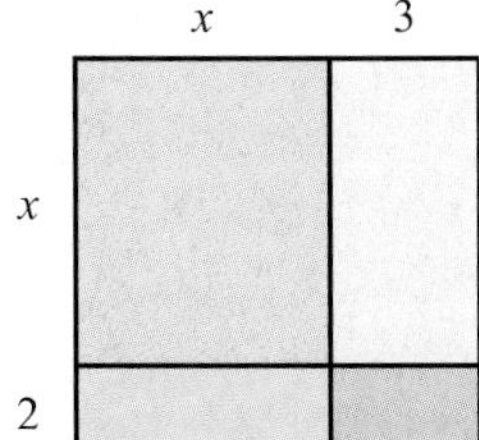

42.

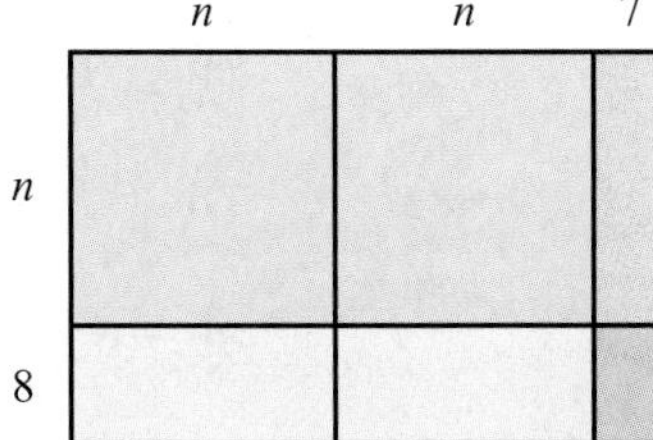

For Exercises 43–54, multiply the polynomials.

43. $(3x - 1)(4x^2 - 2x + 1)$

44. $(3x + 10)(2x^2 - 2x + 1)$

45. $(7a - 1)(3a^2 - 2a - 2)$

46. $(5x + 1)(3x^2 + 12x + 1)$

47. $(7c + 2)(2c^2 - 4c - 3)$

48. $(2x - 3)(7x^2 + 3x - 5)$

49. $(4p + 10q)(3p^2 + 2pq + q^2)$

50. $(2f + 3g)(f^2 - 5fg - 4g^2)$

51. $(2y - 3z)(6y^2 - 2yz + 4z^2)$

52. $(7m - 3n)(2m^2 - 5mn + 3n^2)$

53. $(3u^2 - 2u - 1)(3u^2 + 2u + 1)$

54. $(2a^2 - 3a + 2)(4a^2 + 2a + 1)$

For Exercises 55–64, multiply using the rules for special products.

55. $(2x - 7)(2x + 7)$

56. $(3y + 8)(3y - 8)$

57. $(2q - 3)(2q + 3)$

58. $(3p + 1)(3p - 1)$

59. $(x^2 + 2y)(x^2 - 2y)$

60. $(a^2 + 3b)(a^2 - 3b)$

61. $[(s + 1) + t][(s + 1) - t]$

62. $[(y + 3) + z]\,[(y + 3) - z]$

63. $[3b + (c + 2)]\,[3b - (c + 2)]$

64. $[6k - (x + 1)][6k + (x + 1)]$

For Exercises 65–72, state the conjugate of the given binomial.

65. $x + 8$

66. $y - 4$

67. $3m + 2n$

68. $4q - 9$

69. $2c - 3d$

70. $a^2 - b^2$

71. $-2j - 5k$

72. $-a - 4b$

For Exercises 73–84, find the product.

73. $(x + y)^2$

74. $(m + n)^2$

75. $(4t + 3w)^2$

76. $(7a + 2b)^2$

77. $(4w - 3)^2$

78. $(3k - 2)^2$

79. $(9 - 5y^3)^2$

80. $(6 - 7n^3)^2$

81. $[(x + 1) - y]^2$

82. $[(a + 3) - b]^2$

83. $[p - (q + 5)]^2$

84. $[w - (v + 4)]^2$

For Exercises 85–88, write an expression for the area in simplest form.

85.

86.

$4x - 2$

$8x - 7$

87.

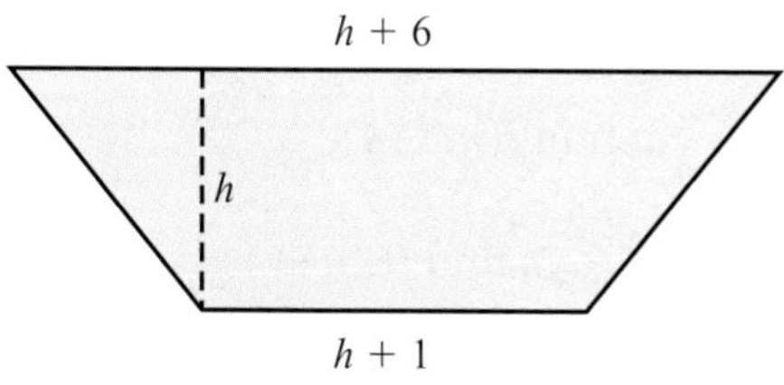

88.

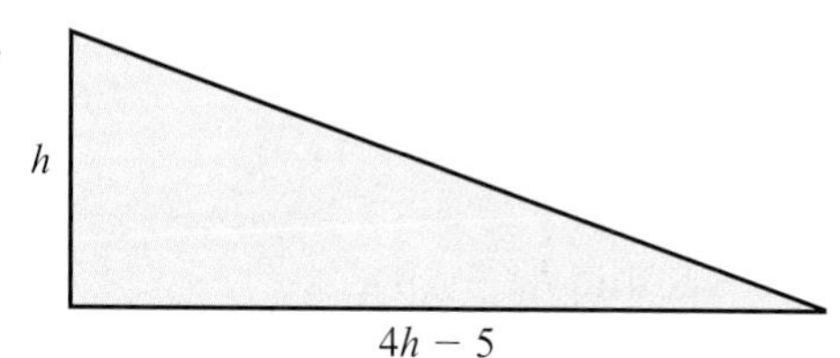

For Exercises 89–92, explain the mistake, then find the correct product.

89. $(3x - 5)(2x - 9) = 6x^2 - 27x - 10x + 45$
$= 6x^2 - 17x + 45$

90. $(t^2 + 3)(t^2 - 1) = t^2 - t^2 + 3t^2 - 3$
$= t^2 + 2t^2 - 3$
$= 3t^2 - 3$

91. $(2x - 7)(2x + 7) = 4x - 14x + 14x - 49$
$= 4x - 49$

92. $(5y + 8)^2 = 25y^2 + 64$

For Exercises 93–112, find the product.

93. $4wv(-w^4v^3 - 4x^2wv^7 + 5xw^2v)$

94. $-x^2y(7x^3y^7 - 5x^2y^2 + 1)$

95. $(8r^2 - 3s)(8r^2 + 3s)$

96. $(4b - 5c^2)(4b + 5c^2)$

97. $(2u^2 + 3v)^2$

98. $(3r - 7s^2)^2$

99. $(2a^2 - 5b^2)(a^2 - 4b^2)$

100. $(7y^2 + 3x^2)(2y^2 + 5x^2)$

101. $(5q^2 - 3t)(3q^2 + 4t)$

102. $(7k^2 - 2j)(2k^2 + 3j)$

103. $3r^2s^3\left(r^4 - \frac{1}{9}r^3s - \frac{1}{6}r^2s^2 + \frac{1}{3}rs - 1\right)$

104. $4x^4y^2\left(x^3 + \frac{1}{8}x^2y - \frac{1}{16}xy^3 + \frac{1}{4}xy - 3\right)$

105. $-0.1t^2r^3(3.5t^2r^3 - 8t^3r^2 + 2.2t^3r - 2tr)$

106. $-0.2a^3b^2(4.3a^4b^5 - 5ab^4 + 3.2a^2b^3 - 4ab)$

107. $[(3m - 4) + n][(3m - 4) - n]$

108. $[(2c - 5) - d][(2c - 5) + d]$

109. $(x^2 + 3xy + y^2)(4x^2 + 2xy - y^2)$

110. $(x^2 + 6x + 9)(x^2 + 4x + 4)$

111. $[(x^2 - 4) + 3]^2$

112. $[(t^2 + 3) - 5]^2$

113. If $f(x) = x^2 - 2x + 9$ and $g(x) = 2x - 1$, find
a. $(f \cdot g)(x)$ **b.** $(f \cdot g)(-4)$ **c.** $(f \cdot g)(2)$

114. If $f(x) = 3x - 4$ and $g(x) = 2x^2 - x - 1$, find
a. $(f \cdot g)(x)$ **b.** $(f \cdot g)(3)$ **c.** $(f \cdot g)(-1)$

115. If $f(x) = 3x + 4$ and $g(x) = 4x^2 - 3x + 2$, find
a. $(f \cdot g)(x)$ **b.** $(f \cdot g)(5)$ **c.** $(f \cdot g)(-3)$

116. If $f(x) = x^2 + 3x - 7$ and $g(x) = 3x + 2$, find $(f \cdot g)(x)$.
a. $(f \cdot g)(x)$ **b.** $(f \cdot g)(-2)$ **c.** $(f \cdot g)(4)$

117. If $f(x) = x^2 + 3x - 1$, find $f(n + 1)$.

118. If $f(x) = 3x^2 - x$, find $f(m - 4)$.

119. If $f(x) = 2x^2 - 1$, find $f(t - 6)$.

120. If $f(x) = 3x^2 - 2x - 5$, find $f(2u + 1)$.

121. A rectangular room has a width that is 2 feet less than the length. Write an expression for the area of the room.

122. A circular metal plate for a machine has radius r. A larger plate with radius $r + 2$ is to be used in another part of the machine. Write an expression in simplest form for the sum of the areas of the two circles.

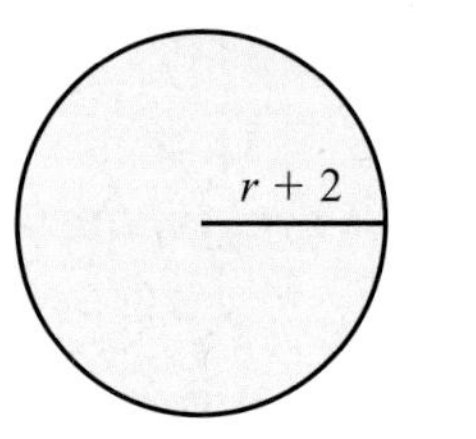

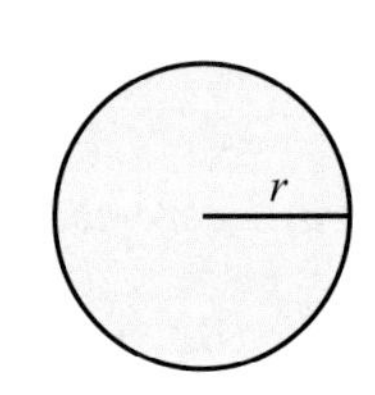

For Exercises 123 and 124, write an expression for the volume in simplest form.

123.

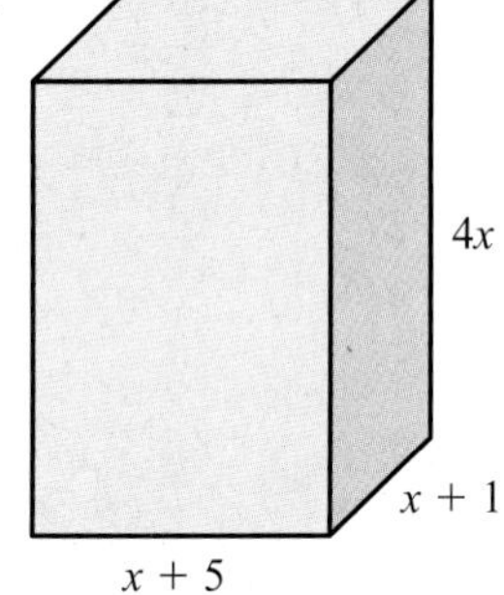

124.

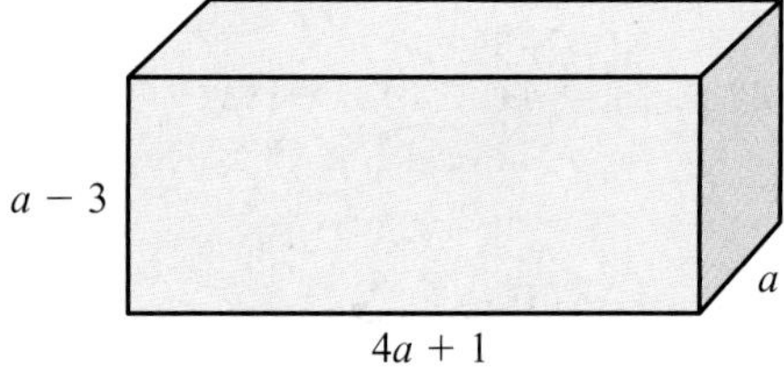

125. A shipping crate's design specifies a length that is triple the width, and a height that is 5 feet more than the length. Let w represent the width. Write an expression for the volume in terms of w.

126. A fish tank's design specifies a length that is 8 inches longer than the width. The height of the tank is to be 5 inches longer than the width. Let w represent the width and write an expression for the volume of the tank in terms of w.

127. A right circular cylinder is to have a height that is 3 inches more than the radius. Write an expression for the volume of the cylinder in terms of the radius.

128. A cone with a circular base is to have a radius that is 4 centimeters less than the height. Write an expression for the volume of the cone in terms of the height.

REVIEW EXERCISES

[2.1] **1.** Solve: $\frac{5}{6}x - \frac{1}{8} = \frac{3}{4}x + \frac{1}{2}$

[2.1] **2.** Solve for t in $d = 30t - n$.

[3.3] **3.** Write the equation of a line passing through the point $(-3, 5)$ and perpendicular to $y = 5x - 2$.

[4.3] **4.** A cereal comes in two box sizes. The smaller box sells for \$3.50 and the larger box sells for \$4.80. In one week a grocer sold a total of 84 boxes of that cereal and made \$349.90 in total revenue from the sale. Use a system of equations to find the number of each size box sold.

[5.1] **5.** Multiply $(-4.26 \times 10^6)(2.1 \times 10^5)$. Write the answer in scientific notation.

[5.1] **6.** Divide $\frac{1.5 \times 10^5}{2.4 \times 10^{-4}}$. Write the answer in scientific notation.

5.4 Dividing Polynomials

OBJECTIVES

1. **Divide a polynomial with multiple terms by a monomial.**
2. **Use long division to divide polynomials.**
3. **Divide polynomial functions.**

In Section 5.1, we learned the quotient rule of exponents. Although we did not say so at that time, we used the quotient rule to divide monomials, as shown here.

$$\frac{28x^5}{4x^2} = \frac{28}{4}x^{5-2} = 7x^3$$

In this section, we will use the quotient rule of exponents and divide polynomials with more terms.

OBJECTIVE 1. Divide a polynomial with multiple terms by a monomial. First, we will consider dividing a polynomial by a monomial, as in $\frac{28x^5 + 20x^3}{4x^2}$. Recall that when fractions with a common denominator are added (or subtracted), the numerators are added and the denominator stays the same.

$$\frac{1}{5} + \frac{2}{5} = \frac{1 + 2}{5}$$

This process can be reversed so that a sum in the numerator of a fraction can be broken into fractions with each addend over the same denominator.

$$\frac{1+2}{5} = \frac{1}{5} + \frac{2}{5}$$

RULE If a, b, and c are real numbers, variables, or expressions with $c \neq 0$, then $\frac{a+b}{c} = \frac{a}{c} + \frac{b}{c}$.

We can apply this rule to $\frac{28x^5 + 20x^3}{4x^2}$.

$$\begin{aligned} \frac{28x^5 + 20x^3}{4x^2} &= \frac{28x^5}{4x^2} + \frac{20x^3}{4x^2} \\ &= 7x^{5-2} + 5x^{3-2} \\ &= 7x^3 + 5x \end{aligned}$$

Note: *We now have a sum of monomial divisions, which we can simplify separately.*

This illustration suggests the following procedure.

PROCEDURE **Dividing a Polynomial by a Monomial**

To divide a polynomial by a monomial, divide each term in the polynomial by the monomial.

EXAMPLE 1 Divide.

a. $\frac{12u^6 - 18u^4 + 42u^2}{6u^2}$

Solution

$$\begin{aligned} \frac{12u^6 - 18u^4 + 42u^2}{6u^2} &= \frac{12u^6}{6u^2} - \frac{18u^4}{6u^2} + \frac{42u^2}{6u^2} && \text{Divide each term in the polynomial by the monomial.} \\ &= 2u^4 - 3u^2 + 7 && \text{Divide.} \end{aligned}$$

Connection In Section 5.1 we learned that $\frac{42u^2}{6u^2} = 7u^{2-2} = 7u^0 = 7 \cdot 1 = 7$.

b. $(32x^5y^3 - 24x^4y^2 - 2xy) \div 8x^2y$

Solution $(32x^5y^3 - 24x^4y^2 - 2xy) \div 8x^2y$

$$\begin{aligned} &= \frac{32x^5y^3 - 24x^4y^2 - 2xy}{8x^2y} \\ &= \frac{32x^5y^3}{8x^2y} - \frac{24x^4y^2}{8x^2y} - \frac{2xy}{8x^2y} && \text{Divide each term in the polynomial by the monomial.} \\ &= 4x^3y^2 - 3x^2y - \frac{1}{4x} \end{aligned}$$

Note: *Using the quotient rule, we have* $\frac{2xy}{8x^2y} = \frac{1}{4}x^{-1} = \frac{1}{4x}$.

YOUR TURN Divide.

a. $\dfrac{54x^5 + 42x^4 - 24x^3}{6x^3}$

b. $(12a^2b^5 - 20a^4b + 36a^2) \div 4a^2b$

OBJECTIVE 2. Use long division to divide polynomials. To divide a polynomial by a polynomial we can use long division. As a reminder of the process of long division, let's divide 157 by 12 using long division.

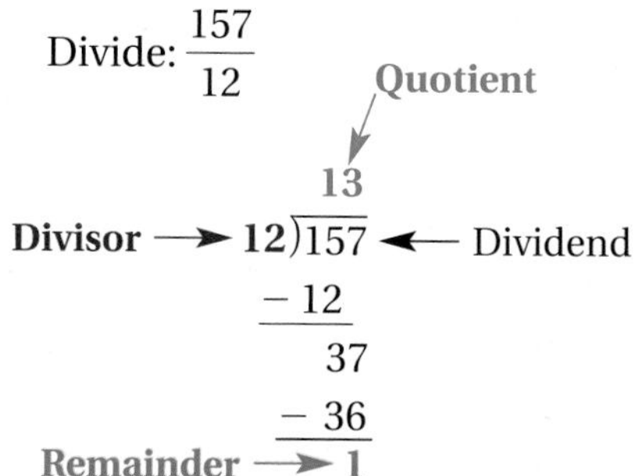

Because the answer has a remainder, we write the result as a mixed number, $13\frac{1}{12}$. Notice we could check the answer by multiplying the divisor by the quotient then adding the remainder. The result should be the dividend.

Now consider polynomial division, which follows the same long division process.

EXAMPLE 2 Divide. $\dfrac{x^2 + 5x + 7}{x + 2}$

Solution

$$x + 2\overline{)x^2 + 5x + 7} \quad \text{quotient: } x$$

Divide the first terms to determine the first term in the quotient: $x^2 \div x = x$.

Next, we multiply the divisor $x + 2$ by the x in the quotient.

$$\begin{array}{r} x \\ x + 2\overline{)x^2 + 5x + 7} \\ x^2 + 2x \end{array}$$

multiply

Next, subtract. Recall that to subtract a polynomial, we first change the signs of its terms. After combining terms, we bring down the next term in the dividend, which is 7.

$$\begin{array}{r} x \\ x + 2\overline{)x^2 + 5x + 7} \\ -(x^2 + 2x) \end{array} \xrightarrow{\text{Change signs.}} \begin{array}{r} x \\ x + 2\overline{)x^2 + 5x + 7} \\ -x^2 - 2x \quad\;\; \\ \hline 3x + 7 \end{array}$$

Combine like terms and bring down the next term.

ANSWERS

a. $9x^2 + 7x - 4$

b. $3b^4 - 5a^2 + \dfrac{9}{b}$

Now we repeat the process with $3x + 7$ as the dividend.

multiply

$$\begin{array}{r} x + 3 \\ x + 2\overline{)x^2 + 5x + 7} \\ \underline{-x^2 - 2x} \\ 3x + 7 \\ 3x + 6 \end{array}$$

Divide the first term of $3x + 7$ by the first term in the divisor, $x + 2$, which gives $3x \div x = 3$. We then multiply the divisor by this 3.

Subtract. As before, we change the signs of the binomial to be subtracted.

$$\begin{array}{r} x + 3 \\ x + 2\overline{)x^2 + 5x + 7} \\ \underline{-x^2 - 2x} \\ 3x + 7 \\ \underline{-(3x + 6)} \end{array} \xrightarrow{\text{Change signs.}} \begin{array}{r} x + 3 \\ x + 2\overline{)x^2 + 5x + 7} \\ \underline{-x^2 - 2x} \\ 3x + 7 \\ \underline{-3x - 6} \\ 1 \end{array}$$

Combine like terms.

Note that we have a remainder, 1. Recall that in the numeric version, we wrote the answer as a mixed number, $13\frac{1}{12}$, which means $13 + \frac{1}{12}$ and is in the form quotient $+ \frac{\text{remainder}}{\text{divisor}}$. With polynomials we write a similar expression:

$$\underbrace{x + 3}_{\text{quotient}} + \frac{1}{x + 2} \begin{array}{l} \leftarrow \text{remainder} \\ \leftarrow \text{divisor} \end{array}$$

To check, we multiply the quotient and the divisor, then add the remainder. The result should be the dividend.

$$(x + 3)(x + 2) + 1 = x^2 + 2x + 3x + 6 + 1$$
$$= x^2 + 5x + 7 \quad \text{It checks.}$$

PROCEDURE **Dividing a Polynomial by a Polynomial**

To divide a polynomial by a polynomial,

1. Use long division.
2. If there is a remainder, write the result in the form quotient $+ \frac{\text{remainder}}{\text{divisor}}$.

EXAMPLE 3 Divide. $\frac{15x^2 - 26x + 17}{5x - 2}$

Solution Begin by dividing the first term in the dividend by the first term in the divisor: $15x^2 \div 5x = 3x$.

$$\begin{array}{r} 3x \\ 5x - 2\overline{)15x^2 - 26x + 17} \\ \underline{-(15x^2 - 6x)} \end{array} \xrightarrow{\text{Change signs.}} \begin{array}{r} 3x \\ 5x - 2\overline{)15x^2 - 26x + 17} \\ \underline{-15x^2 + 6x} \\ -20x + 17 \end{array}$$

Combine like terms and bring down the next term.

To find the next part of the quotient, we divide $-20x$ by $5x$, which is -4. We then repeat the multiplication and subtraction steps.

$$\begin{array}{r} 3x - 4 \\ 5x - 2\overline{)15x^2 - 26x + 17} \\ \underline{-15x^2 + 6x} \\ -20x + 17 \\ \underline{-(-20x + 8)} \end{array} \xrightarrow{\text{Change signs.}} \begin{array}{r} 3x - 4 \\ 5x - 2\overline{)15x^2 - 26x + 17} \\ \underline{-15x^2 + 6x} \\ -20x + 17 \\ \underline{+20x - 8} \\ 9 \end{array} \quad \text{Combine like terms.}$$

Answer $3x - 4 + \frac{9}{5x - 2}$

ANSWERS

a. $x + 6$

b. $6x - 5 + \frac{2}{2x - 3}$

YOUR TURN Divide.

a. $\frac{x^2 + 9x + 18}{x + 3}$

b. $\frac{12x^2 - 28x + 17}{2x - 3}$

Using a Place Marker in Long Division

In polynomial division, it is important to write the terms in descending order of degree. If there is a missing term, then we write that term with a 0 coefficient as a place marker.

EXAMPLE 4 Divide. $\frac{9x^4 - 5 - 7x^2 - 10x}{3x - 1}$

Solution First, we write the dividend in descending order of degree. Because the degree-3 term is missing, we write it with a 0 coefficient as a place holder.

$$3x - 1\overline{)9x^4 + 0x^3 - 7x^2 - 10x - 5}$$

Now we divide as we did in Examples 2 and 3. We show the completed long division next.

$$\begin{array}{r} 3x^3 + x^2 - 2x - 4 \\ 3x - 1\overline{)9x^4 + 0x^3 - 7x^2 - 10x - 5} \\ \underline{-9x^4 + 3x^3} \qquad\qquad\qquad \\ 3x^3 - 7x^2 \qquad\qquad \\ \underline{-3x^3 + x^2} \qquad\qquad \\ -6x^2 - 10x \qquad \\ \underline{+6x^2 - 2x} \qquad \\ -12x - 5 \\ \underline{+12x - 4} \\ -9 \end{array}$$

Note: For simplicity, we will use this second approach to write negative remainders.

Answer $3x^3 + x^2 - 2x - 4 + \frac{-9}{3x - 1}$ or $3x^3 + x^2 - 2x - 4 - \frac{9}{3x - 1}$

Your Turn

Divide. $\dfrac{-12x + 8x^3 - 8}{2x + 1}$

Answer

$4x^2 - 2x - 5 - \dfrac{3}{2x + 1}$

OBJECTIVE 3. Divide polynomial functions. We can find the quotient of two functions using the following rule.

RULE **Dividing Functions**

The quotient of two functions, f/g, is found by $(f/g)(x) = \dfrac{f(x)}{g(x)}$, where $g(x) \neq 0$.

Example 5 a. If $f(x) = x^3 - x + 7$ and $g(x) = x + 2$, find $(f/g)(x)$.

b. Find $(f/g)(4)$.

Solution a.

$$\begin{array}{r} x^2 - 2x + 3 \\ x + 2\overline{)x^3 + 0x^2 - x + 7} \\ \underline{-x^3 - 2x^2 } \\ -2x^2 - x \\ \underline{2x^2 + 4x } \\ 3x + 7 \\ \underline{-3x - 6} \\ 1 \end{array}$$

Answer $(f/g)(x) = x^2 - 2x + 3 + \dfrac{1}{x + 2}$

b. $(f/g)(4) = (4)^2 - 2(4) + 3 + \dfrac{1}{4 + 2}$ In $(f/g)(x)$, replace x with 4.

$= 16 - 8 + 3 + \dfrac{1}{6}$ Simplify.

$= 11\dfrac{1}{6}$, or $\dfrac{67}{6}$

Your Turn

a. If $f(x) = 6x^3 - 13x^2 - 1$ and $g(x) = 2x + 1$, find $(f/g)(x)$.

b. Find $(f/g)(-1)$.

Answers

a. $(f/g)(x) = 3x^2 - 8x + 4 - \dfrac{5}{2x + 1}$

b. 20

5.4 Exercises

For Extra Help

MyMathLab

Videotape/DVT

InterAct Math

Math Tutor Center

Math XL.com

1. Explain how to divide a polynomial by a monomial.

2. Explain why dividing two exponential forms with the same base and the same exponent, such as $\frac{x^2}{x^2}$, results in a quotient of 1.

3. To divide $6x^2 + 8x + 7$ by $2x + 1$ using long division, explain how to determine the first term in the quotient.

4. When dividing polynomials using long division, after determining the first term in the quotient, what is the next step?

5. After performing long division with a divisor of $6x - 1$, a quotient of $3x^2 - 5x + 2$ is found with a remainder of 7. How do you write the final expression with the quotient and remainder?

6. When do you use a zero placeholder in the process of dividing two polynomials by long division?

For Exercises 7–20, divide the polynomial by the monomial.

7. $\dfrac{28a^4 - 7a^3 + 21a^2}{7a^2}$

8. $\dfrac{8k^3 - 4k^2 + 2k}{2k}$

9. $(12u^5 - 6u^4 - 15u^3 + 3u^2) \div 3u^2$

10. $(15m^5 - 10m^4 + 20m^3 - 5m^2) \div 5m^2$

11. $\dfrac{24a^5b^4 - 8a^4b^2 + 16a^2b}{8ab}$

12. $\dfrac{12m^4n^2 - 4m^3n + 8m^2}{4mn}$

13. $\dfrac{36u^3v^4 + 12uv^5 - 15u^2v^2}{3uv^9}$

14. $\dfrac{16hk^4 - 25hk - 4h^2k^2}{4hk}$

15. $\dfrac{30x^3y^2 - 45xy^3 - xy}{-5xy}$

16. $\dfrac{18a^5b^2 - 24a^4b + 2a^3b}{-6a^2b}$

17. $(12t^6u^5v - 8t^4u^4 - 16t^3u^2 + 3tu^2) \div (-4t^3u^2)$

18. $(40x^7y^4z - 4x^4y^5 - 16x^3y^4 + 5xy^2) \div (-8x^2y^4)$

19. $\dfrac{18a^3b^2c^2 + 12a^2b^3c - 3a^2b}{9a^2bc}$

20. $\dfrac{4m^4n^3p^2 - 9m^2n^3p + 6m^2p}{12m^2np}$

21. The area of the parallelogram shown is described by the monomial $25xy^2$. Find the height.

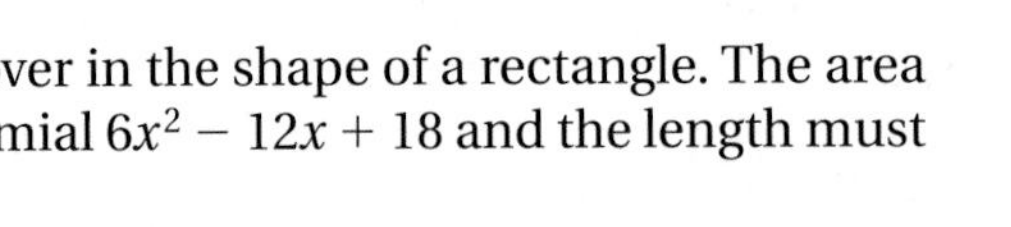

22. The area of the triangle shown is described by the monomial $30m^2n$. Find the base.

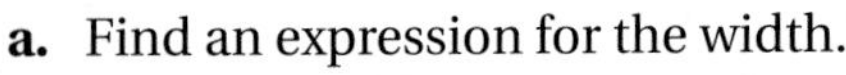

23. A pool company is designing a pool cover in the shape of a rectangle. The area of the cover is described by the polynomial $6x^2 - 12x + 18$ and the length must be $3x$.

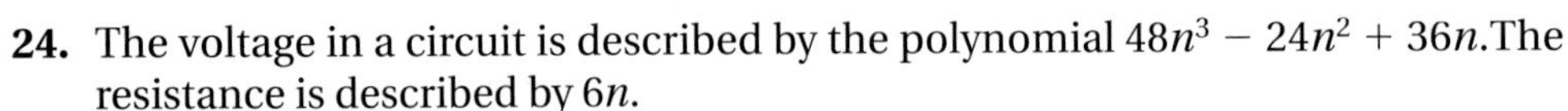

a. Find an expression for the width.

b. Find the area, length, and width if $x = 5$.

24. The voltage in a circuit is described by the polynomial $48n^3 - 24n^2 + 36n$.The resistance is described by $6n$.

a. Find an expression for the current (voltage = current $\times$ resistance).

b. Find the voltage, current, and resistance if $n = 2$.

For Exercises 25–54, use long division to divide the polynomials.

25. $\dfrac{x^2 + 9x + 20}{x + 4}$ **26.** $\dfrac{c^2 - 11c + 24}{c - 8}$ **27.** $\dfrac{3p^2 + 4p + 1}{p + 1}$

28. $\dfrac{3m^2 - 17m + 10}{m - 5}$ **29.** $(5n^2 - 3n - 8) \div (n + 1)$ **30.** $(3q^2 - 10q + 3) \div (q - 3)$

31. $\dfrac{15x^2 + 2x - 8}{3x - 2}$ **32.** $\dfrac{4x^2 + 4x - 3}{2x - 1}$ **33.** $\dfrac{6y^2 + 5y - 4}{3y + 4}$

34. $\dfrac{-13r + 1 + 6r^2}{3r - 2}$ **35.** $\dfrac{2x^3 - 13x^2 + 27x - 18}{2x - 3}$ **36.** $\dfrac{3m^3 + 10m^2 + 9m + 2}{3m + 1}$

37. $\dfrac{x^3 + 8}{x + 2}$ **38.** $\dfrac{a^3 + 125}{a + 5}$ **39.** $\dfrac{y^4 - 16}{y - 2}$

40. $\dfrac{q^4 - 81}{q + 3}$ **41.** $\dfrac{8z^3 + 125}{2z + 5}$ **42.** $\dfrac{3b^3 - 81}{b - 3}$

43. $(2v^3 + 7v^2 + 7v + 2) \div (2v + 1)$ **44.** $(2w^3 + 8w^2 + 2w - 12) \div (2w + 4)$ **45.** $\dfrac{-20a^2 + 13a + 21a^3 - 6}{3a - 2}$

46. $\dfrac{14w - 19w^2 + 21w^3 + 24}{3w + 2}$ **47.** $\dfrac{13q + 12q^3 + 4q^2 + 9}{2q - 1}$ **48.** $\dfrac{4s^2 - 13s + 9 + 12s^3}{2s - 1}$

49. $\dfrac{3x^4 - 6x^3 + 17x^2 - 24x + 20}{x^2 + 4}$

50. $\dfrac{2x^4 + 7x^3 - 14x^2 - 21x + 24}{x^2 - 3}$

51. $\dfrac{-10c - 8 + 3c^4 - 4c^3 - c^2}{3c + 2}$

52. $\dfrac{2x^4 - 28 - x + x^3 - 19x^2}{2x + 7}$

53. $\dfrac{-u^2 + 5u^3 + 10u + 2}{u + 2}$

54. $\dfrac{6w^4 - 11w^3 - 8 - w^2 + 16w}{2w - 3}$

55. The volume of a rectangular spa is described by $(6t^3 + 30t^2 + 12t - 48)$ cubic feet. The width is to be 6 feet and the height (depth) is described by $(t + 2)$ feet.

a. Find an expression for the length of the spa.

b. Find the length, height, and volume of the spa if $t = 2$.

56. On a blueprint, the specifications for a rectangular room call for the area to be $(3y^4 + 11y^3 + 22y^2 + 23y + 5)$ square feet. It is decided that the length should be described by the binomial $(3y + 5)$ feet.

a. Find an expression for the width.

b. Find the length, width, and area of the room if $y = 2$.

57. If $f(x) = 24x^4 - 42x^3 - 30x^2$ and $g(x) = 6x^2$, find

a. $(f/g)(x)$ **b.** $(f/g)(4)$ **c.** $(f/g)(-2)$

58. If $r(x) = 54x^6 - 12x^4 - 27x^3$ and $t(x) = -9x^3$, find

a. $(r/t)(x)$ **b.** $(r/t)(3)$ **c.** $(r/t)(-1)$

59. If $h(x) = 6x^3 + 7x^2 - 9x + 2$ and $k(x) = 2x - 1$, find

a. $(h/k)(x)$ **b.** $(h/k)(3)$ **c.** $(h/k)(-1)$

60. If $f(x) = 18x^3 - 39x^2 + 30x - 8$ and $g(x) = 3x - 2$, find

a. $(f/g)(x)$ **b.** $(f/g)(5)$ **c.** $(f/g)(-4)$

61. If $n(x) = x^3 - x^2 - 7x + 4$ and $p(x) = x - 3$, find

a. $(n/p)(x)$ **b.** $(n/p)(4)$ **c.** $(n/p)(-2)$

62. If $c(x) = 2x^4 + 11x^3 - 40x - 20$ and $d(x) = x + 4$, find

a. $(c/d)(x)$ **b.** $(c/d)(2)$ **c.** $(c/d)(-2)$

Collaborative Exercises DEMO CD

A band decides to produce their own demo CD. They rent recording equipment for \$200 and purchase the CDs for 50 cents each.

1. Write a function $C(x)$ that represents the total cost to produce x CDs.

2. Find the cost to produce 50 CDs.

3. Divide your answer to Problem 2 by 50. The result will be the average cost to produce each CD if 50 CDs are made.

4. Find the average cost of producing 100 CDs.

5. Write a function that can be used to calculate the average cost of producing x CDs.

6. As x gets larger, what happens to the average cost?

REVIEW EXERCISES

For Exercises 1 and 2, solve.

[2.1] **1.** $9x - (6x + 2) = 5x - 2[4x - (6 - x)]$

[2.3] **2.** $7x - 5 \geq 8 - (2x + 1)$

[3.3] **3.** Find the equation of a line perpendicular to the graph of $5x - 2y = 10$ and passing through the point $(3, -1)$.

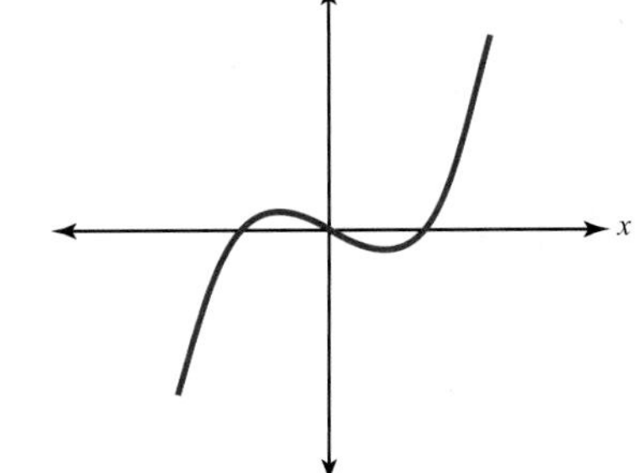

[3.5] **4.** Is the relation graphed to the right a function? Explain.

[4.2] **5.** Solve the following system using elimination: $\begin{cases} 2x + 3y - z = 7 \\ x + y - 5z = -14 \\ -3x - 5y + z = -12 \end{cases}$

[4.6] **6.** Solve the system of inequalities: $\begin{cases} x - y > 3 \\ 2x + y \leq 4 \end{cases}$

5.5 Synthetic Division and the Remainder Theorem

OBJECTIVES

1. **Use synthetic division to divide a polynomial by a binomial in the form $x - c$.**
2. **Use the remainder theorem to evaluate polynomials.**

OBJECTIVE 1. Use synthetic division to divide a polynomial by a binomial in the form $x - c$. We saw in Section 5.4 that long division can be a tedious process. When a polynomial is divided by a binomial in the form $x - c$, where c is a constant, we can streamline the process by focusing on just the coefficients in the polynomials using a method called *synthetic division*.

To get a sense of how synthetic division looks, let's look again at the long division for $\frac{x^2 + 5x + 7}{x + 2}$, which was Example 2 in Section 5.4. To the right of that long division we gradually remove bits so that we are left with the bare essentials for finding the quotient.

Long Division	Variables Removed	Bare Essentials
$x + 3$	$1 + 3$	$1 + 3$
$x + 2\overline{)x^2 + 5x + 7}$	$1 + 2\overline{)1 + 5 + 7}$	$1 - (-2)\overline{)1 + 5 + 7}$
$-x^2 - 2x$	$-1 - 2$	$- 2$
$3x + 7$	$3 + 7$	$+7$
$-3x - 6$	$-3 - 6$	-6
1	1	1

Note: *In polynomial long division, the terms in the positions that we've highlighted in blue are always additive inverses. Since they'll always cancel, we can remove them to get the "bare-essentials" version, as seen to the right.*

Note: *We have rewritten $x + 2$ as $x - (-2)$ so that we have the form $x - c$.*

We can place these "bare-essential" values in an even more compact form, which is the form for synthetic division.

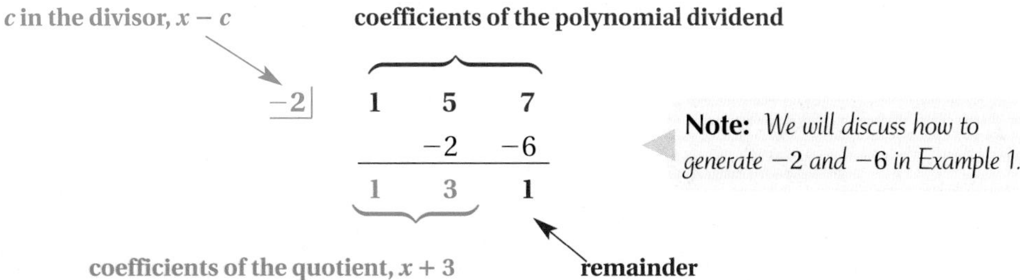

Let's now examine the steps of synthetic division in Example 1.

EXAMPLE 1 Use synthetic division to divide.

a. $\dfrac{x^2 + 5x + 7}{x + 2}$

Solution First, we need the divisor to be in the form $x - c$, so we rewrite $x + 2$ as $x - (-2)$ and we see that $c = -2$. We then set up the synthetic division with c and the coefficients of the dividend, as shown.

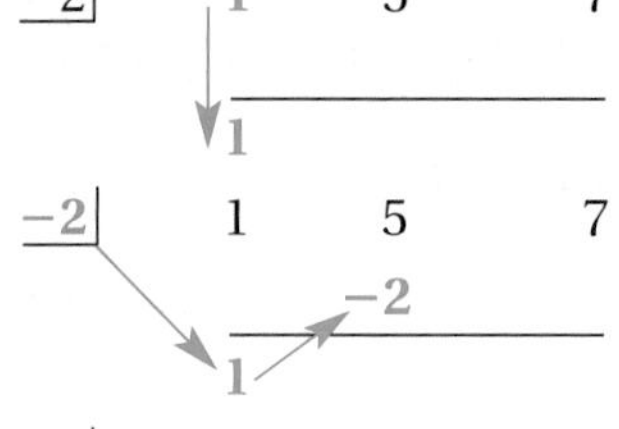

Our first step is to bring down the leading coefficient.

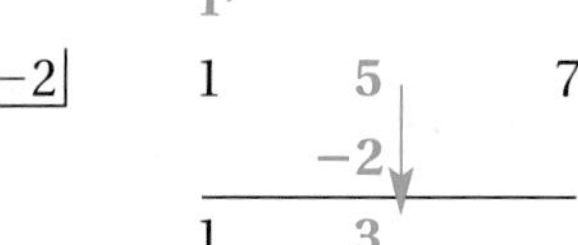

Multiply 1 by −2 and place the result underneath the next coefficient, 5.

Combine: $5 - 2 = 3$.

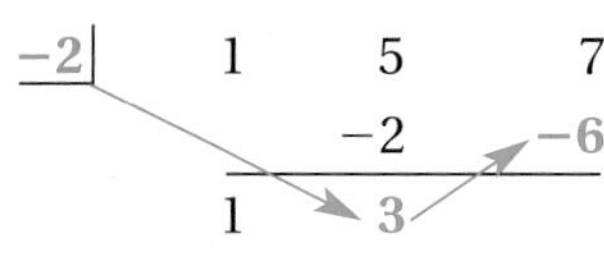

Multiply 3 by −2 and place the result underneath the next coefficient, 7.

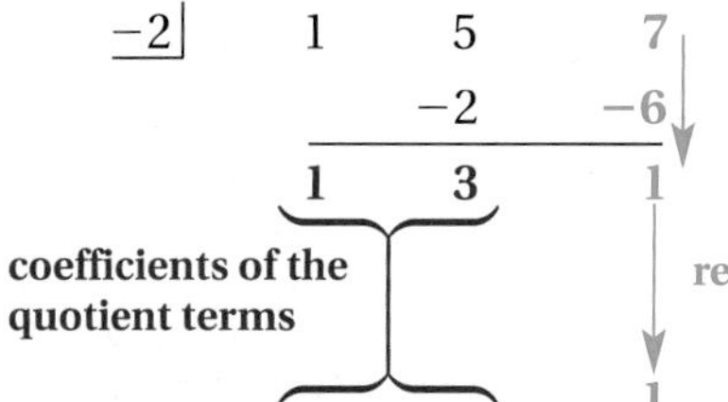

Combine: $7 - 6 = 1$, which is the remainder.

coefficients of the quotient terms

remainder

Answer: $1x + 3 + \dfrac{1}{x + 2}$ or $x + 3 + \dfrac{1}{x + 2}$

Note: *In synthetic division, the degree of the quotient polynomial is always one less than the degree of the dividend polynomial. Since the degree of the dividend $x^2 + 5x + 7$ is 2, the degree of the quotient, $x + 3$, is 1.*

b. $(x^4 - 6x^3 + 35x - 17) \div (x - 4)$

Solution Looking at the divisor, $x - 4$, we see that $c = 4$.

Note: *Since the polynomial dividend has no x^2 term, we place a 0 in the degree 2 position as a place holder.*

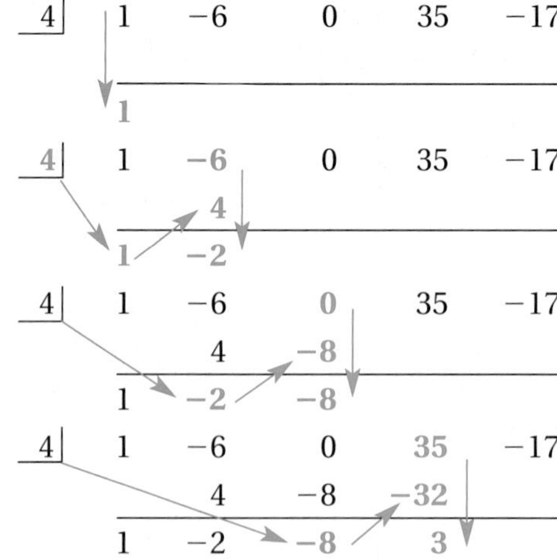

List the coefficients in the dividend and bring down the leading coefficient.

Multiply 1 by 4 and place the result under the next coefficient, −6. Then combine.

Multiply −2 by 4 and place the result underneath the next coefficient, 0. Then combine.

Multiply −8 by 4 and place the result underneath the next coefficient, 35. Then combine.

4	1	−6	0	35	−17
		4	−8	−32	12
	1	−2	−8	3	−5

Multiply 3 by 4 and place the result underneath the next coefficient, −17. Then combine.

Answer: $x^3 - 2x^2 - 8x + 3 - \frac{5}{x - 4}$

Your Turn Use synthetic division to divide.

a. $\frac{4x^3 + 6x^2 - 9x + 27}{x + 3}$ **b.** $(x^4 + x^3 - 26x^2 + 27x - 15) \div (x - 4)$

Objective 2. Use the remainder theorem to evaluate polynomials. There is a helpful connection between evaluating a polynomial using a value c and dividing that polynomial by $x - c$. In Example 1 we used synthetic division to divide $x^2 + 5x + 7$ by $x + 2$, which means $c = -2$, and we found the remainder to be 1. If we evaluate the polynomial by substituting -2 for x, we will get that remainder value, 1.

Answers

a. $4x^2 - 6x + 9$

b. $x^3 + 5x^2 - 6x + 3 - \frac{3}{x - 4}$

$$(-2)^2 + 5(-2) + 7 = 4 - 10 + 7 = 1$$

This suggests the following rule, which is called the *remainder theorem*. The proof of this theorem is beyond the scope of this text.

Rule The Remainder Theorem

Given a polynomial $P(x)$, the remainder of $\frac{P(x)}{x - c}$ is equal to $P(c)$.

Example 2 For $P(x) = x^5 - 8x^3 - 5x^2 + x + 11$, use the remainder theorem to find $P(3)$.

Solution To use the remainder theorem to find $P(3)$, we need to divide $P(x)$ by $x - 3$. We will use synthetic division.

Note: *Using synthetic division and the remainder theorem avoids the big numbers that we often get when evaluating exponential terms.*

3	1	0	−8	−5	1	11
		3	9	3	−6	−15
	1	3	1	−2	−5	−4

The remainder is -4, so $P(3) = -4$.

We can verify our answer by finding $P(3)$ using substitution.

$$P(3) = (3)^5 - 8(3)^3 - 5(3)^2 + (3) + 11 = 243 - 216 - 45 + 3 + 11 = -4$$

Your Turn For $P(x) = x^4 + 4x^3 + 23x - 2$, use the remainder theorem to find $P(-5)$.

Answer

8

5.5 Exercises

For Extra Help

MyMathLab

Videotape/DVT

InterAct Math

Math Tutor Center

Math XL.com

1. What form does the divisor need to have in order to use synthetic division?

2. Suppose you are to use synthetic division to divide $\dfrac{x^3 - 8x^2 - 5x + 9}{x + 6}$. What is the value of c?

3. If you were to use synthetic division for $\dfrac{x^4 - 5x^2 - 3x + 11}{x - 2}$, what list of coefficients would you write?

4. Explain how to finish the synthetic division shown.

$$\begin{array}{r|rrrr} 2 & 1 & 4 & -8 & -9 \\ & & 2 & 12 & \\ \hline & 1 & 6 & 4 & \end{array}$$

5. Where does the remainder appear in synthetic division?

6. When using synthetic division, how do you determine the degree of the first term in the quotient expression?

7. In your own words, explain what the remainder theorem says.

8. What is the advantage of using the remainder theorem as opposed to substitution when evaluating a polynomial?

For Exercises 9–14: ***a. Write the binomial divisor.***
b. Write the polynomial dividend.
c. Write the polynomial quotient with its remainder (if there is a remainder).

9.
$$\begin{array}{r|rrr} -3 & 1 & 7 & 12 \\ & & -3 & -12 \\ \hline & 1 & 4 & 0 \end{array}$$

10.
$$\begin{array}{r|rrr} 9 & 1 & -14 & 45 \\ & & 9 & -45 \\ \hline & 1 & -5 & 0 \end{array}$$

11.
$$\begin{array}{r|rrrr} 3 & 1 & 4 & -25 & 7 \\ & & 3 & 21 & -12 \\ \hline & 1 & 7 & -4 & -5 \end{array}$$

12.

-2	1	-2	-6	7
		-2	8	-4
	1	-4	2	3

13.

4	2	-8	-5	17	10
		8	0	-20	-12
	2	0	-5	-3	-2

14.

-5	3	16	0	-27	-9
		-15	-5	25	10
	3	1	-5	-2	1

For Exercises 15–40, divide using synthetic division.

15. $\dfrac{x^2 - 5x + 6}{x - 3}$

16. $\dfrac{x^2 + 8x - 9}{x - 1}$

17. $\dfrac{2x^2 - x - 5}{x - 4}$

18. $\dfrac{3x^2 - 11x - 7}{x - 5}$

19. $\dfrac{x^3 - x^2 - 5x + 6}{x - 2}$

20. $\dfrac{2x^3 - 6x^2 - 5x + 15}{x - 3}$

21. $\dfrac{3x^3 + 10x^2 + 9x + 2}{x + 2}$

22. $\dfrac{5x^3 + 12x^2 - 36x - 16}{x - 2}$

23. $(3x^3 - x^2 - 22x + 24) \div (x + 3)$

24. $(8x^3 - 10x^2 - x + 3) \div (x - 1)$

25. $\dfrac{2x^3 - 5x^2 - x + 6}{x + 2}$

26. $\dfrac{3x^3 - 5x^2 - 16x + 12}{x - 2}$

27. $(x^3 - 2x^2 + x - 3) \div (x - 2)$

28. $(x^3 + 5x^2 - 4x - 3) \div (x + 6)$

29. $\dfrac{2x^3 + x^2 - 4}{x + 3}$

30. $\dfrac{4x^3 - 2x + 5}{x - 2}$

31. $\dfrac{3x^3 - 4x^2 + 2}{x - 3}$

32. $\dfrac{x^3 + 4x^2 - 7}{x + 3}$

33. $(x^3 - 7x + 6) \div (x - 2)$

34. $(x^3 + 2x + 135) \div (x + 5)$

35. $\dfrac{x^3 + 27}{x - 3}$

36. $\dfrac{x^3 - 8}{x + 2}$

37. $\dfrac{x^4 + 16}{x + 2}$

38. $\dfrac{x^4 + 81}{x - 3}$

39. $\dfrac{6x^3 - x + 1}{x + \frac{1}{3}}$

40. $\dfrac{2x^3 - 13x^2 + 27x - 18}{x - \frac{3}{2}}$

For Exercises 41–52, use the remainder theorem to find P(c).

41. For $P(x) = 3x^3 - 8x^2 + 8x - 3$, find $P(2)$.

42. For $P(x) = 8x^3 + 2x^2 - 13x + 3$, find $P(1)$.

43. For $P(x) = 6x^3 - 13x^2 + x + 2$, find $P(2)$.

44. For $P(x) = x^3 + x^2 - 2x + 12$, find $P(-3)$.

45. For $P(x) = 3x^3 - 11x - 7$, find $P(2)$.

46. For $P(x) = 2x^3 - 26x - 24$, find $P(-3)$.

47. For $P(x) = x^4 - x^3 - 2x^2 + x - 2$, find $P(1)$.

48. For $P(x) = x^4 + 3x^2 - 15x + 7$, find $P(2)$.

49. For $P(x) = 2x^4 - 6x^2 - 5x + 4$, find $P(2)$.

50. For $P(x) = 3x^4 + x^3 + 5x - 1$, find $P(-1)$.

51. For $P(x) = x^5 - x^4 + 3x^2 - 3x - 3$, find $P(1)$.

52. For $P(x) = x^5 + 5x^4 - x - 6$, find $P(-5)$.

53. The area of the parallelogram shown is described by $9x^2 - 10x - 16$.

a. Find an expression for the length of the base.

b. Find the base, height, and area of the parallelogram if $x = 12$ inches.

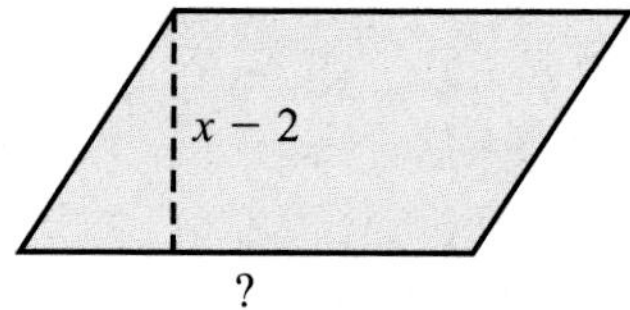

54. The area of the rectangle shown is $12n^2 + 65n - 42$.

a. Find the length.

b. Find the length, width, and area of the rectangle if $n = 9$ centimeters.

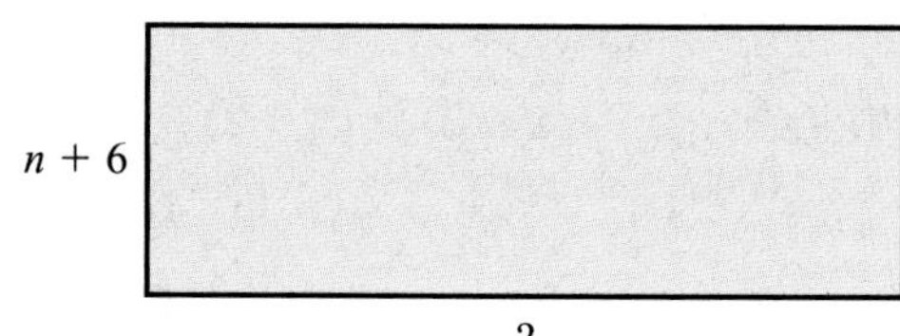

55. The volume of the walk-in freezer shown is described by $2y^3 - 7y^2 - 38y + 88$.

a. Find an expression for the height of the room.

b. Find the length, width, height, and volume of the room if $x = 10$ feet.

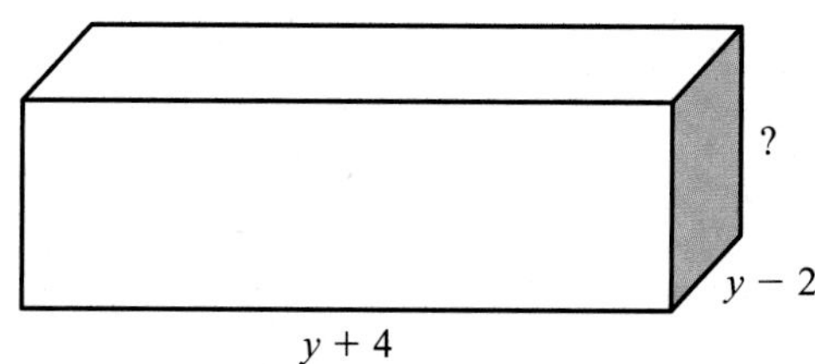

56. The volume of the storage box shown is described by $2x^3 - 168x - 320$.

a. Find an expression for the length of the box.

b. Find the length, width, height, and volume of the box if $x = 16$ inches.

REVIEW EXERCISES

[1.1] **1.** Write a set containing the first five prime numbers.

For Exercises 2 and 3, solve. a. Write the solution set in set-builder notation.
b. Write the solution set in interval notation.
c. Graph the solutions set.

[2.4] **2.** $2x - 3 < 1$ or $2x - 3 > -5$

[2.6] **3.** $|3x - 1| \leq 8$

[3.2] **4.** Find the slope of a line passing through $(3, -5)$ and $(-2, 5)$.

[3.4] **5.** Graph $y > 2$.

[4.3] **6.** The perimeter of an enclosure is to be 100 feet and the length needs to be twice the width. Write a system of equations describing the situation, then solve the system to find the length and width.

Chapter 5 Summary

Defined Terms

Section 5.1
Scientific notation (p. 317)

Section 5.2
Monomial (p. 326)
Coefficient (p. 326)
Degree of a monomial (p. 326)
Polynomial (p. 327)
Polynomial in one variable (p. 327)
Binomial (p. 327)
Trinomial (p. 327)
Degree of a polynomial (p. 327)
Polynomial function (p. 331)
Constant function (p. 331)
Linear function (p. 331)
Quadratic function (p. 332)
Cubic function (p. 332)

Section 5.3
Conjugates (p. 346)

Procedures, Rules, and Key Examples

Procedures/Rules	Key Examples
Section 5.1 Exponents and Scientific Notation	
If n is a natural number and a is a real number, then $a^n = \underbrace{a \cdot a \cdot \cdots \cdot a}_{n \text{ factors of } a}$.	**Example 1:** Evaluate. **a.** $(0.2)^3 = (0.2)(0.2)(0.2) = 0.008$
If $a < 0$ and n is even, then a^n is positive. If $a < 0$ and n is odd, then a^n is negative.	**b.** $(-2)^4 = (-2)(-2)(-2)(-2) = 16$ **c.** $(-2)^5 = (-2)(-2)(-2)(-2)(-2) = -32$
If a and b are real numbers with $b \neq 0$ and n is a whole number, then $\left(\frac{a}{b}\right)^n = \frac{a^n}{b^n}$.	**d.** $\left(\frac{2}{5}\right)^3 = \frac{2^3}{5^3} = \frac{8}{125}$
If a is a real number and $a \neq 0$, then $a^0 = 1$.	**e.** $5^0 = 1$
If a is a real number where $a \neq 0$ and n is a natural number, then $a^{-n} = \frac{1}{a^n}$.	**f.** $3^{-4} = \frac{1}{3^4} = \frac{1}{81}$
If a is a real number where $a \neq 0$ and n is a natural number, then $\frac{1}{a^{-n}} = a^n$.	**g.** $\frac{1}{4^{-3}} = 4^3 = 64$
If a and b are real numbers where $a \neq 0$ and $b \neq 0$ and n is a natural number, then $\left(\frac{a}{b}\right)^{-n} = \left(\frac{b}{a}\right)^n$.	**h.** $\left(\frac{2}{3}\right)^{-4} = \left(\frac{3}{2}\right)^4 = \frac{81}{16}$

continued

Procedures/Rules	Key Examples
Section 5.1 Exponents and Scientific Notation (Continued) If a is a real number and m and n are natural numbers, then $a^m \cdot a^n = a^{m+n}$.	**Example 2:** Simplify. **a.** $x^3 \cdot x^4 = x^{3+4} = x^7$
If m and n are integers and a is a real number where $a \neq 0$, then $\frac{a^m}{a^n} = a^{m-n}$.	**b.** $\frac{y^6}{y^2} = y^{6-2} = y^4$
If a is a real number and m and n are integers, then $(a^m)^n = a^{mn}$.	**c.** $(t^3)^4 = t^{3 \cdot 4} = t^{12}$
If a and b are real numbers and n is an integer, then $(ab)^n = a^n b^n$.	**d.** $(5mn^2)^3 = 5^{1 \cdot 3}m^{1 \cdot 3}n^{2 \cdot 3} = 125m^3n^6$
To change a number from scientific notation, $a \times 10^n$, where $1 \leq \lvert a \rvert < 10$ and n is an integer, if $n > 0$, move the decimal point to the right n places. If $n < 0$, move the decimal point to the left $\lvert n \rvert$ places.	**Example 3:** Write in standard form. **a.** $4.8 \times 10^6 = 4{,}800{,}000$ **b.** $3.1 \times 10^{-4} = 0.00031$
To write a number in scientific notation, **1.** Locate the new decimal position, which will be to the right of the first nonzero digit in the number. **2.** Determine the power of 10. **a.** If the number's absolute value is greater than 1, the power is the number of digits between the old decimal position and new decimal position expressed as a positive power. **b.** If the number's absolute value is less than 1, the power is the number of digits between the old decimal position and new decimal position expressed as a negative power. **3.** Delete unnecessary 0s. **a.** If the number's absolute value is greater than 1, delete the zeroes to the right of the last nonzero digit. **b.** If the number's absolute value is less than 1, delete the zeroes to the left of the first nonzero digit.	**Example 4:** Write in scientific notation. **a.** $943{,}000 = 9.43 \times 10^5$ **b.** $0.00000082 = 8.2 \times 10^{-7}$
Section 5.2 Polynomials and Polynomial Functions To add polynomials, combine like terms.	**Example 1:** Add. $(5x^3 + 12x^2 - 9x + 1) + (7x^2 - x - 13)$ $= 5x^3 + 12x^2 + 7x^2 - 9x - x + 1 - 13$ $= 5x^3 + 19x^2 - 10x - 12$
To subtract polynomials, **1.** Write the subtraction statement as an equivalent addition statement. **a.** Change the operation symbol from a minus sign to a plus sign. **b.** Change the subtrahend (second polynomial) to its additive inverse. To get the additive inverse, we change the sign of each term in the polynomial. **2.** Combine like terms.	**Example 2:** Subtract. $(6y^3 - 5y + 18) - (y^3 - 5y + 7)$ $= (6y^3 - 5y + 18) + (-y^3 + 5y - 7)$ $= 6y^3 - y^3 - 5y + 5y + 18 - 7$ $= 5y^3 + 11$

continued

Procedures/Rules

Key Examples

Section 5.2 Polynomials and Polynomial Functions (Continued)

The graph of a constant function, $f(x) = c$, is a horizontal line through $(0, c)$.

Graphs of linear functions, which have the form $f(x) = mx + b$, are lines with slope m and y-intercept $(0, b)$.

Graphs of quadratic functions, which have the form $f(x) = ax^2 + bx + c$, are parabolas with y-intercept at $(0, c)$.

Graphs of cubic functions, which have the form $f(x) = ax^3 + bx^2 + cx + d$, resemble an S-shape with y-intercept at $(0, d)$.

Example 3: Graph $f(x) = x^2 - 1$.

x	y
−2	3
−1	0
0	−1
1	0
2	3

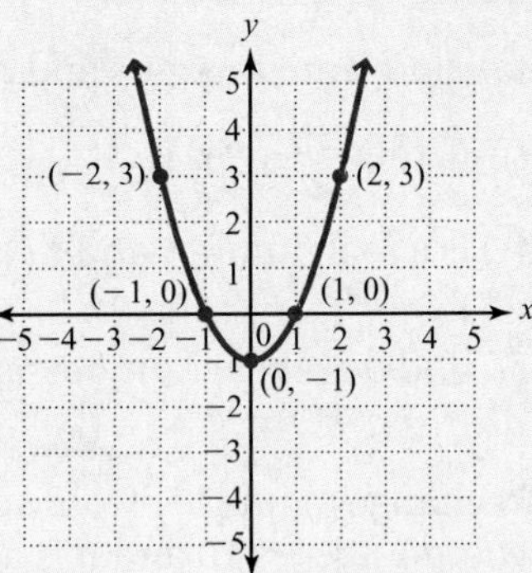

The sum of two functions, $f + g$, is found by $(f + g)(x) = f(x) + g(x)$.

The difference of two functions, $f - g$, is found by $(f - g)(x) = f(x) - g(x)$.

Example 4: If $f(x) = 5x^2 - 2x + 4$ and $g(x) = 3x^2 - 8x - 12$, find $(f + g)(x)$ and $(f - g)(x)$.

$$(f + g)(x) = (5x^2 - 2x + 4) + (3x^2 - 8x - 12)$$
$$= 8x^2 - 10x - 8$$
$$(f - g)(x) = (5x^2 - 2x + 4) - (3x^2 - 8x - 12)$$
$$= (5x^2 - 2x + 4) + (-3x^2 + 8x + 12)$$
$$= 2x^2 + 6x + 16$$

Section 5.3 Multiplying Polynomials

To multiply a polynomial by a monomial, use the distributive property to multiply each term in the polynomial by the monomial.

To multiply two polynomials,

1. Multiply every term in the second polynomial by every term in the first polynomial.
2. Combine like terms.

Special Products:
If a and b are real numbers, variables, or expressions, then

$$(a + b)^2 = a^2 + 2ab + b^2$$
$$(a - b)^2 = a^2 - 2ab + b^2$$
$$(a + b)(a - b) = a^2 - b^2$$

Example 1: Multiply.

a. $5xy(7x^2 + 3x - 6)$
$= 5xy \cdot 7x^2 + 5xy \cdot 3x + 5xy \cdot (-6)$
$= 35x^3y + 15x^2y - 30xy$

b. $(4x + 7)(2x - 1)$
$= 4x \cdot 2x + 4x \cdot (-1) + 7 \cdot 2x + 7 \cdot (-1)$
$= 8x^2 - 4x + 14x - 7$
$= 8x^2 + 10x - 7$

c. $(5n + 6)^2 = 25n^2 + 60n + 36$

d. $(3x - 4y)^2 = 9x^2 - 24xy + 16y^2$

e. $(t^2 + 9)(t^2 - 9) = t^4 - 81$

continued

Procedures/Rules	Key Examples
Section 5.3 Multiplying Polynomials (Continued) The product of two functions, $\boldsymbol{f \cdot g}$, is found by $(f \cdot g)(x) = f(x)\,g(x)$.	**Example 2:** If $f(x) = x - 7$ and $g(x) = x^2 - x + 1$, find $(f \cdot g)(x)$ and $(f \cdot g)(-2)$. Solution: $(f \cdot g)(x) = (x - 7)(x^2 - x + 1)$ $= x^3 - x^2 + x - 7x^2 + 7x - 7$ $= x^3 - 8x^2 + 8x - 7$ $(f \cdot g)(-2) = (-2)^3 - 8(-2)^2 + 8(-2) - 7 = -63$
Section 5.4 Dividing Polynomials If a, b, and c are real numbers, variables, or expressions with $c \neq 0$, then $\frac{a+b}{c} = \frac{a}{c} + \frac{b}{c}$. To divide a polynomial by a monomial, divide each term in the polynomial by the monomial. To divide a polynomial by a polynomial, **1.** Use long division. **2.** If there is a remainder, write the result in the form $\text{quotient} + \frac{\text{remainder}}{\text{divisor}}$.	**Example 1:** Divide. **a.** $\frac{28x^6 + 36x^3 - 8x^2}{4x^2} = \frac{28x^6}{4x^2} + \frac{36x^3}{4x^2} - \frac{8x^2}{4x^2}$ $= 7x^4 + 9x - 2$ **b.** $\frac{2x^3 - 14x - 5}{x + 2}$ $x + 2\overline{)2x^3 + 0x^2 - 14x - 5}$ with quotient $2x^2 - 4x - 6$ $\underline{-2x^3 - 4x^2}$ $-4x^2 - 14x$ $\underline{4x^2 + 8x}$ $-6x - 5$ $\underline{6x + 12}$ 7 Answer: $2x^2 - 4x - 6 + \frac{7}{x + 2}$
The quotient of two functions, $\boldsymbol{f/g}$, is found by $(f/g)(x) = \frac{f(x)}{g(x)}$, where $g(x) \neq 0$.	**Example 2:** If $f(x) = 5x^2 - 13x - 10$ and $g(x) = x - 3$, find $(f/g)(x)$ and $(f/g)(4)$. $x - 3\overline{)5x^2 - 13x - 10}$ with quotient $5x + 2$ $\underline{-5x^2 + 15x}$ $2x - 10$ $\underline{-2x + 6}$ -4 Answers: $(f/g)(x) = 5x + 2 - \frac{4}{x - 3}$ $(f/g)(4) = 5(4) + 2 - \frac{4}{4 - 3} = 22 - 4 = 18$

continued

Procedures/Rules	Key Examples
Section 5.5 Synthetic Division and the Remainder Theorem When a polynomial is divided by a binomial in the form $x - c$, where c is a constant, we can use synthetic division.	**Example 1:** Divide using synthetic division. $\dfrac{2x^3 - 14x - 5}{x + 2}$ $\begin{array}{r\|rrrr} -2 & 2 & 0 & -14 & -5 \\ & & -4 & 8 & 12 \\ \hline & 2 & -4 & -6 & 7 \end{array}$ Answer: $2x^2 - 4x - 6 + \dfrac{7}{x + 2}$
Given a polynomial $P(x)$, the remainder of $\dfrac{P(x)}{x - c}$ is equal to $P(c)$.	For $P(x) = 5x^3 - 18x^2 + x - 6$, use the remainder theorem to find $P(4)$. $\begin{array}{r\|rrrr} 4 & 5 & -18 & 1 & -6 \\ & & 20 & 8 & 36 \\ \hline & 5 & 2 & 9 & 30 \end{array}$ Answer: $P(4) = 30$

Chapter 5 Review Exercises

For Exercises 1–6, answer true or false.

[5.1] **1.** The simplification $n^5 \cdot n^7 = n^{5+7} = n^{12}$ illustrates the product rule for exponents.

[5.5] **2.** When dividing using synthetic division, the divisor must be in the form $x - k$.

[5.2] **3.** The degree of the monomial $9x^3y$ is 3.

[5.3] **4.** FOIL can be used to multiply any two polynomials.

[5.3] **5.** Conjugates have the form $(ax - b)$ and $(cx - d)$, where b and d are positive real numbers.

[5.1] **6.** $a^0 = 0$

For Exercises 7–10, complete the rule.

[5.4] **7.** To divide a polynomial by a monomial, divide each term in the polynomial by the ________.

[5.1] **8.** If m and n are integers and a is a real number, where $a \neq 0$, then $\dfrac{a^m}{a^n} = a^?$.

[5.1] **9.** $(n^a)^b = n^?$, where n is a real number and a and b are integers.

[5.3] **10.** To multiply a polynomial by a monomial, use the ________ property to multiply each term in the polynomial by the monomial.

[5.1] ***For Exercises 11–14, evaluate the exponential expression.***

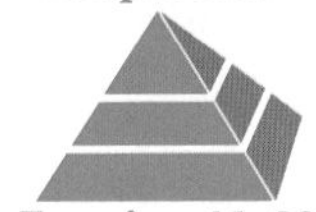

11. $\left(\dfrac{2}{3}\right)^{-3}$

12. -4^2

13. 2^{-5}

14. $14x^0$

[5.1] ***For Exercises 15 and 16, write the number in standard form.***

15. The mass of a hydrogen atom is 1.6736×10^{-24} grams.

16. It is speculated that the universe is about 1.65×10^{10} years old.

[5.1] ***For Exercises 17 and 18, write in scientific notation.***

17. The mass of a dust particle is 0.000000000753 kilograms.

18. The speed of light is 300,000,000 meters per second.

[5.1] ***For Exercises 19 and 20, simplify using the product rule of exponents.***

19. $(4x^2)(5xy^3)$

20. $\left(-\dfrac{2}{3}m^2n^4\right)(3mn^5)$

[5.1] ***For Exercises 21 and 22, simplify using the quotient rule of exponents.***

21. $\dfrac{6x^4}{2x^9}$

22. $-\dfrac{7a^3bc^9}{3abc^{15}}$

[5.1] *For Exercises 23 and 24, simplify using the power rule for exponents.*

23. $(3m^5n)^2$

24. $\dfrac{(6m^2n^3p)^{-1}}{(2m^6n^2)^3}$

[5.1] *For Exercises 25–28, use the rules for exponents to simplify.*

25. $(3j^2k^5)^2(0.1jk^3)^3$

26. $\dfrac{(2s^3t)^{-1}}{(3st)^2\,(s^4t)^8}$

27. $\dfrac{(5m^7n^3)^{-2}}{(3m)^{-1}(3m)^2}$

28. $\left(\dfrac{2a}{b^4}\right)^{-3}$

[5.1] *For Exercises 29 and 30, simplify.*

29. $(5.1 \times 10^4)(-2 \times 10^6)$

30. $\dfrac{8.12 \times 10^{-8}}{2 \times 10^{-5}}$

[5.2] *For Exercises 31–34, identify the degree of each polynomial, then indicate whether the polynomial is a monomial, binomial, trinomial, or none of these names.*

31. $-\dfrac{1}{4} + 2c^2 + 8.7c^3 + \dfrac{2}{5}c$

32. $7m^2 + 1$

33. $-8xy^4$

34. $9h^5 + 4h^3 - h^2 + 1$

[5.2] *For Exercises 35 and 36, perform the indicated operation and write the resulting polynomial in descending order.*

35. $(3c^3 + 2c^2 - c - 1) + (8c^2 + c - 10)$

36. $(y^2 + 3y + 6) - (-5y^2 + 3y - 8)$

[5.2] *For Exercises 37 and 38, add or subtract the polynomials.*

37. $(x^2y^3 - xy^2 + 2x^2y - 4xy + 7y^2 - 9) + (-3x^2y^2 + 2xy^2 + 4x^2y - xy + 6y^2 + 4)$

38. $(4hk - 8k^3) - (5kh + 3k - 2k^3)$

Expressions

Exercises 39–47

[5.2] *For Exercises 39–42, indicate whether the function is a constant function, a linear function, a quadratic function, or a cubic function.*

39.

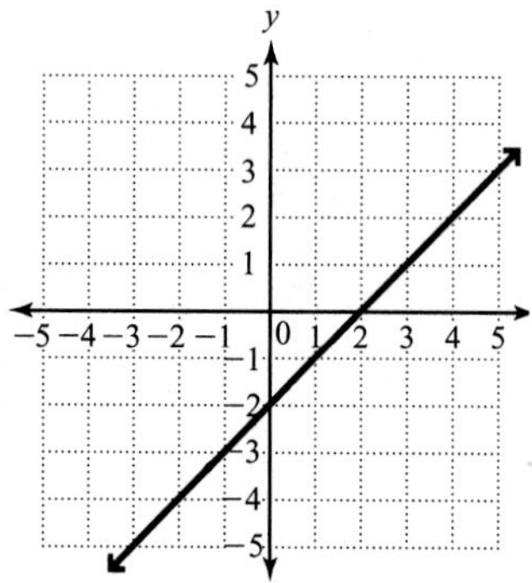

40.

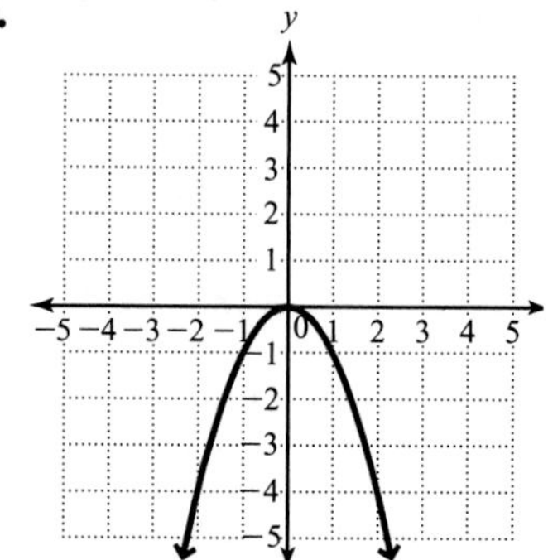

41.

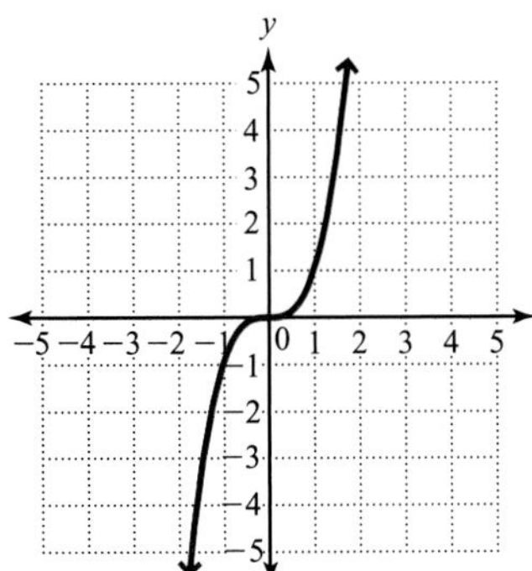

42.

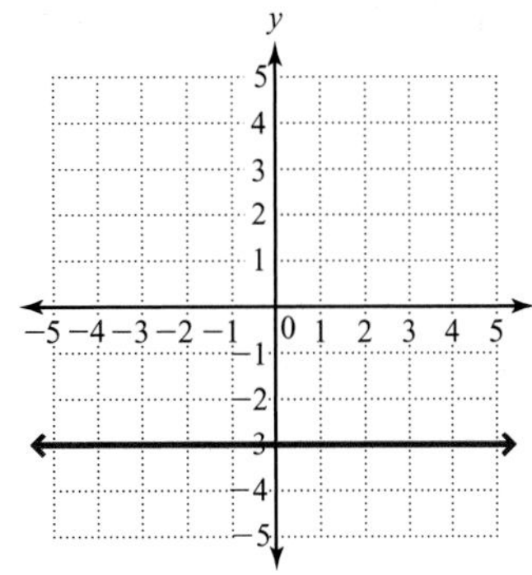

[5.2] ***For Exercises 43–46, graph.***

43. $f(x) = x^2 - 3$

44. $g(x) = -\frac{1}{3}x + 2$

45. $h(x) = x^3 - 1$

46. $w(x) = 7$

[5.2] 47. If $f(x) = x^3 + 2x + 5$ and $g(x) = -2x^3 + 2$, find

a. $(f + g)(x)$

b. $(f + g)(2)$

c. $(f - g)(x)$

d. $(f - g)(-4)$

[5.3] ***For Exercises 48–59, multiply.***

Expressions

Exercises 48–60

48. $3x^2y^3(2x^2 - 3x + 1)$

49. $-pq(2p^2 - 3pq + 3q^2)$

50. $(x - 3)(2x + 8)$

51. $(9w - 1)(3w + 2)$

52. $(3r - s)(6r + 7s)$

53. $(x^2 - 1)(x^2 + 2)$

54. $(x - 1)(2x^3 - 3x^2 + 4x + 7)$

55. $(3t^2 - t + 1)(4t^2 + t - 1)$

56. $(3a - 5)(3a + 5)$

57. $(2p - 1)^2$

58. $(8k + 3)^2$

59. $(4h^2 + 7)(4h^2 - 7)$

[5.3] 60. What is the conjugate of $(5m - 2)$?

[5.3] 61. If $f(x) = 2x + 7$ and $g(x) = x^2 + 4x - 1$, find

a. $(f \cdot g)(x)$

b. $(f \cdot g)(-1)$

[5.3] 62. If $f(x) = 4x^2 - x$, find $f(x - 4)$.

[5.4] ***For Exercises 63 and 64, divide the polynomial by the monomial.***

63. $\dfrac{20m^5 - 5m^4 + 15m^3 - 5m^2}{5m^2}$

64. $\dfrac{8x^2 - 2x + 3}{2x}$

[5.4] ***For Exercises 65–67, divide using long division.***

65. $\dfrac{x^2 + 5x + 7}{x + 2}$

66. $(x^2 - 4x + 4) \div (x - 2)$

67. $\dfrac{4x^2 + 4x - 3}{2x - 1}$

[5.4] 68. If $f(x) = 9x^4 - 7x^2 - 10x - 5$ and $g(x) = 3x - 1$, find $(f/g)(x)$.

[5.5] ***For Exercises 69–72, divide using synthetic division.***

69. $\dfrac{5x^2 - 3x + 2}{x - 1}$

70. $\dfrac{2x^3 - 6x^2 - 5x + 15}{x - 3}$

71. $(4x^3 - 2x + 5) \div (x - 2)$

72. $\dfrac{x^3 - 8}{x - 2}$

[5.5] ***For Exercises 73 and 74, use the remainder theorem to find P(c).***

73. For $P(x) = 5x^3 - x^2 + 8x + 3$, use the remainder theorem to find $P(1)$.

74. For $P(x) = 7x^3 - 2x + 1$, use the remainder theorem to find $P(-2)$.

For Exercises 75–80, solve.

[5.1] 75. The energy in joules of a single photon of light can be determined by $E = hf$, where h is a constant 6.626×10^{-34} joule-seconds and f is the frequency of the light in hertz. Find the energy of blue light with a frequency of 6.1×10^{14} Hz.

[5.2] 76. If we neglect air resistance, the polynomial function $h(x) = -16t^2 + 200$ describes the height of a falling object after falling for t seconds from an initial height of 200 feet. What is its height after 3 seconds?

[5.2] 77. a. Write a polynomial in simplest form that describes the perimeter of the shape shown.

b. Find the perimeter if $a = 12$ centimeters and $b = 9$ centimeters.

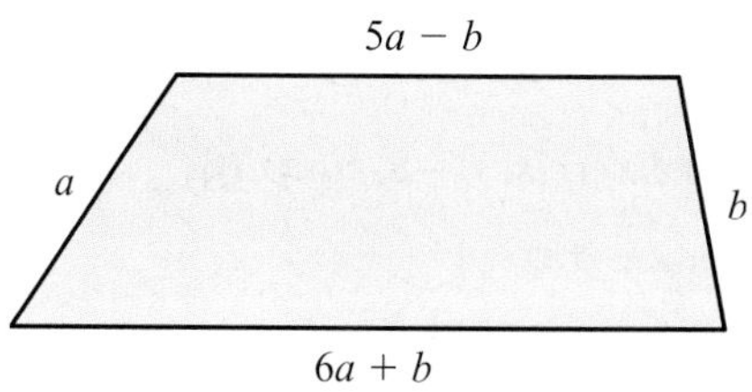

[5.2] 78. The function $R(x) = 0.2x^2 + x + 1000$ describes the revenue a software company makes from the sale of one of its programs, where x represents the number of copies sold. The function $C(x) = 10x + 2000$ describes the cost of producing the software.

a. Find a function, $P(x)$, that describes the profit.

b. Find the profit if the company sells 200 copies of the program.

[5.3] 79. A fish tank's design specifies a length that is 10 inches longer than the width. The height of the tank is to be 5 inches longer than the width. Let w represent the width and write an expression for the volume of the tank in terms of w.

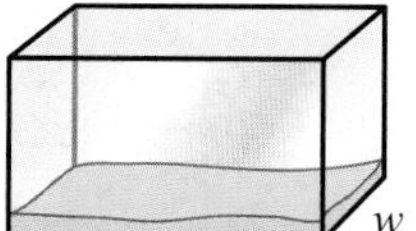

[5.4] 80. The voltage across a component in an amplifier circuit is described by the expression $(3y^4 + 11y^3 + 22y^2 + 23y + 5)$ volts. The resistance in the component is described by $(3y + 5)$ ohms.

a. Find an expression that describes the current. (The formula for voltage is voltage = current × resistance.)

b. Find the current, which is measured in amps, if $y = 3$.

Chapter 5 Practice Test

1. Write 7.2×10^{-3} in standard form.

2. Write 0.00357 in scientific notation.

For Exercises 3–8, simplify.

3. $5x^3(3x^2y)$

4. $(3x^4y)^{-2}$

5. $\dfrac{8u^7v}{2u^4v^5}$

6. $\left(-\dfrac{2}{3}t^5u^2v\right)^4$

7. $(6 \times 10^5)(2.1 \times 10^4)$

8. $\dfrac{8.4 \times 10^{10}}{4.2 \times 10^3}$

For Exercises 9 and 10, indicate whether the function is a constant function, a linear function, a quadratic function, or a cubic function.

9.

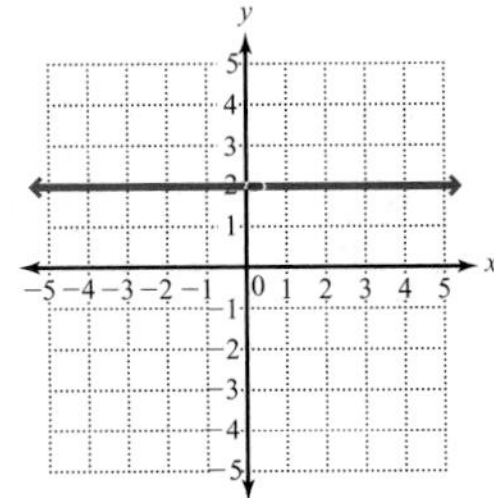

10.

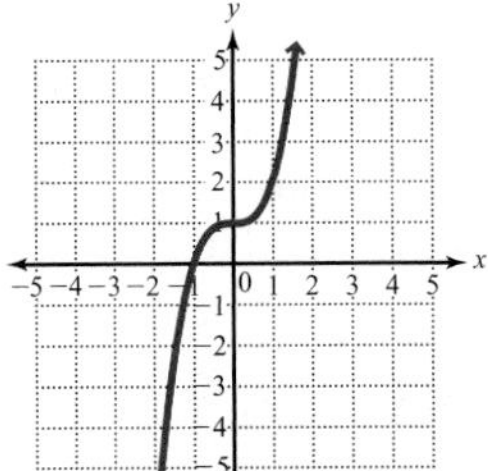

11. Graph $f(x) = x^2 - 1$.

For Exercises 12–18 perform the indicated operation and write the resulting polynomial in descending order.

12. $(a^2 - 4ab + b^2) - (a^2 + 4ab + b^2)$

13. $(6r^5 - 9r^4 - 2r^2 + 8) + (5r^5 + 2r^4 + 7r^3 - 8r^2 - r - 9)$

14. $(3x - 4y)(5x + 2y)$

15. $4m^3n^9(3m^2 - 2mn + 5n^2)$

16. $(7k - 2j)^2$

17. $(2x - 3)(7x^2 + 3x - 5)$

18. $(4h - 3)(4h + 3)$

For Exercises 19 and 20, use $f(x) = 5x - 8$ and $g(x) = 3x + 2$.

19. Find $(f \cdot g)(x)$.

20. Find $(f \cdot g)(2)$.

For Exercises 21 and 22, divide.

21. $\dfrac{8k^3 - 4k^2 + 2k}{2k}$

22. $\dfrac{3m^3 + 10m^2 + 9m + 2}{3m + 1}$

23. Divide $\dfrac{2x^3 + 5x^2 - 8}{x + 3}$ using synthetic division.

24. For $P(x) = x^3 - 5x + 4$, use the remainder theorem to find $P(-3)$.

25. A chemical company produces a respiratory medication. The function $R(x) = 0.1x^2 + x + 5000$ describes the revenue a company makes from sales, where x represents the number of vials sold. The function $C(x) = 15x + 3000$ describes the cost of producing those vials.

a. Find a function, $P(x)$, that describes the profit.

b. Find the profit if the company sells 1000 vials.

CHAPTER 6

Factoring

"We live in succession, in division, in parts, in particles."

—Ralph Waldo Emerson, U.S. essayist, poet, philosopher

In Chapter 5, we discussed a class of expressions called polynomials. We learned that polynomials can be added, subtracted, multiplied, and divided in much the same way as numbers. In this chapter, we will learn how to *factor* polynomial expressions. Factoring is one way to rewrite an expression and is one of the most important skills you will acquire in algebra. It is used extensively in simplifying the more complex expressions that we will study in future chapters. We will also see in Section 6.4 that factoring can be used to solve certain equations that contain polynomial expressions.

6.1 Greatest Common Factor and Factoring by Grouping

OBJECTIVES

1. Find the greatest common factor of a set of terms.
2. Factor a monomial GCF out of the terms of a polynomial.
3. Factor polynomials by grouping.

Often in mathematics we need a number or expression to be in **factored form.**

DEFINITION ***Factored form:*** A number or expression written as a product of factors.

For example, the following polynomials have been rewritten in factored form.

$$2x + 8 = \underbrace{2(x + 4)}_{\textbf{Factored form}} \qquad x^2 + 5x + 6 = \underbrace{(x + 2)(x + 3)}_{\textbf{Factored form}}$$

Notice that we can check an expression's factored form by multiplying the factors to see if their product is the original expression. Writing factored form is called *factoring.*

OBJECTIVE 1. Find the greatest common factor of a set of terms. When factoring polynomials, the first step is to determine if there is a monomial factor that is common to all the terms in the polynomial. Additionally, we want that monomial factor to be the **greatest common factor** of the terms.

DEFINITION ***Greatest common factor (GCF) of a set of terms:*** A monomial with the greatest coefficient and degree that evenly divides all the given terms.

For example, the greatest common factor of $12x^2$ and $18x^3$ is $6x^2$ because 6 is the greatest numerical value that evenly divides both 12 and 18 and x^2 is the highest power of x that evenly divides both x^2 and x^3. Notice that x^2 has the *smaller* exponent, which provides a clue as to how to determine the GCF. We can use the prime factorizations to help us find GCFs.

PROCEDURE **Finding the GCF**

To find the greatest common factor of two or more monomials,

1. Write the prime factorization in exponential form for each monomial. Treat variables like prime factors.
2. Write the GCF's factorization by including the prime factors (and variables) common to all the factorizations, each raised to its smallest exponent in the factorizations.
3. Multiply.

Note: If there are no common prime factors, then the GCF is 1.

Learning Strategy

If you are an auditory learner, you might find it easier to remember the procedure by thinking about the words in the procedure: To find the **greatest common** factor, you use the **common** primes raised to their **smallest** exponent. Notice the words "greatest" and "smallest" are opposites.

EXAMPLE 1 Find the GCF of $24x^2y$ and $60x^3$.

Solution Write the prime factorization of each monomial, treating the variables like prime factors.

$$24x^2y = 2^3 \cdot 3 \cdot x^2 \cdot y$$
$$60x^3 = 2^2 \cdot 3 \cdot 5 \cdot x^3$$

The common prime factors are 2, 3, and x. The smallest exponent for the factor of 2 is 2. The smallest exponent of 3 is 1. The smallest exponent of the x is 2. The GCF will be the product of 2^2, 3, and x^2.

$$\text{GCF} = 2^2 \cdot 3 \cdot x^2 = 12x^2$$

YOUR TURN Find the GCF.

a. $32r^2t$ and $48r^3ts$ **b.** $35a^2$ and $9b$

OBJECTIVE 2. Factor a monomial GCF out of the terms of a polynomial. Earlier we mentioned that $2(x + 4)$ is the factored form of $2x + 8$. Notice that 2 is the GCF of the terms $2x$ and 8. This suggests the following procedure for factoring a monomial GCF out of the terms of a polynomial.

PROCEDURE **Factoring a Monomial GCF out of a Polynomial**

To factor a monomial GCF out of the terms of a polynomial,

1. Find the GCF of the terms that make up the polynomial.
2. Rewrite the polynomial as a product of the GCF and the result of dividing the given polynomial by the GCF.

$$\text{Given polynomial} = \text{GCF}\left(\frac{\text{Given polynomial}}{\text{GCF}}\right)$$

EXAMPLE 2 Factor.

a. $18x^2 + 24x$

Solution The GCF of $18x^2$ and $24x$ is $6x$.

$$18x^2 + 24x = 6x\left(\frac{18x^2 + 24x}{6x}\right) \quad \text{Rewrite using the form GCF}\left(\frac{\text{Given polynomial}}{\text{GCF}}\right).$$
$$= 6x\left(\frac{18x^2}{6x} + \frac{24x}{6x}\right) \quad \text{Separate the terms.}$$
$$= 6x(3x + 4) \quad \text{Divide the terms by the GCF.}$$

Note: *6 is the largest number that divides both 18 and 24 evenly and x has the smaller exponent of the x^2 and x.*

Check We can check using the distributive property to see if the product of the two factors is the given polynomial.

$$6x(3x + 4) = 6x \cdot 3x + 6x \cdot 4 \quad \text{Distribute } 6x.$$
$$= 18x^2 + 24x \quad \text{The product is the given polynomial.}$$

ANSWERS

a. $16r^2t$ **b.** 1

b. $16x^3y - 40x^2$

Solution The GCF of $16x^3y$ and $40x^2$ is $8x^2$.

Note: *From this point on, when we factor out a monomial GCF, we will not show the division steps.*

$$16x^3y - 40x^2 = 8x^2\left(\frac{16x^3y - 40x^2}{8x^2}\right) \quad \text{Rewrite using the form GCF}\left(\frac{\text{Given polynomial}}{\text{GCF}}\right).$$

$$= 8x^2\left(\frac{16x^3y}{8x^2} - \frac{40x^2}{8x^2}\right) \quad \text{Separate the terms.}$$

$$= 8x^2(2xy - 5) \quad \text{Divide the terms by the GCF.}$$

YOUR TURN Factor.

a. $12xy - 20x$

b. $15m^3n - 21mn^2 + 27mnp$

Factoring When the First Term Is Negative

Generally, we prefer the first term inside parentheses to be positive. To avoid a negative first term in the parentheses, factor out the negative of the GCF.

EXAMPLE 3 Factor.

a. $-20xy^3 + 36y$

Solution Because the first term is negative, we will factor out the negative of the GCF, which is $-4y$.

$$-20xy^3 + 36y = -4y\left(\frac{-20xy^3 + 36y}{-4y}\right)$$

$$= -4y(5xy^2 - 9)$$

Note: *Factoring out 4y instead of −4y does give an equivalent expression, but the first term in the parentheses is negative.*

$$-20xy^3 + 36y = 4y(-5xy^2 + 9)$$

We could have a positive first term by rearranging the terms in the parentheses.

$$= 4y(9 - 5xy^2)$$

Although the first term is positive in $9 - 5xy^2$, *the terms are no longer in descending order, which is why factoring out the negative of the GCF is usually the most desirable approach.*

b. $-18x^4y^3 + 9x^2y^2z - 12x^3y$

Solution We will factor out the negative of the GCF, $-3x^2y$.

$$-18x^4y^3 + 9x^2y^2z - 12x^3y = -3x^2y\left(\frac{-18x^4y^3 + 9x^2y^2z - 12x^3y}{-3x^2y}\right)$$

$$= -3x^2y(6x^2y^2 - 3yz + 4x)$$

ANSWERS

a. $4x(3y - 5)$
b. $3mn(5m^2 - 7n + 9p)$

YOUR TURN Factor.

a. $-40ab - 35b$ **b.** $-30t^4u^5 - 24t^3u^4 + 48tu^2v$

Before we discuss additional techniques of factoring, it is important to state that no matter what type of polynomial we are asked to factor, we will always first consider whether a monomial GCF (other than 1) can be factored out of the polynomial.

Factoring When the GCF Is a Polynomial

Sometimes, when factoring, the GCF is a polynomial.

EXAMPLE 4 Factor. $a(c + 5) + b(c + 5)$

Solution Notice that this expression is a sum of two products, $a(c + 5)$ and $b(c + 5)$. Further, note that $(c + 5)$ is the GCF of the two products.

Note: *The parentheses are filled in the same way, by dividing the original expression by the GCF.*

$$a(c + 5) + b(c + 5) = (c + 5)\left(\frac{a(c + 5) + b(c + 5)}{c + 5}\right)$$
$$= (c + 5)(a + b)$$

YOUR TURN Factor. $6n(m - 3) - 7(m - 3)$

OBJECTIVE 3. Factor polynomials by grouping. Factoring out a polynomial GCF as we did in Example 4 is an intermediate step in a process called *factoring by grouping,* which is a technique that we try when factoring a four-term polynomial such as $ac + 5a + bc + 5b$. The method is called *grouping* because we group pairs of terms together and look for a common factor within each group. We begin by pairing the first two terms together as one group and the last two terms as a second group.

$$ac + 5a + bc + 5b = (ac + 5a) + (bc + 5b) \quad \text{Group pairs of terms.}$$

Notice that the first two terms have a common factor of a and the last two terms have a common factor of b. If we factor the a out of the first two terms and the b out of the last two terms, we have the same expression that we factored in Example 4:

$$= a(c + 5) + b(c + 5) \quad \text{Factor out } a \text{ and } b.$$
$$= (c + 5)(a + b) \quad \text{Factor out } c + 5.$$

Our example suggests the following procedure.

ANSWERS

a. $-5b(8a + 7)$
b. $-6tu^2(5t^3u^3 + 4t^2u^2 - 8v)$

ANSWER

$(m - 3)(6n - 7)$

PROCEDURE **Factoring by Grouping**

To factor a four-term polynomial by grouping,

1. Factor out any monomial GCF (other than 1) that is common to all four terms.
2. Group together pairs of terms and factor the GCF out of each pair.
3. If there is a common binomial factor, then factor it out.
4. If there is no common binomial factor, then interchange the middle two terms and repeat the process. If there is still no common binomial factor, then the polynomial cannot be factored by grouping.

EXAMPLE 5 Factor.

a. $32xy^2 - 48xy + 20y^2 - 30y$

Solution There is a monomial GCF, $2y$, common to all four terms.

$32xy^2 - 48xy + 20y^2 - 30y = 2y(16xy - 24x + 10y - 15)$ — **Factor $2y$ out of all four terms.**

Because the polynomial in the parentheses has four terms, we try to factor by grouping.

$= 2y[(16xy - 24x) + (10y - 15)]$ — **Group pairs of terms.**

$= 2y[8x(2y - 3) + 5(2y - 3)]$ — **Factor $8x$ out of $16xy - 24x$ and factor 5 out of $10y - 15$.**

$= 2y(2y - 3)(8x + 5)$ — **Factor out $2y - 3$.**

b. $24m^2n + 6mn - 60m^2 - 15m$

Solution There is a monomial GCF, $3m$, common to all four terms.

$24m^2n + 6mn - 60m^2 - 15m = 3m(8mn + 2n - 20m - 5)$ — **Factor $3m$ out of all four terms.**

$= 3m[(8mn + 2n) + (-20m - 5)]$ — **Group pairs of terms.**

$= 3m[2n(4m + 1) + (-5)(4m + 1)]$ — **Factor $2n$ out of $8mn + 2n$ and -5 out of $-20m - 5$.**

$= 3m(4m + 1)(2n - 5)$ — **Factor out $4m + 1$.**

Note: *Because the first term in $-20m - 5$ is negative, we factored out -5. Also, remember that when factoring by grouping, we need a common binomial factor. Factoring out -5 gives us that common binomial factor, $4m + 1$, whereas factoring out 5 does not.*

YOUR TURN Factor.

a. $3m^2 + 6m + 4mn + 8n$ **b.** $40x^2y - 60x^2 - 8xy + 12x$

ANSWERS

a. $(3m + 4n)(m + 2)$
b. $4x(5x - 1)(2y - 3)$

Calculator **TIPS**

If a factorable polynomial expression is in terms of one variable, we can check its factored form with a graphing calculator by verifying that the graph of the original expression and the factored form are the same.

For example, to verify that $4x^2 + 8x = 4x(x + 2)$ *using a TI-83 Plus, press the* Y= *key, enter* $4x^2 + 8x$ *as* Y1 *and* $4x(x + 2)$ *as* Y2, *and then press* GRAPH. *The graphs of* $4x^2 + 8x$ *and* $4x(x + 2)$ *are identical, so only one parabola appears in the window indicating that the two expressions are equivalent.*

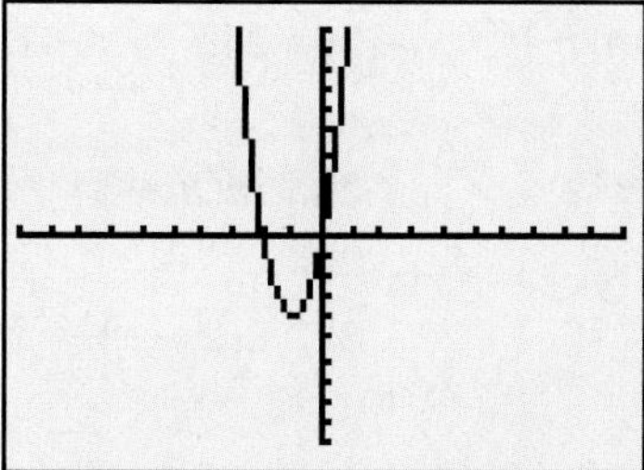

6.1 Exercises

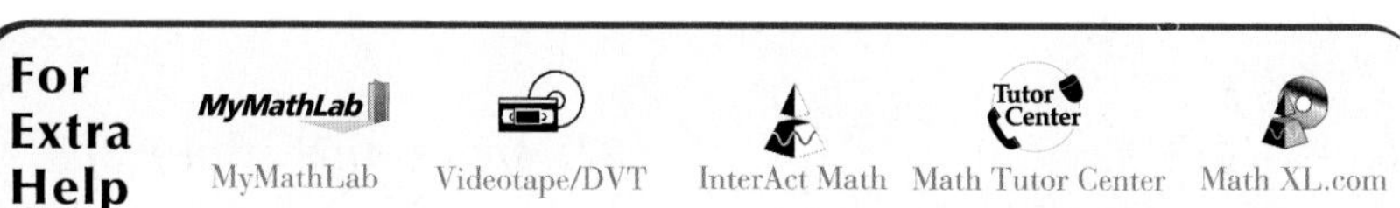

1. Is $4x + 8y$ in factored form? Explain.

2. Explain how to factor a monomial GCF out of the terms in a polynomial.

3. What types of polynomials are factored by grouping?

4. What are the steps in factoring by grouping?

For Exercises 5–12, find the GCF.

5. $4x^3y^2, 24x^2y$

6. $6m^4n^9, 15mn^5$

7. $12u^3v^6, 28u^8v^8$

8. $45g^7h^6, 35g^4h^7$

9. $10a^3b^2c^8, 14abc^8, 20a^7b^3c^4$

10. $35u^8v^2w^3, 40u^3v^7w, 25u^6v^3w^4$

11. $5(a + b), 7(a + b)$

12. $8(m - n), 14(m - n)$

For Exercises 13–32, factor out the GCF.

13. $15c^4d - 20c^2$

14. $8y^5 - 4xy^7$

15. $x^5 - x^3 + x^2$

16. $t^6 - t^4 - t^3$

17. $25xy - 50xz + 100x^2$

18. $60ab + 80ac - 20a^2$

19. $-14u^2v^2 - 7uv^2 + 7uv$

20. $-2x^2y^2 - 8x^3y + 2xy$

21. $9a^7b^3 + 3a^4b^2 - 6a^2b$

22. $12m^6n^8 - 18m^3n^3 + 9m^2n$

23. $3w^3v^4 + 39w^2v + 18wv^2$

24. $15a^6b + 3a^4b^3 + 9ab^2$

25. $18ab^3c - 36a^2b^2c + 24a^5b^2c^8$

26. $24a^2b^5c - 18a^5b^8c + 12ab^3c^2$

27. $-8x^2y + 16xy^2 - 12xy$

28. $-20p^2q - 24pq + 16pq^2$

29. $m(n - 3) + 4(n - 3)$

30. $7(8 + 3b) + a(8 + 3b)$

31. $6(b + 2c) - a(b + 2c)$

32. $5(2 - x) - x(2 - x)$

For Exercises 33–52, factor by grouping.

33. $ax + ay + bx + by$

34. $xy + by + cx + bc$

35. $u^3 + 3u^2 + 3u + 9$

36. $x^3 + 5x^2 + 2x + 10$

37. $4mn + 4np - 3m - 3p$

38. $am - an - 5m + 5n$

39. $cd + d + c + 1$

40. $pq + p + q + 1$

41. $2a^2 - a + 2a - 1$

42. $d^3 - 3d^2 + d - 3$

43. $3ax + 6ay + 8by + 4bx$

44. $ac + 2ad + 2bc + 4bd$

45. $5h^2 + 40h - hk - 8k$

46. $t^2 + 4t - tu - 4u$

47. $3x^2 + 3y^2 - ax^2 - ay^2$

48. $ax^2 + ay^2 - 5x^2 - 5y^2$

49. $3p^3 - 6p^2q + 2pq^2 - 4q^3$

50. $15c^3 - 5c^2d + 6cd^2 - 2d^3$

51. $2x^3 - 8x^2y - 3xy^2 + 12y^3$

52. $8a^3 - 12a^2b - 6ab^2 + 9b^3$

For Exercises 53–64, factor completely.

53. $2ab + 2bx - 2ac - 2cx$

54. $3am - 3an + 3bm - 3bn$

55. $12xy + 8y + 30x + 20$

56. $30ab + 40a + 15b + 20$

57. $3a^2y - 12a^2 + 9ay - 36a$

58. $10uv^2 - 5v^2 + 30uv - 15v$

59. $3m^3 + 6m^2n - 10m^2 - 20mn$

60. $5ab + 5b^2 - ab^2 - b^3$

61. $15st^2 - 5t^2 - 30st + 10t$

62. $12ac + 12cx - 3ac^2 - 3c^2x$

63. $5x^3y - 20x^2y + 5xy - 20y$

64. $10a^2b^2 - 10b^3 + 15a^2b - 15b^2$

For Exercises 65–68, explain the mistake then write the correct factored form.

65. $24x^2y^3 + 36x^3y = 6x^2(4y + 6xy)$

66. $8m^4n - 40m^3 = 8m^3n(m - 5)$

67. $9a^2b^3c - 18a^4b = 9a^2b(bc - 2a^2)$

68. $15h^3k + 12h^3 - 3h^2k = 3h^2k(5h^2 + 4h - 1)$

For Exercises 69–72, use a graphing calculator to verify that the factoring is correct. If it is not, then write the correct factored form.

69. $3x^4 - 6x^3 = 3x^3(x - 2)$

70. $-3x^2 - 6x^3 + 9x = -3x(2x^2 + x - 3)$

71. $4x^3 + 14x^2 + 8x = 2x(x^2 + 7x + 4)$

72. $2x^5 - 6x^3 = 2x^3(2x - 1)$

For Exercises 73 and 74, write an expression for the area of the shaded region, then factor completely.

73.

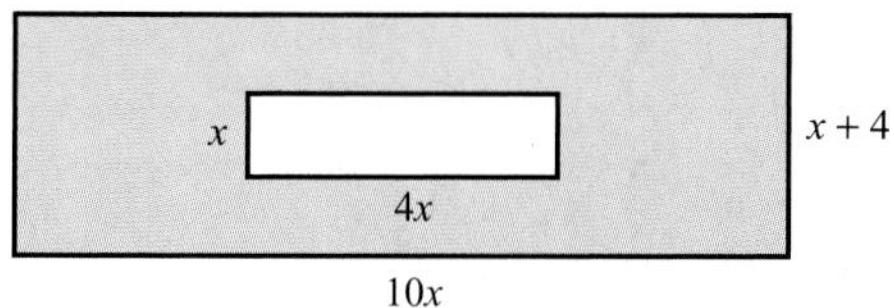

74.

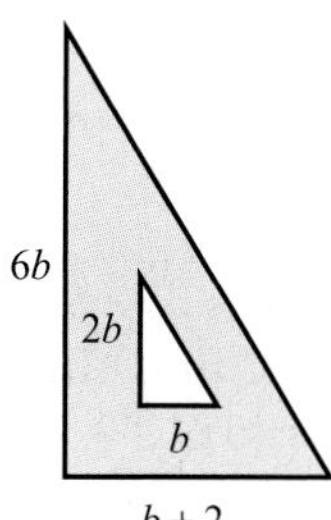

For Exercises 75 and 76: ***a. Write an expression for the area of the blue shaded region.***
b. Write the expression in factored form.
c. Using the factored expression, find the area if x = 3 feet.

75.

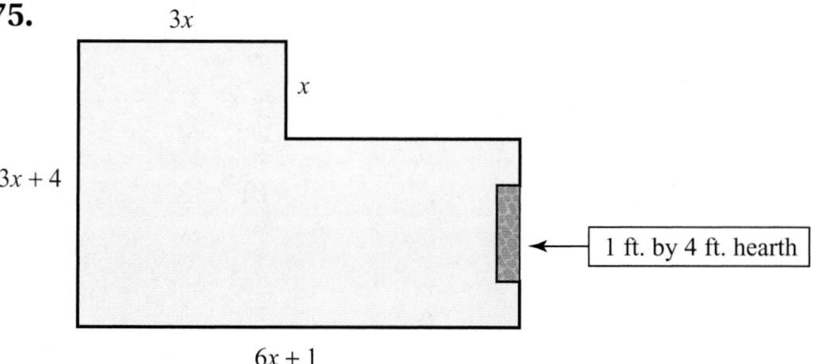

76.

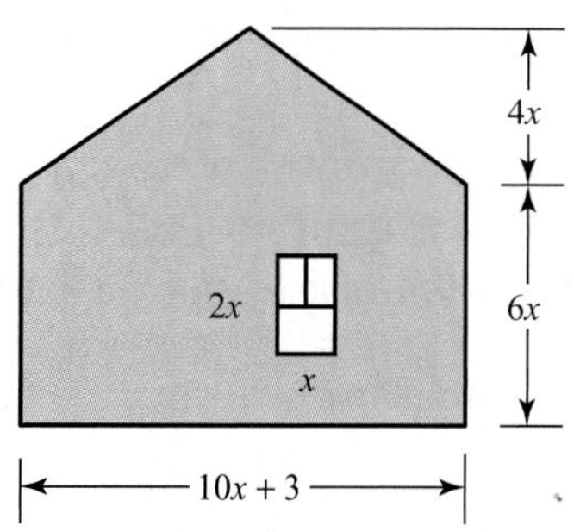

77. a. Write an expression for the volume of the water tower shown. (Treat the base as a cylinder, even though the sphere dips into its top slightly).

b. Write the expression in factored form.

c. Find the volume of a tank if $r = 5$ feet. Round to the nearest tenth.

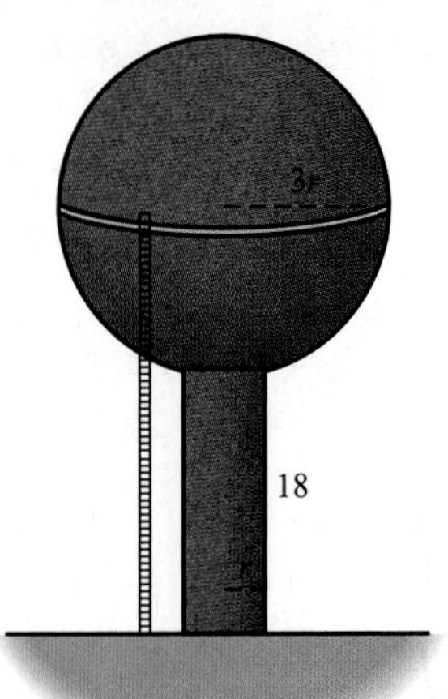

78. a. Write an expression for the volume of the seed spreader shown.

b. Write the expression in factored form.

c. Find the volume of the spreader if $r = 2.5$ feet and $h = 1.5$ feet. Round to the nearest tenth.

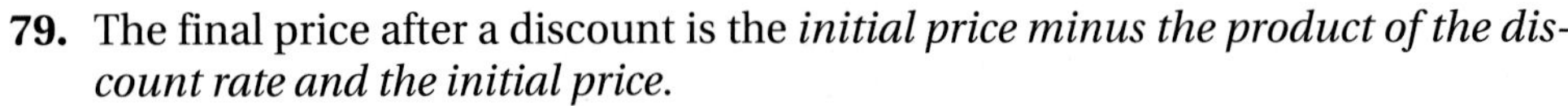

79. The final price after a discount is the *initial price minus the product of the discount rate and the initial price.*

a. Translate the italicized phrase to an expression.

b. Write the expression in factored form.

c. Use the factored expression to find the final price if the initial price is \$54.95 and the discount rate is 40%.

80. The price a store charges after marking up its inventory to make a profit is *the sum of the initial price and the product of the markup rate and the initial price.*

a. Translate the italicized phrase to an expression.

b. Write the expression in factored form.

c. Use the factored expression to find the price the store charges if the initial price is \$24.95 and the markup rate is 30%.

81. The final balance in an account that earns simple interest is *the sum of the principal and the product of the principal, interest rate, and time.*

a. Translate the italicized phrase to an expression.

b. Write the expression in factored form.

c. Use the factored expression to find the final balance in an account containing a principal of \$850 at an interest rate of 3% for half of a year.

82. The distance an object travels after accelerating from an initial velocity is *the product of velocity and the time added to one-half of the product of the acceleration and the square of the time.*

a. Translate the italicized phrase to an expression.

b. Write the expression in factored form.

c. Use the factored expression to find the distance, in feet, that a car travels if it accelerates 7 feet per second per second for 3 seconds from an initial velocity of 58 feet per second.

83. The surface area of a cylinder is *the product of 2, π, and the square of the radius all added to the product of 2, π, the radius, and the height.*

a. Translate the italicized phrase to an expression.

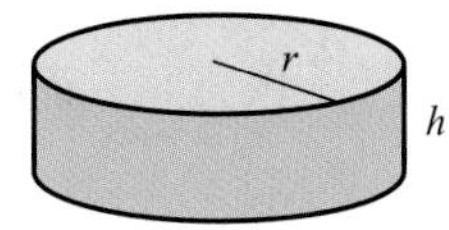

b. Write the expression in factored form.

c. Use the factored expression to find the surface area of a cylinder with a radius of 15 inches and a height of 5 inches. Round to the nearest hundredth.

84. The volume of a capsule is *the product of $\frac{4}{3}$, π, and the cube of the radius of the capsule, all added to the product of π, the square of the radius, and the height of the cylindrical portion of the capsule.*

a. Translate the italicized phrase to an expression.

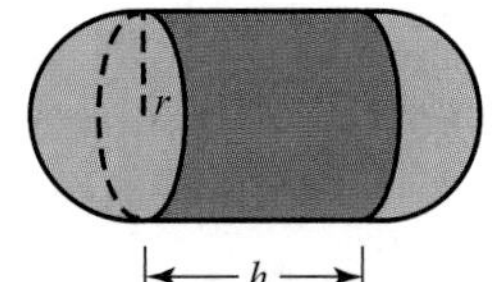

b. Write the expression in factored form.

c. Use the factored expression to find the volume of a capsule with a radius of 5 feet and a height of 8 feet. Round to the nearest hundredth.

REVIEW EXERCISES

[5.1] **1.** Write -2.45×10^7 in standard form.

[5.1] **2.** Write 0.000092 in scientific notation.

[5.3] ***For Exercises 3–6, multiply.***

3. $(x + 3)(x + 5)$

4. $(x - 6)(x - 4)$

5. $(2x + 7)(3x - 1)$

6. $2x(4x - 5)(x + 3)$

6.2 Factoring Trinomials

OBJECTIVES

1. **Factor trinomials of the form $x^2 + bx + c$.**
2. **Factor trinomials of the form $ax^2 + bx + c$, where $a \neq 1$, by trial.**
3. **Factor trinomials of the form $ax^2 + bx + c$, where $a \neq 1$, by grouping.**
4. **Factor trinomials of the form $ax^2 + bx + c$ using substitution.**

In this section, we learn to factor trinomials like $x^2 + 5x + 6$ and $3x^2 - x - 10$.

OBJECTIVE 1. Factor trinomials of the form $x^2 + bx + c$. First we consider trinomials of the form $x^2 + bx + c$ in which the coefficient of the squared term is 1. If a trinomial in this form is factorable, it will have two binomial factors. For example, $x^2 + 5x + 6 = (x + 2)(x + 3)$. Let's use FOIL to multiply $(x + 2)(x + 3)$ and look for patterns that will help us factor.

$$\begin{aligned}(x+2)(x+3) &= \overset{F}{x^2} + \overset{O}{3x} + \overset{I}{2x} + \overset{L}{6} \\ &= x^2 + \underbrace{5x}_{\text{Sum of 2 and 3}} + \underbrace{6}_{\text{Product of 2 and 3}}\end{aligned}$$

Note: *The product of these two numbers is the last term and their sum is the coefficient of the middle term.*

This suggests the following procedure.

PROCEDURE Factoring $x^2 + bx + c$

To factor a trinomial of the form $x^2 + bx + c$,

1. Find two numbers with a product equal to c and a sum equal to b.
2. The factored trinomial will have the form: $(x + \square)(x + \square)$, where the second terms are the numbers found in step 1.

Note: The signs of b and c may cause one or both signs in the binomial factors to be minus signs.

EXAMPLE 1 Factor.

a. $x^2 - 9x + 20$

Solution We must find a pair of numbers whose product is 20 and whose sum is -9. Note that if two numbers have a positive product and negative sum, they must both be negative. We list the possible products and sums next.

Product	Sum
$(-1)(-20) = 20$	$-1 + (-20) = -21$
$(-4)(-5) = 20$	$-4 + (-5) = -9$
$(-2)(-10) = 20$	$-2 + (-10) = -12$

This is the correct combination, so -4 and -5 are the second terms in the binomials of the factored form.

Answer $x^2 - 9x + 20 = (x + (-4))(x + (-5))$
$= (x - 4)(x - 5)$

Note: *We will omit this step in future examples.*

Check Multiply the factors to verify that their product is the original trinomial.

$$(x - 4)(x - 5) = x^2 - 5x - 4x + 20 = x^2 - 9x + 20$$ **It checks.**

b. $x^2 + 2x - 15$

Solution We must find a pair of numbers whose product is -15 and whose sum is 2. Since the product is negative, the two numbers must have different signs.

Product	Sum
$(-1)(15) = -15$	$-1 + 15 = 14$
$(-3)(5) = -15$	$-3 + 5 = 2$

This is the correct combination, so -3 and 5 are the second terms in each binomial of the factored form.

Answer $x^2 + 2x - 15 = (x - 3)(x + 5)$

YOUR TURN Factor.

a. $n^2 - 4n - 12$

b. $y^2 + 10y - 24$

Now, let's consider cases in which we first factor out a monomial GCF.

EXAMPLE 2 Factor.

a. $4x^3 - 20x^2 + 24x$

Solution $4x^3 - 20x^2 + 24x = 4x(x^2 - 5x + 6)$ **Factor out the monomial GCF, $4x$.**

$= 4x(x - 2)(x - 3)$ **Factor $x^2 - 5x + 6$ by finding two numbers, -2 and -3, whose product is 6 and whose sum is -5.**

b. $x^4 - 9x^3 + 15x^2$

Solution $x^4 - 9x^3 + 15x^2 = x^2(x^2 - 9x + 15)$ **Factor out the monomial GCF, x^2.**

To factor $x^2 - 9x + 15$, we need a pair of numbers whose product is 15 and whose sum is -9.

Product	Sum
$(1)(15) = 15$	$1 + 15 = 16$
$(3)(5) = 15$	$3 + 5 = 8$
$(-1)(-15) = 15$	$-1 + (-15) = -16$
$(-3)(-5) = 15$	$-3 + (-5) = -8$

Note: *There are no factors of 15 whose sum is -9.*

The trinomial $x^2 - 9x + 15$ has no binomial factors with integer terms, so $x^2(x^2 - 9x + 15)$ is the final factored form.

ANSWERS

a. $(n + 2)(n - 6)$
b. $(y - 2)(y + 12)$

A polynomial like $x^2 - 9x + 15$ that cannot be factored is like a prime number in that its only factors are 1 and the expression itself. We say such an expression is *prime.*

YOUR TURN Factor.

a. $3n^3 - 18n^2 + 24n$ **b.** $y^5 + 3y^4 - 18y^3$

Let's consider some cases of trinomials that contain two variables.

EXAMPLE 3 Factor. $3mn^3 + 12mn^2 - 96mn$

Solution $3mn^3 + 12mn^2 - 96mn = 3mn(n^2 + 4n - 32)$ Factor out the monomial GCF, $3mn$.

$= 3mn(n - 4)(n + 8)$ Factor $n^2 + 4n - 32$.

In Example 3, we saw that factoring out the monomial GCF left us with a trinomial factor of the form $x^2 + bx + c$. Let's now examine some cases where the trinomial's factors contain both variables.

EXAMPLE 4 Factor.

a. $x^2 - 8xy + 12y^2$

Solution This trinomial has the form $x^2 + bx + c$ if we view b as $-8y$ and c as $12y^2$. We must find a pair of terms whose product is $12y^2$ and whose sum is $-8y$. These terms must be $-2y$ and $-6y$.

$$x^2 - 8xy + 12y^2 = (x - 2y)(x - 6y)$$

b. $2t^2u + 4tu^2 - 30u^3$

Solution This trinomial has a monomial GCF, $2u$.

$2t^2u + 4tu^2 - 30u^3 = 2u(t^2 + 2tu - 15u^2)$ Factor out the monomial GCF, $2u$.

In the remaining trinomial factor, we can view the coefficient of t in the middle term as $2u$. We must find a pair of terms whose product is $-15u^2$ and whose sum is $2u$. These terms would have to be $-3u$ and $5u$.

$$= 2u(t - 3u)(t + 5u)$$

ANSWERS

a. $3n(n - 2)(n - 4)$
b. $y^3(y - 3)(y + 6)$

YOUR TURN Factor.

a. $x^2 + 9xy + 20y^2$ **b.** $x^3 + 4x^2y - 21xy^2$

ANSWERS

a. $(x + 4y)(x + 5y)$
b. $x(x - 3y)(x + 7y)$

OBJECTIVE 2. Factor trinomials of the form $ax^2 + bx + c$, where $a \neq 1$, by trial. Now let's factor trinomials of the form $ax^2 + bx + c$, where $a \neq 1$, such as $5x^2 + 13x + 6$ or $6y^2 - y - 12$. To factor such trinomials, we must consider the factors of the ax^2 term and the factors of the c term. First we will develop a trial-and-error method for finding the factored form.

EXAMPLE 5 Factor. $5x^2 + 13x + 6$

Solution Since all the terms are positive, we know that the terms in the binomial factors will all be positive.

The *first* terms must multiply to equal $5x^2$. These terms must be $5x$ and x.

$$5x^2 + 13x + 6 = (\square + \square)(\square + \square)$$

The *last* terms must multiply to equal 6. These factors could be 1 and 6, or 2 and 3.

We now multiply various combinations of these *first* and *last* terms in binomial factors to see which pair gives the original trinomial. More specifically, we look at the sum of the *inner* and *outer* products to see which combination gives the correct middle term.

$$(5x + 1)(x + 6) = 5x^2 + 30x + x + 6 = 5x^2 + 31x + 6$$
$$(5x + 6)(x + 1) = 5x^2 + 5x + 6x + 6 = 5x^2 + 11x + 6$$
$$(5x + 2)(x + 3) = 5x^2 + 15x + 2x + 6 = 5x^2 + 17x + 6$$

Incorrect combinations.

$$(5x + 3)(x + 2) = 5x^2 + 10x + 3x + 6 = 5x^2 + 13x + 6$$

Correct combination.

Answer $5x^2 + 13x + 6 = (5x + 3)(x + 2)$

Learning Strategy

If you are a visual learner, you may find it helpful to draw lines to connect each product.

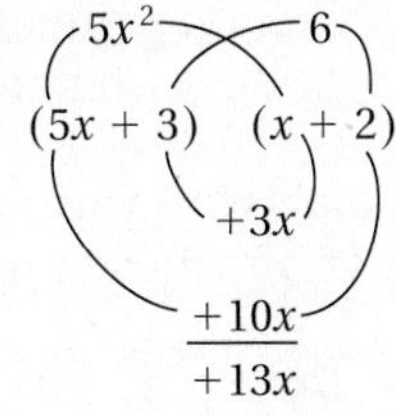

The sum of the *inner* and *outer* products verifies that this is the correct combination.

Our example suggests the following procedure.

PROCEDURE **Factoring by Trial and Error**

To factor a trinomial of the form $ax^2 + bx + c$, where $a \neq 1$, by trial and error,

1. Look for a monomial GCF in all the terms. If there is one, factor it out.
2. Write a pair of *first* terms whose product is ax^2.

$$\overbrace{(\square + \quad)(\square}^{ax^2} + \quad)$$

3. Write a pair of *last* terms whose product is c.

$$(\quad + \overbrace{\square)(\quad + \square}^{c})$$

4. Verify that the sum of the *inner* and *outer* products is bx (the middle term of the trinomial).

$$(\square + \square)(\square + \square)$$

Inner
+ Outer
bx

If the sum of the inner and outer products is not bx, then try the following:

a. Exchange the first terms of the binomials from step 3, then repeat step 4.
b. Exchange the last terms of the binomials from step 3, then repeat step 4.
c. Repeat steps 2–4 with a different combination of first and last terms.

EXAMPLE 6 Factor.

a. $8x^2 - 10x - 7$

Solution The product of the *first* terms must be $8x^2$. These terms could be x and $8x$, or $2x$ and $4x$.

$$8x^2 - 10x - 7 = (\quad + \quad)(\quad - \quad)$$

The product of the *last* terms must be -7, which means they must be 1 and -7 or -1 and 7. We have already written in the appropriate signs.

Now multiply binomials with various combinations of these first and last terms until we find a combination whose inner and outer products combine to equal $-10x$.

$$(x + 1)(8x - 7) = 8x^2 - 7x + 8x - 7 = 8x^2 + x - 7$$
$$(x + 7)(8x - 1) = 8x^2 - x + 56x - 7 = 8x^2 + 55x - 7$$

Incorrect combinations.

$$(2x + 1)(4x - 7) = 8x^2 - 14x + 4x - 7 = 8x^2 - 10x - 7$$

Correct combination.

Answer $8x^2 - 10x - 7 = (2x + 1)(4x - 7)$

Note: *We did not show every possible combination.*

b. $54x^3 - 99x^2 + 36x$

Solution $54x^3 - 99x^2 + 36x = 9x(6x^2 - 11x + 4)$ **Factor out the monomial GCF, $9x$.**

Now we factor $6x^2 - 11x + 4$. Since 4 is positive and $-11x$ is negative, we know that both last terms in the binomial factors must have minus signs.

The product of the *first* terms must be $6x^2$. These terms could be x and $6x$, or $2x$ and $3x$.

$$9x(6x^2 - 11x + 4) = 9x(\square - \square)(\square - \square)$$

The product of the *last* terms must be 4. These terms could be 1 and 4, or 2 and 2.

Now find a combination whose inner and outer products combine to equal $-11x$.

$$9x(6x - 1)(x - 4) = 9x(6x^2 - 24x - x + 4) = 9x(6x^2 - 25x + 4)$$
$$9x(6x - 4)(x - 1) = 9x(6x^2 - 6x - 4x + 4) = 9x(6x^2 - 10x + 4)$$
$$9x(6x - 2)(x - 2) = 9x(6x^2 - 12x - 2x + 4) = 9x(6x^2 - 14x + 4)$$
$$9x(3x - 1)(2x - 4) = 9x(6x^2 - 12x - 2x + 4) = 9x(6x^2 - 14x + 4)$$

Incorrect combinations.

$$9x(3x - 4)(2x - 1) = 9x(6x^2 - 3x - 8x + 4) = 9x(6x^2 - 11x + 4)$$

Correct combination.

Answer $54x^3 - 99x^2 + 36x = 9x(3x - 4)(2x - 1)$

c. $15x^2y + 18xy^2 - 24y^3$

Solution $15x^2y + 18xy^2 - 24y^3 = 3y(5x^2 + 6xy - 8y^2)$ **Factor out the monomial GCF, $3y$.**

Now we factor $5x^2 + 6xy - 8y^2$. Since $-8y^2$ is negative, we know the signs of the last terms in the binomial factors must be different.

The product of the *first* terms must be $5x^2$, so they must be x and $5x$.

$$3y(5x^2 + 6xy - 8y^2) = 3y(\square + \square)(\square - \square)$$

The product of the *last* terms must be $8y^2$. These terms could be y and $8y$, or $2y$ and $4y$.

Now find a combination whose inner and outer products combine to equal $6xy$.

$$3y(x + y)(5x - 8y) = 3y(5x^2 - 8xy + 5xy - 8y^2)$$
$$= 3y(5x^2 - 3xy - 8y^2)$$
$$3y(x + 8y)(5x - y) = 3y(5x^2 - xy + 40xy - 8y^2)$$
$$= 3y(5x^2 + 39xy - 8y^2)$$

Incorrect combinations.

$$3y(x + 2y)(5x - 4y) = 3y(5x^2 - 4xy + 10xy - 8y^2)$$
$$= 3y(5x^2 + 6xy - 8y^2)$$

Correct combination.

Answer $15x^2y + 18xy^2 - 24y^3 = 3y(x + 2y)(5x - 4y)$

YOUR TURN Factor.

a. $6t^2 - 19t + 10$ **b.** $56n^4 - 70n^3 + 21n^2$ **c.** $6x^3 - 28x^2y - 48xy^2$

OBJECTIVE 3. Factor trinomials of the form $ax^2 + bx + c$, where $a \neq 1$, by grouping. Because trial and error can be tedious, an alternative method is to factor by grouping, which we learned in Section 6.1. Recall that we grouped pairs of terms in a four-term polynomial, then factored out the GCF from each pair. Since a trinomial of the form $ax^2 + bx + c$ has only three terms, we split its bx term into two like terms to create a four-term polynomial that we can factor by grouping.

To split the bx term, we use the fact that if $ax^2 + bx + c$ is factorable, b will equal the sum of a pair of factors of the product of a and c. For example, in $5x^2 + 13x + 6$ from Example 5, the product of a and c is $(5)(6) = 30$, so we need to find two factors of 30 whose sum is b, 13. It is helpful to list the factor pairs and their sums in a table.

Factors of *ac*	Sum of Factors of *ac*
(1)(30) = 30	1 + 30 = 31
(2)(15) = 30	2 + 15 = 17
(3)(10) = 30	3 + 10 = 13
(5)(6) = 30	5 + 6 = 11

Notice that 3 and 10 form the only factor pair of 30 whose sum is 13.

Now we can write $13x$ as $3x + 10x$ or $10x + 3x$, and then factor by grouping.

$$\begin{aligned} 5x^2 + 13x + 6 &= 5x^2 + 3x + 10x + 6 \\ &= x(5x + 3) + 2(5x + 3) \\ &= (5x + 3)(x + 2) \end{aligned}$$

Every trinomial factorable by grouping has only one factor pair of ac whose sum is b, which suggests the following procedure.

PROCEDURE **Factoring $ax^2 + bx + c$, Where $a \neq 1$, by Grouping**

To factor a trinomial of the form $ax^2 + bx + c$, where $a \neq 1$, by grouping,

1. Look for a monomial GCF in all the terms. If there is one, factor it out.
2. Find two factors of the product ac whose sum is b.
3. Write a four-term polynomial in which bx is written as the sum of two like terms whose coefficients are the two factors you found in step 2.
4. Factor by grouping.

EXAMPLE 7 Factor by grouping.

a. $6x^2 - 11x + 4$

Solution For this trinomial, $a = 6$, $b = -11$, and $c = 4$, so $ac = (6)(4) = 24$. We must find two factors of 24 whose sum is -11. Since 4 is positive and -11 is negative, the two factors must be negative.

ANSWERS

a. $(2t - 5)(3t - 2)$
b. $7n^2(2n - 1)(4n - 3)$
c. $2x(3x + 4y)(x - 6y)$

Note: *You do not need to list all possible combinations as we did here. We listed them to illustrate that only one combination is correct.*

Factors of ac	Sum of the Factors of ac	
$(-1)(-24) = 24$	$-1 + (-24) = -25$	
$(-2)(-12) = 24$	$-2 + (-12) = -14$	
$(-3)(-8) = 24$	$-3 + (-8) = -11$	← Correct
$(-4)(-6) = 24$	$-4 + (-6) = -10$	

$$6x^2 - 11x + 4 = 6x^2 - 3x - 8x + 4 \quad \text{Write } -11x \text{ as } -3x - 8x.$$

$$= 3x(2x - 1) - 4(2x - 1) \quad \text{Factor } 3x \text{ out of } 6x^2 - 3x\text{; factor } -4 \text{ out of } -8x + 4.$$

$$= (2x - 1)(3x - 4) \quad \text{Factor out } 2x - 1.$$

b. $40y^4 + 50y^3 - 15y^2$

Solution $40y^4 + 50y^3 - 15y^2 = 5y^2(8y^2 + 10y - 3)$ Factor out the monomial GCF, $5y^2$.

For the trinomial factor, note that $a = 8$, $b = 10$, and $c = -3$, so $ac = (8)(-3) = -24$. We must find two factors of -24 whose sum is 10. Since -3 is negative and 10 is positive, the two factors will have different signs and the factor with the greater absolute value must be positive.

Factors of ac	Sum of the Factors of ac	
$(-1)(24) = -24$	$-1 + 24 = 23$	
$(-2)(12) = -24$	$-2 + 12 = 10$	← Correct
$(-3)(8) = -24$	$-3 + 8 = 5$	

$$5y^2(8y^2 + 10y - 3) = 5y^2(8y^2 - 2y + 12y - 3) \quad \text{Write } 10y \text{ as } -2y + 12y.$$

$$= 5y^2[2y(4y - 1) + 3(4y - 1)] \quad \text{Factor } 2y \text{ out of } 8y^2 - 2y\text{; factor 3 out of } 12y - 3.$$

$$= 5y^2(4y - 1)(2y + 3) \quad \text{Factor out } 4y - 1.$$

YOUR TURN Factor by grouping.

a. $10x^2 - 19x + 6$

b. $36x^4y + 3x^3y - 60x^2y$

OBJECTIVE 4. Factor trinomials of the form $ax^2 + bx + c$ using substitution. Some rather complicated-looking polynomials are actually in the form $ax^2 + bx + c$ and can be factored using a method called *substitution.*

EXAMPLE 8 Factor using substitution.

a. $6(y - 3)^2 + 17(y - 3) + 5$

Solution If we substitute another variable, like u, for $y - 3$, we can see that the polynomial is actually in the form $ax^2 + bx + c$.

$$6(y - 3)^2 + 17(y - 3) + 5$$

$$6u^2 + 17u + 5 \quad \text{Substitute } u \text{ for } y - 3.$$

ANSWERS

a. $(5x - 2)(2x - 3)$

b. $3x^2y(3x + 4)(4x - 5)$

Now we can factor $6u^2 + 17u + 5$ by trial or grouping.

$$6u^2 + 17u + 5 = (2u + 5)(3u + 1)$$

Remember that u was substituted for $y - 3$, so we must substitute $y - 3$ back for u to have the factored form for our original polynomial.

$$= (2u + 5)(3u + 1)$$
$$= [2(y - 3) + 5][3(y - 3) + 1] \quad \text{Substitute } y - 3 \text{ for } u.$$

This resulting factored form can be simplified.

$$= [2y - 6 + 5][3y - 9 + 1] \quad \text{Distribute 2 and 3.}$$
$$= (2y - 1)(3y - 8) \quad \text{Combine like terms.}$$

b. $8x^4 + 2x^2 - 15$

Solution Notice that the degree of the middle term, 2, is half of the degree of the first term, 4, and the last term has no variable at all. If we substitute u for x^2, we can see that the polynomial has the form $ax^2 + bx + c$.

$$8(x^2)^2 + 2x^2 - 15 \quad \textbf{Note: } x^4 = (x^2)^2$$
$$8u^2 + 2u - 15 \quad \text{Substitute } u \text{ for } x^2.$$

Now we can factor $8u^2 + 2u - 15$ by trial or grouping.

$$8u^2 + 2u - 15 = (4u - 5)(2u + 3)$$

To finish, we substitute x^2 for u so that we have the factored form for the original polynomial.

$$= (4u - 5)(2u + 3)$$
$$= (4x^2 - 5)(2x^2 + 3) \quad \text{Substitute } x^2 \text{ for } u.$$

Your Turn Factor using substitution.

a. $24(t + 2)^2 - 22(t + 2) + 3$ **b.** $15n^4 - 19n^2 + 6$

Answers

a. $(4t + 5)(6t + 11)$
b. $(5n^2 - 3)(3n^2 - 2)$

6.2 Exercises

For Extra Help

MyMathLab

Videotape/DVT

InterAct Math

Math Tutor Center

Math XL.com

1. If given a trinomial in the form $x^2 + bx + c$, where $c > 0$ and $b > 0$, what can you conclude about the signs of the numbers in the binomial factors?

2. If given a trinomial in the form $x^2 + bx + c$, where $c < 0$ and $b > 0$, what can you conclude about the signs of the numbers in the binomial factors?

3. If given a trinomial in the form $x^2 + bx + c$, where $c > 0$ and $b < 0$, what can you conclude about the signs of the numbers in the binomial factors?

4. If given a trinomial in the form $x^2 + bx + c$, where $c < 0$ and $b < 0$, what can you conclude about the signs of the numbers in the binomial factors?

5. When using grouping to factor a trinomial in the form $ax^2 + bx + c$, where $a \neq 1$, how do you use the product ac?

6. What is the first step when using substitution to factor a trinomial?

For Exercises 7–18, factor completely.

7. $r^2 + 8r + 7$ **8.** $y^2 + 12y + 11$ **9.** $w^2 - 2w - 3$ **10.** $x^2 + 4x - 5$

11. $a^2 + 8a + 12$ **12.** $y^2 + 14y + 45$ **13.** $y^2 - 13y + 36$ **14.** $x^2 - 13x + 30$

15. $m^2 + 2m - 8$ **16.** $b^2 - 7b - 18$ **17.** $b^2 - 6b - 40$ **18.** $n^2 - n - 30$

For Exercises 19–24, factor completely.

19. $3st^2 + 24st + 21s$ **20.** $4x^2y + 24xy + 20y$ **21.** $5y^3 - 65y^2 + 60y$ **22.** $7mn^2 - 35mn + 28m$

23. $6au^3 + 6au^2 - 36au$ **24.** $5k^2lm - 15klm - 50lm$

For Exercises 25–32, factor the trinomials containing two variables.

25. $p^2 + 10pq + 9q^2$ **26.** $r^2 + 8rs + 15s^2$ **27.** $u^2 - 13uv + 42v^2$ **28.** $a^2 - 10ab + 24b^2$

29. $x^2 - 5xy - 14y^2$ **30.** $m^2 + 2mn - 15n^2$ **31.** $a^2 - ab - 42b^2$ **32.** $h^2 - 4hk - 21k^2$

For Exercises 33–62, factor completely.

33. $3a^2 + 10a + 7$ **34.** $5p^2 - 16p + 3$ **35.** $2w^2 - 3w - 2$ **36.** $2y^2 + 5y - 3$

37. $4r^2 - r + 2$

38. $2a^2 - 13a + 1$

39. $4q^2 - 9q + 2$

40. $3c^2 + 13c + 4$

41. $6b^2 + 7b - 3$

42. $8b^2 - 6b - 5$

43. $16m^2 + 24m + 9$

44. $6d^2 - 19d + 10$

45. $4x^2 + 5x - 6$

46. $12p^2 - 13p - 4$

47. $2w^2 + 15wv + 7v^2$

48. $3r^2 + 8rv + 5v^2$

49. $5x^2 - 16xy + 3y^2$

50. $3a^2 - 10ab + 7b^2$

51. $16x^2 - 14xy + 3y^2$

52. $16h^2 + 10hk + k^2$

53. $3t^2 + 19tu - 14u^2$

54. $3x^2 - xy - 14y^2$

55. $3m^2 - 10mn - 8n^2$

56. $5a^2 + 26a - 24$

57. $2a^2 - 13ab + 18b^2$

58. $6u^2 + 13uv + 6v^2$

59. $22m^3 + 200m^2 + 18m$

60. $8x^2y + 60xy + 28y$

61. $4u^2v + 2uv^2 - 30v^3$

62. $24m^4 - 80m^3 - 64m^2$

For Exercises 63–74, factor by grouping.

63. $3y^2 + 16y + 5$

64. $5r^2 - 16r + 3$

65. $6c^2 + 11c + 6$

66. $4b^2 + 15b + 6$

67. $3t^2 - 17t + 10$

68. $3k^2 + 14k + 8$

69. $6x^2 - x - 15$

70. $10u^2 + u - 2$

71. $32x^2y + 24xy - 36y$

72. $12m^3 + 2m^2 - 2m$

73. $15a^2b - 25ab^2 - 10b^3$

74. $2x^2y + 6xy^2 - 8y^3$

For Exercises 75–86, factor using substitution.

75. $x^4 + x^2 - 6$

76. $m^4 - 2m^2 - 3$

77. $8r^4 + 2r^2 - 3$

78. $6w^4 + w^2 - 1$

79. $15x^4 - 11x^2 + 2$

80. $12r^4 - 23r^2 + 10$

81. $y^6 - 16y^3 + 48$

82. $3h^6 + 14h^3 - 5$

83. $7(x + 1)^2 + 8(x + 1) + 1$

84. $2(m - 3)^2 + 7(m - 3) + 6$

85. $3(a + 2)^2 - 10(a + 2) - 8$

86. $8(x - 3)^2 + 6(x - 3) - 9$

For Exercises 87–90, identify the mistake in the factored form, then give the correct factored form.

87. $x^2 - x - 6 = (x - 2)(x + 3)$

88. $x^2 - 5x + 6 = (x - 6)(x + 1)$

89. $4x^2 + 16x + 16 = (2x + 4)(2x + 4)$

90. $2x^2 - 11x + 12 = (2x - 4)(x - 3)$

For Exercises 91–94, find all natural-number values of b that make the trinomial factorable.

91. $x^2 + bx + 16$

92. $x^2 + bx + 20$

93. $x^2 + bx - 63$

94. $x^2 - bx - 35$

For Exercises 95–98, find a natural number c that makes the trinomial factorable.

95. $x^2 + 9x + c$

96. $x^2 - 8x + c$

97. $x^2 + x - c$

98. $x^2 + 4x - c$

PUZZLE PROBLEM

A census taker came to a house where a man lived with his three daughters.

"What are the ages of your three daughters?" asked the census taker.

"The product of their ages is 72 and the sum of their ages is my house number."

"But that clearly is not enough information," insisted the census taker.

"All right," said the man. "The oldest in years loves chocolate milk." The census taker was then able to determine the age of each daughter. Find the ages of the three daughters. (Source: Michael A. Stueben, Discover *magazine; November, 1983)*

REVIEW EXERCISES

[5.3] ***For Exercises 1–3, multiply.***

1. $(2x + 3)(2x + 3)$

2. $(4y - 1)(4y + 1)$

3. $(n - 2)(n^2 + 2n + 4)$

[5.4] **4.** Divide using long division: $\dfrac{6x^3 - 5x^2 + 3x - 2}{x - 1}$

For Exercises 5 and 6, solve, then: a. Write the solution set using set-builder notation.
b. Write the solution set using interval notation.
c. Graph the solution set.

[2.3] **5.** $5(x - 6) + 2x \geq 3x - 4$

[2.6] **6.** $|2x + 7| < 5$

6.3 Factoring Special Products and Factoring Strategies

OBJECTIVES

1. Factor perfect square trinomials.
2. Factor a difference of squares.
3. Factor a difference of cubes.
4. Factor a sum of cubes.
5. Use various strategies to factor polynomials.

In Section 5.3, we explored some special products found by squaring binomials or multiplying conjugates. In this section, we see how to factor those special products.

OBJECTIVE 1. Factor perfect square trinomials. A perfect square trinomial is the product resulting from squaring a binomial. Recall the rules for squaring binomials that we developed in Section 5.3:

$$(a + b)^2 = a^2 + 2ab + b^2$$
$$(a - b)^2 = a^2 - 2ab + b^2$$

Note: *In these perfect square trinomials, the first and last terms are squares and the middle term is twice the product of their square roots.*

To use these rules when factoring, we simply reverse them.

RULES Factoring Perfect Square Trinomials

$$a^2 + 2ab + b^2 = (a + b)^2$$
$$a^2 - 2ab + b^2 = (a - b)^2$$

EXAMPLE 1 Factor.

a. $16x^2 + 40x + 25$

Solution This trinomial is a perfect square because it has the form $a^2 + 2ab + b^2$, where $a = 4x$ and $b = 5$.

$$16x^2 + 40x + 25$$

$a^2 = (4x)^2 = 16x^2$

$b^2 = 5^2 = 25$

$2ab = 2(4x)(5) = 40x$

Using the rule $a^2 + 2ab + b^2 = (a + b)^2$, where $a = 4x$ and $b = 5$, we have

$$16x^2 + 40x + 25 = (4x + 5)^2$$

b. $9y^2 - 42y + 49$

Solution This trinomial is a perfect square because it has the form $a^2 - 2ab + b^2$, where $a = 3y$ and $b = 7$.

$$9y^2 - 42y + 49 = (3y - 7)^2$$ Use the rule $a^2 - 2ab + b^2 = (a - b)^2$, where $a = 3y$ and $b = 7$.

c. $25x^2 - 20xy + 4y^2$

Solution Perfect square trinomials can have two variables. In this case, we have the form $a^2 - 2ab + b^2$, where $a = 5x$ and $b = 2y$.

$$25x^2 - 20xy + 4y^2 = (5x - 2y)^2$$ Use $a^2 - 2ab + b^2 = (a - b)^2$, where $a = 5x$ and $b = 2y$.

d. $m^3n + 8m^2n + 16mn$

Solution Remember to first factor out any monomial GCF.

$$m^3n + 8m^2n + 16mn = mn(m^2 + 8m + 16)$$ Factor out the monomial GCF, mn.

$$= mn(m + 4)^2$$ Factor the perfect square trinomial using $a^2 + 2ab + b^2 = (a + b)^2$, where $a = m$ and $b = 4$.

YOUR TURN Factor.

a. $4x^2 + 28x + 49$

b. $9n^2 - 48n + 64$

c. $16h^2 + 72hk + 81k^2$

d. $2t^3u^2 - 20t^2u^2 + 50tu^2$

OBJECTIVE 2. Factor a difference of squares. Another special product we considered in Section 5.3 was that of conjugates, which are binomials that differ only in the sign separating the terms, as in $3x + 2$ and $3x - 2$. Note that the product of conjugates is a *difference of squares.*

$$(3x + 2)(3x - 2) = 9x^2 - 4$$

This term ($9x^2$) is the square of $3x$. This term (4) is the square of 2.

The rule for multiplying conjugates that we developed in Section 5.3 is

$$(a + b)(a - b) = a^2 - b^2$$

Reversing this rule tells us how to factor a difference of squares.

RULE Factoring a Difference of Squares

$$a^2 - b^2 = (a + b)(a - b)$$

Warning: A *sum* of squares, $a^2 + b^2$, is prime and cannot be factored.

ANSWERS

a. $(2x + 7)^2$ **b.** $(3n - 8)^2$
c. $(4h + 9k)^2$ **d.** $2tu^2(t - 5)^2$

EXAMPLE 2 Factor.

a. $9x^2 - 16y^2$

Solution This binomial is a difference of squares because $9x^2 - 16y^2 = (3x)^2 - (4y)^2$.

$$a^2 - b^2 = (a + b)(a - b)$$

$$9x^2 - 16y^2 = (3x)^2 - (4y)^2 = (3x + 4y)(3x - 4y)$$ Use $a^2 - b^2 = (a + b)(a - b)$ with $a = 3x$ and $b = 4y$.

b. $25x^4 - 64$

Solution This binomial is a difference of squares because $25x^4 - 64 = (5x^2)^2 - (8)^2$.

$$25x^4 - 64 = (5x^2)^2 - (8)^2 = (5x^2 + 8)(5x^2 - 8)$$ Use $a^2 - b^2 = (a + b)(a - b)$ with $a = 5x^2$ and $b = 8$.

c. $5n^4 - 20n^6$

Solution This binomial has a monomial GCF, $5n^4$.

$$5n^4 - 20n^6 = 5n^4(1 - 4n^2)$$ Factor out the monomial GCF, $5n^4$.

$$= 5n^4(1 + 2n)(1 - 2n)$$ Factor $1 - 4n^2$ using $a^2 - b^2 = (a + b)(a - b)$, where $a = 1$ and $b = 2n$.

d. $x^4 - 81$

Solution This binomial is a difference of squares because $x^4 - 81 = (x^2)^2 - (9)^2$.

$$x^4 - 81 = (x^2)^2 - (9)^2 = (x^2 + 9)(x^2 - 9)$$ Use $a^2 - b^2 = (a + b)(a - b)$ with $a = x^2$ and $b = 9$.

Notice the factor $x^2 - 9$ is another difference of squares, with $a = x$ and $b = 3$, so we can factor further.

Note: *The factor $x^2 + 9$ is a sum of squares, which is prime.*

$$= (x^2 + 9)(x + 3)(x - 3)$$

YOUR TURN Factor.

a. $x^2 - 36$ **b.** $16h^4 - 49k^6$ **c.** $24y^5 - 54y^3$ **d.** $t^4 - 16$

OBJECTIVE 3. Factor a difference of cubes. Another special form that we can factor is $a^3 - b^3$, which is a *difference of cubes.* A difference of cubes is the product of a binomial in the form $a - b$ and a trinomial in the form $a^2 + ab + b^2$.

$$(a - b)(a^2 + ab + b^2) = a^3 + a^2b + ab^2 - a^2b - ab^2 - b^3$$ Multiply.

$$= a^3 - b^3$$ Simplify.

Our multiplication suggests the following rule for factoring a difference of cubes.

ANSWERS

a. $(x + 6)(x - 6)$
b. $(4h^2 + 7k^3)(4h^2 - 7k^3)$
c. $6y^3(2y + 3)(2y - 3)$
d. $(t^2 + 4)(t + 2)(t - 2)$

RULE Factoring a Difference of Cubes

$$a^3 - b^3 = (a - b)(a^2 + ab + b^2)$$

Warning: The trinomial $a^2 + ab + b^2$ is not a perfect square and cannot be factored. Remember, a perfect square trinomial has the form $a^2 + 2ab + b^2$.

EXAMPLE 3 Factor.

a. $8x^3 - 27$

Solution This binomial is a difference of cubes because $8x^3 - 27 = (2x)^3 - (3)^3$. To factor, we use the rule $a^3 - b^3 = (a - b)(a^2 + ab + b^2)$ with $a = 2x$ and $b = 3$.

$$a^3 \quad - \quad b^3 \quad = (a \quad - b)(\ a^2 \quad + \quad a \quad b \quad + \ b^2)$$

$$8x^3 - 27 = (2x)^3 - (3)^3 = (2x - 3)[(2x)^2 + (2x)(3) + (3)^2]$$ Substitute $2x$ for a and 3 for b.

$$= (2x - 3)(4x^2 + 6x + 9)$$ Simplify.

b. $64m^5 - 125m^2n^3$

Solution This binomial has a monomial GCF, m^2.

$$64m^5 - 125m^2n^3 = m^2(64m^3 - 125n^3)$$ Factor out the monomial GCF, m^2.

$$= m^2(4m - 5n)[(4m)^2 + (4m)(5n) + (5n)^2]$$ Factor $64m^3 - 125n^3$ using $a^3 - b^3 = (a - b)(a^2 + ab + b^2)$, with $a = 4m$ and $b = 5n$.

$$= m^2(4m - 5n)(16m^2 + 20mn + 25n^2)$$ Simplify.

YOUR TURN Factor.

a. $1 - t^3$

b. $54xy^3 - 128x$

OBJECTIVE 4. Factor a sum of cubes. A *sum of cubes* has the form $a^3 + b^3$ and can be factored using a pattern similar to that for the difference of cubes. A sum of cubes is the product of a binomial in the form $a + b$ and a trinomial in the form $a^2 - ab + b^2$.

$$(a + b)(a^2 - ab + b^2) = a^3 - a^2b + ab^2 + a^2b - ab^2 + b^3$$ Multiply.

$$= a^3 + b^3$$ Simplify.

Our multiplication suggests the following rule for factoring a sum of cubes.

ANSWERS

a. $(1 - t)(1 + t + t^2)$

b. $2x(3y - 4)(9y^2 + 12y + 16)$

RULE Factoring a Sum of Cubes

$$a^3 + b^3 = (a + b)(a^2 - ab + b^2)$$

Warning: In the rule for the sum of cubes, the trinomial factor $a^2 - ab + b^2$ cannot be factored.

EXAMPLE 4 Factor.

a. $27x^3 + 125$

Solution This binomial is a sum of cubes because $27x^3 + 125 = (3x)^3 + (5)^3$. To factor, we use $a^3 + b^3 = (a + b)(a^2 - ab + b^2)$ with $a = 3x$ and $b = 5$.

$$\begin{aligned} a^3 \;+\; b^3 \;&=\; (a + b)\;(a^2 \;-\; a \quad b \;+\; b^2\;) \\ 27x^3 + 125 = (3x)^3 + (5)^3 &= (3x + 5)((3x)^2 - (3x)(5) + (5)^2) \end{aligned}$$

Substitute $3x$ for a and 5 for b.

$$= (3x + 5)(9x^2 - 15x + 25)$$

Simplify.

b. $64tu^5 + 27t^4u^2$

Solution This binomial has a monomial GCF, tu^2.

$$64tu^5 + 27t^4u^2 = tu^2(64u^3 + 27t^3)$$

Factor out the monomial GCF, tu^2.

$$= tu^2(4u + 3t)[(4u)^2 - (4u)(3t) + (3t)^2]$$

Factor $64u^3 + 27t^3$ using $a^3 + b^3 = (a + b)(a^2 - ab + b^2)$, with $a = 4u$ and $b = 3t$.

$$= tu^2(4u + 3t)(16u^2 - 12tu + 9t^2)$$

Simplify.

YOUR TURN Factor.

a. $64 + y^3$

b. $81n^5 + 24n^2$

OBJECTIVE 5. Use various strategies to factor polynomials. We conclude our exploration of factoring by mixing up the various types of factorable expressions. The challenge will be in determining which of the various factoring techniques is appropriate for the given expression. Use the following outline as a general guide.

ANSWERS

a. $(4 + y)(16 - 4y + y^2)$
b. $3n^2(3n + 2)(9n^2 - 6n + 4)$

PROCEDURE **Factoring a Polynomial**

To factor a polynomial, first factor out any monomial GCF, then consider the number of terms in the polynomial.

I. Four terms: Try to factor by grouping.

II. Three terms: Determine if the trinomial is a perfect square.

A. If the trinomial is a perfect square, then consider its form.

1. If it is in the form $a^2 + 2ab + b^2$, then the factored form is $(a + b)^2$.

2. If it is in the form $a^2 - 2ab + b^2$, then the factored form is $(a - b)^2$.

B. If the trinomial is not a perfect square, then consider its form.

1. If it is in the form $x^2 + bx + c$, then find two factors of c whose sum is b and write the factored form as (x + first number)(x + second number).

2. If it is in the form $ax^2 + bx + c$, where $a \neq 1$, then use trial and error. Or, find two factors of ac whose sum is b, write these factors as coefficients of two like terms that, when combined, equal bx, then factor by grouping.

III. Two terms: Determine if the binomial is a difference of squares, sum of cubes, or difference of cubes.

A. If it is a difference of squares, $a^2 - b^2$, then the factors are conjugates and the factored form is $(a + b)(a - b)$. Note that a sum of squares cannot be factored.

B. If it is a difference of cubes, $a^3 - b^3$, then the factored form is $(a - b)(a^2 + ab + b^2)$.

C. If it is a sum of cubes, $a^3 + b^3$, then the factored form is $(a + b)(a^2 - ab + b^2)$.

Note: Always look to see if any of the factors can be factored.

EXAMPLE 5 Factor completely.

a. $12x^2 + 25x - 7$

Solution This trinomial has no monomial GCF. It is of the form $ax^2 + bx + c$, where $a \neq 1$, so we will try using trial and error.

The product of the *first* terms must be $12x^2$. The factors could be x and $12x$, $2x$ and $6x$, or $3x$ and $4x$.

$$12x^2 + 25x - 7 = (\quad + \quad)(\quad - \quad)$$

The product of the *last* terms must be -7. We have already written in the appropriate signs, so these factors are 1 and 7.

After trying various combinations of the preceding terms, we find the following correct combination.

$$= (3x + 7)(4x - 1)$$

b. $15x^5 + 60x^3$

Solution $15x^5 + 60x^3 = 15x^3(x^2 + 4)$ Factor out the monomial GCF, $15x^3$.

The binomial $x^2 + 4$ is a sum of squares, which cannot be factored using real numbers.

c. $5y^4 - 10y^3 - 75y^2$

Solution $5y^4 - 10y^3 - 75y^2 = 5y^2(y^2 - 2y - 15)$ **Factor out the monomial GCF, $5y^2$.**

The trinomial $y^2 - 2y - 15$ is not a perfect square and is in the form $x^2 + bx + c$, so we look for two numbers whose product is -15 and whose sum is -2. Those two numbers are 3 and -5.

$= 5y^2(y + 3)(y - 5)$ **Factor $y^2 - 2y - 15$.**

d. $8m^3n^2 - 24m^2n^3 + 18mn^4$

Solution $8m^3n^2 - 24m^2n^3 + 18mn^4 = 2mn^2(4m^2 - 12mn + 9n^2)$ **Factor out the monomial GCF, $2mn^2$.**

The trinomial $4m^2 - 12mn + 9n^2$ is a perfect square.

$= 2mn^2(2m - 3n)^2$ **Factor $4m^2 - 12mn + 9n^2$ using $a^2 - 2ab + b^2 = (a - b)^2$, where $a = 2m$ and $b = 3n$.**

e. $9(3y - 1)^2 - 36y^2$

Solution $9(3y - 1)^2 - 36y^2 = 9[(3y - 1)^2 - 4y^2]$ **Factor out the monomial GCF, 9.**

The expression $(3y - 1)^2 - 4y^2$ is a difference of squares, where $a = 3y - 1$ and $b = 2y$.

$= 9[(3y - 1) + 2y)][(3y - 1) - 2y]$ **Factor $(3y - 1)^2 - 4y^2$ using $a^2 - b^2 = (a + b)(a - b)$.**

$= 9(5y - 1)(y - 1)$ **Simplify by combining like terms.**

f. $t^5 - 4t^3 - 8t^2 + 32$

Solution We have a four-term polynomial with no monomial GCF, so we try to factor by grouping.

$t^5 - 4t^3 - 8t^2 + 32 = t^3(t^2 - 4) - 8(t^2 - 4)$ **Factor t^3 out of $t^5 - 4t^3$ and -8 out of $-8t^2 + 32$.**

$= (t^2 - 4)(t^3 - 8)$ **Factor out $t^2 - 4$.**

> **Note:** *The binomial $t^2 - 4$ is a difference of squares and $t^3 - 8$ is a difference of cubes, so we can factor these factors.*

$= (t + 2)(t - 2)(t - 2)(t^2 + 2t + 4)$ **Factor $t^2 - 4$ using $a^2 - b^2 = (a + b)(a - b)$, where $a = t$ and $b = 2$. Factor $t^3 - 8$ using $a^3 - b^3 = (a - b)(a^2 + ab + b^2)$, where $a = t$ and $b = 2$.**

$= (t + 2)(t - 2)^2(t^2 + 2t + 4)$ **Simplify by writing $(t - 2)(t - 2)$ as $(t - 2)^2$.**

g. $h^2 + 2h + 1 - 25k^2$

Solution $h^2 + 2h + 1 - 25k^2 = (h + 1)^2 - 25k^2$ **Factor the perfect square trinomial $h^2 + 2h + 1$.**

The expression $(h + 1)^2 - 25k^2$ is a difference of squares, where $a = h + 1$ and $b = 5k$.

$= (h + 1 + 5k)(h + 1 - 5k)$ **Factor $(h + 1)^2 - 25k^2$ using $a^2 - b^2 = (a + b)(a - b)$.**

ANSWERS

a. $2t(5u - 1)(3u - 1)$
b. $x^2y(6 - x)^2$ or $x^2y(x - 6)^2$
c. $6m^3(n + 3)(n - 1)$
d. $(x - 3)(x + 3)^2(x^2 - 3x + 9)$

YOUR TURN Factor completely.

a. $30tu^2 - 16tu + 2t$

b. $36x^2y - 12x^3y + x^4y$

c. $6m^3(n + 1)^2 - 24m^3$

d. $x^5 - 9x^3 + 27x^2 - 243$

6.3 Exercises

For Extra Help

MyMathLab Videotape/DVT InterAct Math Math Tutor Center Math XL.com

1. Explain in your own words how to recognize a perfect square trinomial.

2. The factors of a difference of squares are binomials known as ________.

3. There is only one minus sign in the factors of a difference of cubes. Where is that minus sign placed in the factors?

4. There is only one minus sign in the factors of a sum of cubes. Where is that minus sign placed in the factors?

5. How do you determine the a and b terms in a sum or difference of cubes?

6. What are three different types of factorable binomials?

7. Describe one type of binomial that cannot be factored.

8. When do you try factoring by grouping?

For Exercises 9–20, factor the perfect square.

9. $x^2 + 10x + 25$ **10.** $y^2 + 8y + 16$ **11.** $b^2 - 4b + 4$ **12.** $m^2 - 12m + 36$

13. $25u^2 - 30u + 9$ **14.** $9m^2 - 24m + 16$ **15.** $n^2 + 24mn + 144m^2$ **16.** $w^2 + 2wv + v^2$

17. $9q^2 - 24pq + 16p^2$ **18.** $4r^2 - 20rs + 25s^2$ **19.** $4p^2 - 28pq + 49q^2$ **20.** $36a^2 + 60ab + 25b^2$

For Exercises 21–32, factor the difference of squares.

21. $16a^2 - y^2$ **22.** $a^2 - 121b^2$ **23.** $25x^2 - 4$ **24.** $16m^2 - 49$

25. $100u^2 - 49v^2$ **26.** $121x^2 - 144y^2$ **27.** $9(x - 3)^2 - 16$ **28.** $25(a + 2)^2 - 9$

29. $9x^2 - 36$ **30.** $2x^2 - 50y^2$ **31.** $x^4 - 16$ **32.** $81 - y^4$

For Exercises 33–44, factor the difference of cubes.

33. $27x^3 - 64$

34. $8y^3 - 125$

35. $m^3 - 27$

36. $27k^3 - 8$

37. $64a^3 - 8b^3$

38. $2x^3 - 16y^3$

39. $64 - c^3d^3$

40. $27 - a^3b^3c^3$

41. $27m^3 - 125m^6n^3$

42. $1000a^3b^6 - 27b^3$

43. $27 - (a + b)^3$

44. $64 - (x - y)^3$

For Exercises 45–56, factor the sum of cubes.

45. $125x^3 + 27$

46. $27y^3 + 8$

47. $u^3 + 125v^3$

48. $27k^3 + 8m^3$

49. $8p^3 + q^3z^3$

50. $27u^3 + v^3w^3$

51. $8x^3 + 64y^3$

52. $2a^3 + 128b^3$

53. $8m^3y^3 + 27m^6z^3$

54. $1000u^6 + t^3u^3$

55. $(u + 3)^3 + 8$

56. $27 + (x - 2)^3$

For Exercises 57–64, use the rules for a difference of squares, sum of cubes, or difference of cubes to factor completely.

57. $(2a - b)^2 - 9$

58. $(b - 3)^2 - 16$

59. $16z^2 - 9(x - y)^2$

60. $100a^2 - 9(b + c)^2$

61. $64d^3 - (x + y)^3$

62. $(t + u)^3 + 64$

63. $64x^3 + 27(y + z)^3$

64. $27m^3 - 8(y + b)^3$

For Exercises 65–68, find a natural number b that makes the expression a perfect square trinomial.

65. $16x^2 + bx + 25$

66. $25x^2 + bx + 64$

67. $4x^2 - bx + 81$

68. $36x^2 - bx + 25$

For Exercises 69–72, find the natural number c that completes the perfect square trinomial.

69. $x^2 + 8x + c$

70. $x^2 + 12x + c$

71. $9x^2 - 24x + c$

72. $4x^2 - 28x + c$

For Exercises 73–92, factor completely.

73. $12a^3b^2c + 3a^2b^2c^2 + 9abc^3$

74. $20x^2y^3z^4 - 12x^5yz^2 + 8x^3y^3z$

75. $x^2 + 8x + 15$

76. $p^2 + p - 30$

77. $2x^2 - 32$

78. $2x^2 - 2y^2$

79. $ax - xy - ay + y^2$

80. $2ab - 2ac - 3bd + 3cd$

81. $6b^2 + b - 2$

82. $2a^2 - 5a - 12$

83. $b^3 + 125$

84. $3y^3 - 24$

85. $15x^2 + 7x - 2$

86. $x^2 + x + 2$

87. $ab - 36ab^3$

88. $5m^2 - 45$

89. $x^4 - 16$

90. $t^4 - 81$

91. $7u^2 - 14uv - 105v^2$

92. $ax^2 + 4ax + 4a$

For Exercises 93–104, factor completely. These may require more than one factoring rule or additional simplification.

93. $27 + (x - 1)^3$

94. $2x(4x - 1)^3 - 54x^4$

95. $5m^5 + 10m^3n^2 + 5mn^4$

96. $72x^6y - 24x^3y^3 + 2y^5$

97. $(n - 3)^2 + 6(n - 3) + 9$

98. $(2t + 1)^2 - 10(2t + 1) + 25$

99. $64t^6 - t^3u^3$

100. $(x + y)^3 + (x - y)^3$

101. $2x^3 + 3x^2 - 2xy^2 - 3y^2$

102. $3x^4 - x^3y - 24x + 8y$

103. $36y^2 - (x^2 - 8x + 16)$

104. $24(m - n)^2 - 54n^2$

For Exercises 105 and 106, write a polynomial for the area of the shaded region, then factor completely.

105.

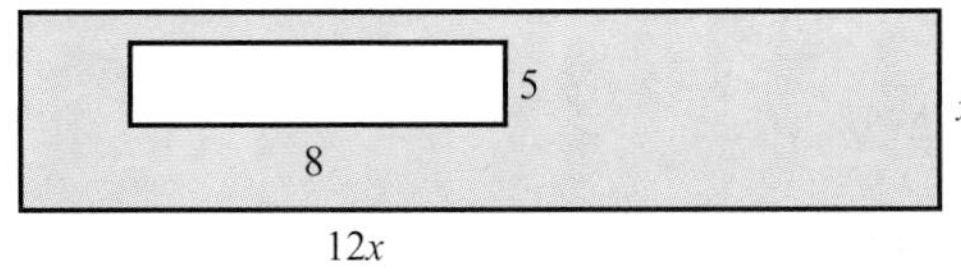

106.

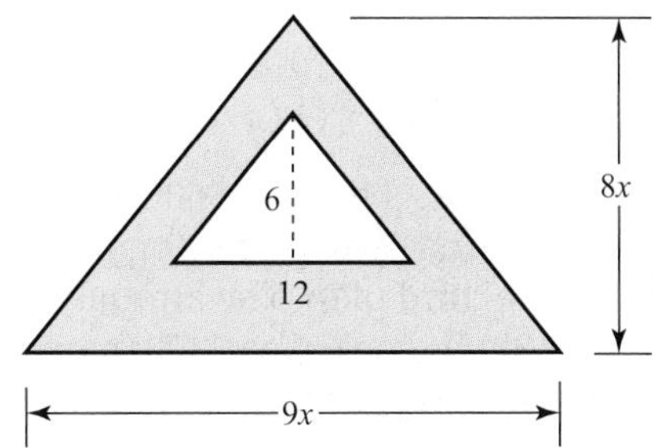

For Exercises 107 and 108, write a polynomial for the volume of the shaded region, then factor completely.

107.

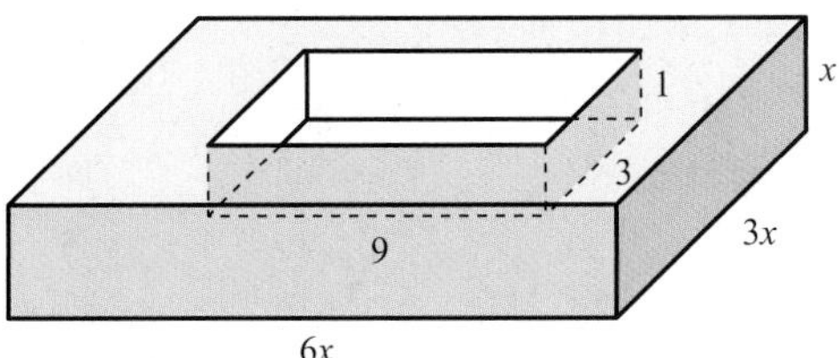

108.

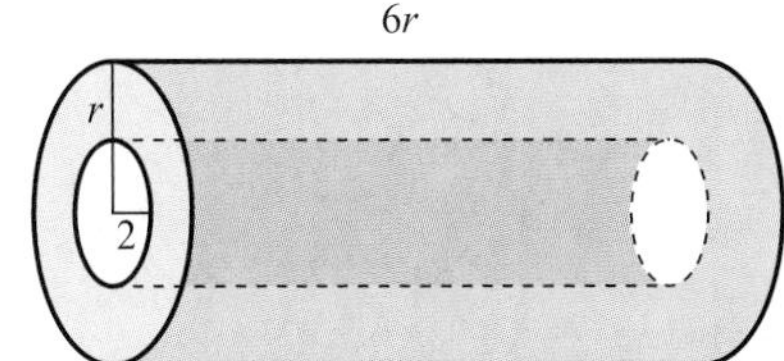

For Exercises 109 and 110, given an expression for the area of each rectangle, find their length and width. Assume the lengths and widths are polynomial factors of the area.

109.

Area $= 6x^2 - 11x + 3$

110.

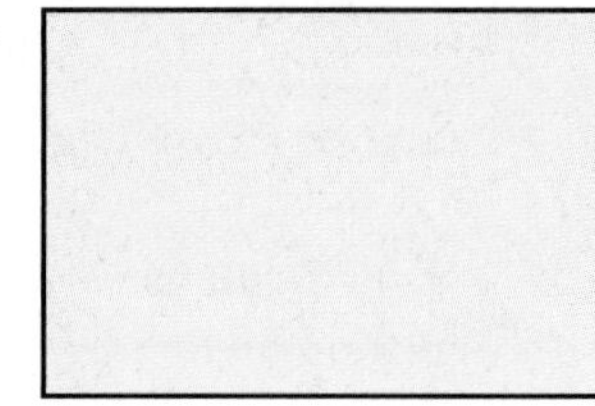

Area $= 16n^2 - 25$

For Exercises 111 and 112, given an expression for the volume of each box, find the length, width, and height. Assume the lengths, widths, and heights are polynomial factors of the volume.

111.

Volume $= 15x^3 + 55x^2 + 30x$

112.

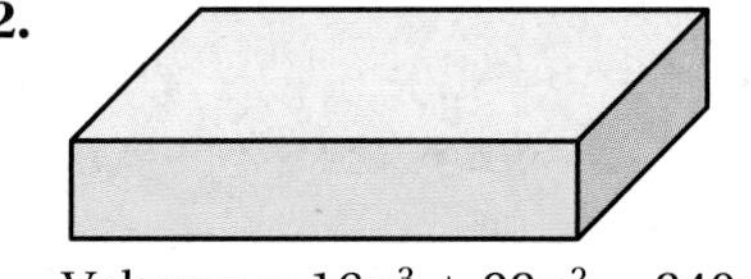

Volume $= 16x^3 + 32x^2 - 240x$

REVIEW EXERCISES

For Exercises 1 and 2, solve.

[2.1] **1.** $9x - 5(3x + 2) = 12x - (4x - 3)$

[2.5] **2.** $|2x - 5| = 7$

[4.3] **3.** An investor invests a total of $6000 in three different plans. The total invested in the first two plans together is equal to twice the amount invested in the third plan. The annual interest rates of the three plans are 4%, 6%, and 8%. If she earns $370 in one year, use a system of equations to find the amount she invested in each plan.

For Exercises 4 and 5, graph.

[3.5] **4.** $f(x) = -\frac{3}{4}x - 2$

[3.4] **5.** $y > 2x - 3$

[3.3] **6.** Write the equation of a line that passes through the points $(3, -2)$ and $(-4, 1)$ in standard form and in slope–intercept form.

6.4 Solving Equations by Factoring

OBJECTIVES

1. **Use the zero-factor theorem to solve equations by factoring.**
2. **Solve problems involving quadratic equations.**
3. **Find the intercepts of quadratic and cubic functions.**

OBJECTIVE 1. Use the zero-factor theorem to solve equations by factoring. We are now ready to move up the Algebra Pyramid from polynomial expressions to **polynomial equations**, which we will solve using factoring.

DEFINITION ***Polynomial equation:*** An equation that equates two polynomials.

Note: *Inequalities are also in the top level of the Algebra Pyramid, but we will not be exploring polynomial inequalities at this point.*

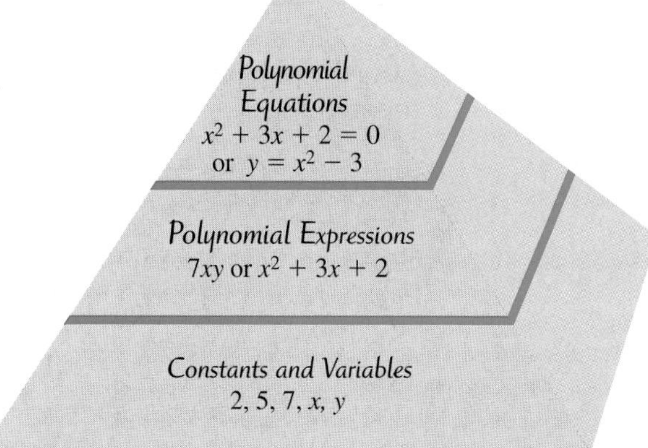

To solve these polynomial equations by factoring, we use the *zero-factor theorem*, which states that if the product of two or more factors is 0, then at least one of the factors is equal to 0.

RULE **Zero-Factor Theorem**

If a and b are real numbers and $ab = 0$, then $a = 0$ or $b = 0$.

EXAMPLE 1 Solve. $(x + 5)(x - 2) = 0$

Solution Since we have a product of two factors, $x + 5$ and $x - 2$, equal to 0, one or both of the factors must equal 0. When using the zero-factor theorem, we set each factor equal to 0 and then solve each of those equations.

$x + 5 = 0$ or $x - 2 = 0$ Use the zero-factor theorem.

$x = -5$ $\quad x = 2$ Solve each equation.

Check Verify that -5 and 2 satisfy the original equation, $(x + 5)(x - 2) = 0$.

For $x = -5$: $(-5 + 5)(-5 - 2) \stackrel{?}{=} 0$
$(0)(-7) = 0$ True

For $x = 2$: $(2 + 5)(2 - 2) \stackrel{?}{=} 0$
$(7)(0) = 0$ True

Both -5 and 2 check, therefore they are both solutions.

YOUR TURN Solve.

$(n - 6)(n + 1) = 0$

To use the zero-factor theorem to solve a polynomial equation, we need the equation to be in **standard form**, which means one side is equal to 0.

DEFINITION ***Polynomial equation in standard form:*** $P = 0$, where P is a polynomial in terms of one variable.

Also, the polynomial needs to be in factored form so that we can use the zero-factor theorem.

EXAMPLE 2 Solve. $3x^2 - 11x = 4$

Note: *The polynomial $3x^2 - 11x - 4$ has two different factors, $3x + 1$ and $x - 4$, which is why the equation $3x^2 - 11x - 4 = 0$ has two different solutions. In general, the number of solutions a polynomial equation in standard form has depends on the number of different factors the polynomial has.*

Solution First, we need the equation in standard form. Then, to use the zero-factor theorem, we will factor the polynomial.

$3x^2 - 11x - 4 = 0$ Subtract 4 from both sides to get standard form.

$(3x + 1)(x - 4) = 0$ Factor $3x^2 - 11x - 4$.

$3x + 1 = 0$ or $x - 4 = 0$ Use the zero-factor theorem.

$3x = -1$ $x = 4$ Solve each equation.

$x = -\frac{1}{3}$

The solutions are $-\frac{1}{3}$ and 4. We will leave the check to the reader.

The equation $3x^2 - 11x = 4$ from Example 2 is a special type of polynomial equation called a **quadratic equation in one variable**.

DEFINITION ***Quadratic equation in one variable:*** An equation that can be written in the form $ax^2 + bx + c = 0$, where a, b, and c are all real numbers and $a \neq 0$.

Notice that $a \neq 0$ means that in a quadratic equation, the degree of the polynomial will always be 2. Also note that if the equation is in the form $ax^2 + bx + c = 0$, it is in standard form. For example, $3x^2 - 12x + 4 = 0$ and $6x^2 - 24 = 0$ are quadratic equations in standard form.

ANSWER

$n = 6$ or -1

Connection The equation $(x+5)(x-2)=0$ from Example 1 is a quadratic equation. We can see why if we multiply the factors so that the equation becomes $x^2+3x-10=0$, which is clearly a quadratic equation.

YOUR TURN Solve.

$12y^2 - 2 = 5y$

Although we have focused on quadratic equations so far, we can use the zero-factor theorem to solve polynomial equations of degree greater than 2.

EXAMPLE 3 Solve. $x^3 - 4x^2 = 12x$

Note: The polynomial $x^3 - 4x^2 - 12x$ has three different factors, so the polynomial equation $x^3 - 4x^2 - 12x = 0$ has three different solutions.

Solution

$$x^3 - 4x^2 - 12x = 0 \quad \text{Subtract } 12x \text{ from both sides to write in standard form.}$$

$$x(x^2 - 4x - 12) = 0 \quad \text{Factor out the monomial GCF, } x.$$

$$x(x-6)(x+2) = 0 \quad \text{Factor } x^2 - 4x - 12.$$

$$x = 0 \quad \text{or} \quad x - 6 = 0 \quad \text{or} \quad x + 2 = 0 \quad \text{Use the zero-factor theorem to solve.}$$

$$x = 6 \qquad\qquad x = -2$$

The solutions are 0, 6, and −2. We will leave the check to the reader.

The equation $x^3 - 4x^2 = 12x$ is called a **cubic equation in one variable**.

DEFINITION ***Cubic equation in one variable:*** An equation that can be written in the form $ax^3 + bx^2 + cx + d = 0$, where a, b, c, and d, are real numbers and $a \neq 0$.

Notice in the definition that $a \neq 0$ means that in a cubic equation the degree-3 term, ax^3, must be present.

YOUR TURN Solve.

$12x^3 + 20x^2 - 8x = 0$

Connection In Examples 2 and 3, notice that the degree of the polynomial in each equation corresponds to the number of solutions the equation has. In general, the degree of a polynomial equation indicates the *maximum* number of real solutions the equation could have. The actual number of real solutions depends upon the number of factors the polynomial has. Since a degree-2 polynomial can have at most two different factors, a quadratic equation can have at most two different solutions. A degree-3 polynomial can have at most three different factors, so a cubic equation could have at most three solutions.

We can now summarize the process of solving polynomial equations using the zero-factor theorem.

ANSWER

$y = \frac{2}{3}$ or $-\frac{1}{4}$

ANSWER

$x = 0, \frac{1}{3}$, or -2

PROCEDURE **Solving Polynomial Equations Using Factoring**

To solve a polynomial equation using factoring,

1. Write the equation in standard form (set one side equal to 0).
2. Write the polynomial in factored form.
3. Use the zero-factor theorem to solve.

EXAMPLE 4 Solve.

a. $3x(2x - 7) = 12$

Solution

$6x^2 - 21x = 12$ Distribute $3x$.

$6x^2 - 21x - 12 = 0$ Subtract 12 from both sides to get standard form.

$3(2x^2 - 7x - 4) = 0$ Factor out the monomial GCF, 3.

$3(2x + 1)(x - 4) = 0$ Factor $2x^2 - 7x - 4$.

$2x + 1 = 0$ or $x - 4 = 0$ Use the zero-factor theorem.

$2x = -1$ $\quad x = 4$ Solve each equation.

$x = -\frac{1}{2}$

Note: *When using the zero-factor theorem, we disregard constant factors like 3 because they have no variable and, therefore, cannot be 0.*

The solutions are $-\frac{1}{2}$ and 4.

b. $\frac{1}{2}(x^2 - 3) + \frac{1}{12}x = \frac{1}{3}(x^2 - 2)$

Solution

$12\left[\frac{1}{2}(x^2 - 3) + \frac{1}{12}x\right] = 12\left[\frac{1}{3}(x^2 - 2)\right]$ Multiply both sides by the LCD, 12, to clear the fractions.

$12 \cdot \frac{1}{2}(x^2 - 3) + 12 \cdot \frac{1}{12}x = 12 \cdot \frac{1}{3}(x^2 - 2)$ Distribute 12.

$6(x^2 - 3) + x = 4(x^2 - 2)$ Simplify.

$6x^2 - 18 + x = 4x^2 - 8$ Distribute 6 on the left side and 4 on the right side.

$2x^2 + x - 10 = 0$ Subtract $4x^2$ from and add 8 to both sides.

$(2x + 5)(x - 2) = 0$ Factor.

$2x + 5 = 0$ or $x - 2 = 0$ Use the zero-factor theorem.

$2x = -5$ $\quad x = 2$ Solve each equation.

$x = -\frac{5}{2}$

The solutions are $-\frac{5}{2}$ and 2.

c. $9y(y^2 + 1) = 4y(6y - 1) - 3y$

Solution

$9y^3 + 9y = 24y^2 - 4y - 3y$	Distribute $9y$ and $4y$.
$9y^3 + 9y = 24y^2 - 7y$	Combine like terms.
$9y^3 - 24y^2 + 16y = 0$	Subtract $24y^2$ from and add $7y$ to both sides.
$y(9y^2 - 24y + 16) = 0$	Factor out the monomial GCF, y.
$y(3y - 4)^2 = 0$	Factor $9y^2 - 24y + 16$, which is a perfect square.
$y = 0$ or $3y - 4 = 0$	Use the zero-factor theorem.
$3y = 4$	Solve each equation.
$y = \frac{4}{3}$	

The solutions are 0 and $\frac{4}{3}$.

d. $t^3 + 3t^2 - 13 = 7t - (3t + 1)$

Solution

$t^3 + 3t^2 - 13 = 7t - 3t - 1$	Distribute.
$t^3 + 3t^2 - 13 = 4t - 1$	Combine like terms.
$t^3 + 3t^2 - 4t - 12 = 0$	Subtract $4t$ from and add 1 to both sides to get standard form.
$t^2(t + 3) - 4(t + 3) = 0$	Factor by grouping.
$(t + 3)(t^2 - 4) = 0$	Factor out $t + 3$.
$(t + 3)(t + 2)(t - 2) = 0$	Factor $t^2 - 4$, which is a difference of squares.
$t + 3 = 0$ or $t + 2 = 0$ or $t - 2 = 0$	Use the zero-factor theorem.
$t = -3$ $\quad t = -2$ $\quad t = 2$	Solve each equation.

The solutions are -3, -2, and 2.

YOUR TURN Solve.

a. $(y - 4)^2 = 1$ **b.** $\frac{1}{4}(x^2 + 8) + 2x = \frac{1}{8}(x - 2)$

c. $n(n^2 - 13) = 3n$ **d.** $h^2(2h + 1) - 3 = 18h + 6$

OBJECTIVE 2. Solve problems involving quadratic equations. Many applications can be solved using polynomial equations.

EXAMPLE 5 The equation $h = -16t^2 + v_0t + h_0$ describes the height, h, in feet of an object t seconds after being thrown upwards with an initial velocity of v_0 feet per second from an initial height of h_0 feet. Suppose a rock is thrown upwards with an initial velocity of 12 feet per second from a 40-foot tower. How many seconds does it take the rock to reach the ground?

Understand We are given a formula, the initial velocity of 12 feet per second, and the initial height of 40 feet. We are to find the time it takes to reach the ground, which is at 0 feet.

ANSWERS

a. 3, 5 **b.** $-6, -\frac{3}{2}$ **c.** 0, -4, 4

d. $-\frac{1}{2}$, 3, -3

Plan Replace the variables in the formula with the given values, then solve for t.

Execute $0 = -16t^2 + 12t + 40$ — In $h = -16t^2 + v_0t + h_0$, substitute 0 for h, 12 for v_0, and 40 for h_0.

$0 = -4(4t^2 - 3t - 10)$ — Factor out the monomial GCF, -4.

$0 = -4(4t + 5)(t - 2)$ — Factor $4t^2 - 3t - 10$.

$4t + 5 = 0$ or $t - 2 = 0$ — Use the zero-factor theorem to solve.

$t = -\frac{5}{4}$ $\quad$ $t = 2$

Note: *Remember, we disregard the constant factor, -4.*

Answer Our answer must describe the amount of time *after* the rock is thrown, so only the positive value, 2, makes sense. This means the rock takes 2 seconds to hit the ground.

Check Verify that after 2 seconds, the rock hits the ground.

$$h = -16(2)^2 + 12(2) + 40 = -64 + 24 + 40 = 0$$

YOUR TURN An object is thrown upwards with an initial velocity of 16 feet per second and an initial height of 60 feet. How many seconds does it take the object to reach the ground?

One of the most well-known theorems in mathematics is the Pythagorean theorem, named after the Greek mathematician Pythagoras. The theorem relates the side lengths of all right triangles. Recall that in a right triangle, the two sides that form the 90° angle are the *legs* and the side directly across from the 90° angle is the *hypotenuse.*

RULE The Pythagorean Theorem

Given a right triangle where a and b represent the lengths of the legs and c represents the length of the hypotenuse, then $a^2 + b^2 = c^2$.

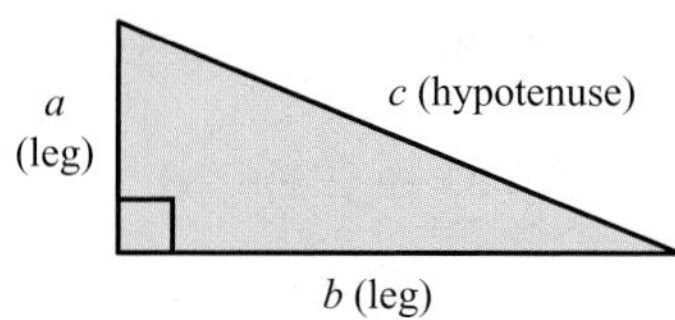

We can use the Pythagorean theorem to find a missing length in a right triangle if we know the other two lengths.

EXAMPLE 6 The figure shows a portion of a roof frame to be constructed. Find the length, in feet, of each side.

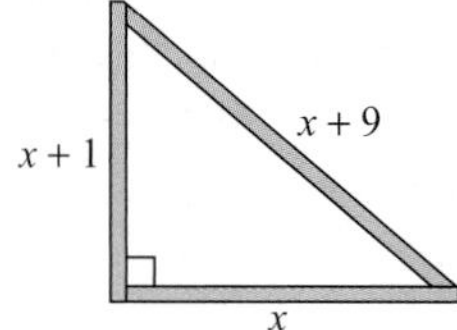

ANSWER

$2\frac{1}{2}$ sec.

Understand We are given expressions for the lengths of the three sides of a right triangle and we are to find those lengths.

Plan and Execute Use the Pythagorean theorem.

$(x)^2 + (x + 1)^2 = (x + 9)^2$ In $a^2 + b^2 = c^2$, substitute x for a, $x + 1$ for b, and $x + 9$ for c.

$x^2 + x^2 + 2x + 1 = x^2 + 18x + 81$ Square the binomials.

$x^2 - 16x - 80 = 0$ Subtract x^2, $18x$, and 81 from both sides to get standard form.

$(x - 20)(x + 4) = 0$ Factor.

$x - 20 = 0$ or $x + 4 = 0$ Use the zero-factor theorem.

$x = 20$ $\quad x = -4$

Answer Because x describes a length in feet, only the positive solution is sensible, so x must be 20 feet. This means $x + 1$ is 21 feet and $x + 9$ is 29 feet.

Check Verify that $20^2 + 21^2 = 29^2$. We will leave this to the reader.

YOUR TURN A sail is to be designed as shown. Find its dimensions in feet.

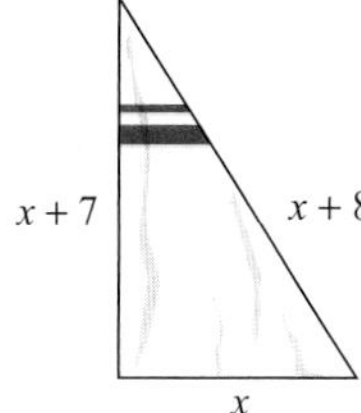

Of Interest

Pythagoras (*c.* 569–500 B.C.) was a Greek mathematician who lived in the town of Croton in what is now Italy. The theorem that now bears his name was known and used in Egypt and elsewhere before Pythagoras's time. It bears his name, however, because he was the first to prove that the relationship is true for all right triangles. In fact, the Greeks were the first culture to value mathematical proofs as opposed to using mathematics solely to solve problems.

OBJECTIVE 3. Find the intercepts of quadratic and cubic functions. In Section 5.2, we graphed some basic polynomial functions. Now we will see a connection between solutions of polynomial equations and the x-intercepts of their corresponding polynomial functions.

Look at the following graphs. Notice that quadratic functions with their parabolic graph could have 0, 1, or 2 x-intercepts. Similarly, cubic functions with their "sideways" S-shaped graphs could have 1, 2, or 3 x-intercepts.

ANSWER

5 ft., 12 ft., 13 ft.

Quadratic Functions

Note: *For simplicity, we have shown only cases where $a > 0$. However, the same number of x-intercepts are possible if $a < 0$.*

$$f(x) = ax^2 + bx + c$$

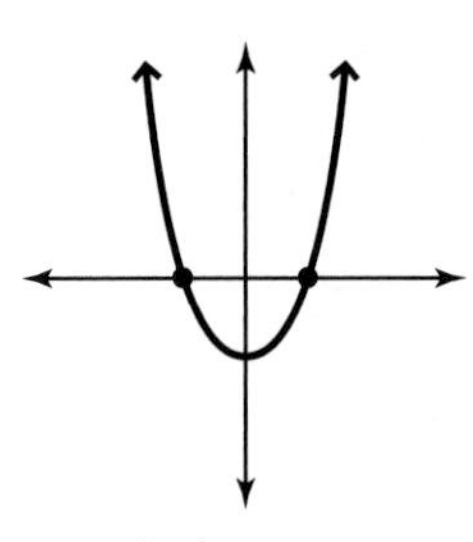

2 x-intercepts

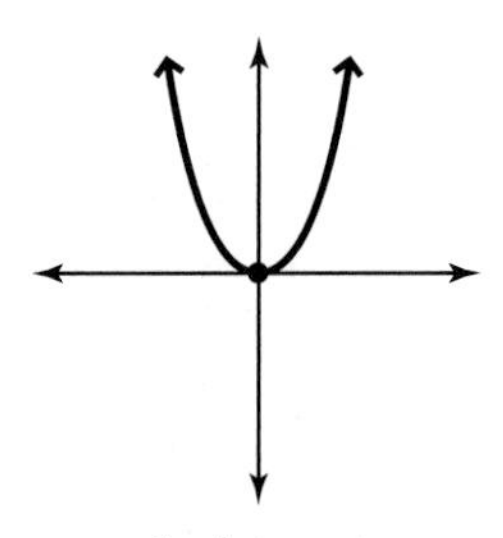

1 x-intercept

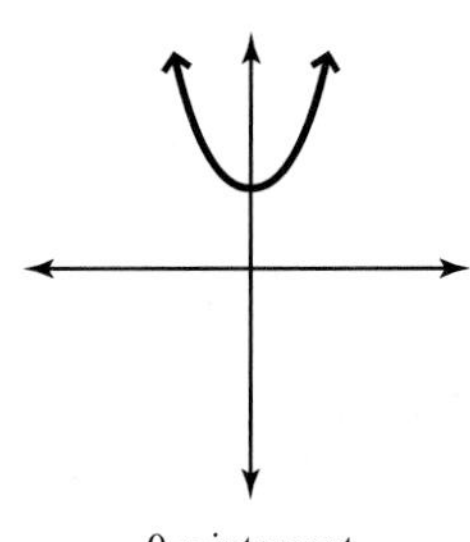

0 x-intercept

Cubic Functions

$$f(x) = ax^3 + bx^2 + cx + d$$

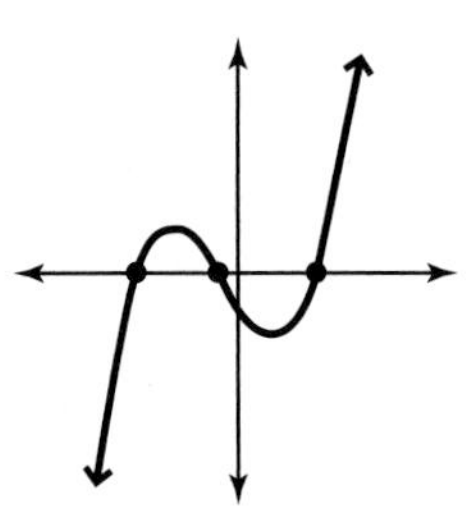

3 x-intercepts

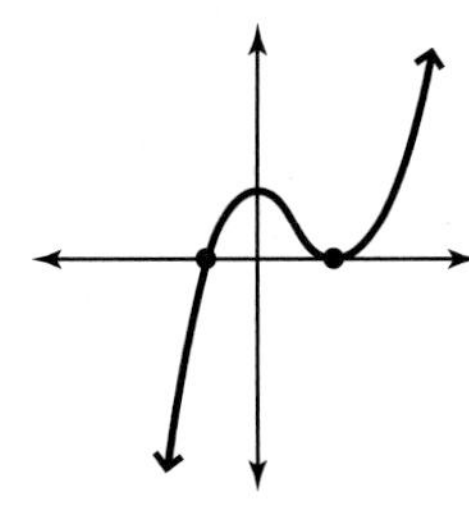

2 x-intercepts

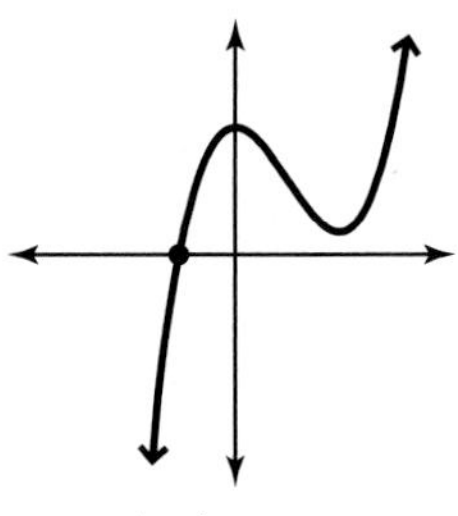

1 x-intercept

Recall that to find x-intercepts, we set $f(x) = 0$ and solve for x. Setting a polynomial function equal to 0 creates a polynomial equation in standard form, so the real solutions to that equation are the x-coordinates of the function's x-intercepts. Consequently, if the polynomial is factorable, we should be able to find those x-intercepts using the zero-factor theorem.

EXAMPLE 7 Find the x-intercept(s) and then sketch the graph.

Note: *Because x-intercepts occur when $f(x) = 0$, they are often called the* zeros *of a function.*

a. $f(x) = x^2 + x - 6$

Solution To find the x-intercepts, we let $f(x) = 0$ and then solve for x.

$$x^2 + x - 6 = 0 \quad \text{Set } f(x) = 0.$$
$$(x + 3)(x - 2) = 0 \quad \text{Factor.}$$
$$x + 3 = 0 \quad \text{or} \quad x - 2 = 0 \quad \text{Use the zero-factor theorem.}$$
$$x = -3 \qquad\qquad x = 2 \quad \text{Solve each equation.}$$

The x-intercepts are $(-3, 0)$ and $(2, 0)$.

To sketch the graph, we will also find the y-intercept by finding $f(0)$.

$$f(0) = (0)^2 + (0) - 6 = -6$$

Connection Earlier in this section, we noted that the degree of the polynomial in a polynomial equation indicates the maximum number of solutions the equation could have. We noted that the actual number of solutions depends on the number of factors the polynomial has. Those connections also apply to the number of x-intercepts a polynomial function has. For example, the polynomial $x^2 + x - 6$ has two factors, $x + 3$ and $x - 2$, so the polynomial equation $x^2 + x - 6 = 0$ has two solutions, -3 and 2. Consequently, the polynomial function $f(x) = x^2 + x - 6$ has two x-intercepts, $(-3, 0)$ and $(2, 0)$.

The y-intercept is $(0, -6)$.

Finding a few more points is helpful.

x	$f(x)$
−4	6
−2	−4
−1	−6
1	−4
3	6

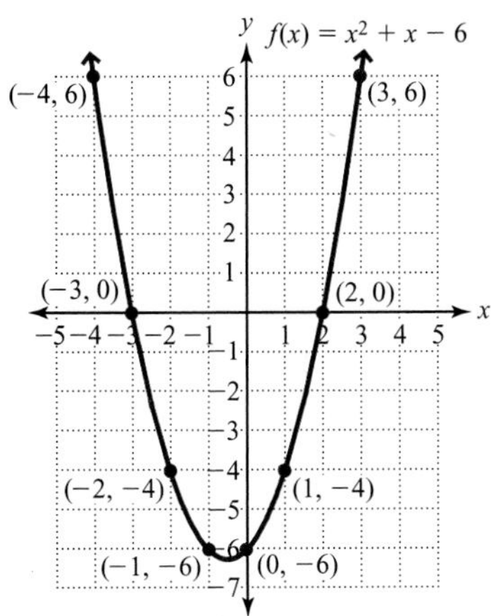

b. $f(x) = (x + 2)(x - 1)(x - 4)$

Solution $(x + 2)(x - 1)(x - 4) = 0$ — Set $f(x) = 0$.

$x + 2 = 0$ or $x - 1 = 0$ or $x - 4 = 0$ — Use the zero-factor theorem.

$x = -2$ $\quad$ $x = 1$ $\quad$ $x = 4$ — Solve each equation.

The x-intercepts are $(-2, 0)$, $(1, 0)$, and $(4, 0)$.

Multiplying the factors in the function reveals this function to be a cubic function.

$$f(x) = (x^2 + x - 2)(x - 4) \quad \text{Multiply } x + 2 \text{ and } x - 1.$$
$$= x^3 - 3x^2 - 6x + 8 \quad \text{Multiply } x^2 + x - 2 \text{ and } x - 4.$$

To sketch the graph, we will also find the y-intercept by finding $f(0)$.

$$f(0) = (0)^3 - 3(0)^2 - 6(0) + 8 = 8$$

The y-intercept is $(0, 8)$.

Finding a few more points is helpful.

x	$f(x)$
−1	10
2	−8
3	−10

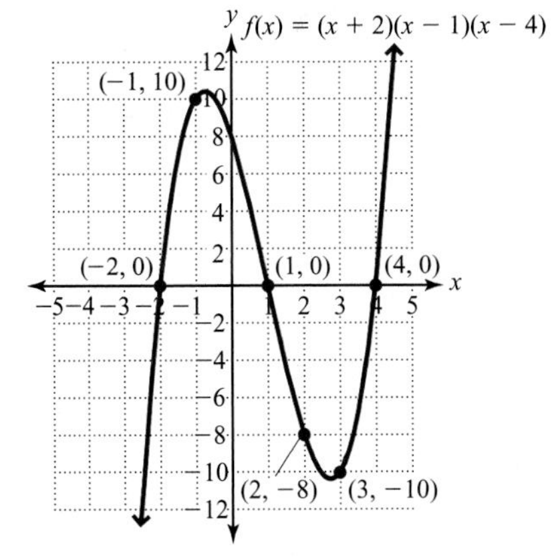

Your Turn Find the x-intercepts and then sketch the graph.

a. $f(x) = x^2 - 4$ $\qquad$ **b.** $f(x) = x(x + 2)(x - 3)$

Answers

a. $(2, 0)$, $(-2, 0)$

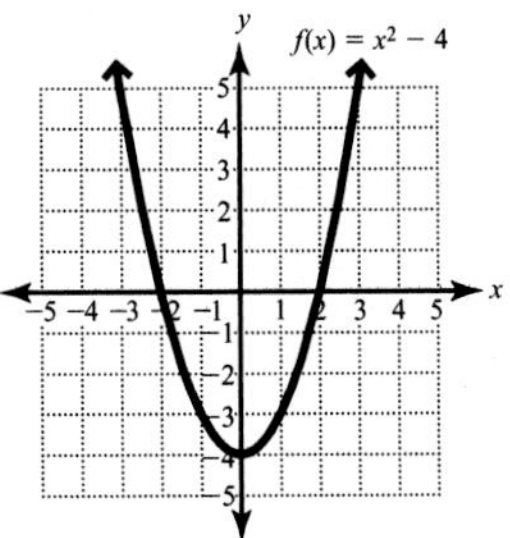

b. $(0, 0)$, $(-2, 0)$, $(3, 0)$

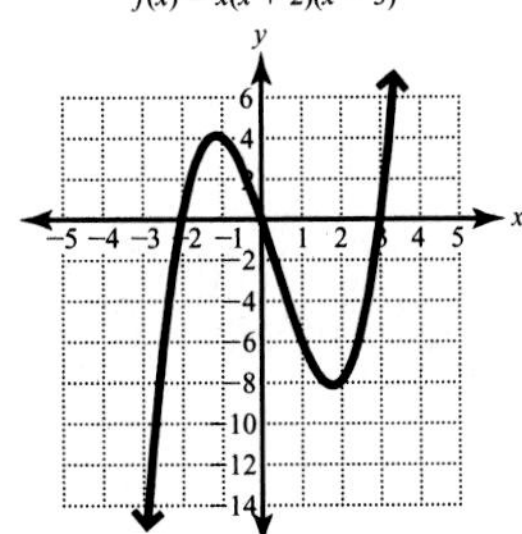

Calculator
TIPS

You can find x-intercepts of a function on most graphing calculators. On a TI-83 Plus, for example, we use the ***ZERO*** *function. After graphing the function, select* ***ZERO*** *from the* ***CALC*** *menu. Enter the left and right bound of the x-intercept that you want to identify, then press* ENTER *one more time at the Guess prompt to get the coordinates of that intercept.*

We can extend this feature to find the solutions to a polynomial equation written in standard form. We simply graph the corresponding polynomial function and then find its x-intercepts (or zeros). Let's solve the equation from Example 3, $x^3 - 4x^2 = 12x$, *using this technique. Written in standard form, the equation is* $x^3 - 4x^2 - 12x = 0$. *The corresponding polynomial function is* $f(x) = x^3 - 4x^2 - 12x$. *If we graph that function, we can see its three x-intercepts.*

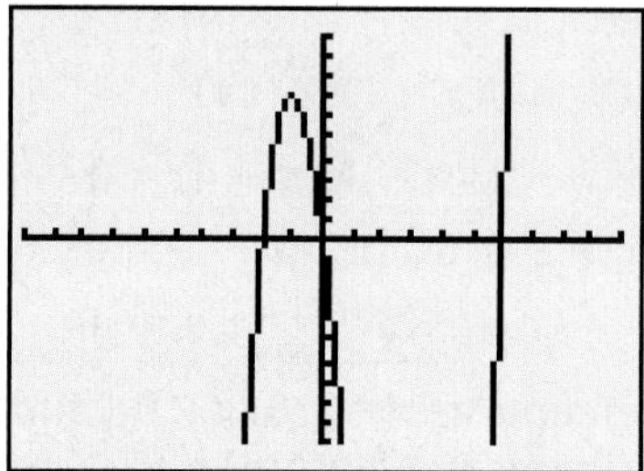

Note: *In the graph shown, we have our window set with* Xmin = −10, Xmax = 10, Ymin = −10, *and* Ymax = 10. *To view the missing bend, use* Ymin = −50.

Using the ***ZERO*** *function for each x-intercept gives us* (0, 0), (6, 0), *and* (−2, 0). *The solutions we found in Example 3 for the equation* $x^3 - 4x^2 = 12x$ *were, in fact, 0, 6, and −2.*

We can also use these x-intercepts and the zero-factor theorem in reverse to backtrack to the factored form. Since $x = 0$, $x = 6$, *and* $x = -2$, *we can rewrite those equations to equal 0, giving us* $x = 0$, $x - 6 = 0$, *and* $x + 2 = 0$. *Those are then the factors in the factored form, so we could write* $x^3 - 4x^2 - 12x = x(x - 6)(x + 2)$.

6.4 Exercises

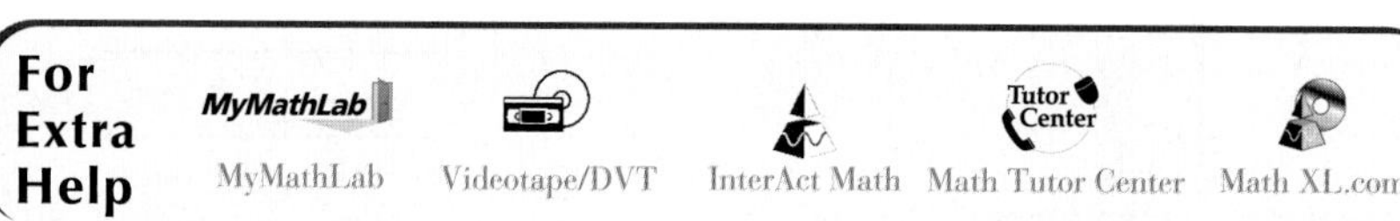

1. Explain what the zero-factor theorem states in your own words.

2. To use the zero-factor theorem, in what form must a polynomial equation be written?

3. What is the degree of the polynomial in a quadratic equation?

4. What is the degree of the polynomial in a cubic equation?

5. Sketch the graph of a general cubic function with two x-intercepts.

6. Sketch the graph of a general quadratic function with one x-intercept.

For Exercises 7–18, solve.

7. $x(x + 4) = 0$

8. $y(y - 8) = 0$

9. $(x - 1)(x + 2) = 0$

10. $(y + 2)(y + 3) = 0$

11. $(3b - 2)(2b + 5) = 0$

12. $(4a - 3)(2a - 5) = 0$

13. $x(x - 3)(2x + 5) = 0$

14. $a(3a - 4)(a + 5) = 0$

15. $(x + 2)(x - 3)(x + 7) = 0$

16. $(y - 3)(y + 4)(y - 6) = 0$

17. $(2x + 3)(3x - 1)(4x + 7) = 0$

18. $(2a - 5)(3a + 2)(5a - 1) = 0$

For Exercises 19–56, solve.

19. $d^2 - 4d = 0$

20. $r^2 + 6r = 0$

21. $x^2 - 9 = 0$

22. $a^2 - 121 = 0$

23. $m^2 + 14m + 45 = 0$

24. $r^2 - 2r - 3 = 0$

25. $2x^2 + 3x - 2 = 0$

26. $6a^2 + 13a - 5 = 0$

27. $3c^2 - c - 2 = 0$

28. $8d^2 - 14d + 3 = 0$

29. $x^2 + 6x + 9 = 0$

30. $b^2 - 10b + 25 = 0$

31. $6y^2 = 3y$

32. $15v^2 = 5v$

33. $x^2 = 25$

34. $u^2 = 64$

35. $n^2 + 6n = 27$

36. $p^2 - 8p = 20$

37. $p^2 = 3p - 2$

38. $m^2 = 11m - 24$

39. $2v^2 + 5 = -7v$

40. $6k^2 + 1 = -7k$

41. $a(a - 5) = 14$

42. $c(2c - 11) = -5$

43. $4x(x + 7) = -49$

44. $m(m + 6) = -9$

45. $4s^2 + 15 = -16s$

46. $6k^2 + 3 = -11k$

47. $x^3 + x^2 - 6x = 0$

48. $2h^3 + 4h^2 - 16h = 0$

49. $9q^4 + 9q^3 - 18q^2 = 0$

50. $d^4 - 8d^3 + 12d^2 = 0$

51. $12x(x+1)+3=5(2x+1)+2$

52. $6-4x(2+x)=8(1-x)-3$

53. $(2v+1)(v-2)=-v(v+2)$

54. $2+4(2m-1)^2=9(2m-1)$

55. $m(m-6)+3(m-3)=-11$

56. $x(x+4)-2(x+3)=9$

For Exercises 57–62, write a polynomial equation in standard form with the given solutions.

57. $-3, 2$

58. $1, 5$

59. $-\frac{2}{3}, 4$

60. $-2, -\frac{1}{4}$

61. $-1, 0, 3$

62. $-2, 0, 4$

For Exercises 63–68, match the graph to the function.

63. $f(x)=(x+3)(x-1)$

64. $f(x)=x(x+1)(x-2)$

65. $f(x)=2x^2-7x+3$

66. $f(x)=x^2-4$

67. $f(x)=x^3+2x^2-3x$

68. $f(x)=x^3+x^2-6x$

a.

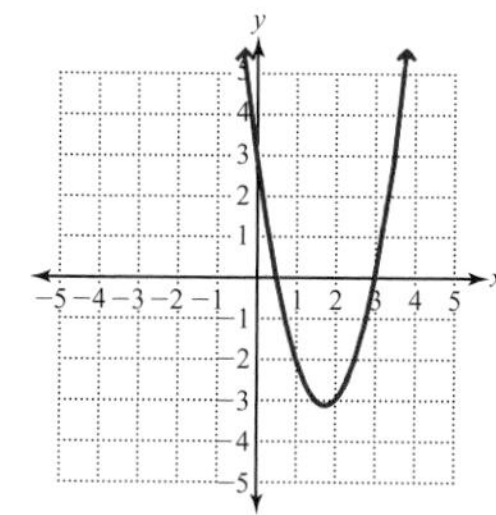

b.

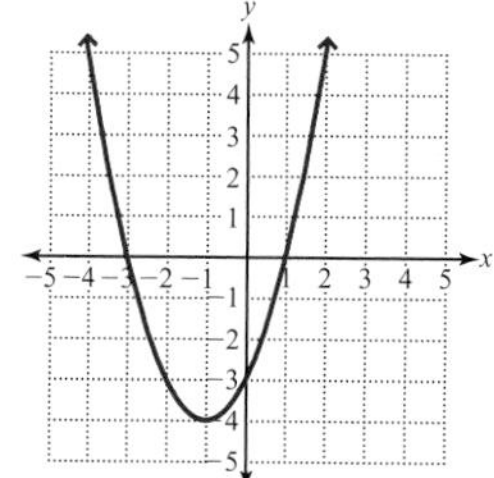

c.

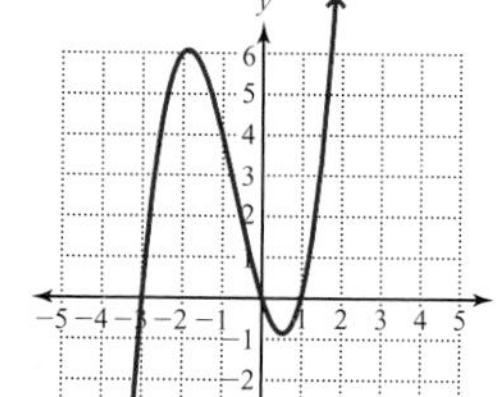

d.

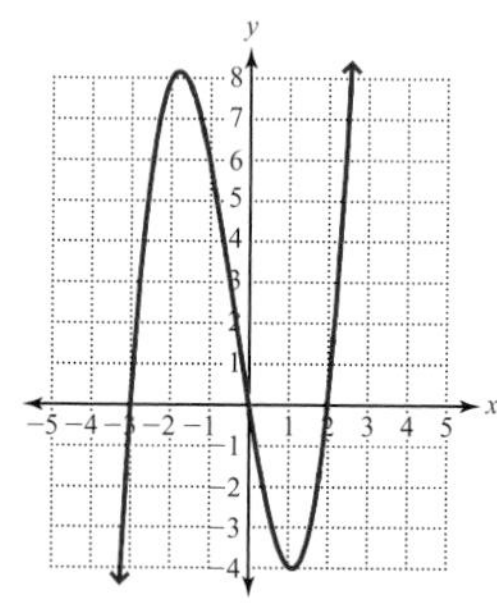

e.

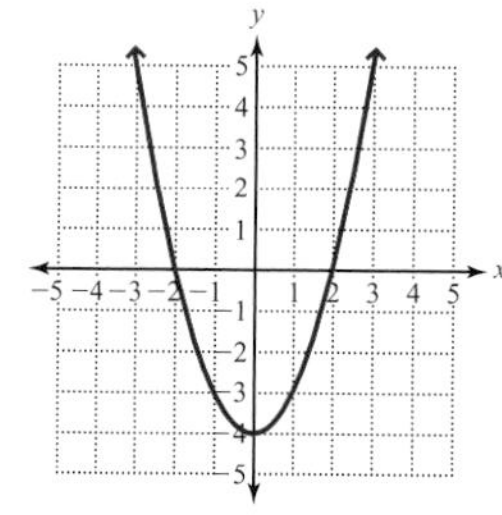

f.

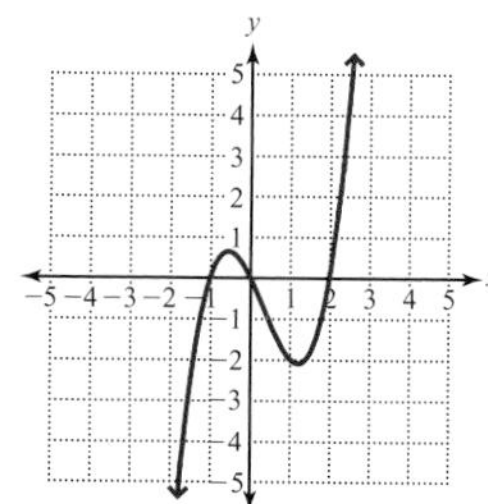

For Exercises 69–76, find the x-intercepts and then sketch the graph.

69. $f(x)=x^2-25$

70. $f(x)=x^2-4$

71. $f(x)=x^2-6x+5$

72. $f(x)=x^2-9x+14$

73. $f(x) = x(x + 4)(x - 1)$ **74.** $f(x) = x(x + 1)(x - 6)$ **75.** $f(x) = x^3 - x^2 - 6x$ **76.** $f(x) = x^3 - x^2 - 2x$

For Exercises 77–82, use a graphing calculator to determine the x-intercepts.

77. $f(x) = x^2 - 9x + 18$ **78.** $f(x) = x^2 + 7x + 12$ **79.** $f(x) = 6x^2 + 7x - 5$

80. $f(x) = 10x^2 - 27x + 5$ **81.** $f(x) = x^3 - x^2 - 2x$ **82.** $f(x) = x^3 - 9x$

For Exercises 83–88, use a graphing calculator to solve the equation. (Hint: Find the x-intercepts of the corresponding polynomial function.)

83. $x^2 = 3x + 40$ **84.** $28 - 3x = x^2$ **85.** $8x^2 - 12 = 18x - 7$

86. $20x^2 + 9 = 15 - 7x$ **87.** $x^3 + x^2 = 12x$ **88.** $x^3 + 10 = 4x^2 + 7x$

For Exercises 89–92, find the length of each side of the right triangle shown.

89.

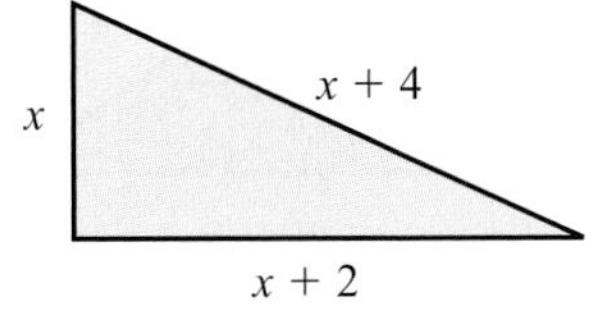

90.

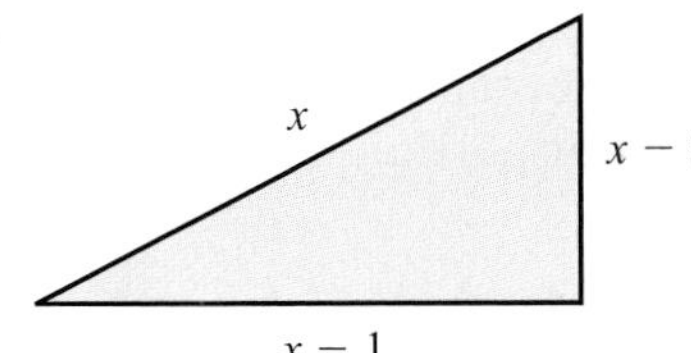

91.

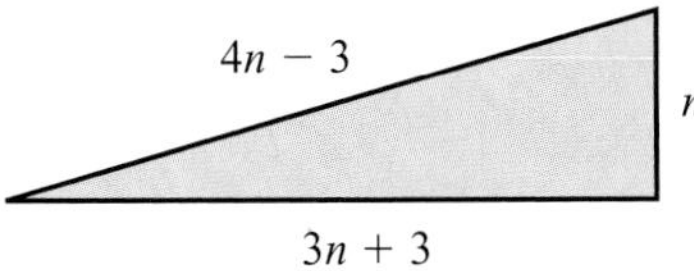

92.

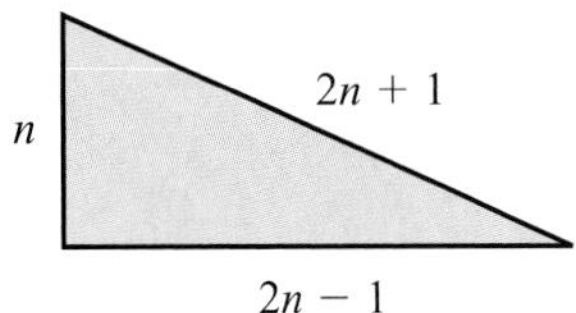

For Exercises 93–104, solve.

93. Find a number such that 55 plus the square of the number is the same as 16 times the number.

94. One natural number is four times another. The product of the two numbers is 100. Find both numbers.

95. The product of two consecutive natural numbers is 342. Find the numbers.

96. The sum of the squares of three consecutive positive even integers is 200. Find the integers.

97. The length of a small rectangular garden is 4 meters more than the width. If the area is 320 square meters, find the dimensions of the garden.

98. A design on the front of a marketing brochure calls for a triangle with a base that is 6 centimeters less than the height. If the area of the triangle is to be 216 square centimeters, what are the lengths of the base and height?

99. The design of a small building calls for a rectangular shape with dimensions of 22 feet by 28 feet. The architect decides to change the shape of the building to a circle but wants it to have the same area. If we use $\frac{22}{7}$ to approximate π, what would be the radius of the circular building?

100. The design of a rectangular room calls for the perimeter to be 40 feet and the area to be 91 square feet. Find the dimensions of the room.

101. A steel plate used in a machine is in the shape of a trapezoid and has an area of 85.5 square inches. The dimensions are shown. Note that the length of the base is equal to the height. Calculate the height.

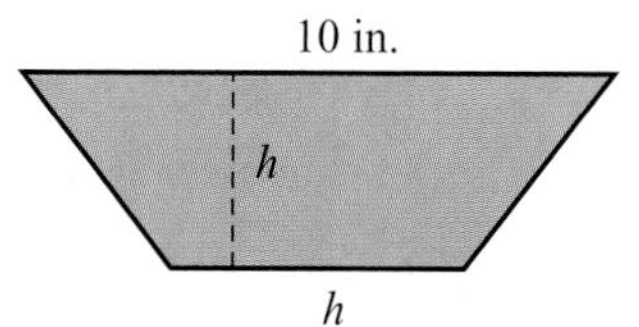

102. The front elevation of one wing of a house is shown. Because of budget constraints, the total area of the front of this wing must be 352 square feet. The height of the triangular portion is 14 feet less than the base. Find the base length.

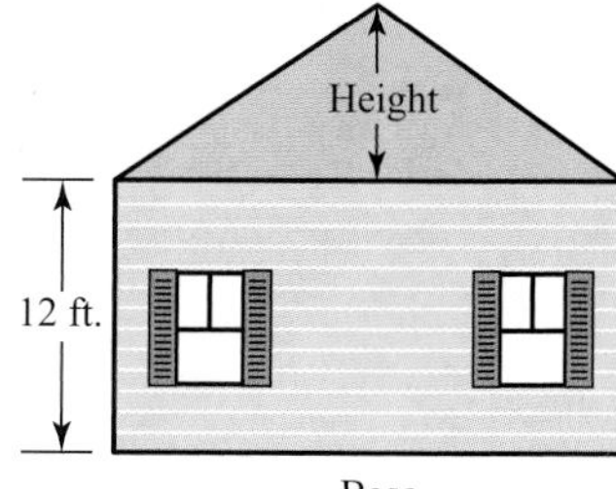

103. A support wire attached to a power pole is to be replaced. The wire needs to be 2 feet longer than the height up the pole. The wire will be staked into the ground 10 feet from the base of the pole. Find the length of the wire and the height at which it is attached to the pole.

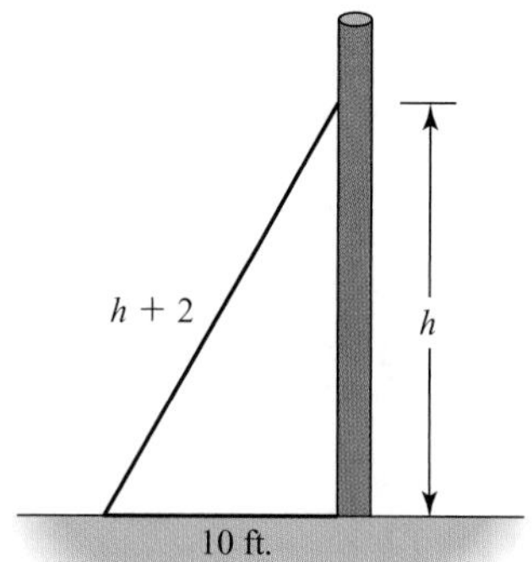

104. Beams in a steel-frame roof structure are configured as shown. The length of the horizontal joist is 2 feet less than the length of the roof support beam. Find the length of the joist and the roof support beam.

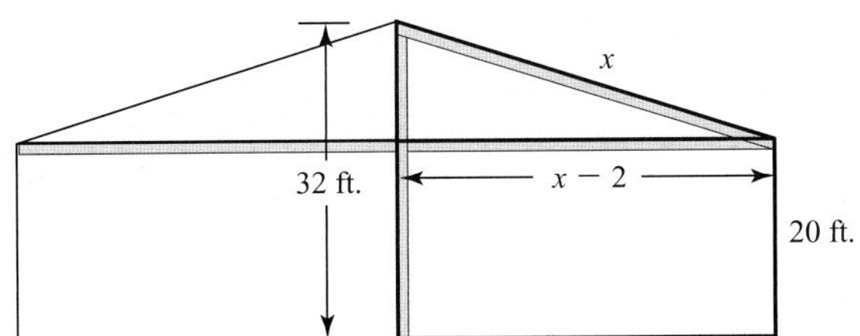

For Exercises 105 and 106, use the formula $h = -16t^2 + v_0 t + h_0$, where h is the final height in feet, t is the time of travel in seconds, v_0 is the initial velocity in feet per second, and h_0 is the initial height in feet of an object traveling upwards.

105. A ball is thrown upward at 4 feet per second from a building 29 feet high. When will the ball be 9 feet above the ground?

106. A rocket is fired from the ground with an upward velocity of 200 feet per second. How many seconds will it take to return to the ground?

For Exercises 107 and 108, use the formula $B = P\left(1 + \frac{r}{n}\right)^{nt}$, which is used to calculate the final balance of an investment or a loan after being compounded. Following is a list of what each variable represents:

B represents the final balance.
P represents the principal, which is the amount invested.
r represents the annual percentage rate.
t represents the time in years in which the principal is compounded.
n represents the number of times the principal is compounded in a year.

107. LaQuita invests \$4000 in an account that is compounded annually. If, after two years, her balance is \$4840, what was the interest rate of the account?

108. David invests \$1000 in an account that is compounded semiannually (every six months). If, after one year, his balance is \$1210, what is the interest rate of the account?

Collaborative Exercises THE SANDBOX

In a daycare center's backyard, there is a 13-foot-by-18-foot rectangular area that would be perfect for a large sandbox. Enough sand to cover a 150-square-foot play area to a sufficient depth has been purchased. The center would like the sandbox to be rectangular and have a uniform wooden border around the play area for seating. How wide should the border be?

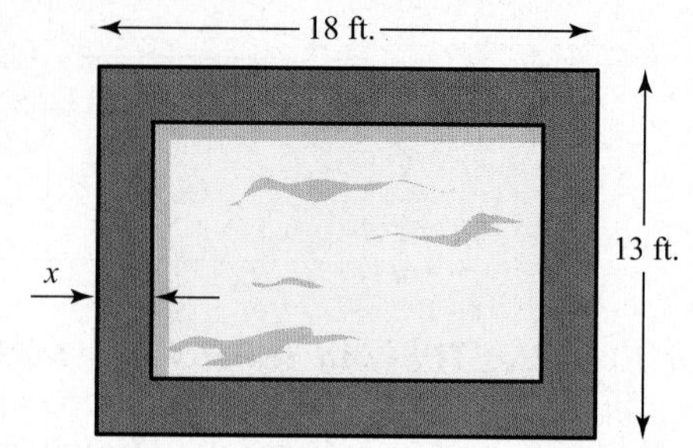

1. The figure to the right is a rough sketch of the desired setup. If x represents the width of the border, write expressions for the length and width of the inside of the box where the sand is placed.

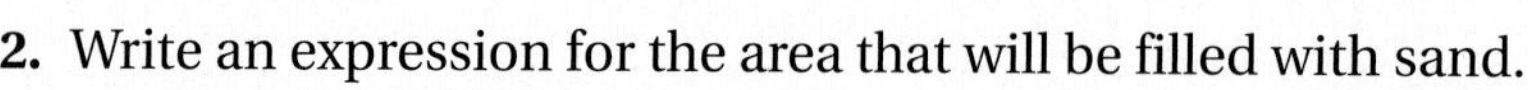

2. Write an expression for the area that will be filled with sand.

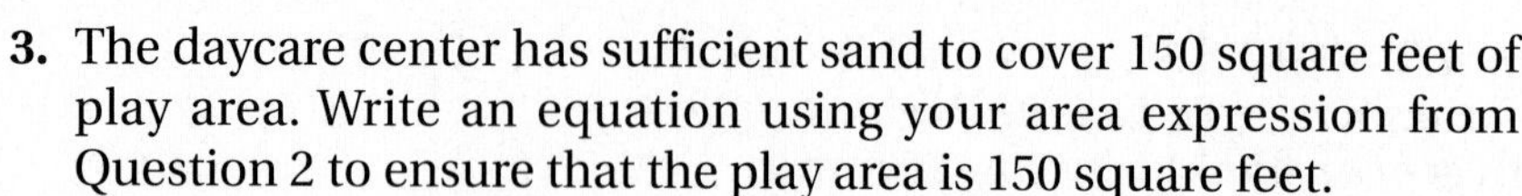

3. The daycare center has sufficient sand to cover 150 square feet of play area. Write an equation using your area expression from Question 2 to ensure that the play area is 150 square feet.

4. Solve your equation from question 3. Which of the two solutions is reasonable in this situation? Why?

5. What are the dimensions of the area to be filled with sand?

PUZZLE PROBLEM

The sum of Gary's age and the square of Rene's age is 62. However, if you add the square of Gary's age to Rene's age, the result is 176. Find their ages. (Source: Henry Ernest Dudney, **The Best of Discover Magazine's Mind Benders,** *1984)*

REVIEW EXERCISES

For Exercises 1 and 2, graph the solution set.

[2.6] **1.** $|2x - 5| \geq 1$

[4.6] **2.** $\begin{cases} x + y < 5 \\ 2x - y \geq 4 \end{cases}$

[5.1] **3.** Write 0.0004203 in scientific notation.

[5.4] **4.** Divide: $\dfrac{x^3 + x^2 - 10x + 11}{x + 4}$

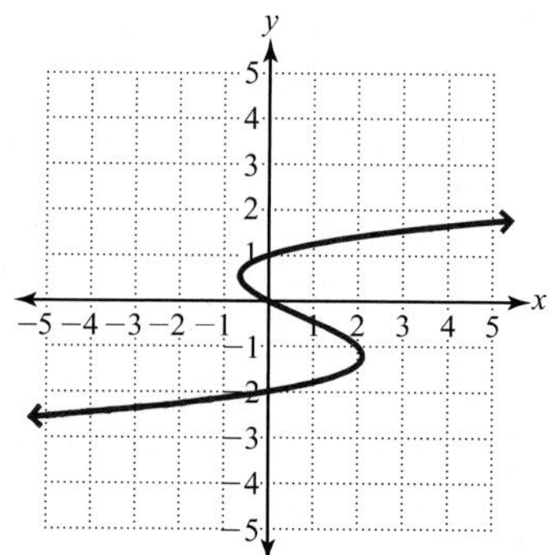

[3.5] **5.** Is the relation graphed to the right a function?

[5.2] **6.** If $f(x) = 5x^3 + 6x - 19$ and $g(x) = 3x^2 - 6x - 8$, find $(f + g)(x)$.

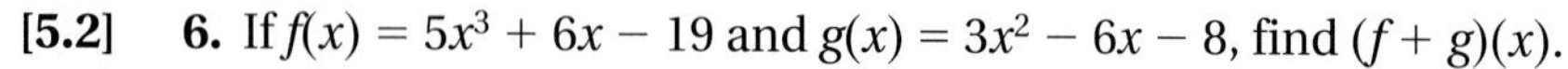

Chapter 6 Summary

Defined Terms

Section 6.1
Factored form (p. 384)
Greatest common factor (p. 384)

Section 6.4
Polynomial equation (p. 417)
Polynomial equation in standard form (p. 418)
Quadratic equation in one variable (p. 418)
Cubic equation in one variable (p. 419)

Procedures, Rules, and Key Examples

Procedures/Rules	Key Examples

Section 6.1 Greatest Common Factor and Factoring by Grouping

To find the greatest common factor of two or more monomials

1. Write the prime factorization in exponential form for each monomial. Treat variables like prime factors.
2. Write the GCF's factorization by including the prime factors (and variables) common to all the factorizations, each raised to its smallest exponent in the factorizations.
3. Multiply.

Note: If there are no common prime factors, then the GCF is 1.

Example 1: Find the GCF of $54x^3y$ and $180x^2yz$ using prime factorization.

$$54x^3y = 2 \cdot 3^3 \cdot x^3 \cdot y$$
$$180x^2yz = 2^2 \cdot 3^2 \cdot 5 \cdot x^2 \cdot y \cdot z$$
$$\text{GCF} = 2 \cdot 3^2 \cdot x^2 \cdot y = 18x^2y$$

To factor a monomial GCF out of the terms of a polynomial,

1. Find the GCF of the terms that make up the polynomial.
2. Rewrite the polynomial as a product of the GCF and the result of dividing the given polynomial by the GCF.

$$\text{Given polynomial} = \text{GCF}\left(\frac{\text{Given polynomial}}{\text{GCF}}\right)$$

Example 2: Factor $54x^3y - 180x^2yz$.

$$54x^3y - 180x^2yz = 18x^2y(3x - 10z)$$

To factor a four-term polynomial by grouping,

1. Factor out any monomial GCF (other than 1) that is common to all four terms.
2. Group together pairs of terms and factor the GCF out of each pair.
3. If there is a common binomial factor, then factor it out.
4. If there is no common binomial factor, then interchange the middle two terms and repeat the process. If there is still no common binomial factor, then the polynomial cannot be factored by grouping.

Example 3: Factor by grouping $12x^3 + 15x^2 - 8x - 10$.

$$\begin{aligned} 12x^3 + 15x^2 - 8x - 10 &= 3x^2(4x + 5) - 2(4x + 5) \\ &= (4x + 5)(3x^2 - 2) \end{aligned}$$

continued

Procedures/Rules	Key Examples
Section 6.2 Factoring Trinomials To factor a trinomial of the form $x^2 + bx + c$, **1.** Find two numbers with a product equal to c and a sum equal to b. **2.** The factored trinomial will have the form $(x + \blacksquare)(x + \blacksquare)$, where the second terms are the numbers found in step 1. Note: The signs of b and c may cause one or both signs in the binomial factors to be minus signs.	**Example 1:** Factor $x^2 - 7x + 12$. **Product** / **Sum** $(-1)(-12) = 12$ / $-1 + (-12) = -13$ $(-3)(-4) = 12$ / $-3 + (-4) = -7$ — Correct combination. $x^2 - 7x + 12 = (x - 3)(x - 4)$
To factor a trinomial of the form $ax^2 + bx + c$, where $a \neq 1$, by trial and error, **1.** Look for a monomial GCF in all the terms. If there is one, factor it out. **2.** Write a pair of *first* terms whose product is ax^2. **3.** Write a pair of *last* terms whose product is c. **4.** Verify that the sum of the *inner* and *outer* products is bx (the middle term of the trinomial). If the sum of the inner and outer products is not bx, then try the following: **a.** Exchange the first terms of the binomials from step 3, then repeat step 4. **b.** Exchange the last terms of the binomials from step 3, then repeat step 4. **c.** Repeat steps 2–4 with a different combination of first and last terms.	**Example 2:** Factor $36x^3 - 42x^2 - 8x$. First, factor out the monomial GCF, $2x$. $= 2x(18x^2 - 21x - 4)$ Factors of 18 are 1 and 18, 2 and 9, or 3 and 6. Factors of 4 are 1 and 4, or 2 and 2. We try various combinations of those factors in the first and last positions of binomial factors, checking each combination using FOIL. The correct combination is $(3x - 4)(6x + 1)$ because it gives a middle term of $-21x$. $= 2x(3x - 4)(6x + 1)$
To factor a trinomial of the form $ax^2 + bx + c$, where $a \neq 1$, by grouping, **1.** Look for a monomial GCF in all the terms. If there is one, factor it out. **2.** Find two factors of the product ac whose sum is b. **3.** Write a four-term polynomial in which bx is written as the sum of two like terms whose coefficients are the two factors you found in step 2. **4.** Factor by grouping.	**Example 3:** Factor $15xy^3 + 54xy^2 - 24xy$. $= 3xy(5y^2 + 18y - 8)$ — Factor out the monomial GCF, $3xy$. Multiply a and c: $5(-8) = -40$. Now we find two factors of -40 whose sum is 18. Note that 20 and -2 work, so we write the middle term, $18y$, as $20y - 2y$ and then factor by grouping. $= 3xy(5y^2 + 20y - 2y - 8)$ $= 3xy[5y(y + 4) - 2(y + 4)]$ $= 3xy(y + 4)(5y - 2)$
If a trinomial is not exactly in the form $ax^2 + bx + c$ but it resembles that form, then we can try substitution to see if it is in fact of the form $ax^2 + bx + c$. If so, then we can factor using trial and error or grouping.	**Example 4:** Factor $6(x - 4)^2 - 19(x - 4) - 7$. This polynomial resembles the form $ax^2 + bx + c$, so we try substitution, substituting u for $x - 4$. $= 6u^2 - 19u - 7$ — Substitute u for $x - 4$. $= (3u + 1)(2u - 7)$ — Factor. $= [3(x - 4) + 1][2(x - 4) - 7]$ — Substitute $x - 4$ for u. $= [3x - 12 + 1][2x - 8 - 7]$ — Distribute. $= (3x - 11)(2x - 15)$ — Simplify. ***continued***

Procedures/Rules	Key Examples

Section 6.3 Factoring Special Products and Factoring Strategies

Rules for factoring special products:

Perfect square trinomials: $a^2 + 2ab + b^2 = (a + b)^2$

$a^2 - 2ab + b^2 = (a - b)^2$

Difference of squares: $a^2 - b^2 = (a + b)(a - b)$

Difference of cubes: $a^3 - b^3 = (a - b)(a^2 + ab + b^2)$

Sum of cubes: $a^3 + b^3 = (a + b)(a^2 - ab + b^2)$

To factor a polynomial, first factor out any monomial GCF, then consider the number of terms in the polynomial.

I. Four terms: Try to factor by grouping.

II. Three terms: Determine if the trinomial is a perfect square.

A. If the trinomial is a perfect square, then consider its form.

1. If it is in the form $a^2 + 2ab + b^2$, then the factored form is $(a + b)^2$.
2. If it is in the form $a^2 - 2ab + b^2$, then the factored form is $(a - b)^2$.

B. If the trinomial is not a perfect square, then consider its form.

1. If it is in the form $x^2 + bx + c$, then find two factors of c whose sum is b and write the factored form as (x + first number)(x + second number).
2. If it is in the form $ax^2 + bx + c$, where $a \neq 1$, then use trial and error. Or, find two factors of ac whose sum is b, write these factors as coefficients of two like terms that, when combined, equal bx, then factor by grouping.

III. Two terms: Determine if the binomial is a difference of squares, sum of cubes, or difference of cubes.

A. If it is a difference of squares, $a^2 - b^2$, then the factors are conjugates and the factored form is $(a + b)(a - b)$. Note that a sum of squares cannot be factored.

B. If it is a difference of cubes, $a^3 - b^3$, then the factored form is $(a - b)(a^2 + ab + b^2)$.

C. If it is a sum of cubes, $a^3 + b^3$, then the factored form is $(a + b)(a^2 - ab + b^2)$.

Example 1: Factor.

a. $25x^2 + 40x + 16 = (5x + 4)^2$

b. $49t^2 - 28tu + 4u^2 = (7t - 2u)^2$

c. $9y^2 - 25 = (3y + 5)(3y - 5)$

d. $8x^3 - 125 = (2x - 5)(4x^2 + 10x + 25)$

e. $n^3 + 125 = (n + 5)(n^2 - 5n + 25)$

Example 2: Factor completely.

a. $4y(2x + 1)^2 - 36x^2y$

$= 4y[(2x + 1)^2 - 9x^2]$ **Factor out the monomial GCF, $4y$.**

$= 4y[(2x + 1) + 3x][(2x + 1) - 3x]$ **Factor the difference of squares.**

$= 4y(5x + 1)(1 - x)$ **Simplify.**

b. $x^5 - 9x^3 + 8x^2 - 72$

$= x^3(x^2 - 9) + 8(x^2 - 9)$ **Factor by grouping.**

$= (x^2 - 9)(x^3 + 8)$ **Factor out $x^2 - 9$.**

$= (x + 3)(x - 3)(x + 2)(x^2 - 2x + 4)$ **Factor the difference of squares and the sum of cubes.**

continued

Procedures/Rules	Key Examples
Section 6.4 Solving Equations by Factoring Zero-factor theorem: If a and b are real numbers and $ab = 0$, then $a = 0$ or $b = 0$. To solve a polynomial equation using factoring, **1.** Write the equation in standard form (set one side equal to 0). **2.** Write the polynomial in factored form. **3.** Use the zero-factor theorem to solve.	**Example 1:** Solve. **a.** $2x^2 = 4 - 7x$ $2x^2 + 7x - 4 = 0$ **Write in standard form.** $(2x - 1)(x + 4) = 0$ **Factor.** $2x - 1 = 0$ or $x + 4 = 0$ **Use the zero-factor theorem.** $2x = 1$ $\quad x = -4$ **Solve the equations.** $x = \frac{1}{2}$ **b.** $n^3 + n(n - 1) = 3n^2 + 2(4n - 9)$ $n^3 + n^2 - n = 3n^2 + 8n - 18$ **Distribute.** $n^3 - 2n^2 - 9n + 18 = 0$ **Write in standard form.** $n^2(n - 2) - 9(n - 2) = 0$ **Factor by grouping.** $(n - 2)(n^2 - 9) = 0$ **Factor out $n - 2$.** $(n - 2)(n + 3)(n - 3) = 0$ **Factor the difference of squares.** $n - 2 = 0$ or $n + 3 = 0$ or $n - 3 = 0$ $n = 2$ $\quad n = -3$ $\quad n = 3$
The Pythagorean theorem: Given a right triangle where a and b represent the lengths of the legs and c represents the length of the hypotenuse, then $a^2 + b^2 = c^2$. 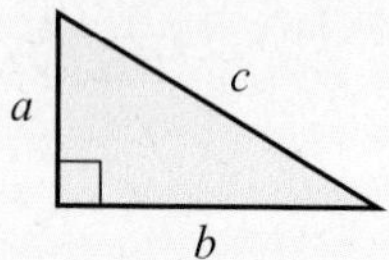	**Example 2:** Find the unknown length in the following right triangle. 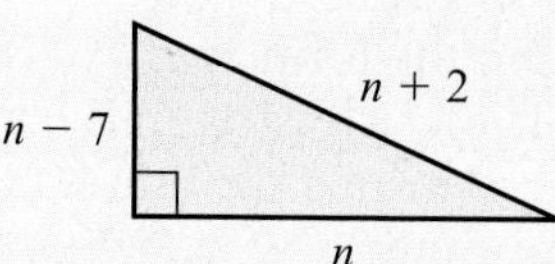 Solution: Use the Pythagorean theorem, $a^2 + b^2 = c^2$. $(n - 7)^2 + (n)^2 = (n + 2)^2$ $2n^2 - 14n + 49 = n^2 + 4n + 4$ $n^2 - 18n + 45 = 0$ **Write in standard form.** $(n - 15)(n - 3) = 0$ **Factor.** $n - 15 = 0$ or $n - 3 = 0$ **Use the zero-factor theorem.** $n = 15$ $\quad n = 3$ Answer: If $n = 3$, then $n - 7 = -4$, which makes no sense as a length, so n must be 15. Therefore, $n - 7 = 8$ and $n + 2 = 17$. *continued*

Procedures/Rules

Section 6.4 Solving Equations by Factoring (continued)

Given a function $f(x)$, we can find its x-intercepts by solving the equation $f(x) = 0$.

Key Examples

Example 3: Find the x-intercepts of the function $f(x) = x^2 + x - 6$ and then graph the function.

$x^2 + x - 6 = 0$		Set $f(x) = 0$.
$(x + 3)(x - 2) = 0$		Factor.
$x + 3 = 0$	or $x - 2 = 0$	Use the zero-factor theorem.
$x = -3$	$x = 2$	Solve each equation.

The x-intercepts are $(-3, 0)$ and $(2, 0)$.

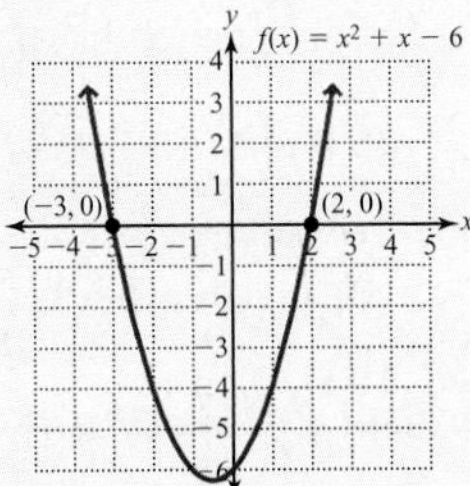

Chapter 6 Review Exercises

For Exercises 1–6, answer true or false.

[6.1] **1.** Factoring by grouping is used when the polynomial contains two terms.

[6.1] **2.** When using prime factorization to find the GCF of a set of monomials, if there are no common prime factors, then the GCF is 0.

[6.2] **3.** $x^2 + x + 1$ is an example of a prime polynomial.

[6.3] **4.** $x^2 - 9 = (x + 3)(x - 3)$

[6.3] **5.** $x^2 + 25 = (x + 5)(x + 5)$

[6.4] **6.** To solve $y(y + 2) = 24$, rewrite the equation as $y = 24$ and $y + 2 = 24$, and then solve each equation.

For Exercises 7–10, complete the rule.

[6.1] **7.** The largest monomial that divides all the terms in a set of terms evenly is the ____________________.

[6.2] **8.** To factor a trinomial of the form $ax^2 + bx + c$, where $a \neq 1$, by grouping,

1. Look for a monomial GCF in all the terms. If there is one, factor it out.
2. Find two factors of ac whose sum is ____.
3. Write a four-term polynomial in which bx is written as the sum of two like terms whose coefficients are the two factors you found in step 2.
4. Factor by grouping.

[6.3] **9.** When factoring, $a^2 + 2ab + b^2 =$ __________.

[6.4] **10.** The zero-factor theorem states that if a and b are real numbers and $ab = 0$, then $a =$ ____ or $b =$ ____.

[6.1] ***For Exercises 11–14, find the GCF.***

11. $16x^4y^2, 24x^3y^9$

12. $20mn^3, 15m^4n^2$

13. $2u^4v^3, 3xy^9$

14. $4(x + 1), 6(x + 1)$

[6.1] ***For Exercises 15–22, factor out the GCF.***

15. $u^5 - u^3 - u$

16. $13d^2 - 26de$

17. $4h^2k^6 - 2h^5k$

18. $12cd - 4c^2d^2 + 10c^4d^4$

19. $16p^6q^3 - 12p^8q^5 + 13p^7q^2$

20. $9w^6v^8 + 6wv^6 - 12w^3v^3$

21. $17(w + 3) - m(w + 3)$

22. $2y(x + 3) + (x + 3)$

[6.1] *For Exercises 23–30, factor by grouping.*

23. $3m + mn + 6 + 2n$

24. $a^3 + 3a^2 + 2a + 6$

25. $2x + 2y - ax - ay$

26. $bc^2 + bd^2 - 5c^2 - 5d^2$

27. $x^2y - x^2s - ry + rs$

28. $4k^2 - k + 4k - 1$

29. $8uv^2 - 4v^2 + 20uv - 10v$

30. $5c^2d^2 - 5d^3 + 20c^2d - 20d^2$

[6.2] *For Exercises 31–44, factor the trinomials completely.*

31. $a^2 - 10a + 9$

32. $m^2 + 20m + 51$

33. $y^2 + y - 42$

34. $x^2 - 7x - 30$

35. $3x^2 - x - 14$

36. $16h^2 + 10h + 1$

37. $4u^2 - 2u + 3$

38. $6t^2 + t - 15$

39. $s^2 - 11st + 10t^2$

40. $6u^2 + 13uv + 6v^2$

41. $5m^2 - 16mn + 3n^2$

42. $4x^2 + 5xy - 6y^2$

43. $b^4 - 7b^3 - 18b^2$

44. $2x^3 - 16x^2 - 40x$

[6.2] *For Exercises 45–48, factor by grouping.*

45. $2u^2 + 9u + 10$

46. $3m^2 - 8m + 4$

47. $10u^2 + 7u - 3$

48. $6y^2 - 13y - 5$

[6.2] *For Exercises 49–52, factor using substitution.*

49. $x^4 + 5x^2 + 4$

50. $3c^4 + 13c^2 + 4$

51. $2h^6 + 9h^3 + 4$

52. $3(k + 1)^2 - 2(k + 1) - 5$

[6.3] *For Exercises 53–58, factor the perfect square.*

53. $x^2 + 6x + 9$

54. $y^2 + 12y + 36$

55. $m^2 - 4m + 4$

56. $w^2 - 14w + 49$

57. $9d^2 + 42d + 49$

58. $4c^2 - 28c + 49$

[6.3] *For Exercises 59–66, factor the difference of squares.*

59. $h^2 - 9$

60. $p^2 - 64$

61. $9d^2 - 4$

62. $81k^2 - 100$

63. $2w^2 - 50$

64. $4q^2 - 36$

65. $25y^2 - 9z^2$

66. $49c^2d^2 - 100b^2$

[6.3] *For Exercises 67–70, factor the difference of cubes.*

67. $c^3 - 27$

68. $v^3 - 8$

69. $8b^3 - 1$

70. $64d^3 - 8c^3$

[6.3] *For Exercises 71–74, factor the sum of cubes.*

71. $27 + m^3$

72. $x^3 + 1000$

73. $27b^3 + 8a^3$

74. $2v^3 + 54$

[6.3] *For Exercises 75–82, factor completely.*

75. $9d^2 - 6d + 1$

76. $3m^2 - 3n^2$

77. $2a^2 - 5a - 12$

78. $x^2 + 9$

79. $3p^4 + 3p^3 - 90p^2$

80. $x^4 - 16$

81. $15a^3b^2c^7 + 3a^2b^4c^2 + 5a^9bc^3$

82. $w^3 - (2 + y)^3$

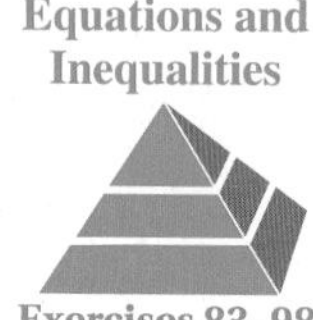

[6.4] *For Exercises 83–88, solve.*

83. $(x + 4)(x - 1) = 0$

84. $x^2 - 64 = 0$

85. $w^2 - 2w - 3 = 0$

86. $2d^2 + 3d - 5 = 0$

87. $6y^2 = 7y - 1$

88. $4x(x + 7) + 9 = -40$

[6.4] *For Exercises 89 and 90, find the x-intercepts and then sketch the graph.*

89. $f(x) = x^2 - 4$

90. $f(x) = x(x + 3)(x - 2)$

[6.4] *For Exercises 91–98, solve.*

91. The product of two natural numbers is 56. If the second number is 10 more than the first, find the two numbers.

92. Find two consecutive positive integers whose product is 110.

93. Find two consecutive positive even integers whose product is 288.

94. A rectangular room is 3 feet longer than it is wide. If the area of the room is 88 square feet, find the dimensions of the room.

95. A stage is to be constructed in the shape of a trapezoid with a base length that is three times the height. If the area is to be 672 square feet, find the height and the length of the base.

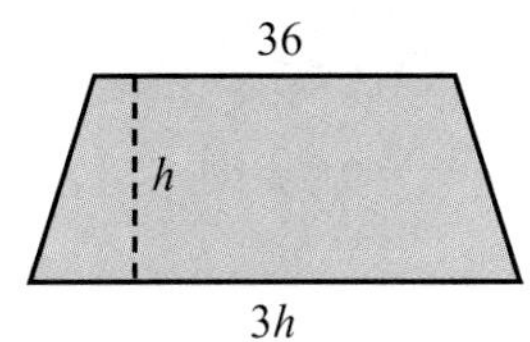

96. Find the length of each side of the right triangle shown.

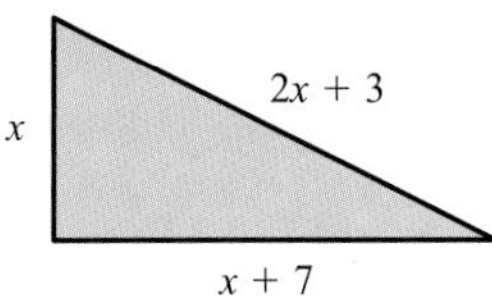

97. A sound designer is trying to find optimal positions for speakers in a room. If she places a speaker on a wall as shown, find the height of the speaker and its distance from her ear.

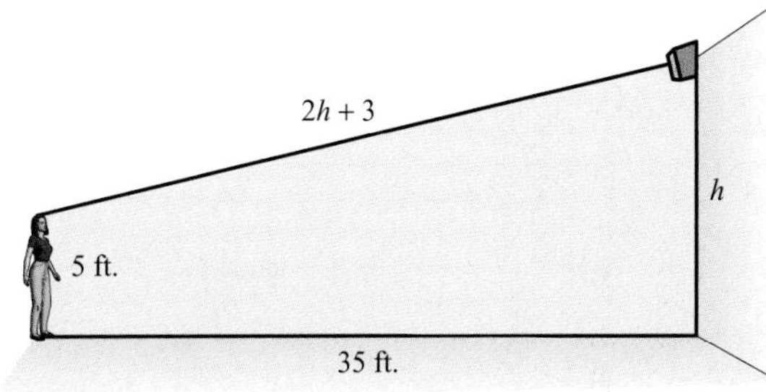

98. A ball is thrown upward at 12 feet per second from a building 60 feet high. Find the time it takes for the ball to be 6 feet above the ground. Use the formula $h = -16t^2 + v_0t + h_0$, where h is the final height in feet, t is the time of travel in seconds, v_0 is the initial velocity in feet per second, and h_0 is the initial height in feet of an object traveling upwards.

Chapter 6 Practice Test

1. Find the GCF of $14m^7n^2$, $21m^3n^9$.

Factor completely.

2. $3m + 6m^3 - 9m^6$

3. $3m + 3n - m - n$

4. $9n^2 - 16$

5. $8x^3 - 27$

6. $y^2 + 14y + 49$

7. $q^2 - 2q - 48$

8. $3ab^2 - 30ab + 24a$

9. $5 - 125t^2$

10. $6d^2 + d - 2$

11. $8c^3 + 8d^3$

12. $w^3 + 2w^2 + 3w + 6$

13. $s^4 - 81$

14. $5p^2 - 7pq - 12q^2$

Solve.

15. $8x^2 - 14x + 5 = 0$

16. $x(x + 3) - 6 = 12$

17. $2x^3 + x^2 = 15x$

18. Find the length of each side of the right triangle shown.

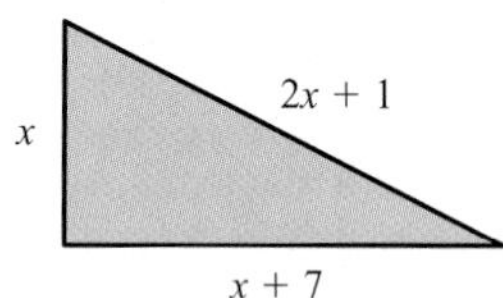

19. A rock is thrown upward at 4 feet per second from a building 56 feet high. Find the time it takes for the rock to hit the ground. Use the formula $h = -16t^2 + v_0t + h_0$, where h is the final height in feet, t is the time of travel in seconds, v_0 is the initial velocity in feet per second, and h_0 is the initial height in feet of an object traveling upwards.

Find the x-intercepts and graph the function.

20. $f(x) = x^2 - 9$

Chapters 1–6 Cumulative Review Exercises

For Exercises 1–6, answer true or false.

[1.1] **1.** The set of irrational numbers is a subset of the real numbers.

[3.5] **2.** The graph of $y = 2x - 1$ is the same as the graph of $f(x) = 2x - 1$.

[3.5] **3.** $f(x) = 3x - x^2$ is an example of a linear function.

[4.4] **4.** A 3×2 matrix has 3 rows and 2 columns.

[5.1] **5.** When writing a number n in scientific notation where $0 < |n| < 1$, the exponent will be negative.

[6.3] **6.** The expression $a^2 + b^2$ can be factored to $(a + b)(a + b)$.

For Exercises 7–11, fill in the blank.

[3.3] **7.** The slopes of parallel lines are ________.

[3.5] **8.** To graph a quadratic equation:

a. Find ordered pair solutions, and plot them in the coordinate plane. Continue finding and plotting solutions until the shape of the ______ can be clearly seen.

b. Connect the points to form a ________.

[5.1] **9.** To multiply monomials: 1. Multiply ________. 2. ____ the exponents of the like bases. 3. Write any unlike variable bases unchanged in the product.

[6.3] **10.** Complete the rule for factoring a sum of cubes: $a^3 + b^3 =$ ________________.

[6.4] **11.** To solve an equation in which two or more factors are equal to 0, use the zero-factor theorem: 1. Set each _____ equal to zero. 2. Solve each of these equations.

[1.4] **12.** Evaluate $xy - 2y^2$ when $x = 2$ and $y = -1$.

For Exercises 13–19, simplify.

[1.3] **13.** $\dfrac{4(3^2 - 10) - 4}{3 - \sqrt{25 - 16}}$

[1.3] **14.** $(3 - 4)^5 + (1 - 2)^4$

[5.2] **15.** $(2x^3 + 5x^2 - 19x + 14) - (17x^2 + 2x - 6)$

[5.3] **16.** $(2x + 1)(4x^2 - 3x + 2)$

[5.3] **17.** $(3x - 1)(3x + 1)$

[5.1] **18.** $17x^0 - 3(2x)^0$

[5.1] **19.** $\left(\dfrac{2}{x}\right)^{-3}$

[5.4] **20.** Use long division to divide $\dfrac{10x^3 + x^2 - 7x + 3}{2x - 1}$.

For Exercises 21–28, factor.

[6.1] **21.** $4m - 16m^2$

[6.2] **22.** $8x^2 + 29x - 12$

[6.2] **23.** $3r^2 + 5rs + 2s^2$

[6.3] **24.** $w^3 - 27$

[6.3] **25.** $49a^2 + 84a + 36$

[6.3] **26.** $64p^3 + 8$

[6.1] **27.** $3x - 6y + ax - 2ay$

[6.3] **28.** $c^4 - 3c^2 - 4$

For Exercises 29–32, solve.

[2.1] **29.** $9x - [14 + 2(x - 3)] = 8x - 2(3x - 1)$

[2.5] **30.** $|2x - 3| - 9 = 16$

[6.4] **31.** $5x^2 + 26x = 24$

[6.4] **32.** $x(2x - 1)(x + 3) = 0$

Equations and Inequalities

Exercises 29–42, 44–50

For Exercises 33–36, solve: a. Write the solution set in set-builder notation.
b. Write the solution set in interval notation.
c. Graph the solution set.

[2.3] **33.** $\frac{3}{4}x + \frac{2}{3} \le \frac{5}{6}x + \frac{1}{2}$

[2.4] **34.** $2x + 1 \le -5$ or $2x + 1 > 9$

[2.4] **35.** $7 > -2x - 1 > 5$

[2.6] **36.** $2|x - 3| - 2 < 10$

[4.1] **37.** Determine whether (1, 1) is a solution for $\begin{cases} 5x + 6y = 11 \\ 2x - 4y = -2 \end{cases}$.

For Exercises 38 and 39, solve.

[4.1] **38.** $\begin{cases} 5x - 2y = -1 \\ 3x + 2y = -7 \end{cases}$

[4.2] **39.** $\begin{cases} x - 2y + z = 8 \\ x + y + z = 5 \\ 2x + y - 2z = -5 \end{cases}$

[4.5] **40.** Find x. $\begin{vmatrix} 3 & -2 \\ x & 4 \end{vmatrix} = 16$

[4.5] **41.** Find the minor of 3 in $\begin{bmatrix} 3 & -5 & 0 \\ 2 & -4 & 1 \\ -2 & -1 & -3 \end{bmatrix}$.

[4.6] **42.** Graph the solution set for the system of inequalities. $\begin{cases} x + y < 6 \\ y \geq 2x \end{cases}$

[3.5] **43.** The table shows the average height of some of the world's most famous walls. Determine if the relation is a function, and identify the domain and range.

Berlin Wall (Germany)	12 ft.
Hadrian's Wall (Britain)	15 ft.
West Bank Wall (Israel)	25 ft.
Great Wall (China)	26 ft.

Source: *USA Today* research

[3.3] **44.** Find the slope of a line perpendicular to the graph of $2x + 4y = 8$.

[6.4] ***For Exercises 45 and 46, find the x- and y-intercepts, then graph.***

45. $f(x) = x^2 + x - 6$

46. $f(x) = x^3 - x^2 - 2x$

For Exercises 47–50, solve.

[2.4] **47.** A mail-order DVD club offers one bonus DVD if the total of the order is from \$50 up to \$80 and two bonus DVDs if the total order is \$80 or more. Lyle decides to buy the lowest-priced DVDs at \$10 each.

a. In what range would the number of DVDs he orders have to be in order to receive one bonus DVD?

b. In what range would the number of DVDs he orders have to be in order to receive two bonus DVDs?

[2.2] **48.** Adults aged 20–24 spend 2 minutes longer driving than twice the time of teenagers aged 15–19. If the combined minutes per day that both groups spend driving is 77 minutes, how many minutes does each group spend driving? (*Source:* U.S. Department of Transportation)

[2.2] **49.** Two angles are complementary. The larger angle is six less than twice the smaller angle. Find the measure of the two angles.

[6.4] **50.** A metal lid is in the shape of a trapezoid as shown. If the area of the metal used is to be 60 square inches, find each length represented in the figure.

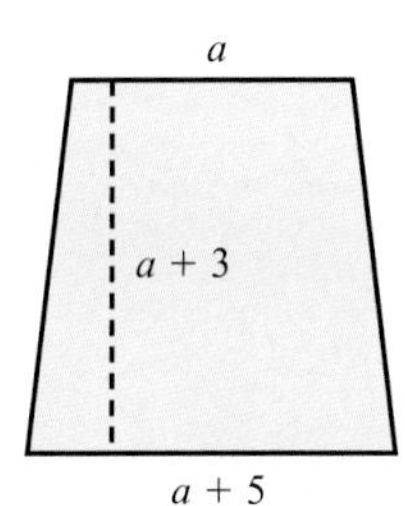

CHAPTER 7

Rational Expressions and Equations

"Nothing is ever enough. Images split the truth in fractions."

—Denise Levertov, U.S. poet

This chapter follows the same pattern of development that we have seen in earlier chapters. We will first define a class of expressions called *rational expressions*. We will then learn to evaluate, rewrite, and perform arithmetic operations with these rational expressions. Finally, we will solve equations that contain rational expressions.

7.1 Simplifying, Multiplying, and Dividing Rational Expressions

OBJECTIVES

1. Simplify rational expressions.
2. Multiply rational expressions.
3. Divide rational expressions.
4. Evaluate rational functions.
5. Find the domain of a rational function.
6. Graph rational functions.

OBJECTIVE 1. Simplify rational expressions. Now that we have explored polynomials, we are ready to learn about **rational expressions**.

DEFINITION ***Rational expression:*** An expression that can be written in the form $\frac{P}{Q}$, where P and Q are polynomials and $Q \neq 0$.

Some rational expressions are

$$\frac{4x^2y^3}{3xy^2}, \quad \frac{3x-5}{x^2-16}, \quad \text{and} \quad \frac{2x^2-3x+6}{x^2-x-6}$$

Connection Notice that the definition of a rational expression is like the definition of a rational number. Recall that a rational number can be expressed in the form $\frac{a}{b}$, where a and b are integers and $b \neq 0$.

Simplifying a rational expression is like simplifying a fraction. To simplify a fraction to lowest terms, we use the rule $\frac{an}{bn} = \frac{a \cdot 1}{b \cdot 1} = \frac{a}{b}$ when $b \neq 0$ and $n \neq 0$. We can rewrite this rule so that it applies to rational expressions.

RULE **Fundamental Principle of Rational Expressions**

$\frac{PR}{QR} = \frac{P \cdot 1}{Q \cdot 1} = \frac{P}{Q}$, where P, Q, and R are polynomials and Q and R are not 0.

The rule indicates that a factor common to both the numerator and denominator can be divided out of a rational expression. Like rational numbers, rational expressions are in lowest terms if the greatest common factor of the numerator and denominator is 1.

Consider the following comparison between simplifying a fraction and simplifying a similar rational expression:

$$\frac{14}{21} = \frac{2 \cdot 7}{3 \cdot 7} = \frac{2 \cdot 1}{3 \cdot 1} = \frac{2}{3} \qquad \frac{2a}{3a} = \frac{2 \cdot a}{3 \cdot a} = \frac{2 \cdot 1}{3 \cdot 1} = \frac{2}{3}$$

7 is the common factor here.

a is the common factor here.

Note: *This simplification is true when $a \neq 0$. If $a = 0$, the original expression is indeterminate. We must avoid variable values that make an expression undefined or indeterminate.*

PROCEDURE **Simplifying Rational Expressions to Lowest Terms**

1. Write the numerator and denominator in factored form.
2. Divide out all common factors in the numerator and denominator.
3. Multiply the remaining factors in the numerator and the remaining factors in the denominator.

EXAMPLE 1 Simplify.

a. $\frac{18x^4}{21x^2}$

Solution Write the numerator and denominator in factored form, then divide out all common factors.

Note: *To the right of each answer in Example 1, we have excluded those values that cause the original expression to be undefined or indeterminate. In future examples we will not have these statements because we will assume variables would not be replaced with such values.*

$$\frac{18x^4}{21x^2} = \frac{2 \cdot 3 \cdot 3 \cdot x \cdot x \cdot x \cdot x}{3 \cdot 7 \cdot x \cdot x}$$

Note: *The common factors are a single 3 and two x's. These form the GCF that we divide out, which is $3x^2$.*

Note: *We will no longer show the step with the 1s and simply highlight the common factors that are eliminated.*

$$= \frac{2 \cdot 1 \cdot 3 \cdot 1 \cdot x \cdot x}{1 \cdot 7 \cdot 1}$$

$$= \frac{6x^2}{7} \quad \text{Answer if } x \neq 0.$$

There are different styles for showing the process of dividing out common factors. For example, some people prefer using cancel marks.

$$\frac{18x^4}{21x^2} = \frac{2 \cdot \cancel{3} \cdot 3 \cdot \cancel{x} \cdot \cancel{x} \cdot x \cdot x}{\cancel{3} \cdot 7 \cdot \cancel{x} \cdot \cancel{x}} = \frac{6x^2}{7}$$

b. $\frac{4x^2 + 16x}{3x^2 + 12x}$

Solution $\frac{4x^2 + 16x}{3x^2 + 12x} = \frac{4 \cdot x \cdot (x + 4)}{3 \cdot x \cdot (x + 4)}$ Factor the numerator and denominator, then divide out the common factors, x and $x + 4$.

$= \frac{4}{3}$ Answer if $x \neq 0, -4$.

c. $\dfrac{x^2 - 9}{x^2 - x - 12}$

Solution $\dfrac{x^2 - 9}{x^2 - x - 12} = \dfrac{(x + 3)(x - 3)}{(x + 3)(x - 4)}$ Factor the numerator and denominator, then divide out the common factor, $x + 3$.

$= \dfrac{x - 3}{x - 4}$ Answer if $x \neq 4, -3$.

d. $\dfrac{x^3 + 27}{3x^2 + 5x - 12}$

Solution $\dfrac{x^3 + 27}{3x^2 + 5x - 12} = \dfrac{(x + 3)(x^2 - 3x + 9)}{(x + 3)(3x - 4)}$ Factor the numerator and denominator, then divide out the common factor, $x + 3$.

$= \dfrac{x^2 - 3x + 9}{3x - 4}$ Answer if $x \neq -3, \frac{4}{3}$.

e. $\dfrac{2x^2 - 11x + 15}{x^4 - 81}$

Solution $\dfrac{2x^2 - 11x + 15}{x^4 - 81} = \dfrac{(2x - 5)(x - 3)}{(x^2 + 9)(x^2 - 9)}$

$= \dfrac{(2x - 5)(x - 3)}{(x^2 + 9)(x + 3)(x - 3)}$ Factor the numerator and denominator completely and divide out the common factor, $x - 3$.

$= \dfrac{2x - 5}{(x^2 + 9)(x + 3)}$ Answer if $x \neq 3, -3$.

Warning: A common error is to divide out common terms instead of common factors, as in $\dfrac{6 + 3}{3} = \dfrac{6 + \cancel{3}}{\cancel{3}} = 6 + 1 = 7$. Using the order of operations, we see the correct answer is $\dfrac{6 + 3}{3} = \dfrac{9}{3} = 3$. The mistake in the first simplification is that we divided out a *term* and not a *factor*. Remember, terms are separated by + and − signs and factor implies multiplication. Therefore, to simplify a rational expression, you must first write the numerator and the denominator in *factored* form and then divide out the common factors. Examples with rational expressions follow.

Correct: $\dfrac{x^2 - 6x + 8}{x^2 + 2x - 8} = \dfrac{(x - 4)\cancel{(x - 2)}}{(x + 4)\cancel{(x - 2)}} = \dfrac{x - 4}{x + 4}$

$x - 2$ is a factor common to both the numerator and the denominator, so we can divide both the numerator and the denominator by $x - 2$.

Incorrect: $\dfrac{x^2 - 6x + 8}{x^2 + 2x - 8} = \dfrac{\cancel{x^2} - 6x + \cancel{8}}{\cancel{x^2} + 2x - \cancel{8}} = \dfrac{1 - 6x + 1}{1 + 2x - 1} = \dfrac{2 - \cancel{6}x}{\cancel{2}x} = \dfrac{1 - 3x}{x}$

x^2 and 8 are terms common to the numerator and the denominator. Since they are not factors, it is wrong to divide them out.

YOUR TURN Simplify.

a. $\dfrac{24a^3}{18a^5}$ **b.** $\dfrac{3x^2 + 6x}{5x^2 + 10x}$ **c.** $\dfrac{x^2 - 4}{x^2 + 9x + 14}$ **d.** $\dfrac{x^3 - 8}{x^2 - 4}$ **e.** $\dfrac{x^4 - 16}{3x^2 + 2x - 8}$

ANSWERS

a. $\dfrac{4}{3a^2}$ b. $\dfrac{3}{5}$

c. $\dfrac{x - 2}{x + 7}$

d. $\dfrac{x^2 + 2x + 4}{x + 2}$

e. $\dfrac{(x^2 + 4)(x - 2)}{3x - 4}$

Rational expressions containing binomial factors that are additive inverses require an extra step. Let's look at binomial factors that are additive inverses in Example 2.

EXAMPLE 2 Simplify. $\dfrac{3x^2 - 7x - 6}{9 - 3x}$

Solution $\dfrac{3x^2 - 7x - 6}{9 - 3x}$

$= \dfrac{(3x + 2)(x - 3)}{3(3 - x)}$ Factor the numerator and denominator.

It appears that there are no common factors. However, $x - 3$ and $3 - x$ are additive inverses because $(x - 3) + (3 - x) = 0$. We can get common factors by factoring -1 out of either expression. We will use the expression in the numerator.

Note: *We could have factored the -1 out of the denominator.*

$$\frac{(3x + 2)(x - 3)}{3(-1)(x - 3)} = \frac{3x + 2}{-3}$$

$= \dfrac{(3x + 2)(-1)(3 - x)}{3(3 - x)}$ Factor -1 out of $x - 3$. $(x - 3) = -1(-x + 3) = (-1)(3 - x)$

$= \dfrac{-3x - 2}{3}$ or $-\dfrac{3x + 2}{3}$ Divide out the common factor $3 - x$, then multiply the remaining factors.

Example 2 also illustrates the rule of sign placement in a fraction or rational expression. Remember that with a negative fraction (or rational expression), the minus sign can be placed in the numerator, denominator, or aligned with the fraction line.

RULE $-\dfrac{P}{Q} = \dfrac{-P}{Q} = \dfrac{P}{-Q}$, where P and Q are polynomials and $Q \neq 0$.

Warning: A rational expression like $\dfrac{4 + x}{4 - x}$ looks like it could be simplified by factoring -1 out of $4 + x$ or $4 - x$. Although $4 + x$ and $4 - x$ appear to be additive inverses, they are not because $(4 + x) + (4 - x) = 8$, not 0. Consequently, factoring -1 out of the numerator or denominator as in $\dfrac{4 + x}{(-1)(-4 + x)}$ does not give common factors. In fact, $\dfrac{4 + x}{4 - x}$ is in simplest form.

YOUR TURN

Simplify. $\dfrac{x^2 - 4}{2x - x^2}$

ANSWER

$-\dfrac{x + 2}{x}$

OBJECTIVE 2. Multiply rational expressions. To multiply fractions, we use the rule $\frac{a}{b} \cdot \frac{c}{d} = \frac{ac}{bd}$, where $b \neq 0$ and $d \neq 0$. We can rewrite this rule so that it applies to multiplying rational expressions.

RULE $\frac{P}{Q} \cdot \frac{R}{S} = \frac{PR}{QS}$, where P, Q, R, and S are polynomials and Q and $S \neq 0$.

We use the same procedure for multiplying rational expressions as for multiplying rational numbers.

PROCEDURE Multiplying Rational Expressions

1. Write each numerator and denominator in factored form.
2. Divide out factors common to both the numerator and denominator.
3. Multiply numerator by numerator and denominator by denominator.
4. Simplify as needed.

Connection Multiplying rational expressions is very much like reducing rational expressions.

EXAMPLE 3 Multiply.

a. $\frac{8a^2b}{7a^3b} \cdot \frac{21ab^2}{4a^2b}$

Solution $\frac{8a^2b}{7a^3b} \cdot \frac{21ab^2}{4a^2b} = \frac{2 \cdot 2 \cdot 2 \cdot a \cdot a \cdot b}{7 \cdot a \cdot a \cdot a \cdot b} \cdot \frac{3 \cdot 7 \cdot a \cdot b \cdot b}{2 \cdot 2 \cdot a \cdot a \cdot b}$ **Factor the numerators and denominators.**

$= \frac{2}{1} \cdot \frac{3 \cdot b}{a \cdot a}$ **Divide out the common factors of 2, 7, *a*, and *b*.**

$= \frac{6b}{a^2}$ **Multiply the remaining factors.**

b. $-\frac{2a^2}{3a - 12} \cdot \frac{5a - 20}{30a^3}$

Solution $-\frac{2a^2}{3a - 12} \cdot \frac{5a - 20}{30a^3} = -\frac{2 \cdot a \cdot a}{3 \cdot (a - 4)} \cdot \frac{5 \cdot (a - 4)}{2 \cdot 3 \cdot 5 \cdot a \cdot a \cdot a}$ **Factor the numerators and denominators.**

$= -\frac{1}{3} \cdot \frac{1}{3a}$ **Divide out the common factors of 2, 5, *a*, and *a* − 4.**

$= -\frac{1}{9a}$ **Multiply the remaining factors.**

c. $\dfrac{x^2 - x - 6}{x^2 + 2x - 15} \cdot \dfrac{x^2 + 4x - 12}{x^2 - 4}$

Solution $\dfrac{x^2 - x - 6}{x^2 + 2x - 15} \cdot \dfrac{x^2 + 4x - 12}{x^2 - 4}$

$= \dfrac{(x+2)(x-3)}{(x+5)(x-3)} \cdot \dfrac{(x-2)(x+6)}{(x+2)(x-2)}$ Factor the numerators and denominators.

$= \dfrac{1}{x+5} \cdot \dfrac{x+6}{1}$ Divide out the common factors of $x + 2, x - 3$, and $x - 2$.

$= \dfrac{x+6}{x+5}$ Multiply the remaining factors.

d. $\dfrac{3 - a}{a^2 + 3a + 2} \cdot \dfrac{a^2 - 4}{a^2 + 2a - 15}$

Solution $\dfrac{3 - a}{a^2 + 3a + 2} \cdot \dfrac{a^2 - 4}{a^2 + 2a - 15}$

$= \dfrac{3 - a}{(a+2)(a+1)} \cdot \dfrac{(a+2)(a-2)}{(a-3)(a+5)}$ Factor the numerators and denominators.

$= \dfrac{-1(a-3)}{(a+2)(a+1)} \cdot \dfrac{(a+2)(a-2)}{(a-3)(a+5)}$ Since $3 - a$ and $a - 3$ are additive inverses, factor -1 from $3 - a$ and get $-1(a - 3)$.

$= \dfrac{-1}{a+1} \cdot \dfrac{a-2}{a+5}$ Divide out the common factors of $a - 3$ and $a + 2$.

$= \dfrac{2 - a}{(a+1)(a+5)}$ Multiply the remaining factors.

Note: *We could write the solution as* $-\dfrac{a-2}{(a+1)(a+5)}$*, which is often the preferred form.*

e. $\dfrac{x^2 + x - 20}{15x^3} \cdot \dfrac{30x^2 - 20x}{3x^2 - 14x + 8}$

Solution $\dfrac{x^2 + x - 20}{15x^3} \cdot \dfrac{30x^2 - 20x}{3x^2 - 14x + 8}$

$= \dfrac{(x+5)(x-4)}{3 \cdot 5 \cdot x \cdot x \cdot x} \cdot \dfrac{2 \cdot 5 \cdot x \cdot (3x-2)}{(3x-2)(x-4)}$ Factor the numerators and denominators.

$= \dfrac{x+5}{3 \cdot x \cdot x} \cdot \dfrac{2}{1}$ Divide out the common factors of $x - 4$, $3x - 2$, 5, and x.

$= \dfrac{2x + 10}{3x^2}$ Multiply the remaining factors.

Your Turn Multiply.

a. $\dfrac{6r^3s}{5r^2s^2} \cdot \dfrac{15rs^2}{2r^3s^2}$ **b.** $-\dfrac{6a^2}{9a + 27} \cdot \dfrac{7a + 21}{28a^4}$ **c.** $\dfrac{a^2 + 6a + 8}{a^2 + 5a - 6} \cdot \dfrac{a^2 + 3a - 18}{a^2 + a - 12}$

d. $\dfrac{y^2 + 3y - 10}{y^2 - 25} \cdot \dfrac{y + 4}{2 - y}$ **e.** $\dfrac{x^2 - 2x - 8}{14} \cdot \dfrac{16x - 40}{2x^2 - x - 10}$

Answers

a. $\dfrac{9}{rs}$ b. $-\dfrac{1}{6a^2}$
c. $\dfrac{a+2}{a-1}$ d. $-\dfrac{y+4}{y-5}$
e. $\dfrac{4x - 16}{7}$

OBJECTIVE 3. Divide rational expressions. To divide fractions, we use the rule $\frac{a}{b} \div \frac{c}{d} = \frac{a}{b} \cdot \frac{d}{c}$, where b, c, and d are not 0. In other words, we multiply by the reciprocal of the divisor. We can rewrite this rule so that it applies to dividing rational expressions.

PROCEDURE Dividing Rational Expressions

1. Write an equivalent multiplication statement using $\frac{P}{Q} \div \frac{R}{S} = \frac{P}{Q} \cdot \frac{S}{R}$, where P, Q, R, and S are polynomials and Q, R, and $S \neq 0$.
2. Simplify using the procedure for multiplying rational expressions.

EXAMPLE 4 Divide.

a. $\frac{4x^2 + 8x}{6x^2} \div \frac{5x + 10}{9x^3}$

Solution $\frac{4x^2 + 8x}{6x^2} \div \frac{5x + 10}{9x^3} = \frac{4x^2 + 8x}{6x^2} \cdot \frac{9x^3}{5x + 10}$ Write an equivalent multiplication statement.

$= \frac{2 \cdot 2 \cdot x \cdot (x + 2)}{2 \cdot 3 \cdot x \cdot x} \cdot \frac{3 \cdot 3 \cdot x \cdot x \cdot x}{5(x + 2)}$ Factor the numerators and denominators.

$= \frac{6x^2}{5}$ Divide out the common factors of 2, 3, x, and $x + 2$ and multiply the remaining factors.

Learning Strategy

When simplifying rational expressions with lots of factors, write larger and more spread out than you usually write. This will reduce clutter, making the common factors easier to see.

b. $\frac{2x^2 + 7x + 3}{x^2 - 9} \div \frac{2x^2 + 11x + 5}{x^2 - 3x}$

Solution $\frac{2x^2 + 7x + 3}{x^2 - 9} \div \frac{2x^2 + 11x + 5}{x^2 - 3x}$

$= \frac{2x^2 + 7x + 3}{x^2 - 9} \cdot \frac{x^2 - 3x}{2x^2 + 11x + 5}$ Write an equivalent multiplication statement.

$= \frac{(2x + 1)(x + 3)}{(x + 3)(x - 3)} \cdot \frac{x(x - 3)}{(2x + 1)(x + 5)}$ Factor the numerators and denominators.

$= \frac{x}{x + 5}$ Divide out the common factors of $2x + 1$, $x + 3$, and $x - 3$ and multiply the remaining factors.

c. $\frac{15b^2 + b - 6}{6b^2 - 11b - 10} \div \frac{10b^2 + 4b - 6}{2b^2 - 13b + 20}$

Solution $\frac{15b^2 + b - 6}{6b^2 - 11b - 10} \div \frac{10b^2 + 4b - 6}{2b^2 - 13b + 20}$

$= \frac{15b^2 + b - 6}{6b^2 - 11b - 10} \cdot \frac{2b^2 - 13b + 20}{10b^2 + 4b - 6}$ Write an equivalent multiplication statement.

$= \frac{(5b - 3)(3b + 2)}{(3b + 2)(2b - 5)} \cdot \frac{(2b - 5)(b - 4)}{2(5b - 3)(b + 1)}$ Factor the numerators and denominators.

$= \frac{b - 4}{2b + 2}$ Divide out the common factors of $5b - 3$, $3b + 2$, and $2b - 5$ and multiply the remaining factors.

d. $\dfrac{2x^2 + 5x - 12}{8 - 4x} \div \dfrac{20 + 5x}{6x^3 - 12x^2}$

Solution $\dfrac{2x^2 + 5x - 12}{8 - 4x} \div \dfrac{20 + 5x}{6x^3 - 12x^2}$

$= \dfrac{2x^2 + 5x - 12}{8 - 4x} \cdot \dfrac{6x^3 - 12x^2}{20 + 5x}$ Write an equivalent multiplication statement.

$= \dfrac{(2x - 3)(x + 4)}{4(2 - x)} \cdot \dfrac{6x^2(x - 2)}{5(4 + x)}$ Factor the numerators and denominators.

$= \dfrac{(2x - 3)(x + 4)}{2 \cdot 2 \cdot (2 - x)} \cdot \dfrac{2 \cdot 3x^2(-1)(2 - x)}{5(4 + x)}$ Continue factoring.

$= \dfrac{2x - 3}{2} \cdot \dfrac{3x^2(-1)}{5}$ Divide out the common factors of 2, $x + 4$, and $2 - x$.

$= -\dfrac{6x^3 - 9x^2}{10}$ Multiply the remaining factors.

Note: *The binomials $x - 2$ and $2 - x$ are additive inverses, so we factored -1 from $x - 2$. Also, $x + 4 = 4 + x$.*

YOUR TURN Divide.

a. $\dfrac{8x^4}{2x^3 + 6x^2} \div \dfrac{6x}{4x + 12}$

b. $\dfrac{2x^2 + 7x + 6}{x^2 - 4} \div \dfrac{2x^2 - 3x - 9}{x^2 - 2x}$

c. $\dfrac{8a^2 - 6a - 9}{2a^2 - 13a + 15} \div \dfrac{12a^2 + a - 6}{3a^2 + 10a - 8}$

d. $\dfrac{8x^2 - 6x - 9}{24x - 4x^2} \div \dfrac{24 + 32x}{6x^3 - 36x^2}$

OBJECTIVE 4. Evaluate rational functions. We can now define a **rational function**.

DEFINITION ***Rational function:*** A function expressed in terms of rational expressions.

For example, $f(x) = \dfrac{3x + 2}{x^2 - 9}$ is a rational function because $\dfrac{3x + 2}{x^2 - 9}$ is a rational expression. Rational functions are evaluated in the same manner as other functions.

EXAMPLE 5 Given $f(x) = \dfrac{x^2 + 2x - 8}{x - 3}$, find $f(4)$.

Solution $f(4) = \dfrac{4^2 + 2(4) - 8}{4 - 3}$ Replace x with 4.

$= \dfrac{16 + 8 - 8}{1}$ Simplify.

$= 16$

ANSWERS

a. $\dfrac{8x}{3}$ b. $\dfrac{x}{x - 3}$

c. $\dfrac{a + 4}{a - 5}$

d. $-\dfrac{3x(2x - 3)}{16}$

YOUR TURN Given $f(x) = \frac{x + 6}{x^2 - 4}$, find the following:

a. $f(3)$ **b.** $f(-6)$

ANSWERS

a. $\frac{9}{5}$ **b.** 0

OBJECTIVE 5. Find the domain of a rational function. Recall from Section 3.5 that the *domain* of a function is the set of all input values for which the function is defined. With a rational function, we must be careful to identify values that cause the function to be undefined and eliminate those values from its domain. Notice the function $f(x) = \frac{x^2 + 2x - 8}{x - 3}$ is undefined if $x = 3$.

$$f(3) = \frac{3^2 + 2(3) - 8}{3 - 3} = \frac{9 + 6 - 8}{0}$$

which is undefined. Because any real number other than 3 can be used as an input value, the domain of $f(x) = \frac{x^2 + 2x - 8}{x - 3}$ is $\{x \mid x \neq 3\}$. Our example suggests the following procedure.

PROCEDURE **Finding the Domain of a Rational Function**

1. Write an equation that sets the denominator of the rational function equal to 0.
2. Solve the equation.
3. Exclude the value(s) found in step 2 from the function's domain.

EXAMPLE 6 Find the domain of $f(x) = \frac{x - 6}{3x^3 + x^2 - 4x}$.

Solution

$3x^3 + x^2 - 4x = 0$ Set the denominator equal to 0, then solve.

$x(3x^2 + x - 4) = 0$ Solve by first factoring out the GCF, x.

$x(3x + 4)(x - 1) = 0$ Factor $3x^2 + x - 4$ by trial.

$x = 0$ or $3x + 4 = 0$ or $x - 1 = 0$ Use the zero-factor theorem.

$x = -\frac{4}{3}$ $\qquad$ $x = 1$

The function is undefined if x is replaced by 0, $-\frac{4}{3}$, or 1, so the domain is $\left\{x \mid x \neq 0, -\frac{4}{3}, 1\right\}$.

Note: *We do not consider the numerator because the rational expression is undefined only if the denominator is 0.*

YOUR TURN Find the domain of each rational function.

a. $f(x) = \frac{2x + 3}{x^2 - 2x - 15}$ **b.** $f(x) = \frac{3x + 4}{2x^3 + 5x^2 + 2x}$

ANSWERS

a. $\{x \mid x \neq -3, 5\}$

b. $\left\{x \mid x \neq -\frac{1}{2}, 0, -2\right\}$

OBJECTIVE 6. Graph rational functions. Let's graph the function $f(x) = \frac{1}{x-2}$ by making a table of ordered pairs. Note that 2 is not in its domain, so there will be no point on the graph whose x value is 2. We say the graph is *discontinuous* at $x = 2$. To complete our table, we will use x-values in the domain that are very close to 2.

x	-1	0	1	1.5	1.9	1.99	2.01	2.1	2.5	3	4	5
y	$-\frac{1}{3}$	$-\frac{1}{2}$	-1	-2	-10	-100	100	10	2	1	$\frac{1}{2}$	$\frac{1}{3}$

Using the table, we can draw the graph shown. Note that we indicate the discontinuity at $x = 2$ with a dashed vertical line called a *vertical asymptote.* Also, notice when $x < 2$, the y-values become smaller as x gets closer to 2, and when $x > 2$, the y-values become greater as x gets closer to 2. Graphs of rational functions usually have this type of shape with vertical asymptotes through x-values that make the function undefined.

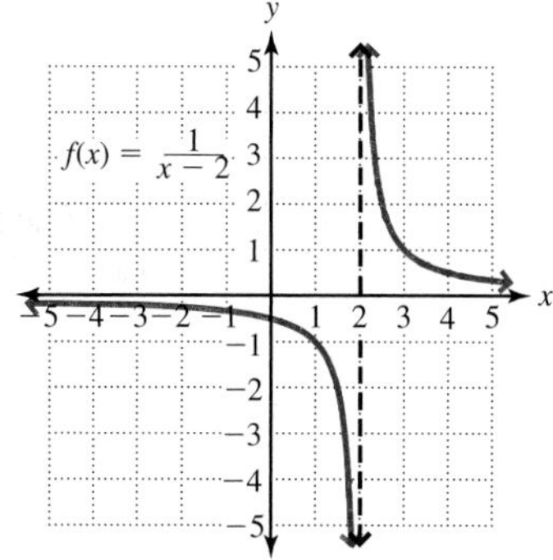

EXAMPLE 7 Graph. $f(x) = \frac{-2}{x+1}$

Solution Note that $x \neq -1$, so the graph has a vertical asymptote at $x = -1$. We find ordered pairs around that asymptote and then graph.

x	-4	-3	-2	-1.5	-1.1	-1.01	-0.99	-0.9	-0.5	0	1	2
y	$\frac{2}{3}$	1	2	4	20	200	-200	-20	-4	-2	-1	$-\frac{2}{3}$

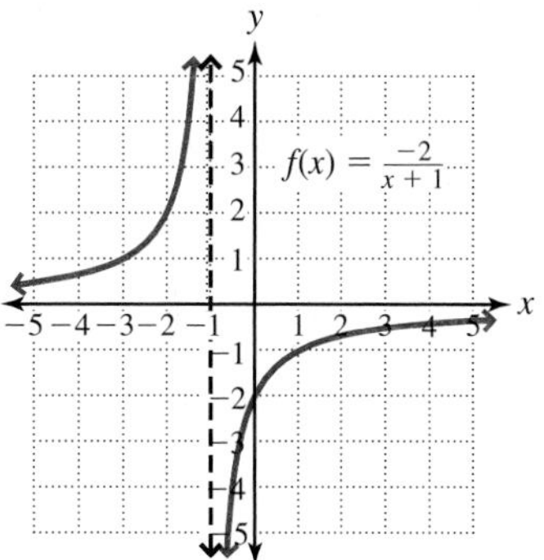

ANSWER

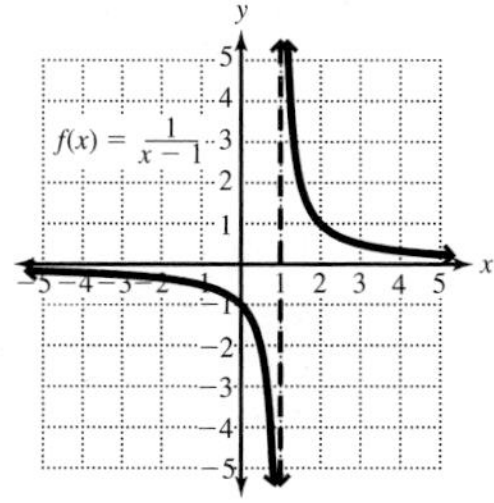

YOUR TURN

Graph. $f(x) = \frac{1}{x-1}$

Calculator **TIP**

We can check a simplification by graphing Y_1 = the original expression and Y_2 = the simplified expression and verifying that the graphs are the same. In Example 1c, we simplified $\frac{x^2 - 9}{x^2 - x - 12}$ to $\frac{x - 3}{x - 4}$. Graphing $Y_1 = \frac{x^2 - 9}{x^2 - x - 12}$ and $Y_2 = \frac{x - 3}{x - 4}$ in the window $[-10, 10]$ for x and $[-10, 10]$ for y on the TI–83 Plus gives the following graphs:

$Y_1 = \frac{x^2 - 9}{x^2 - x - 12}$

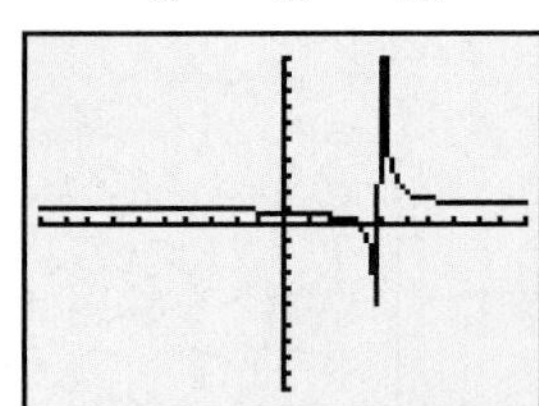

$Y_2 = \frac{x - 3}{x - 4}$

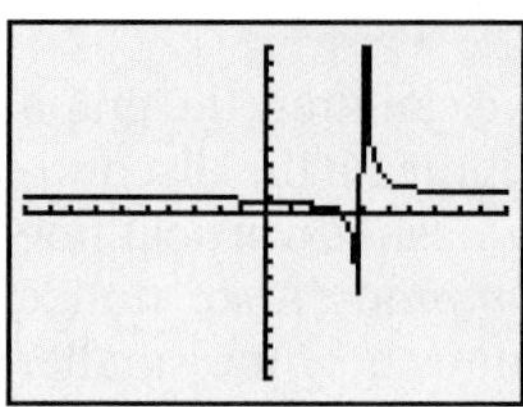

Note the graphs appear to be the same. However, there is a subtle difference resulting from a difference in the domains of the functions. Factoring the denominator of Y_1 we have $Y_1 = \frac{x^2 - 9}{(x + 3)(x - 4)}$ and we can see that its domain is $\{x|x \neq -3, 4\}$, whereas the domain of Y_2 is $\{x|x \neq 4\}$. The number -3 is excluded from the domain of Y_1 because $x + 3$ is a factor in its denominator. During simplification, $x + 3$ divides out, so -3 is in the domain of Y_2. This means Y_1 has a tiny "hole" in its graph at $x = -3$, whereas Y_2 does not. The hole in Y_1 cannot be seen with our current window setting. To see the hole in Y_1, clear Y_2 so that only Y_1 is graphed, press ZOOM, *select ZDecimal, and then press* ENTER. *The graph of Y_1 now looks like this:*

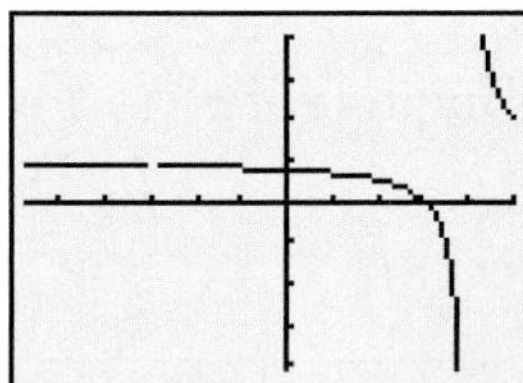

Notice the tiny blank space at $x = -3$. If we use the ZDecimal function on Y_2 we would see no such hole because $x = -3$ is in its domain. In general, holes occur with factors that divide out, whereas asymptotes occur with factors in a denominator that cannot be divided out.

Note: *ZDecimal makes the window*
Xmin = −4.7, *Xmax* = 4.7,
Ymin = −3.1, *Ymax* = 3.1.

7.1 Exercises

For Extra Help

MyMathLab | Videotape/DVT | InterAct Math | Math Tutor Center | Math XL.com

1. If $f(x) = \frac{x^2 + 3x + 2}{x - 5}$, is $f(5)$ defined? Why or why not?

2. What does it mean for a rational expression to be simplified (reduced to lowest terms)?

3. When simplifying $\frac{x^2 + 4x - 5}{x^2 + 2x - 3}$, why can you not divide the x^2's?

4. What is the difference between a term and a factor?

5. Explain how to multiply rational expressions.

6. What is the first step when dividing rational expressions?

For Exercises 7–36, simplify each rational expression.

7. $\dfrac{-28m^3n^5}{16m^7n^6}$

8. $\dfrac{32c^2d^6}{-24c^4d^7}$

9. $\dfrac{-32x^4y^3z^2}{24x^2y^2z^2}$

10. $\dfrac{42w^3x^4y^3}{-28w^2x^3y}$

11. $\dfrac{2x+12}{3x+18}$

12. $\dfrac{3x+6}{5x+10}$

13. $\dfrac{8x^2+12x}{20x^2+30x}$

14. $\dfrac{24a^2-16a}{36a^2-24a}$

15. $\dfrac{-3x-15}{4x+20}$

16. $\dfrac{-5x-25}{6x+30}$

17. $\dfrac{a^2+3a-10}{a^2+8a+15}$

18. $\dfrac{c^2+3c-18}{c^2-c-6}$

19. $\dfrac{x^2-6x}{x^2-4x-12}$

20. $\dfrac{x^2-2x}{x^2-10x+16}$

21. $\dfrac{4x^2-9y^2}{6x^2-xy-15y^2}$

22. $\dfrac{9x^2-25y^2}{6x^2-5xy-25y^2}$

23. $\dfrac{4a^2-4ab-3b^2}{6a^2-ab-2b^2}$

24. $\dfrac{6a^2+13ab+6b^2}{3a^2-7ab-6b^2}$

25. $\dfrac{x^3-8}{3x^2-2x-8}$

26. $\dfrac{x^3+1}{3x^2-4x-7}$

27. $\dfrac{x^3+27}{3x^2-27}$

28. $\dfrac{4x^3+4}{6x^2-8x-14}$

29. $\dfrac{xy-4x+3y-12}{xy-6x+3y-18}$

30. $\dfrac{ab+5a-3b-15}{bc-2b+5c-10}$

31. $\dfrac{6a^3+4a^2+9a+6}{6a^2+13a+6}$

32. $\dfrac{8a^3+6a^2+20a+15}{8a^2+10a+3}$

33. $\dfrac{2b-a}{a-2b}$

34. $\dfrac{3m-n}{n-3m}$

35. $\dfrac{x^2+4x-21}{6-2x}$

36. $\dfrac{20-5x}{x^2-9x+20}$

For Exercises 37–62, find the product.

37. $\frac{8x^2y^4}{12x^3y^2} \cdot \frac{4x^3y^3}{6x^4y^3}$

38. $\frac{10a^2b^2}{18a^2b^3} \cdot \frac{12a^3b}{15a^2b^4}$

39. $-\frac{24a^2bc^3}{16ab^3c^2} \cdot \frac{32a^3b^2c^3}{18a^4b^3c^3}$

40. $-\frac{30x^3y^2z}{42x^2yz^2} \cdot \frac{15x^2y^2z^2}{25x^4yz}$

41. $\frac{3x^2y^4}{15x + 10y} \cdot \frac{9x + 6y}{12xy^2}$

42. $\frac{12p + 16q}{10r^2s} \cdot \frac{15r^3s^4}{15p + 20q}$

43. $\frac{8x + 12}{18x - 24} \cdot \frac{12 - 9x}{10x + 15}$

44. $\frac{24x + 16}{30x - 18} \cdot \frac{27 - 45x}{30x + 20}$

45. $\frac{4a - 16}{3a - 3} \cdot \frac{a^2 - 1}{a^2 - 16}$

46. $\frac{x^2 - 9}{6x + 18} \cdot \frac{5x - 25}{x^2 - 25}$

47. $\frac{4x^2y^3}{2x - 6} \cdot \frac{12 - 4x}{6x^3y}$

48. $\frac{6b^3c^3}{70 - 10b} \cdot \frac{5b - 35}{8b^2c^5}$

49. $\frac{x^2 - 4}{x^2 - 3x - 10} \cdot \frac{x^2 - 8x + 15}{x^2 - 9}$

50. $\frac{x^2 + 3x - 10}{x^2 - 16} \cdot \frac{x^2 - 9x + 20}{x^2 - 25}$

51. $\frac{6x^2 + x - 12}{2x^2 - 5x - 12} \cdot \frac{3x^2 - 14x + 8}{9x^2 - 18x + 8}$

52. $\frac{2x^2 + x - 15}{4x^2 + 11x - 3} \cdot \frac{4x^2 - 9x + 2}{2x^2 - 9x + 10}$

53. $\frac{3x^2 - 5x - 2}{4x^2 - 11x + 6} \cdot \frac{4x^2 + 5x - 6}{3x^2 - 8x - 3}$

54. $\frac{8x^2 + 10x - 3}{3x^2 + 4x - 4} \cdot \frac{x^2 + 6x + 8}{2x^2 + 11x + 12}$

55. $\frac{2x^2 - 11x + 12}{2x^2 + 11x - 21} \cdot \frac{3x^2 + 20x - 7}{4 - x}$

56. $\frac{3x^2 - 17x + 10}{3x^2 + 7x - 6} \cdot \frac{2x^2 + x - 15}{5 - x}$

57. $\frac{ac + 3a + 2c + 6}{ad + a + 2d + 2} \cdot \frac{ad - 5a + 2d - 10}{bc + 3b - 4c - 12}$

58. $\frac{mn + 2m + 4n + 8}{np - 3n + 2p - 6} \cdot \frac{pq + 5p - 3q - 15}{mq + 4m + 4q + 16}$

59. $\frac{4x^3 - 3x^2 + 4x - 3}{4x^3 - 3x^2 - 8x + 6} \cdot \frac{2x^3 + x^2 - 4x - 2}{3x^3 + 2x^2 + 3x + 2}$

60. $\frac{3x^3 + 2x^2 + 12x + 8}{2x^3 - 5x^2 - 8x + 20} \cdot \frac{2x^3 - 5x^2 - 6x + 15}{3x^3 + 2x^2 - 9x - 6}$

61. $\frac{a^3 - b^3}{a^2 + ab + b^2} \cdot \frac{2a^2 + ab - b^2}{a^2 - b^2}$

62. $\frac{3x^2 + 4x - 4}{x^2 - 4} \cdot \frac{x^3 - 8}{x^2 + 2x + 4}$

For Exercises 63–84, find the quotient.

63. $\frac{8x^7g^3}{15x^2g^4} \div \frac{3xg^2}{4x^2g^3}$

64. $\frac{10c^5g^5}{12c^2g^4} \div \frac{5cg^2}{4c^2g^3}$

65. $\frac{4a^3 - 8a^2}{15a} \div \frac{3a^2 - 6a}{5a^4}$

66. $\frac{6b^2 + 24b}{7b^2} \div \frac{7b^3 + 28b^2}{14b^4}$

67. $\frac{10x^3 + 15x^2}{18x - 6} \div \frac{20x + 30}{9x^2 - 3x}$

68. $\frac{12y + 8}{4y - 10} \div \frac{48y + 32}{8y^3 - 20y^2}$

69. $\frac{3a - 4b}{6a - 9b} \div \frac{12b - 9a}{4a - 6b}$

70. $\frac{4x - 3y}{12x - 3y} \div \frac{6y - 8x}{8x - 2y}$

71. $\frac{d^2 - d - 12}{2d^2} \div \frac{3d^2 + 13d + 12}{d}$

72. $\frac{y^2 + y - 20}{7y^2} \div \frac{3y^2 + 19y + 20}{y}$

73. $\frac{m^2 - 25n^2}{2m - 12n} \div \frac{3m + 15n}{m^2 - 36n^2}$

74. $\frac{h^2 - 36j^2}{3h - 21j} \div \frac{4h + 24j}{h^2 - 49j^2}$

75. $\frac{3x + 24}{4x - 32} \div \frac{x^2 + 16x + 64}{x^2 - 16x + 64}$

76. $\frac{5w + 15}{3w - 9} \div \frac{w^2 + 6w + 9}{w^2 - 6w + 9}$

77. $\frac{x^2 - 3xy + 2y^2}{x^2 + 3xy + 2y^2} \div \frac{x^2 - y^2}{x^2 + 2xy + y^2}$

78. $\frac{9y^2 - 4z^2}{3y^2 + 7yz - 6z^2} \div \frac{3y^2 - 7yz - 6z^2}{y^2 + 6yz + 9z^2}$

79. $\frac{2x^2 - 7x - 4}{3x^2 - 14x + 8} \div \frac{2x^2 + 7x + 3}{3x^2 - 8x + 4}$

80. $\frac{8x^2 + 6x - 9}{8x^2 + 14x - 15} \div \frac{2x^2 + 11x + 12}{4x^2 + 19x + 12}$

81. $\frac{27a^3 - 8b^3}{9a^2 - 4b^2} \div \frac{2a^2 - 5ab - 12b^2}{6a^2 + 13ab + 6b^2}$

82. $\frac{x^3 - 64y^3}{x^2 + 2xy - 15y^2} \div \frac{x^2 - 16y^2}{x^2 + xy - 12y^2}$

83. $\frac{ab + 3a + 2b + 6}{bc + 4b + 3c + 12} \div \frac{ac - 3a + 2c - 6}{bc + 4b - 4c - 16}$

84. $\frac{xy - 3x + 4y - 12}{xy + 6x - 3y - 18} \div \frac{xy + 5x + 4y + 20}{xy + 5x - 3y - 15}$

For Exercises 85–88, perform the indicated operations.

85. $\dfrac{x^2 - 3x - 18}{2x^2 + 13x + 20} \cdot \dfrac{3x^2 + 10x - 8}{4x^2 - 24x} \div \dfrac{3x^2 + 7x - 6}{6x^2 + 11x - 10}$

86. $\dfrac{4x^2 - 31x + 21}{12x^2 - 13x - 4} \cdot \dfrac{8x^2 + 14x + 3}{4x^2 - 19x + 12} \div \dfrac{2x^2 - 11x - 21}{9x^2 - 12x}$

87. $\dfrac{5x^2 - 17x - 12}{12x^2 - 29x + 15} \cdot \left(\dfrac{6x^2 + x - 12}{3x^2 + 14x - 24} \div \dfrac{2x^2 - 5x - 12}{3x^2 + 13x - 30}\right)$

88. $\dfrac{6x^3 - 14x^2}{12x^2 + 4x - 5} \cdot \left(\dfrac{4x^2 + 4x - 3}{9x^2 - 27x + 14} \div \dfrac{6x^2 + 9x}{18x^2 + 3x - 10}\right)$

For Exercises 89 and 90, explain the mistake then work the problem correctly.

89. $\dfrac{a^2 + b^2}{a^2 - b^2} \cdot \dfrac{a^2 + 2ab - 3b^2}{a^2 + ab - 6b^2} = \dfrac{(a + b)(a + b)}{(a + b)(a - b)} \cdot \dfrac{(a - b)(a + 3b)}{(a + 3b)(a - 2b)} =$

$\dfrac{\cancel{(a + b)}(a + b)}{\cancel{(a + b)}\cancel{(a - b)}} \cdot \dfrac{\cancel{(a - b)}\cancel{(a + 3b)}}{\cancel{(a + 3b)}(a - 2b)} = \dfrac{a + b}{a - 2b}$

90. $\dfrac{a^2 + b^2}{a^2 - b^2} \cdot \dfrac{a^2 + 2ab - 3b^2}{a^2 + ab - 6b^2} = \dfrac{\cancel{a}^2 + \cancel{b}^2}{\cancel{a}^2 - \cancel{b}^2} \cdot \dfrac{\cancel{a}^2 + 2ab - \cancel{3}\cancel{b}^2}{\cancel{a}^2 + ab - \cancel{6}\cancel{b}^2} = \dfrac{2ab}{ab - 2}$

For Exercises 91–98, find the indicated value of the given functions.

91. $f(x) = \dfrac{x + 3}{x - 4}$,
a. $f(3)$
b. $f(-3)$
c. $f(4)$

92. $f(x) = \dfrac{x + 5}{x - 2}$,
a. $f(1)$
b. $f(-5)$
c. $f(2)$

93. $f(x) = \dfrac{2x - 4}{3x + 2}$,
a. $f(2)$
b. $f(-2.3)$
c. $f(1.2)$

94. $f(x) = \dfrac{3x + 6}{4x - 2}$,
a. $f(-2)$
b. $f(2)$
c. $f(-1.3)$

95. $f(x) = \dfrac{3x - 4}{2x^2 - x - 6}$,
a. $f(3)$
b. $f\left(\dfrac{4}{3}\right)$
c. $f(2)$

96. $f(x) = \dfrac{3x^2 + 11x - 4}{4x - 6}$,
a. $f(-4)$
b. $f(1.5)$
c. $f(4)$

97. $f(x) = \dfrac{x^2 + x - 6}{x^2 + x - 12}$,
a. $f(-3)$
b. $f(5)$
c. $f(3)$

98. $f(x) = \dfrac{x^2 - 4x - 5}{x^2 + 5x + 6}$,
a. $f(5)$
b. $f(2)$
c. $f(-2)$

For Exercises 99–110, find the domain of each rational function.

99. $f(x) = \dfrac{x + 4}{2x + 8}$

100. $f(x) = \dfrac{2x - 5}{3x + 9}$

101. $f(x) = \dfrac{3x - 6}{x^2 - 25}$

102. $g(a) = \dfrac{5a + 15}{a^2 - 64}$

103. $g(c) = \dfrac{3c - 9}{4c^2 - 81}$

104. $h(n) = \dfrac{2n^2 - 3n + 6}{25n^2 - 49}$

105. $h(t) = \dfrac{2t^2 + 4t - 8}{6t^2 + 11t - 10}$

106. $f(a) = \dfrac{3a + 5}{8a^2 + 6a - 9}$

107. $g(x) = \dfrac{4x - 8}{x^3 + 4x^2 - 21x}$

108. $h(x) = \dfrac{3x^2 - 27}{2x^3 - 7x^2 - 4x}$

109. $f(x) = \dfrac{x - 7}{x^3 - 8}$

110. $f(x) = \dfrac{2x^2 - 8}{x^3 + 64}$

For Exercises 111–114, match each function with its graph.

111. $f(x) = \dfrac{1}{x - 3}$

112. $f(x) = \dfrac{-1}{x + 3}$

113. $f(x) = \dfrac{-2}{x + 2}$

114. $f(x) = \dfrac{3}{x - 4}$

a.

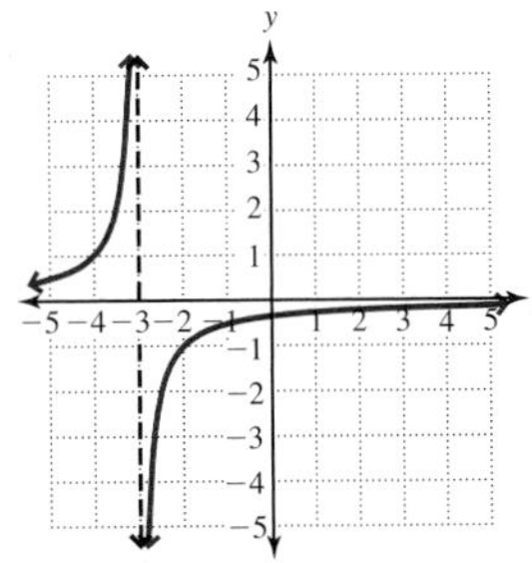

b.

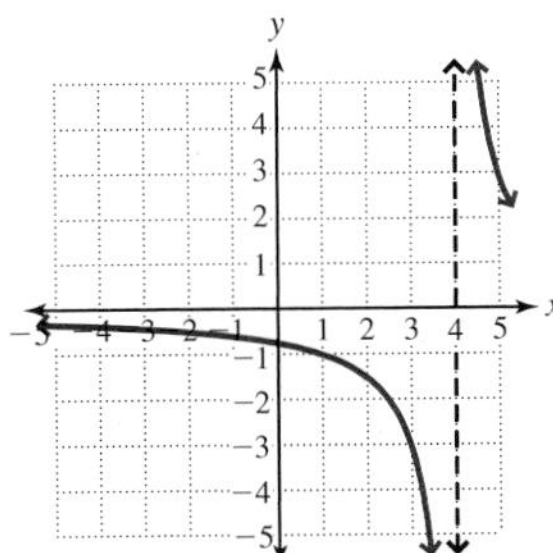

c.

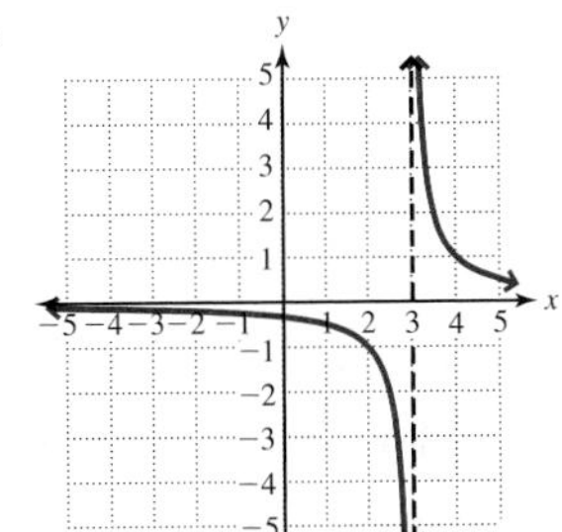

d.

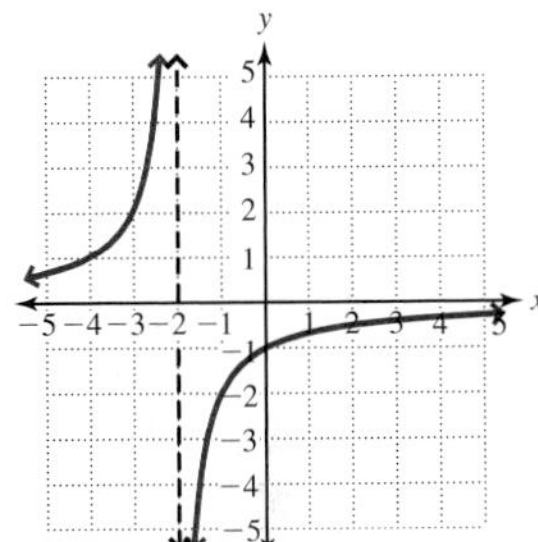

For Exercises 115–118, graph.

115. $f(x) = \dfrac{1}{x - 2}$

116. $f(x) = \dfrac{2}{x - 1}$

117. $f(x) = \dfrac{-2}{x + 1}$

118. $f(x) = \dfrac{-1}{x - 3}$

For Exercises 119–126, solve.

119. The electrical resistance R in ohms of a wire is given by $R = \frac{5}{d^2}$, where d is the diameter of the wire in millimeters. Find the electrical resistance of a wire whose diameter is 2 millimeters.

120. The weight of an object depends on its distance from the center of the Earth. If a person weighs 180 pounds on the surface of the Earth, this person's weight is given by $w = \frac{2{,}880{,}000{,}000}{d^2}$, where w is the weight in pounds and d is the distance from the center of the Earth in miles. Find the weight of this person 2000 miles above the surface of Earth if the radius of Earth is 4000 miles.

121. If the start-up cost of a small manufacturing business is \$1500 per day and the manufacturing costs are \$12 per unit, then the average cost per unit is $C = \frac{12x + 1500}{x}$, where x is the number of units. Find the average cost per unit to produce 100 units.

122. An automobile's velocity starting from rest is $v = \frac{90t}{2t + 18}$, where v is in feet per second and t is in seconds. Find the velocity after 6 seconds.

123. The ordering and transportation cost for the components used in manufacturing a product is given by

$$C = 200\left(\frac{300}{x^2} + \frac{x}{x + 50}\right)$$

where C is the cost in thousands of dollars and x is the size of the order in hundreds. Find the ordering and transportation cost for an order of 5000 components.

124. If 500 bacteria are introduced into a culture, the equation

$$N = 500\left(1 + \frac{4t}{50 + t^2}\right)$$

gives the number, N, of bacteria after t hours. Find the number of bacteria after 5 hours.

125. If a fire truck approaches a stationary observer at a velocity of v miles per hour, the frequency, F, of the truck siren is given by

$$F = \frac{130{,}000}{330 - v}$$

Find the frequency if the truck is approaching at 70 miles per hour.

Of Interest

When an object in motion produces sound, the sound waves in front of the object are compressed by its motion so that they have a higher frequency (pitch). The waves that trail the object are stretched so that they have a lower frequency. You experience this phenomenon, called the Doppler effect, when a train or car passes by blowing its horn and you hear the pitch change from higher to lower.

126. The concentration of a medication in the bloodstream t hours after injection into muscle tissue is given by

$$C = \frac{3t}{27 + t^3}, \quad \text{where} \quad t \geq 0$$

Find the concentration 3 hours after the injection.

REVIEW EXERCISES

For Exercises 1–4, add or subtract.

[1.2] **1.** $\frac{3}{11} + \frac{4}{11}$

[1.2] **2.** $\frac{3}{5} - \frac{5}{6}$

[1.4] **3.** $\frac{5}{12}y + \frac{7}{30}y$

[1.4] **4.** $\frac{11}{15}x + 4 - \frac{2}{15}x - 3$

For Exercises 5 and 6, factor.

[6.1] **5.** $6x^2 - 15x$

[6.3] **6.** $x^2 + x - 20$

7.2 Adding and Subtracting Rational Expressions

OBJECTIVES

1. Add or subtract rational expressions with the same denominator.
2. Find the least common denominator (LCD).
3. Add or subtract rational expressions with unlike denominators.

OBJECTIVE 1. Add or subtract rational expressions with the same denominator. Recall that when adding or subtracting fractions with the same denominator, we add or subtract the numerators and keep the same denominator. We add and subtract rational expressions in the same way.

$$\frac{4a}{13} + \frac{6a}{13} = \frac{4a + 6a}{13} = \frac{10a}{13} \quad \text{or} \quad \frac{b}{2b-3} - \frac{4}{2b-3} = \frac{b-4}{2b-3}$$

PROCEDURE Adding or Subtracting Rational Expressions with the Same Denominator

1. Add or subtract the numerators and keep the same denominator.
2. Simplify to lowest terms.

EXAMPLE 1 Add or subtract.

a. $\dfrac{3a}{a-3} - \dfrac{9}{a-3}$

Solution $\dfrac{3a}{a-3} - \dfrac{9}{a-3} = \dfrac{3a-9}{a-3}$ Subtract the numerators and keep the same denominator.

$= \dfrac{3(a-3)}{a-3}$ Factor the numerator.

$= 3$ Divide out the common factor, $a - 3$.

b. $\dfrac{x^2}{x^2+3x-18} + \dfrac{5x-6}{x^2+3x-18}$

Solution $\dfrac{x^2}{x^2+3x-18} + \dfrac{5x-6}{x^2+3x-18} = \dfrac{x^2+(5x-6)}{x^2+3x-18}$ Add the numerators and keep the same denominator.

$= \dfrac{x^2+5x-6}{x^2+3x-18}$ Remove the parentheses.

$= \dfrac{(x+6)(x-1)}{(x+6)(x-3)}$ Factor the numerator and denominator.

$= \dfrac{x-1}{x-3}$ Divide out the common factor, $x + 6$.

c. $\dfrac{n^2+3n}{n^2+4n+4} - \dfrac{5n+8}{n^2+4n+4}$

Solution $\dfrac{n^2+3n}{n^2+4n+4} - \dfrac{5n+8}{n^2+4n+4} = \dfrac{(n^2+3n)-(5n+8)}{n^2+4n+4}$ Subtract the numerators and keep the same denominator.

$= \dfrac{(n^2+3n)+(-5n-8)}{n^2+4n+4}$ Write the equivalent addition.

$= \dfrac{n^2-2n-8}{n^2+4n+4}$ Combine like terms.

$= \dfrac{(n-4)(n+2)}{(n+2)(n+2)}$ Factor the numerator and denominator.

$= \dfrac{n-4}{n+2}$ Divide out the common factor of $n + 2$.

d. $\frac{p^2 - 1}{2p^2 - 2} - \frac{4p + 4}{2p^2 - 2} + \frac{2p^2 + 2p}{2p^2 - 2}$

Solution $\frac{p^2 - 1}{2p^2 - 2} - \frac{4p + 4}{2p^2 - 2} + \frac{2p^2 + 2p}{2p^2 - 2}$

$= \frac{(p^2 - 1) - (4p + 4) + (2p^2 + 2p)}{2p^2 - 2}$ Subtract and add the numerators and keep the same denominator.

$= \frac{(p^2 - 1) + (-4p - 4) + (2p^2 + 2p)}{2p^2 - 2}$ Write the equivalent addition.

$= \frac{3p^2 - 2p - 5}{2p^2 - 2}$ Combine like terms.

$= \frac{(3p - 5)(p + 1)}{2(p - 1)(p + 1)}$ Factor the numerator and denominator.

$= \frac{3p - 5}{2(p - 1)}$ Divide out the common factor, $p + 1$.

Calculator Note: *To verify your answer, you can graph Y_1 = the original problem and Y_2 = the answer and note that the graphs are the same.*

YOUR TURN Add or subtract.

a. $\frac{4a}{a - 4} - \frac{16}{a - 4}$ **b.** $\frac{2x^2 + 3x}{x^2 - 9} + \frac{4x + 3}{x^2 - 9}$ **c.** $\frac{2x^2 + 4x - 3}{x^2 - x - 6} - \frac{x^2 + 6x + 5}{x^2 - x - 6}$

d. $\frac{3m^2 - 8}{m^2 - 5m + 6} + \frac{m^2 + 4m - 4}{m^2 - 5m + 6} - \frac{2m^2 - 2m + 8}{m^2 - 5m + 6}$

OBJECTIVE 2. Find the least common denominator (LCD). To add fractions with unlike denominators, we have to rewrite each fraction as an equivalent fraction with the **least common denominator (LCD)** as its denominator. The least common denominator is the smallest number divisible by all the denominators. Thus, the LCD contains all the denominators as factors. Likewise, the first step in adding rational expressions with unlike denominators is to find the least common denominator, which is the polynomial of least degree that has all the denominators as factors.

PROCEDURE Finding the LCD

1. Factor each denominator into prime factors.
2. For each unique factor, compare the number of times it appears in each factorization. The LCD is the product of all unique factors the greatest number of times each appears in the denominator factorizations. So, if a factor appears in more than one factorization, use the greatest power of that factor.

ANSWERS

a. 4 **b.** $\frac{2x + 1}{x - 3}$

c. $\frac{x - 4}{x - 3}$ **d.** $\frac{2m + 10}{m - 3}$

EXAMPLE 2 Find the LCD.

a. $\frac{5}{6x^3y}$ and $\frac{7}{8x^2y^2}$

Solution First factor the denominators $6x^3y$ and $8x^2y^2$ by writing their prime factorizations.

$$6x^3y = 2 \cdot 3 \cdot x^3 \cdot y$$
$$8x^2y^2 = 2^3 \cdot x^2 \cdot y^2$$

The unique factors are 2, 3, x, and y. The greatest power of 2 is 3, the greatest power of 3 is 1, the greatest power of x is 3, and the greatest power of y is 2.

$$\text{LCD} = 2^3 \cdot 3 \cdot x^3 \cdot y^2 = 24x^3y^2$$

b. $\frac{y}{y-6}$ and $\frac{y+3}{y^2+5y}$

Solution Factor the denominators.

$$y - 6 \text{ is prime}$$
$$y^2 + 5y = y(y + 5)$$

The unique factors are $y - 6$, y, and $y + 5$ and the highest power of each is 1.

$$\text{LCD} = y(y - 6)(y + 5)$$

c. $\frac{x+3}{x^2+2x-15}$ and $\frac{x^2-4}{x^2-5x+6}$

Solution Factor the denominators.

$$x^2 + 2x - 15 = (x - 3)(x + 5)$$
$$x^2 - 5x + 6 = (x - 3)(x - 2)$$

The unique factors are $x - 3$, $x + 5$, and $x - 2$ and the highest power of each is 1.

$$\text{LCD} = (x - 3)(x + 5)(x - 2)$$

d. $\frac{a+3}{3a^4+15a^3+12a^2}$ and $\frac{a^2+3a+2}{2a^2+4a+2}$

Solution Factor the denominators.

$$3a^4 + 15a^3 + 12a^2 = 3a^2(a + 1)(a + 4)$$
$$2a^2 + 4a + 2 = 2(a + 1)^2$$

The unique factors are 2, 3, a, $a + 1$, and $a + 4$. The highest power of 2, 3, and $a + 4$ is 1 and the highest power of a and $a + 1$ is 2.

$$\text{LCD} = 2 \cdot 3 \cdot a^2 \cdot (a + 1)^2 \cdot (a + 4) = 6a^2(a + 1)^2(a + 4)$$

YOUR TURN Find the LCD.

a. $\frac{7}{15a^2b^4}$ and $\frac{11}{12a^3b^4}$

b. $\frac{b}{b+3}$ and $\frac{b+2}{b^2+3b}$

c. $\frac{y-5}{y^2-7y+10}$ and $\frac{2y-4}{y^2+3y-10}$

d. $\frac{x-5}{4x^2+16x+16}$ and $\frac{3x+2}{3x^2-9x-30}$

ANSWERS

a. $60a^3b^4$
b. $b(b + 3)$
c. $(y - 2)(y - 5)(y + 5)$
d. $12(x + 2)^2(x - 5)$

OBJECTIVE 3. Add or subtract rational expressions with unlike denominators. To add or subtract rational expressions with unlike denominators, we use the same process that we use to add or subtract fractions.

PROCEDURE **Adding or Subtracting Rational Expressions with Different Denominators**

1. Find the LCD.
2. Write each rational expression as an equivalent expression with the LCD.
3. Add or subtract the numerators and keep the LCD as the denominator.
4. Simplify.

EXAMPLE 3 Add or subtract.

a. $\dfrac{a-3}{3a^2} - \dfrac{2a-3}{2a}$

Solution The LCD is $6a^2$, so multiply the numerator and denominator of $\dfrac{a-3}{3a^2}$ by 2 and the numerator and denominator of $\dfrac{2a-3}{2a}$ by $3a$.

$$\frac{a-3}{3a^2} - \frac{2a-3}{2a} = \frac{(a-3)(2)}{3a^2(2)} - \frac{(2a-3)(3a)}{(2a)(3a)}$$ Write each rational expression with the LCD, $6a^2$.

$$= \frac{2a-6}{6a^2} - \frac{6a^2-9a}{6a^2}$$ Distribute.

$$= \frac{(2a-6)-(6a^2-9a)}{6a^2}$$ Combine numerators.

$$= \frac{(2a-6)+(-6a^2+9a)}{6a^2}$$ Write the equivalent addition.

$$= \frac{-6a^2+11a-6}{6a^2}$$ Simplify the numerator.

b. $\dfrac{5}{b-2} + \dfrac{3}{b+4}$

Solution Both denominators are prime, so the LCD is $(b-2)(b+4)$.

$$\frac{5}{b-2} + \frac{3}{b+4} = \frac{5(b+4)}{(b-2)(b+4)} + \frac{3(b-2)}{(b+4)(b-2)}$$ Multiply $\dfrac{5}{b-2}$ by $\dfrac{b+4}{b+4}$, and $\dfrac{3}{b+4}$ by $\dfrac{b-2}{b-2}$.

$$= \frac{5b+20}{(b+4)(b-2)} + \frac{3b-6}{(b+4)(b-2)}$$ Distribute.

$$= \frac{(5b+20)+(3b-6)}{(b+4)(b-2)}$$ Combine numerators.

$$= \frac{5b+20+3b-6}{(b+4)(b-2)}$$ Remove parentheses.

$$= \frac{8b+14}{(b+4)(b-2)}$$ Simplify the numerator.

c. $\dfrac{2x-3}{3x^2+14x+8}+\dfrac{3x-1}{3x^2-13x-10}$

Solution First, find the LCD by factoring the denominators.

$$3x^2+14x+8=(3x+2)(x+4) \quad\text{and}\quad 3x^2-13x-10=(3x+2)(x-5)$$
$$\text{LCD}=(3x+2)(x+4)(x-5)$$

$$\frac{2x-3}{3x^2+14x+8}+\frac{3x-1}{3x^2-13x-10}$$

$$=\frac{2x-3}{(3x+2)(x+4)}+\frac{3x-1}{(3x+2)(x-5)}$$ Factor the denominators.

$$=\frac{(2x-3)(x-5)}{(3x+2)(x+4)(x-5)}+\frac{(3x-1)(x+4)}{(3x+2)(x-5)(x+4)}$$ Write equivalent rational expressions with the LCD, $(3x+2)(x+4)(x-5)$.

$$=\frac{2x^2-13x+15}{(3x+2)(x+4)(x-5)}+\frac{3x^2+11x-4}{(3x+2)(x+4)(x-5)}$$ Multiply the numerators.

$$=\frac{(2x^2-13x+15)+(3x^2+11x-4)}{(3x+2)(x+4)(x-5)}$$ Combine the numerators.

$$=\frac{2x^2-13x+15+3x^2+11x-4}{(3x+2)(x+4)(x-5)}$$ Remove parentheses.

$$=\frac{5x^2-2x+11}{(3x+2)(x+4)(x-5)}$$ Simplify the numerator.

d. $\dfrac{x}{x+3}-\dfrac{21}{x^2-x-12}$

Solution First, find the LCD by factoring $x+3$ and x^2-x-12.

$$x+3 \text{ is prime}$$
$$x^2-x-12=(x-4)(x+3)$$
$$\text{LCD}=(x+3)(x-4)$$

$$\frac{x}{x+3}-\frac{21}{x^2-x-12}=\frac{x}{x+3}-\frac{21}{(x-4)(x+3)}$$ Factor the denominator.

$$=\frac{x(x-4)}{(x+3)(x-4)}-\frac{21}{(x-4)(x+3)}$$ Write equivalent rational expressions with the LCD, $(x-4)(x+3)$

$$=\frac{x^2-4x}{(x+3)(x-4)}-\frac{21}{(x-4)(x+3)}$$ Distribute.

$$=\frac{x^2-4x-21}{(x+3)(x-4)}$$ Combine numerators.

$$=\frac{(x+3)(x-7)}{(x+3)(x-4)}$$ Factor the numerator.

$$=\frac{x-7}{x-4}$$ Divide out the common factor, $x+3$.

e. $\dfrac{a+b}{a-b} - \dfrac{2a-3b}{b-a}$

Solution The expressions $a - b$ and $b - a$ are additive inverses, so we can obtain the LCD by multiplying the numerator and denominator of one rational expression by -1.

$$\frac{a+b}{a-b} - \frac{2a-3b}{b-a} = \frac{a+b}{a-b} - \frac{(2a-3b)(-1)}{(b-a)(-1)}$$ To get the LCD, $a - b$, multiply $\frac{2a-3b}{b-a}$ by $\frac{-1}{-1}$.

$$= \frac{a+b}{a-b} - \frac{-2a+3b}{a-b}$$ Distribute.

$$= \frac{(a+b)-(-2a+3b)}{a-b}$$ Combine numerators.

$$= \frac{(a+b)+(2a-3b)}{a-b}$$ Write the equivalent addition.

$$= \frac{3a-2b}{a-b}$$ Simplify the numerator.

Answers

a. $\dfrac{9x^2+21x-12xy+16y}{24x^2y^2}$

b. $\dfrac{1}{12}$

c. $\dfrac{3x^2+3x-13}{(2x-1)(x+3)(x-2)}$

d. $\dfrac{x-7}{x-2}$

e. $\dfrac{4x-3y}{x-y}$ or $\dfrac{3y-4x}{y-x}$

Your Turn Add or subtract.

a. $\dfrac{3x+7}{8xy^2} - \dfrac{3x-4}{6x^2y}$

b. $\dfrac{x+6}{4x+16} - \dfrac{x+7}{6x+24}$

c. $\dfrac{x+2}{2x^2+5x-3} + \dfrac{2x-3}{2x^2-5x+2}$

d. $\dfrac{x}{x+5} - \dfrac{35}{x^2+3x-10}$

e. $\dfrac{3x+2y}{x-y} - \dfrac{x-5y}{y-x}$

7.2 Exercises

For Extra Help

MyMathLab

Videotape/DVT

InterAct Math

Math Tutor Center

Math XL.com

1. When adding or subtracting rational numbers or rational expressions with common denominators, why do we add or subtract the numerators but we do not add or subtract the denominators?

2. If you are asked to add or subtract rational expressions whose denominators contain multi-term polynomials, explain how to find the LCD.

3. Suppose you are adding two rational expressions with unlike denominators and the LCD is $(2x + 3)(x - 4)(3x - 2)$. If the denominator of one of the fractions is $6x^2 + 5x - 6$, by what factor would you multiply the rational expression in order to write it as an equivalent rational expression with the LCD? Explain how you found this factor.

4. If the denominators of two rational expressions that are to be added or subtracted are additive inverses, by what factors do you multiply one of the rational expressions so that the denominators are the same?

5. Two students were adding the same two rational expressions. One got the answer $\frac{6}{a - 3}$ and the other got the answer $\frac{-6}{3 - a}$. Explain how it is possible for both students to be correct.

6. Suppose you are subtracting two rational expressions. You have found the LCD and you are at the following point in the solution: $\frac{(3a^2 - 2a + 4) - (a^2 + 5a - 1)}{(2a - 3)(a + 3)}$. What is the next step?

For Exercises 7–28, add or subtract. Simplify your answers to lowest terms.

7. $\frac{a}{b^2} + \frac{3a}{b^2}$

8. $\frac{3x}{y} + \frac{2x}{y}$

9. $\frac{3c}{2d} + \frac{5c}{2d}$

10. $\frac{5y}{4x} + \frac{3y}{4x}$

11. $\frac{4a + 3b}{2a - 5b} + \frac{3a - 4b}{2a - 5b}$

12. $\frac{4p - 3q}{3p - 2q} + \frac{2p + 5q}{3p - 2q}$

13. $\frac{2x - 3y}{x + 2y} - \frac{4x + 2y}{x + 2y}$

14. $\frac{3c + 2d}{2c + d} - \frac{5c - 3d}{2c + d}$

15. $\frac{m}{m^2 - 16} - \frac{4}{m^2 - 16}$

16. $\frac{r^2}{r - 7} - \frac{49}{r - 7}$

17. $\frac{a^2}{a^2 - a - 12} + \frac{5a + 6}{a^2 - a - 12}$

18. $\frac{x^2}{x^2 - 9} + \frac{2x - 15}{x^2 - 9}$

19. $\frac{c^2}{c^2 - 5c + 4} - \frac{-2c + 24}{c^2 - 5c + 4}$

20. $\frac{a^2}{a^2 + 3a - 10} - \frac{-a + 20}{a^2 + 3a - 10}$

21. $\frac{y^2 - 2y}{y^2 - 12y + 36} + \frac{-5y + 6}{y^2 - 12y + 36}$

22. $\dfrac{r^2 - 3r}{r^2 - 7r + 12} + \dfrac{5r - 15}{r^2 - 7r - 12}$

23. $\dfrac{h^2 - 4h}{h^2 + 10h + 21} - \dfrac{-6h + 3}{h^2 + 10h + 21}$

24. $\dfrac{w^2 + 6}{w^2 - 3w - 10} - \dfrac{2w + 14}{w^2 - 3w - 10}$

25. $\dfrac{2x^2 + x - 3}{x^2 + 6x + 5} + \dfrac{x^2 - 2x + 4}{x^2 + 6x + 5} - \dfrac{2x^2 + x + 4}{x^2 + 6x + 5}$

26. $\dfrac{3x^2 - 2x + 3}{x^2 + x - 12} - \dfrac{x^2 + x - 2}{x^2 + x - 12} - \dfrac{x^2 + 4x - 7}{x^2 + x - 12}$

27. $\dfrac{x^3}{x^2 - xy + y^2} + \dfrac{y^3}{x^2 - xy + y^2}$

28. $\dfrac{a^3}{a^2 - 2a + 4} + \dfrac{8}{a^2 - 2a + 4}$

For Exercises 29–44, find the LCD for each group.

29. $\dfrac{4}{15a^6b^8}, \dfrac{9}{20a^4b^5}$

30. $\dfrac{11}{12s^6v^3}, \dfrac{13}{18s^4v^5}$

31. $\dfrac{x}{x + 7}, \dfrac{x}{x - 5}$

32. $\dfrac{y}{y - 3}, \dfrac{y}{y - 4}$

33. $\dfrac{6r}{9r + 18}, \dfrac{8r}{15r + 30}$

34. $\dfrac{11x}{8x - 32}, \dfrac{17x}{12x - 48}$

35. $\dfrac{3b}{b^2 - 1}, \dfrac{b}{b^2 + 3b - 4}$

36. $\dfrac{4c}{c^2 - 9}, \dfrac{2c}{c^2 + c - 6}$

37. $\dfrac{c - 5}{c^2 - 2c - 3}, \dfrac{3 + c}{c^2 - 5c + 6}$

38. $\dfrac{w + 6}{w^2 + 6w + 5}, \dfrac{2 - w}{w^2 + w - 20}$

39. $\dfrac{n + 1}{n^2 + 8n + 16}, \dfrac{n - 4}{n^2 + 5n + 4}$

40. $\dfrac{n - 3}{n^2 - 10n + 25}, \dfrac{n - 1}{n^2 - 2n - 15}$

41. $\dfrac{x + 5}{x^3 + x^2 - 6x}, \dfrac{x - 3}{x^4 + 7x^3 + 12x^2}$

42. $\dfrac{y - 2}{y^5 - 16y^3}, \dfrac{y - 1}{y^3 + 5y^2 + 4y}$

43. $\dfrac{3}{x}, \dfrac{5x - 3}{x^2 - 2x - 8}, \dfrac{6}{2x^4 + 4x^2}$

44. $\dfrac{3}{a^2}, \dfrac{6}{a^2 - 16}, \dfrac{5a}{3a + 12}$

For Exercises 45–90, add or subtract as indicated. Simplify your answers to lowest terms.

45. $\dfrac{5}{8u} - \dfrac{7}{12u}$

46. $\dfrac{6}{15v} - \dfrac{4}{9v}$

47. $\dfrac{3z}{10x} + \dfrac{5z}{4x}$

48. $\dfrac{10w}{21y} + \dfrac{7w}{18y}$

49. $\dfrac{y + 6}{4y} + \dfrac{2y - 3}{3y}$

50. $\dfrac{4r - 2}{5r} + \dfrac{3r - 4}{2r}$

51. $\dfrac{m+8}{2m} - \dfrac{m-7}{3m}$

52. $\dfrac{n-9}{7n} - \dfrac{n-2}{4n}$

53. $\dfrac{3p+1}{10p^3q^2} + \dfrac{9p-2}{6p^2q}$

54. $\dfrac{5t-8}{16s^5t^2} + \dfrac{2t-7}{12s^2t^3}$

55. $\dfrac{2a+b}{8a^2b^4} - \dfrac{a-3b}{12a^3b^2}$

56. $\dfrac{3r-2s}{6rs^3} - \dfrac{4r-3s}{8r^2s^2}$

57. $\dfrac{4}{k} - \dfrac{6}{k+2}$

58. $\dfrac{5}{j-5} - \dfrac{9}{j}$

59. $\dfrac{4}{w-3} + \dfrac{5}{w+7}$

60. $\dfrac{6}{c+4} + \dfrac{11}{c-1}$

61. $\dfrac{x-4}{3x+9} - \dfrac{x+5}{6x+18}$

62. $\dfrac{y-3}{2y-8} - \dfrac{y-7}{6y-24}$

63. $\dfrac{2t}{t^2-8t+16} + \dfrac{6}{t-4}$

64. $\dfrac{4}{s^2+14s+49} + \dfrac{5s}{s+7}$

65. $\dfrac{u^2-3u}{u^2-6u+9} + \dfrac{8}{5u-15}$

66. $\dfrac{v^2+4v}{v^2+8v+16} + \dfrac{3}{3v+12}$

67. $\dfrac{x}{x+2} - \dfrac{16}{x^2-4x-12}$

68. $\dfrac{x}{x+4} - \dfrac{36}{x^2-x-20}$

69. $\dfrac{x+1}{x^2-x-6} + \dfrac{2x+8}{x^2+6x+8}$

70. $\dfrac{y+3}{y^2-3y+2} + \dfrac{2y+4}{y^2+y-2}$

71. $\dfrac{2v+5}{v^2-16} - \dfrac{2v+6}{v^2-v-12}$

72. $\dfrac{3t+9}{t^2-9} - \dfrac{3t-2}{t^2-8t+15}$

73. $\dfrac{z-3}{z^2+6z+9} + \dfrac{z+1}{z^2+z-6}$

74. $\dfrac{w-2}{w^2+2w-24} + \dfrac{w-8}{w^2-8w+16}$

75. $\dfrac{x+4}{x^3-36x} - \dfrac{1}{x^2+3x-18}$

76. $\dfrac{3}{x^2+3x-4} - \dfrac{x-2}{x^3-16x}$

77. $\dfrac{3x}{x-2} - \dfrac{4}{x+2} - \dfrac{3}{x^2-4}$

78. $\dfrac{5}{x-4} + \dfrac{2x}{x+4} - \dfrac{2}{x^2-16}$

79. $\dfrac{5}{a^2-4a} + \dfrac{6}{a} + \dfrac{4}{a-4}$

80. $\dfrac{7}{z} - \dfrac{3}{z^2+2z} + \dfrac{4}{z+2}$

81. $\dfrac{3x+1}{x-4} - \dfrac{7}{x} + \dfrac{2}{x^2-4x}$

82. $\dfrac{5}{y} - \dfrac{2y-5}{y+5} - \dfrac{3y-1}{y^2+5y}$

83. $\dfrac{3r}{r+3} - \dfrac{5r}{r-3} + \dfrac{2}{r^2-9}$

84. $\dfrac{2w}{w+1} - \dfrac{3}{w+1} + \dfrac{w^2-3}{w^2-1}$

85. $\dfrac{v}{u-v} - \dfrac{3v}{v-u}$

86. $\dfrac{2x}{x-y} - \dfrac{-3x}{y-x}$

87. $\dfrac{2m-n}{m-n} - \dfrac{m-3n}{n-m}$

88. $\dfrac{p-5q}{p-q} - \dfrac{4p-6q}{q-p}$

89. $\dfrac{2a-3b}{3a-b} + \dfrac{3a+2b}{b-3a}$

90. $\dfrac{4x+3y}{2x-5y} + \dfrac{2x-y}{5y-2x}$

For Exercises 91–96, explain the mistake and then find the correct sum or difference.

91. $\dfrac{4}{a} + \dfrac{a}{5} = \dfrac{4+a}{a+5}$

92. $\dfrac{4}{3x} - \dfrac{5}{12x} = \dfrac{1}{3x} - \dfrac{5}{3x}$
$= -\dfrac{4}{3x}$

93. $\dfrac{6v}{2x} + \dfrac{3v}{2x} = \dfrac{9v}{4x}$

94. $\dfrac{4}{w} + \dfrac{6}{w+z} = \dfrac{4(+z)}{w(+z)} + \dfrac{6}{w+z}$
$= \dfrac{4+z+6}{w+z}$
$= \dfrac{10+z}{w+z}$

95. $\dfrac{5c+2}{3c-5} - \dfrac{2c+3}{3c-5} = \dfrac{5c+2-2c+3}{3c-5}$
$= \dfrac{3c+5}{3c-5}$

96. $\dfrac{x-5}{x^2-1} - \dfrac{x+3}{x-1} = \dfrac{x-5-x-3}{x^2-1} = \dfrac{-8}{x^2-1}$

97. If $f(x) = \dfrac{9x+7}{6x}$ and $g(x) = \dfrac{x-3}{6x}$, find

a. $(f+g)(x)$ **b.** $(f-g)(x)$ **c.** $(f+g)(4)$ **d.** $(f-g)(-2)$

98. If $f(x) = \dfrac{x^2 - 2x + 4}{x - 4}$ and $g(x) = \dfrac{5x - 8}{x - 4}$, find

a. $(f + g)(x)$ **b.** $(f - g)(x)$ **c.** $(f + g)(0)$ **d.** $(f - g)(-5)$

99. If $f(x) = \dfrac{x - 14}{x^2 - 4}$ and $g(x) = \dfrac{x + 1}{x - 2}$, find

a. $(f + g)(x)$ **b.** $(f - g)(x)$ **c.** $(f + g)(1)$ **d.** $(f - g)(-1)$

100. If $f(x) = \dfrac{x + 2}{3x - 18}$ and $g(x) = \dfrac{2}{3x}$, find

a. $(f + g)(x)$ **b.** $(f - g)(x)$ **c.** $(f + g)(-2)$ **d.** $(f - g)(4)$

For Exercises 101–108, solve.

101. The portion of a house that Frank paints after t hours is represented by $\dfrac{t}{5}$, and the portion that Jose paints is represented by $\dfrac{t}{3}$. Find the portion of the house painted by Frank and Jose together after t hours.

102. A contractor is building a highway and the work is divided among three teams. After t months the portion of the highway completed by team A is represented by $\dfrac{t}{2}$, the portion by team B is $\dfrac{t}{4}$, and the portion by team C is $\dfrac{t}{5}$. Find the portion of the highway constructed by the three teams together after t months.

103. In designing a building, an architect describes the length of a rectangular building as $\dfrac{3}{x + 4}$ and the width as $\dfrac{1}{x - 2}$. Find the perimeter of the building.

104. The thickness of a board is given by $\dfrac{4}{x + 3}$. The board is run through a planer, which removes a layer whose thickness is $\dfrac{2}{x}$. What is the new thickness of the board?

105. If x represents a number, write and simplify an expression for the sum of three times the number and two times its reciprocal.

106. Write and simplify an expression for the sum of the reciprocals of two consecutive even integers if x represents the smaller.

107. Find the average of $\frac{x}{5}$ and $\frac{x}{3}$.

108. Write an expression in simplest form for the area of the trapezoid shown.

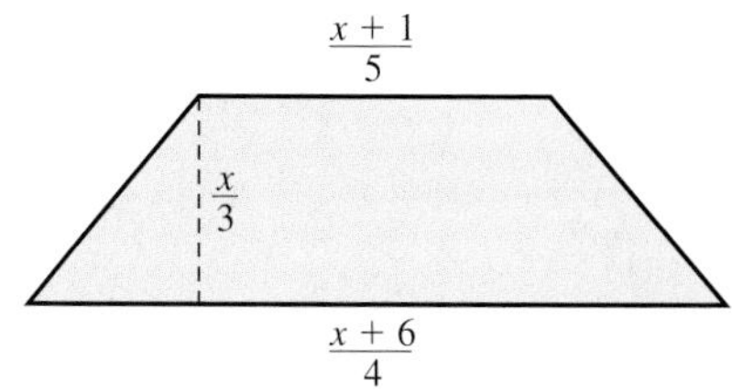

PUZZLE PROBLEM

A 7-minute hourglass and a 5-minute hourglass can be used together to measure exactly 16 minutes. One rule must be followed. As soon as an hourglass finishes leaking, it must immediately be flipped. Explain the steps required to measure 16 minutes.

REVIEW EXERCISES

For Exercises 1–4, perform the indicated operations.

[1.2] **1.** $\frac{1}{3} + \frac{3}{4}$

[1.2] **2.** $\frac{5}{12} \div \frac{3}{8}$

[1.3] **3.** $\left(\frac{2}{3} + \frac{1}{2}\right) \div \left(\frac{3}{4} - \frac{1}{3}\right)$

[1.3] **4.** $12\left(\frac{2}{3} + \frac{5}{6}\right)$

[7.1] ***For Exercises 5 and 6, simplify.***

5. $\frac{x^2 - 9x + 8}{x^2 - 6x - 16}$

6. $\frac{3y + 2}{27y^3 + 8}$

7.3 Simplifying Complex Rational Expressions

OBJECTIVES

1. **Simplify complex rational expressions.**
2. **Simplify rational expressions with negative exponents.**

Complex rational expressions, or *complex fractions*, have parts that are themselves rational expressions.

DEFINITION ***Complex rational expression:*** A rational expression that contains rational expressions in the numerator and/or denominator.

Note: *We will use a slightly larger fraction line to separate the numerator and denominator of complex rational expressions.*

Some complex rational expressions are

$$\frac{\frac{2}{5}}{15} \qquad \frac{2x}{\frac{4}{2x+5}} \qquad \frac{1+\frac{3}{x}-\frac{6}{x^2}}{2+\frac{3}{x}-\frac{4}{x^2}} \qquad \frac{\frac{x+2}{x-4}+\frac{x-5}{x+2}}{\frac{x+1}{x+2}-\frac{x+3}{x-3}}$$

OBJECTIVE 1. Simplify complex rational expressions. A complex rational expression is simplified when the numerator and denominator are polynomials with no common factors. There are two common methods for simplifying complex rational expressions.

PROCEDURE **Simplifying Complex Rational Expressions**

Method 1:

1. Simplify the numerator and denominator if needed.
2. Rewrite as a horizontal division problem.

Method 2:

1. Multiply the numerator and denominator of the complex rational expression by their LCD.
2. Simplify.

For example, using method 1, the complex fraction $\frac{\frac{2}{3}}{\frac{5}{6}}$ can be expressed as the horizontal division problem, $\frac{2}{3} \div \frac{5}{6}$ and then simplified.

$$\frac{\frac{2}{3}}{\frac{5}{6}} = \frac{2}{3} \div \frac{5}{6}$$ Write as division.

$$= \frac{2}{\cancel{3}} \cdot \frac{\cancel{6}^{2}}{5}$$ Rewrite as multiplication.

$$= \frac{4}{5}$$ Simplify.

Using method 2, we multiply the numerator and denominator of the complex fraction by the LCD of the numerator's and denominator's fractions. In this case, the denominators are 3 and 6, so the LCD is 6.

$$\frac{\frac{2}{3}}{\frac{5}{6}} = \frac{\frac{2}{3} \cdot 6}{\frac{5}{6} \cdot 6}$$ Multiply the numerator and denominator by 6.

$$= \frac{4}{5}$$ Simplify.

EXAMPLE 1 Simplify.

a. $\dfrac{\frac{a^2b}{c^2}}{\frac{ab^3}{c^3}}$

Method 1: $\dfrac{\frac{a^2b}{c^2}}{\frac{ab^3}{c^3}} = \dfrac{a^2b}{c^2} \div \dfrac{ab^3}{c^3}$ Write as a horizontal division problem.

$= \dfrac{a^2b}{c^2} \cdot \dfrac{c^3}{ab^3}$ Rewrite as a multiplication problem.

$= \dfrac{ac}{b^2}$ Simplify.

Method 2: $\dfrac{\frac{a^2b}{c^2}}{\frac{ab^3}{c^3}} = \dfrac{\frac{a^2b}{\cancel{c^2}} \cdot \cancel{c^3}^{c}}{\frac{ab^3}{\cancel{c^3}} \cdot \cancel{c^3}}$ Multiply the numerator and denominator by the LCD of the fractions in the numerator and denominator, c^3. Then divide out the common factors.

$= \dfrac{a^2bc}{ab^3}$ Multiply.

$= \dfrac{ac}{b^2}$ Simplify.

b. $\dfrac{1 - \dfrac{9}{x^2}}{1 + \dfrac{3}{x}}$

Method 1: $\dfrac{1 - \dfrac{9}{x^2}}{1 + \dfrac{3}{x}} = \dfrac{1 \cdot \dfrac{x^2}{x^2} - \dfrac{9}{x^2}}{1 \cdot \dfrac{x}{x} + \dfrac{3}{x}}$ Write the numerator fractions as equivalent fractions with their LCD, x^2, and write the denominator fractions with their LCD, x.

$= \dfrac{\dfrac{x^2}{x^2} - \dfrac{9}{x^2}}{\dfrac{x}{x} + \dfrac{3}{x}}$ Simplify.

$= \dfrac{\dfrac{x^2 - 9}{x^2}}{\dfrac{x + 3}{x}}$ Subtract in the numerator and add in the denominator.

$= \dfrac{x^2 - 9}{x^2} \div \dfrac{x + 3}{x}$ Write the complex fraction as a horizontal division problem.

$= \dfrac{x^2 - 9}{x^2} \cdot \dfrac{x}{x + 3}$ Rewrite as multiplication.

$= \dfrac{\cancel{(x + 3)}(x - 3)}{\cancel{x^2}_{x}} \cdot \dfrac{\cancel{x}}{\cancel{x + 3}}$ Factor $x^2 - 9$ and divide out the common factors.

$= \dfrac{x - 3}{x}$ Multiply the remaining factors.

Method 2: $\dfrac{1 - \dfrac{9}{x^2}}{1 + \dfrac{3}{x}} = \dfrac{\left(1 - \dfrac{9}{x^2}\right) \cdot x^2}{\left(1 + \dfrac{3}{x}\right) \cdot x^2}$ Multiply the numerator and denominator by the LCD of the fractions in the numerator and denominator, x^2.

$= \dfrac{1 \cdot x^2 - \dfrac{9}{x^2} \cdot x^2}{1 \cdot x^2 + \dfrac{3}{x} \cdot x^2}$ Distribute x^2.

$= \dfrac{x^2 - 9}{x^2 + 3x}$ Simplify.

$= \dfrac{(x + 3)(x - 3)}{x(x + 3)}$ Factor the numerator and the denominator.

$= \dfrac{x - 3}{x}$ Divide out the common factor, $x + 3$.

c. $\dfrac{x+\dfrac{9}{x-6}}{1+\dfrac{3}{x-6}}$

Method 1: $\dfrac{x+\dfrac{9}{x-6}}{1+\dfrac{3}{x-6}} = \dfrac{x\cdot\dfrac{x-6}{x-6}+\dfrac{9}{x-6}}{1\cdot\dfrac{x-6}{x-6}+\dfrac{3}{x-6}}$ Simplify the numerator and denominator by writing them as equivalent fractions with the LCD, $x-6$.

$= \dfrac{\dfrac{x^2-6x}{x-6}+\dfrac{9}{x-6}}{\dfrac{x-6}{x-6}+\dfrac{3}{x-6}}$ Simplify.

$= \dfrac{\dfrac{x^2-6x+9}{x-6}}{\dfrac{x-3}{x-6}}$ Add in numerator and denominator.

$= \dfrac{x^2-6x+9}{x-6} \div \dfrac{x-3}{x-6}$ Write the complex fraction as a horizontal division problem.

$= \dfrac{x^2-6x+9}{x-6} \cdot \dfrac{x-6}{x-3}$ Write an equivalent multiplication problem.

$= \dfrac{(x-3)^2}{x-6} \cdot \dfrac{x-6}{x-3}$ Factor and divide out the common factors, $x-3$ and $x-6$.

$= x-3$

Method 2: $\dfrac{x+\dfrac{9}{x-6}}{1+\dfrac{3}{x-6}} = \dfrac{\left(x+\dfrac{9}{x-6}\right)(x-6)}{\left(1+\dfrac{3}{x-6}\right)(x-6)}$ Multiply the numerator and denominator by the LCD of the fractions in the numerator and denominator, $x-6$.

$= \dfrac{x(x-6)+\left(\dfrac{9}{\cancel{x-6}}\right)(\cancel{x-6})}{1(x-6)+\left(\dfrac{3}{\cancel{x-6}}\right)(\cancel{x-6})}$ Distribute $x-6$.

$= \dfrac{x^2-6x+9}{x-6+3}$ Multiply.

$= \dfrac{(x-3)^2}{x-3}$ Factor numerator and simplify denominator.

$= x-3$ Divide out the common factor, $x-3$.

Method 2 is always at least as easy as method 1 and is usually the preferred method. Consequently, the remaining examples of this section will be done with method 2 only.

EXAMPLE 2 Simplify.

$$\frac{\dfrac{x+3}{x-3}-\dfrac{x-3}{x+3}}{\dfrac{x+3}{x-3}+\dfrac{x-3}{x+3}}$$

Solution

$$\frac{\left(\dfrac{x+3}{x-3}-\dfrac{x-3}{x+3}\right)(x-3)(x+3)}{\left(\dfrac{x+3}{x-3}+\dfrac{x-3}{x+3}\right)(x-3)(x+3)}$$

Multiply the numerator and denominator by the LCD of the fractions in the numerator and denominator, $(x - 3)(x + 3)$.

$$=\frac{\dfrac{x+3}{\cancel{x-3}}\cdot\cancel{(x-3)}(x+3)-\dfrac{x-3}{\cancel{x+3}}\cdot(x-3)\cancel{(x+3)}}{\dfrac{x+3}{\cancel{x-3}}\cdot\cancel{(x-3)}(x+3)+\dfrac{x-3}{\cancel{x+3}}\cdot(x-3)\cancel{(x+3)}}$$

Distribute.

$$=\frac{(x+3)(x+3)-(x-3)(x-3)}{(x+3)(x+3)+(x-3)(x-3)}$$

Simplify.

$$=\frac{(x^2+6x+9)-(x^2-6x+9)}{(x^2+6x+9)+(x^2-6x+9)}$$

Multiply each pair of binomials.

$$=\frac{(x^2+6x+9)+(-x^2+6x-9)}{x^2+6x+9+x^2-6x+9}$$

Write the equivalent addition.

$$=\frac{12x}{2x^2+18}$$

Combine like terms.

$$=\frac{6x}{x^2+9}$$

Divide out the common factor, 2.

YOUR TURN Simplify.

a. $\dfrac{\dfrac{4cd^3}{3a}}{\dfrac{2c^4d}{6a^2}}$ b. $\dfrac{1-\dfrac{4}{x}-\dfrac{12}{x^2}}{1+\dfrac{6}{x}+\dfrac{8}{x^2}}$ c. $\dfrac{x+\dfrac{16}{x-8}}{1+\dfrac{4}{x-8}}$ d. $\dfrac{\dfrac{a-4}{a+4}+\dfrac{a+4}{a-4}}{\dfrac{a-4}{a+4}-\dfrac{a+4}{a-4}}$

ANSWERS

a. $\dfrac{4ad^2}{c^3}$ b. $\dfrac{x-6}{x+4}$

c. $x-4$ d. $-\dfrac{a^2+16}{8a}$

OBJECTIVE 2. Simplify rational expressions with negative exponents. Rational expressions containing terms with negative exponents often become complex rational expressions when those terms are rewritten with positive exponents.

EXAMPLE 3 Simplify. $\dfrac{x^{-1} + y^{-1}}{x^{-2} - y^{-2}}$

Solution Using $x^{-n} = \dfrac{1}{x^n}$, rewrite the expression with positive exponents only.

$$\frac{x^{-1} + y^{-1}}{x^{-2} - y^{-2}} = \frac{\dfrac{1}{x} + \dfrac{1}{y}}{\dfrac{1}{x^2} - \dfrac{1}{y^2}}$$ Rewrite with positive exponents only.

$$= \frac{\left(\dfrac{1}{x} + \dfrac{1}{y}\right) \cdot x^2y^2}{\left(\dfrac{1}{x^2} - \dfrac{1}{y^2}\right) \cdot x^2y^2}$$ Multiply numerator and denominator by the LCD, x^2y^2.

$$= \frac{xy^2 + x^2y}{y^2 - x^2}$$ Distribute x^2y^2 and simplify the results.

$$= \frac{xy(y + x)}{(y - x)(y + x)}$$ Factor the numerator and denominator.

$$= \frac{xy}{y - x}$$ Divide out the common factor, $y + x$.

YOUR TURN

Simplify. $\dfrac{2a^{-2} + b^{-1}}{a^{-1} - 3b^{-2}}$

ANSWER

$\dfrac{2b^2 + a^2b}{ab^2 - 3a^2}$

7.3 Exercises

For Extra Help

MyMathLab |  Videotape/DVT |  InterAct Math |  Math Tutor Center |  Math XL.com

1. By what would you multiply the numerator and denominator of $\dfrac{\dfrac{x + 3}{x - 2} + 5}{6 - \dfrac{x + 6}{x - 3}}$ to simplify using method 2?

2. Given the complex rational expression $\dfrac{\dfrac{3a + b}{2a - b}}{a - 4b}$, is the numerator $3a + b$ or $\dfrac{3a + b}{2a - b}$? How could the expression be written so that the numerator and denominator are easily identified?

3. Write $\dfrac{\dfrac{x+2}{x-4}}{\dfrac{x-3}{x+5}}$ as a horizontal division problem.

4. Write $\left(\dfrac{c+2d}{2c+3d}-4\right)\div\left(5+\dfrac{3c-d}{c+4d}\right)$ as a complex rational expression.

5. Which of the two methods discussed in this section would you use to simplify $\dfrac{x-3+\dfrac{x+2}{x-4}}{x+2-\dfrac{x-3}{x+4}}$? Why?

6. Which of the two methods discussed in this section would you use to simplify $\dfrac{\dfrac{3a^2b^4}{4c^2d^3}}{\dfrac{9ab^2}{8c^4d^2}}$? Why?

For Exercises 7–54, simplify each complex fraction.

7. $\dfrac{\dfrac{5}{21}}{\dfrac{3}{14}}$

8. $\dfrac{\dfrac{11}{12}}{\dfrac{3}{8}}$

9. $\dfrac{\dfrac{u}{v^3}}{\dfrac{w}{v^4}}$

10. $\dfrac{\dfrac{r}{s^5}}{\dfrac{t}{s^2}}$

11. $\dfrac{\dfrac{ac}{b^4}}{\dfrac{ac}{b^3}}$

12. $\dfrac{\dfrac{rs}{t^2}}{\dfrac{rs}{t^5}}$

13. $\dfrac{\dfrac{u^7v^2}{w^5}}{\dfrac{u^3v^4}{w}}$

14. $\dfrac{\dfrac{m^6p^5}{n^7}}{\dfrac{m^5p^3}{n^4}}$

15. $\dfrac{\dfrac{15}{8}}{20}$

16. $\dfrac{9}{\dfrac{12}{5}}$

17. $\dfrac{a}{\dfrac{b}{c}}$

18. $\dfrac{\dfrac{m}{n}}{p}$

19. $\dfrac{\dfrac{16}{2x-2}}{\dfrac{8}{x-3}}$

20. $\dfrac{\dfrac{x+6}{15}}{\dfrac{x-3}{3}}$

21. $\dfrac{\dfrac{x+4}{6}}{\dfrac{2x+8}{18}}$

22. $\dfrac{\dfrac{b-7}{4}}{\dfrac{3b-21}{16}}$

23. $\dfrac{\dfrac{3x^2}{x+2}}{\dfrac{2x}{x+2}}$

24. $\dfrac{\dfrac{5a^3}{3a-2}}{\dfrac{2a}{3a-2}}$

25. $\dfrac{\dfrac{4}{9}-\dfrac{1}{3}}{\dfrac{7}{12}-\dfrac{5}{18}}$

26. $\dfrac{\dfrac{3}{4}+\dfrac{1}{12}}{\dfrac{5}{8}-\dfrac{7}{20}}$

27. $\dfrac{8-\dfrac{9}{4}}{\dfrac{7}{8}+\dfrac{9}{10}}$

28. $\dfrac{\dfrac{7}{9}+\dfrac{5}{12}}{3-\dfrac{7}{3}}$

29. $\dfrac{1+\dfrac{x}{2}}{1-\dfrac{x}{2}}$

30. $\dfrac{3-\dfrac{y}{3}}{5+\dfrac{y}{3}}$

31. $\dfrac{\dfrac{1}{t}-1}{\dfrac{1}{t^3}-1}$

32. $\dfrac{\dfrac{8}{s^3}-1}{\dfrac{2}{s}-1}$

33. $\dfrac{1+\dfrac{1}{v}-\dfrac{2}{v^2}}{1-\dfrac{4}{v}+\dfrac{3}{v^2}}$

34. $\dfrac{2+\dfrac{13}{w}+\dfrac{6}{w^2}}{1+\dfrac{11}{w}+\dfrac{30}{w^2}}$

35. $\dfrac{\dfrac{1}{u}-\dfrac{2}{u^2}}{1+\dfrac{4}{u}-\dfrac{12}{u^2}}$

36. $\dfrac{\dfrac{4}{p}+\dfrac{4}{p^2}}{3-\dfrac{2}{p}-\dfrac{5}{p^2}}$

37. $\dfrac{\dfrac{2}{x}-\dfrac{7}{x^2}-\dfrac{30}{x^3}}{2+\dfrac{11}{x}+\dfrac{15}{x^2}}$

38. $\dfrac{1-\dfrac{3}{y}-\dfrac{4}{y^2}}{\dfrac{3}{y}-\dfrac{13}{y^2}+\dfrac{4}{y^3}}$

39. $\dfrac{5+\dfrac{4}{r+8}}{r+8}$

40. $\dfrac{\dfrac{7}{s-15}-6}{s-15}$

41. $\dfrac{t+\dfrac{9}{t-7}}{11-\dfrac{3}{t-7}}$

42. $\dfrac{\dfrac{2}{q+4}-q}{\dfrac{13}{q+4}+5}$

43. $\dfrac{\dfrac{x+4}{x^2-9}}{2+\dfrac{1}{x+3}}$

44. $\dfrac{\dfrac{c+3}{c^2-25}}{4+\dfrac{2}{c-5}}$

45. $\dfrac{\dfrac{5}{a+3}-\dfrac{4}{a-3}}{\dfrac{3}{a+3}-\dfrac{2}{a-3}}$

46. $\dfrac{\dfrac{6}{n+4}-\dfrac{3}{n-4}}{\dfrac{2}{n+4}+\dfrac{7}{n-4}}$

47. $\dfrac{\dfrac{r+3}{r-3}-\dfrac{r-3}{r+3}}{\dfrac{r+3}{r-3}+\dfrac{r-3}{r+3}}$

48. $\dfrac{\dfrac{m-7}{m+7}+\dfrac{m+7}{m-7}}{\dfrac{m-7}{m+7}-\dfrac{m+7}{m-7}}$

49. $\dfrac{\dfrac{3x-9}{x-3}-\dfrac{x-3}{x-5}}{\dfrac{x+5}{x-5}+\dfrac{2x-6}{x-3}}$

50. $\dfrac{\dfrac{2x+8}{x+4}+\dfrac{x-4}{x-2}}{\dfrac{3x+12}{x+4}-\dfrac{x+2}{x-2}}$

51. $\dfrac{\dfrac{2}{y^2}-\dfrac{5}{xy}-\dfrac{3}{x^2}}{\dfrac{2}{y^2}+\dfrac{5}{xy}+\dfrac{2}{x^2}}$

52. $\dfrac{\dfrac{3}{y^2}-\dfrac{5}{xy}-\dfrac{2}{x^2}}{\dfrac{3}{y^2}+\dfrac{10}{xy}+\dfrac{3}{x^2}}$

53. $\dfrac{\dfrac{2a}{a+6}+\dfrac{1}{a}}{9a-\dfrac{3}{a+6}}$

54. $\dfrac{8b-\dfrac{7}{b-1}}{\dfrac{4b}{b-1}+\dfrac{21}{b}}$

For Exercises 55–68, rewrite each as a complex fraction and simplify.

55. $\dfrac{x^{-1}}{3x^{-1}+1}$

56. $\dfrac{a^{-1}-1}{2a^{-1}}$

57. $\dfrac{x^{-1}+y^{-1}}{x^{-1}-y^{-1}}$

58. $\dfrac{x^{-1}y^{-1}}{x^{-1}+y^{-1}}$

59. $\dfrac{1-x^{-2}}{1+x^{-1}}$

60. $\dfrac{1-a^{-1}}{1-a^{-2}}$

61. $\dfrac{3x^{-2}+5y^{-1}}{x^{-1}+y^{-1}}$

62. $\dfrac{a^{-2}-b^{-1}}{2a^{-1}+3b^{-2}}$

63. $\dfrac{x^{-2}y^{-2}}{4x^{-1}+3y^{-1}}$

64. $\dfrac{3r^{-1}+4s^{-1}}{r^{-2}s^{-2}}$

65. $\dfrac{36a^{-2}-25b^{-2}}{6a^{-1}+5b^{-1}}$

66. $\dfrac{3x^{-1}+2y^{-1}}{9x^{-2}-4y^{-2}}$

67. $\dfrac{(4a)^{-1}+2b^{-2}}{2a^{-1}+b^{-2}}$

68. $\dfrac{3p^{-1}+q^{-2}}{3p^{-1}+(3q)^{-1}}$

For Exercises 69–72, explain the mistake then find the correct answer.

69. $\dfrac{a+\dfrac{1}{3}}{b+\dfrac{1}{3}} = \dfrac{3a}{3b} = \dfrac{a}{b}$

70. $\dfrac{\dfrac{1}{a}-\dfrac{1}{b}}{a-b} = \dfrac{\dfrac{a}{1}\cdot\dfrac{1}{a}-\dfrac{b}{1}\cdot\dfrac{1}{b}}{a\cdot a-b\cdot b}$

$= \dfrac{1-1}{a^2-b^2}$

$= \dfrac{0}{a^2-b^2}$

$= 0$

71. $\dfrac{n+\dfrac{3}{n}}{\dfrac{3}{n}} = \dfrac{n+\dfrac{n}{1}\cdot\dfrac{3}{n}}{\dfrac{n}{1}\cdot\dfrac{3}{n}}$

$= \dfrac{n+3}{3}$

72. $\dfrac{\dfrac{1}{m}-\dfrac{1}{n}}{\dfrac{1}{m}+\dfrac{1}{n}} = \dfrac{\dfrac{mn}{1}\cdot\dfrac{1}{m}-\dfrac{mn}{1}\cdot\dfrac{1}{n}}{\dfrac{mn}{1}\cdot\dfrac{1}{m}+\dfrac{mn}{1}\cdot\dfrac{1}{n}}$

$= \dfrac{n-m}{n+m}$

$= -1$

For Exercises 73 and 74, average rate can be found using the formula

$$\textbf{Average rate} = \frac{\textbf{Total distance}}{\textbf{Total time}}$$

73. Jamel went on a short trip to visit a friend who lives 20 miles away. The trip there took $\frac{1}{3}$ of an hour and the trip back home took $\frac{1}{4}$ of an hour. What was his average rate for the trip?

74. Ellen took a trip and traveled the first 30 miles in $\frac{1}{2}$ of an hour. She was in heavy traffic the next 10 miles, which took her $\frac{1}{4}$ of an hour. After the traffic cleared up, she traveled the next 50 miles in $\frac{2}{3}$ of an hour. What was her average speed for the trip?

75. Given the area, A, and length, l, of a rectangle, the width can be found using the formula $w = \frac{A}{l}$. Suppose the area of a rectangle is $\frac{3x^2 - 11x - 4}{18}$ square inches and the length is $\frac{x - 4}{2}$ inches. Find the width.

76. If the area, A, and base, b, of a triangle are known, the height can be found using the formula $h = \frac{2A}{b}$. If the area of a triangle is $\frac{10}{4x + 8}$ square inches and the base is $\frac{5}{3x + 6}$ inches, find the height.

77. In electrical circuits, if two resistors with resistance R_1 and R_2 ohms are wired in parallel, the resistance of the circuit is found using the complex rational expression $\frac{1}{\frac{1}{R_1} + \frac{1}{R_2}}$.

a. Simplify this complex rational expression.

b. If a 60-ohm and 40-ohm resistor are wired in parallel, find the resistance of the circuit.

c. If two 40-ohm speakers are wired in parallel, find the resistance of the circuit.

78. Suppose three resistors with resistance R_1, R_2, and R_3 ohms are wired in parallel. The resistance of the circuit is found using the complex rational expression

$$\frac{1}{\frac{1}{R_1} + \frac{1}{R_2} + \frac{1}{R_3}}.$$

a. Simplify this complex rational expression.

b. If a 60-ohm, a 40-ohm, and a 20-ohm resistor are wired in parallel, find the resistance of the circuit.

c. If a 200-ohm, a 400-ohm, and a 600-ohm resistor are wired in parallel, find the resistance of the circuit.

Collaborative Exercises AVERAGE RATES

1. A person drives for 10 minutes at 50 miles per hour and then drives 15 minutes at 60 miles per hour. What is her average rate?
2. A person drives 15 miles at 50 miles per hour and then travels another 20 miles at 40 miles per hour. What is his average rate?
3. A person travels a certain distance at 60 miles per hour and then travels that same distance at 70 miles per hour. What is her average rate?

REVIEW EXERCISES

For Exercises 1–4, solve.

[2.1] **1.** $3x + 5 = 2$ [2.1] **2.** $4x - 6 = 2x - 14$ [6.4] **3.** $3x^2 + 7x = 0$ [6.4] **4.** $2x^2 - 7x - 15 = 0$

[2.2] **5.** Gretchen walks 2.6 miles in 45 minutes. Find Gretchen's average rate in miles per hour. (Use $d = rt$.)

[2.2] **6.** A northbound car and a southbound car meet each other on a highway. The northbound car is traveling 40 mph and the southbound car is traveling 60 mph. How much time elapses from the time they pass each other until they are 20 miles apart?

7.4 Solving Equations Containing Rational Expressions

OBJECTIVE **1. Solve equations containing rational expressions.**

To solve equations containing rational expressions, we multiply both sides of the equation by the LCD. Since all the denominators divide evenly into the LCD, multiplying both sides of the equation by the LCD will eliminate all the denominators of all the rational expressions in the equation. We then solve the resulting equation.

PROCEDURE Solving Equations Containing Rational Expressions

1. Eliminate the denominators of the rational expressions by multiplying both sides of the equation by the LCD.
2. Solve the resulting equation using the methods in Chapter 2 (for linear equations) and Chapter 6 (for quadratic equations).
3. Check your solution(s) in the original equation.

EXAMPLE 1 Solve. $\frac{5}{x} - \frac{3}{4} = \frac{1}{2}$

Solution To solve, eliminate the rational expressions by multiplying both sides by the LCD, $4x$.

Note: *If $x = 0$, then $\frac{5}{x}$ is undefined, so the solution cannot be 0. We will say more about why it is important to note such values after Example 1.*

$$4x\left(\frac{5}{x} - \frac{3}{4}\right) = 4x\left(\frac{1}{2}\right) \quad \text{Multiply both sides by } 4x.$$

$$4x\left(\frac{5}{x}\right) - 4x\left(\frac{3}{4}\right) = 2x \quad \text{Distribute on the left, multiply on the right.}$$

$$20 - 3x = 2x \quad \text{Multiply.}$$

$$20 = 5x \quad \text{Add } 3x \text{ to both sides.}$$

$$4 = x \quad \text{Divide both sides by 5. We see that 4 is a possible solution.}$$

Check $\frac{5}{4} - \frac{3}{4} \stackrel{?}{=} \frac{1}{2}$ Substitute 4 for x.

$\frac{1}{2} = \frac{1}{2}$ The equation is true, so 4 is the solution.

Extraneous Solutions

If we multiply both sides of an equation by an expression that contains a variable, we might obtain a solution that, when substituted into the original equation, makes one of its denominators equal to 0 (and therefore makes one of its expressions undefined). We call such an apparent solution an *extraneous solution*, and we discard it.

By inspecting the denominator of each rational expression, you can determine the value(s) that would cause the expression to be undefined before solving the equation.

EXAMPLE 2 Solve. $\frac{3y}{y-4} = 6 + \frac{12}{y-4}$

Solution Notice that if $y = 4$, then $\frac{3y}{y-4}$ and $\frac{12}{y-4}$ are undefined, so 4 cannot be a solution.

Connection Because the equation in Example 2 has no solution, if we were to write its solution set, we would write an empty set, which is denoted by { } or ∅.

$$(y-4)\left(\frac{3y}{y-4}\right) = (y-4)\left(6 + \frac{12}{y-4}\right) \quad \text{Multiply both sides by the LCD, } y-4.$$

$$3y = (y-4)(6) + (y-4)\frac{12}{y-4} \quad \text{Multiply on the left, distribute on the right.}$$

$$3y = 6y - 24 + 12 \quad \text{Multiply on the right.}$$

$$-3y = -12 \quad \text{Subtract } 6y \text{ from both sides and simplify on the right.}$$

$$y = 4 \quad \text{Divide both sides by } -3\text{, so 4 is a possible solution.}$$

Check We already noted that 4 causes expressions in the equation to be undefined, so 4 is extraneous. Since 4 was the only possible solution, this equation has no solution.

YOUR TURN Solve.

a. $\frac{4}{5} - \frac{1}{x} = \frac{2}{3}$

b. $\frac{5n}{n-2} - 4 = \frac{10}{n-2}$

Proportions

If an equation has the form $\frac{a}{b} = \frac{c}{d}$, which is called a *proportion*, multiplying both sides by the LCD, *bd*, gives the following:

$$\cancel{b}d \cdot \frac{a}{\cancel{b}} = b\cancel{d} \cdot \frac{c}{\cancel{d}}$$

$$ad = bc$$

A faster way to reach that same conclusion is to *cross multiply*.

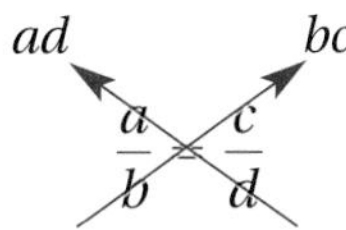

Warning: Cross multiplication can be used *only* when the equation is of the form $\frac{a}{b} = \frac{c}{d}$.

RULE Proportions and Their Cross Products

If $\frac{a}{b} = \frac{c}{d}$, where $b \neq 0$ and $d \neq 0$, then $ad = bc$.

EXAMPLE 3 Solve. $\frac{x}{x+4} = \frac{1}{x-2}$

Solution Note that x cannot equal -4 or 2. Since the equation is in the form $\frac{a}{b} = \frac{c}{d}$, we can cross multiply.

$x(x-2) = (x+4)(1)$	Cross multiply.
$x^2 - 2x = x + 4$	Distribute.
$x^2 - 3x - 4 = 0$	Write the quadratic equation in standard form ($ax^2 + bx + c = 0$).
$(x-4)(x+1) = 0$	Factor.
$x - 4 = 0$ or $x + 1 = 0$	Use the zero-factor theorem.
$x = 4$ or $x = -1$	Solve each equation, so 4 and -1 are possible solutions.

Since $x = 4$ and $x = -1$ are not extraneous solutions, we need only to check to see if our work is correct. We leave the check to the reader.

ANSWERS

a. $\frac{15}{2}$ b. no solution (2 is extraneous)

Connection Cross multiplication is just a shortcut for multiplying both sides by the LCD. Suppose we solved the equation in Example 3 by multiplying both sides by the LCD:

$$(x+4)(x-2)\frac{x}{x+4} = (x+4)(x-2)\frac{1}{x-2}$$
$$(x-2)x = (x+4)1$$
$$x^2 - 2x = x + 4$$

This equation is the same equation we had in the second step of the solution.

YOUR TURN Solve. $\dfrac{y}{y-2} = \dfrac{10}{y+1}$

EXAMPLE 4 Solve.

a. $\dfrac{x}{x+4} - \dfrac{1}{x+2} = \dfrac{8}{x^2+6x+8}$

Note: *Inspecting the denominators, we see that neither −4 nor −2 can be solutions.*

Solution $\dfrac{x}{x+4} - \dfrac{1}{x+2} = \dfrac{8}{(x+4)(x+2)}$ Factor the denominator $x^2 + 6x + 8$.

$(x+4)(x+2)\left(\dfrac{x}{x+4} - \dfrac{1}{x+2}\right) = (x+4)(x+2)\dfrac{8}{(x+4)(x+2)}$ Multiply both sides by the LCD, $(x + 4)(x + 2)$.

$(x+4)(x+2)\dfrac{x}{x+4} - (x+4)(x+2)\dfrac{1}{x+2} = 8$ Distribute on the left side and multiply on the right.

$(x+2)\cdot x - (x+4)\cdot 1 = 8$ Multiply.

$x^2 + 2x - x - 4 = 8$ Simplify the left side.

$x^2 + x - 12 = 0$ Write the quadratic equation in standard form.

$(x+4)(x-3) = 0$ Factor the left side.

$x + 4 = 0$ or $x - 3 = 0$ Use the zero-factor theorem.

$x = -4$ or $x = 3$

Check We have already noted that −4 cannot be a solution, so 3 is the only possible solution. We will leave the check for 3 to the reader.

b. $\dfrac{4}{x^2-x-6} - \dfrac{2}{x^2+3x+2} = \dfrac{4}{x^2-2x-3}$

Solution $\dfrac{4}{(x-3)(x+2)} - \dfrac{2}{(x+2)(x+1)} = \dfrac{4}{(x-3)(x+1)}$ Factor the denominators to determine the LCD.

$(x-3)(x+2)(x+1)\left(\dfrac{4}{(x-3)(x+2)} - \dfrac{2}{(x+2)(x+1)}\right)$ Multiply both sides by the LCD, $(x - 3)(x + 2)(x + 1)$.

$= (x-3)(x+2)(x+1)\left(\dfrac{4}{(x-3)(x+1)}\right)$

ANSWERS

4, 5

$$(x + 1)4 - (x - 3)2 = (x + 2)4 \quad \text{Distribute and simplify.}$$
$$4x + 4 - 2x + 6 = 4x + 8 \quad \text{Simplify.}$$
$$2x + 10 = 4x + 8 \quad \text{Combine like terms on the left side.}$$
$$2 = 2x \quad \text{Subtract } 2x \text{ and 8 from both sides.}$$
$$1 = x \quad \text{Divide both sides by 2.}$$

Check Since 1 does not make any denominator equal to 0, it is not an extraneous solution. We will leave the check to the reader.

YOUR TURN Solve.

a. $\dfrac{12}{x^2 - 4} = \dfrac{3}{x - 2} + 1$ **b.** $\dfrac{7}{n^2 + 3n - 10} = \dfrac{6}{n^2 + 4n - 5} + \dfrac{3}{n^2 - 3n + 2}$

Warning: Be sure you understand the difference between performing operations with rational expressions and solving equations containing rational expressions.

For example, an *expression*, like $\dfrac{2}{x} + \dfrac{3}{x + 1}$, can be *evaluated* or *rewritten* (not solved). To rewrite this expression, we add the two rational expressions.

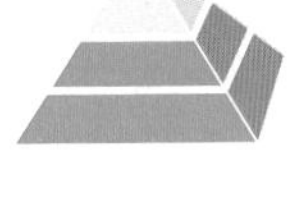

$$\frac{2}{x} + \frac{3}{x + 1} = \frac{2(x + 1)}{x(x + 1)} + \frac{3(x)}{(x + 1)(x)} \quad \text{Rewrite each rational expression with the LCD } x(x + 1).$$
$$= \frac{2x + 2}{x(x + 1)} + \frac{3x}{x(x + 1)} \quad \text{Multiply each numerator.}$$
$$= \frac{5x + 2}{x(x + 1)} \quad \text{Add the numerators and keep the LCD.}$$

An *equation*, like $\dfrac{2}{x} + \dfrac{3}{x + 1} = \dfrac{17}{12}$, can be *solved*. We first eliminate the rational expressions by multiplying both sides of the equation by the LCD, $12x(x + 1)$.

Equations and Inequalities

$$12x(x + 1)\left(\frac{2}{x} + \frac{3}{x + 1}\right) = 12x(x + 1)\frac{17}{12} \quad \text{Multiply both sides by the LCD.}$$
$$12(x + 1)(2) + 12x(3) = x(x + 1)(17) \quad \text{Simplify.}$$
$$24x + 24 + 36x = 17x^2 + 17x \quad \text{Simplify.}$$
$$0 = 17x^2 - 43x - 24 \quad \text{Write in standard form.}$$
$$0 = (17x + 8)(x - 3) \quad \text{Factor.}$$
$$x = -\frac{8}{17} \quad \text{or} \quad x = 3 \quad \text{Use the zero-factor theorem.}$$

ANSWERS

a. -5 (2 is extraneous)
b. no solution (-5 is extraneous)

7.4 Exercises

For Extra Help

MyMathLab Videotape/DVT InterAct Math Math Tutor Center Math XL.com

1. Explain the difference between solving an equation involving rational expressions and adding rational expressions. For example, how is solving $\frac{y}{3y+6} + \frac{3}{4y+8} = \frac{3}{4}$ different from adding $\frac{y}{3y+6} + \frac{3}{4y+8} + \frac{3}{4}$?

2. When solving equations that contain rational expressions, why should you multiply both sides of the equation by the LCD?

3. If you were to solve the equation $\frac{3}{x^2-1} + \frac{2}{x^2+3x+2} = \frac{6}{x^2+x-2}$, by what expression would you multiply both sides of the equation?

4. What is an extraneous solution? What can cause extraneous solutions of rational equations to be introduced?

5. What are the possible extraneous solutions of the equation in Exercise 3? Why?

6. Is $y = -3$ an extraneous solution of $\frac{4y}{y+3} + \frac{12}{y+3} = 3$? Why or why not?

For Exercises 7–12, without solving the equations, find the value(s) of the variable that could result in an extraneous solution.

7. $\frac{3}{x-4} = \frac{1}{2}$

8. $\frac{7}{5} = \frac{4}{y+2}$

9. $\frac{4}{x+6} - \frac{3}{x-2} = \frac{2}{2}$

10. $\frac{6}{a+4} + \frac{2}{a-1} = \frac{3}{4}$

11. $\frac{4x}{x^2-4} + \frac{3}{x^2+5x+6} = \frac{7x}{x^2+x-6}$

12. $\frac{3y}{y^2+2y-8} - \frac{2}{y^2+3y-10} = \frac{6y}{y^2+9y+20}$

For Exercises 13–66, solve and check.

13. $\frac{4}{3u} - \frac{1}{2u} = \frac{5}{6}$

14. $\frac{3}{x} - \frac{4}{3x} = \frac{5}{12}$

15. $\frac{5}{4w} - \frac{3}{5w} = -\frac{13}{40}$

16. $\frac{7}{2z} + \frac{11}{6z} = \frac{16}{9}$

17. $\frac{7}{x-3} = 8 - \frac{1}{x-3}$

18. $\frac{5}{y+2} + 9 = -\frac{4}{y+2}$

19. $\frac{t}{t+4} = 3 - \frac{12}{t+4}$

20. $\frac{8}{z-5} - 2 = \frac{z}{z-5}$

21. $\frac{4a}{a+6} = \frac{10}{a+6} + 2$

22. $\frac{5b}{b-8} - 4 = \frac{3}{b-8}$

23. $\frac{4x}{2x-3} = 3 + \frac{6}{2x-3}$

24. $\frac{9a}{3a-2} + 2 = \frac{6}{3a-2}$

25. $\frac{3y}{y-1} = \frac{-2}{y+1}$

26. $\frac{2n}{n+1} = \frac{-5}{n+3}$

27. $\frac{m}{m-1} = \frac{3m}{4m-3}$

28. $\frac{5p}{p-5} = \frac{p}{p-1}$

29. $\frac{2y}{3y-6} + \frac{3}{4y-8} = \frac{1}{4}$

30. $\frac{2y}{3y+6} - \frac{3}{4y+8} = \frac{1}{4}$

31. $\frac{a^2}{a+2} - 3 = \frac{a+6}{a+2} - 4$

32. $\frac{3b+4}{b-4} = \frac{b^2}{b-4} + 3$

33. $x - \frac{12}{x} = 4$

34. $x - \frac{28}{x} = 3$

35. $\frac{x+1}{x+2} + \frac{5}{2x} = \frac{3x+2}{x+2}$

36. $\frac{x+2}{x+4} - \frac{3}{2x} = \frac{2x-5}{x+4}$

37. $\frac{2x+1}{x+2} - \frac{x+1}{3x-2} = \frac{x}{x+2}$

38. $\frac{2x-2}{x+1} - \frac{x-2}{2x-3} = \frac{x}{2x-3}$

39. $\frac{4}{x+5} - \frac{2}{x-5} = \frac{4x-20}{x^2-25}$

40. $\frac{6}{x+2} - \frac{4}{x-2} = \frac{3x-22}{x^2-4}$

41. $\frac{x}{x-2} - \frac{4}{x-1} = \frac{2}{x^2-3x+2}$

42. $\frac{w}{w+5} - \frac{2}{w-3} = -\frac{16}{w^2+2w-15}$

43. $\frac{6}{t^2+t-12} = \frac{4}{t^2-t-6} + \frac{1}{t^2+6t+8}$

44. $\frac{7}{v^2-6v+5} - \frac{2}{v^2-4v-5} = \frac{3}{v^2-1}$

45. $\frac{5}{x^2-x-6} - \frac{8}{x^2+2x-15} = \frac{3}{x^2+7x+10}$

46. $\frac{8}{a^2+4a-12} - \frac{5}{a^2+a-6} = \frac{2}{a^2+9a+18}$

47. $\frac{p+1}{p^2+8p+15} - \frac{2}{p^2+7p+10} = \frac{1}{p^2+5p+6}$

48. $\frac{w-8}{w^2+2w-8} + \frac{5}{w^2+w-6} = \frac{3}{w^2+7w+12}$

49. $\frac{a-5}{a+5} + \frac{a+15}{a-5} = 2 - \frac{25}{a^2-25}$

50. $\frac{2b}{b+7} - \frac{b}{b+3} = 1 + \frac{1}{b^2+10b+21}$

51. $\frac{3n}{n^2-2n-15} = \frac{2n}{n-5} + \frac{n}{n+3}$

52. $\dfrac{c^2+4}{c^2+c-2}=\dfrac{2c}{c+2}-\dfrac{2c+1}{c-1}$

53. $\dfrac{x+3}{x-3}-\dfrac{x-3}{x+3}=\dfrac{6}{x^2-9}$

54. $\dfrac{1}{p^2-1}=\dfrac{p-1}{p+1}-\dfrac{p+1}{p-1}$

55. $\dfrac{2}{2x^2-4x-6}=\dfrac{3}{3x^2+6x-3}$

56. $\dfrac{3}{3x^2-3x-28}=\dfrac{5}{5x^2-x-20}$

57. $1=\dfrac{3}{a-2}-\dfrac{12}{a^2-4}$

58. $1-\dfrac{2}{u+1}=\dfrac{4}{u^2-1}$

59. $\dfrac{w+2}{w^2-5w+6}+\dfrac{2w}{w^2-w-2}=-\dfrac{2}{w^2-2w-3}$

60. $\dfrac{2}{q^2-5q+6}=\dfrac{q}{q^2-4}+\dfrac{q-2}{q^2-q-6}$

61. $\dfrac{2p}{2p-3}=\dfrac{15-32p^2}{4p^2-9}+\dfrac{3p}{2p+3}$

62. $\dfrac{3x^2+11x+4}{9x^2-4}+\dfrac{2x}{3x+2}=\dfrac{4x}{3x-2}$

63. $\dfrac{4x}{x^2+4x}-\dfrac{2x}{x^2+2x}=\dfrac{3x-2}{x^2+6x+8}$

64. $\dfrac{9v}{v^2-4v}-\dfrac{3v}{2v^2+v}=\dfrac{v-7}{2v^2-7v-4}$

65. $\dfrac{6}{x^3+2x^2-x-2}=\dfrac{6}{x+2}-\dfrac{3}{x^2-1}$

66. $\dfrac{1}{y-3}=\dfrac{6}{y^2-1}+\dfrac{12}{y^3-3y^2-y+3}$

For Exercises 67–74, equations that you will encounter or have encountered in the exercise sets are given. Solve each for the indicated variable.

67. $C=\dfrac{90{,}000p}{100-p}$, for p

68. $v=\dfrac{50t}{t+5}$, for t

69. $I=\dfrac{2E}{R+2r}$, for r

70. $I=\dfrac{E}{R+r}$, for R

71. $\dfrac{1}{s}+\dfrac{1}{S}=\dfrac{1}{f}$, for f

72. $C=\dfrac{400+3x}{x}$, for x

73. $R=\dfrac{1}{\dfrac{1}{R_1}+\dfrac{1}{R_2}}$, for R_1

74. $F=\dfrac{f_1 f_2}{f_1+f_2-d}$, for d

For Exercises 75 and 76, explain the mistake and then find the correct solution(s).

75.
$$\frac{x}{x-4} = \frac{4}{x-4} + 3$$
$$(x-4)\frac{x}{x-4} = (x-4)\frac{4}{x-4} + (x-4)\cdot 3$$
$$x = 4 + 3x - 12$$
$$-2x = -8$$
$$x = 4 \quad \text{So } x = 4 \text{ is the solution.}$$

76.
$$\frac{x}{2} + \frac{x}{x+5} = \frac{-5}{x+5}$$
$$2(x+5)\frac{x}{2} + 2(x+5)\frac{x}{x+5} = 2(x+5)\frac{-5}{x+5}$$
$$x^2 + 5x + 2x = -10$$
$$x^2 + 7x + 10 = 0$$
$$(x+5)(x+2) = 0$$
$$x = -5 \quad \text{or} \quad x = -2 \quad \text{So } x = -5 \text{ and } x = -2 \text{ are the solutions.}$$

For Exercises 77–88, solve.

77. Many utility companies use coal to generate electricity, which emits pollutants into the atmosphere. Suppose the cost for removing the pollutants is given by

$$C = \frac{90{,}000\,p}{100 - p}$$

where C is the cost in thousands of dollars and p is the percentage of the pollutants removed. Find the percentage of pollutants removed if the cost is \$22,500,000. (*Hint:* First change \$22,500,000 to thousands.)

78. Suppose the velocity of an automobile starting from rest is given by

$$v = \frac{50t}{t + 5}$$

where v is the velocity in miles per hour and t is the time in seconds. Find the number of seconds when the velocity is 25 miles per hour.

79. If a police car is approaching a stationary observer with its siren on, then the frequency, F, of the siren is given by

$$F = \frac{132{,}000}{330 - s}$$

where s is the speed of the police car in meters per second. Find the speed of the car if the frequency is 440 cycles per second.

80. For a small business, the cost, C, to order and store x units of a product is given by

$$C = 5x + \frac{2000}{x}$$

Find the number of units if the cost is \$520.

81. If x is the average speed on the outgoing trip and $x - 20$ is the average speed on the return trip, then the total time for a trip of 80 miles is given by

$$T = \frac{80}{x} + \frac{80}{x - 20}$$

where T is in hours and x is in miles per hour. Find the average speed of the outgoing trip if the total time for the trip is 6 hours.

82. If the start-up costs for a small pizza place are \$400 per day and it costs an average of \$3 per pizza, the average cost per pizza for making x pizzas is given by

$$C = \frac{400 + 3x}{x}$$

where C is in dollars. Find the number of pizzas made if the average cost per pizza is \$7.

83. In optics, the thin lens equation is

$$\frac{1}{s} + \frac{1}{S} = \frac{1}{f}$$

where s is the object distance from the lens, S is the image distance from the lens, and f is the focal length of the lens. If the focal length is 6 centimeters and the image distance is 10 centimeters, find the object distance.

84. A particle is moving along the curve whose equation is

$$\frac{xy^3}{1 + y^2} = \frac{8}{5}$$

If the particle is currently located at the point whose y value is 3, what is the x value?

85. The current, I, in amperes in a circuit including an external resistance of R ohms and a cell of electromotive force of E volts and an internal resistance of r ohms is given by

$$I = \frac{E}{R + r}$$

Find the internal resistance in a circuit whose current is 2 amperes, external resistance is 50 ohms, and electromotive force is 110 volts.

86. If two cells are connected in a series, the current, I, in amperes in a circuit including an external resistance of R ohms and a cell of electromotive force of E volts and an internal resistance of r ohms is given by

$$I = \frac{2E}{R + 2r}$$

Find the external resistance in a circuit if the current is 11 amperes, the electromotive force is 220 volts, and the internal resistance is 5 ohms.

87. If two sources of illuminating power, P_1 and P_2, produce equal illuminating power on a screen at distances of r_1 and r_2, respectively, then

$$\frac{P_1}{r_1^2} = \frac{P_2}{r_2^2}$$

If one source is 5 meters from the screen and has illuminating power of 10 foot-candles, find the illuminating power of the other source if it is 15 meters from the screen.

88. If f_1 and f_2 are the focal lengths of two thin lenses separated by a distance of d, the focal length of the system is

$$F = \frac{f_1 f_2}{f_1 + f_2 - d}$$

If the focal length of the system is 30 centimeters, one lens has a focal length of 10 centimeters, and the other 15 centimeters, find the distance between the lenses.

PUZZLE PROBLEM

Amy has nine apparently identical bricks. However one brick weighs slightly less than the others. Using a scale balance and weighing only twice, she was able to identify the lighter brick. Explain the steps.

REVIEW EXERCISES

[2.2, 4.3] *For Exercises 1–6, solve.*

1. Fred has two more nickels than dimes and the total value of the nickels and dimes is $1.30. How many of each type of coin does he have?

2. The length of a rectangular painting is 2 inches more than the width. If the area of the painting is 80 square inches, find the length and width.

3. One number is 3 more than the other. Find the numbers if the difference of their squares is 39.

4. If the sum of two numbers is 28 and their difference is 4, find the numbers.

5. Jose purchased a saw and a drill. If the drill cost $54 less than the saw and he paid a total of $238 for both, how much did each cost?

6. A recreational vehicle leaves New Orleans heading west on I-10 toward Houston at an average rate of 55 miles per hour. Four hours later, a truck also leaves New Orleans on I-10 toward Houston at an average rate of 75 miles per hour. How many hours will it take the truck to catch up to the recreational vehicle?

7.5 Applications with Rational Expressions; Variation

OBJECTIVES

1. **Use tables to solve problems with two unknowns involving rational expressions.**
2. **Solve problems involving direct variation.**
3. **Solve problems involving inverse variation.**
4. **Solve problems involving joint variation.**
5. **Solve problems involving combined variation.**

OBJECTIVE 1. Use tables to solve problems with two unknowns involving rational expressions. In Section 2.2, we used tables to organize information involving two unknown amounts. Let's now consider some similar problems that lead to equations containing rational expressions.

Problems Involving Work

Tables are helpful in solving problems involving two or more people (or machines) working together to complete a task. In these problems, we are given each person's rate of work and asked to find the time for them to complete the task if they work together. For each person involved,

Person's rate of work · Person's time at work = Amount of the task completed by that person

Since the people are working together, the sum of their individual amounts of the task equals the whole task.

Amount completed by one person + Amount completed by the other person = Whole task

EXAMPLE 1 If Mike can paint his room in 10 hours and Susan can paint the same room in 7 hours, how long would it take them to paint the room working together?

Understand Mike paints at a rate of 1 room in 10 hours or $\frac{1}{10}$ of a room per hour.

Susan paints at a rate of 1 room in 7 hours or $\frac{1}{7}$ of a room per hour.

People	Rate of Work (rooms per hour)	Time at Work (number of hours)	Amount of Task Completed
Mike	$\frac{1}{10}$	t	$\frac{1}{10}t$ or $\frac{t}{10}$ of a room
Susan	$\frac{1}{7}$	t	$\frac{1}{7}t$ or $\frac{t}{7}$ of a room

Because they are working together for the same amount of time, we let t represent that amount of time.

Multiplying the rate of work and the time at work gives an expression of the amount of work completed. For example, if Mike works at a rate of $\frac{1}{10}$ of a room per hour for 20 hours, he can paint $\frac{1}{10} \cdot 20 = 2$ rooms.

The total job in this case is 1 room, so we can write an equation that combines their individual expressions for the task completed and set this sum equal to 1 room.

Plan and Execute Amount Mike completed + Amount Susan completed = 1 room

$$\frac{t}{10} + \frac{t}{7} = 1$$

$$70\left(\frac{t}{10} + \frac{t}{7}\right) = 70(1)$$ **Multiply both sides by the LCD, 70.**

$$\overset{7}{\cancel{70}} \cdot \frac{t}{\underset{1}{\cancel{10}}} + \overset{10}{\cancel{70}} \cdot \frac{t}{\underset{1}{\cancel{7}}} = 70$$ **Distribute, then divide out common factors.**

$$7t + 10t = 70$$

$$17t = 70$$ **Combine like terms.**

$$t = \frac{70}{17}$$ **Divide both sides by 17.**

Answer Working together, it takes Mike and Susan $\frac{70}{17}$, or $4\frac{2}{17}$, hours to paint the room.

Check Mike paints $\frac{1}{10}$ of the room per hour, so if he works alone $\frac{70}{17}$ hours, he paints $\frac{1}{10} \cdot \frac{70}{17} = \frac{7}{17}$ of a room. Susan paints $\frac{1}{7}$ of the room per hour, so in $\frac{70}{17}$ hours, she paints $\frac{1}{7} \cdot \frac{70}{17} = \frac{10}{17}$ of the room. Combining their individual amounts, we see that in $\frac{70}{17}$ hours they paint $\frac{7}{17} + \frac{10}{17} = \frac{17}{17} = 1$ room.

ANSWER

$4\frac{4}{5}$ hr.

YOUR TURN If one pipe can fill a tank in 12 hours and a second pipe can fill a tank in 8 hours, how long would it take to fill the tank with both pipes open?

Motion Problems

Recall that the formula for calculating distance, given the rate of travel and time of travel, is $d = rt$. If we isolate r, we have $r = \frac{d}{t}$. If we isolate t, we have $t = \frac{d}{r}$. These equations suggest that we will encounter rational expressions when describing rate or time.

EXAMPLE 2 An airplane flew 450 miles from Miami to Tallahassee in $1\frac{1}{2}$ hours less time than it took to fly 1200 miles from Miami to Newark. If the airplane flew at the same rate on both trips, find the rate of the airplane.

Understand We are to find the rate of the airplane. This situation involves the same rate between each pair of cities, but the distances and times are different. We will use a table and the fact that $t = \frac{d}{r}$ to organize the distance, rate, and time of each leg of the trip.

Flights	Distance (miles)	Rate (miles/hour)	Time (hours)
Miami to Tallahassee	450	r	$\frac{450}{r}$
Miami to Newark	1200	r	$\frac{1200}{r}$

Since $d = rt$, to describe time, we use $t = \frac{d}{r}$.

We can use the fact that the time from Miami to Tallahassee is $1\frac{1}{2}$ or $\frac{3}{2}$ hours less than the time from Miami to Newark to write the equation.

Plan and Execute Time from Miami to Tallahassee = Time from Miami to Newark $- \frac{3}{2}$

$$\frac{450}{r} = \frac{1200}{r} - \frac{3}{2}$$

$$2r\left(\frac{450}{r}\right) = 2r\left(\frac{1200}{r} - \frac{3}{2}\right)$$ Multiply both sides by the LCD, $2r$.

$$2\cancel{r} \cdot \frac{450}{\cancel{r}} = 2\cancel{r} \cdot \frac{1200}{\cancel{r}} - \cancel{2}r \cdot \frac{3}{\cancel{2}}$$ Distribute and divide out common factors.

$$2 \cdot 450 = 2 \cdot 1200 - r \cdot 3$$ Simplify.

$$900 = 2400 - 3r$$ Multiply.

$$-1500 = -3r$$ Subtract 2400 from both sides.

$$500 = r$$ Divide both sides by -3

Answer The plane is traveling at 500 miles per hour.

Check If the plane flies 450 miles from Miami to Tallahassee at 500 miles per hour, the time required is $\frac{450}{500} = \frac{9}{10}$ of an hour. If the plane flies 1200 miles from Miami to Newark at 500 miles per hour, the required time is $\frac{1200}{500} = \frac{12}{5}$ hours. The time from Miami to Tallahassee is to be $1\frac{1}{2}$ hours less than the time from Miami to Newark and $\frac{12}{5} - \frac{3}{2} = \frac{24}{10} - \frac{15}{10} = \frac{9}{10}$, which is the time from Miami to Tallahassee.

EXAMPLE 3 A river has a current of 5 miles per hour. If a boat can make a trip of 20 miles upstream and 30 miles downstream in 2 hours, find the rate of the boat in still water.

Understand We are looking for the rate of the boat in still water, so we let this rate be x. When the boat is traveling upstream, the current slows the boat by 5 miles per hour. When the boat is traveling downstream, the current speeds the boat by 5 miles per hour. Again, we use a table to organize our information.

Direction	Distance (miles)	Rate (miles/hour)	Time (hours)
Upstream	20	$x - 5$	$\frac{20}{x-5}$
Downstream	30	$x + 5$	$\frac{30}{x+5}$

Since x is the rate of the boat in still water, the rate against the current is $x - 5$ and the rate with the current is $x + 5$.

Since $d = rt$, to describe time, we use $t = \frac{d}{r}$.

We can write the equation using the fact that the boat can go 20 miles upstream and 30 miles downstream in 2 hours.

Plan and Execute Time upstream + Time downstream = 2 hours

$$\frac{20}{x-5} + \frac{30}{x+5} = 2$$

$$(x-5)(x+5)\left(\frac{20}{x-5} + \frac{30}{x+5}\right) = (x-5)(x+5)(2) \quad \text{Multiply both sides by the LCD, } (x-5)(x+5).$$

$$\cancel{(x-5)}(x+5)\left(\frac{20}{\cancel{x-5}}\right) + (x-5)\cancel{(x+5)}\left(\frac{30}{\cancel{x+5}}\right) = (x^2-25)(2) \quad \text{Distribute.}$$

$$\begin{aligned} 20(x+5) + 30(x-5) &= 2x^2 - 50 && \text{Multiply.} \\ 20x + 100 + 30x - 150 &= 2x^2 - 50 && \text{Distribute.} \\ 50x - 50 &= 2x^2 - 50 && \text{Combine like terms.} \\ 0 &= 2x^2 - 50x && \text{Add 50 to and subtract } 50x \text{ from both sides.} \\ 0 &= 2x(x-25) && \text{Factor.} \end{aligned}$$

$$2x = 0 \quad \text{or} \quad x - 25 = 0 \qquad \text{Use the zero-factor theorem.}$$

$$x = 0 \quad \text{or} \quad x = 25$$

Answer Since the speed of the boat cannot be 0, the boat travels at 25 miles per hour in still water.

Check If the boat travels at the rate of 25 miles per hour in still water, then its speed against the current is $25 - 5 = 20$ miles per hour. In order to go 20 miles upstream, it would take $\frac{20}{20} = 1$ hour. The speed of the boat with the current is $25 + 5 = 30$ miles per hour. In order to go 30 miles downstream, it would take $\frac{30}{30} = 1$ hour. The time going upstream plus the time going downstream equals 2, so our solution is correct.

A plane flies 1500 miles against the wind in the same amount of time it took to fly 1800 miles with the wind. If the speed of the wind was 50 miles per hour, find the speed of the plane in still air.

OBJECTIVE 2. Solve problems involving direct variation. The Internal Revenue Service allows a \$0.36 per mile deduction for each mile driven for business purposes. If d is the amount of the deduction and n is the number of miles driven, then $d = 0.36n$. In the following table, we use this formula to determine the deduction for various values of n.

Number of Miles	Deduction ($d = 0.36n$)
1	0.36
2	0.72
3	1.08
4	1.44
5	1.80

From the table we see that as the number of miles increases, so does the amount of the deduction. Or, more formally, as values of n increase, so do values of d. In $d = 0.36n$, the two variables, d and n, are said to be in **direct variation** or are *directly proportional* and 0.36 is the constant of variation.

DEFINITION ***Direct variation:*** Two variables y and x are in direct variation if $y = kx$. If y varies directly as the nth power of x, then $y = kx^n$. In both cases, k is the constant of variation.

In words, direct variation is written as "y varies directly as x" or "y is directly proportional to x," and these phrases translate to $y = kx$. The expression $y = kx^n$ is translated as "y varies directly as the nth power of x" or "y is directly proportional to the nth power of x."

In all the variation problems in this section, we will be given one set of values of the variables, which we use to find the constant of variation, k. We will then use this value of k and the equation to find other values of the variable(s).

550 mph

PROCEDURE Solving Variation Problems

1. Write the equation.
2. Substitute the initial values and find k.
3. Substitute for k in the equation found in step 1.
4. Solve for the unknown.

EXAMPLE 4 Solve the direct variations.

a. Suppose y varies directly as x. If $y = 18$ when $x = 5$, find y when $x = 8$.

Solution Translating "y varies directly as x," we have $y = kx$. We replace y with 18 and x with 5 in $y = kx$, and solve for k.

$$18 = k \cdot 5 \quad \text{Replace } y \text{ with 18 and } x \text{ with 5.}$$
$$3.6 = k \quad \text{Divide both sides by 5.}$$

Now, we can replace k with 3.6 in $y = kx$ so that we have $y = 3.6x$. We use this equation to find y when $x = 8$.

$$y = 3.6(8) = 28.8$$

b. Suppose u varies directly as the square of v. If $u = 54$ when $v = 3$, find u when $v = 5$.

Solution Translating "u varies directly as the square of v," we have $u = kv^2$. We replace u with 54 and v with 3 in $u = kv^2$ and solve for k.

$$54 = k \cdot 3^2 \quad \text{Replace } u \text{ with 54 and } v \text{ with 3.}$$
$$54 = 9k \quad \text{Square 3.}$$
$$6 = k \quad \text{Divide both sides by 9.}$$

Now, we can replace k with 6 in $u = kv^2$ so that we have $u = 6v^2$. We use this equation to find u when $v = 5$.

$$u = 6(5)^2 \quad \text{Substitute 5 for } v.$$
$$u = 6(25) \quad \text{Square 5.}$$
$$u = 150 \quad \text{Multiply.}$$

YOUR TURN Suppose m varies directly as the square of n. If $m = 24$ when $n = 2$, find m when $n = -3$.

ANSWER

54

EXAMPLE 5 In physical science, Hooke's Law states that the distance a spring of uniform material and thickness is stretched varies directly with the force applied to the spring. If a force of 5 newtons stretches a spring 80 centimeters, how much force is required to stretch the spring 128 centimeters?

Understand Translating "the distance a spring is stretched varies directly with the force," we write $d = kF$, where d represents distance and F represents the force.

Plan Use $d = kF$, replacing d with 80 centimeters and F with 5 newtons in order to solve for the value of k. Then use that value in $d = kF$ to solve for the force required to stretch the spring 128 centimeters.

Execute

$$80 = k \cdot 5 \quad \text{Replace } d \text{ with 80 and } F \text{ with 5.}$$

$$16 = k \quad \text{Divide both sides by 5.}$$

Replacing k with 16 in $d = kF$, we have $d = 16F$, which we use to solve for F when d is 128 centimeters.

$$128 = 16F \quad \text{Substitute 128 for } d.$$

$$8 = F \quad \text{Divide both sides by 16.}$$

Answer To stretch the spring 128 centimeters, a force of 8 newtons is applied.

Check A force of 8 newtons stretches the spring $d = 16(8) = 128$ centimeters.

YOUR TURN The pressure exerted by a liquid varies directly as the depth beneath the surface. If an object is submerged at a depth of 3 feet in seawater, the pressure is 192 pounds per square foot. Find the pressure per square foot if an object is submerged in seawater to a depth of 8 feet.

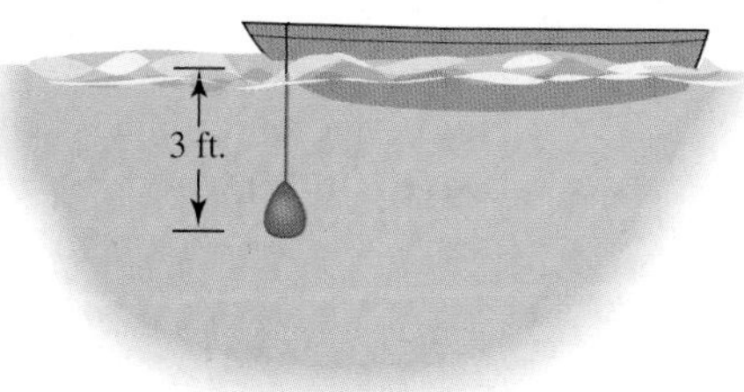

OBJECTIVE 3. Solve problems involving inverse variation. Suppose a campaign worker has to stuff 100 envelopes. If r is the rate at which she can stuff the envelopes and t is the time required, then $t = \frac{100}{r}$. For example, if she can stuff 2 envelopes per minute, then it will take her $\frac{100}{2} = 50$ minutes to stuff the envelopes. In the following table, we

ANSWER

512 lb./ft.^{2}

use $t = \frac{100}{r}$ to see the relationship between the rate at which she can stuff envelopes and the amount of time required to complete the job.

Rate r	Time $\left(t = \frac{100}{r}\right)$
1 per minute	100 minutes
2 per minute	50 minutes
4 per minute	25 minutes

From the table, we see that as the rate at which she stuffs the envelopes increases, the time required decreases. More formally, as values of r increase, values of t decrease. In $t = \frac{100}{r}$, the two variables, t and r, are said to be in **inverse variation**, or are *inversely proportional*, and 100 is the constant of variation.

DEFINITION ***Inverse variation:*** Two variables y and x are in inverse variation if $y = \frac{k}{x}$.

If $y = \frac{k}{x^n}$, then y varies inversely as the nth power of x. The k is the constant of variation.

In words, inverse variation is written as "y varies inversely as x" or "y is inversely proportional to x," and these phrases translate to $y = \frac{k}{x}$. Similarly, "y varies inversely as the nth power of x" or "y is inversely proportional to the nth power of x" translate to $y = \frac{k}{x^n}$.

EXAMPLE 6 Boyle's Law states that if the temperature is held constant, the volume of a gas in a closed container is inversely proportional to the pressure applied to it. If a gas has a volume of 4 cubic feet when the pressure is 10 pounds per square foot, find the volume when the pressure is 2.5 pounds per square foot.

Understand Because the volume and pressure vary inversely, we can write $V = \frac{k}{P}$, where V represents volume and P represents the pressure.

Plan Use the fact that the volume is 4 cubic feet when the pressure is 10 pounds per square foot to find the value of the constant, k. Then use this value of the constant to find the volume when the pressure is 2.5 pounds per square foot.

Execute $4 = \frac{k}{10}$ Substitute for *V* and *P*.

$10 \cdot 4 = 10 \cdot \frac{k}{10}$ Multiply both sides by 10.

$40 = k$

Replacing *k* with 40 in $V = \frac{k}{P}$, we have $V = \frac{40}{P}$, which we use to solve for *V* when *P* is 2.5 pounds per square foot.

$V = \frac{40}{2.5}$ Substitute 40 for *P*.

$V = 16$

Answer With a pressure of 2.5 pounds per square foot, the volume is 16 cubic feet.

YOUR TURN The intensity of a light varies inversely as the square of the distance from the light source. If the intensity is 150 foot-candles when the distance is 2 meters, find the intensity if the distance is 4 meters.

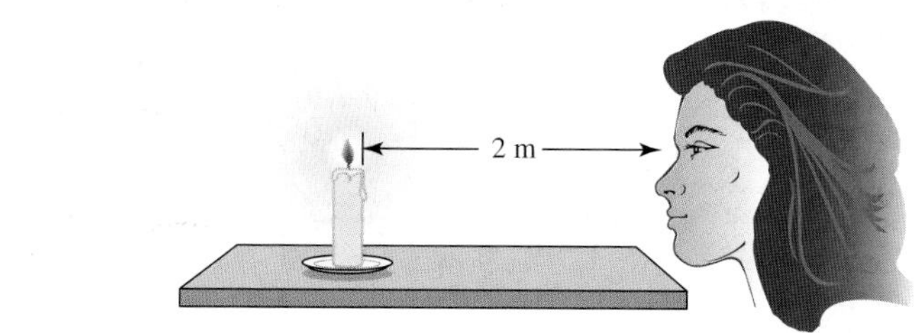

OBJECTIVE 4. Solve problems involving joint variation. Often one quantity will vary as the product of two or more quantities. This is called **joint variation**.

DEFINITION ***Joint variation:*** Three variables *y*, *x*, and *z* are in joint variation if $y = kxz$, where *k* is a constant.

In words, joint variation is written as "*y* varies jointly as *x* and *z*" or "*y* is jointly proportional to *x* and *z*," and these phrases translate to $y = kxz$.

EXAMPLE 7 Suppose *y* varies jointly with *x* and *z*. If $y = 72$ when $x = 4$ and $z = 6$, find *y* when $x = 2$ and $z = 5$.

Solution Translating "*y* varies jointly with *x* and *z*," we have $y = kxz$. We replace *y* with 72, *x* with 4, and *z* with 6 in $y = kxz$ and solve for *k*.

$72 = k \cdot 4 \cdot 6$ Replace *y* with 72, *x* with 4, and *z* with 6.

$72 = 24k$ Multiply 4 and 6.

$3 = k$ Divide both sides by 24.

ANSWER

37.5 foot-candles

Now, we can replace k with 3 in $y = kxz$ so that we have $y = 3xz$. We use this equation to find y when $x = 2$ and $z = 5$.

$y = 3 \cdot 2 \cdot 5$ Substitute for x and z.

$y = 30$ Multiply.

YOUR TURN Suppose n varies jointly with p and the square of q. If $n = 90$ when $p = 5$ and $q = 3$, find n when $p = 2$ and $q = 5$.

OBJECTIVE 5. Solve problems involving combined variation. Problems involving more than one type of variation are called *combined variation*.

EXAMPLE 8 Coulomb's Law states that the force, F, between two charges q_1 and q_2 in a vacuum varies jointly with the charges and inversely with the square of the distance, d, between them. If the force is 5 dynes when q_1 is 9 electrostatic units and q_2 is 20 electrostatic units and the distance between the charges is 6 centimeters, find the force between the particles if q_1 is 16 electrostatic units, q_2 is 12 electrostatic units, and the distance between the particles is 8 centimeters.

Solution Translating "the force, F, between two charges q_1 and q_2 varies jointly with the charges and inversely with the square of the distance, d, between them," we have $F = \frac{kq_1q_2}{d^2}$. Replace F with 5, q_1 with 9, q_2 with 20, and d with 6 in $F = \frac{kq_1q_2}{d^2}$ and solve for k.

$5 = \frac{k \cdot 9 \cdot 20}{6^2}$ Replace F with 5, q_1 with 9, q_2 with 20, and d with 6.

$5 = \frac{180k}{36}$ Simplify.

$5 = 5k$ Simplify.

$1 = k$ Divide both sides by 5.

Now, we can replace k with 1 in $F = \frac{kq_1q_2}{d^2}$ so that we have $F = \frac{q_1q_2}{d^2}$. We use this equation to find F when $q_1 = 16$, $q_2 = 12$, and $d = 8$.

$F = \frac{16 \cdot 12}{8^2}$ Substitute for q_1, q_2, and d.

$F = \frac{192}{64}$ Simplify.

$F = 3$ dynes Simplify.

ANSWER

$n = 100$

YOUR TURN The maximum height, h, obtained by an object that is launched vertically upward varies directly with the square of the velocity, v, and inversely as the acceleration due to gravity, g. If an object that is launched vertically upward with a velocity of 64 feet per second and acceleration due to gravity of 32 feet per second per second obtains a maximum height of 64 feet, find the maximum height obtained by an object launched vertically upward with a velocity of 128 feet per second and acceleration due to gravity of 32 feet per second per second.

ANSWER

256 ft.

7.5 Exercises

1. If a person can complete a task in x hours, what portion of the task can he complete in 1 hour?

2. If a represents the number of hours for one person to complete a task and b represents the number of hours for a second person to complete the same task, represent the portion of the task completed in 1 hour while working together.

3. If a vehicle travels a distance of 100 miles at a rate of r, write an expression for the time, t, it takes the vehicle to travel the 100 miles. Which type of variation is this?

4. If Fred can row a canoe x miles per hour in still water and the current in the Wekiva River is 3 miles per hour, write an expression for Fred's rate rowing upstream and an expression for his rate rowing downstream in the river.

5. If p varies directly as q, then as q decreases, what happens to the value of p? (Assume $k > 0$).

6. If m is inversely proportional to n, then as the n quantity increases, what happens to the value of m? (Assume $k > 0$).

For Exercises 7–22, use a table to organize the information, then solve.

7. Jason can wash and wax his car in 4 hours. His younger sister can wash and wax the same car in 6 hours. Working together, how fast can they wash and wax the car?

8. Jon can clean the house in 7 hours and Linda can clean the house in 5 hours. How long would it take if they worked together?

9. Alicia and Geraldine have volunteered to make quilts for a charity auction. If Alicia can make a quilt in 25 days and Geraldine can make a quilt in 35 days, in how many days can they make a quilt working together?

10. It takes Alice 90 minutes to put a futon frame together and it takes Maya 60 minutes to put the same type of frame together. If they worked together, how long would it take to put a frame together?

11. Working together, it takes two roofers 4 hours to put a new roof on a portable classroom. If the first roofer can do the job by himself in 6 hours, how many hours would it take for the second roofer to do the job by himself?

12. Working together, Rita and Tiffany can cut and trim a lawn in 2 hours. If it takes Rita 5 hours to do the lawn by herself, how long would it take Tiffany to do the lawn by herself?

13. With both the cold water and the hot water faucets open, it takes 9 minutes to fill a bathtub. The cold water faucet alone takes 15 minutes to fill the tub. How long would it take to fill the tub with the hot water faucet alone?

14. The cargo hold of a ship has two loading pipes. Used together, the two pipes can fill the cargo hold in 6 hours. If the larger pipe alone can fill the hold in 8 hours, how many hours would it take the smaller pipe to fill the hold by itself?

15. Two buses leave Kansas City traveling in opposite directions, one going east and one going west. The eastbound bus travels at 40 miles per hour and the westbound bus travels at 65 miles per hour. If both buses leave at 10:00 A.M., at what time will they be 525 miles apart?

16. Sailing in opposite directions, an aircraft carrier and a destroyer leave their base in Hawaii at 5:00 A.M. If the destroyer sails at 30 miles per hour and the aircraft carrier sails at 20 miles per hour, at what time will the two ships be 300 miles apart?

17. Rapid City is 360 miles from Sioux Falls. At 6:00 A.M., a freight train leaves Rapid City for Sioux Falls and at the same time a passenger train leaves Sioux Falls for Rapid City. The two trains meet at 9:00 A.M. If the freight train travels $\frac{3}{5}$ of the speed of the passenger train, how fast does the passenger train travel?

18. Houston and Calgary are about 2100 miles apart. At 2:00 P.M., an airplane leaves Houston for Calgary flying at 250 miles per hour. At the same time an airplane leaves Calgary for Houston flying at 450 miles per hour. How long will it be before the two airplanes meet?

19. An automobile makes a trip of 270 miles in $2\frac{1}{2}$ hours less than it takes the same auto to make a trip of 420 miles traveling at the same speed. What is the speed of the automobile?

20. A ship leaves port traveling at 15 miles per hour. Two hours later, a speedboat leaves the same port traveling at 40 miles per hour. How long will it take for the speedboat to overtake the ship?

21. An airliner flies against the wind from Washington, D.C., to San Francisco in 5.5 hours. It flies back to Washington, D.C., with the wind in 5 hours. If the average speed of the wind is 21 miles per hour, what is the speed of the airliner in still air?

22. A river has a current of 3 miles per hour. A boat goes 40 miles upstream in the same time as it goes 50 miles downstream. Find the speed of the boat in still water.

For Exercises 23–32, solve the direct variations.

23. Suppose q varies directly as p. If $q = 36$ when $p = 9$, find q when $p = 3$.

24. Suppose r varies directly as s. If $r = 6$ when $s = 9$, find r when $s = 15$.

25. Suppose y varies directly as the square of x. If $y = 100$ when $x = 5$, find y when $x = 3$.

26. Suppose t varies directly as the square of u. If $t = 45$ when $u = 3$, find t when $u = -2$.

27. Suppose m varies directly as n. If $m = 6$ when $n = 8$, what is the value of n when $m = 9$?

28. Suppose x varies directly as y. If $x = 12$ when $y = 15$, what is the value of y when $x = 4$?

29. The price of salmon at a fish market is constant, so the cost increases with the quantity purchased. Tamika notes 2.5 pounds cost $16.25. If she plans to buy 6 pounds of salmon, how much will it cost?

30. The cost per cucumber at a farmer's market is constant, so the cost increases with the number purchased. Nashir notes that 5 cucumbers cost \$0.89. If she plans to buy 3 cucumbers, how much will they cost?

31. According to Charles's Law, if the pressure is held constant, the volume of a gas varies directly with the temperature measured on the Kelvin scale. If the volume of a gas is 288 cubic centimeters when the temperature is 80 K, find the volume when the temperature is 50 K.

32. According to Ohm's Law, the current, I, which is measured in amperes, in an electrical circuit varies directly with the voltage, V. If the current is 0.64 amperes when the voltage is 24 volts, find the current when the voltage is 48 volts.

In Exercises 33–38, use the fact that, ignoring air resistance, the distance a free-falling body falls is directly proportional to the square of the time it has been falling.

33. On Earth, if an object falls 144 feet in 3 seconds, how many seconds will it take for the object to fall 400 feet?

Of Interest

Without air resistance, all objects fall at the same rate. This was demonstrated by astronaut David Scott in 1971 when he dropped a hammer and a feather from the same height on the Moon and they landed on the Moon's surface at the same time.

34. On Earth, if an object falls 88.2 meters in 3 seconds, how far will it fall in 6 seconds?

35. On the Moon, an object falls 6.48 meters in 2 seconds. How far will it fall in 5 seconds?

36. On the Moon, an object falls 14.58 meters in 3 seconds. How long will it take for the object to fall 58.32 meters?

37. The circumference of a circle varies directly with its diameter. If the circumference is 12.56 feet when the diameter is 4 feet, find the diameter when the circumference is 21.98 feet.

38. The pressure on an object submerged in a liquid varies directly as the depth beneath the surface. If an object is submerged in gasoline to a depth of 6 feet, the pressure is 253.8 pounds per square foot. Find the pressure if the object is submerged to 9 feet.

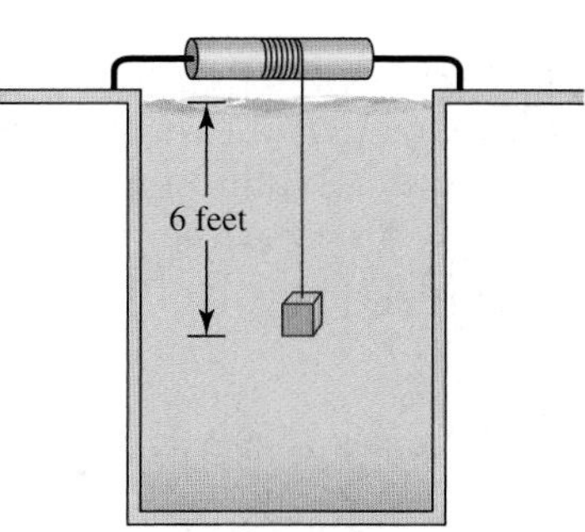

For Exercises 39–46, solve the inverse variations.

39. Suppose y varies inversely as x. If $y = 3.5$ when $x = 6$, what is y when x is 4?

40. Suppose p varies inversely as q. If p is 8 when q is 3.25, what is q when p is 4?

41. Suppose m varies inversely as n. If n is 6 when m is 11, what is n when m is 8?

42. Suppose x varies inversely as y. If $y = 12$ when $x = 3$, what is y when x is 12?

43. By Charles's Law, if the temperature is held constant, the pressure that a gas exerts against the walls of a container is inversely proportional to the volume of the container. A gas is inside a cylinder with a piston at one end that can vary the volume of the cylinder. When the volume of the cylinder is 20 cubic inches, the pressure inside is 40 psi (pounds per square inch). Find the pressure of the gas if the piston compresses the gas to a volume of 16 cubic inches.

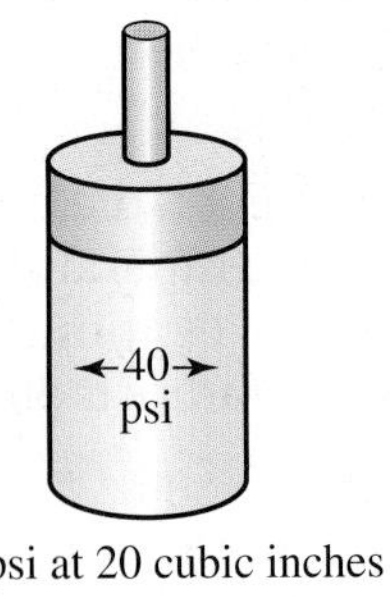

40 psi at 20 cubic inches

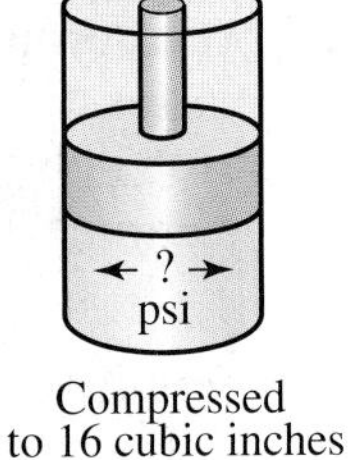

Compressed to 16 cubic inches

44. Find the volume in the cylinder from Exercise 43 if the piston compresses the gas to a pressure of 30 psi.

45. In an electrical conductor, the current, I, which is measured in amperes, varies inversely as the resistance, R, which is measured in ohms. If the current is 10 amperes when the resistance is 15 ohms, find the resistance when the current is 25 amperes.

46. In Exercise 45, find the current when the resistance is 15 ohms.

In Exercises 47 and 48, use the fact that the length of a radio wave varies inversely with its frequency.

47. If the length of a radio wave is 400 meters when the frequency is 900 kilohertz, find the wavelength when the frequency is 600 kilohertz.

48. If the frequency of a radio wave is 800 kilohertz when the wavelength is 200 meters, find the frequency when the radio wavelength is 500 meters.

49. The f-stop setting for a camera lens is inversely proportional to the aperture, which is the size of the opening in the lens. For a particular lens, the f-stop setting is 5.6 when the aperture is 50 mm. Find the aperture when the f-stop is 2.8.

50. The weight of an object varies inversely with the square of the distance from the center of the Earth. At Earth's surface, the weight of the space shuttle is about 4.5 million pounds. It is then 4000 miles from the center of the Earth. How much will the space shuttle weigh when it is in orbit 200 miles above the surface of the Earth?

For Exercises 51–58, solve the joint variations.

51. Suppose y varies jointly with x and z. If $y = 90$ when $x = 3$ and $z = 5$, find y when $x = 2$ and $z = 4$.

52. Suppose m varies jointly with p and q. If $m = 70$ when $p = 5$ and $q = 2$, find m when $p = 6$ and $q = 8$.

53. Suppose a varies jointly as the square of b and c. If $a = 96$ when $b = 2$ and $c = 6$, find a when $b = 3$ and $c = 2$.

54. Suppose x varies jointly with y and the square of z. If $x = 40$ when $y = 1$ and $z = 2$, find x when $y = 2$ and $z = 1$.

55. If the length of a rectangular solid is held constant, the volume varies jointly with the width and the height. If the volume is 192 cubic inches when the width is 8 inches and the height is 4 inches, find the volume when the width is 5 inches and the height is 7 inches.

56. For a fixed amount of principal, the simple interest varies jointly with the rate and the time. If the simple interest is $1000 when the rate is 4% and the time is 5 years, find the simple interest when the rate is 6% and the time is 10 years.

57. The volume of a right circular cylinder varies jointly as the square of the radius and the height. If the volume is 301.6 cubic centimeters when the radius is 4 centimeters and the height is 6 centimeters, find the volume when the radius is 3 centimeters and the height is 6 centimeters.

58. The number of units produced varies jointly with the number of workers and the number of hours worked per worker. If 80 units can be produced by 10 workers who work 40 hours each, how many units can be produced by 15 workers who work 30 hours each?

For Exercises 59–66, solve the combined variations.

59. Suppose y varies directly with x and inversely as z. If $y = 8$ when $x = 4$ and $z = 6$, find y when $x = 5$ and $z = 10$.

60. Suppose m varies directly as the square of n and inversely as the square of p. If $m = 27$ when $n = 6$ and $p = 4$, find m when $n = 9$ and $p = 6$.

61. Suppose y varies jointly with x and z and inversely with n. If $y = 81$ when $x = 4$, $z = 9$, and $n = 8$, find n when $x = 6$, $y = 8$, and $z = 12$.

62. Suppose p varies jointly with q and r and inversely with s. If $p = 15$ when $q = 5$, $r = 3$, and $s = 6$, find p when $q = 5$, $r = 3$, and $s = 4$.

63. The resistance of a wire, R, varies directly with the length and inversely with the square of the diameter. If the resistance is 7.5 ohms when the wire is 6 meters long and the diameter is 0.02 meter, find the resistance in a wire of the same material if the length is 10 meters and the diameter is 0.04 meter.

64. The universal gas law states that the volume of a gas varies directly with the temperature and inversely with the pressure. If the volume of a gas is 1.75 cubic meters when the temperature is 70 K and the pressure is 20 grams per square centimeter, find the volume when the temperature is 80 K and the pressure is 40 grams per square centimeter.

65. Newton's Law of Universal Gravitation states that the force of attraction between two bodies is jointly proportional to their masses and inversely proportional to the square of the distance between them. If the force of attraction between two masses of 4 grams and 6 grams that are 3 centimeters apart is 48 dynes, find the force of attraction between two masses of 2 grams and 12 grams that are 6 centimeters apart.

66. The weight-carrying capacity of a rectangular beam varies jointly with its width and the square of its height and inversely as its length. If a beam is 4 inches wide, 6 inches high, and 10 feet long, it has a carrying capacity of 1400 pounds. Find the carrying capacity of a beam made of the same material if it is 3 inches wide, 5 inches high, and 12 feet long.

67. The following table lists the number of miles a car can drive on the given number of gallons of gas.

Number of Miles	Number of Gallons
135	6
180	8
225	10
270	12

a. Plot the ordered pairs of data as points in the coordinate plane. Connect the points to form a graph.

b. Are the number of miles driven and the number of gallons of gas directly proportional or inversely proportional? Explain.

c. Find the constant of variation. What does it represent?

d. Does the data represent a function? Explain.

68. The following table lists the time required to make a long trip at various average driving speeds.

Speed (in mph)	Driving Time (in hr.)
60	10
50	12
40	15
30	20
20	30

a. Plot the ordered pairs of data as points in the coordinate plane. Connect the points to form a graph.

b. Are speeds and driving times directly proportional or inversely proportional? Explain.

c. Find the constant of variation. What does it represent?

d. Do the data represent a function? Explain.

REVIEW EXERCISES

[1.3] ***For Exercises 1 and 2, evaluate.***

1. 2^3

2. -12^2

[6.4] **3.** Solve: $3x^2 - 13x - 10 = 0$

[5.1] **4.** Multiply: $(6a^3b^2)(-5ab^4)$

[5.1] **5.** Divide: $\dfrac{-48x^3y^6}{-12x^7y^4}$

[1.4] **6.** Combine like terms: $-3(x + 4) - 2(x - 5)$

Chapter 7 Summary

Defined Terms

Section 7.1
Rational expression (p. 448)
Rational function (p. 455)

Section 7.2
Least common denominator (p. 467)

Section 7.3
Complex rational expression (p. 478)

Section 7.5
Direct variation (p. 504)
Constant of variation (p. 504)
Inverse variation (p. 507)
Joint variation (p. 508)

Procedures, Rules, and Key Examples

Procedures/Rules	Key Examples
Section 7.1 Simplifying, Multiplying, and Dividing Rational Expressions To simplify a rational expression to lowest terms, **1.** Write the numerator and denominator in factored form. **2.** Divide out all common factors in the numerator and denominator. **3.** Multiply the remaining factors in the numerator and the remaining factors in the denominator.	**Example 1:** Simplify. **a.** $\frac{12a^3b}{21a^2b^2} = \frac{2 \cdot 2 \cdot 3 \cdot a \cdot a \cdot a \cdot b}{3 \cdot 7 \cdot a \cdot a \cdot b \cdot b} = \frac{4a}{7b}$ **b.** $\frac{2x^2 - 7x - 15}{x^2 - 25} = \frac{(2x + 3)(x - 5)}{(x + 5)(x - 5)} = \frac{2x + 3}{x + 5}$
To multiply rational expressions, **1.** Write each numerator and denominator in factored form. **2.** Divide out factors common to both the numerator and denominator. **3.** Multiply numerator by numerator and denominator by denominator. **4.** Simplify as needed.	**Example 2:** Multiply. **a.** $\frac{6m^2}{9n} \cdot \frac{12n^3}{10m^4} = \frac{2 \cdot 3 \cdot m \cdot m}{3 \cdot 3 \cdot n} \cdot \frac{2 \cdot 2 \cdot 3 \cdot n \cdot n \cdot n}{2 \cdot 5 \cdot m \cdot m \cdot m \cdot m}$ $= \frac{4n^2}{5m^2}$ **b.** $\frac{3x^2 - 15x}{3x^2 - 17x + 10} \cdot \frac{x^2 + 5x + 6}{6x^2 + 12x}$ $= \frac{3x(x - 5)}{(3x - 2)(x - 5)} \cdot \frac{(x + 3)(x + 2)}{2 \cdot 3x(x + 2)}$ $= \frac{1}{3x - 2} \cdot \frac{x + 3}{2}$ $= \frac{x + 3}{2(3x - 2)}$ or $\frac{x + 3}{6x - 4}$

continued

Procedures/Rules	Key Examples

Section 7.1 Simplifying, Multiplying, and Dividing Rational Expressions (continued)

To divide rational expressions,

1. Write an equivalent multiplication statement using $\frac{P}{Q} \div \frac{R}{S} = \frac{P}{Q} \cdot \frac{S}{R}$, where P, Q, R, and S are polynomials and Q, R, and $S \neq O$.
2. Simplify using the procedure for multiplying rational expressions.

Example 3: Divide.

a. $$\frac{12c^2d}{7cd^3} \div \frac{8c^4}{21d^2} = \frac{12c^2d}{7cd^3} \cdot \frac{21d^2}{8c^4}$$
$$= \frac{2 \cdot 2 \cdot 3 \cdot c \cdot c \cdot d}{7 \cdot c \cdot d \cdot d \cdot d} \cdot \frac{3 \cdot 7 \cdot d \cdot d}{2 \cdot 2 \cdot 2 \cdot c \cdot c \cdot c \cdot c}$$
$$= \frac{3}{1} \cdot \frac{3}{2c^3}$$
$$= \frac{9}{2c^3}$$

b. $$\frac{8a^2 + 6a - 9}{8a^2 + 14a - 15} \div \frac{2a^2 + 11a + 12}{4a^2 + 19a + 12}$$
$$= \frac{8a^2 + 6a - 9}{8a^2 + 14a - 15} \cdot \frac{4a^2 + 19a + 12}{2a^2 + 11a + 12}$$
$$= \frac{(4a - 3)(2a + 3)}{(4a - 3)(2a + 5)} \cdot \frac{(4a + 3)(a + 4)}{(2a + 3)(a + 4)}$$
$$= \frac{1}{2a + 5} \cdot \frac{4a + 3}{1}$$
$$= \frac{4a + 3}{2a + 5}$$

To evaluate a rational function, replace the variable with the indicated value and simplify.

Example 4: Given $f(x) = \frac{x^2 + 2x + 3}{x - 4}$, find $f(2)$.

Solution: $$f(2) = \frac{2^2 + 2 \cdot 2 + 3}{2 - 4}$$
$$= \frac{4 + 4 + 3}{-2}$$
$$= -\frac{11}{2}$$

To find the domain of a rational function,

1. Write an equation that sets the denominator equal to zero.
2. Solve the equation.
3. Exclude the values found in step 2 from the function's domain.

Example 5: Find the domain of $f(x) = \frac{x + 5}{x^2 - 2x - 24}$.

Solution:

$x^2 - 2x - 24 = 0$ — Set the denominator equal to 0.

$(x + 4)(x - 6) = 0$ — Factor.

$x + 4 = 0$ or $x - 6 = 0$ — Set each factor equal to 0.

$x = -4$ or $x = 6$ — Solve each equation.

The domain is $\{x | x \neq -4, 6\}$.

continued

Procedures/Rules	Key Examples

Section 7.1 Simplifying, Multiplying, and Dividing Rational Expressions (continued)

Example 6: Graph $f(x) = \dfrac{1}{x-3}$.

Solution: Since the domain of f is $\{x \mid x \neq 3\}$, its graph has an asymptote at $x = 3$.

x	-5	-2	0	2	4	5
y	$-\frac{1}{8}$	$-\frac{1}{5}$	$-\frac{1}{3}$	-1	1	$\frac{1}{2}$

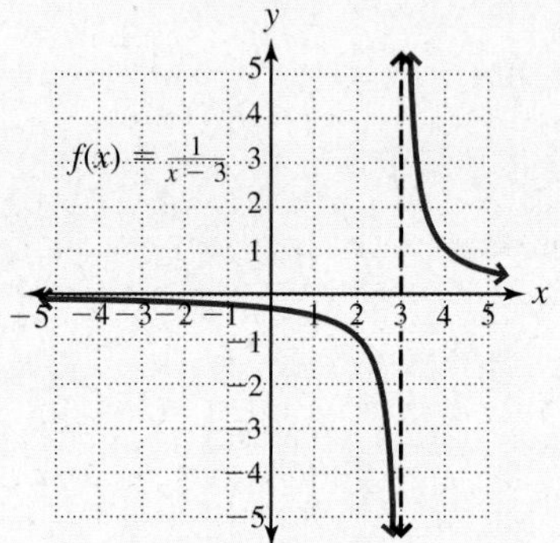

Section 7.2 Adding and Subtracting Rational Expressions

To add or subtract rational expressions that have the same denominator,

1. Add or subtract the numerators and keep the same denominator.
2. Simplify to lowest terms.

Example 1: Add or subtract.

a. $\dfrac{4a}{a+3} + \dfrac{12}{a+3} = \dfrac{4a+12}{a+3}$ Add numerators.

$= \dfrac{4(a+3)}{a+3}$ Factor.

$= 4$ Simplify.

b. $\dfrac{y^2+3y}{y^2+4y+4} - \dfrac{5y+8}{y^2+4y+4}$

$= \dfrac{(y^2+3y)-(5y+8)}{y^2+4y+4}$ Subtract numerators.

$= \dfrac{(y^2+3)+(-5y-8)}{y^2+4y+4}$ Write the equivalent addition.

$= \dfrac{y^2-2y-8}{y^2+4y+4}$ Combine like terms.

$= \dfrac{(y-4)(y+2)}{(y+2)(y+2)}$ Factor.

$= \dfrac{y-4}{y+2}$ Simplify.

continued

Procedures/Rules	Key Examples

Section 7.2 Adding and Subtracting Rational Expressions (continued)

To find the LCD of two or more rational expressions,

1. Factor each denominator.
2. For each unique factor, compare the number of times it appears in each factorization. The LCD is the product of all unique factors the greatest number of times they appear in the denominator factorizations. So if a factor appears in more than one factorization, use the greatest exponent of that factor.

Example 2: Find the LCD for each of the following:

a. $\dfrac{7}{8x^3y^2}$ and $\dfrac{9}{12xy^4}$

$$8x^3y^2 = 2^3 \cdot x^3 \cdot y^2$$
$$12xy^4 = 2^2 \cdot 3 \cdot x \cdot y^4$$
$$\text{LCD} = 2^3 \cdot 3 \cdot x^3 \cdot y^4 = 24x^3y^4$$

b. $\dfrac{x+6}{x^2+4x+4}$ and $\dfrac{x^2-2x+4}{3x^2+6x}$

$$x^2 + 4x + 4 = (x+2)^2$$
$$3x^2 + 6x = 3x(x+2)$$
$$\text{LCD} = 3x(x+2)^2$$

To add or subtract rational expressions with different denominators,

1. Find the LCD.
2. Write each rational expression as an equivalent expression with the LCD.
3. Add or subtract the numerators and keep the LCD.
4. Simplify.

Example 3: Add or subtract.

a. $\dfrac{10x}{21y} + \dfrac{7x}{18y} = \dfrac{10x(6)}{21y(6)} + \dfrac{7x(7)}{18y(7)}$

$$= \frac{60x}{126y} + \frac{49x}{126y}$$
$$= \frac{60x + 49x}{126y}$$
$$= \frac{109x}{126y}$$

b. $\dfrac{x+2}{2x^2+5x-3} + \dfrac{2x-3}{2x^2-5x+2}$

$$= \frac{(x+2)}{(2x-1)(x+3)} + \frac{(2x-3)}{(2x-1)(x-2)}$$
$$= \frac{(x+2)(x-2)}{(2x-1)(x+3)(x-2)} + \frac{(2x-3)(x+3)}{(2x-1)(x-2)(x+3)}$$
$$= \frac{x^2-4}{(2x-1)(x+3)(x-2)} + \frac{2x^2+3x-9}{(2x-1)(x+3)(x-2)}$$
$$= \frac{(x^2-4)+(2x^2+3x-9)}{(2x-1)(x+3)(x-2)}$$
$$= \frac{x^2-4+2x^2+3x-9}{(2x-1)(x+3)(x-2)}$$
$$= \frac{3x^2+3x-13}{(2x-1)(x+3)(x-2)}$$

continued

Procedures/Rules	Key Examples

Section 7.3 Simplifying Complex Rational Expressions

To simplify a complex rational expression, use one of the following methods:

Method 1: **1.** Simplify the numerator and denominator if needed.

2. Rewrite as a horizontal division problem.

Example 1: Simplify $\dfrac{\dfrac{4x^2}{x^2-36}}{\dfrac{2x}{2x^2+9x-18}}$.

$$\frac{\dfrac{4x^2}{x^2-36}}{\dfrac{2x}{2x^2+9x-18}} = \frac{4x^2}{x^2-36} \div \frac{2x}{2x^2+9x-18}$$

$$= \frac{4x^2}{x^2-36} \cdot \frac{2x^2+9x-18}{2x}$$

$$= \frac{\cancel{2} \cdot 2 \cdot \cancel{x} \cdot x}{\cancel{(x+6)}(x-6)} \cdot \frac{(2x-3)\cancel{(x+6)}}{\cancel{2x}}$$

$$= \frac{2x}{x-6} \cdot \frac{2x-3}{1}$$

$$= \frac{4x^2-6x}{x-6}$$

Method 2: **1.** Multiply the numerator and denominator of the complex rational expression by their LCD.

2. Simplify.

Example 2: Simplify $\dfrac{x-\dfrac{8}{x+2}}{1+\dfrac{2}{x+2}}$.

$$\frac{x-\dfrac{8}{x+2}}{1+\dfrac{2}{x+2}} = \frac{\left(x-\dfrac{8}{x+2}\right)(x+2)}{\left(1+\dfrac{2}{x+2}\right)(x+2)}$$

$$= \frac{x(x+2)-\dfrac{8}{\cancel{x+2}}\cdot\cancel{(x+2)}}{1(x+2)+\dfrac{2}{\cancel{x+2}}\cdot\cancel{(x+2)}}$$

$$= \frac{x^2+2x-8}{x+2+2}$$

$$= \frac{\cancel{(x+4)}(x-2)}{\cancel{x+4}}$$

$$= x-2$$

continued

Section 7.4 Solving Equations Containing Rational Expressions

To solve an equation that contains rational expressions,

1. Eliminate the denominators of the rational expressions by multiplying both sides of the equation by the LCD.
2. Solve the resulting equation using the methods we learned in Chapters 2 (for linear equations) and 6 (for quadratic equations).
3. Check your solution(s) in the original equation and discard any extraneous solutions.

Example 1: Solve $\frac{x^2 + 5x - 2}{x + 2} - 3 = \frac{x^2 + 6x}{x + 2} - x$.

Note that x cannot be -2 because it would cause the denominators to be 0, making those expressions undefined.

$$(x + 2)\left(\frac{x^2 + 5x - 2}{x + 2} - 3\right) = (x + 2)\left(\frac{x^2 + 6x}{x + 2} - x\right)$$

$$\cancel{(x + 2)} \cdot \frac{x^2 + 5x - 2}{\cancel{(x + 2)}} - 3(x + 2)$$

$$= \cancel{(x + 2)} \cdot \frac{x^2 + 6x}{\cancel{x + 2}} - x(x + 2)$$

$$x^2 + 5x - 2 - 3x - 6 = x^2 + 6x - x^2 - 2x$$

$$x^2 + 2x - 8 = 4x$$

$$x^2 - 2x - 8 = 0$$

$$(x - 4)(x + 2) = 0$$

$$x - 4 = 0 \quad \text{or} \quad x + 2 = 0$$

$$x = 4 \quad \text{or} \quad x = -2$$

Since -2 is extraneous, the only solution is 4. We will leave the check for 4 to the reader.

Section 7.5 Applications with Rational Expressions; Variation

If y varies directly as x, then we translate to an equation, $y = kx$, where k is a constant.

Example 1: The distance a car can travel varies directly with its number of miles per gallon. On the highway, a Toyota Corolla traveled 280 miles using 8 gallons of gas. How far can the same car travel on 12 gallons of gas?

Solution: Translating "the distance a car can travel varies directly with the number of miles per gallon," we write $d = kn$, where d represents the distance and n is the number of gallons of gas. Use the fact that $d = 280$ when $n = 8$ to find k.

$$280 = k(8) \qquad \text{Substitute 280 for } d \text{ and 8 for } n.$$

$$\frac{280}{8} = \frac{8k}{8} \qquad \text{Divide both sides by 8.}$$

$$35 = k$$

Now use $d = kn$ with $k = 35$ to solve for d when $n = 12$.

$$d = (35)(12) = 420 \text{ miles}$$

continued

Procedures/Rules

Key Examples

Section 7.5 Applications with Rational Expressions; Variation (continued)

If y varies inversely as x, then we translate to an equation, $y = \frac{k}{x}$, where k is a constant.

Example 2: The intensity of light is inversely proportional to the square of the distance from the source. A light meter 6 feet from a lightbulb measures the intensity to be 16 foot-candles. What is the intensity 24 feet from the bulb?

Solution: Translating "the intensity of light is inversely proportional to the square of the distance from the source," we write $I = \frac{k}{d^2}$, where I is the intensity and d is the distance. Use the fact that $I = 16$ when $d = 6$ to find k.

$$16 = \frac{k}{6^2} \qquad \text{Substitute 16 for } I \text{ and 6 for } d.$$

$$36(16) = 36 \cdot \frac{k}{36} \qquad \text{Multiply both sides by 36.}$$

$$576 = k$$

Now use $I = \frac{k}{d^2}$ again with $k = 576$ to solve for I when $d = 24$.

$$I = \frac{576}{24^2}$$

$$I = \frac{576}{576} = 1 \text{ foot-candle}$$

If y varies jointly as x and z, then we translate to an equation, $y = kxz$, where k is a constant.

Example 3: Suppose m varies jointly as n and p. If $m = 288$ when $n = 3$ and $p = 6$, find p when $m = 240$ and $n = 5$.

Solution: Translating "m varies jointly as n and p," we write $m = knp$. Use the fact that $m = 288$ when $n = 3$ and $p = 6$ to find k.

$$288 = k(3)(6)$$
$$288 = 18k$$
$$16 = k$$

Now use $m = knp$ with $k = 16$ to solve for p when $m = 240$ and $n = 5$.

$$240 = 16(5)(p)$$
$$240 = 80p$$
$$3 = p$$

Chapter 7 Review Exercises

For Exercises 1–6, answer true or false.

[7.1] **1.** To simplify a rational expression, we divide out common factors.

[7.2] **2.** In order to add two rational expressions, it is necessary to write each with the GCF as its denominator.

[7.3] **3.** To simplify a complex fraction, multiply the numerator by the LCD of its denominators and multiply the denominator by the LCD of its denominators.

[7.4] **4.** When solving an equation containing rational expressions, it is possible to get extraneous solutions.

[7.4] **5.** To solve an equation containing rational expressions, multiply both sides of the equation by the LCD to clear all fractions.

[7.5] **6.** If x varies inversely with y, then as y increases, x decreases. (Assume $k > 0$.)

[7.1] ***For Exercises 7–18, simplify each rational expression.***

7. $\frac{32x^2y^5}{8xy^3}$

8. $\frac{-42p^6q^2}{27p^3q^4}$

9. $-\frac{18m^3n^2}{54m^5n^5}$

10. $\frac{14x + 63}{10x + 45}$

11. $\frac{4a - 20}{7a - 35}$

12. $\frac{a^2 + 4a - 21}{a^2 + 9a + 14}$

13. $\frac{2x - 3}{6x^2 - x - 12}$

14. $\frac{25x^2 - 9}{10x^2 + x - 3}$

15. $\frac{2c + 3d}{8c^2 + 26cd + 21d^2}$

16. $\frac{5 - 2x}{4x^2 - 25}$

17. $\frac{8ac - 2a + 12c - 3}{6ac - 8a + 9c - 12}$

18. $\frac{8x^3 + 27}{2x^2 - 7x - 15}$

[7.1] *For Exercises 19–28, find each product.*

19. $\dfrac{28m^2n^3}{15p^2q^6} \cdot \dfrac{25p^4q^3}{14mn}$

20. $\dfrac{8m - 12n}{32m^3n} \cdot \dfrac{36mn^3}{6m - 9n}$

21. $\dfrac{14x - 21}{30x - 40} \cdot \dfrac{15x - 20}{8x - 12}$

22. $\dfrac{25x^2 - 16}{16x + 24} \cdot \dfrac{12x + 18}{5x - 4}$

23. $\dfrac{2x - 7}{12} \cdot \dfrac{10}{7 - 2x}$

24. $\dfrac{x^2 - 16}{x^2 + 6x + 8} \cdot \dfrac{x^2 - 3x - 10}{x^2 - 25}$

25. $\dfrac{y^2 - 9}{y^2 - 3y - 18} \cdot \dfrac{y^2 - 4y - 12}{y^2 - 6y + 9}$

26. $\dfrac{8x^2 + 2x - 15}{6x^2 + x - 12} \cdot \dfrac{3x^2 - 13x + 12}{3x^2 - 7x - 6}$

27. $\dfrac{ab + 2ad - 3bc - 6cd}{ab - 4ad + 2bc - 8cd} \cdot \dfrac{ab + 5ad + 2bc + 10cd}{ab + 5ad - 3bc - 15cd}$

28. $\dfrac{8x^3 + y^3}{4x^2 - y^2} \cdot \dfrac{6x^2 + 5xy - 4y^2}{4x^2 - 2xy + y^2}$

[7.1] *For Exercises 29–36, find each quotient.*

29. $\dfrac{39a^2b^4}{27x^3y^2} \div \dfrac{26a^3b}{28xy^4}$

30. $\dfrac{8y^3 - 20y^2}{9y} \div \dfrac{6y^4 - 15y^3}{10y^2}$

31. $\dfrac{21a + 7b}{16a - 24b} \div \dfrac{42a + 14b}{24a - 36b}$

32. $\dfrac{z^2}{z^2 - 2z - 8} \div \dfrac{9z^3 + 3z^4}{z^2 + 5z + 6}$

33. $\dfrac{4p^2 - 9}{24p + 28} \div \dfrac{10p - 15}{36p^2 - 49}$

34. $\dfrac{16p^2 - 8pq - 3q^2}{8p^2 + 22pq + 5q^2} \div \dfrac{8p^2 - 10pq + 3q^2}{10p^2 + pq - 3q^2}$

35. $\dfrac{x^3 - 8y^3}{x^2 - 36y^2} \div \dfrac{x^2 + 2xy - 8y^2}{x^2 - 2xy - 24y^2}$

36. $\dfrac{xz + 2xw - 3yz - 6yw}{4xz - 2xw + 6yz - 3yw} \div \dfrac{xz - 2xw - 3yz + 6yw}{2xz - 4xw + 3yz - 6yw}$

[7.1] **37.** Given $f(x) = \dfrac{2x}{5 - x}$, find

a. $f(3)$ **b.** $f(0)$ **c.** $f(-1)$

[7.1] **38.** Given $g(x) = \dfrac{x + 2}{x^2 - 8x + 15}$, find

a. $g(4)$ **b.** $g(-2)$ **c.** $g(5)$

[7.1] ***For Exercises 39 and 40, find the domain of the rational function.***

39. $f(x) = \dfrac{2x + 4}{3x - 5}$

40. $f(x) = \dfrac{3x - 4}{x^2 + 4x - 12}$

Equations and Inequalities

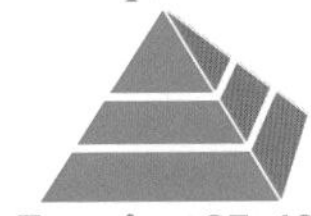

Exercises 37–40

Expressions

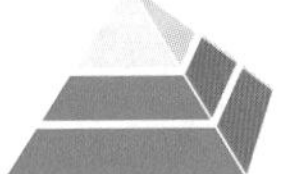

Exercises 41–72

[7.2] ***For Exercises 41–44, add or subtract. Simplify your answers to lowest terms.***

41. $\dfrac{5r}{14x} + \dfrac{3r}{14x}$

42. $\dfrac{3a + 4b}{2a - 3b} - \dfrac{a - 2b}{2a - 3b}$

43. $\dfrac{2p^2 - p + 2}{p^2 - 16} + \dfrac{p^2 + 4p - 8}{p^2 - 16}$

44. $\dfrac{x^2 + 2x - 5}{x^2 - 3x - 18} - \dfrac{3x + 7}{x^2 - 3x - 18}$

[7.2] ***For Exercises 45–52, find the LCD.***

45. $\dfrac{8b}{9p^3q^4}$ and $\dfrac{11c}{12p^2q^5}$

46. $\dfrac{7}{t - 4}$ and $\dfrac{9}{t + 2}$

47. $\dfrac{4u}{8u + 12}$ and $\dfrac{9u}{14u + 21}$

48. $\dfrac{2w}{w^2 - 16}$ and $\dfrac{6w}{w^2 + 5w + 4}$

49. $\dfrac{3a}{a^2 + 6a + 9}$ and $\dfrac{10a}{a^2 - 2a - 15}$

50. $\dfrac{m + 1}{m^2 - 2m - 8}$ and $\dfrac{m - 1}{m^2 - 3m - 4}$

51. $\dfrac{2x - 3}{x^3 + 6x^2 + 8x}$ and $\dfrac{6x + 3}{x^5 + 2x^4 - 8x^3}$

52. $\dfrac{6}{x^3}, \dfrac{4x - 3}{x^2 + 10x + 24}$ and $\dfrac{3x}{4x^3 + 16x^2}$

[7.2] ***For Exercises 53–62, add or subtract as indicated.***

53. $\dfrac{8a}{15x} - \dfrac{4a}{9x}$

54. $\dfrac{y - 3}{4y} - \dfrac{y + 2}{5y}$

55. $\dfrac{2t - 3}{18t^4u^2} + \dfrac{5t + 1}{12t^3u^5}$

56. $\dfrac{8}{w - 3} - \dfrac{-3}{w}$

57. $\dfrac{v + 4}{4v + 8} - \dfrac{v - 2}{2v - 6}$

58. $\dfrac{t^2 - 5t}{t^2 + 8t + 16} + \dfrac{6}{t + 4}$

59. $\dfrac{2w + 5}{w^2 - 25} + \dfrac{6w}{w^2 - 3w - 10}$

60. $\dfrac{z + 9}{4z^2 + 33z + 35} - \dfrac{z - 12}{z^2 + 14z + 49}$

61. $\dfrac{3}{a} - \dfrac{3a + 5}{a + 3} + \dfrac{5a - 3}{a^2 + 3a}$

62. $\dfrac{4x - 3y}{3x - y} - \dfrac{3x - 4y}{y - 3x}$

[7.3] *For Exercises 63–72, simplify.*

63. $\dfrac{\frac{4}{5}}{\frac{3}{10}}$

64. $\dfrac{\frac{4}{3}-\frac{8}{9}}{\frac{5}{6}-\frac{4}{12}}$

65. $\dfrac{\frac{u^4v^2}{w}}{\frac{uv^3}{w^2}}$

66. $\dfrac{\frac{x^6y^3}{t^4}}{\frac{x^2y^2}{t}}$

67. $\dfrac{2w-\frac{w}{4}}{6-\frac{w}{4}}$

68. $\dfrac{\frac{7}{b-15}-6}{b-15}$

69. $\dfrac{1-\frac{1}{y}-\frac{12}{y}}{1-\frac{6}{y}+\frac{8}{y}}$

70. $\dfrac{x+\frac{25}{x+10}}{1-\frac{5}{x+10}}$

71. $\dfrac{\frac{6}{y^2}-\frac{1}{xy}-\frac{12}{x^2}}{\frac{4}{y^2}-\frac{4}{xy}-\frac{3}{x^2}}$

72. $\dfrac{\frac{x+4}{x-4}-\frac{x-4}{x+4}}{\frac{x+4}{x-4}+\frac{x-4}{x+4}}$

[7.4] *For Exercises 73–80, solve and check.*

Equations and Inequalities

Exercises 73–92

73. $\dfrac{5}{4t}-\dfrac{3}{8t}=\dfrac{1}{2}$

74. $\dfrac{12}{m-2}=9+\dfrac{m}{m-2}$

75. $\dfrac{x^2}{x-4}=\dfrac{3x+4}{x-4}-3$

76. $\dfrac{2a-2}{a+1}-\dfrac{a}{2a-3}=\dfrac{a-2}{2a-3}$

77. $\dfrac{1}{q+4}=\dfrac{q}{3q+2}$

78. $\dfrac{5}{v^2+5v+6}-\dfrac{2}{v^2-2v-8}=\dfrac{3}{v^2-16}$

79. $\dfrac{3}{x^2+5x+6}=\dfrac{2}{x+3}+\dfrac{x-3}{x^2+x-2}$

80. $\dfrac{7x}{x^2-2x-8}-\dfrac{4x}{x^2-5x+4}=\dfrac{3}{x-4}$

[7.4] **81.** George can write a chapter for a mathematics textbook in 30 days. Lucille can write a chapter in 45 days. How long would it take them to write a chapter together?

[7.4] **82.** The Gulf Stream is an ocean current off the eastern coast of the United States. A Coast Guard cutter can sail 200 miles against the current in the same amount of time that it can sail 260 miles with the current. If the current is 3 miles per hour, how fast can the cutter sail in still water?

[7.4] **83.** It is 510 road miles from Denver to Salt Lake City. A truck leaves Denver for Salt Lake City at 4:00 A.M. At the same time an automobile leaves Salt Lake City for Denver. If the truck travels at 45 miles per hour and the auto travels at 55 miles per hour, what time will it be when the auto and truck meet? (Give answer to the nearest minute.)

[7.4] ***For Exercises 84–92, solve the variations.***

84. Suppose p varies directly as q. If $p = 24$ when $q = 4$, find p when $q = 7$.

85. Suppose s varies directly as the square of t. If $s = 72$ when $t = 3$, find s when $t = 5$.

86. Suppose y varies inversely as x. If $y = 8$ when $x = 3$, find x when $y = 4$.

87. Suppose m varies inversely as the square of n. If $m = 16$ when $n = 3$, find m when $n = 4$.

88. Suppose y varies jointly with x and z. If $y = 40$ when $x = 2$ and $z = 5$, find y when $x = 4$ and $z = 2$.

89. Suppose m varies directly with n and inversely with p. If $m = 2$ when $n = 3$ and $p = 9$, find m when $n = 6$ and $p = 4$.

90. The distance a car can travel varies directly with the amount of gas it carries. On a trip, a Chevy Corvette travels 156 miles using 6 gallons of fuel. How many gallons are required to travel 234 miles?

91. If the wavelength of a wave remains constant, then the velocity, v, of a wave is inversely proportional to its period, T. In an experiment, waves are created in a fluid so that the period is 7 seconds and the velocity is 4 centimeters per second. If the period is increased to 12 seconds, what is the velocity?

92. The volume of a right circular cylinder varies jointly with the radius squared and the height. If the volume is 62.8 cubic inches when the radius is 2 inches and the height is 5 inches, find the volume when the radius is 4 inches and the height is 2 inches.

Chapter 7 Practice Test

For Exercises 1–4, simplify.

1. $\dfrac{42a^3b^4}{16ab^6}$

2. $\dfrac{6x^2 + 11x - 10}{4x^2 + 4x - 15}$

3. $\dfrac{m^3 - 64n^3}{3m^2 - 14mn + 8n^2}$

4. $\dfrac{2bn - 4bm - 3cn + 6cm}{2b^2 + 7bc - 15c^2}$

For Exercises 5–8, find the products or quotients.

5. $\dfrac{27a^2b^4}{14x^4y} \cdot \dfrac{35xy^3}{18a^5b^2}$

6. $\dfrac{16y^2 - 25}{6y^2 - 17y - 14} \cdot \dfrac{3y^2 + 2y}{8y^2 - 2y - 15}$

7. $\dfrac{6q - 8p}{8p - 2q} \div \dfrac{4p - 3q}{12p - 3q}$

8. $\dfrac{2n^3 - 5n^2 - 6n + 15}{3n^3 + 2n^2 - 9n - 6} \div \dfrac{2n^3 - 5n^2 - 8n + 20}{3n^3 + 2n^2 + 12n + 8}$

9. Given $f(x) = \dfrac{2x - 4}{3x^2 - 7x - 6}$, find the following:

a. $f(-1)$ **b.** $f(2)$ **c.** $f(3)$

10. Find the domain of $f(x) = \dfrac{3x - 4}{2x^2 - x - 10}$.

11. Find the LCD for $\dfrac{3a}{a^2 - 9}$ and $\dfrac{6a}{2a^2 + 13a + 21}$.

For Exercises 12–16, find the sum or difference.

12. $\dfrac{2y^2 + 3}{9y^2} - \dfrac{y - 6}{12y}$

13. $\dfrac{2r^2 - 2r - 5}{r^2 - 16} - \dfrac{-7r + 7}{r^2 - 16}$

14. $\dfrac{t + 3}{t^2 - 10t + 25} - \dfrac{t - 4}{2t^2 - 50}$

15. $\dfrac{3a + 4b}{3a - 5b} + \dfrac{2a - b}{5b - 3a}$

16. $\dfrac{5}{x} - \dfrac{3x - 1}{x^2 + 5x} - \dfrac{2x - 5}{x + 5}$

For Exercises 17–20, simplify.

17. $\dfrac{\dfrac{6a^3b^2}{c^3}}{\dfrac{9a^2b^4}{c^6}}$

18. $\dfrac{6 - \dfrac{1}{x} - \dfrac{15}{x^2}}{4 + \dfrac{4}{x} - \dfrac{3}{x^2}}$

19. $\dfrac{t - \dfrac{14}{t-5}}{2 - \dfrac{4}{t-5}}$

20. $\dfrac{9a^{-2} - 4b^{-2}}{3a^{-1} + 2b^{-1}}$

For Exercises 21–24, solve.

21. $\dfrac{8}{6x} + \dfrac{5}{2x} = \dfrac{13}{9}$

22. $\dfrac{3x}{x-4} = 6 + \dfrac{12}{x-4}$

23. $\dfrac{x}{x+1} + \dfrac{9x}{x^2+3x+2} = \dfrac{12}{x+2}$

24. $\dfrac{n}{n+1} + \dfrac{n+1}{n+2} = \dfrac{6n+5}{n^2+3n+2}$

25. Working together, it takes Elena and Eduardo 10 days to do an architectural project. If Elena can do the project by herself in 15 days, how long would it take Eduardo to do the project working by himself?

26. An air cargo plane and an airliner take off at the same time from the same airport. They fly in opposite directions. The air cargo plane flies at 420 miles per hour and the airliner flies at 530 miles per hour. How long is it before the two airplanes are 1900 miles apart?

For Exercises 27–30, solve the variations.

27. Suppose r varies directly with s. If $r = 14$ when $s = 2$, find s when $r = 42$.

28. Suppose p varies jointly with q and r. If $p = 48$ when $q = 4$ and $r = 6$, find p when $q = 3$ and $r = 5$.

29. The intensity of light is inversely proportional to the square of the distance from the source. A light meter 6 feet from a lightbulb measures the intensity to be 16 foot-candles. What is the intensity at 20 feet from the bulb?

30. The distance a car can travel varies directly with the number of gallons the car uses. On the highway, a Toyota Prius traveled 255 miles using 5 gallons of gas. How far can the same car travel on 8 gallons of gas?

CHAPTER 8

Rational Exponents, Radicals, and Complex Numbers

"Radical simply means 'grasping things at the root.' "

—Angela Davis, U.S. political activist, address June 25, 1987, to Spellman College

In Section 1.3, we reviewed the basic rules of square roots. In this chapter, we will explore square roots and radicals in much more detail. We will begin in Section 8.1 with evaluating and simplifying radical expressions, much like how we started with evaluating and simplifying polynomial expressions in the first part of Chapter 5. Then, in Section 8.2, we will see that exponents can be fractions and that those rational exponents mean roots. In Sections 8.3–8.5, we move to rewriting expressions containing radicals. Once we have explored how to rewrite radical expressions, we will learn to solve equations that contain radicals in Section 8.6. Finally, in Section 8.7, we will discuss a new set of numbers called *complex numbers*, which come from square roots of negative numbers.

8.1 Radical Expressions and Functions

OBJECTIVES

1. **Find the nth root of a number.**
2. **Approximate roots using a calculator.**
3. **Simplify radical expressions.**
4. **Evaluate radical functions.**
5. **Find the domain of radical functions.**
6. **Solve applications involving radical functions.**

In this section through Section 8.5, we focus on the expression portion of our algebra pyramid and explore square root and radical expressions.

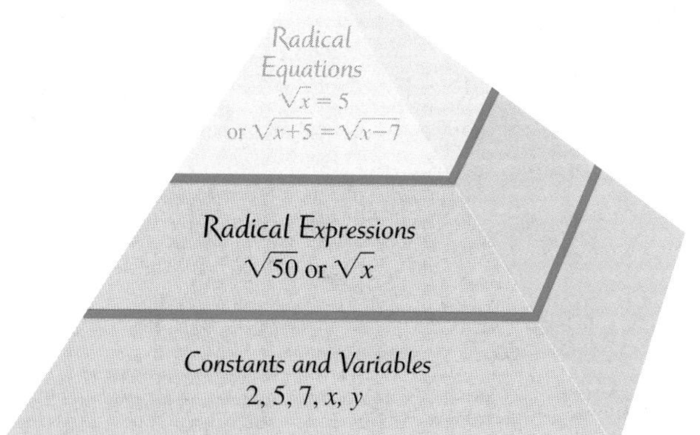

Note: *We will explore radical equations in Section 8.6.*

OBJECTIVE 1. Find the nth root of a number. In Section 1.3, we learned that a square root of a given number is a number that, when squared, equals the given number. For example, a square root of 16 is 4 since $4^2 = 16$. However, another number can be squared to equal 16: $(-4)^2 = 16$, so -4 is also a square root of 16. Consequently, 16 has two square roots: 4 and -4. We can write them more compactly as ± 4.

Similarly, a *cube root* of a given number is a number that, when cubed, equals the given number. For example, the cube root of 8 is 2 because $2^3 = 8$.

The *fourth root* of a number is a number that, when raised to the fourth power, equals the given number. For example, 3 is a fourth root of 81 because $3^4 = 81$. However, $(-3)^4 = 81$ also, so the fourth roots of 81 are ± 3. Our examples suggest the following definition of ***n^{th}* root**.

DEFINITION ***n^{th}* root:** The number b is an nth root of a number a if $b^n = a$.

Recall from Section 1.3 that we used the symbol $\sqrt{}$, called a *radical sign*, to denote the square root of a number. To indicate roots other than square roots we use the symbol $\sqrt[n]{a}$, read "the nth root of a," where n is called the *root index* and indicates which root we are to find. If no root index is given, we assume it is 2, the square root. The number a is called the *radicand* and is the number or expression whose root we are to find. The entire expression is called a *radical* and any expression containing a radical is called a *radical expression*.

We have seen some numbers that have more than one root, such as 16, which has two square roots, 4 and -4. To keep things simple, the symbol $\sqrt[n]{a}$ means to find a single root. If n is even, then $\sqrt[n]{a}$ denotes the nonnegative root only and is called the *principal root*. For example, the principal square root of 16 is 4. Further, for $\sqrt[n]{a}$ to exist as a real number when n is even, a must be nonnegative because there is no real number that can be raised to an even power to equal a negative number. For example, $\sqrt{-4}$ does not exist as a real number because there is no real number whose square is -4.

If the root index is odd, the radicand can be positive or negative since a positive number raised to an odd power is positive and a negative number raised to an odd power is negative.

RULE **Evaluating n^{th} Roots**

When evaluating a radical expression $\sqrt[n]{a}$, the sign of a and the index n will determine possible outcomes.

If a is nonnegative, then $\sqrt[n]{a} = b$, where $b \geq 0$ and $b^n = a$.
If a is negative and n is even, then there is no real-number root.
If a is negative and n is odd, then $\sqrt[n]{a} = b$, where b is negative and $b^n = a$.

EXAMPLE 1 Evaluate the following roots.

a. $\sqrt{225}$

Solution $\sqrt{225} = 15$ Because $15^2 = 225$.

Note: *It can be helpful to think of* $-\sqrt{0.81}$ *as* $-1 \cdot \sqrt{0.81} = -1 \cdot 0.9 = -0.9$.

b. $-\sqrt{0.81}$

Solution $-\sqrt{0.81} = -0.9$ Because $(-0.9)^2 = 0.81$.

c. $\sqrt{-9}$

Solution $\sqrt{-9}$ is not a real number.

Note: *In Section 8.7, we will define the square root of a negative number using a set of numbers called the* imaginary numbers.

d. $\pm\sqrt{121}$

Solution $\pm\sqrt{121} = \pm 11$ Because $(11)^2 = 121$ and $(-11)^2 = 121$.

e. $\sqrt{\frac{4}{9}}$

Solution $\sqrt{\frac{4}{9}} = \frac{2}{3}$ Because $\left(\frac{2}{3}\right)^2 = \frac{4}{9}$.

f. $\sqrt[3]{8}$

Solution $\sqrt[3]{8} = 2$ Because $2^3 = 8$.

g. $\sqrt[3]{-8}$

Solution $\sqrt[3]{-8} = -2$ Because $(-2)^3 = -8$.

h. $\sqrt[4]{16}$

Solution $\sqrt[4]{16} = 2$ Because $2^4 = 16$.

ANSWERS

a. 0.6 b. −2
c. not a real number
d. ±13 e. $\frac{5}{6}$
f. 4 g. −3
h. −3

YOUR TURN Evaluate the following expressions.

a. $\sqrt{0.36}$ b. $-\sqrt{4}$ c. $\sqrt{-64}$ d. $\pm\sqrt{169}$

e. $\sqrt{\frac{25}{36}}$ f. $\sqrt[3]{64}$ g. $\sqrt[3]{-27}$ h. $-\sqrt[4]{81}$

OBJECTIVE 2. Approximate roots using a calculator. All the roots that we have considered so far have been rational. Some roots, like $\sqrt{3}$, are called irrational. As we learned in Section 1.1, we cannot express the exact value of an irrational number using rational numbers. In fact, writing $\sqrt{3}$ with the radical sign is the only way we can express the exact value. However, we may approximate $\sqrt{3}$ with a calculator or a table, such as the one in the endpapers of this text.

For example, a calculator shows that $\sqrt{2} \approx 1.414213562$, which we can round to various decimal places:

Approximating to two decimal places: $\sqrt{2} \approx 1.41$

Approximating to three decimal places: $\sqrt{2} \approx 1.414$

Note: *Recall that the symbol ≈ means "approximately equal to."*

EXAMPLE 2 Approximate the roots using a calculator or the table in the endpapers. Round to three decimal places.

a. $\sqrt{12}$

Answer $\sqrt{12} \approx 3.464$

b. $-\sqrt{38}$

Answer $-\sqrt{38} \approx -6.164$

c. $\sqrt[3]{45}$

Answer $\sqrt[3]{45} \approx 3.557$

Calculator TIPS

To evaluate roots higher than a square root on a calculator, use the [$\sqrt[x]{\ }$] *key. First type the index, then press* [$\sqrt[x]{\ }$], *then the radicand. For example, to evaluate $\sqrt[3]{45}$, type:* [3] [$\sqrt[x]{\ }$] [4] [5] [ENTER].

On some calculators, $\sqrt[x]{\ }$ is a function in a menu accessed by pressing the [MATH] *key.*

YOUR TURN Approximate the following irrational numbers. Round to three decimal places.

a. $\sqrt{19}$ b. $-\sqrt{93}$ c. $\sqrt[3]{63}$

ANSWERS

a. 4.359 b. −9.644 c. 3.979

OBJECTIVE 3. Simplify radical expressions. The definition of a root can also be used to find roots with variable radicands. Recall that with an even index, the principal root is nonnegative, so at first we will assume all variables represent nonnegative values. We will use $(a^m)^n = a^{mn}$ to verify the roots.

Note: *Remember that with an even index, the principal root is nonnegative. Since we are assuming that the variables are nonnegative, our result accurately indicates the principal square root.*

Connection To raise a power to another power using $(a^m)^n = a^{mn}$, we multiply the powers. To find a root of a power, we divide the index into the exponent.

EXAMPLE 3 Find the root. Assume all variables represent nonnegative values.

a. $\sqrt{x^2}$

Solution $\sqrt{x^2} = x$ Because $(x)^2 = x^2$.

b. $\sqrt{a^6}$

Solution $\sqrt{a^6} = a^3$ Because $(a^3)^2 = a^6$.

c. $\sqrt{16x^8}$

Solution $\sqrt{16x^8} = 4x^4$ Because $(4x^4)^2 = 16x^8$.

d. $\sqrt{\dfrac{25x^8}{49y^2}}$

Solution $\sqrt{\dfrac{25x^8}{49y^2}} = \dfrac{5x^4}{7y}$ Because $\left(\dfrac{5x^4}{7y}\right)^2 = \dfrac{25x^8}{49y^2}$.

e. $\sqrt[3]{y^6}$

Solution $\sqrt[3]{y^6} = y^2$ Because $(y^2)^3 = y^6$.

f. $\sqrt[4]{16x^{12}}$

Solution $\sqrt[4]{16x^{12}} = 2x^3$ Because $(2x^3)^4 = 16x^{12}$.

Connection In parts c and f, notice the similarity between finding the root of a product and raising a product to a power. To raise a product to a power, we raise each factor to the power. To find a root of a product, we find the root of each factor.

YOUR TURN Find the root. Assume variables represent nonnegative values.

a. $\sqrt{x^4}$ **b.** $\sqrt{9x^{10}}$ **c.** $\sqrt{36a^{12}}$

d. $\sqrt{\dfrac{100x^4}{81y^6}}$ **e.** $\sqrt[3]{27y^9}$ **f.** $\sqrt[4]{b^8}$

If the variables can represent *any* real number and the index is even, then we must be careful to ensure that the principal root is nonnegative by using absolute value symbols. If the index is odd, however, we do not need absolute value signs because the root can be either positive or negative depending on the radicand's sign.

Note: *To illustrate why $\sqrt{x^2} = |x|$, suppose we were to evaluate $\sqrt{x^2}$ when $x = -3$. We would have: $\sqrt{(-3)^2} = \sqrt{9} = 3$. Notice that the root, 3, is, in fact, the absolute value of -3. If we had incorrectly stated that $\sqrt{x^2} = x$, then $\sqrt{(-3)^2}$ would have to equal -3, which is not true.*

EXAMPLE 4 Find the root. Assume variables represent any real number.

a. $\sqrt{x^2}$

Solution $\sqrt{x^2} = |x|$

b. $\sqrt{a^6}$

Solution $\sqrt{a^6} = |a^3|$

c. $\sqrt{(n+1)^2}$

Solution $\sqrt{(n+1)^2} = |n+1|$

d. $\sqrt{25y^8}$

Solution $\sqrt{25y^8} = 5y^4$

e. $\sqrt[3]{8n^3}$

Solution $\sqrt[3]{8n^3} = 2n$

f. $\sqrt[3]{(t-2)^3}$

Solution $\sqrt[3]{(t-2)^3} = t - 2$

Note: *In part d, we do not need absolute value because y^4, with its even exponent, will always be nonnegative. In parts e and f, we do not use absolute value because the indexes are odd.*

ANSWERS

a. x^2 **b.** $3x^5$ **c.** $6a^6$ **d.** $\dfrac{10x^2}{9y^3}$ **e.** $3y^3$ **f.** b^2

ANSWERS

a. x^2 b. $3|x^5|$
c. $6a^6$ d. $1.1|u^3t|$
e. $\frac{10x^2}{9|y^3|}$ f. $3y^3$
g. b^2

YOUR TURN Find the root. Assume variables represent any real number.

a. $\sqrt{x^4}$ b. $\sqrt{9x^{10}}$ c. $\sqrt{36a^{12}}$ d. $\sqrt{1.21u^6t^2}$

e. $\sqrt{\frac{100x^4}{81y^6}}$ f. $\sqrt[3]{27y^9}$ g. $\sqrt[4]{b^8}$

OBJECTIVE 4. Evaluate radical functions. Now that we have learned about radical expressions, let's examine **radical functions**.

DEFINITION ***Radical function:*** A function of the form $f(x) = \sqrt[n]{P}$, where P is a polynomial.

EXAMPLE 5

a. Given $f(x) = \sqrt{3x - 2}$, find $f(3)$.

Solution To find $f(3)$, substitute 3 for x and simplify.

$$f(3) = \sqrt{3(3) - 2} = \sqrt{9 - 2} = \sqrt{7}$$

b. Given $f(x) = \sqrt{2x - 6}$, find $f(0)$.

Solution To find $f(0)$, substitute 0 for x and simplify. $f(0) = \sqrt{2(0) - 6} = \sqrt{-6}$, which is not a real number.

ANSWER

$f(-1) = \sqrt{3}$

YOUR TURN Given $f(x) = \sqrt{2x + 5}$, find $f(-1)$.

OBJECTIVE 5. Find the domain of radical functions. In Example 5(b), we found that $f(0)$ did not exist because $\sqrt{-6}$ is not a real number. What does this suggest about the domains of functions involving radicals? Consider the graphs of $f(x) = \sqrt{x}$ and $f(x) = \sqrt[3]{x}$, which we can generate by choosing values for x.

x	$f(x) = \sqrt{x}$
−1	Not real
0	0
1	1
4	2

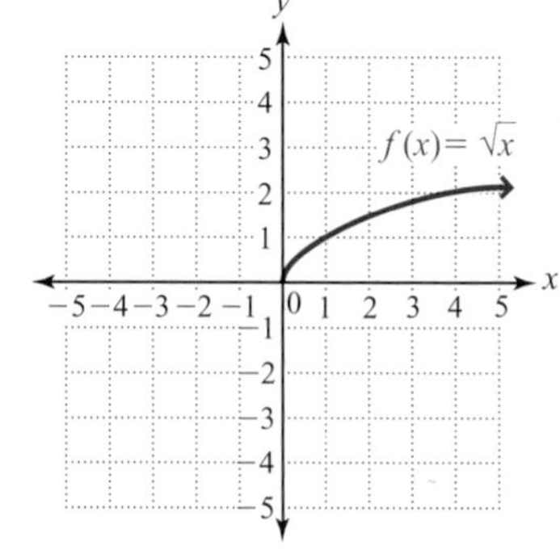

Notice that $\sqrt{x}$ is not a real number if x is negative, which means the domain for $f(x) = \sqrt{x}$ is $\{x|x \geq 0\}$.

x	$f(x) = \sqrt[3]{x}$
−8	−2
−1	−1
0	0
1	1
8	2

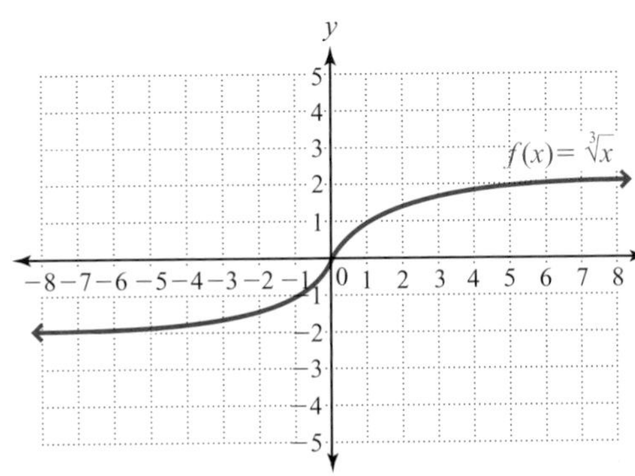

Alternatively, $\sqrt[3]{x}$, with its odd index, is a real number when x is negative, so its domain is all real numbers. Our graphs suggest the following conclusion.

Conclusion The domain of a radical function with an even index must contain values that keep its radicand nonnegative.

Example 6 Find the domain of each of the following.

a. $f(x) = \sqrt{x - 4}$

Solution Since the index is even, the radicand must be nonnegative.

$$x - 4 \geq 0$$
$$x \geq 4 \quad \text{Add 4 to both sides.}$$

Domain: $\{x|x \geq 4\}$, or $[4, \infty)$

b. $f(x) = \sqrt{-2x + 6}$

Solution The radicand must be nonnegative.

$$-2x + 6 \geq 0$$
$$-2x \geq -6 \quad \text{Subtract 6 from both sides.}$$
$$x \leq 3 \quad \text{Divide both sides by } -2.$$

Domain: $\{x|x \leq 3\}$, or $(-\infty, 3]$

Connection Recall that when we divide *both* sides of an inequality by a negative number, we reverse the direction of the inequality.

Your Turn Find the domain of each of the following.

a. $f(x) = \sqrt{2x - 4}$ **b.** $f(x) = \sqrt{-3x - 9}$

Answers

a. $\{x|x \geq 2\}$, or $[2, \infty)$

b. $\{x|x \leq -3\}$, or $(-\infty, -3]$

OBJECTIVE 6. Solve applications involving radical functions. Often, radical functions appear in real-world situations where one variable is a function of another.

Example 7 The velocity of a free-falling object is a function of the distance that it has fallen. Ignoring air resistance, the velocity of an object, v, in meters per second, can be found after falling h meters using the formula $v = -\sqrt{19.6h}$. Find the velocity of a stone that has fallen 30 meters after being dropped from a cliff.

Understand We are to find the velocity of an object after it falls 30 meters.

Plan Use the formula $v = -\sqrt{19.6h}$, replacing h with 30.

Execute $v = -\sqrt{19.6(30)}$ Replace h with 30.

$v = -\sqrt{588}$ Multiply within the radical.

$v \approx -24.2$ Evaluate the square root.

Note: *A negative velocity indicates that the object is traveling downward.*

Answer After falling 30 meters, the stone is traveling at a velocity of -24.2 meters per second.

Check We can verify the calculations, which we will leave to the reader.

YOUR TURN A skydiver jumps from a plane and puts her body into a dive position so that air resistance is minimized. Find her velocity after she falls 100 meters.

EXAMPLE 8 The period of a pendulum is the amount of time it takes the pendulum to swing from the point of release to the opposite extreme then back to the point of release. The period is a function of the length. The period, T, measured in seconds, can be found using the formula $T = 2\pi\sqrt{\frac{L}{9.8}}$, where L represents the length of the pendulum in meters. Find the period of a pendulum that is 0.5 meters long.

Understand We are to find the period of a pendulum that is 0.5 meters long.

Plan Use the formula $T = 2\pi\sqrt{\frac{L}{9.8}}$, approximating π with 3.14 and replacing L with 0.5.

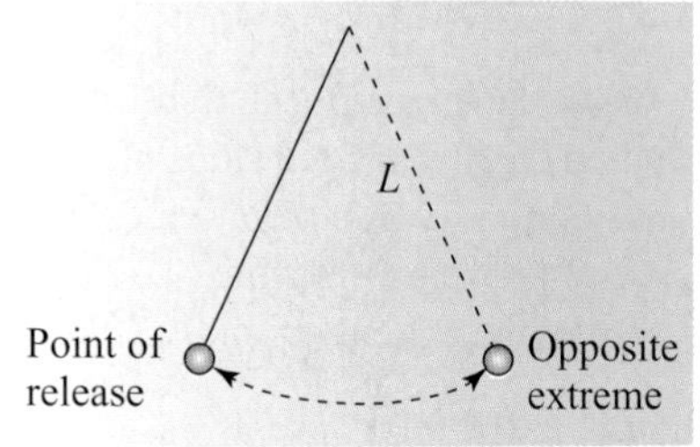

Execute $T \approx 2(3.14)\sqrt{\frac{0.5}{9.8}}$ Simplify.

$T \approx 6.28\sqrt{0.051}$

$T \approx 6.28(0.226)$ Approximate the square root.

$T \approx 1.42$

Answer The period of a pendulum that is 0.5 meters long is 1.42 seconds.

Check We can verify the calculations, which we will leave to the reader.

YOUR TURN Find the period of a pendulum that is 0.2 meters long. Round to the nearest thousandth.

ANSWER

-44.3 m/sec.

ANSWER

≈ 0.898 sec.

8.1 Exercises

For Extra Help

MyMathLab  Videotape/DVT  InterAct Math  Math Tutor Center  Math XL.com

1. Give an example of a number with rational square roots and an example of a number with irrational square roots. What makes the second number's roots irrational?

2. How can we express the exact value of a root that is irrational?

3. Why are there two square roots for every positive real number?

4. If the index is even, what is the principal root of a number?

5. Why is an even root of any negative number not a real number?

6. If x is any real number, explain why $\sqrt{x^2} = |x|$ instead of x (with no absolute value).

For Exercises 7–14, find all square roots of the number given.

7. 36 **8.** 64 **9.** 121 **10.** 81

11. 196 **12.** 400 **13.** 225 **14.** 289

For Exercises 15–56, evaluate the roots, if possible.

15. $\sqrt{25}$ **16.** $\sqrt{49}$ **17.** $\sqrt{-64}$ **18.** $\sqrt{-25}$

19. $-\sqrt{25}$ **20.** $-\sqrt{100}$ **21.** $\pm\sqrt{25}$ **22.** $\pm\sqrt{100}$

23. $\sqrt{1.44}$ **24.** $\sqrt{0.36}$ **25.** $\sqrt{-10.64}$ **26.** $\sqrt{-2.25}$

27. $-\sqrt{0.0121}$ **28.** $-\sqrt{0.0009}$ **29.** $\sqrt{\frac{49}{81}}$ **30.** $\sqrt{\frac{64}{121}}$

31. $-\sqrt{\frac{144}{169}}$ **32.** $-\sqrt{\frac{25}{4}}$ **33.** $\sqrt[3]{27}$ **34.** $\sqrt[3]{64}$

35. $\sqrt[3]{-27}$

36. $\sqrt[3]{-216}$

37. $-\sqrt[3]{-216}$

38. $-\sqrt[3]{-125}$

39. $\sqrt[4]{625}$

40. $\sqrt[4]{81}$

41. $\sqrt[4]{-625}$

42. $\sqrt[4]{-81}$

43. $-\sqrt[4]{16}$

44. $-\sqrt[4]{625}$

45. $\sqrt[5]{32}$

46. $\sqrt[5]{243}$

47. $\sqrt[5]{-32}$

48. $\sqrt[5]{-243}$

49. $-\sqrt[5]{-32}$

50. $-\sqrt[5]{-3125}$

51. $\sqrt[6]{64}$

52. $\sqrt[6]{-64}$

53. $\sqrt[3]{\frac{8}{27}}$

54. $\sqrt[3]{-\frac{64}{125}}$

55. $\sqrt[4]{\frac{16}{81}}$

56. $\sqrt[5]{\frac{1}{32}}$

For Exercises 57–72, use a calculator to approximate each root to the nearest thousandth.

57. $\sqrt{5}$

58. $\sqrt{12}$

59. $-\sqrt{11}$

60. $-\sqrt{41}$

61. $\sqrt[3]{50}$

62. $\sqrt[3]{21}$

63. $\sqrt[3]{-53}$

64. $\sqrt[3]{-83}$

65. $\sqrt[4]{65}$

66. $\sqrt[4]{123}$

67. $-\sqrt[4]{85}$

68. $-\sqrt[4]{77}$

69. $\sqrt[5]{89}$

70. $\sqrt[5]{-62}$

71. $\sqrt[6]{146}$

72. $\sqrt[6]{98}$

For Exercises 73–96, simplify. Assume variables represent nonnegative values.

73. $\sqrt{b^4}$

74. $\sqrt{r^8}$

75. $\sqrt{16x^2}$

76. $\sqrt{81t^6}$

77. $\sqrt{100r^2s^6}$

78. $\sqrt{121x^8y^{10}}$

79. $\sqrt{0.25a^6b^{12}}$

80. $\sqrt{0.36r^4s^{14}}$

81. $\sqrt[3]{m^3}$

82. $\sqrt[3]{n^6}$

83. $\sqrt[3]{27a^9b^6}$

84. $\sqrt[3]{64u^{12}t^9}$

85. $\sqrt[3]{-64a^3b^{12}}$

86. $\sqrt[3]{-27r^{15}s^3}$

87. $\sqrt[3]{0.008x^{18}}$

88. $\sqrt[3]{-0.125r^6}$

89. $\sqrt[4]{a^4}$

90. $\sqrt[4]{x^{12}}$

91. $\sqrt[4]{16x^{16}}$

92. $\sqrt[4]{81t^8}$

93. $\sqrt[5]{32x^{10}}$

94. $\sqrt[5]{243x^{15}}$

95. $\sqrt[6]{x^{12}y^6}$

96. $\sqrt[7]{s^7t^{21}}$

For Exercises 97–108, simplify. Assume variables represent any real number.

97. $\sqrt{36m^2}$

98. $\sqrt{9t^6}$

99. $\sqrt{(r-1)^2}$

100. $\sqrt{(k+3)^2}$

101. $\sqrt[4]{256y^{12}}$

102. $\sqrt[4]{16x^4}$

103. $\sqrt[3]{27y^3}$

104. $\sqrt[3]{125x^6}$

105. $\sqrt{(y-3)^4}$

106. $\sqrt{(x+2)^4}$

107. $\sqrt[3]{(y-4)^3}$

108. $\sqrt[4]{(n+5)^8}$

For Exercises 109–112, find the indicated value of the function.

109. $f(x) = \sqrt{2x+4}$, find $f(0)$

110. $f(x) = \sqrt{3x+2}$, find $f(3)$

111. $f(x) = \sqrt{4x+3}$, find $f(3)$

112. $f(x) = \sqrt{-2x+3}$, find $f(-2)$

For Exercises 113–116, find the domain.

113. $f(x) = \sqrt{2x-8}$

114. $f(x) = \sqrt{3x+12}$

115. $f(x) = \sqrt{-4x+16}$

116. $f(x) = \sqrt{-2x+6}$

For Exercises 117–120: a. Use a graphing calculator to graph the function.
b. Find the domain of the function.

117. $f(x) = \sqrt{x-2}$

118. $f(x) = \sqrt{x+3}$

119. $f(x) = \sqrt[3]{x+1}$

120. $f(x) = \sqrt[4]{x-3}$

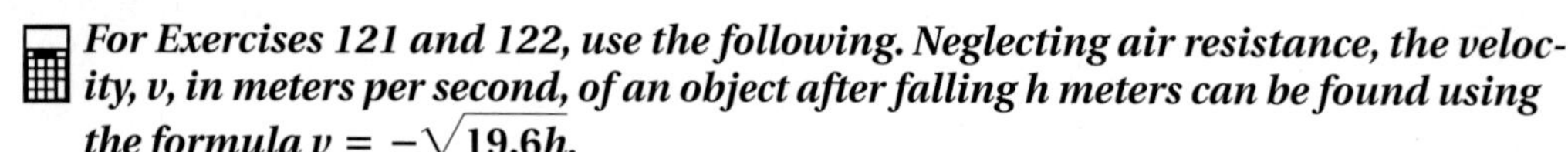

For Exercises 121 and 122, use the following. Neglecting air resistance, the velocity, v, in meters per second, of an object after falling h meters can be found using the formula $v = -\sqrt{19.6h}$.

Of Interest

One of the many topics Galileo Galilei studied was pendulums. His interest in them was piqued when he noticed a swinging light fixture in the cathedral of his hometown of Pisa and timed its period using his pulse.

121. Find the velocity of a rock that has been dropped from a cliff after it falls 16 meters.

122. Find the velocity of a ball that has been dropped from a roof after it falls 5 meters.

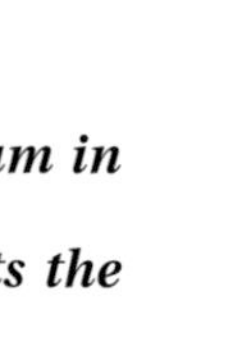

For Exercises 123 and 124, use the following. The period, T, of a pendulum in seconds can be found using the formula $T = 2\pi\sqrt{\frac{L}{9.8}}$, ***where L represents the length of the pendulum in meters.***

123. Find the period of a pendulum that is 3 meters long.

124. Find the period of a pendulum that is 2.5 meters long.

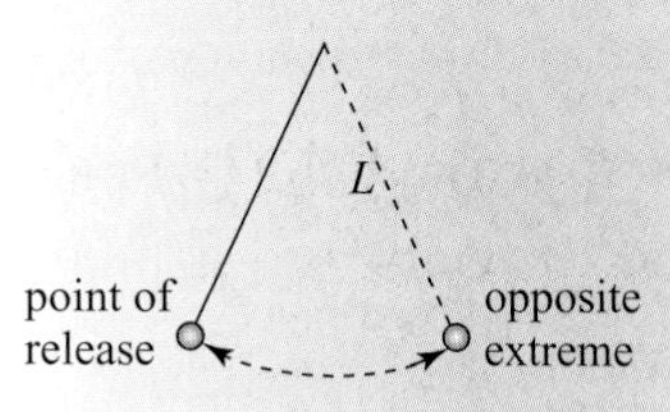

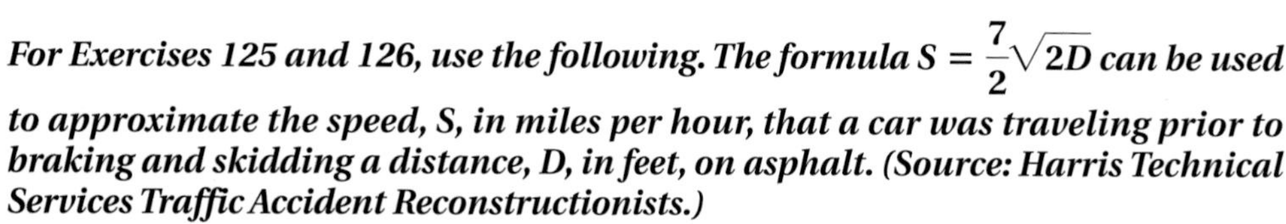

For Exercises 125 and 126, use the following. The formula $S = \frac{7}{2}\sqrt{2D}$ ***can be used to approximate the speed, S, in miles per hour, that a car was traveling prior to braking and skidding a distance, D, in feet, on asphalt. (Source: Harris Technical Services Traffic Accident Reconstructionists.)***

125. Find the speed of a car if the skid distance is 15 feet.

126. Find the speed of a car if the skid distance is 40 feet.

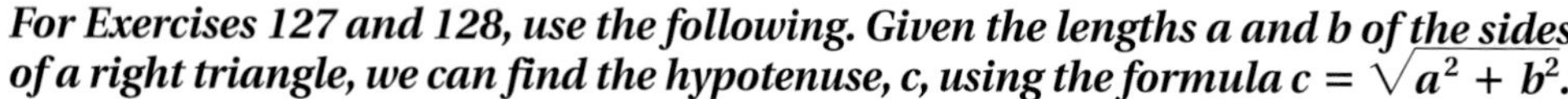

For Exercises 127 and 128, use the following. Given the lengths a and b of the sides of a right triangle, we can find the hypotenuse, c, using the formula $c = \sqrt{a^2 + b^2}$.

127. Three pieces of lumber are to be connected to form a right triangle that will be part of the frame for a roof. If the horizontal piece is to be 12 feet and the vertical piece is to be 5 feet, how long must the connecting piece be?

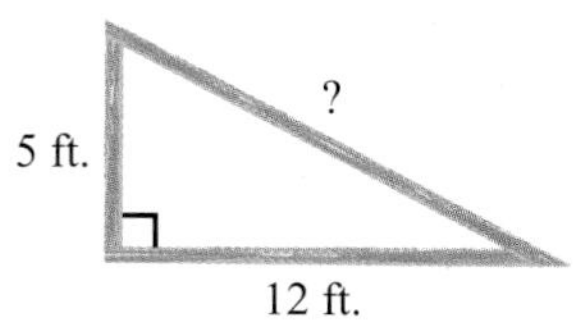

128. A counselor decides to create a ropes course with a zip line. She wants the line to connect from a 40-foot-tall tower to the ground at a point 100 feet from the base of the tower. Assuming the tower and ground form a right angle, find the length of the zip line.

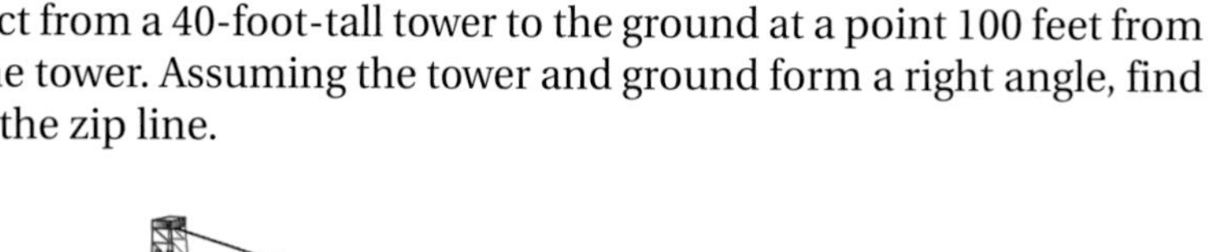

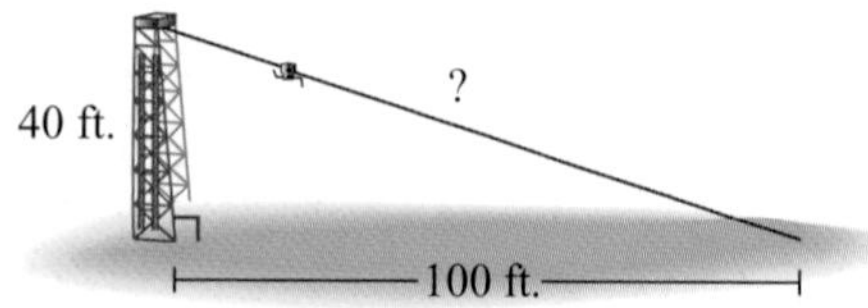

Connection In Section 6.6, we used the Pythagorean theorem, $c^2 = a^2 + b^2$, to find missing side lengths of a right triangle. The formula $c = \sqrt{a^2 + b^2}$ comes from isolating c in that theorem.

For Exercises 129 and 130 use the following. Two forces, F_1 and F_2, acting on an object at a 90° angle will pull the object with a resultant force, R, at an angle in between F_1 and F_2 (see the figure). The value of the resultant force can be found by the formula $R = \sqrt{F_1^2 + F_2^2}$.

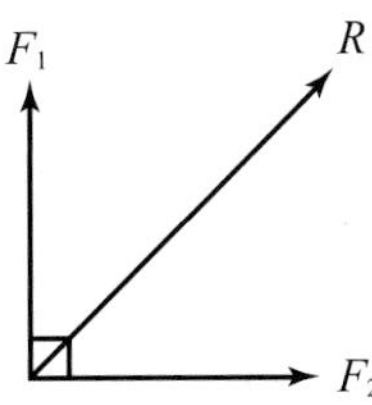

Note: *The unit of force in the metric system is the newton (abbreviated N).*

129. Find the resultant force if $F_1 = 9$ N and $F_2 = 12$ N.

130. Find the resultant force if $F_1 = 6.2$ N and $F_2 = 4.5$ N.

131. The number of earthquakes worldwide for 2000–2004 with a magnitude of 3.0–3.9 can be approximated using the function $f(x) = 1570\sqrt{x} + 4784$, where x represents the number of years after 2000. (*Hint:* The year 2000 corresponds to $x = 0$.) (*Source:* USGS Earthquake Hazards Program)

a. Find the approximate number of earthquakes with a magnitude of 3.0–3.9 that occurred in 2004.

b. Find the approximate number of earthquakes with a magnitude of 3.0–3.9 that occurred in 2002.

132. The number of associate degrees, in thousands, conferred by institutions of higher education for 1970 through 2004 can be approximated using the function $f(x) = 52\sqrt{x} + 252$, where x represents the number of years after 1970. (*Hint:* The year 1970 corresponds to $x = 0$.) (*Source:* National Center for Education Statistics)

a. How many associate degrees were awarded in 1980?

b. How many associate degrees were awarded in 2004?

PUZZLE PROBLEM

A telemarketer calls a house and speaks to the mother of three children. The telemarketer asks the mother, "How old are you?" The woman answers, "41." The marketer then asks, "How old are your children?" The woman replies, "The product of their ages is our house number, 1296, and all of their ages are perfect squares." What are the ages of the children?

REVIEW EXERCISES

[1.3] **1.** Follow the order of operations to simplify.

a. $\sqrt{16 \cdot 9}$

b. $\sqrt{16} \cdot \sqrt{9}$

[5.1, 5.3] ***For Exercises 2–6, multiply.***

2. $x^5 \cdot x^3$ **3.** $(-9m^3n)(5mn^2)$ **4.** $4x^2(3x^2 - 5x + 1)$ **5.** $(7y - 4)(3y + 5)$

6. Write an expression for the area of the figure shown.

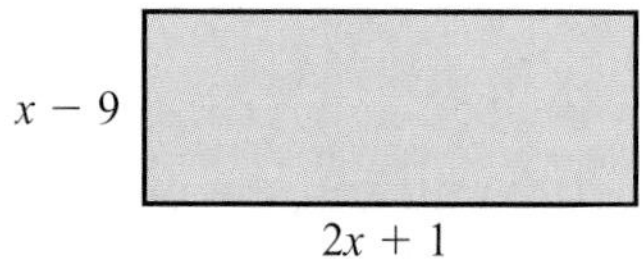

8.2 Rational Exponents

OBJECTIVES

1. Evaluate rational exponents.
2. Write radicals as expressions raised to rational exponents.
3. Simplify expressions with rational number exponents using the rules of exponents.
4. Use rational exponents to simplify radical expressions.

OBJECTIVE 1. Evaluate rational exponents.

Rational Exponents with a Numerator of 1

So far we have only seen integer exponents. However, expressions can have fraction exponents also, as in $3^{1/2}$. The fraction exponent in $3^{1/2}$ is called a **rational exponent.**

DEFINITION ***Rational exponents:*** An exponent that is a fraction.

To discover what rational exponents mean, consider the following.

$$3^{1/2} \cdot 3^{1/2} = 3^{1/2+1/2} = 3^1 = 3$$ **Use $a^m \cdot a^n = a^{m+n}$, which we learned in Section 5.4.**

Notice that multiplying $\sqrt{3}$ by itself gives the same result:

$$\sqrt{3} \cdot \sqrt{3} = 3$$

These two results suggest that a base raised to the 1/2 power and the square root of that same base are equivalent expressions, or $a^{1/2} = \sqrt{a}$. Note that the denominator, 2, of the rational exponent is the index of the root (remember that square roots have an invisible index of 2). This relationship holds for other roots.

RULE **Rational Exponents with a Numerator of 1**

$a^{1/n} = \sqrt[n]{a}$, where n is a natural number other than 1.

Note: *If a is negative and n is odd, then the root will be negative. If a is negative and n is even, then there is no real-number root.*

EXAMPLE 1 Rewrite using radicals, then simplify, if possible. Assume all variables represent nonnegative values.

a. $36^{1/2}$

Solution $36^{1/2} = \sqrt{36} = 6$

b. $81^{1/4}$

Solution $81^{1/4} = \sqrt[4]{81} = 3$

Calculator TIPS

We use the [^] *key to evaluate a number raised to a rational exponent. For example, to evaluate $81^{1/4}$, we type:*

[8] [1] [^] [(] [1] [÷] [4] [)] [ENTER]

On some scientific calculators, we use the [y^x] *key. For example, to evaluate the preceding example, we type:*

[8] [1] [y^x] [1] [$a^b/_c$] [4] [=]

c. $(-125)^{1/3}$

Solution $(-125)^{1/3} = \sqrt[3]{-125} = -5$

d. $(-64)^{1/4}$

Solution $(-64)^{1/4} = \sqrt[4]{-64}$ **There is no real-number answer.**

Note: *In $(-64)^{1/4}$, the parentheses group the minus sign with the radicand, whereas in $-36^{1/2}$, the minus sign is* ***not*** *part of the radicand.*

e. $-36^{1/2}$

Solution $-36^{1/2} = -\sqrt{36} = -6$

f. $x^{1/5}$

Solution $x^{1/5} = \sqrt[5]{x}$

g. $(64x^6)^{1/2}$

Solution $(64x^6)^{1/2} = \sqrt{64x^6} = 8x^3$

h. $12z^{1/4}$

Solution $12\sqrt[4]{z}$

Note: *Only the z is raised to the one-fourth power.*

i. $\left(\dfrac{a^6}{36}\right)^{1/2}$

Solution $\left(\dfrac{a^6}{36}\right)^{1/2} = \sqrt{\dfrac{a^6}{36}} = \dfrac{a^3}{6}$

YOUR TURN Rewrite each of the following using radical notation and evaluate, when possible. Assume all variables represent nonnegative values.

a. $49^{1/2}$ **b.** $625^{1/4}$ **c.** $(-64)^{1/3}$ **d.** $-81^{1/2}$

e. $y^{1/4}$ **f.** $(25a^8)^{1/2}$ **g.** $18n^{1/5}$ **h.** $\left(\frac{49}{x^{10}}\right)^{1/2}$

Rational Exponents with a Numerator Other Than 1

Let's explore how we can rewrite expressions like $8^{2/3}$ in which the numerator of the rational exponent is a number other than 1. Note that we could write the fraction 2/3 as a product:

$$8^{2/3} = 8^{2(1/3)} \quad \text{or} \quad 8^{2/3} = 8^{(1/3)2}$$

Using the rule of exponents $(n^a)^b = n^{ab}$ in reverse, we can write:

$$8^{2/3} = 8^{2(1/3)} = (8^2)^{1/3} \quad \text{or} \quad 8^{2/3} = 8^{(1/3)2} = (8^{1/3})^2$$

Applying the rule $a^{1/n} = \sqrt[n]{a}$, we can write:

$$8^{2/3} = 8^{2(1/3)} = (8^2)^{1/3} = \sqrt[3]{8^2} \quad \text{or} \quad 8^{2/3} = 8^{(1/3)2} = (8^{1/3})^2 = (\sqrt[3]{8})^2$$

Notice that the denominator of the rational exponent becomes the index of the radical. The numerator of the rational exponent can be written either as the exponent of the radicand or as an exponent for the entire radical.

RULE **General Rule for Rational Exponents**

$a^{m/n} = \sqrt[n]{a^m} = (\sqrt[n]{a})^m$, where $a \geq 0$ and m and n are natural numbers other than 1.

EXAMPLE 2 Rewrite using radicals, then simplify, if possible. Assume all variables represent nonnegative values.

a. $8^{2/3}$

Solution

		or		
$8^{2/3} = \sqrt[3]{8^2}$	Rewrite.		$8^{2/3} = (\sqrt[3]{8})^2$	Rewrite.
$= \sqrt[3]{64}$	Square the radicand.		$= (2)^2$	Evaluate the root.
$= 4$	Evaluate the root.		$= 4$	Evaluate the exponential form.

b. $625^{3/4}$

Solution We could rewrite $625^{3/4}$ as $\sqrt[4]{625^3}$, but calculating 625^3 is tedious without a calculator, so we will use the other form of the rule.

$625^{3/4} = (\sqrt[4]{625})^3$	Rewrite.
$= (5)^3$	Evaluate the root.
$= 125$	Evaluate the exponential form.

ANSWERS

a. $\sqrt{49} = 7$
b. $\sqrt[4]{625} = 5$
c. $\sqrt[3]{-64} = -4$
d. $-\sqrt{81} = -9$
e. $\sqrt[4]{y}$
f. $\sqrt{25a^8} = 5a^4$
g. $18\sqrt[5]{n}$
h. $\sqrt{\frac{49}{x^{10}}} = \frac{7}{x^5}$

Warning: In the expression $-9^{5/2}$, the negative sign tells us to find the opposite of the value of the exponential form.

$$-9^{5/2} = -(9^{5/2}) = -(\sqrt{9})^5$$

Or, some people find it helpful to think of $-9^{5/2}$ as $-1 \cdot 9^{5/2}$:

$$\begin{aligned} -9^{5/2} &= -1 \cdot 9^{5/2} \\ &= -1 \cdot (\sqrt{9})^5 \end{aligned}$$

c. $-9^{5/2}$

Solution $-9^{5/2} = -(\sqrt{9})^5 = -(3)^5 = -243$

d. $\left(\frac{1}{9}\right)^{5/2}$

Solution $\left(\frac{1}{9}\right)^{5/2} = \left(\sqrt{\frac{1}{9}}\right)^5 = \left(\frac{1}{3}\right)^5 = \frac{1}{243}$

e. $x^{2/3}$

Solution $x^{2/3} = \sqrt[3]{x^2}$

Note: *Since the base is a variable, we put the exponent beneath the radical sign.*

f. $(3x - 5)^{3/5}$

Solution $(3x - 5)^{3/5} = \sqrt[5]{(3x - 5)^3}$

Note: *The simplification is usually easier if we write $a^{m/n}$ as $(\sqrt[n]{a})^m$ if a is a constant, and as $\sqrt[n]{a^m}$ if a is a variable expression.*

YOUR TURN Simplify.

a. $32^{3/5}$ **b.** $-49^{3/2}$ **c.** $\left(\frac{1}{16}\right)^{3/2}$ **d.** $y^{5/6}$ **e.** $(3a + 4)^{4/5}$

Negative Rational Exponents

Recall from Section 5.1 that $a^{-b} = \frac{1}{a^b}$. For example, $2^{-3} = \frac{1}{2^3}$. This same rule applies to negative rational exponents.

RULE Negative Rational Exponents

$a^{-m/n} = \frac{1}{a^{m/n}}$, where $a \neq 0$ and m and n are natural numbers with $n \neq 1$.

EXAMPLE 3 Rewrite using radicals, then simplify, if possible.

a. $81^{-1/2}$

Solution $81^{-1/2} = \frac{1}{81^{1/2}}$ Rewrite the exponential form with a positive exponent by inverting the base and changing the sign of the exponent.

$= \frac{1}{\sqrt{81}}$ Write the rational exponent in radical form.

$= \frac{1}{9}$ Evaluate the square root.

ANSWERS

a. 8 **b.** -343 **c.** $\frac{1}{64}$ **d.** $\sqrt[6]{y^5}$ **e.** $\sqrt[5]{(3a + 4)^4}$

b. $16^{-3/4}$

Solution $16^{-3/4} = \frac{1}{16^{3/4}}$ Rewrite the exponential form with a positive exponent by inverting the base and changing the sign of the exponent.

$= \frac{1}{(\sqrt[4]{16})^3}$ Write the rational exponent in radical form.

$= \frac{1}{(2)^3}$ Evaluate the radical.

$= \frac{1}{8}$ Simplify the exponential form.

Connection Though we used $a^{-m/n} = \frac{1}{a^{m/n}}$ in part c, we could have used $\left(\frac{a}{b}\right)^{-n} = \left(\frac{b}{a}\right)^{n}$, which we learned in Chapter 5.

c. $\left(\frac{16}{25}\right)^{-1/2}$

Solution $\left(\frac{16}{25}\right)^{-1/2} = \frac{1}{\left(\frac{16}{25}\right)^{1/2}}$ Rewrite the exponential form with a positive exponent by inverting the base and changing the sign of the exponent.

$= \frac{1}{\sqrt{\frac{16}{25}}}$ Write the rational exponent in radical form.

$= \frac{1}{\frac{4}{5}}$ Evaluate the square root.

$= \frac{5}{4}$ Simplify the complex fraction.

d. $(-64)^{-2/3}$

Solution $(-64)^{-2/3} = \frac{1}{(-64)^{2/3}}$ Rewrite the exponential form with a positive exponent.

$= \frac{1}{(\sqrt[3]{-64})^2}$ Write the rational exponent in radical form.

$= \frac{1}{(-4)^2}$ Evaluate the cube root.

$= \frac{1}{16}$ Square -4.

YOUR TURN Simplify.

a. $49^{-1/2}$ **b.** $-27^{-2/3}$ **c.** $\left(\frac{16}{81}\right)^{-3/4}$ **d.** $(-8)^{-5/3}$

ANSWERS

a. $\frac{1}{7}$ b. $-\frac{1}{9}$

c. $\frac{27}{8}$ d. $-\frac{1}{32}$

Of Interest

John Wallis (1616–1703) was one of the first mathematicians to explain rational and negative exponents. Wallis wrote extensively about physics and mathematics but, unfortunately, his work was overshadowed by the work of his countryman Isaac Newton. Newton added to what Wallis began with exponents and roots and popularized the notation that we use today. (*Source:* D. E. Smith, *History of Mathematics*, Dover, 1953.)

OBJECTIVE 2. Write radicals as expressions raised to rational exponents. In upper-level math courses, it is often necessary to change radical expressions into exponential form. To do so, we use the facts that $\sqrt[n]{a^m} = a^{m/n}$ and $(\sqrt[n]{a})^m = a^{m/n}$.

EXAMPLE 4 Write each of the following in exponential form.

a. $\sqrt[4]{x^3}$

Solution $\sqrt[4]{x^3} = x^{3/4}$

b. $\dfrac{1}{\sqrt[3]{x^2}}$

Solution $\dfrac{1}{\sqrt[3]{x^2}} = \dfrac{1}{x^{2/3}} = x^{-2/3}$

c. $(\sqrt[5]{z})^3$

Solution $(\sqrt[5]{z})^3 = z^{3/5}$

d. $\sqrt[6]{(3x-5)^5}$

Solution $\sqrt[6]{(3x-5)^5} = (3x-5)^{5/6}$

YOUR TURN Write each of the following in exponential form.

a. $\sqrt[5]{y^2}$ b. $\dfrac{1}{\sqrt[4]{x^3}}$ c. $(\sqrt[7]{a})^3$ d. $\sqrt[3]{(3y-2)^5}$

OBJECTIVE 3. Simplify expressions with rational number exponents using the rules of exponents. Rational exponents follow the same rules that we established in Chapter 5 for integer exponents, which we review here.

RULE **Rules of Exponents Summary**

(Assume that no denominators are 0, that *a* and *b* are real numbers, and that *m* and *n* are integers.)

Zero as an exponent:	$a^0 = 1$, where $a \neq 0$
	0^0 is indeterminate.
Negative exponents:	$a^{-n} = \dfrac{1}{a^n}$
	$\dfrac{1}{a^{-n}} = a^n$
	$\left(\dfrac{a}{b}\right)^{-n} = \left(\dfrac{b}{a}\right)^n$
Product rule for exponents:	$a^m \cdot a^n = a^{m+n}$
Quotient rule for exponents:	$a^m \div a^n = a^{m-n}$
Raising a power to a power:	$(a^m)^n = a^{mn}$
Raising a product to a power:	$(ab)^n = a^n b^n$
Raising a quotient to a power:	$\left(\dfrac{a}{b}\right)^n = \dfrac{a^n}{b^n}$

ANSWERS

a. $y^{2/5}$ b. $x^{-3/4}$ c. $a^{3/7}$ d. $(3y-2)^{5/3}$

EXAMPLE 5 Use the rules of exponents to simplify. Write answers with positive exponents.

a. $x^{1/5} \cdot x^{3/5}$

Solution $x^{1/5} \cdot x^{3/5} = x^{1/5+3/5}$ Use $a^m \cdot a^n = a^{m+n}$.

$= x^{4/5}$ Add the exponents.

b. $(2a^{1/2})(4a^{1/3})$

Solution $(2a^{1/2})(4a^{1/3}) = 8a^{1/2+1/3}$ Use $a^m \cdot a^n = a^{m+n}$.

$= 8a^{3/6+2/6}$ Rewrite the exponents with a common denominator of 6.

$= 8a^{5/6}$ Add the exponents.

c. $\dfrac{5^{2/7}}{5^{6/7}}$

Solution $\dfrac{5^{2/7}}{5^{6/7}} = 5^{2/7-6/7}$ Use $\dfrac{a^m}{a^n} = a^{m-n}$.

$= 5^{-4/7}$ Subtract the exponents.

$= \dfrac{1}{5^{4/7}}$ Rewrite with a positive exponent.

d. $(-4y^{-3/5})(5y^{4/5})$

Solution $(-4y^{-3/5})(5y^{4/5}) = -20y^{-3/5+4/5}$ Use $a^m \cdot a^n = a^{m+n}$.

$= -20y^{1/5}$ Add the exponents.

e. $(7w^{3/4})^2$

Solution $(7w^{3/4})^2 = 7^2 \cdot (w^{3/4})^2$ Use $(ab)^n = a^n \cdot b^n$.

$= 49 \cdot w^{(3/4)2}$ Use $(a^m)^n = a^{mn}$.

$= 49w^{3/2}$ Multiply the exponents.

f. $(2a^{2/3}b^{3/5})^3$

Solution $(2a^{2/3}b^{3/5})^3 = 2^3(a^{2/3})^3(b^{3/5})^3$ Use $(ab)^n = a^n \cdot b^n$.

$= 8a^{(2/3)3}\, b^{(3/5)3}$ Use $(a^m)^n = a^{mn}$.

$= 8a^2\, b^{9/5}$ Multiply the exponents.

g. $\dfrac{(3x^{4/3})^3}{x^3}$

Solution $\dfrac{(3x^{4/3})^3}{x^3} = \dfrac{3^3(x^{4/3})^3}{x^3}$ Use $(ab)^n = a^n \cdot b^n$ in the numerator.

$= \dfrac{27x^{(4/3)3}}{x^3}$ Use $(a^m)^n = a^{mn}$.

$= \dfrac{27x^4}{x^3}$ Simplify.

$= 27x^{4-3}$ Use $\dfrac{a^m}{a^n} = a^{m-n}$.

$= 27x$ Simplify.

ANSWERS

a. $-6x^{7/6}$ b. $\frac{4}{y^{3/7}}$ c. $16b^{3/2}$
d. $16x^6y^{8/3}$ e. $8z^3$

YOUR TURN Use the rules of exponents to simplify. Write answers with positive exponents.

a. $(3x^{2/3})(-2x^{1/2})$ **b.** $\frac{12y^{2/7}}{3y^{5/7}}$ **c.** $(4b^{3/4})^2$ **d.** $(2x^{3/2}y^{2/3})^4$ **e.** $\frac{(2z^{5/3})^3}{z^2}$

OBJECTIVE 4. Use rational exponents to simplify radical expressions. Often, radical expressions can be simplified by rewriting them with rational exponents, simplifying, and then rewriting as a radical expression.

EXAMPLE 6 Rewrite as a radical with a smaller root index. Assume all variables represent nonnegative values.

a. $\sqrt[4]{49}$

Solution
$$\begin{aligned}\sqrt[4]{49} &= 49^{1/4} && \text{Rewrite in exponential form.}\\ &= (7^2)^{1/4} && \text{Write 49 as } 7^2.\\ &= 7^{2\cdot 1/4} && \text{Use } (a^m)^n = a^{mn}.\\ &= 7^{1/2} && \text{Simplify.}\\ &= \sqrt{7} && \text{Write in radical form.}\end{aligned}$$

b. $\sqrt[6]{x^4}$

Solution
$$\begin{aligned}\sqrt[6]{x^4} &= x^{4/6} && \text{Rewrite in exponential form.}\\ &= x^{2/3} && \text{Simplify to lowest terms.}\\ &= \sqrt[3]{x^2} && \text{Write in radical form.}\end{aligned}$$

c. $\sqrt[6]{a^4b^2}$

Solution
$$\begin{aligned}\sqrt[6]{a^4b^2} &= (a^4b^2)^{1/6} && \text{Rewrite in exponential form.}\\ &= (a^4)^{1/6}(b^2)^{1/6} && \text{Use } (ab)^n = a^nb^n.\\ &= a^{4\cdot 1/6}b^{2\cdot 1/6} && \text{Use } (a^m)^n = a^{mn}.\\ &= a^{2/3}b^{1/3} && \text{Simplify.}\\ &= (a^2b)^{1/3} && \text{Use } (ab)^n = a^nb^n.\\ &= \sqrt[3]{a^2b} && \text{Write in radical form.}\end{aligned}$$

YOUR TURN Rewrite as a radical with a smaller root index. Assume all variables represent nonnegative values.

a. $\sqrt[4]{36}$ **b.** $\sqrt[8]{x^6}$ **c.** $\sqrt[8]{x^6y^2}$

Writing radicals in exponential form also allows us to multiply and divide radical expressions with different root indices.

ANSWERS

a. $\sqrt{6}$ b. $\sqrt[4]{x^3}$ c. $\sqrt[4]{x^3y}$

EXAMPLE 7 Perform the indicated operations. Write the result using a radical.

a. $\sqrt{x} \cdot \sqrt[3]{x^2}$

Solution $\sqrt{x} \cdot \sqrt[3]{x^2} = x^{1/2} \cdot x^{2/3}$ Write in exponential form.

$= x^{1/2+2/3}$ Use $a^m \cdot a^n = a^{m+n}$.

$= x^{3/6+4/6}$ Rewrite the exponents with their LCD.

$= x^{7/6}$ Simplify.

$= \sqrt[6]{x^7}$ Write in radical form.

Note: *In Section 8.3, we will learn to simplify expressions like* $\sqrt[6]{x^7}$ *further.*

b. $\dfrac{\sqrt[4]{x^3}}{\sqrt[3]{x}}$

Solution $\dfrac{\sqrt[4]{x^3}}{\sqrt[3]{x}} = \dfrac{x^{3/4}}{x^{1/3}}$ Write in exponential form.

$= x^{3/4-1/3}$ Use $\dfrac{a^m}{a^n} = a^{m-n}$.

$= x^{9/12-4/12}$ Rewrite the exponents with their LCD.

$= x^{5/12}$ Simplify.

$= \sqrt[12]{x^5}$ Write in radical form.

c. $\sqrt{3} \cdot \sqrt[3]{2}$

Solution $\sqrt{3} \cdot \sqrt[3]{2} = 3^{1/2} \cdot 2^{1/3}$ Write in exponential form.

$= 3^{3/6} \cdot 2^{2/6}$ Write exponents with their LCD.

$= (3^3 \cdot 2^2)^{1/6}$ Use $(ab)^n = a^n b^n$.

$= (27 \cdot 4)^{1/6}$ Evaluate 3^3 and 2^2.

$= 108^{1/6}$ Multiply.

$= \sqrt[6]{108}$ Rewrite as a radical.

YOUR TURN Perform the indicated operations.

a. $\sqrt[3]{x^2} \cdot \sqrt[4]{x}$ **b.** $\dfrac{\sqrt[4]{a^3}}{\sqrt[3]{a^2}}$ **c.** $\sqrt[4]{2} \cdot \sqrt{5}$

ANSWERS

a. $\sqrt[12]{x^{11}}$ b. $\sqrt[12]{a}$ c. $\sqrt[4]{50}$

By writing radical expressions using rational exponents, we can also find the root of a root.

EXAMPLE 8 Write $\sqrt[3]{\sqrt{x}}$ with a single radical.

Solution $\sqrt[3]{\sqrt{x}} = (x^{1/2})^{1/3}$ Write in exponential form.

$= x^{(1/2)(1/3)}$ Apply $(a^m)^n = a^{mn}$.

$= x^{1/6}$ Simplify.

$= \sqrt[6]{x}$ Write as a radical.

YOUR TURN Write $\sqrt[4]{\sqrt[3]{y}}$ with a single radical.

ANSWER

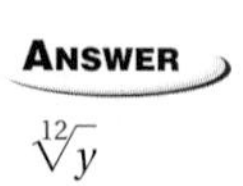

8.2 Exercises

For Extra Help

MyMathLab Videotape/DVT InterAct Math Math Tutor Center Math XL.com

1. If $4^{1/3}$ were written as a radical expression, what would be the radicand?

2. If $4^{3/5}$ were written as a radical expression, what is the root index?

3. If $a^{m/n}$, with n even and m/n reduced, were written as a radical expression, what restrictions would be placed on a? Why?

4. If $a^{m/n}$, with n odd, were written as a radical expression, what restrictions would be placed on a? Why?

5. Does $\sqrt[4]{100} = \sqrt{10}$? Why or why not?

6. Does $\sqrt[3]{3} \cdot \sqrt[5]{2} = \sqrt[15]{1944}$? Why or why not?

For Exercises 7–40, rewrite each of the following using radical notation and evaluate, when possible. Assume all variables represent nonnegative values.

7. $25^{1/2}$

8. $64^{1/2}$

9. $-100^{1/2}$

10. $-64^{1/2}$

11. $27^{1/3}$

12. $125^{1/3}$

13. $(-64)^{1/3}$

14. $(-125)^{1/3}$

15. $y^{1/4}$

16. $w^{1/8}$

17. $(144x^8)^{1/2}$

18. $(9z^{10})^{1/2}$

19. $18r^{1/2}$

20. $22a^{1/2}$

21. $\left(\dfrac{x^4}{121}\right)^{1/2}$

22. $\left(\dfrac{n^8}{36}\right)^{1/2}$

23. $8^{2/3}$

24. $16^{3/4}$

25. $-81^{3/4}$

26. $-16^{5/4}$

27. $(-8)^{4/3}$

28. $(-27)^{5/3}$

29. $16^{-3/2}$

30. $8^{-4/3}$

31. $x^{4/5}$

32. $m^{5/4}$

33. $8n^{2/3}$

34. $6a^{5/6}$

35. $(-32)^{-2/5}$

36. $(-216)^{-2/3}$

37. $\left(\frac{1}{25}\right)^{3/2}$

38. $\left(\frac{1}{32}\right)^{3/5}$

39. $(2a + 4)^{5/6}$

40. $(5r - 2)^{5/7}$

For Exercises 41–56, write each of the following in exponential form.

41. $\sqrt[4]{25}$

42. $\sqrt[8]{33}$

43. $\sqrt[6]{z^5}$

44. $\sqrt[7]{r^5}$

45. $\frac{1}{\sqrt[6]{5^5}}$

46. $\frac{1}{\sqrt[7]{6^2}}$

47. $\frac{5}{\sqrt[5]{x^4}}$

48. $\frac{8}{\sqrt[7]{n^3}}$

49. $(\sqrt[3]{5})^7$

50. $(\sqrt[5]{8})^3$

51. $(\sqrt[7]{x})^2$

52. $(\sqrt[3]{m})^8$

53. $\sqrt[4]{(4a - 5)^7}$

54. $\sqrt[7]{(5w + 3)^5}$

55. $(\sqrt[5]{2r - 5})^8$

56. $(\sqrt[5]{3r - 6})^9$

For Exercises 57–96, use the rules of exponents to simplify. Write answers with positive exponents.

57. $x^{1/5} \cdot x^{3/5}$

58. $n^{1/4} \cdot n^{2/4}$

59. $x^{3/2} \cdot x^{-1/3}$

60. $n^{2/3} \cdot n^{-1/2}$

61. $a^{2/3} \cdot a^{3/4}$

62. $r^{5/2} \cdot r^{5/4}$

63. $(3w^{1/5})(4w^{2/5})$

64. $(5p^{1/9})(3p^{4/9})$

65. $(-3a^{2/3})(4a^{3/4})$

66. $(8c^{4/5})(-4c^{3/2})$

67. $\frac{7^{6/5}}{7^{3/5}}$

68. $\frac{3^{7/9}}{3^{2/9}}$

69. $\frac{x^{2/5}}{x^{4/5}}$

70. $\frac{y^{2/7}}{y^{5/7}}$

71. $\frac{x^{3/4}}{x^{1/2}}$

72. $\frac{x^{5/8}}{x^{1/2}}$

73. $\frac{r^{3/4}}{r^{2/3}}$

74. $\frac{m^{3/5}}{m^{1/2}}$

75. $\frac{x^{-3/7}}{x^{2/7}}$

76. $\frac{v^{-4/5}}{v^{2/5}}$

77. $\frac{a^{3/4}}{a^{-3/2}}$

78. $\frac{b^{4/3}}{b^{-2/3}}$

79. $(5s^{-2/7})(4s^{5/7})$

80. $(6u^{8/9})(-6u^{-5/9})$

81. $(-6b^{-5/4})(4b^{3/2})$

82. $(-6y^{-5/6})(-7y^{5/3})$

83. $(x^{2/3})^3$

84. $(r^{3/4})^4$

85. $(a^{5/6})^2$

86. $(n^{3/8})^4$

87. $(b^{2/3})^{3/5}$

88. $(m^{3/2})^{2/5}$

89. $(2x^{2/3}y^{1/2})^6$

90. $(3a^{1/4}b^{3/2})^4$

91. $(8q^{3/2}t^{3/4})^{1/3}$

92. $(16x^{2/3}y^{1/3})^{3/4}$

93. $\dfrac{(3a^{3/4})^4}{a^2}$

94. $\dfrac{(5v^{5/2})^2}{v^3}$

95. $\dfrac{(9z^{7/3})^{1/2}}{z^{5/6}}$

96. $\dfrac{(36x^{3/4})^{1/2}}{x^{1/8}}$

For Exercises 97–108, represent each of the following as a radical with a smaller root index. Assume all variables represent nonnegative values.

97. $\sqrt[4]{4}$

98. $\sqrt[4]{36}$

99. $\sqrt[6]{25}$

100. $\sqrt[6]{27}$

101. $\sqrt[4]{x^2}$

102. $\sqrt[6]{y^3}$

103. $\sqrt[8]{r^6}$

104. $\sqrt[10]{n^6}$

105. $\sqrt[8]{x^6y^2}$

106. $\sqrt[6]{y^2z^4}$

107. $\sqrt[10]{m^4n^6}$

108. $\sqrt[10]{a^2b^8}$

For Exercises 109–120, perform the indicated operations. Write the result using a radical.

109. $\sqrt[3]{x} \cdot \sqrt{x}$

110. $\sqrt[4]{y} \cdot \sqrt[3]{x^2}$

111. $\sqrt[4]{y^2} \cdot \sqrt[3]{y^2}$

112. $\sqrt[5]{x^4} \cdot \sqrt[3]{x^2}$

113. $\dfrac{\sqrt[3]{x^4}}{\sqrt{x}}$

114. $\dfrac{\sqrt[4]{y^3}}{\sqrt{y}}$

115. $\dfrac{\sqrt[5]{n^4}}{\sqrt[3]{n^2}}$

116. $\dfrac{\sqrt[6]{z^4}}{\sqrt{z}}$

117. $\sqrt{5} \cdot \sqrt[3]{3}$

118. $\sqrt[3]{4} \cdot \sqrt{5}$

119. $\sqrt[4]{6} \cdot \sqrt[3]{2}$

120. $\sqrt[4]{4} \cdot \sqrt[3]{2}$

For Exercises 121–124, write each as a single radical.

121. $\sqrt[3]{\sqrt[3]{x}}$

122. $\sqrt[3]{\sqrt[5]{m}}$

123. $\sqrt{\sqrt[3]{n}}$

124. $\sqrt[4]{\sqrt[5]{z}}$

REVIEW EXERCISES

[1.3] **1.** Rewrite $2 \cdot 2 \cdot 2 \cdot 2 \cdot x \cdot x \cdot x \cdot y \cdot y$ using exponents.

[1.3] ***For Exercises 2 and 3, simplify using the order of operations agreement.***

2. $\sqrt{16} \cdot \sqrt{9}$

3. $\sqrt[3]{27} \cdot \sqrt[3]{125}$

[5.1] ***For Exercises 4 and 5, simplify.***

4. $(2.5 \times 10^6)(3.2 \times 10^5)$

5. $\left(\dfrac{3}{4}x^3y\right)\left(-\dfrac{5}{6}xyz^2\right)$

[5.4] **6.** Use long division to find the quotient: $\dfrac{2x^3 - 2x^2 - 19x + 18}{x + 3}$

8.3 Multiplying, Dividing, and Simplifying Radicals

OBJECTIVES

1. Multiply radical expressions.
2. Divide radical expressions.
3. Use the product rule to simplify radical expressions.

In this section, we explore some ways to simplify expressions that involve multiplication or division of radicals.

OBJECTIVE 1. Multiply radical expressions. Consider the expression $\sqrt{9} \cdot \sqrt{16}$. The usual approach would be to find the roots and then multiply those roots. However, we can also multiply the radicands and then find the root of the product.

Find the roots first:

$$\sqrt{9} \cdot \sqrt{16} = 3 \cdot 4 = 12$$

Multiply the radicands first:

$$\sqrt{9} \cdot \sqrt{16} = \sqrt{9 \cdot 16} = \sqrt{144} = 12$$

Both approaches give the same result, suggesting the following rule:

RULE **Product Rule for Radicals**

If $\sqrt[n]{a}$ and $\sqrt[n]{b}$ are both real numbers, then $\sqrt[n]{a} \cdot \sqrt[n]{b} = \sqrt[n]{a \cdot b}$.

EXAMPLE 1 Find the product and write the answer in simplest form. Assume all variables represent nonnegative values.

a. $\sqrt{3} \cdot \sqrt{27}$

Solution $\sqrt{3} \cdot \sqrt{27} = \sqrt{3 \cdot 27} = \sqrt{81} = 9$

Connection Example 1(a) illustrates that the product of two irrational numbers can be a rational number.

b. $\sqrt{11} \cdot \sqrt{x}$

Solution $\sqrt{11} \cdot \sqrt{x} = \sqrt{11x}$

c. $\sqrt[3]{4} \cdot \sqrt[3]{2}$

Solution $\sqrt[3]{4} \cdot \sqrt[3]{2} = \sqrt[3]{4 \cdot 2} = \sqrt[3]{8} = 2$

d. $\sqrt[3]{3x} \cdot \sqrt[3]{4x}$

Solution $\sqrt[3]{3x} \cdot \sqrt[3]{4x} = \sqrt[3]{3x \cdot 4x} = \sqrt[3]{12x^2}$

e. $\sqrt[4]{5} \cdot \sqrt[4]{7x^2}$

Solution $\sqrt[4]{5} \cdot \sqrt[4]{7x^2} = \sqrt[4]{5 \cdot 7x^2} = \sqrt[4]{35x^2}$

f. $\sqrt{\frac{5}{x}} \cdot \sqrt{\frac{y}{2}}$

Solution $\sqrt{\frac{5}{x}} \cdot \sqrt{\frac{y}{2}} = \sqrt{\frac{5}{x} \cdot \frac{y}{2}} = \sqrt{\frac{5y}{2x}}$

Connection Expressions that have a fraction in a radical, like $\sqrt{\frac{5y}{2x}}$, are not considered to be in simplest form. We will learn how to simplify them in Section 8.5.

g. $\sqrt{x} \cdot \sqrt{x}$

Solution $\sqrt{x} \cdot \sqrt{x} = \sqrt{x \cdot x} = \sqrt{x^2} = x$

YOUR TURN Find the product and write the answer in simplest form.

a. $\sqrt{2} \cdot \sqrt{32}$ **b.** $\sqrt{5} \cdot \sqrt{a}$ **c.** $\sqrt[3]{5x} \cdot \sqrt[3]{2y}$ **d.** $\sqrt{\dfrac{2}{a}} \cdot \sqrt{\dfrac{b}{7}}$

Notice that in Example 1(g), $\sqrt{x} \cdot \sqrt{x} = x$. It is also true that $\sqrt{x} \cdot \sqrt{x} = (\sqrt{x})^2$. Therefore, $(\sqrt{x})^2 = x$. Similarly, $\sqrt[3]{x} \cdot \sqrt[3]{x} \cdot \sqrt[3]{x} = \sqrt[3]{x \cdot x \cdot x} = \sqrt[3]{x^3} = x$. It is also true that $\sqrt[3]{x} \cdot \sqrt[3]{x} \cdot \sqrt[3]{x} = (\sqrt[3]{x})^3$, so $(\sqrt[3]{x})^3 = x$. These examples suggest the following rule.

RULE Raising an nth Root to the nth Power

For any nonnegative real number a, $(\sqrt[n]{a})^n = a$.

This means that $(\sqrt[3]{4})^3 = 4$, $(\sqrt[5]{19})^5 = 19$, and $(\sqrt[4]{3x^2})^4 = 3x^2$.

OBJECTIVE 2. Divide radical expressions. Earlier, we developed the product rule for radicals. Now, we develop a similar rule for quotients like $\dfrac{\sqrt{100}}{\sqrt{25}}$. We can follow the order of operations and divide the roots, or we can divide the radicands, then find the square root of the quotient.

Find the roots first:	Divide the radicands first:
$\dfrac{\sqrt{100}}{\sqrt{25}} = \dfrac{10}{5} = 2$	$\dfrac{\sqrt{100}}{\sqrt{25}} = \sqrt{\dfrac{100}{25}} = \sqrt{4} = 2$

Both approaches give the same result, which suggests the following rule.

RULE Quotient Rule for Radicals

If $\sqrt[n]{a}$ and $\sqrt[n]{b}$ are both real numbers, then $\dfrac{\sqrt[n]{a}}{\sqrt[n]{b}} = \sqrt[n]{\dfrac{a}{b}}$, where $b \neq 0$.

As with all equations, this rule can be used going from left to right or right to left.

EXAMPLE 2 Simplify. Assume variables represent nonnegative values.

a. $\sqrt{\dfrac{7}{36}}$

Solution $\sqrt{\dfrac{7}{36}} = \dfrac{\sqrt{7}}{\sqrt{36}} = \dfrac{\sqrt{7}}{6}$

b. $\dfrac{\sqrt{108}}{\sqrt{3}}$

Solution $\dfrac{\sqrt{108}}{\sqrt{3}} = \sqrt{\dfrac{108}{3}} = \sqrt{36} = 6$

ANSWERS

a. 8 **b.** $\sqrt{5a}$

c. $\sqrt[3]{10xy}$ **d.** $\sqrt{\dfrac{2b}{7a}}$

c. $\sqrt[3]{\dfrac{9}{x^3}}$

Solution $\sqrt[3]{\dfrac{9}{x^3}} = \dfrac{\sqrt[3]{9}}{\sqrt[3]{x^3}} = \dfrac{\sqrt[3]{9}}{x}$

d. $\dfrac{\sqrt[3]{15}}{\sqrt[3]{5}}$

Solution $\dfrac{\sqrt[3]{15}}{\sqrt[3]{5}} = \sqrt[3]{\dfrac{15}{5}} = \sqrt[3]{3}$

e. $\sqrt[4]{\dfrac{y}{81}}$

Solution $\sqrt[4]{\dfrac{y}{81}} = \dfrac{\sqrt[4]{y}}{\sqrt[4]{81}} = \dfrac{\sqrt[4]{y}}{3}$

ANSWERS

a. $\dfrac{\sqrt{x}}{7}$ b. $\sqrt{15}$ c. $\dfrac{\sqrt[4]{6}}{x}$ d. 2

YOUR TURN Simplify. Assume variables represent nonnegative values.

a. $\sqrt{\dfrac{x}{49}}$ **b.** $\dfrac{\sqrt{75}}{\sqrt{5}}$ **c.** $\sqrt[4]{\dfrac{6}{x^4}}$ **d.** $\dfrac{\sqrt[3]{32}}{\sqrt[3]{4}}$

OBJECTIVE 3. Use the product rule to simplify radical expressions. Several conditions exist in which a radical is not considered to be in simplest form. One such condition is if a radicand has a factor that can be written to a power greater than or equal to the index. For example, $\sqrt[3]{81}$ is not in simplest form because the perfect cube 27 is a factor of 81. Our first step in simplifying $\sqrt[3]{81}$ would be to rewrite it as $\sqrt[3]{27 \cdot 3}$ so that we can then use the product rule for radicals.

Note: *We will explore other conditions that require simplification in future sections.*

PROCEDURE Simplifying n^{th} Roots

1. Write the radicand as a product of the greatest possible perfect n^{th} power and a number or expression that has no perfect n^{th} power factors.
2. Use the product rule $\sqrt[n]{ab} = \sqrt[n]{a} \cdot \sqrt[n]{b}$, where a is a perfect n^{th} power.
3. Find the nth root of the perfect n^{th} power radicand.

Connection A list of perfect powers may be helpful.
Perfect squares:
1, 4, 9, 16, 25, 36, 49, 64, 81, 100, ...
Perfect cubes:
1, 8, 27, 64, 125, 216, ...
Perfect fourth powers:
1, 16, 81, 256, 625, ...

EXAMPLE 3 Simplify.

a. $\sqrt{18}$

Solution $\sqrt{18} = \sqrt{9 \cdot 2}$ The greatest perfect square factor of 18 is 9, so we write 18 as $9 \cdot 2$.

$= \sqrt{9} \cdot \sqrt{2}$ Use the product rule of roots to separate the factors into two radicals.

$= 3\sqrt{2}$ Simplify the square root of 9.

b. $5\sqrt{72}$

Solution $5\sqrt{72} = 5 \cdot \sqrt{36 \cdot 2}$ — The greatest perfect square factor of 72 is 36, so we write 72 as $36 \cdot 2$.

$= 5 \cdot \sqrt{36} \cdot \sqrt{2}$ — Use the product rule of roots to separate the factors into two radicals.

$= 5 \cdot 6 \cdot \sqrt{2}$ — Simplify the square root of 36.

$= 30\sqrt{2}$ — Multiply $5 \cdot 6$.

Note: *We can use perfect n^{th} power factors other than the greatest perfect n^{th} power factor. For example, in simplifying $\sqrt{72}$, instead of $\sqrt{72} = \sqrt{36 \cdot 2} = 6\sqrt{2}$, we could write*

$$\sqrt{72} = \sqrt{4 \cdot 18} = 2\sqrt{18} = 2\sqrt{9 \cdot 2} = 2 \cdot 3\sqrt{2} = 6\sqrt{2}$$

Notice that using the greatest perfect n^{th} factor saves steps.

c. $\sqrt[3]{40}$

Solution $\sqrt[3]{40} = \sqrt[3]{8 \cdot 5}$ — The greatest perfect cube factor of 40 is 8.

$= \sqrt[3]{8} \cdot \sqrt[3]{5}$ — Use the product rule of roots.

$= 2\sqrt[3]{5}$ — Simplify the cube root of 8.

d. $4\sqrt[4]{162}$

Solution $4\sqrt[4]{162} = 4\sqrt[4]{81 \cdot 2}$ — The greatest perfect fourth power factor of 162 is 81.

$= 4\sqrt[4]{81} \cdot \sqrt[4]{2}$ — Use the product rule of roots.

$= 4 \cdot 3 \cdot \sqrt[4]{2}$ — Simplify the fourth root of 81.

$= 12\sqrt[4]{2}$

YOUR TURN Simplify.

a. $\sqrt{150}$ **b.** $6\sqrt{80}$ **c.** $\sqrt[3]{108}$ **d.** $3\sqrt[4]{80}$

Using Prime Factorization

If the greatest perfect n^{th} power of a particular radicand is not obvious, then try using the prime factorization of the radicand. Each prime factor that appears twice will have a square root equal to one of the two factors, each prime factor that appears three times will have a cube root that is one of the three factors, and so on. The remaining factors stay in the radical sign.

EXAMPLE 4 Simplify the following radicals using prime factorizations.

a. $\sqrt{375}$

Solution $\sqrt{375} = \sqrt{5 \cdot 5 \cdot 5 \cdot 3}$ — Write 375 as the product of its prime factors.

$= 5\sqrt{3 \cdot 5}$ — The square root of the pair of 5s is 5.

$= 5\sqrt{15}$ — Multiply the prime factors in the radicand.

ANSWERS

a. $5\sqrt{6}$ **b.** $24\sqrt{5}$
c. $3\sqrt[3]{4}$ **d.** $6\sqrt[4]{5}$

b. $\sqrt[3]{324}$

Solution $\sqrt[3]{324} = \sqrt[3]{3 \cdot 3 \cdot 3 \cdot 3 \cdot 2 \cdot 2}$ Write 324 as the product of its prime factors.

$= 3\sqrt[3]{3 \cdot 2 \cdot 2}$ The cube root of the three 3s is 3.

$= 3\sqrt[3]{12}$ Multiply the prime factors in the radicand.

c. $\sqrt[4]{240}$

Solution $\sqrt[4]{240} = \sqrt[4]{2 \cdot 2 \cdot 2 \cdot 2 \cdot 3 \cdot 5}$ Write 240 as the product of its prime factors.

$= 2\sqrt[4]{3 \cdot 5}$ The fourth root of the four 2s is 2.

$= 2\sqrt[4]{15}$ Multiply the prime factors in the radicand.

YOUR TURN Simplify the following radicals using prime factorizations.

a. $\sqrt{294}$ **b.** $\sqrt[3]{324}$ **c.** $\sqrt[4]{486}$

Simplifying Radicals with Variables

We can use either procedure to simplify radicals whose radicands contain variables. Since $\sqrt[n]{a^m} = a^{m/n}$, we can find an exact root if n divides into m evenly. For example, $\sqrt{x^4} = x^{4/2} = x^2$ and $\sqrt[3]{x^{12}} = x^{12/3} = x^4$. Therefore, to find the nth root we rewrite the radicand as a product in which one factor has the greatest possible exponent divisible by the index n.

EXAMPLE 5 Simplify. Assume variables represent nonnegative values.

a. $\sqrt{x^7}$

Solution $\sqrt{x^7} = \sqrt{x^6 \cdot x}$ The greatest number smaller than 7 that is divisible by 2 is 6, so write x^7 as $x^6 \cdot x$.

$= \sqrt{x^6} \cdot \sqrt{x}$ Use the product rule of roots.

$= x^3\sqrt{x}$ Simplify: $\sqrt{x^6} = x^3$.

b. $3\sqrt{24a^5b^9}$

Solution $3\sqrt{24a^5b^9} = 3\sqrt{4 \cdot 6 \cdot a^4 \cdot a \cdot b^8 \cdot b}$ Write 24 as $4 \cdot 6$, a^5 as $a^4 \cdot a$, and b^9 as $b^8 \cdot b$.

$= 3\sqrt{4a^4b^8 \cdot 6ab}$ Regroup the factors so that perfect squares are together.

$= 3\sqrt{4a^4b^8} \cdot \sqrt{6ab}$ Use the product rule of roots.

$= 3 \cdot 2a^2b^4 \cdot \sqrt{6ab}$ Simplify: $\sqrt{4} = 2$, $\sqrt{a^4} = a^2$, and $\sqrt{b^8} = b^4$.

$= 6a^2b^4\sqrt{6ab}$ Multiply: $3 \cdot 2 = 6$.

c. $y^2\sqrt[3]{y^8}$

Solution $y^2\sqrt[3]{y^8} = y^2\sqrt[3]{y^6 \cdot y^2}$ The greatest number smaller than 8 that is divisible by 3 is 6, so write y^8 as $y^6 \cdot y^2$.

$= y^2\sqrt[3]{y^6} \cdot \sqrt[3]{y^2}$ Use the product rule of roots.

$= y^2 \cdot y^2 \cdot \sqrt[3]{y^2}$ Simplify: $\sqrt[3]{y^6} = y^2$.

$= y^4\sqrt[3]{y^2}$ Multiply: $y^2 \cdot y^2 = y^4$.

ANSWERS

a. $7\sqrt{6}$ **b.** $3\sqrt[3]{12}$ **c.** $3\sqrt[4]{6}$

d. $\sqrt[5]{64x^9y^{12}}$

Solution $\sqrt[5]{64x^9y^{12}} = \sqrt[5]{32 \cdot 2 \cdot x^5 \cdot x^4 \cdot y^{10} \cdot y^2}$ — Write 64 as $32 \cdot 2$, x^9 as $x^5 \cdot x^4$, and y^{12} as $y^{10} \cdot y^2$.

$= \sqrt[5]{32x^5y^{10} \cdot 2x^4y^2}$ — Regroup the factors.

$= \sqrt[5]{32x^5y^{10}} \cdot \sqrt[5]{2x^4y^2}$ — Use the product rule of roots.

$= 2xy^2\sqrt[5]{2x^4y^2}$ — Simplify: $\sqrt[5]{32} = 2$, $\sqrt[5]{x^5} = x$, and $\sqrt[5]{y^{10}} = y^2$.

YOUR TURN Simplify. Assume variables represent nonnegative values.

a. $\sqrt{n^{11}}$ **b.** $2\sqrt{45r^7s^3}$ **c.** $\sqrt[4]{m^{13}}$ **d.** $\sqrt[3]{a^8b^{10}}$

After using the product or quotient rules, it is often necessary to simplify the results.

EXAMPLE 6 Find the product or quotient and simplify the results. Assume variables represent nonnegative values.

a. $\sqrt{3} \cdot \sqrt{6}$

Solution $\sqrt{3} \cdot \sqrt{6} = \sqrt{18}$ — Use the product rule of roots to multiply.

$= \sqrt{9 \cdot 2}$ — Write 18 as $9 \cdot 2$.

$= 3\sqrt{2}$ — Simplify: $\sqrt{9} = 3$.

b. $5\sqrt{3x^3} \cdot 3\sqrt{15x^2}$

Solution $5\sqrt{3x^3} \cdot 3\sqrt{15x^2} = 5 \cdot 3\sqrt{3x^3} \cdot \sqrt{15x^2}$ — Regroup the factors.

$= 15\sqrt{45x^5}$ — Multiply.

$= 15\sqrt{9 \cdot 5 \cdot x^4 \cdot x}$ — Write 45 as $9 \cdot 5$ and x^5 as $x^4 \cdot x$.

$= 15 \cdot 3x^2\sqrt{5x}$ — Simplify: $\sqrt{9} = 3$ and $\sqrt{x^4} = x^2$.

$= 45x^2\sqrt{5x}$ — Multiply.

c. $\dfrac{\sqrt{288}}{\sqrt{6}}$

Solution $\dfrac{\sqrt{288}}{\sqrt{6}} = \sqrt{\dfrac{288}{6}}$ — Use the quotient rule of roots.

$= \sqrt{48}$ — Divide the radicand.

$= \sqrt{16 \cdot 3}$ — Write 48 as $16 \cdot 3$.

$= 4\sqrt{3}$ — Simplify: $\sqrt{16} = 3$.

ANSWERS

a. $n^5\sqrt{n}$ **b.** $6r^3s\sqrt{5rs}$
c. $m^3\sqrt[4]{m}$ **d.** $a^2b^3\sqrt[3]{a^2b}$

d. $\dfrac{8\sqrt{756a^8b^5}}{2\sqrt{7a^4b^2}}$

Solution $\dfrac{8\sqrt{756a^8b^5}}{2\sqrt{7a^4b^2}} = 4\sqrt{\dfrac{756a^8b^5}{7a^4b^2}}$ Divide coefficients and use the quotient rule of radicals.

$= 4\sqrt{108a^4b^3}$ Divide the radicand.

$= 4\sqrt{36a^4b^2 \cdot 3b}$ Rewrite the radicand with a perfect square factor.

$= 4 \cdot 6a^2b\sqrt{3b}$ Find the square roots.

$= 24a^2b\sqrt{3b}$ Multiply.

YOUR TURN Find the product or quotient and simplify the results. Assume variables represent nonnegative values.

a. $\sqrt{6} \cdot \sqrt{15}$ **b.** $4\sqrt{14x^3} \cdot 2\sqrt{6x^4}$ **c.** $\dfrac{\sqrt{1296}}{\sqrt{12}}$ **d.** $\dfrac{14\sqrt{315x^{11}y^8}}{2\sqrt{5x^6y^5}}$

ANSWERS

a. $3\sqrt{10}$ **b.** $16x^3\sqrt{21x}$
c. $6\sqrt{3}$ **d.** $21x^2y\sqrt{7xy}$

8.3 Exercises

For Extra Help

MyMathLab

Videotape/DVT

InterAct Math

Math Tutor Center

Math XL.com

1. For $\sqrt{8} \cdot \sqrt{18}$, explain the difference between using the product rule for radicals and multiplying the approximate roots of 8 and 18.

2. Explain why $\sqrt{28}$ is not in simplest form.

3. Describe in your own words how to simplify a cube root containing a radicand with a perfect cube factor.

4. Explain why the expression $3x^2\sqrt[3]{x^5}$ is not in simplest form.

For Exercises 5–32, find the product and write the answer in simplest form. Assume variables represent nonnegative values.

5. $\sqrt{2} \cdot \sqrt{32}$ **6.** $\sqrt{3} \cdot \sqrt{12}$ **7.** $\sqrt{3x} \cdot \sqrt{27x^5}$ **8.** $\sqrt{8y^3} \cdot \sqrt{2y}$

9. $\sqrt{6xy^3} \cdot \sqrt{24xy}$ **10.** $\sqrt{50u^3v^2} \cdot \sqrt{2uv^4}$ **11.** $\sqrt{2} \cdot \sqrt{7}$ **12.** $\sqrt{6} \cdot \sqrt{11}$

13. $\sqrt{15} \cdot \sqrt{x}$

14. $\sqrt{17} \cdot \sqrt{y}$

15. $\sqrt[3]{3} \cdot \sqrt[3]{9}$

16. $\sqrt[3]{4} \cdot \sqrt[3]{16}$

17. $\sqrt[3]{5y} \cdot \sqrt[3]{2y}$

18. $\sqrt[3]{6m} \cdot \sqrt[3]{2m}$

19. $\sqrt[4]{3} \cdot \sqrt[4]{7}$

20. $\sqrt[4]{8} \cdot \sqrt[4]{5}$

21. $\sqrt[4]{12w^3} \cdot \sqrt[4]{6w}$

22. $\sqrt[4]{21r^3} \cdot \sqrt[4]{7r}$

23. $\sqrt[4]{3x^2y} \cdot \sqrt[4]{5xy^2}$

24. $\sqrt[4]{ab^2} \cdot \sqrt[4]{ab}$

25. $\sqrt[5]{6x^3} \cdot \sqrt[5]{5x}$

26. $\sqrt[5]{3m^2} \cdot \sqrt[5]{8m^2}$

27. $\sqrt[6]{4x^2y^3} \cdot \sqrt[6]{2x^3y}$

28. $\sqrt[6]{ab^3} \cdot \sqrt[6]{7a^3b^2}$

29. $\sqrt{\frac{7}{2}} \cdot \sqrt{\frac{3}{5}}$

30. $\sqrt{\frac{5}{2}} \cdot \sqrt{\frac{11}{3}}$

31. $\sqrt{\frac{6}{x}} \cdot \sqrt{\frac{y}{5}}$

32. $\sqrt{\frac{a}{3}} \cdot \sqrt{\frac{7}{b}}$

For Exercises 33–48, use the quotient rule to simplify. Assume variables represent nonnegative values.

33. $\sqrt{\frac{25}{36}}$

34. $\sqrt{\frac{49}{64}}$

35. $\sqrt{\frac{10}{9}}$

36. $\sqrt{\frac{15}{81}}$

37. $\frac{\sqrt{196}}{\sqrt{4}}$

38. $\frac{\sqrt{243}}{\sqrt{3}}$

39. $\frac{\sqrt{15}}{\sqrt{5}}$

40. $\frac{\sqrt{21}}{\sqrt{3}}$

41. $\sqrt[3]{\frac{4}{w^6}}$

42. $\sqrt[3]{\frac{7}{v^3}}$

43. $\sqrt[3]{\frac{5y^2}{27x^9}}$

44. $\sqrt[3]{\frac{5a}{8r^6}}$

45. $\frac{\sqrt[3]{320}}{\sqrt[3]{5}}$

46. $\frac{\sqrt[3]{162}}{\sqrt[3]{6}}$

47. $\sqrt[4]{\frac{3u^3}{16x^8}}$

48. $\sqrt[4]{\frac{3x^2}{81y^4}}$

For Exercises 49–84, simplify. Assume variables represent nonnegative values.

49. $\sqrt{98}$

50. $\sqrt{48}$

51. $\sqrt{128}$

52. $\sqrt{180}$

53. $6\sqrt{80}$

54. $4\sqrt{50}$

55. $5\sqrt{112}$

56. $3\sqrt{245}$

57. $\sqrt{a^3}$

58. $\sqrt{d^5}$

59. $\sqrt{x^2y^4}$

60. $\sqrt{a^6b^2}$

61. $\sqrt{x^6y^8z^{10}}$

62. $\sqrt{p^4q^8r^8}$

63. $rs^2\sqrt{r^9s^5}$

64. $a^2b^3\sqrt{a^{11}b^3}$

65. $3\sqrt{72x^5}$

66. $6\sqrt{75d^3}$

67. $\sqrt[3]{32}$

68. $\sqrt[3]{54}$

69. $\sqrt[3]{x^7}$

70. $\sqrt[3]{b^{11}}$

71. $\sqrt[3]{x^6y^5}$

72. $\sqrt[3]{m^{13}n^9}$

73. $\sqrt[3]{128z^8}$

74. $\sqrt[3]{48h^{14}}$

75. $2\sqrt[3]{24}$

76. $4\sqrt[3]{250}$

77. $\sqrt[4]{80}$

78. $\sqrt[4]{162}$

79. $3x^2\sqrt[4]{243x^9}$

80. $3a^4\sqrt[4]{48a^7}$

81. $\sqrt[5]{486x^{16}}$

82. $\sqrt[5]{160n^{18}}$

83. $\sqrt[6]{x^8y^{14}z^{11}}$

84. $\sqrt[7]{a^{16}b^9c^{12}}$

For Exercises 85–94, find the product and write the answer in simplest form. Assume variables represent nonnegative values.

85. $\sqrt{3} \cdot \sqrt{21}$

86. $\sqrt{5} \cdot \sqrt{15}$

87. $5\sqrt{10} \cdot 3\sqrt{14}$

88. $2\sqrt{6} \cdot 5\sqrt{21}$

89. $\sqrt{y^3} \cdot \sqrt{y^2}$

90. $\sqrt{m^7} \cdot \sqrt{m^4}$

91. $x\sqrt{x^2y^3} \cdot y^2\sqrt{x^4y^4}$

92. $x\sqrt{x^5y^2} \cdot y\sqrt{xy^3}$

93. $4\sqrt{6c^3} \cdot 3\sqrt{10c^5}$

94. $6\sqrt{15c^2} \cdot 2\sqrt{10c^5}$

For Exercises 95 and 96, write an expression in simplest form for the area of the figure.

95.

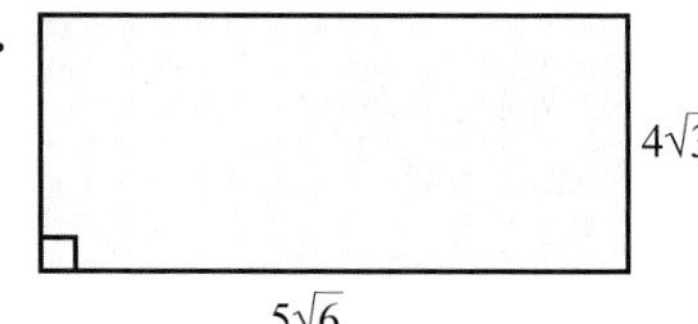

96.

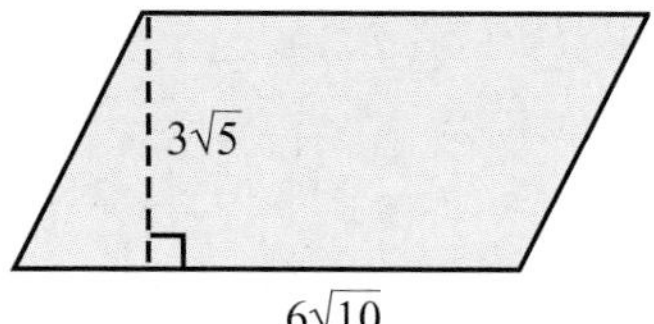

For Exercises 97–108, find the quotient and write the answer in simplest form. Assume variables represent nonnegative values.

97. $\dfrac{\sqrt{48}}{\sqrt{6}}$

98. $\dfrac{\sqrt{54}}{\sqrt{3}}$

99. $\dfrac{9\sqrt{160}}{3\sqrt{8}}$

100. $\dfrac{10\sqrt{280}}{2\sqrt{10}}$

101. $\dfrac{\sqrt{c^5d^4}}{\sqrt{cd^3}}$

102. $\dfrac{\sqrt{m^6n^5}}{\sqrt{m^3n^3}}$

103. $\dfrac{8\sqrt{45a^5}}{2\sqrt{5a}}$

104. $\dfrac{6\sqrt{48n^7}}{3\sqrt{3n^3}}$

105. $\dfrac{12\sqrt{72c^5}}{4\sqrt{6c^2}}$

106. $\dfrac{15\sqrt{48a^7}}{5\sqrt{2a^2}}$

107. $\dfrac{36\sqrt{96x^6y^{11}}}{4\sqrt{3x^2y^4}}$

108. $\dfrac{54\sqrt{240r^9s^{10}}}{9\sqrt{5r^6s^4}}$

For Exercises 109–114, find the products and write the answers in simplest form. Assume variables represent nonnegative values.

109. $\sqrt{\dfrac{3}{7}} \cdot \sqrt{\dfrac{8}{7}}$

110. $\sqrt{\dfrac{8}{5}} \cdot \sqrt{\dfrac{6}{5}}$

111. $\sqrt{\dfrac{a^3}{2}} \cdot \sqrt{\dfrac{a^5}{2}}$

112. $\sqrt{\frac{c^7}{6}} \cdot \sqrt{\frac{c^5}{6}}$

113. $\sqrt{\frac{3x^5}{2}} \cdot \sqrt{\frac{15x^5}{8}}$

114. $\sqrt{\frac{5y^3}{3}} \cdot \sqrt{\frac{10y^3}{27}}$

For Exercises 115 and 116, write an expression in simplest form for the area of the figure.

115.

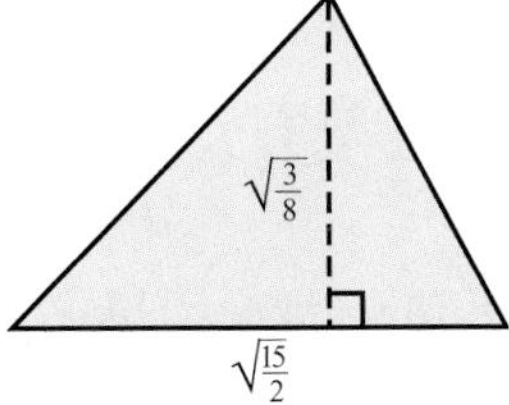

116.

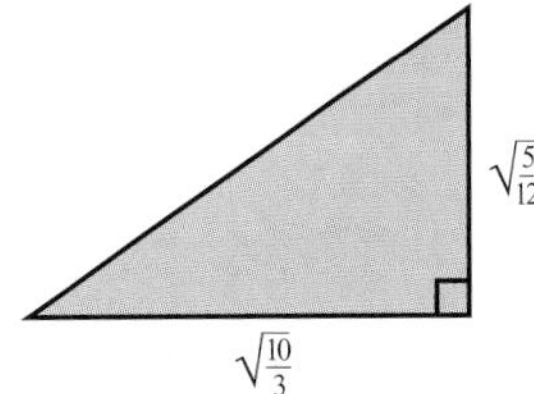

REVIEW EXERCISES

[1.4] ***For Exercises 1 and 2, use a = 6 and b = 8.***

1. Does $\sqrt{a^2 + b^2} = \sqrt{a^2} + \sqrt{b^2} = a + b$?

2. Does $\sqrt{(a + b)^2} = a + b$?

[1.4] **3.** Simplify: $6x^2 - 4x - 3x + 2x^2$

[5.3] ***For Exercises 4–6, multiply.***

4. $(2a - 3b)(4a + 3b)$

5. $(3m + 5n)(3m - 5n)$

6. $(2x - 3y)^2$

8.4 Adding, Subtracting, and Multiplying Radical Expressions

OBJECTIVES

1. **Add or subtract like radicals.**
2. **Use the distributive property in expressions containing radicals.**
3. **Simplify radical expressions that contain mixed operations.**

OBJECTIVE 1. Add or subtract like radicals. Recall that like terms such as $-8a^2$ and $3a^2$ have identical variables and identical exponents. Similarly, **like radicals** have identical radicands and indexes.

DEFINITION ***Like radicals:*** Radical expressions with identical radicands and identical indexes.

The radicals $3\sqrt{2}$ and $5\sqrt{2}$ are like. The radicals $3\sqrt{2}$ and $5\sqrt{3}$ are unlike because their radicands are different. The radicals $3\sqrt{2}$ and $5\sqrt[3]{2}$ are unlike because their indexes are different.

Adding or subtracting like radicals is essentially the same as combining like terms. Remember that the distributive property is at work when we combine like terms because we factor out the common variable, which leaves us a sum or difference of the coefficients.

Like terms:	Like radicals:
$3x + 4x = (3 + 4)x$	$3\sqrt{5} + 4\sqrt{5} = (3 + 4)\sqrt{5}$
$= 7x$	$= 7\sqrt{5}$

Our example suggests the following procedure for adding like radical expressions.

PROCEDURE **Adding Like Radicals**

To add or subtract like radicals, add or subtract the coefficients and keep the radicals the same.

EXAMPLE 1 Simplify.

a. $7\sqrt{5} + 2\sqrt{5}$

Solution $7\sqrt{5} + 2\sqrt{5} = (7 + 2)\sqrt{5}$
$= 9\sqrt{5}$

b. $6\sqrt[3]{4} + \sqrt[3]{4}$

Solution $6\sqrt[3]{4} + \sqrt[3]{4} = (6 + 1)\sqrt[3]{4}$
$= 7\sqrt[3]{4}$

Warning: Never add radicands! Consider $\sqrt{9} + \sqrt{16}$. It might be tempting to add the radicands to get $\sqrt{25}$, but $\sqrt{9} + \sqrt{16} \neq \sqrt{25}$. If we calculate each side separately, we see why they are not equivalent.

$$\sqrt{9} + \sqrt{16} \neq \sqrt{25}$$
$$3 + 4 \neq 5$$
$$7 \neq 5$$

c. $6x\sqrt[4]{3x} - 2x\sqrt[4]{3x}$

Solution $6x\sqrt[4]{3x} - 2x\sqrt[4]{3x} = (6x - 2x)\sqrt[4]{3x}$
$= 4x\sqrt[4]{3x}$

d. $8\sqrt{3} + 6\sqrt{2} - 5\sqrt{3} + 3\sqrt{2}$

Solution $8\sqrt{3} + 6\sqrt{2} - 5\sqrt{3} + 3\sqrt{2} = 8\sqrt{3} - 5\sqrt{3} + 6\sqrt{2} + 3\sqrt{2}$ Regroup the terms.
$= (8 - 5)\sqrt{3} + (6 + 3)\sqrt{2}$
$= 3\sqrt{3} + 9\sqrt{2}$

YOUR TURN Simplify.

a. $5\sqrt{6} + 2\sqrt{6}$

b. $5\sqrt[4]{5x^2} + \sqrt[4]{5x^2}$

c. $7y\sqrt[3]{7} - 12y\sqrt[3]{7}$

d. $9x\sqrt{5x} - 2y\sqrt{3y} - 6x\sqrt{5x} + 7y\sqrt{3y}$

ANSWERS

a. $7\sqrt{6}$
b. $6\sqrt[4]{5x^2}$
c. $-5y\sqrt[3]{7}$
d. $3x\sqrt{5x} + 5y\sqrt{3y}$

In a problem involving addition or subtraction of radicals, if the radicals are not like, it may be possible to simplify one or more of the radicals so that they are like.

EXAMPLE 2 Simplify. Assume variables represent nonnegative quantities.

a. $7\sqrt{3} + \sqrt{12}$

Solution $7\sqrt{3} + \sqrt{12} = 7\sqrt{3} + \sqrt{4 \cdot 3}$ — Factor out the perfect square factor, 4.

$= 7\sqrt{3} + \sqrt{4} \cdot \sqrt{3}$ — Use the product rule to separate the radicals.

$= 7\sqrt{3} + 2\sqrt{3}$ — Simplify.

$= 9\sqrt{3}$ — Combine like radicals.

b. $3\sqrt[3]{24} - 2\sqrt[3]{3}$

Solution $3\sqrt[3]{24} - 2\sqrt[3]{3} = 3\sqrt[3]{8 \cdot 3} - 2\sqrt[3]{3}$ — Rewrite 24 as $8 \cdot 3$.

$= 3\sqrt[3]{8} \cdot \sqrt[3]{3} - 2\sqrt[3]{3}$ — Use the product rule.

$= 3 \cdot 2\sqrt[3]{3} - 2\sqrt[3]{3}$ — Simplify $\sqrt[3]{8}$.

$= 6\sqrt[3]{3} - 2\sqrt[3]{3}$ — Multiply.

$= 4\sqrt[3]{3}$ — Combine like radicals.

c. $\sqrt{48x^3} + \sqrt{12x^3}$

Solution $\sqrt{48x^3} + \sqrt{12x^3} = \sqrt{16x^2 \cdot 3x} + \sqrt{4x^2 \cdot 3x}$ — Rewrite $48x^3$ as $16x^2 \cdot 3x$ and $12x^3$ as $4x^2 \cdot 3x$.

$= \sqrt{16x^2} \cdot \sqrt{3x} + \sqrt{4x^2} \cdot \sqrt{3x}$ — Use the product rule.

$= 4x\sqrt{3x} + 2x\sqrt{3x}$ — Find $\sqrt{16x^2}$ and $\sqrt{4x^2}$.

$= 6x\sqrt{3x}$ — Combine like radicals.

d. $3\sqrt[4]{32x^5} + \sqrt[4]{162x^9}$

Solution $3\sqrt[4]{32x^5} + \sqrt[4]{162x^9} = 3\sqrt[4]{16 \cdot 2 \cdot x^4 \cdot x} + \sqrt[4]{81 \cdot 2 \cdot x^8 \cdot x}$ — Write 32 as $16 \cdot 2$, x^5 as $x^4 \cdot x$, 162 as $81 \cdot 2$ and x^9 as $x^8 \cdot x$.

$= 3 \cdot 2x\sqrt[4]{2x} + 3x^2\sqrt[4]{2x}$ — Find the roots.

$= 6x\sqrt[4]{2x} + 3x^2\sqrt[4]{2x}$ — Multiply.

$= (6x + 3x^2)\sqrt[4]{2x}$ — Combine like radicals.

Note: *Although the radicals are like, we cannot add the coefficients because they are not like.*

YOUR TURN Simplify. Assume variables represent nonnegative quantities.

a. $4\sqrt{24} - 6\sqrt{54}$ **b.** $4\sqrt{50x^5} - 2\sqrt{18x^5}$ **c.** $6a\sqrt[3]{54a^4} - 2\sqrt[3]{128a^7}$

ANSWERS

a. $-10\sqrt{6}$
b. $14x^2\sqrt{2x}$
c. $10a^2\sqrt[3]{2a}$

OBJECTIVE 2. Use the distributive property in expressions containing radicals. Products involving sums and differences of radicals are found in much the same way as products of polynomials. We will use the distributive property, multiply binomials using FOIL, and square a binomial (all from Section 5.3).

EXAMPLE 3 Find the product. Assume variables represent nonnegative values.

a. $\sqrt{3}(\sqrt{3} + \sqrt{15})$

Solution $\sqrt{3}(\sqrt{3} + \sqrt{15}) = \sqrt{3} \cdot \sqrt{3} + \sqrt{3} \cdot \sqrt{15}$ Use the distributive property.

$= \sqrt{3 \cdot 3} + \sqrt{3 \cdot 15}$ Use the product rule.

$= \sqrt{9} + \sqrt{45}$ Multiply.

$= 3 + 3\sqrt{5}$ Find $\sqrt{9}$ and simplify $\sqrt{45}$.

b. $2\sqrt{6}(3 + 5\sqrt{5})$

Solution $2\sqrt{6}(3 + 5\sqrt{5}) = 2\sqrt{6} \cdot 3 + 2\sqrt{6} \cdot 5\sqrt{5}$ Use the distributive property.

$= 2 \cdot 3\sqrt{6} + 2 \cdot 5\sqrt{6 \cdot 5}$ Use the product rule.

$= 6\sqrt{6} + 10\sqrt{30}$ Find the products.

c. $(2 + \sqrt{3})(\sqrt{5} - \sqrt{6})$

Solution $(2 + \sqrt{3})(\sqrt{5} - \sqrt{6}) = 2\sqrt{5} - 2\sqrt{6} + \sqrt{3} \cdot \sqrt{5} - \sqrt{3} \cdot \sqrt{6}$ Use FOIL.

$= 2\sqrt{5} - 2\sqrt{6} + \sqrt{15} - \sqrt{18}$ Use the product rule.

$= 2\sqrt{5} - 2\sqrt{6} + \sqrt{15} - 3\sqrt{2}$ Simplify $\sqrt{18}$.

d. $(3\sqrt{x} + \sqrt{y})(2\sqrt{x} - 5\sqrt{y})$

Solution $(3\sqrt{x} + \sqrt{y})(2\sqrt{x} - 5\sqrt{y}) =$

$3\sqrt{x} \cdot 2\sqrt{x} - 3\sqrt{x} \cdot 5\sqrt{y} + \sqrt{y} \cdot 2\sqrt{x} - \sqrt{y} \cdot 5\sqrt{y}$ Use FOIL.

$= 6x - 15\sqrt{xy} + 2\sqrt{xy} - 5y$ Use the product rule.

$= 6x - 13\sqrt{xy} - 5y$ Combine like radicals.

e. $(5 + \sqrt{3})^2$

Solution $(5 + \sqrt{3})^2 = 5^2 + 2 \cdot 5\sqrt{3} + (\sqrt{3})^2$ Use $(a + b)^2 = a^2 + 2ab + b^2$.

$= 25 + 10\sqrt{3} + 3$ Simplify.

$= 28 + 10\sqrt{3}$ Add 25 and 3.

YOUR TURN Find the product. Assume variables represent nonnegative quantities.

a. $2\sqrt{11}(3 - 3\sqrt{6})$ **b.** $(2\sqrt{a} + 3\sqrt{b})(\sqrt{a} - 3\sqrt{b})$ **c.** $(3 + \sqrt{5})^2$

Radicals in Conjugates

Radical expressions can be conjugates. Like binomial conjugates, conjugates involving radicals differ only in the sign separating the terms. For example, $7 - \sqrt{3}$ and $7 + \sqrt{3}$ are conjugates. Let's explore what happens when we multiply conjugates containing radicals.

ANSWERS

a. $6\sqrt{11} - 6\sqrt{66}$
b. $2a - 3\sqrt{ab} - 9b$
c. $14 + 6\sqrt{5}$

EXAMPLE 4 Find the product.

a. $(2 + \sqrt{3})(2 - \sqrt{3})$

Solution $(2 + \sqrt{3})(2 - \sqrt{3}) = 2^2 - (\sqrt{3})^2$ Use $(a + b)(a - b) = a^2 - b^2$ (from Section 5.5).

$= 4 - 3$ Simplify.

$= 1$

b. $(\sqrt{5} - 3\sqrt{3})(\sqrt{5} + 3\sqrt{3})$

Solution $(\sqrt{5} - 3\sqrt{3})(\sqrt{5} + 3\sqrt{3}) = (\sqrt{5})^2 - (3\sqrt{3})^2$ Use $(a + b)(a - b) = a^2 - b^2$.

$= 5 - 9 \cdot 3$ Simplify.

$= 5 - 27$

$= -22$

Note: *The product of conjugates* always *results in a rational number. This will be useful in Section 8.5 when rationalizing denominators.*

YOUR TURN Find the product.

a. $(4 + \sqrt{10})(4 - \sqrt{10})$

b. $(3\sqrt{5} + 2\sqrt{6})(3\sqrt{5} - 2\sqrt{6})$

OBJECTIVE 3. Simplify radical expressions that contain mixed operations. Now let's use the order of operations to simplify radical expressions that have more than one operation.

EXAMPLE 5 Simplify.

a. $\sqrt{2} \cdot \sqrt{10} + \sqrt{3} \cdot \sqrt{15}$

Solution $\sqrt{2} \cdot \sqrt{10} + \sqrt{3} \cdot \sqrt{15} = \sqrt{2 \cdot 10} + \sqrt{3 \cdot 15}$ Use the product rule.

$= \sqrt{20} + \sqrt{45}$ Multiply.

$= \sqrt{4 \cdot 5} + \sqrt{9 \cdot 5}$ Rewrite 20 as $4 \cdot 5$ and 45 as $9 \cdot 5$.

$= \sqrt{4} \cdot \sqrt{5} + \sqrt{9} \cdot \sqrt{5}$ Use the product rule.

$= 2\sqrt{5} + 3\sqrt{5}$ Find $\sqrt{9}$ and $\sqrt{4}$.

$= 5\sqrt{5}$ Combine like radicals.

b. $\dfrac{\sqrt{54}}{\sqrt{3}} + \sqrt{32}$

Solution $\dfrac{\sqrt{54}}{\sqrt{3}} + \sqrt{32} = \sqrt{\dfrac{54}{3}} + \sqrt{16 \cdot 2}$ Use the quotient rule and rewrite 32 as $16 \cdot 2$.

$= \sqrt{18} + \sqrt{16} \cdot \sqrt{2}$ Divide and use the product rule.

$= \sqrt{9 \cdot 2} + 4\sqrt{2}$ Rewrite 18 as $9 \cdot 2$ and find $\sqrt{16}$.

$= \sqrt{9} \cdot \sqrt{2} + 4\sqrt{2}$ Use the product rule.

$= 3\sqrt{2} + 4\sqrt{2}$ Find $\sqrt{9}$.

$= 7\sqrt{2}$ Combine like radicals.

ANSWERS

a. 6 **b.** 21

Calculator TIPS

We can use a calculator to simplify radical expressions. For Example 5(b), we type:

2nd x^2 5 4) ÷ 2nd x^2 3) + 2nd x^2 3 2) ENTER

The answer, rounded to the nearest thousandth, is 9.899.

YOUR TURN Simplify.

a. $2\sqrt{3} \cdot \sqrt{6} + 4\sqrt{7} \cdot \sqrt{14}$

b. $\sqrt{63} + \dfrac{\sqrt{140}}{\sqrt{5}}$

ANSWERS

a. $34\sqrt{2}$ **b.** $5\sqrt{7}$

8.4 Exercises

For Extra Help MyMathLab Videotape/DVT InterAct Math Math Tutor Center Math XL.com

1. What must be identical in like radicals? What can be different?

2. Add $3x + 2x$ and then add $3\sqrt{2} + 2\sqrt{2}$. Discuss the similarities in the process.

3. Multiply $(x + 3)(x + 2)$ and then multiply $(\sqrt{5} + 3)(\sqrt{5} + 2)$. Discuss the similarities in the process.

4. Why is $\sqrt{a} + \sqrt{b} \neq \sqrt{a + b}$? Use examples if necessary.

For Exercises 5–18, simplify. Assume variables represent nonnegative values.

5. $9\sqrt{6} - 15\sqrt{6}$

6. $2\sqrt{10} - 13\sqrt{10}$

7. $7\sqrt{a} + 2\sqrt{a}$

8. $5\sqrt{y} + 7\sqrt{y}$

9. $4\sqrt{5} - 2\sqrt{6} + 8\sqrt{5} - 6\sqrt{6}$

10. $-6\sqrt{2} + 5\sqrt{7} + 3\sqrt{2} - 2\sqrt{7}$

11. $3a\sqrt{5a} - 4b\sqrt{7b} + 8a\sqrt{5a} + 2b\sqrt{7b}$

12. $12n\sqrt{2n} - 14m\sqrt{5m} - 8n\sqrt{2n} + 18m\sqrt{5m}$

13. $6x\sqrt[3]{9} - 3x\sqrt[3]{9}$

14. $4y\sqrt[3]{3} - y\sqrt[3]{3}$

15. $6x^2\sqrt[4]{5x} + 12x^2\sqrt[4]{5x}$

16. $3y^3\sqrt[4]{8y} - 9y^3\sqrt[4]{8y}$

17. $3x\sqrt{5x} + 4x\sqrt[3]{5x}$

18. $4z\sqrt[4]{2z} - 7z\sqrt{2z}$

For Exercises 19–38, simplify the radicals and then find the sum or difference. Assume variables represent nonnegative values.

19. $\sqrt{48} - \sqrt{75}$

20. $\sqrt{80} - \sqrt{20}$

21. $\sqrt{80y} - \sqrt{125y}$

22. $\sqrt{27x} + \sqrt{75x}$

23. $\sqrt{80} - 4\sqrt{45}$

24. $\sqrt{20} - 2\sqrt{180}$

25. $3\sqrt{96} - 2\sqrt{54}$

26. $5\sqrt{63} + 2\sqrt{28}$

27. $6\sqrt{48a^3} - 2\sqrt{75a^3}$

28. $4\sqrt{98y^5} - 7\sqrt{128y^5}$

29. $\sqrt{150} - \sqrt{54} + \sqrt{24}$

30. $\sqrt{20} + \sqrt{125} - \sqrt{80}$

31. $2\sqrt{8} - 3\sqrt{48} + 2\sqrt{98} - \sqrt{75}$

32. $3\sqrt{216} - \sqrt{147} - 4\sqrt{96} - \sqrt{108}$

33. $\sqrt[3]{16} + \sqrt[3]{54}$

34. $\sqrt[3]{24} + \sqrt[3]{81}$

35. $4\sqrt[3]{135x^5} - 6x\sqrt[3]{320x^2}$

36. $3a^2\sqrt[3]{500a^4} + 6a\sqrt[3]{108a^7}$

37. $-4\sqrt[4]{x^9} + 2\sqrt[4]{16x^7}$

38. $6y\sqrt[4]{81y^5} - 2y\sqrt[4]{16y^6}$

For Exercises 39–46 use the distributive property. Assume variables represent nonnegative values.

39. $\sqrt{2}(3 + \sqrt{2})$

40. $\sqrt{5}(4 - \sqrt{5})$

41. $\sqrt{3}(\sqrt{3} - \sqrt{15})$

42. $\sqrt{6}(\sqrt{6} + \sqrt{2})$

43. $\sqrt{5}(\sqrt{3} + 2\sqrt{15})$

44. $\sqrt{7}(\sqrt{5} - 3\sqrt{14})$

45. $4\sqrt{3x}(2\sqrt{3x} - 4\sqrt{6x})$

46. $6\sqrt{2y}(3\sqrt{2y} + 2\sqrt{10y})$

For Exercises 47–66, multiply (use FOIL). Assume variables represent nonnegative values.

47. $(3+\sqrt{5})(4-\sqrt{2})$

48. $(3+\sqrt{7})(7-\sqrt{3})$

49. $(3+\sqrt{x})(2+\sqrt{x})$

50. $(5-\sqrt{a})(2+\sqrt{a})$

51. $(2+3\sqrt{3})(3+5\sqrt{2})$

52. $(7-3\sqrt{5})(2-2\sqrt{10})$

53. $(\sqrt{2}+\sqrt{3})(\sqrt{3}+\sqrt{5})$

54. $(\sqrt{5}+\sqrt{2})(\sqrt{2}+\sqrt{7})$

55. $(\sqrt{x}+\sqrt{y})(\sqrt{x}-2\sqrt{y})$

56. $(2\sqrt{a}+\sqrt{b})(\sqrt{a}+\sqrt{b})$

57. $(4\sqrt{2}+2\sqrt{5})(3\sqrt{7}-3\sqrt{3})$

58. $(8\sqrt{2}-2\sqrt{3})(2\sqrt{5}+3\sqrt{10})$

59. $(2\sqrt{a}+3\sqrt{b})(4\sqrt{a}-\sqrt{b})$

60. $(\sqrt{m}-4\sqrt{n})(2\sqrt{m}-3\sqrt{n})$

61. $(\sqrt[3]{4}+5)(\sqrt[3]{4}-8)$

62. $(\sqrt[3]{9}+5)(\sqrt[3]{9}-2)$

63. $(\sqrt[3]{9}+\sqrt[3]{4})(\sqrt[3]{3}-\sqrt[3]{2})$

64. $(\sqrt[3]{5}+\sqrt[3]{9})(\sqrt[3]{25}-\sqrt[3]{3})$

65. $(\sqrt[3]{x}+2)(\sqrt[3]{x^2}-2\sqrt[3]{x}+4)$

66. $(\sqrt[3]{r}-3)(\sqrt[3]{r^2}+3\sqrt[3]{r}+9)$

For Exercises 67–74, find the product.

67. $(4+\sqrt{6})^2$

68. $(6+\sqrt{3})^2$

69. $(1-\sqrt{2})^2$

70. $(5-\sqrt{2})^2$

71. $(2+2\sqrt{3})^2$

72. $(3+2\sqrt{5})^2$

73. $(2\sqrt{3}+3\sqrt{2})^2$

74. $(4\sqrt{2}-5\sqrt{6})^2$

For Exercises 75–88, multiply the conjugates. Assume variables represent nonnegative values.

75. $(2+\sqrt{3})(2-\sqrt{3})$

76. $(3+\sqrt{5})(3-\sqrt{5})$

77. $(\sqrt{2}+4)(\sqrt{2}-4)$

78. $(\sqrt{7}-6)(\sqrt{7}+6)$

79. $(6+\sqrt{x})(6-\sqrt{x})$

80. $(5+\sqrt{y})(5-\sqrt{y})$

81. $(\sqrt{3}+\sqrt{2})(\sqrt{3}-\sqrt{2})$

82. $(\sqrt{5}-\sqrt{3})(\sqrt{5}+\sqrt{3})$

83. $(\sqrt{x}+\sqrt{y})(\sqrt{x}-\sqrt{y})$

84. $(\sqrt{a} - \sqrt{b})(\sqrt{a} + \sqrt{b})$

85. $(4 + 2\sqrt{3})(4 - 2\sqrt{3})$

86. $(5 + 3\sqrt{3})(5 - 3\sqrt{3})$

87. $(3\sqrt{7} + \sqrt{13})(3\sqrt{7} - \sqrt{13})$

88. $(4\sqrt{5} - 3\sqrt{2})(4\sqrt{5} + 3\sqrt{2})$

For Exercises 89–96, simplify.

89. $\sqrt{3} \cdot \sqrt{15} + \sqrt{8} \cdot \sqrt{10}$

90. $\sqrt{6} \cdot \sqrt{8} + \sqrt{5} \cdot \sqrt{15}$

91. $3\sqrt{3} \cdot \sqrt{18} - 4\sqrt{18} \cdot \sqrt{12}$

92. $3\sqrt{2} \cdot 2\sqrt{40} - 5\sqrt{12} \cdot \sqrt{15}$

93. $\dfrac{\sqrt{40}}{\sqrt{5}} + \sqrt{50}$

94. $\dfrac{\sqrt{60}}{\sqrt{5}} + \sqrt{48}$

95. $\dfrac{\sqrt{540}}{\sqrt{3}} - 4\sqrt{125}$

96. $\dfrac{\sqrt{288}}{\sqrt{6}} - 6\sqrt{108}$

For Exercises 97 and 98, find the perimeter of the shape.

97.

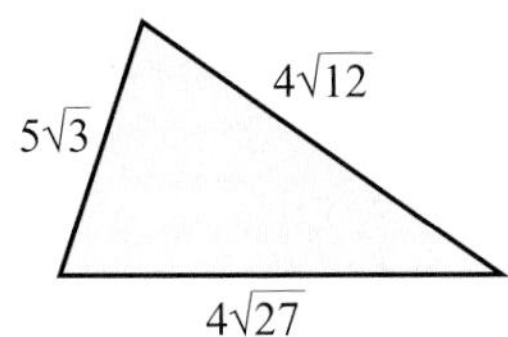

98.

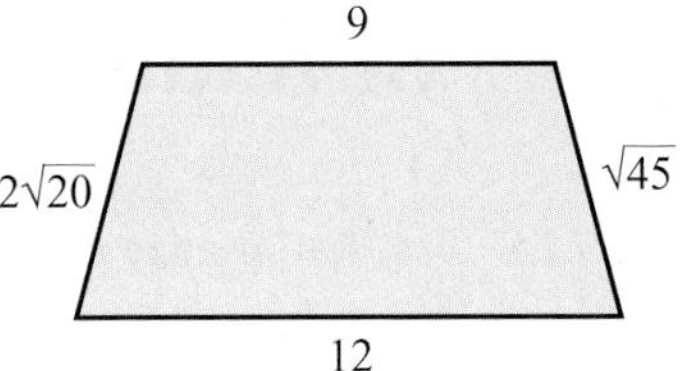

99. Crown molding, which is placed at the top of a wall, is to be installed around the perimeter of the room shown.

a. Write an expression in simplest form for the perimeter of the room.

b. Use a calculator to approximate the perimeter, rounded to the nearest tenth.

c. If crown molding costs \$1.89 per foot length, how much will the crown molding cost for this room?

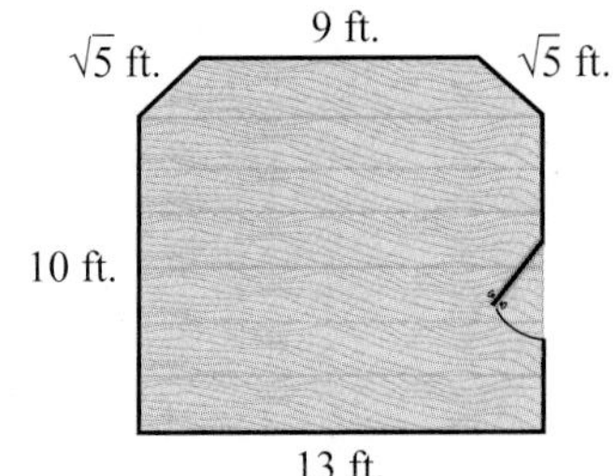

100. A tabletop is to be fitted with veneer strips along the sides.

a. Write an expression in simplest form for the perimeter of the tabletop.

b. Use a calculator to approximate the perimeter, rounded to the nearest tenth.

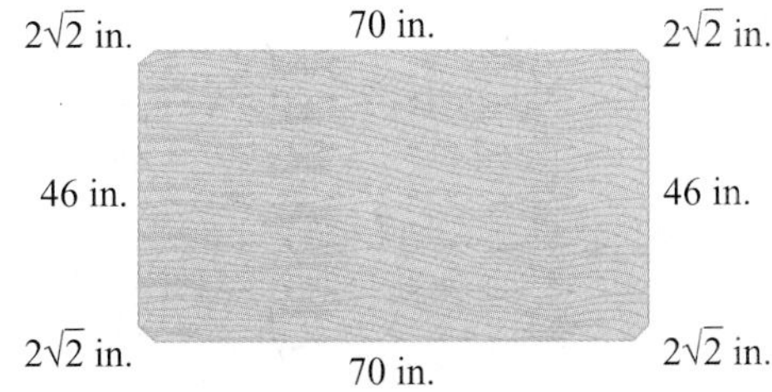

REVIEW EXERCISES

[3.2] **1.** For the equation $5y - 2x = 10$, find the slope and the y-intercept.

[3.1] **2.** Graph: $y = -\frac{1}{3}x + 4$

[5.3] **3.** What is the conjugate of $2x - 5$?

[5.3] **4.** Multiply: $(4x + 3)(4x - 3)$

[8.3] **5.** What factor could multiply $\sqrt{8}$ in order to equal $\sqrt{16}$?

[8.3] **6.** What factor could multiply $\sqrt[3]{2}$ in order to equal $\sqrt[3]{8}$?

8.5 Rationalizing Numerators and Denominators of Radical Expressions

OBJECTIVES

1. **Rationalize denominators.**
2. **Rationalize denominators that have a sum or difference with a square root term.**
3. **Rationalize numerators.**

We are now ready to formalize the conditions for a radical expression that is in simplest form. A radical expression is in simplest form if

1. All rational roots have been found.
2. There are no perfect n^{th} factors of radicands of the form $\sqrt[n]{a^n}$.
3. All possible products, quotients, sums, and differences have been found.
4. There are no radicals in the denominator of a fraction.

In this section, we explore how to simplify expressions that have a radical in the denominator of a fraction, as in $\frac{1}{\sqrt{2}}$.

OBJECTIVE 1. Rationalize denominators. If the denominator of a fraction contains a radical, our goal will be to *rationalize* the denominator, which means to rewrite the expression so that it has a rational number in the denominator. In general, we multiply the fraction by a well-chosen 1 so that the radical is eliminated. We determine that 1 by finding a factor that multiplies the nth root in the denominator so that its radicand is a perfect n^{th} power.

Square Root Denominators

In the case of a square root in the denominator, we multiply it by a factor that makes the radicand a perfect square, which allows us to eliminate the square root. For example, to rationalize $\frac{1}{\sqrt{2}}$, we could multiply by $\frac{\sqrt{2}}{\sqrt{2}}$ because the product's denominator is the square root of a perfect square.

$$\frac{1}{\sqrt{2}} = \frac{1}{\sqrt{2}} \cdot \frac{\sqrt{2}}{\sqrt{2}} = \frac{\sqrt{2}}{\sqrt{4}} = \frac{\sqrt{2}}{2}$$

Note: *We are not changing the value of* $\frac{1}{\sqrt{2}}$ *because we are multiplying it by 1 in the form of* $\frac{\sqrt{2}}{\sqrt{2}}$.

Any factor that produces a perfect square radicand will work. For example, we could have multiplied $\frac{1}{\sqrt{2}}$ by $\frac{\sqrt{8}}{\sqrt{8}}$.

$$\frac{1}{\sqrt{2}} = \frac{1}{\sqrt{2}} \cdot \frac{\sqrt{8}}{\sqrt{8}} = \frac{\sqrt{8}}{\sqrt{16}} = \frac{\sqrt{4 \cdot 2}}{4} = \frac{2\sqrt{2}}{4} = \frac{\sqrt{2}}{2}$$

Notice, however, that multiplying by $\frac{\sqrt{2}}{\sqrt{2}}$ required fewer steps to simplify than multiplying by $\frac{\sqrt{8}}{\sqrt{8}}$.

EXAMPLE 1 Rationalize the denominator. Assume variables represent non-negative values.

a. $\frac{2}{\sqrt{5}}$

Solution $\frac{2}{\sqrt{5}} = \frac{2}{\sqrt{5}} \cdot \frac{\sqrt{5}}{\sqrt{5}}$ Multiply by $\frac{\sqrt{5}}{\sqrt{5}}$.

$= \frac{2\sqrt{5}}{\sqrt{25}}$ Simplify.

$= \frac{2\sqrt{5}}{5}$

b. $\sqrt{\frac{5}{8}}$

Solution $\sqrt{\frac{5}{8}} = \frac{\sqrt{5}}{\sqrt{8}}$ Use the quotient rule of square roots to separate the numerator and denominator into two radicals.

$= \frac{\sqrt{5}}{\sqrt{8}} \cdot \frac{\sqrt{2}}{\sqrt{2}}$ Multiply by $\frac{\sqrt{2}}{\sqrt{2}}$.

$= \frac{\sqrt{10}}{\sqrt{16}}$ Simplify.

$= \frac{\sqrt{10}}{4}$

Note: *We chose to multiply by* $\frac{\sqrt{2}}{\sqrt{2}}$ *because it leads to a smaller perfect square than other choices, like* $\frac{\sqrt{8}}{\sqrt{8}}$. *Multiplying by* $\frac{\sqrt{8}}{\sqrt{8}}$ *produces the same final answer, but requires more steps.*

Warning: Though it may be tempting, we cannot divide out the 4 and 10 because 10 is a radicand, whereas 4 is not. We *never* divide out factors common to a radicand and a number not under a radical.

c. $\frac{3}{\sqrt{2x}}$

Solution $\frac{3}{\sqrt{2x}} = \frac{3}{\sqrt{2x}} \cdot \frac{\sqrt{2x}}{\sqrt{2x}}$ Multiply by $\frac{\sqrt{2x}}{\sqrt{2x}}$.

$= \frac{3\sqrt{2x}}{\sqrt{4x^2}}$ Simplify.

$= \frac{3\sqrt{2x}}{2x}$

YOUR TURN Rationalize the denominator. Assume variables represent nonnegative values.

a. $\frac{1}{\sqrt{7}}$ **b.** $\sqrt{\frac{7}{12}}$ **c.** $\frac{3}{\sqrt{10x}}$

ANSWERS

a. $\frac{\sqrt{7}}{7}$ **b.** $\frac{\sqrt{21}}{6}$ **c.** $\frac{3\sqrt{10x}}{10x}$

n^{th}-Root Denominators

If the denominator contains a higher-order root, such as a cube root, then we multiply appropriately to get a perfect cube radicand in the denominator so that we can eliminate the radical. For example, $\frac{2}{\sqrt[3]{5}} = \frac{2}{\sqrt[3]{5}} \cdot \frac{\sqrt[3]{25}}{\sqrt[3]{25}} = \frac{2\sqrt[3]{25}}{\sqrt[3]{125}} = \frac{2\sqrt[3]{25}}{5}$. We summarize as follows.

PROCEDURE Rationalizing Denominators

To rationalize a denominator containing a single n^{th} root, multiply the fraction by a 1 so that the product's denominator has a radicand that is a perfect n^{th} power.

EXAMPLE 2 Rationalize the denominator. Assume variables represent nonnegative values.

a. $\dfrac{3}{\sqrt[3]{2}}$

Solution $\dfrac{3}{\sqrt[3]{2}} = \dfrac{3}{\sqrt[3]{2}} \cdot \dfrac{\sqrt[3]{4}}{\sqrt[3]{4}}$ Since $\sqrt[3]{2} \cdot \sqrt[3]{4} = \sqrt[3]{8} = 2$, we multiply the fraction by $\dfrac{\sqrt[3]{4}}{\sqrt[3]{4}}$.

$= \dfrac{3\sqrt[3]{4}}{\sqrt[3]{8}}$ Simplify.

$= \dfrac{3\sqrt[3]{4}}{2}$

b. $\dfrac{\sqrt[3]{a}}{\sqrt[3]{b}}$

Solution $\dfrac{\sqrt[3]{a}}{\sqrt[3]{b}} = \dfrac{\sqrt[3]{a}}{\sqrt[3]{b}} \cdot \dfrac{\sqrt[3]{b^2}}{\sqrt[3]{b^2}}$ Since $\sqrt[3]{b} \cdot \sqrt[3]{b^2} = \sqrt[3]{b^3} = b$, we multiply the fraction by $\dfrac{\sqrt[3]{b^2}}{\sqrt[3]{b^2}}$.

$= \dfrac{\sqrt[3]{ab^2}}{\sqrt[3]{b^3}}$ Simplify.

$= \dfrac{\sqrt[3]{ab^2}}{b}$

c. $\sqrt[3]{\dfrac{5}{9a^2}}$

Solution $\sqrt[3]{\dfrac{5}{9a^2}} = \dfrac{\sqrt[3]{5}}{\sqrt[3]{9a^2}}$ Use the quotient rule to separate the numerator and denominator.

$= \dfrac{\sqrt[3]{5}}{\sqrt[3]{9a^2}} \cdot \dfrac{\sqrt[3]{3a}}{\sqrt[3]{3a}}$ Since $\sqrt[3]{9a^2} \cdot \sqrt[3]{3a} = \sqrt[3]{27a^3} = 3a$, multiply the fraction by $\dfrac{\sqrt[3]{3a}}{\sqrt[3]{3a}}$.

$= \dfrac{\sqrt[3]{15a}}{\sqrt[3]{27a^3}}$ Simplify.

$= \dfrac{\sqrt[3]{15a}}{3a}$

d. $\dfrac{5}{\sqrt[4]{3}}$

Solution $\dfrac{5}{\sqrt[4]{3}} = \dfrac{5}{\sqrt[4]{3}} \cdot \dfrac{\sqrt[4]{27}}{\sqrt[4]{27}}$ Since $\sqrt[4]{3} \cdot \sqrt[4]{27} = \sqrt[4]{81} = 3$, multiply the fraction by $\dfrac{\sqrt[4]{27}}{\sqrt[4]{27}}$.

$= \dfrac{5\sqrt[4]{27}}{\sqrt[4]{81}}$ Simplify.

$= \dfrac{5\sqrt[4]{27}}{3}$

YOUR TURN Rationalize the denominator. Assume variables represent nonnegative values.

a. $\dfrac{6}{\sqrt[3]{3}}$ b. $\sqrt[3]{\dfrac{3}{x^2}}$ c. $\dfrac{4}{\sqrt[3]{4y}}$ d. $\dfrac{7}{\sqrt[4]{2}}$

OBJECTIVE 2. Rationalize denominators that have a sum or difference with a square root term. In Example 4 of Section 8.4, we saw that the product of two conjugates containing square roots does not contain any radicals. Consequently, if the denominator of a fraction contains a sum or difference with a square root term, we can rationalize the denominator by multiplying the fraction by a 1 made up of the conjugate of the denominator. For example, to rationalize $\dfrac{5}{7-\sqrt{3}}$, we multiply by $\dfrac{7+\sqrt{3}}{7+\sqrt{3}}$. Because $7-\sqrt{3}$ and $7+\sqrt{3}$ are conjugates, their product will not contain any radicals, so the denominator will be rationalized.

$$\frac{5}{7-\sqrt{3}} = \frac{5}{7-\sqrt{3}} \cdot \frac{7+\sqrt{3}}{7+\sqrt{3}} = \frac{5(7+\sqrt{3})}{(7)^2-(\sqrt{3})^2} = \frac{35+5\sqrt{3}}{49-3} = \frac{35+5\sqrt{3}}{46}$$

PROCEDURE **Rationalizing a Denominator Containing a Sum or Difference**

To rationalize a denominator containing a sum or difference with at least one square root term, multiply the fraction by a 1 whose numerator and denominator are the conjugate of the denominator.

EXAMPLE 3 Rationalize the denominator and simplify. Assume variables represent nonnegative values.

a. $\dfrac{9}{\sqrt{2}+7}$

Solution $\dfrac{9}{\sqrt{2}+7} = \dfrac{9}{\sqrt{2}+7} \cdot \dfrac{\sqrt{2}-7}{\sqrt{2}-7}$ The conjugate of $\sqrt{2}+7$ is $\sqrt{2}-7$, so we multiply by $\dfrac{\sqrt{2}-7}{\sqrt{2}-7}$.

$= \dfrac{9(\sqrt{2}-7)}{(\sqrt{2})^2-(7)^2}$ Multiply. In the denominator, use the rule $(a+b)(a-b) = a^2 - b^2$.

$= \dfrac{9\sqrt{2}-63}{2-49}$ Simplify.

$= \dfrac{9\sqrt{2}-63}{-47}$ We can simplify the negative denominator by factoring out -1 in the numerator and denominator.

$= \dfrac{-1(63-9\sqrt{2})}{-1(47)}$ After factoring out the -1, the signs of the terms change. Since 63 is now positive, we write it first.

$= \dfrac{63-9\sqrt{2}}{47}$ Divide out the common factor -1.

ANSWERS

a. $2\sqrt[3]{9}$ b. $\dfrac{\sqrt[3]{3x}}{x}$ c. $\dfrac{2\sqrt[3]{2y^2}}{y}$ d. $\dfrac{7\sqrt[4]{8}}{2}$

b. $\dfrac{2\sqrt{3}}{\sqrt{6}-\sqrt{2}}$

Solution $\dfrac{2\sqrt{3}}{\sqrt{6}-\sqrt{2}} = \dfrac{2\sqrt{3}}{\sqrt{6}-\sqrt{2}} \cdot \dfrac{\sqrt{6}+\sqrt{2}}{\sqrt{6}+\sqrt{2}}$ The conjugate of $\sqrt{6}-\sqrt{2}$ is $\sqrt{6}+\sqrt{2}$, so we multiply by $\dfrac{\sqrt{6}+\sqrt{2}}{\sqrt{6}+\sqrt{2}}$.

$= \dfrac{2\sqrt{3}(\sqrt{6}+\sqrt{2})}{(\sqrt{6})^2-(\sqrt{2})^2}$

$= \dfrac{2\sqrt{18}+2\sqrt{6}}{6-2}$ Multiply in the numerator and evaluate the exponents in the denominator.

$= \dfrac{2\sqrt{9\cdot 2}+2\sqrt{6}}{4}$ Simplify $\sqrt{18}$ by factoring out a perfect square factor in 18.

$= \dfrac{2\cdot 3\sqrt{2}+2\sqrt{6}}{4}$ Simplify $\sqrt{9\cdot 2}$ by finding the square root of 9.

$= \dfrac{2(3\sqrt{2}+\sqrt{6})}{4}$ Factor out the common 2 factor in the numerator.

$= \dfrac{3\sqrt{2}+\sqrt{6}}{2}$ Divide out the common factor 2.

Note: *We cannot add $3\sqrt{2}$ and $\sqrt{6}$ because their radicands do not match.*

c. $\dfrac{6}{\sqrt{x}-5}$

Solution $\dfrac{6}{\sqrt{x}-5} = \dfrac{6}{\sqrt{x}-5} \cdot \dfrac{\sqrt{x}+5}{\sqrt{x}+5}$ The conjugate of $\sqrt{x}-5$ is $\sqrt{x}+5$, so we multiply by $\dfrac{\sqrt{x}+5}{\sqrt{x}+5}$.

$= \dfrac{6(\sqrt{x}+5)}{(\sqrt{x})^2-(5)^2}$

$= \dfrac{6\sqrt{x}+30}{x-25}$ Multiply in the numerator and evaluate the exponents in the denominator.

Your Turn Rationalize the denominator and simplify. Assume variables represent nonnegative values.

a. $\dfrac{9}{\sqrt{5}+2}$ **b.** $\dfrac{\sqrt{2}}{\sqrt{5}-\sqrt{3}}$ **c.** $\dfrac{3}{\sqrt{x}+4}$

Answers

a. $9\sqrt{5}-18$
b. $\dfrac{\sqrt{10}+\sqrt{6}}{2}$
c. $\dfrac{3\sqrt{x}-12}{x-16}$

OBJECTIVE 3. Rationalize numerators. In later mathematics courses, you may need to rationalize the numerator. We use the same procedure that we use in rationalizing denominators.

EXAMPLE 4 Rationalize the numerator. Assume variables represent nonnegative values.

a. $\dfrac{\sqrt{5x}}{4}$

Solution $\dfrac{\sqrt{5x}}{4} = \dfrac{\sqrt{5x}}{4} \cdot \dfrac{\sqrt{5x}}{\sqrt{5x}}$ To create a perfect square radicand in the numerator, we multiply by $\dfrac{\sqrt{5x}}{\sqrt{5x}}$.

$= \dfrac{\sqrt{25x^2}}{4\sqrt{5x}}$ Simplify.

$= \dfrac{5x}{4\sqrt{5x}}$

b. $\dfrac{3 + \sqrt{2x}}{4}$

Solution $\dfrac{3 + \sqrt{2x}}{4} = \dfrac{3 + \sqrt{2x}}{4} \cdot \dfrac{3 - \sqrt{2x}}{3 - \sqrt{2x}}$ The conjugate of $3 + \sqrt{2x}$ is $3 - \sqrt{2x}$, so we multiply by $\dfrac{3 - \sqrt{2x}}{3 - \sqrt{2x}}$.

$= \dfrac{3^2 - (\sqrt{2x})^2}{4(3 - \sqrt{2x})}$ Simplify.

$= \dfrac{9 - 2x}{12 - 4\sqrt{2x}}$

YOUR TURN Rationalize the numerators. Assume variables represent nonnegative values.

a. $\dfrac{\sqrt{3a}}{7}$ **b.** $\dfrac{5 - \sqrt{3a}}{2}$

ANSWERS

a. $\dfrac{3a}{7\sqrt{3a}}$

b. $\dfrac{25 - 3a}{10 + 2\sqrt{3a}}$

8.5 Exercises

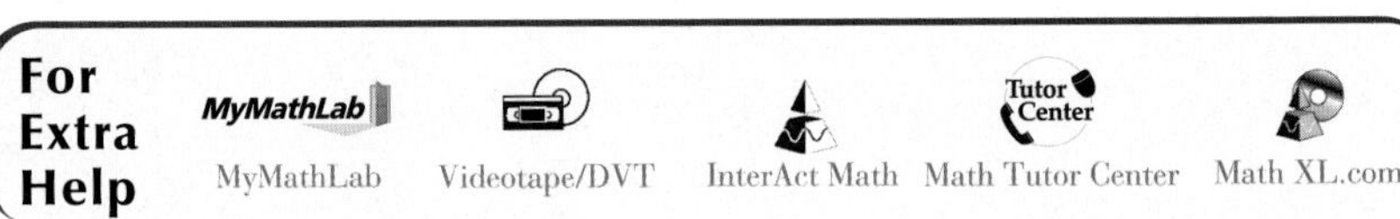

1. Explain why each of the following expressions is not in simplest form.

a. $\sqrt{\dfrac{3}{16}}$

b. $\dfrac{5}{\sqrt{3}}$

2. Although $\frac{1}{\sqrt{3}}$ and $\frac{\sqrt{3}}{3}$ are equal, explain why $\frac{\sqrt{3}}{3}$ is considered simplest form.

3. Explain in your own words how to rationalize a denominator that is the square root of a number $\left(\text{like } \frac{2}{\sqrt{3}} \text{ or } \frac{\sqrt{5}}{\sqrt{7}}\right)$.

4. Explain in your own words how to rationalize a denominator that is a sum or difference with a square root term $\left(\text{like } \frac{3}{5+\sqrt{2}} \text{ or } \frac{2}{\sqrt{x}-\sqrt{y}}\right)$.

For Exercises 5–28, rationalize the denominator. Assume variables represent non-negative values.

5. $\frac{1}{\sqrt{3}}$

6. $\frac{1}{\sqrt{5}}$

7. $\frac{3}{\sqrt{8}}$

8. $\frac{5}{\sqrt{12}}$

9. $\sqrt{\frac{36}{7}}$

10. $\sqrt{\frac{81}{5}}$

11. $\sqrt{\frac{5}{12}}$

12. $\sqrt{\frac{11}{18}}$

13. $\frac{\sqrt{7x^2}}{\sqrt{50}}$

14. $\frac{\sqrt{3x^2}}{\sqrt{32}}$

15. $\frac{\sqrt{8}}{\sqrt{56}}$

16. $\frac{\sqrt{10}}{\sqrt{20}}$

17. $\frac{5}{\sqrt{3a}}$

18. $\frac{11}{\sqrt{7b}}$

19. $\sqrt{\frac{3m}{11n}}$

20. $\sqrt{\frac{5r}{6s}}$

21. $\frac{10}{\sqrt{5x}}$

22. $\frac{20}{\sqrt{10a}}$

23. $\frac{\sqrt{6x}}{\sqrt{32x}}$

24. $\frac{\sqrt{10a}}{\sqrt{18a}}$

25. $\frac{3}{\sqrt{x^3}}$

26. $\frac{5}{\sqrt{b^5}}$

27. $\frac{8x^2}{\sqrt{2x}}$

28. $\frac{14a^3}{\sqrt{7a}}$

For Exercises 29 and 30, explain the mistake, then simplify correctly.

29. $\frac{\sqrt{3}}{\sqrt{2}} = \frac{\sqrt{3}}{\sqrt{2}} \cdot \frac{2}{2} = \frac{2\sqrt{3}}{2}$

30. $\sqrt{\frac{7}{3}} = \frac{\sqrt{7}}{\sqrt{3}} \cdot \frac{\sqrt{3}}{\sqrt{3}} = \frac{\sqrt{21}}{9}$

For Exercises 31–50, rationalize the denominators. Assume variables represent nonnegative values.

31. $\frac{5}{\sqrt[3]{3}}$

32. $\frac{7}{\sqrt[3]{5}}$

33. $\sqrt[3]{\frac{5}{2}}$

34. $\sqrt[3]{\frac{6}{5}}$

35. $\frac{6}{\sqrt[3]{4}}$

36. $\frac{9}{\sqrt[3]{9}}$

37. $\frac{m}{\sqrt[3]{n}}$

38. $\frac{p}{\sqrt[3]{q}}$

39. $\sqrt[3]{\frac{a}{b^2}}$

40. $\sqrt[3]{\frac{m}{n^2}}$

41. $\frac{4}{\sqrt[3]{2x}}$

42. $\frac{9}{\sqrt[3]{3a}}$

43. $\sqrt[3]{\frac{6}{25a^2}}$

44. $\sqrt[3]{\frac{5}{9b^2}}$

45. $\frac{5}{\sqrt[4]{2}}$

46. $\frac{7}{\sqrt[4]{3}}$

47. $\sqrt[4]{\frac{3}{x^2}}$

48. $\sqrt[4]{\frac{5}{y^3}}$

49. $\frac{9}{\sqrt[4]{3x^3}}$

50. $\frac{12}{\sqrt[4]{4x^2}}$

For Exercises 51–72, rationalize the denominator and simplify. Assume variables represent nonnegative values.

51. $\frac{\sqrt{3}}{\sqrt{2}+1}$

52. $\frac{3}{\sqrt{5}+2}$

53. $\frac{4}{2-\sqrt{3}}$

54. $\frac{2}{4-\sqrt{15}}$

55. $\frac{5}{\sqrt{2}+\sqrt{3}}$

56. $\frac{7}{\sqrt{6}+\sqrt{7}}$

57. $\frac{4}{1-\sqrt{5}}$

58. $\frac{6}{1-\sqrt{7}}$

59. $\frac{\sqrt{3}}{\sqrt{3}-1}$

60. $\frac{\sqrt{5}}{1-\sqrt{5}}$

61. $\frac{2\sqrt{3}}{\sqrt{3}-4}$

62. $\frac{2\sqrt{5}}{\sqrt{5}-4}$

63. $\frac{4\sqrt{3}}{\sqrt{7}+\sqrt{2}}$

64. $\frac{\sqrt{8}}{\sqrt{4}+\sqrt{3}}$

65. $\frac{8\sqrt{2}}{4\sqrt{2}-\sqrt{3}}$

66. $\frac{5\sqrt{3}}{\sqrt{2}+4\sqrt{6}}$

67. $\frac{6\sqrt{y}}{\sqrt{y}+1}$

68. $\frac{4\sqrt{x}}{\sqrt{x}+1}$

69. $\frac{3\sqrt{t}}{\sqrt{t}+2\sqrt{u}}$

70. $\frac{2\sqrt{m}}{\sqrt{n}-\sqrt{m}}$

71. $\frac{\sqrt{2y}}{\sqrt{x}-\sqrt{6y}}$

72. $\frac{\sqrt{14h}}{\sqrt{2h}+\sqrt{k}}$

For Exercises 73–84, rationalize the numerator. Assume variables represent non-negative values.

73. $\frac{\sqrt{3}}{2}$ **74.** $\frac{\sqrt{5}}{8}$ **75.** $\frac{\sqrt{2x}}{5}$ **76.** $\frac{\sqrt{7y}}{3}$

77. $\frac{\sqrt{8n}}{6}$ **78.** $\frac{\sqrt{20t}}{8}$ **79.** $\frac{2+\sqrt{3}}{5}$ **80.** $\frac{5-\sqrt{2}}{6}$

81. $\frac{\sqrt{5x}-6}{9}$ **82.** $\frac{\sqrt{2x}+7}{3}$ **83.** $\frac{5\sqrt{n}+\sqrt{6n}}{2n}$ **84.** $\frac{4\sqrt{k}-\sqrt{10k}}{5k}$

85. Given $f(x) = \frac{5\sqrt{2}}{x}$, find each of the following. Express your answer in simplest form.

a. $f(\sqrt{6})$ **b.** $f(\sqrt{10})$ **c.** $f(\sqrt{22})$

86. Given $g(x) = \frac{\sqrt{2}}{x-1}$, find each of the following. Express your answer in simplest form.

a. $g(\sqrt{5})$ **b.** $g(3\sqrt{2})$ **c.** $g(2\sqrt{6})$

87. Graph $f(x) = \frac{1}{\sqrt{x}}$, then graph $g(x) = \frac{\sqrt{x}}{x}$.

a. What do you notice about the two graphs? What does this indicate about the two functions?

b. Simplify $f(x)$ by rationalizing the denominator. What do you notice?

88. Graph $f(x) = -\frac{1}{\sqrt{x}}$, then graph $g(x) = -\frac{\sqrt{x}}{x}$.

a. What do you notice about the two graphs? What does this indicate about the two functions?

b. Simplify $f(x)$ by rationalizing the denominator. What do you notice?

89. In Section 10.1, we used the formula $T = 2\pi\sqrt{\frac{L}{9.8}}$ to determine the period of a pendulum where T is the period in seconds and L is the length in meters.

a. Rewrite the formula so that the denominator is rationalized.

b. Rewrite the formula so that the numerator is rationalized.

90. The formula $t = \sqrt{\frac{h}{16}}$ can be used to find the time, t, in seconds for an object to fall a distance of h feet.

a. Rewrite the formula so that the denominator is rationalized.

b. Rewrite the formula so that the numerator is rationalized.

91. The formula $s = \sqrt{\frac{3V}{h}}$ can be used to find the length, s, of each side of the base of a pyramid having a square base, volume V in cubic feet, and height h in feet.

a. Rationalize the denominator in the formula.

b. The volume of the Great Pyramid at Giza in Egypt is approximately 83,068,742 cubic feet and its height is 449 feet. Find the length of each side of its base.

Of Interest

The Great Pyramid was built for King Khufu from about 2589 to 2566 B.C. The pyramid contains approximately 2,300,000 blocks and weighs about 6.5 million tons, with each block averaging about 2.8 tons.

92. The formula $s = 2\sqrt{\frac{A}{6\sqrt{3}}}$ can be used to find the length, s, of each side of a regular hexagon having an area A.

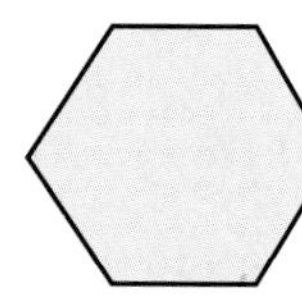

a. Rationalize the denominator in the formula.

b. If A is 100 square meters, write an expression in simplest form for the side length.

c. Use a calculator to approximate the side lengths, rounded to three decimal places.

93. In AC circuits, voltage is often expressed as a *root-mean-square,* or *rms,* value. The formula for calculating the rms voltage, V_{rms}, given the maximum voltage, V_{m}, value is $V_{\text{rms}} = \dfrac{V_{\text{m}}}{\sqrt{2}}$.

a. Rationalize the denominator in the formula.

b. Given a maximum voltage of 163 V, write an expression for the rms voltage.

c. Use a calculator to approximate the rms voltage rounded to the nearest tenth.

94. The velocity, in meters per second, of a particle can be determined by the formula $v = \sqrt{\dfrac{2E}{m}}$, where E represents the kinetic energy, in joules, of the particle and m represents the mass, in kilograms, of the particle.

a. Rationalize the denominator in the formula.

b. A particle with a mass of 1×10^{-6} kilograms has 2.4×10^{7} joules of kinetic energy. Write an expression for its velocity.

c. Use a calculator to approximate the velocity rounded to the nearest tenth.

95. The resistance in a circuit is found to be $\dfrac{5\sqrt{2}}{3 + \sqrt{6}}\ \Omega$. Rationalize the denominator.

96. Two charged particles, q_1 and q_2, are separated by a distance of 8 centimeters. The values of the charges are $q_1 = 3 \times 10^{-6}$ coulombs and $q_2 = 1 \times 10^{-6}$ coulombs. The charged particles each exert an electrical field. At a point in between the two particles x centimeters away from q_1, the electric fields will cancel each other so that the value of the fields at the point x is 0.

a. Use the formula $x = \dfrac{l}{1 + \sqrt{\dfrac{q_2}{q_1}}}$ to find the distance from q_1 at which the electric field is canceled, where l is the distance separating the particles. Write the distance with a rationalized denominator.

b. Use a calculator to approximate the distance rounded to the nearest tenth.

REVIEW EXERCISES

[8.3] **1.** Simplify: $\pm\sqrt{28}$

[6.3] **2.** Factor: $x^2 - 6x + 9$

[2.1] ***For Exercises 3 and 4, solve.***

3. $2x - 3 = 5$

4. $2x - 3 = -5$

[6.4] ***For Exercises 5 and 6, solve and check.***

5. $x^2 - 36 = 0$

6. $x^2 - 5x + 6 = 0$

8.6 Radical Equations and Problem Solving

OBJECTIVE **1. Use the power rule to solve radical equations.**

We now explore how to solve **radical equations**.

DEFINITION ***Radical equation:*** An equation containing at least one radical expression whose radicand has a variable.

The Algebra Pyramid

Radical Equations
$\sqrt{x} = 5$
or $\sqrt{x+5} = \sqrt{x-7}$

Radical Expressions
$\sqrt{50}$ or $\sqrt{x}$

Constants and Variables
2, 5, 7, x, y

Note: *We move upward on the Algebra Pyramid, making equations out of the expressions we've studied.*

OBJECTIVE 1. Use the power rule to solve radical equations. To solve radical equations we will use a new principle of equality called the *power rule.*

RULE **Power Rule**

If $a = b$, then $a^n = b^n$.

Isolated Radicals

First, we consider equations such as $\sqrt{x} = 9$ in which the radical is isolated. When we use the power rule, we raise both sides of the equation to the same power as the root index, then use the principle $(\sqrt[n]{x})^n = x$, which eliminates the radical, leaving us with its radicand.

EXAMPLE 1 Solve.

a. $\sqrt{x} = 9$

Solution $(\sqrt{x})^2 = (9)^2$ Since the root index is 2, we square both sides.

$x = 81$

Check $\sqrt{81} \stackrel{?}{=} 9$

$9 = 9$ True

b. $\sqrt[3]{y} = -2$

Solution $(\sqrt[3]{y})^3 = (-2)^3$ Since the root index is 3, we cube both sides.

$y = -8$

Check $\sqrt[3]{-8} \stackrel{?}{=} -2$

$-2 = -2$ True

Extraneous Solutions

In Section 7.4, we learned that some equations have *extraneous solutions,* which are apparent solutions that actually do not make the original equation true. Using the power rule can sometimes lead to extraneous solutions, so it is important to check solutions. For example, watch what happens when we use the power rule to solve the equation $\sqrt{x} = -9$.

Warning: As we shall see in the check, this result is not a solution. →

$(\sqrt{x})^2 = (-9)^2$ Square both sides.

$x = 81$

By checking 81 in the original equation we see that it is extraneous.

$\sqrt{81} \stackrel{?}{=} -9$

$9 = -9$ This equation is false, so 81 is extraneous.

In fact, $\sqrt{x} = -9$ has no real number solution because if x is a real number, then $\sqrt{x}$ must be positive (or 0).

EXAMPLE 2 Solve.

a. $\sqrt{x-7} = 8$

Solution $(\sqrt{x-7})^2 = (8)^2$ Square both sides.

$x - 7 = 64$

$x = 71$ Add 7 to both sides.

Check $\sqrt{71-7} \stackrel{?}{=} 8$

$\sqrt{64} \stackrel{?}{=} 8$

$8 = 8$ True

b. $\sqrt{2x+1} = -3$

Solution $(\sqrt{2x+1})^2 = (-3)^2$ Square both sides.

$2x + 1 = 9$

$2x = 8$ Subtract 1 from both sides.

$x = 4$ Divide both sides by 2.

Check $\sqrt{2(4)+1} \stackrel{?}{=} -3$

$\sqrt{9} \stackrel{?}{=} -3$

$3 = -3$ False, so 4 is extraneous. This equation has no real-number solution.

Note: *We can see that this equation has no real-number solution because a principal square root cannot be equal to a negative number. However, we will go through the steps to confirm this.*

YOUR TURN Solve.

a. $\sqrt{x-5} = 6$ **b.** $\sqrt[3]{x+2} = 3$ **c.** $\sqrt{x+3} = -7$

Radicals on Both Sides of the Equation

As we will see in Example 3, the power rule can be used to solve equations with radicals on both sides of the equal sign.

EXAMPLE 3 Solve.

a. $\sqrt{6x-1} = \sqrt{x+2}$

Solution $(\sqrt{6x-1})^2 = (\sqrt{x+2})^2$ Square both sides.

$6x - 1 = x + 2$ Subtract x from both sides and add 1 to both sides.

$5x = 3$ Divide both sides by 5.

$x = \frac{3}{5}$

ANSWERS

a. 41
b. 25
c. no real-number solution

Check $\sqrt{6\left(\frac{3}{5}\right) - 1} \stackrel{?}{=} \sqrt{\frac{3}{5} + 2}$

$$\sqrt{\frac{18}{5} - \frac{5}{5}} \stackrel{?}{=} \sqrt{\frac{3}{5} + \frac{10}{5}}$$

$$\sqrt{\frac{13}{5}} = \sqrt{\frac{13}{5}} \qquad \text{True}$$

b. $\sqrt[3]{5x - 2} = \sqrt[3]{3x + 2}$

Solution $(\sqrt[3]{5x - 2})^3 = (\sqrt[3]{3x + 2})^3$ — Cube both sides.

$$5x - 2 = 3x + 2$$

$$2x = 4$$ Subtract $3x$ from both sides and add 2 to both sides.

$$x = 2$$ Divide both sides by 2.

Check $\sqrt[3]{5(2) - 2} \stackrel{?}{=} \sqrt[3]{3(2) + 2}$

$$\sqrt[3]{10 - 2} \stackrel{?}{=} \sqrt[3]{6 + 2}$$

$$\sqrt[3]{8} \stackrel{?}{=} \sqrt[3]{8}$$

$$2 = 2 \qquad \text{True}$$

YOUR TURN Solve.

a. $\sqrt{8x + 5} = \sqrt{2x + 7}$ **b.** $\sqrt[3]{6x + 9} = \sqrt[3]{10x - 3}$

Multiple Solutions

Radical equations may have multiple solutions if, after using the power rule, we are left with a quadratic form.

EXAMPLE 4 Solve. $x + 2 = \sqrt{5x + 16}$

Solution $(x + 2)^2 = (\sqrt{5x + 16})^2$ — Square both sides.

$x^2 + 4x + 4 = 5x + 16$ — Use FOIL on the left-hand side.

$x^2 - x - 12 = 0$ — Since the equation is quadratic, we set it equal to 0 by subtracting $5x$ and 16 from both sides.

$(x + 3)(x - 4) = 0$ — Factor.

$x + 3 = 0$ or $x - 4 = 0$ — Use the zero-factor theorem.

$x = -3$ $\qquad$ $x = 4$

Checks

$x = -3$	$x = 4$
$-3 + 2 \stackrel{?}{=} \sqrt{5(-3) + 16}$	$4 + 2 \stackrel{?}{=} \sqrt{5(4) + 16}$
$-1 \stackrel{?}{=} \sqrt{-15 + 16}$	$6 \stackrel{?}{=} \sqrt{20 + 16}$
$-1 \stackrel{?}{=} \sqrt{1}$	$6 \stackrel{?}{=} \sqrt{36}$
$-1 = 1$ False	$6 = 6$ True

Because -3 does not check, it is an extraneous solution. The only solution is 4.

YOUR TURN Solve.

$$\sqrt{9x + 7} = x + 3$$

ANSWERS

a. $\frac{1}{3}$ **b.** 3

ANSWER

2 and 1

Radicals Not Isolated

Now we consider radical equations in which the radical term is not isolated. In such equations, we must first isolate the radical term.

EXAMPLE 5 Solve.

a. $\sqrt{x+1} - 2x = x + 1$

Solution

$\sqrt{x+1} = 3x + 1$ Add $2x$ to both sides to isolate the radical term.

$(\sqrt{x+1})^2 = (3x+1)^2$ Square both sides.

$x + 1 = 9x^2 + 6x + 1$ Use FOIL on the right-hand side.

$0 = 9x^2 + 5x$ Since the equation is quadratic, we set it equal to 0 by subtracting x and 1 from both sides.

$0 = x(9x + 5)$ Factor.

$x = 0$ or $9x + 5 = 0$ Use the zero-factor theorem.

$9x = -5$

$x = -\frac{5}{9}$

Checks

$\sqrt{0+1} - 2(0) \stackrel{?}{=} 0 + 1$

$\sqrt{1} - 0 \stackrel{?}{=} 1$

$1 = 1$ True

$\sqrt{-\frac{5}{9} + 1} - 2\left(-\frac{5}{9}\right) \stackrel{?}{=} -\frac{5}{9} + 1$

$\sqrt{-\frac{5}{9} + \frac{9}{9}} - 2\left(-\frac{5}{9}\right) \stackrel{?}{=} -\frac{5}{9} + \frac{9}{9}$

$\sqrt{\frac{4}{9}} + \frac{10}{9} \stackrel{?}{=} \frac{4}{9}$

$\frac{2}{3} + \frac{10}{9} \stackrel{?}{=} \frac{4}{9}$

$\frac{6}{9} + \frac{10}{9} \stackrel{?}{=} \frac{4}{9}$

$\frac{16}{9} = \frac{4}{9}$ False

Note: *This false equation indicates that $-\frac{5}{9}$ is an extraneous solution.*

The solution is 0.

b. $\sqrt[4]{3x+4} + 5 = 7$

Solution

$\sqrt[4]{3x+4} = 2$ Subtract 5 from both sides to isolate the radical term.

$(\sqrt[4]{3x+4})^4 = 2^4$ Raise both sides to the fourth power.

$3x + 4 = 16$ Simplify both sides.

$3x = 12$ Subtract 4 from both sides.

$x = 4$ Divide both sides by 3.

Check

$\sqrt[4]{3(4)+4} + 5 \stackrel{?}{=} 7$

$\sqrt[4]{12+4} + 5 \stackrel{?}{=} 7$

$\sqrt[4]{16} + 5 \stackrel{?}{=} 7$

$2 + 5 \stackrel{?}{=} 7$

$7 = 7$ True. The solution is 4.

YOUR TURN Solve.

a. $\sqrt{5x^2 + 6x - 7} + 3x = 5x + 1$ b. $\sqrt[4]{3x + 6} - 7 = -4$

Using the Power Rule Twice

Some equations may require that we use the power rule twice in order to eliminate all radicals.

EXAMPLE 6 Solve. $\sqrt{x + 21} = \sqrt{x} + 3$

Solution $(\sqrt{x + 21})^2 = (\sqrt{x} + 3)^2$ Since one of the radicals is isolated, we square both sides.

$x + 21 = (\sqrt{x} + 3)(\sqrt{x} + 3)$

$x + 21 = x + 3\sqrt{x} + 3\sqrt{x} + 9$ Use FOIL on the right-hand side.

$x + 21 = x + 6\sqrt{x} + 9$ Combine like terms.

$12 = 6\sqrt{x}$ Subtract x and 9 from both sides to isolate the remaining radical expression.

$2 = \sqrt{x}$ Divide both sides by 6.

$(2)^2 = (\sqrt{x})^2$ Square both sides.

$4 = x$

Check $\sqrt{4 + 21} \stackrel{?}{=} \sqrt{4} + 3$

$\sqrt{25} \stackrel{?}{=} 2 + 3$

$5 = 5$ True. The solution is 4.

Our examples suggest the following procedure.

PROCEDURE **Solving Radical Equations**

To solve a radical equation:

1. Isolate the radical. (If there is more than one radical term, then isolate one of the radical terms.)
2. Raise both sides of the equation to the same power as the root index.
3. If all radicals have been eliminated, then solve. If a radical term remains, then isolate that radical term and raise both sides to the same power as its root index.
4. Check each solution. Any apparent solution that does not check is an extraneous solution.

ANSWERS

a. 2 **b.** 25

YOUR TURN Solve.

$\sqrt{2x + 1} = \sqrt{x} + 1$

ANSWER

0 and 4

8.6 Exercises

For Extra Help

MyMathLab  Videotape/DVT  InterAct Math  Math Tutor Center  Math XL.com

1. Explain why we must check all potential solutions to radical equations.

2. What is an extraneous solution?

3. Explain why there is no real-number solution for the radical equation $\sqrt{x} = -6$.

4. Show why $(\sqrt{a})^2 = a$, assuming $a \geq 0$.

5. Given the radical equation $\sqrt{x+2} + 3x = 4x - 1$, what would be the first step in solving the equation? Why?

6. Give an example of a radical equation that would require you to use the power rule twice in solving the equation. Explain why the principle would have to be used twice.

For Exercises 7–28, solve.

7. $\sqrt{x} = 2$

8. $\sqrt{y} = 5$

9. $\sqrt{k} = -4$

10. $\sqrt{x} = -1$

11. $\sqrt[3]{y} = 3$

12. $\sqrt[3]{m} = 4$

13. $\sqrt[3]{z} = -2$

14. $\sqrt[3]{p} = -5$

15. $\sqrt{n-1} = 4$

16. $\sqrt{x+3} = 7$

17. $\sqrt{t-7} = 2$

18. $\sqrt{m+8} = 1$

19. $\sqrt{3x-2} = 4$

20. $\sqrt{2x+5} = 3$

21. $\sqrt{2x+17} = 4$

22. $\sqrt{3y-2} = 5$

23. $\sqrt{2n-8} = -3$

24. $\sqrt{5x-1} = -6$

25. $\sqrt[3]{x-3} = 2$

26. $\sqrt[3]{k+2} = 4$

27. $\sqrt[3]{3y-2} = -2$

28. $\sqrt[3]{2x+4} = -3$

For Exercises 29–38 solve. First isolate the radical term.

29. $\sqrt{u-3}-10=1$

30. $\sqrt{y+1}-4=2$

31. $\sqrt{y-6}+2=9$

32. $\sqrt{r-5}+6=10$

33. $\sqrt{6x-5}-2=3$

34. $\sqrt{8x+4}-2=4$

35. $\sqrt[3]{n+3}-2=-4$

36. $\sqrt[3]{x-4}+2=3$

37. $\sqrt[4]{x-2}-2=-4$

38. $\sqrt[4]{m+3}+2=5$

For Exercises 39–64, solve.

39. $\sqrt{3x-2}=\sqrt{8-2x}$

40. $\sqrt{m+2}=\sqrt{2m-3}$

41. $\sqrt{4x-5}=\sqrt{6x+5}$

42. $\sqrt{3x-4}=\sqrt{5x+2}$

43. $\sqrt[3]{2r+2}=\sqrt[3]{3r-1}$

44. $\sqrt[3]{3h-4}=\sqrt[3]{h+4}$

45. $\sqrt[4]{4x+4}=\sqrt[4]{5x+1}$

46. $\sqrt[4]{2x+4}=\sqrt[4]{3x-2}$

47. $\sqrt{k+4}=k-8$

48. $\sqrt{5n-1}=4-2n$

49. $y-1=\sqrt{2y-2}$

50. $3+x=\sqrt{7+3x}$

51. $\sqrt{3x+10}-4=x$

52. $\sqrt{4y+1}+5=y$

53. $\sqrt{10n+4}-3n=n+1$

54. $\sqrt{6x-1}-6x=2-9x$

55. $\sqrt[3]{5x+2}+2=5$

56. $\sqrt[3]{4x-1}-4=-1$

57. $\sqrt[3]{n^2-2n+5}=2$

58. $\sqrt[3]{y^2+y+7}=3$

59. $1+\sqrt{x}=\sqrt{2x+1}$

60. $\sqrt{t+2}-1=\sqrt{t}$

61. $\sqrt{3x+1}+\sqrt{3x}=2$

62. $\sqrt{x+5}-\sqrt{x}=1$

63. $\sqrt{x+8}-2=\sqrt{x}$

64. $\sqrt{5x-1}-1=\sqrt{x+2}$

For Exercises 65–68, explain the mistake, then solve correctly.

65.
$$\sqrt{x} = -9$$
$$(\sqrt{x})^2 = (-9)^2$$
$$x = 81$$

66.
$$\sqrt{x} = -2$$
$$(\sqrt{x})^2 = (-2)^2$$
$$x = 4$$

67.
$$\sqrt{x+3} = x - 3$$
$$x + 3 = x^2 - 9$$
$$0 = x^2 - x - 12$$
$$0 = (x+3)(x-4)$$
$$x = -3, 4$$

68.
$$\sqrt{x+3} = 4$$
$$(\sqrt{x+3})^2 = 4^2$$
$$x + 9 = 16$$
$$x + 9 - 9 = 16 - 9$$
$$x = 7$$

For Exercises 69–72, given the period of a pendulum, find the length of the pendulum. Use the formula $T = 2\pi\sqrt{\frac{L}{9.8}}$, where T is the period in seconds and L is the length in meters.

69. 2π seconds

70. 6π seconds

71. π seconds

72. $\frac{\pi}{3}$ seconds

For Exercises 73–76, find the distance an object has fallen. Use the formula $t = \sqrt{\frac{h}{16}}$, where t is the time in seconds that it takes an object to fall a distance of h feet.

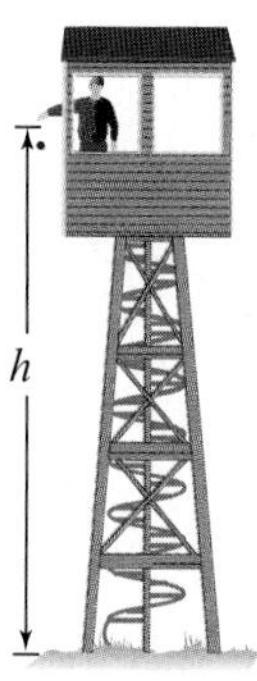

73. 0.3 seconds

74. 0.5 seconds

75. 3 seconds

76. $\frac{1}{4}$ seconds

For Exercises 77–80, find the skid distance after braking hard at a given speed. Use the formula $S = \frac{7}{2}\sqrt{2D}$, where D represents the skid distance in feet on asphalt and S represents the speed of the car in miles per hour. (Source: Harris Technical Services, Traffic Accident Reconstructionists.)

77. 30 miles per hour

78. 60 miles per hour

79. 45 miles per hour

80. 75 miles per hour

For Exercises 81–84, use the following. Two forces, F_1 and F_2, acting on an object at a 90° angle will pull the object with a resultant force, R, at an angle in between F_1 and F_2 (see the figure). The value of the resultant force can be found by the formula $R = \sqrt{F_1^2 + F_2^2}$.

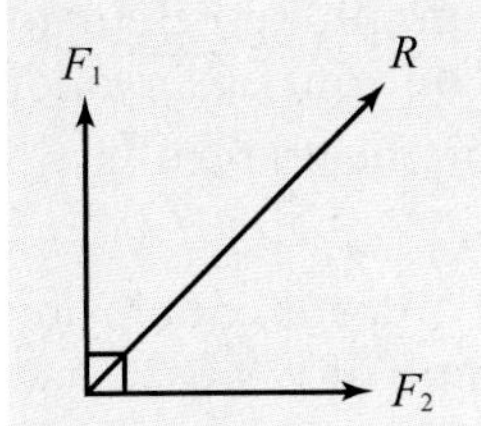

81. Find F_1 if $R = 5$ N and $F_2 = 3$ N.

82. Find F_1 if $R = 10$ N and $F_2 = 8$ N.

83. Find F_2 if $R = 3\sqrt{5}$ N and $F_1 = 3$ N.

84. Find F_2 if $R = 2\sqrt{13}$ N and $F_1 = 6$ N.

85. a. Complete the table of ordered pairs for the equation $y = \sqrt{x}$.

x	0	1	4	9	16	?
y	0	1	2	?	?	5

b. Graph the ordered pairs in the coordinate plane, then connect the points to make a curve.

c. Does the graph extend below or to the left of the origin? Explain.

d. Does the curve represent a function? Explain.

86. a. Complete the table of ordered pairs for the equation $y = \sqrt{x} + 2$.

x	0	1	4	9	16	?
y	2	3	4	?	?	7

b. Graph the ordered pairs in the coordinate plane, then connect the points to make a curve.

c. Does the graph extend below or to the left of the origin? Explain.

d. Does the curve represent a function? Explain.

87. Use a graphing calculator to graph $y = \sqrt{x}$, $y = 2\sqrt{x}$, and $y = 3\sqrt{x}$. Based on your observations, as you increase the size of the coefficient, what happens to the graph?

88. Use a graphing calculator to compare the graph of $y = \sqrt{x}$ to the graph of $y = -\sqrt{x}$. What effect does the negative sign have on the graph?

89. Use a graphing calculator to graph $y = \sqrt{x}$, $y = \sqrt{x} + 2$, and $y = \sqrt{x} - 2$. Based on your observations, what effect does adding or subtracting a constant have on the graph of $y = \sqrt{x}$?

90. Based on your conclusions from Exercises 87–89, without graphing, describe the graph of $y = -2\sqrt{x} + 3$. Use a graphing calculator to graph the equation to confirm your description.

Collaborative Exercises BUILDING TIME

You are constructing a grandfather clock that operates using a pendulum. The period of the pendulum is the time required to complete one full swing. The formula giving the relationship between the length, L (in meters), and the period, T (in seconds), is $T = 2\pi\sqrt{\frac{L}{9.8}}$.

1. Find the length of the pendulum if the period is 1 second.

2. Suppose that you want the pendulum to complete one period in 2 seconds. Would you need to increase or decrease the length of the pendulum found in Exercise 1? Use the same formula to determine the length of the pendulum so that the period is 2 seconds.

3. Based on the results of Exercises 1 and 2, what can you conclude about the required length of the pendulum as the period increases?

REVIEW EXERCISES

[5.1] ***For Exercises 1–3, evaluate.***

1. 3^4

2. $(-0.2)^3$

3. $\left(\frac{2}{5}\right)^{-4}$

[5.1] ***For Exercises 4–6, use the rules of exponents to simplify.***

4. $(x^3)(x^5)$

5. $(n^4)^6$

6. $\frac{y^7}{y^3}$

8.7 Complex Numbers

OBJECTIVES

1. **Write imaginary numbers using i.**
2. **Perform arithmetic operations with complex numbers.**
3. **Raise i to powers.**

We have said that the square root of a negative number is not a real number. In this section, we will learn about the *imaginary number system*, in which square roots of negative numbers are expressed using a notation involving the letter i.

OBJECTIVE 1. Write imaginary numbers using i. Using the product rule of square roots, any square root of a negative number can be rewritten as a product of a real number and an **imaginary unit**, which we express as i.

DEFINITION ***Imaginary unit:*** The number represented by i, where $i = \sqrt{-1}$ and $i^2 = -1$.

A number that can be expressed as a product of a real number and the imaginary unit is called an **imaginary number**.

DEFINITION ***Imaginary number:*** A number that can be expressed in the form bi, where b is a real number and i is the imaginary unit.

Of Interest

When the idea of finding square roots of negative numbers was first introduced, members of the established mathematics community said this type of number existed, but only in the imagination of those finding them. From then on they've been called "imaginary numbers."

Note: *In a product with the imaginary unit, it is customary to write integer factors first, then i, then square root factors.*

EXAMPLE 1 Write each imaginary number as a product of a real number and i.

a. $\sqrt{-9}$

Solution
$$\begin{aligned} \sqrt{-9} &= \sqrt{-1 \cdot 9} && \text{Factor out } -1 \text{ in the radicand.} \\ &= \sqrt{-1} \cdot \sqrt{9} && \text{Use the product rule of square roots.} \\ &= i \cdot 3 \\ &= 3i \end{aligned}$$

b. $\sqrt{-54}$

Solution
$$\begin{aligned} \sqrt{-54} &= \sqrt{-1 \cdot 54} && \text{Factor out } -1 \text{ in the radicand.} \\ &= \sqrt{-1} \cdot \sqrt{54} && \text{Use the product rule of square roots.} \\ &= i\sqrt{9 \cdot 6} && \text{Use the product rule again to simplify further.} \\ &= 3i\sqrt{6} \end{aligned}$$

Example 1 suggests the following procedure.

PROCEDURE **Rewriting Imaginary Numbers**

To write an imaginary number $\sqrt{-n}$ in terms of the imaginary unit i:

1. Separate the radical into two factors, $\sqrt{-1} \cdot \sqrt{n}$.
2. Replace $\sqrt{-1}$ with i.
3. Simplify $\sqrt{n}$.

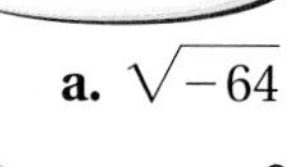

YOUR TURN Write each imaginary number as a product of a real number and i.

a. $\sqrt{-64}$ **b.** $\sqrt{-18}$ **c.** $\sqrt{-48}$

OBJECTIVE 2. Perform arithmetic operations with complex numbers. We now have two distinct sets of numbers, the set of real numbers and the set of imaginary numbers. There is yet another set of numbers, called the set of **complex numbers**, that contains both the real and the imaginary numbers.

DEFINITION ***Complex number:*** A number that can be expressed in the form $a + bi$, where a and b are real numbers and i is the imaginary unit.

Note: *When the coefficient of i is a radical expression as in $-2.1 - i\sqrt{5}$, or $4i\sqrt{3}$, we will write i to the left of the radical sign so that i does not look like it is part of the radicand.*

When written in the form $a + bi$, a complex number is said to be in *standard form*. Following are some examples of complex numbers written in standard form:

$$2 + 3i \qquad 4 - 7i \qquad -2.1 - i\sqrt{5}$$

Note that if $a = 0$, then the complex number is purely an imaginary number, such as:

$$-6i \qquad 4i\sqrt{3}$$

If $b = 0$, then the complex number is a real number.

The following Venn Diagram shows how the set of complex numbers contains both the real numbers and imaginary numbers.

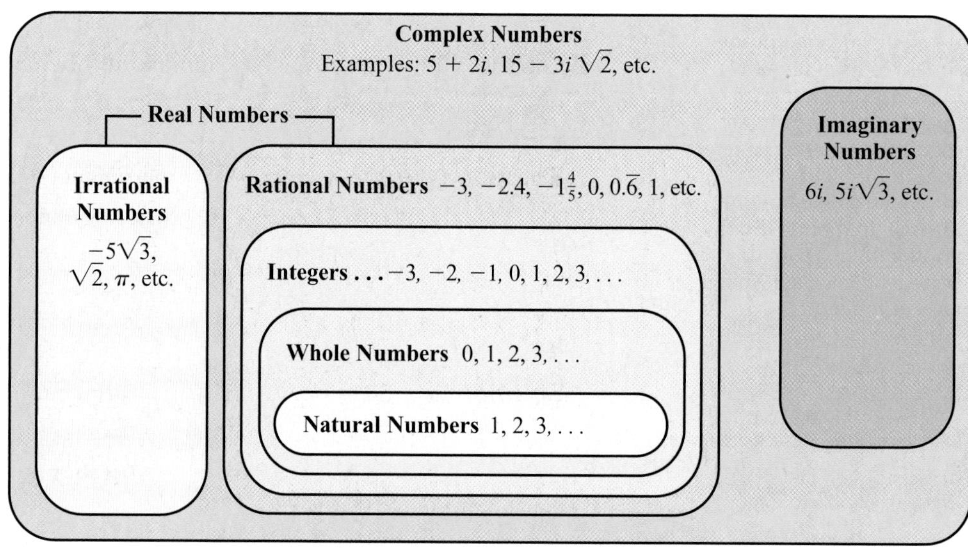

ANSWERS

a. $8i$ **b.** $3i\sqrt{2}$ **c.** $4i\sqrt{3}$

We can perform arithmetic operations with complex numbers. In general, we treat the complex numbers just like polynomials where i is like a variable.

Adding and Subtracting Complex Numbers

EXAMPLE 2 Add or subtract.

a. $(-8 + 7i) + (5 - 19i)$

Solution We add complex numbers just like we add polynomials by combining like terms.

$$(-8 + 7i) + (5 - 19i) = -3 - 12i$$

b. $(-1 + 13i) - (8 - 3i)$

Solution We subtract complex numbers just like we subtract polynomials by writing an equivalent addition and changing the signs in the second complex number.

$$\begin{aligned}(-1 + 13i) - (8 - 3i) &= (-1 + 13i) + (-8 + 3i)\\ &= -9 + 16i\end{aligned}$$

Multiplying Complex Numbers

We multiply complex numbers in the same way that we multiply monomials and binomials. However, we must be careful when simplifying because these products may contain i^2, which is equal to -1.

EXAMPLE 3 Multiply.

a. $(9i)(-8i)$

Solution Multiply in the same way that we multiply monomials.

$$\begin{aligned}(9i)(-8i) &= -72i^2 \\ &= -72(-1) && \text{Replace } i^2 \text{ with } -1.\\ &= 72\end{aligned}$$

b. $(5i)(4 - 7i)$

Solution Multiply in the same way that we multiply a binomial by a monomial.

$$\begin{aligned}(5i)(4 - 7i) &= 20i - 35i^2 && \text{Distribute.}\\ &= 20i - 35(-1) && \text{Replace } i^2 \text{ with } -1.\\ &= 20i + 35\\ &= 35 + 20i && \text{Write in standard form.}\end{aligned}$$

c. $(8 - 3i)(2 + i)$

Solution Multiply in the same way that we multiply binomials.

$$\begin{aligned}(8 - 3i)(2 + i) &= 16 + 8i - 6i - 3i^2 && \text{Use FOIL.}\\ &= 16 + 2i - 3(-1) && \text{Combine like terms and replace } i^2 \text{ with } -1.\\ &= 16 + 2i + 3 && \text{Simplify.}\\ &= 19 + 2i && \text{Write in standard form.}\end{aligned}$$

d. $(6 - 5i)(6 + 5i)$

Solution Note that these complex numbers are conjugates.

$$
\begin{aligned}
(6 - 5i)(6 + 5i) &= 36 + 30i - 30i - 25i^2 && \text{Use FOIL.} \\
&= 36 - (-25) && \text{Combine like terms and replace } i^2 \text{ with } -1. \\
&= 36 + 25 \\
&= 61
\end{aligned}
$$

Note: *The product of these two complex numbers is a real number.*

The complex numbers that we multiplied in Example 3(d) are called **complex conjugates**.

DEFINITION

Complex conjugates: The complex conjugate of a complex number $a + bi$ is $a - bi$.

Some other examples of complex conjugates follow:

$$4 + i \quad \text{and} \quad 4 - i$$
$$9 - 7i \quad \text{and} \quad 9 + 7i$$

Example 3(d) also illustrates the fact that the product of complex conjugates will always be a real number.

YOUR TURN Multiply.

a. $(-4i)(-7i)$ **b.** $(-3i)(6 - i)$ **c.** $(8 - 3i)(5 + 7i)$ **d.** $(7 + 3i)(7 - 3i)$

Dividing Complex Numbers

We have established that the product of complex conjugates is a real number. We can use that fact when the divisor is a complex number. The process is similar to rationalizing denominators.

EXAMPLE 4 Write in standard form.

a. $\dfrac{6 + 5i}{7 - 2i}$

Connection Recall that we rationalize denominators in order to clear undesired square root expressions from a denominator. The imaginary unit i represents a square root expression, $\sqrt{-1}$, which is why we rationalize denominators that contain i.

Solution

$$
\begin{aligned}
\frac{6 + 5i}{7 - 2i} &= \frac{6 + 5i}{7 - 2i} \cdot \frac{7 + 2i}{7 + 2i} && \text{Multiply numerator and denominator by the complex conjugate of the denominator, which is } 7 + 2i. \\
&= \frac{42 + 12i + 35i + 10i^2}{49 - 4i^2} \\
&= \frac{42 + 47i + 10(-1)}{49 - 4(-1)} && \text{Simplify.} \\
&= \frac{42 + 47i - 10}{49 + 4} \\
&= \frac{32 + 47i}{53} \\
&= \frac{32}{53} + \frac{47}{53}i && \text{Write in standard form.}
\end{aligned}
$$

ANSWERS

a. -28 **b.** $-3 - 18i$
c. $61 + 41i$ **d.** 58

b. $\dfrac{9}{2i}$

Solution Think of $2i$ as $0 + 2i$ so the conjugate is $0 - 2i$, which is $-2i$.

$$\frac{9}{2i} = \frac{9}{2i} \cdot \frac{-2i}{-2i} \quad \text{Multiply numerator and denominator by } -2i.$$
$$= \frac{-18i}{-4i^2}$$
$$= \frac{-18i}{-4(-1)} \quad \text{Replace } i^2 \text{ with } -1.$$
$$= \frac{-18i}{4}$$
$$= -\frac{9}{2}i$$

Note: *Remember the goal is to eliminate i from the denominator. Multiplying by $\frac{i}{i}$ would have worked, too.*

$$\frac{9}{2i} = \frac{9}{2i} \cdot \frac{i}{i} = \frac{9i}{2i^2} = \frac{9i}{2(-1)} = -\frac{9}{2}i$$

YOUR TURN Write in standard form.

a. $\dfrac{6 + i}{4 - 5i}$

b. $\dfrac{8}{3i}$

OBJECTIVE 3. Raise i to powers. We have defined i as $\sqrt{-1}$ and seen that $i^2 = -1$. Raising i to other powers leads to an interesting pattern.

$$i^1 = i \qquad i^5 = i^4 \cdot i = 1 \cdot i = i$$
$$i^2 = -1 \qquad i^6 = i^4 \cdot i^2 = 1 \cdot (-1) = -1$$
$$i^3 = i^2 \cdot i = -1 \cdot i = -i \qquad i^7 = i^4 \cdot i^3 = 1 \cdot (-i) = -i$$
$$i^4 = (i^2)^2 = (-1)^2 = 1 \qquad i^8 = (i^4)^2 = 1^2 = 1$$

Note: *If i's exponent is an even number not divisible by 4, then the result is -1. If i's exponent is divisible by 4, then the result is 1. If the exponent is an odd number that precedes a number divisible by 4, then the result is $-i$.*

If we continue the pattern, we get:

$$i^1 = i \qquad i^5 = i \qquad i^9 = i$$
$$i^2 = -1 \qquad i^6 = -1 \qquad i^{10} = -1$$
$$i^3 = -i \qquad i^7 = -i \qquad i^{11} = -i$$
$$i^4 = 1 \qquad i^8 = 1 \qquad i^{12} = 1$$

Notice that i to any integer power is i, -1, $-i$, or 1. This pattern allows us to find i to any integer power. We will use the fact that since $i^4 = 1$, then $(i^4)^n = 1$ for any integer value of n.

EXAMPLE 5 Simplify.

a. i^{25}

Solution $i^{25} = i^{24} \cdot i$ — Write i^{25} as $i^{24} \cdot i$ since 24 is the largest multiple of 4 that is smaller than 25.

$= (i^4)^6 \cdot i$ — Write i^{24} as $(i^4)^6$ since $i^4 = 1$.

$= 1^6 \cdot i$ — Replace i^4 with 1.

$= 1 \cdot i$

$= i$

ANSWERS

a. $\frac{19}{41} + \frac{34}{41}i$

b. $-\frac{8}{3}i$

b. i^{-14}

Solution $i^{-14} = \frac{1}{i^{14}}$ Write i^{-14} with a positive exponent.

$= \frac{1}{i^{12} \cdot i^2}$ Write i^{14} as $i^{12} \cdot i^2$ since 12 is the largest multiple of 4 that is smaller than 14.

$= \frac{1}{(i^4)^3 \cdot (-1)}$ Write i^{12} as $(i^4)^3$ and replace i^2 with -1.

$= \frac{1}{1(-1)}$ $(i^4)^3 = 1^3 = 1$.

$= -1$

YOUR TURN Simplify.

a. i^{43} **b.** i^{-18}

ANSWERS

a. $-i$ **b.** -1

8.7 Exercises

1. As an imaginary number, what does *i* represent?

2. Is every real number a complex number? Explain.

3. Is every complex number an imaginary number? Explain.

4. Explain how to add complex numbers.

5. Explain how to subtract complex numbers.

6. Is the expression $\frac{5 - 4i}{3}$ in standard form for a complex number? Explain.

For Exercises 7–22, write the imaginary number using i.

7. $\sqrt{-36}$ **8.** $\sqrt{-81}$ **9.** $\sqrt{-5}$ **10.** $\sqrt{-7}$

11. $\sqrt{-8}$ **12.** $\sqrt{-12}$ **13.** $\sqrt{-28}$ **14.** $\sqrt{-32}$

15. $\sqrt{-27}$

16. $\sqrt{-72}$

17. $\sqrt{-125}$

18. $\sqrt{-80}$

19. $\sqrt{-63}$

20. $\sqrt{-45}$

21. $\sqrt{-245}$

22. $\sqrt{-810}$

For Exercises 23–38, add or subtract.

23. $(9 + 3i) + (-3 + 4i)$

24. $(5 - 3i) + (4 - 5i)$

25. $(6 + 2i) + (5 - 8i)$

26. $(4 + i) + (7 - 6i)$

27. $(-4 + 6i) - (3 + 5i)$

28. $(8 - 3i) - (-1 - 2i)$

29. $(15 - 3i) - (-7i)$

30. $(19 + 6i) - (-2i)$

31. $(12 + 3i) + (-15 - 13i)$

32. $(14 - 7i) + (-8 - 9i)$

33. $(-5 - 9i) - (-5 - 9i)$

34. $(-4 + 2i) - (-4 + 2i)$

35. $(10 + i) - (2 - 13i) + (6 - 5i)$

36. $(-14 + 2i) + (6 + i) + (19 + 10i)$

37. $(5 - 2i) - (9 - 14i) + (16i)$

38. $(-12i) - (4 - 7i) - (18 + 4i)$

For Exercises 39–54, multiply.

39. $(8i)(3i)$

40. $(4i)(-7i)$

41. $(-8i)(5i)$

42. $(9i)(i)$

43. $2i(6 - 7i)$

44. $6i(9 + i)$

45. $-8i(4 + 9i)$

46. $-7i(5 - 8i)$

47. $(6 + i)(3 - i)$

48. $(5 - 2i)(4 + i)$

49. $(8 + 5i)(5 - 2i)$

50. $(6 - 3i)(7 + 8i)$

51. $(8 + i)(8 - i)$

52. $(4 + 9i)(4 - 9i)$

53. $(10 + i)^2$

54. $(5 - 3i)^2$

For Exercises 55–74, write in standard form.

55. $\frac{2}{i}$

56. $\frac{6}{-i}$

57. $\frac{4}{5i}$

58. $\frac{6}{7i}$

59. $\frac{6}{2i}$

60. $\frac{8}{4i}$

61. $\frac{2 - i}{2i}$

62. $\frac{3 + i}{3i}$

63. $\frac{4 + 2i}{4i}$

64. $\frac{5 - 3i}{4i}$

65. $\frac{7}{2 + i}$

66. $\frac{5}{6 + i}$

67. $\frac{2i}{3 - 7i}$

68. $\frac{-4i}{5 + i}$

69. $\frac{5 - 9i}{1 - i}$

70. $\frac{3 + i}{2 - i}$

71. $\frac{3 + i}{2 + 3i}$

72. $\frac{1 + 3i}{5 + 2i}$

73. $\frac{1 + 6i}{4 + 5i}$

74. $\frac{5 - 6i}{2 - 9i}$

For Exercises 75–90, find the powers of i.

75. i^{19}

76. i^{25}

77. i^{46}

78. i^{51}

79. i^{38}

80. i^{41}

81. i^{60}

82. i^{52}

83. i^{-20}

84. i^{-32}

85. i^{-26}

86. i^{-38}

87. i^{-21}

88. i^{-45}

89. i^{-35}

90. i^{-44}

REVIEW EXERCISES

[6.3] **1.** Factor: $4x^2 - 12x + 9$

[8.3] **2.** Simplify: $\pm\sqrt{48}$

For Exercises 3 and 4, solve.

[2.1] **3.** $3x - 2 = 4$

[6.4] **4.** $2x^2 - x - 6 = 0$

[3.1] **5.** Find the x- and y-intercepts for $2x + 3y = 6$.

[5.2] **6.** Graph: $y = x^2 - 3$

Chapter 8 Summary

Defined Terms

Section 8.1
n^{th} root (p. 534)
Radical function (p. 538)

Section 8.2
Rational exponent (p. 546)

Section 8.4
Like radicals (p. 567)

Section 8.6
Radical equation (p. 588)

Section 8.7
Imaginary unit (p. 599)
Imaginary number (p. 599)
Complex number (p. 600)
Complex conjugate (p. 602)

Procedures, Rules, and Key Examples

Procedures/Rules	Key Examples
Section 8.1 Radical Expressions and Functions If a is positive, then $\sqrt[n]{a} = b$, where $b \geq 0$ and $b^n = a$. If a is negative and n is even, then there is no real-number root. If a is negative and n is odd, then $\sqrt[n]{a} = b$, where b is negative and $b^n = a$.	**Example 1:** Simplify. **a.** $\sqrt{36} = 6$ **b.** $\sqrt{-25}$ is not a real number **c.** $-\sqrt{121} = -11$ **d.** $\sqrt{\frac{49}{100}} = \frac{7}{10}$ **e.** $\sqrt{16x^6} = 4\lvert x^3\rvert$ **f.** $\sqrt[3]{-27} = -3$ **Note:** *If we assume that the variables represent nonnegative numbers, then the solution for part e would be* $\sqrt{16x^6} = 4x^3$.
To evaluate a radical function, replace the variable with the indicated value and simplify.	**Example 2:** If $f(x) = \sqrt{4x + 3}$, find $f(3)$. Solution: $f(3) = \sqrt{4(3) + 3} = \sqrt{12 + 3} = \sqrt{15}$
For radical functions with an even index, the radicand must be nonnegative.	**Example 3:** Find the domain of $f(x) = \sqrt{2x - 4}$. Solution: Since $2x - 4$ must be nonnegative, we solve $2x - 4 \geq 0$. $2x - 4 \geq 0$ $2x \geq 4$ $x \geq 2$
Section 8.2 Rational Exponents $a^{1/n} = \sqrt[n]{a}$, where n is a natural number other than 1. **Note:** If a is negative and n is odd, then the root will be negative. If a is negative and n is even, then there is no real-number solution.	**Example 1:** Evaluate. **a.** $25^{1/2} = \sqrt{25} = 5$ **b.** $(-27)^{1/3} = \sqrt[3]{-27} = -3$ **c.** $(-16)^{1/4} = \sqrt[4]{-16}$, which is not a real number.

continued

Procedures/Rules	Key Examples
Section 8.2 Rational Exponents (continued)	
$a^{m/n} = \sqrt[n]{a^m} = (\sqrt[n]{a})^m$, where $a \geq 0$ and m and n are natural numbers other than 1.	**Example 2:** Write the following in exponential form. **a.** $\sqrt[3]{x^2} = x^{2/3}$ **b.** $(\sqrt[6]{x})^5 = x^{5/6}$
$a^{-m/n} = \dfrac{1}{a^{m/n}}$, where $a \neq 0$ and m and n are natural numbers with $n \neq 1$.	**Example 3:** Evaluate. $16^{-3/4} = \dfrac{1}{16^{3/4}} = \dfrac{1}{(\sqrt[4]{16})^3} = \dfrac{1}{2^3} = \dfrac{1}{8}$ **Example 4:** Write as a radical expression with a smaller root index. **a.** $\sqrt[4]{25} = 25^{1/4} = (5^2)^{1/4}$ $= 5^{2 \cdot 1/4} = 5^{1/2} = \sqrt{5}$ **b.** $\sqrt[8]{x^6} = x^{6/8} = x^{3/4} = \sqrt[4]{x^3}$ **Example 5:** Simplify. **a.** $\sqrt[4]{x} \cdot \sqrt{x} = x^{1/4} \cdot x^{1/2} = x^{1/4+1/2}$ $= x^{1/4+2/4}$ $= x^{3/4} = \sqrt[4]{x^3}$ **b.** $\dfrac{\sqrt[3]{x^2}}{\sqrt[4]{x}} = \dfrac{x^{2/3}}{x^{1/4}} = x^{(2/3)-(1/4)}$ $= x^{(8/12)-(3/12)} = x^{5/12} = \sqrt[12]{x^5}$ **c.** $\sqrt{5} \cdot \sqrt[3]{2} = 5^{1/2} \cdot 2^{1/3} = 5^{3/6} \cdot 2^{2/6}$ $= (5^3 \cdot 2^2)^{1/6} = (125 \cdot 4)^{1/6}$ $= 500^{1/6} = \sqrt[6]{500}$ **Example 6:** Simplify $\sqrt[3]{\sqrt[5]{x}}$ to a single radical. Solution: $\sqrt[3]{\sqrt[5]{x}} = (x^{1/5})^{1/3} = x^{(1/5)\cdot(1/3)}$ $= x^{1/15} = \sqrt[15]{x}$
Section 8.3 Multiplying, Dividing, and Simplifying Radicals Product rule for radicals: If $\sqrt[n]{a}$ and $\sqrt[n]{b}$ are both real numbers, then $\sqrt[n]{a} \cdot \sqrt[n]{b} = \sqrt[n]{a \cdot b}$.	**Example 1:** Simplify. Assume variables represent nonnegative values. **a.** $\sqrt{2} \cdot \sqrt{50} = \sqrt{2 \cdot 50} = \sqrt{100} = 10$ **b.** $\sqrt{3x} \cdot \sqrt{12x^3} = \sqrt{3x \cdot 12x^3}$ $= \sqrt{36x^4} = 6x^2$ **c.** $\sqrt[3]{5x} \cdot \sqrt[3]{3x} = \sqrt[3]{15x^2}$ **d.** $\sqrt[4]{27x^3} \cdot \sqrt[4]{3x} = \sqrt[4]{27x^3 \cdot 3x}$ $= \sqrt[4]{81x^4} = 3x$

continued

Procedures/Rules	Key Examples
Section 8.3 Multiplying, Dividing, and Simplifying Radicals (continued)	
Raising an nth root to the nth power: For any nonnegative real number a, $(\sqrt[n]{a})^n = a$.	**Example 2:** Simplify $(\sqrt[4]{x^3})^4$. Solution: $(\sqrt[4]{x^3})^4 = x^3$
Quotient rule for radicals: If $\sqrt[n]{a}$ and $\sqrt[n]{b}$ are both real numbers, then $\dfrac{\sqrt[n]{a}}{\sqrt[n]{b}} = \sqrt[n]{\dfrac{a}{b}}$, where $b \neq 0$.	**Example 3:** Simplify. **a.** $\dfrac{\sqrt{45}}{\sqrt{5}} = \sqrt{\dfrac{45}{5}} = \sqrt{9} = 3$ **b.** $\dfrac{\sqrt[3]{128x^7}}{\sqrt[3]{2x}} = \sqrt[3]{\dfrac{128x^7}{2x}} = \sqrt[3]{64x^6} = 4x^2$ **c.** $\sqrt{\dfrac{18x^5}{8x^3}} = \sqrt{\dfrac{9x^2}{4}} = \dfrac{\sqrt{9x^2}}{\sqrt{4}} = \dfrac{3x}{2}$ **d.** $\sqrt[4]{\dfrac{8}{x^4}} = \dfrac{\sqrt[4]{8}}{\sqrt[4]{x^4}} = \dfrac{\sqrt[4]{8}}{x}$
Simplifying an nth root. **1.** Write the radicand as the product of the greatest possible n^{th} power and a number or expression that has no perfect n^{th} power factors. **2.** Use the product rule $\sqrt[n]{ab} = \sqrt[n]{a} \cdot \sqrt[n]{b}$, where a is the perfect n^{th} power. **3.** Find the nth root of the perfect n^{th} power radicand.	**Example 4:** Simplify. **a.** $\sqrt{50} = \sqrt{25 \cdot 2} = \sqrt{25} \cdot \sqrt{2} = 5\sqrt{2}$ **b.** $\sqrt{48x^3} = \sqrt{16x^2 \cdot 3x}$ $= \sqrt{16x^2} \cdot \sqrt{3x} = 4x\sqrt{3x}$ **c.** $\sqrt[4]{48a^9} = \sqrt[4]{16a^8 \cdot 3a}$ $= \sqrt[4]{16a^8} \cdot \sqrt[4]{3a} = 2a^2\sqrt[4]{3a}$ **Example 5:** Perform the operation and simplify. $\sqrt{6} \cdot \sqrt{10} = \sqrt{60} = \sqrt{4 \cdot 15}$ $= \sqrt{4} \cdot \sqrt{15} = 2\sqrt{15}$
Section 8.4 Adding, Subtracting, and Multiplying Radical Expressions To add or subtract like radicals, add or subtract the coefficients and keep the radicals the same.	**Example 1:** Simplify. **a.** $2\sqrt{x} - 7\sqrt{x} = (2 - 7)\sqrt{x} = -5\sqrt{x}$ **b.** $4\sqrt[3]{5} + 2\sqrt[3]{5} = (4 + 2)\sqrt[3]{5} = 6\sqrt[3]{5}$ **Example 2:** Simplify the radicals, then combine like radicals. $9\sqrt{3} - \sqrt{12} + 6\sqrt{75}$ $= 9\sqrt{3} - \sqrt{4 \cdot 3} + 6\sqrt{25 \cdot 3}$ $= 9\sqrt{3} - 2\sqrt{3} + 6 \cdot 5\sqrt{3}$ $= 9\sqrt{3} - 2\sqrt{3} + 30\sqrt{3}$ $= 37\sqrt{3}$ **Example 3:** Multiply then simplify. $(\sqrt{3} - 4)(\sqrt{3} + 6)$ $= \sqrt{3} \cdot \sqrt{3} + \sqrt{3} \cdot 6 - 4 \cdot \sqrt{3} - 4 \cdot 6$ **Use FOIL.** $= \sqrt{9} + 6\sqrt{3} - 4\sqrt{3} - 24$ **Multiply.** $= 3 + 6\sqrt{3} - 4\sqrt{3} - 24$ **Simplify $\sqrt{9}$.** $= -21 + 2\sqrt{3}$ **Combine like radicals.** ***continued***

Procedures/Rules	Key Examples

Section 8.4 Adding, Subtracting and Multiplying Radical Expressions (continued)

Example 4: Perform the following operations.

a. $\sqrt{6}\cdot\sqrt{8}+\sqrt{5}\cdot\sqrt{15}$
$$=\sqrt{48}+\sqrt{75}$$
$$=\sqrt{16\cdot 3}+\sqrt{25\cdot 3}$$
$$=4\sqrt{3}+5\sqrt{3}=9\sqrt{3}$$

b. $\dfrac{\sqrt{60}}{\sqrt{5}}+\sqrt{48}=\sqrt{\dfrac{60}{5}}+\sqrt{48}$
$$=\sqrt{12}+\sqrt{48}$$
$$=\sqrt{4\cdot 3}+\sqrt{16\cdot 3}$$
$$=2\sqrt{3}+4\sqrt{3}=6\sqrt{3}$$

Section 8.5 Rationalizing Numerators and Denominators of Radical Expressions

Summary of Simplest Form

A radical expression is in simplest form if:

1. All rational roots have been found.
2. There are no perfect n^{th} factors of radicands of the form $\sqrt[n]{a^n}$.
3. All possible products, quotients, sums, and differences have been found.
4. There are no radicals in the denominator of a fraction.

To rationalize a denominator containing a single n^{th} root, multiply the fraction by a 1 so that the product's denominator has a radicand that is a perfect n^{th} power.

To rationalize a denominator containing a sum or difference with at least one square root term, multiply the fraction by a 1 whose numerator and denominator are the conjugate of the denominator.

Example 1: Rationalize the denominator.

a. $\dfrac{7}{\sqrt{3}}=\dfrac{7}{\sqrt{3}}\cdot\dfrac{\sqrt{3}}{\sqrt{3}}=\dfrac{7\sqrt{3}}{3}$

b. $\dfrac{4}{\sqrt[3]{3}}=\dfrac{4}{\sqrt[3]{3}}\cdot\dfrac{\sqrt[3]{9}}{\sqrt[3]{9}}=\dfrac{4\sqrt[3]{9}}{\sqrt[3]{27}}=\dfrac{4\sqrt[3]{9}}{3}$

c. $\dfrac{6}{4-\sqrt{5}}=\dfrac{6}{4-\sqrt{5}}\cdot\dfrac{4+\sqrt{5}}{4+\sqrt{5}}$
$$=\frac{6\cdot 4+6\cdot\sqrt{5}}{16-5}$$
$$=\frac{24+6\sqrt{5}}{11}$$

The numerator of a rational expression is rationalized exactly the same way a denominator is rationalized.

Example 2: Rationalize the numerator.

a. $\dfrac{\sqrt{6}}{3}=\dfrac{\sqrt{6}}{3}\cdot\dfrac{\sqrt{6}}{\sqrt{6}}=\dfrac{\sqrt{36}}{3\sqrt{6}}$
$$=\frac{6}{3\sqrt{6}}=\frac{2}{\sqrt{6}}$$

b. $\dfrac{2-\sqrt{3}}{5}=\dfrac{2-\sqrt{3}}{5}\cdot\dfrac{2+\sqrt{3}}{2+\sqrt{3}}$
$$=\frac{4-3}{5(2+\sqrt{3})}=\frac{1}{10+5\sqrt{3}}$$

continued

Procedures/Rules	Key Examples

Section 8.6 Radical Equations and Problem Solving

Power Rule: If $a = b$, then $a^n = b^n$.

To solve a radical equation:

1. Isolate the radical. (If there is more than one radical term, then isolate one of the radical terms.)
2. Raise both sides of the equation to the same power as the root index.
3. If all radicals have been eliminated, then solve. If a radical term remains, then isolate that radical term and raise both sides to the same power as its root index.
4. Check each solution. Any apparent solution that does not check is an extraneous solution.

Example 1: Solve $\sqrt{x-5} = 7$.

$(\sqrt{x-5})^2 = 7^2$ **Square both sides.**

$x - 5 = 49$

$x = 54$ **Add 5 on both sides.**

Check: $\sqrt{54-5} = 7$

$\sqrt{49} = 7$ **True**

Example 2: Solve $\sqrt{n+14} = n + 2$.

$(\sqrt{n+14})^2 = (n+2)^2$ **Square both sides.**

$n + 14 = n^2 + 4n + 4$

$0 = n^2 + 3n - 10$ **Subtract n and 14 on both sides.**

$0 = (n+5)(n-2)$ **Factor.**

$n + 5 = 0$ or $n - 2 = 0$ **Use the zero-factor theorem.**

$n = -5 \qquad n = 2$

Check:

$\sqrt{-5+14} = -5 + 2 \qquad \sqrt{2+14} = 2 + 2$

$\sqrt{9} = -3$ **False** $\qquad \sqrt{16} = 4$ **True**

The only solution is 2 (−5 is an extraneous solution).

Example 3: Solve $3 + \sqrt{t} = \sqrt{t+21}$.

$(3+\sqrt{t})^2 = (\sqrt{t+21})^2$ **Square both sides.**

$9 + 6\sqrt{t} + t = t + 21$

$6\sqrt{t} = 12$ **Subtract 9 and t on both sides.**

$\sqrt{t} = 2$ **Divide both sides by 6.**

$(\sqrt{t})^2 = 2^2$ **Square both sides.**

$t = 4$

We will leave the check to the reader.

continued

Procedures/Rules	Key Examples
Section 8.7 Complex Numbers	
To write an imaginary number $\sqrt{-n}$ in terms of the imaginary unit i: **1.** Separate the radical into two factors $\sqrt{-1} \cdot \sqrt{n}$. **2.** Replace $\sqrt{-1}$ with i. **3.** Simplify $\sqrt{n}$.	**Example 1:** Write using the imaginary unit. **a.** $\sqrt{-36} = \sqrt{-1} \cdot \sqrt{36} = 6i$ **b.** $\sqrt{-32} = \sqrt{-1} \cdot \sqrt{32}$ $= i \cdot 4\sqrt{2} = 4i\sqrt{2}$
To add complex numbers, combine like terms.	**Example 2:** Add $(5 - 6i) + (9 + 2i)$. $(5 - 6i) + (9 + 2i) = 14 - 4i$
To subtract complex numbers, change the signs of the second complex numbers, then combine like terms.	**Example 3:** Subtract $(7 - i) - (3 + 5i)$. $(7 - i) - (3 + 5i) = (7 - i) + (-3 - 5i)$ $= 4 - 6i$
To multiply complex numbers, follow the same procedures as for multiplying monomials or binomials (FOIL). Remember that $i^2 = -1$.	**Example 4:** Multiply $(6 + 5i)(2 - 3i)$. $(6 + 5i)(2 - 3i) = 12 - 18i + 10i - 15i^2$ $= 12 - 8i - 15(-1)$ $= 12 - 8i + 15$ $= 27 - 8i$
To divide complex numbers, rationalize the denominator using the complex conjugate.	**Example 5:** Write $\frac{2 - 3i}{4 + 5i}$ in standard form. $\frac{2 - 3i}{4 + 5i} = \frac{2 - 3i}{4 + 5i} \cdot \frac{4 - 5i}{4 - 5i}$ $= \frac{8 - 10i - 12i + 15i^2}{16 - 25i^2}$ $= \frac{8 - 22i + 15(-1)}{16 - 25(-1)}$ $= \frac{-7 - 22i}{41}$ $= -\frac{7}{41} - \frac{22}{41}i$
Powers of i. $i^1 = i$ $i^2 = -1$ $i^3 = -i$ $i^4 = 1$	**Example 6:** Simplify. **a.** $i^{19} = i^{16} \cdot i^3 = (i^4)^4 \cdot (-i)$ $= 1^4(-i) = 1(-i) = -i$ **b.** $i^{-21} = \frac{1}{i^{21}} = \frac{1}{i^{20} \cdot i} = \frac{1}{(i^4)^5 \cdot i} = \frac{1}{1^5 \cdot i}$ $= \frac{1}{i} = \frac{1}{i} \cdot \frac{-i}{-i} = \frac{-i}{-i^2}$ $= \frac{-i}{-(-1)} = \frac{-i}{1} = -i$

Chapter 8 Review Exercises

For Exercises 1–5, answer true or false.

[8.1] **1.** Every positive number has two square roots: a positive root and a negative root.

[8.1] **2.** $\sqrt{x^2} = x$, where x is any real number.

[8.5] **3.** The expression $\frac{3}{\sqrt{2}}$ is in simplest form.

[8.6] **4.** Every radical equation has a single solution.

[8.6] **5.** It is necessary to check all potential solutions to radical equations.

For Exercises 6–10, complete the rule.

[8.4] **6.** Like radicals are two radical expression with identical __________ and indexes.

[8.4] **7.** To add or subtract like radicals, add or subtract the __________ and keep the radicals the same.

[8.5] **8.** To rationalize a denominator that is a single square root,

1. __________ both the numerator and denominator of the fraction by the same square root as appears in the denominator.

2. Simplify.

[8.5] **9.** To rationalize a denominator containing a sum or difference with a square root term, multiply the numerator and denominator by the __________ of the denominator.

[8.6] **10.** An apparent solution to a radical equation that does not solve the equation is considered __________.

[8.1] ***For Exercises 11 and 12, find all square roots of the given number.***

11. 121

12. 49

[8.1] ***For Exercises 13–16, evaluate the square root.***

13. $\sqrt{169}$

14. $-\sqrt{49}$

15. $\sqrt{-36}$

16. $\sqrt{\frac{1}{25}}$

[8.1] *For Exercises 17 and 18, use a calculator to approximate each root to the nearest thousandth.*

17. $\sqrt{7}$

18. $\sqrt{90}$

[8.1] *For Exercises 19–26, simplify. Assume variables represent nonnegative numbers.*

19. $\sqrt{49x^8}$

20. $\sqrt{144a^6b^{12}}$

21. $\sqrt{0.16m^2n^{10}}$

22. $\sqrt[3]{x^{15}}$

23. $\sqrt[3]{-64r^9s^3}$

24. $\sqrt[4]{81x^{12}}$

25. $\sqrt[5]{32x^{15}y^{20}}$

26. $\sqrt[7]{x^{14}y^7}$

[8.1] *For Exercises 27–28, simplify. Assume variables can represent any real number.*

27. $\sqrt{81x^2}$

28. $\sqrt[4]{(x-1)^8}$

[8.2] *For Exercises 29–32, rewrite using radical notation and evaluate. Assume all variables represent nonnegative quantities.*

29. $(-64)^{1/3}$

30. $(24a^4)^{1/2}$

31. $\left(\frac{1}{32}\right)^{3/5}$

32. $(5r-2)^{5/7}$

[8.2] *For Exercises 33–38, write in exponential form.*

33. $\sqrt[8]{33}$

34. $\frac{8}{\sqrt[7]{n^3}}$

35. $(\sqrt[5]{8})^3$

36. $(\sqrt[3]{m})^8$

37. $(\sqrt[4]{3xw})^3$

38. $\sqrt[3]{(a+b)^4}$

[8.2] *For Exercises 39–44, use the rules of exponents to simplify.*

39. $x^{2/3} \cdot x^{4/3}$

40. $(4m^{1/4})(8m^{5/4})$

41. $\frac{y^{3/5}}{y^{4/5}}$

42. $\frac{b^{2/5}}{b^{-3/5}}$

43. $(k^{2/3})^{3/4}$

44. $(2xy^{1/5})^{3/4}$

[8.3] *For Exercises 45–56, simplify. Assume variables represent nonnegative values.*

45. $\sqrt{3} \cdot \sqrt{27}$

46. $\sqrt{5x^5} \cdot \sqrt{20x^3}$

47. $\sqrt[3]{2} \cdot \sqrt[3]{4}$

48. $\sqrt[4]{7} \cdot \sqrt[4]{6}$

49. $\sqrt[5]{3x^2y^3} \cdot \sqrt[5]{5x^2y}$

50. $\sqrt{\frac{49}{121}}$

51. $\sqrt[3]{-\frac{27}{8}}$

52. $\sqrt{72}$

53. $4b\sqrt{27b^7}$

54. $5\sqrt[3]{108}$

55. $2\sqrt[3]{40x^{10}}$

56. $2x^4\sqrt[4]{162x^7}$

[8.3] ***For Exercises 57–60, find the products and write the answer in simplified form. Assume variables represent nonnegative values.***

57. $4\sqrt{6} \cdot 7\sqrt{15}$

58. $\sqrt{x^9} \cdot \sqrt{x^6}$

59. $4\sqrt{10c} \cdot 2\sqrt{6c^4}$

60. $a\sqrt{a^3b^2} \cdot b^2\sqrt{a^5b^3}$

[8.3] ***For Exercises 61–64, find the quotients and write the answers in simplified form. Assume all variables represent nonnegative values.***

61. $\frac{\sqrt{48}}{\sqrt{6}}$

62. $\frac{9\sqrt{160}}{3\sqrt{8}}$

63. $\frac{8\sqrt{45a^5}}{2\sqrt{5a}}$

64. $\frac{36\sqrt{96x^6y^{11}}}{4\sqrt{3x^2y^4}}$

[8.4] ***For Exercises 65–72, find the sum or difference.***

65. $-5\sqrt{n} + 2\sqrt{n}$

66. $3y^3\sqrt[4]{8y} - 9y^3\sqrt[4]{8y}$

67. $\sqrt{45} + \sqrt{20}$

68. $4\sqrt{24} - 6\sqrt{54}$

69. $\sqrt{150} - \sqrt{54} + \sqrt{24}$

70. $4\sqrt{72x^2y} - 2x\sqrt{128y} + 5\sqrt{32x^2y}$

71. $\sqrt[3]{250x^4y^5} + \sqrt[3]{128x^4y^5}$

72. $\sqrt[4]{48} + \sqrt[4]{243}$

[8.4] ***For Exercises 73–82, find the products.***

73. $\sqrt{5}(\sqrt{3} + \sqrt{2})$

74. $\sqrt[3]{7}(\sqrt[3]{3} + 2\sqrt[3]{7})$

75. $3\sqrt{6}(2 - 3\sqrt{6})$

76. $(\sqrt{2} - \sqrt{3})(\sqrt{5} + \sqrt{7})$

77. $(\sqrt[3]{2} - 4)(\sqrt[3]{4} + 2)$

78. $(\sqrt[4]{6x} + 2)(\sqrt[4]{2x} - 1)$

79. $(\sqrt{5a} + \sqrt{3b})(\sqrt{5a} - \sqrt{3b})$

80. $(2\sqrt{3} - \sqrt{5})(2\sqrt{3} + \sqrt{5})$

81. $(\sqrt{2} + 1)^2$

82. $(\sqrt{2} - \sqrt{5})^2$

[8.5] ***For Exercises 83–92, rationalize the denominator and simplify.***

83. $\frac{1}{\sqrt{2}}$ **84.** $\frac{3}{\sqrt[3]{3}}$ **85.** $\sqrt{\frac{4}{7}}$ **86.** $\frac{\sqrt[4]{5x^2}}{\sqrt[4]{2}}$

87. $\sqrt[3]{\frac{17}{3y^2}}$ **88.** $\frac{\sqrt[4]{9}}{\sqrt[4]{27}}$ **89.** $\frac{4}{\sqrt{2}-\sqrt{3}}$ **90.** $\frac{1}{4+\sqrt{3}}$

91. $\frac{1}{2-\sqrt{n}}$ **92.** $\frac{2\sqrt{3}}{3\sqrt{2}-2\sqrt{3}}$

[8.5] ***For Exercises 93–96, rationalize the numerator.***

93. $\frac{\sqrt{10}}{6}$ **94.** $\frac{\sqrt{3x}}{5}$ **95.** $\frac{2-\sqrt{3}}{8}$ **96.** $\frac{2\sqrt{t}+\sqrt{3t}}{5t}$

Equations and Inequalities

Exercises 97–10

[8.6] ***For Exercises 97–108, solve.***

97. $\sqrt{x}=9$ **98.** $\sqrt{y}=-3$ **99.** $\sqrt{w-1}=3$ **100.** $\sqrt[3]{3x-2}=-2$

101. $\sqrt[4]{x-2}-3=-1$ **102.** $\sqrt{y+1}=\sqrt{2y-4}$ **103.** $\sqrt{x-6}=x+2$ **104.** $\sqrt[3]{3x+10}-4=5$

105. $\sqrt[4]{x+8}=\sqrt[4]{2x+1}$ **106.** $\sqrt{5n-1}=4-2n$ **107.** $1+\sqrt{x}=\sqrt{2x+1}$ **108.** $\sqrt{3x+1}=2-\sqrt{3x}$

[8.7] ***For Exercises 109 and 110, write the imaginary number using i.***

109. $\sqrt{-9}$ **110.** $\sqrt{-20}$

[8.7] ***For Exercises 111 and 112, add or subtract.***

111. $(3+2i)+(5-8i)$ **112.** $(7-3i)-(-2+4i)$

[8.7] ***For Exercises 113–116, multiply and simplify.***

113. $(3i)(4i)$ **114.** $2i(4-i)$ **115.** $(6+2i)(4-i)$ **116.** $(5-i)^2$

[8.7] *For Exercises 117–122, write in standard form.*

117. $\frac{5}{i}$

118. $\frac{3}{-i}$

119. $\frac{4}{3i}$

120. $\frac{7+i}{5i}$

121. $\frac{3}{2+i}$

122. $\frac{5+i}{2-3i}$

[8.7] *For Exercises 123 and 124, simplify.*

123. i^{20}

124. i^{15}

[8.3] 125. Write an expression in simplest form for the area of the figure.

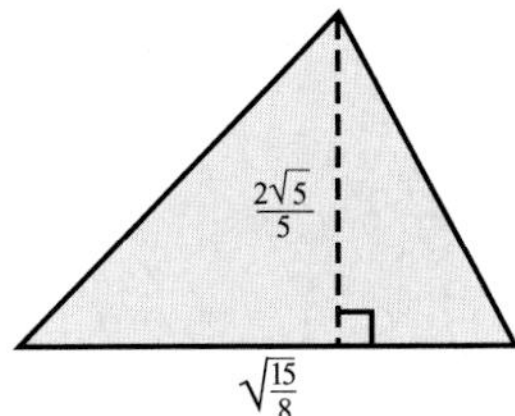

For Exercises 126 and 127, use the following. The speed of a car can be determined by the length of the skid marks using the formula $S = 2\sqrt{2L}$, where L represents the length of the skid mark in feet and S represents the speed of the car in miles per hour. (Source: Harris Technical Services, Traffic Accident Reconstructionists)

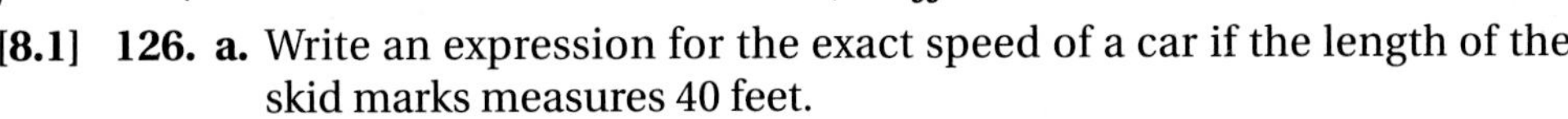

[8.1] 126. a. Write an expression for the exact speed of a car if the length of the skid marks measures 40 feet.

b. Approximate the speed to the nearest tenth.

[8.6] 127. Find the length of skid marks if a driver brakes hard at a speed of 50 miles per hour.

For Exercises 128 and 129, use the formula $T = 2\pi\sqrt{\frac{L}{9.8}}$, where T is the period of a pendulum in seconds and L is the length of the pendulum in meters.

[8.1] 128. Find the period of a pendulum with a length of 2.45 meters.

[8.6] 129. Suppose the period of the pendulum is $\frac{\pi}{3}$ seconds. Find the length.

For Exercises 130–131, use the formula $t = \sqrt{\dfrac{h}{16}}$, where t is the time in seconds that it takes on object to fall a distance of h feet.

[8.1] **130. a.** Write an expression in simplest form of the time that an object takes to fall 40 feet.

b. Approximate the time to the nearest hundredth.

[8.6] **131.** Find the distance an object falls in 0.3 seconds.

[8.1] **132.** Three pieces of lumber are to be connected to form a right triangle that will be part of the roof frame for a small storage building. If the horizontal piece is to be 4 feet and the vertical piece is to be 3 feet, how long must the connecting piece be?

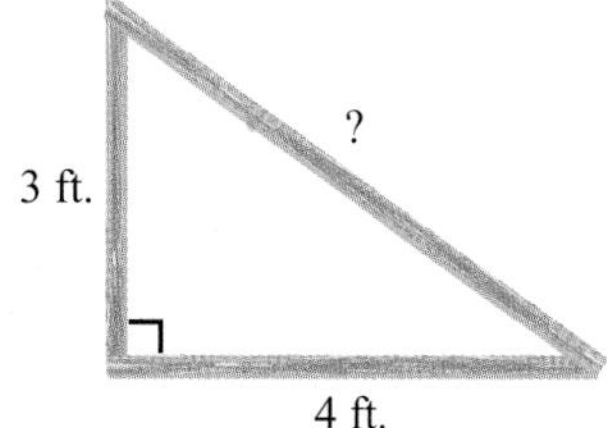

Chapter 8 Practice Test

For Exercises 1 and 2, evaluate the square root.

1. $\sqrt{36}$

2. $\sqrt{-49}$

For Exercises 3–12, simplify. Assume variables represent nonnegative values.

3. $\sqrt{81x^2y^5}$

4. $\sqrt[3]{54}$

5. $\sqrt[4]{4x} \cdot \sqrt[4]{4x^5}$

6. $-\sqrt[3]{-27r^{15}}$

7. $\dfrac{\sqrt{5}}{\sqrt{45}}$

8. $\dfrac{\sqrt[4]{1}}{\sqrt[4]{81}}$

9. $6\sqrt{7} - \sqrt{7}$

10. $(\sqrt{3} - 1)^2$

11. $x^{2/3} \cdot x^{-4/3}$

12. $(\sqrt[3]{2} - 4)(\sqrt[3]{4} + 2)$

For Exercises 13 and 14, write in exponential form.

13. $\sqrt[5]{8x^3}$

14. $\sqrt[3]{(2x + 5)^2}$

For Exercises 15 and 16, rationalize the denominator and simplify. Assume variables represent nonnegative values.

15. $\dfrac{1}{\sqrt[3]{4}}$

16. $\dfrac{\sqrt{x}}{\sqrt{x} + \sqrt{y}}$

For Exercises 17 and 18, solve the equation.

17. $\sqrt{3x - 2} = 8$

18. $\sqrt[4]{x + 8} = \sqrt[4]{2x + 1}$

For Exercises 19–22, simplify and write the answer in standard form ($a + bi$).

19. $(2 - i) - (4 + 3i)$

20. $(2i)(-i)$

21. $(4 - i)(4 + i)$

22. $\dfrac{2}{4 - 3i}$

23. Write an expression in simplest form for the area of the figure.

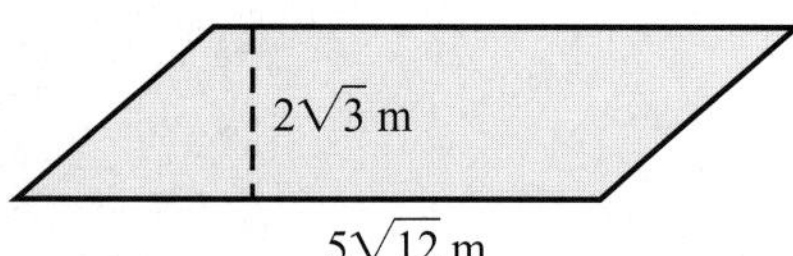

24. The formula $t = \sqrt{\frac{h}{16}}$ describes the amount of time t, in seconds, that an object falls a distance of h feet.

a. Write an expression in simplest form for the exact amount of time an object falls a distance of 12 feet.

b. Find the distance an object falls in 2 seconds.

25. The formula $S = \frac{7}{4}\sqrt{D}$ can be used to approximate the speed, S, in miles per hour, that a car was traveling prior to braking and skidding a distance D, in feet, on ice. (*Source:* Harris Technical Services, Traffic Accident Reconstructionists)

a. Find the speed of a car if the skid length measures 40 feet long.

b. Find the length of the skid marks that a car traveling 30 miles per hour would make if it brakes hard and skids to a halt.

CHAPTER 9

Quadratic Equations

"All our progress is unfolding like the vegetable bud."

—Ralph Waldo Emerson, (1803–1882), U.S. essayist, poet, philosopher

"For beautiful variety no crop can be compared with this. Here is not merely the plain yellow of the grains, but nearly all the colors that we know . . ."

—Henry David Thoreau

In Section 6.4, we introduced quadratic equations and solved them by factoring. However, not all quadratic equations can be solved by factoring, so we need more powerful methods. In this chapter, we explore three other methods for solving quadratic equations, each of which will use square roots. In the first method, we will solve quadratic equations using a new principle of equality, called the *square root principle.* Like factoring, this first method has some limitations, so we will develop a more powerful method called *completing the square.* Although we can complete the square to solve any quadratic equation, it is rather tedious. So we will develop a third method that uses a general formula, called the *quadratic formula,* to solve any quadratic equation. We will then use these methods to solve equations that are quadratic in form and that are quadratic inequalities. Finally, we will add to what we learned in Sections 5.2 and 6.4 and learn more about the graphs of quadratic functions.

9.1 Completing the Square

OBJECTIVES

1. Use the square root principle to solve quadratic equations.
2. Solve quadratic equations by completing the square.

OBJECTIVE 1. Use the square root principle to solve quadratic equations. In Section 6.4, we solved quadratic equations such as $x^2 = 25$ by subtracting 25 from both sides, factoring, and then using the zero-factor theorem. Let's recall that process:

$$x^2 - 25 = 0$$

$$(x - 5)(x + 5) = 0 \quad \text{Factor}$$

$$x - 5 = 0 \quad \text{or} \quad x + 5 = 0 \quad \text{Use the zero-factor theorem.}$$

$$x = 5 \qquad\qquad x = -5 \quad \text{Isolate } x \text{ in each equation.}$$

Another approach to solving $x^2 = 25$ involves square roots. Notice that the solutions to this equation must be numbers that can be squared to equal 25. These numbers are the square roots of 25, which are 5 and -5. This suggests a new rule called the *square root principle.*

RULE **The Square Root Principle**

If $x^2 = a$, where a is a real number, then $x = \sqrt{a}$ or $x = -\sqrt{a}$.

It is common to indicate the positive and negative solutions by writing $\pm\sqrt{a}$.

Note: *The expression $\pm\sqrt{a}$ is read "plus or minus the square root of a."*

For example, if $x^2 = 64$, then $x = \sqrt{64} = 8$ or $x = -\sqrt{64} = -8$. Or, we could simply write $x = \pm\sqrt{64} = \pm 8$.

Solve Equations in the Form $x^2 = a$

The square root principle is especially useful for solving equations in the form $x^2 = a$ when a is not a perfect square.

EXAMPLE 1 Solve. $x^2 = 40$

Connection We learned how to simplify square roots of numbers that have perfect square factors in Section 8.3.

Solution $x^2 = 40$

$$x = \pm\sqrt{40} \quad \text{Use the square root principle.}$$

$$x = \pm\sqrt{4 \cdot 10} \quad \text{Simplify by factoring out a perfect square.}$$

$$x = \pm 2\sqrt{10}$$

Note: *Remember the $\pm$ symbol means that the two solutions are $2\sqrt{10}$ and $-2\sqrt{10}$.*

We can check the two solutions using the original equation:

Check $2\sqrt{10}$: $(2\sqrt{10})^2 \stackrel{?}{=} 40$

$4 \cdot 10 = 40$ True

Check $-2\sqrt{10}$: $(-2\sqrt{10})^2 \stackrel{?}{=} 40$

$4 \cdot 10 = 40$ True

For the remaining examples, the checks will be left to the reader.

YOUR TURN Solve.

a. $x^2 = 81$

b. $x^2 = 50$

Solve Equations in the Form $ax^2 + b = c$

If an equation is in the form $ax^2 + b = c$, we use the addition and multiplication principles of equality to isolate x^2 and then use the square root principle.

EXAMPLE 2 Solve.

a. $x^2 + 2 = 100$

Solution $x^2 + 2 = 100$

$x^2 = 98$ Subtract 2 from both sides to isolate x^2.

$x = \pm\sqrt{98}$ Use the square root principle.

$x = \pm\sqrt{49 \cdot 2}$ Simplify by factoring out a perfect square.

$x = \pm 7\sqrt{2}$

b. $9x^2 = 27$

Solution $9x^2 = 27$

$x^2 = 3$ Divide both sides by 9 to isolate x^2.

$x = \pm\sqrt{3}$ Use the square root principle.

c. $5x^2 + 2 = 62$

Solution $5x^2 + 2 = 62$

$5x^2 = 60$ Subtract 2 from both sides.

$x^2 = 12$ Divide both sides by 5.

$x = \pm\sqrt{12}$ Use the square root principle.

$x = \pm\sqrt{4 \cdot 3}$ Simplify by factoring out a perfect square.

$x = \pm 2\sqrt{3}$

YOUR TURN Solve.

a. $x^2 - 6 = 42$

b. $3x^2 = 60$

c. $2x^2 + 11 = 65$

ANSWERS

a. ± 9 **b.** $\pm 5\sqrt{2}$

ANSWERS

a. $\pm 4\sqrt{3}$ **b.** $\pm 2\sqrt{5}$ **c.** $\pm 3\sqrt{3}$

Solve Equations in the Form $(ax + b)^2 = c$

In an equation in the form $(ax + b)^2 = c$, notice the expression $ax + b$ is squared. We can use the square root principle to eliminate the square.

EXAMPLE 3 Solve.

a. $(x + 6)^2 = 49$

Solution $(x + 6)^2 = 49$

$x + 6 = \pm\sqrt{49}$ — Use the square root principle.

$x + 6 = \pm 7$

$x = -6 \pm 7$ — Subtract 6 from both sides.

$x = -6 + 7$ or $x = -6 - 7$ — Simplify by separating the two solutions.

$x = 1$ $\quad$ $x = -13$

b. $(5x - 1)^2 = 10$

Solution $(5x - 1)^2 = 10$

$5x - 1 = \pm\sqrt{10}$ — Use the square root principle.

$2x = 1 \pm \sqrt{10}$ — Add 1 to both sides to isolate $5x$.

$x = \dfrac{1 \pm \sqrt{10}}{5}$ — Divide both sides by 5.

c. $(8x + 6)^2 = -32$

Note: *The square root of a negative number is an imaginary number, which we can rewrite using i.*

Solution $(8x + 6)^2 = -32$

$8x + 6 = \pm\sqrt{-32}$ — Use the square root principle.

$8x = -6 \pm \sqrt{-16 \cdot 2}$ — Subtract 6 from both sides and simplify the square root.

$8x = -6 \pm 4i\sqrt{2}$ — Rewrite the imaginary number using i notation.

$x = \dfrac{-6 \pm 4i\sqrt{2}}{8}$ — Divide both sides by 8.

$x = -\dfrac{6}{8} \pm \dfrac{4\sqrt{2}}{8}i$ — Write the complex number in standard form.

$x = -\dfrac{3}{4} \pm \dfrac{\sqrt{2}}{2}i$ — Simplify.

YOUR TURN Solve.

a. $(x - 3)^2 = 25$ **b.** $(2x + 1)^2 = 14$ **c.** $(4x - 5)^2 = -12$

ANSWERS

a. 8 or −2

b. $\dfrac{-1 \pm \sqrt{14}}{2}$ **c.** $\dfrac{5}{4} \pm \dfrac{\sqrt{3}}{2}i$

OBJECTIVE 2. Solve quadratic equations by completing the square. To make use of the square root principle, we need one side of the equation to be a perfect square and the other side to be a constant, as in $(x + 2)^2 = 5$. But, suppose we are given an equation like $x^2 + 6x = 2$ whose left-hand side is an "incomplete" square. We can use the addition principle of equality to add an appropriate number (9 in this case) to both sides so that the left-hand side becomes a perfect square. We call this process *completing the square.*

$x^2 + 6x + 9 = 2 + 9$ — Adding 9 to both sides completes the square on the left side.

We can now factor the left-hand side of the equation, then use the square root principle to finish solving the equation.

$(x + 3)^2 = 11$ Write the left-hand side in factored form.

$x + 3 = \pm\sqrt{11}$ Use the square root principle to eliminate the square.

$x = -3 \pm \sqrt{11}$ Subtract 3 from both sides to isolate x.

How do we know what number to add to complete the square?

Notice that squaring half of the coefficient of x gives the constant term in the completed square.

$$x^2 + 6x + 9$$

Half of 6 is 3, which when squared is 9.

Also notice that half of the coefficient of x is the constant in the factored form.

$$x^2 + 6x + 9 = (x + 3)^2$$

Half of 6 is 3, which is the constant in the factored form.

Connection The product of every perfect square in the form $(x + b)^2$ is a trinomial in the form $x^2 + 2bx + b^2$. Notice that half of x's coefficient, $2b$, is b, which is then squared to equal the last term.

Equations in the Form $x^2 + bx = c$

To solve a quadratic equation by completing the square, we need the equation in the form $x^2 + bx = c$, so that the coefficient of x^2 is 1. Once the equation is in the form $x^2 + bx = c$, we complete the square and then use the square root principle.

EXAMPLE 4 Solve by completing the square.

a. $x^2 + 12x + 15 = 0$

Solution We first write the equation in the form $x^2 + bx = c$.

Note: *We found 36 by squaring half of 12.* $\left(\frac{12}{2}\right)^2 = 6^2 = 36$

$x^2 + 12x = -15$ Subtract 15 from both sides to get the form $x^2 + bx = c$.

$x^2 + 12x + 36 = -15 + 36$ Complete the square by adding 36 to both sides.

$(x + 6)^2 = 21$ Factor.

$x + 6 = \pm\sqrt{21}$ Use the square root principle.

$x = -6 \pm \sqrt{21}$ Subtract 6 from both sides to isolate x.

Note: *We found* $\frac{49}{4}$ *by squaring half of* -7.

$$\left(\frac{-7}{2}\right)^2 = \frac{49}{4}$$

b. $x^2 - 7x + 8 = 5$

Solution $x^2 - 7x = -3$ — Subtract 8 from both sides to get the form $x^2 + bx = c$.

$x^2 - 7x + \frac{49}{4} = -3 + \frac{49}{4}$ — Complete the square by adding $\frac{49}{4}$ to both sides.

$\left(x - \frac{7}{2}\right)^2 = \frac{37}{4}$ — Factor the left side and simplify the right.

$x - \frac{7}{2} = \pm\sqrt{\frac{37}{4}}$ — Use the square root principle.

$x = \frac{7}{2} \pm \sqrt{\frac{37}{4}}$ — Add $\frac{7}{2}$ to both sides to isolate x.

$x = \frac{7}{2} \pm \frac{\sqrt{37}}{2}$ — Simplify the square root.

$x = \frac{7 \pm \sqrt{37}}{2}$ — Combine the fractions.

YOUR TURN Solve by completing the square.

a. $x^2 + 8x - 29 = 0$

b. $x^2 - 9x - 6 = 5$

Equations in the Form $ax^2 + bx = c$, where $a \neq 1$

So far our equations have been in the form $x^2 + bx = c$, where the coefficient of the x^2 term is 1. To solve an equation such as $2x^2 + 12x = 3$, we will need to divide both sides of the equation by 2 (or multiply both sides by $\frac{1}{2}$) so that the coefficient of x^2 is 1.

$\frac{2x^2 + 12x}{2} = \frac{3}{2}$ — Divide both sides by 2.

$$x^2 + 6x = \frac{3}{2}$$

We can now solve by completing the square.

$x^2 + 6x + 9 = \frac{3}{2} + 9$ — Add 9 to both sides to complete the square.

$(x + 3)^2 = \frac{21}{2}$ — Factor the left side and simplify the right.

$x + 3 = \pm\sqrt{\frac{21}{2}}$ — Use the square root principle.

$x = -3 \pm \sqrt{\frac{21}{2}}$ — Subtract 3 from both sides to isolate x.

$x = -3 \pm \frac{\sqrt{21}}{\sqrt{2}} \cdot \frac{\sqrt{2}}{\sqrt{2}}$ — Rationalize the denominator.

$$x = -3 \pm \frac{\sqrt{42}}{2}$$

ANSWERS

a. $-4 \pm 3\sqrt{5}$ **b.** $\frac{9 \pm 5\sqrt{5}}{2}$

We can now write a procedure for solving any quadratic equation by completing the square.

PROCEDURE **Solving Quadratic Equations by Completing the Square**

To solve a quadratic equation by completing the square:

1. Write the equation in the form $x^2 + bx = c$.
2. Complete the square by adding $\left(\frac{b}{2}\right)^2$ to both sides.
3. Write the completed square in factored form.
4. Use the square root principle to eliminate the square.
5. Isolate the variable.
6. Simplify as needed.

Connection In Section 9.4, we will use the process of completing the square to rewrite quadratic functions in a form that allows us to easily determine features of the graph.

EXAMPLE 5 Solve by completing the square.

a. $16x^2 - 24x = 1$

Solution

$$\frac{16x^2 - 24x}{16} = \frac{1}{16}$$ Divide both sides by 4.

$$x^2 - \frac{3}{2}x = \frac{1}{16}$$ Simplify.

$$x^2 - \frac{3}{2}x + \frac{9}{16} = \frac{1}{16} + \frac{9}{16}$$ Add $\frac{9}{16}$ to both sides to complete the square.

Note: *To complete the square, we square half of* $\frac{3}{2}$:

$$\left(\frac{1}{2} \cdot \frac{3}{2}\right)^2 = \left(\frac{3}{4}\right)^2 = \frac{9}{16}$$

$$\left(x - \frac{3}{4}\right)^2 = \frac{10}{16}$$ Factor the left side and simplify the right.

$$x - \frac{3}{4} = \pm\sqrt{\frac{10}{16}}$$ Use the square root principle.

$$x = \frac{3}{4} \pm \frac{\sqrt{10}}{4}$$ Add $\frac{3}{4}$ to both sides and simplify the square root.

$$x = \frac{3 \pm \sqrt{10}}{4}$$ Combine the fractions.

b. $2x^2 + 9x + 7 = -5$

Solution

$$2x^2 + 9x = -12$$ Subtract 7 from both sides.

$$\frac{2x^2 + 9x}{2} = \frac{-12}{2}$$ Divide both sides by 2.

$$x^2 + \frac{9}{2}x = -6$$ Simplify.

$$x^2 + \frac{9}{2}x + \frac{81}{16} = -6 + \frac{81}{16}$$ Add $\frac{81}{16}$ to both sides to complete the square.

$$\left(x + \frac{9}{4}\right)^2 = -\frac{15}{16}$$ Factor the left side and simplify the right.

$$x + \frac{9}{4} = \pm\sqrt{-\frac{15}{16}}$$ Use the square root principle.

$$x = -\frac{9}{4} \pm \frac{\sqrt{15}}{4}i$$ Subtract $\frac{9}{4}$ from both sides and write the complex number in standard form.

ANSWERS

a. $\frac{-3 \pm 2\sqrt{2}}{3}$ or $-1 \pm \frac{2\sqrt{2}}{3}$

b. $4 \pm 3i$

YOUR TURN Solve by completing the square.

a. $9x^2 + 18x = -1$

b. $x^2 - 8x + 11 = -14$

9.1 Exercises

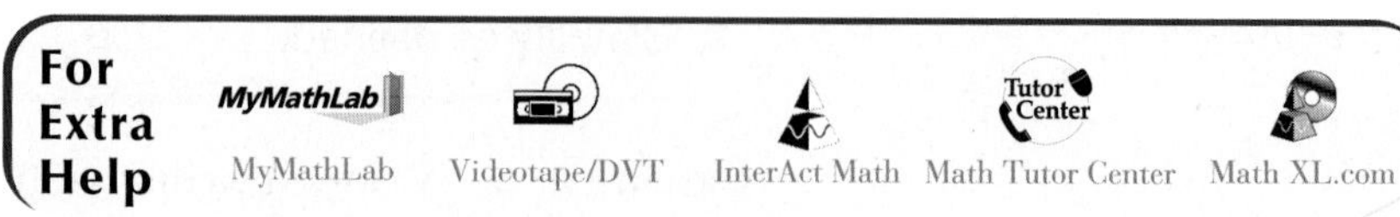

1. Explain why there are two solutions to an equation in the form $x^2 = a$.

2. Write a formula for the solutions of $ax^2 - b = c$ by solving for x.

3. Write a formula for the solutions of $(ax - b)^2 = c$ by solving for x.

4. Given an equation in the form $x^2 + bx = c$, explain how to complete the square.

5. Consider the equation $x^2 - 7x + 12 = 0$. Which is a better method for solving the equation, factoring and then using the zero-factor theorem or completing the square? Explain.

6. Consider the equation $x^2 + 4x + 5 = 0$. Which is a better method for solving the equation, factoring and then using the zero-factor theorem or completing the square? Explain.

For Exercises 7–16, solve and check.

7. $x^2 = 49$

8. $x^2 = 194$

9. $y^2 = \frac{4}{25}$

10. $t^2 = \frac{1}{64}$

11. $n^2 = 0.81$

12. $p^2 = 1.21$

13. $z^2 = 45$

14. $m^2 = 72$

15. $w^2 = -25$

16. $c^2 = -49$

For Exercises 17–36, solve and check. Begin by using the addition or multiplication principles of equality to isolate the squared term.

17. $n^2 - 7 = 42$

18. $y^2 - 5 = 59$

19. $y^2 - 7 = 29$

20. $k^2 + 5 = 30$

21. $4n^2 = 36$

22. $5y^2 = 125$

23. $25t^2 = 9$

24. $16d^2 = 49$

25. $4h^2 = -16$

26. $-3k^2 = 27$

27. $\frac{5}{6}x^2 = \frac{3}{8}$

28. $-\frac{2}{3}m^2 = -\frac{8}{5}$

29. $2x^2 + 5 = 21$

30. $3x^2 + 5 = 80$

31. $5y^2 - 7 = -97$

32. $4n^2 + 20 = -76$

33. $\frac{3}{4}y^2 - 5 = 3$

34. $\frac{25}{9}m^2 - 2 = 6$

35. $0.2t^2 - 0.5 = 0.012$

36. $0.5p^2 + 1.28 = 1.6$

For Exercises 37–52, solve and check. Use the square root principle to eliminate the square.

37. $(x + 8)^2 = 49$

38. $(y + 7)^2 = 144$

39. $(5n - 3)^2 = 16$

40. $(6h - 5)^2 = 81$

41. $(m - 7)^2 = -12$

42. $(t - 5)^2 = -28$

43. $(4k - 1)^2 = 40$

44. $(3x - 7)^2 = 50$

45. $(m - 8)^2 = -1$

46. $(t - 2)^2 = -4$

47. $\left(y - \frac{3}{4}\right)^2 = \frac{9}{16}$

48. $\left(x + \frac{4}{9}\right)^2 = \frac{25}{81}$

49. $\left(\frac{5}{9}d - \frac{1}{2}\right)^2 = \frac{1}{36}$

50. $\left(\frac{3}{4}h + \frac{4}{5}\right)^2 = \frac{1}{100}$

51. $(0.4x + 3.8)^2 = 2.56$

52. $(0.8n - 6.8)^2 = 1.96$

For Exercises 53–56, explain the mistake, then find the correct solutions.

53. $x^2 - 15 = 34$
$x^2 = 49$
$x = \sqrt{49}$
$x = 7$

54. $x^2 = 20$
$x = \sqrt{20}$
$x = 2\sqrt{5}$

55. $(x - 5)^2 = -6$
$x - 5 = \pm\sqrt{6}$
$x = 5 \pm \sqrt{6}$

56. $(x - 1)^2 = -12$
$x - 1 = \pm\sqrt{-12}$
$x = 1 \pm 2\sqrt{3}$

For Exercises 57–62, solve, then use a calculator to approximate the irrational solutions rounded to three places.

57. $x^2 = 96$

58. $t^2 = 56$

59. $y^2 - 15 = 5$

60. $x^2 - 22 = 6$

61. $(n - 6)^2 = 15$

62. $(m + 3)^2 = 10$

For Exercises 63–74: a. Add a term to the expression to make it a perfect square.
b. Factor the perfect square.

63. $x^2 + 14x$

64. $c^2 + 8c$

65. $n^2 - 10n$

66. $a^2 - 4a$

67. $y^2 - 7y$

68. $m^2 - 11m$

69. $s^2 - \frac{2}{3}s$

70. $y^2 - \frac{4}{5}y$

71. $m^2 + \frac{1}{7}m$

72. $v^2 + \frac{1}{5}v$

73. $p^2 + 9p$

74. $z^2 - 15z$

For Exercises 75–90, solve the equation by completing the square.

75. $w^2 + 2w = 15$

76. $p^2 + 8p = 9$

77. $y^2 + 10y = -16$

78. $x^2 + 10x = -24$

79. $r^2 - 2r + 50 = 0$

80. $c^2 - 6c + 45 = 0$

81. $k^2 = 9k - 18$

82. $a^2 = 7a - 10$

83. $n^2 + 9n - 20 = 16$

84. $u^2 - 5u - 9 = -13$

85. $b^2 - 2b - 11 = 5$

86. $z^2 + 2z = 6$

87. $h^2 - 6h + 3 = -26$

88. $j^2 - 4j + 25 = -3$

89. $u^2 + \frac{1}{2}u = \frac{3}{2}$

90. $y^2 + \frac{1}{3}y = \frac{2}{3}$

For Exercises 91–104, solve the equation by completing the square. Begin by writing the equation in the form $x^2 + bx = c$.

91. $4x^2 + 12x = 7$

92. $8m^2 - 10m = 3$

93. $2n^2 - n - 3 = 0$

94. $f^2 - 12f - 45 = 0$

95. $6w^2 - 6 = -5w$

96. $2t^2 - 5 = -3t$

97. $2g^2 + g - 11 = -5$

98. $4l^2 + l - 30 = 30$

99. $2x^2 = 4x + 3$

100. $2x^2 = 6x - 3$

101. $5k^2 + k - 2 = 0$

102. $3s^2 - 4s = 2$

103. $3a^2 - 1 = -6a$

104. $3x^2 - 4 = -8x$

For Exercises 105–106, explain the mistake.

105.

$$\begin{aligned} 3x^2 + 4x &= 2 \\ 3x^2 + 4x + 4 &= 2 + 4 \\ (3x + 2)^2 &= 6 \\ 3x + 2 &= \pm\sqrt{6} \\ 3x &= -2 \pm \sqrt{6} \\ x &= \frac{-2 \pm \sqrt{6}}{3} \end{aligned}$$

106.

$$\begin{aligned} x^2 + 6x &= 7 \\ x^2 + 6x + 9 &= 7 + 9 \\ (x + 3)^2 &= 16 \\ x + 3 &= \sqrt{16} \\ x + 3 &= 4 \\ x &= -3 + 4 \\ x &= 1 \end{aligned}$$

For Exercises 107–122, solve.

107. A square sheet of metal has an area of 196 square inches. What is the length of each side?

108. A severe thunderstorm warning is issued by the National Weather Service for a square area covering 14,400 square miles. What is the length of each side of the square area?

109. A tank is to be made for an aquarium so that the length is 8 feet, the width is twice the height, and the volume is 144 cubic feet. Find the height and width of the tank.

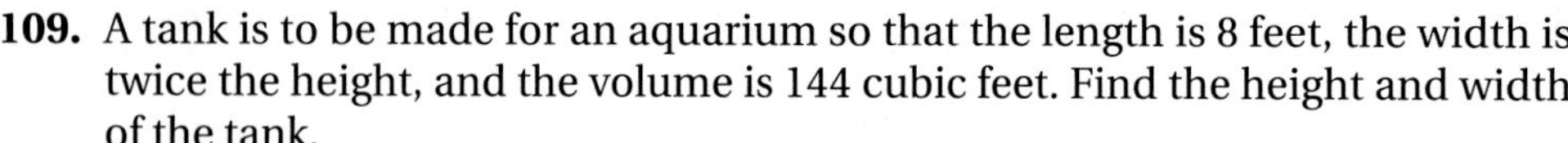

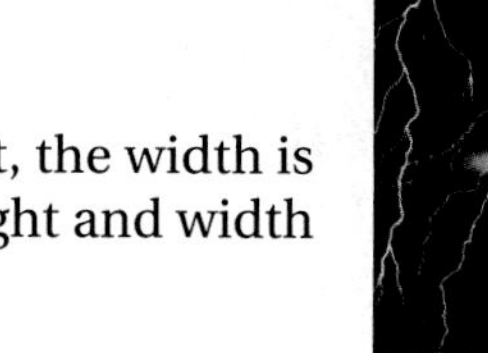

110. The length of a swimming pool is to be three times the width. If the depth is to be a constant 4 feet and the volume is to be 4800 cubic feet, find the length and width.

111. The Arecibo radio telescope is like a giant satellite dish that covers a circular area of approximately 23,256.25π square meters. Find the diameter of the dish.

Of Interest

The Arecibo radio telescope, the world's largest radio telescope, analyzes radiation and signals emitted by objects in space. It was built in a natural depression near Arecibo, Puerto Rico.

112. A field is planted in a circular pattern. A watering device is to be constructed with pipe in a line extending from the center of the field to the edge of the field. The pipe is set upon wheels so that it can rotate around the field and cover the entire field with water. If the area of the field is 7225π square feet, how long will the watering device be?

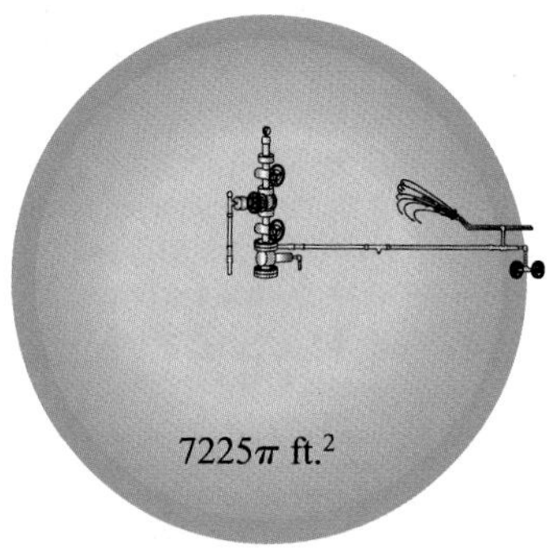

113. The corner of a sheet of plywood has been cut out as shown. If the area of the piece that was removed was 256 square inches, find x.

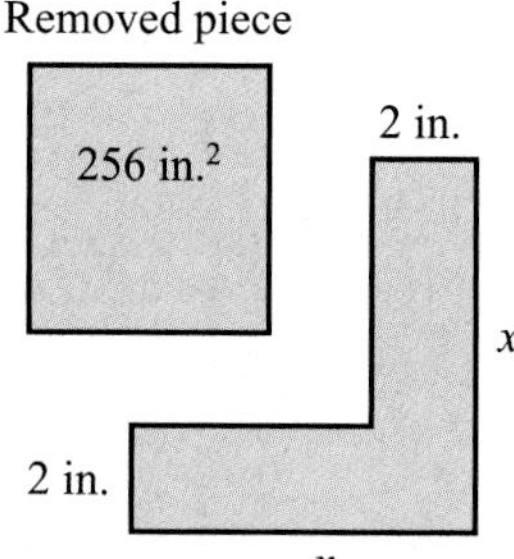

114. The area of the hole in the washer shown is 25π square millimeters. Find r, the radius of the washer.

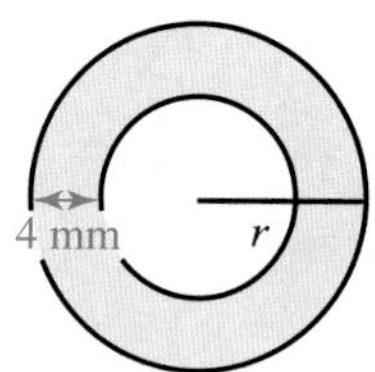

115. A rectangular basement is 6 feet longer than it is wide. If the area of the room is 315 square feet, what are the dimensions of the room?

116. The length of a rectangular garden is 25 feet more than the width. If the area of the garden is 3150 square feet, what are its dimensions?

117. An LCD computer monitor measures 20 inches across its diagonal. If the width is 4 inches less than the length, find the dimensions of the monitor.

118. Two identical right triangles are placed together to form the frame of a roof. If the base of each triangle is 7 feet longer than its height and its hypotenuse is 13 feet, what are the dimensions of the base and height?

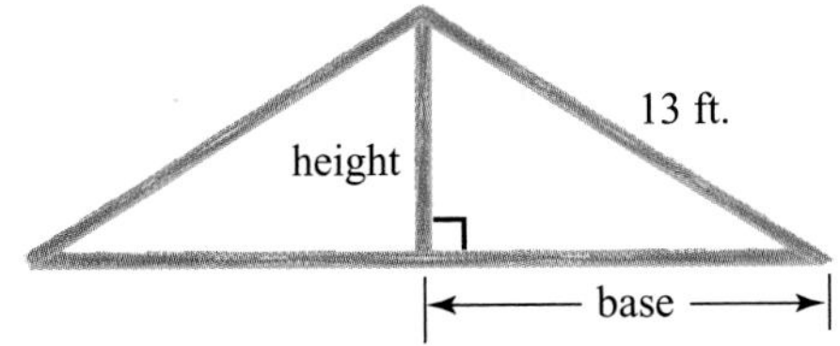

119. The metal panel shown is to have a total area of 1050 square inches. Find the length x.

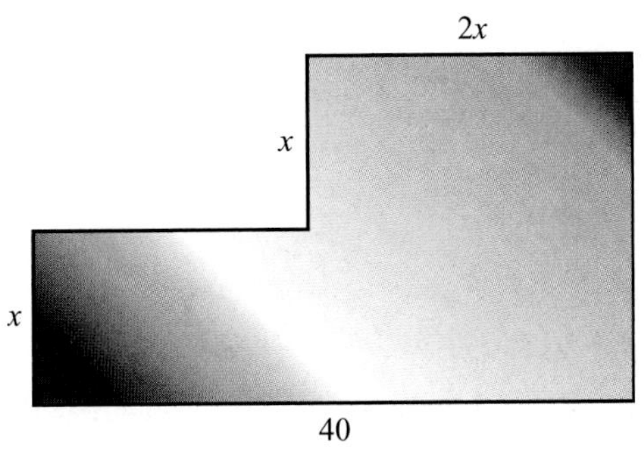

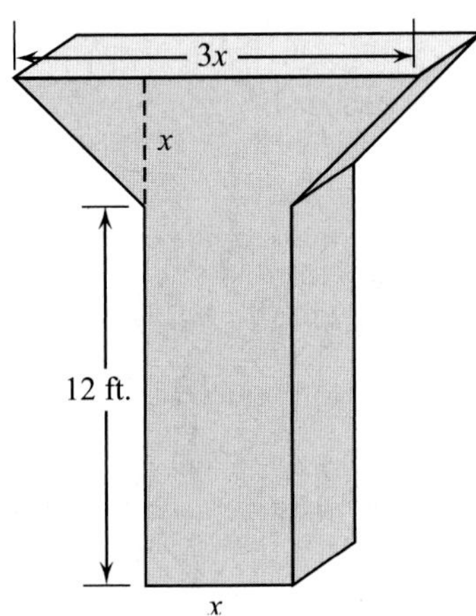

120. A side view of a concrete bridge support footing is shown. The side of the footing is to have a total area of 110 square feet. Find the length of x.

121. A plastic panel is to have a rectangular hole cut as shown.

a. Find l so that the area remaining after the hole is cut is 1230 square centimeters.

b. Find the length and width of the plastic panel.

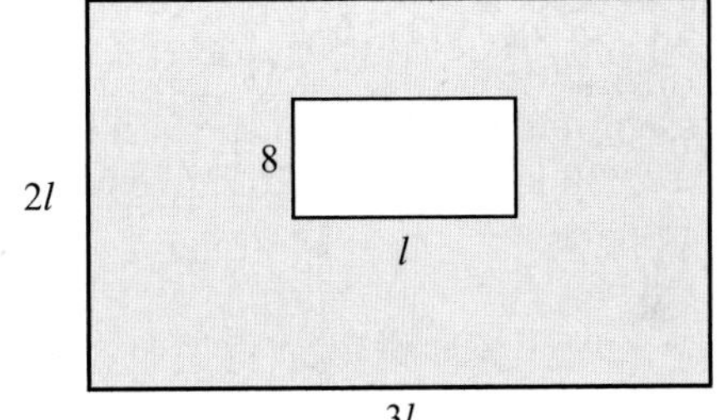

122. A 6-inch-wide groove with a height of h is to be cut into a wood block as shown.

a. Find h so that the volume remaining in the block after the groove is cut is 360 cubic inches.

b. Find the height and width of the block.

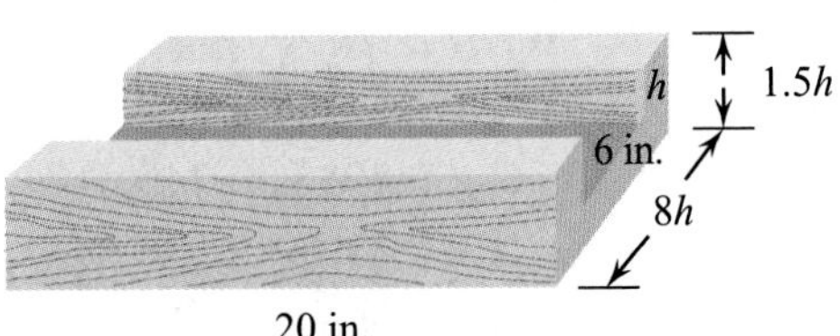

For Exercises 123–126, if an object is dropped, the formula $d = 16t^2$ describes the distance d in feet that the object falls in t seconds.

123. Suppose a cover for a ceiling light falls from a 9-foot ceiling. How long does it take for the cover to hit the floor?

124. A construction worker tosses a scrap piece of lumber from the roof of a house. How long does it take the piece of lumber to reach the ground 25 feet below?

125. A toy rocket is launched straight up and reaches a height of 180 feet in 1.5 seconds then plummets back to the ground. Determine the total time of the rocket's round-trip.

126. A lead weight is dropped from a tower at a height of 150 feet. How much time passes for the weight to be halfway to the ground? Write an exact answer, then approximate to the nearest tenth of a second.

For Exercises 127 and 128, use the following information. In physics, if an object is in motion it has kinetic energy. The formula $E = \frac{1}{2}mv^2$ is used to calculate the kinetic energy E of an object with a mass m and velocity v. If the mass is measured in kilograms and the velocity is in meters per second, then the kinetic energy will be in units called joules (J).

127. Suppose an object with a mass of 50 kilograms has 400 joules of kinetic energy. Find its velocity.

128. In a crash test, a vehicle with a mass of 1200 kilograms is found to have kinetic energy of 117,600 joules just before impact. Find the velocity of the vehicle just before impact.

PUZZLE PROBLEM

Without using a calculator, which of the following four numbers is a perfect square? (Hint: Write a list of smaller perfect squares and look for a pattern)

9,456,804,219,745,618

2,512,339,789,576,516

7,602,985,471,286,543

4,682,715,204,643,182

REVIEW EXERCISES

[6.3] ***For Exercises 1 and 2, factor.***

1. $x^2 - 10x + 25$

2. $x^2 + 6x + 9$

[8.1] ***For Exercises 3 and 4, simplify. Assume variables represent nonnegative values.***

3. $\sqrt{36x^2}$

4. $\sqrt{(2x + 3)^2}$

[1.3, 8.1, 8.7] ***For Exercises 5 and 6, simplify.***

5. $\dfrac{-4 + \sqrt{8^2 - 4(2)(5)}}{2(2)}$

6. $\dfrac{-6 - \sqrt{6^2 - 4(5)(2)}}{2(5)}$

9.2 Solving Quadratic Equations Using the Quadratic Formula

OBJECTIVES

1. Solve quadratic equations using the quadratic formula.
2. Use the discriminant to determine the number of real solutions that a quadratic equation will have.
3. Find the *x*- and *y*-intercepts of a quadratic function.
4. Solve applications using the quadratic formula.

In this section, we will solve quadratic equations using a formula called the *quadratic formula.* This formula is much easier to use than completing the square.

OBJECTIVE 1. Solve quadratic equations using the quadratic formula. To derive the quadratic formula, we begin with the general form of the quadratic equation, $ax^2 + bx + c = 0$, and assume that $a \neq 0$. We will follow the procedure for solving a quadratic equation by completing the square.

$$ax^2 + bx + c = 0$$

$$ax^2 + bx = -c \qquad \text{Subtract } c \text{ from both sides.}$$

$$\frac{ax^2 + bx}{a} = \frac{-c}{a} \qquad \text{Divide both sides by } a \text{ so that the coefficient of } x^2 \text{ will be 1.}$$

$$x^2 + \frac{b}{a}x = -\frac{c}{a} \qquad \text{Simplify.}$$

$$x^2 + \frac{b}{a}x + \frac{b^2}{4a^2} = -\frac{c}{a} + \frac{b^2}{4a^2} \qquad \text{Complete the square.}$$

Note: To complete the square, we square half of $\frac{b}{a}$.

$$\left(\frac{1}{2}\cdot\frac{b}{a}\right)^2 = \left(\frac{b}{2a}\right)^2 = \frac{b^2}{4a^2}$$

$$\left(x + \frac{b}{2a}\right)^2 = -\frac{4ac}{4a^2} + \frac{b^2}{4a^2}$$

On the left side, write factored form. On the right side, rewrite $-\frac{c}{a}$ with the common denominator $4a^2$ so that we can combine the rational expressions.

$$\left(x + \frac{b}{2a}\right)^2 = \frac{b^2 - 4ac}{4a^2}$$

Combine the rational expressions. For simplicity, we rearrange the order of the b^2 and $-4ac$ terms in the numerator.

$$x + \frac{b}{2a} = \pm\sqrt{\frac{b^2 - 4ac}{4a^2}}$$

Use the square root principle to eliminate the square.

$$x + \frac{b}{2a} = \pm\frac{\sqrt{b^2 - 4ac}}{2a}$$

Simplify the square root in the denominator.

$$x = -\frac{b}{2a} \pm \frac{\sqrt{b^2 - 4ac}}{2a}$$

Subtract $\frac{b}{2a}$ from both sides to isolate x.

$$x = \frac{-b \pm \sqrt{b^2 - 4ac}}{2a}$$

Combine rational expressions.

This final equation is the *quadratic formula* and can be used to solve any quadratic equation simply by replacing *a, b,* and *c* with the corresponding numbers from the given equation.

PROCEDURE **Using the Quadratic Formula**

To solve a quadratic equation in the form $ax^2 + bx + c = 0$, where $a \neq 0$, use the quadratic formula:

$$x = \frac{-b \pm \sqrt{b^2 - 4ac}}{2a}$$

Note: *A quadratic equation must be in the form* $ax^2 + bx + c = 0$ *to identify a, b, and c for use in the quadratic formula.*

EXAMPLE 1 Solve.

a. $3x^2 + 10x - 8 = 0$

Solution This equation is in the form $ax^2 + bx + c = 0$, where $a = 3$, $b = 10$, and $c = -8$, so we can use the quadratic formula, $x = \frac{-b \pm \sqrt{b^2 - 4ac}}{2a}$.

$$x = \frac{-10 \pm \sqrt{10^2 - 4(3)(-8)}}{2(3)}$$

Replace a with 3, b with 10, and c with -8.

$$x = \frac{-10 \pm \sqrt{196}}{6}$$

Simplify in the radical and the denominator.

$$x = \frac{-10 \pm 14}{6}$$

Evaluate $\sqrt{196}$.

$$x = \frac{-10 + 14}{6} \quad \text{or} \quad x = \frac{-10 - 14}{6}$$

Now we split up the ± to calculate the two solutions.

$$x = \frac{4}{6} \qquad x = \frac{-24}{6}$$

$$x = \frac{2}{3} \qquad x = -4$$

Connection Notice that the radicand 196 is a perfect square, which causes the solutions to be two rational numbers. As we consider more examples in this section, note how the radicand determines the type of solutions.

b. $x^2 + 8x - 15 = 14$

Solution $x^2 + 8x - 29 = 0$ — Subtract 14 from both sides to get the form $ax^2 + bx + c = 0$.

$$x = \frac{-8 \pm \sqrt{8^2 - 4(1)(-29)}}{2(1)}$$ Use the quadratic formula, replacing a with 1, b with 8, and c with -29.

$$x = \frac{-8 \pm \sqrt{180}}{2}$$ Simplify in the radical and the denominator.

$$x = \frac{-8 \pm \sqrt{36 \cdot 5}}{2}$$ Use the product rule of radicals.

$$x = \frac{-8 \pm 6\sqrt{5}}{2}$$ Simplify the radical.

$$x = \frac{-8}{2} \pm \frac{6\sqrt{5}}{2}$$ Separate into two rational expressions in order to simplify.

$$x = -4 \pm 3\sqrt{5}$$ Simplify by dividing out the 2.

Connection Notice that the radicand 180 is not a perfect square, which causes the two solutions to be irrational numbers.

c. $9x^2 + 4 = -12x$

Solution $9x^2 + 12x + 4 = 0$ — Add $12x$ to both sides to get the form $ax^2 + bx + c = 0$.

$$x = \frac{-12 \pm \sqrt{12^2 - 4(9)(4)}}{2(9)}$$ Use the quadratic formula, replacing a with 9, b with 12, and c with 4.

$$x = \frac{-12 \pm \sqrt{0}}{18}$$ Simplify in the radical and the denominator.

$$x = -\frac{2}{3}$$ Simplify the radical and write the fraction in lowest terms.

Connection Notice that the radicand is 0, which causes this equation to have only one solution.

d. $3x^2 + 9 = -8x + 2$

Solution $3x^2 + 8x + 7 = 0$ — Add $8x$ to and subtract 2 from both sides to get the form $ax^2 + bx + c = 0$.

$$x = \frac{-8 \pm \sqrt{8^2 - 4(3)(7)}}{2(3)}$$ Use the quadratic formula, replacing a with 3, b with 8, and c with 7.

$$x = \frac{-8 \pm \sqrt{-20}}{6}$$ Simplify in the radical and the denominator.

$$x = \frac{-8 \pm 2i\sqrt{5}}{6}$$ Simplify the radical.

$$x = \frac{-4 \pm i\sqrt{5}}{3}$$ Simplify to lowest terms.

Connection Notice that the negative radicand causes the two solutions to be nonreal complex numbers.

Note: *In standard form, the two complex solutions are* $-\frac{4}{3} + \frac{\sqrt{5}}{3}i$ *and* $-\frac{4}{3} - \frac{\sqrt{5}}{3}i$.

YOUR TURN Solve using the quadratic formula.

a. $4x^2 - 5x - 6 = 0$

b. $4x^2 = 15 - 2x$

c. $x^2 - 8x + 16 = 0$

d. $x^2 + 8 = 2 - 4x$

ANSWERS

a. 2 or $-\frac{3}{4}$

b. $\frac{-3 \pm 2\sqrt{6}}{2}$

c. 4

d. $-2 \pm i\sqrt{2}$

Choosing a Method for Solving Quadratic Equations

We have learned several methods for solving quadratic equations. The following table summarizes the methods and the conditions that would make each method the best choice.

Methods for Solving Quadratic Equations

Method	When the Method Is Beneficial
1. Factoring (Section 6.4)	Use when the quadratic equation can be easily factored.
2. Square root principle (Section 9.1)	Use when the quadratic equation can be easily written in the form $ax^2 = c$ or $(ax + b)^2 = c$.
3. Completing the square (Section 9.1)	Rarely the best method, but important for future topics.
4. Quadratic formula (Section 9.2)	Use when factoring is not easy.

OBJECTIVE 2. Use the discriminant to determine the number of real solutions that a quadratic equation will have. In the Connection boxes for Example 1, we pointed out how the radicand in the quadratic formula affects the solutions to a given quadratic equation. The expression $b^2 - 4ac$, which is the radicand, is called the **discriminant**.

DEFINITION ***Discriminant:*** The radicand $b^2 - 4ac$ in the quadratic formula.

We use the discriminant to determine the number and type of solutions to a quadratic equation.

PROCEDURE **Using the Discriminant**

Given a quadratic equation in the form $ax^2 + bx + c = 0$, where $a \neq 0$, to determine the number and type of solutions it has, evaluate the discriminant $b^2 - 4ac$.

If the **discriminant is positive**, then the equation has two real-number solutions. They will be rational if the discriminant is a perfect square and irrational otherwise.

If the **discriminant is 0**, then the equation has one real solution.

If the **discriminant is negative**, then the equation has two nonreal complex solutions.

Note: *When the discriminant is 0, the solution is* $-\frac{b \pm \sqrt{0}}{2a} = -\frac{b}{2a}$. *This real-number solution is a rational number if* $-\frac{b}{2a}$ *is rational.*

EXAMPLE 2 Use the discriminant to determine the number and type of solutions.

a. $3x^2 - 7x = 8$

Solution $3x^2 - 7x - 8 = 0$ Write the equation in the form $ax^2 + bx + c = 0$.

Evaluate the discriminant, $b^2 - 4ac$.

$$(-7)^2 - 4(3)(-8)$$ Replace a with 3, b with -7, and c with -8.

$$= 49 + 96$$

$$= 145$$

Warning: 145 is the value of the discriminant, not a solution for the equation $3x^2 - 7x - 8 = 0$.

Because the discriminant is positive, this equation has two real-number solutions. Since 145 is not a perfect square, the solutions are irrational.

b. $x^2 = \frac{4}{5}x - \frac{4}{25}$

Solution $x^2 - \frac{4}{5}x + \frac{4}{25} = 0$ Write the equation in the form $ax^2 + bx + c = 0$.

It is easier to evaluate the discriminant if the numbers are integers, so we multiply the equation through by the LCD, which is 25.

$$25 \cdot x^2 - \frac{25}{1} \cdot \frac{4}{5}x + \frac{25}{1} \cdot \frac{4}{25} = 25 \cdot 0$$
$$25x^2 - 20x + 4 = 0$$

Note: *Since the discriminant is 0, the solution is* $-\frac{b}{2a} = -\frac{(-20)}{2(25)} = \frac{20}{50} = \frac{2}{5}$ *which is a rational number.*

Now evaluate the discriminant, $b^2 - 4ac$.

$(-20)^2 - 4(25)(4)$ Replace a with 25, b with -20, and c with 4.
$= 400 - 400$
$= 0$

Because the discriminant is zero, this equation has only one real solution.

Note: *We could avoid calculations with decimal numbers by multiplying the original equation through by 10 so that it becomes:*
$5x^2 - 8x + 25 = 0$

c. $0.5x^2 - 0.8x + 2.5 = 0$

Solution $(-0.8)^2 - 4(0.5)(2.5)$ In $b^2 - 4ac$, replace a with 0.5, b with -0.8, and c with 2.5.
$= 0.64 - 5$
$= -4.36$

Because the discriminant is negative, there are two nonreal complex solutions.

YOUR TURN Use the discriminant to determine the number and type of solutions for the equation. If the solutions are real, state whether they are rational or irrational.

a. $5x^2 + 8x - 9 = 0$ **b.** $\frac{5}{2}x^2 + \frac{5}{6} = \frac{4}{3}x$ **c.** $x^2 - 0.12x = -0.0036$

OBJECTIVE 3. Find the *x*- and *y*-intercepts of a quadratic function. In Chapter 3, we learned that x-intercepts are points where a graph intersects the x-axis. Since an x-intercept is on the x-axis, the y-coordinate of the point will be 0. In Sections 5.2 and 6.4, we graphed quadratic functions, which have the form $f(x) = ax^2 + bx + c$ (or $y = ax^2 + bx + c$) and learned their graphs are parabolas. Notice when we replace y with 0 to find the x-intercepts in $y = ax^2 + bx + c$, we have the quadratic equation $ax^2 + bx + c = 0$, which has two, one, or no real-number solutions. As a result, quadratic functions in the form $y = ax^2 + bx + c$ have two, one, or no x-intercepts. The following graphs illustrate the possibilities.

ANSWERS

a. Discriminant = 244; two irrational solutions

b. Discriminant = $-\frac{59}{9}$; two nonreal complex solutions

c. Discriminant = 0; one rational solution

Note: *In Section 5.2, we noted that if $a > 0$, the parabola opens up, and if $a < 0$, the parabola opens down.*

Two x-intercepts: If $0 = ax^2 + bx + c$ has two real-number solutions, which occurs if $b^2 - 4ac > 0$, then the graph of $y = ax^2 + bx + c$ has two x-intercepts and looks like one of the following graphs.

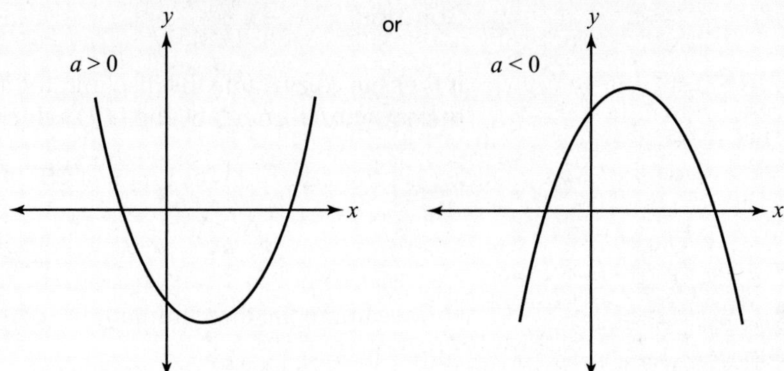

One x-intercept: If $0 = ax^2 + bx + c$ has one real-number solution, which occurs if $b^2 - 4ac = 0$, then the graph of $y = ax^2 + bx + c$ has one x-intercept and looks like one of the following graphs.

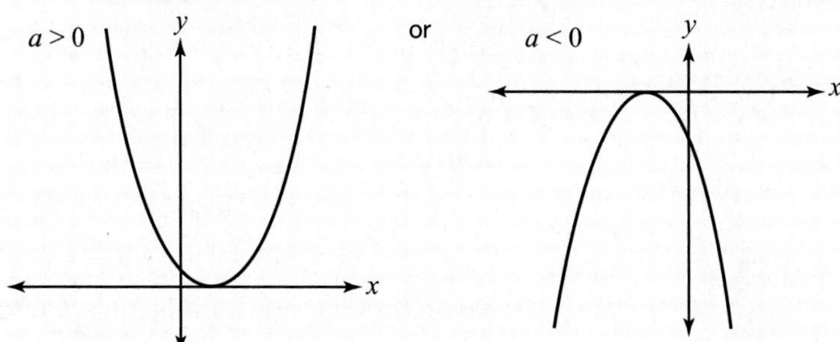

No x-intercepts: If $0 = ax^2 + bx + c$ has no real-number solution, which occurs if $b^2 - 4ac < 0$, then the graph of $y = ax^2 + bx + c$ has no x-intercepts and looks like one of the following graphs.

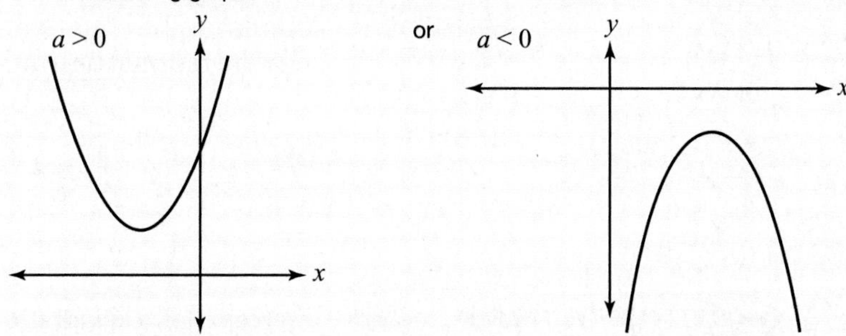

We have also learned that a y-intercept is where a graph intersects the y-axis. To find y-intercepts, we let $x = 0$ and solve for y. Notice when we replace x with 0 in $y = ax^2 + bx + c$, we have $y = a(0)^2 + b(0) + c = c$, so the y-intercept is $(0, c)$.

Connection We could have solved $0 = x^2 - 2x - 8$ by factoring:

$$0 = x^2 - 2x - 8$$

$$0 = (x - 4)(x + 2)$$

$$x - 4 = 0 \quad \text{or} \quad x + 2 = 0$$

$$x = 4 \qquad\qquad x = -2$$

EXAMPLE 3 Find the x- and y-intercepts of $y = x^2 - 2x - 8$, then graph.

Solution For the x-intercepts, letting $y = 0$ gives the equation $0 = x^2 - 2x - 8$, which we will solve using the quadratic formula.

$$x = \frac{(-2) \pm \sqrt{(-2)^2 - 4(1)(-8)}}{2(1)} = \frac{2 \pm \sqrt{36}}{2} = 1 \pm 3 = 4 \quad \text{or} \quad -2$$

x-intercepts: $(4, 0)$ and $(-2, 0)$

Calculator TIPS

The ZERO function on a graphing calculator can be used to find the coordinates of x-intercepts. After graphing the parabola on the calculator, select ZERO from the CALC menu. You will then be prompted to enter the left and right bound of one of the intercepts. Move the cursor along the parabola until it is to the left of one of the intercepts and press ENTER*. Then move the cursor along the parabola to the right of the same intercept and press* ENTER *again. When prompted for the guess, press* ENTER *once more. The coordinates of the intercept will appear at the bottom of the screen.*

Since $c = -8$, the y-intercept is $(0, -8)$.

Note: *We could also have calculated the y-intercept:* $y = (0)^2 + 2(0) - 8 = -8$

Now we can graph. We know the graph opens upwards because $a = 1$, which is positive.

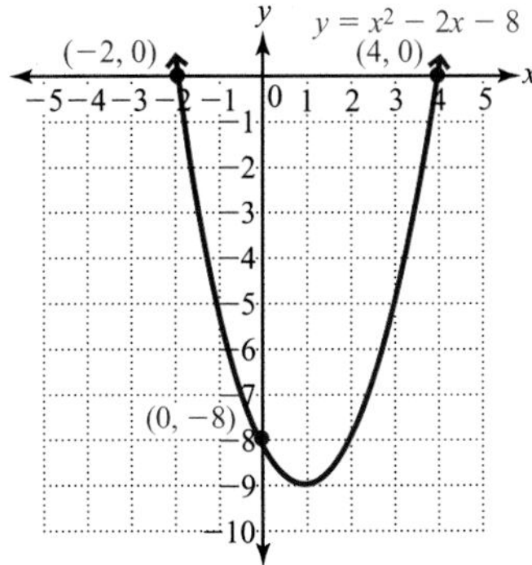

Note: *Knowing the intercepts and whether the parabola opens up or down gives us a pretty good sense of its position. However, if we knew the exact location of the vertex, we could get a more accurate graph. We will learn how to find the coordinates of the vertex in Section 9.4.*

YOUR TURN Find the x- and y-intercepts. Verify on a graphing calculator.

a. $y = 3x^2 - 5x + 1$ **b.** $y = x^2 - 8x + 16$ **c.** $y = 2x^2 - 3x + 5$

OBJECTIVE 4. Solve applications using the quadratic formula. In physics, the general formula for describing the height of an object after it has been thrown upwards is $h = \frac{1}{2}gt^2 + v_0t + h_0$, where g represents the acceleration due to gravity, t is the time in flight, v_0 is the initial velocity, and h_0 is the initial height. For Earth, the acceleration due to gravity is -32.2 ft./sec.^{2} or -9.8 m/sec.^{2}

EXAMPLE 4 In an extreme-games competition, a motorcyclist jumps with an initial velocity of 70 feet per second from a ramp height of 25 feet, landing on a ramp with a height of 15 feet. Find the time the motorcyclist is in the air.

Understand We are given the initial velocity, initial height, and final height and we are to find the time the bike is in the air.

Plan Use the formula $h = \frac{1}{2}gt^2 + v_0t + h_0$ with $h = 15$, $v_0 = 70$, and $h_0 = 25$. Since the units are in feet, we will use -32.2 ft./sec.^{2} for g.

Execute

$$15 = \frac{1}{2}(-32.2)t^2 + 70t + 25$$ Substitute the values.

$$15 = -16.1t^2 + 70t + 25$$

$$0 = -16.1t^2 + 70t + 10$$ Subtract 15 from both sides to get the form $ax^2 + bx + c = 0$.

$$x = \frac{-70 \pm \sqrt{70^2 - 4(-16.1)(10)}}{2(-16.1)}$$ Use the quadratic formula, replacing a with -16.1, b with 70, and c with 10.

$$x = \frac{-70 \pm \sqrt{5544}}{-32.2}$$ Simplify in the radical and the denominator.

$$x \approx -0.138 \quad \text{or} \quad 4.486$$ Approximate the two irrational solutions.

ANSWERS

a. x-intercepts: $\left(\frac{5 + \sqrt{13}}{6}, 0\right)$, $\left(\frac{5 - \sqrt{13}}{6}, 0\right)$; y-intercept: $(0, 1)$

b. x-intercept: $(4, 0)$; y-intercept: $(0, 16)$

c. no x-intercepts; y-intercept: $(0, 5)$

Answer Since the time cannot be negative, the motorcycle is in the air approximately 4.486 seconds.

Check We can check by evaluating the original formula using $t = 4.486$ seconds to see if the motorcycle indeed lands at a height of 15 feet. We will leave this check to the reader.

A ball is thrown from an initial height of 1.5 meters with an initial velocity of 8 meters per second. Find the time for the ball to land on the ground ($h = 0$). Approximate the time to the nearest thousandth.

ANSWER

1.802 sec.

9.2 Exercises

For Extra Help

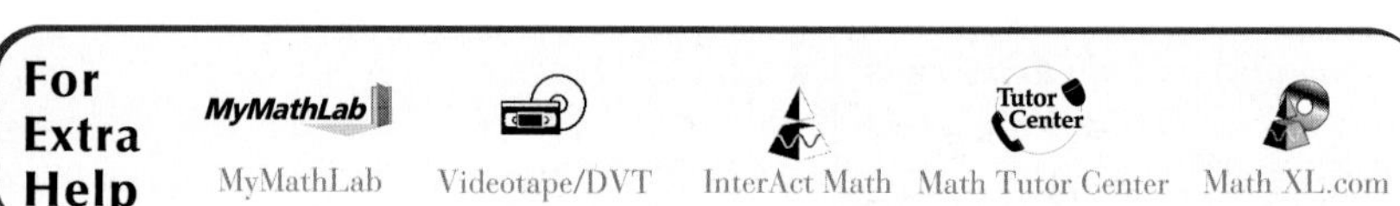

MyMathLab   Videotape/DVT   InterAct Math   Math Tutor Center   Math XL.com

1. Explain the general plan for deriving the quadratic formula.

2. Discuss the advantages and disadvantages of using the quadratic formula to solve quadratic equations.

3. Are there quadratic equations that cannot be solved using factoring? Explain.

4. What part of the quadratic formula is the discriminant?

5. Under what conditions will a quadratic equation have only one solution? Modify the quadratic formula to describe this single solution.

6. If x-intercepts for a quadratic function are imaginary numbers, what does that indicate about its graph?

For Exercises 7–14, rewrite each quadratic equation in the form $ax^2 + bx + c = 0$, then identify a, b, and c.

7. $x^2 - 3x + 7 = 0$

8. $x^2 + 6x - 15 = 0$

9. $3x^2 - 9x = 4$

10. $6x^2 + 10x = -15$

11. $1.5x^2 = x - 0.2$

12. $x = 0.8x^2 + 4.5$

13. $\frac{3}{4}x = -\frac{1}{2}x^2 + 6$

14. $\frac{5}{6}x^2 = \frac{1}{4}x - 6$

For Exercises 15–38, solve using the quadratic formula.

15. $x^2 + 9x + 20 = 0$ **16.** $x^2 - 4x - 21 = 0$ **17.** $x^2 - x - 1 = 0$ **18.** $x^2 + 3x - 5 = 0$

19. $4x^2 + 5x = 6$ **20.** $2x^2 - x = 3$ **21.** $x^2 - 9x = 0$ **22.** $x^2 + 2x = 0$

23. $x^2 - 8x = -16$ **24.** $x^2 + 14x = -49$ **25.** $x^2 + 2 = 2x$ **26.** $x^2 + 5 = 4x$

27. $4x^2 - 3x = 1$ **28.** $5x^2 - 4x = 1$ **29.** $3x^2 + 10x + 5 = 0$ **30.** $3x^2 - x - 3 = 0$

31. $-4x^2 + 6x = 5$ **32.** $-5x^2 = 3 - 4x$ **33.** $18x^2 + 2 = -15x$ **34.** $10x^2 - 12 = -7x$

35. $3x^2 - 4x = -3$ **36.** $-3x^2 + 6x = 8$ **37.** $6x^2 - 3x = 4$ **38.** $6x^2 - 6 - 13x = 0$

For Exercises 39–50, solve using the quadratic formula. If the solutions are irrational, give an exact answer and an approximate answer rounded to the nearest thousandth. (Hint: You might first clear the fractions or decimals by multiplying both sides by an appropriately chosen number.)

39. $2x^2 + 0.1x = 0.03$ **40.** $4x^2 + 8.6x - 2.4 = 0$ **41.** $x^2 + \frac{1}{2}x - 3 = 0$

42. $x^2 + \frac{1}{3}x - \frac{2}{3} = 0$ **43.** $x^2 - \frac{49}{36} = 0$ **44.** $x^2 - 0.0144 = 0$

45. $\frac{1}{2}x^2 + \frac{3}{2} = x$ **46.** $\frac{1}{5}x^2 + 2 = \frac{1}{2}x$ **47.** $x^2 - 0.5 = -0.06x$

48. $\frac{1}{3}x^2 - \frac{1}{4}x - \frac{1}{24} = 0$ **49.** $1.2x^2 - 0.6x = -0.5$ **50.** $2.4x^2 + 4.5 = 6.3x$

For Exercises 51–54, explain the mistake, then solve correctly.

51. Solve $3x^2 - 7x + 1 = 0$ using the quadratic formula.

$$\frac{-7 \pm \sqrt{(-7)^2 - (4)(3)(1)}}{2(3)} = \frac{-7 \pm \sqrt{49 - 12}}{6} = \frac{-7 \pm \sqrt{37}}{6}$$

52. Solve $2x^2 - 6x - 5 = 0$ using the quadratic formula.

$$\frac{6 \pm \sqrt{(-6)^2 - (4)(2)(5)}}{2(2)} = \frac{6 \pm \sqrt{36 - 40}}{4} = \frac{3}{2} \pm \frac{1}{2}i$$

53. Solve $x^2 - 2x + 3 = 0$ using the quadratic formula.

$$\frac{-(-2) \pm \sqrt{(-2)^2 - (4)(1)(3)}}{2(1)} = \frac{2 \pm \sqrt{4 - 12}}{2}$$
$$= \frac{2 \pm \sqrt{-8}}{2}$$

54. Solve $x^2 - 8 = 0$ using the quadratic formula.

$$\frac{-(-8) \pm \sqrt{(-8)^2 - (4)(1)(0)}}{2(1)} = \frac{8 \pm \sqrt{64 - 0}}{2}$$
$$= \frac{8 \pm \sqrt{64}}{2}$$
$$= \frac{8 \pm 8}{2}$$
$$= 8, 0$$

For Exercises 55–64, use the discriminant to determine the number and type of solutions for the equation. If the solution(s) are real, state whether they are rational or irrational.

55. $x^2 + 10x = -25$

56. $2x^2 - 8x + 8 = 0$

57. $\frac{1}{4}x^2 - 4x = -4$

58. $\frac{1}{2}x^2 + 18 = 6x$

59. $x^2 + 4x + 9 = 0$

60. $2x^2 - 3x - 2 = 0$

61. $x^2 - x + 3 = 0$

62. $x^2 + 4x = -5$

63. $x^2 - 6x + 6 = 0$

64. $3x^2 = 13x - 8$

For Exercises 65–74, indicate which of the following methods is the best choice for solving the given equation: factoring, using the square root principle, or using the quadratic formula. Then solve the equation.

65. $x^2 - 81 = 0$

66. $x^2 = 44$

67. $4x^2 + 48 = 0$

68. $2x^2 = 5x - 2$

69. $x^2 - 6x + 13 = 0$

70. $(x + 7)^2 = 40$

71. $x^2 + 6x = 0$

72. $x^2 + 14x + 45 = 0$

73. $x^2 = 8x - 19$

74. $x^2 - 2x = 149$

For Exercises 75–82, find the x- and y-intercepts.

75. $y = x^2 - x - 2$

76. $y = x^2 - 3x + 2$

77. $y = -x^2 - 2x + 8$

78. $y = x^2 - x - 30$

79. $y = 2x^2 + 15x - 8$

80. $y = -15x^2 + x + 6$

81. $y = -2x^2 + 3x - 6$

82. $y = -3x^2 - 2x - 5$

For Exercises 83–90, translate to a quadratic equation then solve using the quadratic formula.

83. A positive integer squared plus five times its consecutive integer is equal to 71. Find the integers.

84. The square of a positive integer minus half of its consecutive integer is equal to 162. Find the integers.

85. A right triangle has side lengths that are three consecutive integers. Use the Pythagorean theorem to find the lengths of those sides. (Remember that the hypotenuse in a right triangle is always the longest side.)

86. A right triangle has side lengths that are consecutive even integers. Use the Pythagorean theorem to find the lengths of those sides.

87. The length of a rectangular fence gate is 3.5 feet less than three times the width. Find the length and width of the gate if the area is 34 square feet.

88. A small access door for an attic storage area is designed so that the width is 3.25 feet more than the length. Find the length and width if the area is 10.5 square feet.

89. An architect is experimenting with two different shapes of a room, as shown.

a. Find x so that the rooms have the same area.

b. Complete the dimensions for the L-shaped room.

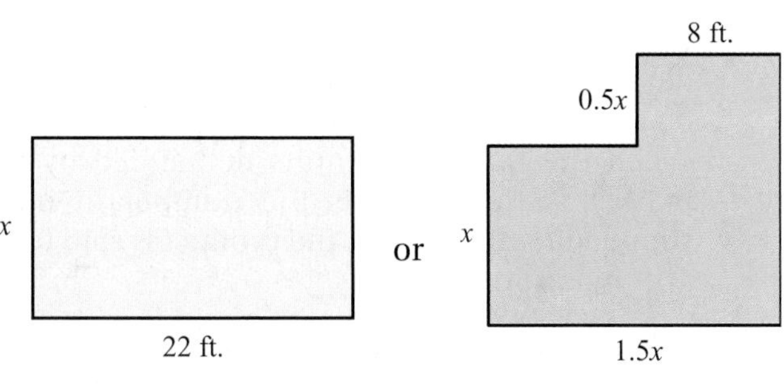

90. A cylinder is to be made so that its volume is equal to that of a sphere with a radius of 9 inches. If a cylinder is to have a height of 4 inches, find its radius.

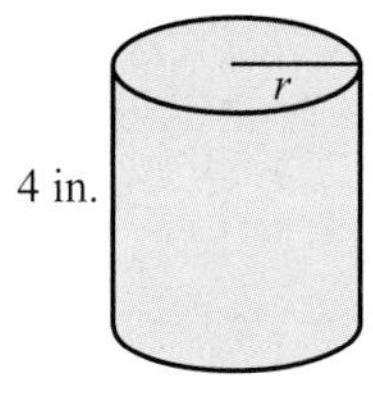

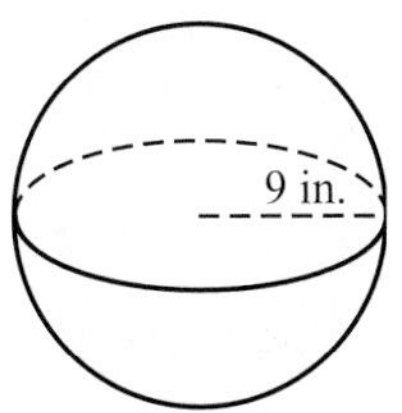

For Exercises 91–94, use the formula $h = \frac{1}{2}gt^2 + v_0t + h_0$, where g represents the acceleration due to gravity, t is the time in flight, v_0 is the initial velocity, and h_0 is the initial height. For Earth, the acceleration due to gravity is -32.2 ft./sec.^{2} or -9.8 m/sec.^{2} Approximate irrational answers to the nearest hundredth.

91. In 2001, Robby Knievel jumped the Grand Canyon on a motorcycle. Suppose that on take off, his motorcycle had a vertical velocity of 48 feet per second and at the end of the launch ramp his altitude was 485 feet above sea level. If he landed on a ramp that was 460 feet above sea level, how long was he in flight?

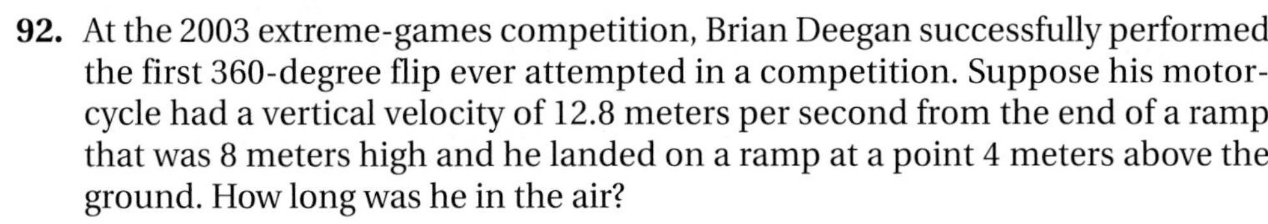

92. At the 2003 extreme-games competition, Brian Deegan successfully performed the first 360-degree flip ever attempted in a competition. Suppose his motorcycle had a vertical velocity of 12.8 meters per second from the end of a ramp that was 8 meters high and he landed on a ramp at a point 4 meters above the ground. How long was he in the air?

93. A platform diver dives from a platform that is 10 meters above the water.

a. Write the equation that describes his height during the dive. (Assume his initial velocity is 0.)

b. Find the time it takes for the diver to be at 5 meters above the water.

c. Find the time for the diver to enter the water.

94. In a cliff-diving championship in Acapulco, a diver dives from a cliff at a height of 70 feet.

a. Write the equation that describes his height during the dive. (Assume his initial velocity is 0.)

b. Find the time it takes for the diver to be at 50 feet above the water.

c. Find the time for the diver to enter the water.

95. The expression $0.5n^2 + 2.5n$ describes the gross income from the sale of a particular software product, where n is the number of units sold in thousands. The expression $4.5n + 16$ describes the cost of producing the n units. Find the number of units that must be produced and sold for the company to break even. (To *break even* means that the gross income and cost are the same.)

96. An economist and marketing manager discover that the expression $2n^2 + 5n$ models the price of a CD based on the demand for it, where n is the number of units (in millions) that the market demands. The expression $-0.5n^2 + 17.1$ describes the price of the CD based on the number of units (also in millions) supplied to the market.

a. Find the number of units that need to be demanded and supplied so that the price based on demand is equal to the price based on supply. When the number of units demanded by the market is the same as the number of units supplied to the market, the product is said to be at equilibrium.

b. What is the price of the CD at equilibrium?

97. For the equation $2x^2 - 5x + c = 0$,

a. Find c so that the equation has only one rational number solution.

b. Find the range of values of c for which the equation has two real-number solutions.

c. Find the range of values of c for which the equation has no real-number solution.

98. For the equation $4x^2 + 6x + c = 0$,

a. Find c so that the equation has only one rational number solution.

b. Find the range of values of c for which the equation has two real-number solutions.

c. Find the range of values of c for which the equation has no real-number solution.

99. For the equation $ax^2 + 12x + 8 = 0$,
   a. Find a so that the equation has only one rational number solution.
   b. Find the range of values of a for which the equation has two real-number solutions.
   c. Find the range of values of a for which the equation has no real-number solution.

100. For the equation $3x^2 + bx + 8 = 0$,
   a. Find b so that the equation has only one rational number solution.
   b. Find all positive values of b for which the equation has two real-number solutions.
   c. Find all positive values of b for which the equation has no real-number solution.

PUZZLE PROBLEM

In the equation shown, A, B, C, D, and E are five consecutive positive integers where $A < B < C < D < E$. What are they?

$$A^2 + B^2 + C^2 = D^2 + E^2$$

REVIEW EXERCISES

[6.2] *For Exercises 1 and 2, factor.*

1. $u^2 - 9u + 14$

2. $3u^2 - 2u - 16$

For Exercises 3 and 4, solve.

[6.4] 3. $5u^2 + 13u = 6$

[7.4] 4. $\dfrac{7}{3u} = \dfrac{5}{u} - \dfrac{1}{u-5}$

For Exercises 5 and 6, simplify.

[5.1] 5. $(x^2)^2$

[8.2] 6. $(x^{1/3})^2$

9.3 Solving Equations That Are Quadratic in Form

OBJECTIVES

1. Solve equations by rewriting them in quadratic form.
2. Solve equations that are quadratic in form by using substitution.
3. Solve applications problems using equations that are quadratic in form.

Many equations that are not quadratic equations are **quadratic in form** and can be solved using the methods for solving quadratic equations.

DEFINITION An equation is **quadratic in form** if it can be rewritten as a quadratic equation $au^2 + bu + c = 0$, where $a \neq 0$ and u is a variable or an expression.

OBJECTIVE 1. Solve equations by rewriting them in quadratic form.

Equations with Rational Expressions

In Section 7.4, we solved equations containing rational expressions by multiplying both sides of the equation by the LCD. Those rewritten equations are often quadratic. Remember that equations containing rational expressions sometimes have extraneous solutions.

Connection Recall that a solution for an equation with rational expressions is extraneous if it causes one or more of the denominators to equal 0.

EXAMPLE 1 Solve. $\frac{3}{x+1} = 1 - \frac{3}{x(x+1)}$

Solution

$x(x+1) \cdot \frac{3}{x+1} = x(x+1)\left(1 - \frac{3}{x(x+1)}\right)$ Multiply both sides by the LCD $x(x+1)$.

$x(x+1) \cdot \frac{3}{x+1} = x(x+1) \cdot 1 - x(x+1) \cdot \frac{3}{x(x+1)}$ Distribute $x(x+1)$.

$3x = x^2 + x - 3$ Simplify both sides.

$0 = x^2 - 2x - 3$ Subtract $3x$ from both sides to get the quadratic form $ax^2 + bx + c = 0$.

$0 = (x-3)(x+1)$ Factor.

$x - 3 = 0$ or $x + 1 = 0$ Use the zero-factor theorem.

$x = 3$ $\quad x = -1$ Solve each equation.

Checks $x = 3$

$$\frac{3}{3+1} = 1 - \frac{3}{3(3+1)}$$

$$\frac{3}{4} = 1 - \frac{3}{12}$$

$$\frac{3}{4} = \frac{3}{4} \quad \text{True}$$

$x = -1$

$$\frac{3}{-1+1} = 1 - \frac{3}{-1(-1+1)}$$

$$\frac{3}{0} = 1 - \frac{3}{0}$$

Note: *The expression* $\frac{3}{0}$ *is undefined, so* -1 *is extraneous.*

The solution is 3 (-1 is extraneous).

YOUR TURN Solve. $\frac{x}{x-2} = \frac{6}{x} + \frac{4}{x(x-2)}$

Equations Containing Radicals

In Section 8.6, we found that after using the power rule on equations containing radicals the result was often a quadratic equation. Remember that these radical equations sometimes have extraneous solutions.

ANSWER

4 (2 is extraneous)

EXAMPLE 2 Solve.

a. $\sqrt{x} + x = 6$

> **Connection** Recall that a solution for an equation containing radicals is extraneous if it makes the original equation false.

Solution

$$\sqrt{x} = 6 - x \quad \text{Subtract } x \text{ from both sides to isolate the radical.}$$
$$(\sqrt{x})^2 = (6 - x)^2 \quad \text{Square both sides.}$$
$$x = 36 - 12x + x^2 \quad \text{Simplify both sides.}$$
$$0 = x^2 - 13x + 36 \quad \text{Subtract } x \text{ from both sides to write in quadratic form.}$$
$$0 = (x - 4)(x - 9) \quad \text{Factor.}$$
$$x - 4 = 0 \quad \text{or} \quad x - 9 = 0 \quad \text{Use the zero-factor theorem.}$$
$$x = 4 \qquad x = 9 \quad \text{Solve each equation.}$$

Checks

$x = 4$: $\sqrt{4} + 4 \stackrel{?}{=} 6$; $2 + 4 = 6$ True

$x = 9$: $\sqrt{9} + 9 \stackrel{?}{=} 6$; $3 + 9 = 6$ False, so 9 is extraneous.

The solution is 4 (9 is extraneous).

b. $\sqrt{x - 1} = 2x - 1$

Solution

$$(\sqrt{x - 1})^2 = (2x - 1)^2 \quad \text{Square both sides of the equation.}$$
$$x - 1 = 4x^2 - 4x + 1 \quad \text{Simplify both sides.}$$
$$0 = 4x^2 - 5x + 2 \quad \text{Subtract } x \text{ and add 1 on both sides to write in quadratic form.}$$

We cannot factor $4x^2 - 5x + 2$, so we will use the quadratic formula.

$$x = \frac{-(-5) \pm \sqrt{(-5)^2 - 4(4)(2)}}{2(4)} = \frac{5 \pm \sqrt{-7}}{8} = \frac{5 \pm i\sqrt{7}}{8}$$

> **Note:** *Written in standard form, these solutions are* $\frac{5}{8} + \frac{\sqrt{7}}{8}i$ *and* $\frac{5}{8} - \frac{\sqrt{7}}{8}i$.

Check Because the solutions are complex, we will not check them.

YOUR TURN Solve.

a. $x - \sqrt{x} - 2 = 0$

b. $\sqrt{6x - 11} = 3x - 1$

ANSWERS

a. 4 (1 is extraneous)

b. $\frac{2 \pm 2i\sqrt{2}}{3}$

OBJECTIVE 2. Solve equations that are quadratic in form using substitution. Recall that an equation that is quadratic in form can be written as $au^2 + bu + c = 0$, where $a \neq 0$ and u can be an *expression*. We now explore a method for solving these equations where we substitute u for an expression. To use substitution, it is important to note a pattern with the exponents of the terms in a quadratic equation. Notice that the degree of the first term, au^2, is 2, the degree of the middle term, bu, is 1, and the third term is a constant. If a trinomial has one term with an expression raised to the second power, a second term with that same expression raised to the first power, and a third term that is a constant, then the equation is quadratic in form and we can use substitution to solve it.

Consider the equation $x^4 - 13x^2 + 36 = 0$. Notice we can rewrite the equation as $(x^2)^2 - 13(x^2) + 36 = 0$, so it is quadratic in form. By substituting u for each x^2,

Note: *Given an equation containing a trinomial, if the degree of the first term is twice the degree of the middle term, and the third term is a constant, then try using substitution.*

we have a "friendlier" form of quadratic equation, which we can then solve by factoring, completing the square, or the quadratic formula.

$$(x^2)^2 - 13(x^2) + 36 = 0$$

$$u^2 - 13u + 36 = 0 \quad \text{Substitute } u \text{ for } x^2.$$

$$(u - 9)(u - 4) = 0 \quad \text{Factor.}$$

$$u - 9 = 0 \quad \text{or} \quad u - 4 = 0 \quad \text{Use the zero-factor theorem.}$$

$$u = 9 \qquad u = 4$$

Note that these solutions are for u, not x. We must substitute x^2 back in place of u to finish solving for x.

Note: *The equation* $x^4 - 13x^2 + 36 = 0$ *has four solutions:* $3, -3, 2,$ *and* -2.

$$x^2 = 9 \qquad x^2 = 4 \quad \text{Substitute } x^2 \text{ for } u.$$

$$x = \pm\sqrt{9} \qquad x = \pm\sqrt{4}$$

$$x = \pm 3 \qquad x = \pm 2$$

Our example suggests the following procedure.

PROCEDURE **Using Substitution to Solve Equations That Are Quadratic in Form**

To solve equations that are quadratic in form using substitution,

1. Rewrite the equation so that it is in the form $au^2 + bu + c = 0$.
2. Solve the quadratic equation for u.
3. Substitute for u and solve.
4. Check the solutions.

EXAMPLE 3 Solve.

a. $(x + 2)^2 - 2(x + 2) - 8 = 0$

Solution If we substitute u for $x + 2$, we have an equation in the form $au^2 + bu + c = 0$.

$$(x + 2)^2 - 2(x + 2) - 8 = 0$$

$$u^2 - 2u - 8 = 0 \quad \text{Substitute } u \text{ for } x + 2.$$

$$(u - 4)(u + 2) = 0 \quad \text{Factor.}$$

$$u - 4 = 0 \quad \text{or} \quad u + 2 = 0 \quad \text{Use the zero-factor theorem.}$$

$$u = 4 \qquad u = -2 \quad \text{Solve each equation for } u.$$

$$x + 2 = 4 \qquad x + 2 = -2 \quad \text{Substitute } x + 2 \text{ for } u.$$

$$x = 2 \qquad x = -4 \quad \text{Solve each equation for } x.$$

Check Verify that 2 and -4 make $(x + 2)^2 - 2(x + 2) - 8 = 0$ true. We will leave this check to the reader.

b. $x^{2/3} - x^{1/3} - 6 = 0$

Solution Since $x^{2/3} - x^{1/3} - 6 = 0$ can be written as $(x^{1/3})^2 - x^{1/3} - 6 = 0$, it is quadratic in form. We will substitute u for $x^{1/3}$.

$(x^{1/3})^2 - x^{1/3} - 6 = 0$		Rewrite in quadratic form.
$u^2 - u - 6 = 0$		Substitute u for $x^{1/3}$.
$(u - 3)(u + 2) = 0$		Factor.
$u - 3 = 0$ or	$u + 2 = 0$	Use the zero-factor theorem.
$u = 3$	$u = -2$	Solve each equation for u.
$x^{1/3} = 3$	$x^{1/3} = -2$	Substitute $x^{1/3}$ for u.
$(x^{1/3})^3 = 3^3$	$(x^{1/3})^3 = (-2)^3$	Cube both sides of the equations.
$x = 27$	$x = -8$	Simplify.

Checks $x = 27$

$$27^{2/3} - 27^{1/3} - 6 \stackrel{?}{=} 0$$
$$(\sqrt[3]{27})^2 - \sqrt[3]{27} - 6 \stackrel{?}{=} 0$$
$$3^2 - 3 - 6 \stackrel{?}{=} 0$$
$$9 - 3 - 6 \stackrel{?}{=} 0$$
$$0 = 0 \quad \text{True}$$

$x = -8$

$$(-8)^{2/3} - (-8)^{1/3} - 6 \stackrel{?}{=} 0$$
$$(\sqrt[3]{-8})^2 - \sqrt[3]{-8} - 6 \stackrel{?}{=} 0$$
$$(-2)^2 - (-2) - 6 \stackrel{?}{=} 0$$
$$4 + 2 - 6 \stackrel{?}{=} 0$$
$$0 = 0 \quad \text{True}$$

YOUR TURN Solve.

a. $x^4 - 10x^2 + 9 = 0$

b. $(n - 2)^2 + 4(n - 2) - 12 = 0$

c. $x^{2/3} - x^{1/3} - 2 = 0$

ANSWERS

a. $\pm 3, \pm 1$

b. ± 4

c. $8, -1$

OBJECTIVE 3. Solve applications problems using equations that are quadratic in form.

EXAMPLE 4 The average speed of a car is 10 miles per hour more than the average speed of a bus. The bus takes 1 hour longer than the car to travel 200 miles. Find how long it takes the car to travel 200 miles.

Understand Both the car and the bus travel 200 miles, but it takes the bus 1 hour longer than the car. The rate of the car is 10 miles per hour more than the rate of the bus.

Plan We will use a table to organize the information, then write an equation, which we can solve.

Execute We will let x represent the time for the car to travel 200 miles. Since the bus travels 1 more hour, $x + 1$ describes the time for the bus to travel 200 miles.

Vehicle	d	t	r
bus	200	$x + 1$	$\frac{200}{x + 1}$
car	200	x	$\frac{200}{x}$

We use $r = \frac{d}{t}$ to describe each rate.

Since the rate of the car is 10 miles per hour faster than the rate of the bus, we can say (the rate of the car) = (the rate of the bus) + 10.

$$\frac{200}{x} = \frac{200}{x+1} + 10$$

$$x(x+1)\left(\frac{200}{x}\right) = x(x+1)\left(\frac{200}{x+1} + 10\right)$$ **Multiply both sides by the LCD $x(x + 1)$.**

$$(x+1)(200) = x(x+1)\left(\frac{200}{x+1}\right) + x(x+1)(10)$$ **Simplify both sides.**

$$200x + 200 = 200x + 10x^2 + 10x$$ **Continue simplifying.**

$$0 = 10x^2 + 10x - 200$$ **Subtract $200x$ and 200 from both sides to get quadratic form.**

$$0 = x^2 + x - 20$$ **Divide both sides by 10.**

$$0 = (x+5)(x-4)$$ **Factor.**

$$x + 5 = 0 \quad \text{or} \quad x - 4 = 0$$ **Use the zero-factor theorem.**

$$x = -5 \qquad\qquad x = 4$$ **Solve each equation.**

Answer Since time cannot be negative, it takes the car 4 hours to travel 200 miles.

Check The rate of the car is $\frac{200}{x} = \frac{200}{4} = 50$ miles per hour, so in 4 hours the car travels $50(4) = 200$ miles. The bus travels 200 miles in $4 + 1 = 5$ hours at a rate of $\frac{200}{4+1} = \frac{200}{5} = 40$ miles per hour, so in 5 hours the bus travels $5(40) = 200$ miles.
Notice also that the car's rate is 10 miles per hour more than the bus's.

EXAMPLE 5 Bobby and Pam manufacture and install blinds. Working together, they can install blinds in every window of an average-sized house in $1\frac{1}{3}$ hours. Working alone, Bobby takes 2 hours longer than Pam to do the same installation. How long does each take to install blinds in an average-sized house working alone?

Understand We are given the time it will take Bobby and Pam working together and are asked to find how long it would take each individually. We also know that it will take Bobby 2 hours longer than it would take Pam.

Plan We will use a table to organize the information, write an equation, and then solve.

Execute We will let x represent the number of hours for Pam to install working alone. Since Bobby takes 2 more hours, his time working alone is represented by $x + 2$.

Worker	Time to Complete the Job Alone	Rate of Work	Time at Work	Portion of Job Completed
Pam	x	$\frac{1}{x}$	$\frac{4}{3}$	$\frac{4}{3x}$
Bobby	$x + 2$	$\frac{1}{x+2}$	$\frac{4}{3}$	$\frac{4}{3(x+2)}$

Multiplying the rate of work and the time at work gives an expression of the amount of the job completed.

The total job in this case is 1 average-sized house, so we can write an equation that combines their individual expressions for work completed and set this sum equal to 1.

(Portion Pam does) + (Portion Bobby does) = 1 (the entire job)

$$\frac{4}{3x} + \frac{4}{3(x+2)} = 1$$

$$3x(x+2)\left(\frac{4}{3x} + \frac{4}{3(x+2)}\right) = 3x(x+2)(1)$$ **Multiply both sides by the LCD $3x(x+2)$.**

$$3x(x+2)\left(\frac{4}{3x}\right) + 3x(x+2)\left(\frac{4}{3(x+2)}\right) = 3x^2 + 6x$$ **Simplify both sides.**

$$4(x+2) + 4x = 3x^2 + 6x$$ **Continue simplifying.**

$$4x + 8 + 4x = 3x^2 + 6x$$ **Distribute.**

$$8x + 8 = 3x^2 + 6x$$ **Combine like terms.**

$$0 = 3x^2 - 2x - 8$$ **Subtract $8x$ and 8 from both sides.**

$$0 = (3x+4)(x-2)$$ **Factor.**

$$3x + 4 = 0 \quad \text{or} \quad x - 2 = 0$$ **Use the zero-factor theorem.**

$$3x = -4 \qquad x = 2$$ **Solve each equation.**

$$x = -\frac{4}{3}$$

Answer Since negative time makes no sense in the context of this problem, it takes Pam 2 hours to install working alone and Bobby $x + 2 = 2 + 2 = 4$ hours working alone.

Check Since Pam can install in 2 hours, she can install $\frac{1}{2}$ of an average-sized house in 1 hour.

Since Bobby takes 4 hours to do the same work, he can install $\frac{1}{4}$ of an average-sized house in 1 hour. Together, they can install $\frac{1}{2} + \frac{1}{4} = \frac{2}{4} + \frac{1}{4} = \frac{3}{4}$ of an average-sized house in 1 hour, so in $1\frac{1}{3}$ hours they can install blinds in $1\frac{1}{3} \cdot \frac{3}{4} = \frac{4}{3} \cdot \frac{3}{4} = 1$ average-sized house.

YOUR TURN Solve.

a. The average speed of a passenger train is 25 miles per hour more than the average speed of a car. The time required for the car to travel 300 miles is 2 hours more than the time required for the train. Find the average speed of the car.

b. Terri and Tommy run a flea market. It takes Terri 2 hours longer to put the merchandise out than it does Tommy. If they can put the merchandise out together in $1\frac{7}{8}$ hours, how long would it take each working alone?

ANSWERS

a. 50 mph
b. It takes Tommy 3 hr. and Terri 5 hr.

9.3 Exercises

For Extra Help

1. What does it mean to say that an equation is quadratic in form?

2. Is $x^{3/4} - x^{1/4} - 6 = 0$ quadratic in form? Why or why not?

3. If you solved $3\left(\frac{x+2}{3}\right)^2 + 13\left(\frac{x+2}{3}\right) - 10 = 0$ using substitution, what would your substitution be?

4. If you solved $x^4 - 7x^2 + 12 = 0$ using substitution, what would your substitution be?

5. If Alisha can clean her house in x hours, what part of her house can she clean in 1 hour?

6. If a car travels 400 miles in x hours, how many miles does it travel in 1 hour?

For Exercises 7–18, solve the equations with rational expressions.

7. $\frac{1}{x} + \frac{1}{x+2} = \frac{3}{4}$

8. $\frac{1}{x} + \frac{2}{x-3} = \frac{5}{6}$

9. $\frac{60}{x+2} = \frac{60}{x} - 5$

10. $\frac{120}{x+2} = \frac{120}{x} - 5$

11. $\frac{1}{p-4} + \frac{1}{4} = \frac{8}{p^2-16}$

12. $\frac{1}{y-5} - \frac{10}{y^2-25} = -\frac{1}{5}$

13. $\frac{6}{2y+5} = \frac{2}{y+5} + \frac{1}{5}$

14. $\frac{6}{2x+3} = \frac{2}{x-6} + \frac{4}{3}$

15. $1 + 2x^{-1} - 8x^{-2} = 0$ 16. $1 + 5x^{-1} + 6x^{-2} = 0$ 17. $3 + 13x^{-1} - 10x^{-2} = 0$ 18. $2 - x^{-1} - 15x^{-2} = 0$

For Exercises 19–30, solve the equations with radical expressions.

19. $x - 8\sqrt{x} + 15 = 0$
20. $x - 3\sqrt{x} + 2 = 0$
21. $2x - 5\sqrt{x} - 7 = 0$
22. $3x + 4\sqrt{x} - 4 = 0$
23. $\sqrt{2a + 5} = 3a - 3$
24. $\sqrt{3b + 1} = 5b - 3$
25. $\sqrt{2m - 8} - m - 1 = 0$
26. $\sqrt{8 - 12x} - 2x + 3 = 0$
27. $\sqrt{4x + 1} = \sqrt{x + 2} + 1$
28. $\sqrt{3x + 7} = \sqrt{2x + 3} + 1$
29. $\sqrt{2x + 1} - \sqrt{3x + 4} = -1$
30. $\sqrt{2x - 1} - \sqrt{4x + 5} = -2$

For Exercises 31–52, solve using substitution.

31. $x^4 - 10x^2 + 9 = 0$
32. $x^4 - 13x^2 + 36 = 0$
33. $4x^4 - 13x^2 + 9 = 0$
34. $9x^4 - 13x^2 + 4 = 0$
35. $x^4 + 5x^2 - 36 = 0$
36. $x^4 - 3x^2 - 4 = 0$
37. $(x + 2)^2 + 6(x + 2) + 8 = 0$
38. $(x - 3)^2 + 2(x - 3) - 15 = 0$
39. $2(x + 3)^2 - 9(x + 3) - 5 = 0$
40. $3(x - 1)^2 + 4(x - 1) - 4 = 0$
41. $\left(\frac{x - 1}{2}\right)^2 + 8\left(\frac{x - 1}{2}\right) + 15 = 0$
42. $\left(\frac{x + 2}{3}\right)^2 - 5\left(\frac{x + 2}{3}\right) - 6 = 0$
43. $2\left(\frac{x + 2}{2}\right)^2 + \left(\frac{x + 2}{2}\right) - 3 = 0$
44. $3\left(\frac{x - 1}{3}\right)^2 + 10\left(\frac{x - 1}{3}\right) - 8 = 0$
45. $x^{2/3} - 5x^{1/3} + 6 = 0$
46. $x^{2/3} + 3x^{1/3} - 10 = 0$
47. $2x^{2/3} - 3x^{1/3} - 2 = 0$
48. $3x^{2/3} - 4x^{1/3} - 4 = 0$
49. $x^{1/2} - 7x^{1/4} + 12 = 0$
50. $x^{1/2} - 4x^{1/4} + 3 = 0$
51. $5x^{1/2} + 8x^{1/4} - 4 = 0$
52. $2x^{1/2} - x^{1/4} - 3 = 0$

For Exercises 53–64, solve.

53. The average rate of a bus is 15 miles per hour more than the average rate of a truck. The truck takes 1 hour longer than the bus to travel 180 miles. How long does it take the bus to travel 180 miles?

54. A charter business has two types of small planes, a jet and a twin propeller plane. The prop-plane's average air speed is 120 miles per hour less than the jet's. The prop-plane takes 1 hour longer to travel 720 miles than the jet. How long does it take the prop-plane to travel the 720 miles?

55. The average speed of the winner of the Boston Marathon in 2003 was 3.8 miles per hour faster than the person who finished 1 hour behind. Find the time it took the winner to run the 26-mile course.

56. In the 2003 Tour de France, Lance Armstrong's average speed was 0.0081 kilometers per hour faster than the second-place finisher, who finished 1 minute behind Armstrong. Find Armstrong's time to complete the 3415-kilometer race.

57. Suppose the time required for a bus to travel 360 miles is 3 hours more than the time required for a motorcycle to travel 300 miles. If the average rate of the motorcycle is 15 miles per hour more than the average rate of the bus, find the average rate of the bus.

58. Suppose the time required for a truck to travel 400 miles is 3 hours more than the time required for a car to travel 300 miles. If the average rate of the car is 10 miles per hour more than the average rate of the truck, find the average rate of the truck.

59. After training hard for a year, a novice cyclist discovers that she has increased her average rate by 6 miles per hour and can travel 36 miles in 1 less hour than a year ago.

a. What was her old average rate? What is her new average rate?

b. What was her old time to travel 36 miles? What is her new time for 36 miles?

60. A high school track coach determines that his fastest long-distance runner runs 2 miles per hour faster than his slowest runner. The slower runner takes a half-hour longer to run 12 miles than the faster runner.

a. Find the average rates of both runners.

b. Find the time for both runners to run 12 miles.

61. Billy and Jody are commercial fishermen. Working alone it takes Billy 2 hours longer to run the hoop nets than it takes Jody working alone. Together they can run the hoop nets in $2\frac{2}{5}$ hours. How long does it take each working alone?

Of Interest

Lance Armstrong has won the Tour de France a record 7 times each year from 1999 to 2005. This seemingly superhuman feat is all the more incredible given that in 1996, he was diagnosed with advanced testicular cancer, which had already spread to his lungs and brain. After surgery, doctors gave him less than a 50/50 chance of recovery even with intensive chemotherapy. Only five months after being diagnosed, Armstrong began training and he returned to professional racing in 1998, winning several prestigious events that year. (*Source:* www.lancearmstrong.com)

62. Using a riding mower, Fran can mow the grass at the campground in 4 hours less time than it takes Donnie using a push mower. Together they can mow the grass in $2\frac{2}{3}$ hours. How long does it take each working alone?

63. A newspaper has two presses, one of which is older than the other. Working alone, the newer press can print all the copies for a typical day in a half-hour less time than the older press. When running at the same time, they print all of a typical day's copies in 2 hours. How long would it take to print a day's worth if they worked alone?

64. A school copy center prints the school newsletter in a half-hour using two copiers running at the same time. If the center uses only one copier, the faster of the two copiers takes 6 minutes less time than the other copier to print all of the newsletters. How long does each copier take to print the newsletters working alone?

On a guitar, the frequency of the vibrating string is related to the tension on the string. Suppose two strings of the same diameter and length are placed on an instrument and wound to different tensions. The formula $\frac{F_1^2}{F_2^2} = \frac{T_1}{T_2}$ ***describes the relationship between their frequency and tension.***

65. One string is wound on a guitar to a tension of 50 pounds. A second string of the same diameter is wound to 60 pounds. If the string with the greater tension has a frequency that is 40 vibrations per second greater than the other string, what are the frequencies of the two strings? Round to the nearest ten.

66. Two strings of the same diameter are wound on a banjo to different tensions: one at 80 pounds and the other at 90 pounds. The frequency of the string under less tension is 20 vibrations per second less than the other string. What are the frequencies of the two strings? Round to the nearest ten.

Of Interest

In addition to tension, the diameter and length of a string also affect its frequency. Increasing the diameter or length of a string under the same tension decreases its frequency.

REVIEW EXERCISES

[1.4] **1.** Evaluate $-\frac{b}{2a}$ when a is 2 and b is 8.

For Exercises 2 and 3, find the x- and y-intercepts.

[3.1] **2.** $2x + 3y = 6$

[6.4] **3.** $f(x) = 3x^2 - 2x - 8$

[5.2] **4.** If $f(x) = 2x^2 - 3x + 1$, find $f(-1)$.

[5.2, 6.4] ***For Exercises 5 and 6, graph.***

5. $f(x) = x^2 - 3$

6. $f(x) = 3x^2$

9.4 Graphing Quadratic Functions

OBJECTIVE

1. **Graph quadratic functions of the form $f(x) = ax^2$.**
2. **Graph quadratic functions of the form $f(x) = ax^2 + k$.**
3. **Graph quadratic functions of the form $f(x) = a(x - h)^2$.**
4. **Graph quadratic functions of the form $f(x) = a(x - h)^2 + k$.**
5. **Graph quadratic functions of the form $f(x) = ax^2 + bx + c$.**
6. **Solve applications involving parabolas.**

In Sections 5.2 and 6.4, we learned that quadratic functions have the form $f(x) = ax^2 + bx + c$, where a, b, and c are real numbers and $a \neq 0$. By plotting lots of ordered pairs, we found that the graphs of these functions are parabolas that open up if $a > 0$ and down if $a < 0$. By replacing x with 0, we found that the y-intercept is $(0, c)$.

In this section, we will learn about an alternate form of the quadratic function, $f(x) = a(x - h)^2 + k$. We will also learn about two important features of parabolas: the **axis of symmetry** and the **vertex**.

DEFINITIONS ***Axis of Symmetry:*** A line that divides a graph into two symmetrical halves.
Vertex: The lowest point on a parabola that opens up or highest point on a parabola that opens down.

Look at the graphs for $f(x) = x^2$ and $f(x) = -x^2 + 4$. Notice on each graph that the y-axis divides the graph into two halves that are mirror images of each other. Consequently, we say that the y-axis is the axis of symmetry for these graphs. The linear equation that describes the y-axis is $x = 0$. Also, in $f(x) = x^2$, we see that $a = 1$, which is positive, so the graph opens up. Since $(0, 0)$ is the lowest point on the graph, it is the vertex of $f(x) = x^2$. Similarly, in $f(x) = -x^2 + 4$, we see that $a = -1$, which is negative, so the graph opens down and its vertex is $(0, 4)$ because that is the highest point on the graph.

x	$f(x) = x^2$
−3	9
−2	4
−1	1
0	0
1	1
2	4
3	9

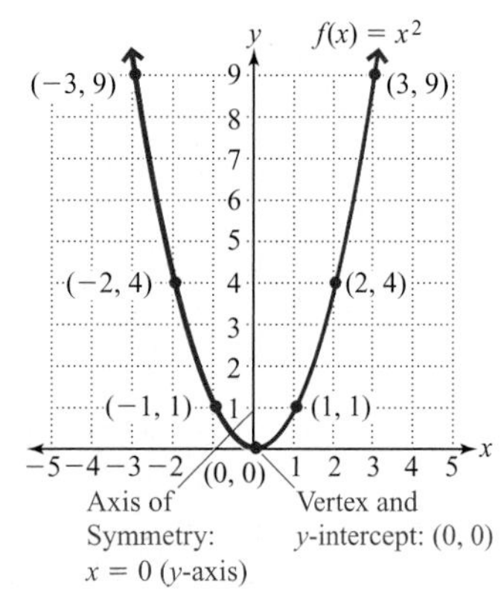

x	$f(x) = -x^2 + 4$
−3	−5
−2	0
−1	3
0	4
1	3
2	0
3	−5

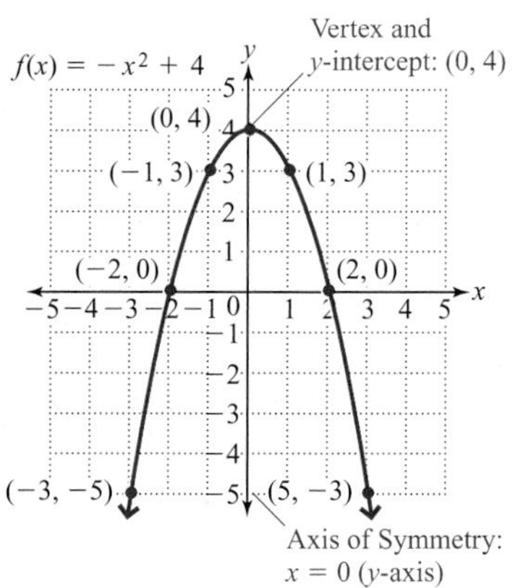

The benefit of the alternate form of the quadratic function $f(x) = a(x - h)^2 + k$ is that we will be able to determine the coordinates of the vertex and the axis of symmetry very easily from the values h and k in the equation. Later in the section, we will revisit the form $f(x) = ax^2 + bx + c$ and learn how to determine a parabola's vertex and axis of symmetry from the values of a and b.

OBJECTIVE 1. Graph quadratic functions of the form $f(x) = ax^2$. First we consider $f(x) = ax^2$ to discover more about how a affects the parabola. We will see that a not only affects whether the parabola opens up or down but also affects the width of the parabola.

EXAMPLE 1 Compare the graphs of each function.

a. $f(x) = \frac{1}{4}x^2, g(x) = \frac{1}{3}x^2, h(x) = \frac{1}{2}x^2, k(x) = x^2$, and $m(x) = 2x^2$

Solution We will graph all five functions on the same grid.

Note: *All these graphs open up because a is positive. Also notice that as a increases, the parabolas appear narrower.*

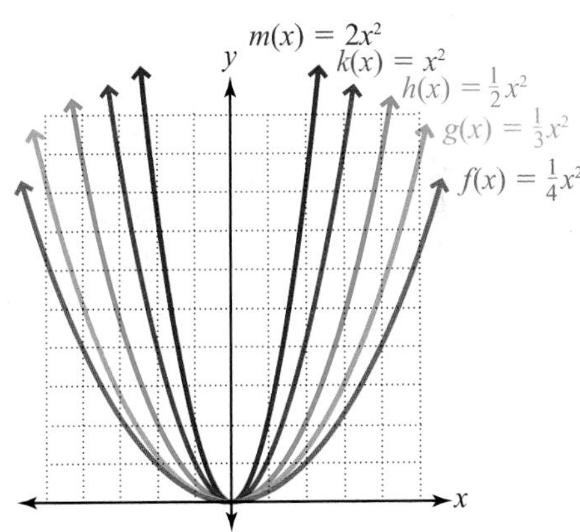

b. $f(x) = -\frac{1}{4}x^2, g(x) = -\frac{1}{3}x^2, h(x) = -\frac{1}{2}x^2, k(x) = -x^2$, and $m(x) = -2x^2$

Note: *All these graphs open down because a is negative. Also notice that as $|a|$ increases, the parabolas appear narrower.*

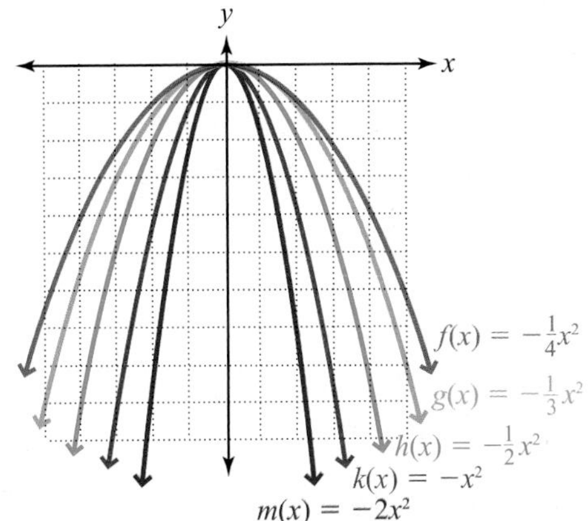

Example 1 suggests the following conclusions about functions of the form $f(x) = ax^2$.

Conclusion Given a function in the form $f(x) = ax^2$, the axis of symmetry is $x = 0$ (the y-axis) and the vertex is (0, 0). Also, the greater the absolute value of a, the narrower the parabola appears (or, the smaller the absolute value of a, the wider the parabola appears).

YOUR TURN Graph the following functions.

a. $f(x) = -4x^2$ **b.** $f(x) = \frac{3}{4}x^2$

OBJECTIVE 2. Graph quadratic functions of the form $f(x) = ax^2 + k$. We have learned that the vertex of a parabola of the form $f(x) = ax^2$ is at (0, 0) and the axis of symmetry is $x = 0$ (the y-axis). Now let's consider quadratic functions of the form $f(x) = ax^2 + k$, where k is a constant. We will see that for the same value of a, the graphs of $f(x) = ax^2 + k$ and $f(x) = ax^2$ have the same width and shape but the graph of $f(x) = ax^2 + k$ is shifted up or down k units on the y-axis from the origin so that the vertex is at $(0, k)$.

EXAMPLE 2 Graph the following functions.

a. $g(x) = 2x^2 + 3$ and $h(x) = 2x^2 - 4$

Solution We will graph both functions on the same grid.

Note: *The vertex of the "basic" function $f(x) = 2x^2$ is the origin (0, 0). The vertex of $g(x) = 2x^2 + 3$ is shifted up 3 and the vertex of $h(x) = 2x^2 - 4$ is shifted down 4 from the origin.*

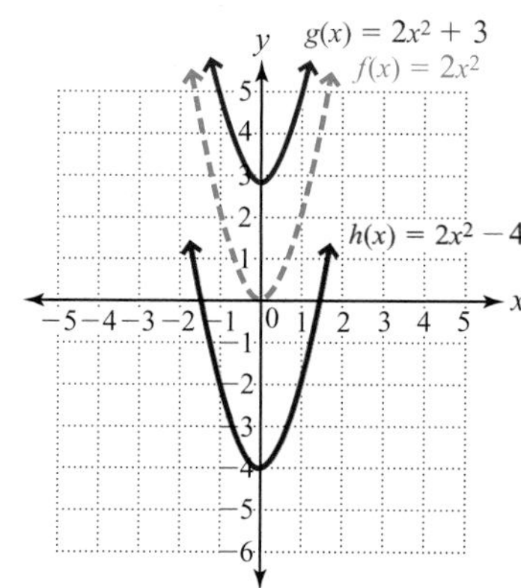

b. $g(x) = -x^2 + 4$ and $h(x) = -x^2 - 2$

Solution

Note: *The vertex of the "basic" function $f(x) = -x^2$ is the origin (0, 0). The vertex of $g(x) = -x^2 + 4$ is shifted up 4 and the vertex of $h(x) = -x^2 - 2$ is shifted down 2 from the origin.*

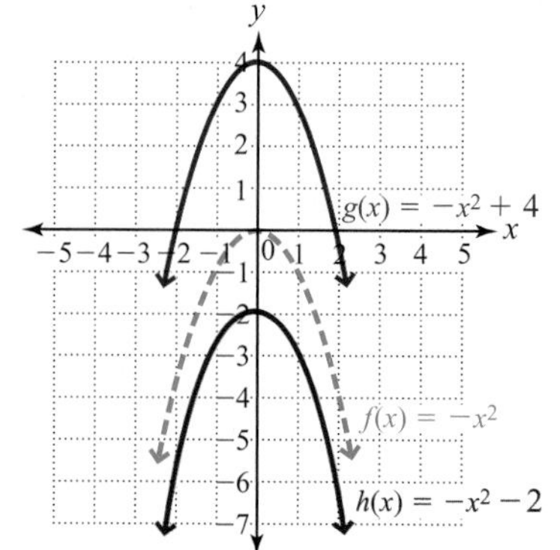

ANSWERS

a.

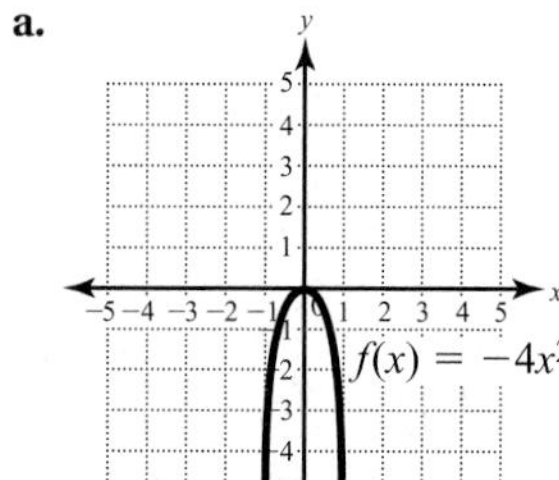

b.

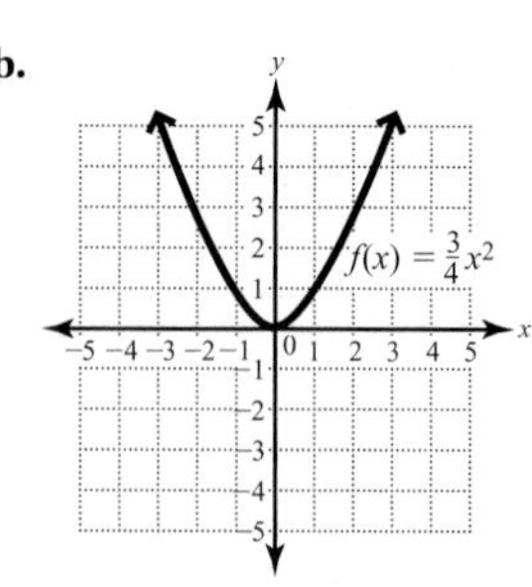

Example 2 suggests the following conclusion about the constant k in functions of the form $f(x) = ax^2 + k$.

Conclusion Given a function in the form $f(x) = ax^2 + k$, if $k > 0$, then the graph of $f(x) = ax^2$ is shifted k units *up* from the origin. If $k < 0$, then the graph of $f(x) = ax^2$ is shifted k units *down* from the origin. The new position of the vertex is $(0, k)$. The axis of symmetry is $x = 0$ (the y-axis).

YOUR TURN Graph the following functions.

a. $f(x) = 3x^2 - 2$ **b.** $f(x) = -\frac{1}{2}x^2 + 4$

OBJECTIVE 3. Graph quadratic functions of the form $f(x) = a(x - h)^2$. We now consider the form $f(x) = a(x - h)^2$ and we will see that the constant h causes the parabola to shift right or left.

EXAMPLE 3 Graph. $m(x) = 2(x - 3)^2$ and $n(x) = 2(x + 4)^2$

Solution

Note: *The vertex of the "basic" function $f(x) = 2x^2$ is $(0, 0)$ and the axis of symmetry is $x = 0$. The graph of $m(x) = 2(x - 3)^2$ has the same shape as $f(x) = 2x^2$, only the vertex and axis of symmetry are shifted* right 3 *units from the origin along the x-axis so that they are $(3, 0)$ and $x = 3$. Similarly, the vertex and axis of symmetry of $n(x) = 2(x + 4)^2$ are shifted* left 4 *units from the origin along the x-axis so that they are $(-4, 0)$ and $x = -4$.*

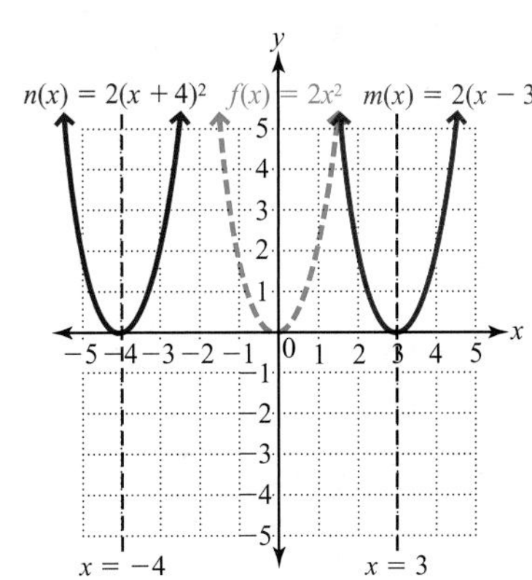

Example 3 suggests the following conclusion about h in $f(x) = a(x - h)^2$.

Conclusion Given a function in the form $f(x) = a(x - h)^2$, if $h > 0$, then the graph of $f(x) = ax^2$ is shifted h units *right* from the origin. If $h < 0$, then the graph of $f(x) = ax^2$ is shifted h units *left* from the origin. The new position of the vertex is $(h, 0)$ and the axis of symmetry is $x = h$.

a.

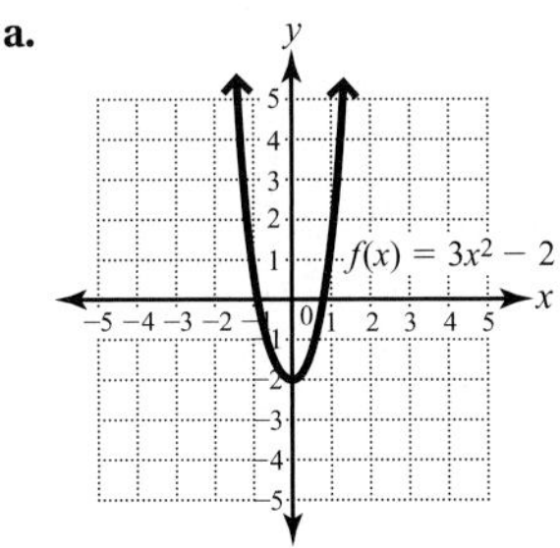

b.

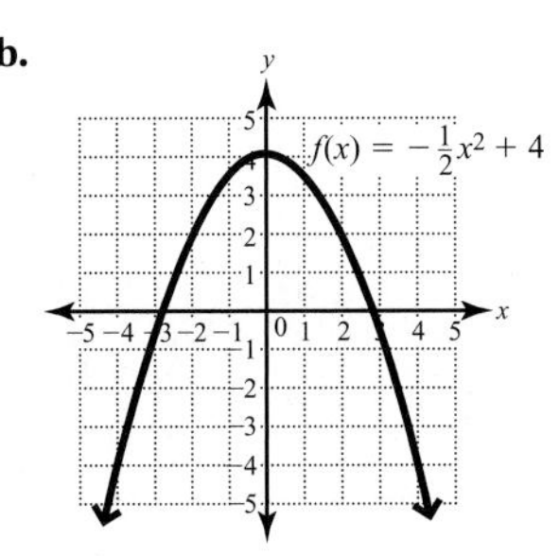

YOUR TURN Graph the following functions.

a. $f(x) = -2(x + 1)^2$ **b.** $g(x) = \frac{1}{3}(x - 4)^2$

OBJECTIVE 4. Graph quadratic functions of the form $f(x) = a(x - h)^2 + k$. Examples 2 and 3 suggest that the graph of a function of the form $f(x) = a(x - h)^2 + k$ has the same shape as $f(x) = ax^2$, but the vertex is shifted from the origin to (h, k) and the axis of symmetry is shifted from $x = 0$ to $x = h$. We can summarize all that we have learned with the following rule.

RULE Parabola with Vertex (h, k).

The graph of a function in the form $f(x) = a(x - h)^2 + k$ is a parabola with vertex at (h, k). The equation of the axis of symmetry is $x = h$. The parabola opens upwards if $a > 0$ and downwards if $a < 0$. The larger the $|a|$, the narrower the graph.

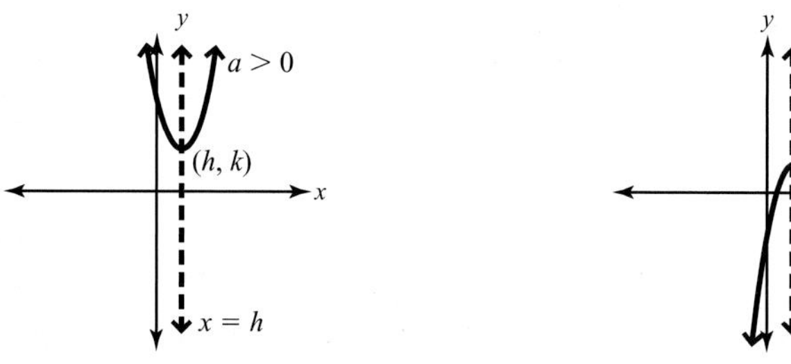

Now let's graph quadratic equations of the form $f(x) = a(x - h)^2 + k$.

EXAMPLE 4 Given $f(x) = 2(x - 3)^2 + 1$, determine whether the graph opens up or down, find the vertex and axis of symmetry, and draw the graph.

Solution We see that $f(x) = 2(x - 3)^2 + 1$ is in the form $f(x) = a(x - h)^2 + k$, where $a = 2$, $h = 3$, and $k = 1$. Since a is positive 2, the parabola opens upwards. The vertex is at (3, 1) and the axis of symmetry is $x = 3$. To complete the graph, we find a few points on either side of the axis of symmetry.

Note: *We find these additional points by choosing x-values on either side of the axis of symmetry and by using the equation to find the corresponding y-values.*

x	y
2	3
1	9
4	3
5	9

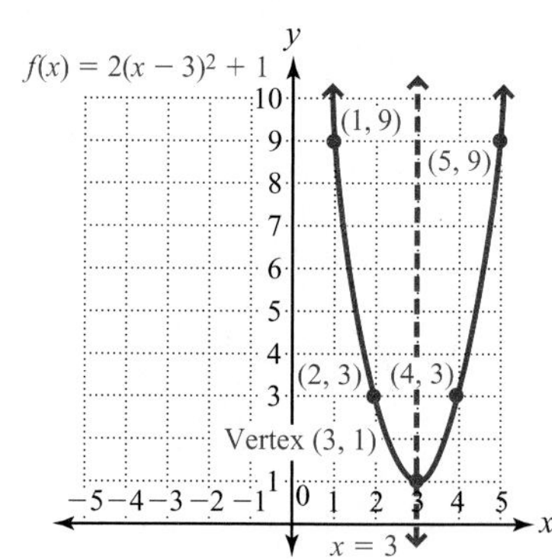

ANSWERS

a.

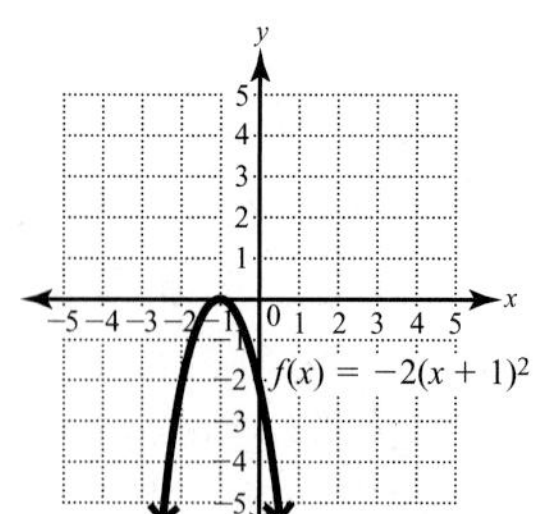

b.

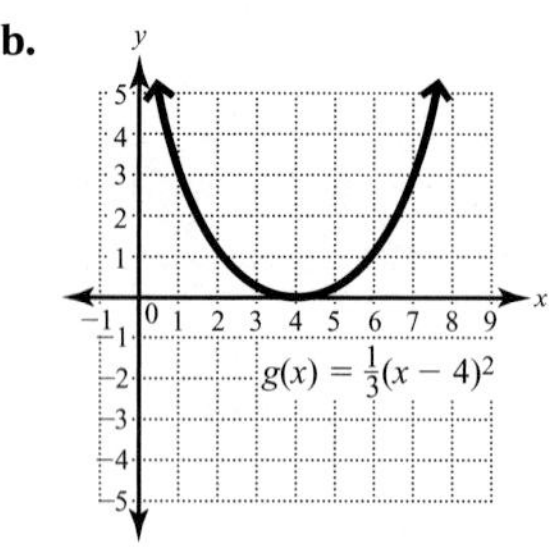

YOUR TURN Given the equation $f(x) = -\frac{1}{2}(x - 1)^2 + 3$, determine whether the graph opens upwards or downwards, find the vertex and axis of symmetry, and draw the graph.

OBJECTIVE 5. Graph quadratic functions of the form $f(x) = ax^2 + bx + c$. The advantage of the form $f(x) = a(x - h)^2 + k$ is that we can "see" the vertex, axis of symmetry, and whether the parabola opens upwards or downwards by looking at the equation. If an equation is in the form $f(x) = ax^2 + bx + c$, it can be transformed into $f(x) = a(x - h)^2 + k$ by completing the square.

ANSWERS

opens downwards, vertex: (1, 3), axis: $x = 1$

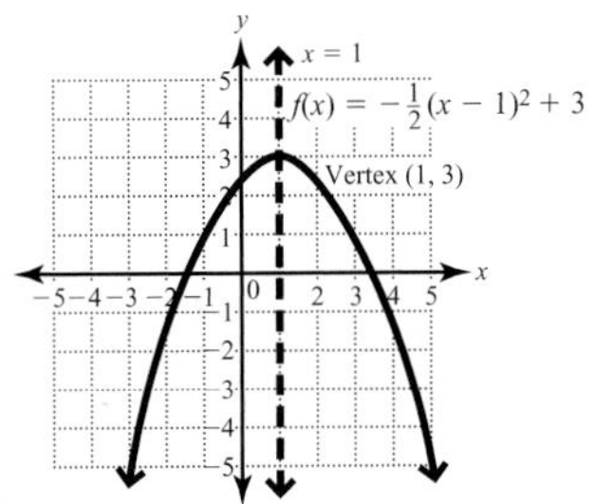

EXAMPLE 5 Write $f(x) = x^2 - 2x - 8$ in the form $f(x) = a(x - h)^2 + k$. Then determine whether the graph opens upwards or downwards, find the vertex and axis of symmetry, and draw the graph.

Solution

$y = x^2 - 2x - 8$ — Replace $f(x)$ with y to make manipulations more workable.

$y + 8 = x^2 - 2x$ — Add 8 to both sides of the equation to isolate the x terms.

$y + 8 + 1 = x^2 - 2x + 1$ — Add 1 to both sides of the equation to complete the square.

$y + 9 = (x - 1)^2$ — Write the right side as a square.

$y = (x - 1)^2 - 9$ — Subtract 9 from both sides to solve for y.

Since $a = 1$ and 1 is positive, the graph opens upwards. The vertex is at $(1, -9)$ and the axis of symmetry is $x = 1$. Plot a few points on either side of the axis of symmetry and graph. We will include the x- and y-intercepts, which we found in Example 3 of Section 9.2

	x	$f(x)$
x-intercept →	−2	0
	−1	−5
y-intercept →	0	−8
vertex →	1	−9
	2	−8
	3	−5
x-intercept →	4	0

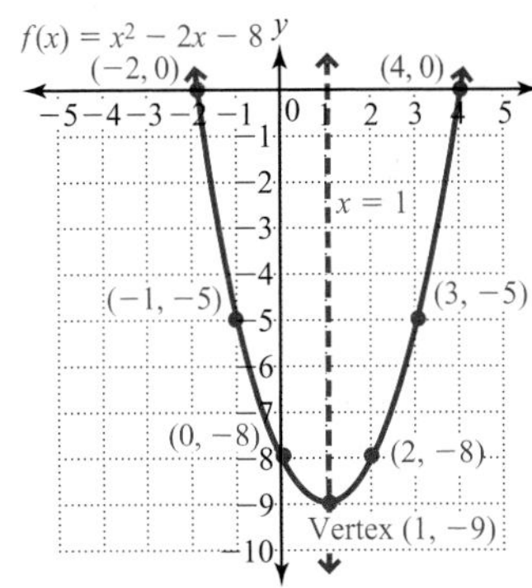

Note: *Remember, to find x-intercepts, we replace $f(x)$ (or y) with 0 and solve for x. To find y-intercepts, we replace x with 0 and solve for y.*

ANSWERS

x-intercepts: $(-1, 0)$, $(3, 0)$
y-intercept: $(0, -3)$
$f(x) = (x - 1)^2 - 4$,
opens upwards, vertex: $(1, -4)$,
axis: $x = 1$

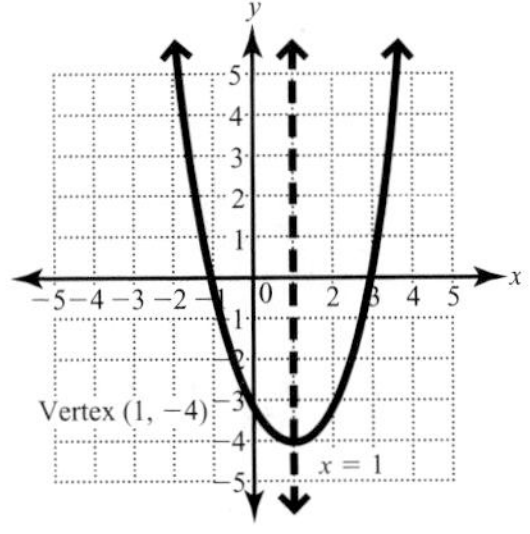

YOUR TURN For $f(x) = x^2 - 2x - 3$, find the x- and y-intercepts, write the equation in the form $f(x) = a(x - h)^2 + k$, determine whether the graph opens upwards or downwards, find the vertex and axis of symmetry, and draw the graph.

We can also find the vertex using a formula, which we can derive by completing the square on the general form $f(x) = ax^2 + bx + c$.

Note: *Adding* $\left[\frac{1}{2}\left(\frac{b}{a}\right)\right]^2$ or $\frac{b^2}{4a^2}$ *completes the square inside the parentheses. Since those parentheses are multiplied by a, we add* $a \cdot \frac{b^2}{4a^2}$ *to the left side to keep the equation balanced.*

$y = ax^2 + bx + c$	Replace $f(x)$ with y.
$y - c = ax^2 + bx$	Isolate the x terms.
$y - c = a(x^2 + \frac{b}{a}x)$	Factor out a.
$y - c + \frac{ab^2}{4a^2} = a\left(x^2 + \frac{b}{a}x + \frac{b^2}{4a^2}\right)$	Complete the square.
$y - \frac{4a^2c - ab^2}{4a^2} = a\left(x + \frac{b}{2a}\right)^2$	Find the LCD of the two terms on the left. Factor on the right.
$y - \frac{a(4ac - b^2)}{4a^2} = a\left(x + \frac{b}{2a}\right)^2$	Remove a common factor of a on the left side.
$y - \frac{4ac - b^2}{4a} = a\left(x + \frac{b}{2a}\right)^2$	Reduce to lowest terms.
$y = a\left(x + \frac{b}{2a}\right)^2 + \frac{4ac - b^2}{4a}$	Add $\frac{4ac - b^2}{4a}$ to both sides.

Notice that if we rewrite the last line above as $y = a\left(x - \left(-\frac{b}{2a}\right)\right)^2 + \frac{4ac - b^2}{4a}$, we see that $h = -\frac{b}{2a}$ and $k = \frac{4ac - b^2}{4a}$ so that the vertex is $\left(-\frac{b}{2a}, \frac{4ac - b^2}{4a}\right)$. Notice this also means that the axis of symmetry is $x = -\frac{b}{2a}$. Also, though the y-coordinate of the vertex is $\frac{4ac - b^2}{4a}$, it is usually easier to find it by substituting the x-coordinate into the original function.

PROCEDURE **Finding the Vertex of a Quadratic Function in the Form $f(x) = ax^2 + bx + c$**

Given an equation in the form $f(x) = ax^2 + bx + c$, to determine the vertex,

1. Find the x-coordinate using the formula $x = -\frac{b}{2a}$.
2. Find the y-coordinate by evaluating $f\left(-\frac{b}{2a}\right)$.

EXAMPLE 6 For the function $f(x) = 2x^2 - 8x + 5$, find the coordinates of the vertex.

Note: *Using* $\frac{4ac - b^2}{4a}$ *also gives the y-coordinate:*

$$\frac{4(2)(5) - (-8)^2}{4(2)} = \frac{-24}{8} = -3$$

However, substituting the x-value, 2, into the original function is easier.

Solution First, find the x-coordinate of the vertex using $-\frac{b}{2a}$.

$$x\text{-coordinate of the vertex: } -\frac{b}{2a} = -\frac{(-8)}{2(2)} = \frac{8}{4} = 2$$

Now find the y-coordinate by evaluating $f(2)$.

$$y\text{-coordinate of the vertex: } f(2) = 2(2)^2 - 8(2) + 5 = 8 - 16 + 5 = -3$$

The vertex is $(2, -3)$.

ANSWER

$(-3, -46)$

YOUR TURN For the equation $f(x) = 3x^2 + 18x - 19$, find the vertex.

OBJECTIVE 6. Solve applications involving parabolas. Because the vertex of a parabola is the highest point in parabolas that open down or the lowest point in parabolas that open up, we can find a maximum or minimum value in applications involving quadratic functions.

EXAMPLE 7 A toy rocket is launched straight up with an initial velocity of 40 feet per second. The equations $h = -16t^2 + 40t$ describes the height, h, of the rocket t seconds after being launched.

a. Find the maximum height that the rocket reaches.

b. Find the amount of time that the rocket is in the air.

Solution

a. Since the graph of $h = -16t^2 + 40t$ is a parabola that opens down ($a = -16$), the maximum height occurs at its vertex.

$$t\text{-coordinate of the vertex: } -\frac{b}{2a} = -\frac{40}{2(-16)} = -\frac{40}{-32} = \frac{5}{4} = 1.25 \text{ seconds}$$

To find the h-coordinate of the vertex we replace t with 1.25 in $h = -16t^2 + 40t$.

$$h = -16(1.25)^2 + 40(1.25) = 25 \text{ feet}$$

Note: *We could have found the vertex by writing the equation in* $y = a(x - h)^2 + k$ *form, which would be* $h = -16(t - 1.25)^2 + 25$.

The vertex is (1.25, 25), so the maximum height is 25 feet, which occurs 1.25 seconds after the rocket is launched.

b. The time the rocket is in the air is from launch until it returns to the ground. At launch and upon returning to the ground, the rocket's height is 0, so we need to find t when $h = 0$.

$$0 = -16t^2 + 40t \quad \text{Replace } h \text{ with 0.}$$

$$0 = -8t(2t - 5) \quad \text{Factor out a common factor of } -8t.$$

$$-8t = 0 \quad \text{or} \quad 2t - 5 = 0 \quad \text{Use the zero-factor theorem.}$$

$$t = 0 \qquad 2t = 5$$

$$t = 2.5$$

This means the height is 0 when $t = 0$ and when $t = 2.5$ seconds, so the rocket is in the air for 2.5 seconds.

Note: *The graph of* $h = -16t^2 + 40t$ *is not the flight path of the rocket. Also, because the situation involves a rocket being launched from the ground (0 feet) and returning to the ground, only the portion of the graph in the first quadrant is realistic.*

The graph of $h = -16t^2 + 40t$ shows that at 0 seconds the rocket is at 0 feet, which is when it is launched. Its height increases to a maximum of 25 feet 1.25 seconds after launch. Then the height decreases back to 0 feet 2.5 seconds after launch.

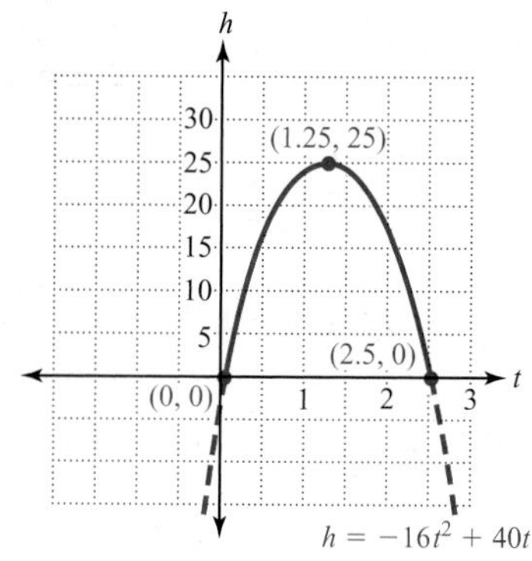

YOUR TURN A soccer player kicks the ball straight up with an initial velocity of 56 feet per second. The equation $h = -16t^2 + 56t$ describes the height, h, of the ball t seconds after being kicked.

a. After how many seconds is the ball at its maximum height?

b. What is the maximum height that the ball reaches?

c. How long is the ball in the air?

ANSWERS

a. 1.75 sec.
b. 49 ft.
c. 3.5 sec.

If an object is thrown or shot upwards and outwards, gravity causes its path, or trajectory, to be in the shape of a parabola.

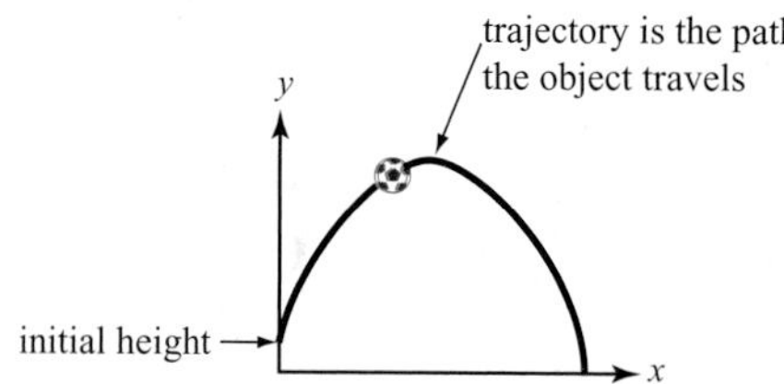

Note: *x represents the horizontal distance the object travels, and y represents the vertical distance the object travels.*

EXAMPLE 8 The equation $y = -0.8x^2 + 2.4x + 6$ when $x \geq 0$ and $y \geq 0$ describes the trajectory of a ball thrown upwards and outwards from an initial height of 6 feet.

a. What is the maximum height that the ball reaches?

b. How far does the ball travel horizontally?

c. Graph the trajectory of the ball.

Solution

a. The maximum height will be the y-coordinate of the vertex.

$$x\text{-coordinate of the vertex: } -\frac{b}{2a} = -\frac{(2.4)}{2(-0.8)} = 1.5$$

$$y\text{-coordinate of the vertex: } y = -0.8(1.5)^2 + 2.4(1.5) + 6 = 7.8$$

The vertex is (1.5, 7.8), so the maximum height is 7.8 feet.

b. To find how far the ball travels, we need to find the x-value when it hits the ground, which is where $y = 0$.

$$0 = -0.8x^2 + 2.4x + 6$$ Substitute 0 for y in the equation.

$$x = \frac{-(2.4) \pm \sqrt{(2.4)^2 - 4(-0.8)(6)}}{2(-0.8)}$$ Substitute $a = -0.8$, $b = 2.4$, and $c = 6$ into the quadratic formula.

$$= \frac{-2.4 \pm \sqrt{24.96}}{-1.6} \approx -1.62 \text{ or } 4.62$$

The negative value does not make sense in the context of this problem, so the distance the ball travels must be approximately 4.62 feet.

Note: *Because the equation describes the trajectory when $x \geq 0$ and $y \geq 0$, we only consider the part of the graph in the first quadrant.*

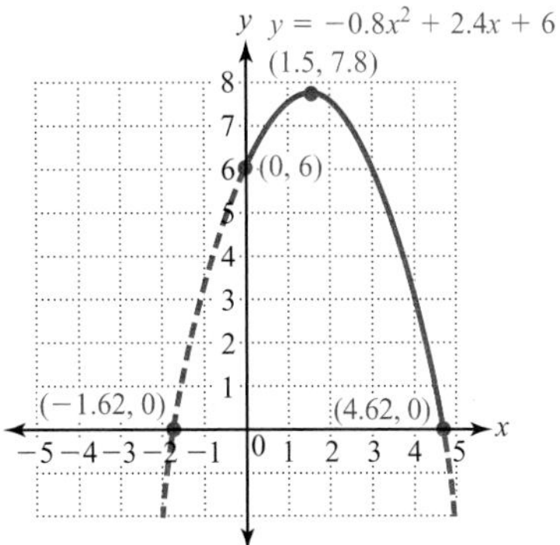

Calculator **TIPS**

Besides being able to graph parabolas on the graphing calculator, you can also find the x-intercepts and the vertex.

Let's work through Example 8 using a graphing calculator. First, press the Y= *key and enter the equation* $y = -0.8x^2 + 2.4x + 6$ *as Y1, then press GRAPH to see the graph.*

Now, to find the x-intercepts, press 2nd TRACE*, which will bring up the CALC menu. Use the arrow keys to highlight ZERO in the CALC menu, then press* ENTER *to select it. This will return you to the graph screen where you will be prompted for the left bound. Use the left-right arrow keys to move the cursor along the parabola until it is to the left of one of the x-intercepts and press* ENTER*. Next you will be prompted for the right bound. Again, use the arrow keys to move the cursor to the right of the x-intercept, then press* ENTER*. You will then be prompted for a guess. Press* ENTER *and you will see the coordinates of this x-intercept. Follow the same procedure for the other x-intercept.*

To find the vertex, we use the MAXIMUM or MINIMUM functions under the CALC menu. Since our graph opens down, the vertex will be a maximum, so we will use the MAXIMUM function. After selecting the MAXIMUM function, you will be prompted for the left bound, then the right bound, then a guess. After entering the left bound, right bound, and pressing ENTER *for the guess, you will see the coordinates of the vertex.*

9.4 Exercises

For Extra Help — MyMathLab | Videotape/DVT | InterAct Math | Math Tutor Center | Math XL.com

1. In an equation in the form $f(x) = a(x - h)^2 + k$, what determines whether the parabola opens upwards or downwards?

2. In an equation in the form $f(x) = a(x - h)^2 + k$, what are the coordinates of the vertex?

3. Describe the axis of symmetry in a parabola that opens up or down. What is its equation?

4. Suppose the y-intercept for a given quadratic equation is $(0, -5)$. If the vertex is at $(3, -2)$, what can you conclude about the x-intercepts?

5. If the solutions for a quadratic equation are imaginary numbers, then what does that indicate about the corresponding quadratic function?

6. Given that the x-coordinate of the vertex is $-\frac{b}{2a}$, what must be true about a and b if the x-coordinate of the vertex is -1?

For Exercises 7–16, state whether the parabola opens upwards or downwards.

7. $f(x) = 2x^2$

8. $g(x) = -3x^2$

9. $h(x) = -x^2 + 3$

10. $k(x) = 4x^2 - 5$

11. $f(x) = 2(x - 1)^2 - 4$

12. $h(x) = -0.8(x + 3)^2 - 1$

13. $g(x) = -\frac{2}{3}x^2 + 4x - 1$

14. $f(x) = -x^2 - 8x - 9$

15. $k(x) = 3x - 0.5x^2$

16. $g(x) = -2x + \frac{3}{4}x^2$

For Exercises 17–34, find the coordinates of the vertex and write the equation of the axis of symmetry.

17. $f(x) = 5(x - 2)^2 - 3$

18. $h(x) = 2(x - 1)^2 + 4$

19. $g(x) = -2(x + 1)^2 - 5$

20. $k(x) = -(x + 4)^2 - 3$

21. $k(x) = x^2 + 2$

22. $g(x) = x^2 - 6$

23. $h(x) = 3(x - 7)^2$

24. $f(x) = 5(x + 1)^2$

25. $f(x) = -0.5x^2$

26. $k(x) = 2.5x^2$

27. $f(x) = x^2 - 4x + 8$

28. $g(x) = x^2 + 6x + 8$

29. $k(x) = 2x^2 + 16x + 27$

30. $f(x) = 3x^2 - 6x + 1$

31. $g(x) = -3x^2 + 2x + 1$

32. $h(x) = 5x^2 - 3x + 2$

33. $k(x) = -0.5x^2 - 0.4x + 1$

34. $f(x) = 0.2x^2 - 0.3x - 5$

For Exercises 35–54: a. State whether the parabola opens upwards or downwards.
b. Find the coordinates of the vertex.
c. Write the equation of the axis of symmetry.
d. Graph.

35. $h(x) = -3x^2$

36. $f(x) = 2x^2$

37. $k(x) = \frac{1}{4}x^2$

38. $g(x) = -\frac{1}{3}x^2$

39. $f(x) = 4x^2 - 3$

40. $h(x) = -x^2 + 4$

41. $g(x) = -0.5x^2 + 2$

42. $k(x) = 0.5x^2 - 3$

43. $f(x) = (x - 3)^2 + 2$

44. $h(x) = (x - 2)^2 - 1$

45. $k(x) = -2(x + 1)^2 - 3$

46. $g(x) = -(x + 3)^2 + 1$

47. $h(x) = \frac{1}{3}(x - 2)^2 - 1$

48. $f(x) = -\frac{1}{4}(x - 3)^2 - 2$

For Exercises 49–54: ***a. Find the x- and y-intercepts.***
b. Write the equation in the form $f(x) = a(x - h)^2 + k$.
c. State whether the parabola opens upwards or downwards.
d. Find the coordinates of the vertex.
e. Write the equation of the axis of symmetry.
f. Graph.

49. $h(x) = x^2 + 6x + 9$

50. $k(x) = 3x^2 - 6x + 1$

51. $f(x) = 2x^2 + 6x + 3$

52. $k(x) = -x^2 - 2x + 3$

53. $g(x) = -3x^2 + 6x - 5$

54. $h(x) = -3x^2 - 6x + 4$

For Exercises 55–66, solve.

55. A toy rocket is launched with an initial velocity of 45 meters per second. The equation $h = -4.9t^2 + 45t$ describes the height, h, of the rocket in meters t seconds after being launched.
   a. After how many seconds does the rocket reach its maximum height?
   b. What is the maximum height the rocket reaches?
   c. How long is the rocket in the air?

56. A ball is drop-kicked straight up with an initial velocity of 36 feet per second. The equation $h = -16t^2 + 36t$ describes the height, h, of the ball in feet t seconds after being kicked.
   a. After how many seconds does the ball reach its maximum height?
   b. What is the maximum height the ball reaches?
   c. How long is the ball in the air?

57. The equation $y = -0.8x^2 + 2.4x + 6$ models the trajectory of a ball thrown upwards and outwards from a height of 6 feet (assume that $x \geq 0$ and $y \geq 0$).
   a. What is the maximum height that the ball reaches?
   b. How far does the ball travel horizontally?
   c. Graph the trajectory of the ball.

58. The javelin toss is an event in track and field in which participants try to throw a javelin the farthest. Suppose the equation $y = -0.02x^2 + 1.3x + 8$ models the trajectory of one particular throw (assume that x and y represent distances in meters and that $x \geq 0$ and $y \geq 0$).
   a. What is the maximum height that the javelin reaches?
   b. How far does the javelin travel horizontally?
   c. Graph the trajectory of the javelin.

59. A record company discovers that the number of CDs sold each week after release follows a parabolic pattern. The function $n(t) = -200t^2 + 4000t$ describes the number, n, of CDs an artist sold each of t weeks after the release of the album.
   a. Which week had the greatest number of CDs sold?
   b. How many CDs sold that week?

60. The function $n(t) = -3t^2 + 42t$ describes the number, n, of tickets sold for a play each of t days after tickets went on sale.
   a. What day had the greatest number of tickets sold?
   b. How many tickets sold that day?

61. A farmer has enough materials to build a fence with a total length of 400 feet. He wants the enclosed space to be rectangular and also wants to maximize the area enclosed. Find the length and width so that the area is maximized.

62. An architect wants the length and width of a rectangular building to be a total of 150 feet long. She also wants to maximize the rectangular area that the building occupies. Find the length and width so that the area is maximized.

63. The function $C(n) = n^2 - 110n + 5000$ describes a company's cost, C, of producing n units of its product. Find the number of units the company should produce to minimize its cost.

64. The function $P(t) = 0.001t^2 - 0.24t + 59.90$ roughly models a particular stock's closing price, P, each of t days of trading during one year.

  a. After how many days is the price at its lowest?

  b. What was the lowest price during that year?

65. One integer is 12 more than another. If their product is minimized, find the integers and their product.

66. The greater of two integers minus the smaller gives a result of 20. Find the two integers so that their product is minimized. Also find their product.

PUZZLE PROBLEM

Given the vertex and the x- and y-intercepts of a parabola, reconstruct the equation.

a. *Vertex* $(-3, -1)$
   Intercepts $(-2, 0), (-4, 0), (0, 8)$

b. *Vertex* $(-1, 0)$
   Intercepts $(-1, 0), (0, 1)$

c. *Vertex* $(-2, 1)$
   Intercept $(0, 5)$

d. *Vertex* $(1, 0)$
   Intercepts $(1, 0), (0, 1)$

Collaborative Exercises ARCH SPAN

The Gateway Arch, located in St. Louis, Missouri, was built from 1963 to 1965 and is the nation's tallest memorial. The equation $h(x) = -0.0063492063x^2 + 630$ can be used to approximate the height of the arch, where x represents the distance from its axis of symmetry and h(x) represents its height above the ground.

1. Using the equation, find the maximum height of the structure.

2. The span of the arch at a given height is the horizontal distance between the two opposing points on the parabola at that height. Find the span of the arch at ground level. (*Hint:* Think of ground level as the x-axis. The height is 0 along the x-axis.)

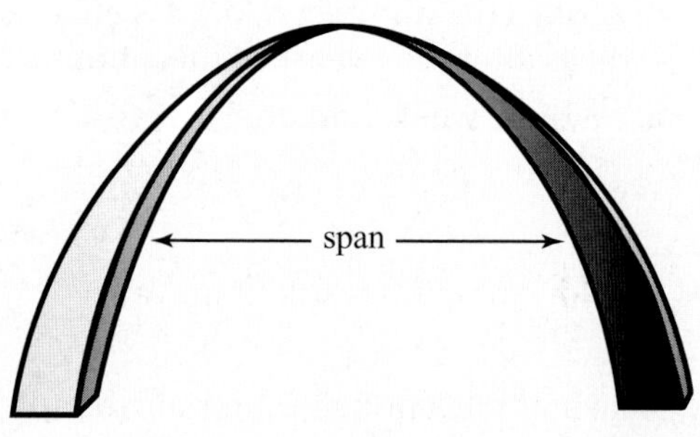

3. How does the span at ground level relate to the maximum height of the arch?

4. The arch has foundations 60 feet below ground level. What is the span of the arch between its foundations?

REVIEW EXERCISES

[1.3] **1.** Evaluate: $-|-2 + 3 \cdot 4| - 4^0$

[6.4] **2.** Solve $x^2 + 2x = 15$ by factoring.

[9.1] **3.** Solve $x^2 = -20$ using the square root principle.

For Exercises 4–6, solve, then graph the solution set.

[2.3] **4.** $4x - 8 \le 2x + 1$

[2.6] **5.** $|x - 3| \ge 8$

[4.6] **6.** $\begin{cases} x + y > 2 \\ 2x - 3y \le 6 \end{cases}$

9.5 Solving Nonlinear Inequalities

OBJECTIVES

1. **Solve quadratic and other inequalities.**
2. **Solve rational inequalities.**

OBJECTIVE 1. Solve quadratic and other inequalities. Now that we have solved quadratic equations, we can learn how to solve **quadratic inequalities**.

DEFINITION ***Quadratic inequality:*** An inequality that can be written in the form $ax^2 + bx + c > 0$ or $ax^2 + bx + c < 0$, where $a \neq 0$.

Note: *In this definition, the symbols $<$ and $>$ can be replaced with $\le$ or $\ge$.*

For example, $2x^2 - x - 3 < 0$ and $2x^2 - x - 3 \ge 0$ are quadratic inequalities. Let's see what we can learn about these inequalities by looking at the graph of the corresponding quadratic function: $y = 2x^2 - x - 3$. Note that by letting $y = 0$, we have the corresponding equation $2x^2 - x - 3 = 0$ and solving this equation gives the x-intercepts $x = -1$ and $x = \frac{3}{2}$. Also notice that those x-intercepts divide the x-axis into three intervals: $(-\infty, -1)$, $\left(-1, \frac{3}{2}\right)$, and $\left(\frac{3}{2}, \infty\right)$.

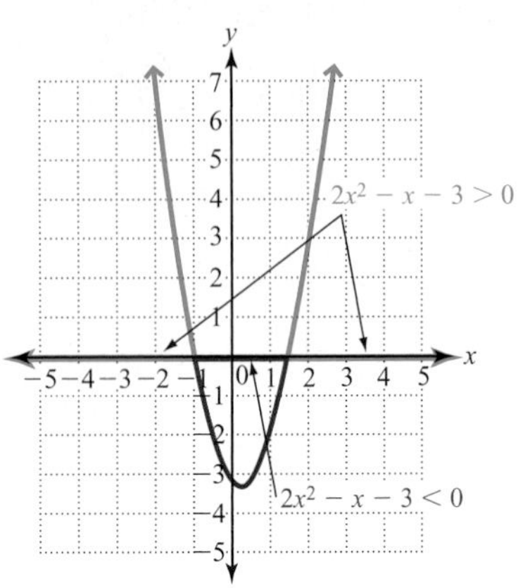

Notice the parabola is above the x-axis in intervals $(-\infty, -1)$ and $\left(\frac{3}{2}, \infty\right)$, shown in blue. Evaluating $y = 2x^2 - x - 3$ using any x-value in these intervals produces a y-value that is greater than 0, so these intervals are the solution sets for $2x^2 - x - 3 > 0$.

The parabola is below the x-axis in the interval $\left(-1, \frac{3}{2}\right)$, shown in red. Evaluating $y = 2x^2 - x - 3$ using any x-value in this interval produces a y-value that is less than 0, so this interval is the solution set for $2x^2 - x - 3 < 0$.

The following number lines indicate various solution sets based on the inequality.

$2x^2 - x - 3 > 0$

Solution set: $(-\infty, -1) \cup \left(\frac{3}{2}, \infty\right)$

$2x^2 - x - 3 < 0$

Solution set: $\left(-1, \frac{3}{2}\right)$

$2x^2 - x - 3 \geq 0$

Solution set: $(-\infty, -1] \cup \left[\frac{3}{2}, \infty\right)$

$2x^2 - x - 3 \leq 0$

Solution set: $\left[-1, \frac{3}{2}\right]$

Based on these observations, we use the following procedure to solve quadratic inequalities.

PROCEDURE **Solving Quadratic Inequalities**

1. Solve the related equation $ax^2 + bx + c = 0$.
2. Plot the solutions of $ax^2 + bx + c = 0$ on a number line. These solutions will divide the number line into intervals.
3. Choose a test number from each interval and substitute the number into the inequality. If the test number makes the inequality *true*, then all numbers in that interval will solve the inequality. If the test number makes the inequality *false*, then no numbers in that interval will solve the inequality.
4. State the solution set of the inequality: It is the union of all the intervals that solve the inequality. If the inequality symbols are $\leq$ or $\geq$, then the values from step 2 are included. If the symbols are $<$ or $>$, they are not solutions.

EXAMPLE 1 Solve $x^2 - x < 6$. Write the solution set using interval notation, then graph the solution set on a number line.

Solution $x^2 - x = 6$ — Write the related equation.

$x^2 - x - 6 = 0$ — Subtract 6 from both sides to get quadratic form.

$(x - 3)(x + 2) = 0$ — Factor.

$x - 3 = 0$ or $x + 2 = 0$ — Use the zero-factor theorem.

$x = 3$ $\quad$ $x = -2$ — These are the x-intercepts of the graph of $y = x^2 - x - 6$.

Plot -2 and 3 on a number line (x-axis), which divides it into three intervals.

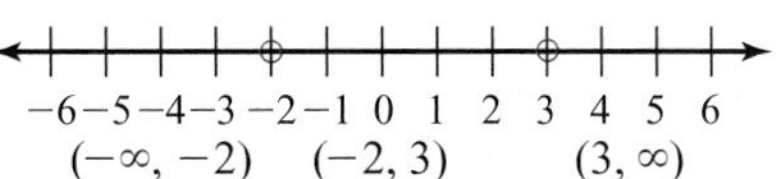

Note: *-2 and 3 are not part of the solution set because the inequality symbol is $<$.*

Connection Notice the solution set for $x^2 - x - 6 < 0$ corresponds to the interval where the graph of $f(x) = x^2 - x - 6$ is below the x-axis and the x-intercepts are the endpoints of the interval.

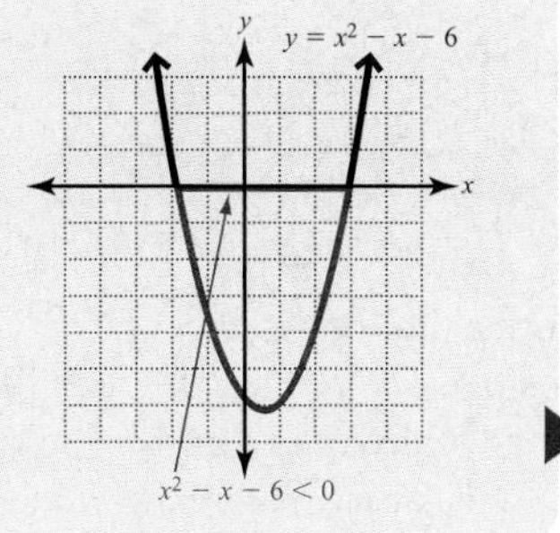

Choose a test number from each interval and substitute that value into $x^2 - x < 6$.

For $(-\infty, -2)$, we choose $x = -3$.

$$(-3)^2 - (-3) < 6$$
$$9 + 3 < 6$$
$$12 < 6$$

This is false, so $(-\infty, -2)$ is not in the solution set.

For $(-2, 3)$, we choose $x = 0$.

$$0^2 - 0 < 6$$
$$0 < 6$$

This is true, so $(-2, 3)$ is in the solution set.

For $(3, \infty)$, we choose $x = 4$.

$$4^2 - 4 < 6$$
$$16 - 4 < 6$$
$$12 < 6$$

This is false, so $(3, \infty)$ is not in the solution set.

Since $(-2, 3)$ is the only interval that has solutions to the inequality, it is the solution set. Following is the graph of the solution set.

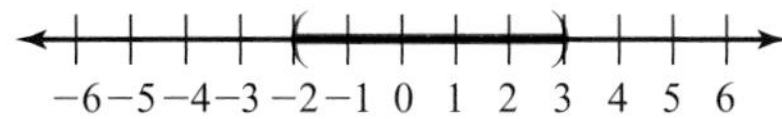

EXAMPLE 2 Solve $x^2 + 3x \geq 0$. Write the solution set using interval notation, then graph the solution set.

Solution

$$x^2 + 3x = 0 \quad \text{Write the related equation.}$$
$$x(x + 3) = 0 \quad \text{Factor.}$$
$$x = 0 \quad \text{or} \quad x + 3 = 0 \quad \text{Use the zero-factor theorem.}$$
$$x = -3$$

Note: *Since we have already determined that -3 and 0 are in the solution set, we will not include them in any of our test intervals, which is why we use parentheses for each interval.*

Plot -3 and 0 on a number line and note the intervals.

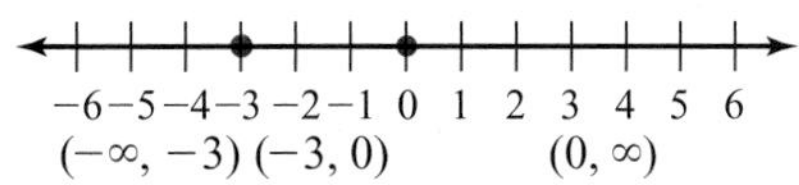

Note: *-3 and 0 are included in the solution set because the inequality symbol is $\geq$.*

Connection Notice that the solution set for $x^2 + 3x \geq 0$ corresponds to the intervals where the graph of $f(x) = x^2 + 3x$ is above the x-axis and the x-intercepts are endpoints in the intervals.

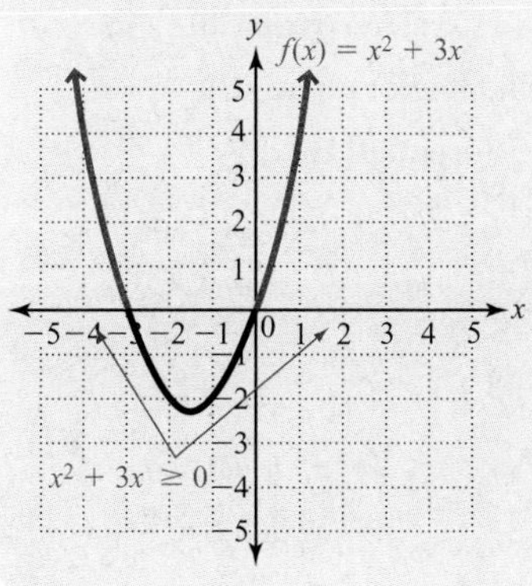

Choose a test number from each interval and test in $x^2 + 3x \geq 0$.

For $(-\infty, -3)$, we choose $x = -4$.

$$(-4)^2 + 3(-4) \geq 0$$
$$16 - 12 \geq 0$$
$$4 \geq 0$$

This is true, so $(-\infty, -3)$ is in the solution set.

For $(-3, 0)$, we choose $x = -1$.

$$(-1)^2 + 3(-1) \geq 0$$
$$1 - 3 \geq 0$$
$$-2 \geq 0$$

This is false, so $(-3,0)$ is not in the solution set.

For $(0, \infty)$, we choose $x = 1$.

$$1^2 + 3(1) \geq 0$$
$$1 + 3 \geq 0$$
$$4 \geq 0$$

This is true, so $(0, \infty)$ is in the solution set.

The solution set is $(-\infty, -3] \cup [0, \infty)$, which is graphed next.

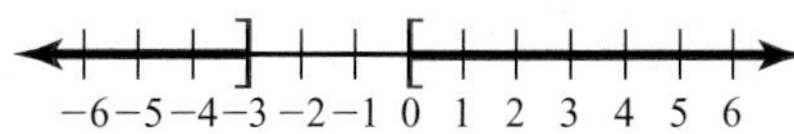

EXAMPLE 3 Solve. Write the solution set using interval notation, then graph the solution set.

a. $(x - 1)^2 > -2$

Solution Since $(x - 1)^2$ is always 0 or positive; it is always greater than -2, so every real number is a solution.

Solution set: $\mathbb{R}$ or $(-\infty, \infty)$

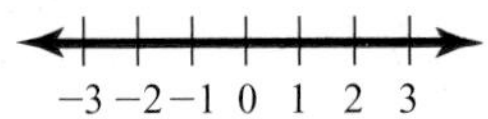

b. $(x - 1)^2 < -2$

Solution Since $(x - 1)^2$ is never negative, its value can never be less than -2, so there are no real solutions for $(x - 1)^2 < -2$.

Solution set: $\varnothing$

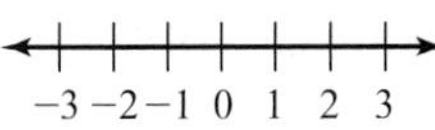

Connection Look at the graph of $f(x) = (x - 1)^2$.

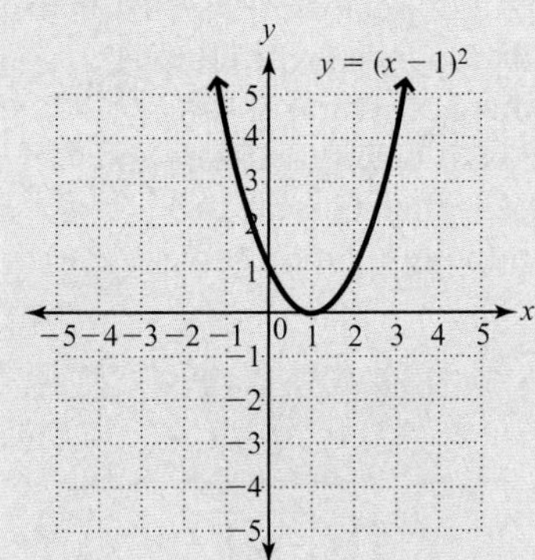

Notice that $f(x)$ is always 0 or positive, so $(x - 1)^2 > -2$ is true for all real x-values and $(x - 1)^2 < -2$ is false for all real x-values.

YOUR TURN Solve. Write the solution set using interval notation, then graph the solution set on a number line.

a. $x^2 - 2x - 8 \leq 0$ **b.** $x^2 - 4x > 0$ **c.** $(3x - 8)^2 < -5$

ANSWERS

a. $[-2, 4]$

b. $(-\infty, 0) \cup (4, \infty)$

c. $\varnothing$

Other Polynomial Inequalities

We use a similar procedure for expressions with more than two factors. In the following example, we condense the procedure to a table.

EXAMPLE 4 Solve $(x + 4)(x + 1)(x - 3) \leq 0$. Write the solution set in interval notation, then graph the solution set on a number line.

Solution $(x + 4)(x + 1)(x - 3) = 0$ — **Write the related equation.**

$x + 4 = 0$ or $x + 1 = 0$ or $x - 3 = 0$ — **Set each factor equal to 0.**

$x = -4$ $\quad x = -1$ $\quad x = 3$ — **Solve the equations.**

Plot -4, -1, and 3 on a number line and note the intervals.

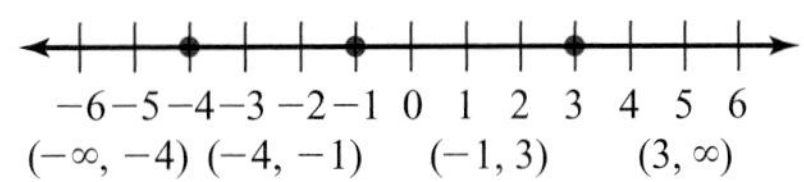

Note: *-4, -1, and 3 are included in the solution set because the inequality symbol is $\le$. Remember that we test intervals around these values, so they are not included in those intervals.*

Interval	$(-\infty, -4)$	$(-4, -1)$	$(-1, 3)$	$(3, \infty)$
Test Number	-5	-2	0	4
Test Results	$-32 \le 0$	$10 \le 0$	$-12 \le 0$	$40 \le 0$
True or False	True	False	True	False

Note: *Because the inequality is $\le$, -4, -1, and 3 are included in the solution set.*

Therefore, the solution set is $(-\infty, -4] \cup [-1, 3]$, which is graphed next.

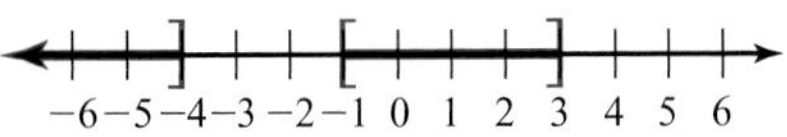

YOUR TURN Solve. Write the solution set using interval notation, then graph the solution set on a number line.

$$(x + 3)(x - 2)(x + 5) \le 0$$

OBJECTIVE 2. Solve rational inequalities. Now we consider solving **rational inequalities**.

DEFINITION ***Rational inequality:*** An inequality containing a rational expression.

For example, $\frac{x + 2}{x - 3} > 4$ is a rational inequality. Recall that in solving a polynomial inequality, we divided the number line into intervals using x-values we found from solving its related equation. With rational inequalities, we not only find values that solve the related equation but we must also consider values that make the rational expression undefined (denominator $= 0$).

PROCEDURE **Solving Rational Inequalities**

1. Find all values that make any denominator equal to 0. These values must be excluded from the solution set.
2. Solve the related equation.
3. Plot the numbers found in steps 1 and 2 on a number line.
4. Choose a test number from each interval and determine whether it solves the inequality.
5. The solution set is the union of all the regions whose test number solves the inequality. If the inequality symbol is $\le$ or $\ge$, include the values found in step 1. The solution set never includes the values found in step 2 because they make a denominator equal to 0.

ANSWER

$(-\infty, -5] \cup [-3, 2]$

−7 −6 −5 −4 −3 −2 −1 0 1 2 3

EXAMPLE 5 Solve. Write the solution set using interval notation, then graph the solution set on a number line.

a. $\dfrac{x-4}{x+2} \geq 0$

Solution First, we find the values that make the denominator equal to 0.

$$x + 2 = 0$$
$$x = -2$$

Now, solve the related equation.

$$\frac{x-4}{x+2} = 0 \qquad \text{Write the related equation.}$$

$$(x+2)\frac{x-4}{x+2} = (x+2)(0) \qquad \text{Multiply both sides by } x+2.$$

$$x - 4 = 0 \qquad \text{Simplify.}$$

$$x = 4$$

Plot −2 and 4 on a number line and note the intervals.

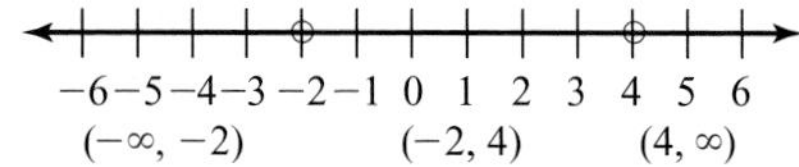

Again, we use a table.

Interval	$(-\infty, -2)$	$(-2, 4)$	$(4, \infty)$
Test Number	−3	0	5
Test Results	$7 \geq 0$	$-2 \geq 0$	$\frac{1}{7} \geq 0$
True or False	True	False	True

Note: *Remember, since −2 makes* $\dfrac{x-4}{x+2}$ *undefined, it is not included in the solution set. Since the inequality is* $\geqslant$, *4 is included in the solution set.*

The solution set is $(-\infty, -2) \cup [4, \infty)$ and is graphed next.

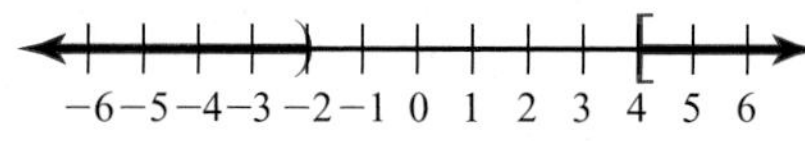

b. $\dfrac{x+5}{x-1} < 4$

Solution Find the values that make the denominator equal to 0.

$$x - 1 = 0$$
$$x = 1$$

Solve the related equation.

$$\frac{x+5}{x-1} = 4 \qquad \text{Write the related equation.}$$
$$(x-1)\frac{x+5}{x-1} = (x-1)(4) \qquad \text{Multiply both sides by the LCD.}$$
$$x + 5 = 4x - 4 \qquad \text{Simplify.}$$
$$-3x = -9 \qquad \text{Subtract } 4x \text{ and 5 from both sides.}$$
$$x = 3$$

Plot 1 and 3 on a number line.

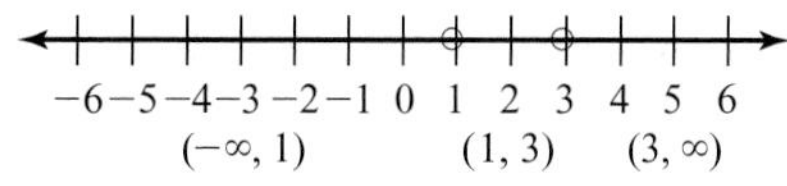

Interval	$(-\infty, 1)$	$(1, 3)$	$(3, \infty)$
Test Number	0	2	4
Test Results	$-5 < 4$	$7 < 4$	$3 < 4$
True or False	True	False	True

The solution set is $(-\infty, 1) \cup (3, \infty)$ and is graphed next.

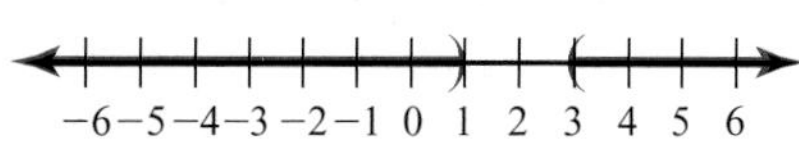

Note: *Since 1 makes* $\dfrac{x+5}{x-1}$ *undefined, it is not included in the solution set. Also, 3 is not in the solution set because* $\dfrac{x+5}{x-1} = 4$ *when* $x = 3$.

YOUR TURN Solve. Write the solution set using interval notation, then graph the solution set on a number line.

a. $\dfrac{x+3}{x-1} < 0$

b. $\dfrac{x-2}{x+4} \le 3$

ANSWERS

a. $(-3, 1)$

−4 −3 −2 −1 0 1 2

b. $(-\infty, -7] \cup (-4, \infty)$

−9 −8 −7 −6 −5 −4 −3 −2

9.5 Exercises

For Extra Help

MyMathLab

Videotape/DVT

InterAct Math

Math Tutor Center

Math XL.com

1. Explain how you would find the solution set of $x^2 + 2x - 15 \geq 0$.

2. If the graph of $y = ax^2 + bx + c$ intersects the x-axis at -2 and 2, what regions would you check to solve $ax^2 + bx + c > 0$?

3. Is it possible to have a quadratic inequality whose solution set is the empty set? Explain.

4. Is it possible to have a quadratic inequality whose solution set is one number? Explain.

5. The quadratic inequality $(x - 2)(x + 5) < 0$ and the rational inequality $\dfrac{x - 2}{x + 5} < 0$ have the same solution sets. Explain how this is possible.

6. If you were to solve $\dfrac{x + 2}{(x + 5)(x - 3)} \geq 0$, what are the regions that you would need to test?

For Exercises 7–14, the graph of a quadratic function is given. Use the graph to solve each equation and inequality. For solution sets that involve intervals, use interval notation.

7. **a.** $x^2 + 6x + 5 = 0$

   b. $x^2 + 6x + 5 < 0$

   c. $x^2 + 6x + 5 > 0$

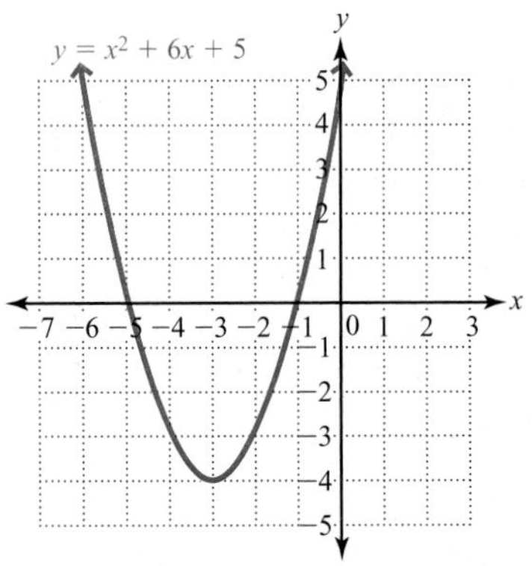

8. **a.** $x^2 - 4 = 0$

   b. $x^2 - 4 > 0$

   c. $x^2 - 4 < 0$

$y = x^2 + 6x + 5$

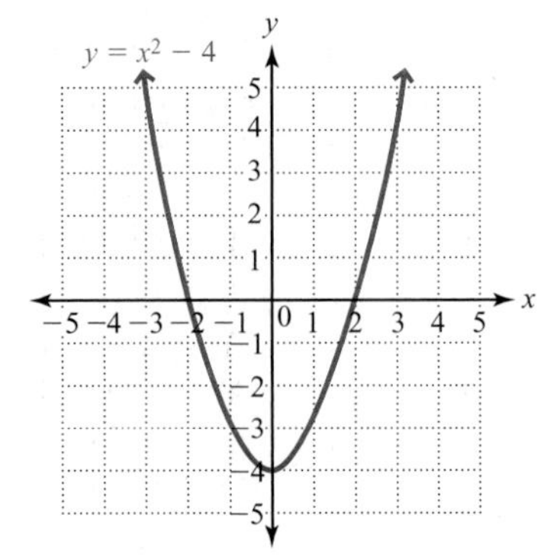

9. **a.** $-x^2 + 2x + 3 = 0$

b. $-x^2 + 2x + 3 \le 0$

c. $-x^2 + 2x + 3 \ge 0$

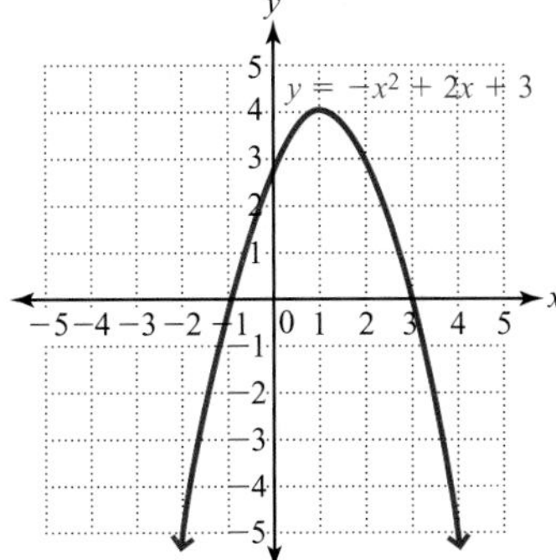

10. **a.** $-x^2 + 5x = 0$

b. $-x^2 + 5x \le 0$

c. $-x^2 + 5x \ge 0$

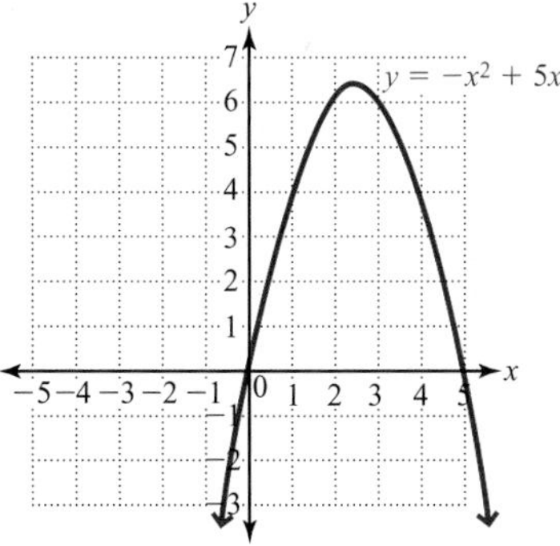

11. **a.** $x^2 + 4x + 4 = 0$

b. $x^2 + 4x + 4 > 0$

c. $x^2 + 4x + 4 < 0$

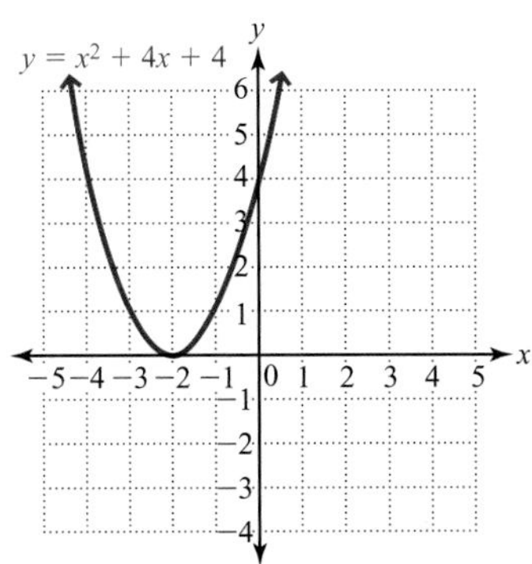

12. **a.** $x^2 - 6x + 9 = 0$

b. $x^2 - 6x + 9 < 0$

c. $x^2 - 6x + 9 > 0$

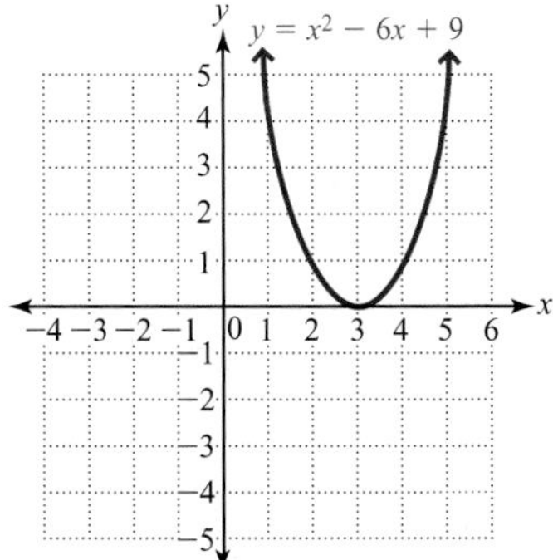

13. **a.** $x^2 + x + 4 = 0$

b. $x^2 + x + 4 \le 0$

c. $x^2 + x + 4 \ge 0$

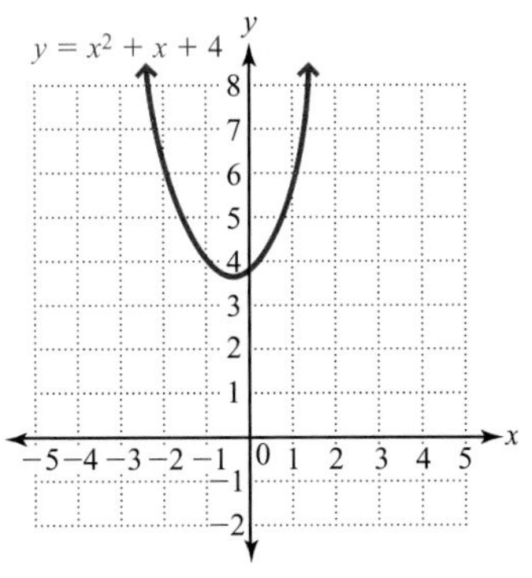

14. **a.** $-x^2 - x - 2 = 0$

b. $-x^2 - x - 2 \le 0$

c. $-x^2 - x - 2 \ge 0$

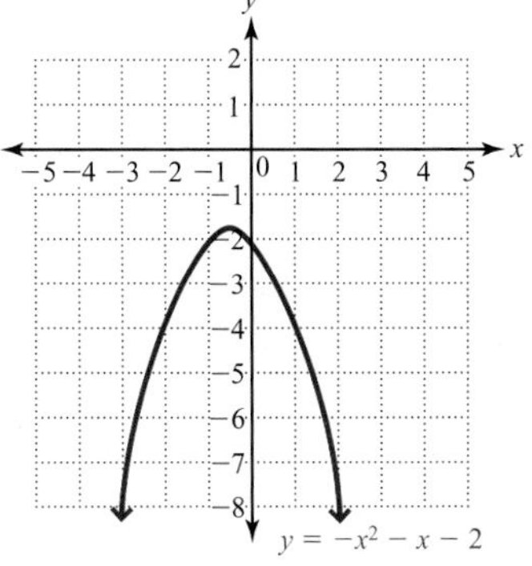

For Exercises 15–44, solve. Write the solution set using interval notation, then graph the solution set on a number line.

15. $(x + 4)(x + 2) < 0$

16. $(x - 3)(x + 1) < 0$

17. $(x - 2)(x - 5) > 0$

18. $(x + 4)(x - 3) > 0$

19. $x^2 + 5x + 4 < 0$

20. $x^2 + 6x + 5 < 0$

21. $x^2 - 4x + 3 > 0$

22. $x^2 - 8x + 7 > 0$

23. $b^2 - 6b + 8 \leq 0$

24. $c^2 - 9c + 14 \leq 0$

25. $y^2 - 5 \geq 4y$

26. $z^2 - 3 \leq 2z$

27. $a^2 - 3a < 10$

28. $b^2 + 4b \geq 21$

29. $y^2 + 6y + 9 \geq 0$

30. $x^2 + 10x + 25 \geq 0$

31. $c^2 - 4c + 5 \leq 0$

32. $2a^2 - 3a + 5 \leq 0$

33. $4r^2 + 21r + 5 > 0$

34. $4s^2 + 5s - 6 < 0$

35. $x^2 - 5x > 0$

36. $x^2 + 3x > 0$

37. $x^2 \leq 6x$

38. $x^2 \geq 3x$

39. $(x - 3)^2 \geq -1$

40. $(2x - 1)^2 > -4$

41. $(x - 4)(x + 2)(x + 4) \geq 0$

42. $(x + 5)(x - 3)(x + 1) \geq 0$

43. $(x + 2)(x + 6)(x - 1) < 0$

44. $(x - 3)(x + 3)(x + 1) < 0$

For Exercises 45–50, solve using the quadratic formula.

45. $3c^2 + 4c - 1 < 0$

46. $5a^2 - 10a + 2 < 0$

47. $4r^2 + 8r - 3 \geq 0$

48. $-2y^2 + 6y + 5 < 0$

49. $-0.2a^2 - 1.6a - 2 \leq 0$

50. $-0.1x^2 - 1.2x + 4 > 0$

For Exercises 51–68, solve the rational inequalities. Write the solution set using interval notation, then graph the solution set on a number line.

51. $\dfrac{a + 4}{a - 1} > 0$

52. $\dfrac{m - 6}{m + 2} \geq 0$

53. $\dfrac{n + 1}{n + 5} \leq 0$

54. $\dfrac{b - 2}{b + 3} < 0$

55. $\dfrac{6}{x + 4} > 0$

56. $\dfrac{3}{x - 3} < 0$

57. $\frac{c}{c+3} < 3$

58. $\frac{m}{m-2} < 2$

59. $\frac{a+5}{a-4} > 4$

60. $\frac{j+4}{j-4} > 5$

61. $\frac{p+2}{p-3} \geq 4$

62. $\frac{c+1}{c-4} \leq 6$

63. $\frac{(k+3)(k-2)}{k-5} \leq 0$

64. $\frac{(m+1)(m-3)}{m+4} \geq 0$

65. $\frac{(2x-1)^2}{x} \geq 0$

66. $\frac{(4x+3)^2}{x} < 0$

67. $\frac{x^2-7x+10}{x+1} < 0$

68. $\frac{x^2-2x-8}{x-1} \geq 0$

For Exercises 69–72, solve.

69. If a ball is thrown upward with an initial velocity of 80 feet per second from the top of a building 96 feet high, then the height, h, above the ground after t seconds is given by $h = -16t^2 + 80t + 96$, where h is in feet.

a. After how many seconds will the ball hit the ground? (*Hint:* Think about the value of h when the ball is on the ground.)

b. After how many seconds is the ball 192 feet above the ground?

c. Find the interval of time when the ball is more than 192 feet above the ground.

d. Use the answers to parts a, b, and c to find the intervals of time when the ball is less than 192 feet above the ground.

70. If an object is dropped from the top of a cliff that is 144 feet high, the equation giving the height, h, above the ground is $h = 144 - 16t^2$, where h is in feet and t is in seconds.

a. After how many seconds will the object hit the ground?

b. After how many seconds is the object 80 feet above the ground?

c. Find the interval of time when the object is more than 80 feet above the ground.

d. Find the interval of time when the ball is less than 80 feet above the ground.

71. In the parallelogram shown, the height is to be 2 inches less than x. The base is to be 6 inches more than x.

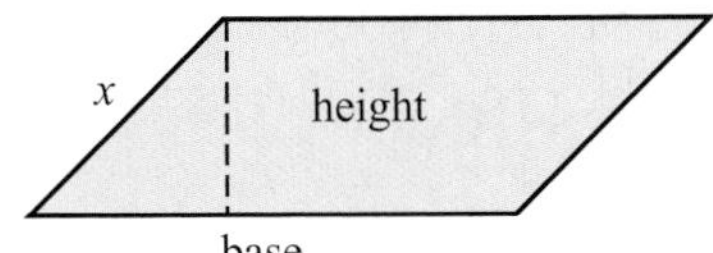

a. Find the range of values for x so that the area of the parallelogram is at least 20 square inches.

b. Find the range of values for the base and the height.

72. An auditorium is to be designed roughly in the shape of a box. The height is set to be 40 feet. It is desired that the length of the space be 20 feet more than the width.

a. Find the range of values for the width so that the volume of the space is at least 140,000 cubic feet.

b. Find the range of values for the length.

For Exercises 73 and 74, use the formula for the slope of a line: $m = \dfrac{y_2 - y_1}{x_2 - x_1}$.

73. Suppose a line is to be drawn in the coordinate plane so that it passes through the point at (2, 5). Find the range of values for the second point (x_2, y_2) so that $x_2 = y_2$ and the slope of the line is at most $\frac{1}{2}$.

74. Suppose a line is to be drawn in the coordinate plane so that it passes through the point at (−3, 2). Find the range of values for the second point (x_2, y_2) so that $x_2 = y_2$ and the slope of the line is at least $\frac{1}{4}$.

REVIEW EXERCISES

[8.6] **1.** Solve: $\sqrt{x-2} = 4$

[6.3] ***For Exercises 2 and 3, add a term to the expression to make it a perfect square then factor the perfect square.***

2. $x^2 - 6x$

3. $x^2 + 5x$

[5.2] **4.** Does the graph of $f(x) = -x^2 + 2x - 3$ open upwards or downwards? Why?

[6.4] **5.** Find the x-intercepts for $y = x^2 + 4x - 12$.

[9.4] **6.** What are the coordinates of the vertex of the graph of $f(x) = x^2 - 6x + 5$?

Chapter 9 Summary

Defined Terms

Section 9.2
Discriminant (p. 638)

Section 9.3
Equations quadratic in form (p. 647)

Section 9.4
Vertex (p. 658)
Axis of Symmetry (p. 658)

Section 9.5
Quadratic Inequality (p. 673)
Rational Inequality (p. 677)

Procedures, Rules, and Key Examples

Procedures/Rules	Key Examples
Section 9.1 Completing the Square The square root principle: If $x^2 = a$, where a is a real number, then $x = \sqrt{a}$ or $x = -\sqrt{a}$. It is common to indicate the positive and negative solutions by writing $\pm\sqrt{a}$.	**Example 1:** Solve. **a.** $x^2 = 36$ Solution: $x = \pm\sqrt{36}$ $x = \pm 6$ **b.** $(x - 3)^2 = 12$ Solution: $x - 3 = \pm\sqrt{12}$ $x = 3 \pm 2\sqrt{3}$
To solve a quadratic equation by completing the square, **1.** Write the equation in the form $x^2 + bx = c$. **2.** Complete the square by adding $\left(\frac{b}{2}\right)^2$ to both sides. **3.** Write the completed square in factored form. **4.** Use the square root principle to eliminate the square. **5.** Isolate the variable. **6.** Simplify as needed.	**Example 2:** Solve by completing the square. **a.** $x^2 + 6x - 7 = 8$ $x^2 + 6x = 15$ **Add 7 to both sides.** $x^2 + 6x + 9 = 15 + 9$ **Complete the square.** $(x + 3)^2 = 24$ **Factor.** $x + 3 = \pm\sqrt{24}$ **Use the square root principle.** $x = -3 \pm 2\sqrt{6}$ **Subtract 3 from both sides and simplify the square root.** **b.** $3x^2 - 9x - 2 = 10$ $3x^2 - 9x = 12$ **Add 2 to both sides.** $x^2 - 3x = 4$ **Divide both sides by 3.** $x^2 - 3x + \frac{9}{4} = 4 + \frac{9}{4}$ **Complete the square.** $\left(x - \frac{3}{2}\right)^2 = \frac{25}{4}$ **Factor.** $x - \frac{3}{2} = \pm\sqrt{\frac{25}{4}}$ **Use the square root principle.** $x = \frac{3}{2} \pm \frac{5}{2}$ **Add $\frac{3}{2}$ to both sides and simplify the square root.** $x = \frac{3}{2} + \frac{5}{2} = 4$ or $x = \frac{3}{2} - \frac{5}{2} = -1$ *continued*

Procedures/Rules	Key Examples
Section 9.2 Solving Quadratic Equations Using the Quadratic Formula To solve a quadratic equation in the form $ax^2 + bx + c = 0$, where $a \neq 0$, use the quadratic formula: $$x = \frac{-b \pm \sqrt{b^2 - 4ac}}{2a}$$	**Example 1:** Solve: $3x^2 - 4x + 2 = 0$ Solution: In the quadratic formula, replace a with 3, b with -4, and c with 2. $$x = \frac{-(-4) \pm \sqrt{(-4)^2 - 4(3)(2)}}{2(3)}$$ $$= \frac{4 \pm \sqrt{16 - 24}}{6} = \frac{4 \pm \sqrt{-8}}{6}$$ $$= \frac{4 \pm 2\sqrt{-2}}{6} = \frac{2}{3} \pm \frac{\sqrt{2}}{3}i$$
Given a quadratic equation in the form $ax^2 + bx + c = 0$, where $a \neq 0$, to determine the number and type of solutions it has, evaluate the discriminant $b^2 - 4ac$. If the **discriminant is positive**, then the equation has two real-number solutions. They will be rational if the discriminant is a perfect square and irrational otherwise. If the **discriminant is 0**, then the equation has one real solution. The solution is rational if $-\frac{b}{2a}$ is a rational number. If the **discriminant is negative**, then the equation has two nonreal complex solutions.	**Example 2:** Use the discriminant to determine the number and type of solutions for $5x^2 - 7x + 8 = 0$. Solution: In the discriminant, replace a with 5, b with -7, and c with 8. $$(-7)^2 - 4(5)(8) = 49 - 160 = -111$$ Since the discriminant is negative, the equation has two nonreal complex solutions.
Section 9.3 Solving Equations That Are Quadratic in Form If the equation involves rational expressions, multiply both sides by the LCD. Be sure to check for extraneous solutions.	**Example 1:** Solve $\frac{6}{x-2} + \frac{6}{x-1} = 5$. Solution: First, multiply both sides by the LCD, $(x-2)(x-1)$. $6(x-1) + 6(x-2) = 5(x-2)(x-1)$ $12x - 18 = 5x^2 - 15x + 10$ **Simplify.** $0 = 5x^2 - 27x + 28$ **Write in $ax^2 + bx + c = 0$ form.** $0 = (5x - 7)(x - 4)$ **Factor.** $5x - 7 = 0$ or $x - 4 = 0$ **Set each factor equal to 0, then solve each equation.** $x = \frac{7}{5}$ $\quad x = 4$ Check: Verify that $\frac{7}{5}$ and 4 solve the original equation. ***continued***

Procedures/Rules

Key Examples

Section 9.3 Solving Equations That Are Quadratic in Form (continued)

If an equation has a radical, isolate the radical and then square both sides. Continue the process as needed until all radicals have been eliminated. Solve the resulting equation and check for extraneous solutions.

Example 2: Solve $\sqrt{4x + 4} = 2x - 2$.

$4x + 4 = 4x^2 - 8x + 4$ Square both sides.

$0 = 4x^2 - 12x$ Write in $ax^2 + bx + c = 0$ form.

$0 = 4x(x - 3)$ Factor.

$4x = 0$ or $x - 3 = 0$ Set each factor equal to 0, then solve each equation.

$x = 0$ $x = 3$

Check:

$x = 0$: $\sqrt{4(0) + 4} = 2(0) - 2$; $\sqrt{0 + 4} = 0 - 2$; $2 = -2$ False

$x = 3$: $\sqrt{4(3) + 4} = 2(3) - 2$; $\sqrt{12 + 4} = 6 - 2$; $4 = 4$ True

$x = 3$ is the only solution.

To solve equations that are quadratic in form using substitution,

1. Rewrite the equation so that it is in the form $au^2 + bu + c = 0$.
2. Solve the quadratic equation for u.
3. Substitute for u and solve.
4. Check the solutions.

Example 3: Solve $(a + 2)^2 + 7(a + 2) + 12 = 0$.

$u^2 + 7u + 12 = 0$ Substitute u for $a + 2$.

$(u + 3)(u + 4) = 0$ Factor.

$u + 3 = 0$ or $u + 4 = 0$ Set each factor equal to 0, then solve each equation.

$u = -3$ $u = -4$

$a + 2 = -3$ or $a + 2 = -4$ Substitute $a + 2$ for u, solve for a.

$a = -5$ $a = -6$

Check: Verify that -5 and -6 satisfy the original equation.

Section 9.4 Graphing Quadratic Functions

The graph of a function in the form $f(x) = a(x - h)^2 + k$ is a parabola with vertex at (h, k). The equation of the axis of symmetry is $x = h$. The parabola opens up if $a > 0$ and down if $a < 0$.

Example 1: For $f(x) = -3(x + 1)^2 - 2$,

a. Determine whether the graph opens upwards or downwards.

b. Find the vertex.

c. Write the equation of the axis of symmetry.

d. Graph.

Answers:

a. downwards **b.** $(-1, -2)$ **c.** $x = -1$

d.

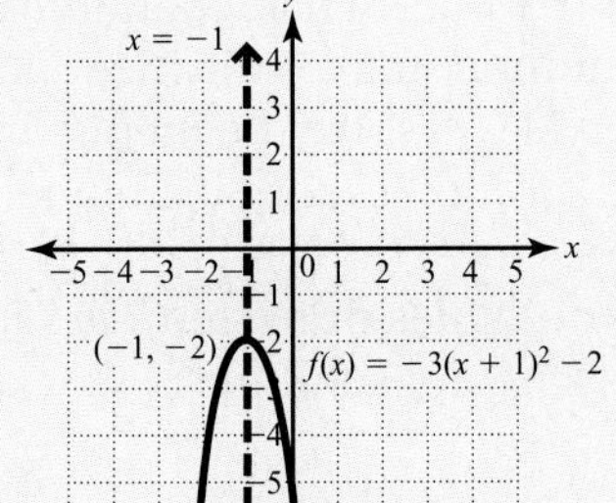

continued

Procedures/Rules

Section 9.4 Graphing Quadratic Functions (continued)

Given an equation in the form $f(x) = ax^2 + bx + c$, to determine the vertex of the corresponding parabola,

1. Find the x-coordinate using the formula $x = -\dfrac{b}{2a}$.
2. Find the y-coordinate by evaluating $f\left(-\dfrac{b}{2a}\right)$.

Note: *As an alternative approach to Example 2, we could have transformed* $f(x) = 2x^2 - 12x + 19$ *to* $f(x) = 2(x - 3)^2 + 1$ *by completing the square.*

Key Examples

Example 2: For $f(x) = 2x^2 - 12x + 19$,

a. Determine whether the graph opens upwards or downwards.
b. Find the vertex.
c. Write the equation of the axis of symmetry.
d. Graph.

Answers:

a. upwards
b. For the x-coordinate, use $-\dfrac{b}{2a}$, replacing a with 2 and b with -12.

$$x = -\frac{(-12)}{2(2)} = 3$$

For the y-coordinate, evaluate $f(3)$.

$$f(3) = 2(3)^2 - 12(3) + 19$$
$$= 18 - 36 + 19 = 1$$

Vertex: $(3, 1)$

c. $x = 3$

d.

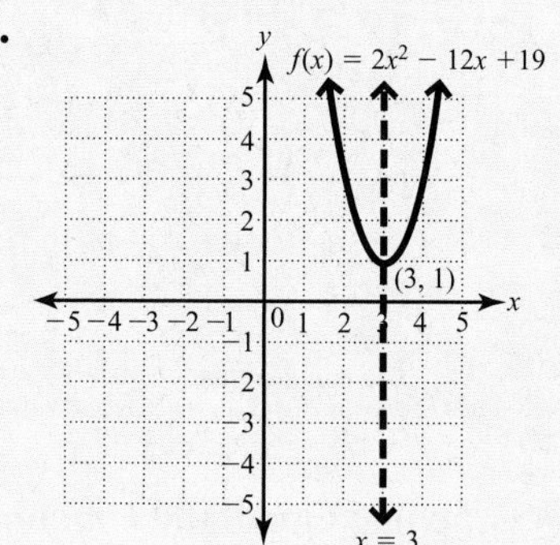

Section 9.5 Solving Nonlinear Inequalities

Solving quadratic inequalities.

1. Solve the related equation $ax^2 + bx + c = 0$.
2. Plot the solutions of $ax^2 + bx + c = 0$ on a number line. These solutions will divide the number line into intervals.
3. Choose a test number from each interval and substitute the number into the inequality. If the test number makes the inequality true, then all numbers in that interval will solve the inequality. If the test number makes the inequality false, then no numbers in that interval will solve the inequality.
4. State the solution set of the inequality: It is the union of all the intervals that solve the inequality. If the inequality symbols are $\le$ or $\ge$, then the values from step 2 are included. If the symbols are $<$ or $>$, they are not solutions.

Example 1: Solve $x^2 + 2x - 15 \ge 0$.

$x^2 + 2x - 15 = 0$ **Write the related equation.**

$(x + 5)(x - 3) = 0$ **Factor.**

$x + 5 = 0$ or $x - 3 = 0$ **Set each factor equal to 0, then solve each equation.**

$x = -5$ $\quad x = 3$

Plot -5 and 3 on a number line and label the intervals.

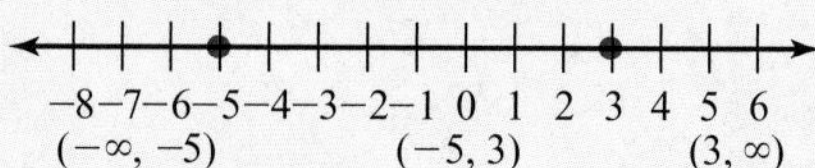

continued

Procedures/Rules

Key Examples

Section 9.5 Solving Nonlinear Inequalities (continued)

Choose a number from each interval and test it in the original inequality.

Interval	$(-\infty, -5)$	$(-5, 3)$	$(3, \infty)$
Test value	-6	0	4
Result	$9 \geq 0$	$-15 \geq 0$	$9 \geq 0$
True/False	True	False	True

The solution set is $(-\infty, -5] \cup [3, \infty)$.

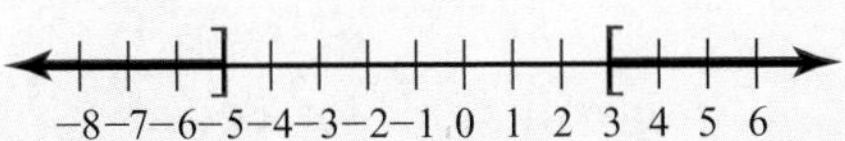

Solving rational inequalities.

1. Find all values that make any denominator equal to 0. These values must be excluded from the solution set.
2. Solve the related equation.
3. Plot the numbers found in steps 1 and 2 on a number line and label the regions.
4. Choose a test number from each interval and determine whether it solves the inequality.
5. The solution set is the union of all the regions whose test number solves the inequality. If the inequality symbol is $\leq$ or $\geq$, include the values found in step 1. The solution set never includes the values found in step 2 because they make a denominator equal to 0.

Example 2: Solve $\frac{x+5}{x-1} > 4$.

$\frac{x+5}{x-1} = 4$ Write the related equation.

$(x-1)\frac{x+5}{x-1} = (x-1)(4)$ Multiply both sides by the LCD.

$x + 5 = 4x - 4$ Simplify.

$9 = 3x$ Subtract x and add 5 on both sides.

$3 = x$ Divide by 3.

Find the value(s) that make any denominator equal to 0.

$$x - 1 = 0$$
$$x = 1$$

Plot 1 and 3 on the number line and label the regions.

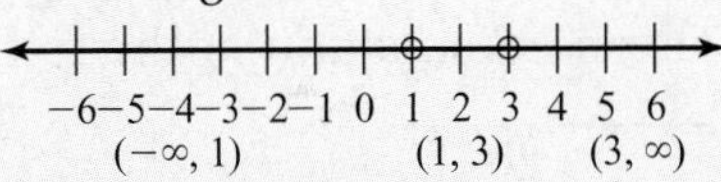

Choose a number from each interval and test it in the original inequality.

Interval	$(\infty, 1)$	$(1, 3)$	$(3, \infty)$
Test value	0	2	4
Result	$-5 > 4$	$7 > 4$	$3 > 4$
True/False	False	True	False

The solution set is $(1, 3)$.

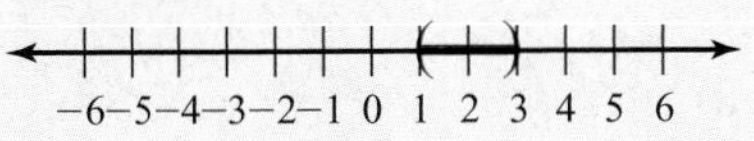

Formulas

The quadratic formula: Given an equation in the form $ax^2 + bx + c = 0$, $x = \frac{-b \pm \sqrt{b^2 - 4ac}}{2a}$

Given an equation in the form $f(x) = ax^2 + bx + c$, the x-coordinate of the vertex of a parabola is $-\frac{b}{2a}$ and the equation of the axis of symmetry is $x = -\frac{b}{2a}$.

Chapter 9 Review Exercises

For Exercises 1–5, answer true or false.

[9.1] **1.** The notation $\sqrt{a}$ has the same meaning as the notation $\pm\sqrt{a}$.

[9.3] **2.** Equations that are quadratic in form never have extraneous solutions.

[9.1, 9.2] **3.** Every quadratic equation has a real-number solution.

[9.2] **4.** If the solutions for a quadratic equation in the form $ax^2 + bx + c = 0$ are complex, the discriminant can be used to find those solutions.

[9.4] **5.** Given an equation in the form $y = ax^2 + bx + c$, if $a > 0$, then the parabola opens up.

For Exercises 6–10, complete the rule.

[9.1] **6.** If $x^2 = a$ and $a \geq 0$, then $x =$ ____________ or ____________.

[9.1] **7.** Given an equation in the form $x^2 + bx = c$, to complete the square we add the constant term ____________ to both sides of the equation.

[9.2] **8.** Given an equation in the form $ax^2 + bx + c = 0$, the quadratic formula is $x =$ ____________.

[9.2] **9.** Given an equation in the form $ax^2 + bx + c = 0$, the discriminant is ____________.

[9.4] **10.** Given a function in the form $f(x) = a(x - h)^2 + k$, the coordinates of the vertex are ____________ and the axis of symmetry is ____________.

Equations and Inequalities

Exercises 11–66

[9.1] ***For Exercises 11–18, solve and check.***

11. $x^2 = 16$

12. $y^2 = \frac{1}{36}$

13. $k^2 + 2 = 30$

14. $3x^2 = 42$

15. $5h^2 + 24 = 9$

16. $(x + 7)^2 = 25$

17. $(x - 9)^2 = -16$

18. $\left(m + \frac{3}{5}\right)^2 = \frac{16}{25}$

[9.1] *For Exercises 19–22, solve by completing the square.*

19. $m^2 + 8m = -7$

20. $u^2 - 6u - 12 = 100$

21. $2b^2 - 6b + 7 = 0$

22. $u^2 + \frac{1}{4}u = \frac{3}{4}$

[9.2] *For Exercises 23–26, solve using the quadratic formula.*

23. $p^2 - 5 = -2p$

24. $3x^2 - 2x + 1 = 0$

25. $2t^2 + t - 5 = 0$

26. $2x^2 + 0.1x - 0.03 = 0$

[9.2] *For Exercises 27–30, use the discriminant to determine the number and type of solutions for the equation.*

27. $b^2 - 4b - 12 = 0$

28. $6z^2 - 7z + 5 = 0$

29. $k^2 + 6k + 9 = 0$

30. $0.8x^2 + 1.2x + 0.3 = 0$

[9.3] *For Exercises 31–34, solve.*

31. $\frac{1}{y} + \frac{1}{y + 3} = \frac{2}{3}$

32. $\frac{1}{u} + \frac{1}{u - 5} = \frac{10}{u^2 - 25}$

33. $6 - 5x^{-1} + x^{-2} = 0$

34. $2 - 3x^{-1} - x^{-2} = 0$

[9.3] *For Exercises 35–38, solve.*

35. $14\sqrt{x} + 45 = 0$

36. $\sqrt{4m} = 3m - 1$

37. $\sqrt{6r + 13} = 2r + 1$

38. $\sqrt{21t + 2} + t = 2 + 4t$

[9.3] *For Exercises 39–44, solve the equations using substitution.*

39. $x^4 - 5x^2 + 6 = 0$

40. $2m^4 - 3m^2 + 1 = 0$

41. $6(x + 5)^2 - 5(x + 5) + 1 = 0$

42. $\left(\frac{x - 1}{3}\right)^2 + 10\left(\frac{x - 1}{3}\right) + 9 = 0$

43. $p^{2/3} - 11p^{1/3} + 24 = 0$

44. $5a^{1/2} + 13a^{1/4} - 6 = 0$

[9.4] ***For Exercises 45–50:*** ***a. Find the x- and y-intercepts.***
b. State whether the parabola opens upwards or downwards.
c. Find the coordinates of the vertex.
d. Write the equation of the axis of symmetry.
e. Graph.

45. $f(x) = -2x^2$

46. $g(x) = \frac{1}{2}x^2 + 1$

47. $h(x) = -\frac{1}{3}(x - 2)^2$

48. $k(x) = 4(x + 3)^2 - 2$

For Exercises, 49 and 50: ***a. Write the equation in the form f(x) = a(x − h)² + k.***
b. Find the x- and y-intercepts.
c. State whether the parabola opens upwards or downwards.
d. Find the coordinates of the vertex.
e. Write the equation of the axis of symmetry.
f. Graph.

49. $m(x) = x^2 + 2x - 1$

50. $p(x) = -0.5x^2 + 4x - 6$

[9.5] ***For Exercises 51–54, solve the following inequalities.***

51. $(x + 5)(x - 3) > 0$

52. $n^2 - 6n \leq -8$

53. $x^2 + 9x + 14 < 0$

54. $(x + 3)(x - 1)(x - 2) \geq 0$

[9.5] ***For Exercises 55–58, solve the rational inequalities.***

55. $\frac{a + 3}{a - 1} \geq 0$

56. $\frac{r}{r + 2} < 2$

57. $\frac{n - 3}{n - 4} \leq 5$

58. $\frac{(k + 2)(k - 3)}{k - 5} < 0$

[9.1–9.2] ***For Exercises 59–66, solve.***

59. A crop circle appeared July 7, 2003 in Windham Hill, England, and covered $22{,}500\pi$ square feet. Find the radius of the circle.

60. Using the formula $E = \frac{1}{2}mv^2$, where E represents the kinetic energy in joules of an object with a mass of m kilograms and a velocity of v meters per second, find the velocity of an object with a mass of 50 kilograms and 400 joules of kinetic energy.

61. The length of a small rectangular shed is 4 feet more than its width. If the area is 285 square feet, what are the dimensions of the shed?

62. A right circular cylinder is to be constructed so that its volume is equal to that of a sphere with a radius of 9 inches. If the cylinder is to have a height of 4 inches, find the radius of the cylinder.

63. A ramp is constructed so that it is a right triangle with a base that is 9 feet longer than its height. If the hypotenuse is 17 feet, find the dimensions of the base and height.

64. An acrobat is launched upwards from one end of a lever with an initial velocity of 24 feet per second. The function $h = -16t^2 + 24t$ describes the height, h, of the acrobat t seconds after being launched.

a. After how many seconds does the acrobat reach maximum height?

b. What is the maximum height that the acrobat reaches?

c. How long is the acrobat in the air?

d. Graph the function.

65. The longest punt on record in the NFL was by Steve O'Neal in a game between the New York Jets and Denver Broncos on September 21, 1969. The function $y = -0.03x^2 + 2.16x$ models the trajectory of the punt. (Note that x and y are distances in yards.)

a. Find the maximum height the punt reached.

b. Find the distance of the punt.

Of Interest

The yardage you found is how far the punt carried in the air. It was actually recorded as a 98-yard punt, which includes the amount of roll after the punt landed and was downed.

66. In a triangle, the height is to be 2 inches less than x. The base is to be 4 inches more than x.

a. Find the range of values for x so that the area of the triangle is at least 56 square inches.

b. Find the range of values for the base and the height.

Chapter 9 Practice Test

For 1 and 2, use the square root principle to solve and check.

1. $x^2 = 81$

2. $(x - 3)^2 = 20$

For 3 and 4, solve by completing the square.

3. $x^2 - 8x = -4$

4. $3m^2 - 6m = 5$

For 5 and 6, solve using the quadratic formula.

5. $2x^2 + x - 6 = 0$

6. $x^2 - 8x + 15 = 0$

For 7–12, solve using any method.

7. $u^2 - 16 = -6u$

8. $x^2 = 81$

9. $4w^2 + 6w + 3 = 0$

10. $2x^2 + 4x = 0$

11. $x^2 + 16 = 0$

12. $3k^2 = -5k$

For 13–16, solve.

13. $\frac{1}{x + 2} + \frac{1}{x} = \frac{5}{12}$

14. $3 - x^{-1} - 2x^{-2} = 0$

15. $9\sqrt{x} + 8 = 0$

16. $\sqrt{x + 8} - x = 2$

For 17 and 18, solve the equations using substitution.

17. $9a^4 + 26a^2 - 3 = 0$

18. $(x + 1)^2 + 3(x + 1) - 4 = 0$

19. For $f(x) = -x^2 + 6x - 4$,

   a. Find the x- and y-intercepts.

   b. Write the equation in the form $f(x) = a(x - h)^2 + k$.

   c. State whether the parabola opens upwards or downwards.

   d. Find the coordinates of the vertex.

   e. Write the equation of the axis of symmetry.

   f. Graph.

For 20 and 21: a. Solve the inequality.
b. Graph the solution set on a number line.

20. $(x + 1)(x - 4) \leq 0$

21. $\dfrac{x + 2}{x - 1} > 0$

For 22–25, solve.

22. A ball is thrown downward from a window in a tall building. The distance, d, traveled by the ball is given by the equation $d = 16t^2 + 32t$, where t is the time traveled in seconds. How long will it take the ball to fall 180 feet?

23. A rectangular parking lot needs to have an area of 400 square feet. The length is to be 20 feet more than the width. Find the dimensions of the parking lot.

24. An archer shoots an arrow in a field. Suppose the equation $y = -0.02x^2 + 1.3x + 8$ models the trajectory of the arrow (assume that x and y represent distances in meters and that $x \geq 0$ and $y \geq 0$).

   a. What is the maximum height that the arrow reaches?

   b. How far does the arrow travel horizontally?

   c. Graph the trajectory of the arrow.

25. A zoo is planning to install a new aquarium tank. The tank is to be in the shape of a box with a height of 12 feet. It is desired that the length be 15 feet more than the width.

   a. Find the range of values for the width so that the volume of the space is at most 12,000 cubic feet.

   b. Find the range of values for the length.

Chapters 1–9 Cumulative Review Exercises

For Exercises 1–4, answer true or false.

[2.1] **1.** An identity is an equation in which all real numbers (for which the expressions in the equation are defined) is a solution.

[3.1] **2.** The point with coordinates $(45, -12)$ is in quadrant II.

[4.1] **3.** A system of equations that has no solution is said to be consistent.

[5.2] **4.** The degree of the monomial $5x^3y$ is 4.

For Exercises 5–18, fill in the blank.

[5.2] **5.** The graph of a polynomial function of the form $f(x) = ax^2 + bx + c$ is a ____________ that opens up if ____________.

[8.3] **6.** If $\sqrt[n]{a}$ and $\sqrt[n]{b}$ are both real numbers, then $\sqrt[n]{a} \cdot \sqrt[n]{b} =$ ____________.

[8.6] **7.** If $a = b$, then $a^n =$ ____________.

[9.1] **8.** If $x^2 = a$ and $a \geq 0$, then $x =$ ____________ or $x =$ ____________.

For Exercises 9–12, simplify.

[5.2] **9.** $(-3x - 4) - (3x + 2)$

[5.1] **10.** $(8n^2)(-7mn^3)$

[5.3] **11.** $(x - 3)(4x^2 - 2x + 1)$

[5.4] **12.** $(15m^2 - 22m + 14) \div (3m - 2)$

[6.3] *For Exercises 13 and 14, factor completely.*

13. $m^3 + 8$

14. $x^2 + 10x + 25$

For Exercises 15–20, simplify.

[5.1] **15.** $-\dfrac{12x^3y}{30x^2z}$

[7.1] **16.** $\dfrac{4u^2 + 4u + 1}{u + 2u^2} \cdot \dfrac{u}{2u^2 - u - 1}$

[7.1] **17.** $\dfrac{a^2 - b^2}{x^2 - y^2} \div \dfrac{a + b}{x - y}$

[7.2] **18.** $\dfrac{3}{x - 2} - \dfrac{2}{x + 2}$

[8.2] **19.** $x^{3/4} \cdot x^{-1/4}$

[8.7] **20.** $(4 + 7i)(5 - 3i)$

[8.5] ***For Exercises 21 and 22, rationalize the denominator.***

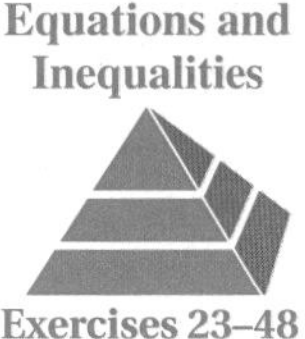

21. $-\frac{2}{\sqrt{5}}$

22. $\frac{\sqrt{n}}{\sqrt{m}-\sqrt{n}}$

For Exercises 23–30, solve.

[2.1] **23.** $3n - 5 = 7n + 9$

[2.1] **24.** $6x - 8 = 2(3x - 4)$

[2.6] **25.** $|3x + 4| + 3 = 7$

[6.4] **26.** $2x^2 + 7x = 15$

[9.1] **27.** $x^2 - 36 = 0$

[7.4] **28.** $\frac{5}{x-2} - \frac{3}{x} = \frac{11}{3x}$

[8.6] **29.** $\sqrt{5x - 4} = 9$

[9.3] **30.** $10x^4 - x^2 - 3 = 0$

[9.2] **31.** Solve $x^2 - 6x + 11 = 0$ using the quadratic formula.

[2.1] **32.** Solve for m in the formula $E = \frac{1}{2}mv^2$.

For Exercises 33 and 34, a. Graph the solution set on a number line.
b. Write the solution set in set-builder notation.
c. Write the solution set in interval notation.

[2.4] **33.** $-2 < x + 4 < 5$

[9.5] **34.** $x^2 + x \leq 12$

For Exercises 35 and 36, graph.

[3.1] **35.** $f(x) = -2x$

[5.2] **36.** $f(x) = x^3 + 2$

[3.3] **37.** Write the equation of a line in standard form that passes through the point $(-2, 4)$ and is perpendicular to the line $3x - 2y = 6$.

[9.4] **38.** For $f(x) = -2(x - 1)^2 + 3$,

a. State whether the parabola opens up or down.

b. Find the x- and y-intercepts.

c. Find the coordinates of the vertex.

d. Write the equation of the axis of symmetry.

e. Graph.

For Exercises 39–42, solve the systems.

[4.1] **39.** $\begin{cases} 4x - 3y = -2 \\ 6x - 7y = 7 \end{cases}$

[4.2] **40.** $\begin{cases} x + y + z = 5 \\ 2x + y - 2z = -5 \\ x - 2y + z = 8 \end{cases}$

[4.5] **41.** Use Cramer's Rule to solve the system: $\begin{cases} 2x + 3y = -5 \\ 3x - y = 9 \end{cases}$

[4.6] **42.** Solve and graph the solution set: $\begin{cases} x + y < 3 \\ x - y \ge 4 \end{cases}$

For Exercises 43–50, solve.

[2.2] **43.** A trucking company finds that the average speed of experienced drivers is 10 miles per hour more than the average speed of inexperienced drivers. If inexperienced drivers take 1 hour longer than experienced drivers to drive 300 miles, find how long it takes the experienced driver to drive 300 miles.

[2.2, 4.3] **44.** A chemist has a bottle containing 50 ml of 15% saline solution and a bottle of 40% saline solution. She wants a 30% solution. How much of the 40% solution must be added to the 15% solution so that a 30% concentration is created?

[4.3] **45.** Joanna purchased two pens, two erasers, and one ream of paper for \$6. Willa purchased four pens, three erasers, and two reams of paper for \$11.50. On the same day, Scott purchased three pens, one eraser, and three reams of paper for \$9.50. Find the cost of each.

[5.1] **46.** If light travels about 3×10^8 meters per second, how long does it take the light from the Sun to travel to Pluto, which is a distance of about 5.9×10^{12} meters. Convert your answer to hours.

[7.5] **47.** If the wavelength of a wave remains constant, then the velocity, v, of a wave is inversely proportional to its period, T. In an experiment, waves are created in a pool of water so that the period is 8 seconds and the velocity is 6 centimeters per second. If the period is increased to 15 seconds, what is the velocity?

48. The formula $t = \sqrt{\frac{h}{16}}$ describes the amount of time t, in seconds, that an object falls a distance of h feet.

[8.1] **a.** Write an expression in simplest form for the exact amount of time an object falls a distance of 24 feet.

[8.6] **b.** Find the distance an object falls in 3 seconds.

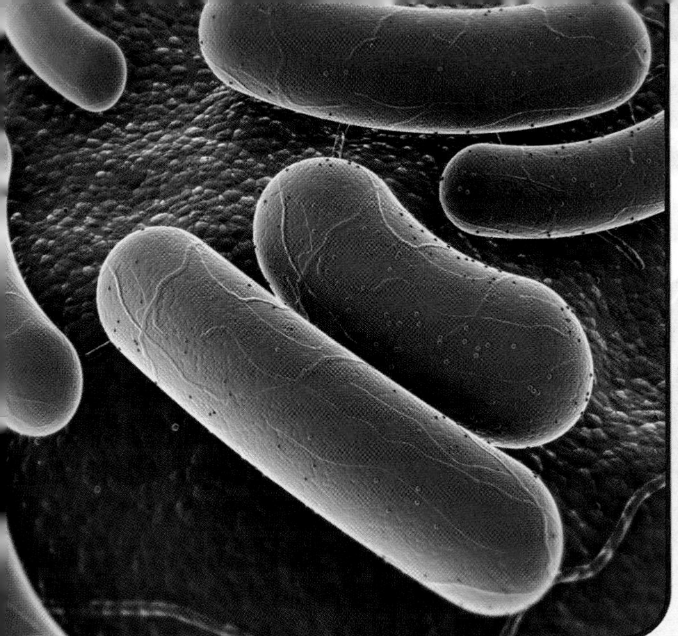

CHAPTER 10

Exponential and Logarithmic Functions

"The mathematics of uncontrolled growth are frightening. A single cell of the bacterium *E. coli* would, under ideal circumstances, divide every twenty minutes. . . . one cell becomes two, two become four, four become eight and so on. In this way it can be shown that in a single day, one cell of *E. coli* could produce a super-colony equal in size and weight to the entire planet Earth."

—Michael Crichton, *The Andromeda Strain* (1971)

We begin this chapter with composite and inverse functions. Exponential and logarithmic functions, the subject of the rest of the chapter, are inverses of each other. These functions are used in modeling population growth, electrical circuits, sound decibels, pH, and the Richter scale.

10.1 Composite and Inverse Functions

OBJECTIVES

1. Find the composition of two functions.
2. Show that two functions are inverses.
3. Show that a function is one-to-one.
4. Find the inverse of a function.
5. Graph a given function's inverse function.

OBJECTIVE 1. Find the composition of two functions. In Section 5.2, we learned to perform the basic operations with functions. We now explore a new operation. Recall that a function pairs elements from an "input" set called the *domain* with elements in an "output" set called the *range*. If the output elements of one function are used as input elements in a second function, the resulting function is the **composition** of the two functions. Composition occurs when one quantity depends on a second quantity, which, in turn, depends on a third quantity.

We can visualize the composition of two functions as a machine with two stages, which are the two functions. For example, some coffee machines have two stages. We put in roasted coffee beans, which are ground in the first stage (first function). The grounds are then fed into the second stage (second function), which runs hot water over the grounds to produce coffee.

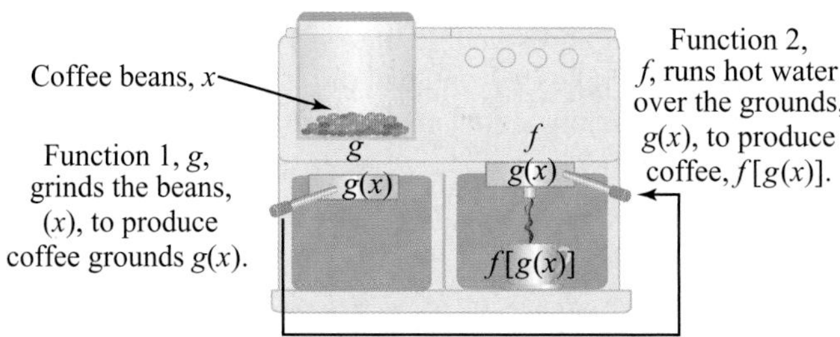

DEFINITION ***Composition of functions:*** If f and g are functions, then the composition of f and g is defined as $(f \circ g)(x) = f[g(x)]$ for all x in the domain g for which $g(x)$ is in the domain of f. The composition of g and f is defined as $(g \circ f)(x) = g[f(x)]$ for all x in the domain of f for which $f(x)$ is in the domain of g.

Note: *The notation $(g \circ f)(x)$ is read "g composed with f of x," or simply "g of f of x."*

Warning: Do not confuse $(f \circ g)(x)$ with $(f \cdot g)(x)$.

For example, suppose a pebble is dropped into a calm lake causing ripples to form in concentric circles. If the radius of the outer ripple increases at a rate of 2 feet per second, then we can describe the length of the radius as a function of t by using $r(t) = 2t$, where t is in seconds. The area of the circle formed by the outer ripple is also a function described by $A(r) = \pi r^2$. If we substitute the output of $r(t)$, which is $2t$ for r, into $A(r)$, then we form the composite function $(A \circ r)(t)$.

$$\begin{aligned}(A \circ r)(t) &= A[r(t)] \\ &= \pi (2t)^2 && \text{Substitute } 2t \text{ for } r \text{ in } A(r) = \pi r^2. \\ &= 4\pi t^2 && \text{Simplify.}\end{aligned}$$

The composite function $(A \circ r)(t)$ gives the area of the circle as a function of time.

EXAMPLE 1 If $f(x) = 2x - 3$ and $g(x) = x^2 + 1$, find the following.

a. $(f \circ g)(2)$

Solution $(f \circ g)(2) = f[g(2)]$ Definition of composition.

$= f(5)$ $g(2) = 2^2 + 1 = 4 + 1 = 5.$

$= 2(5) - 3$ In $f(x)$, replace x with 5.

$= 7$ Simplify.

Note: *We first find $g(2)$, which is 5, then substitute that result into f.*

b. $(g \circ f)(2)$

Solution $(g \circ f)(2) = g[f(2)]$ Definition of composition.

$= g(1)$ $f(2) = 2(2) - 3 = 4 - 3 = 1.$

$= 1^2 + 1$ In $g(x)$, replace x with 1.

$= 2$ Simplify.

c. $(f \circ g)(x)$

Solution $(f \circ g)(x) = f[g(x)]$ Definition of composition.

$= f(x^2 + 1)$ Replace $g(x)$ with $x^2 + 1$.

$= 2(x^2 + 1) - 3$ In $f(x)$, replace x with $x^2 + 1$.

$= 2x^2 + 2 - 3$ Simplify.

$= 2x^2 - 1$

Note: *We could have found $(f \circ g)(2)$ from part a by first finding $(f \circ g)(x) = 2x^2 - 1$, then substituting 2 for x:*

$(f \circ g)(2) = 2(2)^2 - 1$
$= 8 - 1$
$= 7$

d. $(g \circ f)(x)$

Solution $(g \circ f)(x) = g[f(x)]$ Definition of composition.

$= g(2x - 3)$ Replace $f(x)$ with $2x - 3$.

$= (2x - 3)^2 + 1$ In $g(x)$, replace x with $2x - 3$.

$= 4x^2 - 12x + 9 + 1$ Simplify.

$= 4x^2 - 12x + 10$

Note: *We could have found $(g \circ f)(2)$ from part b by first finding $(g \circ f)(x) = 4x^2 - 12x + 10$ then substituting 2 for x:*

$(g \circ f)(2) = 4(2)^2 - 12(2) + 10$
$= 16 - 24 + 10$
$= 2$

Note: *Generally, $(f \circ g)(x) \neq (g \circ f)(x)$ as in Examples 1(c) and (d), but there are special functions for which $(f \circ g)(x) = (g \circ f)(x)$.*

YOUR TURN If $f(x) = x^2 + 2$ and $g(x) = 3x + 5$, find the following:

a. $f[g(-3)]$ **b.** $g[f(2)]$ **c.** $f[g(x)]$

OBJECTIVE 2. Show that two functions are inverses. Operations, such as addition and subtraction, that undo each other are called *inverse operations*. For example, if we begin with a number x and add a second number y, we have $x + y$. Now subtract y (the number we just added) from that result and we have $x + y - y = x$, which is the beginning number.

Likewise, two functions that undo each other under composition are called **inverse functions**.

ANSWERS

a. $f[g(-3)] = 18$
b. $g[f(2)] = 23$
c. $f[g(x)] = 9x^2 + 30x + 27$

DEFINITION ***Inverse functions:*** Two functions f and g are inverses if and only if $(f \circ g)(x) = x$ for all x in the domain of g and $(g \circ f)(x) = x$ for all x in the domain of f.

Loosely speaking, f and g are inverse functions if you evaluate f for a value x in its domain, substitute that result into g and evaluate, and you get x again and vice versa.

PROCEDURE Inverse Functions

To determine whether two functions f and g are inverses of each other,

1. Show that $f[g(x)] = x$ for all x in the domain of g.
2. Show that $g[f(x)] = x$ for all x in the domain of f.

EXAMPLE 2 Verify that f and g are inverses.

a. $f(x) = 3x + 2, g(x) = \dfrac{x - 2}{3}$

Solution We need to show that $f[g(x)] = x$ and $g[f(x)] = x$.

$$f[g(x)] = f\left(\frac{x-2}{3}\right) \quad \text{Substitute } \frac{x-2}{3} \text{ for } g(x).$$

$$= 3\left(\frac{x-2}{3}\right) + 2 \quad \text{Substitute } \frac{x-2}{3} \text{ for } x \text{ in } f(x).$$

$$= x - 2 + 2 \quad \text{Simplify.}$$

$$= x \quad \text{So } f[g(x)] = x.$$

$$g[f(x)] = g(3x + 2) \quad \text{Substitute } 3x + 2 \text{ for } f(x).$$

$$= \frac{3x + 2 - 2}{3} \quad \text{Replace } x \text{ with } 3x + 2 \text{ in } g(x).$$

$$= \frac{3x}{3} \quad \text{Simplify.}$$

$$= x \quad \text{So } g[f(x)] = x.$$

Since $f[g(x)] = x$ and $g[f(x)] = x$, f and g are inverses.

b. $f(x) = x^3 + 5$ and $g(x) = \sqrt[3]{x - 5}$

Solution We need to show that $f[g(x)] = x$ and $g[f(x)] = x$.

$$f[g(x)] = f(\sqrt[3]{x-5}) \quad \text{Replace } g(x) \text{ with } \sqrt[3]{x-5}.$$

$$= (\sqrt[3]{x-5})^3 + 5 \quad \text{Replace } x \text{ with } \sqrt[3]{x-5} \text{ in } f(x).$$

$$= x - 5 + 5 \quad \text{Simplify.}$$

$$= x \quad \text{So } f[g(x)] = x.$$

$$g[f(x)] = g(x^3 + 5) \quad \text{Replace } f(x) \text{ with } x^3 + 5.$$

$$= \sqrt[3]{x^3 + 5 - 5} \quad \text{Replace } x \text{ with } x^3 + 5 \text{ in } g(x).$$

$$= \sqrt[3]{x^3} \quad \text{Simplify.}$$

$$= x \quad \text{So } g[f(x)] = x.$$

Since $f[g(x)] = x$ and $g[f(x)] = x$, f and g are inverses.

YOUR TURN Verify that $f(x) = 5x + 6$ and $g(x) = \dfrac{x - 6}{5}$ are inverse functions.

Calculator TIPS

We can verify that $f[g(x)] = g[f(x)] = x$ without actually finding the composition functions by using a graphing utility. We illustrate using $f(x) = \dfrac{x + 3}{x}$ and $g(x) = \dfrac{3}{x - 1}$, which are inverse functions if $x \neq 0$ and $x \neq 1$. Enter $f(x) = Y_1 = \dfrac{x + 3}{x}$, $g(x) = Y_2 = \dfrac{3}{x - 1}$. Since $f[g(x)] = Y_1(Y_2)$, enter $Y_3 = Y_1(Y_2)$. To enter Y_1 and Y_2 in the Y= *menu on the TI-83 plus, put the cursor at Y_3, press* VARS*, select Y-VARS, then select Function and press* ENTER*. Select your choice and press* ENTER*. Graph Y_3 without graphing Y_1 or Y_2. To graph Y_3 only, press* Y=*, put the cursor on the = sign of Y_1, and press* ENTER*. Repeat for Y_2. Press* GRAPH*. Trace to verify that the graph is $y = x$.*

$Y_1 = (X + 3)/X$
$Y_2 = 3/(X - 1)$
$Y_3 = Y_1(Y_2)$
$Y_4 =$
$Y_5 =$
$Y_6 =$
$Y_7 =$
$Y_8 =$

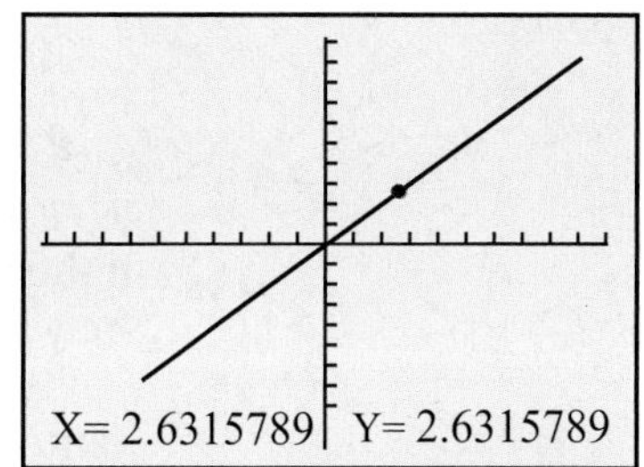

Repeat for $g[f(x)]$.

Before finding inverse functions, we need to determine what types of functions have inverses. Consider the following:

$f = \{(1, 2), (3, 4), (5, 6), (x, y)\}$; Domain $= \{1, 3, 5, x\}$ and range $= \{2, 4, 6, y\}$

$g = \{(2, 1), (4, 3), (6, 5), (y, x)\}$; Domain $= \{2, 4, 6, y\}$ and range $= \{1, 3, 5, x\}$

Note the ordered pairs in g are the ordered pairs of f with x and y interchanged. Since $(1, 2)$ is in f, by definition $f(1) = 2$. Since $(2, 1)$ is in g, $g(2) = 1$. Therefore, $f[g(2)] = f(1) = 2$ and $g[f(1)] = g(2) = 1$. Similarly, $f[g(4)] = f(3) = 4$ and $g[f(3)] = g(4) = 3$. In particular, $f[g(y)] = f(x) = y$ and $g[f(x)] = g(y) = x$ for all real numbers x and y. Consequently, f and g are inverses.

ANSWER

$$f[g(x)] = f\left[\frac{x - 6}{5}\right] = 5\left(\frac{x - 6}{5}\right) + 6 = x - 6 + 6 = x$$

$$g[f(x)] = g[5x + 6] = \frac{5x + 6 - 6}{5} = \frac{5x}{5} = x$$

Since $f[g(x)] = x$ and $g[f(x)] = x$, f and g are inverses.

Warning: Do not confuse f^{-1} with raising an expression to the negative 1 power. It is usually clear from the context that f^{-1} means an inverse function.

In general, to find the inverse of a function, we reverse the ordered pairs by interchanging x and y. To indicate the inverse of the function f, we use a special notation f^{-1} instead of g. The following figure illustrates inverse functions.

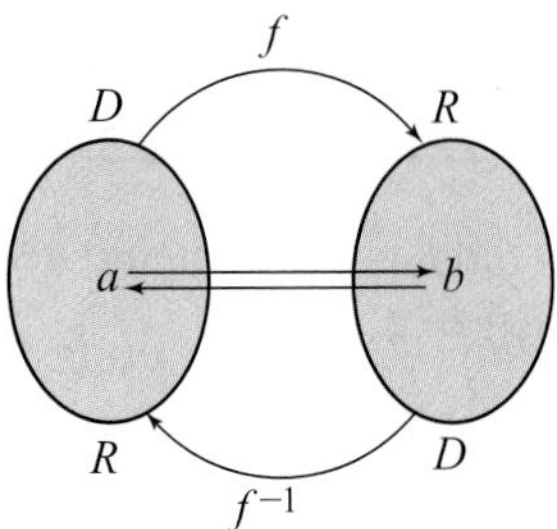

Note: *Notice that f sends a to b and f^{-1} sends b back to a. Hence, $f^{-1}[f(a)] = a$. Also, f^{-1} sends b to a and f sends a back to b. Hence, $f[f^{-1}(b)] = b$. We also see that the domain of f is the range of f^{-1} and the range of f is the domain of f^{-1}.*

Before we can find inverse functions, we need to explore one more concept.

OBJECTIVE 3. Show that a function is one-to-one. If we merely reverse the ordered pairs of a function, we do not always get an inverse function. Consider the following sets in which the ordered pairs of B are the ordered pairs of A with x and y interchanged:

$$A = \{(1, 2), (2, 4), (3, 2), (-2, 5)\}$$
$$B = \{(2, 1), (4, 2), (2, 3), (5, -2)\}$$

Note that A represents a function, but B does not because 2 in the domain of B is paired with 1 and 3. How did this happen? Since the ordered pairs of B are those of A with x and y interchanged, two ordered pairs of A having the same y-value of 2 [(1, 2) and (3, 2)] became two ordered pairs with the same x-value in B. Therefore, if the inverse of a function is to be a function, no two ordered pairs of the function may have the same y-value. Such a function is called a **one-to-one function**.

A function is one-to-one if each value in the domain corresponds to only one value in the range and each value in the range corresponds to only one value in the domain. In terms of x and y, two different x-values must result in two different y-values, which suggests the following formal definition.

DEFINITION ***One-to-one function:*** A function f is one-to-one if for any two numbers a and b in its domain, when $f(a) = f(b)$, $a = b$ and when $a \neq b$, $f(a) \neq f(b)$.

It is possible to determine if a function is one-to-one by looking at its graph. If two ordered pairs have the same y-value, then the corresponding points lie on the same horizontal line. Consequently, if a function is one-to-one, then the graph cannot be intersected by any horizontal line in more than one point.

RULE **Horizontal Line Test for One-to-One Functions**

Given a function's graph, the function is one-to-one if every horizontal line that can intersect the graph does so at one and only one point.

EXAMPLE 3 Determine whether the following are graphs of one-to-one functions.

a. $f(x) = x^3$

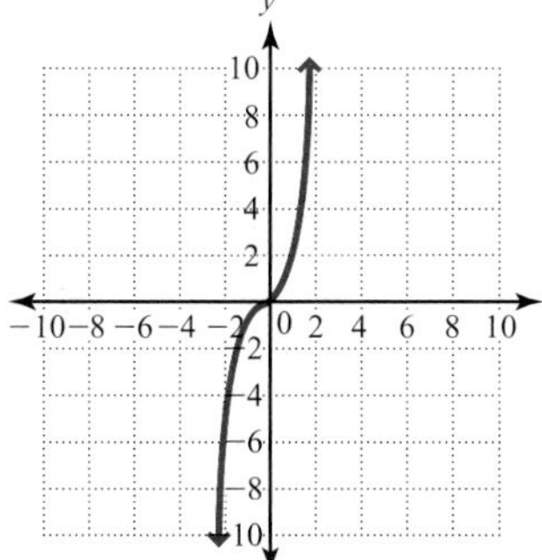

b. $f(x) = x^2 + 2$

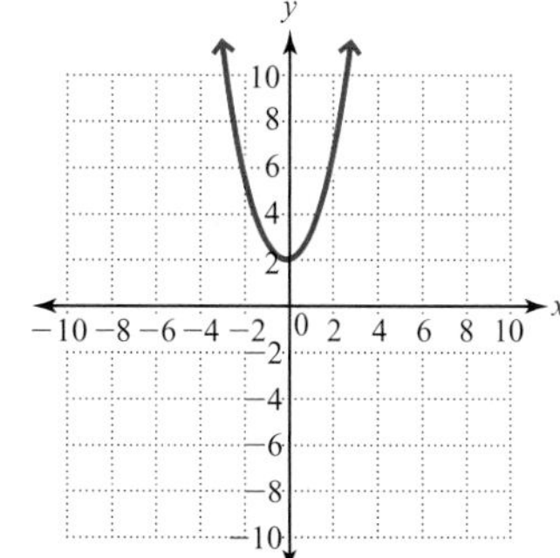

c. $f(x) = \sqrt{36 - x^2}$

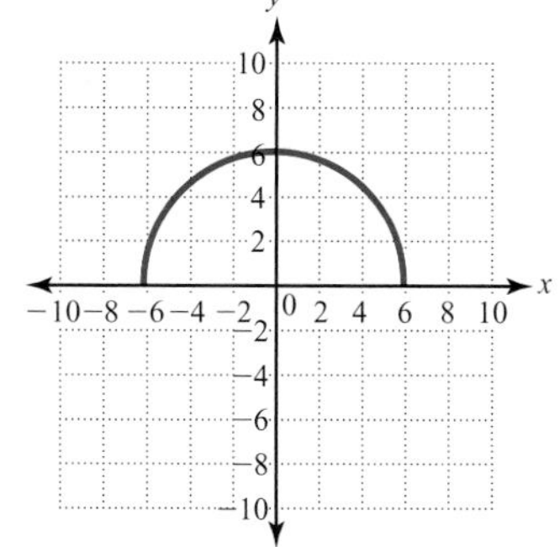

Solution

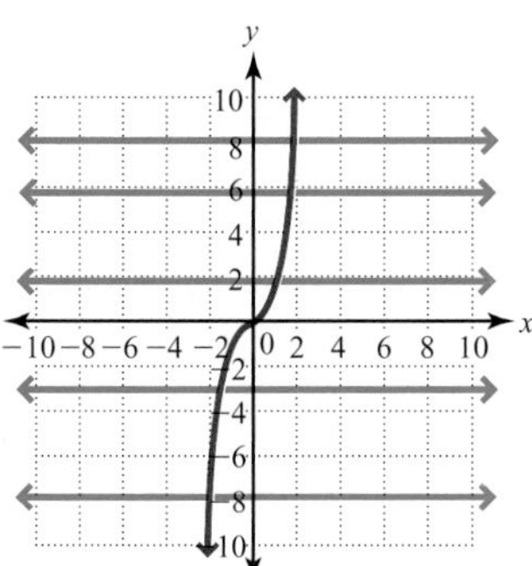

a. Every horizontal line that can intersect this graph does so at one and only one point, so the function is one-to-one.

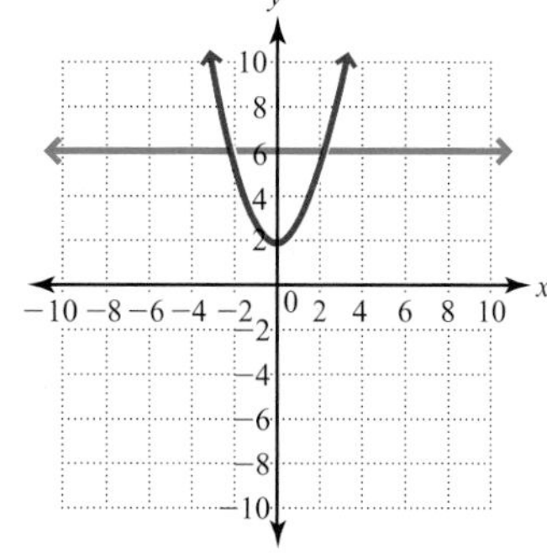

b. A horizontal line can intersect this graph in more than one point, so the function is not one-to-one.

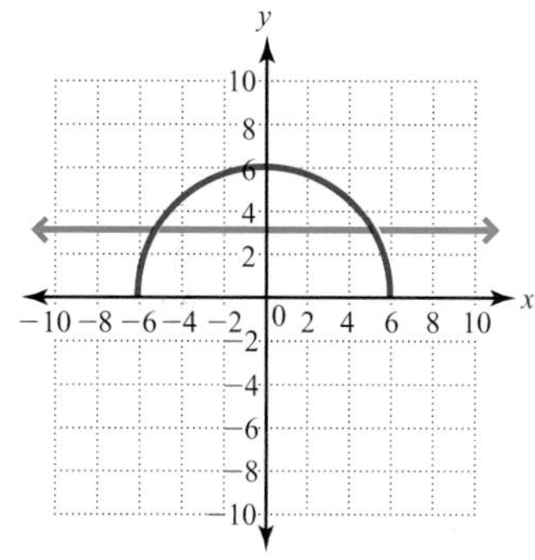

c. A horizontal line can intersect this graph in more than one point, so the function is not one-to-one.

YOUR TURN Determine whether the following are graphs of one-to-one functions.

a.

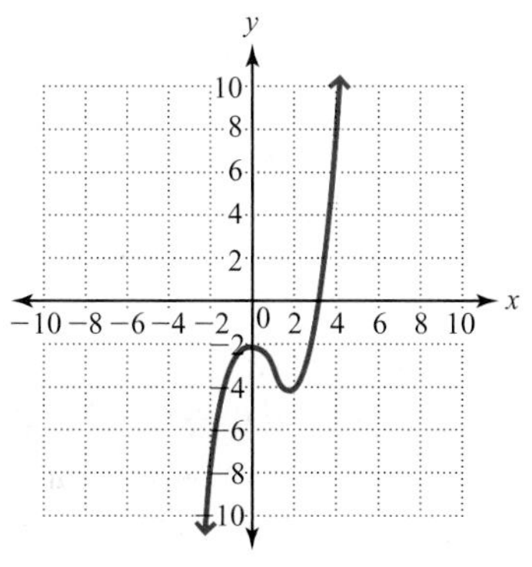

b.

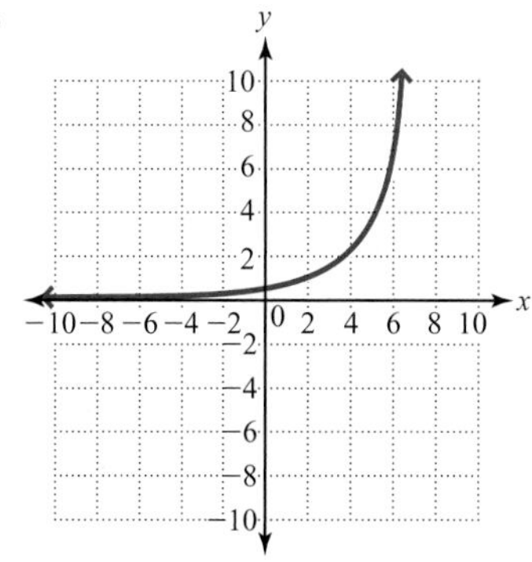

ANSWERS

a. no **b.** yes

OBJECTIVE 4. Find the inverse of a function. Interchanging the ordered pairs of a function that is not one-to-one results in a relation that is not a function. Consequently, we have the following.

RULE **Existence of Inverse Functions**

A function has an inverse function if and only if the function is one-to-one.

We have already seen that in order to find the inverse of a function, we need to interchange the x- and y-values of all the ordered pairs of f. That is, we replace x with y and y with x in the equation defining the function. However, we are not finished. We now have the inverse in the form $f(y) = x$, so we must solve this equation for y in order to have y as a function of x. The steps are summarized as follows.

PROCEDURE **Finding the Inverse Function of a One-to-One Function**

1. If necessary, replace $f(x)$ with y.
2. Replace all x's with y's and y's with x's.
3. Solve the equation from step 2 for y.
4. Replace y with $f^{-1}(x)$.

EXAMPLE 4 Find $f^{-1}(x)$ for each of the following one-to-one functions.

a. $f(x) = 2x + 4$

The domain and range of f is the set of all real numbers, so the domain and range of f^{-1} will also be the set of all real numbers.

Solution

$y = 2x + 4$ Replace $f(x)$ with y.

$x = 2y + 4$ Replace x with y and y with x.

$\frac{x-4}{2} = y$ Solve for y.

$f^{-1}(x) = \frac{x-4}{2}$ Replace y with $f^{-1}(x)$.

In order to verify that we have found the inverse, we need to show that $f[f^{-1}(x)] = x$ and $f^{-1}[f(x)] = x$.

$$f[f^{-1}(x)] = f\left(\frac{x-4}{2}\right) = 2\left(\frac{x-4}{2}\right) + 4 = x - 4 + 4 = x$$

$$f^{-1}[f(x)] = f^{-1}(2x+4) = \frac{(2x+4)-4}{2} = \frac{2x+4-4}{2} = \frac{2x}{2} = x$$

Since $f[f^{-1}(x)] = x$ and $f^{-1}[f(x)] = x$, they are inverses.

b. $f(x) = x^3 + 2$

The domain and range of f is the set of all real numbers, so the domain and range of f^{-1} will also be the set of all real numbers.

Solution $y = x^3 + 2$ — Replace $f(x)$ with y.

$x = y^3 + 2$ — Replace x with y and y with x.

$x - 2 = y^3$ — Begin solving for y by subtracting 2 from both sides.

$\sqrt[3]{x-2} = y$ — Take the cube root of each side to solve for y.

$f^{-1}(x) = \sqrt[3]{x-2}$ — Replace y with $f^{-1}(x)$.

We can verify that f and f^{-1} are inverses as in part a.

c. $f(x) = \sqrt{x-3}$

The domain of f is $[3, \infty)$ and the range is $[0, \infty)$. Therefore, the domain of f^{-1} is $[0, \infty)$ and the range of f^{-1} is $[3, \infty)$.

Solution $y = \sqrt{x-3}$ — Replace $f(x)$ with y.

$x = \sqrt{y-3}$ — Replace x with y and y with x.

$x^2 = y - 3$ — Begin solving for y by squaring both sides.

$x^2 + 3 = y$ — Add 3 to both sides to solve for y.

The domain of $y = x^2 + 3$ is all real numbers, but the domain of f^{-1} is $[0, \infty)$. Therefore, we write $f^{-1}(x) = x^2 + 3, x \geq 0$.

In order to verify that these are inverses, we need to show that $f[f^{-1}(x)] = x$ and $f^{-1}[f(x)] = x$.

$$f[f^{-1}(x)] = f(x^2+3) = \sqrt{(x^2+3)-3} = \sqrt{x^2+3-3} = \sqrt{x^2}$$

Since the domain of f^{-1} is $[0, \infty)$, $\sqrt{x^2} = x$. So, $f[f^{-1}(x)] = x$.

$$f^{-1}[f(x)] = f^{-1}(\sqrt{x-3}) = (\sqrt{x-3})^2 + 3 = x - 3 + 3 = x$$

Since $f[f^{-1}(x)] = x$ and $f^{-1}[f(x)] = x$, they are inverses.

YOUR TURN Find $f^{-1}(x)$ for each of the following one-to-one functions.

a. $f(x) = 5x + 2$ **b.** $f(x) = \dfrac{x+3}{x}$

OBJECTIVE 5. Graph a given function's inverse function. In the following figure, we have plotted pairs of points whose coordinates are interchanged and graphed the line $y = x$.

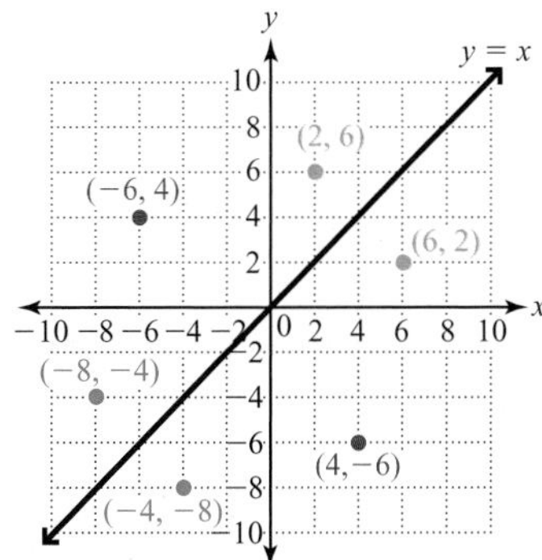

If the graph were folded along the line $y = x$, the points whose coordinates are interchanged would fall on top of each other. These points, therefore, are symmetric with respect to the line $y = x$. It can be shown that the graphs of any two points of the form (a, b) and (b, a) are symmetric with respect to the graph of $y = x$. Similarly, since the ordered pairs of f and f^{-1} have interchanged coordinates, the graphs of f and f^{-1} are symmetric with respect to the line $y = x$.

ANSWERS

a. $f^{-1}(x) = \dfrac{x-2}{5}$

b. $f^{-1}(x) = \dfrac{3}{x-1}$

RULE **Graphs of Inverse Functions**

The graphs of f and f^{-1} are symmetric with respect to the graph of $y = x$.

Learning Strategy

If you are a tactile learner, imagine placing a mirror on the line $y = x$. The graphs of f and f^{-1} are reflections in the line $y = x$.

Following are the graphs of f and f^{-1} for Example 4 along with the graph of $y = x$.

a. $f(x) = 2x + 4$
$f^{-1}(x) = \dfrac{x - 4}{2}$

b. $f(x) = x^3 + 2$
$f^{-1}(x) = \sqrt[3]{x - 2}$

c. $f(x) = \sqrt{x - 3}$
$f^{-1}(x) = x^2 + 3, \quad x \geq 0$

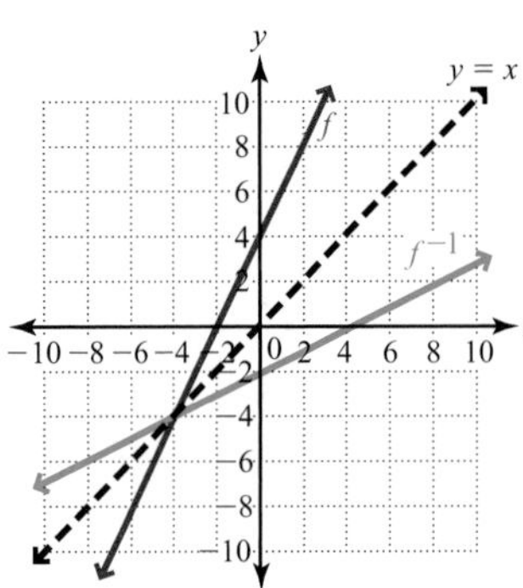

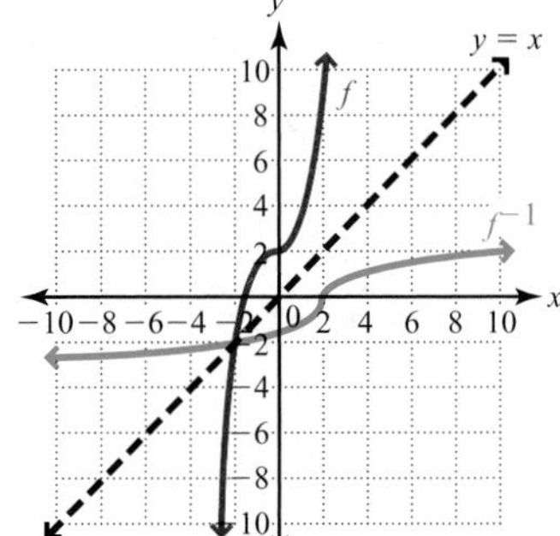

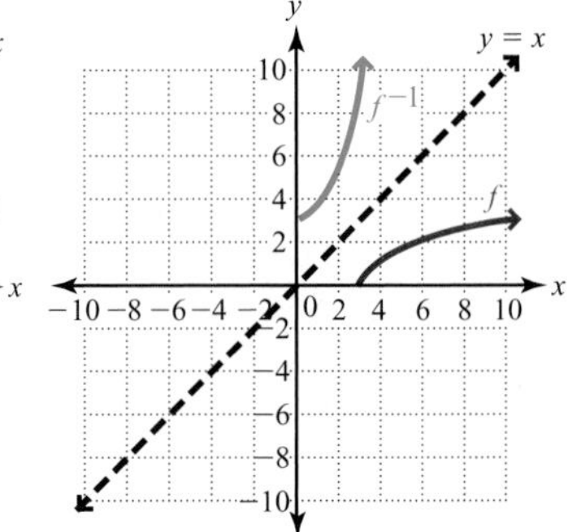

EXAMPLE 5 Sketch the inverse of the functions whose graphs are shown in parts a and b.

a.

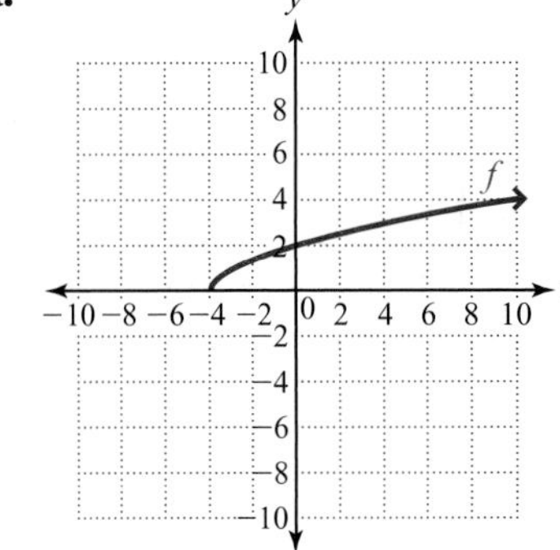

Solution Draw the line $y = x$ and reflect the graph in the line.

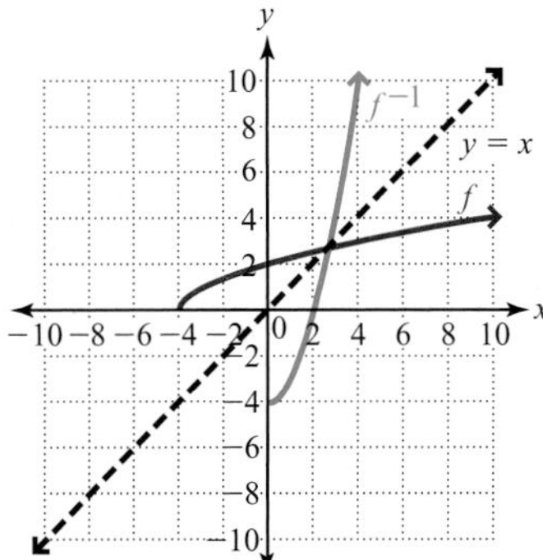

b.

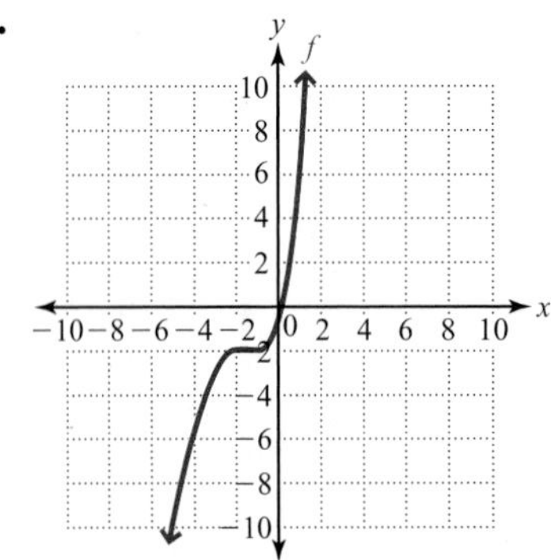

Solution Draw the line $y = x$ and reflect the graph in the line.

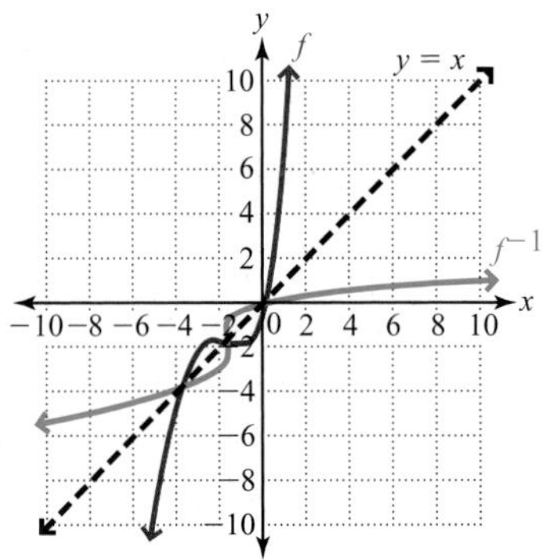

ANSWERS

a.

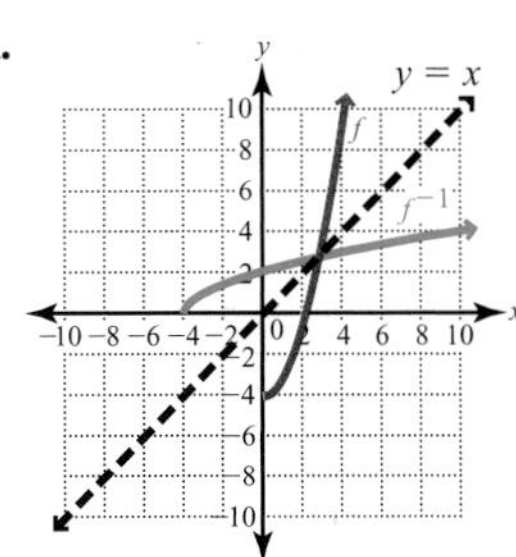

b.

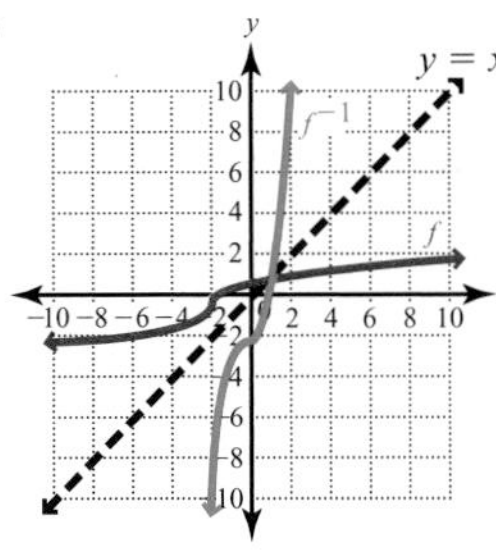

YOUR TURN Sketch the graph of the inverse of the functions whose graphs are given.

a.

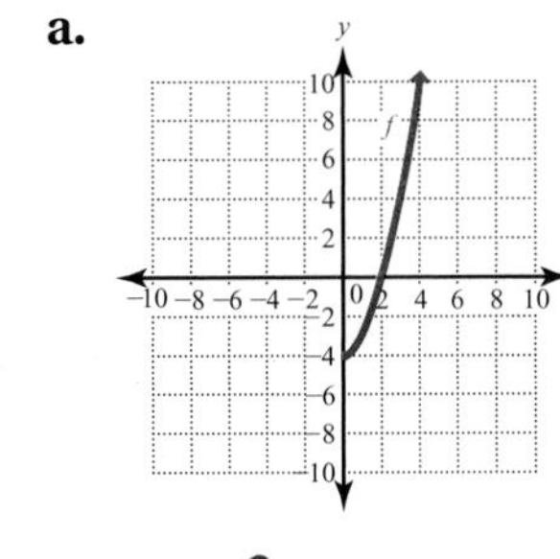

b.

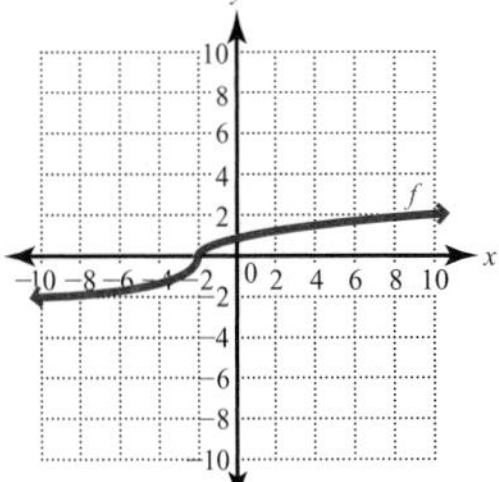

10.1 Exercises

For Extra Help MyMathLab Videotape/DVT InterAct Math Math Tutor Center Math XL.com

1. Does $f \circ g = g \circ f$ for any two functions f and g? Why or why not?

2. What must be shown to prove that two functions, f and g, are inverses?

3. How are the domains and ranges of inverse functions related? Why?

4. If the coordinates of the ordered pairs of a function are interchanged, is the result always the inverse function? Why or why not?

5. How are horizontal lines used to determine whether or not a function is one-to-one?

6. How is the graph of $(-2, 3)$ related to the graph of $(3, -2)$?

For Exercises 7–18, if $f(x) = 3x + 5$, $g(x) = x^2 + 3$, and $h(x) = \sqrt{x + 1}$, find each composition.

7. $(f \circ g)(0)$

8. $(g \circ f)(0)$

9. $(h \circ f)(1)$

10. $(h \circ g)(3)$

11. $(f \circ g)(-2)$
12. $(g \circ f)(-1)$
13. $(f \circ g)(x)$
14. $(g \circ f)(x)$

15. $(f \circ h)(x)$
16. $(h \circ g)(x)$
17. $(h \circ f)(0)$
18. $(g \circ h)(0)$

For Exercises 19–28, find $(f \circ g)(x)$ and $(g \circ f)(x)$.

19. $f(x) = 2x - 2, g(x) = 3x + 4$
20. $f(x) = 4x + 7, g(x) = 3x - 2$
21. $f(x) = x + 2, g(x) = x^2 + 1$
22. $f(x) = x^2 - 3, g(x) = x + 5$
23. $f(x) = x^2 + 3x - 4, g(x) = 3x$
24. $f(x) = -2x, g(x) = x^2 - 2x + 4$
25. $f(x) = \sqrt{x + 2}, g(x) = 2x - 5$
26. $f(x) = 5x + 2, g(x) = \sqrt{x - 5}$
27. $f(x) = \dfrac{x + 1}{x}, g(x) = \dfrac{x - 3}{x}$
28. $f(x) = \dfrac{x + 4}{x}, g(x) = \dfrac{2 - x}{x}$

For Exercises 29–30, answer each question.

29. If the domain of f is $[3, \infty)$ and the range is $[0, \infty)$, what are the domain and range of f^{-1}?

30. If f and g are inverse functions and $f(2) = 5$, then $g(5) =$ __________.

For Exercises 31–34, determine whether the following functions f and g are inverses.

31. $f = \{(1, 2), (-1, -3), (3, 4), (2, -5)\}, g = \{(2, 1), (-3, -1), (4, 3), (-5, 2)\}$

32. $f = \{(4, -3), (1, 4), (5, 2), (-3, 1), (-1, 3)\}, g = \{(-3, 4), (4, 1), (2, 5), (1, -3), (3, -1)\}$

33. $f = \{(-2, -2), (3, -3), (-4, 4), (-6, -6)\}, g = \{(-2, -2), (-3, 3), (-4, 4), (-6, -6)\}$

34. $f = \{(5, 5), (-4, 4), (2, -2), (-7, -7)\}, g = \{(-5, -5), (4, -4), (-2, 2), (7, 7)\}$

For Exercises 35–48, determine if f and g are inverses by determining whether $(f \circ g)(x) = x$ and $(g \circ f)(x) = x$.

35. $f(x) = x + 5, g(x) = x - 5$
36. $f(x) = x - 1, g(x) = x + 1$
37. $f(x) = 6x, g(x) = \dfrac{x}{6}$
38. $f(x) = -\dfrac{x}{3}, g(x) = -3x$

39. $f(x) = 2x - 3, g(x) = \dfrac{x - 3}{2}$

40. $f(x) = \dfrac{x - 4}{5}, g(x) = 5x + 4$

41. $f(x) = x^3 - 4, g(x) = \sqrt[3]{x + 4}$

42. $f(x) = x^5 + 3, g(x) = \sqrt[5]{x - 3}$

43. $f(x) = x^2, g(x) = \sqrt{x}$

44. $f(x) = x^4, g(x) = \sqrt[4]{x}$

45. $f(x) = x^2, x \geq 0; g(x) = \sqrt{x}$

46. $f(x) = x^2 + 2, x \geq 0; g(x) = \sqrt{x - 2}$

47. $f(x) = \dfrac{3}{x + 5}, g(x) = \dfrac{3 - 5x}{x}$

48. $f(x) = \dfrac{x}{x - 3}, g(x) = \dfrac{3x}{x - 1}$

For Exercises 49–50, answer each question.

49. Is $f(x) = \dfrac{1}{x}$ its own inverse? Explain.

50. Is $f(x) = x$ its own inverse? Explain.

For Exercises 51–62, sketch the graph of the inverse of each of the following functions.

51.

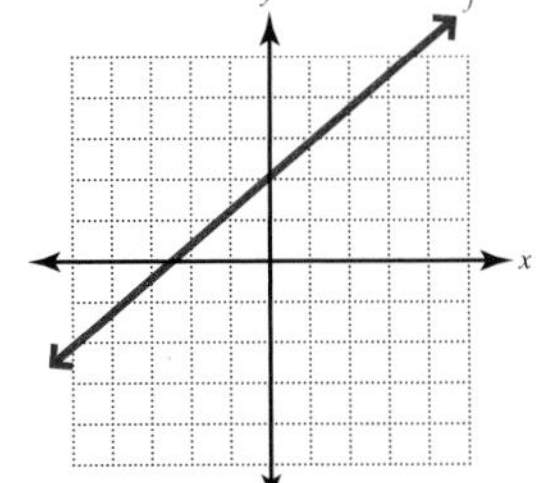

52.

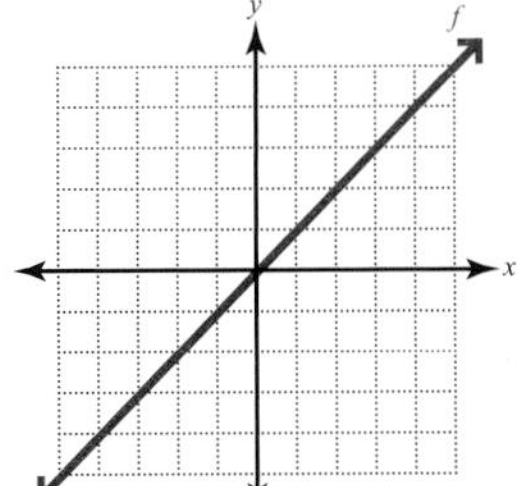

53.

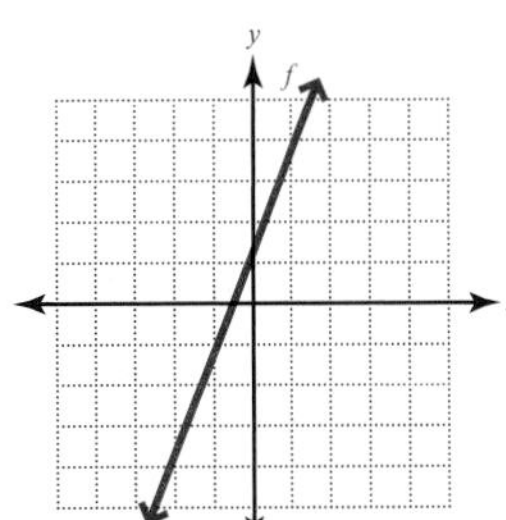

54.

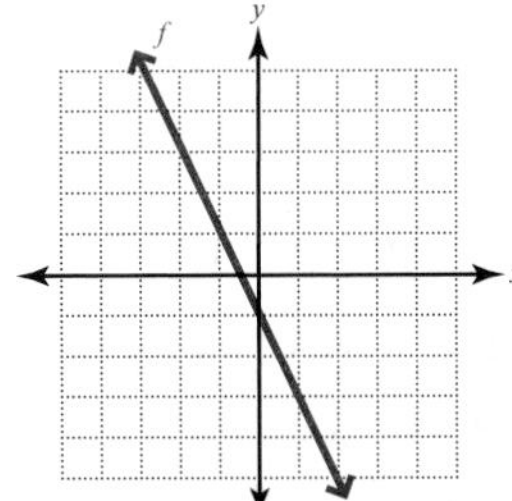

55.

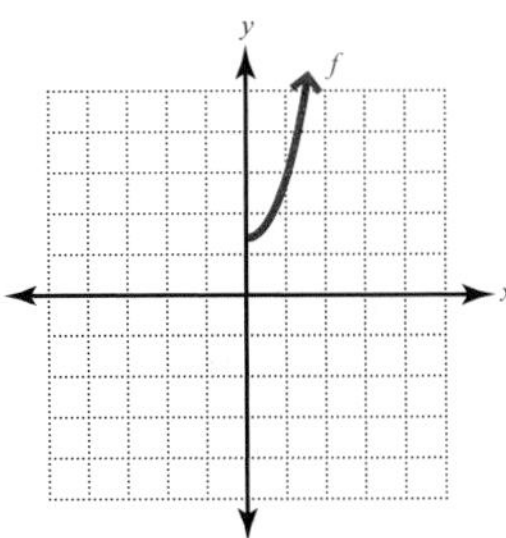

56.

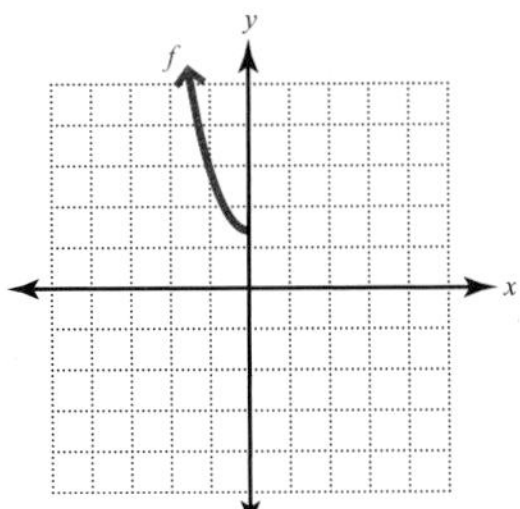

57.

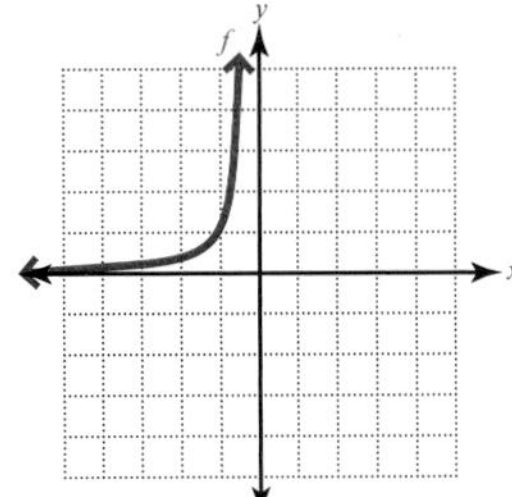

58.

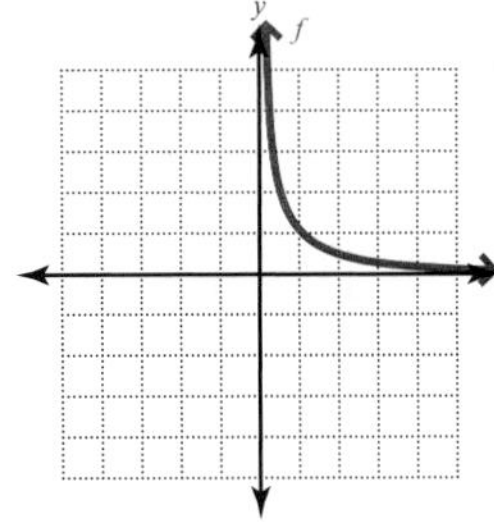

59.

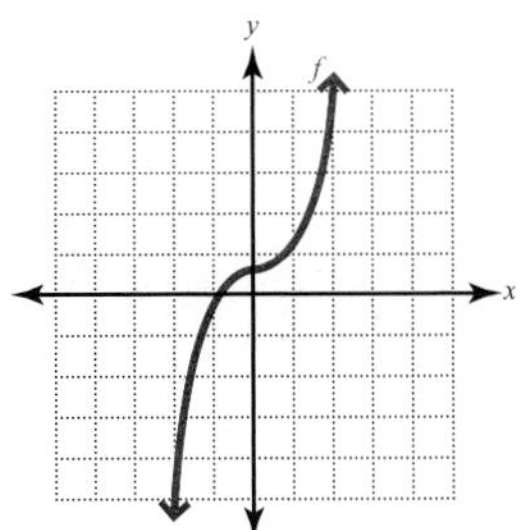

60.

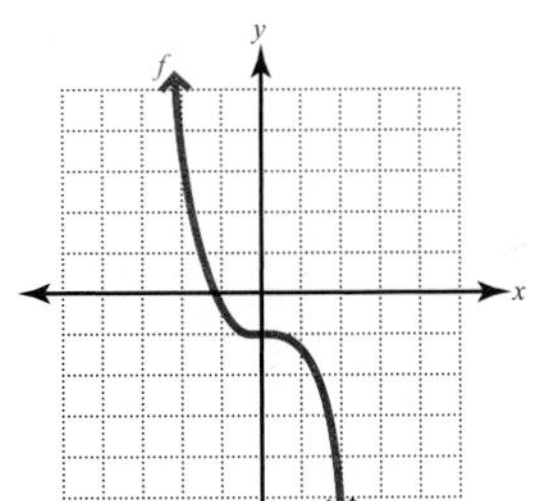

61.

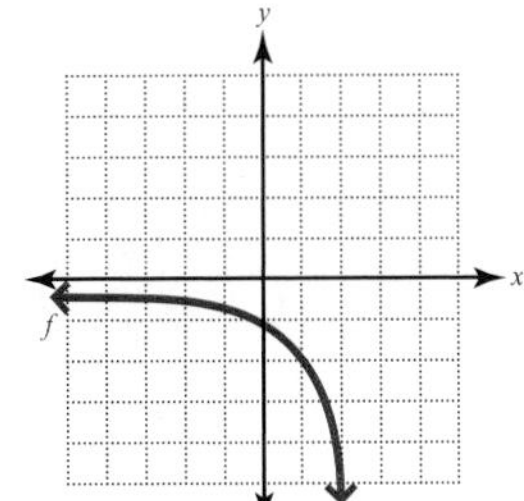

62.

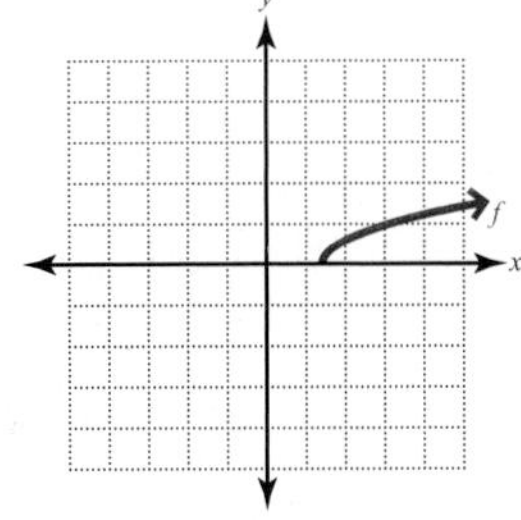

For Exercises 63–66, determine whether the function is one-to-one.

63.

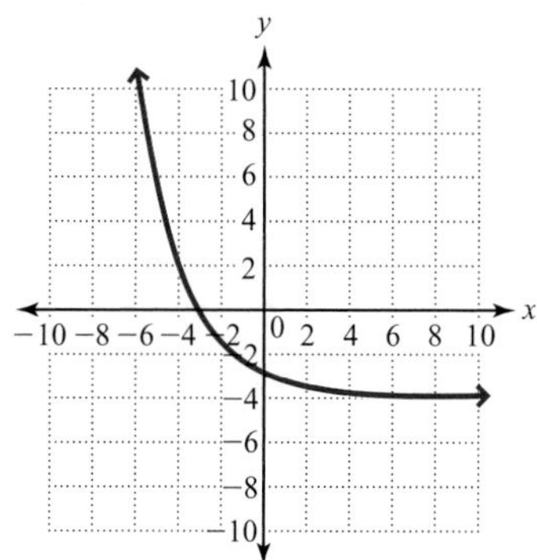

64.

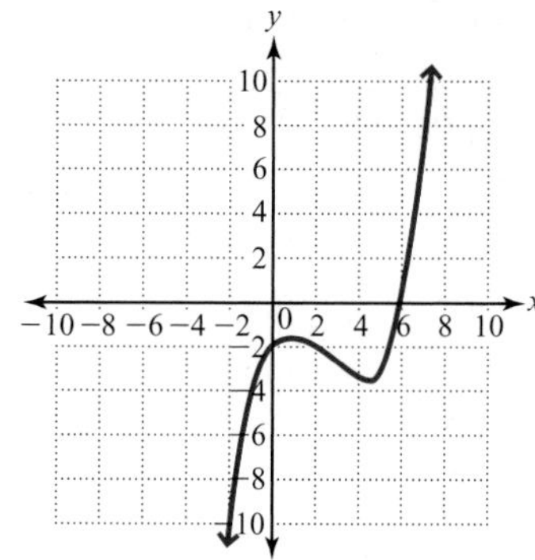

65.

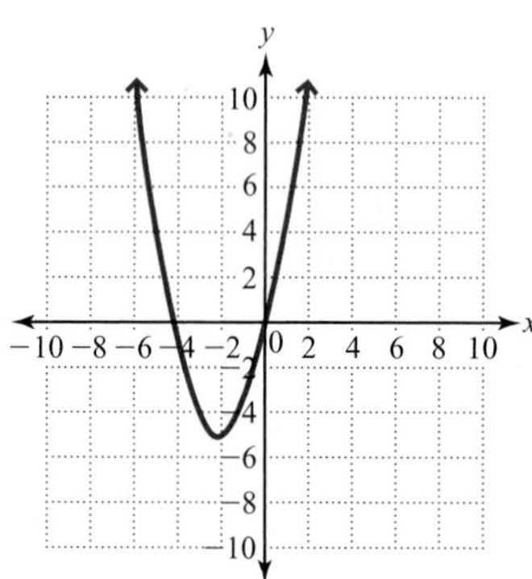

66.

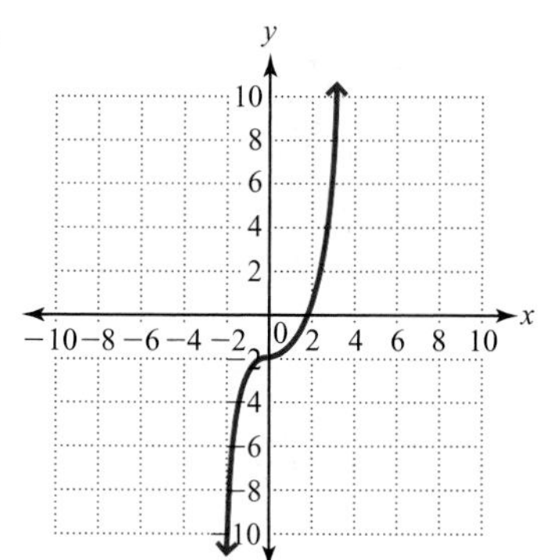

For Exercises 67–88, find $f^{-1}(x)$ for each of the following one-to-one functions f.

67. $f = \{(-3, 2), (-1, -3), (0, 4), (4, 6)\}$

68. $f = \{(-4, 2), (-1, -3), (2, 3), (5, 7)\}$

69. $f = \{(7, -2), (9, 2), (-4, 1), (3, 3)\}$

70. $f = \{(1, 2), (4, 5), (-3, 2), (-9, 0)\}$

71. $f(x) = x + 6$

72. $f(x) = x - 4$

73. $f(x) = 2x + 3$

74. $f(x) = -3x + 2$

75. $f(x) = x^3 + 2$

76. $f(x) = x^3 - 3$

77. $f(x) = \dfrac{2}{x + 2}$

78. $f(x) = \dfrac{-3}{x - 3}$

79. $f(x) = \dfrac{x + 2}{x - 3}$

80. $f(x) = \dfrac{x - 4}{x + 2}$

81. $f(x) = \sqrt{x - 2}$

82. $f(x) = \sqrt{2x - 4}$

83. $f(x) = 2x^3 + 4$

84. $f(x) = 4x^3 - 5$

85. $f(x) = \sqrt[3]{x + 2}$

86. $f(x) = \sqrt[3]{x - 5}$

87. $f(x) = 2\sqrt[3]{2x + 4}$

88. $f(x) = 3\sqrt[3]{4x - 3}$

For Exercises 89–92, solve each problem.

89. If a salesperson worked for \$100 per week plus 5% commission on sales, the weekly salary is $y = 0.05x + 100$, where y represents the salary and x represents the sales.

a. Find the inverse function.

b. What does each variable of the inverse function represent?

c. Use the inverse function to find the sales for a week in which the salary was \$350.

90. A painting contractor purchased 24 gallons of paint for the interior and exterior of a house. If he paid \$12 per gallon for the interior paint and \$18 per gallon for the exterior paint, then the amount that he paid is $y = 12x + 18(24 - x)$, where y represents the amount paid and x represents the number of gallons of interior paint.

a. Find the inverse function.

b. What does each variable of the inverse function represent?

c. Use the inverse function to find the number of gallons of interior paint if he paid \$372 total for the 24 gallons of paint.

91. An office building installed 105 fans, some of which measure 36 inches and the remainder measure 54 inches. If the 36-inch fans cost \$45 each and the 54-inch fans cost \$65 each, then the cost of the fans is $y = 45x + 65(105 - x)$, where y represents the cost and x represents the number of 36-inch fans.

a. Find the inverse function.

b. What does each variable of the inverse function represent?

c. Use the inverse function to find the number of 36-inch fans if the total cost was \$5225.

92. If a Toyota Prius averages 50 miles per hour on a trip, then $y = 50x$, where y represents the number of miles traveled and x represents the number of hours.

a. Find the inverse function.

b. What does each variable of the inverse function represent?

c. Use the inverse function to find the number of hours to travel 210 miles.

For Exercises 93–98, answer the question.

93. If $(-4, 5)$ is an ordered pair on the graph of g, what are the coordinates of an ordered pair on the graph of g^{-1}?

94. If $f(2) = 4$ and $f^{-1}(a) = 2$, find a.

95. A linear function is of the form $f(x) = ax + b, a \neq 0$. Find $f^{-1}(x)$.

96. The square root function is defined by $f(x) = \sqrt{x}$. Find $f^{-1}(x)$. Be careful!

97. The graph of an even function is always symmetric with respect to the y-axis. Is the inverse of an even function also a function? Explain.

98. The graph of an odd function is always symmetric with respect to the origin. Is the inverse of an odd function also a function? Explain.

REVIEW EXERCISES

[1.3] **1.** Write 32 as 2 to a power.

For Exercises 2–5, evaluate each expression.

[1.3] **2.** 2^3

[1.3] **3.** $\left(-\frac{1}{3}\right)^3$

[5.1] **4.** 4^{-2}

[8.2] **5.** $4^{3/2}$

[9.4] **6.** What are the coordinates of the vertex of $g(x) = (x - 3)^2 + 1$?

10.2 Exponential Functions

OBJECTIVES

1. **Define and graph exponential functions.**
2. **Solve equations of the form $b^x = b^y$ for x.**
3. **Use exponential functions to solve application problems.**

OBJECTIVE 1. Define and graph exponential functions. Previously we defined rational number exponents. For example, we know that $b^3 = b \cdot b \cdot b$, $b^{2/3} = \sqrt[3]{b^2}$, $b^0 = 1$, and $b^{-3} = \frac{1}{b^3}$. It can be shown that irrational number exponents have meaning as well, and we can approximate expressions like $2^{\sqrt{3}}$ and 5^{π} by using rational approximations for the exponents. Therefore, the exponential expression b^x has meaning if x is any real number (rational or irrational), so we can define the **exponential function** as follows.

DEFINITION ***Exponential function:*** If $b > 0$, $b \neq 1$, and x is any real number, then the exponential function is $f(x) = b^x$.

Note: The definition of the exponential function has two restrictions on b. If $b = 1$, then $f(x) = b^x = 1^x = 1$, which is a linear function. If $b < 0$, then we could get values for which the function is not defined as a real number. For example, if $f(x) = (-4)^x$, then $f\left(\frac{1}{2}\right) = (-4)^{1/2} = \sqrt{-4}$, which is not a real number.

We graph exponential functions by plotting enough points to determine the graph's shape. We will find that the graph has one typical shape if $b > 1$ and a different typical shape if $0 < b < 1$.

EXAMPLE 1 Graph.

a. $f(x) = 2^x$ and $g(x) = 4^x$

Solution Choose some values of x and find the corresponding values of $f(x)$ and $g(x)$ (which are the y-values).

x	−3	−2	−1	0	1	2	3
f(x)	$\frac{1}{8}$	$\frac{1}{4}$	$\frac{1}{2}$	1	2	4	8
g(x)	$\frac{1}{64}$	$\frac{1}{16}$	$\frac{1}{4}$	1	4	16	64

Plotting the ordered pairs for each function gives smooth curves typical of the graphs of $f(x) = b^x$ with $b > 1$. Comparing the graphs of $f(x) = 2^x$ and $g(x) = 4^x$, we can see that the greater the value of b, the steeper the graph.

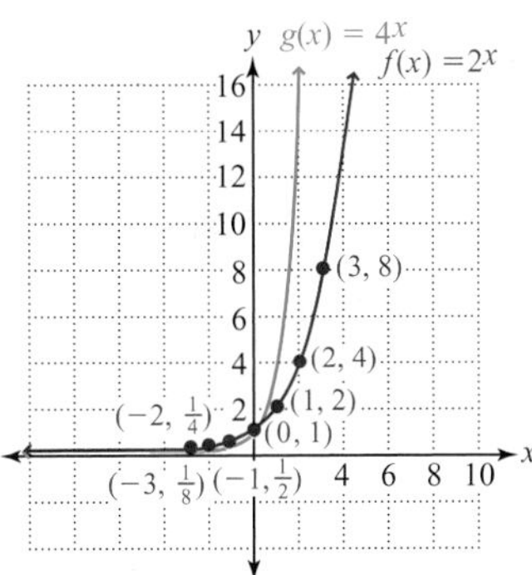

b. $h(x) = \left(\frac{1}{2}\right)^x$ and $k(x) = \left(\frac{1}{4}\right)^x$

Solution Choose some values of x and find the corresponding values of $h(x)$ and $k(x)$.

x	−3	−2	−1	0	1	2	3
h(x)	8	4	2	1	$\frac{1}{2}$	$\frac{1}{4}$	$\frac{1}{8}$
k(x)	64	16	4	1	$\frac{1}{4}$	$\frac{1}{16}$	$\frac{1}{64}$

After plotting the ordered pairs for each function, we see graphs that are typical of exponential functions in the form $f(x) = b^x$ with $0 < b < 1$. Comparing the graphs of $h(x) = \left(\frac{1}{2}\right)^x$ and $k(x) = \left(\frac{1}{4}\right)^x$, we see that the smaller the value of b, the steeper the graph.

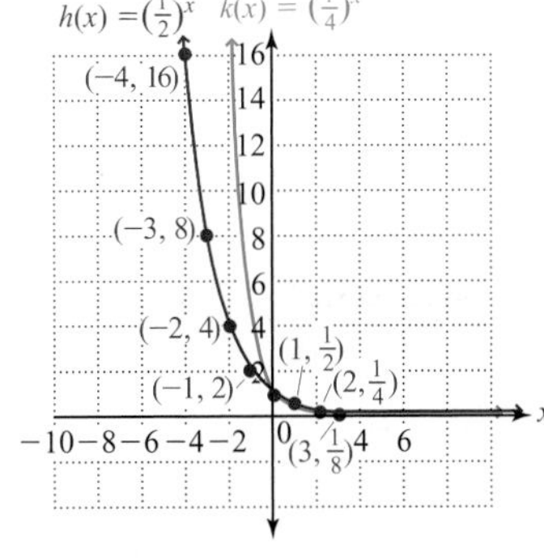

Note: We left the points and coordinate labels off the graphs of $g(x)$ and $k(x)$ to avoid additional clutter. From all the graphs in Example 1, we see that b^x is never negative.

We summarize the graphs of $f(x) = b^x$ as follows.

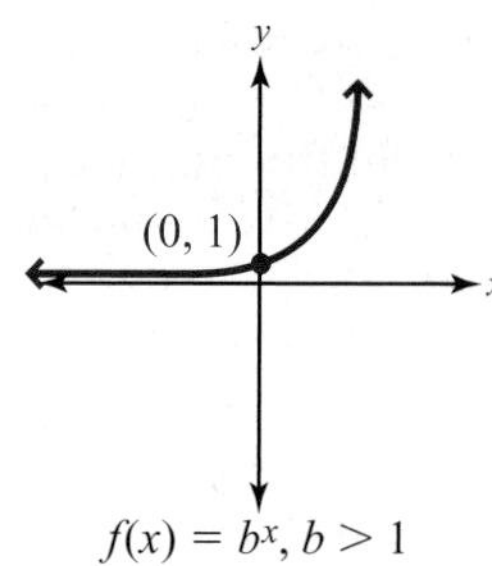

$f(x) = b^x, b > 1$

The function is increasing from left to right and one-to-one. The graph always passes through (0, 1). For negative values of x, the graph approaches the x-axis but never touches it. The larger the values of b, the steeper the graph. The domain is $(-\infty, \infty)$ and the range is $(0, \infty)$.

$f(x) = b^x, 0 < b < 1$

The function is decreasing from left to right and one-to-one. The graph always passes through (0, 1). For positive values of x, the graph approaches the x-axis but never touches it. The smaller the values of x, the steeper the graph. The domain is $(-\infty, \infty)$ and the range is $(0, \infty)$.

Note: *Any function that increases or decreases along its entire domain is one-to-one because no horizontal line will intersect its graph at more than one point.*

More complicated exponential functions are graphed in the same manner.

EXAMPLE 2 Graph. $f(x) = 3^{2x-1}$

Solution Find some ordered pairs, plot them, and draw the graph.

x	y = f(x)
−1	$3^{-3} = \frac{1}{27}$
0	$3^{-1} = \frac{1}{3}$
1	$3^1 = 3$
2	$3^3 = 27$

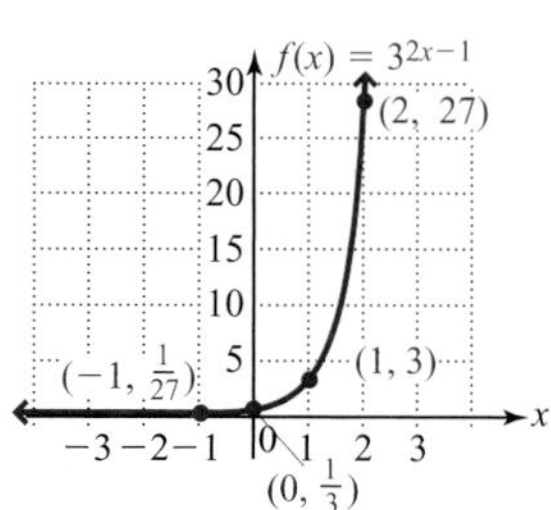

YOUR TURN Graph. $f(x) = 2^{x+1}$

ANSWER

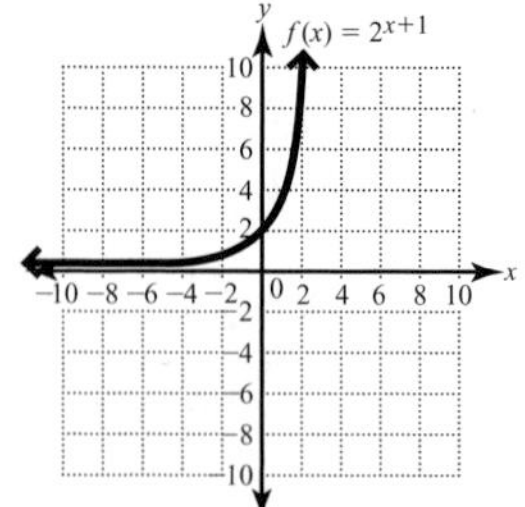

Calculator TIPS

To graph $y = 2^x$ on a graphing calculator, we enter $Y_1 = 2^x$. Set the window to $[-5, 5]$ for x and $[-10, 10]$ for y and press GRAPH. *We graph $y = \left(\frac{1}{2}\right)^x$ in the same window by entering $Y_2 = \left(\frac{1}{2}\right)^x$. Notice that if we folded the graph along the y-axis, the graphs would lie on top of each other, indicating they are symmetric with respect to the y-axis.*

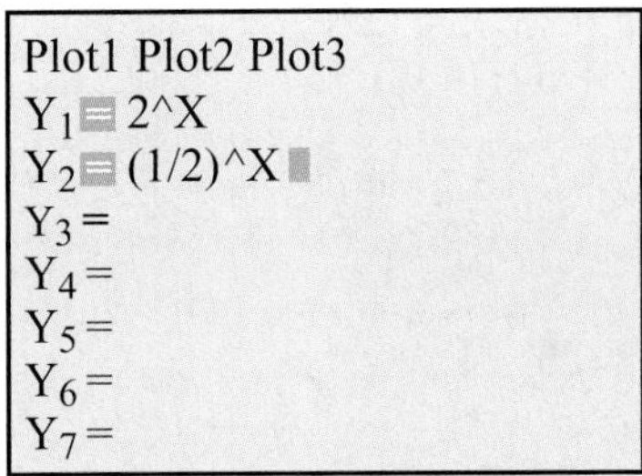

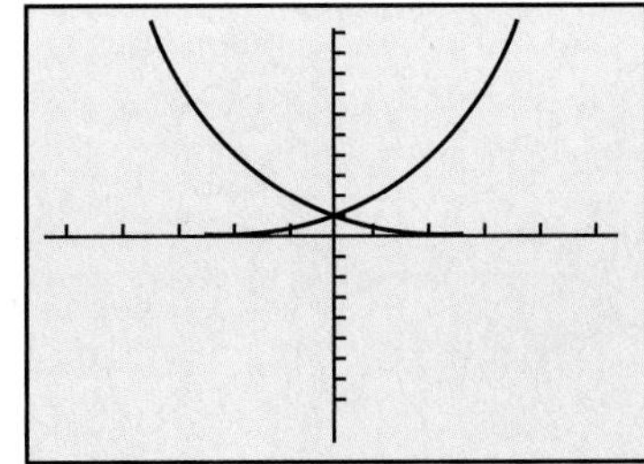

OBJECTIVE 2. Solve equations of the form $b^x = b^y$ for x. Earlier we solved equations containing expressions having a variable base and a constant exponent, like $x^2 = 4$. We will now solve equations that contain expressions having a constant base and a variable exponent, like $2^x = 16$. We need a new method to solve these equations. Previously we noted that the exponential function is one-to-one, so for each x-value there is a unique y-value and vice versa. Consequently, we have the following rule.

RULE **The One-to-One Property of Exponentials**

Given $b > 0$ and $b \neq 1$, if $b^x = b^y$, then $x = y$.

To solve some types of exponential equations, we use the following procedure.

PROCEDURE **Solving Exponential Equations**

1. If necessary, write both sides of the equation as a power of the same base.
2. If necessary, simplify the exponents.
3. Set the exponents equal to each other.
4. Solve the resulting equation.

EXAMPLE 3 Solve.

a. $3^x = 81$

Solution $3^x = 3^4$ Write 81 as 3^4 so both sides have the same base.

$x = 4$ Set the exponents equal to each other.

Check $3^x = 81$

$3^4 = 81$ Replace x with 4.

$81 = 81$ True, so $x = 4$ is the solution.

b. $4^x = 32$

Solution $(2^2)^x = 2^5$ Write 4 as 2^2 and 32 as 2^5 so both sides have the same base.

$2^{2x} = 2^5$ Simplify $(2^2)^x$ by applying $(a^m)^n = a^{mn}$.

$2x = 5$ Set the exponents equal to each other.

$x = \frac{5}{2}$ Solve for x.

Note: *We leave the checks for parts b–d to the reader.*

c. $8^{x+4} = 4^{2x+3}$

Solution $(2^3)^{x+4} = (2^2)^{2x+3}$ Write 8 as 2^3 and 4 as 2^2 so both sides have the same base.

$2^{3x+12} = 2^{4x+6}$ Simplify the exponents by applying $(a^m)^n = a^{mn}$.

$3x + 12 = 4x + 6$ Set the exponents equal to each other.

$6 = x$ Solve for x.

d. $\left(\frac{1}{5}\right)^x = 25$

Solution $(5^{-1})^x = 5^2$ Write $\frac{1}{5}$ as 5^{-1} and 25 as 5^2 so both sides have the same base.

$5^{-x} = 5^2$ Simplify $(5^{-1})^x$ by applying $(a^m)^n = a^{mn}$.

$-x = 2$ Set the exponents equal to each other.

$x = -2$ Solve for x.

YOUR TURN Solve.

a. $2^{x+1} = 8$ **b.** $27^{x-2} = 9^x$ **c.** $\left(\frac{1}{2}\right)^x = 16$

OBJECTIVE 3. Use exponential functions to solve application problems.

Compound Interest Formula

Exponential functions occur in many areas, especially the sciences and business. If P dollars are invested at an annual interest rate of r (written as a decimal) compounded n times per year for t years, the accumulated amount A in the account is given by the formula $A = P\left(1 + \frac{r}{n}\right)^{nt}$.

EXAMPLE 4 Find the accumulated amount in an account if \$5000 is deposited at 6% compounded quarterly for 10 years. Round the answer to the nearest cent.

Understand We are asked to find A given that $P = \$5000$, $r = 0.06$, $n = 4$, and $t = 10$.

Plan Use the formula $A = P\left(1 + \frac{r}{n}\right)^{nt}$.

ANSWERS

a. 2 **b.** 6 **c.** −4

Execute $A = P\left(1 + \frac{r}{n}\right)^{nt}$

$A = 5000\left(1 + \frac{0.06}{4}\right)^{4(10)}$ Substitute 5000 for P, **0.06** for r, 4 for n, and **10** for t.

$A = 5000(1 + 0.015)^{40}$ Simplify.

$A = 5000(1.015)^{40}$

$A = 9070.09$ Evaluate using a calculator and round to the nearest cent (hundredths place).

Answer After 10 years, the accumulated amount in the account is \$9070.09

Check Verify that the principal is \$5000 if the accumulated amount is \$9070.09 after the principal is compounded quarterly for 10 years.

$$9070.09 = P\left(1 + \frac{0.06}{4}\right)^{4(10)}$$

$$9070.09 = P(1.015)^{40}$$

$$\frac{9070.09}{(1.015)^{40}} = P$$

$$4999.998874 = P$$

Since the accumulated amount, 9070.09, was rounded, it is reasonable to expect our calculated value for the principal to be slightly different from \$5000.

Half-Life

The *half-life* of a radioactive substance is the amount of time it takes until only half the original amount of the substance remains. Suppose we begin with 100 grams of a substance that has a half-life of 10 days. After 10 days (one half-life) 50 grams will remain, after 20 days (two half-lives) 25 grams will remain, after 30 days (three half-lives) 12.5 grams will remain, and so forth. The formula $A = A_0\left(\frac{1}{2}\right)^{t/h}$ gives the amount remaining, where A_0 is the initial amount, t is the time, and h is the half-life.

EXAMPLE 5 The isotope ^{45}Ca has a half-life of 165 days. How many grams of a 50-gram sample of ^{45}Ca will remain after 825 days? (*Source: CRC Handbook of Chemistry and Physics*, 62nd edition, CRC Press, 1981)

Understand Given a 50-gram sample of ^{45}Ca, we are to find the amount remaining after 825 days.

Plan Use the formula $A = A_0\left(\frac{1}{2}\right)^{t/h}$.

Execute $A = A_0\left(\frac{1}{2}\right)^{t/h}$

$A = 50\left(\frac{1}{2}\right)^{825/165}$ Substitute 50 for A_0, and **825** for t and 165 for h.

$A = 50\left(\frac{1}{2}\right)^{5}$ Simplify.

$A = 1.5625$ Evaluate using a calculator.

Answer 1.5625 grams of ^{45}Ca will remain after 825 days.

Check Verify that the sample was 50 grams if 1.5625 grams remain after 825 days. We will leave this check to the reader.

Aging

EXAMPLE 6 The number of people in the United States age 65 or over (in millions) is given in the following table.

Year	Number 65 or over (in millions)
1900	3.1
1910	4.0
1920	4.9
1930	6.7
1940	9.0
1950	12.4
1960	16.7
1970	20.1
1980	25.5
1990	31.4
2000	35.6

Source: U.S. Bureau of the Census

The data can be approximated by the function $y = 3.17(1.026)^x$, where x is the number of years after 1900 and y is the number of people in millions. Use the model to estimate the number of people in the United States age 65 or over in the year 2020. Round the answer to the nearest tenth of a million.

Understand We are given the function $y = 3.17(1.026)^x$, which approximates the data in the table. We are to find the number of people age 65 or over in the year 2020.

Plan Since x represents the number of years after 1900, we first subtract 1900 from 2020 to find the value of x that corresponds to the year 2020. We can then use $y = 3.17(1.026)^x$.

Execute $x = 2020 - 1900 = 120$ Subtract 1900 from 2020 to find the value of x that corresponds to the year 2020.

$y = 3.17(1.026)^{120}$ Substitute 120 for x in $y = 3.17(1.026)^x$.

$y = 69.0$ million Evaluate using a calculator and round to the nearest tenth.

Answer According to the function, about 69 million people will be age 65 or over in the year 2020.

Check Verify that if $y = 69$ in the function $y = 3.17(1.026)^x$, then $x = 120$.

$69 = 3.17(1.026)^x$ Substitute 69 for y in $y = 3.17(1.026)^x$.

To solve the above equation, we need to use logarithms, which we have not learned yet. We will learn about logarithms and how to solve equations like the one above in the rest of this chapter.

Connection Following is a graph of the function $y = 3.17(1.026)^x$ with the point corresponding to the solution of Example 6 indicated.

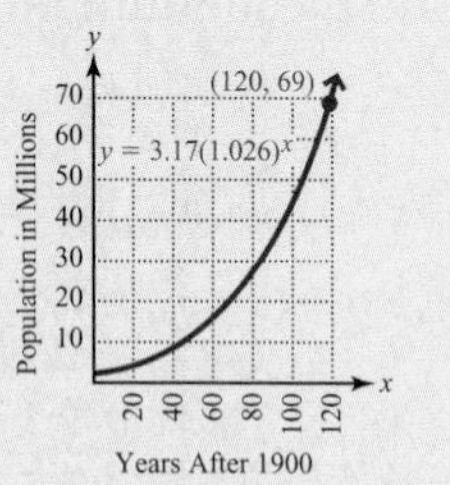

YOUR TURN

a. If \$3000 is invested at 4% compounded semiannually (twice per year), how much money is in the account at the end of 8 years? Round to the nearest cent.

b. The radioactive isotope ^{61}Cr has a half-life of 26 days. How much of a 10-gram sample would remain after 208 days? Give the answer to the nearest thousandth of a gram. (*Source: CRC Handbook of Chemistry and Physics*, 62nd edition, CRC Press, 1981)

c. The following table shows the number of computers (in millions) in use in the United States for selected years since 1984:

Year	Number of computers (in millions)
1985	21.5
1988	40.8
1989	47.6
1991	62.0
1992	68.2
1993	76.5
1994	85.8
1995	96.2
2000	160.5

The data can be approximated by $y = 23.14(1.138)^x$, where x is the number of years after 1984 and y is the number of computers (in millions). Use the model to approximate the number of computers in use in the United States in 2010.

ANSWERS

a. \$4118.36 **b.** 0.039 g
c. 666.9 million

10.2 Exercises

For Extra Help MyMathLab Videotape/DVT InterAct Math Math Tutor Center Math XL.com

1. Are the graphs of $f(x) = 2^x$ and $g(x) = \left(\frac{1}{2}\right)^x$ symmetric with respect to the y-axis? Explain.

2. As x gets larger, which graph is steeper: $f(x) = 3^x$ or $g(x) = 1.5^x$? Why?

3. Find the domain and range of $f(x) = 2^{x-4}$.

4. Is $f(x) = (-2)^x$ an exponential function? Why or why not?

5. For $b > 0$ and $b \neq 1$, if $b^x = b^y$, why does $x = y$?

6. If we have 10 grams of a substance that has a half-life of 50 days, how many grams will be present after 200 days?

For Exercises 7–24 graph.

7. $f(x) = 3^x$

8. $f(x) = 4^x$

9. $f(x) = 4^x - 3$

10. $f(x) = 3^x + 1$

11. $f(x) = \left(\frac{1}{3}\right)^x$

12. $f(x) = \left(\frac{1}{4}\right)^x$

13. $f(x) = \left(\frac{2}{3}\right)^x + 2$

14. $f(x) = \left(\frac{3}{2}\right)^x - 1$

15. $f(x) = -2^x$

16. $f(x) = -3^x$

17. $f(x) = 2^{x-2}$

18. $f(x) = 3^{x+1}$

19. $f(x) = 3^{-x}$

20. $f(x) = 2^{-x}$

21. $f(x) = 2^{2x-3}$

22. $f(x) = 3^{2x+1}$

23. $f(x) = 3^{-x+2}$

24. $f(x) = 2^{-x-1}$

For Exercises 25–42, solve each equation.

25. $2^x = 8$

26. $3^x = 81$

27. $8^x = 32$

28. $27^x = 81$

29. $16^x = 4$

30. $36^x = 216$

31. $5^x = \dfrac{1}{25}$

32. $3^x = \dfrac{1}{27}$

33. $\left(\dfrac{1}{3}\right)^x = 9$

34. $\left(\dfrac{1}{5}\right)^x = 125$

35. $\left(\dfrac{2}{3}\right)^x = \dfrac{8}{27}$

36. $\left(\dfrac{3}{2}\right)^x = \dfrac{9}{4}$

37. $\left(\dfrac{1}{2}\right)^x = 16$

38. $\left(\dfrac{1}{3}\right)^x = 27$

39. $25^{x+1} = 125$

40. $9^{x+2} = 81$

41. $8^{2x-1} = 32^{x-3}$

42. $5^{2x+1} = 125^{2x}$

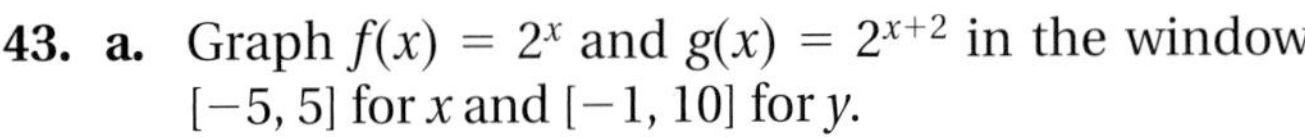

For Exercises 43–46, use a graphing utility.

43. a. Graph $f(x) = 2^x$ and $g(x) = 2^{x+2}$ in the window $[-5, 5]$ for x and $[-1, 10]$ for y.
b. How does the graph of g compare with the graph of f?

44. a. Graph $f(x) = 3^x$ and $g(x) = 3^x - 3$ in the window $[-3, 3]$ for x and $[-3, 9]$ for y.
b. How does the graph of g compare with the graph of f?

45. a. Graph $f(x) = 2^x$ and $g(x) = 2^{-x}$ in the window $[-5, 5]$ for x and $[-1, 10]$ for y.
b. How does the graph of g compare with the graph of f?

46. a. Graph $f(x) = 2^x$ and $g(x) = -2^x$ in the window $[-5, 5]$ for x and $[-5, 5]$ for y.
b. How does the graph of g compare with the graph of f?

For Exercises 47–48, use the following.

Under ideal conditions a culture of *E. coli* bacteria doubles in size every 20 minutes. If A_0 is the initial amount and t is the number of minutes passed, the amount present is A, where $A = A_0(2)^{t/20}$.

47. If a culture of *E. coli* began with 100 cells, how many cells would be present after 120 minutes?

48. If a culture of *E. coli* currently has 500 cells, how many cells were present 90 minutes earlier? (*Hint:* Let $A_0 = 500$ and the time will be negative.)

49. Under ideal conditions, human beings could double their population every 50 years. If A_0 is the initial population and t is the number of years passed, the current population is A, where $A = A_0(2)^{t/50}$. If there were 6 billion humans on Earth in the year 2000, how many would there be in the year 2500 if their growth is uncontrolled?

50. Using the information from Exercise 49, how many human beings were on the Earth in 1900?

For Exercises 51–52, use the formula $A = P\left(1 + \frac{r}{n}\right)^{nt}$.

51. If \$10,000 is deposited into an account paying 8% interest compounded quarterly, how much would be in the account after 12 years?

52. If \$15,000 is deposited into an account paying 6% interest compounded semiannually, how much would be in the account after 9 years?

For Exercises 53–54, use the formula $A = A_0\left(\frac{1}{2}\right)^{t/h}$ _from Example 6._

53. Einsteinium (^{254}Es) has a half-life of 270 days. How much of a 5-gram sample would remain after 2160 days? Give the answer to the nearest thousandth of a gram. (*Source: CRC Handbook of Chemistry and Physics*, 62nd Edition, CRC Press, 1981)

54. Nobelium (^{257}No) has a half-life of 23 seconds. How much of a 100-gram sample would remain after 275 seconds? Give the answer to the nearest thousandth of a gram. (*Source: CRC Handbook of Chemistry and Physics*, 62nd Edition, CRC Press, 1981)

55. Since the first Super Bowl in 1967, ticket prices for the game have risen exponentially. Ticket prices can be approximated by the function $P(t) = 9.046(1.114)^t$, where $t = 1$ is the year 1967 and ticket prices are in dollars. Estimate the price of a Super Bowl ticket in the year 2010. (*Source: Orlando Sentinel*, January 21, 2003, p. A8)

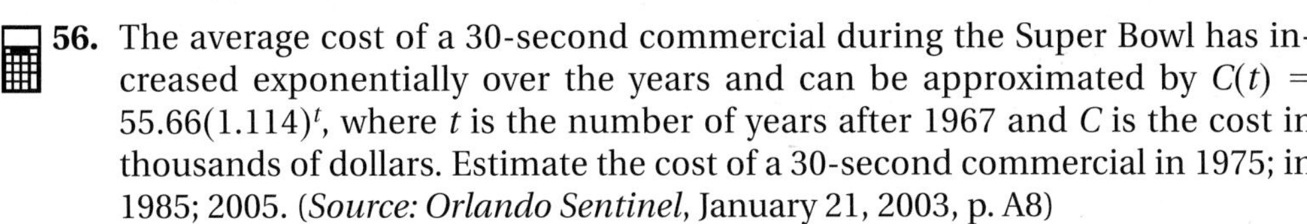

56. The average cost of a 30-second commercial during the Super Bowl has increased exponentially over the years and can be approximated by $C(t) = 55.66(1.114)^t$, where t is the number of years after 1967 and C is the cost in thousands of dollars. Estimate the cost of a 30-second commercial in 1975; in 1985; 2005. (*Source: Orlando Sentinel*, January 21, 2003, p. A8)

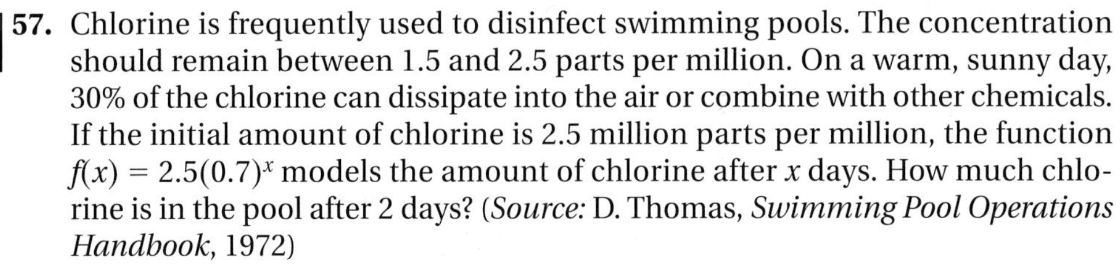

57. Chlorine is frequently used to disinfect swimming pools. The concentration should remain between 1.5 and 2.5 parts per million. On a warm, sunny day, 30% of the chlorine can dissipate into the air or combine with other chemicals. If the initial amount of chlorine is 2.5 million parts per million, the function $f(x) = 2.5(0.7)^x$ models the amount of chlorine after x days. How much chlorine is in the pool after 2 days? (*Source:* D. Thomas, *Swimming Pool Operations Handbook*, 1972)

58. It is estimated that the value of a car depreciates 20% per year for the first five years. If the original price of a car is P, the value, A, of a car after t years is given by $A = P(.8)^t$. If a car originally cost $25,960, find the value of the car after 3 years to the nearest dollar.

59. Between 1971 and 2004 the number of transistors that can be placed on a single chip has grown significantly, as indicated in the table shown.

The data can be approximated by the function $T(x) = 0.001757(1.39)^x$, where x is the number of years after 1970 and $T(x)$ is the number of transistors in millions. If the current trend continues, estimate the number of transistors that could be put on a single chip in 2010. (*Source:* Intel)

Year	Chip	Transistors (millions)
1971	4004	0.0023
1986	386DX	0.275
1989	486DX	1.2
1993	Pentium	3.3
1995	P6	5.5
1997	Pentium II	7.5
1999	Pentium III	9.5
2000	Pentium IV	42
2004	Pentium IV Prescott	125

60. Suppose that Dave fixes a cup of coffee with cream and places it on the counter to cool. The temperature of the coffee at various times is given in the table shown.

The data can be approximated by the function $T(t) = 146.9(.989)^t$, where t is time in seconds and T is the temperature in °F. Estimate the temperature of the coffee after 1 minute.

t (seconds)	T (°F)
0.2	155.8
8.4	133.2
16.6	117.9
24.8	107.9
33	100.7
41.1	94.9
49.3	90.5

PUZZLE PROBLEM

What is the greatest number that can be written using three numerals?

REVIEW EXERCISES

[5.1] ***For Exercises 1–3, evaluate.***

1. 5^3

2. $\left(\frac{1}{3}\right)^{-2}$

3. 4^{-3}

[8.2] **4.** Write $5^{2/3}$ in radical notation.

[10.1] **5.** If $f(x) = 3x + 4$, find $f^{-1}(x)$.

[10.1] **6.** Graph the inverse of the function whose graph is shown.

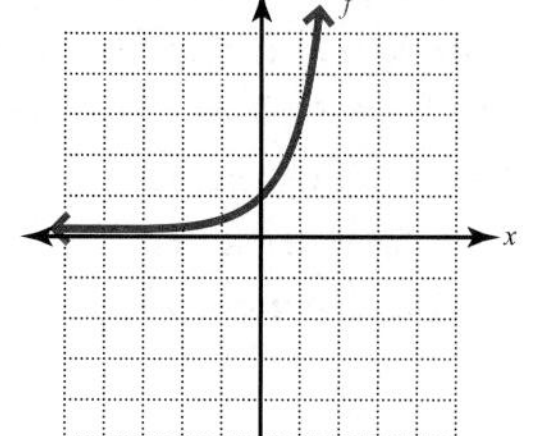

10.3 Logarithmic Functions

OBJECTIVES

1. **Convert between exponential and logarithmic forms.**
2. **Solve logarithmic equations by changing to exponential form.**
3. **Graph logarithmic functions.**
4. **Solve applications involving logarithms.**

OBJECTIVE 1. Convert between exponential and logarithmic forms. In Section 10.2, we defined the exponential function as $f(x) = b^x$ with $b > 0$ and $b \neq 1$. Since the exponential function is a one-to-one function, it has an inverse. Following is the graph of $f(x) = 2^x$ and its inverse, which we find by reflecting the graph about the line $y = x$.

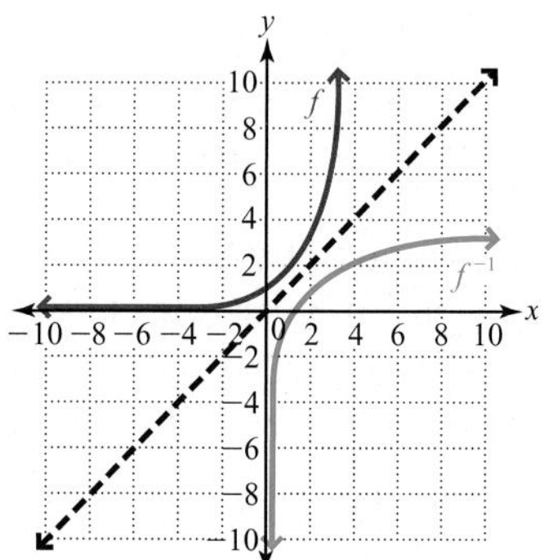

To find the inverse of $f(x) = 2^x$, we use the procedure from Section 10.1.

$$f(x) = 2^x$$

$$y = 2^x \quad \text{Replace } f(x) \text{ with } y.$$

$$x = 2^y \quad \text{Replace } y \text{ with } x \text{ and } x \text{ with } y.$$

The next step is to solve for y, but we haven't learned how to isolate a variable that is an exponent. To rewrite $x = 2^y$, we define a **logarithm**.

DEFINITION ***Logarithm:*** If $b > 0$ and $b \neq 1$, then $y = \log_b x$ is equivalent to $x = b^y$.

If we apply this definition to $x = 2^y$ in the preceding equation, we get $y = \log_2 x$. Replacing y with $f^{-1}(x)$, we get $f^{-1}(x) = \log_2 x$, so the inverse function for $f(x) = 2^x$ is $f^{-1}(x) = \log_2 x$. To generalize, exponential functions and *logarithmic functions* are inverses. Consequently, if $y = \log_b x$, the domain is $(0, \infty)$ and the range is $(-\infty, \infty)$ with $b > 0$ and $b \neq 1$.

The expression $\log_b x$ is read "the logarithm base b of x" and *is the exponent to which b must be raised to get x.* Compare the two forms.

The exponent is the logarithm.

$x = b^y \qquad y = \log_b x$

The base is the base of the logarithm.

Note: *$y = \log_b x$ means y is the power to which we raise b to get x.*

The definition of a logarithm allows us to convert from one form to another. The following table contains pairs of equivalent forms.

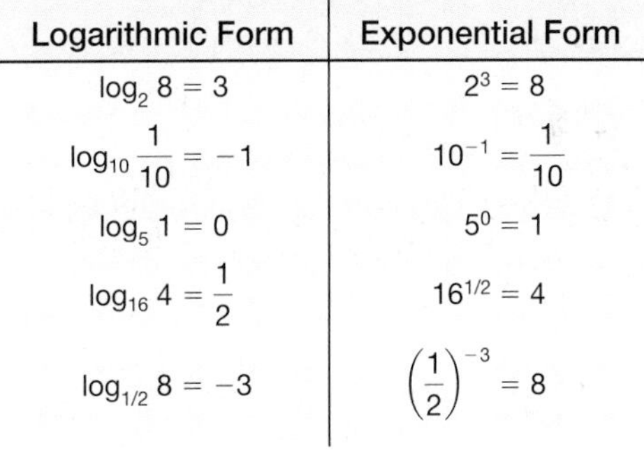

Logarithmic Form	Exponential Form
$\log_2 8 = 3$	$2^3 = 8$
$\log_{10} \frac{1}{10} = -1$	$10^{-1} = \frac{1}{10}$
$\log_5 1 = 0$	$5^0 = 1$
$\log_{16} 4 = \frac{1}{2}$	$16^{1/2} = 4$
$\log_{1/2} 8 = -3$	$\left(\frac{1}{2}\right)^{-3} = 8$

Note: *The logarithmic equations are in the form $\log_b x = y$. The values of x are positive numbers only and the values of y are both positive and negative.*

EXAMPLE 1 Write in logarithmic form.

a. $3^4 = 81$

Solution $\log_3 81 = 4$ — The base of the exponent is the base of the logarithm, and the exponent is the logarithm.

b. $\left(\frac{1}{2}\right)^{-3} = 8$

Solution $\log_{1/2} 8 = -3$ — The base of the exponent is the base of the logarithm, and the exponent is the logarithm.

c. $9^{1/2} = 3$

Solution $\log_9 3 = \frac{1}{2}$ — The base of the exponent is the base of the logarithm, and the exponent is the logarithm.

YOUR TURN Write in logarithmic form.

a. $4^3 = 64$ **b.** $2^{-3} = \frac{1}{8}$ **c.** $27^{1/3} = 3$

ANSWERS

a. $\log_4 64 = 3$ **b.** $\log_2 \frac{1}{8} = -3$

c. $\log_{27} 3 = \frac{1}{3}$

Example 2 Write in exponential form.

a. $\log_6 36 = 2$

Solution $6^2 = 36$ — The base of the logarithm is the base of the exponent, and the logarithm is the exponent.

b. $\log_{16} 2 = \frac{1}{4}$

Solution $16^{1/4} = 2$ — The base of the logarithm is the base of the exponent, and the logarithm is the exponent.

c. $\log_{1/2} 16 = -4$

Solution $\left(\frac{1}{2}\right)^{-4} = 16$ — The base of the logarithm is the base of the exponent, and the logarithm is the exponent.

Learning Strategy

If you are a visual learner, try visualizing the following "loop" for rewriting logarithms in exponential form.

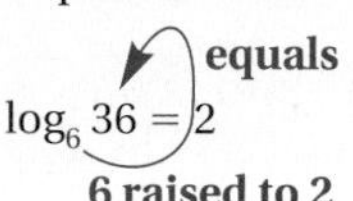

Your Turn Write in exponential form.

a. $\log_4 64 = 3$ **b.** $\log_{16} 2 = \frac{1}{4}$ **c.** $\log_5 \frac{1}{125} = -3$

OBJECTIVE 2. Solve logarithmic equations by changing to exponential form. A logarithmic equation in the form $\log_b x = y$ could have b, x, or y as an unknown.

PROCEDURE Solving Logarithmic Equations

To solve an equation of the form $\log_b x = y$, where b, x, or y is a variable, write the equation in exponential form, $b^y = x$, and then solve for the variable.

Example 3 Solve.

a. $\log_b 16 = 2$

Solution

$b^2 = 16$ — Write in exponential form.

$b = \pm 4$ — Find the positive and negative square roots of 16.

$b = 4$ — The base must be positive and not 1, so $b = 4$.

b. $\log_3 \frac{1}{27} = y$

Solution

$3^y = \frac{1}{27}$ — Write in exponential form.

$3^y = \frac{1}{3^3}$ — Write 27 as 3^3.

$3^y = 3^{-3}$ — Write $\frac{1}{3^3}$ as 3^{-3}.

$y = -3$ — Set the exponents equal to each other.

Answers

a. $4^3 = 64$ **b.** $16^{1/4} = 2$

c. $5^{-3} = \frac{1}{125}$

c. $\log_{25} x = \frac{1}{2}$

Solution $25^{1/2} = x$ Write in exponential form.

$5 = x$ $25^{1/2} = \sqrt{25} = 5.$

d. $\log_{36} \sqrt[4]{6} = y$

Solution $36^y = \sqrt[4]{6}$

$(6^2)^y = 6^{1/4}$ Write 36 as 6^2 and $\sqrt[4]{6}$ as $6^{1/4}$.

$6^{2y} = 6^{1/4}$ $(6^2)^y = 6^{2y}$.

$2y = \frac{1}{4}$ Set the exponents equal to each other.

$y = \frac{1}{8}$ Solve for y.

YOUR TURN Solve.

a. $\log_b \frac{1}{9} = -2$ **b.** $\log_5 \frac{1}{25} = y$ **c.** $\log_{27} x = \frac{1}{3}$

If an equation containing a logarithm is not in the form $\log_b a = c$, then try using the addition or multiplication principles of equality to rewrite the equation in that form.

EXAMPLE 4 Solve. $5 - 3\log_2 x = -7$

Solution Use the addition principle of equality and multiplication principle of equality to write the equation in the form $\log_b a = c$. We can then change the equation to exponential form to solve for x.

$-3 \log_2 x = -12$ Subtract 5 from both sides.

$\log_2 x = 4$ Divide both sides by -3 to isolate the logarithm.

$x = 2^4$ Write in exponential form.

$x = 16$ Simplify.

YOUR TURN Solve.

a. $9 + \log_3 x = 13$ **b.** $5 \log_n 36 = 10$ **c.** $12 - 7 \log_4 t = -9$

The definition of logarithms leads to the following two properties.

RULE For any real number b, where $b > 0$ and $b \neq 1$,

1. $\log_b b = 1$ **2.** $\log_b 1 = 0$

Based on the definition of a logarithm, $\log_b b = 1$ because $b^1 = b$, and $\log_b 1 = 0$ because $b^0 = 1$.

ANSWERS

a. 3 **b.** 2 **c.** 3

ANSWERS

a. 81 **b.** 6 **c.** 64

EXAMPLE 5 Find the value.

a. $\log_5 5$

Solution $\log_5 5 = 1$

b. $\log_e e$

Solution $\log_e e = 1$

c. $\log_{10} 1$

Solution $\log_{10} 1 = 0$

YOUR TURN Find the value.

a. $\log_{10} 10$

b. $\log_{\sqrt{3}} 1$

OBJECTIVE 3. Graph logarithmic functions. To graph logarithmic functions, which have the form $f(x) = \log_b x$, where $b > 0$ and $b \neq 1$, we will first change to exponential form so it will be easier to find ordered pairs.

PROCEDURE **Graphing Logarithmic Functions**

To graph a function in the form $f(x) = \log_b x$,

1. Replace $f(x)$ with y and then write the logarithm in exponential form $x = b^y$.
2. Find ordered pairs that satisfy the equation by assigning values to y and finding x.
3. Plot the ordered pairs and draw a smooth curve through the points.

EXAMPLE 6 Graph.

a. $f(x) = \log_2 x$

Solution $y = \log_2 x$ Replace $f(x)$ with y.

$2^y = x$ Write in exponential form.

Choose values for y and find x.

y	0	1	2	3	−1	−2
x	$2^0 = 1$	$2^1 = 2$	$2^2 = 4$	$2^3 = 8$	$2^{-1} = \frac{1}{2}$	$2^{-2} = \frac{1}{4}$

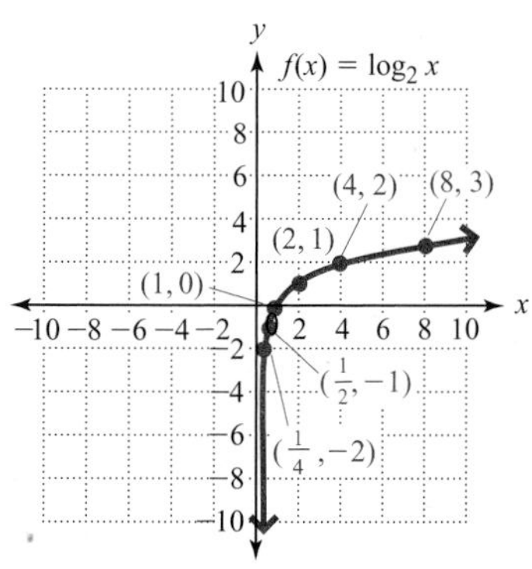

ANSWERS

a. 1 **b.** 0

b. $f(x) = \log_{1/2} x$

Solution $y = \log_{1/2} x$ **Replace $f(x)$ with y.**

$x = \left(\frac{1}{2}\right)^y$ **Write in exponential form.**

Choose values for y and find x.

y	0	1	2	3	−1	−2	−3
x	$\left(\frac{1}{2}\right)^0 = 1$	$\left(\frac{1}{2}\right)^1 = \frac{1}{2}$	$\left(\frac{1}{2}\right)^2 = \frac{1}{4}$	$\left(\frac{1}{2}\right)^3 = \frac{1}{8}$	$\left(\frac{1}{2}\right)^{-1} = 2$	$\left(\frac{1}{2}\right)^{-2} = 4$	$\left(\frac{1}{2}\right)^{-3} = 8$

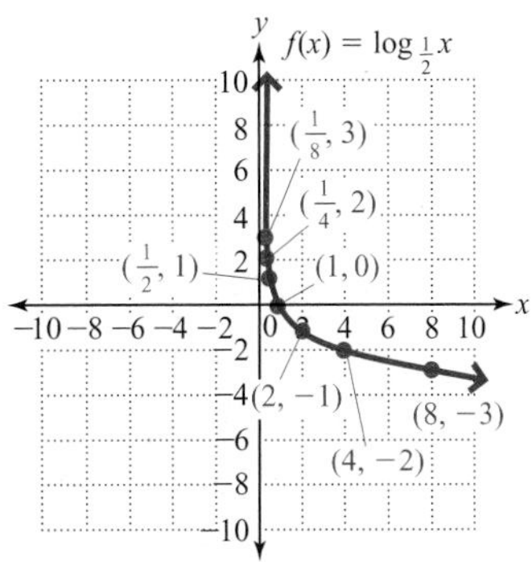

Following is a summary of the key features of the graphs of logarithmic functions.

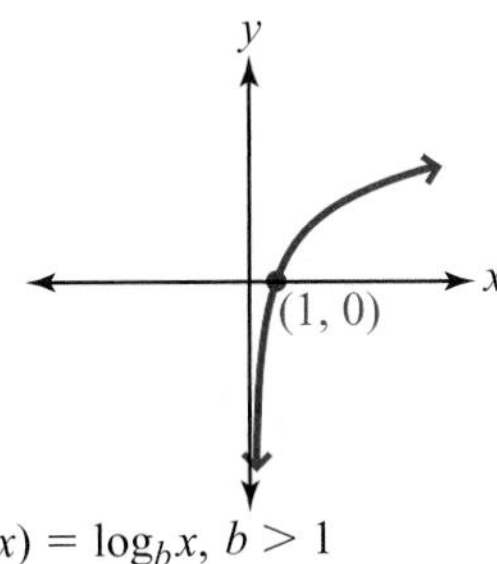

$f(x) = \log_b x,\ b > 1$

The graph passes through (1, 0), approaches the y-axis, and increases. The domain is $(0, \infty)$. The range is $(-\infty, \infty)$.

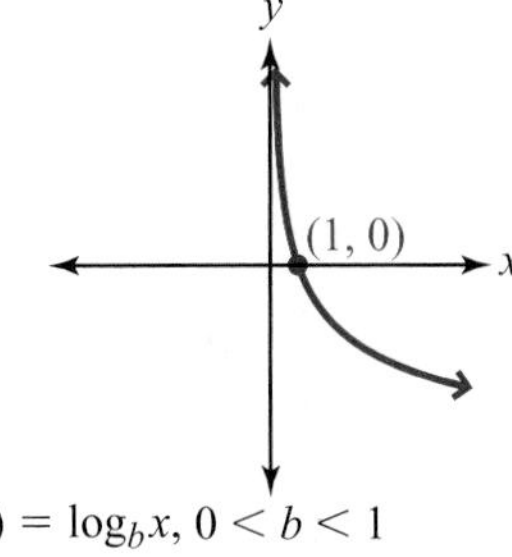

$f(x) = \log_b x,\ 0 < b < 1$

The graph passes through (1, 0), approaches the y-axis, and decreases. The domain is $(0, \infty)$. The range is $(-\infty, \infty)$.

Answer

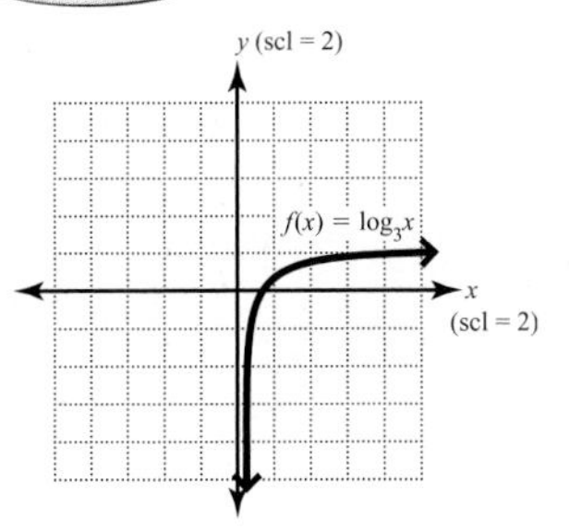

Your Turn Graph.

$f(x) = \log_3 x$

OBJECTIVE 4. Solve applications involving logarithms.

EXAMPLE 7 The function $P = 95 - 30 \log_2 x$ models the percent, P, of students who recall the important features of a classroom lecture over time, where x is the number of days that have elapsed since the lecture was given. What percent of the students recall the important features of a lecture 8 days after it was given? (*Source: Psychology for the New Millennium*, 8th Edition, Spencer A. Rathos, Thomson Publishing Company)

Understand We are given a function that models the percent, P, of students who recall the important features of a lecture x days after it is given. We are to find the percent of students who recall the important features of a lecture 8 days after it was given.

Plan Evaluate $P = 95 - 30 \log_2 x$ where $x = 8$.

Execute

$P = 95 - 30 \log_2 8$ Substitute 8 for x in $P = 95 - 30 \log_2 x$.

$P = 95 - 30(3)$ $\log_2 8 = 3$. Since $2^3 = 8$.

$P = 95 - 90$

$P = 5$ Simplify.

Answer 5% of the students remember the important features of a lecture 8 days after it is given.

Check Verify that 8 days after a lecture is given when 5% of the students recall the important features of the lecture.

$5 = 95 - 30 \log_2 x$ Replace P with 5 in $P = 95 - 30 \log_2 x$.

$-90 = -30 \log_2 x$ Subtract 95 from both sides.

$3 = \log_2 x$ Divide both sides by -30.

$2^3 = x$ Write in exponential form.

$8 = x$ Simplify. It checks.

YOUR TURN Refer to Example 7.

a. Find the percent of students who remember the important features of a lecture 2 days after it was given.

b. When the number of days was decreased by a third (from 8 to 2), was the amount retained also decreased by one-third?

ANSWERS

a. 65% **b.** no

10.3 Exercises

For Extra Help

MyMathLab Videotape/DVT InterAct Math Math Tutor Center Math XL.com

1. If $f(x) = 3^x$, what is $f^{-1}(x)$? Why?

2. If $f(x) = \log_b x$, what are the restrictions on b?

3. If $f(x) = \log_b x$, what are the restrictions on x?

4. What is the relationship between the graphs of $f(x) = 5^x$ and $g(x) = \log_5 x$? Why?

5. Why is a logarithm an exponent?

6. The $\log_b x$ is the ____________ to which ____________ must be raised to get ____________.

For Exercises 7–22, write in logarithmic form.

7. $2^5 = 32$

8. $4^3 = 64$

9. $10^3 = 1000$

10. $10^4 = 10{,}000$

11. $e^4 = x$

12. $e^{-2} = z$

13. $5^{-3} = \frac{1}{125}$

14. $6^{-3} = \frac{1}{216}$

15. $10^{-2} = \frac{1}{100}$

16. $10^{-4} = \frac{1}{10000}$

17. $625^{1/4} = 5$

18. $343^{1/3} = 7$

19. $\left(\frac{1}{4}\right)^2 = \frac{1}{16}$

20. $\left(\frac{3}{4}\right)^3 = \frac{27}{64}$

21. $7^{1/2} = \sqrt{7}$

22. $9^{1/3} = \sqrt[3]{9}$

For Exercises 23–38, write in exponential form.

23. $\log_3 81 = 4$

24. $\log_4 64 = 3$

25. $\log_4 \frac{1}{16} = -2$

26. $\log_3 \frac{1}{243} = -5$

27. $\log_{10} 100 = 2$

28. $\log_{10} \frac{1}{100} = -2$

29. $\log_e a = 5$

30. $\log_e y = -4$

31. $\log_e \frac{1}{e^4} = -4$

32. $\log_e \frac{1}{e} = -1$

33. $\log_{1/8} \frac{1}{64} = 2$

34. $\log_{1/4} \frac{1}{256} = 4$

35. $\log_{1/5} 25 = -2$

36. $\log_{1/3} 81 = -4$

37. $\log_7 \sqrt{7} = \frac{1}{2}$

38. $\log_6 \sqrt[4]{6} = \frac{1}{4}$

For Exercises 39–58, solve.

39. $\log_2 x = 3$

40. $\log_3 x = 4$

41. $\log_5 x = -2$

42. $\log_6 x = -1$

43. $\log_3 81 = y$

44. $\log_2 32 = y$

45. $\log_5 \frac{1}{5} = y$

46. $\log_4 \frac{1}{16} = y$

47. $\log_b 1000 = 3$

48. $\log_b 81 = 4$

49. $\log_m \frac{1}{16} = -4$

50. $\log_n \frac{1}{36} = -2$

51. $\log_{1/2} x = 2$

52. $\log_{1/5} x = 3$

53. $\log_{1/3} h = -5$

54. $\log_{1/6} k = -2$

55. $\log_{1/3} \frac{1}{9} = y$

56. $\log_{1/4} \frac{1}{64} = y$

57. $\log_{1/2} 64 = t$

58. $\log_{1/5} 125 = u$

For Exercises 59–70, solve.

59. $\log_2 x + 4 = 8$

60. $\log_3 x + 2 = -1$

61. $\log_{1/4} h - 2 = 1$

62. $\log_{1/2} k - 3 = -1$

63. $3 \log_b 16 = 12$

64. $-2 \log_b 81 = -8$

65. $\frac{1}{3} \log_5 c = 1$

66. $\frac{1}{4} \log_2 d = 2$

67. $3 \log_t 9 + 6 = 12$

68. $2 \log_u 125 + 8 = 14$

69. $\frac{1}{2} \log_4 m - 2 = -3$

70. $\frac{1}{3} \log_5 n - 4 = -5$

For Exercises 71–74, graph.

71. $f(x) = \log_4 x$

72. $f(x) = \log_5 x$

73. $f(x) = \log_{1/3} x$

74. $f(x) = \log_{1/4} x$

75. Why does $\log_b b = 1$ for any value of $b > 0$ and $b \neq 1$?

76. In the definition of a logarithm, $y = \log_b x$, why must $b \neq 1$?

77. Why does $\log_b 1 = 0$?

78. If $f(x) = b^x$, the domain is $(-\infty, \infty)$ and the range is $(0, \infty)$. What are the domain and range of $f(x) = \log_b x$? Why?

79. The percent of adult height attained by a 5- to 15-year-old girl can be approximated by $f(x) = 62 + 35 \log_{10}(x - 4)$, where x is the age in years and $f(x)$ is the percent. At age 14, what percent of her adult height has a girl reached?

80. The percent of adult height attained by a 5- to 15-year-old boy can be approximated by $f(x) = 29 + 48.8 \log_{10}(x + 1)$, where x is the age in years and $f(x)$ is the percent. At age 9, what percent of his adult height has a boy reached?

81. In Example 7 we learned that the percent of students who recall the important features of a lecture is given by $P = 95 - 30 \log_2 x$, where P is the percent and x is the number of days that have elapsed since the lecture was given. After how many days will 35% of the students recall the important features?

82. Using the formula from Exercise 81, find the percent of the students who recall the important features of a lecture 4 days after it was given.

Collaborative Exercises — EXPLORING GRAPHS OF LOGARITHMS

Using a graphing calculator, draw the graph of $f(x) = \log x$ ***and*** $g(x) = \log(10x)$ ***in the window*** $[0, 100]$ ***for*** x ***and*** $[-2, 4]$ ***for*** y.

1. Press the TRACE key and using up or down arrow keys move from one graph to the other and observe the y-values of several points with the same x-values. What do you observe?

2. How is the graph of $g(x) = \log(10x)$ related to the graph of $f(x) = \log x$? Why?

3. In general, how is the graph of $g(x) = \log(kx)$ for $k > 0$ related to the graph of $f(x) = \log x$?

4. Repeat step 1 using $f(x) = \log x$ and $g(x) = \log \dfrac{x}{10}$.

5. How is the graph of $g(x) = \log \dfrac{x}{10}$ related to the graph of $f(x) = \log x$? Why?

6. In general, how is the graph of $g(x) = \log \dfrac{x}{k}$ for $k > 0$ related to the graph of $f(x) = \log x$?

REVIEW EXERCISES

[5.1] **1.** Write $\frac{1}{x^6}$ as x to a negative power.

[5.1] ***For Exercises 2–5, simplify using the rules of exponents.***

2. $x^4 \cdot x^2$

3. $(x^3)^5$

4. $\frac{x^6}{x^3}$

5. $\frac{(x^3)^2 \cdot x^4}{x^5}$

[8.2] **6.** Write $\sqrt[4]{x^3}$ in exponential form.

10.4 Properties of Logarithms

OBJECTIVES

1. Apply the inverse properties of logarithms.

2. Apply the product, quotient, and power properties of logarithms.

OBJECTIVE 1. Apply the inverse properties of logarithms. Earlier, we developed two properties of logarithms, $\log_b b = 1$ and $\log_b 1 = 0$, which were based on the definition of a logarithm. The fact that $f[f^{-1}(x)] = x$ and $f^{-1}[f(x)] = x$ gives us the following properties of logarithms.

RULE **Inverse Properties of Logarithms**

For any real numbers b and x, where $b > 0$, $b \neq 1$ and $x > 0$,

1. $b^{\log_b x} = x$

2. $\log_b b^x = x$

To prove $b^{\log_b x} = x$, let $f(x) = b^x$ so $f^{-1}(x) = \log_b x$.

$$f[f^{-1}(x)] = x \quad \text{Composition of a function with its inverse is } x.$$

$$f[\log_b x] = x \quad \text{Replace } f^{-1}(x) \text{ with } \log_b x.$$

$$b^{\log_b x} = x \quad \text{Replace } x \text{ with } \log_b x \text{ in } f(x).$$

To prove $\log_b b^x = x$, use the definition of a logarithm. $\text{Log}_a b = c$ means $a^c = b$, so $\log_b b^x = x$, because $b^x = b^x$.

For example, using $b^{\log_b x} = x$ we have $3^{\log_3 8} = 8$, $6^{\log_6 x} = x$, and $e^{\log_e 4} = 4$. Using $\log_b b^x = x$, we have $\log_3 3^6 = 6$, $\log_5 5^a = a$, and $\log_e e^{x^2} = x^2$.

YOUR TURN Find the value.

a. $8^{\log_8 4}$

b. $\log_3 3^{-2}$

ANSWERS

a. 4 **b.** −2

OBJECTIVE 2. Apply the product, quotient, and power properties of logarithms. Logarithms were invented to perform operations on very large or very small numbers. With the invention of handheld calculators, they are no longer used for this purpose. However, we still use the properties of logarithms, which are based on the fact that logarithms are exponents.

Warning: There is no rule for the logarithms of sums or differences. The $\log_b(x + y) \neq \log_b x + \log_b y$.

Note: *When using the product and quotient rules, all the bases of all the logarithms* **must** *be the same.*

RULE Further Properties of Logarithms

For real numbers x, y, and b, where $x > 0$, $y > 0$, $b > 0$, and $b \neq 1$,

Product Rule of Logarithms: $\log_b xy = \log_b x + \log_b y$

(The logarithm of the product of two numbers is equal to the sum of the logarithms of the numbers.)

Quotient Rule of Logarithms: $\log_b \frac{x}{y} = \log_b x - \log_b y$

(The logarithm of the quotient of two numbers is equal to the difference of the logarithms of the numbers.)

Power Rule of Logarithms: $\log_b x^r = r \log_b x$

(The logarithm of a number raised to a power is equal to the exponent times the logarithm of the number.)

To prove $\log_b xy = \log_b x + \log_b y$, let $M = \log_b x$ and $N = \log_b y$.

$x = b^M$ and $y = b^N$	Write each logarithmic equation in exponential form.
$xy = b^M \cdot b^N$	Multiply the left and right sides of the exponential forms.
$xy = b^{M+N}$	Add the exponents.
$\log_b xy = M + N$	Write in logarithmic form.
$\log_b xy = \log_b x + \log_b y$	Substitute $\log_b x$ for M and $\log_b y$ for N and the proof is complete.

The proof of $\log_b \frac{x}{y} = \log_b x - \log_b y$ is left as an exercise. The proof of $\log_b x^r = r \log_b x$ will be a collaborative exercise.

EXAMPLE 1 Use the product rule of logarithms to write the expression as a sum of logarithms.

a. $\log_{10} xyz$

Solution $\log_{10} xyz = \log_{10} x + \log_{10} y + \log_{10} z$

b. $\log_b x(x + 3)$

Solution $\log_b x(x + 3) = \log_b x + \log_b(x + 3)$ **Note:** $\log_b(x + 3) \neq \log_b x + \log_b 3$.

EXAMPLE 2 Use the product rule of logarithms in the form $\log_b x + \log_b y = \log_b xy$ to write the expression as a single logarithm.

a. $\log_3 8 + \log_3 2$

Solution $\log_3 8 + \log_3 2 = \log_3 8 \cdot 2 = \log_3 16$

b. $\log_8 5 + \log_8 x + \log_8(2x - 3)$

Solution $\log_8 5 + \log_8 x + \log_8(2x - 3) = \log_8 5x(2x - 3)$

$= \log_8(10x^2 - 15x)$

YOUR TURN Use the product rule of logarithms to write the expression as a sum of logarithms.

a. $\log_a 6x$ **b.** $\log_2 x(3x + 2)$

Use the product rule of logarithms to write each of the following as a single logarithm.

c. $\log_6 7 + \log_6 9$ **d.** $\log_7 2 + \log_7 x + \log_7(2x + 4)$

EXAMPLE 3 Use the quotient rule of logarithms to write the expression as a difference of logarithms.

a. $\log_5 \frac{5}{11}$

Solution $\log_5 \frac{5}{11} = \log_5 5 - \log_5 11$

$= 1 - \log_5 11$ **Remember, $\log_5 5 = 1$.**

b. $\log_4 \frac{x}{x - 5}$

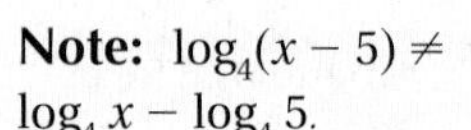

Note: $\log_4(x - 5) \neq \log_4 x - \log_4 5$.

Solution $\log_4 \frac{x}{x - 5} = \log_4 x - \log_4 (x - 5)$

EXAMPLE 4 Use the quotient rule of logarithms in the form $\log_b x - \log_b y = \log_b \frac{x}{y}$ to write the expression as a single logarithm.

a. $\log_9 3 - \log_9 x$

Solution $\log_9 3 - \log_9 x = \log_9 \frac{3}{x}$

b. $\log_{10} x - \log_{10}(x^2 + 4)$

Solution $\log_{10} x - \log_{10}(x^2 + 4) = \log_{10} \frac{x}{x^2 + 4}$

ANSWERS

a. $\log_a 6 + \log_a x$
b. $\log_2 x + \log_2(3x + 2)$
c. $\log_6 63$
d. $\log_7(4x^2 + 8x)$

YOUR TURN Use the quotient rule of logarithms to write the expression as a difference of logarithms.

a. $\log_9 \frac{9}{10}$ **b.** $\log_5 \frac{x}{x + 2}$

Use the quotient rule of logarithms to write each of the following as a single logarithm.

c. $\log_3 15 - \log_3 5$ **d.** $\log_4(x^2 + 2) - \log_4(x + 1)$

ANSWERS

a. $1 - \log_9 10$
b. $\log_5 x - \log_5(x + 2)$
c. 1
d. $\log_4 \frac{x^2 + 2}{x + 1}$

EXAMPLE 5 Use the power rule of logarithms to write the expression as a multiple of a logarithm.

a. $\log_4 a^6$

Solution $\log_4 a^6 = 6 \log_4 a$

b. $\log_b \sqrt[4]{x^3}$

Solution $\log_b \sqrt[4]{x^3} = \log_b x^{3/4}$ Write $\sqrt[4]{x^3}$ as $x^{3/4}$.

$= \frac{3}{4} \log_b x$

c. $\log_b \frac{1}{x^3}$

Solution $\log_b \frac{1}{x^3} = \log_b x^{-3}$ Write $\frac{1}{x^3}$ as x^{-3}.

$= -3 \log_b x$

EXAMPLE 6 Use the power rule of logarithms in the form $r \log_b x = \log_b x^r$ to write the expression as a logarithm of a quantity to a power. Leave answers in simplest form without negative or fractional exponents.

a. $5 \log_4 x$

Solution $5 \log_4 x = \log_4 x^5$

b. $-2 \log_b 5$

Solution $-2 \log_b 5 = \log_b 5^{-2}$

$= \log_b \frac{1}{5^2}$ Write 5^{-2} as $\frac{1}{5^2}$.

$= \log_b \frac{1}{25}$ $5^2 = 25$.

c. $\frac{2}{3} \log_7 y$

Solution $\frac{2}{3} \log_7 y = \log_7 y^{2/3}$

$= \log_7 \sqrt[3]{y^2}$ Write $y^{2/3}$ as $\sqrt[3]{y^2}$.

YOUR TURN Use the power rule of logarithms to write the expression as a multiple of a logarithm.

a. $\log_a z^4$ **b.** $\log_a \sqrt[3]{x^2}$

Use the power rule of logarithms to write the expression as a logarithm of a quantity to a power.

c. $5 \log_7 x$ **d.** $-3 \log_b z$

ANSWERS

a. $4 \log_a z$ **b.** $\frac{2}{3} \log_a x$

c. $\log_7 x^5$ **d.** $\log_b \frac{1}{z^3}$

Often it is necessary to use more than one rule to simplify a logarithmic expression.

EXAMPLE 7 Write the expression as a sum or difference of multiples of logarithms.

a. $\log_b \frac{z^3}{yz}$

Solution $\log_b \frac{z^3}{yz} = \log_b z^3 - \log_b yz$ **Use the quotient rule.**

$= 3 \log_b z - (\log_b y + \log_b z)$ **Use the power rule and product rule. Note the use of parentheses.**

$= 3 \log_b z - \log_b y - \log_b z$ **Remove the parentheses.**

b. $\log_3 \sqrt{\frac{a^3}{b}}$

Solution $\log_3 \sqrt{\frac{a^3}{b}} = \log_3 \left(\frac{a^3}{b}\right)^{1/2}$ **Write** $\sqrt{\frac{a^3}{b}}$ **as** $\left(\frac{a^3}{b}\right)^{1/2}$.

$= \frac{1}{2} \log_3 \left(\frac{a^3}{b}\right)$ **Use the power rule.**

$= \frac{1}{2}(\log_3 a^3 - \log_3 b)$ **Use the quotient rule.**

$= \frac{1}{2}(3 \log_3 a - \log_3 b)$ **Use the power rule.**

$= \frac{3}{2} \log_3 a - \frac{1}{2} \log_3 b$ **Distribute** $\frac{1}{2}$.

c. $\log_5 5^2 b^3$

Solution $\log_5 5^2 b^3 = \log_5 5^2 + \log_5 b^3$ **Use the product rule.**

$= 2 \log_5 5 + 3 \log_5 b$ **Use the power rule.**

$= 2 + 3 \log_5 b$ $\log_5 5 = 1$.

EXAMPLE 8 Write the expression as a single logarithm. Leave answers in simplest form without negative or fractional exponents.

a. $4 \log_b 2 - 2 \log_b 3$

Solution $4 \log_b 2 - 2 \log_b 3 = \log_b 2^4 - \log_b 3^2$ **Use the power rule.**

$= \log_b \frac{2^4}{3^2}$ **Use the quotient rule.**

$= \log_b \frac{16}{9}$ **Simplify.**

b. $\frac{1}{2}(\log_2 5 - \log_2 b)$

Solution $\frac{1}{2}(\log_2 5 - \log_2 b) = \frac{1}{2}\log_2 \frac{5}{b}$ **Use the quotient rule.**

$= \log_2\left(\frac{5}{b}\right)^{1/2}$ **Use the power rule.**

$= \log_2 \sqrt{\frac{5}{b}}$ **Write $\left(\frac{5}{b}\right)^{1/2}$ as $\sqrt{\frac{5}{b}}$.**

c. $\log_a(x + 2) + \log_a(x - 3)$

Solution $\log_a(x + 2) + \log_a(x - 3) = \log_a(x + 2)(x - 3)$ **Apply $\log_b xy = \log_b x + \log_b y$.**

$= \log_a(x^2 - x - 6)$ **Multiply.**

YOUR TURN Write the expression as a single logarithm.

a. $2 \log_a 4 - 3 \log_a x$ **b.** $\frac{1}{3}(\log_a x + 2 \log_a y)$ **c.** $\log_6 x + \log_6(x - 2)$

ANSWERS

a. $\log_a \frac{16}{x^3}$ **b.** $\log_a \sqrt[3]{xy^2}$

c. $\log_6(x^2 - 2x)$

10.4 Exercises

1. Why does $b^{\log_b x} = x$?

2. How can you use the facts that $\log_b x^r = r \log_b x$ and $\log_b b = 1$ to show that $\log_3 81 = 4$?

3. $\log_b(x + y) =$ __________.

4. Why does $\log_a \sqrt[3]{x^4} = \frac{4}{3}\log_a x$?

For Exercises 5–16, find the value.

5. $8^{\log_8 2}$
6. $5^{\log_5 6}$
7. $a^{\log_a r}$
8. $b^{\log_b a}$
9. $a^{\log_a 4x}$
10. $b^{\log_b 5a}$
11. $\log_3 3^5$
12. $\log_7 7^3$
13. $\log_b b^a$
14. $\log_c c^x$
15. $\log_a a^{7x}$
16. $\log_b b^{6y}$

For Exercises 17–24, use the product rule to write the expression as a sum of logarithms.

17. $\log_2 5y$ **18.** $\log_3 4z$ **19.** $\log_a pq$ **20.** $\log_b rs$

21. $\log_4 mnp$ **22.** $\log_4 pqr$ **23.** $\log_a x(x - 5)$ **24.** $\log_b x(2x + 6)$

For Exercises 25–36, use the product rule to write the expression as a single logarithm.

25. $\log_3 5 + \log_3 8$ **26.** $\log_6 4 + \log_6 7$ **27.** $\log_4 2 + \log_4 8$

28. $\log_9 3 + \log_9 27$ **29.** $\log_a 7 + \log_a m$ **30.** $\log_b 2 + \log_b n$

31. $\log_4 a + \log_4 b$ **32.** $\log_6 r + \log_6 m$ **33.** $\log_a 2 + \log_a x + \log_a(x + 5)$

34. $\log_a 4 + \log_a y + \log_a (y - 5)$ **35.** $\log_4(x + 1) + \log_4(x + 3)$ **36.** $\log_6(x - 3) + \log_6(x + 3)$

For Exercises 37–46, use the quotient rule to write the expression as a difference of logarithms.

37. $\log_2 \frac{7}{9}$ **38.** $\log_4 \frac{5}{12}$ **39.** $\log_a \frac{x}{5}$ **40.** $\log_b \frac{y}{3}$

41. $\log_a \frac{a}{b}$ **42.** $\log_b \frac{a}{b}$ **43.** $\log_a \frac{x}{x - 3}$ **44.** $\log_b \frac{y}{2y + 5}$

45. $\log_4 \frac{2x - 3}{4x + 5}$ **46.** $\log_6 \frac{4x - 1}{x^2 + 3}$

For Exercises 47–58, use the quotient rule to write the expression as a single logarithm.

47. $\log_6 24 - \log_6 3$ **48.** $\log_8 12 - \log_8 3$ **49.** $\log_2 24 - \log_2 12$

50. $\log_3 48 - \log_3 16$ **51.** $\log_a x - \log_a 3$ **52.** $\log_a r - \log_a 5$

53. $\log_4 p - \log_4 q$ **54.** $\log_a m - \log_a n$ **55.** $\log_b x - \log_b(x - 4)$

56. $\log_b y - \log_b(y - 5)$ **57.** $\log_x(x^2 - x) - \log_x(x - 1)$ **58.** $\log_a(a^2 + 2a) - \log_a(a + 2)$

For Exercises 59–70, use the power rule to write the expression as a multiple of a logarithm.

59. $\log_7 4^3$ **60.** $\log_3 5^4$ **61.** $\log_a x^7$ **62.** $\log_b y^8$

63. $\log_a \sqrt{3}$ **64.** $\log_a \sqrt{6}$ **65.** $\log_3 \sqrt[3]{x^2}$ **66.** $\log_4 \sqrt[5]{a^3}$

67. $\log_a \frac{1}{6^2}$ **68.** $\log_a \frac{1}{5^4}$ **69.** $\log_a \frac{1}{y^2}$ **70.** $\log_a \frac{1}{x^5}$

For Exercises 71–82, use the power rule to write the expression as a logarithm of a quantity to a power. Leave answers in simplest form without negative or fractional exponents.

71. $4 \log_3 5$ **72.** $5 \log_3 4$ **73.** $-3 \log_2 x$ **74.** $-4 \log_3 y$

75. $\frac{1}{2} \log_5 4$ **76.** $\frac{1}{3} \log_3 8$ **77.** $\frac{3}{4} \log_a x$ **78.** $\frac{5}{6} \log_b y$

79. $\frac{2}{3} \log_a 8$ **80.** $\frac{3}{4} \log_a 16$ **81.** $-\frac{1}{2} \log_3 x$ **82.** $-\frac{1}{3} \log_2 y$

For Exercises 83–94, write the expression as the sum or difference of multiples of logarithms.

83. $\log_a \frac{x^3}{y^4}$ **84.** $\log_a \frac{x^6}{y^5}$ **85.** $\log_3 a^4b^2$

86. $\log_7 m^3n^5$ **87.** $\log_a \frac{xy}{z}$ **88.** $\log_a \frac{pq}{r}$

89. $\log_x \frac{a^2}{bc^3}$ **90.** $\log_x \frac{c^2}{m^2n}$ **91.** $\log_4 \sqrt{\frac{x^3}{y}}$

92. $\log_5 \sqrt[3]{\frac{a^5}{b^2}}$ **93.** $\log_a \sqrt{\frac{x^2y}{z^3}}$ **94.** $\log_3 \sqrt[3]{\frac{c^2d^3}{m^4}}$

For Exercises 95–108, write the expression as a single logarithm.

95. $3 \log_3 2 - 2 \log_3 4$ **96.** $2 \log_4 5 - 3 \log_4 2$ **97.** $4 \log_b x + 3 \log_b y$

98. $5 \log_b a + 4 \log_b c$ **99.** $\frac{1}{2}(\log_a 5 - \log_a 7)$ **100.** $\frac{1}{2}(\log_4 6 - \log_4 5)$

101. $\frac{2}{3}(\log_a x^2 + \log_a y)$

102. $\frac{3}{4}(\log_a m^3 + \log_a n^2)$

103. $\log_b x + \log_b(3x - 2)$

104. $\log_a y + \log_a(2y - 4)$

105. $3 \log_a(x - 2) - 4 \log_a(x + 1)$

106. $2 \log_b(x + 4) - 3 \log_b(2x - 1)$

107. $2 \log_a x + 4 \log_a z - 3 \log_a w - 6 \log_a u$

108. $5 \log_a b + 2 \log_a c - 4 \log_a d - \log_a e$

Collaborative Exercises PROVING THE QUOTIENT AND POWER RULES

1. Using the proof of $\log_b xy = \log_b x + \log_b y$ as a guide, prove $\log_b \frac{x}{y} = \log_b x - \log_b y$.

2. The proof that $\log_b x^r = r \log_b x$ was not given. For r a positive integer, the following use of the product rule suggests (but does not prove) this rule:

$$\log_b x^3 = \log_b x \cdot x \cdot x = \log_b x + \log_b x + \log_b x = 3 \log_b x$$

Using the preceding as a guide, prove that $\log_b x^5 = 5 \log_b x$.

3. Complete the following steps to prove $\log_b x^r = r \log_b x$ for all values of r:
Let $N = \log_b x$.
Rewrite in exponential form.
Raise both sides to the r power.
Simplify the exponents, if necessary.
Rewrite in logarithmic form.
Substitute for N.

REVIEW EXERCISES

[5.1] ***For Exercises 1 and 2, simplify.***

1. $(10^{9.5})(10^{-12})$

2. $\frac{10^{-3}}{10^{-12}}$

[10.2] **3.** If \$10,000 is deposited at 6% compounded quarterly for five years, how much will be in the account?

[10.3] **4.** Write $10^{1.6990} = 50$ in logarithmic form.

[10.3] ***For Exercises 5 and 6, write in exponential form.***

5. $\log_{10} 45 = 1.6532$

6. $\log_e 0.25 = -1.3863$

10.5 Common and Natural Logarithms

OBJECTIVES

1. **Define common logarithms and evaluate them using a calculator.**
2. **Solve applications using common logarithms.**
3. **Define natural logarithms and evaluate them using a calculator.**
4. **Solve applications using natural logarithms.**

OBJECTIVE 1. Define common logarithms and evaluate them using a calculator. Of all the possible bases of logarithms, two are the most useful. As previously mentioned, logarithms were invented to do computations on very large and very small numbers. Since ours is a base-10 number system, base-10 logarithms were commonly used for this purpose. Consequently, base-10 logarithms are called **common logarithms** and $\log_{10} x$ is written as $\log x$, where the base 10 is understood. Base-10 logarithms are found in engineering, economics, social sciences, and in the natural sciences.

DEFINITION ***Common logarithms:*** Logarithms with a base of 10. $\text{Log}_{10}\, x$ is written as $\log x$.

Connection Since logarithmic and exponential functions are inverse functions, if $f(x) = 10^x$, then $f^{-1}(x) = \log x$ and if $g(x) = \log x$, then $g^{-1}(x) = 10^x$.

To evaluate common logarithms, we use a calculator. We will round all results to four places.

Calculator TIPS

On most calculators, to evaluate common logarithms, press the LOG *key, enter the number, and then press* ENTER.

EXAMPLE 1 Use a calculator to approximate each common logarithm to four decimal places.

a. $\log 23$

Solution $\log 23 \approx 1.3617$

b. $\log 0.00236$

Solution $\log 0.00236 \approx -2.6271$

Connection Remember that $\log 23 \approx 1.3617$ means $10^{1.3617} \approx 23$, which provides a way to check. Because we rounded the decimal value, evaluating $10^{1.3617}$ may not give exactly 23.

YOUR TURN Use a calculator to approximate each common logarithm to four decimal places.

a. $\log 436$ **b.** $\log 0.0724$

ANSWERS

a. 2.6395 **b.** −1.1403

OBJECTIVE 2. Solve applications using common logarithms. Common logarithms can be used to calculate sound intensity and runway length.

EXAMPLE 2

a. Sound intensity can be measured in watts per unit of area or, more commonly, in decibels. The function $d = 10 \log \frac{I}{I_0}$ is used to calculate sound intensity, where d represents the intensity in decibels, I represents the intensity in watts per unit of area, and I_0 represents the faintest audible sound to the average human ear, which is 10^{12} watts per square meter. A motorcycle has a sound intensity of about $10^{-2.5}$ watts per square meter. Find the decibel reading for the motorcycle.

Understand We are given the function $d = 10 \log \frac{I}{I_0}$, values for I and I_0, and we are to find the decibel reading, which is d.

Plan Using $d = 10 \log \frac{I}{I_0}$, substitute $10^{-2.5}$ for I and 10^{-12} for I_0 and then solve for d.

Execute

$d = 10 \log \frac{10^{-2.5}}{10^{-12}}$ Substitute for I and I_0 in $d = 10 \log \frac{I}{I_0}$.

$d = 10 \log 10^{9.5}$ Subtract exponents $[-2.5 - (-12) = 9.5]$.

$d = 10(9.5)$ Use $\log_b b^x = x$.

$d = 95$ Simplify.

Answer The motorcycle has a decibel reading of 95 decibels.

Note: *The abbreviation for decibels is dB.*

Check If the sound intensity is 95 decibels, verify that I is $10^{-2.5}$.

$95 = 10 \log \frac{I}{10^{-12}}$ Substitute 95 for d.

$9.5 = \log \frac{I}{10^{-12}}$ Divide both sides by 10 to isolate the logarithm.

$10^{9.5} = \frac{I}{10^{-12}}$ Write in exponential form.

$10^{9.5} \cdot 10^{-12} = I$ Multiply both sides by 10^{-12}.

$10^{-2.5} = I$ Simplify by adding the exponents. It checks.

b. The minimum length of airport runway needed for a plane to take off is related to the weight of the plane. For some planes, the minimum runway length may be modeled by the function $y = 3 \log x$, where x is the plane's weight in thousands of pounds and y is the length of the runway in thousands of feet. Find the minimum length of a runway needed by a Boeing 737 whose maximum takeoff weight is 174,200 pounds.

Understand We are given the function $y = 3 \log x$, where y represents the runway length in thousands of feet required for a plane with a takeoff weight of x, in thousands of pounds. We are given that a plane's takeoff weight is 174,200 pounds and must find the runway length required for that plane.

Plan Since x is the plane's weight in thousands of pounds, we first need to divide 174,200 by 1000 to find the value of x corresponding to 174,200 pounds. We can then use $y = 3 \log x$.

Execute $x = \dfrac{174{,}200}{1{,}000} = 174.2$

$y = 3 \log 174.2$ Substitute 174.2 for x in $y = 3 \log x$.

$y = 6.723$ Evaluate using a calculator.

Answer Since y is in thousands of feet, the minimum runway length is $(6.723)(1000) = 6723$ ft.

Check Verify that a runway length of 6723 feet corresponds to a plane whose takeoff weight is 174,200. Since y is in thousands of feet, we first divide 6723 by 1000 to find the value of y: $y = 6723/1000 = 6.723$.

$6.723 = 3 \log x$ Substitute 6.723 for y in $y = 3 \log x$.

$2.241 = \log x$ Divide both sides by 3 to isolate the logarithm.

$10^{2.241} = x$ Write in exponential form.

$174.2 \approx x$ Simplify. We rounded to the nearest tenth.

Since x is in thousands of feet, $174.2(1000) = 174{,}200$ ft., which checks.

Your Turn

a. The sound intensity of a rock band often exceeds $10^{-0.5}$ watts per square meter. Find the decibel rating of this band.

b. Find the minimum runway length needed for a B-52 Stratofortress whose maximum takeoff weight is 488,000 pounds.

OBJECTIVE 3. Define natural logarithms and evaluate them using a calculator. The number e is an irrational number whose approximate value is 2.7182818285. It is a universal constant like π. In the natural sciences, base-e logarithms are much more prevalent than base-10. Since base-e logarithms occur in so many "natural" situations, they are called **natural logarithms**. The notation for $\log_e x$ is $\ln x$, which is read "el en x."

DEFINITION ***Natural logarithms:*** Base-e logarithms are called natural logarithms and $\log_e x$ is written as $\ln x$. Note that $\ln e = 1$.

Calculator TIPS

On most calculators, to evaluate natural logarithms, press the ln *key, enter the number, and then press* ENTER.

To find natural logarithms, we use a calculator.

Example 3 Use a calculator to approximate each natural logarithm to four decimal places.

a. $\ln 83$

Solution $\ln 83 \approx 4.4188$

b. $\ln 0.0055$

Solution $\ln 0.0055 \approx -5.2030$

Connection In Example 3, $\ln 83 \approx 4.4188$ means that $e^{4.4188} \approx 83$. Similarly, $\ln 0.0055 = -5.2030$ means that $e^{-5.2030} \approx 0.0055$.

Answers

a. $d = 115$ **b.** 8065 ft.

YOUR TURN Use a calculator to approximate each natural logarithm to four decimal places.

a. $\ln 102$

b. $\ln 0.0573$

OBJECTIVE 4. Solve applications using natural logarithms. If money is deposited into an account and the interest is compounded continuously, then the time t (in years) that it would take an investment of P dollars to grow into A dollars at an interest rate r (written as a decimal) is given by $t = \frac{1}{r}\ln\frac{A}{P}$.

EXAMPLE 4 An amount of \$5000 is deposited into an account earning 5% annual interest compounded continuously. How many years would it take until the account has reached \$10,000?

Understand We are to find the time it takes for \$5000 to grow to \$10,000 if it is compounded continuously at 5%.

Plan In $t = \frac{1}{r}\ln\frac{A}{P}$, replace P with 5000, r with 0.05, A with 10,000, and then simplify.

Execute

$$t = \frac{1}{0.05}\ln\frac{10{,}000}{5000}$$ **Substitute for P, r, and A.**

$$t = 20\ln 2$$ **Divide**

$$t \approx 13.86$$ **Multiply using a calculator.**

Answer It would take about 13.86 years for \$5000 to grow to \$10,000 if it is compounded continuously.

Check We can use the formula to verify that if \$5000 is compounded continuously for 13.86 years, the amount will be \$10,000.

$$13.86 = \frac{1}{0.05}\ln\frac{A}{5000}$$ **Substitute for t, r, and P.**

$$0.693 = \ln\frac{A}{5000}$$ **Multiply both sides by 0.05.**

$$0.693 = \ln A - \ln 5000$$ **Use $\log_b\frac{x}{y} = \log_b x - \log_b y$.**

$$0.693 + \ln 5000 = \ln A$$ **Add ln 5000 to both sides.**

$$9.21 \approx \ln A$$ **Add 0.693 and the approximate value of ln 5000.**

$$e^{9.21} \approx A$$ **Write in exponential form.**

$$9996.6 \approx A$$ **Calculate using a calculator.**

Because 13.86 is not the exact time and we rounded the sum of 0.693 and ln 5000, it is reasonable to expect that the amount would be very close to but not exactly \$10,000, so \$9996.60 is reasonable.

YOUR TURN How long would it take \$2000 to grow into \$5000 at 8% interest if the interest is compounded continuously?

ANSWERS

a. 4.6250 **b.** −2.8595

ANSWER

11.5 yr.

10.5 Exercises

For Extra Help

MyMathLab  Videotape/DVT  InterAct Math  Math Tutor Center  Math XL.com

1. If $y = \log_b x$, what values are allowed for b? Why?

2. Is $\log x$ positive or negative for $0 < x < 1$? Why?

3. Is $\ln x$ positive or negative for $x > 1$? Why?

4. Give the symbol and meaning for **a.** common logarithms, **b.** natural logarithms.

5. If $e^{0.91629} = 2.5$, find ln 2.5.

6. Without using a calculator, find the exact value of $\log_{10} 10^{\sqrt{5}}$.

For Exercises 7–26, use a calculator to approximate each logarithm to four decimal places.

7. log 64

8. log 82

9. log 0.0067

10. log 0.00087

11. log 247.8

12. log 785.4

13. $\log(1.5 \times 10^4)$

14. $\log(5.7 \times 10^7)$

15. $\log(1.6 \times 10^{-6})$

16. $\log(7.5 \times 10^{-5})$

17. ln 9.34

18. ln 5.33

19. ln 79.2

20. ln 765.4

21. ln 0.034

22. ln 0.923

23. $\ln(5.4 \times e^4)$

24. $\ln(7.3 \times e^6)$

25. $\log e$

26. ln 10

27. Use your calculator to find log 0. What happened? Why?

28. Use your calculator to find ln −1. What happened? Why?

For Exercises 29–40, find the exact value of each logarithm using $\log_b b^x = x$.

29. log 100

30. log 1000

31. $\log \frac{1}{10}$

32. $\log \frac{1}{1000}$

33. $\log \sqrt[3]{10}$

34. $\log \sqrt{10}$

35. $\log 0.001$

36. $\log 0.00001$

37. $\ln e^3$

38. $\ln e^5$

39. $\ln \sqrt{e}$

40. $\ln \sqrt[5]{e}$

For Exercises 41–44, use the formula $d = 10 \log \frac{I}{I_0}$***, where*** $I_0 = 10^{-12}$ **watts/m^2.**

41. The sound intensity of a firecracker is 10^{-3} watts per square meter. What is the decibel reading for the firecracker?

42. The sound intensity of a race car is 10^{-1} watts per square meter. What is the decibel reading for the race car?

43. If a noisy office has a decibel reading of 60, what is the sound intensity?

44. If loud thunder has a decibel reading of 90, what is the sound intensity?

For Exercises 45–48, use the following:

In chemistry, the pH of a substance determines whether it is a base (pH > 7) or an acid (pH < 7). To find the pH of a solution, we use the formula pH $= -\log [H_3O^+]$, where $[H_3O^+]$ is the hydronium ion concentration in moles per liter. Note that the pH is unitless.

45. Find the pH of vinegar if $[H_3O^+] = 1.6 \times 10^{-3}$ moles per liter. (*Source: CRC Handbook of Chemistry and Physics*, 62nd Edition, CRC Press, 1981)

46. Find the pH of maple syrup if $[H_3O^+] = 2.3 \times 10^{-7}$ moles per liter. (*Source: CRC Handbook of Chemistry and Physics*, 62nd Edition, CRC Press, 1981)

47. Find the hydronium ion concentration of sauerkraut, which has a pH of 3.5.

48. Find the hydronium ion concentration of blood, which has a pH of 7.4.

For Exercises 49 and 50, use the formula $t = \frac{1}{r} \ln \frac{A}{P}$.

49. Find how long it will take \$2000 to grow to \$5000 at 4% interest if the interest is compounded continuously.

50. Find how long it will take \$5000 to grow to \$8000 at 5% interest if the interest is compounded continuously.

51. The magnitude of an earthquake is given by the Richter scale, whose formula is $R = \log \frac{I}{I_0}$, where I is the intensity of the earthquake and I_0 is the intensity of a minimal earthquake and is used for comparison purposes. The 1906 San Francisco earthquake had a magnitude of 7.8, and the 1964 Alaska earthquake had a magnitude of 8.4. Compare the intensity of the two earthquakes. (*Hint:* Express the intensity of each in terms of I_0.)

52. Using the formula from Exercise 51, compare the intensity of the 1949 Queen Charlotte Islands earthquake, whose magnitude was 8.1, with the 1700 earthquake at Cascadia, whose magnitude was estimated at 9.0.

Of Interest

The photograph above shows the destruction of San Francisco that occurred as a result of the 1906 earthquake.

53. During an earthquake, energy is released in various forms. The amount of energy radiated from the earthquake as seismic waves is given by $\log E_s = 11.8 + 1.5\text{ M}$, where E_s is measured in ergs and M is the magnitude of the earthquake as given by the Richter scale. Vancouver Island had an earthquake whose magnitude was 7.3. How much energy was released in the form of seismic waves?

54. Using the formula from Exercise 53, find the energy released in the form of seismic waves from the Double Springs Flat earthquake of 1994, whose magnitude was 6.1 on the Richter scale.

55. Since 1995, the number of adult deaths from AIDS in the United States has been declining and can be approximated by $y = 49{,}971.5 - 21{,}298.87 \ln x$, where y is the number of deaths and $x = 1$ corresponds to 1995. (*Source:* CDC Division of HIV/AIDS Prevention, *Surveillance Report,* Vol. 13, No. 2)

a. Estimate the number of adult AIDS deaths in 2000.

b. Find the year when the number of adult deaths by AIDS was 26,572.

56. The number of adult deaths by AIDS in the United States as a percent of the number of diagnosed cases peaked in 1995 at 73.4%. Since then, the percent has declined and can be approximated by $y = (0.7216 - 0.1911 \ln x)100$, where y is the percent of deaths and $x = 1$ corresponds to 1995. (*Data Source:* CDC Division of HIV/AIDS Prevention, *Surveillance Report,* Vol. 13, No. 2)

a. Find the percent of deaths among diagnosed cases in 1999.

b. In what year did 35% of the diagnosed cases of AIDS result in death?

57. Using the formula from Example 2b, $y = 3 \log x$, find the minimum runway length needed for a Boeing 717 whose maximum weight is 110,000 pounds at takeoff.

58. Walking speeds in various cities is a function of the population, since as populations increase, so does the pace of life. Average walking speeds can be modeled by the function $W = 0.35 \ln P + 2.74$, where P is the population, in thousands, and W is the walking speed in feet per second. Find the average walking speed in Chicago, whose population was approximately 8,500,000 in 2002.

REVIEW EXERCISES

For Exercises 1–4, solve.

[2.1] **1.** $3x - (7x + 2) = 12 - 2(x - 4)$

[6.4] **2.** $x^2 + 2x = 15$

[7.4] **3.** $\frac{5}{x} + \frac{3}{x+1} = \frac{23}{3x}$

[8.6] **4.** $\sqrt{5x - 1} = 7$

[10.4] **5.** Write $\log_3(2x + 1) - \log_3(x - 1)$ as a single logarithm.

[10.3] **6.** Write $\log_3(2x + 5) = 2$ in exponential form.

10.6 Exponential and Logarithmic Equations with Applications

OBJECTIVES

1. **Solve equations that have variables as exponents.**
2. **Solve equations containing logarithms.**
3. **Solve applications involving exponential and logarithmic functions.**
4. **Use the change-of-base formula.**

In Section 10.2, we solved equations that could be put in the form $b^x = b^y$. For example, if $3^x = 81$, then $3^x = 3^4$, so $x = 4$. In order to use this form, we had to write both sides of the equation as the same base raised to a power. In this section, we will solve equations like $3^x = 16$, which is the same as $3^x = 2^4$, where the bases are not the same. To solve this and other exponential and logarithmic equations, we need the following properties.

RULE **Properties for Solving Exponential and Logarithmic Equations**

For any real numbers b, x, and y, where $b > 0$ and $b \neq 1$,

1. If $b^x = b^y$, then $x = y$.
2. If $x = y$, then $b^x = b^y$.
3. For $x > 0$ and $y > 0$, if $\log_b x = \log_b y$, then $x = y$.
4. For $x > 0$ and $y > 0$, if $x = y$, then $\log_b x = \log_b y$.
5. For $x > 0$, if $\log_b x = y$, then $b^y = x$.

These properties are true because the exponential and logarithmic functions are one-to-one.

OBJECTIVE 1. Solve equations that have variables as exponents. To solve equations that have variables as exponents, we use property 4, which says that if two positive numbers are equal, then so are their logarithms.

EXAMPLE 1 Solve. $5^x = 16$

Solution $\log 5^x = \log 16$ Use if $x = y$, then $\log_b x = \log_b y$ (property 4).

$x \log 5 = \log 16$ Use $\log_b x^r = r \log_b x$.

$x = \dfrac{\log 16}{\log 5}$ Divide both sides of the equation by log 5.

The exact solution is $x = \dfrac{\log 16}{\log 5}$. Using a calculator, we find $x \approx 1.7227$ correct to four decimal places.

Check $5^x = 16$

$5^{1.7227} = 16$ Substitute 1.7727 for x.

$15.9998 \approx 16$ The answer is correct.

In Example 1, we took the common logarithm of both sides, but we could have used natural logarithms (or any other base) instead. If one side of the equation contains a power of e, natural logs are preferred so that we can use the fact that $\log_e e^x = x$ or, more simply, $\ln e^x = x$.

EXAMPLE 2 Solve. $e^{4x} = 23$

Solution $\ln e^{4x} = \ln 23$ Use if $x = y$, then $\log_b x = \log_b y$.

$4x = \ln 23$ Use $\log_b b^x = x$.

$x = \dfrac{\ln 23}{4}$ Divide both sides by 4.

$x \approx 0.7839$

We will leave the check to the reader.

YOUR TURN Solve the following for x. Round answers to four decimal places.

a. $6^{2x} = 42$ **b.** $e^{3x} = 5$

OBJECTIVE 2. Solve equations containing logarithms. Now that we have explored additional properties of logarithms, we can modify the procedure for solving equations with logarithms that we learned in Section 10.3.

ANSWERS

a. 1.0430 **b.** 0.5365

PROCEDURE **Solving Equations Containing Logarithms**

To solve equations containing logarithms, use the properties of logarithms to simplify each side of the equation and then use one of the following.

If the simplification results in an equation in the form $\log_b x = \log_b y$, use the fact that $x = y$, and then solve for the variable.

If the simplification results in an equation in the form $\log_b x = y$, write the equation in exponential form, $b^y = x$, and then solve for the variable (as we did in Section 10.3).

EXAMPLE 3 Solve.

a. $\log_5 x + \log_5(x + 3) = \log_5 4$

Solution $\log_5 x(x + 3) = \log_5 4$ — Use $\log_b xy = \log_b x + \log_b y$ to simplify the left side.

$x(x + 3) = 4$ — The equation is in the form $\log_b x = \log_b y$, so $x = y$.

$x^2 + 3x = 4$ — Simplify.

$x^2 + 3x - 4 = 0$ — Write in $ax^2 + bx + c = 0$ form.

$(x + 4)(x - 1) = 0$ — Factor.

$x + 4 = 0$ or $x - 1 = 0$ — Set each factor equal to 0.

$x = -4$ or $x = 1$ — Possible solutions.

If -4 is substituted into the original equation, we have $\log_5(-4) + \log_5(-1) = \log_5 4$, but logarithms are defined for positive numbers only. So $x = -4$ is not a solution. A check will show that $x = 1$ is a solution.

b. $\log_3(2x + 5) = 2$

Solution $3^2 = 2x + 5$ — The equation is in the form $\log_b x = y$, so write it in exponential form, $b^y = x$.

$9 = 2x + 5$

$4 = 2x$

$2 = x$ — Solve for x.

We will let the reader check that $x = 2$ is a solution.

c. $\log_2(5x + 1) - \log_2(x - 1) = 3$

Solution $\log_2 \dfrac{5x + 1}{x - 1} = 3$ — Use $\log_b \dfrac{x}{y} = \log_b x - \log_b y$ to simplify the left side.

$\dfrac{5x + 1}{x - 1} = 2^3$ — The equation is in the form $\log_b x = y$, so $b^y = x$.

$5x + 1 = 8x - 8$ — Multiply both sides by $x - 1$.

$9 = 3x$ — Isolate the x term.

$x = 3$ — Solve for x.

We will let the reader check that $x = 3$ is a solution.

YOUR TURN Solve.

a. $\ln x + \ln(x + 2) = \ln 8$

b. $\log_3(4x + 2) - \log_3(x - 2) = 2$

ANSWERS

a. 2 **b.** 4

OBJECTIVE 3. Solve applications involving exponential and logarithmic functions. A wide variety of problems from business and the sciences can be solved with exponential or logarithmic functions. Earlier, we used the formula for compound interest, $A = P\left(1 + \frac{r}{n}\right)^{nt}$, to find A when given P, r, n, and t. Using logarithms, it is also possible to find t when given A, P, r, and n.

EXAMPLE 4 How long will it take \$6000 invested at 6% interest compounded quarterly to grow to \$10,000?

Understand We are given $A = 10{,}000$, $P = 6000$, $r = 0.06$, and $n = 4$.

Plan Use $A = P\left(1 + \frac{r}{n}\right)^{nt}$.

Execute $10{,}000 = 6000\left(1 + \frac{.06}{4}\right)^{4t}$ Substitute for A, P, r, and n.

$\frac{5}{3} = (1.015)^{4t}$ Divide both sides by 6000 and simplify inside the parentheses.

$\log\frac{5}{3} = \log 1.015^{4t}$ Use if $x = y$, then $\log_b x = \log_b y$.

$\log\frac{5}{3} = 4t \log 1.015$ Use $\log_b x^r = r\log_b x$.

$\frac{\log\frac{5}{3}}{4\log 1.015} = t$ Divide both sides by 4 log 1.015

$8.58 \approx t$ Evaluate using a calculator.

Answer It would take about 8.58 years.

Check We can use the formula to verify that if \$6000 is compounded quarterly for 8.58 years, the amount will be \$10,000.

$A = 6000\left(1 + \frac{0.06}{4}\right)^{4(8.58)}$ Substitute for P, r, n, and t.

$A = 10{,}001.52$ Calculate.

Because 8.58 is not the exact time, it is reasonable to expect that the amount would be very close to, but not exactly, \$10,000, so \$10,001.52 is reasonable.

Many banks compound interest continuously. The formula for interest compounded continuously is $A = Pe^{rt}$, where A is the amount in the account, P is the amount deposited, r is the interest rate as a decimal, and t is the time in years.

EXAMPLE 5 If \$5000 is deposited into an account at 5% interest compounded continuously, how much will be in the account after 9 years?

Understand We are given $P = \$5000$, $r = 0.05$, and $t = 9$ and are asked to find A.

Plan Use $A = Pe^{rt}$.

Execute $A = 5000e^{0.05(9)}$ Substitute for P, r, and t.

$A = 5000e^{0.45}$

$A = \$7841.56$ Evaluate using a calculator.

Answer There will be \$7841.56 in the account.

Check Use the formula $t = \frac{1}{r} \ln \frac{A}{P}$ that we learned in Example 4 of Section 10.5 to verify that it takes 9 years for \$5000 to grow to \$7841.56 if it is compounded continuously at 5%.

$$t = \frac{1}{0.05} \ln \frac{7841.56}{5000} \qquad \textbf{Substitute for } r, A, \textbf{ and } P.$$

$$t \approx 9 \qquad \textbf{Calculate. It checks.}$$

Connection Solving $A = Pe^{rt}$ for t gives the formula $t = \frac{1}{r} \ln \frac{A}{P}$.

$$\frac{A}{P} = e^{rt} \qquad \textbf{Divide both sides by } P \textbf{ to isolate } e^{rt}.$$

$$\log_e \frac{A}{P} = rt \qquad \textbf{Write in log form.}$$

$$\frac{1}{r} \ln \frac{A}{P} = t \qquad \textbf{Divide both sides by } r \textbf{ and by definition, } \log_e \frac{A}{P} \textbf{ is } \ln \frac{A}{P}.$$

EXAMPLE 6 Since the 1970s, Orlando, Florida, has been one of the fastest-growing cities in the United States. The following table shows the population of Orlando for selected years.

Year	Population in millions
1975	0.6896
1980	0.805
1985	0.996
1990	1.225
1995	1.428
2000	1.645
2003	1.803

The data can be approximated by $y = 0.6965e^{0.0344t}$, where y is the population in millions and t is the number of years after 1975. (*Source:* U.S. Bureau of the Census)

a. Assuming the population continues to grow in the same manner, use the model to estimate the population of Orlando in 2010.

Understand We are given the function $y = 0.6965e^{0.0344t}$, where y represents the population of Orlando t years after 1975.

Plan Since t represents the number of years after 1975, we first subtract 1975 from 2010 to determine the value of t that corresponds to 2010. We can then substitute that value into the function.

Execute $t = 2010 - 1975 = 35$.

$$y = 0.6965e^{0.0344t} \qquad \textbf{Formula.}$$

$$y = 0.6965e^{0.0344(35)} \qquad \textbf{Substitute 35 for } t.$$

$$y = 2.322 \qquad \textbf{Evaluate using a calculator.}$$

Answer Since y is in millions, the estimated population is 2,322,000 in 2010.

Check Use $y = 0.6965e^{0.0344t}$ to verify that if the population is 2,322,000 ($y = 2.322$), then the year is 2010 ($t = 35$).

$$2.322 = 0.6965e^{0.0344t} \quad \text{Substitute 2.322 for } y.$$

$$\frac{2.322}{0.6965} = e^{0.0344t} \quad \text{Divide both sides by 0.6965.}$$

$$\ln\frac{2.322}{0.6965} = \ln e^{0.0344t} \quad \text{Use if } x = y, \text{ then } \log_b x = \log_b y.$$

$$\ln\frac{2.322}{0.6965} = 0.0344t \quad \text{Use } \log_b b^x = x.$$

$$\frac{\ln\frac{2.322}{0.6965}}{0.0344} = t \quad \text{Divide both sides by 0.0344.}$$

$$35 \approx t \quad \text{Calculate.}$$

Note: *Our manipulations suggest the following formula for calculating t given y:*

$$t = \frac{1}{0.0344}\ln\frac{y}{0.6965}$$

b. Find the year in which the population will be 2,500,000.

Understand We are to find t given the population.

Plan Since y represents the population in millions, we must first divide 2,500,000 by 1,000,000 to find the value of y that corresponds to 2,500,000. We can then use $y = 0.6965e^{0.0344t}$ and follow the same steps as in our check for part a. Or, as we will do here, use the formula $t = \frac{1}{0.0344}\ln\frac{y}{0.6965}$ that we discovered as a result of checking part a.

Execute $y = \frac{2{,}500{,}000}{1{,}000{,}000} = 2.5$ — Divide 2,500,000 by 1,000,000 to find the value of y that corresponds to 2,500,000.

$$t = \frac{1}{0.0344}\ln\frac{2.5}{0.6965} \approx 37.2$$

Substitute 2.5 for y in $t = \frac{1}{0.0344}\ln\frac{y}{0.6965}$ and then calculate.

Answer Since t represents the number of years after 1975, the population will reach 2,500,000 in $1975 + 37.2 = 2012.2$, which means during the year 2012.

Check Use $y = 0.6965e^{0.0344t}$ to verify that in the year 2012 ($t = 37.2$), the population will be 2,500,000 ($y = 2.5$). We will leave this check to the reader.

YOUR TURN

a. How long will it take $8000 invested at 4% annual interest compounded semiannually to grow to $12,000?

b. Between April 2, 2003, and April 8, 2003, the reported number of SARS (severe acute respiratory syndrome) cases was increasing at a rate of 3.1% per day. As of April 2, there were 2671 reported cases of SARS. If the reported number of SARS cases continues to increase at this rate, the function $A = 2671e^{0.031t}$ models the spread of SARS, where A represents the number of cases and t is the number of days after April 2. How many cases would there be on April 10?

c. How many days after April 2 would it take for there to be 8000 cases?

ANSWERS

a. 10.24 yr.
b. 3423 cases
c. Approximately 35.39 days.

Exponential growth and decay can be represented by the equation $A = A_0e^{kt}$, where A is the amount present, A_0 is the initial amount, t is the time, and k is a constant that is determined by the substance. There is exponential growth indicated if $k > 0$ and exponential decay if $k < 0$. Recall that the half-life of a substance is the amount of time until only one-half the original amount is present.

Plutonium-239 is frequently used as fuel in nuclear reactors to generate electricity. One of the problems with using plutonium is disposing of the radioactive waste, which is extremely dangerous for a very long period of time.

EXAMPLE 7 A nuclear reactor contains 10 kilograms of radioactive plutonium ^{239}P. Plutonium disintegrates according to the formula $A = A_0e^{-0.0000284t}$.

a. How much will remain after 10,000 years?

Understand We are given $A_0 = 10$ kg and $t = 10{,}000$ years and asked to find A.

Plan Use $A = A_0e^{-0.0000284t}$

Execute

$$A = 10e^{-0.0000284(10{,}000)} \quad \text{Substitute 10 for } A_0 \text{ and 10,000 for } t.$$

$$A = 10e^{-0.284} \quad \text{Simplify.}$$

$$A \approx 7.53 \quad \text{Evaluate using a calculator.}$$

Answer About 7.53 kilograms (or about $\frac{3}{4}$ of the original amount) will remain after 10,000 years.

Check Use $A = A_0e^{-0.0000284t}$ to verify that it takes 10,000 years for 10 kilograms of ^{239}P to disintegrate to 7.53 kg.

$$7.53 = 10e^{-0.0000284t} \quad \text{Substitute 7.53 for } A \text{ and 10 for } A_0.$$

$$\frac{7.53}{10} = e^{-0.0000284t} \quad \text{Divide both sides by 10.}$$

$$\ln\frac{7.53}{10} = \ln e^{-0.0000284t} \quad \text{Use if } x = y\text{, then } \log_b x = \log_b y.$$

$$\ln\frac{7.53}{10} = -0.0000284t \quad \text{Use } \log_b b^x = x.$$

$$\frac{\ln\frac{7.53}{10}}{-0.0000284} = t \quad \text{Divide both sides by } -0.0000284.$$

$$9989.1 \approx t \quad \text{Calculate.}$$

Note: *Our manipulations suggest the following formula for calculating t given A and A_0:*

$$t = \frac{1}{-0.0000284}\ln\frac{A}{A_0}$$

Because 7.53 is not the exact amount remaining, we should expect that the corresponding time calculation would not be exactly 10,000 years, so 9989.1 years is reasonable.

b. Find the half-life of ^{239}P.

Understand After one half-life, 5 kg will remain, so we must find t when $A = 5$.

Plan Use $A = A_0e^{-0.0000284t}$ when $A = 5$ and $A_0 = 10$ and follow the same steps as in our check of part a. Or, as we will do, use the formula $t = \frac{1}{-0.0000284}\ln\frac{A}{A_0}$, which we discovered as a result of checking part a.

Execute $t = \frac{1}{-0.0000284}\ln\frac{5}{10} \approx 24{,}406.59$ Substitute 5 for A and 10 for A_0 and then calculate.

Answer The half-life of ^{239}P is about 24,400 years.

Check Use $A = A_0e^{-0.0000284t}$ to verify that in 24,406.59 years, 5 kg out of an initial 10 kg of ^{239}P will remain. We will leave this check to the reader.

YOUR TURN Carbon-14 is a radioactive form of carbon that is present in all living things. Archaeologists and paleontologists frequently use carbon-14 dating in estimating the age of organic fossils. Carbon-14 disintegrates according to the formula $A = A_0e^{-0.000121t}$.

a. If a sample contains 5 grams of carbon-14, how much will be present after 1500 years?

b. What is the half-life of carbon-14?

ANSWERS

a. 4.17 g
b. About 5700 yr.

OBJECTIVE 4. Use the change-of-base formula. Sometimes applications involve logarithms other than common or natural logarithms. For example, earlier we were given the formula $P = 95 - 30\log_2 x$, where P is the percent of students who recall the important features of a lecture after x days. To find the percent after 5 days, we need to calculate $\log_2 5$. Most calculators have only base-10 and base-e logarithms, so to calculate $\log_2 5$ using a calculator, we need to write $\log_2 5$ in terms of common or natural logarithms using the change-of-base formula. To derive the change-of-base formula, we let $y = \log_a x$.

$$a^y = x \quad \text{Write } y = \log_a x \text{ in exponential form.}$$

$$\log_b a^y = \log_b x \quad \text{Take } \log_b \text{ of both sides.}$$

$$y\log_b a = \log_b x \quad \text{Use } \log_b x^r = r\log_b x.$$

$$y = \frac{\log_b x}{\log_b a} \quad \text{Divide both sides by } \log_b a.$$

$$\log_a x = \frac{\log_b x}{\log_b a} \quad \text{Substitute } \log_a x \text{ for } y.$$

RULE **Change-of-Base Formula**

In general, if $a > 0$, $a \neq 1$, $b > 0$, $b \neq 1$, and $x > 0$, then $\log_a x = \frac{\log_b x}{\log_b a}$.

In terms of common and natural logarithms, $\log_a x = \frac{\log x}{\log a} = \frac{\ln x}{\ln a}$.

EXAMPLE 8 Use the change-of-base formula to calculate $\log_5 19$. Round the answer to four decimal places.

Note: *We could have used* ln *rather than* log.

$$\log_5 19 = \frac{\ln 19}{\ln 5} \approx 1.8295$$

Solution $\log_5 19 = \frac{\log 19}{\log 5} \approx 1.8295$ Use $\log_a x = \frac{\log_b x}{\log_b a}$, then evaluate using a calculator.

Check $5^{1.8295} = 19.0005 \approx 19$, so the answer is correct.

EXAMPLE 9 Use $P = 95 - 30 \log_2 x$ and the change-of-base formula to find the percent of students who retained the main points of a lecture after 5 days.

Solution $P = 95 - 30 \log_2 5$ — Substitute 5 for x in $P = 95 - 30 \log_2 x$.

$P = 95 - 30\dfrac{\ln 5}{\ln 2}$ — Use $\log_a x = \dfrac{\log_b x}{\log_b a}$.

$P \approx 95 - 30(2.3219)$ — Evaluate $\dfrac{\ln 5}{\ln 2}$ using a calculator.

$P \approx 25.34$ — Simplify.

About 25% of the students remember the main points of a lecture 5 days later.

YOUR TURN

a. Find $\log_6 25$.

b. Use the formula from Example 9 to find the percent of students who retain the main points of a lecture 8 days later.

ANSWERS

a. 1.7965 **b.** 5%

10.6 Exercises

For Extra Help

MyMathLab Videotape/DVT InterAct Math Math Tutor Center Math XL.com

1. If $\log_a m = \log_a n$, then ______.

2. When solving $10^{x+2} = 45$, would natural or common logarithms be the better choice? Why?

3. What principle is used to solve $\log(x - 3) = \log(3x - 13)$?

4. What principle is used to solve $100 = (5)^{2n}$?

5. In solving $\log x + \log(x + 2) = \log 15$, we get possible solutions of $x = 3$ and $x = -5$. Why must $x = -5$ be rejected?

6. Suppose you were solving an equation that contained $\log(5 - x)$. You did all the algebraic work correctly and you obtained $x = -3$ as an apparent solution. Would you have to reject $x = -3$ as a solution? Why or why not?

For Exercises 7–18, solve. Round answers to four decimal places.

7. $2^x = 9$ **8.** $3^x = 20$ **9.** $5^{2x} = 32$ **10.** $4^{3x} = 13$

11. $5^{x+3} = 10$ **12.** $7^{x+1} = 41$ **13.** $8^{x-2} = 6$ **14.** $5^{x-3} = 12$

15. $4^{x+2} = 5^x$ **16.** $6^{x+4} = 10^x$ **17.** $2^{x+1} = 3^{x-2}$ **18.** $5^{x-3} = 3^{x+1}$

For Exercises 19–26, solve. Round answers to four decimal places.

19. $e^{3x} = 5$ **20.** $e^{2x} = 7$ **21.** $e^{0.03x} = 25$ **22.** $e^{0.002x} = 12$

23. $e^{-0.022x} = 5$ **24.** $e^{-0.0032x} = 8$ **25.** $\ln e^{4x} = 24$ **26.** $\ln e^{5x} = 35$

For Exercises 27–60, solve. Give exact answers.

27. $\log_2(x - 3) = 3$ **28.** $\log_3(x + 4) = 2$ **29.** $\log_4(4x - 8) = 2$ **30.** $\log_3(3x + 9) = 3$

31. $\log_4 x^2 = 2$ **32.** $\log_2 x^2 = 6$ **33.** $\log_6(x^2 + 5x) = 2$ **34.** $\log_4(x^2 + 6x) = 2$

35. $\log(3x - 2) = \log(2x + 5)$ **36.** $\log(4x + 1) = \log(2x + 7)$ **37.** $\ln(4x + 6) = \ln(2x - 8)$

38. $\ln(5x + 6) = \ln(3x - 8)$ **39.** $\log_9(x^2 + 4x) = \log_9 12$ **40.** $\log_8(x^2 + x) = \log_8 30$

41. $\log_4 x + \log_4 8 = 2$ **42.** $\log_6 x + \log_6 4 = 2$ **43.** $\log_5 x - \log_5 2 = 1$

44. $\log_5 x - \log_5 3 = 2$ **45.** $\log_3 x + \log_3(x + 6) = 3$ **46.** $\log_2 x + \log_2(x - 3) = 2$

47. $\log_3(2x + 15) + \log_3 x = 3$ **48.** $\log_2(3x - 2) + \log_2 x = 4$ **49.** $\log_2(7x + 3) - \log_2(2x - 3) = 3$

50. $\log_3(3x + 3) - \log_3(x - 3) = 2$ **51.** $\log(x + 2) - \log(2x - 2) = 0$ **52.** $\log_3(2x + 1) - \log_3(x - 1) = 1$

53. $\log_8 2x + \log_8 6 = \log_8 10$ **54.** $\log_5 3x + \log_5 2 = \log_5 17$ **55.** $\ln x + \ln(2x - 1) = \ln 10$

56. $\ln x + \ln(3x - 5) = \ln 12$ **57.** $\log x - \log(x - 5) = \log 6$ **58.** $\log x - \log(x - 2) = \log 3$

59. $\log_6(3x + 4) - \log_6(x - 2) = \log_6 8$ **60.** $\log_7(5x + 2) - \log_7(x - 2) = \log_7 9$

For Exercises 61–80, solve.

61. How long will it take \$5000 invested at 5% compounded quarterly to grow to \$8000? Round your answer to the nearest tenth of a year.

62. How long will it take \$7000 invested at 6% compounded monthly to grow to \$12,000? Round your answer to the nearest tenth of a year.

63. If \$8000 is deposited into an account at 6% annual interest compounded continuously:

a. How much money will be in the account after 15 years?

b. How long will it take the \$8000 to grow into \$14,000?

64. If \$4000 is deposited into an account at 5% annual interest compounded continuously:

a. How much money will be in the account after 10 years?

b. How long will it take the \$4000 to grow into \$10,000?

65. The probability that a person will have an accident while driving at a given blood alcohol level is approximated by $P(b) = e^{21.5b}$, where b is the blood alcohol level ($0 \leq b \leq 0.4$) and P is the percent probability of having an accident.

a. What is the probability of an accident if the blood alcohol level is 0.08, which is legally drunk in many states?

b. Estimate the blood alcohol level when the probability of an accident is 50%.

66. Atmospheric pressure (in pounds per square inch, psi) is a function of the altitude above sea level and can be modeled by $P(a) = 14.7e^{-0.21a}$, where P is the pressure and a is the altitude above sea level in miles.

a. Find the atmospheric pressure at the peak of Mt. McKinley, Alaska, which is 3.85 miles above sea level.

b. If the atmospheric pressure at the peak of Mt. Everest in Nepal, is 4.68 pounds per square inch, find the height of Mt. Everest.

67. The population of a mosquito colony increases at a rate of 4% per day. If the initial number of mosquitoes is 500, the number present, A, at the end of t days is given by $A = 500e^{0.04t}$.

a. How many mosquitoes are present after 2 weeks?

b. Find the number of days until 10,000 mosquitoes are present.

68. A cake is removed from the oven at a temperature of 210°F and is left to cool on a counter where the room temperature is 70°F. The cake cools according to the function $T = 70 + 140e^{-0.0231t}$, where T is the temperature of the cake and t is in minutes.

a. What is the temperature of the cake after 40 minutes?

b. After how many minutes will the cake's temperature be 75°F?

69. The world population in 2005 was about 6.4 billion and is increasing at a rate of 1% per year. The world population after 2005 can be approximated by the equation $A = A_0e^{0.01t}$, where A_0 is the world population in 2005 and t is the number of years after 2005.

 a. Assuming the growth rate follows the same trend, find the world population in 2020.

 b. In what year will the world population reach 7 billion?

70. In 2004, Africa had a population of 885 million and a natural growth rate of 2.4% (2.4 times the world's growth rate). The population growth can be approximated by $A = A_0e^{0.024t}$, where A_0 is the population in millions in 2004 and t is the number of years after 2004.

 a. Excluding immigration, what would the population of Africa be in 2020?

 b. Excluding immigration, in what year would Africa's population reach 1 billion?

71. The barometric pressure x miles from the eye of a hurricane is approximated by the function $P = 0.48 \ln(x + 1)$, where P is inches of mercury. (*Source:* A. Miller and R. Anthes, *Meteorology*, 4th edition, Merrill Publishing)

 a. Find the barometric pressure 50 miles from the center of the hurricane.

 b. Find the distance from the center where the pressure is 1.5 inches of mercury.

72. The first two-year college, Joliet Junior College in Chicago, was founded in 1901. The number of two-year colleges in the United States grew rapidly, especially during the 1960s, but growth has tapered off. The total number of two-year colleges in the United States since 1960 can be approximated by the function $y = 175.6 \ln x + 513$, where x is the number of years after 1960 and y is the number of two-year colleges. Use the model to estimate the number of two-year colleges in the United States in 2010. (*Source:* American Association of Two-Year Colleges)

73. The percent, $f(x)$, of adult height attained by a boy who is x years old is modeled by $f(x) = 29 + 48.8 \log(x + 1)$, where $5 \le x \le 15$.

 a. At age 10, about what percent of his adult height has a boy reached?

 b. At what age will a boy attain 75% of his adult height?

74. The annual depreciation rate r of a car purchased for P dollars and worth A dollars after t years can be found by the formula $\log(1 - r) = \frac{1}{t}\log\frac{A}{P}$. Find the depreciation rate of a car that is purchased for \$22,500 and is sold 3 years later for \$10,000.

75. The following table shows the number of registered cocaine users in the United Kingdom from 1987 to 1995.

Year	Number of Users
1987	431
1988	462
1989	527
1990	633
1991	882
1992	1131
1993	1375
1994	1636
1995	1809

The data can be approximated by the exponential function $y = 387.7(1.222)^t$, where y is the number of registered users and t is the number of years after 1987. If the current trend continues, use the model to estimate the number of users in 2010.

76. The number of females working in automotive repair has been increasing, as shown in the following table.

Year	Number of Female Technicians
1988	556
1989	614
1990	654
1991	737
1992	849
1993	1086
1994	1329
1995	1592

The data can be approximated by $y = 507(1.116)^t$, where y is the number of technicians and t is the number of years after 1988. If the trend continues in the same manner, use the equation to estimate the number of female technicians in 2012. (*Source:* National Institute for Automotive Service Excellence)

77. Housing prices in Volusia County, Florida, have risen exponentially. The average sales price of existing houses sold in Volusia County can be approximated by $S(t) = 40.2(1.1)^t$, where t is the number of years after 1990 and S is the price in thousands of dollars.

a. Estimate the average price of existing homes sold in 2005.

b. Find the number of years after 1990 when the average price of an existing home will be \$250,000.

78. Since the first Super Bowl in 1967, ticket prices have risen exponentially according to the formula $P(t) = 9.046(1.114)^t$, where $t = 1$ is the year 1967 and ticket prices are in dollars. (*Source: Orlando Sentinel,* January 21, 2003, p. A8)

a. Estimate the price of a Super Bowl ticket in 2012.

b. Find the year when Super Bowl prices were \$230.

79. The thickness of a runway's pavement is determined by the weight of the planes that will be using it. The relationship between runway thickness, x, in inches and gross airplane weight, y, in thousands of pounds can be approximated by $y = 18.29(1.279)^x$. If a partially loaded B-52 Stratofortress weighs 350,000 pounds, how thick should a runway be for it to safely use the runway? (*Source:* Federal Aviation Administration)

80. The gap between available organs for liver transplants and people who need them has widened in recent years. The equation $y = 2329(1.2406)^x$, where y is the number of persons waiting and x is the number of years after 1988, approximated the gap. Assuming the same trend continues, use the equation to estimate the year in which the number of people waiting for a transplant is 200,000.

For Exercises 81–88, use the change-of-base formula to find the logarithms. Round answers to four decimal places.

81. $\log_4 12$

82. $\log_5 23$

83. $\log_8 3$

84. $\log_9 5$

85. $\log_{1/2} 5$

86. $\log_{1/3} 4$

87. $\log_{1/4} \frac{3}{5}$

88. $\log_{1/3} \frac{4}{7}$

REVIEW EXERCISES

For Exercises 1 and 2, simplify.

[5.2] **1.** $(3x + 2) - (2x - 1)$

[8.4] **2.** $(3\sqrt{5} + 2)(2\sqrt{5} - 1)$

For Exercises 3–6, use $f(x) = x^2 + 4$ and $g(x) = 2x - 1$.

[5.2] **3.** Find $f(-5)$.

[5.2] **4.** Find $f + g$.

[10.1] **5.** Find $(f \circ g)$

[10.1] **6.** Find $(f \circ g)(2)$.

Chapter 10 Summary

Defined Terms

Section 10.1
Composition of functions (p. 704)
Inverse functions (p. 706)
One-to-one function (p. 708)

Section 10.2
Exponential function (p. 719)

Section 10.3
Logarithm (p. 731)

Section 10.5
Common logarithms (p. 749)
Natural logarithms (p. 751)

Procedures, Rules, and Key Examples

Procedures/Rules	Key Examples
Section 10.1 Composite and Inverse Functions	
Composition of Functions $(f \circ g)(x) = f[g(x)]$ for all x in the domain of g such that $g(x)$ is in the domain of f. $(g \circ f)(x) = g[f(x)]$ for all x in the domain of f such that $f(x)$ is in the domain of g.	**Example 1:** If $f(x) = x^2 - 3$ and $g(x) = 3x + 4$, find (a) $(f \circ g)(x)$ and (b) $(g \circ f)(x)$. **a.** $(f \circ g)(x) = f[g(x)]$ $= f(3x + 4)$ **Substitute $3x + 4$ for $g(x)$.** $= (3x + 4)^2 - 3$ **Replace x with $3x + 4$ in $f(x)$.** $= 9x^2 + 24x + 13$ **Square $3x + 4$ and add like terms.** **b.** $(g \circ f)(x) = g[f(x)]$ $= g(x^2 - 3)$ **Substitute $x^2 - 3$ for $f(x)$.** $= 3(x^2 - 3) + 4$ **Replace x with $x^2 - 3$ in $g(x)$.** $= 3x^2 - 5$ **Distribute 3 and add.**
Inverse Functions To determine whether two functions f and g are inverses of each other, **1.** Show that $f[g(x)] = x$ for all x in the domain of g. **2.** Show that $g[f(x)] = x$ for all x in the domain of f.	**Example 2:** Show that $f(x) = 5x - 6$ and $g(x) = \dfrac{x+6}{5}$ are inverse functions. $f[g(x)] = f\left(\dfrac{x+6}{5}\right) = 5\left(\dfrac{x+6}{5}\right) - 6 = x + 6 - 6 = x$ $g[f(x)] = g(5x - 6) = \dfrac{5x - 6 + 6}{5} = \dfrac{5x}{5} = x$ Since $f[g(x)] = g[f(x)] = x$, f and g are inverse functions. *continued*

Procedures/Rules	Key Examples

Section 10.1 Composite and Inverse Functions (continued)

Horizontal Line Test for One-to-One Functions

Given a function's graph, the function is one-to-one if every horizontal line that can intersect the graph does so at one and only one point.

Note: Only one-to-one functions have inverses that are functions.

Example 3: Determine whether the functions whose graphs follow are one-to-one.

a.

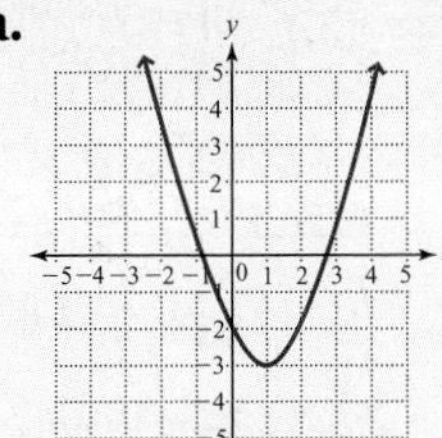

b.

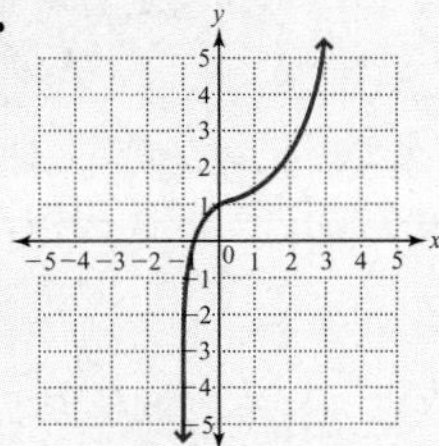

Solution: A horizontal line can intersect this graph in more than one point, so the function is not one-to-one.

Solution: Every horizontal line that can intersect this graph does so at one and only one point, so the function is one-to-one.

Finding the Inverse of a One-to-One Function

1. If necessary, replace $f(x)$ with y.
2. Replace all x's with y's and all y's with x's.
3. Solve the equation from step 2 for y.
4. Replace y with $f^{-1}(x)$.

Example 4: If $f(x) = 3x - 7$, find $f^{-1}(x)$.

Solution:

$$f(x) = 3x - 7$$

$$y = 3x - 7 \quad \text{Replace } f(x) \text{ with } y.$$

$$x = 3y - 7 \quad \text{Interchange } x \text{ and } y.$$

$$\frac{x+7}{3} = y \quad \text{Solve for } y.$$

$$f^{-1}(x) = \frac{x+7}{3} \quad \text{Replace } y \text{ with } f^{-1}(x).$$

Verify by showing that $f[f^{-1}(x)] = x$ and $f^{-1}[f(x)] = x$.

Graphs of Inverse Functions

The graphs of f and f^{-1} are symmetric with respect to the graph of $y = x$.

Example 5: Sketch the inverse of the function whose graph follows.

Solution: Draw the line $y = x$ and reflect the graph in the line.

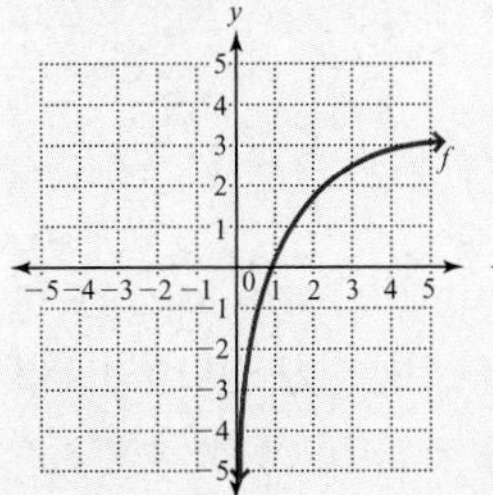

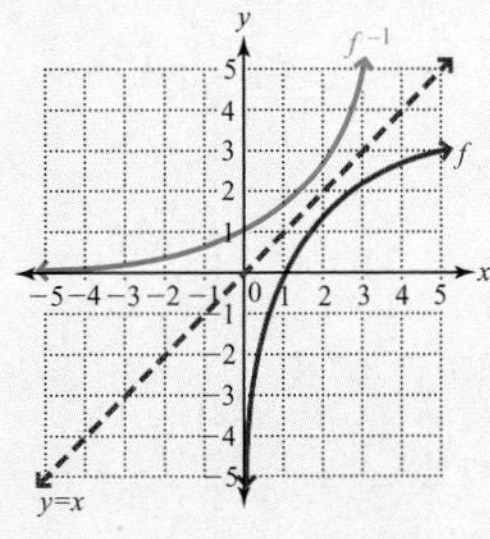

continued

Procedures/Rules	Key Examples

Section 10.2 Exponential Functions

Graphs of Exponential Functions

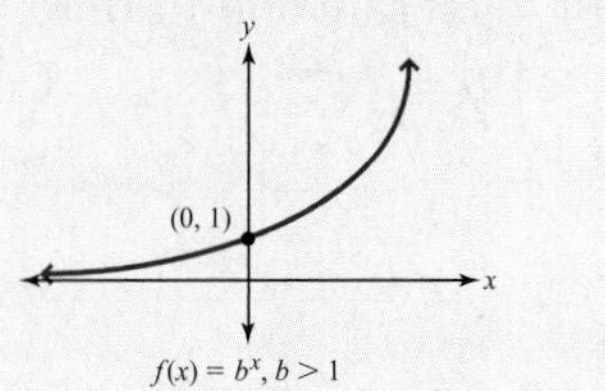

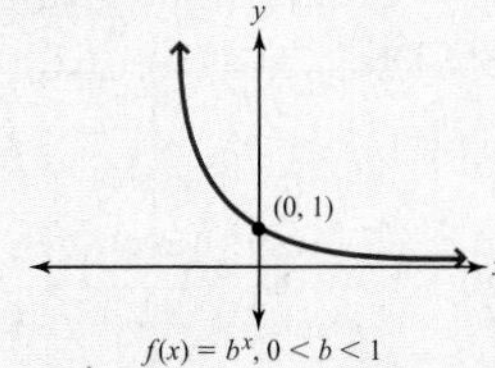

Example 1: Graph $f(x) = 2^x$.

Choose some values for x and find the corresponding values of $f(x)$, which are the y-values.

x	−2	−1	0	1	2
f(x)	$\frac{1}{4}$	$\frac{1}{2}$	1	2	4

Plot the points and draw the graph.

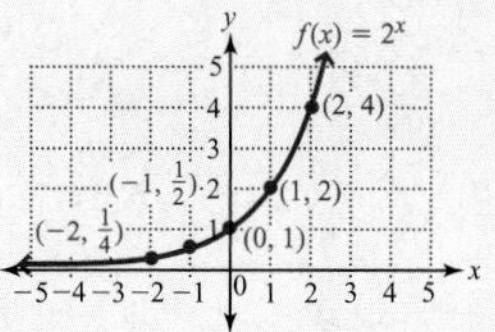

The One-to-One Property of Exponentials

Given $b > 0$ and $b \neq 1$, if $b^x = b^y$, then $x = y$.

Solving Exponential Equations

1. If necessary, write both sides as a power of the same base.
2. If necessary, simplify the exponents.
3. Set the exponents equal to each other.
4. Solve the resulting equation.

Example 2: Solve $8^{x-3} = 16^{2x}$.

Solution: $8^{x-3} = 16^{2x}$

$(2^3)^{x-3} = (2^4)^{2x}$ — Rewrite 8 and 16 as powers of 2.

$2^{3x-9} = 2^{8x}$ — Multiply exponents.

$3x - 9 = 8x$ — Set exponents equal.

$-\frac{9}{5} = x$ — Solve for x.

Section 10.3 Logarithmic Functions

Definition of Logarithm

If $b > 0$ and $b \neq 1$, then $y = \log_b x$ is equivalent to $x = b^y$.

Example 1: Write in logarithmic form.

a. $3^4 = 81$

Solution:
$\log_3 81 = 4$

b. $\left(\frac{1}{4}\right)^{-3} = 64$

Solution:
$\log_{1/4} 64 = -3$

Example 2: Write in exponential form.

a. $\log_5 125 = 3$

Solution: $5^3 = 125$

b. $\log_3 \frac{1}{27} = -3$

Solution: $3^{-3} = \frac{1}{27}$

To solve an equation of the form $\log_b x = y$, where b, x, or y is a variable, write the equation in exponential form, $b^y = x$ and then solve for the variable.

Example 3: Solve.

a. $\log_2 \frac{1}{16} = x$

continued

Procedures/Rules	Key Examples

Section 10.3 Logarithmic Functions (continued)

Solution: $2^x = \frac{1}{16}$ **Change into exponential form.**

$2^x = \frac{1}{2^4}$ $16 = 2^4$

$2^x = 2^{-4}$ $\frac{1}{2^4} = 2^{-4}$

$x = -4$

b. $\log_x 49 = 2$

Solution: $x^2 = 49$ **Change into exponential form.**

$x = \pm 7$ **The square roots of 49 are ±7.**

$x = 7$ **The base must be positive.**

Two Properties of Logarithms
For any real number b, where $b > 0$ and $b \neq 1$,
1. $\log_b b = 1$
2. $\log_b 1 = 0$

Example 4: Find the following logarithms:
a. $\log_8 8$ **b.** $\log_5 1$
Solution: $\log_8 8 = 1$ Solution: $\log_5 1 = 0$

To graph a function in the form $f(x) = \log_b x$,
1. Replace $f(x)$ with y and then write the logarithm in exponential form, $x = b^y$.
2. Find ordered pairs that satisfy the equation by assigning values to y and finding x.
3. Plot the ordered pairs and draw a smooth curve through the points.

Example 5: Graph $y = \log_3 x$.
Solution: Write as $3^y = x$ and let y have values.

x	y
$\frac{1}{9}$	−2
$\frac{1}{3}$	−1
1	0
3	1
9	2

Section 10.4 Properties of Logarithms
Inverse Properties of Logarithms
For any real numbers b and x, where $b > 0$, $b \neq 1$, and $x > 0$.
1. $b^{\log_b x} = x$
2. $\log_b b^x = x$

Example 1: Find the value of each.
a. $6^{\log_6 5}$ **b.** $3^{\log_3 8}$
Solution: $6^{\log_6 5} = 5$ Solution: $3^{\log_3 8} = 8$

Further Properties of Logarithms
For real numbers x, y, and b, where $x > 0$, $y > 0$, $b > 0$, and $b \neq 1$,
Product Rule: $\log_b xy = \log_b x + \log_b y$

Quotient Rule: $\log_b \frac{x}{y} = \log_b x - \log_b y$

Power Rule: $\log_b x^r = r \log_b x$

Example 2: Use the product rule for the following:
a. Write $\log_4 4x$ as the sum of logarithms.
Solution:
$\log_4 4x = \log_4 4 + \log_4 x$ **Use $\log_b xy = \log_b x + \log_b y$.**
$= 1 + \log_4 x$ **$\log_4 4 = 1$**

b. Write $\log_7 4 + \log_7 2$ as a single logarithm.
Solution:
$\log_7 4 + \log_7 2 = \log_7 4 \cdot 2$ **Use $\log_b xy = \log_b x + \log_b y$.**
$= \log_7 8$

continued

Section 10.4 Properties of Logarithms (continued)

Example 3: Use the quotient rule for the following:

a. Write $\log_a \frac{w}{7}$ as the difference of logarithms.

Solution:

$\log_a \frac{w}{7} = \log_a w - \log_a 7$ **Use $\log_b \frac{x}{y} = \log_b x - \log_b y$.**

b. Write $\log_7(x + 5) - \log_7(x - 3)$ as a single logarithm.

Solution:

$\log_7(x + 5) - \log_7(x - 3) = \log_7 \frac{x + 5}{x - 3}$

Use $\log_b \frac{x}{y} = \log_b x - \log_b y$.

Example 4: Use the power rule for the following:

c. Write $\log_b \sqrt[5]{x^3}$ as a multiple of a logarithm.

Solution:

$\log_b \sqrt[5]{x^3} = \log_b x^{3/5} = \frac{3}{5} \log_b x$ **Use $\log_b x^r = r \log_b x$.**

d. Write $-3 \log_2 y$ as a logarithm of a quantity to a power.

Solution:

$$\begin{aligned} -3 \log_2 y &= \log_2 y^{-3} \\ &= \log_2 \frac{1}{y^3} \end{aligned}$$

Use $\log_b x^r = r \log_b x$.

Example 5: Write $\log_a \frac{x^2 y}{z^4}$ as the sum or difference of multiples of logarithms.

Solution:

$\log_a \frac{x^2 y}{z^4} = \log_a x^2 y - \log_a z^4$ **Quotient rule.**

$= \log_a x^2 + \log_a y - \log_a z^4$ **Product rule.**

$= 2 \log_a x + \log_a y - 4 \log_a z$ **Power rule.**

Example 6: Write $\frac{1}{4}(2 \log_5 x - 3 \log_5 y)$ as a single logarithm.

Solution: $\frac{1}{4}(2 \log_5 x - 3 \log_5 y)$

$= \frac{1}{4}(\log_5 x^2 - \log_5 y^3)$ **Power rule.**

$= \frac{1}{4} \log_5 \frac{x^2}{y^3}$ **Quotient rule.**

$= \log_5 \left(\frac{x^2}{y^3}\right)^{1/4}$ **Power rule.**

$= \log_5 \sqrt[4]{\frac{x^2}{y^3}}$

continued

Procedures/rules	Key Examples

Section 10.5 Common and Natural Logarithms

Base-10 logarithms are common logarithms and $\log_{10} x$ is written as $\log x$.

Common logarithms can be evaluated using the LOG key on a calculator.

Base-e logarithms are called natural logarithms and $\log_e x$ is written as $\ln x$.

Natural logarithms can be evaluated using the ln key on a calculator.

Example 1: Evaluate using a calculator and round to 4 decimal places.

a. $\log 356$

Solution: $\log 356 \approx 2.5514$

b. $\log 0.0059$

Solution: $\log 0.0059 \approx -2.2291$

Example 2: Evaluate using a calculator and round to four decimal places.

a. $\ln 72$

Solution: $\ln 72 \approx 4.2767$

b. $\ln 0.097$

Solution: $\ln 0.097 \approx -2.3330$

Section 10.6 Exponential and Logarithmic Equations with Applications

Properties for Solving Exponential and Logarithmic Equations

For any real numbers b, x, and y, where $b > 0$ and $b \neq 1$,

1. If $b^x = b^y$, then $x = y$.

2. If $x = y$, then $b^x = b^y$.

3. For $x > 0$ and $y > 0$, if $\log_b x = \log_b y$, then $x = y$.

4. For $x > 0$ and $y > 0$, if $x = y$, then $\log_b x = \log_b y$.

5. For $x > 0$, if $\log_b x = y$, then $b^y = x$.

Example 1: Solve $3^x = 4$.

Solution:

$\log 3^x = \log 4$ — Use if $x = y$, then $\log_b x = \log_b y$.

$x \log 3 = \log 4$ — Power rule.

$x = \dfrac{\log 4}{\log 3}$ — Divide by log 3.

$x \approx 1.2619$ — Evaluate using a calculator.

Example 2: Solve $e^{3x} = 12$.

Solution: Since the base of the exponent is e, take the natural logarithm of both sides.

$\ln e^{3x} = \ln 12$ — Use if $x = y$, then $\log_b x = \log_b y$.

$3x = \ln 12$ — Use $\log_b b^x = x$.

$x = \dfrac{\ln 12}{3} \approx 0.8283$ — Divide by 3 and approximate.

Solving Equations Containing Logarithms

To solve equations containing logarithms, use the properties of logarithms to simplify each side of the equation and then use one of the following.

If the simplification results in an equation in the form $\log_b x = \log_b y$, use the fact that $x = y$, and then solve for the variable.

If the simplification results in an equation in the form $\log_b x = y$, write the equation in exponential form, $b^y = x$, and then solve for the variable (as we did in Section 10.3).

Example 3: Solve $\log_3 x + \log_3(x - 3) = \log_3 10$.

Solution:

$\log_3 x(x - 3) = \log_3 10$ — Use the product rule.

$x(x - 3) = 10$ — Use if $\log_b x = \log_b y$, then $x = y$.

$x^2 - 3x - 10 = 0$ — Multiply then subtract 10.

$(x - 5)(x + 2) = 0$ — Factor.

$x - 5 = 0$ or $x + 2 = 0$ — Set each factor equal to 0.

$x = 5$ or $x = -2$ — Solve each equation.

We must reject -2 because it results in $\log_3(-2)$ and $\log_3(-5)$ in the original equation, which do not exist.

continued

Procedures/Rules	Key Examples

Section 10.6 Exponential and Logarithmic Equations with Applications (continued)

Example 4: Solve $\log_2(2x + 2) - \log_2(x - 4) = 2$.

Solution:

$$\log_2 \frac{2x+2}{x-4} = 2 \quad \textbf{Use the quotient rule.}$$

$$\frac{2x+2}{x-4} = 2^2 \quad \textbf{Use if } \log_b x = y \textbf{, then } b^y = x.$$

$$\frac{2x+2}{x-4} = 4 \quad 2^2 = 4.$$

$$2x + 2 = 4x - 16 \quad \textbf{Multiply by } x - 4.$$

$$18 = 2x$$

$$9 = x \quad \textbf{Solve for } x.$$

Example 5: How long will it take \$12,000 invested at 4% compounded quarterly to grow to \$15,000?

Solution: Use $A = P\left(1 + \frac{r}{n}\right)^{nt}$ with $A = 15{,}000$, $P = 12{,}000$, $r = 4\% = 0.04$, and $n = 4$. Find t.

$$A = P\left(1 + \frac{r}{n}\right)^{nt}$$

$$15{,}000 = 12{,}000\left(1 + \frac{0.04}{4}\right)^{4t} \quad \textbf{Substitute.}$$

$$\frac{5}{4} = (1.01)^{4t} \quad \textbf{Divide by 12,000.}$$

$$\log \frac{5}{4} = \log 1.01^{4t} \quad \textbf{Apply if } x = y \textbf{, then } \log_b x = \log_b y.$$

$$\log \frac{5}{4} = 4t \log 1.01 \quad \textbf{Power rule.}$$

$$\frac{\log \frac{5}{4}}{4 \log 1.01} = t \quad \textbf{Divide by 4 log 1.01.}$$

$$5.61 \approx t \quad \textbf{Evaluate.}$$

It would take about 5.6 years.

continued

Procedures/Rules

Key Examples

Section 10.6 Exponential and Logarithmic Equations with Applications (continued)

Exponential Growth and Decay
Exponential growth and decay can be represented by the equation $A = A_0e^{kt}$, where A_0 is the initial amount, t is the time, and k is a constant that is determined by the substance. If $k > 0$, there is exponential growth, and if $k < 0$, there is exponential decay. The half-life of a substance is the amount of time that must pass until one-half of the original substance remains.

Example 6: If \$3000 is invested in an account paying 4.5% compounded continuously:

a. How much will be in the account after 10 years?

Solution: Use $A = Pe^{rt}$ with $P = 3000$, $r = 0.045$, and $t = 10$.

$A = Pe^{rt}$
$A = 3000e^{(0.045)(10)}$ **Substitute.**
$A = 3000e^{0.45}$ **$10(0.045) = 0.45$.**
$A = 4704.94$ **Evaluate using a calculator.**

b. How long would it take for the \$3000 to grow to \$5000?

Solution: Use $A = Pe^{rt}$ with $A = 5000$, $P = 3000$, and $r = 0.045$. Find t.

$A = Pe^{rt}$
$5000 = 3000e^{0.045t}$ **Substitute**
$\frac{5}{3} = e^{0.045t}$ **Divide by 3000.**
$\ln\frac{5}{3} = \ln e^{0.045t}$ **Use if $x = y$, then $\log_b x = \log_b y$.**
$\ln\frac{5}{3} = 0.045t$ **Use if $x = y$, then $\log_b x = \log_b y$.**
$\frac{\ln\frac{5}{3}}{0.045} = t$ **Divide by 0.045.**
$11.35 \approx t$ **Evaluate using a calculator.**

It would take about $11\frac{1}{3}$ years.

Example 7: The element bismuth has an isotope, ^{200}Bi, that disintegrates according to the formula $A = A_0e^{-0.0198t}$, where t is in minutes.

a. How much of a 100-gram sample remains after 120 minutes?

Solution: We are given $A_0 = 100$ and $t = 120$.

$A = A_0e^{-0.0198t}$
$A = 100e^{-0.0198(120)}$ **Substitute for A_0 and t.**
$A = 9.29$ **Evaluate using a calculator.**

There would be about 9.29 grams left after 120 minutes.

continued

Procedures/Rules	Key Examples

Section 10.6 Exponential and Logarithmic Equations with Applications (continued)

b. Find the half-life of ^{200}Bi.

Solution: After one half-life, 50 of the original 100 grams remain, so $A = 50$.

$$A = A_0 e^{-0.0198t}$$

$$50 = 100e^{-0.0198t} \quad \text{Substitute for } A \text{ and } A_0.$$

$$\frac{1}{2} = e^{-0.0198t} \quad \text{Divide by 100.}$$

$$\ln\frac{1}{2} = \ln e^{-0.0198t} \quad \text{Use if } x = y, \text{ then } \log_b x = \log_b y.$$

$$\ln\frac{1}{2} = -0.0198t \quad \text{Use } \log_b b^x = x.$$

$$\frac{\ln\frac{1}{2}}{-0.0198} = t \quad \text{Divide by } -0.0198$$

$$35 = t \quad \text{Evaluate using a calculator.}$$

The half-life is 35 minutes.

Change-of-Base Formula

In general, if $a > 0$, $a \neq 1$, $b > 0$, $b \neq 1$, and $x > 0$, then $\log_a x = \frac{\log_b x}{\log_b a}$.

In terms of common and natural logarithms, $\log_a x = \frac{\log x}{\log a} = \frac{\ln x}{\ln a}$.

Example 8: Find $\log_5 16$.

Solution: Use $\log_a x = \frac{\log x}{\log a} = \frac{\ln x}{\ln a}$, then evaluate using a calculator.

$$\log_5 16 = \frac{\log 16}{\log 5} \approx 1.7227$$

or

$$\log_5 16 = \frac{\ln 16}{\ln 5} \approx 1.7227$$

Chapter 10 Review Exercises

For Exercises 1–6, answer true or false.

[10.1] **1.** If f and g are inverse functions and $f(2) = 5$, then $g(5) = 2$.

[10.2] **2.** If b, x, and y are real numbers and $b^x = b^y$, then $x = y$.

[10.3] **3.** If $y = \log_a(x - 4)$, it is possible for x to equal 3.

[10.3] **4.** If $\log_a b = c$, then $c^a = b$.

[10.3] **5.** Logarithms are exponents.

[10.5] **6.** Base-e logarithms are called common logarithms.

Equations and Inequalities

Exercises 7–62

[10.1] ***For Exercises 7–10, find each composition if $f(x) = 3x + 4$ and $g(x) = x^2 - 2$.***

7. $(f \circ g)(3)$ **8.** $(g \circ f)(3)$ **9.** $f[g(0)]$ **10.** $g[f(0)]$

[10.1] ***For Exercises 11–14, find $(f \circ g)(x)$ and $(g \circ f)(x)$.***

11. $f(x) = 3x - 6,\ g(x) = 2x + 3$

12. $f(x) = x^2 + 4,\ g(x) = 3x - 7$

13. $f(x) = \sqrt{x - 3},\ g(x) = 2x - 1$

14. $f(x) = \dfrac{x + 3}{x},\ g(x) = \dfrac{x - 4}{x}$

[10.1] ***For Exercises 15 and 16, determine if each is the graph of a one-to-one function. If the function is one-to-one, sketch the graph of the inverse function.***

15.

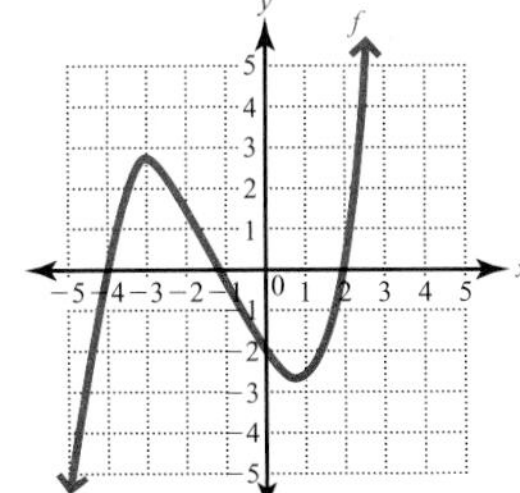

16.

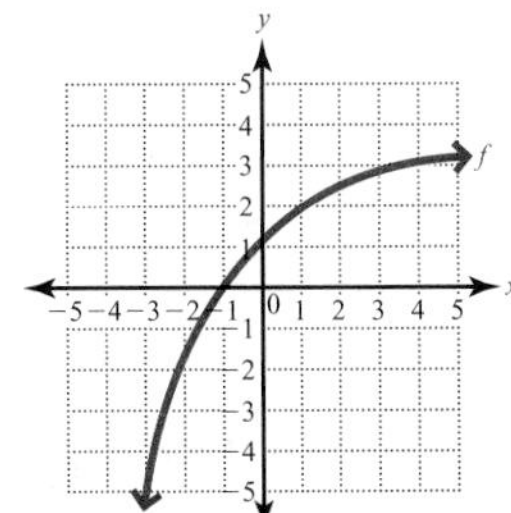

[10.1] ***For Exercises 17–20, determine whether f and g are inverse functions.***

17. $f(x) = 2x - 9,\ g(x) = \dfrac{x+9}{2}$

18. $f(x) = x^3 + 6,\ g(x) = \sqrt[3]{x-6}$

19. $f(x) = x^2 - 3,\ x \geq 0;\ g(x) = \sqrt{x+2}$

20. $f(x) = \dfrac{x}{x+4},\ g(x) = \dfrac{-4x}{x+1}$

[10.1] ***For Exercises 21–24, find $f^{-1}(x)$ for each of the following one-to-one functions.***

21. $f(x) = 5x + 4$

22. $f(x) = x^3 + 6$

23. $f(x) = \dfrac{4}{x+5}$

24. $f(x) = \sqrt[3]{3x+2}$

[10.1] **25.** If $(4, -6)$ is an ordered pair on the graph of f, what ordered pair is on the graph of f^{-1}?

[10.1] **26.** Fill in the blanks. If $f(a) = b$, then $f^{-1}(\_\_) = \_\_\_\_$.

[10.2] ***For Exercises 27–30, graph.***

27. $f(x) = 3^x$

28. $f(x) = \left(\dfrac{1}{2}\right)^x$

29. $f(x) = 2^{x-3}$

30. $f(x) = 3^{-x+2}$

[10.2] ***For Exercises 31–38, solve each equation.***

31. $5^x = 625$

32. $9^x = 27$

33. $6^x = \dfrac{1}{36}$

34. $\left(\dfrac{3}{4}\right)^x = \dfrac{16}{9}$

35. $3^{x-2} = 9$

36. $5^{x+2} = 25^x$

37. $\left(\dfrac{1}{3}\right)^{-x} = 27$

38. $8^{3x-2} = 16^{4x}$

[10.2] **39.** The median doubling time for a malignant tumor is about 100 days. If there are 500 cells initially, then the number of cells, A, after t days is given by $A = A_0 2^{t/100}$. How many cells are present after one year?

[10.2] **40.** If \$25,000 is deposited into an account paying 6% interest compounded monthly, how much will be in the account after eight years?

[10.2] **41.** The radioactive isotope ^{82}R has a half-life of 107 days. How much of a 50-gram sample remains after 300 days? (Use $A = A_0\left(\frac{1}{2}\right)^{t/h}$.)

[10.2] **42.** Cocaine use in the United Kingdom skyrocketed between 1987 and 1995 and can be approximated by the exponential function $N(t) = 387.7(1.222)^t$, where $N(t)$ is the number of addicts and t is the number of years after 1987. Estimate the number of addicts in 1991.

[10.3] ***For Exercises 43–46, write in logarithmic form.***

43. $7^3 = 343$

44. $4^{-3} = \frac{1}{64}$

45. $\left(\frac{3}{2}\right)^4 = \frac{81}{16}$

46. $11^{1/3} = \sqrt[3]{11}$

[10.3] ***For Exercises 47–50, write in exponential form.***

47. $\log_9 81 = 2$

48. $\log_{1/5} 125 = -3$

49. $\log_a 16 = 4$

50. $\log_e c = b$

[10.3] ***For Exercises 51–58, solve.***

51. $\log_3 x = -4$

52. $\log_{1/2} x = -2$

53. $\log_2 32 = x$

54. $\log_{1/4} 16 = x$

55. $\log_x 81 = 4$

56. $\log_x \frac{1}{1000} = 3$

57. $\log_{3/4} \frac{9}{16} = x$

58. $\log_{121} x = \frac{1}{2}$

[10.3] ***For Exercises 59–60, graph.***

59. $f(x) = \log_4 x$

60. $f(x) = \log_{1/3} x$

[10.3] **61.** If $f(x) = \log_b x$, the domain is $(0, \infty)$, and the range is $(-\infty, \infty)$, what are the domain and range of $g(x) = b^x$? Why?

[10.3] **62.** The formula for the number of decibels in a sound is $d = 10 \log \frac{I}{I_0}$. Find the decibel reading of a sound whose intensity is $I = 1000\, I_0$.

Expressions

Exercises 63–92

[10.4] ***For Exercises 63 and 64, find the value.***

63. $3^{\log_3 8}$

64. $\log_9 9^6$

[10.4] ***For Exercises 65 and 66, use the product rule to write the expression as a sum of logarithms.***

65. $\log_6 6x$

66. $\log_4 x(2x - 5)$

[10.4] ***For Exercises 67 and 68, use the product rule to write the expression as a single logarithm.***

67. $\log_3 4 + \log_3 8$

68. $\log_5 3 + \log_5 x + \log_5(x - 2)$

[10.4] ***For Exercises 69 and 70, use the quotient rule to write the expression as a difference of logarithms.***

69. $\log_b \frac{x}{5}$

70. $\log_a \frac{3x - 2}{4x + 3}$

[10.4] ***For Exercises 71 and 72, use the quotient rule to write the expression as a single logarithm.***

71. $\log_5 24 - \log_5 4$

72. $\log_2(x + 5) - \log_2(2x - 3)$

[10.4] ***For Exercises 73–76, use the power rule to write the expression as a multiple of a logarithm.***

73. $\log_3 7^4$

74. $\log_a \sqrt[3]{x}$

75. $\log_4 \frac{1}{a^4}$

76. $\log_a \sqrt[5]{a^4}$

[10.4] ***For Exercises 77 and 78, use the power rule to write the expression as the logarithm of a quantity to a power. Simplify the answer, if possible.***

77. $4 \log_a x$

78. $\frac{3}{5} \log_a y$

[10.4] ***For Exercises 79–82, write the expression as the sum or differences of multiples of logarithms.***

79. $\log_a x^2y^3$

80. $\log_a \frac{c^4}{d^3}$

81. $\log_a \frac{x^2y^3}{z^4}$

82. $\log_a \sqrt{\frac{a^3}{b^4}}$

[10.4] ***For Exercises 83–86, write the expression as a single logarithm.***

83. $2 \log_a b + 4 \log_a c$

84. $3 \log_a 4 - 2 \log_a 3$

85. $\frac{1}{4}(2 \log_a x + 3 \log_a y)$

86. $4 \log_a(x + 5) + 2 \log_a(x - 3)$

[10.5] *For Exercises 87–90, use a calculator to approximate each logarithm to four decimal places.*

87. $\log 326$

88. $\log 0.0035$

89. $\ln 0.043$

90. $\ln 92$

[10.5] *For Exercises 91 and 92, find the exact value without using a calculator.*

91. $\log 0.00001$

92. $\ln \sqrt[4]{e}$

[10.5] 93. The sound intensity of a clap of thunder was $10^{-3.5}$ watts per square meter. What is the decibel reading? (Use $d = 10 \log \frac{I}{I_0}$, where $I_0 = 10^{-12}$ watts/m^2.)

[10.5] 94. Using $\text{pH} = -\log [H_3O^+]$ (the hydronium ion concentration), find the pH of an apple whose $[H_3O^+]$ is 0.001259.

[10.5] 95. How long will it take \$6000 to grow to \$10,000 if it is invested at 3% annual interest compounded continuously? (Use $A = Pe^{rt}$.)

[10.5] 96. Using $R = \log \frac{I}{I_0}$, for Richter scale readings, compare the intensity of an earthquake whose Richter scale reading was 7.8 with one whose reading was 6.8. What do you notice? (*Hint:* Solve for I in terms of I_0.)

[10.6] *For Exercises 97–102, solve. Round answers to four decimal places.*

97. $9^x = 32$

98. $3^{5x} = 19$

99. $6^{2x-1} = 22$

100. $4^{2x-3} = 5^{x+1}$

101. $e^{4x} = 11$

102. $e^{-0.003x} = 5$

[10.6] *For Exercises 103–110, solve. Give exact answers.*

103. $\log_3(x - 4) = 2$

104. $\log_2(x^2 + 2x) = 3$

105. $\log(3x - 8) = \log(x - 2)$

106. $\log 5 + \log x = 2$

107. $\log_3 x - \log_3 4 = 2$

108. $\log_2 x + \log_2(x - 6) = 4$

109. $\log_4 x + \log_4(6x - 9) = \log_4 15$

110. $\log_3(5x + 2) - \log_3(x - 2) = 2$

[10.6] 111. How long will it take \$15,000 invested at 3% compounded monthly to grow to \$18,000? Round the answer to the nearest tenth of a year. (Use $A = P\left(1 + \frac{r}{n}\right)^{nt}$.)

[10.6] **112.** If \$7000 is deposited at 7% annual interest compounded continuously:

a. How much will be in the account after eight years? (Use $A = Pe^{rt}$.)

b. How long will it take until there is \$12,000 in the account?

[10.6] **113.** The population of an ant colony is 400 and is increasing at the rate of 3% per month. Use $A = A_0e^{0.03t}$ to answer the following:

a. How many ants will be in the colony after one year?

b. After how many months will there be 800 ants in the colony?

[10.6] **114.** The number of diagnosed AIDS cases among adults in the United States since 1992 has declined and can be approximated by the function $y = 87{,}419 - 24{,}647.9 \ln x$, where y is the number of diagnosed cases and x is the number of years after 1992. (*Data Source:* CDC Division of HIV/AIDS Prevention, *Surveillance Report*, Vol. 13, No. 2)

a. Estimate the number of cases of diagnosed AIDS in 2000.

b. In what year were 43,256 new cases of AIDS diagnosed?

[10.6] ***For Exercises 115 and 116, use the change-of-base formula to approximate each logarithm to four decimal places.***

115. $\log_4 15$

116. $\log_{1/2} 6$

Chapter 10 Practice Test

1. If $f(x) = x^2 - 6$ and $g(x) = 3x - 5$, find $f[g(x)]$.

2. Use $f(x) = 4x - 3$, to answer a–c.
   a. Find $f^{-1}(x)$.

   b. Verify that $f[f^{-1}(x)] = x$ and $f^{-1}[f(x)] = x$.

   c. What is the relationship between the graphs of $f(x)$ and $f^{-1}(x)$?

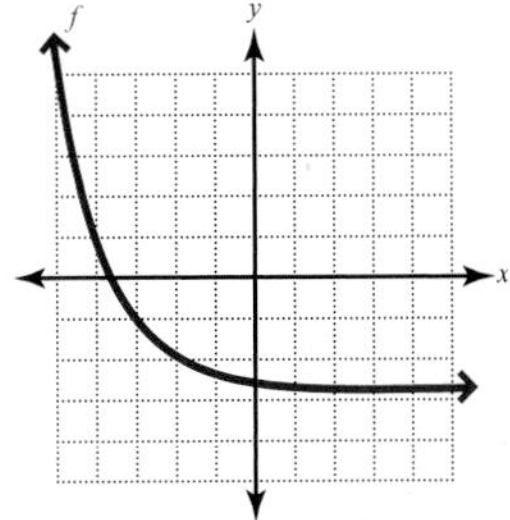

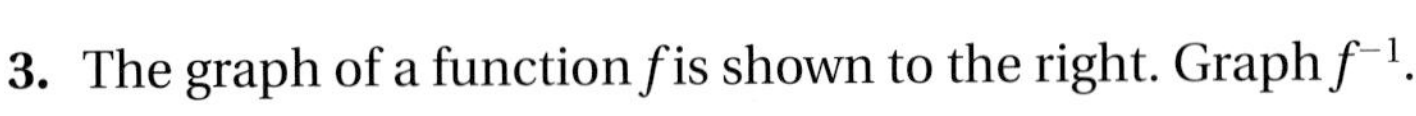

3. The graph of a function f is shown to the right. Graph f^{-1}.

4. Graph $f(x) = 2^{x-1}$.

5. Solve $32^{x-2} = 8^{2x}$.

6. A sum of $20,000 is invested at 5% compounded quarterly.
   a. Find the amount in the account after 15 years.

   b. Find the number of years until there is $30,000 in the account.

7. The isotope ^{98}Nb has a half-life of 30 minutes.
   a. How much of a 100-gram sample remains after 3.6 hours?

   b. How long will it take until there are only 20 grams remaining?

8. The number of people in the United States aged 65 or over has increased rapidly since 1900 and can be approximated by $y = 3.17(1.026)^x$, where y is the number in millions aged 65 or older and x is the number of years after 1900. (*Source:* U.S. Bureau of the Census)
   a. Find the number aged 65 or older in 1960. In 2005.

   b. In what year were there 31.4 million people aged 65 or older?

9. Write $\log_{1/3} 81 = -4$ in exponential form.

For Exercises 10 and 11, solve. Give exact answers.

10. $\log_6 \frac{1}{216} = x$

11. $\log_x 625 = 4$

12. Graph $f(x) = \log_2 x$.

For Exercises 13 and 14, write as the sum or difference of multiples of logarithms.

13. $\log_b \frac{x^4y^2}{z}$

14. $\log_b \sqrt[4]{\frac{x^5}{y^7}}$

15. Write $\frac{3}{4}(2 \log_b x + 3 \log_b y)$ as a single logarithm.

16. The isotope ^{119}Sn disintegrates according to the function $A = A_0e^{-0.0028t}$, where t is the time in days.

a. How much of a 300-gram sample remains after 500 days?

b. What is the half-life of ^{119}Sn?

For Exercises 17–19, solve. If necessary, use a calculator to approximate to four decimal places.

17. $6^{x-3} = 19$

18. $\log_3 x + \log_3(x + 6) = 3$

19. $\log(4x + 2) - \log(3x - 2) = \log 2$

20. The number of children younger than 13 diagnosed with AIDS has declined since 1992 and can be modeled by $y = 1141.7 - 428.64 \ln x$, where y is the number of diagnosed cases and $x = 1$ is 1992. (*Data Source:* CDC Divisions of HIV/AIDS Prevention, *Surveillance Report,* Vol. 13, no. 2)

a. Find the number of diagnosed cases in 1998.

b. In what year were 155 cases diagnosed?

CHAPTER 11

Conic Sections

"The universe . . . is written in the language of mathematics, and its characters are triangles, circles, and other geometrical figures."

—Galileo Galilei (1564–1642), Italian mathematician, astronomer, physicist, and philosopher. *Il Saggiatore* (1623).

"Mankind is not a circle with a single center but an ellipse with two focal points of which facts are one and ideas the other."

—Victor Hugo (1802–1883), French poet, dramatist, novelist. *Les Miserables* (1862)

In this chapter, we expand our understanding of equations, inequalities, and graphs and learn about conic sections. We will see that the conic sections are parabolas, circles, ellipses, and hyperbolas. These shapes are called conic sections because they are formed by the intersection of a cone with a plane. Later, we will study the equations of conic sections, nonlinear systems, and nonlinear inequalities.

11.1 The Parabola and Circle

OBJECTIVES

1. Graph parabolas of the form $x = a(y - k)^2 + h$.
2. Find the distance between two points.
3. Graph circles of the form $(x - h)^2 + (y - k)^2 = r^2$.
4. Find the equation of a circle with a given center and radius.
5. Graph circles of the form $x^2 + y^2 + dx + ey + f = 0$.

The intersection of a plane with a cone will be a circle, ellipse, parabola, or hyperbola. For that reason, these curves are called **conic sections** or **conics**.

DEFINITION ***Conic Section:*** A curve in a plane that is the result of intersecting the plane with a cone. More specifically, a circle, ellipse, parabola, or hyperbola.

Circle

Ellipse

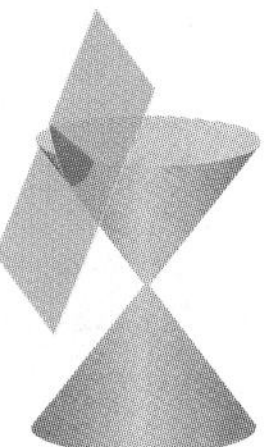
Parabola

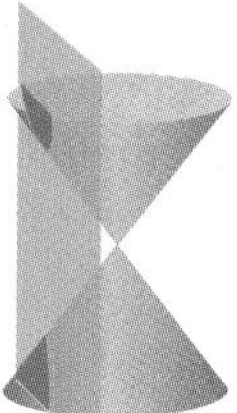
Hyperbola

Recall from Section 9.4 that we graphed parabolas in the form $y = a(x - h)^2 + k$. The graph opened upwards if $a > 0$, downwards if $a < 0$, had a vertex at (h, k), and $x = h$ as the axis of symmetry.

EXAMPLE 1  For $y = 2(x - 3)^2 + 1$, determine whether the graph opens upwards or downwards, find the vertex and axis of symmetry, and draw the graph.

Solution The graph opens upwards since $a = 2$ and 2 is positive. We compare the equation with the form $y = a(x - h)^2 + k$ and observe that the vertex is at the point with coordinates (3, 1) and the axis of symmetry is $x = 3$. Plot a few points on either side of the axis of symmetry by letting x have values on either side of 3 and finding y.

x	y
2	3
1	9
4	3
5	9

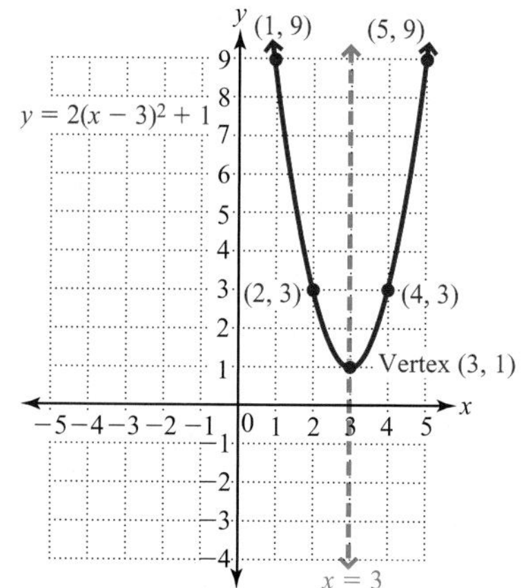

YOUR TURN For $y = -2(x - 1)^2 + 3$, determine whether the graph opens upwards or downwards, find the vertex and axis of symmetry, and draw the graph.

OBJECTIVE 1. Graph parabolas of the form $x = a(y - k)^2 + h$. If we interchange x and y in the equations of parabolas that open upwards and downwards, we get the equations of parabolas that open to the left or right. To keep the vertex at (h, k), we also interchange h and k.

RULE Equations of Parabolas Opening Left or Right

The graph of an equation in the form $x = a(y - k)^2 + h$ is a parabola with vertex at (h, k). The parabola opens to the right if $a > 0$ and to the left if $a < 0$. The equation of the axis of symmetry is $y = k$.

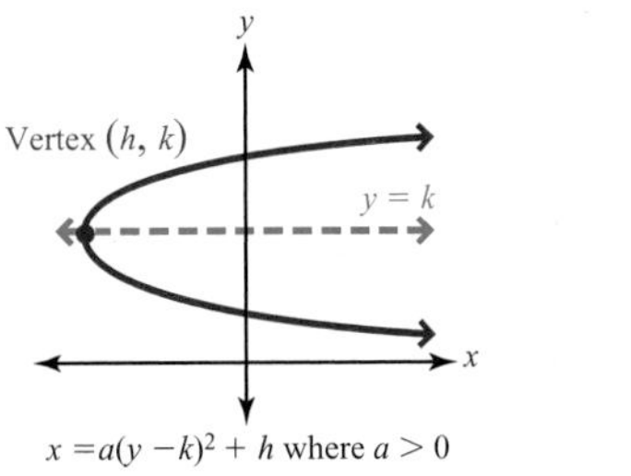

$x = a(y - k)^2 + h$ where $a > 0$

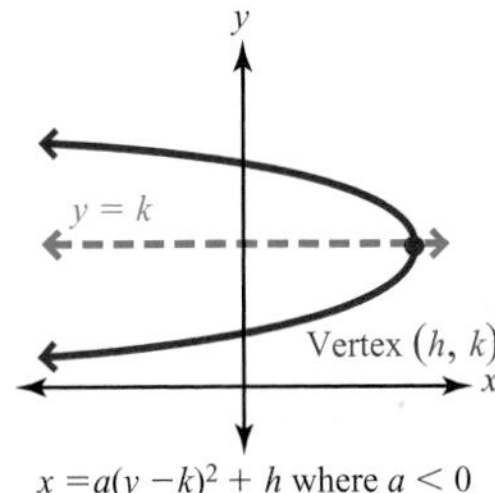

$x = a(y - k)^2 + h$ where $a < 0$

EXAMPLE 2 For each equation, determine whether the graph opens left or right, find the vertex and axis of symmetry, and draw the graph.

a. $x = -2(y + 3)^2 - 2$

Solution This parabola opens to the left because $a = -2$, which is negative. Rewrite the equation as $x = -2(y - (-3))^2 - 2$. Comparing this equation with $x = a(y - k)^2 + h$, we see that $h = -2$ and $k = -3$. The vertex is at the point with coordinates $(-2, -3)$ and the axis of symmetry is $y = -3$. To graph, plot a few points on either side of the axis of symmetry by letting y equal values on either side of -3 and finding x.

x	y
−4	−2
−10	−1
−4	−4
−10	−5

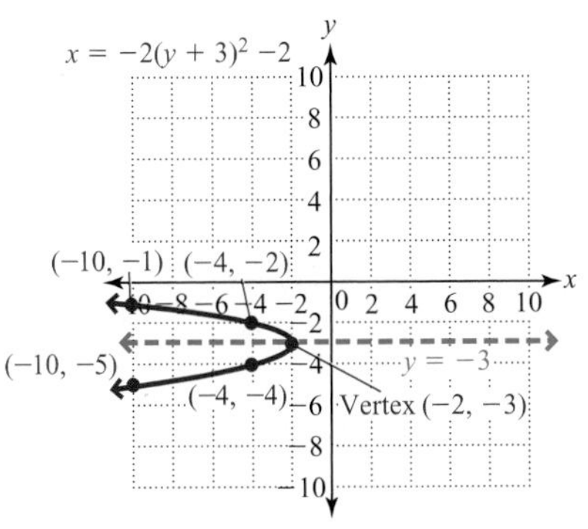

opens downwards; vertex: (1, 3); axis of symmetry: $x = 1$

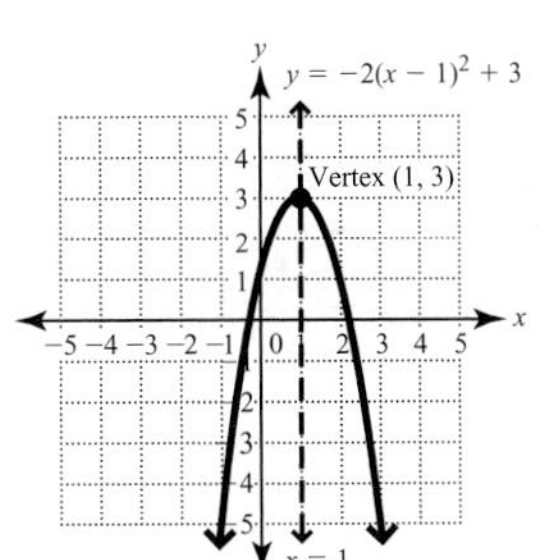

b. $x = 3y^2 + 12y + 8$

Solution This parabola opens to the right because $a = 3$, which is positive. To find the vertex and axis of symmetry we need to write the equation in the form $x = a(y - k)^2 + h$.

$x = 3y^2 + 12y + 8$	Original equation.
$x - 8 = 3y^2 + 12y$	Subtract 8 from both sides.
$x - 8 = 3(y^2 + 4y)$	Factor out the common factor, 3.
$x - 8 + 12 = 3(y^2 + 4y + 4)$	Complete the square. Note that we added $3 \cdot 4 = 12$ to both sides of the equation.
$x + 4 = 3(y + 2)^2$	Simplify the left side and factor the right.
$x = 3(y + 2)^2 - 4$	Subtract 4 from both sides of the equation.

The vertex is at the point with coordinates $(-4, -2)$ and the axis of symmetry is $y = -2$.

To complete the graph, let y equal values on either side of -2 and find x.

x	y
−1	−1
8	0
−1	−3
8	−4

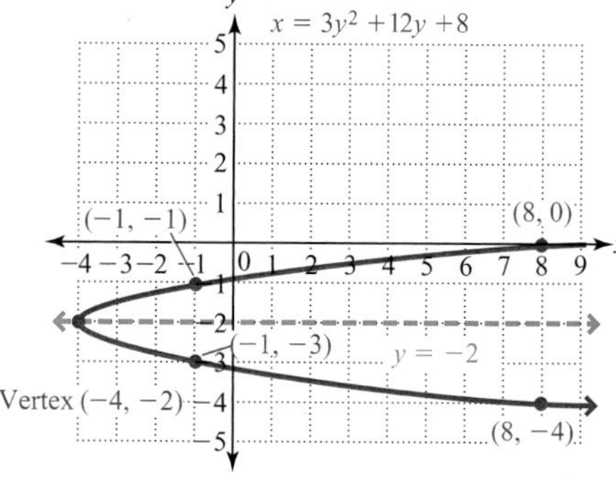

ANSWERS

a. opens left; vertex: (1, 2); axis of symmetry: $y = 2$

b. opens right; vertex: $(-1, -2)$; axis of symmetry: $y = -2$

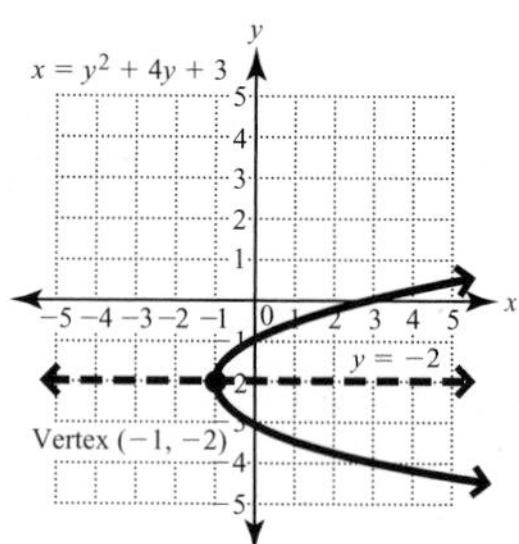

YOUR TURN For each equation, determine whether the graph opens left or right, find the vertex and axis of symmetry, and draw the graph.

a. $x = -(y - 2)^2 + 1$

b. $x = y^2 + 4y + 3$

OBJECTIVE 2. Find the distance between two points. To derive the other conic's general equations, we need to be able to find the distance between any two points in the coordinate plane. Consider the two points (x_1, y_1) and (x_2, y_2) shown in the following graph.

Note: *If y_2 were 6 and y_1 were 2, the distance between would be $6 - 2 = 4$. Therefore, to calculate the length of the vertical leg we calculate $y_2 - y_1$.*

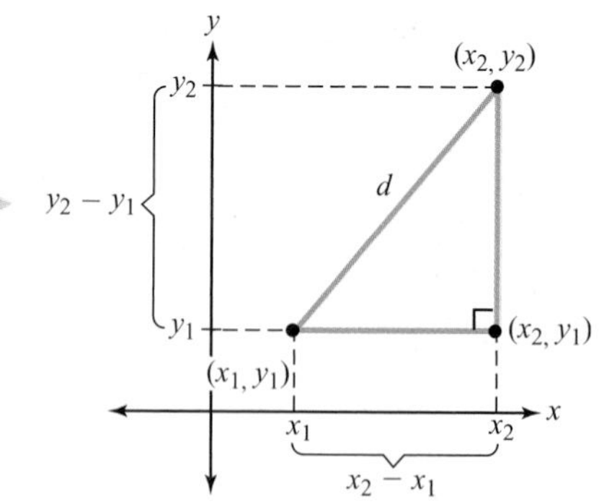

Since the distance between the two points is measured along a line segment that is the hypotenuse of a right triangle, we can use the Pythagorean theorem (see below) to find the distance.

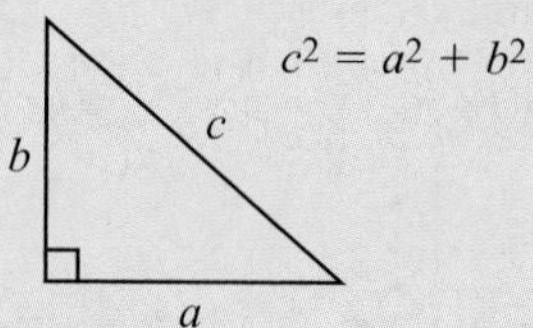

The lengths of the legs of the triangle are $x_2 - x_1$ and $y_2 - y_1$, as illustrated.

Note: *If x_2 were 4 and x_1 were 1, the distance between would be $4 - 1 = 3$. Therefore, to calculate the length of the horizontal leg we calculate $x_2 - x_1$.*

Now we can use the Pythagorean theorem, replacing a with $x_2 - x_1$, b with $y_2 - y_1$, and c with d.

$$d^2 = (x_2 - x_1)^2 + (y_2 - y_1)^2$$

$$d = \pm\sqrt{(x_2 - x_1)^2 + (y_2 - y_1)^2}$$ Use the square root principle to isolate d.

Because d is a distance, it must be positive, so we use only the positive value.

RULE **The Distance Formula**

The distance, d, between two points with coordinates (x_1, y_1) and (x_2, y_2) can be found using the formula

$$d = \sqrt{(x_2 - x_1)^2 + (y_2 - y_1)^2}$$

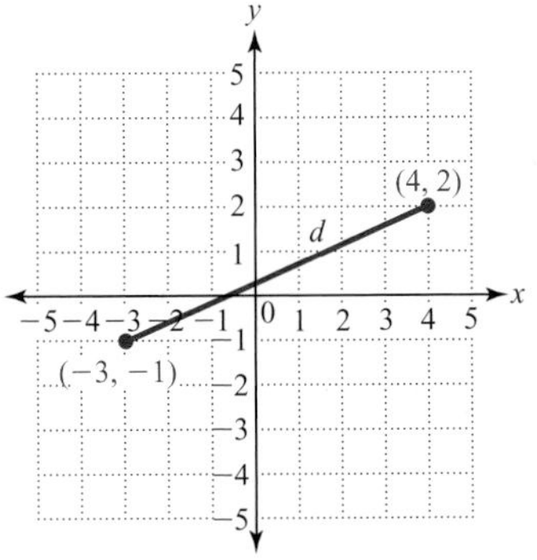

Note: *The distance formula holds no matter what quadrants the points are in.*

EXAMPLE 3 Find the distance between (4, 2) and (−3, −1). If the distance is an irrational number, also give a decimal approximation rounded to three places.

Solution $d = \sqrt{(x_2 - x_1)^2 + (y_2 - y_1)^2}$ Use the distance formula.

$d = \sqrt{(-3 - 4)^2 + (-1 - 2)^2}$ Let $(4, 2) = (x_1, y_1)$ and $(-3, -1) = (x_2, y_2)$.

$d = \sqrt{(-7)^2 + (-3)^2}$

$d = \sqrt{49 + 9}$

$d = \sqrt{58}$

$d \approx 7.616$

Note: *It does not matter which ordered pair is (x_1, y_1) and which is (x_2, y_2). Consider Example 3 again with $(-3, -1)$ as (x_1, y_1) and $(4, 2)$ as (x_2, y_2):*

$$d = \sqrt{(4 - (-3))^2 + (2 - (-1))^2}$$
$$= \sqrt{(7)^2 + (3)^2} = \sqrt{49 + 9}$$
$$= \sqrt{58} \approx 7.616$$

YOUR TURN Determine the distance between the given points. If the distance is an irrational number, also give a decimal approximation rounded to three places.

a. (8, 2) and (3, −4) **b.** (6, −5) and (0, −1)

ANSWERS

a. $\sqrt{61} \approx 7.810$ **b.** $2\sqrt{13} \approx 7.211$

OBJECTIVE 3. Graph circles of the form $(x - h)^2 + (y - k)^2 = r^2$. The second conic section that we will consider is the **circle** with **radius** r.

DEFINITIONS ***Circle:*** A set of points that are equally distant from a central point. The central point is the center.
Radius: The distance from the center of a circle to any point on the circle.

If the center of a circle is (h, k) and the radius is r, we can use the distance formula to derive the equation of the circle. If (x, y) is any point on the circle, the distance between (x, y) and (h, k) must be the radius, r.

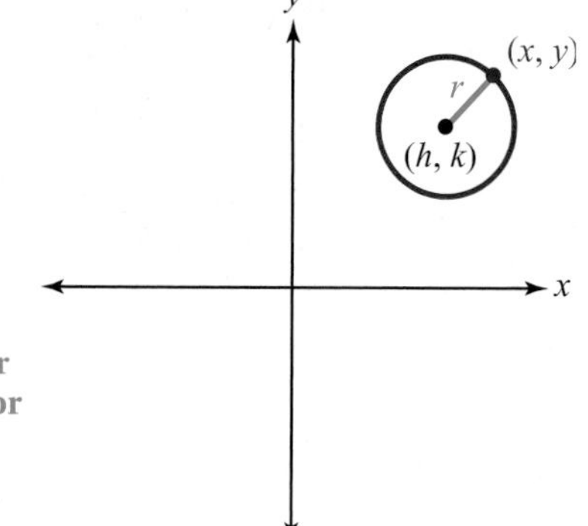

$$\sqrt{(x_2 - x_1)^2 + (y_2 - y_1)^2} = r$$

$$\sqrt{(x - h)^2 + (y - k)^2} = r \quad \text{Substitute } (x, y) \text{ for } (x_2, y_2) \text{ and } (h, k) \text{ for } (x_1, y_1).$$

$$(x - h)^2 + (y - k)^2 = r^2 \quad \text{Square both sides.}$$

RULE **Standard Form of the Equation of a Circle**

The equation of a circle with center at (h, k) and radius r is $(x - h)^2 + (y - k)^2 = r^2$.

Note: *If the center of a circle is at the origin, then $(h, k) = (0, 0)$ and the equation of the circle becomes $(x - 0)^2 + (y - 0)^2 = r^2$, which simplifies to $x^2 + y^2 = r^2$.*

EXAMPLE 4 Find the center and radius of each circle and draw the graph.

a. $(x - 3)^2 + (y + 2)^2 = 36$

Solution $(x - 3)^2 + (y - (-2))^2 = 6^2$ Write in the form $(x - h)^2 + (y - k)^2 = r^2$.

Since $h = 3$ and $k = -2$, the center is $(3, -2)$.
Since $36 = 6^2$, the radius is 6.

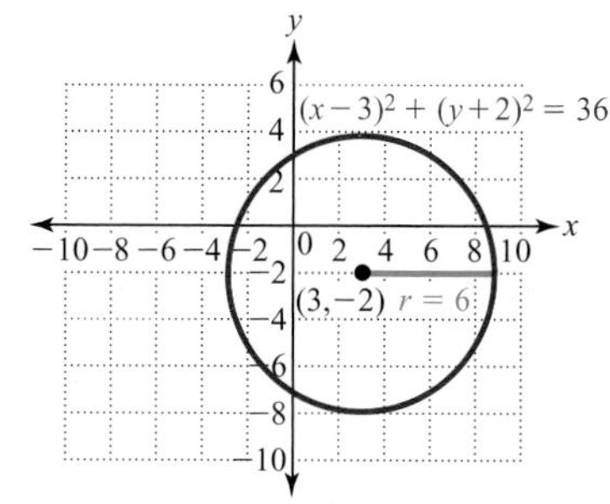

Note: *To find the radius, we evaluate the square root of 36. Because the radius is a distance, we give only the principal square root.*

$$r = \sqrt{36} = 6$$

b. $(x + 4)^2 + (y + 1)^2 = 28$

Solution $(x - (-4))^2 + (y - (-1))^2 = (\sqrt{28})^2$ Write in the form $(x - h)^2 + (y - k)^2 = r^2$.

Since $h = -4$ and $k = -1$, the center is $(-4, -1)$.
For this radius, $r = \sqrt{28} = 2\sqrt{7} \approx 5.292$.

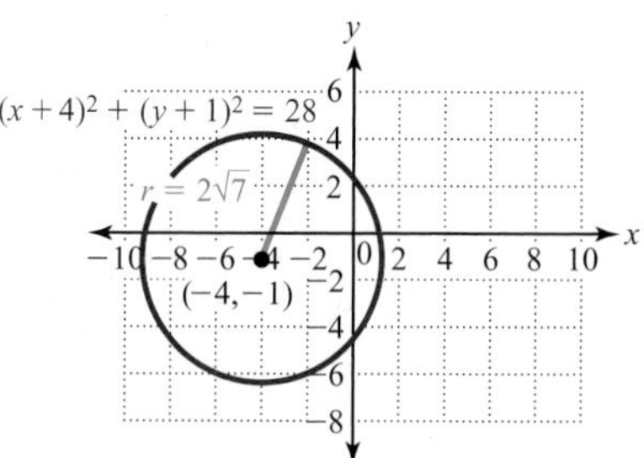

YOUR TURN Find the center and radius of each circle and draw the graph.

a. $(x - 3)^2 + (y + 5)^2 = 4$ **b.** $(x + 1)^2 + (y - 1)^2 = 18$

ANSWERS

a. center: $(3, -5)$; radius: 2

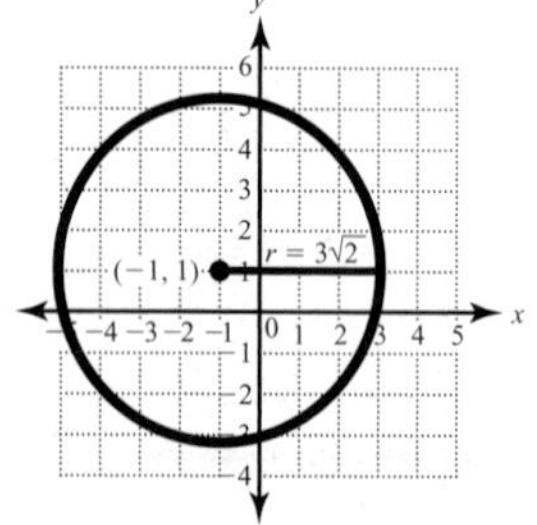

b. center: $(-1, 1)$; radius: $3\sqrt{2}$

(−1, 1)
$r = 3\sqrt{2}$

OBJECTIVE 4. Find the equation of a circle with a given center and radius. We can also use the standard form of a circle to write equations of circles.

EXAMPLE 5 Write the equation of each circle in standard form.

Learning Strategy

Note how the signs of the numbers in the parentheses of the general equation are the opposite of the signs of the center coordinates.

$(x + 4)^2 + (y - 2)^2 = 64$

Center: $(-4, 2)$

a. center: $(-4, 2)$; radius: 8

Solution Since the center is at $(-4, 2)$, $h = -4$ and $k = 2$.

$$(x - h)^2 + (y - k)^2 = r^2$$

$$(x - (-4))^2 + (y - 2)^2 = 8^2 \quad \text{Substitute for } h, k, \text{ and } r.$$

$$(x + 4)^2 + (y - 2)^2 = 64 \quad \text{Simplify}$$

b. center: $(0, 0)$; radius: 2

Solution Since the center is at $(0, 0)$, the standard form of the equation is $x^2 + y^2 = r^2$.

$$x^2 + y^2 = 2^2 \quad \text{Substitute for } r.$$

$$x^2 + y^2 = 4$$

YOUR TURN Write the equation of each circle in standard form.

a. center: $(4, -2)$; radius: 5 **b.** center: $(0, 0)$; radius: 9

ANSWERS

a. $(x - 4)^2 + (y + 2)^2 = 25$
b. $x^2 + y^2 = 81$

OBJECTIVE 5. Graph circles of the form $x^2 + y^2 + dx + ey + f = 0$. If the equation of a circle is not given in standard form, we complete the square to write the equation in the form $(x - h)^2 + (y - k)^2 = r^2$.

EXAMPLE 6 Find the center and radius of the circle whose equation is $x^2 + y^2 - 6x + 8y + 9 = 0$ and draw the graph.

Solution $x^2 + y^2 - 6x + 8y = -9$ — Subtract 9 from both sides to isolate the variable terms.

$$(x^2 - 6x) + (y^2 + 8y) = -9 \quad \text{Group the terms in } x \text{ and } y.$$

$$(x^2 - 6x + 9) + (y^2 + 8y + 16) = -9 + 9 + 16 \quad \text{Complete the square for } x \text{ and } y \text{ by adding 9 and 16 to both sides of the equation.}$$

$$(x - 3)^2 + (y + 4)^2 = 16 \quad \text{Factor and simplify.}$$

The center is at $(3, -4)$ and the radius is 4.

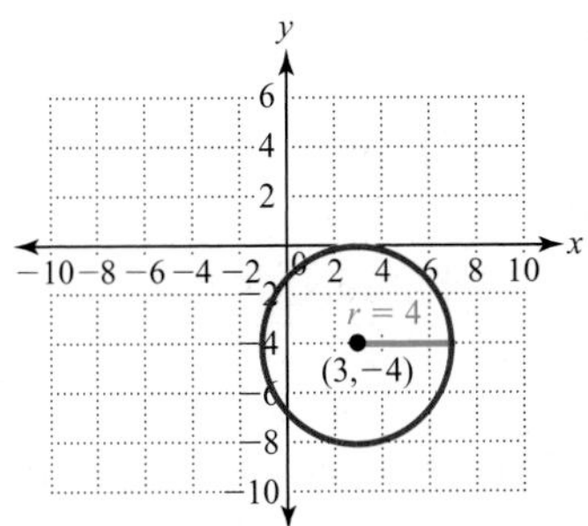

Calculator TIPS

The graph of a circle fails the vertical line test, so the equation of a circle is not a function. To graph equations that are not functions, we solve for y and graph the two resulting functions on the same screen.

EXAMPLE Graph $(x - 2)^2 + (y + 1)^2 = 16$.

Solution We solve $(x - 2)^2 + (y + 1)^2 = 16$ for y.

$$(x - 2)^2 + (y + 1)^2 = 16$$

$$(y + 1)^2 = 16 - (x - 2)^2 \quad \text{Subtract } (x-2)^2 \text{ from both sides.}$$

$$y + 1 = \pm\sqrt{16 - (x - 2)^2} \quad \text{Apply the square root principle.}$$

$$y = -1 \pm \sqrt{16 - (x - 2)^2} \quad \text{Subtract 1 from both sides.}$$

This equation defines two functions. On your graphing calculator define $Y_1 = -1 + \sqrt{16 - (x - 2)^2}$ and $Y_2 = -1 - \sqrt{16 - (x - 2)^2}$ and graph both in the window $[-8, 8]$ for x and $[-8, 8]$ for y. The resulting graph is labeled "**a**" below; note that the graph does not look like the graph of a circle. To make the graph look like a circle, select ZSquare from the ZOOM menu. This results in the graph labeled "**b**" below.

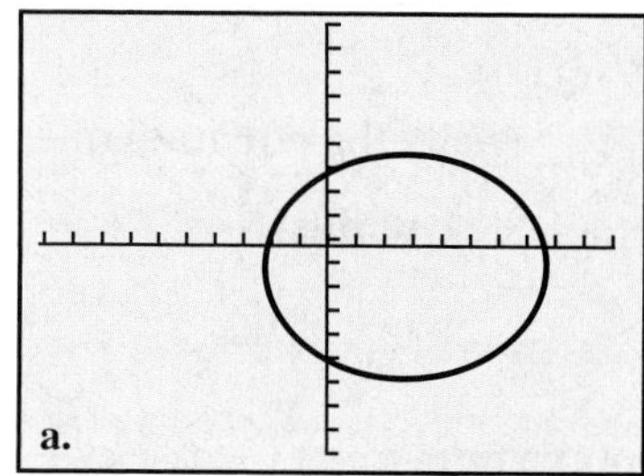
a.

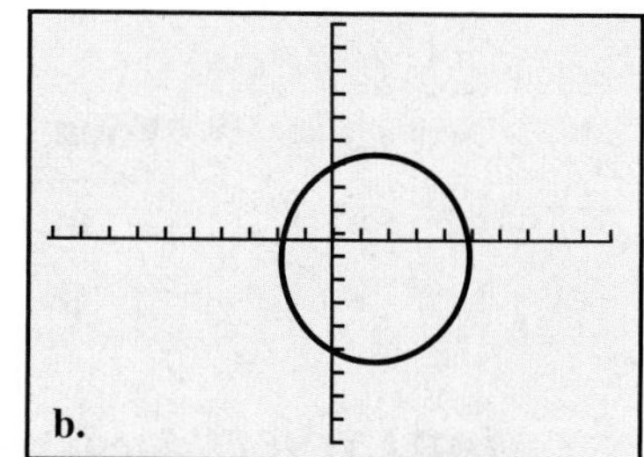
b.

ANSWER

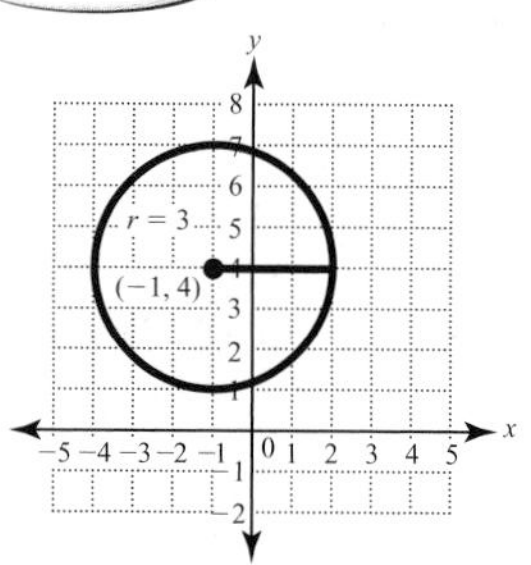

center: $(-1, 4)$; radius: 3

YOUR TURN Find the center and radius of the circle whose equation is $x^2 + y^2 + 2x - 8y + 8 = 0$ and draw the graph.

11.1 Exercises

For Extra Help

MyMathLab

Videotape/DVT

InterAct Math

Math Tutor Center

Math XL.com

1. Is $x^2 + 2x - 3 + y = 0$ the equation of a circle or a parabola? Explain.

2. Find the coordinates of the vertex of the parabola $x = 3y^2 - 12y + 2$.

3. In which direction does the graph of the parabola in Exercise 2 open? Why?

4. Explain how the Pythagorean theorem is used to derive the distance formula.

5. Complete the following: The set of all points that are 4 units from the point whose coordinates are $(-3, 2)$ is a ______________ with center at ______________ and radius of ______________ .

6. How is the distance formula related to the equation of a circle?

For Exercises 7–28, find the direction the parabola opens, the coordinates of the vertex, the equation of the axis of symmetry, and draw the graph.

7. $y = (x - 1)^2 + 2$

8. $y = (x + 2)^2 - 3$

9. $y = -x^2 - 2x + 3$

10. $y = -x^2 + 4x - 1$

11. $x = (y + 2)^2 - 2$

12. $x = (y + 3)^2 + 2$

13. $x = -(y - 1)^2 + 3$

14. $x = -(y - 3)^2 + 3$

15. $x = 2(y + 2)^2 - 4$

16. $x = 3(y + 3)^2 - 4$

17. $x = -3(y + 2)^2 - 5$

18. $x = -2(y - 4)^2 + 1$

19. $x = y^2 + 4y + 3$

20. $x = y^2 - 2y - 8$

21. $x = -y^2 + 6y - 5$

22. $x = -y^2 - 4y - 3$

23. $x = 2y^2 + 8y + 3$

24. $x = 2y^2 - 4y + 1$

25. $x = 3y^2 - 6y + 3$

26. $x = 3y^2 - 12y + 9$

27. $x = -2y^2 + 4y + 5$

28. $x = -3y^2 - 6y - 2$

For Exercises 29–32, match the equation with the correct graph.

29. $x = (y + 3)^2 - 2$

30. $y = (x + 3)^2 - 2$

31. $y = 2x^2 - 8x + 5$

32. $x = 2y^2 - 8y + 5$

a.

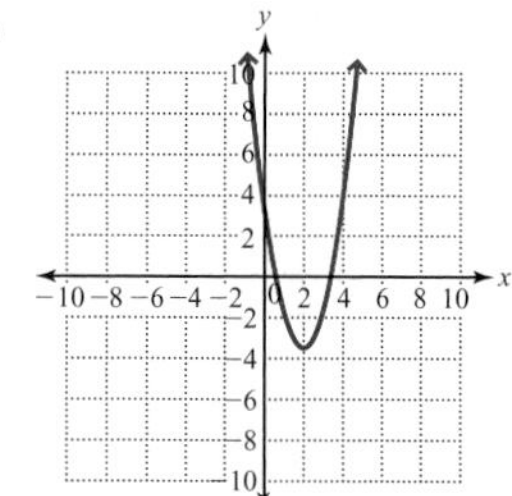

b.

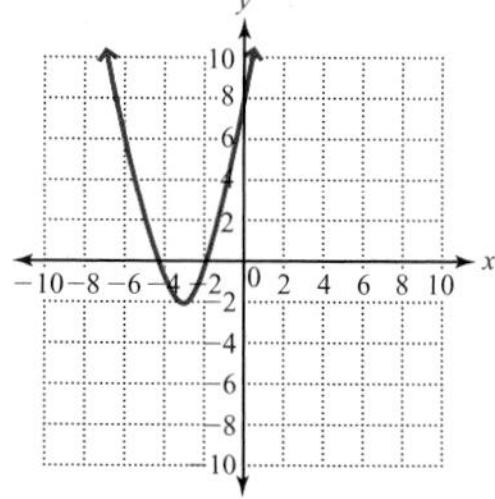

c.

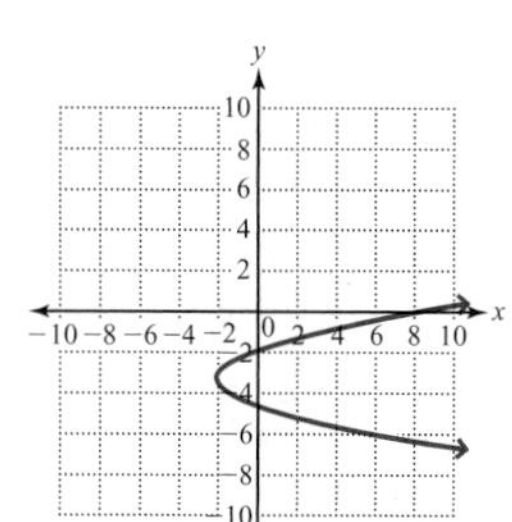

d.

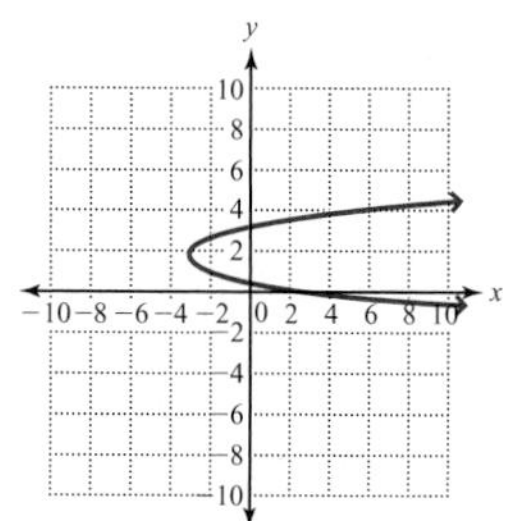

For Exercises 33–44, find the distance between the two points.

33. $(-4, 2)$ and $(-1, 6)$

34. $(5, -1)$ and $(1, 2)$

35. $(-8, -4)$ and $(-3, 8)$

36. $(-3, 2)$ and $(3, -6)$

37. $(-8, -10)$ and $(4, -5)$

38. $(4, -6)$ and $(10, 2)$

39. $(2, 4)$ and $(4, 8)$

40. $(-3, 2)$ and $(1, -4)$

41. $(-5, 2)$ and $(3, -2)$

42. $(3, -4)$ and $(7, 2)$

43. $(6, -2)$ and $(1, -5)$

44. $(3, -6)$ and $(-2, 1)$

For Exercises 45–48, the coordinates of the center of a circle and a point on the circle are given. Find the radius of the circle.

45. center: $(4, 2)$, point on the circle: $(8, -1)$

46. center: $(-4, 6)$, point on the circle: $(2, -2)$

47. center: $(2, -6)$, point on the circle: $(10, -1)$

48. center: $(3, -4)$, point on the circle: $(6, 8)$

For Exercises 49–64, find the center and radius and draw the graph.

49. $(x-2)^2+(y-1)^2=4$

50. $(x-1)^2+(y-3)^2=25$

51. $(x+3)^2+(y+2)^2=81$

52. $(x+5)^2+(y+4)^2=36$

53. $(x-5)^2+(y+3)^2=49$

54. $(x+6)^2+(y-2)^2=1$

55. $(x+1)^2+(y-3)^2=18$

56. $(x-6)^2+(y+1)^2=12$

57. $(x+4)^2+(y+2)^2=32$

58. $(x-6)^2+(y+5)^2=8$

59. $x^2+y^2-2x-6y-39=0$

60. $x^2+y^2+8x-6y+16=0$

61. $x^2+y^2+10x-4y-35=0$

62. $x^2+y^2+12x+10y+60=0$

63. $x^2 + y^2 + 14x - 4y + 49 = 0$

64. $x^2 + y^2 + 8x - 10y + 16 = 0$

For Exercises 65–68, match the equation with the correct graph.

65. $(x - 2)^2 + (y + 3)^2 = 25$

66. $(x + 2)^2 + (y - 3)^2 = 25$

67. $x^2 + y^2 + 8x + 2y - 8 = 0$

68. $x^2 + y^2 + 2x - 8y - 8 = 0$

a.

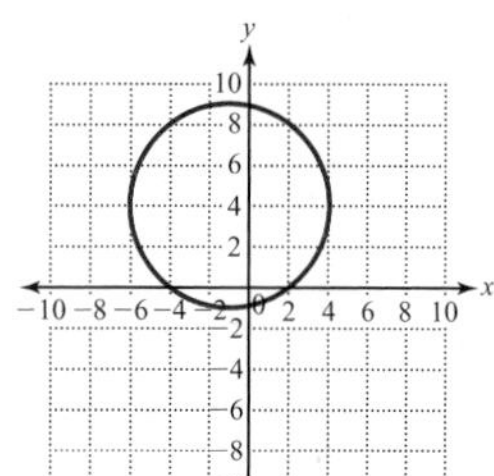

b.

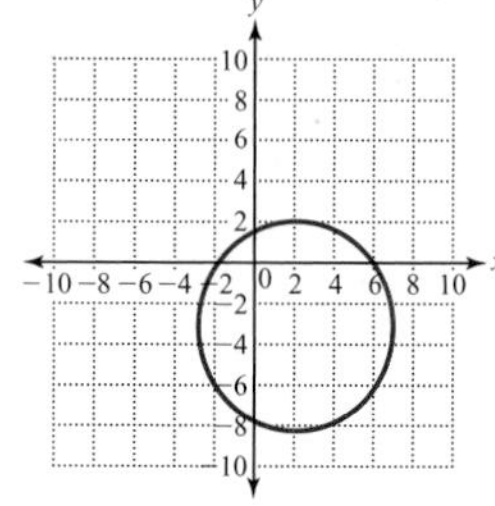

c.

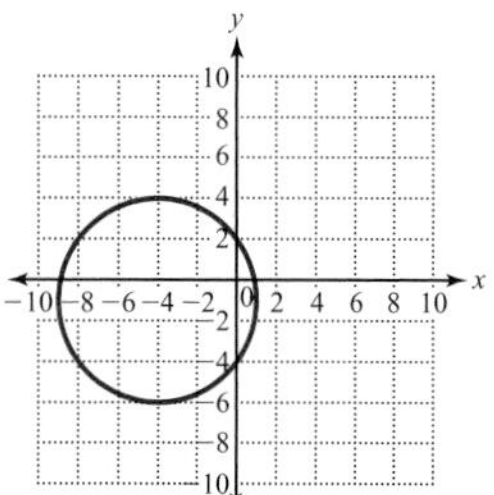

d.

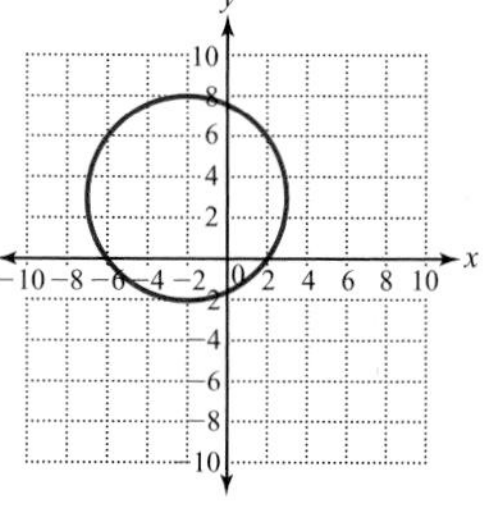

For Exercises 69–72, graph using a graphing calculator.

69. $x^2 + y^2 = 49$

70. $x^2 + y^2 = 36$

71. $(x - 2)^2 + (y + 3)^2 = 25$

72. $(x + 4)^2 + (y - 1)^2 = 9$

For Exercises 73–80, the center and radius of a circle are given. Write the equation of the circle in standard form.

73. center: $(8, -1)$; radius: 6

74. center: $(-3, 2)$; radius: 6

75. center: $(-4, -3)$; radius: 5

76. center: $(-6, -5)$; radius: 2

77. center: $(6, -2)$; radius: $\sqrt{14}$

78. center: $(-3, -3)$; radius: $\sqrt{26}$

79. center: $(-5, 2)$; radius: $3\sqrt{5}$

80. center: $(6, 2)$; radius: $2\sqrt{6}$

For Exercises 81–84, the center of a circle and a point on the circle are given. Write the equation of the circle in standard form.

81. center: $(2, 4)$, point on the circle: $(5, 8)$

82. center: $(-4, 3)$, point on the circle: $(4, 9)$

83. center: $(2, 4)$, point on the circle: $(7, 16)$

84. center: $(-6, 8)$, point on the circle: $(-12, 0)$

85. Write the equation of the set of all points that are a distance of 8 units from $(2, -5)$.

86. Write the equation of the set of all points that are a distance of 10 units from $(-3, -7)$.

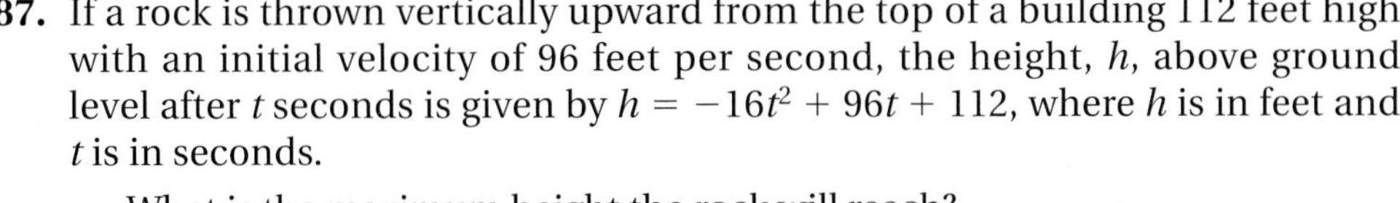

87. If a rock is thrown vertically upward from the top of a building 112 feet high with an initial velocity of 96 feet per second, the height, h, above ground level after t seconds is given by $h = -16t^2 + 96t + 112$, where h is in feet and t is in seconds.

a. What is the maximum height the rock will reach?

b. How many seconds will it take the rock to reach its maximum height?

c. How many seconds will it take the rock to hit the ground?

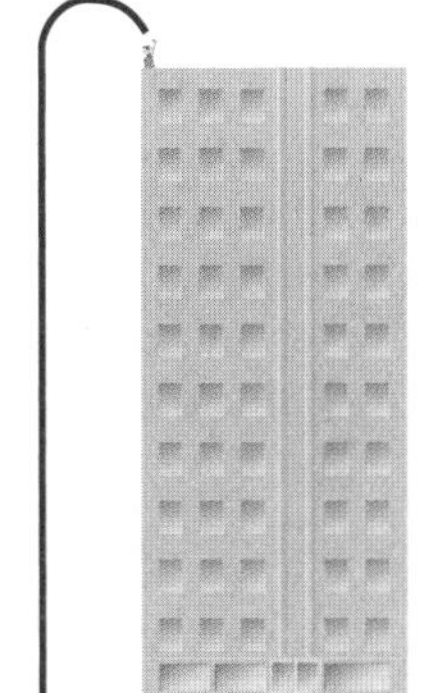

88. The path of a shell fired from ground level is in the shape of the parabola $y = 4x - x^2$, where x and y are given in kilometers.

a. How high does the shell go?

b. How many seconds does it take the shell to reach its maximum height?

c. How far from its firing point does the shell land?

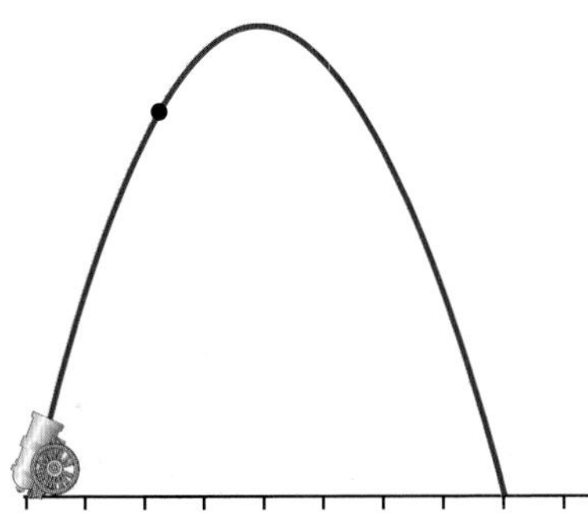

89. Arches in the shapes of parabolas are often used in construction. Find the equation of a parabolic arc that is 18 feet high at its highest point and 30 feet wide at the base, as illustrated in the following figure. Place the origin at the midpoint of the base.

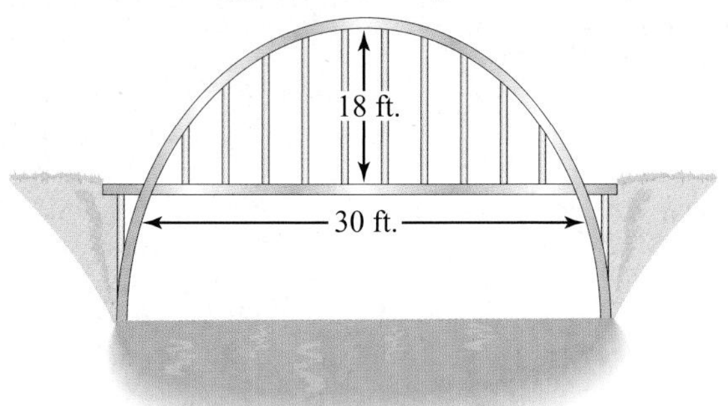

90. The cross sections of satellite dishes are in the shape of parabolas. Find the equation of a dish that is 6 feet across and 1 foot deep if the vertex is at the origin and the parabola is opening upward.

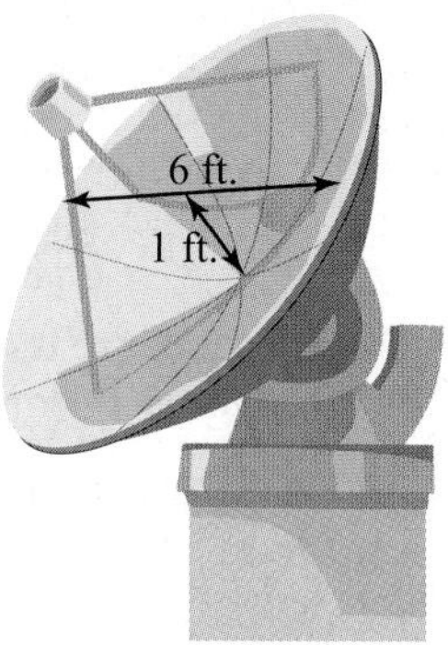

91. The percent of deaths by age per million miles driven can be approximated by the equation $y = 0.0038x^2 - 0.3475x + 8.316$, where x is the age and y is the percent. Find the percent of deaths per million miles for drivers 17 years old. (*Source:* National Highway Traffic Safety Administration)

92. The number of drivers involved in fatal accidents for a given blood alcohol content (BAC) can be approximated by $y = -8862.5x^2 + 26622.6x + 332$, where x is the BAC and y is the number of drivers involved in fatal accidents. Find the number of drivers involved in fatal accidents who had a BAC of 0.20. (*Source:* National Highway Traffic Safety Administration)

93. Bill and Don are fishing in the Gulf of Mexico in separate boats that are equipped with radios with a range of 20 miles. If we put Bill's radio at the origin of a coordinate system, what is the equation of all possible locations of Don's boat where the radios would be at their maximum range.

94. A toy plane is attached to a string pinned to the ceiling so that the plane flies in a circle. If the string is 4 feet long, write an equation that describes the path of the plane if the pin is at the origin.

95. A Ferris wheel has a diameter of 200 feet and the bottom of the Ferris wheel is 10 feet above the ground. Find the equation of the wheel if the origin is placed on the ground directly below the center of the wheel, as illustrated.

96. The Fermilab Tunnel houses the world's largest superconducting synchrotron. A cross section of the tunnel is a circle with radius of 1000 meters. Find the equation of a cross section of the tunnel if the center of the circle is at the center of the tunnel.

97. If a satellite is placed in a circular orbit of 230 kilometers above the Earth, what is the equation of the path of the satellite if the origin is placed at the center of the Earth (the radius of the Earth is approximately 6370 kilometers)?

98. The minute hand of Big Ben is 14 feet long. What is the equation of the circle swept out by the tip of the hand as it makes one complete revolution?

PUZZLE PROBLEM

Using only a pencil, a rough circle and its center can both be drawn without the point of the pencil losing contact with the paper resulting in a picture like the one shown. Explain how.

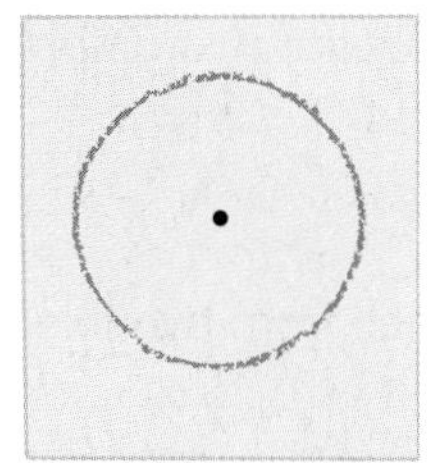

REVIEW EXERCISES

[3.1] **1.** Following is a coordinate system with the point (2, 3) plotted. Plot the points that are 4 units to the left and right of (2, 3) and give their coordinates.

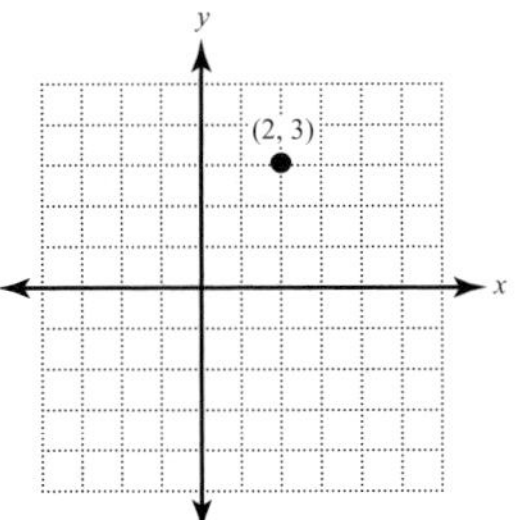

[3.1] **2.** Graph $y = \frac{2}{3}x$ and $y = -\frac{2}{3}x$ on the same set of axes.

[6.3] **3.** Factor: $25x^2 - 9y^2$

[9.1] **4.** Solve: $\frac{x^2}{16} = 1$

[3.1, 9.1] **5.** Find the x- and y-intercepts of $9x^2 + 16y^2 = 144$.

[3.1, 9.1] **6.** Find the x- and y-intercepts of $25x^2 + 9y^2 = 225$.

11.2 Ellipses and Hyperbolas

OBJECTIVES

1. Graph ellipses.
2. Graph hyperbolas.

OBJECTIVE 1. Graph ellipses. Suppose you drive two nails in a board and tie a string to the two nails. Now take a pencil, pull the string taut, and draw a figure around the two nails.

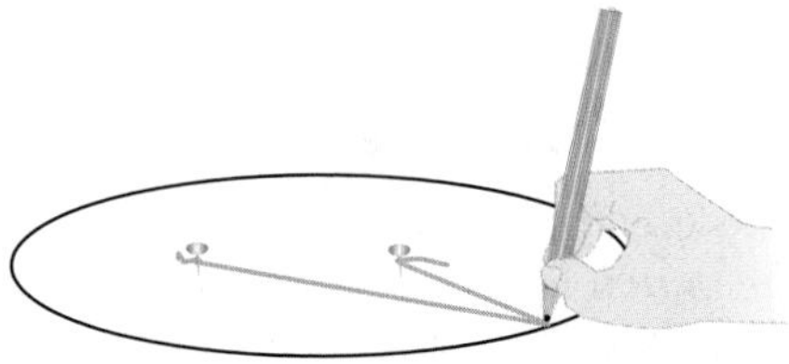

The figure you've drawn is called an **ellipse**. The locations of the two nails are the *focal points*.

DEFINITION ***Ellipse:*** The set of all points the sum of whose distance from two fixed points is constant.

Ellipses occur in many situations. The orbits of the planets about the Sun are elliptical with the Sun at one focal point. The orbits of satellites about the Earth are also elliptical. The cams of compound bows are elliptical, which allows a decrease in the amount of effort required to hold the bow at full draw.

In the definition of an ellipse, the two fixed points are the *foci* (plural of *focus*) and the point halfway between the foci is the *center*. The following figure shows the graph of an ellipse with foci at $(c, 0)$ and $(-c, 0)$, x-intercepts at $(a, 0)$ and $(-a, 0)$, and y-intercepts at $(0, b)$ and $(0, -b)$. Consequently, the center is at the origin, $(0, 0)$. It can be shown that $c^2 = a^2 - b^2$ for ellipses in which $a > b$.

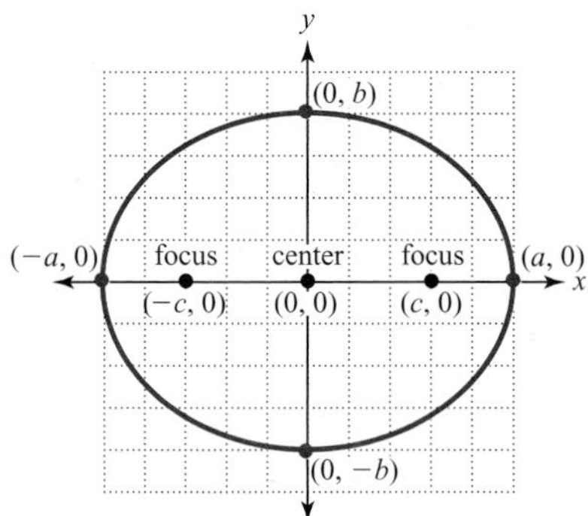

Using the distance formula, it can be shown that an ellipse with these characteristics has the following equation.

RULE **The Equation of an Ellipse Centered at (0, 0)**

The equation of an ellipse with center (0, 0), x-intercepts $(a, 0)$ and $(-a, 0)$, and y-intercepts $(0, b)$ and $(0, -b)$ is

$$\frac{x^2}{a^2} + \frac{y^2}{b^2} = 1$$

EXAMPLE 1 Graph each ellipse and label the x- and y-intercepts.

a. $\dfrac{x^2}{16} + \dfrac{y^2}{9} = 1$

Solution The equation can be rewritten as $\dfrac{x^2}{4^2} + \dfrac{y^2}{3^2} = 1$, so $a = 4$ and $b = 3$. Consequently, the x-intercepts are (4, 0) and (−4, 0) and the y-intercepts are (0, 3) and (0, −3).

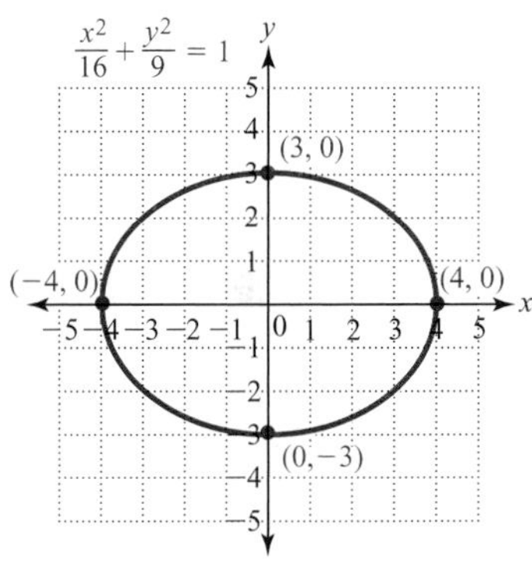

b. $25x^2 + 9y^2 = 225$

Solution We first need to write the equation in standard form: $\dfrac{x^2}{a^2} + \dfrac{y^2}{b^2} = 1$.

$$\frac{25x^2}{225} + \frac{9y^2}{225} = \frac{225}{225} \quad \text{Divide both sides by 225.}$$

$$\frac{x^2}{9} + \frac{y^2}{25} = 1 \quad \text{Simplify both sides.}$$

We now see that this is an equation of an ellipse with $a = 3$ and $b = 5$, so the x-intercepts are (3, 0) and (−3, 0) and the y-intercepts are (0, 5) and (0, −5).

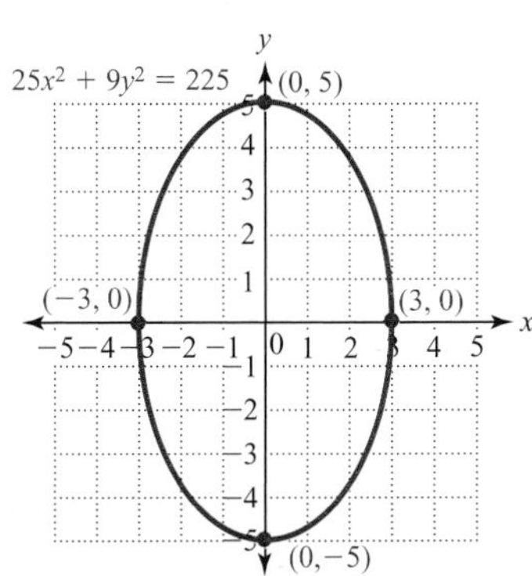

If the center of an ellipse is not at the origin, it can be shown that the equation has the following form.

RULE General Equation for an Ellipse

The equation of an ellipse with center (h, k) is $\frac{(x-h)^2}{a^2} + \frac{(y-k)^2}{b^2} = 1$. The ellipse passes through two points that are a units to the left and right of the center, and two points that are b units above and below the center.

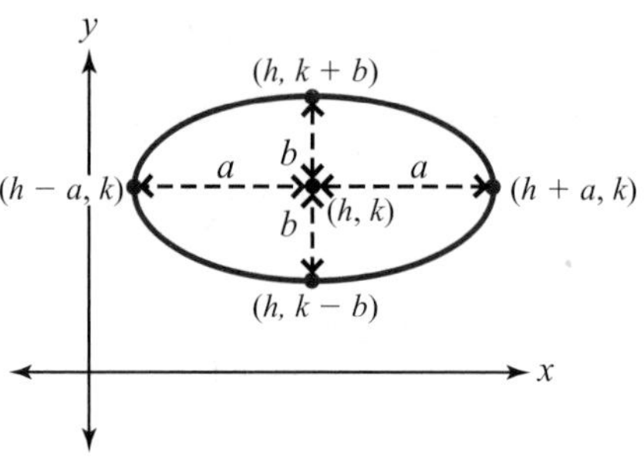

ANSWERS

a. $\frac{x^2}{36} + \frac{y^2}{9} = 1$

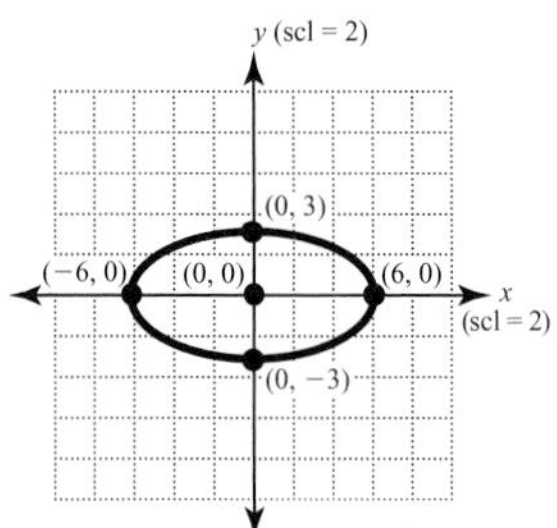

b. $\frac{(x+1)^2}{25} + \frac{(y-2)^2}{36} = 1$

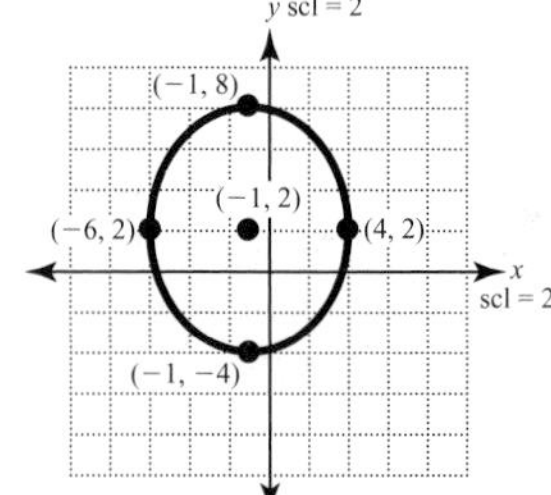

EXAMPLE 2 Graph the ellipse. Label the center and the points above, below, to the left, and to the right of the center. $\frac{(x-2)^2}{36} + \frac{(y+3)^2}{16} = 1$

Solution Since $h = 2$, and $k = -3$, the center of the ellipse is $(2, -3)$. Also, we see that $a = 6$, which means the ellipse passes through two points 6 units to the right and left of $(2, -3)$. These points are $(8, -3)$ and $(-4, -3)$. Since $b = 4$, the ellipse passes through two points 4 units above and below the center. These points are $(2, 1)$ and $(2, -7)$.

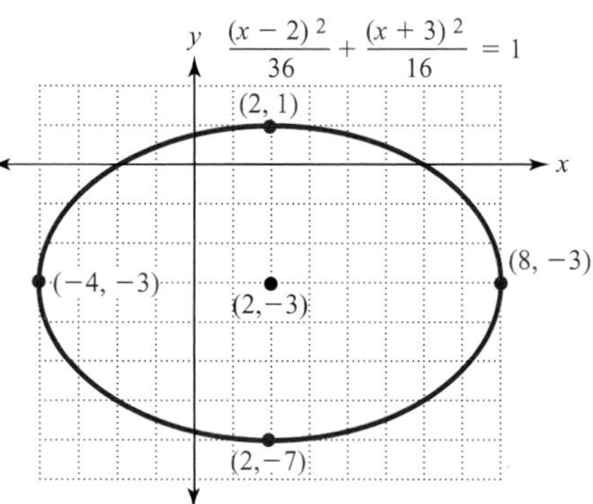

YOUR TURN Graph each ellipse. Label the center and the points above, below, to the left, and to the right of the center.

a. $\frac{x^2}{36} + \frac{y^2}{9} = 1$

b. $\frac{(x+1)^2}{25} + \frac{(y-2)^2}{36} = 1$

OBJECTIVE 2. Graph hyperbolas. The last conic is the **hyperbola**. Applications of hyperbolas include the LORAN tracking system and the orbits of some comets.

DEFINITION ***Hyperbola:*** The set of all points the difference of whose distances from two fixed points remains constant.

Just as with the ellipse, the two fixed points are the *foci* and the point halfway between the foci is the *center.*

RULE **Equations of Hyperbolas in Standard Form**

The equation of a hyperbola with center (0, 0), x-intercepts $(a, 0)$ and $(-a, 0)$, and no y-intercepts is $\frac{x^2}{a^2} - \frac{y^2}{b^2} = 1$.

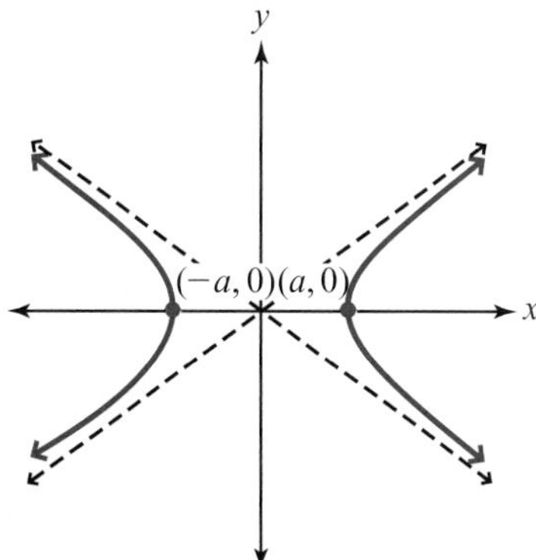

The equation of a hyperbola with center (0, 0), y-intercepts $(0, b)$ and $(0, -b)$, and no x-intercepts is $\frac{y^2}{b^2} - \frac{x^2}{a^2} = 1$.

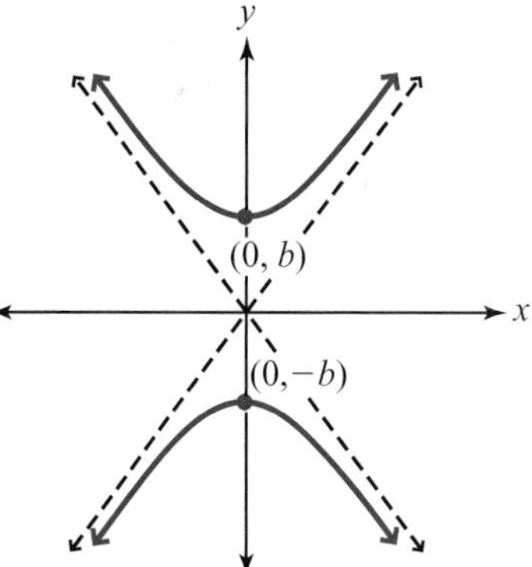

Learning Strategy

Noting which variable term is positive in the Standard form equation of a hyperbola can help you remember how to orient the graph. If the x^2 term is positive, the hyperbola's intercepts are on the x-axis. If the y^2 term is positive, the hyperbola's intercepts are on the y-axis.

The dashed lines that intersect at the center of a hyperbola are *asymptotes* and are not a part of the graph, but are used as an aid in graphing. An asymptote is a line that the graph approaches, but does not cross, as the graph goes away from the origin. The rectangle whose vertices are (a, b), $(-a, b)$, $(a, -b)$, and $(-a, -b)$ is called the *fundamental rectangle*, and the asymptotes are the extended diagonals of the fundamental rectangle, as in the following illustration.

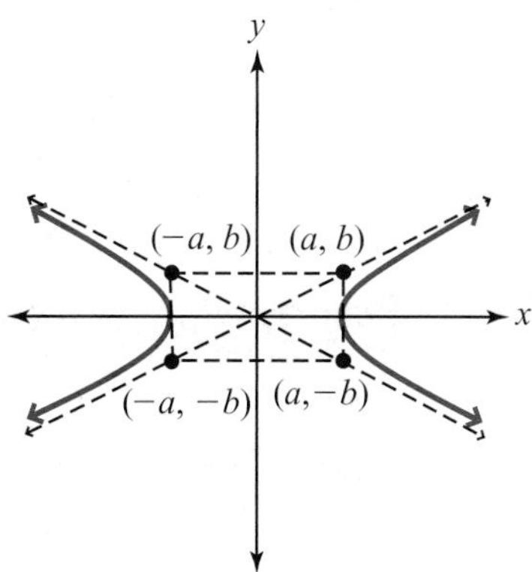

PROCEDURE **Graphing a Hyperbola in Standard Form**

1. Find the intercepts. If the x^2 term is positive, the x-intercepts are $(a, 0)$ and $(-a, 0)$ and there are no y-intercepts. If the y^2 term is positive, the y-intercepts are $(0, b)$ and $(0, -b)$ and there are no x-intercepts.
2. Draw the fundamental rectangle. The vertices are (a, b), $(-a, b)$, $(a, -b)$, and $(-a, -b)$.
3. Draw the asymptotes, which are the extended diagonals of the fundamental rectangle.
4. Draw the graph so that each branch passes through an intercept and approaches the asymptotes the farther they are from the origin.

EXAMPLE 3 Graph each hyperbola. Also show the fundamental rectangle with its corner points labeled, the asymptotes, and the intercepts.

a. $\dfrac{x^2}{16} - \dfrac{y^2}{9} = 1$

Solution

1. This equation can be written as $\dfrac{x^2}{4^2} - \dfrac{y^2}{3^2} = 1$, so $a = 4$ and $b = 3$. Since the x^2 term is positive, the graph has x-intercepts at $(4, 0)$ and $(-4, 0)$.
2. The fundamental rectangle has vertices at $(4, 3)$, $(-4, 3)$, $(4, -3)$, and $(-4, -3)$.
3. Draw the asymptotes.
4. Sketch the graph so it passes through the x-intercepts and then approaches the asymptotes.

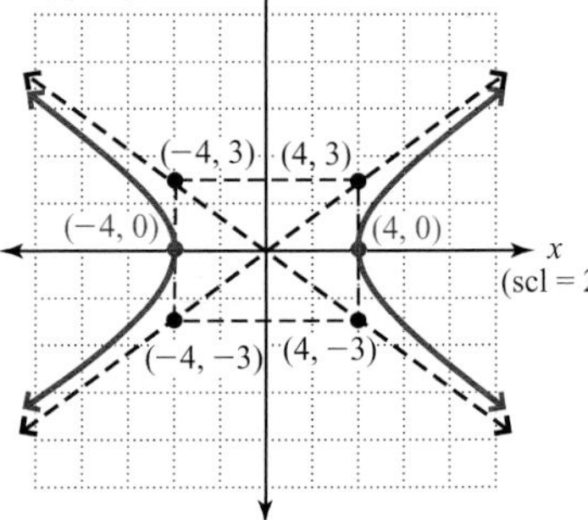

b. $\dfrac{y^2}{4} - \dfrac{x^2}{9} = 1$

Solution

1. This equation can be written as $\dfrac{y^2}{2^2} - \dfrac{x^2}{3^2} = 1$, so $b = 2$ and $a = 3$. Since the y^2 term is positive, the graph has y-intercepts at $(0, 2)$ and $(0, -2)$.
2. The fundamental rectangle has vertices at $(3, 2)$, $(-3, 2)$, $(3, -2)$, and $(-3, -2)$.
3. Draw the asymptotes.
4. Sketch the graph so it passes through the y-intercepts and then approaches the asymptotes.

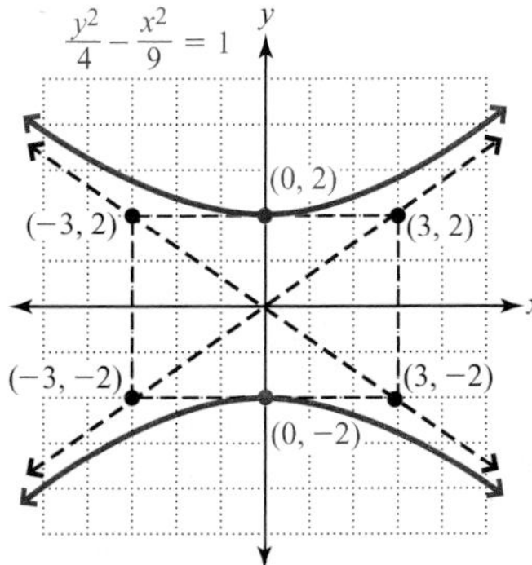

YOUR TURN Graph the hyperbola. Also show the fundamental rectangle with its corner points labeled, the asymptotes, and the intercepts.

$$\frac{x^2}{25} - \frac{y^2}{16} = 1$$

ANSWER

$\dfrac{x^2}{25} - \dfrac{y^2}{16} = 1$

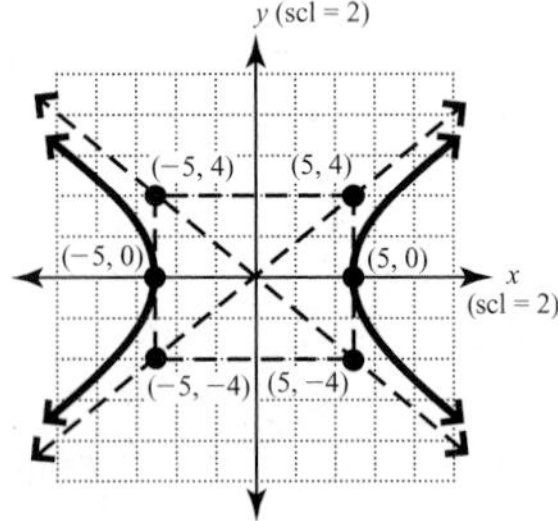

Calculator **TIPS**

Graph the ellipse $\frac{x^2}{9} + \frac{y^2}{16} = 1$. *We solve for y just as when graphing a circle.*

$$\frac{x^2}{9} + \frac{y^2}{16} = 1$$

$$\frac{y^2}{16} = 1 - \frac{x^2}{9} \quad \text{Subtract } \frac{x^2}{9} \text{ from both sides.}$$

$$y^2 = 16\left(1 - \frac{x^2}{9}\right) \quad \text{Multiply both sides by 16.}$$

$$y = \pm 4\sqrt{1 - \frac{x^2}{9}} \quad \text{Apply the square root principle.}$$

Define $Y_1 = 4\sqrt{1 - \frac{x^2}{9}}$ *and* $Y_2 = -4\sqrt{1 - \frac{x^2}{9}}$ *and graph in a window* $[-5, 5]$ *for x and* $[-5, 5]$ *for y and square the window.*

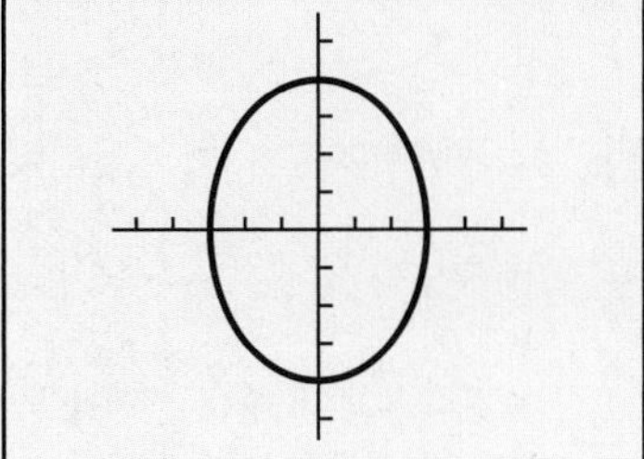

Following is a summary of the conic sections.

Standard Forms

Parabola

$y = a(x - h)^2 + k, a > 0$

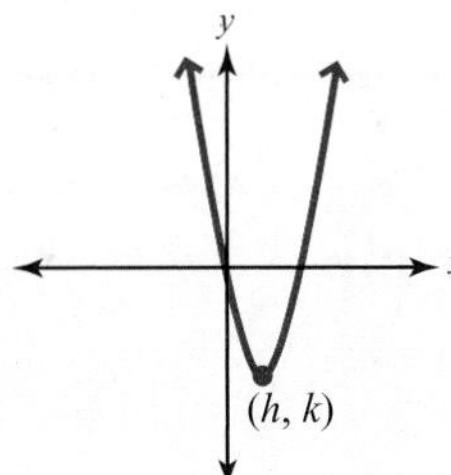

$y = a(x - h)^2 + k, a < 0$

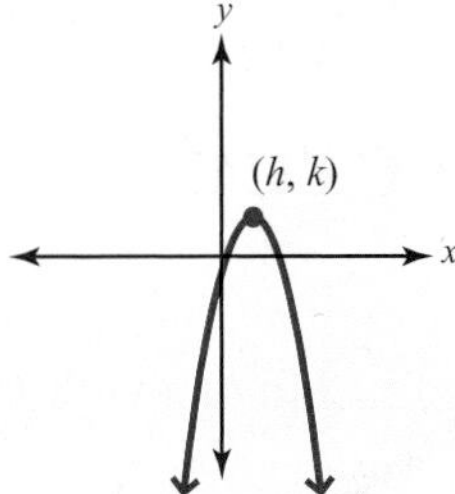

$x = a(y - k)^2 + h, a > 0$

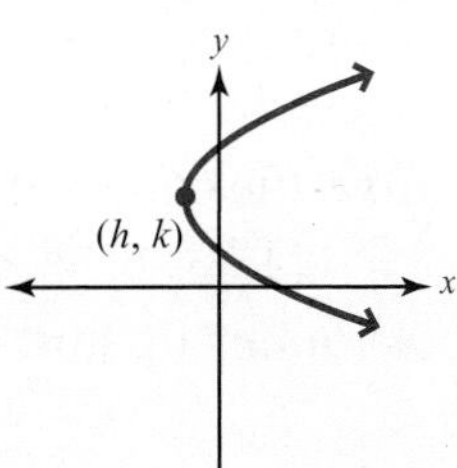

$x = a(y - k)^2 + h, a < 0$

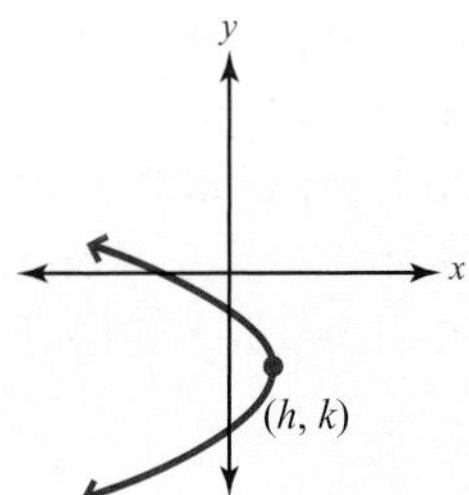

continued

Standard Forms continued		
Circle	$x^2 + y^2 = r^2$	$(x - h)^2 + (y - k)^2 = r^2$
	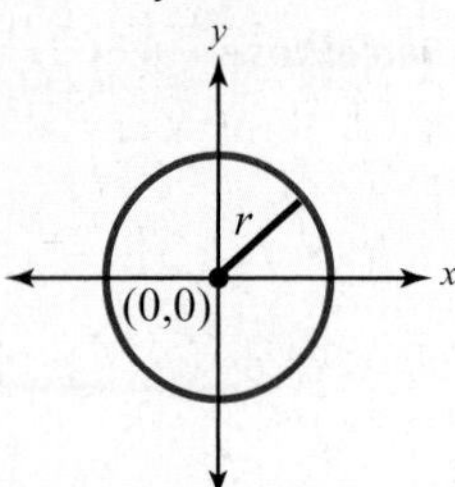	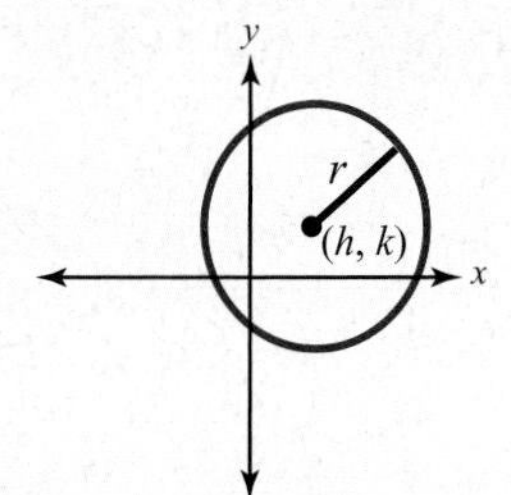
Ellipse	$\frac{x^2}{a^2} + \frac{y^2}{b^2} = 1$	$\frac{(x - h)^2}{a^2} + \frac{(y - k)^2}{b^2} = 1$
	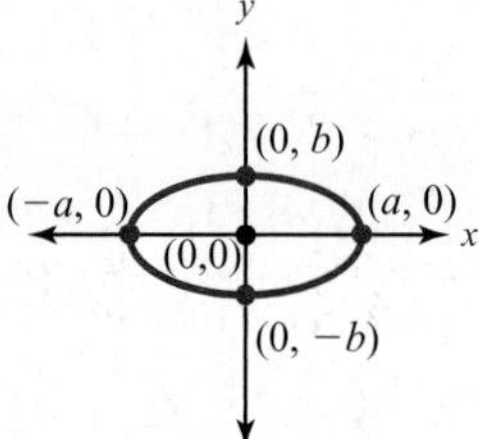	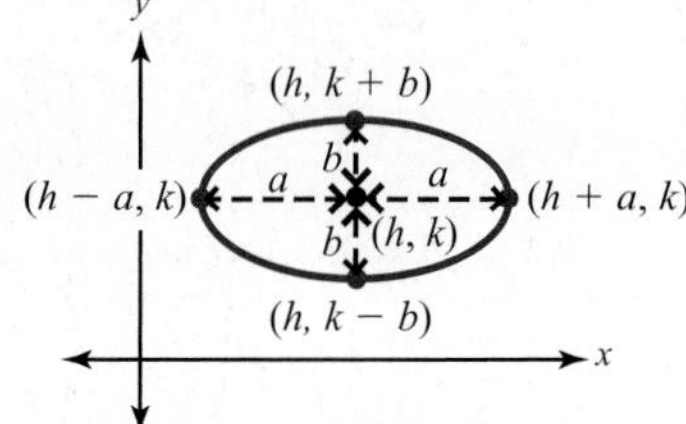
Hyperbola	$\frac{x^2}{a^2} - \frac{y^2}{b^2} = 1$	$\frac{y^2}{b^2} - \frac{x^2}{a^2} = 1$
		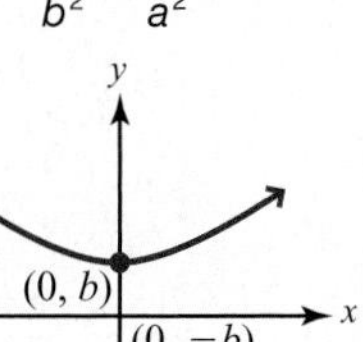

11.2 Exercises

For Extra Help

MyMathLab |  Videotape/DVT | InterAct Math | Math Tutor Center | Math XL.com

1. What is the definition of an ellipse?

2. In the definitions of the ellipse and hyperbola, two fixed points are mentioned. What are these two points called?

3. How can you tell the equation of an ellipse from the equation of a hyperbola?

4. How do you determine whether a hyperbola opens up and down or left and right?

5. Which one conic has an equation that has either x^2 or y^2, but not both?

6. What do the graphs of $\frac{x^2}{9} - \frac{y^2}{16} = 1$ and $\frac{x^2}{9} + \frac{y^2}{16} = 1$ have in common?

For Exercises 7–18, graph each ellipse. Label the center and the points above, below, to the left, and to the right of the center.

7. $\frac{x^2}{25} + \frac{y^2}{4} = 1$

8. $\frac{x^2}{81} + \frac{y^2}{64} = 1$

9. $\frac{x^2}{36} + y^2 = 1$

10. $\frac{x^2}{9} + \frac{y^2}{16} = 1$

11. $4x^2 + 9y^2 = 36$

12. $25x^2 + 4y^2 = 100$

13. $36x^2 + 4y^2 = 144$

14. $x^2 + 9y^2 = 36$

15. $\frac{(x-1)^2}{49} + \frac{(y+3)^2}{25} = 1$

16. $\frac{(x+3)^2}{25} + \frac{(y-2)^2}{64} = 1$

17. $\frac{(x-4)^2}{4} + \frac{(y+3)^2}{36} = 1$

18. $\frac{(x+5)^2}{9} + \frac{(y+3)^2}{25} = 1$

For Exercises 19–22, graph using a graphing calculator.

19. $\frac{x^2}{25} + \frac{y^2}{4} = 1$

20. $\frac{y^2}{12} + \frac{x^2}{8} = 1$

21. $\frac{(y-2)^2}{36} + \frac{(x+4)^2}{25} = 1$

22. $\frac{(x+1)^2}{16} + \frac{(y+2)^2}{4} = 1$

23. If the x-intercepts of an ellipse are $(-3, 0)$ and $(3, 0)$ and the y-intercepts are $(0, 4)$ and $(0, -4)$, what is the equation of the ellipse?

24. If the x-intercepts of an ellipse are $(-6, 0)$ and $(6, 0)$ and the y-intercepts are $(0, 3)$ and $(0, -3)$, what is the equation of the ellipse?

For Exercises 25–32, graph each hyperbola. Also show the fundamental rectangle with its corner points labeled, the asymptotes, and the intercepts.

25. $\frac{x^2}{9} - \frac{y^2}{4} = 1$

26. $\frac{x^2}{16} - \frac{y^2}{25} = 1$

27. $\frac{y^2}{36} - \frac{x^2}{9} = 1$

28. $\frac{y^2}{4} - \frac{x^2}{25} = 1$

29. $9x^2 - y^2 = 36$

30. $x^2 - 4y^2 = 16$

31. $16y^2 - 4x^2 = 64$

32. $9y^2 - 25x^2 = 225$

For Exercises 33–34, graph using a graphing calculator.

33. $\dfrac{x^2}{36} - \dfrac{y^2}{4} = 1$

34. $\dfrac{y^2}{9} - \dfrac{x^2}{25} = 1$

35. If a hyperbola opens left and right and the vertices of the fundamental rectangle are (3, 2), (3, −2), (−3, 2) and (−3, −2), what is the equation of the hyperbola?

36. If a hyperbola opens up and down and the vertices of the fundamental rectangle are (5, 3), (5, −3), (−5, 3) and (−5, −3), what is the equation of the hyperbola?

For Exercises 37–40, match the equation with the graph.

37. $\dfrac{x^2}{9} + \dfrac{y^2}{25} = 1$

38. $\dfrac{x^2}{9} - \dfrac{y^2}{25} = 1$

39. $\dfrac{x^2}{25} + \dfrac{y^2}{9} = 1$

40. $\dfrac{x^2}{25} - \dfrac{y^2}{9} = 1$

a.

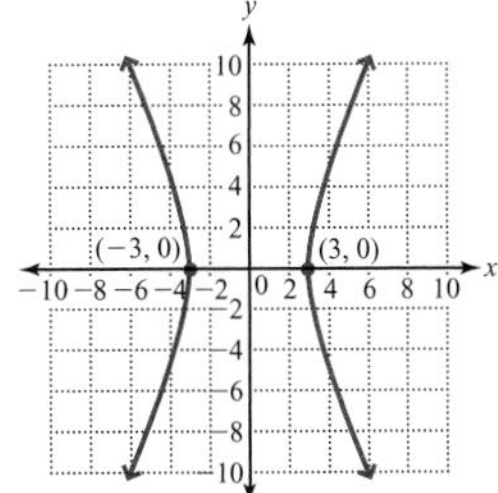

b.

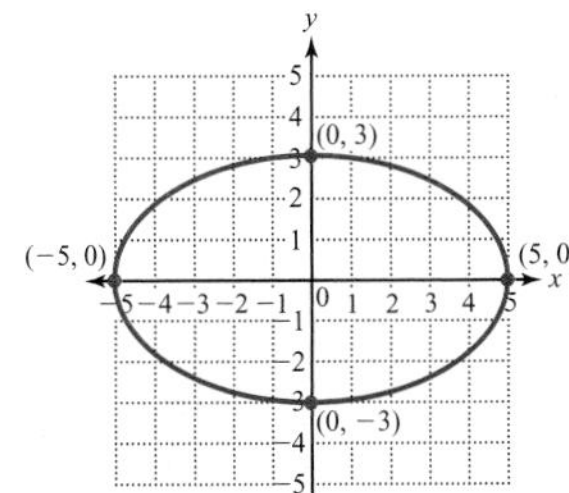

c.

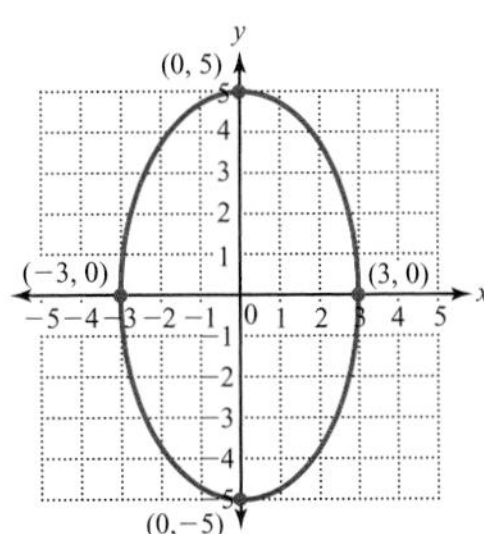

d.

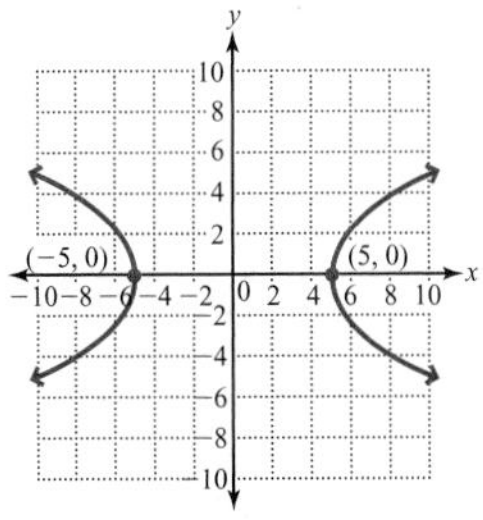

For Exercises 41–44, determine whether the graph of the equation is a circle, parabola, ellipse, or hyperbola. Do not draw the graph.

41. $9x^2 + 16y^2 = 144$

42. $16y^2 - 9x^2 = 144$

43. $x^2 + y^2 - 6x + 8y - 75 = 0$

44. $2x^2 - 12x + 23 - y = 0$

For Exercises 45–58, indicate whether the graph of the given equation is a circle, parabola, ellipse, or hyperbola, then draw the graph.

45. $(x - 2)^2 + (y + 2)^2 = 49$

46. $x^2 + y^2 = 81$

47. $y = 2(x + 1)^2 + 3$

48. $x = (y + 3)^2 - 4$

49. $\frac{x^2}{36} + \frac{y^2}{16} = 1$

50. $\frac{x^2}{36} + \frac{y^2}{81} = 1$

51. $\frac{x^2}{4} - \frac{y^2}{25} = 1$

52. $\frac{y^2}{49} - \frac{x^2}{4} = 1$

53. $y = 2x^2 + 8x + 6$

54. $x = y^2 + 6$

55. $\frac{x^2}{16} - \frac{y^2}{16} = 1$

56. $x = -2y^2 + 4y - 3$

57. $\dfrac{(x+3)^2}{9} + \dfrac{(y+2)^2}{36} = 1$

58. $\dfrac{(x-1)^2}{4} + (y+3)^2 = 1$

For Exercises 59–66, solve.

59. A bridge over a waterway has an arch in the form of half an ellipse. The equation of the ellipse is $400x^2 + 256y^2 = 102{,}400$.

a. A sailboat, the top of whose mast is 18 feet above the water, is approaching the arch. Will the mast clear the bridge? Why or why not?

b. How wide is the base of the arch?

60. A highway passes beneath an overpass that is in the shape of half an ellipse. The overpass is 15 feet high at the center and 40 feet wide at the base.

a. What is the equation of the ellipse?

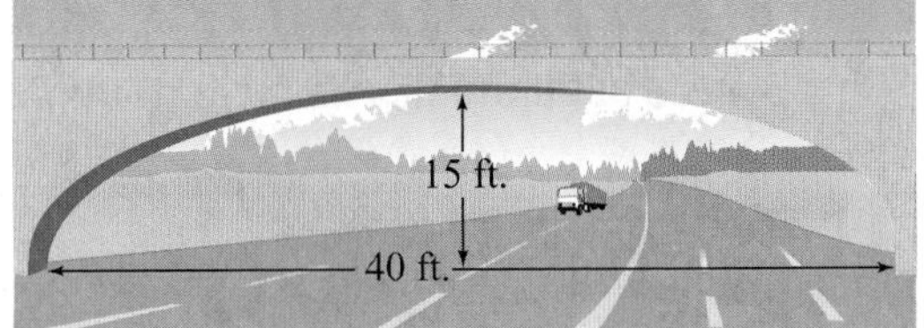

b. A truck that is 10 feet wide and carrying a load that is 14 feet high is approaching the bridge. If the truck goes down the middle of the road, will the load clear the bridge? Why or why not?

61. The comet Epoch has an orbit that is in the shape of an ellipse with the Sun at one of the foci. The equation is approximately $\dfrac{x^2}{3.6^2} + \dfrac{y^2}{2.88^2} = 1$, where x and y are in astronomical units (an astronomical unit is 93,000,000 miles). Sketch the graph of the comet Epoch. (*Source: Orbital Motion,* A. E. Roy, Institute of Physics Publishing, London)

62. The planet Pluto has an orbit that is in the shape of an ellipse with the Sun at one of the foci. The equation is approximately $\dfrac{x^2}{39.4^2} + \dfrac{y^2}{38.2^2} = 1$, where x and y are measured in astronomical units. Sketch the graph of the orbit of Pluto. (*Hint:* Make each unit on the x- and y-axes equal to 10.) (*Source: Orbital Motion,* A. E. Roy, Institute of Physics Publishing, London)

63. Compound bows have elliptical cams that decrease the amount of effort required to hold the bow at full draw. If a cam on a bow is 4 inches from top to bottom and 3 inches across, what is the equation of the ellipse if the origin is at the center of the cam?

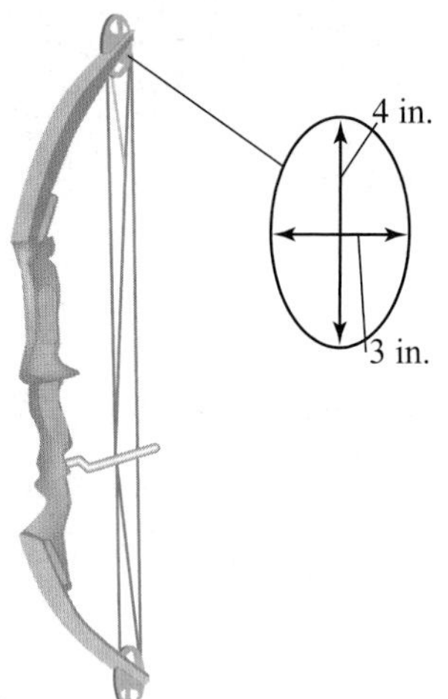

64. One of the most popular exercise machines is an elliptical trainer in which your foot moves in an elliptical path. On a typical machine, the length of the stride is about 19 inches and the height varies from 3 to 5 inches depending on the settings. Write the equation for the path that your foot takes if the total length of the stride is 19 inches and the total height is 4 inches. (*Source:* Precor National Headquarters)

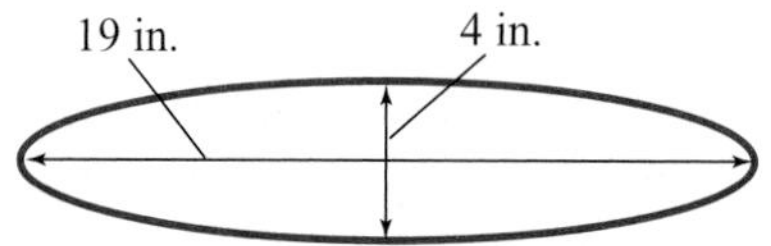

65. If a source of light or sound is placed at one focal point of an elliptic reflector, the light or sound is reflected through the other focal point. This principle is used in a lithotripter that uses sound waves to crush kidney stones by placing a source of sound at one focal point and the kidney stone at the other. If the elliptic reflector is based on the ellipse $\frac{x^2}{25} + \frac{y^2}{9} = 1$, how many units from the center should the kidney stone be placed? (*Hint:* $c^2 = a^2 - b^2$.)

66. The same principle used in Exercise 65 is also used in whispering rooms. The room is in the shape of an elliptic reflector where one person speaks at one focal point and the other places his or her ear at the other and can hear sounds as faint as a whisper. If a whispering room is based on the ellipse $\frac{x^2}{169} + \frac{y^2}{144} = 1$, how many units from the center of the ellipse would the two people stand? (*Hint:* $c^2 = a^2 - b^2$.)

Collaborative Exercises THE ELLIPTICAL TABLECLOTH

Recall that an ellipse can be drawn by fixing the ends of a string to the foci and then tracing out the ellipse. By considering the following figures we can determine an expression for the length of the string and also a relationship between a, b, and c.

1. Use the figure to write a formula for the length of the string.

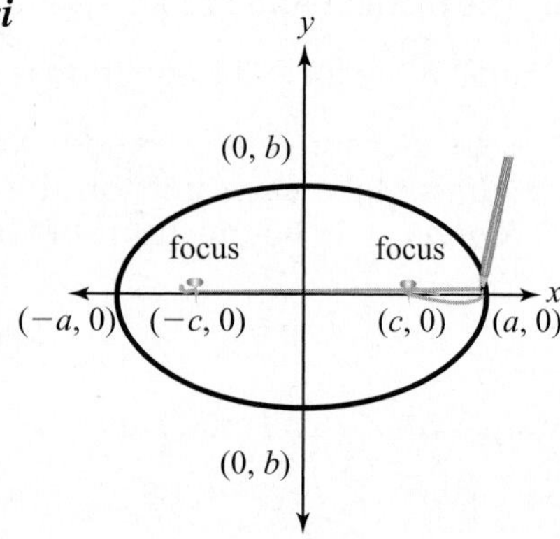

2. In the following figure, notice that the string forms an isosceles triangle and the y-axis splits that triangle into two identical right triangles.

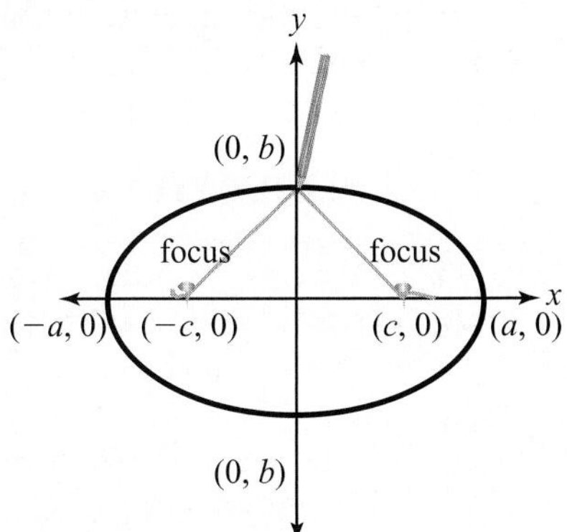

a. Find the length of the hypotenuse of each of those right triangles.

b. What expression describes the length of the string?

c. Use the Pythagorean theorem to write a formula relating a, b, and c.

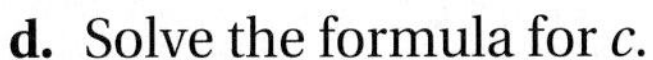

d. Solve the formula for c.

Suppose we are to make an elliptical tablecloth for an elliptical table that is 76 inches long and 58 inches wide. The tablecloth is to drape 6 inches over the edge of the table all the way around plus we need an additional inch for the hem. We have a large rectangular piece of cloth to make the tablecloth. To trace the ellipse on the cloth, we need to know the string length and the location of the foci.

3. Find the dimensions of the elliptical tablecloth taking into account the amount it needs to drape and also the hem.

4. How long must the string be in order to trace the ellipse?

5. How far from the center are the foci located?

PUZZLE PROBLEM

Suppose an ellipse is drawn using the method described on p. 802. Assuming the ellipse is centered at the origin, if the nails are 8 inches apart and the string is 10 inches in length, what are the x- and y-intercepts?

REVIEW EXERCISES

[2.1] **1.** Solve $3x^2 + y = 6$ for y.

[4.1] **2.** Solve the following system using the graphical method. $\begin{cases} x + y = 3 \\ 2x + y = 4 \end{cases}$

[4.1] **3.** Solve the following system using the substitution method. $\begin{cases} 2x + y = 1 \\ 3x + 4y = -6 \end{cases}$

[4.1] **4.** Solve the following system using the elimination method. $\begin{cases} 2x + 3y = 6 \\ 3x - 4y = -25 \end{cases}$

For Exercises 5 and 6, solve.

[6.4] **5.** $3x^2 + 10x - 8 = 0$

[9.1] **6.** $4x^2 = 36$

11.3 Nonlinear Systems of Equations

OBJECTIVES

1. **Solve nonlinear systems of equations using substitution.**
2. **Solve nonlinear systems of equations using elimination.**

We solved systems of linear equations in Chapter 4. Now we will solve **nonlinear systems** of equations.

DEFINITION ***Nonlinear system of equations:*** A system of equations that contains at least one nonlinear equation.

The types of equations in a nonlinear system determine the number of solutions that are possible for the system. For example, a system containing a quadratic equation and a linear equation can have 0, 1, or 2 points of intersection and therefore 0, 1, or 2 solutions, as shown by the following figures.

Connection In Chapter 4, we learned that the number of solutions for a system of linear equations depended on the relative positions of the graphs of each of the equations. We see a similar relationship between the graphs and the number of solutions here.

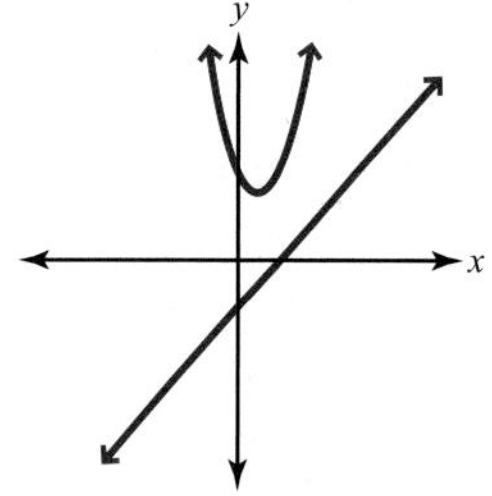

No points of intersection: no solutions

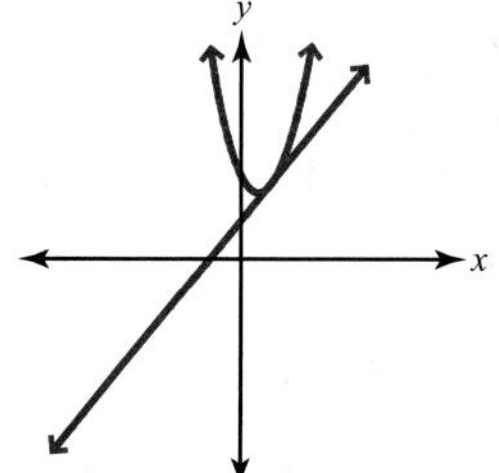

One point of intersection: one solution

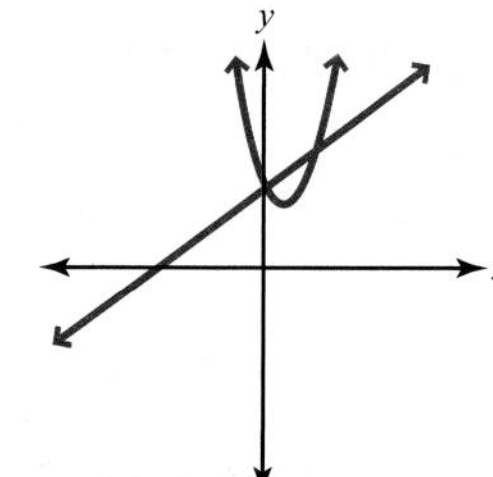

Two points of intersection: two solutions

OBJECTIVE 1. Solve nonlinear systems of equations using substitution. When solving nonlinear systems, if one equation is linear, or one of the variables in an equation is isolated, the substitution method is usually preferred.

EXAMPLE 1 Solve using the substitution method.

a. $\begin{cases} y = 2(x + 2)^2 - 2 \\ 2x + y = -2 \end{cases}$

Note: *The graphs are a parabola and a line, so there will be 0, 1, or 2 solutions.*

Solution It is easier to solve the linear equation for one of its variables and substitute into the nonlinear equation. Since the coefficient of y is 1, we will solve $2x + y = -2$ for y.

$y = -2x - 2$	Subtract $-2x$ from both sides.
$-2x - 2 = 2(x + 2)^2 - 2$	Substitute $-2x - 2$ for y in $y = 2(x + 2)^2 - 2$.
$-2x - 2 = 2(x^2 + 4x + 4) - 2$	To solve this quadratic equation, we need to write it in the form $ax^2 + bx + c = 0$. First, we square $x + 2$.
$-2x - 2 = 2x^2 + 8x + 6$	Distribute 2, then combine like terms.
$0 = 2x^2 + 10x + 8$	Add $2x$ and 2 to both sides.
$0 = x^2 + 5x + 4$	Divide both sides by 2.
$0 = (x + 1)(x + 4)$	Factor.
$0 = x + 1$ or $0 = x + 4$	Use the zero-factor theorem.
$-1 = x$ or $-4 = x$	Solve each equation.

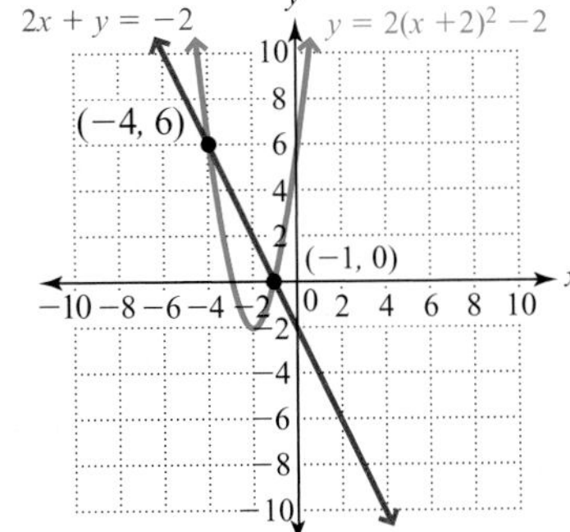
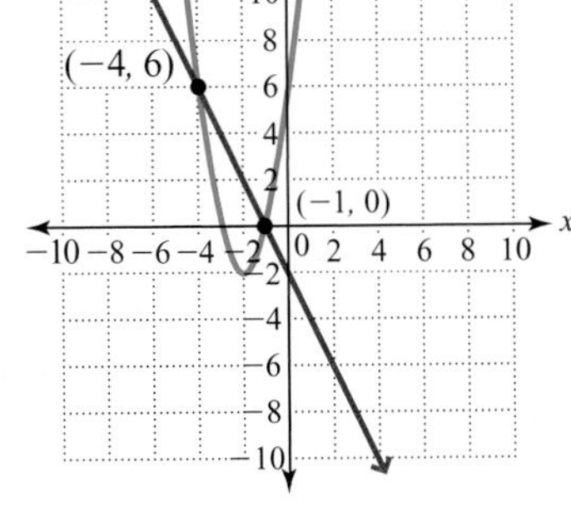

To find y, substitute -1 and -4 for x in $y = -2x - 2$.

$$y = -2(-1) - 2 \qquad y = -2(-4) - 2$$
$$y = 2 - 2 \qquad y = 8 - 2$$
$$y = 0 \qquad y = 6$$

The solutions are $(-1, 0)$ and $(-4, 6)$, as verified by the graph to the left.

b. $\begin{cases} y = \sqrt{x + 2} \\ x^2 + y^2 = 8 \end{cases}$

Solution Since $y = \sqrt{x + 2}$ is already solved for y, substitute $\sqrt{x + 2}$ for y in $x^2 + y^2 = 8$.

$x^2 + (\sqrt{x + 2})^2 = 8$	Substitute $\sqrt{x + 2}$ for y in $x^2 + y^2 = 8$.
$x^2 + x + 2 = 8$	Square $\sqrt{x + 2}$.
$x^2 + x - 6 = 0$	Since the equation is now quadratic, we subtract 8 from both sides to get the form $ax^2 + bx + c = 0$.
$(x + 3)(x - 2) = 0$	Factor
$x + 3 = 0$ or $x - 2 = 0$	Use the zero-factor theorem.
$x = -3$ or $x = 2$	Solve each equation.

To find y, substitute -3 and 2 for x in $y = \sqrt{x + 2}$.

$$y = \sqrt{-3 + 2} \qquad y = \sqrt{2 + 2}$$
$$y = \sqrt{-1} \qquad y = \sqrt{4}$$
$$y = 2$$

Since $\sqrt{-1}$ is imaginary, the only real solution is $(2, 2)$, as verified by the graph to the left.

Note: *A system containing a root function and a circle, as in Example 1b, can have 0, 1, or 2 solutions, as shown.*

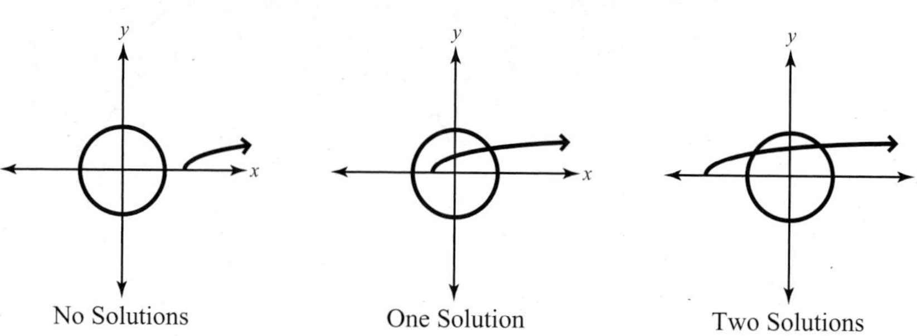

No Solutions   One Solution   Two Solutions

YOUR TURN Solve using the substitution method.

a. $\begin{cases} 2x - y = 1 \\ x^2 + y^2 = 1 \end{cases}$

b. $\begin{cases} xy = 3 \\ 4x - y = 1 \end{cases}$

ANSWERS

a. $\left(\frac{4}{5}, \frac{3}{5}\right)$, $(0, -1)$

b. $\left(-\frac{3}{4}, -4\right)$, $(1, 3)$

OBJECTIVE 2. Solve nonlinear systems of equations using elimination. If neither equation contains a radical expression or both equations contain the same powers of the variables, the elimination method can be used.

EXAMPLE 2 Solve using the elimination method. $\begin{cases} 9x^2 - 4y^2 = 20 \\ x^2 + y^2 = 8 \end{cases}$

Solution $\begin{cases} 9x^2 - 4y^2 = 20 & \text{(Equation 1)} \\ x^2 + y^2 = 8 & \text{(Equation 2)} \end{cases}$

To eliminate y^2, multiply equation 2 by 4 and add the equations.

$$\begin{aligned} 9x^2 - 4y^2 &= 20 \\ x^2 + y^2 &= 8 \end{aligned} \xrightarrow{\text{Multiply by 4}} \begin{aligned} 9x^2 - 4y^2 &= 20 \\ 4x^2 + 4y^2 &= 32 \end{aligned}$$

$$13x^2 = 52 \quad \text{Add the equations.}$$

$$x^2 = 4 \quad \text{Divide both sides by 13.}$$

$$x = \pm 2 \quad \text{Find the square roots of 4.}$$

To find y, substitute 2 and -2 for x in one of the original equations. We will use $x^2 + y^2 = 8$.

$$\begin{aligned} 2^2 + y^2 &= 8 \\ 4 + y^2 &= 8 \\ y^2 &= 4 \\ y &= \pm 2 \end{aligned}$$

Therefore, (2, 2) and (2, −2) are solutions.

$$\begin{aligned} (-2)^2 + y^2 &= 8 \\ 4 + y^2 &= 8 \\ y^2 &= 4 \\ y &= \pm 2 \end{aligned}$$

Therefore, (−2, 2) and (−2, −2) are solutions.

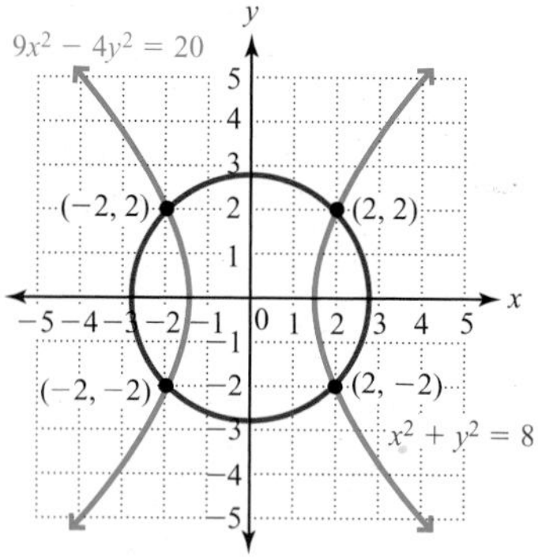

This system has four solutions: (2, 2), (2, −2), (−2, 2), and (−2, −2), as verified by the graph to the left.

Note: *If the graphs are a hyperbola and a circle both centered at the origin as in Example 2, there will be 0, 2, or 4 solutions, as shown.*

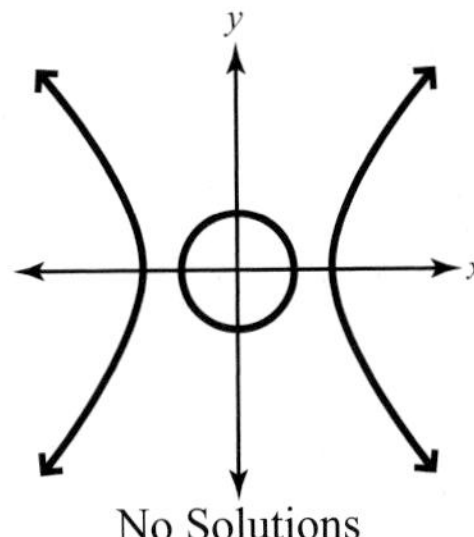

No Solutions

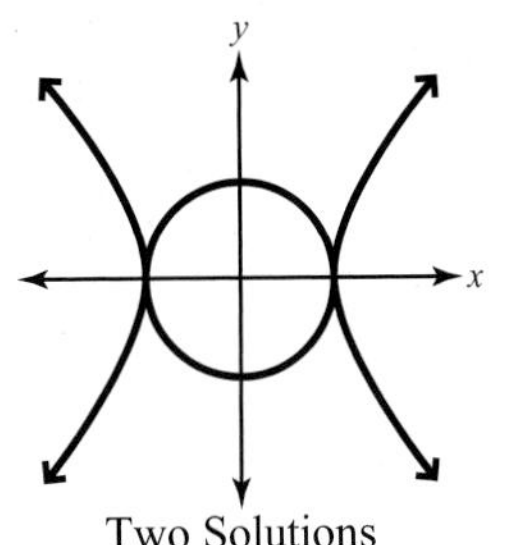

Two Solutions

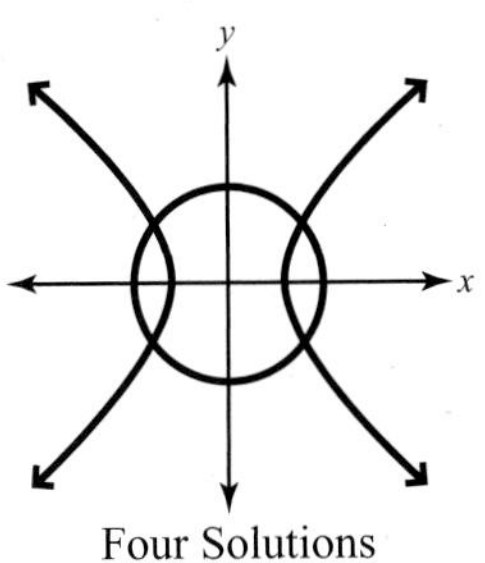

Four Solutions

YOUR TURN Solve using the elimination method. $\begin{cases} 4x^2 + 9y^2 = 72 \\ x^2 + y^2 = 13 \end{cases}$

ANSWERS

(3, 2), (3, −2), (−3, 2), (−3, −2)

Calculator TIPS

A system of equations can be solved using the intersect *function on a graphing calculator.*

Let's solve $\begin{cases} x^2 + y^2 = 10 \\ 3x + y = 6 \end{cases}$ *using a T1-83 Plus.*

First, solve each equation for y:

$$x^2 + y^2 = 10 \qquad 3x + y = 6$$
$$y^2 = 10 - x^2 \qquad y = 6 - 3x$$
$$y = \pm\sqrt{10 - x^2}$$

Enter $Y_1 = \sqrt{10 - x^2}$, $Y_2 = -\sqrt{10 - x^2}$, *and* $Y_3 = 6 - 3x$ *then press* GRAPH.

To find the intersection of y_1 *and* y_3 *and also* y_2 *and* y_3, *select* 5: intersect *from the* CALC *menu. You will then be prompted to indicate the two graphs you want to find the intersection of and the appropriate point of intersection. We show the results of each point of intersection next.*

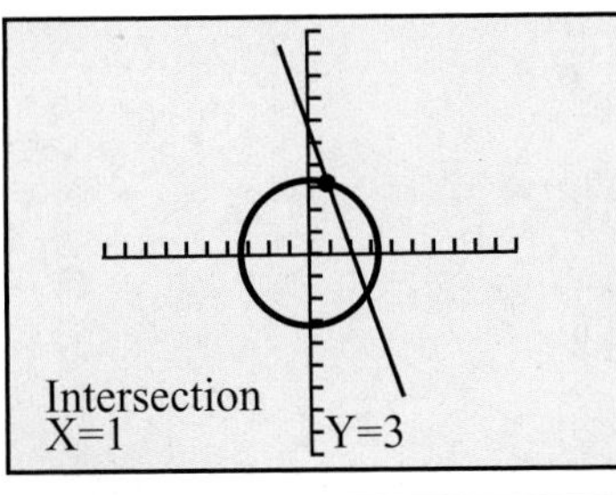

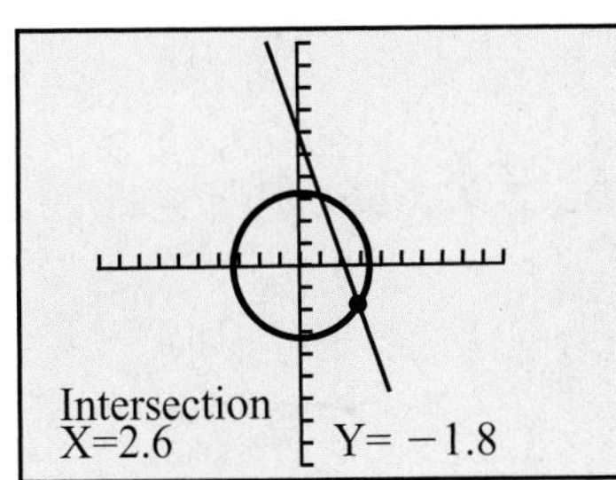

11.3 Exercises

For Extra Help

MyMathLab    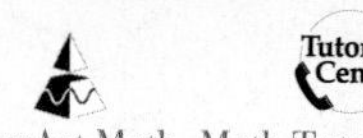 

MyMathLab Videotape/DVT InterAct Math Math Tutor Center Math XL.com

1. a. How many real solutions are possible for a system of two equations whose graphs are a circle and a parabola?

   b. Draw a figure illustrating such a system with two solutions.

2. a. How many real solutions are possible for a system of two equations whose graphs are a line and a hyperbola?

   b. Draw a figure illustrating such a system with one solution.

3. Which method (substitution or elimination) would you use to solve the system $\begin{cases} 2x + y = 8 \\ 4x^2 + 3y^2 = 24 \end{cases}$? Why? (Do not attempt to solve the system.)

4. Which method (substitution or elimination) would you use to solve the system $\begin{cases} 3x^2 - 4y^2 = 12 \\ 2x^2 + 3y^2 = 24 \end{cases}$? Why? (Do not attempt to solve the system.)

5. Without solving the system, what is the number of possible solutions of the system $\begin{cases} 3x - y = 6 \\ 4x^2 + 9y^2 = 36 \end{cases}$?

6. Is it possible for a system of two equations whose graphs are an ellipse and a hyperbola (both centered at the origin) to have three solutions? Why or why not?

For Exercises 7–42, solve.

7. $\begin{cases} y = 2x^2 \\ 2x + y = 4 \end{cases}$

8. $\begin{cases} x^2 + 2y = 1 \\ 2x + y = 2 \end{cases}$

9. $\begin{cases} y = x^2 + 4x + 4 \\ 3x - y = -6 \end{cases}$

10. $\begin{cases} y = 6x - x^2 \\ 2x - y = -3 \end{cases}$

11. $\begin{cases} x^2 + y^2 = 25 \\ x - y = -1 \end{cases}$

12. $\begin{cases} x^2 + y^2 = 25 \\ x - 7y = -25 \end{cases}$

13. $\begin{cases} x^2 + y^2 = 13 \\ 2x + y = 7 \end{cases}$

14. $\begin{cases} x^2 + y^2 = 10 \\ 3x + y = 6 \end{cases}$

15. $\begin{cases} x^2 + 2y^2 = 4 \\ x + y = 5 \end{cases}$

16. $\begin{cases} 3x^2 + y^2 = 9 \\ 2x + y = 11 \end{cases}$

17. $\begin{cases} y = 2x^2 + 3 \\ 2x - y = -3 \end{cases}$

18. $\begin{cases} y = -3x^2 - 2 \\ 3x + y = -2 \end{cases}$

19. $\begin{cases} y = (x-3)^2 + 2 \\ y = -(x-2)^2 + 3 \end{cases}$

20. $\begin{cases} y = 2(x+4)^2 + 2 \\ y = -2(x+3)^2 + 4 \end{cases}$

21. $\begin{cases} x^2 + y^2 = 20 \\ x^2 - y^2 = 12 \end{cases}$

22. $\begin{cases} x^2 + y^2 = 48 \\ x^2 - y^2 = 24 \end{cases}$

23. $\begin{cases} 9x^2 + 4y^2 = 145 \\ x^2 + y^2 = 25 \end{cases}$

24. $\begin{cases} 4x^2 + 9y^2 = 72 \\ x^2 + y^2 = 13 \end{cases}$

25. $\begin{cases} 4x^2 - y^2 = 15 \\ x^2 + y^2 = 5 \end{cases}$

26. $\begin{cases} 9x^2 - 4y^2 = 32 \\ x^2 + y^2 = 5 \end{cases}$

27. $\begin{cases} x^2 + 3y^2 = 36 \\ x = y^2 - 6 \end{cases}$

28. $\begin{cases} 4x^2 + y^2 = 16 \\ y = x^2 - 4 \end{cases}$

29. $\begin{cases} 9x^2 - 16y^2 = 144 \\ 4x^2 + 9y^2 = 36 \end{cases}$

30. $\begin{cases} 9x^2 + 4y^2 = 36 \\ 4x^2 - 9y^2 = 36 \end{cases}$

31. $\begin{cases} y = x^2 \\ x^2 + y^2 = 20 \end{cases}$

32. $\begin{cases} 16x^2 + y^2 = 128 \\ y = 2x^2 \end{cases}$

33. $\begin{cases} 25x^2 - 16y^2 = 400 \\ x^2 + 4y^2 = 16 \end{cases}$

34. $\begin{cases} 9y^2 - 25x^2 = 225 \\ 4y^2 + 25x^2 = 100 \end{cases}$

35. $\begin{cases} xy = 2 \\ 4x^2 + y^2 = 8 \end{cases}$

36. $\begin{cases} xy = 4 \\ 2x^2 - y^2 = 4 \end{cases}$

37. $\begin{cases} y = x^2 - 2x - 3 \\ y = -x^2 + 6x + 7 \end{cases}$

38. $\begin{cases} y = \frac{1}{3}x^2 - 2 \\ 3x^2 + 9y^2 = 36 \end{cases}$

39. $\begin{cases} 4x^2 + 5y^2 = 36 \\ 4x^2 - 3y^2 = 4 \end{cases}$

40. $\begin{cases} 4x^2 + 7y^2 = 64 \\ 4x^2 - 3y^2 = 24 \end{cases}$

41. $\begin{cases} x = -y^2 + 2 \\ x^2 - 5y^2 = 4 \end{cases}$

42. $\begin{cases} x = -y^2 + 2 \\ 9x^2 - 45y^2 = 36 \end{cases}$

43. Create a system of two equations whose graphs are a circle and a line for which there is no solution. Include the graphs.

44. Create a system of two equations whose graphs are a circle and a hyperbola for which there are exactly two solutions. Include the graphs.

45. The sum of the squares of two integers is 34 and the difference of their squares is 16. Find the integers.

46. The difference of the squares of two integers is 32 and their product is 12. Find the integers.

47. A computer keyboard has an area of 144 square inches and a perimeter of 52 inches. Find the length and width.

48. A rectangular living room has an area of 48 square meters and a perimeter of 28 meters. Find the length and width.

49. If p is in dollars and x is in hundreds of units, the demand function for a certain style of chair is given by $p = -3x^2 + 120$ and the supply function is given by $p = 11x + 28$. The *market equilibrium* occurs when the number produced is equal to the number demanded. Find the number of chairs and the price per chair when market equilibrium is reached.

50. If y is in dollars and x is the number of cell phones manufactured (in thousands), a cell phone manufacturer has determined that the cost y to manufacture x cell phones is given by $y = 5x^2 + 30x + 50$ and the revenue from the sales is given by $y = 13x^2$. The *break-even point* is the point (x, y) for which the cost equals the revenue. Find the number of units necessary to break even by solving the system.

For Exercises 51–54, use a graphing calculator to verify the results of the exercise given.

51. Exercise 17

52. Exercise 18

53. Exercise 19

54. Exercise 20

REVIEW EXERCISES

[3.4] ***For Exercises 1 and 2, determine if the ordered pair is a solution for $x + 3y \leq 6$.***

1. $(0, 0)$

2. $(3, 4)$

[3.4] ***For Exercises 3–6, graph.***

3. $y \geq 3$

4. $y < 2x + 3$

5. $2x + 3y < -6$

6. $x \geq -2$

11.4 Nonlinear Inequalities and Systems of Inequalities

OBJECTIVES

1. Graph nonlinear inequalities.
2. Graph the solution set of a system of nonlinear inequalities.

OBJECTIVE 1. Graph nonlinear inequalities. In Section 3.4, we graphed linear inequalities like $x + 2y > 6$ by first graphing the line corresponding to $x + 2y = 6$ and then shading the appropriate region on one side of that boundary line. Recall that we used a dashed line for $<$ and $>$ and a solid line for $\le$ and $\ge$. We determined which side to shade by using a test point on one side of the boundary.

We use a similar procedure to graph nonlinear inequalities such as $x^2 + y^2 \le 25$. The boundary is the graph of $x^2 + y^2 = 25$, which is a circle with center at (0, 0) and radius of 5. The solution set of $x^2 + y^2 \le 25$ contains all ordered pairs on the circle (boundary) along with all ordered pairs inside it or all ordered pairs outside it. To determine which of those two regions is correct, we choose an ordered pair from one of the regions to test in the inequality. Let's test (0, 0).

$0^2 + 0^2 \le 25$ **Substitute (0, 0) into $x^2 + y^2 \le 25$.**

$0 \le 25$ **Simplify. The inequality is true.**

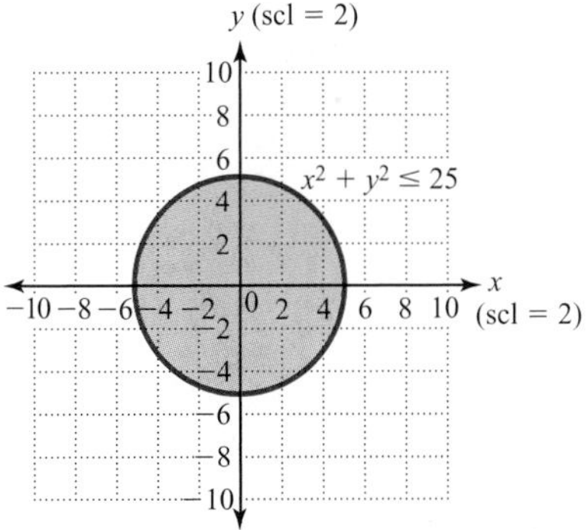

Since $0 \le 25$ is true, all ordered pairs in the region containing (0, 0) are in the solution set, so we shade that region.

Ordered pairs in the region outside the circle are not in the solution set because they do not solve the inequality. To illustrate, let's test (6, 8).

$6^2 + 8^2 \le 25$ **Substitute (6, 8) into $x^2 + y^2 \le 25$.**

$100 \le 25$ **Simplify. The inequality is false.**

Note: *If the inequality had been $x^2 + y^2 < 25$, we would have drawn a dashed circle.*

Since $100 \le 25$ is false, the region containing (6, 8) is not in the solution set, so we do not shade that region.

Our example suggests the following procedure for graphing nonlinear inequalities.

PROCEDURE Graphing Nonlinear Inequalities

1. Graph the related equation. If the inequality symbol is $\le$ or $\ge$, draw the graph as a solid curve. If the inequality symbol is $>$ or $<$, draw a dashed curve.
2. The graph divides the coordinate plane into at least two regions. Test an ordered pair from each region by substituting it into the inequality. If the ordered pair satisfies the inequality, then shade the region containing that ordered pair.

EXAMPLE 1 Graph the inequality.

a. $y > 2(x - 1)^2 + 3$

Solution We first graph the related equation, $y = 2(x - 1)^2 + 3$.

Note: *The parabola is dashed because the inequality is $>$. Also, notice that the graph divides the coordinate plane into two regions: Region 1 above the boundary and Region 2 below.*

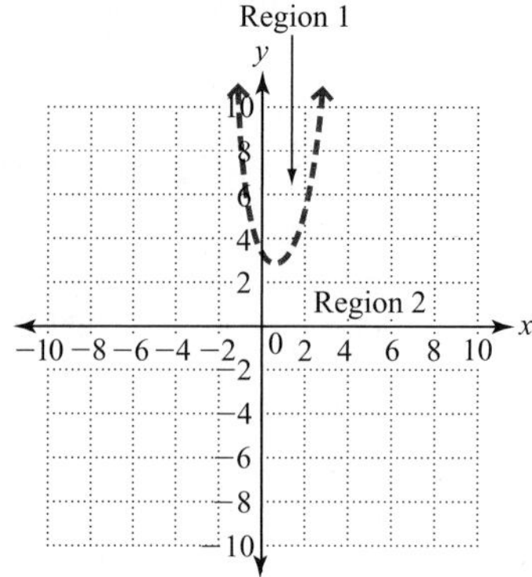

Note: *To make computations easier, choose ordered pairs with 0 for at least one of the coordinates.*

We now test an ordered pair from each region.

Region 1: We choose (0, 6).

$6 > 2(0 - 1)^2 + 3$ **Substitute.**

$6 > 2(-1)^2 + 3$ **Simplify.**

$6 > 5$

True, so Region 1 is in the solution set.

Region 2: We choose (0, 0).

$0 > 2(0 - 1)^2 + 3$ **Substitute.**

$0 > 2(-1)^2 + 3$ **Simplify.**

$0 > 5$

False, so Region 2 is not in the solution set.

Since ordered pairs in only Region 1 solve the inequality, we shade only that region. The solution set contains all ordered pairs in Region 1. Ordered pairs on the parabola are not in the solution set.

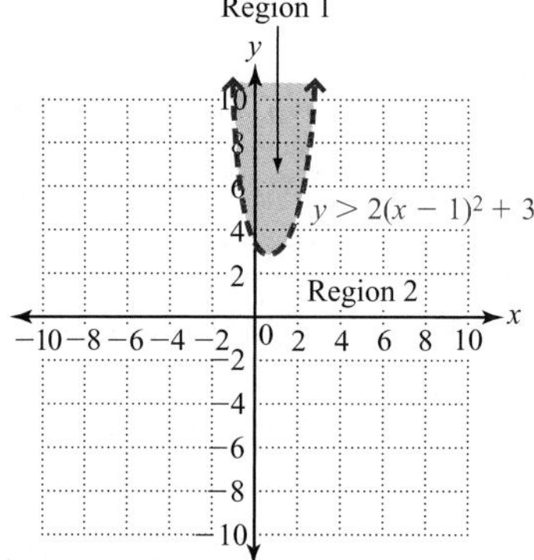

b. 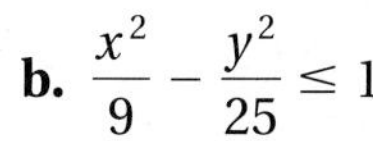 $\dfrac{x^2}{9} - \dfrac{y^2}{25} \le 1$

Solution Graph the related equation, $\dfrac{x^2}{9} - \dfrac{y^2}{25} = 1$.

Note: *The hyperbola is solid because the inequality is $\le$. Also, notice that the hyperbola divides the coordinate plane into three regions.*

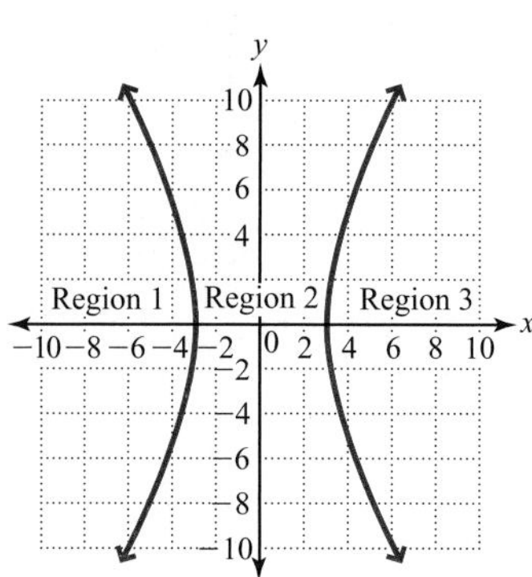

We test an ordered pair from each region.

Region 1: We choose $(-5, 0)$.

$$\frac{(-5)^2}{9} - \frac{0^2}{25} \le 1 \quad \text{Substitute.}$$

$$\frac{25}{9} \le 1 \quad \text{Simplify.}$$

False, so Region 1 is not in the solution set.

Region 2: We choose $(0, 0)$.

$$\frac{0^2}{9} - \frac{0^2}{25} \le 1 \quad \text{Substitute.}$$

$$0 \le 1 \quad \text{Simplify.}$$

True, so Region 2 is in the solution set.

Region 3: We choose $(5, 0)$.

$$\frac{(5)^2}{9} - \frac{0^2}{25} \le 1 \quad \text{Substitute.}$$

$$\frac{25}{9} \le 1 \quad \text{Simplify.}$$

False, so Region 3 is not in the solution set.

Since Region 2 is the only region containing ordered pairs that solve the inequality, we shade only that region. The solution set contains all ordered pairs on the hyperbola along with all ordered pairs in Region 2.

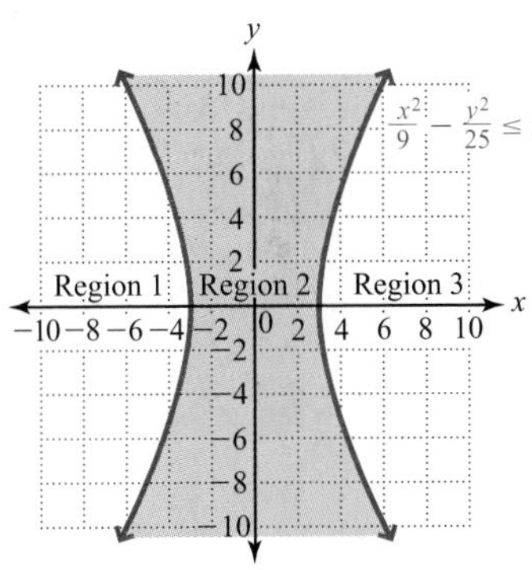

Your Turn Graph the inequality.

a. $y \ge (x + 1)^2 - 3$ **b.** $\frac{x^2}{25} + \frac{y^2}{9} < 1$

Answers

a. $y \ge (x + 1)^2 - 3$

b.

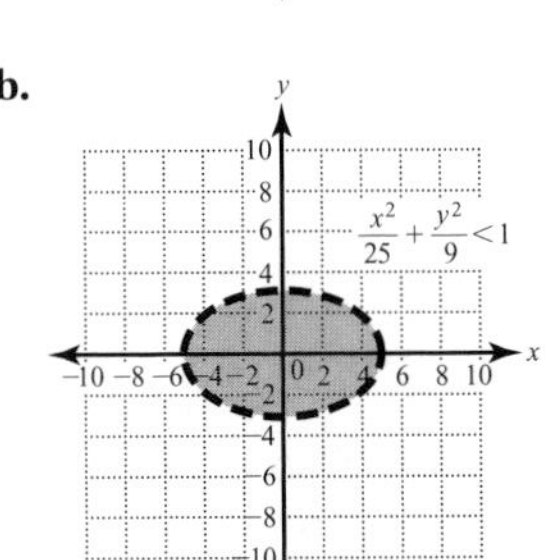

OBJECTIVE 2. Graph the solution set of a system of nonlinear inequalities. In Section 4.6, we solved systems of linear inequalities by graphing each inequality on the same grid. The solution set contained all ordered pairs in the region where the inequalities' solution sets overlapped together with ordered pairs on the portion of any solid line touching the region of overlap. We use a similar procedure for systems of nonlinear inequalities.

Example 2 Graph the solution set of the system of inequalities.

a. $\begin{cases} y \ge x^2 - 4 \\ 2x - y < 2 \end{cases}$

Solution We begin by graphing $y \ge x^2 - 4$ and $2x - y < 2$.

Note: *The boundary graph, $y = x^2 - 4$, is a parabola opening up with vertex at $(0, -4)$. The test point $(0, 0)$ gives the true statement $0 \geq -4$, so we shade the region containing $(0, 0)$.*

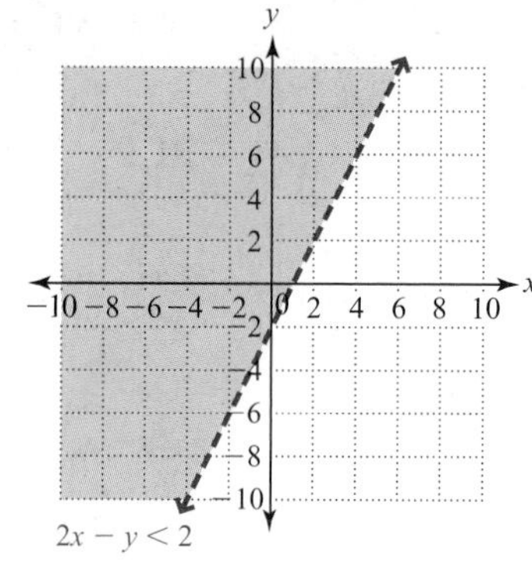

Note: *The boundary graph, $2x - y = 2$, is a dashed line. The test point $(0, 0)$ gives a true statement $0 < 2$, so we shade the region containing $(0, 0)$.*

Note: *Remember, ordered pairs on dashed lines or curves are not in the solution set for a system of inequalities.*

If we place both graphs on the same grid, their intersection (purple shading) is the solution region for the system. In addition to ordered pairs in the purple shaded region, the solution set also contains all ordered pairs on the portions of the parabola that touch the purple shaded region.

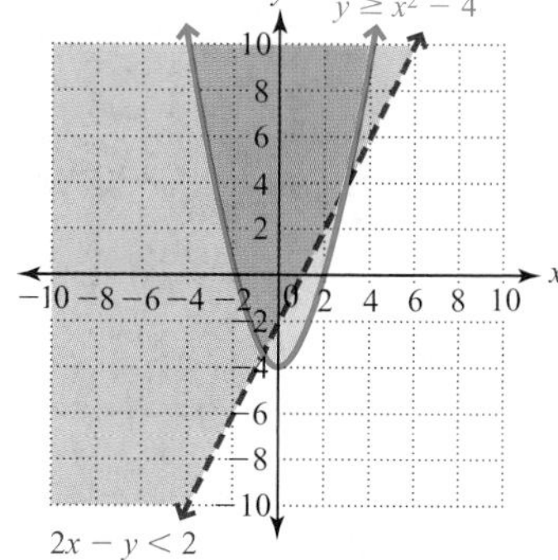

b. $\begin{cases} x^2 + y^2 \leq 49 \\ \dfrac{x^2}{16} - \dfrac{y^2}{9} < 1 \\ y \geq 2x + 2 \end{cases}$

Solution Graph each inequality on the same coordinate system.

The graph of $x^2 + y^2 \leq 49$ is a circle and its interior. The graph of $\dfrac{x^2}{16} - \dfrac{y^2}{9} < 1$ is the region between the branches of the hyperbola with the curve dashed. The graph of $y \geq 2x + 2$ is a solid line and the region above the line. The solution set for the system contains all ordered pairs in the region where the three graphs overlap (purple shaded region) together with all ordered pairs on the portion of the circle and the line that touches the purple shaded region.

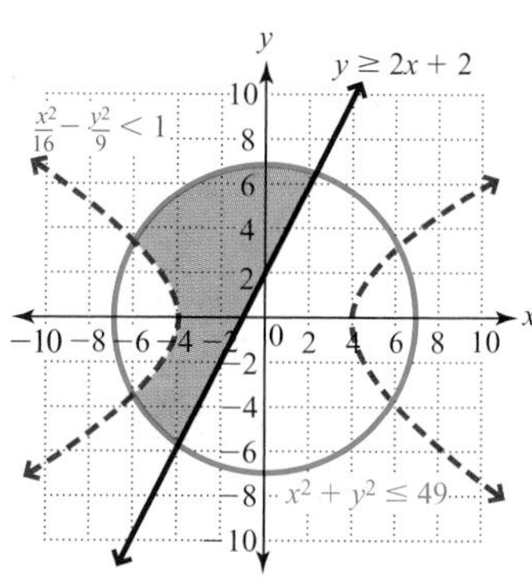

YOUR TURN Graph the solution set of the system of inequalities.

a. $\begin{cases} y > x^2 + 2 \\ x + y < 4 \end{cases}$

b. $\begin{cases} \dfrac{x^2}{9} + \dfrac{y^2}{16} \leq 1 \\ y \geq x^2 \\ -x + y > 2 \end{cases}$

ANSWERS

a.

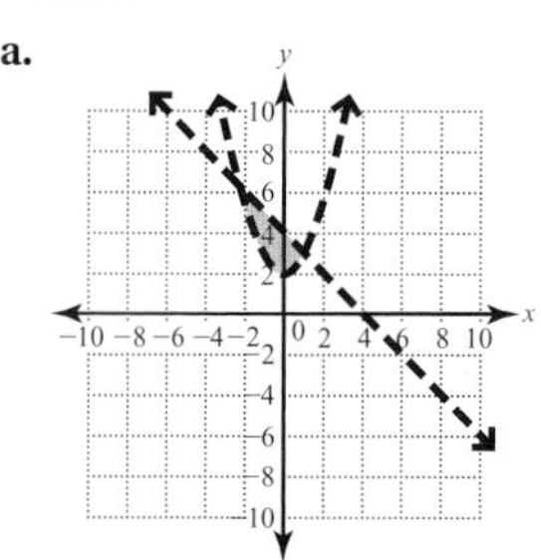

b.

11.4 Exercises

For Extra Help

MyMathLab

Videotape/DVT

InterAct Math

Math Tutor Center

Math XL.com

1. Do the ordered pairs that lie on the boundary curve solve the inequality $9x^2 + 4y^2 > 36$? Why or why not? Would the graph be drawn solid or dashed? Why?

2. Discuss how graphing linear inequalities like $x + 2y \geq 4$ is similar to graphing a nonlinear inequality like $y \geq x^2 + 3$.

3. Draw the graph of a system of nonlinear inequalities for which there is no solution and explain why there is no solution.

4. Describe the procedure you would use to graph $\frac{x^2}{9} - \frac{y^2}{16} > 1$.

For Exercises 5–24, graph the inequality.

5. $x^2 + y^2 \geq 4$

6. $x^2 + y^2 \leq 9$

7. $y > x^2$

8. $y < -x^2$

9. $y < (x - 1)^2 + 2$

10. $y > 2(x + 2)^2 - 3$

11. $\frac{x^2}{16} + \frac{y^2}{9} \leq 1$

12. $\frac{x^2}{4} + \frac{y^2}{9} \geq 1$

13. $\frac{x^2}{25} - \frac{y^2}{4} > 1$

14. $\frac{x^2}{36} - \frac{y^2}{16} < 1$

15. $x^2 + y^2 < 16$

16. $x^2 + y^2 > 25$

17. $y \geq -2(x-3)^2 + 3$

18. $y < -(x+3)^2 - 2$

19. $\frac{x^2}{49} + \frac{y^2}{25} > 1$

20. $\frac{x^2}{16} + \frac{y^2}{9} \leq 1$

21. $\frac{y^2}{25} - \frac{x^2}{4} \leq 1$

22. $\frac{y^2}{36} - \frac{x^2}{25} > 1$

23. $y < x^2 + 4x - 5$

24. $y > x^2 - 6x - 6$

For Exercises 25–50, graph the solution set of each system of inequalities.

25. $\begin{cases} y \geq x^2 \\ x + y \leq 3 \end{cases}$

26. $\begin{cases} y < -x^2 \\ 2x - y < 4 \end{cases}$

27. $\begin{cases} x - 2y < -4 \\ x^2 + y^2 < 16 \end{cases}$

28. $\begin{cases} 3x + 2y \geq -6 \\ x^2 + y^2 \leq 25 \end{cases}$

29. $\begin{cases} x^2 + y^2 \geq 9 \\ x^2 + y^2 \leq 25 \end{cases}$

30. $\begin{cases} x^2 + y^2 > 4 \\ x^2 + y^2 > 9 \end{cases}$

31. $\begin{cases} y > x^2 + 1 \\ 2x + y < 3 \end{cases}$

32. $\begin{cases} 3x - y \leq 2 \\ y \leq -x^2 + 2 \end{cases}$

33. $\begin{cases} y < -x^2 + 3 \\ y > x^2 - 2 \end{cases}$

34. $\begin{cases} y > -x^2 + 2 \\ y < x^2 + 5 \end{cases}$

35. $\begin{cases} \dfrac{x^2}{25} + \dfrac{y^2}{9} \leq 1 \\ x^2 + y^2 \geq 4 \end{cases}$

36. $\begin{cases} \dfrac{x^2}{9} + \dfrac{y^2}{4} \leq 1 \\ x^2 + y^2 \leq 4 \end{cases}$

37. $\begin{cases} \dfrac{x^2}{25} + \dfrac{y^2}{9} < 1 \\ y > x^2 + 1 \end{cases}$

38. $\begin{cases} \dfrac{x^2}{9} + \dfrac{y^2}{4} < 1 \\ y < -x^2 + 2 \end{cases}$

39. $\begin{cases} \dfrac{x^2}{9} - \dfrac{y^2}{4} \leq 1 \\ \dfrac{x^2}{25} + \dfrac{y^2}{9} \leq 1 \end{cases}$

40. $\begin{cases} \dfrac{x^2}{16} - \dfrac{y^2}{9} \geq 1 \\ \dfrac{x^2}{36} + \dfrac{y^2}{16} \leq 1 \end{cases}$

41. $\begin{cases} \dfrac{x^2}{4} - \dfrac{y^2}{4} > 1 \\ y > 2 \end{cases}$

42. $\begin{cases} \dfrac{x^2}{9} - \dfrac{y^2}{9} > 1 \\ x > 3 \end{cases}$

43. $\begin{cases} 3x + 2y \leq 6 \\ x - y > -3 \\ x + 6y \geq 2 \end{cases}$

44. $\begin{cases} 2x + 3y < 6 \\ x - 2y > -4 \\ x + 5y \geq -4 \end{cases}$

45. $\begin{cases} \dfrac{x^2}{16} + \dfrac{y^2}{4} \leq 1 \\ x^2 + y^2 \leq 9 \\ y \leq x \end{cases}$

46. $\begin{cases} \dfrac{x^2}{9} + \dfrac{y^2}{16} < 1 \\ x^2 + y^2 < 9 \\ y > x + 1 \end{cases}$

47. $\begin{cases} \dfrac{x^2}{49} - \dfrac{y^2}{16} \leq 1 \\ \dfrac{x^2}{64} + \dfrac{y^2}{36} \leq 1 \\ 2x - y \leq -3 \end{cases}$

48. $\begin{cases} \dfrac{x^2}{9} + \dfrac{y^2}{49} \leq 1 \\ y \geq x^2 + 3 \\ x + y \leq 6 \end{cases}$

49. $\begin{cases} y < 2x^2 + 8 \\ 2x + y > 3 \\ x - y < 4 \end{cases}$

50. $\begin{cases} y < x^2 + 2 \\ 2x + y < 4 \\ 2x - y > -5 \end{cases}$

REVIEW EXERCISES

[1.3] **1.** Simplify: $(3 \cdot 1 + 2) + (3 \cdot 2 + 2) + (3 \cdot 3 + 2) + (3 \cdot 4 + 2)$

[1.4] **2.** Evaluate: $\frac{n}{2}(a_1 + a_n)$ if $n = 12$, $a_1 = -8$, and $a_n = 60$

[1.4] **3.** Evaluate: $a_1 + (n - 1)d$ if $a_1 = -12$, $n = 25$, and $d = -3$

[1.4] ***For Exercises 4 and 5, evaluate the expression for n = 1, 2, 3, and 4.***

4. $2n^2 - 3$

5. $\dfrac{(-1)^n}{3n - 2}$

[2.2] **6.** The sum of three consecutive odd intgers is 207. Find the intgers.

Chapter 11 Summary

Defined Terms

Section 11.1
Conic Section (p. 788)
Circle (p. 791)
Radius (p. 791)

Section 11.2
Ellipse (p. 802)
Hyperbola (p. 804)

Section 11.3
Nonlinear system of equations (p. 816)

Procedures, Rules, and Key Examples

Procedures/Rules

Section 11.1 The Parabola and Circle

The graph of an equation in the form $x = a(y - k)^2 + h$ is a parabola with vertex at (h, k). The parabola opens to the right if $a > 0$ and to the left if $a < 0$. The equation of the axis of symmetry is $y = k$.

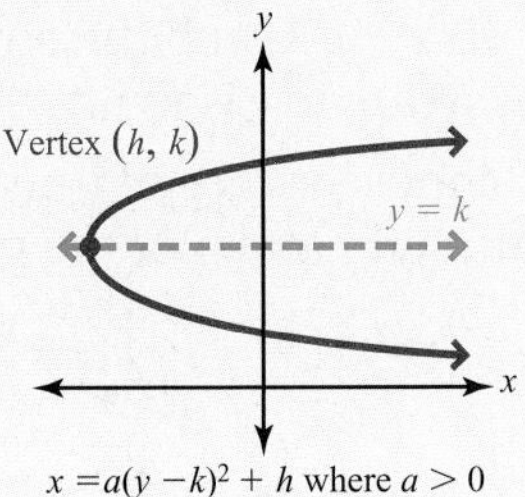

$x = a(y - k)^2 + h$ where $a > 0$

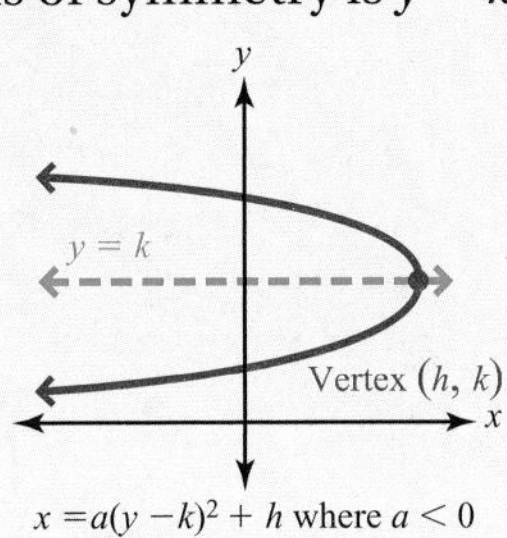

$x = a(y - k)^2 + h$ where $a < 0$

Key Examples

Example 1: For $x = -2(y - 3)^2 - 2$, determine whether the graph opens left or right, find the vertex and the axis of symmetry, and draw the graph.

Solution: Since $a = -2$ and $-2 < 0$, the graph opens to the left. The vertex is $(-2, 3)$ and the equation of the axis of symmetry is $y = 3$.

Choose values for y near the axis of symmetry and plot the graph.

x	y
−10	1
−4	2
−4	4
−10	5

continued

Section 11.1 The Parabola and Circle (continued)

Example 2: For $x = y^2 - 4y - 5$, determine whether the graph opens left or right, find the vertex and axis of symmetry, and draw the graph.

Solution: Since $a = 1$ and $1 > 0$ the graph opens to the right. To find the vertex and axis of symmetry, we will complete the square.

$$x = y^2 - 4y - 5$$

$$x + 5 = y^2 - 4y \quad \text{Add 5 to both sides.}$$

$$x + 5 + 4 = y^2 - 4y + 4 \quad \text{Add 4 to both sides.}$$

$$x + 9 = (y - 2)^2 \quad \text{Simplify both sides.}$$

$$x = (y - 2)^2 - 9 \quad \text{Subtract 9 from both sides.}$$

The vertex is $(-9, 2)$ and the equation of the axis of symmetry is $y = 2$.

To graph, choose values of y near the axis of symmetry and plot the graph.

x	y
−5	0
−8	1
−8	3
−5	4

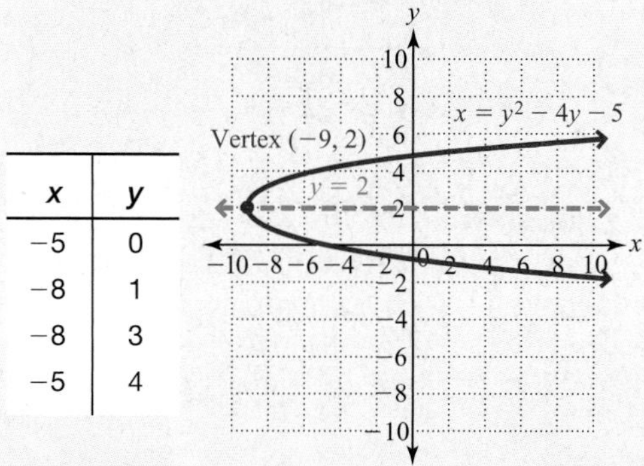

The distance, d, between two points with coordinates (x_1, y_1) and (x_2, y_2), can be found using the formula

$$d = \sqrt{(x_2 - x_1)^2 + (y_2 - y_1)^2}$$

Example 3: Find the distance between the points whose coordinates are $(-4, 3)$ and $(2, -1)$.

Solution: Let $(x_1, y_1) = (-4, 3)$ and $(x_2, y_2) = (2, -1)$.

$$d = \sqrt{(x_2 - x_1)^2 + (y_2 - y_1)^2}$$

$$d = \sqrt{(2 - (-4))^2 + (-1 - 3)^2} \quad \text{Substitute.}$$

$$d = \sqrt{52} \quad \text{Simplify.}$$

$$d = \sqrt{4 \cdot 13} \quad \text{Rewrite 52 as } 4 \cdot 13.$$

$$d = 2\sqrt{13}$$

continued

Procedures/Rules

Key Examples

Section 11.1 The Parabola and Circle (continued)

The equation of a circle with center at (h, k) and radius r is $(x - h)^2 + (y - k)^2 = r^2$.

Example 4: Find the center and radius of the circle whose equation is $(x - 2)^2 + (y + 4)^2 = 25$ and draw the graph.

Solution: Rewrite the equation as

$(x - 2)^2 + (y - (-4))^2 = 5^2$.

Since $h = 2$ and $k = -4$, the center is $(2, -4)$ and $r = 5$.

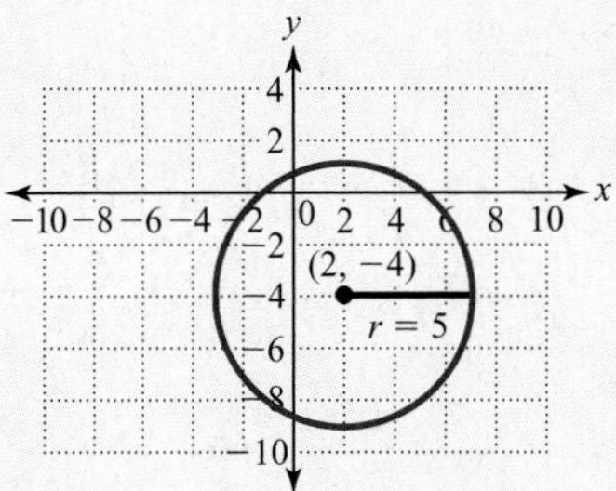

Example 5: Write the equation of the circle whose center is $(5, -6)$ and whose radius is 8.

Solution: Substitute for h, k, and r in

$(x - h)^2 + (y - k)^2 = r^2$.

$(x - 5)^2 + (y - (-6))^2 = 8^2$

$(x - 5)^2 + (y + 6)^2 = 64$ **Simplify.**

To graph circles in the form $x^2 + y^2 + dx + ey + f = 0$, complete the square in x and y to put the equation in the form $(x - h)^2 + (y - k)^2 = r^2$.

Example 6: Find the center and radius of the circle whose equation is $x^2 + y^2 + 8x - 2y + 8 = 0$ and draw the graph.

Solution: Complete the square in x and y.

$x^2 + 8x + y^2 - 2y = -8$ **Group the terms in x and y and subtract 8.**

$x^2 + 8x + 16 + y^2 - 2y + 1 = -8 + 16 + 1$ **Complete the square.**

$(x + 4)^2 + (y - 1)^2 = 9$ **Factor on the left and simplify on the right.**

The center is $(-4, 1)$ and the radius is 3.

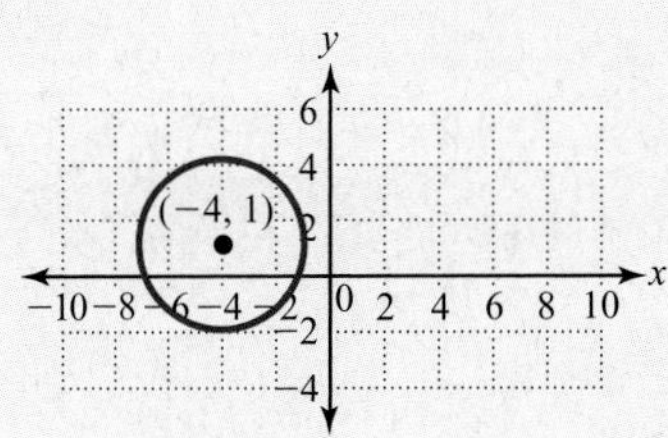

continued

Procedures/Rules

Key Examples

Section 11.2 Ellipses and Hyperbolas

The equation of an ellipse with center (0, 0), x-intercepts $(a, 0)$ and $(-a, 0)$, and y-intercepts $(0, b)$ and $(0, -b)$, is $\frac{x^2}{a^2} + \frac{y^2}{b^2} = 1$.

Example 1: Graph $\frac{x^2}{49} + \frac{y^2}{36} = 1$.

Solution: The equation can be rewritten as $\frac{x^2}{7^2} + \frac{y^2}{6^2} = 1$, so $a = 7$ and $b = 6$. Consequently, the x-intercepts are (7, 0) and (−7, 0) and the y-intercepts are (0, 6) and (0, −6).

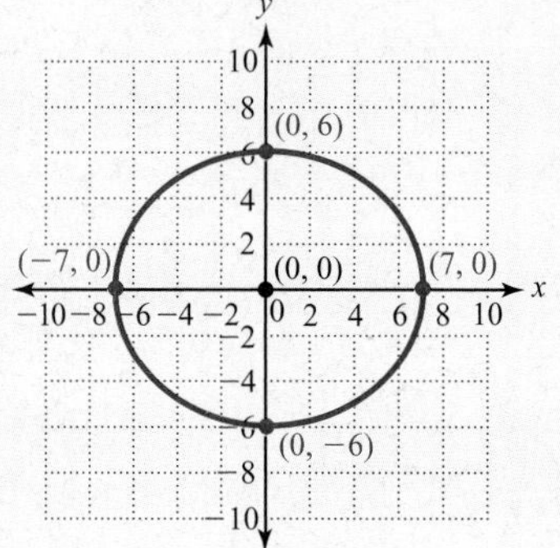

The equation of an ellipse with center (h, k) is $\frac{(x-h)^2}{a^2} + \frac{(y-k)^2}{b^2} = 1$. The ellipse passes through two points that are a units to the left and right of the center, and two points that are b units above and below the center.

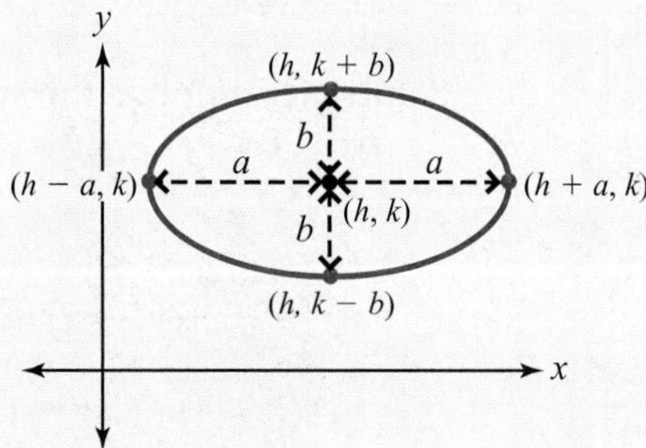

Example 2: Graph $\frac{(x+2)^2}{9} + \frac{(y-1)^2}{25} = 1$.

Solution: The equation can be rewritten as $\frac{(x-(-2))^2}{3^2} + \frac{(y-1)^2}{5^2} = 1$, so $h = -2$, $k = 1$, $a = 3$, and $b = 5$. The center is (−2, 1). To find other points on the ellipse, go 3 units left and right of the center and 5 units up and down from the center.

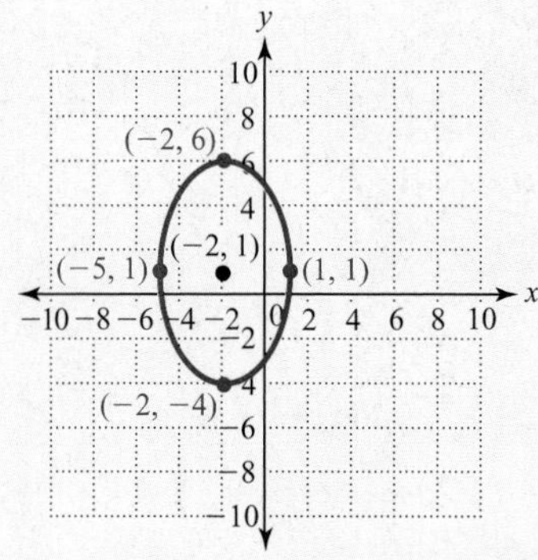

continued

Section 11.2 Ellipses and Hyperbolas (continued)

The equation of a hyperbola with center (0, 0), x-intercepts $(a, 0)$ and $(-a, 0)$, and no y-intercepts, is $\frac{x^2}{a^2} - \frac{y^2}{b^2} = 1$.

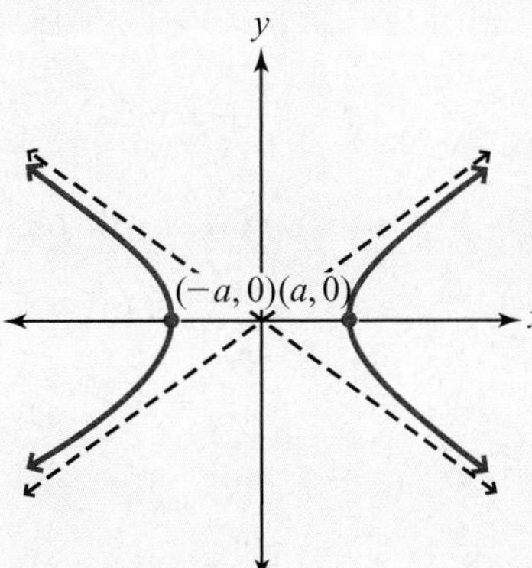

The equation of a hyperbola with center (0, 0), y-intercepts $(0, b)$ and $(0, -b)$, and no x-intercepts, is $\frac{y^2}{b^2} - \frac{x^2}{a^2} = 1$.

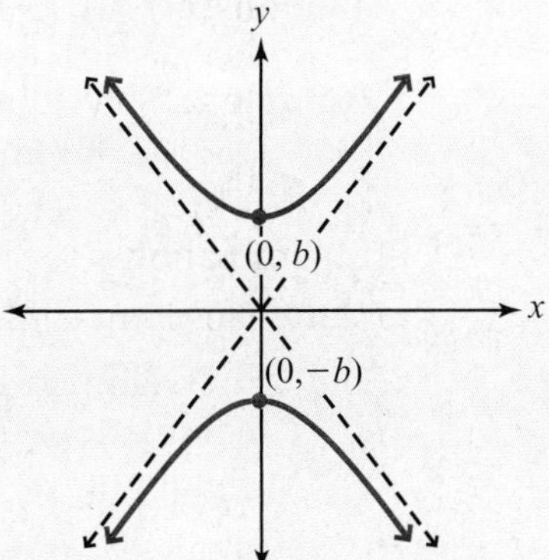

To graph a hyperbola:

1. Find the intercepts. If the x^2 term is positive, the x-intercepts are $(a, 0)$ and $(-a, 0)$ and there are no y-intercepts. If the y^2 term is positive, the y-intercepts are $(0, b)$ and $(0, -b)$ and there are no x-intercepts.
2. Draw the fundamental rectangle. The vertices are (a, b), $(-a, b)$, $(a, -b)$, and $(-a, -b)$.
3. Draw the asymptotes, which are the extended diagonals of the fundamental rectangle.
4. Draw the graph so that each branch passes through an intercept and approaches the asymptotes the farther they are from the origin.

Example 3: Graph $\frac{x^2}{36} - \frac{y^2}{16} = 1$.

Solution: The equation can be rewritten as $\frac{x^2}{6^2} - \frac{y^2}{4^2} = 1$, so $a = 6$ and $b = 4$. Since the x^2 term is positive, the x-intercepts are (6, 0) and (−6, 0) and there are no y-intercepts. The fundamental rectangle has vertices (6, 4), (−6, 4), (6, −4), and (−6, −4).

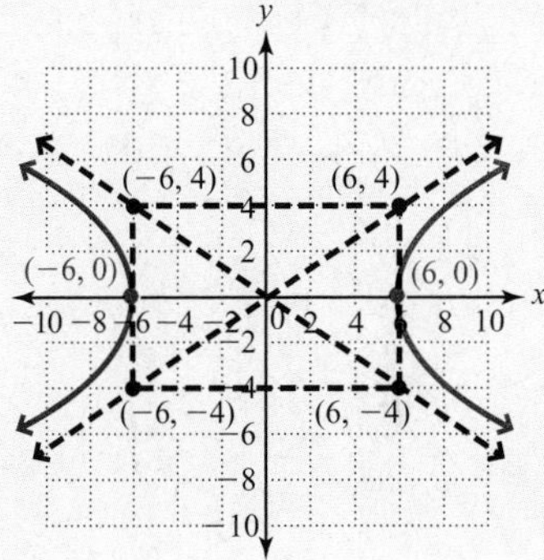

continued

Procedures/Rules	Key Examples
Section 11.3 Nonlinear Systems of Equations If a nonlinear system has a linear equation, the substitution method is usually preferred. Solve the linear equation for one of its variables and substitute for that variable in the other equation.	**Example 1:** Solve the system $\begin{cases} y = (x-3)^2 + 2 \\ 2x - y = -4 \end{cases}$ by substitution. (**Note:** This is a parabola and a line.) Solution: $y = 2x + 4$ **Solve $2x - y = -4$ for y.** $2x + 4 = (x-3)^2 + 2$ **Substitute $2x + 4$ for y in $y = (x-3)^2 + 2$.** $2x + 4 = x^2 - 6x + 11$ **Simplify the right side.** $0 = x^2 - 8x + 7$ **Write in the form $ax^2 + bx + c = 0$.** $0 = (x-7)(x-1)$ **Factor.** $x - 7 = 0$ or $x - 1 = 0$ **Set each factor equal to 0.** $x = 7$ or $x = 1$ **Solve each equation.** To find y, substitute 7 and 1 into $y = 2x + 4$. $y = 2(7) + 4$ $y = 2(1) + 4$ $y = 18$ $y = 6$ Solutions: (7, 18), (1, 6)
If neither equation is linear, the elimination method is often preferred.	**Example 2:** Solve the system $\begin{cases} 4x^2 + 5y^2 = 36 \text{ (Equation 1)} \\ x^2 + y^2 = 8 \text{ (Equation 2)} \end{cases}$ by elimination. (**Note:** This is an ellipse and a circle.) Solution: To eliminate x^2, multiply Equation 2 by -4 and add the equations. $4x^2 + 5y^2 = 36$ $4x^2 + 5y^2 = 36$ $x^2 + y^2 = 8$ → (**Multiply by -4**) $-4x^2 - 4y^2 = -32$ $y^2 = 4$ $y = \pm 2$ Substitute 2 and -2 for y into either equation to find x. Let's use $x^2 + y^2 = 8$. $x^2 + 2^2 = 8$ $x^2 + (-2)^2 = 8$ $x^2 + 4 = 8$ $x^2 + 4 = 8$ $x^2 = 4$ $x^2 = 4$ $x = \pm 2$ $x = \pm 2$ Solutions: (2, 2), (2, −2), (−2, 2), and (−2, −2)

Section 11.4 Nonlinear Inequalities and Systems of Inequalities

Graphing nonlinear inequalities.

1. Graph the related equation. If the inequality symbol is $\leq$ or $\geq$, draw the graph as a solid curve. If the inequality symbol is $<$ or $>$, draw a dashed curve.
2. The graph divides the coordinate plane into at least two regions. Test an ordered pair from each region by substituting it into the inequality. If the ordered pair satisfies the inequality, then shade the region that contains that ordered pair.

Example 1: Graph $y \geq (x - 2)^2 - 1$.

Solution: The related equation is a parabola with vertex $(2, -1)$ and opening upward. Since the inequality is $\geq$, all ordered pairs on the parabola are in the solution set, so we draw it with a solid curve.

We will choose (0, 0) as a test point.

$0 \geq (0 - 2)^2 - 1$

$0 \geq 3$ which is false, so (0, 0) is not in the solution set.

Choose (2, 0) as a test point.

$0 \geq (2 - 2)^2 - 1$

$0 \geq -1$ which is true, so (2, 0) is in the solution set. Shade the region that contains (2, 0).

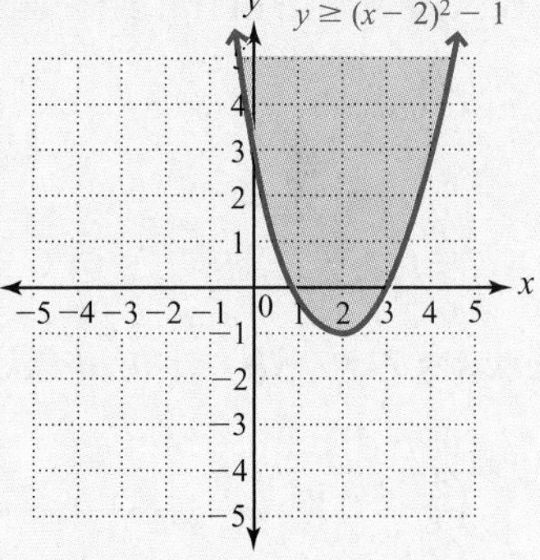

To graph a system of inequalities, graph the solution sets of the individual inequalities on the same grid. The solution set of the system is the intersection of the solution sets of the individual inequalities.

Example 2: Graph the solution set of

$$\begin{cases} \dfrac{x^2}{4} + \dfrac{y^2}{16} \leq 1 \\ x^2 + y^2 \geq 9 \end{cases}.$$

Solution: The graph of $\dfrac{x^2}{4} + \dfrac{y^2}{16} \leq 1$ is an ellipse and all points inside the ellipse. The graph of $x^2 + y^2 \geq 9$ is a circle and all points outside the circle. So, the solution set of the system is the set of all points inside the ellipse and outside the circle.

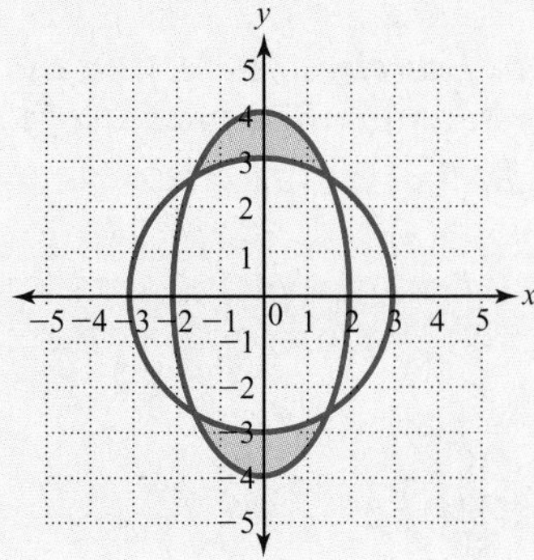

Chapter 11 Review Exercises

For Exercises 1–6, answer true or false.

[11.1] **1.** The graph of $x = -3(y - 2)^2 + 5$ opens down.

[11.1] **2.** The graph of $(x + 2)^2 + (y - 3)^2 = 36$ is a circle with center at $(-2, 3)$ and radius 6.

[11.2] **3.** A parabola has asymptotes.

[11.2] **4.** Choose any two points on an ellipse. The sum of their distances from the foci is the same.

[11.3] **5.** It is possible for a circle and a parabola to intersect in three points.

[11.4] **6.** The graph of the system $\begin{cases} x^2 + y^2 > 36 \\ 2x + y > 3 \end{cases}$ lies inside the circle and above the line.

For Exercises 7–10, fill in the blanks.

[11.1] **7.** The distance between the points whose coordinates are $(-3, 4)$ and $(3, -4)$ is ______________.

[11.1] **8.** The graph of $x^2 + y^2 + 8x - 2y + 13 = 0$ is a circle with center at ______________ and radius of ______________.

[11.2] **9.** The graph of $\frac{x^2}{16} + \frac{y^2}{36} = 1$ is a(n) ______________ with x-intercepts at ______________.

[11.2] **10.** The fundamental rectangle for the graph of $\frac{x^2}{25} - \frac{y^2}{16} = 1$ has vertices at ______________________.

Equations and Inequalities

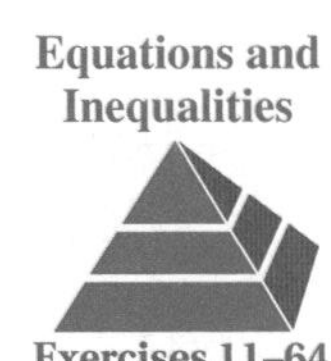

Exercises 11–64

[11.1] ***For Exercises 11–14, find the direction the parabola opens, the coordinates of the vertex, and the equation of the axis of symmetry. Draw the graph.***

11. $y = 2(x - 3)^2 - 5$

12. $y = -2(x + 2)^2 + 3$

13. $x = -(y - 2)^2 + 4$

14. $x = 2(y + 3)^2 - 2$

[11.1] ***For Exercises 15–18, find the direction the parabola opens, the coordinates of the vertex, the equation of the axis of symmetry, and draw the graph.***

15. $x = y^2 + 6y + 8$

16. $x = 2y^2 - 8y - 6$

17. $x = -y^2 - 2y + 3$

18. $x = -3y^2 - 12y - 9$

[11.1] ***For Exercises 19 and 20, find the distance between the two points.***

19. $(-1, -2)$ and $(-5, 1)$

20. $(2, -5)$ and $(-2, 3)$

[11.1] 21. The center of a circle is at $(6, -4)$ and the circle passes through $(-2, 2)$. What is the radius of the circle?

[11.1] 22. The center of a circle is at $(-6, 8)$ and the circle passes through $(3, -4)$. What is the equation of the circle?

[11.1] ***For Exercises 23–26, find the center and radius and draw the graph.***

23. $(x - 3)^2 + (y + 2)^2 = 25$

24. $(x + 5)^2 + (y - 1)^2 = 4$

25. $x^2 + y^2 - 4x + 8y + 11 = 0$

26. $x^2 + y^2 + 10x + 2y + 22 = 0$

[11.1] ***For Exercises 27–28, the center and radius of a circle are given. Write the equation of each circle in standard form.***

27. Center: $(6, -8)$, $r = 9$

28. Center: $(-3, -5)$, $r = 10$

[11.1] **29.** If a heavy object is thrown vertically upward with an initial velocity of 32 feet per second from the top of a building 128 feet high, its height above the ground after t seconds is given by $h = -16t^2 + 32t + 128$, where h is the height in feet and t is the time in seconds.

a. What is the maximum height the object will reach?

b. How many seconds will it take the object to reach its maximum height?

c. How many seconds will it take for the object to strike the ground?

[11.1] **30.** To lay out the border of a circular flower bed, sticks are tied to each end of a rope that is 20 feet long. One person holds one of the sticks stationary while another person uses the other stick to trace out a circular path while keeping the rope taut. If the center of the circle is at the stationary stick, what is the equation of the circle?

[11.2] ***For Exercises 31–36, graph each ellipse. Label the center and the points above, below, to the left, and to the right of the center.***

31. $\frac{x^2}{49} + \frac{y^2}{25} = 1$

32. $\frac{x^2}{9} + \frac{y^2}{25} = 1$

33. $4x^2 + 9y^2 = 36$

34. $25x^2 + 9y^2 = 225$

35. $\frac{(x+2)^2}{16} + \frac{(y-3)^2}{4} = 1$

36. $\frac{(x-1)^2}{9} + \frac{(y+4)^2}{25} = 1$

[11.2] ***For Exercises 37–40, graph each hyperbola. Also show the fundamental rectangle with its corner points labeled, the asymptotes, and the intercepts.***

37. $\frac{x^2}{25} - \frac{y^2}{16} = 1$

38. $\frac{x^2}{36} - \frac{y^2}{9} = 1$

39. $y^2 - 9x^2 = 36$

40. $25y^2 - 9x^2 = 225$

[11.2] 41. A bridge over a canal is in the shape of half an ellipse. If the highest point of the bridge is 15 feet above the water and the base of the bridge is 50 feet across, what is the equation of the ellipse, half of which forms the bridge?

[11.2] 42. The cam of a compound bow is elliptical and is 4 inches long and 3.25 inches wide, as shown in the figure. What is the equation of the ellipse forming the shape of the cam?

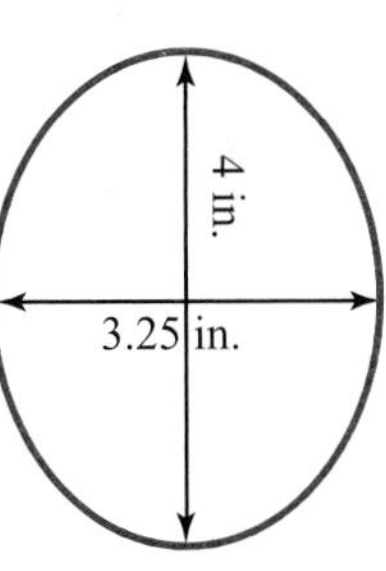

[11.3] ***For Exercises 43–50, solve.***

43. $\begin{cases} y = 2x^2 - 3 \\ 2x - y = -1 \end{cases}$

44. $\begin{cases} x^2 + y^2 = 17 \\ x - y = 3 \end{cases}$

45. $\begin{cases} 25x^2 + 3y^2 = 100 \\ 2x - y = -3 \end{cases}$

46. $\begin{cases} x^2 + y^2 = 64 \\ x^2 - y^2 = 64 \end{cases}$

47. $\begin{cases} x^2 + y^2 = 25 \\ 25y^2 - 16x^2 = 256 \end{cases}$

48. $\begin{cases} x^2 + 5y^2 = 36 \\ 4x^2 - 7y^2 = 36 \end{cases}$

49. $\begin{cases} y = x^2 - 2 \\ 4x^2 + 5y^2 = 36 \end{cases}$

50. $\begin{cases} y = x^2 - 1 \\ 4y^2 - 5x^2 = 16 \end{cases}$

[11.3] 51. The sum of the squares of two integers is 89 and the difference of their squares is 39. Find the integers.

[11.3] 52. A rectangular rug has a perimeter of 36 feet and an area of 80 square feet. Find the length and width of the rug.

[11.3] ***For Exercises 53 and 54, use a graphing calculator to verify the results of the exercise given.***

53. Exercise 43.

54. Exercise 49.

[11.4] ***For Exercises 55–58, graph the inequality.***

55. $x^2 + y^2 \leq 64$

56. $y < 2(x - 3)^2 + 4$

57. $\dfrac{y^2}{25} + \dfrac{x^2}{49} > 1$

58. $\dfrac{x^2}{25} - \dfrac{y^2}{36} < 1$

[11.4] ***For Exercises 59–64, graph the solution set of the system of inequalities.***

59. $\begin{cases} y \geq x^2 - 3 \\ 2x + y < 2 \end{cases}$

60. $\begin{cases} x + 2y < 4 \\ x^2 + y^2 \leq 25 \end{cases}$

61. $\begin{cases} y > x^2 - 2 \\ y < -x^2 + 1 \end{cases}$

62. $\begin{cases} \dfrac{x^2}{9} + \dfrac{y^2}{25} \leq 1 \\ x^2 + y^2 \geq 9 \end{cases}$

63. $\begin{cases} \dfrac{y^2}{4} - \dfrac{x^2}{9} \leq 1 \\ \dfrac{x^2}{9} + \dfrac{y^2}{25} \leq 1 \end{cases}$

64. $\begin{cases} \dfrac{y^2}{4} + \dfrac{x^2}{16} \leq 1 \\ x^2 + y^2 \leq 9 \\ y \leq x + 1 \end{cases}$

Chapter 11 Practice Test

For Exercises 1–3, find the direction the parabola opens, the coordinates of the vertex, the equation of the axis of symmetry, and draw the graph.

1. $y = 2(x + 1)^2 - 4$

2. $x = -2(y - 3)^2 + 1$

3. $x = y^2 + 4y - 3$

4. Find the distance between the points whose coordinates are $(2, -3)$ and $(6, -5)$.

For Exercises 5 and 6, find the center and radius, and draw the graph.

5. $(x + 4)^2 + (y - 3)^2 = 36$

6. $x^2 + y^2 + 4x - 10y + 20 = 0$

7. Write the equation of the circle with center $(2, -4)$ and that passes through the point $(-4, 4)$.

For Exercises 8–11, graph the equation and label relevant points. If the graph is a hyperbola, show the fundamental rectangle with its corner points labeled, the asymptotes, and the intercepts.

8. $16x^2 + 36y^2 = 576$

9. $\dfrac{(x + 2)^2}{4} + \dfrac{(y - 1)^2}{25} = 1$

10. $\dfrac{x^2}{49} - \dfrac{y^2}{25} = 1$

11. $\dfrac{y^2}{9} - \dfrac{x^2}{16} = 1$

For Exercises 12–15, solve.

12. $\begin{cases} y = (x+1)^2 + 2 \\ 2x + y = 8 \end{cases}$

13. $\begin{cases} 3x - y = 4 \\ x^2 + y^2 = 34 \end{cases}$

14. $\begin{cases} x^2 + y^2 = 13 \\ 3x^2 + 4y^2 = 48 \end{cases}$

15. $\begin{cases} x^2 - 2y^2 = 1 \\ 4x^2 + 7y^2 = 64 \end{cases}$

For Exercises 16 and 17, graph.

16. $y \leq -2(x+3)^2 + 2$

17. $\frac{x^2}{4} + \frac{y^2}{9} > 1$

18. Graph the solution set. $\begin{cases} y \geq x^2 - 4 \\ \frac{x^2}{9} + \frac{y^2}{16} \leq 1 \end{cases}$

19. An arch is in the shape of a parabola as shown in the figure. If we place the origin, O, as indicated, find the height of the arch and the distance across the base if the equation is $y = -\frac{1}{2}x^2 + 18$.

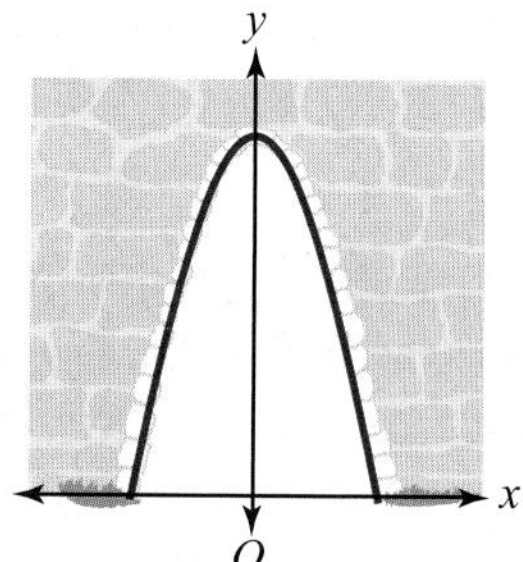

20. The path of a person's foot using an elliptical exercise machine on a particular setting is an ellipse. If the length of the stride is 19 inches and the height of the stride is 2 inches, what is the equation of the path of the foot?

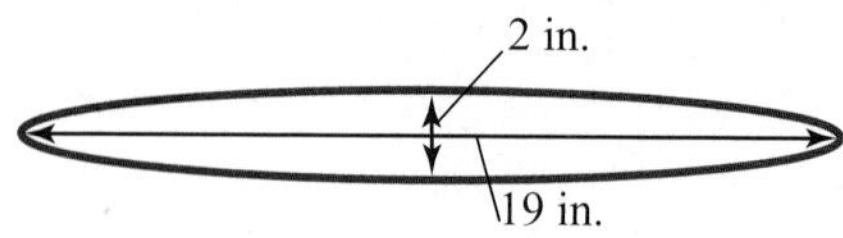

Chapters 1–11 Cumulative Review Exercises

For Exercises 1–6, answer true or false.

[3.5] **1.** The vertical line test can help determine if a graphed relation is a function.

[5.1] **2.** A positive base raised to a negative exponent simplifies to a negative number.

[7.4] **3.** If $\frac{a}{b} = \frac{c}{d}$, then $ad = bc$ where $b \neq 0$ and $d \neq 0$.

[10.1] **4.** A function must be one-to-one to have an inverse that is a function.

[10.3] **5.** Logarithms are exponents.

[11.1] **6.** A circle has asymptotes.

For Exercises 7–10, fill in the blank.

[4.5] **7.** If $A = \begin{bmatrix} a_1 & b_1 \\ a_2 & b_2 \end{bmatrix}$, then $\det(A) = \begin{vmatrix} a_1 & b_1 \\ a_2 & b_2 \end{vmatrix} =$ ______________.

[8.7] **8.** $\sqrt{-1} =$ ______________

[11.1] **9.** The formula for finding the distance between two points (x_1, y_1) and (x_2, y_2) is ________________________.

[11.1] **10.** The equation of a circle with center at (h, k) and radius of r is ___________________.

Expressions

Exercises 11–18

For Exercises 11–16, simplify.

[1.3] **11.** $6 - (-3) + 2^{-4}$

[5.1] **12.** $(5m^3n^{-2})^{-3}$

[7.1] **13.** $\dfrac{x - y}{4y - 4x}$

[7.1] **14.** $7x \cdot \dfrac{8}{21x^2}$

[7.3] **15.** $\dfrac{\frac{1}{x}}{3 + \frac{1}{x^2}}$

[8.7] **16.** $\sqrt{-20}$

For Exercises 17 and 18, factor completely.

[6.1] **17.** $ax + bx + ay + by$

[6.3] **18.** $k^4 - 81$

For Exercises 19–28, solve.

[2.5] **19.** $|2x + 1| = |x - 3|$

[4.1] **20.** $\begin{cases} 2x + y = 3 \\ 2x - y = 5 \end{cases}$

[8.6] **21.** $\sqrt{x - 3} = 7$

[9.2] **22.** $4x^2 - 2x + 1 = 0$

[9.3] **23.** $u^4 - 3u^2 - 4 = 0$

[9.3] **24.** $9 + 24x^{-1} + 16x^{-2} = 0$

[10.2] **25.** $9^x = 27$

[10.3] **26.** $\log_3 x = -2$

[10.3] **27.** $\log_3(x + 1) - \log_3 x = 2$

[11.3] **28.** $\begin{cases} x + y = 5 \\ x^2 + y^2 = 4 \end{cases}$

For Exercises 29 and 30, solve the inequality, then: a. Graph the solution set.
b. Write the solution set in set-builder notation.
c. Write the solution set in interval notation.

[2.6] **29.** $|3x + 1| > 4$

[9.5] **30.** $5m^2 - 3m < 0$

[3.3] **31.** Write the equation of a line perpendicular to $3x + 4y = 8$ and passing through the point $(3, -8)$. Write the equation in slope-intercept form and in standard form.

[5.2, 10.1] **32.** Given $f(x) = 2x + 1$ and $g(x) = x^2 - 1$, find

a. $(f + g)(x)$ **b.** $(f - g)(x)$ **c.** $(f \cdot g)(x)$

d. $(f/g)(x)$ **e.** $(f \circ g)(x)$ **f.** $(g \circ f)(x)$

[10.1] **33.** Determine whether $f(x) = \dfrac{x - 4}{5}$ and $g(x) = 5x + 4$ are inverses.

[10.3] **34.** Write $5^3 = 125$ in logarithmic form.

[10.4] **35.** Write as a sum or difference of multiples of logarithms: $\log_5 x^5 y$

[11.1] **36.** Write the equation of a circle in standard form with a center at (2, 4) passing through the point (7, 16).

For Exercises 37–41, graph.

[4.6] **37.** $\begin{cases} y < -x + 2 \\ y \geq x - 4 \end{cases}$

[10.3] **38.** $f(x) = \log_3 x$

[11.2] **39.** $\frac{x^2}{4} + \frac{y^2}{9} = 1$

[11.2] **40.** $\frac{x^2}{16} - \frac{y^2}{36} = 1$

[11.4] **41.** $\begin{cases} y < x + 2 \\ x^2 + y^2 < 9 \end{cases}$

For Exercises 42–50, solve.

[2.1] **42.** The volume of a cylinder can be found using the formula $V = \pi r^2 h$.

a. Solve the formula for h.

b. Suppose the volume of a cylinder is 15π cubic inches. Find its height.

[2.2 or 4.3] **43.** During 2001, shrimp overtook tuna as the top-selling seafood in the United States. It is estimated that each person consumed a combined total of 6.3 pounds of these two seafoods and that each person ate $\frac{1}{2}$ pound more of shrimp than tuna. How many pounds of each were consumed? (*Source: Good Housekeeping,* March 2003)

[4.3] **44.** A total of \$5000 is invested in three funds. The money market fund pays 5%, the income fund pays 6%, and the growth fund pays 3%. The total annual interest from the three accounts is \$255 and the amount in the growth fund is \$500 less than the amount invested in the money market account. Find the amount invested in each of the funds.

[6.4] **45.** A light-weight frame made of metal rods is in the shape of a right triangle as shown. Find the length of each side.

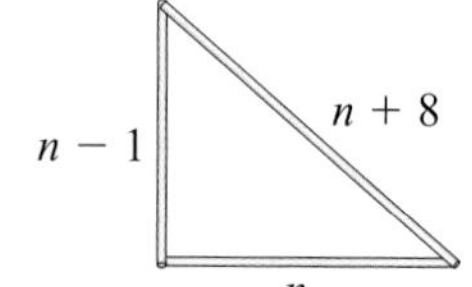

[7.5] **46.** Karen and Eric operate a landscaping company. Karen can mow, edge, and trim hedges on a quarter acre lot in 2 hours. Eric takes 1.5 hours to do the same work. How much time would it take them working together?

[10.2] **47.** If \$15,000 is deposited into an account paying 4% interest compounded monthly, how much will be in the account after 3 years?

[10.5] **48.** What is the pH of a solution whose concentration of hydrogen ions is 4.23×10^{-8} moles per liter?

[10.5] **49.** The magnitude of an earthquake was defined in 1935 by Charles Richter as $M = \log\frac{I}{I_0}$. We let I be the intensity of the earthquake measured by the amplitude of a seismograph reading taken 100 km from the epicenter of the earthquake, and we let I_0 be the intensity of a "standard earthquake" whose amplitude is 10^{-4} cm. On January 26, 2001, an earthquake in Rann of Kutch, Gujarat was measured with an intensity of $10^{2.9}$. What was the magnitude of this earthquake?

[11.2] **50.** A bridge over a canal is in the shape of half an ellipse. If the highest point of the bridge is 20 feet above the water and the base of the bridge is 150 feet across, what is the equation of the ellipse, half of which forms the bridge?

Appendix A Arithmetic Sequences and Series

OBJECTIVES

1. **Find the terms of a sequence when given the general term.**
2. **Define and write arithmetic sequences, find their common difference, and find a particular term.**
3. **Define and write series, find partial sums, and use summation notation.**
4. **Write arithmetic series and find their sums.**

The word **sequence** is used in mathematics much as it is in everyday life. For example, the classes that you attend on a given day occur in a particular order or sequence. As such, a sequence is an ordered list. Suppose we have a bacteria colony that has an initial population of 10,000 and increases at a rate of 10% each day.

On the second day, we would have
$10{,}000 + 0.10(10{,}000) = 10{,}000 + 1000 = 11{,}000$.
On the third day we have $11{,}000 + 0.10(11{,}000) = 11{,}000 + 1100 = 12{,}100$.
On the fourth day we have $12{,}100 + 0.10(12{,}100) = 12{,}100 + 1210 = 13{,}310$.
On the fifth day we have $13{,}310 + 0.10(13{,}310) = 13{,}310 + 1331 = 14{,}641$.

The number of bacteria present after each day forms a sequence that we can summarize as follows.

Days	1	2	3	4	5
Number of Bacteria	10,000	11,000	12,100	13,310	14,641

Based on this, we make the following definition.

DEFINITION ***Sequence:*** A function list whose domain is 1, 2, 3, . . . , *n*.

The *domain* of our bacteria example is the numbers of the days {1, 2, 3, 4, 5}. Each number in the *range* of a sequence is called a *term*, so the terms of our bacteria sequence are the numbers of bacteria: 10,000, 11,000, 12,100, 13,310, 14,641. Sequences can be finite or infinite depending on the number of terms. Our bacteria example is a **finite sequence** because it has a finite number of terms. The sequence 2, 4, 6, 8, 10, . . . is an **infinite sequence** because it has an infinite number of terms.

DEFINITION ***Finite sequence:*** A function with a domain that is the set of natural numbers from 1 to *n*.
Infinite sequence: A function with a domain that is the set of natural numbers.

OBJECTIVE 1. Find the terms of a sequence when given the general term. Since a sequence is a function, we could describe sequences with functional notation. Instead, we use a different notation that emphasizes the fact that the domain is a subset of the natural numbers. We think of the terms of a sequence as $a_1, a_2, a_3, \ldots, a_n$, where the subscript gives the number of the term. Thus, a_n is the nth or *general term* of the sequence and we represent a sequence by giving a formula for a_n. Consider the following sequence:

Term	a_1	a_2	a_3	a_4	a_5	a_n
Term of Sequence	1	4	9	16	25	n^2

We represent this sequence by writing $a_n = n^2$, which means $a_1 = 1^2$, $a_2 = 2^2$, etc.

EXAMPLE 1 Find the first three terms of the following sequences and the 25th term.

a. $a_n = 3n - 1$

Solution We let $n = 1, 2, 3$, and 25 and evaluate.

$a_1 = 3(1) - 1 = 3 - 1 = 2$ Let $n = 1$ and evaluate.

$a_2 = 3(2) - 1 = 6 - 1 = 5$ Let $n = 2$ and evaluate.

$a_3 = 3(3) - 1 = 9 - 1 = 8$ Let $n = 3$ and evaluate.

$a_{25} = 3(25) - 1 = 75 - 1 = 74$ Let $n = 25$ and evaluate.

The first three terms of the sequence are 2, 5, and 8. The 25th term is 74.

b. $a_n = \dfrac{(-1)^n}{n^2 + 1}$

Solution We let $n = 1, 2, 3$, and 25 and evaluate.

$a_1 = \dfrac{(-1)^1}{1^2 + 1} = \dfrac{-1}{1 + 1} = -\dfrac{1}{2}$ Let $n = 1$ and evaluate.

$a_2 = \dfrac{(-1)^2}{2^2 + 1} = \dfrac{1}{4 + 1} = \dfrac{1}{5}$ Let $n = 2$ and evaluate.

$a_3 = \dfrac{(-1)^3}{3^2 + 1} = \dfrac{-1}{9 + 1} = -\dfrac{1}{10}$ Let $n = 3$ and evaluate.

$a_{25} = \dfrac{(-1)^{25}}{25^2 + 1} = \dfrac{-1}{625 + 1} = -\dfrac{1}{626}$ Let $n = 25$ and evaluate.

The first three terms of the sequence are $-\dfrac{1}{2}, \dfrac{1}{5}$, and $-\dfrac{1}{10}$. The 25th term is $-\dfrac{1}{626}$.

OBJECTIVE 2. Define and write arithmetic sequences, find their common difference, and find a particular term. In the sequence $-3, 1, 5, 9, 13, \ldots,$ notice that each term after the first is found by adding 4 to the previous term. This is an example of an **arithmetic sequence** or *arithmetic progression*. Any two successive terms of an arithmetic sequence differ by the same amount, which is called the **common difference** and is denoted by d. To find d, choose any term (except the first) and subtract the previous term.

DEFINITION ***Arithmetic sequence:*** A sequence in which each term after the first is found by adding the same number to the previous term.
Common difference of an arithmetic sequence: The value d found by $d = a_n - a_{n-1}$ where a_n is any value in the sequence and a_{n-1} is the previous value.

EXAMPLE 2 Write the first four terms of the following arithmetic sequences.

a. The first term is 2 and the common difference is 5.

Solution Begin with 2 and find each successive term by adding 5 to the previous term.

$$a_1 = 2, a_2 = 2 + 5 = 7, a_3 = 7 + 5 = 12, a_4 = 12 + 5 = 17$$

The first four terms of the sequence are 2, 7, 12, 17.

b. $a_1 = -1, d = -2$

Solution Begin with -1 and find each successive term by adding -2 to the previous term.

$$a_1 = -1, a_2 = -1 - 2 = -3, a_3 = -3 - 2 = -5, a_4 = -5 - 2 = -7$$

The first four terms of the sequence are $-1, -3, -5, -7$.

EXAMPLE 3 Find the common difference, d, for the following arithmetic sequence.

a. $-4, -1, 2, 5, 8, \ldots$

Solution Pick any term (except the first) and subtract the term before it.

$$d = -1 - (-4) = -1 + 4 = 3$$

or we could use $d = 5 - 2 = 3$, etc.

b. $8, 6, 4, 2, 0, \ldots$

Solution Pick any term (except the first) and subtract the term before it.

$$d = 6 - 8 = -2$$

or we could use $d = 2 - 4 = -2$, etc.

If the first term of an arithmetic sequence is a_1 and the common difference is d, then the arithmetic sequence can be written as $a_1, a_1 + d, a_1 + 2d, a_1 + 3d, a_1 + 4d, \ldots$. Note that the coefficient of d is one less than the number of the term, so we have the following rule.

RULE **The nth Term of an Arithmetic Sequence**

The formula for finding the nth term of an arithmetic sequence is $a_n = a_1 + (n - 1)d$, where a_1 is the first term and d is the common difference.

EXAMPLE 4 Find the 23rd term and an expression for the nth term of an arithmetic sequence in which $a_1 = -8$ and $d = 3$.

Solution

$a_n = a_1 + (n - 1)d$

$a_{23} = -8 + (23 - 1)(3)$ — Substitute **23** for n, -8 for a_1, 3 for d.

$a_{23} = 58$ — Evaluate.

n^{th} term: $a_n = -8 + (n - 1)3$ — Substitute for a_1 and d.

$a_n = -11 + 3n$ — Simplify.

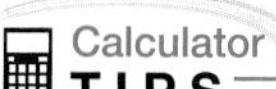

Some graphing calculators have a sequence mode that allows you to display the terms of a sequence in a table. To generate the sequence $a_n = n^2$ on a TI-83, press MODE *and select seq then follow the steps below.*

Press Y= *and enter as below.*

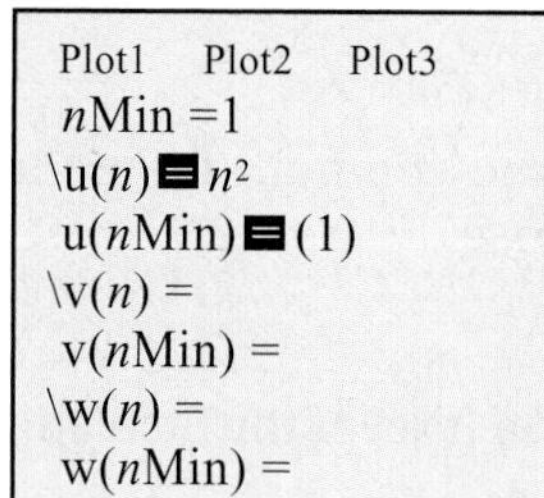

Press 2nd WINDOW *to set the table as below.*

TABLE SETUP
TblStart=1
ΔTbl=1
Indpnt: Auto Ask
Depend: Auto Ask

Press 2nd GRAPH *to display the table.*

n	u(n)
1	1
2	4
3	9
4	16
5	25
6	36
7	49

n=1

The numbers of the terms are in the n column and the terms of the sequence are in the u(n) column.

If we know the first term and one other term, we can use the formula for the nth term of an arithmetic sequence to find the common difference. Consequently, we can find the sequence.

EXAMPLE 5 The first term of an arithmetic sequence is 1 and the 20th term (a_{20}) is 58. Find the common difference and the first four terms of the sequence.

Solution We will use the fact that we know a_{20} and the formula for the nth term to find d.

$$a_n = a_1 + (n - 1)d$$

$$a_{20} = 1 + (20 - 1)d \quad \text{Substitute 1 for } a_1 \text{ and 20 for } n.$$

$$58 = 1 + 19d \quad \text{Substitute 58 for } a_{20} \text{ and solve.}$$

$$3 = d$$

The first four terms are 1, 1 + 3 = 4, 4 + 3 = 7, 7 + 3 = 10.

OBJECTIVE 3. Define and write series, find partial sums, and use summation notation. When the terms of a sequence are added, the sum is called a **series**.

DEFINITION ***Series:*** The sum of the terms of a sequence.

Given a sequence $a_1, a_2, a_3, \ldots, a_n$, then $a_1 + a_2 + a_3 + \cdots + a_n$ is the corresponding series. Finite series correspond to finite sequences and infinite series correspond to infinite sequences. If an expression for the nth term is known, we can write the series in *summation notation* using the capital Greek letter *sigma* (Σ) as follows:

$$a_1 + a_2 + a_3 + \cdots + a_n = \sum_{i=1}^{n} a_i$$

The i is called the *index of summation*; 1 is the *lower limit* of i and n is the *upper limit* of i. If the upper limit is a natural number, the series is finite, and if the upper limit is ∞, the series is infinite. To find a finite series from summation notation, first replace the index of summation (usually i) with its lower limit and evaluate that term, replace i with the lower limit plus 1 and evaluate that term, and so on until you reach the upper limit. This gives the series. To find the sum, add the terms of the series.

EXAMPLE 6

Warning: Do not confuse the use of i in sigma notation, as in $\sum_{i=1}^{5}(3i + 2)$, with its use as the imaginary unit, as in $7 \pm 5i$. In sigma notation, i is a variable representing natural numbers, whereas as the imaginary unit it represents $\sqrt{-1}$.

a. Write the terms of $\sum_{i=1}^{5}(3i + 2)$ and find the sum of the series.

Solution Replace i with 1, 2, 3, 4, and 5. Then evaluate.

$$\sum_{i=1}^{5}(3i + 2) = (3 \cdot 1 + 2) + (3 \cdot 2 + 2) + (3 \cdot 3 + 2) + (3 \cdot 4 + 2) + (3 \cdot 5 + 2)$$

$$= 5 + 8 + 11 + 14 + 17$$

$$= 55$$

b. Write the first five terms of the infinite series $\sum_{i=1}^{\infty} 2i^2$.

Solution Replace i with 1, 2, 3, 4, and 5. Then evaluate.

$$\sum_{i=1}^{\infty} 2i^2 = 2 \cdot 1^2 + 2 \cdot 2^2 + 2 \cdot 3^2 + 2 \cdot 4^2 + 2 \cdot 5^2 + \cdots$$

$$= 2 + 8 + 18 + 32 + 50 + \cdots$$

OBJECTIVE 4. Write arithmetic series and find their sums. If the terms of an arithmetic sequence are added, it is called an **arithmetic series**.

DEFINITION ***Arithmetic series:*** The sum of the terms in an arithmetic sequence.

Consequently, an arithmetic series has the form $a_1 + (a_1 + d) + (a_1 + 2d) + (a_1 + 3d) + \ldots + [a_1 + (n - 1)d]$.

Adding a finite number of terms of an infinite series gives a *partial sum.* The symbol S_n is used to indicate the sum of the first n terms. For example, S_5 means add the first five terms. Let's derive a formula for S_n for an arithmetic series as follows:

$$S_n = a_1 + (a_1 + d) + (a_1 + 2d) + (a_1 + 3d) + \ldots + a_n$$

We also need to include the terms between a_1 and $3d$ and a_n, so we write another version of S_n beginning with the last term, a_n, and subtracting the common difference, d, from the previous term.

$$S_n = a_n + (a_n - d) + (a_n - 2d) + (a_n - 3d) + \ldots + a_1$$

To describe the entire sum, we add our two versions of S_n together.

$$\begin{array}{rl} S_n = a_1 & + (a_1 + d) + (a_1 + 2d) + (a_1 + 3d) + \cdots + a_n \\ + S_n = a_n & + (a_n - d) + (a_n - 2d) + (a_n - 3d) + \cdots + a_1 \\ \hline 2S_n = (a_1 + a_n) & + (a_1 + a_n) + (a_1 + a_n) + (a_1 + a_n) + \cdots + (a_1 + a_n) \end{array}$$

Since S_n has n terms, there are n terms of $(a_1 + a_n)$, so

$$2S_n = n(a_1 + a_n), \text{ or}$$

$$S_n = \frac{n}{2}(a_1 + a_n)$$

RULE **Partial Sum S_n of an Arithmetic Series**

The sum of the first n terms of an arithmetic series, S_n, called the nth partial sum, is given by

$$S_n = \frac{n}{2}(a_1 + a_n)$$

where n is the number of terms, a_1 is the first term, and a_n is the nth term.

EXAMPLE 7 Find the sum of the first 20 terms (S_{20}) of the arithmetic series $-6 - 2 + 2 + 6 + \cdots$.

Solution To find the S_{20}, we first need to find the 20th term. Since $d = 4$,

$a_{20} = -6 + (20 - 1)(4) = 70$

$S_n = \frac{n}{2}(a_1 + a_n)$ Rule for n^{th} partial sum.

$S_{20} = \frac{20}{2}(-6 + 70)$ Substitute 20 for n, -6 for a_1, and 70 for a_n.

$S_{20} = 10(64) = 640$ Evaluate.

Appendix A Exercises

For Extra Help

MyMathLab

Videotape/DVT

InterAct Math

Math Tutor Center

Math XL.com

1. What set of numbers makes up the domain of a sequence?

2. A sequence with an unlimited number of terms is called a(n) ________ sequence.

3. What is a series?

4. What is an arithmetic sequence? How do you find the common difference?

5. The series $-10 - 5 + 0 + 5 + 10 + \cdots$ is an example of a(n) ________ series with a common difference of ________.

6. What does S_n represent?

For Exercises 7–14, write the first four terms of the sequence and the indicated term.

7. $a_n = 2n + 1$, 20th term

8. $a_n = 3n - 4$, 18th term

9. $a_n = n^2 + 2$, 15th term

10. $a_n = n^2 - 3$, 12th term

11. $a_n = \dfrac{n}{n + 2}$, 22nd term

12. $a_n = \dfrac{2n}{n + 3}$, 10th term

13. $a_n = \dfrac{(-1)^n}{n^2 + 1}$, 15th term

14. $a_n = \dfrac{(-1)^n}{n^2 - 3}$, 26th term

For Exercises 15–20, find the common difference, d, for each arithmetic sequence.

15. 2, 7, 12, 17, . . .

16. 3, 11, 19, 27, . . .

17. 25, 22, 19, 16, . . .

18. 42, 36, 30, 24, . . .

19. −12, −5, 2, 9, . . .

20. −24, −27, −30, −33, . . .

For Exercises 21–26, find the indicated term and an expression for the nth term of the given arithmetic sequence.

21. a_{14} if $a_1 = 14$ and $d = 4$

22. a_{24} if $a_1 = 16$ and $d = 6$

23. a_{28} if $a_1 = -8$ and $d = -3$

24. a_{30} if $a_1 = -7$ and $d = -6$

25. a_{34} of −5, −1, 3, 7, . . .

26. a_{21} of 8, 15, 22, 29, . . .

For Exercises 27–32, write the first four terms of the arithmetic sequence with the given characteristics.

27. $a_1 = -6, d = 7$

28. $a_1 = -2, d = 6$

29. $a_1 = -5, d = -3$

30. $a_1 = -13, d = -5$

31. $a_1 = 7, a_{18} = 75$

32. $a_1 = 6, a_{22} = 153$

33. Find the first term of an arithmetic sequence if $a_{45} = 143$ and $d = 3$.

34. Find the first term of an arithmetic sequence if $a_{39} = 181$ and $d = 6$.

35. Find the common difference of an arithmetic sequence if the first term is -110 and the 29th term is 2.

36. Find the common difference of an arithmetic sequence if the first term is 78 and the 15th term is -76.

For Exercises 37–40, write the first four terms of the arithmetic sequence with the given d and a_n.

37. $d = 7, a_8 = 41$

38. $d = 4, a_{11} = 27$

39. $d = -3, a_7 = 9$

40. $d = -6, a_{10} = -36$

For Exercises 41–48, write the series and find the sum.

41. $\sum_{i=1}^{6} i^2$

42. $\sum_{i=1}^{3} 3i^2$

43. $\sum_{i=1}^{4} (2i - 5)$

44. $\sum_{i=1}^{5} (4i + 1)$

45. $\sum_{i=1}^{3} (3i^2 - 4)$

46. $\sum_{i=1}^{4} (-2i^2 + 5)$

47. $\sum_{i=3}^{6} (4i - 3)$

48. $\sum_{i=2}^{5} (4i - 2)$

For Exercises 49–56, find the given S_n for the arithmetic series.

49. If $a_1 = 10$ and $d = 4$, find S_{25}.

50. If $a_1 = -8$ and $d = 2$, find S_{30}.

51. If $a_1 = 12$ and $d = -3$, find S_{22}.

52. If $a_1 = 18$ and $d = -2$, find S_{30}.

53. $3 + 9 + 15 + 21 + \cdots$. Find S_{15}.

54. $5 + 9 + 13 + 17 + \cdots$. Find S_{25}.

55. $54 + 46 + 38 + 30 + \cdots$. Find S_{12}.

56. $44 + 39 + 34 + 29 + \cdots$. Find S_{18}.

For Exercises 57–64, answer each question.

57. Find the sum of the first 100 natural numbers.

58. Find the sum of the even integers 2 through 200.

59. A concert hall has 60 seats in the first row, 64 in the second, 68 in the third, and so on.

a. How many seats are in the 22nd row?

b. How many seats are in the concert hall if there are 35 rows?

60. During the year 2003, many car dealers were offering loans at 0% interest. Johanna bought a car for $24,000 and made monthly payments of $400. Using 24,000 as the first term, write the first five terms of an arithmetic sequence that gives the amount that she still owes at the end of each month. How much will she owe at the end of the twenty-fifth month?

61. Fence posts are arranged in a triangular stack with 25 on the bottom row, 24 on the next, 23 on the next, and so forth until there is a single post on the top.

a. How many posts are on the 10th row from the bottom?

b. How many posts are in the stack?

62. Carlos is doing sit-ups to flatten his stomach. He did 25 the first night and plans to add 1 each night.

a. Write the first five terms of an arithmetic sequence that gives the number of sit-ups that he does each night.

b. Find the number he does on the 30th night.

c. Find the total number of sit-ups that he has done after 30 nights.

63. Tanisha takes a job that pays $28,000 the first year and a raise of $1500 per year.

a. Write the first five terms of an arithmetic sequence that gives her salary at the end of each year.

b. What will her salary be for the 10th year?

c. What are her total earnings for her first 10 years?

64. You have a choice of two jobs. Job A has a starting salary of $25,000 with raises of $900 per year, and job B has a starting salary of $28,000 with raises of $600 per year.

a. Which job will pay the most during the 10th year?

b. Which job will pay the most total amount during the first 15 years?

Appendix B Geometric Sequences and Series

OBJECTIVES

1. Write a geometric sequence and find its common ratio and a specified term.
2. Find partial sums of geometric series.
3. Find the sums of infinite geometric series.
4. Solve applications using geometric series.

OBJECTIVE 1. Write a geometric sequence and find its common ratio and a specified term. In Appendix A, we generated an arithmetic sequence by adding the same number to each term to get the next term. In this section, we will multiply each term by the same number to generate a **geometric sequence**. The number we multiply by is called the **common ratio** and is denoted as r. We can find r by dividing any term (except the first) by the term before it.

DEFINITION ***Geometric sequence:*** A sequence in which every term after the first is found by multiplying the previous term by the same number, called the common ratio, r, where $r = \frac{a_n}{a_{n-1}}$.

Note: *In geometric sequences, the first term is usually denoted as a rather than a_1.*

If the first term is a and the common ratio is r, a geometric sequence has the form $a, ar, ar^2, ar^3, ar^4, \ldots, ar^{n-1}$. Thus, the general term of a geometric series is $a_n = ar^{n-1}$.

We notice that the exponent of r is one less than the number of the term, so we have the following rule.

RULE **The nth Term of a Geometric Sequence**

The formula for finding the nth term of a geometric sequence is $a_n = ar^{n-1}$, where a is the first term and r is the common ratio.

EXAMPLE 1 A geometric sequence has a first term of 3 and a common ratio of 4.

a. Write the first five terms.

Solution To find each term, multiply the term before it by 4. So,

$$a_1 = 3, a_2 = 3(4) = 12, a_3 = 12(4) = 48, a_4 = 48(4) = 192, a_5 = 192(4) = 768$$

The first five terms of the sequence are 3, 12, 48, 192, 768.

b. Find the 10th term.

Solution $a_n = ar^{n-1}$ Formula for a_n.

$a_{10} = 3(4)^{10-1}$ Substitute **10** for n, 3 for a, and 4 for r.

$a_{10} = 786{,}432$ Evaluate.

c. Find the general term.

Solution $a_n = ar^{n-1}$ Formula for a_n.

$a_n = 3(4)^{n-1}$ Substitute for a and r.

EXAMPLE 2 Given the geometric sequence 32, −16, 8, −4, . . . , find the common ratio r.

Solution To find r, divide any term except the first by the term before it.

$$r = \frac{-16}{32} = -\frac{1}{2} \quad \text{or} \quad r = \frac{8}{-16} = -\frac{1}{2}, \text{ etc.}$$

Geometric sequences often occur in populations and other applications.

EXAMPLE 3 The population of rabbits in a large pen increases at a rate of 12% per month. If there are currently 50 rabbits, find the population after 15 months.

Solution Let P_0 be the initial number of rabbits, P_1 the number after 1 month, P_2 the number after 2 months, etc. The number of rabbits at the end of each month is the number at the beginning of the month plus an increase of 12% of that number. So the number at the end of each month can be found as follows:

$$P_1 = P_0 + 0.12P_0 = 1.12P_0$$
$$P_2 = P_1 + 0.12P_1 = 1.12P_1 = 1.12(1.12P_0) = (1.12)^2P_0$$
$$P_3 = P_2 + 0.12P_2 = 1.12P_2 = 1.12[(1.12)^2P_0] = (1.12)^3P_0$$

The number of rabbits at the end of each month can be represented by the sequence $P_0, P_1, P_2, P_3, \ldots P_n = P_0, 1.12P_0, (1.12)^2P_0, (1.12)^3P_0, \ldots, (1.12)^{n-1}P_0$, which is a geometric sequence whose first term is P_0 and $r = 1.12$. Consequently, the nth term is $P_0(1.12)^{n-1}$. The number of rabbits at the end of 15 months is the 16th term of the sequence, so

$$P_{16} = P_0(1.12)^{16-1}$$
$$P_{16} = 50(1.12)^{15}$$
$$P_{16} = 273.68$$

Answer There are about 274 rabbits at the end of 15 months.

Note: *It can be shown using this procedure that if a population is growing at p% per unit time and the initial population is P_0, then the population at the end of each unit of time forms a geometric sequence whose first term is P_0 and whose common ratio is $(1 + p)$, where p is p% written as a decimal. So the geometric sequence is $P_0, P_0(1 + p), P_0(1 + p)^2, P_0(1 + p)^3, \ldots, P_0(1 + p)^{n-1}$. The population after n time periods is the $(n + 1)^{\text{st}}$ term of the sequence, which is $P_0(1 + p)^n$.*

OBJECTIVE 2. Find partial sums of geometric series. If we add the terms of an arithmetic sequence, we get an arithmetic series. Likewise, if we add the terms of a geometric sequence, we get a **geometric series.**

DEFINITION ***Geometric series:*** The sum of the terms of a geometric sequence.

A geometric series is of the form $a + ar + ar^2 + \ldots + ar^{n-1}$ if the series is finite, and $a + ar + ar^2 + \ldots$ if the series is infinite.

In the geometric series $3 + 6 + 12 + 24 + \cdots$, we have $a = 3$ and $r = 2$. In $243 - 81 + 27 - 9 + \cdots$, we have $a = 243$ and $r = -\frac{1}{3}$. Just as with arithmetic series, we can find a formula for partial sum, S_n. Begin with

$$S_n = a + ar + ar^2 + \cdots + ar^{n-1}$$

$$-rS_n = \quad - ar - ar^2 - \cdots - ar^{n-1} - ar^n \qquad \text{Multiply both sides of the equation by } -r.$$

$$S_n - rS_n = a - ar^n \qquad \text{Add the equations.}$$

$$S_n(1 - r) = a(1 - r^n) \qquad \text{Factor both sides.}$$

$$S_n = \frac{a(1 - r^n)}{1 - r} \qquad \text{Divide both sides by } 1 - r.$$

RULE **The Partial Sum, S_n, of a Geometric Series**

The sum of the first n terms of a geometric series, S_n, called the nth partial sum, is given by

$$S_n = \frac{a(1 - r^n)}{1 - r}$$

where n is the number of terms, a is the first term, and r is the common ratio ($r \neq 1$).

EXAMPLE 4 Find the sum of the first 10 terms of the geometric series. $1 - 3 + 9 - 27 + \cdots$

Solution We first find r: $r = \frac{-3}{1} = -3$.

$$S_n = \frac{a(1 - r^n)}{1 - r} \qquad \text{Formula for } S_n.$$

$$S_{10} = \frac{1(1 - (-3)^{10})}{1 - (-3)} \qquad \text{Substitute 10 for } n, 1 \text{ for } a, \text{ and } -3 \text{ for } r.$$

$$S_{10} = -14{,}762 \qquad \text{Evaluate.}$$

OBJECTIVE 3. Find the sums of infinite geometric series. If the common ratio satisfies $|r| > 1$, the partial sums become infinitely large as n becomes infinitely large. However, if $|r| < 1$, the partial sums approach a value as n becomes infinitely large. This value is called the *limit* of the partial sums and is the sum of the infinite series.

For example, for the series $2 + 1 + \frac{1}{2} + \frac{1}{4} + \cdots$, we have

$$S_5 = \frac{2\left(1 - \left(\frac{1}{2}\right)^5\right)}{1 - \frac{1}{2}} = 3.875$$

$$S_{10} = \frac{2\left(1 - \left(\frac{1}{2}\right)^{10}\right)}{1 - \frac{1}{2}} \approx 3.996$$

$$S_{15} = \frac{2\left(1 - \left(\frac{1}{2}\right)^{15}\right)}{1 - \frac{1}{2}} \approx 3.9998779$$

Notice that the greater the value of n, the closer the sum is to 4. The partial sums approach 4 because as n gets larger, $\left(\frac{1}{2}\right)^n$ approaches 0. In the preceding example, $\left(\frac{1}{2}\right)^5 = 0.03125$, $\left(\frac{1}{2}\right)^{10} \approx 0.000977$, and $\left(\frac{1}{2}\right)^{15} \approx 0.0000305$. In general, if $|r| < 1$, then r^n approaches 0 as n becomes large. Consequently, if $|r| < 1$ the formula $S_n = \frac{a(1 - r^n)}{1 - r}$ becomes $S_\infty = \frac{a(1 - 0)}{1 - r} = \frac{a}{1 - r}$ as n becomes infinitely large. We denote the sum as S_∞ rather than S_n.

RULE **The Sum of an Infinite Geometric Series**

If $|r| < 1$, the sum of an infinite geometric series, S_∞, is given by the formula

$$S_\infty = \frac{a}{1 - r}$$

where a is the first term and r is the common ratio. If $|r| \geq 1$, S_∞ does not exist.

EXAMPLE 5 Find the sum of the infinite geometric series.

$$2 + 1 + \frac{1}{2} + \frac{1}{4} + \cdots$$

Solution We know that $a = 2$ and $r = \frac{1}{2}$. Since $|r| < 1$, the sum exists.

$$S_\infty = \frac{a}{1 - r}$$

$$S_\infty = \frac{2}{1 - \frac{1}{2}} \quad \text{Substitute 2 for } a \text{ and } \frac{1}{2} \text{ for } r.$$

$$S_\infty = 4 \quad \text{Evaluate.}$$

Infinite geometric series also provides us with a method of changing a repeating decimal into a fraction.

EXAMPLE 6 Write as a fraction. $0.\overline{37}$

Solution First write $0.\overline{37}$ as an infinite geometric series as follows:

$$0.\overline{37} = 0.37 + 0.0037 + 0.000037 + \cdots$$

$$0.\overline{37} = \frac{37}{100} + \frac{37}{10000} + \frac{37}{1000000} + \cdots$$

The last line is an infinite geometric series with $a = \frac{37}{100}$ and $r = \frac{1}{100}$. Since $|r| < 1$, the sum of this series exists.

$$S_\infty = \frac{a}{1 - r}$$

$$S_\infty = \frac{\frac{37}{100}}{1 - \frac{1}{100}} \qquad \text{Substitute } \frac{37}{100} \text{ for } a \text{ and } \frac{1}{100} \text{ for } r.$$

$$S_\infty = \frac{\frac{37}{100}}{\frac{99}{100}} \qquad \text{Simplify.}$$

$$S_\infty = \frac{37}{99}$$

Answer $0.\overline{37} = \frac{37}{99}$

OBJECTIVE 4. Solve applications using geometric series.

EXAMPLE 7 To save for their child's college education, the McBride family put \$1000 into a savings account the first year, and each year thereafter deposited 10% more than the previous year.

a. Write the first five terms of the geometric sequence that gives the amount of money deposited into the account each year.

Solution From the note following Example 3, this is a geometric series in which $a = 1000$ and $r = (1 + 0.10) = 1.1$

$$a_1 = 1000,\ a_2 = 1000(1.1) = 1100,\ a_3 = 1000(1.1)^2 = 1210,$$
$$a_4 = 1000(1.10)^3 = 1331,\ a_5 = 1000(1.10)^4 = 1464.10$$

The first five terms of the sequence are 1000, 1100, 1210, 1331, 1464.10.

b. How much money will have been deposited into the account at the end of 15 years?

Solution The series 1000 + 1100 + 1210 + 1331 + 1464.10 + $\cdots$ is geometric with $a = 1000$ and $r = 1.1$.

$$S_n = \frac{a(1 - r^n)}{1 - r}$$

$$S_{15} = \frac{1000(1 - 1.1^{15})}{1 - 1.1} \qquad \text{Substitute 15 for } n\text{, 1000 for } a\text{, and 1.1 for } r.$$

$$S_{15} = 31{,}772.48 \qquad \text{Evaluate.}$$

Answer They will have deposited $31,772.48 in the account after 15 years.

Appendix B Exercises

1. If you divide a term of a geometric sequence by the term before it, what do you get?

2. What is the form of a geometric sequence?

3. Can you find the sum of the first n terms of the geometric series $2 + 6 + 18 + \ldots$ using the formula $S_n = \frac{a(1 - r^n)}{1 - r}$? Why or why not?

4. Can you find the sum of the infinite geometric series $2 + 6 + 18 + \cdots$ using the formula $S_\infty = \frac{a}{1 - r}$? Why or why not?

5. The fourth partial sum, S_4, of $2 + 6 + 18 + \ldots$ is ___+___ +___ + ___ = ___.

6. Write $0.\overline{23}$ as an infinite geometric series. What are a and r?

For Exercises 7–14: a. Find the common ratio, r, for the given geometric sequence.
b. Find the indicated term.
c. Find an expression for the general term, a_n.

7. 1, 3, 9, 27, . . . ; 9th term

8. 7, 14, 28, 56, . . . ; 11th term

9. −2, 4, −8, 16, . . . ; 15th term

10. −8, 24, −72, 216, . . . ; 9th term

11. 243, 81, 27, 9, . . . ; 10th term

12. 8, 4, 2, 1, . . . ; 8th term

13. 128, −32, 8, −2, . . . ; 7th term

14. 125, −25, 5, −1, . . . ; 9th term.

For Exercises 15–18: a. Write the first five terms of the geometric sequence satisfying the given conditions.
b. Find the indicated term.

15. $a = -4, r = -3$; 8th term

16. $a = -6, r = -2$; 10th term

17. $a = 243, r = \frac{1}{3}$; 7th term

18. $a = 256, r = -\frac{1}{2}$; 10th term

For Exercises 19–24: a. Use the formula for the nth term to find r.
b. Write the first four terms of the geometric sequence.

19. $a = 1, r > 0, a_5 = 16$

20. $a = 3, a_4 = -24$

21. $a = -6, r > 0, a_5 = -96$

22. $a = -4, a_4 = -32$

23. $a = 128, a_4 = 2$

24. $a = -243, a_6 = 1$

25. Find the first term of a geometric sequence in which $r = 3$ and the fifth term is −486.

26. Find the first term of a geometric sequence in which $r = -\frac{1}{2}$ and the fifth term is $-\frac{1}{8}$.

For Exercises 27–32, find the sum of the first n terms of each geometric series for the given value of n.

27. $3 + 9 + 27 + 81 + \cdots, n = 11$

28. $-1 - 2 - 4 - 8 - \cdots, n = 9$

29. $32 + 16 + 8 + 4 + \cdots, n = 9$

30. $81 + 27 + 9 + 3 + \cdots, n = 7$

31. $128 - 32 + 8 - 2 + \cdots, n = 9$

32. $625 - 125 + 25 - 5 + \cdots, n = 8$

For Exercises 33–38, find the sum of the infinite geometric series, if possible. If it is not possible, explain why.

33. $27 + 9 + 3 + \cdots$

34. $8 + 4 + 2 + \cdots$

35. $15 - 9 + \frac{27}{5} - \cdots$

36. $15 - 10 + \frac{20}{3} - \cdots$

37. $9 + 12 + 16 + \cdots$

38. $16 + 20 + 25 + \cdots$

For Exercises 39–42, write each repeating decimal as a fraction.

39. $0.\overline{4}$

40. $0.\overline{7}$

41. $0.\overline{17}$

42. $0.\overline{25}$

For Exercises 43 and 44, answer each question.

43. If $a_n = 500(1.04)^n$:

a. Find the first five terms of the sequence.

b. Find the 10th term of the sequence.

44. If $a_n = 350(1.06)^n$:

a. Find the first five terms of the sequence.

b. Find the 8th term of the sequence.

For Exercises 45–50, solve.

45. A population of mink is increasing at a rate of 8% per month. The current mink population is 100.

a. Using 100 as the first term, find the first four terms of the geometric sequence that gives the number of mink at the beginning of each month.

b. Find the number of mink present at the beginning of the eighth month.

c. Find the expression for the general term, a_n.

46. The generation time (the time required for the number present to double) for a particular bacteria is 1 hour. Suppose initially one bacteria was present.

a. Using 1 as the first term, write the first five terms of the geometric sequence giving the number of bacteria present after each hour.

b. Find an expression for the number present after the nth hour.

c. How many bacteria are present after 1 day?

47. Suppose you took a job for a month (20 working days) that paid \$0.01 the first day and your salary doubled each day.

a. Write the first five terms of the geometric sequence that gives your salary each day.

b. Find an expression for the amount earned on the nth day.

c. How much would you earn on the 20th day?

d. What are your total earnings for the month?

48. Damarys deposits \$200 in the bank and each month thereafter deposits 5% more than the month before. She does this for 1 year.

a. Find an expression for the amount she deposits on the nth month.

b. Write the first four terms of the geometric sequence that gives the amount of her deposit each month.

c. How much did she deposit on the 10th month?

d. How much does she deposit for the year?

49. A new boat costs \$20,000 and depreciates by 7% each year. What will the boat be worth in 8 years?

50. The isotope ${}_{15}P^{33}$ has a half-life of 25 days. A sample has 400 grams.

a. Find the first five terms of the geometric sequence that gives the amount present at the end of each half-life.

b. Find an expression for the amount present after the nth half-life.

c. Find the amount present after the 10th half-life.

Appendix C The Binomial Theorem

OBJECTIVES

1. **Expand a binomial using Pascal's triangle.**
2. **Evaluate factorial notation and binomial coefficients.**
3. **Expand a binomial using the binomial theorem.**
4. **Find a particular term of a binomial expansion.**

In Section 5.5 we learned to square binomials. In this section, we will learn to raise binomials to natural-number powers.

OBJECTIVE 1. Expand a binomial using Pascal's triangle. We begin by writing out $(a + b)^n$, where n is a natural number, and look for patterns. These products are called *binomial expansions.*

$$(a + b)^0 = 1$$
$$(a + b)^1 = a + b$$
$$(a + b)^2 = a^2 + 2ab + b^2$$
$$(a + b)^3 = a^3 + 3a^2b + 3ab^2 + b^3$$
$$(a + b)^4 = a^4 + 4a^3b + 6a^2b^2 + 4ab^3 + b^4$$
$$(a + b)^5 = a^5 + 5a^4b + 10a^3b^2 + 10a^2b^3 + 5ab^4 + b^5$$

Conclusions: Several patterns can be observed from the preceding expansions.

1. The first term in the expansion, a, is raised to the same power as the binomial and the power of a decreases by 1 in each successive term. Note that the last term does not contain an a.
2. The exponent of b is 0 in the first term and increases by 1 on each successive term.
3. The sum of the exponents of the variables of each term equals the exponent of the binomial.
4. The number of terms in the expansion is one more than the exponent of the binomial.

Now consider the coefficients of the terms in these expansions.

Coefficients of Expansions

$(a + b)^0$	1
$(a + b)^1$	1 1
$(a + b)^2$	1 2 1
$(a + b)^3$	1 3 3 1
$(a + b)^4$	1 4 6 4 1
$(a + b)^5$	1 5 10 10 5 1

If we arrange the coefficients of each expansion in a triangular array, we see an interesting pattern. Each row begins and ends with 1. Each number inside a row is the sum of the two numbers in the row above it. For example, each 10 in the bottom row comes from adding the 4 and 6 directly above.

This triangular array of numbers is called *Pascal's triangle* in honor of the French mathematician Blaise Pascal. Using these observations, the expansion of $(a + b)^6$ would have seven terms and the variable portions of the terms would be $a^6, a^5b, a^4b^2, a^3b^3, a^2b^4, ab^5, b^6$. By continuing the pattern in Pascal's triangle, we can find the coefficients for each of those terms.

$(a+b)^5$	1 5 10 10 5 1
$(a+b)^6$	1 6 15 20 15 6 1

Using the coefficients from the last line of Pascal's triangle and the variables previously listed gives us

$$(a+b)^6 = a^6 + 6a^5b + 15a^4b^2 + 20a^3b^3 + 15a^2b^4 + 6ab^5 + b^6$$

OBJECTIVE 2. Evaluate factorial notation and binomial coefficients. Although Pascal's triangle is easy to use, it isn't practical, especially for binomials raised to large powers. Consequently, another method called the *binomial theorem* is often used. Before introducing the binomial theorem, we need **factorial notation**.

DEFINITION ***Factorial notation:*** For any natural number n, the symbol $n!$ (read "n factorial") means $n(n-1)(n-2)\ldots 3 \cdot 2 \cdot 1$. $0!$ is defined to be 1, so $0! = 1$.

EXAMPLE 1 Evaluate the following factorials.

a. $5!$

Solution $5! = 5 \cdot 4 \cdot 3 \cdot 2 \cdot 1 = 120$

b. $7!$

Solution $7! = 7 \cdot 6 \cdot 5 \cdot 4 \cdot 3 \cdot 2 \cdot 1 = 5040$

Sometimes we may not write all the factors of a factorial. In such cases, the last desired factor is written as a factorial. Below are some alternate ways to write 7!

$$7! = 7 \cdot 6 \cdot 5 \cdot 4 \cdot 3 \cdot 2 \cdot 1 = 7 \cdot 6! \quad \text{or} \quad 7! = 7 \cdot 6 \cdot 5! \quad \text{or} \quad 7! = 7 \cdot 6 \cdot 5 \cdot 4!$$

The coefficients of a binomial expansion can be expressed in terms of factorials using a special notation called the **binomial coefficient**. We will see exactly how the binomial coefficient is used later.

DEFINITION ***Binomial coefficient:*** A number written as $\binom{n}{r}$ and defined as $\dfrac{n!}{r!(n-r)!}$.

EXAMPLE 2 Evaluate the following binomial coefficients.

a. $\binom{6}{2}$

Solution $\binom{6}{2} = \dfrac{6!}{2!(6-2)!}$ — Substitute 6 for n and 2 for r in $\dfrac{n!}{r!(n-r)!}$.

$= \dfrac{6 \cdot 5 \cdot 4 \cdot 3 \cdot 2 \cdot 1}{(2 \cdot 1)(4 \cdot 3 \cdot 2 \cdot 1)}$ — Expand the factorials.

$= 15$ — Simplify.

Note: *These factorials could have been evaluated as follows:*

$$\frac{6!}{2! \cdot 4!} = \frac{6 \cdot 5 \cdot \cancel{4!}}{2 \cdot 1 \cdot \cancel{4!}} = \frac{6 \cdot 5}{2} = \frac{30}{2} = 15$$

b. $\binom{8}{5}$

Solution $\binom{8}{5} = \dfrac{8!}{5!(8-5)!}$ — Substitute 8 for n and 5 for r in $\dfrac{n!}{r!(n-r)!}$.

$= \dfrac{8!}{5! \cdot 3!}$ — Simplify.

$= \dfrac{8 \cdot 7 \cdot 6 \cdot 5 \cdot 4 \cdot \cancel{3!}}{5 \cdot 4 \cdot 3 \cdot 2 \cdot 1 \cdot \cancel{3!}}$ — Rewrite 8! as $8 \cdot 7 \cdot 6 \cdot 5 \cdot 4 \cdot 3!$

$= 56$ — Evaluate.

Calculator TIPS

Factorials can be found on most graphing calculators. To find 10! on the TI-83 Plus, enter [1] [0]*, go to the Math menu, select PRB, and press* [4].

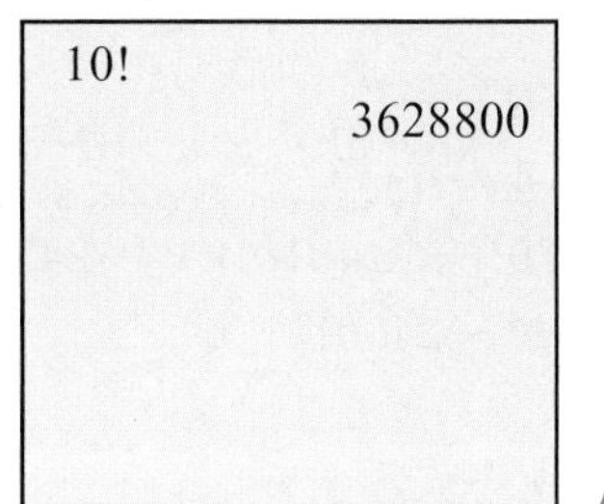

Using the formula for evaluating binomial coefficients, we can prove two special cases: $\binom{n}{0} = 1$ and $\binom{n}{n} = 1$.

OBJECTIVE 3. Expand a binomial using the binomial theorem. Let's look at the expansion of $(a+b)^6$ and make another observation.

$$(a+b)^6 = a^6 + 6a^5b + 15a^4b^2 + 20a^3b^3 + 15a^2b^4 + 6ab^5 + b^6$$

Look at the third term of the expansion, $15a^4b^2$. The coefficient is 15 and from Example 2a, we see that $\binom{6}{2} = 15$. The coefficient of the fourth term is 20 and $\binom{6}{3} = 20$. In both cases, the n value of the binomial coefficient, $\binom{n}{r}$, is the exponent of the binomial and the r value is the exponent of b. This observation leads to the **binomial theorem**.

DEFINITION ***The Binomial theorem:*** For any positive integer n,

$$(a+b)^n = \binom{n}{0}a^n + \binom{n}{1}a^{n-1}b + \binom{n}{2}a^{n-2}b^2 + \binom{n}{3}a^{n-3}b^3 + \cdots + \binom{n}{n}b^n$$

EXAMPLE 3 Expand each of the following binomials using the binomial theorem.

a. $(a+b)^6$

Solution

$$(a+b)^6 = \binom{6}{0}a^6 + \binom{6}{1}a^5b + \binom{6}{2}a^4b^2 + \binom{6}{3}a^3b^3 + \binom{6}{4}a^2b^4 + \binom{6}{5}ab^5 + \binom{6}{6}b^6$$

$$= \frac{6!}{0!6!}a^6 + \frac{6!}{1!5!}a^5b + \frac{6!}{2!4!}a^4b^2 + \frac{6!}{3!3!}a^3b^3 + \frac{6!}{4!2!}a^2b^4 + \frac{6!}{5!1!}ab^5 + \frac{6!}{6!0!}b^6$$

$$= a^6 + 6a^5b + 15a^4b^2 + 20a^3b^3 + 15a^2b^4 + 6ab^5 + b^6$$

Note: *This is the same result we got using Pascal's triangle earlier. Also note that* $\binom{6}{3} = \frac{6!}{3!3!} = \frac{6\cdot5\cdot4\cdot3!}{3!3!} = \frac{6\cdot5\cdot4\cdot\cancel{3!}}{3\cdot2\cdot1\cdot\cancel{3!}} = \frac{\cancel{6}\cdot5\cdot4}{\cancel{3}\cdot\cancel{2}\cdot1} = 20$, *etc.*

b. $(a-3b)^5$

Solution Write $(a-3b)^5$ as $[a+(-3b)]^5$

$$[a+(-3b)]^5 = \binom{5}{0}a^5 + \binom{5}{1}a^4(-3b) + \binom{5}{2}a^3(-3b)^2 + \binom{5}{3}a^2(-3b)^3 + \binom{5}{4}a(-3b)^4 + \binom{5}{5}(-3b)^5$$

$$= a^5 + 5a^4(-3b) + 10a^3(9b^2) + 10a^2(-27b^3) + 5a(81b^4) + (-243b^5)$$

$$= a^5 - 15a^4b + 90a^3b^2 - 270a^2b^3 + 405ab^4 - 243b^5$$

Note: $\binom{5}{2} = \frac{5!}{2!3!} = \frac{5\cdot4\cdot3!}{2\cdot1\cdot3!} = \frac{5\cdot4\cdot\cancel{3!}}{2\cdot1\cdot\cancel{3!}} = \frac{5\cdot4}{2} = \frac{20}{2} = 10$, *etc.*

OBJECTIVE 4. Find a particular term of a binomial expansion. Sometimes, it is necessary to find only a specific term of a binomial expansion without writing out the entire expansion. Look again at the binomial expansion. Note that the third term (which we will call the $(2+1)^{st}$ term) is $\binom{n}{2}a^{n-2}b^2$ and the fourth term (which we will call the $(3+1)^{st}$ term) is $\binom{n}{3}a^{n-3}b^3$. Similarly, the $(m+1)^{st}$ term is $\binom{n}{m}a^{n-m}b^m$. These observations lead to the following.

RULE **Finding the $(m+1)^{st}$ Term of a Binomial Expansion**

The $(m+1)^{st}$ term of the expansion $(a+b)^n$ is $\binom{n}{m}a^{n-m}b^m$.

EXAMPLE 4 Find the indicated term of each of the following binomial expansions.

a. $(a + b)^{11}$, seventh term

Solution Use the formula for the $(m + 1)^{st}$ term with $n = 11$ and $m = 6$ (to find the seventh term).

$$\binom{n}{m}a^{n-m}b^m = \binom{11}{6}a^{11-6}b^6 = 462a^5b^6$$

b. $(2x - 5y)^8$, fourth term

Solution Write $(2x - 5y)^8$ as $[2x + (-5y)]^8$. Use the formula for the $(m+1)^{st}$ term with $n = 8$, $m = 3$ (to find the fourth term), $a = 2x$, and $b = -5y$.

$$\binom{n}{m}a^{n-m}b^m = \binom{8}{3}(2x)^{8-3}(-5y)^3 = 56(32x^5)(-125y^3) = -224{,}000x^5y^3$$

Appendix C Exercises

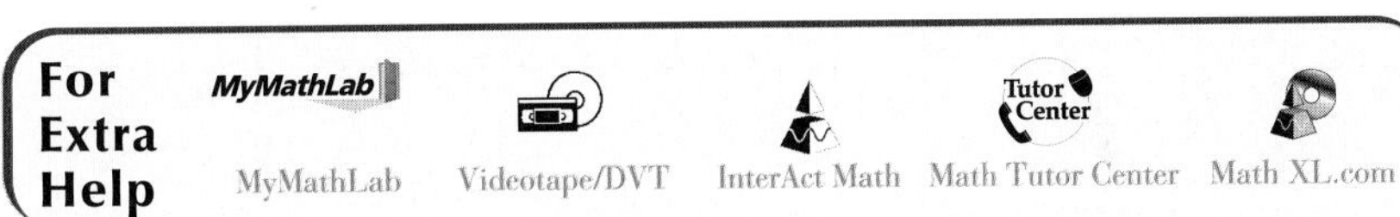

1. How many terms are in the expansion of $(5a + b)^{12}$?

2. What is the sum of the exponents on x and y for any term in the expansion of $(x + y)^9$?

3. How is the symbol 8! read?

4. What is the meaning of 8!? What is its value?

5. What is the exponent of b in the sixth term of the expansion of $(a + b)^{13}$? What is the exponent of a in that term?

6. On a term in the expansion of $(x + y)^n$ the exponent of x is 4. Find the exponent of y.

For Exercises 7–18, evaluate each expression.

7. 4!

8. 7!

9. (4!)(3!)

10. (3!)(2!)

11. (6!)(5!)

12. (4!)(7!)

13. $\frac{8!}{10!}$

14. $\frac{7!}{9!}$

15. $\frac{10!}{9!}$

16. $\frac{12!}{11!}$

17. $\frac{8!}{6!(8-6)!}$

18. $\frac{10!}{6!(10-6)!}$

For Exercises 19–26, evaluate each binomial coefficient.

19. $\binom{7}{3}$ **20.** $\binom{5}{2}$ **21.** $\binom{10}{4}$ **22.** $\binom{6}{5}$

23. $\binom{7}{7}$ **24.** $\binom{4}{4}$ **25.** $\binom{8}{0}$ **26.** $\binom{9}{0}$

For Exercises 27–38, use the binomial theorem to expand each of the following.

27. $(a + b)^5$ **28.** $(a + b)^7$

29. $(x - y)^4$ **30.** $(x - y)^3$

31. $(2a + b)^3$ **32.** $(a + 2b)^3$

33. $(x - 2y)^5$ **34.** $(x - 3y)^4$

35. $(2m + 3n)^6$ **36.** $(3c + 2d)^5$

37. $(3x - 4y)^4$ **38.** $(4a - b)^7$

For Exercises 39–46, find the indicated term of each binomial expansion.

39. $(x + y)^8$, fifth term **40.** $(a + b)^9$, fourth term

41. $(a - b)^{10}$, third term **42.** $(m - n)^7$, second term

43. $(4x + y)^9$, sixth term **44.** $(3a + b)^{11}$, seventh term

45. $(3m - 2n)^7$, fourth term **46.** $(5x - 3y)^{12}$, fifth term

Appendix D Permutations and Combinations

OBJECTIVES

1. **Use the multiplication principle.**
2. **Calculate permutations.**
3. **Calculate combinations.**

In this section, we will be counting the number of different ways in which tasks can be performed. For example, in a race with 8 people, we might count the possible number of first, second, and third place finishers. Or in a club with 25 members, how many different ways can we choose a committee of 5 to plan a party? Before we answer these questions, we need to establish the *multiplication principle.*

OBJECTIVE 1. Use the multiplication principle. Murlene goes into an ice cream shop to buy some fat-free frozen yogurt. For a container, she has a choice of a cup, sugar cone, or waffle cone. Her choices in flavors are vanilla, chocolate, strawberry, or peach. In how many ways can she choose a container and flavor? We use a tree diagram to answer this question.

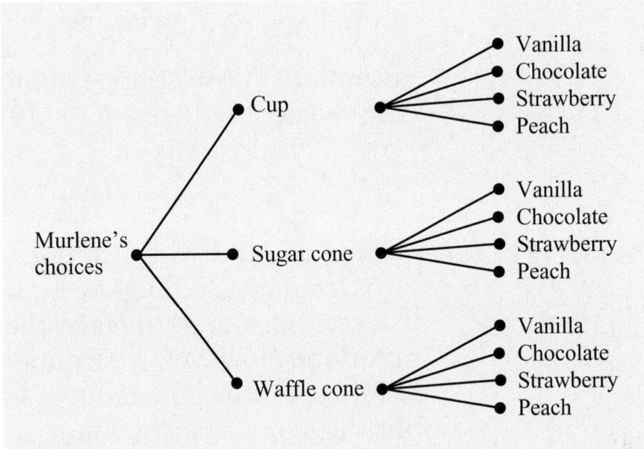

From the diagram, we see that for each of the 3 choices of containers there are 4 choices of flavors, so there are $3 \cdot 4 = 12$ different choices of containers and flavors. Our example suggests the multiplication principle.

RULE The Multiplication Principle

To find the number of ways in which successive tasks can be performed, multiply the number of ways in which each task can be performed. Symbolically, if task A can be performed in m ways and task B can be performed in n ways, then the two tasks can be performed in mn ways.

EXAMPLE 1

a. How many different three-digit numbers can be formed using the digits 2, 3, 5, and 7 if no digit may be repeated?

Solution We have four choices of the first digit (2, 3, 5 or 7), three for the second (can't use the one chosen as the first digit), and two for the third, so the total number of possible different three digit numbers is $4 \cdot 3 \cdot 2 = 24$.

b. In how many different ways (orders) can 6 people sit on a bench?

Solution Six choices are possible for the first person. After that person is seated, 5 choices are possible for the second person. Similarly, there are 4 choices for the third, 3 for the fourth, 2 for the fifth and 1 for the sixth. Therefore, 6 people can sit on a bench in $6 \cdot 5 \cdot 4 \cdot 3 \cdot 2 \cdot 1 = 720$ different ways.

c. A race has 8 participants. How many different first-, second-, and third-place finishers are possible?

Solution Any one of the 8 could finish first. Any one of the remaining 7 could finish second. Similarly, any one of the remaining 6 could finish third, so there are $8 \cdot 7 \cdot 6 = 336$ different first-, second-, and third-place finishers possible.

d. How many different phone numbers are possible for a given area code and prefix such as (920) 423- ___ ___ ___ ___ ?

Solution The telephone numbers have four more digits. Each digit has 10 possible choices, so there are $10 \cdot 10 \cdot 10 \cdot 10 = 10{,}000$ possible phone numbers in area code 920 and prefix 423.

OBJECTIVE 2. Calculate permutations. Examples 1a–c illustrate **permutations**. In a permutation, the *order* in which things occur is important. For example, if points are awarded for first, second, and third place, the order in which the participants finish in a race is important. However, if a troupe of dancers has 15 members and 4 are selected to perform a routine, the order in which dancers are selected is not important. Therefore, this is not a permutation.

DEFINITION ***Permutation:*** An ordered arrangement of objects. The permutation of n things taken r at a time is often denoted as ${}_nP_r$.

In Example 1b, we found ${}_6P_6 = 720$. In Example 1c, we found ${}_8P_3 = 336$. Notice from Example 1b that ${}_6P_6 = 6 \cdot 5 \cdot 4 \cdot 3 \cdot 2 \cdot 1 = 6!$ Instead of finding ${}_8P_3 = 336$ as $8 \cdot 7 \cdot 6$, we could have used $\dfrac{8 \cdot 7 \cdot 6 \cdot 5 \cdot 4 \cdot 3 \cdot 2 \cdot 1}{5 \cdot 4 \cdot 3 \cdot 2 \cdot 1} = \dfrac{8!}{(8-3)!} = \dfrac{8!}{5!}$. This leads to the following formula.

RULE Evaluating ${}_nP_r$.

The formula for the number of permutations of n things taken r at a time is

$$ {}_nP_r = \frac{n!}{(n-r)!} $$

EXAMPLE 2

a. A horse race has 10 horses entered. How many different first-, second-, and third-place finishers are possible?

Solution We need to find the number of permutations of 10 horses taken 3 at a time, ${}_{10}P_3$ so we use the formula ${}_nP_r = \dfrac{n!}{(n-r)!}$.

$$\begin{aligned} {}_{10}P_3 &= \frac{10!}{(10-3)!} && \text{Substitute 10 for } n \text{ and 3 for } r \text{ in } {}_nP_r = \frac{n!}{(n-r)!}. \\ &= \frac{10 \cdot 9 \cdot 8 \cdot \cancel{7!}}{\cancel{7!}} && \text{Replace 10! with } 10 \cdot 9 \cdot 8 \cdot 7! \text{ and } (10-3)! = 7!. \\ &= 10 \cdot 9 \cdot 8 && \text{Divide the 7!s.} \\ &= 720 && \text{Evaluate} \end{aligned}$$

There are 720 different possibilities for first-, second-, and third-place finishers.

b. Johann is going on a short trip and selects 4 CDs to play while driving. In how many different orders can he play the 4 CDs?

Solution We need to find the number of permutations of 4 things taken 4 at a time, ${}_4P_4$. Use ${}_nP_r = \dfrac{n!}{(n-r)!}$ with $n = 4$ and $r = 4$.

$${}_4P_4 = \frac{4!}{(4-4)!} = \frac{4!}{0!} = \frac{4 \cdot 3 \cdot 2 \cdot 1}{1} = 24$$

From Example 2, we can derive the following.

RULE **Three Special Permutation Formulas**

$${}_nP_n = n! \qquad {}_nP_0 = 1 \qquad {}_nP_1 = n$$

Also note that Example 2 could have been done using the multiplication principle.

Calculator TIPS

Most graphing calculators can evaluate ${}_nP_r$. To evaluate ${}_{15}P_5$ on the TI-83 Plus, press 1 5*, MATH, select PRB, select 2, press* 5*, and press* ENTER*. The screen appears as follows.*

```
15 nPr 5
              360360
```

OBJECTIVE 3. Calculate combinations. Suppose a student governance association has 12 members and selects a committee of 4 to organize the upcoming election. The order in which the members of the committee are selected is not important, since an individual is either on the committee or not. So this situation is not a permutation. It is an example of a **combination**.

DEFINITION ***Combination:*** An unordered arrangement of objects. The combination of n things taken r at a time is often denoted as ${}_nC_r$.

To determine the number of different ways the committee of 4 can be formed from the 12 members, let's first calculate the number if order did matter (that is, a permutation).

$$ {}_{12}P_4 = \frac{12!}{(12-4)!} = \frac{12!}{8!} = \frac{12 \cdot 11 \cdot 10 \cdot 9 \cdot \cancel{8!}}{\cancel{8!}} = 12 \cdot 11 \cdot 10 \cdot 9 = 11{,}880 $$

Since order is not important, many of these committees have the same members. For example, a committee made up of Aricellis, Ebony, Hector, and Carol is the same as the committee made up of Ebony, Aricellis, Carol, and Hector. Since each committee has 4 members, there are $4! = 24$ different orders in which the same committee can be written. To eliminate these duplicate committees, divide 11,880 by 24, which gives 495 different committees.

Summarizing, to find the number of combinations of 12 things taken 4 at a time, we found $\frac{{}_{12}P_4}{4!} = \frac{\frac{12!}{(12-4)!}}{4!} = \frac{12!}{4!(12-4)!}$, which suggests the following rule.

PROCEDURE Evaluating ${}_nC_r$

The formula for the number of combinations of n things taken r at a time, is

$$ {}_nC_r = \frac{n!}{r!(n-r)!} $$

Note that ${}_nC_n = 1$, ${}_nC_0 = 1$, and ${}_nC_1 = n$.

Connection ${}_nC_r = \binom{n}{r}$, which is the binomial coefficient.

Calculator TIPS

To evaluate ${}_nC_r$ using a TI-83 Plus, follow the same steps as for evaluating ${}_nP_r$ except choose 3 under the PRB menu instead of 2.

EXAMPLE 3 In a BMX dirt bike race, 4 of the 8 bikes in a preliminary heat can advance to the next heat. How many different groups of 4 can advance?

Solution Since the order in which the top 4 finishers is not important (all 4 go on to the next heat), this is a combination of 8 things taken 4 at a time, or ${}_8C_4$, so we use the formula ${}_nC_r = \frac{n!}{r!(n-r)!}$ with $n = 8$ and $r = 4$.

$$ {}_8C_4 = \frac{8!}{4!(8-4)!} = \frac{8!}{4!(4!)} = \frac{8 \cdot 7 \cdot 6 \cdot 5 \cdot \cancel{4!}}{4 \cdot 3 \cdot 2 \cdot 1(\cancel{4!})} = \frac{8 \cdot 7 \cdot 6 \cdot 5}{4 \cdot 3 \cdot 2 \cdot 1} = 70 $$

There are 70 possible combinations of four riders who can advance to the next heat.

Appendix D Exercises

For Extra Help

MyMathLab Videotape/DVT InterAct Math Math Tutor Center Math XL.com

1. If one task can be performed in x ways and a second task can be performed in y ways, what does $x \cdot y$ represent?

2. How do permutations differ from combinations?

3. A family is going on vacation and can visit any 5 of 12 amusement parks. To determine the number of different groups of 5 parks they could visit, would you use the formula for ${}_nP_r$ or ${}_nC_r$? Why?

4. Fredrica has a choice of 5 pies and a choice of 3 different flavors of ice cream. What formula or principle would you use to determine the total number of different choices of pie and ice cream?

5. The combination to a combination lock consists of 3 different numbers chosen from a possible 50. To determine the number of possible combinations would you use the formula for ${}_nP_r$ or ${}_nC_r$? Why?

6. In deriving the formula ${}_nC_r$, why did we calculate $\frac{{}_nP_r}{r!}$?

For Exercises 7–18, evaluate each.

7. ${}_7P_2$

8. ${}_9P_5$

9. ${}_6P_6$

10. ${}_5P_5$

11. ${}_3P_0$

12. ${}_8P_0$

13. ${}_6C_4$

14. ${}_{10}C_4$

15. ${}_5C_5$

16. ${}_9C_9$

17. ${}_7C_0$

18. ${}_8C_0$

19. John works as a salesperson and has to wear a sports coat and tie to work each day. Assuming all his clothing matches, find the number of possible outfits that he can wear if he has 4 coats, 6 pairs of pants, 5 shirts, and 3 ties.

20. The executive board of a corporation has 8 members. If one member is to be in charge of public relations, a different member in charge of marketing, and a third member in charge of research, how many different ways can these 3 positions be filled?

21. Jorge plans to take one course each in the subject areas of literature, computers, history, and humanities. If he has a choice of 4 literature courses, 3 computer courses, 5 history courses, and 2 humanities courses that fit his schedule, in how many different ways can he build his schedule?

22. At a sandwich shop, you have a choice of 3 breads, one of 3 dressings, with or without lettuce, with or without tomato, with or without pickles, and 5 meats. How many different sandwiches are possible?

23. How many different three-digit numbers can be formed from the digits 4, 6, and 7 (a) if no digit may be repeated? (b) If the digits may be repeated?

24. A club consists of 8 women and 5 men. In how many different ways can the club elect a president and a secretary if the president must be a woman and the secretary must be a man?

25. Some states' license plates have 3 numbers followed by 3 letters. How many different license plates are possible in one of these states?

26. Postal Zip codes have 5 digits. How many Zip codes are possible?

27. A telephone calling card requires a four-digit PIN (personal identification number). How many different PINs are possible? How many different PINs have no repeated digits?

28. A singer has 12 songs to record on a CD. In how many different orders can the songs be recorded?

29. In how many different orders can the 11 members of the starting offensive lineup of a football team be introduced?

30. A television executive has decided on 4 shows to air on Thursday evenings. In how many different orders can the shows be shown?

31. To win a state lottery, you must correctly select 5 numbers, from 1 to 49 inclusive. Since the numbers are drawn using ping-pong balls, no number can be repeated and the order in which the numbers occur is not important. What is the total number of possible winning tickets?

32. Basketball teams in the National Basketball Association have 12 members. How many different starting lineups of 5 players are possible?

33. A singer has been asked to perform 3 songs at a benefit. If she has 18 songs to choose from, how many different groups of 3 songs are possible?

34. A poet is putting together a collection for publication. He has 30 poems to choose from, but can include only 20 in the collection. How many different collections are possible?

Appendix E Probability

OBJECTIVES

1. **Find sample spaces and events.**
2. **Calculate probabilities.**

We often hear the word *probability* in everyday language. For example, a weather forecaster might say that the probability of rain is 40% (0.40). In this usage, probability measures the likelihood that the event will occur. In this section, we will develop a more precise way of defining probability.

OBJECTIVE 1. Find sample spaces and events. To talk about probability, we need to define an **experiment** and **outcomes**.

DEFINITIONS ***Experiment:*** Any act or process whose result is not known in advance.
Outcomes: The possible results of an experiment.

If an experiment is tossing a coin, then an outcome would be getting a head. If an experiment is rolling two dice (plural of *die*), then an outcome would be getting a total of 7. If an experiment is drawing a card from a deck of cards, then getting the King of Hearts is an outcome. Associated with every experiment is the **sample space**.

DEFINITION ***Sample space:*** The set of all possible outcomes of an experiment. The sample space is denoted by S.

EXAMPLE 1 Find the sample spaces for each of the following experiments.

a. Toss a single coin.

Solution $S = \{H, T\}$, where H means heads; T tails

b. Toss two coins (or one coin twice).

Solution $S = \{HH, HT, TH, TT\}$, where HT means heads on the first toss and tails on the second

c. Roll a single die.

Solution $S = \{1, 2, 3, 4, 5, 6\}$

d. Roll two dice.

Solution It is convenient to think of the dice as having different colors. Suppose one is blue and the other red. The sample space contains ordered pairs of the form (B, R), where B represents the number on the blue (first) die and R represents the number on the red (second) die.

$$S = \{(1, 1)\ (1, 2)\ (1, 3)\ (1, 4)\ (1, 5)\ (1, 6)$$
$$(2, 1)\ (2, 2)\ (2, 3)\ (2, 4)\ (2, 5)\ (2, 6)$$
$$(3, 1)\ (3, 2)\ (3, 3)\ (3, 4)\ (3, 5)\ (3, 6)$$
$$(4, 1)\ (4, 2)\ (4, 3)\ (4, 4)\ (4, 5)\ (4, 6)$$
$$(5, 1)\ (5, 2)\ (5, 3)\ (5, 4)\ (5, 5)\ (5, 6)$$
$$(6, 1)\ (6, 2)\ (6, 3)\ (6, 4)\ (6, 5)\ (6, 6)\}$$

Note that there are 36 outcomes in the sample space.

To calculate probabilities, we need to discuss **events**.

DEFINITION ***Event:*** Any subset of a sample space.

EXAMPLE 2 Find the events that correspond with the following outcomes.

a. Having exactly one head on the toss of two coins

Solution $E = \{\text{HT, TH}\}$

b. Rolling a number greater than 2 with a single roll of a die

Solution $E = \{3, 4, 5, 6\}$

c. Rolling two dice for a total of 7

Solution $E = \{(1, 6)\ (2, 5)\ (3, 4)\ (4, 3)\ (5, 2)\ (6, 1)\}$

OBJECTIVE 2. Calculate probabilities. We are now ready to give a more formal definition of probability of an event.

DEFINITION ***Probability of an Event*** **E:** Given an experiment, the probability of an event *E*, written as *P*(*E*), is the number of outcomes in event *E* divided by the number of outcomes in the sample space *S* of the experiment. That is,

$$P(E) = \frac{\text{The number of outcomes in } E}{\text{The number of outcomes in } S}$$

From the definition, we can conclude that the probability of an event must be between 0 and 1 inclusive, that is, $0 \le P(E) \le 1$. If an event cannot occur, then the number of outcomes in the event is 0, so the probability of that event is 0. For example, the probability of rolling a die and getting a 7 is 0 since this cannot occur. The event is {7}, which can occur in 0 ways, and the sample space is $S = \{1, 2, 3, 4, 5, 6\}$, so $P(E) = \frac{0}{6} = 0$. If an event is certain to occur, then the number of outcomes in the event is the same as the number of outcomes in the sample space, so the probability is 1. For example, the probability of rolling a die and getting a number less than 7 is 1 since this event is certain to occur. The event is $E = \{1, 2, 3, 4, 5, 6\}$ and the sample space is $S = \{1, 2, 3, 4, 5, 6\}$, so $P(E) = \frac{6}{6} = 1$.

Many probability problems involve a standard deck of cards, which consists of 52 cards with 4 suits of 13 cards each. The suits are diamonds and hearts, which are red, and spades and clubs, which are black. Each suit has the cards 2, 3, 4, 5, 6, 7, 8, 9, 10, Jack, Queen, King, and Ace. The cards Jack, Queen, and King are called face cards.

EXAMPLE 3 Find the probabilities of each of the following events.

a. Drawing a King from a standard deck on a single draw

Solution Since there are 52 cards in the deck, there are 52 possible outcomes in the sample space. Since there are 4 Kings in the deck, the event has 4 outcomes. Therefore, $P(E) = \frac{4}{52} = \frac{1}{13}$.

b. Rolling a total of 7 on a single roll of two dice

Solution We found in Example 1d that the sample space for rolling two dice has 36 outcomes, and in Example 2c we found 6 different ways of getting a total of 7. Therefore, $P(E) = \frac{6}{36} = \frac{1}{6}$.

c. A club has 20 members, of which 8 are men and 12 are women. If a committee of 5 is chosen to revise the bylaws, what is the probability that all 5 are women?

Solution $P(E) = \dfrac{\text{Number of ways of selecting 5 women from 12}}{\text{Number of ways of selecting 5 members from 20}}$

The number of ways of selecting 5 women from 12:

$$_{12}C_5 = \frac{12!}{5!(12-5)!} = \frac{12!}{5!7!} = 792$$

The number of ways of selecting 5 members from 20:

$$_{20}C_5 = \frac{20!}{5!(20-5)!} = \frac{20!}{5!15!} = 15{,}504$$

Therefore, $P(E) = \frac{792}{15{,}504} = \frac{33}{646}$.

Appendix E Exercises

1. What is the sample space for an experiment?

2. What is an event?

3. The probability of any event lies between what two numbers inclusively?

4. What is the probability of an event that cannot happen?

5. Define the probability of an event.

6. How many face cards are in a standard deck of cards?

For Exercises 7–10, find the sample space for each experiment.

7. Selecting a vowel from the English alphabet

8. A jar contains 1 red, 1 blue, and 1 white ball; draw a ball, replace it, and draw again

9. Tossing 3 coins (or 1 coin 3 times)

10. Jar 1 contains a blue marble and a white marble; jar 2 contains a red marble and a black marble; select a marble from jar 1 and then select a marble from jar 2

For Exercises 11–14, find the number of elements in the sample space of the experiment.

11. Selecting a two-digit number

12. Rolling a die and tossing a coin

13. Choosing a committee of 4 from a club with 16 members

14. Five-card hands using a standard deck of cards

For Exercises 15–18, find the event that corresponds with the following outcomes.

15. Having exactly 1 head if 3 coins are tossed

16. Rolling a total of 4 on a single roll of 2 dice

17. Having a number greater than 6 on a single roll of a die

18. Rolling a total of 13 on a single roll of 2 dice

For Exercises 19 and 20, a single die is rolled. Find the probability of each of the following events.

19. Having a 5

20. Having a number greater than 3

For Exercises 21–24, two coins are tossed. Find the probability of each of the following events.

21. Having exactly 1 head

22. Having exactly 2 tails

23. Having at least 1 tail

24. Having more than 2 heads

For Exercises 25–28, two dice are rolled. Find the probability of each of the following events.

25. Having a total of 4

26. Having a total of 8

27. Having a total less than 4

28. Having a total greater than 7

For Exercises 29–32, a single card is drawn from a standard deck. Find the probability of each of the following events.

29. Drawing a 10

30. Drawing the Jack of spades

31. Drawing a face card

32. Drawing a number less than 5

For Exercises 33–34, use the spinner shown to find the probability of each of the following events.

33. The spinner stopping on red

34. The spinner stopping on blue

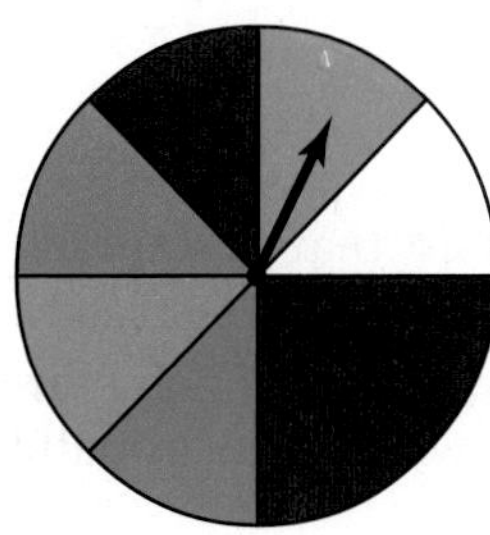

For Exercises 35–40, we have a jar that contains 1 blue, 1 red, and 1 pink ball. Find the probability of each of the following events.

35. Drawing a pink ball on a single draw

36. Drawing a blue ball on a single draw

37. Draw a ball, do not replace it, and then draw a second ball. What is the probability that both balls are blue?

38. Draw a ball, do not replace it, and then draw a second ball. What is the probability that both balls are red?

39. Draw a ball, replace it, and draw again. What is the probability that both balls are red?

40. Draw a ball, replace it, and draw again. What is the probability that the first ball is blue and the second ball is pink?

41. A farm pond is stocked with 80 warmouth perch, 140 bluegills, and 200 shellcrackers. If we assume that each fish is equally likely to be caught, what is the probability that the first fish caught is a shellcracker?

42. The quality-control department of a company that manufactures car batteries has determined that 3% of the batteries produced are defective. If one battery that is manufactured by this company is selected at random, what is the probability that it is defective?

43. A committee has 12 members of which 8 are Democrats and 4 are Republicans. A subcommittee of 3 is appointed. What is the probability that all 3 are Republicans [if each number is equally likely to be chosen]?

44. A club has 20 members of which 13 are women and 7 are men. If a committee of 6 is appointed, what is the probability that all 6 are women [if each number is equally likely to be chosen]?

45. A jar contains 14 balls of which 8 are pink and 6 are purple. If 4 balls are drawn at random, what is the probability that all 4 are pink?

46. A box contains 18 pieces of paper of which 9 are blue and 9 are white. If 4 pieces are selected at random, what is the probability that all 4 are blue?

Answers

Chapter 1

1.1 Exercises **1.** A collection of objects. **3.** If every element of a set B is an element of a set A, then B is a subset of A. **5.** Rational numbers can be expressed as a ratio of two integers; irrational numbers cannot. **7.** {Saturday, Sunday} **9.** {January, June, July} **11.** {New York, New Hampshire, New Mexico, New Jersey} **13.** {0, 1, 2, 3, 4} **15.** {3, 6, 9, . . . } **17.** {−1, 0, 1} **19.** { } **21.** $\{x|x$ is an even number$\}$ **23.** $\{x|x$ is a letter of the alphabet$\}$ **25.** $\{x|x$ is a day of the week$\}$ **27.** $\{x|x$ is a natural-number multiple of 5$\}$ **29.** false **31.** true **33.** true **35.** false **37.** false **39.** true **41.** true **43.** false **45.** true **47.** false **49.** true **51.** true **53.** False, both are infinite sets.

55.

57.

59. **61.**

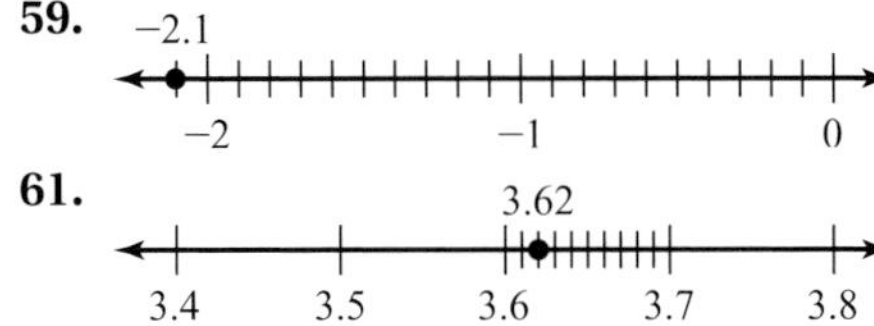

63. 3.6

65. −3.5

67. 2.6 **69.** $1\frac{2}{5}$ **71.** 1 **73.** 8.75 **75.** > **77.** < **79.** > **81.** = **83.** = **85.** < **87.** $-0.6, -0.44, 0, |-0.02|, 0.4, \left|1\frac{2}{3}\right|, 3\frac{1}{4}$ **89.** $-12.6, -9.6, 1, |-1.3|, \left|-2\frac{3}{4}\right|, 2.9$ **91.** {2001, 2002, 2003} **93.** {1994, 1995, 1996, 1999} **95.** {Comedy, Action/Adventure, Family} **97.** {Action/Adventure, Comedy} **99.** {495, 509, 588} **101.** { } or ∅

1.2 Exercises **1.** The order of the addends is changed with the commutative property of addition, whereas the grouping is changed with the associative property of addition. **3.** Their sum is 0. **5.** To add two numbers that have the same sign, add their absolute values and keep the same sign. **7.** To write a subtraction statement as an equivalent addition statement, change the operation symbol from a minus sign to a plus sign and change the subtrahend to its additive inverse. **9.** additive inverse **11.** additive identity **13.** multiplicative identity **15.** multiplicative inverse **17.** multiplicative identity **19.** multiplicative identity **21.** $-8, \frac{1}{8}$ **23.** $7, -\frac{1}{7}$ **25.** $\frac{5}{8}, -\frac{8}{5}$ **27.** $-0.3, \frac{10}{3}$ **29.** commutative property of addition **31.** distributive property **33.** associative property of multiplication **35.** associative property of multiplication **37.** commutative property of addition **39.** commutative property of multiplication **41.** −7 **43.** −40 **45.** −6 **47.** −5 **49.** $-\frac{7}{12}$ **51.** $-\frac{19}{24}$ **53.** 6.52 **55.** −4.38 **57.** 5 **59.** 2 **61.** −4 **63.** 5 **65.** $\frac{13}{10}$ **67.** 0 **69.** 0.36 **71.** −1.6 **73.** −12 **75.** 2 **77.** $-\frac{1}{2}$ **79.** 5 **81.** $-\frac{3}{8}$ **83.** −40 **85.** −6 **87.** 0.54 **89.** 1946.71 **91.** −\$814.66 **93.** −1281.5 N **95.** \$4.632 million **97. a.** −67.65 lb. **b.** Earth **99.** −51.2 V

Review Exercises **1.** {Washington, Adams, Jefferson, Madison} **2.** No, because an element, 6, is in B but not in A. **3.** infinite **4.** −6 can be written as $-\frac{6}{1}$. **5.** 25 **6.** =

1.3 Exercises **1.** To evaluate an exponential form raised to a natural number exponent, write the base as a factor the number of times indicated by the exponent, then multiply. **3.** Because of the sign rules for multiplying two numbers, the square of every real number is positive (or 0). **5.** Multiply 9 and 4 to get 36, then find the principal square root of 36 to get 6. Or, find the principal square roots of 9 and 4, which are 3 and 2, then multiply those roots to get 6. **7.** base = −4; exponent = 3; negative four to the third power **9.** base = 1; exponent = 7; the additive inverse of one raised to the seventh power **11.** 625 **13.** −27 **15.** −36 **17.** 1 **19.** $\frac{9}{64}$ **21.** $-\frac{125}{216}$ **23.** 0.064 **25.** −9.261 **27.** ±15 **29.** ±16 **31.** ±10 **33.** no real-number roots exist **35.** 5 **37.** 2 **39.** 0.6 **41.** −3 **43.** $\frac{2}{3}$ **45.** not a real number **47.** 2 **49.** 5 **51.** −55 **53.** 2 **55.** 8 **57.** 20 **59.** 21 **61.** 26.8 **63.** −3.96 **65.** 18 **67.** $-\frac{253}{300}$ **69.** −4 **71.** undefined **73.** The associative property of multiplication was used to multiply 3 · 3 instead of multiplying 1 · 3 from left to right. **75.** The distributive property was applied instead of adding −1 + 36 in the brackets. **77.** Mistake: Multiplied before division. Correct: −2 **79.** Mistake: Found the square root of the addends 16 and 9 instead of their sum. Correct: 20 **81. a.** 72.9 min. **b.** No, her average time is greater than 72 minutes. **83.** 105,911 **85.** 1206.978 **87.** 2.64 **89.** 35.5 + 0.10 (658 − 500) + 0.12 (45) = \$56.70 **91.** 0.35 (814) + 54.50 + 3 (89.90) + 112.45 = \$721.55 **93.** 4096 **95.** 128 **97.** 270,000

Review Exercises
1. It is an expression because it has no equal sign. **2.** {0,1,2,3,4,5,6,7,8} **3.** commutative property of multiplication **4.** 73 **5.** −60 **6.** −19

1.4 Exercises
1. Addition is commutative. **3.** To evaluate a variable expression: (1) Replace each variable with its corresponding given value. (2) Simplify the resulting numerical expression. **5.** Like terms are variable terms that have the same variable(s) raised to the same exponents, or constant terms. **7.** $5n$ **9.** $2n + 2$ **11.** $5 - p$ **13.** $8n^4$ **15.** $2n - 20$ **17.** $15 \div r^2$ **19.** $\frac{p}{q} - \frac{1}{2}$ **21.** $m - 3(n + 5)$ **23.** $(4 - t)^5$ **25.** $\left(\frac{6}{7}x\right) \cdot 7$ **27.** $(m - n) - (x + y)$ **29.** Mistake: incorrect order. Correct: $y - 6$ **31.** Mistake: incorrect order. Correct: $\sqrt{m} - 4$ **33.** $5w$ **35.** $2 - 3w$ **37.** $2r$ **39.** $17 - n$ **41.** $t + \frac{1}{2}$ **43.** πd **45.** $\frac{1}{2}h(a + b)$ **47.** $\frac{1}{3}\pi r^2 h$ **49.** $\frac{1}{2}mv^2$ **51.** $\frac{Mm}{d^2}$ **53.** $\sqrt{1 - \frac{v^2}{c^2}}$ **55.** −7 **57.** $-\frac{11}{3}$ **59.** 1 **61.** 10 **63. a.** 35.6 **b.** $\frac{16}{3}$ **65. a.** −3 **b.** $\frac{1}{4}$ **67.** 6 **69.** −5, 1 **71.** 0 **73.** $-\frac{1}{4}$ **75.** $27x - 45$ **77.** $-5m - 10$ **79.** $\frac{1}{12}x - 9$ **81.** $-6.3x - 5.04$ **83.** $-11x$ **85.** $-\frac{2}{7}b^2$ **87.** $7x - 8y - 12$ **89.** $-1.3x - 0.4$ **91.** $2.2h^2 + \frac{8}{3}h + 7$ **93.** $14n - 16$ **95.** $-5a + 7b - 12$ **97. a.** $14 + (6x - 8x)$ **b.** $14 - 2x$ **c.** 20

Review Exercises
1. $\{x \mid x \text{ is an integer and } x \geq -2\}$ **2.** commutative property of addition **3.** distributive property **4.** −35 **5.** −84 **6.** −49

Chapter 1 Review Exercises
1. true **2.** false **3.** false **4.** true **5.** true **6.** true **7.** change **8.** positive negative **9.** positive **10.** replace **11.** {Alaska, Hawaii} **12.** {. . ., −3, −1, 1, 3, 5, . . .} **13.** {5, 10, 15, . . .} **14.** {s, i, m, p, l, f, y} **15.** $\{x \mid x \text{ is a natural number multiple of } 3\}$ **16.** $\{x \mid x \text{ is a whole number}\}$ **17.** $\{x \mid x \text{ is a prime number}\}$ **18.** $\{x \mid x \text{ is a day of the week}\}$ **19.** false **20.** false **21.** false **22.** false **23.** = **24.** > **25.** = **26.** > **27.** additive inverse **28.** multiplicative inverse **29.** additive identity **30.** multiplicative identity **31.** distributive property **32.** associative property of multiplication **33.** commutative property of addition **34.** commutative property of multiplication **35.** associative property of addition **36.** distributive property **37.** −1 **38.** 5 **39.** −17 **40.** −7 **41.** −2 **42.** −10 **43.** −13 **44.** 9 **45.** −8 **46.** 15 **47.** −56 **48.** −5 **49.** 2 **50.** −2 **51.** −6 **52.** 25 **53.** base = 2; exponent = 7; the additive inverse of two raised to the seventh power; −128 **54.** base = −1; exponent = 4; negative one raised to the fourth power; 1 **55.** −9 **56.** −8 **57.** 16 **58.** $-\frac{8}{125}$ **59.** 11 **60.** 3 **61.** 3 **62.** 1 **63.** −58 **64.** −61 **65.** −61 **66.** 22 **67.** 625 **68.** −2.6 **69.** $14 - 8n$ **70.** $2(n + 2)$ **71.** $n + \frac{1}{3}(n - 4)$ **72.** $\frac{m}{\sqrt{n}}$ **73.** $\frac{1}{2}(n - 8) - 16$ **74.** $(n + 5) - 20$ **75.** $2w$ **76.** $\frac{1}{3} + t$ **77.** 9 **78.** 0 **79.** 16 **80.** −21 **81.** 3 **82.** $-\frac{5}{3}$ **83.** $-10x - 2$ **84.** $8a + 12b - 16$ **85.** $-x^2 - 2x$ **86.** $m^5 - 2mn^2 + 2mn$ **87.** $-7a^2 + 3ab^2 - 5ab + 3a - 8$ **88.** $3r - 10$ **89.** −\$220.44 **90.** 2.83 **91.** 59 + 0.15(37); \$64.55 **92.** 12,500,000 codes

Chapter 1 Practice Test
1. 8.1 **2.** $-\frac{11}{4}$ **3.** 13 **4.** 5 **5.** $\frac{1}{2}$ **6.** commutative property of addition **7.** associative property of multiplication **8.** 8 **9.** $\frac{11}{12}$ **10.** −7.5 **11.** 25 **12.** $-\frac{4}{25}$ **13.** 2 **14.** 4 **15.** −10 **16.** −6 **17.** 18 **18.** 0 **19.** −8 **20.** −\$389.50 **21.** 5.19 million **22.** −138 **23.** 2 **24.** $-21x - 35$ **25.** $-\frac{23}{5}x + \frac{27}{4}y + 2.7$

Chapter 2

2.1 Exercises
1. A solution for an equation is a number that makes the equation true when it replaces the variable in the equation. **3.** The solution set for an identity contains every real number for which the equation is defined. **5.** Divide both sides of the equation by r. **7.** 6 **9.** −1 **11.** 1 **13.** 3 **15.** 1 **17.** 1 **19.** no solution **21.** 4 **23.** −6.5 **25.** 3 **27.** 7 **29.** −2 **31.** $\frac{32}{7}$ **33.** 30 **35.** $\frac{85}{3}$ **37.** 0.2 **39.** 3 **41.** no solution **43.** all real numbers **45.** Mistake: The distributive property was not used correctly. Correct: $\frac{7}{5}$ **47.** Mistake: Subtracted before distributing into the parentheses. Correct: 3 **49.** $C = R - P$ **51.** $b = \frac{A}{h}$ **53.** $w = \frac{A}{2\pi p}$ **55.** $r^2 = \frac{2A}{\theta}$ **57.** $M = \frac{Fd^2}{km}$ **59.** $s = \frac{A}{\pi(R + r)}$ **61.** $l = \frac{P - 2w}{2}$ **63.** $C = \frac{5}{9}(F - 32)$ **65.** $a = \frac{2(x - vt)}{t^2}$ **67.** Mistake: Subtracted lw instead of dividing by lw. Correct: $h = \frac{V}{lw}$

69. Mistake: Subtracted the coefficient of l instead of dividing by the coefficient. Correct: $l = \frac{P - 2w}{2}$

Review Exercises **1.** $\{1, 3, 5, 7, 9, 11, 13\}$ **2.** 3314 **3.** $7n - 9$ **4.** $-3(n + 8)$ **5.** $-2x - 6y + 5$ **6.** $-54m + 24$

2.2 Exercises **1.** Understand the problem. **2.** Devise a plan. **3.** Execute the plan. **4.** Check results. **3.** Choose a variable for one of the unknowns. Use one of the relationships to describe the other unknown in terms of the chosen variable. Translate the second relationship to an equation. **5.** Two angles are supplementary if the sum of their measures is 180°. **7.** −78.5°C **9.** 180 ft.^{2} **11.** \$3000 **13.** 15 **15.** −6 **17.** 6 **19.** \$88.72 **21.** \$999.93 **23.** 37,500 units **25.** 24, 26, 28 **27.** 53, 55, 57 **29.** 60°, 120° **31.** 23.75°, 66.25° **33.** $0.5(3600 - x); x$ **35.** $50t; 75(t - 3)$ **37.** $x + 2000; 0.20(2000)$ **39.** 850 16-oz. drinks and 2750 12-oz. drinks **41.** 6 hr., 450 mi. **43.** 6000 L **45.** \$1450 per desktop, \$1650 per laptop **47.** 8 Motorola, 16 Nokia **49.** 2 hr. **51.** 2.75 hr. **53.** 16 oz. **55.** 20 gal.

Review Exercises **1.** $-5\frac{3}{8}, -\frac{1}{6}, 0.02, 4.5\%, \sqrt{48}, |-15.8|$

2. 2 **3.** = **4.** > **5.** −6 **6.** $-\frac{15}{8}$

2.3 Exercises **1.** Any value that makes the inequality true. **3.** The graph shows that every real number greater than −1, but not including −1, is in the solution set. **5.** The set of all values x such that x is greater than 2. This means every real number greater than 2 but not including 2 is in the solution set for the variable x **7. a.** $\{x \mid x \geq 5\}$ **b.** $[5, \infty)$

c. −2 −1 0 1 2 3 4 5 6 7 8 9 10

9. a. $\{q \mid q < -1\}$ **b.** $(-\infty, -1)$

c. −6 −5 −4 −3 −2 −1 0 1 2 3 4 5 6

11. a. $\left\{p \mid p < \frac{1}{5}\right\}$ **b.** $\left(-\infty, \frac{1}{5}\right)$

c. $\frac{1}{5}$ −1 0 1 2 3

13. a. $\{r \mid r \leq 1.9\}$ **b.** $(-\infty, 1.9]$

c. 1.9 0 1 2

15. a. $\{r \mid r < -6\}$ **b.** $(-\infty, -6)$

c. −10 −9 −8 −7 −6 −5 −4 −3 −2 −1 0 1 2

17. a. $\{y \mid y \geq 4\}$ **b.** $[4, \infty)$

c. −2 −1 0 1 2 3 4 5 6 7 8 9 10

19. a. $\{p \mid p < 4\}$ **b.** $(-\infty, 4)$

c. −2 −1 0 1 2 3 4 5 6 7 8 9 10

21. a. $\left\{x \mid x < -\frac{23}{5}\right\}$ **b.** $\left(-\infty, -\frac{23}{5}\right)$

c. $-\frac{23}{5}$ −6 −5 −4 −3

23. a. $\{a \mid a < -5\}$ **b.** $(-\infty, -5)$

c. −10 −9 −8 −7 −6 −5 −4 −3 −2 −1 0 1 2

25. a. $\left\{x \mid x > \frac{5}{2}\right\}$ **b.** $\left(\frac{5}{2}, \infty\right)$

c. $\frac{5}{2}$ −1 0 1 2 3 4

27. a. $\{k \mid k < 4\}$ **b.** $(-\infty, 4)$

c. −3 −2 −1 0 1 2 3 4 5 6 7 8 9

29. a. $\{w \mid w \leq 5\}$ **b.** $(-\infty, 5]$

c. −3 −2 −1 0 1 2 3 4 5 6 7 8 9

31. a. $\{y \mid y < 8\}$ **b.** $(-\infty, 8)$

c. −3 −2 −1 0 1 2 3 4 5 6 7 8 9

33. a. $\left\{x \mid x < -\frac{11}{6}\right\}$ **b.** $\left(-\infty, -\frac{11}{6}\right)$

c. $-\frac{11}{6}$ −3 −2 −1 0

35. a. $\left\{m \mid m \leq \frac{9}{4}\right\}$ **b.** $\left(-\infty, \frac{9}{4}\right]$

c. $\frac{9}{4}$ 0 1 2 3 4

37. a. $\{x \mid x \geq -10\}$ **b.** $[-10, \infty)$,

c. −12 −11 −10 −9 −8 −7 −6 −5 −4 −3 −2 −1 0

39. a. $\{z \mid z > 4\}$ **b.** $(4, \infty)$

c. −4 −3 −2 −1 0 1 2 3 4 5 6 7 8

41. $\frac{3}{4}x < -6; x < -8$ **43.** $5x + 1 > 16; x > 3$ **45.** $1 + 4x \leq 25; x \leq 6$ **47.** $6 + 2(x - 5) \leq 12; x \leq 8$ **49.** 84 or higher

51. 80 or less **53.** 13 ft. or less **55.** $\approx$ 23.9 in. or less **57.** at least 70 mph **59.** 4000 or more lamps **61. a.** $t \leq$ 1948.244°F **b.** $t \geq$ 1948.244°F **63.** 2.5 amps or less

Review Exercises

1. yes **2.** commutative property of addition **3.** 10 **4.** 45 **5.** -2 **6.** $\frac{32}{7}$

2.4 Exercises

1. Two inequalities joined by either "and" or "or." **3.** For two sets A and B, the intersection of A and B, symbolized by $A \cap B$, is a set containing only elements that are in both A and B. **5.** We graph the region of overlap of the two inequalities. **7. a.** $\{1, 3, 5\}$ **b.** $\{1, 3, 5, 7, 9\}$ **9. a.** $\{7\}$ **b.** $\{5, 6, 7, 8, 9\}$ **11. a.** $\varnothing$ **b.** $\{a, c, d, g, o, t\}$ **13. a.** $\{x, y, z\}$ **b.** $\{w, x, y, z\}$ **15.** $-4 < x < 5$ **17.** $-2 < y \leq 0$ **19.** $-7 < w < 3$ **21.** $0 \leq u \leq 2$

23.

25.

27.

29.

31. a.

b. $\{x \mid -3 < x < -1\}$ **c.** $(-3, -1)$

33. a.

b. $\{x \mid 3 < x \leq 6\}$ **c.** $(3, 6]$

35. a.

b. { } or $\varnothing$ **c.** no interval notation

37. a.

b. $\{x \mid -5 < x \leq 3\}$ **c.** $[-5, 3)$

39. a.

b. { } or $\varnothing$ **c.** no interval notation

41. a.

b. $\{x \mid -7 < x < -3\}$ **c.** $(-7, -3)$

43. a.

b. $\{x \mid -1 \leq x \leq 2\}$ **c.** $[-1, 2]$

45. a.

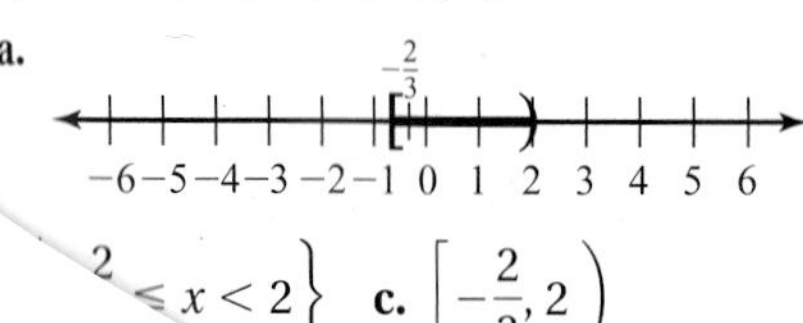

$\frac{2}{} \leq x < 2\}$ **c.** $\left[-\frac{2}{3}, 2\right)$

47. a.

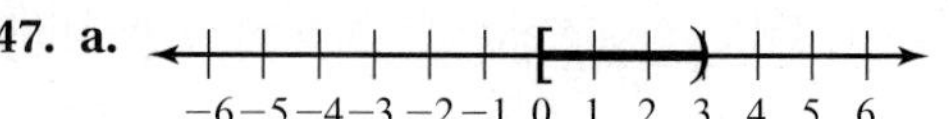

b. $\{x \mid 0 \leq x < 3\}$ **c.** $[0, 3)$

49. a.

b. $\{x \mid 0 \leq x \leq 3\}$ **c.** $[0, 3]$

51.

53.

55.

57.

59. a.

b. $\{y \mid y < -9 \text{ or } y > 5\}$ **c.** $(-\infty, -9) \cup (5, \infty)$

61. a.

b. $\{r \mid r < -2 \text{ or } r > 1\}$ **c.** $(-\infty, -2) \cup (1, \infty)$

63. a.

b. $\{w \mid w \leq -1 \text{ or } w \geq 7\}$ **c.** $(-\infty, -1] \cup [7, \infty)$

65. a.

b. $\{k \mid k \leq 2 \text{ or } k \geq 3\}$ **c.** $(-\infty, 2] \cup [3, \infty)$

67. a.

b. $\left\{x \mid x \leq -\frac{3}{5} \text{ or } x \geq \frac{1}{5}\right\}$ **c.** $\left(-\infty, -\frac{3}{5}\right] \cup \left[\frac{1}{5}, \infty\right)$

69. a.

b. $\{c \mid c > 3\}$ **c.** $(3, \infty)$

71. a.

b. $\left\{m \mid m \leq 1 \text{ or } m \geq \frac{8}{3}\right\}$ **c.** $(-\infty, 1] \cup \left[\frac{8}{3}, \infty\right)$

73. a.

b. $\{x \mid x \leq -4\}$ **c.** $(-\infty, -4)$

75. a.

b. $\{x \mid x \text{ is a real number}\}$ or $\mathbb{R}$ **c.** $(-\infty, \infty)$

77. a.

b. $\{x \mid -7 < x < 3\}$ **c.** $(-7, 3)$

79. a. 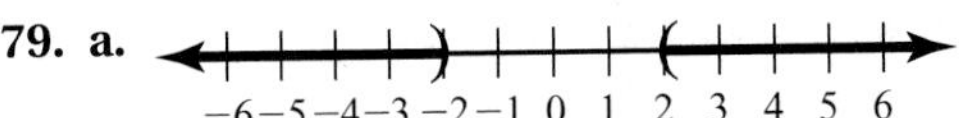

b. $\{x \mid x < -2 \text{ or } x > 2\}$ **c.** $(-\infty, -2) \cup (2, \infty)$

81. a. 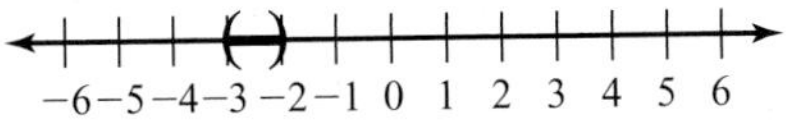 −6 −5 −4 −3 −2 −1 0 1 2 3 4 5 6

b. $\{x \mid -3 < x < -2\}$ **c.** $(-3, -2)$

83. a. 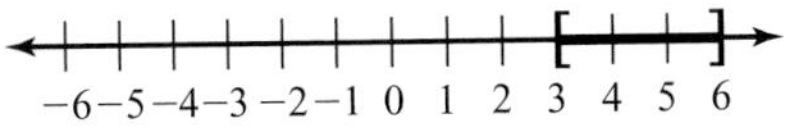 −6 −5 −4 −3 −2 −1 0 1 2 3 4 5 6

b. $\{x \mid 3 \le x \le 6\}$ **c.** $[3, 6]$

85. a. 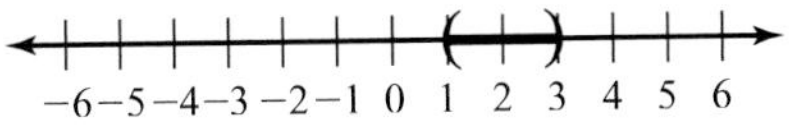 −6 −5 −4 −3 −2 −1 0 1 2 3 4 5 6

b. $\{x \mid 1 < x < 3\}$ **c.** $(1, 3)$

87. a. 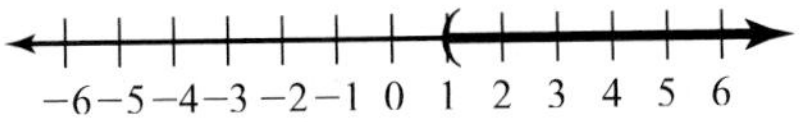 −6 −5 −4 −3 −2 −1 0 1 2 3 4 5 6

b. $\{x \mid x > 1\}$ **c.** $(1, \infty)$

89. a.  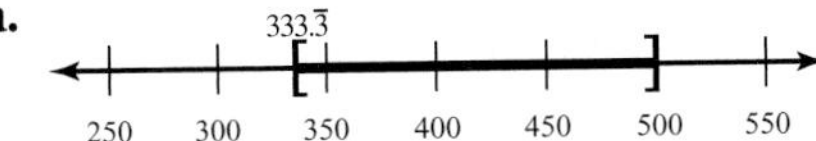

b. $\{x \mid 333.\overline{3} \le x \le 500\}$ **c.** $[333.\overline{3}, 500]$

91. a. 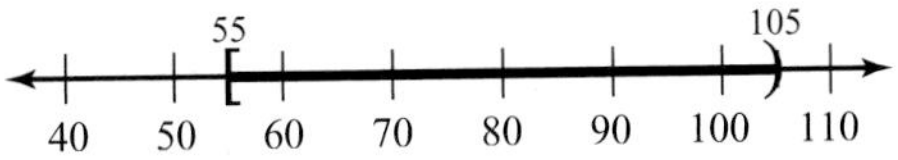 55 105

40 50 60 70 80 90 100 110

b. $\{x \mid 55 \le x < 105\}$ **c.** $[55, 105)$

93. a. 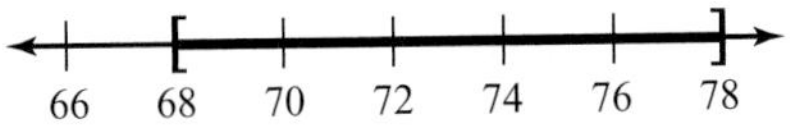 66 68 70 72 74 76 78

b. $\{x \mid 68° \le x \le 78°\}$ **c.** $[68, 78]$

95. a. 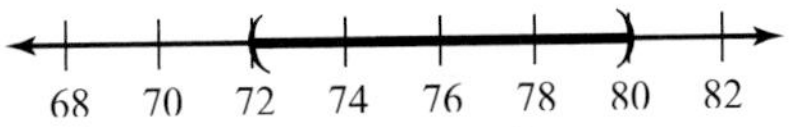 68 70 72 74 76 78 80 82

b. $\{x \mid 72° < x < 80°\}$ **c.** $(72, 80)$

97. a. 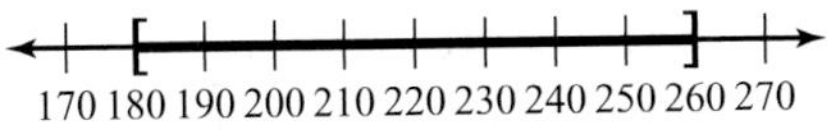 170 180 190 200 210 220 230 240 250 260 270

b. $\{x \mid 180 \text{ ft.} \le x \le 260 \text{ ft.}\}$ **c.** $[180, 260]$

99. $\{x \mid 1990 \le x \le 2002\}$; $[1990, 2002]$ **101.** $\{x \mid 1980 \le x < 1988\}$; $[1980, 1988)$ **103.** $\{x \mid 1989 \le x < 2000 \text{ or } 2001 \le x \le 2002\}$; $[1989, 2000) \cup [2001, 2002]$

Review Exercises

1. No, the absolute value of zero is zero and zero is neither negative or positive. **2.** -11 **3.** -16 **4.** 6 **5.** $\frac{1}{5}$ **6.** -2

2.5 Exercises

1. The units a number is from zero. **3.** If $|n| = a$, where n is a variable or an expression and $a \ge 0$, then $n = a$ or $n = -a$. **5.** Separate the absolute value equation into two equations: $ax + b = cx + d$ and $ax + b = -(cx + d)$. **7.** $-2, 2$ **9.** no solution **11.** $-11, 5$ **13.** $2, 3$ **15.** $1, \frac{7}{5}$ **17.** $-\frac{2}{3}, \frac{10}{3}$ **19.** no solution **21.** $\frac{3}{4}$ **23.** $-4, 4$ **25.** $-1, 3$ **27.** $-12, 4$ **29.** $-\frac{3}{5}, 1$ **31.** $-4, 1$ **33.** $-2, 6$ **35.** $-\frac{9}{2}, \frac{15}{2}$ **37.** $0, 20$ **39.** $-2, 4$ **41.** $\frac{1}{3}, 7$ **43.** $-5, -\frac{3}{5}$ **45.** all real numbers **47.** 1 **49.** $-3, 13$ **51.** $-6, 10$ **53.** $\frac{5}{6}, \frac{11}{6}$ **55.** $-\frac{7}{2}, 2$

Review Exercises

1. $x > -7$ **2.** $x \ge -3$ **3.** $n - 2 < 5$; $x < 7$ **4.** $(-2, 0]$ **5.** −6 −5 −4 −3 −2 −1 0 1 2 3 4 5 6

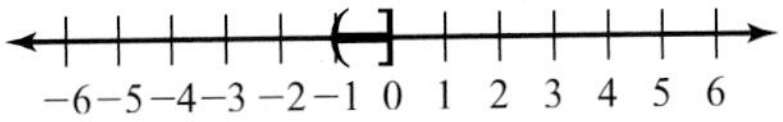

6. $-2 < x < \frac{14}{3}$

2.6 Exercises

1. $x \ge -a$ and $x \le a$ **3.** Shade between the values of $-a$ and a. **5.** when $a \le 0$

7. a. 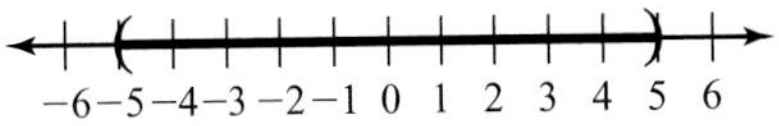 −6 −5 −4 −3 −2 −1 0 1 2 3 4 5 6

b. $\{x \mid -5 < x < 5\}$ **c.** $(-5, 5)$

9. a. 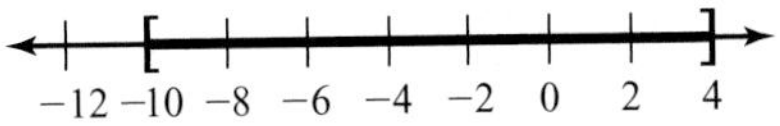 −12 −10 −8 −6 −4 −2 0 2 4

b. $\{x \mid -10 \le x \le 4\}$ **c.** $[-10, 4]$

11. a. 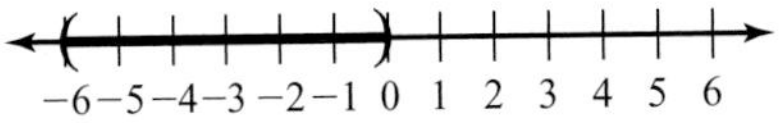 −6 −5 −4 −3 −2 −1 0 1 2 3 4 5 6

b. $\{s \mid -6 < s < 0\}$ **c.** $(-6, 0)$

13. a. 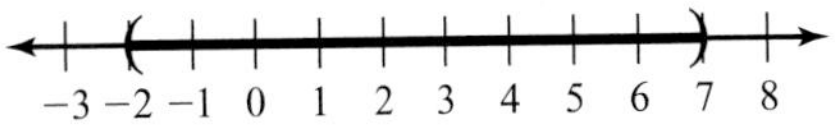 −3 −2 −1 0 1 2 3 4 5 6 7 8

b. $\{m \mid -2 < m < 7\}$ **c.** $(-2, 7)$

15. a. 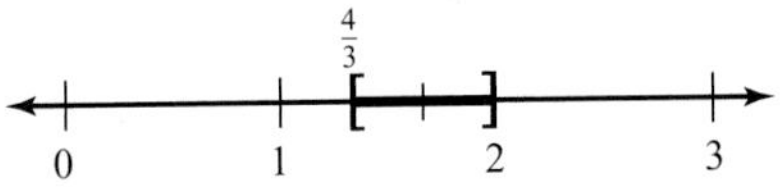 $\frac{4}{3}$

0 1 2 3

b. $\left\{k \mid \frac{4}{3} \le k \le 2\right\}$ **c.** $\left[\frac{4}{3}, 2\right]$

17. a. 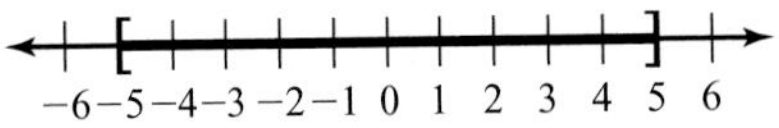 −6 −5 −4 −3 −2 −1 0 1 2 3 4 5 6

b. $\{x \mid -5 \le x \le 5\}$ **c.** $[-5, 5]$

19. a. 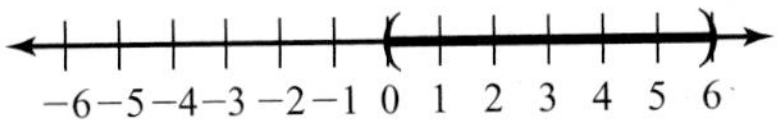 −6 −5 −4 −3 −2 −1 0 1 2 3 4 5 6

b. $\{w \mid 0 < w < 6\}$ **c.** $(0, 6)$

21. a. 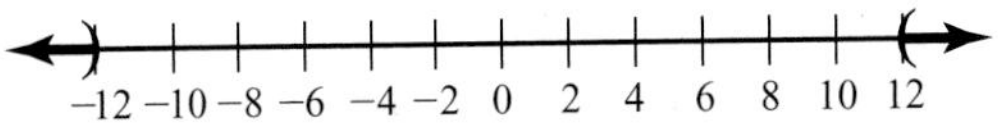 −12 −10 −8 −6 −4 −2 0 2 4 6 8 10 12

b. $\{c \mid c < -12 \text{ or } c > 12\}$ **c.** $(-\infty, -12) \cup (12, \infty)$

23. a. 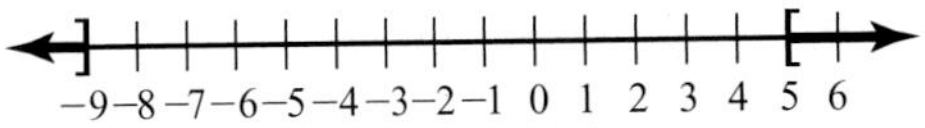 −9 −8 −7 −6 −5 −4 −3 −2 −1 0 1 2 3 4 5 6

b. $\{y \mid y \le -9 \text{ or } y \ge 5\}$ **c.** $(-\infty, -9] \cup [5, \infty)$

25. a. 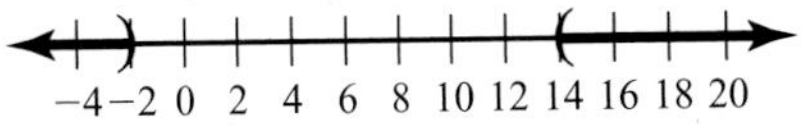 −4 −2 0 2 4 6 8 10 12 14 16 18 20

b. $\{p \mid p < -2 \text{ or } p > 14\}$ **c.** $(-\infty, -2) \cup (14, \infty)$

27. a. 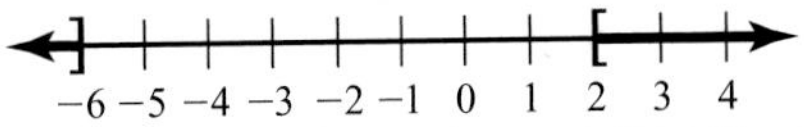 −6 −5 −4 −3 −2 −1 0 1 2 3 4

b. $\{x \mid x \le -6 \text{ or } x \ge 2\}$ **c.** $(-\infty, -6] \cup [2, \infty)$

29. a. 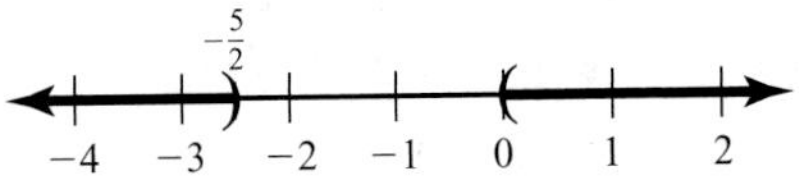 $-\frac{5}{2}$

−4 −3 −2 −1 0 1 2

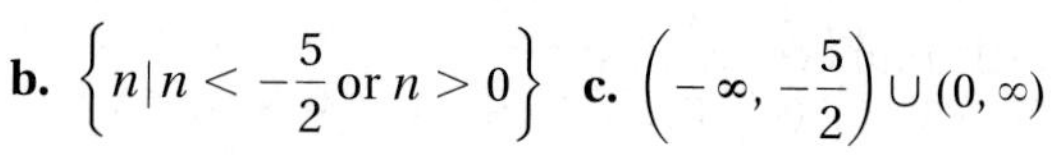

b. $\left\{n \mid n < -\frac{5}{2} \text{ or } n > 0\right\}$ **c.** $\left(-\infty, -\frac{5}{2}\right) \cup (0, \infty)$

31. a. −6 −5 −4 −3 −2 −1 0 1 2 3 4 5 6
b. $\{v \mid v \le -1 \text{ or } v \ge 1\}$ **c.** $(-\infty, -1] \cup [1, \infty)$

33. a. −6 −5 −4 −3 −2 −1 0 1 2 3 4 5 6
b. $\{y \mid y < -3 \text{ or } y > -1\}$ **c.** $(-\infty, -3) \cup (-1, \infty)$

35. a. −6 −5 −4 −3 −2 −1 0 1 2 3 4 5 6
b. $\{m \mid m < -5 \text{ or } m > 1\}$ **c.** $(-\infty, -5) \cup (1, \infty)$

37. a. −6 −5 −4 −3 −2 −1 0 1 2 3 4 5 6
b. $\{x \mid 0 < x < 4\}$ **c.** $(0, 4)$

39. a. −6 −5 −4 −3 −2 −1 0 1 2 3 4 5 6
b. $\{r \mid r$ is a real number$\}$ **c.** $(-\infty, \infty)$

41. a. −6 −5 −4 −3 −2 −1 0 1 2 3 4 5 6
b. $\{x \mid -4 < x < -2\}$ **c.** $(-4, -2)$

43. a. −6 −5 −4 −3 −2 −1 0 1 2 3 4 5 6
b. $\{x \mid 0 < x < 1\}$ **c.** $(0, 1)$

45. a. −14 −12 −10 −8 −6 −4 −2 0 2 4 6 8 10
b. $\varnothing$ **c.** no interval notation

47. a. $\frac{14}{3}$ −6 −5 −4 −3 −2 −1 0 1 2 3 4 5 6
b. $\left\{k \mid -2 \le k \le \frac{14}{3}\right\}$ **c.** $\left[-2, \frac{14}{3}\right]$

49. a. 0 4 8 12 16 20 24 28
b. $\{x \mid x < 4 \text{ or } x > 20\}$ **c.** $(-\infty, 4) \cup (20, \infty)$

51. a. −6.4 12.8 −8 −6 −4 −2 0 2 4 6 8 10 12 14
b. $\{y \mid -6.4 \le y \le 12.8\}$ **c.** $[-6.4, 12.8]$

53. a. −5 −4 −3 −2 −1 0 1 2 3 4 5
b. $\varnothing$ **c.** no interval notation

55. $|x + 1| > 1$ **57.** $|x - 3| \le 2$ **59.** $|x| < 3$
61. $|x| >$ any negative number

Review Exercises

1. true **2.** true **3.** −2 **4.** $\frac{15}{2}$ **5.** $y = \frac{C - Ax}{B}$ **6.** 46, 47, 48

Chapter 2 Review Exercises

1. true **2.** false **3.** true **4.** true **5.** false **6.** true **7.** contradiction **8.** 90° **9.** $ax + b = c$ $ax + b = -c$ **10.** $x > -a$ $x < a$ **11.** 7 **12.** 4 **13.** 0 **14.** −21 **15.** $-\frac{3}{4}$ **16.** $\frac{5}{4}$ **17.** 0 **18.** all real numbers **19.** 6 **20.** no solution **21.** 5 **22.** 8 **23.** $t = \frac{I}{Pr}$ **24.** $w = \frac{P - 2l}{2}$ **25.** $b = \frac{2A}{h}$ **26.** $a = P - b - c$ **27. a.** $\{n \mid n \le 4\}$ **b.** $(-\infty, 4]$
c. −2 −1 0 1 2 3 4 5 6 7

28. a. $\{x \mid x > -2\}$ **b.** $(-2, \infty)$
c. −4 −3 −2 −1 0 1 2 3 4

29. a. $\{m \mid m < -3\}$ **b.** $(-\infty, -3)$ **c.** −7 −6 −5 −4 −3 −2 −1 0 1

30. a. $\{h \mid h \le -5\}$ **b.** $(-\infty, -5]$ **c.** −8 −7 −6 −5 −4 −3 −2 −1 0

31. a. $\left\{t \mid t \le \frac{18}{5}\right\}$ **b.** $\left(-\infty, \frac{18}{5}\right]$
c. 1 2 3 4 5

32. a. $\{u \mid u > 2\}$ **b.** $(2, \infty)$
c. −3 −2 −1 0 1 2 3 4 5

33. $A \cap B = \{2, 4\}$ $A \cup B = \{1, 2, 3, 4, 5, 6\}$ **34.** $A \cap B = \{a\}$ $A \cup B = \{a, b, c, d, e, i, o, u\}$

35. a. −6 −5 −4 −3 −2 −1 0 1 2 3 4 5 6
b. $\{x \mid -3 < x < -2\}$ **c.** $(-3, -2)$

36. a. −8 −7 −6 −5 −4 −3 −2 −1 0 1 2 3 4
b. $\{x \mid -5 \le x \le -3\}$ **c.** $[-5, -3]$

37. a. −8 −7 −6 −5 −4 −3 −2 −1 0 1 2 3 4
b. $\{x \mid -7 < x < 3\}$ **c.** $(-7, 3)$

38. a. −4 −3 −2 −1 0 1 2 3 4 5 6 7 8
b. $\{x \mid 1 < x \le 4\}$ **c.** $(1, 4]$

39. a. −6 −5 −4 −3 −2 −1 0 1 2 3 4 5 6
b. $\{x \mid -2 \le x < 0\}$ **c.** $[-2, 0)$

40. a. −6 −5 −4 −3 −2 −1 0 1 2 3 4 5 6
b. $\{x \mid -1 < x < 3\}$ **c.** $(-1, 3)$

41. a. −8 −7 −6 −5 −4 −3 −2 −1 0 1 2 3 4
b. $\{w \mid w \le -6 \text{ or } w \ge -2\}$ **c.** $(-\infty, -6] \cup [-2, \infty)$

42. a. −6 −5 −4 −3 −2 −1 0 1 2 3 4 5 6
b. $\{w \mid w$ is a real number$\}$ **c.** $(-\infty, \infty)$

43. a.

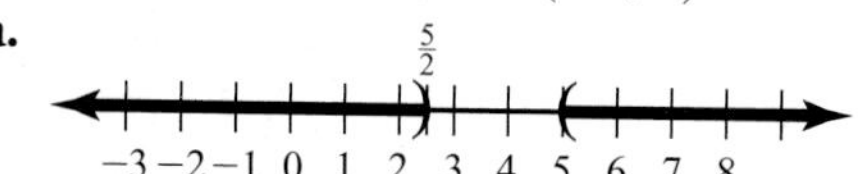

b. $\left\{m \mid m < \frac{5}{2} \text{ or } m > 5\right\}$ **c.** $\left(-\infty, \frac{5}{2}\right) \cup (5, \infty)$

44. a. $-\frac{4}{3}$

−6 −5 −4 −3 −2 −1 0 1 2 3 4 5 6

b. $\left\{x \mid x \le -\frac{4}{3} \text{ or } x \ge 2\right\}$ **c.** $\left(-\infty, -\frac{4}{3}\right] \cup [2, \infty)$

45. a.

−12 −11 −10 −9 −8 −7 −6 −5 −4 −3 −2 −1 0

b. $\{x \mid x \le -9 \text{ or } w \ge -4\}$ **c.** $(-\infty, -9] \cup [-4, \infty)$

46. a.

−6 −5 −4 −3 −2 −1 0 1 2 3 4 5 6

b. $\{w \mid w \le -1 \text{ or } w \ge 1\}$ **c.** $(-\infty, -1] \cup [1, \infty)$

47. $-4, 4$ **48.** $-3, 11$ **49.** $-1, 2$ **50.** $2, \frac{2}{3}$

51. no solution **52.** $-7, 15$ **53.** $-3, 3$ **54.** $\frac{8}{3}, 0$

55. $-\frac{7}{2}, \frac{1}{2}$ **56.** $-1, 11$

57. a.

−6 −5 −4 −3 −2 −1 0 1 2 3 4 5 6

b. $\{x \mid -5 < x < 5\}$ **c.** $(-5, 5)$

58. a.

−6 −5 −4 −3 −2 −1 0 1 2 3 4 5 6

b. $\{p \mid p \le -4 \text{ or } p \ge 4\}$ **c.** $(-\infty, -4] \cup [4, \infty)$

59. a.

−6 −4 −2 0 2 4 6 8 10 12

b. $\{x \mid x < -4 \text{ or } x > 10\}$ **c.** $(-\infty, -4) \cup (10, \infty)$

60. a.

−8 −7 −6 −5 −4 −3 −2 −1 0 1 2 3 4

b. $\{m \mid -5 < m < -1\}$ **c.** $(-5, -1)$

61. a.

−6 −5 −4 −3 −2 −1 0 1 2 3 4 5 6

b. $\{s \mid s \text{ is a real number}\}$ **c.** $(-\infty, \infty)$

62. a.

−6 −5 −4 −3 −2 −1 0 1 2 3 4 5 6

b. $\{b \mid b < -1 \text{ or } b > 1\}$ **c.** $(-\infty, -1) \cup (1, \infty)$

63. a.

−8 −7 −6 −5 −4 −3 −2 −1 0 1 2 3 4

b. $\{m \mid -6 \le m \le 0\}$ **c.** $[-6, 0]$

64. a.

−1 0 1 2 3 4 5 6 7 8 9 10 11

b. $\{t \mid t < 0 \text{ or } t > 10\}$ **c.** $(-\infty, 0) \cup (10, \infty)$

65. a. $-\frac{7}{2}$ $\frac{13}{2}$

−4 −3 −2 −1 0 1 2 3 4 5 6 7

b. $\left\{k \mid k \le -\frac{7}{2} \text{ or } k \ge \frac{13}{2}\right\}$ **c.** $\left(-\infty, -\frac{7}{2}\right] \cup \left[\frac{13}{2}, \infty\right)$

66. a.

−6 −5 −4 −3 −2 −1 0 1 2 3 4 5 6

b. $\{p \mid p \text{ is a real number}\}$ **c.** $(-\infty, \infty)$

67. −320.8°F **68.** $4200 **69.** 12 **70.** 4 **71.** $64.20
72. 34,200 units **73.** 62, 63 **74.** 34, 36, 38 **75.** 21°, 69°
76. 115°, 65° **77.** 320 CDs at $10.50, 80 CDs at $12.00
78. 3 hr **79.** 45 ml **80.** $l \le 25$ ft. **81.** $78 \le x \le 118$
82. a. $32 < t < 212$ **b.** $t \le 32$ or $t \ge 212$ **c.** $0 < t < 100$
d. $t \le 0$ or $t \ge 100$

Chapter 2 Practice Test

1. $A \cap B = \{h, o, e\}$ $A \cup B = \{h, o, m, u, s, e\}$ **2.** 4 **3.** 30 **4.** $-8, 2$ **5.** $-3, 6$ **6.** $-8, \frac{2}{3}$

7. no solution **8.** $b = \frac{2A}{h} - B$

9. a.

−8 −7 −6 −5 −4 −3 −2 −1 0 1 2 3 4

b. $\{x \mid -7 < x \le 3\}$ **c.** $(-7, 3]$

10. a.

−6 −5 −4 −3 −2 −1 0 1 2 3 4 5 6

b. $\{x \mid -3 \le x < -2\}$ **c.** $[-3, -2)$

11. a. −13 5

−14 −12 −10 −8 −6 −4 −2 0 2 4 6

b. $\{x \mid -13 < x < 5\}$ **c.** $(-13, 5)$

12. a.

−6 −5 −4 −3 −2 −1 0 1 2 3 4 5 6

b. $\{x \mid x < -1 \text{ or } x > 3\}$ **c.** $(-\infty, -1) \cup (3, \infty)$

13. a.

−9 −8 −7 −6 −5 −4 −3 −2 −1 0 1 2 3

b. $\{x \mid -7 < x < -1\}$ **c.** $(-7, -1)$

14. a.

−6 −5 −4 −3 −2 −1 0 1 2 3 4 5 6

b. { } or ∅ **c.** no interval notation

15. a.

−6 −5 −4 −3 −2 −1 0 1 2 3 4 5 6

b. $\{t \mid t \text{ is a real number}\}$ **c.** $(-\infty, \infty)$

16. a.

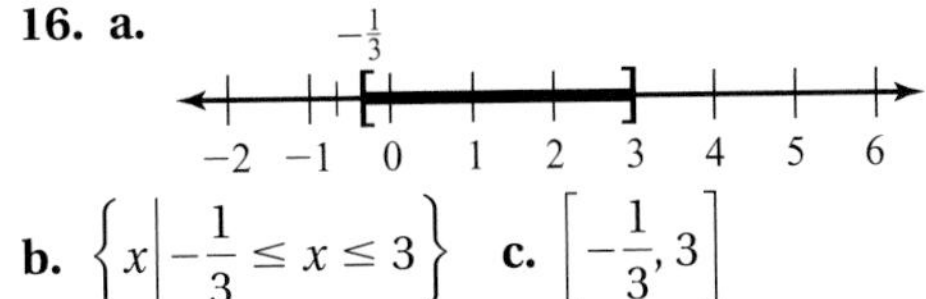

b. $\left\{x \mid -\frac{1}{3} \le x \le 3\right\}$ **c.** $\left[-\frac{1}{3}, 3\right]$

17. −12 **18.** $44.80 **19.** 300 boxes at $250.00, 200 boxes at $225.00 **20.** $5 < x < 6.25$ books; he would have to order 6 books.

Chapter 3

3.1 Exercises

1. Beginning at the origin, move to the left four units along the x-axis, then up three units.
3. Replace the variables in the equation with the corresponding coordinates from the ordered pair. If the resulting equation is true, then the ordered pair is a solution. **5.** The graph of an equation represents the equation's solution set.
7. Plot the solutions as points in the rectangular coordinate system, then connect the points to form a straight line.

9. *A*: (2, 4) *B*: (−2, 1) *C*: (0, −3) *D*: (4, −5)
11. **13.** **15.** II

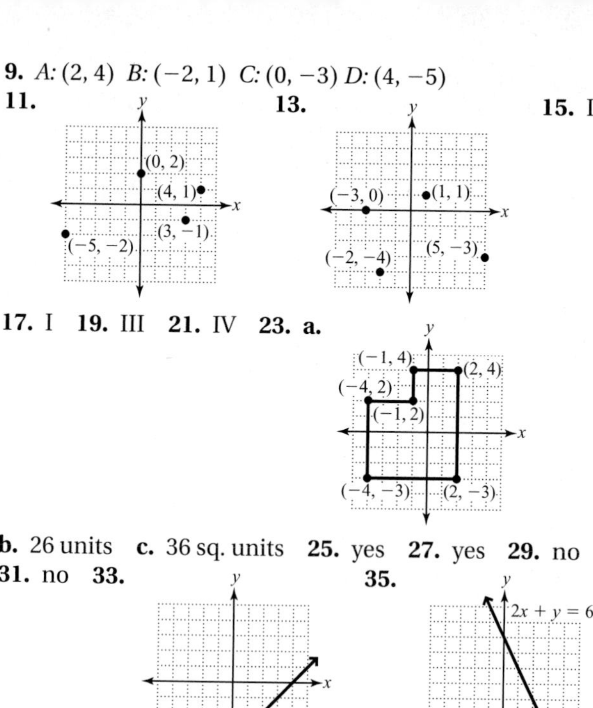

17. I **19.** III **21.** IV **23.** **a.**

b. 26 units **c.** 36 sq. units **25.** yes **27.** yes **29.** no
31. no **33.** **35.**

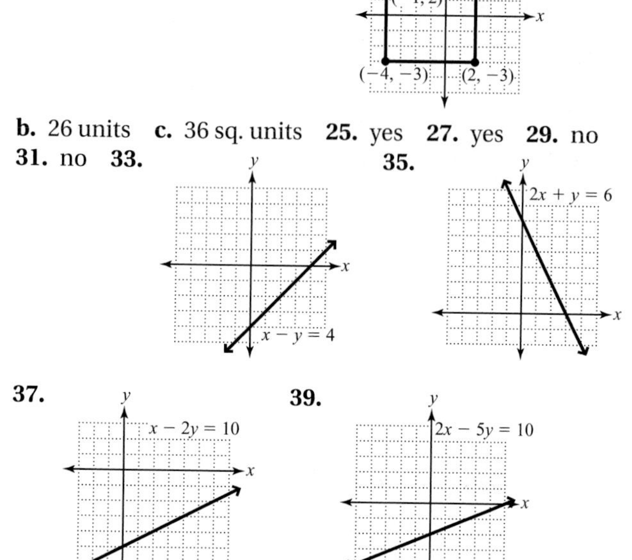

37. **39.**

41. **43.**

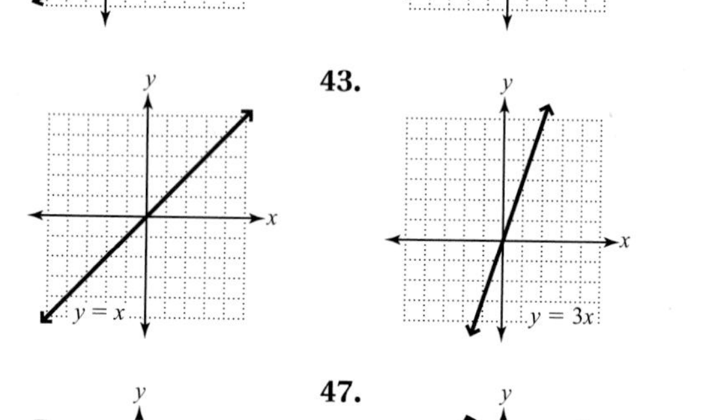

45. **47.**

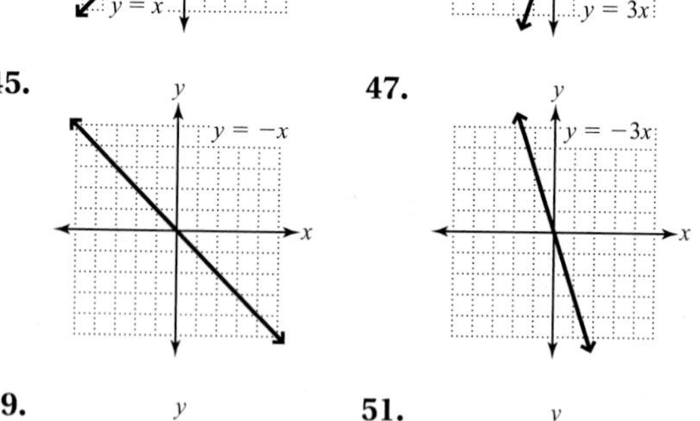

49. **51.**

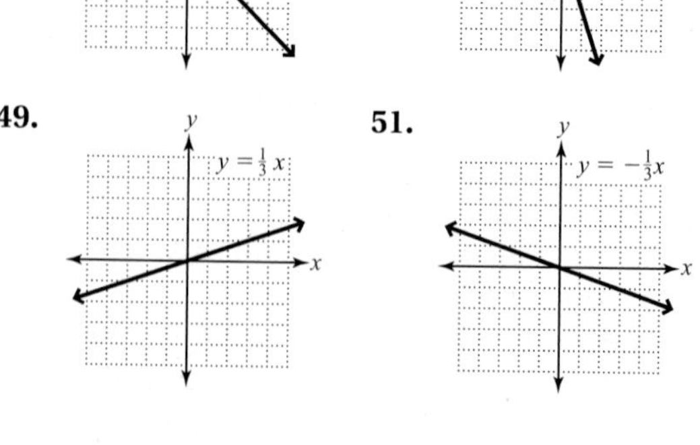

53. **55.** **57.** **59.** **61.** (3, 0) (0, −3) **63.** (2, 0) (0, 2) **65.** (0, 0) **67.** $\left(\frac{4}{3}, 0\right)$, (0, 1) **69.** (2, 0) (0, 4) **71.** (3, 0) (0, −2) **73.** **75.**

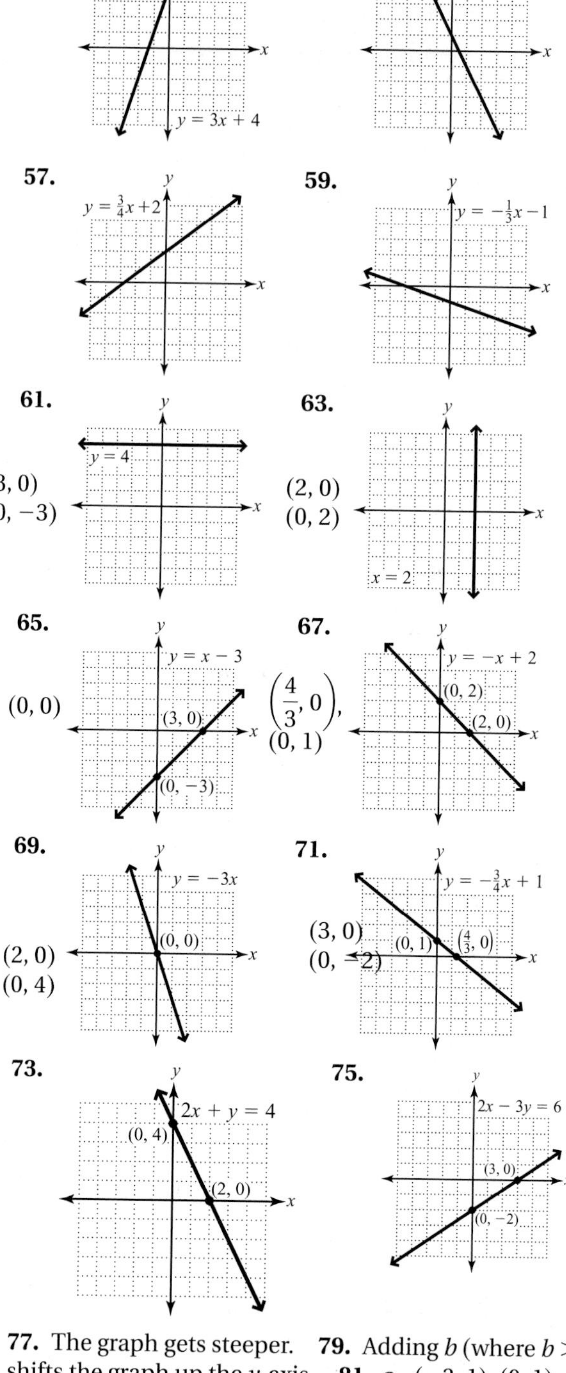

77. The graph gets steeper. **79.** Adding b (where $b > 0$) shifts the graph up the y-axis. **81.** **a.** (−3, 1), (0, 1), (−2, −3), (1, −3) **b.** (1, 2), (4, 2), (2, −2), (5, −2) **c.** $[(x + 4), (y + 1)]$ **83.** **a.** \$95 **b.** 6 hr.

c.

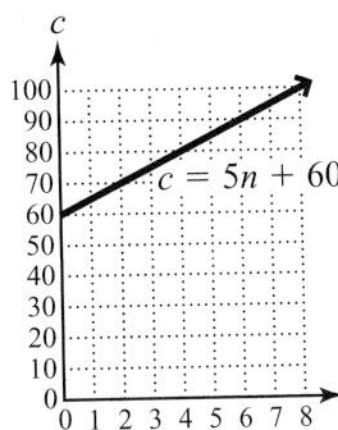

85. a. \$47.50 **b.** 35 min. **c.**

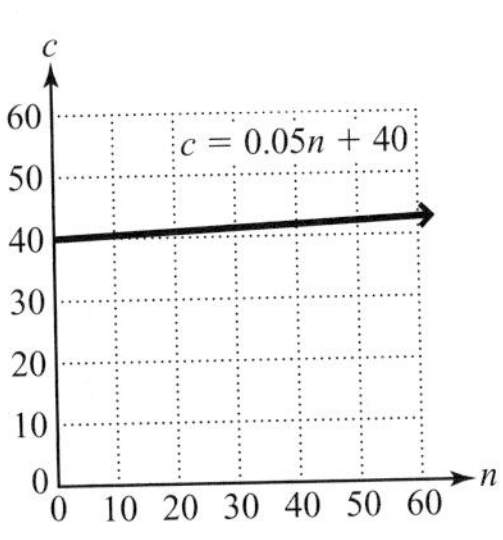

87. a. \$8 **b.** 400 copies

c.

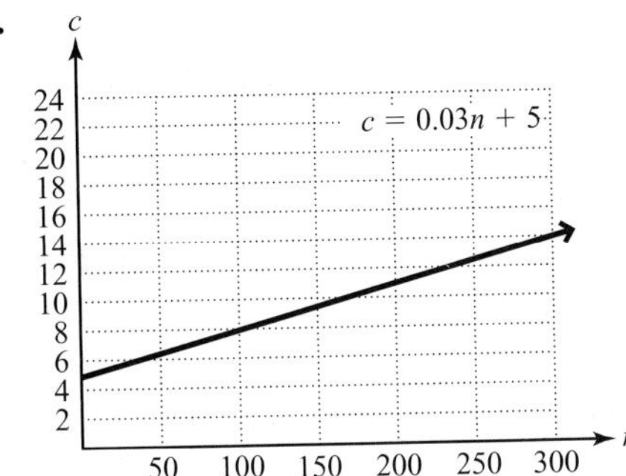

Review Exercises **1.** commutative property of addition **2.** $-\frac{6}{5}$ **3.** -2 **4.** $-\frac{2}{3}$ **5.** $\frac{19}{7}$

6. $\{x|x > 6\}$; $(6, \infty)$; −2 −1 0 1 2 3 4 5 6 7 8 9 10

3.2 Exercises **1.** Slope is the incline of the line. **3.** If $m > 0$, then the graph is a line that slants upward from left to right. **5.** It is a horizontal line because the y-coordinates are the same. **7.** $y = 3x + 1$ **9.** $y = x - 4$ **11.** upward **13.** downward **15.** $m = \frac{2}{3}$; $(0, 1)$ **17.** undefined slope; no y-intercept **19.** e **21.** g **23.** h **25.** d **27.** $m = \frac{2}{3}$; $(0, 5)$

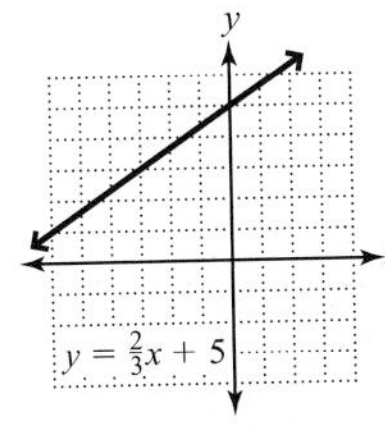

29. $m = -\frac{1}{5}$; $(0, -8)$

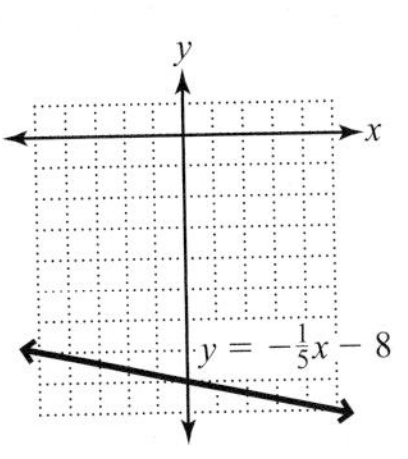

31. $m = 1$; $(0, 4)$

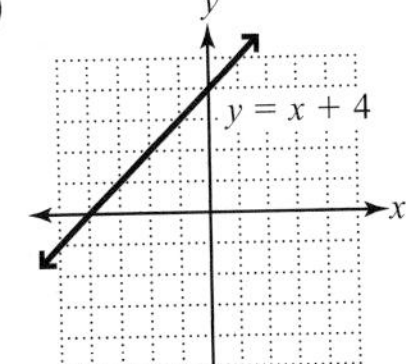

33. $m = -5$; $\left(0, \frac{2}{3}\right)$

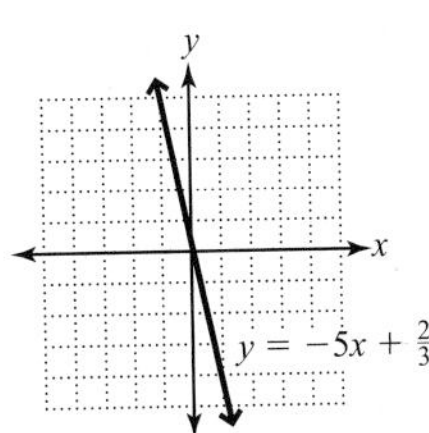

35. $m = -\frac{2}{3}$; $(0, 2)$

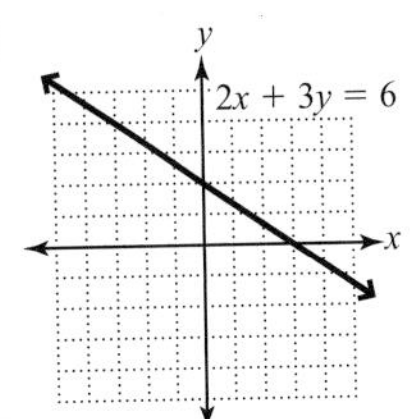

37. $m = -\frac{1}{2}$; $\left(0, -\frac{7}{2}\right)$

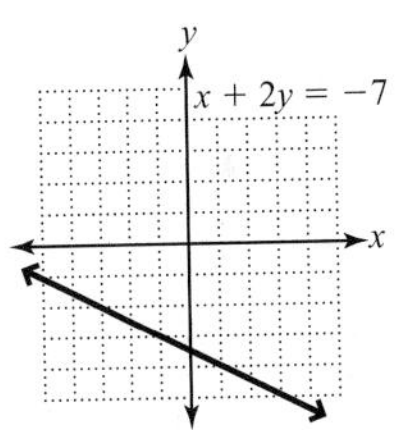

39. $m = \frac{2}{7}$; $\left(0, -\frac{8}{7}\right)$

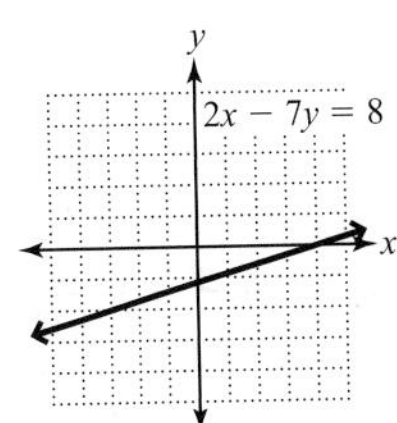

41. $m = 1$; $(0, 0)$

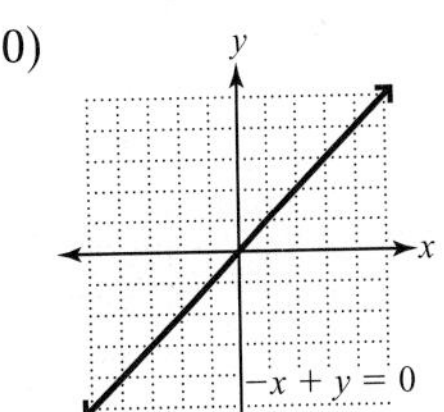

43. -2 **45.** $\frac{9}{5}$

47. $-\frac{3}{2}$ **49.** undefined **51.** 0 **53.** $-\frac{4}{5}$

55. a.

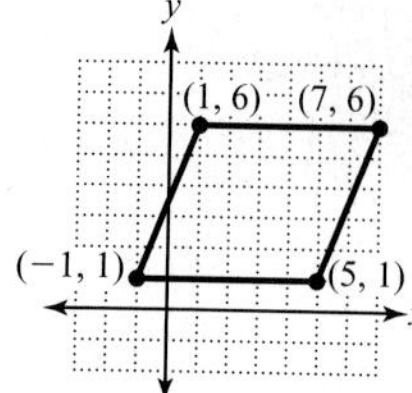

b. Left and right sides: $m = \frac{5}{2}$. Top and bottom sides: $m = 0$.

c. Slopes of parallel sides are equal. **57.** $\frac{1}{12}$ **59.** 0.25

61. a. C: (50, 15), D: (100, 80) **b.** $\frac{13}{10}$ **c.** The coaster climbs 13 ft. vertically for every 10 ft. that it moves horizontally.

63. a.

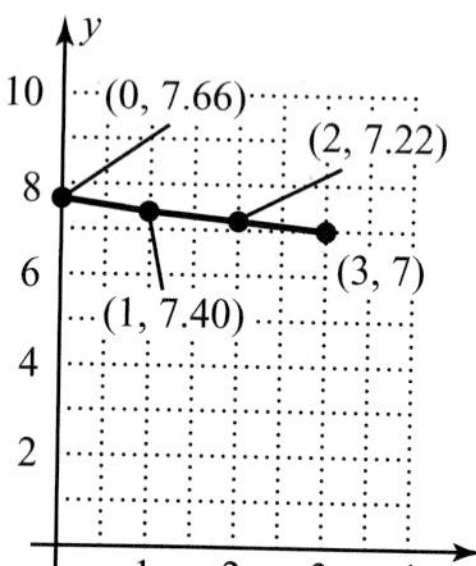

b. −0.22 **c.** \$6.78

65. a.

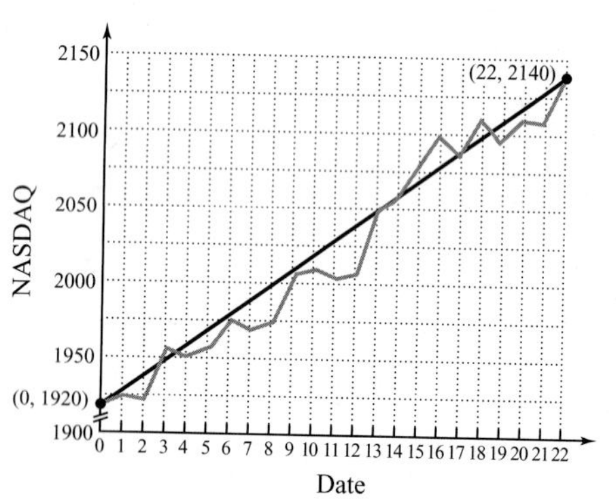

b. 10

c. 2150 **67. a.**

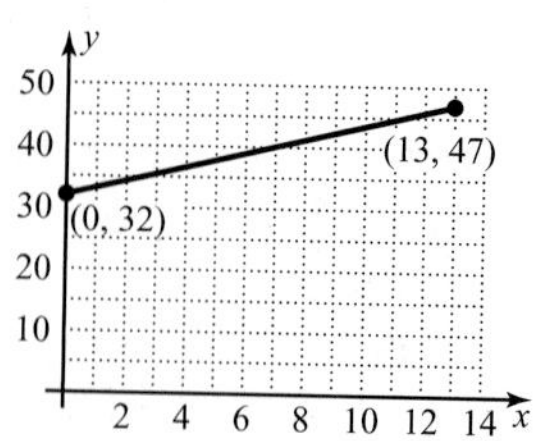

b. 1.15

69. a.

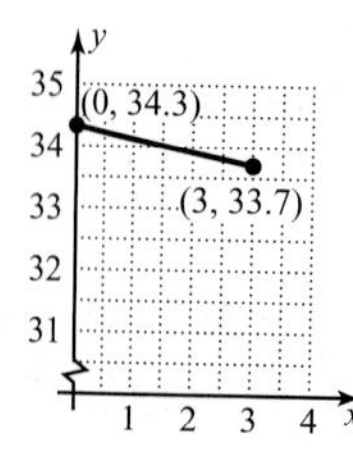

b. −0.2

Review Exercises

1. 16 **2.** 61 **3.** $15x - 27$ **4.** $-15x + 10$ **5.** $h = \frac{V}{lw}$ **6.** $y = \frac{C - Ax}{B}$

3.3 Exercises

1. The slope-intercept form. **3.** Solve the equation for y. **5.** The slopes of parallel lines are equal. **7.** $y = -4x + 3$ **9.** $y = \frac{3}{5}x - 2$ **11.** $y = -0.2x - 1.5$ **13.** $m = \frac{1}{2}$; (0, 2); $y = \frac{1}{2}x + 2$ **15.** $m = -2$; (0, −4); $y = -2x - 4$ **17.** $y = \frac{2}{3}x + 2$ **19.** $y = -\frac{3}{2}x - 6$ **21.** $y = 2x - 5$ **23.** $y = -2x - 10$ **25.** $y = -x - 1$ **27.** $y = \frac{2}{5}x - 2$ **29.** $y = \frac{4}{5}x - \frac{2}{5}$ **31.** $y = -\frac{3}{2}x + \frac{3}{2}$ **33. a.** $y = x - 3$ **b.** $x - y = 3$ **35. a.** $y = -x - 5$ **b.** $x + y = -5$ **37. a.** $y = \frac{2}{3}x - 2$ **b.** $2x - 3y = 6$ **39. a.** $y = \frac{6}{5}x + 6$ **b.** $6x - 5y = -30$ **41. a.** $y = -\frac{1}{13}x + \frac{30}{13}$ **b.** $x + 13y = 30$ **43. a.** $y = -\frac{7}{3}x - \frac{8}{3}$ **b.** $7x + 3y = -8$ **45.** parallel **47.** perpendicular **49.** neither **51.** parallel **53.** perpendicular **55.** perpendicular **57. a.** $y = -5x + 4$ **b.** $5x + y = 4$ **59. a.** $y = 4x + 18$ **b.** $4x - y = -18$ **61. a.** $y = \frac{2}{3}x - 6$ **b.** $2x - 3y = 18$ **63. a.** $y = -\frac{2}{3}x - \frac{1}{3}$ **b.** $2x + 3y = -1$ **65.** $y = -\frac{2}{5}x + \frac{29}{5}$ **b.** $2x + 5y = 29$ **67. a.** $y = 3x - 7$ **b.** $3x - y = 7$ **69. a.** $y = -\frac{5}{2}x - 3$ **b.** $5x + 2y = -6$ **71. a.** $y = \frac{1}{3}x - \frac{25}{3}$ **b.** $x - 3y = 25$ **73. a.** $y = 4x + 5$ **b.** $4x - y = -5$ **75. a.** $y = -\frac{3}{2}x + \frac{11}{2}$ **b.** $3x + 2y = 11$ **77.** $y = -4$ **79.** $y = 0$ **81. a.** 2.62 **b.** $p = 2.62t + 282.4$ **c.** 313.84 million (or 313,840,000) people **83. a.** −130.75 **b.** $b = -130.75t + 6465$ **c.** 4373 thousand (or 4,373,000) barrels

85. a.

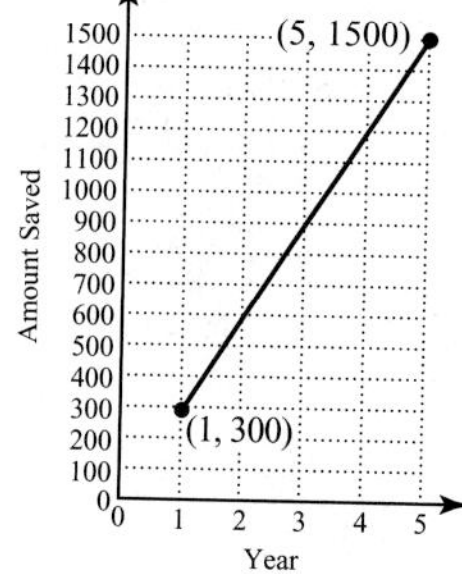

b. 300 **c.** $s = 300t$

d. \$3000

87. a.

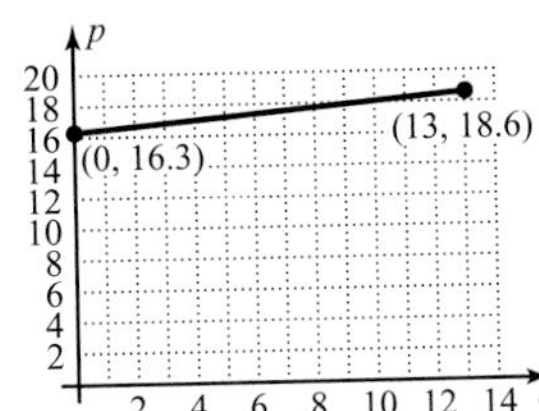

b. ≈ 0.18 **c.** $p = 0.18t + 16.3$ **d.** 19.9% **89. a.**

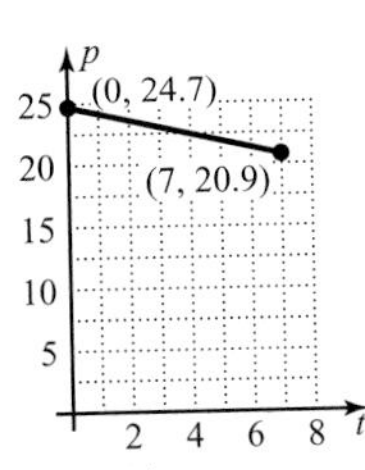

b. ≈ -0.54 **c.** $p = -0.54t + 24.7$ **d.** 14.98%

91. a.

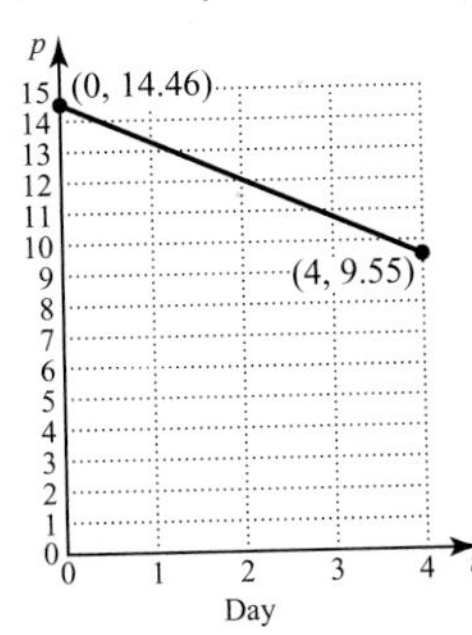

b. -1.2275

c. $p = -1.2275t + 14.46$ **d.** \$5.87

Review Exercises

1. $<$ **2.** -2 **3.** $\frac{2}{15}$

4. $\{x|x \le -5\}$; $(-\infty, -5]$

5. $\{x|x \le 2\}$; $(-\infty, 2]$

6. $w \ge 40$ ft.

3.4 Exercises

1. Begin by graphing the related equation. **3.** If the inequality symbol is $\le$ or $\ge$, draw a solid boundary line. If the inequality symbol is $<$ or $>$, draw a dashed boundary line. **5.** no **7.** yes **9.** no **11.** yes

13.

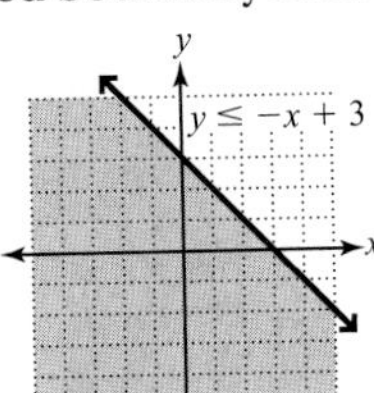

15.

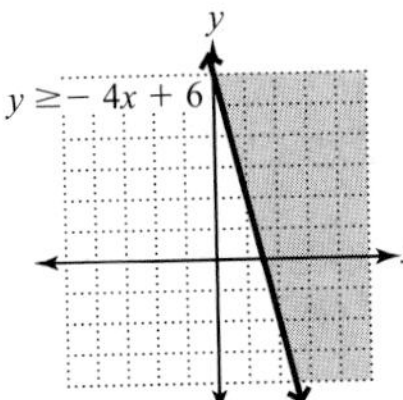

17.

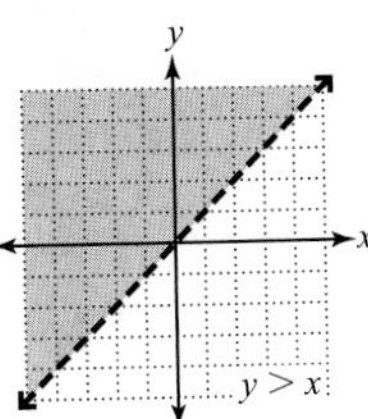

19.

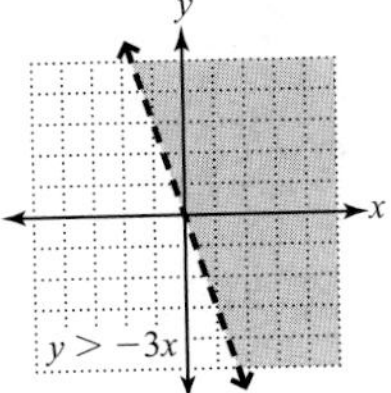

21.

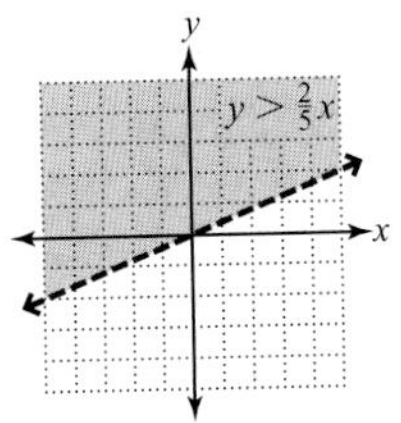

23.

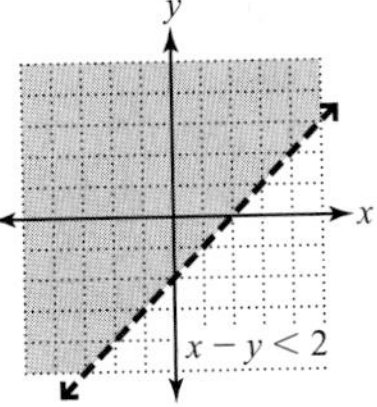

25.

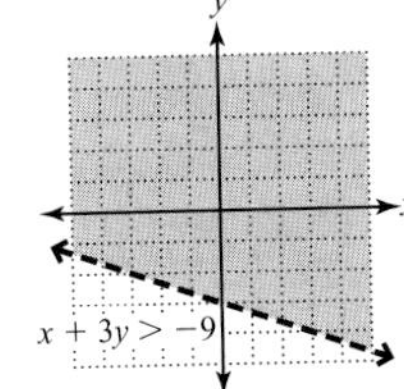

27.

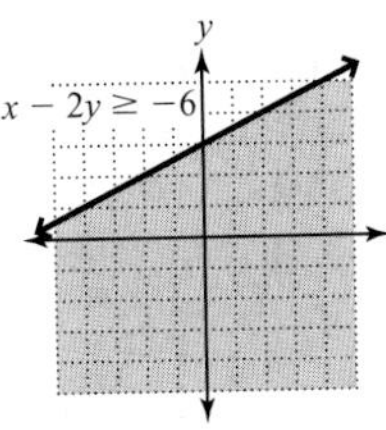

29.

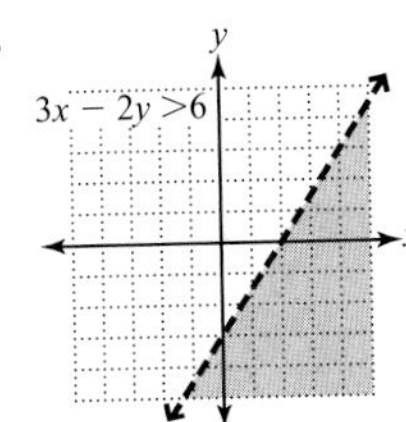

31.

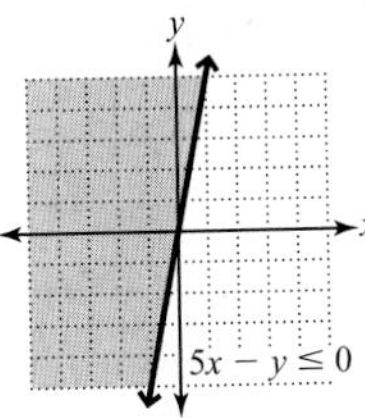

33.

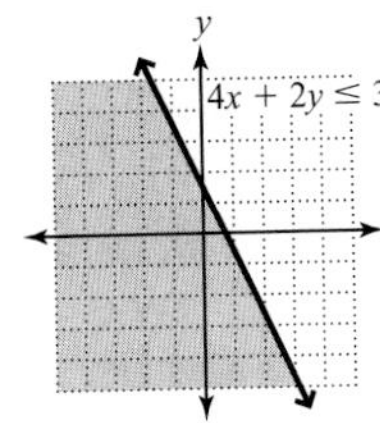

35.

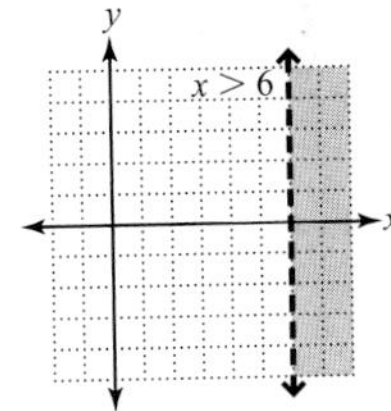

37.

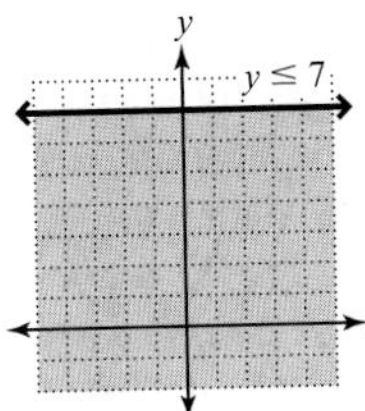

39. a. x represents the number of the board version produced and y represents the number of the video version produced. **b.**

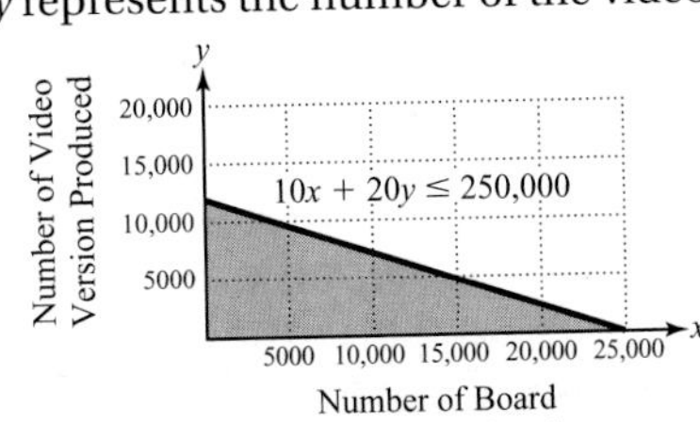

c. All combinations that cost is exactly \$250,000 to produce. **d.** All combinations that cost less than \$250,000 to produce. **e.** Answers may vary. Two examples are (0, 12,500) and (25,000, 0). **f.** Answers may vary. Two examples are (0, 500) and (10,000, 5000). **g.** No, fractions of a game and negative numbers of games are not produced.

41. a. $2l + 2w \leq 200$ **b.**

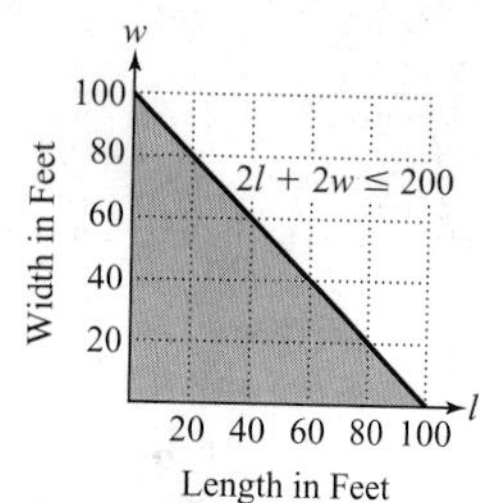

c. Combinations of length and width that make the perimeter exactly 200 ft. **d.** All combinations that make the perimeter less than 200 ft. **e.** Answers may vary. Two examples are (20, 80) and (60, 40). **f.** Answers may vary. Two examples are (20, 40) and (80, 10). **43. a.** $12x + 15y \geq 18{,}000$ **b.**

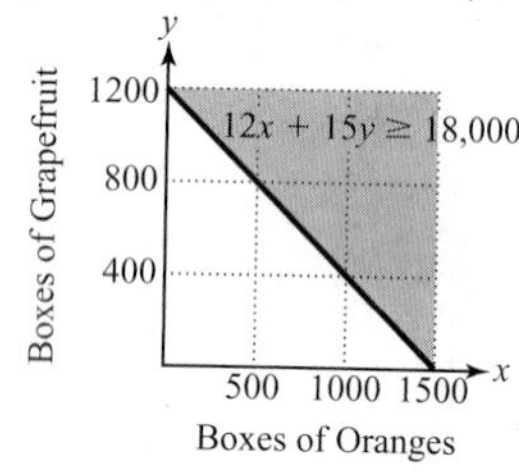

c. All sales combinations that raise exactly $18,000. **d.** All sales combinations that raise more than $18,000. **e.** Answers may vary. Two examples are (500, 800) and (1000, 400). **f.** Answers may vary. Two examples are (500, 1000) and (800, 1200).

Review Exercises

1. 13 **2.** 3 **3.** −8 **4.** (2, 0), $\left(0, -\frac{8}{5}\right)$ **5.** {−2, 0, 3, 4} **6.** {1, 2, 5, 7}

3.5 Exercises

1. The domain is a set containing initial values of a relation, its input values. **3.** A function is a relation in which every value in the domain is assigned to exactly one value in the range. **5.** Given a function $f(x)$, to find $f(a)$, where a is a real number in the domain of f, replace x in the function with a and calculate the value. **7.** Domain: {Landmark College, Sarah Lawrence College, Kenyon College, Trinity College, George Washington University}; Range: {$36,750, $32,416, $32,170, $31,940, $31,710}; It is a function. **9.** Domain: {1, 2, 3, 4, 5} Range: {San Francisco 49ers, Dallas Cowboys, Pittsburgh Steelers, Green Bay Packers, New England Patriots, Oakland/L. A. Raiders, Washington Redskins, New York Giants, Miami Dolphins, Denver Broncos, Baltimore Ravens, Chicago Bears, New York Jets, Tampa Bay Buccaneers, Baltimore Colts, Kansas City Chiefs, St. Louis/L. A. Rams}; It is not a function **11.** Domain: {41%, 16%, 5%, 5%, 4%}; Range: {Savings, Sale of stock or bonds, Equity from other homes, Financial institution loan, Inheritance; It is not a function. **13.** Domain: {1, 2, 3, 4, 5}; Range: {*Lifeguard, The Historian, The Da Vinci Code, The Interruption of Everything, Until I Find You*}; It is a function. **15.** Domain: {2000, 2001, 2002, 2003} Range: {9984, 12,346, 13,505, 14,563}; It is a function. **17.** Domain: ℝ; Range: $\{y \mid y \leq 4\}$; It is a function. **19.** Domain: $\{x \mid x \geq 0\}$ Range: ℝ; It is not a function.

21. Domain: $\{x \mid -4 \leq x \leq 5\}$ Range: {−3, −2}; It is a function. **23.** Domain: $\{x \mid -4 \leq x \leq 0\}$ Range: $\{y \mid -3 \leq y \leq 1\}$; It is not a function **25. a.** −9 **b.** −11 **c.** −7 **d.** $-2a - 11$ **27. a.** 7 **b.** 8 **c.** 10 **d.** $2a^2 - a + 7$ **29. a.** 2 **b.** not real **c.** $\sqrt{5}$ **d.** $\sqrt{3 - t}$ **31. a.** 1 **b.** −1 **c.** $\frac{3}{5}$ **d.** $\frac{2}{5}r + 1$ **33. a.** −3 **b.** 6.46 **c.** ≈ −3.96 **d.** $a^2 - 2.1a - 3$

35. a. 0 **b.** $\sqrt{21}$ **c.** not real **d.** $\sqrt{n^2 - 4n}$ **37. a.** 1 **b.** $\frac{7}{3}$ **c.** 4 **d.** 13 **39. a.** −2 **b.** undefined **c.** $-\frac{5}{6}$ **d.** 0 **41. a.** 0 **b.** undefined **c.** $\frac{2}{3}$ **d.** $\frac{m}{m^2 - 1}$ **43. a.** $\frac{1}{2}$ **b.** 1 **c.** undefined **d.** not real **45. a.** 2 **b.** 0 **c.** −1 **47. a.** −4 **b.** 0 **c.** −4 **49. a.** −3 **b.** 1 **c.** 2

51.

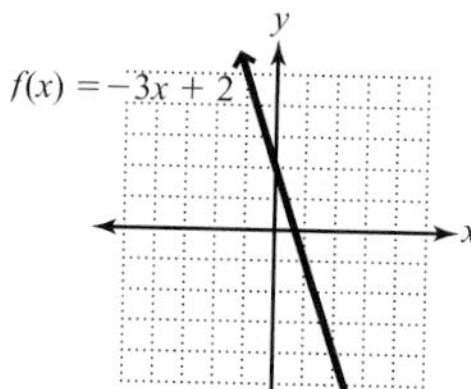

53.

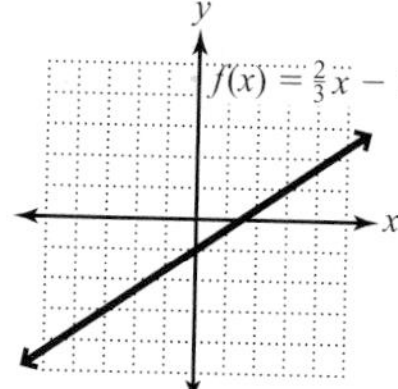

55.

57.

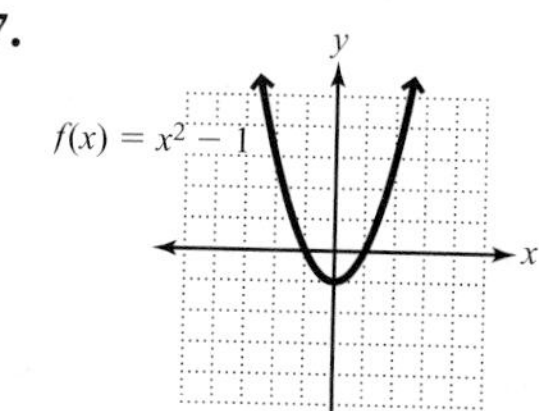

59.

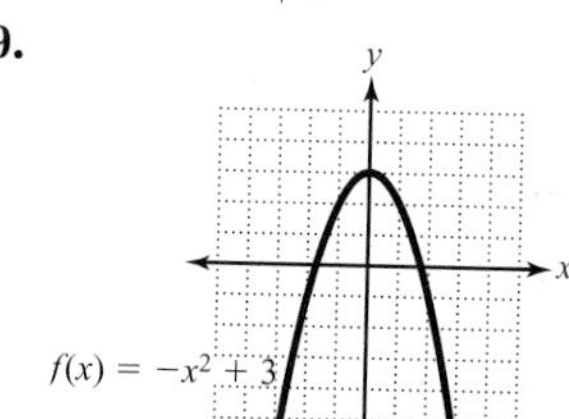

61. a.

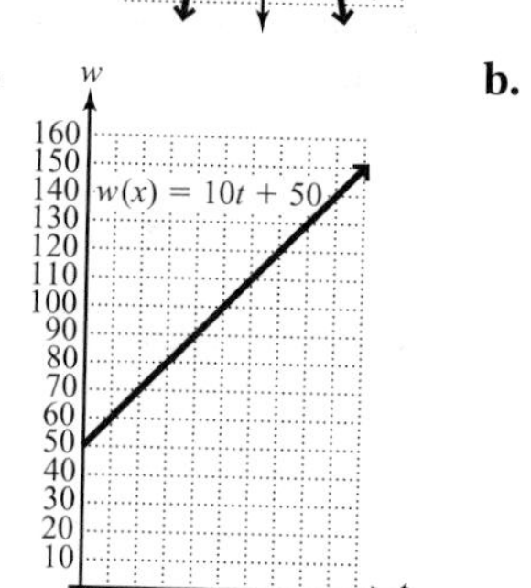

b. $125

63. a. $C(V) = 1.225V + 6$ **b.**

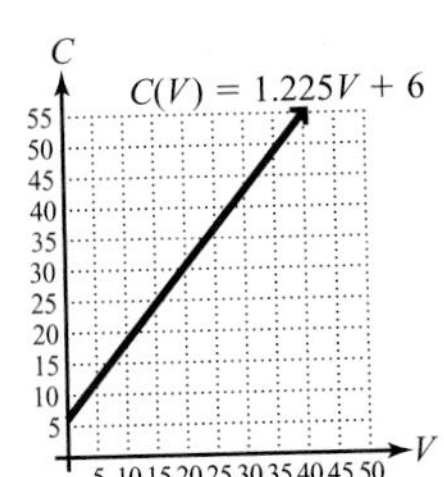

c. \$55

65. a. 

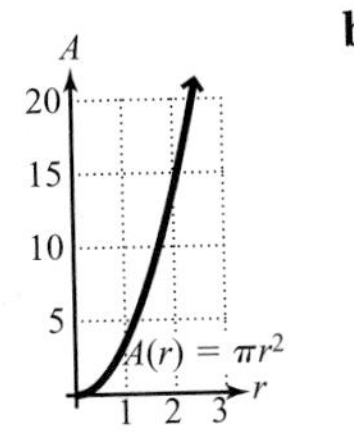

b. no

c. The radius of the circle is 1.5 units. **d.** 7.07 sq. units

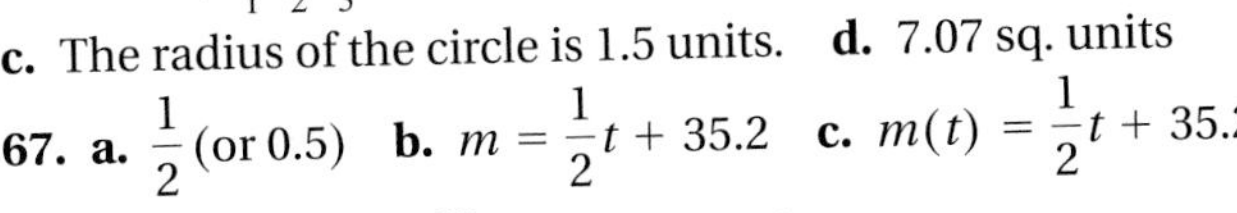

67. a. $\frac{1}{2}$ (or 0.5) **b.** $m = \frac{1}{2}t + 35.2$ **c.** $m(t) = \frac{1}{2}t + 35.2$

d. 45.2 kg/person **69. a.**

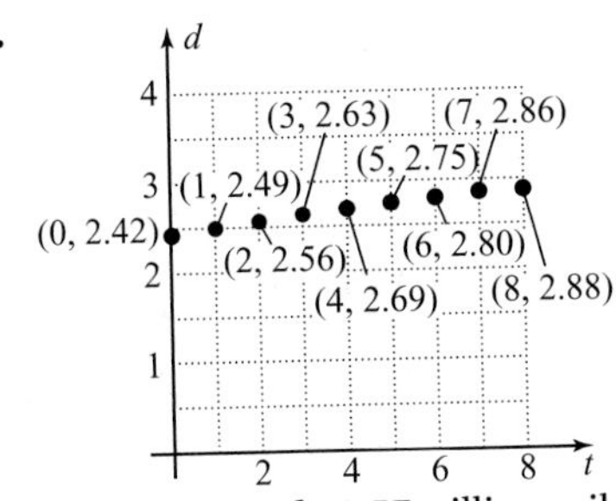

b. yes; 0.0575 **c.** $d(t) = 0.0575t + 2.42$ **d.** 3.57 trillion miles

Review Exercises

1. $-2x + 6y - 2$ **2.** -3 **3.** $\{x|x \leq 2\}$; $(-\infty, 2]$; **4.** 36, 38 **5.** length: 30 in., width: 24 in. **6.** 12 large, 28 small

Chapter 3 Review Exercises

1. false **2.** true **3.** true **4.** true **5.** true **6.** false **7.** $m = \frac{y_2 - y_1}{x_2 - x_1}$ **8.** vertical **9.** upward downward **10.** A function is a relation in which each value in the domain is assigned to exactly one value in the range. **11.** A: (3, 2), B: $(-3, 4)$, C: $(-4, -2)$, D: $(0, -3)$ **12.**

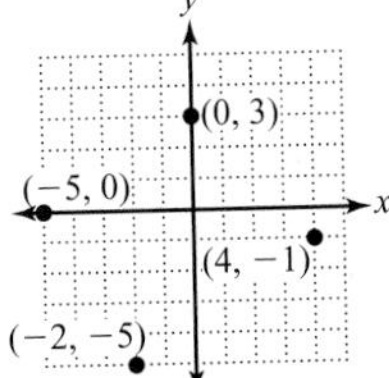

13. II **14.** III **15.** no **16.** yes **17.** no **18.** yes **19.**

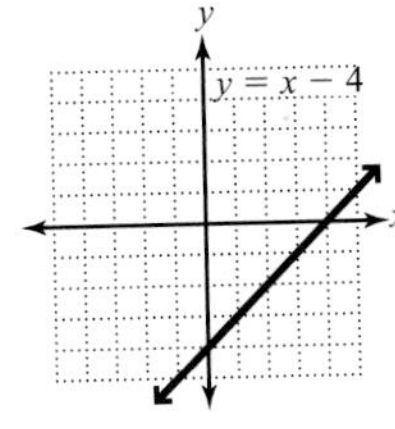

20.

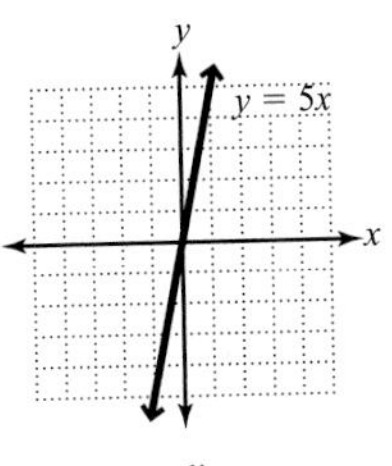

21.

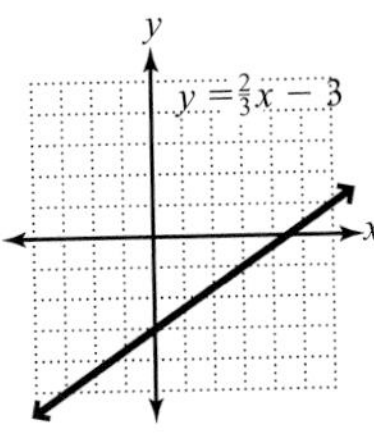

22.

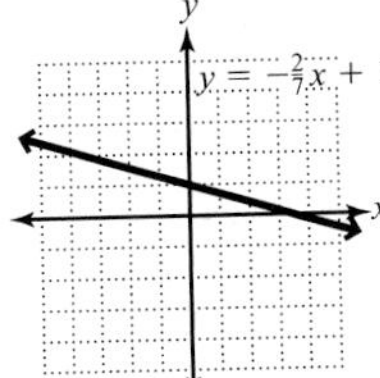

23.

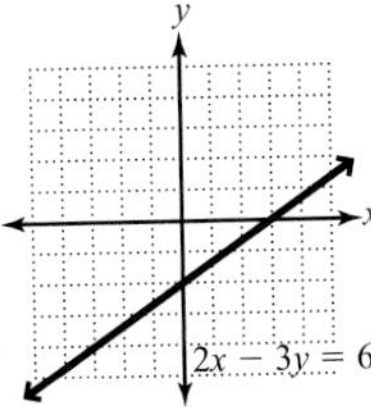

24.

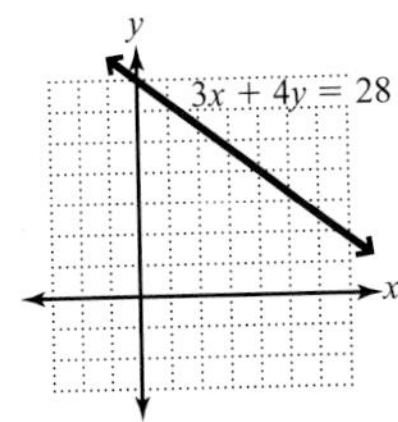

25. $m = -3$; (0, 2)

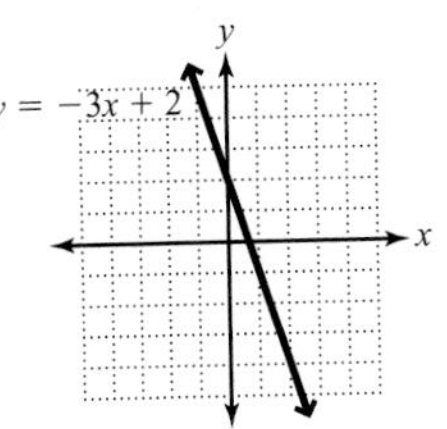

26. $m = \frac{5}{2}$; (0, 3)

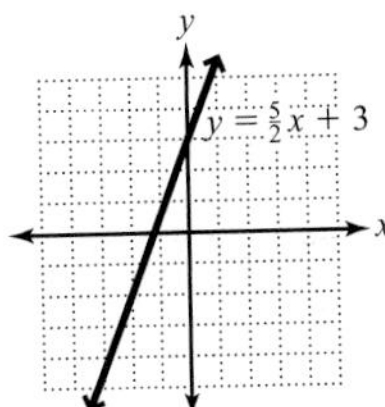

27. $m = -\frac{2}{5}$; (0, 3)

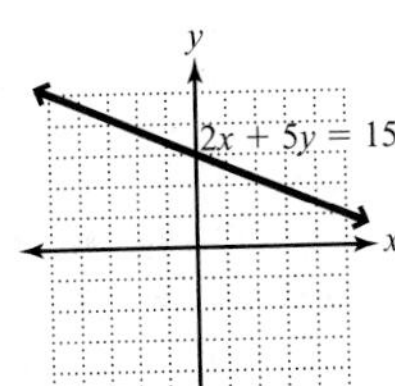

28. $m = 4$; $(0, -3)$

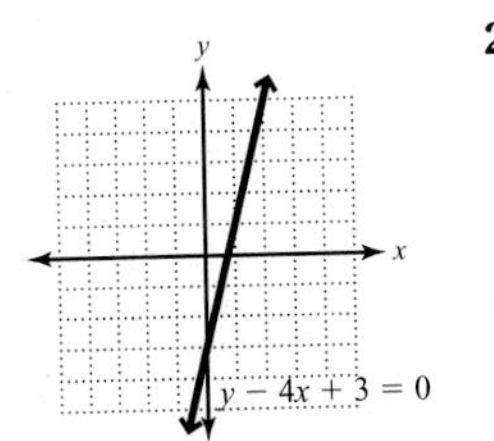

29. -3 **30.** undefined **31.** 1 **32.** $-\frac{3}{4}$ **33.** $y = 2x - 4$

34. $y = -\frac{2}{5}x + 4$ **35.** $y = -0.3x - 1$ **36.** $y = -3x$

37. $m = \frac{2}{5}$; $(0, 2)$; $y = \frac{2}{5}x + 2$ **38.** $m = -2$; $(0, -1)$; $y = -2x - 1$ **39. a.** $y = 3x - 6$ **b.** $3x - y = 6$

40. a. $y = \frac{7}{2}x + 9$ **b.** $7x - 2y = -18$ **41. a.** $y = \frac{1}{13}x - \frac{30}{13}$

b. $x - 13y = 30$ **42. a.** $y = -x + 6$ **b.** $x + y = 6$

43. parallel **44.** neither **45.** perpendicular

46. perpendicular

47. $y = -x + 10$ **48.** $y = 6.2x - 23.6$ **49.** $y = \frac{2}{3}x - \frac{8}{3}$

50. $y = -x - 4$ **51.** $y = -\frac{3}{5}x - 2$ **52.** $y = \frac{1}{5}x + \frac{23}{5}$

53. $y = \frac{3}{2}x - \frac{1}{2}$ **54.** $y = -3x - 14$ **55.** no **56.** yes

57.

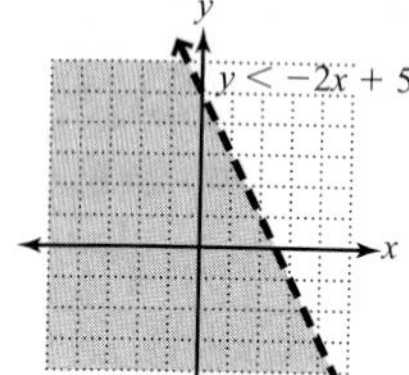

58.

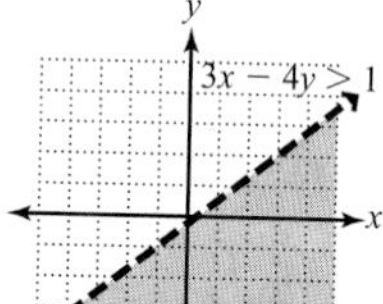

59.

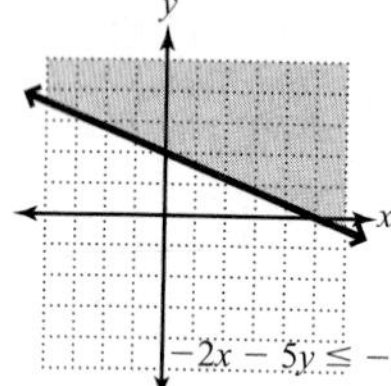

60.

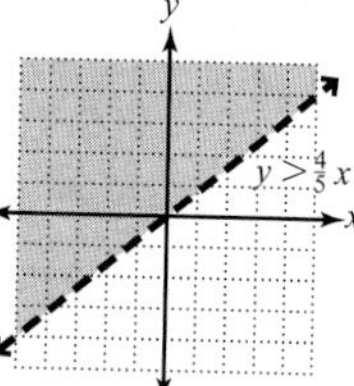

61.

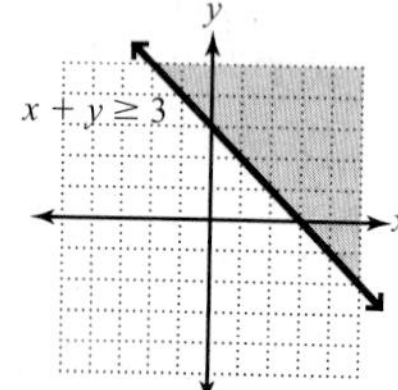

62.

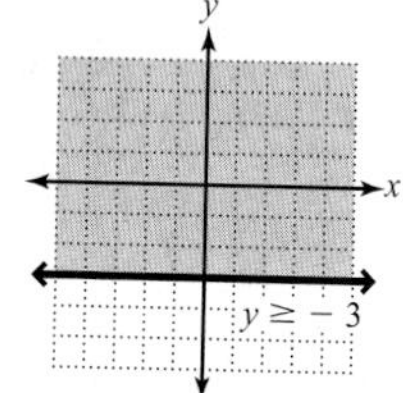

63. Domain: {McKinley, Logan, Pico de Orizaba, St. Elias, Popocatépetl}, Range: {20,320, 19,551, 18,555, 18,008, 17,930}; It is a function. **64.** Domain: {21, 23, 32, 35}, Range: {California, Indiana, New York, Ohio, Pennsylvania}; It is not a function. **65.** Domain: $\{x \mid -4 \le x \le 5\}$, Range: $\{-2, 2, 3\}$; It is a function. **66.** Domain: $\{x \mid -3 \le x \le 3\}$, Range: $\{y \mid 0 \le y \le 3\}$; It is a function. **67.** Domain: $\mathbb{R}$, Range: $\mathbb{R}$; It is a function. **68.** Domain: $\{x \mid x \le 3\}$, Range: **P;** It is not a function. **69. a.** -2 **b.** 3 **c.** 3 **d.** 2 **70. a.** 0 **b.** 3 **c.** 0 **d.** undefined **71. a.** 0 **b.** -4 **c.** 5 **d.** $n^2 - 4$

72. a. $-\frac{1}{7}$ **b.** 0 **c.** undefined **73.** $\frac{5}{6}$ **74. a.** -1.95 **b.** $y = -1.95x + 23.1$ **c.** 1.65%

75. a. $35x + 50y \ge 70{,}000$ **b.**

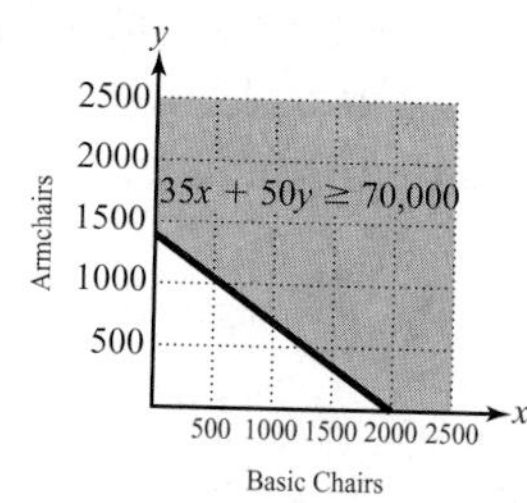

c. All combinations of chair sales that make the company break even. **d.** All combinations of chair sales that make the company a profit. **e.** Answers may vary. Two examples are (1000, 700) and (2000, 0). **f.** Answers may vary. Two examples are (2000, 500) and (2000, 1000). **76. a.** $c(t) = 25t + 75$ **b.** \$112.50 **c.** 3 hr. **d.**

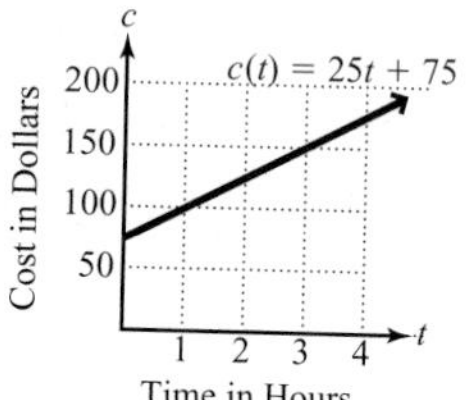

Chapter 3 Practice Test

1. A: (0, 5), B: (3, 3), C: (−2, −3), D: (4, −2) **2.** IV **3.** no

4. $m = -\frac{4}{3}$; $(0, 5)$

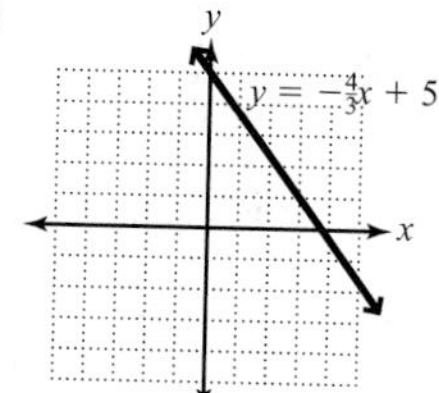

5. $m = \frac{1}{2}$; $(0, 4)$

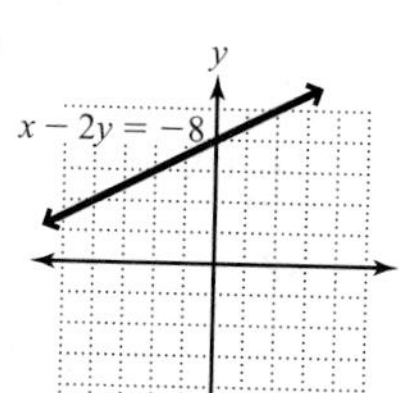

6. 0 **7.** $\frac{8}{7}$

8. $y = \frac{2}{7}x + 5$ **9.** $y = \frac{1}{3}x + \frac{2}{3}$ **10.** $5x - 4y = -11$

11. neither **12.**

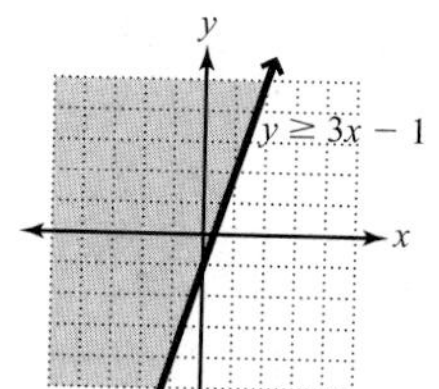

13.

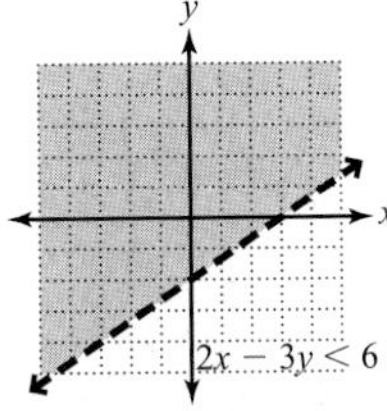

14. Domain: {Spanish, Chinese, French, German, Tagalog}, Range: {28.1, 2.0, 1.6, 1.4, 1.2}; It is a function. **15.** Domain: $\{x|0 \le x \le 3\}$, Range: $\{y|-3 \le y \le 3\}$; It is not a function. **16. a.** 1 **b.** 11 **c.** $2t^2 - 7$ **17. a.** -1 **b.** 2 **c.** undefined **18. a.** Domain: $\{x|-3 \le x \le 3\}$, Range: $\{-2, 1, 3\}$ **b.** 3 **19. a.** $2l + 2w \le 1000$ **b.**

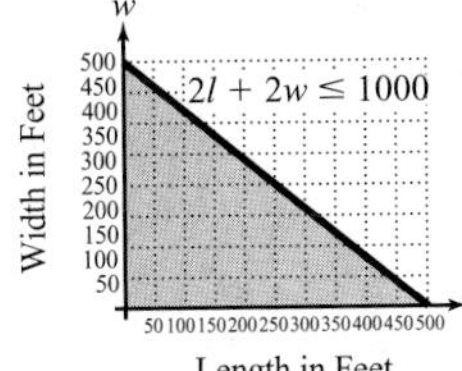

c. Answers may vary. Two possible answers are 300 ft. by 200 ft. or 200 ft. by 200 ft. **20. a.** $c(w) = 0.45w + 3$ **b.**

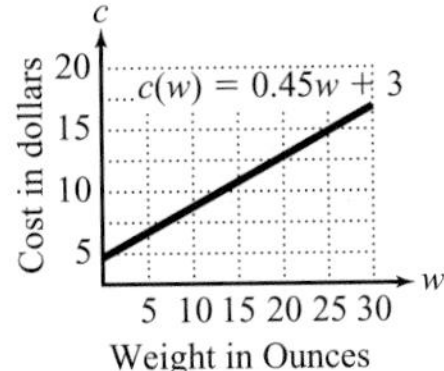

c. $17.40

Chapters 1–3 Cumulative Review Exercises

1. true **2.** false **3.** true **4.** true **5.** true **6.** false **7.** 10 **8.** non negative **9.** We perform operations in the following order: **1.** Grouping symbols: parentheses ( ), brackets [ ], braces { }, absolute value | |, and radicals $\sqrt{\ }$. **2.** Exponents/roots from left to right, in order as they occur. **3.** Multiplication/division from left to right, in order as they occur. **4.** Addition/subtraction from left to right, in order as they occur. **10.** like **11.** {a, e, i, o, u} **12.** $A \cap B = \varnothing$, $A \cup B = \{w,e,l,o,v,m,a,t,h\}$ **13.** 9 **14.** 3 **15.** not a real number **16.** $\frac{2}{3}$ **17.** -135 **18.** 25 **19.** 5 **20.** undefined **21.** $5(6 + n)$ **22. a.** $-4n + 8 - 5n$ **b.** $-9n + 8$ **c.** -28 **23.** multiplicative inverse **24.** distributive property **25.** -29 **26.** all real numbers **27.** $9, -\frac{7}{3}$

28. a. $\left\{x|x \le \frac{5}{6}\right\}$ **b.** $\left(-\infty, \frac{5}{6}\right]$

c.

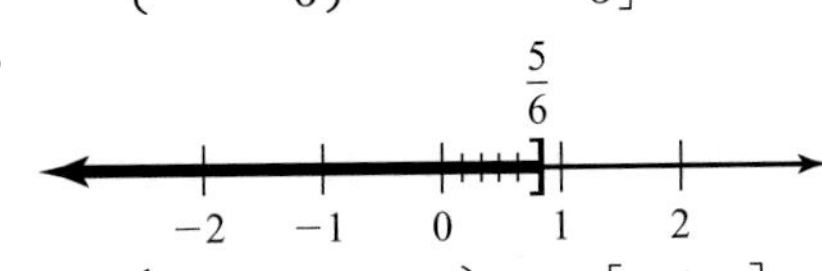

29. a. $\left\{x|-\frac{4}{3} \le x \le 2\right\}$ **b.** $\left[-\frac{4}{3}, 2\right]$

c.

$-\frac{4}{3}$

−5 −4 −3 −2 −1 0 1 2 3 4 5

30. a. $\{x|x < -4 \text{ or } x > 2\}$ **b.** $(-\infty, -4) \cup (2, \infty)$

c.

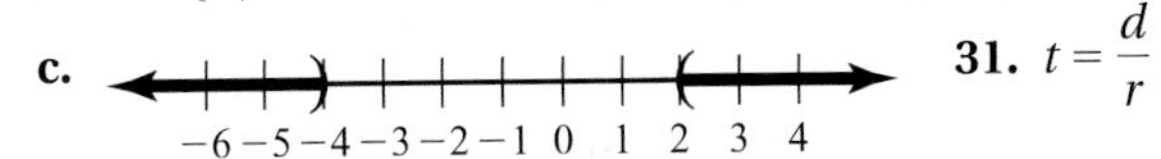

31. $t = \frac{d}{r}$

32. 0 **33.** $m = \frac{3}{4}$; (0, 2) **34.** neither **35.** $y = 2x + 6$

36. no **37.**

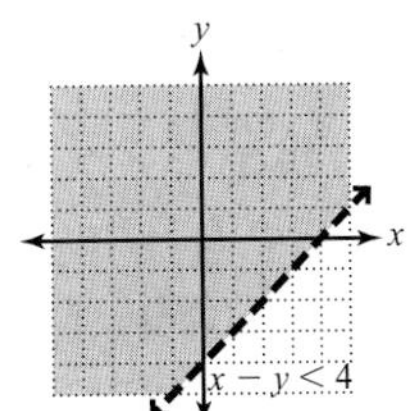

38.

$2x - y \ge 6$

39. yes **40.** 5 **41.**

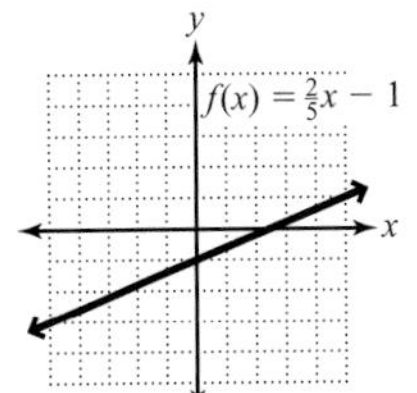

42. 149 million bushels **43.** 3.6 **44.** 92, 94 **45.** 60°, 120° **46.** ≈ 17,913 cell sites **47.** 2 hr. **48.** 4 deluxe, 8 standard **49.** 25 ml **50.** $0 \le t \le 1000$ min.

Chapter 4

4.1 Exercises

1. Replace each variable in each equation with its corresponding value. Verify that each equation is true. **3.** The lines are parallel. **5. a.** y in the second equation **b.** $x + 2$ **7.** yes **9.** no **11.** no **13.** yes **15.** $\begin{cases} x + y = 5 \\ x - y = 3 \end{cases}$

17. $\begin{cases} 2l + 2w = 50 \\ w = l - 2 \end{cases}$ **19. a.** consistent with independent equations **b.** one solution **21. a.** inconsistent with independent equations **b.** no solution **23. a.** consistent with dependent equations **b.** infinite number of solutions **25.** consistent with independent equations **27.** consistent with dependent equations **29.** inconsistent with independent equations

31.

$x + y = 5$, $x - y = 3$, (4, 1)

(4, 1)

33.

$y = 2x + 5$, $y = -x - 4$, (−3, −1)

(−3, −1)

35. 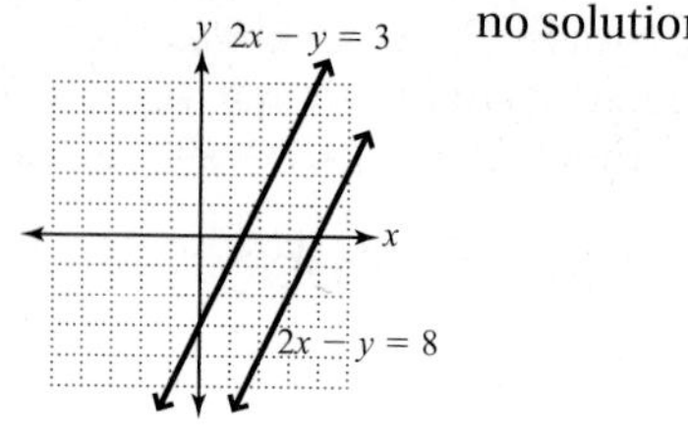

no solution

37. 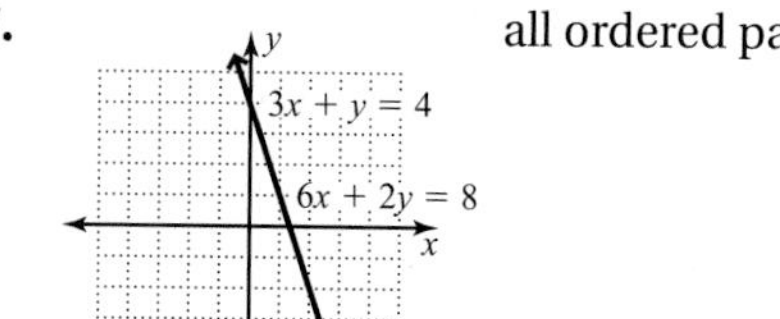

all ordered pairs along $3x + y = 4$

39. 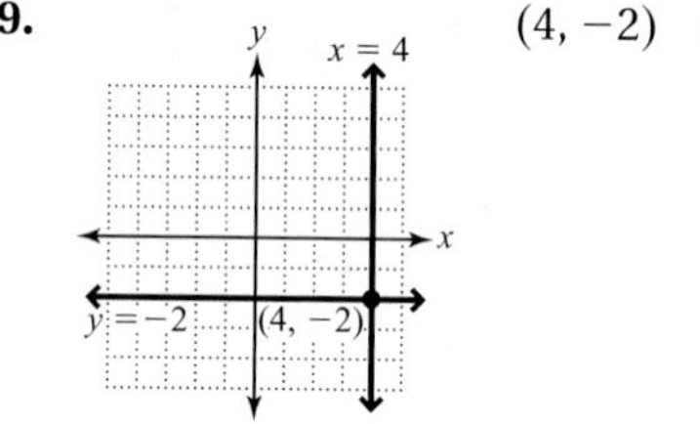

$(4, -2)$ **41.** $(2, 4)$ **43.** $(2, -1)$

45. $\left(-1, \frac{3}{4}\right)$ **47.** $(1, 2)$ **49.** $(-3, -1)$ **51.** $(1, -2)$
53. no solution **55.** all ordered pairs along $x - 2y = 6$
57. Mistake: Did not distribute properly. Correct: $(3, 2)$
59. $(1, 0)$ **61.** $\left(\frac{1}{2}, 1\right)$ **63.** $(3, -4)$ **65.** $\left(1, -\frac{2}{3}\right)$
67. $(1, -2)$ **69.** $\left(\frac{2}{3}, \frac{1}{2}\right)$ **71.** $(-4, 2)$ **73.** $(4, 1)$
75. no solution **77.** all ordered pairs along $x + 2y = 4$
79. Mistake: Subtracted 7 from 8 instead of adding.
Correct: $\left(\frac{15}{2}, \frac{1}{2}\right)$ **81. a.** 5000 **b.** \$4000 **c.** $n > 5000$

83. a. $\begin{cases} c = 45 \\ c = 0.10n + 30 \end{cases}$ **b.**

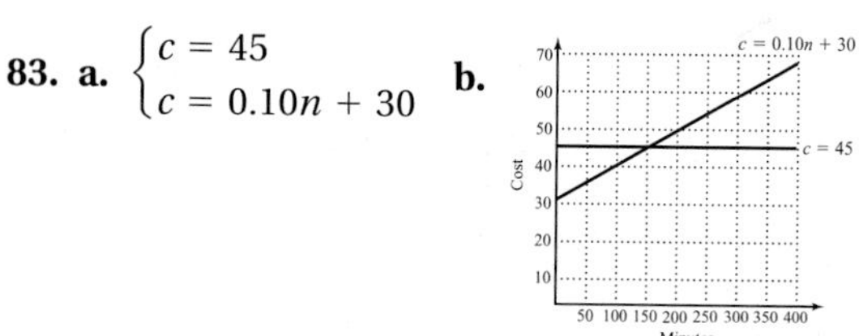

c. 150 min. **d.** Plan 1 **e.** Plan 2

Review Exercises

1. $y = -\frac{1}{3}x + 2$ **2.** $x = 6y + 30$
3. $20x + 28y - 4z$ **4.** $x - 3y$ **5.** $x = 2$ **6.** $y = 0$

4.2 Exercises

1. It does not matter which pair of equations you choose as long as the second pair is different from the first pair. **3.** Answers may vary. A good choice is to eliminate z using equations 1 and 2 and 1 and 3. Multiplying Equation 1 by -3 and then adding this result to both Equations 2 and 3 takes very few steps. **5.** Two parallel planes intersecting a third plane.

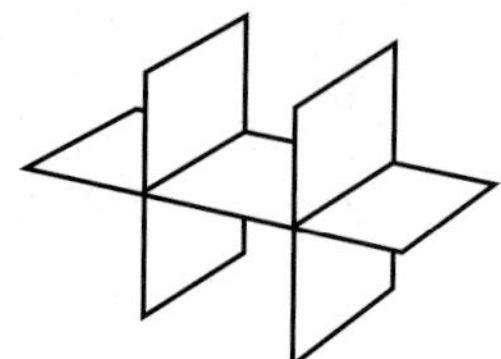

7. no **9.** no

11. yes **13.** $(2, -1, 4)$ **15.** $(-1, 0, 3)$ **17.** $(-1, 1, 3)$
19. no solution **21.** infinite number of solutions (dependent equations)
23. $(2, -3, 1)$ **25.** $\left(-2, -\frac{1}{2}, 1\right)$ **27.** $(2, -1, 1)$
29. $(-3, 7, -4)$ **31.** $(2, -4, 1)$ **33.** $(4, -4, 2)$

Review Exercises

1. 93, 155 **2.** 75 ft. by 125 ft. **3.** 61, 63 **4.** 55°, 125° **5.** 10 min. **6.** 100 ml of 10%, 100 ml of 30%

4.3 Exercises

1. $y = x + 4$ **3.** three **5.** 8 fives, 15 tens
7. *Shrek:* \$268 million, *Shrek II:* \$437 million **9.** 6.7 million
11. Mathematician: \$78,240, Math teacher: \$56,480
13. 85.7% **15.** 10 18-inch wreaths, 36 22-inch wreaths
17. Union: 364,000, Confederacy: 134,000 **19.** width and length: 55.125 ft. **21.** width: 25 yd., length: 40 yd.
23. 60°, 120° **25.** 35°, 55° **27.** 0.2 hr. **29.** 10:00 P.M.
31. boat: 14 mph, current: 2 mph **33.** plane: 600 mph, wind: 50 mph **35.** 5 ml of 5%, 5 ml of 20% **37.** \$14,000 at 9%, \$3500 at 5% **39.** 6, 2, 8 **41.** 105°, 35°, 40° **43.** 8 in., 10 in., and 15 in. **45.** burger: \$2.50, fry: \$1.50, drink: \$1.00
47. 150 children, 250 students, 100 adults **49.** 11 2-point field goals, 2 3-point field goals, 1 free throw **51.** 8 nickels, 7 dimes, 10 quarters **53.** 6 pennies, 12 nickels, 7 dimes
55. zinc: 20 lb., tin: 60 lb., copper: 920 lb. **57.** ham: 3 lb., turkey: 5 lb., beef: 2 lb. **59.** \$2000 at 4%, \$2500 at 6%, \$3500 at 7% **61.** \$30,000 in stocks, \$50,000 in bonds, \$20,000 in certificates **63.** jumbo: 30 oz., extra large: 27 oz., large: 24 oz. **65.** bicycling: 420, walking: 300, climbing stairs: 600 **67.** Angola: 24, Botswana: 25, Mozambique: 26
69. $a = -16, v_0 = 100, h_0 = 150; h = -16t^2 + 100t + 150$

Review Exercises

1. $-6x + 12y + 3z - 27$ **2.** $y = -4$
3. -1 **4.** $(0, 1)$ **5.** $m = -2$ **6.**

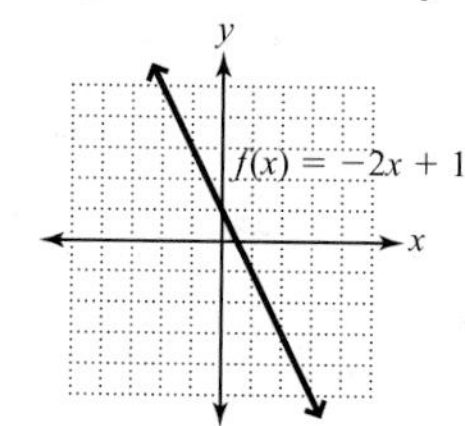

4.4 Exercises

1. 4 rows, 2 columns **3.** The dashed line corresponds to the equal signs in the equations. **5.** A matrix is in echelon form when the coefficient portion of the augmented matrix has 1s on the diagonal from upper left to lower right and 0s below the 1s. **7.** $\left[\begin{array}{cc|c} 14 & 7 & 6 \\ 7 & 6 & 8 \end{array}\right]$ **9.** $\left[\begin{array}{cc|c} 7 & -6 & 1 \\ 8 & -12 & 6 \end{array}\right]$

11. $\left[\begin{array}{ccc|c} 1 & -3 & 1 & 4 \\ 2 & -4 & 2 & -4 \\ 6 & -2 & 5 & -4 \end{array}\right]$ **13.** $\left[\begin{array}{ccc|c} 4 & 6 & -2 & -1 \\ 8 & 3 & 0 & -12 \\ 0 & -1 & 2 & 4 \end{array}\right]$

15. $(4, -2)$ **17.** $(-6, -5, 1)$ **19.** $\left[\begin{array}{cc|c} 1 & 3 & -1 \\ 0 & 11 & 4 \end{array}\right]$

21. $\left[\begin{array}{ccc|c} 1 & -2 & 4 & 6 \\ 0 & 2 & -1 & -5 \\ 0 & 0 & -2 & 17 \end{array}\right]$ **23.** $\left[\begin{array}{cc|c} 1 & 2 & -2.5 \\ -1 & 3 & 2 \end{array}\right]$

25. Replace R_2 with $3R_1 + R_2$. **27.** Replace R_3 with $-2R_2 + R_3$. **29.** $(2, 0)$ **31.** $(1, 2)$ **33.** $(-3, -2)$ **35.** $(-2, 1)$ **37.** $(4, 6)$ **39.** $(4, -4)$ **41.** $(15, -15, 6)$ **43.** $(2, 1, -2)$ **45.** $(2, -3, 1)$ **47.** $(-1, -1, -2)$ **49.** $(-3, 1, 5)$ **51.** $(8, -2, -4)$ **53.** $(3, -4)$ **55.** $(-6, 8)$ **57.** $(4, -3, 5)$ **59.** $(7, -6, 5)$ **61.** Mistake: In the second step, $4R_1 + R_2$ is calculated incorrectly. Correct: The correct calculation is $\left[\begin{array}{cc|c} 1 & 3 & 13 \\ 0 & 11 & 26 \end{array}\right]$. The solution is $(65/11, 26/11)$.
63. chicken sandwich: \$2.50, drink: \$1.20 **65.** 58°, 32° **67.** Nile: 4150 mi., Amazon: 3900 mi. **69.** \$6400 in CD, \$3600 in money market **71.** CD: \$16; tape: \$12, DVD: \$28 **73.** 3 touchdowns, 3 extra points, 1 field goal **75.** $v_1 = 9$, $v_2 = 6$, $v_3 = 2$

Review Exercises

1. 8 **2.** 11 **3.** 2 **4.** 2 **5.** 3 **6.** -1

4.5 Exercises

1. No, because it is not a square matrix. **3.** Eliminate the elements in the same row and column as the element. **5.** 13 **7.** -22 **9.** 24 **11.** -7 **13.** -22 **15.** 15 **17.** 1 **19.** -12 **21.** -14 **23.** -58 **25.** 0.47 **27.** 14.451 **29.** $\frac{13}{30}$ **31.** $-\frac{317}{2400}$ **33.** $-x - 8y - 2$ **35.** $(-4, -1)$ **37.** $(3, 4)$ **39.** $(4, -4)$ **41.** $(-3, -2)$ **43.** $(2, 2)$ **45.** $(2, 4)$ **47.** $(2, 6)$ **49.** $(2, 1, 3)$ **51.** $(-2, 1, -1)$ **53.** $(2, -1, 1)$ **55.** $\left(\frac{2}{3}, -7, \frac{1}{2}\right)$ **57.** $(1, -2, 3)$ **59.** $(2, 2, -2)$ **61.** $x = -4$ **63.** $x = -5$ **65.** 9 **67.** 21.5 **69.** 19 **71.** Hoba is 60 tons, Ahnighito is 30 tons. **73.** 6 lb. of garbanzo, 4 lb. of black turtle **75.** 1% zinc, 4% tin, 95% copper

Review Exercises

1. $x < 3$;

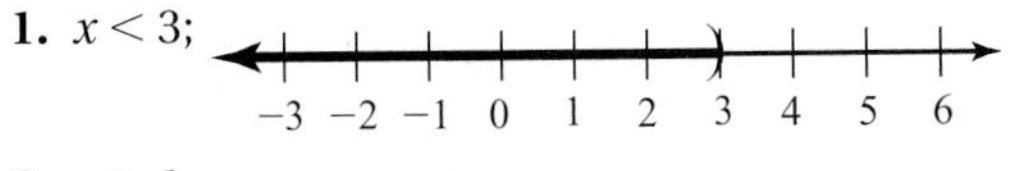

2. $x \geq 1$; **3.** 19, 20

4. **5.**

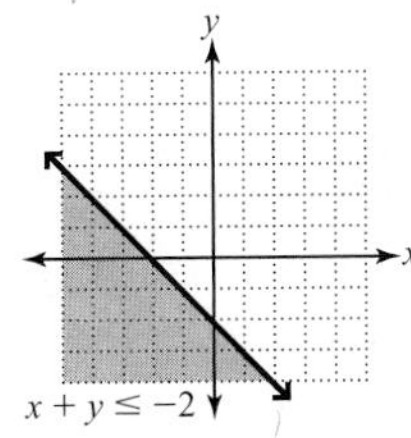

6. $(-2, 5)$

4.6 Exercises

1. Substitute the coordinates of the ordered pair into each inequality of the system. If it makes every inequality in the system true, then it is a solution for the system. **3.** The boundary lines must be parallel. **5.** $\begin{cases} x < 0 \\ y < 0 \end{cases}$

7. **9.**

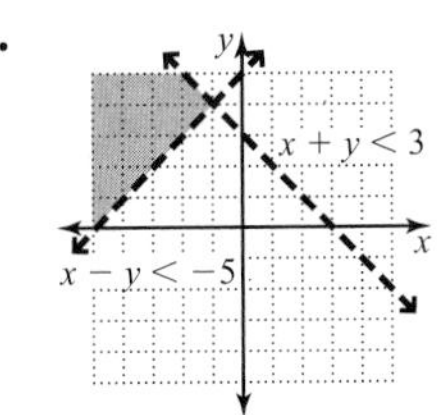

11. **13.**

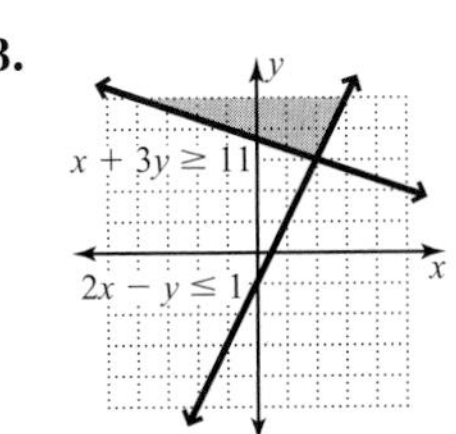

15. **17.**

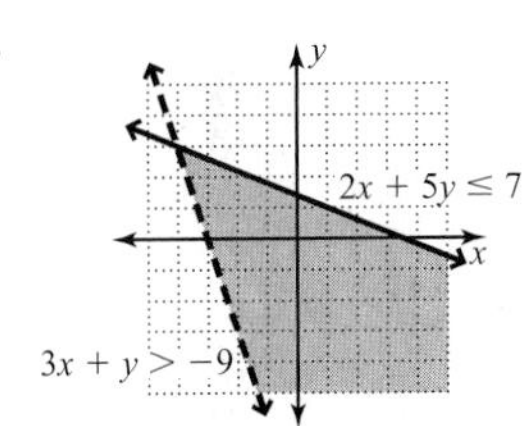

19. **21.**

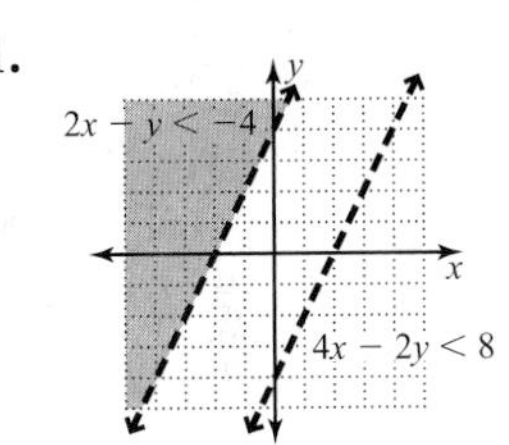

23. no solution

25.

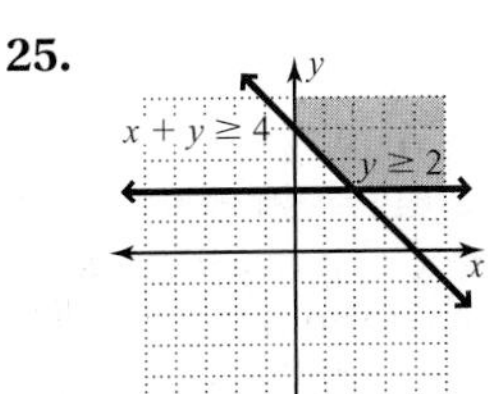

27.

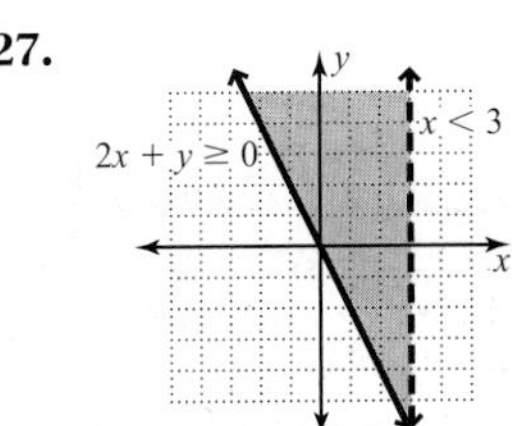

29.

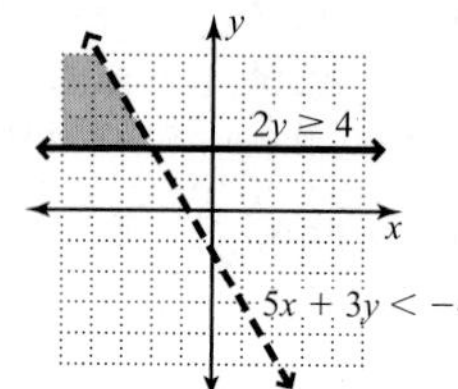

31.

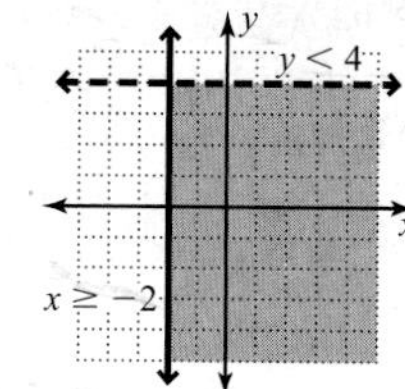

33.

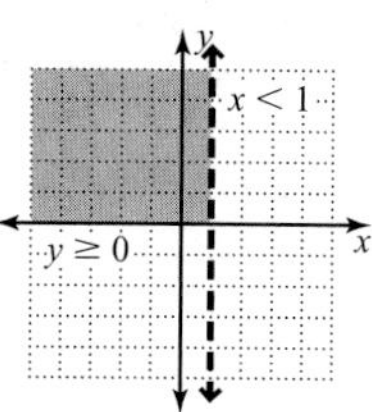

35.

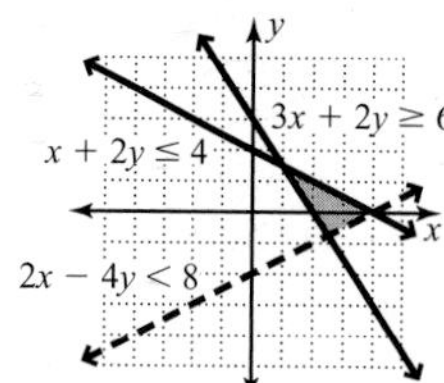

37.

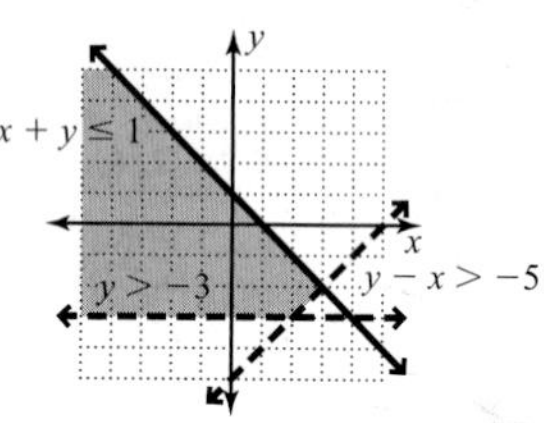

39. The boundary line for $x - y > -1$ should be a dashed line.

41. The wrong region is shaded.

43. a. $\begin{cases} S + G \geq 75 \\ 150S + 200G \geq 12{,}000 \end{cases}$

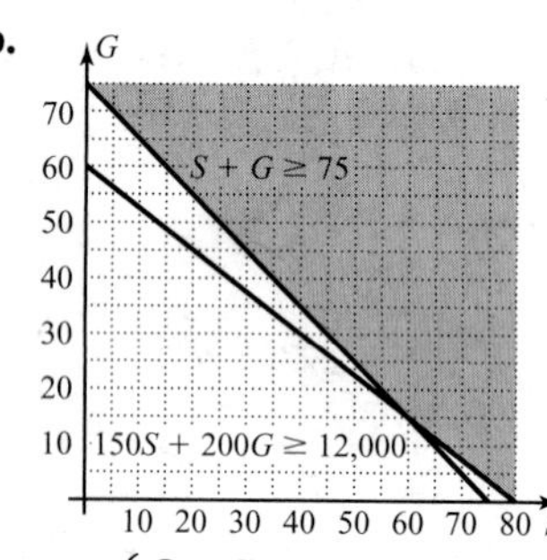

c. Answers may vary. Two examples are (20, 70) and (40, 50).

45. a. $\begin{cases} C + B \leq 5000 \\ 0.04C + 0.05B \geq 100 \end{cases}$

b.

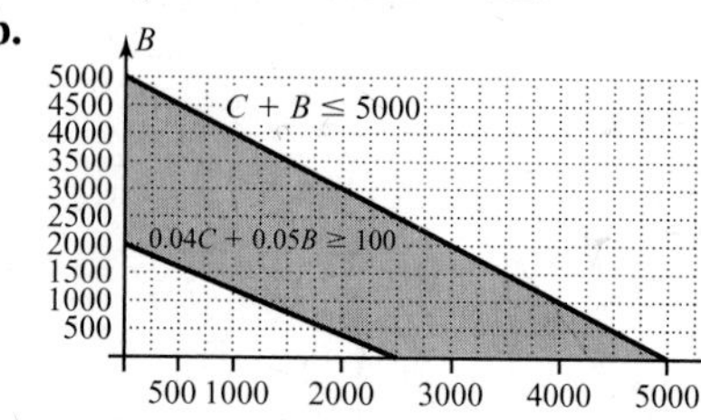

c. Answers may vary. Two examples are (2000, 3000) and (3500, 1000).

Review Exercises

1. 81 **2.** −100 **3.** $-30x + 24$ **4.** $-x - 14$ **5.** 4 **6.** 6

Chapter 4 Review Exercises

1. false **2.** true **3.** true **4.** true **5.** false **6.** Replace **7.** slope, y-intercepts **8.** the intersection (overlap) of the two solution sets **9.** the coefficient portion of the augumented matrix has 1s on the diagonal from upper left to lower right and 0s below the 1s **10.** determinant **11.** yes **12.** no **13. a.** consistent with independent equations **b.** one solution **14. a.** inconsistent with independent equations **b.** no solution **15. a.** consistent with independent equations **b.** one solution **16. a.** consistent with dependent equations **b.** infinite number of solutions **17.** (−1, −4) **18.** (3, −1) **19.** no solution **20.** all ordered pairs along $y = \frac{3}{2}x - 3$ **21.** (4, −1) **22.** $\left(5, -\frac{2}{5}\right)$ **23.** all ordered pairs along $y = \frac{4}{3}x + 2$ **24.** no solution **25.** (1, 3) **26.** (2, −1) **27.** (1, 5) **28.** (5, −3) **29.** (1, 0, −3) **30.** no solution **31.** (3, −3, 2) **32.** (4.2, 0, 1.8) **33.** (4, −1) **34.** (−3, −2) **35.** (−2, 3, 1) **36.** (−1, 2, −4) **37.** (−2, 9) **38.** no solution **39.** (2, 4, −1) **40.** (−2, 4, −2) **41.**

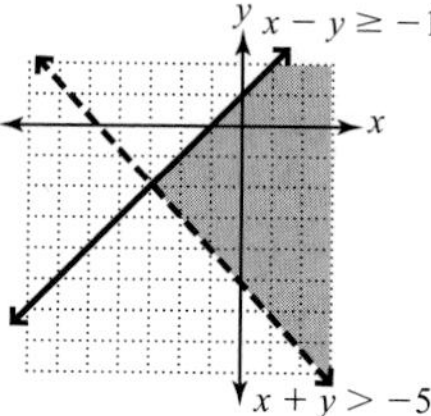

42.

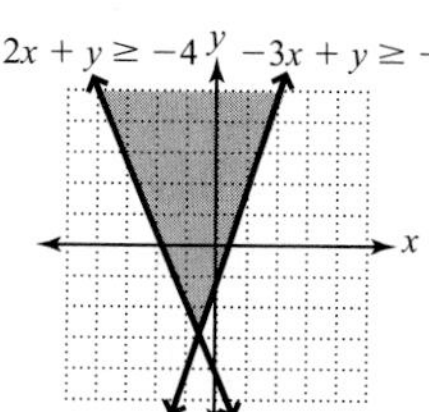

43.

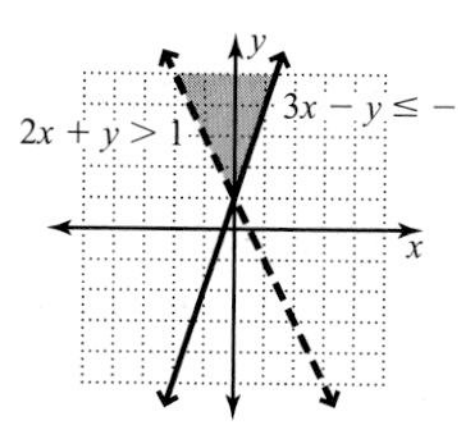

44.

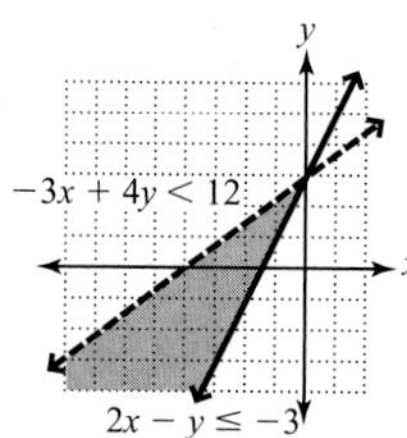

45. 10, 6 **46.** 47 adult, 92 children

47. domestic: \$490 billion, international: \$66 billion **48.** Western Europe: $15\frac{2}{3}\%$, North America: $27\frac{1}{3}\%$ **49.** 75°, 15° **50.** \$3500 at 7% and \$1500 at 9% **51.** 5:45 P.M. **52.** 120 miles per hour **53.** 30 ml **54.** 1 vitamin supplement, 3 rolls of film, 4 bags of candy **55.** 4 first place, 3 second place, 9 third place **56. a.**

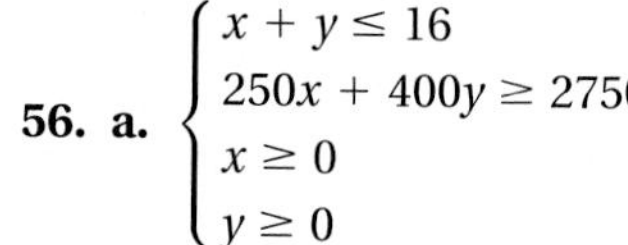

$$\begin{cases} x + y \leq 16 \\ 250x + 400y \geq 2750 \\ x \geq 0 \\ y \geq 0 \end{cases}$$

b.

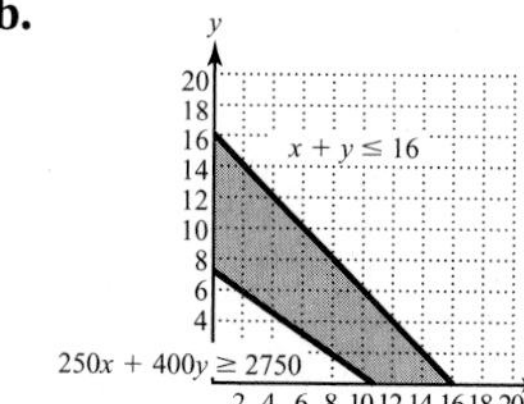

c. Answers may vary. One example is (8, 4).

Chapter 4 Practice Test **1.** yes **2.** no **3.** $(-2, 1)$ **4.** $(8, -1)$ **5.** $(3, 1)$ **6.** $(-4, -2)$ **7.** all ordered pairs along $2x + 3y = 1$ **8.** $(-2, 3, -2)$ **9.** $(-4, 2, 3)$ **10.** $(2, -4)$ **11.** $(-1, 5, 2)$ **12.** $(-2, 5)$ **13.** $(-4, 0, 3)$

14.

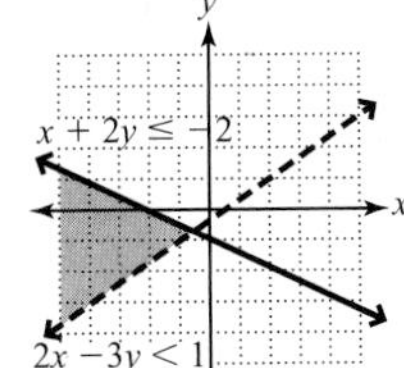

15. 86 accountants; 77 waiters/waitresses

16. 180 **17.** 7 mph **18.** \$4000 at 6%, \$8000 at 8% **19.** children: 150, student: 300, adult: 350

20. a. $\begin{cases} 2l + 2w \le 200 \\ l \ge w + 10 \\ l \ge 0 \\ w \ge 0 \end{cases}$ **b.**

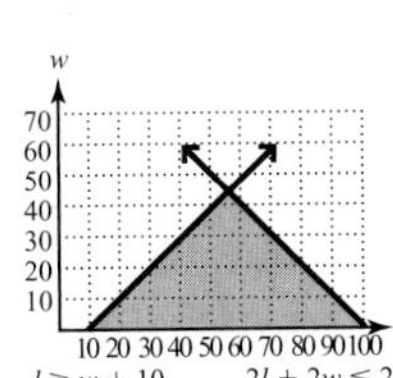

c. Answers may vary. One example is (60, 20).

Chapter 5

5.1 Exercises

1. Keep the same base and add the exponents. **3.** The result will be positive because $5^{-2} = \frac{1}{5^2}$, which is positive. **5. 1.** Locate the new decimal position, which will be to the right of the first nonzero digit in the number. **2.** Determine the power of 10. The power will be the number of digits between the old decimal position and the new decimal position expressed as a positive power. **7.** 1 **9.** 4 **11.** 1 **13.** -16 **15.** $\frac{1}{8}$ **17.** $-\frac{1}{81}$ **19.** 216 **21.** $4x^3$ **23.** $\frac{a^6}{4}$ **25.** Mistake: Assigned the minus sign to the base. Correct: $-(2 \cdot 2 \cdot 2 \cdot 2) = -16$ **27.** Mistake: multiplied 4 by -1. Correct: $\frac{1}{4}$ **29.** m^4n^3 **31.** 3^{12} **33.** $(-3)^8$ or 3^8 **35.** $-12p^6q^5$ **37.** $-18r^6s^4t^{12}$ **39.** $3.72u^7t^{11}$ **41.** h^2 **43.** $\frac{1}{a^5}$ **45.** $2xy^3$ **47.** $-\frac{3s^3}{2r}$ **49.** $-\frac{5w^6}{7u^{11}}$ **51.** $\frac{4b^7c^3}{a^5}$ **53.** x^{12} **55.** $16x^{20}$ **57.** $25x^6y^2$ **59.** $\frac{27}{64}a^6b^{12}$ **61.** $-0.027r^6t^{12}u^3$ **63.** $64x^6y^{14}$ **65.** $\frac{9q^2}{125p^2}$ **67.** $\frac{9}{h^2t^2}$ **69.** $\frac{1}{144u^{19}v^{10}}$ **71.** $\frac{9}{2u^4v^{12}}$ **73.** $-\frac{1}{54a^{18}b^{20}c}$ **75.** 2,900,000 **77.** 200,000,000,000 **79.** \$46,500,000,000 **81.** 0.00000017 **83.** 0.000000000000000000000001675 **85.** 0.00000000000000000000000000006645 **87.** 5.015×10^7 **89.** 7.879×10^{12} **91.** 6×10^{11} **93.** 5.5×10^{-7} **95.** 1×10^{-7} **97.** 7.3×10^{-15} **99.** $6 \times 10^5, 7.4 \times 10^6, 8.3 \times 10^6, 1.2 \times 10^7, 2.4 \times 10^8$ **101.** 6.3×10^7 **103.** -1.312×10^{10} **105.** 1.782×10^2 **107.** 4×10^4 **109.** -3.1×10^8 **111.** 6×10^2 **113.** 2.7×10^{13} **115.** 8×10^{21} **117.** 2.5×10^{-7} **119.** 6.25×10^{10} **121.** 2.78×10^{-19} J **123.** 2.11×10^{14} Hz **125.** 3.78×10^5 J

Review Exercises **1.** Change the operation from subtraction to addition and the second number, -6, to its additive inverse, 6. **2.** Associative property of addition. **3.** $-8n + 32$ **4.** $3x + 21y - 17$ **5.** $13x^2 - 13x + 9$ **6.** $-3u + 12$

5.2 Exercises

1. Add the exponents of all the variables in the monomial. **3.** The degree of a polynomial is the greatest degree of any of the terms in the polynomial. **5.** To add polynomials, combine like terms. **7.** The degree of a quadratic function is 2. The graph of a quadratic function is a parabola. **9.** $d = 4$; monomial **11.** $d = 2$; binomial **13.** $d = 3$; trinomial **15.** $d = 4$; none of these names **17.** $d = 3$; trinomial **19.** $d = 0$; monomial **21.** $d = 3$; binomial **23.** $7x^2 + 4x - 2$ **25.** $2p^4 - 5p^3 + 6$ **27.** $-u^3 - 9u^2 + 2u + 6$ **29.** $\frac{4}{3}u^4 + \frac{1}{2}u^3 + 2u^2 + 3u + 2$ **31.** $7.3t^4 + 1.5t^3 - t^2 + 2.7t + 7$ **33.** $-a^3 - a^2 + 2a - 1$ **35.** $3m^4 - 4m^3 + m^2 + m - 2$ **37.** $-2r^3 - 2r^2 - r + 1$ **39.** $\frac{3}{5}y^3 - y^2 + \frac{25}{4}y - \frac{1}{7}$ **41.** $-8.3w^3 - 3.7w^2 - 0.7w + 6.7$ **43.** $-8w^4 + 3w^3 - 7w^2 + 13w - 9$ **45.** $8x^4 + 4x^3 + 5x^2 - 12x + 13$ **47.** $-g^4 + \frac{4}{5}g^3 - \frac{2}{3}g^2 + \frac{9}{5}$ **49.** $3y^2 + 4y + 1$ **51.** $-19a^3 - 12a^2 + 9a + 9$ **53.** $14a^2 - 5ab + 6b^2$ **55.** $2x^3y^4 + 5x^2y^3 + 7xy - 8$ **57.** $4x^2y^2 + 6xy^2 - 16x^2y - 3xy + y^2 - 13$ **59.** $4x + 5$ **61.** $8a + 4$ **63.** constant **65.** quadratic **67.** linear **69.** cubic

71.

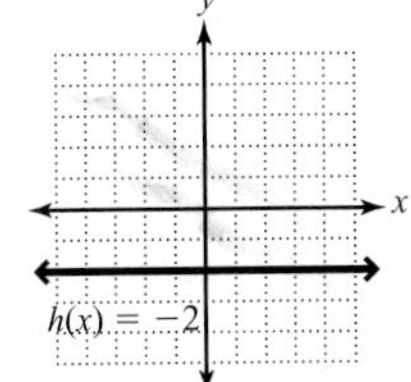

73.

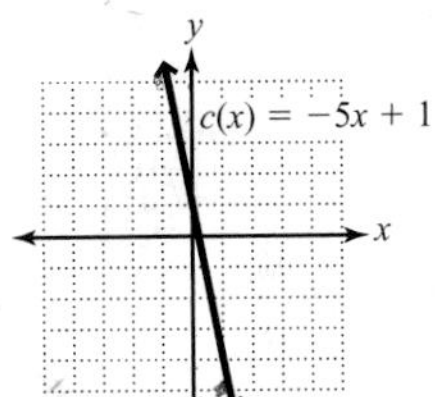

75.

y
x
g(x) = x² + 3x − 4

77.

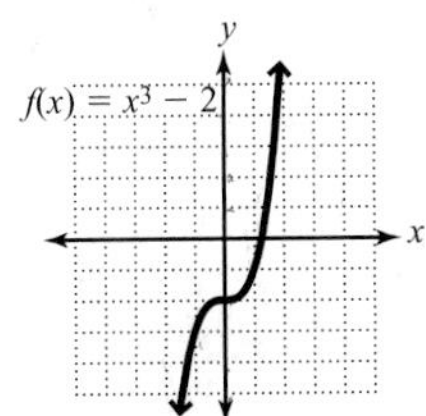

79. a. -1 **b.** -7 **81. a.** $-5x + 2$ **b.** $-3x + 4$ **83. a.** $4x^2 + 1$ **b.** $-2x^2 - 3$ **85. a.** $-x^3 + 2x + 7$ **b.** $3x^3 + 2x + 3$ **87. a.** $P(x) = 0.2x^2 - 17x + 1000$ **b.** \$184,000 **89.** 196 ft. **91.** 58.7 V **93. a.**

0	33.9
2	26.4
4	22.0
6	20.6
8	22.2
10	26.9
12	34.6
14	45.4

b.

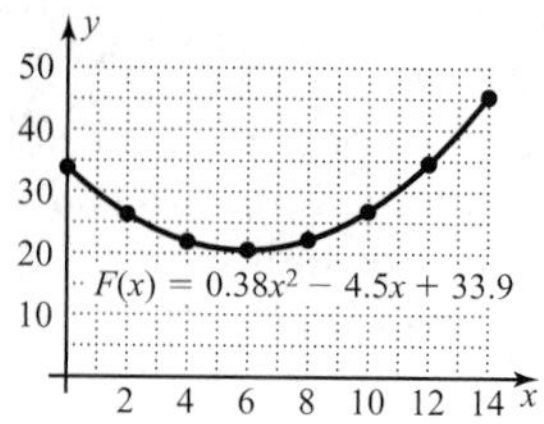

c. 95.9

95. a.

0	915.0
10	823.5
20	713.0
30	604.5
40	519.0
50	477.5

b.

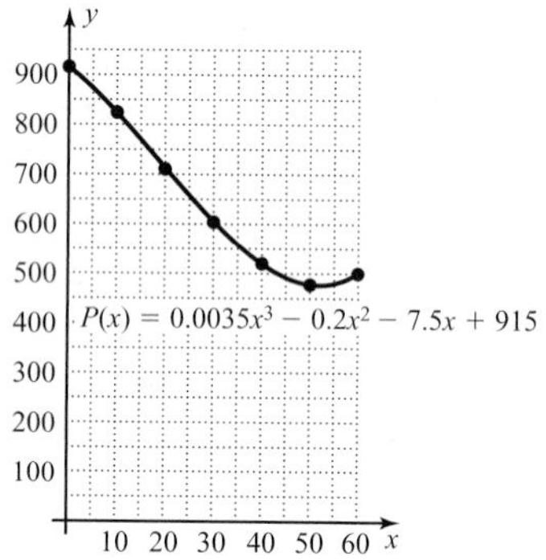

c. 501 million

Review Exercises

1. $24x - 72$ **2.** $-12y - 42$ **3.** $\frac{2}{5}$ **4.** $x > -\frac{29}{4}$ **5. a.** $\{x \mid -4 < x \leq 6\}$ **b.** $(-4, 6]$

c.

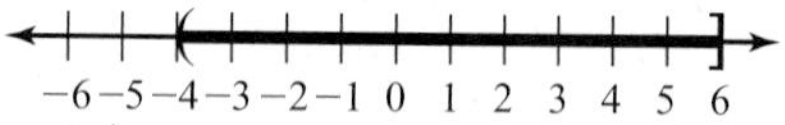

6. a. $\{x \mid x \leq 1 \text{ or } x \geq 6\}$ **b.** $(-\infty, 1] \cup [6, \infty)$

c.

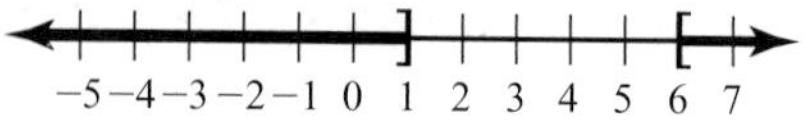

5.3 Exercises

1. We apply the distributive property. **3.** Multiply both terms in the second binomial by both terms in the first binomial (FOIL), then combine like terms. **5.** The product is a difference of squares. **7.** $5x^5 + 15x^4 - 10x^3$ **9.** $-12x^5 - 24x^4 + 6x^3$ **11.** $4n^7 + 28n^6 - 8n^5 - 12n^4$ **13.** $27a^6b^6 - 18a^3b^{12}$ **15.** $\frac{1}{2}m^3n^6p^3 - \frac{5}{4}m^4n^3p^3 + 2m^4n^4p^5$ **17.** $-2.4p^5q^9 + 0.6p^2q^3 - 2.1p^8q^2$ **19.** $-20x^7y^3 - 10x^6y^6 + 25x^{13}y^7$ **21.** $-6r^6s^4 + 18r^4s^3 + 9r^3s^2 - 9r^2s$ **23.** $-0.42a^3b^2 + 1.2a^3b^3 - 0.6a^4b^3 + 0.2a^4b^4$ **25.** $3a^3b^3c^8 - \frac{4}{3}a^3b^2c^7 + 4a^2bc^6$ **27. a.** $x + 5$ **b.** $2x$ **c.** $(x + 5)(2x)$ **d.** $2x^2 + 10x$ **e.** They are equivalent because they both describe the area of the shape. **29.** $6x^2 + 17x + 12$ **31.** $15x^2 + x - 2$ **33.** $6y^2 - 19xy + 10x^2$ **35.** $15m^2 + 11mn - 12n^2$ **37.** $t^4 - 7t^2 + 10$ **39.** $a^4 + 5a^2b^2 - 6b^4$ **41. a.** $x + 3$ **b.** $x + 2$ **c.** $(x + 3)(x + 2)$ **d.** $x^2 + 5x + 6$ **e.** They are equivalent because they both describe the area of the shape. **43.** $12x^3 - 10x^2 + 5x - 1$ **45.** $21a^3 - 17a^2 - 12a + 2$ **47.** $14c^3 - 24c^2 - 29c - 6$ **49.** $12p^3 + 38p^2q + 24pq^2 + 10q^3$ **51.** $12y^3 - 22y^2z + 14yz^2 - 12z^3$ **53.** $9u^4 - 4u^2 - 4u - 1$ **55.** $4x^2 - 49$ **57.** $4q^2 - 9$ **59.** $x^4 - 4y^2$ **61.** $s^2 - t^2 + 2s + 1$ **63.** $9b^2 - c^2 - 4c - 4$ **65.** $x - 8$ **67.** $3m - 2n$ **69.** $2c + 3d$ **71.** $-2j + 5k$ **73.** $x^2 + 2xy + y^2$ **75.** $16t^2 + 24tw + 9w^2$ **77.** $16w^2 - 24w + 9$ **79.** $25y^6 - 90y^3 + 81$ **81.** $x^2 + y^2 - 2xy + 2x - 2y + 1$ **83.** $p^2 + q2 - 2pq - 10p + 10q + 25$ **85.** $3xyz + y^2z$ **87.** $h^2 + \frac{7}{2}h$ **89.** Mistake: Combined like terms incorrectly. Correct: $6x^2 - 37x + 45$. **91.** Mistake: Multiplied $2x$ by $2x$ incorrectly. Correct: $4x^2 - 49$. **93.** $-4v^4w^5 - 16v^8w^2x^2 + 20v^2w^3x$ **95.** $64r^4 - 9s^2$ **97.** $4u^4 + 12u^2v + 9v^2$ **99.** $2a^4 - 13a^2b^2 + 20b^4$ **101.** $15q^4 + 11tq^2 - 12t^2$ **103.** $3r^6s^3 - \frac{1}{3}r^5s^4 - \frac{1}{2}r^4s^5 + r^3s^4 - 3r^2s^3$ **105.** $-0.35t^4r^6 + 0.8t^5r^5 - 0.22t^5r^4 + 0.2t^3r^4$ **107.** $9m^2 - n^2 - 24m + 16$ **109.** $4x^4 + 14x^3y + 9x^2y^2 - xy^3 - y^4$ **111.** $x^4 - 2x^2 + 1$ **113. a.** $(f \cdot g)(x) = 2x^3 - 5x^2 + 20x - 9$ **b.** -297 **c.** 27 **115. a.** $(f \cdot g)(x) = 12x^3 + 7x^2 - 6x + 8$ **b.** 1653 **c.** -235 **117.** $f(n + 1) = n^2 + 5n + 3$ **119.** $f(t - 6) = 2t^2 - 24t + 71$ **121.** $l^2 - 2l$ **123.** $4x^3 + 24x^2 + 20x$ **125.** $9w^3 + 15w^2$ **127.** $\pi r^3 + 3\pi r^2$

Review Exercises

1. $\frac{15}{2}$ **2.** $t = \frac{d + n}{30}$ **3.** $y = -\frac{1}{5}x + \frac{22}{5}$ **4.** 41 small boxes, 43 large boxes. **5.** -8.946×10^{11} **6.** 6.25×10^8

5.4 Exercises

1. To divide a polynomial by a monomial, divide each term in the polynomial by the monomial. **3.** Divide $6x^2$ by $2x$. **5.** Write a sum of the quotient and the remainder over the divisor, as in $3x^2 - 5x + 2 + \frac{7}{6x - 1}$. **7.** $4a^2 - a + 3$ **9.** $4u^3 - 2u^2 - 5u + 1$ **11.** $3a^4b^3 - a^3b + 2a$ **13.** $\frac{12u^2}{v^5} + \frac{4}{v^4} - \frac{5u}{v^7}$ **15.** $-6x^2y + 9y^2 + \frac{1}{5}$ **17.** $-3t^3u^3v + 2tu^2 + 4 - \frac{3}{4t^2}$ **19.** $2abc + \frac{4}{3}b^2 - \frac{1}{3c}$ **21.** $\frac{5y^2}{2}$ **23. a.** $2x - 4 + \frac{6}{x}$ **b.** area: 108, length: 15, width: 7.2 **25.** $x + 5$ **27.** $3p + 1$ **29.** $5n - 8$ **31.** $5x + 4$ **33.** $2y - 1$ **35.** $x^2 - 5x + 6$ **37.** $x^2 - 2x + 4$ **39.** $y^3 + 2y^2 + 4y + 8$ **41.** $4z^2 - 10z + 25$ **43.** $v^2 + 3v + 2$ **45.** $7a^2 - 2a + 3$ **47.** $6q^2 + 5q + 9 + \frac{18}{2q - 1}$ **49.** $3x^2 - 6x + 5$ **51.** $c^3 - 2c^2 + c - 4$ **53.** $5u^2 - 11u + 32 - \frac{62}{u + 2}$ **55. a.** $t^2 + 3t - 4$ **b.** $h = 4$ ft., $l = 6$ ft., $V = 144$ ft.^{3} **57. a.** $(f/g)(x) = 4x^2 - 7x - 5$ **b.** 31 **c.** 25 **59. a.** $(h/k)(x) = 3x^2 + 5x - 2$ **b.** 40 **c.** -4 **61. a.** $(n/p)(x) = x^2 + 2x - 1 + \frac{1}{x - 3}$ **b.** 24 **c.** $-\frac{6}{5}$

Review Exercises **1.** $\frac{7}{4}$ **2.** $x \geq \frac{4}{3}$ **3.** $2x + 5y = 1$ or $y = -\frac{2}{5}x + \frac{1}{5}$ **4.** Yes, because every value in the domain corresponds to only one value in the range (it passes the vertical line test) **5.** $(7, -1, 4)$ **6.**

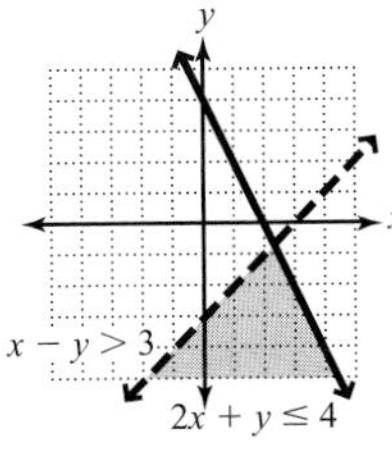

5.5 Exercises

1. $x - c$ **3.** 1, 0, −5, −3, 11 **5.** It is the last number in the last row. **7.** Given a polynomial $P(x)$, the remainder of $\frac{P(x)}{x - c}$ is equal to $P(c)$. **9. a.** $x + 3$ **b.** $x^2 + 7x + 12$ **c.** $x + 4$ **11. a.** $x - 3$ **b.** $x^3 + 4x^2 - 25x + 7$ **c.** $x^2 + 7x - 4 - \frac{5}{x - 3}$ **13. a.** $x - 4$ **b.** $2x^4 - 8x^3 - 5x^2 + 17x + 10$ **c.** $2x^3 - 5x - 3 - \frac{2}{x - 4}$ **15.** $x - 2$ **17.** $2x + 7 + \frac{23}{x - 4}$ **19.** $x^2 + x - 3$ **21.** $3x^2 + 4x + 1$ **23.** $3x^2 - 10x + 8$ **25.** $2x^2 - 9x + 17 - \frac{28}{x + 2}$ **27.** $x^2 + 1 - \frac{1}{x - 2}$ **29.** $2x^2 - 5x + 15 - \frac{49}{x + 3}$ **31.** $3x^2 + 5x + 15 + \frac{47}{x - 3}$ **33.** $x^2 + 2x - 3$ **35.** $x^2 + 3x + 9 + \frac{54}{x - 3}$ **37.** $x^3 - 2x^2 + 4x - 8 + \frac{32}{x + 2}$ **39.** $6x^2 - 2x - \frac{1}{3} + \frac{\frac{10}{9}}{x + \frac{1}{3}}$ **41.** 5 **43.** 0 **45.** −5 **47.** −3 **49.** 2 **51.** −3 **53. a.** $9x + 8$ **b.** length: 116 in., height: 10 in., area: 1160 in.^{2} **55. a.** $2y - 11$ **b.** length: 14 ft., width: 8 ft., height: 9 ft., volume: 1008 ft.^{3}

Review Exercises **1.** {2, 3, 5, 7, 11} **2. a.** $\{x | x \text{ is a real number}\}$ **b.** $(-\infty, \infty)$ **c.**

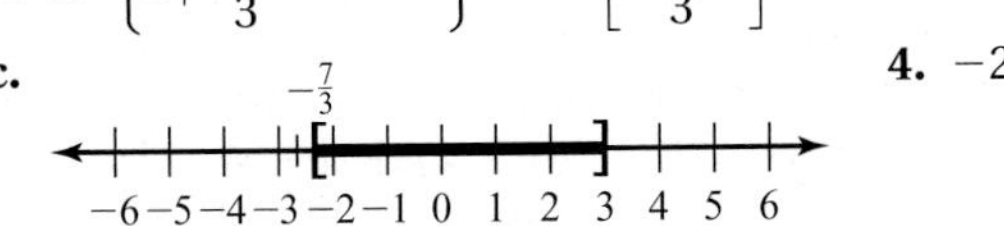

3. a. $\left\{x \middle| -\frac{7}{3} \leq x \leq 3\right\}$ **b.** $\left[-\frac{7}{3}, 3\right]$ **c.**

$-\frac{7}{3}$

−6 −5 −4 −3 −2 −1 0 1 2 3 4 5 6

4. −2 **5.**

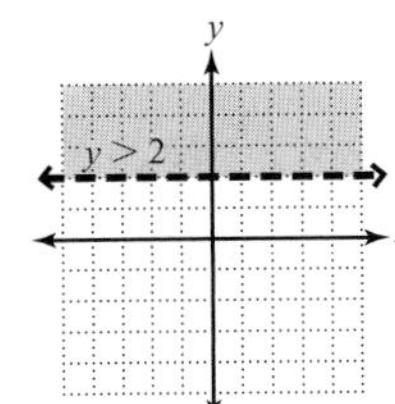

6. $\begin{cases} l = 2w \\ 100 = 2l + 2w \end{cases}$ $w = 16\frac{2}{3}$ ft., $l = 33\frac{1}{3}$ ft.

Chapter 5 Review Exercises **1.** true **2.** true **3.** false **4.** false **5.** false **6.** false **7.** monomial **8.** $m - n$ **9.** $a \cdot b$ **10.** distributive **11.** $\frac{27}{8}$ **12.** −16 **13.** $\frac{1}{32}$ **14.** 14 **15.** 0.0000000000000000000000016736 **16.** 16,500,000,000 **17.** 7.53×10^{-10} **18.** 3×10^8 **19.** $20x^3y^3$ **20.** $-2m^3n^9$ **21.** $\frac{3}{x^5}$ **22.** $-\frac{7a^2}{3c^6}$ **23.** $9m^{10}n^2$ **24.** $\frac{1}{48m^{20}n^9p}$ **25.** $0.009j^7k^{19}$ **26.** $\frac{1}{18s^{37}t^{11}}$ **27.** $\frac{1}{75m^{15}n^6}$ **28.** $\frac{b^{12}}{8a^3}$ **29.** -1.02×10^{11} **30.** 4.06×10^{-3} **31.** $d = 3$; none of these **32.** $d = 2$; binomial **33.** $d = 5$; monomial **34.** $d = 5$; none of these **35.** $3c^3 + 10c^2 - 11$ **36.** $6y^2 + 14$ **37.** $x^2y^3 - 3x^2y^2 + xy^2 + 6x^2y - 5xy + 13y^2 - 5$ **38.** $-hk - 3k - 6k^3$ **39.** linear **40.** quadratic **41.** cubic **42.** constant **43.**

$f(x) = x^2 - 3$

44.

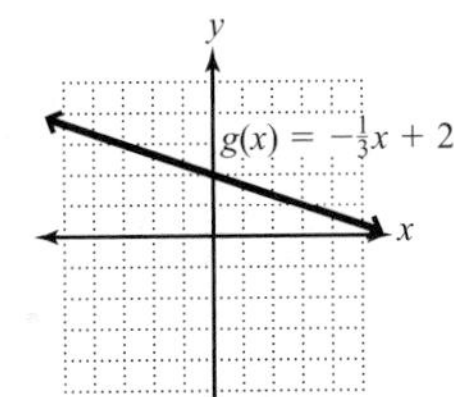

45.

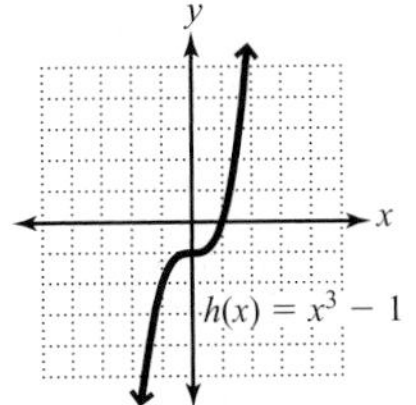

46.

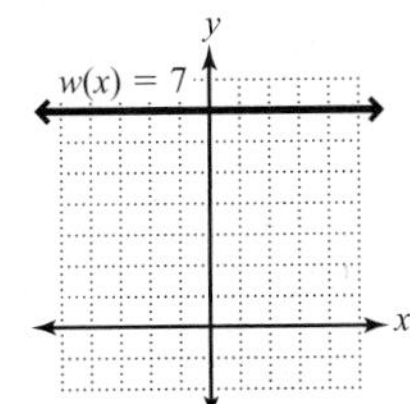

47. a. $(f + g)(x) = -x^3 + 2x + 7$ **b.** $(f + g)(2) = 3$ **c.** $(f - g)(x) = 3x^3 + 2x + 3$ **d.** $(f - g)(-4) = -197$ **48.** $6x^4y^3 - 9x^3y^3 + 3x^2y^3$ **49.** $-2p^3q + 3p^2q^2 - 3pq^3$ **50.** $2x^2 + 2x - 24$ **51.** $27w^2 + 15w - 2$ **52.** $18r^2 + 15rs - 7s^2$ **53.** $x^4 + x^2 - 2$ **54.** $2x^4 - 5x^3 + 7x^2 + 3x - 7$ **55.** $12t^4 - t^3 + 2t - 1$ **56.** $9a^2 - 25$ **57.** $4p^2 - 4p + 1$ **58.** $64k^2 + 48k + 9$ **59.** $16h^4 - 49$ **60.** $5m + 2$ **61. a.** $(f \cdot g)(x) = 2x^3 + 15x^2 + 26x - 7$ **b.** $(f \cdot g)(-1) = -20$ **62.** $f(x - 4) = 4x^2 - 33x + 68$ **63.** $4m^3 - m^2 + 3m - 1$ **64.** $4x - 1 + \frac{3}{2x}$ **65.** $x + 3 + \frac{1}{x + 2}$ **66.** $x - 2$ **67.** $2x + 3$

68. $3x^3 + x^2 - 2x - 4 - \frac{9}{3x - 1}$ **69.** $5x + 2 + \frac{4}{x - 1}$
70. $2x^2 - 5$ **71.** $4x^2 + 8x + 14 + \frac{33}{x - 2}$ **72.** $x^2 + 2x + 4$
73. 15 **74.** −51 **75.** 4.04186×10^{-19} J **76.** 56 ft.
77. a. $12a + b$ **b.** 153 cm **78. a.** $P(x) = 0.2x^2 - 9x - 1000$
b. \$5200 **79.** $w^3 + 15w^2 + 50w$ **80. a.** $y^3 + 2y^2 + 4y + 1$
b. 58 amps

Chapter 5 Practice Test

1. 0.0072 **2.** 3.57×10^{-3}
3. $15x^5y$ **4.** $\frac{1}{9x^8y^2}$ **5.** $\frac{4u^3}{v^4}$ **6.** $\frac{16}{81}t^{20}u^8v^4$ **7.** 1.26×10^{10}
8. 2×10^7 **9.** constant **10.** cubic
11.

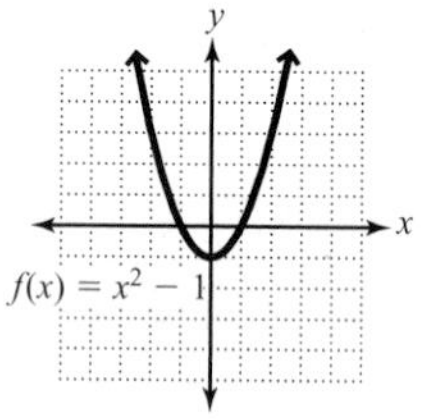

12. $-8ab$ **13.** $11r^5 - 7r^4 + 7r^3 - 10r^2 - r - 1$
14. $15x^2 - 14xy - 8y^2$ **15.** $12m^5n^9 - 8m^4n^{10} + 20m^3n^{11}$
16. $49k^2 - 28jk + 4j^2$ **17.** $14x^3 - 15x^2 - 19x + 15$
18. $16h^2 - 9$ **19.** $(f \cdot g)(x) = 15x^2 - 14x - 16$
20. $(f \cdot g)(2) = 16$ **21.** $4k^2 - 2k + 1$ **22.** $m^2 + 3m + 2$
23. $2x^2 - x + 3 - \frac{17}{x + 3}$ **24.** −8 **25. a.** $P(x) = 0.1x^2 - 14x + 2000$ **b.** \$88,000

Chapter 6

6.1 Exercises

1. No, factored form is a number or expression written as a product of factors. **3.** Four-term polynomials **5.** $4x^2y$ **7.** $4u^3v^6$ **9.** $2abc^4$ **11.** $a + b$
13. $5c^2(3c^2d - 4)$ **15.** $x^2(x^3 - x + 1)$ **17.** $25x(y - 2z + 4x)$
19. $-7uv(2uv + v - 1)$ **21.** $3a^2b(3a^5b^2 + a^2b - 2)$
23. $3wv(w^2v^3 + 13w + 6v)$ **25.** $6ab^2c(3b - 6a + 4a^4c^7)$
27. $-4xy(2x - 4y + 3)$ **29.** $(n - 3)(m + 4)$
31. $(b + 2c)(6 - a)$ **33.** $(x + y)(a + b)$ **35.** $(u + 3)(u^2 + 3)$
37. $(m + p)(4n - 3)$ **39.** $(c + 1)(d + 1)$ **41.** $(2a - 1)(a + 1)$
43. $(x + 2y)(3a + 4b)$ **45.** $(h + 8)(5h - k)$
47. $(x^2 + y^2)(3 - a)$ **49.** $(p - 2q)(3p^2 + 2q^2)$ **51.** $(x - 4y)(2x^2 - 3y^2)$ **53.** $2(b - c)(a + x)$ **55.** $2(3x + 2)(2y + 5)$
57. $3a(a + 3)(y - 4)$ **59.** $m(3m - 10)(m + 2n)$ **61.** $5t(t - 2)(3s - 1)$ **63.** $5y(x^2 + 1)(x - 4)$ **65.** Mistake: Did not factor out the GCF, which is $12x^2y$. Correct: $12x^2y(2y^2 + 3x)$
67. Mistake: Incorrect power of b in the parentheses. Correct: $9a^2b(b^2c - 2a^2)$ **69.** It is correct. **71.** Correct form: $2x(2x^2 + 7x + 4)$ **73.** $2x(3x + 20)$ **75. a.** $15x^2 + 26x$ **b.** $x(15x + 26)$
c. 213 ft.^{2} **77. a.** $36\pi r^3 + 18\pi r^2$ **b.** $18\pi r^2(2r + 1)$
c. ≈15,550.9 ft.^{3} **79. a.** $p - rp$ **b.** $p(1 - r)$ **c.** \$32.97
81. a. $p + prt$ **b.** $p(1 + rt)$ **c.** \$862.75 **83. a.** $2\pi r^2 + 2\pi rh$
b. $2\pi r(r + h)$ **c.** ≈1884.96 in.^{2}

Review Exercises

1. −24,500,000 **2.** 9.2×10^{-5}
3. $x^2 + 8x + 15$ **4.** $x^2 - 10x + 24$ **5.** $6x^2 + 19x - 7$
6. $8x^3 + 14x^2 - 30x$

6.2 Exercises

1. Both will be positive. **3.** Both will be negative. **5.** Find two factors of the product ac whose sum is b. **7.** $(r + 7)(r + 1)$ **9.** $(w + 1)(w - 3)$
11. $(a + 6)(a + 2)$ **13.** $(y - 9)(y - 4)$ **15.** $(m + 4)(m - 2)$
17. $(b - 10)(b + 4)$ **19.** $3s(t + 7)(t + 1)$
21. $5y(y - 12)(y - 1)$ **23.** $6au(u + 3)(u - 2)$ **25.** $(p + 9q)(p + q)$ **27.** $(u - 7v)(u - 6v)$ **29.** $(x - 7y)(x + 2y)$
31. $(a + 6b)(a - 7b)$ **33.** $(3a + 7)(a + 1)$ **35.** $(2w + 1)(w - 2)$ **37.** prime **39.** $(4q - 1)(q - 2)$ **41.** $(3b - 1)(2b + 3)$
43. $(4m + 3)^2$ **45.** $(4x - 3)(x + 2)$ **47.** $(2w + v)(w + 7v)$
49. $(5x - y)(x - 3y)$ **51.** $(2x - y)(8x - 3y)$
53. $(3t - 2u)(t + 7u)$ **55.** $(3m + 2n)(m - 4n)$ **57.** $(2a - 9b)(a - 2b)$ **59.** $2m(11m + 1)(m + 9)$ **61.** $2v(2u - 5v)(u + 3v)$ **63.** $(3y + 1)(y + 5)$ **65.** prime **67.** $(3t - 2)(t - 5)$
69. $(3x - 5)(2x + 3)$ **71.** $4y(4x - 3)(2x + 3)$
73. $5b(3a + b)(a - 2b)$ **75.** $(x^2 + 3)(x^2 - 2)$
77. $(4r^2 + 3)(2r^2 - 1)$ **79.** $(5x^2 - 2)(3x^2 - 1)$
81. $(y^3 - 12)(y^3 - 4)$ **83.** $(7x + 8)(x + 2)$
85. $(3a + 8)(a - 2)$ **87.** Mistake: The signs are incorrect. Correct: $(x + 2)(x - 3)$ **89.** Mistake: The GCF monomial was not factored out. Correct: $4(x + 2)^2$ **91.** 8, 10, 17 **93.** 2, 18, 62 **95.** Answers may vary (8, 14, 18, or 20 are possible answers). **97.** Answers may vary (2, 6, 12, and 20 are a few of the possible answers).

Review Exercises

1. $4x^2 + 12x + 9$ **2.** $16y^2 - 1$
3. $n^3 - 8$ **4.** $6x^2 + x + 4 + \frac{2}{x - 1}$ **5. a.** $\left\{x \middle| x \geq \frac{13}{2}\right\}$
b. $\left[\frac{13}{2}, \infty\right)$ **c.**

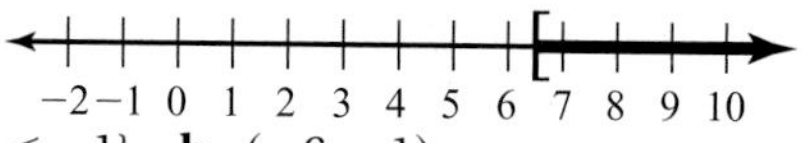

6. a. $\{x \mid -6 < x < -1\}$ **b.** $(-6, -1)$
c.

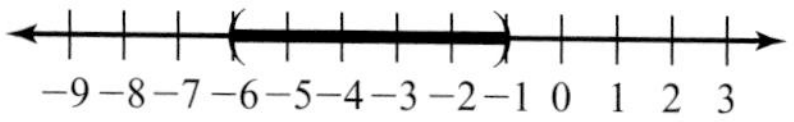

6.3 Exercises

1. The pattern in these perfect square trinomials is that the first and last terms are squares and the middle term is twice the product of the square roots of the first and last terms. **3.** The minus sign is placed in the binomial factor. **5.** They are the cube roots of a^3 and b^3. **7.** A sum of squares cannot be factored. **9.** $(x + 5)^2$ **11.** $(b - 2)^2$
13. $(5u - 3)^2$ **15.** $(n + 12m)^2$ **17.** $(3q - 4p)^2$ **19.** $(2p - 7q)^2$ **21.** $(4a + y)(4a - y)$ **23.** $(5x + 2)(5x - 2)$ **25.** $(10u + 7v)(10u - 7v)$ **27.** $(3x - 5)(3x - 13)$ **29.** $9(x + 2)(x - 2)$
31. $(x^2 + 4)(x + 2)(x - 2)$ **33.** $(3x - 4)(9x^2 + 12x + 16)$
35. $(m - 3)(m^2 + 3m + 9)$ **37.** $8(2a - b)(4a^2 + 2ab + b^2)$
39. $(4 - cd)(16 + 4cd + c^2d^2)$ **41.** $m^3(3 - 5mn)(9 + 15mn + 25m^2n^2)$ **43.** $(3 - a - b)(9 + 3a + 3b + a^2 + 2ab + b^2)$
45. $(5x + 3)(25x^2 - 15x + 9)$ **47.** $(u + 5v)(u^2 - 5uv + 25v^2)$
49. $(2p + qz)(4p^2 - 2pqz + q^2z^2)$ **51.** $8(x + 2y)(x^2 - 2xy + 4y^2)$ **53.** $m^3(2y + 3mz)(4y^2 - 6ymz + 9m^2z^2)$
55. $(u + 5)(u^2 + 4u + 7)$ **57.** $(2a - b + 3)(2a - b - 3)$
59. $(4z + 3x - 3y)(4z - 3x + 3y)$ **61.** $(4d - x - y)(16d^2 + 4dx + 4dy + x^2 + 2xy + y^2)$ **63.** $(4x + 3y + 3z)(16x^2 - 12xy - 12xz + 9y^2 + 18yz + 9z^2)$ **65.** 40 **67.** 36 **69.** 16 **71.** 16
73. $3abc(4a^2b + abc + 3c^2)$ **75.** $(x + 5)(x + 3)$

77. $2(x+4)(x-4)$ **79.** $(a-y)(x-y)$ **81.** $(3b+2)(2b-1)$ **83.** $(b+5)(b^2-5b+25)$ **85.** $(3x+2)(5x-1)$ **87.** $ab(1+6b)(1-6b)$ **89.** $(x^2+4)(x+2)(x-2)$ **91.** $7(u+3v)(u-5v)$ **93.** $(x+2)(x^2-5x+13)$ **95.** $5m(m^2+n^2)^2$ **97.** n^2 **99.** $t^3(4t-u)(16t^2+4tu+u^2)$ **101.** $(x-y)(x+y)(2x+3)$ **103.** $(6y+x-4)(6y-x+4)$ **105.** $4(3x^2-10)$ **107.** $9(2x^3-3)$ **109.** length: $3x-1$, width: $2x-3$ **111.** length: $3x+2$, width: $x+3$, height: $5x$

Review Exercises

1. $-\frac{13}{14}$ **2.** $-1, 6$ **3.** \$1500, \$2500, \$2000 **4.**

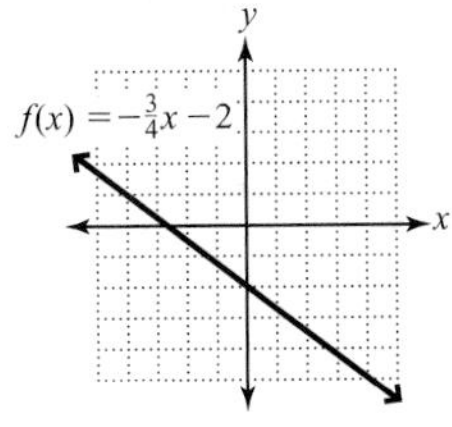

5.

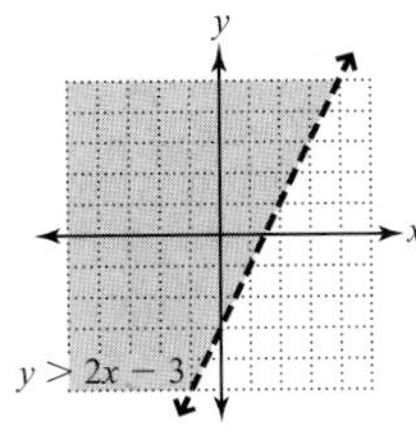

6. $3x+7y=-5$, $y=-\frac{3}{7}x-\frac{5}{7}$

6.4 Exercises

1. If a and b are real numbers and $ab = 0$, then $a = 0$ or $b = 0$. **3.** 2

5. (Answers may vary.)

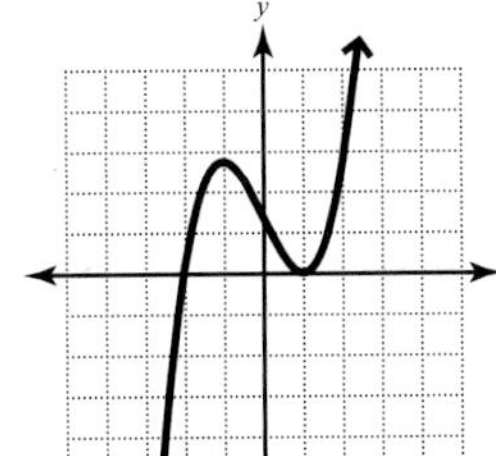

7. $-4, 0$ **9.** $-2, 1$ **11.** $-\frac{5}{2}, \frac{2}{3}$ **13.** $-\frac{5}{2}, 0, 3$ **15.** $-7, -2, 3$ **17.** $-\frac{7}{4}, -\frac{3}{2}, \frac{1}{3}$ **19.** $0, 4$ **21.** $-3, 3$ **23.** $-9, -5$ **25.** $-2, \frac{1}{2}$ **27.** $-\frac{2}{3}, 1$ **29.** -3 **31.** $0, \frac{1}{2}$ **33.** $-5, 5$ **35.** $-9, 3$ **37.** $1, 2$ **39.** $-\frac{5}{2}, -1$ **41.** $-2, 7$ **43.** $-\frac{7}{2}$ **45.** $-\frac{5}{2}, -\frac{3}{2}$ **47.** $-3, 0, 2$ **49.** $-2, 0, 1$ **51.** $-\frac{2}{3}, \frac{1}{2}$ **53.** $-\frac{2}{3}, 1$ **55.** $1, 2$ **57.** $x^2+x-6=0$ **59.** $3x^2-10x-8=0$ **61.** $x^3-2x^2-3x=0$ **63.** b **65.** a **67.** c **69.** $(-5, 0), (5, 0)$

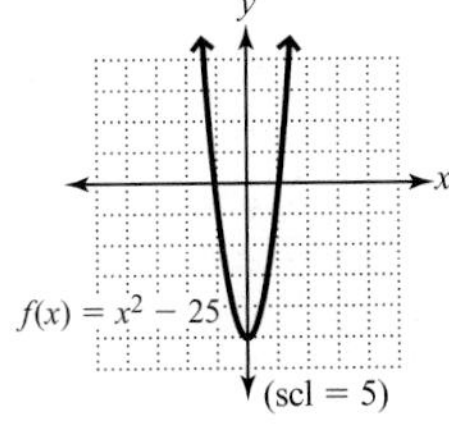

71. $(1, 0), (5, 0)$

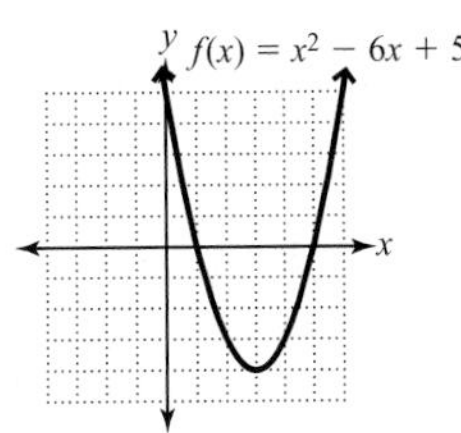

73. $(-4, 0), (0, 0), (1, 0)$

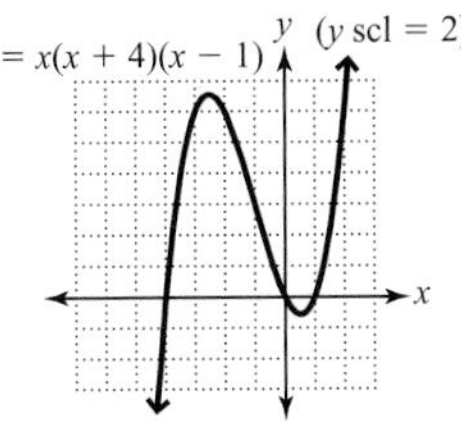

75. $(-2, 0), (0, 0), (3, 0)$

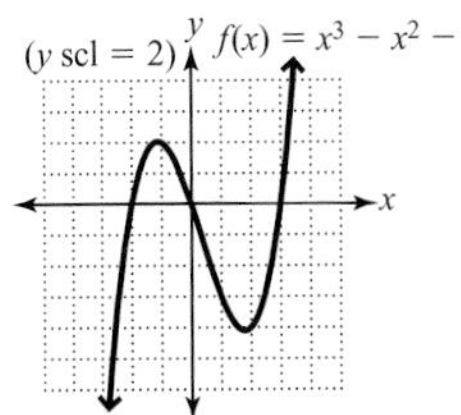

77. $(3, 0), (6, 0)$ **79.** $\left(-\frac{5}{3}, 0\right), \left(\frac{1}{2}, 0\right)$ **81.** $(-1, 0), (0, 0), (2, 0)$ **83.** $-5, 8$ **85.** $-0.25, 2.5$ **87.** $-4, 0, 3$ **89.** 6, 8, 10 **91.** 7, 24, 25 **93.** 5, 11 **95.** 18, 19 **97.** 16 m by 20 m **99.** 14 ft. **101.** 9 in. **103.** height: 24 ft., wire length: 26 ft. **105.** 1.25 sec. **107.** 10%

Review Exercises

1. [number line: −6 −5 −4 −3 −2 −1 0 1 2 3 4 5 6]

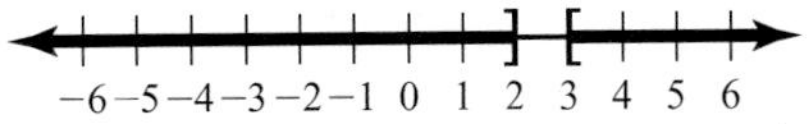

2.

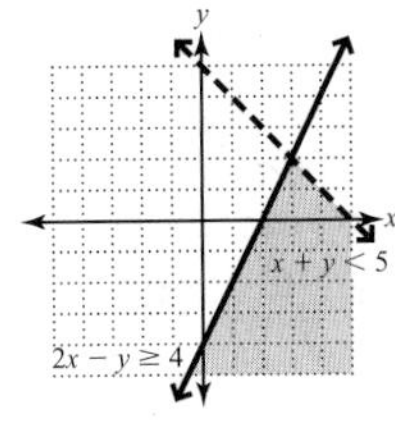

3. 4.203×10^{-4}

4. $x^2 - 3x + 2 + \frac{3}{x+4}$ **5.** no **6.** $(f+g)(x) = 5x^3 + 3x^2 - 27$

Chapter 6 Review Exercises

1. false **2.** false **3.** true **4.** true **5.** false **6.** false **7.** GCF (greatest common factor) **8.** b **9.** $(a+b)^2$ **10.** 0, 0 **11.** $8x^3y^2$ **12.** $5mn^2$ **13.** 1 **14.** $2(x+1)$ **15.** $u(u^4-u^2-1)$ **16.** $13d(d-2e)$ **17.** $2h^2k(2k^5-h^3)$ **18.** $2cd(6-2cd+5c^3d^3)$ **19.** $p^6q^2(16q-12p^2q^3+13p)$ **20.** $3wv^3(3w^5v^5+2v^3-4w^2)$ **21.** $(w+3)(17-m)$ **22.** $(x+3)(2y+1)$ **23.** $(n+3)(m+2)$ **24.** $(a+3)(a^2+2)$ **25.** $(x+y)(2-a)$ **26.** $(c^2+d^2)(b-5)$ **27.** $(y-s)(x^2-r)$ **28.** $(4k-1)(k+1)$ **29.** $2v(2u-1)(2v+5)$ **30.** $5d(c^2-d)(d+4)$ **31.** $(a-9)(a-1)$ **32.** $(m+3)(m+17)$ **33.** $(y+7)(y-6)$ **34.** $(x-10)(x+3)$ **35.** $(3x-7)(x+2)$

36. $(8h + 1)(2h + 1)$ **37.** prime **38.** $(3t + 5)(2t - 3)$ **39.** $(s - 10t)(s - t)$ **40.** $(3u + 2v)(2u + 3v)$ **41.** $(5m - n)(m - 3n)$ **42.** $(4x - 3y)(x + 2y)$ **43.** $b^2(b - 9)(b + 2)$ **44.** $2x(x - 10)(x + 2)$ **45.** $(2u + 5)(u + 2)$ **46.** $(3m - 2)(m - 2)$ **47.** $(u + 1)(10u - 3)$ **48.** $(3y + 1)(2y - 5)$ **49.** $(x^2 + 4)(x^2 + 1)$ **50.** $(3c^2 + 1)(c^2 + 4)$ **51.** $(2h^3 + 1)(h^3 + 4)$ **52.** $(3k - 2)(k + 2)$ **53.** $(x + 3)^2$ **54.** $(y + 6)^2$ **55.** $(m - 2)^2$ **56.** $(w - 7)^2$ **57.** $(3d + 7)^2$ **58.** $(2c - 7)^2$ **59.** $(h + 3)(h - 3)$ **60.** $(p + 8)(p - 8)$ **61.** $(3d + 2)(3d - 2)$ **62.** $(9k + 10)(9k - 10)$ **63.** $2(w + 5)(w - 5)$ **64.** $4(q + 3)(q - 3)$ **65.** $(5y + 3z)(5y - 3z)$ **66.** $(7cd + 10b)(7cd - 10b)$ **67.** $(c - 3)(c^2 + 3c + 9)$ **68.** $(v - 2)(v^2 + 2v + 4)$ **69.** $(2b - 1)(4b^2 + 2b + 1)$ **70.** $8(2d - c)(4d^2 + 2cd + c^2)$ **71.** $(3 + m)(9 - 3m + m^2)$ **72.** $(x + 10)(x^2 - 10x + 100)$ **73.** $(3b + 2a)(9b^2 - 6ab + 4a^2)$ **74.** $2(v + 3)(v^2 - 3v + 9)$ **75.** $(3d - 1)^2$ **76.** $3(m + n)(m - n)$ **77.** $(2a + 3)(a - 4)$ **78.** prime **79.** $3p^2(p + 6)(p - 5)$ **80.** $(x^2 + 4)(x + 2)(x - 2)$ **81.** $a^2bc^2(15abc^5 + 3b^3 + 5a^7c)$ **82.** $(w - 2 - y)(w^2 + 2w + wy + 4 + 4y + y^2)$ **83.** $-4, 1$ **84.** $-8, 8$ **85.** $-1, 3$

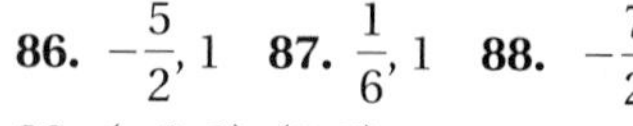

86. $-\frac{5}{2}, 1$ **87.** $\frac{1}{6}, 1$ **88.** $-\frac{7}{2}$

89. $(-2, 0), (2, 0)$

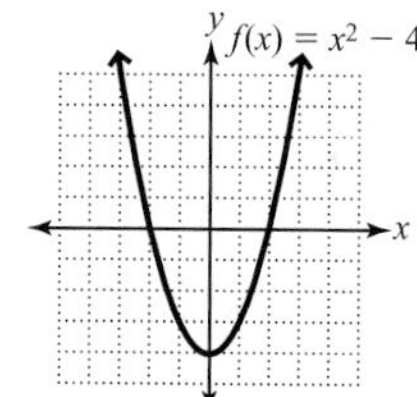

90. $(-3, 0), (0, 0), (2, 0)$

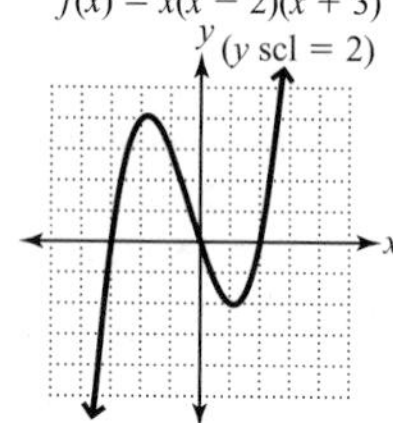

91. 4, 14 **92.** 10, 11 **93.** 16, 18 **94.** 8 ft. by 11 ft. **95.** height: 16 ft., base: 48 ft. **96.** 5, 12, 13 **97.** height: 17 ft., distance from ear: 37 ft. **98.** $2\frac{1}{4}$ sec.

Chapter 6 Practice Test

1. $7m^3n^2$ **2.** $3m(1 + 2m^2 - 3m^5)$ **3.** $2(m + n)$ **4.** $(3n + 4)(3n - 4)$ **5.** $(2x - 3)(4x^2 + 6x + 9)$ **6.** $(y + 7)^2$ **7.** $(q - 8)(q + 6)$ **8.** $3a(b^2 - 10b + 8)$ **9.** $5(1 + 5t)(1 - 5t)$ **10.** $(3d + 2)(2d - 1)$ **11.** $8(c + d)(c^2 - cd + d^2)$ **12.** $(w + 2)(w^2 + 3)$ **13.** $(s^2 + 9)(s + 3)(s - 3)$ **14.** $(5p - 12q)(p + q)$ **15.** $\frac{1}{2}, \frac{5}{4}$ **16.** $-6, 3$ **17.** $-3, 0, \frac{5}{2}$ **18.** 8, 15, 17 **19.** 2 sec. **20.** $(-3, 0), (3, 0)$

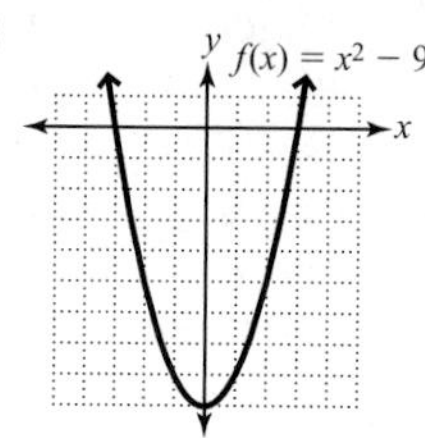

Chapters 1–6 Cumulative Review Exercises

1. true **2.** true **3.** false **4.** true **5.** true **6.** false **7.** the same **8. a.** curve **b.** parabola **9.** coefficients, Add **10.** $(a + b)(a^2 - ab + b^2)$ **11.** factor **12.** -4 **13.** undefined **14.** 0 **15.** $2x^3 - 12x^2 - 21x + 20$ **16.** $8x^3 - 2x^2 + x + 2$ **17.** $9x^2 - 1$ **18.** 14 **19.** $\frac{x^3}{8}$ **20.** $5x^2 + 3x - 2 + \frac{1}{2x - 1}$ **21.** $4m(1 - 4m)$ **22.** $(8x - 3)(x + 4)$ **23.** $(3r + 2s)(r + s)$ **24.** $(w - 3)(w^2 + 3w + 9)$ **25.** $(7a + 6)^2$ **26.** $8(2p + 1)(4p^2 - 2p + 1)$ **27.** $(x - 2y)(3 + a)$ **28.** $(c + 2)(c - 2)(c^2 + 1)$ **29.** 2 **30.** $-11, 14$ **31.** $\frac{4}{5}, -6$ **32.** $-3, 0, \frac{1}{2}$ **33. a.** $\{x \mid x \geq 2\}$ **b.** $[2, \infty)$ **c.**

−6 −5 −4 −3 −2 −1 0 1 2 3 4 5 6

34. a. $\{x \mid x \leq -3 \text{ or } x > 4\}$ **b.** $(\infty, -3] \cup (4, \infty)$ **c.**

−6 −5 −4 −3 −2 −1 0 1 2 3 4 5 6

35. a. $\{x \mid -4 < x < -3\}$ **b.** $(-4, -3)$ **c.**

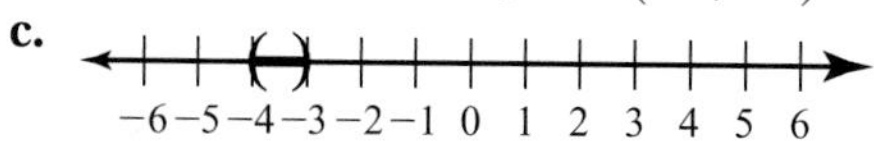

36. a. $\{x \mid -3 < x < 9\}$ **b.** $(-3, 9)$ **c.**

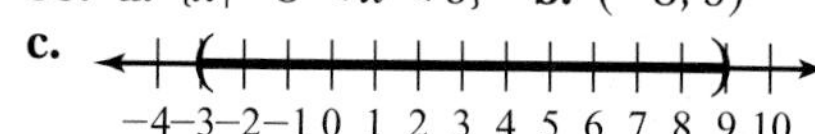

37. yes **38.** $(-1, -2)$ **39.** $(2, -1, 4)$ **40.** 2 **41.** 13 **42.**

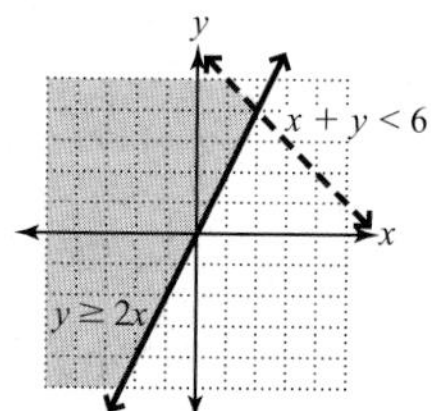

43. Domain: {Berlin Wall, Hadrian's Wall, West Bank Wall, Great Wall}; Range: {12, 15, 25, 26}; It is a function. **44.** 2 **45.** $(-3, 0), (2, 0)$

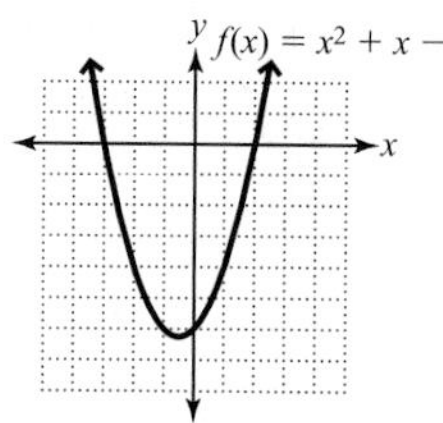

46. $(-1, 0), (0, 0), (2, 0)$

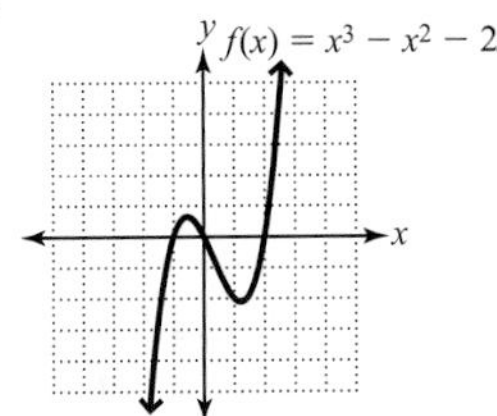

47. a. $5 \le n < 8$ **b.** $n \ge 8$ **48.** teens: 25 min., adults: 52 min. **49.** 32°, 58° **50.** $a = 5$ in., $a + 3 = 8$ in., $a + 5 = 10$ in.

Chapter 7

7.1 Exercises

1. No, since $f(5)$ yields 0 in the denominator, $x = 5$ is not in the domain of the function. **3.** They are not factors of the numerator and denominator. **5.** Multiply the numerators and multiply the denominators. Write both in factored form and reduce to lowest terms.

7. $-\frac{7}{4m^4n}$ **9.** $-\frac{4x^2y}{3}$ **11.** $\frac{2}{3}$ **13.** $\frac{2}{5}$ **15.** $-\frac{3}{4}$ **17.** $\frac{a-2}{a+3}$ **19.** $\frac{x}{x+2}$ **21.** $\frac{2x-3y}{3x-5y}$ **23.** $\frac{2a-3b}{3a-2b}$ **25.** $\frac{x^2+2x+4}{3x+4}$ **27.** $\frac{x^2-3x+9}{3(x-3)}$ **29.** $\frac{y-4}{y-6}$ **31.** $\frac{2a^2+3}{2a+3}$ **33.** -1 **35.** $-\frac{x+7}{2}$ **37.** $\frac{4y^2}{9x^2}$ **39.** $-\frac{8c}{3b^3}$ **41.** $\frac{3xy^2}{20}$ **43.** $-\frac{2}{5}$ **45.** $\frac{4(a+1)}{3(a+4)}$ **47.** $-\frac{4y^2}{3x}$ **49.** $\frac{x-2}{x+3}$ **51.** 1 **53.** $\frac{x+2}{x-3}$ **55.** $-(3x-1)$ **57.** $\frac{(a+2)(d-5)}{(d+1)(b-4)}$ **59.** $\frac{2x+1}{3x+2}$ **61.** $2a-b$ **63.** $\frac{32x^6}{45}$ **65.** $\frac{4a^4}{9}$ **67.** $\frac{x^3}{4}$ **69.** $-\frac{2}{9}$ **71.** $\frac{d-4}{2d(3d+4)}$ **73.** $\frac{(m-5n)(m+6n)}{6}$ **75.** $\frac{3(x-8)}{4(x+8)}$ **77.** $\frac{x-2y}{x+2y}$ **79.** $\frac{x-2}{x+3}$ **81.** $\frac{9a^2+6ab+4b^2}{a-4b}$ **83.** $\frac{b-4}{c-3}$ **85.** $\frac{3x-2}{4x}$ **87.** $\frac{5x+3}{4x-3}$ **89.** Mistake: $a^2 + b^2$ does not factor. Correct: $\frac{a^2+b^2}{(a+b)(a-2b)}$ **91. a.** -6 **b.** 0 **c.** undefined **93. a.** 0 **b.** $\frac{86}{49}$ **c.** $-\frac{2}{7}$ **95. a.** $\frac{5}{9}$ **b.** 0 **c.** undefined **97. a.** 0 **b.** $\frac{4}{3}$ **c.** undefined **99.** $\{x \mid x \ne -4\}$ **101.** $\{x \mid x \ne 5, -5\}$ **103.** $\left\{c \mid c \ne \frac{9}{2}, -\frac{9}{2}\right\}$ **105.** $\left\{t \mid t \ne -\frac{5}{2}, \frac{2}{3}\right\}$ **107.** $\{x \mid x \ne 0, -7, 3\}$ **109.** $\{x \mid x \ne 2\}$ **111.** c **113.** d

115.

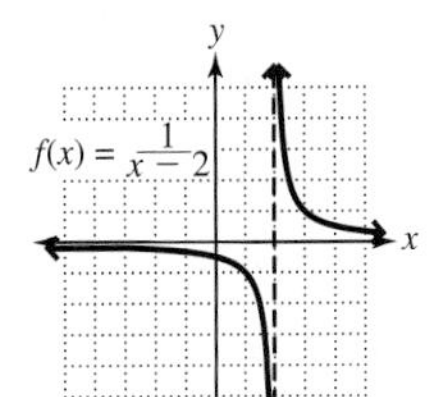

117.

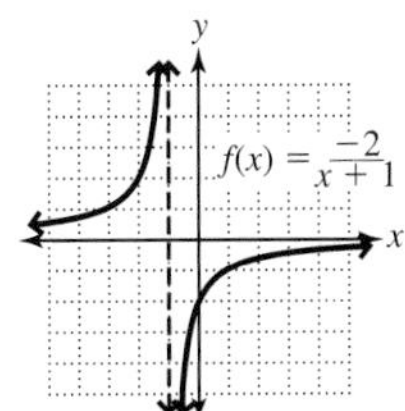

119. $\frac{5}{4}$ ohms **121.** \$27 **123.** \$124,000 **125.** 500

Review Exercises

1. $\frac{7}{11}$ **2.** $-\frac{7}{30}$ **3.** $\frac{13}{20}y$ **4.** $\frac{3}{5}x + 1$ **5.** $3x(2x-5)$ **6.** $(x+5)(x-4)$

7.2 Exercises

1. The units remain the same, only the number of units changes. **3.** Factoring $6x^2 + 5x - 6 = (3x - 2)(2x + 3)$ shows that you must multiply by $(x - 4)$ to obtain the LCD. **5.** Both are correct since $\frac{-6}{3-a} = \frac{6}{-(3-a)} = \frac{6}{a-3}$.

7. $\frac{4a}{b^2}$ **9.** $\frac{4c}{d}$ **11.** $\frac{7a-b}{2a-5b}$ **13.** $\frac{-2x-5y}{x+2y}$ **15.** $\frac{1}{m+4}$ **17.** $\frac{a+2}{a-4}$ **19.** $\frac{c+6}{c-1}$ **21.** $\frac{y-1}{y-6}$ **23.** $\frac{h-1}{h+7}$ **25.** $\frac{x-3}{x+5}$ **27.** $x + y$ **29.** $60a^6b^8$ **31.** $(x+7)(x-5)$ **33.** $45(r+2)$ **35.** $(b+1)(b-1)(b+4)$ **37.** $(c-3)(c+1)(c-2)$ **39.** $(n+4)^2(n+1)$ **41.** $x^2(x+3)(x-2)(x+4)$ **43.** $2x^2(x-4)(x+2)(x^2+2)$ **45.** $\frac{1}{24u}$ **47.** $\frac{31z}{20x}$ **49.** $\frac{11y+6}{12y}$ **51.** $\frac{m+38}{6m}$ **53.** $\frac{9p+3+45p^2q-10pq}{30p^3q^2}$ **55.** $\frac{6a^2+3ab-2ab^2+6b^3}{24a^3b^4}$ **57.** $\frac{8-2k}{k(k+2)}$ **59.** $\frac{9w+13}{(w-3)(w+7)}$ **61.** $\frac{x-13}{6(x+3)}$ **63.** $\frac{8t-24}{(t-4)^2}$ **65.** $\frac{5u+8}{5(u-3)}$ **67.** $\frac{x-8}{x-6}$ **69.** $\frac{3x-5}{(x-3)(x+2)}$ **71.** $\frac{-3}{(v-4)(v+4)}$ **73.** $\frac{2z^2-z+9}{(z+3)^2(z-2)}$ **75.** $\frac{7x-12}{x(x+6)(x-6)(x-3)}$ **77.** $\frac{3x^2+2x+5}{x^2-4}$ **79.** $\frac{10a-19}{a(a-4)}$ **81.** $\frac{3x^2-6x+30}{x(x-4)}$ **83.** $\frac{-2r^2-24r+2}{r^2-9}$ **85.** $\frac{4v}{u-v}$ **87.** $\frac{3m-4n}{m-n}$ **89.** $\frac{-a-5b}{3a-b}$ **91.** Mistake: Added denominators instead of finding the LCD. Correct: $\frac{a^2+20}{5a}$ **93.** Mistake: Added denominators. Correct: $\frac{9v}{2x}$ **95.** Mistake: Did not distribute the subtraction sign to both terms in $2c + 3$.Correct: $\frac{3c-1}{3c-5}$ **97. a.** $\frac{5x+2}{3x}$ **b.** $\frac{4x+5}{3x}$ **c.** $\frac{11}{6}$ **d.** $\frac{1}{2}$ **99. a.** $\frac{x+6}{x+2}$ **b.** $-\frac{x^2+2x+16}{x^2-4}$ or $\frac{x^2+2x+16}{4-x^2}$ **c.** $\frac{7}{3}$ **d.** 5 **101.** $\frac{8t}{15}$ **103.** $\frac{8x-4}{(x+4)(x-2)}$ **105.** $\frac{3x^2+2}{x}$ **107.** $\frac{4x}{15}$

Review Exercises

1. $\frac{13}{12}$ **2.** $\frac{10}{9}$ **3.** $\frac{14}{5}$ **4.** 18 **5.** $\frac{x-1}{x+2}$ **6.** $\frac{1}{9y^2-6y+4}$

7.3 Exercises

1. $(x-2)(x-3)$ **3.** $\frac{x+2}{x-4} \div \frac{x-3}{x+5}$
5. Method 2 because the numerator and denominator are not monomials.
7. $\frac{10}{9}$ **9.** $\frac{uv}{w}$ **11.** $\frac{1}{b}$ **13.** $\frac{u^4}{w^4v^2}$ **15.** $\frac{3}{32}$ **17.** $\frac{ac}{b}$ **19.** $\frac{x-3}{x-1}$ **21.** $\frac{3}{2}$ **23.** $\frac{3x}{2}$ **25.** $\frac{4}{11}$ **27.** $\frac{230}{71}$ **29.** $\frac{2+x}{2-x}$ **31.** $\frac{t^2}{1+t+t^2}$ **33.** $\frac{v+2}{v-3}$ **35.** $\frac{1}{u+6}$ **37.** $\frac{x-6}{x(x+3)}$ **39.** $\frac{5r+44}{(r+8)^2}$ **41.** $\frac{t^2-7t+9}{11t-80}$ **43.** $\frac{x+4}{2x^2+x-21}$ **45.** $\frac{a-27}{a-15}$ **47.** $\frac{6r}{r^2+9}$ **49.** $\frac{2(x-6)}{3x-5}$ **51.** $\frac{x-3y}{x+2y}$ **53.** $\frac{2a^2+a+6}{9a^3+54a^2-3a}$ **55.** $\frac{\frac{1}{x}}{\frac{3}{x}+1} = \frac{1}{3+x}$ **57.** $\frac{\frac{1}{x}+\frac{1}{y}}{\frac{1}{x}-\frac{1}{y}} = \frac{y+x}{y-x}$ **59.** $\frac{1-\frac{1}{x^2}}{1+\frac{1}{x}} = \frac{x-1}{x}$ **61.** $\frac{\frac{3}{x^2}+\frac{5}{y}}{\frac{1}{x}+\frac{1}{y}} = \frac{3y+5x^2}{xy+x^2}$ **63.** $\frac{\frac{1}{x^2y^2}}{\frac{4}{x}+\frac{3}{y}} = \frac{1}{4xy^2+3x^2y}$ **65.** $\frac{\frac{36}{a^2}-\frac{25}{b^2}}{\frac{6}{a}+\frac{5}{b}} = \frac{6b-5a}{ab}$ **67.** $\frac{\frac{1}{4a}+\frac{2}{b^2}}{\frac{2}{a}+\frac{1}{b^2}} = \frac{b^2+8a}{8b^2+4a}$
69. Mistake: The $+1$ was omitted when multiplying $\frac{1}{3} \cdot 3$. Correct: $\frac{3a+1}{3b+1}$ **71.** Mistake: Did not multiply n by n. Correct: $\frac{n^2+3}{3}$ **73.** ≈ 68.6 mph **75.** $\frac{3x+1}{9}$ in. **77. a.** $\frac{R_1 R_2}{R_2+R_1}$ **b.** 24 ohms **c.** 20 ohms

Review Exercises

1. -1 **2.** -4 **3.** $0, -\frac{7}{3}$ **4.** $5, -\frac{3}{2}$ **5.** $\frac{52}{15}$ or $3.4\overline{6}$ mph **6.** 0.2 hr. or 12 min.

7.4 Exercises

1. Adding rational expressions yields another rational expression. Solving an equation involving a rational expression yields a numerical value for the variable. **3.** Multiply by the LCD, $(x+1)(x-1)(x+2)$. **5.** Possible extraneous solutions are $1, -2, -1$. These are the values of x that cause expressions in the equation to be undefined. **7.** 4 **9.** $2, -6$ **11.** $2, -2, -3$ **13.** 1 **15.** -2 **17.** 4 **19.** 0 **21.** 11 **23.** no solution $\left(\frac{3}{2}\text{ is extraneous}\right)$ **25.** $-2, \frac{1}{3}$ **27.** 0 **29.** -3 **31.** 2 (-2 is extraneous) **33.** $6, -2$ **35.** $2, -\frac{5}{4}$ **37.** $2, -1$ **39.** no solution (-5 is extraneous) **41.** 3 (2 is extraneous) **43.** 1 **45.** no solution (3 is extraneous) **47.** 3 (-3 is extraneous) **49.** -17.5 **51.** $0, \frac{2}{3}$ **53.** $\frac{1}{2}$ **55.** $-\frac{1}{2}$ **57.** 1 (2 is extraneous) **59.** $-\frac{2}{3}, 1$ **61.** $\frac{1}{2}, -1$ **63.** 2 (0 is extraneous) **65.** $-\frac{3}{2}, 2$ **67.** $p = \frac{100C}{9000+C}$ **69.** $r = \frac{2E-IR}{2I}$ **71.** $f = \frac{sS}{S+s}$ **73.** $R_1 = \frac{RR_2}{R_2-R}$ **75.** 4 is extraneous. There is no solution. **77.** 20% **79.** 30 m/sec. **81.** 40 mph **83.** 15 cm **85.** 5 ohms **87.** 90 foot-candles

Review Exercises

1. 8 dimes, 10 nickels **2.** length 10 in., width 8 in. **3.** 5 and 8 **4.** 16 and 12 **5.** The saw cost \$146 and the drill cost \$92. **6.** 11 hr.

7.5 Exercises

1. $\frac{1}{x}$ **3.** $\frac{100}{r}$, inverse **5.** p decreases. **7.** $2\frac{2}{5}$ hr. **9.** $14\frac{7}{12}$ days **11.** 12 hr. **13.** 22.5 min. **15.** 3 P.M. **17.** 75 mph **19.** 60 mph **21.** 441 mph **23.** 12 **25.** 36 **27.** 12 **29.** \$39 **31.** 180 cm^3 **33.** 5 sec. **35.** 40.5 m **37.** 7 ft. **39.** 5.25 **41.** 8.25 **43.** 50 psi **45.** 6 ohms **47.** 600 m **49.** 100 mm **51.** 48 **53.** 72 **55.** 210 in.^{3} **57.** 169.65 cm^3 **59.** 6 **61.** 162 **63.** 3.125 ohms **65.** 12 dynes **67. a.**

b. They are directly proportional. As the number of miles increases, so does the number of gallons. **c.** $k = 22.5$, which represents the miles per gallon the car gets. **d.** Yes, the data represent a function because for any given number of miles, there will be one quantity of gasoline (assuming that the miles per gallon stays constant).

Review Exercises **1.** 8 **2.** −144 **3.** $-\frac{2}{3}, 5$ **4.** $-30a^4b^6$ **5.** $\frac{4y^2}{x^4}$ **6.** $-5x - 2$

Chapter 7 Review Exercises

1. true **2.** false **3.** false **4.** true **5.** true **6.** true **7.** $4xy^2$ **8.** $-\frac{14p^3}{9q^2}$ **9.** $-\frac{1}{3m^2n^3}$ **10.** $\frac{7}{5}$ **11.** $\frac{4}{7}$ **12.** $\frac{a-3}{a+2}$ **13.** $\frac{1}{3x+4}$ **14.** $\frac{5x-3}{2x-1}$ **15.** $\frac{1}{4c+7d}$ **16.** $-\frac{1}{2x+5}$ **17.** $\frac{4c-1}{3c-4}$ **18.** $\frac{4x^2-6x+9}{x-5}$ **19.** $\frac{10mn^2p^2}{3q^3}$ **20.** $\frac{3n^2}{2m^2}$ **21.** $\frac{7}{8}$ **22.** $\frac{3(5x+4)}{4}$ **23.** $-\frac{5}{6}$ **24.** $\frac{x-4}{x+5}$ **25.** $\frac{y+2}{y-3}$ **26.** $\frac{4x-5}{3x+2}$ **27.** $\frac{b+2d}{b-4d}$ **28.** $3x+4y$ **29.** $\frac{14y^2b^3}{9x^2a}$ **30.** $\frac{40}{27}$ **31.** $\frac{3}{4}$ **32.** $\frac{1}{3z(z-4)}$ **33.** $\frac{(2p+3)(6p-7)}{20}$ **34.** $\frac{5p+3q}{2p+5q}$ **35.** $\frac{x^2+2xy+4y^2}{x+6y}$ **36.** $\frac{z+2w}{2z-w}$ **37. a.** 3 **b.** 0 **c.** $-\frac{1}{3}$ **38. a.** −6 **b.** 0 **c.** undefined **39.** $\left\{x \middle| x \neq \frac{5}{3}\right\}$ **40.** $\{x|x \neq -6, 2\}$ **41.** $\frac{4r}{7x}$ **42.** $\frac{2a+6b}{2a-3b}$ **43.** $\frac{3p^2+3p-6}{p^2-16}$ **44.** $\frac{x-4}{x-6}$ **45.** $36p^3q^5$ **46.** $(t-4)(t+2)$ **47.** $28(2u+3)$ **48.** $(w+4)(w-4)(w+1)$ **49.** $(a+3)^2(a-5)$ **50.** $(m-4)(m+2)(m+1)$ **51.** $x^3(x+4)(x+2)(x-2)$ **52.** $4x^3(x+6)(x+4)$ **53.** $\frac{4a}{45x}$ **54.** $\frac{y-23}{20y}$ **55.** $\frac{4tu^3-6u^3+15t^2+3t}{36t^4u^5}$ **56.** $\frac{11w-9}{w(w-3)}$ **57.** $\frac{-v^2+v-4}{4(v+2)(v-3)}$ **58.** $\frac{t^2+t+24}{(t+4)^2}$ **59.** $\frac{8w^2+39w+10}{(w+5)(w-5)(w+2)}$ **60.** $\frac{-3z^2+59z+123}{(4z+5)(z+7)^2}$ **61.** $\frac{-3a^2+3a+6}{a(a+3)}$ **62.** $\frac{7x-7y}{3x-y}$ **63.** $\frac{8}{3}$ **64.** $\frac{8}{9}$ **65.** $\frac{u^3w}{v}$ **66.** $\frac{x^4y}{t^3}$ **67.** $\frac{7w}{24-w}$ **68.** $\frac{97-6b}{(b-15)^2}$ **69.** $\frac{y-13}{y+2}$ **70.** $x+5$ **71.** $\frac{3x+4y}{2x+y}$ **72.** $\frac{8x}{x^2+16}$ **73.** $\frac{7}{4}$ **74.** 3 **75.** −4 (4 is extraneous) **76.** 1, 4 **77.** 1, −2 **78.** $-\frac{122}{29}$ **79.** $2, -\frac{5}{3}$ **80.** $\frac{1}{3}$ **81.** 18 days **82.** 23 mph **83.** 9:06 A.M. **84.** 42 **85.** 200 **86.** 6 **87.** 9 **88.** 32 **89.** 9 **90.** 9 gal. **91.** $\frac{7}{3}$ cm/sec. **92.** 100.48 in.^{3}

Chapter 7 Practice Test

1. $\frac{21a^2}{8b^2}$ **2.** $\frac{3x-2}{2x-3}$ **3.** $\frac{m^2+4mn+16n^2}{3m-2n}$ **4.** $\frac{n-2m}{b+5c}$ **5.** $\frac{15b^2y^2}{4a^3x^3}$ **6.** $\frac{y(4y-5)}{(2y-7)(2y-3)}$ **7.** −3 **8.** $\frac{n^2+4}{n^2-4}$ **9. a.** $-\frac{3}{2}$ **b.** 0 **c.** undefined **10.** $\left\{x \middle| x \neq -2, \frac{5}{2}\right\}$ **11.** $(a+3)(a-3)(2a+7)$ **12.** $\frac{5y^2+18y+12}{36y^2}$ **13.** $\frac{2r-3}{r-4}$ **14.** $\frac{t^2+25t+10}{2(t+5)(t-5)^2}$ **15.** $\frac{a+5b}{3a-5b}$ **16.** $\frac{-2x^2+7x+26}{x(x+5)}$ **17.** $\frac{2ac^3}{3b^2}$ **18.** $\frac{3x-5}{2x-1}$ **19.** $\frac{t+2}{2}$ **20.** $\frac{3b-2a}{ab}$ **21.** $\frac{69}{26}$ **22.** no solution **23.** −3, 4 **24.** 2 (−1 is extraneous) **25.** 30 days **26.** 2 hr. **27.** 6 **28.** 30 **29.** 1.44 foot-candles **30.** 408 mi.

Chapter 8

8.1 Exercises

1. Answers will vary. For example, 9 has rational square roots whereas 17 has irrational square roots. The square roots of 17 are irrational because they cannot be expressed in the form $\frac{a}{b}$ where a and b are integers and $b \neq 0$. **3.** Squaring a number or its additive inverse results in the same positive number. **5.** You cannot raise a number to an even power and get a negative value. **7.** ±6 **9.** ±11 **11.** ±14 **13.** ±15 **15.** 5 **17.** not a real number **19.** −5 **21.** ±5 **23.** 1.2 **25.** not a real number **27.** −0.11 **29.** $\frac{7}{9}$ **31.** $-\frac{12}{13}$ **33.** 3 **35.** −3 **37.** 6 **39.** 5 **41.** not a real number **43.** −2 **45.** 2 **47.** −2 **49.** 2 **51.** 2 **53.** $\frac{2}{3}$ **55.** $\frac{2}{3}$ **57.** 2.236 **59.** −3.317 **61.** 3.684 **63.** −3.756 **65.** 2.839 **67.** −3.036 **69.** 2.454 **71.** 2.295 **73.** b^2 **75.** $4x$ **77.** $10rs^3$ **79.** $0.5a^3b^6$ **81.** m **83.** $3a^3b^2$ **85.** $-4ab^4$ **87.** $0.2x^6$ **89.** a **91.** $2x^4$ **93.** $2x^2$ **95.** x^2y **97.** $6|m|$ **99.** $|r-1|$ **101.** $4|y^3|$ **103.** $3y$ **105.** $(y-3)^2$ **107.** $y-4$ **109.** 2 **111.** $\sqrt{15}$ **113.** $\{x|x \geq 4\}$ or $[4, \infty)$ **115.** $\{x|x \leq 4\}$ or $(-\infty, 4]$

117. a.

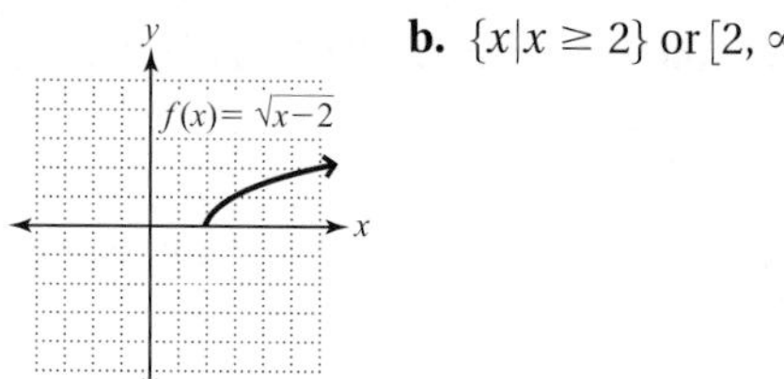

b. $\{x|x \geq 2\}$ or $[2, \infty)$

119. a.

$f(x)=\sqrt[3]{x+1}$

b. $\mathbb{R}$ or $(-\infty, \infty)$

121. ≈ -17.709 m/sec. **123.** ≈ 3.476 sec. **125.** ≈ 19.170 mph **127.** 13 ft. **129.** 15 N **131. a.** 7924 **b.** ≈ 7004

Review Exercises

1. a. 12 **b.** 12 **2.** x^8 **3.** $-45m^4n^3$ **4.** $12x^4 - 20x^3 + 4x^2$ **5.** $21y^2 + 23y - 20$ **6.** $2x^2 - 17x - 9$

8.2 Exercises

1. 4 **3.** a must be positive, because with an even index the radicand must be positive. **5.** Yes, because $100^{1/4} = (10^2)^{1/4} = 10^{1/2}$. **7.** $\sqrt{25} = 5$ **9.** $-\sqrt{100} = -10$ **11.** $\sqrt[3]{27} = 3$ **13.** $\sqrt[3]{-64} = -4$ **15.** $\sqrt[4]{y}$ **17.** $\sqrt{144x^8} = 12x^4$ **19.** $18\sqrt{r}$ **21.** $\sqrt{\dfrac{x^4}{121}} = \dfrac{x^2}{11}$ **23.** $\sqrt[3]{8^2} = 4$ **25.** $(-\sqrt[4]{81})^3 = -27$ **27.** $(\sqrt[3]{-8})^4 = 16$ **29.** $\dfrac{1}{(\sqrt{16})^3} = \dfrac{1}{64}$ **31.** $\sqrt[5]{x^4}$ **33.** $8\sqrt[3]{n^2}$ **35.** $\dfrac{1}{\sqrt[5]{-32^2}} = \dfrac{1}{4}$ **37.** $\left(\sqrt{\dfrac{1}{25}}\right)^2 = \dfrac{1}{125}$ **39.** $\sqrt[6]{(2a+4)^5}$ **41.** $25^{1/4}$ **43.** $z^{5/6}$ **45.** $5^{-5/6}$ **47.** $5x^{-4/5}$ **49.** $5^{7/3}$ **51.** $x^{2/7}$ **53.** $(4a-7)^{7/4}$ **55.** $(2r-5)^{8/5}$ **57.** $x^{4/5}$ **59.** $x^{7/6}$ **61.** $a^{17/12}$ **63.** $12w^{3/5}$ **65.** $-12a^{17/12}$ **67.** $7^{3/5}$ **69.** $\dfrac{1}{x^{2/5}}$ **71.** $x^{1/4}$ **73.** $r^{1/12}$ **75.** $\dfrac{1}{x^{5/7}}$ **77.** $a^{9/4}$ **79.** $20s^{3/7}$ **81.** $-24b^{1/4}$ **83.** x^2 **85.** $a^{5/3}$ **87.** $b^{2/5}$ **89.** $64x^4y^3$ **91.** $2q^{1/2}t^{1/4}$ **93.** $81a$ **95.** $3z^{1/3}$ **97.** $\sqrt{2}$ **99.** $\sqrt[3]{5}$ **101.** $\sqrt{x}$ **103.** $\sqrt[4]{r^3}$ **105.** $\sqrt[4]{x^3y}$ **107.** $\sqrt[5]{m^2n^3}$ **109.** $\sqrt[6]{x^5}$ **111.** $\sqrt[6]{y^7}$ **113.** $\sqrt[6]{x^5}$ **115.** $\sqrt[15]{n^2}$ **117.** $\sqrt[6]{1125}$ **119.** $\sqrt[12]{3456}$ **121.** $\sqrt[9]{x}$ **123.** $\sqrt[6]{n}$

Review Exercises

1. $2^4x^3y^2$ **2.** 12 **3.** 15 **4.** 8×10^{11} **5.** $-\dfrac{5}{8}x^4y^2z^2$ **6.** $2x^2 - 8x + 5 + \dfrac{3}{x+3}$

8.3 Exercises

1. Multiplying the approximate roots gives $\sqrt{8} \cdot \sqrt{18} \approx 2.828 \cdot 4.243 = 11.999204$, which is tedious and inexact. Using the product rule for radicals gives $\sqrt{8} \cdot \sqrt{18} = \sqrt{8 \cdot 18} = \sqrt{144} = 12$, which is fast and exact. **3.** Rewrite the expression as a product of two radicals, the first containing the perfect cube and the second containing no perfect cubes. Then simplify the first radical. For example, $\sqrt[3]{54x^8}$ can be rewritten as $\sqrt[3]{27x^6} \cdot \sqrt[3]{2x^2}$, then simplified to $3x^2 \cdot \sqrt[3]{2x^2}$. **5.** 8 **7.** $9x^3$ **9.** $12xy^2$ **11.** $\sqrt{14}$ **13.** $\sqrt{15x}$ **15.** 3 **17.** $\sqrt[3]{10y^2}$ **19.** $\sqrt[4]{21}$ **21.** $w\sqrt[4]{72}$ **23.** $\sqrt[4]{15x^3y^3}$ **25.** $\sqrt[5]{30x^4}$ **27.** $\sqrt[6]{8x^5y^4}$ **29.** $\sqrt{\dfrac{21}{10}}$ **31.** $\sqrt{\dfrac{6y}{5x}}$ **33.** $\dfrac{5}{6}$ **35.** $\dfrac{\sqrt{10}}{3}$ **37.** 7 **39.** $\sqrt{3}$ **41.** $\dfrac{\sqrt[3]{4}}{w^2}$ **43.** $\dfrac{\sqrt[3]{5y^2}}{3x^3}$ **45.** 4 **47.** $\dfrac{\sqrt[4]{3u^3}}{2x^2}$ **49.** $7\sqrt{2}$ **51.** $8\sqrt{2}$ **53.** $24\sqrt{5}$ **55.** $20\sqrt{7}$ **57.** $a\sqrt{a}$ **59.** xy^2 **61.** $x^3y^4z^5$ **63.** $r^5s^4\sqrt{rs}$ **65.** $18x^2\sqrt{2x}$ **67.** $2\sqrt[3]{4}$ **69.** $x^2\sqrt[3]{x}$ **71.** $x^2y\sqrt[3]{y^2}$ **73.** $4z^2\sqrt[3]{2z^2}$ **75.** $4\sqrt[3]{3}$ **77.** $2\sqrt[4]{5}$ **79.** $9x^4\sqrt[4]{3x}$ **81.** $3x^3\sqrt[5]{2x}$ **83.** $xy^2z\sqrt[6]{x^2y^2z^5}$ **85.** $3\sqrt{7}$ **87.** $30\sqrt{35}$ **89.** $y^2\sqrt{y}$ **91.** $x^4y^5\sqrt{y}$ **93.** $24c^4\sqrt{15}$ **95.** $60\sqrt{2}$ **97.** $2\sqrt{2}$ **99.** $6\sqrt{5}$ **101.** $c^2\sqrt{d}$ **103.** $12a^2$ **105.** $6c\sqrt{3c}$ **107.** $36x^2y^3\sqrt{2y}$ **109.** $\dfrac{2\sqrt{6}}{7}$ **111.** $\dfrac{a^4}{2}$ **113.** $\dfrac{3x^5\sqrt{5}}{4}$ **115.** $\dfrac{3}{8}\sqrt{5}$

Review Exercises

1. no **2.** yes **3.** $8x^2 - 7x$ **4.** $8a^2 - 6ab - 9b^2$ **5.** $9m^2 - 25n^2$ **6.** $4x^2 - 12xy + 9y^2$

8.4 Exercises

1. Like radicals have the same index and the same radicand, but their coefficients may be different. **3.** $(x+3)(x+2) = x^2 + 2x + 3x + 6 = x^2 + 5x + 6$, $(\sqrt{5}+3)(\sqrt{5}+2) = \sqrt{5^2} + 2\sqrt{5} + 3\sqrt{5} + 6 = 5 + 5\sqrt{5} + 6 = 11 + 5\sqrt{5}$. In both cases we use the FOIL method and then simplify the expansion. **5.** $-6\sqrt{6}$ **7.** $9\sqrt{a}$ **9.** $12\sqrt{5} - 8\sqrt{6}$ **11.** $11a\sqrt{5a} - 2b\sqrt{7b}$ **13.** $3x\sqrt[3]{9}$ **15.** $18x^2\sqrt[4]{5x}$ **17.** Cannot combine because the radicals are not like. **19.** $4\sqrt{3} - 5\sqrt{3} = -\sqrt{3}$ **21.** $4\sqrt{5y} - 5\sqrt{5y} = -\sqrt{5y}$ **23.** $4\sqrt{5} - 12\sqrt{5} = -8\sqrt{5}$ **25.** $12\sqrt{6} - 6\sqrt{6} = 6\sqrt{6}$ **27.** $24a\sqrt{3a} - 10a\sqrt{3a} = 14a\sqrt{3a}$ **29.** $5\sqrt{6} - 3\sqrt{6} + 2\sqrt{6} = 4\sqrt{6}$ **31.** $4\sqrt{2} - 12\sqrt{3} + 14\sqrt{2} - 5\sqrt{3} = 18\sqrt{2} - 17\sqrt{3}$ **33.** $2\sqrt[3]{2} + 3\sqrt[3]{2} = 5\sqrt[3]{2}$ **35.** $12x\sqrt[3]{5x^2} - 24x\sqrt[3]{5x^2} = -12x\sqrt[3]{5x^2}$ **37.** $-4x^2\sqrt[4]{x} + 4x\sqrt[4]{x^3}$ **39.** $3\sqrt{2} + 2$ **41.** $3 - 3\sqrt{5}$ **43.** $\sqrt{15} + 10\sqrt{3}$ **45.** $24x - 48x\sqrt{2}$ **47.** $12 - 3\sqrt{2} + 4\sqrt{5} - \sqrt{10}$ **49.** $6 + 5\sqrt{x} + x$ **51.** $6 + 10\sqrt{2} + 9\sqrt{3} + 15\sqrt{6}$ **53.** $\sqrt{6} + \sqrt{10} + 3 + \sqrt{15}$ **55.** $x - \sqrt{xy} - 2y$ **57.** $12\sqrt{14} - 12\sqrt{6} + 6\sqrt{35} - 6\sqrt{15}$ **59.** $8a + 10\sqrt{ab} - 3b$ **61.** $2\sqrt[3]{2} - 3\sqrt[3]{4} - 40$ **63.** $1 - \sqrt[3]{18} + \sqrt[3]{12}$ **65.** $x + 8$ **67.** $22 + 8\sqrt{6}$ **69.** $3 - 2\sqrt{2}$ **71.** $16 + 8\sqrt{3}$ **73.** $30 + 12\sqrt{6}$ **75.** 1 **77.** -14 **79.** $36 - x$ **81.** 1 **83.** $x - y$ **85.** 4 **87.** 50 **89.** $7\sqrt{5}$ **91.** $-15\sqrt{6}$ **93.** $7\sqrt{2}$ **95.** $-14\sqrt{5}$ **97.** $25\sqrt{3}$ **99. a.** $42 + 2\sqrt{5}$ ft. **b.** 46.5 ft. **c.** \$87.89

Review Exercises

1. Slope is $\frac{2}{5}$, y-intercept is $(0, 2)$.

2.

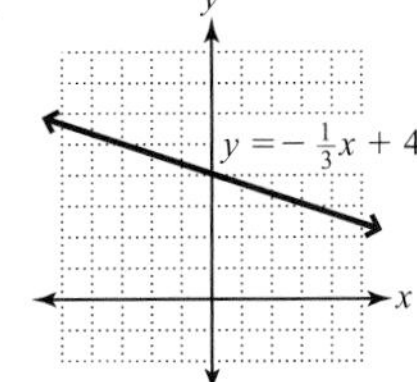

3. $2x + 5$ **4.** $16x^2 - 9$ **5.** $\sqrt{2}$

6. $\sqrt[3]{4}$

8.5 Exercises

1. a. $\sqrt{16}$ is rational. **b.** There is a radical, $\sqrt{3}$, in the denominator. **3.** Multiply the fraction by a 1 so that the product's denominator has a radicand that is a perfect square. **5.** $\frac{\sqrt{3}}{3}$ **7.** $\frac{3\sqrt{2}}{4}$ **9.** $\frac{6\sqrt{7}}{7}$ **11.** $\frac{\sqrt{15}}{6}$ **13.** $\frac{x\sqrt{14}}{10}$ **15.** $\frac{\sqrt{7}}{7}$ **17.** $\frac{5\sqrt{3a}}{3a}$ **19.** $\frac{\sqrt{33mn}}{11n}$ **21.** $\frac{2\sqrt{5x}}{x}$ **23.** $\frac{\sqrt{3}}{4}$ **25.** $\frac{3\sqrt{x}}{x^2}$ **27.** $4x\sqrt{2x}$ **29.** Mistake: The product of $\sqrt{2}$ and 2 is not 2. Correct: $\frac{\sqrt{6}}{2}$. **31.** $\frac{5\sqrt[3]{9}}{3}$ **33.** $\frac{\sqrt[3]{20}}{2}$ **35.** $3\sqrt[3]{2}$ **37.** $\frac{m\sqrt[3]{n^2}}{n}$ **39.** $\frac{\sqrt[3]{ab}}{b}$ **41.** $\frac{2\sqrt[3]{4x^2}}{x}$ **43.** $\frac{\sqrt[3]{30a}}{5a}$ **45.** $\frac{5\sqrt[4]{8}}{2}$ **47.** $\frac{\sqrt[4]{3x^2}}{x}$ **49.** $\frac{3\sqrt[4]{27x}}{x}$ **51.** $\sqrt{6} - \sqrt{3}$ **53.** $8 + 4\sqrt{3}$ **55.** $5\sqrt{3} - 5\sqrt{2}$ **57.** $-1 - \sqrt{5}$ **59.** $\frac{3 + \sqrt{3}}{2}$ **61.** $\frac{-6 - 8\sqrt{3}}{13}$ **63.** $\frac{4\sqrt{21} - 4\sqrt{6}}{5}$ **65.** $\frac{64 + 8\sqrt{6}}{29}$ **67.** $\frac{6y - 6\sqrt{y}}{y - 1}$ **69.** $\frac{3t - 6\sqrt{tu}}{t - 4u}$ **71.** $\frac{\sqrt{2xy} + 2y\sqrt{3}}{x - 6y}$ **73.** $\frac{3}{2\sqrt{3}}$ **75.** $\frac{2x}{5\sqrt{2x}}$ **77.** $\frac{2n}{3\sqrt{2n}}$ **79.** $\frac{1}{10 - 5\sqrt{3}}$ **81.** $\frac{5x - 36}{9\sqrt{5x} + 54}$ **83.** $\frac{19n}{10n\sqrt{n} - 2n\sqrt{6n}}$ **85. a.** $\frac{5\sqrt{3}}{3}$ **b.** $\sqrt{5}$ **c.** $\frac{5\sqrt{11}}{11}$ **87. a.** The graphs are identical. The functions are identical. **b.** $f(x) = g(x)$ **89. a.** $T = 2\pi\sqrt{9.8L}/9.8$ **b.** $T = 2\pi L/\sqrt{9.8L}$ **91. a.** $s = \frac{\sqrt{3Vh}}{h}$ **b.** 745 ft. **93. a.** $V_{\text{rms}} = \frac{\sqrt{2}}{2}V_{\text{m}}$ or $\frac{\sqrt{2}V_{\text{m}}}{2}$ **b.** $V_{\text{rms}} = \frac{163\sqrt{2}}{2}$ **c.** ≈ 115.3 **95.** $\frac{15\sqrt{2} - 10\sqrt{3}}{3}\,\Omega$

Review Exercises

1. $\pm 2\sqrt{7}$ **2.** $(x - 3)^2$ **3.** 4 **4.** -1 **5.** $-6, 6$ **6.** 2, 3

8.6 Exercises

1. Some of the answers may be extraneous. **3.** The principal square root of a number cannot equal a negative. **5.** Subtract $3x$ from both sides to isolate the radical. This allows us to use the squaring principle of equality to eliminate the radical. **7.** 4 **9.** no real-number solution **11.** 27 **13.** -8 **15.** 17 **17.** 11 **19.** 6 **21.** $-\frac{1}{2}$ **23.** no real-number solution **25.** 11 **27.** -2 **29.** 124 **31.** 55 **33.** 5 **35.** -11 **37.** no real-number solution **39.** 2 **41.** no real-number solution **43.** 3 **45.** 3 **47.** 12 (5 is an extraneous solution) **49.** 1, 3 **51.** $-3, -2$ **53.** $\frac{1}{2}$ ($-\frac{3}{8}$ is an extraneous solution) **55.** 5 **57.** $-1, 3$ **59.** 0, 4 **61.** $\frac{3}{16}$ **63.** 1 **65.** Mistake: You cannot take the principal square root of a number and get a negative. Correct: No real-number solution. **67.** Mistake: The binomial $x - 3$ was not squared correctly. Correct: $x = 6$ with $x = 1$ an extraneous solution. **69.** 9.8 m **71.** 2.45 m **73.** 1.44 ft. **75.** 144 ft. **77.** 36.73 ft. **79.** 82.65 ft. **81.** 4 N **83.** 6 N **85. a.** 3, 4, 25 **b.**

y
y = √x
(4, 2) (9, 3) (16, 4) (25, 5)
(1, 1)
(0, 0)
x

c. No. The x-values must be 0 or positive because real square roots exist only when $x \geq 0$. The y-values must be 0 or positive because by definition the principal square root is either 0 or positive. **d.** Yes, because it passes the vertical line test. **87.** The graph becomes steeper from left to right. **89.** The graph rises or lowers according to the value of the constant.

Review Exercises

1. 81 **2.** -0.008 **3.** 39.0625 or $\frac{625}{16}$ **4.** x^8 **5.** n^{24} **6.** y^4

8.7 Exercises

1. $\sqrt{-1}$ **3.** No. The set of complex numbers contains both the real and the imaginary numbers. **5.** We subtract complex numbers just like we subtract polynomials by writing an equivalent addition and changing the signs in the second complex number. **7.** $6i$ **9.** $i\sqrt{5}$ **11.** $2i\sqrt{2}$ **13.** $2i\sqrt{7}$ **15.** $3i\sqrt{3}$ **17.** $5i\sqrt{5}$ **19.** $3i\sqrt{7}$ **21.** $7i\sqrt{5}$ **23.** $6 + 7i$ **25.** $11 - 6i$ **27.** $-7 + i$ **29.** $15 + 4i$ **31.** $-3 - 10i$ **33.** 0 **35.** $14 + 9i$ **37.** $-4 + 28i$ **39.** -24 **41.** 40 **43.** $14 + 12i$ **45.** $72 - 32i$ **47.** $19 - 3i$ **49.** $50 + 9i$ **51.** 65 **53.** $99 + 20i$ **55.** $-2i$ **57.** $-\frac{4i}{5}$ **59.** $-3i$ **61.** $-\frac{1}{2} - i$ **63.** $\frac{1}{2} - i$ **65.** $\frac{14}{5} - \frac{7}{5}i$ **67.** $-\frac{7}{29} + \frac{3}{29}i$ **69.** $7 - 2i$ **71.** $\frac{9}{13} - \frac{7}{13}i$ **73.** $\frac{34}{41} + \frac{19}{41}i$

75. $-i$ **77.** -1 **79.** -1 **81.** 1 **83.** 1 **85.** -1 **87.** $-i$ **89.** i

Review Exercises

1. $(2x-3)^2$ **2.** $\pm 4\sqrt{3}$ **3.** $x = 2$ **4.** $x = -\frac{3}{2}, 2$ **5.** $(0, 2), (3, 0)$ **6.**

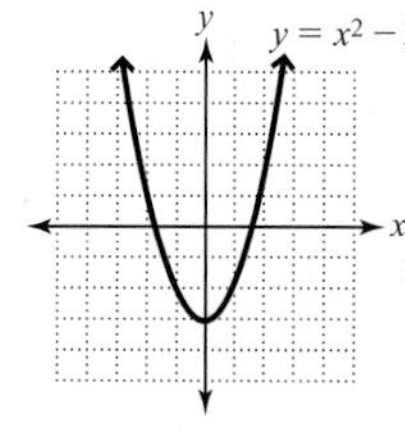

Chapter 8 Review Exercises

1. true **2.** false **3.** false **4.** false **5.** true **6.** radicands **7.** coefficients **8.** 1. Multiply **9.** conjugate. **10.** extraneous **11.** ± 11 **12.** ± 7 **13.** 13 **14.** -7 **15.** $6i$ **16.** $\frac{1}{5}$ **17.** 2.646 **18.** 9.487 **19.** $7x^4$ **20.** $12a^3b^6$ **21.** $0.4mn^5$ **22.** x^5 **23.** $-4r^3s$ **24.** $3x^3$ **25.** $2x^3y^4$ **26.** x^2y **27.** $9|x|$ **28.** $(x-1)^2$ **29.** -4 **30.** $2a^2\sqrt{6}$ **31.** $\frac{1}{8}$ **32.** $\sqrt[7]{(5r-2)^5}$ **33.** $33^{1/8}$ **34.** $8n^{-3/7}$ **35.** $8^{3/5}$ **36.** $m^{8/3}$ **37.** $(3xw)^{3/4}$ **38.** $(a+b)^{4/3}$ **39.** x^2 **40.** $32m^{3/2}$ **41.** $y^{-1/5}$ **42.** b **43.** $k^{1/2}$ **44.** $2^{3/4}x^{3/4}y^{3/20}$ **45.** 9 **46.** $10x^4$ **47.** 2 **48.** $\sqrt[4]{42}$ **49.** $\sqrt[5]{15x^4y^4}$ **50.** $\frac{7}{11}$ **51.** $-\frac{3}{2}$ **52.** $6\sqrt{2}$ **53.** $12b^4\sqrt{3b}$ **54.** $15\sqrt[3]{4}$ **55.** $4x^3\sqrt[3]{5x}$ **56.** $6x^5\sqrt[4]{2x^3}$ **57.** $84\sqrt{10}$ **58.** $x^7\sqrt{x}$ **59.** $16c^2\sqrt{15c}$ **60.** $a^5b^4\sqrt{b}$ **61.** $2\sqrt{2}$ **62.** $6\sqrt{5}$ **63.** $12a^2$ **64.** $36x^2y^3\sqrt{2y}$ **65.** $-3\sqrt{2}$ **66.** $-6y^3\sqrt[4]{8y}$ **67.** $5\sqrt{5}$ **68.** $-10\sqrt{6}$ **69.** $4\sqrt{6}$ **70.** $28x\sqrt{2y}$ **71.** $9xy\sqrt[3]{2xy^2}$ **72.** $5\sqrt[4]{3}$ **73.** $\sqrt{15}+\sqrt{10}$ **74.** $\sqrt[3]{21}+2\sqrt[3]{49}$ **75.** $6\sqrt{6}-54$ **76.** $\sqrt{10}+\sqrt{14}-\sqrt{15}-\sqrt{21}$ **77.** $-6+2\sqrt[3]{2}-4\sqrt[3]{4}$ **78.** $\sqrt[4]{12x^2}-\sqrt[4]{6x}+2\sqrt[4]{2x}-2$ **79.** $5a-3b$ **80.** 7 **81.** $3+2\sqrt{2}$ **82.** $7-2\sqrt{10}$ **83.** $\frac{\sqrt{2}}{2}$ **84.** $\sqrt[3]{9}$ **85.** $\frac{2\sqrt{7}}{7}$ **86.** $\frac{\sqrt[4]{40x^2}}{2}$ **87.** $\frac{\sqrt[3]{153y}}{3y}$ **88.** $\frac{\sqrt[4]{27}}{3}$ **89.** $-4\sqrt{2}-4\sqrt{3}$ **90.** $\frac{4-\sqrt{3}}{13}$ **91.** $\frac{2+\sqrt{n}}{4-n}$ **92.** $\sqrt{6}+2$ **93.** $\frac{5}{3\sqrt{10}}$ **94.** $\frac{3x}{5\sqrt{3x}}$ **95.** $\frac{1}{16+8\sqrt{3}}$ **96.** $\frac{1}{10\sqrt{t}-5\sqrt{3t}}$ **97.** $x = 81$ **98.** no real-number solution **99.** $w = 10$ **100.** $x = -2$ **101.** $x = 18$ **102.** $y = 5$ **103.** no real-number solution **104.** $\frac{719}{3}$ **105.** 7 **106.** $1\left(\frac{17}{4}\text{ is extraneous}\right)$ **107.** 0, 4 **108.** $\frac{3}{16}$ **109.** $3i$ **110.** $2i\sqrt{5}$ **111.** $8-6i$ **112.** $9-7i$ **113.** -12 **114.** $2+8i$ **115.** $26+2i$ **116.** $24-10i$ **117.** $-5i$ **118.** $3i$ **119.** $-\frac{4i}{3}$ **120.** $\frac{1}{5}-\frac{7}{5}i$ **121.** $\frac{6}{5}-\frac{3}{5}i$ **122.** $\frac{7}{13}+\frac{17}{13}i$ **123.** 1 **124.** $-i$ **125.** $\frac{\sqrt{6}}{4}$ ft. **126. a.** $8\sqrt{5}$ mph **b.** 17.9 mph **127.** 312.5 ft. **128.** π sec. **129.** 0.272 m **130. a.** $\frac{\sqrt{10}}{2}$ sec. **b.** 1.58 sec. **131.** 1.44 ft. **132.** 5 ft.

Chapter 8 Practice Test

1. 6 **2.** $7i$ **3.** $9xy^2\sqrt{y}$ **4.** $3\sqrt[3]{2}$ **5.** $2x\sqrt[4]{x^2}$, or $2x\sqrt{x}$ **6.** $3r^5$ **7.** $\frac{1}{3}$ **8.** $\frac{1}{3}$ **9.** $5\sqrt{7}$ **10.** $4-2\sqrt{3}$ **11.** $x^{-2/3}$ **12.** $-6+2\sqrt[3]{2}-4\sqrt[3]{4}$ **13.** $8^{1/5}x^{3/5}$ **14.** $(2x+5)^{2/3}$ **15.** $\frac{\sqrt[3]{2}}{2}$ **16.** $\frac{x-\sqrt{xy}}{x-y}$ **17.** 22 **18.** 7 **19.** $-2-4i$ **20.** $2+0i$, or 2 **21.** $17+0i$, or 17 **22.** $\frac{8}{25}+\frac{6}{25}i$ **23.** 60 m² **24. a.** $\frac{\sqrt{3}}{2}$ **b.** 64 ft. **25. a.** $3.5\sqrt{10}$ **b.** 293.88 ft., or 14,400/49 ft.

Chapter 9

9.1 Exercises

1. $x^2 = a$ has two solutions because squaring $\sqrt{a}$ and $-\sqrt{a}$ gives a for every real number a. **3.** $x = \frac{b \pm \sqrt{c}}{a}$ **5.** Since $x^2 - 7x + 12$ is easy to factor, factoring is better as it will require fewer steps than completing the square. **7.** ± 7 **9.** $\pm\frac{2}{5}$ **11.** ± 0.9 **13.** $\pm 3\sqrt{5}$ **15.** $\pm 5i$ **17.** ± 7 **19.** ± 6 **21.** ± 3 **23.** $\pm\frac{3}{5}$ **25.** $\pm 2i$ **27.** $\pm\frac{3\sqrt{5}}{10}$ **29.** ± 4 **31.** $\pm 3i\sqrt{2}$ **33.** $\pm\frac{4\sqrt{6}}{3}$ **35.** ± 1.6 **37.** $-15, -1$ **39.** $-\frac{1}{5}, \frac{7}{5}$ **41.** $7 \pm 2i\sqrt{3}$ **43.** $\frac{1 \pm 2\sqrt{10}}{4}$ **45.** $8 \pm i$ **47.** $0, \frac{3}{2}$ **49.** $\frac{3}{5}, \frac{6}{5}$ **51.** $-13.5, -5.5$ **53.** Mistake: Gave only the positive solution. Correct: ± 7 **55.** Mistake: Changed -6 to 6. Correct: $5 \pm i\sqrt{6}$ **57.** $\pm 4\sqrt{6} \approx \pm 9.798$ **59.** $\pm 2\sqrt{5} \approx \pm 4.472$ **61.** $6 \pm \sqrt{15} \approx 9.873, 2.127$ **63. a.** $x^2 + 14x + 49$ **b.** $(x+7)^2$ **65. a.** $n^2 - 10n + 25$ **b.** $(n-5)^2$ **67. a.** $y^2 - 7y + \frac{49}{4}$ **b.** $\left(y - \frac{7}{2}\right)^2$ **69. a.** $s^2 - \frac{2}{3}s + \frac{1}{9}$ **b.** $\left(s - \frac{1}{3}\right)^2$ **71. a.** $m^2 + \frac{1}{7}m + \frac{1}{196}$ **b.** $\left(m + \frac{1}{14}\right)^2$ **73. a.** $p^2 + 9p + \frac{81}{4}$ **b.** $\left(p + \frac{9}{2}\right)^2$ **75.** $-5, 3$ **77.** $-8, -2$ **79.** $1 \pm 7i$ **81.** 3, 6 **83.** $-12, 3$ **85.** $1 \pm \sqrt{17}$

87. $3 \pm 2i\sqrt{5}$ **89.** $-\frac{3}{2}, 1$ **91.** $-\frac{7}{2}, \frac{1}{2}$ **93.** $-1, \frac{3}{2}$
95. $-\frac{3}{2}, \frac{2}{3}$ **97.** $-2, \frac{3}{2}$ **99.** $\frac{2 \pm \sqrt{10}}{2}$ **101.** $\frac{-1 \pm \sqrt{41}}{10}$
103. $\frac{-3 \pm 2\sqrt{3}}{3}$ **105.** Did not divide by 3 so that x^2 has a coefficient of 1. Then wrote an incorrect factored form.
107. 14 in. **109.** height: 3 ft., width: 6 ft. **111.** 305 m
113. 18 in. **115.** 15 ft. by 21 ft. **117.** 16 ft. by 12 ft.
119. 15 in. **121. a.** 15 cm **b.** length: 45 cm, width: 30 cm
123. $\frac{3}{4}$ sec. **125.** 4.9 sec. **127.** 4 m/sec.

Review Exercises

1. $(x-5)^2$ **2.** $(x+3)^2$ **3.** $6x$
4. $2x+3$ **5.** $\frac{-2+\sqrt{6}}{2}$ or $-1+\frac{\sqrt{6}}{2}$ **6.** $\frac{-3-i}{5}$ or $-\frac{3}{5}-\frac{1}{5}i$

9.2 Exercises

1. Follow the procedure for solving a quadratic equation by completing the square on $ax^2 + bx + c = 0$. **3.** Yes, if they have irrational or nonreal complex solutions. **5.** When the discriminant is zero. The quadratic formula becomes $x = -\frac{b}{2a}$. **7.** $a = 1, b = -3, c = 7$
9. $a = 3, b = -9, c = -4$ **11.** $a = 1.5, b = -1, c = 0.2$ or $a = -1.5, b = 1, c = -0.2$ **13.** $a = -\frac{1}{2}, b = -\frac{3}{4}, c = 6$ or $a = \frac{1}{2}, b = \frac{3}{4}, c = -6$ **15.** $-5, -4$ **17.** $\frac{1 \pm \sqrt{5}}{2}$
19. $-2, \frac{3}{4}$ **21.** 0, 9 **23.** 4 **25.** $1 \pm i$ **27.** $-\frac{1}{4}, 1$
29. $\frac{-5 \pm \sqrt{10}}{3}$ **31.** $\frac{3 \pm i\sqrt{11}}{4}$ **33.** $-\frac{1}{6}, -\frac{2}{3}$ **35.** $\frac{2 \pm i\sqrt{5}}{3}$
37. $\frac{3 \pm \sqrt{105}}{12}$ **39.** $-0.15, 0.1$ **41.** $-2, \frac{3}{2}$ **43.** $\pm\frac{7}{6}$
45. $1 \pm i\sqrt{2}$ **47.** $\frac{-3 \pm \sqrt{5009}}{100} \approx -0.738, 0.678$
49. $\frac{3 \pm i\sqrt{51}}{12}$ **51.** Mistake: Did not evaluate $-b$. Correct: $\frac{7 \pm \sqrt{37}}{6}$ **53.** Mistake: The result was not completely simplified. Correct: $1 \pm i\sqrt{2}$ **55.** one rational
57. one rational **59.** two nonreal complex **61.** two nonreal complex **63.** two irrational **65.** square root principle or factoring; ± 9 **67.** square root principle; $\pm 2i\sqrt{3}$
69. quadratic formula; $3 \pm 2i$ **71.** factoring; $0, -6$
73. quadratic formula; $4 \pm i\sqrt{3}$ **75.** $(2, 0), (-1, 0), (0, -2)$
77. $(-4, 0), (2, 0), (0, 8)$ **79.** $\left(\frac{1}{2}, 0\right), (-8, 0), (0, -8)$
81. no x-intercepts, $(0, -6)$ **83.** $x^2 + 5(x+1) = 71$; 6, 7
85. $x^2 + (x+1)^2 = (x+2)^2$; 3, 4, 5 **87.** $w(3w - 3.5) = 34$; width: 4 ft., length: 8.5 ft. **89. a.** $22x = 1.5x^2 + 4x$; 12 ft.
b. 18 ft., 12 ft., 10 ft., 6 ft., 8 ft., 18 ft. (from bottom clockwise)
91. ≈ 3.43 sec. **93. a.** $h = -4.9t^2 + 10$ **b.** ≈ 1.01 sec.
c. ≈ 1.43 sec. **95.** 8000 units **97. a.** $\frac{25}{8}$ **b.** $c < \frac{25}{8}$
c. $c > \frac{25}{8}$ **99. a.** $\frac{9}{2}$ **b.** $a < \frac{9}{2}$ **c.** $a > \frac{9}{2}$

Review Exercises

1. $(u-7)(u-2)$
2. $(3u-8)(u+2)$ **3.** $\frac{2}{5}, -3$ **4.** 8 **5.** x^4 **6.** $x^{2/3}$

9.3 Exercises

1. It can be rewritten as a quadratic equation. **3.** $u = \frac{x+2}{3}$ **5.** $\frac{1}{x}$ **7.** $-\frac{4}{3}, 2$ **9.** $-6, 4$
11. -8 (4 is extraneous) **13.** $-7.5, 5$ **15.** $-4, 2$ **17.** $-5, \frac{2}{3}$
19. 9, 25 **21.** $\frac{49}{4}$ (1 is extraneous) **23.** $2\left(\frac{2}{9}\text{ is extraneous}\right)$
25. $\pm 3i$ **27.** $2\left(-\frac{2}{9}\text{ is extraneous}\right)$ **29.** 0, 4 **31.** $\pm 3, \pm 1$
33. $\pm 1.5, \pm 1$ **35.** $\pm 2, \pm 3i$ **37.** $-6, -4$ **39.** $-3.5, 2$
41. $-9, -5$ **43.** $-5, 0$ **45.** 8, 27 **47.** $-\frac{1}{8}, 8$ **49.** 81, 256
51. $\frac{16}{625}$ (16 is extraneous) **53.** 3 hr. **55.** ≈ 2.16 hr.
57. 45 mph **59. a.** Old rate = 12 mph; new rate = 18 mph
b. Old time = 3 hr.; new time = 2 hr. **61.** Jody: 4 hr.; Billy: 6 hr. **63.** ≈ 4.27 hr., 3.77 hr. **65.** 420 vps; 460 vps

Review Exercises

1. -2 **2.** $(3, 0), (0, 2)$ **3.** 6
4.

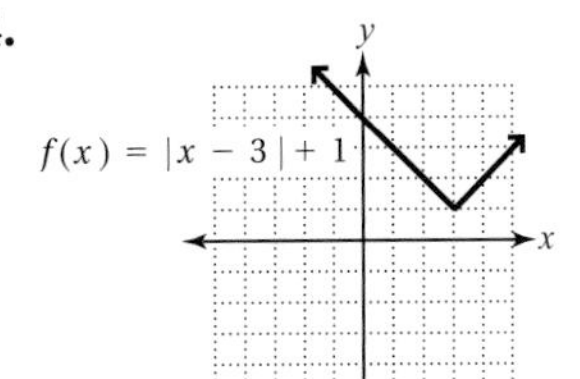

5.

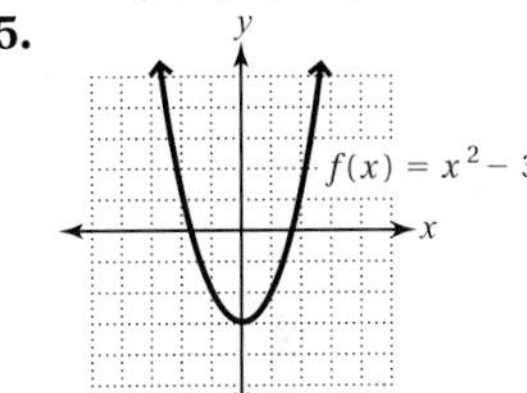

6.

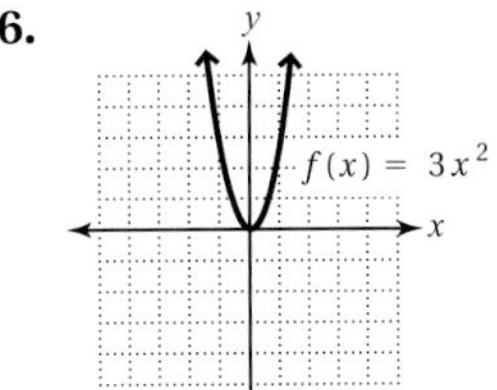

9.4 Exercises

1. The sign of a. **3.** The axis of symmetry is the vertical line through the vertex. If $y = ax^2 + bx + c$, then the equation of symmetry is $x = -\frac{b}{2a}$.
5. There are no x-intercepts. **7.** upwards **9.** downwards
11. upwards **13.** downwards **15.** downwards
17. V: $(2, -3)$; Axis: $x = 2$ **19.** V: $(-1, -5)$; Axis: $x = -1$
21. V: $(0, 2)$; Axis: $x = 0$ **23.** V: $(7, 0)$; Axis: $x = 7$

25. V: $(0, 0)$; Axis: $x = 0$ **27.** V: $(2, 4)$; Axis: $x = 2$

29. V: $(-4, -5)$; Axis: $x = -4$ **31.** V: $\left(\frac{1}{3}, \frac{4}{3}\right)$; Axis: $x = \frac{1}{3}$

33. V: $(-0.4, 1.08)$; Axis: $x = -0.4$ **35. a.** downwards

b. $(0, 0)$ **c.** $x = 0$

d.

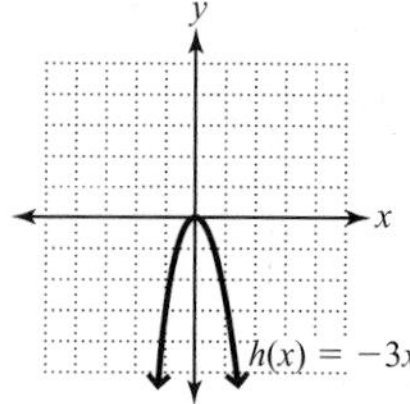

37. a. upwards **b.** $(0, 0)$ **c.** $x = 0$

d. 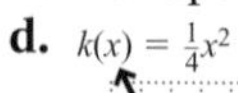

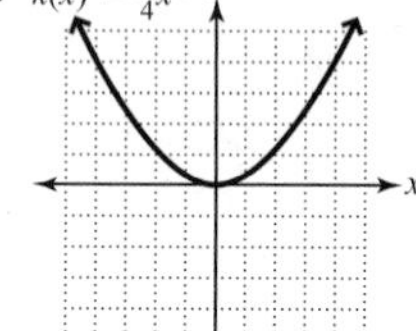

39. a. upwards **b.** $(0, -3)$ **c.** $x = 0$

d.

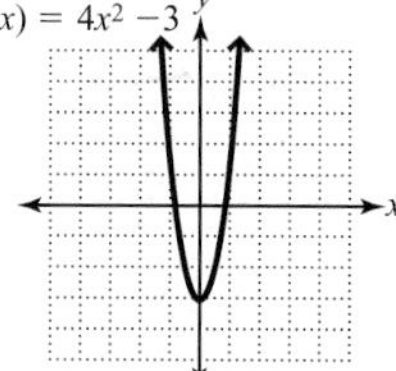

41. a. downwards **b.** $(0, 2)$ **c.** $x = 0$

d.

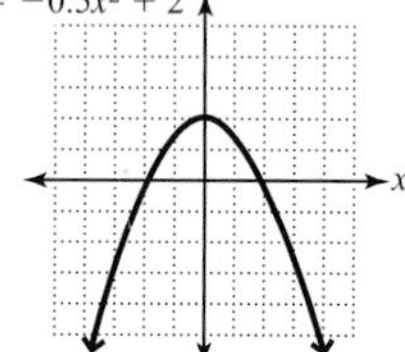

43. a. upwards **b.** $(3, 2)$ **c.** $x = 3$

d.

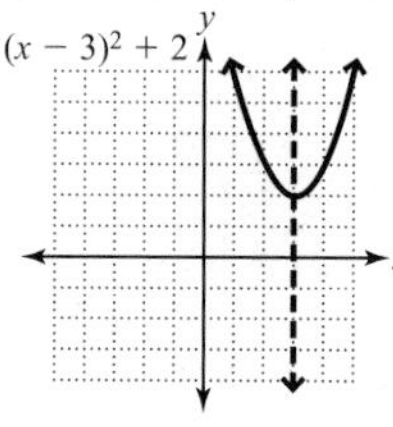

45. a. downwards **b.** $(-1, -3)$ **c.** $x = -1$

d.

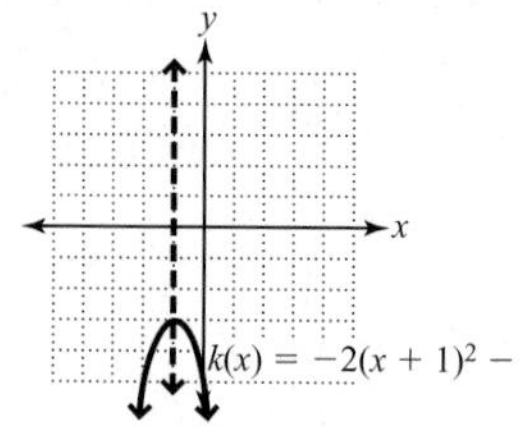

47. a. upwards **b.** $(2, -1)$ **c.** $x = 2$

d.

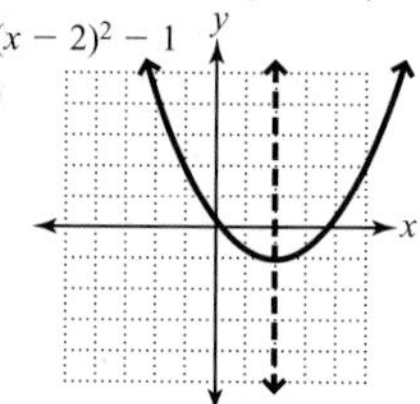

49. a. $(-3, 0), (0, 9)$ **b.** $h(x) = (x + 3)^2$ **c.** upwards

d. $(-3, 0)$ **e.** $x = -3$

f.

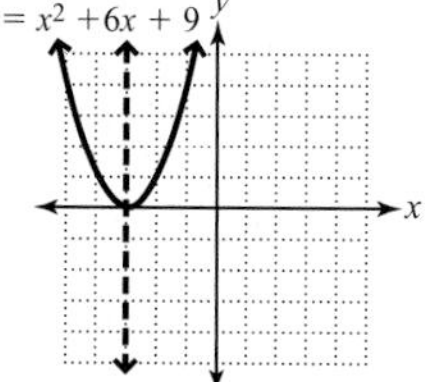

51. a. $\left(\frac{-3 + \sqrt{3}}{2}, 0\right), \left(\frac{-3 - \sqrt{3}}{2}, 0\right), (0, 3)$

b. $f(x) = 2\left(x + \frac{3}{2}\right)^2 - \frac{3}{2}$ **c.** upwards **d.** $\left(-\frac{3}{2}, -\frac{3}{2}\right)$

e. $x = -\frac{3}{2}$ **f.**

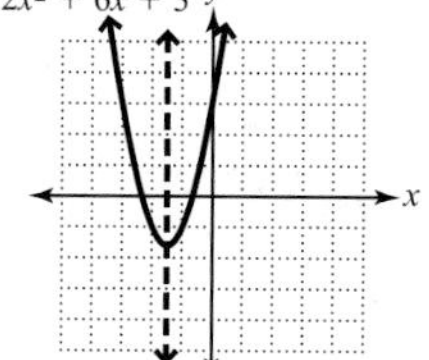

53. a. no x-intercepts, $(0, -5)$ **b.** $g(x) = -3(x - 1)^2 - 2$

c. downwards **d.** $(1, -2)$ **e.** $x = 1$

f.

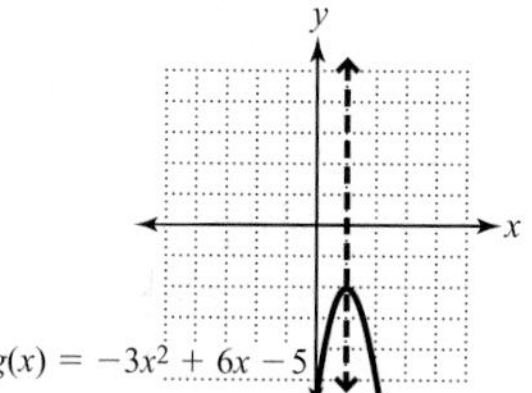

55. a. 4.59 sec. **b.** 103.32 m **c.** 9.18 sec.
57. a. 7.8 ft. **b.** ≈ 4.62 ft. **c.**

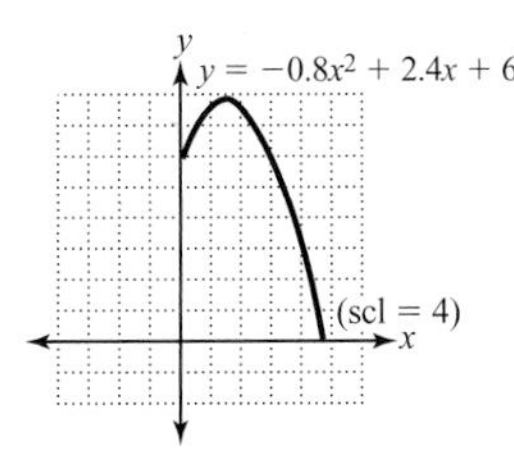

59. a. The greatest number of CDs were sold in the tenth week. **b.** 20,000 **61.** The area is maximized if the length and width are 100 ft. **63.** 55 units **65.** $-6, 6$; The product is -36.

Review Exercises

1. -11 **2.** $-5, 3$
3. $\pm 2i\sqrt{5}$
4. $x \leq 4.5$

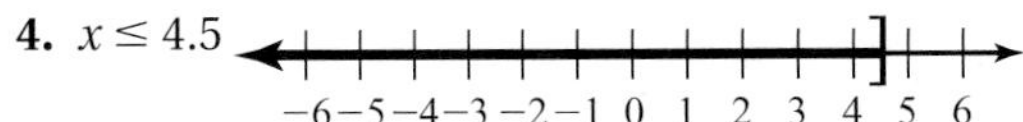

5. $x \leq -5$ or $x \geq 11$

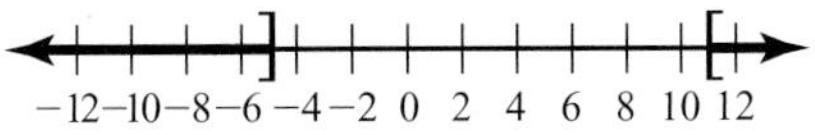

6.

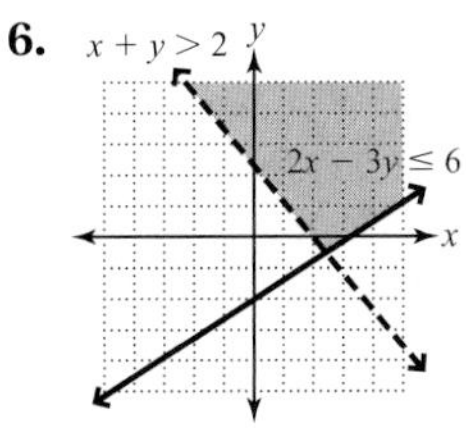

9.5 Exercises

1. Solve the related equation, then determine which of the three intervals satisfy the inequality by examining one value in the interval. If this one value satisfies the inequality, then all points in the interval satisfy the inequality. **3.** Yes, for example, $x^2 + 2 \leq 0$ has an empty solution set since $x^2 + 2$ is always positive. **5.** In both cases, the intervals to check are $(-\infty, -5)$, $(-5, 2)$, and $(2, \infty)$ and in both cases only x-values in $(-5, 2)$ satisfy the original inequalities. **7. a.** $x = -5, -1$ **b.** $(-5, -1)$ **c.** $(-\infty, -5) \cup (-1, \infty)$ **9. a.** $x = -1, 3$ **b.** $(-\infty, -1] \cup [3, \infty)$ **c.** $[-1, 3]$ **11. a.** $x = -2$ **b.** $(-\infty, -2) \cup (-2, \infty)$ **c.** $\varnothing$ **13. a.** $\varnothing$ **b.** $\varnothing$ **c.** $\mathbb{R}$ or $(-\infty, \infty)$

15. $(-4, -2)$

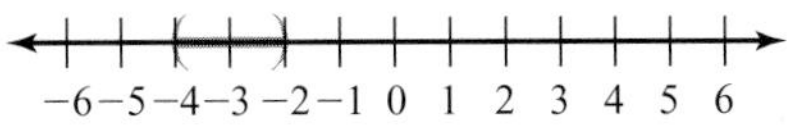

17. $(-\infty, 2) \cup (5, \infty)$

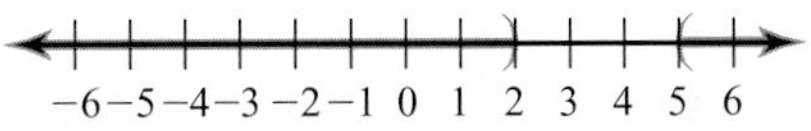

19. $(-4, -1)$

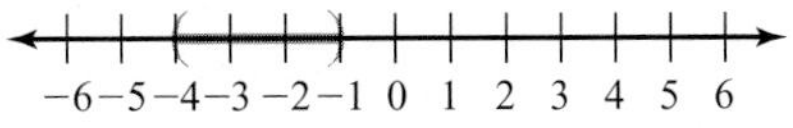

21. $(-\infty, 1) \cup (3, \infty)$

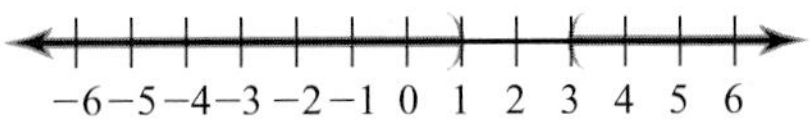

23. $[2, 4]$

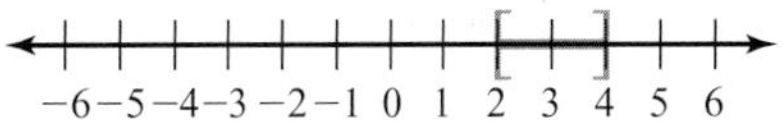

25. $(-\infty, -1] \cup [5, \infty)$

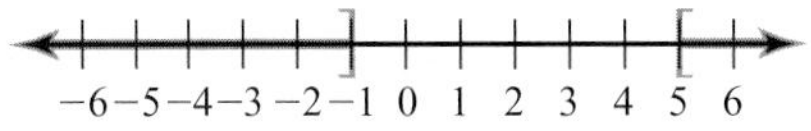

27. $(-2, 5)$

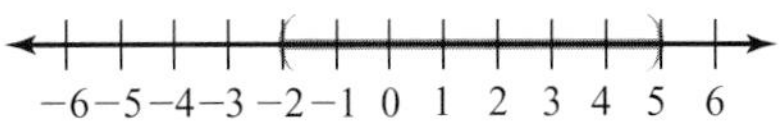

29. $(-\infty, \infty)$, or $\mathbb{R}$

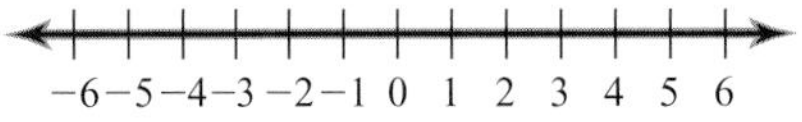

31. no solution, $\varnothing$

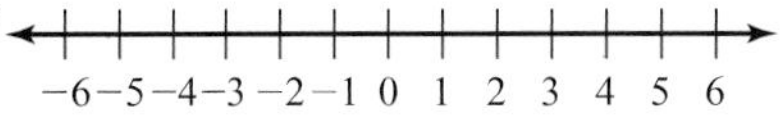

33. $(-\infty, -5) \cup (-0.25, \infty)$

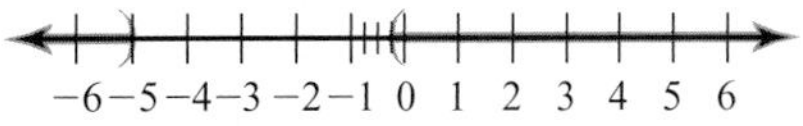

35. $(-\infty, 0) \cup (5, \infty)$

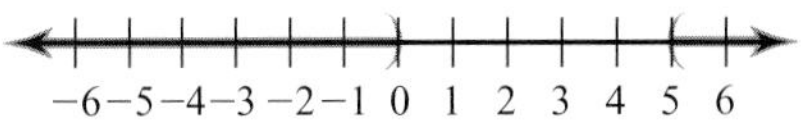

37. $[0, 6]$

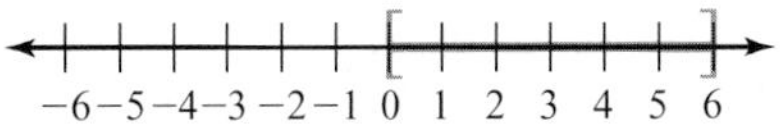

39. $(-\infty, \infty)$, or $\mathbb{R}$

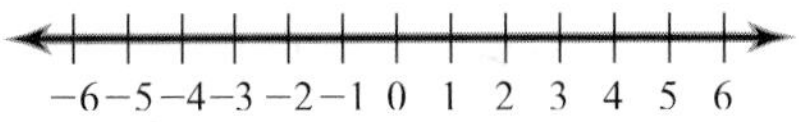

41. $[-4, -2] \cup [4, \infty)$

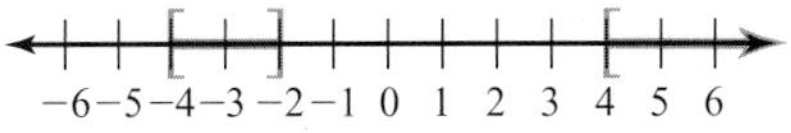

43. $(-\infty, -6) \cup (-2, 1)$

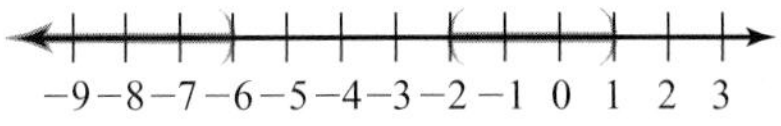

45. $\left(\dfrac{-2-\sqrt{7}}{3}, \dfrac{-2+\sqrt{7}}{3}\right)$

47. $\left(-\infty, \dfrac{-2-\sqrt{7}}{2}\right] \cup \left[\dfrac{-2+\sqrt{7}}{2}, \infty\right)$

49. $(-\infty, -4-\sqrt{6}] \cup [-4+\sqrt{6}, \infty)$

51. $(-\infty, -4) \cup (1, \infty)$

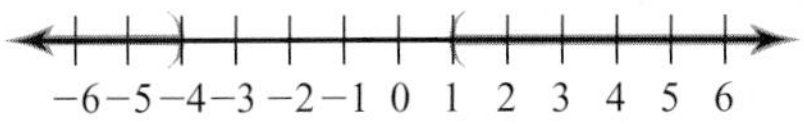

53. $(-5, -1]$

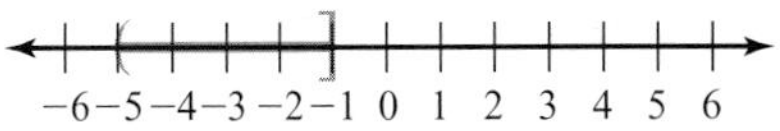

55. $(-4, \infty)$

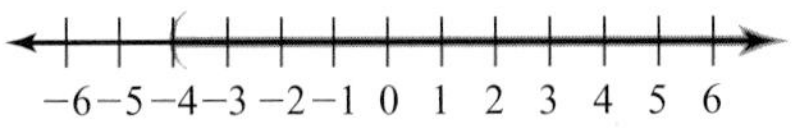

57. $\left(-\infty, -\dfrac{9}{2}\right) \cup (-3, \infty)$

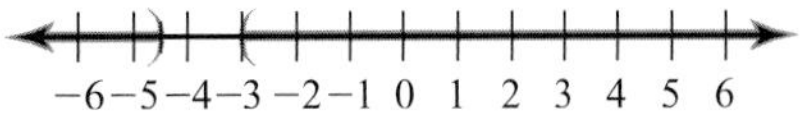

59. $(4, 7)$

61. $\left(3, \frac{14}{3}\right]$

63. $(-\infty, -3] \cup [2, 5)$

65. $(0, \infty)$

67. $(-\infty, -1) \cup (2, 5)$

69. a. 6 sec. **b.** At 2 sec. and again at 3 sec. **c.** $(2, 3)$ sec. **d.** $[0, 2) \cup (3, 6]$ **71. a.** $x \geq 4$ **b.** base ≥ 10, height ≥ 2 **73.** $2 < x \leq 8$

Review Exercises

1. 18 **2.** 9, $(x - 3)^2$ **3.** $\frac{25}{4}, \left(x + \frac{5}{2}\right)^2$ **4.** Downwards, because the coefficient of x^2 is negative. **5.** $(-6, 0), (2, 0)$ **6.** $(3, -4)$

Chapter 9 Review Exercises

1. false **2.** false **3.** false **4.** false **5.** true **6.** $+\sqrt{a}, -\sqrt{a}$ **7.** $\left(\frac{b}{2}\right)^2$ **8.** $\frac{-b \pm \sqrt{b^2 - 4ac}}{2a}$ **9.** $b^2 - 4ac$ **10.** $(h, k); x = h$ **11.** ± 4 **12.** $\pm\frac{1}{6}$ **13.** $\pm 2\sqrt{7}$ **14.** $\pm\sqrt{14}$ **15.** $\pm i\sqrt{3}$ **16.** $-12, -2$ **17.** $9 \pm 4i$ **18.** $-\frac{7}{5}, \frac{1}{5}$ **19.** $-7, -1$ **20.** $-8, 14$ **21.** $\frac{3 \pm i\sqrt{5}}{2}$ **22.** $-1, \frac{3}{4}$ **23.** $-1 \pm \sqrt{6}$ **24.** $\frac{1 \pm i\sqrt{2}}{3}$ **25.** $\frac{-1 \pm \sqrt{41}}{4}$ **26.** $-\frac{3}{20}, \frac{1}{10}$ **27.** $D = 64$; two rational **28.** $D = -71$; two nonreal complex **29.** $D = 0$; one rational **30.** $D = 0.48$; two irrational **31.** $\pm\frac{3\sqrt{2}}{2}$ **32.** $-\frac{5}{2}$ (5 is extraneous) **33.** $\frac{1}{3}, \frac{1}{2}$ **34.** $\frac{3 \pm \sqrt{17}}{4}$ **35.** no solution **36.** 1 $\left(\frac{1}{9}\text{ is extraneous}\right)$ **37.** 2 $\left(-\frac{3}{2}\text{ is extraneous}\right)$ **38.** $\frac{1}{3}, \frac{2}{3}$ **39.** $\pm\sqrt{2}, \pm\sqrt{3}$ **40.** $\pm 1, \pm\frac{\sqrt{2}}{2}$ **41.** $-\frac{9}{2}, -\frac{14}{3}$ **42.** $-26, -2$ **43.** $27, 512$ **44.** $\frac{16}{625}$ (1 is extraneous) **45. a.** $(0, 0)$ **b.** downwards **c.** $(0, 0)$ **d.** $x = 0$ **e.**

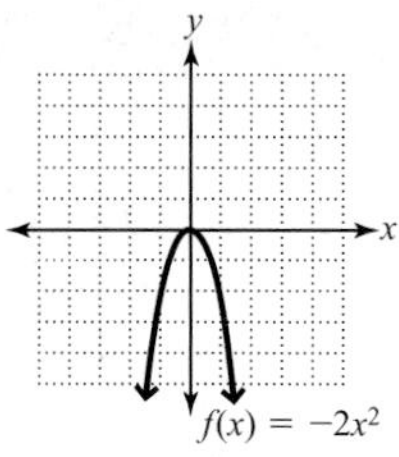

46. a. no x-intercepts, $(0, 1)$ **b.** upwards **c.** $(0, 1)$ **d.** $x = 0$ **e.**

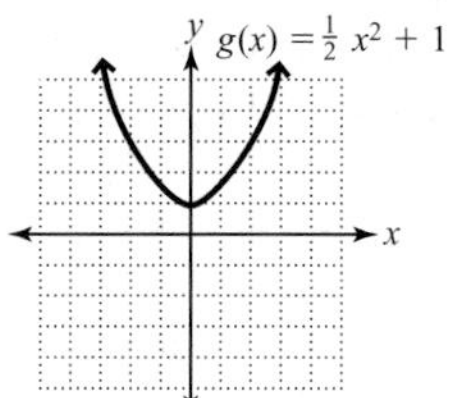

47. a. $(2, 0), \left(0, -\frac{4}{3}\right)$ **b.** downwards **c.** $(2, 0)$ **d.** $x = 2$ **e.**

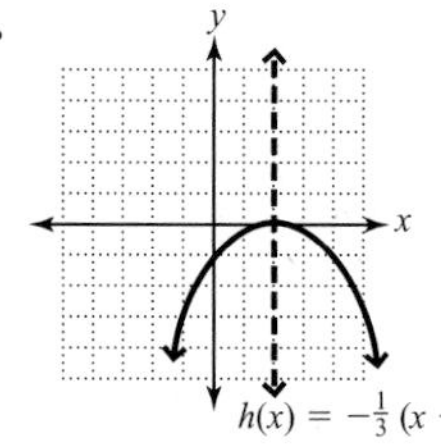

48. a. $\left(-3 + \frac{\sqrt{2}}{2}, 0\right), \left(-3 - \frac{\sqrt{2}}{2}, 0\right)$, $(0, 34)$ **b.** upwards **c.** $(-3, -2)$ **d.** $x = -3$ **e.**

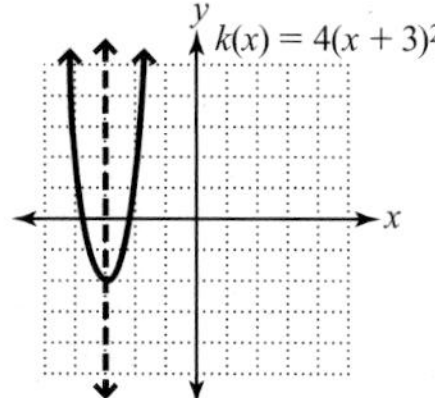

49. a. $(-1 + \sqrt{2}, 0), (-1 - \sqrt{2}, 0), (0, -1)$ **b.** $m(x) = (x + 1)^2 - 2$ **c.** upwards **d.** $(-1, -2)$ **e.** $x = -1$ **f.**

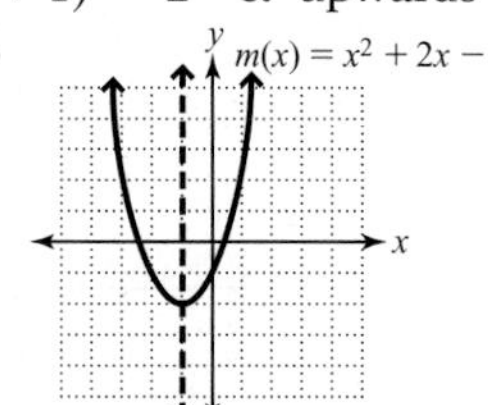

50. a. $(2, 0), (6, 0), (0, -6)$ **b.** $p(x) = -0.5(x + 4)^2 + 2$ **c.** downwards **d.** $(4, 2)$ **e.** $x = 4$ **f.**

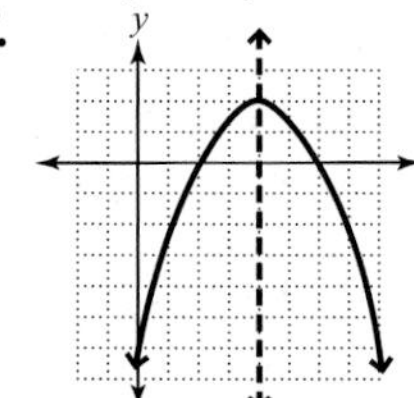

51. $(-\infty, -5) \cup (3, \infty)$ **52.** $[2, 4]$ **53.** $(-7, -2)$
54. $[-3, 1] \cup [2, \infty)$ **55.** $(-\infty, -3] \cup (1, \infty)$
56. $(-\infty, -4) \cup (-2, \infty)$ **57.** $(-\infty, 4) \cup \left[\frac{17}{4}, \infty\right)$
58. $(-\infty, -2) \cup (3, 5)$ **59.** 150 ft. **60.** 4 m/sec.
61. 15 ft. by 19 ft. **62.** $9\sqrt{3}$ **63.** 6.65 ft. and 15.65 ft.
64. a. 0.75 sec. **b.** 9 ft. **c.** 1.5 sec.
d.

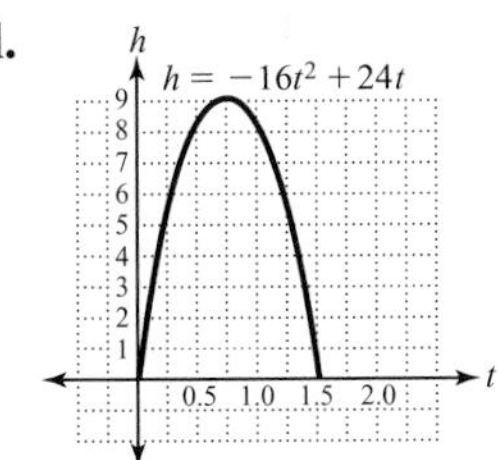

65. a. 38.88 yd. **b.** 72 yd. **66. a.** $x \geq 10$ in. **b.** height $\geq$ 8 in., base $\geq$ 14 in.

Chapter 9 Practice Test

1. ± 9 **2.** $3 \pm 2\sqrt{5}$
3. $4 \pm 2\sqrt{3}$ **4.** $\frac{3 \pm 2\sqrt{6}}{3}$ **5.** $-2, \frac{3}{2}$ **6.** 3, 5 **7.** $-8, 2$
8. ± 9 **9.** $\frac{-3 \pm i\sqrt{3}}{4}$ **10.** $0, -2$ **11.** $\pm 4i$ **12.** $0, -\frac{5}{3}$
13. $-\frac{6}{5}, 4$ **14.** $-\frac{2}{3}, 1$ **15.** no solution **16.** 1 (-4 is extraneous) **17.** $\pm\frac{1}{3}, \pm i\sqrt{3}$ **18.** $-5, 0$
19. a. $(3 + \sqrt{5}, 0)(3 - \sqrt{5}, 0), (0, -4)$
b. $f(x) = -(x - 3)^2 + 5$ **c.** downwards **d.** $(3, 5)$
e. $x = 3$ **f.**

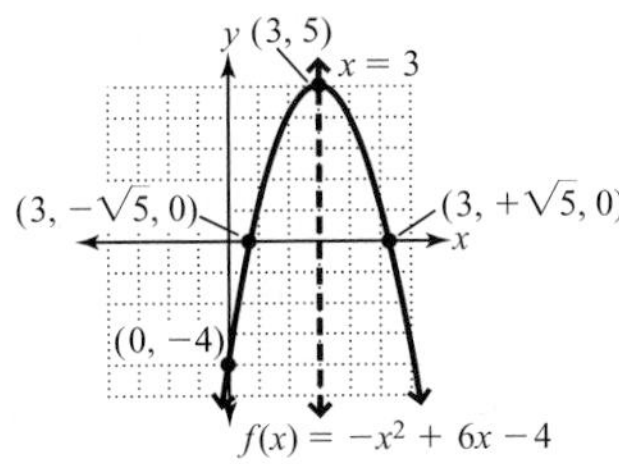

20. $[-1, 4]$
21. $(-\infty, -2) \cup (1, \infty)$
22. 2.5 sec. **23.** 12.36 ft. by 32.36 ft. **24. a.** 29.125 m
b. 70.66 m **c.**

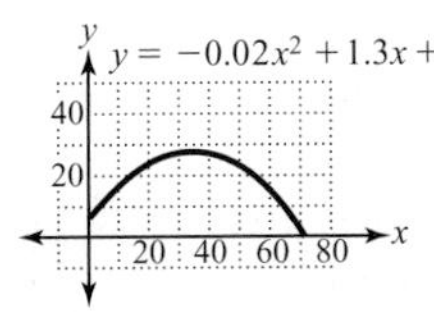

25. a. $0 < w \leq 25$ **b.** $15 < l \leq 40$

Chapters 1–9 Cumulative Review Exercises

1. true **2.** false **3.** false **4.** true **5.** parabola, $a > 0$
6. $\sqrt[n]{a \cdot b}$ **7.** b^n **8.** $\sqrt{a}, -\sqrt{a}$ **9.** $-6x - 6$ **10.** $-56mn^5$
11. $4x^3 - 14x^2 + 7x - 3$ **12.** $5m - 4 + \frac{6}{3m - 2}$
13. $(m + 2)(m^2 - 2m + 4)$ **14.** $(x + 5)^2$ **15.** $-\frac{2xy}{5z}$
16. $\frac{1}{u - 1}$ **17.** $\frac{a - b}{x + y}$ **18.** $\frac{x + 10}{(x - 2)(x + 2)}$
19. $x^{1/2}$ **20.** $41 + 23i$ **21.** $-\frac{2\sqrt{5}}{5}$ **22.** $\frac{\sqrt{mn} + n}{m - n}$
23. $-\frac{7}{2}$ **24.** all real numbers **25.** $-\frac{8}{3}, 0$ **26.** $\frac{3}{2}, -5$
27. ± 6 **28.** 8 **29.** 17 **30.** $\pm\frac{\sqrt{15}}{5}, \pm\frac{\sqrt{2}}{2}i$
31. $3 \pm i\sqrt{2}$ **32.** $m = \frac{2E}{v^2}$
33. a.
b. $\{x \mid -6 < x < 1\}$ **c.** $(-6, 1)$
34. a.
b. $\{x \mid -4 \leq x \leq 3\}$ **c.** $[-4, 3]$
35.

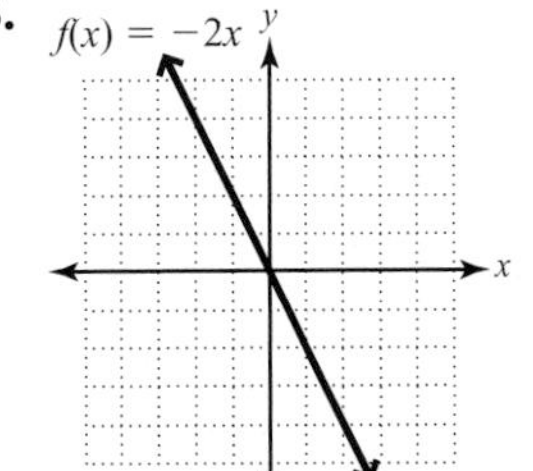

36.

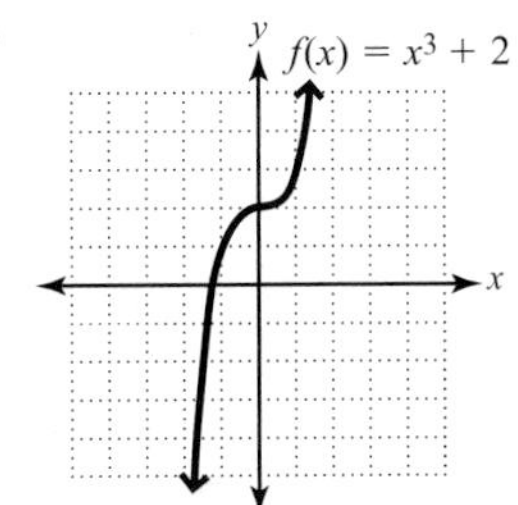

37. $2x + 3y = 8$ **38. a.** down
b. $\left(1 + \frac{\sqrt{6}}{2}, 0\right), \left(1 - \frac{\sqrt{6}}{2}, 0\right), (0, 1)$ **c.** $(1, 3)$ **d.** $x = 1$
e.

$f(x) = -2(x - 1)^2 + 3$

39. $(-3.5, -4)$ **40.** $(2, -1, 4)$ **41.** $(2, -3)$

42.

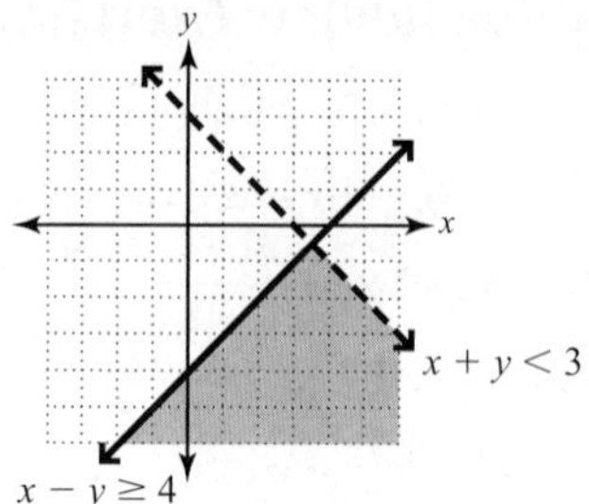

43. 5 hr. **44.** 75 ml **45.** pens: \$2.00, erasers: \$0.50, reams of paper: \$1.00 **46.** ≈ 5.5 hr. **47.** 3.2 cm/sec.

48. a. $\dfrac{\sqrt{6}}{2}$ **b.** 144 ft.

Chapter 10

10.1 Exercises

1. No, $f \circ g = g \circ f$ if f and g are inverses of each other and for a few other functions. **3.** The domain of a function is the range of its inverse; the range of a function is the domain of its inverse. This occurs because the ordered pairs of f and f^{-1} have x and y interchanged. **5.** If any horizontal line intersects the graph in more than one point, it is not one-to-one. **7.** 14 **9.** 3 **11.** 26 **13.** $3x^2 + 14$ **15.** $3\sqrt{x+1} + 5$ **17.** $\sqrt{6}$ **19.** $(f \circ g)(x) = 6x + 6$; $(g \circ f)(x) = 6x - 2$ **21.** $(f \circ g)(x) = x^2 + 3$; $(g \circ f)(x) = x^2 + 4x + 5$ **23.** $(f \circ g)(x) = 9x^2 + 9x - 4$; $(g \circ f)(x) = 3x^2 + 9x - 12$ **25.** $(f \circ g)(x) = \sqrt{2x-3}$; $(g \circ f)(x) = 2\sqrt{x+2} - 5$ **27.** $(f \circ g)(x) = \dfrac{2x-3}{x-3}$; $(g \circ f)(x) = \dfrac{1-2x}{x+1}$ **29.** Domain is $[0, \infty)$ and range is $[3, \infty)$. **31.** yes **33.** no **35.** yes **37.** yes **39.** no **41.** yes **43.** no **45.** yes **47.** yes **49.** Yes, since $(f \circ g)(x) = (g \circ f)(x) = x$.

51.

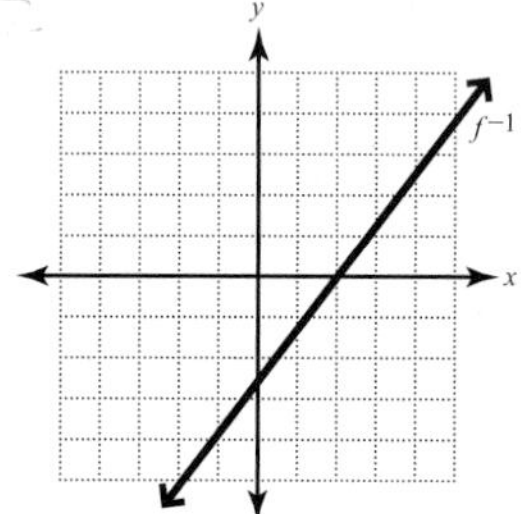

53.

55.

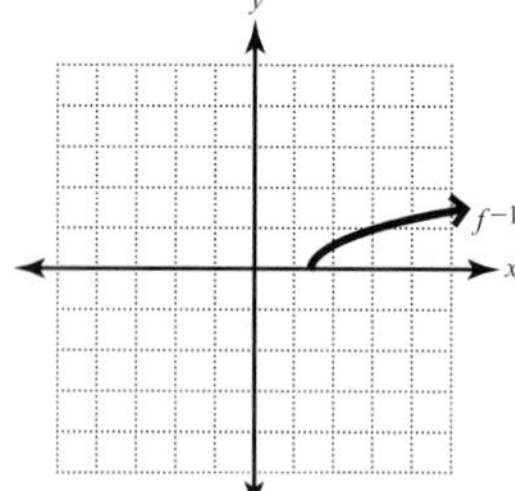

57.

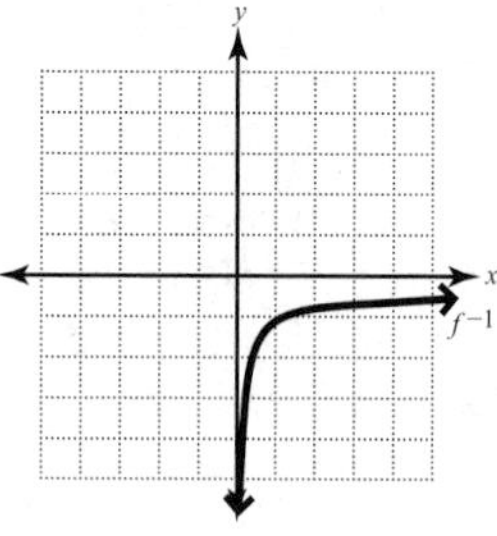

59.

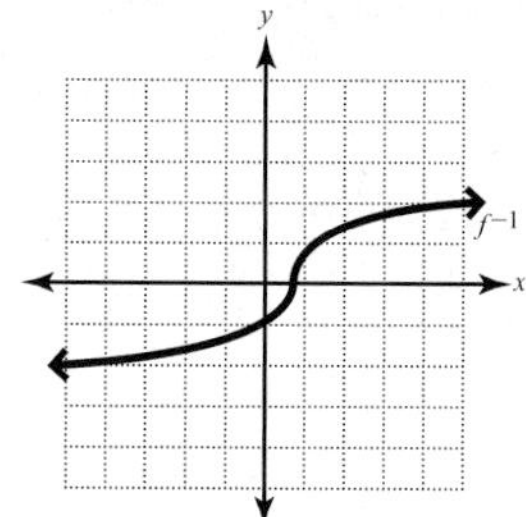

61.

63. yes **65.** no **67.** $f^{-1} = \{(2, -3), (-3, -1), (4, 0), (6, 4)\}$ **69.** $f^{-1} = \{(-2, 7), (2, 9), (1, -4), (3, 3)\}$ **71.** $f^{-1}(x) = x - 6$ **73.** $f^{-1}(x) = \dfrac{x-3}{2}$ **75.** $f^{-1}(x) = \sqrt[3]{x-2}$ **77.** $f^{-1}(x) = \dfrac{2-2x}{x}$ **79.** $f^{-1}(x) = \dfrac{3x+2}{x-1}$ **81.** $f^{-1}(x) = x^2 + 2, x \geq 0$ **83.** $f^{-1}(x) = \sqrt[3]{\dfrac{x-4}{2}}$ **85.** $f^{-1}(x) = x^3 - 2$ **87.** $f^{-1}(x) = \dfrac{x^3}{16} - 2$ **89. a.** $y = \dfrac{x - 100}{0.05}$ **b.** x represents the salary, y represents the sales **c.** \$5000 **91. a.** $y = \dfrac{6825 - x}{20}$ **b.** x represents the cost, y represents the number of 36-in. fans **c.** 80 36-in. fans **93.** $(5, -4)$ **95.** $f^{-1}(x) = \dfrac{x-b}{a}$ **97.** No, the inverse of an even function is not a function. If the graph of a function is symmetric with respect to the y-axis, the function is not one-to-one.

Review Exercises

1. $32 = 2^5$ **2.** 8 **3.** $-\dfrac{1}{27}$ **4.** $\dfrac{1}{16}$ **5.** 8 **6.** (3, 1)

10.2 Exercises

1. Yes, the graphs are symmetric. The point (1, 2) on $f(x)$ corresponds to the point $(-1, 2)$ on $g(x)$, etc. **3.** The domain is all real numbers, the range is the interval $(0, \infty)$. **5.** Because the exponential function is one-to-one.

7.

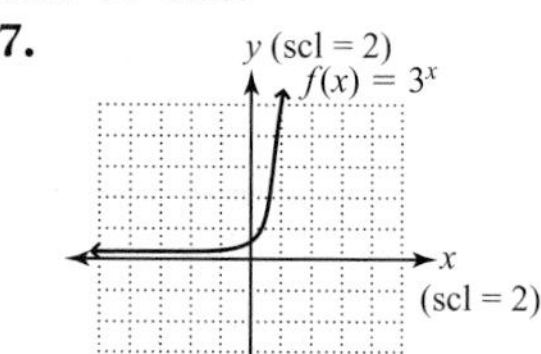

9.

11.

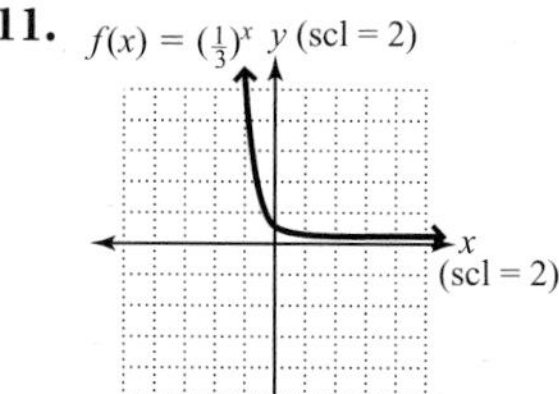

13.

15.

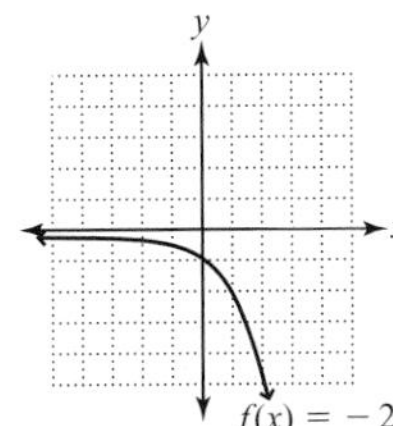

17.

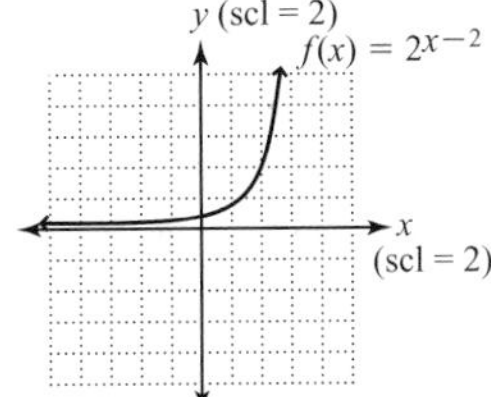

19.

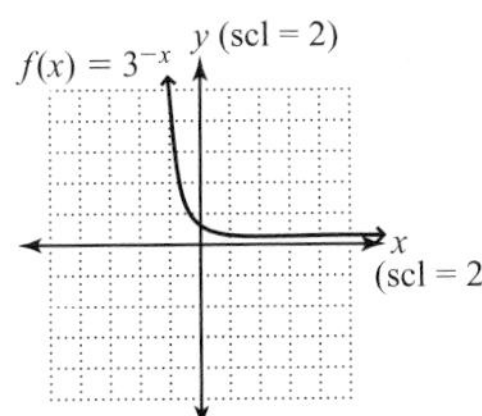

21.

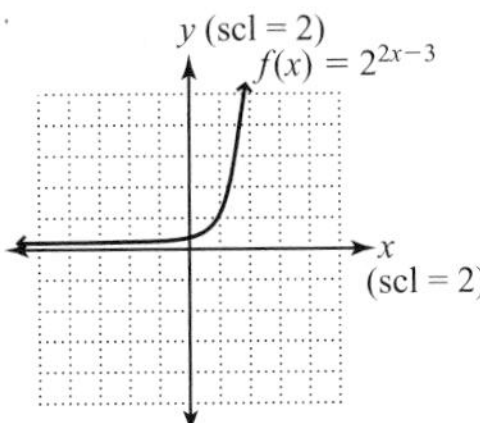

23.

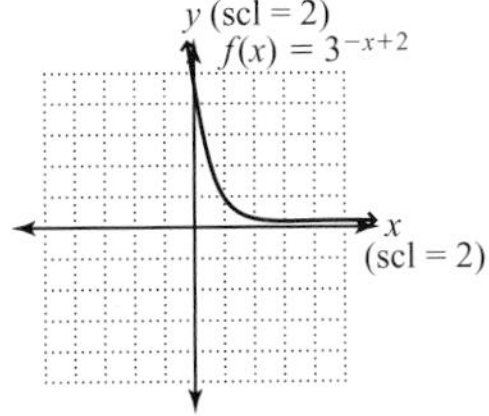

25. 3 **27.** $\frac{5}{3}$ **29.** $\frac{1}{2}$ **31.** -2 **33.** -2 **35.** 3 **37.** -4

39. $\frac{1}{2}$ **41.** -12 **43. a.**

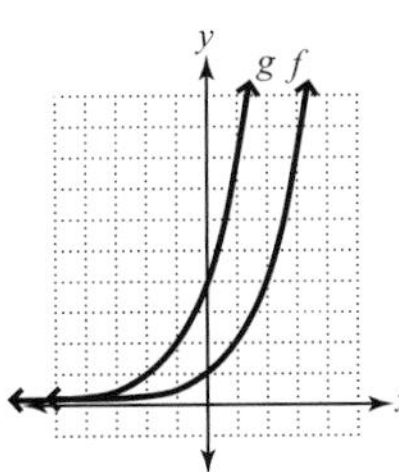

b. The graph of *g* is the graph of *f* shifted 2 units to the left.

45. a.

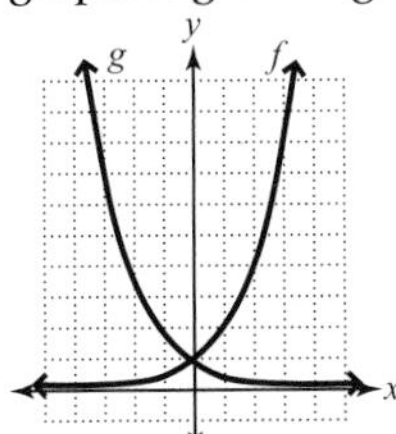

b. The graph of *g* is the graph of *f* reflected in the *y*-axis.
47. 6400 cells **49.** 6144 billion people **51.** \$25,870.70
53. 0.020 g **55.** \$1046 **57.** 1.225 parts per million
59. $\approx$ 923 million transistors

Review Exercises

1. 125 **2.** 9 **3.** $\frac{1}{64}$ **4.** $\sqrt[3]{5^2}$ or $(\sqrt[3]{5})^2$ **5.** $f^{-1}(x) = \frac{x-4}{3}$ **6.**

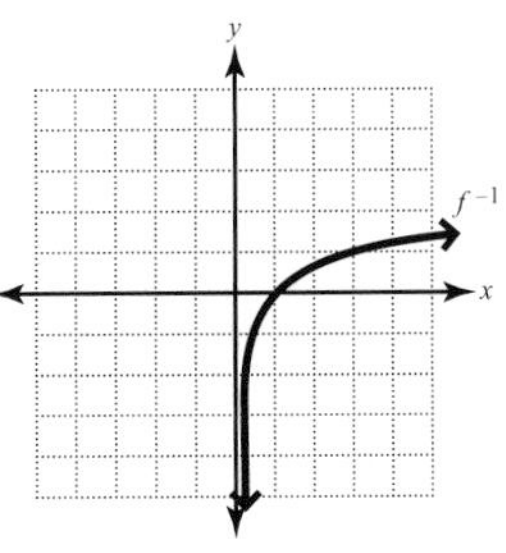

10.3 Exercises

1. $f^{-1}(x) = \log_3 x$, because logarithmic and exponential functions are inverses. **3.** $x > 0$
5. Logarithms are exponents because logarithms are inverses of exponential functions. **7.** $\log_2 32 = 5$ **9.** $\log_{10} 1000 = 3$
11. $\log_e x = 4$ **13.** $\log_5 \frac{1}{125} = -3$ **15.** $\log_{10} \frac{1}{100} = -2$
17. $\log_{625} 5 = \frac{1}{4}$ **19.** $\log_{1/4} \frac{1}{16} = 2$ **21.** $\log_7 \sqrt{7} = \frac{1}{2}$
23. $3^4 = 81$ **25.** $4^{-2} = \frac{1}{16}$ **27.** $10^2 = 100$ **29.** $e^5 = a$
31. $e^{-4} = \frac{1}{e^4}$ **33.** $\left(\frac{1}{8}\right)^2 = \frac{1}{64}$ **35.** $\left(\frac{1}{5}\right)^{-2} = 25$
37. $7^{1/2} = \sqrt{7}$ **39.** 8 **41.** $\frac{1}{25}$ **43.** 4 **45.** -1 **47.** 10
49. 2 **51.** $\frac{1}{4}$ **53.** 243 **55.** 2 **57.** -6 **59.** 16 **61.** $\frac{1}{64}$
63. 2 **65.** 125 **67.** 3 **69.** $\frac{1}{16}$
71.

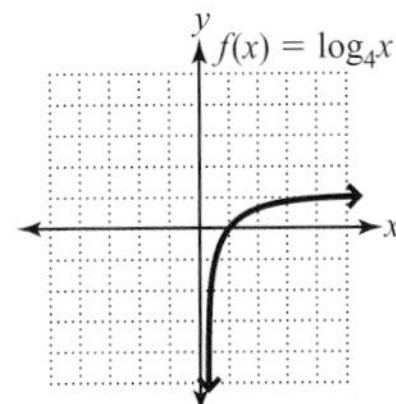

73.

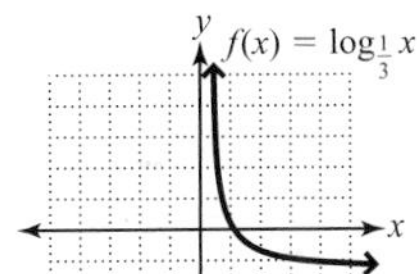

75. $\log_b b = 1$ because $b^1 = b$ **77.** $\log_b 1 = 0$ because $b^0 = 1$.
79. 97% **81.** after 4 days`

Review Exercises

1. x^{-6} **2.** x^6 **3.** x^{15} **4.** x^3 **5.** x^5 **6.** $x^{3/4}$

10.4 Exercises

1. Because the exponential and logarithmic functions are inverses. **3.** $\log_b(x + y)$; there is no rule for the logarithm of a sum. **5.** 2 **7.** r **9.** $4x$
11. 5 **13.** a **15.** $7x$ **17.** $\log_2 5 + \log_2 y$
19. $\log_a p + \log_a q$ **21.** $\log_4 m + \log_4 n + \log_4 p$
23. $\log_a x + \log_a(x - 5)$ **25.** $\log_3(40)$ **27.** $\log_4(16)$
29. $\log_a 7m$ **31.** $\log_4 ab$ **33.** $\log_a(2x^2 + 10x)$
35. $\log_4(x^2 + 4x + 3)$ **37.** $\log_2 7 - \log_2 9$

39. $\log_a x - \log_a 5$ **41.** $1 - \log_a b$ **43.** $\log_a x - \log_a(x - 3)$ **45.** $\log_4(2x - 3) - \log_4(4x + 5)$ **47.** $\log_6 8$ **49.** $\log_2 2 = 1$ **51.** $\log_a \frac{x}{3}$ **53.** $\log_4 \frac{p}{q}$ **55.** $\log_b \frac{x}{x - 4}$ **57.** $\log_x x = 1$ **59.** $3 \log_7 4$ **61.** $7 \log_a x$ **63.** $\frac{1}{2}\log_a 3$ **65.** $\frac{2}{3}\log_3 x$ **67.** $-2 \log_a 6$ **69.** $-2 \log_a y$ **71.** $\log_3 5^4$ **73.** $\log_2 \frac{1}{x^3}$ **75.** $\log_5 2$ **77.** $\log_a \sqrt[4]{x^3}$ **79.** $\log_a 4$ **81.** $\log_3 \frac{1}{\sqrt{x}}$ **83.** $3 \log_a x - 4 \log_a y$ **85.** $4 \log_3 a + 2 \log_3 b$ **87.** $\log_a x + \log_a y - \log_a z$ **89.** $2 \log_x a - \log_x b - 3 \log_x c$ **91.** $\frac{3}{2}\log_4 x - \frac{1}{2}\log_4 y$ **93.** $\log_a x + \frac{1}{2}\log_a y - \frac{3}{2}\log_a z$ **95.** $\log_3 \frac{1}{2}$ **97.** $\log_b x^4 y^3$ **99.** $\log_a \sqrt{\frac{5}{7}}$ **101.** $\log_a \sqrt[3]{(x^2y)^2}$ **103.** $\log_b(3x^2 - 2x)$ **105.** $\log_a \frac{(x - 2)^3}{(x + 1)^4}$ **107.** $\log_a \frac{x^2 z^4}{w^3 u^6}$

Review Exercises

1. $10^{-2.5}$ **2.** 10^9 **3.** $13,468.55 **4.** $\log_{10} 50 = 1.6990$ **5.** $10^{1.6532} = 45$ **6.** $e^{-1.3863} = 0.25$

10.5 Exercises

1. $b > 0, b \neq 1$. If $y = \log_b x$, then $b^y = x$. If $b < 0$ and $y = \frac{1}{2}$ or any other fraction whose denominator is even, then x is imaginary. If $b = 1$, then $x = 1$ for all values of y and the graph is a vertical line. **3.** Positive. If $y = \ln x$, then $e^y = x$. If $x > 1$, then $e^y > 1$, which is true if $y > 0$. **5.** 0.91629 **7.** 1.8062 **9.** −2.1739 **11.** 2.3941 **13.** 4.1761 **15.** −5.7959 **17.** 2.2343 **19.** 4.3720 **21.** −3.3814 **23.** 5.6864 **25.** 0.4343 **27.** Error results, because the domain of $\log_a x$ is $(0, \infty)$, so log 0 is undefined. **29.** 2 **31.** −1 **33.** $\frac{1}{3}$ **35.** −3 **37.** 3 **39.** $\frac{1}{2}$ **41.** 90 dB **43.** 10^{-6} watts/m^2 **45.** 2.796 **47.** $10^{-3.5}$ moles/L **49.** 22.9, or ≈ 23 yr. **51.** The 1964 Alaska earthquake was about four times as severe as the 1906 San Francisco earthquake. **53.** $10^{22.75}$ ergs **55. a.** ≈ 11,809 deaths **b.** 1997 **57.** 6124 ft.

Review Exercises

1. −11 **2.** −5, 3 **3.** 8 **4.** 10 **5.** $\log_3 \frac{2x + 1}{x - 1}$ **6.** $3^2 = 2x + 5$

10.6 Exercises

1. $m = n$ **3.** If $\log_b x = \log_b y$, then $x = y$. **5.** The solution $x = -5$ is rejected because substituting into the equation gives $\log(-5)$ and $\log(-3)$, which are both undefined. **7.** 3.1699 **9.** 1.0767 **11.** −1.5693 **13.** 2.8617 **15.** 12.4251 **17.** 7.1285 **19.** 0.5365 **21.** 107.2959 **23.** −73.1563 **25.** 6 **27.** 11 **29.** 6 **31.** 4 **33.** 4, −9 **35.** 7 **37.** no solution **39.** −6, 2 **41.** 2 **43.** 10 **45.** 3 **47.** $\frac{3}{2}$ **49.** 3 **51.** 4 **53.** $\frac{5}{6}$ **55.** $\frac{5}{2}$ **57.** 6 **59.** 4 **61.** 9.5 yr. **63. a.** $19,676.82 **b.** 9.3 yr. **65. a.** 5.58% **b.** 0.18 **67. a.** 875 mosquitoes **b.** 75 days **69. a.** ≈ 7.44 billion **b.** ≈ 9 yr. after 2005, in 2014. **71. a.** 1.89 in. of mercury **b.** 21.8 mi. **73. a.** 79.8% **b.** 7.76, so about age 8. **75.** ≈ 39,006 **77. a.** $167,925 **b.** ≈ 19.1 after 1990, in 2009. **79.** ≈ 12 in. **81.** 1.7925 **83.** 0.5283 **85.** −2.3219 **87.** 0.3685

Review Exercises

1. $x + 3$ **2.** $28 + \sqrt{5}$ **3.** 29 **4.** $x^2 + 2x + 3$ **5.** $4x^2 - 4x + 5$ **6.** 13

Chapter 10 Review Exercises

1. true **2.** false **3.** false **4.** false **5.** true **6.** false **7.** 25 **8.** 167 **9.** −2 **10.** 14 **11.** $(f \circ g)(x) = 6x + 3, (g \circ f)(x) = 6x - 9$ **12.** $(f \circ g)(x) = 9x^2 - 42x + 53, (g \circ f)(x) = 3x^2 + 5$ **13.** $(f \circ g)(x) = \sqrt{2x - 4}, (g \circ f)(x) = 2\sqrt{x - 3} - 1$ **14.** $(f \circ g)(x) = \frac{4x - 4}{x - 4}, (g \circ f)(x) = \frac{3 - 3x}{x + 3}$ **15.** no **16.** yes

17. yes **18.** yes **19.** no **20.** no **21.** $f^{-1}(x) = \frac{x - 4}{5}$ **22.** $f^{-1}(x) = \sqrt[3]{x - 6}$ **23.** $f^{-1}(x) = \frac{4 - 5x}{x}$ **24.** $f^{-1}(x) = \frac{x^3 - 2}{3}$ **25.** (−6, 4) **26.** $f^{-1}(b) = a$

27.

28.

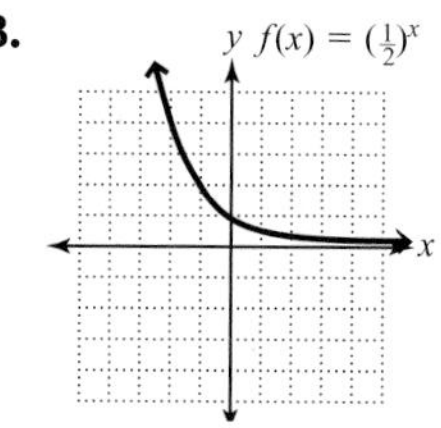

29.

30.

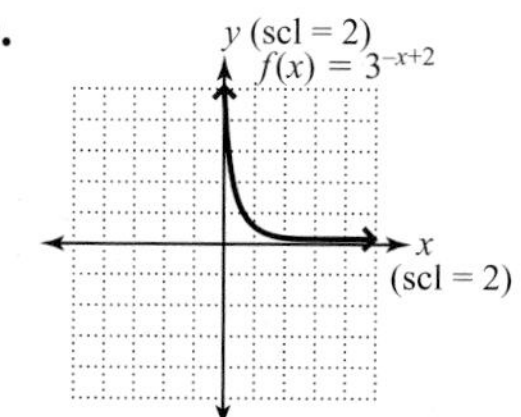

31. 4 **32.** $\frac{3}{2}$ **33.** −2 **34.** −2 **35.** 4 **36.** 2 **37.** 3 **38.** $-\frac{6}{7}$ **39.** 6277 cells **40.** $40,353.57 **41.** 7.16 g **42.** ≈ 865 addicts. **43.** $\log_7 343 = 3$ **44.** $\log_4 \frac{1}{64} = -3$

45. $\log_{3/2}\frac{81}{16} = 4$ **46.** $\log_{11}\sqrt[3]{11} = \frac{1}{3}$ **47.** $9^2 = 81$
48. $\left(\frac{1}{5}\right)^{-3} = 125$ **49.** $a^4 = 16$ **50.** $e^b = c$ **51.** $\frac{1}{81}$ **52.** 4
53. 5 **54.** -2 **55.** 3 **56.** $\frac{1}{10}$ **57.** 2 **58.** 11
59.

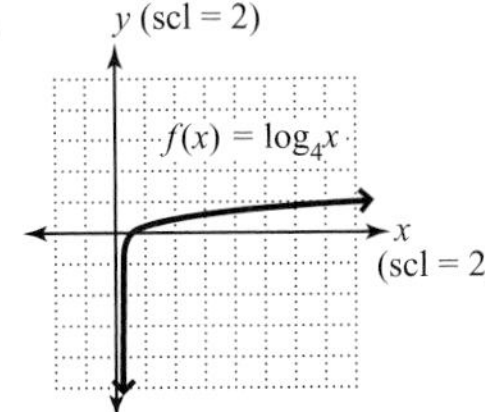

60.

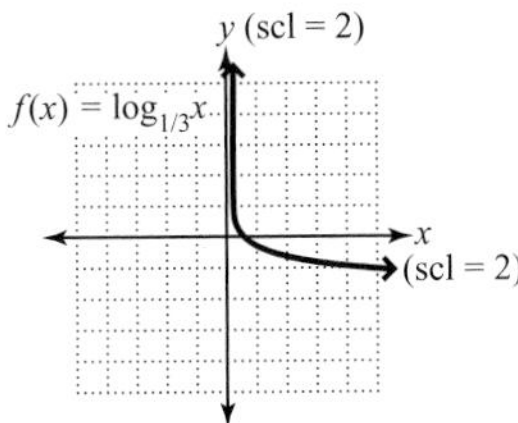

61. Domain is $(-\infty, \infty)$ and range is $(0, \infty)$ because $f(x) = \log_b x$ and $g(x) = b^x$ are inverses. **62.** 30 dB **63.** 8
64. 6 **65.** $1 + \log_6 x$ **66.** $\log_4 x + \log_4(2x - 5)$
67. $\log_3(32)$ **68.** $\log_5(3x^2 - 6x)$ **69.** $\log_b x - \log_b 5$
70. $\log_a(3x - 2) - \log_a(4x + 3)$ **71.** $\log_5 6$
72. $\log_2\frac{x + 5}{2x - 3}$ **73.** $4\log_3 7$ **74.** $\frac{1}{3}\log_a x$ **75.** $-4\log_4 a$
76. $\frac{4}{5}$ **77.** $\log_a x^4$ **78.** $\log_a\sqrt[5]{y^3}$ **79.** $2\log_a x + 3\log_a y$
80. $4\log_a c - 3\log_a d$ **81.** $2\log_a x + 3\log_a y - 4\log_a z$
82. $\frac{3}{2} - 2\log_a b$ **83.** $\log_a b^2c^4$ **84.** $\log_a\frac{4^3}{3^2}$ **85.** $\log_a\sqrt[4]{x^2y^3}$
86. $\log_a(x + 5)^4(x - 3)^2$ **87.** 2.5132 **88.** -2.4559
89. -3.1466 **90.** 4.5218 **91.** -5 **92.** $\frac{1}{4}$ **93.** 85 dB
94. pH = 2.9 **95.** ≈ 17 yr. **96.** The 7.8 earthquake is 10 times as severe. **97.** 1.5773 **98.** 0.5360 **99.** 1.3626
100. 4.9592 **101.** 0.5995 **102.** -536.4793 **103.** 13
104. $-4, 2$ **105.** 3 **106.** 20 **107.** 36 **108.** 8 **109.** $\frac{5}{2}$
110. 5 **111.** 6.1 yr. **112. a.** \$12,254.71 **b.** 7.7 yr.
113. a. 573 ants **b.** ≈ 23 months **114. a.** 36,165 cases
b. 1998 **115.** 1.9534 **116.** -2.5850

Chapter 10 Practice Test

1. $f[g(x)] = 9x^2 - 30x + 19$ **2. a.** $f^{-1}(x) = \frac{x + 3}{4}$

b. $f[f^{-1}(x)] = f\left[\frac{x + 3}{4}\right] = 4\left(\frac{x + 3}{4}\right) - 3 = x + 3 - 3 = x$

$f^{1}[f(x)] = f^{-1}(4x - 3) = \frac{4x - 3 + 3}{4} = \frac{4x}{4} = x$

c. The graphs are symmetric about the graph of $y = x$.

3.

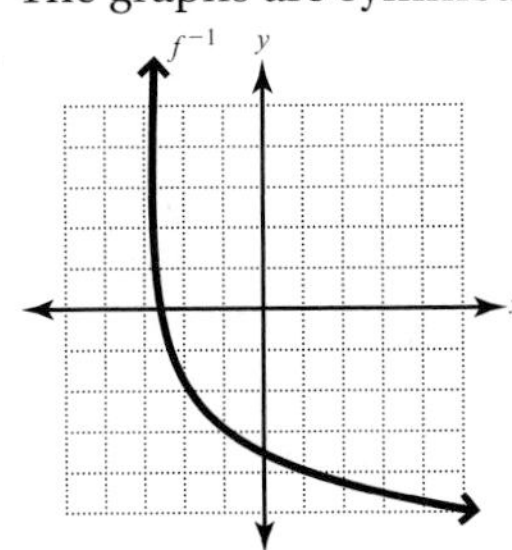

4.

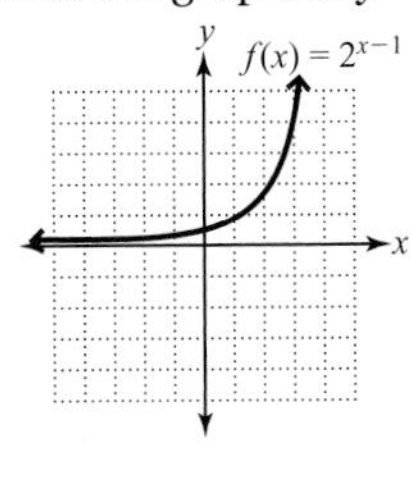

5. -10 **6. a.** \$42,143.63 **b.** ≈ 8.2 yr. **7. a.** 0.68 g **b.** ≈ 1.16 hr. **8. a.** 14.8 million people, 46.9 million people **b.** 1989 **9.** $\left(\frac{1}{3}\right)^{-4} = 81$ **10.** -3 **11.** 5

12.

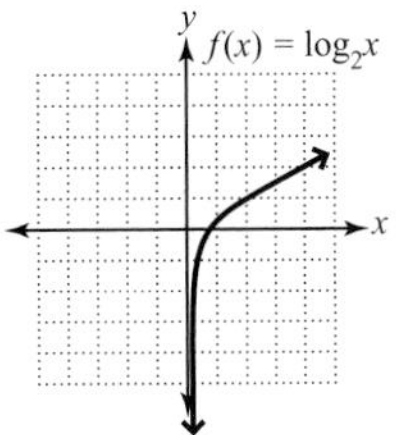

13. $4\log_b x + 2\log_b y - \log_b z$ **14.** $\frac{5}{4}\log_b x - \frac{7}{4}\log_b y$
15. $\log_b\sqrt[4]{x^6y^9}$ **16. a.** ≈ 74 g **b.** ≈ 247.6 days.
17. 4.6433 **18.** 3 **19.** 3 **20 a.** 307 cases **b.** 2001

Chapter 11

11.1 Exercises

1. It is a parabola, because only one variable is squared. **3.** Parabola opens to the right, because the variable y is squared and $a = 3$, which is positive.
5. circle; $(-3, 2)$; 4 **7.** opens upwards; vertex: $(1, 2)$; axis of symmetry: $x = 1$

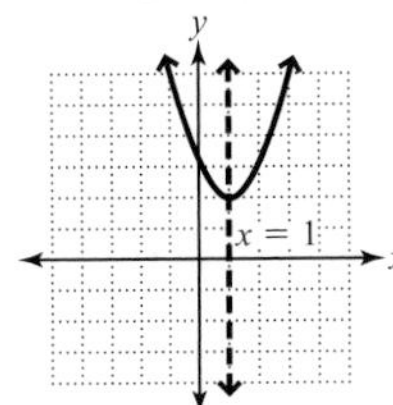

9. opens downwards; vertex: $(-1, 4)$; axis of symmetry: $x = -1$

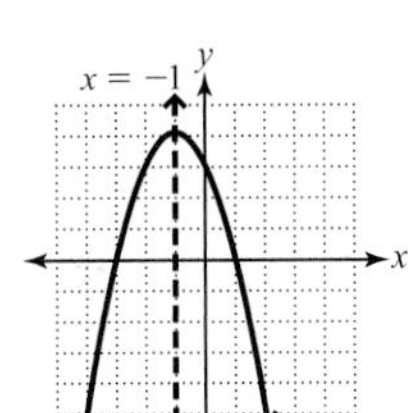

11. opens right; vertex: $(-2, -2)$; axis of symmetry: $y = -2$

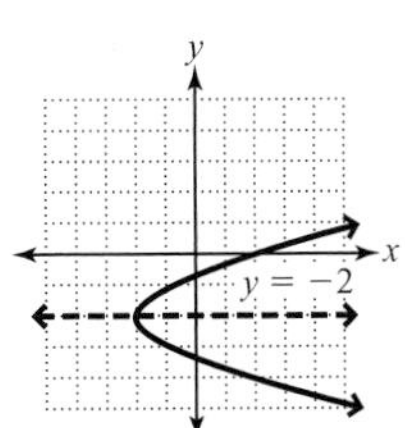

13. opens left; vertex: $(3, 1)$; axis of symmetry: $y = 1$

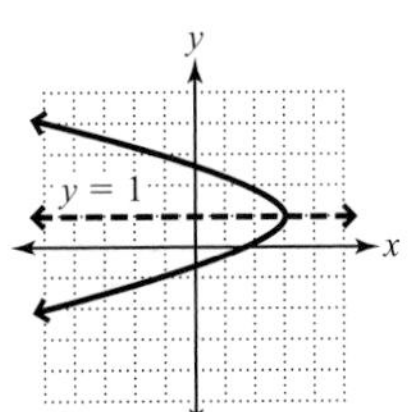

15. opens right; vertex: $(-4, -2)$; axis of symmetry: $y = -2$

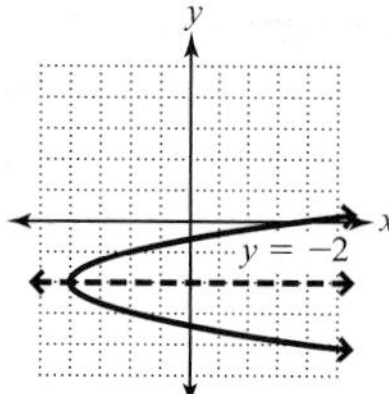

17. opens left; vertex: $(-5, -2)$; axis of symmetry: $y = -2$

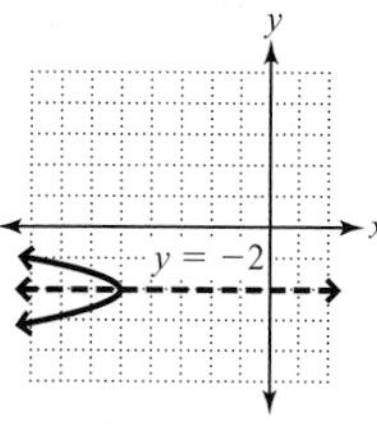

19. opens right; vertex: $(-1, -2)$; axis of symmetry: $y = -2$

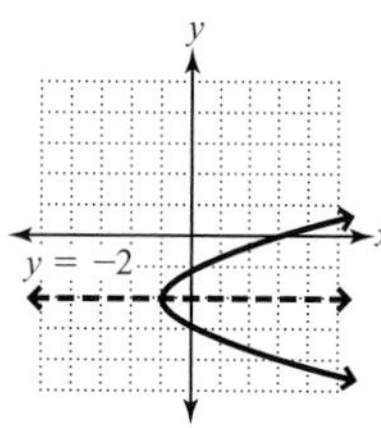

21. opens left; vertex: $(4, 3)$; axis of symmetry: $y = 3$

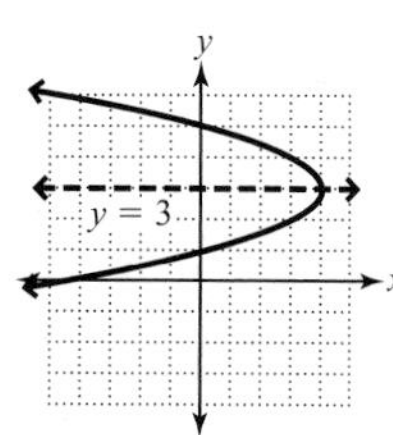

23. opens right; vertex: $(-5, -2)$; axis of symmetry: $y = -2$

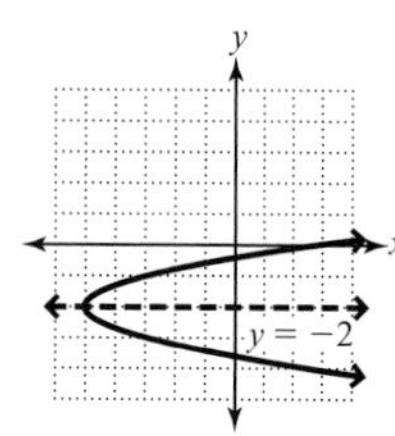

25. opens right; vertex: $(0, 1)$; axis of symmetry: $y = 1$

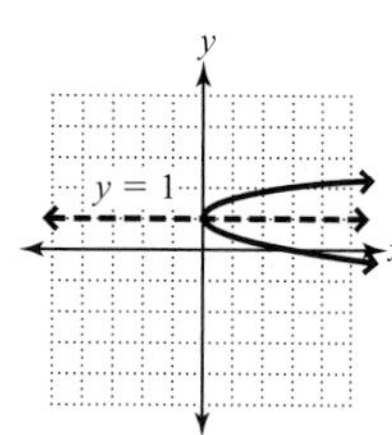

27. opens left; vertex: $(7, 1)$; axis of symmetry: $y = 1$

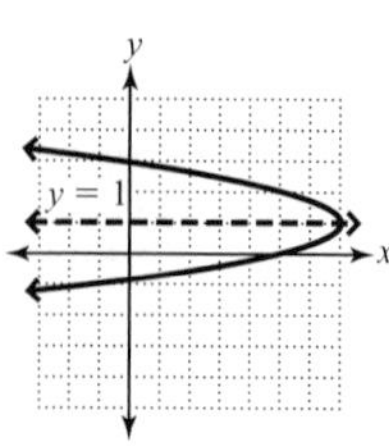

29. c **31.** a **33.** 5 **35.** 13 **37.** 13 **39.** $2\sqrt{5}$ **41.** $4\sqrt{5}$

43. $\sqrt{34}$ **45.** 5 **47.** $\sqrt{89}$

49. center: (2,1); radius: 2

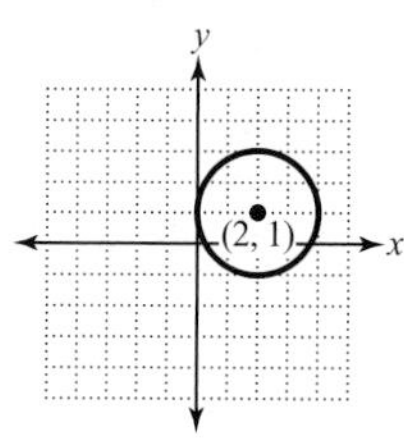

51. center: $(-3, -2)$; radius: 9

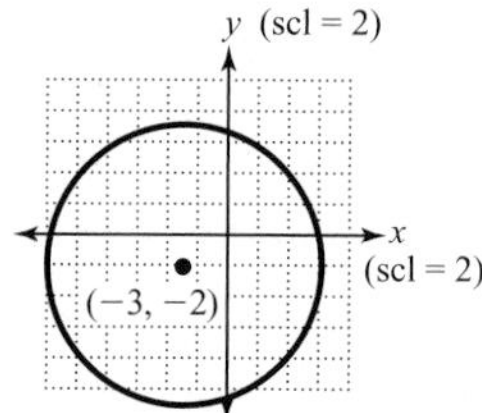

53. center: $(5, -3)$; radius: 7

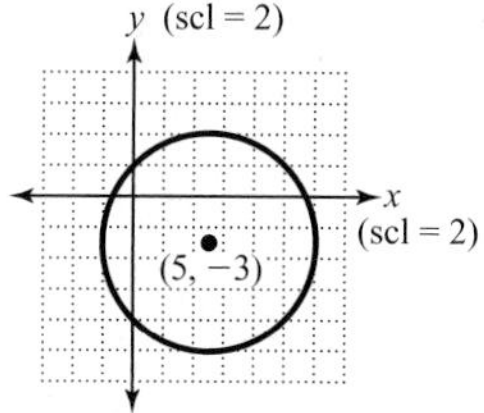

55. center: $(-1, 3)$; radius: $3\sqrt{2}$

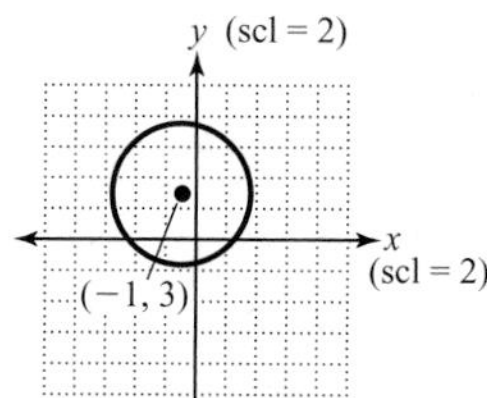

57. center: $(-4, -2)$; radius: $4\sqrt{2}$

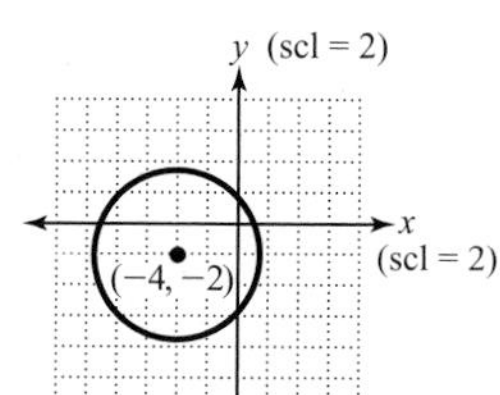

59. center: $(1, 3)$; radius: 7

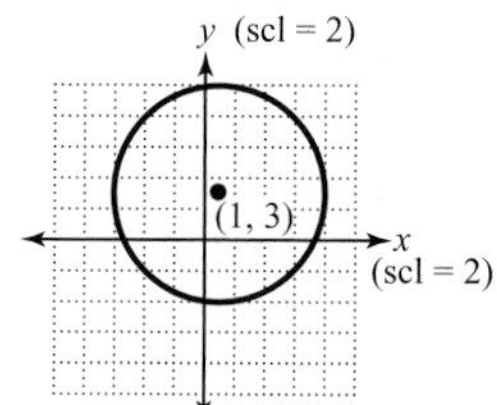

61. center: $(-5, 2)$; radius: 8

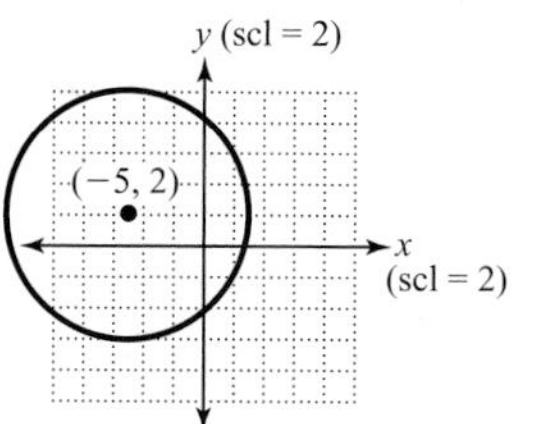

63. center: $(-7, 2)$; radius: 2

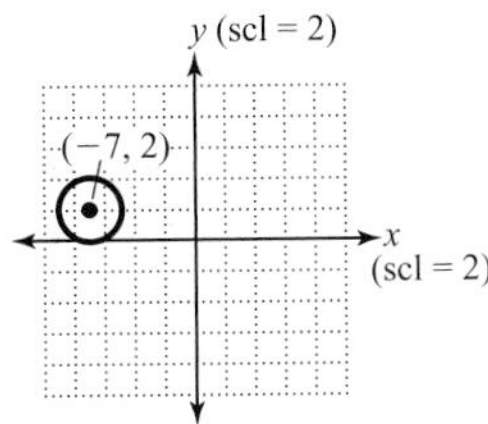

65. b **67.** c **69.**

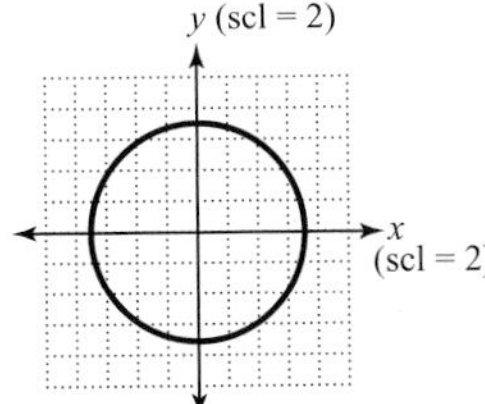

71.

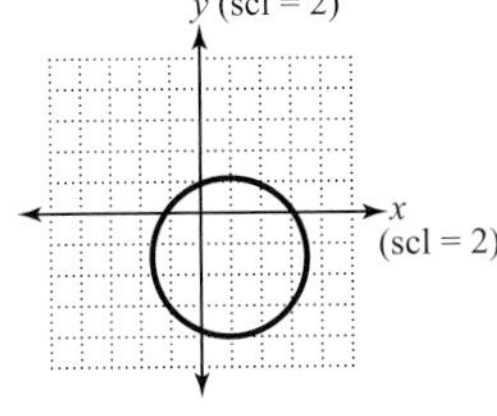

73. $(x - 8)^2 + (y + 1)^2 = 36$ **75.** $(x + 4)^2 + (y + 3)^2 = 25$
77. $(x - 6)^2 + (y + 2)^2 = 14$ **79.** $(x + 5)^2 + (y - 2)^2 = 45$
81. $(x - 2)^2 + (y - 4)^2 = 25$ **83.** $(x - 2)^2 + (y - 4)^2 = 169$
85. $(x - 2)^2 + (y + 5)^2 = 64$ **87. a.** 256 ft. **b.** 3 sec.
c. 7 sec. **89.** $y = -\frac{2}{25}x^2 + 18$ or $y = -0.08x^2 + 18$
91. 3.5% **93.** $x^2 + y^2 = 400$ **95.** $x^2 + (y - 110)^2 = 10{,}000$
97. $x^2 + y^2 = 43{,}560{,}000$

Review Exercises

1.

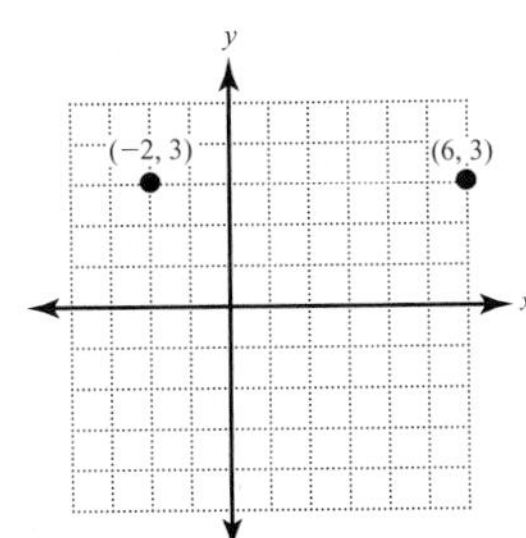

2.

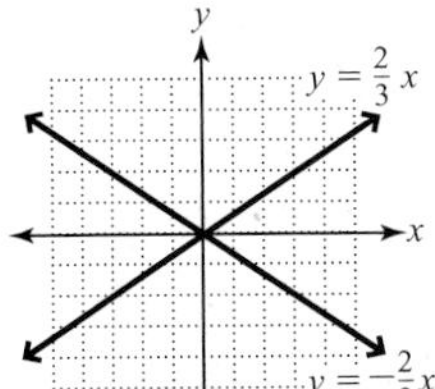

3. $(5x + 3y)(5x - 3y)$ **4.** ± 4

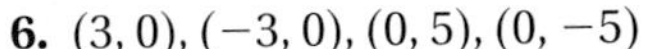

5. $(4, 0), (-4, 0), (0, 3), (0, -3)$ **6.** $(3, 0), (-3, 0), (0, 5), (0, -5)$

11.2 Exercises

1. An ellipse is the set of all points the sum of whose distances from two fixed points is constant. **3.** The equation of an ellipse is a sum; the equation of a hyperbola is a difference. **5.** The parabola has only one squared term.

7.

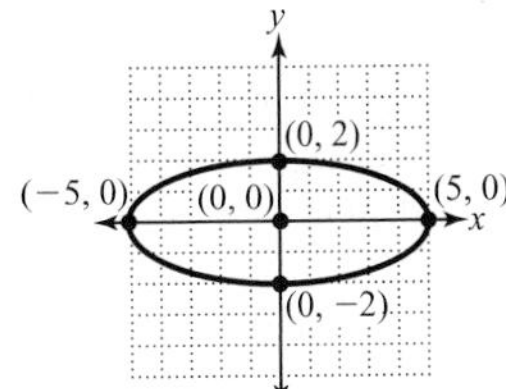

9.

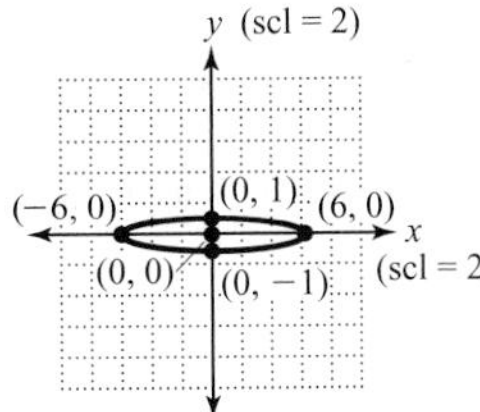

11.

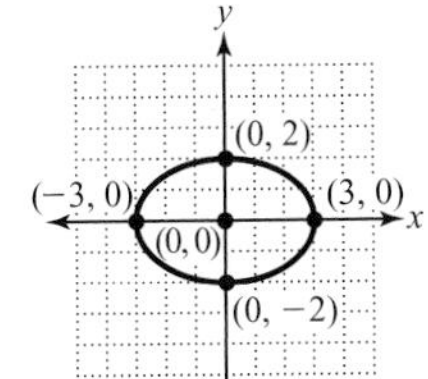

13.

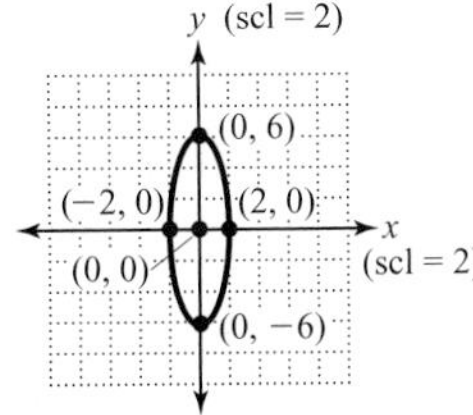

15.

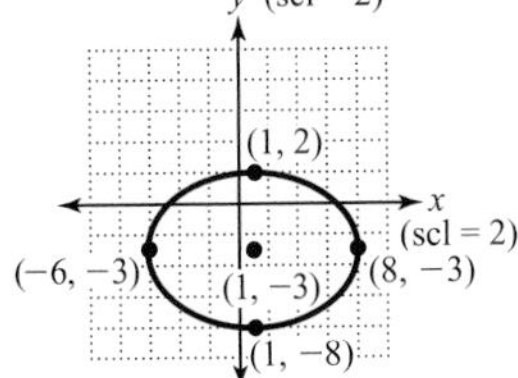

17.

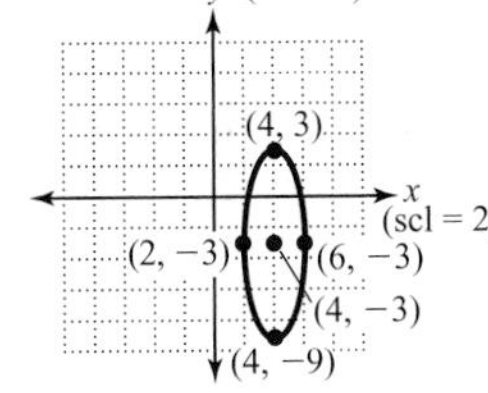

19.

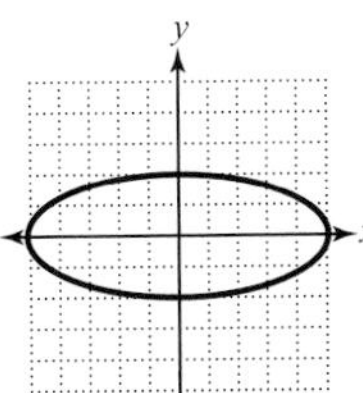

21.

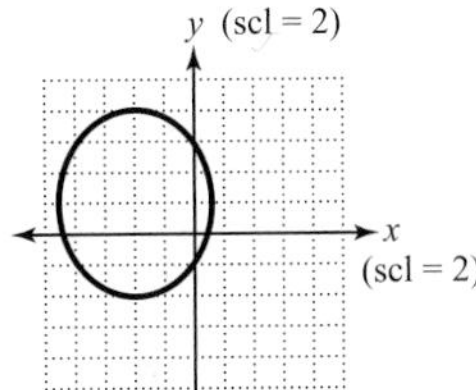

23. $\frac{x^2}{9} + \frac{y^2}{16} = 1$

25.

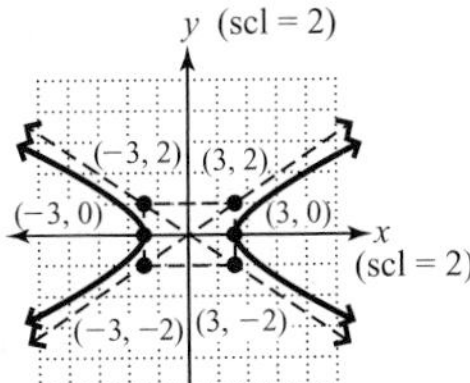

27.

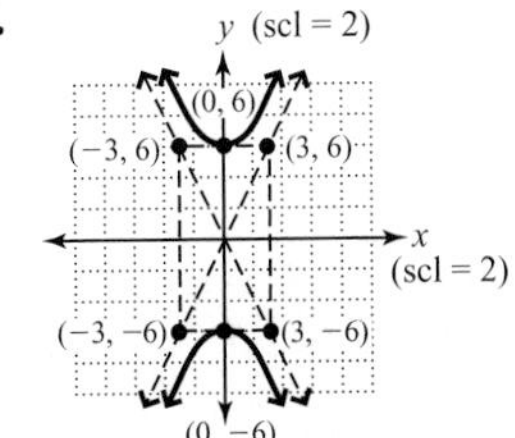

29.

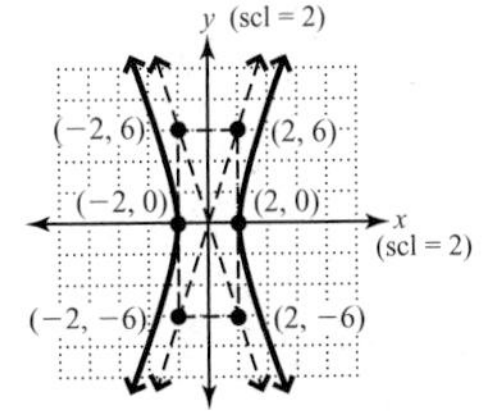

31.

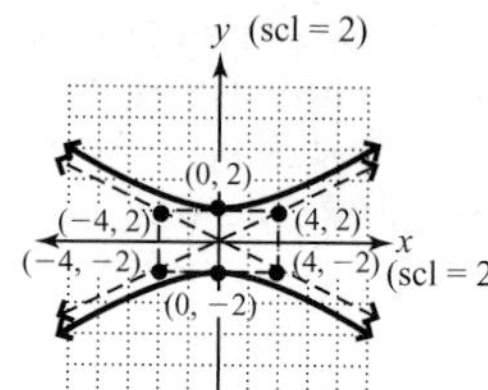

33.

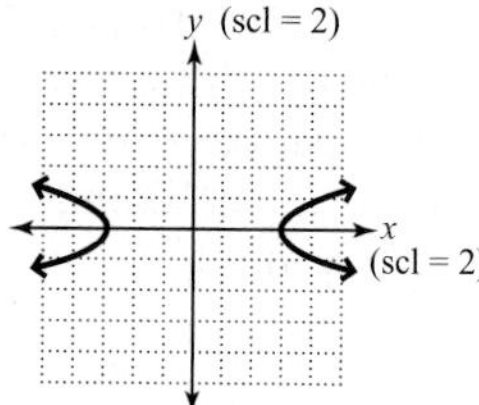

35. $\dfrac{x^2}{9} - \dfrac{y^2}{4} = 1$ **37.** c **39.** b **41.** ellipse **43.** circle

45. circle

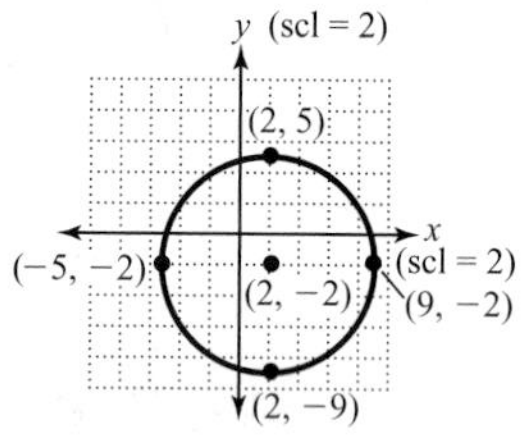

47. parabola

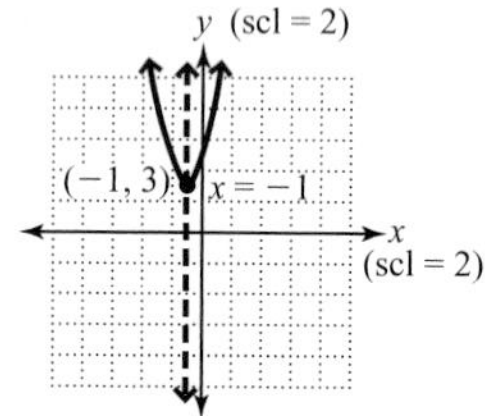

49. ellipse

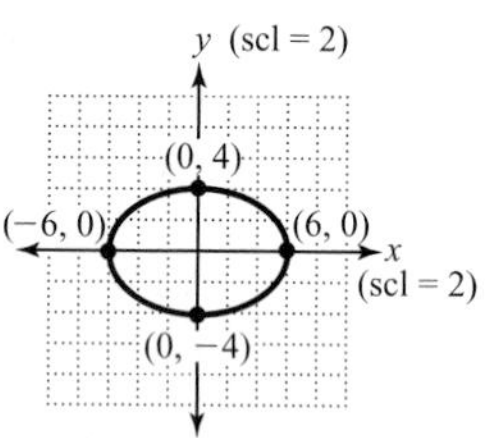

51. hyperbola

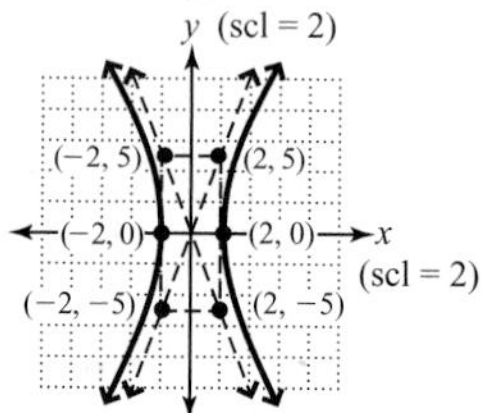

53. parabola

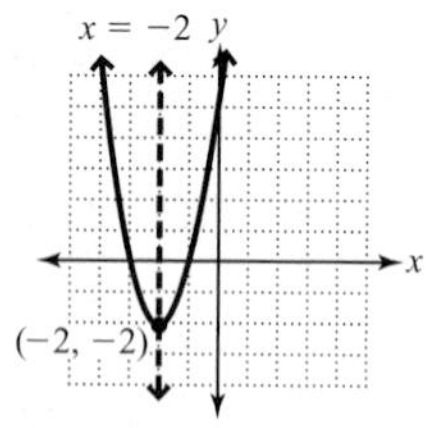

55. hyperbola

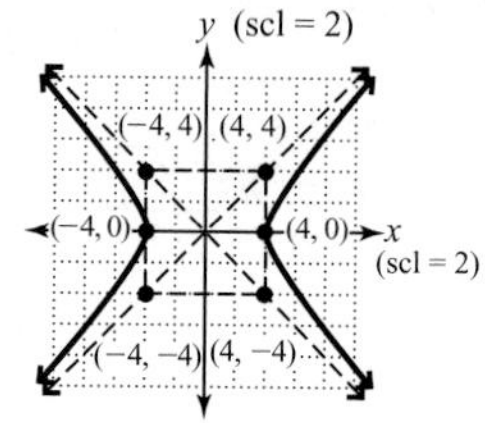

57. ellipse

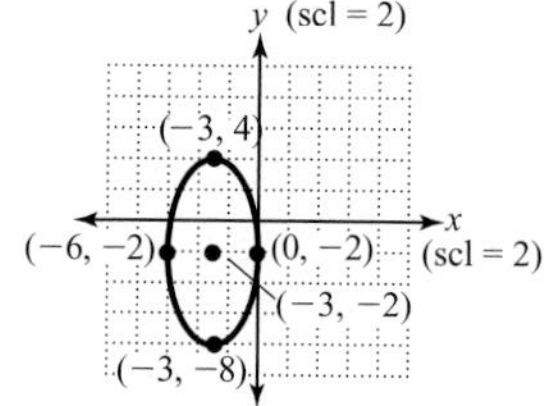

59. a. Yes, the sailboat will clear the bridge. The height of the bridge at the center is 20 feet and the boat's mast is only 18 feet above the water. **b.** The bridge is 32 feet wide at the base of the arch. **61.**

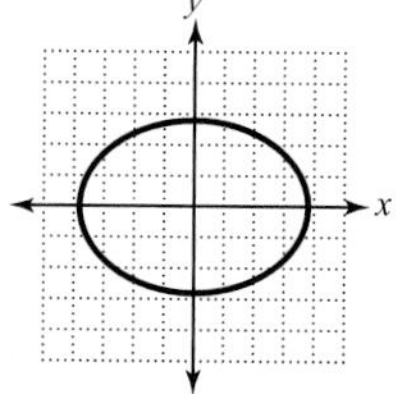

63. $\dfrac{x^2}{2.25} + \dfrac{y^2}{4} = 1$ **65.** 4 units

Review Exercises

1. $y = -3x^2 + 6$ **2.** $(1, 2)$ **3.** $(2, -3)$ **4.** $(-3, 4)$ **5.** $-4, \dfrac{2}{3}$ **6.** ± 3

11.3 Exercises

1. a. 0, 1, 2, 3, or 4 solutions are possible. **b.** Answers may vary.

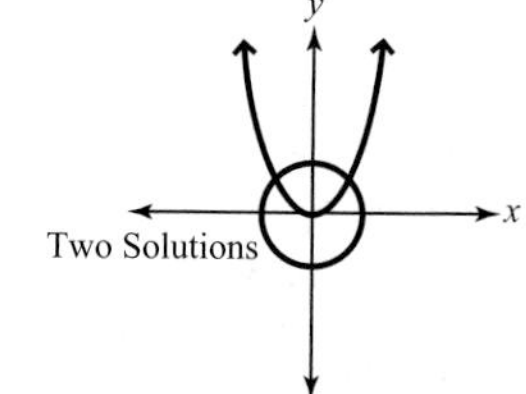

3. Substitution, because the first equation has a linear term and is easy to solve for *y*. **5.** The graph of the first equation is a line and the second is an ellipse, so the system could have 0, 1, or 2 solutions. **7.** $(-2, 8), (1, 2)$ **9.** $(-2, 0), (1, 9)$ **11.** $(-4, -3), (3, 4)$ **13.** $(3.6, -0.2), (2, 3)$ **15.** no solution **17.** $(0, 3), (1, 5)$ **19.** $(3, 2), (2, 3)$ **21.** $(4, 2), (4, -2), (-4, -2), (-4, 2)$ **23.** $(3, 4), (-3, 4), (-3, -4), (3, -4)$ **25.** $(2, 1), (2, -1), (-2, -1), (-2, 1)$ **27.** $(-6, 0), (3, 3), (3, -3)$ **29.** no solution **31.** $(2, 4), (-2, 4)$ **33.** $(4, 0), (-4, 0)$ **35.** $(1, 2), (-1, -2)$ **37.** $(5, 12), (-1, 0)$ **39.** $(2, 2), (2, -2), (-2, -2), (-2, 2)$ **41.** $(-7, 3), (-7, -3), (2, 0)$

43. Answers may vary, but one possible system is $\begin{cases} x + y = 4 \\ x^2 + y^2 = 1 \end{cases}$.

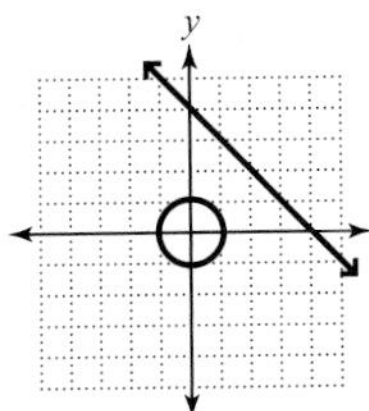

45. (5, 3), (5, −3), (−5, −3), (−5, 3) **47.** length 18 in., width: 8 in.

49. At equilibrium there should be 400 chairs at $72 per chair.

51.

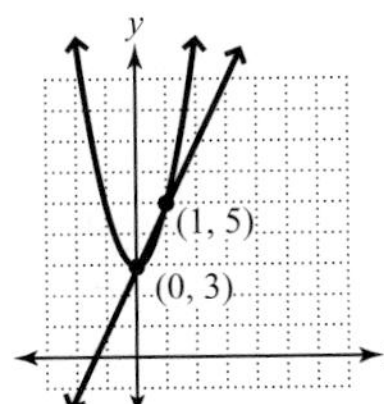

53.

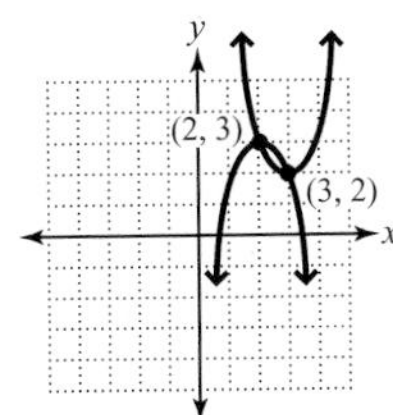

Review Exercises

1. yes **2.** no

3.

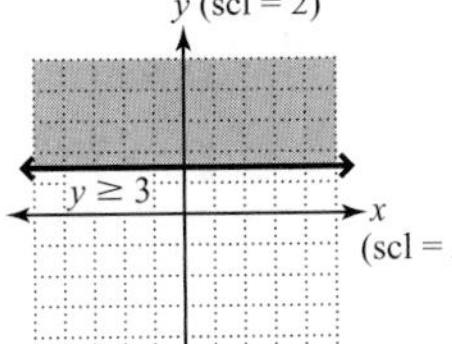

4.

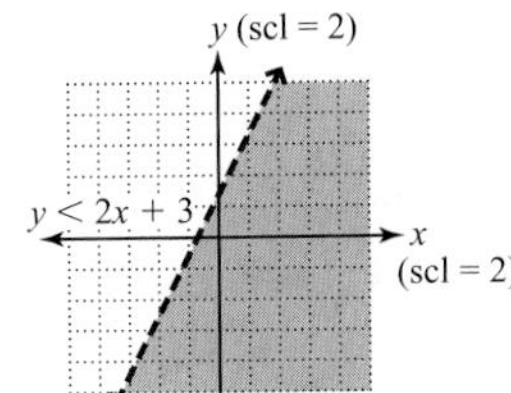

5.

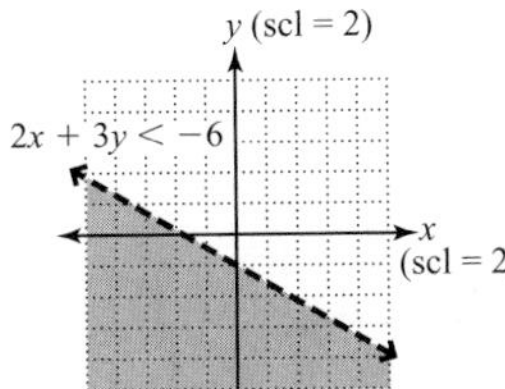

6.

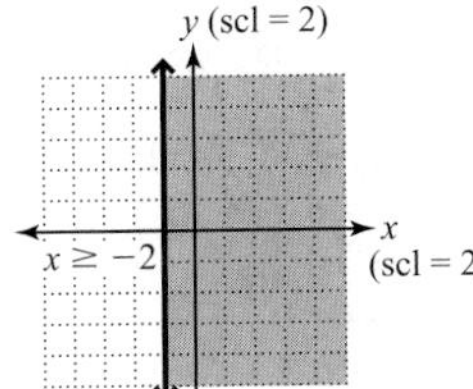

11.4 Exercises

1. No, the coordinates of the points on the curve make $9x^2 + 4y^2 = 36$. The curve would be a dashed curve because the points are not included in the solution.

3. One possible example is the system $\begin{cases} x^2 + y^2 < 1 \\ x^2 + y^2 > 4 \end{cases}$. The graph of $x^2 + y^2 < 1$ is the region inside the circle with radius 1. The graph of $x^2 + y^2 > 4$ is the region outside the circle with radius 2. Thus, there is no common region of solution.

5.

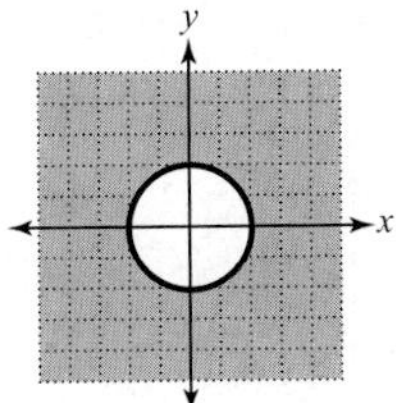

7.

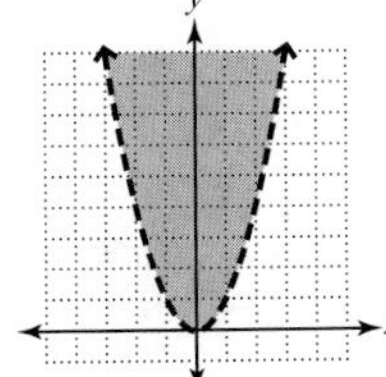

9.

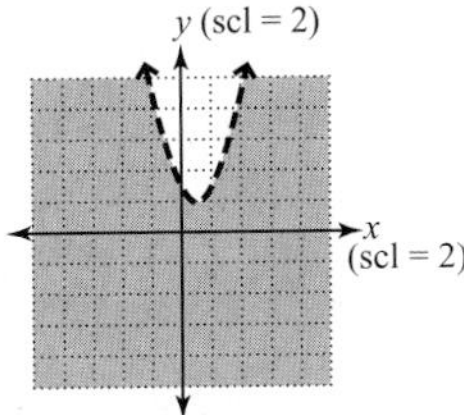

11.

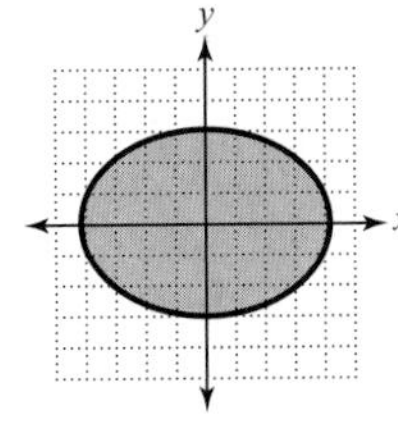

13.

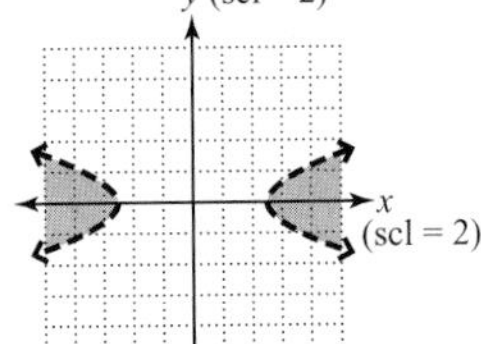

15.

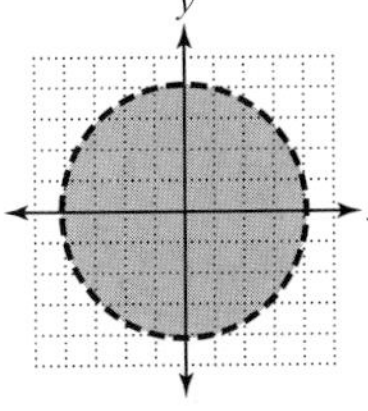

17.

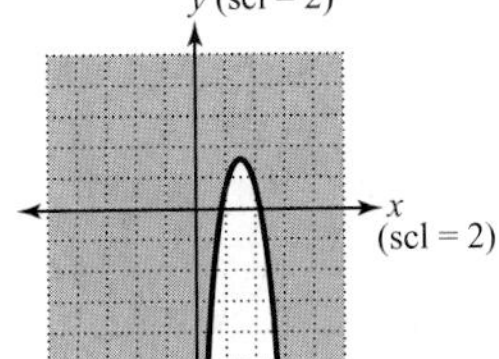

19.

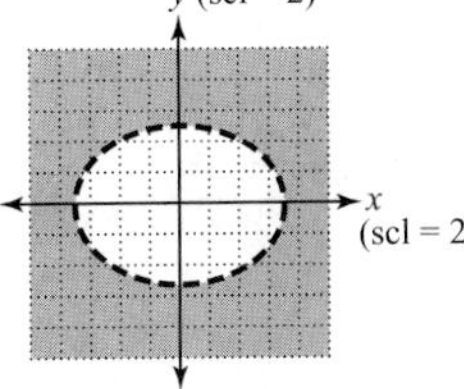

21.

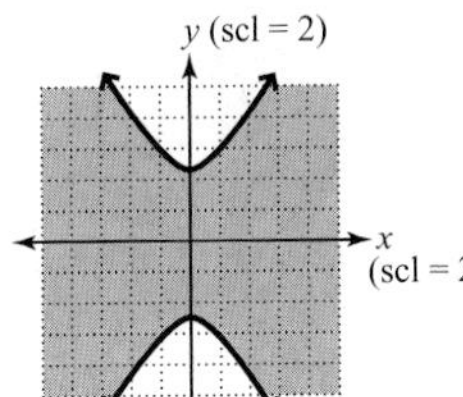

23.

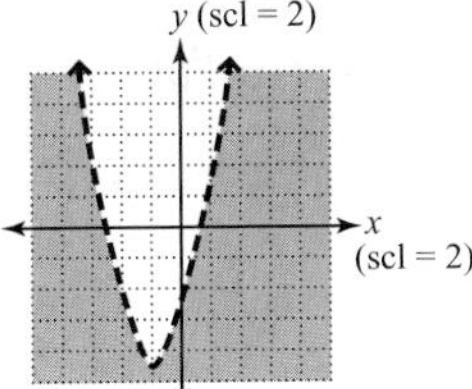

25.

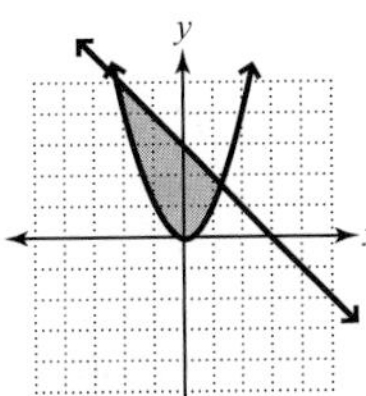

27.

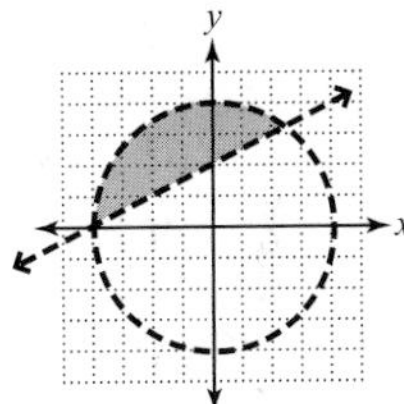

29.

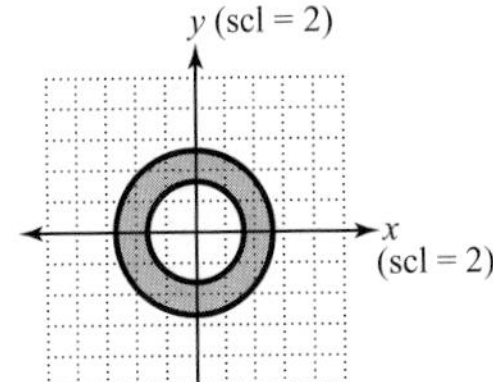

31.

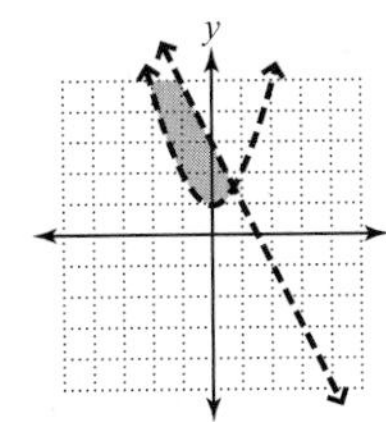

33.

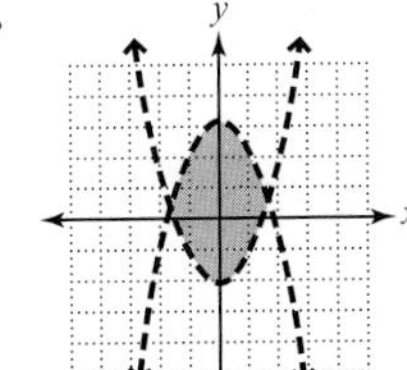

35.

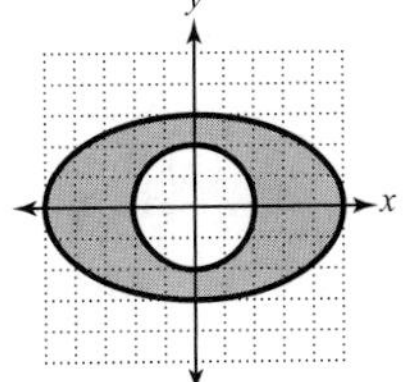

37.

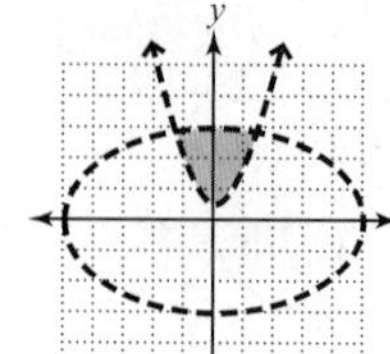

39.

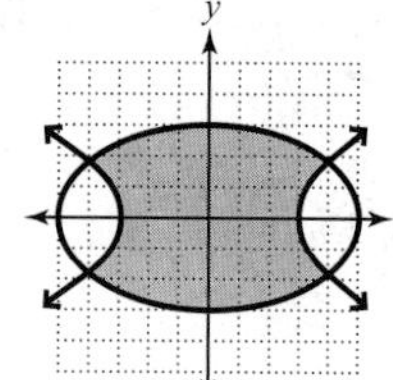

41.

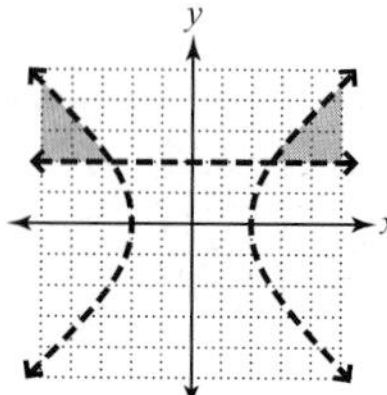

43.

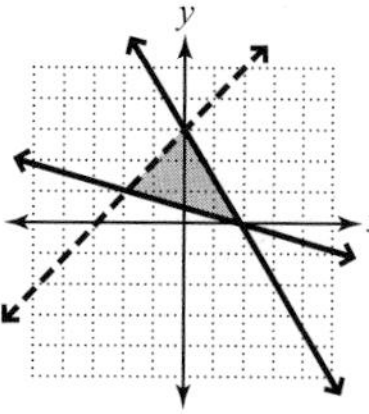

45.

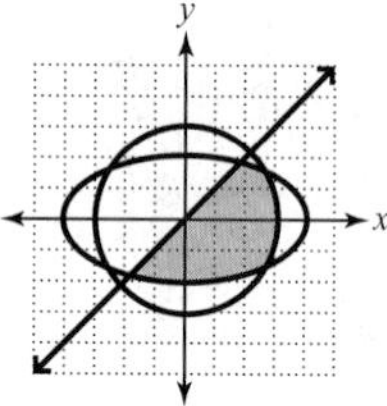

47.

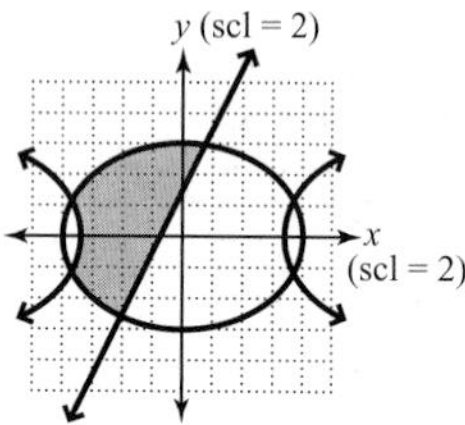

49.

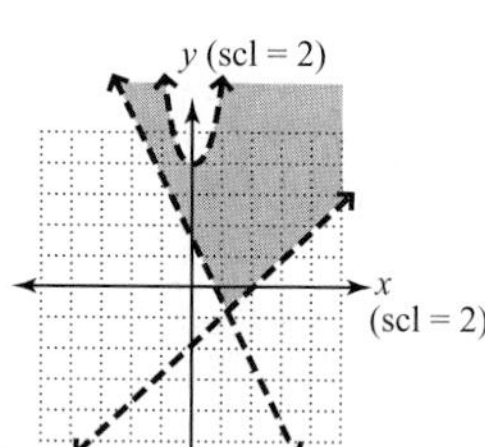

Review Exercises

1. 38 **2.** 312 **3.** −84 **4.** −1, 5, 15, 29 **5.** $-1, \frac{1}{4}, -\frac{1}{7}, \frac{1}{10}$ **6.** 27.5

Chapter 11 Review Exercises

1. false **2.** true **3.** false **4.** true **5.** true **6.** false **7.** 10 **8.** $(-4, 1)$; $r = 2$ **9.** ellipse; $(4, 0)$ and $(-4, 0)$ **10.** $(5, 4)$, $(5, -4)$, $(-5, -4)$, $(-5, 4)$ **11.** opens upwards; vertex: $(3, -5)$; axis of symmetry: $x = 3$

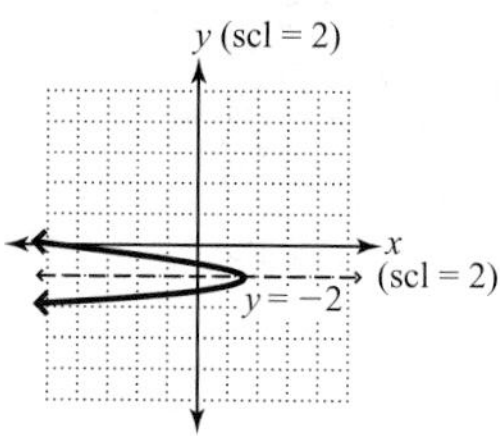

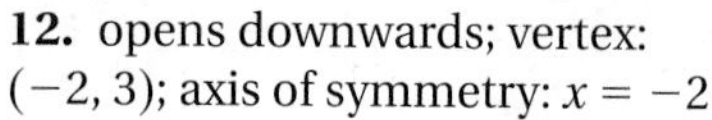

12. opens downwards; vertex: $(-2, 3)$; axis of symmetry: $x = -2$

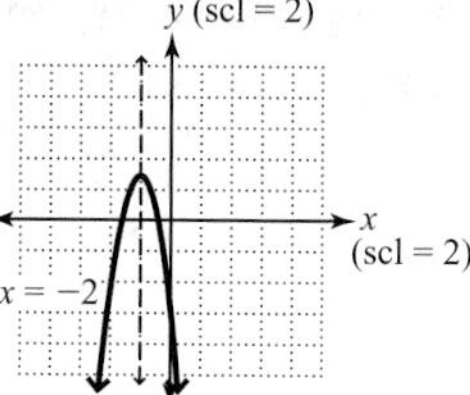

13. opens left; vertex: $(4, 2)$; axis of symmetry: $y = 2$

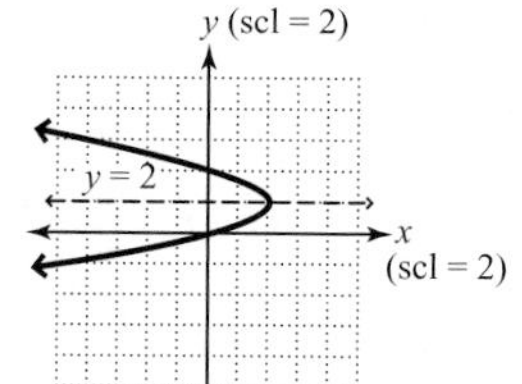

14. opens right; vertex: $(-2, -3)$; axis of symmetry: $y = -3$

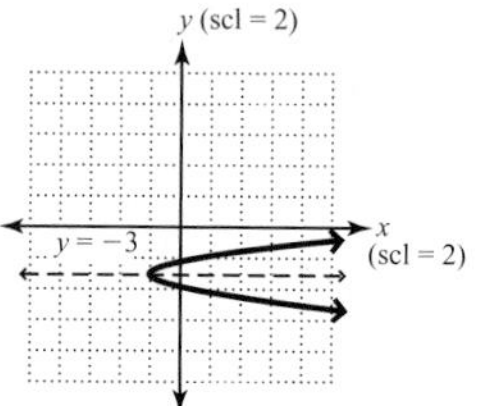

15. opens right; vertex: $(-1, -3)$; axis of symmetry: $y = -3$

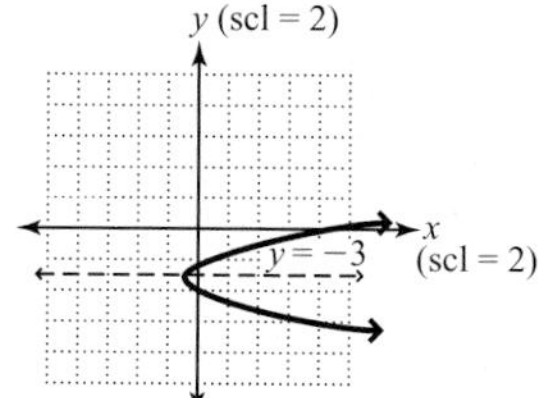

16. opens right; vertex: $(-14, 2)$; axis of symmetry: $y = 2$

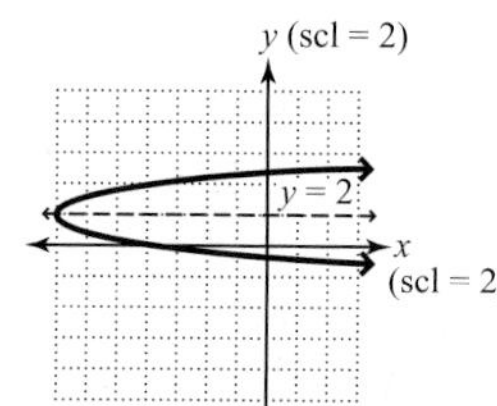

17. opens left; vertex: $(4, -1)$; axis of symmetry: $y = -1$

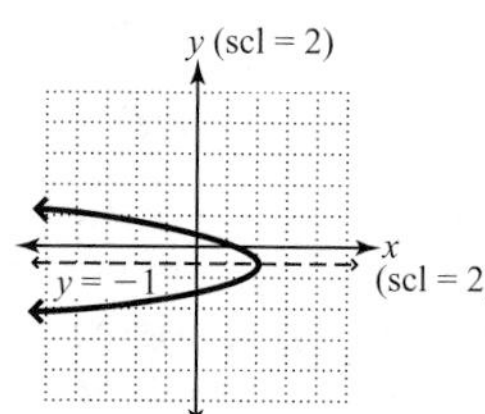

18. opens left; vertex: $(3, -2)$; axis of symmetry: $y = -2$

19. 5 **20.** $4\sqrt{5}$ **21.** 10 **22.** $(x + 6)^2 + (y - 8)^2 = 225$
23. center: (3, −2); radius: 5

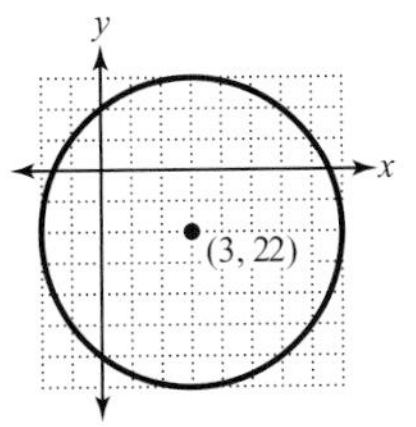

24. center: (−5, 1); radius: 2

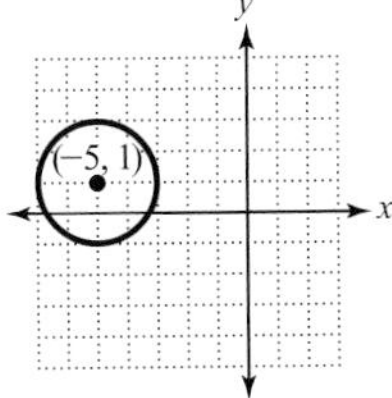

25. center: (2, −4); radius: 3

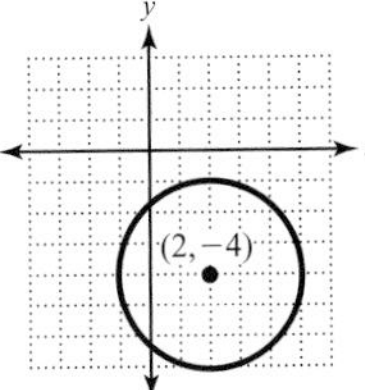

26. center: (−5, −1); radius: 2

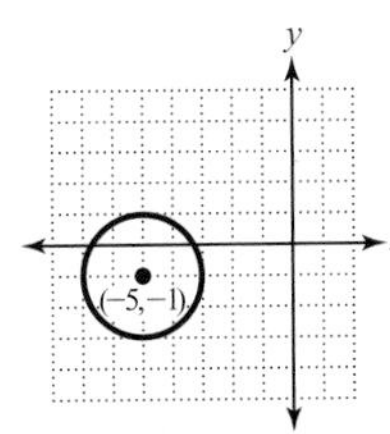

27. $(x - 6)^2 + (y + 8)^2 = 81$ **28.** $(x + 3)^2 + (y + 5)^2 = 100$ **29. a.** 144 ft. **b.** 1 sec. **c.** 4 sec. **30.** $x^2 + y^2 = 400$

31.

32.

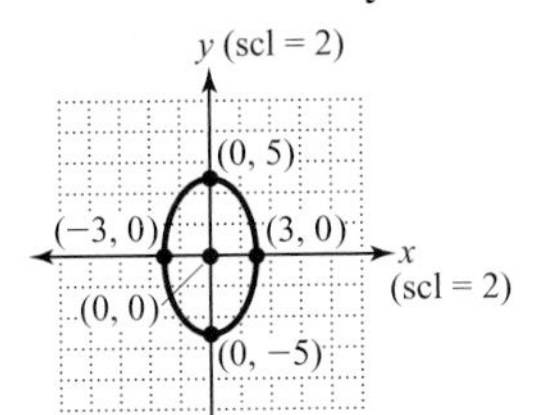

33.

34.

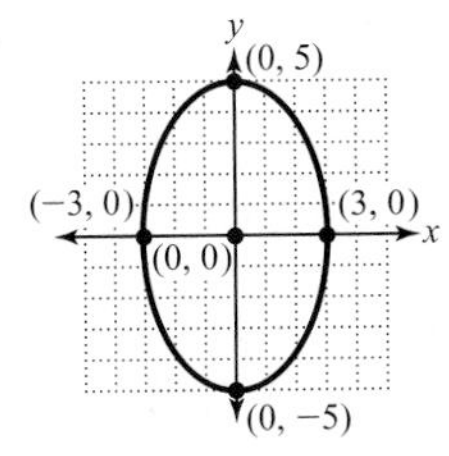

35.

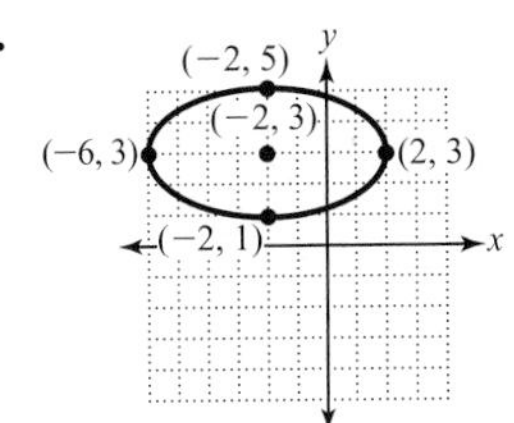

36.

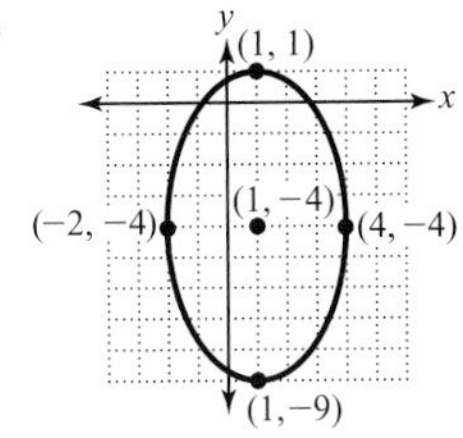

37.

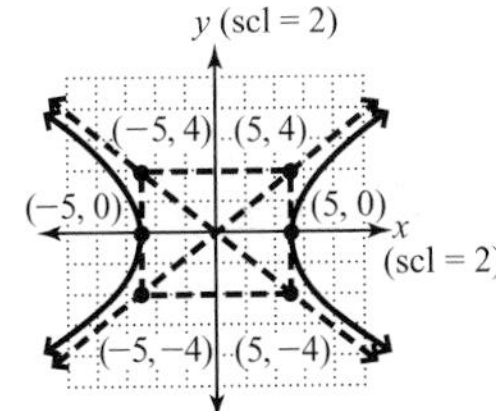

38.

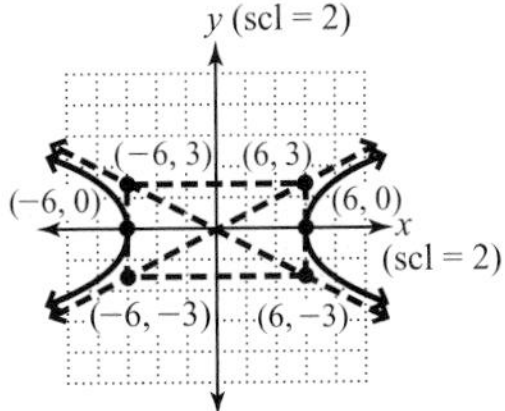

39.

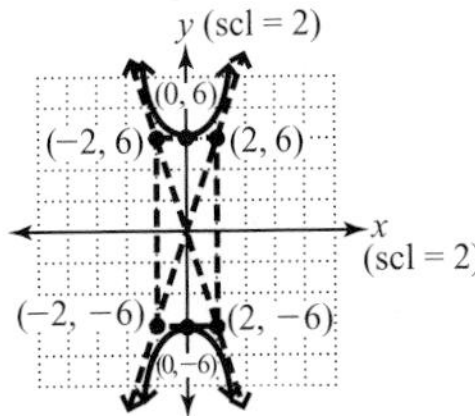

40.

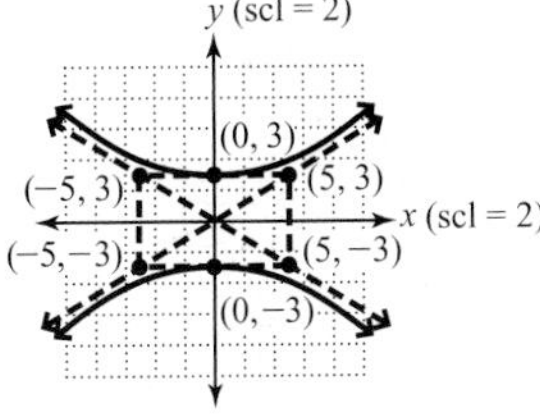

41. $\dfrac{x^2}{625} + \dfrac{y^2}{225} = 1$ **42.** $\dfrac{x^2}{1.625^2} + \dfrac{y^2}{4} = 1$ **43.** (2, 5), (−1, −1) **44.** (−1, −4), (4, 1) **45.** (−73/37, −35/37), (1, 5)
46. (8, 0), (−8, 0) **47.** (3, 4), (3, −4), (−3, −4), (−3, 4)
48. (4, 2), (4, −2), (−4, −2), (−4, 2) **49.** (2, 2), (−2, 2)
50. (2, 3), (−2, 3) **51.** (8, 5), (8, −5), (−8, −5), (−8, 5)
52. 8 ft. by 10 ft. **53.**

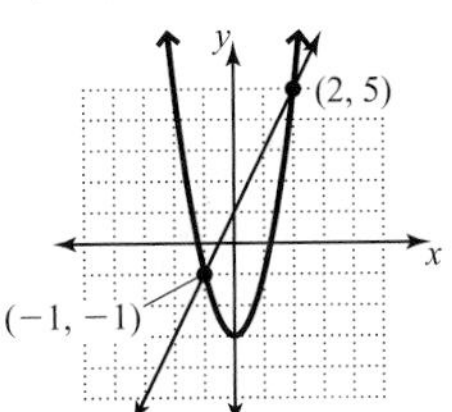

54.

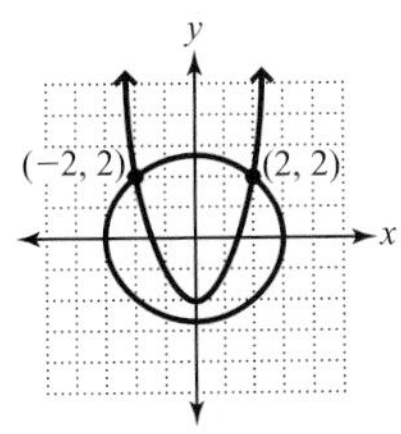

55.

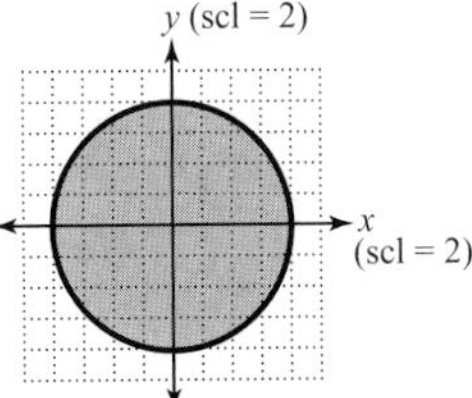

56.

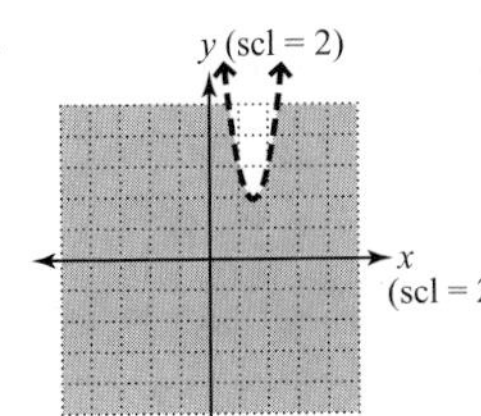

57.

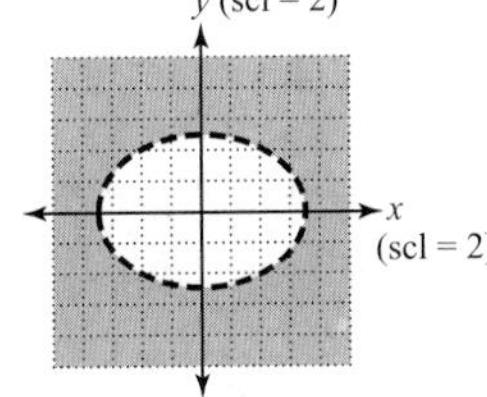

58.

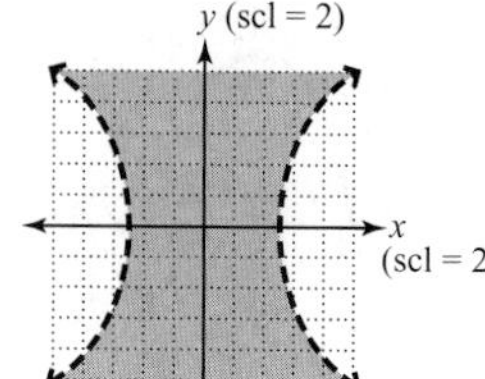

59.

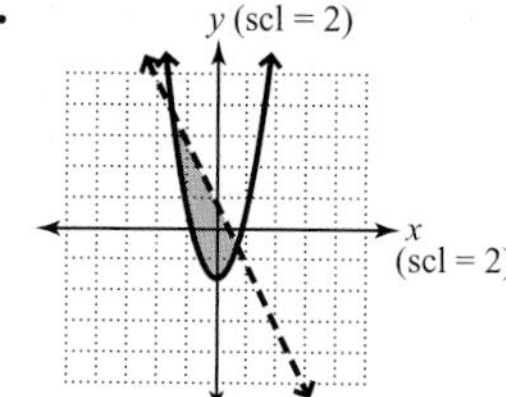

60.

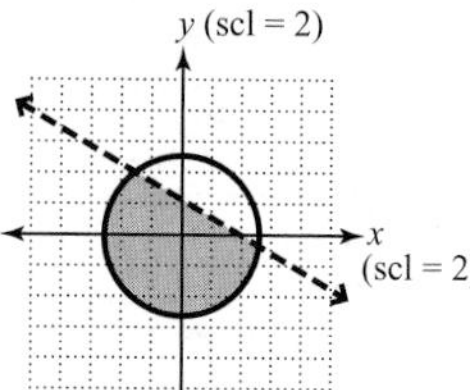

61.

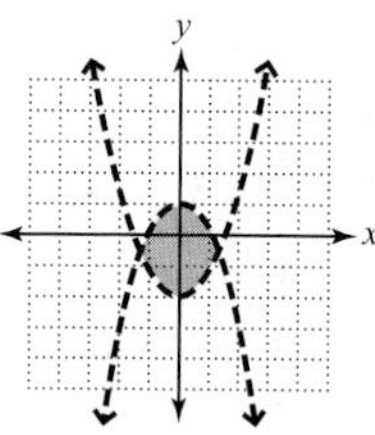

62.

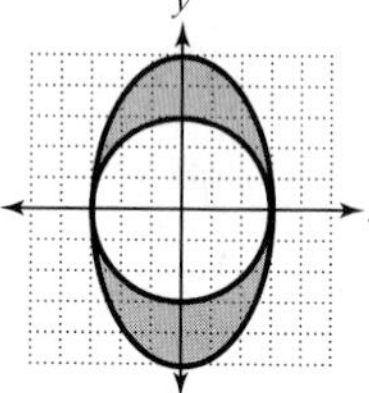

63.

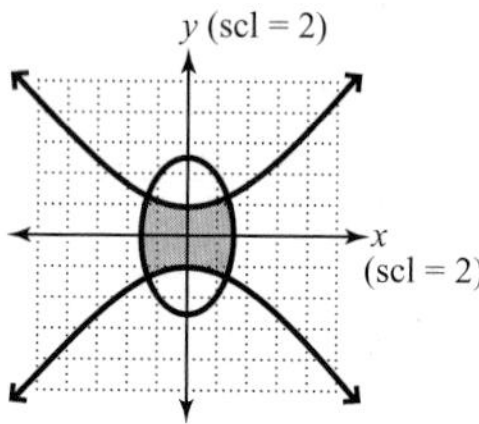

64.

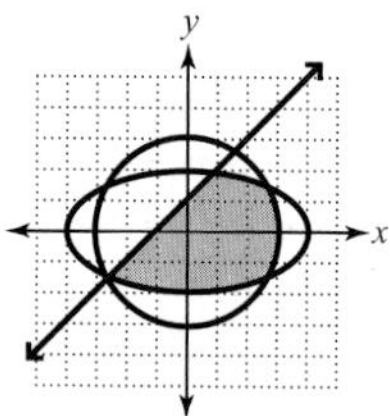

Chapter 11 Practice Test

1. opens upwards; vertex: $(-1, -4)$; axis of symmetry: $x = -1$

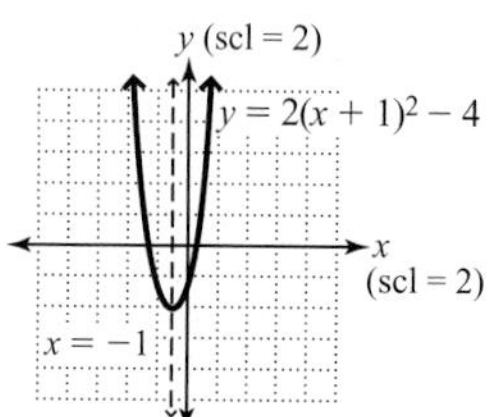

2. opens left; vertex: $(1, 3)$; axis of symmetry: $y = 3$

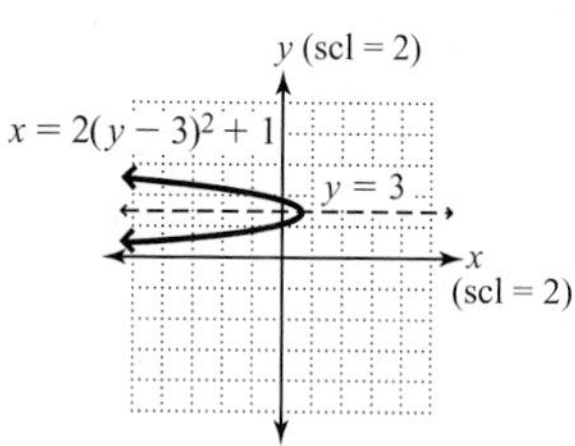

3. opens right; vertex: $(-7, -2)$; axis of symmetry: $y = -2$

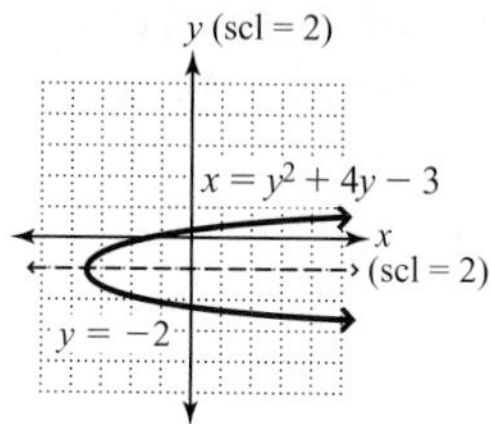

4. $2\sqrt{5}$ **5.** center: $(-4, 3)$; radius: 6

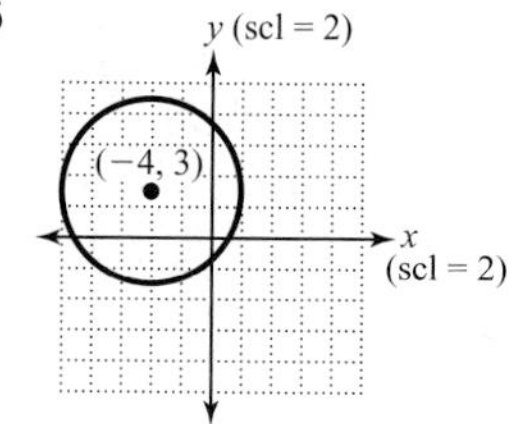

6. center: $(-2, 5)$, radius: 3

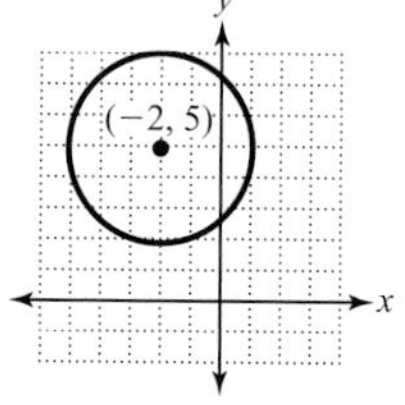

7. $(x - 2)^2 + (y + 4)^2 = 100$ **8.**

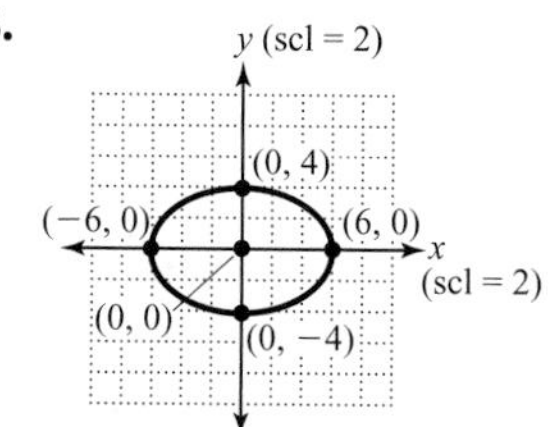

9.

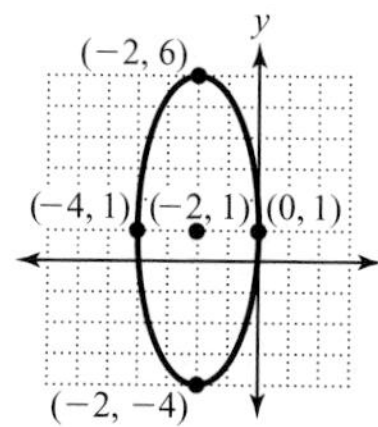

10.

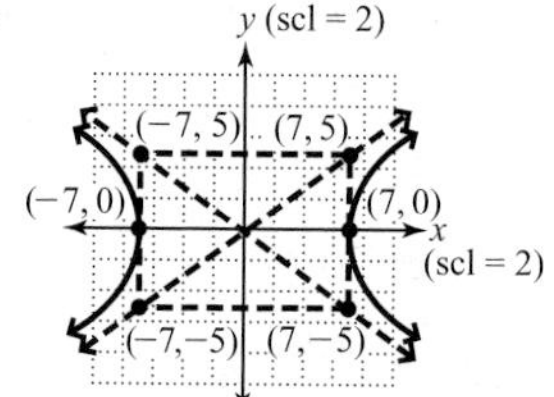

11.

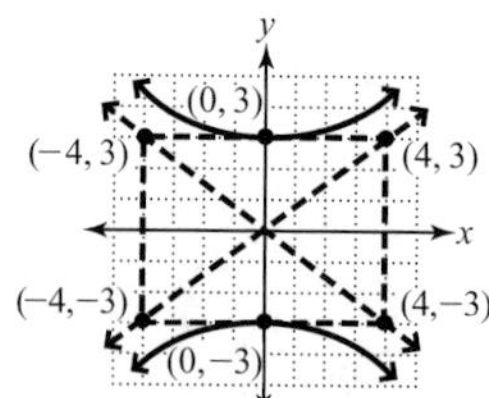

12. $(-5, 18)$, $(1, 6)$

13. $(3, 5)$, $(-3/5, -29/5)$ **14.** $(2, 3)$, $(2, -3)$, $(-2, -3)$, $(-2, 3)$ **15.** $(3, 2)$, $(3, -2)$, $(-3, -2)$, $(-3, 2)$

16.

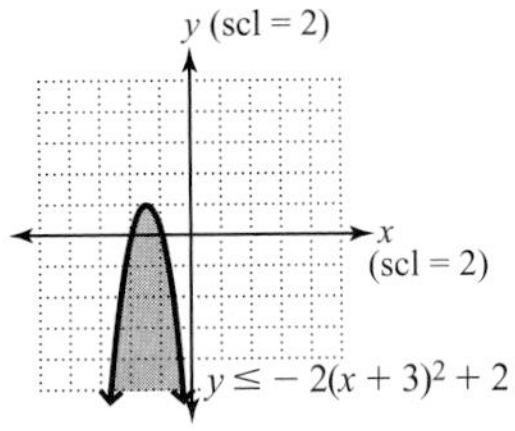

17.

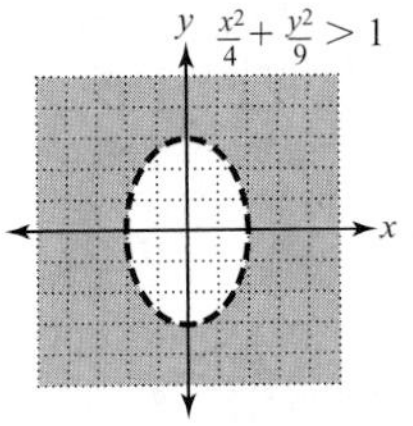

18.

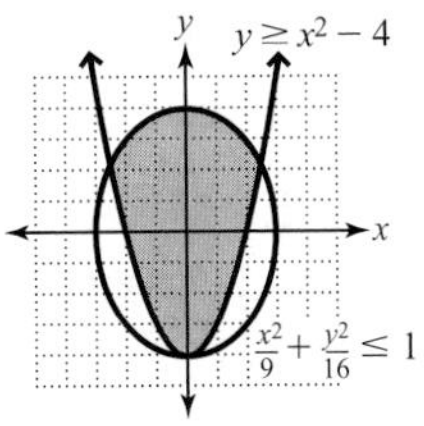

19. Height is 18, distance across the base is 12.

20. $\dfrac{x^2}{9.5^2} + y^2 = 1$

Chapters 1–11 Cumulative Review Exercises

1. true **2.** true **3.** true **4.** true **5.** false **6.** false **7.** $(x - h)^2 + (y - k)^2 - r^2$ **8.** i **9.** $d = \sqrt{(x_2 - x_1)^2 + (y_2 - y_1)^2}$ **10.** $a_1b_2 - a_2b_1$ **11.** $9\frac{1}{16}$ **12.** $2i\sqrt{5}$ **13.** $\dfrac{8}{3x}$ **14.** $\dfrac{x}{3x^2 + 1}$ **15.** $-\dfrac{1}{4}$ **16.** $\dfrac{n^6}{125m^9}$ **17.** $(k^2 + 9)(k + 3)(k - 3)$ **18.** $(x + y)(a + b)$ **19.** $(2, -1)$ **20.** $\dfrac{1 \pm i\sqrt{3}}{4}$ **21.** $-4, \dfrac{2}{3}$ **22.** 52 **23.** $\pm 2, \pm i$ **24.** $-\dfrac{4}{3}$ **25.** 1.5 **26.** no solution **27.** $\dfrac{1}{9}$ **28.** $\dfrac{1}{8}$ **29. a.**

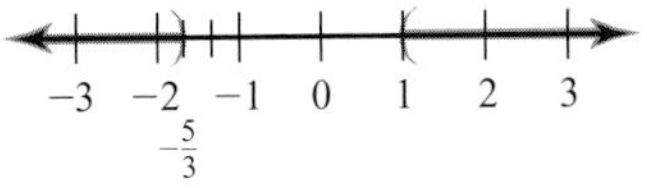

b. $\left\{x \mid x < -\dfrac{5}{3} \text{ or } x > 1\right\}$ **c.** $\left(-\infty, -\dfrac{5}{3}\right) \cup (1, \infty)$

30. a.

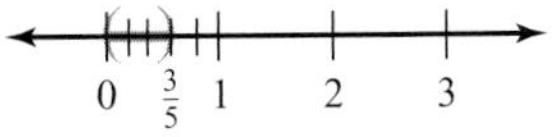

b. $\left\{m \mid 0 < m < \dfrac{3}{5}\right\}$

c. $\left(0, \dfrac{3}{5}\right)$ **31. a.** $x^2 + 2x$ **b.** $-x^2 + 2x + 2$ **c.** $2x^3 + x^2 - 2x - 1$ **d.** $\dfrac{2x + 1}{x^2 - 1}; x \neq \pm 1$ **e.** $2x^2 - 1$ **f.** $4x^2 + 4x$ **32.** yes **33.** $y = \dfrac{4}{3}x - 12; 4x - 3y = 36$ **34.** $\log_5 125 = 3$ **35.** $5 \log_5 x + \log_5 y$ **36.** $(x - 2)^2 + (y - 4)^2 = 13^2$ **37.**

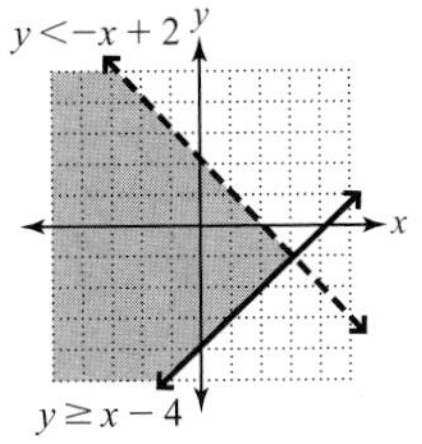

38.

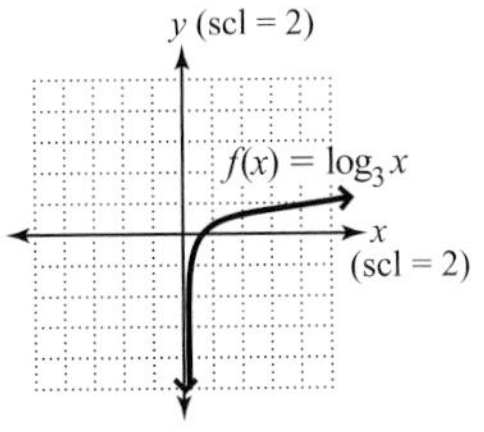

39.

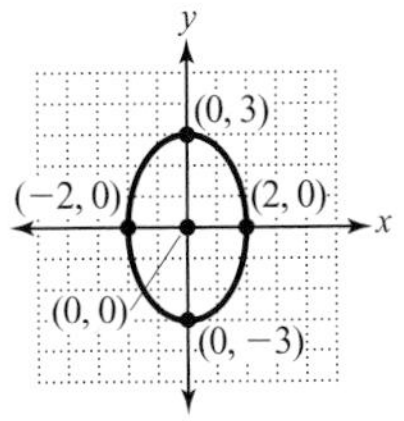

40.

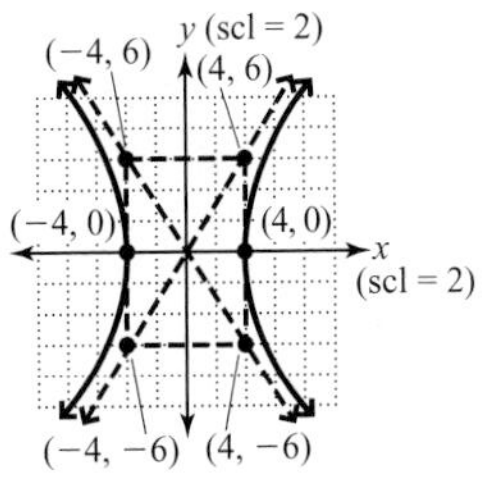

41.

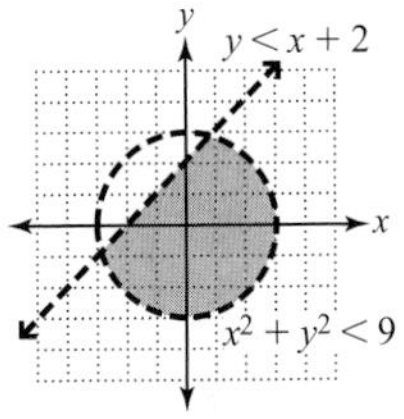

42. 7.3% **43. a.** $h = \dfrac{V}{\pi r^2}$ **b.** $\dfrac{15}{r^2}$ **44.** 2.9 lb. of tuna and 3.4 lb. of shrimp **45.** 0.857 hr. or approximately 51 min. **46.** \$1500 at 5%, \$2500 at 6%, \$1000 at 3% **47.** $\dfrac{x^2}{5625} + \dfrac{y^2}{400} = 1$ **48.** 7.37 **49.** \$16,909.08 **50.** 6.9

Appendix A

Appendix A Exercises

1. The domain of a sequence is the set of natural numbers 1 to n. **3.** A series is the sum of the terms of a sequence. **5.** arithmetic; 5 **7.** 3, 5, 7, 9, 41 **9.** 3, 6, 11, 18, 227 **11.** $\dfrac{1}{3}, \dfrac{1}{2}, \dfrac{3}{5}, \dfrac{2}{3}, \dfrac{11}{12}$ **13.** $-\dfrac{1}{2}, \dfrac{1}{5}, -\dfrac{1}{10}, \dfrac{1}{17}, -\dfrac{1}{226}$ **15.** 5 **17.** −3 **19.** 7 **21.** $a_{n;} = 14 + (n - 1)(4), a_{14} = 66$ **23.** $a_n = -8 + (n - 1)(-3), a_{28} = -89$ **25.** $a_n = -5 + (n - 1)(4), a_{34} = 127$ **27.** −6, 1, 8, 15 **29.** −5, −8, −11, −14 **31.** 7, 11, 15, 19 **33.** $a_1 = 11$ **35.** $d = 4$ **37.** −8, −1, 6, 13 **39.** 27, 24, 21, 18 **41.** $1 + 4 + 9 + 16 + 25 + 36 = 91$ **43.** $-3 - 1 + 1 + 3 = 0$ **45.** $-1 + 8 + 23 = 30$ **47.** $9 + 13 + 17 + 21 = 60$ **49.** 1450 **51.** −429 **53.** 675 **55.** 120 **57.** 5050 **59. a.** 144 seats **b.** 4480 seats **61. a.** 16 posts **b.** 325 posts **63. a.** \$28,000 \$29,500 \$31,000 \$32,500 \$34,000 **b.** \$41,500 **c.** \$347,500

Appendix B

Appendix B Exercises

1. The common ratio, r. **3.** Yes. The first n terms form a finite geometric series.

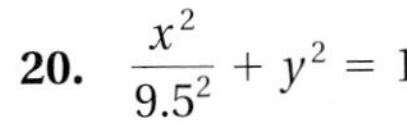

5. $2 + 6 + 18 + 54 = 80$ **7. a.** 3 **b.** $a_9 = 6561$ **c.** $a_n = 1(3)^{n-1}$ **9. a.** -2 **b.** $a_{15} = -32{,}768$ **c.** $a_n = -2(-2)^{n-1}$ **11. a.** $\frac{1}{3}$ **b.** $a_{10} = \frac{1}{81}$ **c.** $a_n = 243\left(\frac{1}{3}\right)^{n-1}$ **13. a.** $-\frac{1}{4}$ **b.** $a_7 = \frac{1}{32}$ **c.** $a_n = 128\left(-\frac{1}{4}\right)^{n-1}$ **15. a.** $-4, 12, -36, 108, -324$ **b.** 8748 **17. a.** 243, 81, 27, 9, 3 **b.** $\frac{1}{3}$ **19. a.** 2 **b.** 1, 2, 4, 8 **21. a.** 2 **b.** $-6, -12, -24, -48$ **23. a.** $\frac{1}{4}$ **b.** 128, 32, 8, 2 **25.** -6 **27.** 265,719 **29.** 63.875 **31.** 102.4 **33.** 40.5 **35.** 9.375 **37.** The sum does not exist, because $|r| \geq 1$. **39.** $\frac{4}{9}$ **41.** $\frac{17}{99}$ **43. a.** 520, 540.8, 562.432, 584.92928, 608.3264512 **b.** 740.122142459 **45. a.** 100, 108, 117, 126 **b.** 171 mink **c.** $a_n = 100(1.08)^{n-1}$ **47. a.** \$0.01, \$0.02, \$0.04, \$0.08, \$0.16 **b.** $a_n = 0.01(2)^{n-1}$ **c.** \$5242.88 **d.** \$10,485.75 **49.** \$11,191.64

Appendix C

Appendix C Exercises

1. 13 **3.** 8 factorial **5.** 5, 8 **7.** 24 **9.** 144 **11.** 86,400 **13.** $\frac{1}{90}$, or $0.0\overline{1}$ **15.** 10 **17.** 28 **19.** 35 **21.** 210 **23.** 1 **25.** 1 **27.** $a^5 + 5a^4b + 10a^3b^2 + 10a^2b^3 + 5ab^4 + b^5$ **29.** $x^4 - 4x^3y + 6x^2y^2 - 4xy^3 + y^4$ **31.** $8a^3 + 12a^2b + 6ab^2 + b^3$ **33.** $x^5 - 10x^4y + 40x^3y^2 - 80x^2y^3 + 80xy^4 - 32y^5$ **35.** $64m^6 + 576m^5n + 2160m^4n^2 + 4320m^3n^3 + 4860m^2n^4 + 2916mn^5 + 729n^6$ **37.** $81x^4 - 432x^3y + 864x^2y^2 - 768xy^3 + 256y^4$ **39.** $70x^4y^4$ **41.** $45a^8b^2$ **43.** $32{,}256x^4y^5$ **45.** $22{,}680m^4n^3$

Appendix D

Appendix D Exercises

1. The number of ways the two tasks can be performed one after the other. **3.** ${}_nC_r$ because the order of selection does not matter. **5.** ${}_nP_r$ because the order matters. **7.** 42 **9.** 720 **11.** 1 **13.** 15 **15.** 1 **17.** 1 **19.** 360 outfits **21.** 120 ways **23. a.** 6 **b.** 27 **25.** 17,576,000 **27.** 10,000; 5040 **29.** 39,916,800 **31.** 1,906,884 **33.** 816

Appendix E

Appendix E Exercises

1. The sample space is the set of all possible outcomes. **3.** 0 and 1 **5.** The probability of an event is the ratio of the number of ways the event can occur to the number of outcomes in the sample space. **7.** {a, e, i, o, u} **9.** {HHH, HHT, HTH, THH, THT, HTT, TTH, TTT} **11.** 90 **13.** 1820 **15.** {HTT, THT, TTH} **17.** { } or $\varnothing$ **19.** $\frac{1}{6}$ **21.** $\frac{1}{2}$ **23.** $\frac{3}{4}$ **25.** $\frac{1}{12}$ **27.** $\frac{1}{12}$ **29.** $\frac{1}{13}$ **31.** $\frac{3}{13}$ **33.** $\frac{3}{8}$ **35.** $\frac{1}{3}$ **37.** 0 **39.** $\frac{1}{9}$ **41.** $\frac{10}{21}$ **43.** 0.018 **45.** 0.070

Glossary

Absolute value: A number's distance from zero on a number line.

Additive inverses: Two numbers whose sum is 0.

Arithmetic sequence: A sequence in which each term after the first is found by adding the same number to the previous term.

Arithmetic series: The sum of an arithmetic sequence.

Augmented matrix: A matrix made up of the coefficients and the constant terms of a system. The constant terms are separated from the coefficients by a dashed vertical line.

Axis of symmetry: A line that divides a graph into two symmetrical halves.

Binomial: A polynomial containing two terms.

Binomial coefficient: A number written as $\binom{n}{r}$ and defined as $\frac{n!}{r!(n-r)!}$.

Binomial theorem: For any positive integer n,

$$(a+b)^n = \binom{n}{0}a^n + \binom{n}{1}a^{n-1}b + \binom{n}{2}a^{n-2}b^2 + \binom{n}{3}a^{n-3}b^3 + p + \binom{n}{n}b^n.$$

Circle: The set of points that is equally distant from a central point. The central point is the center.

Coefficient: The numerical factor in a monomial.

Combination: An unordered arrangement of objects. The combination of n things taken r at a time is often denoted as ${}_nC_r$.

Common logarithms: Logarithms with a base of 10. $\log_{10} x$ is written as $\log x$.

Complementary angles: Two angles are complementary if the sum of their measures is 90°.

Complex conjugates: The complex conjugate of a complex number $a + bi$ is $a - bi$.

Complex number: A number that can be expressed in the form $a + bi$, where a and b are real numbers and i is the imaginary unit.

Complex rational expression: A rational expression that contains rational expressions in the numerator and/or denominator.

Composition of functions: If f and g are functions, then the composition of f and g is defined as $(f \circ g)(x) = f[g(x)]$ for all x in the domain g for which $g(x)$ is in the domain of f. The composition of g and f is defined as $(g \circ f)(x) = g[f(x)]$ for all x in the domain of f for which $f(x)$ is in the domain of g.

Compound inequality: Two inequalities joined by either "and" or "or."

Conditional linear equation in one variable: An equation with exactly one solution.

Conjugates: Binomials that differ only in the sign separating the terms.

Conic section: A curve in a plane that is the result of intersecting the plane with a cone. More specifically, a circle, ellipse, parabola, or hyperbola.

Consistent system of equations: A system of equations that has at least one solution.

Constant: A symbol that does not vary in value.

Constant function: A function of the form $f(x) = c$, where c is a real number.

Contradiction: An equation that has no solution.

Cramer's Rule: The solution to the system of linear equations $\begin{cases} a_1x + b_1y = c_1 \\ a_2x + b_2y = c_2 \end{cases}$ is

$$x = \frac{\begin{vmatrix} c_1 & b_1 \\ c_2 & b_2 \end{vmatrix}}{\begin{vmatrix} a_1 & b_1 \\ a_2 & b_2 \end{vmatrix}} = \frac{D_x}{D} \quad \text{and} \quad x = \frac{\begin{vmatrix} a_1 & c_1 \\ a_2 & c_2 \end{vmatrix}}{\begin{vmatrix} a_1 & b_1 \\ a_2 & b_2 \end{vmatrix}} = \frac{D_y}{D}.$$

The solution to the system of linear equations

$\begin{cases} a_1x + b_1y = c_1z = d_1 \\ a_2x + b_2 = c_2z = d_2 \\ a_3x + b_3y = c_3z = d_3 \end{cases}$ is

$$x = \frac{\begin{vmatrix} d_1 & b_1 & c_1 \\ d_2 & b_2 & c_2 \\ d_3 & b_3 & c_3 \end{vmatrix}}{\begin{vmatrix} a_1 & b_1 & c_1 \\ a_2 & b_2 & c_2 \\ a_3 & b_3 & c_3 \end{vmatrix}} = \frac{D_x}{D}, \; y = \frac{\begin{vmatrix} a_1 & d_1 & c_1 \\ a_2 & d_2 & c_2 \\ a_3 & d_3 & c_3 \end{vmatrix}}{\begin{vmatrix} a_1 & b_1 & c_1 \\ a_2 & b_2 & c_2 \\ a_3 & b_3 & c_3 \end{vmatrix}} = \frac{D_y}{D}, \text{ and } z = \frac{\begin{vmatrix} a_1 & b_1 & d_1 \\ a_2 & b_2 & d_2 \\ a_3 & b_3 & d_3 \end{vmatrix}}{\begin{vmatrix} a_1 & b_1 & c_1 \\ a_2 & b_2 & c_2 \\ a_3 & b_3 & c_3 \end{vmatrix}} = \frac{D_z}{D}.$$

Note: Each denominator, D, is the determinant of a matrix containing only the coefficients in the system. To find D_x, we replace the column of x-coefficients in the coefficient matrix with the constants from the system. To find D_y, we replace the column of y-coefficients in the coefficient matrix with the constant terms, and do likewise to find D_z.

Cubic equation: An equation that can be written in the form $ax^3 + bx^2 + cx + d = 0$, where a, b, c, and d are real numbers and $a \neq 0$.

Cubic function: A function of the form: $f(x) = ax^3 + bx^2 + cx + d$, where a, b, c, and d are real numbers and $a \neq 0$.

Degree of a monomial: The sum of the exponents of all variables in the monomial.

Degree of a polynomial: The greatest degree of any of the terms in the polynomial.

Dependent linear equations in two unknowns: Equations with identical graphs.

Direct variation: Two variables, y and x, are in direct variation if $y = kx$. If y varies directly as the nth power of x, then $y = kx^n$. In both cases, k is the constant of variation.

Discriminant: The radicand $b^2 - 4ac$ in the quadratic formula.

Domain: A set containing initial values of a relation; its input values; first coordinates in ordered pairs.

Echelon form: An augmented matrix whose coefficient portion has 1's on the diagonal from upper left to lower right and 0's below the 1's.

Ellipse: The set of all points the sum of whose distance from two fixed points is constant.

Equation: Two expressions set equal to each other.

Equation quadratic in form: An equation is quadratic in form if it can be rewritten as a quadratic equation $au^2 + bu + c = 0$ where $a \neq 0$ and u is a variable or an expression.

Event: Any subset of a sample space.

Experiment: Any act or process whose result is not known in advance.

Exponential function: If $b > 0$, $b \neq 1$,and x is any real number, then the exponential function is $f(x) = b^x$.

Expression: A collection of constants, variables, and arithmetic symbols.

Factored form: A number or expression written as a product of factors.

Factorial notation: For any natural number n, the symbol $n!$ (read "n factorial") means $n(n - 1)(n - 2)\ldots 3 \cdot 2 \cdot 1$. 0! Is defined to be 1, so $0! = 1$.

Finite sequence: A function with a domain that is the set of natural numbers from 1 to n.

Formula: An equation that describes a mathematical relationship.

Function: A relation in which every value in the domain is assigned to exactly one value in the range.

Geometric sequence: A sequence in which every term after the first is found by multiplying the previous term by the same number, called the common ratio, r, where $r = \dfrac{a_n}{a_{n-1}}$.

Geometric series: The sum of the terms of a geometric sequence.

Greatest common factor (GCF) of a set of terms: A monomial with the greatest coefficient and degree that evenly divides all the given terms.

Hyperbola: The set of all points the difference of whose distances from two fixed points remains constant.

Identity: An equation in which every real number (for which the equation is defined) is a solution.

Imaginary number: A number that can be expressed in the form bi, where b is a real number and i is the imaginary unit.

Imaginary unit: The number represented by i, where $i = \sqrt{-1}$, and $i^2 = -1$.

Inconsistent system of equations: A system of equations that has no solution.

Independent linear equations in two unknowns: Equations that have different graphs.

Inequality: Two expressions separated by $\neq$, $<$, $>$, $\leq$, or $\geq$.

Infinite sequence: A function with a domain that is the set of natural numbers.

Intersection: For two sets A and B, the intersection of A and B, symbolized by $A \cap B$, is a set containing only elements that are in both A and B.

Inverse functions: Two functions f and g are inverses if and only if $(f \circ g)(x) = x$ for all x in the domain of g and $(g \circ f)(x) = x$ for all x in the domain of f.

Inverse variation: Two variables, y and x, are in inverse variation if $y = \dfrac{k}{x}$. If $y = \dfrac{k}{x^n}$, then y varies inversely as the nth power of x. The k is the constant of variation.

Irrational number: Any real number that is not rational.

Joint variation: Three variables y, x, and z are in joint variation if $y = kxz$, where k is a constant.

Like radicals: Radical expressions with identical radicands and identical indexes.

Like terms: Variable terms that have the same variable(s) raised to the same exponents, or constant terms.

Linear equation in one variable: An equation that can be written on the form $ax + b = c$, where a, b, and c are real numbers and $a \neq 0$.

Linear equation in two variables: An equation that can be written in the form $Ax + By = C$, where A and B are not both 0.

Linear function: A function in the form $f(x) = mx + b$, where m and b are real numbers and $m \neq 0$.

Linear inequality in one variable: An inequality that can be written in the form $ax + b \square c$, where $\square$ is an inequality symbol.

Linear inequality in two variables: An inequality that can be written in the form $Ax + By > C$, where the inequality could also be $<$, $\leq$, or $\geq$.

Logarithm: If $b > 0$ and $b \neq 0$, then $y = \log_b x$ is equivalent to $x = b^y$.

Matrix: A rectangular array of numbers.

Minor: The determinant of the remaining matrix when the row and column in which the element is located are ignored.

Monomial: An expression that is a constant or a product of a constant and variables that are raised to whole-number powers.

Multiplicative inverses: Two numbers whose product is 1.

***n*th root:** The number b is the nth root of a number a if $b^n = a$.

Natural logarithms: Base-e logarithms are called natural logarithms and $\log_e x$ is written as $\ln x$. Note that $\ln e = 1$.

Nonlinear system of equations: A system of equations that contains at least one nonlinear equation.

Numerical coefficient: The numerical factor in a term.

One-to-one function: A function f is one to one if for any two numbers a and b in its domain, when $f(a) = f(b)$, $a = b$ and when $a \neq b$, $f(a) \neq f(b)$.

Outcomes: The possible results of an experiment.

Permutation: An ordered arrangement of objects. The permutation of n things taken r at a time is often denoted as ${}_nP_r$.

Polynomial: A monomial or an expression that can be written as a sum of monomials.

Polynomial equation: An equation that equates two polynomials.

Polynomial equation in standard form: $P = 0$, where P is a polynomial in terms of one variable.

Polynomial function: A function of the form $f(x) = ax^m + bx^n + \ldots$, where each coefficient is a real number and each exponent is a whole number.

Polynomial in one variable: A polynomial in which every variable term has the same variable.

Prime number: A natural number with exactly two different factors, 1 and the number itself.

Probability of an event *E*: Given an experiment, the probability of an event E, written as $P(E)$, is the number of outcomes in event E divided by the number of outcomes in the sample space S of the experiment. That is,

$$P(E) = \frac{\text{The number of outcomes in } E}{\text{The number of outcomes in } S}$$

Quadratic equation: An equation that can be written in the form $ax^2 + bx + c = 0$, where a, b, and c are all real numbers and $a \neq 0$.

Quadratic function: A function of the form $f(x) = ax^2 + bx + c$, where a, b, and c are real numbers and $a \neq 0$.

Quadratic inequality: An inequality that can be written in the form $ax^2 + bx + c > 0$ or $ax^2 + bx + c < 0$, where $a \neq 0$.

Radical equation: An equation containing at least one radical expression whose radicand has a variable.

Radical function: A function of the form $f(x) = \sqrt[n]{P}$, where P is a polynomial.

Radius: The distance from the center of a circle to any point on the circle.

Range: A set containing all values that are paired to domain values in a relation; its output values; second coordinates in ordered pairs.

Ratio: A comparison of two quantities using a quotient.

Rational exponents: An exponent that is a fraction.

Rational expression: An expression that can be written in the form $\frac{P}{Q}$, where P and Q are polynomials and $Q \neq 0$.

Rational function: A function expressed in terms of rational expressions.

Rational inequality: An inequality containing a rational expression.

Rational number: Any real number that can be expressed in the form $\frac{a}{b}$, where a and b are integers and $b \neq 0$.

Relation: A set of ordered pairs.

Sample space: The set of all possible outcomes of an experiment. The sample space is denoted by S.

Scientific notation: A number expressed in the form $a \times 10^n$, where a is a decimal number with $1 \leq |a| < 10$ and n is an integer.

Sequence: A function list whose domain is 1, 2, 3, ..., n.

Series: The sum of the terms of a sequence.

Set: A collection of objects.

Slope: The ratio of the vertical change between any two points on a line to the horizontal change between these points.

Solution: A number that makes an equation true when it replaces the variable in the equation.

Solution set: A set containing all the solutions for a given equation.

Solution for a system of equations: An ordered set of numbers that makes all equations in the system true.

Square matrix: A matrix that has the same number of rows and columns.

Subset: If every element of a set B is an element of set A, then B is a subset of A.

Supplementary angles: Two angles are supplementary if the sum of their measures is 180°.

System of equations: A group of two or more equations.

Term: An expression that is separated by addition.

The set of integers: $\{\ldots, -3, -2, -1, 0, 1, 2, 3, \ldots\}$

The set of natural numbers: $\{1, 2, 3, \ldots\}$

The set of whole numbers: $\{0, 1, 2, 3, \ldots\}$

Trinomial: A polynomial containing three terms.

Union: For two sets A and B, the union of A and B, symbolized by $A \cup B$, is a set containing every element in A or B.

Variable: A symbol varying in value.

***x*-intercept:** A point where a graph intersects at the x-axis.

***y*-intercept:** A point where a graph intersects at the y-axis.

Index

D

E

M

N

O

P

S

T

U

V

W

X

Y

Z

Applications Index

Astronomy/Aerospace

Automotive

Biology

Business

Chemistry

Construction

Consumer

Economics

Education

Engineering

Environment

Finance

Geometry

Government

Health

Labor

Medical

Miscellaneous

Physics

Sports/Entertainment

Statistics/ Demographics

Transportation

Photograph Credits

1 Beth Anderson; **30** Peter Dazeley/Getty Images; **43** PFG/Getty Images; **55** Corb; **106** © Japack Company/CORBIS; **107** epicimages.us; **135** Silver Burdett Ginn (45817); **136** Hulton/Archive by Getty Images; **223** NASA Goddard Laboratory for Atmospheres_AACEFCW0; **254** StockTrek/Getty Images; **255** Rudi Von Briel/PhotoEdit; **258** © Chuck Savage/CORBIS; **261** James P Blair/Getty Images; **262** © Sam Sharpe/Corbis; **276** PhotoDisc BS14085; **277** AP Wideworld Photos; **286** Jonathan Blair/Corbis; **309**, **533** PhotoDisc; **322** Kitt Peak National Observatory_AADWNSB0; **323** PhotoDisc vol 14; **383**, **646** Digital Vision; **447** PhotoDisc vol 72; **550** © Bettmann/CORBIS; **621** PhotoDisc Red; **632** Don Farrall/Getty Images; **641** Mike Erhmann/ESPN; **645** AFP/Getty Images Editorial; **656** **(t)** © Reuters NewMedia Inc; **(b)** © Royalty-Free/Corbis; **787** Brand X Pictures; **801** Andrew Ward/Life File/Getty Images; **813** © Kevin Fleming/CORBIS.

POWERS AND ROOTS

n	n^2	n^3	$\sqrt{n}$	$\sqrt[3]{n}$	$\sqrt{10n}$
1	1	1	1.000	1.000	3.162
2	4	8	1.414	1.260	4.472
3	9	27	1.732	1.442	5.477
4	16	64	2.000	1.587	6.325
5	25	125	2.236	1.710	7.071
6	36	216	2.449	1.817	7.746
7	49	343	2.646	1.913	8.367
8	64	512	2.828	2.000	8.944
9	81	729	3.000	2.080	9.487
10	100	1,000	3.162	2.154	10.000
11	121	1,331	3.317	2.224	10.488
12	144	1,728	3.464	2.289	10.954
13	169	2,197	3.606	2.351	11.402
14	196	2,744	3.742	2.410	11.832
15	225	3,375	3.873	2.466	12.247
16	256	4,096	4.000	2.520	12.649
17	289	4,913	4.123	2.571	13.038
18	324	5,832	4.243	2.621	13.416
19	361	6,859	4.359	2.688	13.784
20	400	8,000	4.472	2.714	14.142
21	441	9,261	4.583	2.759	14.491
22	484	10,648	4.690	2.802	14.832
23	529	12,167	4.796	2.844	15.166
24	576	13,824	4.899	2.884	15.492
25	625	15,625	5.000	2.924	15.811
26	676	17,576	5.099	2.962	16.125
27	729	19,683	5.196	3.000	16.432
28	784	21,952	5.292	3.037	16.733
29	841	24,389	5.385	3.072	17.029
30	900	27,000	5.477	3.107	17.321
31	961	29,791	5.568	3.141	17.607
32	1,024	32,768	5.657	3.175	17.889
33	1,089	35,937	5.745	3.208	18.166
34	1,156	39,304	5.831	3.240	18.439
35	1,225	42,875	5.916	3.271	18.708
36	1,296	46,656	6.000	3.302	18.974
37	1,369	50,653	6.083	3.332	19.235
38	1,444	54,872	6.164	3.362	19.494
39	1,521	59,319	6.245	3.391	19.748
40	1,600	64,000	6.325	3.420	20.000
41	1,681	68,921	6.403	3.448	20.248
42	1,764	74,088	6.481	3.476	20.494
43	2,849	79,507	6.557	3.503	20.736
44	2,936	85,184	6.633	3.530	20.976
45	2,025	91,125	6.708	3.557	21.213
46	2,116	97,336	6.782	3.583	21.148
47	2,209	103,823	6.856	3.609	21.679
48	2,304	110,592	6.928	3.534	21.909
49	2,401	117,649	7.000	3.659	22.136
50	2,500	125,000	7.071	3.684	22.361

n	n^2	n^3	$\sqrt{n}$	$\sqrt[3]{n}$	$\sqrt{10n}$
51	2,601	132,651	7.141	3.708	22.583
52	2,704	140,608	7.211	3.733	22.804
53	2,809	148,877	7.280	3.756	23.022
54	2,916	157,464	7.348	3.780	23.238
55	3,025	166,375	7.416	3.803	23.452
56	3,136	175,616	7.483	3.826	23.664
57	3,249	185,193	7.550	3.849	23.875
58	3,364	195,112	7.616	3.871	24.083
59	3,481	205,379	7.681	3.893	24.290
60	3,600	216,000	7.746	3.915	24.495
61	3,721	226,981	7.810	3.936	24.698
62	3,844	238,328	7.874	3.958	24.900
63	3,969	250,047	7.937	3.979	25.100
64	4,096	262,144	8.000	4.000	25.298
65	4,225	274,625	8.062	4.021	25.495
66	4,356	287,496	8.124	4.041	25.690
67	4,489	300,763	8.185	4.062	25.884
68	4,624	314,432	8.246	4.082	26.077
69	4,761	328,509	8.307	4.102	26.268
70	4,900	343,000	8.367	4.121	26.458
71	5,041	357,911	8.426	4.141	26.646
72	5,184	373,248	8.485	4.160	26.833
73	5,329	389,017	8.544	4.179	27.019
74	5,476	405,224	8.602	4.198	27.203
75	5,625	421,875	8.660	4.217	27.386
76	5,776	438,976	8.718	4.236	27.568
77	5,929	456,533	8.775	4.254	27.749
78	6,084	474,552	8.832	4.273	27.928
79	6,241	493,039	8.888	4.291	28.107
80	6,400	512,000	8.944	4.309	28.284
81	6,561	531,441	9.000	4.327	28.460
82	6,724	551,368	9.055	4.344	28.636
83	6,889	571,787	9.110	4.362	28.810
84	7,056	592,704	9.165	4.380	28.983
85	7,225	614,125	9.220	4.397	29.155
86	7,396	636,056	9.274	4.414	29.326
87	7,569	658,503	9.327	4.431	29.496
88	7,744	981,472	9.381	4.448	29.665
89	7,921	704,969	9.434	4.465	29.833
90	8,100	729,000	9.487	4.481	30.000
91	8,281	753,571	9.539	4.498	30.166
92	8,464	778,688	9.592	4.514	30.332
93	8,649	804,357	9.644	4.531	30.496
94	8,836	830,584	9.695	4.547	30.659
95	9,025	857,375	9.747	4.563	30.882
96	9,216	884,736	9.798	4.579	30.984
97	9,409	912,673	9.849	4.595	31.145
98	9,604	941,192	9.899	4.610	31.305
99	9,801	970,299	9.950	4.626	31.464
100	10,000	1,000,000	10.000	4.642	31.623

Linear Functions

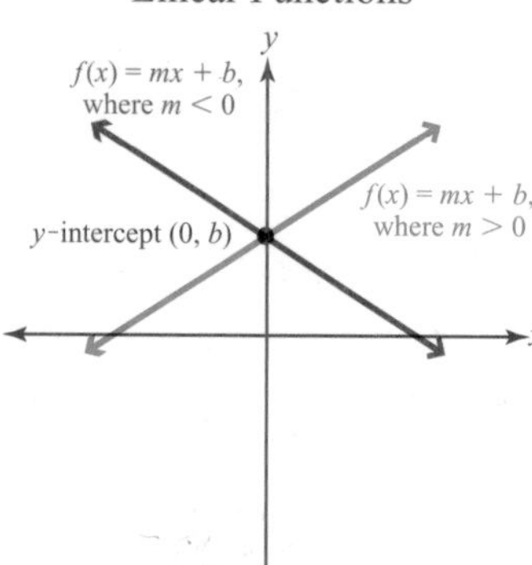

Quadratic Functions

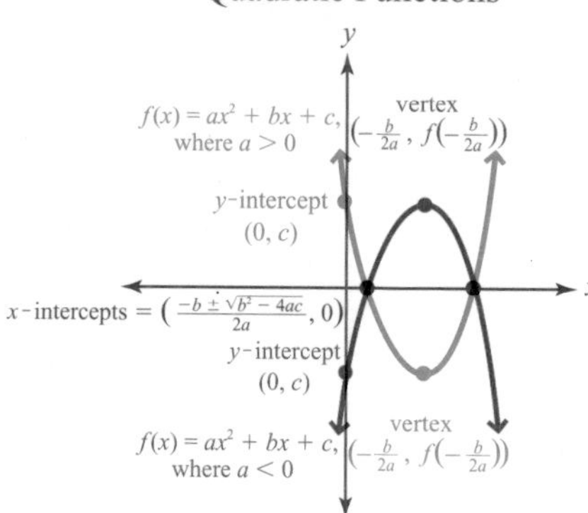

Cubic Functions

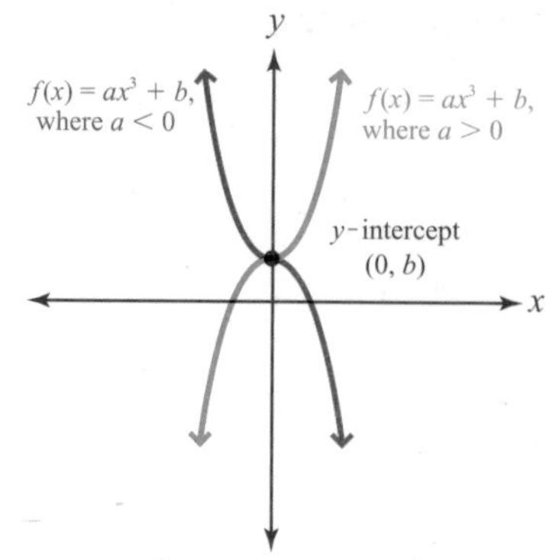

Absolute Value Functions

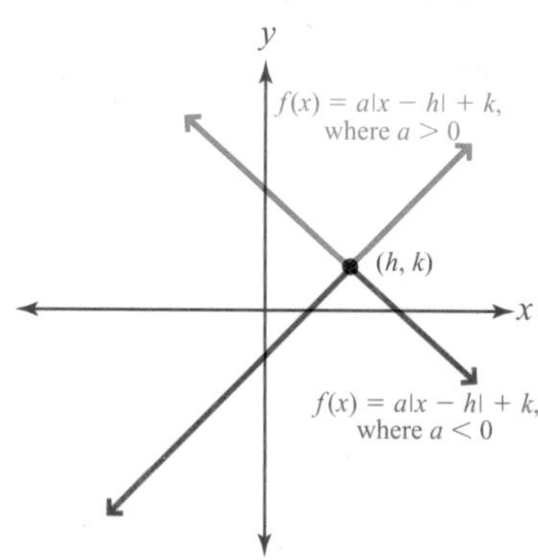

Radical Functions

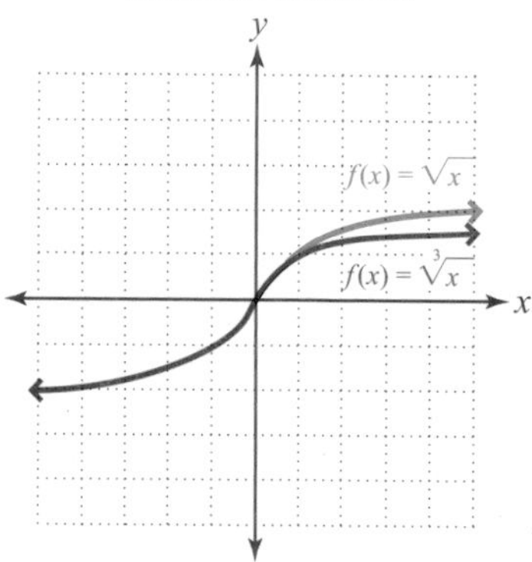

Exponential and Logarithmic Functions

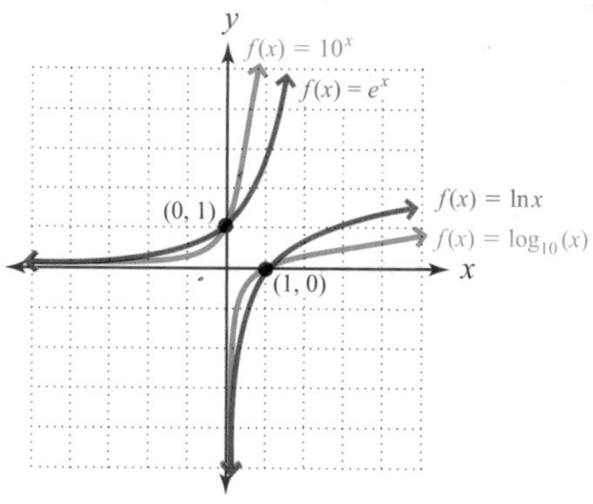

Conic Sections Parabola

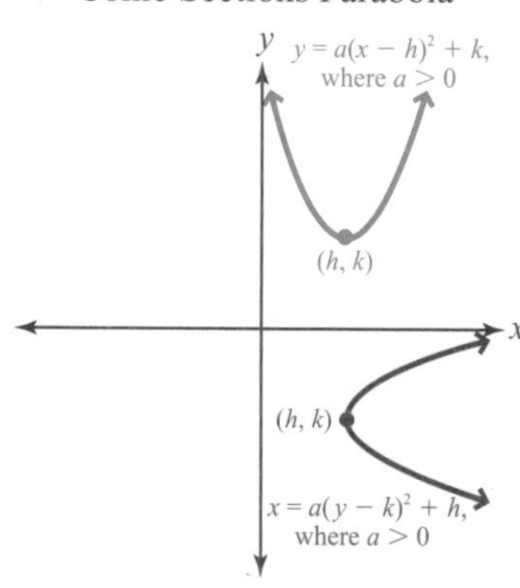

Circle

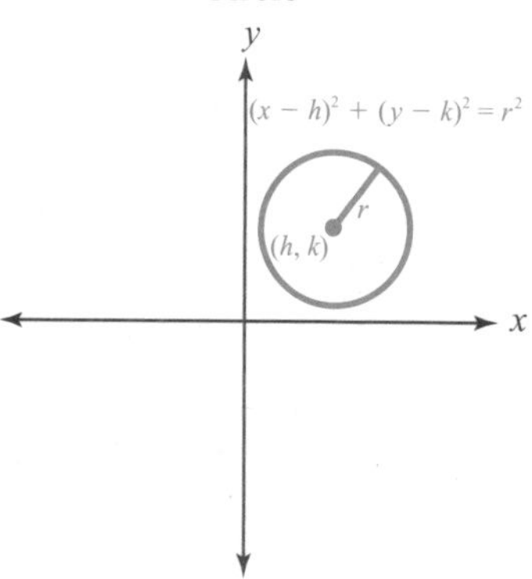

Ellipse

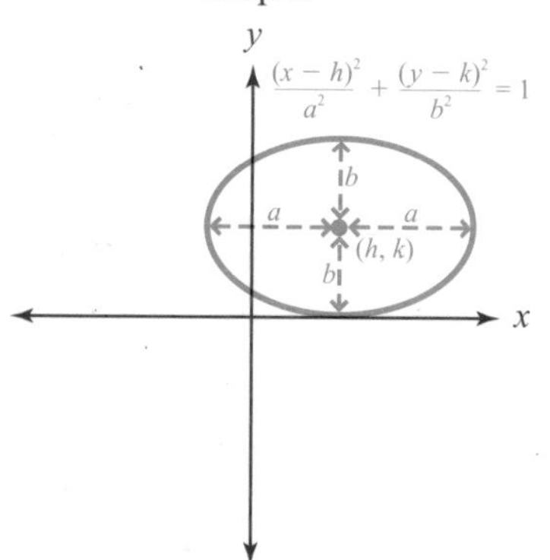

Hyperbola with center (0, 0)

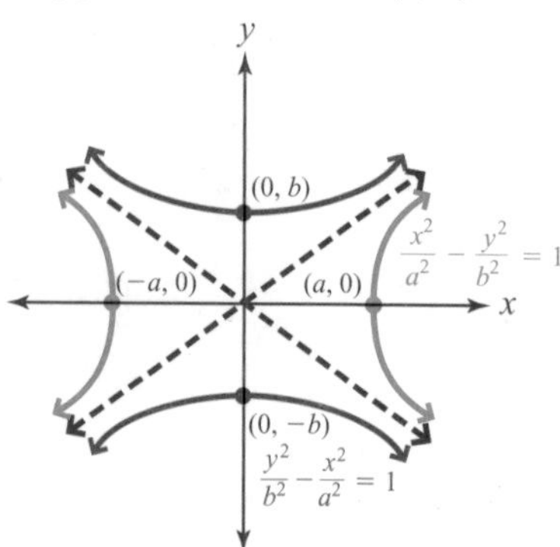

Echolocation by Bats

In 1773, the Italian scientist Lazzaro Spallanzani observed that bats were able to fly in a darkened room in which owls were helpless (cited in Au, 1993). Spallanzani found that disabling the bats' sense of sight didn't affect this ability, but the Swiss scientist Charles Jurine (1798) found that if he plugged a bat's ears it collided with objects. The bat apparently used its sense of hearing for navigation. But how? One problem with this idea was that the bat appeared to be totally silent.

This mystery was solved in 1938 when Pierce and Griffin used an ultrasonic detector to show that bats emit sounds in the 30–70-kHz range, far above the 20-kHz upper limit of human hearing. In addition, Robert Galambos (1941, 1942a,b) showed that bats can hear sounds in the 30–90-kHz range, a finding suggesting that they avoid obstacles in the dark by using a biological form of sonar, the system used in World War II to detect underwater objects such as submarines and mines. *Sonar,* which stands for *so*und *na*vigation and *r*anging, works by sending out pulses of sound and using information contained in the echoes of this sound to determine the location of objects. Griffin (1944) coined the term *echolocation* to describe the biological sonar system used by bats to avoid objects in the dark.

There are many species of bats, each with its own distinctive sounds for echolocation. The mustached bat, which has been the subject of a large amount of research, emits a signal that consists of a 5–50-msec-long pulse at 60 kHz, followed by a rapid 2–4-msec sweep to lower frequencies, which has been likened to a chirp (Suga, 1990). As the bat approaches a target it decreases the duration of the pulses and increases their rate to about 100 pulses per second. The pulses bounce off the target and are reflected back to the bat, which uses the information in the returning echo both to recognize the target and to locate it precisely. For example, the echo reflected by a flying moth (a favorite food of mustached bats) contains the following sources of information: (1) irregularly spaced echoes caused by the beating of the moth's wings, which identifies the object as a moth, and (2) the time between the bat's pulse and the returning echo, which indicates the moth's distance (Figure 8.40).

The bat's ability to identify objects and to measure their distances based on these echoes is extremely impressive. For example, behavioral experiments show that, when a mealworm and several plastic discs with the same dimensions as the mealworm are tossed into the air, a bat will catch the mealworm (Griffin, 1967; Webster & Durlach, 1963) (Figure 8.41). Bats can also detect differences in distance as small as about a centimeter, which corresponds to a difference in pulse

Figure 8.40
When a bat is chasing a moth, it sends out a series of pulses, indicated by the regularly spaced waves. The echo reflected by the beating wings of the moth is a series of irregularly spaced waves that are repeated periodically. This echo pattern, which is determined by the moth's beating wings, identifies the object as a moth. (From Pollak & Casseday, 1989.)

Figure 8.41
A big brown bat uses echolocation to locate and identify a mealworm suspended in the air. (Photograph courtesy of Steven Dear.)

duration of about a 10th of a millisecond (Simmons, 1973, 1979).

What is the physiological basis of these impressive feats? The bat's auditory system contains neurons that are tuned to respond to a specific pulse–echo delay. Just as neurons in the cat's or the monkey's auditory system can be classified as having a "characteristic frequency," neurons in the bat's auditory system have a "best delay" (BD). The best delay of neurons in the mustached bat range from 0.4 to 18 msec, which corresponds to target ranges of 7–310 cm (Riquimaroux, Gaioni, & Suga, 1991; Suga, 1990). These neurons are arranged in an organized way in the bat's cortex; neurons with the same BD are found in columns that are arranged so that BDs change systematically across the cortex.

So far we have been picturing a bat pursuing a single moth. But bats navigate in complex environments, and the pulses they send out reflect off many different objects and create a complicated pattern of echoes returning at different delays. We can compare the situation facing the bat to the situation facing a person viewing the same scene in the light. Both the bat and the human receive reflected energy—sound energy for the bat and light energy for the human. In both cases, the structure of this energy is extremely complex, since it is reflected by many different surfaces located at different distances and having different orientation and reflection properties. This complexity presents a problem in perceptual organization that humans solve by making use of a number of cues for distances, as well as principles such as the Gestalt laws of organization.

What information does a bat use that would enable it to perceptually organize the jumble of reflected energy it receives from the scene? Stephen Dear, James Simmons, and Jonathan Fritz (1993) have proposed an answer to this question, based on their research on the big brown bat. They determined tuning curves for neurons in the bat's cortex and found neurons with BDs ranging from 2 to 28 msec, which correspond to distances between about 30 and 500 cm, the range that contains this bat's biologically important stimuli. These neurons serve as "range finders," the short BD neurons indicating the presence of close objects and the longer BD neurons indicating the presence of farther objects

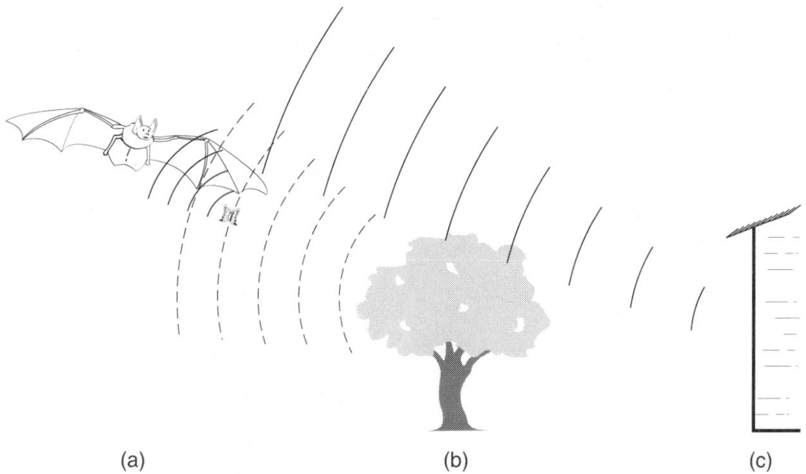

Figure 8.42

When a bat sends out its pulses, it receives echoes from a number of objects in its environment. In the situation in this figure, which shows the echoes that are received by the bat, the bat is receiving echoes (a) from a moth, located about half a meter away; (b) from a tree, located about 2 meters away; and (c) from a house, located about 4 meters away. The echoes from each object return to the bat at different times, with echoes from more distant objects taking longer to return. According to the results of Dear, Simmons, and Fritz, the bat senses each of these echoes by neurons that respond to different delays. They suggest that the bat constructs a perception of the scene based on the discharges of these neurons.

(Figure 8.42). Thus, at any instant, the firing of these neurons simultaneously indicates objects at different distances. Dear et al. described the information from all of the neurons taken together as creating an "acoustic scene."

How does the bat experience this acoustic scene? Does the bat "hear" a complex array of sounds and interpret them as a layout of objects and surfaces? Does the bat actually experience something similar to human vision? After all, both human vision and bat audition are based on the patterns of energy reflected by objects in the scene. Is it far-fetched to wonder if the pattern of acoustic energy received by the bat is transformed into a visually experienced scene? Alvin Liberman and Ignatius Mattingly (1989) speculate that the bat perceives returning echoes not as an echoing bat cry but as the *distance* of the reflecting object.

Even though there is no way we can know what the bat actually experiences, posing questions like the ones above is one of the things that makes studying the "other worlds" of perception experienced by other species so interesting. Of course, we can ask questions like these about the perceptions of any species, but the case of the bat is particularly interesting since bats use sound energy to gather information that humans associate with seeing. We can also ask similar questions about dolphins and whales, which, although different from the bat in many ways, also use sonar to help them navigate (Au, 1993).

Plasticity of Frequency Representation in the Cochlea and the Brain

The cochlea is organized by frequency; high frequencies stimulate the base of the cochlea most strongly, and low frequencies stimulate the apex. Thus, when it was discovered that the base of the cochlea develops before the apex, it was natural to expect that very young animals might be more sensitive to high-frequency sounds than to low-frequency sounds (King & Moore, 1991). However, the opposite is true: Many young animals are most sensitive to low-frequency sounds, a fact that initially puzzled researchers.

The explanation for this puzzling fact was provided by an experiment by Edwin Rubel and Brenda Ryals (1983), who exposed chicks to high-intensity (125 dB SPL) pure tones for 12 hours and observed where on the cochlea their hair cells were being lost and damaged. They found that for a given frequency of tone, hair cells were lost near the base in the youngest chicks (which were exposed to the tone before hatching), and that the site of hair cell loss moved toward the apex in the older chicks (which were exposed to the tone either 10 or 30 days after hatching). The conclusion from this and other studies is that the frequency organization of the cochlea changes during the early stages of development. In very young animals, low frequencies are represented near the base of the cochlea, and as development proceeds, these low frequencies begin moving toward the apex, making the base available for the higher frequencies. This movement is shown in the upper part of Figure 8.43, which indicates that, in the embryonic stage of development, only the part of the basilar membrane near the base vibrates, and that the peak vibration for a 200-Hz tone falls about halfway between the base and the apex (Figure 8.43a). Figures 8.43b and 8.43c show that, as the chick develops toward adulthood, the peak vibration for the 200-Hz tone shifts toward the apex, and higher frequencies take over near the base.

This change in the way frequencies are represented in the cochlea also effects neurons in the central auditory system. This effect is shown in the lower part of Figure 8.43. In a young developing animal, hair cells near the base of the cochlea that respond best to an 800-Hz tone connect to a neuron in the central auditory system and cause it to respond best to an 800-Hz tone (Figure 8.43a). However, as the cochlea develops and this cochlear neuron begins to respond best to higher frequencies, the central neuron with which it connects also shifts its response to higher frequencies (Figures 8.43b and 8.43c). Thus, the frequency responses of both peripheral and central auditory neurons change during the process of early development.

Are further changes in the frequency organization of the cochlea and the higher centers possible once the organization is established in the adult animal? Research has shown that the answer to this question is yes. Donald Robertson and Dexter Irvine (1989) showed that lesioning the guinea pig cochlea changes the tonotopic organization of the auditory cortex. They lesioned a small area in the guinea pig's cochlea with a chemical that destroyed the hair cells. They then mapped the cortex either within hours after making the lesion or 35 days or more after making the lesion. They found that, immediately after the lesion, the map of frequencies in the auditory cortex looked similar to the map of the unlesioned animal, except that the cortical area that normally received signals from the lesioned area

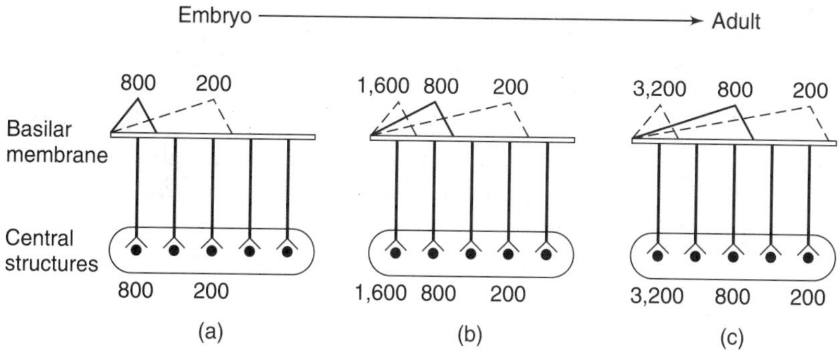

Figure 8.43
Top: The pattern of basilar membrane vibration as a chicken develops from embryo to adult. Bottom: The best frequencies of neurons in the chicken's central auditory system as the chicken develops. The vertical lines between the basilar membrane and the central structures indicate connections between neurons in the cochlea and neurons in the central structures. (Adapted from King & Moore, 1991, based on data in Rubel, 1984.)

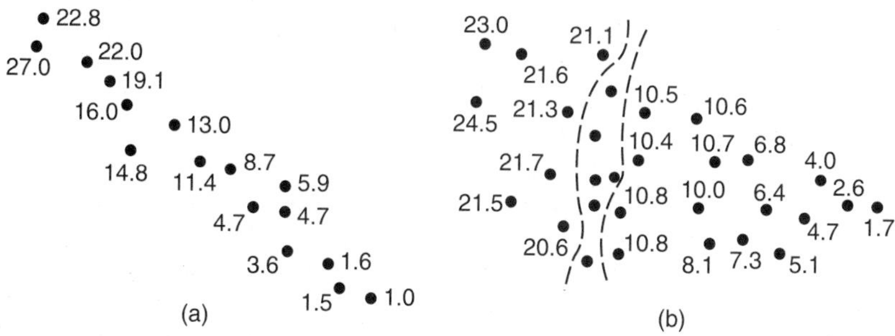

Figure 8.44
Tonotopic mapping in the guinea pig auditory cortex. (a) A map of the normal guinea pig. (b) A map recorded 35 days after the lesioning of an area of the cochlea that responded to frequencies between about 11,000 and 20,000 Hz. Each dot indicates a neuron, and the number beside the dot indicates the best frequency of that neuron in KHz (thousands of Hertz). The neurons between the dashed lines either wouldn't respond or were difficult to classify. (Data from Robertson & Irvine, 1989.)

of the cochlea had elevated thresholds at the original frequencies and also responded to other frequencies but only at high sound intensities. If, however, they waited for 35 days or more to measure the cortical map, they found that neurons that had originally received signals from the lesioned area now responded to new frequencies, even at low sound intensities. We can see this effect by comparing Figures 8.44a and 8.44b. Figure 8.44a shows the map of the cortex before

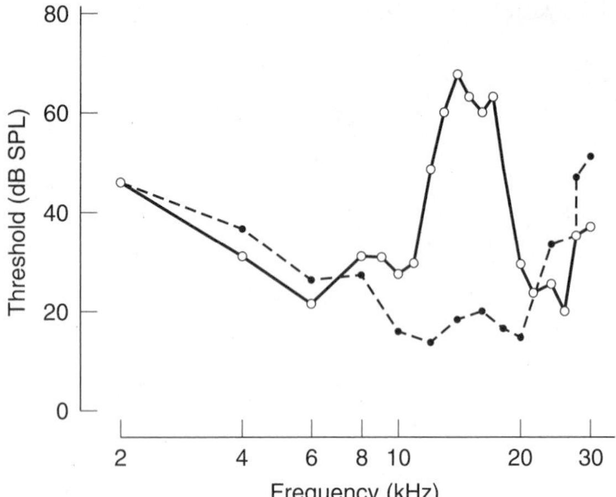

Figure 8.45

Threshold curves for the response of the auditory nerve in the normal guinea pig (filled circles and dashed line) and in the lesioned guinea pig whose cortical map is shown in Figure 8.44b (open circles and solid line). The threshold, plotted on the vertical axis, is the intensity necessary to elicit a compound action potential from the auditory nerve. The compound action potential is the response of the whole auditory nerve and therefore represents the activity of all of the nerve fibers leaving the cochlea. (Adapted from Robertson & Irvine, 1989.)

lesioning. Notice how the best frequencies of the neurons change in an orderly way from high-frequency neurons on the left to low-frequency neurons on the right. The map in Figure 8.44b was determined from another guinea pig 35 days after lesioning of the area of the cochlea that signaled frequencies between about 11,000 and 20,000 Hz. This map contains many neurons that responded best to frequencies above 20,000 Hz (to the left of the dashed line), many that responded best below about 11,000 Hz (to the right of the dashed line), and some that either didn't respond at all or were difficult to classify (the small area between the dashed lines). This result shows that the guinea pig's cortex had been reorganized during the weeks following the lesion so that cortical areas that had originally responded to frequencies between 11,000 and 20,000 Hz now responded to frequencies above 20,000 Hz or below 11,000 Hz.

Confirmation that the cortex was, in fact, receiving stimulation from the cochlea only at frequencies above 20,000 Hz and below 11,000 Hz was provided by measurements of auditory nerve thresholds. The dashed line in Figure 8.45 shows the threshold curve for the normal guinea pig, and the solid line shows the curve for the lesioned guinea pig with the cortical map in Figure 8.44b. The solid curve shows that the threshold of the auditory nerve was greatly elevated between about 11,000 and 20,000 Hz, exactly the same frequencies for which cortical neurons were absent.

Similar types of reorganization occur in other senses. For example, Jon Kaas and his co-workers (1990) did an experiment analogous to Robertson and Irvine's, but on the visual system. They lesioned small areas of an adult cat's retina, waited for two to six months, and then mapped the cat's visual cortex. They found that cortical neurons that had previously received input from the lesioned areas of the retina now responded to stimulation of other areas of the retina. These results from the retina, the ones we described from the auditory system, and similar results from the system serving touch sensations (Kaas, Merzenich, & Killackey, 1983; Merzenich & Kaas, 1982; Wall, 1988) show that sensory systems remain dynamic and open to change in response to damage, even in adult animals.

Review

The Functions of Hearing

Outline

- Hearing informs us of events in our environment, many of which we would be unaware of if we had to rely on vision alone.

Questions

1. What are some specific functions of audition? What was Helen Keller's argument regarding the relative importance of audition and vision? (335)

The Stimulus for Hearing

- Our perception of sound depends on the vibration of objects. We perceive these vibrations indirectly, through the effect of these vibrations on the air or water surrounding these objects.

2. Describe how the vibration of a speaker diaphragm creates pressure changes in the air. (338)
3. What is a pure tone? Describe the following characteristics of pure tones: frequency, amplitude, decibels, and sound pressure level (SPL). Why is the decibel scale used? (338)

- The physical characteristics of sound, such as frequency and amplitude, create varying perceptual experiences.

4. Describe the following perceptual dimensions of sound and the physical characteristics associated with each one: loudness, pitch, localization, and timbre. (340)

- Most of our auditory experience is not with pure tones, but with complex sounds such as speech and music.

5. What is Fourier analysis? What happens when we apply Fourier analysis to a complex tone? Define Fourier frequency spectrum, fundamental frequency, and harmonics. (340)

- According to Ohm's acoustic law, the ear analyzes a complex sound into its frequency components.

6. What is the relationship between Ohm's acoustic law and Fourier analysis? How does the ear analyze a complex sound stimulus into simple frequency components? (342)

Structure and Functioning of the Auditory System

- The goal of the initial stage of auditory processing is to transmit the sound vibrations in the environment to the hair cell receptors in the ear.

7. Describe the overall structures of the outer ear, the middle ear, and the inner ear. Be sure you understand how sound vibrations are transmitted through each one. (343)
8. What are two functions of the outer ear? What are the ossicles, and what is their function? (344)

- The inner ear, as the site of the receptors for hearing, is the place where the sound stimulus is transduced into electrical signals.

9. Describe the following structures of the inner ear: the cochlea, the organ of Corti, the basilar membrane, the tectorial membrane, and the hair cells. How do each of these structures cause the hair cells to bend in response to sound? (346)

- The auditory pathways conduct signals from the cochlea to the auditory cortex.

10. Describe the specific structures that conduct signals from the cochlea to the auditory cortex. (349)

The Code for Frequency

- One of the central questions of auditory research is how the firing of auditory neurons signals different pitches.

11. Describe two general answers that have been proposed to the question: How does the firing of neurons signal frequency? (350)

- The basic idea behind the place code for frequency is that frequency is signaled by individual hair cells that fire to specific frequencies.

12. What two methods did Békésy use to find evidence supporting the place code for frequency? What did he discover using these two methods? (351)
13. Describe the traveling wave. What is the envelope of the traveling wave, and what is its significance? How does the envelope change as frequency changes? Describe the tonotopic map in the cochlea. (351)

- One problem with Békésy's results was that his broad picture of basilar membrane vibration, based on his observations of the membrane's response to high-intensity stimulation, indicated that a single frequency could cause a large portion of the basilar membrane to vibrate.

14. Why is the broad vibration of the basilar membrane reported by Békésy a problem? Describe the two findings that have provided a solution to this problem. (352)

- The timing or rate of nerve firing also provides information about the frequency of the stimulus.

15. Describe Rutherford's proposal linking sound frequency and firing rate. Why can't this proposal work? What was Wever's proposed solution? What is phase locking? (354)

- In order for information about the stimulus frequency that originates in the cochlea to be used for perception, it must be transmitted to central auditory structures.

16. What is the evidence that place and timing information about stimulus frequency are transmitted to central auditory structures? (356)
17. What problem does increasing the stimulus intensity pose for the place theory of frequency coding? (357)

The Physiology of Sound Localization

- Sounds can be localized better with two ears than with one, since each ear receives different sound stimulation and these differences provide cues that indicate a sound's location.
- Auditory localization has been studied at the neural level through recordings from cells that respond best to specific interaural time differences or to specific locations in space.

18. Describe the interaural time difference and the interaural intensity difference cues. Interaural intensity difference works best at certain frequencies. What are they, and why is this so? (358)
19. What is an interaural time difference detector? Describe the receptive field and the map of space associated with cells in the owl's MLD. Describe similarities and differences between the mapping of the owl's MLD cells and the mapping of auditory and visual system neurons. (360)
20. Describe neurons in the cat's superior colliculus and cortex that respond to specific locations. Why are some of the cortical neurons called panoramic? (362)

Comparing the Senses: Audition and Vision

- Although there are large differences between the auditory system and the visual system, they share many mechanisms.

21. Describe seven mechanisms that are shared by the auditory and visual systems. (363)

Echolocation by Bats

- As early as the 1700s, scientists observed that bats can avoid objects while flying in a darkened room.
- The bat's echolocation system enables the bat to identify objects and precisely locate them.

22. Describe the signals that bats emit to navigate. What is the physiological mechanism for bat echolocation? (365)

23. Describe the idea proposed by Dear et al. concerning how the bat uses neural information to perceptually organize a scene. (366)

Plasticity of Frequency Representation

- The fact that the base of the cochlea develops before the apex led researchers to expect that very young animals might be more sensitive to the higher frequencies that are usually represented at the base. However, contrary to this expectation, young animals are most sensitive to low-frequency sounds.

24. Describe Rubel and Ryals' experiment that solved the mystery of why young animals are sensitive to low-frequency sounds even though only the base of their cochlea is developed. (368)

25. Describe Robertson and Irwin's experiment that showed that lesioning the cochlea affects the tonotopic organization of the auditory cortex. (368)

Audition II: Hearing

Sensitivity and Loudness

Perceiving Pitch

Sound Quality

Psychophysical Aspects of Auditory Processing

Auditory Scene Analysis

Auditory Localization

The Psychophysical Approach to Hearing

At the beginning of his book *Auditory Scene Analysis,* Albert Bregman (1990) poses the following problem:

> Imagine that you are on the edge of a lake and a friend challenges you to play a game. The game is this: Your friend digs two narrow channels up from the side of the lake. Each is a few feet long and a few inches wide and they are spaced a few feet apart. Halfway up each one, your friend stretches a handkerchief and fastens it to the sides of the channel. As waves reach the side of the lake they travel up the channels and cause the two handkerchiefs to go into motion. You are allowed to look only at the handkerchiefs and from their motions to answer a series of questions: How many boats are there on the lake and where are they? Which is the most powerful one? Which one is closer? Is the wind blowing? Has any large object been dropped suddenly into the lake? (pp. 5–6) (Figure 9.1)

This seems like an impossible problem. How can we determine how many boats there are, where they are located, and which one is the most powerful simply by observing how two handkerchiefs at the side of the lake respond to the waves created by boats on the lake? Although Bregman's boat problem appears to be an extremely difficult one, we aren't at all surprised that we can identify how many people are talking in a room, where they are located, and which one is loudest, based on how two small membranes on the sides of our head respond to the pressure changes created by people's voices in the air.

Bregman's clever analogy between the moving handkerchiefs and our vibrating eardrums helps us appreciate how amazing it is that the information from these vibrations is transformed into information about the identities and locations of sounds in our environment. But what is even more amazing is that our vibrating membranes not only inform us about the identities and locations of sounds but also help create auditory experiences as varied as hearing music, the rustle of leaves in the wind, or people talking.

In this chapter, we will consider the relationship between pressure changes in the air, vibra-

Figure 9.1
Is it possible to determine what is happening on the lake by watching how the handkerchiefs react to the waves? Since we are able to make judgements about the identities and locations of people talking in a room based on vibrations of our eardrums, perhaps it isn't so far fetched to suggest that we can determine information about the boats based on motions of the handkerchiefs in response to water waves.

tions of our auditory structures, and our various auditory abilities. Chapter 8 considered this relationship at the physiological level by describing how the auditory system transforms pressure changes in the environment first into vibrations and then into neural responses (Figure 9.2a). In this chapter we focus on studying the relationship between these pressure changes and auditory experience (Figure 9.2b). As we determine the relationships between stimuli in our environment and our auditory experience, we will also return to some of the physiological mechanisms that we described in Chapter 8. Continuing in the tradition established in the chapters on vision, we will be illustrating a number of connections between psychophysics and physiology. We begin our consideration of the psychophysics of hearing by looking at two characteristics of hearing: detecting sounds near threshold and perceiving the loudness of sounds that are above the threshold.

Sensitivity and Loudness

The ear is such an exquisitely sensitive instrument that it can detect sounds so faint that they cause the cat's eardrum to move only 10^{-11} cm, a dimension that is less than the diameter of a hydrogen atom (Tonndorf & Khanna, 1968). The ear is so sensitive that the air pressure at thresh-

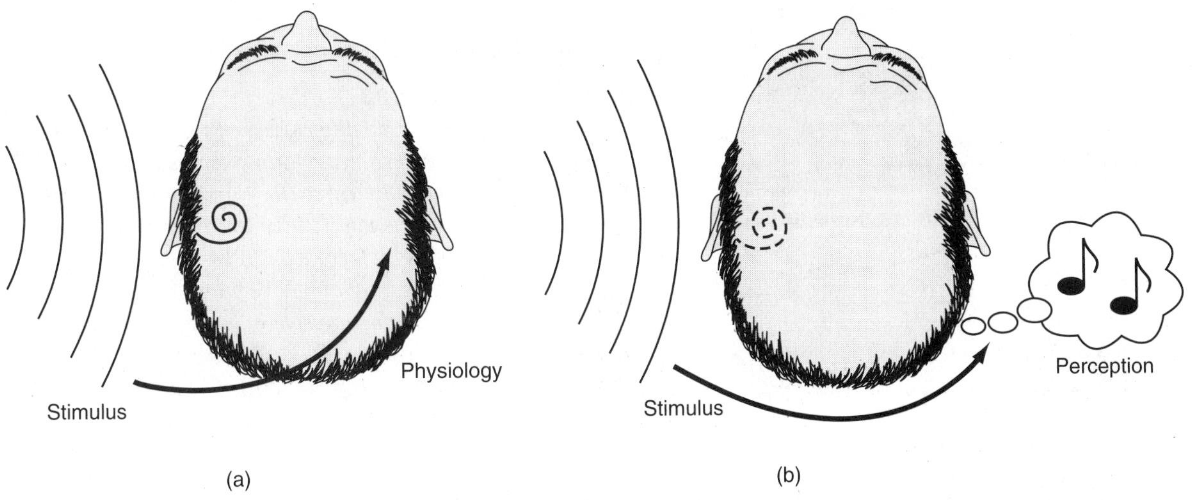

Figure 9.2
(a) In Chapter 8, our emphasis was on the relationships between the stimulus and physiology. (b) In this chapter, our emphasis is on the relationships between the stimulus and perception and on relating some of these stimulus–perception relationships back to the underlying physiology.

old in the most sensitive range of our hearing is only 10–15 dB above the air pressure generated by the random movement of air molecules. Thus, if our hearing were much more sensitive than it is now, we would hear the background hiss of colliding air molecules!

An important fact about auditory functioning is that we are not equally sensitive to all frequencies. This relationship between sensitivity and frequency is depicted by the **audibility curve,** a plot of the minimum sound pressure level in decibels needed to just detect a tone, as a function of the tone's frequency.

The Audibility Curve

The audibility curve is analogous to the spectral sensitivity curve for vision, which we described in Chapter 2. We determined the spectral sensitivity curve by measuring the light intensity necessary to just see a light, at wavelengths across the visible spectrum. We determine the audibility curve by measuring the sound pressure level necessary to just hear a tone, at frequencies across the range of hearing. The resulting curve (Figure 9.3) indicates that the threshold is high at very low frequencies: A 20-Hz tone must have an intensity of 80 dB SPL (sound pressure level) to be just barely audible. However, as we move to higher frequencies the threshold decreases and reaches a minimum threshold (maximum sensitivity) at about 2,000–4,000 Hz. It is interesting that the range of frequencies that is most important for understanding speech—between about 400 and 3,000 Hz—falls within the frequency range to which we are most sensitive. At frequencies above 4,000 Hz, the threshold increases, until 60 dB SPL are required to hear a 20,000-Hz tone. The range of human hearing is usually regarded as being between 20 and 20,000 Hz, since frequencies below or above this range can be heard

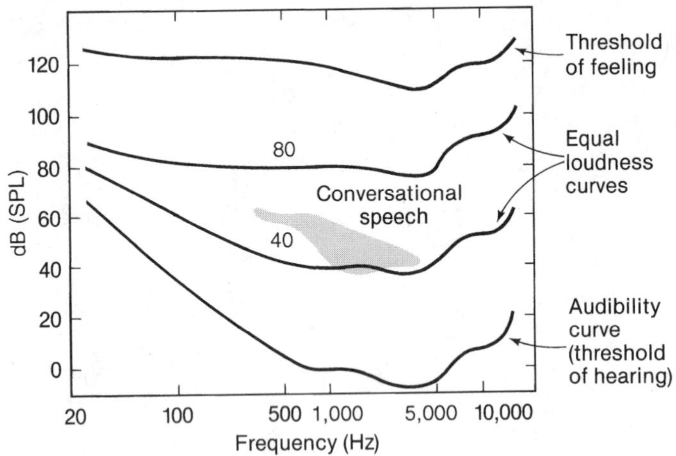

Figure 9.3
The audibility curve and the auditory response area. Hearing occurs between the audibility curve (the threshold for hearing) and the upper curve (the threshold for feeling). Tones with SPLs below the threshold for hearing cannot be heard; tones with SPLs above the threshold of feeling result in pain and eventually cause damage to the ear. The shaded area indicates the frequency and intensity range of conversational speech. (From Fletcher & Munson, 1933.)

only at extremely high pressures or can't be heard at all.

The U-shaped audibility curve defines the threshold for hearing, but what about the large area contained within this U-shaped function? Our above-threshold auditory experience occurs at frequencies and intensities contained within this space, which is called the **auditory response area.** The auditory response area includes all sounds between the audibility curve and the upper curve in Figure 9.3, which defines the threshold for feeling. At intensities below the audibility curve, we can't hear a tone, and at intensities above the threshold for feeling, tones become painful. Thus, the useful range of hearing is contained within the auditory response area between these two curves.

Loudness, Intensity, and Frequency

What happens to our perception of a tone if we hold its frequency constant and increase its intensity? From the connection between decibels and loudness, we already know that increasing the intensity of a tone increases its loudness (Table 8.1). S. S. Stevens used the method of magnitude estimation (see Chapter 1) to measure the relationship between loudness and sound pressure shown in Figure 9.4. The unit for loudness, the **sone,** is plotted versus intensity in decibels. One sone is the loudness of a 1,000-Hz tone at 40 dB. The relation between loudness and decibels is a power function with an exponent of 0.6. The fact that the exponent is less than 1 means that large changes in sound pressure cause smaller changes in loudness, so increasing the sound pressure by a factor of 10 (which is the same as increasing the SPL by 20 dB) increases the loudness by a factor of about 4.0.

Loudness is a function not only of sound pressure, but also of frequency. We can appreciate this relationship by noting that saying, "The tone is 40 dB SPL" tells us nothing about its loudness unless we also know its frequency. A 40-dB-SPL tone can be just above threshold if its frequency is 100 Hz, but it is far above threshold if its frequency is 1,000 Hz. Another way to understand the relationship between loudness and frequency is by measuring equal loudness curves.

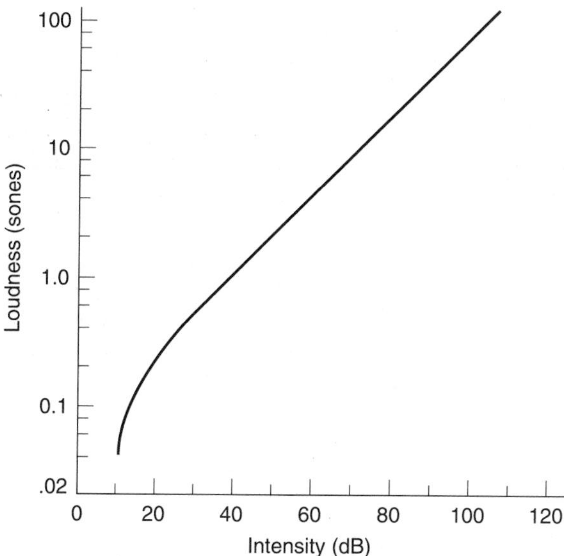

Figure 9.4

The growth in loudness in sones for a 100-Hz tone as a function of intensity. Above about 25 dB SPL, the growth of loudness with intensity is a power function with an exponent of 0.6. (Adapted from Gulick, Gescheider, & Frisina, 1989.)

Equal Loudness Curves

The curves marked **equal loudness curves** in Figure 9.3 indicate the number of decibels that create the same perception of loudness at frequencies across the range of hearing. We measure an equal loudness curve by designating one tone as a standard and matching the loudness of all other tones to it. For example, the curve marked 40 in Figure 9.3 was determined by matching the loudness of frequencies across the range of hearing to a 1,000-Hz 40-dB-SPL tone. Similarly, the curve marked 80 was determined by matching the loudness of different frequencies to a 1,000-Hz 80-dB-SPL tone.

Notice that the equal loudness curve marked 40 looks similar to the audibility function, in that it curves up at high and low frequencies. However, the equal loudness curve marked 80 is considerably flattened, which means that at 80 dB SPL all tones from 30 to 5,000 Hz have about the same loudness. The difference between the flat 80 curve and the upward-curving 40 curve and the audibility function explains something that happens as you adjust the volume control on your stereo system. If you are playing music at a fairly loud level—say, 80 dB SPL—you should be able to easily hear each of the frequencies in the music because, as the equal loudness curve for 80 indicates, all frequencies between about 20 Hz and 5,000 Hz sound equally loud at this intensity. What happens, however, when you turn the loudness down so that the sound pressure is very low—say, 10 dB SPL? Now all frequencies don't sound equally loud. In fact, you can see from the audibility curve in Figure 9.3 that frequencies below about 400 Hz and above about 10,000 Hz are inaudible at 10 dB SPL. (Be sure you understand how to determine the frequencies below and above which a 10-dB-SPL tone is inaudible. Find 10 dB on the SPL scale, and draw a horizontal line across to the right. The place where the 10-dB line intersects the audibility curve indicates that frequencies under about 400 Hz and over 10,000 Hz are not audible at 10 dB SPL.)

Being unable to hear very low and very high frequencies creates a bad situation, because it means that when you play music softly you won't hear the very low or very high pitches. Fortunately most stereo receivers have a button labeled "loudness" that selectively boosts the level of very high and very low frequencies so that you can hear them, even when you are playing your stereo very softly.

Clinical Aspects of Loudness

The perception of loudness is particularly important to those with hearing problems. People who

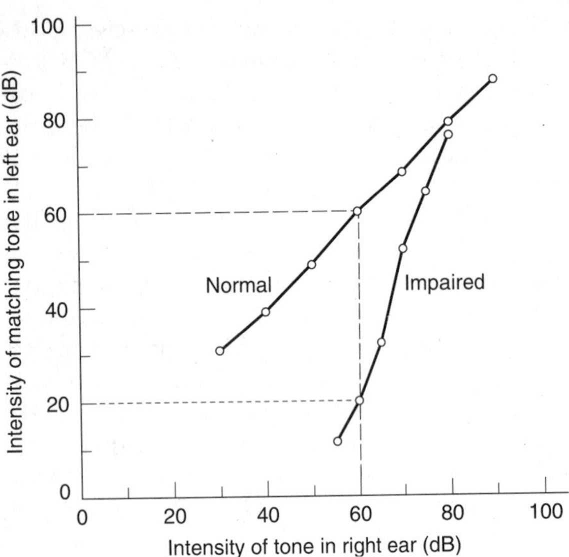

Figure 9.5
How loudness grows with increasing intensity for a person with normal hearing and a person with impaired hearing in one ear. To determine loudness, the listener adjusts a matching tone in one ear to match the loudness of a tone in the other ear. (Adapted from Moore et al., 1985.)

have difficulty hearing perceive some or all frequencies as having a much lower than normal loudness. This perception is dramatically illustrated by the data in Figure 9.5, which compares the perception of loudness by a person with normal hearing in both ears with that of a person with decreased hearing in one ear. The curve marked "Normal" shows how a person with normal hearing matches a tone in one ear with a tone in the other. For example, the person with normal hearing matches a 60-dB-SPL tone in the right ear with a 60-dB-SPL tone in the left ear (dashed line). However, to the person with impaired hearing in the right ear, who can't hear any sound below about 50 dB SPL, the 60-dB tone sounds very soft, so it takes only a 20-dB-SPL tone in the left ear to match it (dotted line).

The steepness of the "impaired" curve in Figure 9.5 illustrates a phenomenon called **loudness recruitment**—an abnormally rapid increase in loudness with increases in intensity. Thus, even though hearing in the impaired ear is far below normal at an intensity of 60 dB SPL, loudness increases rapidly with increasing intensity, until by 80 dB there is no difference between loudness perception in the normal and the impaired ears. Loudness functions such as those in Figure 9.5 are useful for diagnosing hearing disorders since functions like the one marked "Impaired" are usually caused by defects in the hair cells in the cochlea.

Physiological Aspects of Loudness

In addition to its importance clinically, loudness raises important questions related to our basic understanding of auditory functioning. The fact that loudness increases over a range of nearly 120 dB for frequencies in the middle of the range of hearing, poses a difficult question for physiologists: How can nerve firing signal loudness over this large range of intensities? The reason that this is a difficult question is that, as we saw in Figure 8.33, most nerve fibers increase their firing rate over only about a 40-dB range of intensities. At intensities above 40–50 dB SPL, most fibers reach saturation: that is, their rate of firing stops increasing with further increases in intensity. One possible solution to this problem is the fact that some fibers have been discovered that don't begin firing until the middle of the intensity range and that continue to increase their firing at high intensities (Lieberman, 1978; Palmer & Evans, 1979).

Another solution to the problem of the neural coding of loudness is based on place theory: the fact that a tone of a specific frequency causes maximum neural activity at a particular place on the basilar membrane. Thus, as shown in Figure 9.6, at low and moderately low intensities, most of the neural activity is centered on one place on the basilar membrane. However, when intensity

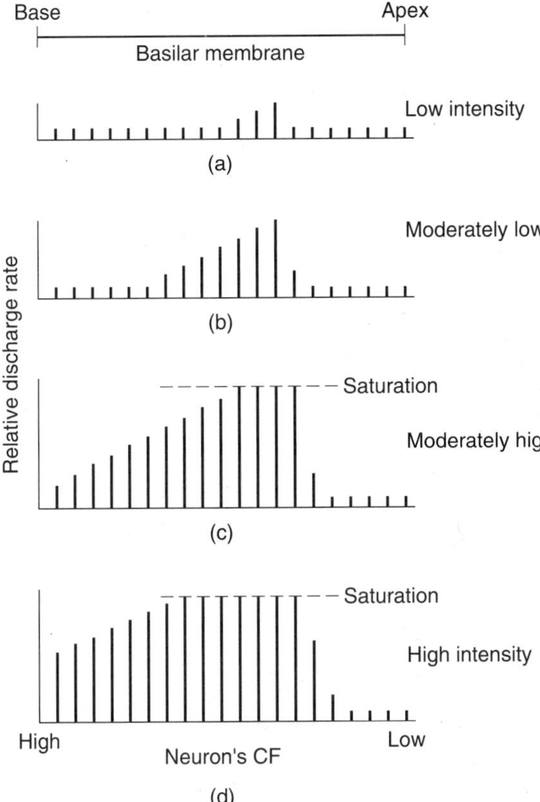

Figure 9.6
How increasing stimulus intensity affects the discharge rate of neurons along the basilar membrane. (a) At low intensity, firing is low and is localized at one place on the membrane; (b) increasing intensity increases firing at that place, and fibers some distance away begin firing; (c) at moderately high intensities, some fibers begin to saturate; (d) more saturation occurs at high intensities. It is possible that the broader pattern of basilar membrane excitation that occurs at high intensities provides information regarding the intensity of the stimulus. (Adapted from Gulick et al., 1989.)

is increased further, two things happen: (1) Neurons that were already firing rapidly begin to saturate, and (2) neurons at other places on the basilar membrane begin to fire more rapidly. The result is that, at higher intensities, the pattern of auditory firing becomes broader, and this may provide the neural information that signals loudness (Gulick, Gescheider, & Frisina, 1989).

Perceiving Pitch

In Chapter 8, we noted that a tone's pitch is related to its frequency. Let's now look at the relationship between pitch and frequency in more detail.

Pitch and Frequency

The relationship between pitch and frequency has been measured by the method of fractionation, in which subjects adjust the frequency of a comparison tone to sound half as high in pitch as a standard tone. By using this method at a number of frequencies, Stevens and Volkman (1940) obtained the curve in Figure 9.7. They called the psychological unit of pitch the **mel** and arbitrarily set the pitch of a 1,000-Hz 40-dB-SPL tone equal to 1,000 mels. The curve indicates that, although pitch and frequency are related, they do not have a one-to-one relationship, so doubling the frequency does not double the pitch.

Not only do the data in Figure 9.7 provide a quantitative relationship between our perception of pitch and the frequency of the stimulus, but when combined with physiological measurements, this curve tells us something about the relationship between our perception of pitch and the vibration of the basilar membrane. In Chapter 8, we saw that there is a relationship between frequency and the place of maximum vibration on the basilar membrane. This relationship is indicated by the circles in Figure 9.7, based on measurements in the guinea pig. Each circle represents a frequency, and the place on the basilar membrane corresponding to that frequency is indicated on the right axis. The almost exact cor-

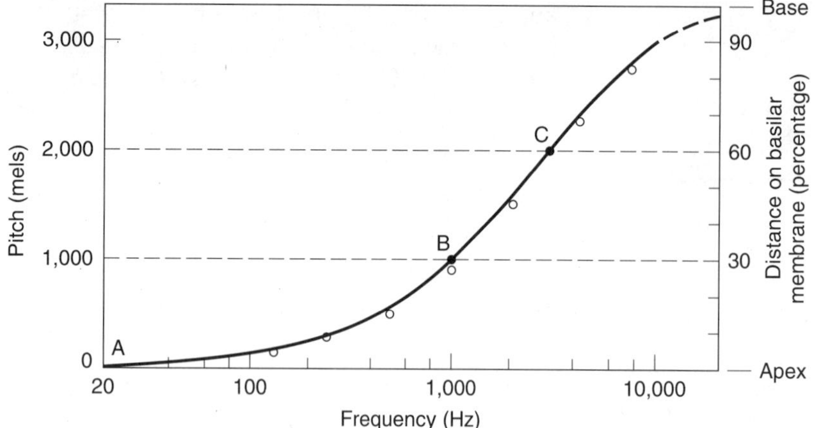

Figure 9.7
The solid curve indicates the relationship for humans between pitch in mels (left axis) and frequency. The open circles give the location along the guinea pig's basilar membrane for seven different frequencies (right axis). (Data from Stevens & Volkmann, 1940; and Stevens, Davis, & Lurie, 1935.)

respondence between our psychophysical curve and the physiological measurements demonstrates a remarkable connection between the experience of pitch and the place of maximum vibration on the basilar membrane.

We can appreciate the nature of this connection between psychophysically measured pitch perception and physiologically measured vibration of the basilar membrane by starting at point A, at the lower end of the curve in Figure 9.7, and moving to point B, which is 1,000 mels higher. Notice from the scale on the right that, by moving up 1,000 mels, we have traversed 30 percent of the length of the basilar membrane. Now do this again, this time moving from point B to point C, another 1,000 mels. Again, increasing the pitch by 1,000 mels causes us to traverse another 30 percent of the basilar membrane. Thus, every time we increase the frequency enough to cause a 1,000-mel rise in pitch, we move the same distance along the basilar membrane—an impressive demonstration of the connection between perception and physiology.

Although the curve in Figure 9.7 provides important information about the relationship between pitch and frequency, there is another psychological dimension of pitch that is not obvious from this curve. We can appreciate this dimension by distinguishing between two properties of pitch: tone height and tone chroma.

Tone Height and Tone Chroma

We can appreciate what we mean by *tone height* and *tone chroma* by starting at the low end of the piano keyboard and moving toward the right. As we do this, the pitch becomes higher. This increase in pitch with increasing frequency reflects an increase in **tone height** as we go up the scale.

But in addition to this increase in tone height that occurs as we move toward higher frequencies on the piano keyboard, something else happens: The letters of the notes A, B, C, D, E, F, and G repeat themselves. This repetition of notes reflects a quality called **tone chroma.** Notes with the same letters sound similar, even though they have different fundamental frequencies. These similar-sounding tones are separated by intervals of one or more **octaves,** so that a tone one octave above another tone has a fundamental frequency twice that of the lower tone. Thus, A_5 on the piano keyboard has a fundamental frequency of 440 Hz, A_6 has a frequency of 880 Hz, A_7 has a fundamental frequency of 1,760 Hz, and so on. Tones in this octave relation to one another have the same chroma.

The concepts of tone height and tone chroma are combined in the spiral of Figure 9.8. Moving up the spiral increases tone height (i.e., the pitch gets higher), and the places where the spiral crosses vertical lines have the same tone chroma

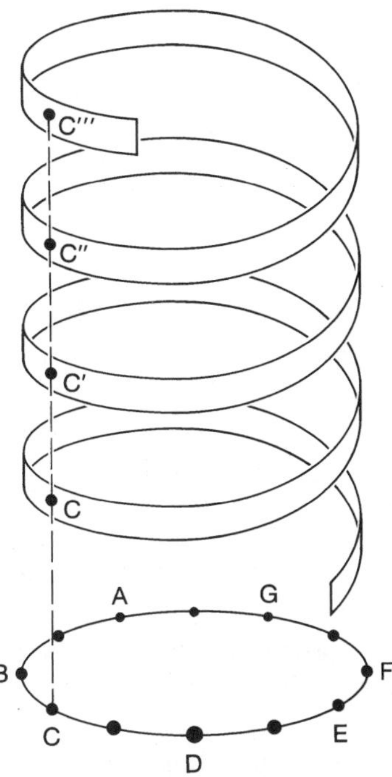

Figure 9.8

Representing the notes of the scale on an ascending spiral graphically depicts the perceptions of tone height and tone chroma. As we move up along a vertical line, we encounter notes with the same letter. These notes, which are separated by octaves and therefore sound similar to each other, have the same tone chroma. As we move from one note to another in ascending the spiral, each note sounds higher than the note preceding it. This property is called tone height. *Thus, two C's separated by an octave or more would have different tone heights but the same tone chroma.*

(i.e., these tones are indicated by the same letter). Tone chroma is important because two notes with the same chroma are psychologically similar. Thus, a male and a female can be regarded as singing "in unison" even if their voices are separated by an octave or more. This similarity between the same notes in different octaves also makes it possible for a singer on the verge of overreaching his or her voice range to shift the melody down to a lower octave.

The perceptual similarity of tones separated by an octave was demonstrated in an experiment by Diana Deutsch (1973). As shown in Figure 9.9, she presented a standard tone, followed by a series of six intervening tones, followed by a test tone. The listeners' task was to indicate whether the standard and test tones were the same or different. Even though Deutsch's subjects were told to ignore the intervening tones, their ability to detect the difference between the standard and the test tones decreased if one of the six intervening tones was either the same as the test tone or one octave higher or lower than the test tone. The fact that intervening tones that were one octave higher or lower than the test tone had the same effect as an intervening tone that was identical to the test tone illustrates **octave generalization:** Tones separated by octaves are perceptually similar and so have similar effects.

Periodicity Pitch (the Effect of the Missing Fundamental)

We saw in Chapter 8 that musical tones have a fundamental frequency and harmonics that are multiples of the fundamental frequency. The fundamental frequency determines the tone's pitch, so the tone in Figure 9.10a with a 400-Hz fundamental frequency sounds lower than the tone in Figure 9.10b with an 800-Hz fundamental frequency. But what happens if we eliminate the 400-Hz tone's fundamental frequency and leave

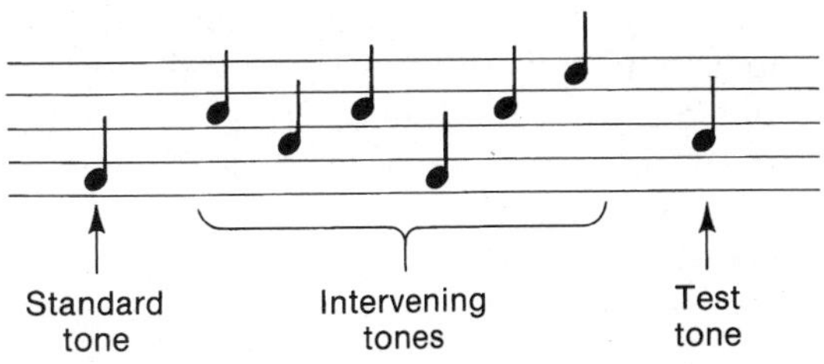

Figure 9.9

Stimuli for Deutsch's (1973) octave generalization experiment.

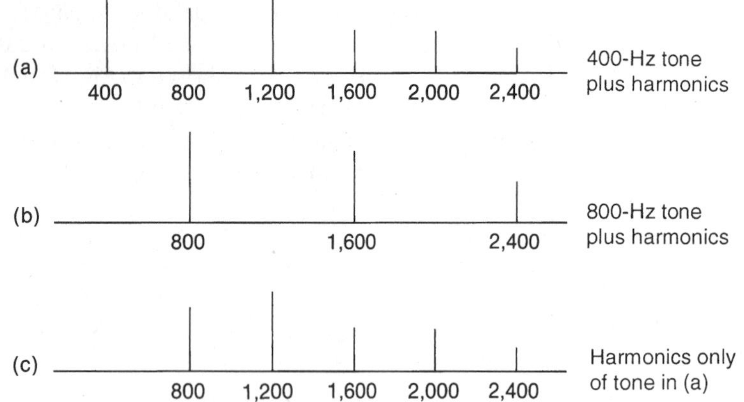

Figure 9.10
Fourier spectra of (a) a 400-Hz tone plus its harmonics, (b) an 800-Hz tone plus its harmonics, and (c) the 400-Hz tone in (a) without its 400-Hz fundamental. See text for details.

the 800-, 1,200-, and 1,600-Hz harmonics, as in Figure 9.10c? We might expect that the tone's pitch might now correspond to the pitch of the 800-Hz tone, since its lowest frequency is now 800 Hz. However, this is not what happens. Removing the 400-Hz fundamental changes the tone's timbre slightly but has no effect on the tone's pitch. The fact that pitch does not change when we remove the fundamental is called **periodicity pitch,** or the **effect of the missing fundamental.**

The fact that tones with different Fourier spectra can have the same pitch poses problems for the place theory of pitch perception, because even though the tones in Figures 9.10a and 9.10c cause different patterns of vibration on the basilar membrane, they are perceived as having the same pitch. Therefore most explanations of periodicity pitch do not focus on the basilar membrane but look to activity in more central structures of the auditory system. One piece of evidence that supports the idea of looking more centrally is the finding that periodicity pitch is perceived even if the harmonics are presented to different ears. Presenting a 1,600-Hz tone to one ear and a 1,700-Hz tone to the other results in a perception of pitch that corresponds to 100 Hz (Houtsma & Goldstein, 1972). (Since harmonics are always multiples of the fundamental frequency, the *difference* between two adjacent harmonics corresponds to the fundamental frequency.) The perception of periodicity pitch when the tones are presented to separate ears means that periodicity pitch must be determined somewhere in the auditory system where signals from both ears are combined. The first place this occurs is the superior olivary nucleus (see Figure 8.19).

The results of many psychophysical experiments have led to the proposal that periodicity pitch is the result of a **central pitch processor**—a central mechanism that analyzes the *pattern* of harmonics and selects the fundamental frequency that is most likely to have been part of that pattern (Evans, 1978; Getty & Howard, 1981; Goldstein, 1978; also see Wightman, 1973, for another type of central processor). This idea that a central mechanism is important for the perception of the pitch of complex tones is supported by the finding that patients with damage to a specific area of the auditory cortex in the right hemisphere are not able to hear the pitch of the missing fundamental (Zatorre, 1988).

Thus, periodicity pitch suggests the following possibility about how the auditory system analyzes the sound stimulus: As we saw in Chapter 8, the firing of auditory nerve fibers contains place information about the various frequency components in the auditory stimulus. However, we perceive a sound's pitch only after the firing caused by these components has been analyzed by more central mechanisms. Thus, the same pitch can be signaled by a large amount of activity in neurons that respond to a tone's fundamental frequency, or by the pattern of neural activity in neurons that fire to a number of different frequencies.

The phenomenon of periodicity pitch not only gives us some insight into how the auditory system may operate at a neural level but also has a number of practical consequences. Consider, for example, what happens as you listen to music on a cheap radio that can't reproduce frequencies below 300 Hz. If you are listening to music that contains a tone with a fundamental frequency of 100 Hz, your radio can't reproduce the fundamental frequency of the 100-Hz tone or of the 200-Hz second harmonic of that tone. Even though the radio reproduces only the third (300-Hz) and higher (400, 500, 600, etc.) harmonics of this tone, periodicity pitch comes to the rescue and causes you to perceive a pitch equivalent to that produced by a 100-Hz tone. Similarly, even though the telephone doesn't reproduce the fundamental frequency of the human voice, we are usually able to identify people's voices on the phone (Truax, 1984).

Periodicity pitch has also been used to overcome the following problem in the construction of pipe organs: An organ designer wants to produce a pitch corresponding to a 55-Hz tone. However, the longest organ pipe he can use is 1.5 m long, a length that produces a pitch with a 110-Hz fundamental frequency. A longer pipe would be needed to produce the 55-Hz tone. The solution: Use the 1.5-m pipe, with its 110-Hz fundamental and a 1.0-m pipe, which produces a 165-Hz fundamental. Since 110 Hz and 165 Hz are the second and third harmonics of a 55-Hz tone ($55 \times 2 = 110$; $55 \times 3 = 165$), these two pipes, when sounded together, produce a tone with a pitch corresponding to 55 Hz (Dowling & Harwood, 1986).

Sound Quality

In Chapter 8, we said that, if two sounds sound different even if they have the same loudness and pitch and are presented in the same way, they differ in a quality called *timbre.* As an example, we noted that even though a trumpet and an oboe are playing the same note, we might describe the sound of the trumpet as *bright* or *brassy* and the sound of the oboe as *nasal* or *reedy.* These differences in timbre illustrate that sounds can have different qualities in addition to pitch and loudness. Tones can differ in other qualities as well. For example, we can describe tones as sounding *flat* or *resonant,* or one tone can sound as if it is reaching us directly from one location, whereas another tone can appear to surround us. We will now consider why some of these differences in sound quality occur.

Timbre

What causes the same note to sound different when played on different strings of a violin or on different instruments? One factor that causes these differences in timbre is the relative strengths of the harmonics. Figure 9.11 shows the heights of the harmonics for the note A_4 (fundamental frequency = 440 Hz) played on the A and the D strings of the violin. Note that, while the frequencies at which the harmonics occur are identical, the relative height of each harmonic is different.

Figure 9.12 compares the harmonics of the guitar, the bassoon, and the alto saxophone playing the note G_3 with a fundamental frequency of 196 Hz. Both the relative heights of the harmonics and the number of harmonics are different in these instruments. For example, at high frequencies, the guitar has more harmonics than either the bassoon or the alto saxophone. Although the frequencies of the harmonics are always multiples of the fundamental frequency, harmonics may be absent, as is true of some of the high-frequency harmonics of the bassoon and the alto saxophone.

The best example of an instrument that produces a tone with few harmonics is the flute. You can see from Figure 9.13 that for a 1,568-Hz tone (G_6) the flute has only one harmonic in addition to the fundamental, and this harmonic has very little energy. Because the fundamental contains

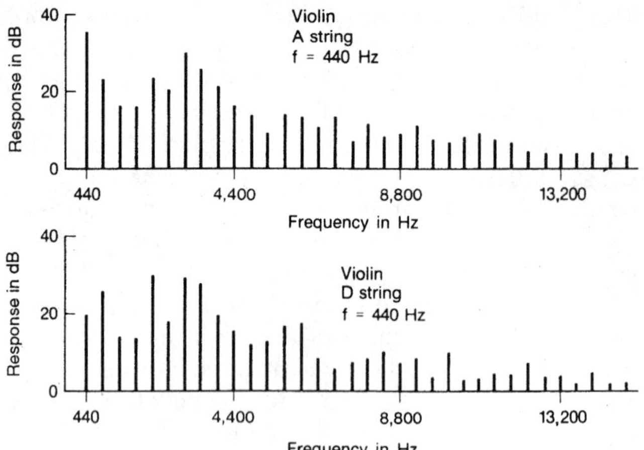

Figure 9.11

Fourier spectra for the A and D strings of a violin for a tone with a fundamental frequency of 440 Hz. Since both tones have the same fundamental frequencies, their pitches are the same; however, differences in the heights of the harmonics cause the tones to have different timbres (Olson, 1967).

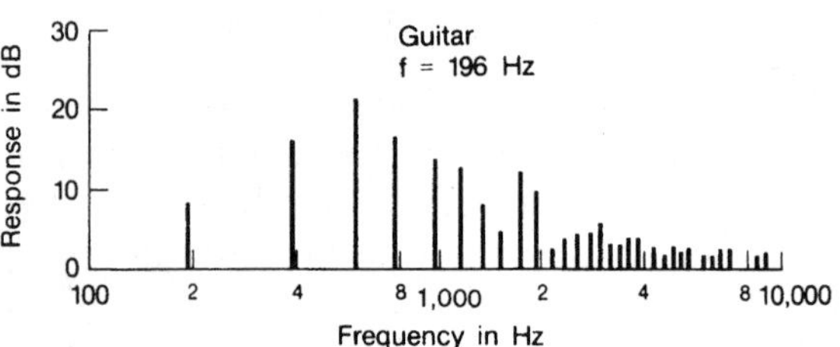

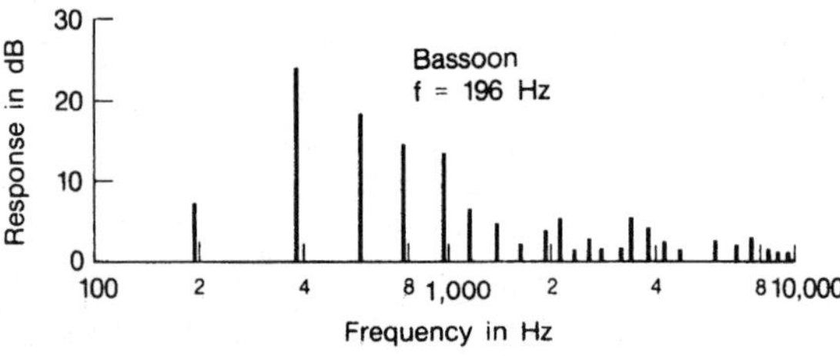

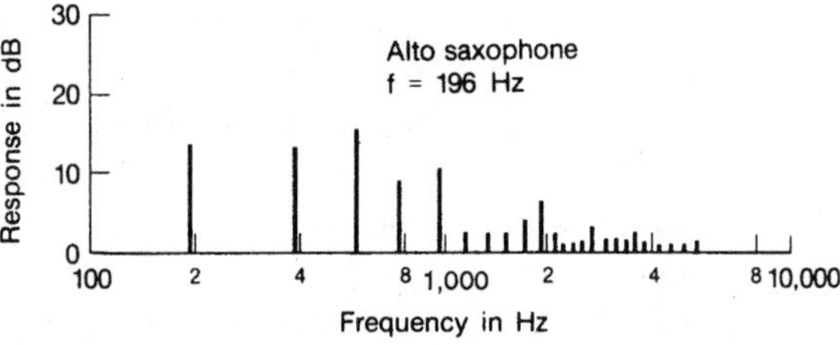

Figure 9.12

Fourier spectra for a guitar, a bassoon, and an alto saxophone playing a tone with a fundamental frequency of 196 Hz. The frequency scale is compressed compared to that in Figure 9.11 (Olson, 1967).

most of the energy of this tone, the flute has the thinnest and purest tone of all of the musical instruments; the tone of the flute is the closest a musical instrument comes to producing a pure tone. At the other extreme, instruments like the guitar and the lower notes on the piano, which have many harmonics, have much fuller, richer tones than the flute.

Timbre does not, however, depend only on harmonic structure. Jean Claude Risset and Max Mathews (1969) synthesized tones with a computer using Fourier spectra such as those shown in Figure 9.14 and found that the resulting tones do not resemble the appropriate musical instruments. For example, when Risset and Mathews synthesized a tone with the harmonic content of a trumpet and presented this tone to a number of people, the tone was unanimously judged *not* to sound like a trumpet.

This result can be explained by studies that show that timbre is affected not only by the harmonic content of the tone but also by the time course of the tone's **attack** (the buildup of sound at the beginning of the tone) and by the time course of the tone's **decay** (the decrease in sound at the end of the tone). This effect is illustrated by the fact that it is easy to tell the difference between a tape recording of a high note played on the clarinet and a recording of the same note

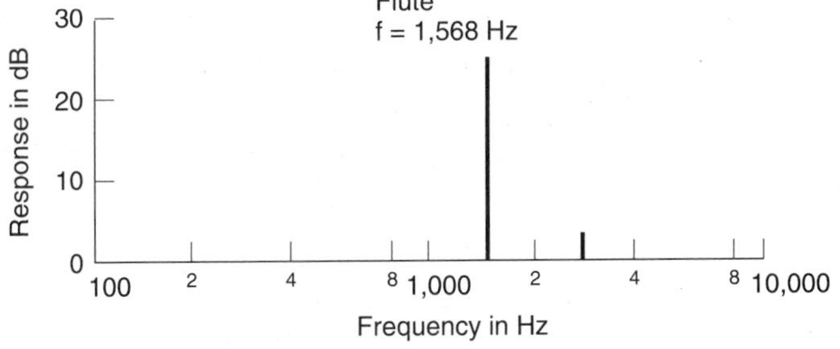

Figure 9.13
Fourier spectrum of a flute playing a tone with a fundamental frequency of 1,568 Hz (Olson, 1967).

played on the flute if the attack, the decay, and the sustained portion of the tone are heard, but if the tone's attack and decay are eliminated by erasing the first and last one-half second of the recording, it becomes very difficult to distinguish between the two instruments (Berger, 1964). Another way to make it difficult to distinguish one instrument from another is to play a tape of an instrument's tone backward. This changes the tone's original decay into the attack and the original attack into the decay but does not affect the harmonic structure of the tone. Even though the harmonic structure is unaffected, a piano tone played backward does not sound like a piano (Berger, 1964; Erickson, 1975).

With these results in mind, Risset and Mathews decided to determine not only the harmonic structure of a trumpet tone but also the way the harmonic structure changes with time. They did this by playing a recording of a trumpet tone into a computer, which analyzed the tone's harmonic structure. The result of this analysis of a 0.20-second tone is shown in Figure 9.14. In this figure, the buildup and decay of each harmonic is indicated by a line, and we can see that low harmonics build up faster than high harmonics. For example, the first harmonic reaches peak intensity in about 0.02 seconds, whereas the fifth harmonic reaches peak intensity in 0.04 seconds. Differences in the decay rates of the various harmonics also occur, with lower harmonics having longer decays. Risset and Mathews also analyzed

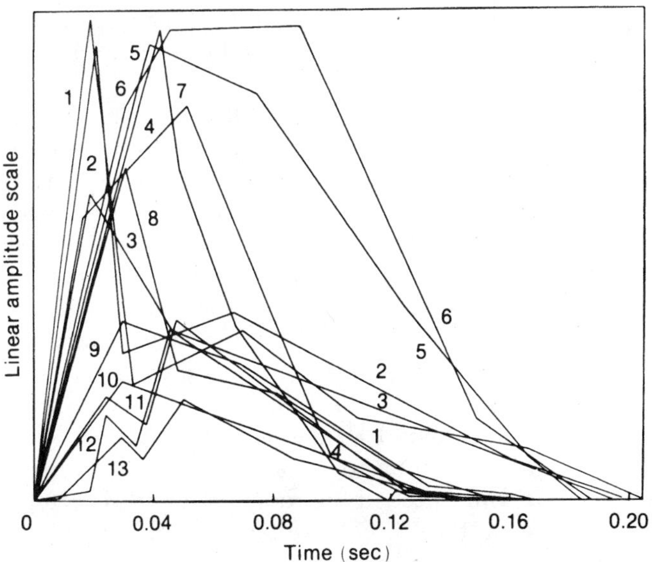

Figure 9.14
Buildup and decay of 13 harmonics of a 0.20-second trumpet tone with a fundamental frequency of 294 Hz. (From Risset & Mathews, 1969.)

longer tones and found that the harmonic structure is relatively constant during the sustained portion of the tone.

By using the data in Figure 9.14, Risset and Mathews synthesized a tone that sounded like a trumpet. When they interspersed five of their synthesized trumpet tones with five actual trumpet tones, nonmusicians could pick the actual tones with only chance accuracy, and even highly trained musicians performed at only slightly-better-than-chance accuracy.

Thus, timbre, the quality of a tone that enables us to distinguish between two tones having the same pitch and loudness, depends both on the tone's steady-state harmonic structure and on the time course of the attack and decay of the tone's harmonics.

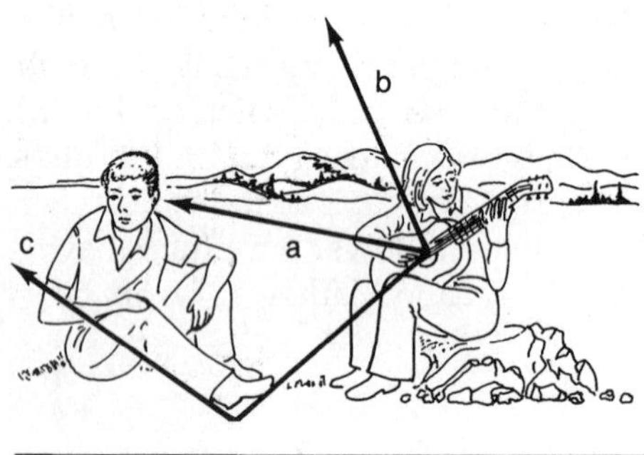

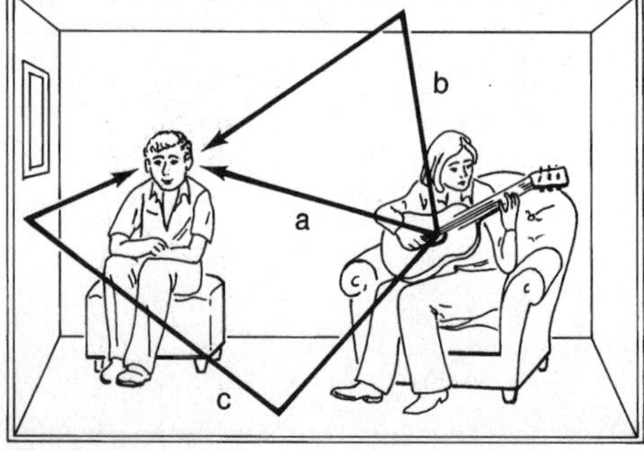

Figure 9.15
Top: When you hear a sound outside, you hear mainly direct sound (path a). Bottom: When you hear a sound inside a room, you hear both direct sound (a) and indirect sound (b and c) that is reflected from the walls, floor, and ceiling of the room.

Reverberation Time

When we studied vision, we saw that our perception of light depends not only on the nature of the light source but also on what happens to the light between the time it leaves its source and the time it enters our eyes. If light passes through haze on its way from an object to our eyes, the object may seem bluer or fuzzier than it would if the haze were not there. Similarly, our perception of sound also depends not only on the sound produced at the source, but also on what happens to the sound between the time it leaves the source and the time it enters our ears.

What happens to a sound between the time it is produced and the time it reaches your ears? We can see from Figure 9.15 that the answer to this question depends on the environment in which you hear the sound. If you are sitting outdoors next to someone playing a guitar, most of the sound you hear travels in a straight line and reaches your ears directly. If, however, you are listening to the same guitar in an enclosed room, then only some of the sound you hear travels directly to your ears; the rest of the sound you hear bounces off of the room's walls, ceiling, and floor before reaching your ears. The sound reaching your ears directly, along path a, is called **direct sound,** and the sound reaching your ears later, along paths like b and c, is called **indirect sound.**

The science of architectural acoustics is largely concerned with how this indirect sound changes the quality of the sounds we hear in rooms. The major factor affecting indirect sound is the amount of sound absorbed by the walls, ceilings, and floors of the room. If most of the sound is absorbed, then there are few sound reflections, and we hear little indirect sound. If little of the sound is absorbed, however, there are many sound reflections, and we hear much indirect sound. The amount of indirect sound produced by a room is expressed as the **reverberation time** of the room, where reverberation time is the time it takes for the sound to decrease to one-thousandth its original pressure.

What is the relationship between reverberation time and our perception of music? If the reverberation time is short, music will sound "dead," because most of the sound is absorbed by the room and it is difficult to produce sounds of very high intensity. If the reverberation time is long, music sounds "muddled," because the sound reflected by the room causes the sounds to overlap each other. Thus, the job of the acoustical engineer is to design a room in which the reverberation time is neither too short nor too long. The optimal reverberation time for a room depends on the size of the room, with most average-sized concert halls needing a reverberation time of about 1.5–2.0 seconds.

However, reverberation time is apparently not the only factor that affects our perception of music in concert halls. This is illustrated by the problems associated with the design of New York's Philharmonic Hall. When it opened in 1962, Philharmonic Hall had a reverberation time of close to 2.0 seconds, a value comparable to the reverberation times of many of the most successful concert halls in the world. Even so, the hall was criticized for sounding as though it had a short reverberation time, and musicians in the orchestra complained that they could not hear each other. These criticisms resulted in a series of alterations to the hall, made over many years, until eventually, when none of the alterations proved satisfactory, the entire interior of the hall was destroyed and the hall was completely rebuilt. The new hall was renamed Avery Fisher Hall and was reopened in 1976, 14 years after Philharmonic Hall's original opening, but this time with good acoustics. Based on experiences like this one, it is safe to say that determining the optimal acoustics of concert halls is not an exact science (Backus, 1977).

The Precedence Effect

Another aspect of the reverberation that occurs when you listen to music or to someone talking in an enclosed room is that the direct sound reaches your ears before the indirect sound. Since your ears receive a sequence of sounds coming from many directions, why do you perceive the sound as coming from only one location? The answer to this question is that, for reasons still not completely understood, perception of the sound's location depends on the sound that reaches the ears first. This effect, which is called the **precedence effect,** was extensively studied by Wallach, Newman, and Rosenzweig (1949).

Wallach et al. demonstrated the precedence effect by having a subject listen to music coming from two speakers located an equal distance away. When the music came from both speakers simultaneously, the subject heard the music coming from a point between the two speakers. However, if Wallach caused the sound from one speaker to precede the sound from the other by a fraction of a second, the subject reported that the sound appeared to come only from the speaker that had first produced the sound. On the basis of this result and those of other experiments, Wallach concluded that the first sound reaching our ears is heard, and that sounds arriving within about 70 msec of this first sound are suppressed. Since Wallach made this proposal, other researchers have suggested that a mechanism other than suppression is responsible for the auditory system's favoring initial sounds over later ones. One reason for this suggestion is that the later sounds do have some effect on perception, as we will see below (Perrott, 1989).

D E M O N S T R A T I O N

Experiencing the Precedence Effect

To demonstrate the precedence effect to yourself, turn your stereo system to monaural (or "mixed"), so that both speakers play the same sounds, and position yourself between the speakers, so that you hear the sound coming from both speakers or from a point between them. Then move a small distance to the left or right. When you do this, the sound suddenly appears to be coming from only the nearer speaker. ●

You might think that the precedence effect occurs because moving toward one speaker makes the sound from that speaker louder. However, Wallach showed that the small increase in loudness that would occur from moving slightly closer to one speaker cannot cause you to hear only that speaker; you hear only the nearer speaker because the sound from that speaker reaches your ears first.

When you hear the sound coming from the near speaker, do you no longer hear the far speaker? You can answer this question by positioning yourself closer to one speaker and having a friend disconnect the other speaker. When this happens, you will notice a difference in the quality of the sound. Even when you think you are listening to only the near speaker, because that is where the sound appears to originate, sounds from the far speaker are also affecting your perception, giving the sound a fuller, more expansive quality. A sound's location is usually determined by the sound that reaches the ears first, but its quality can be influenced by sounds reaching the ears later (Green, 1976).

Psychophysical Aspects of Auditory Processing

So far in this chapter, we have used psychophysics to describe a number of relationships between aspects of the sound stimulus and our perception of sound. We've shown how threshold changes with frequency, how loudness depends on intensity, and how pitch and tone chroma depend on frequency. In addition, we have seen that a tone's timbre depends on the nature of its harmonic components and that the "liveness" or "deadness" of a sound is influenced by how sound reverberates inside a room.

We now consider some psychophysical experiments that focus on what psychophysics can tell us about the basic mechanisms of hearing. We will do this by describing three different experiments, each of which demonstrates a parallel between auditory perception and a physiological mechanism we described in Chapter 8.

Masking

We know from the work of Békésy and others that the basilar membrane vibrates with a traveling wave motion, and that the vibration of the membrane is asymmetrical. That is, low frequencies cause most of the membrane to vibrate, but there is more vibration at the apex, and high frequencies cause vibration only toward the base. Békésy reached these conclusions by constructing a model of the cochlea and by observing the actual vibration of the basilar membrane. Psychologists have reached a similar conclusion by measuring an effect called **masking,** which occurs when the presence of one sound reduces our ability to hear another sound.

J. P. Egan and H. W. Hake (1950) conducted a masking experiment in which they first measured thresholds at frequencies between 100 and 4,000 Hz and then measured these thresholds in the presence of a masking noise. Their procedure for measuring thresholds during masking is diagrammed in Figure 9.16. The masking stimulus is called a **noise** because it consists of a large number of frequencies presented simultaneously. Such a combination of frequencies sounds something like the "shhhhhh" sound you can make by separating your teeth slightly and blowing air across your lower teeth. Egan and Hake used a noise stimulus containing frequencies ranging

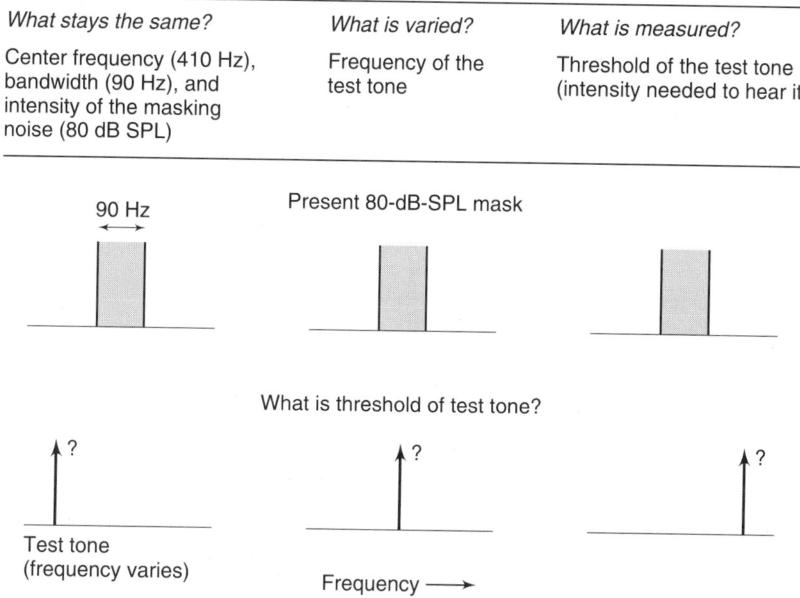

Figure 9.16
The procedure for Egan and Hake's (1950) masking experiment. In this experiment, the threshold of a test tone presented at frequencies across the range of hearing is measured in the presence of a constant masking noise.

from 365 to 455 Hz. The frequency in the center of this frequency range of the noise stimulus is called the **center frequency,** so in this case, the center frequency is 410 Hz. The size of the frequency range, which is 455 Hz – 365 Hz = 90 Hz is called the **bandwidth** of the noise. Egan and Hake held the center frequency, the bandwidth, and the intensity of the masking noise constant, and while presenting the masking noise, they determined the thresholds of single-frequency test tones at a number of frequencies between 100 and 4,000 Hz.

The results of this experiment are shown in Figure 9.17, which indicates how the masking noise affected the threshold at each frequency. An important property of this curve is that it is asymmetrical. That is, the effect of the masking tone spreads more to higher frequencies than to lower frequencies.

The reason that the masking curve is asymmetrical is related to how the basilar membrane vibrates in response to both the masking noise and the test tones. Figure 9.18 shows the vibration patterns caused by 200-Hz, 400-Hz, and 800-Hz tones, taken from Figure 8.25. We can see how a 400-Hz masking tone would affect the 200- and 800-Hz tones by noting how their vibration patterns overlap. Notice that the pattern for the 400-Hz tone, which is shaded, almost totally overlaps the pattern for the higher-frequency 800-Hz tone, but that it does not overlap the place of peak vibration of the lower-frequency 200-Hz tone. We would therefore expect the masking tone to have a large effect on the 800-Hz tone but a smaller effect on the 200-Hz tone, exactly what happened in Egan and Hake's masking experiment. Thus, the asymmetrical masking function in Figure 9.18 reflects the asymmetrical vibration pattern of the basilar membrane.

Masking has also been used in another way to reveal a connection between psychophysics and physiology. One of the major principles from physiological research is the idea that the components of complex sounds are signaled by activity in narrowly tuned nerve fibers. We will now see how we can use masking to determine a function

Masking: Results and Interpretation

Result	*Basilar membrane effect*	*Conclusion*
Test tones with frequencies above the masking frequency are more likely to be affected by the masking noise.	The high-intensity masking noise causes greater vibration toward the high-frequency (base) end of the basilar membrane.	Vibration of the basilar membrane is not symmetrical.

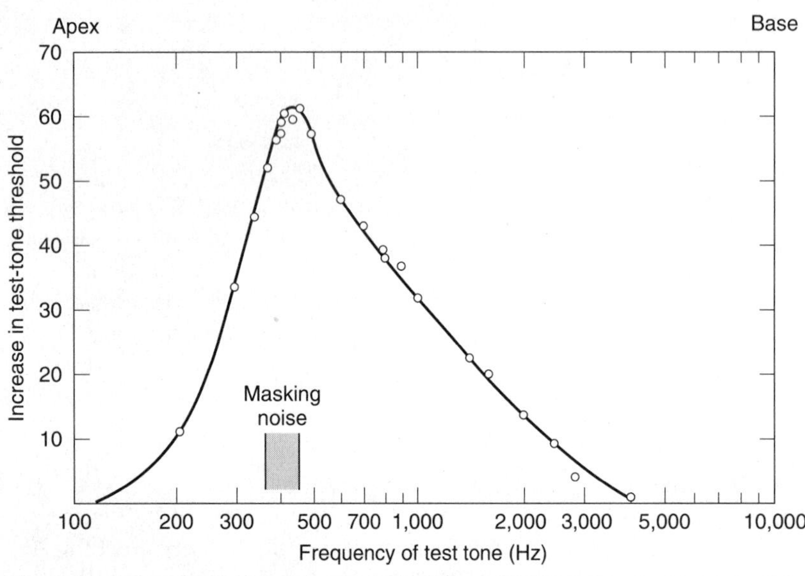

Figure 9.17
Results of Egan and Hake's (1950) masking experiment. The asymmetrical increase in the test-tone threshold across frequencies reflects the asymmetrical vibration of the basilar membrane.

called the *psychophysical tuning curve,* which closely resembles the tuning curves of auditory nerve fibers.

Psychophysical Tuning Curves

The procedure for determining a **psychophysical tuning curve** is shown in Figure 9.19. We present a low-intensity (10-dB-SPL) tone with a frequency that remains constant throughout the experiment. We then present masking tones at frequencies across the range of hearing and determine how intense each of these maskers must be to make the low-intensity tone just barely detectable. The results of this experiment, shown in Figure 9.20a, indicate that masking is achieved

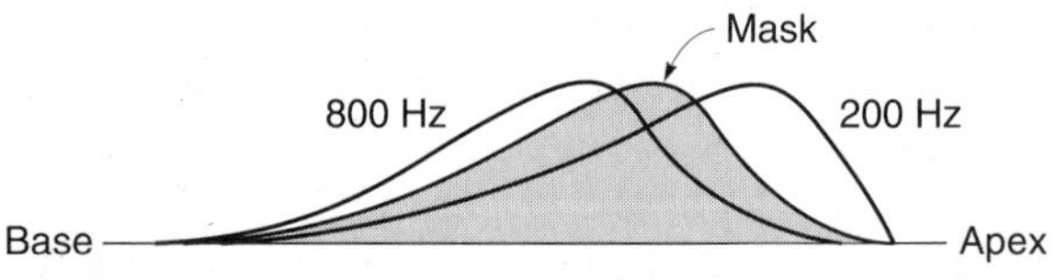

Figure 9.18
Vibration patterns caused by 200- and 800-Hz test tones, and the 410-Hz mask, taken from basilar membrane vibration patterns in Figure 8.25. The pattern for a 400-Hz tone is used for the masking pattern (shaded). Since the mask actually contains a band of frequencies, the actual pattern would be wider than is shown here. It would, however, still be asymmetrical and would overlap the 800-Hz vibration more than the 200-Hz vibration.

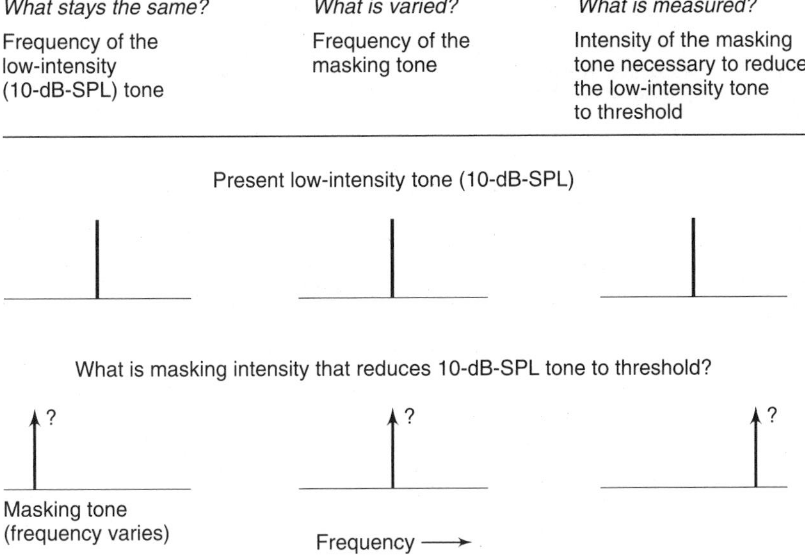

Figure 9.19
Procedure for measuring a psychophysical tuning curve. The intensity of a variable-frequency masking tone needed to reduce a constant-frequency, low-intensity tone to threshold is determined.

with a low-intensity tone for masking frequencies at or near the frequency of the low-intensity tone, but that higher intensities are required above and below that frequency. The psychophysical tuning curve closely resembles the tuning curve for cat auditory nerve fibers shown in Figure 9.20b.

If we repeat our measurement of psychophysical tuning curves using a number of different test frequencies, we obtain a family of tuning functions like those shown with the audibility curve in Figure 9.21. These tuning curves show psychophysically what had previously been demonstrated physiologically: Our perception of sound depends on a number of narrowly tuned fibers that operate across the range of frequencies.

The Critical Band for Loudness

Another connection between psychophysics and physiology is illustrated by measurement of the **critical band for loudness**—the width of the frequency band for which loudness remains constant for a given intensity (dB SPL), no matter which frequencies or combinations of frequencies are presented within the band. We can understand what this means by describing an experiment done by E. Zwicker, G. Flottorp, and S. S. Stevens (1957). As shown in Figure 9.22, they presented a band of noise with a center frequency of 1,000 Hz and an intensity of 60 dB SPL. They measured the loudness of the noise band by having subjects adjust the intensity of a 1,000-Hz test tone to match the noise band's loudness. They then repeated this procedure for wider noise bands, while keeping the intensity constant at 60 dB SPL.

The result of this experiment, shown in Figure 9.23, was that the loudness of the band of noise remained the same as long at the bandwidth of the noise was below 160 Hz. However, when the bandwidth was increased above 160 Hz, the loudness increased. The fact that the loudness remained constant within the 160-Hz band means that, for a center frequency of 1,000 Hz, the size of

Psychophysical tuning curve: Results and interpretation

Result (see (a), below)	*Basilar membrane effect*	*Conclusion*
The masking tone needs to be least intense when it has the same frequency as the low-intensity tone. The intensity needed to mask the tone increases at frequencies above and below the frequency of the low-intensity tone.	The low-intensity tone affects a small area on the basilar membrane.	There are narrow processing units along the basilar membrane.

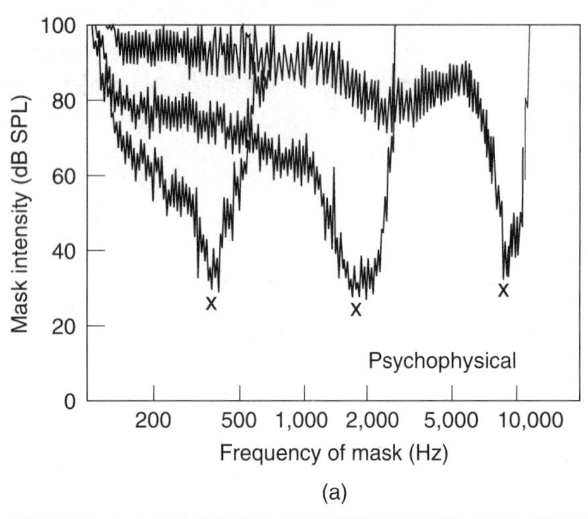

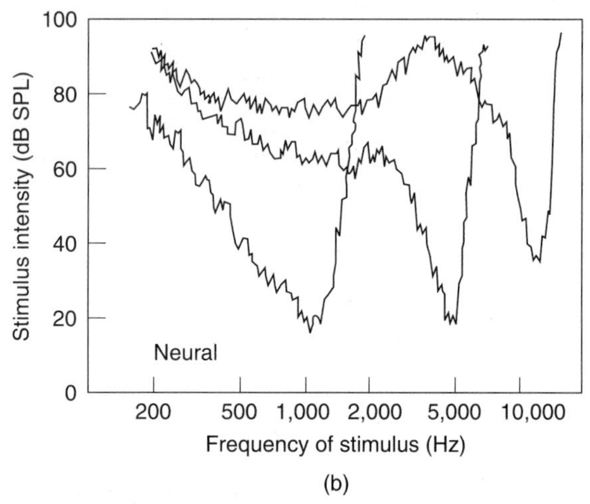

Figure 9.20

(a) Human psychophysical tuning curves determined as described in Figure 9.19. The X's show the frequency of the test-tone. (b) Neural tuning curves for the cat, showing the stimulus intensity needed to generate a constant response. Each curve represents a different auditory nerve fiber. (Adapted from Zwicker, 1974.)

the critical band is 160 Hz. The finding that loudness remains constant within a restricted band of frequencies has led to the conclusion that there are small processing regions on the basilar membrane where neighboring frequencies can add their effects to create loudness. As long as the overall intensity of the frequencies stimulating this region remains constant, the loudness remains the same no matter what frequencies are stimulating this region.

Another finding that supports the idea of small processing regions on the basilar membrane is that the critical band becomes larger at higher frequencies, and that the frequency range of all critical bands takes up the same distance along the basilar membrane. For example, the 160-Hz critical band at a center frequency of 1,000 Hz represents about 1.3 mm along the basilar membrane. If we increase the center frequency to 4,000 Hz, the critical band becomes 600 Hz, but this frequency range also takes up 1.3 mm on the basilar membrane. Based on this result, Roederer (1975) described this 1.3-mm-long stretch of basilar membrane as an "information collection and integration unit."

The idea that sound frequencies are analyzed within small units along the basilar membrane is consistent with the narrow psychophysical and neural tuning curves in Figure 9.20 and with the idea we introduced in Chapter 8 that the auditory

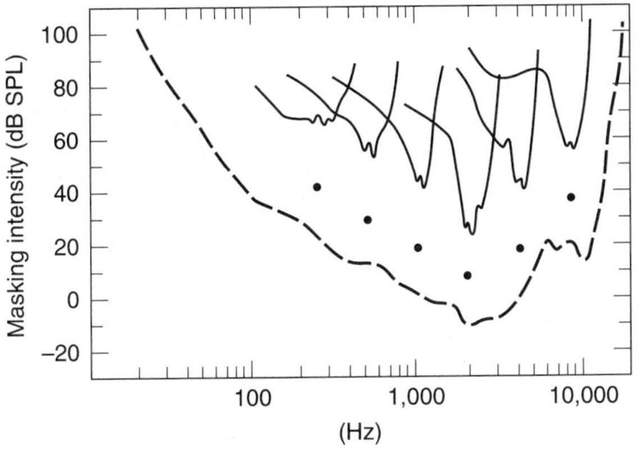

Figure 9.21
Psychophysical tuning curves for a number of test-tone frequencies (dots). Notice how the minimum masking intensities for the curves match the shape of the audibility curve (dashed line). (Based on Vogten, 1974.)

system breaks a sound stimulus down into frequency components so that, when a complex sound consisting of a number of frequency components is presented, each of these components is analyzed within its own small area of the basilar membrane.

When we describe hearing as being the result of processing by a number of narrowly tuned regions on the basilar membrane, we are dealing with very basic mechanisms with origins at the beginning of the auditory system. To fully understand the process of hearing, we also need to consider events that occur more centrally in the auditory system. One approach is to consider how we perceive auditory "scenes."

Auditory Scene Analysis

Sounds in the environment create auditory scenes—images of things happening at various locations in space. When you close your eyes and become aware of the various events that are sig-

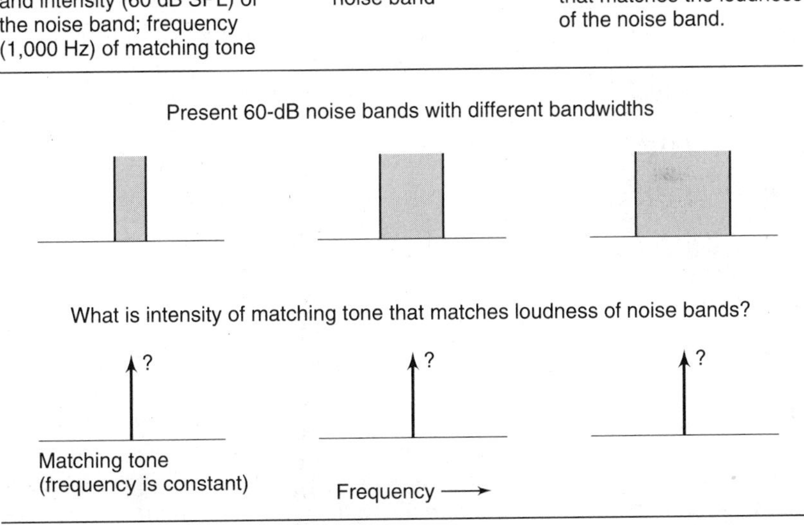

Figure 9.22
Procedure used by Zwicker, Flottorp, and Stevens (1957) to measure the critical band for loudness. The matching tone is adjusted to match the loudness of noise bands with different bandwidths.

Critical band: Results and interpretation

Result	*Basilar membrane effect*	*Conclusion*
The matching tone's intensity remains the same until the bandwidth of the noise exceeds a critical value.	There is a small region on the basilar membrane where neighboring frequencies summate their effects to create loudness.	There are narrow processing units along the basilar membrane.

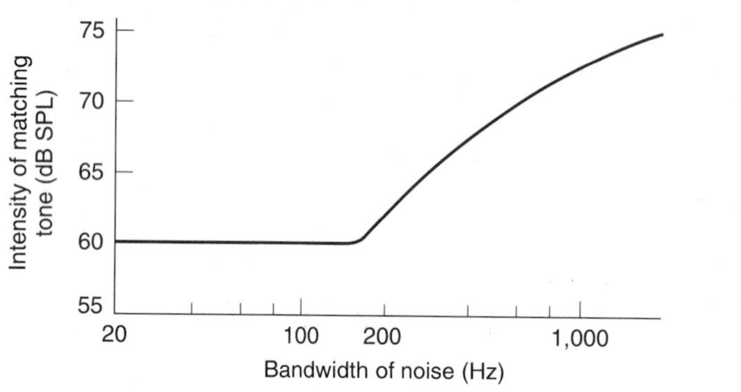

Figure 9.23

Results of Zwicker, Flottorp, and Stevens' (1957) measurement of the critical bandwidth for a noise band with a center frequency of 1,000 Hz. The intensity of the matching tone remains constant until the bandwidth of the noise reaches 160 Hz. This means that the width of the critical band for a 1,000-Hz tone is 160 Hz.

naled by the sounds around you, you are experiencing an auditory scene. This experience is, like most other perceptual experiences, achieved with little conscious effort on your part. However, the effortlessness with which you experience auditory scenes conceals the fact that the auditory system's creation of such scenes is an extremely complex process.

We can understand this complexity by considering what happens when the sounds of a person talking, a symphony recording, a person whistling, and a distant lawnmower simultaneously enter your ears. All of them together set your tympanic membranes into vibration, and these vibrations are eventually transmitted to the basilar membrane and the hair cells. Based only on the vibrations of the tympanic membrane, the basilar membrane, or the hair cells, it is difficult or impossible to tell which parts of this vibration are caused by the person talking, the symphony, the person whistling, or the lawnmower.

The process by which we somehow sort these superimposed vibrations into separate sounds is called **auditory scene analysis** (Bregman, 1990). This process is analogous to the process of perceptual organization in vision that occurs when we perceive different objects as being separate from each other. Bregman describes the analogous process in hearing as one of separating sound stimuli into **auditory streams,** where a stream is a perceptual unit that represents a single happening. Thus, a stream in audition is analogous to an object in vision.

Although we are far from being able to understand totally how we separate our environment into auditory streams, researchers have made some progress by treating stream analysis as a problem in perceptual organization.

Auditory Perceptual Organization

Just as visual stimuli are perceptually organized so that certain elements of a scene appear to "belong together," so are tones. Some of the rules for grouping tones are similar to the Gestalt laws we described in Chapter 5 for the grouping of visual stimuli. Let's now consider how we can apply these principles to the grouping of tones.

Figure 9.24
Four measures of a composition by J. S. Bach (Choral Prelude on Jesus Christus unser Heiland, *1739). When played rapidly, the upper notes become perceptually grouped and the lower notes become perceptually grouped, a phenomenon called* auditory stream segregation.

Similarity "Tones that are similar to one another tend to be perceived as belonging together." This is a statement of the law of similarity for hearing. One dimension of tones that contributes to perceptual grouping by similarity is pitch: Tones more similar in pitch are more likely to be perceived as belonging together. The importance of this principle is reflected in the fact that music of all cultures uses small pitch separations between the notes of melodies (Dowling & Harwood, 1986).

Grouping by similarity was used by composers long before psychologists began studying it. Composers in the Baroque period (1600–1750) knew that, if a single instrument plays notes that alternate rapidly between high and low tones, the listener perceives two separate melodies, with the high tones perceived as being played by one instrument and the low tones as being played by another. An excerpt from a composition by J. S. Bach that uses this device is shown in Figure 9.24. When this passage is played rapidly, the low notes sound as if they are a melody played by one instrument, and the high notes sound like a different melody played by another instrument. This effect, which has been called *implied polyphony* or *compound melodic line* by musicians, is the result of the perceptual phenomenon that psychologists call **auditory stream segregation.**

Bregman and Jeffrey Campbell (1971) demonstrated auditory stream segregation experimentally by alternating high and low tones, as shown in the sequence in Figure 9.25. When the high-pitched tones are slowly alternated with the low-pitched tones, as in Figure 9.25a, the tones

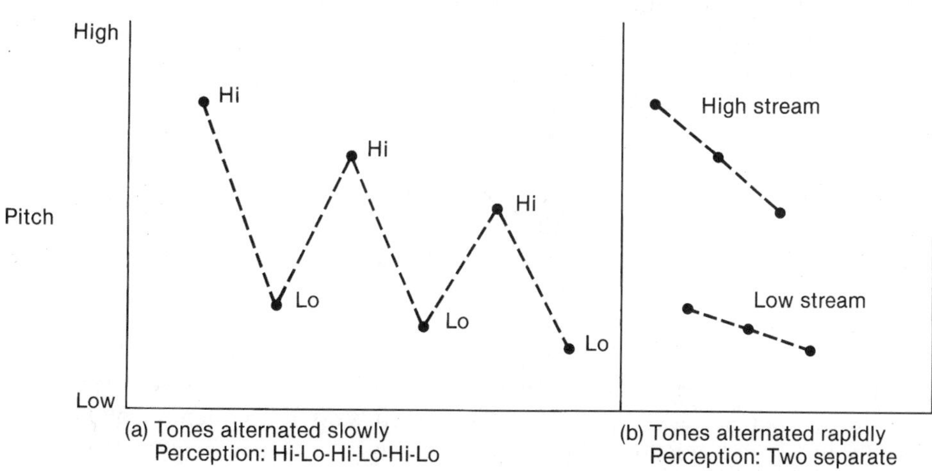

Figure 9.25
(a) When high and low tones are alternated slowly, auditory stream segregation does not occur, so the listener perceives alternating high and low tones. (b) Faster alternation results in segregation into high and low streams.

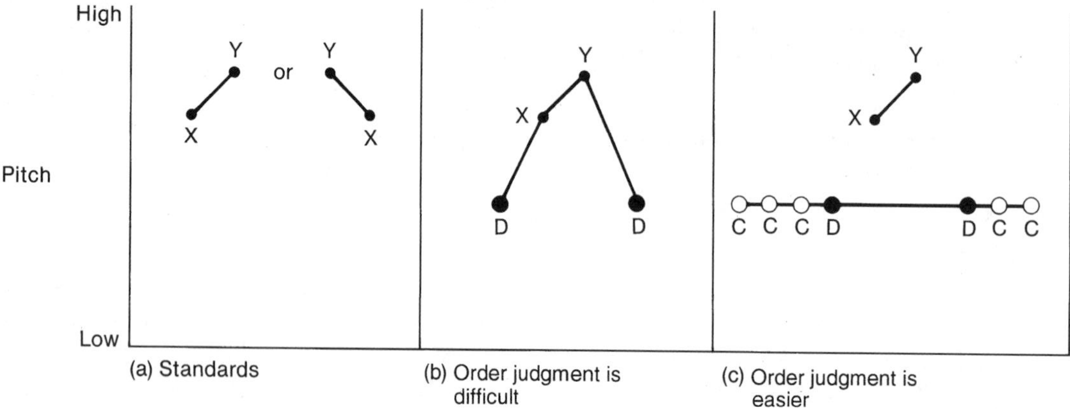

Figure 9.26

Bregman and Rudnicky's (1975) experiment. (a) The standard tones X and Y have different pitches. (b) Test 1: The distractor (D) tones group with X and Y, making it difficult to judge the order of X and Y. (c) Test 2: The addition of captor (C) tones with the same pitch as the distractor tones causes the distractor tones to form a separate stream (law of similarity) and makes it easier to judge the order of tones X and Y. (Based on Bregman and Rudnicky, 1975.)

are heard in one stream, one after another: hi-lo-hi-lo-hi-lo, as indicated by the dashed line. If, however, the tones are alternated very rapidly, as in Figure 9.25b, the high and low tones become perceptually grouped into two auditory streams so that the listener perceives two separate streams of sound, one high-pitched and one low-pitched, occurring simultaneously.

This grouping of tones into streams by similarity of pitch is also demonstrated by an experiment done by Bregman and Alexander Rudnicky (1975), diagrammed in Figure 9.26. The listener is first presented with two standard tones, X and Y (9.26a). When these tones are presented alone, it is easy to perceive their order (XY or YX). However, when these tones are sandwiched between two distractor (D) tones (9. 26b), it becomes very hard to judge their order. The name *distractor tones* is well taken: They distract the listener, making it difficult to judge the order of tones X and Y.

But the distracting effect of the D tones can be eliminated by using the law of similarity. Bregman and Rudnicky accomplished this by adding a series of "captor" tones (C) with the same pitch as the distractors (Figure 9.26c). Since these captor tones have the same pitch as the distractors, they "capture" the distractors and form a stream that separates the distractors from tones A and B. The result is that X and Y are perceived as belonging to a separate stream, and it is much easier to perceive X and Y's order.

A final example of how similarity of pitch affects perception is an effect called the **scale illusion** or **melodic channeling.** Diana Deutsch (1975) demonstrated this effect by presenting two scales simultaneously, one ascending and one descending (Figure 9.27a). The subjects listened to these scales through earphones that presented successive notes from each scale alternately to the left and right ears, as shown in Figure 9.27b. From Figure 9.27b, you can see that, if we focus just on the right ear, the notes alternate from high to low to high. Similarly, if we focus on the left ear, the notes alternate from low to high to low. But this was not what the subjects perceived. They perceived smooth sequences of notes in each ear; the higher notes were perceived in the right ear, and lower ones were perceived in the

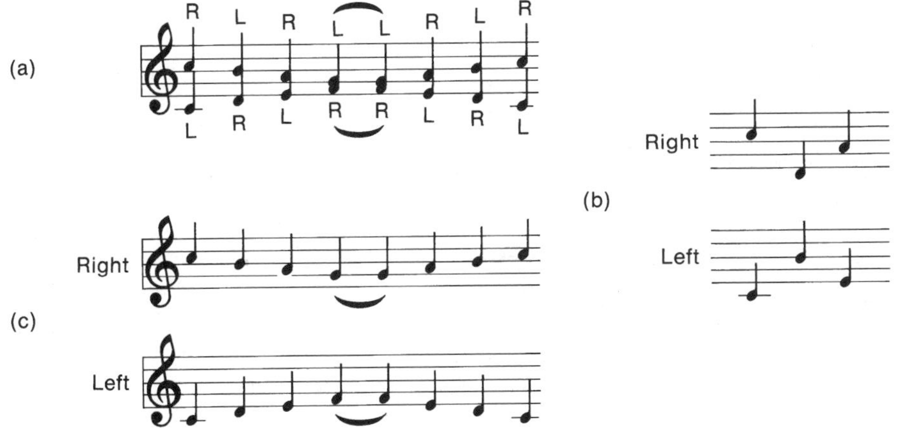

Figure 9.27

(a) These stimuli were presented to the subject's left and right ears in Deutsch's (1975) "scale illusion" experiment. (b) The first three notes presented to the left and right ears in Deutsch's experiment. The notes presented to each ear do not form a scale; they jump up and down. (c) What the subject hears. Although the notes in each ear jump up and down, the subject perceives a smooth sequence of notes in each ear. This effect is called the scale illusion *or* melodic channeling. *(Based on Deutsch, 1975.)*

left ear (Figure 9.26c). The scale illusion also illustrates grouping by similarity: High-pitched tones become perceptually grouped in one ear, and low-pitched tones become perceptually grouped in the other ear.

This illusion also illustrates an important function of perceptual grouping: Perceptual grouping helps us to interpret the environment effectively. It is most effective to perceive similar sounds as coming from the same source, because this is what usually happens in the environment. When psychologists create abnormal stimuli, as in Deutsch's experiment, the perceptual system applies the rule of grouping by similarity and is fooled into assigning similar pitches to the same ear. But in the normal environment (when psychologists aren't controlling the stimuli), this rule of similarity helps us to correctly perceive where sounds are coming from.

Proximity Auditory stream segregation provides an example of how proximity in time affects our perception of tones. Stream segregation requires not only that one series of tones be high-pitched and the other low-pitched but also that these tones occur close together in time. This is the law of proximity: Tones must follow each other rapidly to be perceived together. If the tones are too far apart in time, as in Figure 9.25a, segregation will not occur, even if the tones are similar in pitch.

Good Continuation Auditory good continuation was demonstrated by Richard Warren, C. J. Obuseck, and J. M. Acroff (1972). When they presented bursts of tone interrupted by gaps of silence, as in Figure 9.28a, listeners perceived both the tones and the silence between the tones. If, however, the silent gaps were filled in with noise (Figure 9.28b), the listeners perceived the tone as continuous (Figure 9.28c). This demonstration is analogous to the demonstration of visual good continuation in Pissarro's painting in Figure 5.20. Just as the smokestack in the painting is perceived as continuous even though it is twice interrupted by smoke, an interrupted tone can be

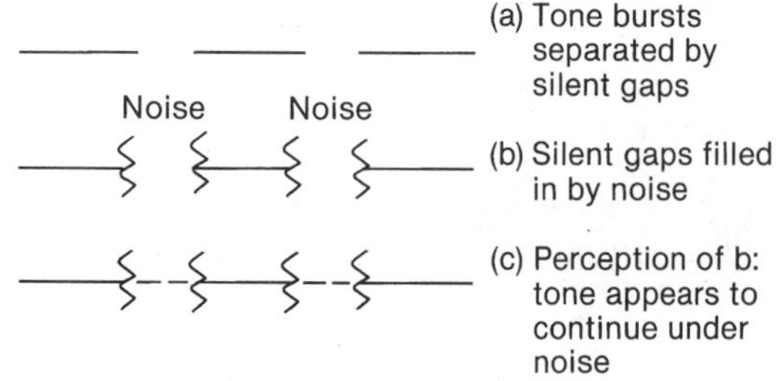

Figure 9.28

A demonstration of good continuation, using tones.

perceived as continuous even though it is interrupted by bursts of noise.

Melody Schema Just as vision is influenced by our past experiences and by our expectations caused by these experiences, so is hearing. The effect of expectation on our perception of melodies is demonstrated by an experiment done by W. Jay Dowling (1973). Dowling had his subjects listen to two **interleaved melodies,** as shown in Figure 9.29a; notes of "Three Blind Mice" are alternated with notes of "Mary Had a Little Lamb." When the subjects listened to these combined melodies, they reported hearing a meaningless jumble of notes. However, when they were told the names of the songs, they were able to hear either melody, depending on which one they paid attention to.

What the listeners were doing, according to Dowling and Dane Harwood (1986), was applying a **melody schema** to the interleaved melodies. A melody schema is a representation of a familiar melody that is stored in a person's memory. When people don't know that a melody is present, they have no access to the schema and therefore have nothing to compare the unknown melody to. But if they are told which melody is present, they compare what they hear to their "Three Blind Mice" or "Mary Had a Little Lamb" schemas and perceive the melodies.

I demonstrate to my classes how schemas affect hearing words by playing a Rolling Stones recording and asking the class to try to identify the lyrics. This task is extremely difficult, because of both the loud instrumental backing and Mick Jagger's less-than-precise enunciation. But when I play the song a second time, accompanied by the words projected onto a screen, the previously difficult-to-understand words become easy to perceive. In the chapter on speech, we will describe additional examples of how our expectations influence our perception of words.

Figure 9.29

(a) "Three Blind Mice," (b) "Mary Had a Little Lamb," and (c) the two melodies interleaved ("Three Blind Mice": stems up; "Mary Had a Little Lamb": stems down).

Auditory Localization

Standing on a hill overlooking a neighborhood in Pittsburgh, at the beginning of our discussion of depth perception in Chapter 6, we were able to see from the cars and houses directly in front of us to the houses and trees lining a ridge in the distance. Our ability to see objects in the visual space around us is something that we are usually much more aware of than the other kind of space that surrounds us: the auditory space created by trees overhead rustling in the wind, the sound of cars approaching from the right or left, and the sound of a dog barking a block away. We can think of these sounds as making up an auditory

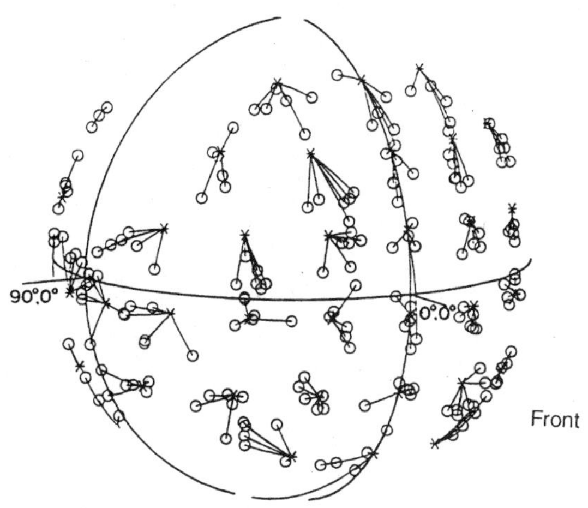

Figure 9.30
A subject's ability to localize sounds. The asterisks are the actual sound locations, and the circles are the subject's estimates of their location. Longer lines connecting the asterisks and circles indicate less accurate localization. (From Makous & Middlebrooks, 1990.)

space because, when we hear them, we can usually tell where they are coming from.

Figure 9.30 illustrates an observer's ability to localize sounds that are in different positions in space (Makous & Middlebrooks, 1990; Middlebrooks & Green, 1991). This figure shows an imaginary sphere that surrounds the observer's head. The asterisks indicate sound locations and the circles, the subject's responses. Localization errors are smallest (about 2–3.5 degrees) for sounds directly in front of the observer, and largest (as high as 20 degrees) for locations behind the observer's head.

Observers can use a number of different sources of information to help localize the source of a sound. In Chapter 8, we introduced the binaural cues of interaural time difference and interaural intensity difference as two sources of information that depend on comparing the patterns of sound reaching the left and right ears.

Interaural Time Difference

We can measure how the interaural time difference changes with a sound's location by placing microphones in a person's ears and measuring the arrival time of sounds originating from different positions in space. The results of these measurements indicate that the interaural time difference is zero when the source is directly in front of or directly behind the listener; however, as the source is moved to the side, there is a delay between the time the sound reaches the near ear and the time it reaches the far ear, this difference reaching a maximum of about 600 microseconds (6/10,000 sec) when the sound is located directly opposite one of the ears, as in B of Figure 8.34 (Feddersen et al., 1957). Although time differences on the order of microseconds are very small, it has been shown that we can detect differences in arrival time as short as 10 microseconds (Durlach & Colburn, 1978).

While the interaural time difference provides some information about the location of a tone, this information is not unambiguous, because a number of points in space can result in the same interaural time difference. Any point lying on a surface called the **cone of confusion,** shown in Figure 9.31, results in the same interaural time

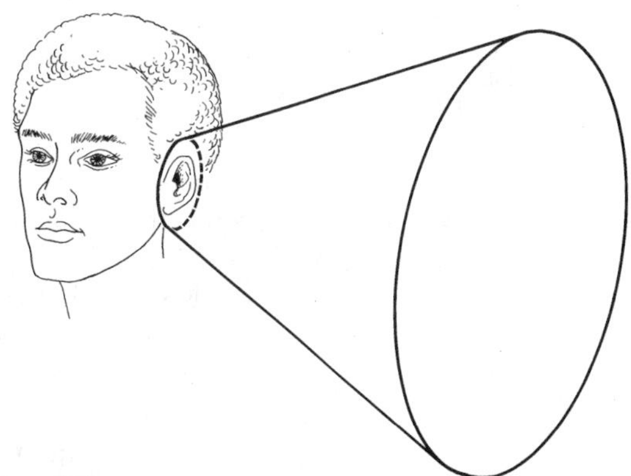

Figure 9.31
The cone of confusion. All points on the surface of this cone result in the same interaural time difference. Thus, without moving one's head, it is difficult to tell the difference in position of two tones located on this cone.

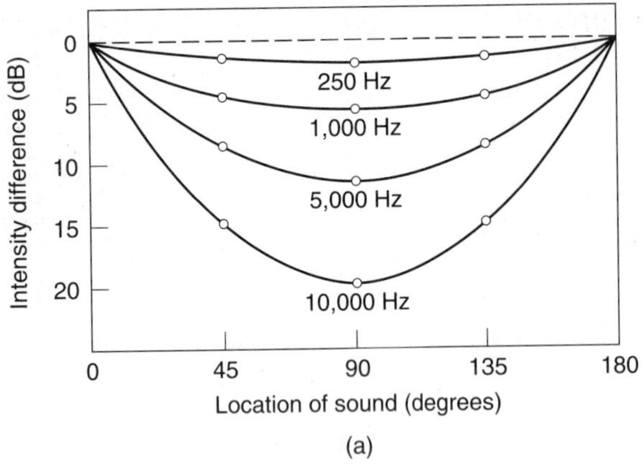

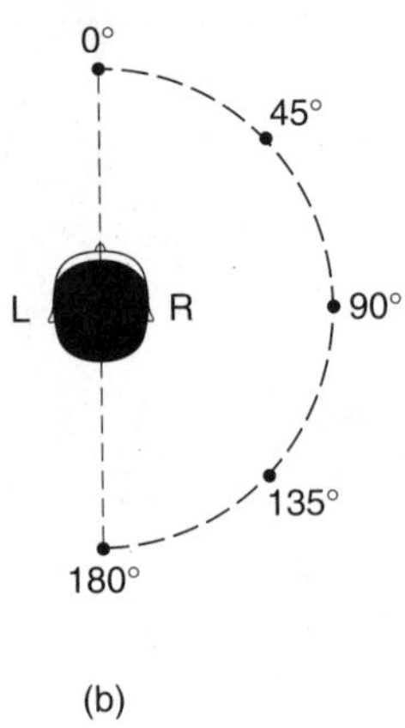

Figure 9.32
(a) The difference in intensity between the two ears for a 70-dB-SPL tone of different frequencies located at different places around the head. (From Gulick, Gescheider, & Frisina, 1989.) (b) The dots indicate the locations of the tones used for the intensity measurement in the graph in (a). The tones were located 2 m from the center of the head.

difference, so it is difficult to tell exactly where on the cone a sound is coming from. The solution to this problem is simple: When the observer moves his or her head, tones that were difficult to localize with the head stationary can be localized much more easily. These head movements play an important role in our ability to localize tones in our everyday experience.

Interaural Intensity Difference

In Chapter 8, we saw that the cue of interaural intensity difference occurs because the head casts a "shadow," which decreases the intensity of the sound reaching the far ear. We also saw that this shadow occurs for high-frequency tones but not for low-frequency tones. Lawrence Gulick, George Gescheider, and Robert Frisina (1989) describe the differences between high- and low-frequency tones by drawing on an analogy between sound waves and ocean waves. Waves such as ocean swells (long wavelengths or low frequencies) ignore the pilings of a pier and proceed uninterrupted toward the shore. However, smaller ripples (short wavelengths or high frequencies) are reflected from these pilings. Similarly, low-frequency sound waves ignore the head and proceed uninterrupted to the ear on the far side of the head, whereas high-frequency sound waves are reflected from the head and cast a shadow that reduces the sound intensity reaching the far ear. This effect of frequency on the interaural intensity difference has been measured by using small microphones to record the intensity of the sound reaching each ear in response to a moveable sound source. The results show that there is little difference in intensity for frequencies below approximately 1,000 Hz, but that quite sizable differences in intensity occur for higher frequencies (Figure 9.32).

Sound Reflections by the Pinna

The **pinnae** (Figure 9.33) also provide information about localization. The pinnae's effect on sound localization was demonstrated in an ex-

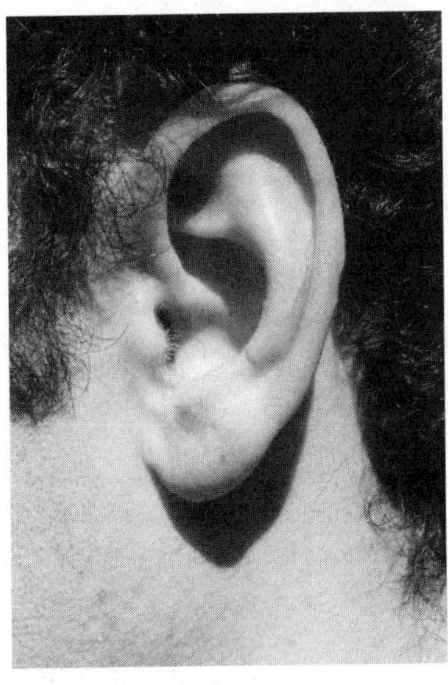

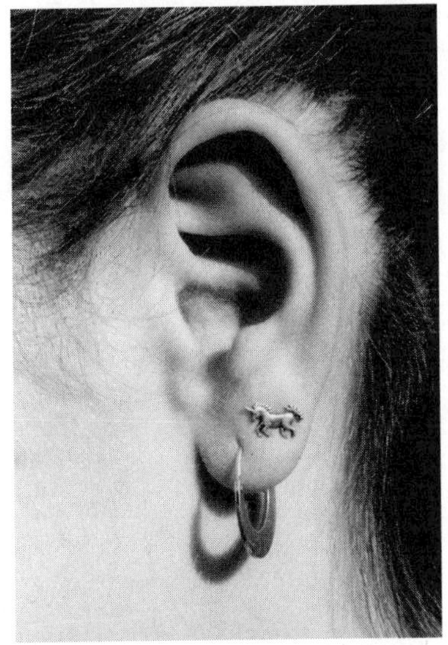

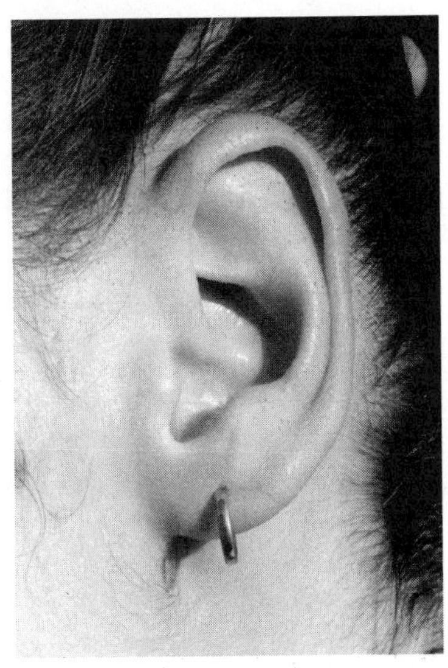

Figure 9.33
Different people have pinnae with very different shapes. These shapes affect how the sound bounces around in the pinna. This "bouncing" provides information for sound localization by changing the frequency composition of the sounds.

periment by Gardner and Gardner (1973), who inserted plugs of various shapes into people's ears. As Gardner and Gardner increasingly smoothed out the pinna by inserting different plugs, the listeners had more and more difficulty in localizing sounds. You can demonstrate how the pinnae affect localization by placing earphones over your ears that still enable you to hear sounds in the environment. When you do this, you will notice that sounds become more difficult to localize than when you are wearing no earphones. This effect becomes particularly evident if you try to localize sounds with your eyes closed.

What is the nature of the localization information provided by the pinnae? The answer appears to be that various frequency components of the complex sounds in our environment are reflected back and forth inside the pinnae (Batteau, 1967; Butler & Belendiuk, 1977; Oldfield & Parker, 1984; Scharf, 1975). Thus, when a sound moves to a different position relative to a listener's head, these reflections could decrease the intensity of some frequency components and amplify others. These different patterns of frequencies then become cues to the location of the sound and provide information that supplements the information gained from head movements and the binaural cues to localization.

Sound as Information for the Visually Impaired

For people who have poor vision or are totally blind, sound provides an important source of information about the environment. Sound provides information about the locations of objects in two ways: (1) The person can create sounds and note the echo created by objects that reflect the sound, or (2) when objects create sounds, the person can use sound cues like those described above to locate the object.

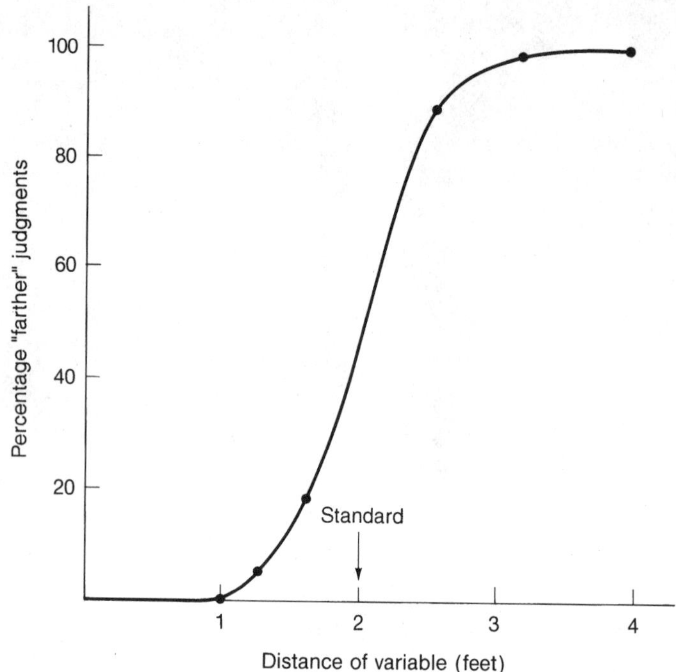

Figure 9.34
The results of an experiment in which a blind observer was presented with two 1-foot-diameter plywood disks, a standard disk that was always 2 feet away and a variable disk that was either closer or farther than the standard. The observer judged whether each variable was closer or farther than the standard, making 100 judgments for each disk. That this blind observer was able to judge distance based on echo ranging is indicated by the fact that changing the distance of the variable by less than a foot (from 1.59 to 2.55 feet) resulted in a change from 19 to 89 percent "farther" responses. (Data from Kellogg, 1962.)

Using Echoes to Locate Objects One way that sound provides information about the environment is through echoes, which can indicate the distance of objects. This effect has been demonstrated in experiments showing that sightless people can judge the distance of objects by paying attention to how the sounds of their footsteps or vocalizations change as they approach these objects. For example, Ammons, Worchel, and Dallenbach (1953) had blindfolded subjects walk toward an obstacle positioned at a distance of 6, 12, 18, 24, or 30 feet, until they perceived it (and stopped) or collided with it. With some practice, all these subjects were able to stop before colliding with the obstacle; however, if they wore earplugs in addition to being blindfolded, they collided with the object more frequently.

Perhaps the most impressive demonstration of how sound can be used to detect objects is an experiment inspired by the way bats and dolphins use sonar, or "echo ranging," to determine the distances and sizes of objects, as we described in the Other Worlds of Perception, "Echolocation by Bats," in Chapter 8. It was reasoned that, if a bat can judge an object's distance by emitting a sound and sensing the echo reflected from the object, perhaps people can do the same thing. Two blind college students were asked to judge which of two objects was farther away. They were told that, to aid them in making this judgment, they could produce any sound they wished. The subjects produced sounds in a variety of ways. They snapped their fingers, hissed, whistled, and, most often, repeated words, such as "now, now, now . . . " By judging the echo produced by these sounds, they were able to tell which of the two objects was closer. In fact, their performance was so impressive that it was possible to vary the separation between the two objects and, using the method of constant stimuli (see Chapter 1), to determine the psychophysical function shown in Figure 9.34 (Kellogg, 1962; also see Rice, 1967).

Using echoes to estimate the distance of objects is an impressive feat, but it is not very accurate and is extremely difficult to use in noisy environments. However, devices such as glasses that locate objects by bouncing sonar signals off of them are now being developed (Goleman, 1994).

Personal Guidance System Based on Binaural Cues The idea of using glasses that send out sonar signals borrows the sonar-ranging technique of bats and dolphins. However, another technique is being developed that locates objects based on the binaural cues we discussed above. The Personal Guidance System being developed by researchers at the University of California at Santa Barbara and Carnegie-Mellon University works as shown in Figure 9.35 (Loomis, Hebert, & Cicinelli, 1990: Loomis et al., 1994).

The person is fitted with an electronic compass and a transmitter and receiver that communicate with Global Positioning Satellites that are currently orbiting the earth. Signals from these satellites determine the person's location, within 1 meter, relative to a map that has been programmed into the computer and that indicates the locations of objects in the scene such as the telephone booth and bus stop. Based on this information, a computer generates messages, such as "Telephone booth here" or "Bus stop 20 feet ahead," that the person hears through earphones. These messages are presented to the earphones so that the loudness and timing of the sounds reaching the left and right ears corresponds to how the sound would be perceived if the objects themselves were making the sounds. Thus, not only does the person hear "Telephone booth here," but the sound appears to come from the location of the telephone booth.

This system has been successfully tested in a prototype form that is quite a bit bulkier than the one shown in the figure, and further work is making such a system practical for actually helping blind people navigate through their environment. Note that this system will work only when an accurate map of the environment can be fed into the computer, so this system can't locate objects, such as cars and people, that are not on

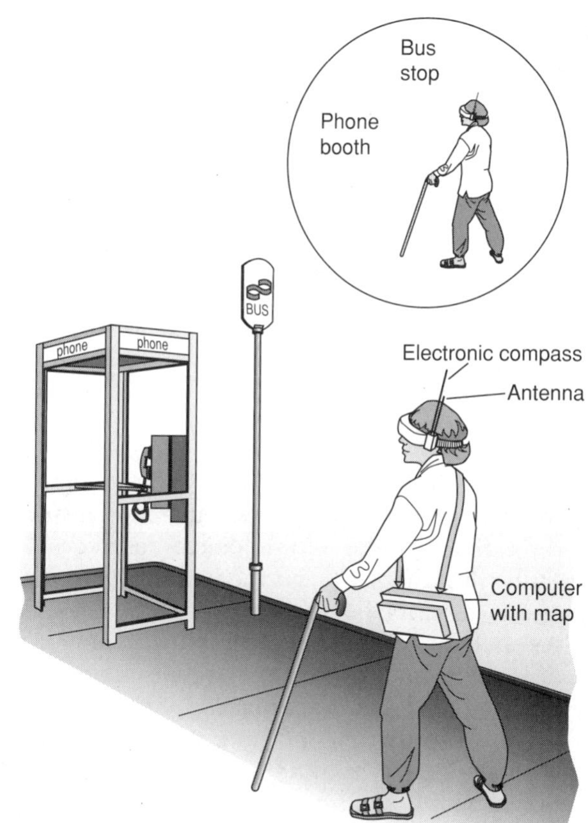

Figure 9.35
A blind person using an electronic navigation system. As the person walks through the environment, her position relative to the objects in the environment is determined by communication with Global Positioning Satellites combined with information provided by the computer, which contains a map of the area within which the person is walking. The circle indicates that the person hears the words bus stop *and* phone booth, *and that these sounds appear to originate from the position in auditory space where these objects are located. The system shown here is a proposed miniaturized version of a larger system that is now being tested. (Adapted from Loomis et al., 1994.)*

the map. Although the availability of actual models of this system for everyday use is at least a decade away, even the creation of the prototypes that now exist is an impressive demonstration of how basic knowledge about hearing mechanisms can be used for practical applications.

The Psychophysical Approach to Hearing

To close this chapter, let's look back at some of the things we have discovered about hearing through both the physiological research described in Chapter 8 and the psychophysical research described in this chapter. Physiological research has provided us with a basic blueprint of how the auditory system operates. For example, we know that the basilar membrane vibrates asymmetrically in a traveling wave; that place, with some help from timing, provides information about a tone's frequency; and that each of the receptors and nerve fibers that line the basilar membrane responds best to a narrow range of frequencies.

Psychophysics has also contributed to our knowledge by showing that this physiology is relevant to perceiving. Thus, the asymmetrical vibration of the basilar membrane creates an asymmetrical masking function, and the narrowly tuned auditory nerve fibers create narrowly tuned psychophysical tuning curves.

In addition to illuminating the parallels between physiology and perception, psychophysical research has also shown that our physiological blueprint is fuzzy in places. For example, the phenomenon of periodicity pitch raises questions about how the mechanisms of pitch perception operate in structures beyond the basilar membrane. In addition, psychophysical phenomena such as the various determinants of auditory stream segregation still need to be explained physiologically. The mystery of how we could possibly determine what is happening on a lake based on the movements of two handkerchiefs or how we could possibly experience the richness of our auditory environment based on the vibration of various membranes, bones, and hair cells inside the ear, still has many components that remain to be solved through physiological research, psychophysical research, and the collaboration of the two.

The Auditory World of Animals

When we studied the visual system, we saw that some animals are able to see wavelengths that are outside of the human visible spectrum. For example, bees and birds can detect ultraviolet light—short wavelengths below 400 nm—that are invisible to humans. A similar situation exists for hearing, but perhaps to an even greater extent.

Humans are sensitive to frequencies between about 20 and 20,000 Hz. However, some animals are sensitive to frequencies below and above the boundaries of human hearing. For example, at the low end of the frequency range, the elephant can hear frequencies below 20 Hz, and the homing pigeon can detect frequencies as low as 0.05 Hz (Heffner & Heffner, 1985; Kreithen & Quine, 1979).

But just saying that some animals have the capacity to hear sounds above or below the frequency range of human hearing doesn't tell the whole story. One way to define more precisely the nature of a particular animal's auditory capacities is to measure its audibility curve. This curve not only indicates the range of frequencies the animal can hear but also indicates the frequency to which the animal is most sensitive. An animal's audibility curve specifies that animal's "window" on the auditory world, and gathering a number of these functions together, as in Figure 9.36, helps us appreciate the variability in different animals' ranges of hearing.

Why do animals have different ranges of hearing? One reason may be that the range of hearing is tailored to the sounds that the animals emit. Thus, bats and dolphins, which use high-frequency sonar pulses to navigate (see Other Worlds of Perception, "Echolocation by Bats," in Chapter 8), can hear extremely high frequencies. However, this explanation does not hold for all

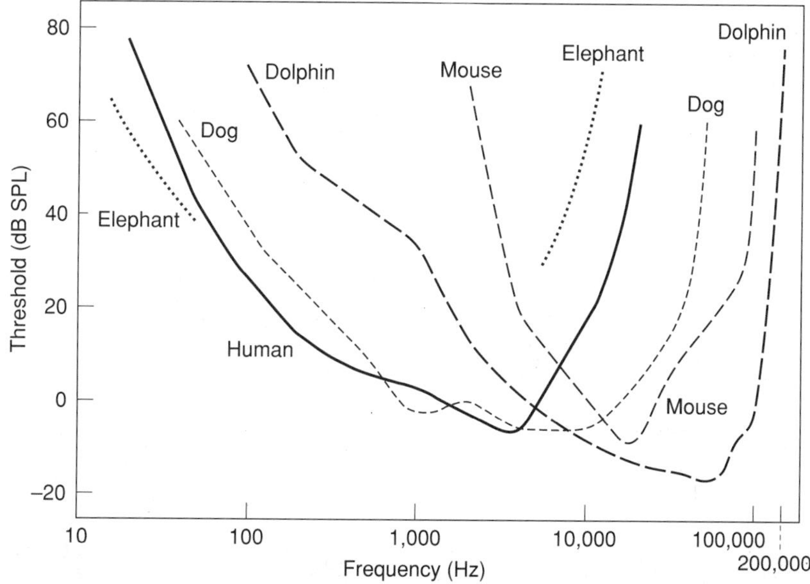

Figure 9.36
Audibility curve for a few animals. Notice that the frequency scale is logarithmic, so high frequencies are allotted less distance than low frequencies. Only the low and high ends of the elephant curve are shown to prevent overlap with the other curves. (Based on data from Au, 1993; Heffner, 1983; Heffner & Heffner, 1980, 1985; Heffner & Masterton, 1980.)

animals. For example, although the cow's vocalizations are concentrated below about 1,500 Hz, cows can hear sounds as high as 35,000 Hz (Heffner & Heffner, 1985).

Rickye Heffner and Henry Heffner (1985) pointed out that there is a relationship between mammals' ability to hear high frequencies and the time it takes for sound to travel the distance between the two ears (Figure 9.37). This relationship indicates that animals with smaller heads generally have better high-frequency hearing. An apparent exception to this relationship between head size and the upper limit of hearing is the dolphin, which has better high-frequency hearing than the smaller mouse and gerbil. In fact, the dolphin's excellent ability to hear high frequencies compares most closely with the excellent high-frequency hearing of the much smaller bat. This apparent discrepancy is explained, however, when we take into account the fact that the dolphin's hearing takes place primarily underwater and that under these conditions sound is transmitted mainly through the head (Heffner & Heffner, 1985). Since sound travels faster through solids than air, sound travels more rapidly through the dolphin's head than if it had to traverse the distance between the dolphin's left and right ears through the air, as it does in animals such as the mouse, gerbil, and bat.

Another ability that varies in different animals is the ability to localize the source of a sound. Dolphins, elephants, and humans can localize sounds to within about 1 degree of visual angle, but other animals, such as horses, gerbils, and cows, can localize sounds only within about 25–30 degrees. These differences in localization ability are certainly not related to head size, as is high-frequency hearing, since animals with very different head sizes, such as the cow and the gerbil, have similar localization abilities. A clue to a possible reason for these localization differences is provided by Figure 9.38, which is a plot of the localization threshold versus the width of the animal's field of best vision. We can understand what this relationship between localization threshold and width of the field of best vision means by considering two examples. In humans, who have good sound localization, the fovea is the area of sharpest vision. This area is very small, accounting for less than 1 degree of visual angle. In the cow, which has poor sound localization, an area called the

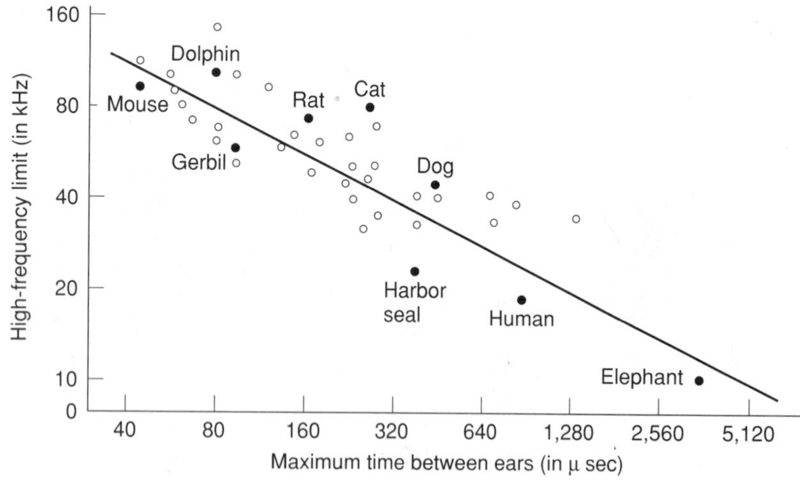

Figure 9.37

Relationship between the time it takes sound to travel between the left and right ears and the high-frequency threshold for nine animals. The open circles indicate data for additional animals. (Adapted from Heffner & Heffner, 1985.)

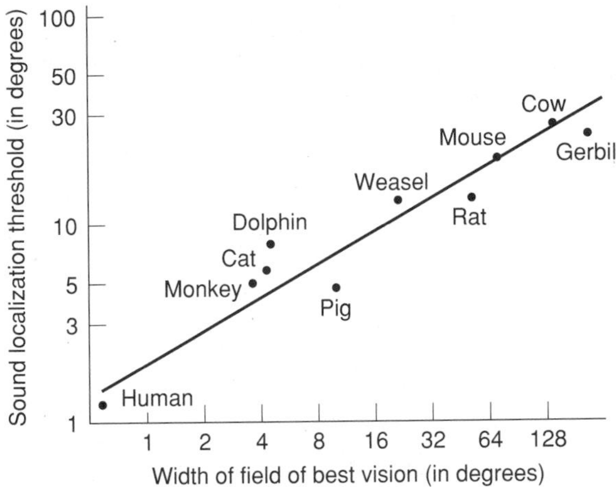

Figure 9.38

Relationship between the accuracy of sound localization and the width of the field of best vision. (Adapted from Heffner & Heffner, 1992.)

Figure 9.39

Locations of four tracking stations that picked up the ultra-low-frequency sound generated by a severe thunderstorm (square). This figure illustrates the ability of very-low-frequency sounds to travel over long distances. (From Kreithen & Quine, 1979.)

visual streak is the area of sharpest vision. This is an elongated area on the retina that stretches across most of the cow's field of view.

The reason for this connection between auditory localization and the width of best vision is that sound localization and vision are behaviorally linked. When an animal hears a sound, it moves its eyes or body to identify the location and the source of that sound. If the animal has a small area of best vision, it is important that its auditory system be able to localize the sound source precisely so that the animal can aim this small area of its retina directly at the sound source. If, on the other hand, the animal's area of best vision is more spread out, the animal does not have to aim its retina precisely at the sound source, and precise auditory localization is not necessary.

We have seen that the auditory ability of animals has evolved in order meet their perceptual needs. Animals with precise areas of good vision need to have good sound localization so that they can quickly aim their area of best vision at the sound source. We can also draw other links between auditory capacities and the animal's behavior. We have already mentioned that, in some animals such as dolphins and bats, the range of high-frequency hearing is related to the high-frequency sounds they emit. At the other end of the frequency range, the elephant's ability to hear at frequencies below the human range of hearing matches the low-frequency calls that they use for communication with each other (Payne, et al., 1986). Birds such as the homing pigeon can hear very low frequencies, such as those generated by weather phenomena like thunderstorms. These low frequencies travel over hundreds of miles because very low frequencies are attenuated only slightly by the atmosphere (Gossard & Hooke, 1975) (Figure 9.39). Although we don't actually know if pigeons use this low-frequency information to detect weather patterns, it is fascinating to think that a bird flying far overhead may, as it flies, be listening to a thunderstorm hundreds of miles away.

The Auditory World of Human Infants

What do newborn infants hear, and how does their hearing change over time? Although some early psychologists felt that newborns were functionally deaf, recent research has shown that newborns do have some auditory capacities and has also demonstrated how this capacity improves as the child gets older (Werner & Bargones, 1992). We begin by considering the question: What can newborns hear?

A simple way to find out if infants can hear is to determine if they will orient toward the source of a sound. Darwin Muir and Jeffry Field (1979) presented infants with a loud (80-dB) rattle sound 20 cm from either their right or their left ear and found that the infants turned toward the sound (Figure 9.40). Newborns can therefore hear and are capable of at least crude sound localization.

Another approach to determining newborns' capacity to hear shows that they can identify sounds they have heard before. Anthony DeCasper and William Fifer (1980) demonstrated this capacity by showing that newborns will modify their sucking on a nipple in order to hear the sound of their mother's voice. These researchers first observed that infants usually suck on a nipple in bursts separated by pauses. They fitted a 2-day-old infant with earphones and let the length of the pause in the infant's sucking determine whether the infant heard a recording of its mother's voice or a recording of a stranger's voice (Figure 9.41). For half of the infants, long pauses activated the tape of the mother's voice, and short pauses activated the tape of the stranger's voice. For the other half, these conditions were reversed.

DeCasper and Fifer found that the babies regulated the pauses in their sucking so that they heard their mother's voice more than the

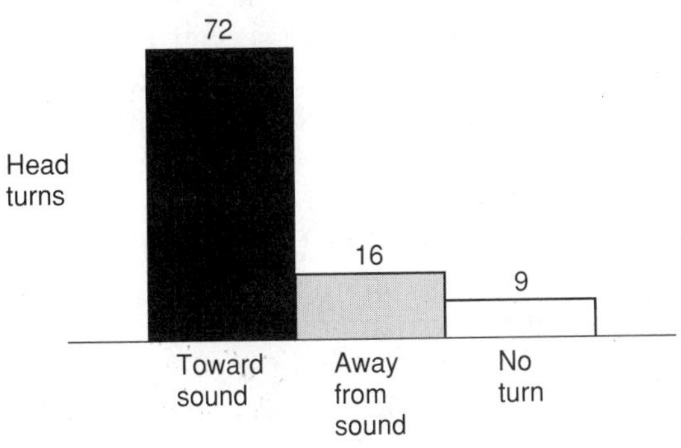

Figure 9.40
Number of head turns made by newborns in response to an 80-dB-SPL sound. Most of the turns were toward the sound, a response indicating that the infants could localize the sound. (Based on data from Muir & Field, 1979.)

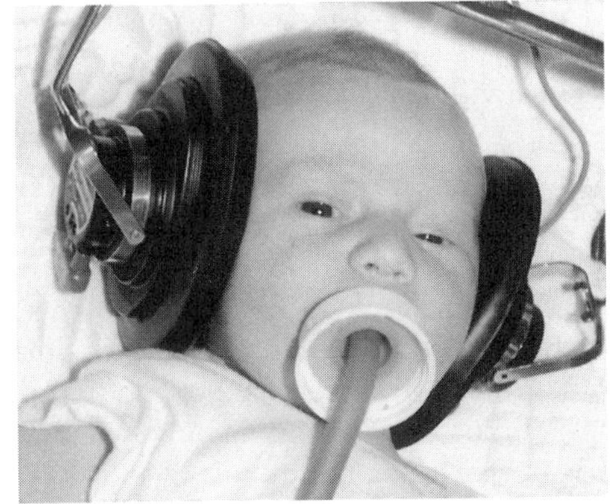

Figure 9.41
This baby, a subject in DeCasper and Fifer's (1980) study, could control whether she heard a recording of her mother's voice or a stranger's voice by the way she sucked on the nipple.

stranger's voice. This is a remarkable accomplishment for a 2-day-old, especially when we recognize that most of them had been with their mothers for only a few hours between the time they were born and the time they were tested.

Why did the newborns prefer their mother's voice? DeCasper and Fifer suggested that newborns recognize their mother's voice because they heard the mother talking while they were developing in the mother's womb. This suggestion is supported by the results of another experiment, in which DeCasper and M. J. Spence (1986) had one group of pregnant women read from Dr. Seuss's book *The Cat in the Hat* and another group read the same story with the words *cat* and *hat* replaced with *dog* and *fog*. Newborns regulated their sucking pattern to hear the version of the story that had been read before they were born. In another experiment, 2-day-old infants regulated their sucking to hear a recording of their native language rather than of a foreign language (Moon, Cooper, & Fifer, 1993). Apparently, even when in the womb, the fetus becomes familiar with the intonation and rhythm of the mother's voice, and also with the sounds of specific words. (Also see DeCasper et al., 1994.)

More precise measurements of infants' capacities have been achieved with older infants, who have a wider repertoire of responses to sound. Lynne Werner Olsho and her co-workers (1988) used a technique called the **observer-based psychoacoustic procedure** to determine infants' audibility curves. This procedure works as follows: An infant is fitted with earphones and sits on the parent's lap. An observer, sitting out of view of the infant, watches the infant through a window. A light blinks on, indicating that a trial has begun, and the infant is either presented with a tone or is not. The observer's task is to decide whether or not the infant heard the tone (Olsho et al., 1987)

How can observers tell whether the infant has heard a tone? They decide by looking for responses such as eye movements, changes in facial expression, a wide-eyed look, a turn of the head, or changes in activity level. Although this determination is not an easy task for observers, we can see from the data in Figure 9.42 that it works (Olsho et al., 1988). Observers only occasionally indicated that the 3-month-old infants heard the tone when the 2,000-Hz tone was presented at low intensities or not at all. But as the tone's intensity was increased, the observers were more likely to say that the infant had heard the tone. The infant's threshold at 2,000 Hz was determined from this curve, and the results from a number of frequencies were combined to create audibility curves, such as those in Figure 9.43. This figure, which shows curves for 3- and 6-month-olds and adults, indicates that infant and adult audibility functions look similar and that, by 6 months of age, the infant's threshold is within about 10–15 dB of the adult threshold.

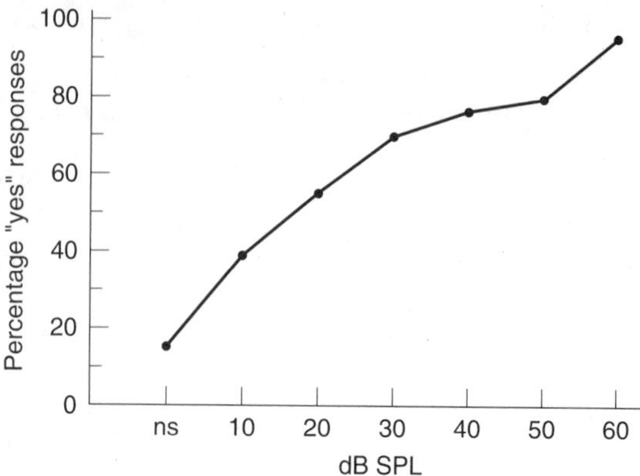

Figure 9.42
Data obtained from the observer-based psychoacoustic procedure, which indicates the percentage of trials on which the observer indicated that a 3-month-old infant heard tones presented at different intensities. ns indicates no sound. (From Olsho et al., 1988.)

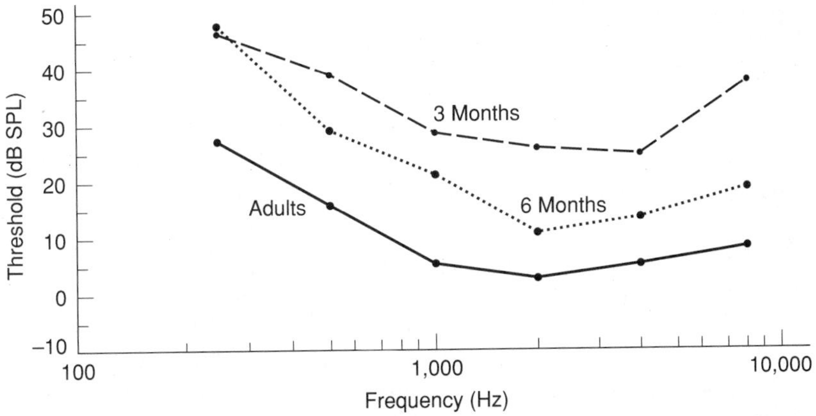

Figure 9.43
Audibility curves for 3- and 6-month-old infants determined from the observer-based psychoacoustic procedure. The curve for 12-month-olds is similar to the curve for 6-month-olds. The adult curve is shown for comparison. (Adapted from Olsho, 1988.)

From the studies we have described, it is clear that human infants are definitely not functionally deaf. Not only do very young infants have many auditory capabilities, but by about 6 months of age, their capacities have reached nearly adult levels (Werner & Bargones, 1992).

Review

Sensitivity and Loudness

Outline

- The ear is so sensitive that it can detect sounds at its most sensitive frequencies that are just 10–15 dB above the air pressure generated by the random movement of air molecules.
- A tone's loudness is related to its intensity, the unit of loudness being the sone.
- An equal loudness curve indicates the number of decibels that create a particular level of loudness across the range of hearing. Equal loudness curves can be determined for any intensity above the threshold.
- People with hearing disorders normally perceive some or all frequencies as having lower than normal loudness.
- Loudness can increase over an intensity range of 100 dB for frequencies in the most sensitive range of hearing.

Questions

1. What is the audibility curve? How is it determined, and what does it tell us about the sensitivity of the ear? What is the auditory response area? (377)
2. Define sone. Describe how the relationship between loudness in sones and intensity was determined. What is the nature of the relationship that was measured? What do we mean by the following statement: "Saying, 'The tone is 40 dB SPL' tells us nothing about the tone's loudness"? (378)
3. How is the equal loudness curve for 80 dB different from the curve for 40 dB? What problem related to loudness perception occurs when you turn down the volume of your stereo system? (379)
4. What is loudness recruitment? How does it help with the diagnosis of hearing disorders? (380)
5. Why does the fact that loudness can increase over a large intensity range pose a problem for physiological explanations of loudness? What solutions have been proposed for this problem? (380)

Perceiving Pitch

- Pitch is related to frequency, the unit of pitch being the mel.

6. Is there a one-to-one relationship between pitch and frequency? What does the relationship between pitch and frequency, combined with the relationship between position on the basilar membrane and frequency, indicate about the relationship between pitch perception and the vibration of the basilar membrane? (381)

- Tone height is a perception of pitch that increases as frequency increases. Tone chroma refers to the similarity in the sound of tones that are separated by octaves.

7. Describe Deutsch's experiment that demonstrated octave generalization. (383)

- Periodicity pitch (also known as the effect of the missing fundamental) refers to the fact that a complex tone's pitch does not change when its fundamental frequency is removed. This means that tones with different Fourier spectra can have the same pitch.

8. What problem does periodicity pitch pose for the place theory of pitch perception? What is the evidence that periodicity pitch occurs more centrally than the basilar membrane? Describe the idea of a central pitch processor. What are some practical consequences of periodicity pitch? (383)

Sound Quality

- In addition to differing in pitch or loudness, tones also differ in other qualities, such as timbre and how "flat" or "resonant" they sound.

9. What is timbre? Describe the factors that cause tones to differ in timbre. (385)

- Hearing a sound in a room affects its quality because of the reverberation caused by the room's acoustics. Even though this reverberation may cause a listener to receive a number of sounds at different times and from different directions, the listener usually perceives the correct location of the sound source.

10. What is the relationship between reverberation time and the perception of music? What is direct sound? Indirect sound? The precedence effect? What causes the precedence effect? (388)

Psychophysical Aspects of Auditory Processing

- Masking, psychophysical tuning curves, and the critical band for loudness are phenomena that can be demonstrated psychophysically and that illustrate parallels between auditory perception and physiological mechanisms.

11. Define: noise, center frequency, bandwidth, and masking. Describe the procedure and results of Egan and Hake's masking experiment. How are the results of this experiment related to the vibration of the basilar membrane? (390)
12. Describe the procedure for measuring the psychophysical tuning curve and the critical band for loudness. Understand the procedure, the results, and how these results relate to the underlying physiology. (392)

Auditory Scene Analysis

- Sounds in the environment create auditory scenes—images of various events that are happening at different locations in space. This sorting of superimposed stimuli into separate sounds, which is called auditory scene analysis, is analogous to perceptual organization in vision.

13. Apply the following principles to the perceptual organization of auditory stimuli: similarity, stream segregation, the scale illusion, proximity, good continuation, and melody schema. Understand the experiments that are described in connection with these principles. (396)

Auditory Localization

- Our ability to hear events at different locations creates auditory space. Our ability to localize sounds in this auditory space depends on information provided by binaural cues, sound reflections within the pinnae, and movements of the head.
- For people who have poor vision or are totally blind, sound provides an important source of information about the environment.

14. How well can people localize sounds located at different places? Describe the psychophysical results that show the operation of interaural time and intensity differences. How do the pinnae provide information for localization? (400)

15. Give examples of two ways that sound can be used to provide information about the locations of objects. Describe the new device that uses technology combined with binaural cues to help blind people navigate through the environment. (403)

The Auditory World of Animals

- There is a great amount of variation in animals' auditory abilities. Animals are sensitive to different frequency ranges and have different localization abilities.

16. Describe the relationship between (a) high-frequency hearing and head size and (b) the ability to localize sounds and an animal's vision? (407)

The Auditory World of Human Infants

- Although some early psychologists felt that newborns were functionally deaf, recent research has demonstrated auditory capacities in newborns and has traced the course of improvement in these capacities as the child gets older.

17. What do the experiments of Muir and Field, DeCasper and Fifer, and DeCasper and Spence tell us about hearing in young infants? Describe the observer-based psychoacoustic procedure. What do audibility curves measured by this technique indicate about the development of audition in infants? (441)

Speech Perception

Our discussion of hearing in the last two chapters focused on how the auditory system responds to pure tones, musical tones, and noise. Studying how the auditory system responds to these stimuli has taught us much about the basic mechanisms that enable us to perceive qualities such as pitch, loudness, and location. However, we are now about to embark on the study of a more complex stimulus—the speech stimulus—and our concerns will be not with perceiving pitch, loudness, and location, but with perceiving the sounds and meanings that are created when people speak.

The complexity of speech perception makes studying it challenging and at the same time provides a good reason to study it—speech perception presents an opportunity to learn more about perception in general by studying how our perceptual system deals with an extremely complex stimulus. One way to illustrate the complexity of the speech stimulus is to try to approach it in the same way we approached the perception of other qualities, such as color and pitch. The first step we took when studying those qualities was to identify the physical stimulus dimensions associated with each quality, establishing relationships between the wavelength of light and our perception of color, and between the frequency of sound and our perception of pitch.

However, when we try to apply this approach to speech perception, we immediately encounter a problem: It has been extremely difficult to discover a straightforward relationship between the sound stimulus and the perception of speech. We know that the pattern of frequency-over-time in the speech stimulus causes us to perceive speech, but what specific parts of this stimulus cause us to perceive specific sounds? As we describe research on speech perception we will begin to appreciate why this seemingly straightforward question has been so difficult to answer. We will first describe the nature of the speech stimulus, and we will then look at why it is so difficult to determine the relationship between this stimulus and our perception of speech.

THE SPEECH STIMULUS

The first hint that the speech stimulus is complex is that we can specify it in three different ways: (1) *phonemes,* which are the shortest segments of speech that change the meaning of a word; (2) *phonetic features,* which are the physical movements of the vocal tract that accompany the production of a phoneme; and (3) the *acoustic signal,* which is the energy in the speech stimulus itself, described in terms of a pattern of frequencies and intensities over time. We will look at each of these ways of specifying speech in turn.

The Phoneme: Sounds and Meaning

Our first task in studying speech perception is to separate speech sounds into units that are small enough to work with. What are these units? The flow of a sentence? A particular word? A syllable? The sound of a letter? A sentence is too large a unit for easy analysis, and some letters have no sounds at all. Much speech research has been based on a unit called the *phoneme.* A **phoneme** is the shortest segment of speech that, if changed, would change the meaning of a word. The phonemes of English, listed in Table 10.1, are represented by phonetic symbols that stand for speech sounds; 13 phonemes have vowel sounds, and 24 phonemes have consonant sounds.

Your first reaction to this table may be that there are more vowels than the standard set you learned in grade school (*a, e, i, o,* and *u*). The reason is that some vowels can have more than one pronunciation, so there are more vowel sounds than vowel letters. For example, the vowel *o* sounds different in *boat* and *hot* and the vowel *e* sounds different in *heed* and *head.* Phonemes, therefore, refer not to letters but to *sounds.*

To illustrate our definition of phoneme, consider the word *bit,* which contains the phonemes /b/, /I/, and /t/. We know that /b/, /I/, and /t/ are phonemes, because we can change the meaning of the word by changing each phoneme individually. Thus, *bit* becomes *pit* if the /b/ is changed to /p/, it becomes *bat* if the /I/ is changed to /æ/, and it becomes *bid* if the /t/ is changed to /d/.

Notice that phonemes are defined not in terms of physical dimensions such as wavelength or frequency, but as a *sound* that creates *meaning.* One reason we define a phoneme in this way is that different languages use different sounds, so the number of phonemes varies in different languages. For example, there are only 11 phonemes in Hawaiian, 48 in English, and as many as 60 in some African dialects. Thus, phonemes are defined in terms of the sounds that are used in a specific language. Another way to specify the speech stimulus is by how each phoneme is produced.

Phonetic Features: The Production of Speech

We can describe how different phonemes are produced by specifying each phoneme's **pho-**

Table 10.1
Major consonants and vowels of English and their phonetic symbols

Consonants				Vowels	
p	*p*ull	s	*s*ip	i	h*ee*d
b	*b*ull	z	*z*ip	I	h*i*d
m	*m*an	r	*r*ip	e	b*ai*t
w	*w*ill	š	*sh*ould	ɛ	h*ea*d
f	*f*ill	ž	plea*s*ure	æ	h*a*d
v	*v*et	č	*ch*op	u	wh*o*'d
θ	*th*igh	ǰ	*g*yp	U	p*u*t
ð	*th*y	y	*y*ip	ʌ	b*u*t
t	*t*ie	k	*k*ale	o	b*oa*t
d	*d*ie	g	*g*ale	ɔ	b*ou*ght
n	*n*ear	h	*h*ail	a	h*o*t
l	*l*ear	ŋ	si*ng*	ə	sof*a*
				ɨ	man*y*

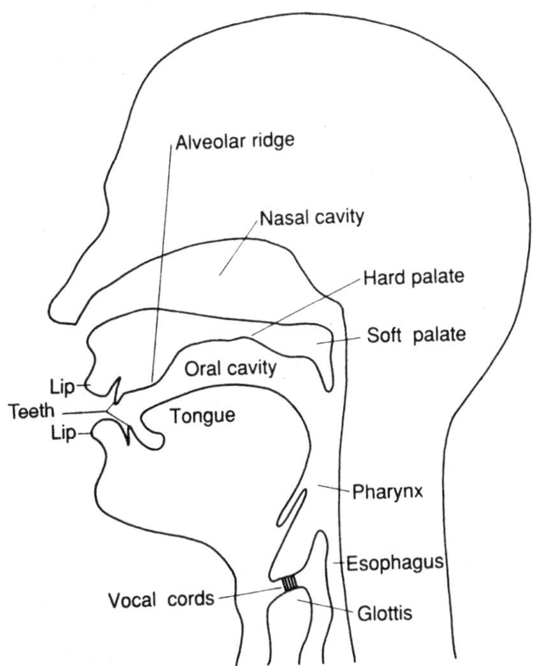

Figure 10.1
The vocal tract includes the nasal and oral cavities and the pharynx, as well as components that move, such as the tongue, lips, and vocal cords.

netic features—physical movements in the vocal tract that accompany the production of a phoneme. We can understand these physical movements by looking at the beginning of the sound stimulus: a pressure wave that is pushed up from the lungs and through the vocal tract (Figure 10.1). The sound that results depends on the state of the vocal tract as the air is pushed through it (Figure 10.2). Vowels are produced with the vocal tract in a relatively open state, and the shape of the opening determines the specific sound produced. You can demonstrate this relation between the shape of the vocal tract and specific vowels by noticing how the opening in your mouth changes as you say different vowels. (Try it: *a, e, i, o, oo.*)

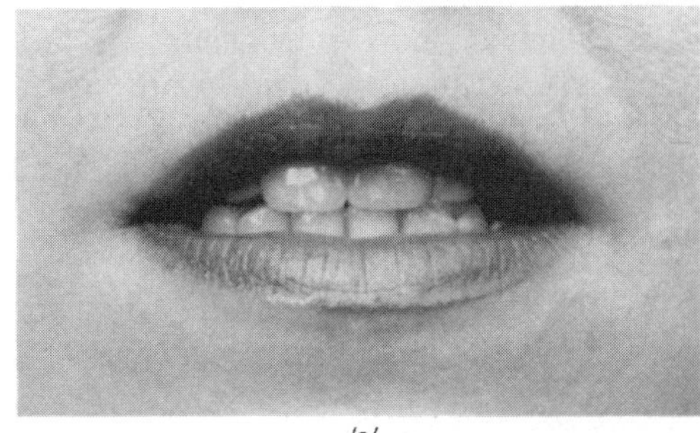

/s/

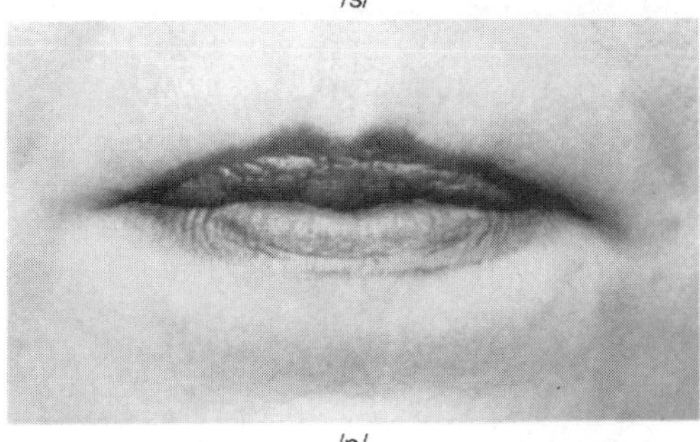

/p/

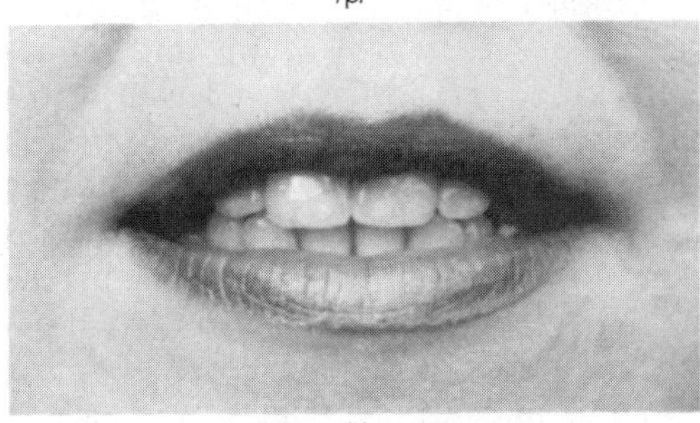

/i/

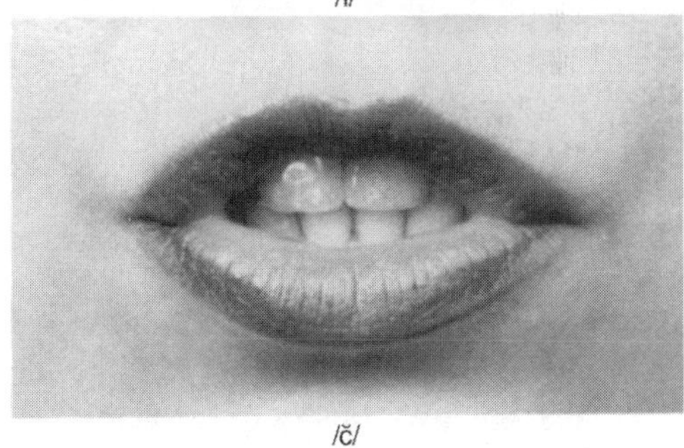

/č/

Figure 10.2
The shape of the lips while producing the word speech. *The differences in these shapes are one indication of the differences in the configuration of the vocal tract during the production of different phonemes.*

The consonants, on the other hand, are produced by a constriction or closing of the vocal tract. Instead of the continuous stream of energy characteristic of vowels (notice that you can hold a vowel until you run out of breath), consonants are characterized by rapidly changing bursts of energy and periods of silence. To illustrate the differences between how different consonants are produced, let's focus on the sounds /d/ and /f/. Make these sounds, and notice what your tongue, lips, and teeth are doing. As you produce the sound /d/, you place your tongue against the ridge above your upper teeth (the alveolar ridge of Figure 10.1) and then release a slight rush of air as you move your tongue away from the alveolar ridge (try it). As you produce the sound /f/, you place your bottom lip against your upper front teeth and then push air between the lips and the teeth.

From these examples, you can see that there are large differences in the way /d/ and /f/ are produced. In fact, when we consider all of the phonemes, we find that *no two phonemes are produced in exactly the same way.* (That's why they sound different!) We can appreciate the differences in how consonants are produced by describing their production in terms of three types of phonetic features: (1) voicing, (2) place of articulation, and (3) manner of articulation. Let's look at each, in turn.

Voicing refers to whether or not producing the sound causes the vocal cords (Figure 10.1) to vibrate. Place your fingers on either side of your neck and say the words *dead* and *fed*. The vibration you feel when you say *dead* is due to vibration of the vocal cords. You feel no vibration at the beginning of *fed* because the vocal cords are not vibrating. Consonants that cause vibration of the vocal cords are *voiced,* whereas consonants that cause no vibration are **unvoiced.** The consonant /d/ is voiced and the consonant /f/ is unvoiced.

The **place of articulation** is the place at which the airstream is obstructed during the production of a sound. For the sound /d/, the place of articulation is the alveolar ridge; therefore, the place of articulation for /d/ is called *alveolar.* The place of articulation for the sound /f/—the bottom lip and the upper front teeth—is called *labiodental.*

The **manner of articulation** is the mechanical means by which the consonants are produced and the way air is pushed through an opening. To produce the sound /d/, we close off the place of articulation by placing the tongue on the alveolar ridge, and we then release a slight rush of air. This manner of articulation is called *stop.* To produce the sound /f/, we push air between the small space between the lips and the upper teeth. This manner of articulation is called *fricative.*

The three types of phonetic features—voicing, place of articulation, and manner of articulation—provide a way to distinguish among the different phonemes. In English, there are two types of voicing, seven different places of articulation, and six different manners of articulation, so there are enough combinations of features to enable each phoneme to have a unique set of phonetic features. For example, we can describe /d/ as a *voiced* (voicing) *alveolar* (place of articulation) *stop* (manner of articulation), and /f/ as an *unvoiced labiodental fricative.* Table 10.2 shows the pattern of phonetic features of six phonemes. If we consider all of the other phonemes, we see that each one has a unique set of phonetic fea-

Table 10.2

Phonetic features of six phonemes (No two sets of phonetic features are the same)

	d	t	m	w	f	v
Place						
Alveolar	d	t				
Bilabial			m	w		
Labiodental					f	v
Manner						
Stop	d	t				
Fricative					f	v
Nasal			m			
Semivowel				w		
Voicing						
Voiced	d		m	w		v
Unvoiced		t			f	

tures. Thus, if we can identify a phoneme's phonetic features, we will have identified the phoneme.

Defining phonemes in terms of phonetic features specifies the stimulus in terms of how it is produced. But what kinds of physical signals are associated with phonemes? We answer this question by describing the acoustic signal and acoustic cues.

The Acoustic Signal and Acoustic Cues

The **acoustic signal** is the pattern of frequencies and intensities over time of the sound stimulus. This pattern of frequencies and intensities can be displayed on a plot called the **sound spectrogram.** Figure 10.3 is the spectrogram for the words *shoo cat.* The vertical axis indicates frequency, the horizontal axis indicates time, and the degree of darkness indicates the intensity. An **acoustic cue** is the specific part of the acoustic signal associated with a particular phoneme. We can identify some acoustic cues in the spectrogram by reading it from left to right. Starting at the left of Figure 10.3, we first encounter the

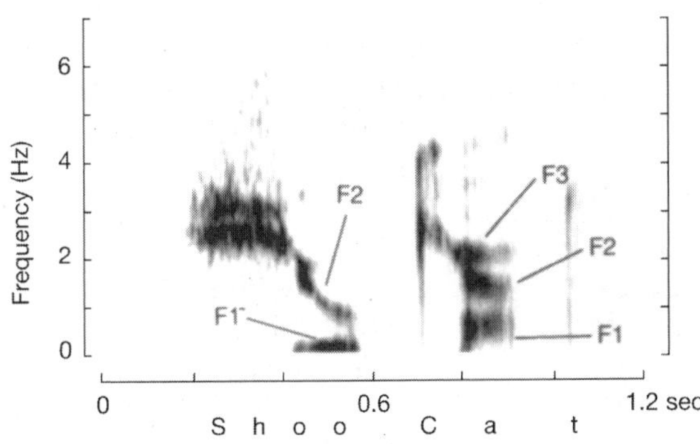

Figure 10.3

A sound spectrogram of the words shoo cat. *The formants (horizontal bands of energy characteristic of vowels) are marked. F1 is the first formant (lowest frequency); F2, the second formant; and F3, the third. (From Kiang, 1975.)*

energy associated with the /s/ (sh) sound of the word *shoo.* The /s/ lasts about 0.2 seconds, and its acoustic cue is a burst of energy between 2,000 and 3,500 Hz (indicated by the dark area on the spectrogram) and also some energy between 3,500 and 6,000 Hz.

The /u/ (oo) sound that follows is lower pitched in this speaker, so the spectrogram indicates energy at lower frequencies. The acoustic cue for /u/ is in the two bands: one (marked F1) below 500 Hz at the very bottom of the spectrogram, and the other (marked F2) at about 1,000 Hz. These horizontal (or nearly horizontal) bands of energy are called **formants.** These formants, which are always associated with vowels, depend on the shape of the vocal tract. The band with the lowest frequency is called the *first formant;* the next highest frequency, the *second formant;* and so on. Notice that three formants (marked F1, F2, and F3) are visible for the /æ/ sound of *cat.*

Now that we have seen that the speech stimulus can be separated into phonemes and that these phonemes can be described in terms of phonetic features or their acoustic cues, we are ready to tackle one of the central questions of speech perception: What is the relationship between the speech stimulus and the sounds that we hear? This question is more challenging than it might at first appear.

The Challenge of Understanding Speech Perception

Figure 10.4 illustrates a way speech perception might work. Our speaker utters a phrase, "Hi Fred," which consists of a series of phonemes. Each phoneme creates a pattern of pressure changes at different frequencies, and when Fred receives these pressure changes, they are de-

Figure 10.4

According to this idea of how speech perception might work, patterns of pressure changes created by the speaker are transformed directly into phonemes by the listener's auditory system, and the listener perceives what the speaker said. Speech perception is, however, much more complex than this.

coded by his auditory system to create each phoneme and then the perception of the phrase "Hi Fred."

Unfortunately, speech perception isn't that simple. The problem is that there is not a constant relationship between the pattern of pressure changes in the air and each phoneme, so "decoding" the speech stimulus into phonemes is not as straightforward as is indicated in Figure 10.4. We can appreciate why by considering two problems: the variability problem and the segmentation problem.

The Variability Problem

The variability problem refers to the fact that a specific phoneme is not always associated with the same acoustic cues. We can illustrate this variability of a phoneme's acoustic cues by considering how they are changed by the context in which a phoneme appears and by variability in the way different people speak.

Context Creates Variable Acoustic Cues We can appreciate how the context within which a phoneme occurs can influence its acoustic cues by looking at Figure 10.5, which shows spectrograms for the sounds /di/ and /du/. These spectrograms are smoother than the one in Figure 10.3 because they are drawn by hand. These hand-drawn spectrograms do, however, accurately show the two most important characteristics of the sounds /di/ and /du/: the formants and the formant transitions. The formants are the horizontal bands of energy at about 200 and 2,600

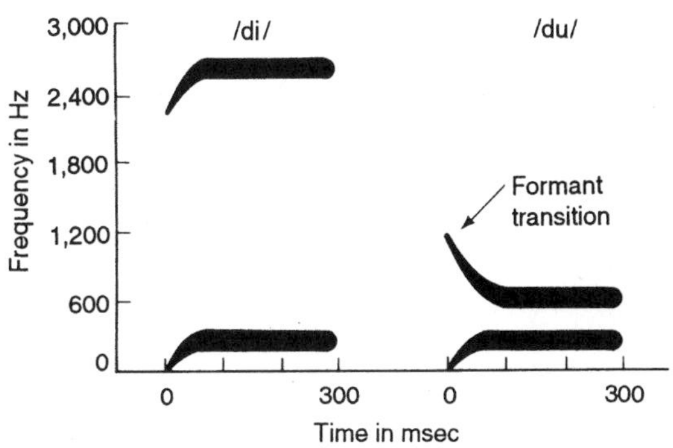

Figure 10.5
Hand-drawn spectrograms for /di/ and /du/. (From Liberman et al., 1967.)

Hz for /di/ and at 200 and 600 Hz for /du/. Knowing that formants are characteristic of vowels, we conclude that the formants at 200 and 2,600 Hz are the acoustic cue for the vowel /i/, and that the formants at 200 and 600 Hz are the acoustic cue for the vowel /u/.

The **formant transitions** are the rapid shifts in frequency preceding each formant. Since we have already identified the formants as the acoustic cues for the vowels, the formant transitions must be the acoustic cue for the consonant /d/. But notice that, even though the /d/ sound is the same for both /di/ and /du/, the formant transitions associated with the second (higher-frequency) formants are different. This difference illustrates the variability of the acoustic cue for /d/. The acoustic cue associated with /d/ depends on the vowel following the /d/, and the /d/ sound is associated with different acoustic cues when followed by /i/ than when followed by /u/.

This context effect, which causes differences in acoustic cues in different contexts, is a result of **coarticulation**—the overlapping of articulation of different phonemes. You can experience this effect for yourself by noticing how you produce phonemes in different contexts. For example, say *bat* and *boot*. Notice that when you say *bat* your lips are unrounded, and that when you say *boot* your lips are rounded, even during the initial /b/ sound. Thus, even though the /b/ is the same in both words, you articulate them differently. In this example, the articulation of /oo/ in *boot* overlaps the articulation of /b/, causing the lips to be rounded even before the /oo/ sound is actually produced.

The fact that we perceive phonemes in the same way, even though the acoustic signal is changed by coarticulation, is called **perceptual constancy.** This term may be familiar to you from our observations of constancy phenomena in the sense of vision, such as color constancy (we perceive an object's chromatic color as constant even when the wavelength distribution of the illumination changes) and size constancy (we perceive an object's size as constant even when the size of its image changes on our retina). Perceptual constancy in speech perception parallels these examples of constancy: We perceive a particular phoneme as constant even when this phoneme appears in changing contexts that cause it to be associated with different acoustic cues. In addition to the context effect, another factor that causes variability in the acoustic signal is that there is a wide variation in the way different people speak.

"Real-World" Speech Creates Variable Acoustic Cues How people speak in normal conversation introduces further variability in the acoustic signal. People have voices that are high-pitched or low-pitched; they speak with accents that are southern, midwestern, down east, or foreign; and they talk at speeds ranging from extremely rapid to extremely slow.

These wide variations in speech between speakers means that, for different speakers, a particular phoneme or word can have very different acoustic cues. Speakers also introduce variability by their sloppy pronunciation. For example, say the following sentence at the speed you would use in talking to a friend: "This was a best buy." How did you say "best buy"? Did you pronounce the /t/ of *best,* or did you say "bes buy"? What about "She is a bad girl"? While saying this rapidly, notice whether your tongue

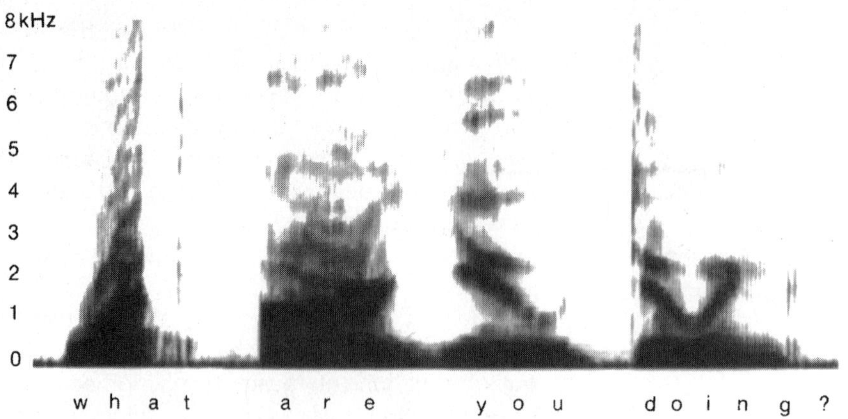

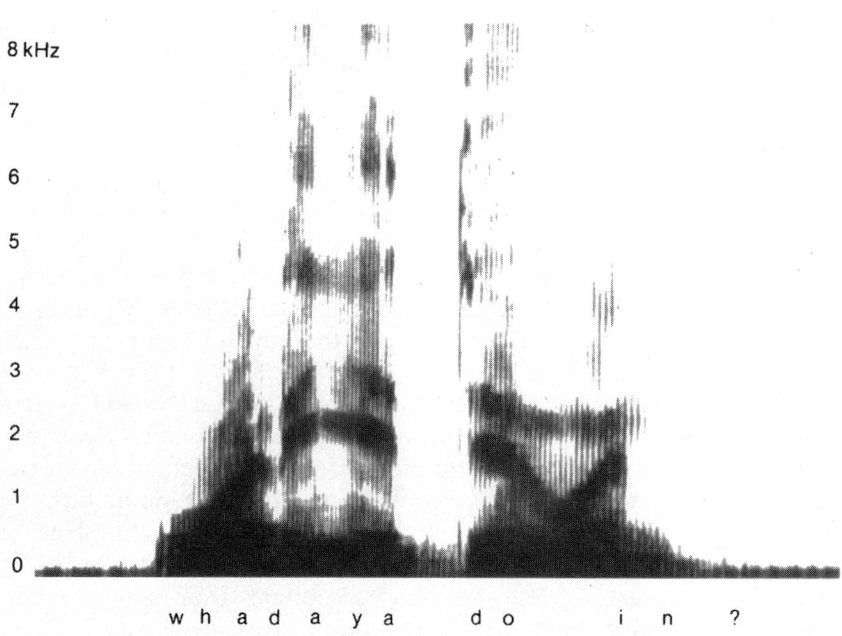

Figure 10.6
(a) Spectrogram of "What are you doing?" pronounced slowly and distinctly. (b) Spectrogram of "What are you doing?" as pronounced in conversational speech. Note that (b) has been enlarged so it is almost as wide as (a). However, (a) lasts about 2.1 seconds, and (b) lasts only about 0.9 seconds. (Spectrograms courtesy of Ron Cole.)

hits the top of your mouth as you say the /d/ in *bad*. Many people omit the /d/ and say "ba girl." Finally, what about "Did you go to the store?" Did you say "did you" or "dijoo"?

That people do not usually enunciate each word individually in conversational speech is reflected in the spectrograms in Figure 10.6. The spectrogram in Figure 10.6a is for the question "What are you doing?" spoken slowly and distinctly, whereas the spectrogram in Figure 10.6b is for the same question taken from conversational speech, in which "What are you doing?" becomes "Whad'aya doin?" This difference shows up clearly in the spectrogram, which indicates that, although the first and last words (*what* and *doing*) create similar patterns in the two spectrograms, the pauses between words are absent or are much less obvious in the spectrogram of Figure 10.6b, and the middle of this spectrogram is completely changed, with a number of speech sounds missing. The absence of pauses between the words in conversational speech creates the *segmentation problem:* How can we perceive individual words if there are no pauses between words?

The Segmentation Problem

Though it may seem that there is a slight break between each word in a conversation, spectrograms like the one in Figure 10.6b show that it is often difficult to tell where one word ends and the next begins. Another example of this lack of physical boundaries between words is shown in the spectrogram of "I owe you a yo-yo" in Figure 10.7. From this spectrogram, it is not at all obvious where one word ends and the other begins, and even an experienced spectrogram reader would find it extremely difficult to locate the individual words.

The fact that there are usually no spaces between words becomes obvious when you consider how a foreign language sounds. To someone unfamiliar with a language, the sounds of that language seem to speed by in an unbroken string; however, to a speaker of that language, the words seem separated, just as the words of English seem separated to you.

The amazing thing about speech perception is that, despite the variability caused by coarticulation, different speakers, and sloppy speech, and despite the lack of actual physical breaks between words, we are hardly aware of any of these difficulties. We experience perceptual constancy and simply hear words that have meanings. In fact, most people are surprised to know that at normal rates of speaking, we are experiencing 12–14 sounds per second, and that we can understand speech at rates as high as 50–60 sounds per second (Werker & Tees, 1992). We are unaware that we are accomplishing amazing feats and instead just go about our business of perceiving words and conversations.

How do we perceive speech even in the face of what might seem to be extreme difficulties? Although we still can't fully answer this question, speech perception research has brought us

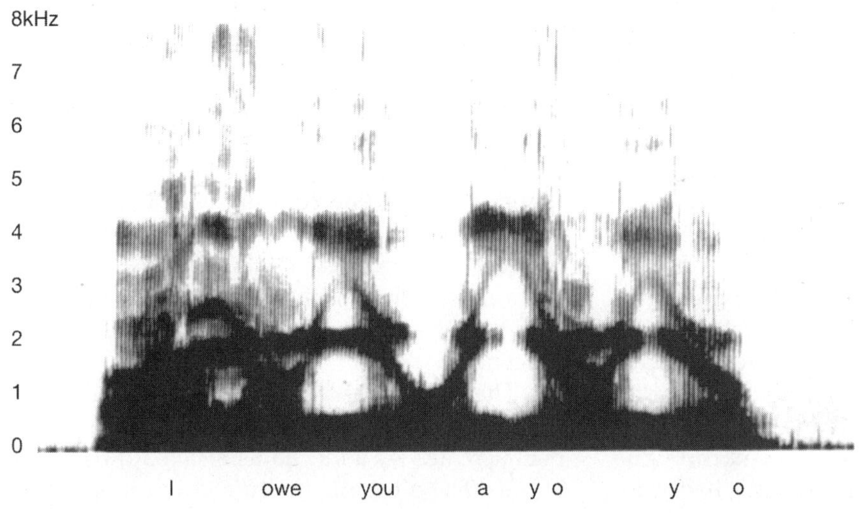

Figure 10.7
Spectrogram of "I owe you a yo-yo." (Spectrogram courtesy of Ron Cole.)

much closer to the answer by asking questions such as the following:

1. Is there a special mechanism for speech perception?
2. Is there information in the acoustic signal that indicates which sounds are present?
3. How do listeners take variations in speakers' voices into account?
4. How do knowledge and meaning influence speech perception?

Is There a Special Mechanism for Speech Perception?

Do we perceive speech because the speech signal activates a mechanism designed specifically to process speech signals, or do we perceive speech because the speech signal activates the same auditory mechanisms responsible for the perception of pitch and loudness that we described in Chapters 8 and 9? These two possibilities have been called the **special speech mechanism** and the **general auditory mechanism.**

The question of whether our perception of speech depends on a special speech mechanism or on a general auditory mechanism is one of the most controversial questions in speech research, and it has therefore generated a great deal of theorizing and research. As we describe the arguments on either side of this debate, we will show how evidence in favor of a special speech mechanism has been presented by some researchers and then has been questioned by others. In addition, we will consider evidence supporting the idea that speech perception is achieved through the operation of a general auditory mechanism.

The idea of a special speech mechanism has its appeal, especially when considered in the context of our knowledge of how the visual system operates. We have seen that there are nuclei in the visual system that are specialized to process information about different qualities, such as color, depth, and movement (Casagrande, 1994). In addition, we know that there are neurons that are specialized to respond to complex visual stimuli such as faces (Perrett et al., 1992). This evidence, along with the fact that, for most people, language is processed in specific areas in the left hemisphere of the cortex (Geshwind, 1979), makes it seem reasonable that there is a specialized mechanism that exists especially to process speech stimuli. This idea of a special speech mechanism was proposed in 1967 in a paper by Alvin Liberman and his co-workers, which described a theory called the *motor theory of speech perception.*

The Motor Theory of Speech Perception

Liberman and co-workers (1967) argued that one reason that speech is special is that it is a stimulus that we both *produce* and *perceive.* Liberman et al. suggested that, since speaking and listening are so closely connected, it would make little sense for us to have two separate and independent mechanisms for producing speech and for perceiving it. Rather, it is more likely that the mechanisms responsible for these two functions are closely linked. This link between production and perception is the basic idea behind the **motor theory of speech perception,** which is called the "motor" theory because it postulates a close link between movements of the vocal tract, such as vibrations of the vocal cords, and movements of the mouth and tongue, and the perception of speech.

The motor theory describes this link between movements of the vocal tract and perception as being created by a mechanism called the *phonetic module.* The **phonetic module** is a proposed neu-

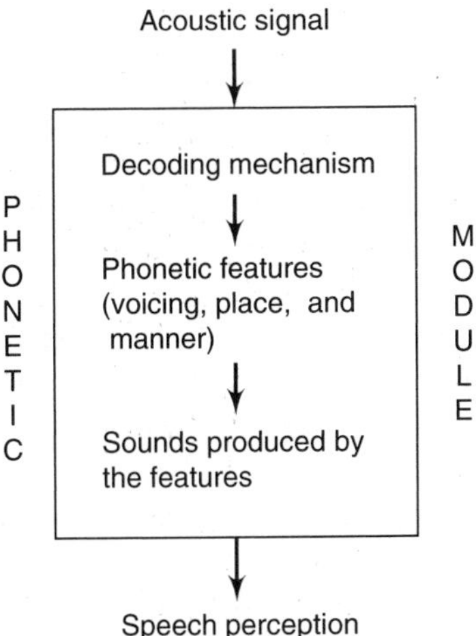

Figure 10.8
The phonetic module proposed by motor theory somehow decodes the incoming acoustic signal to deduce which phonetic features produced it. According to motor theory, speech perception is based on these phonetic features, since each phoneme is associated with a particular set of features.

ral structure that operates separately from other auditory mechanisms, and that serves a very important function: Based on its analysis of the incoming acoustic signals, it re-creates the activity of the vocal tract that created those signals (Figure 10.8). Another way to describe the operation of the phonetic module is to say that it determines which phonetic features (voicing, place and manner of articulation) created the acoustic signal. Since each phoneme has its own set of phonetic features, knowing the features enables the module to identify the phoneme, and we then perceive it.

From this description we can appreciate two major characteristics of the motor theory: (1) It is based on the idea that speech is special, as illustrated by the proposal of a phonetic module that is specialized to analyze speech stimuli, and (2) by hypothesizing that the module determines phonetic features, motor theory avoids the variability problem. Remember that, although a particular phoneme can have a number of different acoustic cues, each phoneme can be described by just one set of phonetic features.

What is the evidence for motor theory? As we will see below, the evidence consists of a series of demonstrations that have been interpreted by proponents of motor theory either as demonstrating the special nature of speech or as illustrating the connection between speech production and speech perception. One of these demonstrations, called *categorical perception,* was described in Liberman et al.'s 1967 paper.

Categorical Perception

Categorical perception refers to the fact that we perceive speech sounds as being in a limited number of perceptual categories. We can understand what this means by describing how categorical perception is measured. To do this, we need to describe a characteristic of the acoustic signal called **voice onset time (VOT).** Voice onset time is the time delay between the beginning of a sound and the beginning of the vibration of the vocal cords that accompanies voicing. We can illustrate this delay by considering the consonants /d/, which is voiced, and /t/, which is unvoiced. This difference between /d/ and /t/ causes a difference in VOT that is illustrated by the spectrograms in Figure 10.9.

We can see from these spectrograms that the time between the beginning of the sound /da/ and the beginning of voicing (indicated by the presence of vertical striations in the spectrogram) is 17 msec for /da/ and 91 msec for /ta/. Thus, the voiced /d/ causes /da/ to have a short VOT, and the unvoiced /t/ causes /ta/ to have a long VOT.

There are two parts to a categorical perception experiment: (1) categorization and (2) discrimination. In the categorization part of the experiment, we use a computer to synthesize a sound with VOT = 0 msec that is perceived as

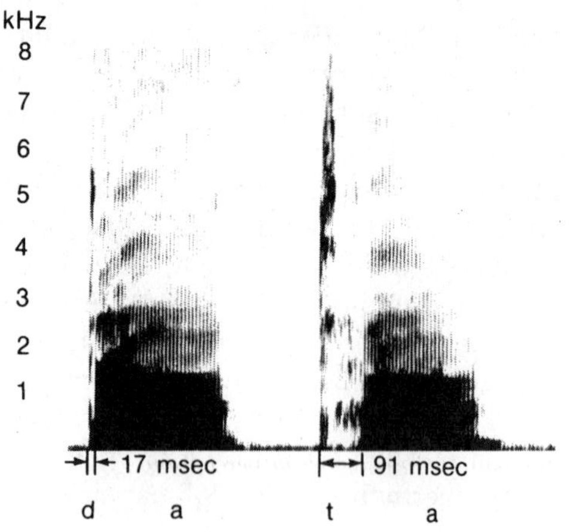

Figure 10.9
Spectrograms for /da/ and /ta/. The voice onset time—the time between the beginning of the sound and the onset of voicing—is indicated at the beginning of the spectrogram for each sound. (Spectrogram courtesy of Ron Cole.)

/da/, and we then increase the VOT in small steps until at longer VOTs the sound is perceived as /ta/. The listeners' task is to indicate whether they hear a /da/ or a /ta/ at each step. The results of such an experiment are shown by the solid line in Figure 10.10 (Eimas & Corbit, 1973). All of the stimuli with VOT = 0 are heard as /da/, and increasing the VOT has no effect until, at about 35 msec, listeners suddenly begin hearing /ta/, and when the VOT is increased just a little more to 40 msec, most of the stimuli are identified as /ta/. The VOT when the perception changes from /da/ to /ta/ is called the **phonetic boundary,** and the key result of the categorical perception experiment is that, even though the VOT is changed in many steps, between 0 and 80 msec, the listener perceives only two categories: /da/ on one side of the phonetic boundary and /ta/ on the other side.

In the discrimination part of the categorical perception experiment, we determine how well a listener can discriminate between two stimuli that differ in VOT. When we do this, we find that a listener typically cannot tell the difference between two stimuli that are in the same category but can tell the difference between two stimuli that are in different categories. For example, a listener can't tell the difference between two stimuli with VOTs of 10 and 30 msec that are both to the left of the phonetic boundary and are therefore perceived as /da/, but the listener can tell the difference between two stimuli with VOTs of 30 and 50 msec, which are on opposite sides of the phonetic boundary, so one is perceived as /da/ and the other as /ta/. Thus, categorical perception occurs when we perceive stimuli in terms of categories so that we can discriminate between stimuli from different categories but can't discriminate between stimuli in the same category (Figure 10.11).

The discovery of categorical perception was important for two reasons. First, categorical perception shows that the auditory system deals with the variability problem by simplifying our perception of speech sounds. Consider, for example, the large number of different sounds that would exist if our perception changed every time we changed the VOT even slightly. But this does not occur. Instead, we experience only two different sounds, even though the VOT changes over a wide range. Categorical perception therefore helps us achieve perceptual constancy (Figure 10.12).

The other reason that the discovery of categorical perception was important was that categorical perception was initially observed only for speech sounds, a finding that supported the idea that speech perception may involve its own special mechanism. For example, although speech sounds are perceived in categories, this kind of perception does not occur for the pitch of pure tones. If we start with a low-frequency pure tone and slowly increase its frequency, our perception of pitch changes continuously, so we perceive hundreds of different pitches across the range of hearing. A similar absence of categorical perception of other nonspeech sounds led researchers to put forth the categorical perception results to support the idea of a special speech mechanism.

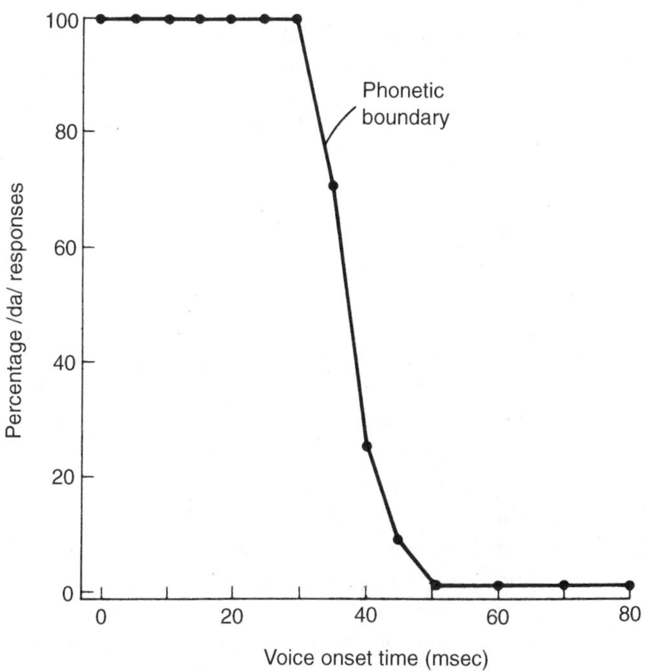

Figure 10.10
The results of a categorical perception experiment indicate that /da/ is perceived for VOTs to the left of the phonetic boundary, and that /ta/ is perceived at VOTs to the right of the phonetic boundary. (From Eimas & Corbit, 1973.)

Later research, however, showed that categorical perception also occurs for nonspeech sounds like buzzes, noises, and musical stimuli (Miller et al., 1976; Pisoni, 1977; Zatorre & Halpern, 1979) and that categorical perception also occurs in monkeys (May, Moody, & Stebbins, 1989), chinchillas (Kuhl, 1986), and quail (Kluender, Diehl, & Killeen, 1987), which don't possess the capacity for speech. These results weakened the special speech argument based on categorical perception, but other evidence soon appeared, in the form of demonstrations that supported the idea of a special speech mechanism. One of these demonstrations was a phenomenon called *audiovisual speech perception,* or the *McGurk effect,* named after Harry McGurk, who described a way of influencing the perception of speech by having a person listen to speech while viewing a videotaped display of a person speaking (McGurk & MacDonald, 1976; MacDonald & McGurk, 1978; Summerfield, 1979).

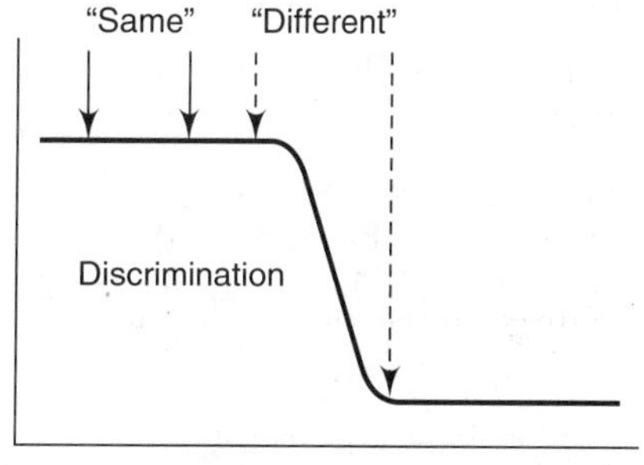

Figure 10.11
In the discrimination part of a categorical perception experiment, two stimuli are presented, and the subject is asked to indicate whether they are the same or different. The typical result is that two stimuli with VOTs on the same side of the phonetic boundary are judged to be the same (solid arrows), and that two stimuli on different sides of the phonetic boundary are judged to be different (dashed arrows).

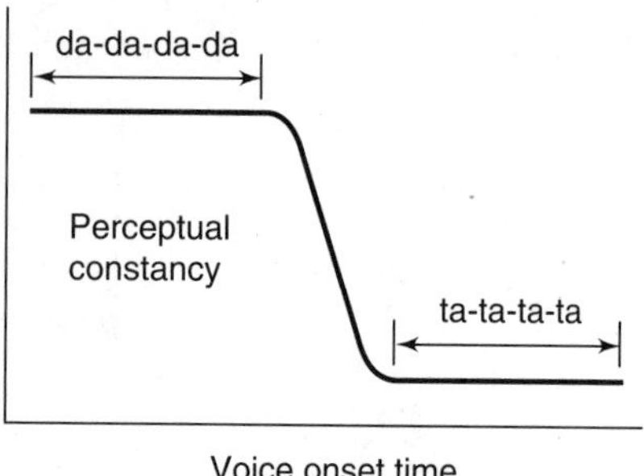

Figure 10.12

Perceptual constancy *occurs when all stimuli on one side of the phonetic boundary are perceived to be in the same category even though their VOT is changed over a substantial range. This diagram symbolizes the constancy observed in the Eimas and Corbit (1973) experiment, in which /da/ was heard on one side of the boundary and /ta/ on the other side.*

The McGurk Effect

McGurk and his co-workers achieved the effect that now bears his name by using the following procedure:

1. They videotaped a person repeating the sounds /ga-ga/.
2. They showed this tape to subjects but substituted the sound /ba-ba/ for the original /ga-ga/. Thus, the visual image on the screen was a woman making the lip movements for /ga-ga/, but the sound from the loudspeakers was /ba-ba/ (Figure 10.13).
3. The subjects reported that they *heard* the sound /da-da/. Thus, presenting an acoustic signal that didn't match the speaker's lip movements, changed the listeners' perception of the sound.

This demonstration that visual observation of a speaker's lip movements can influence what a listener hears has been cited as evidence to support the special speech mechanism, because it demonstrates a link between the production of sound and the perception of speech. Motor theory explains the McGurk effect by stating that the phonetic module uses any available information to determine how the acoustic signal was produced. In the McGurk situation, this information consists of both the acoustic signal *and* the speaker's lip movements. Lip movements are particularly important to the phonetic module, because its task is to determine how the acoustic signal was produced. Since lip movements are closely related to production, they strongly influence the phonetic module and thus alter the listener's perception of the sound.

In response to this motor theory explanation of the McGurk effect, Helena Saldaña and Lawrence Rosenblum (1993) created a McGurk effect using nonspeech stimuli. A subject watched a video in which a person plucked a string on a cello. However, the sound presented to the subject was of a note created by *bowing* the string. Other subjects *saw* the cello being bowed but

Figure 10.13

The McGurk effect. The woman's lips are moving as if she is saying /ga-ga/, but the actual sound being presented is /ba-ba/. The listener, however, reports hearing the sound /da-da/. If the listener closes his eyes, so that he no longer sees the woman's lips, he hears /ba-ba/. Thus, seeing the lips moving influences what the listener hears.

heard a note created by plucking the string. This procedure created a McGurk-type effect: The subjects who were presented with the bowing sound but saw the cello being plucked tended to rate the sound as more plucklike than when they saw the string being bowed. Although the "pluck-bow" McGurk effect was weaker than the effect achieved with speech sounds, Saldaña and Rosenblum suggest that their results cast doubt on a purely special-speech-mechanism interpretation of the McGurk effect.

Yet another phenomenon that has been offered to support the special speech mechanism is called **duplex perception,** a situation in which one stimulus causes a person to hear both a speech sound and a nonspeech sound simultaneously.

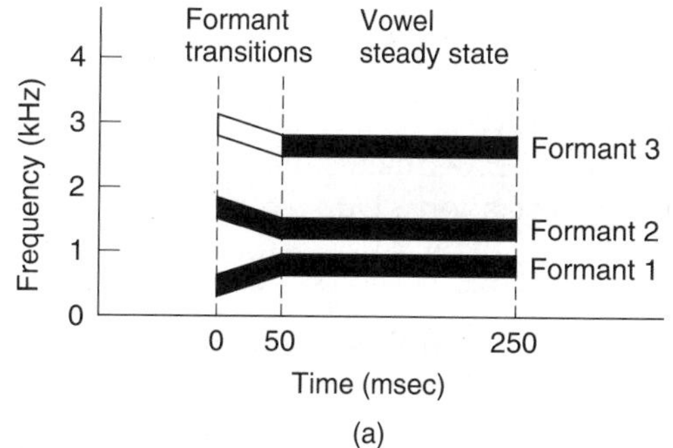

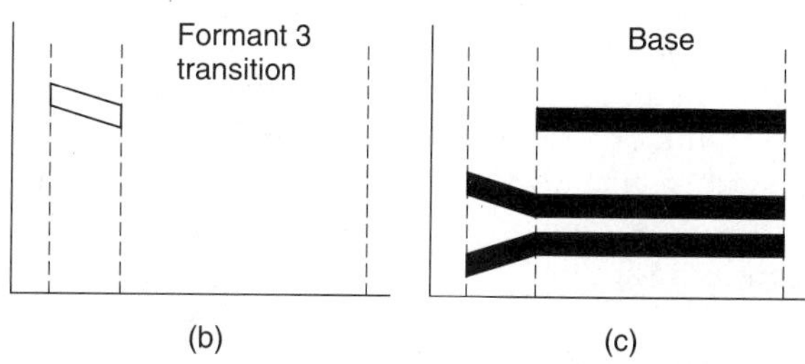

Figure 10.14
(a) The spectrogram for the sound /da/. Duplex perception occurs when the transition for the third formant (b) is presented to the left ear and the rest of the signal, called the base (c), is presented to the right ear. The listener perceives the sound /da/ in the right ear and a "chirp" in the left ear. See text for further details.

Duplex Perception

Duplex perception is created by splitting the acoustic signal for a sound into two parts and presenting one part to each ear. For example, Figure 10.14a shows the speech spectrogram for the sound /da/, which consists of three formant transitions and their formants. To create duplex perception, the transition for the third formant is presented alone to the left ear (Figure 10.14b), and the rest of the signal, which is called the *base*, is presented to the right ear (Figure 10.14c). In response to this stimulus, the listener hears the speech sound /da/, associated with the overall acoustic signal, in the right ear and simultaneously perceives a nonspeech "chirp," associated with the transition alone, in the left ear.

According to proponents of the special speech mechanism, this experiment illustrates two kinds of perception: (1) the speech mode, which occurs when the phonetic module combines the formant transition and the rest of the signal to create the speech sound, and (2) the auditory mode, which occurs when a general auditory mechanism creates the chirp sound from the high frequencies in the formant transition.

Duplex perception has also been demonstrated by presenting the entire stimulus for /da/ to the same ear (Whalen & Liberman, 1987). When the formant transition and the formant have the same intensity, the listener hears the speech sound /da/. But increasing the intensity of the transition causes the listener to hear its chirplike quality. Whalen and Liberman interpret this result to mean that at normal intensities the speech mode, which is created by the phonetic module, dominates the auditory mode. It is only when the transition is much more intense that the auditory mode overcomes the dominance of the speech mode and the listener hears the chirp (also see Werker & Tees, 1992).

Our story so far has been one of demonstrations that can be interpreted as support for the idea of a special speech mechanism, followed by the results of experiments that suggest other possible interpretations. Duplex perception follows this pattern as well, since Carol Fowler and Lawrence Rosenblum (1990) showed that duplex perception can be achieved not only with speech sounds, but with nonspeech sounds as well. This result suggests, according to Fowler and Rosenblum, that duplex perception may not be created by a phonetic module, as claimed by proponents of the motor theory of speech perception.

The long-running debate between researchers who explain speech in terms of a special speech mechanism and those who explain it in terms of a general auditory mechanism is still continuing, almost 30 years after Liberman et al.'s 1967 paper. Another contribution to this debate has been provided by researchers favoring the general auditory mechanism, who have been focusing their attention on looking for information in the auditory signal that may indicate which sounds are present.

Is There Information in the Acoustic Signal That Indicates Which Sounds Are Present?

Near the beginning of this chapter, we introduced the finding that has posed a major challenge to our understanding of speech perception: Coarticulation creates a variable acoustic signal for each phoneme. This variability of the acoustic signal has been accepted by many researchers as a fact that must be dealt with when explaining speech perception. However, what if we could find some aspect of the acoustic signal that would provide information that enables us to identify each phoneme even when it appears in different contexts or is spoken by different speakers. Researchers have been searching for this information, which they call **invariant acoustic cues**—aspects of the acoustic signal associated with a specific phoneme that remain invariant, or constant, in different contexts. The main tactic they have used in this search has been to find new ways of displaying and analyzing the acoustic signal.

One new way of displaying the acoustic signal, which is called the **short-term spectrum,** is shown in Figure 10.15, along with the sound spectrogram for /ga/. This short-term spectrum indicates which frequencies occur during the first 26 msec of the sound /ga/. Notice that the short-term spectrum in Figure 10.15a indicates a peak of energy just below 2,000 Hz. This peak in the short-term spectrum corresponds to the dark area below 2,000 Hz in the narrow band of sound energy indicated by the arrow at the beginning of the spectrogram in Figure 10.15b. Also notice that there is a minimum in the short-term spectrum, just below 3,000 Hz, which is visible as the light area below 3,000 Hz in the first 26 msec of the spectrogram. One advantage of the short-term spectrum over the spectrogram is that it indicates much more precisely exactly how much energy is present at each frequency. The disadvantage of the short-term spectrum is that it gives us information about the frequencies present during only a very brief period of time. However, by combining a sequence of short-term spectra to create a **running spectral display,** as shown in Figure 10.16, we can see not only which frequencies are present at one point in time but can see how these frequencies change as time progresses.

By using short-term spectra and running spectral displays, researchers have identified some invariant acoustic cues. The invariance of these cues has been demonstrated by showing that people can identify phonetic features based on these cues, even in different contexts (Blumstein & Stevens, 1979; Kewley-Port, 1983;

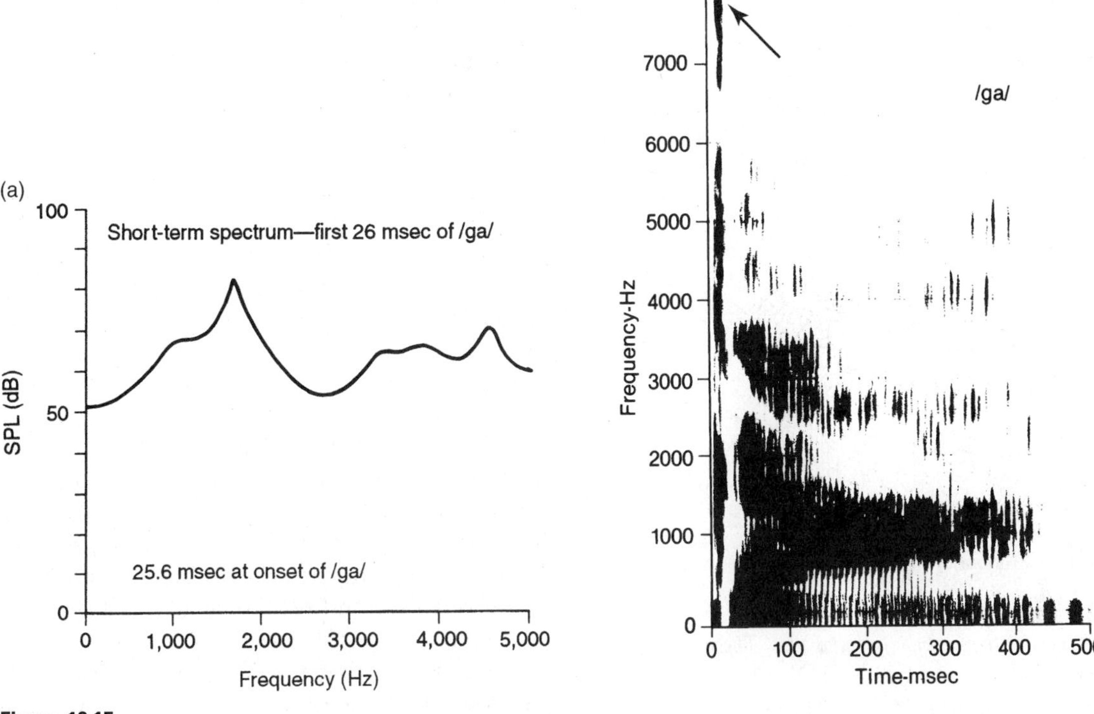

Figure 10.15

(a) A short-term spectrum of the acoustic energy in the first 26 msec of the phoneme /ga/. (b) Sound spectrogram of the same phoneme. The arrow on this spectrogram indicates the top of the narrow band of energy represented by the short-term spectrum. (Courtesy of James Sawusch.)

Kewley-Port & Luce, 1984; Searle, Jacobson, & Rayment, 1979; Stevens & Blumstein, 1978, 1981). For example, Diane Kewley-Port and Paul Luce (1984) found that the voicing and place of articulation of a number of phonemes can be identified correctly about 80 to 90 percent of the time based on running spectral displays. Two examples of cues identified by Kewley-Port and Luce in the running spectral displays in Figure 10.16 are as follows:

1. The prominent low-frequency peak that continues in succeeding frames (marked with V) signals the presence of voicing. Notice that it occurs for /da/ but not for /pi/. From this information, we can conclude that /d/ is voiced but that /p/ is not.
2. The high-frequency peaks are tilted up for the /da/ spectrum (indicated by R). This tilt indicates an alveolar place of articulation.

Although using running spectral displays to find invariant acoustic cues has yielded some encouraging results, this approach has not been completely successful. One problem is that the information in these displays can't explain how listeners can correctly identify speech stimuli virtually 100 percent of the time. In addition, this approach does not work for all speech sounds.

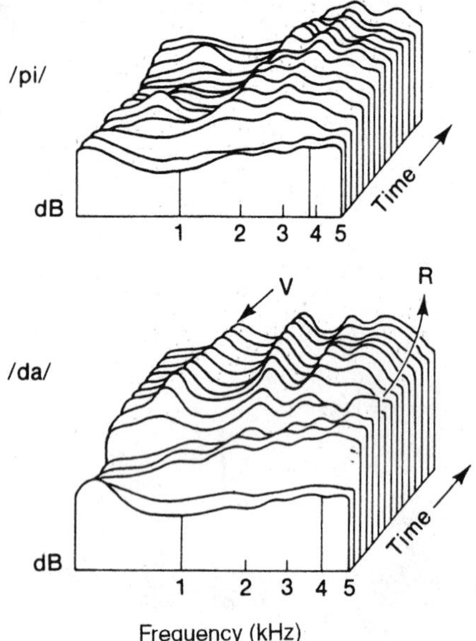

Figure 10.16
Running spectral displays for /pi/ and /da/. These displays are made up of a sequence of short-term spectra like the one in Figure 10.15. Each of these spectra is displaced 5 msec on the time axis, so that each step we move along this axis indicates the frequencies present in the next 5 msec. The low-frequency peak (V) in the /da/ display is a cue for voicing, and the rising high-frequency peak (R) is a cue for alveolar place of articulation. (From Kewley-Port & Luce, 1984.)

Perhaps one reason that running spectral displays don't provide adequate information for predicting human speech perception performance is that they make use of information at only the very beginning of the speech signal. An approach that uses both this early information and information contained later in the acoustic signal is being pursued by Harvey Sussman and his co-workers (Sussman, Hoemeke, & Ahmed, 1993; Sussman, McCaffrey, & Matthews, 1991). Sussman's approach is based on measurements of spectrograms created by a number of speakers who repeated a series of "CVC" syllables—a consonant followed by a vowel followed by a consonant. Examples of some of his stimuli for the consonant /d/ are *deat, debt, dat, doot, dote, daught, dut, dot, date,* and *dit*.

The spectrograms for two of these stimuli, shown in Figure 10.17, illustrate the classic problem of speech perception: The formant transition for the consonant (/d/in this example) varies depending on the sounds that follow it. The formant transition that corresponds to /d/ in *daught* starts at about 1,600 Hz (marked "Transition onset" in the figure) and slopes down, whereas the transition for the /d/ in *deet* starts at about 2,000 Hz and slopes up slightly. However, Sussman went beyond simply looking at these formant transitions, by measuring two things about each spectrogram: (1) the frequency where the transition to the second formant starts, which he calls the onset frequency (marked "Transition onset" in Figure 10.17) and (2) the steady frequency of the second formant (marked "Vowel").

After measuring these characteristics of the spectrogram for a number of words, Sussman plotted the frequency of transition onset versus the frequency of the vowel's second formant for each word, as shown in Figure 10.18, for words beginning with /d/ and /b/. When we look at individual data points in this figure, we can see that the consonants have a range of onset frequencies. However, when we look at the graph as a whole, we see a striking consistency: For each consonant, all of the points fall on a straight line. Based on measurements on hundreds of words, spoken by a wide range of speakers, Sussman found that each consonant creates a line of a particular slope and a particular intercept on the vertical axis (marked I in Figure 10.18). Sussman called this result **relational invariance** since what seems to be important is the *relation* between the transition frequency and the formant frequency.

What this result means is that there is information in the acoustic signal that stands for each consonant. However, by comparing the curves for /b/ and /d/, we can see that some of the data points are the same for the two consonants, so the

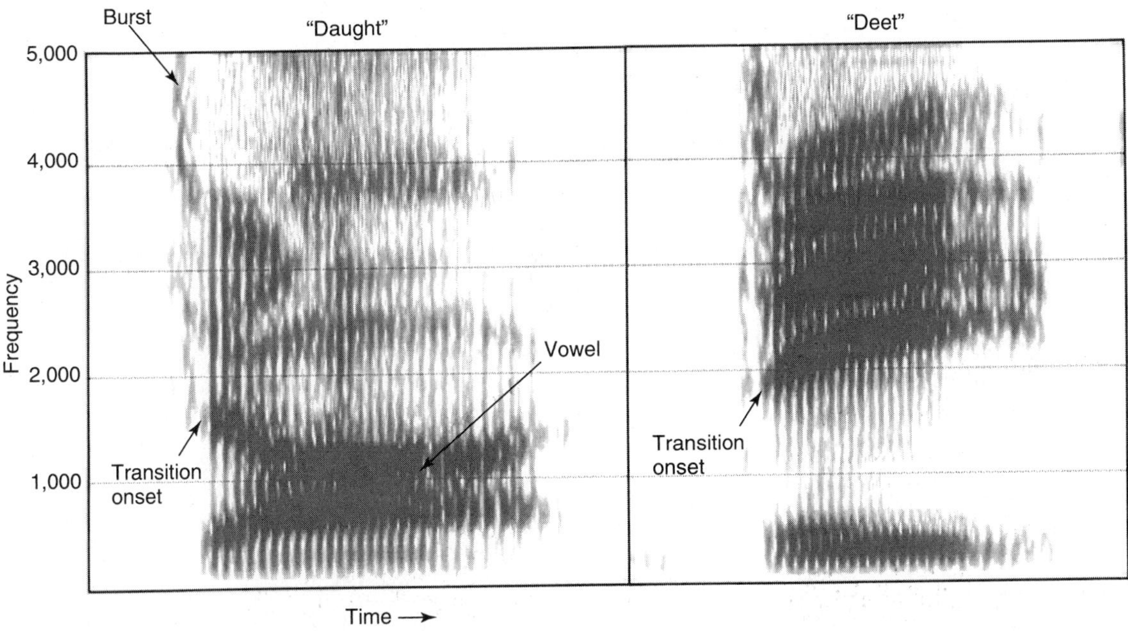

Figure 10.17
Spectrograms for (left) daught *and (right)* deet. *"Transition onset" indicates the place where the formant transition for the second formant begins. "Vowel" indicates the center of the second formant for the vowel, and "Burst" indicates the band of energy preceding the onset of the formant transition. Notice that the formant transitions for* daught *and* deet *are different. (Spectrograms courtesy of Harvey Sussman.)*

same combination of transition frequency and formant frequency could occur for both /b/ and /d/, and for other consonants as well. Sussman proposes that the auditory system deals with this problem by using the information that is contained in the early part of the acoustic spectrum marked "Burst" in Figure 10.17. Information in this burst, which differs for different consonants, combined with the information later in the signal, can, according to Sussman, be used to distinguish between two consonants with the same combination of transition frequency and formant frequency.

Our brief survey of the "arguing points" proposed by researchers in the special-speech-mechanism and the general-auditory-mechanism camps may leave you wondering which idea is correct. The answer is that the debate is still continuing, and that it has yet to be resolved. Remember that science abounds with examples of debates between two or more camps, each of which feeling that it has proposed the correct answer to a particular problem. One such example, which we discussed in Chapter 4, is the debate between followers of Young and Helmholtz's trichromatic theory of color vision and the followers of Hering's opponent-process theory of color vision that began in the 1800s and wasn't resolved until many years later. As we saw in that case, the eventual resolution of the debate was that both explanations were found to explain correctly some of the phenomena of color vision.

Perhaps a similar combination of mechanisms will eventually explain speech perception.

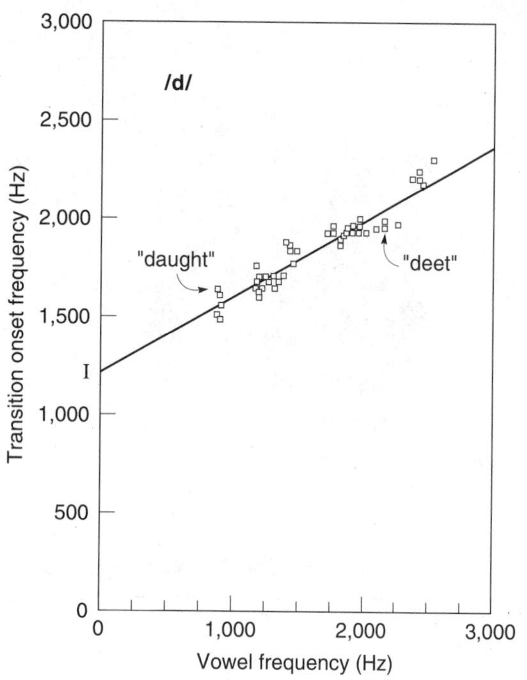

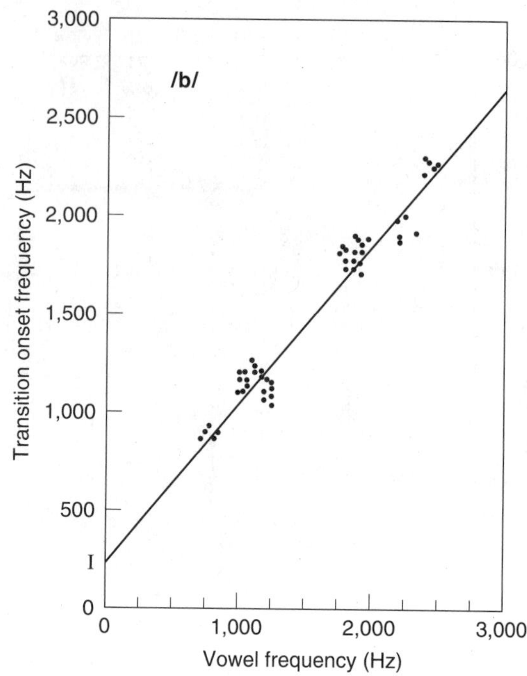

Figure 10.18

The relationship between the transition onset frequency and the frequency of the vowel for a number of words beginning with /d/ and /b/ for an individual subject. Each data point represents measurements on a single word. The points for the words daught *and* deet *from Figure 10.17 are indicated. (Data courtesy of Harvey Sussman.)*

For example, speech perception may be the result of a general auditory mechanism that works in conjunction with the same higher-level cognitive abilities that enable humans to deal not only with complex stimuli such as speech, but with other higher-order processes such as thinking and problem solving. Or perhaps the general auditory mechanism we described in Chapter 8 provides information to a special speech processor like the proposed phonetic module.

Although the "special" versus "general" debate has generated a great deal of research, many researchers are more concerned about solving specific problems posed by speech perception than about whether there is a specialized speech mechanism. Let's now consider some of these problems, beginning with research that focuses on determining how listeners take into account differences in various speakers' voices.

How Do Listeners Take Variations in Speakers' Voices into Account?

Variation in the way different people speak is one of the major sources of variability in the speech signal, and it is a problem that the perceptual system must solve so that listeners can perceive speech across different speakers. In this section, we consider a mechanism called *normalization*, which has been proposed to explain how the perceptual system solves this problem of differences between speakers. **Normalization** is the process by which the auditory system compensates for different speakers' ways of speaking and especially the differences in the frequencies of their voices (Johnson, 1990; Mullennix & Pisoni, 1990; Strange, 1989). Just as we continue to perceive the same melody when a song is transposed into different keys or octaves, this normalization mechanism causes us to continue to perceive the same sounds and meanings of speech whether a person is speaking in a high or a low register.

Kerry Green and his co-workers (1991) illustrated a possible effect of normalization by using the McGurk procedure that we described earlier (Figure 10.13). They repeated McGurk's procedure (page 430), with one modification: In step 2, when the acoustic signal was changed from /ga/ to /ba/, they also changed the gender of the speaker's voice. Thus, the listeners saw a female face making the lip movements for /ga/ paired with a male voice saying /ba/. (Half of the subjects saw a male speaker paired with a female voice.)

What does the listener hear when confronted with this situation? We know that if the gender of the voice matches the gender of the speaker, the McGurk effect occurs, so the speaker's lip movements affect the listener's perception. It is not obvious, however, that this will occur when the voice and the face do not match. Consider, for example, the **ventriloquism effect,** which occurs when a sound produced at one location (for example, from the ventriloquist) appears to come from another location (the ventriloquist's dummy). This effect also occurs in films, in which an actor's voice appears to be coming from his mouth when it is really being produced by a loudspeaker located some distance away. The result that is important for our purposes is that the ventriloquism effect becomes less convincing if there is a poor match between the auditory stimulus and the possible source of that stimulus. Thus, the ventriloquism effect works less well if the dummy's lip movements are not synchronized with the voice, or if it seems unlikely that a particular source could be making the sound that is heard (Jack & Thurlow, 1973; Jackson, 1953; Warren, Welch, & McCarthy, 1981).

Despite these results of the ventriloquism effect, Green and his co-workers found that the McGurk effect still occurred even when there was a poor match between the speaker who was seen and the voice that was presented. They interpreted this result as an effect of normalization and suggested that the reason the poor match between what the listener sees and hears does not eliminate the McGurk effect is that the speaker's voice is normalized by a mechanism that operates early in speech processing. According to this idea, early in the process of speech perception, the speaker's voice is recoded into another form, so that its pitch doesn't matter anymore and we are therefore able to process the speech signals in the same way whether it has a high pitch or a low pitch. The idea of normalization has been proposed recently, so the details of how this recoding of the speaker's voice occurs has not yet been determined.

Most of the experiments we have described in this chapter have approached speech perception by considering how listeners might process the acoustic signal. We have considered the possibility that the signal is processed by a specialized speech mechanism, or that the auditory system uses invariant acoustic cues, or that there is a mechanism that, early in the process of perception, normalizes the speaker's voice. These approaches are called *bottom-up approaches* because they approach speech perception by focusing on the acoustic signal. Remember from Chapter 5 that for perceiving visual objects, bottom-up processing involves mechanisms that focus on how the distribution of light energy on the retina is processed.

We also saw in Chapter 5 that we can also take a *top-down approach* to perceiving visual ob-

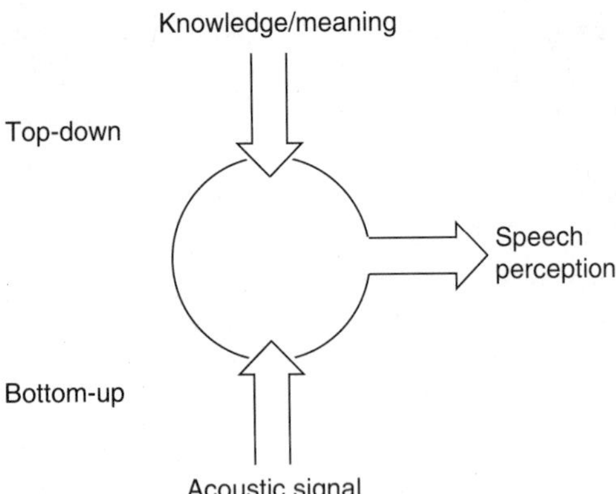

Figure 10.19
Speech perception is the result of top-down processing (based on knowledge and meaning) and bottom-up processing (based on the acoustic signal) working together.

jects by considering how things such as a person's knowledge or experience influence his or her perceptions. We can take a top-down approach to speech perception by considering how the speaker's knowledge or the meanings of words in a sentence influence perception (Figure 10.19). We will now consider some examples of how knowledge and meaning can influence speech perception.

How Do Knowledge and Meaning Influence Speech Perception?

We can demonstrate how knowledge and meaning influence speech perception in a number of ways. Let's begin by returning to the segmentation problem—the fact that, even though there are no physical breaks in words, we perceive words in conversations as if breaks were there.

Segmentation and Meaning

To help you appreciate the role of meaning in achieving segmentation, do the following demonstration.

DEMONSTRATION

Segmenting Strings of Sounds

1. Read the following words: *Anna Mary Candy Lights Since Imp Pulp Lay Things.* Now that you've read the words, what do they mean? If you think that this is a list of unconnected words beginning with the names of two women, Anna and Mary, you're right; but if you read this series of words a little faster, ignoring the spaces between the words on the page, you may hear a connected sentence that does not begin with the names Anna and Mary. (For the answer see the bottom of page 440—but don't peek until you've tried reading the words rapidly.)
2. Read the following phrase fairly rapidly to a few people: "In mud eels are, in clay none are," and ask them to write the phrase. ●

If you succeeded in creating a new sentence from the series of words in the first demonstration, you did so by changing the segmentation, and this change was achieved by your knowledge of the meaning of the sounds. When Raj Reddy (1976) asked subjects to write their perception of "In mud eels are, in clay none are," he obtained responses like "In muddies, sar, in clay nanar"; "In may deals are, en clainanar"; and "In madel sar, in claynanar." In the absence of any context, Reddy's listeners clearly had difficulty figuring out what the phrase meant and therefore forced their own interpretation on the sounds they heard. Had the listeners known that the passage was taken from a book about where amphibians live, their knowledge about possible

Figure 10.20
Archie is experiencing a problem in speech segmentation. (TM © ACP 1988 Archie Comic Publications, Inc.)

meanings of the words would have facilitated segmentation and increased their probability of decoding the sentence correctly.

Pairs of words that flow together in speech also exemplify how segmentation results from meaning: "Big girl" can be interpreted as "big Earl," and the interpretation you pick will probably depend on the overall meaning of the sentence in which these words appear. This example is similar to the familiar "I scream, you scream, we all scream for ice cream" that many people learn as children. The different segmentations of "I scream" and "ice cream" are put there by the meaning of the sentence (Figure 10.20). In the next section, we will look at another example of how meaning influences perception.

Semantics, Syntax, and Speech Perception

Some experiments have shown that words are more intelligible when heard in the context of a grammatical sentence than when presented as items in a list of unconnected words. An experiment by George Miller and Stephen Isard (1963) investigated how semantics and syntax affect the perception of speech. **Semantics** specifies whether it is appropriate to use a word in a sentence based on the word's meaning, and **syntax** specifies grammatical rules that indicate which structures of a sentence are allowed. Let's illustrate what we mean by semantics and syntax with some examples. "The boy spoke a word" is a perfectly good sentence semantically and syntactically. If, however, we change the sentence to "The boy spoke a triangle," it makes no sense. The use of the word *triangle* is semantically incorrect, because the meaning of the word *triangle* does not fit into this sentence; therefore, we call this a *semantically anomalous sentence.* Note that a semantically anomalous sentence is still syntactically correct; however, we can change the sentence further to make it syntactically incorrect by changing the order of the words. Changing the sentence to "Spoke triangle boy the a" turns it into a syntactically incorrect string of words, because the various parts of speech (nouns, verbs, etc.) are not in the correct order.

Miller and Isard constructed sentences that fell into three classes: (1) normal grammatical sentences, (2) anomalous sentences (semantically incorrect), and (3) ungrammatical strings of words (syntactically incorrect). Five examples of normal sentences they used are

Gadgets simplify work around the house.

Accidents kill motorists on the highways.

Trains carry passengers across the country.

Bears steal honey from the hive.

Hunters shoot elephants between the eyes.

Miller and Isard then changed these normal grammatical sentences into the following five semantically anomalous sentences, by taking one word from each normal sentence:

Gadgets kill passengers from the eyes.

Accidents carry honey between the house.

Trains steal elephants around the highways.

Bears shoot work on the country.

Hunters simplify motorists across the hive.

Finally, they produced five ungrammatical strings of words, which were therefore syntactically incorrect, by haphazardly rearranging the words as follows:

Around accidents country honey the shoot.

On trains have elephants the simplify.

Across bears eyes work the kill.

From hunters house motorists the carry.

Between gadgets highways passengers the steal.

Miller and Isard constructed 50 sentences of each type and recorded the resulting 150 sentences in a random order on a tape. They presented these sentences to subjects through earphones and asked them to repeat aloud what they heard as they were hearing it. This technique, which is called **shadowing,** showed that the subjects were able to exactly repeat 89 percent of the normal sentences, 79 percent of the anomalous sentences, and 56 percent of the ungrammatical strings. This result indicated that the subjects must be making use of semantic *and* syntactic information because, when semantic information was lost, as was the case in the anomalous sentences, performance dropped, and when syntactic information was lost, as was the case in the ungrammatical strings of words, performance dropped even further. In fact, these differences among the three types of stimuli became even greater when the subjects heard the stimuli in the presence of a background noise. For example, at a moderately high level of background noise, 63 percent of the normal sentences, 22 percent of the anomalous sentences, and only 3 percent of the ungrammatical strings were perceived correctly. This result is impressive evidence that listeners use both semantic and syntactic information in their perception of sentences. We will now describe some experiments that show that the meaning of a sentence can cause us to hear sounds that aren't even present.

The Phonemic Restoration Effect

Richard Warren (1970) conducted an experiment in which subjects listened to a tape recording of the sentence: "The state governors met with their respective legislatures convening in the capital city." Warren replaced the first /s/ in "legislatures" with the sound of a cough and told his subjects that they should indicate where in the sentence the cough occurred. The result of this experiment was that no subject identified the correct position of the cough, and what is even more interesting, none of the subjects noticed that the /s/ in "legislatures" was missing. This effect, which Warren called the **phonemic restoration effect,** was even experienced by students and staff in the psychology department who knew that the /s/ was missing.

Warren not only demonstrated the phonemic restoration effect but also showed that it can be influenced by the meaning of words *following* the missing phoneme. For example, the last word of the phrase "There was time to •ave . . . " (where the • indicates the presence of a cough or some other sound) could be *shave, save, wave,* or *rave,* but subjects heard the word *wave* if the remainder

Answer: An American delights in simple play things.

of the sentence had to do with saying good-bye to a departing friend.

The phonemic restoration effect was used by Arthur Samuel (1981) to show that speech perception is determined by a person's *expectations* (top-down processing) combined with the nature of the *acoustic signal* (bottom-up processing). Instead of a cough, Samuel used a "white noise"—a hiss-like sound produced by a TV set turned to a nonbroadcasting channel—to mask various phonemes (see Chapter 9 for a description of the masking of pure tones). He found that phonemic restoration is influenced by the length of the word in which the masked phoneme appears and by whether the word is meaningful. The phonemic restoration effect is more likely to occur when the phoneme appears in a long word, because subjects are able to use the additional context provided by the long word to help identify the masked phoneme. The idea that context is important is also supported by Samuel's finding that more restoration occurs for a real word like *prOgress* (where the capital letter indicates the masked phoneme) than for a similar "pseudoword" like *crOgress* (Samuel, 1990).

But speech perception also depends on the nature of the acoustic signal. Thus, Samuel found that restoration is better if the masking sound and the masked phoneme sound similar. What's happening in phonemic restoration, according to Samuel, is that we use the context to develop some expectation of what a sound will be. But before we actually perceive the sound, its presence must be *confirmed* by the presence of a sound that is similar to it. If the white noise mask contains frequencies that make it sound similar to the phoneme we are expecting, phonemic restoration occurs, and we are likely to hear the phoneme. If the mask does not sound similar, phonemic restoration is less likely to occur.

From the results of the experiments discussed above, we can conclude that speech perception depends both on the bottom-up information provided by the acoustic signal and on the top-down information provided by the meanings of words and sentences. We can appreciate the interdependence of the acoustic and contextual units of speech when we realize that, although we use the context to help us to understand the acoustic signal, the acoustic signal is the starting point for determining the context. Look at it this way: There may be enough information in my sloppy handwriting so that a person using bottom-up processing can decipher it solely on the basis of the squiggles on the page, but my handwriting is much easier to decipher if, by using top-down processing, the person takes the meanings of the words into account. Speech perception apparently works in a similar way. Although most of the information is contained in the acoustic signal, taking meaning into account makes understanding speech much easier.

The Physiology of Speech Perception

Researchers have investigated the physiology of speech perception in a number of different ways. They have used psychophysical selective adaptation experiments to determine whether specialized feature detectors may be involved in speech perception; they have recorded neural responses to both natural speech stimuli and stimuli resembling parts of the speech signal; and they have identified areas in the brain that respond to speech.

Selective Adaptation and Feature Detectors

We saw in our study of both the visual and the auditory systems that we can often draw conclusions about physiology from the results of psychophysical experiments. For example, adapting a person to a visual grating stimulus of a particular orientation dims her or his perception of gratings

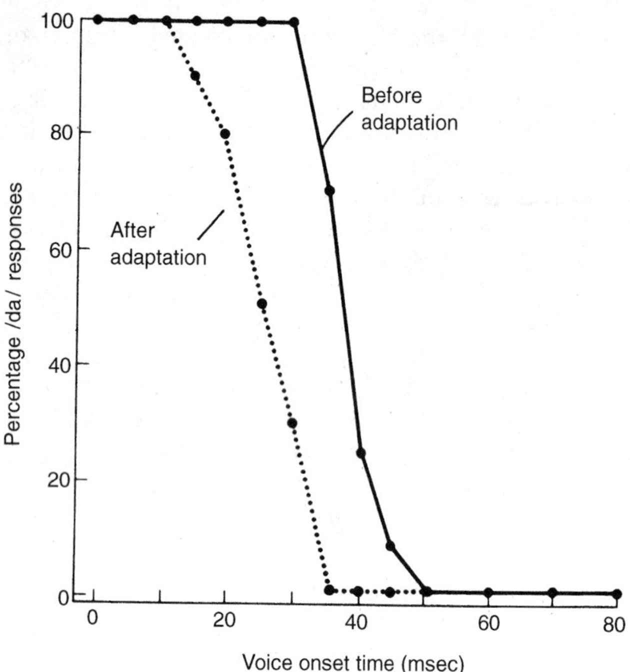

Figure 10.21
Selective adaptation to the voiced syllable /ba/ pushes the phonetic boundary to the left so that the voiced syllable /da/ is heard less. See text for details.

with that orientation but has little affect on gratings with other orientations. This dimming effect has been explained by assuming that adaptation to a particular orientation fatigues feature detectors sensitive to that orientation but does not affect feature detectors sensitive to other orientations.

Peter Eimas and John Corbit (1973) applied this procedure to speech perception by using a selective adaptation procedure with speech stimuli. Their goal was to use selective adaptation to demonstrate the existence of detectors that respond to particular phonetic features. We have already described the first part of Eimas and Corbit's experiment: They are the researchers who determined the categorical perception results shown in Figure 10.10, which are replotted as the solid curve in Figure 10.21. That curve indicates that the phonetic boundary between /da/ and /ta/ is at 35–45 msec. At VOTs shorter than 35 msec, Eimas and Corbit's subjects always perceived /da/, and at VOTs longer than 50 msec, their subjects always perceived /ta/.

Having determined the solid curve in Figure 10.21, Eimas and Corbit adapted their subjects to the voiced syllable /ba/ by repeating it for 2 minutes. They reasoned that if this repetition fatigues a phonetic detector that responds to voicing, then after adaptation subjects should be less able to perceive voiced syllables such as /da/, a result that would be indicated by a shift of the phonetic boundary to the left. This is exactly what Eimas and Corbit found. The dashed curve in Figure 10.21 indicates that, when they asked their adapted subjects to identify what they heard at different VOTs, the phonetic boundary had shifted so that the VOT now had to be less than 10 msec in order for the subject to hear /da/on every trial.

Results such as this seemed to indicate that there are detectors that respond to phonetic features such as voicing. However, later experiments led to a reinterpretation of Eimas and Corbit's selective adaptation results. One of these experiments, by Roberts and Quentin

Summerfield (1981), made use of the McGurk effect described earlier. They first determined a curve like the one in Figure 10.21 with stimuli that caused subjects to perceive /bɛ/ (as in *bed*) on one side of the phonetic boundary and /dɛ/ (as in *dead*) on the other side. Then, they adapted their subjects with the acoustical signal for /bɛ/, while their subjects watched a TV monitor of a person making the lip movements for /gɛ/ (as in *get*). Remember that in the McGurk effect the visual stimulus influences what the listener hears. In this case, seeing the lip movements for /gɛ/ caused the listeners to hear /dɛ/. If detectors that respond to phonetic features are responsible for speech perception, then the detector or detectors for the phonetic features of /d/ should be firing when the subject hears /dɛ/. If this detector is firing during adaptation, the phonetic boundary for /dɛ/ should shift so that /dɛ/ would be heard less following the adaptation and /bɛ/ would be heard more.

This result did not, however, occur. The phonetic boundary did shift, but in the opposite direction from the one predicted, so that adaptation *increased* the perception of /dɛ/ and decreased the perception of /bɛ/. This result, which has been repeated in other experiments (Saldaña and Rosenblum, 1994; Sawusch & Jusczyk, 1981), means that it is unlikely that adaptation fatigues a detector specialized for hearing /d/. What appears to be adapting in these experiments is not a detector associated with what people *hear*, but neurons associated with the *acoustic signal* to which they are being adapted. Thus, when subjects in Roberts and Summerfield's experiment were adapted with the acoustic signal usually associated with /bɛ/, they became less sensitive to /bɛ/ even though they never heard the sound /bɛ/ during the adaptation. Apparently, adaptation shifts the phonetic boundary not by fatiguing a specialized speech detector, but by fatiguing neurons responsible for hearing sounds in general, which are sensitive to the frequencies contained in the adapting stimulus.

Neural Responses to Speech and Complex Sounds

Most of the research explaining the neural response to speech stimuli has focused on how neurons in the auditory nerve respond to speech sounds. For example, Figure 10.22 shows the short-term spectrum for the sound /da/ and the firing pattern for a representative population of cat auditory nerve fibers with low, medium, and high characteristic frequencies. (Remember from Chapter 8 that the neuron's characteristic frequency is the frequency to which this neuron responds best.) The match between the speech stimulus and the firing of a number of neurons in the auditory nerve indicates that information in

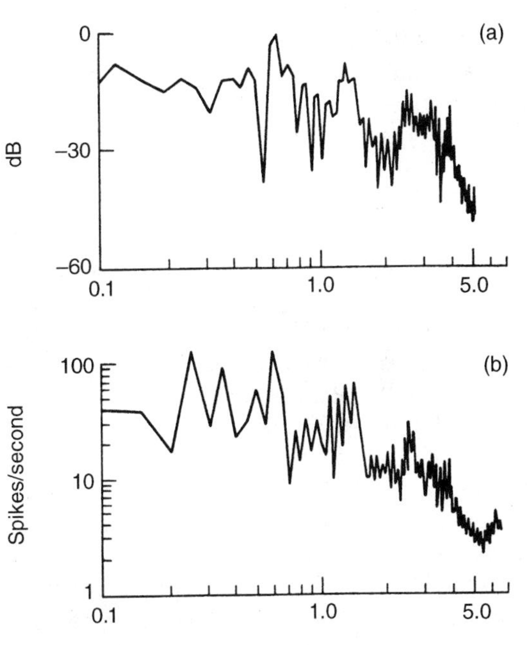

Figure 10.22

(a) Short-term spectrum for /da/. This curve indicates the energy distribution in /da/ between 20 and 40 msec after the beginning of the signal. (b) Nerve firing of a population of cat auditory nerve fibers to the same stimulus. (From Sachs, Young, & Miller, 1981.)

the speech stimulus is represented by the pattern of firing of auditory nerve fibers (also see Delgutte & Kiang, 1984a, 1984b).

Another approach to studying the physiology of speech perception is to look for neurons that respond to parts of the speech stimulus, such as the formant transitions and formants shown in Figure 10.17. Neurons have been found that fire to stimuli like these. For example, we saw in Chapter 8 that Whitfield and Evans (1965) found "frequency sweep detectors" in the cat's cortex that respond when frequencies sweep up or down, as occurs for formant transitions. In addition, neurons have been found in a number of animals, including bats, birds, frogs, and cats, that respond to combinations of tones with specific timing and frequencies (Fuzessery & Feng, 1983; Margoliash, 1983; Nelson, Erulkar, & Bryan, 1966; Olsen & Suga, 1991a, 1991b). If such neurons exist in humans, they would be ideally suited to detecting stimuli like the formant transitions and their formants found in the acoustic signal for speech.

Lateralization of Function

It has been known for over 100 years that some functions are *lateralized* in the brain: They are processed more strongly in either the left or the right hemisphere. For most people, speech is lateralized in specific areas in the left hemisphere of the brain (Figure 10.23). Damage to Broca's area, in the frontal lobe, causes difficulty in speaking, and damage to Wernicke's area, in the temporal lobe, causes difficulty in understanding speech (Geschwind, 1979).

Research using the PET (positron emission tomography) scan on humans, which measures changes in blood flow in the brain during the perception of pitches and the perception of speech stimuli, shows that pitch stimuli activate areas in the right hemisphere, but that speech stimuli activate areas in the left hemisphere (Zatorre et al., 1992). In addition to physiological evidence of the lateralization of speech perception, psychophysical experiments indicate that speech stimuli are more easily processed when presented through earphones to the right ear than when they are presented through earphones to the left ear. Since signals from the right ear are sent preferentially to the left hemisphere, this right-ear preference indicates that speech stimuli are processed in the left hemisphere (Kimura, 1961).

Speech Perception: Complex Phenomena— General Theories

Now that we have come to the end of the chapter, we can better appreciate the statement we made at the beginning that speech perception is complex. Speech perception is complex in a number of ways, including the pattern of frequencies in the acoustic signal, the variability of the acous-

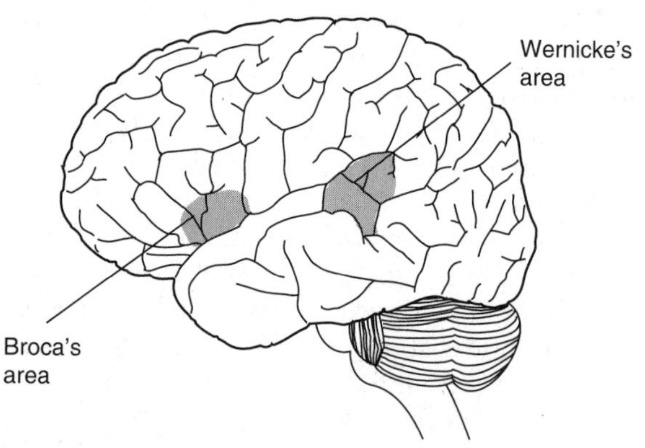

Figure 10.23
Broca's and Wernicke's areas, which are specialized for language production and comprehension, are located in the left hemisphere of the brain in most people.

tic signal, and the complexity of the mechanisms responsible for transforming the acoustic signal into our perception of speech. It is because of this complexity that the mechanisms that have been proposed to explain speech perception are general and lack specifics. Thus, when researchers say that speech perception is determined by a specialized phonetic module, or by a general auditory mechanism, they are specifying a general mechanism but not how that mechanism operates. Although it would be nice if we had more information so that we could fill in the details and also perhaps choose between alternative explanations, or even combine them, it is important to remember that one reason we don't know these details is that the study of speech perception is still in its infancy.

Although the study of speech perception may be in its infancy compared to research in areas such as color perception, depth perception, or pitch perception, we should not forget that we are just beginning to understand the mechanisms responsible for our perception of stimuli that rival the complexity of speech stimuli in both vision and audition. Consider, for example, what is involved in perceiving a visual scene as familiar as the street where you live. Although we have accumulated a great deal of knowledge about how the eye works and how the visual system responds to simple stimuli such as lines, corners, colors, and simple movements, we are far from being able to explain something as complex as even a simple visual scene. Similarly, we know a lot about how the ear processes simple auditory stimuli, but we are just beginning to understand the complexities involved in perceiving complex sounds, whether they be speech sounds or other sounds in our environment.

Perception of "Calls" by Monkeys

Many animals emit sounds to communicate with other members of their species. Deer and prairie dogs emit snorts and high-pitched screeches, respectively, to warn of possible danger. Birds communicate with distinctive songs, and elephants emit low-frequency sounds that are below the range of human hearing, but that can be heard by potential mates many miles away.

The Japanese macaque monkey (Figure 10.24) has been used as a subject in numerous experiments on perceiving vocalizations, because it is a species that is closely related to humans, in terms of both evolution and shared perceptual abilities. In addition, the Japanese macaque's communication signals are graded like humans'; that is, the signal associated with a particular type of vocalization can vary over a range, rather than being strictly limited to one narrow version, as are many bird songs. Our goal in looking at research on the monkey's perception of vocalizations is twofold: (1) to better understand the nature of the monkey's communication system and (2) to provide insights into the mechanisms responsible for human speech perception.

The first question to ask about an animal's communication is: What signals do they produce and what do these signals mean? In 1975, S. Green determined the characteristics of Japanese macaque communication in a classic field study in which he recorded 60,000 vocalizations over 14 months. Green observed a close relationship between the acoustic structure of monkey vocalizations, or "calls," and the circumstances under which the calls occurred. This relationship is shown in Figure 10.25 for one particular type of call, the "coo," which in the Japanese macaque exists in seven basic forms. Coos are an affiliative call used to establish contact. Two of these calls in particular, the smooth early high (SE) and the smooth late high (SL), are the ones that have been studied most frequently. SE calls have a high-frequency peak that occurs during the first two thirds of the vocalization; SL calls have a high-frequency peak that occurs during the last third of the vocalization (see Figure 10.25). The two calls occur in very different social situations. SE calls occur when young monkeys

Figure 10.24
Japanese macaque monkey.

	Type of coo vocalization						
	Low			High			
				Early		*Late*	
Situation	Double	Long	Short	Smooth	Dip	Dip	Smooth
Separated male	xxxx xxxxxxx xxxxxxx		xx	xx		xx	xx
Female minus infant	xxxx	xxx					
Nonconsorting female	xx	xxxxxx	x				
Female at young		xxx xxxxxxx	x				
Dominant at subordinate			xxx xxxxxxx			xx	
Young alone				x xxxxxxx xxxxxxx	xxxxxxx	x	
Dispersal				x xxxxxxx	xx	xxxxx	xx
Young to mother				xxxx	xxx xxxxxxx	xxxx	x xxxxxxx
Subordinate to dominant				xxx	xxxx xxxxxxx	xxx xxxxxxx xxxxxxx xxxxxxx	xxxxx
Estrus female					x	xxxx xxxxxxx xxxxxxx	xxxxx xxxxxxx xxxxxxx xxxxxxx

Figure 10.25

This chart shows when Japanese macaque monkeys made "coo" sounds in different situations. The seven different types of coos are indicated along the top, and the situation is indicated along the left side. Each x represents an observation of a particular sound. Notice that only a few sounds are associated with each situation. The outlined squares indicate the smooth early and smooth late sounds, which have been the subject of many experiments. (From Green, 1975.)

become separated and call to establish contact; SL calls occur in adult females during the early stages of courtship.

The SE and SL calls have been widely used in speech research both because they occur in different social situations and because, by changing the position of the high-frequency peak, researchers can create a series of signals that range between the SE and SL calls. With these stimuli in hand, researchers have asked a number of questions similar to those asked about human speech perception. Brad May, David Moody, and William Stebbins (1988, 1989) asked two questions: (1) What characteristic of the call is used by the monkey to identify the call? (2) Do these monkeys experience categorical perception?

May and his co-workers determined that monkeys use the position of the high-frequency peak to

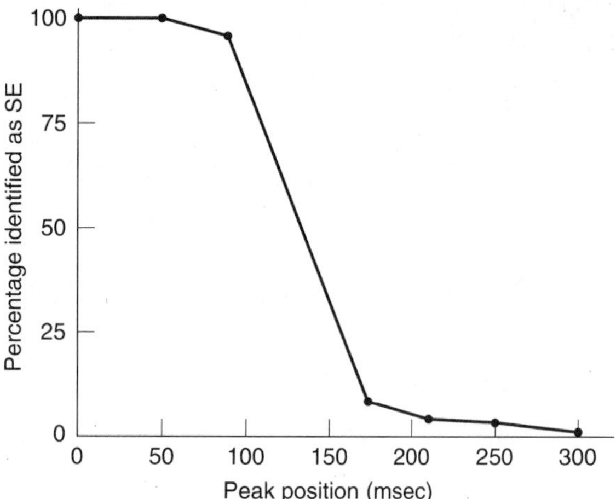

Figure 10.26

The results of a categorical perception experiment on a Japanese macaque monkey. See text for details. (Data from May et al., 1988.)

distinguish between SE and SL calls. They then did a categorical perception experiment in which they varied the position of the high-frequency peak and used a psychophysical technique to determine whether the monkeys perceived the SE or the SL call for each stimulus. The result was a demonstration of categorical perception (Figure 10.26). At peak positions below 125 msec, the monkeys classified most of the signals as SE, and at peak positions above 125 msec, the monkeys classified most of the signals as SL. The key finding is the steep border between the two perceptions, a result similar to the phonetic boundary found in human categorical perception experiments.

This demonstration of categorical perception means that these monkeys experience perceptual constancy. They perceive all signals on the same side of the phonetic boundary as being in the same category, even when they are faced with a large variety of different sounds. This perceptual constancy serves an important function because within a monkey troop different monkeys produce a great number of versions of the same call (Petersen, 1982). Constancy makes it possible for monkeys to understand the wide variety of signals they hear as meaning the same thing.

Japanese macaque monkeys have another similarity to humans: The perception of vocalizations appears to be localized in the left hemisphere. This localization has been demonstrated by the fact that the monkeys are better at discriminating between two coos when they are presented to the right ear (Beecher et al., 1979; Petersen, 1982). Remember from our description of human studies that a right-ear advantage indicates strong left-hemisphere involvement, since signals from the right ear travel preferentially to the left hemisphere.

Interestingly, when another species of monkeys was trained to make the same discrimination, they were able to make the discrimination but did not show a right-ear preference (Petersen et al., 1984). The reason? They were being trained to discriminate Japanese macaque coos, which were not a part of their own vocal repertoire. Apparently, lateralization occurred in the Japanese macaques because the sounds were part of their "language," whereas it didn't occur in the other monkeys because, to them, the coos were simply meaningless sounds.

Do the similarities we have described between humans and monkeys mean that monkeys and humans perceive speech sounds in the same way? Not necessarily. Even though monkeys and humans perform similarly on categorical perception tasks and process vocalizations in the left hemisphere, they do not perform identically on all speech perception tasks. For example, in psychophysical experiments on humans, Patricia Kuhl (1991) showed that humans perceive a certain version of a vowel to be a **prototype**—a particularly good example of that vowel. We can understand what this means by describing re-

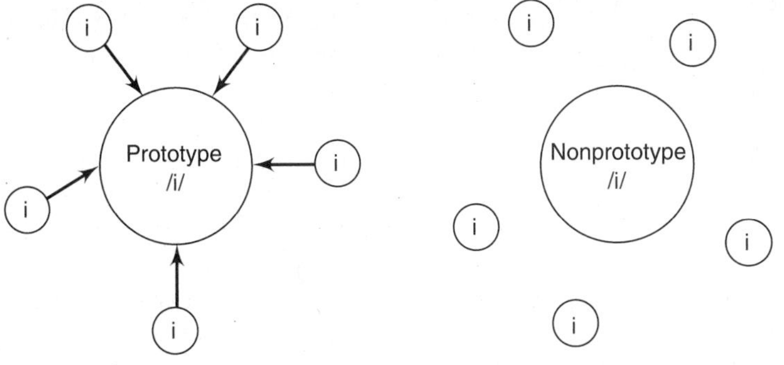

Figure 10.27
When subjects were asked by Kuhl (1991) to make similarity judgments between one example of the vowel /i/ and other examples of the vowel, they tended to judge the various examples of /i/ to be more similar to a prototype /i/ (one judged to be a "good" example), as in (a) than to a nonprototype /i/, as in (b). This is called the perceptual magnet effect.

search in cognitive psychology in which people were asked to rate how good an example various fruits were of the category "fruit." Oranges and apples were considered the best examples of fruits and were therefore the prototypes of the fruit category. Fruits such as olives and coconuts were, however, considered poor examples of fruits (Mervis & Rosch, 1981).

Kuhl showed that a similar situation exists in human speech perception: People were able to pick a particular version of the vowel /i/ as the prototype /i/—the best example of that vowel. After determining the identity of the prototype, Kuhl demonstrated what she called a **perceptual magnet effect.** People judged all of the sounds they classified as /i/ to be more similar to the prototype than to an /i/ that was a nonprototype (Figure 10.27). It was as if the prototype "attracted" the other sounds, so that they were judged to be similar to the prototype, whereas the nonprototype didn't have this effect.

What does all this have to do with monkey speech perception? After Kuhl demonstrated the perceptual magnet effect for humans, she tested monkeys under the same conditions and found that the monkeys did not exhibit this effect. Perhaps, Kuhl suggests, effects like categorical perception, which are found in humans, monkeys, and other animals, are caused by the same general auditory mechanisms, but the existence of speech prototypes and the perceptual magnet effect may represent a higher level of processing that is special to humans.

Infant Speech Perception

Do infants have the capacity to process the complex patterns of the speech stimulus? Peter Eimas and co-workers (1971) opened the modern era of research on infant speech perception with a paper that provided the first indication that the answer to this question might be yes. They accomplished this by showing that infants as young as 1 month old perform similarly to adults in categorical perception experiments. Eimas et al. used the fact that an infant will suck on a nipple in order to hear a series of brief speech sounds. If, however, the same speech sounds are repeated over and over, the infant eventually habituates to these sounds and decreases his or her rate of sucking. By presenting a new stimulus after the rate of sucking has decreased, we can determine whether the infant perceives the new stimulus as sounding the same as or different from the old one. If the new stimulus sounds different, dishabituation occurs, and the rate of sucking increases. If the new stimulus sounds the same, habituation continues, and the rate of sucking either stays the same or decreases further.

The results of Eimas et al.'s experiment are shown in Figure 10.28. The number of sucking responses when no sound was presented is indicated by the point at B. When a sound with VOT = +20 msec (sounds like /ba/ to an adult) is presented when the infant sucks, the sucking increases to a high level and then begins to decrease. When the VOT is changed to +40 msec (sounds like /pa/ to an adult), sucking increases, as indicated by the points to the right of the

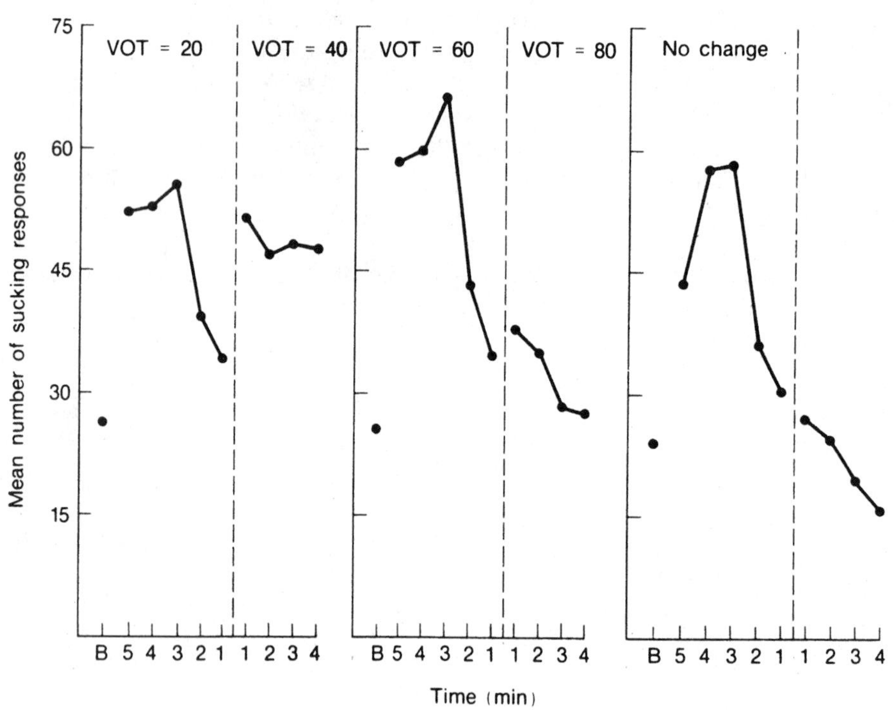

Figure 10.28
Results of a categorical perception experiment on infants, using the habituation procedure. In the left panel, VOT is changed from 20 to 40 msec (across the phonetic boundary). In the center panel, VOT is changed from 60 to 80 msec (not across the phonetic boundary). In the right panel, the VOT was not changed. See text for details.

dashed line. This result means that the infant perceives a difference between sounds with VOTs of +20 and +40 msec. The center graph, however, shows that changing the VOT from +60 to +80 msec (both sound like /pa/ to an adult) has only a small effect on sucking, indicating that the infants perceive little, if any, difference between the two sounds. Finally, the results for a control group (the right graph) show that, when the sound is not changed, the number of sucking responses decreases throughout the experiment.

These results show that, when the VOT is shifted across the phonetic boundary (left graph), the infants perceive a change in the sound, and when the VOT is shifted on the same side of the phonetic boundary (center graph), the infants perceive little or no change in the sound. That infants as young as 1 month old are capable of categorical perception is particularly impressive, especially since these infants have had virtually no experience in producing speech sounds and only limited experience in hearing them.

In another experiment, Eimas and Joanne Miller (1992) extended their finding of early infant speech perception abilities to include the integration of speech signals between the two ears that occurs in duplex perception in adults. In adults, duplex perception occurs when the two stimuli depicted in Figure 10.14 are presented to the left and right ears. Adults receiving these stimuli can combine the third formant transition (Figure 10.14b) and the base (Figure 10.14c) to perceive the sound /da/ associated with the combination of the two (Figure 10.14a). By presenting these stimuli to infants and using a habituation procedure similar to the one used to demonstrate categorical perception in infants, Eimas and Miller were able to show that 3- to 4-month-old infants can perceptually combine the stimuli presented to their left and right ears to create the perception of a speech sound. Eimas and Miller interpret this result as showing that infants, like adults, possess a specialized speech mechanism that combines the sounds presented to the two ears.

Another parallel between infant and adult speech perception is Patricia Kuhl's (1983, 1989) finding that infants have an ability called **equivalence classification**—the ability to classify vowel sounds as belonging to a class, even if the speakers are different. Adults have this ability, so they can classify an /a/ sound as the vowel *a* whether it is spoken by a male or a female. By using a conditioning procedure (Figure 10.29), Patricia Kuhl (1983, 1989) showed that 6-month-old infants are capable of equivalence classification. The infant sits on a parent's lap and hears a speech sound such as /a/ repeated over and over. At some point, the sound changes from /a/ to /i/, and when this change happens, a bear playing a drum is activated, and the child looks toward the bear. After this procedure is repeated a number of times, the child learns to look toward the bear anytime the sound changes from /a/ to /i/. Thus, when the child looks at the bear, the experimenter knows that the child has perceived a change.

Kuhl asked, "Does the infant perceive all /a/sounds as belonging to the same category, even when other characteristics of the sound are changed?" She answered this question by presenting a male voice saying /a/ and then changing it to a female voice saying /a/. Even though the voice changed, no head turn occurred, a result indicating that the child perceived the /a/ sounds as the same. However, when Kuhl changed the stimulus from a male voice saying /a/, to either a male's voice, a female's voice, or a child's voice saying /i/, head turning did occur, a result indicating that the child reacted in the same way to the /i/ stimulus no matter which voice was saying it.

From these results, Kuhl concluded that infants do achieve classification: They classify all

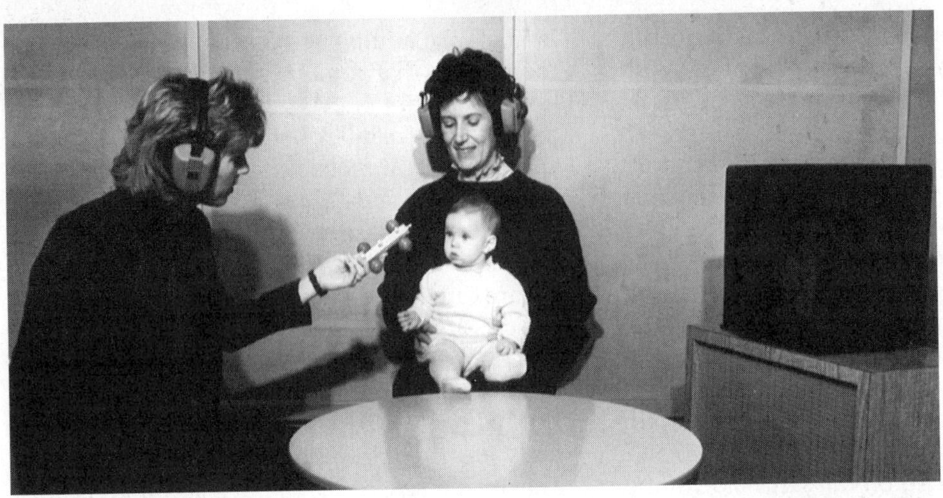

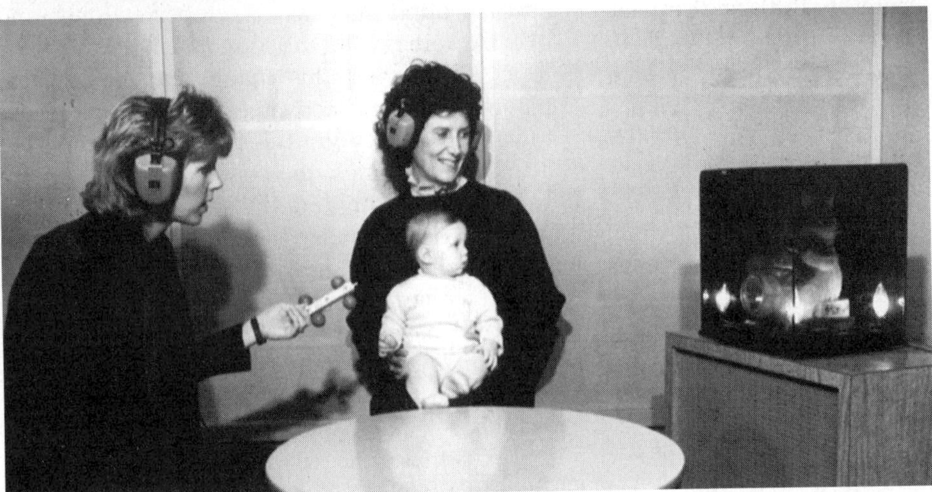

Figure 10.29

In the conditioning procedure used by Kuhl (1983, 1989), infants are trained to turn their heads toward the loudspeaker on their left when they hear a change from one speech category to another.

/a/ sounds as the same and all /i/ sounds as the same, no matter what the quality of the voice saying them (also see Marean, Werner, & Kuhl, 1992, for similar results for 2-month-olds). Equivalence classification is important to infants because, when they eventually begin imitating adult speech, their imitations will be much higher pitched than the speech sounds made by most of the adults in their environment. They need to be able to tell that their production is equivalent to the adults' even though the pitch is different.

The results we have been describing, which indicate that mechanisms for speech perception are in place at a very early age, do not mean that speech perception does not develop as a child gets older. One indication that development occurs is provided by experiments that show that infants lose the ability to distinguish between some pairs of sounds as they get older. For example, there are some speech stimuli that infants can distinguish from each other, but that adults cannot distinguish. Thus, a Japanese infant might respond differently to /r/and /l/, two sounds that Japanese adults have difficulty distinguishing, since they are in the same category in the Japanese language. However, by the time the children are 6 months old, they are no longer able to make distinctions between all pairs of sounds. Their experience in listening to other people talk has changed their speech perception so that they are sensitive only to distinctions between sounds that are important in their native language (Kuhl et al., 1992; Werker & Tees, 1984). Apparently, infants possess mechanisms for perceiving speech fairly early in their development, and these mechanisms become tuned to the specific language that the child experiences during the first six months of life.

Review

The Speech Stimulus

Outline

- One reason speech perception is challenging is that it has been difficult to discover a straightforward relationship between the sound stimulus and the perception of speech.
- The speech stimulus can be specified in terms of how sounds affect the meaning of a word, the way sounds are produced, and the pattern of sound energy.

Questions

1. What is a phoneme? Why are there more phonemes than letters in the alphabet? (418)
2. What is a phonetic feature? Describe each of the three types of phonetic features. How can they be used to distinguish between different phonemes? (418)
3. What is the acoustic signal? How is it displayed on the spectrogram? What are acoustic cues? What are formants? (421)

The Challenge of Understanding Speech Perception

- One way to explain speech perception is to say that the auditory system identifies each phoneme based on the pressure changes associated with each one, and that once the phonemes are identified, speech perception occurs. This explanation is incorrect because of the variability problem: A specific phoneme is not always associated with the same acoustic cues.

4. Describe how context affects acoustic cues. What is coarticulation? What is perceptual constancy in speech? (422)
5. Describe how "real-world" speech contributes to the variability problem. What is the segmentation problem? (423)

Is There a Special Mechanism for Speech Perception?

- There are two competing explanations of speech perception: (1) the special speech mechanism, which states that speech perception is determined by a mechanism designed specifically to process speech signals, and (2) the general auditory mechanism, which states that speech perception is determined by the same auditory mechanisms that are responsible for the perception of pitch and other qualities of pure tones and nonspeech sounds.

6. What are some reasons that the idea of a special speech mechanism has some appeal? (426)
7. What is the basic rationale behind the motor theory of speech perception? What is the phonetic module, and what is it supposed to do? (426)

- Discovery of a phenomenon called *categorical perception* provided early evidence cited by some researchers in favor of the special speech mechanism.

8. What is categorical perception, and how is it measured? Be sure you understand both the categorization and the discrimination parts of the categorical perception experiment. How does categorical perception help explain how the auditory system deals with stimulus variability? (427)
9. How was categorical perception used to support the special speech mechanism? What objections have been raised to this idea? (428)

- Although categorical perception did not live up to its initial promise as evidence supporting the special speech mechanism, two other phenomena, the McGurk effect and duplex perception, have also been proposed as evidence supporting the special speech mechanism.

10. Describe how to achieve the McGurk effect. How have these results been used to support the special speech mechanism? What counterevidence has been provided to argue against the idea that these results support the special speech mechanism? (430)
11. Describe duplex perception. How is it achieved, and why has it been interpreted as supporting the special speech mechanism? What evidence argues against the idea that duplex perception is evidence of a special speech mechanism? (431)

Is There Information in the Acoustic Signal That Indicates Which Sounds Are Present?

- Researchers who support the idea of the general auditory mechanism have attempted to find invariant acoustic cues, which are characteristics of the acoustic signal that remain constant despite the variability of speech.

12. Describe attempts to find invariant acoustic cues. What is a short-term spectrum? A running spectral display? Describe Sussman's approach. What is relational invariance? (432)
13. What is the current status of the debate over a special speech mechanism versus general auditory mechanism? (435)

How Do Listeners Take Variations in Speakers' Voices into Account?

- Differences between speakers are one of the major sources of variability in the speech signal that the perceptual system must solve so that listeners can perceive speech across different speakers.

14. What is normalization of the speech signal? Describe Green's experiment and its relationship to normalization. (437)

How Do Knowledge and Meaning Influence Speech Perception?

- Speech perception is affected not only by bottom-up processing, which depends only on the physical energy in the speech signal, but also by top-down influences, which make use of a listener's knowledge and the meaning of the speech signal.

15. Describe how meaning helps achieve segmentation. How do semantics and syntax affect speech perception? (438)

16. What is the phonemic restoration effect? What is the evidence that it is determined by both top-down and bottom-up processing? (440)

The Physiology of Speech Perception

- The physiological basis of speech perception has been studied by psychophysical selective adaptation experiments, by measuring neural response to speech and complex auditory stimuli, and by determining areas in the brain that respond to speech.

17. Describe Eimas and Corbit's selective adaptation experiment. How did researchers initially interpret this experiment? What later experiment caused a revision of that interpretation? In the final analysis, do the selective adaptation experiments support the idea that speech perception is served by phonetic feature detectors or by general auditory neurons? (441)

18. What relationship has been demonstrated between auditory nerve responses and the speech signal? Describe neurons that respond to complex stimuli that resemble speech stimuli. What is the evidence that speech is lateralized in the brain? (443)

Speech Perception: Complex Phenomena—General Theories

- Because of the great complexity of speech perception, most of the mechanisms that have been proposed to explain speech perception are general and lack specifics.

19. Compare our knowledge of the mechanisms of speech perception to our knowledge of other auditory and visual mechanisms. (444)

Perception of "Calls" by Monkeys

- Many animals emit sounds to communicate with other members of their species. One of the animals that has been the subject of a great deal of experimentation is the Japanese macaque monkey.

20. Describe the signals produced by the Japanese macaque and what they mean. What characteristic of SE and SL calls are used by monkeys to identify the call? Describe the results of categorical perception experiments on these monkeys. (446)

21. Describe research on brain lateralization of speech sounds in the Japanese macaque. Describe experiments on speech prototypes that have been done on humans. How do the results differ in monkeys? (447)

Infant Speech Perception

- The modern era of infant speech perception research began with Eimas and co-workers' demonstration of categorical perception in infants.

22. Describe Eimas et al.'s categorical perception experiment. How have the results been interpreted? What is equivalence classification, and how does it apply to adults and infants? What changes occur in infant speech production and perception as a function of age? (450)

The Somatic Senses

Anatomy of the Somatosensory System

The Psychophysics and Physiology of Tactile Perception

Neural Responses to Temperature and Potentially Damaging Stimuli

Neural Processing

Active Touch

Central Influences on Pain Perception

The **somatic senses** include **cutaneous sensations,** sensations based on the stimulation of receptors in the skin; **proprioception,** the sense of the position of the limbs; and **kinesthesis,** the sense of movement of the limbs. In this chapter we will survey these senses by describing the **somatosensory system,** which includes the receptors for the somatic senses and the pathways that transmit signals generated by these receptors up the spinal cord toward the brain. Our main focus will be on the following cutaneous sensations: (1) tactile perception, the term we will use to refer to all perceptions, except pain, that are caused by mechanical displacement of the skin; (2) the perception of temperature, caused by heating or cooling of the skin; and (3) the perception of pain, caused by stimuli that are potentially damaging to the skin.

We often take our cutaneous sensations for granted. But imagine how your ability to write might change if your hand were anesthetized. Would you know how firmly to grasp your pen if you had no feeling in your hand? Consider, also, all of the other things you do with your hands. How would losing feeling in your hand affect your ability to do these things? We know that a complete loss of our ability to feel with the skin is dangerous, as demonstrated by people who, because of this problem, suffer constant bruises, burns, and broken bones in the absence of the warnings provided by touch and pain (Melzack & Wall, 1983; Rollman, 1991; Wall & Melzack, 1994). And consider for a moment what sex would be like without the sense of touch. Or perhaps a better way to put this is to ask if sex without the sense of touch is something people would care about at all.

When we recognize that the perceptions we experience through our skin are crucial for protecting ourselves from injury and for motivating sexual activity, we can see that these perceptions are crucial to our survival, and to the survival of our species. We could, in fact, make a good case for the idea that perceptions felt through the skin are as important for survival as those provided by vision and hearing.

One of the purposes of this chapter is to describe some of the special properties of the skin and its receptors. But we are also interested in cutaneous sensations for another reason: Studying the perceptions sensed through the skin provides an opportunity to demonstrate a number of basic sensory principles that are also important in vision, audition, and other senses. Studying the cutaneous sensations therefore enables us to illustrate a number of similarities and parallels between the operation of the different senses.

Anatomy of the Somatosensory System

We look at the anatomy of the somatosensory system by first focusing on the periphery: the skin and its receptors. When we do this, we notice an important parallel with the visual system: Both systems have a number of different kinds of receptors.

The Skin and Its Receptors

Comel (1953) called the skin the "monumental facade of the human body" for good reason. It is the heaviest organ in the human body, and if not the largest (the surface areas of the gastrointestinal tract or of the alveoli of the lungs exceed the surface area of the skin), it is certainly the most obvious, especially in humans, whose skin is not obscured by fur or large amounts of hair (Montagna & Parakkal, 1974).

But the real reason that skin is important is the functions it serves. It prevents body fluids from escaping and at the same time protects us by keeping bacteria, chemical agents, and dirt from penetrating our bodies. And as skin maintains the integrity of what's inside and protects us from what's outside, it also provides us with information about the various stimuli that contact it. The sun's rays heat our skin and we feel *warmth;* a pinprick is *painful,* and when someone touches us, we experience sensations such as *touch* or *pressure.*

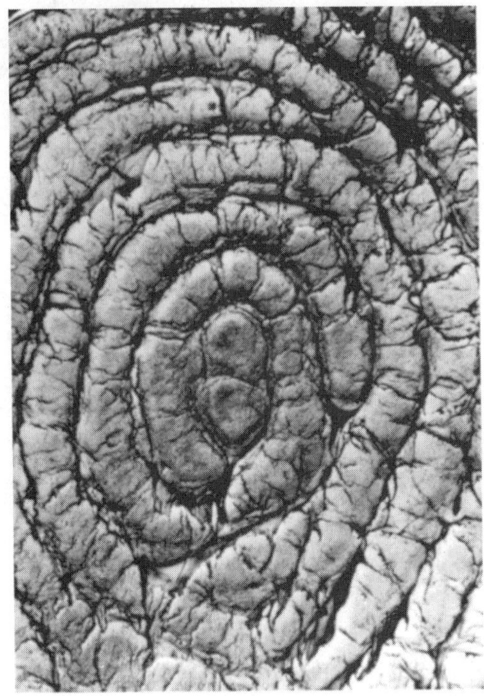

Figure 11.1
A scanning electron micrograph of the glabrous (hairless) skin on the tip of the finger, magnified 27 times. (From Shih & Kessell, 1982.)

The staging ground for the creation of these experiences is the outer layer of the skin, the **epidermis,** which is actually several layers of cells, the outermost one (the surface of the skin) being tough dead skin cells. (Stick a piece of cellophane tape onto your hand and pull it off. The material that sticks to the tape is dead skin cells.) Hairy skin covers most of the body, but hairless (glabrous) skin is found on the fingers and palms of the hands, and on the toes and soles of the feet (Figure 11.1).

One of our main concerns here is with the receptors that are located in the epidermis and the layer directly under the epidermis, the **dermis** (Figures 11.2 and 11.3). We are interested in the following four receptor structures because, as we will see below, each responds best to specific kinds of stimulation and is associated with specific perceptions.

- **Merkel receptor**—a disk-shaped receptor located near the border between the epidermis and the dermis.
- **Meissner corpuscle**—a stack of flattened cells located in the dermis just below the epidermis; a nerve fiber wends its way through these cells.
- **Ruffini cylinder**—many-branched fibers inside a roughly cylindrical capsule.
- **Pacinian corpuscle**—a layered, onionlike capsule that surrounds a nerve fiber; located deep in the skin, the Pacinian corpuscle can also be found in many other places, including the intestines and the joints.

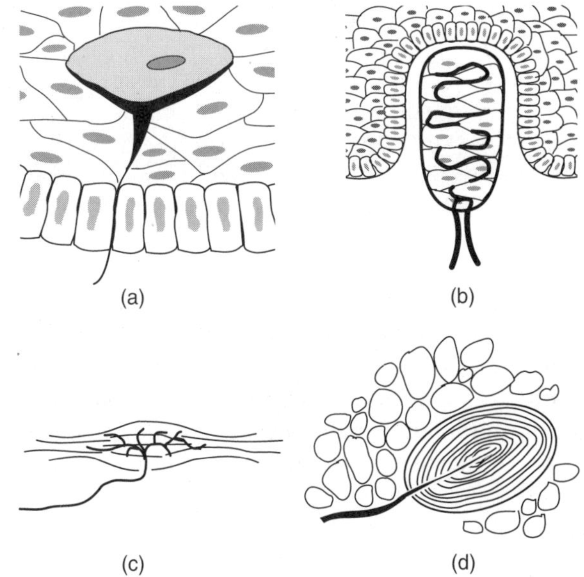

Figure 11.3
The four major receptors for tactile perception: (a) Merkel receptor; (b) Meissner corpuscle; (c) Ruffini cylinder; and (d) Pacinian corpuscle.

As we will see, the differences in these receptors—their physical properties as well as their location in the skin and the sizes of their receptive fields—cause the fibers associated with them to respond best to different types of stimuli and therefore to result in different cutaneous perceptions.

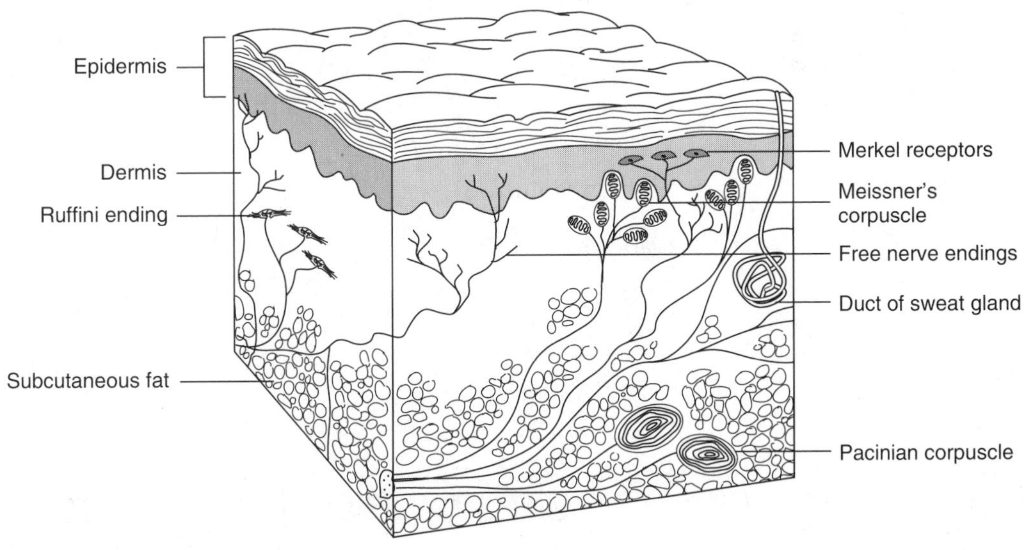

Figure 11.2
A cross section of glabrous skin, showing the layers of the skin and some of its receptors.

Central Structures

Modern anatomical studies have identified a complex web of connections that transmit signals from the skin to the brain that appears to rival that of the visual system. We will focus on a few of the basic characteristics of the system. Nerve fibers from receptors in the skin travel in bundles called *peripheral nerves* that enter the spinal cord through the **dorsal root** (Figure 11.4).

Once entering the spinal cord, the afferent nerve fibers go up the spinal cord in two major pathways: the **medial lemniscal pathway** and the **spinothalamic pathway.** Just as parallel pathways in the visual system serve different perceptual functions, so it is with the cutaneous system. The lemniscal pathway has large fibers that carry signals related to touch perceptions and sensing the positions of the limbs (proprioception). The spinothalamic pathway consists of smaller fibers that transmit signals related to temperature and pain. Fibers from both pathways cross over to the other side of the body at some point during their upward journey and enter the thalamus. Most of these fibers synapse in the **ventral posterior nucleus** in the thalamus, but some synapse in other thalamic nuclei. From the thalamus, signals travel to the **somatosensory receiving area** in the parietal lobe of the cortex (Figure 11.5). Since these signals have crossed over to the opposite

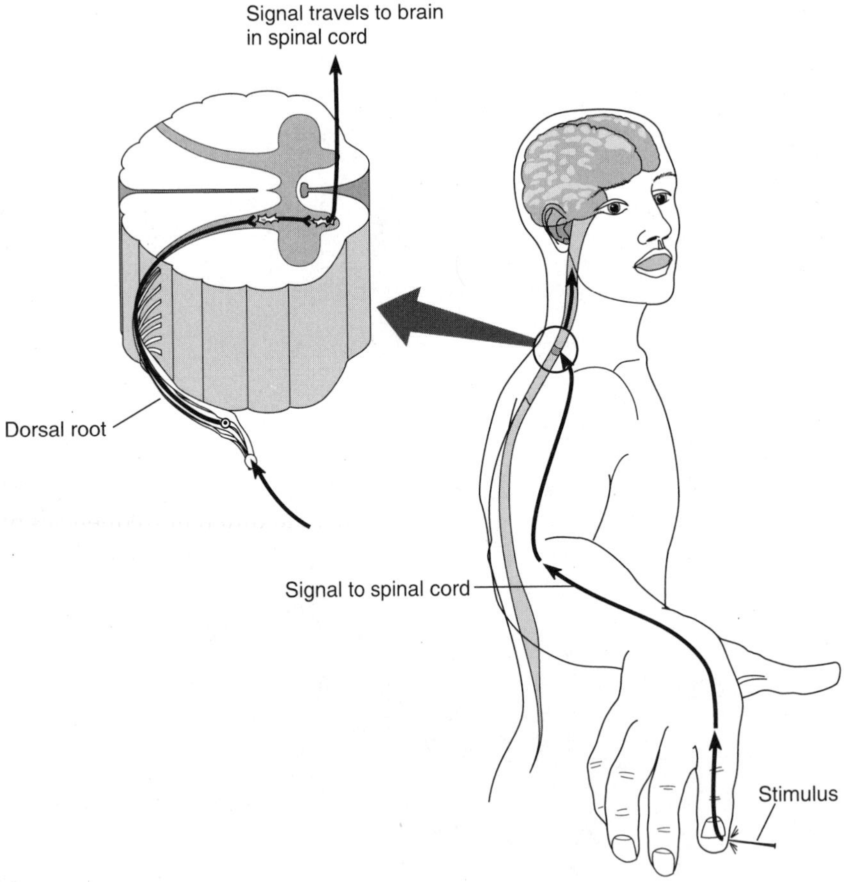

Figure 11.4
The pathway from receptors in the skin to the spinal cord. The fiber carrying signals from a receptor in the finger enters the spinal cord through the dorsal root and then travels up the spinal cord toward the thalamus. This picture is greatly simplified, since there are a number of somatosensory pathways traveling up the spinal cord, and some of them cross over to the opposite side at various points along the way.

side of the body on their way to the thalamus, signals originating from the left side of the body reach the right hemisphere, and signals from the right side of the body reach the left hemisphere.

Now that we have described the overall layout of the somatosensory system, we will focus our attention on how information about different tactile qualities is transmitted from the receptors to the brain. We do this by first looking at psychophysical evidence that indicates that there are four channels for tactile information that correspond to each of the receptors in Figure 11.3.

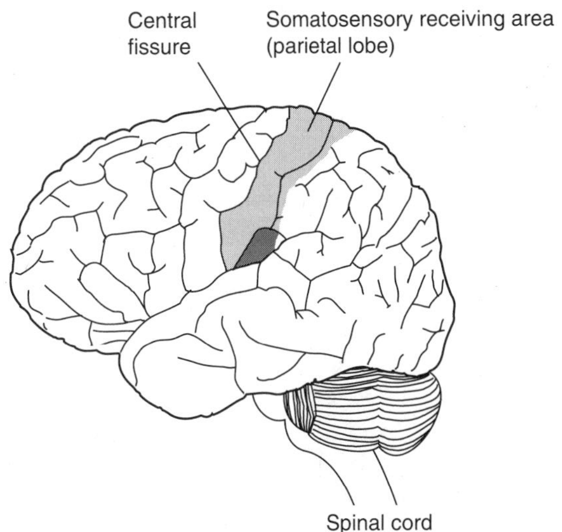

Figure 11.5

The somatosensory cortex in the parietal lobe. The somatosensory system, like the visual and auditory systems, has a number of different areas. The primary somatosensory area, S I (dark shading), receives inputs from the ventral lateral posterior nucleus of the thalamus and then sends fibers to the secondary somatosensory area, S II (light shading), part of which is hidden behind the temporal lobe.

The Psychophysics and Physiology of Tactile Perception

We sense the many types of stimuli that mechanically displace the skin through the operation of separate channels that we can describe both physiologically and psychophysically. We can describe these channels physiologically by noting that there are four different types of receptors that are sensitive to specific aspects of the stimuli presented to the skin, and that there are four types of mechanoreceptive fibers associated with these receptors. We can describe these channels psychophysically by noting that each channel responds to different frequencies of tactile stimulation and results in different tactile perceptions.

Psychophysical Channels for Tactile Perception

When we described the functioning of the visual system in Chapter 2, we saw that there are two different types of receptors: rods and cones. We were able to tell the difference between them because they were distributed differently on the retina, with the fovea containing only cones and the peripheral retina containing both rods and cones. We can also distinguish between different receptors in the skin, but it is more difficult to separate them because there is no area of the skin, analogous to the fovea, that contains only one type of receptor. The various receptors in the skin are, however, sensitive to different frequencies of vibration, and based on these differences, S. J. Bolanowski and his co-workers were able to differentiate four different channels for information about tactile perception in glabrous skin (Bolanowski et al., 1988), and three channels on hairy skin (Bolanowski, Gescheider, & Verrillo, 1994).

In their research on glabrous skin, Bolanowski and his co-workers isolated the different channels by using a number of methods, including cooling the skin (some channels are sensitive to temperature and some are not) and presenting masking

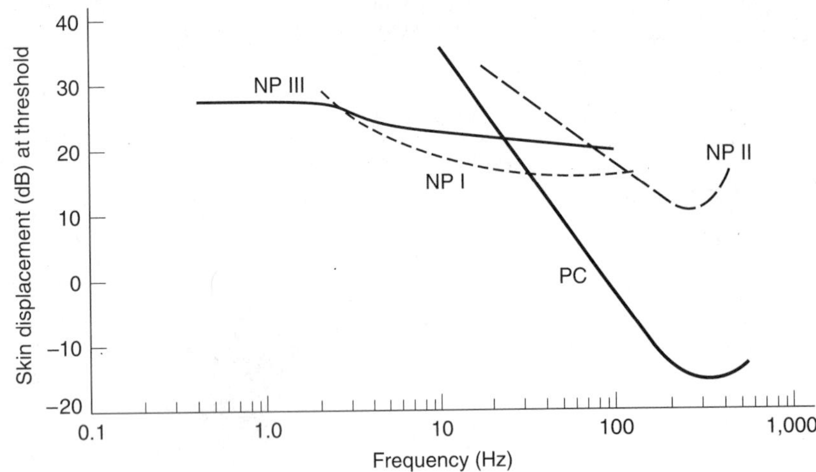

Figure 11.6

Frequency response of psychophysical channels on glabrous skin. These plots indicate the threshold for feeling for each channel as a function of the stimulation presented to the skin. Stimulation at the low-frequency end of the scale is sustained pressure lasting 1 second or longer. Stimulation in the middle of the scale is slow or rapid tapping, and stimulation at the high-frequency end of the scale is vibration. See Table 11.1 for the receptors and properties associated with each of these channels.

vibrations that selectively decreased the sensitivity of one tactile channel so another one could be studied. Rather than describe the specific details of how Bolanowski and his co-workers isolated each of the psychophysical channels, we will look at their final result in Figure 11.6, which shows how each channel responds to the frequency of tactile stimulation. These curves are analogous to the audibility function for hearing, in which the threshold for hearing was plotted versus the frequency of a pure tone stimulus (Figure 9.3). These curves for the cutaneous channels are plots of the threshold for feeling sensations on the skin versus the frequency of tactile stimulation presented to the skin. These channels are described in Table 11.1, by noting (1) the names of the channels; (2) the frequency range over which each channel responds; (3) the receptor associated with each channel; and (4) the perception associated with each channel.

From Figure 11.6 and Table 11.1 we can see that, among them, the four channels respond to frequencies ranging from 0.4 Hz to over 500 Hz, with some overlap, and that each channel is associated with a particular type of tactile perception. For example, the NP III channel responds to low frequencies and is associated with the perception of *pressure,* and the PC channel responds to high frequencies and is associated with the perception of *vibration.* We can also see that each of these channels is associated with one of the receptors in Figure 11.3. Thus, these four receptors provide the neural basis for the four psychophysical channels.

Neural Channels for Tactile Perception

Four separate neural channels for tactile perception have been described by recording from nerve fibers in the skin of both animals and humans; human recording has been accomplished by the use of a technique called **microneurography,** in which a very fine recording electrode is inserted under the skin to record from a single nerve fiber in a person's hand (Vallbo &

Table 11.1

Psychophysical channels for tactile perception

Channel	Best Frequencies	Receptor Structure	Perception
NP III	0.3–3 Hz	Merkel receptors	Pressure
NP I	3–40 Hz	Meissner corpuscle	Flutter
NP II	15–400 Hz	Ruffini cylinder	Buzzing
PC	10 > 500 Hz	Pacinian corpuscle	Vibration

PC stands for "Pacinian corpuscle"; NP stands for "non-Pacinian."

Hagbarth, 1967). The four types of fibers that have been identified are called **mechanoreceptive fibers** because they respond to mechanical displacement of the skin.

These mechanoreceptive fibers are described in Table 11.2 by noting (1) the name of the fiber, which indicates whether its response adapts rapidly (RA = **rapidly adapting fiber**) or slowly (SA = **slowly adapting fiber**); (2) the size of the fiber's receptive fields; (3) the receptors that are associated with each type of fiber; and (4) the stimuli to which each type of fiber responds. This table indicates that there are two types of rapidly adapting fibers (RA I and PC) and two types of slowly adapting fibers (SA I and SA II), that the RA I and SA I fibers have small receptive fields and the SA II and PC fibers have large receptive fields (Figure 11.7), and that the receptors and best stimuli of these fibers are related to the receptors and perceptions of the four psychophysical channels. To appreciate the connection between the psychophysical and physiological channels, note the similarities between the last two columns of Tables 11.1 and 11.2, and also look at Figure 11.8, which combines much of the information in these two tables.

Table 11.2
Types of mechanoreceptive fibers

Type of Fiber	Receptive Field Size	Receptor Structure	Best Stimulus
SA I (slowly adapting)	Small	Merkel receptors	Pressure
RA I (rapidly adapting)	Small	Meissner corpuscle	Taps on skin
SA II (slowly adapting)	Large	Ruffini cylinder	Stretching of skin or movements of joints
PC (rapidly adapting)	Large	Pacinian corpuscle	Rapid vibration

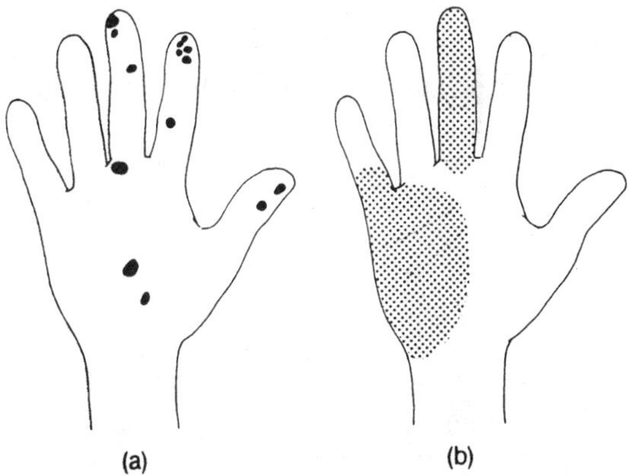

Figure 11.7
Receptive fields of single neurons in the human that are sensitive to touch. (a) Receptive fields of 15 RA I fibers, showing the small receptive fields typical of these fibers; (b) receptive fields of two PC fibers, showing the larger receptive fields typical of these fibers. (From Vallbo & Johansson, 1978.)

Why does each type of fiber respond to a different type of stimulation? One thing that determines their response is the type of receptor ending, which can modify the pressure reaching the fiber. This modification of the stimulus by a receptor ending has been most thoroughly demonstrated for the Pacinian corpuscle, which has been studied extensively because of its distinctive elliptical shape (which makes it easy to identify), its large size (about 1 mm long and 0.6 mm thick), and its accessibility (it is found in the skin, muscles, tendons, and joints, and also in the cat's mesentery, a readily accessible membrane attached to the intestine, from which Pacinian corpuscles can easily be removed).

By taking into account the physical properties of the corpuscle, Werner Lowenstein and R. Skalak (1966) were able to calculate how a push on the corpuscle at A (Figure 11.9) is translated into pressure inside the corpuscle near the nerve fiber at B. According to their calculations, pressing at A should cause brief pressure pulses at B when the stimulus starts and ends, but no pressure at B during the constant part of the stimulus (Figure 11.10).

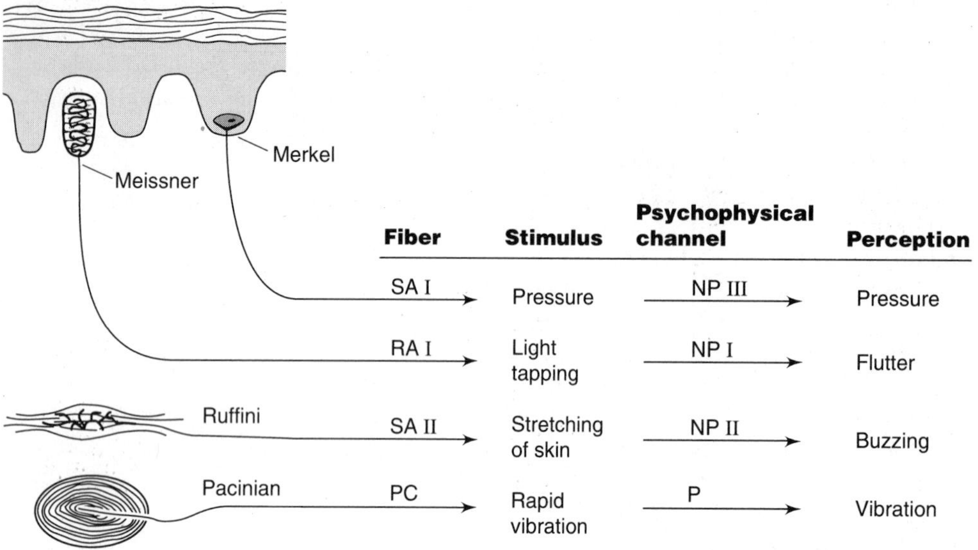

Figure 11.8
The relationship between the four receptor endings (left), the four types of mechanoreceptive fibers (center), and the four psychophysical channels (right). Also note the relationship between the stimulus that best excites each fiber and the perception associated with each psychophysical channel.

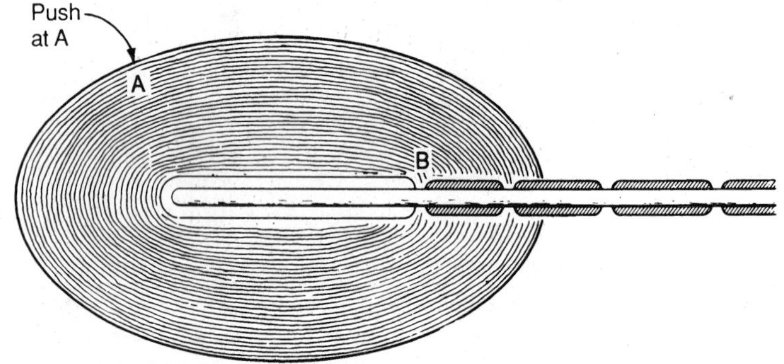

Figure 11.9
A Pacinian corpuscle. (From Lowenstein, 1960.)

To check their prediction, Lowenstein and Skalak measured the response of the nerve fiber under two conditions: (1) when the corpuscle was partially dissected away, so that pressure could be applied closer to the nerve fiber, and (2) when the corpuscle remained, so that pressure could be applied to the corpuscle. Their results were exactly as predicted in their calculations: When most of the corpuscle was removed, the fiber responded when the pressure was first applied, as the pressure continued, and when the pressure was removed. However, when pressure was applied to the intact corpuscle, the fiber responded only when the pressure was first applied and when it was removed. Thus, the presence of the Pacinian corpuscle gives its nerve

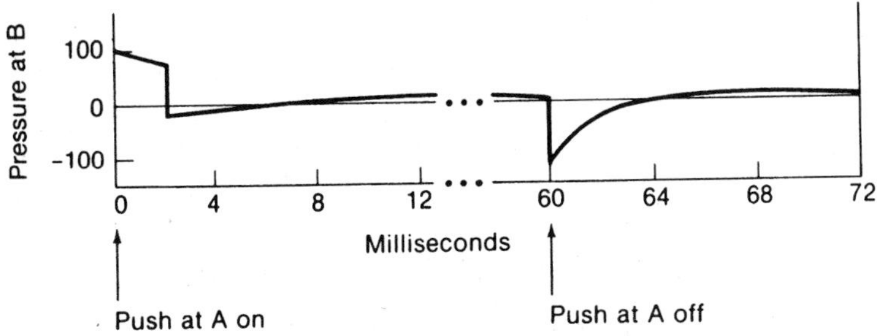

Figure 11.10

Calculated pressure at B inside the Pacinian corpuscle of Figure 11.9 that is caused by pushing the corpuscle at A. Notice that the pressure changes at the onset and offset of stimulation but remains constant as stimulation is applied. (From Lowenstein & Skalak, 1966.)

fiber the properties of a PC fiber, so that it rapidly adapts to constant stimulation such as sustained pressure but responds well to rapidly changing stimulation like vibration.

Mechanoreceptive Fibers and Tactile Perception

The four types of receptors and their fibers result in the four psychophysical channels. When the skin is stimulated, which of these channels are activiated? From Figure 11.6 we can see that the answer to this question depends partially on the frequency of the stimulation presented to the skin. For example, Figure 11.6 shows that stimulation at 1 Hz activates only the NP III channel, since it is the only one that responds at such low frequencies, and we perceive light pressure. Another way to activate just one channel is to present very light stimulation that is near threshold. When we do this, only the channel that is most sensitive to the stimulus frequency responds. Thus, light stimulation at a frequency of 70 Hz activates only the PC channel, since it is the channel most sensitive at 70 Hz, and we perceive very light vibration.

Increasing the stimulation to well above threshold can activate a number of channels at once. Thus, a moderately intense 70-Hz stimulus can activate all four of the tactile channels. In this situation, our perception is probably determined by a combination of the information from all of the channels.

We can appreciate what happens when a number of channels are stimulated simultaneously, by looking at the results of physiological experiments in which raised-dot patterns were rolled across the skin. Kenneth Johnson and Graham Lamb (1981) recorded firing from single fibers in the monkey's finger as raised-dot patterns were presented on a rotating drum (Figure 11.11). By rolling the dots over a fiber's receptive field a number of times, and moving the drum down slightly after each scan, Johnson and Lamb were able to create displays called **spatial event plots,** which indicate how well the fibers are able to resolve the details of the stimulus. (See Figure 11.11 for a description of how these spatial event plots are determined.)

Figure 11.12 shows how the spatial event plots for three types of fibers compare to the original raised-dot pattern. We can see that the plot for the SA I fiber looks similar to the dot pattern, whereas the plots for the RA I and PC fibers do not look like the pattern. (S II fibers respond poorly to this type of stimulus.) This result means that SA I fibers are best suited for resolving details, a result that we might expect both from their location near the surface of the skin and from the fact that SA I fibers have small receptive fields, so the fiber responds only to stimulation of a small area of skin.

Showing that SA I fibers respond well to details does not prove that they are actually used for perception. To show a connection between the fiber's response and perception, we

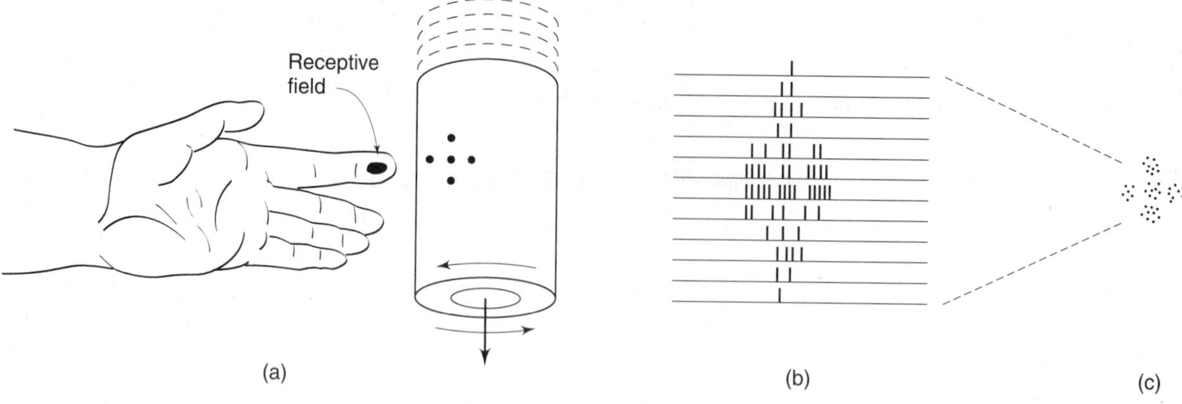

Figure 11.11

(a) Raised-dot stimuli are rolled across the receptive field of a mechanoreceptive fiber on a monkey's fingertip. (b) Sweeping the stimulus across the receptive field generates nerve impulses in the fiber. These nerve impulses are shown on the horizontal lines. Each line represents the impulses generated by one sweep across the receptive field. After each sweep, the drum is moved down slightly, the stimulus is presented again, and a new response is recorded. This process is repeated for a number of positions of the drum. (c) A spatial event plot is created when each nerve impulse is represented by a dot, and the dots are compressed vertically and horizontally so that the pattern of dots has about the same dimensions as the raised-dot stimulus. (Adapted from Martin & Jessell, 1991.)

need to compare the responses of these fibers to psychophysically measured performance. Francisco Vega-Bermudez, Kenneth Johnson, and Steven Hsiao (1991) did this by measuring how well subjects could identify raised letters that were scanned across their fingertips. They found that the subjects identified some of the letters accurately, but that they confused others.

Why were some letters identified accurately and others not? We can answer this question by looking at the spatial event plots in Figure 11.13 for the responses of monkey SA I fibers to the same letters that were used in the psychophysical experiments. First, consider the following psychophysical result: Subjects often identified *C*'s as *O*'s but only rarely said an *O* was a *C*. Now consider the physiological result shown in Figure 11.13. The SA I fiber's response to *C* looks very much like an *O*; however the response to an *O* does not look like a *C*. This parallel between people's psychophysical judgments and physiological responding supports the idea that SA I fibers are important for identifying details. (Also

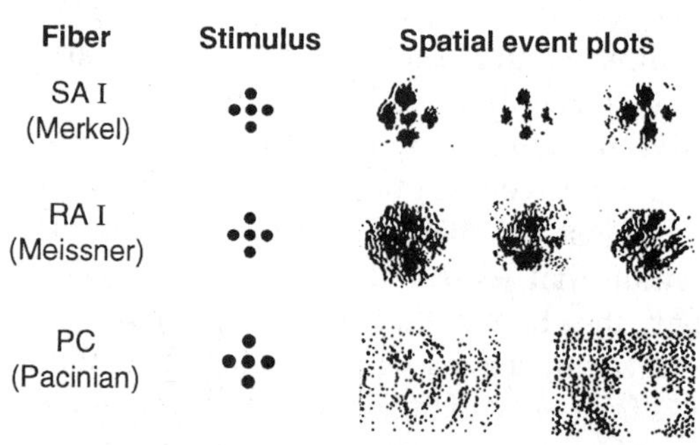

Figure 11.12

Left: Raised-dot stimulus presented as shown in Figure 11.11 to a monkey's fingertip. Right: Some typical spatial event plots generated by this pattern for SA I, RA I, and PC fibers that are associated with Merkel receptors, Meissner corpuscles, and Pacinian corpuscles, respectively. The SA II fiber, which is associated with the Ruffini cylinder, generates only a small response to this type of stimulation. (Adapted from Johnson & Lamb, 1981.)

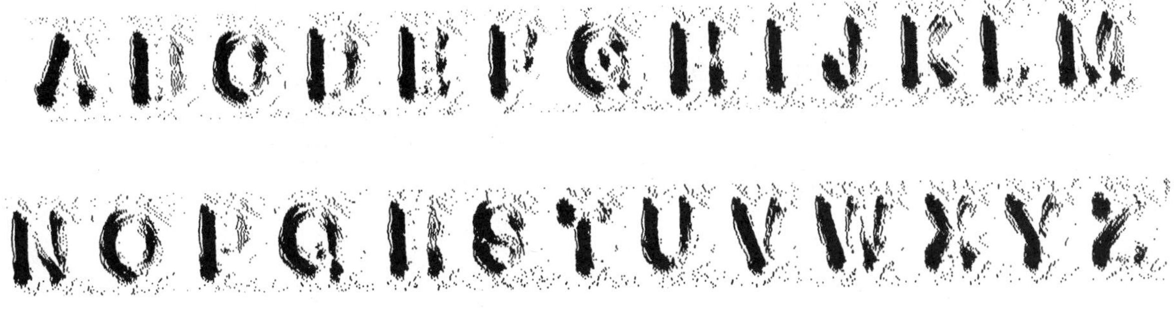

Figure 11.13
Spatial event plots generated by rolling raised letters across the receptive fields of SA I fibers. (From Vega-Bermudez, Johnson, & Hsiao, 1991.)

consider the following psychophysical result: *B* was often identified as *D*, but *D* was rarely identified as *B*. Look at the *B* and *D* responses of the SA I fiber to see the neural parallel to this psychophysical result.)

These results indicate that SA I fibers are probably important for detail perception. However, remember that presenting above-threshold stimuli to the skin activates a number of tactile channels. Thus, the SA I fibers may help us feel the details of a display, but all four types of fibers probably contribute to our overall perceptual experience (Roland, 1992).

Neural Responses to Temperature and Potentially Damaging Stimuli

Stimulation of the skin causes not only tactile sensations, such as pressure and vibration, but also sensations associated with temperature, such as warmth and cold, and sensations associated with potentially damaging stimulation, such as pain. Fibers called *thermoreceptors* and *nociceptors* have been identified in the skin that respond to these types of stimulation.

Thermoreceptors: Temperature

Thermoreceptors respond to specific temperatures and to changes in temperature. There are two kinds of thermoreceptors: warm receptors and cold receptors. Figure 11.14 shows how a **warm fiber** responds to increases in temperature. This neuron illustrates the following properties of a warm fiber: (1) It increases its firing rate when the temperature is increased; (2) it continues to fire as long as the higher temperature is kept on; (3) it decreases its firing rate when the temperature is decreased; and (4) it does not respond to mechanical stimulation (Kenshalo, 1976; Duclaux & Kenshalo, 1980). **Cold fibers,** on the other hand, increase their firing rate when the temperature is decreased and continue to fire at low temperatures. The different ways that cold and warm fibers respond to steady temperatures are shown in Figure 11.15. Cold fibers respond in the 20°C to 45°C range, with the best response at

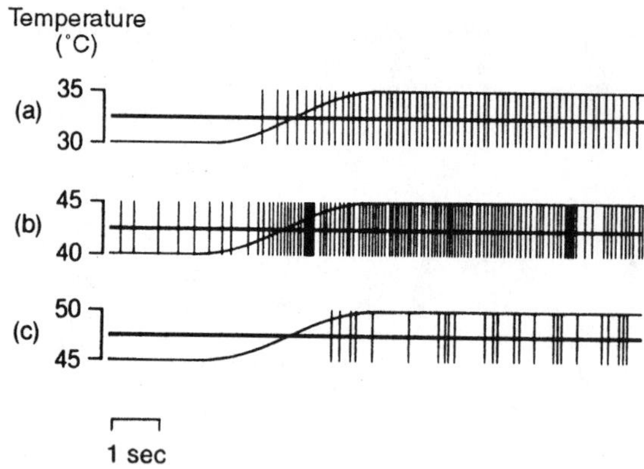

Figure 11.14
The response of a "warm" fiber in the monkey. At the beginning of each record the firing rate is low because the fiber has been at the same temperature for a while. The fiber fires, however, when the temperature is increased (indicated by the sloping line) and continues to fire for a period of time immediately after the increase. Record (b) indicates that this fiber fires best when the temperature is increased from 40° C to 45° C. (From Duclaux & Kenshalo, 1980.)

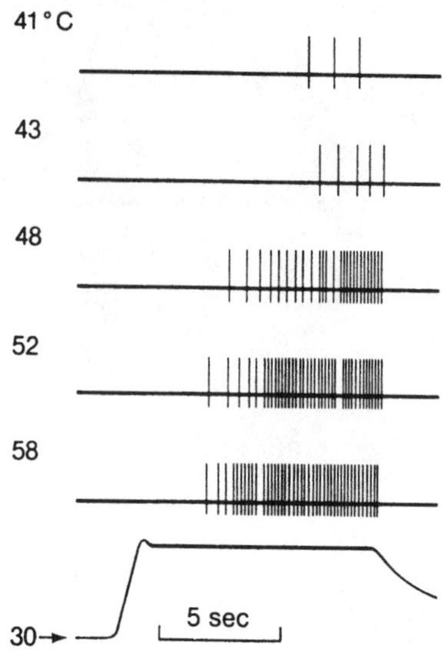

Figure 11.16
The response of a "nociceptor" in the cat to heating of the skin. This fiber begins firing at temperatures above about 45°C and reaches maximum firing rates above about 55°C. (From Beck, Handwerker, & Zimmerman, 1974.)

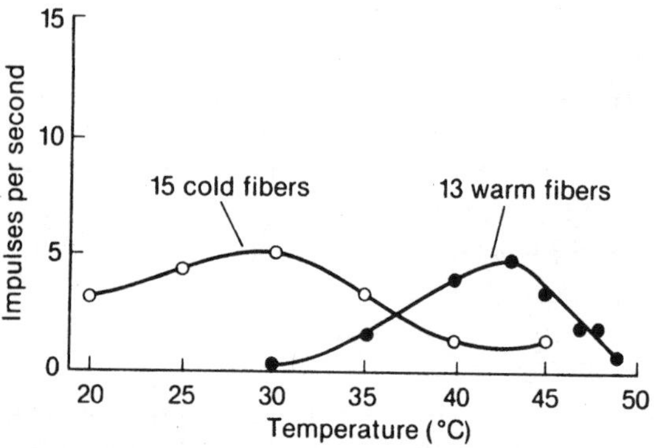

Figure 11.15
Responses of temperature-sensitive fibers in the monkey to constant temperatures. The curve on the left (open circles) shows the average response of a group of 15 "cold" fibers that respond best at about 30°C. The curve on the right (filled circles) shows the average response of a group of 13 "warm" fibers that respond best at about 44°C. (From Kenshalo, 1976.)

about 30°. Warm fibers respond in the 30° to 48° range, with the best response at about 44°.

Nociceptors: Pain

Nociceptors respond to stimulation such as intense pressure, extreme temperature, or burning chemicals that can damage the skin. Figure 11.16 shows the response of a cat's nociceptor, which begins firing when the temperature reaches about 45°C, the same temperature at which a human begins to feel pain (Beck, Handwerker, & Zimmerman, 1974; Zimmerman, 1979). The firing of nociceptors is therefore thought to cause human pain perception, although later in the chapter we will see that the firing of nociceptors does not provide a complete explanation of pain perception, since emotional and cognitive factors also influence people's perception of pain.

Based on our survey of different types of fibers in the skin, we can conclude that receptors and fibers at the very beginning of the somatosensory system have a large influence on what we perceive. The importance of receptors in the skin for perception parallels the situation in the visual system, in which properties of the rod and cone receptors determine our sensitivity to light and our ability to see details and colors. When we studied the visual system, we also saw that the cones in the fovea, which are responsible for the high acuity that enables us to see details, are represented by a disproportionately large area on the visual cortex. We will now see that neural processing creates a similar result in the somatosensory system.

DEMONSTRATION

Comparing Two-Point Thresholds

To measure two-point thresholds on different parts of the body, hold two pencils side by side (or better yet, use a drawing compass) so that their points are about 12 mm (0.5 in.) apart; then touch both points simultaneously to the tip of your thumb and see if you feel two points. If you feel only one, increase the distance between the pencil points until you feel two; then note the distance between the points. Now move the pencil points to the underside of your forearm. With the points about 12 mm apart (or at the smallest separation you felt as two points on your thumb), touch them to your forearm and note whether you feel one point or two. If you feel only one, how much must you increase the separation before you feel two? ●

Neural Processing

We will demonstrate a connection between neural processing and perception in the somatosensory system by making use of a basic fact about tactile perception: Tactile acuity is better on some parts of the body, such as the hands and the fingertips, than on others, such as the limbs, the back, and the trunk. Let's first consider this finding and then consider how it is related to neural processing.

Measuring Tactile Acuity: The Two-Point Threshold

One way to determine tactile acuity is to measure the two-point threshold—the smallest separation between two points on the skin that is perceived as two points. When the two-point threshold is measured on different parts of the body, we find that there are areas on the skin that have higher acuity than others.

If your results from this demonstration match those from the laboratory, you will find that the two-point threshold on your forearm is much larger than the two-point threshold on your thumb. Figure 11.17, which shows how the two-point threshold varies on different parts of the body, indicates that the two-point threshold is over 10 times larger on the forearm than on the thumb. This result, therefore, is analogous to the results for the visual system. Areas that are responsible for making fine spatial discriminations—the fovea for vision and the fingers for touch—have high acuity, whereas other areas have lower acuity. One way to relate these acuity differences to physiology is to compare the sizes of the receptive fields on different parts of the body.

Receptive Fields on the Skin

We have seen that the RA I and SA I fibers have small receptive fields. Vallbo and Johansson (1978) found that the density of these small re-

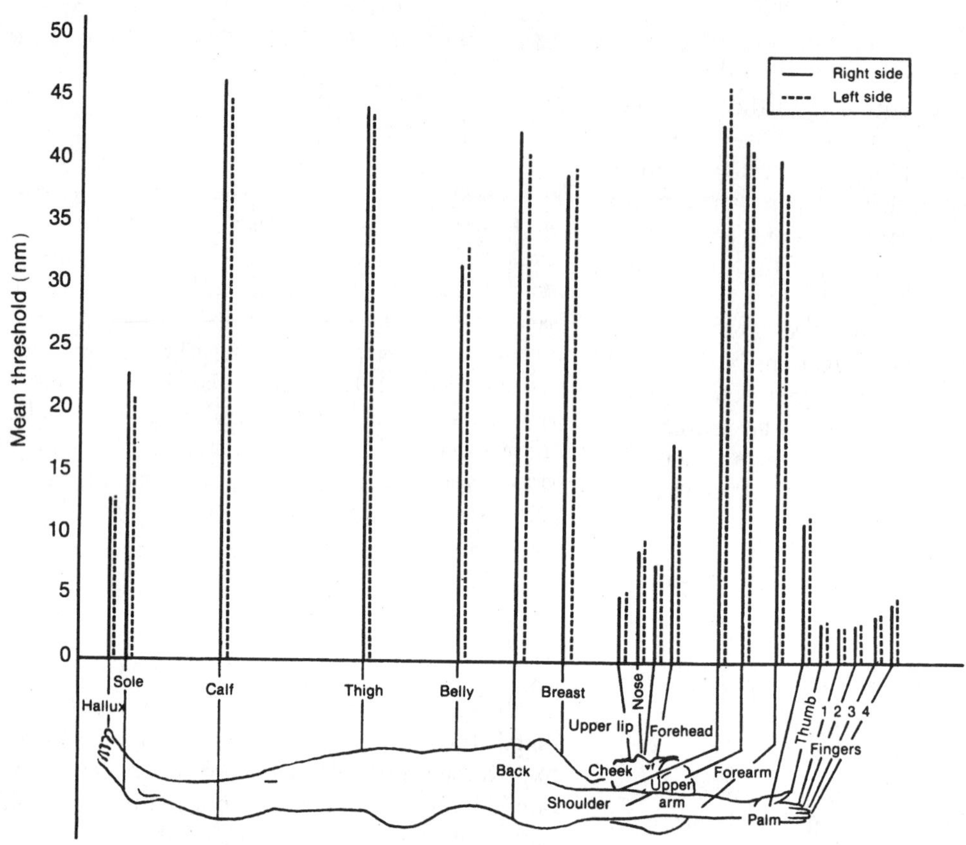

Figure 11.17
Two-point thresholds for females. Two-point thresholds for males follow the same pattern. (From Weinstein, 1968.)

ceptive fields was higher on the fingertips than on the hand—exactly what we would expect from the psychophysical finding that the two-point threshold is smaller on the fingertips than on the hand. In fact, Vallbo and Johansson found a direct relationship between the size of the two-point threshold and the density of the small receptive fields of RA I and SA I fibers (Figure 11.18). Parts of the body with small two-point thresholds had small receptive fields.

We get the same relationship between receptive field size and the area on the body if we record from neurons in the monkey's cortex. Figure 11.19 shows that the size of the receptive field increases as we move from the fingertips toward the arm. Thus, areas that have higher tactile acuity have smaller receptive fields. Another way to relate tactile acuity to physiology is to determine how much space on the somatosensory cortex is allotted to different parts of the body.

Maps of the Body on the Cortex: The Magnification Factor

We can extend the relation between tactile acuity and physiology to the cortex by showing that parts of the body with high acuity are allotted larger areas on the cortex. This relation between acuity and space on the cortex is similar to the *magnification factor* that we described in Chapter 3

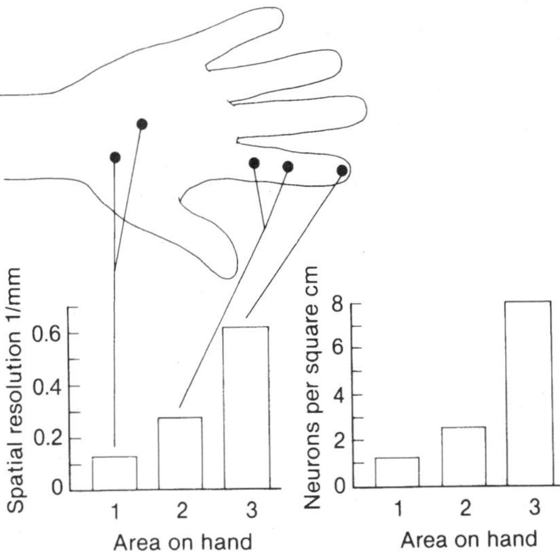

Figure 11.18

Left: The bar graph shows the spatial resolution at one place on the hand and two places on a finger. Spatial resolution, which is determined by taking the inverse of the two-point threshold (so that a small two-point threshold results in high spatial resolution), increases as we move from the hand to the fingertip. Right: The density of small receptive field neurons (RA I and SA I fibers) at the same three locations. These data indicate that areas that have high spatial resolution have high densities of neurons with small receptive fields. (From Vallbo & Johansson, 1978.)

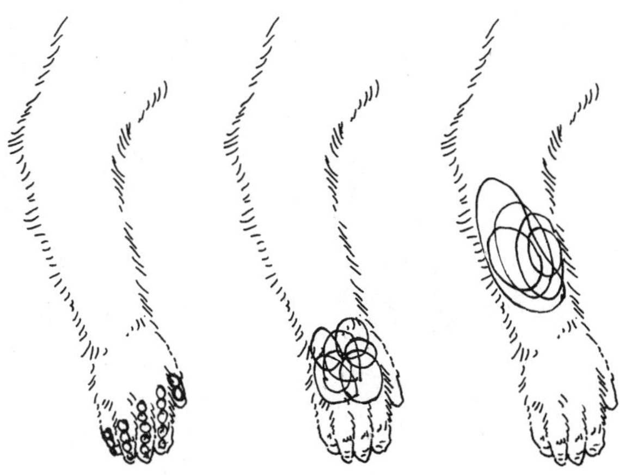

Figure 11.19

Receptive fields of cortical neurons are smallest on the fingers and become larger on the hand and the forearm. (Adapted from Kandel & Jessell, 1991.)

for the visual system. We saw that the fovea, with its densely packed cone receptors and excellent detail vision, is represented by a cortical area far out of proportion to its size: The fovea's cortical area is *magnified* to provide the extra neural processing needed to perceive the fine details of objects imaged on the fovea.

We can show that a similar situation exists in the cutaneous system by measuring the area on the cortex that corresponds to each area of the body. When we do this, we see that, for each area on the skin, there is a corresponding area on the surface of the somatosensory cortex, as shown in Figure 11.20. The strangely shaped cortical representation that results is called a **homunculus,** for "little man." Let's first describe how the homunculus is determined and then relate this strange-looking little man to the two-point threshold.

The relationship between points on the brain and points on the body was determined by applying mild shocks to a small area of a person's cortex and asking the person to describe where he or she felt a sensation. The people who allowed their brains to be electrically stimulated to produce Figure 11.20 were patients of neurosurgeons Wilder Penfield and Theodore Rasmussen (1950), who were operating on these patients to remove brain tumors suspected of causing epileptic seizures. A major problem in this surgery is to remove the tumor without damaging the somatosensory or motor areas of the brain, which are located right next to each other on either side of the central fissure (see Figure 11.5). Damage to the somatosensory area would cause a loss of sensation in a part of the body, and damage to the motor area would cause paralysis. Before removing the tumor, therefore, the surgeons mapped the somatosensory and motor areas of the brain in each patient.

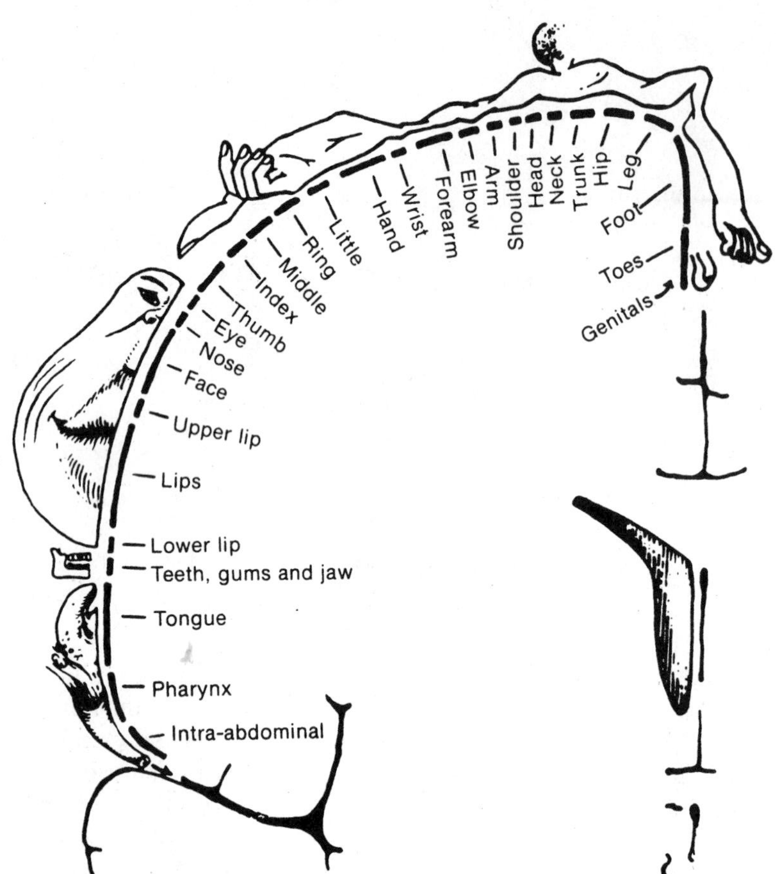

Figure 11.20

The sensory homunculus on the somatosensory cortex. Parts of the body with the highest tactile acuity are represented by larger areas on the cortex. (From Penfield & Rasmussen, 1950.)

When people's somatosensory cortex is shocked they experience a sensation such as tingling, tickling, or numbness at a particular place on the body. Only a local anesthetic is used because the patient must be awake to report the location and type of sensation resulting from stimulation. Since the anesthesia is applied where the skull is cut, and since there are no pain receptors in the brain itself, the patient feels no pain when the brain is stimulated.

The homunculus of Figure 11.20 shows that some areas on the skin are represented by a disproportionately large area of the brain. The area devoted to the thumb, for example, is as large as the area devoted to the entire forearm. As with the magnification factor for foveal vision, the apparent overrepresentation of various body parts on the sensory cortex is related to the functioning of these areas. By comparing Figures 11.17 and 11.20 we can note the close relationship between the two-point threshold on the skin and the cortical representation of the skin: Areas on the skin with small two-point thresholds are represented by large areas on the cortex. Thus, the map of the body on the brain may look distorted, but the distortions provide the extra neural processing that enables us to accurately sense fine details with our fingers and other parts of the body.

Before leaving our discussion of the homunculus, we should note that a number of experiments indicate that there are at least four separate maps of the monkey's body in the somatosensory area of its brain (Kaas & Pons, 1988; Kandel &

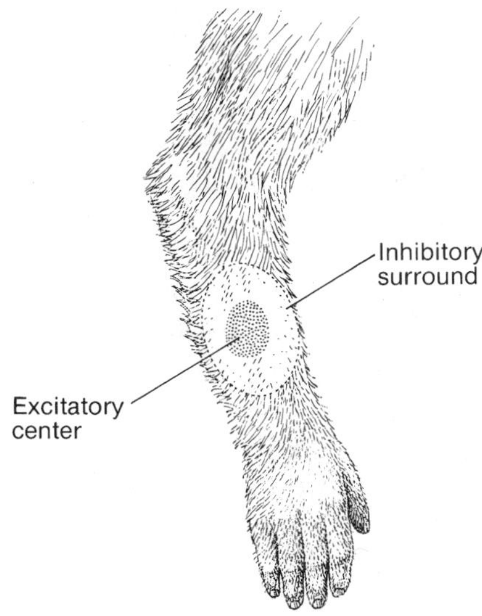

Figure 11.21
An excitatory-center–inhibitory-surround receptive field of a neuron in a monkey's thalamus.

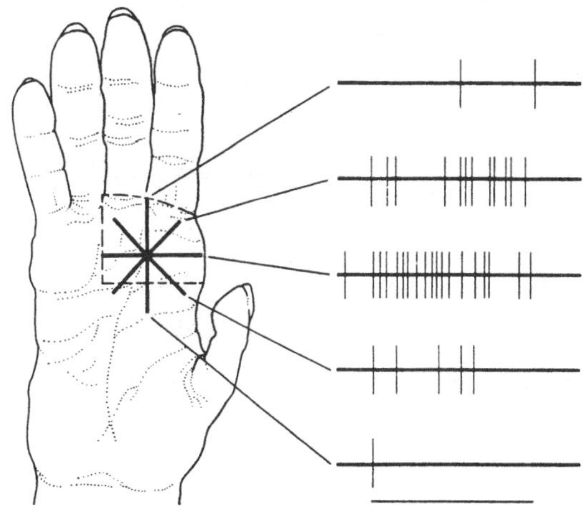

Figure 11.22
The receptive field of a neuron in the monkey's somatosensory cortex that responds when an edge is placed on the hand. This cell responds well when the edge is oriented horizontally but responds less well to other orientations. (From Hyvarinin & Poranen, 1978.)

Jessell, 1991; Nelson et al., 1980). The apparent reason for these multiple representations is that different areas within the sensory cortex have different functions. For example, one area may be specialized for the discrimination of forms and another for the discrimination of textures. Whatever their function, these multiple maps of the body on the brain exhibit the same distortions as our simple homunculus in Figure 11.20: Areas of the body that discriminate fine details are allotted large areas on the cortex.

Receptive Fields

Neural processing in the somatosensory system is reflected not only by the expanded space in the cortex for certain areas of the body, but by changes in neurons' receptive fields as we move from the receptors to the cortex. Early in the system, neurons have center–surround receptive fields, like the one in Figure 11.21, which shows the receptive field of a neuron in the monkey's thalamus that receives signals from receptors in the skin of the monkey's arm. This is an excitatory-center–inhibitory-surround receptive field similar to the center–surround receptive fields of visual neurons in the retina and lateral geniculate nucleus (Mountcastle & Powell, 1959).

If we move up to the cortex, we find some neurons with center–surround receptive fields, and also others that respond to more specialized stimulation of the skin. Figures 11.22 and 11.23 show the receptive fields of neurons in the monkey's somatosensory cortex in the parietal lobe; these are quite similar to the receptive fields of simple and complex cells in the visual cortex. The cell in Figure 11.22 responds to an edge oriented horizontally but responds less well to other orientations. Figure 11.23 shows a cell that responds to movement across the skin in a specified direction (Hyvarinin & Poranen, 1978; see also Costanzo & Gardner, 1980; Warren, Hamalainen, & Gardner, 1986; Whitsel, Roppolo, & Werner, 1972). In Chapter 7, we proposed a neural circuit to explain how visual neurons could respond

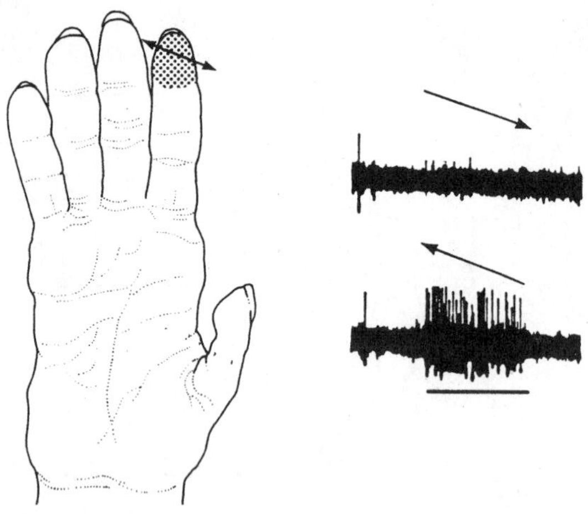

Figure 11.23

The receptive field of a neuron in the monkey's parietal cortex that responds to movement across the fingertip from right to left but does not respond to movement from left to right. (From Hyvarinin & Poranen, 1978.)

selectively to movement in one direction (see Figure 7.12). It is likely that a similar type of circuit is behind the firing of cutaneous neurons like the one in Figure 11.23.

The similarity between these neurons and neurons in the visual system is striking. In both vision and touch, neurons near the receptors respond to a wide range of stimuli, whereas neurons in the cortex respond only to more specific stimuli. Similar principles of information processing apparently operate for both of these senses.

Active Touch

In most of the research we have described so far, experimenters have presented stimuli to passive subjects—humans or monkeys that remain stationary as stimuli are applied to their skin. This state of affairs, in which subjects do not move, is not, however, typical of our everyday experience. In our everyday experience, we move our fingers over an object's surface, we explore its crannies and contours, and we manipulate the object if it is small enough to be held in our hand. This process, which is called **active touch,** has been studied both physiologically and psychophysically. The focus of the physiological approach has been to identify neurons that fire as an object is being actively explored with the fingers or the hands.

The Physiological Approach to Active Touch

Active touch involves a number of different physiological processes. These processes involve sensing the positions and movements of the hand and the fingers as well as cutaneous sensations in the skin. As we noted earlier, sensing limb position is called *proprioception,* and sensing the motions of our limbs is called *kinesthesis.*

Proprioception and Kinesthesis The perception of the position or motion of our limbs and joints depends on receptors in the joints and on **muscle spindle receptors**—slowly adapting mechanoreceptive fibers that are entwined around specialized muscle fibers (Figure 11.24). These muscle spindle receptors are sensitive to changes in the muscle's length, so that, when the muscle stretches, these receptors fire and indicate the position of the limb. The fact that the firing of muscle spindles indicates limb position leads to an interesting tactile illusion: When a limb is vibrated, it feels as if it moves to another position, even though it is actually stationary (Figure 11.25). The reason is that the vibration causes the spindle receptors to fire in the same way that they would if the muscle were stretching and the limb were moving (Jones, 1988).

Just as there are cutaneous neurons that respond to specific types of stimulation of the skin, there are also neurons in the joints that respond to specific limb movements and positions. Esther Gardner and Richard Costanzo (1981) studied

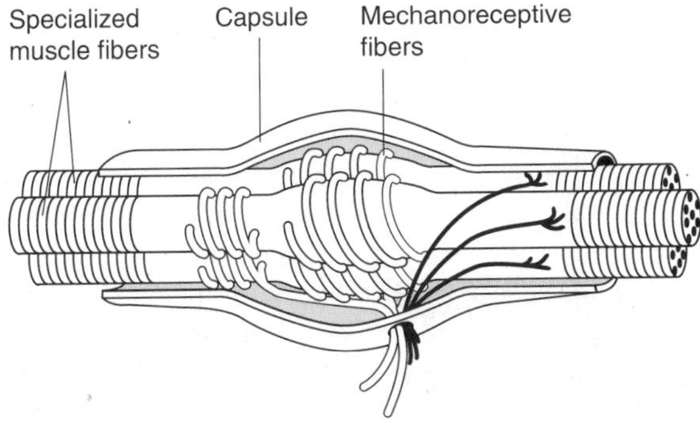

Figure 11.24
A muscle spindle consists of slowly adapting mechanoreceptive fibers wrapped around specialized muscle fibers and enclosed inside a capsule. When these fibers stretch, the mechanoreceptive fibers fire and indicate the position of the limb. (From Martin & Jessell, 1991.)

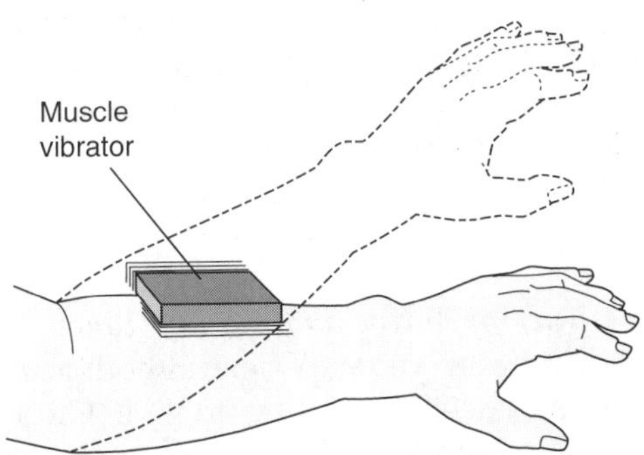

Figure 11.25
Vibration of the arm activates muscle spindles in the arm muscles and causes the person to feel as if the arm has moved to another position.

neurons that help sense limb movement and position by recording from neurons in the monkey's somatosensory cortex that receive inputs from receptors in the joints. They found three different types of neurons:

1. *Rapidly adapting neurons.* These neurons respond only when a limb is moving in a particular direction. Some fire when the limb is flexed, and others fire when the limb is extended. The greater the velocity of movement, the more rapid the firing.
2. *Slowly adapting neurons.* These neurons respond when a limb is moving and also when the limb is held in a specific position. The size of the response to a particular position depends on the limb's state of extension and flexion, with larger responses being generated by greater extensions and flexions.
3. *Neurons that fire maximally to maintained limb positions.* These neurons are called *positional neurons* because they fire best to static position rather than to movement.

Most of the above neurons respond to the movement of a particular joint. In addition, however, Costanzo and Gardner (1981) found neurons that respond best to the flexion of a number of joints. For example, Figure 11.26 shows the firing of a neuron that responds to flexion of the wrist (Figure 11.26a) but responds even better to combined flexion of the wrist and the fingers (Figure 11.26b). This neuron is concerned more with postural information than with movement.

These recordings from single neurons show that information about the positioning and movement of our limbs is signaled by neurons in the parietal cortex. We can imagine the massive amount of information provided by these neurons when we recognize that there are over 60 joints in the human body, all of which must be sending information to the cortex regarding their position and movement.

Tactile Perception and Active Touch Active touch depends on the firing not only of neurons that signal the positions and movements of the fingers and the hands, but also of cutaneous neurons such as the mechanoreceptive fibers we described earlier that sense qualities such as pressure, vibration, and stretching. In addition to these fibers, there are other neurons that are specialized to respond when objects are grasped,

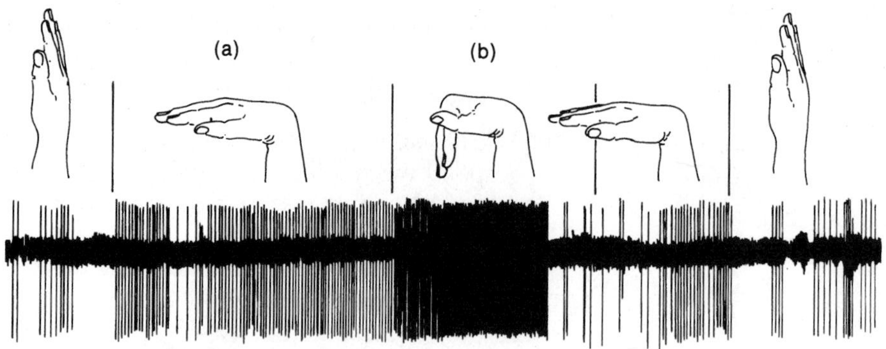

Figure 11.26
These records show the responses of a neuron that responses well to (a) flexion of the wrist, and that responds even better (b) when the fingers are also flexed. (From Costanzo & Gardner, 1981.)

such as the neuron in Figure 11.27, which does not respond when the experimenter stimulates the skin, but which does respond when the monkey grasps certain objects. This cell responds when the monkey grasps a ruler or a block of wood but does not respond if the monkey grasps a cylinder. Neurons like these probably play an important role in haptic perception—the perception of three-dimensional objects by touch, which we will discuss shortly.

The Psychophysical Approach to Active Touch

One of the major concerns of the psychophysical approach to active touch is comparing active and passive touch. Both the *experiences* caused by active and passive touch and people's ability to *identify* objects by using active and passive touch have been compared. Let's begin with a demonstration that illustrates both of these comparisons.

DEMONSTRATION

Comparing Active and Passive Touch

Ask another person to select five or six small objects for you to identify. Close your eyes and have the person place each object in your hand. Your job is to identify the object by touch alone. As you do this, be aware of what you are experiencing: your finger and hand movements, the sensations you are feeling, and what you are thinking. Do this for three objects, and then hold out your hand, keeping it still, with fingers outstretched, and let the person move each of the remaining objects around on your hand, moving their surfaces and contours across your skin. Your task is the same as before: to identify the object and to pay attention to what you are experiencing as the object is moved across your hand. ●

The Experience of Active and Passive Touch In this demonstration, you experienced both active and passive touch. According to J. J. Gibson (1962), who championed the importance of movement in perception (Chapters 6 and 7), this distinction is important because of differences in both what we experience in the two cases and the kinds of information we receive. Gibson and others compared the experience of active and passive touch by noting that we tend to relate active touch to the *object* being touched, whereas we tend to relate passive touch to the *sensation* experienced in the skin. For example, if someone stimulates your skin with a pointed object, you might say, "I feel a prickling sensation on my skin"; however, if you feel the tip of the pointed object yourself, you might say, "I feel a pointed object" (Kruger, 1970). Another example of a difference between active and passive touch is that, when

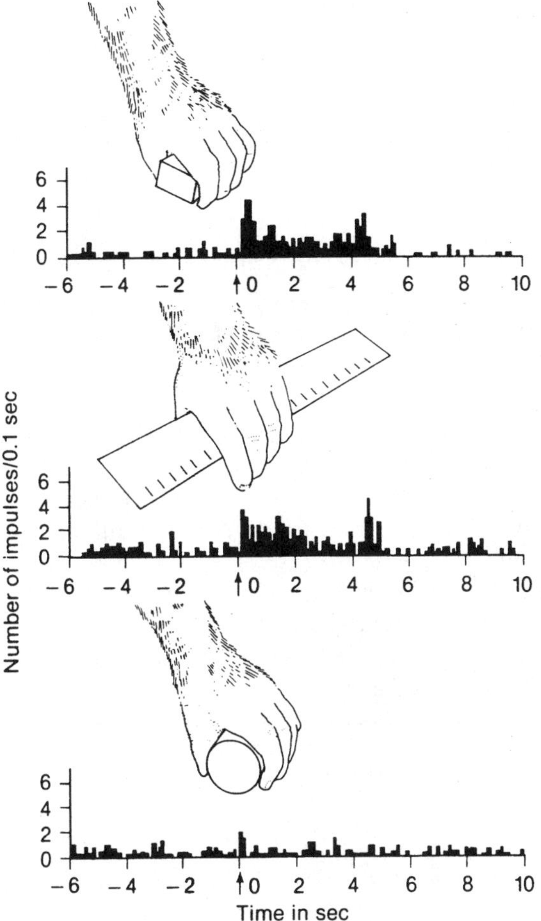

Figure 11.27

The response of a neuron in a monkey's parietal cortex that fires when the monkey grasps a rectangular block or a ruler, but that does not fire when the monkey grasps a cylinder. The neuron's rate of firing is indicated by the height of the bars. The monkey grasps the objects at time = 0. (From Sakata & Iwamura, 1978.)

you move your hand over the edges and surfaces of an object (active touch), you do not perceive the object as moving, even though it is moving relative to the skin. If, however, someone else moves an object across your skin (passive touch), you perceive the object moving across your skin.

Though we do not completely understand why these two types of touch result in such different experiences, a number of properties differentiate active from passive touch. An important property of active touch, according to Gibson, is that it is purposive; that is, when you feel something, especially an object with which you are unfamiliar, your purpose is to determine its shape. Just as you visually scan a scene by looking at its most interesting or important areas, you feel an object by touching the parts that contain information about its shape. This purposiveness may be important in determining the experience of active touch.

In addition to its purposive aspect, active touch includes other properties absent from passive touch. As you feel something, you stimulate receptors not only in the skin but also in the joints and tendons that are activated as you move your fingers or hands over an object, whereas passive touch stimulates only the receptors in the skin. Furthermore, moving your fingers over an object enables you to perceive both the "touch" that occurs when a stimulus is applied to passive skin and the sounds that occur when skin moves actively over a surface (Taylor, Lederman, & Gibson, 1973).

Identifying Objects by Active and Passive Touch According to Gibson (1962), active touch is superior to passive touch for gathering information about objects in our environment. He demonstrated this superiority by showing that, when subjects actively felt cookie cutters of various shapes, they were able to identify the shape correctly 95 percent of the time. However, if the cookie cutters were pushed onto the skin by someone else, the subjects correctly identified the shape only 49 percent of the time.

However, others have found that active touch does not always result in more accurate identification than passive touch. For example, Schwartz, Perez, and Azulaz (1975) repeated Gibson's cookie cutter experiment, with an additional passive condition in which the subjects felt the shapes as the experimenter moved the edges of the cookie cutters across their fingers. Schwartz and his co-workers, like Gibson, found that the subjects had a hard time identifying objects pushed into their palms by the experimenter, but they also found that, when they moved the cookie cutter over the subjects' outstretched (but passive) fingers, the subjects correctly identified the shape 93 percent of the time. Thus, Schwartz et al. argued that, although the subjective experiences of active and passive touch may be different, passive touch can provide information about a stimulus equal to that of active touch, as long as the stimulus is moved across the subject's skin. Movement across the skin, not purposiveness, is what is important in discriminating shapes, according to Schwartz.

Although some studies in addition to Schwartz's have found that passive touch based on moving stimuli can produce identification equal to active touch (Vega-Bermudez et al., 1991), other studies have shown that active touch is superior. For example, Morton Heller (1986) had subjects identify braille dot patterns (Figure 11.28) under three different conditions: The subjects moved their fingers over the dots (active); the experimenters moved the dots over the subjects' stationary fingertips (passive with movement); or the experimenters pushed the dots into the subjects' stationary fingertips (passive with no movement). The results, shown in Figure 11.29, indicate that active touch is better than both types of passive touch. However, in another experiment using braille characters, Grunwald (1965) found that subjects who were experienced braille readers recognized braille characters equally well using passive or active touch. Thus, depending on the conditions and the subjects, active touch is sometimes better than passive touch, and sometimes there is no difference between them. However, when it comes to identifying three-dimensional objects by touch, it seems clear that active touch is the superior mode. This ability to identify three-dimensional objects based on touch is called *haptic perception.*

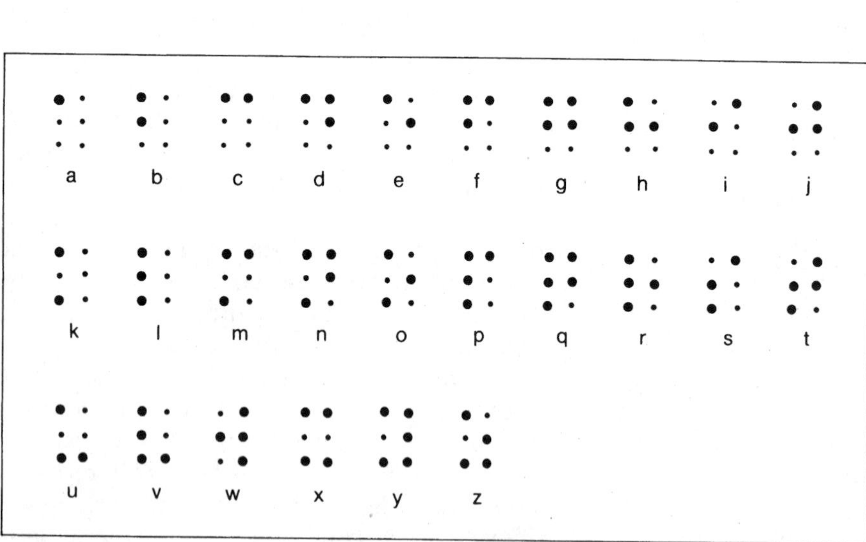

Figure 11.28
The braille alphabet consists of raised dots arranged in the patterns shown here. Blind people read these dots by scanning them with their fingertips, an example of active touch. The pattern above the alphabet says, "Active touch."

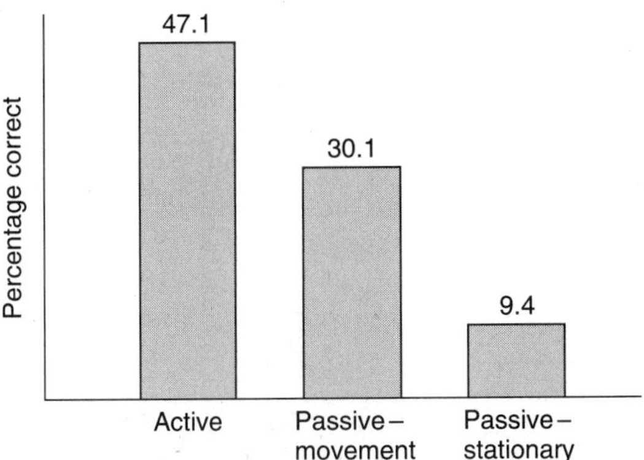

Figure 11.29

The results of Heller's (1986) experiment, in which he had subjects identify braille dot patterns under three different conditions. See text for details.

Haptic Perception: Tactile Perception of Three-Dimensional Objects When you tried to identify objects that were handed to you, you were engaging in **haptic perception,** perception in which three-dimensional objects are explored with the hand. Haptic perception provides a particularly good example of a situation in which a number of different systems are working in coordination with each other. As you manipulated the objects that you were trying to identify, you were using three distinct systems: (1) sensory—sensing the cutaneous sensations such as touch, temperature, and texture and the movements and positions of your fingers and hands; (2) motor—moving your fingers and hands; and (3) cognitive—thinking about the information provided by the sensory and motor systems, in order to identify the object. This is an extremely complex process (although, just as in the rest of perception, the complexity is not always apparent when it is happening) because these three systems must all be coordinated with one another. For example, the motor system's control of finger and hand movements is guided by cutaneous feelings in the fingers and the hands, by the person's sense of the positions of the fingers and hands, and by thought processes that determine what information is needed about the object in order to identify it.

Psychophysical research on this process has shown that people can accurately identify most common objects within one or two seconds, based only on touch (Klatzky, Lederman, & Metzger, 1985). When Susan Lederman and Roberta Klatzky (1987, 1990) observed subjects' hand movements as they made these identifications, they found that people use a number of distinctive movements, which they called **exploratory procedures (EPs),** and that the types of EPs used depend on the object qualities the subjects are asked to judge.

Four of the EPs observed by Lederman and Klatzky are shown in Figure 11.30, and the frequency with which they are used to judge different object qualities are shown in Figure 11.31. Subjects tend to use just one or two EPs to determine a particular quality. For example, people use mainly lateral motion and contour following

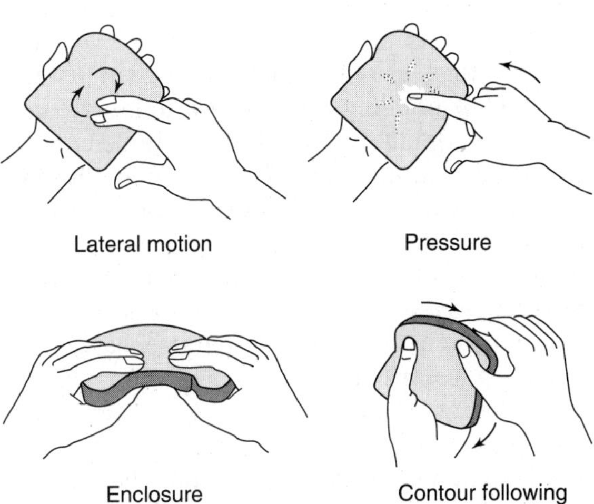

Figure 11.30

Some of the exploratory procedures (EPs) observed by Lederman and Klatzky as subjects identified objects. (From Lederman & Klatzky, 1987.)

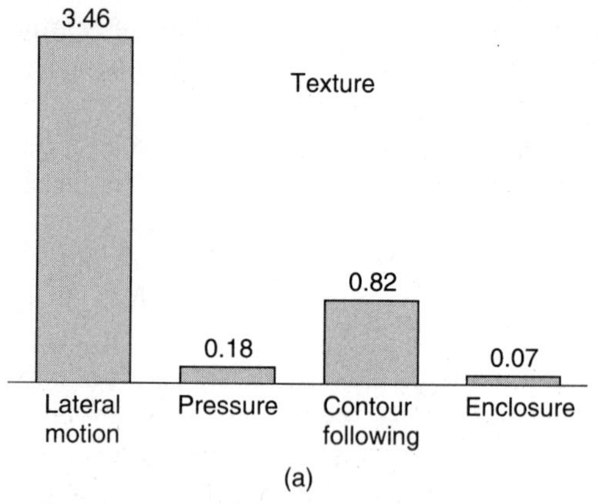

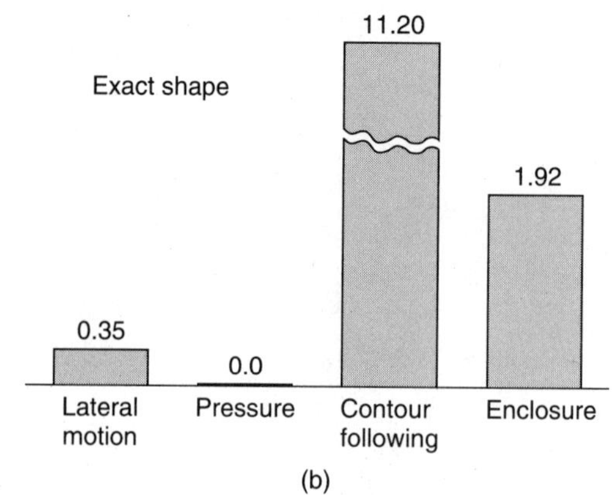

Figure 11.31
Average duration in seconds that subjects used various exploratory procedures (a) to judge texture and (b) to judge an object's exact shape. Subjects use lateral motion and contour following to judge texture. Contour following is the major exploratory procedure used to judge exact shape. (Based on data in Lederman & Klatzky, 1987.)

to judge texture, and they use enclosure and contour following to judge exact shape.

One of the themes of this book is that our perceptions are influenced by the neural activity generated at the beginning of the sensory process, when the receptors are stimulated. But in our brief description of active touch, we have seen that there is more to perception than just the stimulation of receptors. For example, we saw that naive braille readers identify braille characters more accurately by using active touch than by using passive touch, but experienced readers identify characters equally well by using either method. It is likely that the difference between inexperienced and experienced readers has to do both with how they scan the characters and with enhanced cognitive processing in the experienced readers (Foulke, 1991; Kruger, 1982; Pick, Thomas, & Pick, 1966). You may also have noticed, in doing the demonstration, that it was easier to identify familiar objects than unfamiliar objects, and that, if you had trouble identifying a particular object, you were engaging in a good deal of cognitive activity as you were attempting to figure out what the object was. Thus, although our perceptions begin with the stimulation of receptors, they are often influenced by higher-order cognitive factors. To end this chapter, we consider some examples of how people's perceptions are determined not only by signals generated by receptors in the skin but also by higher-level activity, which includes thoughts, emotions, and cultural influences. We will focus our attention on pain perception, since there is a great deal of evidence that pain is affected by factors in addition to stimulation of the receptors.

Central Influences on Pain Perception

We have considered the somatosensory cortex twice in this chapter—first, to show that the homunculus on the cortex is distorted to correspond to the acuity of different areas on the skin, and then, to show that there are central neurons that respond to specific orientations and directions of movement on the skin. In both cases, these properties of the cortex reflect the neural processing that occurs as nerve signals travel from the skin to the cortex and the neural processing that occurs in the cortex itself. In this section we are going to consider not the signals that travel toward the cortex or within the cortex, but the signals that the cortex sends back *toward* the skin. We will see that, far from being only a receiver of signals from "below," the cortex can send out its own signals to influence the neural messages that reach it.

Before considering the neural evidence for this idea, let's discuss some psychophysical observations relevant to this cortical "feedback." These observations show that the perception of pain can be affected by factors other than the stimulation of pain receptors in the skin. Pain can be affected by factors such as the general situation in which the pain stimulus occurs, and by the person's culture and previous experiences. If this sounds familiar, it is because we have encountered similar observations for the sense of vision. Remember the rat–man demonstration of Figure 1.25, an example of how prior experiences can influence our visual perceptions. First seeing Figure 1.25 causes people to perceive Figure 1.28 as a rat, whereas first seeing Figure 1.31 causes people to perceive it as a man. This demonstration, and many others, show that our visual perceptions can be affected by more than simply how our receptors are stimulated. Let's consider evidence that this effect also occurs for the perception of pain.

Culture, Experience, and Pain Perception

The effect of culture on pain perception is most graphically demonstrated by rituals such as the hook-swinging ceremony practiced in some parts of India (Kosambi, 1967). This ritual is described as follows:

> The ceremony derives from an ancient practice in which a member of a social group is chosen to represent the power of the gods. The role of the chosen man (or "celebrant") is to bless the children and crops in a series of neighboring villages during a particular period of the year. What is remarkable about the ritual is that steel hooks, which are attached by strong ropes to the top of a special cart, are shoved under his skin and muscles on both sides of his back [Figure 11.32]. The cart is then moved from

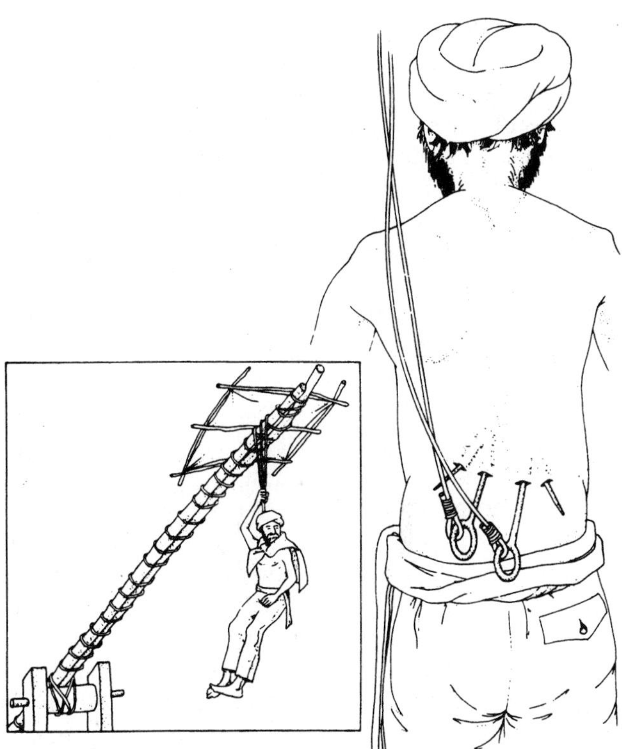

Figure 11.32
Right: Two steel hooks in the back of the "celebrant" of the Indian hook-swinging ceremony. Inset (left): The celebrant hangs onto the ropes as a cart takes him from village to village. After he blesses each child and farm field in the village, he swings freely, suspended by the hooks in his back. (From Kosambi, 1967.)

> village to village. Usually the man hangs on to the ropes as the cart is moved about. But at the climax of the ceremony in each village, he swings free, hanging only from the hooks embedded in his back, to bless the children and crops. Astonishingly, there is no evidence that the man is in pain during the ritual; rather, he appears to be in a "state of exaltation."
> (Melzack & Wall, 1983, p. 28)

It seems unbelievable that a person could feel little or no pain while hanging from steel hooks embedded in his back, but this is apparently what happens during the hook-swinging ceremony. Perhaps the ceremony has some effect on the celebrant's mental state that prevents him from feeling pain. There are, in fact, many examples of situations in which a person's mental state affects pain perception. For example, H. K. Beecher (1972) found that morphine reduces pathologically produced pain that is accompanied by anxiety but doesn't reduce experimentally produced pain that is unaccompanied by anxiety. Results such as these led Beecher to state a new principle of drug action: "Some agents are effective only in the presence of a required mental state."

Further evidence that a person's mental state affects his or her perception of pain is provided by Beecher's observation that 25 percent of men seriously wounded in battle requested a narcotic for pain relief, whereas over 80 percent of civilians about to undergo major surgery requested pain relief. The difference in these percentages can be traced to the mental states of the patients. The civilians were upset about their surgical wounds because they associated these wounds with disturbing health problems. The soldiers, on the other hand, were not upset about their battle wounds because they knew that their wounds would provide escape from a hazardous battlefield to the safety of a behind-the-lines hospital. Later, we will see that the soldier's apparent insensitivity to pain may also be a reaction to the stress of combat.

In a hospital study, when surgical patients were told what to expect and were instructed to relax to alleviate their pain, they requested fewer painkillers following surgery and were sent home 2.7 days earlier than patients who were not given this information. Studies have also shown that up to 35 percent of patients with pathological pain get relief from taking a **placebo,** a pill that they believe contains painkillers but that, in fact, contains no active ingredients (Weisenberg, 1977). The above examples indicate that, just as visual perception is affected by central influences, so is pain perception. Although we have few clues regarding the neural mechanisms responsible for these central effects in vision, we do have some idea of how such neural mechanisms may work for pain. One explanation is called *gate control theory.*

Gate Control Theory

Gate control theory was proposed by Ronald Melzack and Patrick Wall (1965, 1988) to explain how pain perception can be affected both by central influences, such as those described above, and by tactile stimuli, as when rubbing the skin causes a decrease in the perception of pain. The gate control system consists of cells in an area of the dorsal horn of the spinal cord called the **substantia gelatinosa** and transmission cells (T cells) located in the dorsal horn near the substantia gelatinosa (Figure 11.33a). The neural circuit containing these cells is shown in Figure 11.33b. The output of the gate control system, which flows through the T cells, is controlled by two kinds of substantia gelatinosa cells: one (SG+), which opens the pain gate by sending excitation to the T cells, and one (SG–), which closes the pain gate by sending inhibition to the T cells.

If only small-diameter S fibers (which correspond to the nociceptors we discussed earlier) are active, then pain occurs, since SG+ cells open the gate. However, when large-diameter L fibers (which carry information about nonpainful tac-

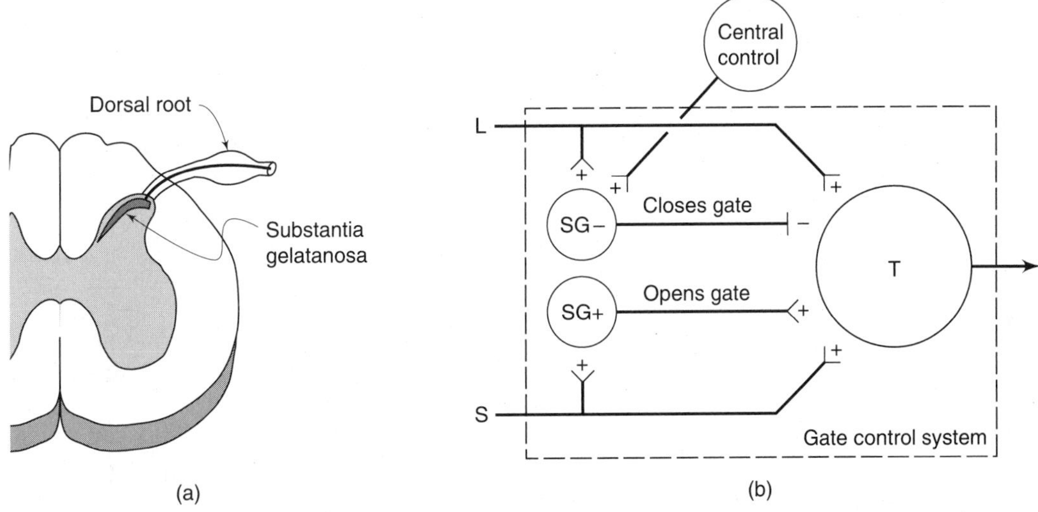

Figure 11.33

(a) Cross section of the spinal cord showing fibers entering through the dorsal root and the substantial gelatinosa (shaded) where some of the cells of Melzack and Wall's (1988) proposed gate control circuit are located. (b) The circuit proposed by Melzack and Wall for their gate control theory of pain perception. See text for details.

tile stimuli) are active, then pain is inhibited, since SG– cells close the gate. Thus, when potentially damaging stimuli cause S fibers to fire, the SG+ cells open the gate and pain increases. But when more gentle stimuli, such as massage, rubbing, and gentle vibration, cause activity in L fibers, the SG– cells close the gate and pain decreases. Another way to close the gate is by sending signals from the structure labeled "Central control," which represents the brain or other higher-order structures. As indicated in the circuit, signals from central control close the gate by activating SG– cells.

The idea that signals from the brain can reduce the perception of pain is supported by the results of an experiment done by David Reynolds (1969). Reynolds first showed that a rat responds vigorously when its tail or paw is pinched. He then electrically stimulated an area in the rat's midbrain and showed that, while the stimulation was on, the animal no longer seemed to mind having its tail or paw pinched. In fact, Reynolds was even able to perform abdominal surgery on rats with no anesthesia other than the electrical stimulation. This effect of brain stimulation, which has been confirmed in many experiments, is called **stimulation-produced analgesia (SPA).**

Gate control theory can also be applied to the pain reduction achieved by **acupuncture.** Acupuncture is the procedure practiced in China in which fine needles are inserted into the skin at certain "acupuncture points" in the body (Figure 11.34). Twirling these needles, or passing electrical current through them, can produce **analgesia** (the elimination of pain without loss of consciousness), which makes it possible to perform major surgery in a totally awake patient (Melzack, 1973). Gate control theory would suggest that the stimulating needles close the gate by activating L fibers or fibers descending from the brain.

Although the specific neural circuit proposed by gate control theory is not accepted by

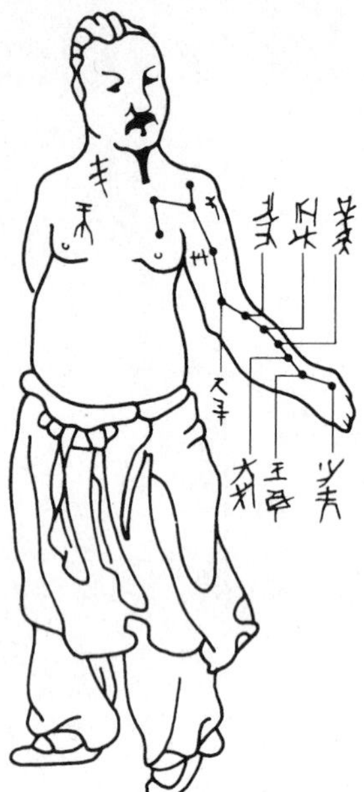

Figure 11.34
A Chinese acupuncture chart showing the sites for the insertion of acupuncture needles.

everyone (Nathan, 1976), the idea that our perception of pain depends not only on input from nociceptors, but also on input from nonpain fibers and central influences, is supported by a large amount of research. Another important development in our understanding of how central factors influence pain perception is the discovery that the nervous system contains endorphins, chemicals produced by the brain that have properties closely related to those of opiates such as morphine.

Endorphins

A family of substances called **endorphins,** endogenous (naturally occurring in the body) morphinelike substances, have been found in the nervous system, and a large amount of evidence indicates that these substances have powerful analgesic effects (Mayer, 1979; Watkins & Mayer, 1982). Some of the evidence supporting this idea is that these endorphins are found at the same sites in the brain where opiates such as morphine act to produce their analgesic effects. Thus, naturally occurring endorphins may reduce pain in the same way as pain-reducing drugs. In fact, stimulation-produced analgesia works best when these endorphin sites are stimulated, a finding suggesting that SPA works by releasing endorphins into the nervous system.

Further evidence linking endorphins with pain relief is provided by the effects of the drug **naloxone,** a substance known to inhibit the activity of opiates. Injection of naloxone decreases the effect of pain-reducing procedures such as acupuncture and SPA. This decreased effect would occur if naloxone were inhibiting the activity of the endorphins responsible for the effects of acupuncture and SPA. This evidence strongly suggests that the brain uses endorphins to control pain.

In addition to decreasing the effect of acupuncture and SPA, naloxone decreases the analgesic effect of placebos. Since placebos contain no active chemicals, their effects have always been thought to be "psychological." Now endorphins provide a physiological explanation of the psychological effect of placebos and, presumably, of some of the other psychological effects we have discussed.

Finally, let's consider another effect that may involve endorphins: **stress-induced analgesia,** a decrease in sensitivity to pain that occurs in stressful situations. There is experimental evidence that animals' pain thresholds are raised by stress (Jessell & Kelly, 1991), and there is anecdotal evidence, such as the reports of soldiers who report feeling little pain in response to serious wounds, that stress decreases pain in humans as well. A possible example of stress-induced analgesia is provided by African explorer David Livingstone's description of a dramatic encounter he had with a lion during one of his early journeys to find the source of the Nile:

> I hear a shout. Starting, and looking half round, I saw the lion just in the act of springing upon

me. I was upon a little height; he caught my shoulder as he sprang, and we both came to the ground below together. Growling horribly close to my ear, he shook me as a terrier does a rat. The shock produced a stupor similar to that which seems to be felt by a mouse after the first shake of the cat. It caused a sort of dreaminess in which there was no sense of pain nor feeling of terror, though quite conscious of all that was happening. It was like what patients partially under the influence of chloroform describe, who see all the operation but feel not the knife. . . . The shake annihilated fear, and allowed no sense of horror in looking round at the beast. This peculiar state is probably produced in all animals killed by the carnivora; and if so, is a merciful provision by our benevolent creator for lessening the pain of death. (David Livingstone, *Missionary Travels,* 1857; quoted in Jessell & Kelly, 1991, p. 398)

This example of a possible central influence on pain perception emphasizes the complexity of the processes involved in the perception of pain. In this chapter, we have only scratched the surface of this complexity, both for pain and for the other somatic senses. Just as we observed for the senses of vision and hearing, for the somatic senses we know quite a bit about how receptors are stimulated, how they respond, and how they send their signals along various pathways on their way to the brain. However, our knowledge of processes later in the system, such as the neural processing that occurs in the cortex, is more limited for all of these senses. As we consider the chemical senses of olfaction and taste in the next chapter, we will experience an analogous situation, as we see that we know far more about responses that occur near the beginning of the system than we do about responses that occur as we move toward the cortex.

"Other Worlds" of Perception in Humans

In each of the Other Worlds of Perception features in this book, we have expressed the idea that many animals may live in different perceptual worlds from humans. But what about the perceptual worlds of individual humans? Are the perceptual worlds of others the same as or different from the one that you experience? In this section we suggest two things: (1) There may be differences between different people's perceptual experiences, and (2) since perceptual experience is private, we have no way of really knowing what other people are experiencing.

From what we have learned about perception so far, we know that it is likely that different people perceive things differently. We know, for example, that some people's vision is color-deficient, because they are missing a cone visual pigment. We also know that even people with normal color vision may perceive colors differently because of subtle differences in the absorption spectra of their visual pigments.

But while evidence such as this allows us to conclude that two people may have different perceptual experiences to the same stimuli, can we ever conclude that two people are having exactly the *same* perceptual experiences? For example, what if two people adjusted the proportion of colors in a color-matching experiment in exactly the same way (see Chapter 4) and described their experience of perceiving these colors in exactly the same way. Can we conclude, based on these observations, that these two people are *experiencing* the colors in the same way? The answer to this question is that we cannot conclude that these two people experience the colors in the same way because, since perceptions are private experiences, we have no way of really knowing what someone else is experiencing. (Also see page 163.)

This idea that perceptions are private experiences is true of all of the senses but is particularly true of pain. Of all of the senses, pain is perhaps the one that is hardest for people to compare. For example, if someone says, "That looks blue to me" and both you and the other person have trichromatic color vision, you have at least an approximate idea of what seeing "blue" means to the other person. If, however, the person says, "That hurts a lot," what do you know about their experience? "Hurting a lot" can describe a wide variety of experiences even for yourself, so how can you begin to know what it means for someone else? Let's consider some of the reasons that people find it difficult to relate to other people's reported pain experiences.

Pain can occur in the absence of outward stimulation. Although outward stimulation is sometimes present, as when a person hits their finger with a hammer, pain can occur even if no external stimulus is present. Headaches, the ache of a bad back, and the pain of terminal cancer all come from within. One of the most dramatic examples of pain from within is provided by the **phantom limb** phenomenon, which is reported by people who have had a limb amputated. These people not only experience the missing limb as still being present but may also report feeling pain in the phantom limb (Melzack, 1992).

Pain is influenced by factors other than tissue damage. We have already seen that pain can be influenced by factors such as a person's emotions, expectations, and level of stress. Thus, when two people are exposed to exactly the same external stimulus, it is possible for one person to experience intense pain while the other experiences little or no pain.

People describe the same stimulus differently. To show that different people describe the same

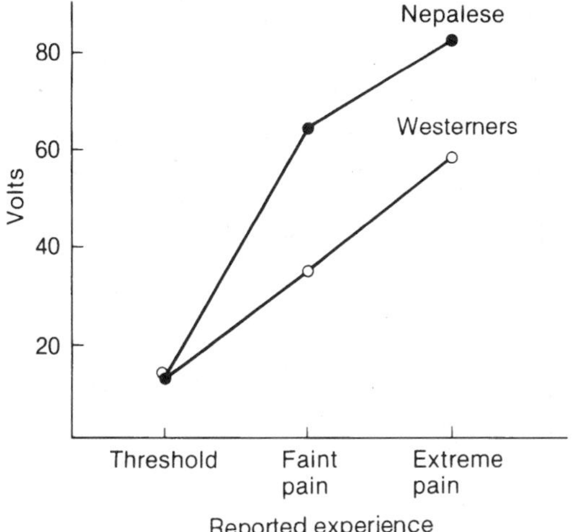

Figure 11.35

Results of the Clark and Clark (1980) experiment, which shows that the threshold for detecting a shock was the same for Western mountain climbers and their native guides (Nepalese porters) but that the Nepalese porters took much higher levels of shock before reporting "faint pain" or "extreme pain." Was this result due to the Nepalese porters' lower sensitivity to the shocks, or did it simply mean that the porters were more conservative responders?

stimulus differently, Clark and Clark (1980) presented electric shocks to a group of Western subjects and to a group of Nepalese subjects and asked them to indicate the intensities at which they experienced "faint pain" and "extreme pain." Although the Westerners and the Nepalese had identical thresholds for experiencing the shock, the Nepalese subjects required much higher stimulus intensities before they said they experienced "faint pain" and "extreme pain" (Figure 11.35).

Analyzing these results by means of a procedure based on signal detection theory (see Appendix 5) indicated that there was probably no difference in the way both groups of subjects experienced the intensity of the shocks. Apparently the Nepalese subjects withstood higher levels of pain before reaching what they considered faint pain or extreme pain. It may be that Nepalese culture teaches people to withstand pain without complaining. A similar result was obtained by Evelyn Hall and Simon Davies (1991), who had female varsity track athletes and female nonathletes rate their level of pain in response to immersion of their hands in freezing ice water. Although both groups were exposed to the same stimulus, the athletes rated their pain as less intense (their average rating was 76 on a 150-point pain scale) than the nonathletes (whose average was 130). Perhaps the athletes had learned, during painful training workouts, to withstand higher levels of pain.

Pain may mean different things to different people. The idea that "pain" may signify different perceptual qualities to different people is supported by the following statement by Katherine Foley, Director of Pain Services at Memorial Sloan-Kettering Cancer Center in New York: "I believe that when a person tells you there is pain, they are telling you there is discomfort. But it may not be a physical pain; it may be a metaphor for anxiety or depression or spiritual suffering. We use 'pain' for physical and emotional distress, and sometimes people don't make the distinction very well" (quoted in Rosenthal, 1992, p. B6).

All of the above examples illustrate why pain is so difficult to describe or to compare between people, and why it is therefore perhaps the most private of our senses. Looking at the privateness of experience in this way, we are saying that, although all sensory experiences are private, some, such as pain, are more difficult to share than others. If, however, you believed in a philosophical position called *radical skepticism*, the special properties of pain we have been describing wouldn't matter to you because you would believe that it is at least as likely as not that other people perceive everything very differently from the way you do. The following statement, taken from a paper on the philosophy of perception, represents how a radical skeptic would view perception:

> You may not have the slightest reason to suppose that visual perception gives other people experiences that are anything like your visual experiences. Perhaps someone else has what would be for you auditory experiences. When he looks at the blue sky, it is like hearing middle C on the piano is for you. There seems to be no way to tell, since he would have been brought up to call that sort of experience the experience of blue. Indeed it is not clear that you have the slightest reason to suppose that others have anything you could recognize as experience. When others see things, their visual experience may be something you could not even imagine.
>
> But then is there any reason to suppose that others have experience at all? The suggestion is that, even if you could know that the people around you were made of flesh and blood, born of women, and nourished by food, they might for all you know be automatons, in the sense that behind their elaborate reactions to the environment there might be no experience.
>
> But the suggestion is not merely that you do not *know* whether other people have any experience but also that you haven't the slightest reason to suppose they do. Similarly, it might be suggested that you haven't the slightest reason to believe you are in the surroundings you suppose you are in, holding a book, reading an article on epistemology. It may look to you and feel as it would look and feel if you were in those surroundings, holding a book and reading an article. But various hypotheses could explain how things look and feel. You might be sound asleep and dreaming, or a playful brain surgeon might be giving you these experiences by stimulating your cortex in a special way. You might really be stretched out on a table in his laboratory with wires running into your head from a large computer. (Harman, 1974, p. 42)

Although the thoughts expressed in this intriguing passage could logically be true, the idea that someone who looks at the blue sky might experience what I hear when I play middle C on the piano seems very unlikely to me. I would instead guess that the quality of seeing or the quality of hearing is probably similar for most people, but that, within this sameness, there is also variation that causes people's perceptions to be slightly different. My guesses notwithstanding, however, remember that we have no way of knowing exactly what other people are experiencing, not only for pain, which we have seen is perhaps the most difficult of the senses to compare between people, but also for all of the other senses. Thus, not only does your dog or cat experience a different perceptual world from yours, but for all you know, other humans also experience a different perceptual world from yours.

Organization of the Somatosensory Cortex

Perceptual development can be studied in a number of ways. We can observe the process of normal development by determining what capacities are present at birth and by then following the changes in these capacities as the organism matures (see the Developmental Dimensions in Chapters 1, 2, 4, 5, 6, 7, 9, and 10). Or we can observe what happens when we interfere with normal development, as when kittens or monkeys are reared with only monocular vision. These monocular rearing experiments showed that a period early in an organism's life, called the *sensitive period*, is important in development (Developmental Dimension, Chapter 3).

Another way of interfering with normal development is to destroy receptors and observe what happens at higher levels in the system. For example, destroying the hair cells of the guinea pig's cochlea causes changes in the tonotopic map in the cortex (Developmental Dimension, Chapter 8). Experiments such as these in hearing and analogous ones in which retinal receptors were destroyed (Kaas et al., 1990) show that, even in adult animals, the organization of the cortex can change in response to changes in the stimulation reaching the cortex.

Do similar effects occur in the somatosensory system? A number of experiments indicate that the answer to this question is yes. Preventing receptor signals from reaching the monkey's somatosensory cortex by surgically removing a finger or by damaging the peripheral nerves conducting impulses from the skin changes the map of the body on the somatosensory cortex, even in fully mature adults (Kaas, 1991; Kaas, Merzenich, & Killackey, 1983; Pettit & Schwark, 1993; Pons et al., 1991).

Maps in the somatosensory cortex can also be changed by *increasing* the stimulation reaching the cortex. William Jenkins and Michael Merzenich (1987) showed that increasing stimulation on a specific area of the skin causes an expansion of the cortical area receiving signals from that area of skin. They demonstrated this effect by training monkeys to carry out a task that involved the heavy use of a particular location on one fingertip. Comparison of the cortical maps of the fingertip measured just before the training and three months later shows that the area representing the stimulated fingertip was greatly expanded after the training (Figure 11.36). Thus, the cortical area representing part of the fingertip, which is large to begin with, becomes even larger if the area receives a large amount of stimulation. The sensory homunculus of Figure 11.20 is not, therefore, a permanent, static map but can be changed by experience.

The brain's capacity to change goes beyond the expansion of a cortical area associated with a heavily used part of the body. Sharon Clark and her co-workers (1988) showed that cortical organization depends not only on how much a particular part of the body is used, but also on how the stimulation of nearby parts of the body is coordinated in time. Clark's experiment was based on the fact that there is normally a sharp dividing line between the areas of cortex representing adjacent fingers, so that, as we move across a monkey's cortex, we encounter neurons that respond only to the stimulation of one finger, and then we suddenly begin encountering neurons that respond to the other finger. Thus, in the normal cortex, there is a discontinuous border between the two fingers, with no neurons responding to both fingers.

What causes this discontinuity between the fingers' representation on the cortex? One possibility is that the discontinuity in the cortex simply mirrors the physical separation between the two fingers. However, Clark suspected that another mechanism might be at work. Perhaps the separation on the cortex reflects the fact that dif-

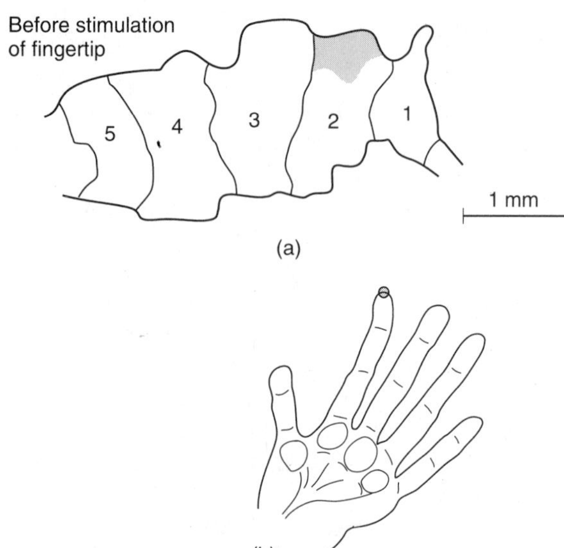

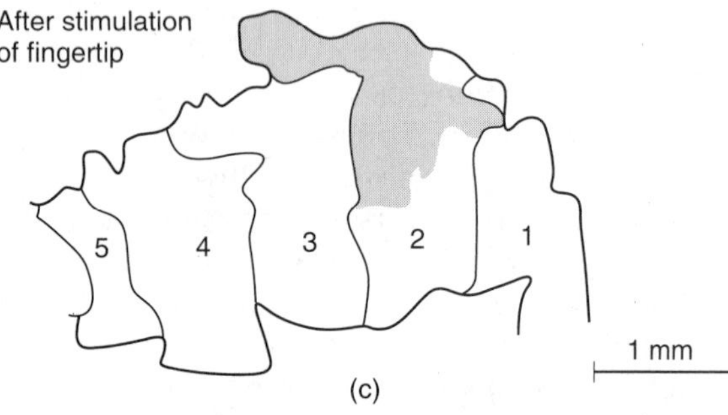

Figure 11.36

(a) Each numbered zone represents the area in the somatosensory cortex that represents one of a monkey's five digits. The shaded area on the zone for digit 2 is the part of the cortex that represents the small area on the tip of the digit shown in (b). (c) The shaded region shows how the area representing the fingertip increased in size after this area was heavily stimulated over a two-month period. (From Merzenich et al., 1988.)

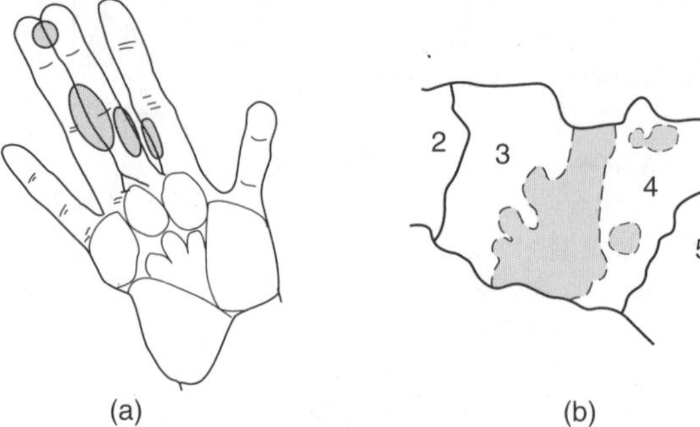

Figure 11.37

(a) A monkey's hand showing how digits 3 and 4 were sutured together by Clark and co-workers (1988). (b) Zones in the somatosensory cortex that represent digits 3 and 4 determined seven months after suturing the digits. Neurons within the shaded area responded to stimulation of both digits 3 and 4, a situation that does not occur when the fingers are kept separated.

ferent fingers do not always experience stimuli that are coordinated in time. If this is so, then presenting stimuli in a coordinated way might eliminate the discontinuity between the maps. Clark achieved this coordination by suturing together an owl monkey's third and fourth fingers, as shown in Figure 11.37a.

When Clark measured the cortical map seven months later, she found that the border between the areas representing these two fingers was no longer abrupt, but that there were a large number of neurons that responded to both fingers (Figure 11.37b). It was as if, for these neurons, two fingers had been changed into one.

Clark's result confirms a conclusion we have reached in other Developmental Dimensions: Cortical organization is not rigid and static but can be shaped by the stimulation reaching the cortex. What Clark's results add to this picture is the idea that cortical organization depends not only on the mere presence of signals reaching the cortex, but also on the relationships between the signals reaching the cortex from different parts of the body.

Review

Outline	Questions
• The somatic senses includes the cutaneous sensations (based on stimulation of receptors in the skin), proprioception (the sense of position of the limbs), and kinesthesis (the sense of movement of the limbs).	1. Why might we be justified in saying that the perceptions we feel through our skin are as important to our survival as those provided by vision and hearing? (459)

Anatomy of the Somatosensory System

• The anatomy of the somatosensory system reveals a parallel with the visual system: Both have a number of different kinds of receptors spread over an area, and both have parallel pathways that ascend toward the cortex.	2. What are some functions of the skin? Describe the epidermis, the dermis, the four major receptors in the skin, and the two kinds of skin. (460) 3. How do nerve fibers from the skin enter the spinal cord? Describe the two pathways in the spinal cord and the functions that they serve. Where do fibers in the somatosensory system synapse in the thalamus and in the cerebral cortex? (462)

The Psychophysics and Physiology of Tactile Perception

• There are four different types of receptors in the skin that are sensitive to different types of stimulation, and four types of mechanoreceptive fibers are associated with these receptors. These physiological mechanisms result in four psychophysical channels that respond to different frequencies of tactile stimulation.	4. Describe the four psychophysical channels for tactile perception, including the following information from Table 11.1 about each channel: the approximate frequency range of stimuli; the receptor structure associated with the channel; and the perception associated with the channel. (463)
• The properties of different neural channels of the transmission of tactile information have been determined by recording from single neurons in animals and by using a technique called microneurography in humans.	5. Describe the four types of mechanoreceptive fibers determined by physiological experiments, including the following information from Table 11.2 about each channel: the name of the fiber; the rate of adaptation; the receptive field size; the receptor structure; and the best stimulus. What is the relationship between the physiological and the psychophysical channels? (464)

- One reason that there are fibers that respond to different kinds of stimuli is that these fibers are associated with different receptors.
- We can determine which psychophysical channels are activated by a specific frequency of stimulation from the curves in Figure 11.6. At some frequencies, only one channel is active. For stimulus intensities near threshold, only the most sensitive channel responds. At intensities above threshold, a number of channels can be activated at once.
- Psychophysical experiments in which subjects identify raised letters presented to the skin, show that some letters are often confused with others, and that others are rarely confused with others.

6. Describe Lowenstein and Skalak's experiments in which they showed how the Pacinian corpuscle receptor influences neural firing. (465)
7. How do different types of fibers respond when raised dots are scanned across the skin? What is a spatial event plot? How closely do spatial event plots generated by each type of fiber correspond to the raised dots that generated these plots? Based on this result, which type of neuron is responsible for perceiving details? (467)
8. Describe the relation between the errors subjects make in identifying raised letters and the neural spatial event plots generated by these letters. (468)

Neural Responses to Temperature and Potentially Damaging Stimuli

- Neurons called thermoreceptors respond best to certain temperatures, and neurons called nociceptors respond best to potentially damaging stimuli.

9. Describe the basic properties of thermoreceptors and nociceptors. (469)

Neural Processing

- A basic fact that enables us to demonstrate a connection between neural processing and perception in the somatosensory system is that tactile acuity, measured psychophysically, is better on some parts of the body than on others.

10. Describe how the two-point threshold varies on different parts of the body. What is the relationship between the size of the two-point thresholds and the sizes of receptive fields on the skin? How does receptive field size change as we move from the fingers to the hand to the arm? (471)

- The magnification factor for vision, described in Chapter 3, refers to the fact that the fovea, the area of the retina with the best acuity, is represented by a cortical area far out of proportion to its size. An analogous situation exists in the somatosensory system.

11. What is the homunculus for tactile perception? How is it determined? How do the sizes of various parts of the homunculus compare to the two-point thresholds of corresponding parts of the body? (472)

- As we described in Chapter 3, there are neurons in the visual system that respond best to specific types of stimuli. A similar situation exists in the somatosensory system.

12. Where are the neurons with center–surround receptive fields located in the somatosensory system? Describe neurons in the cortex that respond to more specialized stimulation. (475)

Active Touch

- Active touch occurs when we actively explore an object by moving our fingers over its surface. Two aspects of active touch that have been studied physiologically are proprioception and kinesthesis.

13. Define proprioception and kinesthesis. What is a muscle spindle? Describe the kinds of neurons in the muscle spindles and the joints that respond to specific limb movements and positions and also those that respond when a number of joints are flexed simultaneously. (476)
14. What is the role of mechanoreceptive fibers in active touch? Describe neurons that respond when objects are grasped. (477)

- One of the major concerns of the psychophysical approach to active touch has been the comparison of (1) the experiences caused by active and passive touch and (2) people's ability to identify objects based on active and passive touch.

15. How did J. J. Gibson differentiate between the experiences of active and passive touch? What is the evidence regarding people's ability to identify objects using active and passive touch? (478)

- Haptic perception occurs during the exploration of a three-dimensional object with the fingers and the hands.

16. What systems are involved in identifying an object by haptic touch? What are the exploratory procedures, and what is their role in haptic perception? (481)

Central Influences on Pain Perception

- A number of psychophysical observations show that the perception of pain is influenced by factors other than the stimulation of pain receptors in the skin.

17. Describe situations that provide evidence that a person's mental state may affect his or her perception of pain. (483)

- Gate control theory is a physiological model proposed by Melzack and Wall to explain how pain perception is affected by signals in addition to those received from nociceptors in the skin.

18. Describe the operation of Melzack and Wall's gate control circuit. Be sure you understand how touch stimuli and thoughts can inhibit the perception of pain, according to this theory. How does gate control theory explain the analgesic effect of acupuncture? What is stimulation-produced analgesia, and how does it relate to gate control theory? (484)

- Endorphins are naturally occurring substances that have been found in the nervous system and that have powerful analgesic effects.

19. Describe the evidence linking endorphins and pain relief. What is stress-induced analgesia? (486)

"Other Worlds" of Perception in Humans

- Just as animals may live in a perceptual world different from that of humans, so humans may experience perceptual worlds different from those experienced by other humans. It is possible that there are differences in different people's perceptual experiences, but since perceptual experience is private, we have no way of really knowing what other people are experiencing.

20. Cite some evidence that people may perceive things differently. If two people exhibit the same behavior with regard to perceptual stimuli and describe their experiences in the same way, can we conclude that they actually experience them in the same way? Explain your answer. (488)

- The idea that perceptions are private experiences is true of all the senses, but it is particularly true of pain.

21. Cite four reasons supporting the statement on the left. How does a radical skeptic view perception? What is the "bottom line" on our ability to know what others are experiencing? (488)

Organization of the Somatosensory Cortex

- There are a number of ways to study perceptual development, including observing the process of normal development, determining what happens when we interfere with normal development early in an animal's life, or determining what happens when we modify the stimulation reaching the cortex of an adult animal by destroying receptors or changing the nature of the stimulation presented to the receptors.

22. What happens to the somatosensory cortex when: (a) A monkey's finger is removed? (b) Stimulation to a monkey's fingertip is increased? (c) A monkey's fingers are sutured together so that stimulation presented to these fingers is coordinated in time? From these results, what overall conclusion can we draw about the stability of cortical organization? (491)

The Chemical Senses

This chapter introduces olfaction (smell) and taste, two senses that respond to chemical stimuli: The sense of smell creates perceptions in response to gaseous molecules that contact receptors in the nose; the sense of taste creates perceptions in response to liquid molecules that contact receptors in the mouth. These senses have been called *molecule detectors* because they endow these gas and liquid molecules with distinctive smells and tastes (Cain, 1988; Kauer, 1987).

In this chapter, as in the ones preceding it, we are interested in looking for commonalities between the senses. But we are also interested in what is unique about olfaction and taste. One of their most unique properties is that the stimuli responsible for tasting and smelling are on the verge of being assimilated into the body (Scott & Giza, 1995). As a result, taste and smell are often seen as "gatekeepers," whose function it is to detect things that would be bad for the body and that should therefore be rejected, and to identify things that the body needs for survival and that should therefore be consumed. This gatekeeper function is aided by the large affective (emotional) component of taste and smell, since things that are bad for us often taste or smell unpleasant, and things that are good for us generally taste good. In addition to assigning "good" and "bad" affect, smelling an odor that you associate with a past place or event can trigger long-lost memories, which in turn may create emotional reactions.

Our plan in this chapter is to consider olfaction and taste separately. We do this because they are served by two separate systems and appear to operate according to different principles. As has been the case throughout this book, our emphasis is on discovering the sensory code, and when we describe how different qualities are signaled in olfaction and taste, we will see that neurons in the taste system tend to be narrowly tuned to specific taste qualities, whereas neurons in the olfactory system are more broadly tuned.

Although we will be considering olfaction and taste separately, we will see that they often interact with each other. Thus, as you savor the "taste" of something you are eating, you may, in fact, be experiencing a combination of taste and smell. We will have more to say about this combined experience, which is called *flavor*, when we begin our discussion of taste in the second part of the chapter.

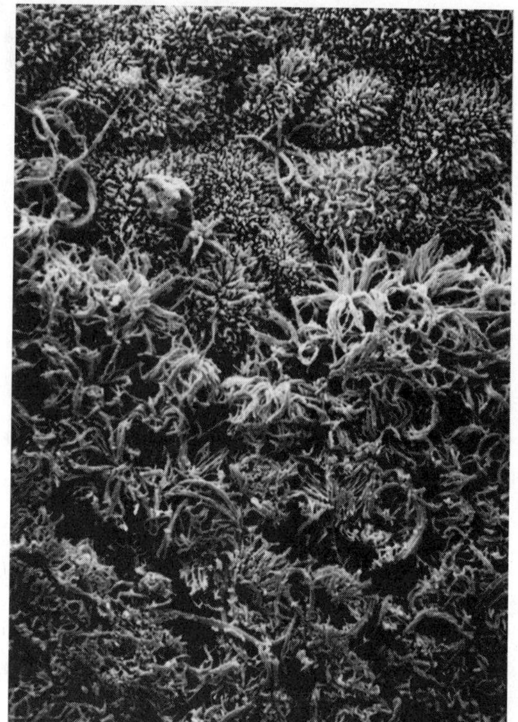

Figure 12.1
A scanning electron micrograph of the surface of the olfactory mucosa. The region in the foreground is densely covered with the cilia of olfactory receptors. (From Morrison & Moran, 1995.)

The Uses of Olfaction

The picture in Figure 12.1 is not an underwater coral reef or vegetation on the forest floor. It is a picture of the surface of the *olfactory mucosa*, the dime-sized region inside the nose that contains the receptors for the sense of smell. As we study the physiological basis of olfaction, we will be interested in how these receptors in the mucosa respond to different types of smell stimuli and how this responding contributes to the neural code for smell. But before we begin our description of the receptors and our search for the neural code, we will consider some of the functions of olfaction and some psychophysical findings about the characteristics of olfaction in humans.

Olfaction is extremely important in the lives of other species, since it is often their main window on the sensory environment (Ache, 1991). One important contrast between human and animal olfaction is captured by the fact that many animals are classified as **macrosmatic** (having a keen sense of smell that is important to their survival), whereas humans are classified as **microsmatic** (having a less keen sense of smell that is not crucial to their survival). The survival value of olfaction for many animals lies in their use of olfaction to provide cues to orient themselves in space, to mark territory, and to guide them to specific places, to other animals, and to food sources (Holley, 1991). Olfaction is also extremely important in reproduction, since it triggers mating behavior in many species (Doty, 1976; Pfeiffer & Johnston, 1994).

Although smell may not be crucial to the survival of humans, the vast sums of money spent yearly on both perfumes and deodorants, as well as the emergence of a new billion-dollar-a-year industry called *environmental fragrancing*, which offers products to add pleasing scents to the air in both homes and businesses, attest to the fact that the role of smell in our daily lives is not inconsequential (Owens, 1994).

But perhaps the most convincing argument for the importance of smell to humans comes from those who suffer from **anosmia**, the loss of the ability to smell due to injury or infection. People suffering from anosmia describe the great void created by their inability to taste many foods because of the close connection between smell and flavor. One woman who suffered from anosmia and then briefly regained her sense of smell stated, "I always thought I would sacrifice smell to taste if I had to choose between the two, but I suddenly realized how much I had missed. We take it for granted and are unaware that *everything* smells: people, the air, my house, my skin" (Birnberg, 1988; quoted in Ackerman, 1990, p. 42). Olfaction is more important in our lives than most of us realize, and while it may not be essential to our survival, life is often enhanced by our ability to smell and becomes a little more dangerous if we lose the olfactory warning system that alerts us to spoiled food, leaking gas, or smoke from a fire.

Some Facts about Human Olfaction

Perhaps because we often take smell for granted and because it is not crucial to our survival, the powers of the human olfactory system have often been underrated, especially when the human sense of smell is compared to the capabilities of other animals. To achieve a more realistic picture of the capabilities of the olfactory system, let's consider some facts that show that our powers of olfaction are greater than we might at first suppose.

Fact #1: Although Humans Are Less Sensitive to Odors Than Other Animals, Our Olfactory Receptors *Are Exquisitely Sensitive*

The human sense of smell is less acute than that of many animals. For example, rats are 8 to 50 times more sensitive to odors than humans, and dogs are from 300 to 10,000 times more sensitive, depending on the odorant (Laing, Doty, & Breipohl, 1991). But even though other animals can detect odors that humans are not aware of, the human's individual olfactory receptors are as sensitive as any animals'. H. deVries and M. Stuiver (1961) demonstrated this by showing that human olfactory receptors can be excited by the action of just one molecule of odorant. Nothing can be more sensitive than one molecule per receptor, so the human's lower sensitivity to odors compared to that of other animals must be due to something else. That something else is the *number* of receptors: only about 10 million in humans compared to about 1 billion in the dog (Moulton, 1977; Dodd & Squirrell, 1980).

Fact #2: Humans Are Surprisingly Good at Detecting Differences in Odor Intensity

The ability to detect differences in intensity is indicated by the difference threshold—the smallest difference in intensity between two stimuli that can just be detected (see Chapter 1). In the past, olfaction has been reputed to have the largest difference threshold of all the senses, with typical values ranging from about 25 to 33 percent (Gamble, 1898; Stone & Bosley, 1965). That is, the concentration of an odorant must be increased by 25 to 33 percent before a person can detect an increase in odor intensity.

When William Cain (1977) carefully measured the difference threshold by placing two odorants of different concentrations on absorbent cotton balls and asking subjects to judge which was more intense, his results were better than those of most other studies, with an average difference threshold of 19 percent and a relatively low difference threshold of 7 percent for n-butyl alcohol. But Cain didn't stop with these measurements because an average difference threshold of 19 percent still seemed high to him. He next analyzed the stimuli he had presented to his human subjects using a **gas chromatograph**, a device that accurately measures the concentration of the vapor given off by each stimulus. Cain found what he had suspected: Stimuli that were supposed to have the same concentration actually varied considerably, apparently because of differences in the airflow pattern through the cotton in different samples.

By eliminating this variability in stimulus concentration, Cain was able to demonstrate that the difference threshold was smaller than had been previously measured. When he presented stimuli to subjects by using an **olfactometer**, a device that presents olfactory stimuli with much greater precision than cotton balls, Cain found an average difference threshold of 11 percent, with n-butyl alcohol having an impressively low threshold of only 5 percent. These figures, which begin to approach the difference thresholds for vision and hearing, show that our ability to detect differences in smell intensity is, in fact, not poor compared to the other senses.

Fact #3: The Olfactory System Is Excellent at Identifying Odors

Early research on odor identification seemed to indicate that our ability to identify odors is poor, because when asked to identify odors people were typically successful only about half the time (Engen & Pfaffmann, 1960). However, later experiments have shown that, under the right conditions, our ability to identify odors is actually quite a bit better than that. For example, J. A. Desor and Gary K. Beauchamp (1974) found that subjects could identify only about half of the smells of familiar substances like coffee, bananas, and motor oil. But when Desor and Beauchamp named the substances when they were first presented and then reminded their subjects of the correct names if they failed to respond correctly on subsequent trials, the subjects could, after some practice, correctly identify 98 percent of the substances.

According to Cain (1979, 1980), the key to the good performance in Desor and Beauchamp's experiment is that their subjects were provided with the correct names, or labels, at the beginning of the experiment. In his own experiments, Cain showed that, if subjects assign a correct label to a familiar object the first time they smell it (for example, labeling an orange "orange"), or if the experimenter provides the correct labels, the subjects usually identify the object correctly the next time it is presented. If, however, subjects assign an incorrect label to an object the first time they smell it (for example, labeling machine oil "cheese"), they usually misidentify it the next time it is presented. Thus, according to Cain, when we have trouble identifying odors, this trouble results not from a deficiency in our olfactory system, but from an inability to retrieve the odor's name from our memory.

The amazing thing about the role memory plays in odor identification is that knowing the correct label for the odor actually seems to transform our perception into that odor. Cain (1980) gives the example of an object initially identified by the subject as "fishy-goaty-oily." When the experimenter tells the subject that the fishy-goaty-oily smell actually comes from leather, the smell is then transformed into that of leather. I recently had a similar experience when a friend gave me a bottle of Aquavit, a Danish drink with a very interesting smell. As I was sampling this drink with some friends, we tried to identify its smell. Many odors were proposed ("anise," "orange," "lemon"), but it wasn't until someone turned the bottle around and read the label on the back that the truth became known: "Aquavit ('Water of Life') is the Danish national drink—a delicious, crystal-clear spirit distilled from grain, with a slight taste of caraway." When we heard the word *caraway*, the previous hypotheses of anise, orange, and lemon were instantly transformed into caraway. Thus, the olfactory system has the information needed to identify specific odors, but it needs an assist from memory to apply that information to the actual naming of these odors.

DEMONSTRATION

Naming and Odor Identification

To demonstrate the effect of naming substances on odor identification, have a friend collect a number of familiar objects for you, and without looking, try to identify the odors. You will find that you can identify some but not others, and when your friend tells you the correct answer for the ones you missed, you will wonder how you could have failed to identify such a familiar smell. But don't blame your misses on your nose; blame your memory. ●

Fact #4: Human Olfaction Has the Potential to Provide Information about Other People

Many animals use their sense of smell to recognize other animals. As McKenzie (1923) remarked about the dog, "He can recognize his master by sight, no doubt, yet, as we know, he is never perfectly satisfied until he has taken stock also of the scent, the more precisely to do so bringing his snout into actual contact with the person he is examining. It is as if his eyes might deceive him, but never his nose." Humans, however, are constrained from behaving like dogs. Except in the most intimate situations, it is considered poor form to smell other people at close range. However, what if we lived in a society that condoned this kind of behavior? Could we identify other people based on their smell? A recent experiment suggests that the answer to this question may be yes.

Michael Russell (1976) had subjects wear undershirts for 24 hours, without showering or using deodorant or perfume. The undershirts were then sealed in a bag and given to the experimenter, who, in turn, presented each subject with three undershirts to smell: One was the subject's own shirt, one was a male's, and one was a female's. About three-quarters of the subjects succeeded in identifying their own undershirt, based on its odor, and also correctly identified which of the other shirts had been worn by males or females (see also McBurney, Levine, & Cavanaugh, 1977, for a similar experiment). Similar results have also been reported for breath odors, which subjects can identify as being produced by a male or by a female (Doty et al., 1982), and it has also been shown that young infants can identify the smell of their mother's breast or armpits (see the Developmental Dimension, "Olfaction and Taste in Infants," at the end of this chapter). These results don't suggest that people can identify other people solely by their smell, but they do suggest that our ability to use smell in such situations may be underrated.

The phenomenon of **menstrual synchrony** also suggests a role for smell in interpersonal relations. Martha McClintock (1971) noted that women who live or work together often report that their menstrual periods begin at about the same time. For example, one group of seven female lifeguards had widely scattered menstrual periods at the beginning of the summer, but by the end of the summer, all were beginning their periods within four days of each other. To investigate this phenomenon, McClintock asked 135 females, aged 17–22, living in a college dormitory, to indicate when their periods began throughout the school year. She found that women who saw each other often (roommates or close friends) tended to have synchronous periods by the end of the school year. After ruling out factors such as awareness of the other person's period, McClintock concluded that "there is some interpersonal physiological process which affects the menstrual cycle" (p. 246).

What might this physiological process be? Michael Russell, G. M. Switz, and K. Thompson (1980) did an experiment that suggests that smell has something to do with this process. He had a "donor" woman wear cotton pads in her armpits for 24 hours, three times a week. The sweat extracted from these pads was then rubbed onto the upper lip of a woman in the "experimental group." A control group of women received the same treatment, but without the sweat. The results for the experimental group showed that, before the experiment there had been an average of 9.3 days between the onset of the donor's and the subject's periods, but after five months, the average time between onsets was reduced to 3.4 days. The control group showed no such synchrony. Since the donors and the subjects in the experimental group never saw each other, Russell concluded that odor must be the factor that causes menstrual synchrony (also see Preti et al., 1986).

From the facts about olfaction presented above, it is clear that the human olfactory system has capacities that are more impressive than it is often given credit for. We will now see that these capacities are created by a system that we are just beginning to understand.

Structure of the Olfactory System

Discovering the structure of the olfactory system poses special problems for researchers for two reasons: (1) Until very recently, the exact sites where odorants act on the receptors have been unknown, and (2) the olfactory system sends signals not only to areas in the brain that result in smell sensations, but also to areas that are responsible for other functions, such as emotions and memory.

The starting point for the complex process of olfaction is receptors contained in the **olfactory mucosa** (Figures 12.2 and 12.3). An odorant molecule entering the nose stimulates one of the 6 to 10 million receptors situated high in the nasal chamber (Doty, 1991). Molecules reach these receptors in two ways: (1) They diffuse through the olfactory mucosa, and (2) they bind to molecules called **olfactory binding proteins**, which are secreted into the nasal cavity, and these proteins transport the odorant molecules to active sites on the olfactory receptors (Pevsner et al., 1985, 1990).

Recent research pictures the active sites on the olfactory receptors as a protein molecule that crosses the receptor membrane seven times (Figure 12.4) (Buck & Axel, 1991). When this protein is stimulated by odorant molecules, it releases active molecules into the olfactory receptor cell, and these molecules cause electrical responses in the receptor. The receptor transmits signals directly to the olfactory bulb in the brain, where the signals are processed before being sent to the **olfactory cortex**, a small area under the temporal lobe, and also to the **orbitofrontal cortex**, located in the frontal lobe, near the eyes (Figure 12.5) (Cinelli, 1993; Dodd & Castellucci, 1991; Frank & Rabin, 1989; McLean & Shipley, 1992; Price et al., 1991; Takagi, 1980).

The Odor Stimulus and Odor Quality

The stimuli for smell are molecules in the air. Thus, when you smell a rose, or anything else, you are sensing molecules that have left the thing

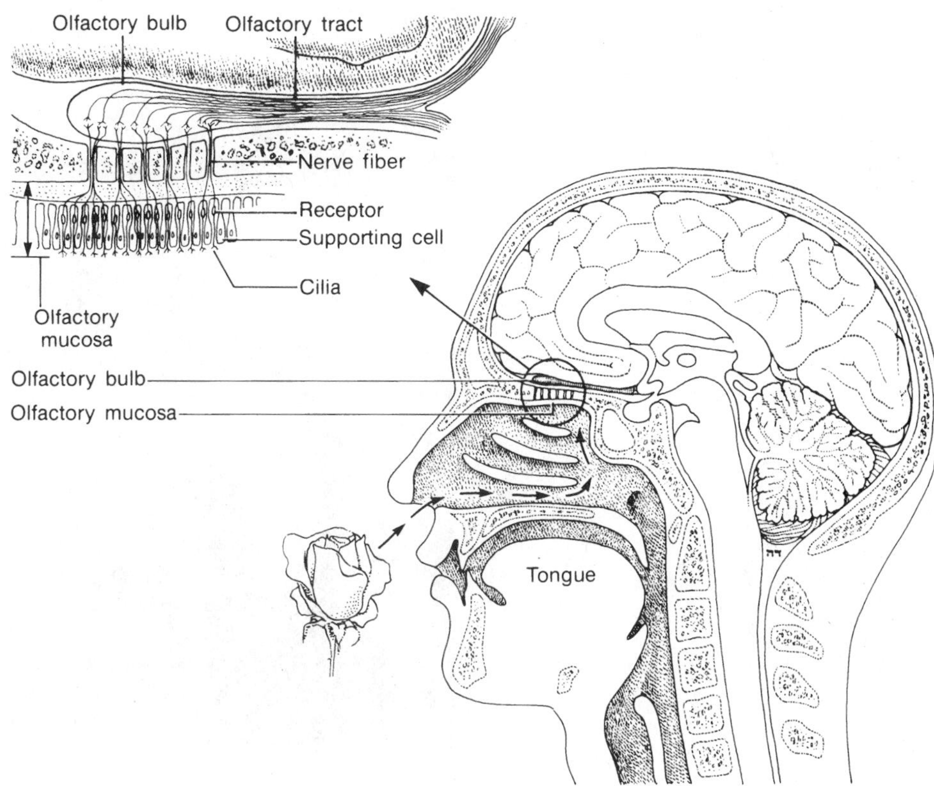

Figure 12.2
The olfactory system, showing the relationship between the olfactory receptors and the olfactory bulb. (Adapted from Amoore et al., 1964.)

you are smelling and have traveled through the air to your nose. The key question in olfaction is: How are the qualities of the thousands of different molecules we can smell indicated by the electrical signals that travel through the olfactory system? In other words, what is the sensory code for odor quality?

In previous chapters, we saw how researchers dealt with the problem of sensory coding by determining which physical property of the stimulus is most closely associated with perception of the stimulus. For example, in the chapter on auditory physiology, we saw that, when researchers identified the relationship between frequency and pitch, the question they asked changed from "What is the code for pitch?" into "What is the code for frequency?" But researchers have not been able to accomplish a similar feat for olfaction because they have not yet identified a relationship between a molecule's smell and a physical property that causes that smell.

Odor Qualities and Physical Properties

One reason it has been difficult to link odor quality with a physical property is that it has been difficult to describe and classify odors. For example, if you asked people to smell the chemical β-ionone, they would probably say that it smells like violets. This description, it turns out, is fairly accurate, but if you compare β-ionone to real violets, they smell different. The perfume industry's solution is to use names such as "woody violet" and "sweet violet" to distinguish between different violet smells, but this labeling hardly solves the problem of determining how olfaction works.

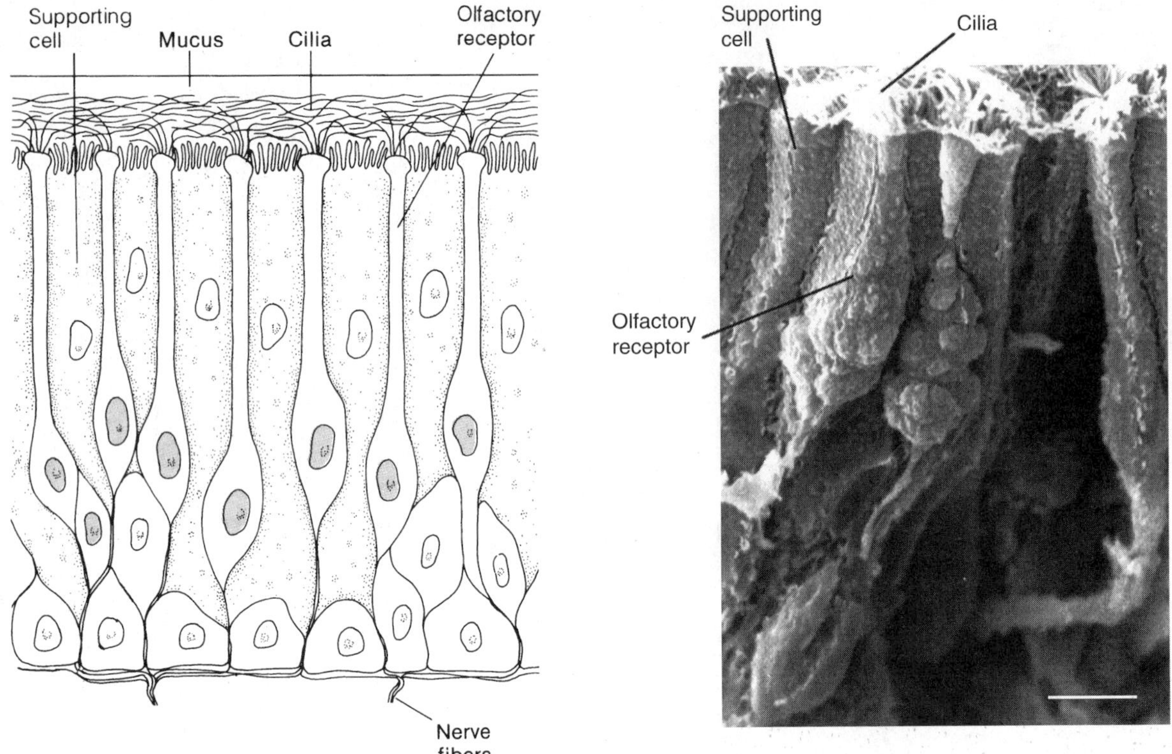

Figure 12.3

Left: The olfactory receptors and mucosa. A smell stimulus is picked up by the mucus layer and stimulates the cilia attached to the olfactory receptors. Note that each receptor is separated from its neighbor by a thick supporting cell. Right: A scanning electron micrograph showing two receptors (arrows and O's) and supporting cells (S). The bar at the lower right represents 1/1,000 of a millimeter. (Photograph courtesy of Edward Morrison.)

A large number of attempts have been made to find a way to classify odors and so to bring some order into this confusing situation. One of the best known attempts is Henning's **odor prism** (Figure 12.6). Henning's prism has six corners, at which the qualities putrid, ethereal, resinous, spicy, fragrant, and burned are located. Odors lying along an edge of the prism resemble the qualities at the corners of the edge, in proportion to their nearness to each corner. For example, Figure 12.7 shows the face of the prism that has fragrant, ethereal, resinous, and spicy at its corners. From its location on the ethereal-resinous edge, we can see that lemon oil has both ethereal and resinous properties, but that it is more like ethereal, since it is closer to the ethereal corner. Although the geometry of Henning's odor prism is aesthetically appealing, this method of classifying odors has proved to be of little use in helping us to understand olfaction, and Henning's prism is now of only historical importance (Bartoshuk, Cain, & Pfaffmann, 1985).

A more modern approach to odor quality is a technique called **odor profiling**, in which trained observers rate the components in odors by comparing them to reference standards such as those

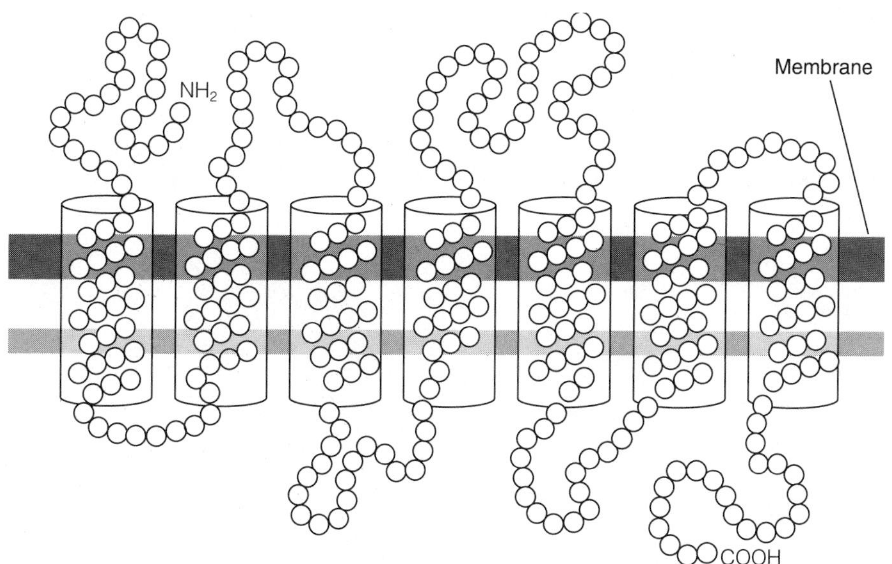

Figure 12.4
An olfactory receptor protein, which creates an active site on the receptor membrane. This molecule consists of strings of amino acids (circles) that cross the membrane of the olfactory receptor cell seven times. The receptor protein responds to odorant molecules by triggering activity inside the cell. There are many different receptor proteins, each made up of different sequences of amino acids. (Adapted from Buck & Axel, 1991.)

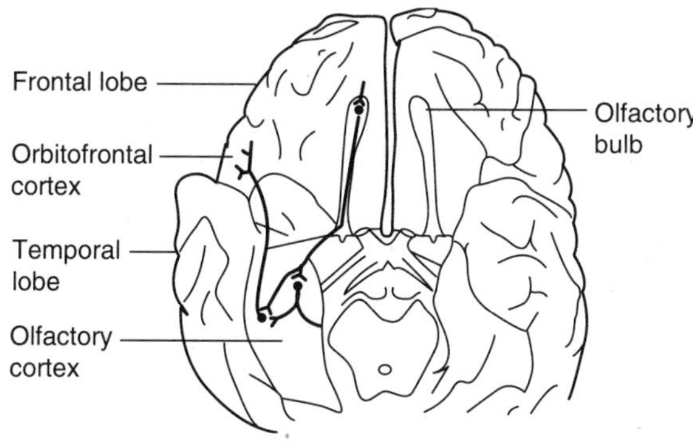

Figure 12.5
The underside of the brain, showing the neural pathways for olfaction. On the left side, the temporal lobe has been deflected to expose the olfactory cortex. See text for details. (From Frank & Rabin, 1989.)

listed in Table 12.1 (Doty, 1991; Mallevialle & Suffet, 1987). Such judgments can result in odor profiles such as the ones in Figure 12.8 for methyl salicylate, which has a wintergreen smell, and methyl disulfide, which has an unpleasant sulfur smell. It is not surprising, considering the difference in their smells, that the odor profiles of these two substances differ considerably.

While odor profiles provide a way to describe odor quality, we are still left with a large number of different odor qualities and no system for classifying them. Other classification schemes

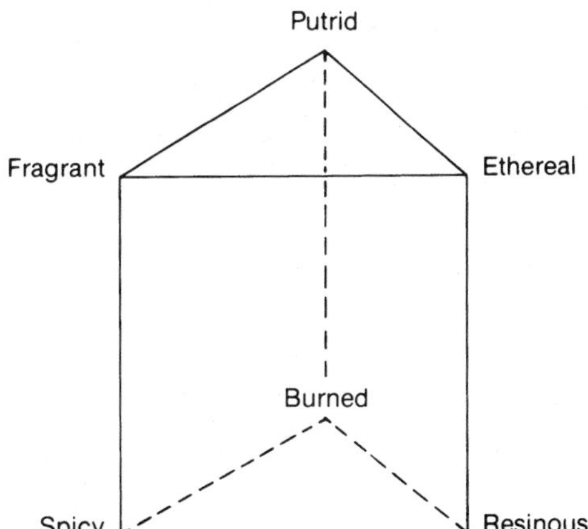

Figure 12.6
Henning's odor prism. (From Woodworth, 1938.)

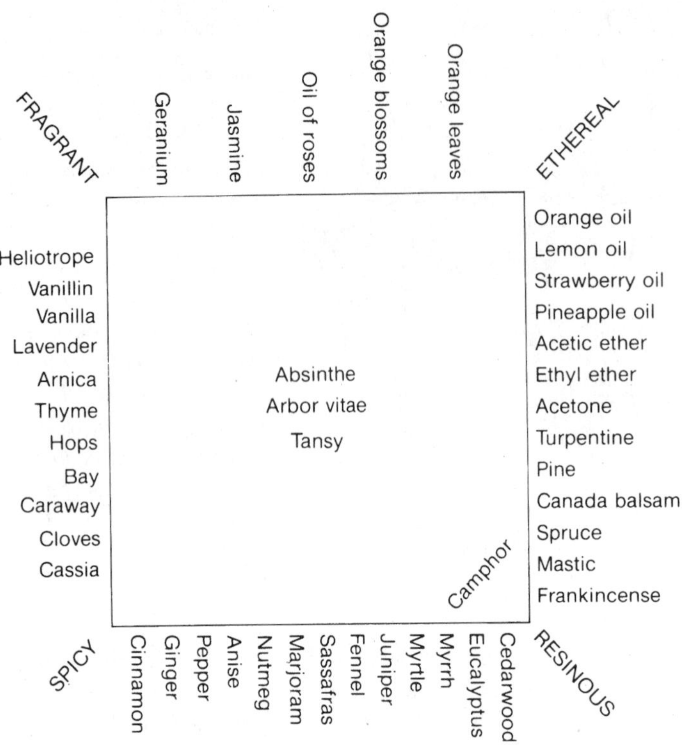

Figure 12.7

The face of Henning's odor prism that is facing us in Figure 12.6. Odors along the edges are related to the two odors at the corners of that edge. Those in the middle are related to all four of the corner odors. (From Woodworth, 1938.)

have also been suggested, but researchers still don't know how to classify odors in a way that will help them study the mechanisms responsible for olfaction. Thus, researchers are still looking for the connection between physical properties and quality.

Table 12.1

Odor reference standards used by the Philadelphia water department

Agent or Compound	Concentration	Descriptor
Benzaldehyde	1,000 μg/L	Almonds
Cumene	100 μg/L	Shoe polish
Styrene	500 μg/L	Model airplane glue
Cloves	3 g in 200 mL	Cloves
Soap	5 g Ivory in 200 mL	Soapy

Source: Adapted from Doty, 1991; originally from Mallevialle & Suffet, 1987.

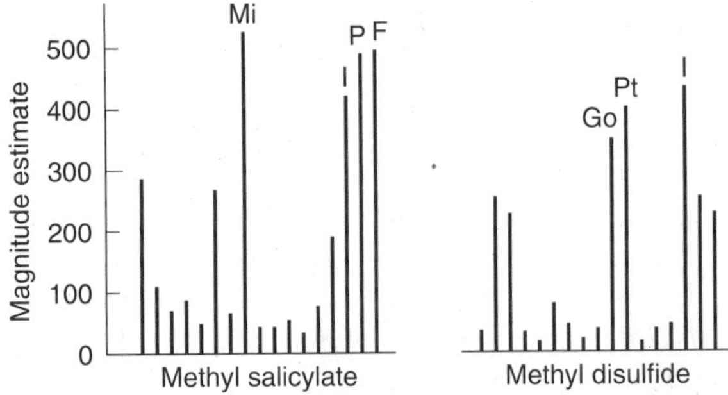

Figure 12.8

Odor profiles of two compounds. Each line indicates the magnitude estimate for different components of each odor. The predominant components of the odor of methyl salicylate are minty (Mi), pleasant (P), and familiar (F). The predominant components of the odor of methyl disulfide are goaty (Go) and putrid (Pt). The line marked "I" represents the judgment of the odor's intensity. (From Doty, 1991.)

To appreciate the problems involved in linking a substance's odor quality and physical properties, consider a situation that might occur as you are cooking breakfast one morning and a friend comments on the smells of coffee, bacon, and toast that are emanating from your kitchen. The fact that your friend can do this may not seem particularly surprising until we recognize that the smell of coffee can arise from more than 100 different chemical constituents, and that toast and bacon smells are also the result of a large number of different molecules that have been released into the air. Somehow, your friend's olfactory system takes this vast array of airborne chemicals with their various physical qualities and groups them into the three classifications: coffee, toast, and bacon (Bartoshuk et al., 1985). This amazing feat of perceptual organization, which is analogous to how we separate a number of sounds coming simultaneously from a number of sources into different perceptual streams (see Bregman, 1990), is far too complex for us to understand at this time.

The complexity of real-life stimuli such as those that create the smells from your kitchen has led researchers to focus their search for connections between physical properties and smell on simple chemicals. Taking this approach, a number of researchers have focused on the idea that a molecule's shape is an important determinant of odor quality. For example, John Amoore's (1970) **stereochemical theory of odor** postulates that there are different molecular shapes for different odor qualities, and that there are "sites" on the olfactory receptors shaped to accept only molecules with the correct shape. As reasonable as this idea may sound, Amoore's theory has not proved useful. One problem is that many molecules with similar shapes have very different odors.

Odor quality is also affected by physical and chemical properties, such as chemical reactivity and the electrical charge of the elements that make up a molecule. Especially in smaller molecules, making a small change in just one molecular group can cause large changes in odor. For example, when the CO group of the molecule on the left in Figure 12.9a is replaced by a CH_2 group, the odor changes from musk to odorless, even though the shapes of the two molecules are almost identical (Beets, 1978). Yet sometimes changing molecular groups has little effect, as in the two compounds in Figure 12.9b, which both have sandalwood odors (Beets, 1982). Thus, we are far from being able to specify a physical property for odor that is analogous to frequency for pitch (Kauer, 1987).

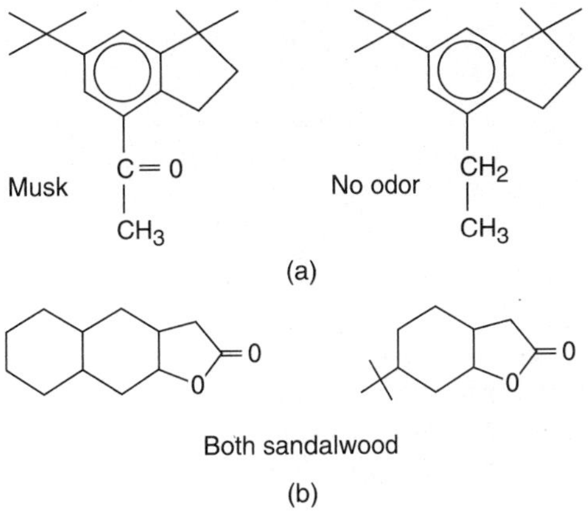

Figure 12.9
(a) Two molecules with similar structures but with different odors. (b) Two molecules with different molecular groups but similar odors.

The Neural Code for Odor Quality

Although we have not yet found a simple way to classify odor quality or to relate odor quality to the physical properties of molecules, researchers have not stopped searching for the neural code for odor quality. We will describe this research by first considering how olfactory receptors respond to various chemicals, and we will then describe how the olfactory bulb and more central structures respond to different chemicals.

Coding at the Level of the Receptors

One approach to investigating olfactory coding at the receptor level has been to look for ways in which different odorants cause the receptors to respond. Researchers have looked at (1) the firing of receptors to different stimuli; (2) the response of areas on the olfactory mucosa to different stimuli; and (3) the way olfactory stimuli physically distribute themselves onto the olfactory mucosa.

Receptor Firing Recordings from a variety of animals show that most olfactory receptors respond to a number of different odorants (Blank, 1974; Holley et al., 1974; Matthews, 1972; Moulton, 1965). Thus, a particular odorant causes a large number of receptors to fire, and these receptors also usually fire to other odorants as well, as illustrated in Figure 12.10 for receptors in the frog's olfactory mucosa (Sicard & Holley, 1984). Notice that each odorant (along the top of the figure) causes a pattern of receptors to fire. Some odorants, like cineole (CIN) and camphor (CAM), cause many of the same receptors to fire, and some, like camphor and heptanol (HEP), cause a different group of receptors to fire, although there is some overlap between the groups. Since single receptors don't respond best to specific odorants, it is likely that odor is coded at the receptor level in terms of the overall pattern of response of many receptors (Cinelli, 1993).

Response of the Mucosa Although individual receptors usually do not respond to just one or a few chemicals, there are areas on the mucosa that are sensitive to some odorants and not as sensitive to others. Alan MacKay-Sim, Paul Shaman, and David Moulton (1982) demonstrated these differences in the responses of different areas of the mucosa, which they called the **regional sensitivity effect**, by presenting chemicals to small areas on the olfactory mucosa and recording responses from the mucosa. By recording the **electro-olfactogram**, a response that indicates the

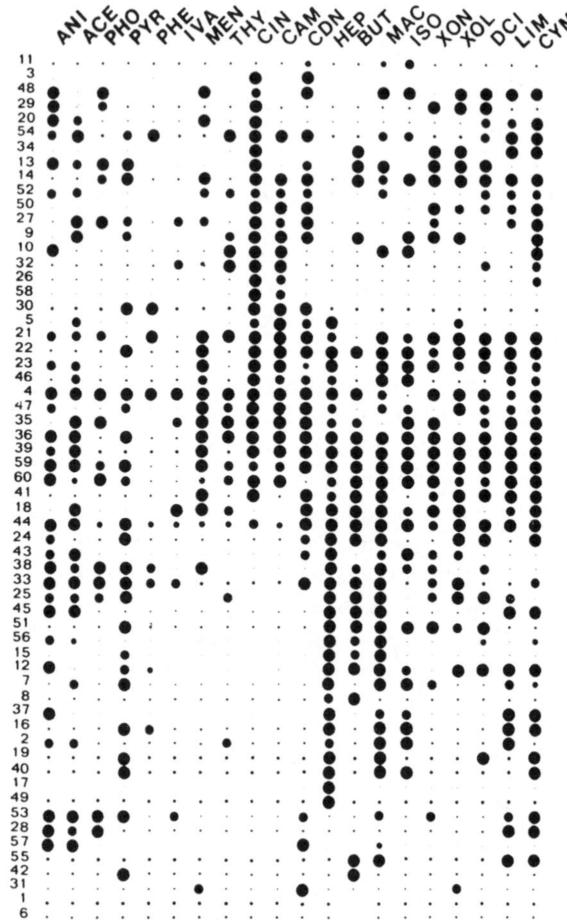

Figure 12.10

Responses of frog olfactory receptors (numbers on the left) to different compounds (letters along the top). The sizes of the spots are roughly proportional to the sizes of the responses. By reading across this diagram horizontally, we can see that most receptors respond to a number of compounds. By reading down from the top, we can see that a given compound causes a large number of receptors to fire. See text for further details. (From Sicard & Holley, 1984.)

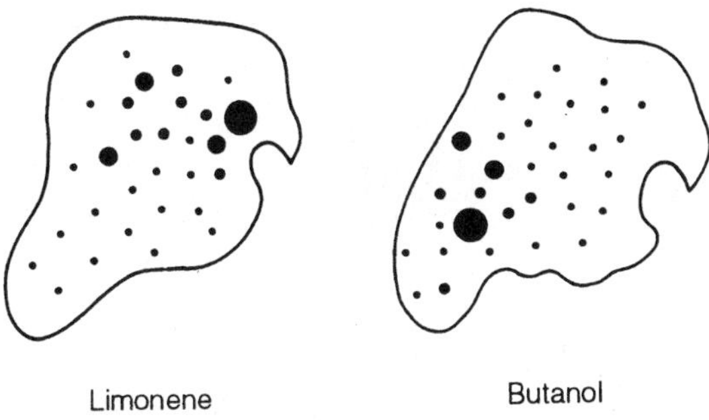

Figure 12.11
Size of the electro-olfactogram (EOG) response of the olfactory mucosa to stimulation of the mucosa by limonene and butanol. The larger the dot, the larger the size of the EOG. (From MacKay-Sim, Shaman, & Moulton, 1982.)

pooled electrical activity of thousands of receptors, they found that the size of the response depends on the chemical used and the site on the mucosa where this chemical is applied. Figure 12.11 shows their results for two chemicals, limonene and butanol, which have very different areas of maximum sensitivity on the mucosa.

Odorant Distribution Maxwell Mozell (1970) showed that, as odorant molecules flow across the olfactory mucosa, different molecules distribute themselves on different areas of the mucosa. Some are strongly attracted to the mucosa, so they flow slowly and are preferentially deposited at the front of the mucosa. Others are more weakly attracted to the mucosa and so flow more rapidly and stimulate receptors more uniformly across the mucosa. According to this idea, each molecule creates a different pattern of receptor stimulation because of differences in how various molecules are deposited on the receptor surface.

David Hornung and Mozell (1981) showed that odorant molecules distribute themselves differently on the mucosa by flowing radioactivity labeled odorants over the mucosa and analyzing the pattern of radioactivity created by different odorant molecules. They found that large amounts of the chemical butanol were deposited on the front of the mucosa and that only a small amount was deposited on the back, whereas octane was distributed fairly evenly over the whole mucosa (Figure 12.12). (Also see Hornung, Lansing, & Mozell, 1975; Hornung & Mozell, 1977; Mozell, 1964, 1966; Mozell & Hornung, 1984, 1985; Mozell & Jagodowiez, 1973.)

The differing patterns of receptor responses to various chemicals, the regional sensitivity ef-

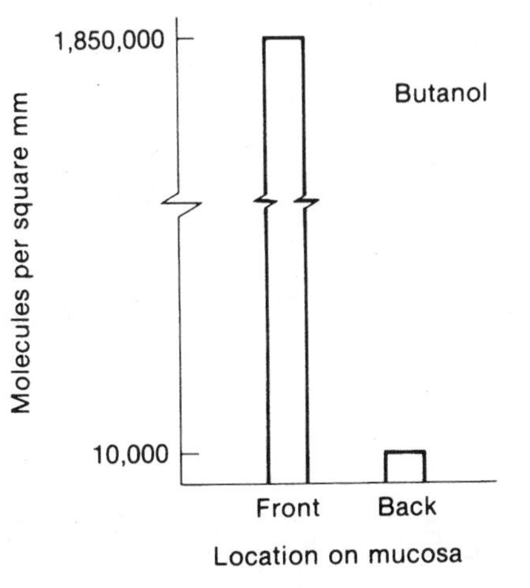

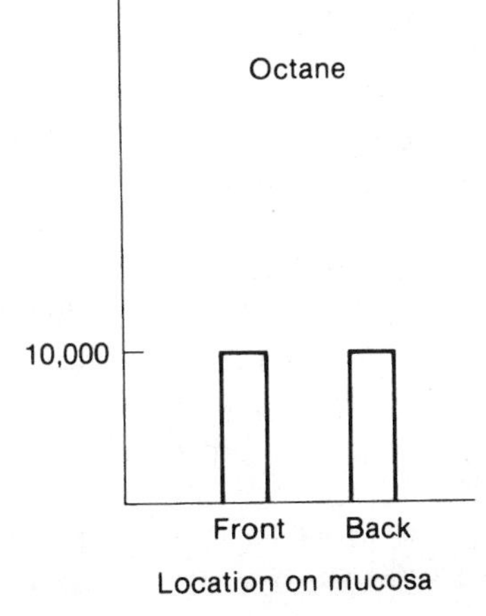

Figure 12.12
How butanol and octane molecules distribute themselves on a frog's olfactory mucosa. Far more butanol is deposited on the front of the mucosa than on the back, but equal amounts of octane are deposited on the front and the back of the mucosa. (Adapted from Hornung & Mozell, 1981.)

fect, and the selective distribution of odorants on different places on the mucosa all work together to create different patterns of receptor activity for different chemicals. We will now see that these differing patterns of receptor activity are translated into differing patterns of activity in the olfactory bulb.

Coding in the Olfactory Bulb and Beyond

If distinctive patterns of receptor firing for different odorants are transmitted from the mucosa, we should be able to observe different patterns of activity in the olfactory bulb. William Stewart, John Kauer, and Gordon Shepherd (1979) demonstrated these patterns by using the 2-deoxyglucose (2-DG) technique that has been used in research on the visual system. When we described this technique on page 100, we saw that 2-deoxyglucose has three important properties: (1) Since its structure is similar to glucose, a primary source of energy for neurons, it is taken up by neurons as though it were glucose; the more active the neuron, the more 2-DG is taken up by the neuron; (2) when 2- DG is taken up by a neuron, it accumulates inside the neuron; and (3) 2-DG can be labeled with a radioactive isotope, carbon 14; therefore, by measuring the amount of radioactivity in the various parts of a structure, we can determine which neurons are most active.

Stewart injected rats with radioactively labeled 2-DG and exposed them to either amyl acetate or camphor for 45 minutes, thereby activating neurons that would be stimulated by amyl acetate or camphor and causing them to take up 2-DG. After this 45-minute exposure, the rats were sacrificed, and their olfactory bulbs were examined to determine the pattern of radioactivity. The results, shown in Figure 12.13, indicate that different patterns of neurons are activated by amyl acetate and camphor. Note, however, that, although the patterns are different, each substance causes activity over a large area, and there is considerable overlap between the areas activated, just as there is overlap in the receptors activated by two different substances (see Figure 12.10).

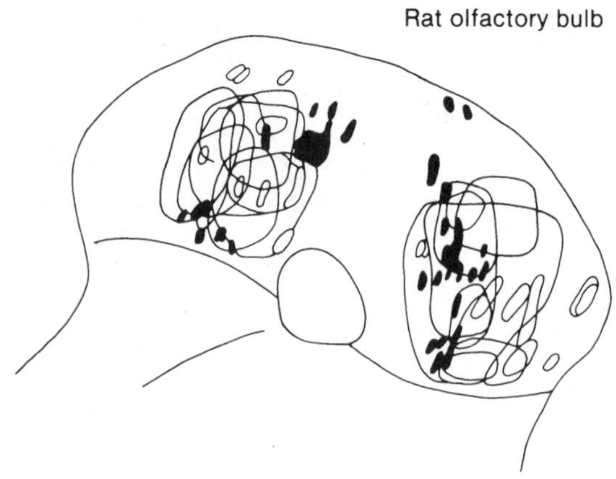

Figure 12.13
Areas of highest activity in the rats' olfactory bulb caused by stimulation with camphor (dark areas) and by stimulation with amyl acetate (open areas), as determined by the 2-DG technique (Stewart, Kauer, & Shepherd, 1979).

The results of both the receptor and the olfactory bulb experiments point to a much more diffuse type of coding in the olfactory system than in the other sensory systems. For example, a touch to your hand activates mechanoreceptors in a small area of skin, and causes activity in a fairly well-defined area of cortex. In contrast, sniffing some amyl acetate activates a fairly diffuse pattern of receptors on your olfactory mucosa and causes widespread activity in your olfactory bulb. This olfactory bulb activity may be different from that generated by sniffing some other substance, but it is not precisely localized, as is the response to having your hand touched. Somehow, even though the coding in the olfactory system appears to be fairly diffuse, we are capable of distinguishing tens of thousands of different odors.

Perhaps we are able to distinguish this large number of odors because of processing in centers higher than the olfactory bulb. We know a little about the nature of this processing from some

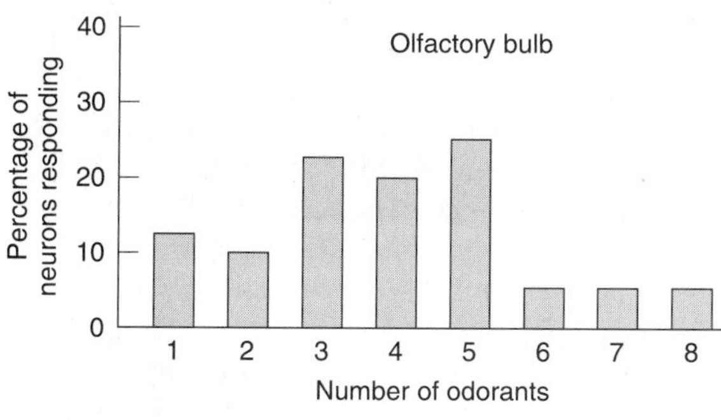

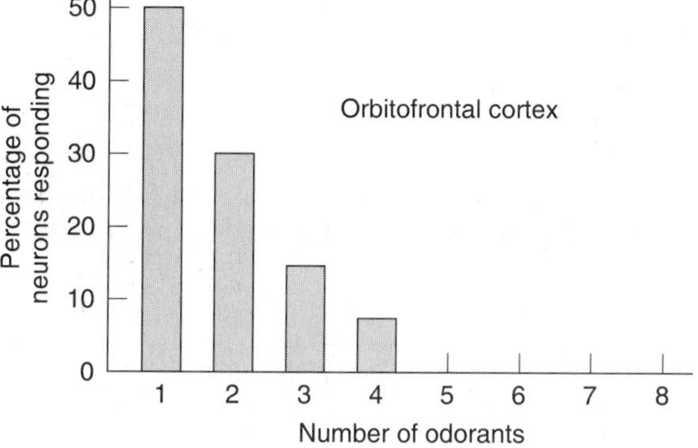

Figure 12.14

Histograms showing the percentage of neurons surveyed in the olfactory bulb and the OFC that responded to a number of odorants, indicated along the horizontal axis. (A total of eight odorants were presented.) (From Tanabe, Iino, & Takagi, 1975.)

behavioral and electrophysiological experiments by T. Tanabe and co-workers (1974; Tanabe, Iino, & Takagi, 1975), which show that the orbitofrontal cortex (OFC), located on the underside of the frontal lobe (see Fig. 12.5), may play an important role in olfaction. The behavioral experiment was simple. Two small pieces of bread were impregnated with different odors, and one piece of bread was given a bitter taste. When each piece was given to a monkey, the monkey first smelled each piece and then ate it, but after determining which smell was associated with the bitter taste, the monkey was able to avoid the bitter piece of bread—solely on the basis of its smell—81 percent of the time. Removal of the OFC, however, reduced this ability to 26 percent. This large decrease in performance did not occur with the removal of other cortical olfactory areas.

Tanabe et al.'s electrophysiological results show the degree to which eight different odorants caused neural responses in the olfactory bulb and the OFC (Figure 12.14). Each bar indicates the percentage of neurons that responded to the number of odorants indicated on the horizontal axis. Few of the cells in the olfactory bulb responded to only one odor, whereas half of the cells in the OFC responded to only one odor. Neurons in the OFC, then, are tuned to respond to much more specific odorants than are the neurons in lower centers. From these results, we can draw an analogy between neurons in the OFC and visually responsive neurons in the extrastriate cortex that respond only to very specific visual stimuli. In both cases, neurons near the beginning of the system respond to a range of stimuli, and neurons at higher levels respond best to more specific stimuli.

The Perception of Flavor

Consider the following description of a dining experience, from a restaurant review:

> It is a surprise to dip your spoon into this mild-mannered soup and experience an explosion of flavor. Mushroom is at the base of the taste sensation, but it is haunted by citric tones—lemongrass, lime perhaps—and high at the top, a resonant note of sweetness. What is it? . . . No single flavor ever dominates a dish. At first you find yourself searching for flavors in this complex tapestry, fascinated by the way they are woven together. In the end, you just give in and allow yourself to be seduced. . . . Each meal is a roller coaster of sensations. (Reichl, 1994)

When I first read this restaurant review, my initial reaction was that I wouldn't mind eating at that restaurant if only I could afford it (it is a very expensive four-star restaurant in New York City), and my next reaction was that such superlatives, when applied to food, did not surprise me at all. We usually do much more than simply notice different flavors as we eat. We often evaluate them, savoring those that we find exceptionally pleasing, and dismissing others as inadequate or even disgusting. Many of these sensory experiences are the result of the combination of taste and olfaction to create **flavor**. You can demonstrate how smell affects flavor with the following demonstration.

DEMONSTRATION

"Tasting" With and Without the Nose

While holding your nostrils shut, drink a beverage with a distinctive taste, such as grape juice, cranberry juice, or coffee. Notice both the quality and the intensity of the taste as you are drinking it. (Take just one or two swallows, as swallowing with your nose closed can cause a build up of pressure in your ears.) After one of the swallows, open your nostrils and notice whether you perceive a flavor. Finally, just drink the beverage normally with nostrils open, and notice the flavor. You can also carry out this demonstration with foods, such as fruits or cooked foods. ●

During this demonstration you probably noticed that, when your nostrils were closed, it was difficult to identify the substance you were drinking or eating, but as soon as you opened your nostrils, the flavor became obvious. This occurred because odor stimuli from the food reached the olfactory mucosa through the **retronasal route**, through the **nasal pharynx**, as shown in Figure 12.15. Although pinching the nostrils shut does not close the nasal pharynx, it prevents vapors from reaching the olfactory receptors by eliminating the circulation of air through this channel (Murphy & Cain, 1980).

The importance of olfaction in the sensing of flavor has been demonstrated experimentally by using both chemical solutions and typical foods. For example, when Maxwell Mozell and coworkers (1969) asked subjects to identify common foods with the nose opened or with the nostrils pinched shut, every substance they tested was easier to identify in the nostrils-open condition than in the nostrils-closed condition (Figure 12.16). In a more recent experiment, Thomas Hettinger, Walter Myers, and Marion Frank (1990) had subjects rate the strength of various qualities experienced when solutions

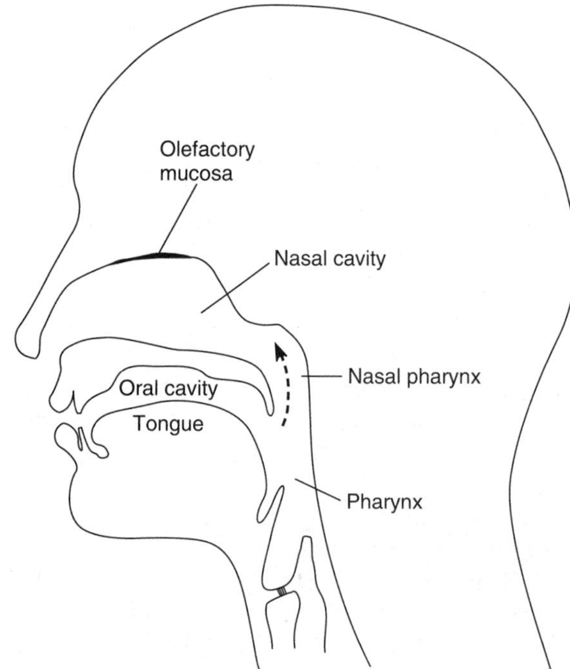

Figure 12.15

Odorant molecules released by food in the oral cavity and pharynx can travel through the nasal pharynx (dashed arrow) to the olfactory mucosa in the nasal cavity. This is the retronasal route to the olfactory receptors.

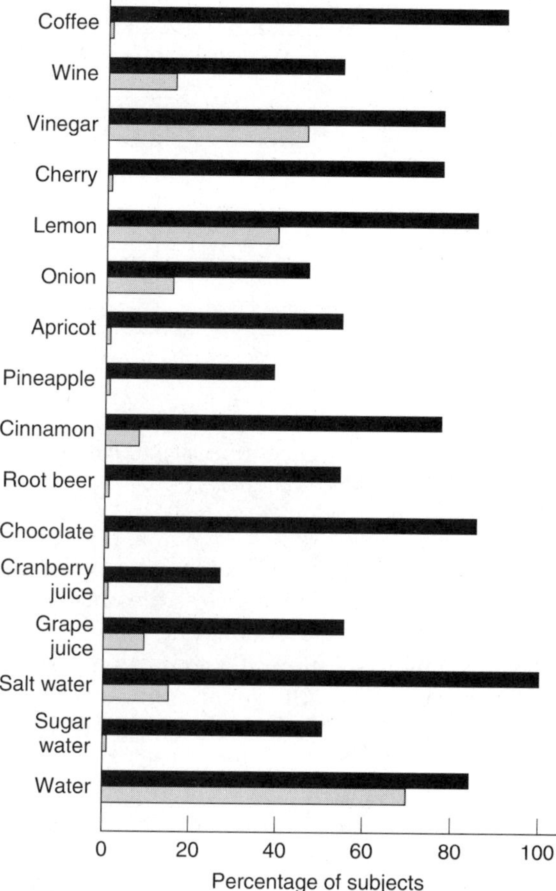

Figure 12.16

Percentage of subjects correctly identifying each flavor listed on the left, with nostrils open (solid bars) and nostrils pinched shut (shaded bars). (Adapted from Mozell et al., 1969.)

were applied to the tongue with the nostrils either open or clamped shut. The results for two of the chemicals, sodium oleate and ferrous sulfate, (in Figure 12.17a and b) show that the oleate had a strong soapy flavor when the nostrils were open but was judged tasteless when they were closed. Similarly, the ferrous sulfate normally has a metallic flavor but was judged predominantly tasteless when the nostrils were closed. Some compounds are not, however, as influenced by olfaction. For example, monosodium glutamate (MSG) had about the same flavor whether or not the nose was clamped (Figure 12.17c). Thus, in this case, the sense of taste predominated.

The results of these experiments indicate that many of the sensations that we call taste, and that we assume are caused only by stimulation of the tongue, are greatly influenced by stimulation of the olfactory receptors. Apparently, we often mislocate the source of our sensations as being in the mouth, partially because the stimuli physically enter the mouth and partially because we experience the tactile sensations associated with chewing and swallowing (Murphy & Cain, 1980; Rozin, 1982).

We will now consider a number of factors that influence our intake of food, an activity that is intimately associated with both taste and smell. As we do this, we will use the term *flavor* when appropriate to indicate that we are describing the joint operation of taste and olfaction. After describing some of the factors involved in determining food preferences, we will describe research on taste as a unique sensation separate from olfaction. Thus, when we describe this research on the mechanisms responsible for taste, we will focus on sensations due to stimulation only of the tongue or of the mouth. With this focus, we will see that taste is a sense in its own right, with its own properties, many of which are different from the properties of the sense of smell.

Factors Influencing Food Preferences

Taste has been called the social sense because we often experience it while dining with other people (Ackerman, 1990). One thing that people often do during these shared eating experiences is to share their taste experiences as they eat: "This chicken would taste better with a little more

	Sodium oleate		Ferrous sulfate		MSG	
	Clamped	Open	Clamped	Open	Clamped	Open
Sweet	x		x			
Salty	x				xxxxxxxxx	xxxxxxxx
Sour			xx	x	xxx	xxx
Bitter		x	x		xx	x
Soapy	xx	xxxxxxxxx	x	xx		
Metallic		xx	x	xxxxxxxxx		
Sulfurous				x	x	xx
Tasteless	xxxxxxxx		xxxxxx	x		
Other	x	x	x		x	xxx
	(a)		(b)		(c)	

Figure 12.17

How subjects described the flavors of three different compounds when they tasted them with nostrils clamped shut and with nostrils open. Each X represents the judgment of one subject. (From Hettinger, Myers, & Frank, 1990.)

salt," "I can't stand liver," "Your apple pie is fantastic, Mom." When we share our taste experiences in this way, we are emphasizing the *affective* aspect of taste; that is, we usually label tastes as pleasant or unpleasant, and these labels affect which foods we choose to eat and which foods we avoid.

Many factors influence how foods taste to us: how much we've already eaten, our past experiences with different foods, our genetic makeup, the food we've just finished eating, and our nutritional state, among others. As we have for the other senses in this book, we want to consider the physiological workings of the sense of taste, but first, let's consider some of the factors that influence which foods we seek out and which we avoid.

Internal State of the Organism

A restaurant near my house has fantastic banana cream pie. When the pie arrives at my table, I usually wish they had given me a bigger piece, but before I've finished, I've usually had enough. This is a phenomenon that Cabanac (1971, 1979) calls **alliesthesia**, changed sensation. Our reaction to a taste stimulus may be positive when we first taste it, but this positive response may become negative after we've eaten for a while. To illustrate this effect, Cabanac describes an experiment in which he had people rate the pleasantness of a sugar solution. If they taste a sample and then spit it out, the sugar continues to get positive ratings over many samplings. If, however, people drink the sample each time they taste it, the originally pleasant sugar solution becomes unpleasant, and eventually the people refuse to drink any more. According to Cabanac, this experiment, and many others with similar results, illustrate that a given stimulus can be pleasant or unpleasant, depending on signals from inside the body.

Past Experiences with Food

Although there is little experimental evidence that early experience with food results in permanent attachments to a particular food or flavor (Beauchamp & Maller, 1977), it is a common observation that people in certain cultures develop tastes for certain foods that are typically associated with that culture. Howard Moskowitz and co-workers (1975) investigated the idea that the perception of flavor can be influenced by experience by asking a group of Indian medical students and a group of Indians from the Karnataka province to rate the pleasantness of a number of compounds. The medical students described citric acid as having an unpleasant sour taste and quinine as having an unpleasant bitter taste. However, the Karnataka Indians described both of these compounds as being pleasant-tasting. Moskowitz felt that past experience was the most likely explanation for the Karnataka Indians' unusual preference for sour and bitter compounds. Their diet consisted of many sour foods, with the tamarind, a particularly sour fruit, making up a large portion of it. Being poor, they ate the tamarind out of necessity and, from constant exposure to this fruit, probably acquired a taste for sour foods. We don't know if the Karnataka Indians' long experience with sour and bitter compounds changed the *flavor* they experienced or simply increased their liking for sour and bitter flavors. It may be that the effect observed in the Karnataka Indians is analogous to the observation that children, who often avoid unfamiliar foods, come to prefer these foods if they are repeatedly exposed to them (Birch, 1979; also see Capretta, Petersik, & Steward, 1975).

Conditioned Flavor Aversion

Food preferences may be affected not only by the exposure to foods that results from long-term eating patterns, as appears to be the case for the Karnataka Indians, but also by a single pairing of food with sickness. The first experiments to illustrate this effect were done on rats by John Garcia and co-workers (Garcia, Ervin, & Koelling, 1966; Garcia & Koelling, 1966). In a typical experiment, Garcia et al. first determined that a rat would drink a sugar solution that tastes sweet to humans. The rats were fed sugar water and were then injected with a chemical that made them sick. After recovering from the resulting sickness, the rats would drink little or none of the sugar solution, which, before it was paired with sickness, they drank in large quantities. This avoidance of a flavor after it is paired with sickness is called **conditioned flavor aversion**. This mechanism also works in humans, as demonstrated by the observation that children made sick by chemotherapy treatments for cancer will avoid eating the ice cream they had eaten just before their chemotherapy injection (Bernstein, 1978; also see Garcia, Hawkins, & Rusiniak, 1974).

Specific Hungers

Conditioned flavor aversion has adaptive value. For example, if an animal survives after eating poison, conditioned flavor aversion prevents the animal from making the same mistake again. The adaptive value of flavor is also illustrated by built-in preferences that cause most people to like sweet foods, to dislike very bitter and very sour foods, and to like salty foods at low concentrations but to dislike them at high concentrations. There are good reasons for these built-in preferences. Sweetness usually indicates the presence of sugar, an important source of energy, whereas sourness or bitterness often indicates the presence of dangerous substances, such as poisons that result in sickness or death. In addition to these preferences, there is evidence that rats and people have a **specific hunger** for sodium, a necessary component of their diets (Beauchamp, 1987). Rats deprived of sodium will increase their sodium intake to make up the deficit, and they use taste to recognize foods that contain sodium (Rozin, 1976).

A dramatic demonstration that taste regulates this hunger for sodium is an experiment by M. Nachman (1963) in which he created a need for salt in a rat by performing an adrenalectomy (removal of the adrenal gland), which caused the rat to eat large amounts of lithium chloride, even though this substance is toxic and makes the rat ill. The rat ate the lithium chloride because it tastes identical to sodium chloride, the substance the rat actually needed to correct the salt deficit created by the adrenalectomy.

Specific hunger for salt analogous to that found in Nachman's adrenalectomized rat has been reported in people who suffer from various diseases. In the 1930s, before the condition of specific salt hunger was widely recognized, a child whose adrenal cortex was diseased craved large amounts of salt. In an attempt to find out what caused this craving, the child was hospitalized, only to die after being placed on a standard hospital diet, which didn't satisfy the child's increased need for salt. Addison's disease also increases hunger for salt. A 34-year-old man with this disease was reported to routinely half-fill a glass with salt before adding tomato juice and to cover his steak with a one-inch-thick layer of salt (Liphovsky, 1977).

The Genetics of Taste Experience

We end our survey of some of the factors that influence food preference by noting yet another reason that people might have different food preferences: The same foods may taste different to different people because of genetic differences. One of the most well-documented genetic effects in taste involves people's ability to taste the bitter substance phenylthiocarbamide (PTC). Linda Bartoshuk (1980) describes the discovery of this PTC effect:

> The different reactions to PTC were discovered accidentally in 1932 by Arthur L. Fox, a chemist working at the E. I. DuPont deNemours Company in Wilmington, Delaware. Fox had prepared some PTC, and when he poured the compound into a bottle, some of the dust escaped into the air. One of his colleagues complained about the bitter taste of the dust, but Fox, much closer to the material, noticed nothing. Albert F. Blakeslee, an eminent geneticist of the era, was quick to pursue this observation. At a meeting of the American Association for the Advancement of Science (AAAS) in 1934, Blakeslee prepared an exhibit that dispensed PTC crystals to 2,500 of the conferees. The results: 28 percent of them described it as tasteless, 66 percent as bitter, and 6 percent as having some other taste. (p. 55)

People who can taste PTC are described as **tasters,** and those who cannot are called **nontasters** (or **taste-blind** to PTC). Molly Hall and coworkers (1975) have found that most people who can taste PTC also perceive a bitter taste in caffeine at much lower concentrations than do nontasters, and Bartoshuk (1979) reported a similar result for the artificial sweetener saccharin. That some people are much more sensitive to the bitter tastes of caffeine and saccharin than others is particularly interesting, since caffeine is found in many common foods, and saccharin was a popular sugar substitute. Caffeine makes coffee taste bitter to tasters but has little effect on nontasters, and saccharin, in the concentrations that used to be added to soft drinks before saccharin was replaced by other artificial sweeteners, tastes two to three times more bitter to tasters than to nontasters.

Recently, additional experiments have been done with a substance called 6-n-propylthiouracil, or PROP, that has properties similar to those of PTC. Researchers have found that about one-third of people report that PROP is tasteless and that two-thirds can taste it. The people who can taste PROP (tasters) also report more bitterness in caffeine, Swiss cheese, and cheddar cheese than nontasters. This result, which is similar to that found for PTC, becomes more interesting when combined with anatomical measurements using a new technique called **video microscopy**, which by combining video technology and microscopy enables researchers to count structures called *taste buds* on people's tongues that contain the receptors for tasting.

The key result of these studies is that people who could taste the PROP had higher densities of taste buds than those who couldn't taste it (Figure 12.18) (Bartoshuk & Beauchamp, 1994). Thus, the next time you disagree with someone else about the taste of a particular food, don't automatically assume that your disagreement is sim-

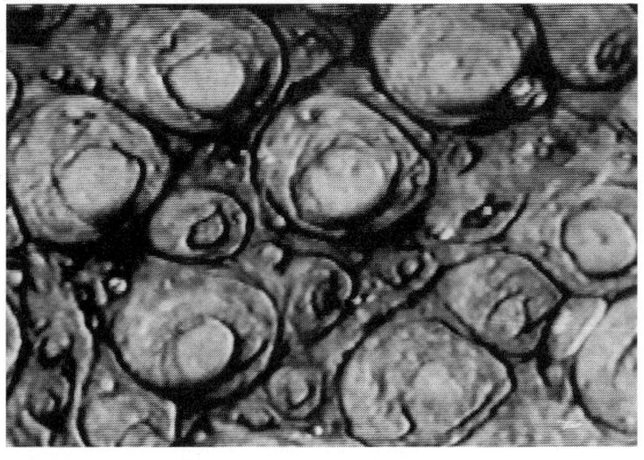
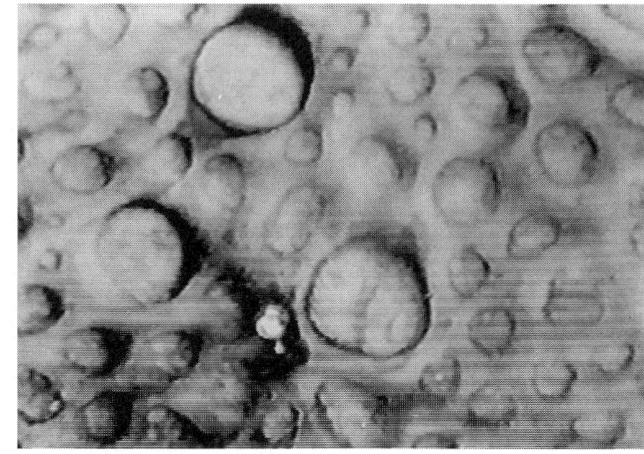

Figure 12.18
Left: Videomicrograph of the tongue showing the fungiform papillae of a "supertaster"—a person who is very sensitive to the taste of PROP. Right: Papillae of a "nontaster," who cannot taste PROP. The supertaster has both more papillae and more taste buds than the nontaster. (Photographs courtesy of Linda Bartoshuk.)

ply a reflection of the fact that you prefer different tastes. It may reflect not a difference in preference (you *like* sweet things more than John does), but a difference in taste *experience* (you *experience* more intense sweet tastes than John does) that could be caused by differences in the number of taste receptors on your tongue.

From our discussion, it is clear that a person's preference for particular foods results from a complex interaction of genetically determined, experiential, and physiological influences (Beauchamp, 1987; Rozin, 1976), and because of its role in food selection, taste is of great importance in our day-to-day lives. We will now consider the physiological workings of the sense of taste, beginning at the logical starting place: the tongue.

The Taste System

To introduce the taste system, we will first describe the anatomy of the tongue and the process of transduction that generates electrical signals from chemical stimuli. We will then describe the central destinations of these signals.

The Tongue and Transduction

The process of tasting begins in the tongue, when receptors are stimulated by taste stimuli. To understand the structures involved in this process, we begin with the tongue in Figure 12.19a. The surface of the tongue contains many ridges and valleys due to the presence of structures called **papillae**, of which there are four kinds: (1) filiform papillae, which are shaped like cones and are found over the entire surface of the tongue, giving it its rough appearance; (2) fungiform papillae, which are shaped like mushrooms and are found at the tip and sides of the tongue; (3) foliate papillae, which are a series of folds along the sides of the tongue; and (4) circumvallate papillae, which are shaped like flat mounds surrounded by a trench and are found at the back of the tongue (also see Figure 12.20).

All of the papillae except the filiform papillae contain taste buds (Figure 12.19b), and the whole tongue contains about 10,000 taste buds (Bartoshuk, 1971). Since the filiform papillae contain no taste buds, stimulation of the central part of the tongue, which contains only these papillae, causes no taste sensations. However, stimulation of the back or perimeter of the tongue results in a broad range of taste sensations.

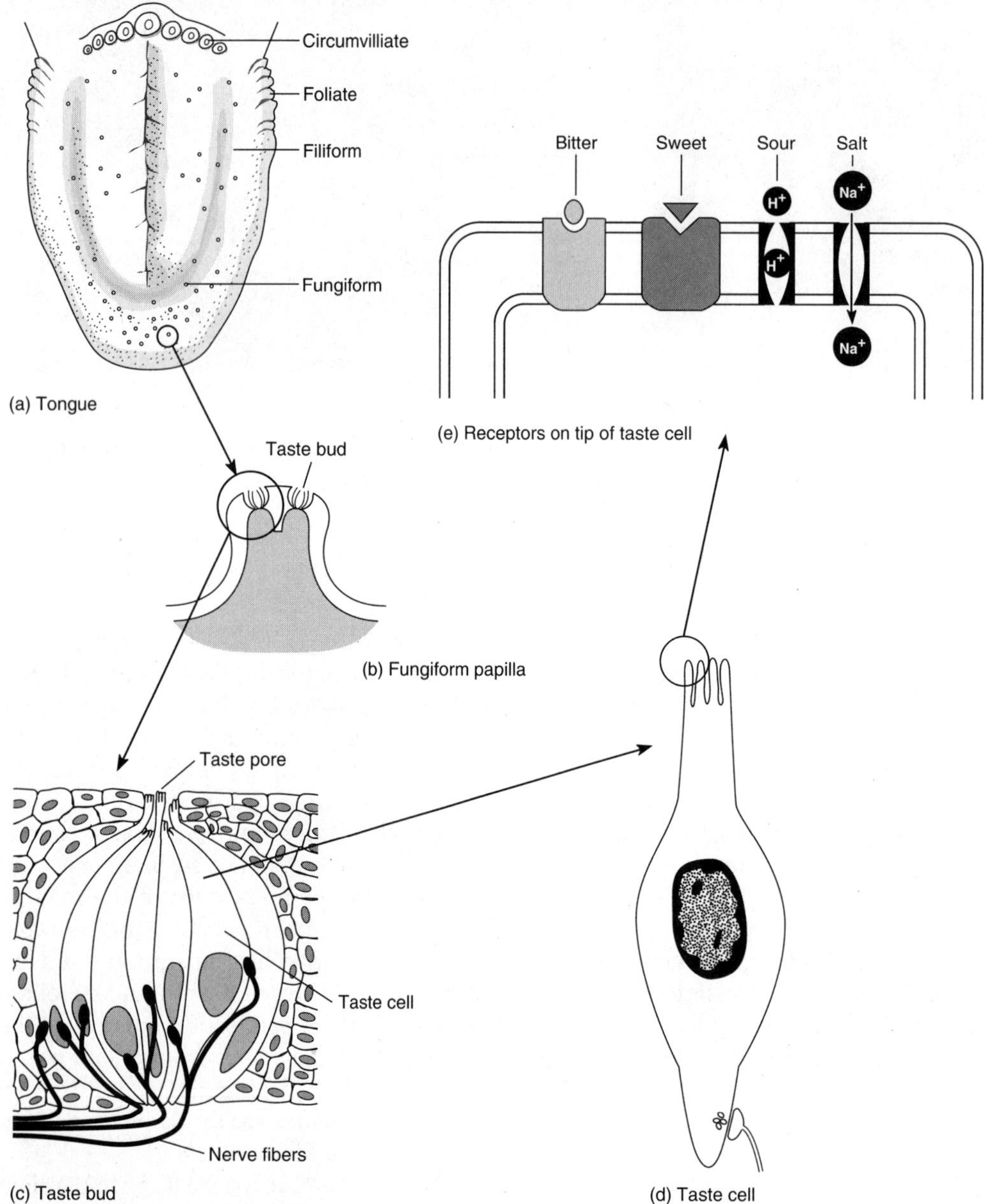

Figure 12.19

(a) The tongue, showing the four different types of papillae. (b) A fungiform papilla on the tongue; each papilla contains a number of taste buds. (c) Cross section of a taste bud showing the taste pore where the taste stimulus enters. (d) The taste cell; the tip of the taste cell is positioned just under the pore. (e) Close-up of the membrane at the tip of the taste cell, showing the receptor sites for bitter, sour, salty, and sweet substances. Stimulation of these receptor sites, as described in the text, triggers a number of different reactions within the cell (not shown) that lead to movement of charged molecules across the membrane, which creates an electrical signal in the receptor.

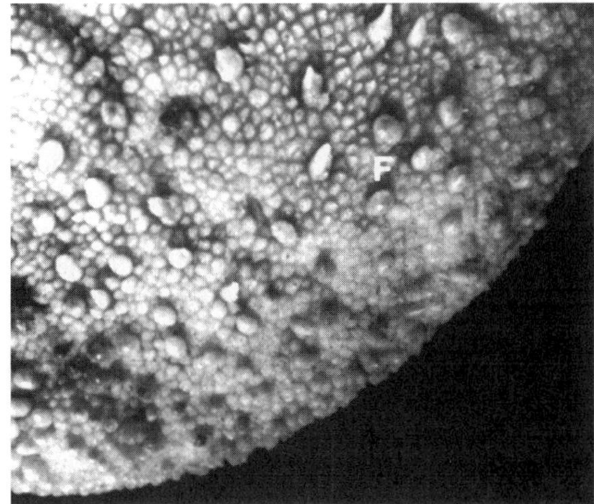

Figure 12.20
The surface of the tongue, showing fungiform (F) papillae. (From Miller, 1995.)

A **taste bud** (Figure 12.19c) contains a number of **taste cells**, which have tips that protrude into the **taste pore** (Figure 12.19d). Transduction occurs when chemicals contact receptor sites or channels located on the tips of these taste cells (12.19e). The details of this transduction process, which have just recently been discovered, involve complex chemical events within the taste cell (Kinnamon, 1988; Ye, Heck, & DeSimone, 1991). In simple terms, however, we can describe transduction as occurring when taste substances affect ion flow across the membrane of the taste cell. Different types of substances affect the membrane in different ways. As shown in Figure 12.19e, molecules of bitter and sweet substances bind to receptor sites, which release other substances into the cell, whereas sour substances are composed of H^+ ions that block channels in the membrane, and the sodium in salty substances becomes sodium ions (Na^+) in solution that flow through channels directly into the cell. Each of these mechanisms affects the cell's electrical charge by affecting the flow of ions into the cell.

Central Destinations of Taste Signals

Electrical signals generated in the taste cells are transmitted from the tongue in two pathways: the **chorda tympani nerve**, which conducts signals from the front and sides of the tongue, and the **glossopharyngeal nerve**, which conducts signals from the back of the tongue. Signals from taste receptors in the mouth and the larynx are transmitted in the **vagus nerve**. The fibers in these three nerves synapse in the brain stem in the **nucleus of the solitary tract** (NST), and from there, the signals travel to the thalamus and then to two areas in the frontal lobe—the **insula** and the **frontal operculum cortex**—that are partially hidden behind the temporal lobe (Figure 12.21) (Frank & Rabin, 1989; Finger, 1987).

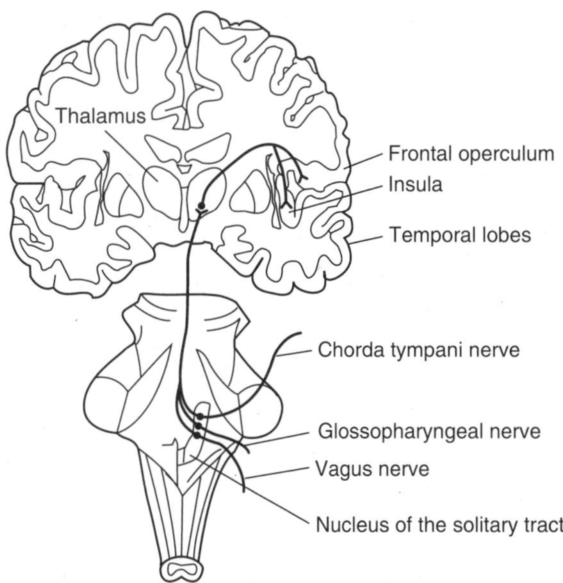

Figure 12.21
The central pathway for taste signals, showing the nucleus of the solitary tract (NST), where nerve fibers from the tongue and the mouth synapse in the medulla at the base of the brain. From the NST, these fibers synapse in the thalamus and the frontal lobe of the brain. (From Frank & Rabin, 1989.)

Taste Quality

When considering taste quality we are in a much better position than we were for olfaction. Although we have not been able to fit the many olfactory sensations into a small number of categories or qualities, taste researchers generally agree (with some exceptions—see Schiffman & Erickson, 1993) that there are four basic taste sensations: salty, sour, sweet, and bitter, and possibly a fifth basic taste called *umami*, the taste of the taste-enhancing chemical monosodium glutamate (MSG) (Scott, 1987). We will focus on salty, sour, bitter, and sweet since most research has used these four qualities.

From the results of psychophysical experiments, it appears that people can describe most of their taste experiences on the basis of the four basic taste qualities. For example, Donald McBurney (1969) presented taste solutions to subjects and asked them to give magnitude estimates of the intensity of each of the four taste qualities for each solution. He found that some substances have a predominant taste, and that other substances result in combinations of the four tastes. For example, sodium chloride (salty), hydrochloric acid (sour), sucrose (sweet), and quinine (bitter) are compounds that come the closest to having only one of the four basic tastes, but the compound potassium chloride (KCl) has substantial salty and bitter components (Figure 12.22). Similarly, sodium nitrate ($NaNO_3$) results in a taste made up of a combination of salty, sour, and bitter. Subjects, therefore, can describe their taste sensations based on these four qualities, and even if given the option of using additional qualities of their own choice, they will usually stay with the four basic tastes. Thus, salty, sour, sweet, and bitter appear to be adequate to describe the majority of our taste experiences.

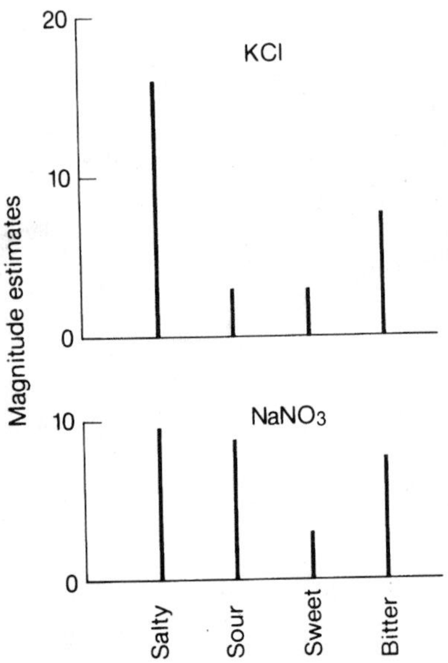

Figure 12.22
The contribution of each of the four basic tastes to the tastes of KCl and $NaNO_3$, determined by the method of magnitude estimation. The height of the line indicates the size of the magnitude estimate for each basic taste (McBurney, 1969).

The Neural Code for Taste Quality

What is the neural code for taste quality? In Chapter 1, when we introduced the idea of sensory coding, we distinguished between *specificity theory*, the idea that quality is signaled by the activity in neurons that are tuned to respond to specific qualities, and *across-fiber pattern theory*, the idea that quality is signaled by the pattern of activity in many fibers. In that discussion, and in others throughout the book, we have favored an across-fiber pattern theory of sensory coding. The situation for taste, however, is not clear-cut, and there are good arguments in favor of both types of coding.

Across-Fiber Pattern Coding

Let's consider some evidence for across-fiber pattern coding. Robert Erickson (1963) carried out one of the first experiments that demonstrated this type of coding by presenting a number of different taste stimuli to a rat's tongue and recording the response of the chorda tympani nerve. Figure 12.23 shows how 13 nerve fibers responded to ammonium chloride (NH_4Cl), potassium chloride (KCl), and sodium chloride (NaCl). The solid and dashed lines show that the across-fiber patterns of response of these 13 neurons to ammonium chloride and potassium chloride are similar to each other but are different from the pattern for sodium chloride, indicated by the open circles.

Erickson reasoned that, if the rat's perception of taste quality depends on the across-fiber pattern, then two substances with similar across-fiber patterns should taste similar. Thus, the electrophysiological results would predict that ammonium chloride and potassium chloride should taste similar and that both should taste different from sodium chloride. To test this hypothesis, Erickson shocked rats for drinking potassium chloride and then gave them a choice between ammonium chloride and sodium chloride. If potassium chloride and ammonium chloride taste similar, the rats should avoid the ammonium chloride when given a choice, and this is exactly what they did. And when the rats were shocked for drinking ammonium chloride, they then avoided the potassium chloride, as predicted by the electrophysiological results.

But what about the perception of taste in humans? When Susan Schiffman and Robert Erickson (1971) had humans make similarity judgments between a number of different solutions, they found that substances that were perceived to be similar were related to the across-fiber patterns for these same substances in the rat. Solutions judged more similar psychophysically had similar across-fiber patterns, as the across-fiber pattern theory would predict.

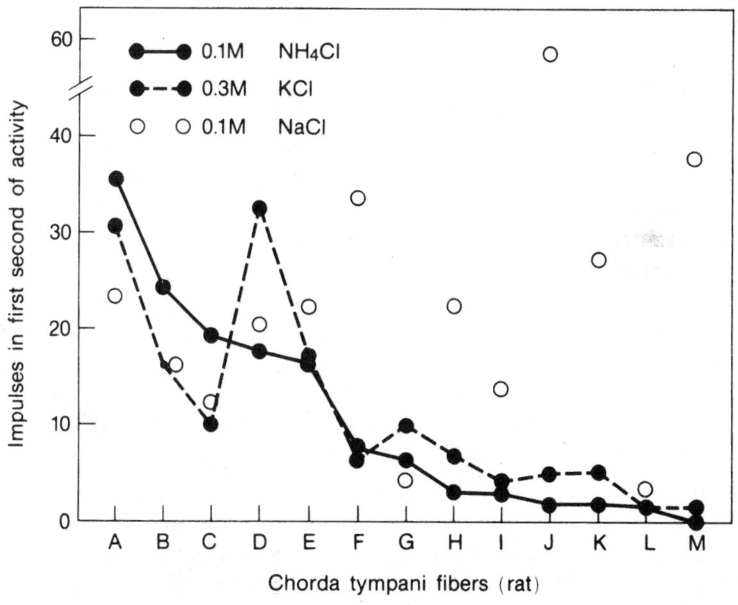

Figure 12.23
Across-fiber patterns of the response of fibers in the rat's chorda tympani nerve to three salts. Each letter on the horizontal axis indicates a different single fiber. (From Erickson, 1963.)

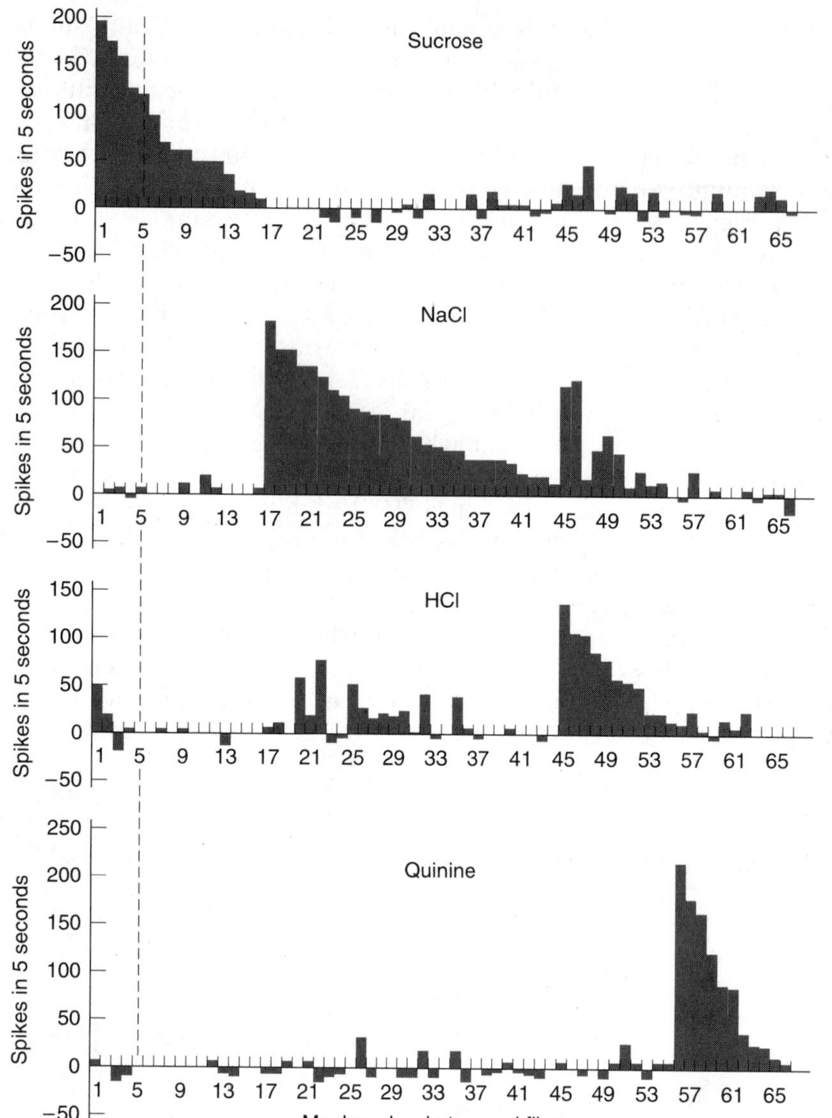

Figure 12.24
Responses of 66 different fibers in the monkey's chorda tympani nerve, showing how they responded to four types of stimuli: sucrose (sweet); sodium chloride (salty); hydrogen chloride (sour); and quinine (bitter). To determine the response of a particular fiber, pick its number and note the height of the bars for each compound. For example, fiber 5 (dashed line) fired well to sucrose but didn't fire to each of the other compounds. Fiber 5 is therefore called a sucrose-best fiber. *(From Sato & Ogawa, 1994.)*

Specificity Coding

Evidence in favor of the possibility that taste quality is coded by the activity in single neurons comes from experiments that show that there are four different types of fibers in the monkey's chorda tympani nerve (Sato, Ogawa, & Yamashita, 1994). Figure 12.24 shows how 66 fibers in the monkey's chorda tympani responded to four substances, each representing one of the basic tastes: sucrose (sweet); salt (NaCl, salty); hydrogen chloride (HCl, sour); and quinine (bitter). We can see that some fibers responded well to sucrose but poorly to almost all other compounds. For example, look at how fiber 5 responded to each substance by noticing the responses where the dashed line crosses the record for each substance. Fibers 1–16 are called *sucrose-best* since they fire best to sucrose. Figure 12.25 shows an example of firing records from one of the sucrose-best fibers (Sato, Ogawa, & Yamashita, 1975). A similar situation exists for the quinine-best fibers (numbers 56–66), most of which respond only to quinine. The NaCl- and HCl-best fibers fire predominantly to one solution, but some fire to both NaCl and HCl (also see Frank, Bieber, & Smith, 1988, for similar results in the hamster).

Another way to test for the operation of specificity coding is, first, to classify neurons in terms of their best response (as in Figure 12.24) and then to expose an animal to a situation that affects neural responding and see if just one type of fiber is affected. This result—only one type of fiber being affected, is what specificity theory would predict. Kimberle Jacobs, Gregory Mark, and Thomas Scott (1988) carried out such an experiment by recording from NST neurons in the rat's medulla before and after depriving the rats of sodium. Before sodium deprivation, about 60 percent of the neurons in this area responded vigorously to sodium. However, after deprivation, these formerly sodium-active neurons fell silent. Apparently, the lack of stimulation caused by the sodium deprivation caused these neurons to become insensitive to sodium.

These results are in line with specificity theory, since sodium deprivation causes only sodium-best neurons to decrease their sensitivity. But the most interesting finding of this experiment is that, after deprivation, sodium caused firing in sugar-best fibers. Apparently, deprivation causes the coding for sodium to be shifted to sugar fibers, a situation that might be expected to cause salt to take on the properties of sugar (Jacobs et al., 1988; Scott & Plata-Salaman, 1991). This is apparently what happens because, when the rats are given access to salt after sodium deprivation, they consume it with the gusto they usually reserve for sugar.

Another finding in line with specificity theory is the recently discovered transduction mechanisms, which indicate that salty, sweet, sour, and bitter substances each use different mechanisms to change the taste cell's membrane properties. Thus, the presentation of a substance called *amiloride*, which blocks sodium channels in membranes, causes a decrease in the responding of rat NST neurons that respond best to salt (Figure 12.26a) but has little effect on neurons that respond best to a combination of salty and bitter tastes (Figure 12.26b) (Scott & Giza, 1990). As we would expect from these results, applying amiloride to a human's tongue causes subjects to describe previously salty substances as "tasteless" (McCutcheon, 1992).

What does all of this evidence mean? Some experiments support across-fiber pattern coding, and others support specificity coding. It is difficult to choose between the two because both can explain most of the data (Scott & Plata-Salaman, 1991). For example, David Smith and co-workers (1983) point out that, while similar-tasting com-

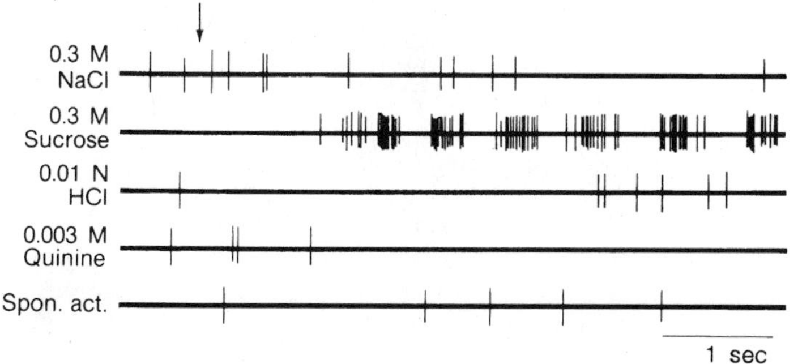

Figure 12.25
Response of a single fiber in the monkey's chorda tympani nerve to salt, sucrose, hydrochloric acid, and quinine. The bottom record indicates the rate of spontaneous activity. The arrow above the record for NaCl indicates the time of application of the stimuli (Sato, Ogawa, & Yamashita, 1975).

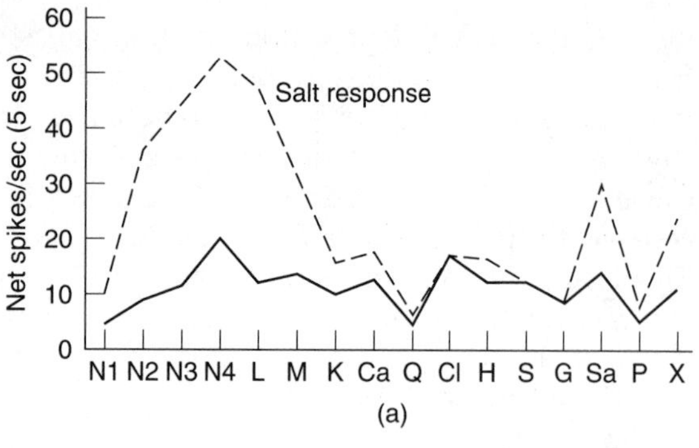

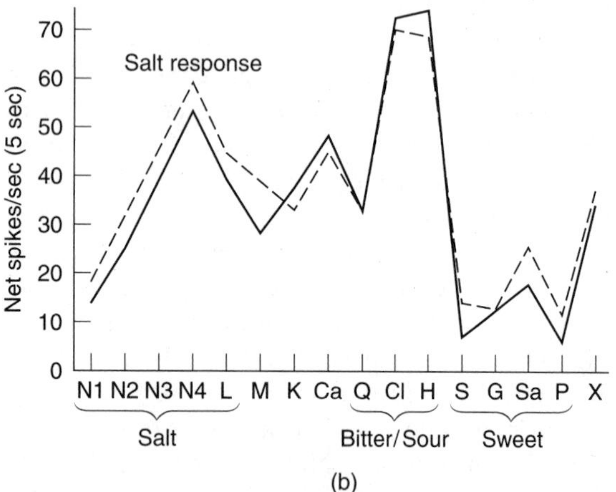

Figure 12.26

The dashed lines show how two neurons in the rat NST respond to a number of different taste stimuli (along the horizontal axis). The neuron in (a) responds strongly to compounds associated with salty tastes. The neuron in (b) responds to a wide range of compounds. The solid lines show how these two neurons fire after the sodium-blocker amiloride is applied to the tongue. This compound inhibits the responses to salt of neuron (a) but has little effect on neuron (b). (From Scott & Giza, 1990.)

pounds have highly similar across-fiber patterns, these patterns are dominated by activity in particular neurons. Thus, sweet-tasting compounds produce highly similar across-fiber patterns in the hamster, primarily because of the high firing rates contributed by the sucrose-best neurons.

Perhaps taste uses both specificity coding and across-fiber patterning at the same time. One possibility is that basic qualities are identified by the firing of specific neurons, and that more subtle differences between substances are then determined by the pattern of firing of larger groups of neurons (Pfaffmann, 1974; Scott & Plata-Salaman, 1991).

It is fitting to end this chapter on this note of uncertainty because it illustrates a problem common to all of the senses, which we can describe as follows: Researchers can, using their electrodes, record neural responses to various kinds of stimuli. Based on how these neurons respond, the researchers might be able to show that the firing of certain neurons is associated with a particular quality. But even after demonstrating this association between neural responding and a sensory quality, researchers are still faced with the following questions: What information is *actually being used by the brain* to determine our experience of that particular quality? Is the key information contained in individual neurons? In the firing patterns of larger populations? In the timing of nerve impulses? As we have seen in this text, we are closer to answering these questions for some of the senses than for others. But even for the sensory qualities we know most about, such as perceiving color or pitch, many questions still remain to be answered by the next generation of researchers, some of whom, hopefully, will include some of the readers of this book.

Similarities in Chemical Sensing across Species

There are many differences in how different animals sense chemicals. For example, receptors for sensing chemicals are located in different places in different animals: Humans smell with their nose, flies detect chemicals with receptors in their feet, and moths trap airborne chemicals in tiny hairs in their antennae. Despite differences such as these, animals that are widely separated from each other on the phylogenetic scale show many striking similarities in how they sense chemicals. We will now describe some of these similarities.

Receptor Cell Structure

Although looking at the olfactory receptors of various species reveals many differences, they all have a common trait: At the receiving end, they have extensions, called *cilia*, that exist in a fluid medium, and at the sending end, they have a single axon that projects directly into the central nervous system without synapsing (Figure 12.27). This lack of synapses between the receptors and the brain contrasts with the situation in vision, in which there are many synapses within the retina, and hearing, in which the receptors synapse with auditory nerve fibers which then travel into the central nervous system.

Receptor Cell Turnover

The receptors in the olfactory system and the taste system of all vertebrates undergo a cycle of

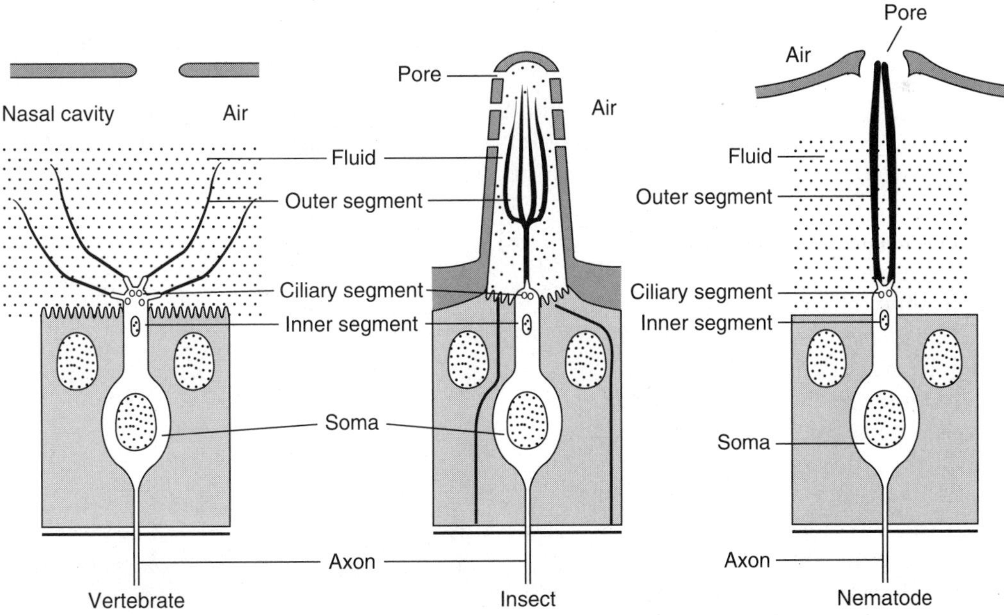

Figure 12.27 *Chemoreceptor cells in three widely different species. Although they at first appear different, they all have similar components. (From Ache, 1991; based on Steinbrecht, 1969, and Ward et al., 1975.)*

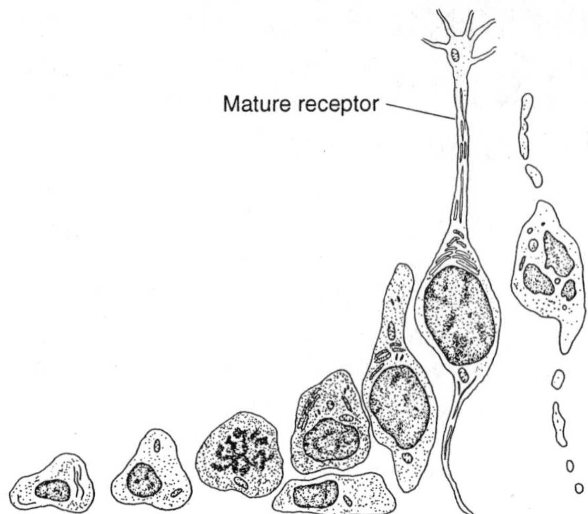

Figure 12.28
Stages in the development and decline of an olfactory receptor. The mature receptor develops from the cell on the left and then dies, as shown on the right. (From Graziadei, 1976.)

birth, development, and death over an average period of five to seven weeks (Figure 12.28), and this turnover has also been observed in an invertebrate, the snail (Chase & Rieling, 1986). This phenomenon of renewal and death, which is called **neurogenesis**, is unusual in the nervous system but serves an important function for chemical-sensing receptors because they are constantly open to the elements. They are bombarded not only by the chemicals they are designed to sense, but also by irritants such as bacteria and dirt and are under the constant threat of drying that comes from being exposed to the air. When we compare this situation to the situation in vision, hearing, and the cutaneous senses, in which the receptors are safely protected inside structures such as the eye and the inner ear, and under the skin, it becomes understandable that the receptors for taste and smell would need a mechanism for renewal.

Receptor Stereospecificity

Humans can distinguish between *enantiomeric forms* of chemicals—molecules that have the same molecular structure but are mirror images of each other. For example, L-proline tastes bitter to most people, and its mirror image D-proline tastes sweet (Figure 12.29). (Schiffman, Sennewald, & Gagnon, 1981). This ability to distinguish among enantiomers means that the taste receptors are stereospecific: They can differentiate between molecules based on their shapes. This property of human receptors is shared by a wide range of other organisms, including mice (Kasahara, Iwasaki, & Sato, 1987), crabs (Case, 1964), fish (Brand et al., 1987), and even bacteria, which will congregate around some forms of chemicals but will ignore the mirror images of the same chemicals (Adler, 1969).

Intermittent Stimulation of Receptors

Olfactory stimuli are typically presented intermittently as a natural consequence of the intake and expulsion of air during breathing in many animals. However, this way of producing intermittent stimulation is often supplemented by sniffing. If we are particularly interested in de-

D-proline

L-proline

Figure 12.29
The enantiomeric molecules L-proline and D-proline have mirror-image structures. L-proline tastes bitter, and D-proline tastes sweet.

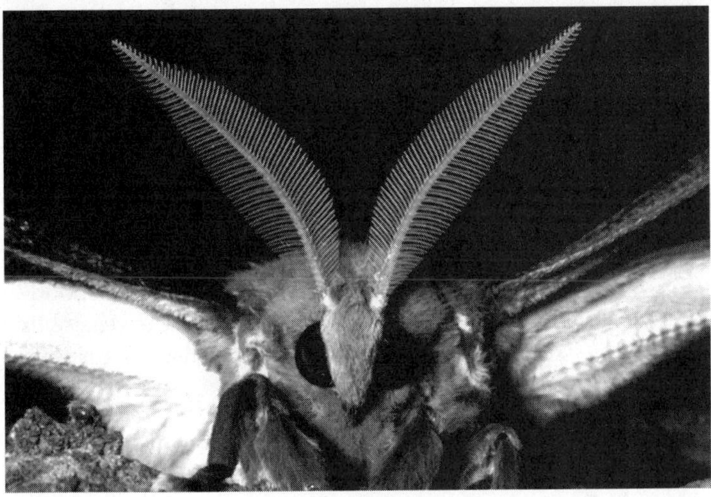

Figure 12.30
The male silkworm moth, showing the large antennae that serve as taste receptors. (From Michael Fogden; Oxford Scientific Films/Animals Animals.)

tecting or experiencing an odor, we sniff (Laing, 1985), an activity especially used by animals with good olfactory sensitivity, such as dogs and rats.

By presenting bursts of intermittent stimuli, sniffing not only helps bring more stimuli into contact with the receptors but also helps overcome the fact that continuous exposure to a chemical causes adaptation that slows or stops the receptor's response. The alternation of air and odorant delays this adaptation and therefore enhances the perception of odors (Dethier, 1987).

Sniffing is how mammals cause intermittent stimulation to forestall adaptation, but other animals have their own ways of achieving this result. Lobsters, insects, and snails flick their olfactory organs (Schmitt & Ache, 1979). For example, snakes flick their tongues to deliver odorants to olfactory receptors in the roof of the mouth (Kubie & Halperin, 1975), moths begin flicking their antennae to help deliver airborne olfactory stimuli to their receptors (Dethier, 1987) (Figure 12.30), fish such as the flounder move stimuli through their nasal chambers by pumping motions (Doving & Thommesen, 1977), and the octopus reacts to the presence of chemicals in the water by blowing water through its siphon (Chase & Wells, 1986).

As we have described the different perceptual worlds inhabited by different animals throughout this book, we have observed vast differences in the sensory capabilities of these animals. It is perhaps appropriate, therefore, that we end our survey of other worlds of perception by showing how vastly different animals, with their different perceptual experiences, still share many mechanisms of sensory functioning. After all, despite their differences, they all have the same task: to detect and discriminate the various forms of environmental energy that will enable them to locomote through their environment, to find food, and, in many animals, to avoid becoming someone else's meal.

Olfaction and Taste in Infants

Do newborn infants perceive odors and tastes? Early researchers, noting that a number of olfactory stimuli elicited responses such as body movements and facial expressions from newborns, concluded that newborns can smell (Kroner, 1881; Peterson & Rainey, 1910–1911). However, some of the stimuli used by these early researchers may have been irritating, so the infants may have been responding to irritation rather than smell (Beauchamp, Cowart, & Schmidt, 1991; Doty, 1991). Modern studies using nonirritating stimuli have, however, provided evidence that newborns can smell and can discriminate between different olfactory stimuli. Studies by J. E. Steiner (1974, 1979), using nonirritating stimuli, showed that infants respond to banana extract or vanilla extract with sucking and facial expressions that are similar to smiles, and that they respond to concentrated shrimp odor and an odor resembling rotten eggs with rejection or disgust responses (Figure 12.31).

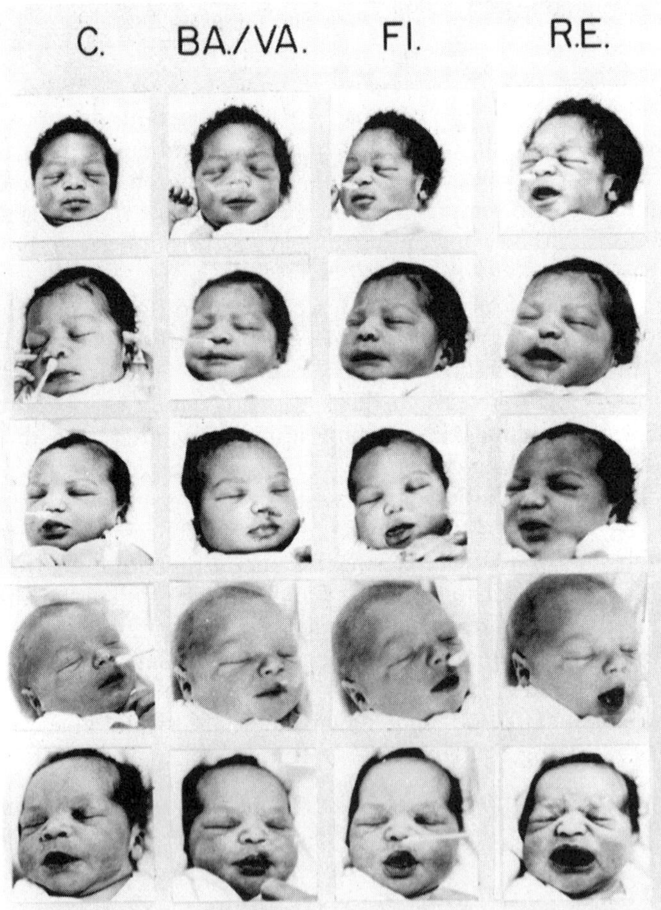

Figure 12.31
The facial expressions of 3- to 8-hour-old infants in response to some food-related odors. In each of the horizontal rows, the reactions of the same infant can be seen to the following stimulation: C = control, odorless cotton swab; BA/VA = artificial solution of banana or vanilla; FI = artificial fish or shrimp odor; R.E. = artificial rotten egg odor. The infants were tested prior to the first breast- or bottle-feeding. (Photographs courtesy of J. E. Steiner, The Hebrew University, Jerusalem.)

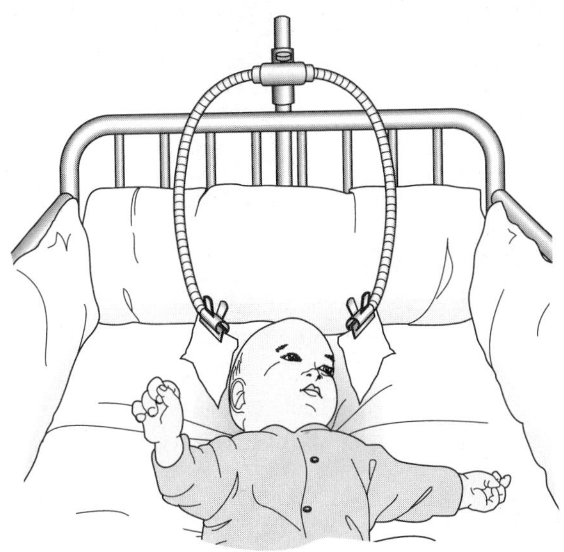

Figure 12.32
Device used to test infants' response to odorized pads. (From Porter & Schaal, 1995.)

It is perhaps not surprising that newborns would respond to strong odors such as rotten eggs. But recent research has also shown that infants can differentiate between subtle qualities in olfactory stimuli. These studies have shown that nursing infants can differentiate the odor of their mother's breasts from the odor from the breast of another mother who is also nursing her infant. A. Macfarlane (1975) demonstrated this preference for the mother's breast odor by presenting breast pads from the mother and the other lactating female to infants in their cribs, and by noting that the infants spent significantly more time turning toward their mother's pads than toward the other woman's pad (Figure 12.32).

More recent research has shown that this preference for the mother's odor is probably learned. Jennifer Cernoch and Richard Porter (1985) demonstrated that 2-week-old breast-fed infants turn toward their mother's axillary (armpit) odors in preference to the axillary odors of either unfamiliar lactating or nonlactating women or their father. Infants who were bottle-fed did not, however, prefer their mother's odor. Cernoch and Porter concluded that the basis for the infants' recognition was that they had learned to recognize their mother's unique olfactory signature while feeding (also see Porter et al., 1992).

Further evidence that learning is important comes from an experiment by Rene Balogh and Richard Porter (1986), in which they exposed 1-day-old infants to either an artificial cherry odor or an artificial ginger odor in the bassinet for two days. After this exposure, the infants were tested by the presentation of two pads, one with cherry odor and the other with ginger odor, on either side of the head for two minutes. The results indicated that female infants preferentially turned toward the odor to which they had previously been exposed, for twice as long as to the other odor. The males showed no preference, mainly because they tended to stay turned in one direction, without changing position, during the two-minute test. The result for females supports the idea that infants can perceive odors and that

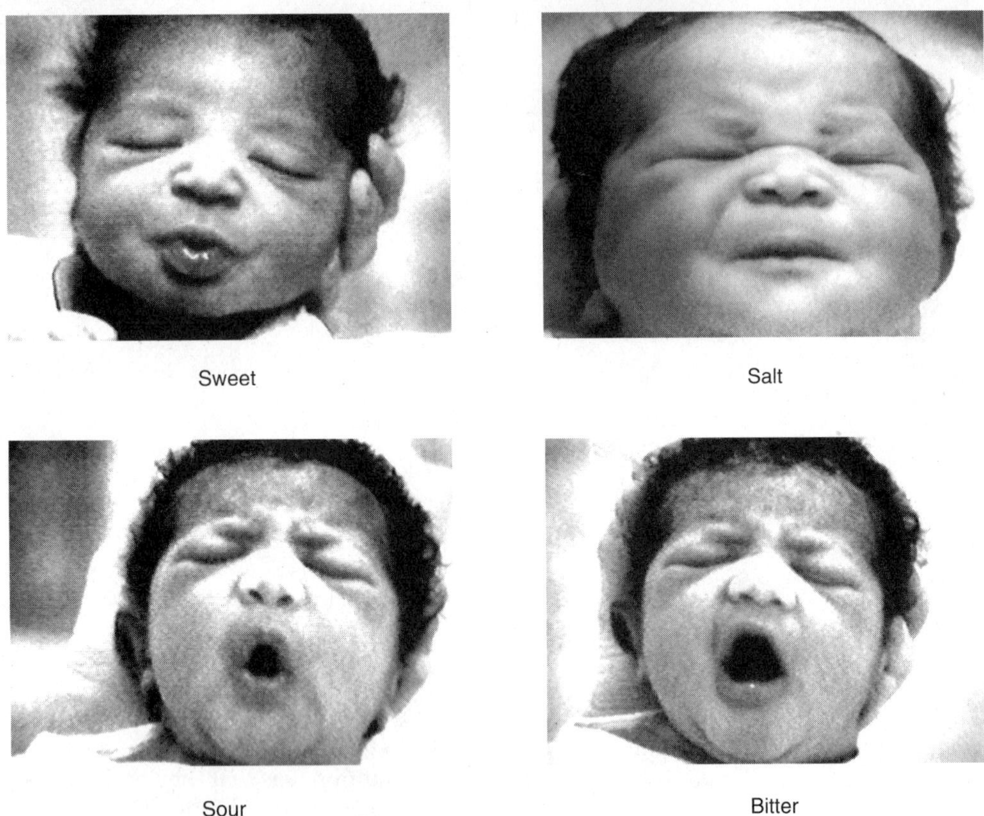

Figure 12.33
The facial expressions of 4- to 10-hour-old infants to sweet, sour, salty, and bitter tastes. (From Rosenstein & Oster, 1988.)

they can develop preferences for odors through learning.

Research on infants' reactions to taste has included numerous studies showing that newborns can discriminate sweet, sour, and bitter stimuli (Beauchamp, Cowart, & Schmidt, 1991). For example, newborns react with different facial expressions to sweet, sour, and bitter stimuli but show little or no response to salty stimuli (Figure 12.33) (Ganchrow, 1995; Ganchrow, Steiner, & Daher, 1983; Rosenstein & Oster, 1988).

Although responses to taste and olfaction do show some changes as the infant grows into childhood (for example, young infants are indifferent to the taste of salt but develop a response to salty stimuli as they get older; Beauchamp, Cowart, & Moran, 1986; Beauchamp, Bertino, & Engelman, 1991), we could argue that taste and olfaction are the most highly developed of all of the senses at birth.

Review

Outline	Questions
• The chemical senses have been called molecule detectors since they create specific smells and tastes in response to specific molecules.	**1.** What two properties do the chemical senses have in common? Describe how taste and smell interact with each other. (499)

The Uses of Olfaction

• Olfaction is extremely important in the lives of many animals, since it is their main window on the sensory environment.	**2.** Define microsmatic and macrosmatic. What are some of the functions of smell in animals? What is a convincing argument for the importance of smell to humans? (500)

Some Facts about Human Olfaction

• One reason the powers of the human olfactory system have often been underrated is that humans' capacity to detect odors is not as acute as other animals'. Our powers of olfaction are, however, actually better than we might at first suppose.	**3.** How does the sensitivity of humans to odors compare to the sensitivity of rats and dogs? How sensitive are the human's olfactory receptors? Why is the dog more sensitive to odors than the human? (501)
• The difference threshold is the ability to detect differences in intensity. Smell has been reputed to have the largest difference threshold of all the senses.	**4.** Describe the procedure that made it possible to show that the olfactory system has a smaller difference threshold than was previously reported. (501)
• Early research showed that people can successfully identify odors only about half the time.	**5.** What conditions cause an increase in the percentage of odors that people can identify? (502)
• Olfaction plays a major role in communication for many animals. However, except in intimate situations, humans are constrained from smelling other people at close range.	**6.** Describe the undershirt experiment and menstrual synchrony. What do each of these tell us about the human sense of smell? (503)

Structure of the Olfactory System

- Odorant molecules entering the nose stimulate receptors in the olfactory mucosa, which send their processes to the olfactory bulb.

7. Describe the two ways in which odorant molecules can reach receptors in the mucosa. Describe the active site on the olfactory receptors. Where do signals go after traveling from the receptors to the olfactory bulb? (504)

The Odor Stimulus and Odor Quality

- When you smell something, you are sensing molecules that have left the thing you are smelling and have traveled through the air to your nose.

8. What two problems make it difficult to determine the code for olfactory quality? (505)

- Many attempts have been made to devise a classification system for odors, although none have proved successful.

9. Describe Henning's odor prism. What is odor profiling? What have these two approaches to odor classification failed to do? (506)

- The complexity of the olfactory stimuli actually found in the environment has caused researchers to focus their search for connections between physical properties and smell on simpler chemicals.

10. Explain why identifying the smells of coffee, toast, and bacon coming from the kitchen is a feat of perceptual organization. (509)

- In their search for physical properties correlated with odor quality, researchers have considered molecular shape and chemical composition.

11. What are some problems associated with drawing connections between odor quality and molecular shape and between quality and chemical composition? (509)

The Neural Code for Odor Quality

- In investigating olfactory coding at the receptor level, researchers have considered how individual receptors fire to stimuli, how different areas of the olfactory mucosa respond to stimuli, and how odorant molecules are deposited on the mucosa as they flow over it.

12. Are there olfactory receptors that are tuned to respond to just one type of odorant? Describe how a population of receptors would respond to a single odorant. (510)

13. What is the regional sensitivity effect? Describe how different molecules distribute themselves on the surface of the mucosa. (510)

- Coding of odor quality at the level of the olfactory bulb occurs when different odorants cause different patterns of activity in the bulb.
- We know little about olfactory coding at centers higher than the olfactory bulb.

14. Describe the 2-DG experiment that measured how odorants stimulate neurons in the olfactory bulb. Compare coding in the olfactory system to coding in other sensory systems. (512)

15. Describe the behavioral and electrophysiological evidence that the OFC may be important in olfaction. (513)

The Perception of Flavor

- Many of the sensations we call taste are actually flavor, a combination of taste and smell.

16. What is the retronasal route for stimulation of the olfactory receptors? What evidence demonstrates the importance of olfaction in sensing flavor? Why do we sometimes mislocate the source of flavor sensations? (514)

Factors Influencing Food Preferences

- Taste has been called the social sense because we often eat together and share our taste experiences as we eat.
- A number of factors influence our food preferences. Some have to do with past behaviors, and many are based on physiology.

17. What does it mean to say that we often share the affective aspects of taste with other people? (515)

18. Give examples of how each of the following influences food preferences and eating behavior: the internal state of the organism; past experiences with food; pairing food with sickness; and specific hungers. (516)

19. What is the evidence that food may taste different to different people because of genetic differences between the people? What are tasters? Nontasters? What is a physiological difference between them? (518)

The Taste System

- The process of tasting begins when receptors in the tongue are stimulated by taste stimuli. The electrical signals generated in the tongue are transmitted via the chorda tympani and glossopharyngeal nerves to the nucleus of the solitary tract (NST) and then to the thalamus and to areas in the frontal cortex. Signals from taste receptors in the mouth and the larynx are transmitted in the vagus nerve to the NTS.

20. Describe the following structures in the tongue: papillae; taste buds; taste pores; taste cells; receptor sites or channels on the taste cells. (519)

21. When does transduction occur in the taste system? (521)

Taste Quality

- Researchers have established that there are four basic taste sensations: salty, sour, sweet, and bitter.

22. Describe how magnitude estimation has been used to show that subjects can describe their taste experience based on the four basic tastes. (522)

The Neural Code for Taste Quality

- The neural code for taste quality has been explained in terms of across-fiber pattern theory, which states that quality is signaled by the firing pattern of large groups of neurons.

23. Describe Erickson's experiment in the rat that supports an across-fiber pattern code for taste quality. How do these data relate to human perception? (523)

- The neural code for taste quality has also been explained in terms of specificity theory, which states that taste quality is signaled by the firing of neurons tuned to respond to specific qualities.

24. What is the evidence that there are neurons that respond best to specific taste qualities? (524)

25. How does depriving rats of sodium affect the response of their sodium-best fibers? Which fibers fire to sodium after deprivation has had its effect? Describe the experiments in which amiloride was applied to the tongue. (525)

- A problem that is common to the study of sensory coding in all of the senses is determining whether neural responding that is correlated with a particular quality is actually used by the brain to determine the experience associated with that quality.

26. What have researchers concluded about the roles of specificity theory and across-fiber pattern theory in determining taste quality? How might the two approaches work together to determine taste quality? (525)

Similarities in Chemical Sensing across Species

- Although there are many differences in how various animals sense chemicals, there are also many similarities, even in animals that are widely separated on the phylogenetic scale.

27. Give examples of how different species share the following properties: structure of the receptor cells; turnover of the receptor cells; receptor stereospecificity; intermittent stimulation of the receptors. (527)

Olfaction and Taste in Infants

- Early researchers' conclusion that infants can smell may have been compromised by the fact that some of the stimuli they used may have caused irritation. Modern research, using nonirritating stimuli, has shown that newborn and very young infants can taste and smell and can differentiate different stimuli from one another.

28. Describe the evidence from observing facial expressions that indicates that newborns can smell and taste. Describe the experiments that showed that nursing infants can distinguish between their mother's breast and axillary odors and the odors of other people. What is the evidence that the preference for the mother's odor is learned? (530)

Clinical Aspects of Vision and Hearing

Although it is obvious that the man in Figure 13.1 is examining the woman's eye, most people do not understand exactly what he is seeing or what he is looking for. Even though most Americans have had their eyes and ears examined because of problems with either vision or hearing, or just as part of a routine physical examination, few people understand, except in the most general sense, what is going on during these examinations. One of the purposes of this chapter is to demystify what goes on during examinations of the eye and the ear.

Before we can understand what eye and ear specialists look for during an examination, we must understand the major problems that can cause impairments in vision and hearing. We therefore begin this chapter by describing a number of the most common visual problems and how they are treated to improve or restore vision. After we understand the nature of the most common causes of visual problems, we will describe how a routine eye examination detects these problems. Following our discussion of vision, we cover the same material for hearing.

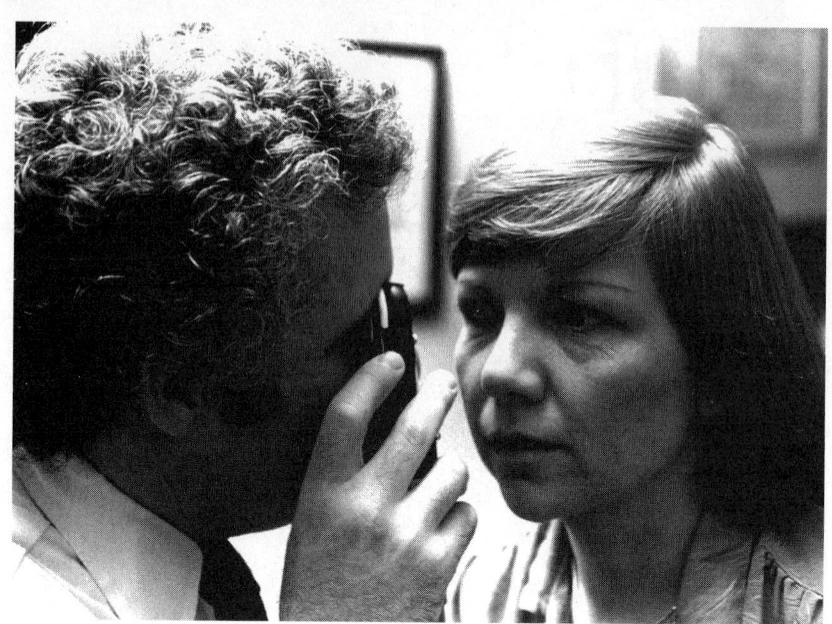

Figure 13.1
Ophthalmologist examining patient.

VISUAL IMPAIRMENT

How Can Vision Become Impaired?

Four major types of problems can cause poor vision (Figure 13.2):

1. Light is not focused clearly on the retina. Problems in focusing light can occur because the eyeball is too short or too long or because the cornea or the lens does not function properly. We will describe the following specific problems: myopia (nearsightedness), hyperopia (farsightedness), presbyopia ("old eye"), and astigmatism.
2. Light is blurred as it enters the eye. Scarring of the cornea or clouding of the lens blurs light as it enters the eye. Specific problems: corneal injury or disease, cataract.
3. There is damage to the retina. The retina can be damaged by disruption of the vessels that supply it with blood, by its separation from the blood supply, and by diseases that attack its receptors. Specific problems: macular degeneration, diabetic

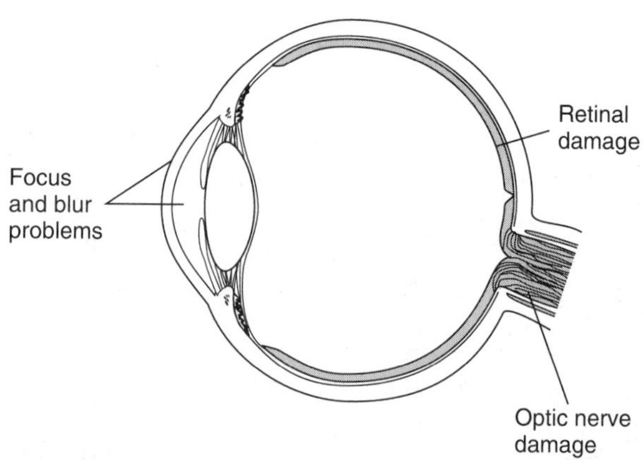

Figure 13.2
Places in the eye where visual problems can occur.

retinopathy, detached retina, hereditary retinal degeneration.

4. There is damage to the optic nerve. The optic nerve can degenerate. When this degeneration is due to a pressure buildup inside the eyeball, the cause is glaucoma. In addition, degeneration can be caused by poor retinal circulation, toxic substances, or the presence of a tumor. We will focus on glaucoma in our discussion.

We begin by considering a problem that affects more people than all the others combined: an inability to adequately focus incoming light onto the retina.

Focusing Problems

In Chapter 2, we described the optical system of the eye—the cornea and the lens—which, if everything is working properly, brings light entering the eye to a sharp focus on the retina. We also described the process of accommodation, which adjusts the focusing power of the eye to bring both near and far objects into focus.

We will now consider the conditions myopia, hyperopia, presbyopia, and astigmatism, four problems that affect a person's ability to focus an image on the retina.

Myopia

Myopia, or nearsightedness, is an inability to see distant objects clearly. The reason for this difficulty, which affects over 70 million Americans, is illustrated in Figure 13.3a: In the myopic eye, parallel rays of light are brought to a focus in front of the retina so that the image reaching the retina is blurred. This problem can be caused by either of two factors: (1) **refractive myopia,** in which the cornea and/or the lens bends the light too much, or (2) **axial myopia,** in which the eyeball is too long. Either way, light comes to a focus in front of the retina, so that the image on the retina is out of focus, and far objects look blurred.

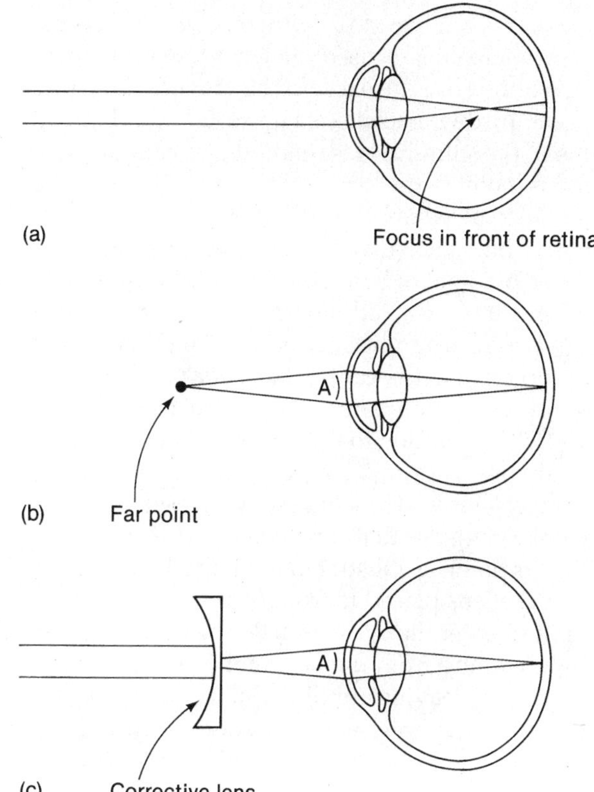

Figure 13.3
Focusing of light by the myopic (nearsighted) eye. (a) Parallel rays from a distant spot of light are brought to a focus in front of the retina, so distant objects appear blurred. (b) As the spot of light is moved closer to the eye, the focus point is pushed back until, at the far point, the rays are focused on the retina, and vision becomes clear. Vision is blurred beyond the far point. (c) A corrective lens, which bends light so that it enters the eye at the same angle as light coming from the far point, brings light to a focus on the retina. Angle A is the same in (b) and (c).

How can we deal with this problem? One way to create a focused image on the retina is to move the stimulus closer. This pushes the focus point further back (see Figure 2.7), and if we move the stimulus close enough, we can push the focus point onto the retina (Figure 13.3b). The distance at which the spot of light becomes focused on the retina is called the **far point,** and when our spot of light is at the far point, a myope can see it clearly. Although a person with myopia can see nearby objects clearly (which is why a myopic person is called *nearsighted*), objects beyond the far point are still out of focus (see Table 13.1). The solution to this problem is well known to anyone with myopia: corrective eyeglasses or contact lenses. These corrective lenses bend incoming light so that it is focused as if it were at the far point, as illustrated in Figure 13.3c. Notice that the lens placed in front of the eye causes the light to enter the eye at exactly the same angle as light coming from the far point in Figure 13.3b.

Table 13.1
Comparison of focusing problems associated with the far point and the near point

Far Point (Farthest Distance for Clear Vision)	Near Point (Closest Distance for Clear Vision)
Problem: In myopia, the far point is close to the eye, and vision is blurred beyond the far point.	Problem: In presbyopia, the near point moves away from the eye, and vision is blurred closer than the near point.

Before leaving our discussion of myopia, let's consider the following question: How strong must a corrective lens be to give the myope clear far vision? To answer this question, we have to keep in mind what is required of a corrective lens: It must bend parallel rays so that light enters the eye at the same angle as a spot of light positioned at the far point. Figure 13.4 shows what this means for two different locations of the far point. When the far point is close, as in Figure 13.4a, we need a powerful corrective lens to bend

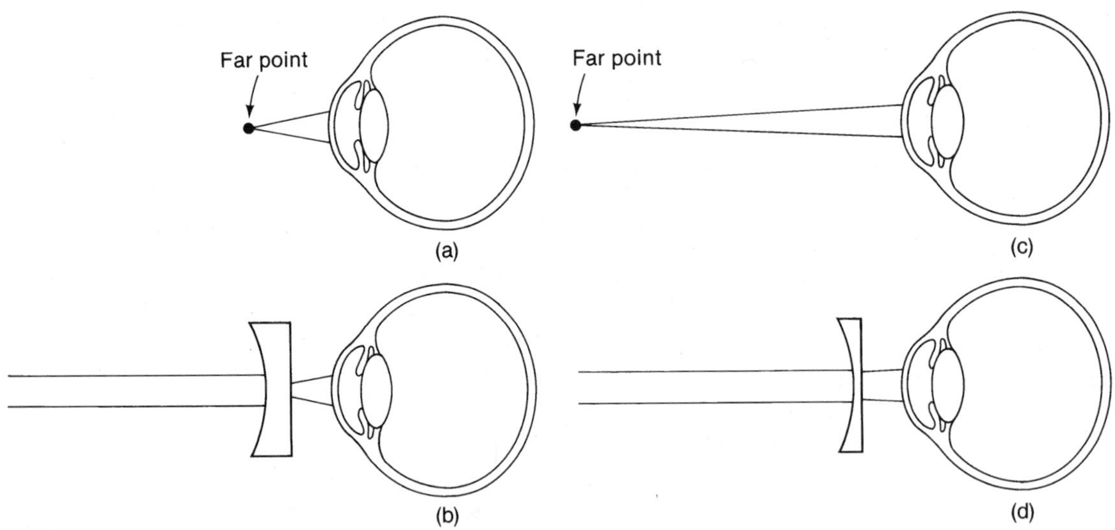

Figure 13.4
The strength of a lens required to correct myopic vision depends on the location of the far point. (a) A close far point requiring (b) a strong corrective lens. (c) A distant far point requiring (d) a weak corrective lens.

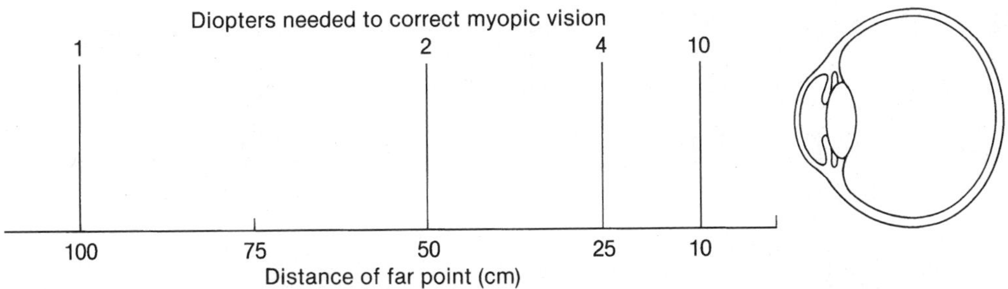

Figure 13.5
The number of diopters of lens power needed to correct myopic vision for different far points. Without a corrective lens, vision is blurred at distances greater than the far point. A far point of 10 cm represents severe myopia, and a far point of 100 cm represents mild myopia.

the light in the large angle shown in Figure 13.4b. However, when the far point is distant, as in Figure 13.4c, we need only a weak corrective lens to bend the light in the small angle shown in Figure 13.4d. Thus, the strength of the corrective lens depends on the location of the far point: A powerful lens is needed to correct vision when the far point is close, and a weak lens is needed to correct vision when the far point is distant.

When ophthalmologists or optometrists write a prescription for corrective lenses, they specify the strength of the lens in **diopters,** using the following relationship: number of diopters = l/far point in meters. Thus, a slightly myopic person with a far point at 1 meter (100 cm) requires a 1-diopter correction (diopters = 1/1 = 1.0). However, a very myopic person with a far point at 2/10 of a meter (20 cm) requires a 5-diopter correction (diopters = 1/0.2 = 5.0). This relationship between the distance of the far point and the required number of diopters of correction is shown in Figure 13.5.

Although glasses or contact lenses are the major route to clear vision for the myope, surgical procedures have recently been introduced that, for some people, may provide good vision without corrective lenses. In a procedure called **radial keratotomy,** four or eight cuts are made on the cornea (Figure 13.6a). These cuts cause the cornea to become flatter, which reduces its power to bend light and causes the focus point of the image to move toward the retina (Figure 13.6b). Most people who have had this operation experience an improvement of vision, although sometimes their vision is either undercorrected or overcorrected, and some people's vision drifts back toward myopia over time. Because of these problems and other undesirable side effects, some ophthalmologists are cautious about recommending this operation to their patients.

A newer technique, which provides more precise control over the shape of the cornea, is

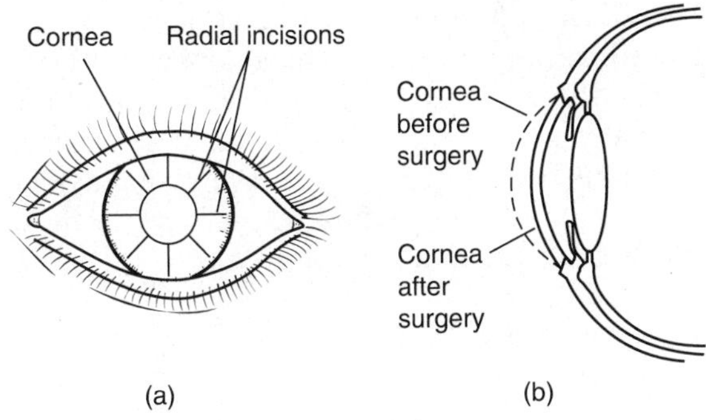

Figure 13.6
The radial keratotomy operation. (a) Four or eight incisions are made in the cornea. (b) The incisions flatten the cornea, weakening its focusing power and causing the focus point to move back toward the retina.

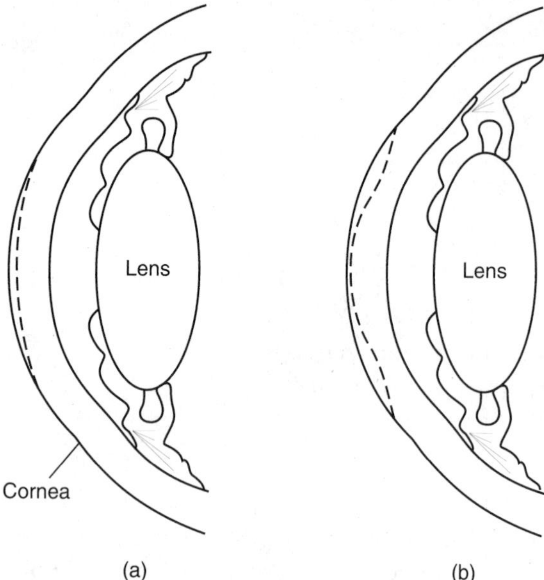

Figure 13.7
In the laser photorefractive keratotomy operation, an excimer laser is used to reshape the cornea, as shown by the dashed lines. (a) Reducing the curvature of the cornea on the myopic eye reduces the focusing power of the cornea so that the focus point moves back. (b) Increasing the curvature of the cornea in the hyperopic eye increases the focusing power of the cornea so that the focus point moves forward.

called **laser photorefractive keratotomy.** In this procedure, a type of laser called an *excimer laser,* which does not heat tissue, is used to sculpt the cornea to give it either less power (for myopia; Figure 13.7a) or more power (for hyperopia; Figure 13.7b). This procedure appears to be most effective for myopia.

Hyperopia

A person with **hyperopia,** or **farsightedness,** can see distant objects clearly but has trouble seeing nearby objects (Figure 13.8a). In the hyperopic eye, the focus point for parallel rays of light is located behind the retina, usually because the eyeball is too short. By accommodating to bring the focus point back to the retina, people with hyperopia are able to see distant objects clearly.

Nearby objects, however, are more difficult for the hyperope to deal with because moving an object closer pushes the focus point farther back. The hyperope's focus point, which is behind the retina for far objects, is pushed even farther back for nearby objects, so the hyperope must exert a great deal of accommodation to return the focus point to the retina. The hyperope's constant need to accommodate when looking at nearby objects (as in reading or doing close-up work) results in eyestrain and, in older people, headaches. Headaches do not usually occur in young people since

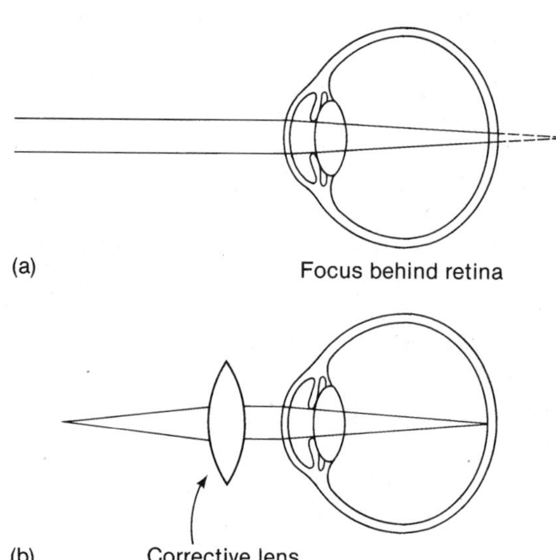

Figure 13.8
Focusing of light by the hyperopic (farsighted) eye. (a) Parallel rays from a distant spot of light are brought to a focus behind the retina, so that, without accommodation, far objects are blurred. Hyperopes can, however, achieve clear vision of distant objects by accommodating. (b) If hyperopia is severe, the constant accommodation needed for clear vision may cause eyestrain, and a corrective lens is required.

they can accommodate easily, but older people, who have more difficulty accommodating because of a condition called *presbyopia,* which we will describe next, are more likely to experience headaches and may therefore require a corrective lens that brings the focus point forward onto the retina (Figure 13.8b).

Presbyopia

A decrease in the ability to accommodate due to old age is called **presbyopia,** or "old eye." This decrease in accommodation affects the location of the **near point,** the closest distance at which a person can still see an object in focus (Table 13.1). As a person ages, the near point moves farther and farther away, as shown in Figure 13.9. The near point for most 20-year-olds is at about 10 cm, but it increases to 14 cm by age 30, 22 cm at 40, and 100 cm at 60. This loss in the ability to accommodate occurs because the lens hardens with age, and the ciliary muscles, which control accommodation, become weaker. These changes make it more difficult for the lens to change its shape for vision at close range. Though this gradual decrease in accommodative ability poses little problem for most people before the age of 45, at around that age the ability to accommodate begins to decrease rapidly, and the near point moves beyond a comfortable reading distance. This is the reason you may have observed older people holding their reading material at arm's length. But the real solution to this problem is a corrective lens that provides the necessary focusing power to bring light to a focus on the retina.

Astigmatism

Imagine what it would be like to see everything through a pane of old-fashioned wavy glass, which causes some things to be in focus and others to be blurred. This describes the experience of a person with a severe **astigmatism;** an astigmatic person sees through a misshapen cornea, which correctly focuses some of the light reaching the retina but distorts other light. The normal cornea is spherical, curved like a round kitchen bowl, but an astigmatic cornea is somewhat elliptical, curved like the inside of a teaspoon. Because of this elliptical curvature, a person with astigmatism will see the astigmatic fan in Figure 13.10 partially in focus and partially out of focus. As in hyperopia, eyestrain is a symptom of astigmatism, because no matter how much the person accommodates to try to achieve clear vision, something is always out of focus.

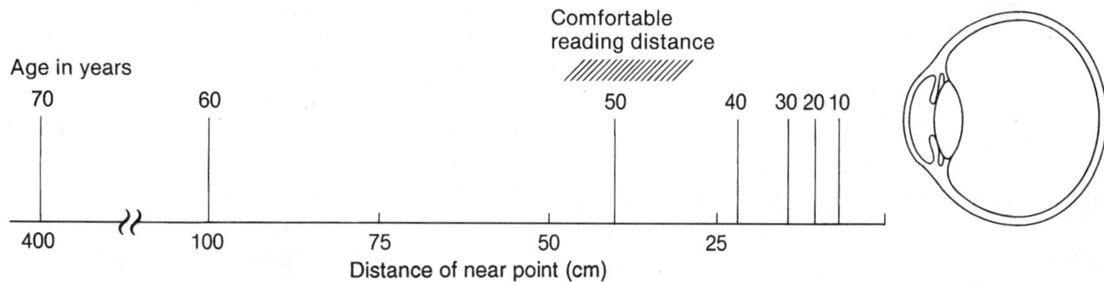

Figure 13.9

The near point as a function of age. The distance of the near point in centimeters is indicated on the scale at the bottom, and various ages are indicated by the vertical lines. Objects closer than the near point cannot be brought into focus by accommodation. Thus, as age increases, the ability to focus on nearby objects becomes poorer and poorer; eventually, past the age of about 50, reading becomes impossible without corrective lenses.

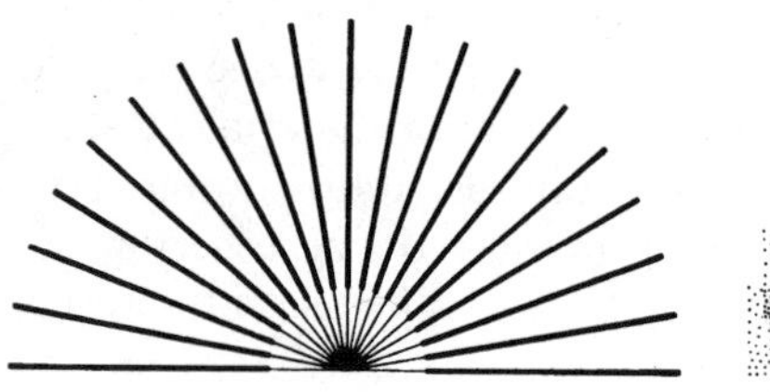
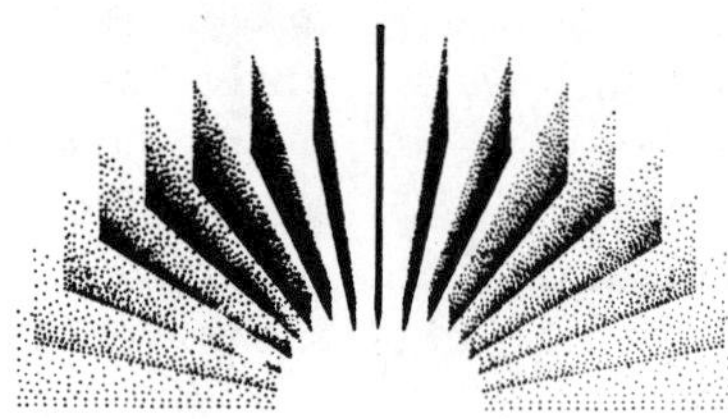

Figure 13.10
Left: Astigmatic fan chart used to test for astigmatism. Right: An astigmatic patient will perceive the lines in one orientation (in this case vertical) as sharp and the lines in the other orientation as blurred. (From Trevor-Roper, 1970.)

Fortunately, astigmatism can be corrected with the appropriate lens.

Decreased Transmission of Light

The focusing problems described above are the most prevalent visual problems, as evidenced by the large number of people who wear glasses or contact lenses. Because these problems can usually be corrected, most people with focusing problems see normally or suffer only mild losses of vision. We will now consider situations in which disease or physical damage causes severe visual losses or, in some cases, blindness. But before we begin to discuss these problems, we will define what we mean by **blindness.**

What Is Blindness?

It is a common conception that a person who is blind lives in a world of total darkness or formless diffuse light. While this description is true for some blind people, many people who are classified as legally blind do have some vision, and many can read with the aid of a strong magnifying glass. According to the definition accepted in most states, a person is considered legally blind if, after correction with glasses or contact lenses, he or she has a visual acuity of 20/200 or less in the better eye. A visual acuity of 20/20 means that a person can see at 20 feet what a person with normal vision can see at 20 feet. However, a person with an acuity of 20/200 needs to be at a distance of 20 feet to see what a person with normal vision can see from a distance of 200 feet.

When we define blindness in terms of visual acuity, we are evaluating a person's ability to see with his or her fovea (which, as we saw in Chapter 2, is the cone-rich area of the retina that is responsible for detail vision). While poor foveal vision is the most common reason for legal blindness, a person with good foveal vision but little peripheral vision may also be considered legally blind. Thus, a person with normal (20/20) foveal vision but little or no peripheral vision may be legally blind. This situation, which is called **tunnel vision,** results from diseases that affect the retina, such as advanced glaucoma or retinitis pigmentosa (a form of retinal degeneration), which affect peripheral vision but leave the foveal cones unharmed.

We begin our discussion of problems caused by disease or injury by considering some conditions that affect both peripheral and central vision because they affect the perception of light at the beginning of the visual process, as light enters the eye through the cornea and the lens.

Corneal Disease and Injury

The **cornea,** which is responsible for about 70 percent of the eye's focusing power (Lerman, 1966), is the window to vision because light first passes through this structure on its way to the retina. In order for a sharp image to be formed on the retina, the cornea must be transparent, but this transparency is occasionally lost when injury, infection, or allergic reactions cause the formation of scar tissue on the cornea. This scar tissue decreases visual acuity and sometimes makes lights appear to be surrounded by a halo, which looks like a shimmering rainbow. In addition, **corneal disease and injury** can also cause pain. Drugs, which often bring the cornea back to its transparent state, are the first treatment for corneal problems. If drugs fail, however, clear vision can often be restored by a **corneal transplant** operation.

The basic principle underlying a corneal transplant operation is shown in Figure 13.11. The scarred area of the cornea, usually a disk about 6–8 mm in diameter, is removed and replaced by a piece of cornea taken from a donor. For best results, this donor should be a young adult who died of an acute disease or of an injury that left the corneal tissue in good condition. In the past, a major problem with this operation was the necessity of transplanting the donor cornea within a few hours after the donor's death. Now, however, donor corneas are preserved by low-temperature storage in a specially formulated solution.

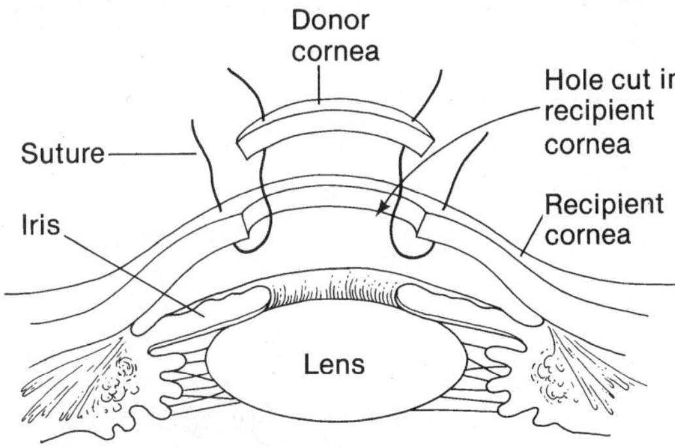

Figure 13.11
Corneal transplant operation. The scarred part of the cornea has been removed, and the donor cornea is about to be sutured in place.

Of the over 10,000 corneal transplants performed every year, 85 percent are successful. Remember, however, that a corneal transplant operation involves only a small piece of the eye—there is no such thing as an eye transplant. Indeed, the problems involved in transplanting a whole eye are overwhelming. For one thing, the optic nerve and the retina are sensitive to lack of oxygen, so that, once the circulation is cut off, irreversible damage occurs within minutes, just as is the case for the brain. Thus, keeping the donor's eye alive presents a serious problem. And even if it were possible to keep an eye alive, there is the problem of connecting the 1 million optic nerve fibers of the donor's eye to the corresponding nerve fibers of the patient's optic nerve. At this point, whole eye transplants are purely science fiction.

Clouding of the Lens (Cataract)

Like the cornea, the **lens** is transparent and is important for focusing a sharp image on the retina. Clouding of the lens, which is called a **cataract**, is sometimes present at birth **(congenital cataract)**, may be caused by an eye disease **(secondary cataract)**, or may be caused by injury **(traumatic cataract)**, but the most common cause of cataract is old age **(senile cataract)**. Cataracts develop, for reasons as yet unknown, in 75 percent of people over 65 and in 95 percent of people over 85.

Although millions of people have cataracts, in only about 15 percent of the cases does the cataract interfere with a person's normal activities, and only 5 percent of cataracts are serious enough to require surgery—the only treatment. The basic principle underlying a cataract operation is illustrated in Figure 13.12a. A small opening is made in the eye, and the surgeon removes the lens while leaving in place the *capsule*, the structure that helps support the lens. A method for removing the lens that has the advantage of requiring a small incision in the eye is **phacoemulsification** (Figure 13.12b). In this procedure, a hollow tubelike instrument that emits ultrasound vibrations up to 40,000 times per second is inserted through a small incision in the cornea. The vibrations break up the lens, and the resulting pieces are sucked out of the eye through the tube.

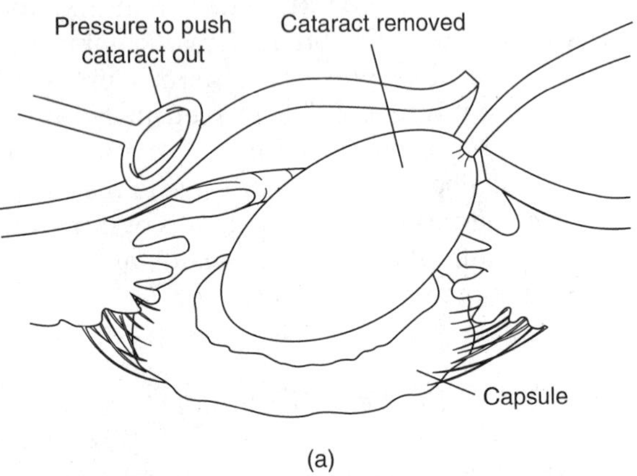

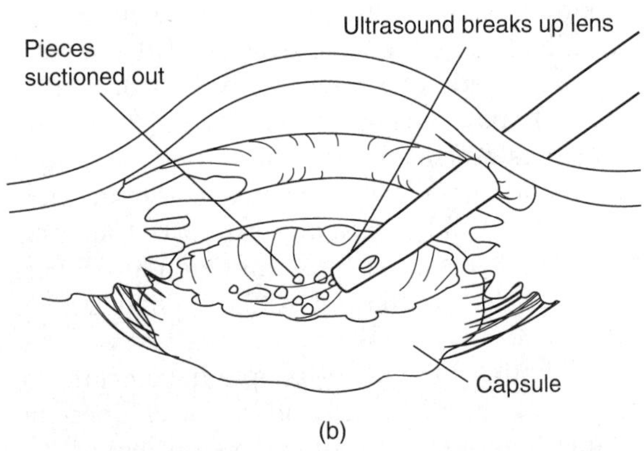

Figure 13.12
A cataract operation. (a) The cataract (the clouded lens) is removed through an incision in the cornea. (b) The phacoemulsification procedure for removing the cataract. High-frequency sound vibrations break up the lens and the pieces are sucked into the tube. After the lens is removed, an intraocular lens is inserted.

Removal of the clouded lens clears a path so that light can reach the retina unobstructed, but in removing the lens, the surgeon has also removed some of the eye's focusing power. (Remember that the cornea accounts for 70 percent of the eye's focusing power; the lens is responsible for the remaining 30 percent.) Although the patient can be fitted with glasses, these create problems of their own, because glasses enlarge the image falling on the retina by as much as 20 to 35 percent. If one eye receives this enlarged image and the other receives a normal image, the brain cannot combine the two images to form a single, clear perception. The **intraocular lens,** a plastic lens which is placed inside the eye where the original lens used to be, is the solution to this problem.

The idea of implanting a lens inside the eye goes back 200 years, but the first workable design for an intraocular lens was not proposed until 1949. Although lenses introduced in the 1950s were not very successful, recent developments in plastics have resulted in small ultralightweight lenses, like the one shown in Figure 13.13, and installing an intraocular lens is now a routine part of most cataract operations. Notice that the lens is placed in the same location as the clouded lens that was removed, just above the capsule, which the surgeon was careful to leave in place when removing the cataract. The presence of the capsule helps hold the intraocular lens in place.

Damage to the Retina

The retina receives nourishment from the retinal circulation and from the **pigment epithelium** on which it rests. All four conditions described below cause a loss of vision because of their effects on the retinal circulation and on the relationship between the retina and the pigment epithelium.

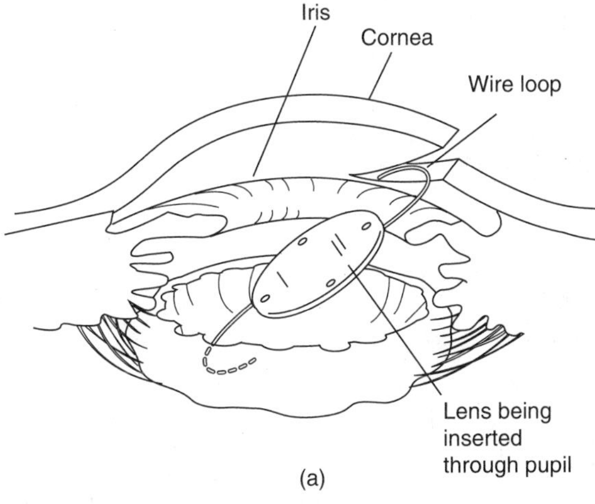

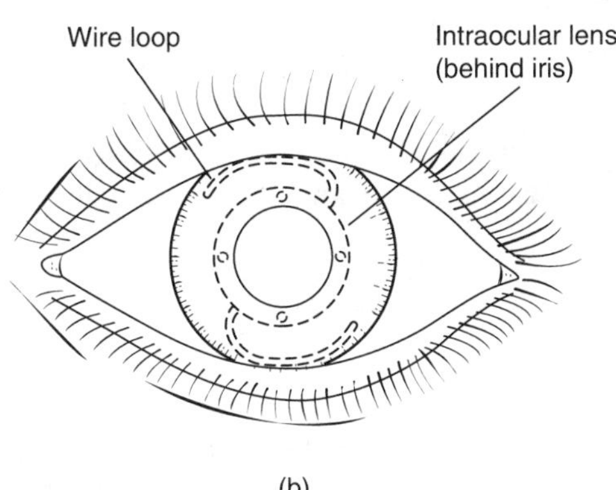

Figure 13.13
Installing an intraocular lens in the eye after the cataract has been removed. (a) The lens is inserted through an incision in the cornea. Notice that it is being inserted through the pupil so that it will be positioned where the original lens was, behind the iris and just above the capsule. (b) Frontal view, showing the lens in place behind the iris. The small wire loops hold the lens in place.

Diabetic Retinopathy

Before the isolation of insulin in 1922, most people with severe **diabetes,** a condition in which the body doesn't produce enough insulin, had a life expectancy of less than 20 years. The synthesis of insulin (which won the 1923 Nobel prize for its discoverers) greatly increased the life expectancy of diabetics, but one result of this greater life expectancy has been a great increase in an eye problem called **diabetic retinopathy.** Of the 10 million diabetics in the United States, about 4 million show some signs of this problem.

Figure 13.14 shows what happens as the disease progresses. At first, the capillaries swell, as shown in Figure 13.14a. Most cases of diabetic retinopathy stop here; even so, a large number of diabetics suffer vision losses even when the disease stops at this point. The disease's further progression, which occurs in a small percentage of patients, involves a process called **neovascularization.** Abnormal new blood vessels are formed (Figure 13.14b), which do not supply the retina with adequate oxygen and which are fragile and so bleed into the **vitreous humor** (the jellylike substance that fills the eyeball); this bleeding interferes with the passage of light to the retina. Neovascularization can also cause scarring of the retina and retinal detachment (see below).

One technique for stopping neovascularization is called **laser photocoagulation,** in which a

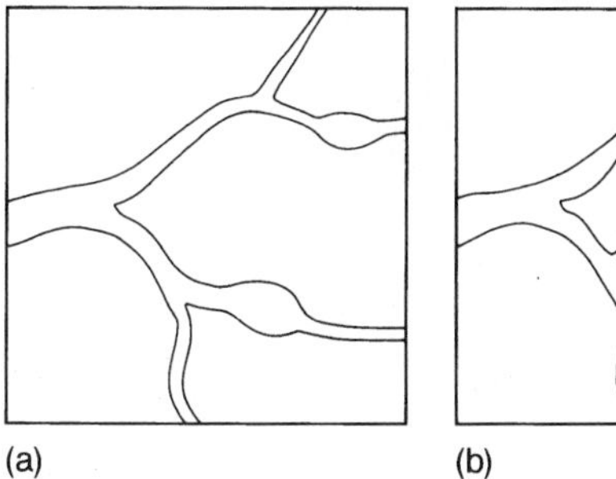

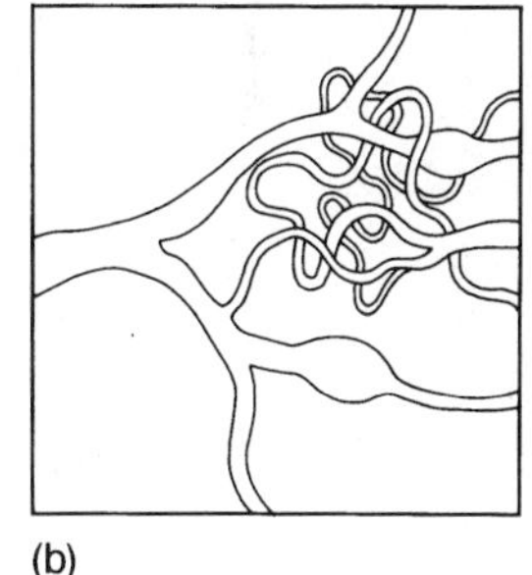

Figure 13.14
Blood vessels in diabetic retinopathy. (a) In early stages of the disease, the blood vessels swell and leak slightly. (b) In later stages, in a process called neovascularization, abnormal new blood vessels grow on the surface of the retina.

laser beam of high-energy light is aimed at leaking blood vessels. The laser "photocoagulates," or seals off, these vessels and stops the bleeding. A procedure called **panretinal photocoagulation** has been used with some success. In this technique, the laser scatters 2,000 or more tiny burns on the retina, as shown in Figure 13.15. The burns do not directly hit the leaking blood vessels, but by destroying part of the retina, they decrease the retina's need for oxygen, so that the leaking blood vessels dry up and go away.

If laser photocoagulation is not successful in stopping neovascularization, a procedure called a **vitrectomy,** shown in Figure 13.16, is used to eliminate the blood inside the eye. In this operation, which is done only as a last resort, a hollow tube containing a guillotinelike cutter takes in the vitreous humor and chops it into pieces small enough to be sucked out of the eye through the tube. When the vitreous humor and blood are removed, they are replaced with a salt solution. While this procedure removes the blood inside the eye, it does not stop the neovascularization that caused the bleeding in the first place.

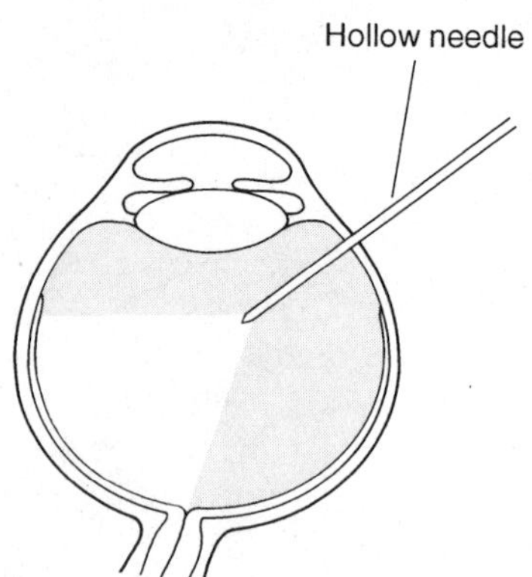

Figure 13.16
Vitrectomy. The hollow needle inserted into the eyeball first sucks out the liquid inside the eye and then fills the eyeball with a salt solution.

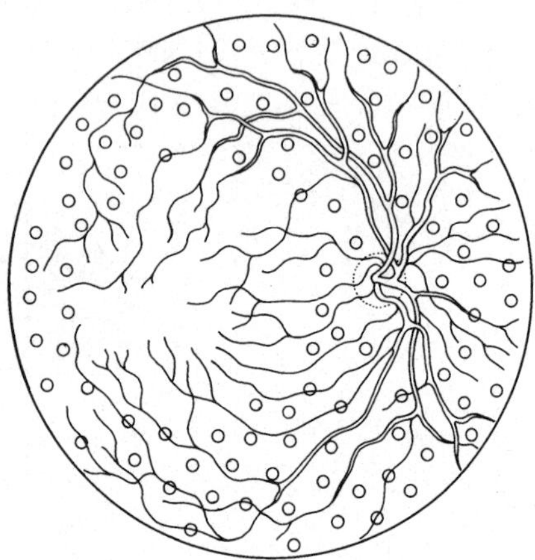

Figure 13.15
Laser photocoagulation in the treatment of diabetic retinopathy. The picture illustrates the technique of panretinal photocoagulation. Each dot represents a small laser burn on the retina.

Macular Degeneration

Imagine your frustration if you could see everywhere *except* where you were looking, so that every time you looked at something you lost sight of it. That is exactly what happens if a region of the retina called the **macula** is damaged. The macula is an area about 5 mm in diameter that surrounds and includes the cone-rich fovea (itself only slightly larger than one of the periods on this page). If the macula degenerates, blindness results in the center of vision (Figure 13.17). This condition is extremely debilitating because, although peripheral vision remains intact, the elimination of central vision makes reading impossible.

There are a number of forms of **macular degeneration,** but the most common is called **age-related macular degeneration** because it occurs, without obvious reason, in older people. In its mild form, there is a slight thinning of the cone receptors and the formation of small

Figure 13.17
Macular degeneration causes a loss of central vision.

white or yellow lumps on the retina. This form of macular degeneration usually progresses slowly and may not cause serious visual problems. In 5 to 20 percent of the cases, however, small new blood vessels, similar to those in diabetic retinopathy, grow underneath the macular area of the retina. These new blood vessels form very rapidly—over a period of only one or two months—and leak fluid into the macula, killing the cone receptors.

Until recently, there was no treatment for age-related macular degeneration. However, a recent study by the National Eye Institute indicates that, if the problem can be caught at an early stage in some patients with the more severe form of the disease, laser photocoagulation can stop or greatly reduce leakage of the newly formed vessels.

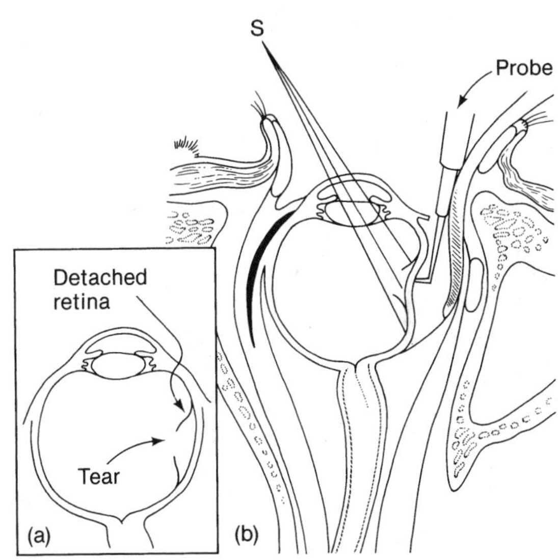

Figure 13.18
(a) A detached retina. (b) Procedure for reattaching the retina. To locate the site of detachment, a probe pushes the eyeball from outside while the surgeon, at S, looks into the eye. Once the site of the detachment is located, the outside of the eye is marked, and a cooling or heating probe is applied at the marked point.

Detached Retina

Detached retina, a condition in which the retina becomes separated from the underlying pigment epithelium (Figure 13.18), has occurred in a number of athletes because of traumatic injuries to the eye or the head. Tony Conigliaro, a star shortstop of the Boston Red Sox in the 1960s, was hit by a pitch and began complaining that he could not see well: He had fuzzy vision and found it difficult to catch what would have been routine fly balls. Tests revealed that his retina had become detached, and although attempts were made to reattach his retina, Conigliaro's vision never returned to normal and his baseball career was ended. More recently, the detached retina suffered by Sugar Ray Leonard, the welterweight boxing champion, caused him to retire temporarily from boxing. He returned to boxing a number of years later amid much discussion about whether returning to the ring was worth the risk of losing his sight in one eye. As it turned out, Leonard won both the fight and the gamble with his sight, apparently escaping without further damaging his eye.

A detached retina affects vision for two reasons: (1) For good image formation, the retina must lie smoothly on top of the pigment epithelium, and (2) when the retina loses contact with the pigment epithelium, the visual pigments in the detached area are separated from enzymes in the epithelium necessary for pigment regeneration. When the visual pigment can no longer regenerate, that area of the retina becomes blind.

The treatment for a detached retina is an operation to reattach it. The basic idea behind this operation is to cause the formation of scar tissue inside the eye that will attach itself to the retina and anchor it in place. This process is accomplished by applying either a cooling or a heating probe to exactly the right place on the outside of the eyeball. Figure 13.18b shows the procedure used to determine where to apply the probe. While looking into the eye with a special viewing device, the surgeon presses on the outside of the eyeball, which causes an indentation that can be seen inside the eye. The surgeon presses at a number of points, until the indentation inside the eyeball matches the location of the tear or hole in the retina, where the detachment originated.

Once the point where the detachment has occurred is located, it is marked on the outside of the eyeball, and that point is cooled or heated to create an inflammatory response. The retina must then be pushed flush with the wall of the eyeball. This is accomplished by placing a band around the outside of the eyeball that creates a dumbbell-shaped eye. Then, with the retina pressed against the wall of the eye, the inflammation causes scarring that "welds" the retina back onto the pigment epithelium. If the area of detached retina is not too big, there is a 70 to 80 percent chance that this procedure will work. In most cases, it restores vision, although vision is sometimes not restored even though the retina is successfully reattached. The larger the detached area, the less likely it is that this operation (or others, which we will not describe here) will work. Sometimes, if a tear is present, this procedure is used to weld the retina down and prevent detachment from occurring.

Hereditary Retinal Degeneration

The most common form of hereditary retinal degeneration is a disease called **retinitis pigmentosa,** a degeneration of the retina that is passed from one generation to the next (although not always affecting everyone in a family). We know little about what actually causes the disease, although one hypothesis is that it is caused by a problem in the pigment epithelium.

A person with retinitis pigmentosa usually shows no signs of the disease until reaching adolescence. At this time, the person may begin to notice some difficulty in seeing at night, since the disease first attacks the rod receptors. As the person gets older, the disease slowly progresses, causing further losses of vision in the peripheral retina. Then, in its final stages, which may occur as early as a person's 30s or as late as the 50s or 60s (depending on the strain of the disease), retinitis pigmentosa also attacks the cones, and the result is complete blindness.

Optic Nerve Damage

A leading cause of blindness in the United States is **glaucoma**, which causes nerve fibers in the optic nerve to degenerate and therefore prevents the nerve impulses generated by the retina from being transmitted to the brain.

Glaucoma

Although the end result of glaucoma is damage to the optic nerve, the source of the problem is at the front of the eye. We can understand how damage to the front of the eye affects the optic nerve by looking at the cross section of the eye in Figure 13.19a. Under normal conditions, the aqueous humor (the liquid found in the space between the cornea and the lens), which is continuously produced at A, passes between the iris and the lens following the path indicated by the arrows; it then drains from the eye at B. In glaucoma, the drainage of aqueous humor is partially blocked. **Closed-angle glaucoma** is a rare form of glaucoma in which a pupillary block (Figure 13.19b) constricts the opening between the iris and the lens and causes a pressure buildup that pushes the iris up, thereby "closing the angle" between the cornea and the iris and blocking the area at B where the aqueous humor leaves the eye.

In **open-angle glaucoma,** which is the most common form of the disease, the eye looks normal, as in Figure 13.19a, but the drainage area at B is partially blocked, so that it is more difficult for the aqueous humor to leave the eye. The blocks that occur in both closed- and open-angle glaucoma result in a large resistance to the outflow of aqueous humor, and since the aqueous humor continues to be produced inside the eye, the **intraocular pressure**—the pressure inside the eyeball—rises. This increase in intraocular pressure has two effects: (1) It compresses the blood vessels that provide nourishment to the retina,

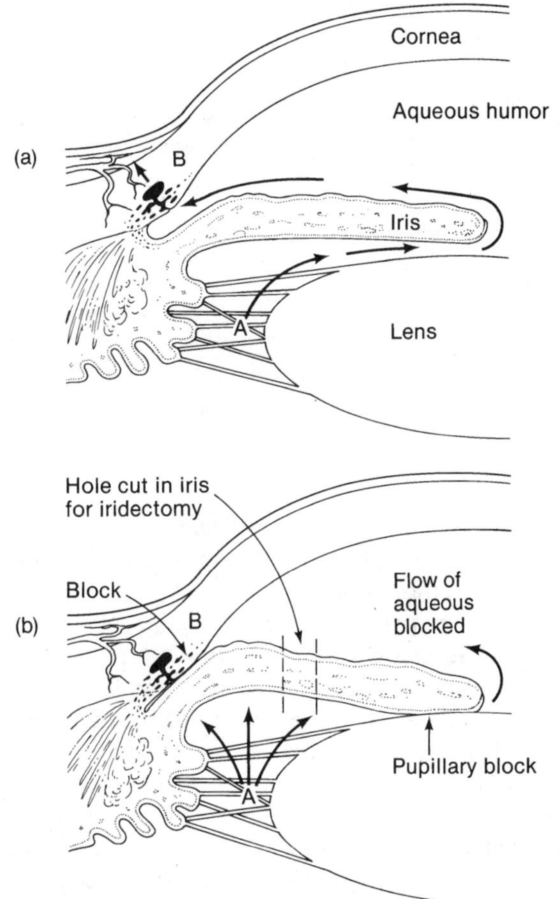

Figure 13.19
(a) Arrows indicate the flow of aqueous humor in the normal eye. The aqueous humor is produced at A and leaves the eye at B. In open-angle glaucoma, the aqueous humor cannot leave the eye because of a blockage at B. (b) In closed-angle glaucoma, the raised iris hinders the flow of aqueous humor from the eye. An iridectomy—cutting a hole in the iris—can provide a way for the aqueous humor to reach B.

and (2) it presses on the head of the optic nerve at the back of the eye; this pressure sometimes causes an effect called *cupping of the optic disk*, in which the head of the optic nerve is pushed in

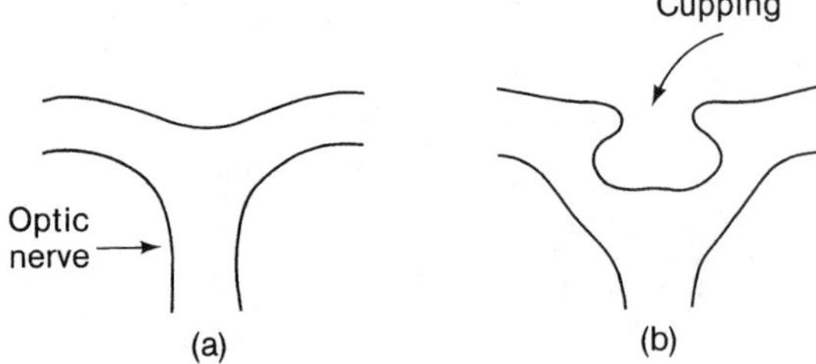

Figure 13.20
Side view of the optic nerve where it leaves the eye. (a) Normal optic nerve. (b) Optic nerve in eye with glaucoma, which shows "cupping" of the optic disk caused by excessive pressure inside the eyeball.

(Figure 13.20). These two effects result in degeneration of the optic nerve fibers, and this degeneration results in blindness.

The increase in pressure that occurs in closed-angle glaucoma usually happens very rapidly and is accompanied by pain. The treatment for this type of glaucoma is an operation called an **iridectomy,** in which a small hole is cut in the iris, as shown in Figure 13.19b. This hole opens a channel through which the aqueous humor can flow and releases the pressure on the iris. With the pressure gone, the iris flattens out and uncovers the area at B so that aqueous humor can flow out of the eye.

Intraocular pressure increases more slowly in open-angle glaucoma, so the patient may be unaware of any symptoms. In many cases, visual loss is so gradual that much of the patient's peripheral vision is gone before its loss is noticed. For that reason, ophthalmologists strongly recommend that people over 40 have their eyes checked regularly for glaucoma, since early detection greatly enhances the chances of effective treatment by medication. In 5 to 10 percent of the cases of open-angle glaucoma, medications do not decrease the pressure, and an operation becomes necessary. The goal of this operation is to cut an opening at B in Figure 13.19b that enables the aqueous humor to flow out.

The Eye Examination

So far, we have described some of the things that can go wrong with the eye and how these problems are treated. In this part of the chapter, we will describe the procedures used to uncover some of these problems. Before describing the eye examination, however, we will consider who examines the eyes.

Who Examines Eyes?

Three types of professionals are involved in eye care: ophthalmologists, optometrists, and opticians. We will consider each in turn.

1. An **ophthalmologist** is an M.D. who has completed undergraduate school and four years of medical school, which provide general medical training. In order to become an ophthalmologist, a person needs four or more years of training after graduation from medical school to learn how to treat eye problems medically and surgically. Some ophthalmologists receive even further training and then specialize in specific areas, such as pediatric ophthalmology (practice limited to children), diseases of the cornea, retinal diseases, or glaucoma. Most ophthalmologists, however, treat all eye problems, as well as prescribing glasses and fitting contact lenses.
2. An **optometrist** has completed undergraduate school and, after four years of additional study, has received a doctor of optometry (O.D.) degree. Optometrists can examine eyes and fit and prescribe glasses or contact lenses. In some states, optometrists have won the right to include medical treatment using drugs, for some

eye conditions. Surgery, however, is still done exclusively by ophthalmologists.

3. An **optician** is trained to fabricate and fit glasses and, in some states, contact lenses, on the prescription of an ophthalmologist or an optometrist.

What Happens during an Eye Exam?

The basic aims of an eye exam are (1) to determine how well the patient can see, (2) to correct vision if it is defective, (3) to determine the causes of defective vision by examining the optics of the eye and checking for eye diseases, and (4) to diagnose diseases that the patient may not even be aware of. To accomplish these aims, an examination by an eye specialist usually includes the following.

Medical History The first step in an eye exam is to take a medical history. This history focuses on any eye problems that the patient may have had in the past, on any current eye problems, and on any general medical problems that may be related to the patient's vision.

Visual Acuity This is the familiar part of the eye exam, in which you are asked to read letters on an eye chart like the one in Figure 13.21. The *E* at the top of the chart is usually the 20/400 line, which means that a person with normal vision should be able to see the *E* from a distance of 400 feet. Since the eye chart is usually viewed from about 20 feet, people with normal vision see the *E* easily. When asked to read the smallest line he or she can see, the patient usually picks a line that is easily read. With a little encouragement, however, most patients find that they can see lines smaller than the one they originally picked, and the examiner has the patient read smaller and smaller lines until letters are missed. The smallest line a person can read indicates his or her visual acuity, with normal vision defined as an acuity of

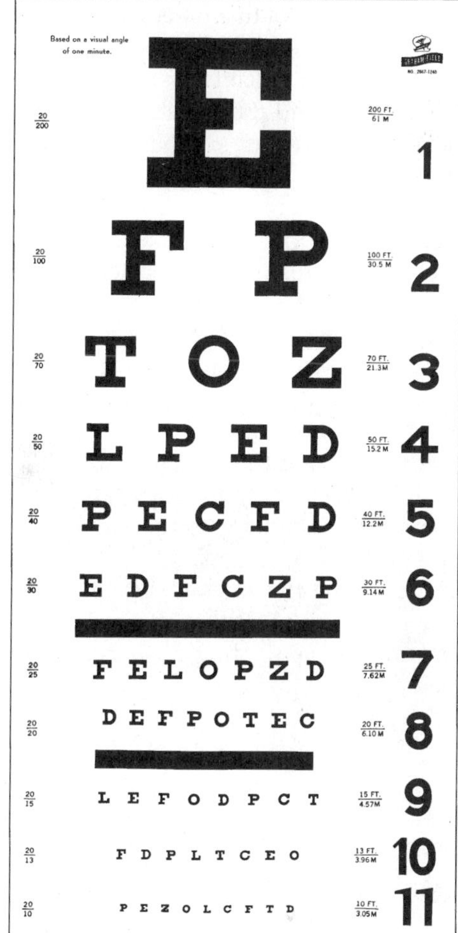

Figure 13.21
The Snellen eye chart used to test visual acuity.

20/20. A person with worse than normal acuity—say, 20/40—must view a display from a distance of 20 feet to see what a person with normal acuity can see at 40 feet. A person with better than normal acuity—say, 20/10—can see from a distance of 20 feet what a person with normal vision can see only at 10 feet.

It is important to realize that the visual acuity test described above tests only foveal vision. When you read an eye chart, you look directly at each letter, so the image of that letter falls on your fovea. Thus, as mentioned earlier, a person who scores 20/20 on a visual acuity test may still be classified as legally blind if he or she has little or no peripheral vision. Testing peripheral vision is usually not part of a routine eye exam, but when peripheral vision problems are suspected, a technique called **visual perimetry** is used, in which the patient is asked to indicate the location of small spots of light presented at different locations in the periphery. This test locates blind spots (called *scotomas*) that may be caused by retinal degeneration, detachment of the retina, or diseases such as glaucoma.

In addition to using the eye chart to test far vision, it is also customary to test near vision, especially in older patients who may be experiencing the effects of presbyopia. This testing is done by determining the smallest line of a card like the one in Figure 13.22 that the patient can see from a comfortable reading distance.

No. 1

In short-sighted persons the eye-ball is too long, and the light rays come to a focus in front of the retina,

No. 2

while in far-sighted persons the eye-ball is too short, and the focal point, therefore, falls be-

No. 3

hind the retina. In either case a blurred image is received upon the retina. In order

No. 4

to overcome this blurring, and thus correct the optical defect, the eye un-

No. 6

consciously makes an effort by which the ciliary muscle acts on

No. 8

the lens. This effort explains why eye-strain may cause

No. 10

pain and discomfort. An optical correction for

No. 12

the refractive error is found in specta-

Figure 13.22

A card for testing close vision. The patient's close vision is determined by the smallest line that he or she can read from a comfortable reading distance.

Refraction A score of 20/60 on a visual acuity test indicates worse than normal acuity but does not indicate what is causing this loss of acuity. Acuity could be decreased by one of the diseases described earlier or by a problem in focusing: myopia, hyperopia, presbyopia, or astigmatism. If the problem lies in the focusing mechanism of the eye, it is usually easily corrected by glasses or contact lenses. **Refraction** is the procedure used to determine the power of the corrective lenses needed to achieve clear vision.

The first step in refraction is a **retinoscopy exam,** an examination of the eye with a device called a *retinoscope*. This device projects a streak of light into the eye that is reflected into the eye of the examiner. The examiner moves the retinoscope back and forth and up and down across the eye, noticing what the reflected light looks like. If the patient's eye is focusing the light correctly, the examiner sees the whole pupil filled with light, and no correction is necessary (in this case, the patient will usually have tested at 20/20 or better in the visual acuity test). If, however, the patient's eye is not focusing the light correctly, the examiner sees a streak of light move back and forth across the pupil as the streak of light from the retinoscope is moved across the eye.

To determine the correction needed to bring the patient's eye to 20/20 vision, the examiner places corrective lenses in front of the eye while still moving the streak of light from the retinoscope back and forth. One way of placing these lenses in front of the eye is to use a device like the one shown in Figure 13.23. This device contains a variety of lenses that can be changed by turning a dial. The examiner's goal is to find the lens that causes the whole pupil to fill up with light when the retinoscope is moved back and forth. This lens brings light to a focus on the retina and is usually close to the one that will be prescribed to achieve 20/20 vision.

The retinoscopy exam results in a good first approximation of the correct lens to prescribe for a patient, but the ultimate test is what the patient sees. To determine this, the examiner has the

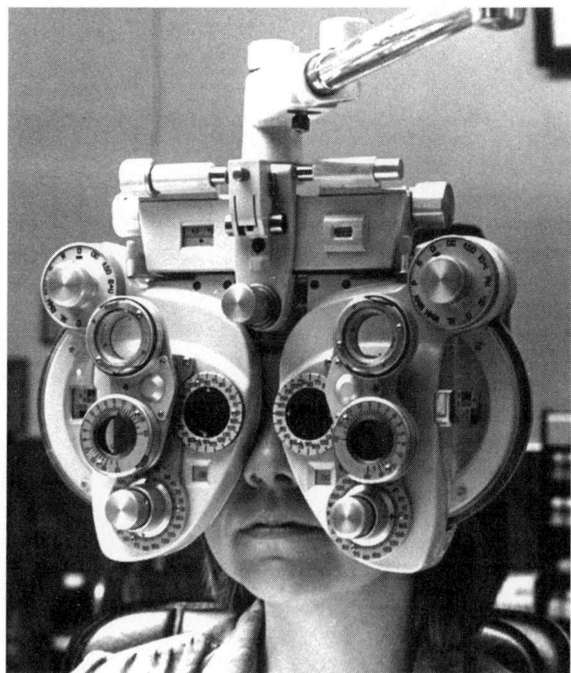

Figure 13.23
A device for placing different corrective lenses in front of the patient's eyes. Different lenses are placed in front of the eye during the retinoscopy exam and again as the patient looks at the eye chart.

patient look at the eye chart, and places lenses in front of the patient's eyes to determine which one results in the clearest vision. When the examiner determines which lens results in 20/20 vision, he or she writes a prescription for glasses or contact lenses. To fit contact lenses after determining the prescription, the examiner must match the shape of the contact lens to the shape of the patient's cornea.

Refraction is used to determine the correction needed to achieve clear far vision. Using a procedure we will not describe here, the examiner also determines whether a correction is needed to achieve clear near vision. This determination is particularly important for patients over 45 years old, who may experience reading difficulties due to presbyopia.

External Eye Exam In an **external eye exam,** the examiner uses a variety of tests to check the condition of the external eye. She or he checks pupillary reaction by shining light into the eye, to see if the pupil responds by closing when the light is presented and by opening when the light is removed. The examiner also checks the color of the eye and the surrounding tissues. "Red eye" may indicate that an inflammation is present. The examiner checks eye movement by having the patient follow a moving target, and checks the alignment of the eyes by having the patient look at a target. If the eyes are aligned correctly, both eyes will look directly at the target, but if the eyes are misaligned, one eye will look at the target, and the other will veer off to one side.

Slit-Lamp Examination The **slit-lamp examination** checks the condition of the cornea and the lens. The slit lamp, shown in Figure 13.24, projects a narrow slit of light into the patient's eye. This light can be precisely focused at different places inside the eye, and the examiner views this sharply focused slit of light through a binocular magnifier. This slit of light is like the sharp edge of a knife that cuts through the eye.

What does the examiner see when looking at the "cutting edge" of light from the slit lamp? By focusing the light at different levels inside the cornea and lens, the examiner can detect small imperfections—places where the cornea or the lens is not completely transparent—that cannot be seen by any other method. These imperfections may indicate corneal disease or injury or the formation of a cataract.

Tonometry **Tonometry** measures intraocular pressure, the pressure inside the eye, and is therefore the test for glaucoma. Nowadays, an instrument called a **tonometer** is used to measure intraocular pressure, but before the development of this device, it was known that large increases

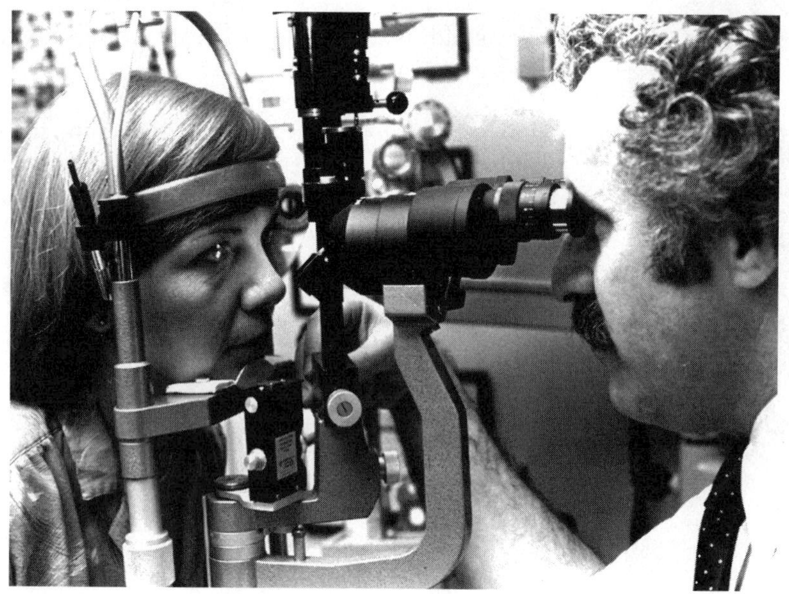

Figure 13.24
A patient being examined with a slit lamp. The examiner is checking the condition of the lens and the cornea by viewing the slit of light through a binocular magnifier.

of intraocular pressure, which accompany severe cases of glaucoma, cause the eye to become so hard that this hardness could be detected by pushing on the eyeball with a finger.

There are several types of tonometers, which measure the intraocular pressure by pushing on the cornea. The Schiotz tonometer is a hand-held device that consists of a small plunger attached to a calibrated weight. The weight pushes the plunger and indents the cornea. If the intraocular pressure is high, the plunger causes a smaller indentation than if the intraocular pressure is normal. Thus, intraocular pressure is determined by measuring the indentation of the cornea. (Though this procedure may sound rather painful, it is not, because the examiner applies a few drops of anesthetic to the cornea before applying the tonometer.)

The applanation tonometer, shown being applied to a patient's cornea in Figure 13.25, is a more sophisticated and accurate instrument than the Schiotz tonometer. After a few drops of anesthetic are applied to the cornea, the flat end of a cylindrical rod, called an **applanator,** is slowly moved against the cornea by the examiner, who watches the applanator's progress through the same magnifiers used for the slit-lamp exam (Figure 13.24). The examiner pushes the end of the applanator against the cornea until enough pressure is exerted to flatten a small area on the cornea's curved surface. The greater the force that must be exerted to flatten the cornea, the greater the intraocular pressure.

Ophthalmoscopy So far, we have looked at the outside of the eye (external eye exam), examined the lens and cornea (slit-lamp exam), and measured the intraocular pressure (tonometry), but we have yet to look at perhaps the most important structure of all: the retina. Since there is a hole (the pupil) in the front of the eye, it should be simple to see the retina; we only have to look into the hole. Unfortunately, it's not that simple; if you've ever looked into a person's pupil, you realize that it's dark in there. In order to see the retina, we must find some way to light up the inside of the eye.

You might try placing a light at L, as shown in Figure 13.26. This seems like a good idea until you try to look into the patient's eye. If you place your eye at A, your head blocks the light from L, and if you place your eye at B, you are blinded by

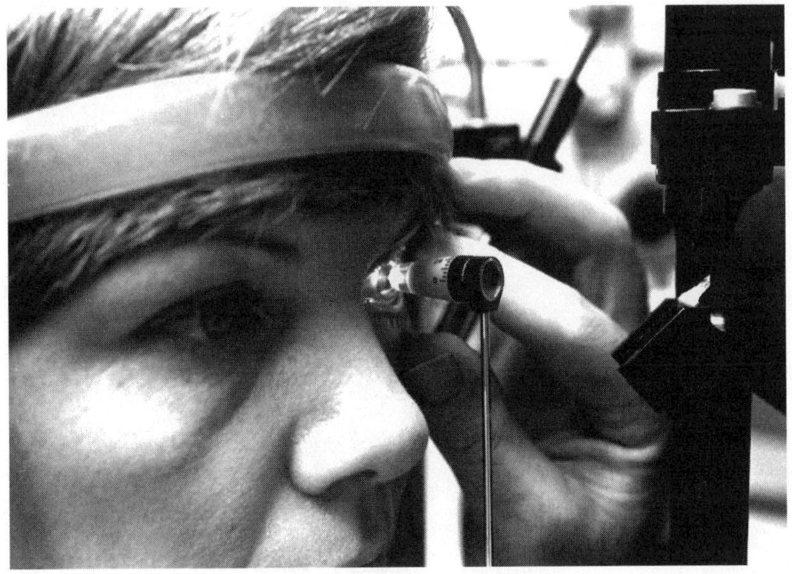

Figure 13.25
An applanation tonometer being applied to a patient's cornea.

the light at L. Clearly, neither of these methods will work.

It was not until 1850 that Hermann von Helmholtz, of the Young–Helmholtz theory of color vision, invented a device called the **ophthalmoscope.** The principle underlying Helmholtz's ophthalmoscope is shown in Figure 13.27. Helmholtz solved the problem of the blocked or blinding light of Figure 13.26 by placing the light off to the side and directing it into the patient's eye with a half-silvered mirror. The important property of a half-silvered mirror is that it reflects some of the light and transmits the rest, so that an examiner positioned as shown in Figure 13.27 can see through the mirror and into the patient's eye. Actual ophthalmoscopes are much more complicated than the one diagrammed here, since they include numerous lenses, mirrors, and filters, but the basic principle remains the same as that of the original ophthalmoscope designed by Helmholtz in 1850.

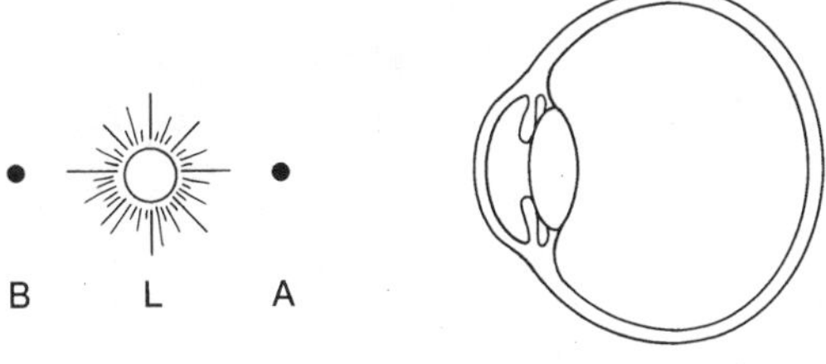

Figure 13.26
Ways to light up the inside of the eye that won't work. The light at L is blocked if the examiner's eye is positioned at A, and it blinds the examiner's eye positioned at B.

Figure 13.28 is a patient's-eye view of an examination with an ophthalmoscope, although the examiner is actually very close, as shown in Figure 13.1. Figure 13.29 shows a close-up of what the ophthalmologist sees if the patient has a normal retina. The most prominent features of this view of the retina are the place where the optic nerve leaves the eye (called the **optic disk,** or blind spot) and the arteries and veins of the retina. In this examination, the ophthalmologist focuses on these features, noting any abnormalities in the appearance of the optic disk and the retinal circulation. For example, the ophthalmologist may detect the presence of diabetic retinopathy by noticing a number of very small blood

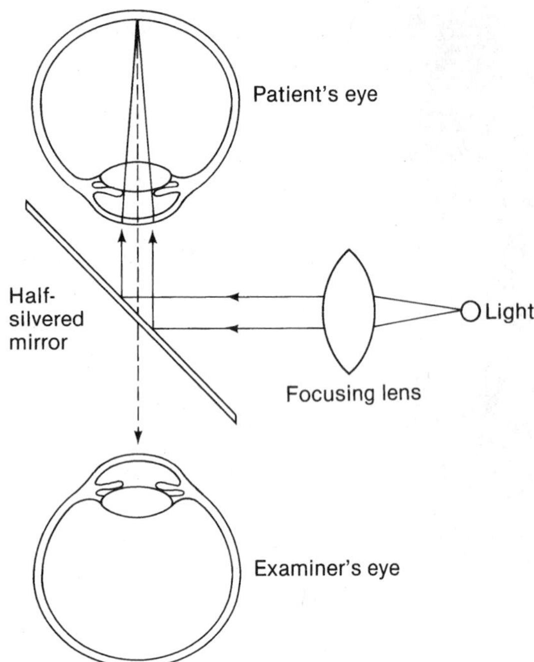

Figure 13.27
The principle behind the ophthalmoscope. Light is reflected into the patient's eye by the half-silvered mirror. Some of this light is then reflected into the examiner's eye (along the dashed line), allowing the examiner to see the inside of the patient's eye.

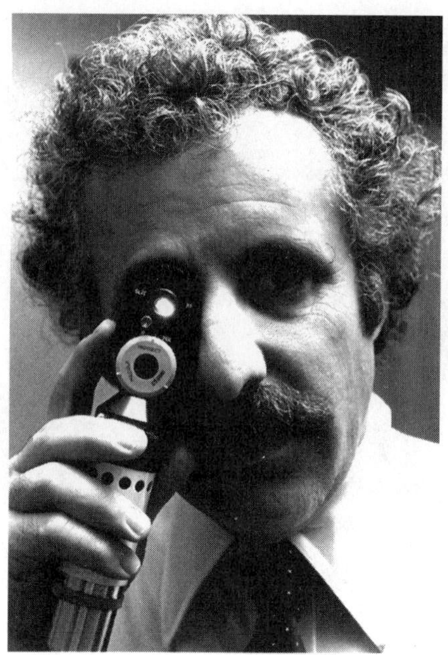

Figure 13.28
Patient's-eye view of an ophthalmoscopy exam.

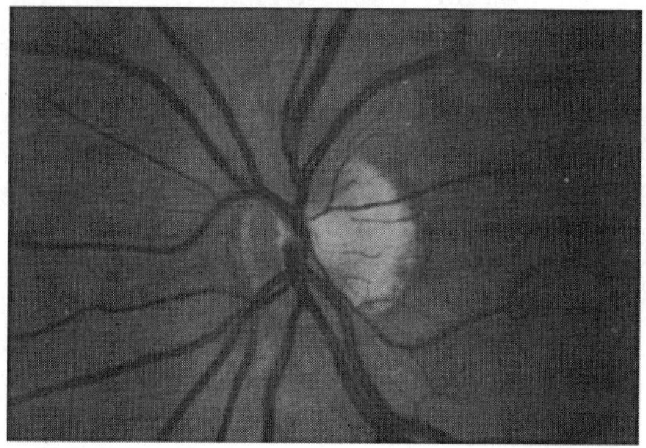

Figure 13.29
Close-up view of the head of the optic nerve and the retinal circulation as seen through an ophthalmoscope.

vessels (neovascularization). In fact, all the retinal injuries and diseases described above cause some change in the appearance of the retina, which can be detected by looking at the retina with an ophthalmoscope.

Our description of an eye examination has covered most of the tests included in a routine exam. The examiner may decide to carry out other tests if a problem is suggested by the routine tests. For example, a technique called **fluorescein angiography** is used to examine more closely the retinal circulation in patients with diabetic retinopathy. A fluorescent dye is injected intravenously into the arm, and when this dye reaches the retina, it sharply outlines the retinal arteries and veins, as shown in Figure 13.30. Only by this technique can we clearly see the small arteries formed by the neovascularization that accompanies diabetic retinopathy.

Other tests, which we will not describe here, include the **electroretinogram**, which measures

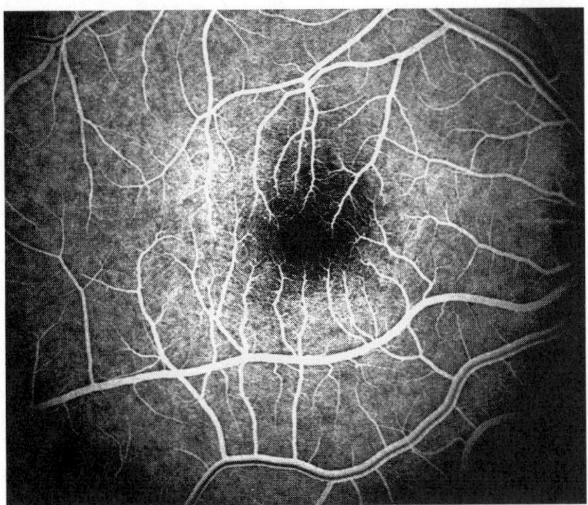

Figure 13.30
Fluorescein angiograph of a normal eye. In this view, the head of the optic nerve is on the far right, just outside the picture. The fovea is in the dark space near the middle of the picture. In the normal eye, the blood vessels stand out in sharp contrast to the background. (Photograph courtesy of Eye and Ear Hospital of Pittsburgh.)

the electrical response of the rod and cone receptors and is therefore useful in diagnosing such retinal degeneration as retinitis pigmentosa, and the cortical evoked potential, which measures the electrical response of the visual cortex and is useful for diagnosing vision problems caused by head injuries or tumors.

HEARING IMPAIRMENT

In our consideration of the clinical aspects of vision, we saw that visual functioning can be impaired because of problems in delivering the stimulus to the receptors, because of damage to the receptors, and because of damage to the system that transmits signals from the receptors toward the brain. An analogous situation exists in hearing, as we will see by considering the various causes of hearing impairment.

How Can Hearing Become Impaired?

In considering the question "How can hearing become impaired?" it is important to distinguish between impairments in the auditory system and what effects these impairments have on a person's hearing. A **hearing impairment** is a deviation or change for the worse in either the structure or the functioning of the auditory system. A **hearing handicap** is the disadvantage that a hearing impairment causes in a person's ability to communicate or in the person's daily living (American Speech and Hearing Association, 1981). The distinction between an impairment and a handicap means that a hearing impairment does not always cause a large hearing handicap. For example, although a person who has lost the ability to hear all sounds above 6,000 Hz has lost a substantial portion of his or her range of hearing, this particular hearing loss has little effect on the person's ability to hear and understand speech. We can appreciate this when we realize that, even though telephones transmit frequencies only between about 500 and 3,000 Hz, most people have no trouble using the telephone for communication. Such a hearing impairment would, however, change a person's perception of music, which often contains frequencies above 6,000 Hz, and would therefore have some impact on the person's quality of life.

Problems can develop in the auditory system for the following reasons: (1) problems in delivering the sound stimulus to the receptors, (2) damage to the receptors, (3) damage to the transmission system, and (4) damage to the auditory cortex (Figure 13.31). The following is a list of the types of things that can go wrong

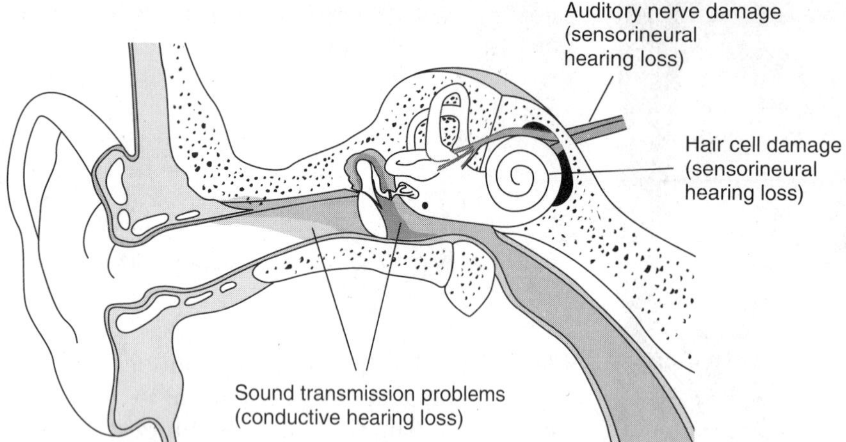

Figure 13.31
Places in the ear where hearing problems can occur.

in the auditory system within each of these categories:

1. Sound is not properly transmitted to the receptors. Problems in delivering sound to the receptors can occur because of problems such as blockage of the outer ear or damage to the system that transmits vibrations through the middle ear. These types of problems result in *conductive hearing losses.*
2. The hair cells are damaged, so they can't generate electrical signals. This problem and the one below result in *sensorineural hearing losses.*
3. There is damage to the auditory nerve or the brainstem that keeps signals that are generated from being transmitted to the auditory area of the brain. Damage at the brainstem level can interfere with the listener's ability to integrate the signals coming from the left and right ears.
4. There is damage at the auditory cortex, so when the signal reaches the cortex, it is not processed properly.

We will describe some of the major ways in which these problems occur in the auditory system, focusing on the first two categories above: conductive hearing loss and sensorineural hearing loss.

Conductive Hearing Loss

A conductive hearing loss is one in which the vibrations that would normally be caused by a sound stimulus are not conducted from the outer ear into the cochlea. This kind of loss can occur in either the outer ear or the middle ear.

Outer-Ear Disorders

Sound can be blocked at the ear canal by the buildup of excessive cerumen (ear wax), or by the insertion of objects, as might occur when children decide it would be fun to put beans or wads of paper into their ears. A more serious problem occurs in children who are born with outer- or middle-ear malformations that prevent sound from traveling down the outer-ear canal and through the middle ear. Blockage may also occur because of a swelling of the canal caused by infection by microorganisms, a situation that often occurs in swimmers when water is trapped in the ear, hence the name "swimmer's ear." Another problem occurs if the tympanic membrane at the end of the outer ear is ruptured either by a very loud noise such as an explosion or by the insertion of a sharp object too far into the ear. Such a rupture may allow microorganisms into the middle ear that may cause infection. Also, once the tympanic membrane is ruptured, it does not efficiently set the ossicles into vibration, which causes hearing loss. Problems of the outer ear are generally treated with medication or surgery. Normal hearing is often restored after these treatments.

Middle-Ear Disorders

Most people have experienced **otitis media,** middle-ear infection, at some time. Middle-ear infections are caused by bacteria that also cause swelling of the eustachian tube, the passageway that leads from the middle ear to the pharynx, which normally opens when a person swallows. This natural opening allows the pressure in the middle ear space to equalize with the pressure in the environment. However, if the eustachian tube is blocked, the pressure in the middle ear starts to decrease. With the eustachian tube closed, the bacteria have a nice, warm home to grow inside the middle ear space, and this growth eventually produces fluid in the middle ear, which prevents the tympanic membrane and the ossicles from vibrating properly.

Repeated exposure to middle-ear infections may cause a tissue buildup in the middle ear called a *cholesteatoma*. This growth interferes with the vibrations of the tympanic membrane and the ossicles and must be surgically removed. If a person does not seek treatment for a middle-ear infection, the fluid may build up until the tympanic membrane ruptures in order to release the pressure. An infection that is left untreated may also diffuse through the porous mastoid bone, which creates the middle-ear cavity. This is a very serious condition and must be treated immediately before the infection is allowed to spread to the brain. Luckily, diffusion through the mastoid bone rarely occurs if middle-ear infections are promptly treated by antibiotics.

Otosclerosis is a hereditary condition in which there is a growth of bone in the middle ear. Usually, the stapes becomes fixed in place, so it can't transmit vibrations to the inner ear. This was the condition that caused Beethoven to become so deaf that, late in his career, he was unable to hear his own music. Today, this condition can be successfully treated by a surgical procedure called *stapedectomy*, in which the stapes is replaced with an artificial strut.

Sensorineural Hearing Loss

Sensorineural hearing loss is caused by a number of factors, which have in common their site of action in the inner ear.

Presbycusis

The most common form of sensorineural hearing loss is called **presbycusis,** which means "old hearing" (remember that the equivalent term for vision is *presbyopia*, for "old eye"). This loss of sensitivity, which is greatest at higher frequencies, accompanies aging and affects males more severely than females. Figure 13.32 shows the progression of loss as a function of age. The most common complaint of people with presbycusis is

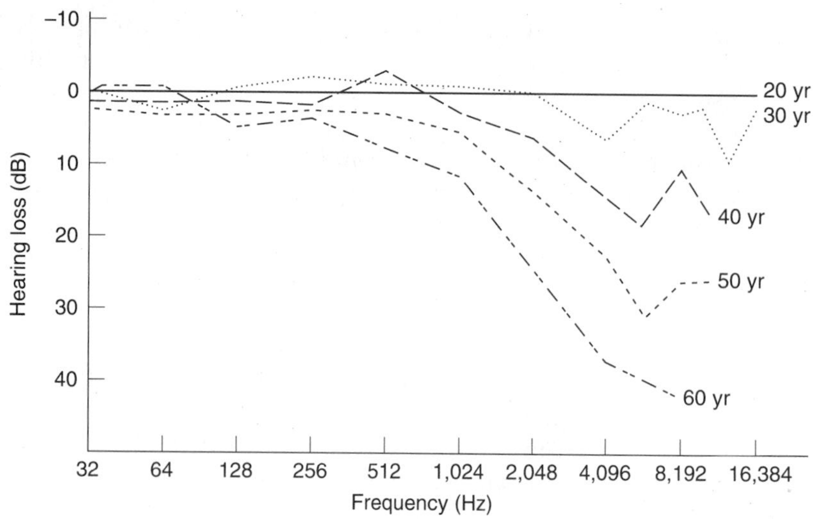

Figure 13.32
Hearing loss in presbycusis as a function of age. All of the curves are plotted relative to the 20-year curve, which is taken as the standard. (Adapted from Bunch, 1929.)

that they have difficulty hearing people talking when there is noise or when other people are talking at the same time. Presbycusis is treated by the amplification provided by hearing aids and by teaching people more effective communication strategies.

Unlike the visual problem of presbyopia, which is an inevitable consequence of aging, presbycusis is apparently caused by factors in addition to aging, since people in preindustrial cultures, who have not been exposed to the noises that accompany industrialization, often do not experience a decrease in high-frequency hearing in old age. This may be why males, who are exposed to more workplace noise than females, as well as to noises associated with hunting and wartime, experience a greater presbycusis effect. Because of its link to environmental conditions, presbycusis is also called *sociocusis*.

Noise-Induced Hearing Loss

Noise-induced hearing loss occurs when loud noises cause degeneration of the hair cells. This degeneration has been observed in examinations of the cochleas of people who have worked in noisy environments and have willed their ear structures to medical research. Damage to the organ of Corti is often observed in these cases. For example, examination of the cochlea of a man who worked in a steel mill indicated that his organ of Corti had collapsed and no receptor cells remained (Miller, 1974). Apparently, this kind of damage also occurs in people who have exposed themselves to loud music for extended periods of time. Because of this exposure to loud music, rock musicians such as Steven Stills and Peter Townsend have become partially deaf and have urged musicians and concertgoers to wear earplugs (Ackerman, 1995). In fact, members of many symphony orchestras, including the Chicago Symphony, wear ear protection to preserve their hearing.

Acoustic trauma caused by implosive noises, such as explosions or machines that create a loud impact, can also result in sensorineural hearing loss. An example is a 21-year-old college student who was in the process of raiding a rival fraternity house when a firecracker exploded in his hand, 15 inches from his right ear. The result was a hearing loss of over 50 dB at frequencies above 3,000 Hz. In addition, the student also experienced a ringing sensation in his ear that was still present two years after the accident (Ward & Glorig, 1961).

Tinnitus

Ringing in the ears, which is known as **tinnitus** (ti-NYE-tus or TIN-ni-tus, from the Latin for "tinkling"), affects more than 36 million Americans, nearly 8 million of them severely. The most common cause of tinnitus is exposure to loud sounds, although this condition can also be caused by certain drugs, ear infections, or food allergies.

Whatever causes tinnitus, it is an extremely debilitating condition. According to Jack Vernon, director of the Kresge Hearing Research Laboratory at the University of Oregon, tinnitus is the third worst thing that can happen to a person, ranking only below intractable severe pain and intractable severe dizziness. In its most serious form, the constant noise of tinnitus is totally incapacitating, making it impossible for people to maintain their concentration long enough to complete a task and, in some cases, even driving people to suicide.

Is there a cure for tinnitus? Unfortunately, for most people the answer to this question is no. Some people, however, can gain relief by using a device called a **tinnitus masker.** The masker, which is worn in the ear like a hearing aid, produces noise that sounds like a waterfall. This externally produced noise masks the internal noise of tinnitus, making life bearable for some tinnitus sufferers. Also, tinnitus sufferers who use a hearing aid to compensate for a loss of hearing sometimes find that they are unaware of the tinnitus while using the hearing aid, and for several hours after taking the hearing aid off.

Meniere's Disease

Another cause of sensorineural hearing loss is **Meniere's disease,** a debilitating condition that is caused by an excessive buildup of the liquid that fills the cochlea and the semicircular canals. The symptoms of the disease include fluctuating hearing loss, tinnitus, and severe vertigo (dizziness) that is often accompanied by nausea and vomiting. By the end of the disease, the vertigo subsides, but some people are left with a sensorineural hearing loss. Physicians attempt a variety of treatments to relieve the symptoms and to treat the increase in fluid, but no one treatment is effective for all patients. The fluctuating hearing loss can be helped by a flexible hearing aid that can be reprogrammed as the hearing loss changes.

Neural Hearing Loss

All of the conditions described above have their effects primarily in the inner ear and on the hair cells. A type of sensorineural hearing loss called *neural hearing loss* may be caused by tumors on the auditory nerve along the auditory pathways in the brainstem. These tumors generally grow slowly and are benign. However, when they are surgically removed, the patient is often left with some hearing loss. In addition, neural hearing loss can also be caused by tumors or damage further along the auditory pathway.

The Ear Examination and Hearing Evaluation

We begin our description of the ear examination and hearing evaluation by considering the types of professionals involved in the care of the ear and in helping people maintain their hearing.

Who Examines Ears and Evaluates Hearing?

A number of types of professionals examine the ear and test hearing. The two main categories are otorhinolaryngologists and audiologists.

1. An **otorhinolaryngologist** is an M.D. who has specialized in the treatment of diseases

and disorders affecting the ear, nose, and throat, and so the name of this specialty is often abbreviated *ENT*, for "ear, nose, and throat." ENT specialists carry out physical examinations of the ear, nose, and throat and provide treatment through drugs and surgery. Some physicians with ENT training specialize in one area. For example, an **otologist** is an otorhinolaryngologist whose practice is limited to problems involving the auditory and vestibular (balance) system.

2. An **audiologist** is a professional with a master's degree or Ph.D. who measures the hearing ability of children and adults and identifies the presence and severity of any hearing problems. If a hearing loss is identified, the audiologist can fit the person with a hearing aid to make sound audible and may also work with the person on a long-term basis to teach communication strategies such as speech reading (also called *lipreading*) and other techniques for more effective communication. When hearing loss is found in children, the audiologist works with other professionals to make sure that the child develops a communication system (speech or sign language) and has access to appropriate schooling.

ENT specialists and audiologists often work together in dealing with hearing problems and the ear. The physician treats ear diseases that may or may not be causing a hearing loss. If an audiologist sees a patient complaining of hearing loss and finds that the cause is a medical problem, the audiologist refers the patient to the physician for care. On the other hand, if a person with a hearing loss first sees a physician, who determines that nothing is medically treatable, the physician refers the patient to an audiologist, who can help by fitting the patient with a hearing aid. Sometimes, a person has a problem that needs medical treatment by a physician and at the same time sees an audiologist who helps deal with the hearing loss.

What Happens during an Ear Examination and Hearing Evaluation?

The basic aims of the ear examination and hearing evaluation are to assess hearing and to determine the cause of defective hearing so it can be treated. The basic components of the examination are the following:

Medical History The medical history focuses on hearing problems that the patient now has or may have had in the past, on general medical problems that could affect the person's hearing, and on medications that may be responsible for a hearing loss.

Otoscopy The purpose of otoscopy is to examine the tympanic membrane. To do this, the physician looks into the ear using an **otoscope,** which, much in the manner of the ophthalmoscope used to see the inside of the eye, illuminates the ear and makes it possible to view the illuminated area. The physician inspects the ear canal for foreign objects and signs of disease, notes the color of the tympanic membrane, and inspects it for evidence of tears.

Hearing Evaluation A person's hearing is typically measured in two ways: (1) by *pure-tone audiometry*, which determines an *audiogram*, the function relating hearing loss to frequency, and (2) by *speech audiometry*, which determines a person's ability to recognize words as a function of the intensity of the speech stimulus.

Pure-tone audiometry is typically measured by a device called an **audiometer,** which can present pure tone stimuli at different frequencies and intensities. The audiologist varies the intensity of the test tone and instructs the patient to indicate when he or she hears it. When the per-

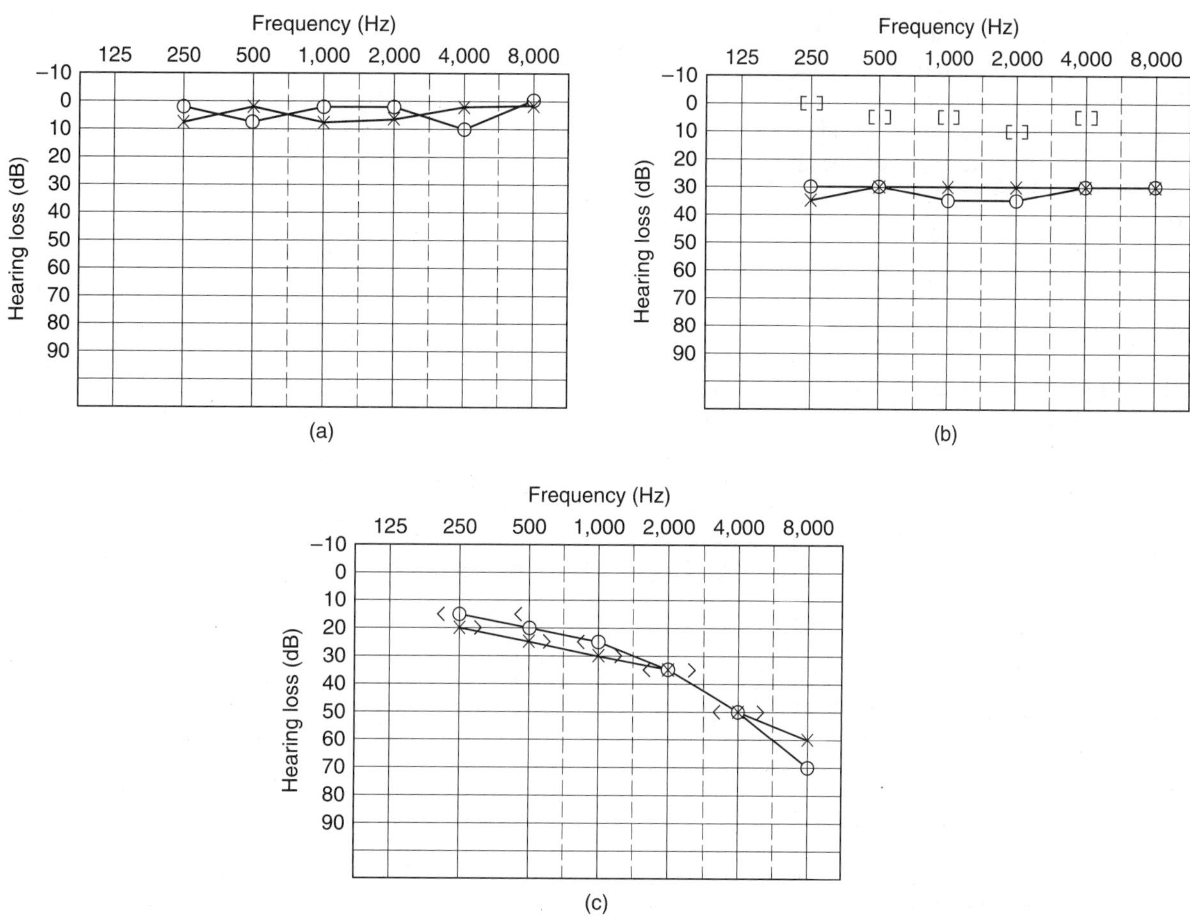

Figure 13.33

Audiograms for people with (a) normal hearing, (b) conductive hearing loss, and (c) sensorineural hearing loss. Symbols: O = right ear; X = left ear; [and] indicate bone conduction for the left and right ears for (b); < and > indicate bone conduction for the left and right ears for (c). See text for details.

son's threshold has been determined at a number of frequencies, the audiometer creates an **audiogram,** a plot of degree of hearing loss (compared to normal) versus frequency. Figure 13.33a shows the audiogram of a patient with normal hearing, and the audiograms in Figures 13.33b and 13.33c are of a patient with about a 30-dB loss of hearing at all frequencies and a patient with high-frequency loss, respectively. Audiograms are plotted so the curve for a person with normal hearing falls along the zero line for all frequencies, and any hearing loss is indicated by symbols below the zero line.

The pattern of hearing loss sometimes provides information regarding the nature of the patient's problem. For example, the audiogram in Figure 13.33b, in which the hearing loss is approximately the same across the range of hearing, is typical of a patient with a conductive hearing loss. The record in Figure 13.33c, in which hearing becomes progressively worse at high frequencies, is typical of sensorineural hearing loss. Hearing loss due to exposure to noise typically shows a maximum loss at about 4,000 Hz.

Another way to differentiate between sensorimotor and conductive hearing losses is to compare a person's hearing when the stimulation is presented through the air, as when sound is heard from a loudspeaker or through earphones, and when it is presented by vibrating the mastoid bone in back of the ear. Bone conduction is measured by means of an audiometer connected to an electronic vibrator that presents vibrations of different frequencies to the mastoid. The person responds to the bone-vibrated signal in the same way as to air-conducted signals, and thresholds for bone conduction hearing are plotted on the audiogram. If air conduction hearing is worse than bone conduction hearing, this result indicates that something must be blocking sound in the outer or middle ear. If the bone conduction and air conduction results are the same, the problem must be beyond the outer and middle ear. The bracket symbols ("[" and "]") in Figure 13.33b indicate normal bone conduction in the patient with conductive hearing loss.

Another way to measure hearing, which is particularly important in people with sensorineural loss, is to measure their word recognition ability. To do this, the audiologist presents a series of tape-recorded words at different intensities, and the patient is asked to identify the words that are spoken. The result is an **articulation function** like the one in Figure 13.34, which plots the percentage of words identified correctly versus the intensity of the sound. Patients with conductive losses tend to have articulation functions that are shifted to higher intensities, but if the intensity is high enough, they can understand the words. Patients with sensorineural loss have functions that are also shifted to higher intensities, but they often fail to reach 100 percent performance. Thus, in sensorineural hearing loss, many words cannot be consistently identified, no matter what their intensity. This result is most likely caused by the poor frequency and intensity discrimination, which is associated with sensorineural hearing loss.

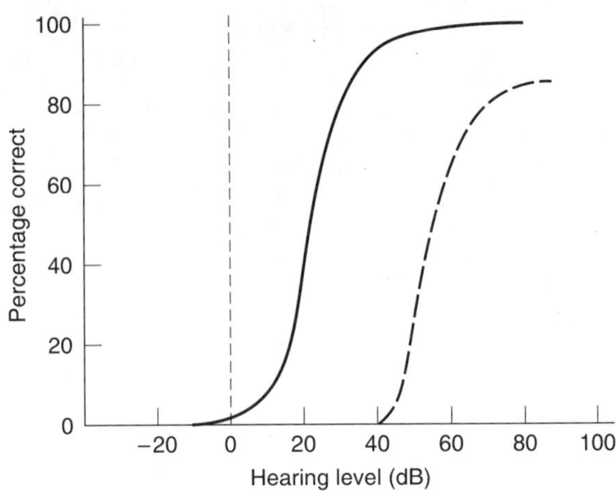

Figure 13.34

Articulation functions for a patient with conductive hearing loss (solid line) and a patient with sensorineural hearing loss (dashed line). Both curves are shifted to the right compared to normal, indicating that the patient requires greater than normal intensity to recognize the words. Also, notice that the curve for sensorineural hearing loss never reaches 100 percent, no matter what the intensity.

Tympanometry An important indication of auditory functioning is how well the tympanic membrane and the middle-ear bones are responding to sound vibrations. This response is measured by a device called a **tympanometer,** which consists of a snug-fitting ear probe with a sound source, a pressure generator, and a microphone. This probe is placed in the outer ear, and as the probe varies the pressure in the ear canal, it presents sound and the microphone measures how much of this sound is bounced off the tympanic membrane. If the tympanic membrane is immobilized because the middle ear is full of fluid or for some other reason, a large proportion of the sound hitting it bounces back. If the tympanic membrane is operating normally, it is flexible and so absorbs most of the sound striking it. The tympanometer uses its measurement of the amount of sound reflected to assess the functioning of the tympanic membrane.

Measurement of the Acoustic Reflex Threshold The acoustic reflex is the activation of the middle-ear muscles in response to high-intensity sounds. This activation stiffens the chain of ossicles and dampens their vibration in order to protect the inner ear from being overstimulated. The acoustic reflex can be measured with the tympanometer. A loud sound is produced in the ear canal, and a graph is produced that indicates the movement of the tympanic membrane. If the acoustic reflex is normal and the muscles contract in response to the sound, the graph will show a normal level of movement.

In addition to these diagnostic techniques that access hearing and the functioning of structures in the ear, there are also tests that measure the electrical response of the auditory nerve and the auditory cortex, which are important if there is a hearing loss even though the structures of the ear seem to be operating normally. These electrophysiological measures can therefore be useful in determining the location of sensorineural hearing loss.

Managing Hearing Loss

Hearing loss covers a very large continuum, from the mildly hearing-impaired individual to someone who is totally deaf and cannot make use of sound for the purpose of communication (Figure 13.35). Different individuals need different types of technology and strategies in order to communicate effectively.

The majority of individuals with conductive hearing loss receive medical treatment for their condition, and the hearing loss is eliminated as soon as the disease process is eliminated. Patients who cannot or do not choose to take advantage of surgical or medical procedures may use an amplification system such as a hearing aid to help them hear. As we noted when describing word recognition tests, as long as the sound is loud enough, these patients can hear quite well.

Individuals with sensorineural hearing loss generally use some kind of amplification. Hearing aids have changed drastically since the beginning of the 1990s, with the development of smaller hearing aids that can be automatically programmed to work effectively in different listening situations. Almost all individuals with sensorineural hearing loss can benefit from a hearing aid, which is fitted by an audiologist.

Hearing-impaired individuals may also receive training in speech reading (often called *lipreading*) and communication strategies. This type of training is often called **aural rehabilitation** and is conducted by an audiologist. Some individuals have so much hearing loss that they cannot benefit from amplification. Many of these individuals consider themselves part of the deaf culture and are happy to communicate using sign language. However, most people who lose their hearing after being able to hear for the majority of their lives wish to continue to be connected to the hearing world. For these people, a new technology called the *cochlear implant* is available.

A **cochlear implant** is a device in which electrodes are inserted in the cochlea to create hearing by electrically stimulating the auditory nerve fibers. This device offers the hope of regaining some hearing to some people who have lost their hearing because of damaged hair cells, so that hearing aids, which can amplify sound, but which can't cause that sound to be translated into electrical signals in the hair cells, are ineffective.

The cochlear implant bypasses the damaged hair cells and stimulates auditory nerve fibers directly. The following are the basic components of a cochlear implant (Figure 13.36):

- The *microphone* (1), which is worn behind the person's ear, receives the speech signal, transforms it into electrical signals, and sends these signals to the speech processor.

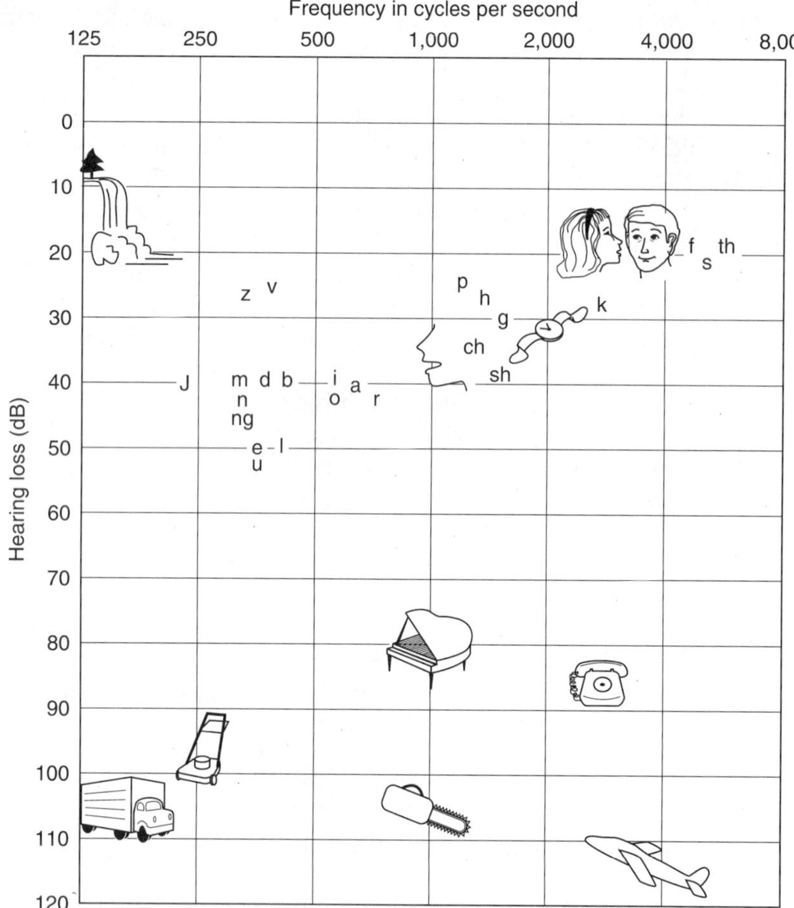

Figure 13.35
Audiogram showing the frequencies and degree of hearing loss that cause difficulties in the perception of various environmental sounds. The letters represent spoken sounds. For example, a person with a hearing loss of 100 dB around 250 Hz will not be able to hear the lawn mower. (From Northern & Downs, 1978.)

- The *speech processor* (2), which looks like a small transistor radio, shapes the signal generated by the microphone to emphasize the information needed for the perception of speech by splitting the range of frequencies received by the microphone into a number of frequency bands. These signals are sent, in the form of an electrical code, from the processor to the transmitter.
- The *transmitter* (3), mounted on the mastoid bone, just behind the ear, transmits the coded signals received from the processor, through the skin, to the receiver.
- The *receiver* (4) is surgically mounted on the mastoid bone, beneath the skin. It picks up the coded signals from the transmitter and converts the code into signals that are sent to electrodes implanted inside the cochlea (5).

The implant makes use of Békésy's observation, which we described in Chapter 8, that there is a tonotopic map of frequencies on the cochlea, low frequencies being represented by activity near the base of the cochlea, and high frequencies being represented by activity at the

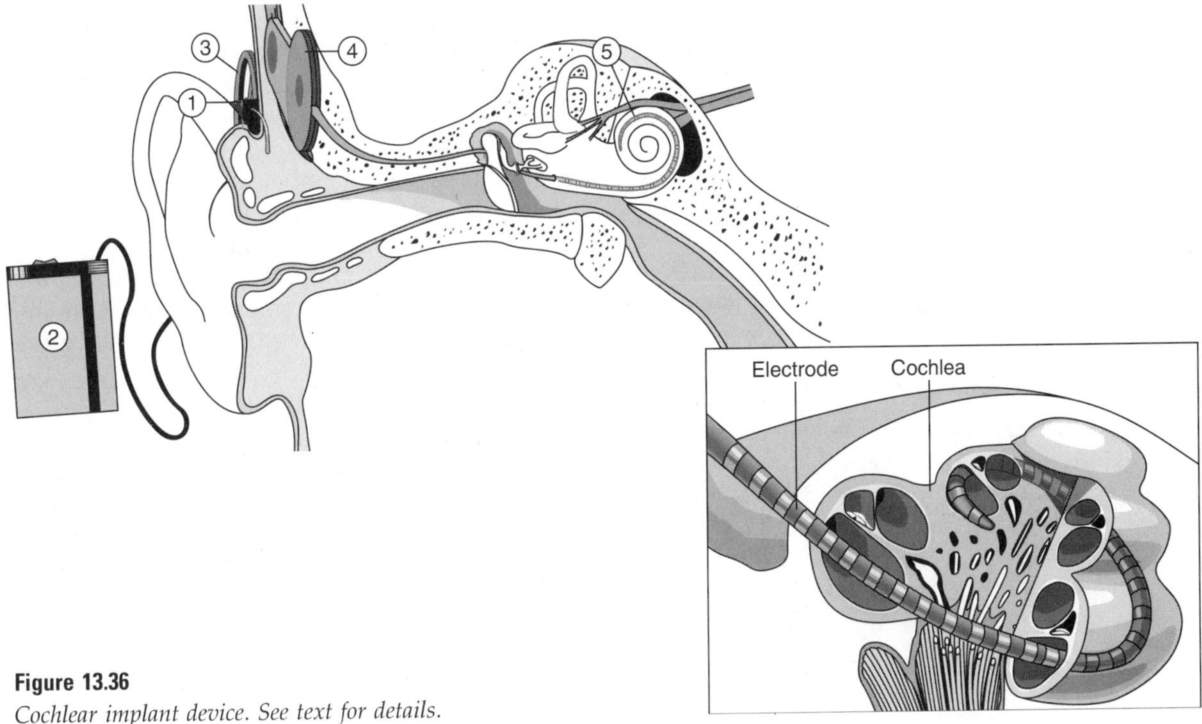

Figure 13.36
Cochlear implant device. See text for details.

apex of the cochlea. The most widely used implants therefore have a multichannel design, which uses a number of electrodes to stimulate the cochlea at different places along its length, depending on the frequencies in the stimuli received by the microphone. This electrical stimulation of the cochlea then causes signals to be sent to the auditory area of the cortex, and hearing results.

What does a person using this system hear? The answer to this question depends on the person. Most patients are able to recognize a few everyday sounds, such as horns honking, doors closing, and water running. In addition, many patients are able to perceive speech. In the best cases, patients can perceive speech on the telephone, but it is more common for cochlear-implant patients to use the sounds perceived from their implant in conjunction with speech reading. In one test, 24 patients scored 54 percent on a test of speech reading alone, and 83 percent when speech reading was combined with sound from the implant. In addition, the implant enabled patients to track speech much more rapidly—16 words per minute using speech reading alone, and 44 words per minute with speech reading plus the implant (Brown, Dowell, & Clark, 1987; Owens, 1989).

As of 1995, over 10,000 people have cochlear implants. The best results occur for *postlingually deaf* individuals—people who were able to perceive speech before they became deaf. These people are most likely to be able to understand speech with the aid of the implant because they already know how to connect the sounds of speech with specific meanings. Thus, these people's ability to perceive speech often improves with time, as they again learn to link sounds with meanings.

To place a more human face on the effects of the cochlear implant operation, we will close this chapter, and the book, by relating the story of Gil McDougald, the standout New York Yankee infielder who played in eight World Series in the 1950s. McDougald, who in 1995 was 66 years old, had gradually gone deaf in both ears after an accident in which he was hit by a line drive during batting practice in 1955. He was almost completely deaf for about 20 years, being able to make out some sounds, but no intelligible words. His deafness cut him off from other people. He could no longer talk on the telephone, and he stopped attending "old-timers'" functions with ex-teammates like Yogi Berra and Mickey Mantle, because he was unable to communicate with them. At family functions, he would leave the table because of his frustration at not being able to hear what was going on.

McDougald heard about the implant operation and, in 1995, called implant specialist Noel Cohen at New York University Medical Center. After testing had determined that he was a good candidate for an implant, he underwent the operation. Six weeks after the operation, he went to see Betsy Bromberg, an audiologist, to have the apparatus programmed and activated (Figure 13.37). The following is an excerpt from a newspaper account of what transpired in her office (Berkow, 1995):

> In the office, McDougald sat at a desk with a computer on it. Bromberg sat across from him. His wife, Lucille, and daughter, Denise, sat within arm's length.

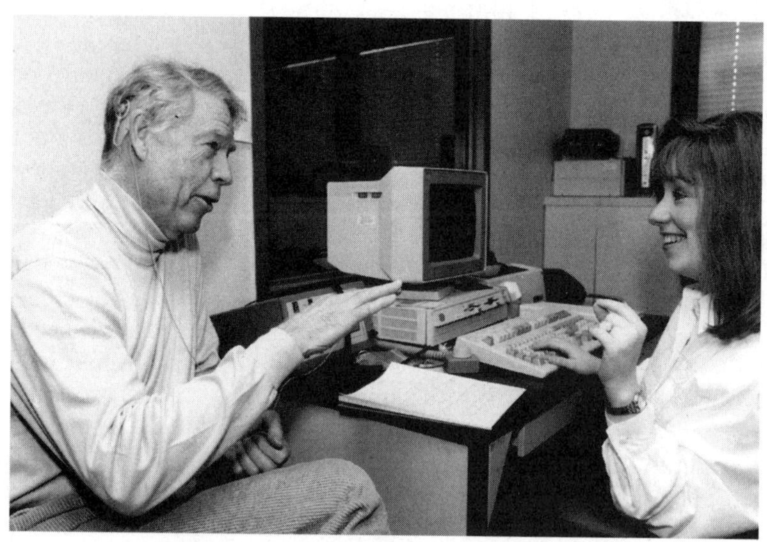

Figure 13.37
Gil McDougald in his playing days and trying out his cochlear implant for the first time, with audiologist Betsy Bromberg.

A small microphone was set behind his ear, and a transmitter with a magnet was placed over the site of the implant. A cable was extended from the microphone to a speech processor the size of a hand calculator that can be worn on a belt or placed in a breast pocket.

Then Bromberg began the test that would determine how much McDougald's hearing had improved.

Bromberg covered her mouth with a sheet of paper so he could not lip read.

"Tell me," she said, "what you hear."

She said, "aah." He hesitated. "Aah," he answered. She went, "eeeh." He said, "eeeh."

"Hello," she said. "Hello," he said. "I'm going to count to five," she said. "Do you hear me?"

"Oh yeah!" he said. "Wow! This is exciting!"

His wife and daughter stared, hardly moving.

Bromberg wrote down four words on a pad of paper and said them: "football," "sidewalk," "cowboy," and "outside." "Now Gil," she said, "I'm going to mix up the word order and cover my mouth and you tell me the word I say."

"Cowboy," she said. "Cowboy," he said. "Outside," she said. "Outside," he said. And then he began to flush. Tears welled in his eyes.

"This is the first time in . . . " Lucille said and then choked up, unable to finish her sentence. "It's unbelievable."

"It's a miracle," said Denise.

Both began crying.

Bromberg said, "It's O.K. Everybody cries at times like this." And then mother and daughter embraced. And they hugged Gil. And they hugged Bromberg, and hugged the director of the unit, Susan Waltzman, who had been observing.

. . .

Last night, the McDougald household was bursting with children and grandchildren. "Everyone," said Lucille McDougald, "has come to watch grandpa hear." (p. B8)

Cochlear implantation is an impressive demonstration of how basic research yields practical benefits. Advances such as this, which influence people's lives, are the end result of discoveries that began in perception or physiology laboratories many years earlier. In this case, it was George von Békésy's research on how the vibration of the basilar membrane depends on the frequency of the sound that provided the knowledge about the operation of the ear that made implants possible. Systems have also been proposed for restoring vision to people who are blind, by stimulating their visual cortex (Dobelle, 1977; Dobelle et al., 1974, 1976), and for guiding blind people through the environment by applying the principles of auditory localization to the design of a "personal guidance system" (see Chapter 9, page 405).

It is clear that the study of both the psychophysics and the physiology of perception has yielded not only knowledge about how our senses operate, but also ways to apply this knowledge to create new perceptual worlds for people like Gil McDougald and countless others.

Review

How Can Vision Become Impaired?

Outline

- Four major types of problems can cause poor vision: (1) Light isn't focused clearly on the retina; (2) light is blurred as it enters the eye; (3) the retina is damaged; and (4) the optic nerve is damaged.

Questions

1. On a picture of the eye, be able to indicate the site of each of the four major problems that can cause poor vision. (540)

Focusing Problems

- Myopia (nearsightedness) is an inability to see far objects clearly.

2. What are the two kinds of myopia, and why do they occur? Why does moving an object closer eventually result in clear vision of the object? (541)

- Although people with myopia can see objects clearly when they are near, the only way they can see far objects clearly is by using corrective lenses or by a surgical change in the focusing power of the cornea.

3. How does a lens have to bend light to create a sharp image on a myope's retina? Define diopter. Describe the surgical procedures radial keratotomy and laser refractive keratotomy. How do these surgical techniques improve vision? (541)

- Hyperopia (farsightedness) is a difficulty in focusing the eye to see near objects clearly.

4. Describe how the hyperope's eye focuses light. Why can accommodation help a hyperope see clearly? (544)

- Presbyopia ("old eye") makes it difficult for older people to adjust their vision to see nearby objects.

5. Why does presbyopia occur? What is the near point? What is the solution to presbyopia? (545)

- Astigmatism is a condition that causes blurred vision in some parts of the visual field.

6. Why does astigmatism cause blurred vision? (545)

Decreased Transmission of Light

- Disease or physical damage to the eye can cause severe visual losses and, in some cases, blindness.

7. What is the definition of blindness? How can a person have 20/20 vision and still be legally classified as being blind? (548)

- Clouding of the cornea due to disease or injury blocks the transmission of light as it enters the eye.
- A cataract is a clouding of the lens, which may be present at birth, or may be caused by disease or trauma; however, it is most commonly associated with old age.

8. Describe the corneal transplant operation. Why isn't there such as thing as an "eye transplant"? (547)

9. Describe a cataract operation. What is an intraocular lens, and how is it installed? (547)

Damage to the Retina

- Diabetic retinopathy occurs in about 40 percent of diabetics as they get older.

10. What happens to the circulation and the retina in diabetic retinopathy? Describe a technique for dealing with this problem. (549)

- Macular degeneration is a degeneration of the fovea and the area around it. It is most commonly associated with old age.

11. Describe what happens to the retina in macular degeneration. Is there a treatment for it? (550)

- Detached retina occurs when the retina becomes separated from the pigment epithelium.

12. Why is detached retina a problem? How is it treated? (551)

- Retinitis pigmentosa is the most common hereditary retinal degeneration.

13. What happens to the receptors in retinitis pigmentosa? Which receptors are attacked first? (552)

Optic Nerve Damage

- The leading cause of blindness in the United States is glaucoma, a condition that causes degeneration of fibers in the optic nerve.

14. Describe how problems at the front of the eye can lead to degeneration of the optic nerve. Describe closed-angle and open-angle glaucoma. How are they treated? (553)

The Eye Examination

- Ophthalmologists, optometrists, and opticians are the major professionals involved in eye care.

15. Distinguish between ophthalmologists, optometrists, and opticians. (554)

- The eye examination consists of the following main components: (1) medical history; (2) visual acuity measurement; (3) refraction; (4) external eye examination; (5) slit-lamp examination; (6) tonometry; and (7) ophthalmoscopy.

16. Describe the reasons for each of the components of the eye exam and the procedures used for each one. What does 20/40 vision mean? Why is near vision tested? What are the steps in determining the prescription for a patient's corrective lenses? Understand the design of the ophthalmoscope. What is fluorescein angiography? (555)

How Can Hearing Become Impaired?

- Four major problems can cause an impairment in hearing: (1) Sound is not transmitted to the receptors; (2) the hair cells are damaged; (3) the auditory nerve is damaged; and (4) the auditory cortex is damaged.

17. What is the difference between a hearing impairment and a hearing handicap? Be able to indicate, on a diagram of the ear, the site of each of the problems that can cause hearing impairment. (561)

Conductive Hearing Loss

- A conductive hearing loss is one in which the vibrations that would normally be caused by a sound stimulus are not conducted from the outer ear into the cochlea.

18. Describe the following outer- and middle-ear disorders that cause conductive hearing loss and indicate treatments, if they exist: swimmer's ear, otitis media, cholesteatoma, otosclerosis. (562)

Sensorineural Hearing Loss

- Sensorineural hearing loss can be caused by tumors that damage the auditory nerve or by damage to the hair cells.

19. What is presbycusis? What causes it, and which frequencies are usually affected? How does noise-induced hearing loss affect hearing? What is tinnitus? Is there any treatment for it? What are Meniere's disease and neural hearing loss? (563)

The Ear Examination and Hearing Evaluation

- Otorhinolaryngologists and audiologists are the major professionals involved in ear care.

20. Describe the training and responsibilities of otorhinolaryngologists and audiologists. (565)

- The ear examination consists of the following components: (1) medical history; (2) otoscopy; (3) hearing assessment; (4) tympanometry; and (5) measurement of the acoustic reflex.

21. Describe the reasons for each of the components of the ear exam and the procedures used for each one. What is pure-tone audiometry? Describe the audiograms typical of conductive and sensorineural hearing loss. Describe speech audiometry and the articulation function that results. What is tympanometry? How is the acoustic reflex measured? (566)

Managing Hearing Loss

- Hearing loss covers a very large continuum, from mild impairment to total deafness.

22. What is the generally expected outcome for people with conductive and sensorineural hearing losses? Describe the use of hearing aids in conductive and sensorineural hearing loss. (569)

- An operation called the *cochlear implant operation* has been developed to provide hearing to people who have lost their hearing because of damage to their hair cells.

23. Describe the components of the cochlear implant. What is the basic auditory principle behind the implant? What is the typical outcome of the implant operation? Who are the best candidates for the implant operation? (569)

APPENDIX A
Synopses of Developmental Dimensions

The following are summaries of the material in each of the end-of-chapter Developmental Dimension (DD) features. See Appendix B for suggested ways to create topical units by combining these DDs.

DD 1: Infant Psychophysics. The preferential looking and habituation techniques for measuring visual perception.

DD 2: Visual Acuity and the Newborn's Cones. Measurement of acuity using preferential looking and visual evoked potential; acuity as a function of age; poor development of the cortex and the foveal cone development as reasons for poor newborn acuity.

DD 3: Sensitive Periods in Perceptual Development. The effects of monocular deprivation on ocular dominance and binocularity in cats and monkeys; the sensitive period for monocular deprivation effects; the effect of image misalignment on binocularity; stimulus deprivation amblyopia in children as caused by patching or strabismus; use of interocular transfer of the tilt aftereffect to measure binocularity.

DD 4: Infant Color Vision. Infant perception of color categories; psychophysical evidence for short-, medium-, and long-wavelength receptors.

DD 5: Textons and Infant Perception. Psychophysical evidence that infants use textons in perception; pop-out vision in infants.

DD 6: The Emergence of Depth Perception in Infants. The development of binocular disparity, binocular fixation, and stereoacuity; the physiological mechanism of disparity development; the development of sensitivity to pictorial cues such as interposition and familiar size.

DD 7: Infants' Perception of Movement. Basic movement perception capabilities; perceiving movement behind an occluding object; perceiving the biological motion of point-light walkers; what discrimination tells us about perception.

DD 8: Plasticity of Frequency Representation in the Cochlea and the Brain. Change in the cochlear frequency map in chickens as a function of age; effect of the cochlear changes on the cortical map; effect of lesioning the adult guinea pig's cochlea on the cortical representation of frequency; effect of lesioning the adult cat's retina on the representation of the retina in the cortex.

DD 9: The Auditory World of Human Infants. Measuring infant auditory capability by head turning; recognition of the mother's voice by fetus and newborn; measuring infant audibility curves by the observer-based psychoacoustic procedure.

DD 10: Infant Speech Perception. Categorical perception, duplex perception, and equivalence classification in infants; effect of experience on infants' ability to discriminate sounds.

DD 11: Organization of the Somatosensory Cortex. Ways to study perceptual development; reorganization of somatosensory maps in response to elimination of input from the receptors; reorganization of somatosensory maps in response to high levels of receptor stimulation; evidence that cortical organization is based on temporal coordination of stimulation to the monkey's fingers.

DD 12: Olfaction and Taste in Infants. Infant capacity to detect smell and taste stimuli as measured by facial expressions; infant ability to detect maternal odors.

APPENDIX B

UNITS ON PERCEPTUAL DEVELOPMENT

In the third edition of *Sensation and Perception,* perceptual development appeared in a single chapter on the development of visual perception. In this edition, perceptual development appears in the Developmental Dimension (DD) feature at the end of each chapter. This arrangement expands the coverage of development to cover all of the senses and also provides two ways to cover perceptual development: (1) Each quality can be considered separately following the material in the relevant chapter, and (2) development can be covered in larger units by grouping a number of DDs together. The following are suggested outlines for some possible units:

Unit 1: *Visual Development/Psychophysical Approach*

DD 1: Infant Psychophysics

DD 2: Visual Acuity and the Newborn's Cones

DD 4: Infant Color Vision

DD 5: Textons and Infant Perception

DD 6: The Emergence of Depth Perception in Infants

DD 7: Infants' Perception of Movement

OWP 5: Contrast Sensitivity in Cats and Human Infants

Unit 2: *Visual Development/Physiological Approach*

DD 2: Visual Acuity and the Newborn's Cones (relates perception and physiology)

DD 3: Sensitive Periods in Perceptual Development

DD 6: The Emergence of Depth Perception in Infants (relates perception and physiology)

Unit 3: *Auditory Development*

PSYCHOPHYSICS:

DD 9: The Auditory World of Human Infants

DD 10: Infant Speech Perception

PHYSIOLOGY:

DD 8: Plasticity of Frequency Representation in the Cochlea and the Brain

Unit 4: *Plasticity of Development*

DD 3: Sensitive Periods in Perceptual Development

DD 8: Plasticity of Frequency Representation in the Cochlea and the Brain

DD 11: Organization of the Somatosensory Cortex

APPENDIX C
Synopses of Other Worlds of Perception

The following are summaries of the material in each of the end-of-chapter Other Worlds of Perception (OWP) features. See Appendix D for suggested ways to create topical units by combining these OWPs.

OWP 1: What Do Other Species Experience? How the perceptual worlds of animals compare to the perceptual world of humans; how we can know what animals perceive; the privateness of experience; "extraordinary" animal capacities (compared to human capacities): elephants' low-frequency hearing, praying mantises' high-frequency neurons, snakes' detection of infrared radiation, duck-billed platypuses' detection of electrical fields, birds' detection of magnetic fields; the nature of animal experience.

OWP 2: Animal Eyes. Single-lens eyes and multiple-lens eyes; diurnal and nocturnal eyes: lizards, ground squirrels, birds, owls, opossums; eyes for underwater vision: the "four-eyed" fish *Anableps;* retinas with areas for detail vision: eagle, falcon, human foveas; turtle area centralis; visual pigments and habitat; animals with frontal and lateral eyes.

OWP 3: Orientation Detectors in Insects. Bees' detection of orientation separate from movement; feature-detecting neurons in the dragonfly; bees' preference for flower-like patterns; possible characteristics of feature-detecting neurons in bees.

OWP 4: Color Experience in Animals. Creation of experience by the nervous system; animals with color vision different from humans': pigeons' and honeybees' color vision (what do they perceive?); requirements for animal color vision tests; pigeon wavelength discrimination; physiological evidence for color vision in the goldfish; ecological evidence for animal color vision; color vision of birds, cats, dogs, and monkeys.

OWP 5: Contrast Sensitivity in Cats and Human Infants. Determining the cat's contrast sensitivity function; comparison of cats' and humans' spatial frequency channels; determining the human infant's contrast sensitivity function; development of humans' contrast sensitivity as a function of age.

OWP 6: The Depth Information Used by Animals. Behavioral evidence for depth perception in animals; animals that use stereopsis: cats, monkeys, pigeons; binocular and monocular depth perception in frogs; use of the relative size cue by the backswimmer *Notonecta;* use of movement parallax by locusts and honeybees.

OWP 7: The Cat's Perception of Biological Motion. The difference between discrimination and perception; research on the cat's ability to discriminate "point-light cat walkers"; do cats perceive biological motion in the same way humans do?

OWP 8: Echolocation by Bats. Early research on bat echolocation; behavioral demonstrations of echolocation in the mustached bat; echolocation tuning curves in the cortex of the mustached and big brown bats; how bats experience an acoustic scene.

OWP 9: The Auditory World of Animals. Animals sensitive to frequencies above and below human sensitivity: elephant, homing pigeon; audibility curves of various animals; relationship between high-frequency hearing and interaural time difference; sound localization in various animals; relationship between sound localization and width of field of best vision; behavioral reason for link between sound localization and vision; ecological factors in animal hearing.

OWP 10: Perception of "Calls" by Monkeys. Signals produced by the Japanese macaque monkey; characteristics of the stimulus that the monkey uses to identify calls; a similarity between monkey and human perception: categorical perception in monkeys; a difference between monkey and human perception: lack of a perceptual magnet effect in monkeys.

OWP 11: "Other Worlds" of Perception in Humans. How can we tell what other people are experiencing? Reasons it is difficult to relate to other people's reported pain experiences; the radical skeptic's view of perception.

OWP 12: Similarities in Chemical Sensing across Species. Similarities in receptor cell structure, receptor cell turnover, receptor stereospecificity, and intermittent receptor stimulation across many different species.

APPENDIX D
UNITS ON COMPARATIVE PERCEPTION

The Other Worlds of Perception (OWP) features at the end of each chapter can also be covered individually or can be combined to create units. Some possible units are as follows:

Unit 1: *Visual Perception in Animals*

OWP 1: What Do Other Species Experience?

OWP 2: Animal Eyes

OWP 3: Orientation Detectors in Insects

OWP 4: Color Experience in Animals

OWP 5: Contrast Sensitivity in Cats and in Human Infants

OWP 6: The Depth Information Used by Animals

OWP 7: The Cat's Perception of Biological Motion

Unit 2: *Auditory Perception in Animals*

OWP 8: Echolocation by Bats

OWP 9: The Auditory World of Animals

OWP 10: Perception of "Calls" by Monkeys

Unit 3: *Adaptations to the Environment*

OWP 2: Animal Eyes

OWP 9: The Auditory World of Animals

Unit 4: *Extraordinary Powers of Animal Perception Compared to That of Humans*

OWP 1: What Do Other Species Experience?

OWP 4: Color Experience in Animals

OWP 8: Echolocation by Bats

OWP 9: The Auditory World of Animals

Unit 5: *Comparison of the Perception of Animals and Human Infants*

OWP 4: Color Experience in Animals (in conjunction with DD 4: Infant Color Vision)

OWP 5: Contrast Sensitivity in Cats and in Human Infants

OWP 7: The Cat's Perception of Biological Motion (in conjunction with DD 7: Infants' Perception of Movement)

Unit 6: *The Nature of Perceptual Experience*

OWP 1: What Do Other Species Experience? (humans and animals live in different perceptual worlds)

OWP 4: Color Experience in Animals (creation of experience by the nervous system)

OWP 7: The Cat's Perception of Biological Motion (discrimination vs. experience)

DD 7: Infants' Perception of Movement (what discrimination tells us about perceptual experience)

OWP 8: Echolocation by Bats (how bats experience echoes)

OWP 11: "Other Worlds" of Perception in Humans (the privateness of experience)

APPENDIX E
Signal Detection: Procedure and Theory

In Chapter 1, we saw that various psychophysical methods can be used to determine an observer's absolute threshold. For example, using the method of constant stimuli to randomly present tones of different intensities, we can determine the intensity to which the subject reports "I hear it" 50 percent of the time. But can the experimenter be confident that this intensity truly represents the subject's sensory threshold? For a number of reasons, which we will discuss below, the answer to this question is "Maybe not." Many researchers feel that the idea of an *absolute* measure of sensitivity, called the *threshold,* which can be measured by the classic psychophysical methods, is not valid. In this appendix, we will first discuss the question of whether there is an absolute threshold, and we will then describe a way of looking at this problem, called *signal detection theory,* that takes into account both the characteristics of the sensory system and the characteristics of the observer.

Is There an Absolute Threshold?

To understand the position of researchers who question the idea of an absolute threshold, let's consider a hypothetical experiment. In this experiment we use the method of constant stimuli to measure two subjects' thresholds for hearing a tone. We pick five different tone intensities, present them in random order, and ask our subjects to say "yes" if they hear the tone and "no" if they don't hear it. Our first subject, Laurie, thinks about these instructions and decides that she wants to appear supersensitive to the tones; since she knows that tones are being presented on every trial, she will answer "yes" if there is even the slightest possibility that she hears the tone. We could call Laurie a *liberal responder:* She is more willing to say "Yes, I hear the tone" than to report that no tone was present. Our second subject, Chris, has read the same instructions, but Chris is different from Laurie; she doesn't care about being supersensitive. In fact, Chris wants to be totally sure that she hears the tone before saying "yes." We could call Chris a *conservative responder:* She is not willing to report that she hears the tone unless it is very strong.

The results of this hypothetical experiment are shown in Figure E.1. Laurie gives many more "yes" responses than Chris and therefore ends up with a lower threshold. But given what we know about Laurie and Chris, should we conclude that Laurie is more sensitive to the tones than Chris? It could be that their actual sensitivity to the tones is exactly the same, but Laurie's apparently lower threshold is simply due to her being more willing than Chris to report that she heard a tone. A way to describe this difference between the two subjects is that each has a different **response criterion.** Laurie's response criterion is low (she will say "yes" if there is the slightest chance there is a tone present), whereas Chris's response criterion is high (she says "yes"

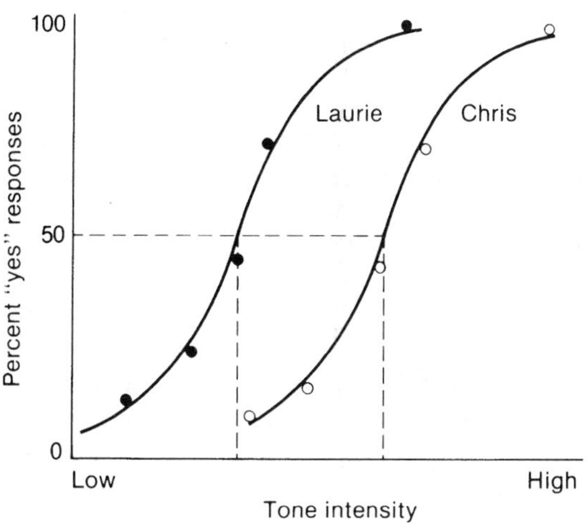

Figure E.1

Data from experiments in which the threshold for hearing a tone is determined for Laurie and Chris by means of the method of constant stimuli. These data indicate that Laurie's threshold is lower than Chris's. But is Laurie really more sensitive to the tone than Chris, or does she just appear to be more sensitive because she is a more liberal responder? Signal detection theory helps provide an answer to this question.

only when she is sure that she heard the tone). That factors other than the subject's sensitivity to the signal may influence the results of a psychophysical experiment has caused many researchers to doubt the validity of the absolute threshold, as determined by these psychophysical experiments, and to create new procedures based on a theory called **signal detection theory (SDT).**

In the next section, we will describe the basic procedure of a signal detection experiment and will show how we can tell whether Chris and Laurie are, in fact, equally sensitive to the tone even though their response criteria are very different. After describing the signal detection experiment, we will look at the theory on which the experiment is based.

A Signal Detection Experiment

Remember that, in a psychophysical procedure such as the method of constant stimuli, at least five different tone intensities are presented, and that a stimulus is presented on every trial. In a signal detection experiment, however, we use only a single low-intensity tone that is difficult to hear, and we present this tone on some of the trials and present no tone at all on the rest of the trials. Thus, a signal detection experiment differs from a classical psychophysical experiment in two ways: In a signal detection experiment (1) only one stimulus intensity is presented, and (2) on some of the trials, no stimulus is presented. Let's consider the results of such an experiment, using Laurie as our subject. We present the tone for 100 trials and no tone for 100 trials, mixing the tone and no-tone trials at random. Laurie's results are as follows:

When the tone is presented, Laurie

- Says "yes" on 90 trials. This is a correct response and, in signal detection terminology, is called a **hit.**
- Says "no" on 10 trials. This is an incorrect response and is called a **miss.**

When no tone is presented, Laurie

- Says "yes" on 40 trials. This is an incorrect response and is called a **false alarm.**
- Says "no" on 60 trials. This is a correct response and is called a **correct rejection.**

These results are not very surprising, given that we know Laurie has a low criterion and likes to say "yes" a lot. This gives her a high hit rate of 90 percent but also causes her to say "yes" on many trials when no tone is presented at all, so her 90 percent hit rate is accompanied by a 40 percent false-alarm rate. If we do a similar experiment on Chris, who has a higher criterion and therefore says "yes" much less often, we find that she has a lower hit rate (say, 60 percent) but also a lower false-alarm rate (say, 10 percent). Note that, although Laurie and Chris say "yes" on numerous trials on which no stimulus is presented, that result would not be predicted by classical threshold theory. Classical theory would say "no stimulus—no response," but that is clearly not the case here.

By adding a new wrinkle to our signal detection experiment, we can obtain another result that would not be predicted by classical threshold theory. Without changing the tone's intensity at all, we can cause Laurie and Chris to change their percentages of hits and false alarms. We do this by manipulating each subject's motivation by means of **payoffs.** Let's look at how payoffs might influence Chris's responding. Remember that Chris is a conservative responder who is hesitant to say "yes." But being clever experimenters, we can make Chris say "yes" more frequently by adding some financial inducements to the experiment. "Chris," we say, "we are going to reward you for making correct responses and are going to penalize you for making incorrect responses by using the following payoff scale:

Hit	Win $100
Correct rejection	Win $10
False alarm	Lose $10
Miss	Lose $10

What would you do if you were in Chris's position? Being smart, you analyze the payoffs and realize that the way to make money is to say "yes" more. You can lose $10 if a "yes" response results in a false alarm, but this small loss is more than counterbalanced by the $100 you can win for a hit. While you don't decide to say "yes" on every trial—after all, you want to be honest with the experimenter about whether or not you heard the tone—you do decide to stop being so conservative. *You decide to change your criterion for saying "yes."* The results of this experiment are interesting. Chris becomes a more liberal responder and says "yes" a lot more, responding with 98 percent hits and 90 percent false alarms.

This result is plotted as data point L (for "liberal" response) in Figure E.2, a plot of the percentage of hits versus the percentage of false alarms. The solid curve going through point L is called a **receiver operating characteristic (ROC) curve.** We will see why the ROC curve is important in a moment, but first let's see how we determine the other points on the curve. Determining the other points on the ROC curve is simple: All we have to do is to change the payoffs. We can make Chris raise her criterion and therefore respond more conservatively by means of the following payoffs:

Hit	Win $10
Correct rejection	Win $100
False alarm	Lose $10
Miss	Lose $10

This schedule of payoffs offers a great inducement to respond conservatively since there is a big reward for saying "no" when no tone is presented. Chris's criterion is therefore shifted to a much higher level, so Chris now returns to her conservative ways and says "yes" only if she is quite certain that a tone is presented; otherwise she says "no." The result of this new-found conservatism is a hit rate of only 10 percent and a minuscule false-alarm rate of 1 percent, indicated by point C (for "conservative" response) on the ROC curve. We should note that, although Chris hits on only 10 percent of the trials in which a tone is presented, she scores a phenomenal 99 percent correct rejections on trials in which a tone is not presented. (This result follows from the fact that, if there are 100 trials in which no tone is presented, then correct rejections + false alarms = 100. Since there was one false alarm, there must be 99 correct rejections.)

Chris, by this time, is rich and decides to go buy the Miata she's been dreaming about. (So far she's won $8,980 in the first experiment and $9,090 in the second experiment, for a total of $18,070! To be sure you understand how the payoff system works, check this calculation yourself. Remember that the signal was presented on 100 trials and was not presented on 100 trials.) However, we point out that she may need a little extra cash to have a tape deck installed in her car, so she agrees to stick around for one more experiment. We now use the following neutral schedule of payoffs:

Hit	Win $10
Correct rejection	Win $10
False alarm	Lose $10
Miss	Lose $10

and obtain point N on the ROC curve: 15 percent hits and 20 percent false alarms. Chris wins $1,000 more and becomes the proud owner of a Miata, and we are the proud owners of the world's most expensive ROC curve. (Do not, at this point, go to the psychology department in search of the nearest signal detection experiment. In real life, the payoffs are quite a bit less than in our hypothetical example.)

Chris's ROC curve shows that factors other than sensitivity to the stimulus determine the subject's response. Remember that in all of our experiments the intensity of the tone has remained constant. The only thing we have changed is the subject's criterion, but in doing this, we have succeeded in drastically changing the subject's responses.

What does the ROC curve tell us in addition to demonstrating that subjects will change how they respond to an unchanging stimulus? Remember, at the beginning of this discussion, we said that a signal detection experiment can tell us whether or not Chris and Laurie are equally sensitive to the tone. The beauty of signal detection theory is that the subject's sensitivity is indicated by the *shape* of the ROC curve, so if experiments on two subjects result in identical ROC curves, their sensitivities must be equal. (This conclusion is not obvious from our discussion so far. We will explain below why the shape of the ROC curve is related to the subject's sensitivity.) If we repeat the above experiments on Laurie, we get the following results (data points L′, N′, and C′ in Figure E.2):

Liberal payoff:

Hits = 99 percent

False alarms = 95 percent

Neutral payoff:

Hits = 92 percent

False alarms = 50 percent

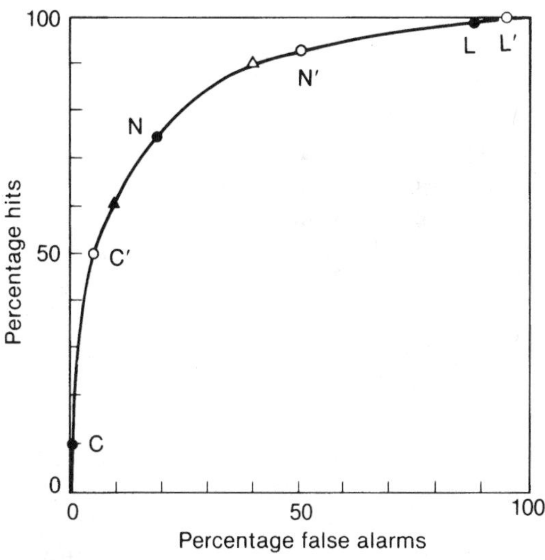

Figure E.2

A receiver operating characteristic (ROC) curve. The fact that Chris's and Laurie's data points all fall on this curve means that they have the same sensitivity to the tone. See text for details.

Conservative payoff:

Hits = 50 percent

False alarms = 6 percent

The data points for Laurie's results are shown by the open circles in Figure E.2. Note that although these points are different from Chris's, they fall on the same ROC curve as do Chris's. We have also plotted the data points for the first experiments we did on Laurie (open triangle) and Chris (filled triangle) before we introduced payoffs. These points also fall on the ROC curve.

That Chris's and Laurie's data both fall on the same ROC curve indicates their equal sensitivity to the tones, thus confirming our suspicion that the method of constant stimuli misled us into thinking that Laurie is more sensitive, when the real reason for her apparently greater sensitivity is her lower criterion for saying "yes."

Before we leave our signal detection experiment, it is important to note that signal detection procedures can be used without the elaborate payoffs that we described for Chris and Laurie. Much briefer procedures, which we will describe below, can be used to determine whether differences in the responses of different subjects are due to differences in threshold or to differences in the subjects' response criteria.

What does signal detection theory tell us about functions such as the spectral sensitivity curves in Figure 2.25 and the audibility function in Figure 9.3, which were determined by one of the classical psychophysical methods? When the classical methods are used to determine functions such as the spectral sensitivity curve and the audibility function, it is usually assumed that the subject's criterion remains constant throughout the experiment, so that the function measured is due not to changes in the subject's criterion but to changes in the wavelength or some other physical property of the stimulus. This is a good assumption, since changing the wavelength of the stimulus probably has little or no effect on factors such as motivation, which would shift the subject's criterion. Furthermore, experiments such as the one for determining the spectral sensitivity curve usually use highly practiced subjects who are trained to give stable results. Thus, even though the idea of an "absolute threshold" may not be strictly correct, classical psychophysical experiments run under well-controlled conditions have remained an important tool for measuring the relationship between stimuli and perception.

Signal Detection Theory

We will now discuss the theoretical basis for the signal detection experiments we have just described. Our purpose is to explain the theoretical bases underlying two ideas: (1) The percentage of hits and false alarms depends on the subject's criterion, and (2) a subject's sensitivity to a stimulus is indicated by the shape of the subject's ROC curve. We will begin by describing two of the key concepts of signal detection theory (SDT): signal and noise. (See Swets, 1964.)

Signal and Noise

The **signal** is the stimulus presented to the subject. Thus, in the signal detection experiment we described above, the signal is the tone. The **noise** is all the other stimuli in the environment, and since the signal is usually very faint, noise can sometimes be mistaken for the signal. Seeing what appears to be a flash of light in a completely dark room is an example of visual noise. Seeing a flash of light when there is none is what we have been calling a false alarm, which, according to signal detection theory, is caused by the noise. In the experiment we described above, hearing a tone on a trial in which no tone was presented is an example of auditory noise.

Let's now consider a typical signal detection experiment, in which a signal is presented on some trials and no signal is presented on the other trials. Signal detection theory describes this procedure not in terms of presenting a signal or no signal, but in terms of presenting signal plus noise (S + N) or noise (N). That is, the noise is always present, and on some trials, we add a signal. Either condition can result in the perceptual effect of hearing a tone. A false alarm occurs if the subject says "yes" on a noise trial, and a hit occurs if the subject says "yes" on a signal-plus-noise trial. Now that we have defined signal and noise, we introduce the idea of probability distributions for noise and signal plus noise.

Probability Distributions

Figure E.3 shows two probability distributions. The probability distribution on the left represents the probability that a given perceptual effect will be caused by noise (N), and the one on the right represents the probability that a given perceptual effect will be caused by signal plus noise (S + N). The key to understanding these distributions is to realize that the value labeled "Perceptual effect" on the horizontal axis is what the subject experiences on each trial. Thus, in an experiment in which the subject is asked to indicate whether or not a tone is present, the perceptual effect is the perceived loudness of the tone. Remember that in an SDT experiment the tone always has the same intensity. The loudness of the tone, however, can vary from trial to trial. The subject perceives different loudnesses on different trials, because of either trial-to-trial changes in attention or changes in the state of the subject's auditory system.

The probability distributions tell us what the chances are that a given loudness of tone is due to (N) or to (S + N). For example, let's assume that a subject hears a tone with a loudness of 10 on one of the trials of a signal detection experiment. By extending a vertical dashed line up from 10 on the "Perceptual effect" axis in Figure E.3, we see that the probability that a loudness of 10 is due to (S + N) is extremely low, since the distribution for (S + N) is essentially zero at this loudness. There is, however, a fairly high probability that a loudness of 10 is due to (N), since the (N) distribution is fairly high at this point.

Let's now assume that, on another trial, the subject hears a tone with a loudness of 20. The probability distributions indicate that, when the tone's loudness is 20, it is equally probable that this loudness is due to (N) or to (S + N). We can also see from Figure E.3 that a tone with a perceived loudness of 30 would have a high probability of being caused by (S + N) and only a small probability of being caused by (N).

Now that we understand the curves of Figure E.3, we can appreciate the problem confronting the subject. On each trial, she has to decide whether no tone (N) was present or whether a tone (S + N) was present. However, the overlap in the probability distributions for (N) and (S + N) makes this judgment difficult. As we

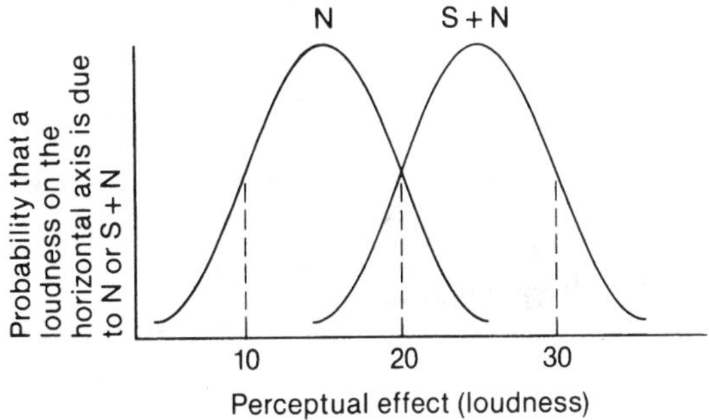

Figure E.3
Probability distributions for noise alone (N), on the left, and for signal plus noise (S + N), on the right. The probability that any given perceptual effect is caused by the noise (no signal is presented) or by the signal plus noise (signal is presented) can be determined by finding the value of the perceptual effect on the horizontal axis and extending a vertical line up from that value. The place where that line intersects the (N) and (S + N) distributions indicates the probability that the perceptual effect was caused by (N) or by (S + N).

saw above, it is equally probable that a tone with a loudness of 20 is due to (N) or to (S + N). So, on a trial in which the subject hears a tone with a loudness of 20, how does she decide whether or not the signal was presented? According to signal detection theory, the subject's decision depends on the location of her criterion.

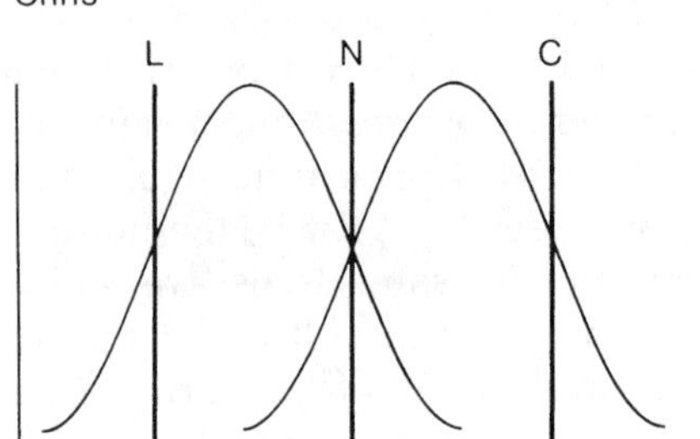

Figure E.4
The same probability distributions from Figure E.3, showing three criteria: liberal (L), neutral (N), and conservative (C). When a subject adopts a criterion, he or she uses the following decision rule: Respond "yes" ("I detect the stimulus") if the perceptual effect is greater than the criterion. Respond "no" ("I do not detect a stimulus") if the perceptual effect is less than the criterion. See text for details.

The Criterion

We can see how the criterion affects the subject's response by looking at Figure E.4. In this figure, we have labeled three different criteria: liberal (L), neutral (N), and conservative (C). Remember that we can cause subjects to adopt these different criteria by means of different payoffs. According to signal detection theory, once the subject adopts a criterion, he or she uses the following rule to decide how to respond on a given trial: If the perceptual effect is greater than (to the right of) the criterion, say, "Yes, the tone was present"; if the perceptual effect is less than (to the left of) the criterion, say, "No, the tone was not present." Let's consider how different criteria influence the subject's hits and false alarms.

Liberal Criterion To determine how criterion L will affect the subject's hits and false alarms, let's consider what happens when we present (N) and when we present (S + N):

1. Present (N): Since most of the probability distribution for (N) falls to the right of the criterion, the chances are good that presenting (N) will result in a loudness to the right of the criterion. This means that the probability of saying "yes" when (N) is presented is high; therefore, the probability of a false alarm is high.
2. Present (S + N): Since the entire probability distribution for (S + N) falls to the right of the criterion, the chances are excellent that presenting (S + N) will result in a loudness to the right of the criterion. Thus, the probability of saying "yes" when the signal is presented is high; therefore, the probability of a hit is high. Since criterion L results in high false alarms and high hits, adopting that criterion will result in point L on the ROC curve in Figure E.5.

Neutral Criterion

1. Present (N): The subject will answer "yes" only rarely when (N) is presented, since only a small portion of the (N) distribution falls to the right of the criterion. The false-alarm rate, therefore, will be fairly low.
2. Present (S + N): The subject will answer "yes" frequently when (S + N) is presented, since most of the (S + N) distribution falls to the right of the criterion. The hit rate, therefore, will be fairly high (but not as high as for the L criterion). Criterion N results in point N on the ROC curve in Figure E.5.

Conservative Criterion

1. Present (N): False alarms will be very low, since none of the (N) curve falls to the right of the criterion.
2. Present (S + N): Hits will also be low, since only a small portion of the (S + N) curve falls to the right of the criterion. Criterion C results in point C on the ROC curve in Figure E.5.

You can see that applying different criteria to the probability distributions generates the ROC curve in Figure E.5. But why are these probability distributions necessary? After all, when we described the experiment with Chris and Laurie, we determined the ROC curve simply by plotting the results of the experiment. The reason the (N) and (S + N) distributions are important is that, according to signal detection theory, the subject's sensitivity to a stimulus is indicated by the distance (d′) between the peaks of the (N) and (S + N) distributions and that this distance affects the shape of the ROC curve. We will now consider how the subject's sensitivity to a stimulus affects the shape of the ROC curve.

The Effect of Sensitivity on the ROC Curve

We can understand how the subject's sensitivity to a stimulus affects the shape of the ROC curve by considering what the probability distributions would look like for Alice, a subject with super-sensitive hearing. Alice's hearing is so good that a tone barely audible to Chris sounds very loud to Alice. If presenting (S + N) causes Alice to hear a loud tone, this means that Alice's (S + N) distribution should be far to the right, as shown in Figure E.5. In signal detection terms, we would say that Alice's high sensitivity is indicated by the large separation (d′) between the (N) and the (S + N) probability distributions. To see how this greater separation between the probability distributions will affect Alice's ROC curve, let's see how she would respond when adopting liberal, neutral, and conservative criteria.

Liberal Criterion

1. Present (N): high false alarms.
2. Present (S + N): high hits.

The liberal criterion, therefore, results in point L′ on the ROC curve of Figure E.5.

Neutral Criterion

1. Present (N): low false alarms. It is important to note that Alice's false alarms for the neutral criterion will be lower than Chris's false alarms for the neutral criterion, because only a small portion of Alice's (N) distribution falls to the right of the criterion, whereas some of Chris's (N)

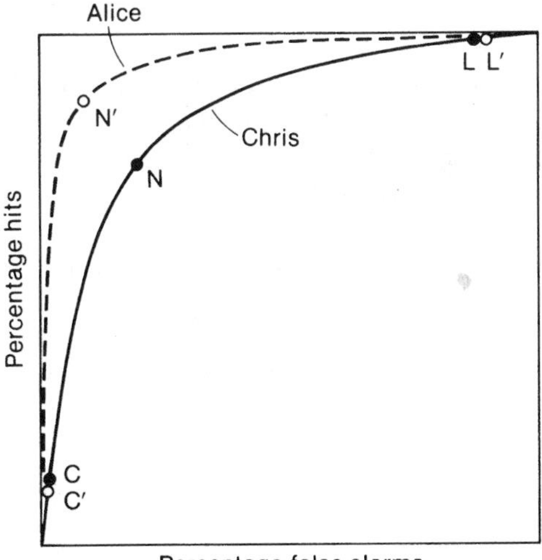

Figure E.5
ROC curves for Chris (solid) and Alice (dashed). See text for details.

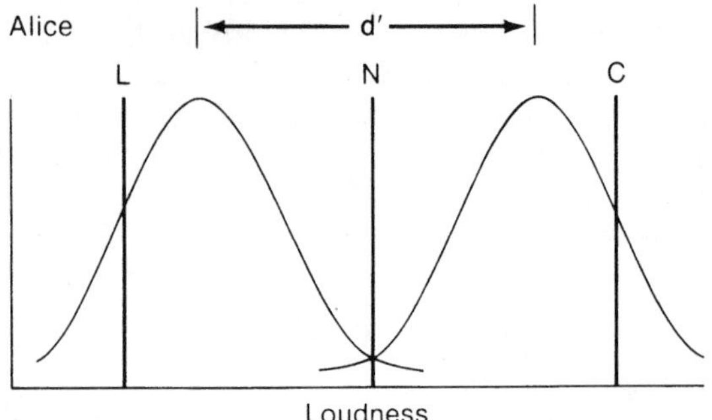

Figure E.6

Probability distributions for a subject who is extremely sensitive to the signal. The noise distribution remains the same, but the (S + N) distribution moves to the right. See text for details.

distribution falls to the right of the neutral criterion (Figure E.4).

2. Present (S + N): high hits. In this case Alice's hits will be higher than Chris's, because all of Alice's (S + N) distribution falls to the right of the neutral criterion, whereas not all of Chris's does (Figure E.4).

The neutral criterion, therefore, results in point N′ on the ROC curve in Figure E.6.

Conservative Criterion

1. Present (N): low false alarms.
2. Present (S + N): low hits.

The conservative criterion, therefore, results in point C′ on the ROC curve.

The difference between the two ROC curves in Figure E.6 is obvious, since Alice's curve is more "bowed." But before you conclude that the difference between these two ROC curves has anything to do with where we positioned Alice's L, N, and C criteria, see if you can get an ROC curve like Alice's from the two probability distributions of Figure E.4. You will find that, no matter where you position the criteria, there is no way that you can get a point like point N′ (with very high hits and very low false alarms) from the curves of Figure E.4. In order to achieve very high hits and very low false alarms, the two probability distributions must be spaced far apart, as in Figure E.5.

Thus, increasing the distance (d′) between the (N) and the (S + N) probability distributions changes the shape of the ROC curve. When the subject's sensitivity (d′) is high, the ROC curve is more bowed. In practice, d′ can be determined by comparing the experimentally determined ROC curve to standard ROC curves (see Gescheider, 1976), or d′ can be calculated from the proportions of hits and false alarms that occur in an experiment, by means of a mathematical procedure we will not discuss here. This mathematical procedure for calculating d′ enables us to determine a subject's sensitivity by determining only one data point on an ROC curve. Thus, this mathematical procedure makes it possible to use the signal detection procedure without running a large number of trials.

Glossary

The number in parentheses after each item indicates the chapter in which the term is introduced. DD stands for Developmental Dimensions; OWP stands for Other Worlds of Perception. See the index to determine other places where these terms are discussed.

Absolute threshold. See *Threshold, absolute.*

Absorption spectrum. A plot of the amount of light absorbed by a visual pigment versus the wavelength of light. (2)

Accommodation (depth cue). A possible depth cue. Muscular sensations that occur when the eye accommodates to bring objects at different distances into focus may provide information regarding the distance of that object. (6)

Accommodation (focus). The eye's ability to bring objects located at different distances into focus by changing the shape of the lens. (2)

Accretion. The uncovering of the farther of two surfaces due to observer movement. (6)

Achromatic color. Colors without hue; white, black, and all the grays between these two extremes. (4)

Acoustic cues. The sound energy associated with a particular phoneme. (10)

Acoustic signal. The pattern of frequencies and intensities of the sound stimulus. (10)

Acoustic trauma. Damage to the inner ear caused by implosive noises such as those created by explosions or machines. (13)

Across-fiber-pattern coding. The theory that sensory quality is signaled by the pattern of nerve activity across a large number of nerve fibers. (1)

Active touch. Touch in which the observer plays an active role in touching and exploring an object, usually with his or her hands. (11)

Acupuncture. A procedure in which fine needles are inserted into the skin at specific points. Twirling these needles or passing electrical current through them can cause analgesia. (11)

Adapting field. A field of light presented to adapt the receptors. (2)

Additive color mixture. See *Color mixture, additive.*

Adjustment, method of. A psychophysical method in which the experimenter or the observer slowly changes the stimulus until the observer detects the stimulus. (1)

Aerial perspective. See *Atmospheric perspective.*

Afferent fibers. Fibers conducting signals from the receptors toward the brain. (11)

Aftereffect of movement. Illusory movement of a stationary object that occurs after one views a moving inducing stimulus for 30 to 60 seconds. The spiral aftereffect and the waterfall illusion are examples of aftereffects of movement. (7)

Afterimage. An image that is perceived after the original source of stimulation is removed. A visual afterimage usually occurs after one sees a high-contrast stimulus for 30 to 60 seconds. (4)

Age-related macular degeneration. Degeneration of the macular area of the retina associated with old age. (13)

Alberti's window. Used to draw a picture in perspective. Alberti's window is a transparent surface on which an artist traces the scene viewed through the surface. (6)

Alliesthesia. "Changed sensation." The change in reaction to a stimulus, which may be positive when we first experience it but, after repeated presentations, becomes more negative. (12)

All-or-none response. The nerve impulse either fires or it doesn't. When it does fire, it has just one size no matter what the intensity of the stimulus that generated the response. (1)

Amacrine cell. A neuron that transmits signals laterally in the retina. Amacrine cells synapse with bipolar cells and ganglion cells. (2)

Ambient optic array. The way the light of the environment is structured by the presence of objects, surfaces, and textures. (6)

Amblyopia. A large reduction in the acuity in one eye. (3)

Ames room. A distorted room, first built by Adelbert Ames, that creates an erroneous perception of the sizes of people in the room. The room is constructed so that two people at the far wall of the room appear to stand at the same distance from an observer. In actuality, one of the people is much farther away than the other. (6)

Amplitude. In the case of a repeating sound wave, such as the sine wave of a pure tone, amplitude represents the pressure difference between atmospheric pressure and the maximum pressure of the wave. (8)

Analgesia. The elimination of pain without loss of consciousness. (11)

Angle of disparity. The visual angle between the images of an object on the two retinas. If the images of an object fall on corresponding points, the angle of disparity is zero. If the images fall on noncorresponding points, the angle of disparity indicates the degree of noncorrespondence. (6)

Angular expansion, rate of. The rate at which an object's visual angle expands as it gets closer to an observer. (7)

Angular size-contrast theory. An explanation of the moon illusion that states that the perceived size of the moon is determined by the sizes of the objects that surround it. According to this idea, the moon appears small when it is surrounded by large objects, such as the expanse of the sky when the moon is overhead. (6)

Anomalous trichromat. A person who needs to mix a minimum of three wavelengths to match any other wavelength in the spectrum but mixes these wavelengths in different proportions from a trichromat. (4)

Aperture problem. The ambiguity of the direction of movement that occurs when a moving stimulus is seen through an aperture small enough so that only part of the object is visible. (7)

Apex of the basilar membrane. The end of the basilar membrane farthest from the middle ear. (8)

Apparent distance theory. An explanation of the moon illusion that is based on the idea that the horizon moon, which is viewed across the filled space of the terrain, should appear farther away than the zenith moon, which is viewed through the empty space of the sky. This theory states that, since the horizon and zenith moons have the same visual angle, the farther-appearing horizon moon should appear larger. (6)

Apparent movement (or stroboscopic movement). An illusion of movement that occurs between two objects separated in space when the objects are flashed rapidly on and off, one after another, separated by a brief time interval. (5)

Applanator. The part of an applanation tonometer that is pushed against the patient's cornea to determine the intraocular pressure. (13)

Aqueous humor. The liquid in the space between the cornea and the lens of the eye. (13)

Artichoke effect. The effect in which some people taste water as sweet after eating artichokes. (12)

Articulation function. A plot of the number of words identified correctly versus the intensity of the words. (13)

Astigmatism. A condition in which vision is blurred in some orientations because of a misshapen cornea. (13)

Atmospheric perspective. A depth cue. Objects that are farther away look more blurred and bluer than objects that are closer, because we must look through more air and particles to see them. (6)

Attack. The buildup of sound at the beginning of a tone. (9)

Audibility curve. A curve that indicates the sound pressure level (SPL) at threshold for frequencies across the audible spectrum. (9)

Audiogram. A plot of the threshold for hearing pure tones versus the frequencies of the tones. (13)

Audiologist. A professional with a master's or doctoral degree who measures the hearing ability of children and adults to identify the presence and severity of any hearing problems. Audiologists also fit hearing-impaired people with hearing aids and teach them strategies for more effective communication. (13)

Audiometer. A device for measuring an audiogram. (13)

Audio-visual speech perception. A perception of speech that is affected by both auditory and visual stimulation, as when a person sees a tape of someone saying /ga/ with the sound /ba/ substituted and perceives /da/. (10)

Auditory canal. The canal through which air vibrations travel from the environment to the tympanic membrane. (8)

Auditory receiving area. The area of the cortex, located in the temporal lobe, that is the primary receiving area for hearing. (8)

Auditory response area. The area that defines the frequencies and sound pressure levels over which hearing functions. This area extends between the audibility curve and the curve for the threshold of feeling. (9)

Auditory scene analysis. The process by which listeners sort superimposed vibrations into separate sounds (9)

Auditory stream. A perceptual unit that represents a single auditory event. A stream in audition is analogous to an object in vision. (9)

Auditory stream segregation. The effect that occurs when a series of tones that differ in pitch are played so that the high- and low-pitched tones alternate rapidly, and the high and low pitches become perceptually separated into simultaneously occurring independent streams of sound. (9)

Aural rehabilitation. Training for hearing-impaired people that consists of training in speech reading and other communication strategies. (13)

Autostereogram. A stereogram that creates a three-dimensional perception within a single image. (6)

Axial myopia. See *Myopia, axial.*

Axon. The part of the neuron that conducts nerve impulses over distances. Also called the *nerve fiber.* (1)

Bandwidth. The difference between the highest and lowest frequencies in a stimulus consisting of a number of frequencies presented simultaneously, such as an auditory noise stimulus. (9)

Base of the basilar membrane. The part of the basilar membrane nearest the middle ear. (8)

Basilar membrane. A membrane that stretches the length of the cochlea and controls the vibration of the cochlear partition. (8)

Binaural cells. Neurons in the auditory system that receive inputs from both ears. (8)

Binding problem. The problem of how neural activity in many separated areas in the brain is combined to create a perception of a coherent object. (3)

Binocular depth cell. A neuron in the visual cortex that responds best to stimuli that fall on points separated by a specific degree of disparity on the two retinas. (6)

Binocular depth cue. A depth cue that requires the participation of both eyes. Binocular disparity is the major binocular depth cue. (6)

Binocular disparity. The result when the retinal images of an object fall on disparate points on the two retinas. (6)

Binocular fixation. Fixation on an object with both eyes simultaneously. (DD 6)

Binocularity. A neuron is binocular if it responds to stimulation of both the left and the right eyes. (DD 3)

Biological movement. Motion produced by biological organisms. Most of the experiments on biological motion have used walking humans with lights attached to their joints and limbs as stimuli. (7)

Bipolar cell. A neuron that is stimulated by the visual receptors and sends electrical signals to the retinal ganglion cells. (2)

Blindness. A visual acuity of 20/200 or less after correction or little peripheral vision (the legal definition of blindness). (13)

Blindsight. A situation in which a person can't see a light at a particular place in the visual field but can indicate that a stimulus is at that location by pointing or by other means. (2)

Blind spot. The small area where the optic nerve leaves the back of the eye; there are no visual receptors in this area. (2)

Blobs. Cells found in areas of the visual cortex that take up a stain that selectively colors areas that contain the enzyme cytochrome oxidase. Many of these cells have double color-opponent receptive fields. (4)

Bottom-up processing. Processing in which a person constructs a perception by first analyzing small units such as primitives. Treisman's preattentive stage of processing, in which a stimulus is analyzed into parts, is an example of bottom-up processing. (1)

Braille. A system of raised dots in which different patterns of dots stand for different letters of the alphabet. Blind people can read by feeling these dots with their fingertips. (11)

Bruce effect. An effect in which a female mouse that has just mated and is exposed to the smell of a strange male mouse within 24 hours will not become pregnant. (OWP 12)

Cataract. A lens that is clouded. (13)

Cataract, congenital. A cataract present at birth. (13)

Cataract, secondary. A cataract caused by another eye disease. (13)

Cataract, senile. A cataract due to old age. This is the most common form of cataract. (13)

Cataract, traumatic. A cataract caused by injury. (13)

Categorical perception. In speech perception, perceiving one sound at short voice onset times and another sound at longer voice onset times. The listener perceives only two categories across the whole range of voice onset times. (10)

Cell assembly. A group of neurons that fire together in response to a particular stimulus. (3)

Cell body. The part of a neuron that receives stimulation from other neurons. (1)

Center frequency. The frequency positioned midway between the lower and upper frequencies in a complex sound stimulus such as noise. (9)

Center–surround antagonism. The competition between the center and surround regions of a center–surround receptive field. (2)

Center–surround receptive field. A receptive field that consists of a roughly circular excitatory area surrounded by an inhibitory area, or a circular inhibitory center surrounded by an excitatory area. (2)

Central pitch processor. A hypothetical central mechanism that analyzes the pattern of a tone's harmonics and selects the fundamental frequency that is most likely to have been part of that pattern. (9)

Characteristic frequency. The frequency at which a neuron in the auditory system has its lowest threshold. (8)

Chorda tympani nerve. A nerve that transmits signals from receptors on the front and sides of the tongue. (12)

Chromatic adaptation. The adaptation of the eye to chromatic light. Chromatic adaptation is selective adaptation to wavelengths in a particular region of the visible spectrum. (4)

Chromatic color. Colors with hue, such as blue, yellow, red, and green. (4)

Chromatographic effect. An effect in which different molecules distribute themselves differently on the surface of the olfactory mucosa. (12)

Classical psychophysical methods. The methods of limits, adjustment, and constant stimuli, described by Fechner. (1)

Classical threshold theory. The idea that, if a mental event is to be experienced, it has to be stronger than some critical amount. (1)

Coarticulation. The overlapping articulation of different phonemes. (10)

Cochlea. The snail-shaped, liquid-filled structure that contains the structures of the inner ear, the most important of which are the basilar membrane, the tectorial membrane, and the hair cells. (8)

Cochlear implant. A device in which electrodes are inserted into the cochlea to create hearing by electrically stimulating the auditory nerve fibers. This device is used to restore hearing in people who have lost their hearing because of damaged hair cells. (13)

Cochlear nucleus. The nucleus where nerve fibers from the cochlea first synapse. (8)

Cochlear partition. A partition in the cochlea, extending almost its full length, that separates the scala tympani and the scala vestibuli. (8)

Cognitive approach to perception. The approach to perception that focuses on how perception is affected by mental processing, stimulus meaning, and the subject's knowledge and expectations. (1)

Cognitive psychology. An approach to psychology that emphasizes the study of mental processes. (1)

Coherent motion. The perception that an entire pattern is moving in one direction. In the case of plaid patterns used in movement research, coherent motion occurs even when the two plaid components are moving in different directions. (7)

Cold fiber. A nerve fiber that responds to decreases in temperature or to steady low temperatures. (11)

Color. A perceptual response to objects and lights that causes them to possess qualities such as redness, greenness, whiteness, and grayness. (4)

Color constancy. The effect in which the perception of an object's hue remains constant even when the wavelength distribution of the illumination is changed. Approximate color constancy means that our perception of hue usually changes a little when the illumination changes, though not as much as we might expect from the change in the wavelengths of light reaching the eye. (4)

Color deficiency. People with color deficiency (sometimes incorrectly called *color blindness*) see fewer colors than people with normal color vision and need to mix fewer wavelengths to match any other wavelength in the spectrum. (4)

Color-matching experiment. A procedure in which observers are asked to match the color in one field by mixing two or more lights in another field. (4)

Color mixing. Combining two or more different colors to create a new color. See *Color mixture, additive*, and *Color mixture, subtractive*. (4)

Color mixture, additive. The result when lights of different colors are superimposed. (4)

Color mixture, subtractive. The result when paints of different colors are mixed together. (4)

Columnar arrangement. The arrangement of neurons with similar properties in columns perpendicular to the surface of the cortex. For example, there are location, orientation, and ocular dominance columns in the visual system and frequency columns in the auditory system. (3)

Common fate, law of. Gestalt law: Things that are moving in the same direction appear to be grouped together. (5)

Comparator. A structure hypothesized by the corollary discharge theory of movement perception. The corollary discharge signal and the sensory movement signal meet at the comparator. (7)

Complex cell. A neuron in the visual cortex that responds best to moving bars with a particular orientation. (3)

Component cells. Neurons that respond to the direction of movement of single gratings (see *Pattern cells*). (7)

Componential recovery, principle of. A principle stating that, if an object's geons can be identified, then the object can be rapidly and correctly recognized. (5)

Computational approach. An approach to explaining object perception that treats perception as the end result of a mathematical analysis of the retinal image. (5)

Conditioned flavor aversion. Avoidance of a flavor after it has been paired with sickness. (12)

Conditioned taste aversion. An aversion to the taste of a food due to a pairing of the food's taste with sickness. (12)

Conductive hearing loss. Hearing loss that occurs when the vibrations of a sound stimulus are not conducted normally from the outer ear into the cochlea. (13)

Cone of confusion. The cone-shaped surface that is defined by all of the points in space that are the same distance from both ears. (9)

Cones. Cone-shaped receptors in the retina that are primarily responsible for vision in high levels of illumination, and for color vision and detail vision. (2)

Constant stimuli, method of. A psychophysical method in which a number of stimuli with different intensities are presented repeatedly in a random order. (1)

Constructivist approach to perception. An approach to perception that focuses on how perceptions are constructed by the mind. This approach, which emphasizes mental processing, is associated with Herman von Helmholtz and Richard Gregory, among others. (5)

Contralateral eye. The eye on the opposite side of the head from a particular structure. (2)

Contrast. For a visual grating stimulus, the contrast is the amplitude of the grating divided by its mean intensity. (5)

Contrast sensitivity. Sensitivity to the difference in the light intensities in two adjacent areas. Contrast sensitivity is usually measured by taking the reciprocal of the minimum intensity difference between two bars of a grating necessary to see the bars. (3)

Contrast sensitivity function (CSF). A plot of contrast sensitivity versus the spatial frequency of a grating stimulus. (5)

Convergence (depth cue). A possible depth cue. Muscular sensations that occur when the eyes move inward (convergence) or outward (divergence) to view objects at different distances may provide information regarding the depth of that object. (6)

Convergence (neural). The process of many neurons synapsing onto fewer neurons. (2)

Cornea. The transparent focusing element of the eye that is the first structure through which light passes as it enters the eye and that is the eye's major focusing element. (2)

Corneal disease and injury. Any disease or injury that damages the cornea, causing a loss of transparency. (13)

Corneal transplant. The replacement of a damaged piece of cornea with a piece of healthy cornea taken from a donor. (13)

Corollary discharge. A copy of the signal sent from the motor area of the brain to the eye muscles. The corollary discharge is sent not to the eye muscles, but to the hypothetical comparator of corollary discharge theory. (7)

Corollary discharge theory. According to the corollary discharge theory of motion perception, the corollary discharge signal is sent to a structure called the *comparator*, where the information in the corollary discharge is compared to the sensory movement signal. If the corollary discharge signal and the sensory movement signal do not cancel each other, movement is perceived. (7)

Correct rejection. In a signal detection experiment, saying, "No, I don't detect a stimulus" on a trial in which the stimulus is not presented (a correct response). (Appendix E)

Correspondence problem (depth). The visual system's matching of points on one image with similar points on the other image in order to determine disparity. (6)

Correspondence problem (movement). The visual system's matching of elements in one frame of a stimulus with elements in succeeding frames in order to create apparent movement. (7)

Corresponding retinal points. The points on each retina that would overlap if one retina were slid on top of the other. (6)

Cortical malleability. The ability of neurons in the visual cortex to change their properties in response to environmental changes. (DD 3)

Covering and uncovering heuristic. The "rule of thumb" that a moving object progressively covers and uncovers parts of the background. This "rule" has been applied to describing apparent motion phenomena. (7)

Critical band for loudness. The width of the frequency band for which loudness remains constant for a given intensity. (9)

Cross adaptation. In taste, flowing one substance over the tongue affects a person's perception of another substance with the same quality. For example, adaptation with one salt might raise the threshold for detecting another salt. Adaptation across qualities, such as salt versus bitter, does not usually occur. (12)

Crossed disparity. Binocular disparity in which objects are located in front of the horopter. (6)

Cryopreservation. A technique in which a donor cornea is preserved by being slowly brought down to a temperature of –190°F. (13)

Cue theory. The approach to depth perception that focuses on identifying information in the retinal image that is correlated with depth in the world. (6)

Cutaneous sensations. Sensations based on the stimulation of receptors in the skin. (11)

Cutaneous senses. The senses that serve the qualities sensed through the skin. (11)

Cutaneous sensitivity. Sensitivity of the skin to qualities such as touch, temperature, and pain. (11)

Dark adaptation. Visual adaptation that occurs in the dark, in which the sensitivity to light increases. (2)

Dark adaptation curve. The function that traces the time course of the increase in visual sensitivity that occurs during dark adaptation. (2)

Dark adapted sensitivity. The sensitivity of the eye after it has adapted to the dark. (2)

Decay. The decrease in sound at the end of a tone. (9)

Decibel (dB). A unit that indicates the presence of a tone relative to a reference pressure: $dB = 20 \log (p/p_o)$ where p is the pressure of the tone and p_o is the reference pressure. (8)

Decision rule. The rule usually adopted by subjects in signal detection experiments, which states, "Respond yes if the perceptual effect is greater than the response criterion. Respond no if the effect is less than the response criterion." (Appendix E)

Degree of disparity. If the two retinas are stimulated by images that fall on disparate (noncorresponding) points, the degree of disparity is the distance that one of the images must be moved so that both images will fall on corresponding points. (6)

Deletion. The covering of the farther of two surfaces due to observer movement. (6)

Dendrites. Nerve processes on the cell body that receive stimulation from other neurons. (1)

Deprivation experiments. Experiments in which an animal is deprived of some aspect of visual experience during rearing, followed by an assessment of the effect of the deprivation on the animal's behavior and/or physiology. (DD 3)

Depth cues. Two-dimensional information on the retina that is correlated with depth in the scene. (6)

Dermatome. The area of skin served by one dorsal root. (11)

Dermis. The inner layer of skin that contains nerve endings and receptors. (11)

Detached retina. A condition in which the retina is detached from the back of the eye. (13)

Deuteranopia. A form of red–green color dichromatism caused by lack of the middle-wavelength cone pigment. (4)

Diabetes. A condition in which the body doesn't produce enough insulin. One side effect of diabetes is a loss of vision due to diabetic retinopathy. (13)

Diabetic retinopathy. Damage to the retina that is a side effect of diabetes. This condition causes neovascularization—the formation of abnormal blood vessels that do not supply the retina with adequate oxygen and that bleed into the vitreous humor. (13)

Dichromat. A person who has a form of color deficiency. Dichromats can match any wavelength in the spectrum by mixing two other wavelengths. Deuteranopes, protanopes, and tritanopes are all dichromats. (4)

Diopter. The strength of a lens. Diopters = 1/far point in meters. See Figure 13.5. (13)

Direct perception. J. J. Gibson's idea that we pick up the information provided by invariants directly and that perceptions result from this information without the need of any further processing. (6)

Direct sound. Sound that is transmitted to the ears directly from a sound source. (9)

Disparate points. See *Noncorresponding points.*

Diurnal vision. Vision specialized to function during the day. (OWP 2)

Divergence. Rotation outward of the eyes to fixate on far objects. (2)

Doctrine of specific nerve energies. A principle stating that the brain receives environmental information from sensory nerves and that the brain distinguishes between the different senses by monitoring the activity in these sensory nerves. (1)

Dorsal root. The pathway through which fibers from the skin enter the spinal cord. (11)

Double color-opponent cell. A cell with a center–surround receptive field that responds in an opponent manner to stimulation of the field's center with a reversed opponent response to stimulation of the surround. For example, if the center response is R+G– the surround response will be R–G+. (4)

Duplex perception. The result when one stimulus causes a person to hear both speech and nonspeech sounds simultaneously. (10)

Duplicity theory of vision. The idea that the rod and cone receptors in the retina operate under different conditions and have different properties. (2)

Eardrum. Another term for the tympanic membrane, the membrane located at the end of the auditory canal that vibrates in response to sound. (8)

Earlobe. See *Pinna.*

Echolocation. Locating objects by sending out high-frequency pulses and sensing the echo created when these pulses are reflected from objects in the environment. Echolocation is used by bats and dolphins. (OWP 8)

Ecological approach. The approach to perception that emphasizes studying perception as it occurs in natural settings, particularly emphasizing the role of observer movement. (6)

Effect of the missing fundamental. See *Periodicity pitch.* (9)

Efferent fibers. Fibers carrying signals from the brain toward the periphery. (11)

Electro-olfactogram. An electrical response recorded from the pooled activity of thousands of receptors in the olfactory mucosa. (12)

Electroretinogram. An electrical response of the visual receptors that is used in diagnosing retinal degenerations. (13)

Emmert's law. A law stating that the size of an afterimage depends on the distance of the surface against which the afterimage is viewed. The farther away the surface against which an afterimage is viewed, the larger the afterimage appears. (6)

Endorphins. Chemicals that are naturally produced in the brain and that cause analgesia. (11)

End-stopped cell. A cortical neuron that responds best to lines of a specific length that are moving in a particular direction. (3)

Envelope of the traveling wave. A curve that indicates the maximum displacement at each point along the basilar membrane caused by a traveling wave. (8)

Epidermis. The outer layers of the skin, including a layer of dead skin cells. (11)

Equal loudness curve. A curve that indicates the sound pressure levels that result in a perception of the same loudness at frequencies across the audible spectrum. (9)

Equiluminance. The result when the intensities of two areas of a field are adjusted so that the fields appear equally bright. (3)

Equivalence classification. In speech perception, the ability to classify vowel sounds as belonging to the same class, even if the speakers are different. (DD 10)

Error of perseveration. In the method of limits, the tendency to continue resonding in the same way, even after the threshold has been crossed. (1)

Evoked potential. See *Visual evoked potential.*

Excitation. A condition that facilitates the generation of nerve impulses. (1)

Excitatory. Referring to the type of neurotransmitter associated with increases in the rate of nerve firing. Can also refer to neural responses that are associated with increases in firing rate. (1)

Excitatory-center–inhibitory-surround receptive field. A center–surround receptive field in which stimulation of the center area causes an excitatory response and stimulation of the surround causes an inhibitory response. (2)

Excitatory response. The response of a nerve fiber in which the firing rate increases. (2)

Exploratory procedures (EPs). People's movements of their hands and fingers while they are identifying three-dimensional objects by touch. (11)

External eye exam. Examination of the condition of the outer eye. This exam includes, among other things, examination of the reaction of the pupil to light, the color of the eye, and the alignment of the eyes. (13)

Extrastriate visual areas. Areas in the cortex that are activated by visual stimuli but are outside the striate cortex. (2)

Eye elevation hypothesis. An explanation of the moon illusion, hypothesizing that the zenith moon looks smaller than the horizon moon because we move our eyes upward to observe the zenith moon. (6)

Eye movements. Movements of the eyes that occur as an observer views a stimulus. Voluntary eye movements are rapid movements of the eyes that last 10 to 80 msecs and are separated by fixations. (5)

False alarm. In a signal detection experiment, saying, "Yes, I detect the stimulus" on a trial in which the stimulus is not presented (an incorrect response). (Appendix E)

Familiarity, law of. Gestalt law: Things are more likely to form groups if the groups appear familiar or meaningful. (5)

Familiar size. A depth cue. Our knowledge of an object's actual size sometimes influences our perception of an object's distance. (6)

Far point. The distance at which the rays from a spot of light are focused on the retina of the unaccommodated eye. For a person with normal vision, the far point is at infinity. For a person with myopic vision, the spot must be moved closer to the eye to bring the rays to a focus on the retina. (13)

Farsightedness. See *Hyperopia.*

Feature detector. A neuron that responds selectively to a specific feature of the stimulus. (3)

Feature integration theory. A sequence of steps proposed by Treisman to explain how objects are broken down into primitives and how these primitives are recombined to result in a perception of the object. (5)

Figure–ground segregation. Segregation of a pattern into a figure and a background. The figure is seen as being in front of the ground, which extends behind the figure. (5)

Fixations. Pauses made by the eye as an observer views a stimulus. During these pauses, which last fractions of a second or more, the observer takes in information about the stimulus. (5)

Flavor. The perception that occurs from the combination of taste and olfaction. (12)

Fluorescein angiography. A technique in which a fluorescent dye is injected into a person's circulation. The outline of the retinal arteries and veins produced by this dye gives information about the condition of the retinal circulation. (13)

Focused-attention stage of processing. The stage of processing in which the primitives are combined. This stage requires conscious attention. (5)

Focusing power. The degree to which a structure such as the lens or the cornea bends light. The greater the focusing power, the more the light passing through the structure is bent. (2)

Focus of expansion (F.O.E.) The point in the flow pattern caused by observer movement in which there is no expansion. According to Gibson, the focus of expansion always remains centered on the observer's destination. (7)

Formants. Horizontal bands of energy in the speech spectrogram that are associated with vowels. (10)

Formant transitions. In the speech stimulus, the rapid shifts in frequency that preceed formants. (10)

Fourier analysis. A mathematical technique that analyzes complex periodic waveforms into a number of sine-wave components. (5, 8)

Fourier spectrum. The sine-wave components that make up a periodic waveform. Fourier spectra are usually depicted by a line for each sine-wave frequency, the height of the line indicating the amount of energy at that frequency. (8)

Fourier synthesis. The process of constructing a repeating waveform by adding the sine-wave components of the waveform. (5)

Fourier's theorem. A theorem stating that any periodic waveform can be reproduced by the addition of a number of sine-wave components. (5, 8)

Fovea. A small area in the human retina that contains only cone receptors. The fovea is located on the line of sight, so that when a person looks at an object, its image falls on the fovea. (2)

Free nerve endings. Nerve endings in the skin without an end receptor. (11)

Frequency. In the case of a sound wave that repeats itself like the sine wave of a pure tone, frequency is the number of times per second that the wave repeats itself. (8)

Frequency sweep detector. A neuron in the auditory cortex that fires only when frequencies are smoothly increased or decreased. (8)

Frequency theory of pitch perception. A theory of pitch perception that says that pitch is signaled by the frequency of nerve firing. An early version of frequency theory, proposed by Rutherford, hypothesized that the basilar membrane vibrates as a whole and that the rate of this vibration matches the frequency of the stimulus. (8)

Frontal operculum cortex. An area in the frontal lobe of the cortex that receives signals from the taste system. (12)

Fundamental frequency. Usually the lowest frequency in the Fourier spectrum of a complex tone. The tone's other components, called *harmonics*, have frequencies that are multiples of the fundamental frequency. (8)

Ganglion cell. A neuron in the retina that receives inputs from bipolar and amacrine cells. The axons of the ganglion cells are the fibers that travel toward the lateral geniculate nucleus of the thalamus in the optic nerve. (2)

Gas chromatograph. A device that accurately measures the concentration of the vapor given off by a chemical stimulus. (12)

Gate control theory. Melzak and Wall's idea that our perception of pain is controlled by a neural circuit that takes into account the relative amount of activity in large (L) fibers and small (S) fibers. (11)

General auditory mechanism. A proposed mechanism for speech perception. The proposal is that we perceive speech because the speech signal activates the same auditory mechanisms responsible for our perception of other types of sound stimuli. (10)

Geon. "Geometric ion"; Volumetric primitives proposed by Biederman. (5)

Gestalt psychology. A school of psychology that has focused on developing principles of perceptual organization, proposing that "the whole is different from the sum of its parts." (5)

Glaucoma. A disease of the eye that usually results in an increase in intraocular pressure. (13)

Glaucoma, closed-angle. A rare form of glaucoma in which the iris is pushed up so that it closes the angle between the iris and the cornea and blocks the area through which the aqueous humor normally drains out of the eye. (13)

Glaucoma, open-angle. A form of glaucoma in which the area through which the aqueous humor normally drains out of the eye is blocked. In this form of glaucoma, the iris remains in its normal position so that the angle between the iris and the cornea remains open. (13)

Glossopharyngeal nerve. A nerve that transmits signals from receptors located at the back of the tongue. (12)

Good continuation, law of. Gestalt law: Points that, when connected, result in straight or smoothly curving lines are seen as belonging together, and lines tend to be seen in such a way as to follow the smoothest path. (5)

Good figure, law of. Gestalt law: Every stimulus pattern is seen so that the resulting structure is as simple as possible. (5)

Grating. A stimulus pattern consisting of alternating bars with different lightnesses or colors. (3)

Habituation. The result when the same stimulus is presented repeatedly. For example, infants look at a stimulus less and less on each succeeding trial. (DD1)

Hair cells. Small hairs, or cilia, that are the receptors for hearing. There are two kinds of hair cells: inner and outer (8)

Haptic perception. The perception of three-dimensional objects by touch. (11)

Harmonics. Fourier components of a complex tone with frequencies that are multiples of the fundamental frequency. (8)

Hearing handicap. The disadvantage that a hearing impairment causes in a person's ability to communicate or in the person's daily living. (13)

Hearing impairment. A deviation or change for the worse in either the structure or the functioning of the auditory system (see *Hearing handicap*). (13)

Height in the field of view. A depth cue. Objects that rest on a surface below the horizon and are higher in the field of view are usually seen as being more distant. (6)

Hereditary retinal degeneration. A degeneration of the retina that is inherited. Retinitis pigmentosa is an example of a hereditary retinal degeneration. (13)

Hermann grid. A grid of black squares separated by white areas that causes an illusion in which there appear to be gray spots at the intersections of the white areas. It is thought that this effect is due to lateral inhibition. (2)

Hertz (Hz). The unit for designating the frequency of a tone. One Hertz equals one cycle per second. (8)

Hit. In a signal detection experiment, saying, "Yes, I detect a stimulus" on a trial in which the stimulus is present (a correct response). (Appendix E)

Homunculus. "Little man," a term referring to the map of the body in the somatosensory cortex. (11)

Horizon ratio. The proportion of an object that is above the horizon divided by the proportion that is below the horizon. (6)

Horizon ratio principle. A principle stating that, if a person is standing on flat terrain, a point on an object that intersects the horizon will be one eye-height above the ground, and that if two objects that are in contact with the ground are the same size, the proportions of the objects above and below the horizon will be the same. (6)

Horizontal cell. A neuron that transmits signals laterally across the retina. Horizontal cells synapse with receptors and bipolar cells. (2)

Horopter. An imaginary surface that passes through the point of fixation. Objects falling on this surface result in images that fall on corresponding points on the two retinas. (6)

Hue. The experience of a chromatic color such as red, green, yellow, or blue or combinations of these colors. (4)

Hypercolumn. A column of cortex about 1 mm on a side that contains a location column for a particular area of the retina, left and right ocular dominance columns, and a complete set of orientation columns. A hypercolumn can be thought of as a processing module for a particular location on the retina. (3)

Hyperopia (farsightedness). The inability to see near objects clearly because the focus point for parallel rays of light is behind the retina. (13)

Hypothesis testing. The idea that sensory stimulation provides data for hypotheses about the world. According to this idea, perceiving involves testing different hypotheses about what is causing the stimulation. (5)

Illumination contour. A contour created at the border between two areas with different light intensities. (4)

Illusion. A situation in which an observer's perception of a stimulus does not correspond to the physical properties of the stimulus. For example, in the Muller–Lyer illusion, two lines of equal length are perceived to be of different lengths. (12)

Illusions of movement. The perception of movement in situations in which there is actually no movement in the physical stimulus. Examples of illusions of movement are the waterfall illusion, induced movement, and apparent movement. (7)

Illusory conjunctions. Illusory combinations of primitives that are perceived when stimuli containing a number of primitives are presented briefly. (5)

Illusory conjunction technique. A technique developed by Treisman showing that primitives exist independently of one another in the preattentive stage of processing. (5)

Illusory contours. Contours that are perceived even though they are not present in the physical stimulus. (3)

Impossible object. An "object" that can be represented by a two-dimensional picture but cannot exist in three-dimensional space. (5)

Incus. The second of the three ossicles of the middle ear. It transmits vibrations from the malleus to the stapes. (8)

Indirect sound. Sound that reaches the ears after being reflected from a surface such as a room's walls. (9)

Induced movement. The illusory movement of one object that is caused by the movement of another object that is nearby. (7)

Inducing stimulus. A moving stimulus, such as a waterfall or a moving belt, that, after being viewed, causes an aftereffect of movement. (7)

Inertia heuristic. The "rule of thumb" that objects in motion continue along the same path. This "rule" has been applied in describing apparent motion. (7)

Inferior colliculus. A nucleus in the hearing system along the pathway from the cochlea to the auditory cortex. The inferior colliculus receives inputs from the superior olivary nucleus. (8)

Infrasound. Vibrations below 10 Hertz. (OWP 1)

Inhibition. A condition that decreases the likelihood that nerve impulses will be generated. (1)

Inhibitory-center–excitatory-surround receptive field. A center–surround receptive field in which stimulation of the center causes an inhibitory response and stimulation of the surround causes an excitatory response. (2)

Inhibitory response. The response of a nerve fiber in which the firing rate decreases. (1)

Inner ear. The innermost division of the ear, containing the cochlea and the receptors for hearing. (8)

Insula. An area in the frontal lobe of the cortex that receives signals from the taste system. (12)

Interaural intensity difference. The greater intensity of a sound at the closer ear when a sound source is positioned closer to one ear than to the other. This effect is most pronounced for high-frequency tones. (8)

Interaural time difference. The effect that, when a sound source is positioned closer to one ear than to the other, the sound reaches the close ear slightly before reaching the far ear. (8)

Interaural time difference detector. A neuron that fires only when a stimulus is presented first to one ear and then to the other, with a specific delay between the stimulation of the two ears. (8)

Interleaved melodies. The stimulus produced when the notes of two different melodies are presented alternately. (9)

Interocular transfer. The aftereffect in one eye when an adaptation stimulus is presented to the other eye. (DD 3)

Interstimulus interval (ISI). The time interval between two flashes of light in an apparent movement display. (7)

Intraocular lens. A plastic or silicone lens that is inserted into the eye after the removal of a cataract. This lens partially compensates for the loss of focusing power caused by removal of the patient's lens. (13)

Intraocular pressure. Pressure inside the eyeball. (13)

Invariant acoustic cues. In speech perception, aspects of an auditory signal that remain constant even in different contexts. (10)

Invariant information. Environmental properties that do not change as the observer moves. For example, the spacing, or texture, of the elements in a texture gradient does not change as the observer moves on the gradient. The texture of the gradient therefore supplies invariant information for depth perception. (7)

Ions. Charged molecules found floating in the water that surrounds nerve fibers. (1)

Ipsilateral eye. The eye on the same side of the head of the structure to which the eye sends inputs. (2)

Iridectomy. A procedure used to treat closed-angle glaucoma, in which a small hole is cut in the iris. This hole opens a channel through which aqueous humor can flow out of the eye. (13)

Ishihara plate. A display made up of colored dots used to test for the presence of color deficiency. The dots are colored so that people with normal (trichromatic) color vision can perceive numbers in the plate, but people with color deficiency cannot perceive these numbers or perceive different numbers from someone with trichromatic vision. (4)

Just noticeable difference (JND). The smallest difference in intensity that results in a noticeable difference between two stimuli. (1)

Kinematogram, random-dot. See *Random-dot kinematogram.*

Kinesthesis. The sense that enables us to feel the motions and positions of the limbs and body. (11)

Kinetic depth effect. The resulting effect when a stimulus's three-dimensional structure becomes apparent from viewing a two-dimensional image of the stimulus as it rotates. (7)

Krause end bulb. A receptor in the skin. (11)

Labeled lines. An application of specificity coding to taste that suggests that different taste qualities are signaled by activity in specific nerve fibers. (12)

Large-diameter fiber (L-fiber). According to gate control theory, activity in L-fibers closes the gate control mechanism and therefore decreases the perception of pain. (11)

Laser photocoagulation. A procedure in which a laser beam is aimed at blood vessels that are leaking because of neovascularization. This laser beam photocoagulates—seals off—the blood vessels and stops the leaking. (13)

Laser photorefractive keratotomy. A surgical procedure in which an excimer laser is used to change the shape of the cornea to improve the vision of people with myopia or hyperopia. (13)

Lateral geniculate nucleus (LGN). The nucleus in the thalamus that receives inputs from the optic nerve and sends fibers to the cortical receiving area for vision. (2)

Lateral inhibition. Inhibition that is spread laterally across a nerve circuit. In the retina, lateral inhibition is spread by the horizontal and amacrine cells. (2)

Lateral plexus. A structure that transmits nerve impulses laterally in the *Limulus* eye. (2)

Lateral postero-orbital frontal cortex (LPOF). An area on the underside of the frontal lobe of the cortex that may be involved in processing olfactory information. (12)

Law of size constancy. A principle stating that we correctly perceive an object's physical size no matter what its distance from us or the size of its image on the retina. This "law" is generally true when there is good depth information. (6)

Law of visual angle. A principle stating that our perception of an object's size is determined solely by its visual angle. This "law" generally holds when there is little or no depth information. (6)

Laws of organization. A number of rules developed by the Gestalt psychologists that describe how small elements are grouped into larger configurations. (5)

Lens. The transparent focusing element of the eye through which light passes after passing through the cornea and the aqueous humor. The lens's change in shape to focus at different distances is called *accommodation*. (2)

Light adaptation. Visual adaptation that occurs in the light, in which the sensitivity to light decreases. (2)

Light adapted sensitivity. The sensitivity of the light adapted eye. (2)

Lightness. Perception of reflectance. The perception of lightness is usually associated with the achromatic colors: white, gray, and black. (4)

Lightness constancy. The constancy of our perception of an object's lightness under different intensities of illumination. (4)

Likelihood principle. A principle proposed by Helmholtz stating that we will perceive the object that is most likely to be the cause of our sensory stimulation. (5)

Limits, method of. A psychophysical method in which the experimenter presents stimuli in alternating ascending and descending series. (1)

***Limulus*.** A primitive animal, also called the horseshoe crab, which has large visual receptors that make it especially well suited to the study of the electrophysiology of vision. (2)

Linear circuit. A circuit in which one neuron synapses with another, with no convergence. (2)

Linear perspective (depth cue). The visual effect that parallel lines (like railroad tracks) converge as they get farther away. This convergence of parallel lines is a depth cue, with greater convergence indicating greater distance. (6)

Linear perspective (drawing system). A method of representing three-dimensional space on a two-dimensional surface. (6)

Local depth information. Information at a localized place on a figure that indicates depth. (5)

Localization. In audition, the capacity to locate a sound in space. (8)

Localization of function. The principle that specific areas of the brain serve specific functions. (1)

Location column. A column in the visual cortex that contains neurons with the same receptive field locations on the retina. (3)

Locomotor flow line. The flow line that passes directly under a moving observer. (7)

Long-wavelength pigment. A cone visual pigment that absorbs light maximally at the long-wavelength end of the spectrum. In humans, this pigment absorbs maximally at 558 nm. (2)

Loudness. The quality of sound that ranges from soft to loud. For a tone of a particular frequency, loudness usually increases with increasing decibels. (8)

Loudness recruitment. An abnormally rapid increase in loudness with increases in intensity, usually associated with hearing impairment. (9)

Mach bands. A perceptual effect that causes a thin dark band on the dark side of a light–dark border and a thin light band on the light side of the border. (2)

Macrosmatic. Having a keen sense of smell that is important to an animal's survival. (12)

Macula. An area about 5 mm in diameter, that surrounds and includes the fovea. (13)

Macular degeneration. A degeneration of the macula area of the retina. (13)

Macular degeneration, senile. The most common form of macular degeneration, occurring in older people. (13)

Magnification factor. The apportioning of proportionally more space on the cortex to the representation of specific areas of sensory receptors. For example, a small area on the retina in or near the fovea receives more space on the cortex than the same area of peripheral retina. Similarly, the fingertips receive more space on the somatosensory cortex than the forearm or leg. (3)

Magnitude estimation. A psychophysical method in which the subject assigns numbers to a stimulus that are proportional to the subjective magnitude of the stimulus. (1)

Magnocellular (or magno). Neurons in layers 1 and 2 of the lateral geniculate nucleus that receive inputs from the M ganglion cells. (2)

Malleus. The first of the ossicles of the middle ear. Receives vibrations from the tympanic membrane and transmits these vibrations to the incus. (8)

Manner of articulation. The mechanical means by which consonants are produced and the way air is pushed through openings in the vocal tract. (10)

Masking. The presence of a sound that reduces a person's ability to hear another sound. (9).

Mechanoreceptor fibers. Fibers that respond to mechanical displacements of the skin. See *Rapidly adapting fiber* and *Slowly adapting fiber*. (11)

Medial geniculate nucleus. A nucleus in the auditory system along the pathway from the cochlea to the auditory cortex. The medial geniculate nucleus receives inputs from the inferior colliculus. (8)

Medial lemniscal pathway. A pathway in the spinal cord that transmits signals from the skin toward the thalamus. (11)

Medium-wavelength pigment. A cone visual pigment that absorbs light maximally in the middle of the spectrum. In humans, this pigment absorbs maximally at 531 nm. (2)

Meissner corpuscle. A receptor in the skin, associated with RA I mechanoreceptors, that responds best to taps on the skin. (11)

Mel. The psychological unit of pitch. A 1,000-Hz tone has a pitch of 1,000 mels. (9)

Melodic channeling. See *Scale illusion*.

Melody schema. A representation of a familiar melody that is stored in a person's memory. (9)

Memory color. The idea that an object's characteristic color influences our perception of that object's color. (4)

Meniere's disease. A form of sensorineural hearing loss caused by an excessive buildup of the liquid that fills the cochlea and the semicircular canals. (13)

Menstrual synchrony. The effect that women who live together often have menstrual periods that begin at approximately the same time. (12)

Merkel disk receptors. Receptors in the skin, associated with SA I mechanoreceptors, that respond best to light pressure and are sensitive to details. (11)

Merkel receptor. A disk-shaped receptor in the skin associated with slowly adapting fibers, small receptive fields, and the perception of pressure. (11)

Mesencephalicus lateralus dorsalis (MLD). A nucleus in the owl that contains cells that respond only when the sound source is located in a particular area of space relative to the owl's head. (8)

Metamers. Two lights that have different wavelength distributions but are perceptually identical. (4)

Method of adjustment. See *Adjustment, method of*.

Method of constant stimuli. See *Constant stimuli, method of.*

Method of limits. See *Limits, method of.*

Microelectrode. A thin piece of wire or glass that is small enough to record electrical signals from single nerve fibers. (1)

Microneurography. A procedure for recording the activity of single neurons in the skin of awake humans. (11)

Microsmatic. Having a weak sense of smell that is not crucial to an animal's survival. (12)

Microspectrophotometry. A procedure for determining pigment absorption spectra that involves shining light through single receptors or through small numbers of receptors. (2)

Microstimulation. The stimulation of neurons by passing electrical current through the tip of a microelectrode. (3)

Middle ear. The small air-filled space between the auditory canal and the cochlea that contains the ossicles. (8)

Middle-ear muscles. Muscles attached to the ossicles in the middle ear. The smallest skeletal muscles in the body, they contract in response to very intense sounds and dampen the vibration of the ossicles. (8)

Mind–body problem. The problem of how physical processes such as nerve impulses cause mental processes such as perceptual experience. (1)

Misapplied size constancy. A principle, proposed by Gregory, that when mechanisms that help maintain size constancy in the three-dimensional world are applied to two-dimensional pictures, an illusion of size sometimes results. (6)

Miss. In a signal detection experiment, saying, "No, I don't detect a stimulus" on a trial in which the stimulus is present (an incorrect response). (Appendix E)

Missing fundamental, effect of. See *Periodicity pitch.*

Monochromat. A person who is completely colorblind and therefore sees everything as black, white, or shades of gray. A monochromat can match any wavelength in the spectrum by adjusting the intensity of any other wavelength. (4)

Monocular deprivation. Rearing an animal with one eye occluded so that the animal receives visual input through only the unoccluded eye. (DD 3)

Monocular depth cues. Depth cues, such as overlap, relative size, relative height, familiar size, linear perspective, movement parallax, and accommodation that work if we use only one eye. (6)

Moon illusion. An illusion in which the moon appears to be larger when it is on or near the horizon than when it is high in the sky. (6)

Mossbauer technique. A technique that has been used to determine how the basilar membrane vibrates. This technique involves placing on the basilar membrane a small radioactive source that emits gamma radiation. (8)

Motion agnosia. A condition in which the ability to perceive motion is disrupted. (3)

Motion parallax. A depth cue. As an observer moves, nearby objects appear to move rapidly whereas far objects appear to move slowly. (6)

Motion-produced cues. Cues that depend on the movement of the observer, or on the movement of objects in the environment. (6)

Motor theory of speech perception. The theory of speech perception that proposes that there is a close link between the perception of speech and movements of the vocal tract that occur during the production of speech. This theory also proposes that speech is perceived through the operation of a mechanism that is specialized for the perception of speech. (10)

Movement aftereffect. An illusion of movement that occurs after a person views an inducing stimulus such as a waterfall. (7)

Muller–Lyer illusion. An illusion consisting of two lines of equal length that appear to be different lengths because of the addition of "fins" to the ends of the lines. (6)

Multimodal neurons. Neurons that respond to stimulation by more than one sense modality. (8)

Multiple-joint neurons. Kinesthetic neurons that respond best to the simultaneous flexion of a number of joints. (11)

Muscle spindle receptors. Slowly adapting mechanoreceptors that are entwined around specialized muscle fibers. They are sensitive to changes in the muscle's length and so provide information about limb position. (11)

Myopia (nearsightedness). The inability to see distant objects clearly because parallel rays of light are brought to a focus in front of the retina. (13)

Myopia, axial. Myopia caused by the fact that the eyeball is too long. (13)

Myopia, refractive. Myopia caused by the fact that the cornea and the lens bend light too much (they have too much focusing power). (13)

Naloxone. A substance that inhibits the activity of opiates. It is hypothesized that naloxone also inhibits the activity of endorphins. (11)

Nasal pharynx. A passageway that connects the mouth cavity and the nasal cavity. (12)

Near point. The distance at which the lens can no longer accommodate to bring close objects into focus. Objects nearer than the near point can be brought into focus only by corrective lenses. (2)

Neovascularization. The formation of abnormal small blood vessels that occurs in patients with diabetic retinopathy. (13)

Nerve. A group of nerve fibers traveling together. (1)

Nerve axon. See *Nerve fiber*.

Nerve fiber. In most sensory neurons, the long part of the neuron that transmits electrical impulses from one point to another. (1)

Nerve impulse. A rapid change in electrical potential that travels down a nerve fiber. (1)

Neural circuit. A number of neurons that are connected by synapses. (2)

Neurogenesis. The cycle of birth, development, and death of a neuron. This process occurs for the receptors for olfaction and taste. (12)

Neuron. A cell in the nervous system that generates and transmits electrical impulses. (1)

Neurotransmitter. A chemical stored in synaptic vesicles that is released in response to a nerve impulse and has an excitatory or inhibitory effect on another neuron. (1)

Neutral point. The wavelength at which a dichromat perceives gray. (4)

Nociceptor. A fiber that responds to stimuli that are damaging to the skin. (11)

Nocturnal vision. Vision specialized to function at night. (2)

Noise. All stimuli in the environment other than the signal. Noise can also be generated within a person's nervous system. The subject's perception of noise in a signal detection experiment sometimes causes the subject to think mistakenly that a signal has been presented. (Appendix E)

Noise-induced hearing loss. A form of sensorineural hearing loss that occurs when loud noises cause degeneration of the hair cells. (13)

Noncorresponding (disparate) points. Two points, one on each retina, that would not overlap if the retinas were slid onto each other. (6)

Nontasters. People who cannot taste the compound phenylthiocarbamide (PTC). (12)

Normalization. The process in speech perception that takes differences between speakers into account. (10)

Nuclei. Small areas in the nervous system at which many synapses occur. (1)

Nucleus of the solitary tract (NST). The nucleus in the brainstem that receives signals from the tongue, the mouth, and the larynx transmitted by the chorda tympani, glossopharyngeal, and vagus nerves. (12)

Observer-based psychoacoustic procedure. A psychophysical procedure in which an observer determines whether an infant has heard a sound from observing the infant's behavior. (DD 9)

Occipital lobe. A lobe at the back of the cortex that is the site of the cortical receiving area for vision. (1)

Octave. Tones that have frequencies that are binary multiples of each other (x2, x4, etc.). For example, an 800-Hz tone is one octave above a 400-Hz tone. (9)

Octave generalization. An effect observed in an experiment by Deutsch in which two tones separated by an interval of an octave had perceptual effects similar to the effects of two identical tones. These similar effects are caused by the perceptual similarity of tones separated by octaves. (9)

Ocular dominance. The degree to which a neuron is influenced by stimulation of each eye. A neuron has a large amount of ocular dominance if it responds only to stimulation of one eye. There is no ocular dominance if the neuron responds equally to stimulation of both eyes. (3)

Ocular dominance column. A column in the visual cortex that contains neurons with the same ocular dominance. (3)

Ocular dominance histogram. A histogram that indicates the degree of ocular dominance of a large population of neurons. (3)

Oculomotor cues. Depth cues that depend on our ability to sense the position of our eyes and the tension in our eye muscles. (6)

Odor prism. Henning's system for classifying odors. (12)

Odor profiling. A technique for describing an olfactory stimulus, in which trained observers rate components in odors by comparing them to reference standards. (12)

Ohm's acoustic law. A law formulated by G. S. Ohm stating that the ear is an analyzing instrument that breaks tones down into simple components. (8)

Olfactometer. A device that presents olfactory stimuli with great precision. (12)

Olfactory binding proteins. Proteins contained in the olfactory mucosa that are secreted into the nasal cavity and bind to olfactory stimuli to transport them to active sites on the olfactory receptors. (12)

Olfactory bulb. The structure that receives signals directly from the olfactory receptors. (12)

Olfactory cortex. A small area under the temporal lobe of the cortex that receives signals that originate in the olfactory receptors. (12)

Olfactory mucosa. The region inside the nose that contains the receptors for the sense of smell. (12)

Ommatidium. A structure in the eye of the *Limulus* that contains a small lens, located directly over a visual receptor. The *Limulus* eye is made up of hundreds of these ommatidia. (2)

On response. The response of a nerve fiber in which there is an increase in the firing rate when the stimulus is turned on; the same as an excitatory response. (2)

Ophthalmologist. A person who has specialized in the medical treatment of the eye by completing four or more years of training after receiving the M.D. degree. (13)

Ophthalmoscope. A device that enables an examiner to see the retina and the retinal circulation inside the eye. (13)

Ophthalmoscopy. The use of an ophthalmoscope to visualize the retina and the retinal circulation. (13)

Opponent cell. A neuron that has an excitatory response to wavelengths in one part of the spectrum and an inhibitory response to wavelengths in the other part of the spectrum. (4)

Opponent-process theory of color vision. A theory stating that our perception of color is determined by the activity of two opponent mechanisms: a blue–yellow mechanism and a red–green mechanism. The responses to the two colors in each mechanism oppose each other, one being an excitatory response and the other an inhibitory response. (This theory also includes a black–white mechanism, which is concerned with the perception of brightness.) (4)

Opsin. The protein part of the visual pigment molecule, to which the light-sensitive retinal molecule is attached. (2)

Optic array. See *Ambient optic array.*

Optic disk. The disk-shaped area at the back of the eye where the optic nerve leaves the eye. (2)

Optic flow pattern. The flow pattern that occurs when an observer moves relative to the environment. Forward movement causes an expanding optic flow pattern, whereas backward movement causes a contracting optic flow pattern. The term *optic flow field* is used by some researchers to refer to this flow pattern. (7)

Optician. A person who is trained to fit glasses and, in some cases, contact lenses. (13)

Optometrist. A person who has received the doctor of optometry (O.D.) degree by completing four years of postgraduate study in optometry school. (13)

Orbitofrontal cortex. An area in the frontal lobe, near the eyes, that receives signals originating in the olfactory receptors. (12)

Organization, laws of. See *Laws of organization.*

Organ of Corti. The major structure of the cochlear partition, containing the basilar membrane, the tectorial membrane, and the receptors for hearing. (8)

Orientation. The angle of a stimulus relative to vertical. (5)

Orientation column. A column in the visual cortex that contains neurons with the same orientation preference. (3)

Orientation tuning curve. A function relating the firing rate of a neuron to the orientation of the stimulus. (3)

Oscillator. An electronic device used to produce pure tones. (8)

Ossicles. Three small bones in the middle ear that transmit vibrations from the outer to the inner ear. (8)

Otitis media. An infection of the middle ear. (13)

Otologist. An otorhinolaryngologist whose practice is limited to problems involving the auditory and vestibular systems. (13)

Otorhinolaryngologist. A medical doctor who has specialized in the treatment of diseases and disorders affecting the ear, nose, and throat. More commonly known as an ENT (ear, nose, and throat) specialist. (13)

Otosclerosis. A hereditary condition in which there is a growth of bone in the middle ear. (13)

Otoscope. A device used to see the tympanic membrane. (13)

Outer ear. The pinna and the external auditory meatus. (8)

Oval window. A small membrane-covered hole in the cochlea that receives vibrations from the stapes. (8)

Overlap. A depth cue. If object A covers object B, then object A is seen as being in front of object B. (6)

Pacinian corpuscle. A receptor with a distinctive elliptical shape associated with RA II mechanoreceptors. It transmits pressure to the nerve fiber inside it only at the beginning or end of the pressure stimulus. (11)

Panretinal photocoagulation. A procedure in which a laser is used to create many small burns on the retina. This procedure has been successful in treating the neovascularization associated with diabetic retinopathy. (13)

Papillae. Ridges and valleys on the tongue, some of which contain taste buds. There are four types of papillae: filiform, fungiform, foliate, and circumvallate. (12)

Parietal lobe. A lobe at the top of the cortex that is the site of the cortical receiving area for touch. (2)

Parietal pathway. A visual pathway from the occipital to the parietal lobe that processes information about where an object is. (3)

Parvocellular (or parvo). Neurons in layers 3, 4, 5, and 6 of the lateral geniculate. These neurons receive inputs from the P ganglion cells. (2)

Passive touch. The presentation of a tactile stimulus to stationary skin. (11)

Pattern cells. Neurons that respond to the overall direction of the movement of a pattern. (7)

Payoffs. A system of rewards and punishments used to influence a subject's motivation in a signal detection experiment. (Appendix E)

Perceptual constancy. The perception of a particular stimulus property, such as size, shape, or color, as remaining the same even when the conditions of stimulation are changed. (10)

Perceptual grouping. The process of grouping or organizing small elements into larger configurations; essentially the same thing as perceptual organization. (5)

Perceptual magnet effect. The effect when people judge a particular sound to be more similar to a sound that is a prototype than to a sound that is a nonprototype. (OWP 10)

Perceptual organization. The perceptual grouping of small units into larger forms. (5)

Perceptual processing. Mental or neural processing that occurs during the process of perception. (5)

Perceptual segregation. Perceptual organization in which one object is seen as separate from other objects. (4)

Periodicity. The repetition of a sound wave's pattern. (8)

Periodicity pitch. The effect in which a complex tone's pitch remains the same even if we eliminate the fundamental frequency. This is also called the *effect of the missing fundamental*. (9)

Peripheral retina. All of the retina except the fovea and a small area surrounding the fovea. (2)

Permeability. A property of a membrane that refers to the ability of molecules to pass through the membrane. If the permeability to a molecule is high, the molecule can easily pass through the membrane. (1)

Phacoemulsification. A technique for removing a cataract by breaking up the lens with ultrasonic vibrations and then sucking the pieces of lens out of the eye through a hollow needle. (13)

Phantom limb. A person's continued perception of a limb, such as an arm or a leg, even though that limb has been amputated. (11)

Phase. The position of a grating stimulus relative to a landmark. (5)

Phase locking. Auditory neurons' firing in synchrony with the phase of an auditory stimulus. (8)

Phenomenal identity. A stroboscopic movement effect that occurs when a group of lights appears to move as a whole to a new position. (7)

Phenomenological observations. A method in which the subject describes his or her perceptions (or "experience"). (4)

Phenylthiocarbamide (PTC). A substance that some people (tasters) can taste and others (nontasters) can't. (12)

Pheromone. A chemical released by an animal that causes specific physiological or behavioral reactions in another animal of the same species. (12)

Phoneme. The shortest segment of speech that, if changed, would change the meaning of a word. (10)

Phonemic restoration effect. An effect that occurs in speech perception when listeners perceive a phoneme in a word even though the acoustic signal of that phoneme is obscured by another sound, such as white noise or a cough. (10)

Phonetic boundary. The voice onset time when perception changes from one speech category to another in a categorical perception experiment. (10)

Phonetic features. Physical movements of the vocal tract that accompany the production of a phoneme. (10)

Phonetic module. A proposed mechanism that creates the special speech mode. (10)

Physical size. The size of an object as measured by a device such as a ruler or tape measure. (6)

Physiological approach to perception. As used in this book, this term refers to explanations of perceptual processes based on the relationship between physiological processes and perception. Related to the physiological approach to perception is the study of "pure" physiology, in which physiological processes of sensory systems are studied but are not directly related to perception. (1)

Pictorial cues. Depth cues, such as overlap, relative height, and relative size, that can be depicted in pictures. (6)

Pigment bleaching. The process that begins when a visual pigment molecule absorbs light. The molecule changes shape, and the color of the rod visual pigment changes from red to transparent. Sometime, early in this process, visual transduction takes place. (2)

Pigment epithelium. A layer of cells that lines the inside of the eyeball under the retina. (2)

Pigment regeneration. The reconstruction of the visual pigment molecule from its bleached state to its original unbleached state. (2)

Pinna. The part of the ear that is visible on the outside of the head. (8)

Pitch. The quality of sound, ranging from low to high, that is most closely associated with the frequency of a tone. (8)

Pit organs. Small heat-sensitive organs near the snake's mouth. (OWP 1)

Placebo. A substance that a person believes will relieve symptoms such as pain but that contains no chemicals that actually act on these symptoms. (11)

Place code for frequency. The idea that the frequency of a tone is signaled by the place in the auditory system that is maximally stimulated. (8)

Place of articulation. The place where the airstream is obstructed during the production of a sound. (10)

Ponzo illusion. An illusion of size in which two rectangles of equal length that are drawn between two converging lines appear to be different in length. Also called the *railroad track illusion.* (6)

Pop-out boundaries. Boundaries between areas in a display that are seen almost immediately because they "pop out." (5)

Positional neurons. Kinesthetic neurons that fire best to the static position of a limb. (11)

Postsynaptic neuron. A neuron on the receiving side of a synapse that receives neurotransmitter from the presynaptic neuron. (1)

Power function. A mathematical function of the form $P = KS^n$, where P is perceived magnitude, K is a constant, S is the stimulus intensity, and n is an exponent. (1)

Pragnanz, law of. A gestalt law that is also called the *law of good figure* or the *law of simplicity.* It states that every stimulus pattern is seen in such a way that the resulting structure is as simple as possible. (5)

Preattentive stage of processing. An automatic and rapid stage of processing, during which a stimulus is decomposed into small units called *primitives.* (5)

Precedence effect. The effect that occurs when two identical or very similar sounds reach a listener's ears separated by a time interval of less than about 50–100 msec, and the listener hears the sound that reaches his or her ears first. (9)

Preferential looking (PL) technique. A technique used to measure perception in infants. Two stimuli are presented, and the infant's looking behavior is monitored for the amount of time the infant spends viewing each stimulus. (DD1)

Presbycusis. A form of sensorineural hearing loss that occurs as a function of age and is usually associated with a decrease in the ability to hear high frequencies. Since this loss also appears to be related to exposure to environmental sounds, it is also called *sociocusis*.(13)

Presbyopia ("old eye"). The inability of the eye to accommodate due to the hardening of the lens and a weakening of the ciliary muscles. It occurs as people get older. (13)

Presynaptic neuron. A neuron on the sending side of the synapse, which releases neurotransmitter onto the postsynaptic neuron. (1)

Primary receiving area. The area of the cerebral cortex that first receives most of the signals initiated by a sense's receptors. (1)

Primary somatosensory receiving area (SI). The area in the parietal lobe that receives signals of the somatic system, and that signals perceptions such as touch, temperature, and pain. (11)

Primer. A pheromone that triggers a chain of long-lasting physiological (usually hormonal) effects in an animal. (12)

Primitives. Basic properties of a stimulus. For example, Treisman proposed primitives such as color, line tilt, and curvature. Biederman's primitives are volumetric shapes. (5)

Principle of componential recovery. Biederman's principle stating that we can identify an object if we can perceive its individual geons. (5)

Problem of sensory coding. The problem of determining how characteristics of nerve impulses represent properties of the environment. (1)

Processing module. A term used in Chapter 3 to refer to hypercolumns, each of which processes information from a small area of the retina. (3)

Propagated response. A response, such as a nerve impulse, that travels all the way down the nerve fiber without decreasing in amplitude. (1)

Proprioception. The sensing of the position of the limbs. (11)

Proprioceptive senses. The senses responsible for sensing the location, orientation, and movement of the body. (11)

Prosopagnosia. A form of visual agnosia in which the person can't recognize faces. (3)

Protanopia. A form of red–green dichromatism caused by a lack of the long-wavelength cone pigment. (4)

Prototype. A version of a particular vowel sound that listeners rate as being a particularly good example of that vowel. (OWP 10)

Proximity, law of. Gestalt law: Things that are near to each other appear to be grouped together. Also called the *law of nearness*. (5)

Psychophysical approach to perception. The approach to perception that uses psychophysical methods to determine the relationship between the stimulus and its perception. (1)

Psychophysical tuning curve. A function that indicates the intensity of masking tones of different frequencies that cause a low-intensity pure tone to become just barely detectable. (9)

Psychophysics. Traditionally, this term has referred to a number of methods of quantitative measurement of thresholds, but it is used more broadly in this book to refer to an approach to perception in which the relationship between properties of the stimulus and the subject's experience is determined. (1)

Pupil. The small opening at the front of the eye. (2)

Pupillary block. A blockage that constricts the opening between the iris and the lens of the eye, making it difficult for aqueous humor to leave the eye. It is caused by the pushed-up iris characteristic of closed-angle glaucoma. (13)

Pure tone. A tone with pressure changes that can be described by a single sine wave. (8)

Pure-tone audiometry. Measurement of the threshold for hearing as a function of the frequency of a pure tone. (13)

Purkinje shift. The shift from cone spectral sensitivity to rod spectral sensitivity that takes place during dark adaptation. (2)

RA I mechanoreceptors. See *Rapidly adapting mechanoreceptors*.

RA II mechanoreceptors. See *Rapidly adapting mechanoreceptors*.

Radial keratotomy. A surgical procedure in which four to eight cuts are placed radially around the cornea. When successful, this operation decreases the focusing power of the cornea and improves the vision of people with myopia. (13)

Radical skepticism. A philosophical position stating that it is just as likely as not that other people perceive things very differently from you. (OWP 12)

Random-dot kinematogram. A stimulus that occurs when the left and right parts of a random-dot stereogram are flashed, one after the other, to the same eye. (7)

Random-dot stereogram. A stereogram in which the stimuli are pictures of random dots. If one section of this pattern is shifted slightly in one direction, the resulting disparity causes the perception of depth when the patterns are viewed in a stereoscope. (6)

Rapidly adapting (RA) fiber. A mechanoreceptive fiber that adapts rapidly to continuous stimulation of the skin. Rapidly adapting fibers are associated with Meissner corpuscle and Pacinian corpuscle receptors. (11)

Rapidly adapting (RA) mechanoreceptors. Fibers that respond as pressure indents the skin, generally firing more rapidly to faster velocities of indentation. These neurons do not fire to constant pressure. RA I mechanoreceptors are associated with Meissner corpuscles; RA II with Pacinian corpuscles. (11)

Rate of angular expansion. See *Angular expansion, rate of.*

Ratio principle. A principle stating that two areas that reflect different amounts of light will look the same if the ratios of their intensities to the intensities of their surrounds are the same. (4)

Rat–man demonstration. The demonstration in which presentation of a "ratlike" or "manlike" picture influences an observer's perception of a second picture, which can be interpreted either as a rat or as a man. (1)

Raw primal sketch. In Marr's computational approach to object perception, the raw primal sketch consists of an object's primitives and edges. (5)

Rayleigh scattering. The scattering of sunlight by small particles in the earth's atmosphere, the amount of scatter being inversely proportional to the fourth power of the light's wavelength. This means that short-wavelength light is scattered more than long-wavelength light. (6)

Real movement. The physical movement of a stimulus. (7)

Receiver-operating-characteristic (ROC) curve. A graph in which the results of a signal detection experiment are plotted as the proportion of hits versus the proportion of false alarms for a number of different response criteria. (Appendix E)

Receptive field. A neuron's receptive field is the area on the receptor surface (the retina, for vision; the skin, for touch) that, when stimulated, affects the firing of that neuron. An exception is the case of receptive fields for auditory space perception in the owl. In this case, the receptive fields are locations in space rather than areas on the receptor surface. (2)

Receptor. A sensory receptor is a neuron sensitive to environmental energy that changes this energy into electrical signals in the nervous system. (1)

Recognition by components (RBC). A mechanism of object perception proposed by Biederman, in which we recognize objects by decomposing them into primitives called *geons.* (5)

Recording electrode. A small shaft of metal or glass that, when connected to appropriate electronic equipment, records electrical activity in nerves or nerve fibers. (1)

Reflectance. The percentage of light reflected from a surface. (4)

Reflectance curve. A plot showing the percentage of light reflected from an object versus wavelength. (4)

Refraction. A procedure used to determine the power of the corrective lenses needed to achieve clear vision. (13)

Refractive myopia. See *Myopia, refractive.*

Refractory period. The time period of about 1/1,000 second that a nerve fiber needs to recover from conducting a nerve impulse. No new nerve impulses can be generated in the fiber until the refractory period is over. (1)

Regional sensitivity effect. In the olfactory system, the fact that different areas on the mucosa are sensitive to some odorants and are not as sensitive to others. (12)

Relational invariance. A straight-line function is created when a consonant is paired with different vowels and the formant transition onset frequency of the vowels' second formant is plotted versus the frequency of the vowels' second formant. (10)

Releaser. A pheromone that triggers an immediate reaction in another animal. (12)

Resonance. A mechanism that enhances the intensity of certain frequencies because of the reflection of sound waves in a closed tube. Resonance occurs in the auditory canal. (8)

Resonance theory. Helmholtz's theory of pitch perception, which proposes that the basilar membrane is made up of a series of transverse fibers, each tuned to resonate to a specific frequency. (8)

Resonant frequency. The frequency that is most strongly enhanced by resonance. The resonance frequency of a closed tube is determined by the length of the tube. (8)

Response compression. The result when doubling the physical intensity of a stimulus less than doubles the subjective magnitude of the stimulus. (1)

Response criterion. In a signal detection experiment, the subjective magnitude of a stimulus above which the subject will indicate that the stimulus is present. (Appendix E)

Response expansion. The result when doubling the physical intensity of a stimulus more than doubles the subjective magnitude of the stimulus. (1)

Response profile. A graph that shows a neuron's response to each of the four basic taste qualities. (12)

Resting potential. The difference in charge between the inside and the outside of the nerve fiber when the fiber is not conducting electrical signals. (1)

Retina. A complex network of cells that covers the inside back of the eye. These cells include the receptors, which generate an electrical signal in response to light, as well as the horizontal, bipolar, amacrine, and ganglion cells. (6)

Retinal. The light-sensitive part of the visual pigment molecule. (2)

Retinal densitometry. A procedure for measuring the concentration of visual pigment in the living eye that involves projecting a dim beam of light into the eye and measuring the fraction of this beam that is reflected back out of the eye. (2)

Retinal size. The size of an image on the retina. (6)

Retinitis pigmentosa. A retinal disease that causes a gradual loss of vision. (13)

Retinoscopy exam. Examination with a device called a *retinoscope* that indicates the power of the corrective lenses needed to achieve normal vision. (13)

Retinotopic map. A map on a structure in the visual system, such as the lateral geniculate nucleus or the cortex, that indicates locations on the structure that correspond to locations on the retina. In retinotopic maps, locations adjacent to each other on the retina are usually represented by locations that are adjacent to each other on the structure. (3)

Retronasal route. The route through the nasal pharynx by which olfactory stimuli from food in the mouth reach the olfactory mucosa. (12)

Reverberation time. The time it takes for a sound produced in an enclosed space to decrease to 1/1000 of its original pressure. (9)

Reversible figure–ground. A figure–ground pattern that perceptually reverses as it is viewed, so that the figure becomes the ground and the ground becomes the figure. (5)

Rigidity heuristic. The "rule of thumb" that parts of objects are rigid and linked together so that they move in synchrony. This "rule" has been applied to describing apparent movement phenomena. (7)

Rod monochromat. A person who has a retina in which the only functioning receptors are rods. (2)

Rod spectral sensitivity curve. A graph showing the rod system's sensitivity to light as a function of the light's wavelength. (2)

Rod–cone break. The point on the dark adaptation curve at which vision shifts from cone vision to rod vision. (2)

Rods. Rod-shaped receptors in the retina that are primarily responsible for vision at low levels of illumination. The rod system is extremely sensitive in the dark but cannot resolve fine details. (2)

Round window. A small membrane-covered opening at the end of the scala tympani in the cochlea of the ear. (8)

Ruffini cylinder. A receptor structure in the skin associated with slowly adapting fibers, large receptive fields, and the perception of "buzzing," stretching of the skin, and limb movements. (11)

Ruffini ending. A receptor in the skin, associated with SA II mechanoreceptors. (11)

Running spectral display. A way of representing the speech stimulus in which a number of short-term spectra are arranged to show how the frequencies in the speech stimulus change as time progresses. (10)

SA I mechanoreceptors. See *Slowly adapting mechanoreceptors.*

SA II mechanoreceptors. See *Slowly adapting mechanoreceptors.*

Saturation (nerve firing). The intensity at which a nerve fiber reaches its maximum response. Once the fiber is saturated, further increases in intensity cause no further increase in the fiber's firing rate. (8)

Saturation (color). The relative amount of whiteness in a chromatic color. The less whiteness a color contains, the more saturated it is. (4)

Scale illusion. An illusion that occurs when successive notes of a scale are presented alternately to the left and the right ears. Even though each ear receives notes that jump up and down in frequency, smoothly ascending or descending scales are heard in each ear. (9)

Secondary somatosensory receiving area (S II). The area in the parietal lobe next to the primary somatosensory area (S I) that processes neural signals related to touch, temperature, and pain. (11)

Selective adaptation. A procedure in which a person or animal is selectively exposed to one stimulus and then the effect of this exposure is assessed by testing with a wide range of stimuli. Exposing a person to vertical bars and then testing a person's sensitivity to bars of all orientations is an example of selective adaptation to orientation. Selective adaptation can also be carried out for spatial frequency, wavelength, and speech sounds. (3)

Selective reflection. When an object reflects some wavelengths of the spectrum more than others. (4)

Semantics. A system that specifies whether it is appropriate to use a word in a sentence based on its meaning. (10)

Sensitive period. A period of time, usually early in an organism's life, during which changes in the environment have a large effect on the organism's physiology or behavior. (DD 3)

Sensitivity. 1.0 divided by the threshold for detecting a stimulus. Thus, lower thresholds correspond to higher sensitivities. (2)

Sensory code. A code, formed by electrical signals in the nervous system, that represents the properties of stimuli in the environment. (1)

Sensory movement signal. The electrical signal generated by the movement of an image across the retina. This is one of the signals that plays a role in the corollary discharge theory of movement perception. (7)

Shadowing. Subjects' repetition aloud of what they hear as they are hearing it. (10)

Shape constancy. The constancy of the perception of an object's shape that is maintained even when the object is viewed from different angles. (4)

Shortest-path constraint. The principle that apparent movement occurs along the shortest path between two stimuli that cause apparent movement when flashed on and off with the appropriate timing. (7)

Short-term spectrum. A plot that indicates the frequencies in a sound stimulus during a short period, usually at the beginning of the stimulus. (10)

Short-wavelength pigment. The cone visual pigment that absorbs maximally at short wavelengths. In the human, this pigment absorbs maximally at about 419 nm. (2)

Signal. The stimulus presented to a subject. (Appendix E)

Signal detection theory (SDT). A theory stating that the detection of a stimulus depends both on the subject's sensitivity to the stimulus and on the subject's response criterion. (Appendix E)

Similarity, law of. A Gestalt law stating that similar things appear to be grouped together. (5)

Simple cortical cell. A neuron in the visual cortex that responds best to bars of a particular orientation. (3)

Simplicity, law of. See *Good figure, law of.*

Simultaneous contrast. The effect that occurs when surrounding one color with another changes the appearance of the surrounded color. (2)

Sine-wave grating. A grating stimulus with a sine-wave intensity distribution. (5)

Size constancy. The constancy of the perception of the size of a stimulus that is maintained even when the object is viewed from different distances. (6)

Size constancy, law of. A law stating that our perception of an object's size remains constant, no matter what its distance from us. (6)

Size–distance scaling. A hypothesized mechanism that helps maintain size constancy by taking an object's distance into account. (6)

Size in the field of view. A depth cue. Objects that take up a small part of the field of view are, everything else being equal, perceived as farther away than objects that take up a large part of the field of view. (6)

Slit lamp examination. An examination that checks the condition of the cornea and lens. (13)

Slowly adapting (SA) fiber. A mechanoreceptive fiber in the skin that adapts slowly to continuous stimulation of the skin. Slowly adapting fibers are associated with Merkel receptors and Ruffini cylinders. (11)

Slowly adapting (SA) mechanoreceptors. Fibers that respond to indentation of the skin and continue to respond during constant indentation. SA I mechanoreceptors are associated with Merkel disks and SA II mechanoreceptors are associated with Ruffini endings. (11)

Small-diameter fiber (S-fiber). According to gate control theory, activity in S-fibers opens the gate control mechanism and therefore increases the perception of pain. (11)

Somatic senses. The senses that are responsible for cutaneous sensations, proprioception, and kinesthesis. (11)

Somatosensory area. An area in the parietal lobe of the cortex that receives inputs from the skin and the viscera that are associated with somatic senses such as touch, temperature, and pain. There are a number of such areas, including S I and S II. (11)

Somatosensory receiving area. The area in the parietal lobe of the cortex that receives input from the somatic senses. (11)

Sone. Unit of loudness. One sone is the loudness of a 1,000-Hz tone at 40 dB. (9)

Sound pressure level (SPL). A designation used to indicate that the reference pressure used for calculating a tone's decibel rating is set at 2×10^{-5} Newtons/m^2, near the threshold in the most sensitive frequency range for hearing. (8)

Sound spectrogram. A plot showing the pattern of intensities and frequencies of a speech stimulus. (10)

Sound waves. Pressure changes in a medium. Most of the sounds we hear are due to pressure changes in the air. (8)

Spatial event plots. Plots showing the pattern of response generated by a neuron to a touch stimulus. (11)

Spatial frequency. For a grating stimulus, *spatial frequency* refers to the frequency with which the grating repeats itself per degree of visual angle. For more natural stimuli, high spatial frequencies are associated with fine details, and low spatial frequencies are associated with grosser features. (5)

Spatial frequency channels. Hypothesized channels in the visual system that are sensitive to narrow ranges of spatial frequencies. (5)

Spatial summation. The summation, or accumulation, of the effect of stimulation over a large area. (2)

Special speech mechanism. A proposed mechanism of speech perception, that hypothesizes that the speech stimulus activates a mechanism specifically designed to process speech signals. (10)

Specific hunger. A genetically programmed taste preference that helps organisms seek out food that meets specific nutritional needs. (12)

Specificity theory. The theory that different qualities are signaled to the brain by the activity in specific nerve fibers that fire only to that quality. (1)

Spectral sensitivity. The sensitivity of visual receptors to different parts of the visible spectrum. See *Spectral sensitivity curve.* (2)

Spectral sensitivity curve. The function relating a subject's sensitivity to light to the wavelength of the light. (2)

Spinothalamic pathway. One of the nerve pathways in the spinal cord that conducts nerve impulses from the skin to the somatosensory area of the thalamus. (11)

Spiral aftereffect. An aftereffect of movement. Objects that are viewed immediately after one views a rotating spiral for 30 to 60 seconds appear to shrink or expand. (7)

Spontaneous activity. Nerve firing that occurs in the absence of environmental stimulation. (1)

Spontaneous looking preference. Infants' preferring to look at certain types of stimuli. This property of infant behavior is the basis of the preferential looking technique. (DD 1)

S-potential. An electrical response with opponent properties that has been recorded from cells in the fish retina. (4)

Square-wave grating. A grating stimulus with a square-wave intensity distribution. (3)

Stapes. The last of the three ossicles in the middle ear. It receives vibrations from the incus and transmits these vibrations to the oval window of the inner ear. (8)

Stereoacuity. The ability to resolve small differences in disparity. (DD 6)

Stereochemical theory of odor. The theory of olfaction, proposed by Amoore, postulating that there are different molecular shapes for different odor qualities and that there are sites on the olfactory receptors that accept only molecules with a particular shape. (12)

Stereopsis. The impression of depth that results from differences in the images on the retinas of the two eyes. (6)

Stereoscope. A device that presents pictures to the left and the right eyes so that the binocular disparity a person would experience when viewing an actual scene is duplicated. The result is a convincing illusion of depth. (6)

Stevens's power law. A law concerning the relationship between the physical intensity of a stimulus and the perception of the subjective magnitude of the stimulus. The law states that $P = KS^n$, where P is perceived magnitude, K is a constant, S is the stimulus intensity, and n is an exponent. (1)

Stimulation deafness. Deafness caused by intense auditory stimulation that damages structures in the cochlea. (8)

Stimulation-produced analgesia (SPA). Brain stimulation that eliminates or strongly decreases the perception of pain. (11)

Stimulus deprivation amblyopia. Amblyopia due to early closure of one eye. (DD 3)

Strabismus. A condition in which an imbalance in the eye muscles upsets the coordination between the two eyes. (DD 3)

Stress-induced analgesia. Analgesia that occurs when an organism experiences a stressful situation. (11)

Striate cortex. The visual receiving area of the cortex, located in the occipital lobe. (2)

Stroboscopic movement. See *Apparent movement.*

Structuralism. The approach to psychology, prominent in the late 19th and early 20th centuries, that postulated that perceptions result from the summation of many elementary sensations. (5)

Subjective contour. The perception of a contour when no contour is physically present. (OWP 3)

Substantia gelatinosa. A nucleus in the spinal cord that, according to gate control theory, receives inputs from S-fibers and L-fibers and sends inhibition to the T-cell. (11)

Subtractive color mixture. See *Color mixture, subtractive.*

Superior colliculus. A structure at the base of the brain that is important in controlling eye movement. A small proportion of the nerve fibers in the optic nerve synapse in the superior colliculus. (2)

Superior olivary nucleus. A nucleus along the auditory pathway from the cochlea to the auditory cortex. The superior olivary nucleus receives inputs from the cochlear nucleus. (8)

Synapse. A small space between the end of one neuron and the cell body of another neuron. (1)

Syntax. Rules that specify which structures of a sentence are allowed. (10)

Tapetum. A reflecting surface located behind the retina in some animals. (OWP 2)

Taste-blind. A person who is taste-blind cannot taste phenylthiocarbamide (PTC) and also tends to be less sensitive to certain other tastes than someone who is not taste-blind. (12)

Taste bud. A structure located within papillae on the tongue that contains the taste cells. (12)

Taste cells. Cells located in taste buds that cause the transduction of chemical to electrical energy when chemicals contact receptor sites or channels located at the tips of these cells. (12)

Taste pore. An opening in the taste bud through which the tips of taste cells protrude. When chemicals enter a taste pore, they stimulate the taste cells and result in transduction. (12)

Tasters. People who can taste the compound phenylthiocarbamide (PTC). (12)

Tectorial membrane. A membrane that stretches the length of the cochlea and is located directly over the hair cells. Vibrations of the cochlear partition cause the tectorial membrane to stimulate the hair cells by rubbing against them. (8)

Temporal code for frequency. A code in which frequency is signaled by the timing of nerve impulses. One example of a temporal code is one in which high rates of nerve firing signal high frequencies and low rates of nerve firing signal low frequencies. (8)

Temporal lobe. A lobe on the side of the cortex that is the site of the cortical receiving area for hearing. (1)

Temporal pathway. A visual pathway from the occipital lobe to the temporal lobe that processes information concerned with what an object is. (3)

Textons. Primitives for object perception and texture segregation proposed by Julesz. (5)

Texture gradient. The pattern formed by a regularly textured surface that extends away from the observer. The elements in a texture gradient appear smaller as distance from the observer increases. (6)

Texture segregation. The perceptual separation of fields with different textures. (5)

Thalamus. A nucleus in the brain where neurons from all of the senses, except smell, synapse on their way to their cortical receiving areas. (2)

Thermoreceptor. Receptors in the skin that responds to specific temperatures or changes in temperature. (11)

3-D representation. The end result of Marr's computational process; the perception of the three-dimensional stimulus. (5)

Threshold, absolute. The minimum stimulus energy necessary for an observer to detect a stimulus. (1)

Threshold, difference. The minimal detectable difference between two stimuli. (1)

Threshold, relative. The amount of stimulus energy that can just be detected, expressed relative to another threshold. For example, "The amount of energy needed to detect a 500-nm light is twice as high as the amount of energy needed to detect a 540-nm light." (2)

Tilt aftereffect. The result when staring at an adapting field of tilted lines and then looking at vertical lines causes the vertical lines to appear to be tilted in a direction opposite to the tilt of the adapting field. (DD 3)

Timbre. The quality of a tone. Different musical instruments have different timbres, so when we play the same note on different instruments, the notes have the same pitch but sound different. (8)

Timing code of frequency. The sensory code for the frequency of an auditory stimulus in which stimulus frequency is signaled by the timing of nerve impulses in nerve fibers or groups of nerve fibers. (8)

Tinnitus. A condition caused by damage in the inner ear in which a person experiences ringing in the ears. (13)

Tinnitus masker. A unit that generates white noise to mask the ringing in the ears associated with tinnitus. (13)

Tone chroma. The perceptual similarity of notes separated by one or more octaves. (9)

Tone height. The increase in pitch that occurs as frequency is increased. (9)

Tonometer. A device for measuring the eye's intraocular pressure. (13)

Tonometry. An examination that determines the pressure inside the eye. (13)

Tonotopic map. The frequency map that is formed on an auditory structure when neurons with the same characteristic frequency are grouped together and neurons with nearby characteristic frequencies are found near each other. (8)

Top-down processing. Processing that starts with the analysis of high-level information, such as the context in which a stimulus is seen. (1)

Transduction. In the senses, the transformation of environmental energy into electrical energy. For example, the retinal receptors transduce light energy into electrical energy. (2)

Transmission cell (T-cell). According to gate control theory, the cell that receives input from the L- and S-fibers. Activity in the T-cell determines the perception of pain. (11)

Traveling wave. In the auditory system, vibration of the basilar membrane in which the peak of the vibration travels from the base of the membrane to its apex. (8)

Trichromat. A person with normal color vision. Trichromats can match any wavelength in the spectrum by mixing three other wavelengths in various proportions. (4)

Trichromatic theory of color vision. A theory postulating that our perception of color is determined by the ratio of activity in three cone receptor mechanisms with different spectral sensitivities. (4)

Tritanopia. A form of dichromatism thought to be caused by a lack of the short-wavelength cone pigment. (4)

Tungsten light. Light produced by a tungsten filament. Tungsten light has a wavelength distribution that has relatively more intensity at long wavelengths than at short wavelengths. (4)

Tuning curve. In the auditory system, a curve that indicates the intensity necessary to elicit a threshold response from a neuron at different frequencies along the audible range. (8)

Tuning curve, orientation. See *Orientation tuning curve.*

Tunnel vision. Vision that results when there is little peripheral vision. (13)

2½-D sketch. The second stage of Marr's computational process. This stage is the result of processing the primitives. The resulting 2½-D sketch is then transformed into the 3-D representation. (5)

Two-color threshold method. A method used by Stiles in which the thresholds for different wavelengths are measured with a test flash that is superimposed on an adapting field. (2)

2-Deoxyglucose technique. An anatomical technique that makes it possible to see which neurons in a structure have been activated. For example, this technique was used to visualize the orientation columns in the visual cortex. (3)

Two-point threshold. The smallest separation between two points on the skin that is perceived as two points; a measure of acuity on the skin. (11)

Tympanic membrane (eardrum). A membrane at the end of the auditory canal that vibrates in response to vibrations of the air and transmits these vibrations to the ossicles in the middle ear. (8)

Tympanometer. A device for measuring how well the tympanic membrane and the middle-ear bones respond to sound vibrations. (13)

Uncrossed disparity. Binocular disparity that occurs when objects are located beyond the horopter. (6)

Unilateral dichromat. A person who has dichromatic vision in one eye and trichromatic vision in the other eye. (4)

Unvoiced. Referring to consonants that do not cause vibration of the vocal chords. (10)

Vagus nerve. A nerve that conducts signals from taste receptors in the mouth and larynx. (12)

Ventral posterior nucleus. A nucleus in the thalamus that receives inputs from the somatosensory system, primarily from the spinothalamic and lemniscal pathways. (11)

Ventriloquism effect. The effect when cues such as lip movements cause a sound produced at one location (such as a ventriloquist's voice) to appear to come from another location (such as the ventriloquist's dummy). (10)

Video microscopy. A technique that has been used to take pictures of papillae and taste buds on the tongue. (12)

Visible spectrum. The range of wavelengths in the electromagnetic spectrum, from about 350 to 700 nm, that humans can see. (2)

Visual acuity. The ability to resolve small details. (2)

Visual agnosia. A condition in which a person can see clearly but has difficulty recognizing what he or she sees. This condition, which is often caused by brain injuries, makes it difficult for people to synthesize parts of an object into an integrated whole. (5)

Visual angle. The angle between two lines that extend from the observer's eye, one line extending to one end of an object and the second to the other end of the object. An object's visual angle is always determined relative to an observer; therefore, an object's visual angle changes as the distance between the object and the observer changes. (5)

Visual angle, law of. A law stating that our perception of an object's size is determined solely by its visual angle. (6)

Visual evoked potential (VEP). An electrical response to visual stimulation recorded by the placement of disk electrodes on the back of the head. This potential reflects the activity of a large population of neurons in the visual cortex. (DD 2)

Visual field test. A test that measures a person's ability to see both in the fovea and in the periphery. (13)

Visual perimetry. A procedure for testing vision that tests a person's ability to detect small spot stimuli presented at various locations in the person's visual field. (13)

Visual pigment. A light-sensitive molecule contained in the rod and cone outer segments. The reaction of this molecule to light results in the generation of an electrical response in the receptors. (2)

Visual receiving area (or visual cortex). The area in the occipital lobe, also called the *striate cortex*, that receives inputs from the lateral geniculate nucleus. (2)

Vitrectomy. A procedure in which a needle placed inside the eye removes vitreous humor and replaces it with a salt solution. This procedure is used if the vitreous humor is filled with blood, usually because of neovascularization. (13)

Vitreous humor. The jellylike substance that fills the eyeball. (13)

Voice onset time. In speech production, the time delay between the beginning of a sound and the beginning of the vibration of the vocal chords. (10)

Voicing. A property of the speech stimulus associated with the vibration of the vocal cords. A voiced phoneme is one that causes the vocal cords to vibrate. (10)

Volley principle. Wever's idea that groups of nerve fibers fire in volleys, some fibers firing while others are refractory. In this way, groups of fibers can effect high rates of nerve firing. (8)

Warm fiber. A nerve fiber that responds to increases in temperature or to steady high temperatures. (11)

Waterfall illusion. An aftereffect of movement. Objects that are viewed immediately after one views the downward flow of a waterfall for 30 to 60 seconds appear to move up. (7)

Waveform. A waveform describes functions in which stimulus intensity or amplitude is plotted versus time or space. For example, we can describe the waveform of a pure tone stimulus (a plot of pressure change vs. time) as a sine wave, or we can refer to the waveform of a grating stimulus (a plot of light intensity vs. distance) as a sine wave or a square wave. (5)

Wavelength. For light energy, the distance between one peak of a light wave and the next peak. (2)

Wavelength distribution. The amount of energy in a light at each of the wavelengths in the spectrum. (4)

Weber's law. A law stating that the just noticeable difference (JND) equals a constant (K), called the Weber fraction, times the size of the stimulus (S). This law is usually expressed in the form $K = JND/S$. (1)

White light. Light that contains an equal intensity of each of the visible wavelengths. (4)

Young–Helmholtz theory of color vision. See *Trichromatic theory of color vision.* (4)

Zenith. The sky directly overhead. (6)

REFERENCES

Abeles, M., & Goldstein, M. H. (1970). Functional architecture in cat primary auditory cortex: Columnar organization and organization according to depth. *Journal of Neurophysiology, 33,* 172–187.

Able, K. P., & Able, M. A. (1993). Daytime calibration of magnetic orientation in a migratory bird requires a view of skylight polarization. *Nature, 364,* 523–525.

Abramov, I., & Gordon, J. (1994). Color appearance: On seeing red, or yellow, or green, or blue. *Annual Review of Psychology, 45,* 451–485.

Abramov, I., Gordon, J., Hendrickson, A., Hainline, L., Dobson, V., & LaBossiere, E. (1982). The retina of the newborn human infant. *Science, 217,* 265–267.

Ache, B. W. (1991). Phylogeny of smell and taste. In T. V. Getchell et al. (Eds.), *Smell and taste in health and disease* (pp. 3–18). New York: Raven.

Ackerman, D. (1990). *A natural history of the senses.* New York: Vintage.

Ackerman, D. (1995). *Mystery of the senses: Taste.* Boston: WGBH and Washington, DC: WETA.

Adelson, E. H. (1993). Perceptual organization and the judgment of brightness. *Science, 262,* 2042–2044.

Adelson, E. H., & Movshon, J. A. (1982). Phenomenal coherence of moving visual patterns. *Nature, 300,* 523–525.

Adler, J. (1969). Chemoreceptors in bacteria. *Science, 166,* 1588–1597.

Adler, S. A., & Rovee-Collier, C. (1994). The memorability and discriminability of primitive perceptual units in infancy. *Vision Research, 34,* 449–459.

Aitkin, L. M. (1986). Sensory processing in the mammalian auditory system—Some parallels and contrasts with the visual system. In. J. D. Pettigrew, K. J. Sanderson, & W. R. Levick (Eds.), *Visual neuroscience* (pp. 223–235). Cambridge: Cambridge University Press.

Ali, M. A., & Klyne, M. A. (1985). *Vision in vertebrates.* New York: Plenum.

Alpern, M., Kitahara, K., & Krantz, D. H. (1983). Perception of colour in unilateral tritanopia. *Journal of Physiology, 335,* 683–697.

Ammons, C. H., Worchel, P., & Dallenbach, K. M. (1953). Facial vision: The perception of obstacles out of doors by blindfolded and blindfolded-deafened subjects. *American Journal of Psychology, 66,* 519–553.

Amoore, J. E. (1970). *Molecular basis of odor.* Springfield, IL: Charles C Thomas.

Amoore, J. E., Johnston, J. W., Jr., & Rubin, M. (1964). The stereochemical theory of odor. *Scientific American, 210*(2), 42–49.

Anstis, S. M., & Gregory, R. L. (1964). The aftereffect of seen motion: The role of retinal stimulation and eye movements. *Quarterly Journal of Experimental Psychology, 17,* 173–174.

Arnheim, R. (1974). *Art and visual perception* (2nd ed.). Berkeley: University of California Press.

Aslin, R. N. (1977). Development of binocular fixation in human infants. *Journal of Experimental Child Psychology, 23,* 133–150.

Aslin, R. N. (1981a). Development of smooth pursuit in infants. In D. F. Fisher, R. A. Monty, & J. W. Senders (Eds.), *Eye movements: Cognition and visual perception.* Hillsdale, NJ: Erlbaum.

Aslin, R. N. (1981b). Experiential influences and sensitive periods of perceptual development: A unified model. In R. N. Aslin, J. Alberts, & M. J. Petersen (Eds.), *Development of perception* (Vol. 2, pp. 45–93). New York: Academic.

Aslin, R. N., & Banks, M. S. (1978). Early experience in humans: Evidence for a critical period in the development of binocular vision. In Sten-Schneider, H. Liebowitz, H. Pick, & H. Stevenson (Eds.), *Psychology: From basic research to practice.* New York: Plenum.

Aslin, R. N., & Pisoni, D. B. (1980). Some developmental processes in speech perception. In G. Yeni-Konishian, J. F. Kavanagh, & C. A. Ferguson (Eds.), *Child phonology: Vol. 2. Perception.* New York: Academic Press.

Au, W. W. L. (1993). *The sonar of dolphins.* New York: Springer-Verlag.

Aubert, H. (1886). Die Bewegungsempfindung. *Archiv für die gesamte Physiologie des Menschen und der Tiere, 39,* 347–370.

Awaya, S., Miyake, Y., Imayuni, Y., Shiose, Y., Kanda, T., & Komuro, K. (1973). Amblyopia in man, suggestive of stimulus deprivation amblyopia. *Japanese Journal of Ophthalmology, 17,* 69–82.

Azzopardi, P., & Cowey, A. (1993). Preferential representation of the fovea in the primary visual cortex. *Nature, 361,* 719–721.

Backus, J. (1977). *The acoustical foundations of music* (2nd ed.). New York: Norton.

Baird, J. C., Wagner, M., & Fuld, K. (1990). A simple but powerful theory of the moon illusion. *Journal of Experimental Psychology: Human Perception and Performance, 16,* 675–677.

Balogh, R. D., & Porter, R. H. (1986). Olfactory preferences resulting from mere exposure in human neonates. *Infant Behavior and Development, 9,* 395–401.

Banks, M. S. (1982). The development of spatial and temporal contrast sensitivity. *Current Eye Research, 2,* 191–198.

Banks, M. S. (1992). Optics, receptors, and spatial vision in human infants. In. L. A. Werner & E. W. Rubel (Eds.), *Developmental psychoacoustics.* Washington, DC: American Psychological Association.

Banks, M. S., Aslin, R. N., & Letson, R. D. (1975). Sensitive period for the development of human binocular vision. *Science, 190,* 675–677.

Banks, M. S., & Bennett, P. J. (1988). Optical and photoreceptor immaturities limit the spatial and chromatic vision of human neonates. *Journal of the Optical Society of America, A5,* 2059–2079.

Banks, M. S., & Salapatek, P. (1978). Acuity and contrast sensitivity in 1-, 2-, and 3-month-old human infants. *Investigative Ophthalmology and Visual Science, 17,* 361–365.

Banks, M. S., & Salapatek, P. (1981). Infant pattern vision: A new approach based on the contrast sensitivity function. *Journal of Experimental Child Psychology, 31,* 1–45.

Barinaga, M. (1994). Neurons tap out a code that may help locate sounds. *Science, 264,* 775.

Barlow, H. B., Blakemore, C., & Pettigrew, J. D. (1967). The neural mechanism of binocular depth discrimination. *Journal of Physiology, 193,* 327–342.

Barlow, H. B., & Brindley, G. S. (1963). Intraocular transfer of movement after-effects during pressure blinding of the stimulated eye. *Nature, 200,* 1347.

Barlow, H. B., & Hill, R. M. (1963). Evidence for a physiological explanation of the waterfall illusion and figural after-effects. *Nature, 200,* 1345–1347.

Barlow, H. B., & Mollon, J. D. (Eds.). (1982). *The senses.* Cambridge: University of Cambridge Press.

Barrow, H. G., & Tannenbaum, J. M. (1986). Computational approaches to vision. In K. R. Boff, L. Kaufman, & J. P. Thomas (Eds.), *Handbook of perception and human performance* (Chapter 35). New York: Wiley.

Bartoshuk, L., Cain, W. S., & Pfaffmann, C. (1985). Taste and olfaction. In G. A. Kimble & K. Schlesinger (Eds.), *Topics in the history of psychology* (Vol. 1, pp. 221–260). Hillsdale, NJ: Erlbaum.

Bartoshuk, L. J., & Beauchamp, G. K. (1994). Chemical senses. *Annual Reviews of Psychology, 45,* 419–449.

Bartoshuk, L. M. (1971). The chemical senses I. Taste. In J. W. Kling & L. A. Riggs (Eds.), *Experimental psychology* (3rd ed.). New York: Holt, Rinehart & Winston.

Bartoshuk, L. M. (1978). Gustatory system. In R. B. Masterson (Ed.), *Handbook of behavioral neurobiology: Vol. 1. Sensory integration* (pp. 503–567). New York: Plenum.

Bartoshuk, L. M. (1979). Bitter taste of saccharin: Related to the genetic ability to taste the bitter substance propylthioural (PROP). *Science, 205,* 934–935.

Bartoshuk, L. M. (1980, September). Separate worlds of taste. *Psychology Today,* pp. 48–56.

Batteau, D. W. (1967). The role of the pinna in human localization. *Proceedings of the Royal Society of London, 168B,* 158–180.

Beauchamp, G. K. (1987). The human preference for excess salt. *American Scientist, 75,* 27–33.

Beauchamp, G. K., Bertino, M., & Engelman, K. (1991). Human salt appetite. In M. I. Friedman, M. G.

Tordoff, & M. R. Kare. (Eds.), *Chemical senses* (Vol. 4, pp. 85–108). New York: Dekker.

Beauchamp, G. K., Cowart, B. J., & Moran, M. (1986). Developmental changes in salt acceptability in human infants. *Developmental Psychobiology, 19,* 17–25.

Beauchamp, G. K., Cowart, B. J., & Schmidt, H. J. (1991). Development of chemosensory sensitivity and preference. In T. V. Getchell et al. (Eds.), *Smell and taste in health and disease* (pp. 405–416). New York: Raven.

Beauchamp, G. K., & Maller, O. (1977). The development of flavor preferences in humans: A review. In M. R. Kare & O. Maller (Eds.), *Chemical senses and nutrition* (pp. 291–311). New York: Academic.

Beck, J. (1966). Effect of orientation and shape similarity in perceptual grouping. *Perception and Psychophysics, 1,* 300–302.

Beck, J. (1972). *Surface color perception.* Ithaca, NY: Cornell University Press.

Beck, J. (1982). Textural segmentation. In J. Beck (Ed.), *Organization and representation in perception.* Hillsdale, NJ: Erlbaum.

Beck, J. (1993). The British aerospace lecture: Visual processing in texture segregation. In D. Brogan, A. Gale, & K. Carr (Eds.), *Visual search* (Vol. 2, pp. 1–35). London: Taylor & Francis.

Beck, J., Hope, B., & Resenfeld, A. (Eds.). (1983). *Human and machine vision.* New York: Academic.

Beck, P. W., Handwerker, H. O., & Zimmerman, M. (1974). Nervous outflow from the cat's foot during noxious radiant heat stimulation. *Brain Research, 67,* 373–386.

Beecher, H. K. (1972). The placebo effect as a nonspecific force surrounding disease and the treatment of disease. In R. Janzen, W. D. Kerdel, A. Herz, C. Steichele, J. P. Payne, & A. P. Burt (Eds.), *Pain: Basic principles, pharmacology, and therapy.* Stuttgart, West Germany: Georg Thiene.

Beecher, M. D., Petersen, M. R., Zoloth, S. R., Moody, D. B., & Stebbins, W. C. (1979). Perception of conspecific vocalizations by Japanese macaques. *Brain Behavior and Evolution, 16,* 443–460.

Beets, M. G. J. (1978). Odor and stimulant structure. In E. C. Carterette & M. P. Friedman (Eds.), *Handbook of perception* (Vol. 6A, pp. 245–255). New York: Academic.

Beets, M. G. J. (1982). Odor and stimulant structure. In E. T. Theimer (Ed.). *Fragrance chemistry: The science of the sense of smell* (pp. 77–122). New York: Academic Press.

Békésy, G. von (1960). *Experiments in hearing.* New York: McGraw-Hill.

Berger, K. W. (1964). Some factors in the recognition of timbre. *Journal of the Acoustical Society of America, 36,* 1881–1891.

Berkley, M. A. (1990). Behavioral determination of spatial selectivity of contrast adaptation in cats: Some evidence for a common plan in the mammalian visual system. *Visual Neuroscience, 4,* 413–426.

Berkow, I. (1995, January 5). The sweetest sound of all. *New York Times,* B8.

Berlin, B., & Kay, P. (1969). *Basic color terms: Their universality and evolution.* Berkeley: University of California Press.

Bernstein, I. (1978). Learned taste aversions in children receiving chemotherapy. *Science, 200,* 1302–1303.

Bertenthal, B. I., Proffitt, D. R., & Kramer, S. J. (1987). Perception of biomechanical motions by infants: implementation of various processing constraints. *Journal of Experimental Psychology: Human Perception and Performance, 13,* 577–585.

Bertenthal, B. I., Proffitt, D. R., Spetner, N. B., & Thomas, M. A. (1985). The development of infant sensitivity to biomechanical motions. *Child Development, 56,* 531–543.

Biederman, I. (1981). On the semantics of a glance at a scene. In M. Kubovy & J. Pomerantz (Eds.), *Perceptual organization.* Hillsdale, NJ: Erlbaum.

Biederman, I. (1987). Recognition-by-components: A theory of human image understanding. *Psychological Review, 94,* 115–147.

Biederman, I., & Cooper, E. E. (1991). Priming contour-deleted images: Evidence for intermediate representations in visual object recognition. *Cognitive Psychology, 23,* 393–419.

Biederman, I., Cooper, E. E., Hummel, J. E., & Fiser, J. (1993). Geon theory as an account of shape recognition in mind, brain, and machine. In J. Illingworth (Ed.), *Proceedings of the Fourth British Machine Vision Conference* (pp. 175–186). Guildford, Surrey, U.K.: BMVA Press.

Birch, L. (1979). Dimensions of preschool children's food preferences. *Journal of Nutrition Education, 11,* 77–80.

Bisti, S., & Maffei, L. (1974). Behavioral contrast sensitivity of the cat in various visual meridians. *Journal of Physiology, 241,* 201–210.

Blake, R. (1988). Cat spatial vision. *Trends in Neurosciences, 1,* 78–83.

Blake, R. (1993). Cats perceive biological motion. *Psychological Science, 4,* 54–57.

Blake, R. (1994). Gibson's inspired but latent prelude to visual motion perception. *Psychological Review, 101,* 324–328.

Blake, R., Cool, S. J., & Crawford, M. L. J. (1974). Visual resolution in the cat. *Vision Research, 14,* 1211–1217.

Blake, R., & Hirsch, H. V. B. (1975). Deficits in binocular depth perception in cats after alternating monocular deprivation. *Science, 190,* 1114–1116.

Blake, R., & Wilson, H. R. (1991). Neural models of stereoscopic vision. *Trends in Neuroscience, 14,* 445–452.

Blakemore, C., & Tobin, E. A. (1972). Lateral inhibition between orientation detectors in the cat's visual cortex. *Experimental Brain Research, 15,* 439–440.

Blank, D. L. (1974). Mechanism underlying the analysis of odorant quality at the level of the olfactory mucosa: II. Receptor selective sensitivity. *Annals of the New York Academy of Sciences, 237,* 91–101.

Bloom, F., Lazerson, A., & Hofstadter, L. (1985). *Brain, mind, and behavior.* New York: Freeman.

Blumstein, S. E., & Stevens, K. N. (1979). Acoustic invariance in speech production: Evidence from measurements of the spectral characteristics of stop consonants. *Journal of the Acoustical Society of America, 66,* 1001–1007.

Bolanowski, S. J., Gescheider, G. A., & Verrillo, R. T. (1994). Hairy skin: Psychophysical channels and their physiological substrates. *Somatosensory and Motor Research, 11,* 279–290.

Bolanowski, S. J., Jr., Gescheider, G. A., Verrillo, R. T., & Checkosky, C. M. (1988). Four channels mediate the mechanical aspects of touch. *Journal of the Acoustical Society of America, 84,* 1680–1694.

Borg, G., Diamant, H., Strom, C., & Zotterman, Y. (1967). The relation between neural and perceptual intensity: A comparative study of neural and psychophysical responses to taste stimuli. *Journal of Physiology, 192,* 13–20.

Bornstein, M. H., Kessen, W., & Weiskopf, S. (1976). Color vision and hue categorization in young human infants. *Journal of Experimental Psychology: Human Perception and Performance, 2,* 115–119.

Bough, E. W. (1970). Stereoscopic vision in the macaque monkey: A behavioural demonstration. *Nature, 225,* 42–44.

Boussourd, D., Ungerleider, L. G., & Desimone, R. (1990). Pathways for motion analysis: Cortical connections of the medial superior temporal and fundus of the superior temporal visual areas in the macaque. *Journal of Comparative Neurology, 296,* 462–495.

Bowmaker, J. K., & Dartnall, H. J. A. (1980). Visual pigments of rods and cones in a human retina. *Journal of Physiology, 298,* 501–511.

Boynton, R. M. (1979). *Human color vision.* New York: Holt, Rinehart & Winston.

Boynton, R. M. (1990). Human color perception. In K. N. Leibovic (Ed.), *Science of vision* (pp. 211–253). New York: Springer-Verlag.

Boynton, R. M., & Olson, C. X. (1987). Locating basic colors in the OSA space. *Color Research Applications, 12,* 94–105.

Boynton, R. M., & Olson, C. X. (1990). Salience of basic chromatic color terms confirmed by three measures. *Vision Research, 30,* 1311–1317.

Bradley, D. R., & Petry, H. M. (1977). Organizational determinants of subjective contour: The subjective Necker cube. *American Journal of Psychology, 90,* 253–262.

Brainard, D. H., & Wandell, B. A. (1992). Asymmetric color matching: how color appearance depends on the illuminant. *Journal of the Optical Society of America, A9,* 1433–1448.

Brand, J. G., Bryant, B. P., Cagan, R. H., & Kalinoski, D. L. (1987). Biochemical studies of taste sensation: 13. Enantiomeric specificity of alanine taste receptor sites in catfish, *Ictalurur punctatus. Brain Research, 416,* 119–128.

Bregman, A. (1981). Asking the "what for" question in auditory perception. In M. Kubovy & J. R. Pomerantz (Eds.), *Perceptual organization* (pp. 99–119). Hillsdale, NJ: Erlbaum.

Bregman, A. S. (1990). *Auditory scene analysis.* Cambridge: MIT Press.

Bregman, A. S., & Campbell, J. (1971). Primary auditory stream segregation and perception of order in rapid sequence of tones. *Journal of Experimental Psychology, 89,* 244–249.

Bregman, A. S., & Rudnicky, A. I. (1975). Auditory segregation: Stream or streams? *Journal of Experimental Psychology: Human Perception and Performance, 1,* 263-267.

Bridgeman, B., & Stark, L. (1991). Ocular proprioception and efference copy in registering visual direction. *Vision Research, 31,* 1903–1913.

Britten, K. H., Shadlen, M. N., Newsome, W. T., & Movshon, J. A. (1993). Responses of neurons in macaque MT to stochastic motion signals. *Visual Neuroscience, 10,* 1157–1169.

Brown, A. A., Dowell, R. C., & Clark, G. M. (1987). Clinical results for postlingually deaf patients implanted with multichannel cochlear prosthetics. *Annals of Otology, Rhinology, and Laryngology, 96* (Suppl. 128), 127–128.

Brown, C. M. (1984). Computer vision and natural constraints. *Science, 224,* 1299–1305.

Brown, K. T. (1969). A linear area centralis extending across the turtle retina and stabilized to the horizon by non-visual cues. *Vision Research, 9,* 1053–1062.

Brown, P. K., & Wald, G. (1964). Visual pigments in single rods and cones of the human retina. *Science, 144,* 45–52.

Bruce, C., Desimone, R., & Gross, C. G. (1981). Visual properties of neurons in a polysensory area in the superior temporal sulcus of the macaque. *Journal of Neurophysiology, 46,* 369–384.

Brugge, J. F., & Merzenich, M. M. (1973). Responses of neurons in auditory cortex of the macaque monkey to monaural and binaural stimulation. *Journal of Neurophysiology, 36,* 1138–1158.

Bruner, J. S., & Postman, L. (1949). On the perception of incongruity: A paradigm. *Journal of Personality, 18,* 206–228.

Buck, L., & Axel, R. (1991). A novel multigene family may encode odorant receptors: A molecular basis for odor recognition. *Cell, 65,* 175–187.

Bugelski, B. R., & Alampay, D. A. (1961). The role of frequency in developing perceptual sets. *Canadian Journal of Psychology, 15,* 205–211.

Bunch, C. C. (1929). Age variations in auditory acuity. *Archives of Otolaryngology, 9,* 625–626.

Buswell, G. T. (1935). *How people look at pictures.* Chicago: University of Chicago Press.

Butler, R. A., & Belendiuk, K. (1977). Spectral cues utilized in the localization of sound in the median sagittal plane. *Journal of the Acoustical Society of America, 61,* 1264–1269.

Cabanac, M. (1971). Physiological role of pleasure. *Science, 173,* 1103–1107.

Cabanac, M. (1979). Sensory pleasure. *Quarterly Review of Biology, 54,* 1–29.

Cain, W. S. (1977). Differential sensitivity for smell: "Noise" at the nose. *Science, 195,* 796–798.

Cain, W. S. (1979). To know with the nose: Keys to odor identification. *Science, 203,* 467–470.

Cain, W. S. (1980). Sensory attributes of cigarette smoking. *Branbury Report: 3. A safe cigarette?* Cold Spring Harbor, NY: Cold Spring Harbor Laboratory, pp. 239–249.

Cain, W. S. (1988). Olfaction. In R. A. Atkinson, R. J. Herrnstein, G. Lindzey, & R. D. Luce (Eds.), *Stevens' handbook of experimental psychology: Vol. 1. Perception and motivation* (rev. ed., pp. 409–459). New York: Wiley.

Cameron, E. L., Baker, C. L., & Boulton, J. C. (1992). Spatial frequency selective mechanisms underlying the motion aftereffect. *Vision Research, 32,* 561–568.

Campbell, F. W., & Robson, J. G. (1968). Application of Fourier analysis to the visibility of gratings. *Journal of Physiology, 197,* 551–566.

Capretta, P. J., Petersik, J. T., & Steward, D. J. (1975). Acceptance of novel flavours is increased after early exposure of diverse taste. *Nature, 254,* 689–691.

Casagrande, V. A. (1994). A third parallel visual pathway to primate area V1. *Trends in Neuroscience, 17,* 305–310.

Casagrande, V. A., & Norton, T. T. (1991). Lateral geniculate nucleus: A review of its physiology and function. In A. G. Leventhal (Ed.), *The neural basis of visual function* (pp. 41–84), Vol. 4 of J. R. Cooney-Dillon (Ed.), *Vision and visual dysfunction.* London: Macmillan.

Case, J. F. (1964). Properties of the dactyl chemoreceptors of *Cancer antennarius* (L.) and *C. productus* (Randall). *Biological Bulletin, 127,* 428–446.

Cavanagh, P., Tyler, C. W., & Favreau, O. E. (1984). Perceived velocity of moving chromatic gratings. *Journal of the Optical Society of America A, 1,* 893–899.

Cernoch, J. M., & Porter, R. H. (1985). Recognition of maternal axillary odors by infants. *Child Development, 56,* 1593–1598.

Chapanis, A. (1947). The dark adaptation of the color anomalous. *Journal of General Physiology, 30,* 423–437.

Chase, R., & Rieling, J. (1986). Autoradiographic evidence for receptor cell renewal in the olfactory epithelium of a snail. *Brain Research, 384,* 232–239.

Chase, R., & Wells, M. J. (1986). Chemotactic behavior in *Octopus. Journal of Comparative Physiology A, 158,* 375–381.

Cholewiak, R. W. (1991). Sensory and physiological bases of touch. In M. A. Heller & W. Schiff (Eds.), *The psychology of touch* (pp. 23–60). Hillsdale, NJ: Erlbaum.

Cinelli, A. R. (1993). Review of *Science of olfaction. Trends in Neurosciences, 16,* 123–124.

Clark, S. A., Allard, T., Jenkins, W. M., & Merzenich, M. M. (1988). Receptive fields in the body-surface map in adult cortex defined by temporally correlated inputs. *Nature, 332,* 444–445.

Clark, W. C., & Clark, S. B. (1980). Pain responses in Nepalese porter. *Science, 209,* 410–412.

Clulow, F. W. (1972). *Color: Its principles and their applications.* New York: Morgan & Morgan.

Cohen, L. H. (1994). *Train go sorry.* Boston: Houghton Mifflin.

Collett, T. S. (1978). Peering—A locust behavior pattern for obtaining motion parallax information. *Journal of Experimental Biology, 76,* 237–241.

Collett, T. S. (1993). Orientation detectors in insects. *Nature, 362,* 494.

Collett, T. S., & Harkness, L. I. K. (1982). Depth vision in animals. In D. J. Ingle, M. A. Goodale, & R. J. W. Mansfield (Eds.), *Analysis of visual behavior* (pp. 111–176). Cambridge: MIT Press.

Coltheart, M. (1970). The effect of verbal size information upon visual judgments of absolute distance. *Perception and Psychophysics, 9,* 222–223.

Comel, M. (1953). *Fisiologia normale e patologica della cute umana.* Milan: Fratelli Treves Editori.

Conel, J. L. (1939). *The postnatal development of the cerebral cortex* (Vol. 1). Cambridge: Harvard University Press.

Conel, J. L. (1947). *The postnatal development of the cerebral cortex* (Vol. 2). Cambridge: Harvard University Press.

Conel, J. L. (1951). *The postnatal development of the cerebral cortex* (Vol. 3). Cambridge: Harvard University Press.

Connolly, M., & Van Essen, D. (1984). The representation of the visual field in parvocellular and magnocellular layers of the lateral geniculate nucleus in the macaque monkey. *Journal of Comparative Neurology, 226,* 544–564.

Costanzo, R. M., & Gardner, E. B. (1980). A quantitative analysis of responses of direction-sensitive neurons in somatosensory cortex of awake monkeys. *Journal of Neurophysiology, 43,* 1319–1341.

Costanzo, R. M., & Gardner, E. B. (1981). Multiple-joint neurons in somatosensory cortex of awake monkeys. *Brain Research, 24,* 321–333.

Cowey, A., & Stoerig, P. (1991). The neurobiology of blindsight. *Trends in Neuroscience, 14,* 140–145.

Crawford, M. L. J., & von Noorden, G. K. (1980). Optically induced concomitant strabismus in monkey. *Investigative Ophthalmology and Visual Science, 19,* 1105–1109.

Culler, E. A., Coakley, J. D., Lowy, K., & Gross, N. (1943). A revised frequency-map of the guinea-pig cochlea. *American Journal of Psychology, 56,* 475–500.

Cutting, J. E. (1986). *Perception with an eye for motion.* Cambridge: MIT Press.

Cynader, M., Timney, B. N., & Mitchell, D. E. (1980). Period of susceptibility of kitten visual cortex to the effects of monocular deprivation extends beyond six months of age. *Brain Research, 191,* 545–550.

Dalton, J. (1948). Extraordinary facts relating to the vision of colour: With observations. In W. Dennis (Ed.), *Readings in the history of psychology* (pp. 102–111). New York: Appleton-Century-Crofts. (Original work published 1798).

Damasio, A. R., Tranel, D., & Damasio, H. (1990). Face agnosia and the neural substrates of memory. *Annual Review of Neuroscience, 13,* 89–109.

Dartnall, H. J. A., Bowmaker, J. K., & Mollon, J. D. (1983). Human visual pigments: Microspectrophometric results from the eyes of seven persons. *Proceedings of the Royal Society of London, 220B,* 115–130.

Dear, S. P., Simmons, J. A., & Fritz, J. (1993). A possible neuronal basis for representation of acoustic scenes in auditory cortex of the big brown bat. *Nature, 364,* 620–623.

DeCasper, A. J., & Fifer, W. P. (1980). Of human bonding: newborns prefer their mothers' voices. *Science, 208,* 1174–1176.

DeCasper, A. J., Lecanuet, J-P., Busnel, M-C., Deferre-Granier, C., & Maugeais, R. (1994). Fetal reactions to recurrent maternal speech. *Infant Behavior and Development, 17,* 159–164.

DeCasper, A. J., & Spence, M. J. (1986). Prenatal maternal speech influences newborn's perception of speech sounds. *Infant Behavior and Development, 9,* 133–150.

Delgutte, B., & Kiang, N. Y. S. (1984a). Speech coding in the auditory nerve: 1. Vowel-like sounds. *Journal of the Acoustical Society of America, 75,* 866–878.

Delgutte, B., & Kiang, N. Y. S. (1984b). Speech coding in the auditory nerve. 3. Voiceless fricative consonants. *Journal of the Acoustical Society of America, 75,* 887–896.

Delk, J. L., & Fillenbaum, S. (1965). Differences in perceived color as a function of characteristic color. *American Journal of Psychology, 78,* 290–293.

DeLucia, P., & Hochberg, J. (1985). Illusions in the real world and in the mind's eye (Abstract). *Proceedings of the Eastern Psychological Association, 56,* 38.

DeLucia, P., & Hochberg, J. (1986). Real-world geometrical illusions: Theoretical and practical implications (Abstract). *Proceedings of the Eastern Psychological Association, 57,* 62.

DeLucia, P. R., & Hochberg, J. (1991). Geometrical illusions in solid objects under ordinary viewing conditions. *Perception and Psychophysics, 50,* 547–554.

Derrington, A. M., Lennie, P., & Krauskopf, J. (1983). Chromatic response properties of parvocellular neurons in the macaque LGN. In J. D. Mollon & L. T. Sharpe (Eds.), *Colour vision* (pp. 245–251). London: Academic Press.

Desor, J. A., & Beauchamp, G. K. (1974). The human capacity to transmit olfactory information. *Perception and Psychophysics, 13,* 271–275.

Dethier, V. G. (1987). Sniff, flick, and pulse: an appreciation of interruption. *Proceedings of the American Philosophical Society, 131,* 159–176.

Deutsch, D. (1973). Octave generalization of specific interference effects in memory for tonal pitch. *Perception and Psychophysics, 13,* 271–275.

Deutsch, D. (1975). Two-channel listening to musical scales. *Journal of the Acoustical Society of America, 57,* 1156–1160.

DeValois, R. L. (1960). Color vision mechanisms in monkey. *Journal of General Physiology, 43,* 115–128.

DeValois, R. L., & DeValois, K. K. (1993). A multistage color model. *Vision Research, 33,* 1053–1065.

DeValois, R. L., & Jacobs, G. H. (1968). Primate color vision. *Science, 162,* 533–540.

DeValois, R. L., & Jacobs, G. H. (1984). Neural mechanisms of color vision. In J. M. Brookhart & V. B. Mountcastle (Eds.), *Handbook of physiology: 3. The nervous system* (pp. 425–456). Bethesda, MD: American Physiological Society.

deVries, H., & Stuiver, M. (1961). The absolute sensitivity of the human sense of smell. In W. A. Rosenblith (Ed.), *Sensory communication.* Cambridge: MIT Press.

DeYoe, E. A., & Van Essen, D. C. (1988). Concurrent processing streams in monkey visual cortex. *Trends in Neurosciences, 11,* 219–226.

Dobelle, W. H. (1977). Current status of research on providing sight to the blind by electrical stimulation of the brain. *Journal of Visual Impairment and Blindness, 71,* 290–297.

Dobelle, W. H., Mladejovsky, J. J., Evans, J. R., Roberts, T. S., & Girvin, J. P. (1976). "Braille" reading by a blind volunteer by visual cortex stimulation. *Nature, 259,* 111–112.

Dobelle, W. H., Mladejovsky, M. J., & Girvin, J. P. (1974). Artificial vision for the blind: Electrical stimulation of visual cortex offers hope for a functional prosthesis. *Science, 183,* 440–444.

Dobson, V., & Teller, D. (1978). Visual acuity in human infants: Review and comparison of behavioral and electrophysiological studies. *Vision Research, 18,* 1469–1483.

Dodd, G. G., & Squirrell, D. J. (1980). Structure and mechanism in the mammalian olfactory system. *Symposium of the Zoology Society of London, 45,* 35–56.

Dodd, J., & Castellucci, V. F. (1991). Smell and taste: The chemical senses. In E. R. Kandel, J. H. Schwartz, & T. M. Jessell (Eds.), *Principles of neural science* (3rd ed., pp. 512–529). New York: Elsevier.

Doty, R. L. (Ed.). (1976). *Mammalian olfaction, reproductive processes and behavior.* New York: Academic Press.

Doty, R. L. (1991). Olfactory system. In T. V. Getchell, R. L. Doty, L. M. Bartoshuk, & J. B. Snow (Eds.), *Smell and taste in health and disease* (pp. 175–203). New York: Raven.

Doty, R. L., Green, P. A., Ram, C., & Yankell, S. L. (1982). Communication of gender from human breath odors: Relationship to perceived intensity and pleasantness. *Hormones and Behavior, 16,* 13–22.

Doving, K. B., & Thommesen, G. (1977). Some properties of the fish olfactory system. In J. LeMagnen & P. MacLeod (Eds.), *Olfaction and taste* (Vol. 6, pp. 175–183). London: Information Retrieval.

Dowling, J. E., & Boycott, B. B. (1966). Organization of the primate retina. *Proceedings of the Royal Society of London, 166B,* 80–111.

Dowling, W. J. (1973). The perception of interleaved melodies. *Cognitive Psychology, 5,* 322–337.

Dowling, W. J., & Harwood, D. L. (1986). *Music cognition.* New York: Academic Press.

Duclaux, R., & Kenshalo, D. R. (1980). Response characteristics of cutaneous warm fibers in the monkey. *Journal of Neurophysiology, 43,* 1–15.

Duffy, C. J., & Wurtz, R. H. (1991). Sensitivity of MST neurons to optic flow stimuli: 2. Mechanisms of response selectivity revealed by small-field stimuli. *Journal of Neurophysiology, 65,* 1346–1359.

Duhamel, J-R., Colby, C. L., & Goldberg, M. E. (1992). The updating of the representation of visual space in parietal cortex by intended eye movements. *Science, 255,* 90–92.

Duncker, D. K. (1929). Über induzierte Bewegung (Ein Beitrag zur Theorie optisch wahrgenommener Bewegung). *Psychologische Forschung, 12,* 180–259. (Summary in W. D. Ellis, 1938, *A sourcebook of Gestalt psychology.* London: Kegan Paul, Trench, Trubner)

Durlach, N. I., & Colburn, H. S. (1978). Binaural phenomena. In E. C. Carterette & M. P. Friedman (Eds.), *Handbook of perception* (Vol. 4, pp. 365–466). New York: Academic.

Durrant, J., & Lovrinic, J. (1977). *Bases of hearing science.* Baltimore: Williams & Wilkins.

Egan, J. P., & Hake, H. W. (1950). On the masking pattern of a simple auditory stimulus. *Journal of the Acoustical Society of America, 22,* 622–630.

Eimas, P. D., & Corbit, J. D. (1973). Selective adaptation of linguistic feature detectors. *Cognitive Psychology, 4,* 99–109.

Eimas, P. D., & Miller, J. L. (1992). Organization in the perception of speech by young infants. *Psychological Science, 3,* 340–345.

Eimas, P. D., Siqueland, E. R., Jusczyk, P., & Vigorito, J. (1971). Speech perception in infants. *Science, 171,* 303–306.

Emmerton, J., & Delius, J. D. (1980). Wavelength discrimination in the "visible" and ultraviolet spectrum by pigeons. *Journal of Comparative Physiology, 141,* 47–52.

Engel, A. K., Konig, P., Kreiter, A. K., Schillen, T. B., & Singer, W. (1992). Temporal coding in the visual cortex: New vistas on integration in the nervous system. *Trends in Neurosciences, 15,* 218–226.

Engel, A. K., Konig, P., Kreiter, A. K., & Singer, W. (1991a). Interhemispheric synchronization of oscillatory neuronal responses in cat visual cortex. *Science, 252,* 1177–1179.

Engel, A. K., Konig, P., & Singer, W. (1991b). Direct physiological evidence for scene segmentation by temporal coding. *Proceedings of the National Academy of Sciences, 88,* 9136–9140.

Engen, T. (1972). Psychophysics, In J. W. Kling & L. A. Riggs (Eds.) *Experimental psychology* (3rd ed., pp. 1–46). New York: Holt, Rinehart and Winston.

Engen, T., & Pfaffmann, C. (1960). Absolute judgments of odor quality. *Journal of Experimental Psychology, 59,* 214–219.

Enns, J. (1986). Seeing textons in context. *Perception and Psychophysics, 39,* 143–147.

Epstein, W. (1965). Nonrelational judgments of size and distance. *American Journal of Psychology, 78,* 120–123.

Epstein, W. (1977). What are the prospects for a higher-order stimulus theory of perception? *Scandinavian Journal of Psychology, 18,* 164–171.

Erickson, R. (1975). *Sound structure in music.* Berkeley: University of California Press.

Erickson, R. P. (1963). Sensory neural patterns and gustation. In Y. Zotterman (Ed.), *Olfaction and taste* (Vol. 1, pp. 205–213). Oxford: Pergamon.

Erickson, R. P. (1982). The across-fiber pattern theory: An organizing principle for molar neural function. In W. D. Neff (Ed.), *Contributions to sensory physiology* (Vol. 6, pp. 79–109). New York: Academic Press.

Erickson, R. P. (1984). On the neural bases of behavior. *American Scientist, 72,* 233–241.

Evans, E. F. (1978). Place and time coding of frequency in the peripheral auditory system: Some physiological pros and cons. *Audiology, 17,* 369–420.

Fagan, J. F. (1976). Infant's recognition of invariant features of faces. *Child Development, 47,* 627–638.

Fantz, R. L. (1965). Visual perception from birth as shown by pattern selectivity. *Annals of the New York Academy of Sciences, 118,* 793–814.

Fantz, R. L., & Nevis, S. (1967). Pattern preferences and perceptual-cognitive development in early infancy. *Merrill-Palmer Quarterly, 13,* 77–108.

Fantz, R. L., Ordy, J. M., & Udelf, M. S. (1962). Maturation of pattern vision in infants during the first six months. *Journal of Comparative and Physiological Psychology, 55,* 907–917.

Farah, M. J. (1990). *Visual agnosia: Disorders of object recognition and what they tell us about normal vision.* Cambridge: MIT Press.

Farah, M. J. (1992). Is an object an object an object? Cognitive and neuropsychological investigations of domain specificity in visual object recognition. *Current Directions in Psychological Science, 1,* 164–169.

Feddersen, W. E., Sandel, T. T., Teas, D. C., & Jeffress, L.A. (1957). Localization of high frequency tones. *Journal of the Acoustical Society of America, 5,* 82–108.

Felleman, D. J., & Van Essen, D. C. (1987). Receptive field properties of neurons in area V3 of macaque monkey extrastriate cortex. *Journal of Neurophysiology, 57,* 889-920.

Felleman, D. J., & Van Essen, D. C. (1991). Distributed hierarchical processing in the primate cerebral cortex. *Cerebral Cortex, 1,* 1–47.

Fendrich, R., Wessinger, M., & Gazzaniga, M. S. (1992). Residual vision in a scotoma: Implications for blindsight. *Nature, 258,* 1489–1491.

Finger, T. E. (1987). Gustatory nuclei and pathways in the central nervous system. In T. E. Finger & W. L. Silver (Eds.), *Neurobiology of taste and smell* (pp. 331–353). New York: Wiley.

Fiorentini, A., & Maffei, L. (1973). Contrast in night vision. *Vision Research, 13,* 73–80.

Fletcher, H., & Munson, W. A. (1933). Loudness: Its definition, measurement, and calculation. *Journal of the Acoustical Society of America, 5,* 82–108.

Fodor, J. A., & Pylyshyn, Z. W. (1981). How direct is visual perception? Some reflections of Gibson's "ecological approach." *Cognition, 9,* 139–196.

Foulke, E. (1991). Braille. In M. A. Heller & W. Schiff (Eds.), *The psychology of touch* (pp. 219–223). Hillsdale, NJ: Erlbaum.

Fowler, C. A., & Rosenblum, L. D. (1990). Duplex perception: A comparison of monosyllables and slamming doors. *Journal of Experimental Psychology: Human Perception and Performance, 16,* 742–754.

Fox, C. R. (1990). Some visual influences on human postural equilibrium: Binocular versus monocular fixation. *Perception and Psychophysics, 47,* 409–422.

Fox, R., Aslin, R. N., Shea, S. L., & Dumais, S. T. (1980). Stereopsis in human infants. *Science, 207,* 323–324.

Fox, R., Lehmukuhle, S. W., & Westendorf, D. H. (1976). Falcon visual acuity. *Science, 192,* 263–265.

Fox, R., & McDaniel, C. (1982). The perception of biological motion by human infants. *Science, 218,* 486–487.

Frank, M. E., Bieber, S. L., & Smith, D. V. (1988). The organization of taste sensibilities in hamster chorda tympani nerve fibers. *Journal of General Physiology, 91,* 861-896.

Frank, M. E., & Rabin, M. D. (1989). Chemosensory neuroanatomy and physiology. *Ear, Nose and Throat Journal, 68,* 291–292, 295–296.

Friedman, M. B. (1975). Visual control of head movements during avian locomotion. *Nature, 225,* 67–69.

Fuld, K., Wooten, B. R., & Whalen, J. J. (1981). Elemental hues of short-wave and spectral lights. *Perception and Psychophysics, 29,* 317–322.

Fuzessery, Z. M., & Feng, A. S. (1983). Mating call selectivity in the thalamus and midbrain of the leopard frog (*Rana p. pipiens*): Single and multiunit analyses. *Journal of Comparative Physiology, 150A,* 333–344.

Galambos, R. (1941). Cochlear potentials from the bat. *Science, 93,* 215.

Galambos, R. (1942a). The avoidance of obstacles by bats: Spallanzani's ideas (1794) and later theories. *Isis, 34,* 132–140.

Galambos, R. (1942b). Cochlear potentials elicited from bats by supersonic sounds. *Journal of the Acoustical Society of America, 14,* 41–49.

Gamble, A. E. McC. (1898). The applicability of Weber's law to smell. *American Journal of Psychology, 10,* 82–142.

Ganchrow, J. R. (1995). Ontogeny of human taste perception. In R. L. Doty (Ed.), *Handbook of olfaction and gustation* (pp. 715–729). New York: Marcel Dekker.

Ganchrow, J. R., Steiner, J. E., & Daher, M. (1983). Neonatal facial expressions in response to different qualities and intensities of gustatory stimuli. *Infant Behavior and Development, 6,* 473–484.

Garcia, J., Ervin, F. R., & Koelling, R. A. (1966). Learning with prolonged delay of reinforcement. *Psychonomic Science, 5,* 121–122.

Garcia, J., Hawkins, W. G., & Rusiniak, K. W. (1974). Behavioral regulation of the milieu interne in man and rat. *Science, 185,* 824–831.

Garcia, J., & Koelling, R. A. (1966). A relation of cue to consequence in avoidance learning. *Psychonomic Science, 4,* 123–124.

Gardner, E. B., & Costanzo, R. H. (1981). Properties of kinesthetic neurons in somatosensory cortex of awake monkeys. *Brain Research, 214,* 301–319.

Gardner, M. B., & Gardner, R. S. (1973). Problem of localization in the median plane: Effect of pinnae cavity occlusion. *Journal of the Acoustical Society of America, 53,* 400–408.

Gazzaniga, M. S., Fendrich, R., & Wessinger, C. M. (1994). Blindsight reconsidered. *Current Directions in Psychological Science, 3,* 93–96.

Gelb, A. (1929). Die "Farbenkoinstanz" der Sehding. *Handbook norm. path. Phys., 12,* 594–678.

Gescheider, G. A. (1976). *Psychophysics: Method and theory.* Hillsdale, NJ: Erlbaum.

Geschwind, N. (1979, September). Specializations of the human brain. *Scientific American,* pp. 108–119.

Getty, D. J., & Howard, J. H. (Eds.). (1981). *Auditory and visual pattern recognition.* Hillsdale, NJ: Erlbaum.

Gibson, J. J. (1950). *The perception of the visual world.* Boston: Houghton Mifflin.

Gibson, J. J. (1962). Observations on active touch. *Psychological Review, 69,* 477–491.

Gibson, J. J. (1966). *The senses considered as perceptual systems.* Boston: Houghton Mifflin.

Gibson, J. J. (1979). *The ecological approach to visual perception.* Boston: Houghton Mifflin.

Gilbert, C. D., & Wiesel, T. N. (1989). Columnar specificity of intrinsic horizontal and corticocortical connections in cat visual cortex. *Journal of Neuroscience, 9,* 2432–2442.

Gilinsky, A. S. (1965). The effect of attitude upon the perception of size. *American Journal of Psychology, 68,* 173–192.

Ginsburg, A. (1983). *Contrast perception in the human infant.* Unpublished manuscript.

Glickstein, M. (1988, September). The discovery of the visual cortex. *Scientific American, 259,* 118–127.

Goldstein, J. L. (1978). Mechanisms of signal analysis and pattern perception in periodicity pitch. *Audiology, 17,* 421–445.

Goleman, D. (1994, September 6). Sonic device for blind may aid navigation. *New York Times,* C1 (late edition, final).

Gombrich, E. H. (1960). *Art and illusion.* Princeton, NJ: Princeton University Press.

Gordon, J., & Abramov, I. (1988). Scaling procedures for specifying color appearance. *Color Research Applications, 13,* 146–152.

Gossard, E. E., & Hooke, W. H. (1975). *Waves in the atmosphere.* Amsterdam: Elsevier.

Gould, J. L. (1984). Magnetic field sensitivity in animals. *Annual Review of Physiology, 46,* 585–598.

Gould, J. L. (1993). Birds lost in the red. *Nature, 364,* 491–492.

Gouras, P. (1991a). Color vision. In E. R. Kandel, J. H. Schwartz, & T. M. Jessell (Eds.), *Principles of neural science* (3rd ed., pp. 467–480). New York: Elsevier.

Gouras, P. (1991b). Cortical mechanisms of colour vision. In P. Gouras (Ed.), *Vision and visual dysfunction* (Vol. 6, pp. 179–197). London: Macmillan.

Gouras, P. (1991c). Precortical physiology of colour vision. In P. Gouras (Ed.), *Vision and visual dysfunction* (Vol. 6, pp. 163–178). London: Macmillan.

Graham, C. H. (1965). Perception of movement. In C. Graham (Ed.), *Vision and visual perception* (pp. 575–588). New York: Wiley.

Graham, C. H., Sperling, H. G., Hsia, Y., & Coulson, A. H. (1961). The determination of some visual functions of a unilaterally color-blind subject: Methods and results. *Journal of Psychology, 51,* 3–32.

Graham, N. (1992). Breaking the visual stimulus into parts. *Current Directions in Psychological Science, 1,* 55–61.

Graham, N., Beck, J., & Sutter, A. (1992). Nonlinear processes in spatial-frequency channel models of perceived texture segregation: Effects of sign and amount of contrast. *Vision Research, 32*, 719–743.

Granrud, C. E. (1987). *Visual size constancy in newborn infants.* Paper presented at 1987 meeting of the Association for Research in Vision and Ophthalmology, Sarasota, Florida.

Granrud, C. E., Haake, R. J., & Yonas, A. (1985). Infants' sensitivity to familiar size: The effect of memory on spatial perception. *Perception and Psychophysics, 37*, 459–466.

Granrud, C. E., & Yonas, A. (1984). Infants' perception of pictorially specified interposition. *Journal of Experimental Child Psychology, 37*, 500–511.

Granrud, C. E., Yonas, A., & Opland, E. A. (1985). Infants' sensitivity to the depth cue of shading. *Perception and Psychophysics, 37*, 415–419.

Gray, C. M., & Singer, W. (1989). Stimulus-specific neuronal oscillations in orientation columns of cat visual cortex. *Proceedings of the National Academy of Sciences, 86*, 1698–1702.

Graziadei, P. P. C. (1976). Functional anatomy of the mammalian chemoreceptor system. In D. Muller-Schwarze & M. Mozell (Eds.), *Chemical signals in vertebrates* (pp. 435–454). New York: Plenum.

Graziano, M. S. A., Andersen, R. A., & Snowden, R. J. (1994). Tuning of MST neurons to spiral motions. *Journal of Neuroscience, 14*, 54–67.

Green, D. M. (1976). *An introduction to hearing.* Hillsdale, NJ: Erlbaum.

Green, K. P., Kuhl, P. K., Meltzoff, A. N., & Stevens, E. B. (1991). Integrating speech information across talkers, gender, and sensory modality: Female faces and male voices in the McGurk effect. *Perception and Psychophysics, 50*, 524–536.

Green, S. (1975). Variation of vocal pattern with social situation in the Japanese monkey (*Macaca fuscata*): A field study. In L. A. Rosenblum (Ed.), *Primate behavior* (Vol. 4, pp. 1–102). New York: Academic Press.

Gregory, R. L. (1966). *Eye and brain.* New York: McGraw-Hill.

Gregory, R. L. (1973). *Eye and brain* (2nd ed.). New York: McGraw-Hill.

Grether, W. F. (1939). Color vision and color blindness in monkeys. *Comparative Psychology Monographs, 15*, 1–38.

Griffin, D. R. (1944). Echolocation by blind men and bats. *Science, 100*, 589–590.

Griffin, D. R. (1967). Discriminative echolocation by bats. In R-G Busnel (Ed.), *Animal sonar systems* (Vol. 1, p. 273). Jouy-en-Josas, France: Lab Physiological Acoustics.

Grunwald, A. P. (1965). A braille reading machine. *Science, 154*, 144–146.

Gulick, W. L. (1971). *Hearing.* New York: Oxford University Press.

Gulick, W. L., Gescheider, G. A., & Frisina, R. D. (1989). *Hearing.* New York: Oxford University Press.

Gyr, J. W. (1972). Is a theory of direct perception adequate? *Psychological Bulletin, 77*, 246–261.

Hagen, M. A. (Ed.). (1979). *The perception of pictures* (Vols. 1, 2). New York: Academic.

Haith, M. M. (1983). Spatially determined visual activity in early infancy. In A. Hein & M. Jeannerod (Eds.), *Spatially oriented behavior.* New York: Springer.

Hall, E. G., & Davies, S. (1991). Gender differences in perceived intensity and affect of pain between athletes and nonathletes. *Perceptual and Motor Skills, 73*, 779–786.

Hall, J. L. (1965). Binaural interaction in the accessory superior-olivary nucleus of the cat. *Journal of the Acoustical Society of America, 37*, 814–823.

Hall, M. J., Bartoshuk, L. M., Cain, W. S., & Stevens, J. C. (1975). PTC taste blindness and the taste of caffeine. *Nature, 253*, 442–443.

Hamer, R. D., Alexander, K. R., & Teller, D. Y. (1982). Rayleigh discriminations in young human infants. *Vision Research, 22*, 575–587.

Hanson, A. R., & Riseman, E. M. (1978). *Computer vision systems.* New York: Academic.

Harman, G. (1974). Epistemology. In E. C. Carterette & M. P. Friedman (Eds.), *Handbook of perception* (Vol. 1, pp. 410–455). New York: Academic.

Harris, L., Atkinson, J., & Braddick, O. (1976). Visual contrast sensitivity of a 6-month-old infant measured by the evoked potential. *Nature, 246*, 570–571.

Harrison, R. V. (1988). *The biology of hearing and deafness.* Springfield, IL: Charles C Thomas.

Hartline, H. K., Wagner, H. G., & Ratliff, F. (1956). Inhibition in the eye of *Limulus*. *Journal of General Physiology, 39*, 651–673.

Haxby, J. V., Grady, C. L., Horwitz, B., Ungerleider, L. G., Mishkin, M., Carson, R. E., Hersovitch, P., Schapiro, M. B., & Rapoport, S. I. (1991). Dissociation of object and spatial visual processing pathways in human extrastriate cortex. *Proceedings of the National Academy of Sciences, 88,* 1621–1625.

Heffner, H. E. (1983). Hearing in large and small dogs: Absolute thresholds and size of the tympanic membrane. *Behavioral Neuroscience, 97,* 310–318.

Heffner, H. E., & Masterton, R. B. (1980). Hearing in glires: Domestic rabbit, cotton rat, feral house mouse, and kangaroo rat. *Journal of the Acoustical Society of America, 68,* 1584–1599.

Heffner, R. S., & Heffner, H. E. (1980). Hearing in the elephant (*Elephas maximus*). *Science, 208,* 518–520.

Heffner, R. S., & Heffner, H. E. (1985). Hearing in mammals: The least weasel. *Journal of Mammalogy, 66,* 745–755.

Heffner, R. S., & Heffner, H. E. (1992). Visual factors in sound localization in mammals. *Journal of Comparative Neurology, 317,* 219–232.

Held, R. (1985). Binocular vision—Behavioral and neural development. In J. Mehler & R. Fox (Eds.), *Neonate cognition: Beyond the blooming, buzzing confusion* (pp. 37–44). Hillsdale, NJ: Erlbaum.

Held, R. (1991). Development of binocular vision and stereopsis. In D. Regan (Ed.), *Vision and visual dysfunction* (Vol. 9, pp. 170–178). New York: Macmillan.

Held, R. (1993). Two stages in the development of binocular vision and eye alignment. In K. Simons (Ed.), *Early visual development: Normal and abnormal* (Chap. 15). New York: Oxford University Press.

Held, R., Birch, E. E., & Gwiazda, J. (1980). Stereoacuity of human infants. *Proceedings of the National Academy of Sciences, 77,* 5572–5574.

Heller, M. A. (1986). Active and passive tactile braille recognition. *Bulletin of the Psychonomic Society, 24,* 201–202.

Helmholtz, H. von (1852). On the theory of compound colors. *Philosophical Magazine, 4,* 519–534.

Helmholtz, H. von (1909–1911). *Treatise on physiological optics* (Vols. 2, 3, 3rd ed.) (J. P. Southall, Ed. and Trans.). Rochester, NY: Optical Society of America. (Original work published 1866)

Helmholtz, H. von (1954). *Die Lehr von den Tonenpfindungen als physiologische Grundlege für die Theorie der Musik* [On the sensations of tone as a physiological basis for the theory of music] (A. J. Ellis, Trans.). New York: Dover. (Original work published 1863)

Helson, H. (1933). The fundamental propositions of Gestalt psychology. *Psychological Review, 40,* 13–32.

Helson, H., Judd, D. B., & Wilson, M. (1956). Color rendition with fluorescent sources of illumination. *Illuminating Engineering, 51,* 329–346.

Hering, E. (1878). *Zur Lehre vom Lichtsinn.* Vienna: Gerold.

Hering, E. (1905). Grundzuge der Lehre vom Lichtsinn. In *Handbuch der gesamter Augenheilkunde* (Vol. 3, Chap. 13). Berlin.

Hering, E. (1964). *Outlines of a theory of the light sense* (L. M. Hurvich & D. Jameson, Trans.). Cambridge: Harvard University Press.

Hershenson, M. (Ed.). (1989). *The moon illusion.* Hillsdale, NJ: Erlbaum.

Hettinger, T. P., Myers, W. E., & Frank, M. E. (1990). Role of olfaction in perception of nontraditional "taste" stimuli. *Chemical Senses, 15,* 755–760.

Hildreth, E. C. (1990). The neural computation of the velocity field. In B. Cohen & I. Bodis-Wollner (Eds.), *Vision and the brain* (pp. 139–164). New York: Raven.

Hiris, E., & Blake, R. (1992). Another perspective on the visual motion aftereffect. *Proceedings of the National Academy of Sciences, 89,* 9025–9028.

Hochberg, J. E. (1970). Attention, organization, and consciousness. In D. I. Mostofsky (Ed.), *Attention: Contemporary theory and analysis* (pp. 99–124). New York: Appleton-Century-Crofts.

Hochberg, J. E. (1971). Perception. In J. W. Kling & L. A. Riggs (Eds.), *Experimental psychology* (3rd ed., pp. 396–550). New York: Holt, Rinehart & Winston.

Hochberg, J. E. (1987). Machines should not see as people do, but must know how people see. *Computer Vision, Graphics and Image Processing, 39,* 221–237.

Hohmann, A., & Creutzfeldt, O. D. (1975). Squint and the development of binocularity in humans. *Nature, 254,* 613–614.

Holley, A. (1991). Neural coding of olfactory information. In T. V. Getchell et al. (Eds.), *Smell and taste in health and disease* (pp. 329–343). New York: Raven.

Holley, A., Duchamp, A., Revial, M. F., Juge, A., & Macleod, P. (1974). Qualitative and quantitative discrimination in the frog olfactory receptors:

Analysis from electrophysiological data. *Annals of the New York Academy of Sciences, 237,* 102–114.

Holway, A. H., & Boring, E. G. (1941). Determinants of apparent visual size with distance variant. *American Journal of Psychology, 54,* 21–37.

Hornung, D. E., Lansing, R. D., & Mozell, M. M. (1975). Distribution of butanol molecules along bullfrog olfactory mucosa. *Nature, 254,* 617–618.

Hornung, D. E., & Mozell, M. M. (1977). Factors influencing the differential sorption of odorant molecules across the olfactory mucosa. *Journal of General Physiology, 69,* 343–361.

Hornung, D. E., & Mozell, M. M. (1981). Accessibility of odorant molecules to the receptors. In R. H. Cagan & M. R. Kare (Eds.), *Biochemistry of taste and olfaction* (pp. 33–45). New York: Academic.

Hothersall, D. (1990). *History of psychology* (2nd ed.). New York: McGraw-Hill.

Houtsma, A. J. M., & Goldstein, J. L. (1972). Perception of musical intervals: Evidence for the central origin of the pitch of complex tones. *Journal of the Optical Society of America, 51,* 520–529.

Hubel, D. H. (1982). Exploration of the primary visual cortex, 1955–1978. *Nature, 299,* 515–524.

Hubel, D. H. (1988). *Eye, brain and vision.* New York: Scientific American Library.

Hubel, D., & Livingston, M. (1990). Color puzzles. *Cold Spring Harbor Symposia on Quantitative Biology, 60,* 643–649.

Hubel, D. H., & Wiesel, T. N. (1959). Receptive fields of single neurons in the cat's striate cortex. *Journal of Physiology, 148,* 574–591.

Hubel, D. H., & Wiesel, T. N. (1961). Integrative action in the cat's lateral geniculate body. *Journal of Physiology, 155,* 385–398.

Hubel, D. H., & Wiesel, T. N. (1965a). Receptive fields and functional architecture in two non-striate visual areas (18 and 19) of the cat. *Journal of Neurophysiology, 28,* 229–289.

Hubel, D. H., & Wiesel, T. N. (1965b). Binocular interaction in striate cortex of kittens reared with artificial squint. *Journal of Neurophysiology, 28,* 1041–1059.

Hubel, D. H., & Wiesel, T. N. (1970). Cells sensitive to binocular depth in area 18 of the macaque monkey cortex. *Nature, 225,* 41–42.

Hubel, D. H., & Wiesel, T. N. (1974). Uniformity of monkey striate cortex: A parallel relationship between field size, scatter, and magnification factor. *Journal of Comparative Neurology, 158,* 295–306.

Hubel, D. H., & Wiesel, T. N. (1977). Functional architecture of macaque monkey cortex. *Proceedings of the Royal Society of London, 198,* 1–59.

Hubel, D. H., Wiesel, T. N., & Stryker, M. P. (1978). Anatomical demonstration of orientation columns in macaque monkey. *Journal of Comparative Neurology, 177,* 361–379.

Hudspeth, A. J. (1983). The hair cells of the inner ear. *Scientific American, 248* (1), 54–64.

Humphrey, A. L., & Saul, A. B. (1994). The temporal transformation of retinal signals in the lateral geniculate nucleus of the cat: Implications for cortical function. In D. Minciacchi, M. Molinari, G. Macchi, & E. G. Jones (Eds.), *Thalamic networks for relay and modulation* (pp. 81–89). New York: Pergamon.

Hurvich, L. (1981). *Color vision.* Sunderland, MA: Sinauer.

Hyvarinin, J., & Poranen, A. (1974). Functions of the association area 7 as revealed from cellular discharges in alert monkeys. *Brain, 97,* 673–692.

Hyvarinin, J., & Poranen, A. (1978). Movement-sensitive and direction and orientation-selective cutaneous receptive fields in the hand area of the postcentral gyrus in monkeys. *Journal of Physiology, 283,* 523–537.

Ilg, U. J., Bridgeman, B., & Hoffmann, K. P. (1989). Influence of mechanical disturbance on oculomotor behavior. *Vision Research, 29,* 545–551.

Inslicht, S. S. (1994). *Search asymmetries in 3-month-old infants.* B.S. thesis, Department of Psychology, Rutgers University, New Brunswick, New Jersey.

Ittleson, W. H. (1952). *The Ames demonstrations in perception.* Princeton, NJ: Princeton University Press.

Jack, C. E., & Thurlow, W. R. (1973). Effects of degree of visual association and angle of displacement on the "ventriloquism" effect. *Perceptual and Motor Skills, 37,* 967–979.

Jackson, C. V. (1953). Visual factors in auditory localization. *Quarterly Journal of Experimental Psychology, 5,* 52–65.

Jacobs, G. H. (1993). The distribution and nature of colour vision among the mammals. *Biological Review, 68,* 413–471.

Jacobs, G. H., Deegan, J. F., Crognale, M. A., & Fenwick, J. A. (1993). Photopigments of dogs and foxes and their implications for canid vision. *Visual Neuroscience, 10,* 173–180.

Jacobs, K. M., Mark, G. P., & Scott, T. R. (1988). Taste responses in the nucleus tractus solitarius of sodium-deprived rats. *Journal of Physiology, 406,* 393–410.

Jacobson, A., & Gilchrist, A. (1988). The ratio principle holds over a million-to-one range of illumination. *Perception and Psychophysics, 43,* 1–6.

James, W. (1981). *The principles of psychology* (rev. ed.). Cambridge: Harvard University Press. (Original work published 1890)

Jameson, D. (1985). Opponent-colors theory in light of physiological findings. In D. Ottoson & S. Zeki (Eds.), *Central and peripheral mechanisms of color vision* (pp. 8–102). New York: Macmillan.

Jenkins, W. M., & Merzenich, M. M. (1987). Reorganization of neocortical representations after brain injury: A neurophysiological model of the bases of recovery from stroke. *Progress in Brain Research, 71,* 249–266.

Jessell, T. M., & Kelly, D. D. (1991). Pain and analgesia. In E. R. Kandel, J. H. Schwartz, & T. M. Jessell (Eds.), *Principles of neural science* (3rd ed., pp. 385–399). New York: Elsevier.

Johansson, G. (1975). Visual motion perception. *Scientific American, 232,* 76–89.

Johnson, K. (1990). The role of perceived speaker identity in F0 normalization of vowels. *Journal of the Acoustical Society of America, 88,* 642–654.

Johnson, K. O., & Lamb, G. D. (1981). Neural mechanisms of spatial tactile discrimination: Neural patterns evoked by braille-like dot patterns in the monkey. *Journal of Physiology, 310,* 117–144.

Johnstone, B. M., & Boyle, A. J. F. (1967). Basilar membrane vibrations examined with the Mossbauer technique. *Science, 158,* 390–391.

Jones, K. R., Spear, P., & Tong, L. (1984). Critical periods for effects of monocular deprivation differences between striate and extrastriate cortex. *Journal of Neuroscience, 4,* 2543–2552.

Jones, L. A. (1988). Motor illusions: What do they reveal about proprioception? *Psychological Bulletin, 103,* 72–86.

Judd, D. B., & Kelly, K. L. (1965). *The ISCC-NBS method of designating colors and a dictionary of color names* (2nd ed.). U. S. National Bureau of Standards Circular 553, Washington, DC: U.S. Government Printing Office.

Judd, D. B., MacAdam, D. L., & Wyszecki, G. (1964). Spectral distribution of typical daylight as a function of correlated color temperature. *Journal of the Optical Society of America, 54,* 1031–1040.

Julesz, B. (1971). *Foundations of cyclopean perception.* Chicago: University of Chicago Press.

Julesz, B. (1978). Perceptual limits of texture discrimination and their implications to figure-ground separation. In E. Leeuwenberg & H. Buffart (Eds.), *Formal theories of perception* (pp. 205–216). New York: Wiley.

Julesz, B. (1981). Textons, the elements of texture perception, and their interactions. *Nature, 290,* 91–97.

Julesz, B. (1984). A brief outline of the texton theory of human vision. *Trends in Neuroscience, 7,* 41–45.

Kaas, J. H. (1991). Plasticity of sensory and motor maps in adult mammals. *Annual Review of Neuroscience, 14,* 137–167.

Kaas, J. H., Krubitzer, L. A., Chino, Y. M., Langston, A L., Polley, E. H., & Blair, N. (1990). Reorganization of retinotopic cortical maps in adult mammals after lesions of the retina. *Science, 248,* 229–231.

Kaas, J. H., Merzenich, M. M., & Killackey, H. P. (1983). The reorganization of somatosensory cortex following peripheral nerve damage in adult and developing mammals. *Annual Review of Neuroscience, 6,* 325–356.

Kaas, J. H., & Pons, T. P. (1988). The somatosensory system of primates. In H. D. Steklis & J. Erwin (Eds.), *Comparative primate biology* (Vol. 4, pp. 421–468). New York: Alan R. Liss.

Kandel, E. R., & Jessell, T. M. (1991). Touch. In E. R. Kandel, J. H. Schwartz, & T. M. Jessell (Eds.), *Principles of neural science* (3rd ed., pp. 367–384). New York: Elsevier.

Kanizsa, G. (1979). *Organization in vision.* New York: Praeger.

Kaplan, E., Mukherjee, P., & Shapley, R. (1993). Information filtering in the lateral geniculate nucleus. In R. Shapley & D. Man-Kit Lam (Eds.), *Contrast sensitivity* (Vol. 5). Cambridge: MIT Press.

Kaplan, G. (1969). Kinetic disruption of optical texture: The perception of depth at an edge. *Perception and Psychophysics, 6,* 193–198.

Kasahara, T., Iwasaki, K., & Sato, M. (1987). Taste effectiveness of some D- and L-amino acids in mice. *Physiology and Behavior, 39,* 619–624.

Kauer, J. S. (1987). Coding in the olfactory system. In T. E. Finger & W. C. Silver (Eds.), *Neurobiology of taste and smell* (pp. 205–231). New York: Wiley.

Kaufman, L., & Rock, I. (1962a). The moon illusion. *Science, 136,* 953–961.

Kaufman, L., & Rock, I. (1962b). The moon illusion. *Scientific American, 207,* 120–132.

Keeton, W. T., Larkin, T. S., & Windson, D. M. (1974). Normal fluctuations in the earth's magnetic field influence pigeon orientation. *Journal of Comparative Physiology, 95,* 95–103.

Kellman, P., & Spelke, E. (1983). Perception of partly occluded objects in infancy. *Cognitive Psychology, 15,* 483–524.

Kellogg, W. N. (1962). Sonar system of the blind. *Science, 137,* 399–404.

Kelly, J. P. (1991). Hearing. In E. R. Kandel, J. H. Schwartz, & T. M. Jessell (Eds.), *Principles of neural science* (3rd ed., pp. 481–499). New York: Elsevier.

Kenshalo, D. R. (1976). Correlations of temperature sensitivity in man and monkey, a first approximation. In Y. Zotterman (Ed.), *Sensory functions of the skin in primates, with special reference to man* (pp. 305–330). New York: Pergamon.

Keverne, E. B. (1982). Olfaction and the reproductive behavior of nonhuman primates. In C. T. Snowdon, C. H. Brown, & M. R. Petersen (Eds.), *Primate communication* (pp. 396–412). Cambridge: Cambridge University Press.

Kewley-Port, D. (1983). Time-varying features as correlates of place of articulation in stop consonants. *Journal of the Acoustical Society of America, 73,* 322–335.

Kewley-Port, D., & Luce, P. A. (1984). Time-varying features of initial stop consonants in auditory running spectra: A first report. *Perception and Psychophysics, 35,* 353-360.

Kiang, N. Y. S. (1975). Stimulus representation in the discharge patterns of auditory neurons. In E. L. Eagles (Ed.), *The nervous system* (Vol. 3, pp. 81–96). New York: Raven.

Kimura, D. (1961). Cerebral dominance and the perception of verbal stimuli. *Canadian Journal of Psychology, 15,* 166–171.

King, A. J., & Moore, D. R. (1991). Plasticity of auditory maps in the brain. *Trends in Neuroscience, 14,* 31–37.

King, W. L., & Gruber, H. E. (1962). Moon illusion and Emmert's law. *Science, 135,* 1125–1126.

Kinnamon, S. C. (1988). Taste transduction: A diversity of mechanisms. *Trends in Neurosciences, 11,* 491–496.

Klatzky, R. L., Lederman, S. J., & Metzger, V. A. (1985). Identifying objects by touch: An "expert system." *Perception and Psychophysics, 37,* 299–302

Kluender, K. R., Diehl, R. L., & Killeen, P. R. (1987). Japanese quail can learn phonetic categories. *Science, 237,* 1195–1197.

Knill, D. C., & Kersten, D. (1991). Apparent surface curvature affects lightness perception. *Nature, 351,* 228–230.

Knudsen, E. I., & Konishi, M. (1978a). A neural map of auditory space in the owl. *Science, 200,* 795–797.

Knudsen, E. I., & Konishi, M. (1978b). Center-surround organization of auditory receptive fields in the owl. *Science, 202,* 778–780.

Konishi, M. (1984). Spatial receptive fields in the auditory system. In L. Bolis, R. D. Keynes, & S. H. Maddrell (Eds.), *Comparative physiology of sensory systems.* Cambridge: Cambridge University Press.

Kosambi, D. D. (1967). Living prehistory in India. *Scientific American, 216,* 105.

Kozlowski, L., & Cutting, J. (1977). Recognizing the sex of a walker from a dynamic point-light display. *Perception and Psychophysics, 21,* 575–580.

Kreithen, M. L., & Quine, D. B. (1979). Infrasound detection by the homing pigeon: A behavioral audiogram. *Journal of Comparative Physiology, 129,* 1–4.

Kremenitzer, J. P., Vaughn, H. G., Kurtzberg, D., & Dowling, K. (1979). Smooth-pursuit eye movements in the newborn infant. *Child Development, 50,* 442–448.

Kroner, T. (1881). Über die Sinnesempfindungen der Neugeborenen. *Breslauer aerzliche Zeitschrift.* (Cited in Peterson, F., & Rainey, L. H. (1910–1911). The beginnings of mind in the newborn. *Bulletin of the Lying-In Hospital, 7,* 99–122.

Kruger, L. M. (1970). David Katz: Der Aufbau der Tastwelt (The world of touch): A synopsis. *Perception and Psychophysics, 7,* 337–341.

Kruger, L. S. (1982). A word-superiority effect with print and braille characters. *Perception and Psychophysics, 31,* 345–352.

Kubie, J., & Halperin, M. (1975). Laboratory observation of trailing behavior in garter snakes. *Journal of Comparative and Physiological Psychology, 89,* 667–674.

Kubovy, M. (1986). *The psychology of perspective and Renaissance art.* Cambridge: Cambridge University Press.

Kuffler, S. W. (1953). Discharge patterns and functional organization of mammalian retina. *Journal of Neurophysiology, 16,* 37–68.

Kuhl, P. K. (1983). Perception of auditory equivalence classes for speech in early infancy. *Infant Behavior and Development, 6,* 263–285.

Kuhl, P. K. (1986). Theoretical contributions of tests on animals to the special-mechanisms debate in speech. *Experimental Biology, 53,* 31–39.

Kuhl, P. K. (1989). On babies, birds, modules, and mechanisms: A comparative approach to the acquisition of vocal communication. In R. J. Dooling & S. H. Hulse (Eds.), *Comparative psychology of audition* (pp. 379–419). Hillsdale, NJ: Erlbaum.

Kuhl, P. K. (1991). Human adults and human infants show a "perceptual magnet effect" for the prototypes of speech categories, monkeys do not. *Perception and Psychophysics, 50,* 93–107.

Kuhl, P. K., Williams, K. A., Lacerda, F., Stevens, K. N., & Lindblom, B. (1992). Linguistic experience alters phonetic perception in infants by 6 months of age. *Science, 255,* 606–608.

Külpe, O. (1904). Versuche über Abstraktion. *Berlin International Congress of Experimental Psychology,* 56–68.

Kunnapas, T. (1957). Experiments on figural dominance. *Journal of Experimental Psychology, 53,* 31–39.

LaBarbera, J. D., Izard, C. E., Vietze, P., & Parisi, S. A. (1976). Four and six-month-old infants' visual responses to joy, anger, and neutral expressions. *Child Development, 47,* 535–538.

Laing, D. D., Doty, R. L., & Breipohl, W. (Eds.). (1991). *The human sense of smell.* New York: Springer.

Laing, D. G. (1985). Optimum perception of odor intensity by humans. *Physiology and Behavior, 34,* 569–574.

Land, E. H. (1983). Recent advances in retinex theory and some implications for cortical computations: Color vision and the natural image. *Proceedings of the National Academy of Sciences, USA, 80,* 5163–5169.

Land, E. H. (1986). Recent advances in retinex theory. *Vision Research, 26,* 7–21.

Land, E. H., & McCann, J. J. (1971). Lightness and retinex theory. *Journal of the Optical Society of America, 61,* 1–11.

Lederman, S. J., & Klatzky, R. L. (1987). Hand movements: A window into haptic object recognition. *Cognitive Psychology, 19,* 342–368.

Lederman, S. J., & Klatzky, R. L. (1990). Haptic classification of common objects: Knowledge-driven exploration. *Cognitive Psychology, 22,* 421–459.

Lee, D. N. (1974). Visual information during locomotion. In R. B. MacLeod & H. L. Pick, Jr. (Eds.), *Perception: Essays in honor of J. J. Gibson* (pp. 250–267). Ithaca, NY: Cornell University Press.

Lee, D. N. (1976). A theory of visual control of braking based on information about time to collision. *Perception, 5,* 437–459.

Lee, D. N. (1980). The optic flow field: The foundation of vision. *Transactions of the Royal Society, 290B,* 169–179.

Lee, D. N., & Aronson, E. (1974). Visual proprioceptive control of standing in human infants. *Perception and Psychophysics, 15,* 529–532.

Lee, D. N., Young, D. S., Reddish, P. E., Lough, S., & Clayton, T. M. H. (1983). Visual timing in hitting an accelerating ball. *Quarterly Journal of Experimental Psychology, 35A,* 333–346.

LeGrand, Y. (1957). *Light, color and vision.* London: Chapman & Hall.

Lehrer, M., Horridge, G. A., Zhang, S. W., & Gadagkar, R. (1995). Shape vision in bees: Innate preference for flower-like patterns. *Philosophical Transactions of the Royal Society of London B, 347,* 123–137.

Lehrer, M., Srinivasan, M. V., Zhang, S. W., & Horridge, G. A. (1988). Motion cues provide the bee's visual world with a third dimension. *Nature, 332,* 356–357.

Lerman, S. (1966). *Basic ophthalmology.* New York: McGraw-Hill.

Lettvin, J. Y., Maturana, H. R., McCulloch, W. S., and Pitts, W. H. (1959). What the frog's eye tells the frog's brain. *Proceedings of the Institute of Radio Engineers, 47,* 1940–1951.

LeVay, S., & Voigt, T. (1988). Ocular dominance and disparity coding in cat visual cortex. *Visual Neuroscience, 1,* 395–414.

Levine, J. (1955). Consensual pupillary reflex in birds. *Science, 122,* 690.

Levine, J. S., & MacNichol, E. F., Jr. (1982, February). Color vision in fishes. *Scientific American,* pp. 140–149.

Lewis, E. R., Zeevi, Y. Y., & Werblin, F. S. (1969). Scanning electron microscopy of vertebrate visual receptors. *Brain Research, 15,* 559–562.

Liberman, A. M., Cooper, F. S., Shankweiler, D. P., & Studdert-Kennedy, M. (1967). Perception of the speech code. *Psychological Review, 74,* 431–461.

Liberman, A. M., & Mattingly, I. G. (1989). A specialization for speech perception. *Science, 243,* 489–494.

Lichten, W., & Lurie, S. (1950). A new technique for the study of perceived size. *American Journal of Psychology, 63,* 280–282.

Lieberman, M. C. (1978). Auditory-nerve response from cats raised in a low-noise chamber. *Journal of the Acoustical Society of America, 63,* 442–455.

Liebowitz, H. W., Shina, K., & Hennessy, H. R. (1972). Oculomotor adjustments and size constancy. *Perception and Psychophysics, 12,* 497–500.

Lindsay, P. H., & Norman, D. A. (1977). *Human information processing* (2nd ed.). New York: Academic Press.

Liphovsky, S. (1977). The role of the chemical senses in nutrition. In M. R. Kare & O. Muller (Eds.), *The chemical senses and nutrition* (pp. 413–428). New York: Academic.

Livingstone, M. (1990). Segregation of form, color, movement and depth processing in the visual system: Anatomy, physiology, art, and illusion. In B. Cohen & I. Bodis-Wollner (Eds.), *Vision and the brain* (pp. 119–138). New York: Raven.

Livingstone, M. S., & Hubel, D. H. (1984). Anatomy and physiology of a color system in the primate visual cortex. *Journal of Neuroscience, 4,* 309–356.

Livingstone, M. S., & Hubel, D. H. (1987). Psychophysical evidence for separate channels for the perception of form, color, movement, and depth. *Journal of Neuroscience, 7,* 3416–3468.

Livingstone, M. S., & Hubel, D. H. (1988). Segregation of form, color, movement, and depth: Anatomy, physiology, and perception. *Science, 240,* 740–749.

Logothetis, N. K., Schiller, P. H., Charles, E. R., & Hurlbert, A. C. (1990). Perceptual deficits and the activity of the color-opponent and broad-band pathways at isoluminance. *Science, 247,* 214–217.

Loomis, J. M., Golledge, R. G., Klatzky, R. L., Speigle, J. M., & Tietz, J. (1994). Personal guidance system for the visually impaired. *Proceedings of the First Annual International ACM/SIGCAPH Conference on Assistive Technologies,* Marina del Rey, California.

Loomis, J. M., Hebert, C., & Cicinelli, J. G. (1990). Active localization of virtual sounds. *Journal of the Acoustical Society of America, 88,* 1757–1764.

Lowenstein, W. R. (1960). Biological transducers. *Scientific American, 203,* 98–108.

Lowenstein, W. R., & Skalak, R. (1966). Mechanical transmission in a Pacinian corpuscle: An analysis and a theory. *Journal of Physiology, 182,* 346–378.

Lythgoe, J. N., & Partridge, J. C. (1989). Visual pigments and the acquisition of visual information. *Journal of Experimental Biology, 146,* 1–20.

MacDonald, J., & McGurk, H. (1978). Visual influences on speech perception processes. *Perception and Psychophysics, 24,* 253–257.

Macfarlane, A. (1975). Olfaction in the development of social preferences in the human neonate. In A. Macfarlane (Ed.), *Ciba Foundation Symposium, 33,* 103–117.

Mach, E. (1959). *The analysis of sensations.* New York: Dover. (Original work published 1914)

MacKay, D. M. (1961). Visual effects of non-redundant stimulation. *Nature, 192,* 739–740.

MacKay-Sim, A., Shaman, P., & Moulton, D. (1982). Topographic coding of olfactory quality: Odorant-specific patterns of epithelial responsivity in the salamander. *Journal of Neurophysiology, 48,* 548–596.

Maffei, L., & Fiorentini, A. (1973). The visual cortex as a spatial frequency analyzer. *Vision Research, 13,* 1255–1267.

Makous, J. C., & Middlebrooks, J. C. (1990). Two-dimensional sound localization by human listeners. *Journal of the Acoustical Society of America, 87,* 2188–2200.

Mallevialle, J., & Suffet, I. H. (Eds.). (1987). *Identification and treatment of tastes and odors in drinking water.* Denver: American Water Works Association.

Marean, G. C., Werner, L. A., & Kuhl, P. K. (1992). Vowel categorization by very young infants. *Developmental Psychology, 28,* 396–405.

Margoliash, D. (1983). Acoustic parameters underlying the responses of song-specific neurons in the white-crowned sparrow. *Journal of Neuroscience, 3,* 1029–1057.

Marks, W. B. (1965). Visual pigments of single goldfish cones. *Journal of Physiology, 178,* 14–32.

Marks, W. B., Dobelle, W. H., & MacNichol, E. F. (1964). Visual pigments of single primate cones. *Science, 143,* 1181–1183.

Marr, D. (1976). Early processing of visual information. *Transactions of the Royal Society of London, 275B,* 483–524.

Marr, D. (1982). *Vision.* San Francisco: W. H. Freeman.

Marr, D., & Hildreth, E. (1980). Theory of edge detection. *Proceedings of the Royal Society of London, 207B,* 187–207.

Marr, D., & Nishihara, H. K. (1978). Representation and recognition of the spatial organization of three-dimensional shapes. *Proceedings of the Royal Society of London, 200B,* 269–294.

Martin, J. H. (1991). Coding and processing of sensory information. In E. R. Kandel, J. H. Schwartz, & T. M. Jessell (Eds.), *Principles of neural science* (3rd ed., pp. 329–340). Norwalk, CT: Appleton & Lange.

Martin, J. H., & Jessell, T. M. (1991). Modality coding in the somatic sensory system. In E. R. Kandel, J. H. Schwartz, & T. M. Jessell (Eds.), *Principles of neural science* (3rd ed., pp. 339–352). Norwalk, CT: Appleton & Lange.

Masland, R. H. (1988). Amacrine cells. *Trends in Neuroscience, 9,* 405–410.

Matin, L., Picoult, E., Stevens, J., Edwards, M., & MacArthur, R. (1982). Oculoparalytic illusion: Visual-field dependent spatial mislocations by humans partially paralyzed with curare. *Science, 216,* 198–201.

Matthews, D. F. (1972). Response patterns of single neurons in the tortoise olfactory epithelium and olfactory bulb. *Journal of General Physiology, 60,* 166–180.

Maunsell, J. H. R., & Newsome, W. T. (1987). Visual processing in monkey extrastriate cortex. *Annual Review of Neuroscience, 10,* 363–401.

May, B., Moody, D. B., & Stebbins, W. C. (1988). The significant features of Japanese macaque coo sounds: A psychophysical study. *Animal Behavior, 36,* 1432–1444.

May, B., Moody, D. B., & Stebbins, W. C. (1989). Categorical perception of conspecific communication sounds by Japanese macaques, *Macaca fuscata. Journal of the Acoustical Society of America, 85,* 837–846.

Mayer, A. M. (1876). Researchers in acoustics (absence of masking of one sound by one higher in pitch). *Philosophical Magazine, 11,* 500–507.

Mayer, D. J. (1979). Endogenous analgesia systems: Neural and behavioral mechanisms. In J. J. Bonica (Ed.), *Advances in pain research and therapy* (Vol. 3, pp. 385–410). New York: Raven.

McArthur, D. J. (1982). Computer vision and perceptual psychology. *Psychological Bulletin, 92,* 283–309.

McBurney, D. H. (1969). Effects of adaptation on human taste function. In C. Pfaffmann (Ed.), *Olfaction and taste* (pp. 407–419). New York: Rockefeller University Press.

McBurney, D. H., Levine, J. M., & Cavanaugh, P. H. (1977). Psychophysical and social ratings of human body odor. *Personality and Social Psychology Bulletin, 3,* 135–138.

McClintock, M. K. (1971). Menstrual synchrony and suppression. *Nature, 229,* 244–245.

McCutcheon, N. B. (1992). Human psychophysical studies of saltiness suppression by amiloride. *Physiology and Behavior, 51,* 1069–1074.

McFadden, S. A. (1987). The binocular depth stereoacuity of the pigeon and its relation to the anatomical resolving power of the eye. *Vision Research, 27,* 1967–1980.

McFadden S. A., & Wild, J. M. (1986). Binocular depth perception in the pigeon. *Journal of the Experimental Analysis of Behavior, 45,* 149–160.

McGurk, H., & MacDonald, T. (1976). Hearing lips and seeing voices. *Nature, 264,* 746–748.

McKee, S. P. (1993). Psychophysics and perception. *Perception, 22,* 505–507.

McKenzie, D. (1923). *Aromatics and the soul: A study of smells.* New York: Hoeber.

McLean, J. H., & Shipley, M. T. (1992). Neuroanatomical substrates of olfaction. In M. J. Serby & K. L. Chobor (Eds.), *Science of olfaction* (pp. 126–171). New York: Springer-Verlag.

McLeod, R. W., & Ross, H. E. (1983). Optic-flow and cognitive factors in time-to-collision estimates. *Perception, 12,* 417–423.

Mello, N. K., & Peterson, N. J. (1964). Behavioral evidence for color discrimination in cats. *Journal of Neurophysiology, 27,* 323–333.

Melzack, R. (1973). *The puzzle of pain.* New York: Basic Books.

Melzack, R. (1992). Phantom limbs. *Scientific American, 266,* 121–126.

Melzack, R., & Wall, P. D. (1965). Pain mechanisms: A new theory. *Science, 150,* 971–979.

Melzack, R., & Wall, P. D. (1983). *The challenge of pain.* New York: Basic Books.

Melzack, R., & Wall, P. D. (1988). *The challenge of pain* (rev. ed.). New York: Penguin.

Menzel, R., & Backhaus, W. (1989). Color vision honey bees: Phenomena and physiological mechanisms. In D. G. Stavenga & R. C. Hardie (Eds.), *Facets of vision* (pp. 281–297). Berlin: Springer-Verlag.

Menzel, R., Ventura, D. F., Hertel, H., deSouza, J., & Greggers, U. (1986). Spectral sensitivity of photoreceptors in insect compound eyes: comparison of species and methods. *Journal of Comparative Physiology, 158A,* 165–177.

Merigan, W. H., & Maunsell, J. H. R. (1993). How parallel are the primate visual pathways? *Annual Review of Neuroscience, 16,* 369–402.

Merker, H. (1994). *Listening.* New York: HarperCollins.

Mervis, C. B., & Rosch, E. (1981). Categorization of natural objects. *Annual Review of Psychology, 32,* 89–115.

Merzenich, M. M., & Kass, J. H. (1980). Principles of organization of sensory-perceptual systems in mammals. In A. N. Epstein & J. M. Sprague (Eds.), *Progress in psychobiology and physiological psychology* (Vol. 9, pp. 1–42). New York: Academic.

Merzenich, M. M., & Kaas, J. H. (1982). Reorganization of mammalian somatosensory cortex following peripheral nerve injury. *Trends in Neuroscience, 5,* 434–436.

Merzenich, M. M., Recanzone, G., Jenkins, W. M., Allard, T. T., & Nudo, R. J. (1988). Cortical representational plasticity. In P. Rakic & W. Singer (Eds.), *Neurobiology of neocortex* (pp. 42–67). Berlin: Wiley.

Meyer, D. R., & Anderson, R. A. (1965). Colour discrimination in cats. In CIBA Foundation Symposium, *Color vision* (pp. 324–344). London: Churchill.

Michael, C. R. (1969). Retinal processing of visual images. *Scientific American, 220*(5), 104–114.

Michael, C. R. (1978). Color vision mechanisms in monkey striate cortex: Dual-opponent cells with concentric receptive fields. *Journal of Neurophysiology, 41,* 572–588.

Michael, C. R. (1986). *Functional and morphological identification of double and single opponent color cells in layer IVCb of the monkey's striate cortex.* Paper presented at the meeting of the Society for Neuroscience.

Middlebrooks, J. C. (1988). Auditory mechanisms underlying a neural code for space in the cat's superior colliculus. In G. M. Edelman, W. E. Gall, & W. M. Cowan (Eds.), *Auditory function: Neurobiological bases of hearing* (pp. 431–455). New York: Wiley.

Middlebrooks, J. C., Clock, A. E., Xu, L., & Green, D. M. (1994). A panoramic code for sound location by cortical neurons. *Science, 264,* 842–844.

Middlebrooks, J. C., & Green, D. M. (1991). Sound localization by human listeners. *Annual Review of Psychology, 42,* 135–159.

Middlebrooks, J. C., & Pettigrew, J. D. (1981). Functional classes of neurons in primary auditory cortex of the cat distinguished by sensitivity to sound location. *Journal of Neuroscience, 1,* 107–120.

Miller, G. A., & Isard, S. (1963). Some perceptual consequences of linguistic rules. *Journal of Verbal Learning and Verbal Behavior, 2,* 212–228.

Miller, I. J. (1995). Anatomy of the peripheral taste system. In R. L. Doty (Ed.), *Handbook of olfaction and gustation* (pp. 521–548). New York: Marcel Dekker.

Miller, J. D. (1974). Effects of noise on people. *Journal of the Acoustical Society of America, 56,* 729–764.

Miller, J. D., Wier, C. C., Pastore, R., Kelly, W. J., & Dooling, R. J. (1976). Discrimination and labeling of noise-buzz sequences with varying noise-lead times: An example of categorical perception. *Journal of the Acoustical Society of America, 60,* 410–417.

Mishkin, M. (1986). *Two visual systems.* Talk presented at Western Psychiatric Institute and Clinic, Pittsburgh, January 24.

Mishkin, M., Ungerleider, L. G., & Macko, K. A. (1983). Object vision and spatial vision: Two central pathways. *Trends in Neuroscience, 6,* 414–417.

Mitchell, D. E., Reardon, J., & Muir, D. W. (1975). Interocular transfer of the motion aftereffect in normal and stereoblind observers. *Experimental Brain Research, 22,* 163–173.

Mitchell, D. E., & Ware, C. (1974). Interocular transfer of a visual aftereffect in normal and stereoblind humans. *Journal of Physiology, 236,* 707–721.

Mollon, J. D. (1989). "Tho' she kneel'd in that place where they grew . . ." *Journal of Experimental Biology, 146,* 21–38.

Mollon, J. D. (1990). The club-sandwich mystery. *Nature, 343,* 16–17.

Mollon, J. D. (1992). Signac's secret. *Nature, 358,* 379–380.

Mollon, J. D. (1993). Mixing genes and mixing colours. *Current Biology, 3,* 82–85.

Montagna, W., & Parakkal, P. F. (1974). *The structure and function of skin* (3rd ed.). New York: Academic.

Moon, C., Cooper, R. P., & Fifer, W. P. (1993). Two-day-olds prefer their native language. *Infant Behavior and Development, 16,* 495–500.

Moore, B. C. J., Glasberg, B. R., Hess, R. F., & Birchall, J. P. (1985). Effects of flanking noise bands on the rate of growth of loudness of tones in normal and recruiting ears. *Journal of the Acoustical Society of America, 77,* 1505–1513.

Moran, J., & Desimone, R. (1985). Selective attention gates visual processing in the extrastriate cortex. *Science, 229,* 782–784.

Morrison, E. E., & Moran, D. T. (1995). Anatomy and ultrastructure of the human olfactory neuroepithelium. In R. L. Doty (Ed.), *Handbook of olfaction and gustation* (pp. 75–101). New York: Marcel Dekker.

Moskowitz, H. R., Kumriach, V., Sharma, H., Jacobs, L., & Sharma, S. D. (1975). Cross-cultural differences in simple taste preference. *Science, 190,* 1217–1218.

Moulton, D. G. (1965). Differential sensitivity to odors. *Cold Spring Harbor Symposium on Quantitative Biology, 30,* 201–206.

Moulton, D. G. (1977). Minimum odorant concentrations detectable by the dog and their implications for olfactory receptor sensitivity. In D. Miller-Schwarze & M. M. Mozell (Eds.), *Chemical signals in vertebrates* (pp. 455–464). New York: Plenum.

Mountcastle, V. B., & Powell, T. P. S. (1959). Neural mechanisms subserving cutaneous sensibility, with special reference to the role of afferent inhibition in sensory perception and discrimination. *Bulletin of the Johns Hopkins Hospital, 105,* 201–232.

Movshon, J. A., Adelson, E. H., Gizzi, M. S., & Newsome, W. T. (1985). The analysis of moving visual patterns. In C. Chagas, R. Gattass, & C. Gross (Eds.), *Pattern recognition mechanisms* (pp. 117–151). New York: Springer-Verlag.

Movshon, J. A., & Newsome, W. T. (1992). Neural foundations of visual motion perception. *Current Directions in Psychological Science, 1,* 35–39.

Mozell, M. M. (1964). Olfactory discrimination: Electrophysiological spatiotemporal basis. *Science, 143,* 1336–1337.

Mozell, M. M. (1966). The spatiotemporal analysis of odorants at the level of the olfactory receptor sheet. *Journal of General Physiology, 50,* 25–41.

Mozell, M. M. (1970). Evidence for a chromatographic model of olfaction. *Journal of General Physiology, 56,* 46–63.

Mozell, M. M., & Hornung, D. E. (1984). Initial events influencing olfactory analysis. In L. Bolis., R. D. Keynes, & S. H. Maddrell (Eds.), *Comparative physiology of sensory systems* (pp. 227–244). Cambridge: Cambridge University Press.

Mozell, M. M., & Hornung, D. E. (1985). Peripheral mechanisms in the olfactory process. In D. W. Pfaff (Ed.), *Taste, olfaction and the central nervous system* (pp. 253–279). New York: Rockefeller University Press.

Mozell, M. M., & Jagodowiez, M. (1973). Chromatographic separation of odorants by the nose: Retention times measured across *in vivo* olfactory mucosa. *Science, 181,* 1247–1249.

Mozell, M. M., Smith, B. P., Smith, P. E., Sullivan, R. L., & Swender, P. (1969). Nasal chemoreception in flavor identification. *Archives of Otolaryngology, 90,* 131–137.

Muir, D., & Field, J. (1979). Newborn infants orient to sounds. *Child Development, 50,* 431–436.

Mullennix, J. W., & Pisoni, D. B. (1989). Speech perception: Analysis of biologically significant signals. In R. J. Dooling & S. H. Hulse (Eds.), *Comparative psychology of audition* (pp. 97–128). Hillsdale, NJ: Erlbaum.

Mullennix, J. W., & Pisoni, D. B. (1990). Stimulus variability and processing dependencies in speech perception. *Perception and Psychophysics, 47,* 379–390.

Müller, J. (1842). *Elements of physiology* (W. Baly, Trans.). London: Tayler & Walton.

Murphy, C., & Cain, W. S. (1980). Taste and olfaction: Independence vs. interaction. *Physiology and Behavior, 24,* 601–606.

N. E. Thing Enterprises. (1993). *Magic eye: A new way of looking at the world.* Kansas City, MO: Andrews & McMeel.

Nachman, M. (1963). Taste preferences for sodium salts by adrenalectomized rats. *Journal of Comparative and Physiological Psychology, 55,* 1124–1129.

Nakamura, H., Gattass, R., Desimone, R., & Ungerleider, L. G. (1993). The modular organization of projections from areas V1 and V2 to areas V4 and TEO in macaques. *Journal of Neuroscience, 13,* 3681–3691.

Nakayama, K. (1990). The iconic bottleneck and the tenuous link between early visual processing and perception. In C. Blakemore (Ed.), *Vision: Coding and efficiency* (pp. 411–422). New York: Cambridge University Press.

Nakayama, K. (1994). James J. Gibson—An appreciation. *Psychological Review, 101,* 329–335.

Nathan, P. W. (1976). The gate control theory of pain. *Brain, 99,* 123–158.

Nathans, J., Thomas, D., & Hogness, D. S. (1986). Molecular genetics of human color vision: The genes encoding blue, green, and red pigments. *Science, 232,* 193–202.

Neisser, U. (1967). *Cognitive psychology.* Englewood Cliffs, NJ: Prentice-Hall.

Neitz, J., Geist, T., & Jacobs, G. H. (1989). Color vision in the dog. *Visual Neuroscience, 3,* 119–125.

Neitz, J., Neitz, M., & Jacobs, G. H. (1993). More than three different cone pigments among people with normal color vision. *Vision Research, 33,* 117–122.

Neitz, M., Neitz, J., & Jacobs, G. H. (1991). Spectral tuning of pigments underlying red-green color vision. *Science, 252,* 971–974.

Nelson, C. A., & Horowitz, F. D. (1987). Visual motion perception in infancy: A review and synthesis. In P. Salapatek & L. Cohen (Eds.), *Handbook of infant perception* (Vol. 2, pp. 123–153). New York: Academic.

Nelson, M. E., & Bower, J. M. (1990). Brain maps and parallel computers. *Trends in Neuroscience, 13,* 403–408.

Nelson, P. G., Erulkar, S. D., & Bryan, S. S. (1966). Responses of units of the inferior colliculus to time-varying acoustic stimuli. *Journal of Neurophysiology, 29,* 834–860.

Nelson, R. J., Sur, M., Felleman, D. J., & Kaas, J. H. (1980). Representations of the body surface in postcentral parietal cortex of *Macaca fasicularis. Journal of Comparative Neurology, 192,* 611–643.

Newman, E. R., & Hartline, P. H. (1982, March). The infrared "vision" of snakes. *Scientific American, 246,* 116–127.

Newsome, W. T., Britten, K. H., & Movshon, J. A. (1989). Neuronal correlates of a perceptual decision. *Nature, 341,* 52–54.

Newsome, W. T., & Paré, E. B. (1988). A selective impairment of motion perception following lesions of the middle temporal visual area (MT). *Journal of Neuroscience, 8,* 2201–2211.

Newton, I. (1704). *Optiks.* London: Smith and Walford.

Nordby, K. (1990). Vision in a complete achromat: A personal account. In R. F. Hess, L. T. Sharpe, & K. Nordby (Eds.), *Night vision* (pp. 290–315). Cambridge: Cambridge University Press.

Northern, J., & Downs, M. (1978). *Hearing in children* (2nd ed.). Baltimore: Williams & Wilkins.

Nothdurft, H. C. (1990). Texton segregation by associated differences in global and local luminance distribution. *Proceedings of the Royal Society of London, B239,* 295–320.

O'Carroll, D. O. (1993). Feature-detecting neurons in dragonflies. *Nature, 362,* 541–543.

Oldfield, S. R., & Parker, S. P. A. (1984). Acuity of sound localisation: A topography of auditory space: 2. Pinna cues absent. *Perception, 13,* 601–617.

Olsen, J. F., & Suga, N. (1991a). Combination sensory neurons in the medial geniculate body of the mustache bat: Encoding of relative velocity information. *Journal of Neurophysiology, 65,* 1254–1273.

Olsen, J. F., & Suga, N. (1991b). Combination sensory neurons in the medial geniculate body of the mustache bat: Encoding of target range information. *Journal of Neurophysiology, 65,* 1275–1296.

Olsho, L. W., Koch, E. G., Carter, E. A., Halpin, C. F., & Spetner, N. B. (1988). Pure-tone sensitivity of human infants. *Journal of the Acoustical Society of America, 84,* 1316–1324.

Olsho, L. W., Koch, E. G., Halpin, C. F., & Carter, E. A. (1987). An observer-based psychoacoustic procedure for use with young infants. *Developmental Psychology, 23,* 627–640.

Olson, C. R., & Freeman, R. D. (1980). Profile of the sensitive period for monocular deprivation in kittens. *Experimental Brain Research, 39,* 17–21.

Olson, H. F. (1967). *Music, physics and engineering* (2nd ed.). New York: Dover.

Orban, G. A., Lagae, L., Verri, A., Raiguel, S., Xiao, D., Maes, H., & Torre, V. (1992). First-order analysis of optical flow in monkey brain. *Proceedings of the National Academy of Sciences, 89,* 2595–2599.

O'Shea, R. P. (1991). Thumb's rule tested: Visual angle of thumb's width is about 2 deg. *Perception, 20,* 415–418.

Owens, E. (1989). Present status of adults with cochlear implants. In E. Owens & D. K. Kessler (Eds.), *Cochlear implants in young deaf children* (pp. 25–52). Boston: Little, Brown.

Owens, M. (1994, June 6). Designers discover the sweet smell of success. *New York Times.*

Oyama, T. (1960). Figure-ground dominance as a function of sector angle, brightness, hue and orientation. *Journal of Experimental Psychology, 60,* 299–305.

Palmer, A. R., & Evans, E. F. (1979). On the peripheral coding of the level of individual frequency components of complex sounds at high sound levels. In O. Creutzfeldt, H. Scheich, & C. Schreiner (Eds.), *Hearing mechanisms and speech.* Berlin: Springer-Verlag.

Palmer, S. E. (1975). The effects of contextual scenes on the identification of objects. *Memory and Cognition, 3,* 519–526.

Pasternak, T. (1990). Vision following loss of cortical directional selectivity. In. M. A. Berkley & W. C. Stebbins (Eds.), *Comparative perception* (Vol. 1, pp. 407–428). New York: Wiley.

Pasternak, T., & Merigan, E. H. (1994). Motion perception following lesions of the superior temporal sulcus in the monkey. *Cerebral Cortex, 4,* 247–259.

Payne, K. B., Langbauer, W. R., & Thomas, E. H. (1986). Infrasonic calls of the Asian elephant (*Elephas maximus*). *Behavioral Ecology and Sociobiology, 18,* 297–301.

Pearlman, A. L., & Daw, N. (1970). Opponent color cells in the cat lateral geniculate nucleus. *Science, 167,* 84–86.

Penfield, W., & Rasmussen, T. (1950). *The cerebral cortex of man.* New York: Macmillan.

Perrett, D. I., Harries, M. H., Benson, P. J., Chitty, A. J., & Mistlin, A. J. (1990). Retrieval of structure from rigid and biological motion: An analysis of the visual responses of neurones in the macaque temporal cortex. In A. Blake & T. Troscianko (Eds.), *AI and the Eye* (pp. 181–200). New York: Wiley.

Perrett, D. I., Hietanen, J. K., Oram, M. W., & Benson, P. J. (1992). Organization and functions of cells responsive to faces in the temporal cortex. *Transactions of the Royal Society of London B, 225,* 23–30.

Petersen, M. R. (1982). The perception of species-specific vocalizations by primates: A conceptual framework. In C. T. Snowdon, C. H. Brown, & M. R. Petersen (Eds.), *Primate communication* (pp. 171–211). Cambridge: Cambridge University Press.

Petersen, M. R., Beecher, M. D., Zoloth, S. R., Green, S., Marler, P. R., Moody, D. B., & Stebbins, W. C. (1984). Neural lateralization of vocalizations by Japanese macaques: Communicative significance is more important than acoustic structures. *Behavioral Neuroscience, 98,* 779–790.

Peterson, F., & Rainey, L. H. (1910–1911). The beginnings of mind in the newborn. *Bulletin of the Lying-in Hospital, 7,* 99–122.

Peterson, M. A. (1994). Object recognition processes can and do operate before figure-ground organization. *Current Directions in Psychological Science, 3,* 105–111.

Peterson, M. A., & Hochberg, J. (1983). Opposed-set measurement procedure: A quantitative analysis of the role of local cues and intention in form perception. *Journal of Experimental Psychology: Human Perception and Performance, 9,* 183–193.

Pettigrew, J. D. (1986). The evolution of binocular vision. In J. D. Pettigrew, K. J. Sanderson, & W. R. Levick (Eds.), *Visual neuroscience* (pp. 208–222). Cambridge: Cambridge University Press.

Pettit, M. J., & Schwark, H. D. (1993). Receptive field reorganization in dorsal column nuclei during temporary denervation. *Science, 262,* 2054–2056.

Pevsner, J., Hou, V., Snowman, A. M., & Snyder, S. H. (1990). Odorant-binding protein: Characterization of ligand binding. *Journal of Biological Chemistry, 265,* 6118–6125.

Pevsner, J., Trifletti, R. R., Strittmatter, S. M., & Snyder, S. H. (1985). Isolation and characterization of an olfactory protein for odorant pyrazines. *Proceedings of the National Academy of Sciences, 82,* 3050–3054.

Pfaffmann, C. (1974). Specificity of the sweet receptors of the squirrel monkey. *Chemical Senses, 1,* 61–67.

Pfeiffer, C. A., & Johnston, R. E. (1994). Hormonal and behavioral responses of male hamsters to females and female odors: Roles of olfaction, the vemeronasal system, and sexual experience. *Physiology and Behavior, 55,* 129–138.

Phillips, R., Johansson, R. S., & Johnson, K. O. (1990). Representation of braille characters in human nerve fibers. *Experimental Brain Research, 81,* 589–592.

Pick, A. D., Thomas, M. L., & Pick, H. L. (1966). The role of grapheme-phoneme correspondences in the perception of braille. *Journal of Verbal Learning and Verbal Behavior, 5,* 298–300.

Pierce, G. W., & Griffin, D. R. (1938). Experimental determination of supersonic notes emitted by bats. *Journal of Mammalogy, 19,* 454–455.

Pirchio, M., Spinelli, D., Fiorentini, A., & Maffei, L. (1978). Infant contrast sensitivity evaluated by evoked potentials. *Brain Research, 141,* 179–184.

Pisoni, D. B. (1977). Identification and discrimination of the relative onset of two-component tones: Implications for voicing perception in stops. *Journal of the Acoustical Society of America, 61,* 1352–1361.

Plug, C., & Ross, H. E. (1994). The natural moon illusion: A multifactor angular account. *Perception, 23,* 321–333.

Poggio, G. F., Motter, B. C., Squatrito, S., & Trotter, Y. (1985). Responses of neurons in visual cortex (V1 and V2) of the alert macaque to dynamic random-dot stereograms. *Vision Research, 25,* 397–406.

Poggio, T. (1984, April). Vision by man and machine. *Scientific American,* 106–116.

Pokorny, J., Shevell, S. K., & Smith, V. C. (1991). Colour appearance and colour constancy. In P. Gouras (Ed.), *The perception of colour: Vol. 6. Vision and visual dysfunction* (pp. 43–61). Boca Raton, FL: CRC.

Pollak, G. D., & Casseday, J. H. (1989). *The neural basis of echolocation in bats.* New York: Springer-Verlag.

Pomerantz, J. R. (1981). Perceptual organization in information processing. In M. Kubovy & J. Pomerantz (Eds.), *Perceptual organization.* Hillsdale, NJ: Erlbaum.

Pons, T. P., Garraghty, P. E., Ommaya, A. K., Kaas, J. H., Taub, E., & Mishkin, M. (1991). Massive cortical reorganization after sensory deafferentiation in adult macaques. *Science, 252,* 1857–1860.

Porter, R. H., Makin, J. W., Davis, L. B., & Christensen, K. M. (1992). Breast-fed infants respond to olfactory cues from their own mother and unfamiliar lactating females. *Infant Behavior and Development, 15,* 85–93.

Porter, R. H., & Schaal, B. (1995). Olfaction and development of social preferences in neonatal organisms. In R. L. Doty (Ed.), *Handbook of olfaction and gustation* (pp. 299–321). New York: Marcel Dekker.

Preti, G., Cutler, W. B., Garcia, C. R., Huggins, G. R., & Lawley, H. J. (1986). Human axillary secretions influence women's menstrual cycles: The role of donor extract from females. *Hormones and Behavior, 20,* 474–482.

Price, J. L., Carmichael, S. T., Carnes, K. M., Clugnet, M. C., Kuroda, M., & Ray, J. P. (1991). Olfactory output to the prefrontal cortex. In J. L. Davis & H. Eichenbaum (Eds.), *Olfaction: A model system for computational neuroscience* (pp. 101–120). Cambridge: MIT Press.

Proffitt, D. R., Cutting, J. E., & Stier, D. M. (1979). Perception of wheel-generated motions. *Journal of Experimental Psychology: Human Perception and Performance, 5,* 289–302.

Prosen, C. A., Moody, D. B., Stebbins, W. C., & Hawkins, J. E., Jr. (1981). Auditory intensity discrimination after selective loss of cochlear outer hair cells. *Science, 212,* 1286–1288.

Purkinje, J. E. (1825). *Neurre Beitrage zur Kenntniss des Sehens in subjectiver Hinsicht.* Berlin: Reimer.

Quinn, P. C., Rosano, J. L., Wooten, B. R. (1988). Evidence that brown is not an elemental color. *Perception and Psychophysics, 43,* 156–164.

Ramachandran, V. S. (1990). Visual perception in people and machines. In A. Blake & T. Troscianko (Eds.), *AI and the Eye* (pp. 21–77). New York: Wiley.

Ramachandran, V. S. (1992, May). Blind spots. *Scientific American,* 86–91.

Ramachandran, V. S., & Anstis, S. M. (1986, May). The perception of apparent motion. *Scientific American,* 102–109.

Ranganathan, R., Harris, W. A., & Zuker, C. S. (1991). The molecular genetics of invertebrate phototransduction. *Trends in Neuroscience, 14,* 486–493.

Ratliff, F. (1965). *Mach bands: Quantitative studies on neural networks in the retina.* New York: Holden-Day.

Ratoosh, P. (1949). On interposition as a cue for the recognition of relative distance. *Proceedings of the National Academy of Sciences, 35,* 257–259.

Rayleigh, L. (1881). Experiments on colour. *Nature, 25,* 64–66.

Reddy, D. R. (1976). Speech recognition by machine: A review. *Proceedings of the Institute of Electrical and Electronic Engineers, 64,* 501–531.

Regan, D. (1986). Luminance contrast: Vernier discrimination. *Spatial Vision, 1,* 305–318.

Regan, D. (1991). Spatial vision for objects defined by color contrast, binocular disparity, and motion parallax. In D. Regan (Ed.), *Spatial vision* (pp. 135–178). London: Macmillan.

Regan, D., & Cynader, M. (1979). Neurons in area 18 of cat visual cortex selectively sensitive to changing size: Nonlinear interactions between responses to two edges. *Vision Research, 19,* 699–711.

Reichardt, W. (1961). Autocorrelation, a principle for the evaluation of sensory information by the central nervous system. In W. A. Rosenblith (Ed.), *Sensory communication* (pp. 303–318). New York: Wiley.

Reichl, R. (1994, March 11). Dining in New York. *New York Times.*

Restle, F. (1970). Moon illusion explained on the basis of relative size. *Science, 167,* 1092–1096.

Reymond, L. (1985). Spatial visual acuity of the eagle *Aquila audax:* A behavioural, optical and anatomical investigation. *Vision Research, 25,* 1477–1491.

Reynolds, D. V. (1969). Surgery in the rat during electrical analgesia induced by focal brain stimulation. *Science, 164,* 444–445.

Rice, C. E. (1967). Human echolocation. *Science, 155,* 656–664.

Riggs, L. A. (1965). Visual acuity. In C. Graham (Ed.), *Vision and visual perception.* New York: Wiley.

Riquimaroux, H., Gaioni, S. J., & Suga, N. (1991). Cortical computational maps control auditory perception. *Science, 251,* 565–568.

Risset, J. C., & Mathews, M. W. (1969). Analysis of musical instrument tones. *Physics Today, 22,* 23–30.

Roberts, J. C., & Summerfield, Q. (1981). Audiovisual adaptation in speech perception. *Perception and Psychophysics, 30,* 309–314.

Robertson, D., & Irvine, D. R. F. (1989). Plasticity of frequency organization in auditory cortex of guinea pigs with partial unilateral deafness. *Journal of Comparative Neurology, 282,* 456–471.

Robinson, D. L., & Wurtz, R. (1976). Use of an extraretinal signal by monkey superior colliculus neurons to distinguish real from self-induced stimulus movement. *Journal of Neurophysiology, 39,* 852–870.

Robson, J. G., Tolhurst, D. J., Freeman, R. D., & Ohzawa, I. (1988). Simple cells in the visual cortex of the cat can be narrowly tuned for spatial frequency. *Visual Neuroscience, 1,* 415–419.

Rock, I., & Brosgole, L. (1964). Grouping based on phenomenal proximity. *Journal of Experimental Psychology, 67,* 531–538.

Rock, I., & Kaufman, L. (1962). The moon illusion, Part 2. *Science, 136,* 1023–1031.

Rock, I., Nijhawan, R., Palmer, S., & Tudor, L. (1992). Grouping based on phenomenal similarity of achromatic color. *Perception, 21,* 779–789.

Rock, I., & Palmer, S. (1990, December). The legacy of Gestalt psychology. *Scientific American,* 84–90.

Roeder, K. (1963). *Nerve cells and insect behavior.* Cambridge: Harvard University Press.

Roederer, J. G. (1975). *Introduction to the physics and psychophysics of music* (2nd ed.). New York: Springer-Verlag.

Roland, P. (1992). Cortical representation of pain. *Trends in Neuroscience, 15,* 3–5.

Rollman, G. B. (1991). Pain responsiveness. In M. A. Heller & W. Schiff (Eds.), *The psychology of touch* (pp. 91–114). Hillsdale, NJ: Erlbaum.

Rolls, E. T. (1992). Neurophysiological mechanisms underlying face processing within and beyond the temporal cortical visual areas. *Transactions of the Royal Society of London B, 335,* 11–21.

Rose, J. E., Brugge, J. F., Anderson, D. J., & Hind, J. E. (1967). Phase locked response to low frequency tones in single auditory nerve fibers of the squirrel monkey. *Journal of Neurophysiology, 30,* 769–793.

Rosenstein, D., & Oster, H. (1988). Differential facial responses to four basic tastes in newborns. *Child Development, 59,* 1555–1568.

Rosenthal, E. (1992, December 29). Chronic pain fells many yet lacks clear cause. *New York Times,* B5, B6.

Rovee-Collier, C., Hankins, E., & Bhatt, R. (1992). Textons, visual pop-out effects, and object recognition in infancy. *Journal of Experimental Psychology: General, 121,* 435–445.

Rozin, P. (1976). The selection of foods by rats, humans, and other animals. In J. S. Rosenblatt, R. A.

Hinde, & C. Beer (Eds.), *Advances in the study of behavior.* New York: Academic.

Rozin, P. (1982). "Taste-smell confusions" and the duality of the olfactory sense. *Perception and Psychophysics, 31,* 397–401.

Rubel, E. W., 1984. Ontogeny of auditory system function. *Annual Review of Physiology, 46,* 213–229.

Rubel, E. W., & Ryals, B. M. (1983). Development of the place principle: Acoustic trauma. *Science, 219,* 512–514.

Rubin, E. (1915). *Synoplevde Figurer.* Copenhagen: Gyldendalske.

Runeson, S. (1977). On the possibility of "smart" perceptual mechanisms. *Scandinavian Journal of Psychology, 18,* 172–179.

Runeson, S., & Frykholm, G. (1981). Visual perception of lifted weights. *Journal of Experimental Psychology: Human Perception and Performance, 7,* 733–740.

Rushton, W. A. H. (1961). Rhodopsin measurement and dark adaptation in a subject deficient in cone vision. *Journal of Physiology, 156,* 193–205.

Rushton, W. A. H. (1964). Colour blindness and cone pigments. *American Journal of Optometry and Archives of the American Academy of Optometry, 41,* 265–282.

Russell, I. J., & Sellick, P. M. (1977). Tuning properties of cochlear hair cells. *Nature, 267,* 858–860.

Russell, M. J. (1976). Human olfactory communication. *Nature, 260,* 520–522.

Russell, M. J., Switz, G. M., & Thompson, K. (1980). Olfactory influence on the human menstrual cycle. *Pharmacology, Biochemistry and Behavior, 13,* 737–738.

Rutherford, W. (1886). A new theory of hearing. *Journal of Anatomy and Physiology, 21,* 166–168.

Sachs, M. B., Young, E. D., & Miller, M. I. (1981). Encoding of speech features in the auditory nerve. In R. Carson & B. Grandstrom (Eds.), *The representation of speech in the peripheral auditory system* (pp. 115–130). New York: Elsevier.

Sakata, H., & Iwamura, Y. (1978). Cortical processing of tactile information in the first somatosensory and parietal association areas in the monkey. In G. Gordon (Ed.), *Active touch* (pp. 55–72). Elmsford, NY: Pergamon.

Salapatek, P., & Banks, M. S. (1978). Infant sensory assessment: Vision. In F. D. Minifie & L. L. Lloyd (Eds.), *Communicative and cognitive abilities: Early behavioral assessment* (pp. 61–106). Baltimore: University Park Press.

Salapatek, P., Bechtold, A. G., & Bushness, E. W. (1976). Infant visual acuity as a function of viewing distance. *Child Development, 47,* 860–863.

Saldaña, H. M., & Rosenblum, L. D. (1993). Visual influences on auditory pluck and bow judgments. *Perception and Psychophysics, 54,* 406–416.

Saldaña, H. M., & Rosenblum, L. D. (1994). Selective adaptation in speech perception using a compelling audiovisual adapter. *Journal of the Acoustical Society of America, 95,* 3658–3661.

Salzman, C. D., Britten, K. H., & Newsome, W. T. (1990). Cortical microstimulation influences perceptual judgments of motion direction. *Nature, 346,* 174–177.

Salzman, C. D., Murasugi, C. M., Britten, K. H., & Newsome, W. T. (1992). Microstimulation in visual area MT: Effects on direction discrimination performance. *Journal of Neuroscience, 12,* 2331–2355.

Samuel, A. G. (1981). Phonemic restoration: Insights from a new methodology. *Journal of Experimental Psychology: General, 110,* 474–494.

Samuel, A. G. (1990). Using perceptual-restoration effects to explore the architecture of perception. In G. T. M. Altmann (Ed.), *Cognitive models of speech processing* (pp. 295–314). Cambridge: MIT Press.

Sary, G., Vogels, R., & Orban, G. A. (1993). Cue-invariant shape selectivity of macaque inferior temporal neurons. *Science, 260,* 995–997

Sato, M., & Ogawa, H. (1994). Neural coding of taste in macaque monkeys. In K. Kurihara, N. Suzuki, & H. Ogawa (Eds.), *Olfaction and taste* (Vol. 11, pp. 388–392). Tokyo: Springer-Verlag.

Sato, M., Ogawa, H., & Yamashita, S. (1975). Response properties of macaque monkey chorda tympani fibers. *Journal of General Physiology, 66,* 781–810.

Sato, M., Ogawa, H., & Yamashita, S. (1994). Gustatory responsiveness of chorda tympani fibers in the cynomolgus monkey. *Chemical Senses, 19,* 381–400.

Sawusch, J. R., & Jusczyk, P. (1981). Adaptation and contrast in the perception of voicing. *Journal of Experimental Psychology: Human Perception and Performance, 7,* 408–421.

Scharf, B. (1975). Audition. In B. Scharf (Ed.), *Experimental sensory psychology* (pp. 112–149). Glenview, IL: Scott Foresman.

Scheich, H., Langner, G., Tidemann, C., Coles, R. B., & Guppy, A. (1986). Electroception and electrolocation in platypus. *Nature, 319,* 401–402.

Schiff, W. (1980). *Perception: An applied approach.* Boston: Houghton Mifflin.

Schiff, W., & Detwiler, M. L. (1979). Information used in judging impending collision. *Perception, 8,* 647–658.

Schiffman, H. R. (1967). Size-estimation of familiar objects under informative and reduced conditions of viewing. *American Journal of Psychology, 80,* 229–235.

Schiffman, S. S., & Erickson, R. P. (1971). A psychophysical model for gustatory quality. *Physiology and Behavior, 7,* 617–633.

Schiffman, S. S., & Erickson, R. P. (1993). Psychophysics: Insights into transduction mechanisms and neural coding. In S. A. Simon & S. D. Roper (Eds.), *Mechanisms of taste transduction* (pp. 395–424). Boca Raton, FL: CRC Press.

Schiffman, S. S., Sennewald, K., & Gagnon, J. (1981). Comparison of taste qualities and thresholds of d- and l-amino acids. *Physiology and Behavior, 27,* 51–59.

Schiller, P. H. (1992). The ON and OFF channels of the visual system. *Trends in Neurosciences, 15,* 86–92.

Schiller, P. H. (1993). The effects of V4 and middle temporal (MT) area lesions on visual performance in the rhesus monkey. *Visual Neuroscience, 10,* 717–746.

Schiller, P. H. (1994). Area V4 of the primate visual cortex. *Current Directions in Psychological Science, 3,* 89–92.

Schiller, P. H., Logothetis, N. K., & Charles, E. R. (1990). Functions of the colour-opponent and broad-band channels of the visual system. *Nature, 343,* 68–70.

Schmitt, B. C., & Ache, B. W. (1979). Olfaction: Responses of a decapod crustacean are enhanced by flicking. *Science, 205,* 204–206.

Schnapf, J. L., Kraft, T. W., & Baylor, D. A. (1987). Spectral sensitivity of human cone photoreceptors. *Nature, 325,* 439–441.

Schubert, E. D. (1980). *Hearing: Its function and dysfunction.* Wien: Springer-Verlag.

Schultze, M. (1872). Aur Anatomie und Physiologie der Retina. *Arch. Mikroskop. Anat., 2,* 165–286.

Schwartz, A. S., Perez, A. J., & Azulaz, A. (1975). Further analysis of active and passive touch in pattern discrimination. *Bulletin of the Psychonomic Society, 6,* 7–9.

Schwind, R. (1978). Visual system of *Notonecta glaucia:* A neuron sensitive to movement in the binocular visual field. *Journal of Comparative Physiology, 123,* 315–328.

Scott, T. R. (1987). Coding in the gustatory system. In T. E. Finger & W. L. Silver (Eds.), *Neurobiology of taste and smell* (pp. 355–378). New York: Wiley.

Scott, T. R., & Giza, B. K. (1990). Coding channels in the taste system of the rat. *Science, 249,* 1585–1587.

Scott, T. R., & Giza, B. K. (1995). Theories of gustatory neural coding. In R. L. Doty (Ed.), *Handbook of olfaction and gustation* (pp. 611–633). New York: Marcel Dekker.

Scott, T. R., & Plata-Salaman, C. R. (1991). In T. V. Getchell et al. (Eds.), *Smell and taste in health and disease* (pp. 345–368). New York: Raven.

Searle, C. L., Jacobson, J. Z., & Rayment, S. G. (1979). Stop consonant discrimination based on human audition. *Journal of the Acoustical Society of America, 65,* 799–809.

Sechzer, J. A., & Brown, J. L. (1964). Color discrimination in the cat. *Science, 144,* 427–429.

Sedgwick, H. A. (1973). The visible horizon: A potential source of visual information for the perception of size and distance. *Dissertation Abstracts International, 34,* 1301B–1302B.

Sedgwick, H. A. (1983). Environment-centered representation of spatial layout: Available visual information from texture and perspective. In A. Rosenthal & J. Beck (Eds.), *Human and machine vision* (pp. 425–458). New York: Academic.

Sekuler, R. W., & Ganz, L. (1963). Aftereffect of seen motion with a stabilized retinal image. *Science, 139,* 419–420.

Shapley, R., & Perry, V. H. (1986). Cat and monkey retinal ganglion cells and their visual functional roles. *Trends in Neuroscience, 9,* 229–235.

Shea, S. L., Fox, R., Aslin, R., & Dumais, S. T. (1980). Assessment of stereopsis in human infants. *Investigative Ophthalmology and Visual Science, 19,* 1400–1404.

Sherrick, C. E., & Cholewiak, R. W. (1986). Cutaneous sensitivity. In K. R. Boff, L. Kaufman, & J. P.

Thomas (Eds.), *Handbook of perception and human performance* (Chap. 12). New York: Wiley.

Shiffrar, M. (1994). When what meets where. *Current Directions in Psychological Science, 3,* 96–100.

Shiffrar, M., & Freyd, J. J. (1990). Apparent motion of the human body. *Psychological Science, 1,* 257–264.

Shih, G., & Kessell, R. (1982). *Living images.* Boston: Jones & Bartlett.

Shiffrar, M., & Freyd, J. J. (1993). Timing and apparent motion path choice with human body photographs. *Psychological Science, 4,* 379–384.

Sicard, G., & Holley, A. (1984). Receptor cell responses to odorants: Similarities and differences among odorants. *Brain Research, 292,* 283–296.

Simmons, J. A. (1973). The resolution of target range by echolocating bats. *Journal of the Acoustical Society of America, 54,* 157–173.

Simmons, J. A. (1979). Perception of echo phase in bat sonar. *Science, 204,* 1334–1338.

Singer, W., Artola, A., Engel, A. K., Konig, P., Kreiter, A. K., Lowel, S., & Schillen, T. B. (1993). Neuronal representations and temporal codes. In T. A. Poggio & D. A. Glaser (Eds.), *Exploring brain functions: Models in neuroscience* (pp. 179–194). New York: Wiley.

Sivak, J. G. (1976). Optics of the eye of the "four-eyed" fish *Anableps anableps. Vision Research, 16,* 513–516.

Slater, A. M., & Findlay, J. M. (1975). Binocular fixation in the newborn baby. *Journal of Experimental Child Psychology, 20,* 248–273.

Slater, A. M., Morison, V., & Rose, D. (1984). Habituation in the newborn. *Infant Behavior and Development, 7,* 183–200.

Sloan, L. L., & Wollach, L. (1948). A case of unilateral deuteranopia. *Journal of the Optical Society of America, 38,* 502–509.

Smith, D., Van Buskirk, R. L., Travers, J. B., & Bieber, S. L. (1983). Coding of taste stimuli by hamster brainstem neurons. *Journal of Neurophysiology, 50,* 541–558.

Smith, K. R. (1947). The problem of stimulation deafness: 2. Histological changes in the cochlea as a function of tonal frequency. *Journal of Experimental Psychology, 37,* 304–317.

Solomon, A. (1994, August 28). Defiantly deaf. *New York Times Magazine,* 39–45, 62, 65–68.

Srinivasan, M. V., Zhang, S. W., & Rolfe, B. (1993). Is pattern vision in insects mediated by "cortical" processing? *Nature, 362,* 539–540.

Stark, L., & Bridgeman, B. (1983). Role of corollary discharge in space constancy. *Perception and Psychophysics, 34,* 371–380.

Stebbins, W. C., Hawkins, J. E., Jr., Johnson, L. G., & Moody, D. B. (1979). Hearing threshold with outer and inner hair cell loss. *American Journal of Otolaryngology, 1,* 15–27.

Steinbrecht, R. A. (1969). Comparative morphology of olfactory receptors. In C. Pfaffmann (Ed.), *Olfaction and taste* (Vol. 3, pp. 1–21). New York: Rockefeller University Press.

Steiner, J. E. (1974). Innate, discriminative human facial expressions to taste and smell stimulation. *Annals of the New York Academy of Sciences, 237,* 229–233.

Steiner, J. E. (1979). Human facial expressions in response to taste and smell stimulation. *Advances in Child Development and Behavior, 13,* 257–295.

Stevens, J. K., Emerson, R. C., Gerstein, G. L., Kallos, T., Neufeld, G. R., Nichols, C. W., & Rosenquist, A. C. (1976). Paralysis of the awake human: Visual perceptions. *Vision Research, 16,* 93–98.

Stevens, K. N., & Blumstein, S. (1978). Invariant cues for place of articulation in stop consonants. *Journal of the Acoustical Society of America, 64,* 1358–1368.

Stevens, K. N., & Blumstein, S. (1981). The search for invariant acoustic correlates of phonetic features. In P. D. Eimas & L. L. Miller (Eds.), *Perspectives on the study of speech.* Hillsdale, NJ: Erlbaum.

Stevens, S. S. (1957). On the psychophysical law. *Psychological Review, 64,* 153–181.

Stevens, S. S. (1961). To honor Fechner and repeal his law. *Science, 133,* 80–86.

Stevens, S. S. (1962). The surprising simplicity of sensory metrics. *American Psychologist, 17,* 29–39.

Stevens, S. S., Davis, H., & Lurie, M. H. (1935). The localization of pitch perception on the basilar membrane. *Journal of General Psychology, 13,* 297–315.

Stevens, S. S., & Volkman, J. (1940). The relation of pitch to frequency: A revised scale. *American Journal of Psychology, 53,* 329–353.

Stewart, W. B., Kauer, J. S., & Shepherd, G. M. (1979). Functional organization of rat olfactory bulb analyzed by the 2-deoxyglucose method. *Journal of Comparative Neurology, 185,* 715–734.

Stiles, W. S. (1953). Further studies of visual mechanisms by the two-color threshold method. *Coloquio sobre problemas opticos de la vision.* Madrid: *Union Internationale de Physique Pure et Appliquée, 1,* 65.

Stone, H., & Bosley, J. J. (1965). Olfactory discrimination and Weber's law. *Perceptual and Motor Skills, 20,* 657–665.

Stone, J. (1965). A quantitative analysis of the distribution of ganglion cells in the cat's retina. *Journal of Comparative Neurology, 124,* 277–352.

Strange, W. (1989). Evolving theories of vowel perception. *Journal of the Acoustical Society of America, 85,* 2081–2087.

Stryker, M. P. (1989). Is grandmother an oscillation? *Nature, 338,* 297–298.

Suga, N. (1990, June). Biosonar and neural computation in bats. *Scientific American,* 60–68.

Summerfield, Q. (1979). Use of visual information for phonetic perception. *Phonetica, 36,* 314–331.

Sussman, H. M., Hoemeke, K. A., & Ahmed, F. S. (1993). A cross-linguistic investigation of locus equations as a phonetic descriptor for place of articulation. *Journal of the Acoustical Society of America, 94,* 1256–1268.

Sussman, H. M., McCaffrey, H. A., & Matthews, S. A. (1991). An investigation of locus equations as a source of relational variance for stop place categorization. *Journal of the Acoustical Society of America, 90,* 1309–1325.

Svaetichin, G. (1956). Spectral response curves from single cones. *Acta Physiologica Scandinavica Supplementum, 134,* 17–46.

Swets, J. A. (1964). *Signal detection and recognition by human observers.* New York: Wiley.

Takagi, S. F. (1980). Dual nervous systems for olfactory functions in mammals. In H. Van derStarre (Ed.), *Olfaction and taste* (Vol. 7, pp. 275–278). London: IRC Press.

Tanabe, T., Iino, M., Oshima, Y., & Takagi, S. F. (1974). An olfactory area in the prefrontal lobe. *Brain Research, 80,* 127–130.

Tanabe, T., Iino, M., & Takagi, S. F. (1975). Discrimination of odors in olfactory bulb, pyriform-amygdaloid areas and orbito-frontal cortex of the monkey. *Journal of Neurophysiology, 38,* 1284–1296.

Tanaka, K. (1993). Neuronal mechanisms of object recognition. *Science, 262,* 684–688.

Tansley, K. (1965). *Vision in vertebrates.* London: Chapman & Hall.

Tarr, M. J. (1994). Visual representation. In V. Ramachandran (Ed.), *Encyclopedia of human behavior* (Vol. 4, pp. 503–512). New York: Academic.

Taylor, M. M., Lederman, S. J., & Gibson, R. H. (1973). Tactual perception of texture. In E. C. Carterette & M. P. Friedman (Eds.), *Handbook of perception* (Vol. 3). New York: Academic.

Teller, D. Y. (1990). The domain of visual science. In L. Spellman, & J. S. Werner, (Eds), *Visual perception: The neurophysiological foundations* (pp. 11–21). San Diego: Academic.

Teller, D. Y., Morse, R., Borton, R., & Regal, D. (1974). Visual acuity for vertical and diagonal gratings in human infants. *Vision Research, 14,* 1433–1439.

Teuber, H. L. (1960). Perception. In J. Field, H. W. Magoun, & V. E. Hall (Eds.), *Handbook of physiology* (Section 1, Vol. 3). Washington, DC: American Physiological Society.

Thimbleby, H. W., Inglis, S., & Witten, I. H. (1994). Displaying 3D images: Algorithms for single image random dot stereograms. *IEEE Computer, 27* (10).

Thomas, B. (1993). *Magic eye.* Kansas City: Andrews & McMeel.

Thompson, R. F. (1985). *The brain.* New York: W. H. Freeman.

Todd, J. T. (1981). Visual information about moving objects. *Journal of Experimental Psychology: Human Perception and Performance, 7,* 795–810.

Todd, J. T., & Norman, J. F. (1991). The visual perception of smoothly curved surfaces from minimal apparent motion sequences. *Perception and Psychophysics, 50,* 509–523.

Tomita, T., Kaneko, A., Murakami, M., & Paulter, E. L. (1967). Spectral response curves of single cones in the carp. *Vision Research, 7,* 519–531.

Tonndorf, J. (1960). Shearing motion in scala media of cochlear models. *Journal of the Acoustical Society of America, 32,* 238–244.

Tonndorf, J., & Khanna, S. M. (1968). Submicroscopic displacement amplitudes of the tympanic membrane (cat) measured by laser interferometer. *Journal of the Acoustical Society of America, 44,* 1546–1554.

Treisman, A. (1986). Features and objects in visual processing. *Scientific American, 255,* 114B–125B.

Treisman, A. (1987). Properties, parts, and objects. In K. R. Boff, L. Kaufman, & F. P. Thomas (Eds.), *Handbook of perception and human performance.* (Chap. 35). New York: Wiley.

Treisman, A. (1993). The perception of features and objects. In A. Baddeley & L. Weiskrantz (Eds.), *Attention: Selection, awareness, and control* (pp. 5–34). Oxford: Clarendon.

Trevor-Roper, P. (1970). *The world through blunted sight.* Indianapolis: Bobbs-Merrill.

Truax, B. (1984). *Acoustic communication.* Norwood, NJ: Ablex.

Ts'o, D. Y. (1989). The functional organization and connectivity of color processing. In D. M-K. Lam & C. Gilbert (Eds.), *Neural mechanisms of visual perception* (pp. 87–115). Woodlands, TX: Portfolio.

Tyler, C. W. (1983). Sensory processing of binocular disparity. In C. M. Schor & K. J. Ciuffreda (Eds.), *Vergence eye movements: Basic and clinical aspects.* London: Butterworths.

Tyler, C. W. (1990). A stereoscopic view of visual processing streams. *Vision Research, 30,* 1877–1895.

Tyler, C. W. (1994). The birth of computer stereograms for unaided stereovision. In S. Horibuchi (Ed.), *Stereogram* (pp. 83–89). San Francisco: Cadence.

Tyler, C. W., & Clarke, M. B. (1990). The autostereogram. In J. O. Merritt & S. S. Fisher (Eds.), *Stereoscopic displays and applications* (pp. 182–197). Bellingham, WA: International Society for Optical Engineering.

Ullman, S. (1980). Against direct perception. *Behavioral and Brain Sciences, 3,* 373–415.

Ungerleider, L. G., & Haxby, J. V. (1994). "What" and "where" in the human brain. *Current Opinion in Neurobiology, 4,* 157–165.

Ungerleider, L. G., & Mishkin, M. (1982). Two cortical visual systems. In D. J. Ingle, M. A. Goodale, & R. J. Mansfield (Eds.), *Analysis of visual behavior* (pp. 549–580). Cambridge: MIT Press.

Vallbo, A. B., & Hagbarth, K. E. (1967). Impulses recorded with microelectrodes in human muscle nerves during stimulation of mechanoreceptors and voluntary contractions. *Electroencephalography and Clinical Neurophysiology, 23,* 392.

Vallbo, A. B., & Johansson, R. S. (1978). The tactile sensory innervation of the glabrous skin of the human hand. In G. Gordon (Ed.), *Active touch* (pp. 29–54). New York: Oxford University Press.

Van Sluyters, R. C., Atkinson, J., Banks, M. S., Held, R. M., Hoffmann, K. P., & Chatz, C. J. (1990). The development of vision and visual perception. In L. Spillman & J. Werner (Eds.), *Visual perception: The neurophysiological foundations* (pp. 349–379). San Diego: Academic.

Varela, F. J., Palacios, A. G., & Goldsmith, T. H. (1993). Color vision of birds. In H. P. Zeigler & H-J. Bishof (Eds.), *Vision, brain and behavior in birds* (pp. 77–98). Cambridge: MIT Press.

Varner, D., Cook, J. E., Schneck, M. E., McDonald, M., & Teller, D. Y. (1985). Tritan discriminations by 1- and 2-month-old human infants. *Vision Research, 25,* 821–831.

Vega-Bermudez, F., Johnson, K. O., & Hsiao, S. S. (1991). Human tactile pattern recognition: Active versus passive touch, velocity effects, and patterns of confusion. *Journal of Neurophysiology, 65,* 531–546.

Verheijen, F. J. (1961). A single afterimage method demonstrating the involuntary multi-directional eye movements during fixation. *Optica Acta, 8,* 309–311.

von Holst, E. (1954). Relations between the central nervous system and the peripheral organs. *British Journal of Animal Behaviour, 2,* 89–94.

von Noorden, G. K., & Maumanee, A. E. (1968). Clinical observations on stimulus deprivation amblyopia (amblyopia ex anopsia). *American Journal of Ophthalmology, 65,* 220–224.

Walcott, C., Gould, J. L., & Kirschvink, J. L. (1979). Pigeons have magnets. *Science, 205,* 1027–1028.

Walcott, C., & Green, R. P. (1974). Orientation of homing pigeons altered by a change in the direction of an applied magnetic field. *Science, 184,* 180–182.

Wald, G. (1964). The receptors of human color vision. *Science, 145,* 1007–1017.

Wald, G., & Brown, P. K. (1958). Human rhodopsin. *Science, 127,* 222–226.

Wald, G., & Brown, P. K. (1965). Human color vision and color blindness. *Cold Spring Harbor Symposia on Quantitative Biology, 30,* 345–359.

Wall, J. T. (1988). Variable organization in cortical maps of the skin as an indication of the lifelong adaptive capacities of circuits in the mammalian brain. *Trends in Neuroscience, 12,* 549–557.

Wall, P. D., & Melzack, R. (Eds.). (1994). *Textbook of pain* (3rd ed.). Edinburgh: Churchill Livingstone.

Wallach, H. (1963). The perception of neutral colors. *Scientific American, 208*(1), 107–116.

Wallach, H., Newman, E. B., & Rosenzweig, M. R. (1949). The precedence effect in sound localization. *American Journal of Psychology, 62,* 315–336.

Wallach, H., & O'Connell, D. N. (1953). The kinetic depth effect. *Journal of Experimental Psychology, 45,* 205–217.

Walls, G. L. (1942). *The vertebrate eye.* New York: Hafner. (Reprinted in 1967)

Walls, G. L. (1953). *The lateral geniculate nucleus and visual histophysiology.* Berkeley: University of California Press.

Ward, S., Thomson, N., White, J. G., & Brenner, S. (1975). Electron microscopical reconstruction of the anterior sensory anatomy of the nematode *Caenorhabditis elegans. Journal of Comparative Neurology, 160,* 313–338.

Ward, W. D., & Glorig, A. (1961). A case of firecracker induced hearing loss. *Laryngoscope, 71,* 1590–1596.

Ware, C., & Mitchell, D. E. (1974). On interocular transfer of various visual aftereffects in normal and stereoblind observers. *Vision Research, 14,* 731–735.

Warren, D. H., Welch, R. B., & McCarthy, T. J. (1981). The role of visual-auditory "compellingness" in the ventriloquism effect: Implications for transivity among the special senses. *Perception and Psychophysics, 30,* 557–564.

Warren, R. M. (1970). Perceptual restoration of missing speech sounds. *Science, 167,* 392–393.

Warren, R. M., Obuseck, C. J., & Acroff, J. M. (1972). Auditory induction of absent sounds. *Science, 176,* 1149.

Warren, S., Hamalainen, H., & Gardner, E. P. (1986). Objective classification of motion- and direction-sensitive neurons in primary somatosensory cortex of awake monkeys. *Journal of Neurophysiology, 56,* 598–622.

Warren, W. H., Jr., Mestre, D. R., Blackwell, A. W., & Morris, M. (1991). Perception of circular heading from optical flow. *Journal of Experimental Psychology: Human Perception and Performance, 17,* 28–43.

Warren, W. H., Morris, M. W., & Kalish, M. (1988). Perception of translational heading from optical flow. *Journal of Experimental Psychology: Human Perception and Performance, 14,* 646–660.

Wassle, H., Grunert, U., Rohrenbeck, J., & Boycott, B. B. (1990). Retinal ganglion cell density and cortical magnification factor in the primate. *Vision Research, 30,* 1897–1911.

Watkins, C. R., & Mayer, D. J. (1982). Organization of endogenous opiate and nonopiate pain control system. *Science, 176,* 1149.

Webster, F. A., & Durlach, N. I. (1963). Echolocation systems of the bat. MIT Lincoln Lab Report No 41-G-3, Lexington, MA.

Weinstein, S. (1968). Intensive and extensive aspects of tactile sensitivity as a function of body part, sex, and laterality. In D. R. Kenshalo (Ed.), *The skin senses* (pp. 195–218). Springfield, IL: Charles C Thomas.

Weisenberg, M. (1977). Pain and pain control. *Psychological Bulletin, 84,* 1008–1044.

Weiskrantz, L. (1986). *Blindsight: A case study and implications.* New York: Oxford University Press.

Weisstein, N., & Wong, E. (1986). Figure-ground organization and the spatial and temporal responses of the visual system. In E. C. Schwab & H. C. Nusbaum (Eds.), *Pattern recognition by humans and machines* (Vol. 2). New York: Academic.

Werker, J. F., & Tees, R. C. (1984). Cross-language speech perception: Evidence for perceptual reorganization during the first year of life. *Infant Behavior and Development, 7,* 49–63.

Werker, J. F., & Tees, R. C. (1992). The organization and reorganization of human speech perception. *Annual Review of Neuroscience, 15,* 377–402.

Werner, L. A., & Bargones, J. Y. (1992). Psychoacoustic development of human infants. In C. Rovee-Collier & L. Lipsett (Eds.), *Advances in infancy research* (Vol. 7, pp. 103–145). Norwood, NJ: Ablex.

Wever, E. G. (1949). *Theory of hearing.* New York: Wiley.

Whalen, D. H., & Liberman, A. M. (1987). Speech perception takes precedence over nonspeech perception. *Science, 237,* 169–171.

White, J. (1968). *The birth and rebirth of pictorial space* (2nd ed.). London: Faber & Faber.

Whitfield, I. C., & Evans, E. F. (1965). Responses of auditory neurons to stimuli of changing frequency. *Journal of Neurophysiology, 28,* 655–672.

Whitsel, B. L., Roppolo, J. R., & Werner, G. (1972). Cortical information processing of stimulus motion on primate skin. *Journal of Neurophysiology, 35,* 691–717.

Wiesel, T. N. (1982). Postnatal development of the visual cortex and the influence of environment. *Nature, 299,* 583–591.

Wiesel, T. N., & Hubel, D. H. (1963). Single cell responses in striate cortex of kittens deprived of vision in one eye. *Journal of Neurophysiology, 26,* 1003–1017.

Wightman, F. L. (1973). Pitch and stimulus fine structure. *Journal of the Acoustical Society of America, 54,* 397–406.

Winston, P. H. (1975). *The psychology of computer vision.* New York: McGraw-Hill.

Wittreich, W. J. (1959). Visual perception and personality. *Scientific American, 200*(4), 56–60.

Woodworth, R. S. (1938). *Experimental psychology.* New York: Holt, Rinehart & Winston.

Wurm, L. H., Legge, G. E., Isenberg, L. M., & Luebker, A. (1993). Color improves object recognition in normal and low vision. *Journal of Experimental Psychology: Human Perception and Performance, 19,* 899–911.

Wyszecki, G., & Stiles, W. S. (1967). *Color science: Concepts and methods, quantitative data and formulas.* New York: Wiley.

Yager, D. D., & Hoy, R. R. (1986). The cyclopean ear: A new sense for the praying mantis. *Science, 231,* 727–729.

Yager, D. D., & Thorpe, S. (1970). Investigation of goldfish color vision. In W. C. Stebbins (Ed.), *Animal psychophysics* (pp. 259–275). New York: Appleton-Century-Crofts.

Yarbus, D. L. (1967). *Eye movements and vision.* New York: Plenum.

Ye, Q., Heck, G., & DeSimone, J. A. (1991). The anion paradox in sodium taste reception: Resolution by voltage-clamp studies. *Science, 254,* 724–726.

Yonas, A., Granrud, C. E., Arterberry, M. E., & Hanson, B. L. (1986). Infant's distance from linear perspective and texture gradients. *Infant Behavior and Development, 9,* 247–256.

Yonas, A., Pettersen, L., & Granrud, C. E. (1982). Infant's sensitivity to familiar size as information for distance. *Child Development, 53,* 1285–1290.

Young, T. (1802). On the theory of light and colours. *Transactions of the Royal Society of London, 92,* 12–48.

Young-Browne, G., Rosenfield, H. M., & Horowitz, F. D. (1977). Infant discrimination of facial expression. *Child Development, 48,* 555–562.

Yuodelis, C., & Hendrickson, A. (1986). A qualitative and quantitative analysis of the human fovea during development. *Vision Research, 26,* 847–855.

Zatorre, R. J. (1988). Pitch perception of complex tones and human temporal-lobe function. *Journal of the Acoustical Society of America, 84,* 566–572.

Zatorre, R. J., Evans, A. C., Meyer, E., & Gjedde, A. (1992). Lateralization of phonetic and pitch discrimination in speech processing. *Science, 256,* 846–849.

Zatorre, R. J., & Halpern, A. R. (1979). Identification, discrimination, and selective adaptation of simultaneous musical intervals. *Perception and Psychophysics, 26,* 384–395.

Zeki, S. (1983a). Color coding in the cerebral cortex: The responses of wavelength-selective and color coded cells in monkey visual cortex to changes in wavelength composition. *Neuroscience, 9,* 767–781.

Zeki, S. (1983b). Colour coding in the cerebral cortex: The reaction of cells in monkey visual cortex to wavelengths and colours. *Neuroscience, 9,* 741–765.

Zeki, S. (1984). The construction of colours by the cerebral cortex. *Proceedings of the Royal Institute of Great Britain, 56,* 231–257.

Zeki, S., Watson, J. D. G., Lueck, C. J., Friston, K. J., Kennard, C., & Frackowiak, R. S. J. (1991). A direct demonstration of functional specialization in human visual cortex. *Journal of Neuroscience, 11,* 641–649.

Zihl, J., von Cramon, D., & Mai, N. (1983). Selective disturbance of movement vision after bilateral brain damage. *Brain, 106,* 313–340.

Zihl, J., Von Cramon, D., Mai, N., & Schmid, Ch. (1991). Disturbance of movement vision after bilateral posterior brain damage. *Brain, 114,* 2235–2252.

Zimmerman, M. (1979). Peripheral and central nervous mechanisms of nociception, pain, and pain therapy: Facts and hypotheses. In J. J. Bonica, J. D. Liebeskind, & D. G. Albe-Fessard (Eds.), *Advances in pain research and therapy* (Vol. 3, pp. 3–32). New York: Raven.

Zrenner, E., Abramov, I., Akita, M., Cowey, A., Livingston, M., & Valberg, A. (1990). Color perception. In L. Spillmann & J. S. Werner (Eds.), *Visual perception: The neurophysiological foundations* (pp. 163–204). San Diego: Academic.

Zwicker, E. (1974). On the psychoacoustic equivalent of tuning curves. In E. Zwicker & E. Terhardt (Eds.), *Facts and models in hearing* (pp. 132–141). Berlin: Springer-Verlag.

Zwicker, E., Flottorp, S., & Stevens, S. S. (1957). Critical bandwidth in loudness summation. *Journal of the Acoustical Society of America, 29,* 548–557.

Author Index

Subject Index

Credits

This page constitutes an extension of the copyright page. We have made every effort to trace the ownership of all copyrighted material and to secure permission from copyright holders. In the event of any question arising as to the use of any material, we will be pleased to make the necessary corrections in future printings. Thanks are due to the following authors, publishers, and agents for permission to use the material indicated.

Illustration and Text Credits

About the cover: The cover picture, a detail from Landforms West 1980, is a wool weaving by fiber artist Lia Cook. This composition, which is based on the idea of fabric draped over a rolling landscape, creates the illusion of three-dimensional shape by varying the curvature, spacing, and width of parallel contours. Lia, whose weavings often relate to perceptual concerns, is professor of art at the California College of Arts and Crafts, Oakland, California. Her work has been exhibited worldwide and is represented in the permanent collections of the Museum of Modern Art and the Metropolitan Museum of Art.

Chapter One: **20**: Figure 1.18, From T. Engen: Figure from *Woodworth & Schlosberg's Experimental Psychology*, Third Edition, by J. W. Kling and Lorrin A. Riggs, copyright © 1971 by Holt, Rinehart and Winston, Inc., reproduced by permission of the publisher. **21**: Figure 1.19, Adapted from "The Surprising Simplicity of Sensory Metrics," by S. S. Stevens, 1962, *American Psychologist, 17*, pp. 29–39. Copyright © 1962 by American Psychological Association. **21**: Figure 1.20, Adapted from "The Surprising Simplicity of Sensory Metrics," by S. S. Stevens, 1962, *American Psychologist, 17*, pp. 29–39. Copyright © 1962 by American Psychological Association. Reprinted by permission. **22**: Figure 1.21, Adapted from "The Surprising Simplicity of Sensory Metrics," by S. S. Stevens, 1962, *American Psychologist, 17*, pp. 29–39. Copyright © 1962 by American Psychological Association. Reprinted by permission. **24**: Figure 1.24, Adapted from "The Role of Frequency in Developing Perceptual Sets," by B. R. Bugelski and D. A. Alampay, 1961, *Canadian Journal of Psychology, 15*, 205–211. Copyright © 1961 by the Canadian Psychological Association. Reprinted by permission. **26**: Figure 1.28, From "The Effects of Contextual Scenes on the Identification of Objects," by S. E. Palmer, 1975, *Memory and Cognition, 3*, 519–526, figure 1. Copyright © 1975 by The Psychonomic Society. Reprinted by permission. **26**: Figure 1.27, Adapted from "The Role of Frequency in Developing Perceptual Sets," by B. R. Bugelski and D. A. Alampay, 1961, *Canadian Journal of Psychology, 15*, 205–211. Copyright © 1961 by the Canadian Psychological Association. Reprinted by permission. **29**: Figure 1.31, Adapted from "The Role of Frequency in Developing Perceptual Sets," by B. R. Bugelski and D. A. Alampay, 1961, *Canadian Journal of Psychology, 15*, 205–211. Copyright © 1961 by the Canadian Psychological Association. Reprinted by permission. **31**: Figure 1.32b, From "The Cyclopean Ear: A New Sense for the Praying Mantis," by D. D. Yager and R. R. Hoy, 1986, *Science, 231*, figure 1a. Copyright © 1986 by the American Association for the Advancement of Science. Reprinted by permission. **35**: Figure 1.37, Courtesy of Velma Dobson. **36**: Figure 1.38, Adapted from "Visual Acuity for Vertical and Diagonal Gratings in Human Infants," by D. Teller, R. Morse, R. Borton, and D. Regal, 1974, *Vision Research, 14*, 1433–1439. Copyright © 1974 with kind permission from Elsevier Science Ltd., The Boulevard, Langford Lane, Kidlington 0X5 1GB, UK.

Chapter Two: **43**: Figure 2.3, Adapted from "Organization of the Primate Retina," by J. E. Dowling and B. B. Boycott, 1966. *Proceedings of the Royal Society of London, 16*, Series B, 80–111. Copyright © 1966 by The Royal Society. Adapted by permission. **44**: Figure 2.4, Adapted from *Human Information Processing*, by P. Lindsay and D. A. Norman, 1977, 2nd ed., p. 126. Copyright © 1977 Academic Press, Inc. Adapted by permission. **47**: Figure 2.9, From "Scanning Electron Microscopy of Vertebrate Visual Receptors," by E. R. Lewis, Y. Y. Zeevi, and F. S. Werblin, 1969, *Brain Research, 15*, 559–562. Copyright © 1969 Elsevier Science Publishers, B.V. Reprinted by permission. **49**: Figure 2.12, Adapted from *Vision and the Eye*, by M. H. Pirenne, 1948, figure 29, p. 30. Copyright © 1948 by Chapman & Hall Ltd., London. Adapted by permission. **50**: Figure 2.16, From a figure by Johnny Johnson on page 88 of "Blind Spots," by Vilaynaur S. Ramachandran, *Scientific American*, May 1992. Copyright © 1992 by Scientific American. All rights reserved. **51**: Figure 2.17, From "Segregation of Form, Color, Movement and Depth: Anatomy, Physiology and Perception," by M. Livingston and D. H. Hubel, 1988, *Science, 240*, 740–749. Copyright © 1988 by the American Association for the Advancement of Science. Reprinted by permission of M. Livingston. **53**: Figure 2.18, From "Functional Architecture of Macaque Monkey Cortex," by D. H. Hubel and T. N. Wiesel, 1977, *Proceedings of the Royal Society of London, 198*, 1–59. Copyright © 1977 by the Royal Society. Reprinted by permission of D. H. Hubel. **56**: Figure 2.22, Partial data from "Rhodopsin Measurement and Dark Adaptation in a Subject Deficient in Cone Vision," by W. A. H. Ruston, 1961, *Journal of Physiology, 156*, 193–205. Copyright © 1961 by the Physiological Society, Cambridge University Press. **59**: Figure 2.24, Adapted from "The Receptors of Human Color Vision," by E. Wald, 1964, *Science, 145*, pp. 1009 and 1011. Copyright © 1964 by the American Association for the Advancement of Science. Adapted by permission. **60**: Figure 2.25, From "Human Rhodopsin," by G. Wald and P. K. Brown, 1958, *Science, 127*, pp. 222–226, figure 6, and "The Receptors of Human Color Vision," by G. Wald, 1964, *Science, 145*, pp. 1007–1017. Copyright © 1964 by the American Association for the Advancement of Science. Reprinted by permission. **60**: Figure 2.26, Reproduced from "The Dark Adaptation of the Color Anomalous," by A. Chapanis, 1947, *Journal of General Physiology, 30*, 423–437, figure 9, by copyright permission of The Rockefeller University Press. **61**: Figure 2.27, Data from "Human Visual Pigments: Microspectrophometric Results from the Eyes of Seven Persons," by H. J. A. Dartnall, J. K. Bowmaker, and J. D. Mollon, 1983, *Proceedings of the Royal Society of London, 220B*, 115–130. The Royal Society. **65**: Figure 2.32, From "Amacrine Cells," by R. H. Masland, 1988, *Trends in Neuroscience, 11* (9), 405–410, figure 2. Copyright © 1988 by Elsevier Science Ltd. Reprinted by permission. **67**: Figure 2.36, From " Integrative Action in the Cat's Lateral Geniculate Body," by D. H. Hubel and T. N. Wiesel, 1961, *Journal of Physiology, 155*, 385–398, figure 1. Copyright © 1961 by The Physiological Society, Cambridge University Press. Reprinted by permission. **70**: Figure 2.39, From "Visual Acuity," by L. A. Riggs, 1965. In C. Graham, Ed., *Vision and Visual Perception*, figure 11.4, p. 324. Copyright © 1965 by John Wiley & Sons., Inc. Reprinted by permission of John Wiley & Sons. **73**: Figure 2.43, From *Mach Bands: Quantitative Studies on Neural Networks in the Retina*, by F. Ratliff, 1965, figure 3.25, p. 107. Copyright © 1965 Holden-Day, Inc. Reprinted by permission. **74**: Figure 2.45, From "A Simple Afterimage Method Demonstrating the Involuntary Multi-Directional Eye Movements During Fixation," by F. J. Verheijen, 1961, *Optica Acta, 8*, 309–311. Copyright © 1961 Taylor & Francis, Ltd. Reprinted by permission. **78**: Figure 2.52, From "Apparent Surface Curvature Affects Lightness Perception," by D. C. Knill and D. Kersten, 1991, *Nature, 351*, 228–230, figure 1. Copyright © 1991 Macmillan Magazines Ltd. Reprinted by permission. **82**: Figure 2.56, Reprinted from *Vision Research, 16*, p. 534, "Optics of the Eye of the 'Four-Eyed Fish' (Anableps Anableps)," by J. G. Sivak, 1976. Copyright © 1976 with kind permission from Elsevier Science Ltd., The Boulevard, Langford Lane, Kidlington OX5 1GB, UK. **84**: Figure 2.59, From "Infant Contrast Sensitivity Evaluated by Evoked Potentials," by M. Pirchio, D. Spinelli, A. Fiorentini, and L. Maffei, 1978, *Brain Research, 141*, 179–184, figure 3C. Copyright © 1978 Elsevier Biomedical Press B.V. Reprinted by permission. **84**: Figure 2.58, Adapted from "Visual Acuity for Vertical and Diagonal Gratings in Human Infants," by D. Teller, R. Morse, R. Borton, and D. Regal, 1974, *Vision Research, 14*, 1433–1439. Copyright © 1974 with kind permission from Elsevier Science Ltd., The Boulevard, Langford Lane, Kidlington 0X5 1GB, UK. **85**: Figure 2.60, From *The Postnatal Development of the Cerebral Cortex*, Vol. 1, 1939 and Vol. 3, 1947, and *The Postnatal Development of the Human Cerebral Cortex*, Vol. 4, 1951, by J. L. Conel, plates LVIII, LXIV and LXIV. Copyright © 1939, 1947, 1951 by Harvard University Press. Reprinted by permission of the publisher. **86**: Figure 2.61, Based on "Optical and Photoreceptor Immaturities Limit the Spatial and Chromatic Vision of Human Neonates" by M. S. Banks and P. J. Bennett, 1988, *Journal of the Optical Society of America, (5)*, 12, p. 2062, figure 2. **86**: Figure 2.62, Based on "Optical and Photoreceptor Immaturities Limit the Spatial and Chromatic Vision of Human Neonates" by M. S. Banks and P. J. Bennett, 1988, *Journal of the Optical Society of America, (5)*, 12, p. 2062, figure 3.

Chapter Three: **93**: Figure 3.3 and Color Plate 2.7, Reprinted with permission from *Nature*, "Functions of the Colour-Opponent and Broad-Band Channels of the Visual System," by P. H. Schiller, N. K. Logthetis, and E. R. Charles, 1990, 343, p. 68, figure 1a & b. Copyright © 1990 by Macmillan Magazines, Ltd. **95**: Figure 3.5, From "Receptive Fields of Single Neurons in the Cat's Striate Cortex," by D. H. Hubel and T. N. Wiesel, 1959, *Journal of Physiology, 148*, 574–591, figure 2. Copyright © 1959 by The Physiological Society, Cambridge University Press. Reprinted by permission. **95**: Figure 3.4, Courtesy of David Hubel and Torsten Wiesel. **96**: Figure 3.7, From "Receptive Fields of Single Neurons in the Cat's Striate Cortex," by D. H. Hubel and T. N. Wiesel, 1959, *Journal of Physiology, 148*, 574–591, figure 8. Copyright © 1959 by The Physiological Society, Cambridge University Press. Reprinted by permission. **96**: Figure 3.8, From "Receptive Fields and Functional Architecture in Two Non-Striate Visual Areas (18 & 19) of the Cat," by D. H. Hubel and T. N. Wiesel, 1965, *Journal of Neurophysiology, 28*, 229–289, figure 8. Copyright © 1965 by The American Physiological Society. Reprinted by permission. **99**: Figure 3.12, Based on data from "Uniformity of Monkey Striate Cortex: A Parallel Relationship Between Field Size, Scatter and Magnification Factor," by D. H. Hubel and T. N. Wiesel, 1974, *Journal of Comparative Neurology, 158*, 295–306. John Wiley & Sons. **101**: Figure 3.14, From "Anatomical Demonstration of Orientation Columns in Macaque Monkey," by D. H. Hubel, T. N. Wiesel, and M. P. Stryker, 1978, *Journal of Comparative Neurology, 177*, 261–379, figure 4. Copyright © 1978 by John Wiley & Sons. Reprinted by permission of John Wiley & Sons. **103**: Figure 3.16, Data from "The Relation Between Neural and Perceptural Intensity: A Comparative Study of Neural and Psychophysical Responses to Taste Stimuli," by G. Borg, H. Diamant, L. Strom, and Y. Zotterman, 1967, *Journal of Physiology, 192*, 13–20, figure 2. The Physiological Society. **105**: Figure 3.18 and Color Plate 2.8, From "Object Vision and Spatial Vision: Two Central Pathways," by M. Mishkin, L. G. Ungerleider, and K. A. Macko, 1983, *Trends in Neuroscience, 6*, 414–417, figure 1. Copyright © 1983 Elsevier Science Publishers B.V. Reprinted by permission. **106**: Figure 3.19, From "Object Vision and Spatial Vision: Two Central Pathways," by M. Mishkin, L. G. Ungerleider, and K. A. Macko, 1983, *Trends in Neuroscience, 6*, 414–417, figure 2. Copyright © 1983 Elsevier Science Publishers B.V. Reprinted by permission. **108**: Figure 3.21, From "A Selective Impairment of Motion Perception Following Lesions of the Middle Temporal Visual Area (MT)," by W. T. Newsome and E. B. Paré, 1988, *Journal of Neuroscience, 8* (6), 2201–2211, figure 1. Copyright © 1988 by Society for Neuroscience. Reprinted by permission of Oxford University Press. **108**: Figure 3.22, Adapted from "How Parallel Are the Primate Visual Pathways," by W. H. Merigan and J. H. R. Maunsell [from data from Felleman & Van Essen, 1987], 1992, *Annual Review of Neuroscience, 16*, 369–402, figure 3. Annual Reviews, Inc. **109**: Figure 3.23, Adapted from "How Parallel Are the Primate Visual Pathways," by W. H. Merigan and J. H. R. Maunsell [from data from Felleman & Van Essen, 1987], 1992, *Annual Review of Neuroscience, 16*, 369–402, figure 3. Annual Reviews, Inc. **109**: Figure 3.24, From "Visual Properties of Neurons in a Polysensory Area in the Superior Temporal Sulcus of the Macaque," by C. Bruce, R. Desimone, and C. G. Gross, 1981, *Journal of Neurophysiology, 46*, 369–384, figure 7. Copyright © 1981 by The American Physiological Society. Reprinted by permission. **110**: Figure 3.25, Reprinted with permission from "Neuronal Mechanisms of Object Recognition," by K. Tanaka, 1993, *Science, 262*, 646–688, figure 2. Copyright © 1993 by the American Association for the Advancement of Science. **114**: Figure 3.28, Adapted by permission from "Selective Attention Gates Visual Processing in the Extrastriate Cortex," by J. Moran and R. Desimone, 1985, *Science, 229*, 782–784. Copyright © 1985 by the American Association for the Advancement of Science. **116**: Figure 3.30, Adapted from "Temporal Coding in the Visual Cortex: New Vistas on Integration in the Nervous System," by A. K. Engel, P. Konig, A. K. Kreiter, T. B. Schillen, and W. Singer, 1992, *Trends in Neuroscience, 15*, 218–226, figure 1. Copyright © 1992 by Elsevier Science Ltd. Adapted by permission. **118**: Figure 3.31, Adapted from "Is Pattern Vision in Insects Mediated by 'Cortical' Processing?," by M. V. Srinivasan, S. W. Zhang, and B. Rolfe, 1993, *Nature, 362*, 539–540. Macmillan Magazines, Ltd. **119**: Figure 3.32, From "Shape Vision in Bees: Innate Preference for Flower-Like Patterns," by M. Lehrer, G. A. Horridge, S. W. Zhang, and R. Gadagkar, 1995, *Philosophical Transactions of the Royal Society of London B, 347*, 123–137. Reprinted by permission of the Royal Society. **121**: Figure 3.33, From "Single Cell Responses in Striate Cortex of Kittens Deprived of Vision in One Eye," by T. N. Wiesel and D. H. Hubel, 1963, *Journal of Neurophysiology, 26*, 1002–1017, figures 1 and 3. Copyright © 1963 by The American Physiological Society. Reprinted by permission. **122**: Figure 3.34, Adapted from "Profile of the Sensitive Period for Monocular Deprivation in Kittens," by C. R. Olson and R. D. Freeman, 1980, *Experimental Brain Research, 39*, 17–21, figure 3. Copyright © 1980 by Springer-Verlag. Adapted by permission. **122**: Figure 3.36, From "Optically Induced Concomitant Strabismus in Monkey," by M. L. J. Crawford and G. K. von Noorden, 1980, *Investigative Ophthalmology and Visual Science, 19*, 1105–1109, figure 1. Copyright © 1980 by The C.V. Mosby Company. Reprinted by permission. **124**: Figure 3.38, From "Sensitive Period for the Development of Human Binocular Vision," by M. S. Banks, R. N. Aslin, and R. D. Letson, 1975, *Science, 190*, 675–677, figure 1C. Copyright © 1975 by the American Association for the Advancement of Science. Reprinted by permission.

Chapter Four: **132**: Figure 4.1 and Color Plate 4.3, From *Color Vision*, by Leo M. Hurvich, 1981. Reprinted by permission of Dr. Leo M. Hurvich. **134**: Figure 4.4, Based on data from "Spectral Distribution of Typical Daylight as a Function of Correlated Color Temperature," by D. B. Judd, D. L. MacAdam, and G. Wyszecki, 1964, *Journal of the Optical Society of America, 54*, 1031–1040. **135**: Figure 4.5, Adapted from *Color: Its Principles and Their Applications*, by F. W. Clulow, 1972, Morgan and Morgan. Copyright © 1972 by F. W. Clulow. Used by permission of the author. **135**: Figure 4.6, Adapted from *Color: Its Principles and Their Applications*, by F. W. Clulow, 1972, Morgan and Morgan. Copyright © 1972 by F. W. Clulow. Used by permission of the author. **142**: Figure 4.14, From "Color Appearance: On Seeing Red, or Yellow, or Green, or Blue," by I. Abramov and J. Gordon, 1994. Reproduced, with permission, from *Annual Review of Psychology, 45*, 451–485, figure 1a. Copyright © 1994 by Annual Reviews, Inc. **145**: Figure 4.18, From "Spectral Response Curves from Single Cones," by G. Svaetichin, 1956, *Acta Physiologica Scandanavica, Suppl. 134*, 17–46, figure 1. Copyright © 1956 Karolinska Institutet. Reprinted by permission. **146**: Figure 4.19, From "Primate Color Vision," by R. L. DeValois and G. H. Jacobs, 1968, *Science, 162*, 533–540, figure 5. Copyright © 1968 by the American Association for the Advancement of Science. Reprinted by permission. **147**: Figure 4.21, From "Anatomy and Physiology of a Color System in the Primate Visual Cortex," by M. S. Livingstone and D. H. Hubel, 1984, *Journal of Neuroscience, 4*, (1), 309–356, figure 34. Copyright © 1984 Society for

Advancement of Science. Reprinted by permission. **209**: Figure 5.58, Courtesy of University of Pittsburgh, Office of University Relations. **212**: Figure 5.63, From "Application of Fourier Analysis to the Visibility of Gratings," by F. W. Campbell and J. G. Robson, 1968, *Journal of Physiology, 197*, 551–556, figure 2. Copyright © 1968 Physiological Society, Cambridge University Press. Reprinted by permission. **214**: Figure 5.65, From "The Visual Cortex as a Spatial Frequency Analyzer," by L. Maffei and A. Fiorentini, 1973, *Vision Research, 13*, 1255–1267, figure 3. Copyright © 1973 with kind permission from Elsevier Science Ltd., The Boulevard, Langford Lane, Kidlington 0X5 1GB, UK. **219**: Figure 5.69, Reprinted from *Vision Research, 14*, 1211–1217, "Visual Resolution in the Cat," by R. Blake, S. J. Cool, and M. L. J. Crawford, 1974, Copyright © 1974 with kind permission from Elsevier Science Ltd., The Boulevard, Langford Lane, Kidlington 0X5 1GB, UK. **220**: Figure 5.71, Adapted from "Behavioral Determination of Spatial Selectivity of Contrast Adaptation in Cats: Some Evidence for a Common Plan in the Mammalian Visual System," by M. A. Berkley, 1990, *Visual Neuroscience, 4*, 413–426. Copyright © 1990 Cambridge University Press. Adapted by permission. **223**: Figure 5.74, Photo courtesy of Carolyn Rovee-Collier. **225**: Figure 5.75, Courtesy of Carolyn Rovee-Collier.

Chapter Six: **234**: Figure 6.3, Paul Signac, *Place des Lices, St. Tropez*, 1893. Museum of Art, Carnegie Institute, Pittsburgh, PA. Reproduced by permission. **235**: Figure 6.4, Reproduced by Special Permission of *Playboy* Magazine. Copyright © 1971 by Playboy. **239**: Figure 6.10, School of Brumante, *A Street with Various Buildings, Colonnades and an Arch*, c. 1500. Museum of Art, Carnegie Institute, Pittsburgh, PA. Reproduced by permission. **242**: Figure 6.14, Copyright © Michael Chikiris, 1977. Reproduced by permission. **246**: Figure 6.22, Figure from *Foundations of Cyclopean Perception*, by B. Julesz, 1971, figures 2.4-1 and 2.4-3. Copyright © 1971 University of Chicago Press. Reprinted by permission. **247**: Figure 6.23, From "Cells Sensitive to Binocular Depth in Area 18 of the Macaque Monkey Cortex, " by D. H. Hubel and T. Wiesel, 1970, *Nature, 225*, 41–42, figure 1. Copyright © 1970 by Macmillan Magazines, Ltd. Reprinted by permission. **248**: Figure 6.24, Adapted from "Ocular Dominance and Disparity Coding in Cat Visual Cortex," by S. LeVay and T. Voigt, 1988, *Visual Neuroscience, 1*, 395–414, figure 5b. Cambridge University Press. **252**: Figure 6.26, From " Displaying 3D Images: Algorithms for Single-Image Random-Dot Stereograms," by H. W. Thimbleby, S. Inglis, and I. H. Witten, 1994, *IEEE Computer, 27*, no. 10, pp. 38–48. Copyright © 1994 by IEEE Computer Society Press. Reprinted by permission. **256**: Figure 6.32, From "Determinants of Apparent Visual Size with Distance Variant," by A. H. Holway and E. G. Boring, 1941, *American Journal of Psychology, 54*, 21–34, figure 2. University of Illinois Press. **256**: Figure 6.33, From "Determinants of Apparent Visual Size with Distance Variant," by A. H. Holway and E. G. Boring, 1941, *American Journal of Psychology, 54*, 21–34, figure 22. University of Illinois Press. **259**: Figure 6.36, 1981 The Far Side cartoon by Gary Larson is reprinted by permission of Chronicle Features, San Francisco, CA. All rights reserved. **260**: Figure 6.37, Photo © 1959 by William Vandivert. Reproduced by permission. First appeared in "Visual Perception and Personality," by W. J. Wittreich, 1959, *Scientific American, 200*, (4) April, p. 58. **267**: Figure 6.48, Photo by Philip Brodatz. **269**: Figure 6.50, Victor Vasarely: *Vega-Nor*, 1969. Oil on canvas, 78-3/4 x 78-3/4". Albright-Knox Art Gallery, Buffalo, New York. Gift of Seymour H. Knox, 1969. **275**: Figure 6.57, From "Depth Vision in Animals," by T. S. Collett and L. I. K. Harkness, 1982. In M. A. Goodale and R. J. W. Mansfield (Eds.), *Analysis of Visual Behavior*, pp. 111–176. Copyright © 1982 by MIT Press. Reprinted by permission. **278**: Figure 6.60, From "Assessment of Stereopsis in Human Infants," by S. L. Shea, R. Fox, R. Aslin, and S. T. Dumais, 1980, *Investigative Ophthalmology and Visual Science, 19*, 1440–1404, figure 1. Copyright © 1980 C.V. Mosby Company, St. Louis, MO. Reprinted by permission. **279**: Figure 6.61, From "Stereoacuity of Human Infants," by R. Held, E. E. Birch, and J. Gwiazda, 1980, *Proceedings of the National Academy of Sciences, 77*, 5572–5574, figure 1. Copyright © 1980 by R. Held, E. E. Birch, and J. Gwiazda. Reprinted by permission. **279**: Figure 6.62, From "Binocular Vision: Behavioral and Neural Development," by R. Held, 1985. In J. Mehler and R. Fox (Eds.), *Neonate Cognition: Beyond the Blooming, Buzzing Confusion*, pp. 37–44. Copyright © 1985 by Lawrence Erlbaum Associates, Inc. Reprinted by permission. **280**: Figure 6.63, From "Infants' Perception of Pictorially Specified Interposition," by C. E. Granrud and A. Yonas, 1984, *Journal of Experimental Child Psychology, 27*, 500–511, figure 1 b and d. Copyright © 1984 Academic Press, Inc. Reprinted by permission. **281**: Figure 6.64, From "Infants' Sensitivity to Familiar Size: The Effect of Memory on Spatial Perception," by C. E. Granrud, R. J. Haake, and A. Yonas, 1985, *Perception and Psychophysics, 37*, 459–466. Copyright © 1985 Psychonomic Society Publications. Reprinted by permission.

Chapter Seven: **288**: Figure 7.1, From *Form in Motion Parallax;* and from "Luminance Contrast: Vernier Discrimination," *Spatial Vision, 1*, 305–318. Reprinted by permission of David Regan. **289**: Figure 7.2, Detail from "A Passing Umbrella," by Kenneth Antol, 1983. Reprinted by permission. **292**: Figure 7.7, Courtesy of Mark Friedman, Pittsburgh, PA. **296**: Figure 7.14, From "Lateral Inhibition Between Orientation Detectors in the Cat's Visual Cortex," by C. Blakemore and E. A. Tobin, 1972, *Experimental Brain Research, 15*, 439–440, figure 1. Copyright © 1972 Springer-Verlag Publishers. Reprinted by permission. **298**: Figure 7.16, From "Evidence for a Physiological Explanation of the Waterfall Illusion and Figural Aftereffects," by H. B. Barlow and R. M. Hill, 1963, *Nature, 200*, 1345–1347, figure 1. Copyright © 1963 by Macmillan Magazines, Ltd. Reprinted by permission. **299**: Figure 7.17, From "A Selective Impairment of Motion Perception Following

Lesions of the Middle Temporal Visual Area (MT)," by W. T. Newsome and E. B. Paré, 1988, *Journal of Neuroscience, 8* (6), 2201–2211, figure 1. Copyright © 1988 by Society for Neuroscience. Reprinted by permission of Oxford University Press. **300**: Figure 7.19, From "Perception," by H. L. Teuber, 1960. In J. Field, H. W. Magoun, and V. E. Hall (Eds.), *Handbook of Physiology, Section 1, Neurophysiology*, Vol. 3, pp. 1595–1668, figure 31. Copyright © 1960 by the American Physiological Society. Reprinted by permission. **304**: Figure 7.24, Adapted from "Segmentation Versus Integration in Visual Motion Processing," by O. Braddick, 1993, *Trends in Neuroscience, 16*, (7), 263–268, figure 1. Copyright © 1993 by Elsevier Science Ltd. Adapted by permission. **305**: Figure 7.25, Based on "The Analysis of Moving Visual Patterns," by J. A. Movshon, E. H. Adelson, M. S. Gizzi, and W. T. Newsome, 1985. In C. Chagas, R. Gattass, and C. Gross (Eds.), *Pattern Recognition*, figure 2, p. 120. Springer-Verlag. **305**: Figure 7.26, Based on "The Analysis of Moving Visual Patterns,", by J. A. Movshon, E. H. Adelson, M. S. Gizzi, and W. T. Newsome, 1985. In C. Chagas, R. Gattass, and C. Gross (Eds.), *Pattern Recognition*, figure 2, p. 120. Springer-Verlag. **306**: Figure 7.27, From "The Neural Computation of the Velocity Field," by E. C. Hildreth, 1990. In B. Cohen and Bodis-Wollner (Eds.), *Vision and the Brain*, pp. 139–164, figure 6. Copyright © 1990 Raven Press. Reprinted by permission. **306**: Figure 7.28, Adapted from "The Neural Computation of the Velocity Field," by E. C. Hildreth, 1990. In B. Cohen and Bodis-Wollner (Eds.), *Vision and the Brain*, pp. 139–164, figure 12. Copyright © 1990 Raven Press. Adapted by permission. **307**: Figure 7.29, From "The Neural Computation of the Velocity Field," by E. C. Hildreth, 1990. In B. Cohen and Bodis-Wollner (Eds.), *Vision and the Brain*, pp. 139–164, figure 13. Copyright © 1990 Raven Press. Reprinted by permission. **311**: Figure 7.34, From "Retrieval of Structure from Rigid and Biological Motion: An Analysis of the Visual Responses of Neurones in the Macaque Temporal Cortex," by D. I. Perrett, M. H. Harries, P. J. Bensen, A. J. Chitty, and A. J. Mistlin, 1990. In A. Blake and T. Troscianko (Eds.), *AI and the Eye*, pp. 181–200, figure 2. Copyright © 1990 by John Wiley & Sons, Ltd. Reprinted by permission of D. I. Perrett and John Wiley & Sons, Ltd. **317**: Figure 7.45, Based on figure from "Visual Proprioceptive Control of Standing in Human Infants," by D. N. Lee and E. Aronson, 1974, *Perception and Psychophysics, 15*, 529–532, figure 2. Psychonomics Society Publications. **322**: Figure 7.53, Adapted from "Tuning MST Neurons to Spiral Motions," by M. S. A. Graziano, R. A. Anderson, and R. J. Snowden, 1994, *Journal of Neuroscience, 14*, 54–67, figures 1 & 2. Society for Neuroscience. **323**: Figure 7.54, From "Cats Perceive Biological Motion," by R. Blake, 1993, *Psychological Science, 4*, 54–57, figure 1. Copyright © 1993 Cambridge University Press. Reprinted with permission of Cambridge University Press. **324**: Figure 7.55, Adapted from "Cats Perceive Biological Motion," by R. Blake, 1993, *Psychological Science, 4*, 54–57, figures 1, 2. Copyright © 1993 Cambridge University Press. Adapted with permission of Cambridge University Press. **327**: Figure 7.57, From "Perception of Partly Occluded Objects in Infancy," by P. J. Kellman and E. S. Spelke, 1983, *Cognitive Psychology, 15*, 483–524, figure 3. Copyright © 1983 by Academic Press. Reprinted by permission.

Chapter Eight: **341**: Figure 8.6, Reprinted with permission of C.G. Conn Ltd., Hatfield Herts, England. **343**: Figure 8.9, Redrawn by permission from *Human Information Processing*, by P. H. Lindsay and D. A. Norman, 2nd ed., 1977, p. 229. Copyright © 1977 by Academic Press, Inc. **344**: Figure 8.10, Adapted from *Hearing: Physiology and Psychophysics*, by W. Lawrence Gulick. Copyright © 1971 by Oxford University Press, Inc. Reprinted by permission. **345**: Figure 8.13, From "Hearing: Its Function and Dysfunction," by E. D. Schubert, 1980, *Disorders of Human Communication*, Vol. 1, pp. 15 and 18. Copyright © 1980 Springer-Verlag, Wien-New York. Reprinted by permission. **348**: Figure 8.17, Adapted from *Hearing: Physiology and Psychophysics*, by W. Lawrence Gulick. Copyright © 1971 by Oxford University Press, Inc. Reprinted by permission. **351**: Figure 8.22, From "Shearing Motion in Scala Media of Cochlear Models," by J. Tonndorf, 1960, *Journal of the Acoustical Society of America*, 32, 238–244, figure 7b. Copyright © 1960 by the American Institute of Physics. Reprinted by permission. **351**: Figure 8.23, From "Hearing: Its Function and Dysfunction," by E. D. Schubert, 1980, *Disorders of Human Communication*, Vol. 1, pp. 15, 18. Copyright © 1980 Springer-Verlag, Wien-New York. Reprinted by permission. **352**: Figure 8.24, From *Experiments in Hearing*, by G. von Békésy, 1960, figure 11.43. Copyright © 1960 by McGraw-Hill Book Company, New York. Reprinted by permission. **353**: Figure 8.25, From *Experiments in Hearing*, by G. von Békésy, 1960, figure 11.59. Copyright © 1960 by McGraw-Hill Book Company, New York. Reprinted by permission. **353**: Figure 8.26, From "A Revised Frequency Map of the Guinea Pig Cochlea," by E. A. Culler, J. D. Coakley, K. Lowy, and N. Gross, 1943, *American Journal of Psychology, 56*, 475–500, figure 11. Copyright © 1943 by the University of Illinois Press. Reprinted by permission. **354**: Figure 8.27, Data from "The Tuning Properties of Cochlea Hair Cells—Addendum" by I. J. Russell and P. M. Selleck, 1977. In E. F. Evans and J. P. Wilson, *Psychophysics and Physiology of Hearing*, p. 81. Academic Press. **355**: Figure 8.28, Adapted from "Stimulus Representation in the Discharge Patterns of Auditory Neurons," by N. Y. S. Kiang, 1975. In E. L. Eagles (Ed.), *The Nervous System*, Vol. 3, pp. 81–96. Copyright © 1975 by Raven Press. Adapted by permission. **360**: Figure 8.36, From "Space and Frequency Are Represented Separately in Auditory Midbrain of the Owl," by E. I. Knudsen and M. Konishi, 1978, *Journal of Neurophysiology*, 41, 870–883, figure 1. Copyright © 1978 by The American Physiological Society. Reprinted by permission. **361**: Figure 8.37, From "A Neural Map of Auditory Space in the Owl," by E. I. Knudsen and M. Konishi, 1978, *Science, 200*, 795–797, figure 1. Copyright © 1978 by the American Association for the Advancement of Science. Reprinted by permission. **363**: Figure 8.39, Adapted from "A Panoramic Code for Sound Location by Cortical Neurons," by J. C. Middlebrooks, A. E. Clock, L. Xu, and D. M. Green, 1994, *Science, 264*, 842–844. Copyright © 1994 by the American Association for the Advancement of Science. Adapted by permission. **365**: Figure 8.40, Based on *The Neural Basis of Echolocation in Bats*, by G. D. Pollak & J. H. Casseday, 1989, figure 1.7. Springer-Verlag, New York. **369**: Figure 8.43, From "Ontogeny of Auditory System Function," by E. W. Rubel, 1984. Reproduced, with permission, from *Annual Review of Physiology, 46*, 213–229, figure 1. Copyright © 1984 by Annual Reviews, Inc. **369**: Figure 8.44, Data from "Plasticity of Frequency Organization in Auditory Cortex of Guinea Pigs with Partial Unilateral Deafness," by D. L. Robertston and D. R. F. Irvine, 1989, *Journal of Comparative Neurology, 282*, 456–471, figures 9a and 14a. Alan R. Liss, Inc. **370**: Figure 8.45, From "Plasticity of Frequency Organization in Auditory Cortex of Guinea Pigs with Partial Unilateral Deafness," by D. L. Robertston and D. R. F. Irvine, 1989, *Journal of Comparative Neurology, 282*, 456–471, figure 7b. Alan R. Liss, Inc.

Chapter Nine: **378**: Figure 9.3, From "Loudness: Its Definition, Measurement and Calculation," by H. Fletcher and W. A. Munson, 1933, *Journal of the Acoustical Society of America, 5*, 82–108, figure 4. Copyright © 1993 by the American Institute of Physics. Reprinted by permission. **379**: Figure 9.4, Adapted from *Hearing: Physiological Acoustics, Neural Coding, and Pyschoacoustics*, by W. Lawrence Gulick, George A. Gescheider, and Robert D. Frisina, 1989, figure 11.9. Oxford University Press. **380**: Figure 9.5, Adapted from Effects of Flanking Noise Bands on the Rate of Growth of Loudness of Tones in Normal and Recruiting Ears," by B. C. J. Moore, B. R. Glasberg, R. F. Hess, and J. P. Birchall, 1985, *Journal of the Acoustical Society of America, 77* (4) 1505–1513. Copyright © 1985 by the American Institute of Physics. Adapted by permission. **381**: Figure 9.6, From *Hearing: Physiological Acoustics, Neural Coding and Psychoacoustics*, by W. Lawrence Gulick, G. A. Gescheider, and R. D. Frisina, 1989, p 235. Copyright © 1989 by Oxford University Press, Inc. Reprinted by permission. **382**: Figure 9.7, Data from "The Relation of Pitch to Frequency: A Revised Scale," by S. S. Stevens and J. Volkman, 1940, *American Journal of Psychology, 53*, 329–353; and from "The Localization of Pitch Perception on the Basilar Membrane," by S. S. Stevens, H. Davis, and M. H. Lurie, 1935, *Journal of General Psychology*, 13, 297–315. **385**: Figure 9.11, From *Music, Physics and Engineering*, 2nd ed., by H. F. Olson, 1967. Copyright © 1967 by Dover Publications, Inc. Reprinted by permission. **386**: Figure 9.12, From *Music, Physics and Engineering*, 2nd ed., by H. F. Olson, 1967. Copyright © 1967 by Dover Publications, Inc. Reprinted by permission. **387**: Figure 9.13, From *Music, Physics and Engineering*, 2nd ed., by H. F. Olson, 1967. Copyright © 1967 by Dover Publications, Inc. Reprinted by permission. **387**: Figure 9.14, From "Analysis of Musical Instrument Tones," by J. C. Risset and M. V. Mathews, 1969, *Physics Today, 22*, (2), 23–30, figure 4. Copyright © 1967 by the American Institute of Physics. Reprinted by permission. **392**: Figure 9.17, Adapted from "On the Masking Pattern of a Simple Auditory Stimulus," by J. P. Egan and H. W. Hake, 1950, *Journal of the Acoustical Society of America, 22*, 622–630. Copyright © 1950 by the American Institute of Physics. Adapted by permission. **394**: Figure 9.20a, From "On a Psychoacoustical Equivalent of Tuning Curves," by E. Zwicker, 1974. In E. Zwicker and E. Terhardt (Eds.), *Facts and Models in Hearing*, p. 134. Copyright © 1974 Springer-Verlag Publishers. Reprinted by permission. **394**: Figure 9.20b, Originally presented by Kiang and Moxon at a Georg von Békésy symposium in 1973. **395**: Figure 9.21, From "Pure Tone Masking; A New Result from a New Method," by L.L. M. Vogten, 1974. In E. Zwicken and E. Terhardt (Eds.), *Facts and Models in Hearing*, Copyright © 1974 by Springer-Verlag Publishers. Reprinted by permission. **395**: Figure 9.22, Based on "Critical Bandwidth in Loudness Summation," by E. Zwicker, S. Flottorp, and S. S. Stevens, 1957, *Journal of Acoustical Society of America, 29*, 548–557. Physics Institute of America. **396**: Figure 9.23, From *Hearing: Physiological Acoustics, Neural Coding and Psychoacoustics*, by W. Lawrence Gulick, G. A. Gescheider, and R. D. Frisina, 1989, p. 276. Copyright © 1989 by Oxford University Press, Inc. Reprinted by permission. **397**: Figure 9.25, Based on "Primary Auditory Stream Segregation and Perception of Order in Rapid Sequence of Tone," by A. S. Bregman and J. Campbell, 1971, *Journal of Experimental Psychology, 89*, 244–249. American Psychological Association. **398**: Figure 9.26, Based on "Auditory Segregation: Stream or Streams?," by A. S. Bregman and A. I. Rudnicky, 1975, *Journal of Experimental Psychology: Human Perception and Performance, I*, 263–267. American Psychological Association. **399**: Figure 9.27, From "Two-Channel Listening to Musical Scales," by D. Deutsch, 1975, *Journal of the Acoustical Society of America, 57*. Copyright © 1975 by the American Institute of Physics. Reprinted by permission. **400**: Figure 9.28, Based on "Auditory Induction of Absent Sounds," by R. M. Warren, C. J. Obuseck, and J. M. Acroff, 1972, *Science, 176*, 1149. Association for the Advancement of Science. **400**: Figure 9.29, From "The Perception of Interleaved Melodies," by W. J. Dowling, 1973, *Cognitive Psychology, 5*, 322–337. Academic Press, Inc. **401**: Figure 9.30, From "Sound Localization by Human Listeners," by J. C. Middlebrooks and David M. Green, 1991, *Annual Review of Psychology, 42*, 135–159, figure 1. Copyright © 1991 Annual Reviews, Inc. Reprinted by permission. **402**: Figure 9.32, From *Hearing: Physiological Acoustics, Neural Coding and Psychoacoustics*, by W. Lawrence Gulick, G. A. Gescheider, R. D. Frisina, 1989, p. 324. Copyright © 1989 by Oxford University Press, Inc. Reprinted by permission. **404**: Figure 9.34, Data from "Sonar System of the Blind," by W. N. Kellogg, 1962, *Science, 137*, 399–505, figure 2. American Association for the Advancement of Science. **408**: Figure 9.37, Adapted from "Hearing in Mammals: The Least Weasel," by R. S. Heffner and H. E. Heffner, 1985, *Journal of Mammology, 66* (4), 745–755, figure 5. **409**: Figure 9.38, Adapted from "Visual Factors in Sound Localization in Mammals," by R. S. Heffner and H. E. Heffner, 1992, *Journal of Comparative Neurology, 317*, 219–232, figure 6. Copyright © 1992 by Wiley-Liss, Inc. Reprinted by permission of John Wiley & Sons. **409**: Figure 9.39, From "Infrasound Detection by the Homing Pigeon: A Behavioral Audiogram," by M. L. Kreithen and D. B. Quinn, 1979, *Journal of Comparative Physiology, 129*, 1–4. Springer-Verlag, Germany. Reprinted by permission of the author. **411**: Figure 9.40, Data from "Newborn Infants Orient to Sounds," by D. Muir and J. Field, 1979, *Child Development, 50*, 431–436. Society for Research in Child Development. **412**: Figure 9.41, Photo by Walter Salinger/Property of Anthony DeCasper. **412**: Figure 9.42, Adapted from "Pure-Tone Sensitivity of Human Infants," by L. W. Olsho, E. G. Koch, E. A. Carter, C. F. Halpin, and N. B. Spetner, 1988, *Journal of the Acoustical Society of America, 84*, 1316–1324. American Institute of Physics. **413**: Figure 9.43, Adapted from "Pure-Tone Sensitivity of Human Infants," by L. W. Olsho, E. G. Koch, E. A. Carter, C. F. Halpin, and N. B. Spetner, 1988, *Journal of the Acoustical Society of America*, 84, 1316–1324, fig. 5. Copyright © 1988 by the American Institute of Physics. Reprinted by permission.

Chapter Ten: **421**: Figure 10.3, From "Stimulus Representation in the Discharge Patterns of Auditory Neurons," by N. Y. S. Kiang, 1975. In E. L. Eagles (Ed.), *The Nervous System*, Vol. 3, fig. 7. Copyright © 1975 Raven Press. Reprinted by permission. **423**: Figure 10.5, From "Perception of the Speech Code," by A. M. Liberman, 1967, *Psychological Review, 74*, 431–461, figure 1. Copyright © 1967 by the American Psychological Association. Reprinted by permission of the author. **424**: Figure 10.6, Courtesy of Ronald A. Cole, Carnegie-Mellon University. **425**: Figure 10.7, Courtesy of Ronald A. Cole, Carnegie-Mellon University. **428**: Figure 10.9, Courtesy of Ronald A. Cole, Carnegie-Mellon University. **429**: Figure 10.10, From "Selective Adaptation of Linguistic Feature Detectors," by P. Eimas and J. D. Corbit, 1973, *Cognitive Psychology, 4*, 99–109, figure 2. Copyright © Academic Press, Inc. Reprinted by permission. **431**: Figure 10.14, Adapted from "Speech Perception Takes Precedence over Nonspeech Perception," by D. H. Whalen and A. M. Liberman, 1987, *Science, 237*, 169–171, 10 July 1987, figure 1. Copyright © 1987 by the American Association for the Advancement of Science. Adapted by permission. **433**: Figure 10.15, Courtesy of James Sawusch. **434**: Figure 10.16, From "Time-Varying Features of Initial Stop Consonants in Auditory Running Spectra: A First Report," by D. Kewley-Port and P. A. Luce, 1984, *Perception and Psychophysics, 35*, 353–360, figure 1. Copyright © 1984 by Psychonomic Society Publications. Reprinted by permission. **435**: Figure 10.17, Courtesy of Harvey M. Sussman. **436**: Figure 10.18, Adapted from "The Investigation of Locus Equations as a Source of Relational Variance for Stop Place Categorization," by H. M. Sussman, H. A. McCaffrey, and Sandra A. Matthews, 1991, *Journal of the Acoustical Society of America, 90* (3), 1309–1317. Copyright © 1991 by the American Institute of Physics. Reprinted by permission. **439**: Figure 10.20, TM © ACP 1988 Archie Comic Publications, Inc. **442**: Figure 10.21, From "Selective Adaptation of Linguistic Feature Detectors," by P. Eimas and J. D. Corbit, 1973, *Cognitive Psychology*, 4, 99–109, figure 2. Copyright © Academic Press, Inc. Reprinted by permission. **443**: Figure 10.22, From "Encoding of Speech Features in the Auditory Nerve," by M. B. Sachs, E. D. Young, and M. I. Miller, 1981. In R. Carlson and B. Granstrom (Eds.), *The Representation of Speech in the Peripheral Auditory System*, pp. 115–130. Copyright © 1981 by Elsevier Science Publishing, New York. Reprinted by permission. **449**: Figure 10.26, Data from "The Significant Features of Japanese Macaque Coo Sounds: A Psychophysical Study," by B. May, D. B. Moody, and W. C. Stebbins, 1988, *Animal Behavior, 36*, 1432–1444. Association for the Study of Animal Behavior, London. **451**: Figure 10.28, From "Speech Perception in Infants," by P. Eimas, E. P. Siqueland, P. Jusczyk, and J. Vigorito, 1971, *Science, 171*, 303–306, figure 2. Copyright © 1971 by the American Association for the Advancement of Science. Reprinted by permission.

Chapter Eleven: **460**: Figure 11.1, From *Living Images*, by G. Shih and R. Kessell, 1982. Jones and Barlett Publishers, Inc. **464**: Figure 11.6, Adapted from "Four Channels Mediate the Mechanical Aspects of Touch," by S. J. Bolanowski Jr., G. A. Gescheider, R. T. Verrillo, and C. M. Checkosky, 1988, *Journal of the Acoustical Society of America, 84*, 1680–1694, figure 8. Copyright © 1988 by the American Institute of Physics. Reprinted by permission. **465**: Figure 11.7, From "The Tactile Sensory Innervation of the Glabrous Skin of the Human Hand," by A. B. Vallbo and R. S. Johansson, 1978. In G. Gordon (Ed.), *Active Touch*, figure 1, p. 33. Copyright © 1978 by Pergamon Press Ltd. Reprinted by permission. **466**: Figure 11.9, From "Biological Transducers," by W. R. Lowenstein, 1960, p 103. Copyright © 1960 by *Scientific American, Inc.* All rights reserved. **467**: Figure 11.10, From "Mechanical Transmission in a Pacinian Corpuscle. An Analysis and a Theory," by W. R. Lowenstein and R. Skalak, 1966, *Journal of Physiology, 182*, 246–378, figure 13. Copyright © The Physiological Society. Reprinted by permission. **468**: Figure 11.11, From "Modality Coding in the Somatic Sensory System," by J. H. Martin and T. M. Jessell, 1991. In E. R. Kandel, J. H. Schwartz, and T. M. Jessell (Eds.), *Principle of Neural Science*, 3rd ed., figure 24.10. Copyright © 1991 by Appleton & Lange, Norwalk, CT. Reprinted by permission. **468**: Figure 11.12, Adapted from "Neural Mechanisms of Spatial Tactile Discrimination: Neural Patterns Evoked by Braille-Like Dot Patterns in the Monkey," by K. O. Johnson and G. D. Lamb, 1981, *Journal of Physiology, 310*, 117–144. Copyright © 1981 by The Physiological Society, Oxford. Adapted by permission. **469**: Figure 11.13, From " Human Tactile Pattern Recognition: Active Versus Passive Touch, Velocity Effects, and Patterns of Confusion," by F. Vega-Bermudez, K. O. Johnson, and S. S. Hsiao, 1991, *Journal of Neurophysiology, 65*, 531–546, figure 12. Copyright © 1991 by The American Physiological Society. Reprinted by permission. **470**: Figure 11.14, From "Response Characteristics of Cutaneous Warm Fibers in the Monkey," by R. Duclaux and D. R. Kenshalo, 1980, *Journal of Neurophysiology, 43*, 1–15, figure 3. Copyright © 1980 by The American Physiological Society. Reprinted by permission. **470**: Figure 11.15, From "Correlations of Temperature Sensitivity in Man and Monkey, A First Approximation," by D. R. Kenshalo, 1976. In Y. Zotterman (Ed.), *Sensory Functions in the Skin of Primates*, p. 309. Copyright © 1976 by Plenum Publishing Group. Reprinted by permission. **470**: Figure 11.16, From "Nervous Outflow from the Cat's Foot During Noxious Radiant Heat Stimulation," by P. W. Beck, H. O. Handwerker, and M. Zimmerman, 1974, *Brain Research, 67*, 373–386, figure 3A. Copyright © 1974 by Elsevier Biomedical Press, B.V. Reprinted by permission. **470**: Figure 11.17, From "Intensive and Extensive Aspects of Tactile Sensitivity as a Function of Body Part, Sex, and Laterality," by S. Weinstein, 1968. In D. R. Kenshalo (Ed.), *The Skin Senses*, pp. 206, 207. Copyright © 1968 by Charles C Thomas. Courtesy of Charles C Thomas, Publishers, Springfield, IL. **473**: Figure 11.18, From "The Tactile Sensory Innervation of the Glabrous Skin of the Human Hand," by A. B. Vallbo and R. S. Johansson, 1978. In G. Gordon (Ed.), *Active Touch*, figure 1, p. 45. Copyright © 1978 by Pergamon Press Ltd. Reprinted by permission. **473**: Figure 11.19, From "Touch," by E. R. Kandel and T. M. Jessell, 1991. In E. R. Kandel, J. H. Schwartz, and T. M. Jessell (Eds.), *Principle of Neural Science*, 3rd ed., figure 26-8a. Copyright © 1991 Appleton & Lange, Norwalk, CT. Reprinted by permission. **474**: Figure 11.20, Reprinted with permission of Simon & Schuster, Inc. from *The Cerebral Cortex of Man*, by Wilder Penfield and Theodore Rasmussen, 1950, p. 214. Copyright © 1950 by Macmillan Publishing Co., renewed 1978 by Theodore Rasmussen. **475**: Figure 11.22, From "Movement-Sensitive and Direction and Orientation Selective Cutaneous Receptive Fields in the Hand Area of the Postcentral Gyrus in Monkeys," by L. Hyvarinen and A. Poranen, 1978, *Journal of Physiology, 283*, 523–537, figure 3. Copyright © 1978 by The Physiological Society, UK. Reprinted by permission. **476**: Figure 11.23, From "Movement-Sensitive and Direction and Orientation-Selective Cutaneous Receptive Fields in the Hand Area of the Postcentral Gyrus in Monkeys," by L. Hyvarinen and A. Poranen, 1978, *Journal of Physiology, 283*, 523–537, figure 7. Copyright © 1978 by The Physiological Society, UK. Reprinted by permission. **477**: Figure 11.24, From "Modality Coding in the Somatic Sensory System," by J. H. Martin and T. M. Jessell, 1991. In E. R. Kandel, J. H. Schwartz, and T. M. Jessell (Eds.), *Principle of Neural Science*, 3rd ed., figure 24.11. Copyright © 1991 by Appleton & Lange, Norwalk, CT. Reprinted by permission. **478**: Figure 11.26, From "Multiple-Joint Neurons in Somatosensory Cortex of Awake Monkeys," by R. M. Costanzo and E. P. Gardner, 1981, *Brain Research, 214*, 321–333. Copyright © 1981 by Elsevier Science Publishers, B.V. Reprinted by permission. **479**: Figure 11.27, From "Cortical Processing of Tactile Information in the First Somatosensory and Parietal Association Areas in the Monkey," by H. Sakata and Y. Iwamura, 1978. In G. Gordon (Ed.), *Active Touch*, p. 61. Copyright © 1978 by Pergamon Press Ltd. Reprinted by permission. **481**: Figure 11.30, From "Hand Movements: A Window into Haptic Object Recognition," by S. J. Lederman and R. L. Klatzky, 1987, *Cognitive Psychology, 19*, 342–368, figure 1. Academic Press, Inc. **482**: Figure 11.31, Data from "Hand Movements: A Window into Haptic Object Recognition," by S. J. Lederman and R. L. Klatzky, 1987, *Cognitive Psychology, 19*, 342–368. Academic Press, Inc. **483**: Figure 11.32, Adapted from "Living Prehistory in India," by D. D. Kosambi, *Scientific American*, February 1967, figures on pp. 210–211. **485**: Figure 11.33, Based on figures in "Pain Responsiveness," by G. B. Rollman, 1991. In M. A. Heller & W. Schiff (Eds.), *The Psychology of Touch*, pp. 91–114, Lawrence Erlbaum & Associates, and in *The Challenge of Pain*, rev. ed. by Ronald Melzack and Patrick D. Wall, 1988. Penquin USA. **489**: Figure 11.35, Data from "Pain Responses in Nepalese Porter," by W. C. Clark and S. B. Clark, 1980, *Science, 209*, 410–412. American Association for the Advancement of Science. **492**: Figure 11.36, From Cortical Representational Plasticity," by M. M. Merzenich, G. Reconzone, W. M. Jenkins, T. T. Allard, and R. J. Nudo, 1988. In P. Rakic and W. Singer (Eds.), *Neurobiology of Neocortex*, pp. 42–67, figure 1. Copyright © 1988 John Wiley & Sons. Reproduced by permission of M. M. Merzenich. **492**: Figure 11.37, From Cortical Representational Plasticity," by M. M. Merzenich, G. Reconzone, W. M. Jenkins, T. T. Allard, and R. J. Nudo, 1988. In P. Rakic and W. Singer (Eds.), *Neurobiology of Neocortex*, pp. 42–67, figure 2. Copyright © 1988 John Wiley & Sons. Reproduced by permission of M. M. Merzenich.

Chapter Twelve: **505**: Figure 12.2, Adapted from "The Stereochemical Theory of Odor," by J. E. Amoore, J. W. Johnston Jr., and M. Rubin, from figures on pp. 43a and 44–45, *Scientific American*, 210, Feb. Copyright © 1964 by Scientific American, Inc. All rights reserved. **507**: Figure 12.4, From "A Novel Multigene Family May Encode Odorant Receptors: A Molecular Basis for Odor Recognition," by L. Buck and Axel, 1991, *Cell, 65*, 175–187, figure 5. Copyright © 1991 by Cell Press. Reprinted by permission. **507**: Figure 12.5, Adapted from "Chemosensory Neuroanatomy and Physiology," by M. E. Frank and M. D. Rabin, 1989, *Ear, Nose and Throat Journal, 68*. **507**: Figure 12.6, Adapted from Y. C. Tsang, 1936, *Comprehensive Psychology Monthly*, Vol. 12, #57. **508**: Figure 12.7, Adapted from Honzik, 1936, *Comprehensive Psychology Monthly*, Vol. 13, #64. **508**: Figure 12.8, From part of figure 10, "Olfactory System," by R. L. Doty, 1991. In T. V. Getchell et. al. (Eds.), *Smell and Taste in Health and Disease*, pp. 175–203. Raven Press, Inc. **510**: Figure 12.10, From "Receptor Cell Responses to Odorants: Similarities and Differences Among Odorants," by G. Sicard and A. Holley, 1984, *Brain Research, 292*, 283–296, figure

4. Copyright © 1984 by Elsevier Science Publishing. Reprinted by permission. **511**: Figure 12.11, Adapted from "Topographic Coding of Olfactory Quality. Ordorant-Specific Patterns of Epithelial Responsivity in the Salamander," by A. MacKay-Sim, P. Shaman, and D. Moulton, 1982, *Journal of Neurophysiology, 48*, 584–596, figure 4. Copyright © 1982 by the American Physiological Society. Adapted by permission. **511**: Figure 12.12, Data from "Accessibility of Ordorant Molecules to the Receptors," by D. E. Hornung and M. M. Mozell, 1981. In R. H. Cagan and M. R. Kare (Eds.), *Biochemistry of Taste and Olfaction*, pp. 33–45. Academic Press. **512**: Figure 12.13, From "Functional Organization of Rat Olfactory Bulb Analysed by the 2-Deoxyglucose Method," by W. B. Stewart, J. S. Kauer, and G. M. Shepherd, 1979, *Journal of Comparative Neurology*, 185, 715–734, figure 9D. Copyright © 1979 by Wiley-Liss, Inc. Reprinted by permission. **513**: Figure 12.14, Adapted from "Discrimination of Odors in Olfactory Bulb, Pyriformamygdaloid Areas and Orbito-Frontal Cortex of the Monkey," by T. Tanabe, M. Iino, and S. F. Takagi, 1975, *Journal of Neurophysiology, 38*, 1284–1296. Copyright © 1975 by The American Physiological Society. Reprinted by permission. **515**: Figure 12.16, Adapted from Nasal Chemoreception in Flavor Identification," by M. M. Mozell, B. P. Smith, P. E. Smith, R. L. Sullivan, and P. Swender, 1969, *Archives of Otolaryngology, 90*, 367–373, figure 3. Copyright © 1969 by the American Medical Association. Adapted by permission of M. M. Mozell. **516**: Figure 12.17, Adapted from "Role of Olfaction in Perception of Non-Traditional 'Taste' Stimuli," by T. P. Hettinger, W. E. Myers, and M. E. Frank, 1990, *Chemical Senses, 15*, 755–760, fig. 2. Copyright © 1990 by Oxford University Press. Adapted by permission of Oxford University Press, UK. **522**: Figure 12.22, From "Effects of Adaptation on Human Taste Function," by D. H. McBurney, 1969. In C. Pfaffmann (Ed.), *Olfaction and Taste*, pp. 407–419. Rockefeller University Press. **523**: Figure 12.23, From "Sensory Neural Patterns and Gustation," by R. Erickson, 1963. In Y. Zotterman (Ed.), *Olfaction and Taste*, Vol. 1, pp. 205–213, figure 4. Copyright © 1963 by Pergamon Press, Ltd. Reprinted by permission. **524**: Figure 12.24, Adapted from "Neural Coding of Taste in Macaque Monkeys," by M. Sato and H. Ogawa, 1993. In K. Kurihara, N. Suzuki, and H. Ogawa (Eds.), *Olfaction and Taste XI*, p. 398. Copyright © 1993 by Springer-Verlag. Adapted by permission. **525**: Figure 12.25, Reproduced from *The Journal of General Physiology*, "Response Properties of Macaque Monkey Chorda Tympani Fibers," by M. Sato, H. Ogawa, and S. Yamashita, *66*, 781–810, figure 4, copyright © 1975 by permission of The Rockefeller University Press. **526**: Figure 12.26, Adapted from "Coding Channels in the Taste System of the Rat," by T. R. Scott and B. K. Giza, 1990, *Science, 249*, 1585–1587, figure 1. Copyright © 1990 by the American Association for the Advancement of Science. Adapted by permission. **527**: Figure 12.27, A & B from "Comparative Morphology of Olfactory Receptors," by R. A. Steinbrecht, p. 17. In C. Pfaffmann (Ed.), *Olfaction and Taste III*. Copyright © 1969 Rockefeller University Press. Reprinted by permission. **531**: Figure 12.32, From "Olfaction and Development of Social Preferences in Neonatal Organisms," by R. H. Porter and B. Schaal, 1995. In R. L. Doty (Ed.), *Handbook of Olfaction and Gustation*, pp. 299–321, figure 1. Copyright © Marcel Dekker, Inc. Reprinted by permission.

Chapter Thirteen: **546**: Figure 13.10, From *The World Through Blunted Sight*, by P. Trevor-Roper, 1970, p 39. Copyright © 1970 by Bobbs-Merrill Company. Reprinted by permission. **562**: Figure 13.31, Adapted from *Human Information Processing*, 2nd ed., by Peter Lindsay and Donald Norman, 1977. Copyright © Academic Press, Inc. Adapted by permission. **564**: Figure 13.32, Adapted from "Age Variations in Auditory Acuity," by C. C. Bunch, 1929, *Archives of Otolaryngology, 9*, 625–636. Copyright © 1929 by American Medical Association. Adapted by permission. **570**: Figure 13.35, Based on a figure in *Hearing in Children*, 2nd ed., by J. Northern and M. Downs, 1978. Williams & Wilkins. **572**: Excerpt from "The Sweetest Sound of All," by I. Berkow, New York Times, January 1, 1995, p. B8.

Photo Credits

About the Author: Christopher Baker.

Chapter One: **2**, Larry Fisher/Masterfile; **10**, The Bettmann Archive; **16**, The Bettmann Archive; **27**, Donald Johnson/Tony Stone; **28**, Edward Hopper/Yale University Art Gallery, New Haven, CN; **31**, Fred Whitehead/Animals Animals; **32**, Breck P. Kent; **32**, Norman Hendricks/FPG; **35**, Courtesy of Velma Dobson.

Chapter Two: **47**, From "Scanning Electron Microscopy of Vertebrate Visual Receptors," By E. R. Lewis, Y. Y. Zeevi, and F. S. Werblin, 1969, *Brain Research, 15*, 559–562, © 1969 Elsevier Science Publishers, B. V. Reprinted by permission; **51**, From "Segregation of Form, Color, Movement and Depth: Anatomy, Physiology and Perception" by M. Livingston and D. H. Hubel, 1988, *Science, 240*, 740–749, © 1988 by the American Association for the Advancement of Science. Reprinted by permission; **53**, From "Functional Architecture of Macaque Monkey Cortex," by D. H. Hubel and T. N. Wiesel, 1977, *Proceedings of the Royal Society of London, 198*, 1–59, © 1977 by the Royal Society. Reprinted by permission.

Chapter Three: **94**, Courtesy of the Francis A. Countway Library of Medicine; **101**, Hubel, Wiesel, and Stryker, 1978; **110**, Grant LeDuc/Monkmeyer Press; **122**, Crawford and von Noorden, 1980.

Chapter Four: **133**, Bodlein Library, MSWC 361 Vol. 2; **164**, Stephen Dalton/Animals Animals.

Chapter Five: **177**, Thomas Macaulay; **179**, Bev Doolittle; **181**, Bettmann Archives; **183**, R. C. James; **185**, Courtesy of the Pittsburgh Ballet Theater; **186**, Carnegie Museum; **187**, Bill Wood/The Montgomery Journal; **188**, Courtesy of the Pittsburgh Ballet Theater; **189**, Bev Doolittle.

Chapter Six: **234**, The Carnegie Museum of Art, Pittsburgh; Acquired through the generosity of the Sarah Mellon Scaife family, 66.24.2; **236**, Photograph by Ansel Adams. Copyright © by the Trustees of the Ansel Adams Publishing Rights Trust; **239**, The Carnegie Museum of Art, Pittsburgh; **242**, Mike Chikiris/Pittsburgh Stereogram Company, Pittsburgh; **260**, Vandervit; **267**, Courtesy of E. Gibson; **267**, Kathleen Olson; **269**, Albright-Knox Gallery.

Chapter Seven: **289**, Kenneth Antol; **290**, Kathleen Olson; **290**, Paul Berger/Tony Stone; **291**, From Edwin S. Porter's *The Great Train Robbery*, 1903; **291**, Kathleen Olson.

Chapter Eight: **366**, Steven Dear.

Chapter Nine: **412**, Walter Salinger/Photo property of Anthony DeCasper.

Chapter Ten: **421**, From "Stimulus Representation in the Discharge Patterns of Auditory Neurons," by N. Y. S. Kiang. In E. L. Eagles (Ed.), *The Nervous System*, Vol. 3, figure 7, © 1975 Raven Press; **424**, Courtesy of Ronald A. Cole, Carnegie-Mellon University; **425**, Courtesy of Ronald A. Cole, Carnegie-Mellon University; **428**, Courtesy of Ronald A. Cole, Carnegie-Mellon University; **433**, Courtesy of James Sawusch; **447**, Esao Hashimoto/Animals Animals; **453**, Kuhl 1986.

Chapter Eleven: **460**, Shih and Kessell, 1982.

Chapter Twelve: **500**, Courtesy of Dr. Edward E. Morrison; **509**, Courtesy of Dr. Edward E. Morrison; **519**, Courtesy of Linda Bartoshuk; **521**, Courtesy of Inglis Miller; **529**, Michael Fogden; Oxford Scientific Films/Animals Animals; **530**, Prof. J. E. Steiner, Hebrew University, Jerusalem, Israel; **532**, D. S. Rosenstein and H. Oster, 1988, Differential facial responses to four basic tastes in newborns. *Child Development, 59*; **561**, Courtesy of Eye and Ear Hospital of Pittsburgh.

Chapter Thirteen: **572**, left, Ernest Sisto/NYT Pictures; **572**, right, Chester Higgins Jr./NYT Pictures.

Color Essays: **Plate 2.3:** Courtesy of Mitchell Glickstein; **Plate 2.4a, b:** F. de Monasterio et al., 1981; **Plate 4, 1a, b:** Phil Degginger/Tony Stone; **Plate 4.2a, b:** Ruth Dixon/Stock, Boston; **Plate 4.5:** Courtesy of Yale University Press; **Plate 4.10:** Courtesy of North Carolina Department of Commerce; **Plate 4.15:** The Carnegie Museum of Art, Pittsburgh; Acquired through the generosity of the Sarah Mellon Scaife family, 66.24.2; **Plate 6.1:** MAGIC EYE © N. E. Thing Enterprises. Reprinted with permission of Andrews and McMeel. All rights reserved.